COMPREHENSIVE
COORDINATION CHEMISTRY
IN 7 VOLUMES

COMPREHENSIVE COORDINATION CHEMISTRY

The Synthesis, Reactions, Properties
& Applications of Coordination Compounds

Volume 2
Ligands

EDITOR-IN-CHIEF

SIR GEOFFREY WILKINSON, FRS
Imperial College of Science & Technology, University of London, UK

EXECUTIVE EDITORS

ROBERT D. GILLARD
University College, Cardiff, UK

JON A. McCLEVERTY
University of Birmingham, UK

PERGAMON PRESS
OXFORD · NEW YORK · BEIJING · FRANKFURT
SÃO PAULO · SYDNEY · TOKYO · TORONTO

U.K.	Pergamon Press, Headington Hill Hall, Oxford OX3 0BW, England
U.S.A.	Pergamon Press, Maxwell House, Fairview Park, Elmsford, New York 10523, U.S.A.
PEOPLE'S REPUBLIC OF CHINA	Pergamon Press, Room 4037, Qianmen Hotel, Beijing, People's Republic of China
FEDERAL REPUBLIC OF GERMANY	Pergamon Press, Hammerweg 6, D-6242 Kronberg, Federal Republic of Germany
BRAZIL	Pergamon Editora, Rue Eça de Queiros, 346, CEP 04011, São Paulo, Brazil
AUSTRALIA	Pergamon Press Australia, P.O. Box 544, Potts Point, NSW 2011, Australia
JAPAN	Pergamon Press, 8th Floor, Matsuoka Central Building, 1-7-1 Nishishinjuku, Shinjuku-ku, Tokyo 160, Japan
CANADA	Pergamon Press Canada, Suite No. 271, 253 College Street, Toronto, Ontario M5T 1R5, Canada

First edition 1987

Library of Congress Cataloging-in-Publication Data
Comprehensive coordination chemistry.
Includes bibliographies
Contents: v. 1. Theory and background – v. 2. Ligands – v. 3. Main group and early transition elements – [etc.]
1. Coordination compounds.
I. Wilkinson, Geoffrey, Sir, 1921–
II. Gillard, Robert D.
III. McCleverty, Jon A.
QD474.C65   1987   541.2'242   86-12319

British Library Cataloguing in Publication Data
Comprehensive coordination chemistry: the synthesis, reactions, properties and applications of coordination compounds.
1. Coordination compounds.
I. Wilkinson, Geoffrey, 1921–
II. Gillard, Robert D.
III. McCleverty, Jon A.
541.2'242   QD474

ISBN 0-08-035945-0 (vol. 2)
ISBN 0-08-026232-5 (set)

Printed in Great Britain by A. Wheaton & Co. Ltd., Exeter

Contents

Multidentate Macrocyclic Ligands

Preface

Since the appearance of water on the Earth, aqua complex ions of metals must have existed. The subsequent appearance of life depended on, and may even have resulted from, interaction of metal ions with organic molecules. Attempts to use consciously and to understand the metal-binding properties of what are now recognized as electron-donating molecules or anions (ligands) date from the development of analytical procedures for metals by Berzelius and his contempories. Typically, by 1897, Ostwald could point out, in his 'Scientific Foundations of Analytical Chemistry', the high stability of cyanomercurate(II) species and that 'notwithstanding the extremely poisonous character of its constituents, it exerts no appreciable poison effect'. By the late 19th century there were numerous examples of the complexing of metal ions, and the synthesis of the great variety of metal complexes that could be isolated and crystallized was being rapidly developed by chemists such as S. M. Jørgensen in Copenhagen. Attempts to understand the 'residual affinity' of metal ions for other molecules and anions culminated in the theories of Alfred Werner, although it is salutary to remember that his views were by no means universally accepted until the mid 1920s. The progress in studies of metal complex chemistry was rapid, perhaps partly because of the utility and economic importance of metal chemistry, but also because of the intrinsic interest of many of the compounds and the intellectual challenge of the structural problems to be solved.

If we define a coordination compound as the product of association of a Brönsted base with a Lewis acid, then there is an infinite variety of complexing systems. In this treatise we have made an arbitrary distinction between coordination compounds and organometallic compounds that have metal–carbon bonds. This division roughly corresponds to the distinction — which most but not all chemists would acknowledge as a real one — between the cobalt(III) ions $[Co(NH_3)_6]^{3+}$ and $[Co(\eta^5\text{-}C_5H_5)_2]^+$. Any species where the number of metal–carbon bonds is at least half the coordination number of the metal is deemed to be 'organometallic' and is outside the scope of our coverage; such compounds have been treated in detail in the companion work, 'Comprehensive Organometallic Chemistry'. It is a measure of the arbitrariness and overlap between the two areas that several chapters in the present work are by authors who also contributed to the organometallic volumes.

We have attempted to give a contemporary overview of the whole field which we hope will provide not only a convenient source of information but also ideas for further advances on the solid research base that has come from so much dedicated effort in laboratories all over the world.

The first volume describes general aspects of the field from history, through nomenclature, to a discussion of the current position of mechanistic and related studies. The binding of ligands according to donor atoms is then considered (Volume 2) and the coordination chemistry of the elements is treated (Volumes 3, 4 and 5) in the common order based on the Periodic Table. The sequence of treatment of complexes of particular ligands for each metal follows the order given in the discussion of parent ligands. Volume 6 considers the applications and importance of coordination chemistry in several areas (from industrial catalysis to photography, from geochemistry to medicine). Volume 7 contains cumulative indexes which will render the mass of information in these volumes even more accessible to users.

The chapters have been written by industrial and academic research workers from many countries, all actively engaged in the relevant areas, and we are exceedingly grateful for the arduous efforts that have made this treatise possible. They have our most sincere thanks and appreciation.

We wish to pay tribute to the memories of Professor Martin Nelson and Dr Tony Stephenson who died after completion of their manuscripts, and we wish to convey our deepest sympathies to their families. We are grateful to their collaborators for finalizing their contributions for publication.

Because of ill health and other factors beyond the editors' control, the manuscripts for the chapters on Phosphorus Ligands and Technetium were not available in time for publication. However, it is anticipated that the material for these chapters will appear in the journal *Polyhedron* in due course as Polyhedron Reports.

We should also like to acknowledge the way in which the staff at the publisher, particularly

Dr Colin Drayton and his dedicated editorial team, have supported the editors and authors in our endeavour to produce a work which correctly portrays the relevance and achievements of modern coordination chemistry.

We hope that users of these volumes will find them as full of novel information and as great a stimulus to new work as we believe them to be.

ROBERT D. GILLARD   JON A. McCLEVERTY
Cardiff   *Birmingham*

GEOFFREY WILKINSON
London

Contributors to Volume 2

Professor G. Anderegg
Laboratorium für Anorganische Chemie, ETHZ, Universitätstrasse 6, CH-8092 Zurich, Switzerland

Dr F. J. Berry
Department of Chemistry, University of Birmingham, PO Box 363, Birmingham B15 2TT, UK

Dr J. Burgess
Department of Chemistry, University of Leicester, Leicester LE1 7RH, UK

Professor M. Calligaris
Dipartimento di Scienze Chimiche, Università Degli Studi di Trieste, Piazzale Europa 1, 34127 Trieste, Italy

Professor M. H. Chisholm
Department of Chemistry, Indiana University, Bloomington, IN 47405, USA

Professor R. H. Crabtree
Department of Chemistry, Yale University, PO Box 6666, New Haven, CT 06511, USA

Dr J. A. Cras
Faculteit der Wiskunde en Natuurwetenschappen, Katholieke Universiteit Nijmegen, Toernooiveld, 6525 ED Nijmegen, The Netherlands

Professor N. F. Curtis
Chemistry Department, Victoria University of Wellington, Private Bag, Wellington, New Zealand

Professor P. A. W. Dean
Chemistry Department, The University of Western Ontario, London, Ontario N6A 5B7, Canada

Dr E. Diemann
Lehrstuhl für Anorganische Chemie I, Universität Bielefeld, Postfach 8640, 4800 Bielefeld 1, Federal Republic of Germany

Professor J. R. Dilworth
Department of Chemistry, University of Essex, Wivenhoe Park, Colchester CO4 3SQ, UK

Professor D. Dolphin
Department of Chemistry, University of British Columbia, 2036 Main Mall, Vancouver BC V6T 1Y6, Canada

Dr A. J. Edwards
Department of Chemistry, University of Birmingham, PO Box 363, Birmingham B15 2TT, UK

Professor Dr H. Endres
Anorganische-Chemisches Institut der Universität, Im Neuenheimer Feld 270, D-6900
Heidelberg 1, Federal Republic of Germany

Dr P. L. Goggin
School of Chemistry, University of Bristol, Bristol BS8 1TS, UK

Dr P. G. Harrison
Department of Chemistry, University of Nottingham, University Park, Nottingham NG7 2RD,
UK

Professor B. J. Hathaway
Chemistry Department, University College, Cork, Republic of Ireland

Professor B. L. Haymore
Monsanto Co, 800 N Lindberg Boulevard, St Louis, MO 63167, USA

Dr H. A. O. Hill
Inorganic Chemistry Laboratory, University of Oxford, South Parks Road, Oxford OX1 3QR,
UK

Dr D. A. House
Department of Chemistry, University of Canterbury, Christchurch, New Zealand

Dr B. F. G. Johnson
University Chemical Laboratory, Lensfield Road, Cambridge CB2 1EW, UK

Dr T. Kikabbai
Department of Chemistry, University of Nottingham, University Park, Nottingham NG7 2RD,
UK

Dr S. H. Laurie
School of Chemistry, Leicester Polytechnic, PO Box 143, Leicester LE1 9BH, UK

Professor J.-M. Lehn
Institut de Chimie, Université Louis Pasteur, BP 296/R8, 1 rue Blaise Pascal, F-67008
Strasbourg Cedex, France

Professor S. E. Livingstone
School of Chemistry, University of New South Wales, PO Box 1, Kensington, New South
Wales 2033, Australia

Dr T. Mashiko
2-2-7 Shiroyama, Otawara, Tochigi 324, Japan

Dr C. A. McAuliffe
Department of Chemistry, University of Manchester Institute of Science and Technology, PO
Box 88, Sackville Street, Manchester M60 1QD, UK

Professor R. C. Mehrotra
Chemistry Department, University of Rajasthan, Jaipur, 302 004, India

Professor K. B. Mertes
Department of Chemistry, University of Kansas, Lawrence, KS 66045, USA

Professor Dr A. Müller
Lehrstuhl für Anorganische Chemie I, Universität Bielefeld, Postfach 8640, 4800 Bielefeld 1,
Federal Republic of Germany

Professor U. T. Mueller-Westerhoff
Department of Chemistry, University of Connecticut, Storrs, CT 06266, USA

Dr C. Oldham
Department of Chemistry, University of Lancaster, Bailrigg, Lancaster LA1 4YA, UK

Dr J. D. Pedrosa de Jesus
Department of Chemistry, University of Aveiro, 3800 Aveiro, Portugal

Professor L. Randaccio
Dipartimento di Scienze Chimiche, Università Degli Studi di Trieste, Piazzale Europa 1, 34127 Trieste, Italy

Professor Dr J. Reedijk
Department of Chemistry, Gorlaeus Laboratories, Leiden University, PO Box 9502, 2300 RA Leiden, The Netherlands

Professor I. Rothwell
Department of Chemistry, Purdue University, West Lafayette, IN 47906, USA

Dr A. G. Sharpe
Jesus College, Cambridge CB5 8BL, UK

Dr A. Shaver
Department of Chemistry, McGill University, 801 Sherbrooke Street West, Montreal, Quebec, H3A 2K6, Canada

Dr A. R. Siedle
3M Central Research, 3M Company, PO Box 33221, St Paul, MN 55144, USA

Dr D. G. Tew
Department of Pharmaceutical Chemistry, School of Pharmacy, University of California, San Francisco, CA 94143, USA

Dr R. S. Vagg
School of Chemistry, Macquarie University, North Ryde, NSW 2109, Australia

Professor G. van Koten
Organisch Chemisch Laboratorium, Rijks Universiteit Utrecht, Postbus 5055, 3584 CH Utrecht, The Netherlands

Professor K. Vrieze
Anorganisch Chemisch Laboratorium, Universiteit van Amsterdam, NWE Achtergracht 166, 1018 WV Amsterdam, The Netherlands

Dr J. Willemse
Faculteit der Wiskunde en Natuurwetenschappen, Katholieke Universiteit Nijmegen, Toernooiveld, 6525 ED Nijmegen, The Netherlands

Contents of All Volumes

11

Mercury as a Ligand

PHILIP A. W. DEAN
University of Western Ontario, London, Ontario, Canada

11.1 INTRODUCTION

When a transition metal is the focus of interest, it is common to regard an attached —HgY moiety, wherein Y can be a wide variety of supporting groups, as a ligand donating through mercury. Section 11.2 deals with transition metal complexes of such ligands, and with other transition metal complexes containing mercury as a coordinated atom.

Mercury forms a unique series of catenated polyatomic cations Hg_n^{2+} that can be considered as complexes having monatomic Hg^0 as a ligand. These cations are the subject of Section 11.3.

References for this chapter have been taken from the literature covered by *Chemical Abstracts* up to the end of vol. 98 (1982).

11.2 COMPOUNDS WITH MERCURY TO TRANSITION METAL BONDS

Mercury–transition metal bonds have been described for all members of Groups V–VIII of the transition series except, apparently, technetium. They commonly involve a low oxidation state of the transition element and are particularly numerous for the chromium, iron and cobalt families.[1] In addition, mercury–titanium bonded species have been postulated as unstable reaction intermediates.[2]

Basically four types of mercury-containing transition metal complexes are recognizable. The first type, **A**, at present represented solely by the complex $(np_3)CoHgHgCo(np_3)$ (**1**; $np_3 = N(CH_2CH_2PPh_2)_3$) and its $P(CH_2CH_2PPh_2)_3$ analog,[3] contains a μ-Hg_2 group with a short Hg—Hg bond comparable in length to those found in more conventional mercury (I) complexes (see Chapter 56.2). In complexes of the second type, **B**, each mercury atom bears an attached group Y (see below) and is bound directly to only one transition metal; $Cp_2Mo(HgSEt)_2$ (**2**)[4] and $(OC)_5MnHg\{N_3(2\text{-}ClC_6H_4)_2\}$ (**3**)[5] are examples of this type. Direct bonding of mercury to two or more separate transition metals characterizes the third type, **C**, exemplified by $[Hg\{Co(CO)_4\}_n]^{2-n}$ ($n = 2$ or 3) (see ref. 1 and references therein) and $Hg_6[Rh(PMe_3)_3]_4$ (**4**).[6] The fourth type, **D**, is similar but here the mercury atom bridges form an edge or face of a transition metal cluster (or clusters); examples are $Hg[Ru_3(CO)_9(C_2Bu^t)]_2$ (**5**),[7] $Pd_4(HgBr)_2(CO)_4(PEt_3)_4$ (**6**)[8] and $[HgPt_3(CO)_3(PPhPr_2^i)_3]_2$ (**7**; $R_3 = PhPr_2^i$).[9] Some complexes are intermediate in type, *e.g.* $[Ge\{Co_2(CO)_7\}\{Co_2(CO)_6[HgCo(CO)_4]\}]^-$ (**8**)[10] shows some features of both type **C** and type **D**.

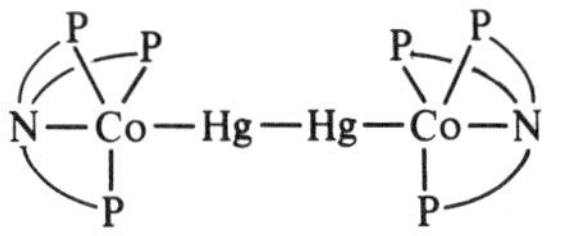

(1)

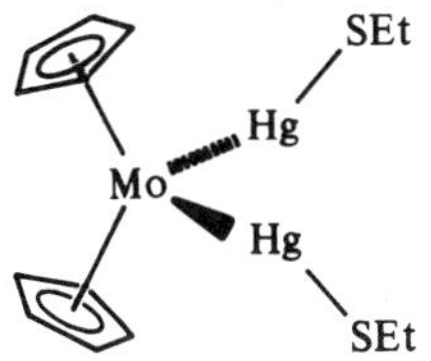

(2)

1

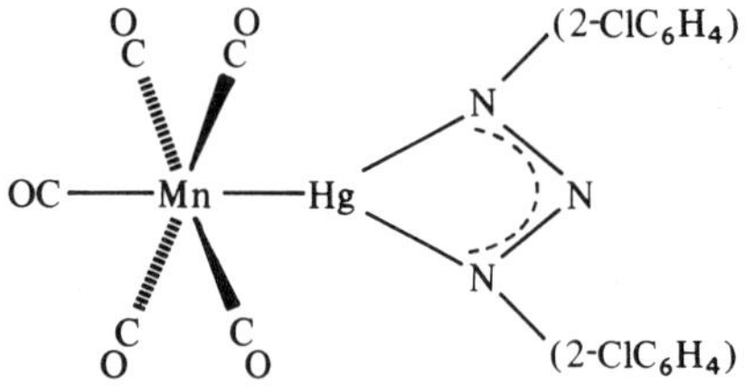

(3)

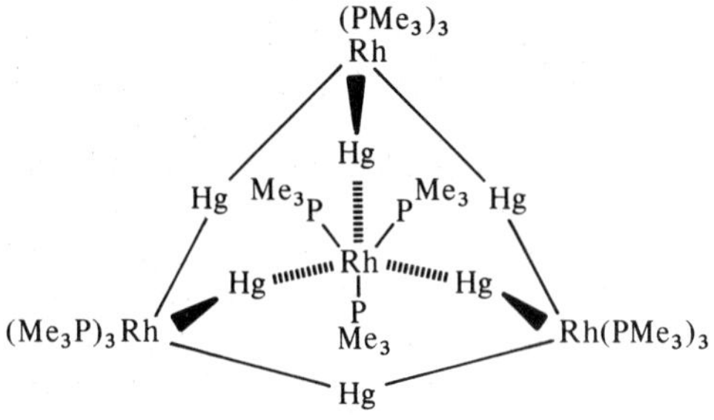

(4)

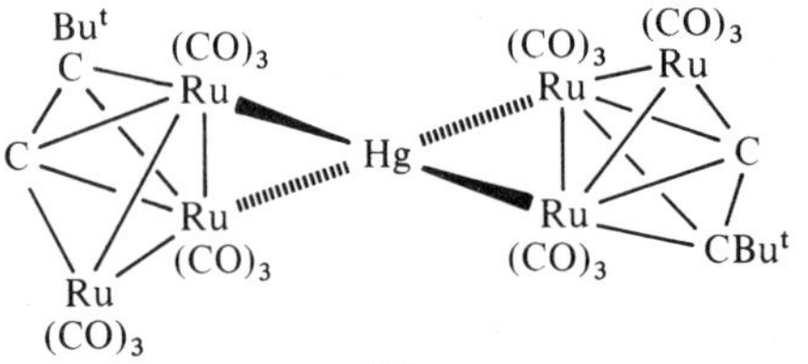

(5)

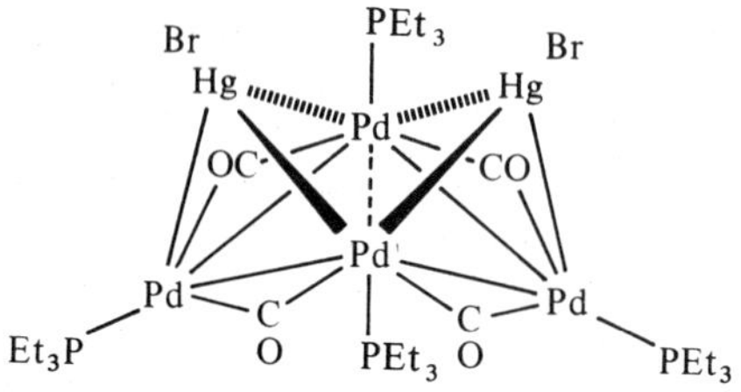

(6)

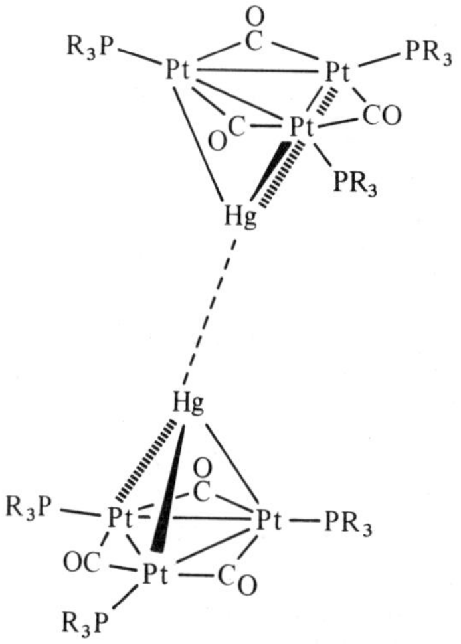

(7)

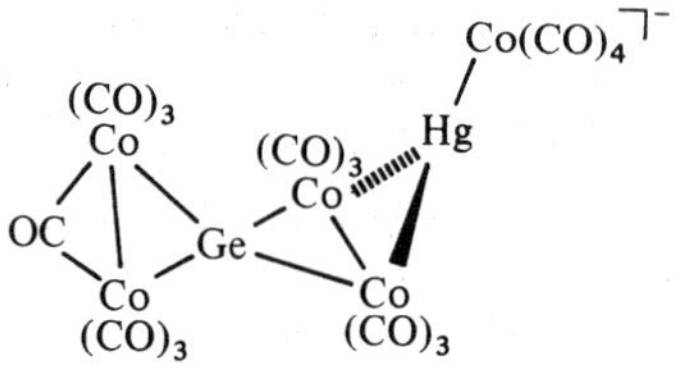

(8)

A recent comprehensive review, with references into 1980, outlines the main synthetic routes to and common properties of complexes of types **B** and **C**.[1] Table 1 gives references to some of the more unusual species that have been described in the interim. New organic syntheses exploiting Pd— and Pt—Hg complexes have been summarized lately.[14]

Table 1   Some Mercury-containing Transition Metal Complexes

Compound	Type[a]	Ref.
LCoHgHgCoL (L = E(CH$_2$CH$_2$PPh$_2$)$_3$; E = N or P)	**A**	3
[Pt$_3$Hg(CO)$_3$(PR$_3$)$_3$]$_2$ (R$_3$ = Pr$^i_3$, PhPr$^i_2$ or PEt$_2$Bu^t)	**D**	9
Pd$_4$(HgBr)$_2$(CO)$_4$(PEt$_3$)$_4$	**D**	8
Rh$_4$Hg$_6$(PMe$_3$)$_{12}$	**D**	6
(NEt$_4$)[Ge{Co$_2$(CO)$_7$}{Co$_2$(CO)$_6$[HgCo(CO)$_4$]}]	**C–D**	10
(Bu^{t}C$_2$)Ru$_3$(CO)$_9$HgMo(Cp)(CO)$_3$	**C–D**	7
[(Bu^{t}C$_2$)Ru$_3$(CO)$_9$(HgX)]$_2$ (X = Br, I, OAc)	**D**	11
Hg[Ru$_3$(CO)$_9$(C$_2$Bu^t)]$_2$	**D**	7
Hg[Pd$_3$(CO)$_3$(PEt$_3$)$_3$]$_2$[b]	**D**	12
Hg[Pt$_3$(2,6-Me$_2$C$_6$H$_3$NC)$_6$]$_2$	**D**	13

[a] See text.
[b] The Pt analog has been reported briefly.[9]

For the purposes of counting electrons, the neutral fragments HgY, Hg and Hg_2 contribute one, two and two skeletal electrons respectively. In terms of oxidation states, (1) has been described as a complex of mercury(I), *i.e.* a mercurous complex,[3] and complexes of class **C** are normally held to be Hg^{II} species.[1,15] There is, however, no general agreement on the oxidation state to be assigned to mercury in complexes of class **B**: some authors describe these as Hg^{II} species,[15] while others regard them as Hg^0 species either explicitly (*e.g.* ref. 16) or implicitly, by describing formation reactions of the general type shown in equation (1) (L = neutral ligand) as redox processes (*e.g.* ref. 1). In the cluster-containing type **D** species, assignment of an oxidation state to mercury has not been attempted.

$$ML_n + XHgY \longrightarrow ML_m(X)(HgY) + (n - m)L \qquad (1)$$

The range of substituents Y that can be attached to mercury in complexes of type **B** is large: Burlitch's review[1] includes examples in which Y is an aliphatic, aromatic, germyl, stannyl or carboxylate group, halide or pseudohalide, and Y in more recent work has included SR with various R groups,[4,17] dissymmetrically bidentate R_2NCS_2[18-20] and RN_3R',[5] and $EtOCS_2$ and $(EtO)_2PS_2$.[20] Also, examples are known of up to three HgY groups attached to a single transition metal;[1,19] as $C(HgY)_4$ are known for various Y,[21-23] there seems no reason to believe that attachment of three HgY groups should be the limit in transition metal chemistry.

In aromatic compounds, HgY is known to be a substituent that is mildly electron-withdrawing through the σ-framework but which exerts little influence on the π-system.[24] In type **B** complexes the donor–acceptor nature of the mercury is expected to be influenced by the particular transition element (M) and its oxidation state, the supporting ligands at M, and the substituent Y at mercury. The effects of the last two have been studied in detail for the series $Cp(OC)_2LMoHgY$;[20] for $Cp(OC)_3MoHgY$, for instance, IR data suggest that the basicity of HgY varies with Y in the order $Cl < Br < I \approx S_2P(OEt)_2 < S_2COEt < S_2CNEt_2$. From the IR data, net electron donation from HgCl in this series seems roughly comparable to that from H, *e.g.* $v(CO)$ are 2029, 1949 and 1945 for $Cp(OC)_3MoH$,[25] and 2024, 1954 and 1936 cm^{-1} for $Cp(OC)_3MoHgCl$.[20] Also, perhaps coincidentally, the ^{95}Mo NMR spectra of $Cp(OC)_3MoHgY$ place the ^{95}Mo NMR substituent effects of various HgY between those for H and Me.[20,26] However, transition metal–mercury bond distances are often significantly less than expected from the appropriate covalent radii, and back-bonding from the filled $d\pi$-orbitals of the transition element into the empty $6p\pi$-orbitals of the mercury is thought to occur (*e.g.* refs. 4, 17, 27, but see also ref. 28). Such π-acceptor character was invoked to account for the high positions of various HgY in the Spectrochemical Series.[29]

The mercury atom is large. Grdenić[30] quotes 1.30 Å (or 1.34 Å) for the digonal covalent radius and 1.54 Å for the van der Waals radius, while others[31] think the latter is at least 1.73 Å. In complexes with multiple HgY groups (*e.g.* **2** and **6**) or mercury atoms (*e.g.* **4** and **7**) the intramolecular (or, perhaps, in **7**, intermolecular) Hg $\cdots$ Hg distances usually fall within the range spanned by the different estimates of r_{vdW}, though Hg $\cdots$ Hg distances less than 3.0 Å are known (*e.g.* in $Cp_2Nb\{Hg(S_2CNEt_2)\}_3$).[19] The interpretation of these interactions is subjective. Thus, for instance, Venanzi and coworkers[9] regard the Hg $\cdots$ Hg distance of 3.225(1) Å in (**7**) as indicating a weak bonding interaction between the two subunits, while Slovokhotov and coworkers[8] reserve judgment on an intramolecular Hg $\cdots$ Hg distance of 3.251(1) Å in (**6**), pointing out that similar distances are observed in $C(HgY)_4$ where there is probably no Hg $\cdots$ Hg bonding interaction. A notable feature of structures of type **D** is the smallness of the angles subtended at mercury by the cluster fragment(s): *e.g.* 60.7(1) and 60.4(1) Å in (**5**), and 55.90(2)–67.58(2) Å in (**6**). This probably results from the small 'bite' of the cluster fragment and the large size of the mercury atom. Small bite angles are observed in various mercury complexes with chelating ligands. For example, the intrachelate angles are 72° in both $Hg(bipy)_2(NO_3)_2 \cdot 2H_2O$[32a] and $Hg(phen)_2(NO_3)_2$;[32b] further, these two bis-(chelate) complexes also exhibit the same twisted square-planar geometry about mercury that is found in (**5**).

11.3 POLYATOMIC CATIONS OF MERCURY

In their 1966 treatise, Phillips and Williams[33] pointed out that the mercury(I) ion, Hg_2^{2+} (**9**), can be regarded formally as a very stable complex of the type $Hg^{II}L$ where L is Hg^0; they also speculated about the possible existence of Hg_3^{2+} (**10**), which by the same formalism is a complex of the type $LHg^{II}L$. The position of equilibrium (2) ($n = 1$) in water does favour (**9**) very strongly; a recent value of log K_1 is 8.28 at 25 °C, $I = 0.1$ M $NaClO_4$,[34] in between log K_1 values reported[35] for the formation of $HgCl^+$ and $HgBr^+$, for instance. For equilibrium (2), ΔH_1° is $-79\,kJ\,mol^{-1}$,[34] similar to the ΔH_1° of $-75\,kJ\,mol^{-1}$ found for formation of HgI^+.[35] Entropically, formation of (9) is disfavoured relative to HgX^+: ΔS_1° is $-108\,J\,mol^{-1}\,K^{-1}$ for equation (2)[34] whereas ΔS_1° is 46, 32 and $-6\,J\,mol^{-1}\,K^{-1}$ for formation of HgX^+ with $X = Cl$, Br and I, respectively.[35] This difference presumably reflects smaller solvation of the reactant ligand and larger solvation of the product cation for Hg^0–Hg_2^{2+} than for X^-–HgX^+. However, whether (10) is regarded as a 1:2 complex of Hg^{2+} and Hg^0 or a 1:1 complex of (9) and Hg^0, its apparent total absence from aqueous media remains a puzzle: in aqueous solution, $K(HgL)/K(Hg_2L) < 100$ for a range of ligands L,[36,37] and also $K_1/K_2 \leqslant 83$ for the series $HgX_n^{2-n}(X = Cl$, Br or I) in water,[35] for example. On this basis, $\log K_2 > 6$ is expected for equation (2) ($n = 2$) in water.

$$Hg_n^{2+}(aq) + Hg^0(aq) \rightleftharpoons Hg_{n+1}^{2+}(aq) \qquad (2)$$

The linear (or, sometimes in the solid state, near linear) yellow Hg_3^{2+} and deep red Hg_4^{2+} (11) ions are now known both in the solid state and as solutions in weakly basic media; liquid SO_2, FSO_3H and $NaCl$–$AlCl_3$ melts have been used as solvents for (10), and liquid SO_2 for (11).[28,38–40] Of note also are the related infinite chain compounds $Hg_{3-x}(MF_6)$ ($M = As$ or Sb)[38], which have interesting electrical conduction properties,[41] but which are not discussed further here.

By comparison with the mercury(I) and mercury(II) ions (Chapter 56.1), the coordination chemistry of (10) and (11) has received little attention. Both ions disproportionate to Hg^0 and Hg_2^{2+}/Hg^{2+} in media more basic than those from which they can be prepared. However, the existence of (11) (which can be regarded as a complex of (10) with Hg^0) and the cation–anion coordination found in solid salts of (10)[38] suggest that this ion, at least, might form stable complexes with suitable weak donors. In addition, the formation of as yet incompletely characterized $Hg_2BrClO_4 \cdot 2SnBr_2$, which may contain both Hg—Hg and Hg—Sn bonds,[42] and the isolation and characterization of $[\{(np_3)Co\}HgHg\{Co(np_3)\}]$ (1; see Section 11.2) may presage a wider occurrence of catenated heterometallic polymercury species. Slow disproportionation of (11) into (10) and $Hg_{3-x}(MF_6)$ occurs even in liquid SO_2.[39] As discussed below, there is evidence for mercury atom transfer between (10) and (11) in liquid SO_2.

The ^{199}Hg NMR chemical shift of $(Hg)Hg(Hg)^{2+}$ (-1968 and -1832 p.p.m. from $HgMe_2$ as reference for $Hg_3(AsF_6)_2$ and $Hg_3(SO_3F)_2$, in SO_2 and FSO_3H, respectiely, at $-70\,^\circ C$[43]) places the ^{199}Hg NMR substituent effect of Hg^0 close to that observed for Br^- in $HgBr_2$ in various solvents.[44a] $^1J(^{199}Hg–^{199}Hg)$ in $(Hg)Hg(Hg)^{2+}$ is enormous: 139.7 and 139.6 kHz in $Hg_3(AsF_6)_2$ and $Hg_3(SO_3F)_2$.[43] In several other ways the coordinated Hg^0 ligand resembles an η_1-carbon donor group. Thus, for example, the formation constants of the carboxylate complexes $HgHgO_2CR^+$ and $MeHgO_2CR$ are remarkably similar.[37] Also, both transfer of phenyl groups and transfer of mercury atoms between mercury(II) ions (equations 3 and 4) are fast second-order processes: from recent ^{199}Hg NMR studies, $k_2 = 1.3 \times 10^4$ and $3.1 \times 10^2\,M^{-1}\,s^{-1}$ for (3) in MeOH and DMSO, respectively, at 300 K,[45] and $k_2 = 3.7 \times 10^5\,M^{-1}\,s^{-1}$ for (4) in water at 303 K.[44b] Rapid mercury atom exchange between (11) and an equilibrium concentration of (10; equation 5) has been postulated to account for inability to observe the slow exchange ^{199}Hg NMR spectrum of (11) in liquid SO_2.[43]

$$Hg^{II} + PhHg^{II} \rightleftharpoons Hg^{II}Ph + Hg^{II} \qquad (3)$$

$$Hg^{2+}(aq) + HgHg^{2+}(aq) \rightleftharpoons HgHg^{2+}(aq) + Hg^{2+}(aq) \qquad (4)$$

$$HgHgHg^{2+} + HgHgHgHg^{2+} \underset{SO_2}{\rightleftharpoons} HgHgHgHg^{2+} + HgHgHg^{2+} \qquad (5)$$

11.4 REFERENCES

1. J. M. Burlitch, in 'Comprehensive Organometallic Chemistry', ed. G. Wilkinson, F. G. A. Stone and E. W. Abel, Pergamon, Oxford, 1982, vol. 6, chap. 42.
2. G. A. Razuvaev, V. N. Latyaeva, L. I. Vishinskaya, V. T. Bytchkov and G. A. Vasil'eva, *J. Organomet. Chem.*, 1975, **87**, 93.
3. F. Cecconi, C. A. Ghilardi, S. Midollini and S. Moneti, *J. Chem. Soc., Dalton Trans.*, 1983, 349.
4. M. M. Kubicki, R. Kergoat, J. E. Guerchais, I. Bkouche-Waksman, C. Bois and P. L'Haridon, *J. Organomet. Chem.*, 1981, **219**, 329.
5. P. E. Jaitner, P. Peringer, G. Huttner and L. Zsolnai, *Transition Met. Chem.*, 1981, **6**, 86.
6. R. A. Jones, F. M. Real, G. Wilkinson, A. M. R. Galas and M. B. Hursthouse, *J. Chem. Soc., Dalton Trans.*, 1981, 126.

7. S. Ermer, K. King, K. I. Hardcastle, E. Rosenberg, A. M. M. Lanfredi, A. Tiripicchio and M. T. Camellini, *Inorg. Chem.,* 1983, **22**, 1339.
8. E. G. Mednikov, V. V. Bashilov, V. I. Sokolov, Yu. L. Slovokhotov and Yu. T. Struchkov, *Polyhedron*, 1983, **2**, 141.
9. A. Albinati, A. Moor, P. S. Pregosin and L. M. Venanzi, *J. Am. Chem. Soc.,* 1982, **104**, 7672.
10. D. N. Duffy, K. M. Mackay, B. K. Nicholson and W. T. Robinson, *J. Chem. Soc., Dalton Trans.,* 1981, 381.
11. E. Rosenberg, R. Fahmy, K. King, A. Tiripicchio and M. T. Camellini, *J. Am. Chem. Soc.,* 1980, **102**, 3626.
12. E. G. Mednikov, N. K. Eremenko, V. V. Bashilov and V. I. Sokolov, *Inorg. Chim. Acta,* 1983, **76**, L31.
13. Y. Yamamoto, H. Yamazaki and T. Sakurai, *J. Am. Chem. Soc.,* 1982, **104**, 2329.
14. V. V. Bashilov, V. I. Sokolov and O. A. Reutov, *Bull. Acad. Sci. USSR, Div. Chem. Sci.,* 1982, **31**, 1825.
15. F. A. Cotton and G. Wilkinson, 'Advanced Inorganic Chemistry', 4th edn., Wiley-Interscience, New York, 1980, chap. 19.
16. M. M. Mickiewicz, C. L. Raston, A. H. White and S. B. Wild, *Aust. J. Chem.,* 1977, **30**, 1685.
17. M. M. Kubicki, R. Kergoat, J. E. Guerchais, C. Bois and P. L'Haridon, *Inorg. Chim. Acta,* 1980, **43**, 17.
18. M. M. Kubicki, R. Kergoat, J. E. Guerchais, R. Mercier and J. Douglade, *J. Cryst. Mol. Struct.,* 1981, **11**, 43.
19. R. Kergoat, M. M. Kubicki, J. E. Guerchais, N. C. Norman and A. G. Orpen, *J. Chem. Soc., Dalton Trans.,* 1982, 633.
20. M. M. Kubicki, R. Kergoat, J.-Y. LeGall, J. E. Guerchais, J. Douglade and R. Mercier, *Aust. J. Chem.,* 1982, **35**, 1543.
21. J. L. Wardell, in 'Comprehensive Organometallic Chemistry', ed. G. Wilkinson, F. G. A. Stone and E. W. Abel, Pergamon, Oxford, 1982, vol. 2, chap. 17, and references therein.
22. D. Grdenić, B. Kamenar, B. Korpar-Colig, M. Sikirica and G. Jovanovski, *Cryst. Struct. Commun.,* 1982, **11**, 568, and references therein.
23. D. K. Breitinger, W. Kress, R. Sendelbeck and K. Ishiwada, *J. Organomet. Chem.,* 1983, **243**, 245, and references therein.
24. O. Exner, in 'Correlation Analysis in Chemistry', ed. N. B. Chapman and J. Shorten, Plenum, New York, 1978, chap. 10, and references therein.
25. W. K. Dean and W. A. G. Graham, *Inorg. Chem.,* 1977, **16**, 1061.
26. J. Y. LeGall, M. M. Kubicki and F. Y. Petillon, *J. Organomet. Chem.,* 1981, **221**, 287.
27. J. P. Oliver, M. J. Albright and M. D. Glick, *J. Organomet. Chem.,* 1978, **161**, 221.
28. M. J. Taylor, 'Metal-to-Metal Bonded States of the Main Group Elements', Academic, London, 1975, chap. 2.
29. R. S. Nyholm and K. Vrieze, *J. Chem. Soc.,* 1965, 5331.
30. D. Grdenić, in 'Structural Studies of Molecules of Biological Interest', ed. G. Dodson, J. P. Glusker and D. Sayre, Clarendon, Oxford, 1981, chap. 21, and references therein.
31. A. J. Canty and G. B. Deacon, *Inorg. Chim. Acta,* 1980, **45**, L225.
32. (a) D. Grdenić, B. Kamenar and A. Hergold-Brundic, *Croat. Chem. Acta,* 1979, **52**, 339; (b) *Cryst. Struct. Commun.,* 1978, **7**, 245.
33. C. S. G. Phillips and R. J. P. Williams, 'Inorganic Chemistry', Oxford, New York, 1966, vol. 2, p. 529.
34. I. Sanemasa, T. Kobayashi, T. Deguchi and H. Nagai, *Bull. Chem. Soc. Jpn.,* 1983, **56**, 1231.
35. S. Ahrland, *Pure Appl. Chem.,* 1979, **51**, 2019, and references therein.
36. R. M. Smith and A. E. Martell, 'Critical Stability Constants', Plenum, New York, 1976.
37. F. J. C. Rossotti and R. J. Whewell, *J. Chem. Soc., Dalton Trans.,* 1977, 1229.
38. B. D. Cutforth and R. J. Gillespie, *Inorg. Synth.,* 1979, **19**, 22, and references therein.
39. B. D. Cutforth, R. J. Gillespie, P. Ireland, J. F. Sawyer and P. K. Ummat, *Inorg. Chem.,* 1983, **22**, 1344.
40. L. N. Balyatinskaya, Deposited Doc., 1982, VINITI 1593-82 (*Chem. Abstr.,* 1983, **98**, 171 793).
41. I. D. Brown, W. R. Datars and R. J. Gillespie, in 'Extended Linear Chain Compounds', ed. J. S. Miller, Plenum, New York, 1983, vol. 3, p. 1.
42. K. Brodersen and J. Hoffman, *Z. Anorg. Allg. Chem.,* 1980, **469**, 32.
43. R. J. Gillespie, P. Granger, K. R. Morgan and G. J. Schrobilgen, *Inorg. Chem.,* 1984, **23**, 887.
44. (a) P. Peringer, *Inorg. Chim. Acta,* 1980, **39**, 67; (b) *J. Chem. Res. (S),* 1980, 194.
45. P. Peringer and P.-P. Winkler, *J. Organomet. Chem.,* 1980, **195**, 249.

12.1

Cyanides and Fulminates

ALAN G. SHARPE
University of Cambridge, UK

12.1.1 INTRODUCTION

In the history of chemistry, cyanides and fulminates have both played important parts. Prussian blue, now known to be an iron(III) hexacyanoferrate(II), has some claim to be considered the first synthetic coordination compound, having been accidentally discovered by Diesbach as long ago as 1704. Silver fulminate, AgCNO, and silver cyanate, AgOCN, were shown to be isomers by Liebig and Gay-Lussac in 1824. In the 19th and early 20th centuries, knowledge of many complex cyanides was put on a firm basis, and the older literature is still very important in this field. The development of this subject up to the end of 1974 has, however, been dealt with in a comprehensive review,[1] and only selected references before that date will be given here. Two other reviews should also be noted; they cover the cyano complexes of the members of the titanium, vanadium, chromium and manganese groups,[2] and of the platinum metals.[3] In contrast, although there is an extensive technical literature on the use of mercury(II) fulminate as an initiatory explosive, the systematic study of fulminic acid and of the fulminate ion as a ligand is of quite recent origin: the structure of the acid was not established as HCNO until 1967, and proofs that carbon is the ligand atom in AgCNO and $Ph_4As[Au(CNO)_2]$ (the only fulminates studied by X-ray diffraction) were obtained only in 1965 and 1980 respectively. The organic chemistry of fulminic acid lies outside the scope of this account, but a good review[4] of its inorganic chemistry and that of complex fulminates covers all but a small amount of the most recent work and is the source of most of the material in Section 12.1.5.

The oxidation of free cyanide ion yields cyanogen or cyanate, never fulminate. It might seem that when the carbon atom is attached to hydrogen or a metal, addition of oxygen to the nitrogen atom should be possible, but there is no record of such an oxidation ever having been achieved. The reverse process, conversion of combined fulminate into combined cyanide, has, however, been demonstrated for platinum(II) and iron(II) complexes.

Most fulminates are very dangerously explosive, but this appears not to be the case for transition metal complexes containing tetraphenylarsonium, tetrapropylammonium or other very large cations. Much recent work on fulminate complexes has involved these compounds, and, as a result of it, the very close analogy with cyano complexes is now established beyond doubt.

12.1.2 GENERAL PROPERTIES OF CYANIDE AS A LIGAND

The cyanide ion is both a pseudohalide and an isoelectronic equivalent of carbon monoxide. Nearly all the complexes which it forms are charged species, prepared in water or other polar solvents, and isolated as involatile salts whose detailed structures must be determined by diffraction methods. There is, however, convincing evidence that, like carbon monoxide, cyanide enters into π- as well as σ-bonding with transition metal atoms or ions; it is next to carbon monoxide in the spectrochemical series, and there are numerous structural analogies between metal carbonyls and cyano complexes. Both of these aspects of the properties of cyanide as a ligand are considered below.

The effective crystallographic radius of the cyanide ion, derived from studies of the cubic high-temperature forms of the alkali metals salts, is 1.92 Å, intermediate between the values for chloride and bromide. The standard enthalpy of formation of the gaseous ion is, at the latest estimate, $+36\,kJ\,mol^{-1}$; its hydration enthalpy is $-329\,kJ\,mol^{-1}$; and the electron affinity of the gaseous CN radical is 380 or $399\,kJ\,mol^{-1}$ according to the value taken for the dissociation energy of cyanogen.[5] The interatomic distance in the ion is 1.16 Å, and the fundamental vibration frequency is $2080\,cm^{-1}$.

Hydrogen cyanide, like hydrogen fluoride, is only slightly dissociated in aqueous solution; its pK values are 9.63, 9.21 and 8.88 at 10, 25 and 40 °C respectively. Aqueous solutions of alkali metal cyanides therefore contain quite high concentrations of hydroxide ions, and in the course of attempts to make cyano complexes these sometimes compete successfully for the metal ion. Further, complex cyano acids persist in aqueous solution only if their anions have extremely high formation constants (*e.g.* $[Au(CN)_2]^-$) or are kinetically inert with respect to substitution (*e.g.* $[Co(CN)_6]^{3-}$). It is widely recognized that water as solvent may be oxidized or reduced under drastic conditions, but it is less generally appreciated that this is also true of the cyanide ion, which can be oxidized to cyanogen (in alkaline media to cyanate) or reduced to methylamine; in the presence of an excess of cyanide, for example, iron(III) and cobalt(II) respectively bring about these changes. Frequently, however, there are kinetic barriers to oxidation or reduction, and it is often possible to effect a change in the oxidation state of a metal without free or combined cyanide being attacked.

The electronic structure of the cyanide ion, most simply described as $(\sigma 1s)^2(\sigma^*1s)^2(\sigma 2s)^2(\sigma^*2s)^2(\sigma 2p_z)^2(\pi 2p_x)^2(\pi 2p_y)^2$, corresponds to a triple bond between carbon and nitrogen, with all bonding orbitals fully occupied. Back-donation from filled metal orbitals must therefore involve occupation of antibonding orbitals of the cyanide ion, with consequent weakening of the carbon–oxygen bond; since back-donation is most likely when the metal has a large number of electrons (*i.e.* is in a low oxidation state), the value of the CN stretching frequency is, especially in the absence of more than a small variation in the bond length, a useful structural characteristic of a complex. A decrease in $v(CN)$ is normally associated with an increase in $v(MC)$. Unfortunately change in oxidation state is often accompanied by change in geometry, and the interpretation of changes in stretching frequencies is then open to some doubt. Being negatively charged, cyanide ion is less effective than carbon monoxide as a π-acceptor, and the range of CN stretching frequencies observed is much smaller than found for CO stretching frequencies. For the same reason, the 18-electron rule does not hold nearly so well for cyano complexes as for metal carbonyls.

Cyanide ion acting as a monodentate ligand always appears to have carbon as the donor atom, *i.e.* to form cyano rather than isocyano complexes.[1] When it acts as a bidentate ligand it usually does so by coordinating at both ends to form linear systems typified by $Fe^{II}CNFe^{III}$ (in Prussian blue) and $Fe^{II}CNBF_3$ (in boron trifluoride adducts of hexacyanoferrates(II)). In $CuCN \cdot NH_3$, however, there are three Cu atoms bonded to each cyanide ion, probably one to the nitrogen and two to the carbon atom in the same way as two metal atoms bond to a bridging carbonyl group in, for example, $Fe_2(CO)_9$.[6] (It is difficult to distinguish the carbon and nitrogen atoms with certainty unless neutron diffraction is used, which was not the case here.) Similar bridging cyanide groups have been suggested for the anion (1), isolated in the form of its tetraethylammonium salt.[7] In the fluxional anion $[(C_5H_5)_2Mo_2(CO)_4(CN)]^-$, a single cyanide ion bridges the molybdenum atoms as in (2); again, however, carbon and nitrogen atoms are not distinguished with certainty, and accurate bond lengths are not available.[8] A rare example of a discrete bridged group containing cyanide as a bidentate ligand is provided by the compound $[Pd_3(Ph_2PCH_2CH_2PPh_2)_3(\mu\text{-}CN)_3](ClO_4)_3$, in which the three palladium atoms and three cyanide groups form a nine-membered ring.[9]

Cyano complexes have been involved in many kinetic studies.[1] The fast electron transfer reactions between $[Fe(CN)_6]^{3-}$ and $[Fe(CN)_6]^{4-}$, and between $[Mo(CN)_8]^{3-}$ and $[Mo(CN)_8]^{4-}$, for example, were important in establishing the outer-sphere mechanism for redox reactions. The kinetics of

$$[(OC)_4(NC)V \cdots V(CN)(CO)_4]^{4-}$$

(1)

$$Mo-C{\equiv}N \rightarrow Mo$$

(2)

water replacement in the complex $[Co(CN)_5H_2O]^{2-}$ have provided evidence for the dissociative mechanism for substitution at a cobalt(III) atom, and the study of Co^{II}-catalyzed substitution of cyanide into $Co^{III}(NH_3)_5X$ complexes has provided insights into reactions in which redox and substitution processes are both involved. In general, however, little work has been done on the formation or decomposition of cyano complexes, and in this review of cyanide coordination chemistry we shall therefore concentrate on preparative methods, structures and thermodynamics, and mention kinetic factors only occasionally.

12.1.3 CYANO COMPLEXES: PREPARATIVE METHODS AND REACTIONS

The reaction most frequently used for the preparation of cyano complexes is substitution in aqueous solution without change of oxidation state, *e.g.* for compounds containing the $[Fe(CN)_6]^{4-}$, $[Co(CN)_5]^{3-}$ and $[Ni(CN)_4]^{2-}$ ions. Although partially substituted complexes can sometimes be detected or isolated (*e.g.* $[Fe(CN)_5H_2O]^{3-}$ in the $[Fe(H_2O)_6]^{2+}/CN^-$ and $[Co(CN)_5Cl]^{3-}$ in the $[Co(NH_3)_5Cl]^{2+}/CN^-$ reactions respectively), the end product of the replacement is usually the only species to be obtained in this way; occasionally, however, the use of a very large excess of cyanide enables another complex to be prepared (*e.g.* $[Ni(CN)_5]^{3-}$ from $[Ni(CN)_4]^{2-}$ and CN^-). The very wide applicability of this substitution procedure is qualitative proof (quite often, indeed, the sole proof) that cyano complexes are amongst the most stable transition metal complexes known, the only reagents to destroy all of them being strong acids that remove the very weak acid HCN either as such or as its hydrolysis products.

Mention has already been made of the fact that cyanide can itself act as an oxidant or, more commonly, as a reductant in aqueous solution. Examples in which the latter possibility is utilized are the preparation of complexes of copper(I), molybdenum(IV), tungsten(II) and rhenium(III) from compounds of these metals in higher oxidation states; in such reactions it may be necessary to work under anaerobic conditions.[10] More commonly, however, substitution is effected in the presence of an oxidizing agent or reducing agent. Air acts as the oxidant in the usual preparations of complexes of manganese(III), cobalt(III) and tungsten(IV); peroxodisulfate has been used in the case of palladium(IV) complexes. In a recent preparation[11] of the elusive $[Ru(CN)_6]^{3-}$ ion, photo-oxidation in aqueous solution has been employed; conversely, photolytic reduction of the complex $[Nb(CN)_8]^{4-}$ gives $[Nb(CN)_8]^{5-}$. Reducing agents which have been found effective in the preparation of low oxidation state complexes in aqueous media include amalgams of alkali metals [for complexes of vanadium(II), manganese(I) and nickel(I)] and of aluminum (for complexes of chromium(II), described fully for the first time only in 1977).[12] Electrolytic oxidation [for complexes of iron(III) and molybdenum(V)] and electrolytic reduction [for complexes of manganese(I) and nickel(I)] have also been used.

The complications which result from the hydrolysis of alkali metal cyanides in aqueous media may be avoided by the use of non-aqueous solvents. The one most often employed is liquid ammonia, in which derivatives of some of the lanthanides and of titanium(III) may be obtained from the metal halides and cyanide.[13] By addition of potassium as reductant, complexes of cobalt(0), nickel(0), titanium(II) and titanium(III) may be prepared; and a complex of zirconium(0) has been obtained in a remarkable disproportion of zirconium(III) into zirconium(IV) and zirconium(0).[14] Other solvents which have been shown to be suitable for halide–cyanide exchange reactions include ethanol, methanol, tetrahydrofuran, dimethyl sulfoxide and dimethylformamide. With their aid, species of different stoichiometry from those isolated from aqueous media can sometimes be made: $[Hg(CN)_3]^-$, for example, is obtained as its cesium salt form CsF, KCN and $Hg(CN)_2$ in ethanol.[15]

There have been reports of syntheses of cyano complexes in fused potassium cyanide, though the characterization of the products has not always been satisfactory.[1] The dry reaction between potassium cyanide and potassium hexaiodoplatinate(IV), however, certainly gives the impure hexacyanoplatinate(IV); and substitution of cycloocta-1,5-diene bonded to platinum(II) may be achieved by the use of solid potassium cyanide in the presence of a crown ether catalyst.[16]

Binuclear cyano complexes are occasionally formed in ordinary preparations; from the green solution containing cobalt(II) in the presence of excess of potassium cyanide, $K_6[Co_2(CN)_{10}] \cdot 4H_2O$ can be precipitated by ethanol, though most salts obtainable from the solution contain the mononuclear $[Co(CN)_5]^{3-}$ anion. In recent years many binuclear complexes containing a single bidentate cyanide group as bridging ligand have been prepared by redox reactions (*e.g.* $[(NC)_5CoNCFe(CN)_5]^{6-}$ from $[Co(CN)_5]^{3-}$ and $[Fe(CN)_6]^{3-}$) or by substitution processes (*e.g.* $[(NC)_5RuCNFe(CN)_5]^{6-}$ from $[Ru(CN)_6]^{4-}$ and $[Fe(CN)_5H_2O]^{2-}$). In such substitution processes it is usually a ligand other than cyanide that is displaced, but this is not always so; the interaction of the ions

$[Fe(CN)_6]^{3-}$ and $[Fe(CN)_5NH_3]^{2-}$ yields the complex $[(NC)_5FeNCFe(CN)_4(NH_3)]$.[17] As an elegant example of synthetic planning we may note the methods employed for the preparation of a pair of linkage isomers containing Cr^{III} and Co^{III}: the slow reaction of $[Cr(H_2O)_6]^{3+}$ and $[Co(CN)_6]^{3-}$ gives $[(H_2O)_5CrNCCo(CN)_5]$, whilst the rapid reaction of $[Co(CN)_5N_3]^{3-}$ with HNO_2 in the presence of $[Cr(H_2O)_5CN]^{2+}$ gives $[(H_2O)_5CrNCCo(CN)_5]$.[18] Where different oxidation states are involved in the same complex, the species involved are of considerable current interest as mixed valence compounds showing charge-transfer spectral bands.[19,20]

In addition to the aqua, ammine and azido complexes mentioned above, many other substituted cyano complexes are known. Most of them, of course, are derivatives of metals in non-labile oxidation states, especially octahedral chromium(III) and low-spin octahedral iron(II), iron(III) and cobalt(III). Ligands capable of replacing cyanide under mild conditions are usually species bearing considerable similarities to it, *e.g.* CO, NO^+, 2,2′-bipyridyl and 1,10-phenanthroline. Monoaqua complexes of iron(II) and cobalt(III) may be obtained by photochemical aquation, and the water molecule may then be replaced by other ligands. Many monosubstituted cobalt(III) complexes, including $[Co(CN)_5H]^{3-}$, $[Co(CN)_5O_2]^{3-}$ and $[Co(CN)_5Br]^{3-}$, may be made by oxidative addition to the anion $[Co(CN)_5]^{3-}$, which is formed when cyanide reacts with solutions of cobalt(II) salts. The analogous addition of excess of halogens to $K_2[Pt(CN)_4]$ leads to the *trans* forms of $K_2[Pt(CN)_4X_2]$, where X = Cl, Br or I; when only limited amounts of chlorine or bromine are used complexes of composition $K_2[Pt(CN)_4]X_{0.3}\cdot2.5H_2O$ result.[21-23]

The above examples illustrate the principal methods used in the synthesis of cyano complexes (especially in recent work), but many others have found occasional use, and details of them may be found in the reviews cited. The recent appearance of a comprehensive article[24] on the preparation of the numerous mononuclear substituted cyano complexes of cobalt(III), in which both the rationale and the details of the methods employed are discussed, is especially noteworthy.

12.1.4 CYANO COMPLEXES: STRUCTURES AND THERMODYNAMICS

There are two aspects of the structural chemistry of cyano complexes that we shall discuss here: the coordination of the metal atom, and relative values of interatomic distances and vibrational stretching frequencies.[1] The correlation between the distances and frequencies and the very limited available thermodynamic data is considered briefly at the end of the section.

Metal ions having coordination number two occur in $Hg(CN)_2$, $[Ag(CN)_2]^-$ and $[Au(CN)_2]^-$, all derivatives of d^{10} species; in each case the structure is linear. In $K[Cu(CN)_2]$ the copper atom is actually three-coordinated, the anion being an infinite chain; each copper has a carbon atom of an unshared cyanide, and a carbon atom and a nitrogen atom of shared cyanides, as nearest neighbours in approximately planar coordination. Three-coordination also occurs in the $[Hg(CN)_3]^-$ ion, but there are weak interactions between two of the nitrogen atoms and mercury atoms of other anions, giving rise to a loosely bonded chain structure.

The isoelectronic complexes $[Cu(CN)_4]^{3-}$, $[Ag(CN)_4]^{3-}$, $[Zn(CN)_4]^{2-}$, $[Cd(CN)_4]^{2-}$ and $[Hg(CN)_4]^{2-}$ are all tetrahedral d^{10} derivatives; for d^8 ions, the tetracyano complexes are low-spin square planar ions. Although the structure of $K_2[Ni(CN)_4]$ is built up from quite separate ions, many tetracyanonickelates and their palladium and platinum analogues have structures in which planar $[M(CN)_4]^{2-}$ ions are so stacked that the M atoms form an infinite chain; the strength of the M—M interaction increases progressively from nickel to platinum. Partially oxidized cyanoplatinates(II), typified by the complex $K_2[Pt(CN)_4]Br_{0.3}\cdot3H_2O$, also contain such chains and are important one-dimensional metallic conductors.[21,22] The distinction between planar diamagnetic complexes of Ni^{II} and the more common tetrahedral or octahedral paramagnetic complexes is, of course, easily accounted for on the basis of simple ligand field theory, and it is always clear-cut.

This is by no means equally true of the distribution of the basic structures for five-coordination, the square pyramid and the trigonal bipyramid. Actual structures are often only approximately described by these designations. In the square pyramidal $[Ni(CN)_5]^{3-}$ anion, for example, the Ni^{2+} ion is 0.34 Å above the basal plane, the average C—Ni—C angle between opposing basal carbon atoms being 160°; the equatorial and axial bond lengths are 1.86 and 2.17 Å respectively. In the trigonal bipyramidal form of the same anion, the average axial bond length is 1.84 Å, and the equatorial bond lengths are 1.91, 1.91 and 1.99 Å; the C(axial)—Ni—C(axial) angle is 173° and the C—Ni—C angle between the equivalent equatorial bonds is 141°. Both of these ions occur in the same crystal of $[Cren_3][Ni(CN)_5]\cdot1.5H_2O$, but on dehydration or on application of pressure all the anions assume the square pyramidal form.[25-27] The square pyramid is also the structure found for the $[Co(CN)_5]^{3-}$ ion. In view of the ease of interconversion of the square pyramid and the trigonal bipyramid by the Berry pseudorotation mechanism, their occurrence may well be determined by very small energy considerations.

All known hexacyano complexes contain octahedral anions. There is, it may be noted, no octahedral cyano complex containing a metal ion having more than six d-electrons (corresponding to all bonding orbitals being filled); since cyano complexes are always low-spin in character the possibility of a substantial Jahn–Teller distortion does not arise. Along the series $[Cr(CN)_6]^{3-}$ to $[Co(CN)_6]^{3-}$ there is a steady decrease in the metal–carbon bond length (from 2.08 to 1.89 Å), quite different from the erratic variation in metal–ligand distance found in analogous high-spin complexes such as the fluorides. Moreover, whereas for high-spin complexes such as $[Fe(H_2O)_6]^{2+}$ and $[Fe(H_2O)_6]^{3+}$, or $[FeCl_6]^{4-}$ and $[FeCl_6]^{3-}$, the iron(II)–ligand distance is greater by 0.10–0.15 Å than the iron(III)–ligand distance, the reverse is true for $[Fe(CN)_6]^{4-}$ and $[Fe(CN)_6]^{3-}$, in which the Fe—C bond lengths are 1.90 and 1.93 Å respectively. Further, the M—C stretching frequency is higher, and the the C≡N stretching frequency slightly lower, in the $[Fe(CN)_6]^{4-}$ ion. These data provide some of the strongest evidence available for the importance of back-donation in lower oxidation state cyano complexes. The small difference in the Fe—C bond lengths is, incidentally, an important factor in accounting for the rapid electron transfer between the ions.

Since the discovery that the $[V(CN)_7]^{4-}$ ion has the pentagonal bipyramidal structure, a number of other seven-coordinated cyano complexes have been prepared, amongst them those of Mo^{II}, Mo^{III}, W^{II} and Re^{III}. In some solid salts at least, all these have been shown crystallographically to be isostructural with $[V(CN)_7]^{4-}$, the presence of one or two extra d-electrons having no stereochemical effect.[28] Electronic and electron spin resonance spectra suggest that the capped trigonal prism, the other seven-coordination structure found in complexes containing only monodentate ligands, may be present in $K_4[Mo(CN)_7]\cdot 2H_2O$,[29] but this conclusion cannot yet be regarded as proved.

The three basic structures for eight-coordination are the cube, the dodecahedron and the square antiprism, of which only the last two are found in cyano complexes. It is again necessary to stress that these are usually only approximate descriptions; some dodecahedra, for example, are part-way towards becoming bicapped trigonal prisms. All octacyano complexes yet made are derivatives of d^1 or d^2 ionic configurations, *i.e.* are 17- or 18-electron systems. The subtlety of the factors determining which structure is adopted may be illustrated for the molybdenum-containing ions: $[Mo(CN)_8]^{4-}$ is dodecahedral in $K_4[Mo(CN)_8]\cdot 2H_2O$ but antiprismatic in $Cd_2[Mo(CN)_8]\cdot 2N_2H_4\cdot 4H_2O$ and $H_4[Mo(CN)_8]\cdot 6H_2O$; $[Mo(CN)_8]^{3-}$ is dodecahedral in $(Bu_4N)_3[Mo(CN)_8]$, antiprismatic in $Na_3[Mo(CN)_8]\cdot 4H_2O$, and approximately bicapped trigonal prismatic in $Cs_3[Mo(CN)_8]\cdot 2H_2O$.[1,30] Similar distributions of structures have also been found for salts containing octacyanotungstate anions, but to date the $[Nb(CN)_8]^{4-}$ and $[Nb(CN)_8]^{5-}$ ions in the solid state are known only in dodecahedral forms.[30,31,32] Vibrational spectroscopic studies indicate, however, that $[Nb(CN)_8]^{5-}$ changes its structure on dissolution in water.[32]

It may be noted that, for d^2, d^4, d^6, d^8 and d^{10} transition metal ions, octa-, hepta-, hexa-, penta- and tetra-cyano complexes represent the species of maximum coordination number. It should thus be possible, so far as electronic factors are concerned, to prepare an ennea complex of a d^0 system (*e.g.* Re^{VII} or La^{III}). No ennea cyano complex has, however, yet been reported.

Many cyano complexes are undoubtedly thermodynamically very stable, but the evaluation of their formation constants is often difficult. Some species are formed only extremely slowly and are not in equilibrium with their constituent ions (so that formation constants must be obtained from enthalpy and entropy data); for others, measurements in aqueous media are suspect owing to the possibility of errors arising from oxidation by air or by the solvent. Even when accurate data are available, their interpretation in molecular terms is complicated by the fact that both metal–carbon and carbon–nitrogen bonds are involved. If we have accurate values for the enthalpies of conversion of M^{3+}(aq) into $[M(CN)_6]^{3-}$(aq) along the series M = Cr to Co, for example, and can make all necessary allowances for ionic solvation terms, we can still discuss the variation in M^{3+}–CN^- interaction from the point of view of the variation in metal–carbon bond strength only if we assume that the bonding in the coordinated ligands is the same in all the complexes. There is in practice a small increase in the Raman-active symmetric CN stretching frequency along this series, from $2132\ cm^{-1}$ for M = Cr to $2151\ cm^{-1}$ for M = Co, but this is much less than the variation in the MC symmetric stretching frequency (from $348\ cm^{-1}$ for M = Cr to $410\ cm^{-1}$ for M = Co), and this fact, coupled with the substantial diminution in the M—C distance whilst the C≡N distance remains constant within experimental error, would seem to justify making the stated assumption. (Rather sophisticated arguments do in fact suggest that ν(CN) should increase slightly as the strength of the M—C bond increases.[33])

A critical survey of the formation constant data available for the hexacyano complexes of the elements vanadium to cobalt inclusive indicates that for the dipositive ions the sequence is probably $Mn^{II} < V^{II} < Cr^{II} < Fe^{II}$ and for the tripositive ions it is probably $Cr^{III} \sim Mn^{III} < Fe^{III} < Co^{III}$, though it is only for $[Fe(CN)_6]^{4-}$ and $[Fe(CN)_6]^{3-}$ that totally reliable values for β_6 (10^{35} and 10^{44} respectively) are available.[1] Along each series it may reasonably be assumed that entropy changes are nearly constant, so the relative values are determined by enthalpy change variations. In the case of the dipositive cations all the aqua complexes are high-spin, all the cyano complexes low-spin; for the tripositive cations, however, $[Co(H_2O)_6]^{3+}$ is, exceptionally, a low-spin complex. With pairing energies, ligand field stabilization energies and hydration energies involved in addition to the basic M^{2+}—CN^- or M^{3+}—CN^- energy terms, a comprehensive discussion is impossible, and we can only draw attention to the fact that, all interpretative doubts notwithstanding, the low-spin d^6 systems, which would maximize back-donation from metal to ligand, do appear to be the most stable.

It is instructive now to consider the $[Fe(CN)_6]^{4-}$ and $[Fe(CN)_6]^{3-}$ ions in more detail. Both ions are kinetically inert, as is shown by the slowness of their exchange with labelled cyanide; but electron transfer between them is rapid. From the $[Fe(CN)_6]^{3-}/[Fe(CN)_6]^{4-}$ and Fe^{3+} (aq)/Fe^{2+} (aq) standard potentials ($+0.36$ and $+0.77$ V respectively), it is easily seen that β_6 for the Fe^{III} complex is greater than that for the Fe^{II} complex by a factor of about 10^7. The difference in standard potentials is, however, markedly temperature-dependent, and calorimetric determinations of the enthalpies of formation and of oxidation of $[Fe(CN)_6]^{4-}$ reveal that the enthalpy of complexing of Fe^{2+} (aq) ($-359 \, kJ \, mol^{-1}$) is substantially more negative than that of Fe^{3+} (aq) ($-293 \, kJ \, mol^{-1}$). The less negative standard free energy of complexing of Fe^{2+} (aq) thus arises from a very much more negative entropy of complexing: the quadruply charged $[Fe(CN)_6]^{4-}$ ion imposes a very high degree of order on the solvent. In enthalpic terms, cyano complexing, like complexing by 2,2'-bipyridyl or 1,10-phenanthroline, favours the lower oxidation state.[1] We should not conclude from this evidence alone that the bond to Fe^{II} is stronger than that to Fe^{III}, since all thermodynamic functions of complex formation in aqueous solution are really those of solvent replacement. Nevertheless, the bond length and spectroscopic data mentioned earlier do indicate the metal–carbon bond to be stronger in the Fe^{II} complex.

For the formation of $[Ni(CN)_4]^{2-}$ ion, β_4 is 10^{30}; the value for the $[Pd(CN)_4]^{2-}$ ion was formerly accepted as 10^{42}, but it has been argued that the electrochemical data on which this value is based are unreliable; a study of the $[PdCl_4]^{2-}/CN^-$ reaction[34] leads to a value of about 10^{63}. Standard enthalpies of formation of $[Ni(CN)_4]^{2-}$ and $[Pd(CN)_4]^{2-}$ from aqua cations have been determined calorimetrically as -181 and $-386 \, kJ \, mol^{-1}$ respectively, but a meaningful comparison of these values is difficult since the aqua cations involved are high-spin octahedral $[Ni(H_2O)_6]^{2+}$ and low-spin planar $[Pd(H_2O)_4]^{2+}$ respectively.

It might seem that comparisons between $[Zn(CN)_4]^{2-}$, $[Cd(CN)_4]^{2-}$ and $[Hg(CN)_4]^{2-}$ would provide the best basis for a study of a triad of transition metals. For these ions the standard enthalpies of formation from aqua ions are -117, -112 and $-250 \, kJ \, mol^{-1}$, and β_4 values are 10^{20}, 10^{18} and 10^{39} respectively. Accurate bond length data are, unfortunately, not available for $[Cd(CN)_4]^{2-}$ and $[Hg(CN)_4]^{2-}$; the symmetric MC stretching frequencies are 340, 318 and $336 \, cm^{-1}$ for M = Zn, Cd and Hg respectively. If the hydrated cations of all three metals are identical in geometry, these results are difficult to reconcile, the more so since MC stretching frequencies for $[Pt(CN)_4]^{2-}$ and $[Au(CN)_2]^-$ are higher than for analogous ions containing other metals in their triads.[1] The correlation of structural and thermodynamic data for complex cyanides is thus clearly only at a very early stage.

12.1.5 FULMINATO COMPLEXES

The only directly accessible metal fulminates are those of mercury(II) and silver(I), very dangerously exposive solids obtained by the action of nitric acid and ethanol on the metals or their salts. Most modern preparations of fulminato complexes involve the conversion of a known amount of mercury fulminate into aqueous sodium fulminate by the action of sodium amalgam and ice-cold water; the sodium fulminate solution is then allowed to react with the appropriate amount of a transition metal salt, and the resulting complex fulminato ion is precipitated as the salt of a large cation, most frequently Ph_4As^+ or R_4N^+; these are not explosive.[4,35] Alkali and alkaline earth metal salts containing complex fulminato anions may be isolated from aqueous solutions, but they are reported to be as exposive as the binary silver and mercury fulminates, and are therefore usually avoided.

Other aqueous preparative methods include aerial oxidation of an alkaline solution of $CoSO_4$ and NaCNO to give the fulminatocobaltate(III) anion $[Co(CNO)_6]^{3-}$, reduction of ruthenate(VI) by excess of fulminate to give $[Ru(CNO)_6]^{4-}$, and displacement of 2,2'-bipyridyl or 1,10-phenanthroline from nickel(II) or cobalt(III) complexes to give $[Ni(CNO)_4]^{2-}$ or $[Co(CNO)_6]^{3-}$. Liquid ammonia may replace water as solvent; $[Ni(NH_3)_6]^{2+}$ and $[Co(NH_3)_6]^{3+}$, for example, react with sodium fulminate in this solvent to form $[Ni(CNO)_4]^{2-}$ and $[Co(CNO)_6]^{3-}$. In all these reactions fulminate behaves very like cyanide; with $[AuCl_4]^-$, however, reduction to form the gold(I) complex $[Au(CNO)_2]^-$ takes place and no gold(III) complex can be isolated.

Mercury fulminate may also be obtained by the action of the sodium salt of nitromethane on mercury(II) chloride according to equation (1).[36] (This should discourage the use of nitromethane as solvent in operations involving mercury or silver salts.) In a modification of this reaction, *trans*-$(Ph_3P)_2Pt(CNO)_2$ has recently been made by the action of nitromethane on $(Ph_3P)_4Pt$; Ph_3P, H_2 and H_2O are the other products.[37] The complex is reduced to *trans*-$(Ph_3P)_2Pt(CN)_2$ by the action of triphenylphosphine (a chemical proof that the carbon atom is bonded to platinum), and undergoes isomerization to *cis*-$(Ph_3P)_2Pt(NCO)_2$ when heated.[38] Only *N*-cyanato complexes are obtained in the course of attempts to extend the reaction to palladium and rhodium complexes.[37] A further reduction of a fulminato complex to the corresponding cyano complex occurs when $[Fe(CNO)_6]^{4-}$ is treated with an alkaline suspension of $Fe(OH)_2$; tracer studies confirm that the iron in the resulting hexacyanoferrate(II) originates in the hexafulminato complex.[39]

$$2MeNO_2 + 2NaOH + HgCl_2 \xrightarrow{H_2O} Hg(CNO)_2 + 4H_2O + 2NaCl \tag{1}$$

The microwave spectrum of fulminic acid vapour shows the molecule to be linear HCNO with bond lengths H—C, 1.03 Å; C≡N, 1.17 Å; N—O, 1.21 Å. There are, unfortunately, no diffraction data for any simple ionic fulminates, and indeed the only X-ray crystallographic studies reported for fulminates are those of $AgCNO$[41,42] and $Ph_4As[Au(CNO)_2]$.[43] Silver fulminate is dimorphic; one form contains infinite AgCAgCAg chains and the other a 12-membered ring of alternate C and Ag atoms; in each case the CNO group is linear. Bond lengths and angles in the chain form are Ag—C, 2.18; C≡N, 1.16; N—O, 1.25 Å; ∠AgCAg, 83°; ∠AgCN, 138°. Values for the cyclic anion are not very different; in each case the long Ag—C distance suggests a bond order of 0.5, *i.e.* a three-centre AgCAg bond. An additional feature of interest is that the distance between nearest silver atoms is almost the same as that in metallic silver. The anion $[Au(CNO)_2]^{2-}$ is, as expected, linear, with Au—C, 2.10; C≡N, 1.10; N—O, 1.25 Å. The vibrational spectra of the ions $[Ni(CNO)_4]^{2-}$ and $[Zn(CNO)_4]^{2-}$ are compatible with planar and tetrahedral coordination respectively and with linear CNO groups.[44] Colour changes on dehydration of alkaline earth metal tetrafulminatoplatinates(II) suggest structural analogies with the tetracyanoplatinates(II), but this early observation[45] has not been followed up.

Fulminato complexes are now known for Fe^II, Ru^II, Co^III, Rh^III, Ir^III, Ni^II, Pd^II, Pt^II, Cu^I, Ag^I, Au^I, Zn^II, Cd^II and Hg^II. The electronic spectra[46] of the diamagnetic ions $[Fe(CNO)_6]^{4-}$ and $[Co(CNO)_6]^{3-}$ are extremely like those of the corresponding cyano complexes, and values for Δ_{oct} (27 000 and 32 200 cm^{-1} for $[Fe(CNO)_6]^{4-}$ and $[Fe(CN)_6]^{4-}$; 26 100 and 33 500 cm^{-1} for $[Co(CNO)_6]^{3-}$ and $[Co(CN)_6]^{3-}$ respectively) show that CNO^- is near CN^- in the spectrochemical series (a further indication that carbon is the ligand atom). The *trans* effect of CNO^- in Pt^II complexes, as inferred from the PtH stretching frequency for $(Et_3P)_2Pt(H)(CNO)$, is also close to that of CN^-. The only formation constant data reported for fulminato complexes are β_2 for $Hg(CNO)_2$ and β_4 for $[Hg(CNO)_4]^{2-}$; both are smaller by a factor of about 10^5 than those for the analogous cyano complexes.[47]

In addition to the phosphineplatinum(II) fulminate complexes mentioned already, many other mixed complexes have been prepared and their vibrational spectra studied.[4,35] Equations (2)–(6) illustrate typical preparations. In each case the fulminato complex formed closely resembles the analogous cyano complex.

$$Zn + 2py + Hg(CNO)_2 \xrightarrow{py} 2Hg + py_2Zn(CNO)_2 \tag{2}$$

$$[Fe(CNO)_6]^{4-} + 2bipy \xrightarrow{H_2O} bipy_2Fe(CNO)_2 + 4CNO^- \tag{3}$$

$$[Fe(CN)_5NH_3]^{3-} + CNO^- \xrightarrow{H_2O} [Fe(CN)_5CNO]^{4-} + NH_3 \tag{4}$$

$$Cr(CO)_6 + 2CNO^- \xrightarrow[THF]{hv} cis\text{-}[Cr(CO)_4(CNO)_2]^{2-} + 2CO \tag{5}$$

$$PhHgOH + MeNO_2 \xrightarrow{20°C} PhHgCNO + 2H_2O \tag{6}$$

In principle the fulminate anion should, like cyanate and thiocyanate, be capable of coordinating to a metal atom in two ways, but although it was until quite recently assumed that oxygen was the donor atom, there is at present no evidence for this mode of bonding in any fulminato compound for which structural information is available. Nor has it yet been shown that fulminate can act as a bridging ligand. Further studies of these often dangerous compounds should yield interesting results.

12.1.6 REFERENCES

1. A. G. Sharpe, 'The Chemistry of Cyano Complexes of the Transition Metals', Academic, London, 1976.
2. W. P. Griffith, *Coord. Chem. Rev.*, 1975, **17**, 177.
3. I. B. Baranovskii, *Russ. J. Inorg. Chem. (Engl. Transl.)*, 1978, **23**, 1429.
4. W. Beck, *Organomet. Chem. Rev., Ser. A*, 1971, **7**, 159.
5. H. D. B. Jenkins, K. F. Pratt and T. C. Waddington, *J. Inorg. Nucl. Chem.*, 1977, **39**, 213.
6. D. T. Cromer, A. C. Larson and R. B. Roof, *Acta Crystallogr.*, 1965, **19**, 192.
7. D. Rehder, *J. Organomet. Chem.*, 1972, **37**, 303.
8. M. D. Curtis, K. R. Han and W. M. Butler, *Inorg. Chem.*, 1980, **19**, 2096.
9. J. A. Davies, F. R. Hartley, S. G. Murray and M. A. Pierce-Butler, *J. Chem. Soc., Dalton Trans.*, 1983, 1305.
10. A.-M. Soares and W. P. Griffith, *J. Chem. Soc., Dalton Trans.*, 1981. 1886.
11. A. Vogler. W. Losse and H. Kunkely, *J. Chem. Soc., Chem. Commun.*, 1979, 187.
12. E. Ljungström, *Acta Chem. Scand., Ser. A*, 1977, **31**, 104.
13. D. Nicholls and T. A. Ryan, *Inorg. Chim. Acta*, 1980, **41**, 233.
14. D. Nicholls and T. A. Ryan, *Inorg. Chim. Acta*, 1977, **21**. L17.
15. B. M. Chadwick, D. A. Long and S. U. Qureshi, *J. Mol. Struct.*, 1980, **63**, 167.
16. M. E. Fakley and A. Pidcock, *J. Chem. Soc., Dalton Trans.*, 1977, 1444.
17. P. Roder, A. Ludi, G. Chapuis, K. J. Schenk, D. Schwarzenbach and K. O. Hodgson, *Inorg. Chim. Acta*, 1979, **34**, 113.
18. D. Gaswick and A. Haim, *J. Inorg. Nucl. Chem.*, 1978, **40**, 437.
19. F. Felix and A. Ludi, *Inorg. Chem.*, 1978, **17**, 1782.
20. S. Bagger and P. Stoltze, *Acta Chem. Scand., Ser. A*, 1983, **37**, 247.
21. A. E. Underhill and D. M. Watkins, *Chem. Soc. Rev.*, 1980, **9**, 429.
22. J. M. Williams, *Adv. Inorg. Chem. Radiochem.*, 1983, **26**, 235.
23. G. S. V. Coles, A. E. Underhill and K. Carneiro, *J. Chem. Soc., Dalton Trans.*, 1983, 1411.
24. M. G. Burnett, *Chem. Soc. Rev.*, 1983, **12**, 267.
25. K. N. Raymond, P. W. R. Corfield and J. A. Ibers, *Inorg. Chem.*, 1968, **7**, 1362.
26. A. Terzis, K. N. Raymond and T. G. Spiro, *Inorg. Chem.*, 1970, **9**, 2415.
27. L. J. Basile, J. R. Ferraro, M. Choca and K. Nakamoto, *Inorg. Chem.*, 1974, **13**, 496.
28. J.-M. Manoli, C. Potvin, J.-M. Brégeault and W. P. Griffith, *J. Chem. Soc., Dalton Trans.*, 1980, 192.
29. G. R. Rossman, F.-D. Tsay and H. B. Gray, *Inorg. Chem.*, 1973, **12**, 824.
30. S. S. Basson, J. G. Leipoldt, L. D. C. Bok, J. S. van Vollenhoven and P. J. Cilliers, *Acta Crystallogr., Sec. B*, 1980, **36**, 1765.
31. M. Laing, G. Gafner, W. P. Griffith and P. M. Kiernan, *Inorg. Chim. Acta*, 1979, **33**, L119.
32. M. B. Hursthouse, A. M. Galas, A. M. Soares and W. P. Griffith, *J. Chem. Soc., Chem. Commun.*, 1980, 1167.
33. L. H. Jones and B. J. Swanson, *Acc. Chem. Res.*, 1976, **9**, 128.
34. R. D. Hancock and A. Evers, *Inorg. Chem.*, 1976, **15**, 995.
35. W. Beck, P. Swoboda, K. Feldl, and E. Schuierer, *Chem. Ber.*, 1970, **103**, 3591.
36. J. U. Nef, *Annalen*, 1894, **280**, 275.
37. K. Schorpp and W. Beck, *Chem. Ber.*, 1974, **107**, 1371.
38. W. Beck, K. Schorpp and C. Oetker, *Chem. Ber.*, 1974, **107**, 1380.
39. W. Beck and F. Lux, *Chem. Ber.*, 1962, **95**, 1683.
40. M. Winnewisser and H. Bodenseh, *Z. Naturforsch., Teil A*, 1967, **22**, 1724.
41. D. Britton and J. D. Dunitz, *Acta Crystallog., Sect. B*, 1965, **19**, 662.
42. J. C. Barrick, D. Canfield and B. C. Giessen, *Acta Crystallogr., Sect. B*, 1979, **35**, 464.
43. U. Nagel, K. Peters, H. G. von Schnering and W. Beck, *J. Organomet. Chem.*, 1980, **185**, 427.
44. W. Beck, C.-J. Oetker and P. Swoboda, *Z. Naturforsch., Teil B*, 1973, **28**, 229.
45. L. Wöhler and A. Berthmann, *Ber. Deutsch. Chem. Ges.*, 1929, **62**, 2748.
46. W. Beck and K. Feldl, *Z. Anorg. Allg. Chem.*, 1965, **341**, 113.
47. B. Klatt and K. Schwabe, *J. Electroanal. Chem.*, 1970, **24**, 61.

12.2
Silicon, Germanium, Tin and Lead

PHILIP G. HARRISON and THAKOR KIKABBAI
University of Nottingham, UK

12.2.1 SILICON, GERMANIUM, TIN AND LEAD SPECIES AS LIGANDS

Complexes of transition metals in which a Group IV metal moiety functions as a donor ligand are quite profuse. Many routes for their synthesis are available, and many types of structure are adopted. The formally tetravalent organometallic $M^{IV}R_xX_{3-x}$ (R = organic group, X = halogen; $x = 1-3$) ligand units function formally as single electron donors, whilst bivalent species $M^{II}X_2$ (X = unidentate or bidentate group) are formally analogous to other more common two-electron donor molecules such as phosphines and arsines. Like carbonyl ligands, however, bivalent molecules of Group IV metals can also function as bridging ligands. The trihalogenometallate ligand may be considered either as an anionic two-electron X_3M^{II-} donor, or as a neutral one-electron X_3M^{IV} ligand, although the similar neutral analogues $MX_2 \cdot L$ (L = a neutral oxygen or nitrogen donor) are formally bivalent. Examples of the structural units which are formed are illustrated in Figure 1.

Complexes of type A are readily obtained by metathesis between a Group IV organometallic

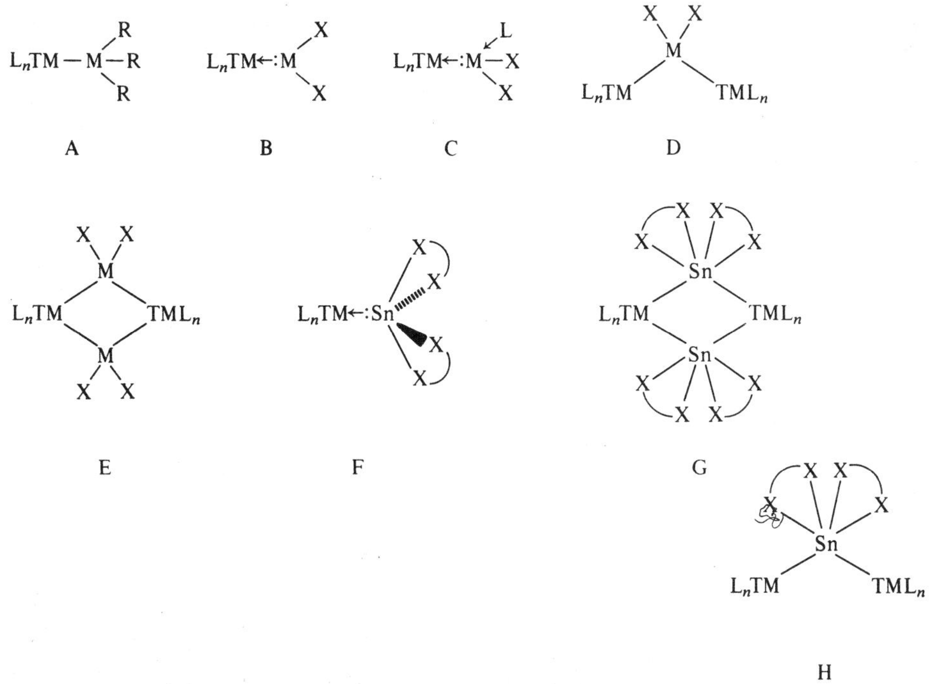

Figure 1 Some structural types adopted by transition metal complexes of Group IV metal ligands

halide and a transition metal carbonyl anion, or by the converse reaction of a triorganometal(IV) lithium, sodium or even mercury reagent and the transition metal halide derivative. These reactions are very versatile, and many such have been characterized, some of which are illustrated in equations (1)–(7).[1-7]

$$Me_3SiCl + Na[Co(CO)_4] \longrightarrow Me_3SiCo(CO)_4 \tag{1}$$

$$2Me_3GeCl + Na_2[Fe(CO)]_4 \longrightarrow cis\text{-}(Me_3Ge)_2Fe(CO)_4 \tag{2}$$

$$3Ph_3SnCl + Na_4[Cr(CO)_4] \longrightarrow Na[(Ph_3Sn)_3Cr(CO)_4] \tag{3}$$

$$Ph_3GeLi + ClAuPPh_3 \longrightarrow Ph_3GeAuPPh_3 \tag{4}$$

$$(Me_3Ge)_2Hg + ClAuPPh_3 \longrightarrow Me_3GeAuPPh_3 + Me_3GeCl + Hg \tag{5}$$

$$Me_3SnCl + M(C_9H_7)(CO)_3 \longrightarrow Me_3SnM(C_9H_7)(CO)_3 \tag{6}$$

$$M = Cr, Mo, W$$

$$Me_3PbCl + Na[FeCp(CO)_2] \longrightarrow Me_3PbFeCp(CO)_2 \tag{7}$$

Oxidative addition of a suitable organo-Group IV compound to low oxidation state transition metal centres has also been employed to produce similar complexes. In particular, triorganosilanes and trichlorosilane have been used widely in this type of reaction (equations 8–11).[8-11] A wide variety of organotin(IV) compounds react similarly by the oxidative addition of the tin–carbon bond to platinum(0) complexes (equation 12).[12] However, addition of the tin–tin bond in hexamethylditin can also participate in this type of reaction, as with $Hf(PhMe)_2(PMe_3)$ (equation 13).[13] In contrast, hexamethyldilead reacts with $Pt(C_2H_4)(PPh_3)_2$ by lead–carbon bond fission to yield cis-$[PtPh(Pb_2Ph_5)(PPh_3)_2]$, which decomposes in solution to cis-$[PtPh(PbPh_3)(PPh_3)_2]$. This latter complex can also be obtained by the reaction of tetraphenyllead. Triphenyllead chloride and bromide similarly react by lead–carbon bond cleavage, but reaction with trimethyllead chloride affords both cis- and trans-$[PtCl(PbMe_3)(PPh_3)_2]$. Trimethyllead(IV) acetate reacts in an entirely different fashion with bis(cyclopentadienyl)-molybdenum and -tungsten dihydrides affording the complexes $[Cp_2HM]_2Pb(O_2CMe)_2$ (M = Mo, W; equation 14).

$$2HSiCl_3 + VCp_2 \longrightarrow VCp_2(SiCl_3)_2 \tag{8}$$

$$Pt(diphos) + HSiCl_3 \longrightarrow Pt(diphos)(H)(SiCl_3) \tag{9}$$

$$2Et_3SiH + [(C_5Me_5)Ir]_2(H)_2Cl_2 \longrightarrow 2(C_5Me_5)Ir(H)_2Cl(SiEt_3) \tag{10}$$

$$[(C_5Me_5)Rh]_2Cl_4 + 4Et_3SiH \longrightarrow 2(C_5Me_5)Rh(H)_2(SiEt_3)_2 \tag{11}$$

$$R_3SnX + Pt(C_2H_4)(PPh_3)_2 \longrightarrow cis\text{-}PtR(SnR_2X)(PPh_3)_2 \tag{12}$$

$$X = Cl, Br, I, NMe_2, O_2CMe, OMe, OSnPh_3, SnPh_3, N(CO)(CH_2)_2(CO)$$

$$Me_3SnSnMe_3 + Hf(PhMe)_2(PMe_3) \longrightarrow (Me_3Sn)_2Hf(PhMe) \tag{13}$$

$$Me_3PbO_2CCMe + MH_2Cp_2 \longrightarrow [CpHM]_2Pb(O_2CMe)_2 + CH_4 + PbMe_4$$
$$+ \text{ other products} \tag{14}$$

Silanediyl complexes (structure type B or C) cannot be prepared directly due to the unavailability of isolable silicon(II) donor molecules. Nevertheless, the diiodo and dimethylsilanediyl complexes (1) and (2) have been obtained by the routes shown in equations (15) and (16).[16,17] These complexes are extremely air sensitive, but complexes in which the $[R_2Si]$ group functions as a bridging ligand between two transition metal centres (structure type G), for example as in (3) and (4), are much more stable.[18] The corresponding germanium, tin and lead heterocycles (5) undergo a base-induced ring cleavage in donor solvents (B) such as tetrahydrofuran and pyridine, in which solvents the equilibrium given in equation (17) involving the metallene complexes (6) (structure type C) is established. The relative propensity for cleavage follows the order Ge > Sn > Pb and pyridine > acetone > tetrahydrofuran > diethyl ether.[19] In the presence of strongly hydridic reagents such as sodium hydride, or by treatment with sodium amalgam, the stannylenes (6; M = Sn) undergo a two-electron reduction to complexes which contain the $[Fe(CO)_4(SnR_2)]^{2-}$ anion.[20]

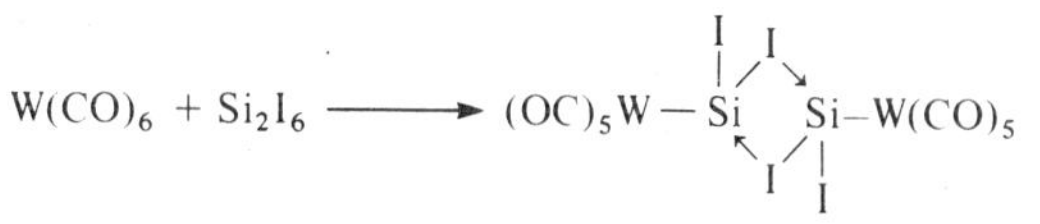

$$W(CO)_6 + Si_2I_6 \longrightarrow (OC)_5W-\underset{\underset{\displaystyle I}{|}}{\overset{\overset{\displaystyle I}{|}}{Si}}\diagdown\underset{\underset{\displaystyle I}{|}}{\overset{\overset{\displaystyle I}{|}}{Si}}-W(CO)_5 \qquad (15)$$

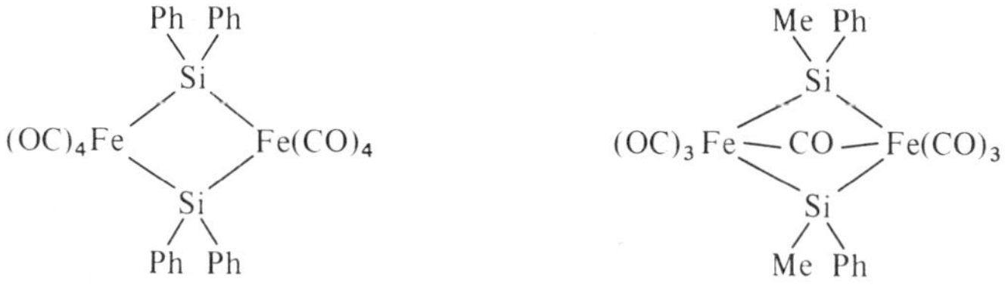

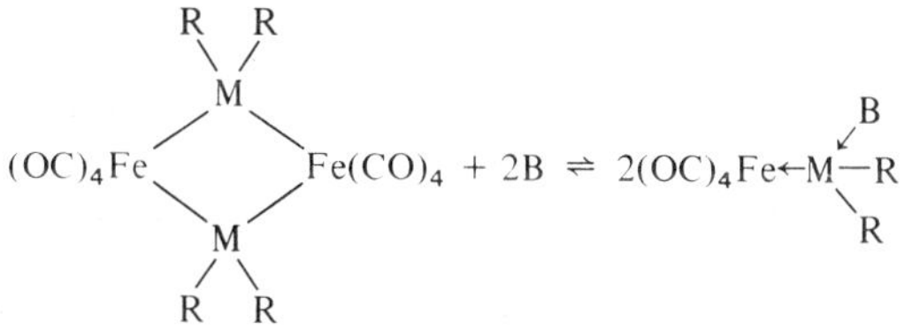

(1)

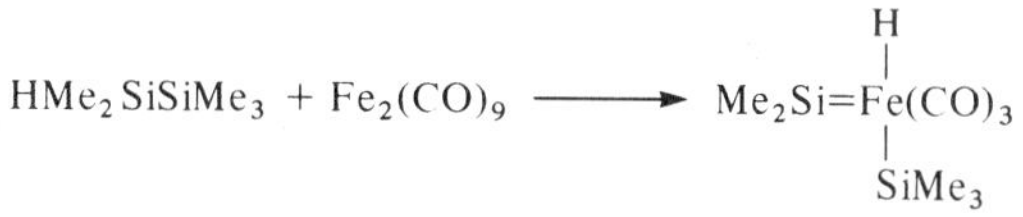

$$HMe_2SiSiMe_3 + Fe_2(CO)_9 \longrightarrow Me_2Si=\underset{\underset{\displaystyle SiMe_3}{|}}{\overset{\overset{\displaystyle H}{|}}{Fe}}(CO)_3$$

(2)

(3)

(4) $\qquad (16)$

(5) M = Ge, Sn, Pb   (6) $\qquad (17)$

Direct synthesis of complexes of structure type B from germylenes, stannylenes and plumbylenes is possible in several cases. Many complexes of the very sterically hindered metallenes M[CH-(SiMe$_3$)$_2$]$_2$ (M = Ge, Sn, Pb) have been obtained (Scheme 1).[21–27]

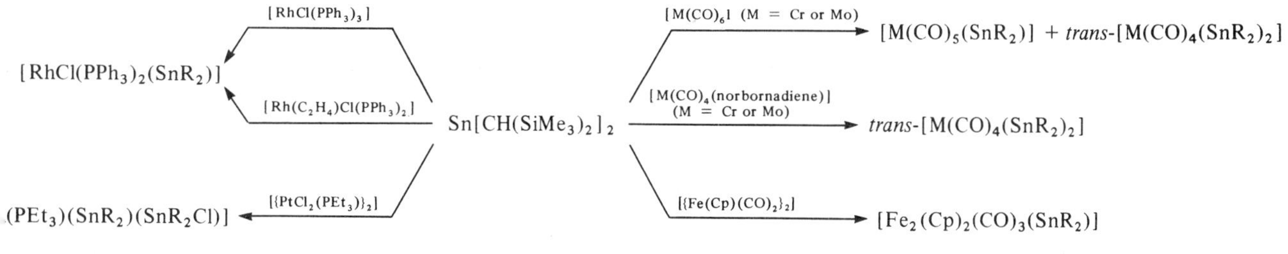

Scheme 1

The structures of one of these complexes, (OC)$_5$Cr ← Sn[CH(SiMe$_3$)$_2$]$_2$,[21] and of the base-stabilized stannylene complex, (OC)$_5$Cr ← SnBu$_2^i$(NC$_5$H$_5$),[28] have been determined and are illustrated in Figure 2. As expected for simple sp^2 hybridization, the former complex has a trigonal planar geometry at tin, but the latter is significantly distorted due to donation of electron density of the nitrogen lone pair into the $5p_x$ orbital at tin. Dicyclopentadienyltin(II) behaves similarly, forming the complexes Cp$_2$Sn → Mn(CO)$_2$(C$_5$H$_4$Me) and Cp$_2$Sn → M(CO)$_5$ (M = Cr, Mo, W),[29,30] and with diiron enneacarbonyl gives the complex [Cp$_2$SnFe(CO)$_4$].[31]

The rather unusual stannylene–platinum complex (7) has been characterized from the reaction of Pt(CO)$_3$(SEt$_2$)(PEt$_3$) with tris(4-tolyl)stannane in methanol. The geometry about the formally

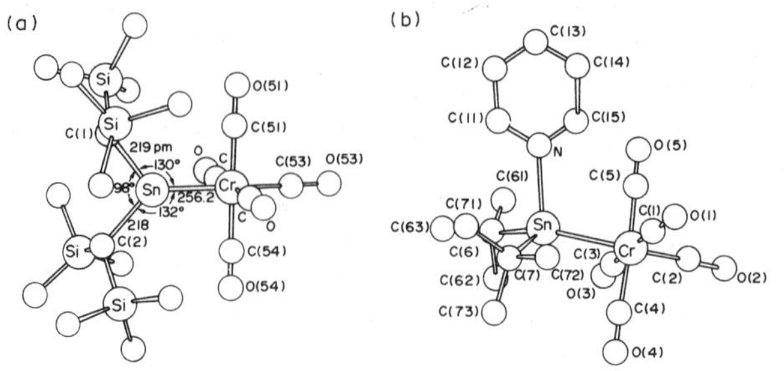

Figure 2 Molecular structures of (a) $Cr(CO)_5[Sn\{CH(SiMe_3)_2\}_2]$ (reproduced by permission from J. D. Cotton, P. J. Davison and M. F. Lappert, *J. Chem. Soc., Dalton Trans.*, 1976, 2275), and (b) $Cr(CO)_5[SnBu_2^t(C_5H_5N)]$ (reproduced by permission from T. J. Marks, *J. Am. Chem. Soc.*, 1971, **93**, 7090)

bivalent tin atom is that of a trigonal bipyramid, with these two tin–oxygen distances longer than the others in the molecule.[32]

Tin(II) bis(β-ketonolates), $Sn[OCRCHCR'(=O)]_2$ (R, R' = Me, CF_3, Ph), are also excellent donor molecules to transition metal carbonyl residues, readily forming the type D structure chromium, molybdenum and tungsten complexes (**8**) under photolysis in tetrahydrofuran,[33] as well as the manganese complexes (**9**).[29] The cobalt complexes (**10**) and (**11**) prepared by the same method are somewhat different, and in these the stannylene functions as a bridging group.[34] The two platinum(0) complexes (**12**) and (**13**) have been obtained from the reaction of tin(II) bis(pentane-2,4-dionate) and $Pt(C_2H_4)(PPh_3)_2$ under different conditions.[35]

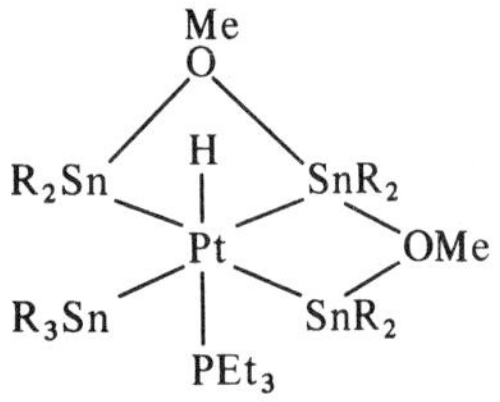

(7)

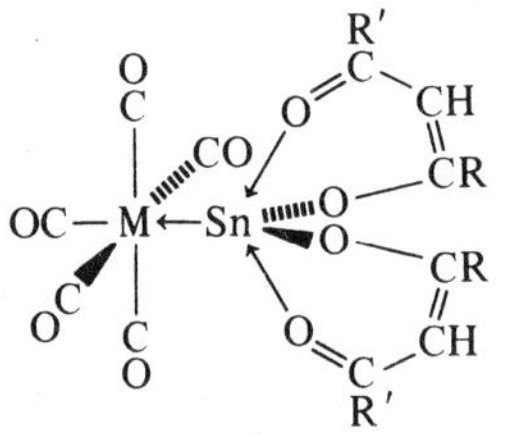

(8)

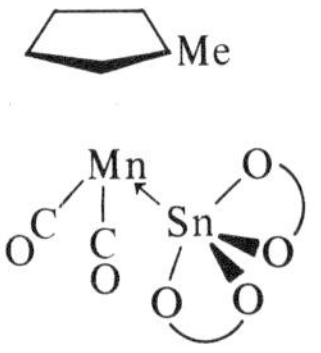

(9)

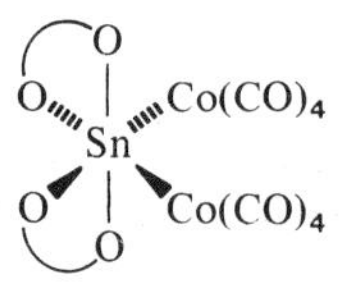

(10)

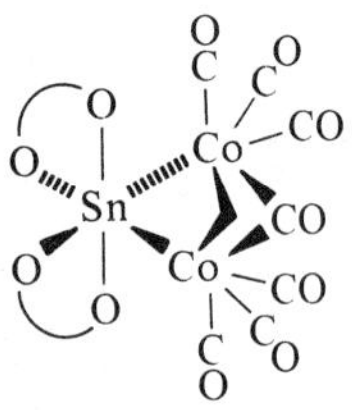

(11)

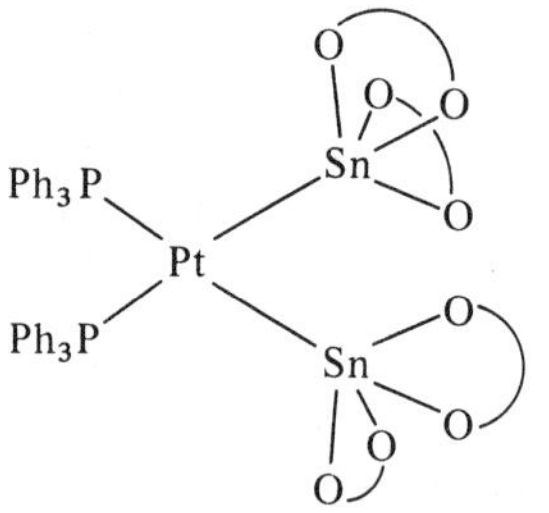

(12)

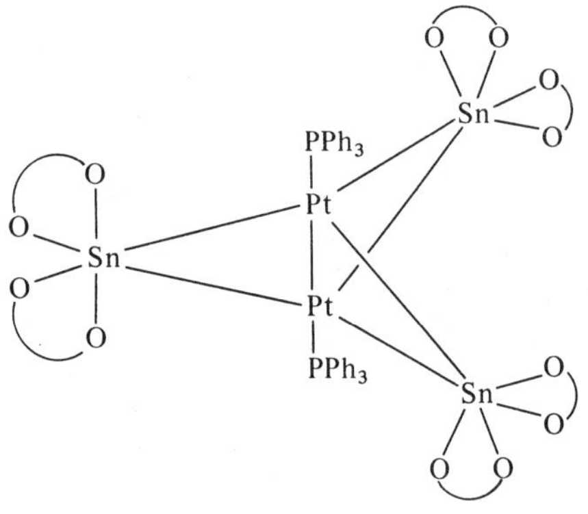

(13)

The reaction of the same tin(II) bis(β-ketoenolates) with diiron enneacarbonyl affords complexes of stoichiometry $[(OC)_4FeSn\{OCRCHCR'(=O)\}_2]$. In chloroform solution, the pentane-2,4-dionate complex is dimeric, presumably with structure type G, but the others are only partially associated, indicating the presence of a dimer $\rightleftharpoons$ monomer equilibrium involving structures G and F similar to that above for the dialkylstannylene–tetracarbonyliron complexes. The extent of dissociation is strongly influenced by the steric bulk of the organic groups R and R', and appears to be essentially complete for the bulky 4-phenylbutane-2,4- and 1,3-diphenylpropane-1,3-dionato complexes. Complete dissociation also occurs upon dissolution in pyridine when base-stabilized monomers are again formed. Varying degrees of dissociation occur in less nucleophilic solvents such as tetrahydrofuran and acetonitrile.[36]

Similar base-stabilized monomers are also formed from the halogenotin complexes

$[X_2SnFe(CO)_4]$ (X = Cl, Br), obtained from tin(II) halides and diiron enneacarbonyl, but have not been isolated. However, analogous halogeno-germanium and -tin complexes of pentacarbonyl-chromium, -molybdenum and -tungsten, $X_2Sn \rightarrow M(CO)_5$ (X = Cl, Br, I; M = Cr, Mo, W), sometimes solvated by tetrahydrofuran or other base at the Group IV metal centre, and the analogous anionic species $[X_3SnM(CO)_5]^-$ have been isolated and characterized.[37-39] Numerous other trihalogenostannate–metal complexes have been synthesized either by substitution or by the insertion of the tin(II) halide into a transition metal–halogen bond. Platinum(II) systems are especially well studied because of their catalytic activity.[40-42] Such solutions contain *inter alia* the anion $[Pt(SnCl_3)_5]^-$, which has a regular trigonal bipyramidal geometry. The structure is retained in solution, but is stereochemically non-rigid due to intramolecular exchange.[43]

Some structural data are listed in Table 1.

Table 1 Structural Data for Compounds with Group IV Metal–Transition Metal Bonds

Compound	CN (M)[a]	CN (M[IV])[b]	M—M[IV] (pm)	Ref.
$[Cp_2TiSiH_2]_2$	4	4	216(1)	c
$Cp_2Nb(CO)SnPh_3$	3	4	282.5(2)	d
$Cr(CO)_5[Sn\{CH(SiMe_3)_2\}_2]$	6	3	256.2(5)	e, f
$Cr(CO)_5[SnBu_2^t(C_5H_5N)]$	6	4	265.4(3)	g
$[Et_4N][(Ph_3Sn)_3Cr(CO)_4]$	7	4	269.5(1)	h
$[Mo(SnCl_3)(CNBu^t)_6]^+[(Ph_3B)_2CN]^-$	7	4	266.3(1)	i
$[W\{C(H)C_6H_4Me-4\}(SnPh_3)(CO)_2Cp]$	5	4	283.7(1)	j
$[Cp_2(H)Mo]_2Pb(O_2CMe)_2$	4	4	280.8(1)	k
$[(Me_3Si)_3Si]Mn(CO)_5$	6	4	256.4(6)	l
$Mn_2(CO)_8(SiPh_2)_2$	7	4	240.2(2)	m
$Me_3SnMn(CO)_5$	6	4	267.4(2)	n
$H_2Sn_2[Mn(CO)_5]_4$	6	4	267	o
			273	
$ClSn[Mn(CO)_5]_3$	6	4	274	p
$(-)-(MeC_5H_4)(CO)_2(1\text{-}NpPhMeSi)(H)Mn$	4	4	246.1(7)	q
$(S)(-)-(1\text{-}NpPhMeGe)Mn(CO)_4[CMe(OEt)]$	6	4	252.4(4)	r
$Cp(CO)_2Mn=Ge[Mn(CO)_2Cp]_2$	5 (×2)	3	236.4(1)	s
	4 (×1)		225.0(1)	
			236.0(1)	
$Cp(CO)_2FeSi(F)Ph_2$	4	4	227.8(1)	t
$Cp(CO)FeSiCl_3$	4	4	221.6(1)	t
$(OC)_4FeSiPh_2CEt=CEtSiPh_2$	6	4 (×2)	240.5(3)	u
			241.8(2)	
$[Me_2SnFe(CO)_4]_2$	6	4	264.7(8)	v
$[Cp_2SnFe(CO)_4]_2$	6	4	265.1(1)	w
			267.0(1)	
$[MeSn\{Fe(CO)_4\}_2]_2Sn$	6	4 (×3)	262.5(8)	x
			274.7(8)	
$Me_2Sn[Fe(CO)_2Cp]_2$		4	260.5(4)	y
$Cl_2Sn[Fe(CO)_2Cp]_2$		4	249.2(8)	z
$Ru(SiMe_3)(CO)_2[C_8H_8(SiMe_3)]$	5	4	242.4(2)	aa
$(-)-(C_6H_6)Ru(SnCl_3)(Me)[Ph_2PNHCH(Me)Ph]$		4	254.3(1)	bb
$Os_3SnH_2(CO)_{10}[CH(SiMe_3)_2]_2$	5	5	264.5(3)	cc
			285.5(3)	
$HOs_3(S)(SCH_2)(CO)_7(PMe_2Ph)(SnMe_3)$	7 (×1)	4	265.3(1)	dd
	6 (×2)			
$[NEt_4]_3[SiCo_9(CO)_{21}][Co(CO)_4]$	8	8	231.4(2)	ee
			228.3(2)	
			252.7(4)	
$1\text{-}NpPhMeGeCo(CO)_4$	5	4	245.8	ff
$[(Me_2Sn)CoCp(CO)]_2$	4	4	253.8(5)	gg
			254.4(5)	
$BrSn[(Co(CO)_4]_3$	5	4	260.2(6)	hh
$[Sn_2Co_5Cl_2(CO)_{19}][CoHB(pz)_3]_2$	5	4	250.8(1)	ii
			266.8(1)	
			265.6(1)	
$(C_5Me_5)Rh(H)(SiEt_3)_2$	5	4	237.9(2)	jj
$(Me_3Ge)(H)_2(CO)(PPh_3)_2Ir \cdot \frac{1}{2}C_6H_6$	4	4	248.4(2)	kk
$(+)-PtCl[SiMe(1\text{-}Np)Ph](PMe_2Ph)_2$	4	4	231.7(4)	ll
$Cp(GeCl_3)(PPh_3)Ni \cdot \frac{1}{2}C_6H_6$	3	4	224.8(1)	mm
trans-$[Pt(SnCl_3)_2\{P(OPh)_3\}_2]$	4	4	259.9(2)	nn

Table 1 Structural Data for Compounds with Group IV Metal–Transition Metal Bonds

Compound	CN (M)[a]	CN (M^IV)[b]	M—M^IV (pm)	Ref.
[Ph₄As][Pt(SnCl₃)₃(1,5-cod)]	5	4	256.8(2)	oo
			254.6(2)	
			264.3(2)	
cis-[PtPh(PbPh₃)(PPh₃)₂]	4	4	269.8(9)	pp

[a] Coordination number at the transition metal.
[b] Coordination number at the Group IV metal.
[c] G. Heneken and E. Weiss, *Chem. Ber.*, 1973, **106**, 1747.
[d] Yu. V. Skripkin, O. G. Volkov, A. A. Pasynskii, A. K. Antsyshkina, I. M. Dikareva, V. N. Ostrikova, M. A. Porai-Koshits, S. I. Davydova and S. G. Sakharov, *J. Organomet. Chem.*, 1984, **263**, 345.
[e] J. D. Cotton, P. J. Davidson, D. E. Goldberg, M. F. Lappert and K. M. Thomas, *J. Chem. Soc., Chem. Commun.*, 1974, 893.
[f] J. D. Cotton, P. J. Davidson and M. F. Lappert, *J. Chem. Soc., Dalton Trans.*, 1976, 2275.
[g] M. D. Brice and F. A. Cotton, *J. Am. Chem. Soc.*, 1973, **95**, 4529.
[h] J. T. Lin, G. P. Hagen and J. E. Ellis, *Organometallics*, 1984, **3**, 1288.
[i] C. M. Giandomenico, J. C. Dewan and S. J. Lippard, *J. Am. Chem. Soc.*, 1981, **103**, 1407.
[j] G. A. Carriedo, D. Hodgson, J. A. K. Howard, K. Marsden, F. G. A. Stone, M. J. Went and P. Woodward, *J. Chem. Soc., Chem. Commun.*, 1982, 1006.
[k] M. M. Kubicki, R. Kergoat and J.-E. Guerchais, *J. Chem. Soc., Dalton Trans.*, 1984, 1791.
[l] B. K. Nicholson, J. Simpson and W. T. Robinson, *J. Organomet. Chem.*, 1973, **47**, 403.
[m] G. L. Simon and L. F. Dahl, *J. Am. Chem. Soc.*, 197, **95**, 783.
[n] R. F. Bryan, *J. Chem. Soc (A)*, 1968, 696.
[o] K. D. Bos, E. J. Bulten, J. G. Noltes and A. L. Spek, *J. Organomet. Chem.*, 1974, **71**, C52.
[p] J. H. Tsai, J. J. Flynn and F. P. Boer, *Chem. Commun.*, 1967, 702.
[q] F. Carre, E. Colomer, R. J. P. Corriu and A. Vioux, *Organometallics*, 1984, **3**, 1272.
[r] F. Carre, G. Cerveau, E. Colomer and R. J. P. Corriu, *J. Organomet. Chem.*, 1982, **229**, 257.
[s] D. Melzer and E. Weiss, *J. Organomet. Chem.*, 1984, **263**, 67.
[t] U. Schubert, G. Kraft and E. Walther, *Z. Anorg. Allg, Chem.*, 1984, **549**, 96.
[u] F. H. Carre and J. J. E. Moreau, *Inorg. Chem.*, 1982, **21**, 3099.
[v] C. J. Gilmore and P. Woodward, *J. Chem. Soc., Dalton Trans.*, 1972, 1387.
[w] P. G. Harrison, T. J. King and J. A. Richards, *J. Chem. Soc., Dalton Trans.*, 1975, 2097.
[x] R. M. Sweet, C. J. Fritchie and R. A. Schunn, *Inorg. Chem.*, 1967, **6**, 749.
[y] B. P. Biryukov and Yu. T. Struchkov, *J. Struct. Chem. (Engl. Transl.)*, 1968, **9**, 412.
[z] J. E. Connor and E. R. Corey, *Inorg. Chem.*, 1967, **6**, 968.
[aa] P. J. Harris, J. A. K. Howard, S. A. R. Knox, R. J. McKinney, R. P. Phillips, F. G. A. Stone and P. Woodward, *J. Chem. Soc., Dalton Trans.*, 1978, 403.
[bb] J. D. Korp and I. Bernal, *Inorg. Chem.*, 1981, **20**, 4065.
[cc] C. J. Cardin, D. J. Cardin, H. E. Parge and J. M. Power, *J. Chem. Soc., Chem. Commun.*, 1984, 609.
[dd] R. D. Adams and D. A. Katahira, *Organometallics*, 1982, **1**, 460.
[ee] K. M. Mackay, B. K. Nicholson, W. T. Robinson and A. W. Sims, *J. Chem. Soc., Chem. Commun.*, 1984, 1276.
[ff] F. Dahan and Y. Jeannin, *J. Organomet. Chem.*, 1977, **135**, 251.
[gg] J. Weaver and P. Woodward, *J. Chem. Soc., Dalton Trans.*, 1973, 1060.
[hh] R. D. Ball and D. Hall, *J. Organomet. Chem.*, 1973, **52**, 293.
[ii] O. J. Curnow and B. K. Nicholson, *J. Organomet. Chem.*, 1984, **267**, 257.
[jj] M.-J. Fernandez, P. M. Bailey, P. O. Bentz, J. S. Ricci, T. F. Koetzle and P. M. Maitlis, *J. Am. Chem. Soc.*, 1984, **106**, 5458.
[kk] N. A. Bell, F. Glockling, M. Schneider, H. M. M. Shearer and M. D. Wilbey, *Acta Crystallogr., Sect. C*, 1984, **40**, 625.
[ll] C. Eaborn, P. B. Hitchcock, D. J. Tune and D. R. M. Walton, *J. Organomet. Chem.*, 1973, **54**, C1.
[mm] N. A. Bell, F. Glockling, A. McGregor, M. L. Schneider and H. M. M. Shearer, *Acta Crystallogr., Sect. C*, 1984, **40**, 623.
[nn] A. Albinati, P. S. Pregosin and H. Ruegger, *Inorg. Chem.*, 1984, **23**, 3223.
[oo] A. Albinati, P. S. Pregosin and H. Ruegger, *Angew. Chem., Int. Ed. Engl.*, 1984, **23**, 78.
[pp] B. Crociani, M. Nicolini, D. A. Clemente and G. Bandoli, *J. Organomet. Chem.*, 1973, **49**, 249.

12.2.2 REFERENCES

1. B. K. Nicholson and J. Simpson, *J. Organomet. Chem.*, 1978, **155**, 237.
2. A. Bonny and K. M. Mackay, *J. Organomet. Chem.*, 1978, **144**, 389.
3. J. T. Lin, G. P. Hagen and J. E. Ellis, *Organometallics*, 1984, **3**, 1288.
4. F. Glockling and K. A. Hooton, *J. Chem. Soc.*, 1962, 2658.
5. F. Glockling and M. D. Wilbey, *J. Chem. Soc. (A)*, 1968, 2168.
6. A. N. Nesmeyanov, N. A. Ustynyuk, L. N. Novikova, T. N. Rybina, Yu. A. Ustynyuk, Yu. F. Oprunenko and O. I. Trifonova, *J. Organomet. Chem.*, 1980, **184**, 63.
7. K. H. Pannell, *J. Organomet. Chem.*, 1980, **198**, 37.
8. A. M. Cardoso, R. J. H. Clark and S. Moorhouse, *J. Organomet. Chem.*, 1980, **186**, 241.
9. R. N. Haszeldine, R. V. Parish and D. J. Parry, *J. Chem. Soc. (A)*, 1969, 683.
10. M.-J. Fernandez, P. M. Maitlis, *J. Chem., Soc., Dalton Trans.*, 1984, 2063.
11. M.-J. Fernandez, P. M. Bailey, P. O. Bentz, J. S. Ricci, T. F. Koetzle and P. M. Maitlis, *J. Am. Chem. Soc.*, 1984, **106**, 5458.
12. G. Butler, C. Eaborn and A. Pidcock, *J. Organomet. Chem.*, 1980, **185**, 367.
13. F. G. N. Cloke, K. P. Cox, M. L. H. Green, J. Bashkin and K. Prout, *J. Chem. Soc., Chem. Commun.*, 1981, 1117.
14. T. A. K. Al-Allaf, G. Butler, C. Eaborn and A. Pidcock, *J. Organomet. Chem.*, 1980, **188**, 335.
15. M. M. Kubicki, R. Kergoat, J.-E. Guerchais and P. L'Haridon, *J. Chem. Soc., Dalton Trans.*, 1984, 1791.
16. G. Schmid and R. Boese, *Chem. Ber.*, 1972, **105**, 3306.
17. H. Sakurai, Y. Kamiyama and Y. Nakadaira, *Angew. Chem., Int. Ed. Engl.*, 1978, **17**, 674.

18. F. H. Carre and J. L. E. Moreau, *Inorg. Chem.*, 1982, **21**, 3099.
19. T. J. Marks and A. R. Newman, *J. Am. Chem. Soc.*, 1973, **95**, 769.
20. B. A. Sosinsky, J. Shelly and R. Strong, *Inorg. Chem.*, 1981, **20**, 1370.
21. J. D. Cotton, P. J. Davidson and M. F. Lappert, *J. Chem. Soc., Dalton Trans.*, 1975, 2275.
22. D. E. Goldberg, D. H. Harris, M. F. Lappert and K. M. Thomas, *J. Chem. Soc., Chem. Commun.*, 1976, 261.
23. P. J. Davidson, D. H. Harris and M. F. Lappert, *J. Chem. Soc., Dalton Trans.*, 1976, 2268.
24. P. B. Hitchcock, M. F. Lappert, S. J. Miles and A. J. Thorne, *J. Chem. Soc., Chem. Commun.*, 1984, 480.
25. J. D. Cotton, P. J. Davidson, M. F. Lappert, J. D. Donaldson and J. Silver, *J. Chem. Soc., Dalton Trans.*, 1976, 2286.
26. J. D. Cotton, P. J. Davidson, D. E. Goldberg, M. F. Lappert and K. M. Thomas, *J. Chem. Soc., Chem. Commun.*, 1974, 893.
27. M. F. Lappert, S. J. Miles, P. P. Power, A. J. Carty and N. J. Taylor, *J. Chem. Soc., Chem. Commun.*, 1977, 458.
28. M. D. Brice and F. A. Cotton, *J. Am. Chem. Soc.*, 1973, **95**, 4529; see also T. J. Marks, *J. Am. Chem. Soc.*, 1972, **93**, 7090.
29. A. B. Cornwell and P. G. Harrison, *J. Chem. Soc., Dalton Trans.*, 1976, 1054.
30. A. B. Cornwell, J. A. Richards and P. G. Harrison, *J. Organomet. Chem.*, 1976, **108**, 47.
31. P. G. Harrison, T. J. King and J. A. Richards, *J. Chem. Soc., Dalton Trans.*, 1975, 2097.
32. J. F. Almeida, K. R. Dixon, C. Eaborn, P. B. Hitchcock and A. Pidcock, *J. Chem. Soc., Chem. Commun.*, 1982, 1315.
33. A. B. Cornwell and P. G. Harrison, *J. Chem. Soc., Dalton Trans.*, 1975, 1486.
34. A. B. Cornwell and P. G. Harrison, *J. Chem. Soc., Dalton Trans.*, 1976, 1608.
35. G. W. Bushnell, D. T. Eadie, A. Pidcock, A. R. Sam, R. D. Holmes-Smith, S. R. Stobart, E. T. Brennan and T. S. Cameron, *J. Am. Chem. Soc.*, 1982, **104**, 583.
36. A. B. Cornwell and P. G. Harrison, *J. Chem. Soc., Dalton Trans.*, 1975, 2017.
37. D. Uhlig, H. Behrens and E. Lindner, *Z. Anorg. Allg. Chem.*, 1973, **401**, 233.
38. T. Kruck and H. Breuer, *Chem. Ber.*, 1974, **107**, 263.
39. T. Kruck, F. J. Becker, H. Breuer, K. Ehlert and W. Rother, *Z. Anorg. Allg. Chem.*, 1974, **405**, 95.
40 G. K. Anderson, H. C. Clark and J. A. Davies, *Inorg. Chem.*, 1983, **22**, 427.
41. G. K. Anderson, H. C. Clark and J. A. Davies, *Inorg. Chem.*, 1983, **22**, 434.
42. G. K. Anderson, C. Billard, H. C. Clark, J. A. Davies and C. S. Wong, *Inorg. Chem.*, 1983, **22**, 439.
43. J. M. Nelson and N. W. Alcock, *Inorg. Chem.*, 1982, **21**, 1196.

13.1
Ammonia and Amines

DONALD A. HOUSE

University of Canterbury, Christchurch, New Zealand

13.1.1   INTRODUCTION

Ligands containing nitrogen donor atoms transcend the conventional boundaries between organic and inorganic chemistry and their classification has often caused difficulties. Mellor's 'Comprehensive Treatise of Inorganic Chemistry'[1] restricts nitrogen ligands to ammonia (ammine complexes) and hydrazine as all other related nitrogen donor systems contain carbon atoms and were regarded as part of organic chemistry. Traditional organic texts regard amines as derivatives of ammonia in which one or more hydrogen atoms are replaced by alkyl or aromatic groups.

To a coordination chemist there is little distinction between ammonia and an organic amine. Both have the all important nitrogen donor atom and thus have the potential to behave similarly. Regrettably, structural information for ammine complexes is not included in the Cambridge Crystallographic Data File,[2] as simple salts lack the necessary carbon atom.

Much of the background chemistry of amines (*e.g.* synthetic methods and physical properties) can be found in standard organic chemistry reference works[3-6] or encyclopedias[7,8] and there is at least one journal[9] devoted to research in this field.

The interaction of the lone pair on the nitrogen atom of an amine with a metal ion spans a considerable range from the very weak 'template assisted' association of alkali and alkaline earth cations[10] to the very robust Co^{III} and Ru^{III} metal nitrogen bonds. In aqueous solution, most amines are basic (Chapter 4) and there will always be competition between the lone pairs on oxygen (OH^- or H_2O) and the lone pair on the nitrogen, for the metal center. This competition can be overcome to a certain extent by using non-aqueous systems or the free amine as a solvent.[11,12]

It is perhaps convenient to classify complexes with nitrogen donors into 'labile' and 'inert' systems, with 'inert' being defined as systems which allow geometric isomers. Another attempt at classification is the hard-acid, soft-base concept[13] with the amine nitrogen being considered as a 'hard' center, as it binds strongly to protons and is thus expected to bind strongly to 'hard' metal centers (good lone pair acceptors). In classifying metal centers in this way, it is usual to compare the accessibility or stability of metal complexes of nitrogen donor ligands with those of analogous phosphorus donor ligands (a typical 'soft' center) (Chapter 14). In both methods of classification, the boundaries are not sharp and the formation of a particular metal amine complex depends to a considerable extent on the nature of the ligand.

Frequently, amine complexes are designed to illustrate some subtle mechanistic or stereochemical effect in coordination chemistry and much ingenuity has been exercised in the design of specific amine ligands. One example, from the many that could be quoted, is the use of $PdCl(1,1,7,7-Et_4-dien)^+$ and $PdCl(1,1,7,7-Et_4-4-Me-dien)^+$ to highlight the importance of a ligand N—H proton in the mechanism of base hydrolysis.[14]

Coordination of the amine nitrogen lone pair to a central metal ion completes the tetrahedral stereochemistry about the nitrogen. Thus, prochiral nitrogen bases (with three different substituents) now become chiral. For labile complexes, inversion about the nitrogen center is generally rapid, but for inert complexes, both geometric and optical isomers can arise.[15,16] While such complexes have not yet been resolved for monodentate amines (suitable aliphatic tertiary amines[17] are not easily coordinated to an inert metal center), the phenomenon is more widespread than is generally recognized with saturated polyamine ligands. In systems such as (1) or (2), inversion (racemization or isomerization) can take place in basic solution *via* a deprotonation–protonation mechanism. These reactions are the simplest type of reaction of a coordinated amine ligand and proton exchange rates have been measured for a number of inert transition-metal amine complexes.[18,19] With suitable central metals, *e.g.* Pt^{IV}, the deprotonated intermediate can be isolated and characterized.[20]

(1a)   Geometric isomers of *mer*-CoCl(en)(3,3-tri)$^{2+}$ [15]   (1b)

(2a) (RR)   Optical isomers of *trans*-MCl$_2$(3,2,3-tet)$^+$ [16]   (2b) (SS)

Other reactions of coordinated amine ligands continue to be discovered, *e.g.* condensation with aldehydes, ketones or nitriles to give new macrocyclic (Chapter 12.2) or linear polyamine ligands,[21] dehydrogenation to give imines (Chapter 13.8),[22,23] the 'capping' of hexamines to give multicompartmental ligands (Chapter 21.3), and *in situ* N-alkylation.[24]

The above examples are ones involving modification of existing amine complexes, but there are also cases where amine ligands can be formed from non-amine precursors: the oxidation of NCS^- coordinated to Co^{III} gives $Co—NH_3$[25] and the reaction of $Pt^{II}—N_3$ with an alkyne gives a coordinated triazole (Chapter 13.5).[26]

Amines containing a chiral carbon atom in the aliphatic residue attached to the nitrogen atom are of considerable interest to the coordination chemist as, when coordinated, these ligands can induce chirality in the metal–ligand chromophore. The recent compilation[27] on the methods of optical resolution of more than 1000 amines and amino alcohols (Chapter 20.3) is an excellent resource.

Although single-crystal X-ray diffraction continues to provide coordination chemists with the most unambiguous structural data in terms of bond lengths and angles, there has been considerable recent use of ^{13}C NMR spectroscopy for isomer characterization of diamagnetic metal amine complexes. Geometric isomerism (Chapter 5) of the *cis/trans* type for diamines, *mer/fac* for tri-amines and *trans/asym-fac/sym-fac* for tetramines can be readily distinguished[28–31] although the assignment of one particular geometry, in the absence of any other isomer, is often more difficult.[32] ^{15}N NMR spectroscopy could also have considerable potential in this aspect of coordination chemistry.[33]

Another approach to the problem of the structural assignment of isomers formed by Co^{III} and linear polyamines is the use of 'structure–reactivity' patterns. In particular it has been observed that in chloropentaaminecobalt(III) complexes containing a 'planar' (meridional)—$NH(CH_2)_x NH(CH_2)_x NH$—moiety, base hydrolysis is (Chapter 7.1) several orders of magnitude more rapid than in isomeric analogs with this feature absent.[15,34]

Traditionally, isomers formed by polyamines and inert metal centres have been separated by fractional crystallization. More recently, the techniques of column chromatography have been used, especially for Co^{III} hexamines.[31,35] Both geometric and optical isomers can be separated by judicious use of eluting agents and under favorable conditions isomers occurring at the 1% level can be isolated.

There are certain classes of coordination complexes that are particularly well characterized with amine ligands. Among these are the μ-peroxo and μ-superoxo Co^{III} complexes with general formula $(N_5)Co(O_2)Co(N_5)^{n+}$, where N_5 can be a variety of amine nitrogen donors[36–42] and the tri-μ-hydroxo Co^{III} systems $(N_3)Co(OH)_3Co(N_3)^{3+}$.[43]

Another particularly interesting class of amine complex that has had relatively little recent attention is the clathrate series $M(X)_2(A) \cdot Y$, where $M = Ni^{II}$ or Cd^{II}, $X = CN^-$ or NCS^-, $A = NH_3$ or $\frac{1}{2}(NH_2(CH_2)_x NH_2)$ and Y is the host, *e.g.* benzene, thiophene, furan, pyrrole, aniline or phenol.[44,45]

This section would not be complete without a brief mention of the use of molecular mechanics calculations in coordination chemistry, as these have been applied mainly to amine complexes.[46,47] While most calculations relate to complexes of Co^{III}, these are now being extended to complexes with other metal centres, *e.g.* Ni^{II}.[48]

13.1.1.1 Nomenclature

The standard organic nomenaclature for di- and poly-amine ligands is not always strictly applied and many trivial names abound, especially for simple diamine ligands, *e.g.* propylenediamine ($NH_2CH(Me)CH_2NH_2$), stilbenediamine ($NH_2CH(Ph)CH(Ph)NH_2$) and isobutylenediamine ($NH_2C(Me)_2CH_2NH_2$). This in turn leads to a variety of ligand abbreviations, *e.g.* $NH_2(CH_2)_3NH_2$ is sometimes tmd (trimethylenediamine) or tn (by analogy with en for ethylenediamine, $NH_2(CH_2)_2NH_2$).

Most polyamine ligands form coordination compounds with five- or six-membered ring systems and this has led to a system of ligand abbreviations, where linear polyamines are described by the number of carbon atoms between linking nitrogen atoms. Thus, triethylenetetramine, $NH_2(CH_2)_2NH(CH_2)_2NH(CH_2)_2NH_2$ (trien) can be described as 2,2,2-tet. Alternatively, if a five-membered ring is related to en and a six-membered diamine ring to tn, then trien can be described as enenen and the polyamine $NH_2(CH_2)_3NH(CH_2)_2NH(CH_2)_3NH_2$ (3,2,3-tet) as tnentn.

The introduction of substituents, either at carbon or nitrogen causes additional complications, and, for example, it is not always clear just where the four methyl substituents are located when a ligand is described as tetramethylethylenediamine.

For ease of systematization, we will regard all chelate diamine ligands with the two nitrogen donors separated by —$(CH_2)_2$— as related to ethylenediamine (en). These generally form five-membered chelate rings and the atoms will be numbered consecutively N(1)—C(2)—C(3)—N(4).

Similarly, all diamine ligands with three aliphatic carbon atoms between the nitrogen donors will be related to $NH_2(CH_2)_3NH_2$ (tn), forming six-membered rings and numbered N(1)—C(2)—C(3)—C(4)—N(5). This will be extended to the basic triamine chains $NH_2(CH_2)_2NH(CH_2)_2NH_2$ (dien), $NH_2(CH_2)_2NH(CH_2)_3NH_2$ (2,3-tri) and $NH_2(CH_2)_3NH(CH_2)_3NH_2$ (3,3-tri) and the tetramine chains $NH_2(CH_2)_2NH(CH_2)_2NH(CH_2)_2NH_2$ (trien), $NH_2(CH_2)_2NH(CH_2)_3NH(CH_2)_2NH_2$ (2,3,2-tet) and $NH_2(CH_2)_3NH(CH_2)_2NH(CH_2)_3NH_2$ (3,2,3-tet).

Where cyclic aliphatic ring systems are fused to the carbon chain, these will be indicated as —$(CH_2)_n$—. Thus $(\pm)$-*trans*-cyclohexane-1,2-diamine [$(\pm)$-chxn] becomes $(\pm)$-2,3-$(CH_2)_4$-en.

For optically active ligands of known absolute configuration, the *R,S* nomenclature of Cahn, Ingold and Prelog[49] will be used. If the absolute configuration is unknown, the sign of rotation at the sodium D line (589.1 nm) is generally sufficient to characterize the chirality. It should be pointed out that, for amines, the sign of rotation is highly pH dependent and amine hydrochlorides will not, in general, have the same sign as the parent amine.

In most cases, ligand abbreviations will be those accepted by custom and usage.

13.1.2 AMMONIA AND AMMINES

Complexes with ammonia as the only nitrogen-containing ligand are called ammines; those with nitrogen donor ligands, which may be regarded as organic derivatives of ammonia, are referred to as amine complexes. The properties of the free ligand have been well described[50] and a recent review on the vibrational modes available for NH_3 has appeared.[51] The early work on ammines has been comprehensively summarized by Mellor[1] and there are few cations which do not show some sort of interaction with the NH_3 molecule.

Coordination chemists are, however, generally interested in those systems with definite coordination numbers. The maximum coordination number is often achieved in a transition-metal ammine complex, especially where synthetic routes involve the use of excess ligand, as the gas, aqueous solution or the anhydrous liquid. Even so, forcing conditions or catalysts are required to form such complexes as $Cu(NH_3)_6^{2+}$ or $Co(NH_3)_6^{3+}$.

Liquid ammonia[50] (b.p. $-33.35\,°C$) is an excellent solvent for many coordination complexes and a number of mechanistic studies are now being conducted in this medium.[52]

There is an extensive literature on ammines and the Chemical Abstracts 10th Collective Index (1977–81) has four pages under this heading. One important parameter for this class of compound is the metal–nitrogen bond length and Table 1 lists some of the available data. Transition metal–NH_3 bond energies have also been calculated.[100]

Ammine complexes often predominate in the diet of undergraduate synthetic preparations and

Table 1 Metal—NH_3 Bond Distances (Å)

		Refs.
Co^III (389.3)[a]		
$[Co(NH_3)_6](CrO_4)Cl\cdot3H_2O$	1.960(4)	53
$[Co(NH_3)_6]CdCl_5$	1.960(6)	54, 55
$[Co(NH_3)_6]I_3$	1.936(15)	56, 57
$[Co(NH_3)_6]ZnCl_4\cdot Cl$	1.968(11)	58
$[Co(NH_3)_6][Co(CN)_6]$	1.972(1)	59
$[Co(NH_3)_6][Cr(CN)_6]$	1.970(3)	60
$[Co(NH_3)_6]SnCl_4\cdot Cl$	1.969(10)	61
$[Co(NH_3)_6][Na(edta-4H)]\cdot3.5H_2O$	1.966(6)	62
$[Co(NH_3)_6]TlCl_6$	2.07	63
$[Co(NH_3)_6]TlBr_6$	2.02	63
$[Co(NH_3)_6]Cl_3$	1.963	64
$[Co(NH_3)_6][Pb_4Cl_{11}]$		65
$[Co(NH_3)_6][HgCl_5]$	1.960(6)	66a, 66b
$[Co(NH_3)_6]CuCl_5$		67
$[Co(NH_3)_6]I_3I_4$	1.968(13)	68
Co^II (186.4)[a]		
$[Co(NH_3)_6]Cl_2$	2.114(9)	69
Cr^III		
$[Cr(NH_3)_6]CuCl_5$	2.0644(25)	70, 71
$[Cr(NH_3)_6][Ni(CN)_5\cdot2H_2O]$	2.080(4)	72
$[Cr(NH_3)_6]CuBr_5$	2.0592(57)	73
$[Cr(NH_3)_6][MnCl_5(OH_2)]$	2.072(4)	74
$[Cr(NH_3)_6][HgCl_5]$		75, 66b

Table 1 (*continued*)

[Cr(NH$_3$)$_6$]ZnCl$_4$·Cl		2.071(3)	76a
[Cr(NH$_3$)$_6$]MnF$_6$		2.067(3)	76b
[Cr(NH$_3$)$_6$]FeF$_6$		2.073(2)	76b
	Ru^{III}		
[Ru(NH$_3$)$_6$](BF$_4$)$_3$		2.104(4)	77
	Ru^{II}		
[Ru(NH$_3$)$_6$]I$_2$		2.144(4)	77
	V^{III}		
(NH$_4$)$_2$[V(NH$_3$)Cl$_5$]		2.138	78
	Pt^{II}		
[Pt(NH$_3$)$_4$][TCNQ]			79
cis-Pt(NH$_3$)$_2$(NO$_3$)$_2$		2.00	80
[(NH$_3$)$_2$Pt(OH)$_2$Pt(NH$_3$)$_2$]CO$_3$		2.02	80
[Pt(NII$_3$)$_4$]PtCl$_4$		2.06	81
	Pt^{II}–Pt^{IV}		
[Pt(NH$_3$)$_4$][Pt(2,6-Me$_2$-py)Cl$_3$]$_2$		2.055	82
[Pt(NH$_3$)$_4$][Pt(NH$_3$)$_4$Br$_2$](HSO$_4$)$_4$		2.059	83
	Pd^{II}		
[Pd(NH$_3$)$_4$]I$_8$			84
	Cu^{II} (213.5)^a		
[Cu(NH$_3$)$_6$]Cl$_2$		eq 2.07(7)	85
		ax 2.62(11)	
[Cu(NH$_3$)$_6$]Br$_2$		eq 2.15(8)	85
		ax 2.45(28)	
K[Cu(NH$_3$)$_5$(PF$_6$)$_3$		eq 2.00, 2.048	86
		ax 2.193	
Cu(NH$_3$)$_4$S$_2$O$_6$		2.05	87
Cu(NH$_3$)$_4$SO$_4$·H$_2$O		2.031(6)	88
Cu(NH$_3$)$_4$SeO$_4$		2.005(9)	88
Cu(NH$_3$)$_4$(NO$_3$)$_2$		2.017(9)	89
[Cu(NH$_3$)$_4$]I$_4$		2.025(8)	90
[Cu(NH$_3$)$_4$]I$_6$		2.04(1)	90
Na$_4$[Cu(NH$_3$)$_4$][Cu(S$_2$O$_3$)$_2$]$_2$		1.994	91
	Cu^I (128.0)^a		
[Cu(NH$_3$)$_2$][Ag(NCS)$_3$]		2.00	92
	Zn^{II} (218.9)^a		
[Zn(NH$_3$)$_4$](I$_3$)$_2$		2.05(1)	93
	Mg^{II} (176.8)^a		
[Mg(NH$_3$)$_6$](ClO$_4$)$_2$			94
	Cd^{II} (191.7)^a		
Cd(NH$_3$)$_6^{2+}$		2.37	95
[Cd(NH$_3$)$_4$]I$_4$		2.341(8)	96
[Cd(NH$_3$)$_4$]I$_6$		2.319(5)	96
	Ag^I (106.6)^a		
[Ag(NH$_3$)$_2$][Ag(NO$_2$)$_2$]		2.114	97
	Fe^{III}		
Fe—N (est)		2.00	98
	Fe^{II}		
Fe—N (est)		2.15	98
	Ni^{II} (197.3)^a		
Ni(NH$_3$)$_6^{2+}$		2.15	99

^{a} Numbers in parentheses are the (M^{n+}—NH$_3$) bond energy in kJ mol^{-1}.[100]

Table 2 Typical Ammine Complexes[101]

$[Ag(NH_3)_2]_2SO_4$	$[Cr(NH_3)_6]Cl_3$
$Hg(NH_3)_2Cl_2$	$[Co(NH_3)_6]Cl_3$
$Co(NH_3)_4Cl_2$	$[Co(NH_3)_5Cl]Cl_2$
$[Co(NH_3)_6]Cl_2$	$Ni(NH_3)_4(NO_2)_2$
$[Cu(NH_3)_4]Cl_2$	$[Ni(NH_3)_6]Cl_2$
$[Zn(NH_3)_4](BF_4)_2$	$[Cr(NH_3)_5(OH_2)](NO_3)_2$
$[Cr(NH_3)_5Cl]Cl_2$	$[Co(NH_3)_4CO_3]NO_3$
trans-$[Co(NH_3)_4Br_2]Br$	$[Cr(NH_3)_4ox]NO_3$
cis-$[Co(NH_3)_4(NO_2)_2]Cl$	$[Cr(NH_3)_3(OH_2)Cl_2]Cl$
trans-$[Co(NH_3)_4(NO_2)_2]NO_3$	$[Co(NH_3)_3(OH_2)Cl_2]Cl$
mer-$Co(NH_3)_3(NO_2)_3$	*trans*-$NH_4[Cr(NH_3)_2(NCS)_4]$
$NH_4[Co(NH_3)_2(NO_2)_4]$	$Na_3[Fe(CN)_5(NH_3)]$
$[(NH_3)_4Co(OH)_2Co(NH_3)_4]Br_4$	

Table 2 lists a selection given by Schlessinger.[101] Naturally, these reflect complexes of relatively inexpensive central metal ions. An increasingly important use of transition metal ammines is in the stabilization and characterization of unusual anions, *e.g.* polyiodides (Table 1).

The $M(NH_3)_5X^{n+}$ systems (M = Co, Cr)[15] have particular utility in providing many examples for the study of substitution and inner-sphere electron transfer reactions[102] (Chapter 7.2) and recent work has shown that the $Co(NH_3)_5$ group has excellent potential as a protecting group in the synthesis of peptides.[103] The ammine chemistry of Co[III],[15] Cr[III],[11,12,15] Ru[III],[104] Ru[II],[105,106] Os[II],[105] Group IIIA[107] and Group IIA[108] has been reviewed.

13.1.3 MONOAMINE LIGANDS

The replacement of six NH_3 ligands about an octahedral metal center by six alkylamine groups apparently results in serious steric interaction, as examples of such complexes, *e.g.* $Ru(NH_2R)_6^{2+}$,[109] are rare. More common are the $MX(NH_2R)_5^{n+}$ systems with M = Co[III],[110–114] Cr[III][115,116] and Rh[III].[113] Steric effects are proposed to account for the very much faster rates of base hydrolysis in $CoCl(NH_2R)_5^{2+}$ when compared to $CoCl(NH_3)_5^{2+}$. A variety of conformational arrangements can be postulated for $MCl(NH_2R)_5^{2+}$[117] or $Pt(NH_2R)_4^{2+}$.[118] Structure (3) shows the 'spiderleg' conformation adopted in the mixed valent Pt[II]–Pt[IV] complex with R = Et,[118] but in $[Pt(NH_2Et)_4][PtCl_4]$ the 'legs' are folded.[119] The structure of *trans*-$[RhI_2(az)_4]I$ is also known (az = aziridine).[120]

$$C-C\overset{}{\underset{C-C}{\diagdown}}N\longrightarrow Pt\overset{N}{\diagup}\diagdown\overset{C-C}{\diagdown}N\diagdown\overset{}{\underset{C-C}{}}$$

(3)[118]

Octahedral systems with only one coordinated alkylamine are well known for Co[III]. These include the $Co(NH_3)_5(NH_2R)^{3+}$[112–123] and *cis*-$CoX(en)_2(NH_2R)^{2+}$ cases.[15] The former can be prepared using $Co(NH_3)_5(DMSO)^{3+}$ and a DMSO solution of the alkylamine, and the latter from the reaction of an aqueous slurry of *trans*-$[CoCl_2(en)_2]Cl$ with the alkylamine (the Meisenheimer reaction).[15] Other *cis*-$CoX(N_4)(NH_2R)^{n+}$ complexes have recently been prepared where N_4 = trien[124] or $(tn)_2$[125] and the *trans* analogues where N_4 = 3,2,3-tet.[126]

Co[III] complexes with two coordinated alkylamine groups are illustrated by $CoX(dien)(NH_2R)_2^{n+}$[127] and $Co(NH_2Me)_2(NO_2)_4^-$,[128] and a series of octahedral Ni[II] complexes of general formula $Ni(dien)(L)(ONO)_2(L = NH_2R)$ have been prepared, but the geometric configuration remains unknown.[129]

In addition to the $[Pt(NH_2R)_4][PtCl_4]$ (R = Me, Et) (Magnus' green analogues[119]) in square planar systems, complexes of the type $M(acac)(L_2)^+$ (M = Pd, Pt) are known for a variety of monoamines.[130,131]

One amine that has potentially four coordination sites is hexamethylenetetramine.[132] In most complexes, however, this acts as a terminally bonded monodentate ligand (Mn[II],[133] Cd[II],[134] Ni[II],[132] Co[II][132]) or a bidentate ligand bridging between two metal atoms (Cu[II], Zn[II], Cd[II]).[132]

Excessive alkyl substitution on the N-donor atoms can result in the isolation of amine complexes with unusually low coordination numbers, *e.g.* Cr[III]$(NPr^i_2)_3$.[135]

Table 3 Diamine Ligands

Formula	Name	Formula	Name	Formula	Name
NH_2NH_2	Hydrazine				
$NH_2CH_2NH_2$	Methanediamine				
$NH_2(CH_2)_2NH_2$	Ethylenediamine (en)	$\begin{array}{cc} Me & Me \\ NH_2-C-C-NH_2 \\ H & H \end{array}$	2,3-Diaminobutane (sbn)		*trans*-Cyclohexane-1,2-diamine (chxn)
$NH_2(CH_2)_3NH_2$	Trimethylenediamine (tn)				
$NH_2(CH_2)_4NH_2$	Putrescine (bn)				
$NH_2(CH_2)_5NH_2$	Cadaverine	$\begin{array}{cc} Ph & Ph \\ NH_2-C-C-NH_2 \\ H & H \end{array}$	Stilbenediamine (stien)		*trans*-Cyclopentane-1,2-diamine (cptn)
$NH_2(CH_2)_6NH_2$	Hexamethylenediamine				

13.1.4 DIAMINE LIGANDS

13.1.4.1 Introduction

The simplest diamine, hydrazine[136] N_2H_4 (Table 3), is normally available as the monohydrate. An X-ray structure of the crystalline solid has been determined[137] (there is some doubt as to the correct space group[138]) as well as an electron diffraction study of the vapor.[139,140] The structures of the di(hydrogen fluoride),[141] di(hydrogen chloride)[141] and monoperchlorate crown ether[142] salts are also known. The N—N distance in hydrazine (1.499 Å) is an important parameter, as one half this distance is used for the covalent radius of the single-bonded nitrogen atom.

Hydrazine can act as either a monodentate or bridging ligand with IR spectroscopy being used as the main method of distinguishing between these bonding modes.[146–148]

Methanediamine ($NH_2CH_2NH_2$) is hardly mentioned in standard textbooks of organic chemistry and at least one casts doubts on its existence. Despite this, salts of the diamine are quite stable and easily prepared (equation 1).[149,150] The potential of this diamine to act as a ligand has yet to be exploited.

$$(CH_2O)_n + HCO(NH_2) \longrightarrow CH_2(NHCHO)_2 \xrightarrow{HCl} NH_2CH_2NH_2 \cdot 2HCl \qquad (1)$$

On the other hand, *N*-tetraalkyl derivatives[151,152] are quite well known (equation 2) but only in one case have ligands of this type been reported to bind in the bidentate chelating mode to a central atom[153] to form a four-membered ring. The more normal use of such compounds, as ligands, is in carbene formation[154] *via* metal–central carbon atom bond.

$$2Et_2NH + CH_2O \longrightarrow Et_2NCH_2NEt_2 \qquad (2)$$

13.1.4.2 en Systems

Ethylenediamine (1,2-diaminoethane), once described as 'God's gift to the coordination chemist', is commercially available either as the anhydrous amine (m.p. 8.5 °C) or the monohydrate (m.p. 10 °C). Both forms are clear fuming liquids at room temperature and these readily absorb carbon dioxde from the atmosphere. The vapor can cause an asthmatic allergic reaction after repetitious exposure.

The free ligand, and its salts, can adopt two staggered configurations, *trans* (4) and *gauche* (*cis*) (5), with an energy difference of about $4 \, kJ \, mol^{-1}$. Structural studies and molecular mechanics calculations show that when the *gauche* form is adopted (Table 4), the dihedral angle varies from 63 to 70°. The (+)-(*R,R*)-tartrate salts of ethylenediamine have been especially well studied, as crystals of $[NH_3(CH_2)_2NH_3][(+)\text{-tartrate}(-2)]$ exhibit a marked piezo-electric effect.

(4) *trans*   (5) *gauche*

Despite the fact that ethylenediamine is by far the most widely studied bidentate amine in the chelating mode, there are a number of examples where it, and other diamines, act in the monodentate or bridging[178,179] forms. One of the earliest examples is in the Pt^{IV} complex[180] $[Pt_2(en)_3Me_6]I_2$ (6) with two bidentate chelates per Pt, one bridging ethylenediamine, and a crystallographic center of symmetry. Other examples are shown in (7)–(10).[181–185]

(6)[180]   (7)[181]

Table 4 Configurations Adopted by Ethylenediamine and the Ethylene-diammonium Cation

Anion	Configuration	Refs.
None ($-60\,^\circ$C)	trans	155
None (gas)	gauche	156
Mol. mech. calcn.	gauche	157, 158
$2Cl^-$	trans	159, 160
$2Br^-$	trans	161
SO_4^{2-}	gauche	162
$PdCl_4^{2-}$	trans	163
$CuCl_4^{2-}$	trans	164, 165
$Pb_2Cl_6^{2-}$	trans	166
$CoCl_4^{2-} \cdot 2Cl^-$	gauche	167
$CuBr_4^{2-} \cdot 2Br^-$	gauche	168
$[Au(SO_3)_2en]^{2-}$	trans	169
Citrate(-2)	gauche	170
($+$)-Tartrate(-2)	trans	171–174
$[(+)$-Tartrate($-1)]_2$	gauche	175
Mol. mech. calcn.[a]	trans	176
Mol. mech. calcn.[b]	trans	177

[a] 1,4-Me_2-en. [b] 1,1,4,4-Me_4-enH_2^{2+}

(8)[182] (9)[183] (10)[183a, 184, 185]

Chelate complexes of ethylenediamine provide many of the examples on which the theories of coordination chemistry have been founded. Co^{III} complexes of this ligand were studied by Alfred Werner and his students[186] and the separation of *cis*-CoCl(en)$_2$(NH$_3$)$^{2+}$ into its optical enantiomers[187] was a key factor in establishing octahedral stereochemistry.

Two nitrogen donors, separated by two carbon atoms bound to a central metal ion, M, form a five-membered puckered ring. This ring has a two-fold axis of symmetry (through M and the mid-point of the C—C bond) as shown in (11). In an undistorted five-membered ring, the C—C bond lies at an angle of $\pm48.8°$ (the dihedral angle) with respect to the N—M—N plane. Two enantiomeric conformations are possible: one labeled λ ($=k$) (left handed helicity) and the other δ ($=k'$) (right handed helicity). The $\lambda \rightleftharpoons \delta$ ring inversion barrier is low[188] ($\sim 20\,\mathrm{kJ\,mol^{-1}}$)[189] and it is not possible to resolve such complexes as Co(en)(NH$_3$)$_4^{3+}$. The barrier does increase, however, with increasing *N*-alkylation.[19]

(11a) δ (11b) λ

Bidentate mono(ethylenediamine) complexes are quite well known in Cr^{III} chemistry,[11] *e.g.* *trans*-CrF$_2$(en)(OH$_2$)$_2^+$.[191]

When the above considerations are extended to bis(ethylenediamine) chelate systems, with the en rings *trans* to each other, three stereochemical possibilities arise: $\delta\delta$, $\lambda\lambda$ and $\delta\lambda$. The first two are enantiomeric and energetically equivalent and, because the —NH$_2$ protons are in the staggered conformation, they are estimated to be about $4\,\mathrm{kJ\,mol^{-1}}$ more stable than the diastereoisomeric $\delta\lambda$ form (with an eclipsed —NH$_2$ proton configuration). However, in the solid state the majority of *trans*-bis(ethylenediamine) complexes adopt the slightly less stable $\delta\lambda$ form, perhaps reflecting the importance of crystal packing forces (Table 5). Crystallographers from Finland have determined the structure of a considerable number of Ni^{II} and Cu^{II} M(diamine)$_2^{2+}$ complexes.

Table 5   Ring Conformations in Some $M(en)_3^{n+}$, *trans*-$M(en)_2XY^{n+}$ and $M(en)_2^{n+}$ Complexes

Complex	Configuration	Refs.
	Co^{III}	
$(\pm)$-$[Co(en)_3]Cl_3 \cdot 2.8H_2O$	*lel lel lel*	a
$(+)$-$[Co(en)_3]Cl_3$	$\Lambda\ \delta\delta\delta$	a
$(\pm)$-$[Co(en)_3]Br_3$	*lel lel lel*	a
$(+)$-$[Co(en)_3]Br_3$	$\Lambda\ \delta\delta\delta$	a
$(\pm)$-$[Co(en)_3]I_3$	*lel ob ob*	192
$(\pm)$-$[Co(en)_3](NCS)_3$	*lel lel lel*	a
$(+)$-$[Co(en)_3](NCS)_3$	Λ	a
$(+)$-$[Co(en)_3](NO_3)_3$	$\Lambda\ \delta\delta\delta$	a
$(+)$-$[Co(en)_3]Cl_3 \cdot 0.5NaCl$	$\Lambda\ \delta\delta\delta$	a
$(\pm)$-$[Co(en)_3][SnCl_3]Cl$	*lel ob ob*	a
$(\pm)$-$[Co(en)_3][Pb_2Cl_9]Cl$	*lel lel ob*	193
$(\pm)$-$[Co(en)_3]_2[CdCl_6]Cl_2$	*lel lel lel*	194
$(\pm)$-$[Co(en)_3]_2[Cu_2Cl_8]Cl_2$	*lel ob ob*	195
$(\pm)$-$[Co(en)_3][Cr(CN)_5NO]$	*lel ob ob*	196
$(+)$-$[Co(en)_3][(+)$-$tartrate(-2)]Br$	$\Lambda\ \delta\delta\delta$	197
$(+)$-$[Co(en)_3][(+)$-$tartrate(-2)]Cl \cdot 5H_2O$	$\Lambda\ \delta\delta\delta$	198
$(\pm)$-$[Co(en)_3][H_3A(-1)][H_2A(-2)] \cdot 2.5H_2O^b$	*lel lel lel*	199
$(\pm)$-$[Co(en)_3][Fe(CN)_6]$	*lel lel lel*	a
$[(+)$-$Co(en)_3 \cdot (-)$-$Cr(en)_3](NCS)_6$	$\Lambda\ \delta\delta\delta \cdot \Lambda\lambda\lambda(0.5\lambda)(0.5\delta)$	a
$[(+)$-$Co(en)_3 \cdot (-)$-$Cr(en)_3]Cl_6$	$\Lambda\ \delta\delta\delta \cdot \Delta\ \lambda\lambda\lambda$	a
$(\pm)$-$[Co(en)_3]_2[HPO_4]_3 \cdot 9H_2O$	*lel lel lel*	200
$[Co(en)_2Cl_2]Cl \cdot HCl \cdot 2H_2O$	$\delta\lambda$	201
$[Co(en)_2Cl_2]NO_3$	$\delta\lambda$	202
$[Co(en)_2Cl_2]TlCl_4$	$\delta\lambda$	203
$[Co(en)_2Br_2]Br \cdot HBr \cdot 2H_2O$	$\delta\lambda$	204
$\overline{Co(en)_2(CN)_2]Cl}$	$\delta\lambda$	205
$[Co(en)_2(NO_2)_2]NO_3$	$\delta\lambda$	206
$[Co(en)_2(NCS)(ONO)]I$	$\delta\delta$	207
$[Co(en)_2(NCS)(ONO)]ClO_4$	$\lambda\lambda$	207
$[Co(en)_2(NCS)(NO_2)]I$	$\lambda\lambda$	207
$[Co(en)_2(NCS)(NO_2)]ClO_4$	$\lambda\lambda$	207
$[Co(en)_2(NCS)(NO_2)]NCS$	$\delta\lambda$	208
$[Co(en)_2(A)(Cl)]Cl^c$	$\delta\lambda$ and $\delta\delta$	209, 210
$[Co(en)_2(imid)(SO_3)]ClO_4$	$\delta\lambda$	211
$[Co(en)_2(NH_3)(SO_3)]ClO_4$	$\delta\lambda$	212
	Cr^{III}	
$(\pm)$-$[Cr(en)_3]Cl_3 \cdot 3H_2O$	*lel lel lel*	a
$(+)$-$[Cr(en)_3]Cl_3 \cdot 2H_2O$	$\Lambda\ \delta\delta\delta$	a
$(\pm)$-$[Cr(en)_3]Br_3$		a
$(+)$-$[Cr(en)_3]Br_3 \cdot 0.6H_2O$	$\Lambda\ \delta\delta\lambda$	a
$(\pm)$-$[Cr(en)_3]I_3$	*lel ob ob*	a
$(\pm)$-$[Cr(en)_3](NCS)_3 \cdot 0.75H_2O$	*lel lel* (0.7 *lel*, 0.3 *ob*)	a
$(+)$-$[Cr(en)_3](NCS)_3$	$\Lambda\ \delta\lambda\lambda$	213
$(-)$-$[Cr(en)_3](NCS)_3$	$\Delta\ \lambda\delta\delta$	a
$(\pm)$-$[Cr(en)_3][Co(CN)_6]$	*ob ob ob*	a
$(\pm)$-$[Cr(en)_3][Ni(CN)_5]$	*lel lel ob, lel ob ob*	214
$Li(+)$-$[Cr(en)_3][(+)$-$tartrate(-2)]_2$	$\Lambda\ \delta\delta\delta$	197
$[(+)$-$Cr(en)_3 \cdot (+)$-$Rh(en)_3]Cl_6$	$\Lambda\ \delta\delta\delta \cdot \Lambda\ \lambda\lambda\lambda$	a
$(\pm)$-$[Cr(en)_3][MoO(OH)(CN)_4]$	*lel lel lel*	215
$[Cr(en)_2(OH_2)_2][Cr(en)_2(OH)F](ClO_4)_5$	$\delta\lambda, \delta\lambda$	216
	Rh^{III}	
$(\pm)$-$[Rh(en)_3]Cl_3 \cdot 3H_2O$	*lel lel lel*	a
$(-)$-$[Rh(en)_3]Cl_3$	$\Lambda\ \delta\delta\delta$	a
$(-)$-$[Rh(en)_3]Br_3$	$\Lambda\ \delta\delta\delta$	217
	Ru^{III}	
$(\pm)$-$[Ru(en)_3]Cl_3$	*lel lel lel*	218a
	Ni^{II}	
$(\pm)$-$[Ni(en)_3][NO_3]_2$	*lel lel lel*	219
$(+)$-$[Ni(en)_3][NO_3]_2$	$\Lambda\ \delta\delta\delta$	220
$(\pm)$-$[Ni(en)_3](SO_4)$	*lel lel lel*	221, 222
$(\pm)$-$[Ni(en)_3][BPh_4]_2 \cdot 3DMSO$	*ob lel lel*	223
$[Ni(en)_2(NCS)_2]$	$\delta\lambda$	224
$[Ni(en)_2(ONO)]_n$	$\delta\lambda$	225

Complex	Configuration	Refs.
[Ni(en)$_2$(ONO)$_2$]	$\delta\lambda$	226
[Ni(en)$_2$][AgI$_2$]$_2$	$\delta\lambda$	227
[Ni(en)$_2$][AgBr$_2$]$_2$	$\delta\lambda$	227
	Cu^{II}	
[Cu(en)$_3$]Cl$_2$ · 0.75en	*lel lel lel*	228
[Cu(en)$_3$]SO$_4$	*lel lel lel*	228, 229
[Cu(en)$_2$](NO$_3$)$_2$	$\delta\lambda$	230
[Cu(en)$_2$](ClO$_4$)$_2$	$\delta\lambda$	231
[Cu(en)$_2$](BF$_4$)$_2$	$\delta\lambda$	232
[Cu(en)$_2$(SCN)$_2$]	$\delta\lambda$	233
[Cu(en)$_2$][Hg(SCN)$_4$]	$\delta\lambda$	234
[Cu(en)$_2$(OH$_2$)Cl]Cl	$\delta\lambda$	235
	Pd^{II}	
[Pd(cn)$_2$]Cl$_2$	$\delta\lambda$	236
	Pt^{II}	
[Pt(en)$_2$(crown ether)]$_n$[PF$_6$]$_{2n}$	$\delta\lambda$	237
[Pt(en)$_2$][R,R-(+)-tartrate(−2)]	$\delta\lambda$	238
	Pt^{II}–Pt^{IV}	
[Pt(en)$_2$][Pt(en)$_2$I$_2$](ClO$_4$)$_4$	$\delta\lambda, \delta\lambda$	239, 240
[Pt(en)$_2$]$_3$[Pt(en)$_2$Cl$_2$]$_3$[CuCl$_4$]$_4$	$\delta\lambda, \delta\lambda$	241, 242
	Pt^{IV}	
[Pt(en)$_2$Br$_2$][C$_4$O$_4$] · 2H$_2$O	$\delta\lambda$	243
[Pt(en)$_2$I$_2$]$_2$[Ag$_2$I$_6$]	$\delta\lambda$	244
[Pt(en)$_2$I$_2$]I$_2$ · 2H$_2$O	$\delta\lambda$	245
	Zn^{II}	
[Zn(en)$_3$]Cl$_2$ · H$_2$O	*ob ob ob*	766
	Ru^{II}	
[Ru(en)$_3$]ZnCl$_4$	*lel lel* (0.74 *ob*, 0.26 *lel*)	218b

[a] References to M(en)$_3^{n+}$ complexes are given in Table 1 of Ref. 246. [b] A = tetrahydrofurantetracarboxylate. [c] A = theophylline(−1).

Chelate complexes with two ethylenediamine rings in a *cis* configuration lack a plane of symmetry and thus have the potential to be separated into enantiomeric (Δ, Λ) (**12**) forms. Inert *cis*-bis(en) complexes of Co^{III},[247] Cr^{III}[11] or Rh^{III}[248] can be resolved by the 'method of racemic modification'[249] or using chromatographic techniques,[35] but labile systems, such as Ni(en)$_3^{2+}$, which occasionally crystallize in one chiral form,[220] rapidly racemize in solution.

(12a) Δ (12b) Λ (13a) *lel*$_3$ (13b) *ob*$_3$

Rather more theoretical work has been done with respect to tris(ethylenediamine) complexes. By considering the two stable chiral (skew) conformations of the five-membered chelate rings (designated λ and δ as above), four unique conformers of M(en)$_3^{n+}$ can be constructed for each absolute configuration (Δ or Λ) (**12**): $\lambda\lambda\lambda$, $\lambda\lambda\delta$, $\lambda\delta\delta$, and $\delta\delta\delta$. In this representation, $\Delta(\lambda\lambda\lambda)$ is energetically equivalent to $\Lambda(\delta\delta\delta)$.

Alternatively, a chirality invariant nomenclature can be used, in which the two possible conformations of an individual chelate ring in M(en)$_3$ are named *lel* and *ob*, respectively, where the central C—C bond of the ring is (approximately) paral*lel* and *ob*lique with respect to the C_3 or pseudo-C_3 symmetry axis in M(en)$_3^{n+}$ (**13**). Thus the four conformers are symbolized as *lel*$_3$, *lel*$_2$*ob*, *lelob*$_2$ and *ob*$_3$, with *lel*$_3$ being equivalent to $\Delta(\lambda\lambda\lambda)$ [or $\Lambda(\delta\delta\delta)$].

With regard to conformation change, the three rings can move independently of each other, and the saddle point of each chelate ring is a symmetrical envelope, with C atoms 0.7 Å from N—M—N plane. The barrier height is 30 kJ mol^{-1} for the *ob* → *lel* process and 35 kJ mol^{-1} for the reverse.[250,251] For $Co(en)_3^{3+}$ the lel_3 form is calculated to be more stable than the ob_3 form by about 7.6 kJ mol^{-1}.[47]

In the solid state, the actual ring conformation adopted (Table 5) in a variety of $M(en)_3^{n+}$ systems depends on the anion used to crystallize the salt and is apparently related to the degree of hydrogen bond formation.[246] Attempts to distinguish the individual ring conformers using IR spectroscopy appear to be most encouraging.[252]

C or N substituents on these five-membered puckered rings can be either axial or equatorial and the bulk of accumulated crystallographic evidence suggests that the equatorial configuration is favored[47] for octahedral systems. Thus $Co(NH_3)_4(R,S\text{-pn})^{3+}$ should consist of equal amounts of $Co(NH_3)_4(R\text{-pn-}\lambda)^{3+}$ and $Co(NH_3)_4(S\text{-pn-}\delta)^{3+}$ (14).[188]

(14a) λ-(*R*)-pn   (14b) δ-(*S*)-pn

Bis(propylenediamine) complexes with the four nitrogen atoms in the square planar configuration usually adopt the *trans* methyl (equatorial) configuration[253-255] (15), but in $Pt(R\text{-pn})_2^{2+}$ one methyl group is axial (16).[256]

(15)[253]   (16)[256]

The introduction of more than one *C*- or *N*-substituted en ring into the coordination sphere considerably increases the isomeric complexity. Many of these ligands are now unsymmetrical and their complexes may exhibit geometric isomerism dependent upon the end-for-end orientation.[257] Thus there are 24 distinct configurational and conformational forms expected for $Co(R,S\text{-pn})_3^{3+}$.[258] Table 6 lists a variety of *C*- and *N*-substituted ethylenediamine type ligands that have been investigated.

In 1969 it was discovered that *cis*-dichlorodiammineplatinium(II) exhibited potent antitumor activity (Chapter 62.2) and could be used in treating human cancers. This has prompted the synthesis of hundreds of platinium(II) compounds in the hopes of finding safer and perhaps more effective analogues.[320] Consequently, many new *C*- and *N*-substituted diamine ligands have been synthesized, including the combination of estrogens (known antitumor agents for cancers) with ethylenediamine, to form $[PtCl_2(\text{estrone-NH(CH}_2)_2\text{NH}_2)]$.[293]

Among the more widely studied 2,3-*C*-disubstituted ligands related to ethylenediamine are *s*bn and stien[518] (Table 3). These ligands can exist in *meso* and *racemo* forms and the properties (*e.g.* solubility) of transition metal complexes (especially Co^{III}) with the two forms of this type of ligand are often sufficiently different as to allow amine isomer separation to be achieved without separation of the amine mixture. *Meso* and *racemic s*bn are prepared together by the reduction of dimethylglyoxime and can be separated by fractional crystallization of the dihydrochloride salts. There are, however, stereospecific synthetic routes available for the *meso* and *racemo* forms of stien (Table 6).

Once the racemic diamine has been resolved,[27] there is the problem of assigning the absolute configuration. This can be achieved by classical organic transformations from starting materials of known absolute configuration, by single-crystal X-ray methods using a transition metal complex and the Bijvoet technique[247] or, less reliably, by comparing the chiroptical properties of a transition metal complex of the chiral ligand with those of a related ligand of known absolute configuration.[519] Thus $(-)_{589}$-stien has been assigned (after some confusion[300]) the *S,S* configuration[313,377] and $(-)_{589}$-*s*bn the *R,R* absolute configuration.[300]

Table 6 *C*-Substituted and *N*-Substituted Diamine Ligands

$en = N\!-\!C\!-\!C\!-\!N$ 1 2 3 4	Refs.
1-Me-en-	
Co^{III}	188, 259–276
Ni^{II}	267, 276, 277
Rh^{III}	278
Cu^{II}	279–281
Pt^{II}	188, 274, 282–285
1-Et-en-	
Co^{III}	265, 269, 276, 286, 287
Rh^{III}	278
Cu^{II}	288, 289
Pt^{II}	274
Ni^{II}	277
1-Pr^n-en-	
Co^{III}	260, 274, 276, 286, 287
Cu^{II}	268, 288
Pt^{II}	274
1-Pr^i-en-	
Co^{III}	287, 290
1-Bu^n-en-	
Co^{III}	287, 290
1-Bu^i-en-	
Co^{III}	287, 290
1-Hx^n-en-	
Co^{III}	287, 290
1-Bz-en-	
Cu^{II}	291, 292
Ni^{II}	291
1-Cyclohex-en-	
Co^{III}	287
1-Ph-en-	
Pt^{II}	274
1-Estrone-en-	
Pd^{II}	293
Pt^{II}	293
2-Me-en-(R,S)- [pn]	
Co^{III}	273, 294–296
Pt^{II}	274, 282, 283
Cu^{II}	253, 268
Cr^{III}	11, 12, 15
Ni^{II}	297
$(-)$-2-Me-en-(R)- [$(-)$-(R)-pn-]	
Co^{III}	296, 298–304
Rh^{III}	305, 306
Fe^{II}	22
Cu^{II}	292, 307
Pt^{II}	256, 308, 309
Pt^{IV}	306, 310
Pd^{II}	309
Pd^{IV}	306
Ni^{II}	267
Cr^{III}	311

Table 6 *(continued)*

$en = N—C—C—N$ $\quad\quad 1 \quad 2 \quad 3 \quad 4$	Refs.
(+)-2-Me-en-(S)- [(+)-(S)-pn-] Co^{III} Rh^{III} Cr^{III}	270, 296, 301, 312–314 314 314
2-Bu^t-en-(R,S)- Pt^{II}	188
(+)-2-Bu^t-en-(S)- Ni^{II} Co^{III} Pt^{II}	267, 315 315–317 315
(−)-2-Bu^t-en-(R)- Co^{III}	188
2-Pr^n-en- Ni^{II}	318
2-Pr^i-en-(S)- Co^{III}	313
2-Bu^i-en-(S)- Co^{III}	313
2-Bu^s-en-(S)- Co^{III}	313
2-Bz-en-(S)- Co^{III}	313
2-Ph-en-(R,S)- Pt^{II}	319 320
(−)-2-Ph-en-(R)- Pt^{II} Pd^{II} Co^{III} Mo^0	321, 322 309 313, 317, 321–323 322
2-Et-en-(R,S)-	319
2-(3-pentane)-en-(R,S)- Pt^{II}	320
2-(Cyclohexylmethyl)-en-(R,S)- Pt^{II}	320
2,4-Me$_2$-en-(S)- Pt^{II} Co^{III} Pt^{IV}	282, 324, 325 270, 326 325
2-Me-4-Et-en-(S)- Co^{III}	325
1,2-Me$_2$-en-(R)- Cu^{II}	292
1,2-Me$_2$-en-(S)- Co^{III}	270, 326

Table 6 *(continued)*

$en = N{-}C{-}C{-}N$ $\quad\;1\quad 2\quad 3\quad 4$	*Refs.*
1-Et-2-Me-en-(R,S)- Co^{III}	21
1-Bz-2-Me-en-(R,S)- Co^{III}	21
1-Bz-2-Me-en-(R)- Cu^{II}	289
1-(4-ChloroBz)-2-Me-(en)-(R,S)- Co^{III}	21
4,4-Et$_2$-2-Me-en-(S)- Pd^{II}	327
1,4-Bz$_2$-2-Me-(R)- Cu^{II}	289
1,2,4-Me$_3$-en-(R)- Cu^{II}	289
1,2,4-Me$_3$-en-(S)- Co^{III}	326
1,2,4-Me$_3$-en-(R,S)- Co^{III}	270
2-Me-2-Bu^i-en-(R,S)- Pt^{II}	320
2,2-(CH$_2$)$_5$-en- Pt^{II} Cu^{II} Cr^{III} Co^{III}	320 318 318 318
2,2-(CH$_2$)$_7$-en- Pt^{II}	320
2,2-[2-Me-(CH$_2$)$_5$]-en-1(R),2(S)- Co^{III} Pt^{II}	328 328
1,1,2,4,4-Me$_5$-en-(R,S)- Ni^{II} Cu^{II} Pr^{III} Pd^{II} Pt^{II} Co^{II} Fe^{II}	267, 315, 329 289, 330 190 331 315 329 330
1,1,2,4,4-Me$_5$-en-(R)- Cu^{II} Pd^{II}	289 332
3,3-d_2-2-Me-en-(S)- Pt^{II}	321
2-d-2-Me-en-(R,S)- Pt^{II}	321

Table 6   (*continued*)

$en = N-C-C-N$ 1 2 3 4	Refs.
1,1-Me$_2$-en-	
Co^{III}	266, 271, 333
Rh^{III}	278
Ni^{II}	226, 267, 277, 334–338
Pd^{II}	327
Cu^{II}	339–341
Pt^{II}	274
Zn^{II}	342
1,1-Et$_2$-en-	
Cu^{II}	268, 289, 339, 341, 343–352
Pd^{II}	327
Ni^{II}	99, 348, 352, 353
Pt^{II}	274
1-Me-1-Et-en-	
Ni^{II}	334
2,3-Me$_2$-en-(R,S)- (*meso*-sbn)	354
Co^{III}	250, 260, 298, 355–357
Rh^{III}	278
Pt^{II}	358–360
Pd^{II}	358, 359
2,3-Me$_2$-en-(RR,SS)- (*rac*-sbn)	354
Co^{III}	260, 295, 361
Cr^{III}	362
Rh^{III}	278
(+)-2,3-Me$_2$-en-(SS)-[(+)-sbn-(SS)-]	
Co^{III}	250, 317, 357, 363
Pt^{II}	321
(−)-2,3-Me$_2$-en-(RR)-[(−)-sbn-(RR)-]	
Co^{III}	250, 300, 357, 364
Rh^{III}	305
2,3-Ph$_2$-en-(R,S)- (*meso*-stien)	365–367
Co^{III}	368
Ni^{II}	369–373
Pd^{II}	327, 374
Pt^{II}	321, 375
2,3-Ph$_2$-en-(RR,SS)- (*rac*-stien)	376
(+)-2,3-Ph$_2$-en-(RR)- [(+)-stien-(RR)-]	
Co^{III}	304
Pd^{II}	309
(−)-2,3-Ph$_2$-en-(SS)- [(−)-stien-(SS)-]	
Co^{III}	305, 377–380
Pt^{II}	321
Cr^{III}	311
2,2-Me$_2$-en- [ibn]	
Co^{III}	15, 259, 295, 355, 381, 382
Ni^{II}	335
Pd^{II}	327, 374
Pt^{II}	274, 375
Cr^{III}	11, 12
2,2-Me$_2$-4-Pr^i-en-	
Cu^{II}	383

Table 6   *(continued)*

en = N—C—C—N 1 2 3 4	*Refs.*
1,1,2,2-Me$_4$-en- Pd^{II}	327
2,2-Ph$_2$-en- Pt^{II}	320
(+)-2,3-(CH$_2$)$_3$-en-(*SS*)- [(+)-*trans*-cptn-(*SS*)-] Cr^{III} Co^{III} Rh^{III}	362 384–389 388
(−)-2,3-(CH$_2$)$_3$-en-(*RR*)- [(−)-*trans*-cptn-(*RR*)-] Fe^{II} Co^{III}	22 390
(−)-2,3-(CH$_2$)$_4$-en-(*RR*)- [(−)-*trans*-chxn-(*RR*)-] Co^{III} Cr^{III} Fe^{II} Pd^{IV}	391–394 298, 300, 312, 363, 391, 395–399 12, 383 22 306
(+)-2,3-(CH$_2$)$_4$-en-(*SS*)- [(+)-*trans*-chxn-(*SS*)-] Cr^{III} Pt^{II} Co^{III} Cu^{II} Ni^{II}	12, 362 282 400–404 299 299
2,3-(CH$_2$)$_7$-en- [*trans*-cotn] Pt^{II}	405
1,4-Me$_2$-en- Co^{III} Ni^{II} Pt^{II} Pr^{III} Ln^{III} Re^{I} Rh^{III} Ru^{III} Cu^{II} Pd^{II} U^{VI}	267, 269–271, 273, 275 267, 337, 406 266, 274, 308, 407 190 408 409 278 410 339–341, 411–415 416 417
1,4-Et$_2$en- Cu^{II} Ni^{II} Pt^{II}	280, 339, 345, 413 277, 334, 337 274
1,4-Pr$_2^n$-en- Cu^{II}	268
1,4-Bu^t-en- Pt^{II}	418, 419
1,4-(1-methylheptyl)-en- Pd^{II}	416
1,4-Bz$_2$-en- Cu^{II} Ni^{II}	291, 292, 420–422 291
1,4-Ph$_2$-en- Ni^{II} Pd^{II} Cu^{II}	423 331, 416 415

Table 6 *(continued)*

$en = N-C-C-N$ 1 2 3 4	Refs.
1,1,4-Me$_3$-en-	
Pt^{II}	308, 424
Rh^{III}	278
Cu^{II}	415
Ni^{II}	425
1,1,4-Et$_3$-en-	
Cu^{II}	280
1,1-Et$_2$-4-Me-en-	
Cu^{II}	280, 349
1,1-Me$_2$-4-Et-en-	
Pt^{II}	274
1,1,1-Me$_3$-en$^+$-	
Co^{III}	266
1,1,4,4-Me$_4$-en-	
Ni^{II}	266, 287, 329, 425–431
Cu^{II}	291, 330, 415, 432–444
Pr^{III}	190
Ln^{III}	408
Re^{I}	445
Rh^{III}	278
Pd^{II}	331, 416, 442, 446
Hg^{II}	447
Co^{III}	448
Co^{II}	329, 449
Fe^{II}	330
Pt^{II}	274, 442, 450, 451
Cd^{II}	442
Ir^{III}	442
Zn^{II}	430
Li^{I}	452–454
1,1,4,4-Et$_4$-en-	
Cu^{II}	291, 349, 434, 455–460
Ni^{II}	291
Pr^{III}	190
Pd^{II}	331
2,2,3,3-d_4-en-	461
Pd^{II}	462
Pt^{II}	462
2,2,3,3-Me$_4$-en-	
Ni^{II}	369, 463–465
Co^{III}	259, 466
Cu^{II}	467
1,1-Me$_2$-4,4-Et$_2$-en-	
Pt^{II}	274
1,4-Me$_2$-1,4-Bz$_2$-en-	
Cu^{II}	291, 468
Ni^{II}	291, 468
1,4-Me$_2$-1,4-(R-α-MeBz)$_2$-en-	
Pt^{II}	469

Table 6 *(continued)*

en = N—C—C—N 1 2 3 4	Refs.
1,1,4,4-Bz$_4$-en- Cu^{II}	470
tn = N—C—C—C—N 1 2 3 4 5	
1-Me-tn- Co^{III} Ni^{II}	471–473 129
1-Et-tn- Ni^{II}	129
1-Estrone-tn- Pd^{II} Pt^{II}	293 293
3-Me-tn- Pt^{II} Pd^{II} Co^{III}	358 359 474
1,1-Me$_2$-tn- Ni^{II} Pd^{II} Hg^{II}	129 327 447
3,3-Me$_2$-tn-(dan) Co^{III} Pd^{II} Pt^{II} Zn^{II} Cd^{II} Hg^{II} Cu^{II}	475, 476 260, 477, 478 358 283, 359 479 479 479 480
1,5-Me$_2$-tn- Pt^{II}	450
2-Me-tn-(*R,S*)- Pt^{II} Co^{III} Cu^{II}	484 283 481 482
(+)-2-Me-tn-(*S*)- Co^{III}	483 304, 485–488
2-Ph-tn-(*S*)- Co^{III}	304, 487, 489
2,4-Me$_2$-tn-(*R,S*)- Pd^{II} Pt^{III} Co^{III}	490–492 309, 358 283, 309, 358 300, 357, 481, 487, 493–498
2,4-Me$_2$-tn-(*RR,SS*)- Pd^{II} Pt^{II} Co^{III}	490–492 359 283, 359 497, 499, 500
(−)-2,4-Me$_2$-tn-(*RR*) Co^{III}	300, 304, 481, 487, 497, 501–506
(+)-2,4-Me$_2$-tn-(*SS*) Co^{III}	300, 497

Nitrogen Ligands

Table 6 *(continued)*

$tn = \underset{1}{N} - \underset{2}{C} - \underset{3}{C} - \underset{4}{C} - \underset{5}{N}$	*Refs.*
Pt^{II}	507
Cr^{III}	311
2,4-Ph$_2$-tn-(SS)-	
Co^{III}	304, 378, 487
2,4-Ph$_2$-tn-(R, S)-	
Co^{III}	304, 487
1,1,5,5-Me$_4$-tn	
Pt^{II}	450
Cu^{II}	508
Fe^{II}	330
Ni^{II}	329
Co^{II}	329
2,2,4-Me$_3$-tn-(R, S)-	
Co^{III}	509
Pt^{II}	509
2,2,4-Me$_3$-tn-(S)-	
Co^{III}	510
2,3-[1,2,2-Me$_3$-(CH$_2$)$_3$]	
Cu^{II}	511
2,3-(CH$_2$)$_4$-tn-(S, R)-	
Pt^{II}	512
2,3-(CH$_2$)$_4$-tn-(R, R)-	
Pt^{II}	512
cis -2,4-(CH$_2$)$_3$-tn-	
Cu^{II}	513, 514
Pt^{II}	515, 516
Ni^{II}	514
2,2-[2-Me-(CH$_2$)$_5$]-tn-	
Co^{III}	328
Pt^{II}	328
$bn = \underset{1}{N} - \underset{2}{C} - \underset{3}{C} - \underset{4}{C} - \underset{5}{C} - \underset{6}{N}$	
3-Me-bn-(R)-	
Co^{III}	517
3,5-Me$_2$-bn-(R, S)-	
Co^{III}	517
3,5-Me$_2$-bn-(RR, SS)-	
Co^{III}	517
(−)-3,5-Me$_2$-bn-	
Co^{III}	517
1,1,6,6-Me$_4$-bn-	
Pr^{III}	190

If two amine groups are located in the 1,2- (or 1,3-) positions in a cyclohexane or cyclopentane ring, the ring system becomes unsymmetrical, and the opportunity exists for both geometric and optical isomerism. The *cis* form can be regarded as *meso* configuration and the *trans* as a racemate. Coordination of the *trans* isomer allows an equatorial orientation of the ring system and complexes of *trans*-chxn and *trans*-cptn are well known.[47] The forced axial orientation for coordination of the *cis* isomers appears to preclude the isolation of complexes, at least in octahedral systems. In the optically active forms of *trans*-chxn and *trans*-cptn the $(-)_{589}$ enantiomers are assigned to the R,R absolute configuration (Table 6).

Both 1- and 1,4-*N*-alkylated ethylenediamine ligands are prochiral and coordination completes the asymmetry at the nitrogen atom. Under favorable conditions, such complexes can be resolved and the absolute configuration determined by X-ray or other methods. Structures (17)–(22) show some representative configurations.

(17) B = (R)-CH(Me)(Ph)[469]   (18)[423]   (19)[326]

(20)[282,324,325]   (21)[308,407]   (22)[308,309]

There has also been considerable interest in the bis(1,1-dialkylethylenediamine) complexes of Cu^{II} and Ni^{II} as these exhibit marked thermochroism.[339] The origin of this effect still remains to be precisely defined, despite approaches using ESR,[350] single-crystal X-ray[289,343] and high-pressure techniques.[335]

Complexes of the fully *N*-alkylated ethylenediamine ligand, *e.g.* 1,1,4,4-Me_4-en and 1,1,4,4-Et_4-en, have not been neglected and (23) shows the arrangements of the Et groups in Cu(1,1,4,4-Et_4-en)-$(NO_3)_2$.[455] This complex shows C_2 symmetry and should be compared with the arrangement adopted for Pt(1,1,7,7-Et_4-dien)X$^+$ (43) which has a mirror plane.

(23)[455]

13.1.4.3 tn Systems

Coordination of 1,3-propanediamine (tn) results in a six-membered ring and chair (c), boat (b) or twist (t) ring conformations are possible. The twist conformation can adopt enantiomeric δ and λ orientations of the C—C axis with respect to the N—M—N plane (24). Simple tn complexes appear to prefer the chair conformation, and although twist conformations are known, no boat conformations have been detected.

(24a) λ   (24b) δ   (25a) p   (25b) a

With an octahedral metal center there are two orientations possible for an isolated tn ring in the chair conformation (**25**) and these become non-equivalent if some of the other four donor atoms are different. There is solid state evidence for such alternative conformations.[520]

Two or three tn rings in the chair conformtion can also have alternative arrangements. A tris(tn) or *cis*-bis(tn) complex in the Λ configuration has a unique rotational direction (clockwise) when viewed down the C_3 (or pseudo C_3) axis (**26**). The chair rings can then be arranged with the central CH_2 group of the ring folded in the same direction as the rotation axis (parallel, p) or in the opposite direction (antiparallel, a) (**25**). Thus the combination of chair and twist arrangements for $M(tn)_3$ can give rise to 16 possible conformations.[521] for a *cis*-bis(tn,c) complex, the pp or aa arrangement results in the chairs folding in the same direction, but in the pa arrangement the rings can fold together, or apart.

(26)

Applying the same arguments to the *trans*-bis(tn,c), the two rings can again be parallel or antiparallel to a rotation axis.[125] Table 7 lists the ring conformations adopted in a variety of tn complexes.

Although twist conformations are rare in simple tn complexes, these become much more available in tn ligand systems with 2- or 2,4-alkyl substituents.[499] In such cases, both axial and equatorial arrangements have been observed, even in octahedral complexes.

In summary, the six-membered tn ring is much more flexible than the corresponding five-membered en ring, with the various ring conformations and substituent orientations being more equivalent in energy.[497,572]

Of the *C*-alkyl substituted tn analogues (Table 6), 2,4-Me$_2$-tn is probably the most widely investigated as this is prepared by reduction of acetylacetone dioxime to give a mixture of *RR,SS*- and *R,S*-2,4-Me$_2$-tn. $(-)$-2,4-Me$_2$-tn has been assigned to the *R,R* absolute configuration.[300]

13.1.4.4 Other Aliphatic Diamines

Although simple chelates of Cu^{II} with putrescine (bn), cadaverine and hexamethylenediamine have been known for some time,[573] it is only recently that octahedral Co^{III} complexes with bn and *C*-substituted bn ligands have been characterized. The crystal structure of $(+)_{589}$-Co(bn)$_3^{3+}$, containing three seven-membered chelate rings, has been determined and the absolute configuration of this enantiomer is established as Λ (**12b**).[47] Formula (**86**) also shows the conformation of a seven-membered diamine ring, although in this case it is part of a polyamine ligand.

It is, of course, not necessary for all chelate rings in an octahedral tris-diamine complex to be of the same size, and the complete series $M(en)_x(tn)_y^{3+}$ (M = Co, Cr;[536,574] $x + y = 3$) and Co(en)$_x$(tn)$_y$(bn)$_z^{3+}$ ($x + y + z = 3$) have been prepared.[575] Pt^{II} complexes containing chelated bn are also known,[576] as well as Co(en)$_2$(L)$^{3+}$ (L = *cis*-NH$_2$CH$_2$CH=CHCH$_2$NH$_2$).[577] The key to the isolation of such complexes has been the use of non-aqueous solvents, especially DMSO.

13.1.5 TRIAMINE LIGANDS

All the linear triamines (**27**)–(**31**) are commerically available and there are others, such as 6,6-tri, which, as yet, do not appear to have been used in the preparation of coordination complexes. These linear polyamine ligands can coordinate in either meridional or facial topology and both bis(triamine) and mono(triamine) complexes are possible for octahedral central metal ions.

(27) dien(2,2-tri)   (28) 2,3-tri   (29) dpt(3,3-tri)   (30) 2,4-tri   (31) 3,4-tri (spermidine)

Table 7 Ring Conformations in 1,3-Propanediamine Complexes

Complex	Configuration	Refs.
	Co^{III}	
(−)-[Co(tn)$_3$]Cl$_3$	Λ(ccc)(ppp)	522
(−)-[Co(tn)$_3$]Br$_3$	Λ(ccc)(ppp)	523
(±)-[Co(tn)$_2$CO$_3$]ClO$_4$	(cc)(pp)	524, 525
(±)-*cis*-[Co(tn)$_2$(NO$_2$)$_2$]Cl·H$_2$O	(cc)(pp)	526
(−)-[Co(tn)$_2$(acac)][As$_2$-(+)-tart$_2$(−2)]	Δ(cc)(aa)	527
(−)-*cis*-[Co(tn)$_2$(NCS)$_2$]$_2$[Sb$_2$-(+)-tart$_2$(−2)]	Λ(cc)(pa)[a]	528
trans-[Co(tn)$_2$Cl$_2$]Cl·HCl·2H$_2$O	(cc)(p)	529
trans-[Co(tn)$_2$(ONO$_2$)$_2$]NO$_3$	(cc)(p)	530
trans-[Co(tn)$_2$(NO$_2$)$_2$]NO$_2$	(cc)	531
(+)-[Co(tn)(en)$_2$]Br$_3$	Λ(c)(a)	532
trans-[Co(tn)(β-ala)(NO$_2$)$_2$]	(c)	533
cis-[Co(tn)(β-ala)(NO$_2$)$_2$]	(c)	534
	Cr^{III}	
(±)-[Cr(tn)$_3$][Ni(CN)$_5$]	(cct) (ap,*lel*)[b]	72, 521, 535
(±)-[Cr(tn)$_2$(en)]I$_3$	(cc)(aa,*ob*)	536
(−)-*cis*-[Cr(tn)$_2$(NCS)$_2$][Sb$_2$-(+)-tart$_2$(−2)]	Λ(cc)(pa)[a]	528
trans-[Cr(tn)$_2$(NH$_3$)F](ClO$_4$)$_2$	(ct)(pδ)	537
trans-[Cr(tn)$_2$(OH$_2$)F](ClO$_4$)$_2$	(cc)(aa)	538
trans, trans-[Cr(tn)$_2$(OH$_2$)F][Cr(tn)$_2$(OH)F](ClO$_4$)$_3$	(cc)(p)	216
(±)-[Cr(tn)(en)$_2$]Br$_3$	(c)(a,*ob*,*lel*)	536
(−)-[Cr(tn)(acac)$_2$]I·H$_2$O	Λ(c)	539
	Cu^{II}	
[Cu(tn)$_2$](ClO$_4$)$_2$	(cc)(p)	482
[Cu(tn)$_2$]SO$_4$·H$_2$O	(cc)(p)	540
[Cu(tn)$_2$]SeO$_4$·H$_2$O	(cc)(p)	540
[Cu(tn)$_2$](NO$_3$)$_2$	(cc)(p)	541
[Cu(tn)$_2$](NO$_2$)$_2$	(cc)(p)	542
[Cu(tn)$_2$(NCS)]ClO$_4$	(cc)	543
[Cu(tn)$_2$(SCN)$_2$]		544
[Cu(tn)$_2$(benzoate)$_2$	(cc)(p)	545
[Cu(tn)$_2$(*m*-chlorobenzoate)$_2$	(cc)(p)	546
[Cu(tn)$_2$(*p*-chlorobenzoate)$_2$	(cc)(p)	547
[Cu(tn)$_2$(*p*-iodobenzoate)$_2$	(cc)(p)	548
[Cu(tn)$_2$(*p*-bromobenzoate)$_2$	(cc)(p)	549
[Cu(tn)$_2$(*p*-fluorobenzoate)$_2$	(cc)(p)	550
[Cu(tn)$_2$(OH)$_2$](*m*-iodobenzoate)$_2$	(cc)(p)	551
[Cu(tn)$_2$(*m*-bromobenzoate)$_2$]	(cc)(p)	552
[Cu(tn)$_2$(*m*-nitrobenzoate)$_2$]	(cc)(p)	553
[Cu(tn)$_2$(*p*-nitrobenzoate)$_2$]	(cc)(p)	554
[Cu(tn)$_2$(*p*-methylbenzoate)$_2$]	(cc)(p)	555
[Cu(tn)$_2$(*m*-methylbenzoate)$_2$]	(cc)(p)	556
[Cu(tn)(malonato)]$_2$	(c)	557
	Cd^{II}	
[Cd(tn)Cl$_2$]$_2$	(c)	558
	Ni^{II}	
(±)-[Ni(tn)$_3$][Ni(tn)$_2$(OH$_2$)$_2$]Cl$_4$·H$_2$O		559
trans-[Ni(tn)$_2$(OH$_2$)$_2$](ClO$_4$)$_2$	(cc)(p)	560
trans-[Ni(tn)$_2$(OH$_2$)$_2$](NO$_3$)$_2$	(cc)(p)	561
trans-[Ni(tn)$_2$(OH$_2$)$_2$]Br$_2$·2H$_2$O		562
[Ni(tn)$_2$](benzoate)$_2$	(cc)(p)	563
[Ni(tn)$_2$(*p*-methylbenzoate)$_2$]	(cc)(p)	564
[Ni(tn)$_2$(*m*-methylbenzoate)$_2$]	(cc)(p)	565
[Ni(tn)$_2$(*p*-nitrobenzoate)$_2$]	(cc)(p)	566
[Ni(tn)$_2$(*m*-nitrobenzoate)$_2$]	(cc)(p)	567
	Pt^{II}	
Na[Pt(tn)(inosine-5′-phosphato)$_2$]-[inosine-5′-phosphate]·H$_2$O		568
[Pt(tn)(guanosine-5′-monophosphate-methylester)]·H$_2$O	disordered	569
	Pt^{II}–Pt^{IV}	
[Pt(tn)$_2$][Pt(tn)$_2$Cl$_2$](BF$_4$)$_4$	(tt)(δλ)	570
[Pt(tn)$_2$][Pt(tn)$_2$Br$_2$](BF$_4$)$_4$	(tt)(δλ)	570
[Pt(tn)$_2$][Pt(tn)$_2$Br$_2$](ClO$_4$)$_4$	(tt)(δλ)	570
[Pt(tn)$_2$][Pt(tn)$_2$I$_2$](ClO$_4$)$_4$		571

[a] The central —CH$_2$— groups of the tn ligands are folded towards each other. [b] The central —CH$_2$— groups of the tn ligands are folded away from each other.

The remaining branched chain ligands (**32**)–(**36**) are restricted to facial coordination and, while ML_2^{n+} systems predominate, Co(tach)X_3 are known,[381,586,587] and there is no apparent reason why other mono(branched chain triamine) complexes cannot be prepared (Section 13.1.5.3).

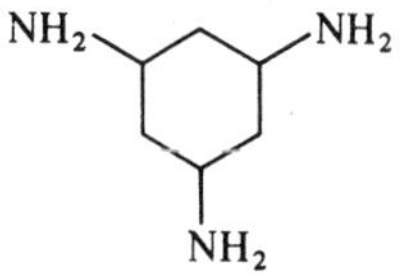

(**32**) *cis,cis*-tach[476,578]

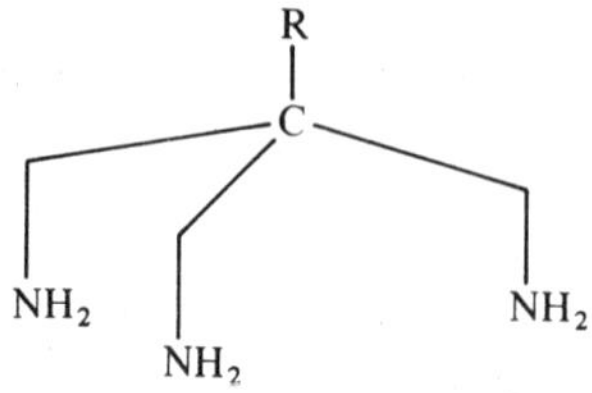

(**33**) tamm (R = H)[476,579]
(**34**) tame (R = Me)[580,581]
(**35**) tamp (R = Et)[582,585]

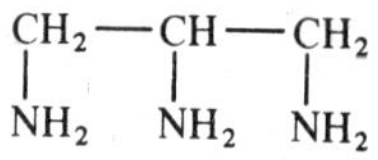

tap (**36**)[583,584]

Martell and Calvin[588] have summarized the pioneering studies of Mann and Pope, using tap (**36**) as a stereochemical probe in Pt^{II} chemistry and ML_2^{3+} (M = Co, Rh; L = tap) have been prepared.[583,584] The other tripodal ligands have been used more frequently in stability constant studies and the results have been reviewed.[589,590]

13.1.5.1 dien Systems

The free ligand, or its protonated form dien-H_n^{n+} ($n = 0$–3), has the potential to adopt the *cis,cis* (**37**), *cis,trans* (**38**) or *trans,trans* (**39**) conformations. Crystal structures of dien-H_n^{n+} ($n = 2, 3$) (Table 8) salts show that both the counter ion and degree of protonation determine which form is adopted.

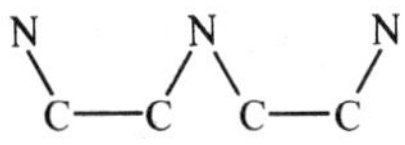

(**37**) *cis,cis*

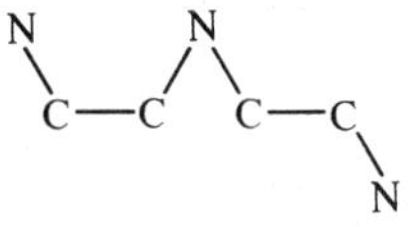

(**38**) *cis,trans*

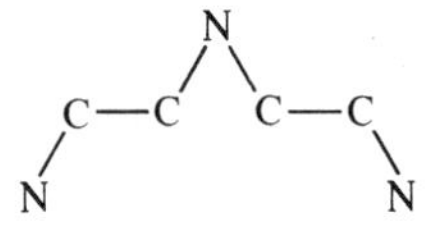

(**39**) *trans,trans*

Coordination compounds using dien (**27**) as a ligand have been widely investigated from both octahedral and square planar complexes and there are numerous references to various aspects of Co^{III}, Cr^{III}, Cu^{II} and Ni^{II} chemistry.

For octahedral bis(dien) complexes there are three potential isomers (**40**)–(**42**). All three have been isolated[597,598] for Co(dien)$_2^{3+}$ and the *mer* form is optically active by virtue of the dissymmetric arrangement of the 'planar' *sec*-NH protons. The equilibrium isometric ratios *sym-fac:asym-fac:mer* are 7:28:65 respectively and crystal structures have been determined for all three isomers.[599–603] Likewise, all three isomeric forms of Co(dien)(4-Me-dien)$^{3+}$ have been isolated;[604–606] in this case the equilibrium isomeric ratios are 81:10:9 respectively.[604] Only the *sym-fac* isomer has been detected for Co(4-Me-dien)$_2^{3+}$.[604–606] The configuration of Cr(dien)$_2^{3+}$ is unknown.[607,608] Structural investigations on the analogous Ni^{II},[609] Cu^{II}[610] and Zn^{II}[611] bis(dien) complexes show these to have the *mer* configuration.

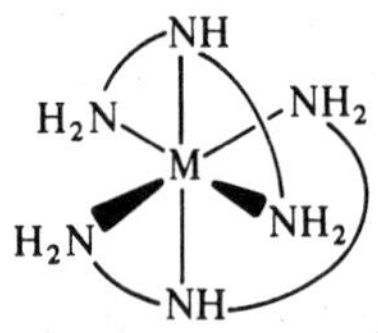

(**40**) *mer*

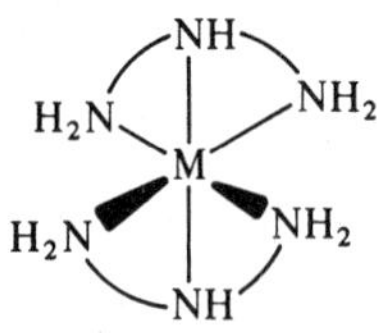

(**41**) *sym-fac*

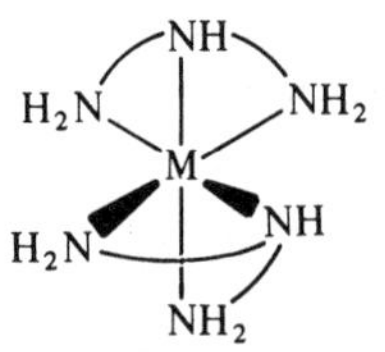

(**42**) *asym-fac*

Table 8 Salts of dien-H_n^{n+} (L^{n+})

	Conformation	Refs.
$L^{2+} \cdot 2Cl^-$	cis-cis	591
$L^{3+} \cdot 3Cl^-$	trans-trans	591
$L^{3+} \cdot CrCl_4^{2-} \cdot Cl^-$	trans-trans	592
$L^{3+} \cdot CuCl_4^{2-} \cdot Cl^-$	trans-trans	593, 594
$(L^{3+})_2 \cdot PtCl_4^{2-} \cdot 4Cl^-$	cis-trans	595
$L^{3+} \cdot CdCl_4^{2-} \cdot Cl^-$	unknown	596

Table 9 Linear Tridentate Polyamine Ligand Configurations in M(tridentate)(XYZ)$^{n+}$ and M(tridentate)X^{n+} Complexes

Complex	Configuration	Refs.
Co(NO$_2$)$_3$(dien)	*mer*	614, 615
CoCl$_3$(dien)	*mer*	616
Co(N$_3$)$_3$(dien)	*mer*	617
Co(*CN*)$_3$(dien)	*fac*	618
Co(dien)(OH$_2$)$_3^{3+}$	*mer, fac*	619
Co(dien)(NH$_3$)$_3^{3+}$	*mer, fac*	620, 621
CrCl$_3$(dien)	*mer, fac*	622, 623
Cr(CO)$_3$(dien)	*fac*	624
Mo(CO)$_3$(dien)	*fac*	625
MoO$_3$(dien)	*fac*	626
RhCl$_3$(dien)	*mer, fac*	627
Co(dien)(L-pen)$^+$	*fac*	628
Cr(dien)$_2^{2+}$	unknown	629
Co(NO$_2$)$_2$(dien)(NH$_3$)$^+$	*sym-fac*	630
	mer (*trans*-NO$_2$)	631, 632
CoCl$_2$(dien)(NH$_3$)$^+$	*mer* (*trans*-Cl)	633
Co(ox)(dien)(NH$_3$)$^+$	*mer*	634, 635
CoBr(dien)(NH$_3$)(OH$_2$)$^{2+}$	*fac*	636
CoX(gly)(dien)$^+$	*mer* (*trans*-O,X)	618, 637
Co(dapo)(dien)$^{3+}$	*fac*	638
Co(dien)(A)(OH$_2$)$_2^{3+}$	*mer, fac*	619, 639
CoCl(dien)(en)$^{2+}$	*asym-fac, sym-fac, mer*	640–642
CoCl(dien)(tn)$^{2+}$	*asym-fac, sym-fac, mer*	520
CoCl(dien)(2-Me-en)$^{2+}$	*mer*	643
CoCl(dien)(NH$_3$)$_2^{2+}$	*mer*	644
CoCl(dien)(1-Me-tn)$^{2+}$	*asym-fac, mer*	471, 473
CoCl(dien)(A)$_2^{2+\ a}$	*asym-fac, mer*	127
CoCl(dien)(bipy)$^{2+}$	*mer*	645
Co$_2$O$_2$(dien)$_2$(en)$_2^{4+}$	*mer*	646
CrCl(dien)(en)$^{2+}$	*sym-fac*	15, 647
Cu(*NCS*)$_2$(dien)	*mer*	648
[Cu(ox)(dien)]·4H$_2$O	*mer*	649
[Cu(dien)(*NCS*)]$_2^{2+}$	*mer*	650
[Cu(dien)(A)$_2$]^a	*mer*	651
[{Cu(dien)}$_2$ox]$^{2+}$	*mer*	652
[Cu(HCO$_2$)$_2$(dien)]$_n$	*mer*	653
[Cd(*NCS*)$_2$(dien)]$_n$	bridging	654
[CdCl$_2$(dien)]$_n$	*mer*	655
AuCl(dien)$^{2+}$	*mer*	656
AuCl(dien-H)$^+$	*mer*	656
PdN$_3$(dien)$^+$	*mer*	657
Cr(dpt)$_2^{3+}$	unknown	658
CoCl$_3$(dpt)	*mer*	616, 659
Co(dpt)$_2^{3+}$	*mer*	660
CoCl(dpt)(en)$^{2+}$	*mer*	661, 662
Co$_2$(OH)$_3$(dpt)$_2^{3+}$	*fac*	30
Ni(dpt)$_2^{2+}$	*mer*	609, 663
[{Zn(dpt)}$_2$ox]$^{2+}$	trigonal bipyramid	652
Cd(*NCS*)$_2$(dpt)	*mer*	664
[CdCl$_2$(dpt)]$_2$	*mer*	655
Cu(*NCS*)$_2$(dpt)	*mer*	655
Cu(OAc)(dpt)$^+$	*mer*	666
Cu(dpt)(AA)$^{2+\ b}$	*mer*	667, 668
Cu$_4$(dpt)$_4$CO$_3^{6+}$	*mer*	669
Cd(*NCS*)$_2$(2,3-tri)	*mer?*	664
CdCl$_2$(2,3-tri)	*mer*	655
CoCl$_3$(2,3-tri)	*mer*	616, 670
PdBr(dien)$^+$	*mer*	671
Pd(NO)$_2$(dien)$^+$	*mer*	672
[PtCl(dien)(NH$_3$)$_2$]$^{2+}$	*mer*	673
[Pt(dien)I][Pt(dien)I$_3$]I$_2$	*mer*	676
[Pt(dien)Cl]Cl	*mer*	675
[Pt(dien)NO$_3$]NO$_3$	*mer*	675
[Pt(dien)Cl$_3$]Cl	*fac*	674
HgCl$_2$(dien)	unknown	447
Cu(dpt)$_2^{2+}$	unknown	677
Zn(dpt)$_2^{2+}$	unknown	677
Hg$_3$(dpt)$_2$Cl$_6$	unknown	447
Cu(3,4-tri)Br$_2$	square pyramid	678
Co(3,4-tri)Cl$_3$	*mer*	679

^{a} A = monodentate amine. ^{b} AA = bidentate diamine.

Both *mer-* and *fac-*M(dien)(XYZ)$^{n+}$ complexes for octahedral metal centers can be formed but the factors determining the relative stabilities have yet to be determined.[612] In diamagnetic systems these configurations can be distinguished using NMR.[29,613] Table 9 summarizes the available information.

Like Co(trien(XY)$^{n+}$ (Section 13.1.6.2), Co(dien)(XYZ)$^{n+}$ has been used to bind peptide residues[612,680] and, in this case, two linked amino acid groups can be simultaneously coordinated.

The Cr^{III}—N bond is much more susceptible to acid hydrolysis than is the Co^{III}—N bond, and Garner and his coworkers have carefully investigated the stepwise dechelation kinetics of Cr(dien)-(OH$_2$)$_3^{3+}$ and other Cr^{III} polyamine complexes[11] in acid solution. Thus, Cr(dien-H)(OH$_2$)$_4^{4+}$ and Cr(dien-H$_2$)(OH$_2$)$_3^{5+}$ have been isolated chromatographically and characterized in solution.[681]

The coordination compounds involving *N-* and *C-*alkylated linear triamines (mainly dien) are summarized in Table 10. In most cases, *N-*alkylation involves the use of the free ligand[722] but alkylation of [Pt(dien)I]I by methyl iodide in liquid NH$_3$ ($-70\,°$C) is reported to produce [Pt(1-Me-dien)I]I.[24,688]

Of particular interest in the chemistry of Pd(1,1,7,7-Et$_4$-dien)X$^+$ systems is the early proposal that these may be 'pseudo-octahedral' by virtue of the alkyl groups 'blocking' the axial positions.[14] Recent X-ray crystals studies[704,708] have now shown that, in the solid state at least, the alkyl groups are not arranged in the 'pseudo-octahedral' sites, and the axial positions are quite open (**43**).

Table 10 *N*-Alkylated and *C*-Alkylated Linear N$_3$ Ligands

	Metal ion	Refs.
dien = N—C—C—N—C—C—N 1 2 3 4 5 6 7		
1-Me-	Pt^{II}	23
3-Me-(*R,S*)-	Co^{III}	682
4-Me-		683
	Co^{III}	604–606, 684, 685
	Cu^{II}	686
	Au^{III}	687
	Pt^{II}	23, 673, 688
2,6-Me$_2$-(*RR,SS*)-	Co^{III}	689
1,4,7-Alkyl-		690
1,4,7-Me$_3$-	Pd^{II}	691
	Ni^{II}	692
1,1,4-Me$_3$-	Au^{III}	693
1,1,7,7-Me$_4$-	Au^{III}	693
	Cu^{II}	694
1,1,7,7-Et$_4$-	Co^{II}	695, 696
	Cu^{II}	432, 697
	Ni^{II}	696, 698, 699
	Au^{III}	700, 701
	Pd^{II}	14, 672, 699, 702–706
	Pt^{II}	671, 699, 707, 708
1,1,4,7,7-Me$_5$-	Cu^{II}	697, 709–714
	Au^{III}	693
	Pd^{II}	704
	Co^{II}	713, 715
	Fe^{II}	713
	Mn^{II}	713
	Ni^{II}	713
	Zn^{II}	713
	Li^{I}	452
1,1,7,7-Et$_4$-4-Me-	Pd^{II}	14, 699, 702–704, 706, 716–718
	Pt^{II}	699
1,1,4,7,7-Et$_5$-	Cu^{II}	697, 719
	Pd^{II}	704
dpt = N—C—C—C—N—C—C—C—N 1 2 3 4 5 6 7 8 9		
5-Me-	Ni^{II}	720
	Cu^{II}	721
	Co^{III}	689
1,5,9-Me$_3$-	Ni^{II}	692
1,1,5,9,9-Me$_5$-		722

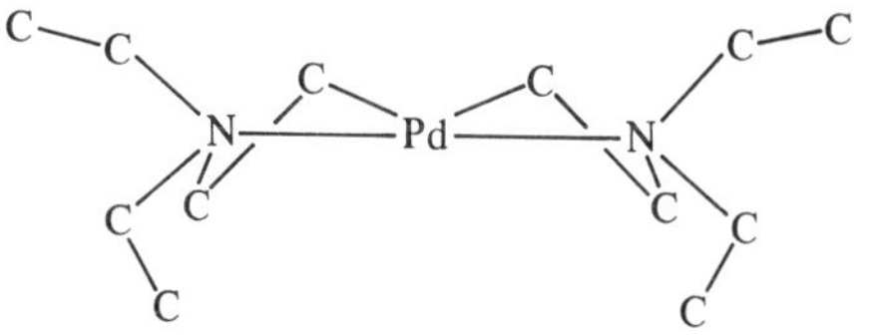

(43)[708,672]

Another intriguing aspect of dien and 4-Me-dien chemistry is the preparation of a '*trans* spanning' Pt[II] complex involving an eight-membered ring (45). The synthetic strategy used to prepare this unusual situation is outlined in Scheme 1.[673] The use of 4-Me-dien in Co[III] chemistry also results in the formation of an unusual byproduct (46), as well as the expected $Co(4\text{-Me-dien})_2^{3+}$ isomers.[604,605]

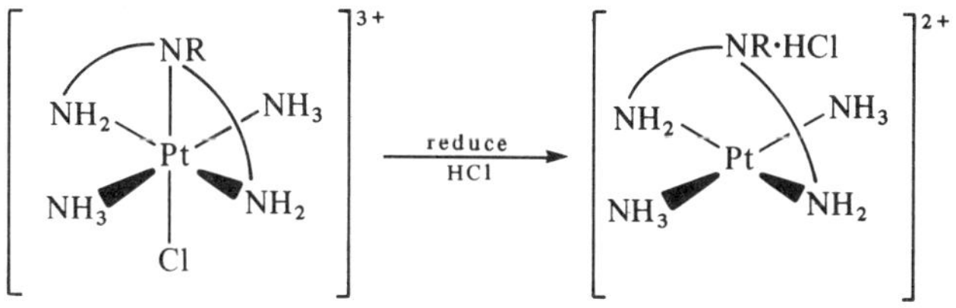

(44) Pt[IV] R = H, Me   (45) Pt[II]

Scheme 1

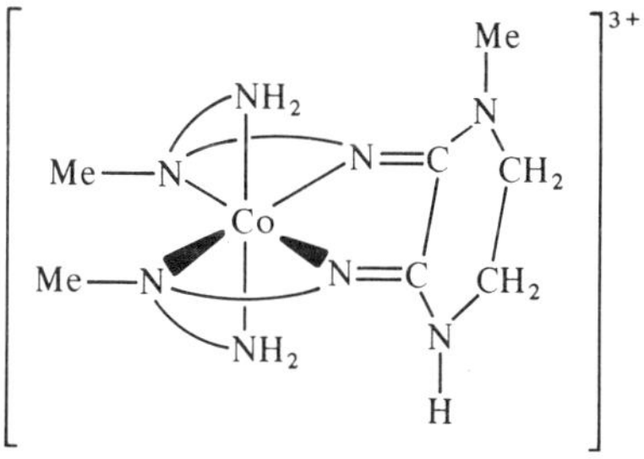

(46)[604,605]

13.1.5.2 Other Linear Triamine Systems

Remarks made with regard to the isomeric arrangements possible for dien are also applicable to the other linear triamines (28)–(31). For 3,3-tri(dpt) (29), with two fused six-membered rings, there is also the possibility (as yet unrealized) that the *sec*-NH proton in a facial configuration may lie *between* (*endo*) (47) the fold of the bent ligand. Such *endo* configurations have been shown to be of much higher energy than the *exo* (48) forms.[660] Indeed, complexes with dpt in a facial configuration are rather uncommon (Table 9). All three isomers (*asym-fac*, *mer* and *sym-fac*) of $Co(dien)(dpt)^{3+}$ have been characterized in a ratio of 4:2:1 respectively, and X-ray crystal structures of *mer*-$Co(dien)(dpt)^{3+}$ and *mer*-$Co(dien)(2,3\text{-tri})^{3+}$ have been determined.[723,724] However, for $Co(dpt)_2^{3+}$, only the *mer* isomer[660] has been detected and this is also the configuration adopted in $Ni(dpt)_2^{2+}$.[609]

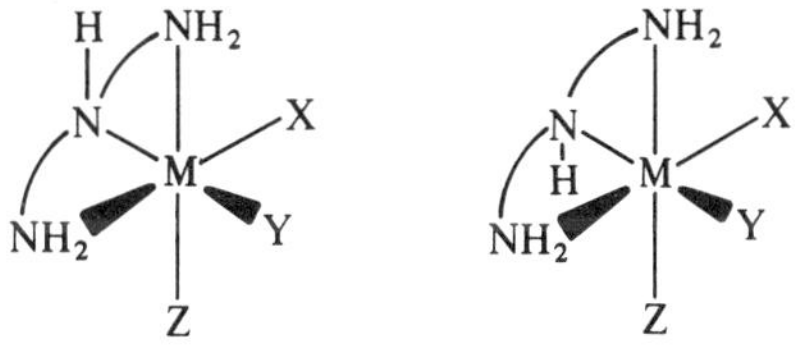

(47) *endo*   (48) *exo*

Mono(2,3-tri) (28) complexes are all potentially chiral, regardless of configuration, as the coordinated *sec*-NH center is now asymmetric. The *mer* configuration in $CoCl(2,3\text{-tri})(AA)^{2+}$ (AA = en, tn) complexes has been assigned[725] from ^{13}C NMR measurements and reactivity properties. Isomeric forms (**1a**) and (**1b**) resulting from the alternative arrangements of the *sec*-NH proton in a *mer* configuration were distinguished, but no attempt was made to establish optical forms. While Cr(2,3-tris)(NCS)$_3$,[726] M(2,3-tri)(OH$_2$)$_2$Cl$_2$ (M = Ni, Cu; distorted octahedral)[727] and Zn(2,3-tri)Cl$_2$ (tetrahedral)[727] are kown, the amine ligand configuration remains speculative.

Apart from the isolation of the Cu^{II} complexes[728] Cu(2,4-tri)$_2^{2+}$, Cu(2,4-tri-H)(2,4-tri)$^{3+}$ and Cu(2,4-tri-H)$_2^{4+}$, this ligand has been used mainly in stability constant measurements.[729,730] In this series of Cu^{II} complexes it is proposed that the 'longer' arm of the ligand is the one that is preferentially protonated.[729]

Despite considerable biochemical interest in the naturally occurring triamine 3,4-tri, spermidine (**31**), surprisingly few coordination compounds have been reported. The crystal structure of Cu(3,4-tri)Br$_2$ (five-coordinate, square pyramidal) has been determined[678] and *mer*-Co(3,4-tri)X$_3$ (X = NO$_2$, Cl) have been prepared.[679] The triaqua complex resulting from the latter is unstable with respect to spontaneous thermal reduction in acid solution (presumably *via* Co—N bond rupture) to give Co^{II}.

13.1.5.3   Branched Chain Triamine Systems

Both Co(tame)$_2^{3+}$ and CoCl(tame)(tame-H)$^{2+}$ have been isolated, with the latter[431] having one arm of the ligand not coordinated (**49**). The isomeric possibilities for both these complexes have been discussed in relation to the crystal structures.[581,731] One interesting feature of the divalent metal ion chemistry of tamm (**33**) and related ligands is that while crystalline solids of the type NiL$_2^{2+}$ can be isolated,[732] there is little evidence that such species are present in solution.[589]

(49)

The *N*-alkylated tripodal ligands Me$_3$-tame[589,733] and Me$_6$-tame[733] have also been used for the preparation of Zn^{II} complexes,[733] and while Me-tame is known,[476] the coordination chemistry of this ligand has yet to be investigated.

The Birch reduction (Na/NH$_3$) of the trioxime of α-phloroglucinol gives[578] a 1:1 mixture of *cis,cis*- and *cis,trans*-tach (**32**), with only the *cis,cis* forming stable M(tach)$_2^{n+}$ (M = Co, Rh, Ni) ions.[734] Treatment of the Ni^{II} complex with HCl allows the isolation of pure *cis,cis*-tach·3HCl[734] and this in turn can be used to prepare tach complexes of Co^{III},[381,586,587,734,714] Rh^{III},[734] Ni^{II},[734–738] Cu^{II},[735,737] Zn^{II},[735] Cd^{II},[735] Co^{II}[735] and Pt^{IV}.[516] Alternatively, the ligand can be prepared according to equation (3),[476] using *cis,cis*-1,3,5-cyclohexanetriol.

$$\text{ROH} \quad \longrightarrow \quad \text{ROSO}_2\text{Ph} \quad \longrightarrow \quad \text{RN}_3 \quad \longrightarrow \quad \text{RNH}_2 \tag{3}$$

13.1.6   TETRAMINE LIGANDS

The skeleton types are shown in (**50**)–(**61**). Symmetrical linear tetramines are normally synthesized by condensation of the appropriate centrally bridging α,ω-dihalide with excess of two moles of the diamine in ethanol. The addition of an inorganic base (NaOH or KOH) is usually required.[739,740] Spermine (**56**) is naturally occurring and most of these polyamines are commerically available.

(50) 2,2,2-tet(trien)   (51) 2,3,2-tet(entnen)   (52) 3,2,3-tet(tnentn)   (53) 2,2,3-tet

(54) 3,3,2-tet   (55) 3,3,3-tet   (56) 3,4,3-tet (spermine)   (57) tren

(58) trtn   (59) trbn

$CH_2-CH-CH-CH_2$
$|\quad\ |\quad\ |\quad\ |$
$NH_2\ NH_2\ NH_2\ NH_2$

(60) tab

$C(CH_2NH_2)_4$

(61) tam

The unsymmetrical linear tetramines [3,3,2-tet (54) and 2,2,3-tet (53)][32,741] have not been widely investigated, presumably because of synthetic difficulties, and the chemistry of the branched chain ligands, tren (57), trtn (58) and trbn (59), has been reviewed by Zipp *et al.*[742]

Pure tren is commercially available, but most samples of commerically available technical trien (50) are contaminated with variable amounts of tren. These isomeric tetramines can be separated by fractional crystallization of the hydrochloride salts (tren·3HCl, trien·4HCl) from ethanol[743,744] and ^{13}C NMR has been used to monitor the ligand composition.[745]

13.1.6.1 tren Systems

Only one isomeric form exists (62) when all four nitrogen donors bind to an octahedral metal center with two other equivalent ligands.[746] If the remaining two ligand sites are occupied by non-equivalent ligands, two isomers can be formed (63 and 64; Scheme 2). Both isomers have been characterized for $CoCl(tren)(NH_3)^{2+}$ [747,748] and the p-form for $CoCl(tren)(py)^{2+}$.[749]

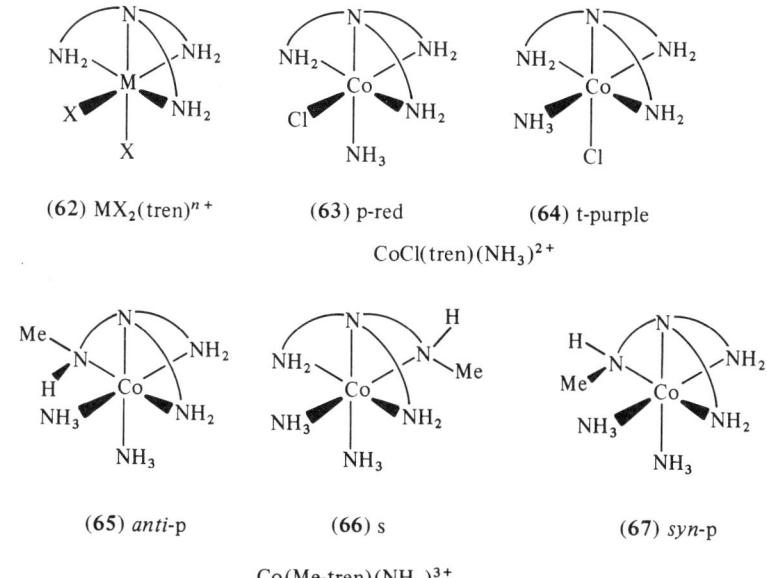

(62) $MX_2(tren)^{n+}$ (63) p-red   (64) t-purple

$CoCl(tren)(NH_3)^{2+}$

(65) *anti*-p   (66) s   (67) *syn*-p

$Co(Me\text{-}tren)(NH_3)_2^{3+}$

Scheme 2 Nomenclature is that suggested by Buckingham *et al.*[748,754] Note that the p and t assignments to the red and purple forms of $CoCl(tren)(NH_3)^{2+}$ are reversed from those originally suggested[747]

In addition to Co[III], octahedral $MCl_2(tren)^+$ complexes are also known for M = Rh,[750] Cr[608,743] and Ti[751] and a Tl[III][752] complex of composition $[Me_2Tl(tren)]BPh_4$ is formed by refluxing Me_2TlNO_3 with tren in ethanol. The Ti[III] complex $Ti(tren)Cl_3$ is non-ionic, indicating NH_2 bonding only. An ion with an exceptionally high charge is postulated in the formation of $Co_3(en)_6(tren)_2^{9+}$.[753]

A Co[III] complex with Me-tren is also known,[754] but details of the ligand synthesis are not given. $Co(Me\text{-}tren)(NH_3)_2^{3+}$ can exist in three potentially chiral forms (Scheme 2; 65–67).

Coordination numbers higher than six are postulated for lanthanide complexes[744] of tren but no X-ray structural data are available. A number of five-coordinate systems have been characterized using divalent central metal ions, *e.g.* Mn(tren)X$^+$ (X = NCS^{755} or NCO^{755}), Cu(tren)X^{n+} (X = Cl,[756] NCO,[756] NCS,[756] NH_3,[86] or CN^{757}) and Zn(tren)X$^+$ (X = NCS,[758] Cl[759]). If a suitable fifth ligand is used in such systems, bridging[710,760-762] can occur to give dinuclear five-coordinate complexes such as $[Ni_2(tren)_2(N_3)_2]^{2+}$.

With tren showing a marked propensity to stabilize five-coordinate M^{II} systems, it was of interest to investigate the Pd^{II} tren system, as a stereochemistry other than square planar is rare for this central ion. Senoff and his coworkers[763] have isolated [Pd(tren)Cl]Cl with a suspected five-coordinate structure, but there is some evidence for Pd—N bond rupture in solution to give a four-coordinate species. Pd^{II} complexes with the fully *N*-methylated ligand Me_6-tren are definitely four-coordinate and structural data are available for both [Pd(Me_6-tren)(NCS)]NCS and Pd(Me_6-trenH)Cl^{2+}, the latter having one arm of the ligand protonated.[764,765]

Complete *N*-alkylation of tren to give Me_6- or Et_6-tren stabilizes the Co^{II} oxidation state with respect to Co^{III} and monomeric N_4 complexes with Cu^{II},[767] Ni^{II}, Zn^{II},[768] Cr^{II}, Co^{II} and Fe^{II} are known.[742]

13.1.6.2 trien Systems

Fully N-bonded complexes have been isolated for most transition metal ions forming stable M—N bonds: Co^{III},[769] Ru^{III},[770,771] Rh^{III},[772,773] Ir^{III},[774] Cr^{III},[775,776] Cu^{II},[777,778] Ni^{II},[778] Pd^{II},[779] Pt^{II}[780] and Hg^{II};[447] an impressive amount of structural and reactivity data has been accumulated.

Schemes 3 and 4 show the topological arrangements for trien in both octahderal and square planar (*trans*) geometries. These stereochemical arrangements are available for any of the linear tetramine ligands (50)–(56), but the *R,S* nomenclature must be assigned for each individual polyamine. The *R,S* nomenclature shown in Scheme 3, applicable to trien and 3,2,3-tet, must be systematically reversed for 2,3,2-tet.

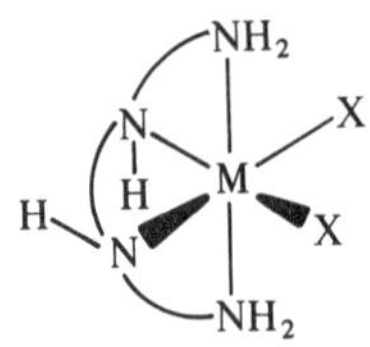

(68) Λ-(*RR*)-*cis-α* 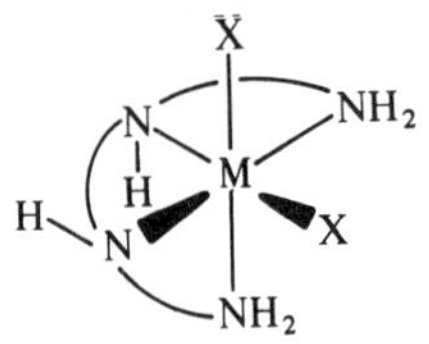   (69) Δ-(*RR*)-*cis-β* 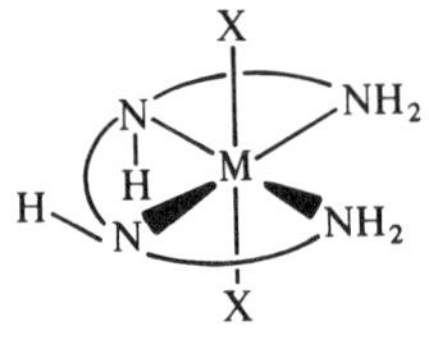   (70) (*RR*)-*trans* 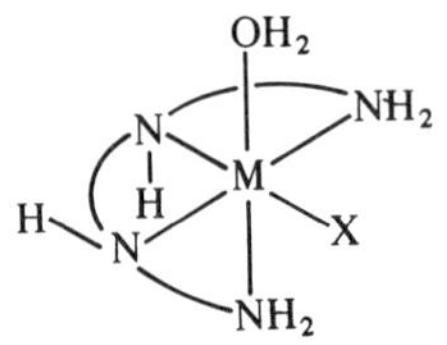   (71) Δ-(*RR*)-*cis-β₁*

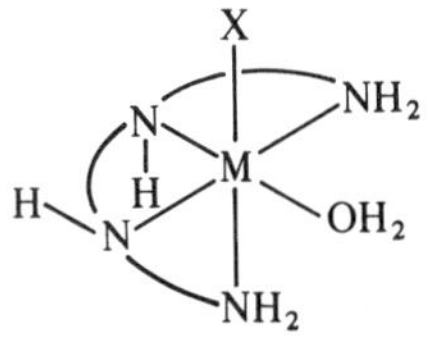

(72) Δ-(*RR*)-*cis-β₂*   (73) Δ-(*R,S*)-*cis-β* 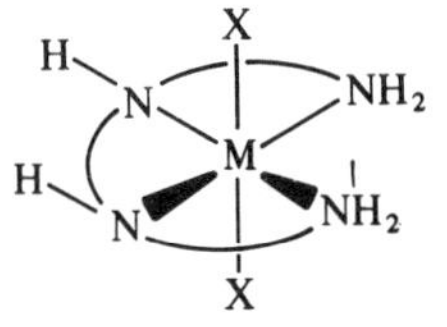   (74) (*R,S*)-*trans*

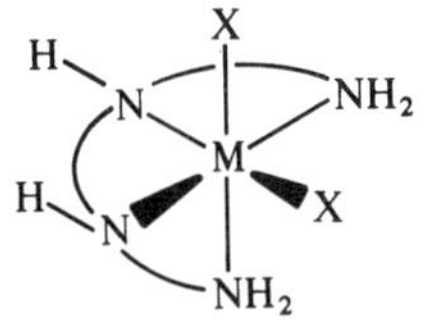

Scheme 3 Topological forms for linear N_4 polyamine ligands in an octrahedral complex

The pioneering work of Sargeson and Searle[769] in establishing the stereochemistry of the Co^{III} trien system on a sound basis has been the cornerstone of polyamine coordination chemistry. They were among the first to demonstrate unambiguously the stereochemical importance of the *sec*-NH proton in such systems, although isomers resulting from a similar source had previously been looked for by Meisenheimer.[781]

All three topological forms (*cis-α*, *cis-β* and *trans*) have been isolated[769] for CoCl$_2$(trien)$^+$ and the formation of the optically active *trans* isomer from the chiral *cis-β* establishes the (*RR,SS*) configuration for the racemic form of the latter. The alternative *R,S* configuration for the chiral *cis-β* would lead to an inactive (*R,S-meso*) *trans* form.

The single-crystal X-ray structure[782] of the major *cis-β*-CoCl(trien)(OH$_2$)$^{2+}$ isomer (formed in a ratio of about 20:1 from *cis-β*-CoCl$_2$(trien)$^+$) shows this to have the $β_2$-(*RR,SS*) configuration (72),

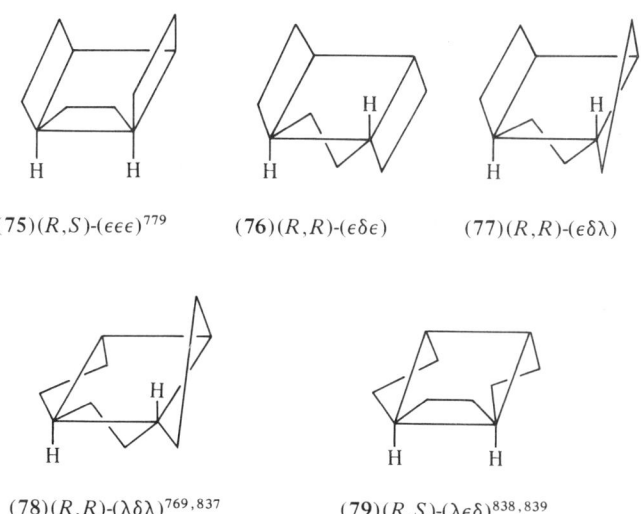

(75)(R,S)-(εεε)[779]   (76)(R,R)-(εδε)   (77)(R,R)-(εδλ)

(78)(R,R)-(λδλ)[769,837]   (79)(R,S)-(λεδ)[838,839]

Scheme 4 Potential ring conformations in square planar and *trans*-octahedral trien complexes

and both *RR,SS* and *R,S* configurations for the chiral *cis-β*-Co(trien)(OH₂)₂³⁺ have been detected in the acid hydrolysis products from *trans-(RR,SS)*-CoCl₂(trien)⁺.[783]

An additional source of isomerism (Scheme 5) can arise when a 5-alkyl-trien adopts the *cis-β* configuration. The single-crystal X-ray structure[784] of Λ-(+)-*cis-β*-(SS)-Co(NO₂)₂{5(R)-Me-trien}⁺ shows that for this system, at least, the 'planar' *sec*-NH is adjacent to the 5-position (**81**).

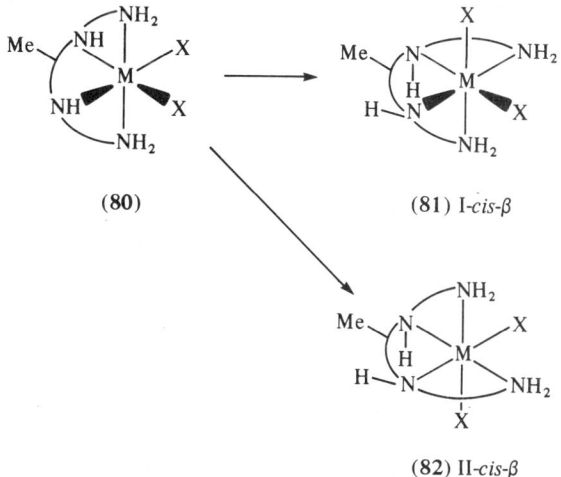

(80)   (81) I-*cis-β*

(82) II-*cis-β*

Scheme 5 Isomers of *cis-β*-MX₂(5-Me-trien)⁺. Λ-(+)-*cis-β*-Co(NO₂)₂{5(R)-Me-trien}⁺ has the (SS)-I configuration[784] while Λ-(+)-*cis-β*-Co(NO₂)₂{5(R)-Me-2,3,2-tet}⁺ has the (SR)-II configuration.[817] However, in both cases, the 'planar' *sec*-NH proton is remote from the coordinated nitro group

A similar situation occurs for the unsymmetrical linear tetramine 2,2,3-tet in the *cis-β* configuration,[741] where the 'fold' can occur at the *sec*-NH adjacent to the 'long' or 'short' arm of the ligand (Scheme 6).

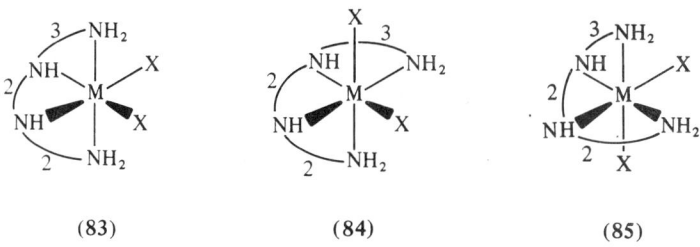

(83)   (84)   (85)

Scheme 6 Topological isomers possible for *cis-β*-MX₂(2,2,3-tet)⁺ [32,741]

It is also evident from Scheme 3 that the cis-$\alpha \rightleftharpoons cis$-$\beta$ isomerization will be accompanied by optical inversion ($\Lambda \rightleftharpoons \Delta$). This has been observed in acid solution[776] for the reaction shown in equation (4) and in basic solution, for the reaction shown in equation (5) although in the latter case[296] the configuration of the sec-NH protons in the resulting cis-β was not established.

$$\Delta\text{-}(RR)\text{-}cis\text{-}\beta\text{-Cr(trien)(ox)}^+ + \text{HCl} \longrightarrow \Lambda\text{-}(RR)\text{-}cis\text{-}\alpha\text{-CrCl}_2\text{(trien)}^+ \tag{4}$$

$$\Lambda\text{-}(RR)\text{-}cis\text{-}\alpha\text{-CoCl}_2\text{(trien)}^+ + \text{en} \longrightarrow \Delta\text{-}cis\text{-}\beta\text{-Co(en)(trien)}^{3+} \tag{5}$$

The stereochemical relationship outlined in Scheme 3, and established for the Co^{III} trien system, provides the basis for much of the subsequent work with other flexible linear tetramine ligands and the information up to 1971 for Co^{III} has been reviewed.[16]

The complete isomeric set (cis-α, cis-β-(SS)-, cis-β-(RS)-, $trans$-(RR,SS)- and $trans$-(R,S)-) of Co(trien)(NH$_3$)$_2^{3+}$ has recently been reported[31] and characterized by ^{13}C NMR.

The situation for octahedral M^{III} trien systems[770-774] other than Co^{III} and Cr^{III} is not quite so satisfactory. At best, only cis-α, cis-β or $trans$ configurations have been assigned, with little cognisance taken of the potential isomers available for the latter two.

One reason for an otherwise apparently excessive interest in Co(trien)X$_2^{n+}$ systems is the use of cis-Co(OH)(trien)(OH$_2$)$^{2+}$ in the hydrolysis of amino acid esters, amino acid amides and peptides[785] to form cis-β_1- and cis-β_2-Co(trien)(aa)$^{2+}$ (aa = amino acid) complexes.[16] In principle, a peptide could be degraded in a stepwise manner and each amino acid residue successively characterized. By the introduction of a chiral center into the backbone of the trien moiety, it was hoped to make such reactions stereoselective. Consequently, while fully N-alkylated trien systems have been widely investigated for M^{II} central ions, the C-alkylated trien systems have been almost exclusively the reserve of the Co^{III} chemist (Table 11).

In addition, it was observed that when complexes such as Λ-cis-β-Co(trien)(R-aa)$^{2+}$ were treated with base, the resultant product contained varying amounts of the chelated S-amino acid. The overall reaction is an epimerization at the α-carbon of the chelated amino acid and a range of selectivities has been achieved using a variety of chiral C-alkylated trien ligands.[824]

Several potential isomers are also possible for the square planar M^{II} trien systems (Scheme 4) and single-crystal X-ray data are available for Ni(trien)$^{2+}$, Pd(trien)$^{2+}$, Cu(trien)(NCS)$^+$ and Cu-(trien)(SCN)$^+$. Although the latter two are five-coordinate,[777,837] The trien adopts the racemic ($\delta\lambda\delta$) configuration (as has been proposed[769] for $trans$-(RR,SS)-CoCl$_2$(trien)$^+$).

The ($\lambda\epsilon\delta$) $meso$ configuration is adopted in the Ni^{II}[838] and $trans$-(R,S)-Co(CN)$_2$(trien)$^+$[839] complexes while the Pd^{II}[779] complex adopts the ($\epsilon\epsilon\epsilon$) $meso$ arrangement. Hg(trien)Cl$_2$ is also believed to be five coordinate.[840]

In addition to the mononuclear Ni(trien)$^{2+}$ ion, Ni^{II} also forms Ni$_2$(trien)$_3^{4+}$ with a bridging trien ligand occupying two sites on each central atom.[778]

Fully N-alkylated trien complexes are known for Cu^{II},[841] Pd^{II}[704] and Li^I[842] and the synthesis of Me$_6$- and Et$_6$-trien have been described.[841]

13.1.6.3   2,3,2-tet Systems

Co^{III}, Cr^{III}, Rh^{III}, Cu^{II} and Ni^{II} complexes with this ligand (51) were first prepared in 1966[843,844] and the developments of an appreciation of the isomeric complexities have paralleled those for trien.[16,845] The (RR,SS)- and (R,S)-$trans$-CoCl$_2$(2,3,2-tet)$^+$ isomers were isolated by Hamilton and Alexander[844] and are rather more easily obtained than the (RR,SS)-trien analogue, as the central six-membered ring allows more flexibility.

However, the preferred chair conformation for this central ring appears to stabilize the $trans$-(R,S) configuration and this is adopted for $trans$-MX$_2$(2,3,2-tet)$^+$ (M = Cr,[740] X = Cl, F, NCS; M = Ru,[410] X = Cl; M = Os,[846] X = Cl).

Most cis-MCl$_2$(2,3,2-tet)$^+$ complexes have been assigned the β-(RR,SS) configuration (M = Co,[847] Rh[848]) and the same configation is adopted by cis-CrX$_2$(2,3,2-tet)$^+$ (X = Cl, $\frac{1}{2}$ox).[849]

The Cu^{II}[850,851] and Ni^{II}[852,853] 2,3,2-tet complexes have been further studied and the single-crystal X-ray structure of [Cu(2,3,2-tet)](ClO$_4$)$_2$ shows this to have the (λ, chair, δ) conformation, with weakly bonded axial ClO$_4^-$ groups.[850]

Both trien[745] and 2,3,2-tet[856] have been used to treat Wilson's disease in patients sensitive to D-penicillamine. Wilson's disease results from an inherited metabolic disorder which leads to the

Table 11 *N*-Alkylated and *C*-Alkylated N$_4$ Linear Polyamines Used in the Formation of Co^{III} Complexes

Substituent	Refs.	Substituent	Refs.
	trien = N—C—C—N—C—C—N—C—C—N		
	1 2 3 4 5 6 7 8 9 10		
1-Me-	786	5,6-(CH$_2$)$_4$-(*R,R*)-	363, 812
5-Me-(*R,S*)-	363, 787	1,2:9,10-bis(CH$_2$)$_3$-(*S,S*)-	813
5-Alkyl-(*R,S*)-	788	1,2:9,10-bis(CH$_2$)$_3$-5-Me-	814
5-Me-(*R*)-	363, 784, 789–792	(2*S*:5*R*:9*S*)- and (2*S*:5*S*:9*S*)-	
5-Me-(*S*)-	789, 790	1,2:9,10-bis(CH$_2$)$_3$-5,6-	
1,10-Me$_2$-	786, 793	(CH$_2$)$_4$-(2*S*:5*R*:6*R*:9*S*)-	815
2,9-Me$_2$-(*S,S*)-	790, 792, 794–801	1,10-Bz$_2$-	816
3,8-Me$_2$-(*RR,SS*)-	787, 802	1,10-Bz$_2$-5-Me-(*R*)-	816
3,8-Me$_2$-(*R,S*)-	802, 507 [Pt^{II}]	5,6-(CH$_2$)$_4$-2,9-Bz-	397, 398
3,8-Me$_2$-(*S,S*)-	786, 787, 791, 792,	(2*S*:5*R*:6*R*:9*S*)-	
	802–806	1,5,10-Me$_3$-(*R*)-	786
4,7-Me$_2$-	34, 807–810	2,5,9-Me$_3$-(*S,R,S*)	790–792, 794
5,6-Me$_2$-(*RR,SS*)-	363	2,5,9-Me$_3$-(*S,S,S*)-	790–792, 794
5,6-Me$_2$-(*R,S*)-	363, 811	1,2,9,10-Me$_4$-(*S,S*)-	786
5,6-Me$_2$-(*S,S*)-	363	1,3,8,10-Me$_4$-(*S,S*)-	786
5,6-(CH$_2$)$_4$-(*RR,SS*)-	363	2,5,6,9-Me$_4$-(*S,R,S,S*)-	811
	2,2,3-tet = N—C—C—N—C—C—N—C—C—C—N		
	1 2 3 4 5 6 7 8 9 10 11		
9-Me-(*R,S*)-	32		
	2,3,2-tet = N—C—C—N—C—C—C—N—C—C—N		
	1 2 3 4 5 6 7 8 9 10 11		
5-Me-(*R*)-	817, 818	2,10-Me$_2$-(*S,S*)-	822, 823
5,7-Me$_2$-(*R,R*)-	819–821	1,2:10,11-bis(CH$_2$)$_3$-(*S,S*)-	814, 824–826
5,7-Me$_2$-(*R,S*)-	820	1,1,5,7-Me$_4$-	827
1,11-Me$_2$-	786	2,5,7,7,10-Me$_5$-(*R,R,R*)-	828 [Ni^{II}]
2,10-Me$_2$-(*RR,SS*)-	822	2,5,7,7,10-Me$_5$-(*R,S,R*)-	828 [Ni^{II}]
		1,5,7,11-Me$_4$-(*R,S*)-	820, 827
	3,2,3-tet = N—C—C—C—N—C—C—N—C—C—C—N		
	1 2 3 4 5 6 7 8 9 10 11 12		
6-Me-(*R,S*)-	829	2,4,9,11-Me$_4$-(*S,R,R,S*)	820, 835
6-Me-(*R*)-	829–831	2,4,9,11-Me$_4$-(*R,S,R,S*)-	820, 835
3,10-Me$_2$-(*RR,SS*)-	832	2,4,9,11-Me$_4$-(*R,R,S,S*)-	834
3,10-Me$_2$-(*R,S*)-	832	2,4,9,11-Me$_4$-(*R,S,S,R*)-	820
2,4,9,11-Me$_4$-(*R,R,R,R*)-	820, 833, 834	2,4,4,9,9,11-Me$_6$-(*RR,SS*)-	836 [Cu^{II}]

accumulation of excess deposits of copper in the body. D-Penicillamine is successfully used in 90% of the cases, but for the remainder, allergic responses require the use of alternative treatments.

The *N*- and *C*-alkylated derivatives of 2,3,2-tet are summarized in Table 11.

13.1.6.4 3,2,3-tet Systems

The preparation of Co$^{III\,854}$ and Cr$^{III\,743,855}$ complexes of 3,2,3-tet (**52**) followed chronologically after the trien and 2,3,2-tet analogues, but the stereochemical subtleties of this ligand in an octahedral environment were first developed by Bosnich and his coworkers.[856,857]

For Co^{III}, at least, the sequence of 6,5,6 in ring size imparts considerable *trans* stabilization, and *trans*-(*RR,SS*)-CoCl$_2$(3,2,3-tet)$^+$ reacts with monoamines to give CoCl(3,2,3-tet)(monoamine)$^{2+}$ with retention of both geometric and *sec*-NH configuration,[858] whereas most other *trans*-CoCl$_2$(N$_4$)$^+$ systems react to form *cis* products.[15]

The *cis-β*-(*RR,SS*) configuration can be forced with bidentate ligands but the *cis*-dichloro is difficult to stabilize.[16,857] For the less stereomobile Cr^{III} system, both *cis-β*-(*RR,SS*)- and *trans*-(*RR,SS*)-CrCl$_2$(3,2,3-tet)$^+$ have been characterized.[855]

Rather less work has been done with other transition metal complexes using 3,2,3-tet as a ligand. The Cu^{II},[836,859] Os$^{III\,846}$ and Ru$^{III\,410}$ complexes have been reported, with *trans* octahedral configurations assigned to the latter two.

13.1.6.5 3,3,3-tet and Other Linear Tetramine Ligands

Crystal structures of the Cu^{II} perchlorate salts of 3,3,3-tet[850] and 3,4,3-tet[860] show that all four N atoms of the ligands are bonding in a square planar arrangement with the six-membered rings adopting chair conformations and the *sec*-NH protons the *meso* configuration (similar to 2,3,2-tet,[850] but different from trien[777,837]). In $Cu(3,4,3\text{-tet})^{2+}$ the central seven-membered ring lies above the CuN_4 plane, almost blocking the axial position, while the two six-membered rings hang below in a 'parachute' arrangement (86). The terminal six-membered rings in $Cu(3,3,3\text{-tet})^{2+}$ are similarly orientated, but the central six-membered ring is not nearly so vertical.[850]

(86)

Co^{III} complexes with these ligands are rather unstable with respect to Co—N bond rupture and reduction to Co^{II}.[861] Nevertheless, *trans*-$CoCl_2(3,3,3\text{-tet})^+$ has been synthesized and characterized[862] by ^{13}C NMR and *trans*-$RuCl_2(3,3,3\text{-tet})^+$ has been reported recently.[410]

13.1.6.6 Miscellaneous Tetramine Ligands

1,2,3,4-Tetraminobutane (tab; 60) is, for spatial reasons, not expected to coordinate in either octahedral or square planar arrangements, with all four N atoms bonded. However, *C*-substituted derivatives of this ligand have been used to prepare Schiff's base complexes (Chapter 20.1) with Cu^{II} as the central atom.[863]

Likewise, the spirotetramine tam (61)[476,864a,864b] cannot bind with all four N atoms to one metal center; nevertheless, polymeric complexes (M = Cu^{II}, Co^{III}, Pt^{II}) using this ligand have been described.[865] It is also possible that the Co^{III} complex could be mononuclear with one arm of the ligand not coordinated, *e.g.* $Co(tamH)_2^{5+}$.[866a]

13.1.7 PENTAMINE LIGANDS

The skeleton types for linear and branched chain N_5 polyamine ligands are shown in (87)–(90). Commerically available technical 'tetraethylenepentamine' is a mixture of linear tetren,[866b] branched chain trenen and other unidentified materials. Thus it is not surprising, in hindsight, that five-coordinate Cu^{II} complexes of trenen[867,868] and octahedral Co^{III} complexes of tetren (*e.g.* $CoCl(tetren)^{2+}$ [869]) have been isolated from the technical polyamine. Both Fe^{III} [870–872] and Cr^{III} also give complexes with the linear tetren, rather than the branched chain isomer.

(87) tetren   (88) trenen   (89) 3,2,2,3-pent   (90) 4-Me-trenen

More reliably, but with more difficulty, Co^{III} complexes of both tetren and trenen have been prepared by Sargeson and his coworkers *via* reactions of coordinated ligands (Schemes 7, 8). In these cases, while the nature of the ligand is not in doubt, there is still a problem in deciding which topological isomer has been formed (Schemes 9, 10). Data for Co^{III} and Cr^{III} complexes with tetren, trenen and 4-Me-trenen, prior to 1977, have been reviewed.[15] In the dinuclear Fe^{III} complex $[Fe_2(tetren)_2O]^{4+}$, the ligand adopts the $\alpha\beta R$ configuration (99).[870,873]

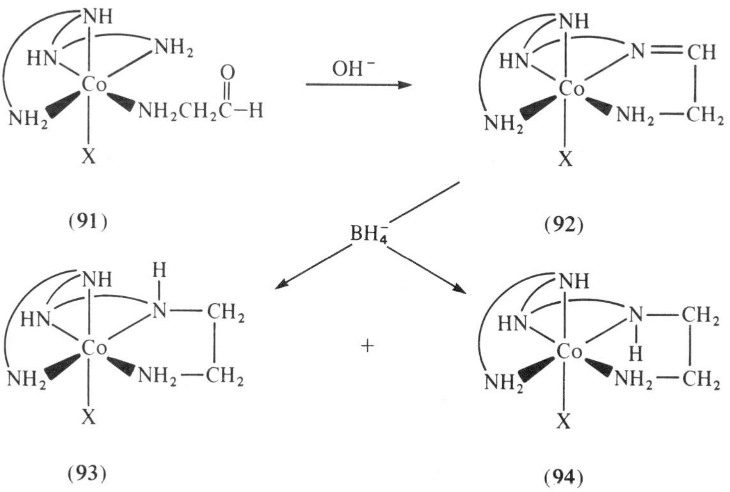

(91)　(92)

(93)　+　(94)

Scheme 7 Synthesis of CoX(tetren)$^{2+}$

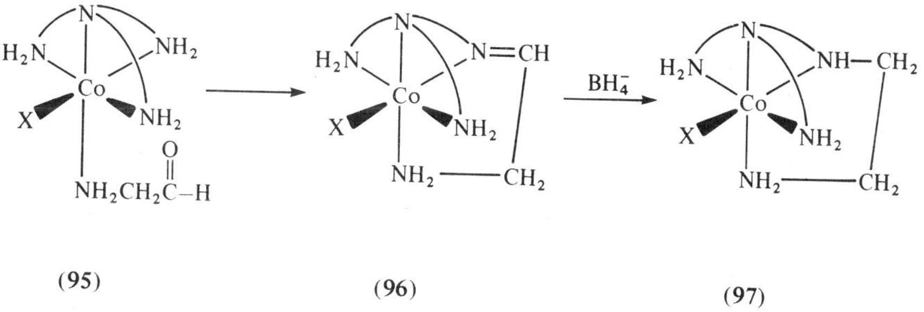

(95)　(96)　(97)

Scheme 8 Synthesis of CoCl(trenen)$^{2+}$

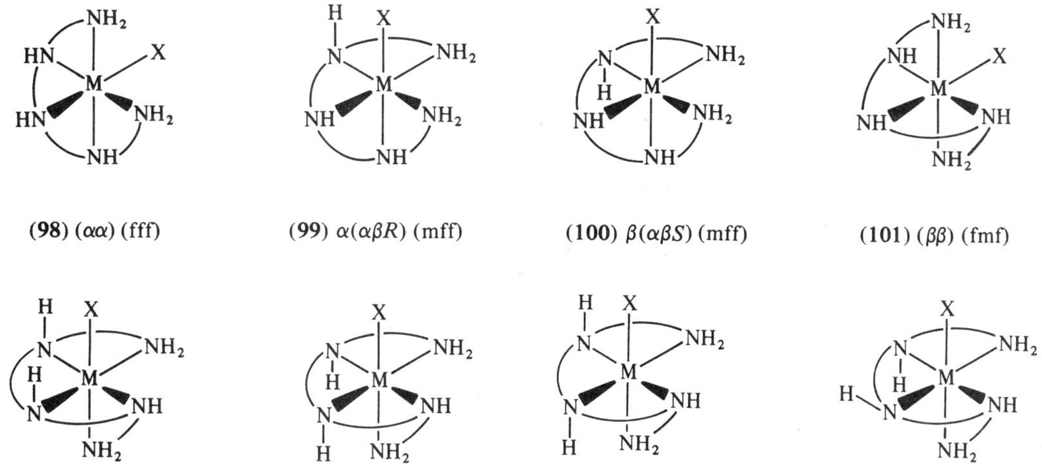

(98) ($\alpha\alpha$) (fff)　(99) $\alpha(\alpha\beta R)$ (mff)　(100) $\beta(\alpha\beta S)$ (mff)　(101) ($\beta\beta$) (fmf)

(102) (β-trans) (mmf)　(103) (mmf)　(104) (mmf)　(105) (mmf)

Scheme 9 Topological arrangements possible for a linear N$_5$ polyamine in an octahedral complex. The α,β nomenclature was originally used by House and Garner[869] and this was superseded by the ($\alpha\alpha$), ($\alpha\beta R$), ($\alpha\beta S$) *etc.* of Snow *et al.* based on the nomenclature used in trien systems.[873] More recently, Saito[47] has proposed the m,f nomenclature based on *me*riodional or *f*acial components of the chelated polyamine

Recently, μ-peroxo Co^{III} complexes with the linear pentamine ligand 3,2,2,3-pent (papd) have been described. Here the ligand adopts configuration (**105**), with three chelate rings in one plane and with the 'planar' *sec*-NH proton closest to the end of the polyamine adjacent to the μ-peroxo bridge. This particular μ-peroxo complex is remarkably resistant to O—O bond fission by acid.[874]

The naturally occurring polyamine 3,3,3,3-pent (caldopentamine) has recently been isolated from high-temperature resistant bacteria (*Thermus thermophilus*).[875]

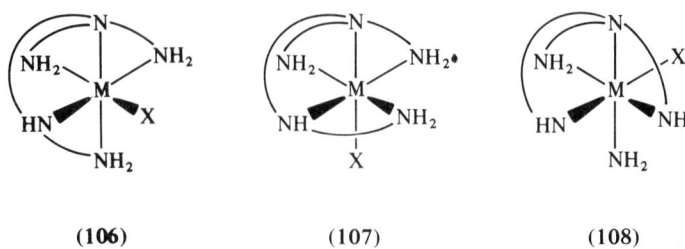

(106)   **(107)**   **(108)**

Scheme 10 Topological arrangements for octahedral trenen complexes

13.1.8 HEXAMINE LIGANDS

Both the linear hexamine ligands **(109)** and **(110)** are commercially available, but only Co[III] complexes with linpen have been described. All eight isomeric forms of Co(linpen)$^{3+}$ (Scheme 11) have been separated by chromatographic techniques[35,876] and the single-crystal X-ray structure of the ffff isomer **(111)** has been determined.[47,877]

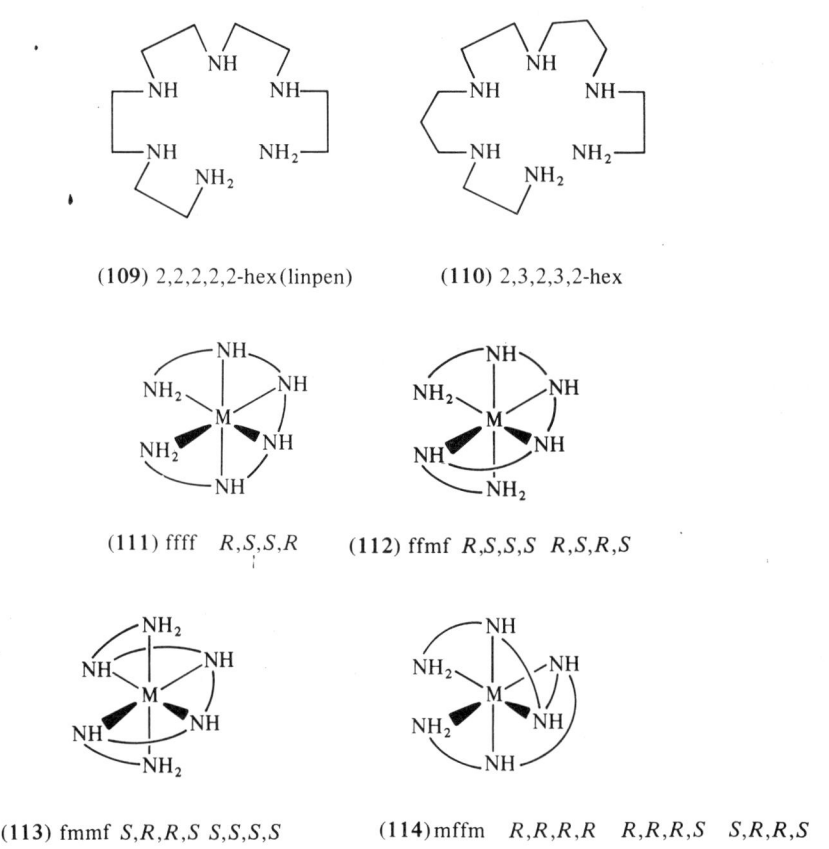

(**109**) 2,2,2,2,2-hex(linpen)   (**110**) 2,3,2,3,2-hex

(**111**) ffff   *R,S,S,R*   (**112**) ffmf *R,S,S,S   R,S,R,S*

(**113**) fmmf *S,R,R,S S,S,S,S*   (**114**) mffm   *R,R,R,R   R,R,R,S   S,R,R,S*

Scheme 11 Topological isomers possible for linpen and 2,3,2,3,2-hex. The *R,S* designation is applicable to linpen and the nomenclature is that used in Scheme 9

The branched chain penten **(115a)** (the amine analogue of edta) was first synthesized by Schwarzenbach and coworkers[878,879] and, more recently, several analogues **(115b–115f)** with differing central bridging systems have been prepared.[492,880–881]

The crystal structure of penten·5HCl[882] shows the ligand to have the central 'en' bridge in the *cis* conformation with the terminal arms adopting a *trans* arrangement (Section 13.1.4) and the crystal structure of Co(penten)$^{3+}$ shows[883] all six N atoms are coordinating, yielding a complex structurally similar to that formed by edta. However, Co[III] complexes of both penten and ttn **(115c)** can also be isolated in the 'pentamine' form, with one arm of the ligand dechelated.[492]

The ligands sen **(116a)**[884] and stn **(116b)**[885] are also formally of the hexamine type.

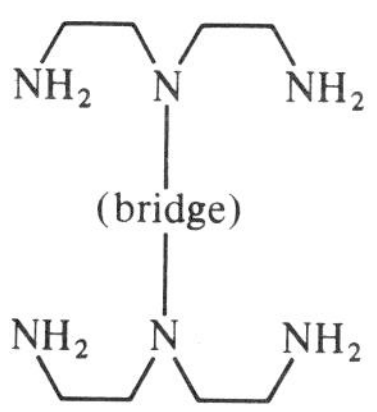

(115)

(bridge)

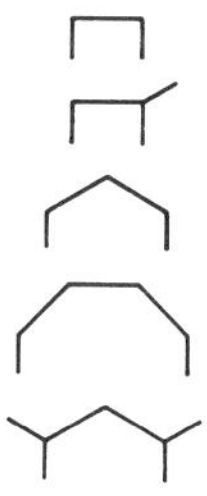

penten, ten (115a)

R-5-Me-penten, tpn (115b)

ttn (115c)

tbn (115d)

R,R-tptn (115e)
R,S-tptn (115f)

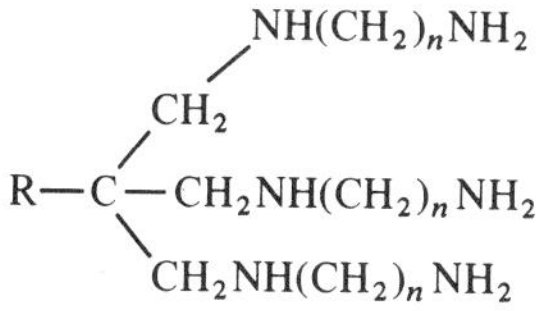

(116a) sen (R = Me, n = 2)

(116b) stn (R = Me, n = 3)

(116)

13.1.9 AMINOBENZENE LIGANDS

Ligands such as aniline (an), 1,2-diaminobenzene (dab, *o*-phenylenediamine) and 2,2′-dia-minobiphenyl[886,887] are classed separately, not because their ability to bind to a central metal is any less than the ligands discussed previously, but because of their potential 'non-innocent' behavior[888] with respect to internal redox reactions. Indeed, the dark blue complex isolated from the air oxidation of Co^{II}/dab in aqueous ethanol (a conventional route to yellow $Co(diamine)_3^{3+}$ systems) has been shown to have structure (117) with five-coordinate Co^{II}.[888] Related diimine complexes have been reported for Ni^{II}[889] as well as the conventional $Ni(dab)_3^{2+}$,[890] $Co(dab)_3^{3+}$[891] and $Pt(dab)_2^{2+}$[892] systems.

The structures of dab[893] and dab·2HBr[894] have been determined, showing that the two N atoms are coplanar with the benzene ring. For the free ligand, the lone pairs adopt a *cis* configuration with respect to the ring plane.[893]

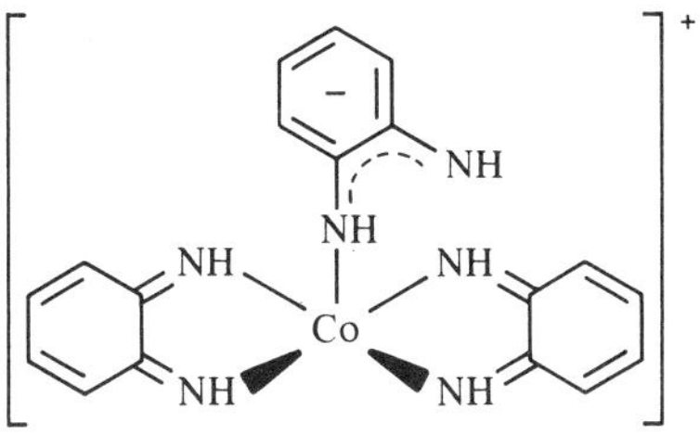

(117)

Attempts have been made to investigate the electrophilic substitution reactivity of coordinated aniline relative to the free ligand. For $CrCl_3(an)_3$, little rate enhancement was observed for bromina-tion reactions, but extensive complex decomposition accompanied the substitution.[895] Complexes such as *cis*-CoCl(en)$_2$(an)$^{2+}$ have abnormally high base hydrolysis rates when compared with their alkylamine analogues.[15]

Polyamine ligands such as (118)–(120) which contain terminal aniline residues have also been reported.[896–898]

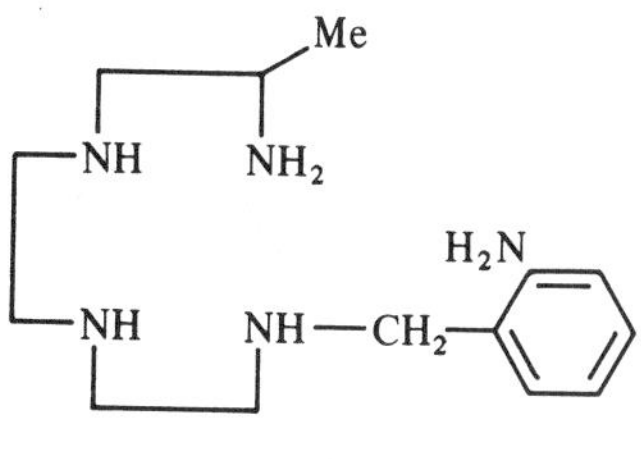

(118)

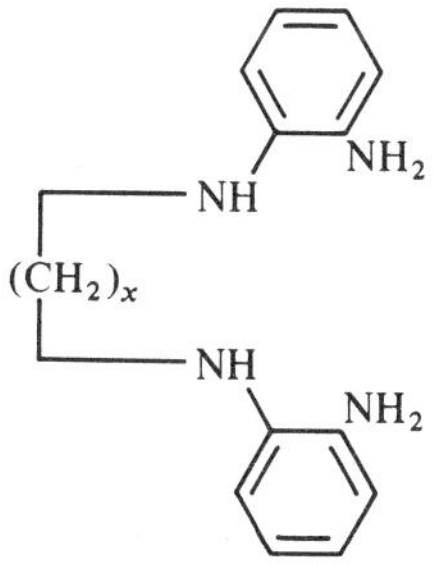

(119) x = 0, 1

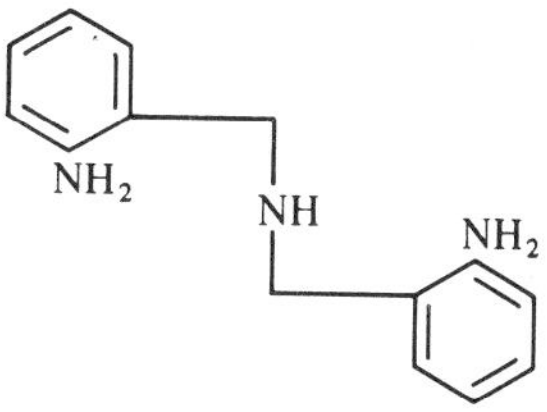

(120)

13.1.10 REFERENCES

1. J. W. Mellor, 'A Comprehensive Treatise on Inorganic and Theoretical Chemistry', Longmans, London, 1931.
2. Cambridge Crystallographic Data File.
3. J. M. Z. Gladych and D. Hartley, in 'Comprehensive Organic Chemistry', ed. D. H. R. Barton and W. D. Ollis, Pergamon, Oxford, 1979, vol. 2, p. 61.
4. J. R. Malpass, in 'Comprehensive Organic Chemistry', ed. D. H. R. Barton and W. D. Ollis, Pergamon, Oxford, 1979, vol. 2, p. 3.
5. G. Kneen, *Gen. Synth. Methods*, 1981, **4**, 172.
6. D. Ginsburg, 'Concerning Amines', Pergamon, Oxford, 1967.
7. E. Muller (ed.), *Methoden Org. Chem. (Houben-Weyl)*, **10**, **11**.
8. R. D. Spitz, in 'Kirk-Othmer Encyclopedia of Chemical Technology', 3rd edn., Wiley, New York, 1979, vol. 7, p. 580.
9. C. Caldarera, V. Zappia and U. Bachrach (eds.), 'Advances in Polyamine Research', Raven, New York.
10. M. G. B. Drew, J. Nelson and S. M. Nelson, *J. Chem. Soc., Dalton Trans.*, 1981, 1678.
11. C. S. Garner and D. A. House, *Transition Met. Chem. (New York)*, 1970, **6**, 59.
12. J. C. Chang, *J. Ind. Chem. Soc.*, 1977, 98.
13. R. G. Pearson, *J. Chem. Educ.*, 1968, **45**, 581, 643.
14. W. H. Baddley and F. Basolo, *J. Am. Chem. Soc.*, 1964, **86**, 2075.
15. D. A. House, *Coord. Chem. Rev.*, 1977, **23**, 223.
16. G. R. Brubaker, D. P. Schaefer, J. H. Worrell and J. I. Legg, *Coord. Chem. Rev.*, 1971, **7**, 161.
17. L. Spialter and J. A. Pappalardo, 'The Acyclic Aliphatic Tertiary Amines', Macmillan, New York, 1965.
18. T. P. Pitner and R. B. Martin, *J. Am. Chem. Soc.*, 1971, **93**, 4400.
19. D. A. House and Othman Nor, *Inorg. Chim. Acta*, 1983, **70**, 13.
20. J. E. Sarneski, A. T. McPhail, K. D. Onan, L. E. Erikson and C. N. Reilley, *J. Am. Chem. Soc.*, 1977, **99**, 7376.
21. A. R. Gainsford and A. M. Sargeson, *Aust. J. Chem.*, 1978, **31**, 1679.
22. M. Goto, M. Takeshita and T. Sakai, *Bull. Chem. Soc. Jpn.*, 1981, **54**, 2491.
23. M. da Costa Ferreira and H. E. Toma, *J. Chem. Soc., Dalton Trans.*, 1983, 2051.
24. I. Mochida and J. C. Bailar, Jr., *Inorg. Chem.*, 1983, **22**, 1834.
25. R. S. Nyholm and M. L. Tobe, *J. Chem. Soc.*, 1956, 1707.
26. T. Kemmerich, J. H. Nelson, N. E. Takach, H. Boehme, B. Jablonski and W. Beck, *Inorg. Chem.*, 1982, **21**, 1226.
27. P. Newman, 'Optical Resolution Procedures for Chiral Compounds', vol. 1, 'Amines and Related Compounds', Optical Resolution Information Center, Manhattan College, Riverdale, NY, 1978.
28. D. A. House and J. W. Blunt, *Inorg. Nucl. Chem. Lett.*, 1975, **11**, 219.
29. J. W. Blunt, Foo Chuk Ha and D. A. House, *Inorg. Chim. Acta*, 1979, **32**, L5.
30. G. H. Searle and T. W. Hambley, *Aust. J. Chem.*, 1982, **35**, 1297.
31. S. Utsuno, Y. Sakai, Y.Yoshikawa and H. Yamatera, *J. Am. Chem. Soc.*, 1980, **102**, 6903.
32. G. R. Brubaker and D. W. Johnson, *Inorg. Chim. Acta*, 1982, **62**, 141.
33. J. Mason, *Chem. Br.*, 1983, 654.
34. R. A. Henderson and M. L. Tobe, *Inorg. Chem.*, 1977, **16**, 2576.
35. H. Nakazawa and H. Yoneda, *J. Chromatogr.*, 1978, **160**, 89.
36. A. G. Sykes and J. A. Weil, *Prog. Inorg. Chem.*, 1970, **13**, 1.
37. Y. Sasaki, J. Fujita and K. Saito, *Bull. Chem. Soc. Jpn.*, 1971, **44**, 3373.
38. C. G. Barraclough, G. A. Lawrance and P. A. Lay, *Inorg. Chem.*, 1978, **17**, 3317.
39. G. A. Lawrance and P. A. Lay, *J. Inorg. Nucl. Chem.*, 1979, **41**, 301.
40. J. D. Ortego and M. Seymour, *Polyhedron*, 1982, **1**, 21.
41. D. L. Duffy, D. A. House and J. A. Weil, *J. Inorg. Nucl. Chem.*, 1969, **31**, 2053.
42. S. Fallab and P. R. Mitchell, *Adv. Inorg. Bioinorg. Mech.*, 1984, **3**, 311.
43. K. Weighardt, W. Schmidt, B. Nuber and J. Weiss, *Chem. Ber.*, 1979, **112**, 2220.
44. M. M. Hagan, 'Clathrate Inclusion Compounds', Van Nostrand-Reinhold, New York, 1962.
45. S. Nishikiori, T. Iwamoto and Y. Yoshino, *Bull. Chem. Soc. Jpn.*, 1980, **53**, 2236.
46. D. A. Buckingham and A. M. Sargeson, *Top. Stereochem.*, 1971, **6**, 219.
47. Y. Saito, *Top Stereochem.*, 1978, **10**, 96.
48. G. J. McDougall, R. D. Hancock and J. C. A. Boeyens, *J. Chem. Soc., Dalton Trans.*, 1978, 1438.
49. R. S. Cahn, C. Ingold and V. Prelog, *Angew. Chem., Int. Ed. Engl.*, 1966, **5**, 385.
50. E. C. Franklin, 'The Nitrogen System of Compounds', Reinhold, New York, 1935.
51. D. Papousek, *J. Mol. Struct.*, 1983, **100**, 179.
52. S. Balt and A. Jelsma, *Inorg. Chem.*, 1981, **20**, 733.
53. B. N. Figgis, B. W. Skelton and A. H. White, *Aust. J. Chem.*, 1979, **32**, 417.
54. E. F. Epstein and I. Bernal, *J. Chem. Soc. (A)*, 1971, 3628.
55. T. V. Long, A. W. Herlinger, A. W. Epstein and I. Bernal, *Inorg. Chem.*, 1970, **9**, 459.
56. N. E. Kime and J. A. Ibers, *Acta Crystallogr., Part B*, 1969, **25**, 168.
57. R. T. Welberry and B. E. Williamson, *Acta Crystallogr., Part C*, 1983, **39**, 513.
58. D. W. Meek and J. A. Ibers, *Inorg. Chem.*, 1970, **9**, 465.
59. M. Iwata and Y. Saito, *Acta Crystallogr., Part B*, 1973, **29**, 822.
60. M. Iwata, *Acta Crystallogr., Part B*, 1977, **33**, 59.
61. H. J. Haupt, F. Huber and H. Preut, *Z. Anorg. Allg. Chem.*, 1976, **422**, 97.
62. E. O. Schlemper, *J. Cryst. Mol. Struct.*, 1977, 7, 81.
63. T. Watanabe, M. Atoji and C. Okazaki, *Acta Crystallogr.*, 1950, **3**, 405.
64. G. J. Kruger and E. C. Reynhardt, *Acta Crystallogr., Part B*, 1978, **34**, 915.
65. P. Mauersberger, H. J. Haupt and F. Huber, *Acta Crystallogr., Part B*, 1979, **35**, 295.
66. (a) A. W. Herlinger, J. N. Brown, M. A. Dwyer and S. F. Pavkovic, *Inorg. Chem.*, 1981, **20**, 2366; (b) W. Clegg, *J. Chem. Soc., Dalton Trans.*, 1982, 593.
67. I. Bernal, N. Elliott, R. A. Lalancette and T. Brennan, in 'Progess in Coordination Chemistry', ed. M. Cais, Elsevier, Amsterdam, 1968, p. 518.

68. K-F. Tebbe, *Acta Crystallogr., Part C*, 1983, **39**, 154.
69. M. T. Barnet, B. M. Craven, H. C. Freeman, N. E. Kime and J. A. Ibers, *Chem. Commun.*, 1966, 307.
70. K. N. Raymond, D. W. Meek and J. A. Ibers, *Inorg. Chem.*, 1968, **7**, 1111.
71. M. Mori, Y. Saito and T. Watanabe, *Bull. Chem. Soc. Jpn.*, 1961, **34**, 295.
72. F. A. Jurnak and K. N. Raymond, *Inorg. Chem.*, 1974, **13**, 2387.
73. S. A. Goldfield and K. N. Raymond, *Inorg. Chem.*, 1971, **10**, 2604.
74. W. Clegg, *Acta Crystallogr., Part B*, 1978, **34**, 3328.
75. W. Clegg, D. A. Greenhalgh and B. P. Straughan, *J. Chem. Soc., Dalton Trans.*, 1975, 2591.
76. (a) W. Clegg, *Acta Crystallogr., Part B*, 1976, **32**, 2907; (b) K. Wieghardt and J. Weiss, *Acta Crystallogr., Part B*, 1972, **28**, 529.
77. H. C. Stynes and J. A. Ibers, *Inorg. Chem.*, 1971, **10**, 2304.
78. M. Weishaupt, H. Bezler and J. Strahle, *Z. Anorg. Allg. Chem.*, 1978, **440**, 52.
79. H. Endres, H. J. Keller, W. Moroni, D. Nothe and Vu Dong, *Acta Crystallogr., Part B*, 1978, **34**, 1703.
80. B. Lippert, C. J. L. Lock, B. Rosenberg and M. Zvagulis, *Inorg. Chem.*, 1977, **16**, 1525; 1978, *17*, 2971.
81. M. Atoji, J. W. Richardson and R. E. Rundle, *J. Am. Chem. Soc.*, 1957, **79**, 3017.
82. F. D. Rochon and R. Melanson, *Acta Crystallogr., Part B*, 1980, **36**, 691.
83. M. Tanaka, I. Tsujikawa, K. Toriumi and T. Ito, *Acta Crystallogr., Part B*, 1982, **38**, 2793.
84. K.-F. Tebbe and B. Freckmann, *Z. Naturforsch., Teil B*, 1982, **37**, 542.
85. T. Distler and P. A. Vaughan, *Inorg. Chem.*, 1967, **6**, 126.
86. M. Duggan, N. Ray, B. Hathaway, G. Tomlinson, P. Brint and K. Pelin, *J. Chem. Soc., Dalton Trans.*, 1980, 1342.
87. M. Leskela and J. Valkonen, *Acta Chem. Scand., Ser. A*, 1978, **32**, 805.
88. B. Morosin, *Acta Crystallogr., Part B*, 1969, **25**, 19.
89. B. Morosin, *Acta Crystallogr., Part B*, 1976, **32**, 1237.
90. K.-F. Tebbe, *Z. Anorg. Allg. Chem.*, 1982, **489**, 93.
91. A. Ferrari, A. Braibanti and A. Tiripicchio, *Acta Crystallogr.*, 1966, **21**, 605.
92. Huang Tin-Ling, Li Tien-Ming and Li Tia-Xi, *Acta Chim. Sinica*, 1966, **32**, 162.
93. K.-F. Tebbe, *Z. Kristallogr.*, 1980, **153**, 297.
94. A. Migdal-Mikuli, E. Mikuli, M. Rachwalska and S. Hodorowicz, *Phys. Status Solidi A*, 1978, **47**, 57 (*Chem. Abstr.*, 1978, **89**, 51 651).
95. T. Yamaguchi and H. Ohtaki, *Bull. Chem. Soc. Jpn.*, 1979, **52**, 1223.
96. K.-F. Tebbe and M. Plewa, *Z. Anorg. Allg. Chem.*, 1982, **489**, 111.
97. H. M. Maurer and A. Weiss, *Z. Kristallogr.*, 1977, **146**, 227.
98. E. Kai, T. Misawa and K. Nishimoto, *Bull. Chem. Soc. Jpn.*, 1980, **53**, 2481.
99. A. B. P. Lever, I. M. Walker, P. J. McCarthy, K. B. Mertes, A. Jircitano and R. Sheldon, *Inorg. Chem.*, 1983, **22**, 2252.
100. C. Glidewell, *J. Coord. Chem.*, 1977, **6**, 189.
101. G. G. Schlessinger, 'Inorganic Laboratory Preparations', Chemical Publishing Company, New York, 1962.
102. K. F. Purcell and J. C. Kotz, 'Inorganic Chemistry', W. B. Saunders, Eastbourne, 1977, p. 659.
103. S. S. Isied, A. Vassilian and J. M. Lyon, *J. Am. Chem. Soc.*, 1982, **104**, 3910.
104. P. C. Ford, *Coord. Chem. Rev.*, 1970, **5**, 75.
105. H. Taube, *Pure Appl. Chem.*, 1979, **51**, 901.
106. D. Waysbort and G. Navon, *Inorg. Chem.*, 1979, **18**, 9.
107. Y. Pauleau, J. J. Hantzpergue and J. C. Remy, *Bull. Soc. Chim. Fr.*, 1978, 246.
108. W. S. Glaunsinger, *J. Phys. Chem.*, 1980, **84**, 1163.
109. W. R. McWhinnie, J. D. Miller, B. J. Watts and D. Y. Wadden, *Inorg. Chim. Acta*, 1973, **7**, 461.
110. R. Mitzner, W. Depkat and P. Blankenburg, *Z. Chem.*, 1970, **10**, 34.
111. D. A. Buckingham, B. M. Foxman and A. M. Sargeson, *Inorg. Chem.*, 1970, **9**, 1790.
112. L. F. Book, K. Y. Hui, O. W. Lau and W.-K. Li, *Z. Anorg. Allg. Chem.*, 1976, **426**, 215, 227.
113. T. W. Swaddle, *Can. J. Chem.*, 1977, **55**, 3166.
114. K. Ohashi, T. Hasegawa, Y. Kurimura and K. Yamamoto, *J. Inorg. Nucl. Chem.*, 1978, **40**, 647.
115. A. Rodgers and P. J. Staples, *J. Chem. Soc.*, 1965, 6834.
116. M. Parris and W. J. Wallace, *Can. J. Chem.*, 1969, **47**, 2257.
117. B. R. Foxman, *Inorg. Chem.*, 1978, **17**, 1932.
118. K. L. Brown and D. Hall, *Acta Crystallogr., Part B*, 1976, **32**, 279.
119. M. E. Cradwick, D. Hall and R. K. Phillips, *Acta Crystallogr., Part B*, 1971, **27**, 480.
120. R. Lussier, J. O. Edwards and R. Eisenberg, *Inorg. Chim. Acta*, 1969, **3**, 468.
121. C. J. Hawkins and G. A. Lawrance, *Inorg. Nucl. Chem. Lett.*, 1973, **9**, 483.
122. M. W. G. de Bolster and J. W. M. Wegener, *J. Inorg. Nucl. Chem.*, 1977, **39**, 1458.
123. S. Balt, M. W. G. de Bolster and H. J. Gamelkoorn, *Inorg. Chim. Acta*, 1979, **35**, L329.
124. A. C. Dash, N. K. Mohanty and R. K. Nanda, *J. Indian Chem. Soc.*, 1977, **54**, 89.
125. B. M. Oulaghan and D. A. House, *Inorg. Chem.*, 1978, **17**, 2197.
126. G. Bombieri, F. Benetollo, A. del Pra, M. L. Tobe and D. A. House, *Inorg. Chim. Acta*, 1981, **50**, 89.
127. Foo Chuk Ha, D. A. House and J. W. Blunt, *Inorg. Chim. Acta*, 1979, **33**, 269.
128. A. G. Gantsev, Sh. Sh. Tukhtaev and Kh. U. Ikramov, *Russ. J. Inorg. Chem. (Engl. Transl.)*, 1972, **17**, 717.
129. R. W. Green and B. Bell, *Aust. J. Chem.*, 1973, **26**, 1663.
130. S. Matsumoto and S. Kawaguchi, *Bull. Chem. Soc. Jpn.*, 1980, **53**, 1577.
131. S. Okeya, Y. Nakamura and S. Kawaguchi, *Bull. Chem. Soc. Jpn.*, 1982, **55**, 1460.
132. I. S. Ahuja, R. Singh and C. L. Yadava, *Spectrochim. Acta, Part A*, 1981, **37**, 407.
133. Y. C. Tang and J. H. Sturdivant, *Acta Crystallogr.*, 1952, **5**, 74.
134. M. Shimoi, A. Ouchi, M. Aikawa, S. Saito and Y. Saito, *Bull. Chem. Soc. Jpn.*, 1982, **55**, 2089.
135. E. C. Alyea, J. S. Basi, D. C. Bradley and M. H. Chisholm, *Chem. Commun.*, 1968, 495.
136. F. Cardulla, *J. Chem. Educ.*, 1983, **60**, 505.
137. R. L. Collin and W. N. Lipscomb, *Acta Crystallogr.*, 1951, **4**, 10.
138. P. A. Giguere and V. Schomaker, *J. Am. Chem. Soc.*, 1943, **65**, 2025.

139. F. G. Baglin, S. F. Bush and J. R. Durig, *J. Chem. Phys.*, 1967, **47**, 2104.
140. K. Kohata, T. Fukuyama and K. Kuchitsu, *J. Phys. Chem.*, 1982, **86**, 602.
141. A. F. Wells, 'Structural Inorganic Chemistry', 2nd edn., Oxford University Press, Oxford, 1950, p. 246.
142. K. N. Trueblood, C. B. Knobler, D. S. Lawrence and R. V. Stevens, *J. Am. Chem. Soc.*, 1982, **104**, 1355.
143. F. A. Cotton and G. Wilkinson, 'Advanced Inorganic Chemistry', 3rd edn., Wiley-Interscience, New York, 1972, p. 350.
144. D. Sellmann, P. Kreutzer, G. Huttner and A. Frank, *Z. Naturforsch., Teil B*, 1978, **33**, 1341 (*Chem. Abstr.*, 1979, **90**, 33 234).
145. J. A. Broomhead, J. Budge, J. H. Enemark, R. D. Feltham, J. I. Gelder and P. L. Johnson, *Adv. Chem. Ser.*, 1977, **162**, 421 (*Chem. Abstr.*, 1977, **87**, 94 725).
146. R. Ya. Aliev, M. N. Guseinov and N. G. Klyuchnikov, *Zh. Neorg. Khim.*, 1977, **22**, 3381 (*Chem. Abstr.*, 1978, **88**, 57 760).
147. A. I. Stetsenko, L. S. Tikhonova, L. I. Shigina and J. M. Ginsburg, *Zh. Neorg. Khim.*, 1978, **23**, 1871 (*Chem. Abstr.*, 1978, **89**, 122 239).
148. R. Ya. Aliev, D. B. Musaev, A. D. Kuliev and N. G. Klyuchnikov, *Zh. Neorg. Khim.*, 1980, **25**, 1801 (*Chem. Abstr.*, 1980, **93**, 140 248).
149. R. Ohme and E. Schmitz, *Ber.*, 1964, **97**, 297.
150. J. H. Grimes, R. G. Hannis and A. J. Huggard, *J. Chem. Soc.*, 1964, 266.
151. R. Kiesel and E. P. Schram, *Inorg. Chem.*, 1973, **12**, 1090.
152. H. Van Der Does, F. C. Mijlhoff and G. H. Reres, *J. Mol. Struct.*, 1981, **74**, 153.
153. S. R. Wade and G. R. Willey, *J. Chem. Soc., Dalton Trans.*, 1981, 1264.
154. W. Petz, *J. Organomet. Chem.*, 1979, **172**, 415.
155. S. Jamet-Delcroix, *J. Chim. Phys.*, 1967, **64**, 601; *Acta Crystallogr., Part B*, 1973, **29**, 977.
156. A. Yokozeki and K. Kuchitsu, *Bull Chem. Soc. Jpn.*, 1970, **43**, 2664.
157. N. Hadjiliadis, A. Diot and T. Theophanides, *Can. J. Chem.*, 1972, **50**, 1005.
158. K. Rasmussen and C. Tosi, *Acta Chem. Scand., Ser. A*, 1983, **37**, 79.
159. T. Ashida and S. Hirokawa, *Bull. Chem. Soc. Jpn.*, 1963, **36**, 704.
160. C. H. Koo, M. I. Kim and C. S. Yoo, *J. Korean Chem. Soc.*, 1963, **1**, 293.
161. I. Søtofte, *Acta Chem. Scand., Ser. A*, 1976, **30**, 309.
162. K. Sakurai, *J. Phys. Soc. Jpn.*, 1961, **16**, 1205.
163. R. W. Berg and I. Søtofte, *Acta Chem. Scand., Ser. A*, 1976, **30**, 843.
164. K. Tichy, J. Benes, W. Halg and H. Arend, *Acta Crystallogr., Part B*, 1978, **34**, 2970.
165. G. B. Birrell and B. Zaslow, *J. Inorg. Nucl. Chem.*, 1972, **34**, 1751.
166. I. Lofving, *Acta Chem. Scand., Ser. A*, 1976, **30**, 715.
167. H. W. Smith and W. J. Stratton, *Inorg. Chem.*, 1977, **16**, 1640.
168. D. N. Anderson and R. D. Willet, *Inorg. Chim. Acta*, 1971, **5**, 41.
169. A. Dunand and R. Gerdil *Acta Crystallogr., Part B*, 1975, **31**, 370.
170. N. Gavrushento, H. L. Carrell, W. C. Stallings and J. P. Glusker, *Acta Crystallogr., Part B*, 1977, **33**, 3936.
171. S. Pérez, J. M. Léger and J. Housty, *Cryst. Struct. Commun.*, 1973, **2**, 303.
172. S. Pérez, *Acta Crystallogr., Part B*, 1976, **32**, 2064.
173. R. A. Palmer and M. F. C. Ladd, *J. Cryst. Mol. Struct.*, 1977, **7**, 123.
174. C. K. Fair and E. O. Schlemper, *Acta Crystallogr., Part B*, 1977, **33**, 1337.
175. S. Pérez, *Acta Crystallogr., Part B*, 1977, **33**, 1083.
176. G. J. Reibnegger and B. M. Rode, *Inorg. Chim. Acta*, 1983, **72**, 47.
177. A. Pajunen and S. Pajunen, *Acta Crystallogr., Part B*, 1980, **36**, 2425.
178. T. Iwamoto and D. F. Shriver, *Inorg. Chem.*, 1971, **10**, 2428.
179. T. Theophanides and P. C. Kong, *Inorg. Chim. Acta*, 1971, **5**, 485.
180. M. R. Truter and E. G. Cox, *J. Chem. Soc.*, 1956, 948.
181. H. Ogino *Bull. Chem. Soc. Jpn.*, 1977, **50**, 2459; *Inorg. Chem.*, 1980, **19**, 1619, 3178.
182. H. Ogino and J. Fujita, *Bull. Chem. Soc. Jpn.*, 1975, **48**, 1836.
183. M. D. Alexander and H. G. Kilcrease, *J. Inorg. Nucl. Chem.*, 1973, **35**, 1583.
183a. R. W. Hay and K. B. Nolan, *J. Chem. Soc., Dalton Trans.*, 1975, 1621.
184. S. C. Chan and S. F. Chan, *J. Inorg. Nucl. Chem.*, 1973, **35**, 1247.
185. M. D. Alexander and C. A. Spillert, *Inorg. Chem.*, 1970, **9**, 2344.
186. 'Werner Centennial', Advances in Chemistry, ACS, 1967, **62**.
187. M. Kuramoto, Y. Kushi and H. Yoneda, *Bull. Chem. Soc. Jpn.*, 1978, **51**, 3196.
188. T. W. Hambley, C. J. Hawkins, J. Martin, J. A. Palmer and M. R. Snow, *Aust. J. Chem.*, 1981, **34**, 2505.
189. J. R. Gollogly, C. J. Hawkins, and J. K. Beattie, *Inorg. Chem.*, 1971, **10**, 317.
190. D. F. Evans and G. C. de Villardi, *J. Chem. Soc., Dalton Trans.*, 1978, 315.
191. J. W. Vaughn, *J. Cryst. Spectrochem. Res.*, 1983, **13**, 231.
192. A. Whuler, P. Spinat and C. Brouty, *Acta Crystallogr.*, 1980, **36**, 1086.
193. H. J. Haupt and F. Huber, *Z. Anorg. Allg. Chem*, 1978, **442**, 31.
194. J. T. Veal and D. J. Hodgson, *Inorg. Chem.*, 1972, **11**, 597.
195. D. J. Hodgson P. K. Hale and W. E. Hatfield, *Inorg. Chem.*, 1971, **10**, 1061; see also *Inorg. Chem.*, 1985, **24**, 1194.
196. J. H. Enemark, M. S. Quinby, L. L. Reed, M. J. Steuck and K. K. Walthers, *Inorg. Chem.*, 1970, **9**, 2397.
197. Y. Kushi, M. Kuramoto and H. Yoneda, *Chem. Lett.*, 1976, 135.
198. D. H. Templeton, A. Zalkin, H. W. Ruben and L. K. Templeton, *Acta Crystallogr., Part B*, 1979, **35**, 1608.
199. J. C. Barnes and J. D. Paton, *Acta Crystallogr., Part B*, 1982, **38**, 1588.
200. E. N. Duesler and K. N. Raymond, *Inorg. Chem.*, 1971, **10**, 1486.
201. A. Nakahara, Y. Saito and H. Kuroya, *Bull. Chem. Soc. Jpn.*, 1952, **25**, 331.
202. S. Ooi and H. Kuroya, *Bull. Chem. Soc. Jpn.*, 1963, **36**, 1083.
203. K. Brodersen, J. Rath and G. Thiele, *Z. Anorg. Allg. Chem.*, 1972, **394**, 13.
204. S. Ooi, Y. Komiyama, Y. Saito and H. Kuroya, *Bull. Chem. Soc. Jpn.*, 1959, **32**, 263.

205. T. Okamato, K. Matsumoto and H. Kuroya, *Nippon Kagaku Zasshi*, 1970, **91**, 650.
206. O. Börtin, *Acta Chem. Scand., Ser. A*, 1976, **30**, 657.
207. I. Grenthe and E. Nordin, *Inorg. Chem.*, 1979, **18**, 1109.
208. O. Börtin, *Acta Chem. Scand., Ser. A*, 1976, **30**, 503.
209. T. J. Kistenmacher, *Acta Crystallogr., Part B*, 1975, **31**, 85.
210. T. J. Kistenmacher and D. J. Szalda, *Acta Crystallogr., Part B*, 1975, **31**, 90.
211. C. L. Raston, A. H. White and J. K. Yandell, *Aust. J. Chem.*, 1978, **31**, 993.
212. C. L. Raston, A. H. White and J. K. Yandell, *Aust. J. Chem.*, 1980, **33**, 1123.
213. K. Akabori and Y. Kushi, *J. Inorg. Nucl. Chem.*, 1978, **40**, 625.
214. K. N. Raymond, P. W. R. Corfield and J. A. Ibers, *Inorg. Chem.*, 1968, **7**, 842.
215. P. R. Robinson, E. O. Schlemper and R. K. Murman, *Inorg. Chem.*, 1975, **14**, 2035.
216. X. Solans, M. Font-Altaba, J. L. Briansó, A. Solans, J. Casabó and J. Ribas, *Cryst. Struct. Commun.*, 1982, **11**, 1199.
217. P. Spinat, A. Whuler and C. Brouty, *J. Appl. Crystallogr.*, 1980, **13**, 616.
218. (a) H. J. Peresie and J. A. Stanko, *Chem. Commun.*, 1970, 1674; (b) R. J. Smolenaers, J. K. Beattie and N. P. Hutchinson, *Inorg. Chem.*, **20**, 2202.
219. L. N. Swink and M. Atoji, *Acta Crystallogr.*, 1960, **13**, 639.
220. J. D. Korp, I. Bernal, R. A. Palmer and J. C. Robinson, *Acta Crystallogr., Part B*, 1980, **36**, 560.
221. Mazhar-Ul-Haque, C. N. Caughlan and K. Emerson, *Inorg. Chem.*, 1970, **9**, 2421.
222. G. B. Jameson, R. Schneider, E. Dubler and H. R. Oswald, *Acta Crystallogr., Part B*, 1982, **38**, 3016.
223. R. E. Cramer and J. T. Huneke, *Inorg. Chem.*, 1978, **17**, 365.
224. B. W. Brown and E. C. Lingafelter, *Acta Crystallogr.*, 1963, **16**, 753.
225. F. J. Llewellyn and J. M. Waters, *J. Chem. Soc.*, 1962, 3845.
226. A. J. Finney, M. A. Hitchman, C. L. Raston, G. L. Rowbottom and A. H. White, *Aust. J. Chem.*, 1981, **34**, 2047.
227. R. Stomberg, *Acta Chem. Scand.*, 1969, **23**, 3498.
228. I. Bertini, P. Dapporto, D. Gatteschi and A. Scozzafava, *J. Chem. Soc., Dalton Trans.*, 1979, 1409.
229. D. L. Cullen and E. C. Lingafelter, *Inorg. Chem.*, 1970, **9**, 1859.
230. Y. Komiyama and E. C. Lingafelter, *Acta Crystallogr.*, 1964, **17**, 1145.
231. A. Pajunen, *Suom. Kemistil. B*, 1967, **40**, 32.
232. D. S. Brown, J. D. Lee, B. G. A. Melsom, B. J. Hathaway, I. M. Procter and A. A. G. Tomlinson, *Chem. Commun.*, 1967, 369.
233. B. W. Brown and E. C. Lingafelter, *Acta Crystallogr.*, 1964, **17**, 254.
234. H. Scouloudi, *Acta Crystallogr.*, 1953, **6**, 651.
235. R. D. Ball, D. Hall, C. E. F. Rickard and T. N. Waters, *J. Chem. Soc. (A)*, 1967, 1435.
236. J. R. Wiesner and E. C. Lingafelter, *Inorg. Chem.*, 1966, **5**, 1770.
237. H. M. Colquhoun, J. F. Stoddart and D. J. Williams, *J. Chem. Soc., Chem. Commun.*, 1981, 851.
238. W. A. Freeman, *Inorg. Chem.*, 1976, **15**, 2235.
239. H. Endres, H. J. Keller, R. Martin, Hae Nam Gung and U. Traeger, *Acta Crystallogr., Part B*, 1979, **35**, 1885.
240. N. Matsumoto, M. Yamashita, S. Kida and I. Ueda, *Acta Crystallogr., Part B*, 1979, **35**, 1458.
241. H. Endres, H. J. Keller, R. Martin and U. Traeger, *Acta Crystallogr., Part B*, 1979, **35**, 2880.
242. F. H. Herbstein and R. E. Marsh, *Acta Crystallogr., Part B*, 1982, **38**, 1051.
243. H. J. Keller and U. Traeger, *Acta Crystallogr., Part B*, 1979, **35**, 1887.
244. H. J. Keller, B. Keppler and H. Pritzkow, *Acta Crystallogr., Part B*, 1982, **38**, 1603.
245. B. Freckmann and K.-F. Tebbe, *Acta Crystallogr., Part B*, 1981, **37**, 1520.
246. P. Spinat, C. Brouty and A. Whuler, *Acta Crystallogr., Part B*, **36**, 1980, 544, 1267.
247. C. J. Hawkins, 'Absolute Configuration of Metal Complexes', Wiley-Interscience, New York, 1971.
248. R. D. Gillard and L. R. H. Tipping, *J. Chem. Soc., Dalton Trans.*, 1977, 1241.
249. F. T. Williams, *J. Chem. Educ.*, 1962, **39**, 211.
250. S. R. Niketic and K. Rasmussen, *Acta Chem. Scand., Ser. A*, 1978, **32**, 391.
251. S. R. Niketic and K. Rasmussen, *Acta Chem. Scand., Ser. A*, 1981, **35**, 213.
252. R. E. Cramer and J. T. Huneke, *Inorg. Chem.*, 1975, **14**, 2565.
253. A. Pujunen and M. Lehtonen, *Suom. Kemistil. B*, 1972, **45**, 43.
254. Y. Saito and H. Iwasaki, *Bull. Chem. Soc. Jpn.*, 1962, **35**, 1131.
255. H. Breer, H. Endres, H. J. Keller and R. Martin, *Acta Crystallogr., Part B*, 1978, **34**, 2295.
256. C. Maeda, K. Matsumoto and S. Ooi, *Bull. Chem. Soc. Jpn.*, 1980, **53**, 1755.
257. R. E. Tapscott, J. D. Mather and T. F. Them, *Coord. Chem. Rev.*, 1979, **29**, 87.
258. S. E. Harnung, S. Kallesøe, A. M. Sargeson and C. E. Schaffer, *Acta Chem. Scand., Ser. A*, 1974, **28**, 385.
259. R. G. Pearson, C. R. Boston and F. Basolo, *J. Am. Chem. Soc.*, 1953, **75**, 3089.
260. R. G. Pearson, R. E. Meeker and F. Basolo, *J. Am. Chem. Soc.*, 1956, **78**, 709.
261. D. A. Buckingham, L. G. Marzilli and A. M. Sargeson, *J. Am. Chem. Soc.*, 1967, **89**, 825, 3428.
262. D. A. Buckingham, L. G. Marzilli and A. M. Sargeson, *J. Am. Chem. Soc.*, 1968, **90**, 6028.
263. W. T. Robinson, D. A. Buckingham, G. Chandler, L. G. Marzilli and A. M. Sargeson, *Chem. Commun.*, 1969, 539.
264. M. Kojima and T. Ishii, *Inorg. Nucl. Chem. Lett.*, 1974, **10**, 1095.
265. G. Daffner, D. A. Palmer and H. Kelm, *Inorg. Chim. Acta*, 1982, **61**, 57.
266. H. Ogino, Y. Orihara and N. Tanaka, *Inorg. Chem.*, 1980, **19**, 3178.
267. C. J. Hawkins and R. M. Peachey, *Acta Chem. Scand., Ser. A*, 1978, **32**, 815.
268. H. Ojima and K. Sone, *Bull. Chem. Soc. Jpn.*, 1962, **35**, 298.
269. H. E. LeMay, Jr. and J. C. Seher, *J. Inorg. Nucl. Chem.*, 1977, **39**, 1965.
270. J. A. Tiethof and D. W. Cooke, *Inorg. Chem.*, 1972, **11**, 318.
271. K. Akamatsu and Y. Shimura, *Bull. Chem. Soc. Jpn.*, 1978, **51**, 2586.
272. D. A. Buckingham, L. G. Marzilli and A. M. Sargeson, *Inorg. Chem.*, 1968, **7**, 915.
273. J. I. Legg and B. E. Douglas, *Inorg. Chem.*, 1968, **7**, 1452.
274. R. J. Mureinik and W. Robb, *Spectrochim. Acta, Part A*, 1968, **24**, 837.
275. J. A. Hearson, S. F. Mason and R. H. Seal, *J. Chem. Soc., Dalton Trans.*, 1977, 1026.

276. R. N. Keller and L. J. Edwards, *J. Am. Chem. Soc.*, 1952, **74**, 215.
277. A. J. Finney, M. A. Hitchman, C. L. Raston, G. L. Rowbottom and A. H. White, *Aust. J. Chem.*, 1981, **34**, 2139.
278. M. P. Hancock, B. T. Heaton and H. Huw Vaughan, *J. Chem. Soc., Dalton Trans.*, 1979, 761.
279. R. Hamalainen, U. Turpeinen and M. Ahlgren, *Acta Crystallogr., Part B*, 1979, **35**, 2408.
280. P. Pfeiffer and H. Glaser, *J. Prakt. Chem.*, 1938, **151**, 134 (*Chem. Abstr.*, 1938, **32**, 8361).
281. A. Pajunen and R. Hamalainen, *Suom. Kemistil. B*, 1972, **45**, 117, 122 (*Chem. Abstr.*, 1972, **76**, 159 562, 159 554).
282. Y. Nakayama, K. Matsumoto, S. Ooi and H. Kuroya, *Bull. Chem. Soc. Jpn.*, 1977, **50**, 2304.
283. T. G. Appleton and J. R. Hall, *Inorg. Chem.*, 1971, **10**, 1717.
284. D. A. Buckingham, L. G. Marzilli and A. M. Sargeson, *J. Am. Chem. Soc.*, 1969, **91**, 5227.
285. J. B. Goddard and F. Basolo, *Inorg. Chem.*, 1969, **8**, 2223.
286. F. Basolo, *J. Am. Chem. Soc.*, 1953, **75**, 227.
287. T. Kitamura, M. Saburi and S. Yoshikawa, *Bull. Chem. Soc. Jpn.*, 1978, **51**, 1563.
288. I. Bertini, C. Luchinat, F. Mani and A. Scozzafava, *Inorg. Chem.*, 1980, **19**, 1333.
289. I. Grenthe, P. Paoletti, M. Sandstrom and S. Glikberg, *Inorg. Chem.*, 1979, **18**, 2687.
290. T. Kitamura, M. Saburi and S. Yoshikawa, *Inorg. Chem.*, 1977, **16**, 585.
291. K. C. Patel, *J. Inorg. Nucl. Chem.*, 1978, **40**, 1631.
292. A. A. Kurganov, V. A. Davankov, L. Ya. Zhuchkova and T. M. Ponomaryova, *Inorg. Chim. Acta*, 1980, **39**, 237, 243.
293. J. M. Fernandez G., M. F. Rubio-Arroyo, C. Rubio-Poo and A. de la Pena, *Monatsh. Chem.*, 1983, **114**, 535.
294. H. E. LeMay, Jr., *Inorg. Chem.*, 1971, **10**, 1990.
295. L. M. Eade, G. A. Rodley and D. A. House, *J. Inorg. Nucl. Chem.*, 1975, **37**, 589.
296. E. Kyuno and J. C. Bailar, Jr., *J. Am. Chem. Soc.*, 1966, **88**, 5447.
297. D. A. Cooper, S. J. Higgins and W. Levason, *J. Chem. Soc., Dalton Trans.*, 1983, 2131.
298. T. Kudo and Y. Shimura, *Bull. Chem. Soc. Jpn.*, 1980, **53**, 1588.
299. R. Saito and Y. Kidani, *Bull. Chem. Soc. Jpn.*, 1978, **51**, 159.
300. B. Bosnich and J. MacB. Harrowfield, *J. Am. Chem. Soc.*, 1972, **94**, 3425.
301. A. J. McCaffery, S. F. Mason, B. J. Norman and A. M. Sargeson, *J. Chem. Soc. (A)*, 1968, 1304.
302. T. E. MacDermott and A. M. Sargeson, *Aust. J. Chem.*, 1963, **16**, 334.
303. B. E. Douglas, *Inorg. Chem.*, 1965, **4**, 1813.
304. K. Kashiwabara, M. Kojima and J. Fujita, *Bull. Chem. Soc. Jpn.*, 1979, **52**, 772.
305. S. K. Hall and B. E. Douglas, *Inorg. Chem.*, 1968, **7**, 533.
306. H. Ito, J. Fujita and K. Saito, *Bull. Chem. Soc. Jpn.*, 1969, **42**, 1286.
307. R. D. Gillard, *J. Inorg. Nucl. Chem.*, 1964, **26**, 1455.
308. K. Matsumoto and S. Ooi, *Z. Kirstallogr.*, 1979, **150**, 139.
309. E. A. Sulivan, *Can. J. Chem.*, 1979, **57**, 67.
310. R. Larsson, G. H. Searle, S. F. Mason and A. M. Sargeson, *J. Chem. Soc. (A)*, 1968, 1310.
311. S. Kaizaki and Y. Shimura, *Bull. Chem. Soc. Jpn.*, 1975, **48**, 3611.
312. C. J. Hawkins, E. Larsen and I. Olsen, *Acta Chem. Scand.*, 1965, **19**, 1915.
313. S. Yano, M. Saburi, S. Yoshikawa and J. Fujita, *Bull. Chem. Soc. Jpn.*, 1976, **49**, 101.
314. P. Andersen, F. Galsbøl and S. E. Harnung, *Acta Chem. Scand.*, 1969, **23**, 3027.
315. C. J. Hawkins and R. M. Peachey, *Aust. J. Chem.*, 1976, **29**, 33.
316. M. Kojima and J. Fujita, *Bull. Chem. Soc. Jpn.*, 1983, **56**, 139.
317. M. Kojima and J. Fujita, *Bull. Chem. Soc. Jpn.*, 1981, **54**, 2691.
318. F. Hein and H. Schade, *Z. Anorg. Allg. Chem.*, 1957, **289**, 90.
319. H. K. J. Powell and N. F. Curtis, *J. Chem. Soc. (B)*, 1966, 1205.
320. L. M. Hall, R. J. Speer, H. J. Ridgway and J. S. Norton, *J. Inorg. Biochem.*, 1979, **11**, 139.
321. S. Yano, T. Tukada, M. Saburi and Y. Yoshikawa, *Inorg. Chem.*, 1978, **17**, 2520.
322. C. J. Hawkins and M. L. McEniery, *Aust. J. Chem.*, 1978, **31**, 1699; 1979, **32**, 1433.
323. N. Bernth and E. Larsen, *Acta Chem. Scand., Ser. A*, 1978, **32**, 545.
324. R. G. Ball, N. J. Bowman and N. C. Payne, *Inorg. Chem.*, 1976, **15**, 1704.
325. B. Bosnich and E. A. Sullivan, *Inorg. Chem.*, 1975, **14**, 2768.
326. M. Saburi, T. Tsujito and S. Yoshikawa, *Inorg. Chem.*, 1970, **9**, 1476.
327. K. Nakayama, T. Komorita and Y. Shimura, *Bull. Chem. Soc. Jpn.*, 1981, **54**, 1056.
328. R. Saito and Y. Kidani, *Bull. Chem. Soc. Jpn.*, 1983, **56**, 449.
329. L. Sacconi, I. Bertini and F. Mani, *Inorg. Chem.*, 1967, **6**, 262.
330. I. Bertini and F. Mani, *Inorg. Chem.*, 1967, **6**, 2032.
331. W. C. Fultz, J. L. Burmeister, C. P. Cheng and T. L. Brown, *Inorg. Chem.*, 1981, **20**, 1734.
332. M. Suzuki and Y. Nishida, *J. Inorg. Nucl. Chem.*, 1977, **39**, 1459.
333. K. Miyoshi, Y. Matsumoto and H. Yoneda, *Inorg. Chem.*, 1981, **20**, 1057.
334. R. Birdy, D. M. L. Goodgame, J. C. McConway and D. Rodgers, *J. Chem. Soc., Dalton Trans.*, 1977, 1730.
335. J. R. Ferraro, L. Fabbrizzi and P. Paoletti, *Inorg. Chem.*, 1977, **16**, 2127.
336. D. M. L. Goodgame and M. A. Hitchman, *Inorg. Chem.*, 1966, **5**, 1303.
337. Y. Ihara, E. Izumi, A. Uehara, R. Tsuchiya, S. Nakagawa and E. Kyuno, *Bull. Chem. Soc. Jpn.*, 1982, **55**, 1028.
338. A. J. Finney, M. A. Hitchman, C. L. Raston, G. L. Rowbottom and A. H. White, *Aust. J. Chem.*, 1981, **34**, 2139.
339. A. B. P. Lever and E. Mantovani, *Inorg. Chem.*, 1971, **10**, 817.
340. A. Pajunen and S. Pajunen, *Suom. Kemistil. B.*, 1971, **44**, 331 (*Chem. Abstr.*, 1972, **76**, 7525).
341. H. Irving and J. M. M. Griffiths, *J. Chem. Soc.*, 1954, 213.
342. A. J. Finney, M. A. Hitchman, C. L. Raston, G. L. Rowbottom and A. H. White, *Aust. J. Chem.*, 1981, **34**, 2061.
343. M. M. Andino, J. D. Curet and M. M. Muir, *Acta Crystallogr., Part B*, 1976, **32**, 3185.
344. H. J. Gysling, *Chem. Abstr.*, 1977, **86**, 16 784.
345. Y. Fukuda, H. Okamura and K. Sone, *Bull. Chem. Soc. Jpn.*, 1977, **50**, 313.
346. A. Pajunen and E. Nasakkala, *Cryst. Struct. Commun.*, 1978, **7**, 299.
347. A. Pajunen and S. Pajunen, *Cryst. Struct. Commun.*, 1979, **8**, 331.
348. J. R. Ferraro, L. J. Basile, L. R. Garcia-Ineguez, P. Paoletti and L. Fabbrizzi, *Inorg. Chem.*, 1976, **15**, 2342.

349. W. E. Hatfield, T. S. Piper and U. Klabunde, *Inorg. Chem.*, 1963, **2**, 629.
350. H. Yokoi, M. Sai and T. Isobe, *Bull. Chem. Soc. Jpn.*, 1969, **42**, 2232.
351. A. B. P. Lever, E. Montovani and J. C. Donini, *Inorg. Chem.*, 1971, **10**, 2424.
352. L. Fabbrizzi, M. Micheloni and P. Paoletti, *Inorg. Chem.*, 1974, **13**, 3019.
353. D. M. L. Goodgame and L. M. Venanzi, *J. Chem. Soc.*, 1963, 616.
354. F. H. Dickey, W. Fickett and H. J. Lucas, *J. Am. Chem. Soc.*, 1952, **74**, 944.
355. S. Bagger, O. Bang and F. Woldbye, *Acta Chem. Scand.*, 1973, **27**, 2663.
356. M. Kojima, H. Funaki, Y. Yoshikawa and K. Yamasaki, *Bull. Chem. Soc. Jpn.*, 1975, **48**, 2801.
357. S. R. Niketic and F. Woldbye, *Acta Chem. Scand.*, 1973, **27**, 621, 3811; *Acta Chem. Scand., Ser. A*, 1974, **28**, 248.
358. T. G. Appleton and J. R. Hall, *Inorg. Chem.*, 1970, **9**, 1800.
359. T. G. Appleton and J. R. Hall, *Inorg. Chem.*, 1970, **9**, 1807.
360. S. Yano, H. Ito, Y. Koike, J. Fujita and K. Saito, *Chem. Commun.*, 1969, 460.
361. E. N. Duesler, M. Fe Gargallow and R. E. Tapscott, *Acta Crystallogr., Part B*, 1982, **38**, 1300.
362. R. Tsuchida, A. Uehara and T. Yoshikuni, *Bull. Chem. Soc. Jpn.*, 1982, **21**, 590.
363. M. Goto, M. Saburi and S. Yoshikawa, *Inorg. Chem.*, 1969, **8**, 358.
364. S. Bagger and O. Bang, *Acta Chem. Scand., Ser. A*, 1976, **30**, 765.
365. S. Trippett, *J. Chem. Soc.*, 1957, 4407.
366. R. E. Tapscott, *Inorg. Chem.*, 1975, **14**, 216.
367. M. N. H. Irving and R. M. Parkins, *J. Inorg. Nucl. Chem.*, 1965, **27**, 270.
368. M. Kojima, M. Ishiguro and J. Fujita, *Bull. Chem. Soc. Jpn.*, 1978, **51**, 3651.
369. A. Larena and E. Bernabeu, *Spectrochim. Acta, Part A*, 1980, **36**, 345.
370. S. C. Nyburg and J. S. Wood, *Inorg. Chem.*, 1964, **3**, 468.
371. W. A. Sadler and D. A. House, *J. Chem. Soc., Dalton Trans.*, 1973, 1937.
372. S. C. Nyburg, J. S. Wood and W. C. E. Higginson, *Proc. Chem. Soc.*, 1961, 297.
373. A. J. Finney, M. A. Hitchman, C. L. Raston, G. L. Rowbottom and A. H. White, *Aust. J. Chem.*, 1981, **34**, 2069.
374. A. G. Lidstone and W. H. Mills, *J. Chem. Soc.*, 1939, 1754.
375. W. M. Mills and T. H. H. Quibell, *J. Chem. Soc.*, 1935, 839.
376. O. F. Williams and J. C. Bailar, Jr., *J. Am. Chem. Soc.*, 1959, **81**, 4464.
377. R. Kuroda and S. F. Mason, *J. Chem. Soc., Dalton Trans.*, 1977, 1016.
378. S. Arakawa, K. Kashiwabara, J. Fujita and K. Saito, *Bull. Chem. Soc. Jpn.*, 1977, **50**, 2331, 2108.
379. G. G. Hawn, C. Maricondi and B. E. Douglas, *Inorg. Chem.*, 1979, **18**, 2542.
380. N. Matsuoka, J. Hidaka and Y. Shimura, *Inorg. Chem.*, 1970, **9**, 719; *Bull. Chem. Soc. Jpn.*, 1975, **48**, 458.
381. B. Bosnich and F. P. Dwyer, *Aust. J. Chem.*, 1966, **19**, 2045, 2051.
382. M. Kojima, Y. Yoshikawa and K. Yamasaki, *Bull. Chem. Soc. Jpn.*, 1973, **46**, 1687.
383. J. Kansikas and R. Hamalainen, *Finn. Chem. Lett.*, 1977, 118.
384. J. H. Dunlop, R. D. Gillard and G. Wilkinson, *J. Chem. Soc.*, 1964, 3160.
385. M. Kunimatsu, H. Kanno, M. Kojima, K. Kashiwabara and J. Fujita, *Bull. Chem. Soc. Jpn.*, 1980, **53**, 1571.
386. H. Toftlund and E. Pedersen, *Acta Chem. Scand.*, 1972, **26**, 4019.
387. M. Ito, F. Marumo and Y. Saito, *Acta Crystallogr., Part B*, 1971, **27**, 2187.
388. F. M. Jaeger and H. B. Blumendal, *Z. Anorg. Allg. Chem.*, 1928, **175**, 161; *Proc. Akad. Amsterdam*, 1928, **37**, 412.
389. J. F. Phillips and D. J. Royer, *Inorg. Chem.*, 1965, **4**, 616.
390. M. Goto, M. Takeshita and T. Sakai, *Bull. Chem. Soc. Jpn.*, 1979, **52**, 2589.
391. R. S. Treptow, *Inorg. Chem.*, 1966, **5**, 1593.
392. K. Kashiwabara, K. Hanaki and J. Fujita, *Bull. Chem. Soc. Jpn.*, 1980, **53**, 2275.
393. F. Galsbøl, P. Steenbøl and B. S. Sørensen, *Acta Chem. Scand.*, 1972, **26**, 3605.
394. F. M. Jaeger and L. Bijkerk, *Z. Anorg. Allg. Chem.*, 1937, **233**, 101.
395. R. G. Asperger and C. F. Liu, *Inorg. Chem.*, 1965, **4**, 1492.
396. W. C. G. Baldwin, *Proc. R. Soc. London, Ser. A*, 1938, **167**, 539.
397. J. W. Turley and R. G. Asperger, *Inorg. Chem.*, 1971, **10**, 558.
398. R. G. Asperger, *Inorg. Chem.*, 1969, **8**, 2127.
399. A. Kobayashi, F. Marumo and Y. Saito, *Acta Crystallogr., Part B*, 1972, **28**, 2709; *Acta Crystallogr., Part C*, 1983, **39**, 807.
400. S. C. Chan and F. K. Chan, *Aust. J. Chem.*, 1970, **23**, 1477.
401. T. Laier and E. Larsen, *Acta Chem. Scand., Ser. A*, 1979, **33**, 257.
402. S. E. Harnung, B. S. Sørensen, I. Creaser, H. Maegaard, U. Pfenninger and C. E. Schäffer, *Inorg. Chem.*, 1976, **15**, 2123.
403. F. Marumo, Y. Utsumi and Y. Saito, *Acta Crystallogr., Part B*, 1970, **26**, 1492.
404. S. Sato and Y. Saito, *Acta Crystallogr., Part B*, 1977, **33**, 860.
405. R. J. Speer, H. Ridgway, L. M. Hall, D. P. Stewart, A. D. Newman and J. M. Hill, *J. Clin. Hematol. Oncol.*, 1979, **9**, 150 (*Chem. Abstr.*, 1979, **90**, 145166).
406. R. W. Green, *Aust. J. Chem.*, 1973, **26**, 1841.
407. K. Matsumoto, S. Ooi, M. Sakuma and H. Kuroya, *Bull. Chem. Soc. Jpn.*, 1976, **49**, 2129.
408. J. E. McDonald and T. Moeller, *J. Inorg. Nucl. Chem.*, 1978, **40**, 253.
409. E. W. Abel, M. M. Bhatti, M. B. Hursthouse, K. M. Abdul Malik and M. A. Mazid, *J. Organomet. Chem.*, 1980, **197**, 345.
410. C. K. Poon and C. M. Che, *J. Chem. Soc., Dalton Trans.*, 1980, 756.
411. R. Nasanen and E. Luukkonen, *Suom. Kemistil. B*, 1968, **41**, 27.
412. R. Nasanen, L. Lemmetti and P. Ilyonen, *Suom. Kemistil. B*, 1968, **41**, 111.
413. R. Nasanen and P. Tilus, *Suom. Kemistil. B*, 1969, **42**, 11 (*Chem. Abstr.*, 1969, **70**, 71562).
414. R. Nasanen, I. Virtamo and P. Suvanto, *Suom. Kemistil. B*, 1966, **39**, 158 (*Chem. Abstr.*, 1967, **66**, 22910).
415. D. W. Meek and S. A. Ehrhardt, *Inorg. Chem.*, 1965, **4**, 584.
416. D. W. Meek, *Inorg. Chem.*, 1965, **4**, 250.
417. J. G. Reynolds, A. Zalkin, D. H. Templeton and N. M. Edelstien, *Inorg. Chem.*, 1977, **16**, 599, 1858.
418. H. van der Poel, G. van Koten, M. Kokkes and C. H. Stam, *Inorg. Chem.*, 1981, **20**, 2941.

419. A. De Renzi, A. Panunzi, A. Saporito and A. Vitagliano, *Gazz. Chim. Ital.*, 1977, **107**, 549 (*Chem. Abstr.*, 1978, **89**, 43 734).
420. M. F. C. Ladd and D. H. G. Perrins, *J. Cryst. Mol. Struct.*, 1977, **7**, 157; *Z. Kristallogr.*, 1981, **154**, 155.
421. L. F. Larkworthy and K. C. Patel, *J. Inorg. Nucl. Chem.*, 1970, **32**, 1263.
422. K. C. Patel and. D. E. Goldberg, *Inorg. Chem.*, 1972, **11**, 759.
423. D. B. Sowerby and I. Haiduc, *Inorg. Nucl. Chem. Lett.*, 1976, **12**, 791.
424. K. Yokoho, K. Matsumoto, S. Ooi and H. Kuroya, *Bull. Chem. Soc. Jpn.*, 1976, **49**, 1864.
425. N. Hoshino, Y. Fukuda and K. Sone, *Bull. Chem. Soc. Jpn.*, 1981, **54**, 420.
426. D. B. Sowerby and I. Haiduc, *Inorg. Chim. Acta*, 1976, **17**, L15.
427. U. Turpeinen, M. Ahlgren and R. Hamalainen, *Finn. Chem. Lett.*, 1977, 246.
428. M. G. B. Drew and D. Rogers, *Chem. Commun.*, 1965, 476.
429. J. G. H. Du Preez, H. E. Rohwer, B. J. Van Brecht and M. R. Caira, *Inorg. Chim. Acta*, 1983, **73**, 67.
430. A. J. Finney, M. A. Hitchman, C. L. Raston, G. L. Rowbottom and A. H. White, *Aust. J. Chem.*, 1981, **34**, 2159.
431. M. Ahlgren and U. Turpeinen, *Acta Crystallogr., Part B*, 1982, **38**, 276.
432. K. Nonoyama, H. Ojima, K. Ohki and M. Nonoyama, *Inorg. Chim. Acta*, 1980, **41**, 155.
433. A. Pajunen and S. Pajunen, *Acta Crystallogr., Part B*, 1979, **35**, 2401.
434. M. Nasakkala, *Ann. Acad. Sci. Fenn. Ser. 2A*, 1977, 181 (*Chem. Abstr.*, 1977, **86**, 181 117).
435. A. Pajunen and S. Pajunen, *Cryst. Struct. Commun.*, 1977, **6**, 413.
436. M. Ahlgren, R. Hamalainen and U. Turpeinen, *Acta Chem. Scand., Ser. A*, 1978, **32**, 57.
437. U Turpeinen, M. Ahlgren and R. Hamalainen, *Cryst. Struct. Commun.*, 1978, **7**, 617.
438. T. J. Greenhough and M. F. C. Ladd, *Acta Crystallogr., Part B*, 1978, **34**, 2744.
439. D. W. Meek and S. A. Ehrhardt, *Inorg. Chem.*, 1965, **4**, 585.
440. Y. Fukuda and K. Sone, *Bull. Chem. Soc. Jpn.*, 1972, **45**, 465.
441. T. P. Mitchell, W. H. Bernard and J. R. Wasson, *Acta Crystallogr., Part B*, 1970, **26**, 2096.
442. F. G. Mann and H. R. Watson, *J. Chem. Soc.*, 1958, 2772.
443. I. Bkouche-Waksman, S. Sikorav and O. Kahn, *J. Crystallogr. Spectrochem. Res.*, 1983, **13**, 303.
444. H. Okawa, M. Mikuriya and S. Kida, *Bull. Chem. Soc. Jpn.*, 1983, **56**, 2142.
445. M. C. Couldwell and J. Simpson, *J. Chem. Soc., Dalton Trans.*, 1979, 1101.
446. D. A. Baldwin and G. J. Leigh, *J. Chem. Soc. (A)*, 1968, 1431.
447. J. Dwyer, W. Levason and C. A. McAuliffe, *J. Inorg. Nucl. Chem.*, 1976, **38**, 1919.
448. R. J. York, W. D. Bonds, Jr., B. P. Cotsoradis and R. D. Archer, *Inorg. Chem.*, 1969 **8**, 789.
449. U. Turpeinen, M. Ahlgren and R. Hamalainen, *Acta Crystallogr., Part B*, 1982, **38**, 1580.
450. T. G. Appleton and J. R. Hall, *Inorg. Chem.*, 1972, **11**, 124.
451. A. Tiripicchio, M. T. Camellini, L. Marescar, G. Natile and G. Rizzardi, *Cryst. Struct. Commun.*, 1979, **8**, 689.
452. M. Walczak and G. Stucky, *J. Am. Chem. Soc.*, 1976, **98**, 5531; *J. Organomet. Chem.*, 1975, **97**, 313.
453. J. J. Brooks and G. D. Stucky, *J. Am. Chem. Soc.*, 1972, **94**, 7333.
454. W. E. Rhine and G. D. Stucky, *J. Am. Chem. Soc.*, 1975, **97**, 737.
455. M. Nasakkala and A. Pajunen, *Cryst. Struct. Commun.*, 1980, **9**, 897.
456. M. R. Churchill, G. Davies, M. A. El-Sayed, M. F. El-Shazly, J. P. Hutchinson and M. W. Rupich, *Inorg. Chem.*, 1980, **19**, 201.
457. Y. Fukuda, Y. Miura and K. Sone, *Bull. Chem. Soc. Jpn.*, 1977, **50**, 143.
458. M. Ahlgren and R. Hamalainen, *Finn. Chem. Lett.*, 1977, **8**, 239.
459. A. Pajunen and S. Pajunen, *Acta Crystallogr., Part B*, 1979, **35**, 460.
460. M. Ahlgren, U. Turpeinen and R. Hamalainen, *Acta Chem. Scand., Ser A.*, 1980, **34**, 67.
461. R. W. Gaver and R. K. Murmann, *J. Inorg. Nucl. Chem.*, 1964, **26**, 881.
462. R. W. Berg and K. Rasmussen, *Spectrochim. Acta, Part A*, 1972, **28**, 2319; 1973, **29**, 37.
463. R. Alcala and J. Fernández, *Cryst. Struct. Commun.*, 1977, **6**, 635.
464. F. Gómez-Beltrán, A. Valero Capilla and R. Alcala Aranda, *Cryst. Struct. Commun.*, 1978, **7**, 153.
465. F. Basolo, Y. T. Chen and R. K. Murmann, *J. Am. Chem. Soc.*, 1954, **76**, 956.
466. F. Gómez-Beltrán, A. Valero Capilla and R. Alcala Aranda, *Cryst. Struct. Commun.*, 1979, **8**, 87.
467. F. Gómez-Beltrán, A. Valero Capilla and R. Alcala Aranda, *Cryst. Struct. Commun.*, 1978, **7**, 255.
468. K. C. Patel and R. R. Patel, *J. Inorg. Nucl. Chem.*, 1977, **39**, 1325.
469. A. De Renzi, D. Di Blasio, A. Saporito, M. Scalone and A. Vitagliano, *Inorg. Chem.*, 1980, **19**, 960.
470. K. C. Patel, S. R. Shirali and D. E. Goldberg, *J. Inorg. Nucl. Chem.*, 1975, **37**, 1659.
471. A. R. Gainsford, G. J. Gainsford and D. A. House, *Cryst. Struct. Commun.*, 1981, **10**, 365.
472. Lim Say Dong, R. A. Gainsford and D. A. House, *Inorg. Chim. Acta*, 1976, **19**, 23.
473. Lim Say Dong and D. A. House, *Inorg. Chim. Acta*, 1978, **30**, 271.
474. J. D. Mather, R. E. Tapscott and C. F. Campana, *Inorg. Chim. Acta*, 1983, **73**, 235.
475. A. Lambert and A. Lowe, *J. Chem. Soc.*, 1947, 1517.
476. E. B. Fleischer, A. E. Gebala, A. Levey and P. A. Tasker, *J. Org. Chem.*, 1971, **36**, 3042.
477. J. C. Bailar, Jr. and J. B. Work, *J. Am. Chem. Soc.*, 1946, **68**, 232.
478. U. Tinner and W. Marty, *Helv. Chim. Acta*, 1977, **60**, 1629.
479. F. Cariati, G. Ciani, L Manabue, G. G. Pellacani, G. Rassu and A. Sironi, *Inorg. Chem.*, 1983, **22**, 1897.
480. L. P. Battaglia, A. B. Corradi, G. Marcotrigiano, L. Menabue and G. C. Pellacani, *J. Chem. Soc., Dalton Trans.*, 1981, 8.
481. F. Mizukami, M. Ito, J. Fujita and K. Saito, *Bull. Chem. Soc. Jpn.*, 1970, **43**, 3973; 1972, **45**, 2129.
482. A. Pajunen, K. Smolander and I. Belinskij, *Suom. Kemistil. B*, 1972, **45**, 317 (*Chem. Abstr.*, 1973, **78**, 49 273).
483. E. Balieu, P. M. Boll and E. Larsen, *Acta Chem. Scand.*, 1969, **23**, 2191.
484. E. Strack and H. Schwaneberg, *Chem. Ber.*, 1934, **67**, 39.
485. M. Kojima and J. Fujita, *Bull. Chem. Soc. Jpn.*, 1977, **50** 3237.
486. C. J. Hawkins and G. A. Lawrance, *Aust. J. Chem.*, 1973, **26**, 2401.
487. K. Kashiwabara, M. Kojima, S. Arakawa and J. Fujita, *Bull. Chem. Soc. Jpn.*, 1979, **52**, 243.
488. M. Parris, L. J. De Hayes and D. H. Busch, *Can. J. Chem.*, 1972, **50**, 3569.

489. M. Kojima and J. Fujita, *Bull. Chem. Soc. Jpn.*, 1982, **55**, 1454.
490. C. J. Dipple, *Recl. Trav. Chim. Pays-Bas*, 1931, **50**, 525.
491. A. H. I. Ben-Bassat and I. Binenboym, *Chem. Analyst*, 1964, **53**, 36.
492. K. Hata, M.-K. Doh, K. Kashiwabara and J. Fujita, *Bull. Chem. Soc. Jpn.*, 1981, **54**, 190.
493. R. G. Ball, R. T. Thurier and N. C. Payne, *Inorg. Chim. Acta*, 1978, **30**, 227.
494. H. Boucher and B. Bosnich, *Inorg. Chem.*, 1976, **15**, 2364.
495. J. R. Gollogly and C. J. Hawkins, *Inorg. Chem.*, 1972, **11**, 156.
496. I. Oonishi, S. Sato and Y. Saito, *Acta Crystallogr., Part B*, 1974, **30**, 2256.
497. S. R. Niketic, K. Rasmussen, F. Woldbye and S. Lifson, *Acta Chem. Scand., Ser. A*, 1976, **30**, 485.
498. M. Kojima and J. Fujita, *Chem. Lett.*, 1976, 429.
499. T. W. Hambley, C. J. Hawkins, J. A. Palmer and M. R. Snow, *Aust. J. Chem.*, 1981, **34**, 45.
500. H. Yoneda, M. Muto and K. Tamaki, *Bull Chem. Soc. Jpn.*, 1971, **44**, 2863.
501. A. Kobayashi, F. Marumo, Y. Saito, J. Fujita and F. Mizukami, *Inorg. Nucl. Chem. Lett.*, 1971, **7**, 777.
502. A. Kobayashi, F. Marumo and Y. Saito, *Acta Crystallogr., Part B*, 1972, **28**, 3591; 1973, **29**, 2443.
503. M. Kojima, M. Fujita and J. Fujita, *Bull. Chem. Soc. Jpn.*, 1977, **50**, 898.
504. F. Mizukami, H. Ito, J. Fujita and K. Saito, *Bull. Chem. Soc. Jpn.*, 1971, **44**, 3051; 1973, **46**, 2410.
505. S. Ohba, H. Miyamae, S. Sato and Y. Saito, *Acta Crystallogr., Part B*, 1979, **35**, 321.
506. H. Boucher and B. Bosnich, *Inorg. Chem.*, 1976, **15**, 1471.
507. Y. Nakayama, S. Ooi and H. Kuroya, *Bull. Chem. Soc. Jpn.*, 1979, **52**, 914.
508. M. R. Churchill, G. Davies, M. A. El-Sayed, M. F. El Shazly, J. P. Hutchinson, M. W. Rupich and K. O. Watkins, *Inorg. Chem.*, 1979, **18**, 2296.
509. T. W. Hambley, C. J. Hawkins, J. A. Palmer and M. R. Snow, *Aust. J. Chem.*, 1981, **34**, 2525.
510. M. Parris, *Inorg. Chim. Acta*, 1978, **31**, L429.
511. P. Pfeiffer, W. Christeleit, T. Hesse, H. Pfitzner and H. Thielert, *J. Prakt. Chem.*, 1938, **150**, 261 (*Chem. Abstr.*, 1938, **32**, 5783).
512. K. Okamoto, M. Noji and Y. Kidani, *Bull. Chem. Soc. Jpn.*, 1981, **54**, 713.
513. K. Kamisawa, K. Matsumoto, S. Ooi, R. Saito and Y. Kidani, *Bull. Chem. Soc. Jpn.*, 1981, **54**, 1072.
514. R. Saito and Y. Kidani, *Bull. Chem. Soc. Jpn.*, 1979, **52**, 57.
515. K. Kamisawa, K. Matsumoto, S. Ooi, H. Kuroya, R. Saito and Y. Kidani, *Bull. Chem. Soc. Jpn.*, 1978, **51**, 2330.
516. J. E. Sarneski, A. T. McPhail, K. D. Onan, L. E. Erickson and C. N. Reilley, *J. Am. Chem. Soc.*, 1977, **99**, 7376.
517. M. Kojima, K. Morita and J. Fujita, *Bull. Chem. Soc. Jpn.*, 1981, **54**, 2947.
518. P. Ivanov and I. Pajarlieff, *J. Mol. Struct.*, 1977, **38**, 269.
519. R. D. Gillard, *Tetrahedron*, 1965, **21**, 503.
520. A. R. Gainsford and D. A. House, *Inorg. Chim. Acta*, 1972, **6**, 227.
521. F. A. Jurnak and K. N. Raymond, *Inorg. Chem.*, 1972, **11**, 3149.
522. R. Nagao, F. Marumo and Y. Saito, *Acta Crystallogr., Part B*, 1973, **29**, 2438.
523. T. Nomura, F. Marumo and Y. Saito, *Bull. Chem. Soc. Jpn.*, 1969, **42**, 1016.
524. R. J. Geue and M. R. Snow, *J. Chem. Soc. (A)*, 1971, 2981.
525. M. Dwyer, R. J. Geue and M. R. Snow, *Inorg. Chem.*, 1973, **12**, 2057.
526. G. Srdanov, R. Herak and B. Prelesnik, *Inorg. Chim. Acta*, 1979, **33**, 23.
527. K. Matsumoto, H. Kawaguchi, H. Kuroya and S. Kawaguchi, *Bull. Chem. Soc. Jpn.*, 1973, **46**, 2424.
528. K. Matsumoto, S. Ooi, H. Kawaguchi, M. Nakano and S. Kawaguchi, *Bull. Chem. Soc. Jpn.*, 1982, **55**, 1840.
529. K. Matsumoto, S. Ooi and H. Kuroya, *Bull. Chem. Soc. Jpn.*, 1970, **43**, 1903.
530. E. Yasaki, I. Oonishi, H. Kawaguchi, S. Kawaguchi and Y. Komiyama, *Bull. Chem. Soc. Jpn.*, 1970, **43**, 1354.
531. I. Krstanovic and M. Celap, *Cryst. Struct. Commun.*, 1977, **6**, 171.
532. H. V. F. Schousboe-Jensen, *Acta Chem. Scand.*, 1972, **26**, 3413.
533. R. M. Herak, M. B. Celap and I. Krstanovic, *Acta Crystallogr., Part B*, 1977, **33**, 3368.
534. R. M. Herak, M. B. Celap and I. Krstanovic, *Acta Crystallogr., Part A*, 1975, **31**, S142, S143.
535. A. Terzis, K. N. Raymond and T. G. Spiro, *Inorg. Chem.*, 1970, **9**, 2415.
536. E. N. Duesler and K. N. Raymond, *Inorg. Chim. Acta*, 1978, **30**, 87.
537. J. W. Vaughn, *Inorg. Chem.*, 1981, **20**, 2397.
538. X. Solans, M. Font-Altaba, M. Montfort and J. Ribas, *Acta Crystallogr., Part B*, 1982, **38**, 2899.
539. K. Matsumoto and S. Ooi, *Bull. Chem. Soc. Jpn.*, 1979, **52**, 3307.
540. B. Morosin and J. Howatson, *Acta Crystallogr., Part B*, 1970, **26**, 2062.
541. A. Pajunen, *Suom. Kemistil. B*, 1969, **42**, 15 (*Chem. Abstr.*, 1969, **71**, 7494).
542. A. Pajunen and I. Belinskij, *Suom. Kemistil. B*, 1970, **43**, 70 (*Chem. Abstr.*, 1970, **72**, 94156).
543. M. Cannas, G. Carta and G. Marongiu, *J. Chem. Soc., Dalton Trans.*, 1974, 550.
544. G. D. Andreetti, L. Cavalca and P. Sgarabotto, *Gazz. Chim. Ital.*, 1971, **101**, 483.
545. R. Uggla and M. Klinga, *Suom. Kemistil. B*, 1972, **45**, 10 (*Chem. Abstr.*, 1972, **76**, 77849).
546. R. Uggla, O. Orama, M. Sundberg, E. Tirronen and M. Klinga, *Finn. Chem. Lett.*, 1974, 185.
547. R. Uggla, O. Orama and M. Klinga, *Suom. Kemistil. B*, 1973, **46**, 43.
548. O. Orama and G. Huttner, *Finn. Chem. Lett.*, 1976, 140.
549. O. Orama, *Finn. Chem. Lett.*, 1976, 151.
550. O. Orama, *Finn. Chem. Lett.*, 1976, 154.
551. O. Orama and A. Pajunen, *Finn. Chem. Lett.*, 1977, 193.
552. O. Orama, G. Huttner, H. Lorenz, M. Marsili and A. Frank, *Finn. Chem. Lett.*, 1976, 137.
553. M. Klinga, *Finn. Chem. Lett.*, 1979, 223.
554. M. Klinga, *Finn. Chem. Lett.*, 1976, 179.
555. M. Klinga, *Finn. Chem. Lett.*, 1976, 71.
556. M. Klinga, *Finn. Chem. Lett.*, 1977, 153.
557. A. Pajunen and E. Nasakkala, *Finn. Chem. Lett.*, 1977, 100 (*Chem. Abstr.*, 1977, **87**, 175917).
558. G. D. Andreetti, L. Cavalca, M. A. Pellin-Ghelli and P. Sgarabotto, *Gazz. Chim. Ital.*, 1971, **101**, 488.
559. G. D. Andreetti, L. Cavalca and P. Sgarabotto, *Gazz. Chim. Ital.*, 1971, **101**, 494.

560. A. Pajunen, *Suom. Kemistil. B*, 1969, **42**, 397 (*Chem. Abstr.*, 1969, **71**, 129 682).
561. A. Pajunen, *Suom. Kemistil. B*, 1968, **41**, 232 (*Chem. Abstr.*, 1968, **69**, 46 833).
562. J. M. Franco, J. Pardo and S. Garcia-Blanco, European Crystallographic Meeting, 1973.
563. A. Pajunen and U. Turpeinen, *Suom. Kemistil. B*, 1973, **46**, 281.
564. M. Klinga, *Cryst. Struct. Commun.*, 1980, **9**, 457.
565. M. Klinga, *Cryst. Stuct. Commun.*, 1980, **9**, 439.
566. M. Klinga, *Cryst. Struct. Commun.*, 1981, **10**, 521.
567. M. Klinga, *Cryst. Struct. Commun.*, 1980, **9**, 567.
568. T. K. Kistenmacher, C. C. Chiang, P. Chalilpoyil and L. G. Marzilli, *Biochem. Biophys. Res. Commun.*, 1978, **84**, 70.
569. L. G. Marzilli, P. Chalilpoyil, C. C. Chiang and T. J. Kistenmacher, *J. Am. Chem. Soc.*, 1980, **102**, 2480.
570. N. Matsumoto, M. Yamashita and S. Kida, *Bull. Chem. Soc. Jpn.*, 1978, **51**, 3514.
571. B. Keppler, B. Muller, M. Cannas and G. Marongiu, *Acta Crystallogr., Part A*, 1981, **37**, C238.
572. L. J. DeHayes and D. H. Busch, *Inorg. Chem.*, 1973, **12**, 1505.
573. A. E. Martell and M. Calvin, 'The Chemistry of the Metal Chelate Compounds', Prentice-Hall, 1956.
574. M. Rancke-Madsen and F. Woldbye, *Acta Chem. Scand.*, 1972, **26**, 3405.
575. M. Kojima, H. Yamada, H. Ogino and J. Fujita, *Bull. Chem. Soc. Jpn.*, 1977, **50**, 2325.
576. R. Romeo, D. Minniti, S. Lanza and M. L. Tobe, *Inorg. Chim. Acta*, 1977, **22**, 87.
577. Y. Soma and F. Mizukami, *Bull. Chem. Soc. Jpn.*, 1978, **51**, 641.
578. F. Lions and K. Martin, *J. Am. Chem. Soc.*, 1957, **79**, 1572.
579. T. A. Geissman, M. J. Schlatter and I. D. Webb, *J. Org. Chem.*, 1946, **11**, 736.
580. H. Stetter and W. Beckmann, *Chem. Ber.*, 1951, **84**, 834.
581. R. J. Geue and M. R. Snow, *Inorg. Chem.*, 1977, **16**, 231.
582. M. Micheloni, A. Sabatini and A. Vacca, *Inorg. Chim. Acta*, 1977, **25**, 41.
583. F. G. Mann and W. J. Pope, *Proc. R. Soc. London, Ser. A*, 1925, **107**, 80.
584. M. S. Okamoto and E. K. Barefield, *Inorg. Chem.*, 1974, **13**, 2611.
585. F. Hein and R. Burkhardt, *Chem. Ber.*, 1957, **90**, 928.
586. G. Schwarzenbach, J. Boesch and H. Egli, *J. Inorg. Nucl. Chem.*, 1971, **33**, 2141.
587. M. Umehara, M. Ishii, T. Nomura, I. Muramatsu and M. Nakahara, *Nippon Kagaku Kaishi*, 1980, 657 (*Chem. Abstr.*, 1980, **93**, 60 131).
588. A. E. Martell and M. Calvin, 'The Chemistry of the Metal Chelate Compounds', Prentice-Hall, 1956, pp. 137, 287, 326.
589. A. Sabatini and A. Vacca, *Coord. Chem. Rev.*, 1975, **16**, 161; *J. Chem. Soc., Dalton Trans.*, 1980, 519.
590. L. Bologni, M. Micheloni, A. Sabatini and A. Vacca, *Inorg. Chim. Acta*, 1983, **70**, 117.
591. S. N. Golubev and Y. D. Kondvashev, *Zh. Strukt. Khim.*, 1981, **22**, 80.
592. M. A. Babar, M. F. C. Ladd, L. F. Larkworthy, G. P. Newell and D. C. Povey, *J. Cryst. Mol. Struct.*, 1979, **8**, 43.
593. T. J. Greenhough and M. F. C. Ladd, *Acta Crystallogr., Part B*, 1977, **33**, 1266.
594. G. L. Ferguson and B. Zaslow, *Acta Crystallogr., Part B*, 1971, **27**, 849.
595. J. Britten and C. J. L. Lock, *Acta Crystallogr., Part B*, 1979, **35**, 3065.
596. A. Daoud and R. Perret, *Bull. Soc. Chim. Fr.*, 1975, 109.
597. F. R. Keene and G. H. Searle, *Inorg. Chem.*, 1972, **11**, 148.
598. Y. Yoshikawa and K. Yamasaki, *Bull. Chem. Soc. Jpn.*, 1972, **45**, 179.
599. F. R. Keene and G. H. Searle, *Inorg. Chem.*, 1974, **13**, 2173.
600. M. Kobayashi, F. Marumo and Y. Saito, *Acta Crystallogr., Part B*, 1972, **28**, 470.
601. M. Konno, F. Marumo and Y. Saito, *Acta Crystallogr., Part B*, 1973, **29**, 739.
602. F. D. Sancilio, L. F. Druding and D. M. Lukaszewski, *Inorg. Chem.*, 1976, **15**, 1626.
603. K. Okiyama, S. Sato and Y. Saito, *Acta Crystallogr., Part B*, 1979, **35**, 2389.
604. G. H. Searle, S. F. Lincoln, S. G. Teague and D. G. Rowe, *Aust. J. Chem.*, 1979, **32**, 519.
605. E. O. Horn, S. F. Lincoln, G. H. Searle and M. R. Snow, *Aust. J. Chem.*, 1980, **33**, 2151, 2159.
606. M. Kojima, M. Iwagaki, Y. Yoshikawa and J. Fujita, *Bull. Chem. Soc. Jpn.*, 1977, **50**, 3216.
607. R. E. Hamm and F. C. Fushimi, *J. Indian Chem. Soc.*, 1973, **50**, 33.
608. E. Pedersen, *Acta Chem. Scand.*, 1970, **24**, 3362.
609. S. Biagini and M. Cannas, *J. Chem. Soc. (A)*, 1970, 2398.
610. F. S. Stephens, *J. Chem. Soc. (A)*, 1969, 883, 2233.
611. M. Zocchi, A. Albinati and G. Tieghi, *Cryst. Struct. Commun.*, 1972, **1**, 135.
612. L. F. Vilas Boas, R. D. Gillard and P. R. Mitchell, *J. Chem. Soc., Dalton Trans.*, 1977, 1215.
613. G. H. Searle, S. F. Lincoln, F. R. Keene, S. G. Teague and D. G. Rowe, *Aust. J. Chem.*, 1977, **30**, 1221.
614. M. R. Churchill, G. M. Harris, T. Inoue and R. A. Lashewycz, *Acta Crystallogr., Part B*, 1981, **37**, 933.
615. Y. Kushi, K. Watanabe and H. Kuroya, *Bull. Chem. Soc. Jpn..*, 1967, **40**, 2985.
616. A. R. Gainsford and D. A. House, *J. Inorg. Nucl. Chem.*, 1970, **32**, 688.
617. L. F. Druding and F. D. Sancillo, *Acta Crystallogr., Part B*, 1974, **30** 2386.
618. H. Kuroya, *Nippon Kagaku Zasshi*, 1971, **92**, 905.
619. D. A. House, *Inorg. Chim. Acta.*, 1978, **30**, 281.
620. I. F. Burshtein, M. D. Mazus, V. N. Biyushkin, E.V. Popa, A. V. Ablov, T. I. Malinovski, N. V. Rannev and B. M. Shchedrin, *Koord. Khim.*, 1978, **4**, 282 (*Chem. Abstr.*, 1978, **88**, 161 969).
621. A. V. Ablov, E. V. Popa, M. D. Mazus, V. N. Biyushkin, A. P. Gulya and T. I. Malinovski, *Koord. Khim.*, 1979, **5**, 287 (*Chem. Abstr.*, 1979, **90**, 145 183).
622. D. A. House, *Inorg. Nucl. Chem. Lett.*, 1967, **3**, 67.
623. A. D. Fowlie, D. A. House, W. T. Robinson and S. Sheat Rumball, *J. Chem. Soc. (A)*, 1970, 803.
624. F. A. Cotton and D. C. Richardson, *Inorg. Chem.*, 1966, **5**, 1851.
625. F. A. Cotton and R. M. Wing, *Inorg. Chem.*, 1965, **4**, 314.
626. F. A. Cotton and R. C. Elder, *Inorg. Chem.*, 1964, **3**, 397.
627. H.-H. Schmidtke, *Z. Anorg. Allg. Chem.*, 1965, **339**, 103.
628. K. Okamoto, K. Wakayama, H. Einaga, M. Ohmasa and J. Hidaka, *Bull. Chem. Soc. Jpn.*, 1982, **55**, 3473.

629. A. Earnshaw, L. F. Larkworthy and K. C. Patel, *J. Chem. Soc. (A)*, 1969, 2276.
630. M. R. Churchill, G. M. Harris, T. Inoue and R. A. Lashewycz, *Acta Crystallogr., Part B*, 1981, **37**, 695.
631. P. H. Crayton, *Inorg. Synth.*, 1963, **7**, 211.
632. M. D. Mazus, V. N. Biyushkin, E. V. Popa, A. V. Ablov, T. I. Malinovskii and V. V. Tkachev, *Dokl. Akad. Nauk SSSR*, 1976, **230**, 613 (*Chem. Abstr.*, 1977, **86**, 10 944).
633. M. C. Couldwell and D. A. House, *Inorg. Chem.*, 1974, **13**, 2949.
634. M. C. Couldwell, D. A. House and B. R. Penfold, *Inorg. Chim. Acta*, 1975, **13**, 61.
635. K.Kobayashi and M. Shibata, *Bull. Chem. Soc. Jpn.*, 1975, **48**, 2561.
636. A. V. Ablov, M. D. Mazus, E. V. Popa, T. I. Malinovski and V. N. Biushkin, *Dokl. Akad. Nauk SSSR*, 1972, **202**, 333 (Russ.); 1972, **202**, 21 (Engl. Transl.).
637. K. Ohkawa, J. Fujita and Y. Shimura, *Bull. Chem. Soc. Jpn.*, 1972, **45**, 161.
638. G. J. Gainsford, D. A. House, W. Marty and P. Comba, *Cryst. Struct. Commun.*, 1982, **11**, 215.
639. Foo Chuk Ha and D. A. House, *Inorg. Chim. Acta*, 1980, **38**, 167.
640. A. R. Gainsford, D. A. House and W. T. Robinson, *Inorg. Chim. Acta*, 1971, **5**, 595.
641. A. V. Ablov, M. D. Mazus, E. V. Popa, T. I. Malinovski and V. N. Biyushkin, *Dokl. Akad. Nauk SSSR*, 1970, **194**, 701 (Engl. Transl.) (*Chem. Abstr.*, 1971, **74**, 16 728).
642. J. H. Johnston and A. G. Freeman, *J. Chem. Soc., Dalton Trans.*, 1975, 2153.
643. B. F. Anderson, J. D. Bell, A. R. Gainsford and D. A. House, *Inorg. Chim. Acta*, 1978, **30**, 59.
644. R. G. Holloway, D. A. House and B. R. Penfold, *Cryst. Struct. Commun.*, 1978, **7**, 139.
645. A. R. Gainsford and D. A. House, *Inorg. Chim. Acta*, 1983, **74**, 205.
646. J. R. Fritch, G. G. Christoph and W. P. Schaefer, *Inorg. Chem.*, 1973, **12**, 2170.
647. B. S. Dawson and D. A. House, *Inorg. Chem.*, 1977, **16**, 1354.
648. M. Cannas, G. Carta and G. Marongiu, *J. Chem. Soc., Dalton Trans.*, 1974, 553.
649. F. S. Stephens, *J. Chem. Soc. (A)*, 1969, 2493.
650. M. Cannas, G. Carta and G. Marongiu, *J. Chem. Soc., Dalton Trans.*, 1974, 556.
651. T. Sorrell, L. G. Marzilli and T. J. Kistenmacher, *J. Am. Chem. Soc.*, 1976, **98**, 2181.
652. N. F. Curtis, I. R. N. McCormick and T. N. Waters, *J. Chem. Soc., Dalton Trans.*, 1973, 1537.
653. G. Davey and F. S. Stephens, *J. Chem. Soc. (A)*, 1971, 103.
654. M. Cannas, G. Carta, A. Cristini and G. Marongiu, *Inorg. Chem.*, 1977, **16**, 228.
655. M. Cannas, G. Marongiu and G. Saba, *J. Chem. Soc., Dalton Trans.*, 1980, 2090.
656. G. Nardin, L. Randaccio, G. Annibale, G. Natile and B. Pitteri, *J. Chem. Soc., Dalton Trans.*, 1980, 220.
657. N. Bresciani, M. Calligaris, L. Randaccio, V. Ricevuto and U. Belluco, *Inorg. Chim. Acta*, 1975, **14**, L17.
658. O. Kling and H. L. Schläfer, *Z. Anorg. Allg. Chem.*, 1961, **313**, 187.
659. E. K. Barefield, A. M. Carrier and D. G. Vanderveer, *Inorg. Chim. Acta*, 1980, **42**, 271.
660. T. W. Hambley, G. H. Searle and M. R. Snow, *Aust. J. Chem.*, 1982, **35**, 1285.
661. D. A. House, P. R. Ireland, I. A. Maxwell and W. T. Robinson, *Inorg. Chim. Acta*, 1971, **5**, 397.
662. P. R. Ireland, D. A. House and W. T. Robinson, *Inorg. Chim. Acta*, 1970, **4**, 137.
663. R. D. Hancock, G. J. McDougall and F. Marsicano, *Inorg. Chem.*, 1979, **18**, 2847.
664. M. Cannas, A. Cristini and G. Marongiu, *Inorg. Chim. Acta*, 1977, **22**, 233.
665. M. Cannas, G. Carta, A. Cristini and G. Marongiu, *J. Chem. Soc., Dalton Trans.*, 1974, 1278.
666. B. W. Skelton, T. N. Waters and N. F. Curtis, *J. Chem. Soc., Dalton Trans.*, 1972, 2133.
667. N. J. Ray and B. J. Hathaway, *Acta Crystallogr., Part B*, 1978, **34**, 3224.
668. G. Ponticelli, *Inorg. Chim. Acta*, 1971, **5**, 461.
669. F. W. B. Einstein and A. C. Willis, *Inorg. Chem.*, 1981, **20**, 609.
670. W. D. Stanley, T. Davies, T. D. Tullins and C. S. Garner, *J. Inorg. Nucl. Chem.*, 1973, **35**, 3857.
671. R. Melanson, J. Hubert and F. D. Rochon, *Can. J. Chem.*, 1975, **53**, 1139.
672. N. Bresciani Pahor, M. Calligaris and L. Randaccio, *J. Chem. Soc., Dalton Trans.*, 1976, 725.
673. I. Mochida, J. A. Mattern and J. C. Bailar, Jr., *J. Am. Chem. Soc.*, 1975, **97**, 3021.
674. J. F. Britten and C. J. L. Lock, *Acta Crystallogr., Part B*, 1980, **36**, 2958.
675. J. F. Britten, C. J. L. Lock and W. M. C. Pratt, *Acta Crystallogr., Part B*, 1982, **38**, 2148.
676. R. J. H. Clark, M. Kurmoo, A. M. R. Galas and M. B. Hursthouse, *J. Chem. Soc., Dalton Trans.*, 1983, 1583.
677. N. F. Curtis, R. W. Hay and Y. M. Curtis, *J. Chem. Soc. (A)*, 1968, 182.
678. R. Barbucci, M. J. M. Campbell, M. Cannas and G. Marongiu, *Inorg. Chim. Acta*, 1980, **46**, 135.
679. D. A. House and R. P. Moon, *J. Inorg. Nucl. Chem.*, 1981, **43**, 2572.
680. Y. Wu and D. H. Busch, *J. Am. Chem. Soc.*, 1972, **94**, 4115.
681. D. K. Lin and C. S. Garner, *J. Am. Chem. Soc.*, 1969, **91**, 6637.
682. A. R. Gainsford and A. M. Sargeson, *Aust. J. Chem.*, 1978, **31**, 1679.
683. G. Riggio, W. H. Hopff, A. A. Hofmann and P. G. Waser, *Helv. Chim. Acta*, 1980, **63**, 488.
684. G. H. Searle, *Aust. J. Chem.*, 1977, **30**, 2625.
685. G. H. Searle, *Aust. J. Chem.*, 1980, **33**, 2159.
686. F. G. Mann, *J. Chem. Soc.*, 1934, **137**, 461, 466.
687. G. Annibale, G. Natile and L. Cattalini, *J. Chem. Soc., Dalton Trans.*, 1976, 1547.
688. G. W. Watt and W. A. Cude, *J. Am. Chem. Soc.*, 1968, **90**, 6382; *Inorg. Chem.*, 1968, **7**, 335.
689. G. C. Schlessinger, *Bull. N. J. Acad. Sci.*, 1967, **42**, 12.
690. G. Gelbard and P. Rumpf, *C.R. Hebd. Seances Acad. Sci.*, 1966, **262**, 1587.
691. E. L. J. Breet and R. Van Eldik, *Inorg. Chim. Acta Lett.*, 1983, **76**, L303.
692. M. T. Halfpenny, W. Levason, C. A. McAuliffe, W. E. Hill and F. P. McCullough, *Inorg. Chim. Acta*, 1979, **32**, 229.
693. W. J. Louw and D. J. A. De Waal, *Inorg. Chim. Acta*, 1978, **28**, 35.
694. C.-L. O'Young, J. C. Dewan, H. R. Lilienthal and S. J. Lippard, *J. Am. Chem. Soc.*, 1978, **100**, 7291.
695. Z. Dori, R. Eisenberg and H. B. Gray, *Inorg. Chem.*, 1967, **6**, 483.
696. Z. Dori and H. B. Gray, *J. Am. Chem. Soc.*, 1966, **88**, 1394.
697. J. T. Reinprecht, J. G. Miller, G. C. Vogel, M. S. Haddad and D. N. Hendrickson, *Inorg. Chem.*, 1980, **19**, 927.
698. L. Campbell and J. J. McGarvey, *J. Am. Chem. Soc.*, 1977, **99**, 5809.

699. J. L. Burmeister, R. L. Hassel, K. A. Johnson and J. C. Lim, *Inorg. Chim. Acta*, 1974, **9**, 23.
700. M. J. Blandamer, J. Burgess, S. J. Hamshere and P. Wellings, *Transition Met. Chem. (Weinheim, Ger.)*, 1979, **4**, 161.
701. R. D. Alexander and P. N. Holper, *Transition Met. Chem.*, 1980, **5**, 108.
702. L. Costanzo, A. Giuffrida, G. Guglielmo and V. Ricevuto, *Inorg. Chim. Acta*, 1979, **33**, 29.
703. F. Basolo, H. B. Gray and R. G. Pearson, *J. Am. Chem. Soc.*, 1960, **82**, 4200.
704. S. N. Bhattacharya, C. V. Sneoff and F. Walker, *Synth. React. Inorg. Metal-Org. Chem.*, 1979, **9**, 5.
705. N. Bresciani, M. Colligaris, L. Randaccio, V. Ricevuto and U. Belluco, *Inorg. Chim. Acta*, 1975, **14**, L17.
706. M. Cusumano, G. Guglielmo, P. Marricchi and V. Ricevuto, *Inorg. Chim. Acta*, 1978, **30**, 29.
707. C. Bartocci, A. Ferri, V. Carassiti and F. Scandola, *Inorg. Chim. Acta*, 1977, **24**, 251.
708. R. C. E. Durley, W. L. Waltz and B. E. Robertson, *Can. J. Chem.*, 1980, **58**, 664.
709. G. Kolks, C. R. Frihart, P. K. Coughlin and S. J. Lippard, *Inorg. Chem.*, 1981, **20**, 2928, 2933.
710. C. G. Pierpont, L. C. Francesconi and D. N. Hendrickson, *Inorg. Chem.*, 1977, **16**, 2367.
711. T. R. Felthouse and D. N. Hendrickson, *Inorg. Chem.*, 1978, **17**, 444.
712. M. S. Haddad, E. N. Duesler and D. N. Hendrickson, *Inorg. Chem.*, 1979, **18**, 141.
713. M. Ciampolini and G. P. Speroni, *Inorg. Chem.*, 1966, **5**, 45.
714. T. F. Brennan, G. Davies, M. A. El-Sayed, M. F. El-Shazly, M. W. Rupich and M. Veidis, *Inorg. Chim. Acta*, 1981, **51**, 45.
715. M. Di Vaira and P. L. Orioli, *Inorg. Chem.*, 1969, **8**, 2729.
716. M. Cusumano, G. Guglielmo, V. Ricevuto, R. Romeo and M. Trozzi, *Inorg. Chim. Acta*, 1976, **16**, 135.
717. M. Cusumano, G. Guglielmo, V. Ricevuto, R. Romeo and M. Trozzi, *Inorg. Chim. Acta*, 1976, **17**, 45.
718. D. A. Palmer, R. Schmidt, R. Van Eldik and H. Kelm, *Inorg. Chim. Acta*, 1978, **29**, 261.
719. T. R. Felthouse, E. J. Laskowski and D. N. Hendrickson, *Inorg. Chem.*, 1977, **16**, 1077.
720. I. Bertini, D. L. Johnston and W. de W. Horrocks, Jr., *Inorg. Chim. Acta*, 1970 **4**, 79.
721. R. D. Keen, D. A. House and H. K. J. Powell, *J. Chem. Soc., Dalton Trans.*, 1975, 688.
722. A. Marxer and K. Miescher, *Helv. Chim. Acta*, 1951, **34**, 924.
723. G. H. Searle and T. W. Hambley, *Aust. J. Chem.*, 1982, **35**, 2399.
724. M. Ishii, S. Sato, Y.Saito and M. Nukahara, *Chem. Lett.*, 1981, 1613.
725. D. A. House, A. R. Gainsford and J. W. Blunt, *Inorg. Chim. Acta*, 1982, **57**, 141.
726. C. L. Sharma and T. K. De, *J. Indian Chem. Soc.*, 1981, **58**, 119.
727. S. K. Srivastava, *Indian J. Chem., Sect. A*, 1979, **17**, 193.
728. R. Barbucci and M. Budini, *J. Chem. Soc., Dalton Trans.*, 1976, 1321.
729. R. Barbucci, P. Paoletti and A. Vacca, *Inorg. Chem.*, 1975, **14**, 302.
730. M. Gold and H. K. J. Powell, *J. Chem. Soc., Dalton Trans.*, 1976, 1418.
731. J. R. Flückiger, C. W. Schläpfer and C. Couldwell, *Inorg. Chem.*, 1980, **19**, 2493.
732. T. G. Spiro and C. J. Ballhausen, *Acta Chem. Scand.*, 1961, **15**, 1707.
733. W. J. Kasowski and J. C. Bailar, Jr., *J. Am. Chem. Soc.*, 1969, **91**, 3212.
734. R. A. D. Wentworth and J. J. Felten, *J. Am. Chem. Soc.*, 1968, **90**, 621.
735. R. A. D. Wentworth, *Inorg. Chem.*, 1968, **7**, 1030.
736. J. C. Huffman, R. A. D. Wentworth, C. C. Tsai, C. J. Huffman and W. E. Streib, *Cryst. Struct. Commun.*, 1981, **10**, 1493.
737. J. H. Ammeter, H. B. Bürgi, E. Gamp, V. Meyer-Sandrin and W. P. Jensen, *Inorg. Chem.*, 1979, **18**, 733.
738. W. Marty, *Synth. React. Inorg. Metal-Org. Chem.*, 1981, **11**, 411.
739. J. Van Alphen, *Recl. Trav. Chim. Pays-Bas*, 1936, **55**, 835. (*Chem. Abstr.*, 1936, **30**, 5992, 7100; 1937, **31**, 1007, 4645, 5361).
740. D. A. House and D. Yang, *Inorg. Chim. Acta*, 1983, **74**, 179.
741. G. R. Brubaker, F. H. Jarke and I. M. Brubaker, *Inorg. Chem.*, 1979, **18**, 2032.
742. S. G. Zipp, A. P. Zipp and S. K. Madan, *J. Indian Chem. Soc.*, 1977, **54**, 149.
743. J. Glerup, J. Josephsen, K. Michelsen, E. Pedersen and C. E. Schaffer, *Acta Chem. Scand.*, 1970, **24**, 247.
744. J. H. Forsberg, T. M. Kubik, T. Moeller and K. Gucwa, *Inorg. Chem.*, 1971, **10**, 2656.
745. A. F. Casey, *J. Pharm. Pharmacol.*, 1981, **33**, 333.
746. G. A. Bottomley, L. G. Glossop, B. W. Skelton and A. H. White, *Aust. J. Chem.*, 1979, **32**, 285.
747. C.-H. L. Yang and M. W. Grieb, *J. Chem. Soc., Chem. Commun.*, 1972, 656.
748. D. A. Buckingham, P. J. Cresswell and A. M. Sargeson, *Inorg. Chem.*, 1973, **14**, 1485.
749. G. Bombieri, F. Benetollo, A. Del Pra, M. L. Tobe and D. A. House, *Inorg. Chim. Acta*, 1981, **50**, 89.
750. E. Martins and P. S. Sheridan, *Inorg. Chem.*, 1978, **17**, 2822.
751. J. Hughes and G. R. Willey, *J. Coord. Chem.*, 1974, **4**, 33.
752. G. C. Stocco and A. Venezia, *Inorg. Chim. Acta*, 1980, **46**, 1.
753. F. M. Jaeger and P. Koets, *Z. Anorg. Allg. Chem.*, 1928, **170**, 347 (*Chem. Abstr.*, 1928, **22**, 3366).
754. D. A. Buckingham, C. R. Clark and T. W. Lewis, *Inorg. Chem.*, 1979, **18**, 2041.
755. E. J. Laskowski and D. N. Hendrickson, *Inorg. Chem.*, 1978, **17**, 457, 460.
756. E. J. Laskowski, D. M. Duggan and D. N. Hendrickson, *Inorg. Chem.*, 1975, **14**, 2449.
757. D. M. Duggan and D. N. Hendrickson, *Inorg. Chem.*, 1974, **13**, 1911.
758. G. D. Andreetti, P. C. Jain and E. C. Lingafelter, *J. Am. Chem. Soc.*, 1969, **91**, 4112.
759. R. J. Sime, R. P. Dodge, A. Zalkin, and D. H. Templeton, *Inorg. Chem.*, 1971, **10**, 537.
760. C. G. Pierpont, D. N. Hendrickson, D. M. Duggan, F. Wagner and E. K. Barefield, *Inorg. Chem.*, 1975, **14**, 604.
761. D. M. Duggan and D. N. Hendrickson, *Inorg. Chem.*, 1974, **13**, 2056.
762. B. G. Segal and S. J. Lippard, *Inorg. Chem.*, 1977, **16**, 1623.
763. S. N. Bhattacharya and C. V. Senoff, *Inorg. Chim. Acta*, 1980, **41**, 67.
764. G. Ferguson and M. Parvez, *Acta Crystallogr., Part B*, 1979, **35**, 2207.
765. G. Ferguson and P. J. Roberts, *Acta Crystallogr., Part B*, 1978, **34**, 3083.
766. C. Muralikrishna, C. Mahaderan, S. Shakuntala Sastry, M. Seshasayee and S. Subramanian, *Acta Crystallog., Part C*, 1983, **39**, 1630.
767. M. Di Vaira and P. L. Orioli, *Acta Crystallogr., Part B*, 1968, **24**, 595.
768. S. F. Lincoln, A. M. Hounslow and J. E. Coates, *Inorg. Chim. Acta Lett.*, 1983, **77**, L7.

769. A. M. Sargeson and G. H. Searle, *Inorg. Chem.*, 1967, **6**, 787.
770. J. A. Broomhead and L. A. P. Kane-Maguire, *Inorg. Chem.*, 1968, **7**, 2519.
771. T. Eliades, R. O. Harris and P. Reinslau, *Can. J. Chem.*, 1969, **47**, 3823.
772. E. Martins and P. S. Sheridan, *Inorg. Chem.*, 1978, **14**, 3631.
773. P. M. Gidney, R. D. Gillard, B. T. Heaton, P. S. Sheridan and D. H. Vaughn, *J. Chem. Soc., Dalton Trans.*, 1973, 1462.
774. J. Burgess, K. W. Bowker, E. R. Gardner and F. M. Mekhail, *J. Inorg. Nucl. Chem.*, 1979, **41**, 1215.
775. D. A. House and C. S. Garner, *J. Am. Chem. Soc.*, 1966, **88**, 2156,
776. D. A. House, *Aust. J. Chem.*, 1982, **35**, 659.
777. G. Marongiu and M. Cannas, *J. Chem. Soc., Dalton Trans.*, 1979, 41.
778. N. F. Curtis and D. A. House, *J. Chem. Soc.*, 1965, 6195.
779. F. Hori, K. Matsumoto, S. Ooi and H. Kuroya, *Bull. Chem. Soc. Jpn.*, 1977, **50**, 138.
780. H. B. Jonassen and N. L. Cull, *J. Am. Chem. Soc.*, 1949, **71**, 4097.
781. J. Meisenheimer, L. Angermann, H. Holsten and E. Kiderlen, *Justus Liebigs Ann. Chem.*, 1924, **438**, 217, 261.
782. H. C. Freeman and I. E. Maxwell, *Inorg. Chem.*, 1969, **8**, 1293.
783. D. A. Buckingham, P. A. Marzilli and A. M. Sargeson, *Inorg. Chem.*, 1967, **6**, 1032.
784. K. Tanaka, F. Marumo and Y. Saito, *Acta Crystallogr., Part B*, 1973, **29**, 733.
785. W. Bentley and E. H. Creaser, *Inorg. Chem.*, 1974, **13**, 1115.
786. M. Goto, A. Okubo, T. Sawai and S. Yoshikawa, *Inorg. Chem.*, 1970, **9**, 1488.
787. S. Yoshikawa, T. Sekihara and M. Goto, *Inorg. Chem.*, 1967, **6**, 169.
788. Hon-Peng Lau and C. D. Gutsche, *J. Am. Chem. Soc.*, 1978, **100**, 1850, 1857.
789. M. M. Muir, R. R. Rechani and J. A. Diaz, *Synth. React. Inorg. Metal-Org. Chem.*, 1981, **11**, 317.
790. M. M. Muir and J. A. Diaz, *Synth. React. Inorg. Metal-Org. Chem.*, 1981, **11**, 333.
791. M. Saburi, M. Homma and S. Yoshikawa, *Inorg. Chem.*, 1973, **12**, 1250.
792. M. Saburi, T. Sawai and S. Yoshikawa, *Bull. Chem. Soc. Jpn.*, 1972, **45**, 1086.
793. R. M. Clay, H. McCormac, M. Micheloni and P. Paoletti, *Inorg. Chem.*, 1982, **21**, 2494.
794. M. Saburi and S. Yoshikawa, *Bull. Chem. Soc. Jpn.*, 1972, **45**, 1086.
795. R. G. Asperger and C. F. Liu, *Inorg. Chem.*, 1965, **4**, 1395; 1967, **6**, 796.
796. R. G. Asperger and C. F. Liu, *J. Am. Chem. Soc.*, 1967, **89**, 708, 1533.
797. J. P. Glusker, H. L. Carrell, R. Job and T. C. Bruice, *J. Am. Chem. Soc.*, 1974, **96**, 5741.
798. R. Job, *J. Am. Chem. Soc.*, 1978, **100**, 5089.
799. R. Job, *Inorg. Chim. Acta*, 1980, **40**, 59.
800. R. Job and T. C. Bruice, *J. Am. Chem. Soc.*, 1974, **96**, 809.
801. M. Saburi and S. Yoshikawa, *Bull. Chem. Soc. Jpn.*, 1972, **45**, 806.
802. T. Shibahara and M. Mori, *Bull. Chem. Soc. Jpn.*, 1972, **45**, 1433.
803. M. Ito, F. Marumo and Y. Saito, *Acta Crystallogr., Part B*, 1972, **28**, 457, 463.
804. M. Saburi, T. Sato and S. Yoshikawa, *Bull. Chem. Soc. Jpn.*, 1976, **49**, 2100.
805. K. Toriumi and Y. Saito, *Acta Crystallogr., Part B*, 1975, **31**, 1247.
806. M. Saburi and S. Yoshikawa, *Inorg. Chem.*, 1968, **7**, 1890.
807. D. J. Francis and G. H. Searle, *Aust. J. Chem.*, 1974, **27**, 269.
808. G. H. Searle, M. Petkovic and F. R. Keene, *Inorg. Chem.*, 1974, **13**, 399.
809. R. W. Hay and J. Bakir, *Transition Met. Chem. (Weinheim, Ger.)*, 1980, **5**, 252.
810. H. Macke, M. Zehnder, U. Thewalt and S. Fallab, *Helv. Chim. Acta*, 1979, **62**, 1804.
811. M. Goto, H. Matsushita, M. Saburi and S. Yoshikawa, *Inorg. Chem.*, 1973, **12**, 1498.
812. M. Saburi and S. Yoshikawa, *Bull. Chem. Soc. Jpn.*, 1974, **47**, 1184.
813. M. J. Jun and C. F. Liu, *Inorg. Chem.*, 1975, **14**, 2310.
814. M. J. Jun and C. F. Liu, *J. Chem. Soc., Dalton Trans.*, 1976, 1031.
815. M. J. Jun and C. F. Liu, *Inorg. Chim. Acta*, 1975, **15**, 111.
816. M. J. Jun and C. F. Liu, *Taehan Hwahak Hoechi*, 1975, **19**, 98 (*Chem. Abstr.*, 1975, **83**, 52 579).
817. P. W. R. Corfield, J. C. Dabrowiak and E. S. Gore, *Inorg. Chem.*, 1973, **12**, 1734.
818. J. J. Fitzgerald and G. R. Brubaker, *Inorg. Chem.*, 1973, **12**, 2988.
819. A. Fjuioka, S. Yano and S. Yoshikawa, *Inorg. Nucl. Chem. Lett.*, 1975, **11**, 341.
820. F. Mizukami, *Bull. Chem. Soc. Jpn.*, 1975, **48**, 472, 1205.
821. S. Yano, A. Fujioka, M. Yamaguchi and S. Yoshikawa, *Inorg. Chem.*, 1978, **17**, 14.
822. P. C. Harrington, S. Linke and M. D. Alexander, *Inorg. Chem.*, 1973, **12**, 168.
823. M. Goto, T. Makino, M. Saburi and S. Yoshikawa, *Bull. Chem. Soc. Jpn.*, 1976, **49**, 1879.
824. S. Yoshikawa, M. Saburi and M. Yamaguchi, *Pure Appl. Chem.*, 1978, **50**, 915.
825. M. J. Jun and C. F. Liu, *J. Coord. Chem.*, 1975, **5**, 1.
826. W. A. Freeman, *Inorg. Chem.*, 1978, **17**, 2982.
827. S. Yano, K. Furuhashi and S. Yoshikawa, *Bull. Chem. Soc. Jpn.*, 1977, **50**, 685.
828. B. Güntert, S. Claude and K. Bernauer, *Helv. Chim. Acta*, 1975, **58**, 780.
829. J. Cragel and G. R. Brubaker, *Inorg. Chem.*, 1972, **11**, 303.
830. L. J. De Hayes and D. H. Busch, *Inorg. Chem.*, 1973, **12**, 2010.
831. M. Saburi, C. Hattori and S. Yoshikawa, *Inorg. Chim. Acta*, 1972, **6**, 427.
832. G. R. Brubaker and D. W. Johnson, *Inorg. Chem*, 1983, **22**, 1422.
833. S. Yaba, S. Yano and S. Yoshikawa, *Inorg. Nucl. Chem. Lett.*, 1976, **12**, 267; 831.
834. B. Bosnich and J. MacB. Harrowfield, *Inorg. Chem.*, 1975, **14**, 853, 861.
835. B. Bosnich and J. MacB. Harrowfield, *Inorg. Chem.*, 1975, **14**, 836, 847.
836. G. R. Hedwig, J. L. Love and H. K. J. Powell, *Aust. J. Chem.*, 1970, **23**, 981.
837. G. Marongiu, E. C. Lingafelter and P. Paoletti, *Inorg. Chem.* 1969, **8**, 2763.
838. A. McPherson, M. G. Rossman, D. W. Margerum and M. R. James, *J. Coord. Chem.*, 1971, **1**, 39.
839. R. K. Wismer and R. A. Jacobson, *Inorg. Chim. Acta*, 1973, **7**, 477.
840. J. Dwyer, W. Levason and C. A. McAuliffe, *J. Inorg. Nucl. Chem.*, 1976, **38**, 1919.

841. R. Barbucci, A. Mastroianni and M. J. M. Campbell, *Inorg. Chim. Acta*, 1978, **27**, 109.
842. G. Helary, L. Lefevre-Jenot and M. Fontanille, *J. Organomet. Chem.*, 1981, **205**, 139.
843. B. Bosnich, R. D. Gillard, E. D. McKenzie and G. A. Webb, *J. Chem. Soc. (A)*, 196, 1331.
844. H. G. Hamilton, Jr. and M. D. Alexander, *J. Am. Chem. Soc.*, 1967, **89**, 5065.
845. E. Ahmed and M. L. Tobe, *Inorg. Chem.*, 1974, **13**, 2956.
846. C. K. Poon, C. M. Che and T. W. Tang, *J. Chem. Soc., Dalton Trans.*, 1981, 1697.
847. R. Niththyananthan and M. L. Tobe, *Inorg. Chem.*, 1969, **8**, 1589.
848. E. Martins, E. B. Kaplan and P. S. Sheridan, *Inorg. Chem.*, 1979, **18**, 2195.
849. D. A. House and B. R. Penfold, unpublished X-ray crystal structure. This implies that the α configurational assignments in ref. 740 are incorrect.
850. T G. Fawcett, S. M. Rudich, B. H. Toby, R. A. Lalancette, J. A. Potenza and H. J. Schugar, *Inorg. Chem.*, 1980, **19**, 940.
851. M. D. Alexander, P. C. Harrington and A. Van Heuvelen, *J. Phys. Chem.*, 1971, **75**, 3355.
852. D. F. Cook and E. D. McKenzie, *Inorg. Chim. Acta*, 1977, **29**, 193; 1978, **31**, 59.
853. J. D. Vitiello and E. J. Billo, *Inorg. Chem.*, 1980, **19**, 3477.
854. M. D. Alexander and H. G. Hamilton, *Inorg. Chem.*, 1969, **8**, 2131.
855. D. Yang and D. A. House, *Inorg. Chem.*, 1982, **21**, 2999.
856. B. Bosnich and J. MacB. Harrowfield, *Inorg. Chem.*, 1975, **14**, 828.
857. B. Bosnich, J. MacB. Harrowfield and H. Boucher, *Inorg. Chem.*, 1975, **14**, 815.
858. D. A. House and J. W. Blunt, *Inorg. Chim. Acta*, 1981, **49**, 193.
859. G. R. Hedwig and H. K. J. Powell, *Anal. Chem.*, 1971, **43**, 1206; *J. Chem. Soc., Dalton Trans.*, 1973, 793.
860. R. Boggs and J. Donohue, *Acta Crystallogr., Part B*, 1975, **31**, 320.
861. G. R. Brubaker and D. P. Schaefer, *Inorg. Chem.*, 1971, **10**, 968.
862. Y. Yamamoto, H. Kudo and E. Toyota, *Bull. Chem. Soc. Jpn.*, 1983, **56**, 1051.
863. T. Izumitani, M. Nakamura, H. Okawa and S. Kida, *Bull. Chem. Soc. Jpn.*, 1982, **55**, 2122.
864. (a) A. Litherland and F. G. Mann, *J. Chem. Soc.*, 1938, 1588; (b) T. S. Cameron, W. J. Chute and O. Knop, *Can. J. Chem.*, 1985, **63**, 586.
865. R. W. Oehmke and J. C. Bailar, Jr., *J. Inorg. Nucl. Chem.*, 1965, **27**, 2199.
866. (a) J. C. Bailar, Jr., D. J. Baker and R. A. Bauer, *US NTIS, AD Rep.*, 1975, AD-AO15508 (*Chem. Abstr.*, 1976, **84**, 115 243); (b) I. W. Stapleton, *Aust. J. Chem.*, 1985, **38**, 633.
867. G. Ponticelli and A. Diaz, *Ann. Chim. (Rome)*, 1976, **61**, 46.
868. M. Cannas, A. Cristini and G. Marongiu, *Inorg. Chim. Acta*, 1976, **19**, 241.
869. D. A. House and C. S. Garner, *Inorg. Chem.*, 1966, **5**, 2097; 1967, **6**, 272.
870. A. Coda, B. Kamenar, K. Prout, J. R. Carruthers and J. S. Rollett, *Acta Crystallogr., Part B*, 1975, **31**, 1438.
871. L. R. Melby, *Inorg. Chem.*, 1970, **9**, 2186.
872. M. V. Hanson, W. E. Marsh and G. O. Carlisle, *Inorg. Nucl. Chem. Lett.*, 1977, **13**, 277.
873. M. R. Snow, D. A. Buckingham, P. A. Marzilli and A. M. Sargeson, *Chem. Commun.*, 1969, 891.
874. M. Zehnder and U. Thewalt, *Z. Anorg. Allg. Chem.*, 1980, **461**, 53.
875. T. Oshima, *J. Biol. Chem.*, 1982, **257**, 9913; 1979, **254**, 8720 (*Chem. Abstr.*, 1979, **91**, 87 019, 189 428; 1982, **97**, 178 421).
876. Y. Yoshikawa and K. Yamasaki, *Bull. Chem. Soc. Jpn.*, 1973, **46**, 3448.
877. S. Sato and Y. Saito, *Acta Crystallogr., Part B*, 1975, **31**, 2456.
878. W. Gauss, P. Moser and G. Schwarzenbach, *Helv. Chim. Acta*, 1952, **35**, 2359; 1953, **36**, 581.
879. F. P. Emmenegger and G. Schwarzenbach, *Helv. Chim. Acta*, 1966, **49**, 625.
880. J. R. Gollogly and C. J. Hawkins, *Aust. J. Chem.*, 1967, **20**, 2395.
881. Y. Yoshikawa, E. Fujii and K. Yamasaki, *Proc. Jpn. Acad.*, 1967, **43**, 495.
882. C. O. Haagensen, *Acta Chem. Scand.*, 1967, **21**, 457.
883. A. Muto, F. Marumo and Y. Saito, *Inorg. Nucl. Chem. Lett.*, 1969, **5**, 85.
884. H. Okazaki, U. Sakaguchi and H. Yoneda, *Inorg. Chem.*, 1983, **22**, 1539.
885. R. E. Hermer and B. E. Douglas, *J. Coord. Chem.*, 1977, **7**, 43.
886. F. McCullough and J. C. Bailar, Jr., *J. Am. Chem. Soc.*, 1956, **78**, 714.
887. T. Habu and J. C. Bailar, Jr., *J. Am. Chem. Soc.*, 1966, **88**, 1128.
888. M. Zehnder and H. Löliger, *Helv. Chim. Acta*, 1980, **63**, 754.
889. G. S. Hall and R. H. Soderberg, *Inorg. Chem.*, 1968, **7**, 2300.
890. D. R. Marks, D. J. Phillips and J. P. Redfern, *J. Chem. Soc. (A)*, 1967, 1464.
891. G. J. Grant and D. J. Royer, *J. Am. Chem. Soc.*, 1981, **103**, 868.
892. S. Miya, K. Kashiwabara and K. Saito, *Bull. Chim. Soc. Jpn.*, 1981, **54**, 2309.
893. C. Stalhandske, *Cryst. Struct. Commun.*, 1981, **10**, 1081.
894. C. Stalhandske, *Acta Chem. Scand.*, 1972, **26**, 3029.
895. J. C. Taft and M. M. Jones, *J. Am. Chem. Soc.*, 1960, **82**, 4196.
896. H. Lau and C. D. Gutsche, *J. Am. Chem. Soc.*, 1978, **100**, 1857.
897. P. A. Tasker and E. B. Fleischer, *J. Am. Chem. Soc.*, 1970, **92**, 7072.
898. D. St. C. Black and N. E. Rothnie, *Aust. J. Chem.*, 1983, **36**, 1141.

13.2

Heterocyclic Nitrogen-donor Ligands

JAN REEDIJK
Leiden University, The Netherlands

13.2.1 INTRODUCTION, SCOPE AND LIMITATIONS

This chapter will deal with heterocyclic nitrogen-donor ligands that are known to coordinate to metal ions, in the following categories: (i) simple monocyclic rings as ligands (monodentate or bridging); (ii) simple monocyclic rings with an additional ligand group as a substituent (only nitrogen-donor substituents are considered); (iii) chelating ligands containing two or more heterocyclic rings; and (iv) condensed, conjugated heterocyclic chelating ligands.

The reader is referred to other chapters for the following heterocyclic systems: ligands with mainly amine-type groups and one heterocyclic substituent (Chapter 13.1); polypyrazolyl compounds (Chapter 13.6); ligands that also contain another donor atom, such as O, S or P (Chapters 14, 20.3 and 20.4); macrocyclic ligands (Chapter 21); and naturally occurring ligands, such as histidine (Chapters 22 and 62).

The four above-mentioned main categories will be treated in different sections. In each of these main categories, subgroups of ligands will be discussed, divided according to ring size and ring atoms. For each subgroup of ligands the following facts will be given or referred to: accessibility or synthesis; basic properties; steric and electronic properties as a ligand; and examples of a few characteristic compounds with this ligand group.

For most heterocyclic ligand groups, extensive reviews on synthesis and physical, spectroscopy and chemical properties have been published. For detailed studies the reader is referred to these excellent compilations.[1-10] This chapter will focus upon the properties of these molecules as ligands for metal ions.

The nomenclature of the ligands will follow IUPAC rules. Ligand abbreviations will be explained in the text, unless generally accepted abbreviations are dealt with. Ligands bearing an acidic hydrogen that may be dissociated upon coordination in certain compounds usually end with a capital H, such as in imH (for imidazole), compared to the anionic ligand im$^-$ (for imidazolato).

13.2.2 WHY ARE HETEROCYCLIC NITROGEN ATOMS GOOD LIGAND DONOR ATOMS?

13.2.2.1 Electronic Aspects

Heterocyclic ligands such as pyridine (py), *N*-methylimidazole (Meim), benzotriazole (btaH) and related molecules are good ligands due to the presence of at least one nitrogen ring atom with a localized pair of electrons. When this nitrogen bears a substituent, such as in pyrrole, the coordinating possibilities are reduced or absent.

A poor indicator of the coordinating ability of such heterocyclic nitrogen atoms is the value of the basic pK_a value (*i.e.* the ease with which a hydrogen ion is added). Values for pK_a(base) have been reported[5] to vary from values as small as -2 to as large as $+14$. One should realize that the values for the pK_a of ligands are not necessarily proportional to the binding strength to metal ions. This is illustrated by comparing the pK_a values of the following three ligands: pyrazole (1) with a pK_a of 2.5, pyridine (2) with a pK_a of 5.2 and imidazole (3) with a pK_a of 7.0 under comparable conditions.[5] However, the affinity for the Ni^{2+} ion in water is almost identical for the three ligands, as deduced from stability constants;[11] also, the ligand-field splitting parameters and the metal–ligand vibrations in the solid state are very close together.[12,13]

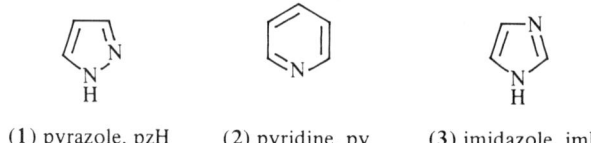

(1) pyrazole, pzH   (2) pyridine, py   (3) imidazole, imH

Probably better indicators of the strength of this type of ligand are the so-called donor numbers (DN) introduced by Gutmann,[14] with reference to reaction with SbCl$_5$.

Summarizing all available data, it is evident that any heterocyclic nitrogen atom with a non-conjugated pair of free electrons can act as a donor atom towards metal ions, provided that steric effects of substituents do not hamper the coordination (see Section 13.2.2.2).

13.2.2.2 Steric Aspects

Steric effects are known to be important in coordination chemistry, dictating geometry and coordination number and greatly influencing the spectroscopic, magnetic and catalytic properties. In heterocyclic nitrogen-donor ligands, the shape of the ligands usually allows octahedral coordination geometry for transition metal ions.

Steric effects, such as those originating from ring substituents, hydrogen bonding and/or ring-stacking interactions, are usually sufficient to explain differences in coordination geometry and coordination number in a comparison of closely related systems.

The following examples may illustrate this. Let us first compare the ligands pyrazole (pzH; 1), imidazole (imH; 3) and pyridine (py; 2) again. Only with pyrazole and imidazole are octahedral species [M(ligand)$_6$](anion)$_2$ with the metal completely solvated by the ligands (so-called solvates) easily formed (anions are, for example, nitrate or perchlorate). With pyridine, however, only

[M(py)$_4$(anion)$_2$] can be isolated,[15] containing square-planar, tetrahedral or tetragonal M^{II} with axially coordinating anionic ligands. The reason for this is thought to be the larger C—N—C angle in pyridine (120° *vs.* 108° in pzH and imH), which results in strong internal steric hindrance of the C—H groups when six pyridines are arranged around the small Ni^{II} ion. On going from Ni^{II} to Fe^{II} and using a large counterion such as [Fe$_4$(CO)$_{13}$]$^{2-}$, the [Fe(py)$_6$]$^{2+}$ ion can indeed be isolated.[16]

Another illustrative example comes from comparison of the species M(ligand)$_n$Cl$_2$. Even with a large excess of pyrazole only [Ni(pzH)$_4$Cl$_2$] has been isolated,[12,17] whereas with imidazole [Ni(imH)$_6$]Cl$_2$ is found. In this case the steric effect due to hydrogen bonding in the solid state is important. In [Ni(pzH)$_4$Cl$_2$], intramolecular hydrogen bonding keeps the Cl$^-$ ions in a coordinated position (see Figure 1), whereas in the case of [Ni(imH)$_6$]Cl$_2$ the intermolecular hydrogen bonding determines the structure and stoichiometry (here the N—H group is at the outside of the coordinated ligand).[12]

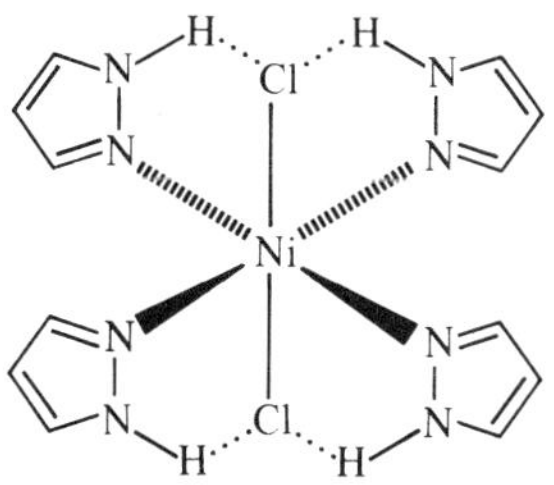

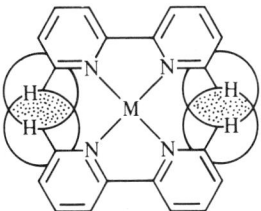

Figure 1 Schematic representation of the molecular structure of [Ni(pzH)$_4$Cl$_2$] and the stabilization by intramolecular hydrogen bonding

A further example comes from the comparison of imidazole and 2-methylimidazole (2-MeimH). As said above, [Ni(imH)$_6$](anion)$_2$ is easily formed. However, with 2-MeimH only species [Ni(2-MeimH)$_4$(anion)$_2$] can be isolated. Apparently the methyl group near the donor site hampers the arrangement of six ligands around Ni^{II}.[18] On going from Ni^{II} to the larger Cd^{II}, on the other hand, octahedral species [Cd(2-MeimH)$_6$](anion)$_2$ are easily obtained.[19]

The last examples come from bidentate ligands of the 2,2'-bipyridyl (bipy) type. Octahedral species of the type [M(bipy)$_3$]$^{2+}$ are easily formed and very stable for most metal ions.[20] When metal ions, *e.g.* Pd^{II}, require a square-planar geometry, the steric effects of the hydrogen atoms (see Figure 2) are quite large and serious distortions occur.[21] In the case of Cu^{II} this usually results in five-coordinate species, or distorted six-coordinate species, as indicated in Figure 3.[22] When, in addition, substituents are added near the donor atoms, a strong preference for tetrahedral geometry results. This has been used to develop selective reagents to coordinate to Cu^I rather than to Cu^{II} (see Figure 4).[23]

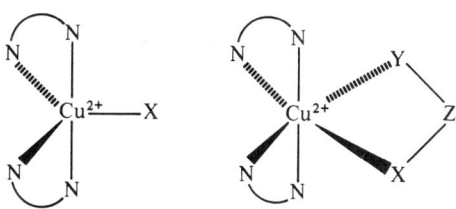

Figure 2 Illustration of the steric effect of hydrogens in position 6 of bipy, preventing the formation of tetragonal and square-planar [M(bipy)$_2$]$^{n+}$ species

Figure 3 Schematic representation of observed five- and six-coordinate Cu(bipy)$_2$(anion)$_2$ compounds

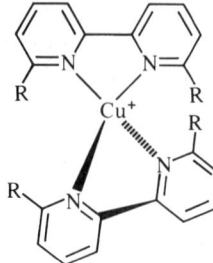

Figure 4 Stabilization of Cu^I by steric effects in substituted bipy ligands, resulting in tetrahedral geometry

13.2.3 LIGANDS CONSISTING OF MONOCYCLIC RINGS

13.2.3.1 Five-membered Rings

13.2.3.1.1 Azoles

This section will only deal with heterocyclic ligands having nitrogen as the donor atom. This means that isoxazoles, phosphazoles and thiazoles are excluded, and that only pyrazoles, imidazoles, triazoles and tetrazoles will be dealt with. These compounds are powerful and very interesting ligands which have been studied only relatively recently. The structures and ring numbering of the ligands are given in Scheme 1. The synthesis and the ligand properties such as pK_a(base), pK_a(acid), dipole moments, NMR and UV spectra have been discussed by Schofield et al.[4] Table 1 summarizes some of these properties.

Scheme 1

Reviews of the coordination chemistry of this type of ligand have been published for pyrazoles[24] and imidazoles,[25] and to a lesser extent for the triazoles[26] and tetrazoles.[27] Many azole ligands are commercially available or are easily synthesized by a variety of condensation and ring-closure reactions.[4-6]

Contrary to the case with pyridines, octahedral species [M(azole)$_6$]$^{2+}$ have been found, both in the solid state and in solution, for all the azoles lacking a substituent next to the donor atom. This appears to be a quite general observation for divalent metal ions with mainly electrostatic interactions. With monovalent ions (e.g. Cu$^+$) or metal ions with a strong preference for a non-octahedral geometry (e.g. Pt^{2+}), species [M(azole)$_p$]$^{n+}$ ($p = 2, 3, 4$) are usually found. With large metal ions such as Cd^{2+}, octahedral coordination may be found even with ligands bearing a bulky substituent.[19]

The azoles do not coordinate only as monodentate ligands; bidentate (or even tridentate) coordination with bridging between two (or more) metal ions has been observed. In the case of the pyrazoles and imidazoles such bridging coordination is only possible after deprotonation of the N—H group. In the case of triazoles and tetrazoles, bridging as a neutral ligand may also occur. Some examples of coordination compounds with bridging azoles are shown in Figures 5–7.

Figure 5 Schematic structure of dinuclear unit Cu(im)Cu with a bridging imidazolato ligand, as found in many copper compounds

When metal ions with unimpaired electrons are used, compounds with interesting magnetic structures may be observed,[28] e.g. dimers,[29] trimers,[30] tetramers,[31] pentamers,[32] linear chains[33] and also two-dimensional layered compounds.[34] An imidazolato bridge is known to be present in the enzyme bovine superoxide dismutase.[35]

Table 1 Values of pK_a for a Selection of Azole Ligands,
Based on Recent Review Papers[5]

Ligands	pK_a(base)	pK_a(acid)
Pyrazoles		
1-H	2.5	14.2
1-Me	2.0	a
3(5)-Me	3.6	b
3,5-diMe	4.4	15.3
4-NO$_2$	−2.0	9.6
Imidazoles		
1-H	7.0	14.4
1-Me	7.3	a
2-Me	7.9	b
4(5)-Me	7.5	b
2-Ph	6.4	13.3
4-NO$_2$	0.0	9.3
2-Br	3.9	11.0
Benzimidazole	5.6	b
1,2,3-Triazoles		
1-H	1.1	9.3
1-Me	1.2	a
Benzotriazole	1.6	8.6
1,2-4-Triazoles		
1-H	2.5	10.1
4-Me	3.4	b
3,5-diMe	3.7	b
Tetrazoles		
1-H	b	4.9
1-Me	4.3	a
2-Me	3.6	5.6

[a] No acid hydrogen available. [b] Not known with accuracy.

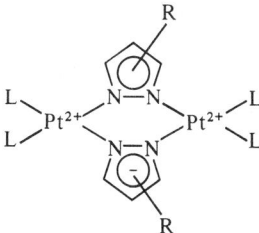

Figure 6 Example of a structure with two bridging pyrazolato ligands, resulting in a dinuclear unit

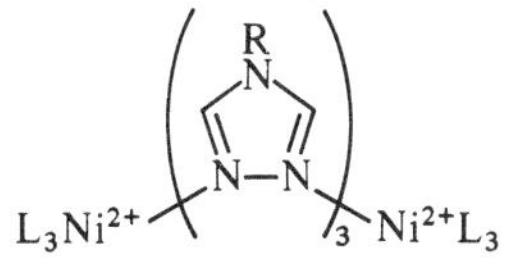

L_3Ni^{2+} ⟨ ⟩$_3$ $Ni^{2+}L_3$

Figure 7 Example of a structural unit with three bridging 4-substituted triazole ligands, as found in many dinuclear and trinuclear compounds[26]

Azoles with bulky substituents near the donor site, such as 1,2-dimethylimidazole and 3,5-dimethylpyrazole, result in lower coordination numbers.[36,37] When a ligand such as 3(5)-methylpyrazole is used, in which the methyl group is adjacent to the donor site in one tautomer and distant in the other tautomer, coordination takes place exclusively through the non-sterically-hindered form of the ligand, *i.e.* the 5-methylpyrazole. This has been proved by X-ray studies for this ligand[38] and for indazole (**4**).[39]

With ligands such as pentamethylenetetrazole[27] or 2-methyltetrazole[40] the tetrazole has three

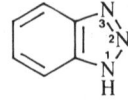   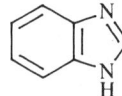

(4) indazole, inH   (5) benzotriazole, btaH   (6) benzimidazole, bimH

nitrogen-donor atoms available for coordination, but only the nitrogen furthest from the substituent (*i.e.* N-3 in pentamethylenetetrazole, and N-4 in 2-methyltetrazole; see also Scheme 1) coordinates to the metal ion when no other coordinating ligands are present. This observation seems quite general for heterocyclic ligands. Thus the coordinating atom always appears to be the nitrogen atom with the smallest substituents on nearby atoms. However, when steric limitations are less important, coordination by a more sterically hindered nitrogen atom is possible, as is observed with, for example, benzotriazole (5), where coordination *via* N-3 has been observed in compounds such as $[ZnCl_2(btaH)_2]$ and $[Ir(CO)(PPh_3)_2(bta)]$.[41,42] In these compounds the coordination number of the metal ions is four and the steric effect of the benzo group is rather small.

Early studies on benzimidazole (6) as a ligand have shown that it behaves as a sterically hindered imidazole, not allowing hexakis solvates with divalent metal ions.[25]

The steric effects of the substituents, such as in 3,5-dialkylpyrazoles, may induce the coordination of small ligands or counterions that usually do not coordinate to transition metal ions. This behaviour is believed to be responsible for the unusual decomposition of transition metal tetrafluoroborates in the presence of such ligands.[43] Otherwise, the coordination number changes, resulting in, for example, square-planar Ni^{II} (low-spin) complexes, even with rather weak ligands such as 1,2-dimethylimidazole.[44]

The number of coordination compounds containing an azole-type ligand for which a crystal structure determination is available is very large. This allows one to make a comparison of the various coordination numbers and valence (or spin) states of the different metal ions. Such a comparison of the M—N distances is shown in Table 2.

Although the majority of azole ligands have been studied as neutral ligands, the deprotonated forms of, for example, pyrazoles and imidazoles have frequently been studied as ligands. The poor solubility of such salt-type compounds has prevented extensive X-ray structural studies. In strongly alkaline media, even salts such as $Na_2[Cu(im)_4]$ may be isolated in which the imidazolato ligand bridges between sodium and copper, but with the strongest bond to copper.[45]

Deprotonation of azoles may also occur at a carbon atom, *e.g.* when the nitrogen is substituted. This protonation may be induced by an already existing M—N bond, as in the case when metallation reactions such as those depicted in Scheme 2 occur.[46] These carbon-bonded systems will not be discussed further.

Another unusual form of deprotonated azoles has been reported for pyrazolate coordinated to U^{IV} in $[U(C_5Me_5)_2(pz)Cl]$. Both N-1 and N-2 are coordinated to the same uranium at a distance of 2.35 Å.[47] This side-on coordination is unique for azoles and in fact resembles the side-on coordination of dioxygen.

Table 2 Observed Ranges of Metal–Nitrogen Distances (in pm) in Azole Coordination Compounds, Without Steric Effects of the Substituents

	Coordination numbers				
Metal ions	2	4	5	6	Other
Mn^{II}	—	—	—	222–226	—
Fe^{II} (HS)[a]	—	200–206[b]	—	216–224	—
Fe^{II} (LS)[a]	—	—	—	200–206	—
Co^{II} (HS)[a]	—	196–204[b]	—	208–219	—
Ni^{II}	—	190–199[c]	—	204–214	—
Cu^{II}	—	195–202[c]	198–208	205–230[d]	—
Zn^{II}	—	190–205[b]	—	215–224	—
Cd^{II}	—	—	—	225–242	—
Pt^{II}	—	199–205[c]	—	—	—
Fe^{III} (HS)[a]	—	—	205–215	205–220	—
Fe^{III} (LS)[a]	—	—	—	198–205	—
Cu^{I}	186–190	198–204[b]	—	—	190–200[e]
Ag^{I}	210–218	226–235[b]	—	—	—
Rh^{I}	—	205–212[c]	—	—	—

[a] HS = high spin; LS = low spin. [b] In case of tetrahedral geometry. [c] In case of square-planar geometry. [d] In fact two Cu—N bonds are elongated. [e] T-shape three-coordination.

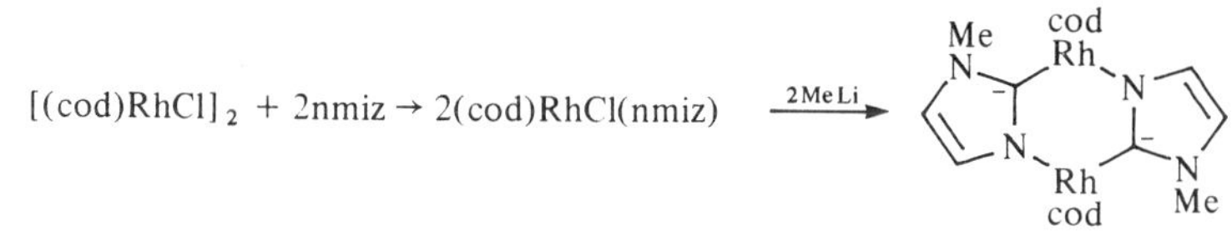

$$[(cod)RhCl]_2 + 2nmiz \rightarrow 2(cod)RhCl(nmiz) \xrightarrow{2\,MeLi}$$

Scheme 2

Finally it should be mentioned that the uncharged azoles could also coordinate principally *via* the N-1 atoms (bearing a hydrogen or a substituent). This is highly unlikely for steric and electronic reasons, and although it has been proposed by a few investigators,[48] this mode has never been proved.

13.2.3.1.2 *Other five-membered rings*

Although the number of non-azole five-membered rings with a nitrogen atom is quite large, the coordination chemistry is poorly developed for the simple rings. Rings such as pyrrole, C_4H_5N, are only important in complex formation after deprotonation and when present in a chelate-forming system such as the tetrapyrroles (see Chapter 21).

Hydrogenated products of the pyrazoles and the imidazoles, *i.e.* the pyrazolines and imidazolines, have some importance as ligands, but have seldom been reported. Only some chelate ligands containing imidazoline have been studied in some detail and will be discussed in Section 13.2.5.3.

13.2.3.2 Six-membered Rings

13.2.3.2.1 *Pyridines*

Undoubtedly, pyridine, C_5H_5N (2), is the best-known heterocyclic nitrogen ligand and its coordination chemistry has been studied in great detail, as have its simple derivatives bearing a non-coordinating substituent. For the physical properties, the reader is referred to the heterocyclic literature.[1–3,5,9] The basic properties of pyridine have been mentioned above. Alkyl-substituted derivatives are slightly more basic [pK_a(base) values of about 5–7].

A commonly found stoichiometry with transition metal salts is $M(py)_4(anion)_2$, in which the anions are often in a *trans* orientation.[15] However, other stoichiometries have also been observed, such as $M(py)_2(anion)_2$ (often with halide anions), $M(py)_3(NO_3)_2$, $MX_3(py)_3$ and $M(py)_2(anion)$. The compounds of formula $M(py)_2X_2$ (X = halide) have been found in at least five different structures, examples of which are presented in Scheme 3. The dimers with a five-coordinate structure are usually found with substituted pyridines, such as 2-picoline.[53]

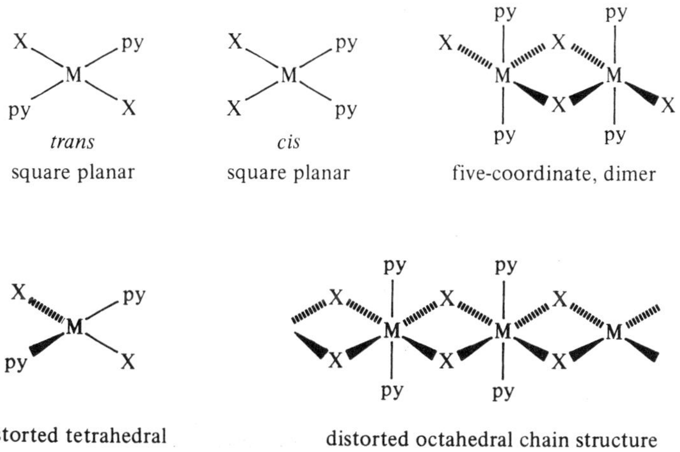

trans
square planar

cis
square planar

five-coordinate, dimer

distorted tetrahedral

distorted octahedral chain structure

Scheme 3

As mentioned above, the hydrogen atoms at the 2- and 6-positions of the pyridine hamper octahedral coordination of pyridine to most metal ions. Only when a large counterion is present, such as $[Fe_4(CO_{13})]^{2-}$ or perhaps PF_6^- or AsF_6^-, do octahedral $[M(py)_6]^{2+}$ species have some stability.[15,16] With a small cation such as Ni^{2+} even a weak ligand such as 4-methylpyridine (4-Mepy) may induce a low-spin square-planar geometry as found in $[Ni(4-Mepy)_4](PF_6)_2$.[54]

Metal–nitrogen distances, as observed in a very large number of crystal structure determinations, are almost the same as those observed for the azole ligands (see Table 2) within experimental error. Comparing pyridine with substituted ligands, it is observed that substituents at position 2 (and 6) next to the donor atom have a dramatic effect upon the stoichiometry and properties of the compounds formed. It appears that a stoichiometry $M(Rpy)_4(anion)_2$ with a *trans* orientation of the anions is not possible, because the methyl group effectively blocks the axial coordination sites. This results in square-planar low-spin compounds for Ni^{II} and tetrahedral geometry for some other metals such as Co^{2+} and Zn^{2+}.[55]

The chain-type compounds $M(py)_2Cl_2$ (see Scheme 3) have been the subject of intensive research by physical chemists and physicists, since the magnetic exchange through the halogen bridges provides interesting pathways with the possibility of subtle variations of bond angles, bond distances and lattice packing.[56,57]

Although the large majority of pyridine compounds use the free electron pair to bind to metal ions in a monodentate fashion, bridging coordination using the nitrogen atom is also possible, although stabilization in a so-called crevice seems to be required, as shown in the compound $[MoO(C_8H_9PS_2)_2OS]_2py$.[58]

Usually, coordinated pyridine ligands are rather stable towards substitution. In the case of *trans*-$[PtCl_2(py)_4]$, however, reversible addition of OH^- to one of the pyridine ligands has been proposed.[59]

13.2.3.2.2 Rings with two nitrogen atoms

This section deals with molecules derived from the diazines $C_4H_4N_2$: pyrazine (1,4-diazine; **7**), pyrimidine (1,3-diazine; **8**) and pyridazine (1,2-diazine; **9**).

(7) pyrazine   (8) pyrimidine   (9) pyridazine

Pyridazine resembles pyrazole and the triazoles with regard to bonding, as it has the ability to bridge between two metal ions. Unsubstituted pyridazine, however, yields few stable compounds, and only when chelating substituents are present at positions 3 and 6 can stable compounds (**10**) be obtained with transition metal ions.[60]

(10)   (11)

Free pyrimidine has been little studied as a ligand, although substituted derivatives (occurring in the nucleobases cytidine and uracil) do bind metals (see Chapter 62).

The most intensively studied diazine is undoubtedly pyrazine (**7**), present in the so-called Creutz–Taube complex[61] $[(H_3N)_5Ru(pyrazine)Ru(NH_3)_5]^{5+}$, a well-known mixed-valence compound. Examples of other bridged compounds (**11**) are the chain compound[62] $Cu(NO_3)_2(pyrazine)$ and the two-dimensional[63] $CoCl_2(pyrazine)$ (see Figure 8). The synthesis of the corresponding Zn compound probably proceeds *via* an intermediate which is immediately converted to the end-product *via* crystallization, resulting in a unique acoustic emission.[64]

The interest in pyrazine and related ligands has increased recently as a result of a new interest in electron-transfer reactions *via* ligand aromatic systems.[65]

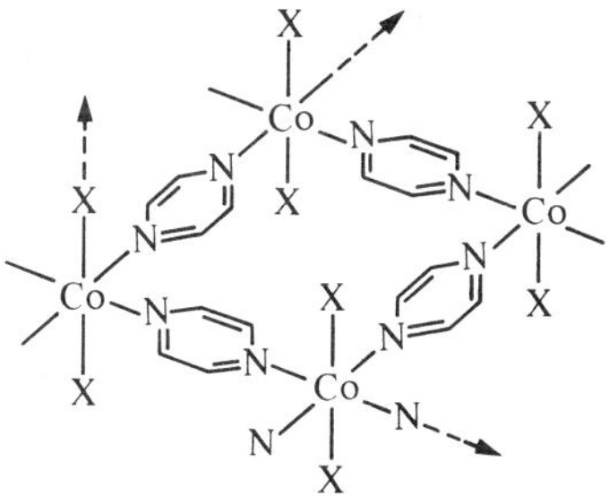

Figure 8 Schematic representation of the structure of the two-dimensional pyrazine-bridged compound with CoX_2[63]

13.2.3.2.3 *Other six-membered rings*

The variety of heterocyclic six-membered nitrogen-donor molecules which may bind to metal ions through one of the ring nitrogen atoms is very large. However, the coordination chemistry of such molecules is not well developed. Some such ring systems are the triazines (**12**), cyclophosphazenes, borazines and azacyclohexane (**13**). In many cases the coordination of such heterocyclic ligands is restricted to substituted derivatives with coordinating substituents and therefore these are not further treated here.

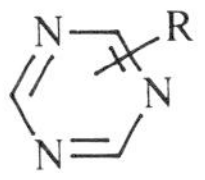

(12)    (13)

13.2.3.3 **Rings of Other Size**

Rings consisting of only three or four atoms are not very stable and have been little used as ligands in coordination chemistry. Examples are azetidine (**14**) and aziridine (**15**). In fact only the coordination chemistry of aziridine has been investigated in some detail. Apart from behaving as a secondary amine (Chapter 13.1), it may be an interesting ligand after deprotonation, as shown in the aziridinyl-gallane trimer shown in Figure 9.[66]

(14)    (15)

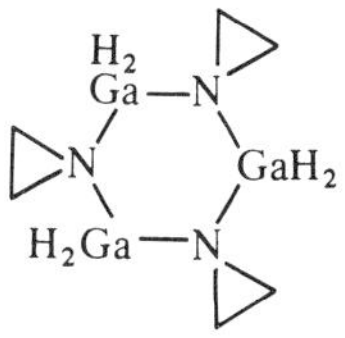

Figure 9 Structure of the six-membered ring form with the anion of aziridine and gallane[6]

Rings consisting of seven atoms or more have also scarcely been studied as ligands. Examples are azepine (**16**), thiadiazepines (**17**) and cyclotetraphosphazenes, (**18**). Substituted thiadiazepines form interesting oligomeric compounds with transition metal ions, with bridging *via* both nitrogens (see Figure 10).[67]

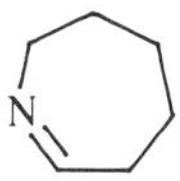

(16) 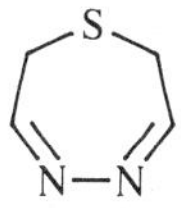   (17)

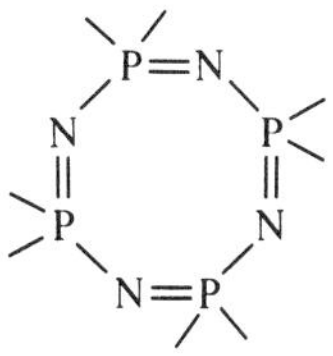

(18)

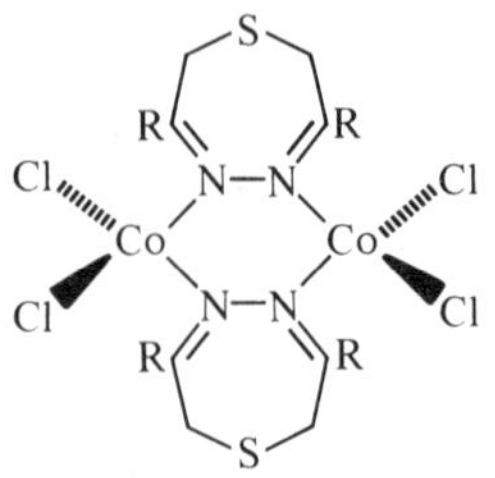

Figure 10   Schematic structure of a dinuclear tetrahedral Co[II] compound with a substituted thiadiazepine[67]

The ligand octamethylcyclotetraphosphazene appears to be able to chelate to a *cis*-[PtCl$_2$] unit (Figure 11).[68] A large variety of other rings occurs, many of these also containing sulfur, *e.g.* N$_4$S$_4$, oxygen or other donor atoms. These are not treated in this chapter, although many of them use only nitrogen as the donor atom (see Figure 10).

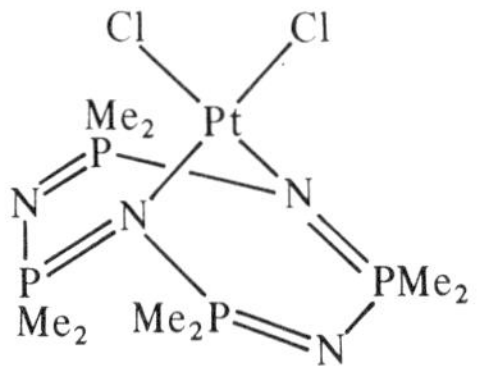

Figure 11   Structure of the *cis*-[PtCl$_2$] adduct of octamethylcyclotetraphosphazene[68]

Finally it should be noted that rings with more than eight atoms, *e.g.* triazacyclononanes,[69] are treated elsewhere (Chapter 21).

13.2.4   LIGANDS CONSISTING OF A MONOCYCLIC RING WITH AN ADDITIONAL LIGAND GROUP AS A SUBSTITUENT

13.2.4.1   Five-membered Rings with Coordinating Substituents

As discussed in Section 13.2.3, substituents in five-membered rings can have a dramatic influence on coordination. When (bulky) substituents near the donor atom of the ligand are present, the coordination number and the stoichiometry of the compound can change dramatically. This in fact implies that when a coordinating substituent is present near the donor atom, coordination may be enhanced by the possibility of chelate formation. Examples of such ligands are presented in formulae (19), (20) and (21).

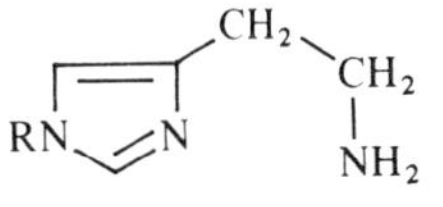

(19) 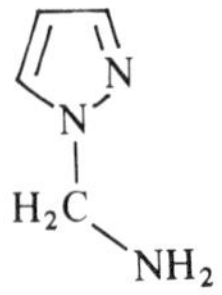   (20) 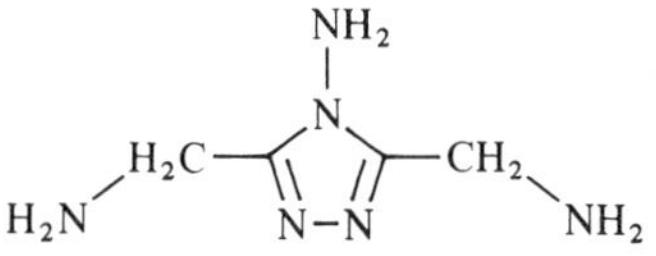   (21) 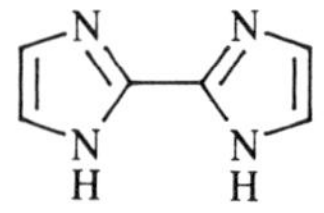   (22) biimH$_2$

Chelation sometimes does not take place automatically, as illustrated by the coordination of histamine, 4-(2-aminoethyl)imidazole (19; R = H). It appears to tautomerize and coordinates with the non-chelating nitrogen, whereas the NH$_2$ group is protonated, so that a monodentate ligand (in fact a 4-substituted imidazole) coordinates to the metal, in this case Ni[II] with NCS$^-$ anions in the compound [Ni(NCS)$_4$(hamH)$_2$] (hamH = histaminium cation).[70]

It should be noted here that when the substituents are conjugated with the heterocyclic ligand, such as in 2,2'-biimidazolyl (biimH$_2$; 22), the ligands are dealt with in Section 13.2.6.

13.2.4.2   Six-membered Rings with Nitrogen-donor Substituents

After the discovery was made that pyridine is a potentially powerful ligand, but with steric problems, several authors synthesized pyridine and pyrazine derivatives with coordinating substituents. These coordinating substituents are most interesting when located at 2 (and 6) positions of the pyridine rings, because in these cases chelates can be formed. Examples of such compounds are shown in formulae (23)–(25). As discussed above for five-membered rings, the steric effects are much smaller in the case of coordinating substituents, and octahedrally coordinated metal ions are often found in 2-substituted pyridine complexes with the general formula $[Fe(LL)_3](anion)_2$. The ligand-field strength, as is usual with such chelating ligands, is much stronger than expected for the individual donor groups. This results, for example, in low-spin Fe^{II} compounds[71] in the case of 2-methylaminopyridine (mapy; 23, $n = 1$), *i.e.* $[Fe(mapy)_3]Cl_2$ (see Figure 12).

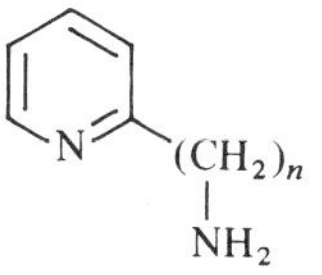

(23) 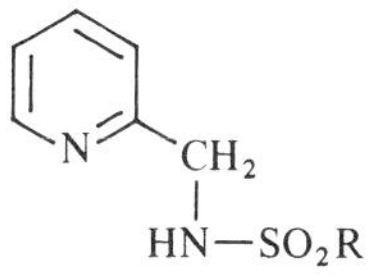   (24) 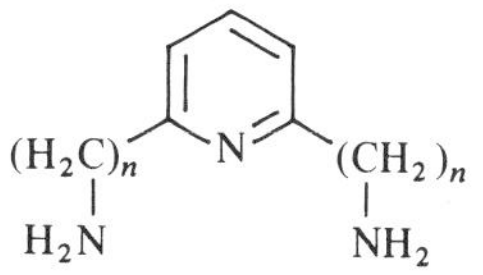   (25)

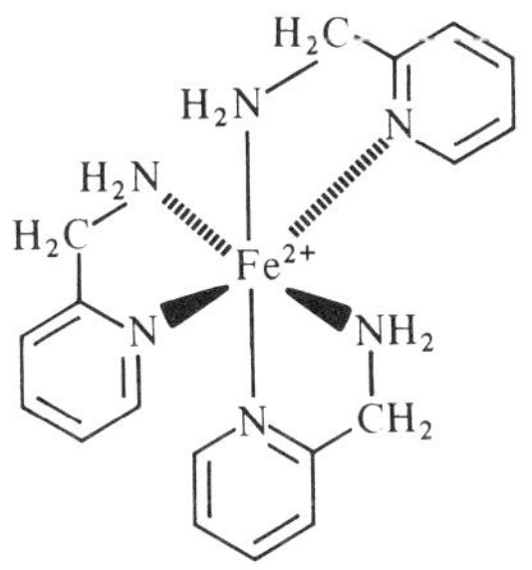

Figure 12   Structure of the cation $[Fe(mapy)_3]^{2+}$, which shows a high- to low-spin transition upon cooling[72]

Many related ligands of this type have been examined, *e.g.* (26) and (27), to study the high-spin/low-spin phenomena.[72] With this type of ligand a large variety of copper(II) dimers has been obtained of general formula $Cu_2(OH)_2(LL)_2(anion)_2$. A schematic structure is shown in Figure 13. These investigations have contributed significantly to a better understanding of the magnetic exchange coupling between the pairs of copper(II) ions.[73]

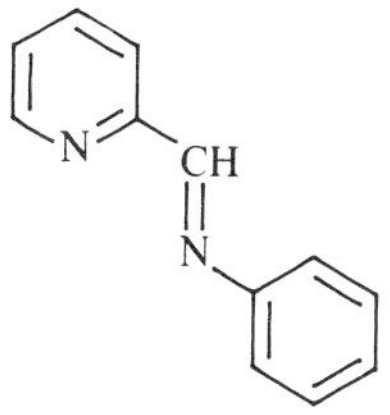

(26) 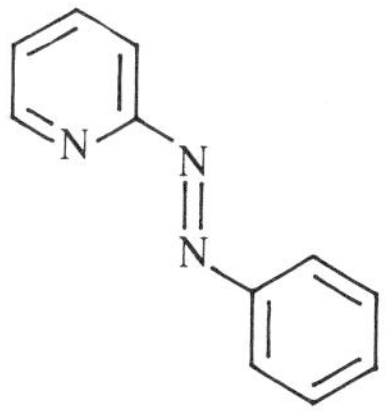   (27)

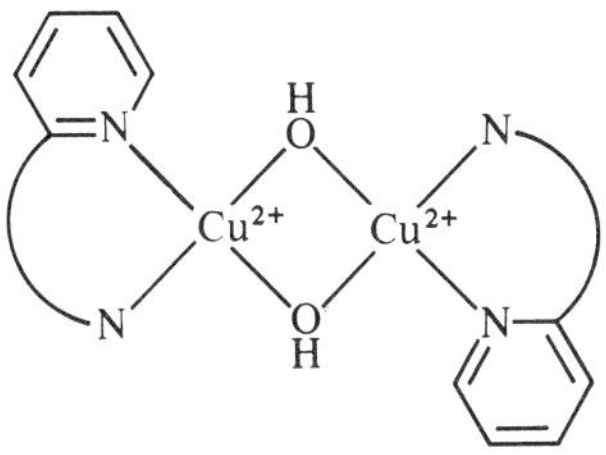

Figure 13   Schematic structure of the cations $[Cu_2(OH)_2(LL)_2]^{2+}$ in coplanar dimeric Cu^{II} compounds with substituted pyridine ligands[73]

When the substituents in the pyridine ring are in a sterically unfavourable position, such as in (28) and (29), either polymeric compounds are formed, with both ends of the molecule coordinating to a different metal ion, or the molecule may use only one side. For example, in 4-cyanopyridine (29) the molecule may bind to metal ions *via* the nitrile group (*i.e.* in the case of Cu^{II}), or *via* the pyridine group (*i.e.* in the case of, for example, Fe^{III} and mercury).[74,75] Although in the case of 2-amino-pyridine (30) chelation might occur, the number of compounds with both the pyridine and the amine group coordinating to the same metal is small.

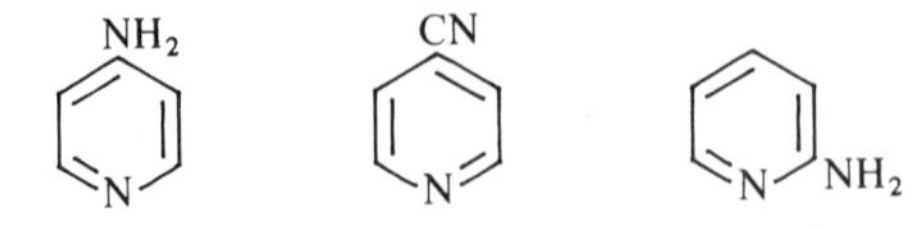

(28)   (29)   (30)

When chelate-forming substituents are added to the diazines, dinucleating or trinucleating reagents may result (see formula **10**). These will not be discussed in detail because of limited space.

13.2.4.3 Other Chelating Ligands with a Heterocyclic Nitrogen-donor Atom

As already concluded above, heterocyclic ligands with a ring size smaller than five or larger than six are rare, and few coordination compounds are known. In principle, however, ligands such as (**31**) and (**32**) could be expected to coordinate to metal ions in a chelating manner.

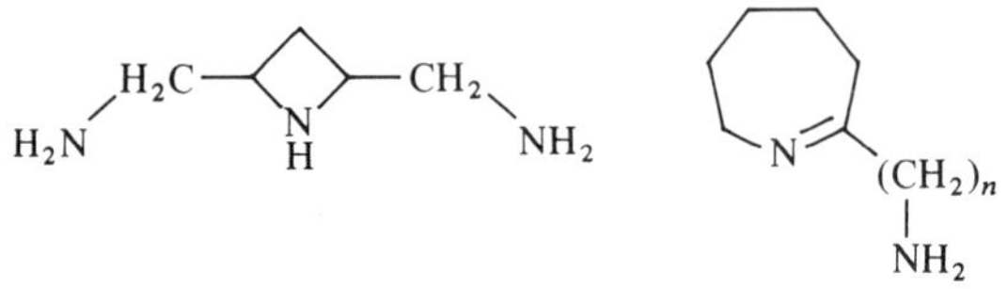

(31)   (32)

13.2.5 FLEXIBLE CHELATING LIGANDS CONTAINING TWO OR MORE HETEROCYCLIC LIGANDS

13.2.5.1 Introduction

In this section we will deal with chelating ligands consisting of two or more heterocyclic nitrogen-donor groups held together by a more or less flexible connecting group. Ths group may, or may not, contain additional donor atoms, such as nitrogen, oxygen, sulfur or phosphorus.

Strictly speaking, the chelating ligands with the amine linkages could also have been dealt with in Chapter 13.1, whereas the ligands with P, O and S groups bridging between the heterocycles should have been dealt with in Chapter 20. Because the ligand properties are mainly determined by the heterocyclic parts, and since such bridging groups sometimes do not, or at best very weakly, bind to the metals, these chelating systems are briefly dealt with in this chapter.

Ligands having two (or more) heterocyclic ligand components connected only *via* a C—C (or N—N) bond are not treated in this section, but in Section 13.2.6. The rings in such systems are in fact conjugated and are therefore best compared with the fused ring systems having more than one donor atom.

The chelating ligands described in this section are often not commercially available, and are usually prepared by condensation reactions. Some examples are given in Schemes 4 and 5. It should be noted that many other condensation reactions, *e.g.* Schiff-base type reactions (Chapter 20), have been reported for the synthesis of novel chelating ligands.

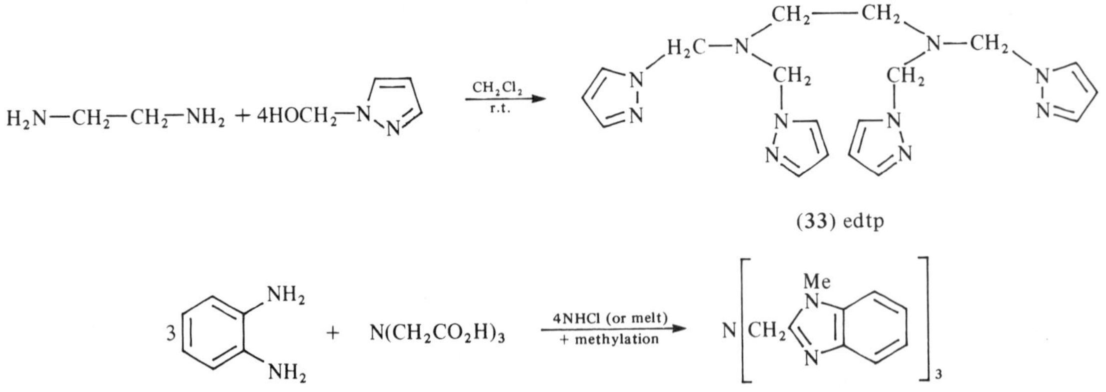

(33) edtp

(34) mntb

Scheme 4

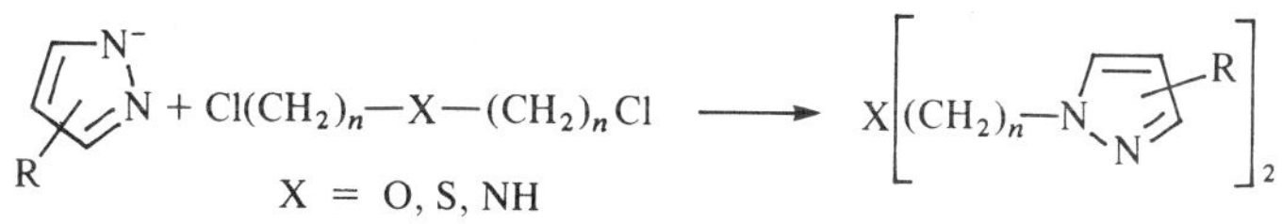

(35)

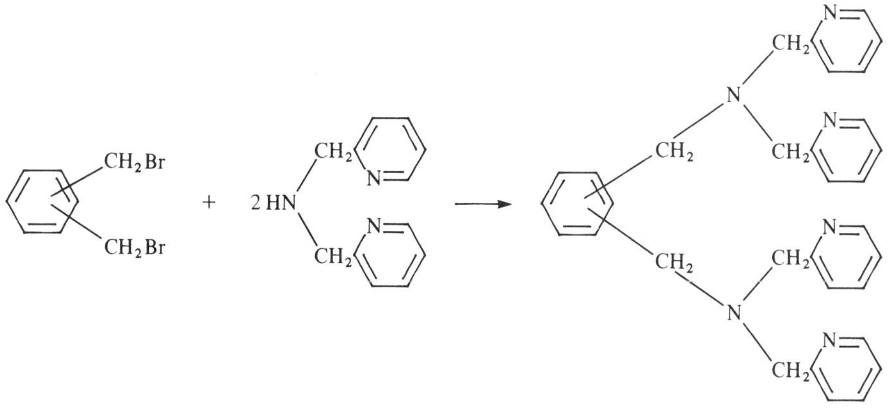

(36)

Scheme 5

13.2.5.2 Chelating Ligands Containing Only Five-membered Rings

During the last decade there has been a large increase in interest in chelating ligands containing two or more five-membered rings. Part of the interest in such ligands originates from the aim of mimicking certain metal sites in metalloproteins. In particular, copper proteins and zinc proteins are known to contain two or more imidazole groups (from histidine side-chains) as ligands for the metal ions.[76,77]

The synthesis of imidazole-containing chelating ligands has been shown to be rather elaborate[78,79] and therefore several investigators have decided to study the benzimidazole-type ligands (see Scheme 4). The use of this ligand type has an additional advantage in that it imposes some steric hindrance that might be an approximation to the protein environment.[80] In Scheme 6 a variety of such benzimidazole-type chelating ligands is shown. As well as the nitrogen-donor atoms of the imidazole part of these molecules, other donor atoms, such as N, O and S, may coordinate to transition metal ions. Because this type of ligand has been designed to mimic the active sites of (mainly) copper and zinc proteins, the coordination chemistry of other metal ions with these ligand systems has been studied only to a limited extent.

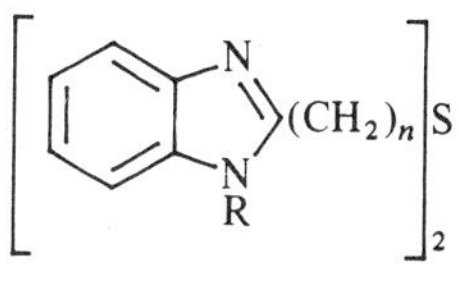

(37)

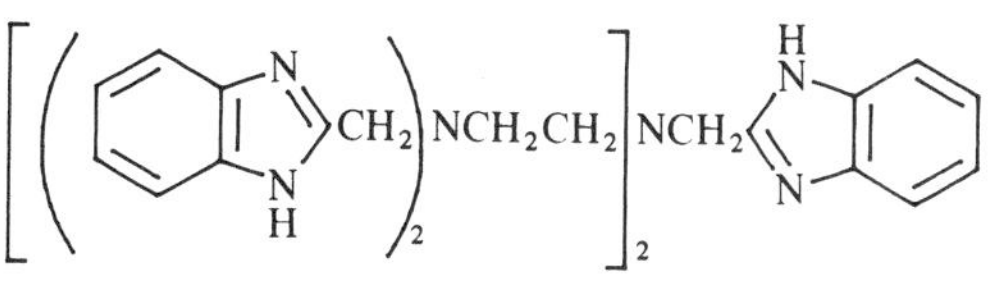

(38)

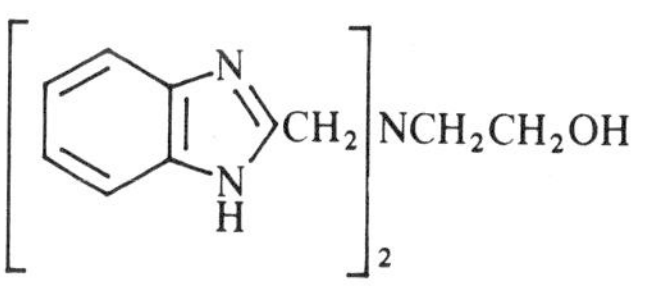

(39)

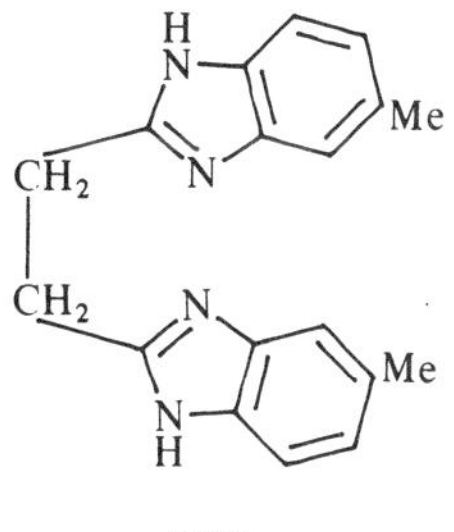

(40)

Scheme 6

More recently, a few groups have started to look at pyrazole-type ligands in chelating systems (Schemes 4 and 5),[81-83] as these are easily formed[84] and also resemble the imidazole ligands in their coordinating behaviour (*vide infra*).

Some of the pyrazole-containing chelating ligands are shown in Scheme 7. It turns out that the presence of methyl groups at positions 3 and 5 in the pyrazole rings causes significant steric effects,[81,84] just as is found for the above-mentioned benzimidazoles.

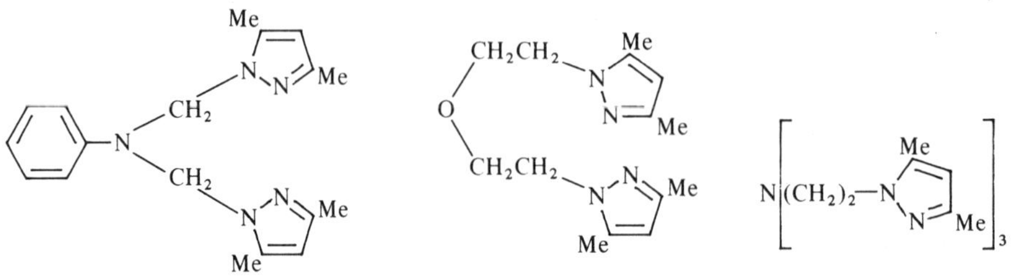

(41)   (42)   (43)

Scheme 7

It is impossible to depict the coordination behaviour of these and many other ligands, and only a few examples will be given as an illustration. The hexadentate ligand edtb can bind to one or two metal ions, as shown in Figures 14 and 15,[85,86] yielding six-coordinate and linear-coordinated copper ions, respectively. A ligand such as edtp (33)[87] also yields distorted six-coordination, with sometimes a seventh ligand added. The ligand mntb (34) yields five-coordinated species, imposed by the ligand's tripodal character.[88] Ligands of structure (35) yield T-shaped Cu^I compounds[89] (see Figure 16), whereas ligands of structure (36) yield dinuclear compounds.[90] For the structures of compounds with the ligands depicted in formulae (37)–(42), the reader is referred to other compilations.[77,91-93] Distorted tetrahedral geometry has been found for ligand (43) in the case of Co^{II}, resulting in very unusual spectral properties.[94]

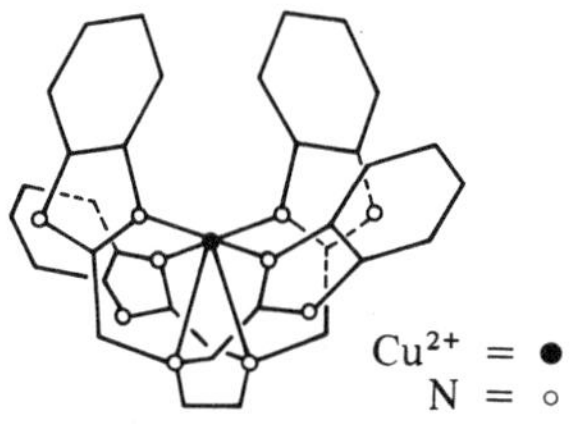

Cu^{2+} = ●
N = ○

Figure 14 Structure of the cation $[Cu(edtb)]^{2+}$, with a bicapped square-pyramidal structure[85]

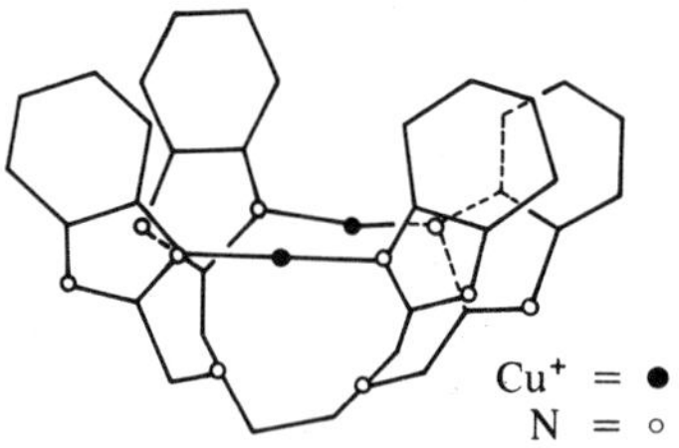

Cu^+ = ●
N = ○

Figure 15 Schematic structure of the dimeric cation $[Cu_2(edtb)]^{2+}$ with a linear coordination for the Cu^I ions. The Cu—Cu distance is 305 pm[86]

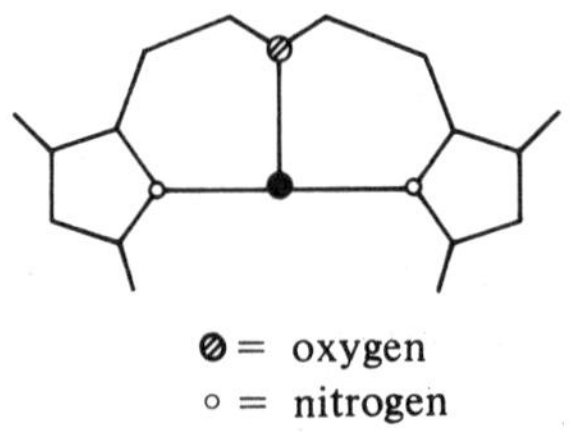

⊘ = oxygen
○ = nitrogen

Figure 16 Schematic structure of T-shaped coordination of Cu^I by a tridentate pyrazole-type ligand[89]

Increasing the size of such ligands may result in dinuclear species (such as in **36**, **38**, **44** and **45**).[95,96] Aims in such studies are often the mimicking of dinuclear copper sites in enzymes or the synthesis of new catalysts.[77]

More recently, imidazole- and pyrazole-type ligands with a phosphorus atom as the bridge have been reported,[97,98] as depicted in formulae (**46**)–(**48**). Also rings with imidazolines (**49**),[99] or bridged by a —C—OH group (**50**),[100] have been reported. Boron-bridged polypyrazolylborates and related systems are dealt with in Chapter 13.6.

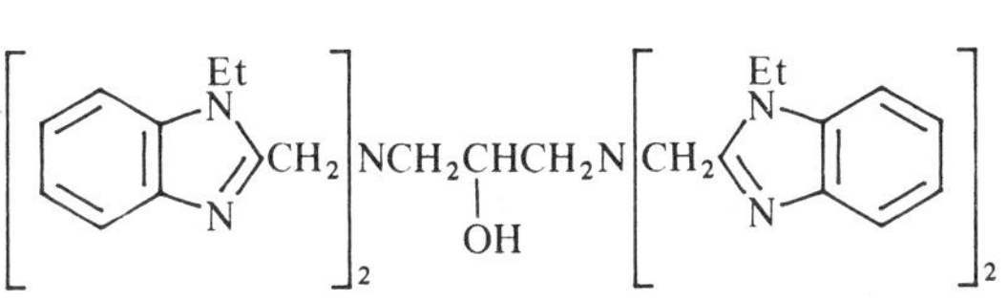

(**44**)

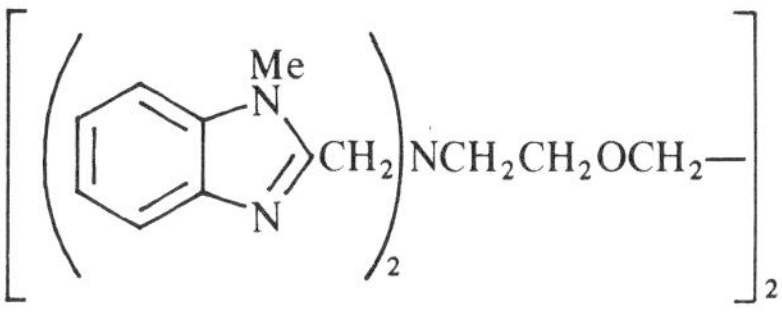

(**45**) nmegtb

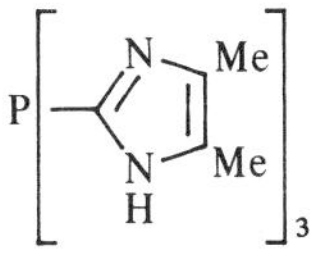

(**46**) 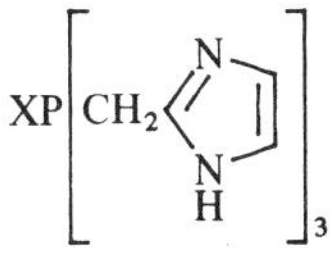   (**47**) 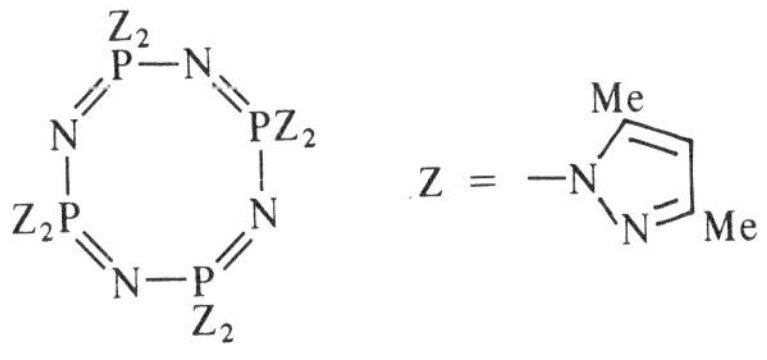   (**48**) $Z = $

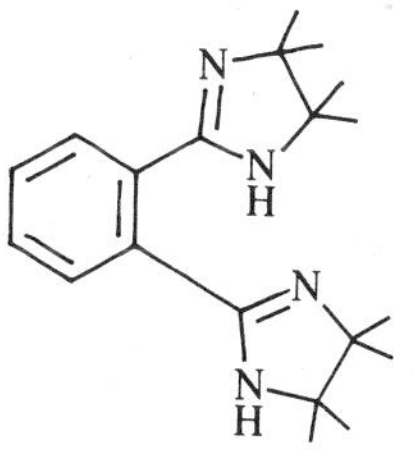

(**49**) 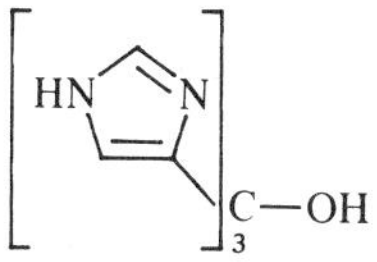   (**50**)

13.2.5.3 Chelating Ligands Containing Only Six-membered Rings

This is undoubtedly the largest class of ligands and is based mainly on pyridine with linkages *via* the C-2 atom to other groups. The linking groups may or may not contain atoms that are able to coordinate, such as in (**36**) or (**51**) (in this case the amine nitrogen cannot reach the metal).[101] A selection of ligands is depicted in formulae (**52**),[102] (**53**) (a non-chelating system),[103] (**54**),[104] (**55**) (a negative tridentate)[105] and the polydentates (**56**) and (**57**).[106,107] A schematic structure of one of these compounds is depicted in Figure 17.[77]

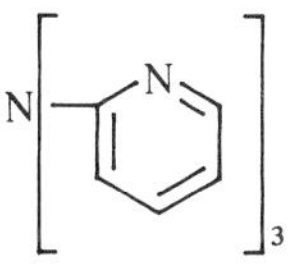

(**51**) 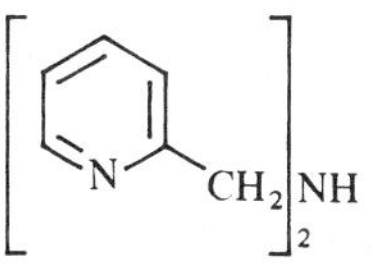   (**52**) 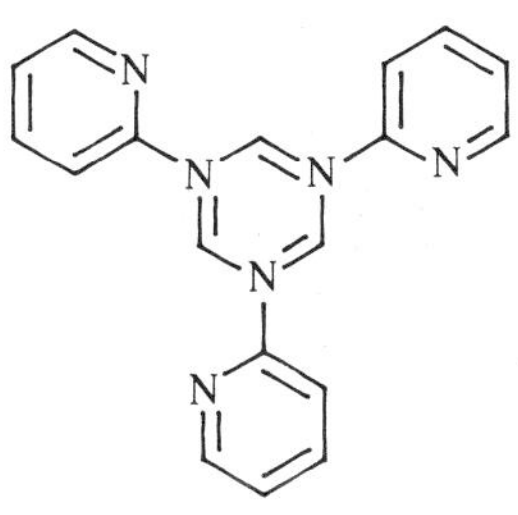   (**53**) 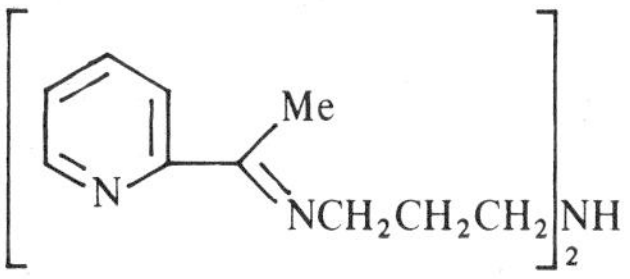   (**54**)

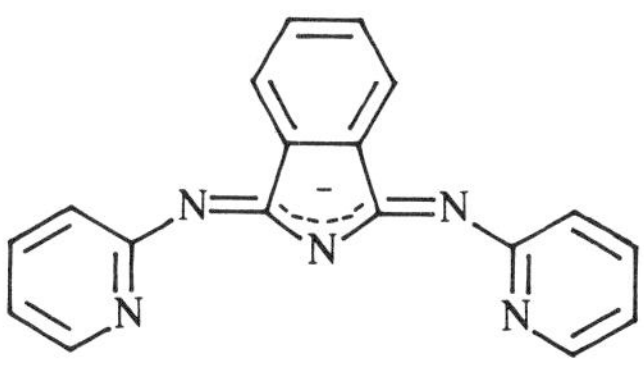

(**55**) 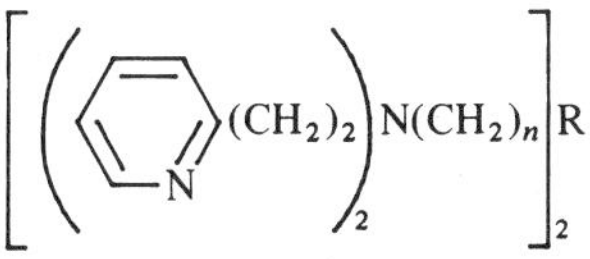   (**56**) 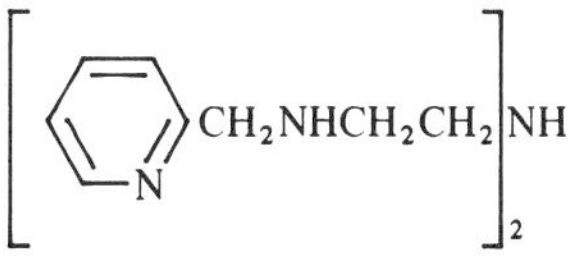   (**57**)

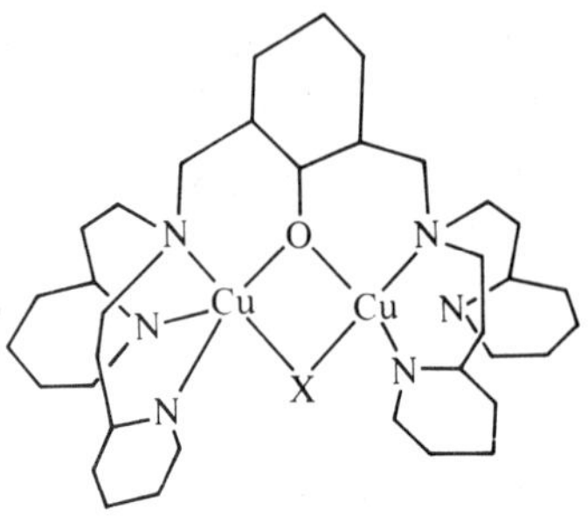

Figure 17 Structure of the dimeric Cu[II] compound formed by oxidation of a Cu[I] species in a pyridine ligand. The phenolic oxygen is introduced upon oxidation[77]

An interesting dinucleating system, formed from a pyrazine, is depicted in (**58**).[108] With saturated combinations of six-membered rings, *e.g.* sparteine,[109] complex formation is much more difficult for steric reasons, as is seen from formula (**59**).

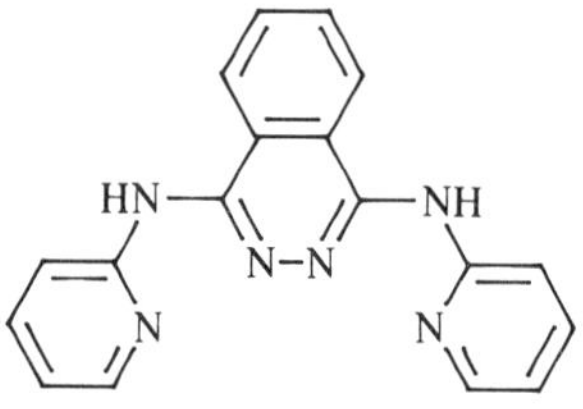

(**58**)

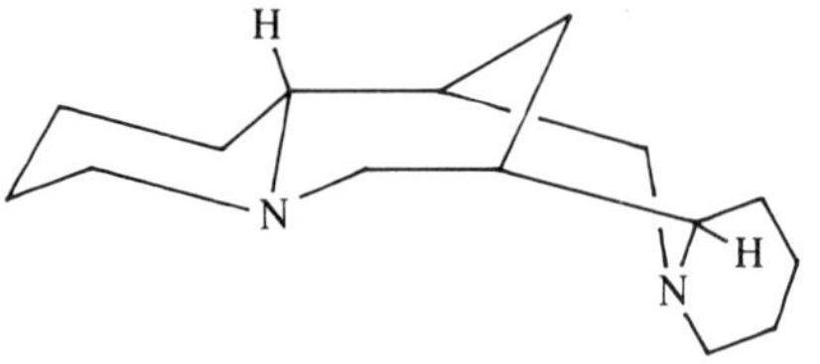

(**59**) sparteine

Crystal structures proving the chelating nature of the ligands shown in formulae (**51**)–(**59**) are numerous and cannot be mentioned here.

13.2.5.4 Chelating Ligands Containing Rings of Different Size

Using somewhat more elaborate synthetic methods, chelating ligands containing, for example, both a pyridine and an imidazole-type ligand can be obtained. Some examples are shown in formulae (**60**),[110] (**61**) (a ligand the Cu[I] complex of which seems to bind dioxygen reversibly),[111] and two smaller systems (**62**) and (**63**).[112,113] Although many metal ions have been reacted with these and related ligands, copper has been studied the most since its compounds with these ligands are used to mimic the active sites in copper proteins.[77]

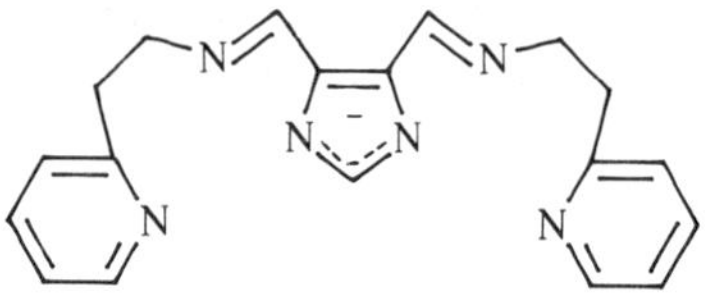

(**60**)

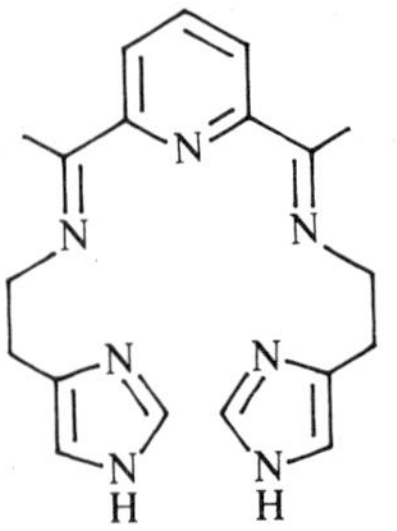

(**61**)

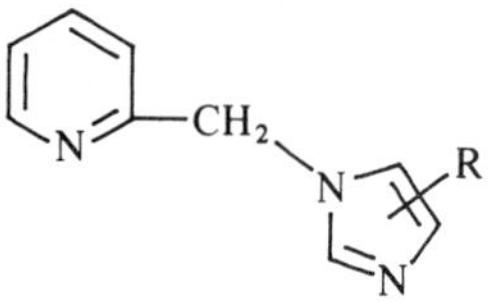

(**62**)

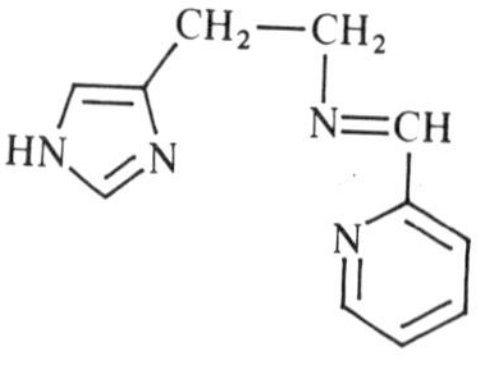

(**63**)

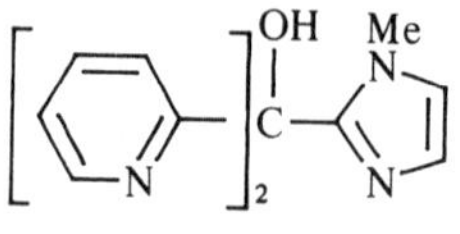

mbpm (**64**)

It should be mentioned that even when a potentially tridentate or tetradentate ligand, such as *N*-methylimidazol-2-yldi(pyrid-2-yl)methanol (**64**) is present, coordination may proceed through only one or two of the ligand donor atoms.[114]

Many other chelating ligands consisting of rings of different size could be designed. Most of the mixed-ring systems produced so far contain azoles and pyridines. Some recent examples are presented in ref. 114a. Metal compounds of other ring sizes are rare and will not be discussed here.

13.2.6 CONDENSED OR CONJUGATED HETEROCYCLIC CHELATING LIGANDS

13.2.6.1 Introduction

In addition to the more or less flexible chelating ligands described in the previous section, a large class of chelating ligands occurs in which two or more heterocyclic ligands are linked together *via* a single bond and conjugation, *e.g.* the ligands 2,2'-bipyridyl (bipy; **65**) and 2,2',2″-terpyridyl (terpy; **66**), or *via* a fused set of ring systems, *e.g.* 1,10-phenanthroline (phen; **67**) and purine (**68**). Of these ligand systems, both bipy and phen have been studied extensively.[115,116,20]

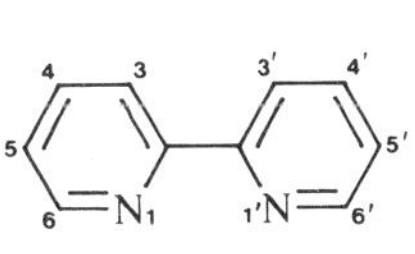

(65) bipy 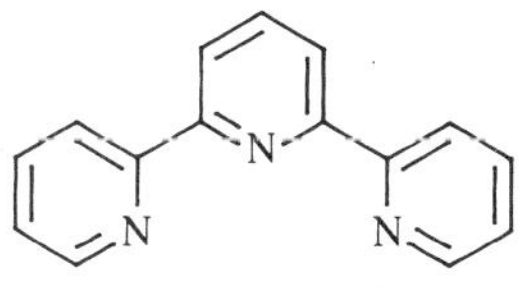   (66) terpy 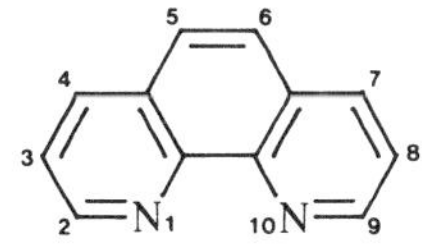   (67) phen 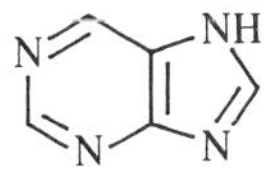   (68) purine

Because of the less-flexible nature of such ligands, the steric effects are usually much more severe, resulting in unusual geometries for the metal ions. The present section will deal with a variety of examples of such ligands and their coordination ability. Ligands of this type are sometimes commercially available. Otherwise syntheses along methods outlined above (see Schemes 4 and 5) may be applied.

13.2.6.2 Ligands Based on Five-membered Rings Only

The various heterocyclic rings of the azole type, discussed above in Section 13.2.3.1, can be combined together to form new multidentate ligands which may or may not act as chelate-forming systems. Well-known examples of such systems are biimH$_2$ (**22**), biim (**69**),[117] 2,2'-bi-2-imidazolinyl (**70**)[118] and 4,4'-bi-1,2,4-triazolyl (**71**).[119] Of these, biim (**69**) has been most frequently studied because it may form interesting dinuclear or polynuclear compounds, in which metal–metal exchange coupling can be studied.[120] Although the synthesis of biimH$_2$ seems simple (from NH$_3$ and glyoxal),[121] the yields of the synthesis of this and related ligands are rather low. The coordinating behaviour of biim has been proved by X-ray single-crystal structure determinations, showing that both mononucleating bidentate coordination and dinucleating tetradentate coordination are possible (see Figures 18 and 19).[122,123]

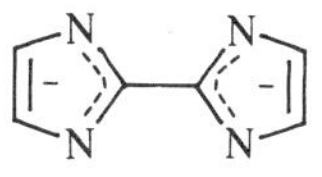

(69) biim 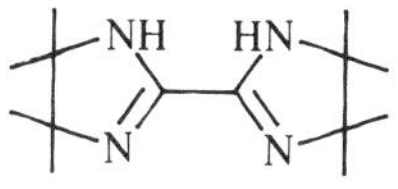   (70) bimd 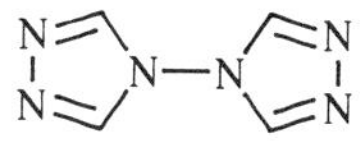   (71) btrz 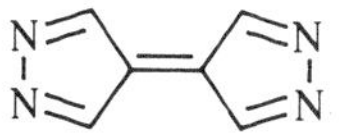   (72) bpz

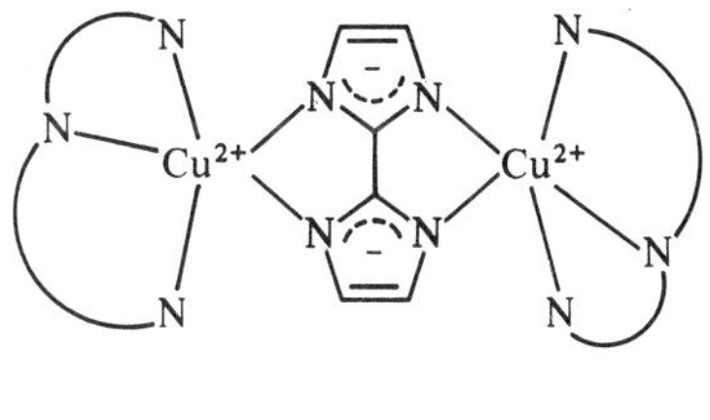

Figure 18 Schematic representation of the structure of a dinuclear Cu^{II} species, with asymmetric bridging of biim. The end ligands come from diethylenetriamine[122]

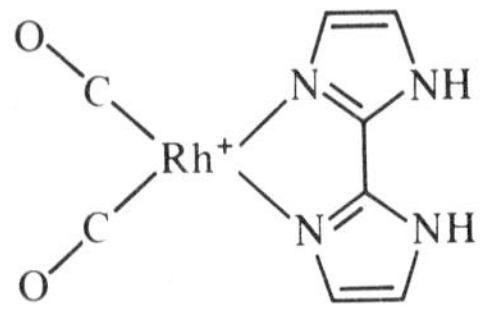

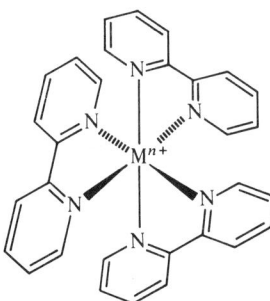

Figure 19 Schematic structure of a square-planar Rh^I cation with a chelating neutral biimH$_2$ ligand[123]

The ligand 4,4'-bi-1,2,4-triazolyl is known to form polymeric species, since the ligand cannot chelate.[119] Similar behaviour is likely for 4,4'-bipyrazolyl (**72**).[124]

The main interest in ligands of this type centres on their magnetochemistry and will not be discussed in further detail here. Another, more recent, interest in these ligands is in the area of electron-transfer reactions and the effect of a metal ion on (de)protonation of heterocyclic ligands.[125]

13.2.6.3 Ligands Based on Six-membered Rings Only

This category of ligand undoubtedly belongs to the most intensively studied class of chelating ligands, with bipy (**65**) and phen (**67**) as the classical examples. Reviews of the coordinating properties of these ligands and derivatives have been available since the last decade.[115,116,126–128,20]

As mentioned above (Section 13.2.2.2) the steric effect of these ligands especially the ligands of the phen type, can be signficant in preventing the formation of square-planar or *trans*-octahedral compounds (Figures 2 and 4). Substituents at positions 2 and 9 may increase the preference for tetrahedrally distorted species; this has also been found in other systems, such as in spartein (**59**), and also in combinations of five- and six-membered rings (Section 13.2.6.4).

In the case of octahedral geometry, the steric effect of the hydrogens at the 6-positions in bipy and the 2,9-positions in phen are less severe (see Figure 20), because the bulky groups are directed to the outside of the molecule. In fact the octahedral species such as [M(bipy)$_3$]$^{n+}$ are very stable and often kinetically inert (especially with d^3 and d^6 low-spin metal ions). With Fe^{2+}, low-spin species are observed,[127] and with the *cis*-octahedral species [Fe(bipy)$_2$(NCS)$_2$], high-spin to low-spin transitions may be observed upon cooling.[72]

Figure 20 Representation of the cation [M(bipy)$_3$]$^{n+}$. In D_3 symmetry each ring hinders coordination of the other only to a small extent

During recent years, interest in the bipy (and phen) compounds has been renewed because of their interesting redox and photoredox properties,[129] with particular attention being given to the ruthenium compounds[130] (see also Chapter 45). In the case of Cu^I compounds of these ligands, many systems have been shown to have strong antimicrobial effects.[131]

Apart from the above-mentioned interest in *cis*-octahedral species [M(bipy)$_2$(anion)$_2$] from the point of view of high-spin to low-spin transitions,[72] this geometry is also of interest for the formation of reactive species, leading to, for example, dinuclear species with catalytic properties, such as the cobalt–peroxo compound shown in Figure 21. This compound is active in oxidative phenol coupling.[132] With Cu^{II} the stoichiometry Cu(bipy)$_2$X$_2$ results in a variety of five-coordinate and distorted six-coordinate structures (see Figure 3).[21]

With the ligand phen, eight-coordinate structures (tetrabidentate) have been proposed for the larger metal ions such as Sr, Ba, Pb and even for Mn^{II}.[133]

Although the polypyridyl ligands are reported to be quite stable after coordination, covalent hydration and pseudo-base formation have been reported, although not proved unequivocally.[134]

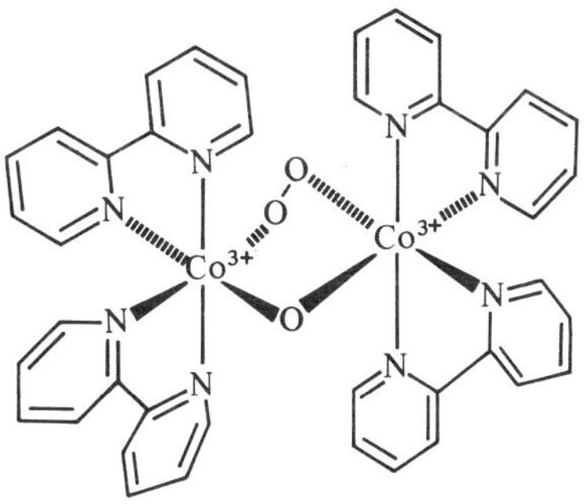

Figure 21 Dinuclear Co^{III} species with bipy ligands and with a peroxo and hydroxo bridge; this compound shows interesting catalytic properties[132]

The ligands of the bipy type can be enlarged in several ways. A classical example is the extension to tripyridyls (*e.g.* terpy; **66**) and even tetrapyridyls. The terpy ligand is an interesting one, because it does not allow the formation of regular octahedral species $[M(terpy)_2]^{n+}$. Instead, approximate D_{2d} symmetry is imposed. In the case of Co^{2+}, interesting high-spin/low-spin transitions have been reported with this ligand, whereas both Cu^{2+} and Co^{2+} result in Jahn–Teller distortions.[135] With tetrapy (**73**), distorted tetragonal geometries with two additional axial ligands have been reported[136] for Co^{II}, Co^{III} and Cu^{II}.

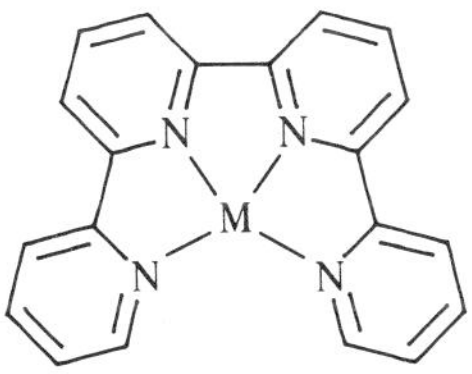

(73) tetrapy 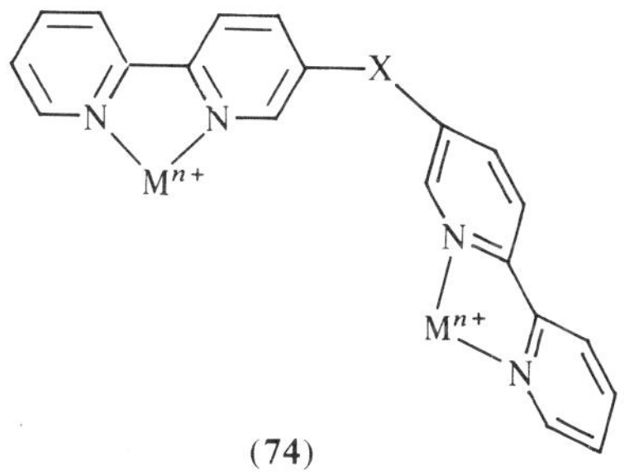   (74)

Dinucleating systems with ligands of the bipy type are also known. Provided that the different units of bipy are connected in a suitable way, dinuclear compounds may be obtained, as shown in (**74**). With certain ligands of sufficient size, however, tetracoordinate mononuclear species may also be formed (Figures 22 and 23).[137,138]

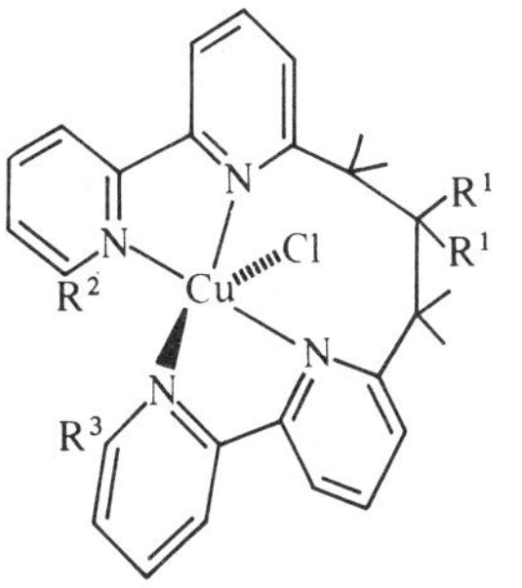

Figure 22 Schematic structure of a mononuclear bis(bipy) chelate with five-coordinate Cu^{II} [137]

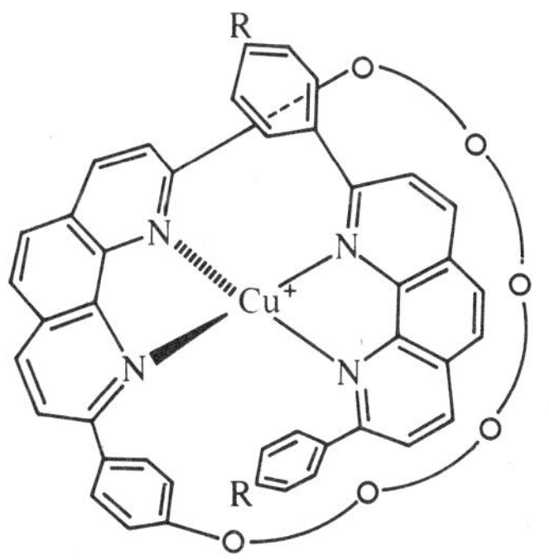

Figure 23 Representation of the structure of an open chain Cu^{I} chelate,[138] with two phen ligands arranged in a distorted tetrahedron around Cu^{I}. Subsequent ring closure results in a very interesting catenane structure

Use of combinations of pyridine and diazines may result in dinuclear species of the type indicated in (75) and (76). Although a variety of such compounds has already been reported by several groups,[60,139,140,141] the magnetic exchange between metal ions in such systems has not yet been studied in great detail.

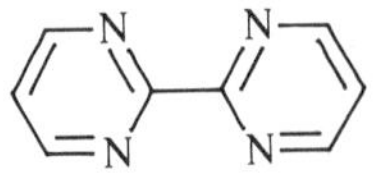

(75) 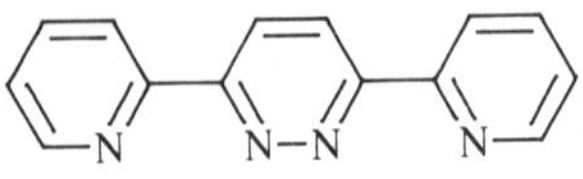   (76)

A very interesting bidentate pyridine-type ligand is 1,8-naphthyridine (napy; 77). The coordination modes that have been observed for this ligand are depicted schematically in Scheme 8. When this ligand coordinates in a bidentate fashion to one metal ion, the coordination number may be unusually high, *i.e.* octacoordination for Fe^{II}[142] and even dodecacoordination for lanthanide(III) compounds.[143] Upon coordination with both nitrogen atoms to different metal ions, short M—M distances may result.[144] With certain metal ions, mixed-valence species may even be stabilized such as in $[Ni_2(napy)_4Br_2]^+$, where the formal valence state of nickel is $+1.5$.[145] Modification of napy with coordinating substituents has also been reported recently.[145a]

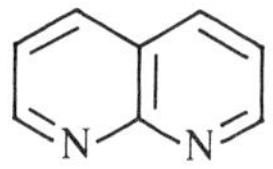

(77) napy

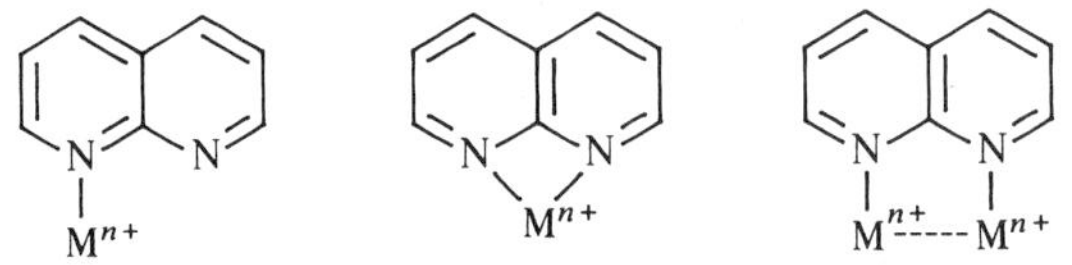

Scheme 8

13.2.6.4 Systems Based on Two Different Ring Types

13.2.6.4.1 Purine ligand systems

The coordination chemistry of the purine-type ligands has been studied on a rather large scale during the last decade, due to its relevance in biological systems. Detailed reviews are available of the coordination chemistry aspects both in the solid state and in solution.[146-149] Recently it has also been proposed that the role of metal-ion binding to purines influences the conversions between, for example, B-DNA and Z-DNA.[150]

As a free ligand, purine (68) in the neutral form has been studied only to a limited extent. The acidic hydrogen on the nitrogen is in fact labile and may also occur on N-9 and N-1. Upon deprotonation the purine anion (78) results, and now four nitrogen atoms are available for metal-ion binding. Bridging between two metal ions has been observed quite frequently for the deprotonated purine ligands.

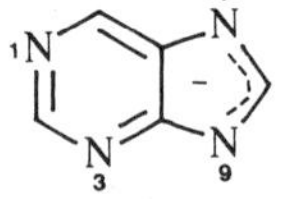

(78)

Bridging in the same way as that discussed for the imidazolato ligand has been reported with Cu^{II},[151] whereas bridging using N-3 and N-9 has been observed for several compounds and proved by X-ray structures (see Figure 24).[152,153]

In the case of adenine (79) and derivatives an external NH_2 adds a new potential donor atom. However, this has only rarely been found to coordinate,[146,147] and in fact binding to metal ions is restricted to N-1 and N-7, since N-9 is blocked in, for example, adenosine and derivatives, thereby also hindering coordination at N-3.

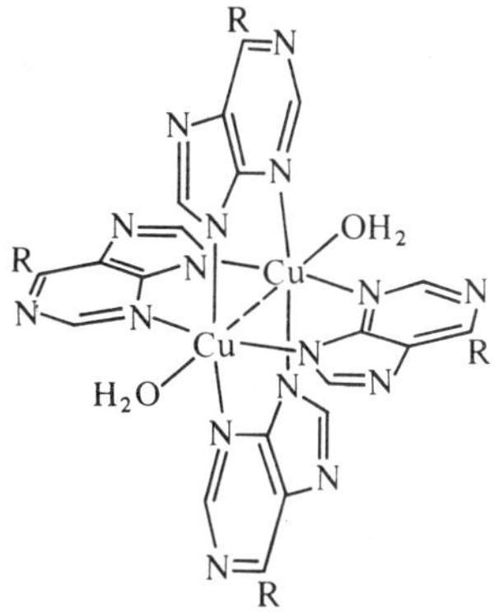

Figure 24 Schematic structure of copper(II) dimeric species held together by four bridging purine ligands (after ref. 28).

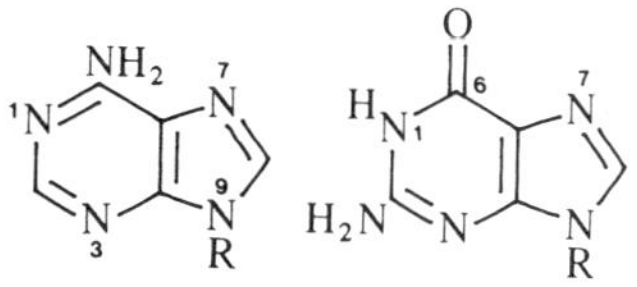

(79)   (80)

In the case of guanine derivatives (**80**), coordination to the O-6 atom may also be possible. In fact this has only rarely been observed.[146,147,154] Coordination at N-3 is sterically hindered, whereas binding at the NH$_2$ group is rare and binding at N-1 can only occur after deprotonation in an alkaline medium.[146,147,155] Coordination of guanines and hypoxanthines (**80**, but with the NH$_2$ replaced by hydrogen) takes place almost universally at the N-7 position, as proved by numerous crystal structures, unless N-7 is blocked.[155]

The binding of guanines *via* N-7 to platinum compounds is of particular interest from the point of view of antitumour chemistry,[156,157] since this binding is generally accepted to play a key role in the mechanism of action of *cis*-[PtCl$_2$(NH$_3$)$_2$] as an anticancer drug. Despite the fact that many other basic sites are available in DNA, these guanine N-7 sites appear to have the highest affinity (probably mainly kinetic) for the interaction with *cis*-[Pt(NH$_3$)$_2$]$^{2+}$. Space limitations prevent a more detailed discussion. For this the reader is referred to the above-mentioned reviews[156,157] and to Chapter 62.

Although not a purine, the ligand properties of 7-azaindole (**81**) strongly resemble those of deprotonated purines. Metal binding of this anionic ligand may occur to one or two metal ions, just as with 1,8-naphthyridine (Scheme 8).[158]

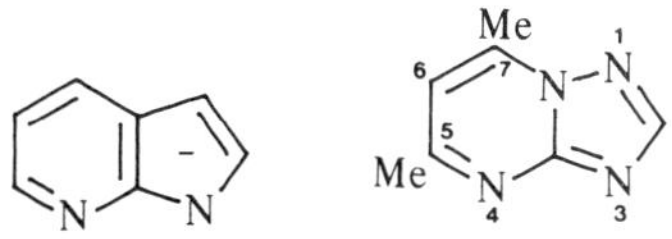

(81)   (82) dmtp

Apart from the normal purines, several derivatives have been reported and their complex-forming behaviour studied, examples being 8-azapurine[148] and dimethyltriazolopyrimidine.[159] The latter compound (**82**) is easily synthesized from acetylacetone and 3-amino-1,2,4-triazole.[159] This ligand coordinates to transition metal ions *via* N-3, since N-1 and N-4 appear to be blocked by the neighbouring methyl groups.[159] It seems that a very interesting area of coordination chemistry remains to be explored using ligands of this type, *i.e.* changes in the positions of the methyl substituents may result in binding types as observed for the other purines described above.

13.2.6.4.2 *Other mixed-ring ligand systems*

Inspired by the good chelating properties of bipy (**65**) and phen (**67**), analytical chemists had already studied their derivatives 30 years ago[160] (see also Chapter 10). Some examples are 2-(2-pyridyl)benzimidazole (**83**), 2-(2-pyridyl)imidazoline (**84**)[160,161] and 1-(2-pyridyl)pyrazoles

(85).[162,163] The coordination chemistry of these ligands was studied at an early date[161,164] and many interesting properties have been revealed, such as isomerism[164] and high-spin to low-spin transitions for the iron compounds.[62,127,164]

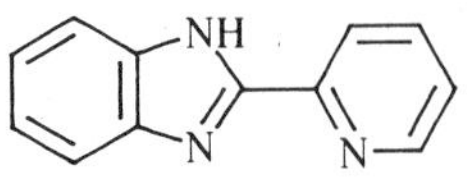

(83) 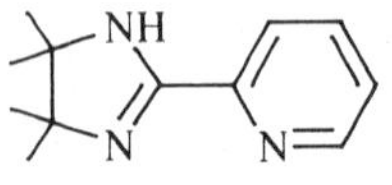   (84) pyim 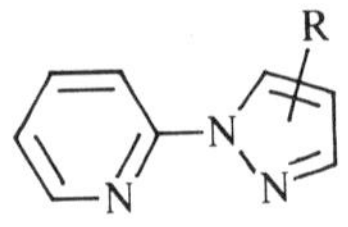   (85)

More recently, many other variations have been made, including extension to three rings, *e.g.* tridentate ligands (86).[165] These and many other derivatives of terpy will not be further discussed here, although the coordination chemistry of such ligands can be very interesting.

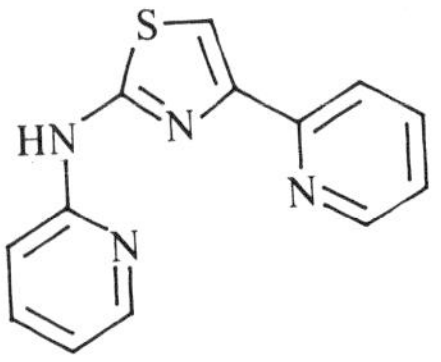

(86) 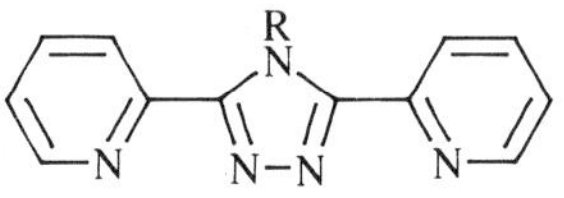   (87)

When the steric requirements of such ligands are severe, interesting dinucleating chelating ligands such as 3,5-di(2-pyridyl)pyrazoles and the corresponding triazoles (87) may be formed.[166,167] Such ligands allow the study of magnetic exchange through a heterocyclic ligand.

13.2.7 CONCLUDING REMARKS

The above sections have given a brief overview of the simple heterocyclic ligands and many derivatives of varying degrees of complexity. The numerous coordinating possibilities of such ligands have been discussed in a detailed manner, without being complete. In organometallic chemistry especially, many coordination modes of heterocyclic ligands, usually including metal to carbon bonds, have been observed. For a review of these aspects the reader is referred to 'Comprehensive Organometallic Chemistry'.[174]

Although the treatment of the subject has been rather systematic in this chapter, it is incomplete since not all heterocyclic ligand types known today are mentioned. On the other hand, the variety of examples presented should give an idea as to how to obtain or synthesize such ligands, and what their coordination behaviour will be with respect to several different metal ions.

The treatment of the subject according to the ligand types has excluded or somewhat neglected other interesting aspects of heterocyclic coordination chemistry. Some of these are mentioned here. First of all the possibility of studying the effect of ligand constraints on stoichiometry and stereochemistry should be mentioned. For instance, in a group of ligands such as is depicted in Figure 25, the coordination polyhedron around copper may be any one of a variety of distorted tetrahedra.[168]

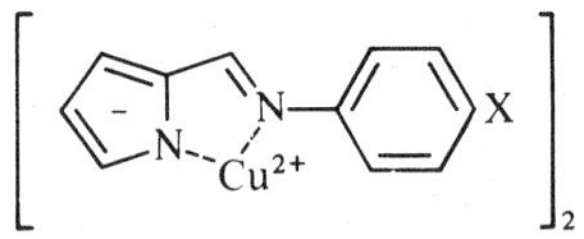

Figure 25 Illustration of the distorted tetrahedral Cu^{II} species with sterically hindered amines[168]

A second less extensively discussed area is the large group of compounds containing two different kinds of heterocyclic ligand coordinated to the same metal ion, *e.g.* Fe(bipy)(pzH)Cl$_3$,[169] and many ruthenium compounds with the general formula [Ru(bipy)(L)$_2$]$^{n+}$ (L = pyrazine, pyrazole, imidazole, *etc.*).[163,170] A very interesting alternating zig-zag chain of this type is presented in Figure 26.[171] When rather large ligands of this type are coordinated to the same metal ion, mutual ligand interaction may stabilize the coordination compound.[172]

Finally, the large group of ligands containing non-coordinating (or at best weakly coordinating) atoms such as sulfur and oxygen is not discussed in great detail. An example of such a compound[173]

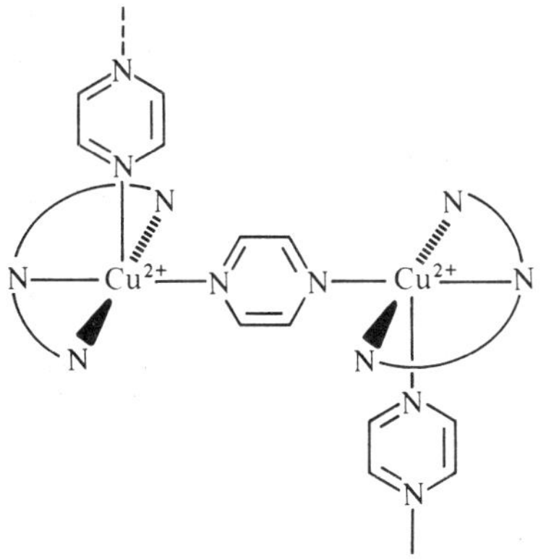

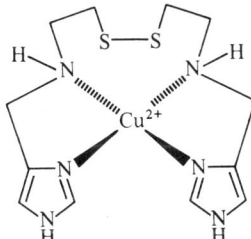

Figure 26 Schematic presentation of the zig-zag structure in the cation [Cu(pyrazine)(terpy)]$^{2+}$ (after ref. 171)

is depicted in Figure 27, where the sulfur atoms are close to copper and, in principle, coordination might occur. Ligands of this type are also discussed in some detail in Chapters 20.3 and 20.4.

Figure 27 A potentially hexadentate ligand, chelating to Cu^{II} in a tetradentate fashion, without significant Cu—S interaction[173]

13.2.8 REFERENCES

1. A. R. Katritzky (ed.), 'Advances in Heterocyclic Chemistry', Academic, New York; a series of over 34 volumes.
2. A. R. Katritzky and J. M. Lagowski, 'The Principles of Heterocyclic Chemistry', Methuen, London, 1967.
3. J. A. Joule and G. F. Smith, 'Heterocyclic Chemistry', 2nd edn., Van Nostrand Reinhold, London, 1978.
4. K. Schofield, M. R. Grimmett and B. R. T. Keene, 'The Azoles', Cambridge University Press, Cambridge, 1976.
5. A. Weissberger and E. C. Taylor (eds.), 'The Chemistry of Heterocyclic Compounds', Wiley, New York; a series with over 40 volumes, some of which are cited below as refs. 6–8.
6. K. Th. Finley, ref. 5, vol. 39, 'Triazoles: 1,2,3', Wiley, New York, 1980.
7. G. B. Barlin, ref. 5, vol. 41, 'The Pyrazines', Wiley, New York, 1982.
8. P. N. Preston, D. M. Smith and G. Tennant, ref. 5, vol. 40, 'Benzimidazoles and Congeneric Tricyclic Compounds', Wiley, New York, 1982.
9. A. R. Katritzky and C. W. Rees (eds.), 'Comprehensive Heterocyclic Chemistry', Pergamon, Oxford, 1984.
10. G. R. Newkome and W. W. Paudler, 'Contemporary Heterocyclic Chemistry', Wiley, New York, 1982.
11. R. M. Smith and A. E. Martell, 'Critical Stability Constants', Chemical Society, London, 1975.
12. J. Reedijk, *Recl. Trav. Chim. Pays-Bas*, 1969, **88**, 1451.
13. J. Reedijk, *Recl. Trav. Chim. Pays-Bas*, 1971, **90**, 117.
14. V. Gutmann, *Angew. Chem., Int. Ed. Engl.*, 1970, **9**, 843.
15. D. W. Herlocker and M. R. Rosenthal, *Inorg. Chim. Acta*, 1970, **4**, 501.
16. R. J. Doedens and L. F. Dahl, *J. Am. Chem. Soc.*, 1966, **88**, 4847.
17. C. W. Reimann, A. D. Mighell and F. A. Mauer, *Acta Crystallogr.*, 1967, **23**, 135.
18. J. Reedijk. *Recl. Trav. Chim. Pays-Bas*, 1972, **91**, 507.
19. J. Reedijk and G. C. Verschoor, *Acta Crystallogr., Sect. B*, 1973, **29**, 721.
20. E. D. McKenzie, *Coord. Chem. Rev.*, 1971, **6**, 187.
21. J. Foley, S. Tyagi and B. J. Hathaway, *J. Chem. Soc., Dalton Trans.*, 1984, 1.
22. W. D. Harrison and B. J. Hathaway, *Acta Crystallogr., Sect. B.*, 1979, **35**, 2910.
23. K. Burger, 'Organic Reagents in Metal Analysis', Pergamon, Oxford, 1973.
24. S. Trofimenko, *Chem. Rev.*, 1972, **72**, 497.
25. R. J. Sundberg and R. B. Martin, *Chem. Rev.*, 1974, **74**, 471.
26. G. Vos, Ph.D. Thesis, State University Leiden, 1983 (abstract in *Recl. Trav. Chim. Pays-Bas*, 1983, **102**, 468).
27. A. I. Popov, *Coord. Chem. Rev.*, 1969, **4**, 463.
28. M. Inoue and M. Kubo, *Coord. Chem. Rev.*, 1976, **21**, 1.
29. G. Kolks, S. J. Lippard, J. V. Waszczak and H. R. Lilienthal, *J. Am. Chem. Soc.*, 1982, **104**, 717.
30. F. B. Hulsbergen, R. W. M. ten Hoedt, G. C. Verschoor, J. Reedijk and A. L. Spek, *J. Chem. Soc., Dalton Trans.*, 1983, 539.

31. F. B. Hulsbergen, R. W. M. ten Hoedt, G. C. Verschoor and J. Reedijk, *Inorg. Chem.*, 1982, **21**, 2369.
32. G. F. Kokoszka, J. Baranowski, C. Goldstein, J. Orsini, A. D. Mighell, V. L. Himes and A. R. Siedle, *J. Am. Chem. Soc.*, 1983, **105**, 5627.
33. J. A. J. Jarvis, *Acta Crystallogr.*, 1962, **15**, 964.
34. D. W. Engelfriet, W. den Brinker, G. C. Verschoor and S. Gorter, *Acta Crystallogr., Sect. B*, 1979, **35**, 2922.
35. J. A. Tainer, E. D. Getzoff, J. S. Richardson and D. C. Richardson, *Nature (London)*, 1983, **306**, 284.
36. M. A. Guichelaar, J. A. M. van Hest and J. Reedijk, *Delft Prog. Rep.*, 1976, **2**, 51.
37. D. M L. Goodgame, M. Goodgame and G. W. Rayner-Canham, *J. Chem. Soc. (A)*, 1971, 1923.
38. J. Reedijk, B. A. Stork-Blaisse and G. C. Verschoor, *Inorg. Chem.*, 1971, **10**, 2594.
39. J. L. Atwood, K. R. Dixon, D. T. Eadie, S. R. Stobart and M. Zaworotko, *Inorg. Chem.*, 1983, **22**, 774.
40. E. J. van den Heuvel, P. L. Franke, G. C. Verschoor and A. P. Zuur, *Acta Crystallogr., Sect. C*, 1983, **39**, 337.
41. I. Sotofte and K. Nielsen, *Acta Chem. Scand., Ser. A*, 1981, **35**, 733.
42. L. D. Brown, J. A. Ibers and A. R. Siedle, *Inorg. Chem.*, 1978, **17**, 3026.
43. J. Reedijk, *Comments Inorg. Chem.*, 1982, **1**, 379.
44. J. Reedijk, *Recl. Trav. Chim. Pays-Bas*, 1972, **91**, 1373.
45. F. Seel and J. Rodrian, *Justus Liebigs Ann. Chem.*, 1974, 1784.
46. J. Müller and R. Stock, *Angew. Chem., Int. Ed. Engl.*, 1983, **22**, 993.
47. C. W. Eigenbrot, Jr. and K. N. Raymond, *Inorg. Chem.*, 1982, **21**, 2653.
48. B. S. Tovrog and R. S. Drago, *J. Am. Chem. Soc.*, 1977, **99**, 2203.
49. L. I. Chudinova and R. M. Klykova, *Russ. J. Inorg. Chem. (Engl. Transl.)*, 1973, **18**, 23.
50. A. F. Cameron, D. W. Taylor and. R. H. Nuttall, *J. Chem. Soc., Dalton Trans.*, 1972, 1603.
51. A. H. Lewin, I. A. Cohen and R. J. Michl, *J. Inorg. Nucl. Chem.*, 1974, **36**, 1951.
52. J. A. C. van Ooijen, J. Reedijk and A. L. Spek, *J. Chem. Soc., Dalton Trans.*, 1979, 1183.
53. W. E. Marsh, W. E. Hatfield and D. J. Hodgson, *Inorg. Chem.*, 1982, **21**, 2679.
54. R. M. Morrison, R. C. Thompson and J. Trotter, *Can. J. Chem.*, 1980, **58**, 238.
55. R. M. Morrison, R. C. Thompson and J. Trotter, *Can. J. Chem.*, 1979, **57**, 135.
56. V. H. Crawford and W. E. Hatfield, *Inorg. Chem.*, 1977, **16**, 1336.
57. P. C. M. Gubbens, A. M. van der Kraan, J. A. C. van Ooijen and J. Reedijk, *J. Magn. Magn. Mater.*, 1980, **15–18**, 636.
58. M. G. B. Drew, P. C. H. Mitchell and A. R. Read, *J. Chem. Soc., Chem. Commun.*, 1982, 238.
59. R. D. Gillard and R. J. Wademan, *J. Chem. Soc., Chem. Commun.*, 1981, 448.
60. G. Bullock, F. W. Hartstock and L. K. Thompson, *Can. J. Chem.*, 1983, **61**, 57.
61. C. Creutz, *Prog. Inorg. Chem.*, 1983, **30**, 1.
62. H. W. Richardson, J. R. Wasson and W. E. Hatfield, *Inorg. Chem.*, 1977, **16**, 484.
63. P. W. Carreck, M. Goldstein, E. M. McPartlin and W. D. Unsworth, *Chem. Commun.*, 1971, 1634.
64. J. A. C. van Ooijen, E. van Tooren and J. Reedijk, *J. Am. Chem. Soc.*, 1978, **100**, 5569.
65. P. H. Citrin and A. P. Ginsberg, *J. Am. Chem. Soc.*, 1981, **103**, 3673.
66. W. Harrison, A. Storr and J. Trotter, *Chem. Commun.*, 1971, 1101.
67. S. S. Sandhu, S. S. Tandon and H. Singh, *Inorg. Chim. Acta*, 1979, **34**, 81.
68. J. P. O'Brien, R. W. Allen and H. R. Allcock, *Inorg. Chem.*, 1979, **18**, 2230.
69. D. E. Bolster, P. Gütlich, W. E. Hatfield, S. Kremer, E. W. Müller and K. Wieghardt, *Inorg. Chem.*, 1983, **22**, 1725.
70. A. Wojtczak, M. Jaskolski and Z. Kosturkiewicz, *Acta Crystallogr., Sect. C*, 1983, **39**, 545.
71. G. A. Renovitsh and W. A. Baker, Jr., *J. Am. Chem. Soc.*, 1967, **89**, 6377.
72. P. Gütlich, *Struct. Bonding (Berlin)*, 1981, **44**, 83.
73. V. H. Crawford, H. W. Richardson, J. R. Wasson, D. J. Hodgson and W. E. Hatfield, *Inorg. Chem.*, 1976, **15**, 2107.
74. J. C. Daran, Y. Jeannin and L. M. Martin, *Inorg. Chem.*, 1980, **19**, 2935.
75. R. E. Clarke and P. C. Ford, *Inorg. Chem.*, 1970, **9**, 495.
76. H. Beinert, *Coord. Chem. Rev.*, 1980, **33**, 55.
77. K. Karlin and J. Zubieta (eds.), 'Copper Coordination Chemistry; Biochemical and Inorganic Perspectives', Adenine Press, Guilderland, NY, 1983.
78. P. J. M. W. L. Birker, E. F. Godefroi, J. Helder and J. Reedijk, *J. Am. Chem. Soc.*, 1982, **104**, 7556.
79. R. Breslow, J. T. Hunt, R. Smiley and T. Tarnowski, *J. Am. Chem. Soc.*, 1983, **105**, 5337.
80. P. J. M. W. L. Birker and J. Reedijk, in ref. 77, p. 205.
81. J. Verbiest, J. A. C. van Ooijen and J. Reedijk, *J. Inorg. Nucl. Chem.*, 1980, **42**, 971.
82. F. Mani and G. Scapacci, *Inorg. Chim. Acta*, 1980, **38**, 151.
83. T. N. Sorrell, M. R. Malachowski and D. L. Jameson, *Inorg. Chem.*, 1982, **21**, 3250.
84. W. L. Driessen, *Recl. Trav. Chim. Pays-Bas*, 1982, **101**, 441.
85. P. J. M W. L. Birker, H. M. J. Hendriks, J. Reedijk and G. C. Verschoor, *Inorg. Chem.*, 1981, **20**, 2408.
86. H. M. J. Hendriks, P. J. M. W. L. Birker, J. van Rijn, G. C. Verschoor and J. Reedijk, *J. Am. Chem. Soc.*, 1982, **104**, 3607.
87. F. B. Hulsbergen, W. L. Driessen, J. Reedijk and G. C. Verschoor, *Inorg. Chem.*, 1984, **23**, 3588.
88. H. M. J. Hendriks, P. J. M. W. L. Birker, G. C. Verschoor and J. Reedijk, *J. Chem. Soc., Dalton Trans.*, 1982, 623.
89. T. N. Sorrell and M. R. Malachowski, *Inorg. Chem.*, 1983, **22**, 1883.
90. K. D. Karlin, Y. Gultneh, J. P. Hutchinson and J. Zubieta, *J. Am. Chem. Soc.*, 1982, **104**, 5240.
91. J. V. Dagdigian, V. McKee and C. A. Reed, *Inorg. Chem.*, 1982, **21**, 1332.
92. K. Takahashi, E. Ogawa, N. Oishi, Y. Nishida and S. Kida, *Inorg. Chim. Acta*, 1982, **66**, 97.
93. A. W. Addison, H. M. J. Hendriks, J. Reedijk and L. K. Thompson, *Inorg. Chem.*, 1981, **20**, 103.
94. F. Mani and C. Mealli, *Inorg. Chim. Acta*, 1981, **54**, L77.
95. J. van Rijn and J. Reedijk, *Recl. Trav. Chim. Pays-Bas*, 1984, **103**, 78.
96. V. McKee, J. V. Dagdigian, R. Bau and C. A. Reed, *J. Am. Chem. Soc.*, 1981, **103**, 7000.
97. R. S. Brown, D. Salmon, N. J. Curtis and S. Kusuma, *J. Am. Chem. Soc.*, 1982, **104**, 3188.
98. K. D. Gallicano, N. L. Paddock, S. J. Rettig and J. Trotter, *Inorg. Nucl. Chem. Lett.*, 1979, **15**, 417.
99. G. C. Wellon, D. V. Bautista, L. K. Thompson and F. W. Hartstock, *Inorg. Chim. Acta.*, 1983, **75**, 271.
100. R. Breslow, J. T. Hunt, R. Smiley and T. Tarnowski, *J. Am. Chem. Soc.*, 1983, **105**, 5337.

101. J. C. Lancaster and W. R. McWhinnie, *Inorg. Chim. Acta*, 1971, **5**, 515.
102. G. Anderegg and F. Wenk, *Helv. Chim. Acta*, 1967, **50**, 2330.
103. D. Sedney, M. Kahjehnassiri and W. M. Reiff, *Inorg. Chem.*, 1981, **20**, 3476.
104. J. C. van Niekerk and L. R. Nassimbeni, *Acta Crystallogr., Sect. B*, 1979, **35**, 1221.
105. R. R. Gagné, R. S. Gall, G. C. Lisensky, R. E. Marsh and L. M. Speltz, *Inorg. Chem.*, 1979, **18**, 771.
106. K. D. Karlin, J. Shi, J. C. Hayes, J. W. McKown, J. P. Hutchinson and J. Zubieta, *Inorg. Chim. Acta*, 1984, **91**, L3.
107. G. Bombieri, E. Forsellini, A. Delpra, M. L. Tobe, C. Chatterjee and C. J. Cooksey, *Inorg. Chim. Acta*, 1983, **75**, 93.
108. J. A. Doull and L. K. Thompson, *Can. J. Chem.*, 1980, **58**, 221.
109. S. N. Choi, R. D. Bereman and J. R. Wasson, *J. Inorg. Nucl. Chem.*, 1975, **37**, 2087.
110. G. Kolks and S. J. Lippard, *J. Am. Chem. Soc.*, 1977, **99**, 5804.
111. J. D. Korp, I. Bernal, C. L. Merrill and L. J. Wilson, *J. Chem. Soc., Dalton Trans.*, 1981, 1951.
112. R. J. Sundberg, I. Yilmaz and D. C. Mente, *Inorg. Chem.*, 1977, **16**, 1470.
113. Y. Nakao, M. Yamazaki, S. Suzuki, W. Mori, A. Nakahara, K. Matsumoto and S. Ooi, *Inorg. Chim. Acta*, 1983, **74**, 159.
114. A. J. Canty, J. M. Patrick and A. H. White, *J. Chem. Soc., Dalton Trans.*, 1983, 1873.
115. W. R. McWhinnie and J. D. Miller, *Adv. Inorg. Chem. Radiochem.*, 1969, **12**, 135.
116. L. F. Lindoy and S. E. Livingstone, *Coord. Chem. Rev.*, 1967, **2**, 173.
117. A. D. Mighell, C. W. Reimann and F. A. Mauer, *Acta Crystallogr., Sect. B*, 1969, **25**, 60.
118. M. G. Burnett, V. McKee and S. M. Nelson, *J. Chem. Soc., Chem. Commun.*, 1980, 599.
119. W. Vreugdenhil, S. Gorter, J. G. Haasnoot and J. Reedijk, *Polyhedron*, 1985, **4**, 1369.
120. M. S. Haddad, E. N. Duesler and D. N. Hendrickson, *Inorg. Chem.*, 1979, **18**, 141.
121. F. Holmes, K. M. Jones and E. G. Torrible, *J. Chem. Soc.*, 1961, 4790.
122. G. Kolks, S. J. Lippard, J. V. Waszczak and H. R. Lilienthal, *J. Am. Chem. Soc.*, 1982, **104**, 717.
123. S. W. Kaiser, R. B. Saillant, W. M. Butler and P. G. Rasmussen, *Inorg. Chem.*, 1976, **15**, 2681.
124. A. Cuadro, J. Elguero, P. Navarro, E. Royer and A. Santos, *Inorg. Chim. Acta*, 1984, **81**, 99.
124a. R. Uson, L. A. Oro, M. Esteban, D. Carmona, R. M. Claramunt and J. Elguero, *Polyhedron*, 1984, **3**, 213.
125. M. A. Haga, *Inorg. Chim. Acta*, 1983, **75**, 29.
126. W. R. McWhinnie, *Coord. Chem. Rev.*, 1970, **5**, 293.
127. E. König, *Coord. Chem. Rev.*, 1968, **3**, 471.
128. J. H. Forsberg, *Coord. Chem. Rev*, 1973, **10**, 195.
129. C. Creutz, M. Chou, T. L. Netzel, M. Okumura and N. Sutin, *J. Am. Chem. Soc.*, 1980, **102**, 1309.
130. W. H. Elfring and G. A. Crosby, *J. Am. Chem. Soc.*, 1981, **103**, 2683.
131. H. Smit, H. van der Goot, W. T. Nauta, P. J. Pijper, S. Balt, M. W. G. de Bolster, A. H. Stouthamer, H. Verheul and R. D. Vis, *Antimicrob. Agents Chemother.*, 1980, **18**, 249.
132. S. A. Bedell and A. E. Martell, *Inorg. Chem.*, 1983, **22**, 364.
133. B. Chiswell and E. J. O'Reilly, *Inorg. Chim. Acta*, 1973, **7**, 707.
134. N. Serpone, G. Ponterini, M. A. Jamieson, F. Bolletta and M. Maestri, *Coord. Chem. Rev.*, 1983, **50**, 209.
135. S. Kremer, W. Henke and D. Reinen, *Inorg. Chem.*, 1982, **21**, 3013.
136. W. Henke, S. Kremer and D. Reinen, *Z. Anorg. Allg. Chem.*, 1982, **491**, 124.
137. G. R. Newkome, V. K. Gupta and F. R. Fronczek, *Inorg. Chem.*, 1983, **22**, 171.
138. C. O. Dietrich-Buchecker, J. P. Sauvage and J. P. Kintzinger, *Tetrahedron Lett.*, 1983, 5095.
139. V. F. Sutcliffe and G. B. Young, *Polyhedron*, 1984, **3**, 87.
140. G. de Munno, G. Denti and P. Dapporto, *Inorg. Chim. Acta*, 1983, **74**, 199.
141. S. Lanza, *Inorg. Chim. Acta*, 1983, **75**, 131.
142. A. Clearfield, P. Singh and I. Bernal, *Chem. Commun.*, 1970, 389.
143. A. Clearfield, R. Gopal and R. W. Olsen, *Inorg. Chem.*, 1977, **16**, 911.
144. A. Tiripicchio, M. Tiripicchio Camellini, R. Uson, L. A. Oro, M. A. Ciriano and F. Viguri, *J. Chem. Soc., Dalton Trans.*, 1984, 125.
145. L. Sacconi, C. Mealli and D. Gatteschi, *Inorg. Chem.*, 1974, **13**, 1985.
145a. W. R. Tikkanen, E. Binamira-Soriaga, W. C. Kaska and P. C. Ford, *Inorg. Chem.*, 1984, **23**, 141.
146. L. G. Marzilli, *Prog. Inorg. Chem.*, 1977, **23**, 255.
147. R. W. Gellert and R. Bau, *Met. Ions Biol. Syst.*, 1979, **8**, 1.
148. D. J. Hodgson, *Prog. Inorg. Chem.*, 1977, **23**, 211.
149. V. Swaminathan and M. Sundaralingam, *CRC Crit. Rev. Biochem.*, 1979, **6**, 245.
150. V. Narasimhan and A. M. Bryan, *Inorg. Chim. Acta*, 1984, **91**, L39.
151. P. I. Vestues and E. Sletten, *Inorg. Chim. Acta*, 1981, **52**, 269.
152. P. de Meester, D. M. L. Goodgame, K. A. Price and A. C. Skapski, *Nature (London)*, 1971, **229**, 191.
153. A. Terzis, A. L. Beauchamp and R. Rivest, *Inorg. Chem.*, 1973, **12**, 1166.
154. D. L. Smith and H. R. Luss, *Photogr. Sci. Eng.*, 1976, **20**, 184.
155. T. J. Kistenmacher, K. Wilkowski, B. de Castro, C. C. Chiang and L. G. Marzilli, *Biochem. Biophys. Res. Commun.*, 1979, **91**, 1521.
156. S. J. Lippard, *Science*, 1982, **218**, 1075.
157. A. T. M. Marcelis and J. Reedijk, *Recl. Trav. Chim. Pays-Bas*, 1983, **102**, 121; J. Reedijk, *Pure Appl. Chem.*, 1987, **59**, 181.
158. R. W. Brookes and R. L. Martin, *Aust. J. Chem.*, 1975, **28**, 1363; R. W. Brookes and R. L. Martin, *Inorg. Chem.*, 1975, **14**, 528.
159. J. Dillen, A. T. H. Lenstra, J. G. Haasnoot and J. Reedijk, *Polyhedron*, 1983, **2**, 195.
160. J. L. Walter and H. Freiser, *Anal. Chem.*, 1954, **26**, 217.
161. T. J. Lane, I. Nakagawa, J. L. Walter and A. J. Kandathil, *Inorg. Chem.*, 1962, **1**, 267.
162. A. J. Canty, C. V. Lee, N. Chaichit and B. M. Gatehouse, *Acta Crystallogr., Sect. B*, 1982, **38**, 743.
163. P. J. Steel, F. Lahousse, D. Lerner and C. Marzin, *Inorg. Chem.*, 1983, **22**, 1488.
164. D. M. L. Goodgame and A. A. S. C. Machado, *Chem. Commun.*, 1969, 1420.
165. S. E. Livingstone and J. D. Nolan, *J. Chem. Soc., Dalton Trans.*, 1972, 218.

166. P. W. Ball and A. B. Blake, *J. Chem. Soc. (A)*, 1969, 1415.
167. F. S. Keij, R. A. G. de Graaff, J. G. Haasnoot and J. Reedijk, *J. Chem. Soc., Dalton Trans.*, 1984, 2093.
168. A. W. Addison and J. H. Stenhouse, *Inorg. Chem.*, 1978, **17**, 2161.
169. W. L. Driessen, R. A. G. de Graaff and J. G. Vos, *Acta Crystallogr., Sect. B*, 1983, **39**, 1635.
170. J. G. Vos, J. G. Haasnoot and G. Vos, *Inorg. Chim. Acta*, 1983, **71**, 155.
171. E. Coronado, M. Drillon and D. Beltran, *Inorg. Chim. Acta*, 1984, **82**, 13.
172. K. H. Scheller and H. Sigel, *J. Am. Chem. Soc.*, 1983, **105**, 5891.
173. O. Yamauchi, H. Seki and T. Shoda, *Bull. Chem. Soc. Jpn.*, 1983, **56**, 3258.
174. G. Wilkinson, F. G. A. Stone and E. W. Abel (eds.), 'Comprehensive Organometallic Chemistry', Pergamon, Oxford, 1982, vols. 1–9.

13.3

Miscellaneous Nitrogen-containing Ligands

BRIAN F. G. JOHNSON
University of Cambridge, UK

BARRY L. HAYMORE
Monsanto Company, St. Louis, MO, USA

and

JON R. DILWORTH
University of Essex, Colchester, UK

13.3.1   INTRODUCTION

This chapter is concerned with the chemistry of transition metal complexes of an important group of π-bonded nitrogen ligands. Transition metal nitrosyls have a long history, have been thoroughly reviewed and are the most extensively studied and most developed of this group. Closely related to the coordination chemistry of NO is that of the electronically equivalent NS, $N{=}CR_2$ and N_2R, and important examples of the synthesis and reactivity of their metal complexes, together with relevant structural, spectroscopic and physico-chemical properties, are illustrated appropriately in this chapter. These species form a closely related group of compounds and the analogies between NO and the other form ligand systems are very close.

Transition metal hydrazido(2−) and (1−) ligands form a distinctly different group of complexes.

Interest in these species stems largely from study of the mechanism of conversion of coordinated dinitrogen to hydrazine and ammonia. It seems abundantly clear that hydrazido(2 −) species are involved in the conversion, in simple molecular systems at least, of N_2 to NH_3, and there is evidence to suggest their involvement enzymatically.

Species not mentioned in this chapter include the coordination chemistry of dinitrogen, and of hydrazines, hydroxylamine and hydroxylaminato anions. The chemistry of coordinated dinitrogen has been extensively reviewed, and its protonation has been discussed in detail.[1]

Hydrazine ligands are well established and behave similarly to ammonia. In addition to being unidentate they can act as bridges between two metals. There are no authenticated cases of 'sideways' or η^2-bound N_2H_4, but two cases of η^2-organohydrazine ligands have been reported, *e.g.* $[Mo(\eta^5\text{-}C_5H_5)(NO)I(NH_2NHPh)]^+$.[2]

Hydroxylamine ligands too can behave like amines as ligands, but their ability to act as a singly and even doubly deprotonated ligand is very interesting and, in many ways, parallel to the behaviour of hydrazido(1 −) and (2 −) species. For example R_2NO^- can serve as an N—O ligand with some similarity to peroxide, while RNO^{2-} forms complexes entirely analogous to those obtained using η^2-peroxo and η^2-disulfido ligands.[3,4] The hydroxylamido(2 −) complexes may be thought of as metallo-oxaziridines.[3] The formation of transition metal nitrosyls using hydroxylamine in basic media may well proceed *via* $NHOH^-$ and NHO^{2-} intermediates.[5]

The chemical and structural relationships between hydrazine and hydrazido complexes and between hydroxylamine and hydroxylamido complexes have been recently reviewed.[510]

13.3.2 NITROSYL COMPLEXES

The chemistry of transition metal nitrosyl complexes has undergone something of a transformation in recent years. This was stimulated by the emergence of two important aspects in the behaviour of nitric oxide as a ligand. The first unequivocal X-ray analysis revealing a bent metal—nitrosyl linkage (angle about 120°) was reported by Ibers[6] in 1968 and confirmed the earlier suggestion[7] that nitric oxide is capable of coordinating to a metal either in a linear or in a bent fashion. Subsequently a large number of structural determinations on nitrosyl complexes were reported and a reappraisal of bonding concepts has taken place. The reactivity of the coordinated nitric oxide group has also received much attention. Until recently this study was largely restricted to one nitrosyl species, the nitroprusside ion[8] but now extends to several other systems resulting in a greater understanding of the chemical behaviour of the NO ligand.

A series of comprehensive reviews[7,9–11] documents the historical growth of nitrosyl chemistry and also shows the recent expansion that has taken place both in the more traditional areas and in the comparatively new organometallic branch[12] of this subject. In this section it is our intention to discuss the structural aspects of nitrosyl coordination, to provide an account of the methods available for the preparation of nitrosyl complexes and to survey the reactions that occur at the nitrosyl group.

13.3.2.1 Structure and Bonding

13.3.2.1.1 Mononuclear nitrosyl complexes

Nitric oxide in its gaseous state exists as a monomeric species which possesses an unpaired electron, rendering the molecule paramagnetic. In terms of simple MO theory this electron is placed in a π-antibonding orbital (Figure 1) so that electron configuration is $(\sigma_1)^2(\sigma_1^*)^2(\sigma_2, \pi)^6(\pi^*)$. Thus, the NO bond order is 2.5, consistent with the interatomic distance of 1.15 Å which is intermediate between the triple bond distance in NO^+ (see below) of 1.06 Å and representative double bond distances of about 1.20 Å.[13]

Loss of the π^* electron occurs relatively easily as indicated by the low ionization potential of nitric oxide (9.23 eV) compared, for example, with that of carbon monoxide (14.1 eV) and nitrogen (14.5 eV).[13,14] In consequence nitrosonium salts ($NO^+ClO_4^-$, $NO^+BF_4^-$, *etc.*) are readily isolable. As expected the IR stretching frequency $v(NO)$ is considerably higher for these species[15] (2300–2200 cm^{-1}) than for free NO (1877 cm^{-1}), consistent with the increased bond strength. The nitrosonium cation is isoelectronic with carbon monoxide and since its mode of coordination to transition metals is potentially similar to that of CO, the convenience of formulating metal nitrosyls as derivatives of NO^+ has been recognized for some time.[16] Essential to metal carbonyl bond formation is the occurrence of π-back donation from the metal which takes place because of the

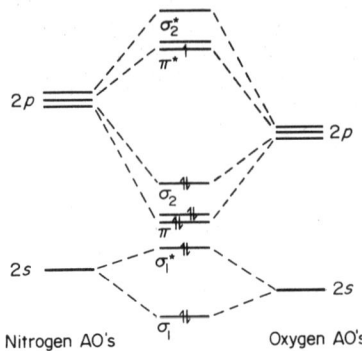

Figure 1   MO diagram for nitric oxide

π-acid nature of the ligand. The 'synergic' mode of bonding that occurs here might also be supposed to operate in all nitrosyl complexes. Resemblance between M—CO and M—NO coordination is suggested in a superficial manner by IR data. For although the regions in which CO and NO stretching frequencies appear are different, the ranges are very similar; for $v(CO)$ 2130–1700 cm^{-1} and for $v(NO)$ 1950–1520 cm^{-1}. However, such similarities are deceptive. It has now become clear that more than one type of coordination can occur in metal nitrosyls. The presence of the additional electron in nitric oxide has a considerable influence on M–NO bond formation and a more complex description is required to accommodate this aspect.

Lewis, Irving and Wilkinson[17,18] provided the first attempt at a complete classification of the different types of NO coordination. Their rationalizations, however, were derived from more limited spectroscopic and structural information than is currently available, much of which was unreliable. For example it has been shown that the red isomer of '[Co(NO)(NH$_3$)$_5$]$^{2+}$' is not a nitrosyl complex at all but is in fact a dimeric species containing a hyponitrite bridge.[19] Similarly '[Co(CN)$_5$NO]$^{3-}$' should now be reformulated as [(CN)$_5$Co(N$_2$O$_2$)Co(CN)$_5$]$^{6-}$.[20] Although laying the basis for present day concepts, the classification above has now become outmoded and is in part incorrect. In one of the proposed types of coordination these authors considered that the nitrosyl could be bonded in a 'side-on' fashion, involving σ-donation from a nitrogen–oxygen π-bonding MO, as is found with ethylene or O$_2$. This mode may be rejected (see below). Evidence available at that time, apparently supporting this arrangement, came from an incomplete X-ray structural determination of Co(S$_2$C-NMe$_2$)$_2$NO.[21] A three-dimensional study,[22] however, indicates an alternative orientation for the ligand here.

The first accurately documented example of a distinctly bent M–N–O linkage in a metal nitrosyl complex was presented in 1968. Ibers[6] showed that the cation [IrCl(CO)NO(PPh$_3$)$_2$]$^+$ (Figure 2) possessed a square pyramidal structure with a metal–nitrosyl bond angle of 124°. Subsequently other species exhibiting similar structural properties have been reported. It has now become clear that there are essentially two distinct orientations in which nitric oxide may be complexed. In one case the ligand is coordinated linearly and in the other it is bent (angle approximately 120°); for structural data available on linear and bent nitrosyls see Tables 1 and 2 respectively.

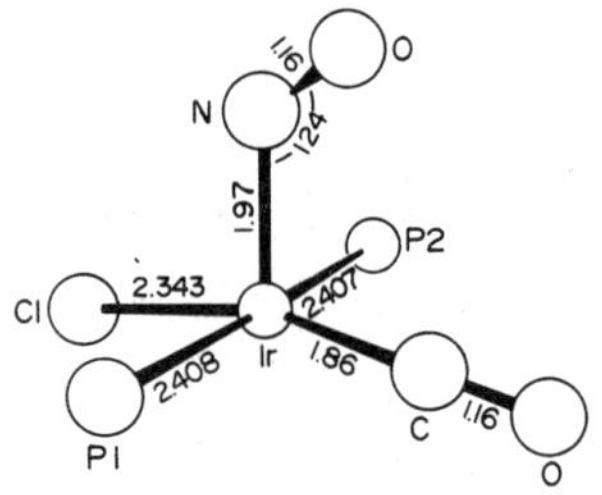

Figure 2   Coordination geometry in [IrCl(CO)NO(PPh$_3$)$_2$][BF]

A description involving two extreme bonding formulations is currently used to explain the nature of the M—N—O linkage in most mononuclear complexes. In the sense of formal oxidation state it is now customary to think of linear nitrosyls as derived from NO$^+$ and bent (120°) nitrosyls as derived from NO$^-$.[23,24] Bond formation between free nitric oxide and a transition metal ion must then be considered to involve either prior donation of an electron from NO (giving NO$^+$) or prior acceptance of an electron by NO (giving NO$^-$);[25] in each case this would of course be followed by lone pair donation from the nitrogen. Using this somewhat simplistic view it is easy to explain the

Table 1 IR and Structural Data on Linear Mononuclear Nitrosyl Complexes

Complex	$v(NO)$ (cm^{-1})	M—N—O (°)	M—N (Å)	N—O (Å)	Ref.
$[Fe(CN)_5NO]^{2-}$	1944	178	1.63	1.13	62
$[Ru(OH)(NO_2)_4NO]^{2-}$	1907	180	1.75	1.13	63
$[RuCl_5(NO)]^{2-}$	1887	177	1.74	1.13	64
$[RuCl(NO)_2(PPh_3)_2]^+$	1845	180	1.74	1.16	24
$[Os(OH)(NO)_2(PPh_3)_2]^+$	1840	179	1.71	1.25	65
$[Ni(Cp)NO]^a$	1820	180	1.68	1.10	66
$[IrH(NO)(PPh_3)_3]^{+\,b}$	1780	175	1.68	1.21	32
$[Mn(CO)_4NO]$	1759	180	1.80	1.15	34
$[Os(CO)_2NO(PPh_3)_2]^+$	1750	177	1.89	1.12	36
$[Mn(CN)_5NO]^{3-}$	1725	174	1.66	1.21	67
$[Mn(CO)_3NO(PPh_3)]$	1713	178	1.78	1.15	68
$[Ru(NO)(dppe)_2]^+$	1673	174	1.74	1.20	35
$[Mn(CO)_2NO(PPh_3)_2]$	1661	174	1.76	1.18	69
$[Cr(CN)_5NO]^{3-}$	1660	176	1.71	1.21	70
$[RuH(NO)(PPh_3)_3]$	1640	176	1.80	1.18	37
$[Mo(S_2CNBu_2^n)_3NO]$	1630	173	1.73	1.15	71
$[MoCp_2(Me)NO]$	1600	178	1.76	1.23	72
$[Ir(NO)(PPh_3)_3]$	1600	180	1.67	1.24	73

^{a} Microwave data. N—O estimated to calculated M—N.
^{b} Black isomer.

Table 2 IR and Structural Data on Bent Mononuclear Nitrosyl Complexes

Complex	$v(NO)$ (cm^{-1})	M—N—O (°)	M—N (Å)	N—O (Å)	Ref.
$[Cr(NCO)Cp(NO)_2]$	1818, 1783	171	1.72	1.16	74
$[Ru(S_2CNEt_2)_3NO]$	1803	170	1.72	1.17	75
$[Ir(NO)_2(PPh_3)_2]^+$	1760, 1715	164	1.77	1.21	76
$[IrH(NO)(PPh_3)_3]^{+\,a}$	1720	167	1.77	1.20	26
$[IrI(CO)NO(PPh_3)_2]^+$	1720	125	1.89	1.17	77
$[Ni(N_3)(NO)(PPh_3)_2]$	1710	153	1.69	1.16	53
$[Fe(S_2CNMe_2)_2NO]$	1690	170	1.72	1.10	56
$[RuCl(NO)_2(PPh_3)_2]^+$	1687	136	1.87	1.17	24
$[IrCl(CO)NO(PPh_3)_2]^+$	1680	124	1.97	1.16	6
$[Co(Ea)NO]^b$	1654	122	1.82		78
$[Co(Eb)NO]^c$	1635	123	1.83		78
$[Co(S_2CNMe_2)_2NO]$	1630	135	1.75	1.07	22
$[Os(OH)(NO)_2(PPh_3)_2]^+$	1630	128	1.98	1.12	65
$[CoCl(NO)(en)_2]^+$	1611	121	1.81	1.11	33
$[Co(NO)(NH_3)_5]^{2+\,d}$	1610	119	1.87	1.15	79
$[IrCl_2(NO)(PPh_3)_2]$	1560	123	1.94	1.03	25
$[IrI(Me)NO(PPh_3)_2]$	1525	120	1.92	1.23	52

^{a} Brown isomer.
^{b} Ea = N,N'-ethylenebis(acetylacetoneiminato).
^{c} Eb = N,N'-ethylenebis(benzoylacetoneiminato).
^{d} Black isomer.

occurrence of linear and bent arrangements by an examination of the σ-framework involved. The nitrogen and oxygen atoms in NO$^+$ are considered to be *sp* hybridized and in NO$^-$ to be *sp^2* hybridized (see Figure 3); the bonding orientations are thus prescribed. Bent nitrosyls with angles between 120° and 180° can be visualized as possessing intermediate hybridization of *s* and *p* ligand atomic orbitals. Here the formalism of oxidation state will have little meaning. Several species of this type have been reported, for example the brown form of $[IrH(NO)(PPh_3)_3]^+$ (angle 167°),[26] also

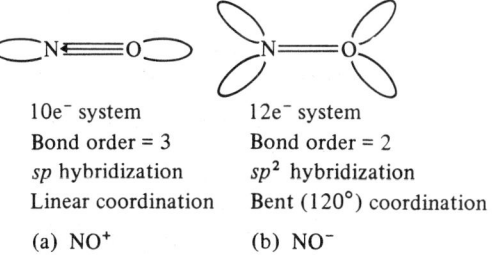

10e$^-$ system	12e$^-$ system
Bond order = 3	Bond order = 2
sp hybridization	*sp^2* hybridization
Linear coordination	Bent (120°) coordination
(a) NO$^+$	(b) NO$^-$

Figure 3

$MoCl_2(NO)_2(PPh_3)_2$ (mean angle 162°),[27] $Co(S_2CNMe_2)_2NO$ (two orientations, mean angle 135°)[22] and $[RuCl(NO)_2(PPh_3)_2]^+$ (one angle is 136°).[24,28]

It might seem strange that NO^- which is isoelectronic with O_2 should bond in an end-on fashion. However, Griffith[29] has argued that the oxygen molecule coordinates to transition metals in a π-bonded manner because the bonding π orbitals $(\pi_{2p_x,2p_y})$ are of higher energy than the σ_{2p_z} bonding orbital, *i.e.* the electronic configuration of O_2 is $(\sigma_{1s})^2(\sigma_{1s}^*)^2(\sigma_{2s})^2(\sigma_{2s}^*)^2(\sigma_{2p_z})^2(\pi_{2p_x,2p_y})^4(\pi_{2p_x,2p_y}^*)^2$. In NO^- a *different* ordering occurs. A σ-bonding orbital is of higher energy than the π-bonding orbitals (see Figure 1), thus the electronic configuration is $(\sigma_{1s})^2(\sigma_{1s}^*)^2(\sigma_{2s})^2(\sigma_{2s}^*)^2(\pi_{2p_x,2p_y})^4(\sigma_{2p_z})^2(\pi_{2p_x,2p_y}^*)^2$. The NO^- ligand is therefore expected to utilize its σ *orbitals* in coordination to a metal. Griffith has also shown that the σ orbitals for the oxygen molecule in its valence state (and presumably also for NO^-) have the same form as filled sp^2 hybrid orbitals on the oxygen atom.[25] This provides correlation with the more rudimentary approach given above and similarly predicts coordination of NO^- at a 120° angle. It is pertinent to consider that the localization of *any* electron density in a non-bonding nitrogen orbital should lead to a bending of the M—N—O linkage from linearity.

In the NO^+ (linear) type of nitrosyl coordination, π-acceptance is considered to play an important role as it does in metal carbonyls. The two extreme forms of bonding involved here are often represented as in Figure 4 (z is the oxidation state of the free metal ion). In the NO^- type of coordination however, it is considered that π-backbonding is *not* very important.[23,24,30] This is ascribed to the electron density already localized at the nitrogen atom, to the orientation of the NO ligand which will grossly restrict π-overlap, and to the availability of only one π^* ligand orbital compared with two in NO^+. Indeed in the case of NO^+ the formation of *two* M—N π bonds could enable triple bond character to be approached, *i.e.* M≡NO.[31] Although the factors determining which ligand orientation is found in a particular nitrosyl complex are not well understood it is clear from the above that one important influence will be the readiness of the metal to undergo π-bonding, *i.e.* the 'hardness' or 'softness' of the metal centre.

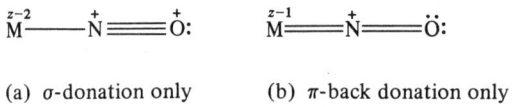

(a) σ-donation only   (b) π-back donation only

Figure 4 Coordination as NO^+

Since the involvement of π-bonding should normally effect a contraction in bond length, M—N distances are expected to be *shorter* in *linear* arrangements.[23] However, the bond order in this type of linkage is somewhat variable (Figure 4) and so differences may not always be very significant. In practice M—(NO^+) is often shorter than M—(NO^-) by 0.1 to 0.3 Å (see Tables 1 and 2). Similarly in a crude sense one might expect $v(NO)$ to be lower for NO^- species than for NO^+ species (Figure 3). This is found to be true in many cases but there are anomalies, some of which could result from electronic of steric hindrance to bending of the M—N—O moiety [*e.g.* crowding in $Rh(NO)(PPh_3)_3$]; however the picture is complex and other factors will be involved. For compounds of NO^+ and of NO^-, $v(NO)$ has been observed in the ranges $1950-1600\,cm^{-1}$ and $1720-1520\,cm^{-1}$ respectively.[10,11] Clearly a sizeable overlap region exists and this restricts the use of IR stretching frequency as a criterion for bond type. In fact no direct correlation can be found between $v(NO)$ and metal–nitrogen bond length.[32] Weaver and Snyder[33] have also noted that at present valuable information cannot be derived from N—O bond lengths because of the difficulties in accurately measuring N—O distances by X-ray methods and the possible insensitivity of bond length to bond change in these systems.

A result of the *restricted* π-overlap in NO^- bonding, and also of the strong σ-donation envisaged, is the observation of very large *trans*-labilizing effects induced by this ligand. Such behaviour[33] has been reported for the coordinatively saturated cobalt III complex $[CoCl(NO)(en)_2]^+ClO_4^-$ (Table 2). Now in most of the crystallographically confirmed NO^- nitrosyls the metal ion has formally a d^6 configuration (*e.g.* Ru^{II} or Ir^{III}). Many of these species are coordinatively unsaturated (five- instead of six-coordination) and this has been explained[23] in terms of the labilizing effect being so great that a ligand *trans* to NO cannot be held by the metal. Collman[23] has found that the unsaturated complex $RhCl_2(NO)(PPh_3)_2$ (Figure 5 shows the structure of the iridium analogue)[25] reacts with carbon monoxide to give $RhCl_2(CO)NO(PPh_3)_2$. In both species the NO stretching frequency is *identical* ($1630\,cm^{-1}$) and in the latter $v(CO)$ is $2080\,cm^{-1}$. The high value of the carbonyl frequency suggests a rhodium(III) and NO^- formulation, and the fact that $v(NO)$ is the same, whether CO is coordinated or not, suggests both strong σ-donation and little π-bonding for the presumed *trans* NO^- group, giving it a large *trans* effect. This is as expected (above).

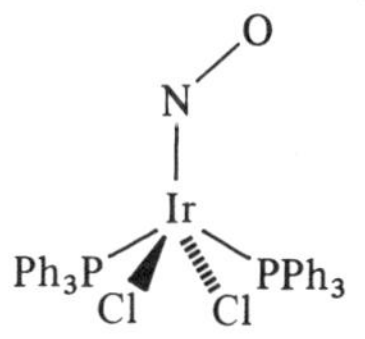

Figure 5 Coordination geometry in $IrCl_2NO(PPh_3)_2$. The NO group lies approximately in the P—Ir—P plane

The preponderance of five-coordination amongst d^6 NO^- complexes has attracted considerable interest. The reason for this is that all these species possess essentially square pyramidal stereo-chemistries with the bent nitrosyl in an apical position (Figures 2 and 5). In five-coordinate species where the NO linkage is linear, however, trigonal bipyramidal geometries with equatorial nitrosyls are invariably found, for example in $Mn(CO)_4NO$,[34] $[Ru(NO)(dppe)_2]^+$ [35] and $[Os(CO)_2NO(PPh_3)_2]^+$;[36] data on these structures are given in Table 1. Only when the species also contains a metal hydride linkage does this distinction break down, a result of the very small spatial requirements of the H atom, *e.g.* in $[IrH(NO)(PPh_3)_3]^+$ [32] and $RuH(NO)(PPh_3)_3$. The two basic stereochemical variations have been ascribed to different oxidation states and electron configurations. Mingos and Ibers[25] consider that five-coordinate d^6 complexes are generally square pyramidal; two structures of such triphenylphosphine complexes are known, $RuCl_2(PPh_3)_3$[38] and $RhI_2Me(PPh_3)_2$.[39] On the other hand five-coordinate d^8 complexes are generally trigonal bipyramidal; examples are $IrCl(CO)_2(PPh_3)_2$,[40] $Os(CO)_3(PPh_3)_2$,[41] *etc.* Therefore, the square pyramidal geometries of the bent nitrosyl complexes are consistent with their formulation as d^6 species; a similar argument implies that linear nitrosyl complexes should have the d^8 configuration. It has also predicted on theoretical grounds that the square pyramidal geometry ought to be more stable than the trigonal bipyramidal geometry for low spin d^6 phosphine complexes of the type ML_5.[42] However, the lower symmetry of the species considered here may somewhat invalidate the use of this argument.

Pierpont *et al.*[28] have reported the preparation and structural determination of a five-coordinate ruthenium dinitrosyl cation, $[RuCl(NO)_2(PPh_3)_2]^+$. They found that it possesses square pyramidal geometry with a bent apical nitrosyl (NO^-) and also a linear nitrosyl (NO^+) situated in the square plane (Figure 6a). Others[23] have repeated the preparation of this species using a ^{15}N-labelled precursor (see equation 1), and have discovered in the product that ^{14}N and ^{15}N were equally distributed in each of the dissimilar nitrosyl ligands. This was taken to indicate equilibration of the two structurally different groups. A plausible pathway has been suggested by which interconversion of these groups may be effected.[24] This involves a change in formal oxidation state of Ru^{II} to Ru^0 accompanied, as should be expected (see above), by a stereochemical interconversion of the type SP to TBP. Scrambling of the labelled nitrosyl ligand therefore occurs through a trigonal bipyramidal intermediate with equivalent linear (or slightly bent) nitrosyl groups in the equatorial plane. The mechanism is essentially an inverse Berry-type rearrangement,[43] see Figure 6. Since the two equilibrating forms (a) and (c) are identical and only the nitrosyls are undergoing redox interexchange the overall result may be described as a hybridization tautomerism. Some support for the mechanism comes from a study of the species $CoCl_2(NO)L_2$ (L = monotertiary phosphine). Collman[23] considers that in solution these compounds also exist in two interconverting forms, each of which contains a different type of nitrosyl group. His argument is based on IR data. In each system two $v(NO)$ bands are observed, their assignment being confirmed by ^{15}N substitution. The higher band has been attributed to linear (NO^+) and the lower to bent (NO^-) coordination (Figure 7). Only two stereochemistries are thought to be involved, corresponding to the similar geometries in Figure 6(a) and (b) and therefore their interchange represents one half of the inverse-Berry rearrangement.

$$RuCl(^{15}NO)(PPh_3)_2 + NOPF_6 \longrightarrow [RuCl(^{15}NO)NO(PPh_3)_2]PF_6 \qquad (1)$$

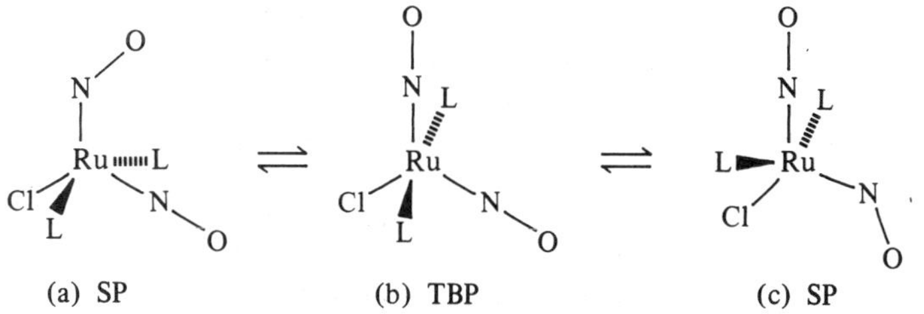

Figure 6 Scrambling of linear and bent NOs in $[RuCl(NO)_2(PPh_3)_2]^+$

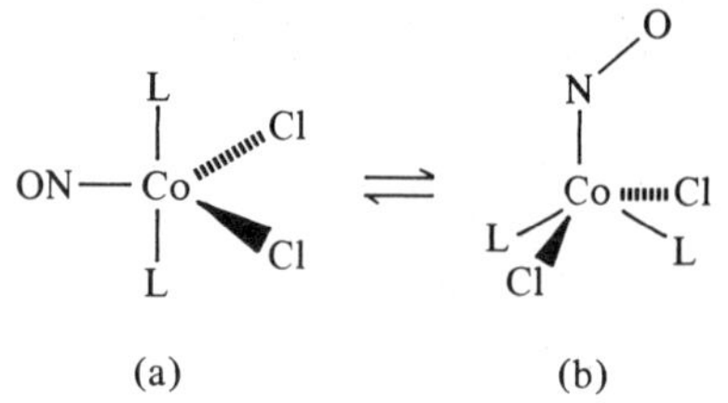

Figure 7

This redox facility possessed by the nitrosyl ligand (Figures 6 and 7 above) and represented by equation (2) may also be considered, in a formally NO^+ species, as the property of either donating or accepting electron pairs [z is the formal oxidation state of $M(NO^+)$]. Any tendency to localize electron density at the ligand, *i.e.* tending towards NO^- coordination, would leave the metal ion somewhat coordinatively unsaturated and thus susceptible to nucleophilic attack.[44] Behaviour of this type can be used to explain the otherwise unexpected results obtained from studies of substitution kinetics in carbonyl nitrosyl complexes. Whereas binary metal carbonyls, *i.e.* $M_x(CO)_y$, generally undergo unimolecular substitution reactions (with ligands such as monotertiary phosphines) following a dissociative process (S_{N1} path), carbonlyl nitrosyls undergo bimolecular displacements (S_{N2} path).[45] Thus under substitution conditions nickel carbonyl [$Ni(CO)_4$] behaves differently from the pseudonickel carbonyl species $Co(NO)(CO)_3$, $Fe(NO)_2(CO)_2$ and $Mn(NO)_3CO$. In nitrosyl complexes the course of substitution may be represented as shown in equation (3), although alternative views have been expressed.[46]

$$M^z(NO^+) \rightleftharpoons M^{z+2}(NO^-) \tag{2}$$

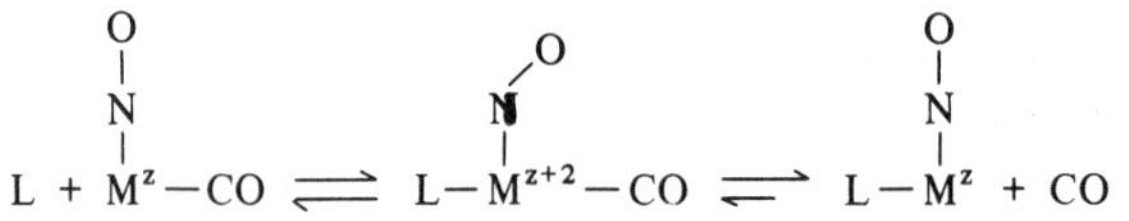

$$L + \overset{\displaystyle \overset{O}{|}}{\underset{\displaystyle |}{N}} M^z - CO \rightleftharpoons L - \overset{\displaystyle N \nearrow^O}{\underset{\displaystyle |}{M^{z+2}}} - CO \rightleftharpoons L - \overset{\displaystyle \overset{O}{|}}{\underset{\displaystyle |}{N}} M^z + CO \tag{3}$$

The ability of the nitrosyl ligand to behave as an 'electron pair reservoir' has also been considered to play an important part in certain catalytically active systems.[44,47] The vacant site provided by isomerization of the ligand could enable an unsaturated organic molecule to enter the transition metal's coordination sphere, thus forming an active intermediate. Examples of catalysis by nitrosyl complexes include the hydrogenation of alkenes by $Rh(NO)L_3$ species[44] and the dimerization of dienes in the presence of $Fe(CO)_2(NO)_2$ or $Fe(\pi\text{-}C_3H_5)(CO)_2NO$.[48] Certain molybdenum dinitrosyl complexes, such as $MoCl_2(NO)_2(PPh_3)_2$, have also been found to provide very efficient alkene dismutation catalysts.[49]

In the above discussion no attempt has been made to rationalize the effects which lead to the adoption of particular orientation by the nitric oxide group on coordination. Several authors have considered factors which might contribute to bending of the M—N—O linkage but it is clear that no all-embracing theory has emerged; at present one cannot predict with certainty the bond type that will be found in a complex. Molecular orbital calculations[50] have given orderings for the orbital levels in pentacyanonitrosyl complexes (where linear coordination is found). Pierpont and Eisenberg[24,57] used the essential order of levels that was obtained for the nitroprusside ion[31] to explain why the NO ligand bends in square pyramidal structures. They considered the hypothetical case of a species with square pyramidal geometry but with a linear nitrosyl group (C_{4v} symmetry); the order of levels here should be similar to that in the six-coordinate $[Fe(CN)_5NO]^{2-}$ except that the d_{z^2} orbital is expected to drop in energy below the $d_{x^2-y^2}$ orbital, a result of the absence of the sixth ligand. As the nitrosyl group bends to 120° the symmetry of the molecule is lowered from C_{4V} to C_S and one N—O and both M—NO π bonds are broken. A new ordering of levels takes place which has been correlated with the C_{4V} levels as in Figure 8. The correlation diagram implies that for a 20-electron system composed of 10 electrons in the ligand σ^b functions, four electrons in the $\pi^b(NO)$ set, and a d^6 metal ion, the C_{4v} configuration in which strong $M \rightarrow NO$ π bonding exists is the geometry of the ground state. However, if an additional electron pair is added to the system the nitrosyl bends and the molecule undergoes an electron shift from metal to ligand. The placement of electrons in a significantly antibonding orbital (with respect to the M—NO bond) in C_{4v} symmetry, that is the d_{z^2} orbital, is thus avoided at the energetic expense of significant metal–nitrosyl π-bonding. Hence, from the correlation it can be concluded that a 22-electron system possessing a square-based pyramidal structure (*e.g.* $IrCl_2NO(PPh_3)_2$ and other species in Table 2) should exhibit a bent nitrosyl coordination as is found.

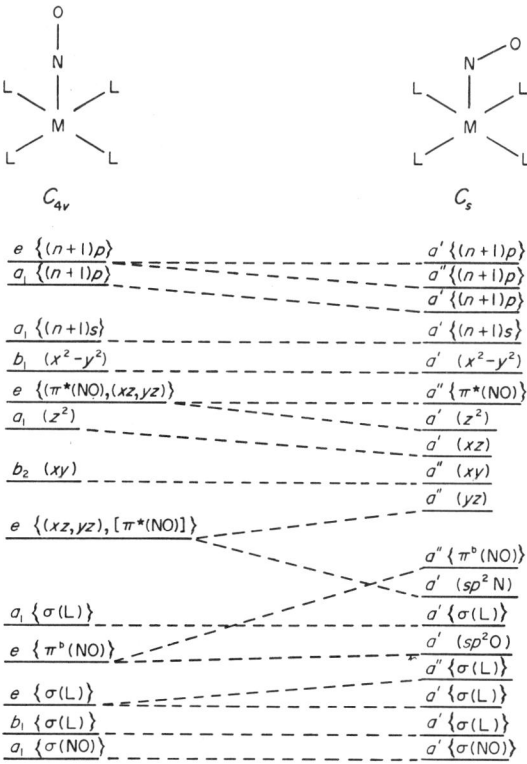

Figure 8 An orbital correlation diagram which shows the principal levels of interest in the bending of nitrosyls in five-coordinate tetragonal systems

Other proposals by Ibers and co-workers[25,32,52] are also based on the filling of molecular orbital levels, with particular reference to the relative energy of the $\pi^*(NO)$ level in the different bonding situations. Enemark,[53] however, takes a quite different view. In interpreting the structure of $Ni(N_3)$-$(NO)(PPh_3)_2$ he considers that the short M—N bond (1.69 Å) indicates a NO^+ type of coordination, and to explain the M—N—O angle of 153° (which is equally distant from 120° and 180°) he refers to Kettle's π-bonding mechanism[54] for the deviation from linear coordination. This states essentially that in most substituted metal carbonyl and nitrosyl (NO^+) complexes the two π^* orbitals on each ligand belong to a different irreducible representation and hence need not interact equally with the metal d orbitals. Such an effect may lead to appreciable distortions from 180°.

In another treatment Mingos[55] has evaluated the influence of a molecule's electronic excited states on the coordinated ligand geometry; this also gives a correlation with the experimental results. Finally M—N—O bending in certain species has been attributed to ligand–ligand interactions. Thus, in the complex $[RuCl(NO)_2(PPh_3)_2]PF_6$ it has been considered possible that a direct but weak linkage could occur which involves one of the π^* orbitals of the basal nitrosyl and one of the oxygen lone pairs of the apical ligand.[24,28] Similarly in $Fe(S_2CNMe_2)_2NO$ the 170° M–N–O angle has been associated with interaction between the oxygen (NO) lone pairs and the dithiocarbamato nitrogen atom.[56]

13.3.2.1.2 Bridging nitrosyl complexes

Relatively few unequivocal X-ray structural determinations of bridging nitrosyl complexes have been reported and therefore the nature of such bonding is not well understood. Several of the species studied possess doubly bridging nitrosyl groups: $Pt_4(MeCO_2)_6(NO)_2$,[57] $Ru_3(CO)_{10}(NO)_2$ (Figure 9a)[58] and $Mn_2(NO_2)(\pi\text{-}Cp)_2(NO)_3$ (Figure 9b).[59] In the first of these the M—N—M angle is 120°, consistent with a sp^2 hybridized nitric oxide group and leading to its formulation as a NO^- derivative.[57] Here both the non-bonding sp^2 nitrogen orbitals of NO^- are utilized in forming a metal to metal bridge. The other two species above have M—N—M angles that are not 120° and rationalizations therefore can only be based on multicentre bonding in a MO description. This also applies to triply bridging nitrosyls. Such a mode of coordination has been structurally confirmed in the species $Mn_3(\pi\text{-}Cp)_3(NO)_4$[60] (Figure 10) but, as a result of crystal disorder, it was not possible to obtain accurate structural parameters. Bridging need not occur symmetrically as is also found in polynuclear carbonyl species, and in the manganese complex (Figure 9b) the distance Mn_1—N is greater than N—Mn_2 by 0.17 Å.

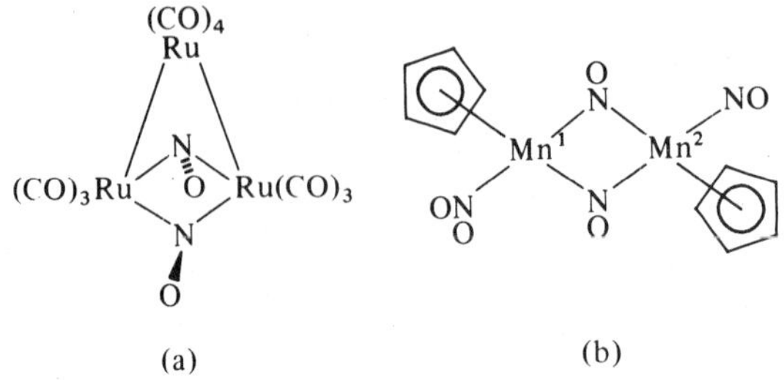

Figure 9

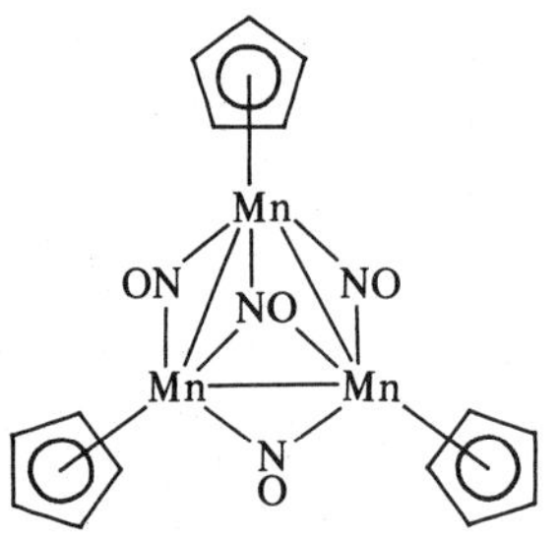

Figure 10

Nitrosyl bridging may also be detected in complexes using IR spectroscopy, for example in $[Co(\pi\text{-}Cp)NO]_2$.[61] The NO stretching frequency is lower in these species than in terminal nitrosyls and as expected it is found to decrease as the extent of bridging increases. This is illustrated by $Mn_3(\pi\text{-}Cp)_3(NO)_4$ which shows two bands due to the doubly bridging groups at 1543 and 1481 cm^{-1} and one from the triply bridging group at 1320 cm^{-1} (see also section 13.3.2.4).[30]

13.3.2.2 Preparative Methods

Nitric oxide gas provides the most direct route to many nitrosyl complexes. It will frequently displace CO groups, three of which must be substituted by two NO groups in order that the effective atomic number rule is obeyed.

The well authenticated binary metal nitrosyl $Co(NO)_3$, is prepared as shown in equation (4).[80]

$$Co(CO)_3NO + 2NO \longrightarrow Co(NO)_3 + 3CO \qquad (4)$$

Metal–metal bonds can also be formed or broken in such displacements, *e.g.* equation (5) or (6).

$$Fe_3(CO)_{12} + 6NO \longrightarrow 3Fe(CO)_2(NO)_2 + 6CO \qquad (5)[81]$$

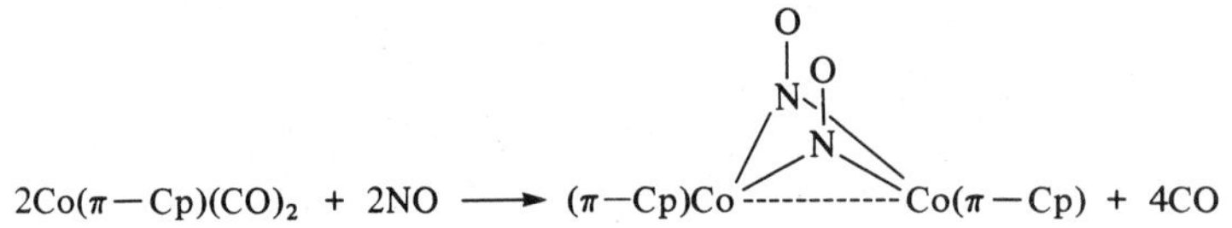

$$2Co(\pi-Cp)(CO)_2 + 2NO \longrightarrow (\pi-Cp)Co\text{-----}Co(\pi-Cp) + 4CO \qquad (6)[61]$$

Sometimes a *formal* disproportionation of NO occurs. For example reaction with $Co(NO)(PPh_3)_3$ yields the nitrito dinitrosyl species $Co(NO_2)(NO)_2(PPh_3)$ and nitrogen gas is evolved.[82] The initial step in this reaction may be described as in equation (7).

$$3NO \longrightarrow NO^+ NO_2^- + \tfrac{1}{2}N_2 \qquad (7)$$

Nitrosyl halides (or dinitrogen tetroxide acting as $NO^+NO_3^-$) can also displace carbonyl groups. The metal nitrosyl halides which result are often polymeric in nature.

$$M(CO)_6 + 2NOCl \xrightarrow{CH_2Cl_2} \frac{1}{n}[MCl_2(NO)_2]_n + 6CO \qquad (8)^{83}$$

$$(M = Mo \text{ or } W)$$

Nitrosonium salts NO^+Y^- ($Y = BF_4$, PF_6 etc.) react with *basic* precursors which are either coordinatively unsaturated or have displaceable groups. The use of these electrophilic reagents has only recently been developed and their sensitivity to conditions, such as the nature of the solvent, has tended to restrict their application (equation 9).

$$IrCl(CO)(PPh_3)_2 + NOBF_4 \longrightarrow [IrCl(CO)NO(PPh_3)_2]BF_4 \qquad (9)^6$$

N-Methyl-*N*-nitrosotoluene-*p*-sulfonamide reacts with metal hydride complexes which possess a labile neutral ligand. Entry of this reagent into the coordination sphere is followed by exchange of H for NO and a concurrent *formal* reduction of the metal ion by two units (equation 10).

$$MH(CO)_5 \xrightarrow{\text{Me}-\langle\rangle-SO_2N\langle^{NO}_{Me}} Mn(CO)_4NO + CO \qquad (10)^{84}$$

It has been known for a long time that nitrosyl complexes can be prepared using nitric acid; the nitroprusside ion $[Fe(CN)_5NO]^{2-}$ was first obtained by its action on $[Fe(CN)_6]^{3-}$.[85] In many cases the production of such species was accidental and the reactions are often quite complex. Several nitrosyls of the platinum metals have been obtained in this way and recently the polymeric species $Pt_4(MeCO_2)_6(NO)_2$ was isolated, together with other products, from the reduction of Pt^{IV} in a mixture of nitric and acetic acids.[57]

A number of reactions have been reported which occur at coordinated nitrogen-containing groups and lead to the isolation of nitrosyl complexes; these will now be considered. Electrophilic attack at a nitro group, leading to nitrosyl formation, has been reported (see equation 11).[86]

$$Ru^{II}Cl(NO_2)(bipy)_2 + 2H^+ \longrightarrow [Ru^{II}(Cl)NO(bipy)_2]^{2+} + H_2O \qquad (11)$$

Alkyl nitrites have also been used in nitrosyl synthesis and their mode of action may be explained by analogy with the above. One can consider that initial coordination of RONO is quickly followed by H^+ attack giving a free alcohol and the nitrosyl complex (equation 12).

$$Mn(CO)_4(PCy_3) + RONO \longrightarrow {}'Mn(CO)_3(PCy_3)N\langle^{O}_{OR}{}' + CO$$

$$R = \text{isoamyl} \qquad\qquad\qquad H^+\downarrow$$

$$Mn(CO)_3NO(PCy_3) + ROH \qquad (12)^{87}$$

In certain cases reduction of NO_2^- and NO_3^- groups can be effected using carbon monoxide or triphenylphosphine. The reducing agent may first coordinate to the metal and then oxygen transfer can occur. Examples are shown in equations (13) and (14).

$$Ni^{II}(NO_3)_2(PEt_3)_2 + 2CO \longrightarrow Ni^0(NO_3)NO(PEt_3)_2 + 2CO_2 \qquad (13)^{88}$$

$$Ru(NO_2)_2(CO)_2(PPh_3)_2 + 2PPh_3 \longrightarrow Ru^0(NO)_2(PPh_3)_2 + 2CO + 2OPPh_3 \qquad (14)^{89}$$

It is possible that in the use of alkali metal nitrites for nitrosyl synthesis, the first formed nitrito complex undergoes rapid ligand reduction similar to that above, as in equations (15) and (16).

$$NiBr_2(PPh_3)_2 + NaNO_2 \xrightarrow{THF} {}'NiBr(NO_2)(PPh_3)_2{}' + NaBr \qquad (15)^{90}$$

$$\downarrow PPh_3$$

$$NiBr(NO)(PPh_3)_2 + OPPh_3$$

$$Fe(CO)_5 + KNO_2 \xrightarrow{MeOH} {}'K[Fe(CO)_4NO_2]{}' + CO \qquad (16)^{91}$$

$$\downarrow$$

$$K[Fe(CO)_3NO] + CO_2$$

Nitrosyl complexes have also been synthesized using hydroxylamine. Although the mechanism of this process is not well understood, hydroxylamine complexes are known.[92] Therefore it is possible that oxidation or deprotonation of the coordinated NH_2OH group could occur, leading to the products obtained in equation (17).

$$FeCl_2(CNPh)_4 + 4NH_2OH \longrightarrow Fe(NO)_2(CNPh)_2 + 2NH_4Cl + 2H_2O + 2CNPh \qquad (17)^{[93]}$$

A mechanism of this type is supported by the observation[94] that an NH_3 group (which is essentially similar to a NH_2OH group) in the complex $[RuCl(NH_3)_5]^{2+}$ is oxidized in dilute perchloric acid leading to a mixture of the nitrosyls $[Ru(OH)NO(NH_3)_4]^{2+}$ and $[Ru(NO)(H_2O)(NH_3)_4]^{3+}$.

13.3.2.3 Reactivity of the Coordinated Nitric Oxide Group and Related Chemistry

Reactions have been known to occur at the nitrosyl group in the nitroprusside ion, $[Fe(CN)_5NO]^{2-}$,[8] for some considerable time; recently similar behaviour has been observed in other systems. Different types of reactivity are exhibited depending on the nature of the nitrosyl complex and the mode of coordination of the NO ligand. For general reviews of this topic see references 9, 11 and 14.

13.3.2.3.1 Nucleophilic addition

Nucleophilic attack can occur at the nitrogen atom in NO^+ complexes where there is strong σ-donation and weak π-acceptance by the ligand. In such species, which are often cationic, electron density is drawn from the nitric oxide group. This results in an increased N—O bond order and is thus reflected in the observation of a high NO stretching frequency. Consequently IR data can be used loosely as a criterion for susceptibility to attack by nucleophiles.

The high $v(NO)$ ($1944\,cm^{-1}$) found in $[Fe(CN)_5NO]^{2-}$ is ascribed to the effect of the electron-withdrawing CN^- ligands, for strong σ-donation by the NO group must take place in order to satisfy the electron-seeking metal ion. This leaves the nitrosyl susceptible to attack. Hence reaction occurs with base leading to the isolation of a nitro complex (equation 18).

$$[Fe(CN)_5NO]^{2-} + 2OH^- \longrightarrow [Fe(CN)_5(NO_2)]^{4-} + H_2O \qquad (18)$$

An analogous product is obtained with SH^-. Alkaline mercaptan solutions give red-purple colourations (equation 19) and similar behaviour is observed for alkaline solutions of ketones or other species containing 'acidic' hydrogen bound to carbon. Here, however, colourations are often transient and fade as the NO-containing ligand is cleaved from the metal by hydrolysis (equation 20). Oximes and related organic products are isolable from such solutions.[8]

$$[Fe(CN)_5NO]^{2-} + SR^- \longrightarrow \left[[Fe(CN)_5N \overset{O}{\underset{SR}{\diagup}} \right]^{3-} \qquad (19)$$

$$[Fe(CN)_5NO]^{2-} + Me\overset{O^-}{\underset{|}{C}}=CH_2 \longrightarrow \left[Fe(CN)_5N \overset{O}{\underset{CHC(O)Me}{\diagup}} \right]^{4-} + H$$

$$\downarrow H_2O$$

$$[Fe(CN)_5(H_2O)]^{3-} + MeC(O)CH(NOH) \qquad (20)$$

Hydroxide ion attack, again leading to a nitro complex, has been observed in a ruthenium system. The process is reversed in the presence of H^+ (equation 21).

$$[RuClNO(L_2)_2]^{2+} + 2OH^- \rightleftharpoons RuCl(NO_2)(L_2)_2 + H_2O \qquad (21)$$

$$L_2 = \text{bipy or diars}$$

Cupric halides in alcohols absorb nitric oxide giving unstable blue or violet species. Decomposition leads to alkyl nitrites, presumably *via* RO^- attack (equation 22).

$$Cu(NO)X_2 + ROH \rightleftharpoons CuX_2 + H^+ + RONO \qquad (22)$$

However, there is only one well-defined example of alkoxide addition. Here the formation of a *coordinated* alkyl nitrite occurs and the reaction may be reversed by treatment with acid (equation 23).

$$[IrCl_3(NO)(PPh_3)_2]^+ \underset{H^+}{\overset{ROH}{\rightleftharpoons}} IrCl_3(RONO)(PPh_3)_2 \tag{23}$$

$$(R = Me, Et \text{ or } Pr)$$

Azide ion also reacts with the dication $[RuCl(NO)(bipy)_2]^{2+}$ (*above*) but leads to complete removal of the nitric oxide group. It is possible that an intermediate is first formed, containing a N_4O group, and this then decomposes. The overall reaction is shown in equation (24).

$$[RuCl(NO)(bipy)_2]^{2+} + N_3^- + \text{solvent} \longrightarrow RuCl(\text{solvent})(bipy)_2 + N_2 + N_2O \tag{24}$$

It has been reported that the dinitrogen species $[Fe(CN)_5N_2]^{2-}$ * can be obtained from the reaction of hydrazine with nitroprusside ion[100] but that a similar reaction with $[RuCl(NO)(diars)_2]^{2+}$ gives an azide[96] (equation 25).

$$[RuCl(NO)(diars)_2]^{2+} + NH_2NH_2 \longrightarrow {}'RuCl(N{\overset{O}{\underset{N-NH_2}{\diagdown}}})(diars)_2{}' + 2H^+$$

$$\downarrow$$

$$RuClN_3(diars)_2 + H_2O \tag{25}$$

With phenylhydrazine the product isolated is analogous to the intermediate above and the N-bonded ligand is formulated as the oxotriazeno group *i.e.* $-N{\overset{\overset{O}{|}}{=}}N-NHPh$.[16]

Hydrazine also reacts with $[Ru(NH_3)_5NO]^{3+}$ giving the N_2 species $[Ru(NH_3)_5N_2]^{2+}$ whereas with hydroxylamine as the nucleophile the nitrous oxide complex $[Ru(NH_3)_5N_2O]^{2+}$ is produced.[101]

13.3.2.3.2 Electrophilic addition

Protonation of the nitric oxide group may occur either at nitrogen or the oxygen atom. In NO^+ species, where this behaviour has been observed, the NO ligand is considered to act as a strong π acceptor and consequently $v(NO)$ values are low. This does not imply, however, that NO^- complexes will necessarily be susceptible to attack. Although electron density is localized in a non-bonding nitrogen orbital in these species, the effects of strong σ-donation and weak π-acceptance may not result in an electron rich nitrosyl group.

The decomposition of nitrosyl complexes has frequently been observed in acid solution, giving nitrous oxide, nitrogen or even ammonia. Here protonation must occur at the oxygen atom which is then removed as water. Further reaction gives gaseous products. An example[9] is shown in equation (26).

$$2[Ni(NO)(CN)_3]^{2-} + 2H^+ \longrightarrow [Ni(CN)_4]^{2-} + Ni(CN)_2 + N_2O + H_2O \tag{26}$$

Protonation of a reduced form of nitroprusside ion gives a stable blue solid; apparently attack also occurs at the O atom here (equation 27).[102]

$$[Fe(CN)_5NO]^{3-} + H^+ \longrightarrow [Fe(CN)_5(NOH)]^{2-} \tag{27}$$

Electrophilic addition of H^+ (from HCl) to $OsCl(CO)NO(PPh_3)_2$ produces a coordinated $:N{\overset{O}{\underset{H}{\diagdown}}}$ moiety equation (28), and the expected decrease in $v(NO)$ is observed ($1565\,cm^{-1}$ to $1410\,cm^{-1}$). Protonation of both N and O atoms occurs in $Os(NO)_2(PPh_3)_2$ giving the species in Figure 11, whereas $Ir(NO)(PPh_3)_3$ reacts with three equivalents of HCl leading to a coordinated hydroxylamine complex $IrCl_3(NHOH)(PPh_3)_2$.[92] The rhodium analogue $Rh(NO)(PPh_3)_3$ behaves similarly. All these reactions are readily reversible, the processes involving formal oxidation or reduction of the metal centre. The fact that the reaction (29) is reversible may shed some light on the function of hydroxylamine as a nitrosating agent (see previous section).

* This formulation is questionable.

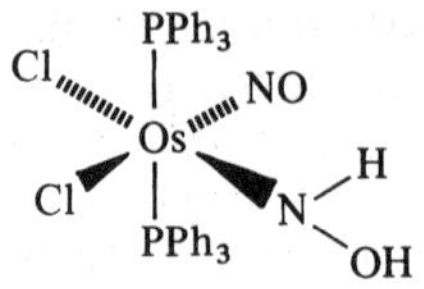

Figure 11

$$OsCl(CO)NO(PPh_3)_2 + HCl \longrightarrow OsCl_2(CO)(HNO)(PPh_3)_2 \tag{28}$$

$$M^{z+3}-NH_2OH \underset{+3H^+}{\overset{-3H^+}{\rightleftharpoons}} M^z-NO \tag{29}$$

The anions $[CoX(CO)_2NO]^-$ (X = halogen) react with alkyl halides giving organic nitroso compounds and finally oximes (equation 30).[104] Here, however, the $(Co-NO)^-$ grouping is acting as a nucleophile, attacking the $R^{\delta+}$ centre. Addition of R^+ to the complex can occur either directly at the nitrosyl or at the metal centre. In the latter case migration must then occur causing insertion of NO into the R—M bond.

$$C_6H_5CH_2-Cl + [CoCl(CO)_2NO]^- \longrightarrow C_6H_5-NO + `[CoCl(CO)_2]` + Cl^- \tag{30}$$

Another related feature is the observation that monomeric organo-lanthanides (Lewis acids) interact with nitrosyl groups. This occurs for example with $CrCl(\pi\text{-}Cp)(NO)_2$ and $(MeCp)_3Sm$. Adduct formation through the oxygen atom is proposed in order to explain the decreases that are found in the NO stretching frequencies.[105]

13.3.2.3.3 Oxidative processes

Nitro and nitrato complexes have been isolated from reactions of coordinated NO groups with oxygen. Thus, the formation of $[Os(NO_2)CO(CNR)_2(PPh_3)_2]^+$ (R = *p*-tolyl) has been reported[36] (equation 31) and others[106] have oxidized $Co(dmgH)_2NO$ in aqueous acetone giving $Co(dmgH)_2(NO_2)(H_2O)$. Similarly $[Co(NO)(en)_2]^{2+}$ in acetonitrile produces $[Co(NO_2)(MeCN)(en)_2]^{2+}$, and $[Ru(NO_3)O_2(NO)_2]_2O$ in aqueous solution gives $[Ru(NO_2)(NO_3)_2O_2(H_2O)]^-$.

$$[Os(CO)_2NO(PPh_3)_2]^+ + 2CNR + \tfrac{1}{2}O_2 \longrightarrow [Os(NO_2)CO(CNR)_2(PPh_3)_2]^+ + CO \tag{31}$$

The above reactions have been generalized by Basolo (in equation 32)[107] who considers that the presence of base is of fundamental importance. From a kinetic study on the oxidation of $CoL_4(NO)$ species (L_4 = quadridentate Schiff base), the same author has proposed a mechanism in which the first step is shown in equation (33). He finds that base strength effects reaction rate. This suggests that electron density, driven onto the nitrosyl group by base addition, facilitates oxidation. Conversion to the nitro group then occurs directly at the NO ligand.

$$M(NO) + B + \tfrac{1}{2}O_2 \longrightarrow BM(NO_2) \tag{32}$$

$$CoL_4(NO) + B \rightleftharpoons BCoL_4(NO) \tag{33}$$

It would seem possible, however, that an alternative mechanism could operate at least in —NO to —ONO_2 conversion. Here prior formation of an O_2 complex is considered. The addition of a coordinating ligand (or a base) would then cause intramolecular ligand oxidation in a second step. Such behaviour has been observed in a ruthenium system (shown in equations 34 and 35).[108–110] In the oxidation of $Ru(NO)_2(PPh_3)_2$,[89] O_2 itself is acting as the coordinating ligand (equation 36).

$$RuX(NO)(PPh_3)_2 + O_2 \longrightarrow RuX(O_2)NO(PPh_3)_2 \tag{34}$$

$$RuX(O_2)NO(PPh_3)_2 + 2CO \longrightarrow RuX(NO_3)(CO)_2(PPh_3)_2 \tag{35}$$

$$Ru(NO)_2(PPh_3)_2 + O_2 \longrightarrow `Ru(O_2)(NO)_2(PPh_3)_2` \xrightarrow{O_2} Ru(NO_3)(O_2)NO(PPh_3)_2 \tag{36}$$

No compelling evidence favouring a particular mechanism for these reactions has been presented. Clearly care must be exercized in comparing the oxidations to —NO_2 and —ONO_2. In the first case the oxidative step may be considered to involve two metal centres (and O_2) whereas in the second case a migration about one metal centre must occur (since a change from N to O coordination is observed).

In one instance, however, clear evidence has been obtained to suggest the formation of a peroxynitrate intermediate. Thus, the nickel complex $Ni(NO)Cl(diphos)$ reacts with O_2 either in dimethylformamide or on irradiation in dichloromethane to form the nitro complex $Ni(NO_2)Cl(diphos)$. Evidence for the formation of a d^9 nickel(I) intermediate came from ESR studies. This intermediate was considered to be the peroxy-bridge species (4) and the reaction sequence was believed to follow the sequence shown in Scheme 1.

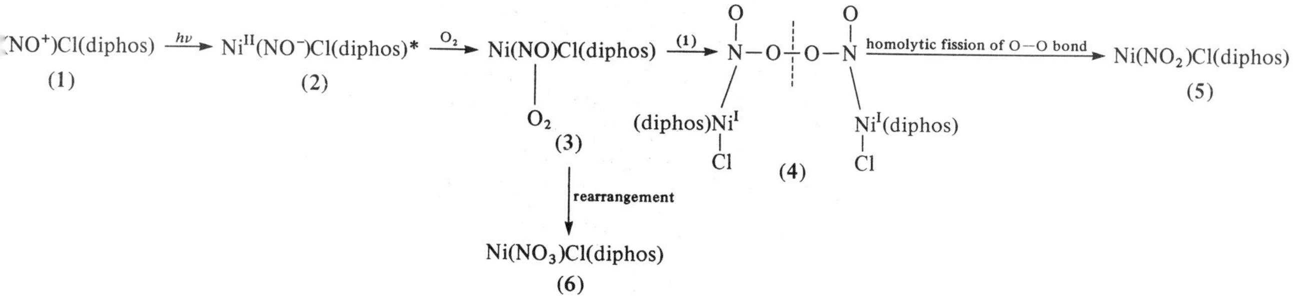

Scheme 1

There is a second possible reaction path. In the absence of an excess of compound (1) it is possible that compounds (3) could undergo an intramolecular rearrangement of the peroxy–nitrate group to the nitrato complex (6).

On the basis of these observations and conclusions it was suggested that the cobalt complexes described above might follow a related pathway. Here, however, it can be argued that the presence of base in, *e.g.* the reactions of $Co(dmgH)_2NO$, leads initially to the *quantitative* formation of $Co^{III}(dmgH)_2(NO)B$. In this case attack of O_2 leads first to the peroxy compounds and then, because of the absence of a suitable NO^+ compound, rearrangement occurs to give the nitrito derivative. Clearly whether or not nitro or nitrato compounds are produced would depend on the concentration of base B. In some cases it is possible to obtain mixtures of such products. These oxidations are also observed for other metals, *e.g.* equations (37) and (38).

$$[Ir(NO)_2(PPh_3)_2]^+ \xrightarrow{O_2} [Ir(NO_2)(NO_3)(PPh_3)_2]^+ \tag{37}$$

$$Pt(NO)(NO_3)(PPh_3)_2 \xrightarrow{O_2} Pt(NO_3)_2(PPh_3)_2 \tag{38}$$

13.3.2.3.4 *The catalytic conversion of NO and CO to N₂O and CO₂*

In 1973 it was shown that the monocation $[Ir(NO)_2(PPh_3)_2]^+$ undergoes reaction with CO to give $[Ir(CO)_3(PPh_3)_2]^+$, CO_2 and N_2O (equation 39), and that the tricarbonyl cation is easily converted back to the dinitrosyl cation by reaction with NO. The reactions were recognized as the basis of a possible homogeneous catalytic O-transfer process.

$$2NO + CO \longrightarrow CO_2 + N_2O \tag{39}$$

Later related reactions of iridium nitrosyl complexes of a slightly different type were reported (equations 40–42).

$$[IrBr(NO)_2(PPh_3)_2] + 3CO \longrightarrow [IrBr(CO)_2(PPh_3)_2] + CO_2 + N_2O \tag{40}$$

$$[Ir_2(NO)_4(PPh_3)_2] + 8CO \longrightarrow [Ir_2(CO)_6(PPh_3)_2] + 7CO_2 + 2N_2O \tag{41}$$

$$[IrCl(NO)(NO_2)(PPh_3)_2] + 4CO \longrightarrow [IrCl(CO)_2(PPh_3)_2] + 2CO_2 + N_2O \tag{42}$$

A key feature of all these reactions is the simultaneous displacement of 3CO groups by 2NO groups. Related rhodium compounds behave similarly and function as good catalysts for the conversion of NO and CO into N_2O and CO_2. The neutral compounds $M(NO)_2(PPh_3)_2$ of ruthenium and osmium, which are isoelectronic with the rhodium and iridium cations, also undergo the same reaction but far more slowly. In contrast, and disappointingly in view of the relative costs of the materials, the cobalt cation $[Co(NO)_2(PPh_3)_2]^+$ and the iron complex $Fe(NO)_2(PPh_3)_2$ do not undergo this reaction. In this case simple substitution to give, *e.g.* $Fe(NO)_2(CO)(PPh_3)$ occurs.

Ethanolic solutions of rhodium trichloride also catalyze the conversion of CO and NO into CO_2 and N_2O. In this case the active catalyst appears to be $[Rh(CO)_2Cl_2]^-$.

Two mechanisms for these reactions have been put forward. The first involve the migration of O from coordinated NO to CO to produce a 'metal nitrene' intermediate.[120,121] Transfer of NO would then produce N_2O. This is now regarded as unlikely but cannot be totally excluded as a viable possibility. The second involves the coupling of the two NO groups before O abstraction takes place.[126] In the first step the initial tetrahedral cation or molecule rearranges to a planar system, formally $M(NO)(NO)L_2$. This formally 16 electron system then undergoes CO addition to give a new 18 electron complex $M(NO)(NO)(CO)L_2$. It has been argued that since the initial step involves a tetrahedral planar interconversion the ability of $[M(NO)(PPh_3)_2]^2$ to catalyze the reduction of NO to N_2O may be correlated with the extent of distortion shown by the catalyst.

The mechanism possibilities of these important reductions have been discussed in detail.[14]

13.3.2.3.5 *Reductive processes*

Common reducing agents have frequently been observed to attack nitrosyl ligands, liberating ammonia. A recently reported example is the formation of $Co(NO)(PPh_3)_3$ and NH_3 by the reaction of $[Co(NO)_2(PPh_3)_2]^+$ with sodium borohydride.[111] In certain cases the product of NO ligand reduction remains coordinated. Thus, the nitroprusside ion is converted to an ammino derivative $[Fe(CN)_5NH_3]^{3-}$ by the action of sodium amalgam.[112] Similarly, a reexamination[113] of the reduction of $CrCl(\pi-Cp)(NO)_2$ (preparation of $[Cr(\pi-Cp)(NO)_2]_2$) led to isolation of the by-product $Cr_2(NH_2)(\pi-Cp)_2(NO)_3$. X-Ray data has confirmed formulation of this complex as a dinuclear species with unsymmetrical bridging by one NO and one NH_2 group (Figure 12).[114] Reduction can also lead to nitride formation. This is observed in reaction of $[RuCl_5(NO)]^{2-}$ with $SnCl_2/HCl$ where the product isolated is the μ-nitrido complex $[Ru_2Cl_8N(H_2O)_2]^{3-}$.[115]

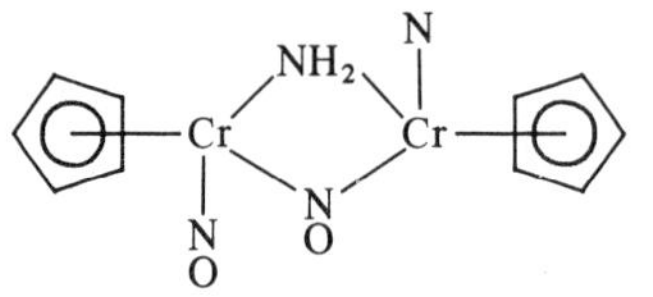

Figure 12

Nitrosyl groups have been observed to react with triphenylphosphine. In these reactions oxygen abstraction occurs and triphenylphosphine oxide, together with nitrous oxide or nitrogen, can be identified amongst the reaction products. For example, see equations (43) and (44).

$$2FeX(NO)_3 + 3PPh_3 \longrightarrow 2FeX(NO)_2PPh_3 + OPPh_3 + N_2O \qquad (43)^{116}$$

$$[CoCl(NO)_2]_2 + 6PPh_3 \xrightarrow{140\,°C} Co(NO)(PPh_3)_3 + CoCl_2(OPPh_3)_2 + OPPh_3 + \tfrac{3}{2}N_2 \qquad (44)$$

It is possible that nitric oxide may be displaced from the metal centre prior to oxidation of the phosphine. Indeed, such a nitrosyl displacement in the complex $CrCl(\pi-Cp)(NO)_2$, leading to derivatives of the type $CrCl(\pi-Cp)NO(L)$, has been reported.[118] However, such behaviour is rare (compared with CO displacements in metal carbonyls) and it seems likely that the above reduction takes place by an alternative process.

It has been proposed[119] that oxygen abstraction at a coordinated nitrosyl group could lead to formation of a metal nitrene*, $M—\ddot{N}:$. The attack of this highly reactive species at another nitrosyl group, or its combination with another nitrene, could result in the gaseous products identified (equations 45 and 46).

$$M—NO + PPh_3 \longrightarrow M—\ddot{N}: + OPPh_3 \qquad (45)$$

$$M—\ddot{N}: + M—NO + 2PPh_3 \longrightarrow 2M(PPh_3) + N_2O \qquad (46)$$

---

* Nitrenes $(R—\ddot{N}:)^{120}$ are well known in organic chemistry as reaction intermediates. They may be generated from nitroso complexes R—NO (organic analogues of metal nitrosyls) by attack at the oxygen atom which is followed by deoxygenation.[121] Several attacking agents have been used, including phosphines and phosphites.

Such a mechanism is also considered to operate in the photochemical reaction of Mo(π-Cp)(CO)$_2$(NO) with triphenylphosphine. Isolation of the isocyanate complex Mo(NCO)(π-Cp)CO(PPh$_3$)$_2$ {together with Mo(π-Cp)(CO)NO(PPh$_3$)} results from trapping of the metal nitrene by carbon monoxide in an intramolecular process.[122] Nitrene formation is also considered to participate in the process of conversion of Mo(NO)$_2$(S$_2$CNR$_2$)$_2$ to Mo(NO)(S$_2$CNR$_2$)$_3$ using triphenylphosphine.[119]

An especially interesting and important observation is the transformation of coordinated NO$_2^-$ to NH$_3$ on the complex [Ru(trpy)(cipy)L]$^{2+}$ (trpy = 2,2',2''-terpyridine and bipy = 2,2'-bipyridine) through the nitrosyl complex [Ru(trpy)NO]$^{3+}$. This models the reaction of the enzyme nitrite reductions.

13.3.2.3.6 *Nitric oxide insertion and other reactions involving coordinated ligands*

The insertion of nitric oxide into transition metal–alkyl bonds has been little studied.[123] Recently, however, an investigation of the reaction of NO with M(Me)$_2$(π-Cp)$_2$ (M = Zr or Ti) and ZrClMe(π-Cp)$_2$ has been undertaken. The products isolated from these reactions are of the type MX(π-Cp)$_2$(ON(Me)NO), X = Me or Cl, where the new monodentate ligand is considered to be derived from *N*-methyl-*N*-nitrosohydroxylamine (*i.e.* —ON(Me)NO).[124] Confirmation of this ligand geometry comes from X-ray diffraction studies on the related complex WMe$_4$(ON(Me)NO)$_2$ obtained by Wilkinson *et al.* from the reaction of NO with hexamethyltungsten. The structure of this complex has been described as intermediate between a square antiprism and a dodecahedron (Figure 13). Here *both* oxygen atoms of the ligand are found to be involved in coordination to the metal.[123] A possible mechanism for NO insertion into a metal–carbon σ-bond is shown in Scheme 2.

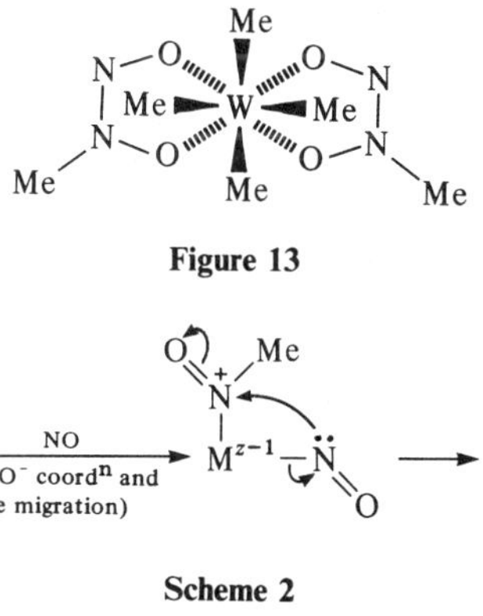

Figure 13

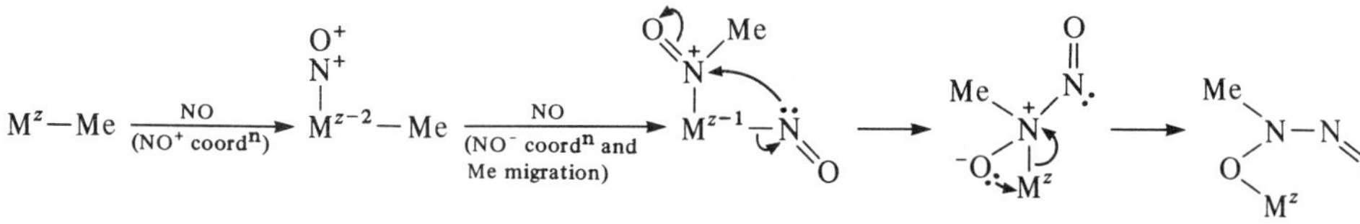

Scheme 2

The reaction of Co(π-Cp)(CO)$_2$ and of [Co(π-Cp)NO]$_2$ with nitric oxide in the presence of norbornene has been reported. In both cases the species shown in Figure 14 may be isolated in high yield.[125] The mechanism of these three component syntheses could well be related to that of the NO insertion reactions (Scheme 2) in that here NO insertion might occur into the metal–carbon π-bond of a cobalt–norbornene intermediate.

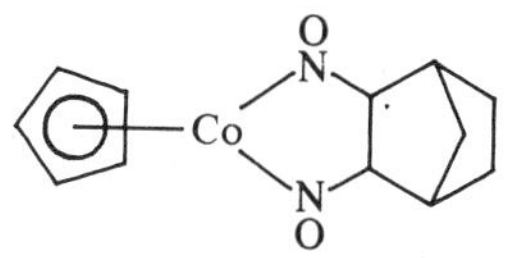

Figure 14

Nitric oxide also reacts with [Ru(NH$_3$)$_6$]$^{3+}$ giving a dinitrogen complex [RuN$_2$(NH$_3$)$_5$]$^{2+}$,[128] as shown in Scheme 3.

Finally, it has been proposed[129] that the nitrosation of [CoN$_3$(NH$_3$)$_5$]$^{2-}$ by NOClO$_4$ occurs through the addition of NO$^+$ at the azide group and leads to the intermediate [Co(NH$_3$)$_5$(NNNNO)]$^{3-}$. This rapidly decomposes to yield N$_2$, N$_2$O and [Co(NH$_3$)$_5$(H$_2$O)]$^{3-}$. Such behaviour is found to occur in other azido species and the same products are also obtained from the attack of azide on a coordinated NO group (as discussed under nucleophilic attack).

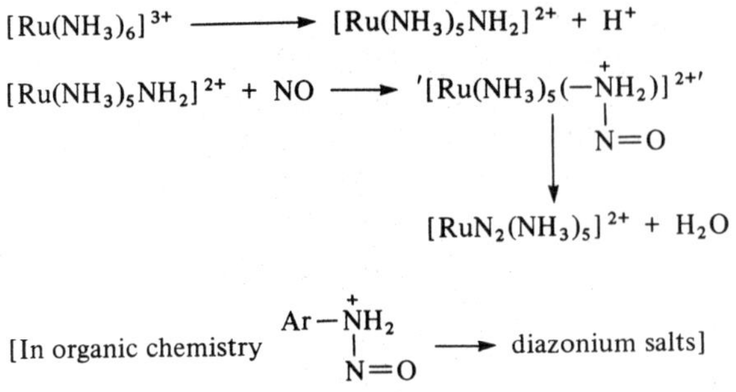

$$[Ru(NH_3)_6]^{3+} \longrightarrow [Ru(NH_3)_5NH_2]^{2+} + H^+$$

$$[Ru(NH_3)_5NH_2]^{2+} + NO \longrightarrow {}'[Ru(NH_3)_5(-\overset{+}{N}H_2)]^{2+\prime}$$

(with branch to)

$$\overset{|}{N}=O$$

$$\downarrow$$

$$[RuN_2(NH_3)_5]^{2+} + H_2O$$

[In organic chemistry $\quad Ar-\overset{+}{N}H_2 \underset{\overset{|}{N}=O}{} \longrightarrow$ diazonium salts]

Scheme 3

13.3.2.4 Nitrosyl Clusters

Two reviews devoted to this topic have appeared.[131,132] It is an area still in its infancy but one which shows considerable promise for the future. To date, emphasis has been placed largely on synthetic methods but increasingly the reactivity of these multimetal systems has come under investigation. In particular the ease by which NO undergoes bond cleavage to produce nitrido species has enabled workers in the field to provide a better understanding of the Haber process.

To some extent much of the previous discussion of synthetic routes and reactivity also applies here. There are, however, notable differences and these will be discussed in this section.

In polymetal (cluster) systems the nitrosyl ligand can adopt one of the several bonding modes. As with the monometal systems described above, when coordinating as a terminal ligand, NO can bind formally as a linear group or as a bent group. The former is by far the dominant mode and although examples of the latter have been proposed in intermediate complexes no structurally characterized example exists. Nitric oxide may also function as a bridging group (μ_2-NO). Here two classes of compound exist, *viz.* those containing an M—M bond (7) and those with no supporting M—M bond present (8). However, it should be realized that the existence or non-existence of M—M interactions is a controversial subject and based on simple electron-counting rules rather than reliable structural information.

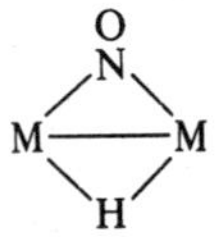

(7) 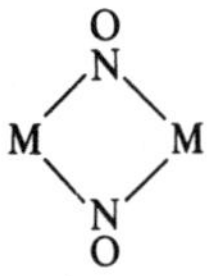   (8)

Several features emerge from a consideration of the N—O bond in its various bonding configurations. First, the N—O bond distance increases from terminal [1.17(2) Å] to doubly bridging types. Secondly, there is a sharp increase in the M—N bond length of *ca.* 0.25 Å in going from a terminal to bridging NO; and, thirdly, clusters with NO bridging two metals with an M—M bond have longer M—N distances by *ca.* 0.07 Å than those without supporting M—M interaction. The M—N—O for terminal systems usually falls around 176(2)°. Deviations do occur especially in compounds containing the M(NO)$_2$ grouping and in several examples *e.g.* [Fe$_4$S$_3$(NO)$_7$]$^-$, values $\sim 166°$ are found. This deviation is usually ascribed to a strong electronic coupling between two NO ligands.

The N—O stretching frequencies are useful for differentiating between the various NO bonding modes in polymetal systems. Terminal NO ligands absorb in the region 1664–1789 cm^{-1}, double bridging nitrosyl absorb in the range 1445–1603 cm^{-1}. Triply bridging NO absorbs at 1340 cm^{-1} in the well-documentated example (η^5-C$_5$H$_5$)$_3$Mn$_3$(NO)$_4$ (see Section 13.3.2.1.3).[127]

Frequently, because of the low intensity of bridging NO stretching frequencies, it can be difficult to observe and assign v_{NO}. This problem arises primarily because of the broad energy profile of the bridging group and can be overcome either by using isotopically enriched samples or by lowering the temperature.

^{15}N NMR spectroscopy has proved to be a reliable means of establishing bonding type.[134] In general, isotopically (^{15}NO) enriched samples are required but this does not usually pose an unduly difficult problem. ^{15}N chemical shift values fall in four general regions, as shown in Table 3.

Table 3 ^{15}N Chemical Shift Values for Different Nitrosyl Systems

Region (p.p.m.)	System
350–450	Terminal, linear, mononitrosyl
(All values downfield from NH$_3$(l), 25 °C)	
510–570	Terminal, linear, dinitrosyl
740–870	Terminal, bent, mononitrosyl
750–815	μ_2-Bridging group

Several points of interest arise from this data. First, dinitrosyl systems appear in a different region to the simpler mononitrosyl systems, indicating that a significant interaction between *cis* NO must occur. Secondly, complexes containing bent NO ligands are substantially deshielded because of the lone pair on the nitrogen. Thirdly, this latter region overlaps with that found for μ_2-bridged systems. Although this is inconvenient it appears to provide no real problems since — as mentioned above — bent NO groups are rare in multimetal systems.

13.3.2.4.1 Reactions with electrophiles

O methylation and O protonation of coordinated NO have been observed[135] for bridged systems.

$$[\eta^5\text{-C}_5\text{H}_5\text{Me}]_3\text{Mn}_3(\text{NO})_4 + \text{HBF}_4 \longrightarrow [\eta^5\text{-C}_5\text{H}_5\text{Me}]_3\text{Mn}_3(\text{NO})_3(\text{NOH})]^+ \tag{47}$$

$$[\text{Ru}_3(\text{CO})_{10}(\text{NO})]^- + \text{CF}_3\text{SO}_3\text{Me} \longrightarrow \text{Ru}_3(\text{CO})_{10}(\text{NOMe}) \tag{48}$$

$$[\text{Ru}_3(\text{CO})_{10}(\text{NO})]^- + \text{CF}_3\text{SO}_3\text{H} \longrightarrow \text{Ru}_3(\text{CO})_{10}(\text{NOH}) \tag{49}$$

In the examples given NO is bridging *three* (equation 47) or *two* metal atoms (equations 48 and 49) and its behaviour towards either Me$^+$ or H$^+$ is similar to that of CO bonded in a related fashion. In these instances the O atom of the bridging group is considered to be more nucleophilic as a result of enhanced back bonding to the metal — this is reflected in the change in v_{NO} (terminal, linear ~ 1800 cm^{-1}, μ_2 ~ 1500 cm^{-1}, μ_3 ~ 1300 cm^{-1}).

The protonation of [Ru$_3$(CO)$_{10}$(NO)]$^-$ shows a dependence on the acid.[131] Whilst CF$_3$SO$_3$H leads to the O-protonated product, CF$_3$CO$_2$H gives the species HRu$_3$(CO)$_{10}$(NO) in which H$^+$ addition has occurred at the metal unit. The two isomeric forms do not interconvert in inert solvents such as hexane, ether and CH$_2$Cl$_2$. However, addition of a catalytic amount of [CF$_3$CO$_2$]$^-$ causes the rapid O—H to M—H rearrangement.

The manganese compound [(η^5-C$_5$H$_5$Me)$_3$Mn$_3$(NO)$_3$(NOH)]$^+$ reacts with further acid to generate [(η^5-C$_5$H$_5$Me)$_3$Mn$_3$(NO)$_3$(NH)]$^+$ and water. The cleavage of the N—O bond and the concomitant formation of N—H bonds using the hydridic reagent Na[H$_2$Al(OCH$_2$CH$_2$OMe)$_2$] has also been observed for the dinuclear complex (η^5-C$_5$H$_5$)$_2$Cr$_2$(NO)$_4$ when the NH$_2$ ligand is produced in the complex (η^5-C$_5$H$_5$)$_2$Cr$_2$(NO)$_3$(NH$_2$).

13.3.2.4.2 Reduction with H$_2$

Molecular hydrogen will also reduce the NO group in certain polynuclear compounds.[136] A typical example is reaction (50).

$$\text{HRu}_3(\text{CO})_{10}(\text{NO}) + \text{H}_2 \xrightarrow{75\,°C} \text{HRu}_3(\text{NH})(\text{CO})_9 + \text{HRu}_3(\text{NH}_2)(\text{CO})_{10}$$
$$(10\%) \qquad (37\%)$$
$$+ \text{H}_4\text{Ru}_4(\text{CO})_{12} + \text{H}_3\text{Ru}_4(\text{NH}_2)(\text{CO})_{12} \tag{50}$$
$$(8\%) \qquad (\text{very small amount})$$

It has been suggested that one possible mechanism for this reduction involves the intermediacy of coordinated NOH. In support of this conclusion the reaction of Ru$_3$(NOMe)(CO)$_{10}$ and H$_2$ yields H$_2$Ru$_3$(NOMe)(CO)$_9$.

The same sequence of reactions takes place with the related cluster HOs$_3$(CO)$_{10}$(NO). In this case, however, higher temperatures (140 °C) are required.

13.3.2.4.3 Substitution reactions

The phosphine substitution of $Ru_3(CO)_{10}(NO)_2$ has been reported to proceed *via* a dissociative mechanism,[137] shown in equations (51) and (52).

$$Ru_3(CO)_{10}(NO)_2 \longrightarrow Ru_3(CO)_9(NO)_2 + CO \tag{51}$$

$$Ru_3(CO)_9(NO)_2 + PR_3 \longrightarrow Ru_3(CO)_9(PR_3)(NO)_2 \tag{52}$$

An osmium cluster of formula $Os_3(CO)_9(NO)_2$ undergoes reaction with donor ligands to two types of product which depend on the ligand employed.[138]

With relatively weak ligands such as NH_3 or $NHEt_2$ step 1 is reversible and the adducts have been identified solely on the basis of their IR spectra (ν_{NO} 1678 and 1617 cm^{-1} compared to 1731 and 1705 cm^{-1} in parent). With $P(OMe)_3$ or PPh_3 no simple adduct is observed. Here substitution occurs rapidly by what is assumed to be an associative step, followed by CO dissociation. Reaction with CO yields a metastable form of $Os_3(CO)_{10}(NO)_2$ which at 80 °C rearranges to $Os_3(CO)_{10}(\mu_2\text{-}NO)_2$. Finally reaction with pyridine gives the green cluster $Os_3(CO)_9(py)(NO)_2$. There is some spectroscopic evidence (IR and ^{13}NMR) to suggest that even toluene forms an unstable adduct with $Os_3(CO)_9(NO)_2$.

13.3.2.4.5 Formation of nitrido species

The removal of the O atom to form a nitride cluster is a relatively common reaction path for cluster nitrosyls. Several examples have been reported and in general the O atom is removed by reaction with CO to produce CO_2 (equations 53 and 54).

$$[Ru(CO)_3(NO)]^- + Ru_3(CO)_{12} \longrightarrow [Ru_4N(CO)_{12}]^- + CO_2 + CO \tag{53}{}^{[139]}$$

$$[Fe(CO)_3(NO)]^- + 3Fe(CO)_{12} \longrightarrow 2[Fe_4N(CO)_{12}]^- + 3Fe(CO)_5 + CO_2 + CO \tag{54}$$

Under the appropriate conditions O transfer to the cluster unit has also been observed (equation 55).[141]

$$M(\eta^5\text{-}C_5H_5)(CO)_2(NO) + [M'(\eta^5\text{-}C_5H_5)(CO)_3]_2 \longrightarrow MM'_2(\eta^5\text{-}C_5H_5)_3(N)(O)(CO)_4 \tag{55}$$

13.3.3 THIONITROSYL COMPLEXES

13.3.3.1 Background

The first thionitrosyl complex was discovered by chance during an attempted synthesis of molybdenum nitrido complexes in the presence of a source of sulfur, tetrathiuram sulfide.[168] This work of Dilworth and Chatt was reported in 1974 and ultimately led to the syntheses of a range of Mo, Re and Os thionitrosyl complexes starting with the respective nitrido complexes.[143,164] Nitrosyl complexes, like carbonyls, had been known for decades, and the more recent syntheses of thiocarbonyl complexes forebode the advent of thionitrosyl complexes. However, convenient synthetic routes based on analogies with syntheses of NO complexes were generally not available because the requisite precursors did not exist or were inconvenient to handle. Sulfiliminato complexes ($M{=}N{=}SR_2$) of transition metals are as yet unknown. The chemistry of thionitrosyl complexes was the subject of a recent review.[179]

13.3.3.2 Nomenclature

Fortunately, the —NS ligand has received only one name, thionitrosyl, which is used here. A radical name, rather than an anionic name, is used which follows the precedent set by the nitrosyl ligand itself.

13.3.3.3 Syntheses of Thionitrosyl Complexes

Most of the known thionitrosyl complexes have been prepared by one of two synthetic techniques. As always, each technique has limitations, and future expansion of this field of coordination chemistry will depend upon the development of new synthetic methods.

13.3.3.3.1 *Sulfur transfer to nitrido complexes*

This method has been used most widely, but it is limited to the availability of reactive nitrido complexes. The molybdenum nitrides simply abstract sulfur from episulfides ($MeCHCH_2S$) or elemental sulfur (equations 56 and 57).

$$MoNBr(diphos)_2 + \overset{S}{\triangle} \longrightarrow MoBr(NS)(diphos)_2 \qquad (56)^{143}$$

$$MoN(S_2CNMe_2)_3 + S_8 \longrightarrow Mo(NS)(S_2CNMe_2)_3 \qquad (57)^{164}$$

Ru, Os and Re nitrido complexes require more reactive sulfur transfer agents, such as S_2Cl_2 or SCl_2, which can also serve to oxidize the metal if used in excess (equations 58–60).

$$OsNCl_3(AsPh_3)_2 + S_2Cl_2 \longrightarrow OsCl_3(NS)(AsPh_3)_2 \qquad (58)^{164}$$

$$OsNCl_4^- + NCS^- \longrightarrow Os(NS)(NCS)_5^{2-} \qquad (59)^{145}$$

$$ReNCl_2(PEt_2Ph)_3 + SCl_2 \longrightarrow ReCl_3(NS)(PEt_2Ph)_2 \qquad (60)^{164}$$

13.3.3.3.2 *Thionitrosylation using thiazylhalides*

Solvated thiazylchloride trimer ($N_3S_3Cl_3$) acts as a source of NSCl and can be used to prepare thionitrosyl complexes. The use of thiazylchloride is limited because it can also function as a simple halogenating or oxidizing agent (equations 61 and 62).

$$CpCr(CO)_3^- + NSCl \longrightarrow CpCr(CO)_2(NS) \qquad (61)^{167}$$

$$RuCl_2(PPh_3)_3 + NSCl \longrightarrow RuCl_3(NS)(PPh_3)_2 \qquad (62)^{422}$$

It can also lead to condensed sulfur–nitrogen species and to NSCl complexes in the presence of metals.[175,176]

13.3.3.3.3 *Halide abstraction from ligated thiazylhalide*

There is only one example of this reaction (equation 63). The scarcity of thiazylhalide complexes will probably prevent this method from becoming widely used.

$$Re(CO)_5(NSF)^+ + AsF_5 \longrightarrow Re(CO)_5(NS)^{2+} + AsF_6^- \qquad (63)^{144}$$

13.3.3.3.4 *Substitution or addition reactions with NS$^+$*

There are not many examples of this reaction, and only mediocre yields were obtained (equations 64 and 65). Reactions of NS$^+$ with transition metal complexes certainly deserve more study.

$$ReBr(CO)_5 + NSSbF_6 \longrightarrow Re(CO)_5(NS)^{2+} \qquad (64)^{144}$$

$$(C_6H_6)Cr(CO)_3 + NSPF_6 \xrightarrow{\text{MeCN}} Cr(MeCN)_5(NS)^{2+} \qquad (65)^{154}$$

A number of metal complexes prepared from S_4N_4, $S_4N_4H_4$, $S_3N_3Cl_3$ or S_2N_2 have been reported. Many of these compounds have S to N ratios of unity and are sometimes called 'thionitrosyl' complexes. With the exception of thiazylchloride, however, these reagents rarely yield ordinary monomeric thionitrosyl ligands. Usually, complexes are produced which contain $S_2N_2H^-$, $S_2N_2^{2-}$ or cyclo-S_2N_2 groups (equations 66–69). It remains to be seen whether S_4N_4 and related species can be used as convenient sources of the —N=S ligand. One recent example has been reported (equation 70).

$$Pt(PPh_3)_3 + S_4N_4H_4 \longrightarrow Pt(S_2N_2)(PPh_3)_2 \qquad (66)^{169}$$

$$Ni(CO)_4 + S_4N_4 \longrightarrow Ni(HS_2N_2)_2 \qquad (67)^{170,171}$$

$$RuCl_2(CO)_2(PPh_3)_2 + S_2N_2 \cdot 2AlCl_3 \longrightarrow RuCl(N_2S_2)(CO)_2(PPh_3)_2^+ \qquad (68)^{173}$$

$$H_2PtCl_6 + N_4S_4 \longrightarrow Pt(HS_2N_2)_2 \qquad (69)^{172}$$

$$ReCl_5 + S_4N_4 + Ph_4AsCl \longrightarrow [Re(NS)Cl_5][Ph_4As] \qquad (70)^{430}$$

13.3.3.4 Structure and Bonding

The structures of several thionitrosyl complexes are known and tabulated in Table 4. All known structures possess linear —NS ligands with short metal nitrogen bond distances indicative of multiple metal–nitrogen bonding (bond orders in the range 2.0–2.5). The N—S bond lengths in the thionitrosyl complexes are in the range 1.51–1.59 Å, which is indicative of partial multiple bonding (N=S ≅ 1.52 Å). It is noteworthy that the Mo—N distances in $Mo(NS)(S_2CNMe_2)_3$, $[Mo(NCPh_3)(S_2CNMe_2)_3]^+$ and $Mo(NO)(S_2CNBu_2)_3$[174] are the same within ±0.01 Å. Correcting for the larger size (0.11 Å longer) of Mo with respect to Cr, the metal–nitrogen bond length in $Mo(NS)(S_2CNMe_2)_3$ is 0.06 Å shorter than that in $CpCr(CO)_2(NS)$; this correlates with a longer N—S distance (0.04 Å longer) in the molybdenum thionitrosyl complex. In the structures where it can be observed, the thionitrosyl ligand exerts little, if any, *trans* influence. Bent and bridging NS ligands (**9, 12, 13**) have not been structurally characterized. The two structures which have *cis* —NS and —NSCl ligands, $ReCl_4(NS)(NSCl)^{2-}$ and $OsCl_4(NS)(NSCl)^-$, have unusually long M—N distances (~1.84 Å) for the thionitrosyl ligand.[180,375]

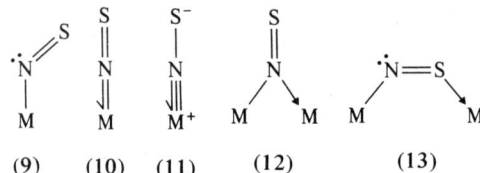

(9)   (10)   (11)   (12)   (13)

Table 4 Structures of Thionitrosyl Complexes

Compound[c]	M—N (Å)	N—S (Å)	M—N—S (°)	Ref.
$CpCr(CO)_2(NS)$	1.694(2)	1.551(2)	1.76.8(1)	163, 167
$Mo(NS)(S_2CNMe_2)_3$[a]	1.738(11)	1.592(11)	172.0(7)	166
$[Ru(NS)Cl_4(H_2O)][Ph_4P]$	1.729(4)	1.504(4)	170.9(3)	201
$[Os(NS)Cl_4(H_2O)][Ph_4P]$	1.731(4)	1.514(5)	174.9(3)	421
$OsCl_3(NS)(PPh_3)_2$	1.779(9)	1.503(10)	180	422
$TcCl_2(NS)(S_2CNEt_2)_2$[b]	1.75	1.52	178	423

[a] Compare with $[Mo(NSO_2Ph)(S_2CNMe_2)_3][PF_6]$, Mo—N = 1.70(2) Å, N—S = 1.71(2) Å, Mo—N —S = 161(1)°; see M. W. Bishop, J. Chatt, J. R. Dilworth, B. D. Neaves, P. Dahlstrom, J. Hyde and J. Zubieta, *J. Organomet. Chem.*, 1981, **213**, 109.
[b] Partially disordered.
[c] Abbreviations, see footnote in Table 6.

13.3.3.5 Spectroscopic Studies

Infrared spectroscopy has been used to great advantage in identifying and characterizing thionitrosyl complexes. All —NS complexes display a strong infrared active band which is associated with $\nu(NS)$. Isotope labeling (^{15}N) of the thionitrosyl ligand has not been carried out. The N—S stretching frequencies occur in the 1050–1400 cm^{-1} region; cationic complexes exhibit bands in the upper part and anionic ones in the lower part of this region. As with nitrosyl complexes, we would expect linear —NS complexes of Mo and Re to have lower values of $\nu(NS)$ than analogous complexes of Os and Ir. Likewise we would expected to observe $\nu(NS)$ for bent —NS ligands at lower energy than for linear —NS ligands given the same metal and similar ligands. A comparison of $v(NO)$ and $v(NS)$ frequencies (Table 5) for analogous complexes shows that the $v(NO)/v(NS)$ ratio falls in the range of 1.40–1.47 for most complexes. Thus, it is reasonable to expect the respective pairs of complexes to be isostructural. The values of $v(CO)$ in thionitrosyl complexes are nearly identical with those found in analogous nitrosyl complexes. This suggests that the thionitrosyl ligand like the nitrosyl ligand is a good π acceptor.[431]

Table 5 Comparison of N—O and N—S Stretching Frequencies in Nitrosyl and Thionitrosyl Complexes

Compound[b]	$v(NS)$ (cm^{-1})	$v(NO)$ (cm^{-1})	$v(NO)/v(NS)$
RuCl$_3$(NX)(PPh$_3$)$_2$	1295–1310	1875	1.440
OsCl$_3$(NX)(PPh$_3$)$_2$	1290–1310	1850	1.423
[OsCl$_5$(NX)]$^-$	1340	1870	1.396
HB(Me$_2$pz)$_3$Mo(CO)$_2$(NX)	1125–1136	1650	1.460
	(2003, 1924)[a]	(1997, 1905)[a]	
HB(Me$_2$pz)$_3$W(CO)$_2$(NX)	1127–1140	1640	1.446
	(1988, 1902)[a]	(1993, 1884)[a]	
ReCl$_3$(NX)(PEt$_2$Ph)$_2$	1230	1735	1.411
ReCl$_2$(NX)(PEt$_2$Ph)$_3$	1167	1680	1.440
CpCr(CO)$_2$(NX)	1180	1713	1.451
	(2033, 1962)[a]	(2028, 1961)[a]	
[Cr(Bu^{t}NC)$_5$(NX)]$^+$	1135	1673	1.474
[Cr(MeCN)$_5$(NX)]$^{2+}$	1245	1797	1.443
MoBr(NX)(diphos)$_2$	1065	1550	1.455

[a] Values of $v(CO)$.
[b] Abbreviations, see footnote in Table 6.

Photoelectron spectra of the core and valence electrons for CpCr(CO)$_2$(NS) have been measured and compared with the spectra for CpCr(CO)$_2$(NO).[149,158,160,429] The changes in the energies when O is replaced by S have been interpreted to suggest that —NS is a better π acceptor and a better σ donor.[424,431] The core binding energy of the electrons at the Cr change very little and so do the C—O stretching frequencies of the carbonyl ligands. There is one report of electrochemical studies on thionitrosyl complex, Mo(NS)(S$_2$CNMe$_2$)$_3$.[165] Electron impact measurements at 70 eV suggest that the thionitrosyl ligand is more tightly bound in CpCr(CO)$_2$(NS) than is the nitrosyl ligand in CpCr(CO)$_2$(NO).[167]

Both ^{95}Mo and ^{14}N NMR spectra have been measured for some thionitrosyl complexes.[424,427] When compared to nitrosyl analogues, the thionitrosyl caused significant downfield shifts in both the Mo and N spectra.

13.3.3.6 Reactions of Thionitrosyl Ligands

There are several reports of thionitrosyl ligands undergoing reactions. In those case where the —NS ligand can be formed from nitrides, phosphines will remove the sulfur to reform the original nitride (equations 71 and 72). There are two reports of thionitrosyl ligands being displaced by nitrosyl ligands from nitrosyl halides (equation 73–75).[159,167] There is one report of an S alkylation of a thionitrosyl ligand,[164] and there is another report of the oxidation of —N═S to form —N═S═O (sulfinylamide, thiazate) as shown in equations 74 and 75 respectively.[148]

$$RuCl_3(NS)(PEt_2Ph)_2 + PEt_2Ph \longrightarrow RuNCl_3(PEt_2Ph)_2 + S{=}PEt_2Ph \qquad (71)^{164}$$

$$Mo(NS)(S_2CNMe_2)_3 + PBu_3 \longrightarrow MoN(S_2CNMe_2)_3 + S{=}PBu_3 \qquad (72)^{164}$$

$$CpCr(CO)_2(NS) + ONCl \longrightarrow CpCr(NO)_2Cl \qquad (73)^{167}$$

$$Mo(NS)(S_2CNMe_2)_3 + Ph_3C^+ \longrightarrow Mo(NSCPh_3)(S_2CNMe_2)_3^+ \qquad (74)^{164}$$

$$IrCl_2(CO)(NS)(PPh_3)_2 + O_2 \longrightarrow IrCl_2(NSO)(CO)(PPh_3)_2 \qquad (75)^{148}$$

13.3.3.7 Survey of Thionitrosyl Complexes

Table 6 gives a listing of thionitrosyl complexes. There are no complexes from the titanium and vanadium groups, largely because adequate synthetic methods have not been developed. Nitrosyl complexes of these metals are also rare, primarily because they are so oxophilic. It is possible that the reduced thiophilicity may allow thionitrosyl complexes to be prepared. Bis(thionitrosyl) complexes are not known for any metal. The coupling of thionitrosyl ligands on a metal and bonding of metals to the sulfur end of the —NS ligand present new opportunities for future work.

Table 6 Listing of Thionitrosyl Complexes

Complex[a]	Ref.	Complex[a]	Ref.
$[Cr(NS)(MeCN)_5]^{2+}$	154	$[Cr(NS)(Bu^tNC)_5]^{2+}$	154
$[Cr(NS)(Bu^tNC)_5]^+$	154	$CpCr(CO)_2(NS)$	149, 158, 160, 162, 152, 167, 424, 429
$[CpCr(CO)(NO)(NS)]^+$	167		
$Mo(NS)(S_2CNR_2)_3$	143, 161, 164–166, 168, 427	$MoX(NS)(diphos)_2$	147
$HB(Me_2pz)_3Mo(CO)_2(NS)$	424, 425		
$WX(NS)(diphos)_2$	147	$HB(Me_2pz)_3W(CO)_2(NS)$	424, 425
$CpMn(CO)_2(NS)^+$	177		
$TcCl_2(NS)(S_2CNEt_2)_2$	423	$TcCl_2(NS)(PMe_2Ph)_3$	178
$TcCl_3(NS)(PMe_2Ph)_2$	178		
$[Re(NS)(CO)_5]^{2+}$	144, 177	$ReX_2(NS)(PR_3)_3$	142, 164
$[ReX(NS)(diphos)_2]^+$	142, 164	$ReX_3(NS)(PR_3)_2$	164
$ReX(NS)(S_2CNR_2)(PR_3)_2$	164	$ReCl_5(NS)^-$	430
$ReCl_4(NS)(NSCl)^{2-}$	430		
$CpFe(CO)_2(NS)^{2+}$	177		
$RuX_3(NS)(PR_3)_2$	159, 181, 422, 426	$RuX_3(NS)(AsPh_3)_2$	159, 181
$RuCl_5(NS)^{2-}$	201, 426	$RuCl_4(NS)(H_2O)^-$	201, 426
$[Os(NS)(NCS)_5]^{2-}$	145	$OsX_3(NS)(bpy)$	164
$OsX_3(NS)(py)_2$	164	$OsX_3(NS)(PR_3)_2$	142, 159, 164, 422, 426
$OsX_3(NS)(AsPh_3)_2$	142, 159, 164, 422, 426	$OsCl_4(NS)(NSCl)^-$	180
$OsCl_4(NS)(H_2O)^-$	421, 426	$OsCl_5(NS)^{2-}$	421, 426
$OsCl_5(NS)^-$	200		
$RhCl_2(NS)(PPh_3)_2$	146, 151–153, 199, 428	$RhCl_2(NS)(AsPh_3)_2$	151, 152, 428
$RhCl_2(CO)(NS)(PPh_3)_2$	150, 153, 428	$RhCl_2(CO)(NS)(AsPh_3)_2$	150, 153, 428
$Rh(CO)(NS)(PPh_3)_2$	150, 428	$[RhCl_2(NS)(PPh_3)]_2$	145, 151, 152, 199
$[RhCl_2(NS)(AsPh_3)]_2$	151, 152	$[RhCl_2(CO)(NS)(PPh_3)]_2$	150
$[RhCl_2(CO)(NS)(AsPh_3)]_2$	150		
$IrCl_2(CO)(NS)(PPh_3)_2$	148	$IrCl_2(NS)(PPh_3)_2$	199

[a] Abbreviations/ R = alkyl or aryl, diphos = $Ph_2PCH_2CH_2PPh_2$, $HB(Me_2pz)_3$ = tris(3,5-dimethylpyrazolyl)borate.

13.3.4 PHOSPHINIMINATO COMPLEXES

13.3.4.1 Background

The first phosphiniminato (—N=PR$_3$) complexes were prepared in 1973 by Pawson and Griffith during their studies with osmium nitrido complexes.[192] Since that time, a number of new synthetic routes to these compounds have been devised. It was early recognized that —N=PR$_3$ could be a three-electron donor like —N=O and that similarities between phosphiniminato and nitrosyl complexes might be expected. Yet the =PR$_3$ moiety is much less electronegative than =O, and σ and π donation to the metal would become more important for the phosphiniminato ligand. While $OsCl_3(NO)(PPh_3)_2$ is diamagnetic (indicative of Os^{II}), $OsCl_3(NPPh_3)(PPh_3)_2$ is paramagnetic (indicative of Os^{IV}).[183] The trivalent phosphorus analogue, —N=PR (phosphazenido), has not been reported as a ligand to a transition metal.

13.3.4.2 Nomenclature

The —N=PPh$_3$ ligand has been named *P,P,P*-triphenylphosphineiminato. The endings phosphaneiminato, phosphineiminido and phosphineimidato are also used. In addition, these names have been (a) spelled without the letter 'e', (b) written as two words and (c) hyphenated, for example, phosphiniminato, phosphine iminato and phosphine-iminato. The name 'phosphiniminato' will be used here.

13.3.4.3 Syntheses of Phosphiniminato Complexes

Several synthetic techniques for the preparation of phosphiniminato complexes have been devised. It is difficult to know if any of these routes possesses wide applicability. The most obvious method of synthesis, use of alkali metal salts of phosphinimides {[Li][NPPh$_3$]}, has not led to generally useful results.[190]

13.3.4.3.1 Reaction of phosphines with metal azides

This method has been used to prepare Ti, Nb and Ta complexes (equations 76 and 77). Sometimes unisolable intermediates have been observed in these reactions. In some cases, the reactions must be photolyzed in order to promote the reaction. One arsiniminato complex [Ph$_4$As][NbCl$_5$(NAsPh$_3$)] has been prepared *via* this route.[194]

$$NbCl_4N_3 + PPh_3 \longrightarrow [NbCl_4(NPPh_3)]_2 + N_2 \qquad (76)^{191}$$

$$[Ph_4P][TaCl_5N_3] + PPh_3 \xrightarrow{h\nu} [Ph_4P][TaCl_5(NPPh_3)] + N_2 \qquad (77)^{195}$$

13.3.4.3.2 Reaction of phosphines with metal nitrides

Most nitrido ligands are not sufficiently electrophilic; however, when this method works, it is particularly convenient (equation 78). While the Re^{VII} nitride, ReNCl$_4$, forms a phosphiminato ligand with triphenylphosphine, Re^{VI} and Re^{V} nitrides (ReNCl$_3$(PPh$_3$)$_2$ and ReNCl$_2$(PPh$_3$)$_2$) will not form these complexes (equation 79).[193] Thionitrosyl and thiazylchloride complexes can sometimes act as sources of incipient nitrido complexes (equation 80).

$$OsNCl_3(PPh_3)_2 + PPh_3 \longrightarrow OsCl_3(NPPh_3)(PPh_3)_2 \qquad (78)^{183}$$

$$ReNCl_4 + PPh_3 \longrightarrow ReCl_4(NPPh_3)(PPh_3) \qquad (79)^{193}$$

$$MoCl_4(NSCl) + PPh_3 \longrightarrow MoCl_4(NPPh_3)(PPh_3) \qquad (80)^{445}$$

13.3.4.3.3 Halide–phosphinimide exchange with N-silylphosphinimines

This reaction may be the most generally useful of all. It seems to work best with acidic halide ligands attached to early transition metals, but the scope and limitations of this reaction are not yet known (equations 81 and 82).

$$CpTiCl_3 + Me_3SiNPPh_3 \longrightarrow CpTiCl_2(NPPh_3) + Me_3SiCl \qquad (81)^{189}$$

$$VOCl_3 + 2Me_3SiNPPh_3 \longrightarrow VOCl(NPPh_3)_2 + 2Me_3SiCl \qquad (82)^{190}$$

13.3.4.3.4 Reaction of tosylazide with phosphine complexes

There is only one example of this reaction (equation 83), but it deserves further study.

$$MoCl_4(PR_3)_2 + ArSO_2N_3 \longrightarrow MoCl_4(NPR_3)(OPR_3) + \tfrac{1}{2}ArSSO_2Ar + N_2 \qquad (83)^{187}$$

13.3.4.3.5 Reaction of phosphines with nitrosyl complexes

There are only two examples of this reaction, and it is probably limited to a small group of reactive nitrosyl complexes of early transition metals (equations 84 and 85).

$$\text{MoCl}_3(\text{NO}) + 2\text{PR}_3 \longrightarrow \text{MoCl}_4(\text{NPR}_3)(\text{OPR}_3) \qquad (84)^{186}$$

$$\text{ReCl}_3(\text{NO})_2 + 2\text{PPh}_3 \longrightarrow \text{ReCl}_3(\text{NO})(\text{NPPh}_3)(\text{OPPh}_3) \qquad (85)^{444}$$

13.3.4.4 Structure and Bonding

Six full structural reports dealing with phosphiniminato complexes have been published along with partial details of one other structure (see Table 7). Generally, the metal nitrogen bonds are short, indicative of multiple bonding, and the M—N—P angles are nearly linear (163°–175°). The short P—N bonds in the range 1.59–1.63 Å are also indicative of multiple bonding (P=N $\cong$ 1.56 Å, P—N $\cong$ 1.77 Å). The two bis(phosphiniminato) structures show significant bending at nitrogen (M—N—P = 157°–162°) and somewhat longer M—N bond lengths. Here, competitive π-donation to the metal by the two phosphiniminato ligands is evident. There is structural evidence for a bent phosphiniminato ligand (14) in ReCl₃(NO)(NPPh₃)(OPPh₃) in which Re—N—P is 139°. The more common ligand geometry is linear, as expected for formalisms (15) and (16).

(14)   (15)   (16)

Table 7 Structures of Phosphiniminato Complexes

Complex[a]	M—N (Å)	N—P (Å)	M—N—P (°)	Ref.
[NbCl₄(NPPh₃)]₂	1.776(8)	1.637(9)	171.1(6)	191
[TaCl₄(NPPh₃)]₂	1.801(8)	1.593(9)	176.8(7)	196
[TaCl₄(NPPh₃)₂]⁻	1.97	1.56	162	191
WF₄(NPPh₃)₂	1.825(6)	1.594(6)	157.2(4)	197
Re(NPPh₃)(SPh)₄	1.743(7)	1.634(9)	163.1(6)	448
ReCl₃(NO)(NPPh₃)(OPPh₃)	1.855(8)	1.630(8)	138.5(5)	444
RuCl₃(NPEt₂Ph)(PEt₂Ph)₂	1.841(3)	1.586(3)	174.9(3)	184, 185

[a] Abbreviations, see footnote in Table 8.

Although a careful analysis of the structural aspects of bonding must await the accumulation of more data, several observations can be made. The metal—nitrogen bond lengths cannot be correlated with those in imido complexes (M≌NR), nor with those in diazenido complexes (M⇐N=NR), but they can be correlated with those in nitrido complexes (M≡N). This may reflect the ionic and dipolar nature of the P—N bond in phosphiniminato ligands. The nature of the bonding in RuCl₃(NPEt₂Ph)(PEt₂Ph)₂ is fundamentally different from that in [MCl₄(NPPh₃)]₂ (M = Nb, Ta) even though both phosphiniminato ligands are linear. It is likely that the steric and electronic effects of the ancillary ligands and the nature of the metal can significantly alter the structural properties of the —NPR₃ ligand. After correction for metal size, the Nb—N distance in NbCl₄(NPPh₃) is 0.15 Å shorter than the Ru—N distance in RuCl₃(NPR₃)(PR₃)₂; this change in distance translates into a change in bond order of about 0.9–1.0 units. There is no evidence for a *trans* influence in RuCl₃(NPEt₂Ph)(PEt₂Ph)₂. The unsymmetric chlorine bridges in [NbCl₄(NPPh₃)]₂ and the Ta analogue indicate a possible *trans* influence there.

The paramagnetism of RuCl₃(NPR₃)(PR₃)₂ and its Os analogues,[183] in contrast to the diamagnetism of RuCl₃(NO)(PR₃)₂, RuCl₃(NS)(PR₃)₂ and RuCl₃(NNAr)(PR₃)₂, shows that the phosphiniminato ligand is a poor π-acceptor ligand. This coupled with the linearity of the Ru—N—P moiety suggests that phosphiniminito ligands should be regarded as a π donor like alkoxy (—OR) and amide (—NR₂) ligands. Phosphiniminato ligands can serve as one-electron donors (14) and as three-electron donors (15).

13.3.4.5 Infrared Spectroscopy

The only spectroscopic technique that has been applied to phosphiniminato complexes is infrared spectroscopy. Virtually all —NPR₃ ligands give rise to a strong absorption in the 1050–1200 cm⁻¹

region, with the majority between 1100 and 1150 cm^{-1}. This band has been assigned to $\nu(PN)$ and has been shown to shift upon ^{15}N isotopic labeling; in $MoCl_4(NPPh_2Pr^n)(OPPh_2Pr^n)$ the 1128 cm^{-1} band shifts to 1120 cm^{-1}, and in $OsCl_3(NPPh_3)(PPh_3)_2$ the 1127 cm^{-1} band shifts to 1104 cm^{-1}.[183]

13.3.4.6 Reactions of Phosphiniminato Ligands

There are a few reports where coordinated —NPR_3 undergoes a reaction. The phosphiniminato ligand can be protonated to form a phosphinimine ligand; this ligand can be further protonated and removed from the metal.[183] Complexes of the type $MX_3(NPR_3)(PR_3)_2$ (M = Ru, Os) react with chlorine to generate nitrido complexes in good yields (Scheme 4).[183] There is some evidence that —NPR_3 ligands can suffer hydrolytic cleavage to form amido complexes (equation 86).[186]

$$OsCl_4(HNPPh_3)(PPh_3) \xleftarrow[-PPh_3]{HCl} OsCl_3(NPPh_3)(PPh_3)_2 \xrightarrow[-[H_2NPPh_3][Cl]]{HCl} OsCl_4(PPh_3)_2$$

$$\downarrow {}^{-OPPh_3} \big| Cl_2$$

$$OsNCl_3(PPh_3)_2$$

Scheme 4

$$MoCl_4(NPMePh_2)(OPMePh_2) \xrightarrow{H_2O} [MoCl_3(NH_2)(OPMePh_2)_2]Cl \qquad (86)$$

13.3.4.7 Survey of Phosphiniminato Complexes

Table 8 gives a listing of phosphiniminato complexes. The most common complexes are those of V, Nb and Ta, and all complexes but $MX_3(NPR_3)(PR_3)_2$ (M = Ru, Os) contain metals from the first half of the transition series (Ti group → Mn group). The azophilic metals of the Ti and V groups readily bind to the phosphiniminato ligand; owing to the shortage of valence electrons, these metals do not readily form discrete nitrido complexes. Metals of the Cr and Mn groups also readily bind phosphiniminato ligands, but they also form nitrido complexes that are often more stable.

Table 8 Listing of Phosphiniminato Complexes

Complex[a]	Ref.	Complex[a]	Ref.
$TiCl_2(NPPh_3)_2$	49	$TiCl(NPPh_3)(S_2CNEt_2)_2$	190
$CpTiCl_2(NPR_3)$	48, 305		
$VOCl(NPPh_3)_2$	49	$VCl_3(NPPh_3)_2$	190
$VOCl_2(NPPh_3)$	49		
$[NbCl_4(NPPh_3)]_2$	50	$[NbX_5(NPPh_3)]^-$	194, 195
$[NbCl_5(NAsPPh_3)]^-$	53		
$[TaCl_4(NPPh_3)_2]^-$	50	$[TaX_5(NPPh_3)]^-$	194, 195
$[TaCl_4(NPPh_3)_2]$	55	$TaCl_4(NPPh_3)(HNPPh_3)$	196
$MoCl_4(NPR_3)(OPR_3)$	45, 46, 47	$MoCl_4(NPPh_3)(PPh_3)$	445
$MoCl_4(NPPh_3)(PPh_3)^-$	304	$[MoCl_5(NPCl_3)]^-$	445
$WF_5(NPPh)_3$	56	$WF_4(NPPh_3)_2$	197
$Re(NPPh_3)(SPh)_4$	307	$ReCl_3(NO)(NPPh_3)(OPPh_3)$	444
$ReCl_4(NPPh_3)(PPh_3)$	52, 306		
$RuCl_3(NPR_3)(PR_3)_2$	41, 42, 43, 44		
$OsX_3(NPR_3)(PR_3)_2$	4, 41, 42, 51	$OsCl_3(NPPh_3)(py)_2$	145
$OsCl_3(NPPh_3)(py)(PPh_3)$	4		

[a] Abbreviations: R = alkyl or aryl.

13.3.5 METHYLENEAMIDO COMPLEXES

13.3.5.1 Background

The first methyleneamido (—$N{=}CR_2$) complex, $Fe_2(CO)_6(NCR_2)_2$, was produced in low yield from the reaction of an azine with $Fe_2(CO)_9$.[239] Subsequent work by Kilner and coworkers[212,216] on $CpMo(CO)_2(NCR_2)$ and related compounds was the beginning of a series of new methyleneamido complexes. Using ketimide salts or *N*-stannyl- or *N*-silyl-ketimines, a variety of new methylene-amido complexes have been prepared. The placement of electron withdrawing groups on carbon lead to the —N=C(CF$_3$)$_2$ ligand which was a better π acceptor and hence a more versatile ligand.[207] Methyleneamido ligands were briefly reviewed in 1972.[235]

13.3.5.2 Nomenclature

The —N=CR$_2$ ligand has been most commonly called alkylideneamido or ketimido. Alternate names such as alkaneiminato, alkylideneimido, alkylideneamino, azomethine and aldimido also appear in literature. The name methyleneamido will be used throughout this review.

13.3.5.3 Synthesis of Methyleneamido Complexes

Several synthetic methods for the synthesis of methyleneamido complexes have been developed. The only method which shows any degree of generality is the use of metalloketimines. The other synthetic methods listed below have not been studied in depth.

13.3.5.3.1 Halide–methyleneamide exchange with N-metalloketimines

The majority of methyleneamido complexes have been prepared using the Li, Si or Sn derivatives of ketimines, M—N=CR$_2$ (equations 87–90). The lithium derivatives have some disadvantages; they have poor solubility in some non-polar organic solvents and are sufficiently nucleophilic to lead to unwanted reaction by-products in some cases. The nature of the organic groups on carbon have some bearing on the outcome of the reactions; sterically bulky groups (—CMe$_3$) and electron withdrawing groups (—CF$_3$) have been used to advantage.

$$CpW(CO)_3Cl + LiN{=}CBu^t_2 \longrightarrow CpW(CO)_2(NCBu^t_2) + LiCl + CO \qquad (87)^{[213]}$$

$$CpMo(CO)_3Cl + Me_3SiN{=}CPh_2 \longrightarrow CpMo(CO)_2(NCPh_2) + Me_3SiCl \qquad (88)^{[216]}$$

$$PtCl_2(PPh_3)_2 + Me_3SnN{=}C(CF_3)_2 \longrightarrow PtCl(NC(CF_3)_2)(PPh_3)_2 + Me_3SnCl \qquad (89)^{[222]}$$

$$MnBr(CO)_5 + Me_3SnN{=}C(CF_3)_2 \longrightarrow \tfrac{1}{2}Mn_2(CO)_7(NC(CF_3)_2)_2 + CO + Me_3SnBr \qquad (90)^{[230]}$$

A variation of this reaction involves the oxidative addition of *N*-stannylketimines to low valent complexes (equation 91).

$$Pt(PPh_3)_4 + Me_3SnN{=}C(CF_3)_2 \longrightarrow Pt(SnMe_3)(NC(CF_3)_2)(PPh_3)_2 \qquad (91)^{[218]}$$

13.3.5.3.2 Addition of metal hydride or metal alkyl to a nitrile

In certain early transition metal complexes and in metal clusters, organic nitriles can insert into metal–hydrogen or metal–carbon bonds (equations 92 and 93). Similarly, nucleophilic attack at coordinated nitrile can yield methyleneamido complexes (equation 94).

$$Cp_2ZrHCl + RC{\equiv}N \longrightarrow Cp_2ZrCl(NCHR) \qquad (92)^{[203]}$$

$$Cp_2TiR + MeC{\equiv}N \longrightarrow Cp_2Ti(NCRMe) \qquad (93)^{[234]}$$

$$Fe_3(NCR)(CO)_9 + BH_4^- \longrightarrow Fe(NCHR)(CO)_9^- \qquad (94)^{[431]}$$

13.3.5.3.3 Reaction of ketimine with metal alkyls

The protolysis of a basic metal–alkyl by the acidic proton in ketimines can lead to methylene-amido complexes (equation 95).

$$(cod)PtPr^i_2 + 2HN{=}C(CF_3)_2 \longrightarrow (cod)Pt\{NC(CF_3)_2\}_2 + 2C_3H_8 \qquad (95)^{[223]}$$

13.3.5.3.4 Condensation of ketone with metal amide

There is one example of an inorganic analogue of ketimine formation (equation 96). If requisite, primary amido complexes were available, this could be a useful synthetic method.

$$\{HB(Me_2pz)_3\}MoI(NO)(NH_2) + Me_2CO \longrightarrow \{HB(Me_2pz)_3\}MoI(NO)(NCMe_2) + H_2O \quad (96)^{227}$$

13.3.5.3.5 Reductive coupling of nitriles

A two metal, two-electron (not two metal, four-electron) reductive coupling of organic nitriles can lead to dimeric methyleneamido complexes (equation 97).

$$2Cp_2TiCl + 2PhC\equiv N \longrightarrow Cp_2ClTiNCPh{-}CPhNTiCp_2Cl \quad (97)^{233}$$

13.3.5.3.6 Oxidation of primary amine complexes

Carefully controlled oxidation of ligated primary amines has been used to prepare methylene-amido complexes (equation 98).

$$Ru(tpy)(bipy)(NH_2CHMe_2)^{2+} + 4Ce^{4+} \longrightarrow Ru(tpy)(bipy)(NCMe_2)^{3+} + 4Ce^{3+} \quad (98)^{231}$$

13.3.5.3.7 Deprotonation of alkylimido complexes

The single known example of a deprotonation of an alkylamido ligand suggests that a potentially convenient synthetic pathway may not be generally useful owing to the low acidity of the aliphatic C—H bonds (equation 99).

$$ReCl_3(NMe)(PMe_2Ph)_2 + PMe_2Ph \xrightarrow{Et_3N} ReCl_2(NCH_2)(PMe_2Ph)_3 + Et_3NHCl \quad (99)^{205}$$

13.3.5.3.8 Other methods

Azines and 2-bromo-2-nitrosoalkanes have been used to produce methyleneamido complexes from low valent carbonyl complexes; yields were mediocre and by-products were generated (equations 100 and 101). Other attempts to use azines or *N*-bromoketimines have failed to give useful products.[216]

$$Fe_2(CO)_9 + Ar_2C{=}NN{=}CAr_2 \longrightarrow Fe_2(CO)_6(NCAr_2)_2 + \text{other products} \quad (100)^{239}$$

$$Na_2Fe(CO)_4 + Me_2BrC(NO) \longrightarrow Fe_2(CO)_6(NCMe_2)_2 + Fe_2(CO)_6(ONCMe_2)(NCMe_2) \quad (101)^{236}$$

13.3.5.4 Structure and Bonding

The structures of six methyleneamido complexes have been determined and are listed in Table 9. The quality of the structural data varies, but the structures can be categorized as linear, bent and bridging. $CpMo(CO)_2(NCBu_2^t)$ has the linear geometry (17); $Rh\{C(NMeCH_2)_2\}\{NC(CF_3)_2\}(PPh_3)_2$ has a bent geometry (18), and the remaining structures have the bridging geometry (19). It appears that the methyleneamido ligand shows a stronger tendency to form bridging structures than the analogous nitrosyl ligand. The C=N double bond is almost invariant at 1.27 ± 0.02 Å which is the normal value in free ketimines. The metal–nitrogen distances in $CpMo(CO)_2(NCBu_2^t)$ (1.89 Å) and $Mn_2(CO)_7\{NC(CF_3)_2\}_2$ (2.01 Å, av.) can be compared with similar distances in $(HBpz_3)Mo(CO)_2(N_2Ph)$ (1.83 Å) and $Mn_2(CO)_8(N_2Ph)_2$ (2.03 Å, av.). The diamagnetism of $CpWI_2(CO)(NCR_2)$ and $ReCl_2(NCH_2)(PMe_2Ph)_3$ show that these complexes are adequately described as having d^6 metal configurations like their nitrosyl analogues. It is not clear whether the linear methyleneamido ligand exerts any significant structural *trans* influence; there is no obvious Mo—C lengthening in $CpMo(CO)_2(NCBu_2^t)$.

Table 9 Structures of Methyleneamido Complexes

Compound[a]	M—N (Å)	N—C (Å)	M—N—C (°)	Ref.
CpMo(CO)$_2$(N=CBu$_2^t$)	1.892(5)	1.259(8)	171.8(4)	211
Rh{C(NMeCH$_2$)$_2$}{N=C(CF$_3$)$_2$}(PPh$_3$)$_2$	2.02(3)	1.27(3)	152	214
Mn$_2$(CO)$_7${N=C(CF$_3$)$_2$}$_2$	1.972(7)	1.259(9)	142.7(6)	237
	1.999(7)	1.258(9)	139.7(6)	
	2.021(7)		141.8(6)	
	2.036(7)		140.3(6)	
Fe$_2$(CO)$_6${N=C(*p*-C$_6$H$_4$Me)$_2$}$_2$	1.91(1)	1.24(2)	—[b]	239
	1.93(1)	1.29(2)	—[b]	
	1.94(1)		—[b]	
	1.95(1)		—[b]	
Fe$_2$(CO)$_6$(ON=CMe$_2$)(N=CMe$_2$)	1.902(4)	1.254(7)	138.8(4)	232
	1.921(4)		139.3(4)	
Fe$_2$(CO)$_6$(N$_2$C$_{23}$H$_{18}$)	1.920(av.)	1.254	—[b]	206
	1.965(av.)	1.263	—[b]	
Re$_2$H(CO)$_6$(N=CHMe)(Ph$_2$PCH$_2$PPh$_2$)	2.08(2)	1.39(6)	—[b]	198
	2.18(3)		—[b]	
Os$_3$H(CO)$_9$(N=CHCF$_3$)(PMe$_2$Ph)	2.075(12)	1.247(15)	142.5(10)	435
(*anti* isomer)	2.119(11)		131.8(10)	
Os$_3$H(CO)$_9$(N=CHCF$_3$)(PMe$_2$Ph)	2.081(6)	1.290(11)	143.5(6)	436
(*syn* isomer)	2.065(16)		130.2(6)	
Os$_3$H(CO)$_{10}$(N=CHCF$_3$)	2.065(4)	1.271(8)	142.2(4)	434
	2.078(5)		132.3(4)	
Fe$_3$H(CO)$_9$(N=CHMe)	1.960(3)	1.321(5)	—[b]	432
	1.877(3)		—[b]	
	1.878(3)			

[a] Abbreviations, see footnote in Table 10.
[b] Data not given.

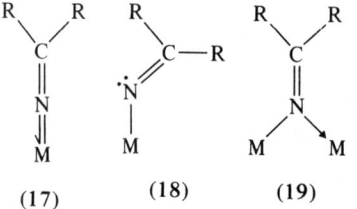

(17)   (18)   (19)

13.3.5.5 Spectroscopic Studies

Infrared spectroscopy has been used to identify and characterize methyleneamido complexes. Most if not all —N=CR$_2$ complexes have a medium to weak band in the infrared spectrum in the region 1525–1725 cm^{-1}. The positions and intensities of the C—N vibrations are such that they can be mistaken for other ligand vibrations which can lead to erroneous assignments of ν(CN). Isotope labeling (^{15}N) of the methylenamido ligand has not been carried out. A listing of some reported ν(CN) bands appears in Table 10. The relatively narrow range of ν(CN) reflects the weaker π-acceptor characteristics of —N=CR$_2$; these bands seem to be less sensitive to the electronic environment of the metal and geometry of the ligand.

X-Ray photon emission spectra of some Mo and W methyleneamido complexes have been measured. The limited data show that the bind energies of N(1S) are insensitive to changes in structure and bonding environment. The binding energies for the metals show some variation (1.0–1.5 eV) which has been correlated with ν(CO) of carbonyl coligands.[220]

13.3.5.6 Reactions of Methyleneamido Ligands

There are several reactions of coordinated methyleneamido ligands cited in the literature. The methylineamido ligands can be displaced by carbon monoxide or by protolysis using strong acids (equations 102 and 103).

Table 10 Listing of Methyleneamido Complexes

Compound[a]	$v(CN)$ cm^{-1}	Ref.
$Cp_2TiCl(N{=}CR_2)$	1634–1715	207, 218
$Cp_2XTi(N{=}CR{-}CR{=}N)TiXCp_2$	1638–1705	233
$Cp_2Ti(N{=}CR_2)_2$	1601–1603	234
$Cp_2ZrCl(N{=}CR_2)$	1640	203, 204, 208
$Cp_2Zr(N{=}CPh_2)_2$	1640, 1660	208
$Cp_2Hf(N{=}CPh_2)_2$	1643, 1665	208
$NbCl_4(NCR_2)(Et_2NH)$	—	229
$TaCl_4(NCR_2)(Et_2NH)$	—	229
$CpMo(CO)(PPh_3)(N{=}CBu^tPh)$	1547	219
$CpMoI_2(CO)(N{=}CR_2)$	1639–1660	213, 219
$CpMo(CO)_2(N{=}CR_2)$	1534–1636	210–213, 215–217, 219, 221
$Cp_2Mo_2(CO)_4(N{=}CR_2)_2$	—	215, 217
$Cp_2Mo_2(CO)_2(N{=}CR_2)_2$	—	215, 216
$\{HB(Me_2pz)_3\}MoI(NO)(N{=}CMe_2)$	—	227, 228
$\{HB(Me_2pz)_3\}WBr(NO)(N{=}CMe_2)$	—	228
$Cp_2W_2(CO)_2(N{=}CR_2)_2$	—	215–217
$CpW(CO)_2(N{=}CR_2)$	—	210, 213, 215–217, 219–221
$CpWI_2(CO)(N{=}CR_2)$	1640–1667	313, 219
$Mn_2(CO)_7\{N{=}C(CF_3)_2\}_2$	—	230, 237, 238
$Mn_2(CO)_6(py)\{N{=}C(CF_3)_2\}_2$	—	230
$Mn(CO)_2\{N{=}C(CF_3)_2\}(PR_3)_2$	—	230
$ReX_2(N{=}CH_2)(PMe_2Ph)_3$	1560	202, 205
$ReX_2(N{=}CHR)(py)(PR_3)_2$	1585–1592	202, 205
$Re_2H(CO)_6(NCHMe)(Ph_2PCH_2PPh_2)$	—	198
$CpFe(CO)(N{=}CR_2)$	—	226
$Fe_2(CO)_6(N{=}CR_2)_2$	—	206, 226, 236, 239
$Fe_2I(CO)_6(N{=}CR_2)$	—	226
$Fe_2(CO)_6(ON{=}CMe_2)(N{=}CMe_2)$	—	232, 236
$Fe_3H(CO)_9(NCHR)$	1325	241, 431, 432
$[Fe_3(CO)_9(NCHR)]^-$	—	241, 431
$FeAr(MeCN)(NCMeAr)$	—	240
$[Ru(terpy)(bipy)(N{=}CMe_2)][PF_6]_3$	—	240
$Os_3H(CO)_9(NCHR)$	—	433
$Os_3H(CO)_{10}(NCHR)$	—	433, 434, 435
$OsH(CO)_9(PMe_2Ph)(NCHR)$	—	435, 436
$Co\{N{=}C(CF_3)_2\}(oep)$	—	225
$Rh\{N{=}C(CF_3)_2\}(PPh_3)_3$	1665	207, 218
$Rh\{C(NR_2)_2\}\{N{=}C(CF_3)_2\}(PPh_3)_2$	—	214
$Ir(CO)\{C(NR_2)_2\}\{N{=}C(CF_3)_2\}(PPh_3)$	—	224
$Ir(CO)\{N{=}C(CF_3)_2\}(PPh_3)_2$	1695	218
$PtX(N{=}CR_2)(PR_3)_2$	1662–1675	207, 218, 222
$Pt\{N{=}C(CF_3)_2\}_2(PMe_2Ph)_2$	1650–1675	207, 218
$(cod)Pt\{N{=}C(CF_3)_2\}_2$	1685	223
$Zn(N{=}CPh_2)_2$	1600	209
$RZn(N{=}CPh_2)$	1600	209
$RZn(N{=}CPh_2)$	1607–1624	209
$MeZn(N{=}CPh_2)(py)_2$	1613	209

[a] Abbreviations: X = halide or pseudohalide, R = alkyl or aryl, oep = octaethylporphyrin, pz = $C_3H_3N_2$, $HB(Me_2pz)_3$ = tris(3,5-dimethylpyrazolyl)borate

$$CpMo(CO)_2(NCPh_2) \xrightarrow{CO} \tfrac{1}{2}Cp_2Mo_2(CO)_6 \qquad (102)^{210}$$

$$PtCl\{NC(CF_3)_2\}(PMe_2Ph)_2 + HCl \longrightarrow PtCl_2(PMe_2Ph)_2 + (CF_3)_2C{=}NH_2Cl \qquad (103)^{218}$$

There is one interesting report of an inner sphere reductive migration in a platinum hydrido complex (equation 104).

$$PtH\{NC(CF_3)_2\}(PPh_3)_2 \xrightarrow{\Delta} Pt\{\eta^2\text{-}HN{=}C(CF_3)_2\}(PPh_3)_2$$
$$\xrightarrow{PPh_3} Pt\{HN{=}C(CF_3)_2\}(PPh_3)_3 \qquad (104)$$

The reduction of methyleneamide ligands to organoimido ligands has also been observed (equation 105).

$$Fe_3H(NCHR)(CO)_9 \begin{array}{c} \xrightarrow{H_2} Fe_3H_2(NCH_2R)(CO)_9 \\ \xrightarrow{CO} Fe_3(NCH_2R)(CO)_{10} \end{array} \qquad (105)^{431}$$

13.3.5.7 Survey of Methyleneamido Complexes

Table 10 gives a listing of methyleneamido complexes. The listing includes several bis(methylene-amido) complexes and representatives from most of the metals in the transition series with the exception of complexes from the copper group.

13.3.6 DIAZENIDO COMPLEXES

13.3.6.1 Background

During the past twenty years since the appearance of $CpMo(CO)_2(N_2Ph)$[249] in 1964 and $PtCl(N_2Ph)(PEt_3)_2$,[250] in 1965, reports dealing with the synthetic, structural and chemical properties of organodiazenido complexes have continued to appear in the open literature. At first, these complexes were novelties in which the RN_2 group was stablized toward N_2 extrusion, but the development of organodiazenido chemistry, which coincided with the renewed interest in nitrosyl chemistry and the discovery of dinitrogen complexes, then became a topic of interest in its own right. It was initially postulated and finally shown that ligated RN_2 groups were intermediates in the chemical reduction of ligated N_2 (equation 106 and 107).

$$W(N_2)_2(diphos)_2 + MeBr \longrightarrow WBr(N_2Me)(diphos)_2 \longrightarrow \text{ further reaction} \qquad (106)^{391,397}$$

$$CpMn(CO)_2(N_2) + LiC_6H_5 \longrightarrow CpMn(CO)_2(PhN_2)Li \longrightarrow \text{ further reaction} \qquad (107)^{333,394}$$

It also was evident to many early workers that NO and N_2R were isoelectronic ligands, and analogies with nitrosyl complexes facilitated development of organodiazenido chemistry. However, during the most recent decade, unique properties of diazenido complexes were discovered which distinguished diazenido ligands from their nitrosyl analogues. As expected, the outer nitrogen atom (N_β) in N_2R is more basic than the oxygen atom in NO. This led to the formation of side-bonded diazenido (M—N=NR) and N_α,N_β bridging diazenido (M—N=$\overline{NR\ M}$) N_2R ligands and to more facile protonation or alkylation of N_β (equations 108 and 109).

$$Mo(NNMe)(S_2CNMe_2)_3 + MeI \longrightarrow [Mo(NNMe_2)(S_2CNMe_2)_3][I] \qquad (108)^{398}$$

$$Mo(NO)(S_2CNMe_2)_3 + MeI \longrightarrow \text{ no reaction} \qquad (109)$$

That NO is a better π-acceptor ligand than N_2R parallels the greater electronegativity of O over NH and the greater Lewis acidity of NO^+ over N_2R^+. However, the organodiazenido ligand renders the metal more susceptible to internal electron transfer (M^n—$N_2R^+ \rightarrow M^{n+2}$—$N_2R^-$) and to attack by electrophiles at N_α (equations 110 and 111).

$$RhCl_2(N_2Ph)(PPh_3)_2 + HCl \longrightarrow RhCl_3(HN=NPh)(PPh_3)_2 \qquad (110)^{261}$$

$$RhCl_2(NO)(PPh_3)_2 + HCl \longrightarrow \text{ no reaction} \qquad (111)$$

In contrast to nitrosyls, the absence of a transferable oxygen atom in N_2R ligands allows the preparation of stable diazenido complexes of oxophilic, early transition metals; see for example $Cp_2TiCl(N_2Ph)$. Furthermore, there are as yet no diazotate ($RN=NO^-$) forming reactions anologous to the nitrite forming reactions in nitrosyl chemistry (see equations 112 and 113).

$$M—NO + O^{2-} \longrightarrow M—NO_2^{2-} \qquad (112)$$

$$M—NO + \tfrac{1}{2}O_2 \longrightarrow N—NO_2 \qquad (113)$$

Diazenido complexes were covered in a review published in 1975.[349]

13.3.6.2 Nomenclature

Widelyl accepted names for $C_6H_5N_2^+$ and MeN_2^+ are benzenediazonium and methanediazonium, although phenyldiazonium and methyldiazonium are occasionally seen. Accepted names for $C_6H_5N_2^-$ and MeN_2^- are phenyldiazenide and methyldiazenide. Names derived from the older designations of the parent molecule (HN=NH) as diimide or diimine have almost disappeared from current literature. Radical names for $C_6H_5N_2$ are phenyldiazenyl or benzeneazo when used as part of another name such as benzeneazobenzene (abbreviated azobenzene, 1,2-diphenyldiazene). Hence, ligand names for $C_6H_5N_2$— span the entire range above and more: benzenediazoniumato, phenyldiazoniumato, benzenediazo, phenyldiazo, phenyldiazenyl, benzeneazo, phenylazo, phenyldiazenato and phenyldiazenido. Fortunately, the nitrosyl ligand was spared such confusion because the neutral radical name has been almost universally accepted. For the sake of simple euphony and following the precedent set in a previous review,[349] the anionic name *aryldiazenido* (or *alkyldiazenido*) is used hereafter for the N_2R ligand regardless of the formal ligand charge, mode of bonding or ligand geometry.

13.3.6.3 Syntheses of Diazenido Complexes

A number of synthetic techniques have been developed for the preparation of organodiazenido complexes. Each method has certain advantages, and several seem to have some degree of generality. One synthetic method may succeed where a similar one may fail. Two of the most useful synthetic techniques used in the preparation of nitrosyl complexes are not available for the preparation of organodiazenido complexes (see equations 114 and 115).

$$M + NO \longrightarrow M-NO \qquad (114)$$

$$M-CO + NO_2^- \longrightarrow M-NO^- + CO_2 \qquad (115)$$

13.3.6.3.1 Addition of arenediazonium ions to low valent transition metal complexes

The metal complex must be coordinatively unsaturated or possess a labile ligand which is easily displaced (equations 116–118).

$$Rh(BuNC)_4^+ + ArN_2^+ \longrightarrow Rh(N_2Ar)(BuNC)_4^{2+} \qquad (116)^{275}$$

$$IrCl(N_2)(PPh_3)_2 + ArN_2^+ \longrightarrow IrCl(N_2Ar)(PPh_3)_2^+ + N_2 \qquad (117)^{271}$$

$$CpW(CO)_3^- + ArN_2^+ \longrightarrow CpW(CO)_2(N_2Ar) + CO \qquad (118)^{291}$$

It is not unusual to observe no reaction between diazonium ions and metal complexes (equation 119). Furthermore, a common side reaction is simple oxidation of the starting complex. It is thus sometimes necessary to carefully choose the ligands on the metal and substituents on the diazonium salt in order to obtain a stable, isolable product (equations 120–123).

$$CpMn(CO)_3 + ArN_2^+ \longrightarrow \text{no reaction} \qquad (119)^{375}$$

$$CpMn(CO)(PPh_3)_2 + ArN_2^+ \longrightarrow CpMn(CO)(PPh_3)_2^+ \qquad (120)^{400}$$

$$(C_6Me_6)Cr(CO)_2(PPh_3) + ArN_2^+ \longrightarrow (C_6Me_6)Cr(CO)_2(PPh_3)^+ \qquad (121)^{283}$$

$$(C_6Me_6)Cr(CO)_3 + ArN_2^+ \longrightarrow (C_6Me_6)Cr(CO)_2(N_2Ar)^+ \qquad (122)^{283}$$

$$
\begin{array}{c}
\qquad\qquad \xrightarrow{\;C_6H_5N_2^+\;} (MeC_5H_4)Mn(CO)_2(N_2C_6H_5)^+ + Ph_3SiH \\
\qquad\qquad\qquad\qquad\qquad \text{unstable} \\
(MeC_5H_4)Mn(CO)_2H(SiPh_3) \qquad\qquad\qquad\qquad\qquad (123)^{375} \\
\qquad\qquad \xrightarrow{\;o\text{-}CF_3C_6H_4N_2^+\;} \\
\qquad\qquad\qquad\qquad (MeC_5H_4)Mn(CO)_2(N_2C_6H_4CF_3)^+ + Ph_3SiH \\
\qquad\qquad\qquad\qquad\qquad \text{stable}
\end{array}
$$

13.3.6.3.2 Deprotonation of diazene complexes

Aryldiazene complexes can be formed from arenediazonium ions and metal hydrides, and subsequent deprotonation leads to aryldiazenido complexes. Mechanistic details of ArN_2^+ insertion into Rh—H and Pt—H bonds have been recently studied (equations 124–126).[453,454]

$$\text{PtHCl(PEt}_3)_2 \xrightarrow{\text{ArN}_2^+} \text{PtCl(HN}_2\text{Ar)(PEt}_3)_2^+ \xrightarrow{\text{MeCO}_2^-} \text{PtCl(N}_2\text{Ar)(PEt}_3)_2 \qquad (124)^{242}$$

$$\text{MnH(CO)}_3(\text{PPh}_3)_2 \xrightarrow{\text{ArN}_2^+} \text{Mn(CO)}_3(\text{HN}_2\text{Ph})(\text{PPh}_3)_2^+$$

$$\xrightarrow{\text{MeO}^-} \text{Mn(CO)}_2(\text{N}_2\text{Ph})(\text{PPh}_3)_2 + \text{CO} \qquad (125)^{322}$$

$$\text{OsH}_2(\text{CO})_2(\text{PPh}_3)_2 \xrightarrow{\text{ArN}_2^+} \text{OsH(CO)}_2(\text{HN}_2\text{Ph})(\text{PPh}_3)_2^+$$

$$\xrightarrow{\text{MeO}^-} \text{OsH(CO)}_2(\text{N}_2\text{Ph})(\text{PPh}_3)_2 \qquad (126)^{310}$$

Sometimes an ancillary ligand is also lost upon deprotonation. Dehydrohalogenation of neutral diazene complexes is generally more difficult but can be accomplished in some cases. The nature of the base is important (equation 127).

$$\text{RhCl}_3(\text{HN}_2\text{Ar})(\text{PPh}_3)_2 \xrightarrow{\text{Bu}_3\text{N}} \text{RhCl}_2(\text{N}_2\text{Ar})(\text{PPh}_3)_2 \qquad (127)^{262}$$

When they can be used, weaker bases such as carboxylate anions or tertiary amines are useful. Sometimes stronger bases such as hydroxide or alkoxide can be used. The reaction with diazonium ions seems to be limited to a select group of transition metal hydrides. Most are d^6 hydrido complexes and a few are d^8 hydrido complexes. There is a large number of higher valent polyhydride complexes which induce rapid decomposition of diazonium ions. The d^4 hydrido complexes, CpMoH(CO)_3, $\text{CpReH}_2(\text{CO})_2$ and $\text{CpReH(CO)}_2(\text{SiPh}_3)$, give aryldiazenido complexes, but aryldiazenes are not detected as intermediates and no bases are required. It is possible that the hydride ligand is lost as H^+ or H_2 prior to reaction of the metal complex with the diazonium ion.

13.3.6.3.3 Halide–diazenide exchange using silyldiazenes

Although monosubstituted diazenes and their anions are too unstable to be synthetically useful, the silyldiazines, $\text{ArN}{=}\text{NSiMe}_3$, are stable and can be used to prepare organodiazenido complexes (equations 128 and 129).

$$\text{MnBr(CO)}_5 + \text{PhN}_2\text{SiMe}_3 \longrightarrow \tfrac{1}{2}[\text{Mn(N}_2\text{Ph})(\text{CO})_4]_2 + \text{Me}_3\text{SiBr} + \text{CO} \qquad (128)^{291}$$

$$\text{CpTiCl}_3 + \text{PhN}_2\text{SiMe}_3 \longrightarrow \text{CpTiCl}_2(\text{N}_2\text{Ph}) + \text{Me}_3\text{SiCl} \qquad (129)^{334}$$

Though its use has been limited by the availability of silyldiazenes, this method of introducing diazenido ligands onto transition metals is an important one. Silyldiazines can be used to prepare bis(aryldiazenido) complexes from polyhalide compounds. Just as PhN_2^+ can be an oxidizing agent, a source of PhN_2^- can be a reducing agent (equations 130 and 131).

$$\text{WCl}_4(\text{PR}_3)_2 + 2\text{PhN}_2\text{SiMe}_3 \longrightarrow \text{WCl}_2(\text{N}_2\text{Ph})_2(\text{PR}_3)_2 + 2\text{Me}_3\text{SiCl} \qquad (130)^{367}$$

$$\text{ReOCl}_3(\text{PPh}_3)_2 + 4\text{PhN}_2\text{SiMe}_3 \longrightarrow \text{ReCl(N}_2\text{Ph})_2(\text{PPh}_3)_2 + 2\text{Me}_3\text{SiCl} + 2\text{N}_2$$

$$+ \text{Me}_3\text{SiOSiMe}_3 \qquad (131)^{402}$$

13.3.6.3.4 Reactions with monosubstituted hydrazines

The reactions of monosubstituted hydrazines with higher valent halo or oxo complexes yields organodiazenido complexes (equations 132 and 133).

$$\text{ReOCl}_3(\text{PPh}_3)_2 + \text{PhC(O)NHNH}_2 \longrightarrow \text{ReCl}_2(\eta^2\text{-N}_2\text{C(O)Ph})(\text{PPh}_3)_2 + \text{H}_2\text{O} + \text{HCl} \qquad (132)^{254}$$

$$\downarrow \text{py}$$

$$\text{ReCl}_2(\text{py})(\eta^1\text{-N}_2\text{C(O)Ph})(\text{PPh}_3)_2$$

$$\text{MoO}_2(\text{S}_2\text{CNMe}_2)_2 + 2\text{MeNHNH}_2 \longrightarrow \text{Mo(N}_2\text{Me})(\text{NHNHMe})(\text{S}_2\text{CNMe}_2)_2 + 2\text{H}_2\text{O} \qquad (133)^{359}$$

Many of the reactions with hydrazines are complicated multistep sequences which sometimes require the presence of an oxidizing agent. Oxidizing capability can come from adventitious O_2, sacrificial reduction of the metal or the hydrazine itself (equations 134–136).

$$ReCl_3(PMe_2Ph)_3 + 2PhNHNH_2 \longrightarrow ReCl_2(NH_3)(N_2Ph)(PMe_2Ph)_2 + PMe_2Ph \qquad (134)^{306}$$

$$\downarrow \text{slow}$$

$$ReCl_2(N_2Ph)(PMe_3Ph)_3 + NH_3$$

$$CpMoH(CO)_3 + PhNHNH_2 \xrightarrow{2[O]} CpMo(CO)_2(N_2Ph) \qquad (135)^{252}$$

$$MoCl_2(tpp) + 2PhNHNH_2 \xrightarrow{2[O]} Mo(N_2Ph)_2(tpp) \qquad (136)^{378}$$

The oxidation of preformed hydrazine or hydrazido($1-$) complexes can also lead to the formation of diazenido ligands (equations 137 and 138).

$$Mo(N_2Me)(NHNHMe)(S_2CNMe_2)_2 \xrightarrow{\frac{1}{2}O_2} Me(N_2Me)_2(S_2CNMe_2)_2 + H_2O \qquad (137)^{359}$$

$$[ReBr_2(N_2Ph)(NHNHPh)(PPh_3)_2]^+ \xrightarrow{Br_2} [ReBr_2(N_2Ph)_2(PPh_3)_2]^+ + 2HBr \qquad (138)^{400}$$

13.3.6.3.5 *Protonation, alkylation or acylation of dinitrogen complexes*

There are many dinitrogen complexes, but only a small group of Re, Mo and W complexes have coordinated N_2 ligands which are sufficiently reactive (equations 139 and 140).

$$ReCl(N_2)(PMe_2Ph)_4 + Ph\overset{O}{\overset{\|}{C}}Cl \longrightarrow ReCl_2(N_2\overset{O}{\overset{\|}{C}}Ph)(PMe_2Ph)_3 + PMe_2Ph \qquad (139)^{312}$$

$$W(N_2)_2(diphos)_2 + EtBr \xrightarrow{h\nu} WBr(N_2Et)(diphos)_2 + N_2 \qquad (140)^{397}$$

13.3.6.3.6 *Nucleophilic attack at coordinated dinitrogen*

There is a single report of this kind of reaction (equation 141). The diazenido complex was not isolated but was characterized spectroscopically and chemically in solution. Tentative spectroscopic data suggested that the inner nitrogen atom was the site of attack.

$$Cp(CO)_2Mn(N_2) + LiPh \longrightarrow Cp(CO)_2Mn\!-\!\underset{\underset{Ph}{|}}{N}\!=\!N\!-\!Li \qquad (141)^{338}$$

13.3.6.3.7 *Nucleophilic attack at coordinated diazoalkane*

There is a single report of this reaction. The number of conveniently accessible diazoalkane complexes is limited (equation 142).

$$WBr(N_2CH_2)(diphos)_2^+ + LiMe \longrightarrow WBr(N_2Et)(diphos)_2 + Li^+ \qquad (142)^{450}$$

13.3.6.3.8 *Reaction of amines with coordinated nitrosyl ligand*

The NO ligand loses much of its electrophilic character (NO^+) upon coordination to a metal; this severely limits the number of metal nitrosyl complexes which can react with amines in this manner (equation 143).

$$RuCl(NO)(bipy)_2^{2+} + p\text{-}MeC_6H_4NH_2 \longrightarrow RuCl(N_2C_6H_4Me)(bipy)_2^{2+} + H_2O \qquad (143)^{281}$$

13.3.6.3.9 *Reactions with triazenes*

The reaction of metal hydrides with *N*-tosyl-*N*-methylnitrosamine is a useful synthetic method for preparing nitrosyl complexes, but this is not true for aryldiazenido complexes. The analogous precursor, 1-phenyl-3-tosyl-3-methyltriazene, can be prepared but is thermally unstable much above room temperature. Stable triazenes such as 1,3-diaryltriazenes usually react with metal hydrides to give stable triazenido complexes. In a few cases, however, aryldiazenido complexes can be isolated directly from triazenes (equations 144 and 145).

$$RuCl_3 + ArMeN-N=NAr + PPh_3 \xrightarrow{EtOH} RuCl_3(N_2Ar)(PPh_3)_2 \qquad (144)^{261}$$

$$RhCl_3 + ArMeN-N=NAr + PPh_3 \xrightarrow{EtOH} RhCl_2(N_2Ar)(PPh_3)_2 \qquad (145)^{261}$$

13.3.6.3.10 Insertion of diazoalkanes into metal–hydride bonds

This reaction offers attractive possibilities for preparing alkyldiazenido complexes. However, this reaction may be limited as only one hydride is known to work well, $CpMH(CO)_3$ (M = Mo, W), see equation 146.

$$CpWH(CO)_3 + CH_2N_2 \longrightarrow CpW(CO)_2(N_2Me) + CO \qquad (146)^{342}$$

13.3.6.4 Structure and Bonding

For the sake of simple structure and bonding considerations, it is convenient to classify the amphoteric organodiazenido ligand as RN_2^+ (three-electron donor, singly bent geometry) or as RN_2^- (one-electron donor, doubly bent geometry). This leads to reasonable oxidation numbers of the metal from which stereochemical information about the metal may be inferred, and it leads to simple bonding notions using valence bond theory from which ligand geometries may be inferred. The limitations and over simplications inherent in the RN_2^+/RN_2^- formalism are well known, but its utility and simplicity are also well demonstrated.

The coordination geometries of diazenido complexes follow clear and distinct patterns based on coordination number and maximum electron count (MEC). This is summarized in Table 11. The MEC is the total count of valence electrons on the metal and those donated by all ligands assuming that the amphoteric diazenido ligand always donates three electrons to the metal. The strong π-bonding capability of the diazenido ligand coupled with other common ancilliary ligands gives low spin configurations to almost all diazenido complexes.

Table 11 Geometry of Diazenido Complexes[a]

Coordination number	14 electrons (RN_2^+)	16 electrons (RN_2^+)	Maximum electron count 18 electrons (RN_2^+)	18 electrons (RN_2^-)	20 electrons (RN
4	d^6	d^8, sq. planar $IrCl(N_2Ph)(PPh_3)_2^+$	d^{10}, tetrahedral $Fe(CO)(NO)(N_2Ph)(PPh_3)$	d^8, sq. planar $PtCl(N_2Ph)(PEt_3)_2$	d^{10}
5	d^4, trig. bipyramid $Mo(N_2Ph)(MeCN)(SAr)_3$	d^6	d^8 trig. bipyramid $Fe(CO)_2(N_2Ph)(PPh_3)_2^+$	d^6 sq. pyramid $RhCl(N_2Ph)(PPP)^+$	d^8
6	d^2	d^4, octahedral $(HBPz_3)Mo(N_2Ph)(SPh)_2$	d^6, octahedral $RuCl_3(N_2Ph)(PPh_3)_2$	d^4	d^6, octahedral $IrCl_2(CO)(N_2Ph)(F$
7	d^0	d^2	d^4, pent. bipyramid $Mo(N_2Ph)(S_2CNMe_2)_3$	d^2	d^4

[a] Abbreviations, see footnote in Table 15.

Four-coordinate complexes can adopt either square planar (low spin d^8) or tetrahedral geometries (d^{10}). Complexes with MEC of 18 can adopt either geometry depending on the demands of the metal and coligands; $Co(CO)_2(N_2Ph)(PPh_3)$ ($\nu(NN) = 1689\,cm^{-1}$, RN_2^+) is probably tetrahedral[272] while $PtCl(N_2Ph)(PEt_2)_2$ ($\nu(NN) = 1455\,cm^{-1}$, RN_2^-) is square planar.[314] There are no known 20 electron, four-coordinate diazenido complexes. *Five-coordinate* complexes can adopt either the square pyramidal (low spin d^6) or trigonal bipyramidal geometries (low spin d^8). Again, complexes with MEC of 18 can adopt either geometry; $Fe(CO)_2(N_2Ph)(PPh_3)_2^+$ ($\nu(NN) = 1723\,cm^{-1}$, RN_2^+) has a trigonal bipyramidal geometry[311] while $RhCl(N_2Ph)\{PhP(CH_2CH_2CH_2PPh_2)_2\}^+$ ($\nu(NN) = 1600\,cm^{-1}$, RN_2^-) is square pyramidal.[292] There are few 16 electron, five-coordinate diazenido complexes. $[RuCl_2(N_2Ph)(PPh_3)_2]^+$ has been reported but it is probably a chloro-bridged dimer.[263] Fourteen electron, five-coordinate complexes such as $Mo(N_2Ph)(MeCN)(SAr)_3$ and $Re(N_2Ph)(SAr)_4$ are known, and the structure of the Mo complex displays a trigonal bipyramidal geometry.[403] *Six-coordinate* complexes possess octahedral geometries. Other geometries such as trigonal prismatic or pentagonal pyramidal are too high in energy relative to octahedral. Octahedral complexes with MEC of 16, 18 and 20 are all known (Table 11). Complexes of the type $(HBpz_3)WBr_2(N_2Ph)$ are probably dimeric and seven-coordinate.[287] The unique, monomeric, 14-electron complex $(C_5H_5)TiCl_2(N_2Ph)$ possesses an unusual 'side-on' bonding mode for the diazenido ligand;[381] the η^5-cyclopentadienyl ligand is considered to occupy three coordination sites. The only *seven-coordinate* diazenido complex complexes which have been structurally characterized are the d^4, Mo^{II} complexes of the type $Mo(N_2R)(S_2CNR_2)_3$ which have pentagonal bipyramidal geometries with apical diazenido ligands.[330] There is no structurally characterized *eight-coordinate* complex. The only report of such a complex is $(C_5H_5)_2TiCl(N_2Ph)$ which contains 18 electrons, a d^2 metal and an RN_2^+ ligand and probably has the typical Cp_2MXY geometry.

The organodiazenido ligand can adopt one of several geometries as described by the valence bond descriptions **(20)**–**(24)**.

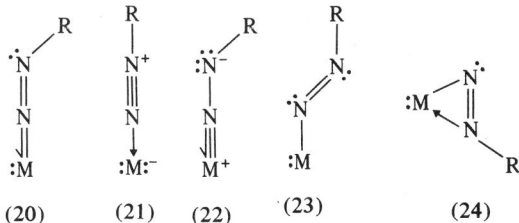

(20)   (21) (22) ·(23)   (24)

Structural data for terminal diazenido ligands are summarized in Table 12. By far, the most prevalent bonding made for RN_2 ligands is the singly bent geometry, **(20)**; typical metric parameters for this ligand are M—N—N = 170°–180°, N—N—C = 118°–125°, N–N = 1.20 Å. The metal–nitrogen bond lengths for six-coordinate complexes are given by the following table which reflects the different inherent bonding radii of the various metals.

Ti	V	Cr	Mn	Fe	Co	Ni
1.85	1.75	1.71	1.70	1.70	1.69	1.68

Zr	Nb	Mo	Tc	Ru	Rh	Pd
1.98	1.87	1.83	1.81	1.78	1.78	1.81

Hf	Ta	W	Re	Os	Ir	Pt
1.97	1.87	1.83	1.81	1.79	1.80	1.83

Lower coordination numbers of four or five should have metal–nitrogen distances about 0.01–0.02 Å smaller. Of the singly bent complexes listed in Table 12, $OsH(CO)(N_2Ph)(PPh_3)_2$ stands out as having a metal–nitrogen bond length about 0.10 Å too long.[315]

There are no good structural examples of the linear geometry, **(21)** or the singly bent geometry, **(22)**. However, two complexes, $RuCl_3(N_2Ar)(PPh_3)_2$ and $[IrCl(N_2Ar)(PPh_3)_2]^+$, have shortened N—N distances (1.16 Å), enlarged N—N—C angles (137°, 127°) and high values of $v(NN)$ (1868, 1881 cm^{-1}), but the geometries are much closer to **(20)** than **(21)**. Two groups of complexes, $Mo(N_2Ar)(S_2CNR_2)_3$ and $ReX_2(N_2Ar)L_3$, have low values of $v(NN)$ (1530, 1600 cm^{-1}) and metal–nitrogen distances which are shortened by about 0.06 Å and 0.03 Å respectively, but again the geometries are much closer to **(20)** than **(22)**. There are three structurally characterized examples of the doubly bent geometry, **(23)**. The angles at both nitrogen atoms are near the ideal values for sp^2 hybridization (115°–135°), and the metal–nitrogen distances are lengthened by 0.15–0.25 Å over that expected for geometry **(20)**. It is likely that there will be significant differences in the structural parameters of doubly bent diazenido ligands in 18-electron and 20-electron complexes (MEC), but the very limited amount of good quality structural data disallow firm conclusions at present. Finally, there is one structurally characterized example of a side-bonded diazenido ligand, **(24)**. In $CpTiCl_2(N_2Ph)$, the diazenido ligand has structural parameters like a double bent ligand which is bent sufficiently accutely at N_α so that N_β also interacts with the same metal. This side-bonding is rare; the diazenido ligand donates three electrons *via* two σ bonds rather than one σ and one π bond as in **(20)**. The filled π-orbital of the N—N double bond is also in a position to interact with the electron deficient metal. Under certain conditions, a diazenido complex can adopt a geometry intermediate between the limiting geometries. One example is $[IrCl(N_2Ph)(PMePh_2)_3]^+$ whose geometry about the metal is part way between square pyramidal and trigonal bipyramidal and which has a diazenido ligand with a structure part way between singly bent and double bent.[319] The singly

Nitrogen Ligands

Table 12 Structures of Diazenido Ligands

Compound[a]	M—N (Å)	N—N (Å)	M—N—N (°)	N—N—C (°)	Ref.
Singly Bent					
Mo(N$_2$Ph)(S$_2$CNMe$_2$)$_3$	1.781(4)	1.233(6)	171.5(4)	120.5(5)	330
Mo(3-N$_2$C$_6$H$_4$NO$_2$)(S$_2$CNMe$_2$)$_3$	1.770(6)	1.262(9)	170.6(6)	117.9(7)	330
Mo(N$_2$CO$_2$Et)(S$_2$CNMe$_2$)$_3$	1.732(5)	1.274(7)	178.9(5)	117.0(6)	329
Mo(N$_2$CS$_2$Et)(NH$_2$N==C(S)SEt)(S$_2$CNMe$_2$)$_2$	1.77(1)	1.21(1)	178(1)	120(1)	380
(HBpz$_3$)Mo(CO)$_2$(N$_2$Ph)	1.825(4)	1.211(6)	174.2(1)	121.1(2)	301
(HBpz$_3$)Mo(p-N$_2$C$_6$H$_4$F)(p-SC$_6$H$_4$Me)$_2$	1.807(8)	1.229(9)	170.8(8)	121.4(9)	382
(MeC$_5$H$_4$)Mo(p-N$_2$C$_6$H$_4$F)$_2$Cl	1.826(2)	1.229(4)	167.4(2)	117.4(3)	389
	1.834(3)	1.218(4)	176.5(2)	122.2(3)	
MoCl(N$_2$C(O)Ph)(NHNC(O)Ph)(PMe$_2$Ph)$_2$	1.770(8)	1.24(1)	174(1)	116(1)	337
	1.793(9)	1.27(1)	175(1)	114(1)	337
MoCl(N$_2$C(O)Ph)(diphos)$_2$	1.813(7)	1.255(10)	172.1(6)	116.7(7)	392
MoI(N$_2$C$_6$H$_{11}$)(diphos)$_2$	1.834(9)	1.155(12)	177(1)	132(1)	387
Mo$_2$(N$_2$Ph)$_4$(acac)$_2$(OMe)$_2$	1.802(14)	1.259(19)	175.9(11)	117.1(12)	438
Isomer A	1.810(9)	1.240(13)	177.7(11)	118.9(13)	
Mo$_2$(N$_2$Ph)$_4$(acac)$_2$(OMe)$_2$	1.832(9)	1.216(14)	173.3(9)	117.6(10)	438
Isomer B	1.859(11)	1.194(15)	176.5(9)	119.7(10)	
[Et$_3$NH]$_2$[Mo$_2$(N$_2$Ph)$_4$(MoO$_4$)$_2$(OMe)$_2$]	1.823(av.)	1.227(av.)	~180	—[b]	405
[Et$_3$NH][Mo$_2$(N$_2$Ph)$_4$(SPh)$_5$]	1.73(1)	1.37(2)	172.4(11)	—[b]	415
	1.83(2)	1.23(3)	165.9(16)	—[b]	
	1.74(1)	1.26(2)	162.5(12)	—[b]	
	1.82(1)	1.30(2)	166.9(15)	—[b]	
[Et$_3$NH]$_2$[Mo$_2$(N$_2$Ph)(NNHPh)(SCH$_2$CH$_2$S)$_3$(SCH$_2$CH$_2$SH)]	1.740(7)	1.35(2)	162.2(6)	—[b]	416
Mo(p-N$_2$C$_6$H$_4$OMe)$_2$(L')	1.81(1)	1.28(2)	170.4(17)	113.8(17)	417
	1.81(2)	1.20(3)	168.3(19)	119.6(17)	
Mo(N$_2$Ph)(MeCN)(SC$_6$H$_2$Pr$_3^i$)$_3$	1.782(12)	1.211(17)	171.2(11)	124.7(13)	403
CpW(CO)$_2$(N$_2$Me)	1.856(3)	1.215(5)	173.3(3)	116.5(4)	
ReCl$_2$(N$_2$Ph)(PMe$_2$Ph)$_3$	1.77(2)	1.23(2)	173(2)	119(2)	393, 276
ReCl$_2$(N$_2$C(O)Ph)(PMe$_2$Ph)$_3$	1.74(2)	1.22(3)	170(2)	124(2)	276
ReBr$_2$(N$_2$Ph)(N$_2$HPh)(PPh$_3$)$_2$	1.793(11)	1.212(16)	172.4(10)	120.2(11)	401
CpRe(CO)(p-N$_2$C$_6$H$_4$OMe)(AuPPh$_3$)	1.78(2)	1.27(2)	171(1)	119(1)	412
[Et$_3$NH][Re$_2$(N$_2$Ph)$_2$(SPh)$_7$]	1.81(2)	1.23(2)	170.0(18)	—[b]	404
	1.81(2)	1.24(3)	171.4(19)	—[b]	
[(MeC$_5$H$_4$)Mn(CO)$_2$(2-N$_2$C$_6$H$_4$CF$_3$)][BF$_4$]	1.693(7)	1.211(8)	171.8(8)	125.6(9)	375
[Fe(CO)$_2$(N$_2$Ph)(PPh$_3$)$_2$][BF$_4$]	1.702(6)	1.201(7)	179.2(5)	124.2(6)	311
RuCl$_3$(p-N$_2$C$_6$H$_4$Me)(PPh$_3$)$_2$	1.796(9)	1.144(10)	171.2(9)	135.9(11)	270
	1.784(5)	1.158(6)	171.9(5)	137.1(5)	816
OsH(CO)(N$_2$Ph)(PPh$_3$)$_2$	1.876(6)	1.211(8)	171.1(6)	118.5(7)	315
[IrCl(N$_2$Ph)(PPh$_3$)$_2$][BF$_4$]	1.800(10)	1.163(11)	175.8(8)	126.9(10)	400
Bridging — N$_\alpha$,N$_\beta$					
CpW(CO)$_2$($\overline{\text{NNMe}}$)Cr(CO)$_5$	N$_\alpha$-1.830(3)	1.247(4)	174.4(3)	113.8(3)	383
	N$_\beta$-2.106(3)				
CpMo(CO)$_2$($\overline{p\text{-NNC}_6\text{H}_4\text{Me}}$)ReCp(CO)$_2$	N$_\alpha$-1.822(4)	1.256(6)	177.7(4)	113.5(4)	379
	N$_\beta$-2.151(4)				
Bridging — N$_\alpha$,N$_\alpha$,N$_\beta$					
Mn$_3$(CO)$_{12}$(N$_2$Me)	N$_\alpha$-1.879(8)	1.22(1)	119.9(7)	112(1)	339
	N$_\alpha$-1.964(8)		146.2(7)		
	N$_\beta$-2.064(8)				
Doubly Bent					
[RhCl(N$_2$Ph)(PPP)][PF$_6$]	1.961(7)	1.172(9)	125.1(6)	118.9(8)	292
PtCl(p-N$_2$C$_6$H$_4$F)(PEt$_3$)$_2$	1.97(3)	1.17(3)	118(2)	118(2)	314
IrCl$_2$(o-N$_2$C$_6$H$_4$NO$_2$)(CO)(PPh$_3$)$_2$	2.05(4)	1.19(4)	115(3)	115(3)	320
Half Doubly Bent					
[IrCl(N$_2$Ph)(PMePh$_2$)$_3$][PF$_6$]	1.835(8)	1.241(11)	155.2(7)	118.8(8)	319
Mo(N$_2$Ph)$_2$(tpp)	2.060(5)	1.133(9)	149.1(9)	128.6(9)	378
Side-bonded					
CpTiCl$_2$(N$_2$Ph)	N$_\alpha$-2.053(5)	1.215(8)	70.4(3)	126.7(5)	381
	N$_\beta$-2.004(4)				
Bridging — N$_\alpha$,N$_\alpha$					
Ir$_2$O(NO)$_2$(o-N$_2$C$_6$H$_4$NO$_2$)(PPh$_3$)$_2$	2.07(6)	1.22(6)	121(4)	111(4)	346
	2.13(5)		140(4)		
[Pt$_2$(N$_2$H)$_2$(PPh$_3$)$_4$][BPh$_4$]$_2$	1.97(5)	1.18(9)	112(5)	—	253
	2.09(5)		137(5)		

$Mn_2(CO)_8(N_2Ph)_2$	2.031(2)	1.233(2)	119.5(1)	119.6(2)	355
	2.023(2)		134.5(1)		
$HOs_3(CO)_{10}(p\text{-}N_2C_6H_4Me)$	2.04(1)	1.24(2)	—[b]	116(1)	377
	2.06(1)		—[b]		
$Mo_2O(N_2C(O)Ph)_2(S_2CNEt_2)_2$	1.90	—[b]	—[b]	—[b]	352
	1.96	—[b]	—[b]	—[b]	
	1.92		—[b]		
	1.97		—[b]		

[a] Abbreviations, see footnote in Table 15.
[b] Value not given.

bent RN_2^+ ligand has a small or negligible structural *trans* influence on *trans* ligands such as halide, cyclopentadienyl or pyrazole. On the other hand, there is a significant *trans* lengthening (~ 0.10 Å) of the metal–chloride bonds in $PtCl(N_2Ar)(PEt_3)_2$ and $IrCl_2(CO)(N_2Ar)(PPh_3)_2$ which are *trans* to doubly bent RN_2^- ligands.[314,320] This *trans* influence is manifest in another way. Many five-coordinate, 18 electron diazenido complexes are reluctant to bind to a sixth ligand; when binding does occur, a cationic diazenido complex binds to an anionic halide, and the halide (or other ligand) in the six-coordinate complex suffers a bond weakening and lengthening by the *trans* RN_2^-.

The presence of three lone pairs of electrons in the diazenido ligand ($R–\ddot{N} = \ddot{N}$) allows it to bind in a bridging fashion to several metals at once. Three bridging modes of binding for diazenido ligands have been observed (**25–27**).

(25) (26) (27)

The only example of structure (**25**) is in $Mn_3(CO)_{12}(N_2Me)$,[339] and there are two examples of structure (**26**), $CpW(CO)_2(NNMe)Cr(CO)_5$[383] and $CpMo(CO)_3(p\text{-}NNC_6H_4Me)ReCp(CO)_2$.[379] Finally, several examples of structure (**27**) are known and are listed in Table 12. In some cases the two metals are engaged in metal–metal bonding, and in others there is no metal–metal bond.

Qualitative molecular orbital calculations have been carried out on diazenido complexes.[356] These results showed that bending at N_β produces a ligand which is both a stronger π acceptor and a better σ donor. Hence, diazenido ligands are likely to be strongly bent at N_β and the linear geometry (**21**) will be rare or nonexistent. Furthermore, metal complexes with poorer σ-donor ligands, higher metal oxidation state and higher positive charges will yield diazenido complexes with larger N_α —N_β—C angles. Calculations show that the singly bent diazendio ligand prefers to lie in the coordination plane with the better π donors. Estimates of charge distributions tend to corroborate what is known about protonation of the nitrogen atoms; diazenido ligands in d^4 and d^6 complexes are protonated at N_β and diazenido ligands in d^8 complexes are generally more easily protonated and usually at N_α.

13.3.6.4 Spectroscopic Properties

Infrared spectroscopy of diazenido complexes provides useful structural information as well as a means of characterizing new compounds. The N—N stretching frequencies of diazenido ligands span a wide range from 1400–2100 cm^{-1}. The $v(NN)$ band usually has medium to strong intensity and can be used as a diagnostic tool to identify the presence of a diazenido ligand. Table 13 lists some values of $v(NN)$ for selected diazenido complexes. The values of $v(NN)$ are usually 25–75 cm^{-1} lower than those of $v(NO)$ for analogous nitrosyl complexes. Bands associated with $v(NN)$ in the lower region (1400–1600 cm^{-1}) are sometimes obscured by other overlapping bands in the complex. A further complication, especially in aryldiazenido ligands, is that the N—N stretching mode is often vibrationally coupled with other vibrational modes of the diazenido ligand when $v(NN)$ is in the region 1400 and 1675 cm^{-1}.[255,284,309] Consequently, isotopic substitution using ^{15}N can lead to the shifting of two or even three bands. By measuring the energies and intensities of bands which shift upon isotopic substitution, it is possible to estimate the uncoupled value of $v(NN)$ to a good degree of accuracy.[309]

Table 13 N—N Stretching Frequencies for Selected Diazenido Complexes

Complex	Observed $v(NN)$ (cm^{-1})	Uncoupled $v(NN)$ (cm^{-1})	Ref.
[(C$_6$Me$_6$)Cr(CO)$_2$(N$_2$Ph)][BF$_4$]	1697	1697	283
[(HBpz$_3$)Mo(N$_2$Ph)I]$_2$	1556	—	287
MoBr(N$_2$Me)(diphos)$_2$	1540	1540	312
Mo(CO)$_2$(bipy)(N$_2$Ph)(PPh$_3$)	1561, 1633	—	336
MoBr$_3$(phen)(N$_2$Ph)	1410	—	336
Mo(N$_2$Ph)(S$_2$CNMe$_2$)$_3$	1481, 1530, 1601	1529	400
MoBr(N$_2$C(O)Ph)(diphos)$_2$	1355	—	353
[CpMo(NO)(N$_2$Ph)(PPh$_3$)][BF$_4$]	1563, 1651	—	255
CpMoCl(NO)(N$_2$Ph)	1559, 1623	—	255
CpMo(CO)$_2$(N$_2$Ph)	1549, 1620	—	255
[CpMo(N$_2$Ph)$_2$(PPh$_3$)][BF$_4$]	1547, 1621, 1655	—	255
WBr(N$_2$Me)(diphos)$_2$	1525	1525	312
CpW(CO)$_2$(N$_2$Me)	1627	1627	396
[CpMn(CO)$_2$(N$_2$Ph)][PF$_6$]	1790	1790	375
Mn(CO)$_2$(N$_2$Ph)(PPh$_3$)$_2$	1476, 1543, 1612	1556	322
MnBr(NO)(N$_2$Ph)(PPh$_3$)$_2$	1476, 1565, 1631	1598	322
[Mn(CO)(N$_2$Ph)$_2$(PPh$_3$)$_2$][PF$_6$]	1574, 1668, 1709	1686	322
[CpRe(CO)$_2$(N$_2$Ar)][BF$_4$]	1760	1760	375
Re(CO)$_2$(N$_2$Ph)(PPh$_3$)$_2$	1471, 1538, 1604	1536	322
[ReBr$_2$(N$_2$Ph)$_2$(PPh$_3$)$_2$][ClO$_4$]	1785, 1853	1819	400
[Fe(CO)$_2$(N$_2$Ph)(PPh$_3$)$_2$][BF$_4$]	1723	1723	273
Fe(CO)(NO)(N$_2$Ph)(PPh$_3$)	1664	1664	255
RuCl$_3$(N$_2$Ph)(PPh$_3$)$_2$	1882	1882	310
OsH(CO)(N$_2$Ph)(PPh$_3$)$_2$	1472, 1541, 1605	1542	309
OsH(CO)$_2$(N$_2$Ph)(PPh$_3$)$_2$	1458, 1440	1455	310
[Os(CO)$_2$(N$_2$Ph)(PPh$_3$)$_2$][PF$_6$]	1577, 1668	1660	310
Co(CO)$_2$(N$_2$Ph)(PPh$_3$)	1689	1689	363
[RhCl(N$_2$Ph)(PPP)][PF$_6$]	1550, 1624	1600	309
[RhCl(N$_2$Ph)(PPh$_2$Me)$_3$][PF$_6$]	1573, 1662	1646	292
[RhCl(N$_2$Ph)(diphos)$_2$][PF$_6$]	1446, 1466, 1494	1488	292
RhCl$_2$(N$_2$Ph)(PPh$_3$)$_2$	1549, 1614	1575	309
[IrCl(N$_2$Ph)(PPh$_3$)$_2$][PF$_6$]	1868	1868	271
[IrCl(N$_2$Ph)(PPh$_2$Me)$_3$][PF$_6$]	1569, 1644	1619	319
[Ir(N$_2$Ph)(diphos)$_2$][PF$_6$]$_2$	1705	1705	292
IrCl$_2$(CO)(N$_2$Ph)(PPh$_3$)$_2$	1444, 1465	1463	309
PtCl(N$_2$Ph)(PEt$_3$)$_2$	1439, 1465	1455	302

^{a} Abbreviations, see footnote in Table 15.

The energy of the N—N stretching frequency has been correlated with ligand geometry. The higher values of $v(NN)$ correspond to the singly bent geometry and lower values correspond to the terminal doubly bent geometry or to the bridging, doubly bent structure (**27**). Other factors, such as the charge on the complex and the electronic nature of the metal and ancillary ligands, also affect $v(NN)$ so intermediate values of $v(NN)$ cannot be directly correlated with a specific ligand geometry. However, an attempt has been made to empirically correct for these factors and obtain a corrected value of the N—N stretching frequency, $v'(NN)$, which primarily reflects ligand geometry.[316] In certain complexes with singly bent diazenido ligands such as W(N$_2$Ph)(S$_2$CNR$_2$)$_3$ ($v(NN) = 1460$ cm^{-1}) where the canonical description (**22**) plays a role, $v(NN)$ is lower than for many complexes which contain double bent diazenido ligands. Furthermore, $v(NN)$ for singly bent acyldiazenido ligands are often low because the carbonyl group can stabilize a negative charge (**28**).[353]

$$^+M\equiv N-N=\overset{\displaystyle O^-}{\overset{\displaystyle |}{C}}-R$$

(28)

The diazenido ligand shows vibrational coupling with nitrosyl and other diazenido ligands in nitrosyl–diazenido complexes and bis(diazenido) complexes.[255,363,322,400] [RuCl(N$_2$Ar)$_2$(PPh$_3$)$_2$]$^+$ in solution is unusual in that only one N—N band is observed.[307] The values of $v(CO)$ in carbonyl–diazenido complexes are usually lower than the same values in analogous carbonyl–nitrosyl complexes indicating the superior π-accepting capability of the nitrosyl ligand.[273] Mössbauer spectra of iron diazenido complexes support the same conclusions.[363] There have been some attempts to obtain Raman spectra of diazenido complexes.[284]

Nuclear magnetic resonance spectra of the ^{15}N nucleus is one of the most important spectroscopic techniques for studying diazenido ligands. Fortunately, in many cases, isotope labeling with highly enriched ^{15}N is not difficult to accomplish. The data which are now available (Table 14) indicate that nitrogen chemical shifts are quite sensitive to geometric and electronic environment of diazenido ligand.[369] Resonances for N_α and N_β occur in the range -250 to $+300$ p.p.m. (MeNO$_2$ standard) with N_α located downfield from N_β by 60–160 p.p.m. and $^1J_{NN} = 8$–19 Hz. The ^{15}N resonances for diazenido ligands fall into three regions.

	δN_α	δN_β
High region, doubly bent	290 to 220	170 to 140
Intermediate region	120 to 40	10 to -40
Low region, singly bent	0 to -90	-120 to -240

Table 14 ^{15}N Chemical Shift Data for Selected Diazenido Complexes[a]

Complex[c]	δN_α (ppm)	δN_β (ppm)	$^1J_{NN}$ (Hz)	Solvent
MoBr(N$_2$Et)(diphos)$_2$	-29.0	-146.8	12	THF
MoCl(N$_2$C(O)Me)(diphos)$_2$	-35.4	-123.7	12	THF
WBr(N$_2$Et)(diphos)$_2$	-28.2	-164.7	12	THF
W(N$_2$Ph)(S$_2$CNMe$_2$)$_3$	-38.2	-138.0	16	CH$_2$Cl$_2$
WCl(N$_2$C(O)Me)(diphos)$_2$	-32.2	-134.5	12	THF
WBr(N$_2$H)(diphos)$_2$	-25.9	-187.1	14	THF
ReCl$_2$(N$_2$C(O)Ph)(C$_5$H$_5$N)(PPh$_3$)$_2$	-55.9	-148.6	15	THF
ReBr$_2$(N$_2$Ph)(N$_2$HPh)(PPh$_3$)$_2$	-3.7	-124.7	13	CH$_2$Cl$_2$
RuCl$_3$(N$_2$Ph)(PPh$_3$)$_2$	-46.8	-185.6	—[b]	CH$_2$Cl$_2$
[IrCl(N$_2$Ph)(PPh$_3$)$_2$][PF$_6$]	-89.9	-236.4	8	CH$_2$Cl$_2$
PhN$_2$PF$_6$	-66.9	-149.7	1.6	Me$_2$CO
[PhN$_2$·18-crown-6][PF$_6$]	-63.4	-155.8	<0.4	Me$_2$CO
N$_2$ (dissolved)	-71.3	-71.3	—	Me$_2$CO
ReCl$_2$(N$_2$C(O)Ph)(PPh$_3$)$_2$	$+157.4$	-72.0	23	CH$_2$Cl$_2$
[Ru(CO)$_2$(N$_2$Ph)(PPh$_3$)$_2$][BF$_4$]	$+116.8$	-25.2	16	CH$_2$Cl$_2$
OsH(CO)(N$_2$Ph)(PPh$_3$)$_2$	$+98.9$	-35.5	17	CH$_2$Cl$_2$
[RhCl(N$_2$Ph)(PPh$_2$Me)$_3$][PF$_6$]	$+109.2$	$+4.8$	16	CH$_2$Cl$_2$
[IrCl(N$_2$Ph)(PPh$_2$Me)$_3$][BF$_4$]	$+64.0$	-23.6	15	CH$_2$Cl$_2$
[Ir(N$_2$Ph)(diphos)$_2$][PF$_6$]$_2$	$+46.4$	-38.2	14	CH$_2$Cl$_2$
[IrCl(N$_2$Ph)(PPh$_3$)$_3$][PF$_6$]	$+58.6$	-22.1	14	CH$_2$Cl$_2$
RhCl$_2$(N$_2$Ph)(PPh$_2$Et)$_2$	$+241.0$	—[b]	16	CH$_2$Cl$_2$
IrCl$_2$(CO)(N$_2$Ph)(PPh$_3$)$_2$	$+240.2$	$+148.4$	—[b]	CH$_2$Cl$_2$
PtCl(N$_2$Ph)(PEt$_3$)$_2$	$+285.0$	$+162.0$	19	THF
[IrBr(N$_2$Ph)(diphos)$_2$][PF$_6$]	$+220.5$	$+158.3$	18	CH$_2$Cl$_2$

[a] Taken from ref. 369 plus unpublised data by same authors, external MeNO$_2$ standard.
[b] Not measured.
[c] Abbreviations, see footnote in Table 15.

The upper region contains resonances for doubly bent diazenido ligands. The lower region contains singly bent diazenido ligands. The free benzenediazonium ion also falls in this region. The complexes with the highest values of ν(NN) (RuCl$_3$(N$_2$Ph)(PPh$_3$)$_2$; IrCl(N$_2$Ph)(PPh$_3$)$_2^+$) also have the lowest ^{15}N chemical shifts (in the region of free PhN$_2^+$) and the smallest values of $^1J_{NN}$. There is an intermediate region which contains resonances for five-coordinate, d^8 diazenido complexes of the second and third transition series. The structures of two members of this group are known (OsH(CO)(N$_2$Ph)(PPh$_3$)$_2$, IrCl(N$_2$Ph)(PPh$_2$Me)$_3^+$), and both display structural distortions away from the trigonal bipyramidal, singly bent geometry toward the square pyrimidal, doubly bent geometry. The protonation at N_α in doubly bent diazenido complexes causes a dramatic upfield shift of this nitrogen by ~300 p.p.m., causing N_α to be located *upfield* from N_β by ~150 p.p.m. It appears that the values of the coupling constants $^2J_{NY}$ (Y = F,P,C,H ligands) are useful in understanding the stereochemistry around the metal and the environment of the diazenido ligand.

X-Ray photoelectron spectroscopy has been used to study diazenido complexes.[264,265,348,304,308,361,410,411] Sample decomposition, problems with sample charging, and poor spectral resolution (range of chemically induced shifts *vs.* spectral line widths) limit the general utility of XPES. The diazenido ligand shows a large positive ligand shifts (RN$_2^+$ = 1.6 eV, RN$_2^-$ = $+0.6$ eV) to the core bending energy of the metal; only the nitrosyl ligand has a larger ligand shift.[361] These observations are in harmony with π-acidic nature of the diazenido ligand. Electronic substituent effects of groups on aromatic rings of aryldiazenido ligands are strongly transmitted to the metal by singly bent ligands (RuX$_3$(N$_2$Ar)L$_2$), but they are only weakly transmitted by doubly bent ligands $(RhX_2(N_2Ar)L_2)$.[308] Nitrogen ($1S$) core bending energies show that singly and doubly bent ligands do not have large differences in effective charge thus showing that the RN_2^+/RN_2^- formalism is somewhat artificial. Notwithstanding, singly bent diazenido ligands seem to be slightly more positively charged than $RN{=}NH$, and doubly bent ligands seem to be more negatively charged than $RN{=}NH$.[308] XPE spectra show that the diazenido ligand strives toward electroneutrality. The two nitrogen atoms show differences in binding energies of only 0.6–$1.7\,eV$ (N_α minus N_β); this difference cannot be used to distinguish singly from doubly bent ligands.[304] In some cases the XPE emissions of the two nitrogen atoms in a diazenido ligand cannot be resolved into two peaks.

13.3.6.6 Reactions of Organodiazenido Ligands

There are several interesting and useful reactions of coordinated —NNR ligands. Diazenido complexes undergo protonation or alkylation on N_β in some d^4 complexes and d^6 complexes (N_2R considered to have positive charge) and protonation on N_α in five-coordinate d^8 complexes and in complexes with doubly bent diazenido ligands (equations 147–151).

$$
\begin{array}{c}
& \xrightarrow{\text{HBF}_4} & Mo(NNHPh)(S_2CNMe_2)_3^+ \\
Mo(N_2Ph)(S_2CNMe_2)_3 & & \\
& \xrightarrow{\text{MeI}} & Mo(NNMePh)(S_2CNMe_2)_3^+
\end{array}
\qquad (147)^{[398]}
$$

$$
WBr(N_2Me)(diphos)_2 \xrightarrow{\text{HBF}_4} W(NNHMe)(diphos)_2^+ \qquad (148)^{[397]}
$$

$$
CpW(CO)_2(N_2Me) + Cr(CO)_5(THF) \longrightarrow Cp(CO)_2W\overline{NNMe}\,Cr(CO)_5 \qquad (149)^{[390]}
$$

$$
Re(CO)_2(N_2Ph)(PPh_3)_2 \xrightarrow{\text{HX}} ReX(HN{=}NPh)(CO)_2(PPh_3)_2 \qquad (150)^{[400]}
$$

$$
IrCl_2(CO)(N_2Ph)(PPh_3)_2 \xrightarrow{\text{HBF}_4} IrCl_2(CO)(HN{=}NPh)(PPh_3)_2^+ \qquad (151)^{[271]}
$$

Other Lewis acids including transiton metals also bind to N_β in these complexes.[386] Hydride sources can attack diazenido ligands at N_α forming the expected diazene complex (equation 152).

$$
\begin{array}{c}
& \xrightarrow{\text{NaBH}_4} & CpRe(CO)_2(HN{=}NAr) \\
CpRe(CO)_2(N_2Ar)^+ & & \\
& \xrightarrow{\text{LiMe}} & CpRe(CO)_2(NNMeAr)
\end{array}
\qquad (152)^{[441]}
$$

However, sources of carbanions attack at N_β in the same complexes forming unusual, doubly bent —NNR$_2$ ligands owing to the greater steric demands of the organic carbanions. There are some cases where the isodiazene (doubly bent —NNHR) form of the ligand is as stable as or more stable than the normal diazene (HN=NR) form (equations 153 and 154). There are several instances of diazenido complexes being converted into dinitrogen complexes which offers the possibility of preparing the isotopomers, M—^{15}N≡^{14}N and M—^{14}N≡^{15}N (equations 155 and 156). In contrast, there are reactions of aryldiazenido complexes which extrude N_2 while retaining the aryl group in the coordination sphere (equation 157).

$$
ReCl(N_2Ph)_2(PPh_3)_2 \xrightarrow{\text{HCl}} ReCl_2(N_2Ph)(NNHPh)(PPh_3)_2 \qquad (153)^{[402]}
$$

$$
ReBr(NO)(N_2Ph)(PPh_3)_2 \xrightarrow{\text{HBr}} ReBr_2(NO)(NNHPh)(PPh_3)_2
$$
$$
+ \; ReBr_2(NO)(HN{=}NPh)(PPh_3)_2 \qquad (154)^{[400]}
$$

$$
(C_5Me_5)Re(CO)_2(N_2Ar)^+ \xrightarrow{\text{KI}} (C_5Me_5)Re(CO)_2(N_2) + PhI \qquad (155)^{[375]}
$$

$$
ReCl_2(N_2C(O)Ph)(PPh_3)_2 \xrightarrow[\text{MeOH}]{\text{CO}} ReCl(N_2)(CO)_2(PPh_3)_2 + PhCO_2Me \qquad (156)^{[299]}
$$

$$Pd(N_2Ar)(PPh_3)_3^+ \quad\underset{KBr}{\overset{h\nu}{\diagup\diagdown}}\quad \begin{array}{l} PdAr(PPh_3)_3^+ + N_2 \\[2ex] PdBr(Ar)(PPh_3)_2 + N_2 \end{array} \qquad (157)^{368}$$

There are fewer documented examples of these N_2-elimination reactions than might be expected. Aryldiazenido complexes are known to react with molecular hydrogen to yield arylhydrazine complexes *via* aryldiazene intermediates (equations 158 and 159).

$$Pt(BF_4)(N_2Ar)(PPh_3)_2 \xrightarrow{H_2} PtH(H_2NNHAr)(PPh_3)_2^+ BF_4^- \qquad (158)^{321}$$

$$RhCl_2(N_2Ar)(PPh_3)_2 \xrightarrow{H_2} [RhCl_3(H_2NNHAr)(PPh_3)]_2 \qquad (159)^{261}$$

Chiral diazenido complexes of the type $CpReX(CO)(N_2Ar)$ and $CpFe(CO)(N_2Ar)(PR_3)$ are known, and diastereomers of iron complexes with chiral PR_3 ligands have been separated.[372] There is little known ligand substitution chemistry involving the displacement of diazenido ligands (equation 160); however, the reductively fragile diazenido ligand has been used to advantage in displacing kinetically inert carbonyl ligands (equations 161 and 162).

$$IrCl(N_2Ph)(PPh_3)_2^+ + NO^+ \longrightarrow IrCl(NO)(PPh_3)_2^+ + PhN_2^+ \qquad (160)^{271}$$

$$Fe(CO)_3(PPh_3)_2 \xrightarrow{PhN_2^+} Fe(CO)_2(N_2Ph)(PPh_3)_2^+ \xrightarrow[LiOEt]{PPh_3} Fe(CO)_2(PPh_3)_3 \qquad (161)^{289,376}$$

$$Ru(CO)_3(PPh_3)_2 \xrightarrow{PhN_2^+} Ru(CO)_2(N_2Ph)(PPh_3)_2^+ \xrightarrow{NaBH_4} RuH_2(CO)_2(PPh_3)_2 \qquad (162)^{313}$$

13.3.6.7 Survey of Diazenido Complexes

Table 15 gives a listing of diazenido complexes. Most transition metals are represented except for the vanadium and copper groups. There are no diazenido complexes of Zr, Hf, Tc and Ni, but it is not unlikely that diazenido complexes of these metals can be prepared under the proper conditions. The reduced *trans* influence of RN_2^- coupled with its greater stability towards aerial oxidation as compared with analogous nitrosyls means that larger numbers of 20-electron (MEC) complexes are known. Several bis(diazenido) complexes have been reported. The ability to modulate the properties of the RN_2 ligand by varying the nature of the organic group attached to nitrogen coupled with newer developments in synthetic pathways to these complexes should further expand the number and kinds of diazenido complexes which can be prepared.

13.3.7 HYDRAZIDO(2−) AND HYDRAZIDO(1−) COMPLEXES

13.3.7.1 Introduction

The interest in these ligands has stemmed largely from their proven involvement in the chemistry of coordinated dinitrogen. However, they are of considerable intrinsic interest as their electronic flexibility permits them to bind to transition metals with a wide variety of bonding modes.

The participation of hydrazido(2−) (NNH_2) and hydrazido(1−) ($NHNH_2$) ligands in the protonation of coordinated N_2 has been extensively discussed elsewhere[445,446] and will not be considered in great detail here. A relatively brief discussion of the chemistry of hydrazide ligands in general was included in a 1983 account of the nitrogenase models,[457] but the area has not been reviewed since.

13.3.7.2 Methods of Synthesis of Hydrazide Complexes

13.3.7.2.1 Hydrazido(2−) complexes

Both protonated and alkyl-substituted hydrazido(2−) complexes are readily available from the protonation or alkylation of coordination dinitrogen (equations 163 and 164).

Table 15 Listing of Organodiazenido Complexes

Diazenido Complex[a]	Ref.	Diazenido Complex[a]	Ref.
$CpTiCl_2(N_2Ar)$	334, 381	$Cp_2TiCl(N_2Ar)$	334
$CpCr(CO)_2(N_2Ar)$	280	$[(C_6Me_6)Cr(CO)_2(N_2Ar)]^+$	283, 340
$(C_6HMe_6)Cr(CO)_2(N_2Ar)$	283		
$CpMo(CO)_2(N_2R)$	247–249, 252, 293, 303, 335, 366, 378 389, 390, 396	$CpMo(CO)(PR_3)(N_2Ar)$	248, 335, 366
$CpMo(CO)(RNC)(N_2Ar)$	293	$CpMo(RNC)_2(N_2Ar)$	293
$[CpMo(SR)(N_2Ar)]_2$	248	$[CpMo(PR_3)(N_2Ar)_2]^+$	335, 366
$CpMoX(N_2Ar)_2$	335, 389	$(RBpz_3)Mo(CO)_2(N_2Ar)$	257, 287, 290, 301, 282
$(RBpz_3)Mo(CO)(PPh_3)(N_2Ar)$	382	$(RBpz_3)MoX(NO)(N_2Ar)$	290
$[(RBpz_3)MoX(N_2Ar)]_2$	287	$(RBpz_3)Mo(SR)_2(N_2Ar)$	382
$[(C_5H_4PPh_3)Mo(CO)_2(N_2Ar)]^+$	285, 288	$[(C_5H_4PPh_3)Mo(CO)(PPh_3)(N_2Ar)]^+$	288
$[(RCpz_3)Mo(CO)_2(N_2Ar)]^+$	296	$Mo_2O(N_2C(O)Ar)_2(S_2CNR_2)_2$	352
$Mo(N_2R)(S_2CNR_2)_3$	165, 329–332, 359, 374, 398	$Mo(N_2R)_2(S_2CNR_2)_2$	332, 335, 343, 359, 374
$Mo(N_2R)(HNNHCS_2R)(S_2CNR_2)_2$	380	$Mo(N_2R)(HNNHR)(S_2CNR_2)_2$	359
$Mo(N_2R)(HNNHCO_2R)(S_2CNR_2)_2$	332, 359	$Mo(N_2Ar)_2(ox)_2$	358, 398
$Mo(N_2Ar)_2(tox)_2$	358	$Mo(N_2Ar)_2(acac)_2$	335
$MoX_2(N_2Ar)_2(PR_3)_2$	367	$MoX(N_2R)(diphos)_2$	312, 338, 551, 387, 388, 391, 395, 397, 439, 449, 451, 437
$Mo(OH)(N_2R)(diphos)_2$	388	$MoX(N_2C(O)R)(diphos)_2$	298, 312, 353, 370, 392, 395
$MoX(N_2C(O)Ar)(HNNC(O)Ar)(PR_3)_2$	337	$[Mo(CO)(PPh_3)_2(N_2Ar)(bipy)]^+$	336
$[Mo(CO)_3(N_2Ar)(bipy)]^+$	336	$Mo(N_2Ar)_2(L')$	417, 418
$[Mo(CO)_2(PPh_3)(N_2Ar)(bipy)]^+$	336	$MoX(CO)_2(N_2Ar)(bipy)$	336
$MoX_3(N_2Ar)(bipy)$	336	$MoX(CO)_2(N_2Ar)(diars)$	343
$Mo(CO)(N_2Ar)(S_2CNR_2)(diars)$	343	$[Mo(CO)_3(N_2Ar)(diars)]^+$	343
$[Mo(CO)_2(PPh_3)(N_2Ar)(phen)]^+$	336	$[Mo(CO)(PPh_3)_2(N_2Ar)(phen)]^+$	336
$MoX(CO)_2(N_2Ar)(phen)$	336	$Mo(OH)(CO)_2(N_2Ar)(phen)$	386
$MoX_3(N_2Ar)(phen)$	336	$[Mo(N_2Ar)_2(phen)_2]^{2+}$	335
$[Mo(N_2Ar)_2(bipy)]^{2+}$	335	$[Mo(N_2Ar)_2\{S_2C=C(CN)\}_2]^{2-}$	335
$Mo(N_2Ar)_2(tpp)$	378	$[Mo(OH)(CO)_2(N_2Ar)]_4$	286
$[Mo(OH)(OPPh_3)(CO)_2(N_2Ar)]_4$	286	$CpMo(CO)_2(N\overline{N}R)Cr(CO)_5$	390
$CpMo(CO)_2(N\overline{N}R)MnCp(CO)_2$	390	$CpMo(CO)_2(N\overline{N}R)ReCp(CO)_2$	379
$Mo(N_2Ar)(MeCN)(SR)_3$	403, 405	$Mo(N_2Ar)(PR_3)(SR)_3$	403
$Mo(N_2C(O)OPh)(HNNHC(O)OPh(L')$	418	$[Mo_2(NPh)_4(OMe)_2(MoO_4)_2]^{2-}$	405
$[Mo_2(N_2Ph)_4(SAr)_5]^-$	405, 415	$Mo_2Cl_2(N_2Ph)_4(OEt)_4$	405
$Mo_2(N_2Ph)_4(acac)_2(OMe)_2$	439	$[Mo_2(N_2Ph)(NNHPh)(SCH_2CH_2S)_3 - (SCH_2CH_2SH)]^{2-}$	416
$[Mo_2(N_2Ph)(NNHPh)(SAr)_7]^{2-}$	416	$[Mo_2(N_2Ph)(NNH-Ph)(OMe)(SAr)_6]^{2-}$	416
$CpW(CO)_2(N_2R)$	246, 247, 252, 335, 342, 366, 393, 396, 397	$CpW(CO)\{P(OPh)_3\}(N_2R)$	246
$CpWX(N_2Ar)_2$	335	$CpW(CO)(PPh_3)(N_2R)$	247
$[(RBpz_3)WX_2(N_2Ar)]_2$	287	$(RBpz_3)W(CO)_2(N_2Ar)$	251, 287, 290 382

$(RBpz_3)W(CO)(PPh_3)(N_2Ar)$	382	$(RBpz_3)WX(NO)(N_2Ar)$	290
$(RBpz_3)WX_2(CO)(N_2Ar)$	287	$[(C_5H_4PPh_3)W(CO)_2(N_2Ar)]^+$	285, 288
$WX(N_2R)(diphos)_2$	312, 395, 439, 450, 451, 437	$WX(N_2C(O)R)(diphos)_2$	298, 312, 370, 395
$WX(N_2C(O)R)(PR_3)_4$	370	$WCl_2(N_2Ar)_2(PR_3)_2$	367
$[W(CO)_3(N_2Ar)(diars)]^+$	343	$W(N_2Ar)(S_2CNR_2)_3$	400
$CpW(CO)_2(NN\overline{R})Cr(CO)_5$	383, 390	$CpW(CO)_2(NN\overline{R})MnCp(CO)_2$	390
$WCl_2(N_2Ph)_2(py)_2$	401		
$[CpMn(CO)_2(N_2Ar)]^+$	375, 419, 442	$[CpMn(CO)_2(N_2R)]^-$	333, 394, 443
$Mn_3(CO)_{12}(N_2Me)$	339, 414	$[Mn(CO)_4(N_2Ar)]_2$	291, 326, 355
$Mn(CO)_3(N_2Ar)(PPh_3)$	326	$Mn(CO)_2(N_2Ar)(PPh_3)_2$	291, 322, 326
$[Mn(CO)(N_2Ar)_2(PPh_3)_2]^+$	322	$[Mn(CO)(NO)(N_2Ar)(PPh_3)_2]^+$	322
$MnX(N_2Ar)_2(PPh_3)_2$	322	$MnX(NO)(N_2Ar)(PPh_3)_2$	322
$CpMn(C(O)NR_2)(CO)(N_2Ar)$	419		
$[CpRe(CO)_2(N_2Ar)]^+$	375, 413, 419, 420, 442, 441, 440	$Re(CO)_2(N_2Ar)(PPh_3)_2$	322
$ReX(N_2Ar)_2(PPh_3)_2$	400, 402, 404	$ReX_2(NNHAr)(N_2Ar)(PPh_3)_2$	400, 402
$[ReX_2(HNNHAr)(N_2Ar)(PPh_3)_2]^+$	400	$[ReX_2(N_2Ar)_2(PPh_3)_2]^+$	400
$ReX_2(N_2Ar)(PR_3)_3$	256, 276, 306	$ReX_2(CO)(N_2C(O)Ar)(PR_3)_2$	384
$ReX_2(N_2C(O)Ar)(PR_3)_3$	254, 260, 276, 299, 312, 384, 385	$ReX_2(N_2C(O)Ar)\{P(OR)_3\}_3$	384, 385, 452
$ReX_2(CO)(N_2Ar)(PR_3)_2$	254, 306	$ReX_2(NH_3)(N_2Ar)(PR_3)_2$	276, 306
$ReX_2(py)(N_2C(O)Ar)(PR_3)_2$	254	$ReX_2(RCN)(N_2C(O)Ar)(PR_3)_2$	254, 259
$ReX_2(N_2C(O)Ar)(PR_3)_2$	254, 259, 299, 300, 384, 385, 452	$ReCl_3(N_2C(O)Ar)(PR_3)_2$	254
$CpReX(CO)(N_2Ar)$	419, 420	$CpRe(C(O)OR)(CO)(N_2Ar)$	419, 420
$[CpRe(CO_2)(CO)(N_2Ar)]^-$	412, 420	$CpRe(C(O)NR_2)(CO)(N_2Ar)$	419
$Re(N_2Ph)(SAr)_4$	404	$CpRe(CO)(N_2Ar)(AuPPh_3)$	412
$[Re_2(N_2Ph)_2(SPh)_7]^-$	404		
$[Fe(CO)_2(N_2Ar)(PR_3)_2]^+$	272–274, 311, 313, 372, 376, 410, 406, 406–408	$Fe(CO)(NO)(N_2Ar)(PR_3)$	272, 363, 372, 406, 409
$FeX(CO)(N_2Ar)(PR_3)_2$	289		
$RuX_3(N_2Ar)(PR_3)_2$	261–263, 270, 292, 307, 309, 316	$[Ru(CO)_2(N_2Ar)(PR_3)_2]^+$	277, 292, 310, 313
$RuH(CO)(N_2Ar)(PPh_3)_2$	310	$RuX(CO)(N_2Ar)(PPh_3)_2$	310
$RuH(CO)_2(N_2Ar)(PPh_3)_2$	310	$RuX(CO)_2(N_2Ar)(PPh_3)_2$	292, 310
$[RuX(N_2Ar)(bipy)_2]^{2+}$	281, 341	$[RuX(N_2Ar)_2(PPh_3)_2]^+$	263, 307
$[RuX_2(N_2Ar)(PPh_3)_2]^+$	263, 307		
$OsX_3(N_2Ar)(PPh_3)_2$	261, 262, 292, 310, 411	$[Os(CO)_2(N_2Ar)(PR_3)_2]^+$	310
$OsX(CO)(N_2Ar)(PPh_3)_2$	310	$OsH(CO)(N_2Ar)(PPh_3)_2$	310, 315
$OsX(CO)_2(N_2Ar)(PPh_3)_2$	310	$OsH(CO)_2(N_2Ar)(PPh_3)_2$	310
$Os_3H(CO)_{10}(N_2Ar)$	377		
$[Co(N_2Ar)(PPh(OEt)_2)_4]^{2+}$	328	$[Co(N_2Ar)(P(OEt)_3)_4]^{2+}$	328
$Co(CO)_2(N_2Ar)(PPh_3)$	272, 363, 366		
$RhX_2(N_2Ar)(PPh_3)_2$	261, 262, 269, 295	$[RhX(N_2Ar)(PPP)]^+$	282, 292, 354
$[RhX(N_2Ar)(PR_3)_3]^+$	292, 400	$[RhX(N_2Ar)(diphos)_2]^+$	292
$[Rh(N_2Ar)(BuNC)_4]^{2+}$	275	$[Rh_2(CO)_2(N_2Ar)(Ph_2PCH_2PPh_2)_2]^+$	399

Table 15 (continued)

Diazenido Complex[a]	Ref.	Diazenido Complex[a]	Ref.
$IrX_2(CO)(N_2Ar)(PR_3)_2$	244, 267, 271, 317, 320	$IrX_2(RNC)(N_2Ar)(PPh_3)_2$	271
$IrX_2(py)(N_2Ar)(PPh_3)_2$	400	$IrX_2(N_2Ar)(PPh_3)_2$	261, 271
$IrX(C_2Ph)(CO)(N_2Ar)(PPh_3)_2$	318	$[IrX(N_2Ar)(PR_3)_3]^+$	271, 319
$IrX(N_2Ar)(AsPh_3)(PPh_3)_2]^+$	271, 319	$[IrX(N_2Ar)(SbPh_3)(PPh_3)_2]^+$	271
$[Ir(N_2Ar)(diphos)_2]^{2+}$	292	$[IrX(N_2Ar)(diphos)_2]^+$	292
$[IrX(C_5Cl_4N_2)(N_2Ar)(PPh_3)_2]^+$	324	$[IrX(N_2Ar)(PR_3)_2]^+$	271
$IrX(CO)(N_2Ar)(PPh_3)_2]^+$	271, 317, 318, 320	$IrX(CO)(N_2C_6H_3R)(PPh_3)_2$	357
$[IrX(RNC)(N_2Ar)(PPh_3)_2]^+$	271, 319	$[IrX(NH_3)(N_2Ar)(PPh_3)_2]^+$	319
$[IrX(PF_3)(N_2Ar)(PPh_3)_2]^+$	319	$[IrX(py)(N_2Ar)(PPh_3)_2]^+$	319
$[Ir_2(CO)_4(SO_2)(N_2Ar)(PPh_3)_2]^+$	373	$Ir_2X(CO)_2(SO_2)(N_2Ar)(PPh_3)_2$	373
$[Ir_2O(NO)_2(N_2Ar)(PPh_3)_2]^+$	279, 346	$Ir_2(CO)_4(N_2C_6H_3R)_2(PPh_3)_2$	327
$[Pd(N_2Ar)(PPh_3)_3]^+$	368	$[Pd_2Cl_2(N_2Ar)(Ph_2PCH_2PPh_2)_2]^+$	350
$PtX(N_2Ar)(PR_3)_2$	242, 250, 257, 258, 266, 278, 314, 325	$[Pt(py)(N_2Ar)(PEt_3)_2]^+$	257
$[Pt(RNC)(N_2Ar)(PEt_3)_2]^+$	257	$[Pt(NH_3)(N_2Ar)(PEt_3)_2]^+$	257
$[Pt(N_2Ar)(PR_3)_3]^+$	257, 125	$[Pt(N_2Ar)(PPh_3)_2]_2^{2+}$	253, 278, 321
$Pt(o\text{-}N_2C_6H_4CO_2)(PPh_3)_2$	245	$Pt(o\text{-}N_2C_6H_4O)(PPh_3)_2$	245
$Pt(o\text{-}N_2C_6H_4SO_2)(PPh_3)_2$	245		

[a] Abbreviations: X = halide or pseudohalide, R = alkyl or aryl, Ar = aryl, pz = $C_3H_3N_2$, ox = oxinate, tox = thiooxinate, oep = octaethylporphyrin, diphos = $Ph_2PCH_2CH_2PPh_2$, diars = $o\text{-}C_6H_4(AsMe_2)_2$, PPP = $Ph_2P(CH_2CH_2CH_2PPh_2)_2$, tpp = tetraphenylporphyrin, L' = $SCH_2CH_2N(Me)CH_2CH_2N(Me)CH_2CH_2S$.

$$[Mo(N_2)_2(dppe)_2] + 2RX \longrightarrow [Mo(NNR_2)X(dppe)_2]X + N_2 \tag{163}$$

$$[W(N_2)_2(PMe_2Ph)_4] + 3HCl \longrightarrow [WCl_2(NNH_2)(PMe_2Ph)_3] + [HPMe_2Ph]Cl + N_2 \tag{164}$$

Protonation or alkylation of aryldiazenido precursors provides a route to aryl-substituted hydrazido(2−) complexes (equations 165, 166 and 167). Diazenido complexes have been covered in Section 13.3.6.

$$[Mo(NNPh)(S_2CNMe_2)_3] + MeI \longrightarrow [Mo(NNMePh)(S_2CNMe_2)_3]I \tag{165}^{458}$$

$$[ReCl(NNPh)_2(PPh_3)_2] + HCl \longrightarrow [ReCl_2(NNHPh)(N_2Ph)(PPh_3)_2] \tag{166}^{459}$$

$$[(\eta^5\text{-}C_5H_5)Re(CO)_2(N_2Ph)]^+ + Me^- \longrightarrow [(\eta^5\text{-}C_5H_5)Re(CO)_2(NNMePh)] \tag{167}^{460}$$

The most widely used methods for the syntheses of monohydrazido(2−) complexes have been from the parent hydrazide and metal oxo or halide complexes with formal elimination of water (equation 168), hydrogen halide (equation 169) or trimethylchlorosilane (equation 170). The preparation of the hydrazido(2−) complexes, shown in equation (169) and (170), shows that the driving force for the formation of metal–nitrogen multiple bonds is such that the N_α hydrogens are lost.

$$[MoO_2(S_2CNMe_2)_2] + Me_2NNH_2 \longrightarrow [MoO(NNMe_2)(S_2CNMe_2)_2] \tag{168}^{461}$$

$$MoCl_5 + MePhNHNH_2 \xrightarrow{CH_2Cl_2} MoCl_4(NNMePh) + HCl \tag{169}^{462}$$

$$[MoCl_4(PPh_3)_2] + 2Me_3SiNHNMePh \longrightarrow [MoCl(NNMePh)_2(PPh_3)_2]Cl \tag{170}^{463}$$

An unusual route to hydrazido(2−) complexes involves the rearrangement of $Me_3Si-N{=}N-SiMe_3$ on reaction with metallocenes (equation 171).

$$[Cp_2V] + Me_3Si-N{=}N-SiMe_3 \longrightarrow [Cp_2V{=}N-N(SiMe_3)_2] \tag{171}^{464}$$

13.3.7.2.2 *Hydrazido(1−) complexes*

These are not accessible from coordinated dinitrogen and the preparative routes involve alkylation (equation 172) or protonation (equation 173) of hydrazido(2−) precursors or direct use of the hydrazine with elimination of hydrogen halide (equation 174).

$$[Mo(NNMePh)_2(S_2CNMe_2)_2] + RI \longrightarrow [Mo(NMeNMePh)(NNMePh)(S_2CNMe_2)_2]I \quad (172)^{465}$$

$$[(\eta^5\text{-}C_5H_5)Re(CO)_2(NNHPh)] + HX \longrightarrow [(\eta^5\text{-}C_5H_5)Re(CO)_2(NHNHPh)]X \quad (173)^{466}$$

$$[(\eta^5\text{-}C_5H_5)TiCl_3] + Me_2NNHMe \xrightarrow{Et_3N} [(\eta^5\text{-}C_5H_5)TiCl_2(\eta^2\text{-}Me_2NNMe)] \quad (174)^{467}$$

13.3.7.3 Structures and Chemistry of Hydrazido(2−) Complexes

These complexes are considered according to the mode of bonding of the hydrazido fragment. The known coordination types are shown in valence bond representations in Figure 15 together with the formal charge, geometry and numbers of valence electrons donated to the metal.

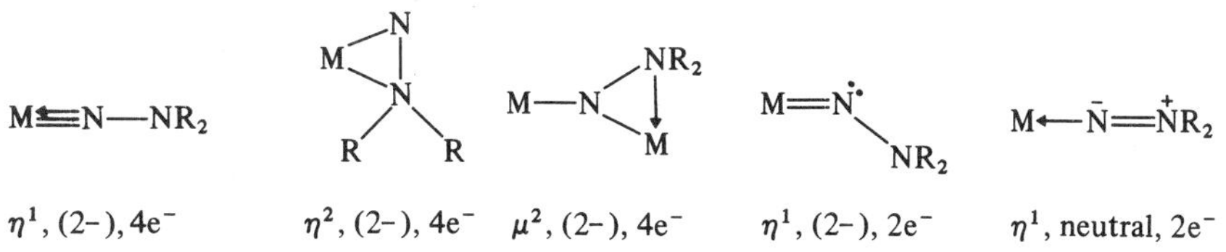

$\eta^1, (2-), 4e^-$ $\eta^2, (2-), 4e^-$ $\mu^2, (2-), 4e^-$ $\eta^1, (2-), 2e^-$ η^1, neutral, $2e^-$

Figure 15

13.3.7.3.1 *Terminal, linear, hydrazido(2−)*

The simplest member of this class is the NNH_2 ligand, which, as mentioned above, is a crucial and persistent intermediate in the protonation of coordinated dinitrogen. Indeed it remains the only coordinated protonated nitrogen ligand identifiable between dinitrogen and ammonia.[455,456] Spectroscopic studies have provided a detailed mechanistic picture of the protonation of $[Mo(N_2)_2(depe)_2]$ (M = Mo, W; depe = $Et_2PCH_2CH_2PEt_2$).[472]

Many NNH_2 complexes have now been structurally characterized and some representative examples are shown in Table 16. The characteristic bond lengths and angles within the hydrazido(2−) ligand are also given. These change little with different metals and coligands and are consistent with the NNH_2 ligand functioning as a four electron donor. ^{15}N NMR has proved to be an invaluable diagnostic tool for the characterization of reduced nitrogen species[473] in general and the NNH_2 ligand in particular. Generally two widely spaced resonances are observed. Thus $[MoF(NNH_2)(dppe)_2]^+$ has resonances at −83 and −243.9 p.p.m. relative to C_2H_3NO; the higher field resonance is assigned to the NH_2 nitrogen.

The cationic $[MX(NNH_2)(dppe)_2]^+$ complexes do not protonate further even under forcing conditions, due to the overall positive charge and the stability of the chelated phosphine ligands. However, the monophosphine complexes $[MX_2(NNH_2)(PMe_2Ph)_3]$ undergo rapid further protonation to give ammonia, hydrazine or dinitrogen. The relative proportions of these products are a complex function of metal, coligands and solvent.[455,474] Despite intensive efforts no intermediates between NNH_2 and NH_3 have been unequivocally identified. Indeed a species initially formulated as $[WCl_3(NHNH_2)(PMe_2Ph)_3]$ was subsequently shown to be the hydrido–hydrazido(2−) complex $[WHCl_3(NNH_2)(PMe_2Ph)_3]$.[475] There is a dramatic difference between the Mo and W NNH_2 complexes, the former giving up to one equivalent of NH_3 whereas the W analogue yields up to two equivalents. This is in part due to the molybdenum completing the reaction in a lower oxidation state than tungsten whose oxidation state rises to six. The additional electrons donated to nitrogen in the tungsten system provide the reduction necessary to generate the addition NH_3. In addition, ^{15}N labelling studies suggest that degradation of the $Mo=NNH_2$ complex proceeds *via* disproportionation reactions where the tungsten system protonates stepwise.[476]

The NNH_2 ligand is the most reactive hydrazido(2−) ligand due to the comparative lability of the NH_2 hydrogens. Thus reaction of $[MX(NNH_2)(dppe)_2]^+$ (M = Mo, W) with one equivalent of Et_3N generates the diazenido complexes $[MX(NNH)(dppe)_2]$ in good yield.[477] Uses of excess Et_3N

Nitrogen Ligands

Table 16 Representative Structural Data for Hydrazido(2−) Complexes

Complex	M–N (Å)	N–N (Å)	M–N̂–M (°)	Ref.
η^1, (2−), linear				
[WCl(NNH$_2$)(dppe)$_2$]BPh$_4$	1.73(1)	1.37(2)	171(1)	14
[Mo(NNH$_2$)(ox)(PMe$_2$Ph)$_3$]Br	1.743(4)	1.37(7)	172.3(5)	15
[WCl$_3$(NNH$_2$)(PMe$_2$Ph)$_2$]	1.752(10)	1.300(17)	178.7(9)	16
[MoCl(NNH$_2$)(triphos)(PPh$_3$)]$^+$	1.694(12)	1.34(2)	172.2(11)	17
[VCl$_2$(NNMePh)(H$_2$NNMePh)$_2$]Cl	1.696(24)	1.295(17)	174.4(19)	35
[MoO(NNMe$_2$)(S$_2$CNMe$_2$)$_2$]	1.799(8)	1.288(10)	168.0(7)	36
[MoO(NNMe$_2$)(ox)$_2$]	1.800(9)	1.28(1)	155.5(9)	37
[MoO(NNMe$_2$)(SPh)$_3$]$^-$	1.821(9)	1.292(14)	152.5(10)	38
[W(NNPh$_2$)(SC$_6$H$_2$Pr$_3^i$)$_4$]	1.779(11)	1.317(16)	173.7(7)	39
[MoCl(NNMe$_2$)$_2$(PPh$_3$)$_2$]$^+$	1.752(5)	1.291(7)	163.8(4)	40
[Mo(NNMe$_2$)$_2$(bipy)$_2$]$^{2+}$	1.80(1)	1.27(2)	168.2(7)	40
[Mo(NNPh$_2$)$_2$(S$_2$CNMe$_2$)$_2$]	1.790(8)	1.31(1)	169.9(8)	40
[Mo$_4$O$_{10}$(OMe)$_2$(NNPh$_2$)$_2$]$^{2-}$	1.780(7)	1.32(1)	173.7(6)	46
[Mo$_8$O$_{16}$(OMe)$_6$(NNMePh)$_6$]$^{2-}$	1.71(1), 1.75(2)	—	174.3(11), 168.8(22)	46
	1.82(2)	—	173.9(18)	
[S$_2$MoS$_2$Mo(NNMe$_2$)$_2$(PPh$_3$)]	1.78(3)	1.30(4)	165.0(3)	47
[(CO)$_6$Fe$_2$S$_2$Mo(NNMe$_2$)$_2$(PPh$_3$)]	[1.758(17)]	[1.344(26)]	[178.2(15)]	45
	[1.825(15)]	[1.300(24)]	[159.4(14)]	
η^1, (2−), bent				
[ReBr$_2$(NNHPh)(NNPh)(PPh$_3$)$_2$	1.922(11)	1.287(15)	131.2(10)	5
[W(η^5-C$_5$H$_5$)$_2$H(NNHC$_6$H$_4$F)]$^+$	1.837(7)	1.315(9)	146.4(5)	48
[Re(η^5-C$_5$H$_5$)(CO)$_2$(NNHC$_6$H$_4$OMe)]	1.937(7)	1.283(10)	138.1(6)	6
η^1, neutral, linear				
[S$_2$MoS$_2$Mo(NNMe$_2$)$_2$S$_2$MoS$_2$]$^{2-}$	2.13(1)	1.16(2)	165.9(16)	49

Abbreviations: dppe = Ph$_2$PCH$_2$CH$_2$PPh$_2$; ox = 8-hydroxyquinoline; triphos = Ph$_2$PCH$_2$CH$_2$PPhCH$_2$CH$_2$PPh$_2$; bipy = 2,2′-bipyridyl.

under dinitrogen leads to reformation of parent bis(nitrogen) complexes [M(N$_2$)$_2$(dppe)$_2$]. The NNH$_2$ group reacts with carbonyl compounds under acid catalysis to give complexes of substituted diazomethane (equation 175).[478,479] These complexes can also be prepared by reaction of the bis(dinitrogen) complexes with the *gem*-dihalides R$_2$CBr$_2$.[480] The condensation reactions also occur for the monophosphine NNH$_2$ complexes and represent the only way to make N—C bonds for these systems. A number of X-ray crystal structures of diazomethane derivatives obtained from dinitrogen have been determined and are generally consistent with the bonding represented in (29).

$$[MoCl(NNH_2)(dppe)_2]^+ + R_2CO \xrightarrow{H^+} [MoCl(NNCR_2)(dppe)_2]^+ + H_2O \qquad (175)$$

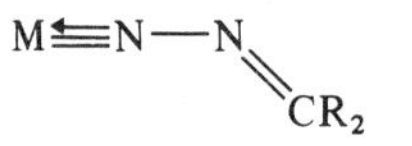

$$M \equiv N - N \diagdown_{CR_2}$$

(29)

An interesting dichlorodiazomethane complex [WBr(NNCCl$_2$)(dppe)$_2$]Br has been obtained by reaction of [WBr(NNH$_2$)(dppe)$_2$]Br with chloroform in the presence of base and diphenyliodonium bromide.[487] A mechanism involving free radicals has been postulated. If the reaction is carried out in organic solvents organonitrogen species are formed by attack of solvent radicals on the coordinated dinitrogen.[482]

$$P_4M(N_2)_2 \xrightarrow{HX} [P_4MX(NNH)] \xrightarrow{HX} P_3MX_2(NNH_2) \xrightarrow{HX} [P_2MX_3(NHNH_2)] \xrightarrow{HX} [P_2X_4(NH)] \xrightarrow{2HX} 'MX_6'$$
$$+ NH_3 \qquad + NH_3$$

P = monodentate phosphine; M = W; X = Cl, Br
Species in brackets postulated

Scheme 5

There is a reasonably extensive reported chemistry for diazomethane complexes[483] obtained by oxidative addition of a diazomethane to relatively low oxidation state metal complexes. A representative example is given in equation (176).[483]

$$[Mo(CO)_3(S_2CNEt_2)_2] + N_2C(Me)Ph \longrightarrow [Mo(CO)(NNC(Me)Ph)(S_2CNEt_2)_2] + 2CO \quad (176)$$

Extensive studies of the mechanism of formation of dialkylhydrazido(2−) complexes from $[Mo(N_2)_2(dppe)_2]$ (M = Mo, W) have revealed a free radial mechanism for addition of the first alkyl group with loss of one N_2 molecule as the rate-determining step.[484] Addition of the second alkyl group proceeds *via* an S_N2 mechanism (see Scheme 6) and the effect of the metal and coligands on reaction rates has been studied in detail.[485] Acoyl and aroyl hydrazido(2−) complexes can be obtained by reaction of the N_2 complexes with acoyl or aroyl halides in the presence of acid (equation 177).[486]

$$[M(N_2)_2(dppe)_2] \xrightarrow{-N_2} [M(N_2)(dppe)_2] \xrightarrow{RX} [M(N_2)(XR)(dppe)_2] \longrightarrow [M(N_2)X(dppe)_2] + R \longrightarrow$$

$$[MX(N_2R)(dppe)_2] \xrightarrow{RX} [MX(NNR_2)(dppe)_2]X$$

$$M = Mo, W; \ X = Cl, Br; \ R = alkyl$$

Scheme 6

$$[W(N_2)_2(dppe)_2] + RCOCl + HCl \longrightarrow [WCl(NNHCOR)(dppe)_2]Cl + N_2 \quad (177)$$

The dialkylhydrazido complexes do not undergo further protonation unless reduced chemically or electrochemically. Addition of two electrons or Bu^tLi to $[MX(NNR_2)(dppe)_2]X$ (X = Cl, Br; M = Mo, W) causes loss of two halide anions and formation of the 16 electron five-coordinate complexes $[M(NNR_2)(dppe)_2]$, presumed to contain a linear four-electron NNR_2 ligand. Further reduction by two electrons in the presence of acid and dinitrogen liberates the disubstituted hydrazine, R_2NNH_2, and with reformation of the parent bis(dinitrogen) complex (Scheme 7).[487]

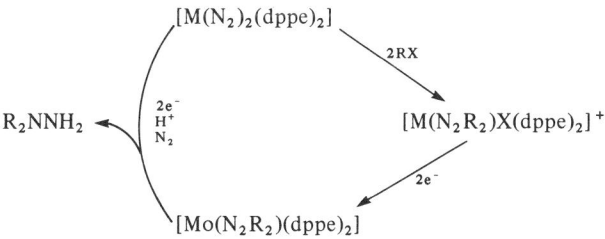

$$M = Mo, W; \ X = Cl, Br, I; \ R = alkyl$$

Scheme 7

Electrochemical reduction of the corresponding NNH_2 complexes $[MX(NNH_2)(dppe)_2]^+$ (X = Cl, Br; M = Mo, W) leads to loss of the NNH_2 hydrogens as H_2 and no further protonation of the N_2 ligand. However, a significant recent development has involved the electrochemical reduction of $[W(tosyl)(NNH_2)(dppe)_2]^+$ where the labile tosylate anion is lost in preference to H_2 evolution, and the parent bis(dinitrogen) complex is formed.[488] Concomitantly 0.2–0.3 moles of NH_3 are formed in the catholyte solution. The cycle may be repeated several times after addition of further toluenesulfonic acid, further increasing the yields of ammonia. This research augurs well for the eventual development of an electrochemically driven, metal mediated process for the reduction of dinitrogen to ammonia.

Dialkylhydrazido complexes cannot be prepared by alkylation of bis(nitrogen) complexes of monotertiary phosphines. However, these have been recently prepared from *N,N*-disubstituted hydrazines (equation 178).[462]

$$WCl_6 + RPhNNH_2 \longrightarrow WCl_4(NNRPh) \xrightarrow{PR_3} [WCl_3(NNRPh)(PR_3)_2]$$

$$\xrightarrow{PR_3, \ Na/Hg} [WCl_2(NNRPh)(PR_3)_3] \qquad R = Me, Ph \quad (178)$$

A wide range of monodisubstituted hydrazido(2−) complexes with other types of coligands have been prepared from the corresponding hydrazines. Some representative examples that have been structurally characterized are shown in Table 16. As anticipated, these complexes are found for those metals which show a pronounced tendency to form metal nitrogen multiple bonds. In general the structural parameters for the hydrazido(2−) ligands are relatively insensitive to the metal, coligands or coordination geometry.

The complexes generally have linear or nearly linear M—N—NR$_2$ systems with the exception of a few complexes such as [MoO(NNMe$_2$)(oxine)$_2$] which has an Mo—N̄—NMe$_2$ angle of 155.5(9)°.[481] The reasons for such relatively small distortions are not clear, but *ab initio* calculations on LiNNH$_2$ suggest that only small energies, of the order available from crystal packing, are necessary to induce angular distortion of up to about 25°.[495] The non-coordinated nitrogen N$_\beta$ is rigorously planar suggesting formal sp^2 hybridization. As a consequence the p orbitals on N$_\alpha$ lie in the same plane as the alkyl substituents and the orientation of the hydrazide substituents provides a useful guide to the metal orbitals involved in forming metal—nitrogen multiple bonds.

Although the hydrazido(2−) ligand has the same formal charge as the oxo group, isoelectronic oxo and hydrazido species exhibit very different redox properties. Thus MoOCl$_4$ is rapidly reduced by bulky aromatic thiolates to the Mo^V species [MoO(SAr)$_4$]$^{−}$ [496] whereas no reduction occurs with MoCl$_4$(NNMePh), the Mo^{VI} species [Mo(NNMePh)(SPh)$_4$] being formed.[497]

There is only one reported mononuclear hydrazido(2−) complex for the titanium group, [Ti(η^5-C$_5$H$_5$)$_2$(NNSiMe$_3$)$_2$].[498] Other hydrazido complexes are known but involve bridging hydrazido groups (see below). For reasons not yet fully understood, titanium is apparently reluctant to form four electron metal–nitrogen multiple bonds.

Vanadium by contrast readily forms such metal–nitrogen multiple bonds and [V(η^5-C$_5$H$_5$)$_2$\{NN(SiMe$_3$)$_2$\}] (see equation 171) and [VCl$_2$(NNMePh)(H$_2$NNMePh)$_2$]Cl[489] have been structurally characterized. The former complex formally has 20 valence electrons as the V–N–N system is virtually linear, implying it is functioning as a four electron donor. However, the C$_5$H$_5$ rings are not symmetrically bonded, thereby achieving a formal eighteen valence electron count. The complex [VCl$_2$(NNMePh)(H$_2$NNMePh)$_2$]$^+$ is prepared by reaction of [VCl$_3$(MeCN)$_3$] directly with the hydrazine.[489] The structure (30) reveals a conventional linear four electron NNMePh ligand but in addition two η^2-coordinated hydrazine ligands, making the vanadium seven-coordinate.

(30)

Hydrazido(2−) complexes are most prevalent within group VI although there are to date no reported hydrazido complexes of chromium. Representative examples of known structure are shown in Table 16. Sulfur coligands predominate reflecting the metal–sulfur coordination in nitrogenase and the participation of hydrazide ligands in dinitrogen chemistry.

The occurrence of mononuclear bis[hydrazido(2−)] complexes is restricted to molybdenum and tungsten and examples also appear in Table 16. Once again the bond distances and angles are generally similar. As with oxo groups the two hydrazido(2−) ligands are invariably *cis* in octahedral complexes to minimize π-bonding competition. In trigonal bipyramidal complexes, such as [MoCl-(NNMe$_2$)$_2$(PPh$_3$)$_2$]$^+$,[494] π bonding is maximized by the hyrazido ligands occupying equatorial sites (31). Complexes such as [Mo(NNMePh)$_2$(S$_2$CNMe$_2$)$_2$][494] and [Mo(NNMe$_2$)$_2$(bipy)$_2$]$^{2+}$ [494] have linear Mo—N—NR$_2$ systems implying each contributes four electrons to the metal. Overall the metal achieves a formal valence electron count of twenty and in contrast to systems discussed below appears stable.

(31)

The high hydrolytic and redox stability of the Mo(NNR$_2$)$_2^{2+}$ core permits the synthesis of a number of sulfido-bridged heteronuclear hydrazido(2−) clusters. The most recent is [(CO)$_6$Fe$_2$S$_2$Mo(NNMePh)(PPh$_3$)][499] obtained as a black crystalline solid by reaction of [MoCl(NNMePh)$_2$(PPh$_3$)$_2$]$^+$ with Li$_2$[Fe$_2$S$_2$(CO)$_6$]. The structure[499] (32) showed that the bridged sulfido group occupies equatorial and axial sites at the trigonal bipyramidally coordinated molyb-

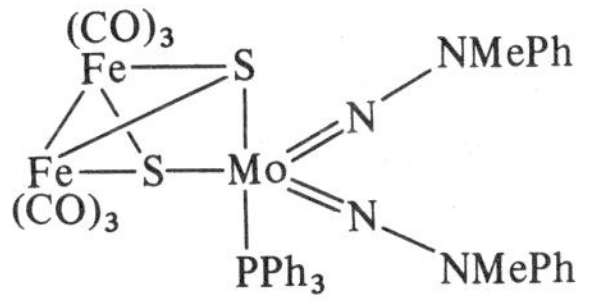

(32)

denum. The small bite angle of the bridging sulfides imposes asymmetric coordination and the NNMe$_2$ groups are not equivalent.

An interesting recent development in hydrazido(2−) chemistry has seen the preparation and structural characterization of polymolybdate clusters, in which the peripheral oxo groups have been replaced by NNR$_2$ groups.[500] Thus [Mo$_4$O$_{10}$(OMe)$_2$(NNPh$_2$)$_2$]$^{2-}$ is prepared by reaction of [Mo$_8$O$_{26}$]$^{4-}$ with Ph$_2$NNH$_2$ and has the structure shown in (33). This and other related anions are useful precursors for the synthesis of a range of complexes containing the (MoO(NNPh$_2$)$^{2+}$ core.

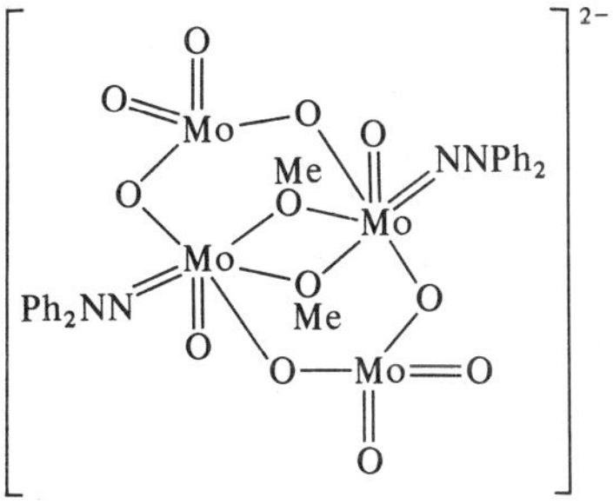

(33)

13.3.7.3.2 Terminal, 'bent', hydrazido(2 −)

The three structurally characterized examples of this type of bonding are shown in Table 16 and all would involve a twenty electron valence count for the metal if the hydrazido(2−) ligand functioned as a four-electron donor. The structure of [Re(η^5-C$_5$H$_5$)(CO)$_2$(NNMeC$_6$H$_4$OMe)]^{460} is typical and is shown in (34). The formal localization of a lone pair on N$_\alpha$ causes a significant increase in the metal nitrogen distance consistent with decreased multiple bonding.

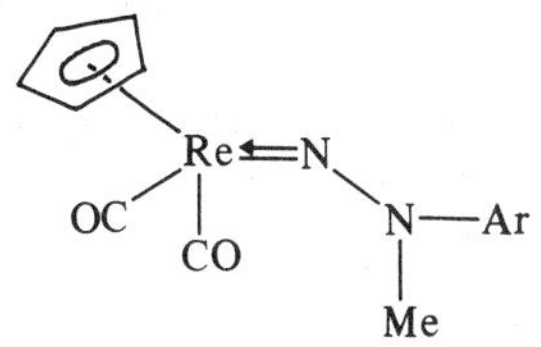

(34)

13.3.7.3.3 Terminal, neutral, hydrazido

The only example of the 'isodiazene' type of bonding for the NNR$_2$ groups is provided by [Mo$_3$S$_8$(NNMe$_2$)$_2$]$^{2-}$ whose structure[503] is represented in (35). Strong electron delocalization within

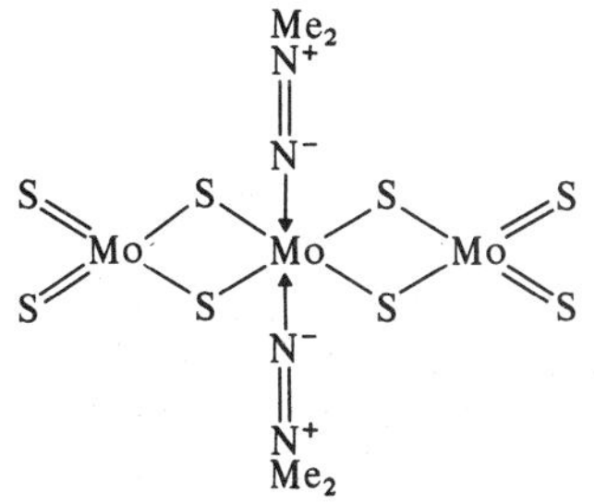

(35)

the planar MoS_2MoS_2Mo system prevents the $NNMe_2$ groups from acting as π donors. This is reflected in long Mo—N distances of 2.13(1) Å and a very short N—N distance of 1.16(2) Å. The eclipsed configuration adopted by the NMe_2 groups is also consistent with a lack of π donation.

13.3.7.3.4 'Side-on' hydrazido(2−)

There are no structurally characterized examples of this type of bonding, but there is clear spectroscopic evidence for its involvement as an intermediate in the deprotonation of $[Mo(\eta^2\text{-}NHNMePh)(NNMePh)(S_2CNMe_2)_2]^+$.[504]

13.3.7.3.5 Bridging hydrazido(2−)

Several interesting examples of this mode of bonding have been structurally characterized and are currently restricted to metal cyclopentadienyl and pyrazolylborate systems. The first example was $[\{(Mo(\eta^5\text{-}C_5H_5)(NO)I\}_2(NNMe_2)]$[505] and an X-ray crystal structure revealed an unusual bridging-η^2 bonding for the $NNMe_2$ ligand (36). A detailed review of the complex protonation behaviour of this system is available.[506] A related titanium complex $[\{Ti(\eta^5\text{-}C_5H_5)Cl_2(NNPh_2)\}_2]$ has recently been reported[507] and contains both bridging and η^2-bridging $NNPh_2$ groups (37). It is suggested from ^1H NMR studies that equilibria between the two bonding modes are present in solution.

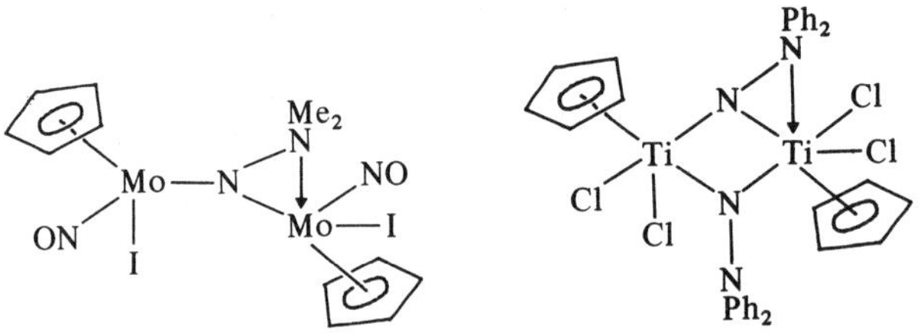

(36)   (37)

13.3.7.4 Structures and Chemistry of Hydrazido(1−) Complexes

In principle there are three possible bonding modes for the $NRNR_2$ ligand shown in Figure 16, and examples of all three are known. The first structurally verified example of η^2-binding was provided by $[W(\eta^5\text{-}C_5H_5)_2(H_2NNAr)]^+$, prepared by reaction of $[W(\eta^5\text{-}C_5H_5)_2H_2]$ with diazonium salts (38).[502] Relevant bonding parameters are summarized in Table 17. The dimensions of the η^2-hydrazide ligand in $[Mo(NHNMePh)(NNMePh)(S_2CNMe_2)_2]^+$[465] are very similar to those in the tungsten complex. A detailed mechanistic study of the protonation of $[Mo(NNMePh)_2(S_2CNMe_2)_2]$ to give $MePhNNH_2$ and $[MoCl_2(NNMePh)(S_2CNMe_2)_2]$ revealed that an η^2-hydrazido(1−) species is *not* an intermediate as its rate of formation is relatively slow.[504]

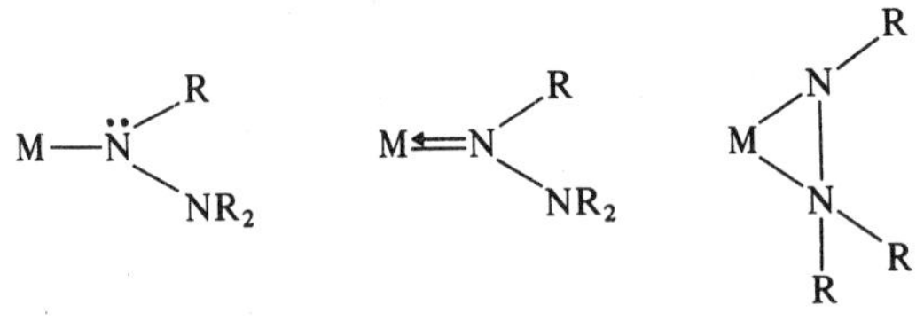

$\eta^1, 1e^-$ $\eta^1, 3e^-$ $\eta^2, 3e^-$

(A)   (B)   (C)

Figure 16

An extensive series of η^2-hydrazido(1−) complexes of titanium have recently been reported[467] together with X-ray crystal structures (Table 17). The bonding of the η^2-hydrazido fragment is perceptibly less symmetrical than for the Mo and W complexes due to increased multiple bonding between Ti and the monosubstituted nitrogen.

(38)   (39)

Table 17 Representative Structural Data for Hydrazido(1 −) Complexes

η^2-*bonded*	*M—NR* (Å)	*M—NR$_2$* (Å)	*N—N* (Å)	*N—M—N* (°)	*Ref.*
[W(η^5-C$_5$H$_5$)$_2$(H$_2$NNC$_6$H$_4$F)]	2.034(9)	2.156(9)	1.43(1)	39.7(3)	50
[Mo(NHNMePh)(NNMePh)(S$_2$CNMe$_2$)$_2$]$^+$	2.069(8)	2.175(9)	1.388(12)	38.1(3)	11
[Ti(η^5-C$_5$H$_5$)Cl$_2$(NHNMe$_2$)]	1.83(1)	2.22(1)	1.41(2)	39.1(2)	13
[Ti(η^5-C$_5$H$_5$)Cl$_2$(H$_2$NNPh)]	1.877(9)	2.14(1)	1.41(2)	40.4(4)	13

η^1-*bonded*	*M—N* (Å)	*N—N*(å)	*M—N—NR* (°)		*Ref.*
[Mo{HB(Me$_2$pz)$_3$}(NO)I(NHNMe$_2$)]	1.980(17)	1.34(3)	140.3(15)		54
[ReCl$_2$(NHNHCOPh)(NNHCOPh)(PPh$_3$)$_2$]	2.211(8)	1.44(1)	116.1(7)		55

pz = pyrazole

In the absence of steric or electronic factors it appears that the η^2-bonding mode is preferred. However, in the complex [Mo{HB(Me$_2$pz)$_3$}(NO)I(NHNMe$_2$)] the steric hindrance provided by the pyrazolylborate ligand imposes η^1-coordination of the NHNMe$_2$ ligand.[508] The Mo—N distance of 1.980(17) Å (Table 17) is indicative of substantial multiple bonding (*i.e.* form B Figure 16). It is also possible to prepare the only example of the analogous NHNH$_2$ complex with the Mo pyrazolylborate system, but this has not been structurally characterized. In the rhenium complex [ReCl$_2$(NHNHCOPh)(NNHCOPh)(PPh$_3$)$_2$] η^2 bonding of the NHNHCOPh ligand would involve a formal valence electron count of 20 (NNHCOPh is a linear 4e$^-$ (donor) and a η^1-bonding mode is found (form A Figure 16).[509] Moreover the long Re—N distance of 2.211(8) Å is consistent with little or no π bonding of the N$_\alpha$ nitrogen to the metal (**38**).

13.3.7.5 Conclusions

Space constraints have prevented a totally comprehensive survey of hydrazide ligands but hopefully the wide range of bonding and chemistry possible has been demonstrated. The continuing intense interest in inorganic nitrogen fixation should ensure further interesting developments in this area of coordination chemistry.

13.3.8 REFERENCES

1. J. R. Dilworth and R. L. Richards, 'Comprehensive Organometallic Chemistry', ed. G. Wilkinson, F. G. A. Stone and E. W. Abel, Pergamon, Oxford, 1982, 1017; J. Chatt, J. R. Dilworth and R. L. Richards, *Chem. Rev.*, 1978, **78**, 589; R. A. Henderson, G. J. Leigh and C. J. Pickett, *Adv. Inorg. Chem. Radiochem.*, 1983, **27**, 197.
2. P. D. Frisch, M. M. Hunt, W. G. Kita, J. A. McCleverty, A. E. Rae, D. Seddon, D. C. Povery, G. W. Smith, J. R. Dilworth and G. J. Leigh, *J. Chem. Soc., Chem. Commun.*, 1986, 1748. D. Swann and J. Williams, *J. Chem. Soc., Dalton Trans.*, 1979, 1819.
3. L. S. Liebeskind, K. B. Sharpless, R. D. Wilson and J. A. Ibers, *J. Am. Chem. Soc.*, 1978, **100**, 8–61; D. A. Muccigrosso, S. E. Jacobson, P. A. Apgar and F. Mares, *J. Am. Chem. Soc.*, 1978, **100**, 7063.
4. K. Wieghardt, M. Hahn, J. Weiss and W. Swiridoff, *Z. Anorg. Allg. Chem.*, 1982, **494**, 164; S. Bristow, D. Collison, C. D. Garner and W. Clegg, *J. Chem. Soc., Dalton Trans.*, 1983, 2495; S. Bristow, J. H. Enemark, C. D. Garner, M. Minelli, G. A. Morris and R. B. Ortega, *Inorg. Chem.*, 1985, **214**, 4070.
5. K. Wieghardt, M. Kleine-Boymann, W. Swiridoff, B. Nuber and J. Wiess, *J. Chem. Soc., Dalton Trans.*, 1985, 2493.
6. D. J. Hodgson, N. C. Payne, J. A. McGinnety, R. G. Pearson and J. A. Ibers, *J. Am. Chem. Soc.*, 1968, **90**, 4486; D. J. Hodgson and J. A. Ibers, *Inorg. Chem.*, 1968, **7**, 2345.
7. J. Lewis, 'Transition Metal—Nitric Oxide Complexes', *Sci. Prog. (Oxford)*, 1959, **47**, 506.
8. For a review on this subject see J. H. Swinehart, *Coord. Chem. Rev.*, 1967, **2**, 385.
9. C. C. Addison and J. Lewis, *Q. Rev., Chem. Soc.*, 1955, **9**, 115.
10. B. F. G. Johnson and J. A. McCleverty, *Prog. Inorg. Chem.*, 1966, **7**, 277.

11. J. A. McCleverty, *Chem. Rev.*, 1979, **79**, 53.
12. W. P. Griffith, *Adv. Organomet. Chem.*, 1968, **7**, 211.
13. F. A. Cotton and G. Wilkinson, 'Advanced Inorganic Chemistry', 3rd edn., Wiley-Interscience, New York, 1972, p. 355.
14. Reference 5, p. 278.
15. D. J. Millen and D. Watson, *J. Chem. Soc.*, 1957, 1369.
16. See N. V. Sidgwick, 'Chemical Elements and their Compounds' Oxford, 1950, vol. 1, p. 685; L. O. Brockway and J. S. Anderson, *Trans. Faraday Soc.*, 1937, **33**, 1233.
17. J. Lewis, R. J. Irving and G. Wilkinson, *J. Inorg. Nucl. Chem.*, 1958, **7**, 32.
18. W. P. Griffith, J. Lewis and G. Wilkinson, *J. Inorg. Nucl. Chem.*, 1958, **7**, 38.
19. R. D. Feltham, *Inorg. Chem.*, 1964, **3**, 1038, and refs. therein.
20. B. Jeżowska-Trzebiatowska, J. Hanuza, M. Ostern and J. Ziółkowski, *Inorg. Chim. Acta*, 1972, **6**, 141.
21. P. R. H. Alderman, P. G. Owston and J. M. Rowe, *J. Chem. Soc.*, 1962, 668.
22. J. H. Enemark and R. D. Feltham, *J. Chem. Soc., Dalton Trans.*, 1972, 718.
23. J. P. Collman, P. Farnham and G. Dolcetti, *J. Am. Chem. Soc.*, 1971, **93**, 1788.
24. C. G. Pierpont and R. Eisenberg, *Inorg. Chem.*, 1972, **11**, 1088.
25. D. M. P. Mingos and J. A. Ibers, *Inorg. Chem.*, 1971, **10**, 1035.
26. W. R. Roper, personal communication.
27. M. O. Visscher and K. G. Caulton, *J. Am. Chem. Soc.*, 1972, **94**, 5923.
28. C. G. Pierpont, D. G. Van Derveer, W. Durland and R. Eisenberg, *J. Am. Chem. Soc.*, 1970, **92**, 4761.
29. J. S. Griffith, *Proc. R. Soc. London, Ser. A*, 1956, **235**, 23.
30. Reference 8, p. 713.
31. P. T. Manoharan and H. B. Gray, *J. Am. Chem. Soc.*, 1965, **87**, 3340.
32. D. M. P. Mingos and J. A. Ibers, *Inorg. Chem.*, 1971, **10**, 1479.
33. D. A. Snyder and D. L. Weaver, *Inorg. Chem.*, 1970, **9**, 2760.
34. B. A. Frenz, J. H. Enemark and J. A. Ibers, *Inorg. Chem.*, 1969, **8**, 1288.
35. C. G. Pierpont, A. Pucci and R. Eisenberg, *J. Am. Chem. Soc.*, 1971, **93**, 3050.
36. G. R. Clark, K. R. Grundy, W. R. Roper, J. M. Waters and K. R. Whittle, *J. Chem. Soc., Chem. Commun.*, 1972, 119.
37. C. G. Pierpont and R. Eisenberg, *Inorg. Chem.*, 1972, **11**, 1094.
38. S. J. La Placa and J. A. Ibers, *Inorg. Chem.*, 1965, **4**, 778.
39. A. C. Skapski and P. G. H. Troughton, *Chem. Commun.*, 1968, 575.
40. N. C. Payne and J. A. Ibers, *Inorg. Chem.*, 1969, **8**, 2714.
41. J. K. Stalick and J. A. Ibers, *Inorg. Chem.*, 1969, **8**, 419.
42. R. G. Pearson, *J. Am. Chem. Soc.*, 1969, **91**, 4947.
43. R. S. Berry, *J. Chem. Phys.*, 1960, **32**, 933.
44. J. P. Collman, N. W. Hoffman and D. E. Morris, *J. Am. Chem. Soc.*, 1969, **91**, 5659.
45. F. Basolo, 'Mechanisms of Substitution Reactions of Metal Carbonyls', *Chem. Br.*, 1969, **5**, 505 and refs. therein.
46. J. P. Day, D. L. Diemente and F. Basolo, *Inorg. Chim. Acta*, 1969, **3**, 363.
47. F. Basolo and R. G. Pearson, 'Mechanisms of Inorganic Reactions', 2nd edn., Wiley, New York, 1967, p. 573.
48. J. P. Candlin and W. H. Janes, *J. Chem. Soc. (C)*, 1968, 1856.
49. See for example E. A. Zuech, W. B. Hughes, D. H. Kubicek and E. T. Kittleman, *J. Am. Chem. Soc.*, 1970, **92**, 529.
50. P. T. Manoharan and H. B. Gray, *Inorg. Chem.*, 1966, **5**, 823.
51. C. G. Pierpont and R. Eisenberg, *J. Am. Chem. Soc.*, 1971, **93**, 4905.
52. D. M. P. Mingos, W. T. Robinson and J. A. Ibers, *Inorg. Chem.*, 1971, **10**, 1043.
53. J. H. Enemark, *Inorg. Chem.*, 1972, **10**, 1952.
54. S. F. A. Kettle, *Inorg. Chem.*, 1965, **4**, 1661.
55. D. M. P. Mingos, *Nature (London)*, 1971, **229**, 193.
56. G. R. Davies, J. A. J. Jarvis, B. J. Kilbourn, R. H. B. Mais and P. G. Owston, *J. Chem. Soc. (A)*, 1970, 1275.
57. P. de Meester, A. C. Skapski and J. P. Heffer, *J. Chem. Soc., Chem. Commun.*, 1972, 1039.
58. J. R. Norton, J. P. Collman, G. Dolcetti and W. T. Robinson, *Inorg. Chem.*, 1972, **11**, 382.
59. J. L. Calderon, F. A. Cotton, B. G. DeBoer and N. Martinez, *Chem. Commun.*, 1971, 1476.
60. R. C. Elder, F. A. Cotton and R. A. Schunn, *J. Am. Chem. Soc.*, 1967, **89**, 3645.
61. H. Brunner, *J. Organometallic Chem.*, 1968, **12**, 517.
62. P. T. Manoharan and W. C. Hamilton, *Inorg. Chem.*, 1963, **2**, 1043.
63. S. H. Simonsen and M. H. Mueller, *J. Inorg. Nucl. Chem.*, 1965, **27**, 309.
64. J. T. Veal and D. J. Hodgson, *Inorg. Chem.*, 1972, **11**, 1420.
65. J. M. Waters and K. R. Whittle, *Chem. Commun.*, 1971, 518.
66. A. P. Cox, L. F. Thomas and J. Sheridan, *Nature (London)*, 1958, **181**, 1157.
67. A. Tullberg, N.-G. Vannerberg, *Acta Chem. Scand.*, 1967, **21**, 1462.
68. J. H. Enemark and J. A. Ibers, *Inorg. Chem.*, 1968, **7**, 2339.
69. J. H. Enemark and J. A. Ibers, *Inorg. Chem.*, 1967, **6**, 1575.
70. J. H. Enemark, M. S. Quinby, L. L. Reed, M. J. Steuck and K. K. Walthers, *Inorg. Chem.*, 1970, **9**, 2397.
71. T. F. Brennan and I. Bernal, *Chem. Commun.*, 1970, 138.
72. F. A. Cotton and G. A. Rusholme, *J. Am. Chem. Soc.*, 1972, **94**, 402.
73. V. G. Albano, P. Bellon and M. Sansoni, *J. Chem. Soc. (A)*, 1971, 2420.
74. M. A. Bush and G. A. Sim. *J. Chem. Soc. (A)*, 1970, 605.
75. A. Domenicano, A. Vaciago, L. Zambonelli, P. L. Loader and L. M. Venanzi, *Chem. Commun.*, 1966, 476.
76. D. M. P. Mingos and J. A. Ibers, *Inorg. Chem.*, 1970, **9**, 1105.
77. D. J. Hodgson and J. A. Ibers, *Inorg. Chem.*, 1969, **8**, 1282.
78. R. Wiest and R. Weiss, *J. Organomet. Chem.*, 1970, **30**, C33.
79. C. S. Pratt, B. A. Coyle and J. A. Ibers, *J. Chem. Soc. (A)*, 1971, 2146.
80. I. H. Sabherwal and A. B. Burg, *Chem. Commun.*, 1970, 1001.

81. R. L. Mond and A. E. Wallis, *J. Chem. Soc.*, 1922, **121**, 32; J. S. Anderson, *Z. Anorg. Allg. Chem.*, 1932, **208**, 238.
82. M. Rossi and A. Sacco, *Chem. Commun.*, 1971, 694.
83. F. A. Cotton and B. F. G. Johnson, *Inorg. Chem.*, 1964, **3**, 1609.
84. P. M. Treichel, E. Pitcher, R. B. King and F. G. A. Stone, *J. Am. Chem. Soc.*, 1961, **83**, 2593.
85. L. Playfair, *Proc. R. Soc.* London, 1849, **5**, 849 and *Philos. Mag.*, 1850, **36**, 197.
86. T. J. Meyer, J. B. Godwin and N. Winterton, *Chem. Commun.*, 1970, 872; J. B. Godwin and T. J. Meyer, *Inorg. Chem.*, 1971, **10**, 471.
87. W. Hieber and H. Tengler, *Z. Anorg. Allg. Chem.*, 1962, **318**, 136.
88. G. Booth and J. Chatt, *J. Chem. Soc.*, 1962, 2099.
89. K. R. Grundy, K. R. Laing and W. R. Roper, *Chem. Commun.*, 1970, 1500.
90. R. D. Feltham, *Inorg. Chem.*, 1964, **3**, 116.
91. R. B. King, 'Organometallic Syntheses', Academic, London, 1965, vol. 1, p. 165 and refs. therein.
92. K. R. Grundy, C. A. Reed and W. R. Roper, *Chem. Commun.*, 1970, 1501.
93. L. Malatesta and A. Sacco, *Z. Anorg. Allg. Chem.*, 1953, **274**, 341.
94. J. A. Broomhead and H. Taube, *J. Am. Chem. Soc.*, 1969, **91**, 1261.
95. J. B. Godwin and T. J. Meyer, *Inorg. Chem.*, 1971, **10**, 2150.
96. P. G. Douglas, R. D. Feltham and H. G. Metzger, *J. Am. Chem. Soc.*, 1971, **93**, 84.
97. Reference 5, p. 311.
98. C. A. Reed and W. R. Roper, *J. Chem. Soc., Dalton Trans.*, 1972, 1243.
99. F. J. Miller and T. J. Meyer, *J. Am. Chem. Soc.*, 1971, **93**, 1294; S. A. Adeyemi, F. J. Miller and T. J. Meyer, *Inorg. Chem.*, 1972, **11**, 995.
100. R. D. Feltham, H. G. Metzger and R. Singler, *Proc. I.C.C.C.* (Sydney), XII, 1969, 225.
101. F. Bottomley and J. R. Crawford, *Chem. Commun.*, 1971, 200.
102. R. Nast and J. Schmidt, *Angew. Chem., Int. Ed. Engl.*, 1969, **8**, 383.
103. C. A. Reed and W. R. Roper, *J. Chem. Soc. (A)*, 1970, 3054.
104. M. Foá and L. Cassar, *J. Organomet. Chem.*, 1971, **30**, 123.
105. A. E. Crease and P. Legzdins, *J. Chem. Soc., Chem. Commun.*, 1972, 268.
106. M. Tamaki, I. Masuda and K. Shinra, *Bull. Chem. Soc. Jpn.*, 1972, **45**, 171.
107. S. G. Clarkson and F. Basolo, *J. Chem. Soc., Chem. Commun.*, 1972, 670.
108. T. Ishimyama and T. Matsumara, *Bull. Chem. Soc. Jpn.*, 1972, **76**, 252.
109. M. H. B. Stiddard and R. E. Townsend, *Chem. Commun.*, 1969, 1372.
110. K. R. Laing and W. R. Roper, *Chem. Commun.*, 1968, 1568.
111. B. F. G. Johnson, S. Bhaduri and N. G. Connelly, *J. Organomet. Chem.*, 1972, **40**, C36.
112. Reference 4, p. 145 and K. A. Hoffman, *Z. Anorg. Allg. Chem.*, 1896, **12**, 146.
113. N. Flitcroft, *J. Organomet. Chem.*, 1968, **15**, 254.
114. L. Y. Y. Chan and F. W. B. Einstein, *Acta Crystallogr., Sect. B*, 1970, **26**, 1899.
115. M. J. Cleare and W. P. Griffith, *J. Chem. Soc. (A)*, 1970, 1117 and refs. therein.
116. W. Hieber and R. Kramolowshi, *Z. Naturforsch., Teil B*, 1961, **16**, 555; *Z. Anorg. Allg. Chem.*, 1963, **321**, 94.
117. W. Hieber and K. Heineck, *Z. Naturforsch., Teil B*, 1961, **16**, 553.
118. E. O. Fischer and H. Strametz, *J. Organomet. Chem.*, 1967, **10**, 323.
119. B. F. G. Johnson, personal communication.
120. W. Lwowski (ed.), 'Nitrenes', Interscience, New York, 1970.
121. J. H. Boyer, 'Deoxygenation of Nitro and Nitroso Groups', references 115, p. 163.
122. A. T. McPhail, G. R. Knox, C. G. Robertson and G. A. Sim, :*J. Chem. Soc. (A)*, 1971, 205.
123. S. R. Fletcher, A. Shortland, A. C. Skapski and G. Wilkinson, *J. Chem. Soc., Chem. Commun.*, 1972, 922.
124. P. C. Wailes, H. Weigold and A. P. Bell, *J. Organomet. Chem.*, 1972, **34**, 155 and refs. therein.
125. H. Brunner and S. Loskot, *Angew. Chem.*, 1971, **10**, 515.
126. J. P. Collman, M. Kubota and J. W. Hosking, *J. Am. Chem. Soc.*, 1967, **89**, 4809 and J. P. Collman, *Acc. Chem. Res.*, 1968, **1**, 136.
127. S. Cenin, R. Ugo, G. La Monica and S. D. Robinson, *Inorg. Chim. Acta*, 1972, **6**, 182.
128. S. Pell and J. N. Armour, *J. Am. Chem. Soc.*, 1972, **94**, 686.
129. R. B. Jordan, A. M. Sargeson and H. Taube, *Inorg. Chem.*, 1966, **5**, 1091.
130. See for example R. H. Reimann and E. Singleton, *J. Organomet. Chem.*, 1971, **32**, C44.
131. W. L. Gladfelter, *Adv. Organomet. Chem.*, 1985, **24**, 41.
132. R. D. Adams and I. T. Horváth, *Prog. Inorg. Chem.*, 1985, **33**, 127.
133. R. C. Elder, *Inorg. Chem.*, 1974, **13**, 1037.
134. J. Mason, *Chem. Rev.*, 1981, **81**, 205; L. K. Bell, J. Mason, D. M. P. Mingos and D. G. Tew, *Inorg. Chem.*, 1983, **22**, 3497; D. H. Evans, D. M. P. Mingos, J. Mason and A. Richards, *J. Organomet. Chem.*, 1983, **249**, 293; J. Mason, *Chem. Bri.*, 1983, 654; R. E. Botto, B. W. S. Kolthammer, P. Legzdins and J. D. Roberts, *Inorg. Chem.*, 1979, **18**, 2049; R. E. Stevens and W. L. Gladfelter, *Inorg. Chem.*, 1983, **22**, 2034.
135. R. E. Stevens and W. L. Gladfelter, *J. Am. Chem. Soc.*, 1982, **104**, 645; P. Legzdins, C. R. Nurse and S. J. Rettig, *J. Am. Chem. Soc.*, 1983, **105**, 3727.
136. B. F. G. Johnson, J. Lewis and J. M. Mace, *J. Chem. Soc., Chem. Commun.*, 1984, 186.
137. J. R. Norton and J. P. Collman, *Inorg. Chem.*, 1973, **12**, 476.
138. S. Bhaduri, B. F. G. Johnson, J. Lewis, P. J. Watson and C. Zuccaro, *J. Chem. Soc., Dalton Trans.*, 1979, 557.
139. T. B. Rauchfuss, T. D. Weatherill, S. R. Wilson and J. P. Zebrowski, *J. Am. Chem. Soc.*, 1983, **105**, 6508.
140. C. T. W. Chu, R. S. Gall and L. F. Dahl, *J. Am. Chem. Soc.*, 1982, **104**, 737.
141. N. D. Feasey, S. A. R. Knox and A. G. Orpen, *J. Chem. Soc., Chem. Commun.*, 1982, 75.
142. M. W. Bishop, J. Chatt, and J. R. Dilworth, *J. Chem. Soc., Chem. Commun.*, 1975, 780.
143. J. Chatt and J. R. Dilworth, *J. Chem. Soc., Chem. Commun.*, 1974, 508.
144. R. Mews and C. Liu, *Angew. Chem., Int. Ed. Engl.*, 1983, **22**, 162. R. Mews and C. Liu, *Angew. Chem.*, 1983, **95**, 156.
145. M. J. Wright and W. P. Griffith, *Transition Met. Chem.*, 1982, **7**, 53.
146. K. K. Pandey and U. C. Agarwala, *Indian J. Chem., Sect. A*, 1982, **21**, 77.

147. P. C. Bevan, J. Chatt, J. R. Dilworth, R. A. Henderson and G. J. Leigh, *J. Chem. Soc., Dalton Trans.*, 1982, 821.
148. K. K. Pandey and U. C. Agarwala, *Indian J. Chem., Sect. A*, 1981, **20**, 906.
149. H. W. Chen, W. L. Jolly and S. F. Xiang, *J. Electron Spectrosc. Relat. Phenom.*, 1981, **24**, 121.
150. K. K. Pandey, K. C. Jain and U. C. Agarwala, *Inorg. Chim. Acta*, 1981, **48**, 23.
151. K. K. Pandey and U. C. Agarwala, *Indian J. Chem., Sect. A*, 1981, **20**, 74.
152. K. K. Pandey and U. C. Agarwala, *Inorg. Chem.*, 1981, **20**, 1308.
153. K. K. Pandey, S. Datta and U. C. Agarwala, *Z. Anorg. Allg. Chem.*, 1980, **468**, 228.
154. M. Heberhold and L. Haumaier, *Z. Naturforsch., Teil B*, 1980, **35**, 1277.
155. M. B. Comisarow, *Adv. Mass Spectrom.*, 1980, **8B**, 1698.
156. G. Parisod and M. B. Comisarow, *Adv. Mass Spectrom.* 1980, **8A**, 212.
157. M. Herberhold, *Nachr. Chem. Tech. Lab.*, 1981, **29**, 269.
158. S. C. Avanzino, A. A. Bakke, H. W. Chen, C. J. Donahue, W. L. Jolly, T. H. Lee and A. J. Ricco, *Inorg. Chem.*, 1980, **19**, 1931.
159. K. K. Pandey and U. C. Agarwala, *Z. Anorg. Allg. Chem.*, 1980, **461**, 231.
160. J. L. Hubbard and D. L. Lichtenberger, *Inorg. Chem.*, 1980, **19**, 1388.
161. M. W. Bishop, J. Chatt, J. R. Dilworth, M. B. Hursthouse and M. Motevalli, *J. Less-Common Met.*, 1977, **54**, 487.
162. B. W. S. Kolthammer and P. Legzdins, *J. Am. Chem. Soc.*, 1978, **100**, 2247.
163. T. J. Greenhough, B. W. S. Kolthammer, P. Legzdins and J. Trotter, *J. Chem. Soc., Chem. Commun.*, 1978, 1036.
164. M. W. Bishop, J. Chatt and J. R. Dilworth, *J. Chem. Soc., Dalton Trans.*, 1979, 1.
165. G. Butler, J. Chatt, G. J. Leigh and C. J. Pickett, *J. Chem. Soc., Dalton Trans.*, 1979, 113.
166. M. B. Hursthouse and M. Motevalli, *J. Chem. Soc., Dalton Trans.*, 1979, 1362.
167. T. J. Greenhough, B. W. S. Kolthammer, P. Legzdins and J. Trotter, *Inorg. Chem.*, 1979, **18**, 3548.
168. J. Chatt, *Pure Appl. Chem.*, 1977, **49**, 815.
169. U. Müller, P. Klingelhofer, U. Kynast and K. Dehnicke, *Z. Anorg. Allg. Chem.* 1985, **520**, 18.
170. M. Goehring and A. Debo, *Z. Anorg. Allg. Chem.*, 1953, **273**, 319.
171. M. Goehring, K. W. Daum and J. Weiss, *Z. Naturforsch., Teil B*, 1955, **10**, 298; J. Weiss and M. Ziegler, *Z. Anorg. Allg. Chem.*, 1963, **322**, 184.
172. E. Fluck, M. Goehring and J. Weiss, *Z. Anorg. Allg. Chem.*, 1956, **287**, 51; I. Lindqvist and J. Weiss, *J. Inorg. Nucl. Chem.*, 1957, **6**, 184.
173. H. W. Roesky, J. Anhaus and W. S. Sheldrick, *Inorg. Chem.*, 1984, **23**, 75.
174. T. F. Brennan and I. Bernal, *Inorg. Chim. Acta*, 1973, **7**, 283.
175. U. Demant, E. Conradi, J. Pebler, U. Müller and K. Dehnick, *Z. Anorg. Allg. Chem.*, 1984, **510**, 180.
176. U. Kynast, U. Muller and K. Dehnicke, *Z. Anorg. Allg. Chem.*, 1984, **508**, 26.
177. G. Hartmann and R. Mews, *Angew. Chem.*, 1985, **97**, 218.
178. L. Kaden, B. Lorenz, R. Kirmse and J. Stach, *Z. Chem.*, 1985, **25**, 29.
179. H. W. Roesky and K. K. Pandey, *Adv. Inorg. Chem. Radiochem.*, 1983, **26**, 337.
180. R. Weber, U. Muller and K. Dehnicke, *Z. Anorg. Allg. Chem.*, 1983, **504**, 13.
181. K. N. Udupa, K. C. Jain, M. I. Kahn and U. C. Agarwala, *Inorg. Chim. Acta*, 1983, **74**, 191.
182. D. Pawson and W. P. Griffith, *Inorg. Nucl. Chem. Lett.*, 1974, **10**, 253.
183. D. Pawson and W. P. Griffith, *J. Chem. Soc., Dalton Trans.*, 1975, 417.
184. F. L. Phillips and A. C. Skapski, *J. Chem. Soc., Dalton Trans.*, 1976, 1448.
185. F. L. Phillips and A. C. Skapski, *J. Chem. Soc., Chem. Commun.*, 1974, 49.
186. F. King and G. J. Leigh, *J. Chem. Soc., Dalton Trans.*, 1977, 429.
187. G. R. Beilharz, D. Scott and A. G. Wedd, *Aust. J. Chem.*, 1976, **29**, 2167.
188. D. Scott and A. G. Wedd, *J. Chem. Soc., Chem. Commun.*, 1974, 527.
189. J. R. Dilworth, H. J. De Liefde Meijer and J. H. Teuben, *J. Organometal. Chem.*, 1978, **159**, 47.
190. R. Choukroun, D. Gervais and J. R. Dilworth, *Transition Met. Chem.*, 1979, **4**, 249.
191. H. Bezler and J. Straehle, *Z. Naturforsch., Teil B*, 1979, **34**, 1199.
192. W. P. Griffith and D. Pawson, *J. Chem. Soc., Chem. Commun.*, 1973, 418.
193. K. Dehnicke, H. Prinz, W. Kafitz and R. Kujanek, *Liebigs Ann. Chem.*, 1981, 20.
194. U. Müller, F. Weller and K. Dehnicke, *Z. Anorg. Allg. Chem.*, 1981, **473**, 115.
195. R. Dübgen, U. Müller, F. Weller and K. Dehnicke, *Z. Anorg. Allg. Chem.*, 1980, **471**, 89.
196. H. Bezler and J. Straehle, *Z. Naturforsch., Teil B*, 1982, **38**, 317.
197. H. W. Roesky, U. Seseke, M. Noltemeyer, P. G. Jones and G. M. Sheldrick, *J. Chem. Soc., Dalton Trans.*, 1986, 1309.
198. M. J. Mays, D. W. Prest and P. R. Raithby, *J. Chem. Soc., Chem. Commun.*, 1980, 171.
199. K. C. Jain and U. C. Agarwala, *Indian J. Chem., Sect. A*, 1983, **22**, 336.
200. R. Weber and K. Dehnicke, *Z. Naturforsch., Teil B*, 1984, **39**, 262.
201. J. W. Bats, K. K. Pandey and H. W. Roesky, *J. Chem. Soc., Dalton Trans.*, 1984, 2081.
202. J. Chatt, R. J. Dosser and G. J. Leigh, *J. Chem. Soc., Chem. Commun.*, 1972, 1243.
203. P. Etievant, B. Gautheron and G. Tainturier, *Bull. Soc. Chim. Fr.*, 1978, II, 292.
204. P. Etievant, G. Tainturier and B. Gautheron, *C. R. Hebd. Seances Acad. Sci., Ser. C*, 1976, **283**, 233.
205. J. Chatt, R. J. Dosser, F. King and G. J. Leigh, *J. Chem. Soc., Dalton Trans.*, 1976, 2435.
206. A. J. Carty, D. P. Madden, M. Matthew, G. J. Palenik and T. Birchall, *J. Chem. Soc., Chem. Commun.*, 1970, 1664.
207. M. F. Lappert, B. Cetinkaya and J. McMeeking, *Chem. Commun.*, 1971, 215.
208. M. R. Colllier, M. F. Lappert and J. McMeeking, *Inorg. Nucl. Chem. Lett.*, 1971, **7**, 689.
209. I. Pattison and K. Wade, *J. Chem. Soc. (A)*, 1968, 57.
210. T. Inglis and M. Kilner, *J. Chem. Soc., Dalton Trans.*, 1976, 562.
211. H. Shearer, M. M. Harrison and J. D. Sowerby, *J. Chem. Soc., Dalton Trans.*, 1973, 2629.
212. M. Kilner and C. Midcalf, *Chem. Commun.*, 1970, 552.
213. M. Kilner and C. Midcalf, *J. Chem. Soc. (A)*, 1971, 292.
214. M. J. Doyle, M. E. Lappert, G. M. McLaughlin and J. McMeeking, *J. Chem. Soc., Dalton Trans.*, 1974, 1494.
215. T. Inglis, H. R. Keable, M. Kilner and E. E. Robertson, *J. Less-Common Met.*, 1974, **36**, 217.
216. K. Farmery, M. Kilner and C. Midcalf, *J. Chem. Soc. (A)*, 1970, 2279.

217. H. R. Keable, M. Kilner and E. E. Robertson, *J. Chem. Soc., Dalton Trans.*, 1974, 639.
218. B. Cetinkaya, M. F. Lappert and J. McMeeking, *J. Chem. Soc., Dalton Trans.*, 1973, 1975.
219. M. Kilner and J. N. Pinkney, *J. Chem. Soc. (A)*, 1971, 2887
220. D. Briggs, D. T. Clark, H. R. Keable and M. Kilner, *J. Chem. Soc., Dalton Trans.*, 1973, 2143.
221. H. R. Keable and M. Kilner, *J. Chem. Soc., Dalton Trans.*, 1972, 153.
222. M. F. Lappert, J. McMeeking and D. E. Palmer, *J. Chem. Soc., Dalton Trans.*, 1973, 151.
223. J. Browning, H. D. Empsall, M. Green and F. G. A. Stone, *J. Chem. Soc., Dalton Trans.*, 1973, 381.
224. B. Cetinkaya, P. Dixneuf and M. F. Lappert, *J. Chem. Soc., Dalton Trans.*, 1974, 1827.
225. M. Cetinkaya, A. W. Johnson, M. F. Lappert, G. M. McLaughlin and K. W. Muir, *J. Chem. Soc., Dalton Trans.*, 1974, 1236.
226. M. Kilner and C. Midcalf, *J. Chem. Soc., Dalton Trans.*, 1974, 1620.
227. J. A. McCleverty, A. E. Rae, I. Wolochowicz, N. A. Bailey and J. M. A. Smith, *J. Chem. Soc., Dalton Trans.*, 1982, 429.
228. J. A. McCleverty, A. E. Rae, I. Wolochowicz, N. A. Bailey and J. M. A. Smith, *J. Organomet. Chem.*, 1979, **168**, C1.
229. E. G. Ilin, G. A. Ermakov, M. M. Ershov, M. A. Glushkova, L. V. Khmelevskaya and Y. A. Buslaev, *Koord. Khim.*, 1984, **10**, 964.
230. E. W. Abel and C. A. Burton, *J. Inorg. Nucl. Chem.*, 1980, **42**, 1697.
231. P. A. Adcock and R. F. Keene, *J. Am. Chem. Soc.*, 1981, **103**, 6494.
232. G. P. Khane and R. J. Doedens, *Inorg. Chem.*, 1976, **15**, 86.
233. E. J. M. DeBoer and J. H. Teuben, *J. Organomet. Chem.*, 1978, **153**, 53.
234. E. Klei, J. H. Teuben, H. J. De Liefde Meijer and E. J. Kwak, *J. Organomet. Chem.*, 1982, **224**, 327.
235. M. Kilner, *Adv. Organomet. Chem.*, 1972, **10**, 115.
236. R. B. King and W. M. Douglas, *Inorg. Chem.*, 1974, **13**, 1339.
237. M. K. Churchill and K. G. Lin, *Inorg. Chem.*, 1975, **14**, 1675.
238. E. W. Abel, C. A. Burton, M. R. Churchill and K. G. Lin, *J. Chem. Soc., Chem. Commun.*, 1974, 917.
239. D. Bright and O. S. Mills, *J. Chem. Soc., Chem. Commun.*, 1967, 245.
240. W. Seidel and K. J. Lattermann, *East Ger. Pat.* 205 437-A1 (1983).
241. M. A. Andrews and H. D. Kaesz, *J. Am. Chem. Soc.*, 1979, **100**, 7238.
242. G. W. Parshall, *J. Am. Chem. Soc.*, 1967, **89**, 1822.
243. M. C. Baird and G. Wilkinson, *J. Chem. Soc. (A)*, 1967, 865.
244. A. J. Deeming and B. L. Shaw, *J. Chem. Soc. (A)*, 1969, 1128.
245. C. D. Cook and G. S. Jauhal, *J. Am. Chem. Soc.*, 1968, **90**, 1464.
246. A. N. Nesmeyanov, Y. A. Chapovskii, N. A. Ustynyuk and L. G. Makarova, *Izv. Akad. Nauk SSSR, Ser. Khim.*, 1968, 449.
247. M. F. Lappert and J. S. Poland, *Chem. Commun.*, 1969, 1061.
248. R. B. King and M. B. Bisnette, *Inorg. Chem.*, 1966, **5**, 300.
249. R. B. King and M. B. Bisnette, *J. Am. Chem. Soc.*, 1964, **86**, 5694.
250. G. W. Parshall, *J. Am. Chem. Soc.*, 1965, **87**, 2133.
251. S. Trofimenko, *Inorg. Chem.*, 1969, **8**, 2675.
252. M. L. H. Green, T. R. Sanders and R. N. Whiteley, *Z. Naturforsch., Teil B*, 1967, **23**, 106.
253. G. C. Dobinson, R. Mason and G. B. Robertson, R. Ugo, F. Conti, D. Morelli, S. Cenini and F. Bonati, *Chem. Commun.*, 1967, 739.
254. J. Chatt, J. R. Dilworth, G. J. Leigh and V. D. Gupta, *J. Chem. Soc. (A)*, 1971, 2631.
255. W. E. Carroll, M. E. Deane and F. J. Lalor, *J. Chem. Soc., Dalton Trans.*, 1974, 1837.
256. V. F. Duckworth, P. G. Douglas, R. Mason and B. L. Shaw, *Chem. Commun.*, 1970, 1083.
257. A. W. B. Garner and M. J. Mays, *J. Organomet. Chem.*, 1974, **67**, 153.
258. G. W. Parshall, *Inorg. Synth.*, 1970, **12**, 26.
259. J. Chatt, J. R. Dilworth, G. J. Leigh and I. A. Zakharova, *Dokl. Akad. Nauk SSSR*, 1971, **199**, 848.
260. J. Chatt, G. A. Heath, N. E. Hooper and G. J. Leigh, *J. Organomet. Chem.*, 1973, **57**, C67.
261. K. R. Lang, S. D. Robinson and M. F. Uttley, *J. Chem. Soc., Dalton Trans.*, 1973, 2713.
262. K. R. Laing, S. D. Robinson and M. F. Uttley, *J. Chem. Soc., Chem. Commun.*, 1973, 176.
263. J. A. McCleverty and R. N. Whiteley, *Chem. Commun.*, 1971, 1159.
264. V. I. Nefedov, M. A. Porai-Koshits, I. A. Zakharova and M. E. Dyatkina, *Dokl. Akad. Nauk SSSR*, 1972, **202**, 605.
265. V. I. Nefedov, I. A. Zakharova, M. A. Porai-Koshits and M. E. Dyatkina, *Izv. Akad. Nauk SSSR, Ser. Khim.*, 1971, 1846.
266. S. Cenini, R. Ugo and G. LaMonica, *J. Chem. Soc. (A)*, 1971, 3441.
267. B. L. Shaw and R. E. Stainbank, *J. Chem. Soc., Dalton Trans*, 1972, 223.
268. G. W. Rayner-Canham and D. Sutton, *Can. J. Chem.*, 1971, **49**, 3994.
269. L. Toniolo, G. Deluca, C. Panattoni and G. Deganello, *Gazz. Chim. Ital.*, 1974, **104**, 961.
270. J. V. McArdle, A. J. Schultz, B. J. Corden and R. Eisenberg, *Inorg. Chem.*, 1973, **12**, 1676.
271. B. L. Haymore and J. A. Ibers, *J. Am. Chem. Soc.*, 1973, **95**, 3052.
272. W. E. Carroll and F. J. Lalor, *J. Organomet. Chem.*, 1973, **54**, C37.
273. W. E. Carroll and F. J. Lalor, *J. Chem. Soc.., Dalton Trans.*, 1973, 1754.
274. D. R. Fisher and D. Sutton, *Can. J. Chem.*, 1973, **51**, 1697.
275. J. W. Dart, M. K. Lloyd, R. Mason and J. A. McCleverty, *J. Chem. Soc., Dalton Trans.*, 1973, 2039.
276. R. Mason, K. M. Thomas, J. A. Zubieta, P. G. Douglas, A. R. Galbraith and B. L. Shaw, *J. Am. Chem. Soc.*, 1974, **96**, 260.
277. S. Cenini, F. Porta and M. Pizzotti, *Inorg. Nucl. Chem. Lett.*, 1974, **10**, 983.
278. L. Toniolo, J. A. McGinnety, T. Boschi and G. Deganello, *Inorg. Chim. Acta*, 1974, **11**, 143.
279. G. S. Brownlee, P. Carty, D. N. Cash and A. Walker, *Inorg. Chem.*, 1974, **14**, 323.
280. M. Herberhold and W. Bernhagen, *Z. Naturforsch., Teil B*, 1974, **29**, 801.
281. W. L. Bowden, W. F. Little and T. J. Meyer, *J. Am. Chem. Soc.*, 1973, **95**, 5084.
282. T. Tatsumi, M. Hidai and Y. Uchida, *Inorg. Chem.*, 1975, **14**, 2530.

283. N. G. Connelly and Z. Demidowicz, *J. Organomet. Chem.*, 1974, **73**, C31.
284. D. Sutton, *Can. J. Chem.*, 1974, **52**, 2634.
285. D. Cashman and F. J. Lalor, *J. Organomet. Chem.*, 1970, **24**, C29.
286. F. J. Lalor and P. L. Pauson, *J. Organomet. Chem.*, 1970, **25**, C51.
287. M. E. Deane and F. J. Lalor, *J. Organomet. Chem.*, 1974, **67**, C19.
288. D. Cashman and F. J. Lalor, *J. Organomet. Chem.*, 1971, **32**, 351.
289. W. E. Carroll, F. A. Deeney and F. J. Lalor, *J. Organomet. Chem.*, 1973, **57**, C61.
290. M. E. Deane and F. J. Lalor, *J. Organomet. Chem.*, 1973, **57**, C61.
291. E. W. Abel, C. A. Burton, M. R. Churchill and K. G. Lin, *J. Chem. Soc., Chem. Commun.*, 1974, 268.
292. A. P. Gaughan, B. L. Haymore, J. A. Ibers, W. H. Myers, T. E. Nappier and D. W. Meek, *J. Am. Chem. Soc.*, 1973, **95**, 6859.
293. R. B. King and M. S. Saran, *Inorg. Chem.*, 1974, **13**, 364.
294. S. D. Ittel and J. A. Ibers, *J. Am. Chem. Soc.*, 1974, **96**, 4804.
295. L. Toniolo, *Inorg. Chim. Acta*, 1972, **6**, 660.
296. S. Trofimenko, *J. Am. Chem. Soc.*, 1970, **92**, 5118.
297. D. F. Gill, B. E. Mann and B. L. Shaw, *J. Chem. Soc., Dalton Trans.*, 1973, 311.
298. J. Chatt, G. A. Heath and G. J. Leigh, *J. Chem. Soc., Chem. Commun.*, 1972, 444.
299. J. Chatt, J. R. Dilworth and G. J. Leigh, *J. Chem. Soc.., Dalton Trans.*, 1973, 612.
300. R. W. Adams, J. Chatt, N. E. Hooper and G. J. Leigh, *J. Chem. Soc., Dalton Trans.*, 1974, 1075.
301. G. Avitabile, P. Ganis and M. Nemiroff, *Acta Crystallogr., Sect. B*, 1971, **27**, 725.
302. G. Caglio and M. Angoletta, *Gazz. Chim. Ital.*, 1972, **102**, 462.
303. M. L. H. Green and J. R. Sanders, *J. Chem. Soc. (A)*, 1971, 1947.
304. P. Brant and R. D. Feltham, *J. Organomet. Chem.*, 1976, **120**, C53.
305. B. L. Haymore and J. A. Ibers, *J. Am. Chem. Soc.*, 1975, **97**, 5369.
306. P. G. Douglas, A. R. Galbraith and B. L. Shaw, *Transition Met. Chem.*, 1975, **1**, 17.
307. J. A. McCleverty, D. Seddon and R. N. Whiteley, *J. Chem. Soc., Dalton Trans.*, 1975, 839.
308. D. T. Clark, I. S. Woolsy, S. D. Robinson, K. R. Laing and J. N. Wingfield, *Inorg. Chem.*, 1977, **16** 1201.
309. B. L. Haymore, J. A. Ibers and D. W. Meek, *Inorg. Chem.*, 1975, **14**, 541.
310. B. L. Haymore and J. A. Ibers, *Inorg. Chem.*, 1975, **14**, 2784.
311. B. L. Haymore and J. A. Ibers, *Inorg. Chem.*, 1975, **14**, 1369.
312. J. Chatt, A. A. Diamantis, G. A. Heath, N. E. Hooper and G. J. Leigh, *J. Chem. Soc., Dalton Trans.*, 1977, 688.
313. S. Cenini, F. Porta, M. Pizzotti, *Inorg. Chim. Acta*, 1976, **20**, 119.
314. S. Krogsrud and J. A. Ibers, *Inorg. Chem.*, 1975, **14**, 2298.
315. M. Cowie, B. L. Haymore and J. A. Ibers, *Inorg. Chem.*, 1975, **14**, 2617.
316. B. L. Haymore and J. A. Ibers, *Inorg. Chem.*, 1975, **14**, 3060.
317. L. Toniolo, *Chem. Ind. (London)*, 1976, 30.
318. L. Toniolo and D. Leonesi, *J. Organomet. Chem.*, 1976, **113**, C73.
319. M. Cowie, B. L. Haymore and J. A. Ibers, *J. Am. Chem. Soc.*, 1976, **98**, 7608.
320. R. E. Cobbledick, F. W. B. Einstein, N. Farrell, A. B. Gilchrist and D. Sutton, *J. Chem. Soc., Dalton Trans.*, 1977, 373.
321. S. Krogsrud, L. Toniolo, U. Croatto and J. A. Ibers, *J. Am. Chem. Soc.*, 1977, **99**, 5277.
322. B. L. Haymore, *J. Organomet. Chem.*, 1977, **137**, C11.
323. N. Farrell and D. Sutton, *J. Chem. Soc., Dalton Trans.*, 1977, 2124.
324. K. D. Schramm and J. A. Ibers *J. Am. Chem. Soc.*, 1978, **100**, 2932.
325. M. Pizzotti, S. Cenini, F. Porta, W. Beck and J. Erbe, *J. Chem. Soc., Dalton Trans.*, 1978, 1155.
326. E. W. Abel and C. A. Burton, *J. Organomet. Chem.*, 1979, **170**, 229.
327. M. Angoletta and G. Caglio, *J. Organomet. Chem.*, 1979, **182**, 425.
328. G. Albertin, E. Bordignon and A. Orio, *Congr. Naz. Chim. Inorg., [Atti], 12th*, 1979, 93.
329. G. Butler, J. Chatt, W. Hussain, G. J. Leigh and D. L. Hughes, *Inorg. Chim. Acta*, 1978, **30**, L287.
330. G. Butler, J. Chatt, W. Hussain, G. J. Leigh and D. L. Hughes, *Inorg. Chim. Acta*, 1978, **28**, L165.
331. E. O. Bishop, G. Butler, J. Chatt, J. R. Dilworth, G. J. Leigh and D. Orchard, *J. Chem. Soc., Dalton Trans.*, 1978, 1654.
332. G. Butler, J. Chatt and G. J. Leigh, *J. Chem. Soc., Chem. Commun.*, 1978, 352.
333. D. Sellman and W. Weiss, *Angew. Chem., Int. Ed. Engl.*, 1977, **16**, 880.
334. J. R. Dilworth, H. J. de Liefde Meijer and J. H. Teuben, *J. Organomet. Chem.*, 1978, **159**, 47.
335. W. E. Carroll, D. Condon, M. E. Deane and F. J. Lalor, *J. Organomet. Chem.*, 1978, **157**, C58.
336. D. Condon, M. E. Deane, F. J. Lalor, N. G. Connelly and A. C. Lewis, *J. Chem. Soc., Dalton Trans.*, 1977, 925.
337. A. V. Butcher, J. Chatt, J. R. Dilworth, G. J. Leigh, M. B. Hursthouse, S. A. A. Jayaweera and A. Quick, *J. Chem. Soc., Dalton Trans.*, 1979, 921.
338. V. S. Day, T. A. George and S. D. A. Iske, *J. Am. Chem. Soc.*, 1975, **97**, 4127.
339. W. A. Herrmann, M. L. Ziegler and K. Weidenhammer, *Angew. Chem., Int. Ed. Engl.*, 1976, **15**, 368.
340. N. G. Connelly, Z. Demidowicz and R. L. Kelly, *J. Chem. Soc., Dalton Trans.*, 1975, 2335.
341. W. L. Bowden, G. M. Brown, E. M. Gupton, W. F. Little and T. J. Meyer, *Inorg. Chem.*, 1977, **16**, 213.
342. W. A. Herrmann, *Angew. Chem., Int. Ed. Engl.*, 1975, **14**, 355.
343. N. G. Connelly and C. Gardener, *J. Organomet. Chem.*, 1978, **159**, 179.
344. U. Croatto, L. Toniolo, A. Immirzi and G. Bombieri, *J. Organomet. Chem*, 1975, **102**, C31.
345. M. Angoletta, L. Malatesta, P. L. Bellon and G. Caglio, *J. Organomet. Chem.*, 1976, **114**, 219.
346. F. W. B. Einstein, D. Sutton and P. L. Vogel, *Inorg. Nucl. Chem. Lett.*, 1976, **12**, 671.
347. J. M. Manriquez, R. D. Sanner, R. E. Marsh and J. E. Bercaw, *J. Am. Chem. Soc.*, 1976, **98**, 3042.
348. P. Brant and R. D. Feltham, *J. Less-Common Met.*, 1977, **54**, 81.
349. D. Sutton, *Chem. Soc. Rev.*, 1975, **4**, 443.
350. A. D. Rattray and D. Sutton, *Inorg. Chim. Acta*, 1978, **27**, L85.
351. V. W. Day, T. A. George, S. D. A. Iske and S. D. Wagner, *J. Organomet. Chem.*, 1976, **112**, C55.

352. M. W. Bishop, J. Chatt, J. R. Dilworth, G. Kaufman, S. Kim and J. Zubieta, *J. Chem. Soc., Chem. Commun.*, 1977, 70.
353. T. Tatsumi, M. Hidai and Y. Uchida, *Inorg. Chem.*, 1975, **14**, 2530.
354. A. P. Graughan and J. A. Ibers, *Inorg. Chem.*, 1975, **14**, 352.
355. M. R. Churchill and K. G. Lin, *Inorg. Chem.*, 1975, **14**, 1133.
356. D. L. DuBois and R. Hoffmann, *Nouv. J. Chim.*, 1977, **1**, 479.
357. A. B. Gilchrist and D. Sutton, *J. Chem. Soc., Dalton Trans.*, 1977, 677.
358. A. Nakamura, M. Nakayama, K. Sugihashi and S. Otsuka, *Inorg. Chem.*, 1979, **18**, 394.
359. J. Chatt, J. R. Dilworth and G. J. Leigh, *J. Chem. Soc., Dalton Trans.*, 1979, 1843.
360. J. Chatt, B. A. L. Chrichton, J. R. Dilworth, P. Dahlstrom, G. Gutkeska and J. A. Zubieta, *Transition Met. Chem.*, 1979, **4**, 271.
361. R. D. Feltham and P. Brant, *J. Am. Chem. Soc.*, 1982, **104**, 641.
362. K. D. Schramm and J. A. Ibers, *Inorg. Chem.*, 1980, **19**, 2435.
363. W. E. Carroll, F. A. Deeney and F. J. Lalor, *J. Organomet. Chem.*, 1980, **198**, 189.
364. T. R. Gaffney and J. A. Ibers, *Inorg. Chem.*, 1982, **21**, 2851.
365. K. D. Schramm and J. A. Ibers, *Inorg. Chem.*, 1980, **19**, 2441.
366. R. A. Michelin and R. J. Angelici, *Inorg. Chem.*, 1980, **19**, 3850.
367. B. A. Crichton, J. R. Dilworth, P. Dahlstrom and J. Zubieta, *Transition Met. Chem.*, 1980, **5**, 316.
368. R. Tamashita, K. Kikukawa, F. Wada and T. Matsuda, *J. Organomet. Chem.*, 1980, **201**, 463.
369. J. R. Dilworth, C. Kan, R. L. Richards, J. Mason and I. A. Stenhouse, *J. Organomet. Chem.*, 1980, **201**, C24.
370. H. M. Colquhoun, *Transition Met. Chem.*, 1981, **6**, 57.
371. M. Angoletta, and G. Caglio, *J. Organomet. Chem.*, 1982, **234**, 99.
372. H. Brunner and W. Miehling, *Angew. Chem., Int. Ed. Engl.*, 1983, **22**, 164.
373. M. Angoletta and G. Caglio, *J. Organomet. Chem.*, 1980, **185**, 105.
374. J. R. Dilworth, B. D. Neaves and C. J. Pickett, *Inorg. Chem.*, 1980, **19**, 2859.
375. C. F. Barrientos-Penna, F. W. B. Einstein, D. Sutton and A. C. Willis, *Inorg. Chem.*, 1980, **19**, 2740.
376. S. Vancheesan, *Proc. Indian. Acad. Sci., Sect. A, Chem. Sci.*, 1982, **91**, 343.
377. M. R. Churchill and H. J. Wasserman, *Inorg. Chem.*, 1981, **20**, 1580.
378. J. Colin, G. Butler and R. Weiss, *Inorg. Chem.*, 1980, **19**, 3828.
379. C. F. Barrientos-Penna, F. W. B. Einstein, T. Jones and D. Sutton, *Inorg. Chem.*, 1983, **22**, 2614.
380. R. Mattes and H. Scholand, *Angew. Chem., Int. Ed. Engl.*, 1983, **22**, 245.
381. J. R. Dilworth, I. A. Latham, G. J. Leigh, G. Huttner and I. Jibril, *J. Chem. Soc., Chem. Commun.*, 1983, 1368.
382. D. Condon, G. Ferguson, F. J. Lalor, M. Parvez and T. Spalding, *Inorg. Chem.*, 1982, **21**, 188.
383. G. L. Hillhouse, B. L. Haymore, S. A. Bistram and W. A. Herrmann, *Inorg. Chem.*, 1983, **22**, 314.
384. G. J. Leigh, R. H. Morris, C. J. Pickett and D. R. Stanley, *J. Chem. Soc., Dalton Trans.*, 1981, 800.
385. F. N. N. Carvalho, A. J. L. Pombeiro, O. Orama, U. Schubert, C. J. Pickett and R. L. Richards, *J. Organomet. Chem.*, 1982, **240**, C18.
386. T. Takahashi, Y. Mizobe, M. Sato, Y. Uchida and M. Hidai, *J. Am. Chem. Soc.*, 1980, **102**, 7461.
387. C. S. Day, V. W. Day, T. A. George and I. Tavanaiepour, *Inorg. Chim. Acta*, 1980, **45**, L54.
388. D. C. Busby, T. A. George, S. D. A. Iske and S. D. Wagner, *Inorg. Chem.*, 1981, **20**, 22.
389. F. J. Lalor, D. Condon, G. Ferguson and M. A. Kahn, *Inorg. Chem.*, 1981, **20**, 2178.
390. W. A. Herrmann and S. A. Bistram, *Chem. Ber.*, 1980, **113**, 2648.
391. G. E. Bossard, D. C. Busby, M. Chang, T. A. Geogre and S. D. A. Iske, *J. Am. Chem. Soc.*, 1980, **102**, 1001.
392. M. Sato, T. Kodama, M. Hidai and Y. Uchida, *J. Organomet. Chem.*, 1978, **152**, 239.
393. G. L. Hillhouse, B. L. Haymore and W. A. Herrmann, *Inorg. Chem.*, 1979, **18**, 2423.
394. D. Sellmann and W. Weiss, *Angew. Chem., Int. Ed. Engl.*, 1978, **17**, 269.
395. J. Chatt, R. A. Head, G. J. Leigh and C. L. Pickett, *J. Chem. Soc., Dalton Trans.*, 1978, 1638.
396. W. A. Herrmann and H. Biersack, *Chem. Ber.*, 1977, **110**, 896.
397. A. A. Diamantis, J. Chatt, G. J. Leigh and G. A. Heath, *J. Organomet. Chem.*, 1975, **84**, C11.
398. M. W. Bishop, J. Chatt and J. R. Dilworth, *J. Organomet. Chem.*, 1974, **73**, C59.
399. C. Woodcock and R. Eisenberg, *Organometallics*, 1985, **4**, 4.
400. B. L. Haymore, unpublished data.
401. J. Chatt, M. E. Fakley, P. B. Hitchcock, R. L. Richards and N. T. Luong-Thi, *J. Chem. Soc., Dalton Trans.*, 1982, 345.
402. J. R. Dilworth, S. A. Harrison, D. R. M. Walton and E. Schweda, *Inorg. Chem.*, 1985, **24**, 2594.
403. P. J. Blower, J. R. Dilworth, J. Hutchinson, T. Nicholson and J. A. Zubieta, *J. Chem. Soc., Dalton Trans.*, 1985, 2639.
404. T. Nicholson and J. Zubieta, *Inorg. Chim. Acta*, 1985, **100**, L35.
405. T. C. Hsieh and J. Zubieta, *Inorg. Chem.*, 1985, **24**, 1287.
406. F. J. Lalor and L. H. Brookes, *J. Organomet. Chem.*, 1983, **251**, 327.
407. H. C. Ashton and A. R. Manning, *Inorg. Chim. Acta*, 1983, **71**, 163.
408. H. C. Ashton and A. R. Manning, *Inorg. Chem.*, 1983, **22**, 1440.
409. H. Brunner and W. Miehling, *Angew. Chem.*, 1983, **95**, 162.
410. P. Brant and R. D. Feltham, *J. Electron Spectrosc. Relat. Phenom.*, 1983, **32**, 205.
411. I. V. Linko, B. E. Zaitsev, A. K. Molodkin, T. M. Ivanov and R. V. Linko, *Zh. Neorg. Khim.*, 1983, **28**, 1520.
412. C. F. Barrientos-Penna, F. W. B. Einstein, T. Jones and D. Sutton, *Inorg. Chem.*, 1985, **24**, 632.
413. C. F. Barrientos-Penna, G. F. Campana, F. W. B. Einstein, T. Jones, D. Sutton and A. S. Tracey, *Inorg. Chem.*, 1984, **23**, 363.
414. W. A. Herrmann, H. Biersack, K. K. Mayer and B. Reiter, *Chem. Ber.*, 1980, **113**, 2655.
415. T. C. Hsieh and J. Zubieta, *Inorg. Chim. Acta*, 1985, **99**, L47.
416. T. C. Hsieh, K. Gebreyes and J. Zubieta, *J. Chem. Soc., Dalton Trans.*, 1984, 1172.
417. P. L. Dahlstrom, J. R. Dilworth, P. Shulman and J. Zubieta, *Inorg. Chem.*, 1982, **21**, 933.
418. C. Pickett, S. Kumar, P. A. Vella and J. Zubieta, *Inorg. Chem.*, 1982, **21**, 908.
419. C. F. Barrientos-Penna, A. H. Klahn-Oliva and D. Sutton, *Organometallics*, 1985, **4**, 367.

420. C. F. Barrientos-Penna, A. B. Gilchrist, A. J. L. Hanlan and D. Sutton, *Organometallics*, 1985, **4**, 478.
421. K. K. Panday, H. W. Roesky, M. Noltemeyer and G. M. Sheldrick, *Z. Naturforsch., Teil B*, 1984, **39**, 5980.
422. H. W. Roesky, K. K. Pandey, W. Clegg, M. Noltemeyer and G. M. Sheldrick, *J. Chem. Soc., Dalton Trans.*, 1984, 719.
423. J. Baldas, J. Bonnyman, M. F. Mackay and G. A. Williams, *Aust. J. Chem.*, 1984, **37**, 751.
424. M. Minelli, J. L. Hubbard, D. L. Lichtenberger and J. H. Enemark, *Inorg. Chem.*, 1984, **23**, 2721.
425. D. L. Lichtenberger and J. L. Hubbard, *Inorg. Chem.*, 1984, **23**, 2718.
426. K. K. Pandey, S. R. Ahuja and M. Goyal, *Indian J. Chem., Sect. A*, 1985, **24**, 1059.
427. M. Minelli, C. G. Young and J. H. Enemark, *Inorg. Chem.*, 1985, **24**, 1111.
428. M. B. Hursthouse, N. P. C. Walker, C. P. Warrens and J. D. Woollins, *J. Chem. Soc., Dalton Trans.*, 1985, 1043.
429. D. L. Lichtenberger and J. L. Hubbard, *Inorg. Chem.*, 1985, **24**, 3825.
430. J. Anhaus, Z. A. Siddigi, H. W. Roesky, J. W. Bats and Y. Elerman, *Z. Naturforsch. Teil B*, 1985, **40**, 740.
431. M. A. Andrews and H. D. Kaesz, *J. Am. Chem. Soc.*, 1979, **101**, 7255.
432. M. A. Andrews, G. Van Bushkirk, C. B. Knobler and H. D. Kaesz, *J. Am. Chem. Soc.*, 1979, **101**, 7245.
433. Z. Dawoodi, M. J. Mays and P. R. Raithby, *J. Organomet. Chem.*, 1981, **219**, 103.
434. Z. Dawoodi, A. G. Orphen and M. J. Mays, *J. Organomet. Chem.*, 1981, **219**, 251.
435. R. D. Adams, D. A. Katahira and L. W. Yang, *J. Organomet. Chem.*, 1981, **219**, 85.
436. R. D. Adams, D. A. Katahira and L. W. Yang, *J. Organomet. Chem.*, 1981, **219**, 241.
437. P. C. Bevan, J. Chatt, R. A. Head, P. B. Hitchcock and G. J. Leigh, *J. Chem. Soc., Chem. Commun.*, 1976, 509.
438. D. Carrillo, P. Gouzerh and Y. Jeannin, *Nouv. J. Chim.*, 1985, **9**, 749.
439. J. Chatt, W. Hussain, G. J. Leigh, H. Neukomm, C. J. Pickett and D. A. Rankin, *J. Chem. Soc., Chem. Commun.*, 1980, 1024.
440. F. W. B. Einstein, T. Jones, A. J. L. Hanlan and D. Sutton, *Inorg. Chem.*, 1982, **21**, 2585.
441. C. F. Barrientos-Penna, F. W. B. Einstein, T. Jones and D. Sutton, *Inorg. Chem.*, 1982, **21**, 2578.
442. C. Barrientos-Penna and D. Sutton, *J. Chem. Soc., Chem. Commun.*, 1980, 111.
443. D. Sellmann and W. Weiss, *J. Organomet. Chem.*, 1978, **160**, 183.
444. N. Mronga, F. Weller and K. Dehnicke, *Z. Anorg. Allg. Chem.*, 1983, **502**, 35.
445. I. Schmidt, U. Kynast, J. Hanich and K. Dehnicke, *Z. Naturforsch., Teil B*, 1984, **39**, 1248.
446. R. Choukroun and D. Gervais, *J. Chem. Soc., Dalton Trans.*, 1980, 1800.
447. K. Dehnicke, H. Prinz, W. Kafitz and R. Kujanek, *Liebigs Ann. Chem.*, 1981, 20.
448. J. R. Dilworth, B. D. Neaves, J. P. Hutchinson and J. A. Zubieta, *Inorg. Chim. Acta*, 1982, **65**, L223.
449. J. Chatt, G. J. Leigh, H. Neukomm, C. J. Pickett and D. R. Stanley, *J. Chem. Soc., Dalton Trans.*, 1980, 121.
450. R. Ben-Shoshan, J. Chatt, W. Hussain and G. J. Leigh, *J. Organomet. Chem.*, 1976, **112**, C9.
451. J. Chatt, R. A. Head, G. J. Leigh and C. J. Pickett, *J. Chem. Soc., Chem. Commun.*, 1977, 299.
452. M. Fernanda, N. N. Carvalho, A. J. L. Pombeiro, U. Schubert, O. Orama, C. J. Pickett and R. L. Richards, *J. Chem. Soc., Dalton Trans.*, 1985, 2079.
453. V. L. Frost and R. A. Henderson, *J. Chem. Soc., Dalton Trans.*, 1985, 2059.
454. R. A. Henderson, *J. Chem. Soc., Dalton Trans.*, 1985, 2067.
455. J. R. Dilworth and R. L. Richards, in 'Comprehensive Organometallic Chemistry', ed. G. Wilkinson, F. G. A. Stone, E. W. Abel, Pergamon, Oxford, 1982, p. 1017.
456. J. Chatt, J. R. Dilworth and R. L. Richards, *Chem. Rev.*, 1978, **78**, 589.
457. R. A. Henderson, G. J. Leigh and C. J. Pickett, *Adv. Inorg. Radiochem.*, 1985, **27**, 197.
458. M. W. Bishop, G. Butler, J. Chatt, J. R. Dilworth and G. J. Leigh, *J. Chem. Soc., Dalton Trans.*, 1979, 1843.
459. J. R. Dilworth, S. A. Harrison, D. R. M. Walton and E. Schweda, *Inorg. Chem.*, 1985, **24**, 2595.
460. C. F. Barientos-Penna, F. W. B. Einstein, T. Jones and D. Sutton, *Inorg. Chem.*, 1982, **21**, 2578.
461. M. W. Bishop, J. Chatt, J. R. Dilworth, M. R. Hursthouse and M. Molevalli, *J. Chem. Soc., Dalton Trans.*, 1979, 1600.
462. J. R. Dilworth and S. Morton, *Transition Met. Chem.*, in press.
463. J. Chatt, B. A. L. Crichton, J. R. Dilworth, P. Dahlstrom and J. A. Zubieta, *J. Chem. Soc.., Chem. Commun.*, 1980, 786.
464. M. Veith, *Angew. Chem., Int. Ed. Engl.*, 1976, **15**, 387.
465. J. Chatt, J. R. Dilworth, P. Dahlstrom and J. A. Zubieta, *J. Chem. Soc., Chem. Commun.*, 1986, 786.
466. C. F. Barrientos-Penna, C. F. Campana, F. W. B. Einstein, T. Jones, D. Sutton and A. S. Tracey, *Inorg. Chem.*, 1984, **23**, 363.
467. I. A. Latham, G. J. Leigh, G. Huttner and I. Jibril, *J. Chem. Soc., Dalton Trans.*, 1986, 385.
468. M. Hidai, T. Kodama, M. Sato, M. Harakawa and Y. Uchida, *Inorg. Chem.*, 1975, **15**, 2694.
469. J. Chatt, A. J. Pearman and R. L. Richards, *J. Chem. Soc., Dalton Trans.*, 1978, 1766.
470. J. Chatt, M. E. Fakley, P. B. Hitchcock, R. L. Richards and N. T. Luong-Thi, *J. Organomet. Chem.*, 1979, **172**, C55.
471. K. Gebreyes, J. A. Zubieta, T. A. George, L. M. Koczon and R. C. Tisdale, *Inorg. Chem.*, 1986, **25**, 407.
472. R. A. Henderson, *J. Chem. Soc., Dalton Trans.*, 1984, 2259.
473. S. Donovan-Mtunzi, R. L. Richards and J. Mason, *J. Chem. Soc., Dalton Trans*, 1984, 1329.
474. S. N. Anderson, M. E. Fakeley, R. L. Richards and J. Chatt, *J. Chem. Soc., Dalton Trans.*, 1981, 1973.
475. J. Chatt, A. J. Pearman and R. L. Richards, *J. Organomet. Chem.*, 1975, **101**, C45.
476. J. A. Baumann and T. A. George, *J. Am. Chem. Soc.*, 1980, **102**, 6153; R. L. Richards, 'Nitrogen Fixation, The Chemical Biochemical Interface', ed. A. Muller and W. E. Newton, Plenum, London, 1983, 275.
477. J. Chatt, G. A. Heath and R. L. Richards, *J. Chem. Soc., Dalton Trans.*, 1974, 2074.
478. M. Hidai, Y. Mizobe, M. Sato, T. Kodama and Y. Uchida, *J. Am. Chem. Soc.*, 1978, **100**, 5740.
479. P. C. Bevan, J. Chatt, M. Hidai and G. J. Leigh, *J. Organomet. Chem.*, 1978, **160**, 165.
480. R. Ben-Shoshan, J. Chatt, G. J. Leigh and W. Hussain, *J. Chem. Soc., Dalton Trans.*, 1980, 771.
481. H. M. Colquhoun, *J. Chem. Res.*, 1981, **9**, 3416.
482. H. M. Colquhoun, *J. Chem. Res.*, 1981, **9**, 3401.
483. G. L. Hillhouse and B. L. Haymore, *J. Am. Chem. Soc.*, 1982, **104**, 1537 and references therein.
484. J. Chatt, R. A. Head, G. J. Leigh and C. J. Pickett, *J. Chem. Soc., Dalton Trans.*, 1986, 1473.

485. W. Hussain, G. J. Leigh, H. Modh-Ali and C. J. Pickett, *J. Chem. Soc., Dalton Trans.*, 1986, 1473.
486. J. Chatt, A. A. Diamantis, G. A. Heath, N. E. Hooper and G. J. Leigh, *J. Chem. Soc., Dalton Trans.*, 1977, 688.
487. C. J. Pickett and G. J. Leigh, *J. Chem. Soc., Chem. Commun.*, 1981, 1033.
488. C. J. Pickett, K. S. Ryder and J. Talarmin, *J. Chem. Soc., Dalton Trans.*, 1986, 1453.
489. J. Bultitude, L. F. Larkworthy, D. C. Povey, G. W. Smith, J. R. Dilworth and G. J. Leigh, *J. Chem. Soc., Chem. Commun.*, 1986, 1748.
490. M. W. Bishop, J. Chatt, J. R. Dilworth, M. B. Hursthouse and M. Motevalle, *J. Chem. Soc., Dalton Trans.*, 1979, 1600.
491. J. Chatt, B. A. L. Crichton, J. R. Dilworth, P. Dahlstrom and J. A. Zubieta, *J. Chem. Soc., Dalton Trans.*, 1982, 1041.
492. R. J. Burt, J. R. Dilworth, J. Hutchinson and J. A. Zubieta, *J. Chem. Soc., Dalton Trans.*, 1982, 2295.
493. P. T. Bishop, J. R. Dilworth and J. A. Zubieta, unpublished results.
494. J. Chatt, J. R. Dilworth, B. A. L. Crichton, P. Dahlstrom, R. Gutkoska and J. A. Zubieta, *Inorg. Chem.*, 1982, **21**, 2283.
495. J. R. Dilworth, A. Garcia-Rodriguez, G. J. Leigh and J. N. Murrell, *J. Chem. Soc., Dalton Trans.*, 1983, 455.
496. P. J. Blower, J. R. Dilworth and J. A. Zubieta, unpublished results.
497. P. T. Bishop, J. R. Dilworth and J. A. Zubieta, unpublished results.
498. N. Wiberg, *Adv. Organomet. Chem.*, 131 1984, **23**, and references therein.
499. J. R. Dilworth and S. Morton, *J. Organomet. Chem.*, 1986, **314**, C25.
500. S. N. Shaikh and J. A. Zubieta, *Inorg. Chem.*, 1986, **25**, 4615.
501. J. R. Dilworth and J. A. Zubieta, *J. Chem. Soc., Chem. Commun.*, 1987, 132.
502. T. Jones, A. J. Hanlan, F. W. B. Einstein and D. J. Sutton, *J. Chem. Soc., Chem. Commun.*, 1980, 1978.
503. J. R. Dilworth, J. A. Zubieta and J. R. Hyde, *J. Am. Chem. Soc.*, 1982, **104**, 365.
504. J. R. Dilworth, R. A. Henderson, P. Dahlstrom, T. Nicholson and J. A. Zubieta, *J. Chem. Soc., Dalton Trans.* in press.
505. W. G. Kita, J. A. McCleverty, B. E. Mann, D. Seddon, G. A. Sim and D. I. Woodhouse, *J. Chem. Soc., Chem. Commun.*, 1974, 132; P. D. Frisch, M. M. Hunt, W. G. Kita, J. A. McCleverty, A. E. Rose, D. Seddon, D. Swann and J. Williams, *J. Chem. Soc., Dalton Trans.*, 1979, 1819.
506. J. A. McCleverty, *Chem. Soc. Rev.*, 1983, **12**, 331.
507. D. L. Hughes, I. A. Latham and G. J. Leigh, *J. Chem. Soc., Dalton Trans.*, 1986, 393.
508. J. A. McCleverty, A. E. Rose, I. Wolochowicz, N. A. Bailey and J. M. A. Smith, *J. Chem. Soc., Dalton Trans.*, 1983, 71.
509. T. Nicholson and J. A. Zubieta, *J. Chem. Soc., Chem. Commun.*, 1985, 367.
510. J. A. McCleverty, *Transition Met. Chem.*, 1987, **12**, Highlights 16, in press.

13.4

Amido and Imido Metal Complexes

MALCOLM H. CHISHOLM
Indiana University, Bloomington, IN, USA

and

IAN P. ROTHWELL
Purdue University, West Lafayette, IN, USA

13.4.1 INTRODUCTION

Compounds included in this chapter are the homoleptic amides and imides of empirical formula $M(NRR')_n$ and $M(NR)_n$, where R, R' = H, alkyl, aryl or trialkylsilyl, and the mixed amido–imido compounds of formula $M(NRR')_x(NR)_y$.[1-4] Brief mention is given to certain compounds of formula $L_nM(NRR')_m$ or $L_nM(NR)_m$, where L_n represents a group of ligands, neutral or anionic, which also share the coordination sphere of the metal, when the chemistry of the M—N bond is of particular note. Metal-containing compounds derived from halogen azides[5] are not discussed.

13.4.2 SYNTHESIS OF METAL AMIDES AND IMIDES[1-4]

13.4.2.1 Metal Amides

13.4.2.1.1 *From the metal*

Only amides of the more electropositive metals can be prepared by the direct interaction of ammonia or a primary or secondary amine with the metal. Hence, the molten alkali metals will react with gaseous ammonia liberating hydrogen (*e.g.* equation 1).[6-8]

$$Na + NH_3 \xrightarrow{\ 300\,^{\circ}C\ } NaNH_2 + \tfrac{1}{2}H_2 \tag{1}$$

The deep blue solutions formed by dissolving alkali metals in ammonia do not rapidly generate the amide unless a catalyst is added.[9] However, a hydrogen acceptor will also initiate the reaction and this forms the basis of the important Birch reduction of aromatic compounds (equation 2).[10]

$$4Na + 4NH_3 + C_{10}H_8 \text{ (naphthalene)} \longrightarrow 4NaNH_2 + C_{10}H_{12} \tag{2}$$

This method can also be applied to both primary and secondary amines.[11,12]

Of the other elements, only magnesium appears to form amides directly, reacting with anilines on heating (equation 3).[13]

$$Mg + 2H_2NAr \longrightarrow Mg(NHAr)_2 + H_2 \tag{3}$$

13.4.2.1.2 *From metal halides*

(*i*) *By direct reaction with amines*

The reaction of halides with the NH function of amines can set up an equilibrium as shown (equation 4).

$$L_nMX + RR'NH \rightleftharpoons L_nM\!-\!NRR' + HX \tag{4}$$

The equilibrium can be easily forced to the right by addition of an excess of amine which will be protonated by HX with overall elimination of the amine salt. Hence, boron halides form the corresponding amides on reaction with excess of amine (equation 5).[14]

$$BCl_3 + 6RNH_2 \longrightarrow B(NHR)_3 + 3RNH_3^+Cl^- \tag{5}$$

With more bulky secondary amines, the reaction may stop before total substitution yielding mixed haloamides.[15,16]

This synthetic method is of great importance in the synthesis of silicon amides. Halides of the type R_xSiCl_{4-x} ($x = 0, 1, 2, 3$) react with ammonia or primary and secondary amines to give the corresponding amides.[17-24] Again, with sterically demanding amines only partial halide displacement is possible (equations 6 and 7).[22,23]

$$SiCl_4 + 8HNMe_2 \longrightarrow Si(NMe_2)_4 + 4Me_2NH_2^+Cl^- \tag{6}$$

$$SiCl_4 + 6H_2NBu^t \longrightarrow Si(NHBu^t)_3Cl + 3BuNH_3^+Cl^- \tag{7}$$

The amides of phosphorus can be similarly obtained, using either PCl_3, $POCl_3$ or PCl_5, for example.[24-27] However, with PCl_5 only secondary amines yield amides,[27] primary amines generally giving phosphazene derivatives.[28]

The higher valent, early transition metal halides will react readily with ammonia, primary or secondary amines to form metal amides. By this method amides of Ti,[29] V,[30] Nb,[31] Ta,[32] Mo[33] and W[34] have been obtained (equation 8).[33]

$$TiCl_4 + 8HNMe_2 \longrightarrow Ti(NMe_2)_4 + 4Me_2NH_2^+Cl^- \tag{8}$$

However, the degree of aminolysis is rarely total, giving mixed haloamides. Furthermore, the products sometimes contain the amine reagent as a simple donor ligand (*e.g.* equations 9,[35] 10[36] and 11[33]).

$$ZrCl_4 + 3RNH_2 \longrightarrow Zr(NHR)Cl_3 \cdot NH_2R + RNH_3^+Cl^- \tag{9}$$

$$(R = Me, Et, Pr, Bu)$$

$$NbCl_5 + 5Me_2NH \longrightarrow Nb(NMe_2)_2Cl_3 \cdot NHMe_2 + 2Me_2NH_2^+Cl^- \tag{10}$$

$$MoCl_5 + 5Me_2NH \longrightarrow Mo(NMe_2)_2Cl_3 \cdot NHMe_2 + 2Me_2NH_2^+Cl^- \tag{11}$$

(ii) By reaction with alkali metal amides

This method, sometimes referred to as transmetallation, is by far the most versatile synthetic route to transition metal amide complexes. The amide, typically lithium or sodium, is reacted with the corresponding transition metal halide either in a hydrocarbon or, more typically, an ether solvent. The method has been applied to virtually all of the transition elements and normally results in complete substitution except for the most bulky amido substituents (equations 12,[37] 13,[38] 14,[39] 15[40] and 16[41,42]).

$$NbCl_5 + 5LiNMe_2 \longrightarrow Nb(NMe_2)_5 + 5LiCl \tag{12}$$

$$MoCl_3 + 3LiNMe_2 \longrightarrow \tfrac{1}{2}Mo_2(NMe_2)_6 + 3LiCl \tag{13}$$

$$ZrCl_4 + 4LiNMe_2 \longrightarrow Zr(NMe_2)_4 + 4LiCl \tag{14}$$

$$MCl_3 + 3Na[N(SiMe_3)_2] \longrightarrow M[N(SiMe_3)_2]_3 + 3NaCl \tag{15}$$

$$(M = Fe, Cr)$$

$$MCl_4 + 3Na[N(SiMe_3)_2] \longrightarrow MCl[N(SiMe_3)_2]_3 + 3NaCl \tag{16}$$

$$(M = Ti, Zr, Hf, Th, U)$$

However, in some cases either disproportionation or reduction of the metal occurs (equations 17,[3] 18[43] and 19[44]).

$$MoCl_5 + 5LiNMe_2 \longrightarrow Mo(NMe_2)_4 + Mo_2(NMe_2)_6 + 5LiCl \tag{17}$$

$$WCl_4 + 4LiNMe_2 \longrightarrow W(NMe_2)_6 + W_2(NMe_2)_6 + 4LiCl \tag{18}$$

$$NbCl_5 + 5LiNEt_2 \longrightarrow Nb(NEt_2)_4 + 5LiCl + \text{other products} \tag{19}$$

13.4.2.1.3 From metal amides by amine exchange

Addition of an amine to a metal amide can set up an equilibrium as shown (equation 20).

$$L_nM{-}NR^1R^2 + HNR^3R^4 \rightleftharpoons L_nM{-}NR^3R^4 + HNR^1R^2 \tag{20}$$

The utility of this reaction for the synthesis of a new amide is influenced by both steric effects and the relative volatilities of the amines. As with amine for HX exchange, total substitution typically does not occur (equations 21,[45] 22[39] and 23[46]).

$$\tfrac{1}{2}[Al(NMe_2)_3]_2 + Pr_2^iNH \longrightarrow \tfrac{1}{2}[Al(NMe_2)_2(NPr_2^i)]_2 + Me_2NH \tag{21}$$

$$Ti(NMe_2)_4 + 3HNPr_2^n \longrightarrow Ti(NPr_2^n)_3(NMe_2) + 3HNMe_2 \tag{22}$$

$$U(NEt_2)_4 + 2dmedH_2 \longrightarrow \tfrac{1}{3}U_3(dmed)_6 + 4HNEt_2 \tag{23}$$

$$(dmedH_2 = MeNH(CH_2)_2NHMe)$$

In the case of niobium, the attempts to substitute with bulky dialkylamines can result in reduction (equation 24).[38]

$$Nb(NMe_2)_5 + HNEt_2 \longrightarrow Nb(NEt_2)_3(NMe_2) + HNMe_2 + \text{other products} \tag{24}$$

Further complications may arise on using primary amines where imido functions can sometimes be generated (see Section 13.4.2.2.1.ii).

13.4.2.1.4 From metal alkyls and hydrides

(i) By treatment with amines

Although of relatively weak acidity, amines will react with either carbanionic metal alkyls or hydridic metal hydrides to form amides with the elimination of alkane or hydrogen, respectively. The easiest and most exploited method for the synthesis of lithium and magnesium amides is to treat lithium alkyls or Grignard reagents (normally commercially available) with the corresponding amine (equations 25,[47] 26[48] and 27[49]).

$$LiBu^n + H_2NAr \xrightarrow{Et_2O} \tfrac{1}{2}[Li(NHAr)(OEt_2)]_2 + Bu^nH \tag{25}$$

$$(Ar = 2,4,6\text{-tri-t-Butylphenyl})$$

$$R_2Mg + 2HNMe_2 \longrightarrow Mg(NMe_2)_2 + 2RH \tag{26}$$

$$RMgCl + HN(SiMe_3)_2 \longrightarrow ClMgN(SiMe_3)_2 + RH \tag{27}$$

Similar reactivity is seen for the alkyls of beryllium,[50,51] calcium[52] and aluminum.[53,54]

The hydrides of the more electropositive metals will react with amines to liberate hydrogen and give the metal amide. Typically these reactions are carried out in hydrocarbon solvents (equations 28[55] and 29[15,45]).

$$MH + HNR^1R^2 \longrightarrow MNR^1R^2 + H_2 \tag{28}$$

$$(M = Na, K)$$

$$AlH_3 + 3HNR^1R^2 \longrightarrow Al(NR^1R^2)_3 + 3H_2 \tag{29}$$

Diborane or one of its tertiary amine adducts will react with primary or secondary amines to give aminoboranes (equation 30).[56]

$$B_2H_6 + 2HNMe_2 \longrightarrow [H_2BNMe_2]_2 + 2H_2 \tag{30}$$

13.4.2.1.5 From metal alkoxides, thioalkoxides or oxides

The alkali metal oxides will deprotonate ammonia to give a mixture of hydroxide and amide (equation 31),[57] while P_4O_{10} reacts with amines as shown in equation (32).[58]

$$Na_2O + NH_3 \longrightarrow NaNH_2 + NaOH \tag{31}$$

$$P_4O_{10} + 2H_2NAr \rightleftharpoons (ArNH)_2P(O)(OH) + \text{other products} \qquad (32)$$

Attempts to displace either alkoxy or alkylthio groups by direct reaction with amines generally fail due to the poor acidity of the NH bond and the unfavorable position of the equilibrium (equation 33).

$$L_nM—OR + HNR^1R^2 \rightleftharpoons L_nM—NR^1R^2 + ROH \qquad (33)$$

Indeed the alcoholysis of metal amides is an excellent method for the synthesis of alkoxides. However, some synthetic utility to this reaction has been found for the synthesis of aminoboranes where the equilibrium can be forced to the right by using either an involatile or a chelating amine.[59,60] Hydrocarbon solutions of $Mo_2Me_2(OBu^t)_4$ also react with $MeNHCH_2CH_2NHMe$ (> 6 equiv.) to give $Mo_2(MeNCH_2CH_2NMe)_3$ with the liberation of methane (2 equiv.) and *t*-butyl alcohol.[61]

13.4.2.1.6 By coupling of iminoacyl groups

The migratory insertion of alkyl and aryl isocyanides into metal alkyl bonds produces an iminoacyl function. The early transition metal iminoacyl, like its oxygen counterpart, has been shown to bond in a dihapto fashion.[62] It has recently been shown that it is possible to couple either an acyl and an iminoacyl or two iminoacyls on the metals titanium or zirconium to produce enamidolate and enediamide ligands, respectively (equation 34).[62,63]

$$(ArO)_2MR_2 + 2xyNC \longrightarrow (ArO)_2M(\eta^2\text{-}xyNCR)_2 \longrightarrow (ArO)_2\overline{M(xyN—CR{=}CR—N}xy)$$

$$(M = Ti, Zr; xy = 2,6\text{-dimethylphenyl}) \qquad (34)$$

13.4.2.2 Metal Imides

Unlike the case of metal amides where it was convenient to discuss synthesis in terms of the metal substrate, it is more informative for organoimido groups to consider synthetic methods as a function of the nitrogen-containing substrate or ligand that ultimately generates the NR groups.

13.4.2.2.1 From primary amines

As mentioned previously, the synthesis of transition metal amides from primary amines can be complicated by the concomitant generation of imido ligands. In general the imido group replaces two uninegative ligands originally coordinated to the metal, these groups being lost with the two protons also generated (equation 35).

$$L_nMX_2 + H_2NR \longrightarrow L_nMNR + 2HX \qquad (35)$$

(i) And metal halides

The reaction of the bulky amine Bu^tNH_2 with WCl_6 in hexane affords the four-coordinate bis-imide as shown (equation 36).[64]

$$WCl_6 + 10Bu^tNH_2 \longrightarrow (Bu^tN)_2W(NHBu^t)_2 + 6Bu^tNH_3^+ Cl^- \qquad (36)$$

A somewhat related reaction makes use of the monolithium salt, $LiNHBu^t$, to give the *t*-butyl imido derivatives of Nb and Ta (equation 37).[65]

$$MCl_5 + 4LiNMe_2 + LiNHBu^t \longrightarrow (Me_2N)_3MNBu^t + 5LiCl + Me_2NH \qquad (37)$$

$$(M = Nb, Ta)$$

(ii) And metal dialkylamides

The attempted amine exchange reaction to prepare metal amides can be complicated if primary amines are used. Hence, titanium dialkylamides will react with many primary amines to produce polymeric materials (equation 38).[66]

$$Ti(NMe_2)_4 + 2RNH_2 \longrightarrow [Ti(NR)_2]_x + 4HNMe_2 \tag{38}$$

With the more bulky amine Bu^tNH_2, only partial substitution leads to discrete mixed amido-imido compounds (equation 39).[67]

$$Ti(NMe_2)_4 + Bu^tNH_2 \longrightarrow \tfrac{1}{2}[(Me_2N)_2Ti(NBu^t)]_2 + 2HNMe_2 \tag{39}$$

Analogous compounds are obtained using $M(NMe_2)_4$ (M = Zr, Hf).[72]

A similar amine exchange on tantalum dialkylamides with primary amines again leads to an imido derivative (equation 40).[65,68]

$$Ta(NMe_2)_5 + RNH_2 \longrightarrow (Me_2N)_3TaNR + 2Me_2NH \tag{40}$$

$$(R = Bu^{t},^{65}Ph^{68})$$

(iii) And metal oxides

The direct aminolysis of metal oxides with primary amines is a useful synthetic route for imido compounds of rhenium and osmium (equation 41).

$$L_nMO + H_2NR \longrightarrow L_nMNR + H_2O \tag{41}$$

A large range of substituted arylimido derivatives of rhenium have been made by this method (equation 42).[69,74] However, the reaction does not proceed with aliphatic amines.[69] Similar reactivity is seen for the dimeric oxydithiocarbamates (equation 43).[71]

$$L_2Cl_3ReO + ArNH_2 \longrightarrow L_2Cl_3ReNAr + H_2O \tag{42}$$

$$(L = \text{a tertiary phosphine})$$

$$Re_2O_3(S_2CNEt_2)_4 + 2PhNH_2 \longrightarrow [Re(NPh)(S_2CNEt_2)_2]_2O + 2H_2O \tag{43}$$

The reaction of OsO_4 with bulky amines provided the first alkylimido transition metal complexes (equation 44).[67,73] A simple amine adduct is initially formed prior to loss of water.[74]

$$OsO_4 + RNH_2 \longrightarrow O_3OsNR + H_2O \tag{44}$$

13.4.2.2.2 *From silylamines*

This very versatile method for the synthesis of organoimido functions makes use of the great thermodynamic strength of Si—O and Si—F bonds. With metal oxides two synthetic strategies can be used leading to either complete removal of the oxygen or its conversion to a siloxy derivative (equations 45 and 46).

$$L_nMO + RN(SiMe_3)_2 \longrightarrow L_nMNR + (Me_3Si)_2O \tag{45}$$

$$L_nMO + RNH(SiMe_3) \longrightarrow L_nM(OSiMe_3)NR^- + H^+ \tag{46}$$

In the latter reaction the proton typically combines with a suitable leaving group or is picked up by excess amine. The silylated amine is also sometimes used as its lithium salt, *e.g.* $LiN(R)SiMe_3$. Examples of the use of this technique are shown in equations (47),[75,76] (48),[77] (49)[77] and (50).[78,79]

$$Cl_3VO + MeN(SiMe_3)_2 \longrightarrow Cl_3VNMe + (Me_3Si)_2O \tag{47}$$

$$(Me_3SiO)_2ClVO + xsHN(SiMe_3)_2 \longrightarrow (Me_3SiO)_3VNSiMe_3 + Cl^- H_2N(SiMe_3)_2^+ \tag{48}$$

$$MO_2Cl_2 + 2Bu^tNH(SiMe_3) \longrightarrow M(OSiMe_3)_2(NBu^t)_2 + 2[HCl] \tag{49}$$

$$(M = Cr, Mo)$$

$$WF_6 + HN(SiMe_3)_2 \xrightarrow{\text{MeCN}} (MeCN)F_4WNH + 2Me_3SiF \tag{50}$$

13.4.2.2.3 From phosphinimines, isocyanates and sulfinylamines

All three of these reagents have been used to prepare metal imido complexes from metal oxides. In particular the phosphinimines are very useful and have in some cases allowed multiple imido functions to be introduced on to one metal center (equation 51).[73]

$$O_3OsNR + 2Ph_3PNR \longrightarrow OOs(NR)_3 + 2Ph_3PO \tag{51}$$

The complexes L_2Cl_3ReO, besides reacting with anilines, can be deoxygenated by phosphinimines as well as isocyanates and sulfinylamines (equations 52,[80] 53[81] and 54[80]).

$$L_2Cl_3ReO + Ph_3PNAr \longrightarrow L_2Cl_3ReNAr + Ph_3PO \tag{52}$$

$$L_2Cl_3ReO + ArNCO \longrightarrow L_2Cl_3ReNAr + CO_2 \tag{53}$$

$$L_2Cl_3ReO + ArNSO \longrightarrow L_2Cl_3ReNAr + SO_2 \tag{54}$$

More recently *p*-tolyl isocyanate has been used to prepare a simple imido complex of vanadium(V) (equation 55).[83]

$$Cl_3VO + p\text{-tolNCO} \longrightarrow Cl_3V(N\text{-}p\text{-tol}) + CO_2 \tag{55}$$

Similar reactivity is seen for the tungsten oxychloride (equation 56)[84,85] and for oxomolybdenum complexes (equation 57).[46]

$$Cl_4WO + PhNCO \longrightarrow Cl_4W(NPh) + CO_2 \tag{56}$$

$$MoOCl_2(S_2CNEt_2)_2 + PhNCO \longrightarrow Mo(NPh)Cl_2(S_2CNEt_2)_2 + CO_2 \tag{57}$$

13.4.2.2.4 From organic azides

This technique makes use of the strong thermodynamic driving force for elimination of N_2 from azides, leaving behind an RN fragment which can bond to metal centers (equations 58[4] and 59[86]).

$$MoO(S_2CNEt_2)_2 + ArN_3 \longrightarrow Mo(O)(NAr)(S_2CNEt_2)_2 + N_2 \tag{58}$$

$$MoCl_4(THF)_2 + ArN_3 \longrightarrow ArNMoCl_4(THF) + N_2 + THF \tag{59}$$

The azide $CF_3CFHCF_2N_3$ provides a route to alkylimido compounds of Pd and Pt (equation 60).[87]

$$M(PMePh_2)_4 + C_3F_6HN_3 \longrightarrow M(NC_3F_6H)(PMePh_2)_2 + N_2 + 2PMePh_2 \tag{60}$$

Similarly perfluoroazides allow the synthesis of analogous complexes for Rh and Ir.[87]

13.4.2.2.5 From nitriles

There are a number of ways that nitriles may be converted to alkylimido ligands. Low valent niobium and tantalum complexes have been shown to reductively couple nitriles to give bridging bis-imido ligands (equation 61).[88,89]

$$MCl_4 + 2MeCN \longrightarrow \tfrac{1}{2}[(MeNC)MCl_3]_2(NCMeCMeN) \tag{61}$$

$$(M = Nb, Ta)$$

Niobium and tantalum alkylidenes have also been shown to react with nitriles to form imido groups as follows:[90]

$$R_3M{=}CHBu^t + R'CN \longrightarrow R_3M(NCR'{=}CHBu^t)$$

13.4.3 PHYSICAL AND SPECTROSCOPIC PROPERTIES

13.4.3.1 Metal Amides

13.4.3.1.1 Molecular complexities and volatilities[1]

As one would expect, there is a strong correlation between solubility in non-polar solvents, volatility and molecular complexity of metal amide compounds. The degree of association can typically be related to empirical formula (ML_x), metal covalent radius and the steric demand of the amide substituents.

Although the amide derivatives of the alkali metals are normally only sparingly soluble in hydrocarbon solvents,[91] the very bulky bistrimethylsilylamide derivatives are soluble in benzene. Cryoscopic molecular weight measurements indicate a dimeric nature for $MN(SiMe_3)_2$ (M = Li, Na and K).[92,93] Both the lithium and sodium derivatives can be distilled under reduced pressure.

In the absence of steric bulk, the bis-amides of magnesium tend to be polymeric in non-polar solvents with amido bridges.[48] In the presence of coordinating ligands or solvents, however, breakdown to smaller oligomers takes place.[48]

Homoleptic boron amides typically exist in a monomer–dimer equilibrium,[94,95] while their aluminum counterparts are dimeric in the absence of coordinating ligands.[96] The use of the bulky ligand $N(SiMe_3)_2$ has allowed a number of strictly mononuclear bis- and tris-amido derivatives to be characterized for a large number of metals.[3,97] Such derivatives tend to be volatile as well as having solution spectral properties consistent with the presence of the low valent, mononuclear metal center. The bis-amide derivatives of a number of metals have been successfully isolated (equation 62),[97] although more typically three-coordinate mono-adducts are obtained.

$$MCl_2 + 2NaN(SiMe_3)_2 \longrightarrow M[N(SiMe_3)_2]_2 + 2NaCl \tag{62}$$

The relatively high volatility of complexes of this type makes their purification by sublimation routine and also lends itself to mass spectrometric characterization. To date a complex of stoichiometry $M[N(SiMe_3)_2]_4$ has not been isolated, even the large metals Th^{4+} and U^{4+} giving only partially substituted derivatives, e.g. $ClTh[N(SiMe_3)_2]_3$.[98]

With the metals molybdenum and tungsten the molecular complexity of the tris-amides is complicated by the ability of these metals to form strong metal–metal bonds. Hence the complexes $M_2(NR_2)_6$ (R = Me, Et; M = Mo, W) are dimeric both in solution and in the gas phase but with no bridging NR_2 groups.[3,43] The dimetal centers are held together by an unbridged metal–metal triple bond. By far the largest class of homoleptic metal amides are those of formula $M(NR_2)_4$ (M = Ti,[39] Zr,[39] Hf,[99] V,[100] Nb,[44] Ta,[101] Cr[102] Mo,[44] U,[103] Th[104]). Typically strict mononuclearity is maintained for these complexes although for uranium there is evidence for dimer formation.[105] All derivatives of this type tend to be soluble in hydrocarbon solvents and can normally be sublimed/distilled under reduced pressure.

Besides the phosphoranes, the homoleptic amides of niobium(V) and tantalum(V) are the only pentamides known.[1] These compounds, exemplified by $M(NMe_2)_5$ (M = Nb, Ta) are volatile, hydrocarbon-soluble solids.[37] All physical and spectroscopic data are consistent with a strictly mononuclear existence for these complexes.

$W(NMe_2)_6$ represents the only isolated hexamide. The deep-red crystalline material is hydrocarbon soluble and volatile.[106] Molecular weight measurements indicate its discrete mononuclearity in solution.

13.4.3.1.2 *NMR spectroscopy of metal amides*

Besides the routine and widespread uses of this technique for the characterization of diamagnetic metal amides, a number of systems demonstrate fluxionality on the NMR time scale (typically ^{1}H) that can often be correlated with the presence of significant nitrogen-p to metal-d π-bonding. The ensuing restricted rotation can generally be detected in systems that possess asymmetry in the metal coordination environment, although in most cases it is difficult to factor out any steric contributions. Initial studies of this type concerned aminoboranes of the form $XYBNR_2$.[54] Restricted rotation about the B—N bond results in non-equivalent R groups with the ground state containing the NR_2 group lying in the same plane as the XYB atoms, (**1**).

$$\begin{array}{c} X \\ \diagdown \\ \diagup \\ Y \end{array} B-N \begin{array}{c} \diagup R \\ \\ \diagdown R \end{array}$$

(1)

Variable temperature NMR measurements have allowed a number of rotational barriers to be quantified.[54] Similar measurements have been made on some phosphoranes.[107]

An interesting example of restricted M—NR_2 rotation is seen in the dinuclear species $M_2(NMe_2)_6$ (M = Mo, W). Here, because of the bonding requirements of the M_2^{6+} core, the NMe_2 units lie coplanar with the M_2 axis in order to maximize M—NR_2 π-bonding. This results in a ground state structure in which two types of methyl group are present, one lying *proximal* and the other *distal* to the metal—metal triple bond.[38] The diamagnetic anisotropy of this latter function results in an impressive separation of close to 2 p.p.m. for the resonances of these methyl groups in the ^{1}H NMR spectrum at low temperature. On warming, collapse and coalescence of the signals occurs so that activational parameters can be obtained. A number of other related molecules have also been studied.[43,108,109]

The series of molecules of stoichiometry $M[N(SiMe_3)_2]_3X$ (M = Ti, Zr, Hf;[41] X = typically a halide) exhibit temperature-dependent NMR spectra involving a ground state non-equivalence of the $SiMe_3$ groups. One set lies closer to the X function than the other. Restricted rotation primarily for steric reasons is hence observable.

13.4.3.1.3 *Infrared spectroscopy of metal amides*

IR spectroscopy is not the most useful tool for probing the structures of metal amide complexes. The assignment of M—N vibrational modes is not a simple task as such bands can be readily coupled to other vibrational modes involving the NR_2 moiety.[1]

13.4.3.1.4 *Electronic spectra of metal amides*

The most extensively studied electronic spectra are those of the homoleptic complexes of the transition elements and lanthanides of stoichiometry $M(NR_2)_4$ and $M[N(SiMe_3)_2]_3$. For the tetramides of the Group V and Group VI metals, in particular V,[110] Cr[102] and Mo,[111] where the d—d transitions can be seen, the spectra are consistent with the non-tetrahedral, D_{2d} ground state structure also implied by IR data. Although $Cr(NEt_2)_4$ is high-spin (paramagnetic), the purple compound $Mo(NMe_2)_4$ is diamagnetic—an unusual situation for a tetrahedral d^2 ion, but again consistent with a D_{2d} ground state symmetry and extensive Me_2N-to-Mo π-bonding (see Section 13.4.5).

The sterically demanding ligand $N(SiMe_3)_2$ has allowed the electronic and EPR spectra of the first row d-block elements in a purely trigonal planar environment to be measured.[112,113] All of the data obtained by Bradley *et al.* on this series were consistent with the theoretical predictions for this geometry, and allowed in some cases values for spin–orbit coupling constants to be evaluated.

The electronic spectra of the lanthanide complexes $M[N(SiMe_3)_2]_3$ have been investigated.[114] Besides the presence of a strong ligand-to-metal charge transfer band at high energies, the electronic spectra of the Pr and Nd complexes indicated considerable splitting of the various single electron $4f$-orbitals. This was attributed to the significant covalent nature of the M—$N(SiMe_3)_2$ bond giving rise to a correspondingly higher ligand field.[114]

13.4.3.2   Metal Imides

Because the organoimido function is normally found associated with a range of other ligation, L_nMNR, the physical properties such as volatility and molecular complexities of such compounds are not always representative of this particular function. However, certain spectroscopic properties are characteristic of organoimido groups.

As could be expected, terminal M—NR groups can give rise to a sometimes characteristic absorption in the IR spectrum. This vibration lies typically in the range of 1000–1200 cm^{-1} and is consistently higher than the corresponding metal–oxo function.[73,115] Probably the most illustrative systems are the mixed oxoimido complexes of osmium(VIII).[73] However, the fact that coupling with other metal–ligand modes as well as the possible obscuration of this part of the spectrum by other functional groups regrettably does not allow IR spectra to be used as a definitive method for detecting organoimide functions.[4]

One of the most extensively used tools in studying diamagnetic metal imide complexes is NMR spectroscopy. The metal–imido bond has been found to have some dramatic effects on the ^{1}H NMR chemical shifts of the hydrogen atoms located in the α-position relative to the nitrogen atoms. In nearly all cases a downfield shift of these protons is observed relative to their expected positions.[116,117] However, in some complexes an equally large upfield shift has also been seen.[73] In the ^{13}C NMR spectra of *t*-butylimido complexes the metal–nitrogen bond has been found to give rise to a downfield shift of the α- and an upfield shift of the β-carbons of the Bu^t substituent.[4] A correlation between the magnitude of these shifts and the reactivity of the metal–nitrogen bond has been made.

Given the correct metal geometry and substituents it has been found possible to observe direct coupling between the ^{14}N nuclei of MNR functions and various nuclei, including ^{1}H and ^{19}F.[118] An interesting example involves the well-resolved 1:1:1 triplet in the ^{51}V NMR spectrum of $(Me_3SiO)_3V(NBu^t)$ with $^1J(^{51}V-^{14}N) = 95$ Hz.[119] A number of related complexes containing the *p*-tolylimido function (which destroys the axial symmetry) result in only broad single resonances in their ^{51}V NMR spectra.[83]

13.4.4   SOLID STATE AND MOLECULAR STRUCTURES

13.4.4.1   Prototypal Structures for Homoleptic Amides

13.4.4.1.1.   *Mononuclear complexes*

(*i*)   *Of formula M(NRR′)*

In the solid state there are no known mononuclear compounds of formula M(NRR′) in the absence of supporting donor molecules. A few of the Group I metal bis(trimethylsilyl)amides have been structurally characterized with O-donor ligands. Examples include (i) $K(NSi_2Me_6)\cdot 2$dioxane[120] which contains a five-coordinate potassium atom and (ii) $Li(NSi_2Me_6)\cdot 12$-crown-4.[121] The compounds $Au(L)(NSi_2Me_6)$, where L = PMe$_3$, PPh$_3$ or AsPh$_3$, most likely have linear coordination at the Au^I center.[122]

(*ii*)   *Of formula M(NRR′)$_2$*

Compounds of this type fall into three groups.

(1) For the combination of an extremely bulky amide ligand and a very small central metal a discrete monomeric linear molecule can be envisaged and this has been seen for $Be(NSi_2Me_6)_2$.[123] In the solid state the Si_2N—Be—NSi_2 unit has D_{2d} symmetry and a pseudo-allenic bonding scheme may be envisaged.

(2) The bulky bis(trimethylsilyl)amide ligand, bulky dialkylamido ligands (NPr$^i_2$ and NBu$^t_2$) and the cyclic amido ligand $\overline{NCMe_2CH_2CH_2CH_2CMe_2}$ are known to allow the formation of bent monomeric M(NR$_2$)$_2$ complexes for the elements M = Ge, Sn and Pb in their divalent states.[124–127] The NMN angles are in the range 90–115°. The origin of the bending may be viewed in terms of the stereochemically active lone pair which is present in these carbene analogues. The compound $[Me_2NSn(\mu-NMe_2)]_2$ has a planar central Sn_2N_2 moiety but the local geometry for each SnN$_3$ unit is pyramidal.[128] Isovalent with these carbene analogues are stabilized phosphenium ions[129] of the type $[(R_2N)_2P]^+$ and the analogous amido-substituted dications of sulfur, $[(R_2N)_2S]^{2+}$.[130]

Somewhat related to these subvalent Group IV amides is the recent discovery of the novel two-coordinate phosphorus radicals $P(NRR')_2$ where R, R' = $SiMe_3$, Bu^t, Pr^i.[131,134] The ESR spectra of these compounds consist of a large doublet due to a nuclear hyperfine interaction with ^{31}P ($I = \frac{1}{2}$, 100% natural abundance). The data are interpretable in terms of a bent N—P—N structure with the unpaired electron in an essentially phosphorus $2p$-orbital.

(3) Some compounds of formula $M(NRR')_2$ are stabilized by donor solvent molecules or neutral donor ligands and these may give rise to unusual coordination geometries. Examples include (i) $Cr(NSi_2Me_6)_2(THF)_2$,[132] which provides a rare example of square planar Cr^{II}, and (ii) the three-coordinate Co^{II} complex $Co(NSi_2Me_6)_2(PPh_3)$.[133]

(iii) Of formula M(NRR')₃

The use of the bulky bis(trimethylsilyl)amide ligand dominates this class of compound, though for small central elements three-coordination is possible for less bulky amide groups as in $B(NMe_2)_3$,[134] and the anion $Be(NH_2)_3^-$.[135] Structurally characterized examples of three-coordinate metal atoms in MX_3 compounds, where X = NSi_2Me_6, are known for M = Al,[136] Sc,[110] Ti,[137] Cr,[138] Fe[139] and the lanthanides Eu[110] and Nd.[140] Several other lanthanide MX_3 compounds are known (M = La, Ce, Pr, Sm, Gd, Ho, Yb and Lu) and based upon spectroscopic properties and cryoscopic molecular weight data there is no doubt that these too contain three-coordinate metal atoms.[134,141] Even $U(NSi_2Me_6)_3$ is monomeric.[142] The bulky diisopropylamido ligand has also been shown to stabilize three-coordination in $Cr(NPr^i_2)_3$.[143] There is little doubt that related three-coordinate metal ions are found for Al, Ti, V and Cr with other bulky dialkylamides or cyclic alkylamides.[134] Also in the lithium salt, $Li(THF)_4Ni(NPh_2)_3$, there is a three-coordinate Ni^{II} ion.[144]

For Al and the transition elements, structural studies reveal common features involving (i) planar MN_3 units of virtual D_{3h} symmetry and (ii) planar MNC_2 or $MNSi_2$ units with dihedral angle between the MN_3 and the NX_2 planes (X = C or Si) being *ca.* 50° (Figure 1). For the compounds containing Sc, Eu and Nd, the metal atoms lie out of the N_3 plane by *ca.* 0.35 Å. This leads to a pyramidal MN_3 unit in the solid state. In solution, however, the compounds do not show a dipole moment nor do vibrational spectra provide evidence for the pyramidal MN_3 unit. Thus it is believed that the distortion from planarity is a solid state effect and for the lanthanides, if not for Sc, this may reflect a greater degree of ionic character in the M—N bonds. Support for the latter hypothesis comes from the somewhat large SiNSi angles (130°) and relatively short Si—N bond distances which are comparable to those in $KNSi_2Me_6 \cdot 2dioxane$.[120]

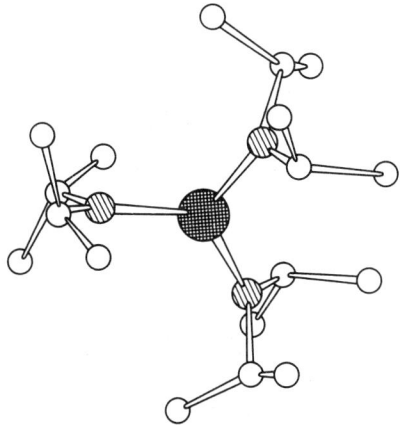

Figure 1 Molecular structure of $Cr(NPr^i_2)_3$ viewed approximately down the virtual C_3 axis of symmetry

Several of the lanthanide complexes MX_3, where X = NSi_2Me_6, have been shown to give 1:1 adducts MX_3L where M = La, Eu and Lu and L = $OPPh_3$.[145] The central MN_3O unit has virtual C_{3v} symmetry. Evidence has also been presented for the formation of unstable 1:2 adducts $MX_3L'_2$ where L' = $OPMe_3$.[146]

Two other classes of $M(NRR')_3$ compounds are known. (1) For M = Ge and Sn, the use of the bulky amides NSi_2Me_6, NGe_2Me_6 and $N(Bu^t)SiMe_3$ has allowed isolation of kinetically stable radical species which, based on ESR data, can reliably be established to be pyramidal at the central element M.[124] (2) The metalloids P, As and Sb form $M(NMe_2)_3$ compounds which are pyramidal at

M. However, unlike the amides of the transition elements and lanthanides the NC_2 units are not orientated in a propeller-like manner. At least one NR_2 group is pyramidal.[147]

(iv)   Of formula M(NRR')₄

A number of M^{IV} dialkyl- or diaryl-amides are known to be monomeric in solution and this knowledge coupled with solution and Nujol mull IR spectroscopy leaves little doubt that they contain pseudo-tetrahedral MN_4 units. Examples include amides of the Group IV metals (Si, Ge, Sn, Pb, Ti, Zr, Hf, V, Cr, Mo and U.[134] It should be noted that four NSi_2Me_6 ligands cannot be accommodated at a metal atom. Only three structural characterizations of this class of compounds are available and these are for $Sn(NMe_2)_4$,[148] $Mo(NMe_2)_4$[111] and $U(NPh_2)_4$.[46] In each of these there is an essentially tetrahedral MN_4 moiety and the coordination about the nitrogen atoms is planar. The central $Mo(NC_2)_4$ skeleton of the $Mo(NMe_2)_4$ molecule has virtual D_{2d} symmetry (Figure 2). The $Mo(NMe_2)_4$ molecule provides a very rare example of a diamagnetic four-coordinate d^2 metal ion, the only other being $Mo(SBu^t)_4$.[149] The diamagnetism and other spectrochemical data can only be explained as a result of Me_2N-to-Mo π-bonding (see Section 13.4.5).

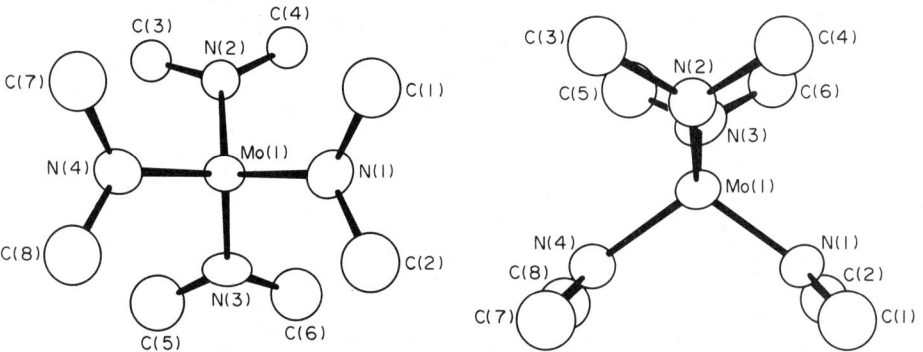

Figure 2   Two views of the $Mo(NC_2)_4$ skeleton of the $Mo(NMe_2)_4$ molecule showing the orientation of the NC_2 units in the symmetry group D_{2d}

(v)   Of formula M(NRR')₅

Certain discrete monomeric compounds of this class are known for M = Nb and Ta, and for M = Nb both the dimethylamide and piperidide have a central square-based pyramidal MN_5 unit distorted toward a trigonal bipyramid (Figure 3).[150] However, the related $Ta(NEt_2)_5$ molecule has a TaN_5 unit more closely approximating the idealized D_{3h} geometry.[151]

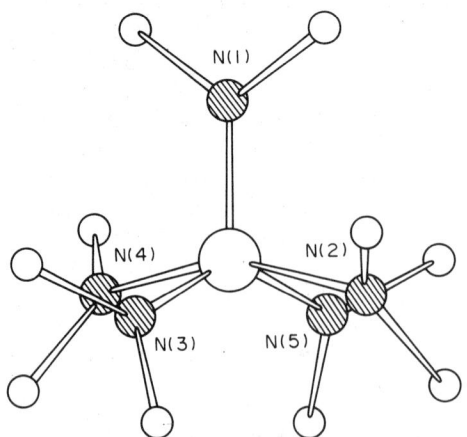

Figure 3   The $Nb(NC_2)_5$ skeleton of the $Nb(NMe_2)_5$ molecule. The NbN_5 moiety corresponds closely to a square-based pyramid with N(1) at the apical position

(vi) *Of formula* $M(NRR')_6$

The only known structurally characterized example of this group is $W(NMe_2)_6$.[43,106,152] The central $W(NC_2)_6$ unit has O_h symmetry (Figure 4). The molecule is of particular interest in terms of Me_2N-to-M π-bonding as is discussed in Section 13.4.5.

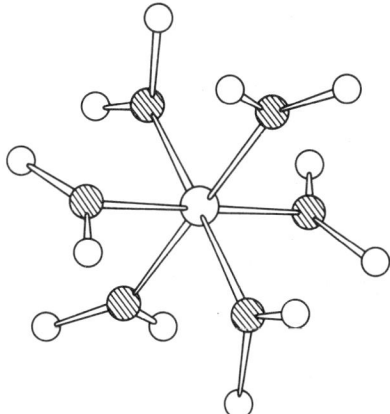

Figure 4 The $W(NC_2)_6$ skeleton of the $W(NMe_2)_6$ molecule has rigorous O_h symmetry

13.4.4.1.2 Polynuclear complexes

(i) *Of formula* $[M(NRR')]_x$

The only structural example of an infinite chain polymer is seen for $NaNSi_2Me_6$.[153] The N—Na—N angle is 150° and the Na—N—Na angle 102°. Lithium forms a cyclic trimer $[Li(NSi_2Me_6)]_3$ having a planar $(LiN)_3$ unit: Li—N = 2.00 Å, N—Li—N = 92° and Li—N—Li = 148°.[154] Cyclic tetramers are found for $[Li(NCMe_2CH_2CH_2CH_2CMe_2)]_4$,[155] $[Cu(NEt_2)]_4$[156] and $[Cu(NSi_2Me_6)]_4$[157] involving planar M_4N_4 rings. The coordination at the metal is close to linear and the M—N—M angles are correspondingly close to 90°. Oxygen donor solvents readily break up these oligomers. A crystalline etherate of $LiNSi_2Me_6$ has been shown in the solid state to be $[Et_2OLi(\mu\text{-}NSi_2Me_6)]_2$.[157] Coordination about lithium is trigonal planar and for the $(LiN)_2$ unit the angles are Li—N—Li = 75° and N—Li—N = 105°. A similar structure was reported for the secondary amide etherate $[(Et_2O)Li(\mu\text{-}NHAr)]_2$, where Ar = 2,4,6-tri-*t*-butyl-phenyl.[158]

(ii) *Of formula* $[M(NRR')_2]_x$

Dimeric species are known for certain divalent transition metal ions with sterically demanding amides: $[XM(\mu\text{-}X)]_2$ where M = Mn[159] and Co[160] and X = NSi_2Me_6; M = Co and Ni where X = NPh_2.[161] These compounds contain a planar M_2N_4 unit and the M—NC_2 or M—NSi_2 blades are arranged so that the central $M_2(NX_2)_4$ skeleton (X = C or Si) deviates little from D_{2d} symmetry (Figure 5). Of particular interest in this class of fused trigonally coordinate metal atoms are the M · · · M distance and the magnetism. The M · · · M distances are 2.84, 2.56 and 2.33 Å for the manganese, cobalt and nickel compounds, respectively. In the manganese compound the magnetic moment is 3.3 BM at 296 K and falls with decreasing temperature, consistent with strong anti-ferromagnetic coupling. The question concerning direct Mn · · · Mn bonding is clearly moot, but in the case of the nickel compound the diamagnetism and extremely short distance are suggestive of strong direct Ni—Ni bonding (formally a double bond).

The molecular structure of $[Be(NMe_2)_2]_3$[161] contains a three-metal-atom chain involving the fusing of two trigonally coordinate Be atoms to a central tetrahedrally coordinated Be atom through the agency of $\mu\text{-}NMe_2$ ligands (Figure 6).

(iii) *Of formula* $[M(NRR')_3]_x$

There do not appear to be any structural characterizations of homoleptic compounds of this class. However, the simple fusing of two tetrahedral units most probably occurs for $[Al(NMe_2)_3]_2$ as is seen

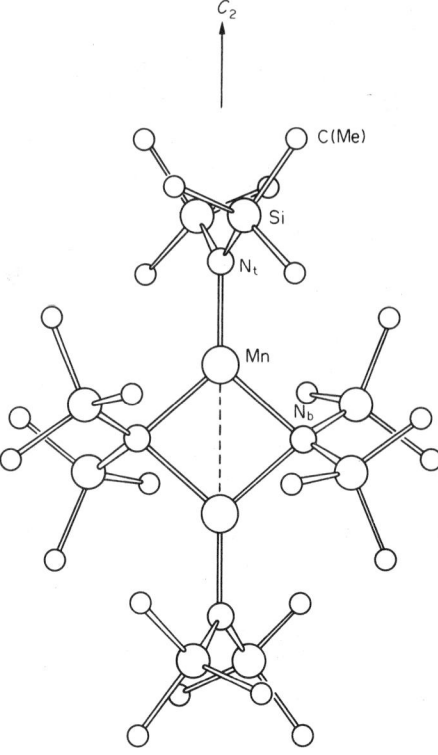

Figure 5 A ball and stick drawing depicting the molecular structure of the $Mn_2(NSi_2Me_6)_4$ molecule

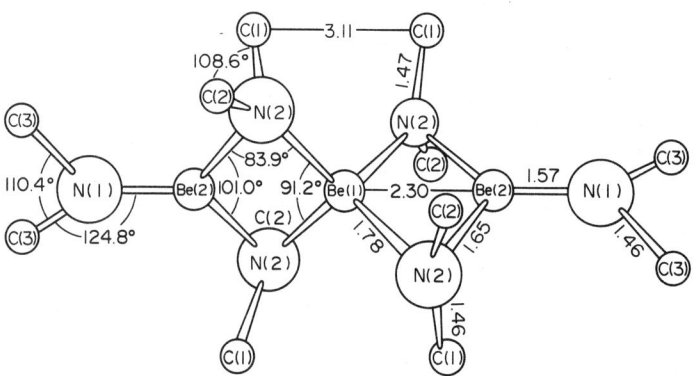

Figure 6 A ball and stick drawing of the trimer $[Be(NMe_2)_2]_3$

in $X_4Al_2(\mu\text{-}NMe_2)_2$ compounds where $X = Cl$,[163] Br[163] and Me.[164] A similar molecular structure for $[Ti(NMe_2)_2(\mu\text{-}NMe_2)]_2$ is also proposed on the basis of low temperature ^1H NMR data.[100]

(iv)   Of formula $[M(NRR')_4]_x$

U(NEt$_2$)$_4$ has been shown to adopt a dimeric structure in the solid state: $[U(NEt_2)_3(\mu\text{-}NEt_2)]_2$.[105] The central U_2N_8 unit involves the fusing of two distorted trigonal bipyramidal UN$_5$ units joined by a common axial–equatorial edge. The U^{IV} amide U(MeNCH$_2$CH$_2$NMe)$_2$,[46] adopts a trimeric structure having a linear arrangement of uranium atoms, each of which is in a distorted octahedral environment: $N_3U(\mu\text{-}N)_3U(\mu\text{-}N)_3UN_3$. In both U^{IV} amides the U to U distances are greater than 3.5 Å indicating a non-bonding distance. The angles subtended by the bridging nitrogen atoms are close to those expected for sp^3 hybridization, while those of the terminal amido nitrogens are typical for trigonal planar M—NC$_2$ moieties.

(v) With M—M bonds

Based on either considerations of magnetism and/or short M—M distances a few amido-bridged compounds may have significant M—M bonds, *e.g.* [Ti(NMe$_2$)$_2$(μ-NMe$_2$)]$_2$,[100] and [Ni(NPh$_2$)(μ-NPh$_2$)]$_2$.[161] However, the most dramatic effect of M—M bonding is seen in the compounds M$_2$(NMe$_2$)$_6$, where M = Mo[108] and W.[43] These have an ethane-like central M$_2$N$_6$ core, M—M—N = 104°, M—M = 2.212(2) Å (M = Mo), 2.292(1) Å (M = W) and the M$_2$(NC$_2$)$_6$ unit has virtual D_{3d} symmetry (Figure 7). They are members of an extensive series of compounds having M≡M triple bonds, $\sigma^2\pi^4$, between molybdenum and tungsten atoms.[165]

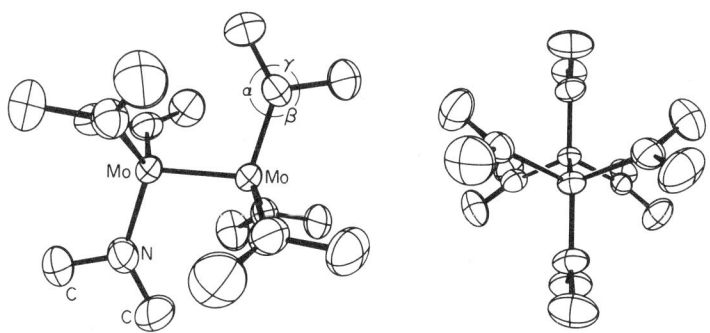

Figure 7 Two views of the central Mo$_2$(NC$_2$)$_6$ skeleton of the Mo$_2$(NMe$_2$)$_6$ molecule

13.4.4.2 Homoleptic Imides

Though imido ligands are isoelectronic with oxo groups, very few homoleptic imido compounds are known and fewer still are structurally characterized. The main group elements P and As form M$_4$(NR)$_6$ compounds where R = alkyl or aryl,[1] and these are analogues of the oxides P$_4$O$_6$ and As$_4$O$_6$. The structure of As$_4$(NMe)$_6$ is adamantane-like.[166] The mean As—N bond distance is 1.87 Å and the average N—As—N angle is 102°. Sulfur forms bisimido compounds S(NR)$_2$,[167] and even a tris-imide is known, S(NSiMe$_3$)$_3$,[168] analogues of SO$_2$ and SO$_3$, respectively.

In the transition series, osmium alone is known to form a homoleptic imide Os(NBu^t)$_3$NSO$_2$Ar,[169] where Ar = mesityl or 2,4,6-triisopropylphenyl, an analogue of OsO$_4$, but it has not been structurally characterized.

13.4.4.3 Mixed Ligand Complexes

13.4.4.3.1 With terminal imido groups

A large number of imido analogues of oxometal-containing compounds are known. Examples include [VCl$_3$(NR)],[83,170] Mo(NR)Cl$_4$(THF),[86] Nb(NR)(S$_2$CNEt$_2$)$_3$,[171] Ta(NR)(NMe$_2$)$_3$,[65] Mo$_2$(NR)$_2$Cp$_2$S$_2$,[172] Mo(NR)$_2$(S$_2$CNEt$_2$)$_2$,[173] Re(NR)Cl$_3$(PEt$_2$Ph),[174] Os(NR)O$_3$[175] and Os(NR)$_2$O$_2$,[175] where R = alkyl, phenyl or aryl. For isoelectronic compounds the metal–oxygen bond distances are shorter by *ca.* 0.05 Å than the metal–nitrogen (imido) distances. The absolute length of the M—N bond is, of course, dependent on several factors including the metal, the coordination number and geometry and the specific nature of the other ligands. Of particular interest is the M—N—C angle. Two limiting valence bond (VB) descriptions can be envisaged as shown in (2) and (3). In (2) there is a linear M—N—C moiety and an M≡N triple bond while in (3) there is an M=N double bond and a lone pair of electrons in a nitrogen sp^2 hybrid orbital. In general M—N bond distances correlate with M—N—C angles: shorter distances go with larger angles. For example, in the compound Mo(NPh)$_2$(S$_2$CNEt$_2$)$_2$[173] the M—N distances of 1.789(4) and 1.754(4) Å correlate with Mo—N—C angles of 139° and 169°, respectively. When a markedly bent M—N—C angle is observed, then an electronic effect at the metal center can generally be identified. See Section 13.4.5.

$$M \stackrel{\text{···}}{\equiv} N - R \qquad M = \overset{\displaystyle ..}{N} \diagdown R$$

(2)   (3)

13.4.4.3.2 *With bridging imido ligands*

Imido groups are known to bridge two and three metal atoms and the metal–nitrogen distances may be symmetric or asymmetric. For example, in the compound $(Me_2N)_2Zr(\mu\text{-}NBu^t)_2Zr(NMe_2)_2$, which involves two pseudo-tetrahedrally coordinated zirconium atoms sharing a common edge, there is a central planar $(ZrN)_2$ unit with essentially four equivalent Zr—N distances: 2.06 Å (averaged) and Zr—N—Zr = 97°, N—Zr—N = 114°.[67] Similar structures are expected for other Group IV mixed amide imides of formula $[M(NR)(NR_2')_2]_2$, where the bridging imido ligand may be viewed in terms of equal contributions from the resonance VB structures (4) and (5).

(4)   (5)

By contrast the compound $[Mo(OBu^t)_2(NAr)(\mu\text{-}NAr)]_2$, where Ar = *p*-tolyl, has a very asymmetric bridge.[176] Each molybdenum is in a distorted trigonal bipyramidal environment with the two halves of the dimer sharing a common equatorial–axial edge. These are terminal imido groups with M—N—C = 174° and M—N = 1.75 Å and the bridging imido ligands form short M—N bonds, 1.84 Å, in the equatorial positions and long bonds, Mo—N = 2.30 Å, in the axial positions *trans* to the terminal Mo—NAr bond. Formally these bond distances correspond to triple, double and weak dative bonds, respectively, leading to the VB description shown in (6).

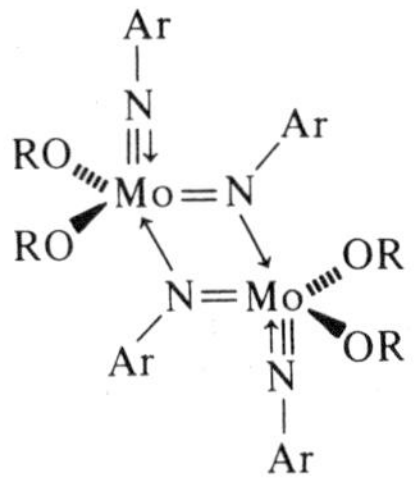

(6)

A special group of bridging amido compounds are seen in the cyclic trimers of B—N-containing compounds and the cage compounds of Al and Ga. The former, the borazines[177] are planar as a result of π-delocalization in a manner somewhat akin to benzene. Examples of cages include the cubane-like tetramer $[PhAl(\mu_3\text{-}NPh)]_4$,[178] $(HAlNPr^i)_6$,[179] which has two essentially planar $(AlN)_3$ six-membered rings joined together by a μ-NPr^i group, and $(MeAlNMe)_7$.[180] Triply bridging imido groups are also not uncommon in transition metal compounds and may cap triangular M_3 units having metal–metal bonds as in $Fe_3(CO)_9(NMe)_2$,[117] $Fe_3(CO)_{10}(NSiMe_3)$[181] and $W_3(NH)(OPr^i)_{10}$.[182] The latter compound is isoelectronic and isostructural with the oxo-capped compound $W_3(O)(OPr^i)_{10}$:[183] the W—O (oxo) and W—N distances are identical.

13.4.5 EVIDENCE FOR AND CONSEQUENCES OF AMIDO AND IMIDO LIGAND-TO-METAL π-BONDING

Evidence for nitrogen-*p*-to-metal-*d* π-bonding can be found in structural, physicochemical and spectroscopic data and may be inferred from the properties of spectator ligands. In some instances the data are open to various interpretations but in others the case for M—N π-bonding is unequivocal. The ability of amides and imides to π-donate is commonly argued as a reason for the stability of amide and imide complexes of the early and middle transition elements in high oxidation states and low coordination numbers. Conversely, the nitrogen *p*-orbitals have the opposite effect in M—N bonds involving the *d*-electron-rich later transition elements, and amides of the platinum group metals, for example, are not common.

13.4.5.1 Structural Data

Relatively short M—N distances and planar M—NC$_2$ units are commonly observed and often quoted as evidence in favor of N-to-M π-bonding. However, before these data can be considered acceptable evidence several other factors need to be examined including electronegativity differences between M and N, the oxidation state and hybridization of the metal and the nature of the other ligands present at the metal center. The symmetry of the molecule must also be examined. These considerations are well exemplified in the compounds $M_2(NMe_2)_6$, where M = Mo[108] and W,[43] and $M_2X_2(NMe_2)_4$, where M = Mo and W and X = Cl,[184,185] Br,[184,185] I,[184,185] Me,[186] Et,[187] Ph,[188] CH_2Ph[188] and o- and p-tolyl.[188]

The $M_2(NMe_2)_6$ compounds have a central $M_2(NC_2)_6$ skeleton with virtual D_{3d} symmetry (Figure 7, p. 175). The M≡M triple bond ($\sigma^2\pi^4$) is formed from metal d_{z^2}, d_{xz} and d_{yz} orbitals.[108,189] The M—N σ- and π-bonds must use metal s, p_x, p_y, d_{xy} and $d_{x^2-y^2}$ atomic orbitals which limits the M—N bond order to one and two-thirds, *i.e.* four N-to-M π-bonds are delocalized over six M—N σ-bonds. The M—N distance is 1.98 Å (averaged) for both M = Mo and W.[43,108] Note that the molecular symmetry limits the valence electron count on the metal atoms to 16 and dictates the alignment of the M—NC$_2$ blades along the M—M axis. For steric reasons one might expect the lower symmetry group S_6 for the $M_2(NC_2)_6$ moiety as seen in C_2Ar_6 where Ar = 2,6-di-t-butyl-4-biphenyl[190] but the twisting of the M—NC$_2$ units would diminish N-to-M π-bonding since the metal d_{xz} and d_{yz} orbitals are not available for ligand-to-metal π-bonding.

On replacing two NMe$_2$ ligands, one at each metal center, by non-π-donating or weakly π-donating ligands such as Me,[186] Ph[188] or a halide[184,185] we arrive at the $M_2X_2(NMe_2)_4$ compounds. These adopt *anti* and *gauche* $M_2X_2N_4$ moieties but in both conformations there are now four NMe$_2$ ligands and four metal orbitals of π symmetry, two at each metal, and the M—N distances are in the range 1.92–1.94 Å. This is roughly 0.2 Å shorter than might be expected based on covalent radii calculated using the M—X distances found in these molecules. The marked preference for alignment of the M—NC$_2$ blades along the M—M axis is seen in all of these compounds, even in the compound $Mo_2(o\text{-tolyl})_2(NMe_2)_4$ where the phenyl blade is aligned perpendicularly to the M—M bond.[188]

To contrast with the aforementioned favorable π-bonding situation, we may examine the structure of $(Ph_2PCH_2CH_2PPh_2)Pt(Me)(N(Me)Ph)$. Here the amide ligand is attached to a d^8 metal centre and there are no vacant metal d-orbitals to receive the nitrogen p-electrons. The Pt—NC$_2$ unit is, however, still planar and one might invoke some nitrogen-p-to-platinum-p orbital interaction, although the orbital energies are not expected to be well suited for bonding. The Pt—C and Pt—N bonds, which are both *trans* to Pt—P bonds and thus experience the same *trans* influence,[192] are of similar length: Pt—C = 2.093(6) Å and Pt—N = 2.080(5) Å. They differ less than the difference anticipated from considerations of the covalent radii of C_{sp^3} and N_{sp^2} which contrasts with Mo—C_{sp^3} = 2.17 Å and Mo—N = 1.94 Å in $Mo_2R_2(NMe_2)_4$ compounds.

The above examples represent extreme cases within the transition series. It is also interesting to compare metals in the main group with those in the transition series. Tin(IV) and molybdenum(IV) have similar covalent radii and the $M(NMe_2)_4$ compounds both contain a pseudo-tetrahedral MN$_4$ unit with planar M—NC$_2$ units arranged such that the $M(NC_2)_4$ moiety has D_{2d} symmetry (Figure 2). However, the Mo—N distance (1.92 Å)[193] and Sn—N distance (2.04 Å)[146] differ by more than 0.1 Å. We might conclude that Me$_2$N-to-M π-bonding is less for Sn than for Mo and this is also supported by PES studies (see Section 13.4.5.3). $Mo(NMe_2)_4$ provides a rare example of a four-coordinate diamagnetic d^2 ion. This implies a sizable splitting of the e-orbitals (e in $T_d \rightarrow a_1(d_{z^2}) + b_1(d_{x^2-y^2})$ in D_{2d}) and this splitting is directly a result of Me$_2$N-to-Mo-d π-bonding.[193]

For imido complexes the length of the M—N bond generally correlates with the valence electron count at the metal center. When the metal center has precisely 18 valence electrons or is deficient of this number, then linear M—N—C groups are found and the shorter bonds correlate with increasing electron deficiency or electrophilicity of the metal center. For compounds where the ligand set could provide more than 18 valence electrons then longer bonds are observed with or without bent M—N—C groups. The excess electrons may reside on localized nitrogen lone pairs (bent M—N—C groups) or may be delocalized over a set of ligands. Assignments of M—N bond order ranging from two to three and correlations with M—N distance and electron count have been tabulated for imido and related compounds.[4]

13.4.5.2 Metal–Nitrogen Bond Rotational Barriers

Variable temperature NMR studies sometimes reveal restricted rotations about M—N bonds in M—NR$_2$-containing compounds, *e.g.* $MX(NR_2)_3$[140] (M = Hf, Zr, X = halide and methyl, and R = SiMe$_3$; M = Cr, R = Pr^i, and X = NO).[130] These may be attributed to π-bonding effects but in many instances steric factors are more important. However, in the $M_2X_2(NMe_2)_4$ compounds, where M = Mo and W, and X = halide,[184,185] alkyl[186,187] or aryl[188] the barriers can reliably be traced to electronic factors. Particularly pertinent are the comparisons of rotational barriers about M—NC_2 and M—C (aryl) bonds.[188] In these compounds the barriers to M—N rotation are *ca.* $62.8\,kJ\,mol^{-1}$, somewhat less than that estimated for rotation about the carbene–metal bond in $Cp_2Ta(Me)(CH_2)$.[194]

13.4.5.3 Photoelectron Spectroscopic Data

Valence shell UV photoelectron spectroscopy has allowed certain trends to be identified. Ionization energies follow in increasing order of energy (i) d^n electrons, *i.e.* metal-centered non-bonding electrons, (ii) nitrogen lone pair electrons or ligand-centered MOs, (iii) M—N π-bonding MOs and (iv) M—N σ MOs.[195-198]

Ionizations from d^n electrons showed a marked difference in energy as a function of NSi_2Me_6 ligands *versus* NR_2 ligands (R = alkyl). This is clearly seen in the PES of $Cr(NSi_2Me_6)_3$ and $Cr(NPr^i_2)_3$ with the first ionization potential of the former being at higher energy.[197,198] This has been interpreted in terms of the poorer electron releasing properties of the NSi_2Me_6 ligand (see Section 13.4.5.4).

In $W(NMe_2)_6$ which has O_h symmetry (Figure 4) there are π-electrons which occupy orbitals belonging to the representations T_1 and T_2. Tungsten can only accommodate six π-electrons using its d_{xz}, d_{yz} and d_{xy} orbitals (T_2) which results in three W—N π-bonds (t_2^6) delocalized over six W—N σ-bonds and six π-electrons being ligand centered in non-bonding MOs, t_1^6. The splitting in energy of the t_2 and t_1 MOs is clearly visible in the PES.[195] Similarly the compounds $M_2(NMe_2)_6$, where M = Mo and W, show the first ionization from NMe_2-centered non-bonding MOs.[199]

A comparison of the PES of $M(NMe_2)_4$ compounds where M = Mo and Sn is also noteworthy since both compounds have a D_{2d} $Mo(NC_2)_4$ skeleton. This gives rise to nitrogen π-orbitals which transform as a_1, b_2 and e. The spectrum for $Sn(NMe_2)_4$ shows little evidence of splitting of the first ionization band at 7.7 eV implying that the energies of a_1, b_2 and e nitrogen π-orbitals are little perturbed by the metal.[195] By contrast the first IE band for $Mo(NMe_2)_4$ is at 5.3 eV corresponding to ionization of the metal-centered d^2 electrons b_1^2 $(d_{x^2-y^2})^2$. This is followed by ionizations at 7.3, 7.7 and 9.0 eV corresponding to ionizations from the M—N π-bonding MOs (e, b_2 and a_1). At 10.7 eV ionizations from Mo—N σ set in. The marked splitting of nitrogen π-MOs is a sure indication of significant Mo—N π-bonding.[198] The same result was inferred from structural data, the diamagnetism of the compound and MO calculations (Section 13.4.5.1).[193]

13.4.5.4 Spectator Ligands

A π-acid ligand such as CO or NO may act as a monitor for electron density at the metal center, particularly π-electron density and shifts in $\nu(NO)$ or $\nu(CO)$ bands may be easily monitored. For example, there is a series of compounds of formula CrX_3NO that have linear Cr—N—O moieties with C_3 molecular symmetry where X = OBu^t, NSi_2Me_6 and NPr^i_2. Defining the Cr—N—O axis as the z axis then the filled Cr d_{xz} and d_{yz} orbitals are the ones which can π-back-bond to NO π^*. These too can interact with ligand π-orbitals on X which, being filled, will raise the energy of the Cr e set and thus promote metal-to-nitrosyl π-bonding. For the compounds $CrX_3(NO)$, $\nu(NO)$ = 1641 (X = NPr^i_2), 1673 (X = 2,6-dimethylpiperidide), 1698 (X = NSi_2Me_6), 1707 (X = OBu^t), $1720\,cm^{-1}$ (X = OPr^i).[200] The lower the N—O stretching frequency the greater the Cr-to-NO π^*-back-bonding which in turn indicates the electron-releasing order of the ligands $NPr^i_2 > NSi_2Me_6 >$ OR. Note the big difference in $\nu(NO)$ for X = NPr^i_2 *versus* X = NSi_2Me_6 (see Section 13.4.5.3).

An extensive series of compounds of formula $Mo(pz)(NO)(X)(Y)$ have been synthesized and characterized, where pz = a substituted tridentate pyrazolylborate ligand and X, Y = various combinations of halides, SR, OR and NHR.[201,202] Here again the values of $\nu(NO)$ may be monitored for the linear Mo—N—O moiety and are indicative of a $(\sigma + \pi)$ donor series RNH > OR > SR.[201,202]

The reactivity of a spectator ligand may also be greatly modified by the presence of the amido π-donor ligands. For example, early transition metal alkyl moieties are prone to β-hydrogen abstraction reactions (also α- and γ-H abstraction). In this regard the compounds $Ta(NMe_2)_4(Bu^t)$[203] and $M_2(R)_2(NMe_2)_4$,[187] where M = Mo and W and R = Et, Pr^n, Pr^i and $Bu^{n,i,t}$, are unusually thermally stable being sublimable at 80–100 °C in vacuum with only slight decomposition. Although the metal atoms in these compounds are formally coordinatively unsaturated, the metal d orbitals that are not used in metal–ligand σ bonding are involved in metal–nitrogen π-bonding and thus less available for M $\cdots$ H—C interactions. The compounds $RM(NSi_2Me_6)_3$, where M = Zr and Hf, and $Et_2Zr(NSi_2Me_6)_2$ are also thermally stable ($< 100\,°C$) but here steric factors may prevent facile β-hydrogen abstraction.[204,205]

13.4.5.5 Metal–Nitrogen Bond Reactivity

The reactivity of the M—N bond is to some extent reflective of the degree of nitrogen-to-metal π-bonding. The more strongly π-donating the amido group then the less basic is the lone pair on nitrogen. Extensive M—N π-bonding reduces the susceptibility of the amido group toward protolysis and electrophilic attack.

A similar situation holds for imido groups and a simple orbital energy picture has been proposed[206] to account for the fact that imido groups sometimes show electrophilic reactivity (equation 63), while in other complexes the same ligands may be nucleophilic (equation 64).

$$M(X)(=NR) + ZnPh_2 \rightarrow M-N\overset{\displaystyle R}{\underset{\displaystyle Ph}{\diagup}} + PhZnX \tag{63}$$

$$M=NR + PhCHO \rightarrow M=O + PhCHNR \tag{64}$$

The orbital picture proposed to account for these limiting differences in reactivity is essentially analogous to that proposed to account for the electrophilic *versus* nucleophilic character of carbene ligands.[207] (A) If the nitrogen p-orbitals are energetically well below the metal d-orbitals then the N—M π-bond will be nitrogen centered and the nitrogen will be nucleophilic. (B) If the metal d-orbital–nitrogen p-orbital energy is inverted such that the N-p is higher in energy than the metal-d then the nitrogen will become electrophilic. Case A is commonly encountered for the early transition elements while case B remains hypothetical but the move toward B is expected as one moves across the transition series from left to right.

13.4.6 REACTIONS OF AMIDE AND IMIDE LIGANDS

Reactions of amides have been extensively reviewed and tabulated.[1] The most commonly encountered classes of reactions involve electrophilic attack on the nitrogen lone pair(s) or metal–nitrogen π-bond.

13.4.6.1 With Molecules Containing Acidic Hydrogen

Amides typically react with amines to set up an equilibrium involving amine which may lead to imide formation (*e.g.* as noted in equations 21, 38, 39 and 40). The reactions are sensitive to the steric bulk of the entering amine and the ability of the metal to expand its coordination number. For example in toluene-d_8 $M(NMe_2)_4$ compounds (M = Ti, Zr and Hf) and $M(NMe_2)_5$ compounds (M = Nb and Ta) show exchange with added $HNMe_2$ which is rapid on the NMR timescale, while the coordinatively saturated tungsten compound $W(NMe_2)_6$ shows no exchange with added $HN(CD_3)_2$ at room temperature.[208] An associative mechanism is implicated involving a four-center transition state (7).

$$\begin{array}{ccc} R_2N & \text{------} & H \\ | & & | \\ L_nM & \text{------} & NR_2 \end{array}$$

$$(7)$$

The bis(imido)pinacolates $(Bu^tN)_2W(OCX_2CX_2O)(H_2NBu^t)$, where X = Me, CF_3 and Ph, have been shown[209] to exist in equilibrium with the bis(amido)imidopinacolate (equation 65).

$$(RN)_2W(OCX_2CX_2O)(H_2NR) \; \rightleftharpoons \; (RN)(RNH)_2W(OCX_2CX_2O) \tag{65}$$

The position of the equilibrium (equation 65) depends upon the nature of X, and the intramolecular hydrogen atom transfer can be monitored by ^1H NMR magnetization transfer. This provides a model for the generalized imide–amide–amine interconversion by proton transfer.

Alcohols and thiols also react rapidly with most amides to give alkoxide or thiolate ligands with liberation of amine. Steric factors can be sufficiently important to limit this exchange (*e.g.* equation 66) and in some instances redox reactions may occur (equation 67).[212] Salt formation is also possible in some cases (equation 68).[213]

$$Mo_2(NMe_2)_6 + \text{excess ArXH} \; \longrightarrow \; Mo_2(NMe_2)_2(XAr)_4 + 4HNMe_2 \tag{66}$$

$$(ArX = 2\text{-Me-6-Bu}^t\text{-phenoxide}^{210} \quad \text{and} \quad 2,4,6\text{-Me}_3\text{-phenylthiolate}^{211})$$

$$2Cr(NEt_2)_4 + 7RCH_2OH \; \longrightarrow \; 2Cr(OCH_2R)_3 + RCHO + 8HNEt_2 \tag{67}$$

$$Mo_2(NMe_2)_6 + \text{excess ArOH} \; \longrightarrow \; [Me_2NH_2]^+ [Mo_2(OAr)_7(HNMe_2)_2]^- + 3HNMe_2 \tag{68}$$

$$(Ar = 4\text{-methylphenyl})$$

In general, imido groups, particularly those that have M—N bonds of order three, are less susceptible to protolysis than amide bonds and this can allow selective replacement of amido groups (equation 69).[214]

$$(Bu^tN)_2W(NHBu^t)_2 + 2ROH \; \longrightarrow \; (Bu^tN)_2W(OR)_2 + 2Bu^tNH_2 \tag{69}$$

$$(R = Bu^t \quad \text{and} \quad Ph_3Si)$$

Reactions between carbon acids such as $RC{\equiv}CH$ and cyclopentadiene, CpH, have been noted for metal amides which have very basic amide ligands (equations 70,[215,216] 71,[215–217] 72[218] and 73[219]).

$$2R_3SnNR_2' + HC{\equiv}CH \; \longrightarrow \; R_3SnC{\equiv}CSnR_3 + 2HNR_2' \tag{70}$$

$$R_{4-n}Sn(NR_2')_n + nR''C{\equiv}CH \; \longrightarrow \; R_{4-n}Sn(C{\equiv}CR'')_n + nHNR_2' \tag{71}$$

$$Ti(NMe_2)_4 + CpH \; \longrightarrow \; CpTi(NMe_2)_3 + HNMe_2 \tag{72}$$

$$U(NEt_2)_4 + 2CpH \; \longrightarrow \; Cp_2U(NEt_2)_2 + 2HNEt_2 \tag{73}$$

Certain metal hydrides that are acidic are also known to react with metal–amide bonds liberating amine and forming metal–metal bonds (equations 74[220,221] and 75[222]).

$$trans\text{-Pt(H)(Cl)(PPh}_3)_2 + Me_3Sn(NMe_2) \; \longrightarrow \; cis\text{-Pt(Me)(SnMe}_2\text{Cl)(PPh}_3)_2 + HNMe_2 \tag{74}$$

$$Zr(NEt_2)_4 + 3Ph_3SnH \; \longrightarrow \; Zr(NEt_2)(SnPh_3)_3 + 3HNEt_2 \tag{75}$$

13.4.6.2 Insertion into Metal–Amide Bonds

The general form of the reaction is shown in equation (76) and may be viewed as an aminometallation of the substrate, un, which must be unsaturated and susceptible to nucleophilic attack.[219]

$$L_nM{-}NR_2 + un \; \longrightarrow \; L_nM{-}un{-}NR_2 \tag{76}$$

Typical examples of this reaction involve the heterocumulenes CO_2, CS_2, COS, RNCO, RNCS and other related molecules such as SO_2, SO_3, $RC{\equiv}N$ and RNC, and activated alkynes bearing strongly electron-withdrawing substituents such as CO_2Me. These reactions find utility in the syntheses of homoleptic carbamates (equation 77),[223] monothiocarbamates and dithiocarbamates from homoleptic amides though redox reactions can occur in some instances (equations 78[102] and 79[224]).

$$M(NMe_2)_n + nCO_2 \; \longrightarrow \; M(O_2CNMe_2)_n \tag{77}$$

$$(M = Ti, Zr, Hf; \quad n = 4; \quad Nb, Ta; \quad n = 5)$$

$$Cr(NEt_2)_4 + 4CS_2 \longrightarrow Cr(S_2CNEt_2)_3 + \tfrac{1}{2}(Et_2NC(S)S-)_2 \tag{78}$$

$$Cr(NEt_2)_4 + \text{excess } CO_2 \longrightarrow Cr_2(O_2CNEt_2)_4 +$$

$$Cr_2(NEt_2)_2(O_2CNEt_2)_4 + HNEt_2 + EtN{=}CHMe \tag{79}$$

The mechanism of these insertion reactions is generally believed to involve nucleophilic attack by the amido group on the unsaturated molecule within the coordination sphere of the metal. A generalized four-center transition state having polar character can be invoked.

Though such a mechanism is eminently plausible, other reaction pathways exist and have in some instances been shown to be operative. For example, the reactions between each of $W_2Me_2(NEt_2)_4$ and $W(NMe_2)_6$ with CO_2 to give $W_2Me_2(O_2CNEt_2)_4$ and $W(NMe_2)_3(O_2CNMe_2)_3$ have been shown to proceed *via* an amine-catalyzed reaction sequence (equation 80).[208]

$$\text{(i) } HNR_2 + CO_2 \rightleftharpoons R_2NCO_2H$$

$$\tag{80}$$

$$\text{(ii) } L_nM{-}NR_2 + R_2NCO_2H \rightleftharpoons L_nMO_2CNR_2 + HNR_2$$

Insertion of carbon monoxide, which is so common for metal–alkyl bonds, is very rare for metal amides but has been documented for amides of the actinides (equations 81 and 82).[225]

$$(\eta^5\text{-}C_5Me_5)_2ThCl(NR_2) + CO \longrightarrow (\eta^5\text{-}C_5Me_5)_2ThCl(\eta^2\text{-}CONR_2) \tag{81}$$

$$(R = Me, Et)$$

$$(\eta^5\text{-}C_5Me_5)_2U(NEt_2)_2 + 2CO \longrightarrow (\eta^5\text{-}C_5Me_5)_2U(\eta^2\text{-}CONEt_2)_2 \tag{82}$$

The driving force for these CO insertions would appear to be the formation of the η^2-carbamoyl ligands, as evidenced by the very low IR C—O stretching frequencies of *ca.* $1500\,cm^{-1}$.[225]

13.4.6.3 Metathetical Exchange Reactions

These reactions can also be called ligand redistribution reactions and may be classified by the general equation (83).

$$L_nMNR_2 + L_nM'X \rightleftharpoons L_nMX + L_nM'NR_2 \tag{83}$$

In general M and M′ may differ or be the same and X may be NR_2, OR, a halide, SR, $\tfrac{1}{2}S$, $\tfrac{1}{2}O$, *etc.* The use of Group I metal amides in the synthesis of other metal amides (equations 12–19) provides one class of this type of reaction. A four-center transition state can in most cases be invoked and in studies of the ligand redistributions between $Ti(NMe_2)_4$ and $Ti(OR)_4$ compounds (R = Bu^t and Pr^i) activation parameters are consistent with this view.[226,227]

However, it is also possible that other reaction pathways are involved. For example, in the reaction between $M_2(NMe_2)_6$ compounds (M = Mo and W) and Me_3SiCl (equation 84)[181] an amine-catalyzed reaction sequence appears to be operative (equation 85).[18]

$$M_2(NMe_2)_6 + 2Me_3SiCl \longrightarrow M_2Cl_2(NMe_2)_4 + 2Me_3SiNMe_2 \tag{84}$$

$$\text{(i) } Me_3SiCl + HNMe_2 \rightleftharpoons Me_3SiNMe_2 + HCl$$

$$\tag{85}$$

$$\text{(ii) } M_2(NMe_2)_6 + HCl \rightleftharpoons M_2Cl(NMe_2)_5 + HNMe_2$$

$$\text{etc.}$$

13.4.6.4 With Metal Carbonyls to Form Aminocarbene–Metal Complexes

The reaction between a lithium amide and $Cr(CO)_6$ was first noted to give attack on a coordinated carbonyl and provide entry into amino-substituted carbene–metal complexes (equation 86).[228,229]

$$\text{(i)} \quad Cr(CO)_6 + LiNEt_2 \xrightarrow{Et_2O} Cr(CO)_5[C(OLi)NEt_2]\cdot Et_2O$$

$$\text{(ii)} \quad Cr(CO)_5[C(OLi)NEt_2]\cdot Et_2O \xrightarrow{Et_3O+} Cr(CO)_5C(OEt)NEt_2$$

$$(86)$$

Other metal amides have subsequently been shown to give similar reactions (equation 87).[230,231]

$$Mo(CO)_6 + Ti(NMe_2)_4 \longrightarrow Mo(CO)_5[C(NMe_2)OTi(NMe_2)_3] \tag{87}$$

13.4.6.5 Amination of Alkenes

There is keen interest in developing nitrogen–carbon bond-forming reactions akin to oxygen atom transfer and selective oxidations in organic chemistry.

Imido selenium compounds $Se(NR)_2$, where $R = Bu^t$ or Ts, were first noted to give allylic amination of alkenes and alkynes.[232] Formally the NR function is inserted into the allylic C—H bond yielding the C—NHR moiety. Related reactivity was also found for the sulfur imides, $S(NR)_2$.[233] Reactions between 1,3-dienes and $Se(NTs)_2$ give [4 + 2] adducts which, in the presence of $TsNH_2$, react to generate 1,2-disulfonamides.[234]

Osmium(VIII) oxo imido compounds of formula $OsO_{4-x}(NR)_x$, where $R = Bu^t$ and Ad (adamantyl), and $x = 1$, 2 and 3, react with alkenes in an analogous manner to OsO_4, which gives *cis* hydroxylation. The compound $OsO_3(NBu^t)$ provides a reagent for oxyamination (equation 88)[235] while $OsO_2(NR)_2$ and $OsO(NR)_3$ compounds provide reagents for diamination (equation 89).[73] In both cases the reactions are stereospecific, delivering the O,N or N,N to carbon bonds *cis* to one another. The 'work-up' involves treatment with $LiAlH_4$ followed by hydrolysis. Of note in reactions (88) and (89) is the preference for NR group transfer relative to O.

$$R'CH{=}CHR'' \xrightarrow[\text{(ii) work-up}]{\text{(i) } OsO_3(NR)} R'CH(OH)C(NR)HR'' \tag{88}$$

$$R'CH{=}CHR'' \xrightarrow[\text{(ii) work-up}]{\text{(i) } OsO_2(NR)_2} R'CH(NHR)CH(NHR)R'' \tag{89}$$

The initial step in the reaction between the $OsO_{4-x}(NR)_x$ compound and the *cis*-alkene generates an osmium(VI) species shown schematically in (**8**)

$$(8) \quad X = O, NR$$

More recently a tetraimido osmium(VIII) compound has been formed by the reaction outlined in equation (90).[236] The osmium complex subsequently formed in the further reaction with dimethylfumarate has been structurally characterized by a single crystal X-ray diffraction study (Figure 8).[237] The structure conforms to the proposed general intermediate (**8**).

$$\text{(i)} \quad Os(O)(NBu^t)_3 + Ph_3P \longrightarrow Os(OPPh_3)(NBu^t)_3$$

$$(90)$$

$$\text{(ii)} \quad Os(OPPh_3)(NBu^t)_3 + ArSO_2NNCl \longrightarrow (Bu^tN)_3Os(NAr)$$

(Ar = 4-Me-phenyl, 2,4,6-Me$_3$-phenyl   and   2,4,6-Pr$_3^i$-phenyl)

The industrial process for the preparation of acrylonitrile from propylene and ammonia uses a bismuth molybdate catalyst (equation 91).

$$CH_2{=}CHCH_3 + NH_3 + \tfrac{3}{2}O_2 \xrightarrow[400^\circ C]{Bi_2O_3/MoO_3} CH_2{=}CHCN + 3H_2O \tag{91}$$

The active site in this ammoxidation of propylene has been proposed[238] to be of the type depicted by (**9**), where L represents a coordinated group which may be NH_3, H_2O or an allyl radical during various points of the catalytic cycle.

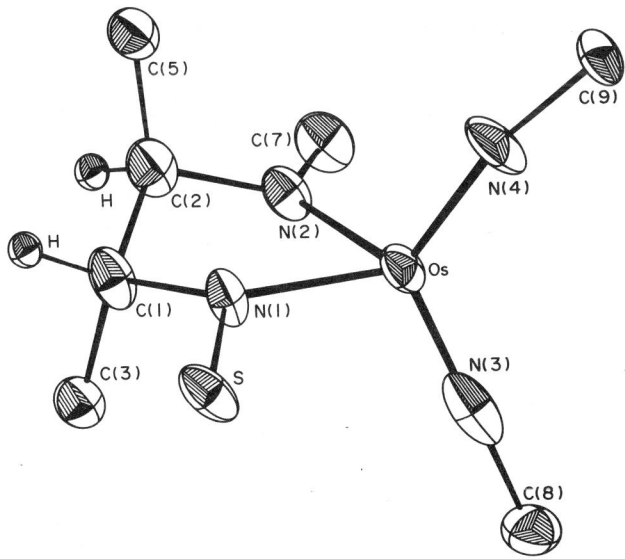

Figure 8 An ORTEP view of the central skeleton of the $(Bu^tN)_2Os[N(SO_2\text{-}2,4,6\text{-}Pr^i_3\text{-}C_6H_2)CH(CO_2Me)CH(CO_2\text{-}Me)NBu^t]$ molecule showing the stereochemistry of the amination of the *trans*-alkene

(9)

DuPont workers[206] have recently provided a homogeneous model for (9) involving glycolate bis-imido Group VI transition metal complexes. The compounds $[OC(CF_3)_2C(CF_3)_2O]W(N\text{-}Bu^t)_2(H_2NBu^t)$ and $(OCPh_2CPh_2O)W(NBu^t)(NHBu^t)_2$ provide structural models[206] for (9), and the reactivity of related chromium, molybdenum and tungsten imido compounds with benzyl and allyl radicals to generate $RCH{=}NBu^t$, where $R = Ph$ and Me, respectively, provides a model for the $C{=}N$ bond-forming part of the reaction.[239]

13.4.6.6 Activation of Carbon–Hydrogen Bonds in Dialkylamido Ligands

Although there exists an extensive body of chemistry dealing with the cyclometallation of amine ligands, particularly by the Group VIII metals,[240–242] analogous reactions involving amido groups are less common. However, a number of reactions involving activation of either the β- or γ-CH bonds of dialkylamido groups have been characterized over the last six years.[243]

Attempts to prepare $Ta(NEt_2)_5$ by metathetic exchange reactions were found to lead to a complex in which activation of a β-CH bond of an NEt_2 ligand occurs with the elimination of diethylamine (equation 92).[244,245] Presumably steric pressure at the tantalum center promotes the cyclometallation reaction to generate the iminomethyl group.

$$TaCl_5 + 5LiNEt_2 \rightarrow (Et_2N)_3Ta\underset{NEt}{\overset{CHMe}{\diagdown|}} + Et_2NH + 5LiCl \qquad (92)$$

Recent work by Nugent *et al.* has indicated that this type of reactivity is common for most dialkylamido complexes although much more forcing conditions are normally required.[246] Hence, he has demonstrated that complexes of the type $M(NMe_2)_x$ ($M = Zr$, $x = 4$; $M = Nb$, Ta, $x = 5$; and $M = W$, $x = 6$) will catalyze deuterium incorporation into the methyl groups of Me_2ND. This observation is consistent with a reaction scheme involving a reversible cyclometallation of the dimethylamido group coupled with facile exchange of free dimethylamine with coordinated NMe_2

groups (reaction 93). It was also found possible to intercept the intermediate azametallo-cyclopropane by observing its aminomethylation of alkenes.[246]

$$M(NMe_2)_2 \xrightarrow{-HNMe_2} \overline{M(CH_2NMe)} \xrightarrow{DNMe_2} M[N(CH_2D)Me](NMe_2) \qquad (93)$$

The complex $Cp^*TaMe_3(NMe_2)$ undergoes loss of methane to generate a cyclometallated complex (equation 94).[247] The reaction is first order with $k_H/k_D = 9.7$ at 34°C when using $N(CD_3)_2$ ligands.[247]

$$Cp^*TaMe_3(NMe_2) \longrightarrow Cp^*\overline{Ta(CH_2NMe)}Me_2 + CH_4 \qquad (94)$$

$$(Cp^* = \eta^5C_5Me_5)$$

Very recently, β-H bond activation of an NMe_2 ligand has been observed[248] at a dinuclear tungsten center generating an azadimetallabicyclobutane core (equation 95).

$$W_2Cl_2(NMe_2)_2(\mu\text{-}Cl)(\mu\text{-}NMe_2)(\mu\text{-}HCCH)L_2 \longrightarrow L(NMe_2) +$$

$$ClW(\mu\text{-}NMe_2)(\mu\text{-}CHCH_2)(\mu\text{-}N(CH_2)Me)WCl_2L \qquad (95)$$

$$(L = PMe_3 \quad \text{and} \quad PMe_2Ph)$$

The reaction is first order in metal complex and the reaction rate is independent of added L. With $k_H/k_D = 6.0$ at 63°C, $\Delta H^\ddagger = 99.2 \pm 8.4\,kJ\,mol^{-1}$ and $\Delta S^\ddagger = -17 \pm 25\,J\,K^{-1}\,mol^{-1}$ for $L = PMe_2Ph$, and $\Delta H^\ddagger = 71.5 \pm 12.6\,kJ\,mol^{-1}$ and $\Delta S^\ddagger = 105 \pm 33\,J\,K^{-1}\,mol^{-1}$ for $L = PMe_3$.

The activation of γ-CH bonds of the ligand $N(SiMe_3)_2$ has been observed in a number of early transition metal and actinide complexes.[243] Attempts to prepare metallocene derivatives of this ligand for the Group IV metals lead to cyclometallation as shown (reactions 96 and 97).[249,250]

$$Cp_2TiCl_2 + 2LiN(SiMe_3)_2 \longrightarrow Cp_2\overline{Ti[CH_2SiMe_2N(SiMe_3)]} + 2LiCl + NH(SiMe_3)_2 \qquad (96)$$

$$Cp_2Zr(H)Cl + LiN(SiMe_3)_2 \longrightarrow Cp_2\overline{Zr[CH_2SiMe_2N(SiMe_3)]} + LiCl + H_2 \qquad (97)$$

In the case of actinide systems it was found possible to effect catalytic deuteration of the $SiMe_3$ groups using D_2 and making use of a hydride leaving group.[251]

Thermolysis of the complexes $R_2M[N(SiMe_3)_2]_2$ (M = Zr, Hf) was found to proceed with double activation of the γ-CH bonds leading to a dimeric complex containing a bridging alkylidene function (equation 98).[252]

$$R_2M[N(SiMe_3)_2]_2 \longrightarrow \tfrac{1}{2}[(Me_3Si)_2N]_2\overline{M_2[\mu\text{-}CHSiMe_2N(SiMe)_3]_2} + RH \qquad (98)$$

Addition of dmpe ($Me_2PCH_2CH_2PMe_2$) leads to a bis-cyclometallated complex.[253] Some insight into how the metal interacts with, and eventually activates, the γ-CH bonds of $N(SiMe_3)_2$ groups has come from the observation of a significant ground state interaction between the metal center and these methyl groups in the complex $Yb[N(SiMe_3)_2]_2(dmpe)$.[254]

13.4.7 REFERENCES

1. M. F. Lappert, P. P. Power, A. R. Sanger and R. C. Srivastava, 'Metal and Metalloid Amides', Wiley, New York, 1980.
2. P. J. Davidson, M. F. Lappert and R. Pearce, *Acc. Chem. Res.*, 1974, **7**, 209.
3. D. C. Bradley and M. H. Chisholm, *Acc. Chem. Res.*, 1976, **9**, 273.
4. W. A. Nugent and B. L. Haymore, *Coord. Chem. Rev.*, 1980, **31**, 123.
5. K. Dehnicke, *Adv. Inorg. Chem. Radiochem.*, 1983, **26**, 169.
6. R. W. Chambers and P. C. Scherer, *J. Am. Chem. Soc.*, 1926, **48**, 1054.
7. F. W. Bergstrom and W. C. Fernelius, *Chem. Rev.*, 1937, **20**, 413.
8. H. Rupe, M. Seiberth and W. Kussmaul, *Helv. Chim. Acta*, 1920 **3**, 50.
9. G. Brauer, 'Handbook of Preparative Inorganic Chemistry', 2nd ed., Academic, New York, 1963, vol. 1, p. 464.
10. F. Fieser and M. Fieser, 'Reagents for Organic Synthesis', Wiley, 1976, vols. 1–6.
11. D. O. DePree, *US Pat.* 2 799 705 (1957) (*Chem. Abstr.*, 1958, **52**, 1202).
12. K. Ziegler, *US Pat.* 2 141 058 (1938) (*Chem. Abstr.*, 1939, **33**, 2538).
13. A. Terentev, *Bull. Soc. Chim. Fr.*, 1924, **35**, 1164.

14. J. L. Atwood and G. D. Stucky, *J. Am. Chem. Soc.*, 1970, **92**, 285.
15. D. W. Aubrey, W. Gerrard and E. F. Mooney, *J. Chem.Soc.*, 1962, 1786.
16. D. W. Aubrey, M. F. Lappert and M. K. Majundar, *J. Chem. Soc.*, 1962, 4088.
17. C. A. Kraus and R. Rosen, *J. Am. Chem. Soc.*, 1925, **47**, 2739.
18. H. J. Emeleus and N. Miller, *Nature (London)*, 1938, **142**, 996.
19. H. Schumann, I. Schumann-Ruidisch and S. Ronecker, *Z. Naturforsch., Teil B*, 1970, **25**, 565.
20. L. Tansjo, *Acta Chem. Scand.*, 1969, **14**, 2097.
21. S. S. Washburne and W. R. Peterson, *Inorg. Nucl. Chem. Lett.*, 1969, **5**, 17.
22. J. Mack and C. H. Yoder, *Inorg. Chem.*, 1969, **8**, 278.
23. M. Wieber and M. Schmidt, *Z. Naturforsch., Teil B*, 1963, **18**, 849.
24. S. Fisher, L. K. Peterson and J. F. Nixon, *Can. J. Chem.*, 1974, **52**, 3981.
25. G. M. Kosolapoff and L. Maier, 'Organic Phosphorus Compounds', Wiley, New York, 1973.
26. P. Sarignac, M. Dreux and G. Ple, *J. Organomet. Chem.*, 1973, **60**, 103.
27. D. D. Poulin and R. G. Cravell, *Inorg. Chem.*, 1974, **13**, 2324.
28. H. A. Klein and H. P. Latschra, *Z. Anorg. Allg. Chem.*, 1974, **406**, 214.
29. R. T. Cowdell and G. W. A. Fowles, *J. Chem. Soc.*, 1960, 2522.
30. G. W. A. Fowles and C. M. Pleass, *J. Chem. Soc.*, 1957, 1674.
31. G. W. A. Fowles and C. M. Pleass, *J. Chem. Soc.*, 1957, 2078.
32. J. C. Fuggle, D. W. A. Sharp and J. M. Winfield, *J. Chem. Soc., Dalton Trans.*, 1972, 1766.
33. D. A. Edwards and G. W. A. Fowles, *J. Chem. Soc.*, 1961, 24.
34. B. J. Brisdon, G. W. A. Fowles and B. P. Osborne, *J. Chem. Soc.*, 1962, 1330.
35. J. E. Drake and G. W. A. Fowles, *J. Chem. Soc.*, 1960, 1498.
36. P. J. H. Garnell and G. W. A. Fowles, *J. Less-Common Met.*, 1962, **4**, 40.
37. D. C. Bradley and I. M. Thomas, *Can. J. Chem.*, 1962, **4**, 449.
38. M. H. Chisholm and W. W. Reichert, *J. Am. Chem. Soc.*, 1974, **96**, 1249.
39. D. C. Bradley and I. M. Thomas, *J. Chem. Soc.*, 1960, 3857.
40. D. H. Harris and M. F. Lappert, *J. Organomet. Chem. Libr.*, 1976, **2**, 13.
41. C. Airold, D. C. Bradley, H. Chudzynska, M. B. Hursthouse, K. M. A. Malik and P. R. Raithby, *J. Chem. Soc., Dalton Trans.*, 1980, 2010.
42. H. W. Turner, R. A. Andersen, A. Zalkin and D. H. Templeton, *Inorg. Chem.*, 1979, **18**, 1221.
43. M. H. Chisholm, M. W. Extine, F. A. Cotton and B. R. Stults, *J. Am. Chem. Soc.*, 1976, **98**, 4477.
44. D. C. Bradley and M. H. Chisholm, *J. Chem. Soc. (A)*, 1971, 1511.
45. E. A. Cohen and R. A. Beaudet, *Inorg. Chem.*, 1973, **12**, 1570.
46. J. G. Reynolds, A. Zalkin, D. H. Templeton and N. M. Edelstein, *Inorg. Chem.*, 1977, **16**, 599.
47. B. Cetinkaya, P. B. Hitchcock, M. F. Lappert, M. C. Misra and A. J. Thorne, *J. Chem. Soc., Chem. Commun.*, 1984, 148.
48. G. E. Coates and D. Ridley, *J. Chem. Soc. (A)*, 1967, 56.
49. U. Wannagat, H. Autzen, H. Kuckertz and H. J. Wismar, *Z. Anorg. Allg. Chem.*, 1972, **394**, 254.
50. G. E. Coates and A. H. Fishwick, *J. Chem. Soc. (A)*, 1967, 1199.
51. N. A. Bell and G. E. Coates, *J. Chem. Soc. (A)*, 1966, 1069.
52. R. Mastoff, G. Krieg and C. Vieroth, *Z. Anorg. Allg. Chem.*, 1969, **364**, 316.
53. C. A. Brown and R. C. Osthoff, *J. Am. Chem. Soc.*, 1952, **74**, 2340.
54. P. A. Barfield, M. F. Lappert and J. Lee, *Trans. Faraday Soc.*, 1968, **64**, 2571.
55. V. Wannagat and H. Niederprum, *Chem. Ber.*, 1961, **94**, 1540.
56. R. Koster, H. Bellut and S. Hattori, *Liebigs Ann. Chem.*, 1969, **720**, 1.
57. G. W. Watt and W. C. Fernelius, *J. Am. Chem. Soc.*, 1939, **61**, 1692.
58. A. C. Buck and H. P. Lankelma, *J. Am. Chem. Soc.*, 1948, **70**, 2398.
59. W. Gerrard, M. F. Lappert and C. A. Pearce, *J. Chem. Soc.*, 1957, 381.
60. R. J. Brotherton and H. Steinberg, *J. Org. Chem.*, 1961, **26**, 4632.
61. M. H. Chisholm and R. J. Tatz, unpublished results.
62. A. K. McMullen, I. P. Rothwell and J. C. Huffman, *J. Am. Chem. Soc.*, 1985, **107**, 1072.
63. A. K. McMullen, S. Latesky and I. P. Rothwell, unpublished results.
64. W. A. Nugent and R. L. Harlow, *Inorg. Chem.*, 1980, **19**, 777.
65. W. A. Nugent and R. L. Harlow, *J. Chem. Soc., Chem. Commun.*, 1978, 579.
66. D. C. Bradley and E. G. Torrible, *Can. J. Chem.*, 1963, **451**, 134.
67. W. A. Nugent and R. L. Harlow, *Inorg. Chem.*, 1979, **18**, 2030.
68. A. Majid, D. W. A. Sharp and J. M. Winfield,, *J. Chem. Soc., Dalton Trans.*, 1972, 87.
69. J. Chatt and G. A. Rowe, *J. Chem. Soc.*, 1962, 4019.
70. J. Chatt, J. D. Garforth, N. P. Johnson and G. A. Rowe, *J. Chem. Soc.*, 1964, 1012.
71. J. F. Rowbottom and G. Wilkinson, *J. Chem. Soc., Dalton Trans.*, 1972, 826.
72. A. F. Clifford and C. S. Kobayashi, *Inorg. Synth.*, 1960, **6**, 207.
73. A. O. Chong, K. Oshima and K. B. Sharpless, *J. Am. Chem. Soc.*, 1977, **99**, 3420.
74. D. W. Patrick, L. K. Truesdale, S. A. Biller and K. B. Sharpless, *J. Org. Chem.*, 1978, **43**, 2628.
75. A. Slawisch, *Z. Anorg. Allg. Chem.*, 1970, **374**, 291.
76. A. Slawisch, *Naturwissenschaften*, 1969, **56**, 369.
77. W. A. Nugent and R. L. Harlow, *J. Chem. Soc., Chem. Commun.*, 1979, 342.
78. M. Harmon, D. W. A. Sharp and J. M. Winfield *Inorg. Nucl. Chem. Lett.*, 1974, **10**, 183.
79. Yu. V. Kokurov, Yu. D. Chuber, V. A. Bochkareva and Yu. A. Buslaev, *Koord. Khim.*, 1973, **1**, 1100 (*Chem. Abstr.*, 1973, **84**, 53 395).
80. J. Chatt and J. R. Dilworth, *J. Chem. Soc., Chem. Commun.*, 1972, 549.
81. I. S. Kolomikov, Yu. D. Koreshkov, T. S. Lobeeva and M. E. Volpin, *J. Chem. Soc., Chem. Commun.*, 1970, 1432.
82. G. La Monica and S. Cenini, *Inorg. Chim. Acta*, 1978, **29**, 183.
83. E. A. Maatta, *Inorg. Chem.*, 1984, **23**, 2561.

84. S. F. Pedersen and R. R. Schrock, *J. Am. Chem. Soc.*, 1982, **104**, 7483.
85. D. C. Bradley, M. B. Hursthouse, K. M. A. Malik and A. J. Nielson, *J. Chem. Soc., Chem. Commun.*, 1981, 103.
86. C. Y. Chou, J. C. Huffman and E. A. Maatta, *J. Chem. Soc., Chem. Commun.*, 1984, 1184.
87. M. J. McGlinchey and F. G. A. Stone, *J. Chem. Soc., Chem. Commun.*, 1970, 1265.
88. P. A. Finn, M. S. King, P. A. Kilty and R. E. McCarley, *J. Am. Chem. Soc.*, 1975, **97**, 220.
89. F. A. Cotton and W. T. Hall, *Inorg. Chem.*, 1978, **17**, 3525.
90. R. R. Schrock and J. D. Fellman, *J. Am. Chem. Soc.*, 1978, **100**, 3359.
91. H. Burger and H. Seyffert, *Angew. Chem., Int. Ed. Engl.*, 1964, **3**, 646.
92. V. Wannagat and H. Seyffert, *Angew. Chem., Int. Ed. Engl.*, 1965. **4**, 438.
93. V. Wannagut and H. Niedesprum, *Angew. Chem.*, 1959, **71**, 574.
94. M. F. Lappert, M. K. Majumdar and B. P. Tilley, *J. Chem. Soc. (A)*, 1966, 1590.
95. H. Noth and H. Varenkamp, *Chem. Ber.*, 1967, **100**, 3353.
96. E. Wiberg and A. May, *Z. Naturforsch., Teil B*, 1955, **10**, 234.
97. D. H. Harris and M. F. Lappert, *J. Organomet. Chem. Libr.*, 1976, **2**, 13.
98. D. C. Bradley, J. S. Ghotra and F. A. Hart, *Inorg. Nucl. Chem. Lett.*, 1974, **10**, 209.
99. D. C. Bradley and M. H. Gitlitz, *J. Chem. Soc. (A)*, 1969, 980.
100. E. C. Alyea, D. C. Bradley, M. F. Lappert and A. R. Sanger, *Chem. Commun.*, 1969, 1064.
101. D. C. Bradley and I. M. Thomas, *Can. J. Chem.*, 1962, **40**, 1355.
102. J. S. Basi, D. C. Bradley and M. H. Chisholm, *J. Chem. Soc. (A)*, 1971, 1433.
103. K. W. Bagwall and E. Yanir, *J. Inorg. Nucl. Chem.*, 1974, **36**, 777.
104. J. G. Reynolds and N. M. Edelstein, *Inorg. Chem.*, 1977, **16**, 2822.
105. J. G. Reynolds, A. Zalkin, D. H. Templeton, N. M. Edelstein and L. K. Templeton, *Inorg. Chem.*, 1976, **15**, 2498.
106. D. C. Bradley, M. H. Chisholm and M. W. Extine, *Inorg. Chem.*, 1977, **16**, 1791.
107. A. H. Cowley, R. W. Braun and J. W. Gilje, *J. Am. Chem. Soc.*, 1975, **97**, 434.
108. M. H. Chisholm, F. A. Cotton, B. A. Frenz, W. W. Reichert, L. W. Shive and B. R. Stults, *J. Am. Chem. Soc.*, 1976, **98**, 4469.
109. M. H. Chisholm, F. A. Cotton, M. W. Extine, M. Millar and B. R. Stults, *J. Am. Chem. Soc.*, 1976, **98**, 4486.
110. J. S. Ghotra, M. B. Hursthouse and A. J. Welch, *J. Chem. Soc., Chem. Commun.*, 1973, 669.
111. M. H. Chisholm, F. A. Cotton and M. W. Extine, *Inorg. Chem.*, 1978, **17**, 1329.
112. E. C. Alyea, D. C. Bradley and R. G. Copperthwaite, *J. Chem. Soc., Dalton Trans.*, 1972, 1580.
113. E. C. Alyea, D. C. Bradley and R. G. Copperthwaite, *J. Chem. Soc., Dalton Trans.*, 1973, 185.
114. D. C. Bradley, J. S. Ghotra and F. A. Hart, *J. Chem. Soc., Dalton Trans.*, 1973, 1021.
115. J. Chatt, J. R. Dilworth and G. J. Leigh, *J. Chem. Soc. (A)*, 1970, 2239.
116. M. Veith, *Angew. Chem., Int. Ed. Engl.*, 1976, **15**, 387.
117. R. J. Doedens, *Inorg. Chem.*, 1969, **8**, 570.
118. O. R. Chambers, M. E. Harmon, D. S. Rycroft, D. W. A. Sharp and J. M. Winfield, *J. Chem. Res. (M)*, 1977, 1849.
119. W. A. Nugent and R. L. Harlow, *J. Chem. Soc., Chem. Commun.*, 1979, 342.
120. A. M. Domingos and G. M. Sheldrick, *Acta Crystallogr., Sect. B*, 1974, **30**, 517.
121. P. P. Power and X. Xiaojie, *J. Chem. Soc., Chem. Commun.*, 1984, 358.
122. H. Schmidbaur and A. Shiotani, *J. Am. Chem. Soc.*, 1970, **92**, 7003.
123. A. H. Clark and A. Haaland, *Chem. Commun.*, 1969, 912.
124. M. J. S. Gynane, D. H. Harris, M. L. Lappert, P. P. Power, P. Riviere and M. Riviere-Baudet, *J. Chem. Soc., Dalton Trans.*, 1977, 2004.
125. M. F. Lappert, P. P. Power, M. J. Slade, L. Hedberg, K. Hedberg and K. Schomaker, *J. Chem. Soc., Chem. Commun.*, 1979, 369.
126. M. Veith, *Z. Naturforsch., Teil B*, 1978, **33**, 7.
127. T. Fjelberg, H. Hope, M. F. Lappert, P. P. Power and A. J. Thorne, *J. Chem. Soc., Chem. Commun*, 1983, 639.
128. M. M. Olmstead and P. P. Power, *Inorg. Chem.*, 1984, **23**, 413.
129. A. H. Cowley and R. A. Kemp, *Chem. Soc. Rev.*, 1985, **86**, 367.
130. A. H. Cowley, D. J. Pagel and M. L. Walker, *J. Am. Chem. Soc.*, 1978, **100**, 7065.
131. M. J. S. Gynane, A. Hudson, M. F. Lappert, P. P. Power and H. Goldwhite, *Chem. Commun.*, 1970, 623; *J. Chem. Soc., Dalton Trans.*, 1980, 2428.
132. D. C. Bradley, M. B. Hursthouse, C. W. Newing and A. J. Welch, *J. Chem. Soc., Chem. Commun.*, 1972, 567.
133. D. C. Bradley, M. B. Hursthouse, R. J. Smallwood and A. J. Welch, *J. Chem. Soc., Chem. Commun.*, 1972, 872.
134. A. H. Clark and A. G. Anderson, *Chem. Commun.*, 1969, 1082.
135. L. Guemas, M. G. B. Drew and J. E. Goulter, *J. Chem. Soc., Chem. Commun.*, 1972, 916.
136. G. M. Sheldrick and W. S. Sheldrick, *J. Chem. Soc. (A)*, 1969, 2279.
137. C. E. Heath and M. B. Hursthouse, personal communication.
138. D. C. Bradley, M. B. Hursthouse and P. F. Rodesiler, *J. Chem. Soc., Dalton Trans.*, 1972, 2100.
139. D. C. Bradley, M. B. Hursthouse and P. F. Rodesilver, *Chem. Commun.*, 1969, 14.
140. R. A. Andersen, D. H. Templeton and A. Zalkin, *Inorg. Chem.*, 1978, **17**, 2317.
141. D. C. Bradley, in 'Coordination Chemistry—20', ed. D. Banerjea, Pergamon, Oxford, 1980, p. 249.
142. R. A. Andersen, *Inorg. Chem.*, 1979, **18**, 1507.
143. D. C. Bradley, M. B. Hursthouse and C. W. Newing, *Chem. Commun.*, 1971, 411.
144. H. Hope, M. M. Olmstead, B. D. Murray and P. P. Power, *J. Am. Chem. Soc.*, 1985, **107**, 712.
145. D. C. Bradley, J. S. Ghotra, F. A. Hart, M. B. Hursthouse and P. R. Raithby, *J. Chem. Soc., Dalton Trans.*, 1977, 1166.
146. D. C. Bradley and Y. C. Gao, *Polyhedron*, 1982, **1**, 307.
147. A. H. Cowley, M. J. S. Dewar, D. W. Goodman and J. R. Schweiger, *J. Am. Chem. Soc.*, 1973, **95**, 6506.
148. L. V. Vilkov, N. A. Tarasenko and A. K. Prokofev, *J. Struct. Chem. (Engl. Transl.)*, 1979, **11**, 114.
149. S. Otsuka, M. Kamata, K. Hirotsu and T. Higuchi, *J. Am. Chem. Soc.*, 1981, **103**, 3011.
150. C. E. Heath and M. B. Hursthouse, *Chem. Commun.*, 1971, 145.

151. C. E. Heath and M. B. Hursthouse, unpublished results.
152. D. C. Bradley, M. H. Chisholm, C. E. Heath and M. B. Hursthouse, *Chem. Commun.*, 1969, 1261.
153. R. Gruning and J. L. Atwood, *J. Organomet. Chem.*, 1977, **137**, 101.
154. R. D. Rogers, J. L. Atwood and R. Gruning, *J. Organomet. Chem.*, 1978, **157**, 299.
155. M. F. Lappert, M. J. Slade, A. Singh, J. L. Atwood, R. D. Rogers and R. Shakir, *J. Am. Chem. Soc.*, 1983, **105**, 302.
156. H. Hope and P. P. Power, *Inorg. Chem.*, 1984, **23**, 936.
157. M. B. Hursthouse, unpublished results.
158. B. Cetinkaya, P. B. Hitchcock, M. F. Lappert, M. C. Misra and A. J. Thorne, *J. Chem. Soc., Chem. Commun.*, 1984, 148.
159. D. C. Bradley, M. B. Hursthouse, K. M. Abdul-Malik and R. Moseler, *Transition Met. Chem.*, 1978, **3**, 253.
160. B. R. Murray and P. P. Power, *Inorg. Chem.*, 1984, **23**, 4584.
161. H. Hope, M. M. Olmstead, B. D. Murray and P. P. Power, *J. Am. Chem. Soc.*, 1985, **107**, 712.
162. J. L. Atwood and G. D. Stucky, *J. Am. Chem. Soc.*, 1969, **91**, 4426.
163. A. Ahmed, W. Schwarz and H. Hess, *Z. Naturforsch., Teil B*, 1978, **33**, 43.
164. H. Hess, A. Hinderer and S., Steinhauser, *Z. Anorg. Allg. Chem.*, 1970, **377**, 1.
165. M. H. Chisholm and F. A. Cotton, *Acc. Chem. Res.*, 1978, **11**, 356.
166. J. Weiss and W. Eisenhuth, *Z. Anorg. Allg. Chem.*, 1967, **350**, 9.
167. R. Mews, *Adv. Inorg. Chem. Radiochem.*, 1976, **19**, 185.
168. O. Glemser and J. Wegener, *Angew. Chem., Int. Ed. Engl.*, 1970, **9**, 309.
169. B. K. Sharpless and S. G. Hentges, personal communication.
170. J. Strahle and H. Barnighamsen, *Z. Anorg. Allg. Chem.*, 1968, **357**, 325.
171. L. S. Tan, G. V. Goeden and B. L. Haymore, *Inorg. Chem.*, 1983, **22**, 1743.
172. L. F. Dahl, P. D. Frisch and G. R. Gust, *J. Less-Common Met.*, 1974, **36**, 255.
173. E. A. Maatta, B. L. Haymore and R. A. D. Wentworth, *J. Am. Chem. Soc.*, 1979, **101**, 2063.
174. D. A. Bright and J. A. Ibers, *Inorg. Chem.*, 1968, **7**, 1099.
175. W. A. Nugent, R. L. Harlow and R. J. McKinney, *J. Am. Chem. Soc.*, 1979, **101**, 7265.
176. M. H. Chisholm, K. Folting, J. C. Huffman and A. L. Ratermann, *Inorg. Chem.*, 1982, **21**, 978.
177. D. F. Gaines and J. Borlin, in 'Boron Hydride Chemistry', ed. E. L. Muetterties, Academic, New York, 1975, p. 241.
178. T. R. R. McDonald and W. S. McDonald, *Acta Crystallogr., Sect. B*, 1972, **28**, 1619.
179. M. Cesari, G. Perego, G. Del Piero, S. Cucinelli and E. Cernia, *J. Organomet. Chem.*, 1974, **78**, 203.
180. P. B. Hitchcock, J. D. Smith and K. M. Thomas, *J. Chem. Soc., Dalton Trans.*, 1976, 1433.
181. B. L. Bardett and C. Kruger, *Angew. Chem., Int. Ed. Engl.*, 1971, **10**, 910.
182. M. H. Chisholm, D. M. Hoffman and J. C. Huffman, *Inorg. Chem.*, 1985, **24**, 796.
183. M. H. Chisholm, E. M. Kober and J. C. Huffman, *Inorg. Chem.*, 1985, **24**, 241.
184. M. H. Chisholm, F. A. Cotton, M. W. Extine, M. Millar and B. R. Stults, *Inorg. Chem.*, 1977, **16**, 320.
185. M. Akiyama, M. H. Chisholm, F. A. Cotton, M. W. Extine and C. A. Murillo, *Inorg. Chem.*, 1977, **16**, 2407.
186. M. H. Chisholm, F. A. Cotton, M. W. Extine and R. L. Kelly, *J. Am. Chem. Soc.*, 1978, **100**, 2256.
187. M. H. Chisholm, D. A. Haitko and J. C. Huffman, *J. Am. Chem. Soc.*, 1981, **103**, 4046.
188. M. J. Chetcuti, M. H. Chisholm, K. Folting, D. A. Haitko, J. C. Huffman and J. Janos, *J. Am. Chem. Soc.*, 1983, **105**, 1163.
189. B. E. Bursten, F. A. Cotton, J. C. Green, E. A. Seddon and G. Stanley, *J. Am. Chem. Soc.*, 1980, **102**, 4579.
190. M. Stein, W. Winter and A. Riecker, *Angew. Chem., Int. Ed. Engl.*, 1978, **17**, 692.
191. H. E. Bryndza, W. C. Fultz and W. Tam, *Organometallics*, 1985, **4**, 939.
192. T. G. Appleton, H. C. Clark and L. E. Manzer, *Coord. Chem. Rev.*, 1973, **10**, 335.
193. M. H. Chisholm, F. A. Cotton and M. W. Extine, *Inorg. Chem.*, 1978, **17**, 2944.
194. L. J. Guggenberger and R. R. Schrock, *J. Am. Chem. Soc.*, 1975, **97**, 6578.
195. S. G. Gibbins, M. F. Lappert, J. B. Pedley and G. J. Sharp, *J. Chem. Soc., Dalton Trans.*, 1975, 72.
196. D. H. Harris, M. F. Lappert, J. B. Pedley and G. J. Sharp, *J. Chem. Soc., Dalton Trans.*, 1976, 945.
197. M. F. Lappert, J. B. Pedley, G. J. Sharp and D. C. Bradley, *J. Chem. Soc., Dalton Trans.*, 1976, 1737.
198. M. H. Chisholm, A. H. Cowley and M. Lattman, *J. Am. Chem. Soc.*, 1980, **102**, 46.
199. F. A. Cotton, G. G. Stanley, B. Kalbacher, J. C. Green, E. Seddon and M. H. Chisholm, *Proc. Natl. Acad. Sci. USA*, 1977, **74**, 3109.
200. D. C. Bradley, C. W. Newing, M. H. Chisholm, R. L. Kelly, D. A. Haitko, D. Little, F. A. Cotton and P. E. Fanwick, *Inorg. Chem.*, 1980, **19**, 3010.
201. J. A. McCleverty, *Chem. Soc. Rev.*, 1983, **12**, 331.
202. A. S. Drane and J. A. McCleverty, *Polyhedron*, 1983, **2**, 53.
203. M. H. Chisholm, L. S. Tan and J. C. Huffman, *J. Am. Chem. Soc.*, 1982, **104**, 4879.
204. R. A. Andersen, *Inorg. Chem.*, 1979, **18**, 2928.
205. R. A. Andersen, *J. Organomet. Chem.*, 1980, **192**, 183.
206. W. A. Nugent, R. J. McKinney, R. V. Kasowski and F. A. Van-Catlidge, *Inorg. Chim. Acta*, 1982, **65**, L91.
207. R. J. Goddard, R. Hoffmann and E. D. Jemmis, *J. Am. Chem. Soc.*, 1980, **102**, 7667.
208. M. H. Chisholm and M. W. Extine, *J. Am. Chem. Soc.*, 1977, **99**, 792.
209. D. M.-T. Chan, W. D. Fultz, W. A. Nugent, D. C. Roe and T. H. Tulip, *J. Am. Chem. Soc.*, 1985, **107**, 251.
210. T. W. Coffinder, J. C. Huffman and I. P. Rothwell, *Inorg. Chem.*, 1985, **24**, 1643.
211. M. H. Chisholm, J. F. Corning and J. C. Huffman, *Polyhedron*, 1985, **4**, 383.
212. E. C. Alyea, J. S. Basi, D. C. Bradley and M. H. Chisholm, *J. Chem. Soc. (A)*, 1971, 772.
213. T. W. Coffinder, I. P. Rothwell and J. C. Huffman, *Inorg. Chem.*, 1983, **22**, 3178.
214. W. A. Nugent, *Inorg. Chem.*, 1983, **22**, 965.
215. K. Jones and M. F. Lappert, *Proc. Chem. Soc.*, 1964, 22.
216. K. Jones and M. F. Lappert, *J. Organomet. Chem.*, 1965, **3**, 295.
217. A. D. Jenkins, M. F. Lappert and R. C. Srivastava, *J. Organomet. Chem.*, 1970, **23**, 165.
218. G. Chandra and M. F. Lappert, *J. Chem. Soc. (A)*, 1968, 1940.

219. J. D. Jameson and J. Takats, *J. Organomet. Chem.*, 1974, **78**, C23.
220. D. J. Cardin, S. A. Keppie and M. F. Lappert, *J. Chem. Soc. (A)*, 1970, 2594.
221. C. Eaborn, A. Pidcock and B. R. Steele, *J. Chem. Soc., Dalton Trans.*, 1976, 767.
222. H. M. J. C. Creemers, F. Verbeek and J. G. Noltes, *J. Organomet. Chem.*, 1968, **14**, 125.
223. M. H. Chisholm and M. W. Extine, *J. Am. Chem. Soc.*, 1977, **99**, 782.
224. M. H. Chisholm, F. A. Cotton, M. W. Extine and D. C. Rideout, *Inorg. Chem.*, 1978, **17**, 3536.
225. P. J. Fagan, J. M. Manriquez, S. H. Vollmer, C. S. Day, V. W. Day and T. J. Marks, *J. Am. Chem. Soc.*, 1981, **103**, 2206.
226. H. Weingarten and J. R. Van Wazer, *J. Am. Chem. Soc.*, 1965, **87**, 724.
227. H. Weingarten and J. R. Van Wazer, *J. Am. Chem. Soc.*, 1966, **88**, 2700.
228. E. O. Fischer and H. J. Kollmeier, *Angew. Chem., Int. Ed. Engl.*, 1970, **9**, 309.
229. E. O. Fischer, E. Winkler, C. G. Kreiter, G. Huttner and B. Krieg, *Angew. Chem., Int. Ed. Engl.*, 1971, **10**, 922.
230. W. Petz, *J. Organomet. Chem.*, 1973, **55**, C42; 1974, **72**, 369; 1975, **90**, 223.
231. W. Petz and A. Jonas, *J. Organomet. Chem.*, 1976, **120**, 423.
232. K. B. Sharpless, T. Hori, L. K. Truesdale and C. O. Dietrich, *J. Am. Chem. Soc.*, 1976, **98**, 269.
233. K. B. Sharpless and T. Hori, *J. Org. Chem.*, 1976, **41**, 176.
234. K. B. Sharpless and S. P. Singer, *J. Org. Chem.*, 1976, **41**, 2504.
235. K. B. Sharpless, D. W. Patrick, L. K. Truesdale and S. A. Biller, *J. Am. Chem. Soc.*, 1975, **97**, 2305.
236. S. G. Hentges and K. B. Sharpless, results to be published.
237. T. H. Tulip, results to be published.
238. J. D. Burrington, C. T. Kartisek and R. K. Grasselli, *J. Catal.*, 1983, **81**, 489.
239. D. M.-T. Chan and W. A. Nugent, *J. Am. Chem. Soc.*, 1985, **107**, 251.
240. J. Dehand and M. Pfeffer, *Coord. Chem. Rev.*, 1976, **18**, 327.
241. M. I. Bruce, *Angew. Chem., Int. Ed. Engl.*, 1977, **16**, 73.
242. G. W. Parshall, *Acc. Chem. Res.*, 1975, **8**, 113.
243. I. P. Rothwell, *Polyhedron*, 1985, **4**, 177.
244. C. Airoldi, D. C. Bradley and G. Vuru, *Transition Met. Chem.*, 1979, **4**, 64.
245. Y. Takahashi, N. Onoyama, Y. Ishikawa, S. Motojima and K. Sugiyana, *Chem. Lett.*, 1978, 525.
246. W. A. Nugent, D. W. Overall and S. J. Holmes, *Organometallics*, 1983, **2**, 161.
247. J. M. Mayer, C. J. Curtis and J. E. Bercaw, *J. Am. Chem. Soc.*, 1983, **105**, 2651.
248. K. J. Ahmed, M. H. Chisholm, K. Folting and J. C. Huffman, *J. Chem. Soc., Chem. Commun.*, 1985, 152.
249. C. R. Bennett and D. C. Bradley, *J. Chem. Soc., Chem. Commun.*, 1974, 29.
250. S. J. Simpson and R. A. Andersen, *Inorg. Chem.*, 1981, **20**, 2991.
251. S. J. Simpson, H. W. Turner and R. A. Andersen, *Inorg. Chem.*, 1981, **20**, 2991.
252. R. P. Planalp, R. A. Andersen and A. Zalkin, *Organometallics*, 1983, **2**, 16.
253. R. P. Planalp and R. A. Andersen, *Organometallics*, 1983, **2**, 1675.
254. T. D. Tilley, R. A. Andersen and A. Zalkin, *J. Am. Chem. Soc.*, 1982, **104**, 3725.

13.5

Sulfurdiimine, Triazenido, Azabutadiene and Triatomic Hetero Anion Ligands

KEES VRIEZE and GERARD VAN KOTEN
University of Amsterdam, The Netherlands

13.5.1 SULFURDIIMINE AND THIONITRITE METAL COMPLEXES

13.5.1.1 Introduction

Sulfurdiimines $RN{=}S{=}NR$ and thionitrites $RN{=}S{=}O$ (R = alkyl, aryl) have been relatively little investigated as ligands in metal complexes. This lack of attention has been little deserved as both types of ligands have interesting ambident coordination properties, since they may bind *via* N, S, O, the π-N$=$S or π-S$=$O bond for example. They may be compared with allenes and naturally also with the isostructural and the isoelectronic sulfur dioxide $O{=}S{=}O$.[1-3] We will first discuss some relevant data concerning RNSNR and RNSO and subsequently the properties of the metal complexes.

13.5.1.2 Properties of RN$=$S$=$NR and RN$=$S$=$O

Compounds RNSNR (R = *p*-tol,[4] *p*-tolylsulfonyl[5]) occur in the *cis,trans* configuration (**1a**) in the solid state. An electron diffraction study of *N,N'*-dimethylsulfurdiimine[6] in the gas phase shows that the *cis,trans* configuration is the most abundant one. The S$=$N bond lengths are of the order of 1.53 Å, which is close to the double bond value of 1.52 Å and much less than the S—N single bond value of 1.69 Å. The N$=$S$=$N bond angle lies between 120 and 140°.[6] In solution the *trans,trans* isomer (**1b**) may be observed by ^{1}H and ^{13}C NMR.[8] The ratio of (**1a**) to (**1b**) mainly depends on the steric properties of R. For example, if R = 2,4,6-mesityl configuration (**1b**) is the only one observed. The energy difference between both configurations is small, as predicted by CNDO/2-SCMO calculations[9] and as borne out by temperature-dependent NMR studies.[8] The interconversion rates of the process (**1a**) $\rightleftharpoons$ (**1b**) decrease in the order Ar < Bu^t < Pr^i $\approx$ Et $\approx$ Me, *i.e.* the interconversion rate slows down with increasing electron donor properties of R. The mechanism of the (**1a**) $\rightleftharpoons$ (**1b**) process may involve a rotation about the N$=$S bond, but is more likely an inversion at the N atoms. IR and Raman data show that ν_{as}(N$=$S), ν_s(N$=$S) and δ(NSN) lie in the ranges of 1190–1220, 1010–1100 and 675–810 cm^{-1} respectively for R = alkyl, while for R = Ar these values are 1250–1300, 940–985 and 775–805 cm^{-1} respectively.[10]

$$R-N{\nearrow}^{S}{\searrow}_{N} \qquad R-N{\nearrow}^{S}{\searrow}_{N}-R$$
$$\qquad\quad\ \ \overset{|}{R}$$

$$\text{(1a)} \qquad\qquad\qquad \text{(1b)}$$

In the case of RN$=$S$=$O it was shown conclusively that the *cis* conformation is the most stable one.[10] The values for ν(S$=$O), ν(N$=$S), δ(NSO) and δ(PhNS) are about 1275–1305, 1145–1165, 720–870 and 640–650 cm^{-1} respectively.

Molecular orbital calculations[11] indicate that in the series RNSNR, RNSO, SO_2 the tendency for S-coordination to a metal should increase in the order RNSNR < RNSO < SO_2, while it is predicted that N-coordination should be favored more by RNSNR than by RNSO. Finally coordination *via* the π-N$=$S bond should be about similar for RNSNR and RNSO, while the tendency to bind *via* the π-S$=$O bond should lessen in going from SO_2 to RNSO.

13.5.1.3 Metal Complexes of RN=S=NR and RN=S=O

In this section we will discuss coordination properties of RNSNR and RNSO towards metal atoms and subsequently the fluxional properties connected with the coordinated ligands and the chemical reactivity of the compounds.

13.5.1.3.1 Nitrogen coordination; monodentate (σ-N; two electron)

The most common coordination mode involves the linkage of RNSNR to a metal atom *via* one N atom (σ-N; two electron). An example is $PtCl_2(Bu^tNSNBu^t)(\eta^2-C_2H_4)$ (**2a**),[8,12] in which the sulfurdiimine is in the (**1a**) configuration while the platinum atom is bonded to the N-donor atom of the *cis* side of the ligand, since this position provides more space for coordination than the *trans* side (**2b**). The N=S bond length at the coordinated side of the ligand is 1.55 Å, while the NSN angle is 113° indicating that the ligand is scarcely perturbed by this type of monodentate N-coordination. Other examples probably having an analogous structure are $trans$-$MCl_2(Bu^tNSNBu^t)L$ (M = Pd^{II},[13] Pt,[8,14] L = PR_3, AsR_3, SR_2 and SeR_2). When R is less bulky than Bu^t it also becomes possible for the metal atom to become coordinated to the *trans* side of the RNSNR ligand (**2b**). Isomers (**2a**) and (**2b**) have been observed for $CDCl_3$ solutions at $-60\,°C$ in the case of $trans$-$PtCl_2(RNSNR)L$ (R = Me, Et, Pr^i, $pent^n$; L = PR_3, AsR_3)[8] [$trans$-$M(CO)(PPh_3)_2$(MeNSNMe)]-ClO_4 (M = Rh^I, Ir^I)[15] and [cis-$PtCl(MeNSNMe)L_2$]ClO_4 (L = PMe_2Ph, $AsMe_2Ph$).[15] A third isomer appears to occur for $trans$-$PtCl_2(ArNSNAr)L$ (Ar = 3,5-xylyl, 2,4,6-mesityl, 4-XC_6H_4 with X = Cl, Me, OMe; L = PR_3).[14] In this type of isomer the sulfurdiimine ligand is in the *trans,trans* configuration.

(2a)   (2b)

Finally monodentate N-coordination has also been observed for metal atoms in a zerovalent oxidation state. Examples are $M(CO)_5(RNSNR)$ (M = Cr, Mo, W; R = Me, Et, Pr^i) with the metal atom bonded to the N atom at the *cis* side of the sulfurdiimine ligand in the *cis,trans* configuration.[16–18]

13.5.1.3.2 Nitrogen coordination; bidentate (σ,σ-N,N'; four electron)

The chelate bonding mode has been found for $M(CO)_4(RNSNR)$ (M = Cr, Mo, W)[16,17] in which R is a bulky group, *e.g.* Bu^t. The crystal structure determination of $W(CO)_4(Bu^tNSNBu^t)$[19] shows an average N=S bond length of 1.58 Å while the NSN angle is very acute (73.4°). The elongation of the N=S bonds with respect to the normal double bond value indicates some electron donation from the metal atoms into the π^*-N=S orbital, as is also shown by resonance Raman experiments.[16,17]

In the case of RNSO compounds the chelated bonding mode has not yet been observed, as far as we are aware, but there are no arguments against such a type of bonding.

13.5.1.3.3 Metal–sulfur (σ-S; two electron) and metal–η^2-N=S (two electron) bonding of RNSNR and RNSO

Although formally metal–sulfur bonding is not treated in this section, some observations will be reported as they are pertinent to the subject at hand.

The crystal structure determination of $trans$-$RhCl(4-MeC_6H_4NSO)(PPr^i_3)_2$ (**3a**) shows that the ArNSO ligand is S-bonded to Rh^I with the Rh atom in the NSO plane.[19] The Rh—S bond is short (2.100(3) Å), analogous to the Rh—S bond in $Rh(\eta^5-C_5H_5)(SO_2)(\eta^2-C_2H_4)$.[1–3] The N=S and S=O bond lengths of 1.520(9) Å and 1.440(9) Å respectively and the NSO angle of 117.2(5)° show

(3a)   (3b)

that the ligand is not appreciably influenced by S-coordination to Rh(I). These values should be compared with the values of 1.525(4) Å, 1.466(4) Å and 117.0(2)° respectively of MeNSO.[6]

In solution there are two isomers present for *trans*-MCl(L)$_2$(RNSO) (L = PPr$_3^i$, P(Cy)$_3$; M = Rh^I, Ir^I; R = Ph, 4-MeC$_6$H$_4$, 4-ClC$_6$H$_4$, 2,4,6-Me$_3$C$_6$H$_2$). Both IR studies and ^{1}H and ^{31}P NMR measurements clearly indicate that in addition to the S-bonded isomer (3a) an η^2-N=S-bonded isomer is present (3b). The equilibrium moves to the side of the S-bonded isomer in the order Ir^I > Rh^I and by increasing the donor properties of the Ar group.[19]

Metal-η^2-N=S bonding has been demonstrated conclusively for Pt(2,4,6-Me$_3$C$_6$H$_2$-NSO)(PPh$_3$)$_2$,[20] as shown by a crystallographic determination. The N=S bond has now lengthened to 1.63(1) Å indicating appreciable backbonding into the π*-N=S orbital. The bond angle of 116.4(7)° and the S=O bond length of 1.455(6) Å have not changed, however, to any appreciable extent. This type of bonding has now also been observed for the analogous complexes M(PPh$_3$)$_2$(RNSO) with M = Ni(O)[21] and M = Pd(O),[22] and Fe(ArNSO)(CO)$_2$(PR$_3$)$_2$.[23]

The solid state features of the platinum complexes are also retained in solution, as shown by ^{31}P and ^{195}Pt NMR measurements of ^{15}N-enriched samples. These types of measurements also made it possible to determine that the rather unstable sulfurdiimine analogue Pt(PPh$_3$)$_2$(RNSNR) has a similar structure with the π-N=S bond being η^2-bonded to Pt.[20]

13.5.1.3.4 Fluxional behavior

Since RN=S=NR and RN=S=O are cumulated double bond systems they may be considered as heteroallene compounds. In the case of metal–allene compounds the allene may rotate about the metal–η^2-allene bond and in addition the metal may jump intramolecularly from one π-C=C bond to the other and *vice versa*.[7]

It is therefore of interest to note that RN=S=NR and RN=S=O ligands may also rotate about the metal–η^2-N=S bond in Pt(PPh$_3$)$_2$(RNSNR).[20] Furthermore, it has been demonstrated for Pt(PPh$_3$)$_2$(RNSNR) that between −30 and +30°C the Pt(PPh$_3$)$_2$ moiety may glide from one π-N=S to the other and *vice versa*, probably *via* an S-bonded intermediate.[20]

A very detailed investigation of the dynamic behaviour of complexes of Pt^{II}, Pd^{II}, Rh^I and Ir$^{I[18,13–15]}$ containing monodentate RNSNR groups, which are N-bonded to the metal atom has been carried out. A survey of the reactions occurring in these systems has been given in Scheme 1.

There are in principle three steps in the various pathways of Scheme 1, *i.e.* (4a) ⇌ (4b) (rotation about the N=S bond), (4b) ⇌ (4c) (N-inversion) and (4b) ⇌ (4d) (formation of a five-coordinate intermediate). The route chosen depends on the relative activation enthalpies. The N-inversion is expected to commence at about the same temperature (~ −45°C) as found for the free ligands. The rotation about the N=S bond has been found experimentally to start at room temperature, while the formation of the five-coordinate intermediate should vary greatly in activation enthalpy, depending on the ease of coordination of a fifth ligand.[7]

In the case of *trans*-PtCl$_2$(RNSNR)L (R = Me, Et, Pr^i, pent^n) the activation enthalpy of step (4b) ⇌ (4d) is apparently high, since the first observable process involves the steps (4a) ⇌ (4b) ⇌ (4c), *i.e.* the platinum atom remains bonded to the same N atom. At higher temperatures the process (4b) ⇌ (4d) ⇌ (4b′) is also observed. For the complexes *trans*-MCl(CO)(MeNSNMe)(PPh$_3$)$_2$ [M = Rh(I), Ir(I)],[15] [*cis*-PtCl(MeNSNMe)L$_2$]ClO$_4$[14] and for *trans*-PdCl$_2$(RNSNR)L (R = Me, Et, Pr^i)[13] the intramolecular pathway (4c) ⇌ (4b) ⇌ (4d) ⇌ (4b′) ⇌ (4c′) is the only one observed, while isomer (4a) is not involved in the fluxional behavior in the NMR time scale. Clearly five-coordination occurs more easily for these compounds than for *trans*-PtCl$_2$(RNSNR)L.[8]

Another alternative for the 'N–N' jump, *i.e.* the process (4b) ⇌ (4d) ⇌ (4b′), may well involve an S-bonded intermediate. This appears less likely, but evidence for such an intermediate has been discussed for the intramolecular movements occurring for another type of compound, *e.g.* M(CO)$_5$(RNSNR) (R = Me, Et, Pr^i; M = Cr^0, W^0).[16,17] The proposed mechanism is shown in Scheme 2.

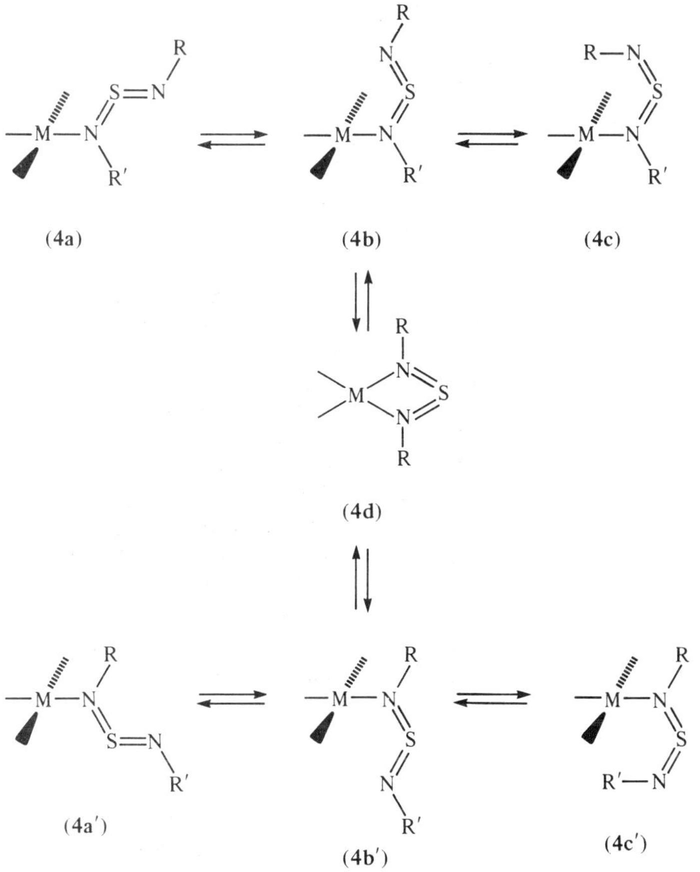

Scheme 1 Reactions occurring for monodentate RNSNR. Complexes of Pt(II), Pd(II), Rh(I) and Ir(I)[8,13–15]

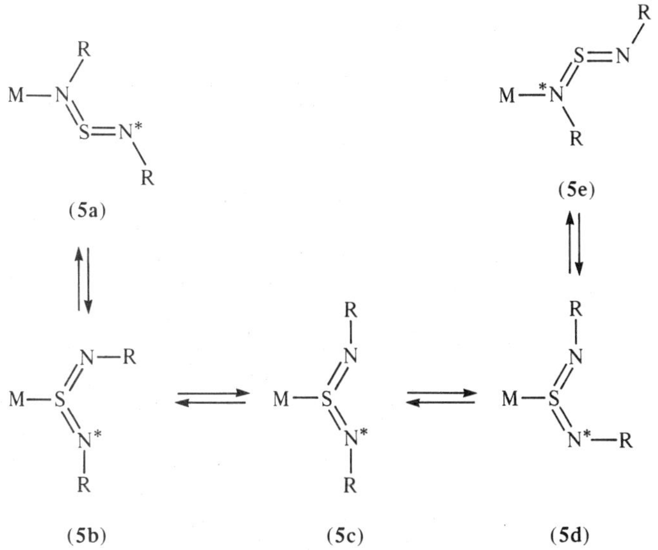

Scheme 2 Proposed fluxional movements of $M(CO)_5(RNSNR)$ [M = Cr(0), W(0)][16,17]

The NMR spectrum shows the presence of (**5a**) and (**5b**), containing a monodentate N-bonded RNSNR group in the *cis,trans* configuration, and the S-bonded isomer for which configurations (**5b**), (**5c**) and (**5d**) are in the limit of fast exchange.

At elevated temperatures the N- and S-bonded isomers interconvert intramolecularly. A chelate-bonded (σ,σ-N,N′) intermediate is not likely, since this would give Cr^0 or W^0 seven-coordination.

13.5.1.3.5 Chemical reactivity of metal–η^2-RNSNR compounds

In Scheme 3 the conversion of $Pt(PPh_3)_2(ArNSNAr)$ (6a; $Ar = 3,5\text{-}Me_2C_6H_3$) into $Pt\{-S-N(2\text{-}NH\text{-}4,6\text{-}Me_2C_6H_2)(3,5\text{-}Me_2C_6H_3)(PPh_3)_2$ (6b) has been shown.[24] The structural features of (6b) indicate a mechanism in which the η^2-bonded N=S group breaks at $T > 30\,^\circ C$. Models show that the Pt atom in (6a) is situated close to the other N atom and to the *ortho* C—H group of one of the aryl groups. It is then understandable that S=N bond rupture of the η^2-linked N=S bond may be accompanied by C—H bond rupture, H migration, N—C bond making and oxidation of Pt^0 to Pt^{II}.

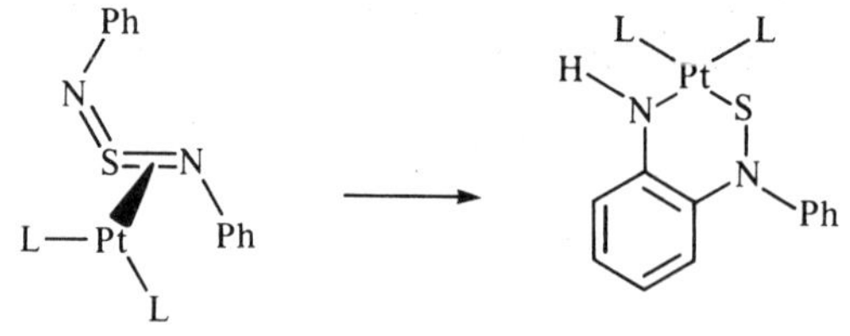

(6a)   (6b)

Scheme 3 Interconversion of $Pt(PPh_3)_2(ArNSNAr)$ (6a) into $Pt[SN(2\text{-}NH\text{-}4,6\text{-}Me_2C_6H_2)(3,5\text{-}Me_2C_6H_3)](PPh_3)_2$ (6b)[24]

Another case of chemical activation is represented by the reaction of $Pt(PPh_3)_2(C_2H_4)$ with 5,6-dimethyl-2,1,3-benzothiadiazole which produced $Pt_2S\{N(6\text{-}\mu\text{-}N\text{-}4,5\text{-}Me_2C_6H_2)\}(\mu\text{-}PPh_2)(PPh_3)_2(Ph)$ (7).[25] It is of interest that N=S bond rupture and insertion of Pt in the S—N bond is accompanied by P—C bond cleavage resulting in PPh_2 and Ph fragments which have both been captured in the complex.

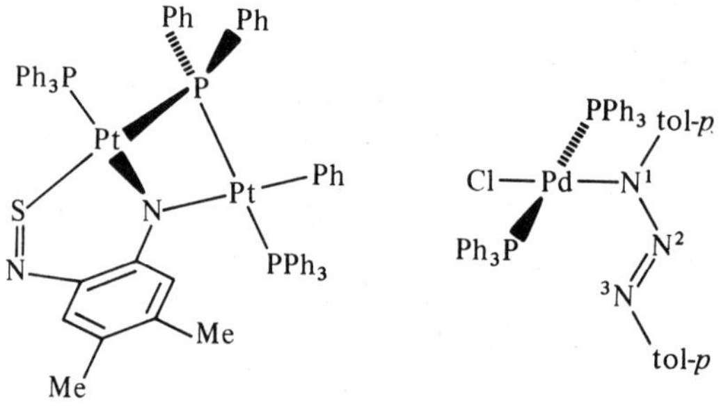

(7)   (8)

Another example of the activation of N=S bonds is provided by the reaction of $Fe_2(CO)_9$ with RNSNR ($R = Bu^t$, $4\text{-}MeC_6H_4$) which produces a number of compounds presented in the general Scheme 4.[26] It is not useful to discuss this reaction scheme in detail, but it is relevant to mention that RNSNR is clearly fragmented by N=S bond rupture in such a way that nitrene (NR), S and RN=S moieties are formed which are captured in clusters.

A different way of activating the RNSNR and RNSO ligands involves reactions with LiR′, MgBrR′ and AlR'_3.[27-29] The resulting products, *e.g.* $Li\{RNS(R')NR\}$ and $MgBr\{RNS(R')NR\}$, are excellent starting materials to transfer the $[RNS(R')NR]^-$ anion to other metal atoms, producing for example $[M\{RNS(R')NR\}]_2$ [$M = Cu(I)$, $Ag(I)$],[27] $Rh(CO)_2\{RNS(R')NR\}$[27] and $Pd(\eta^2\text{-}allyl)\{RNS(R')NR\}$[28] (8) with $R = 2,4,6\text{-}Me_3C_6H_2$, Bu^t, $3,5\text{-}Cl_2C_6H_3$, $4\text{-}XC_6H_4$ ($X = Cl$, Me, MeO) and $R' = Bu^t$, Me, Et, Pr^i. A crystal structure determination of $Rh(CO)_2\{(2,4,6\text{-}Me_3C_6H_2)NS(Bu^t)N(2,4,6\text{-}Me_3C_6H_2)\}$ shows that the $Rh(CO)_2$ and the $-NS(R')N-$ moieties are not coplanar, but that the ligand is bonded in a way reminiscent of metal-η^3-allyl fashion.[27] The average S—N bond length is 1.65 Å indicating virtually single bond character.

The Cu^I, Ag^I, Rh^I and Pd^{II} complexes were found to convert quantitatively and selectively into azo compounds RN=NR, while in some cases the SR′ fragments could be captured (Scheme 5). Bulky R and R′ groups retard the conversion reaction, while the rate increases in the order for $R = 3,5\text{-}Cl_2C_6H_3 < 4\text{-}ClC_6H_4 < 4\text{-}MeC_6H_4$. It was proposed that the plane of the RNS(R′)NR ligand moves in such a way that the S atom approaches the metal atom as the reaction progresses. Electron donating R groups bring more electronic charge in the ligand which, when going into the first LUMO, will bring the two N atoms together and weaken the S—N bonds, since the LUMO

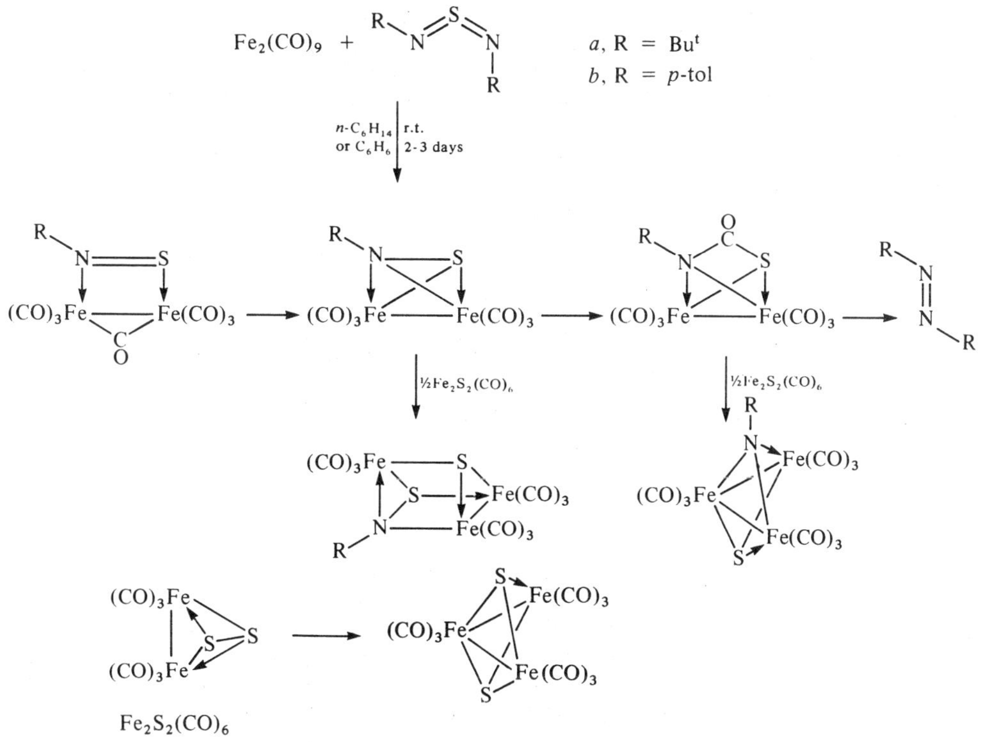

Scheme 4 Schematic representation of the products of the reactions of $Fe_2(CO)_9$ with RNSNR (R = Bu^t, *p*-tol)[26]

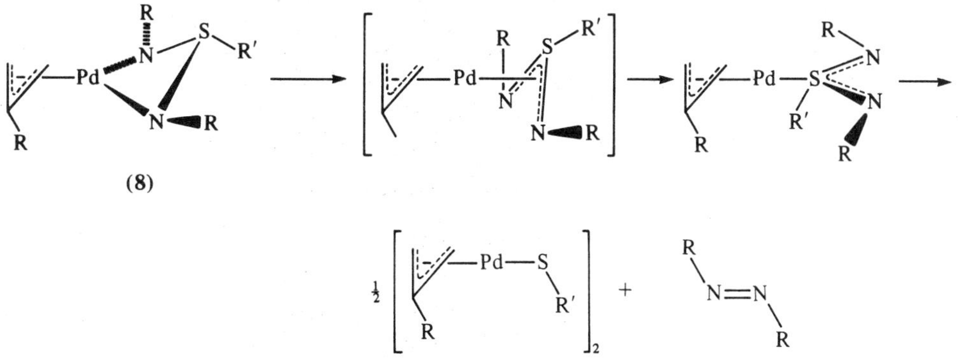

Scheme 5 Interconversion of Pd(η^3-allyl)[RNS(R')NR] (9) into [Pd(SR')(η^3-allyl)]$_2$ and RN = NR[27–29]

has bonding character in the N · · · N interaction and antibonding character in the S—N bonds.[7,27–29]

13.5.2 TRIAZENIDO AND AMIDINO METAL COMPLEXES

13.5.2.1 Introduction

The triazenido [RN$_3$R]$^-$ and the amidino [RNC(R')NR]$^-$ anions are isoelectronic with the carboxylato [OC(R')O]$^-$ anions. It is to be expected that the electron donor properties of the nitrogen ligands are stronger than those of the oxygen ligands. The nitrogen ligands are more versatile, since the ligating properties are influenced by both the electronic *and* steric properties of R and R'.

The preparations of the triazenido and amidino ligands are relatively simple and run generally along similar lines. Some methods are (i) reactions involving Grignard, organozinc or organoaluminum reagents[30,31,32] (equations 1 and 2); (ii) reactions of metal hydrides[31] (equations 3 and 4); (iii) ligand exchange reactions[30] (equation 5) and (iv) metathesis reactions[33,34] (equations 6 and 7).

$$n(RN_3R)MgX + MX_n \longrightarrow M(RN_3R)_n + nMgX_2 \qquad (1)$$

$$(RN_3R)H + MR_n \longrightarrow M(RN_3R)_n + nRH \qquad (2)$$

$$MHL_n + (RN_3R)H \longrightarrow M(RN_3R)L_n + H_2 \qquad (3)$$

$$MHL_n + (RN_3R)H \longrightarrow MH_2(RN_3R)L_{n-1} + L \qquad (4)$$

$$M(RN_3R) + (R'N_3R')H \longrightarrow M(R'N_3R') + (RN_3R)H \qquad (5)$$

$$NiClCp(PPh_3) + Ag(RN_3R) + PPh_3 \longrightarrow Ni(RN_3R)Cp + AgCl(PPh_3) \qquad (6)$$

$$MCl_2(PPh_3)_2 + Li(ArNYNAr) \longrightarrow MCl(ArNYNAr)(PPh_3)_2 + LiCl \qquad (7)$$

In the following we will discuss the complexes in the order of the coordination behavior of the nitrogen ligands, *i.e.* monodentate (σ-N; two electron), bridging (σ-N,σ-N'; four electron) and bidentate (σ,σ-N,N'; four electron).

13.5.2.2 Monodentate Bonded (σ-N; two electron) Triazenido and Amidino Anions

13.5.2.2.1 Structural features

Relatively few complexes have been prepared with monodentate bonded [RNYNR]$^-$ [Y = N, C(R')] groups. It was shown for *trans*-PdCl(p-tolN$_3$tol-p)(PPh$_3$)$_2$ (**8**) that the triazenido group has a *trans* configuration with respect to the double bond N(2)=N(3) and a *trans* conformation about the N(1)—N(2) single bond.[35] The aryl groups are slightly twisted with respect to the N(1)N(2)N(3) plane, *i.e.* by about 4.4 and 15.2° respectively. The Pd $\cdots$ N(3) distance of 2.836(6) Å is relatively small indicating that the lone pair of N(3) is in a favorable orientation for the formation of a five-coordinate species. Some relevant crystallographic details of compounds with monodentate bonded [RNYNR]$^-$ groups are given in Table 1.

Other examples for which the monodentate mode of bonding has been claimed are Rh(ArN$_3$Ar)(CO)(PPh$_3$)$_2$, Rh(ArN$_3$Ar)(cod)(PPh$_3$), *trans*-MX(ArN$_3$Ar)(PPh$_3$)$_2$, *cis*-M(ArN$_3$Ar)$_2$(PR$_3$)$_2$ and *cis*-MCl(ArN$_3$Ar)(PR$_3$)$_2$ (M = Pd^{II}, Pt^{II}; Ar = Ph, p-tol, p-MeOC$_6$H$_4$; X = Cl, Br, I; R = Et, Ph)[31] and furthermore MCp(RN$_3$R')(PPh$_3$) (M = Ni^{II}, Pd^{II}; R = R' = p-tol, p-ClC$_6$H$_4$ and R = p-tol, R' = p-ClC$_6$H$_4$).[33] Monodentate bonding may also occur for M(MeN$_3$Me)$_4$ (M = Ti^{IV}, Zr^{IV}).

In the case of non-transition metal complexes little work has been done. It is certain that monodentate bonding occurs for P(MeN$_3$Me)$_3$, while this may be the case for SiMe$_3$(MeN$_3$Me), [SiMe$_2$(MeN$_3$Me)]$_2$, B(MeN$_3$Me)$_3$ and Al(MeN$_3$Me)$_3$.[36] HgPh(RN$_3$R) also appears to contain monodentate triazenido groups.[37,38]

13.5.2.2.2 Some relevant spectroscopic properties

It is to be expected that rigidly monodentate bonded [RNYNR]$^-$ groups will show the presence of inequivalent R groups in the ^{1}H and ^{13}C NMR spectra. A good example is P(MeN$_3$Me)$_3$, which shows a ^{1}H NMR signal at δ 5.75 p.p.m. (^3J (P–CH) = 4.7 Hz) for the P-bonded side of the ligand and a singlet at 6.20 p.p.m. for the non-bonded side.[36]

IR data have been used as a diagnostic tool for distinguishing monodentate bonded triazenido groups from chelating and bridging groups. The monodentate bonded triazenido groups show absorptions at about 1150, 1190–1210, 1280 and 1360–1380 cm^{-1}, while the chelating groups have absorptions at about 1280 cm^{-1}. The bridging groups absorb at about 1190–1210 and 1360–1380 cm^{-1}. In the case of the amidino systems IR absorptions appear at about 1200,

Table 1 Some Bond Distances (Å) and Bond Angles (°) of Monodentate Bonded Triazenido Groups

Compound	N(1)—N(2)	N(2)—N(3)	M—N(1)	⟨NNN	Ref.
trans-PdCl(p-tolN$_3$tol-p)(PPh$_3$)$_2$	1.336(8)	1.286(7)	2.033(4)	113.0(5)	35a
PtH(p-tolN$_3$tol-p)(PPh$_3$)$_2$	1.31(3)	1.28(3)	2.09(2)	111(2)	35b
PtCl(p-tolN$_3$tol-p)(PPh$_3$)$_2$·CHCl$_3$	1.26(3)	1.26(3)	2.11(2)	116(2)	35c
cis-Pt(PhN$_3$Ph)$_2$(PPh$_3$)$_2$·C$_6$H$_6$	1.336(6)	1.278(6)	2.089(6)	113.9(5)	35d
trans-Ir(CO)(p-tolN$_3$tol-p)(PPh$_3$)$_2$	1.31(1)	1.28(2)	2.16(1)	109.7(5)	35e

1310–1320, 1550–1560 and 1610 cm^{-1} for monodentate bonded [p-tolNC(H)Ntol-p]$^-$ groups. Little is known about the IR absorptions of the other bonding modes.[33,39]

13.5.2.2.3 Dynamic NMR spectroscopy

It has been mentioned that the metal $\cdots$ N(3) distances are relatively short (**8**) and that therefore the situation is favorable for the occurrence of five-coordination. Indeed it has been observed for almost all monodentate bonded [RNYNR]$^-$ groups that both R groups may become magnetically equivalent in the NMR time scale. The temperature-dependent processes are generally intramolecular (Scheme 6) and proceed in the pathway (**9a**) $\rightleftarrows$ (**9b**) *via* a thermally easily attainable chelated intermediate (**9b**).[40,41] In the case of square-planar d^8 complexes the intermediate is five-coordinate. Intermolecular exchange of the [RNYNR]$^-$ group might probably occur for complexes of Li^I, Mg^{II}, Ti^{IV}, Zr^{IV}, Zn^{II} and Hg^{II}, but no detailed studies have been carried out.[36]

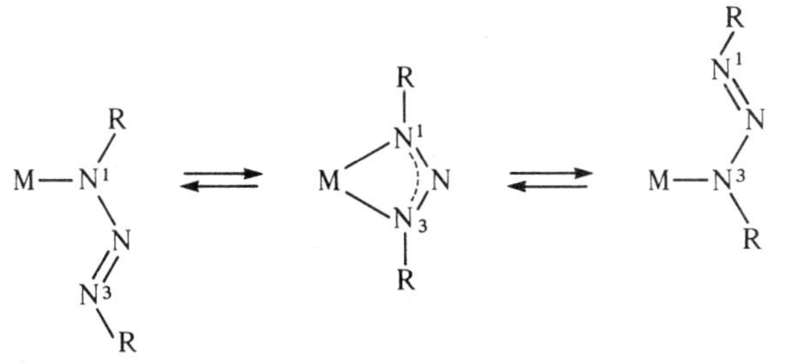

(9a)   (9b)   (9c)

Scheme 6 Intramolecular process involving the chelated intermediate (**9b**)

A rather unusual situation is represented by the complexes NiCp(RN$_3$R)(PPh$_3$) for which there is a fast exchange between the diamagnetic singlet ground state with a monodentate bonded [RN$_3$R]$^-$ group (**10a**) and a paramagnetic intermediate in a triplet state with a chelated group (**10b**).[33]

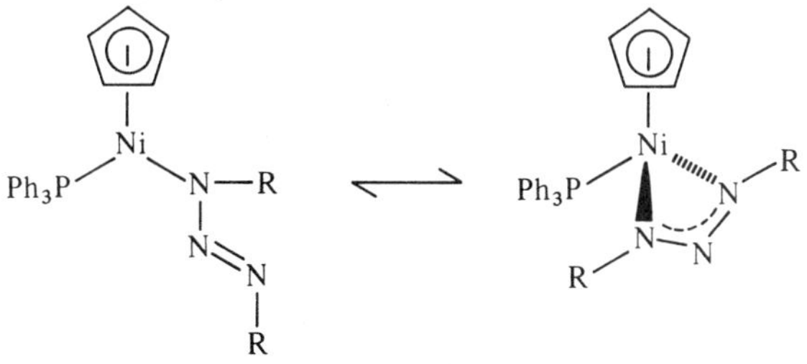

(10a) diamagnetic   (10b) paramagnetic

13.5.2.2.4 Reactivity; insertion reactions

Addition of CO or CNR to compounds with monodentate [RNYNR]$^-$ ligands may lead to insertion and formation of five-membered rings as shown for Ni(Cp){RN—N=N(R)CO} (**11a**) and Ni(Cp){RN—N=N(R)CNR} (**11b**) which have been formed from Ni(Cp)(RN$_3$R) and CO and CNR respectively.[33] Characteristically, the v(CO) and v(CN) stretching frequencies lie at about 1675

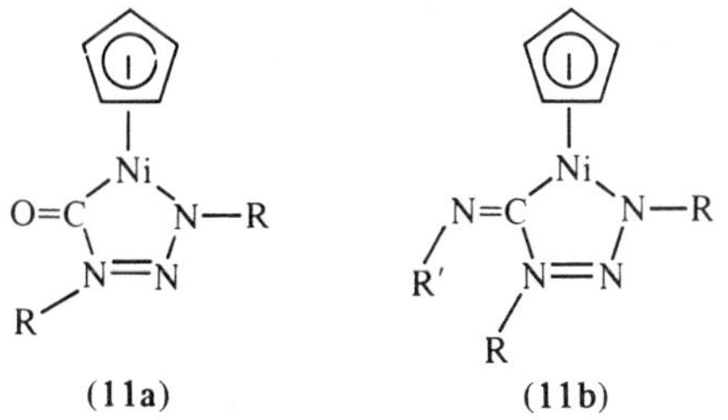

(11a)   (11b)

and 1670 cm^{-1} respectively, indicating inserted carbonyl and isocyanide groups respectively. Similar products of the composition Ir{RN—N=N(R)(CO)}(CO)$_2$(PPh$_3$) were formed by reacting IrAgCl(RN$_3$R)(CO)(PPh$_3$)$_2$ with CO.[39]

On the other hand reaction of *cis*-PtH(*p*-tolN$_3$tol-*p*)(PPh$_3$)$_2$ with CO resulted in reductive elimination of triazene [(*p*-tolN$_3$tol-*p*)H] and zerovalent platinum carbonyl compounds.[42] Finally, reaction of (ArN$_3$Ar)H with CO at 100 atm in the presence of catalytic amounts of PdCl$_2$(PPh$_3$)$_2$ afforded ArNH(CO)Ar.[43] The mechanism probably involves PdCl$_2${(ArN$_3$Ar)H}(PPh$_3$) containing neutral RN=N—N(R)H as a neutral coordinated ligand. This intermediate palladium complex may then give an aryldiazenido group and a diaryldiazonium salt in a decomposition reaction. Insertion of CO and loss of N$_2$ then yield ArNH(CO)Ar. Reaction of Ir(ArN$_3$Ar)(CO)(PPh$_3$)$_2$ with [ArN$_2$]BF$_4$ produces [Ir(ArN$_3$Ar)(N$_2$Ar)(CO)(PPh$_3$)$_2$]BF$_4$ containing a chelated triazenido group and a bent Ir—N=NAr unit. On the other hand treatment of Rh(ArN$_3$Ar)(CO)(PPh$_3$)$_2$ with [ArN$_2$]BF$_4$ afforded [Rh(ArN$_3$ArCO)(N$_2$Ar)(PPh$_3$)$_2$]BF$_4$ containing the carbamoyl moiety with an inserted CO group (v(CO) 1690 cm^{-1}).[44]

In analogy to triazenes it has been shown that neutral formamidines (RNC(H)NR)H may also coordinate, *e.g.* as in [M{2,6-(Me$_2$NCH$_2$)$_2$C$_6$H$_3$}{*p*-tolNC(H)N(H)tol-*p*}](SO$_3$CF$_3$) (M = Ni^{II}, Pd^{II}, Pt^{II}). The structure of the Pt derivative (12) shows that the imine N atom is coordinated to the metal atom.[45] The H atom is situated on top of the metal atom and close to the amino N atom, so that its position is similar to the Ag atom in PtAgCl{(2,6-(Me$_2$NCH$_2$)$_2$C$_6$H$_3$}(*p*-tolNYNtol-*p*).[46]

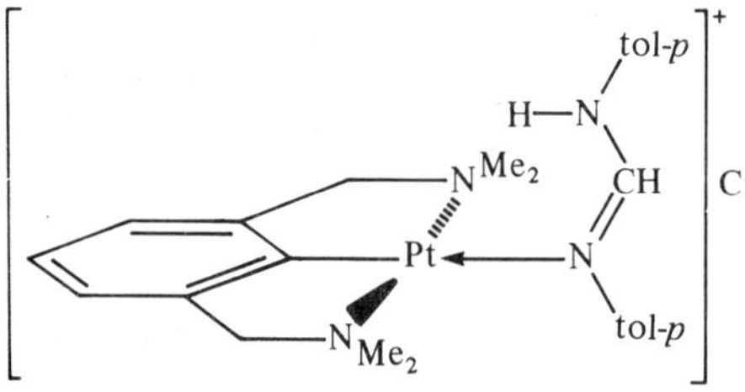

(12)

The complexes show dynamic behavior, *i.e.* the *p*-tol groups change positions indicating movement of both the NYN skeleton and the H atom. The activation decreases in the order Ni^{II} > Pd^{II} > Pt^{II}. In the light of the structure of (12) it seems likely that in PdCl(1,3-η-C$_3$H$_5$){(RNYNR)H} which has been prepared from [PdCl(1,3-η-C$_3$H$_5$)]$_2$ and (RNYNR)H (Y = CH, N) the triazene is also coordinated *via* the imine N atom to Pd and not *via* the amino N atom, as was proposed.[47] Finally complexes of (RNYNR)H and Ag^I, Zn^{II}, Cd^{II}, Hg^{II}, Co^I and Rh^I have been reported.[48]

13.5.2.3 Bridging Triazenido and Amidino Ligands

13.5.2.3.1 Compounds of Cr, Mo and W

Complexes containing only metal–metal bridging triazenido or amidino groups and for which the crystal structure is known are Cr$_2${MeNC(Ph)NMe}$_4$,[49] Cr$_2$(PhN$_3$Ph)$_4$,[50] Mo$_2${PhNC(Ph)NPh}$_4$[51] and Mo$_2$(PhN$_3$Ph)$_4$(0.5C$_7$H$_8$)[50] (see also Table 2) (13). The quadruple metal–metal bonds are of the order of 1.85 Å for Cr and lie in the range of 2.08 and 2.09 Å for Mo. The NYN angles are about 112 to 116° which is appreciably larger than observed for the four-membered ring in the chelating

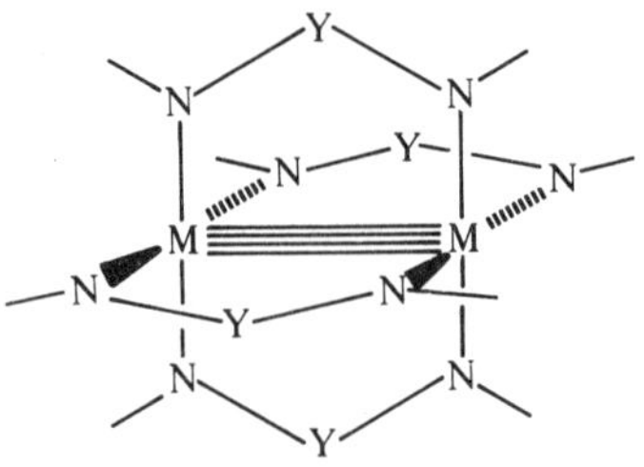

(13)

situation and slightly larger than in the case of the monodentate bonding mode. Of interest is that the R groups on the N atoms make axial coordination to the MM unit difficult. When we consider the electronic influences of the equatorially bonded donor atoms it appears that the stronger electron donor properties of N cause metal–metal shortening, since in $Cr_2(O_2CR)_4$[50] the Cr—Cr bond length is 2.093(1) Å. It should be noted that axial electron donation results in metal–metal bond lengthening.[50]

The amidino compounds $M_2\{RNC(H)NR\}_4$ (M = Cr, Mo) may be prepared from $M(CO)_6$ and $\{RNC(H)NR\}H$ (R = p-tol, o-tol, m-tol, 3,5-xylyl, p-ClC$_6$H$_4$).[52] It was observed that the reactions proceed in several steps. The first steps involve the formation of $M(CO)_5[\{RNC(H)NR\}H]$ and $M(CO)_4[\{RNC(H)NR\}H]_2$ after which $M_2\{RNC(H)NR\}_4$ is produced with loss of CO and hydrogen. The complexes $M_2\{RNC(H)NR\}_4$ are dimeric for R = 3,5-xylyl, but monomeric for R = p-tol and o-tol. Reactions of $M_2\{RNC(H)NR\}_4$ with further $M(CO)_6$ afford unusual complexes $M_2\{RNC(H)NR\}_3\{RNC(H)NR\}Mo(CO)_3$ and $M_2\{RNC(H)NR\}_2\{RNC(H)NRMo(CO)_3\}_2$ in which the $M(CO)_3$ groups are bonded to aromatic groups. These dimeric compounds also dissociate to some degree in CHCl$_3$ and C$_6$H$_6$.[52]

Both Cotton and coworkers and our group have found it difficult to make the analogous tungsten complexes. Indeed, reaction of W(CO)$_6$ and $\{(3,5\text{-xylyl})NC(H)(xylyl-3,5)\}H$ afforded $W_2(\mu\text{-CO})_2\{\mu\text{-}RNC(H)NR\}_2\{RNC(H)NR\}\{RNC(H)NR(CH_2)\}$ (14).[53] In this complex one may distinguish a W—W bond of 2.464(3) Å bridged by two formamidino and two CO groups. One W atom is bonded to a chelating formamidino group, while the other W atom is part of a five-membered ring with a CH$_2$ inserted between N and W (14). This CH$_2$ group is very likely formed by reduction of an inserted CO group.

(14)

The complex $Mo_2Me_2(NMe_2)_4$ reacts with (p-tolN$_3$tol-p)H to give $Mo_2Me_2(NMe_2)_2(p\text{-tolN}_3\text{tol-}p)_2$ containing an Mo≡Mo triple bond of 2.174(1) Å bridged by two *cis*-positioned triazenido groups (Table 2).[54] It is interesting that $Mo_2(NMe_2)_4(p\text{-tolN}_3\text{tol-}p)_2$ contains only chelated triazenido groups. The sterically less demanding Me groups probably allow the triazenido groups to act as bridging groups. A triple Mo≡Mo bond is also very likely present in $Mo_2(OPr^i)_4(PhN_3Ph)_2$, which was formed from $Mo_2(OPr^i)_6$ and $(PhN_3Ph)H$.[55]

13.5.2.3.2 Compounds of Re

Fusing of $[Bu_4N]_2[Re_2Cl_8]$ with $\{PhNC(Ph)NPh\}H$ produces $Re_2\{PhNC(Ph)NPh\}_2Cl_4$ which, when washed with THF and recrystallized from chloroform, gives monoclinic $Re_2\{PhNC(Ph)NPh\}Cl_4$ and triclinic $Re_2\{PhNC(Ph)NPh\}_2Cl_4\cdot$THF.[56] The Re—Re quadruple bond in the THF solvate is 2.209(1) Å which is longer than the Re—Re bond in the compound lacking THF (2.177(2) Å). As pointed out above, this is probably due to the THF molecule being coordinated axially to one of the Re atoms.

However, when $[Bu_4N]_2[Re_2Cl_8]$ was fused with $\{PhNC(Me)NPh\}H$ or with $\{MeNC(Ph)NMe\}H$ two compounds were isolated after recrystallization, *i.e.* $Re_2\{PhNC(Me)NPh\}_2Cl_4$ and $Re_2\{MeNC(Ph)NMe\}_4Cl_2\cdot CCl_4$ respectively (see Table 2).[57] In the first product the two amidino groups are *trans* positioned with the four Cl atoms in equatorial positions forming a planar Re$_2$Cl$_4$ unit. In the second product there are four bridging amidino groups and two axially coordinated Cl atoms. The Re—Re quadruple bond lengths are 2.178(1) and 2.208(2) Å. The bond lengthening in the last compound may be due to the axially coordinated Cl atoms. From the above it is clear that relatively small changes in the amidino ligands may influence greatly the type of product formed.[57]

Table 2 Some Bond Distances (Å) and Bond Angles (°) of Bridging Triazenido and Amidino Compounds

Compound	$M\cdots M$	$M-N(1)$ av.	$M-N(3)$ av.	$Y-N(1)$ av.	$Y-N(2)$ av.	$\langle NYN\rangle$	Ref.
$Cr_2(MeNC(Ph)NMe)_4$	1.843(2)	2.033(5)		1.325(7)	1.345(7)	116.4(5)	49
$Cr_2(PhN_3Ph)_4$	1.858(1)	2.043(5)		1.299(7)		112.9(5)	50
$Mo_2Me_2(p\text{-}MeC_6H_4N_3C_6H_4Me\text{-}p)_2(NMe_2)_2$	2.174(1)	2.220(3)		1.302(3)		113(2)	54
$Mo_2(PhNC(Ph)NPh)_4$	2.090	2.144(10)		1.35(1)		116(1)	51
$Mo_2(PhN_3Ph)_4(0.5C_7H_8)$	2.083	2.15(2)		1.32(1)		112.5(2)	50
$Re_2(PhNC(Ph)NPh)_2Cl_4$	2.177(2)	2.06(2)		1.36(3)		116(2)	56
$Re_2(PhNC(Me)NPh)_2Cl_4$	2.178(1)	2.07(2)		1.35(3)		116.7(19)	57
$Re_2(PhNC(Ph)NPh)_2Cl_4(THF)$	2.209(1)	2.08(1)		1.34(2)		117(2)	56
$Re_2(MeNC(Ph)NMe)_4Cl_2(CCl_4)$	2.208(2)	2.055(9)		1.32(2)		112.5(2)	50
$[Pd(1,3\text{-}\eta^3\text{-}C_3H_5)(p\text{-}MeC_6H_4N_3C_6H_4Me\text{-}p)]_2$	2.856(1)	2.109(6)		1.299(7)		116.8(5)	66
$[Pd(1,3\text{-}\eta^3\text{-}C_4H_7)(MeN_3Me)]_2$	2.97(1)	2.12(1)		1.28(1)	1.34(1)		60
$[Ni(PhN_3Ph)_2]_2$	2.395(3)	1.916(10)		1.31(2)		116.5(8)	64
$[Pd(PhN_3Ph)_2]_2$	2.5626(7)	2.041(6)		1.313(9)		117.6(6)	64
$[Cu(PhN_3Ph)_2]_2$	2.441(2)	2.019(7)		1.296(9)		117.1(6)	64
$[Cu(MeN_3Me)]_4$	2.66(1)	1.87(3)		1.29(4)		116.0(3)	71
$[Cu(PhN_3Ph)]_2$	2.451(8)	1.92(2)		1.30(13)		115.8(2)	70
$RhCuCl(MeN_3Me)(CO)(Ph_3P)_2$	2.730(3)	2.14(1)[a]	1.91(2)[c]	1.25(2)	1.27(2)	119(2)	78
$IrCuCl(MeN_3Me)(CO)(PhMe_2P)_2$	2.686(3)	2.08(2)[b]	1.89(2)[c]	1.29(2)	1.29(2)	116(1)	77
$PtHg\{2,6\text{-}(Me_2NCH_2)_2C_6H_3\}BrCl\cdot\{p\text{-}MeC_6H_4NC(H)NPr^i\}$	2.8331(7)	2.155(9)[d]	2.156(11)[e]	1.32(1)	1.28(2)	125(1)	82
$[IrHgCl\{EtN_3C_6H_4Me\text{-}p\}_2(C_8H_{12})]_2$		2.10(1)[f]	2.42(1)[g]	1.31(2)	1.27(2)	—	84
$Zn_4O(PhN_3Ph)_6$	2.618(1)	2.19(1)[h]	2.06(1)	1.29(1)	1.30(2)	106(1)	84
		2.04(1)		1.31(2)		117(1.5)	73

[a] Rh—N(1). [b] Ir—N(1). [c] Cu—N(3). [d] Pt—N(1). [e] Hg—N(3). [f] Ir—N(1) of bridging ligand. [g] Hg—N(3) of bridging ligand. [h] Ir—N distances of chelating ligands.

13.5.2.3.3 Compounds of Rh and Ir

Complexes $[Rh(\mu\text{-}RNYNR)(CO)_2]_2$ (Y = N, CMe, CH; R = Ar) have been prepared by (i) reaction of $[RhCl(CO)_2]_2$ with M(RNYNR) (M = Li, Na)[58,59] and (ii) reaction of $[RhCl(CO)_2]_2$ with triazene in the presence of triethylamine.[41]

In the case of Ir the analogous $[Ir(\mu\text{-}RNYNR)(CO)_2]_2$ was prepared from $IrCl(CO)_3$ and Li{RNC(H)NR}. The dimeric Rh and Ir complexes probably have structures similar to $[Pd(\eta^3\text{-}allyl)(RN_3R)]_2$ (*vide infra*).[60,61] The carbonyl groups can be substituted by PPh_3 and also by dienes such as cod and nbd giving complexes, *e.g.* of the composition $[Rh\{RN_3R\}nbd]_2$.[58]

It is interesting that oxidative addition of $[Rh(Rh_3R)(CO)_2]_2$ affords $[RhI(RN_3R)(CO)_2]_2$ containing an Rh—Rh bond with the iodine atoms on the axial positions. The Rh_2 pair is bridged by two *cis*-positioned RN_3R ligands and each Rh atom is coordinated by two terminal carbonyl ligands.[58]

13.5.2.3.4 Compounds of Ni, Pd and Pt

Early reports by Dwyer and coworkers[62,63] indicated the existence of dimers of the composition $[M(RN_3R)_2]_2$ (M = Ni^{II}, Pd^{II}; R = Ar). Single crystal X-ray structures on $[Ni(PhN_3Ph)_2]_2$ and $[Pd(PhN_3Ph)_2]$ showed a dimeric structure for these $[M(RN_3R)_2]_2$ compounds (15) consisting of a binuclear pair of metal atoms bridged by four triazenido groups and with M $\cdots$ M distances of 2.395(3) and 2.562(7) Å respectively (Table 2).[64] Analogous amidino complexes $[Pd\{p\text{-}tol\text{-}NC(H)Ntol\text{-}p\}_2]_2$ have been reported by Toniolo *et al.*[65] These were prepared by decomposition of $Pd(RNYNR)_2(PPh_3)_2$ in ethanol or by reacting $PdCl_2(NCCH_3)_2$ with Li{$p\text{-}tolNC(H)Ntol\text{-}p$} in THF.

(15)

Complexes containing two *cis*-positioned bridging RN_3R groups are $[Pd(1,3\text{-}\eta^3\text{-}C_3H_5)(p\text{-}tolN_3\text{-}tol\text{-}p]_2$[66] and $[Pd(1,3\text{-}\eta^3\text{-}C_4H_7)(MeN_3Me)]_2$ (Table 2).[60] The complexes were formed from $[PdCl(\eta^3\text{-}allyl)]_2$ with M(RN_3R) (M = Li,[67] K,[61] Ag[61]) or by reaction of $[Pd(OAc)(1,3\text{-}\eta^3\text{-}C_3H_5)]_2$ with $(RN_3R)H$.[68] The dimeric complexes are very resistent against bridge-splitting reactions. The complexes generally exist only in two isomeric forms, whereas three are possible depending on the orientation of the allyl group. The isomers do not interchange and are rigid on the NMR time scale in the temperature range of -50 to $+115\,^{\circ}C$.[61]

13.5.2.3.5 Compounds of Cu, Ag and Zn

The crystal structure of the copper(II) compound $[Cu(PhN_3Ph)_2]_2$, which has been prepared from copper(II) acetate and $(PhN_3Ph)H$, is similar to $[M(RN_3R)_2]_2$ (M = Ni^{II}, Pd^{II}) (15) and contains four bridging triazenido groups with a Cu^{II}–Cu^{II} distance of 2.441(2) Å (Table 2).[64] Cu^{I} and Ag^{I} derivatives of the general formula $[M(RNYNR)]_n$ (M = Cu^{I}, Ag^{I}; Y = N, CH) were formed from M^{I} complexes and $(RNYNR)H$.[69] The number n may be two or four depending on the R group. According to a crystal structure determination $[Cu(PhN_3Ph)]_2$ is a dimer with two bridging triazenido groups and a Cu^{I}—Cu^{I} distance of 2.45 Å (Table 2) (16).[70] However, $[Cu(MeN_3Me)]_4$ is a tetramer (17).[71] ^1H and ^{13}C NMR studies on $[M\{RNC(H)NR'\}]_n$ (M = Cu^{I}, Ag^{I}; R = p-tol, R' = alkyl) in solution indicated the presence of both dimers and tetramers. Furthermore it was found that the dimer/tetramer ratio increases with (i) increasing temperature, (ii) increasing bulkiness of R' = alkyl, and (iii) when Ag^{I} is substituted by Cu^{I}. It was proposed that the tetramers are reversibly formed by bimolecular reactions of dimers.[72]

(16)   (17)

A rather unusual complex is $Zn_4O(p\text{-tolN}_3\text{tol-}p)_6$ which is formed from $ZnEt_2$ and $(RN_3R)H$ in benzene which is incompletely water-free, probably by hydrolysis of the intermediate $Zn(RN_3R)_2$.[73] The crystal structure determination shows four Zn atoms on the corner of a tetrahedron and an oxygen in the centre of this tetrahedron. The six triazenido groups bridge the six edges of the Zn_4O tetrahedron (Table 2).[73]

13.5.2.3.6 *Heterobinuclear complexes*

From reactions of planar d^8 metal complexes with d^{10} metal compounds a rich chemistry has developed. It is possible to form metal–metal donor bonds for which the d_z^2 orbital of the d^8 metal atom denotes electrons to empty s and p-orbitals of Cu^I, Ag^I and Hg^{II}. Partial electron transfer leads to covalent metal–metal bonds, while complete two-electron transfer also results in ligand transfer. This last aspect will, however, not be treated here.

Complexes with metal-to-metal donor bonds include $L_2(CO)MM'X(RNYNR)$ ($L = PR_3$, AsR_3; $M = Rh^I$, Ir^I; $M' = Cu^I$, Ag^I; X = halide, carboxylate, *e.g.* **(18)**.[39,74-78] Crystal structure determinations of $(PPh_3)_2(CO)RhCuCl(MeN_3Me)$[78] and of $(PPhMe_2)_2(CO)IrCu(MeN_3Me)$[77] showed a metal(I)-to-copper(I) donor bond bridged by a triazenido group. It was assumed that the reaction proceeds *via* an attack of the nucleophilic d^8 metal atom on the d^{10} metal atom, after which the halide migrates from the Rh^I or Ir^I atom to the Cu^I or Ag^I atom with concomitant formation of a bridging $RNYNR$ group. Employing the rigid and strongly electron donating terdentate mono-anionic $2,6\text{-}(Me_2NCH_2)_2C_6H_3$ ligand one may form $\{2,6\text{-}(Me_2NCH_2)_2C_6H_3\}PtAgX(RNYNR)$ X = halide; $Y = N$, CH) by reacting $PtX\{2,6\text{-}(Me_2NCH_2)_2C_6H_3\}$ with $[Ag(RNYNR)]_n$.[79,80] The halide X has migrated from Pt^{II} to Ag^I with concomitant formation of a bridging $RNYNR$ ligand. Extensive NMR studies and in particular ^{109}Ag INEPT NMR measurements of $\{2,6\text{-}(Me_2NCH_2)_2C_6H_3\}PtAgBr\{p\text{-tolNC(H)NR}\}$ showed the presence of a Pt^{II}-to-Ag^I donor bond with relatively small s-character as suggested by a $J(^{195}Pt–^{109}Ag)$ value of only about 540 Hz.[79,80]

Metal-to-metal donor bonds may also be formed with Hg^{II}, as evidenced by $\{2,6\text{-}(Me_2NCH_2)_2C_6H_3\}PtHgBrCl(p\text{-tolNYNR})$ ($Y = N$, CH; R = alkyl), which was prepared from $PtBr\{2,6\text{-}(Me_2NCH_2)_2C_6H_3\}$ and $HgCl(p\text{-tolNYNR})$.[81] The X-ray crystal structure determination of $\{2,6\text{-}(Me_2NCH_2)_2C_6H_3\}PtHgBrCl\{p\text{-tolNC(H)NPr}^i\}$ **(19)** showed a $Pt^{II}–Hg^{II}$ bond of ·2.833(7) Å bridged by the amidino ligand.[82] The Hg^{II} atom is four-coordinated by Br, Cl, Pt and amidino N. It is of interest to note that $HgCl_2$ also affords **(19)** when reacted with $Pt\{2,6\text{-}(Me_2NCH_2)_2C_6H_3\}\{p\text{-tolNC(H)NPr}^i\}$ indicating that this starting Pt compound may be regarded as a bidentate ligand with the Pt and N atoms acting as a chelate to Hg^{II}.[82]

(18)   (19)   (20)

One-electron transfer can be recognized in compounds [(diene)MHgCl(RNYNR)]$_2$ (M = Rh^I, Ir^I; diene = cod, nbd; R = Ar, alkyl), which have been prepared from [MCl(diene)]$_2$ and Hg(RNYNR)$_2$.[83,84] A crystal structure of [(cod)IrHgCl(RN$_3$R)]$_2$ shows that the complex is dimeric with two Cl atoms bridging two Hg atoms.[84] In each half of the dimer (20) the Ir—Hg bond of 2.618(1) Å is bridged by one triazenido group, while the other triazenido group is chelating to the Ir atom. In the case of solutions of the iridium complexes the bridging and chelating RNYNR groups appear to interchange intramolecularly *via* monodentate bonded intermediates. The structural details and the six-coordination around Rh (or Ir) clearly imply that this central metal atom has a d^6-non-bonding electron configuration and that the metal–metal bond may be regarded as an Rh^{II}— (or Ir^{II}—) Hg^I bond.[84] In solution the complexes are monomeric.

13.5.2.4 Chelating Triazenido and Amidino Ligands

The chelating bonding mode is less common for triazenido and amidino ligands than the bridging mode. A factor may be the steric bulk of the ligand, but another factor is clearly the strain involved in the formation of a four-membered ring, which is clear from the acute NYN angle of about 102–105° (Table 3).

13.5.2.4.1 Compounds of Cr, Mo and W

Reaction of W$_2$(NMe$_2$)$_6$ with (PhN$_3$Ph)H affords W$_2$(NMe$_2$)$_4$(PhN$_3$Ph)$_2$.[85] Its X-ray crystal structure demonstrates that the W—W triple bond of 2.314(1) Å is unsupported by bridges. Each W atom is coordinated by one chelating triazenido group and two NMe$_2$ groups (Table 3). It was argued that this is thermodynamically more favorable, since for bridging triazenido groups an eclipsed form would be involved which is sterically less favored.[85]

Mononuclear complexes containing only RNYNR groups are scarce. An example is octahedral Cr(PhN$_3$Ph)$_3$ which is formed from Cr$_2$(Me)$_8^{2-}$ and (PhN$_3$Ph)H (Table 3).[50]

Substituted complexes are more abundant. Complex anions of the type [M(PhN$_3$Ph)(CO)$_4$]$^-$ (M = Cr, Mo, W) have been produced by the reaction of M(CO)$_6$ with Na(PhN$_3$Ph) and can be isolated as the tetraethylammonium salts.[86] More convenient is treatment of [NEt$_4$][MCl(CO)$_5$] with K{RNC(H)NR} (R = Ar, Bu^t). Further reaction of the product [NEt$_4$]-[M{RNC(H)NR}(CO)$_4$] with L (pyridine or PPh$_3$) afforded [NEt$_4$][M{RNC(H)NR}(CO)$_3$L] in which the anion has a *fac* configuration.[87] The chelating amidino groups are strong electron donors, as indicated by the relatively large lowering of the CO stretching frequencies.

It is interesting that for R = Me, carbamoyl-type complexes are obtained with the composition [NEt$_4$][M{MeNC(H)NMeCO}(CO)$_4$] with one CO group inserted between M and N as indicated by the CO stretching frequency of about 1642–1649 cm^{-1} (M = Cr, Mo, W).[87] The complexes are fluxional with *cis/trans* exchange of the terminal CO groups *via* a transition state of C_{4v} or D_{3d} symmetry.

Reactions of MoClCp(CO)$_3$, [MoCp(CO)$_3$]$_2$ or MoMeCp(CO)$_3$ with Na(PhN$_3$Ph) or of MoCl-Cp(CO)$_3$ with SnMe$_3$(ArN$_3$Ar) afforded MoCp(PhN$_3$Ph)(CO)$_2$.[86] This complex can be most conveniently prepared from MoClCp(CO)$_3$ with [Ag(RN$_3$R)]$_n$ (R = *p*-tol, Pr^i; M = Mo, W).[88,89] The asymmetrically substituted compounds MoCp(RN$_3$R')(CO)$_2$ appeared to be fluxional. The process involves an intramolecular exchange of the two carbonyl groups, while the triazenido group remains

Table 3 Some Bond Distances (Å) and Bond Angles (°) of Chelate Bonded Triazenido and Amidino Groups

Compound	M—N(1)	M—N(3)	N(1)—N(2)	N(2)—N(3)	⟨NNN	Ref.
[W(NMe$_2$)$_2$(PhN$_3$Ph)]$_2$	2.215(15)	2.194(14)	1.32	1.34	104.5(1)	85
MoCp{3,5-(CF$_3$)$_2$C$_6$H$_3$N$_3$C$_6$H$_3$(CF$_3$)$_2$-3,5}(CO)$_2$	2.114(9)	2.126(8)	1.326(12)	1.292(11)	100.8(7)	90
Re(*p*-tolN$_3$tol-*p*)(CO)$_2$(PPh$_3$)$_2$	2.21(1)	2.18(1)	1.33(1)	1.31(1)	105(1)	96
RuH(*p*-tolN$_3$tol-*p*)(CO)(PPh$_3$)$_2$	2.149(3)	2.179(4)	1.318(4)	1.310(4)	105.2(3)	101
RuCl{CH$_2$=C(Me)NC(H)NPr^i}(CO)(PPh$_3$)$_2$	2.048(10)	2.242(10)	*a*	*b*	*c*	100
Cr(PhN$_3$Ph)$_3$(0.5C$_7$H$_8$)	2.00(1)	2.00(1)	1.33	1.33	104(1)	50
Co(PhN$_3$Ph)$_3$(C$_7$H$_8$)	1.92(1)	1.92(1)	1.32	1.32	103.6(7)	105
Co(PhN$_3$Ph)$_3$	1.925(5)	1.913(5)	1.325(7)	1.317(7)	102.9(4)	104
Mn(CO)$_5$Hg(2-ClC$_6$H$_4$N$_3$C$_6$H$_4$Cl-2)	2.314(7)	2.435(7)	1.332(11)	1.262(11)	109.19(76)	109

^{a} N(1)—C 1.328 Å. ^{b} N(3)—C 1.300(14) Å. ^{c} ⟨NCN 112.6(10)°.

rigid with respect to the cyclopentadienyl group.[89] An X-ray crystal structure has been carried out for $MoCp\{3,5-(CF_3)_2C_6H_3N_3C_6H_3(CF_3)_2-3,5\}(CO)_2$ (Table 3).[90]

The analogous amidino compounds $MCp\{ArNC(Ph)NAr\}(CO)_2$ (M = Mo, W) can be prepared (i) from $Li\{RNC(Ph)NAr\}$ and $MClCp(CO)_3$,[91] (ii) from $MClCp(CO)_3$ and $M'\{RNC(H)NR\}$ (M' = K, Ag, Cu),[92] (iii) by irradiation of a mixture of $MClCp(CO)_3$ and an amidine and finally (iv) by treating $[MoCp(CO)_3]_2$ with $[M'\{RNC(R)NR\}]_n$ (M' = Cu^I, Ag^I).[92]

It is of interest that reaction of $\{MeNC(H)NMe\}H$ with $MClCp(CO)_3$ afforded a carbamoyl complex $M\{MeNC(H)NMe\}C(O)(CO)_2$ (M = Mo, W), which on irradiation was converted into the complex with the chelated amidino group, *i.e.* $MCp\{MeNC(H)NMe\}(CO)_2$.[92] The Me substituents on the amidino N atoms clearly favor intramolecular attack on a CO group, which may be caused by the strongly electron donating properties of the Mc groups.

Nitrosyl derivatives of the type $[MoCp(NO)(PhN_3Ph)]I$, prepared from $[MoI_2Cp(NO)]_2$ and $(PhN_3Ph)H$, are also known.[86]

13.5.2.4.2 Compounds of Mn and Re

The complex anions $[M(CO)_4(PhN_3Ph)]^-$ have been prepared from $MBr(CO)_5$ (M = Mn, Re) and $Na(PhN_3Ph)$.[86] However, reaction of $MnX(CO)_5$ (X = Cl, Br, I) with $Li\{RNC(R')NR\}$ (R = Ar; R' = Me, Ar) afforded $Mn\{RNC(R')NRC(O)\}(CO)_4$. These carbamoyl complexes can be decarbonylated by UV light to form $Mn\{RNC(R')NR\}(CO)_4$.[93] Abel and Skittral[94] in more recent work showed that $MCl(CO)_5$ and $[MCl(CO)_4]_2$ (M = Mn, Re) undergo reactions with $Li\{ArNC(H)NAr\}$ to give $M\{RNC(H)NRC(O)\}(CO)_4$ ($v(CO)$ of CO inserted between M and N at $\approx 1700\,cm^{-1}$) and $M\{RNC(H)NR\}(CO)_4$ in which the amidino ligand is chelating.

The CO groups in these compounds may be partially substituted by PPh_3, *e.g.* $ReCl(CO)_3(PPh_3)_2$ reacts with $Li(ArNYNAr)$ (Y = CH, N) to yield $Re(ArNYNAr)(CO)_2(PPh_3)_2$.[95] The X-ray crystal structure determination of $Re(p\text{-}tolN_3tol\text{-}p)(CO)_2(PPh_3)_2$ confirms the predictions based on spectroscopic data that the phosphine ligands are *trans* positioned while the triazenido group is *trans* to both CO groups (Table 3).[96] Finally $Mn(ArN_3Ar)(CO)_4$ could also be prepared from $MnBr(CO)_5$ and $SnMe_3(ArN_3Ar)$.[97]

13.5.2.4.3 Compounds of Fe, Ru and Os

Complexes $FeCp(ArN_3Ar)L$ (L = PPh_3, CO) have been prepared from $FeI_2(PPh_3)_2$ with $M(ArN_3Ar)$ (M = Cu^I, Ag^I) in the presence of TlCp or by reaction of $FeICp\{P(OMe)_3\}_2$ with $Cu(RN_3R)$.[41] The treatment of $FeI_2Cp(CO)_2$ with $Na(RN_3R)$, however, did not yield the required product.[86] It was suggested that the formation of the complexes possibly proceeds *via* Fe—Ag bonded intermediates. The compounds $FeCp(L)(ArN_3Ar)$ (L = PPh_3, CO) could be oxidized reversibly in the potential range of 0.25–0.65 V *vs.* an Ag–AgI electrode in CH_2Cl_2.[98]

Treatment of $RuH_2(PPh_3)_4$ and $OsH_4(PPh_3)_3$ with *p*-tolyl isocyanate affords the formamidino complexes $MH\{RNC(H)NR\}(CO)(PPh_3)_2$ (M = Ru, Os).[99]

Similar reactions of $Pr^iN{=}C{=}NPr^i$ with $RuHX(CO)(PPh_3)_2$ (X = Cl, Br) led, *via* insertion of the carbodiimide in the metal–hydride bond, to $RuX(CO)(PPh_3)_2\{CH_2{=}C(Me)NC(H)NPr^i\}$. A single crystal X-ray diffraction study of the chloro complex (21) showed the presence of a chelating formamidino group, while an isopropyl group has been converted to an isopropenyl group (Table 3).[100]

(21)

$OsH_2(CO)(PPh_3)_3$ reacts with $Pr^iN=C=NPr^i$ to give $OsH(CO)(PPh_3)_2\{Pr^iNC(H)NPr^i\}$ which on longer heating results in dehydrogenation and formation of $OsH(CO)(PPh_3)_2\{CH_2=C(Me)NC-(H)NPr^i\}$.[100]

Various other Ru and Os complexes of the type $Ru(ArN_3Ar)_2(PPh_3)_2$, $RuH(ArN_3Ar)-(CO)(PPh_3)_2$, $RuCl(ArN_3Ar)(CO)(PPh_3)_2$, $OsH(ArN_3Ar)(CO)(PPh_3)_2$ and $OsH_3(ArN_3Ar)(PPh_3)_2$ have been prepared by reactions involving complex metal hydrides and triazenes.[31,102] The single crystal X-ray determination of *trans*-$RuH(p\text{-}tolN_3tol\text{-}p)(CO)(PPh_3)_2$ shows an approximately octa-hedral coordination around Ru[II], while the chelating triazenido group is *trans* to the CO group and the hydride ligand (Table 3).[101]

13.5.2.4.4 Compounds of Co, Rh and Ir

The X-ray crystal structure of $Co(PhN_3Ph)_3$ was solved by two groups. Krigbaum and Rubin[103] determined the structure of a monoclinic toluene-solvated form, while Corbett and Hoskins[104] worked on the triclinic solvent-free form (Table 3). The bond distances and bond angles are not appreciably different and are in line with the values of the other compounds in Table 3.

Reactions of $CoCl_2(PPh_3)_2$ with TlCp subsequently followed by treatment with $[M(ArN_3Ar)]_n$ (M = Cu[I], Ag[I]) afforded $CoCp(RN_3R)(PPh_3)$. These complexes are representatives of the very small number of paramagnetic monocyclopentadienyl complexes. ESR spectra indicate the presence of low spin Co[II] in an environment of low symmetry (22).[105]

(22)

Complexes of trivalent cobalt $[CoCp(RN_3R)(L)][PF_6]$ were prepared by reaction of $CoI_2Cp(L)$ with $[Ag(RN_3R)]_n$ followed by anion exchange with $TlPF_6$ (L = PEt_3, PPh_3, P(OMe)_3 and P(OPh)_3).[106] Electrochemical measurements showed that $[CoCp(ArN_3Ar)(PR_3)][PF_6]$ for example can be reversibly reduced in a one-electron step to the neutral Co[II] species at about -0.2 to $+0.1$ V *vs.* an Ag–AgCl electrode in acetone.[98]

In the case of rhodium, complexes have been reported in some variety, *e.g.* $Rh(ArN_3Ar)_3$, $RhH_2(ArN_3Ar)(PPh_3)_2$, $RhCl(ArN_3Ar)_2(PPh_3)$ and $Rh(NO)(ArN_3Ar)_2(PPh_3)$, while for iridium $IrHCl(ArN_3Ar)(PPh_3)_2$ and $IrH_2(ArN_3Ar)(PPh_3)_2$, which all contain chelated triazenido groups and normal coordination geometries, have been reported.[31]

Finally, reaction of $Ir(ArN_3Ar)(CO)(PPh_3)_2$ with $[ArN_2][BF_4]$ produces $[Ir(N_2Ar)(ArN_3Ar)(CO)(PPh_3)_2][BF_4]$ with a chelating triazenido group, a monodentate N_2Ar unit, two phosphines and a terminal carbonyl group.[107]

13.5.2.4.5 Compounds of post transition elements

Only a few elements have been investigated and only data on Hg and Tl complexes are reported.

An interesting type of complex is $Mn(CO)_5Hg(RN_3R)$ (R = 2-XC_6H_4 with X = F, Cl, Br, I), obtained in high yields by the synproportionation of $Hg\{Mn(CO)_5\}_2$ and $Hg(RN_3R)_2$. The single crystal X-ray structure of the 2-chlorophenyl derivative showed the presence of an unsupported Mn—Hg bond and an asymmetric bonded triazenido ligand. The HgN_3 ring is coplanar with both aryl rings. The reason for the asymmetric triazenido bonding to Hg is not clear.[108]

The bonding of the triazenido groups in the compounds $Hg(RN_3R)_2$ and $HgX(RN_3R)$ in the solid and solution has not been clarified sufficiently.[37,38] Possibly both chelating and/or monodentate groups are involved.

Finally, the preparation of $Tl(RN_3R)_3$ has been reported involving the reaction of $Tl(C_6F_5)_3(1,4-$ dioxane) with $(RN_3R)H$.[109] A more convenient preparation has been evolved in our laboratory[110] and involves the reaction of $Tl(OAc)_3$ with $(RN_3R)H$ (R = Ph, *p*-tol) in ethanol.

13.5.3 1,4-DIAZA-1,3-BUTADIENE (α-DIIMINE) METAL COMPLEXES

13.5.3.1 Introduction

Metal complexes of the 1,4-diaza-1,3-butadiene ligands [RN=C(R′)C(R″)=NR = R-DAB{R′, R″}*] have recently been surveyed by van Koten and Vrieze.[111,112] Therefore we will refer extensively to these reviews and to some key references and we will discuss here the main points and new results. The discussion will be restricted to the above ligands, with passing mention of complexes of 2-pyridinecarbaldehyde-*N*-imines, which are discussed in Chapter 13.2.

13.5.3.2 1,4-Diaza-1,3-butadiene Ligands

Very recently the crystal and molecular structure of CyN=C(H)C(H)=NCy was reported,[113] from which it was concluded that the ligand has a flat *E-s-trans-E* structure with C=N bond lengths of 1.258(3) Å and a central C—C′ bond length of 1.457(3) Å. These values and the N=C—C′ angle of 120.8(2)° indicate that the C and N atoms are purely sp^2-hybridized with scarcely any conjugation in the central DAB skeleton. In solution the R-DAB molecules exist predominantly in the *E-s-trans-E* conformation, while in the gas phase according to electron diffraction analyses Bu^t-DAB predominantly has a *gauche* conformation with respect to the central C—C bond with a torsion angle of about 65° from the *s-cis* form. A small amount of the *s-trans* form is probably also present.[114,115]

Calculations indicate that the planar *s-trans* conformation is 20–28 kJ mol^{-1} more stable than the *s-cis* form owing mainly to the interaction of the lone pairs.[112] Furthermore, it appears that the π-acceptor properties of the N=C—C=N skeleton increase in the sequence 2,2′-bipyridine < 2-pyridinecarbaldehyde-*N*-methylimine < R-DAB.[116]

13.5.3.3 Possible Coordination Modes of R-DAB Ligands

In Table 4 examples are given of the various coordination modes of R-DAB ligands, but only for the modes which have been crystallographically ascertained. The R-DAB ligand may be linked *via* one N atom to a metal atom in the σ-N (two electron) bonding mode as in $PdCl_2(PPh_3)(Bu^t$-DAB) (**23**).[113] The Bu^t-DAB ligand is in a planar *E-s-trans-E* conformation with C=N, C—C distances and N=C—C angles not significantly different from the values of the free ligand, indicating little disturbance of the N=C—C=N skeleton by metal coordination. The σ,σ-N,N′ (four electron) donor bonding mode with the planar Bu^t-DAB ligand bridging between two metal atoms has been found for $[PtCl_2(PBu_3)]_2(Bu^t$-DAB) (**25**).[117] The C=N bond lengths of 1.27(3) Å, the C—C bond length of 1.48(2) Å and the C—C—N angle of 118.2(1.3)° again indicate scarcely any disturbance of the R-DAB skeleton.

The σ-N,σ-N′ (four electron) chelating bonding mode has been found for many compounds. The C=N bond lengths vary between 1.26 and 1.34 Å while the central C—C bond varies between 1.44 and 1.50 Å. The C—C—N bond angles lie in the range of 114 to 124°. An example of the chelating bonding mode is given in Table 4 for $PtCl_2(Bu^t$-DAB)(η^2-styrene) (**24**).[118] The above values indicate that the four electron chelating bonding mode may cause some more disturbance of the N=C —C=N skeleton. In particular for electron-rich metal atoms the first N=C—C=N LUMO may become partially populated thereby resulting in C=N bond lengthening and C—C bond shortening, since the LUMO is bonding in the C—C bond and antibonding in the C—N bond. The reader is referred to Table II of our review for relevant crystallographic details.[112]

The σ-N,μ^2-N′,η^2-C=N′ (six electron) donor mode is represented by $Fe_2(CO)_6(Bu^t$-DAB) (**26**) in Table 4.[119] The structure indicates that Bu^t-DAB donates two electrons *via* one N atom of the σ-bonded N=C group (N=C 1.260(5) Å) and four electrons *via* the other N=C group which has a bond length of 1.397(4) Å. In this case this N=C group denotes two electrons through the bridging N atom and two electrons by η^2-bonding to the second Fe atom.

---

* When R′ = R″ = H we will use as notation R-DAB instead of R-DAB{H,H}.

Table 4 Crystallographically Determined Bonding Modes of R-DAB in Metal Complexes

Compound	Coordination mode	Structure	Ref.
dCl$_2$(PPh$_3$)(Bu^t-DAB)	σ-N; two electron	(23)	113
tCl$_2$(η^2-styrene)(Bu^t-DAB)	σ,σ-N,N′; four electron	(24)	118
tCl$_2$(PBu$_3$)]$_2$(Bu^t-DAB)	σ-N,σ-N′; two electron + two electron	(25)	117
$_2$(CO)$_6$(Bu^t-DAB)	σ-N,μ^2-N′,η^2-C=N′; six electron	(26)	119
$_3$(CO)$_8$(pent^n-DAB)	σ-N,σ-N′,η^2-C=N,η^2-C=N′; eight electron	(27)	120, 121

As in the case of the σ,σ-N,N' (four electron) chelating bonding mode (Table 4, **24**), the Bu^t-DAB ligand in (**26**) is in the *E-s-cis-E* conformation. The elongated bond length of the η^2-bonded C=N group is due to π-backbonding from Fe to this C=N group.

Finally the σ-N,σ-N',η^2-C=N,η^2-C=N' (eight electron) bonding type of R-DAB is demonstrated by the structure of Ru$_3$(CO)$_8$(pent^n-DAB) (**27**; Table 4).[120,121] The R-DAB ligand is in the *E-s-cis-E* conformation with four electrons donated by two N atoms and four electrons by the two η^2-C=N bonds. The C=N bonds have an average bond length of 1.39 Å and C—C bond length of 1.39 Å, again indicating population of the first LUMO *via* π-backbonding from the two Fe atoms into the N=C—C=N skeleton.

It should be noted that the η^2-C=N,η^2-C=N' (four electron) bonding mode has been proposed for Fe(CO)$_3$[Me$_2$CN(Me)C(Me)C(Me)N(Me)] in which the N atoms are coupled together *via* a CMe$_2$ bridge.[122]

The reader is referred to Table II of ref. 112 for relevant crystallographic details of compounds containing the R-DAB ligand in the two electron, four electron, six electron and eight electron donor bonding modes. Some data including new information will be discussed in this chapter.

13.5.3.4 1,4-Diaza-1,3-butadiene Metal Complexes; Synthesis, Occurrence and Properties

13.5.3.4.1 *Monodentate bonded 1,4-diaza-1,3-butadiene ligands (σ-N; two electron)*

The monodentate type of bonding for R-DAB groups has been observed fairly recently, while there are relatively few examples. Stable monodentate R-DAB complexes have been reported for the d^8 metals Pd^{II}, Pt^{II} and Rh^I.[111,112] An example is *trans*-PdX$_2$(R-DAB)$_2$ (**28**) (X = halide; R = Bu^t, EtMe$_2$C) which were prepared from PdX$_2$(PhCN)$_2$ and R-DAB.[123] The *E-s-trans-E* conformation for the monodentate type of bonding has been established for *trans*-PdCl$_2$(PPh$_3$)(Bu^t-DAB) (**23**) (Table 4). The H$^\beta$ atom is situated close to Pd and above the coordination plane (calculated distance is about 2.6 Å) resulting in an anomalously low field shift of the H$^\beta$ NMR signal.[113] Analogous complexes could also be prepared for Pt^{II}.[123] It is of interest that at elevated temperatures the complexes are fluxional which involves a process in which the R-DAB changes its point of attachment from one N atom to the other which necessitates an *E* to *Z* inversion at the N atom and a rotation around the central C—C bond.

(28)

In the case of the early transition metals (M = Cr, Mo) reaction of M(CO)$_5$(THF) with Ph-DAB at $-60\,°$C afforded the intermediate M(CO)$_5$(Ph-DAB) in which the R-DAB ligand is σ-N-bonded (two electron) to the voluminous M(CO)$_5$ group.[124]

The monodentate coordination also occurs in the case of AlEt$_3$(R-DAB), which is stable at low temperatures.[125]

13.5.3.4.2 *Bridging 1,4-diaza-1,3-butadiene ligands (σ-N,σ-N'; two electron + two electron)*

In Table 4 the example of [PtCl$_2$(PBu$_3$)]$_2$(Bu^t-DAB) (**25**) has been given as a representative of the bridging R-DAB ligand.[117] In this case the ligand is also in the *E-s-trans-E* conformation, resulting in both imine H atoms being located close to a Pt site. These structural aspects are also retained in solution, whereby these H atoms again show an anomalously low field NMR chemical shift. Other examples containing bridging σ-N,σ-N'-bonded R-DAB groups include [PdCl(η^3-C$_4$H$_7$)]$_2$(Bu^t-DAB), which was prepared by oxidative addition of Pd(η^2-alkene)(Bu^t-DAB) and 2-methallyl chloride.[126] The complex $[Pd(\eta^2\text{-alkene})_2]_2(Bu^t\text{-DAB})$ was prepared from Bu^t-DAB and $Pd(\eta^2\text{-}$ alkene)(Bu^t-DAB) with a chelated Bu^t-DAB.[126]

Bridge-splitting reactions of $Rh(CO)_2(\mu\text{-Cl})_2Rh(CO)_2$ with R-DAB yielded $[RhCl(CO)_2]_2(R\text{-}$ DAB) (R = Bu^t, $EtMe_2C$).[127,128] In solution there appears to be an equilibrium between this dinuclear species with a bridging R-DAB group and the ionic compound $[Rh(CO)_2(R\text{-DAB})][RhCl_2\text{-}(CO)_2]$.

13.5.3.4.3 *Chelating 1,4-diaza-1,3-butadiene ligands (σ,σ-N,N'; four electron)*

Complexes containing the four electron donating σ,σ-N,N' chelating R-DAB ligands (see for example **24** in Table 4) occur in much greater abundance than complexes with monodentate and bridging R-DAB ligands. The general methods of preparation have been discussed fairly extensively in our review.[112] Let it suffice to mention that these complexes may be most commonly prepared by direct addition reactions or by substitution. Less common is the preparation *in situ* of the ligand on the metal atom, *e.g.* by reaction of arylamine with O=C(Me)C(Me)=O and $Mo(CO)_6$ to give $Mo(CO)_4(aryl\text{-DAB}\{Me,Me\})$.[112] In the following we will discuss a number of compounds with some interesting properties.

(i) *Metals in Groups IIIA–VA*

Very few metal compounds in this area of the Periodic System are known. Complexes of the composition $TiCl_4(R\text{-DAB})$ (R = Pr^i, Bu^t, Cy) have been claimed,[129] but not unambiguously established. The complex $V(Pr^i\text{-DAB})_3$ formed by reduction of VCl_3 with sodium in the presence of Pr^i-DAB appears to be better characterized.[129]

(ii) *Metals in Group VIA*

Complexes containing only R-DAB ligands have been reported for Cr.[129,130] Tetracoordinate $Cr(R\text{-DAB})_2$ appears to be formed for R = Pr^i_2CH and Bu^t and hexacoordinate $Cr(R\text{-DAB})_3$ for the less bulky substituent R = Pr^i. Much attention has been paid to the synthesis and properties of $M(CO)_4(R\text{-DAB})$ (M = Cr, Mo, W).[112] The compounds have strong colors and a strong π-back-bonding from metal to the diimine skeleton was deduced.[112] The CO groups on the axial positions may be substituted by ligands such as PPh_3 and acetonitrile with the formation of $M(CO)_n(R\text{-}DAB)L_n$ (n = 1, 2). R-DAB complexes may also be formed with the M atom in divalent oxidation state.[112] Complexes $MX(\eta^3\text{-allyl})(CO)_2(R\text{-DAB})$ may be formed by reaction of $MX(\eta^3\text{-allyl})\text{-}(CO)_2(MeCN)_2$ (M = Mo, W; X = Cl, Br) with R-DAB (R = alkyl, aryl). Structures **(29)** of $MoCl(\eta^3\text{-}C_3H_4Me)(CO)_2(Cy\text{-DAB})$[131] and of $WBr(\eta^3\text{-}C_3H_5)(CO)_2(Cy\text{-DAB})$[131] have been determined and the structure is schematically shown. Complexes $M(CO)_4(R\text{-DAB})$ (M = Cr, Mo, W) are strongly colored owing to an absorption in the visible region, which is due to a metal-to-ligand charge transfer from filled *d*-orbitals into empty π^*-orbitals of the ligand.[132] Further details on spectroscopic properties may be gleaned from the literature, where the solvatochromism of, for example, the charge transfer band for $Mo(CO)_4(R\text{-DAB})$ is discussed in detail.[112] It was also shown that the charge transfer transitions, as investigated by resonance Raman studies, involve both metal-to-R-DAB character and orbitals of the *cis*-carbonyls mixed in the first excited states of the complex.[124]

(29)

The complexes $Mo(CO)_4(R\text{-DAB})$ are not conformationally stable, as indicated by ^{13}C NMR spectroscopy which shows *cis–trans* exchange of the carbonyl ligands, probably *via* a transition state with pseudo C_{4v} symmetry.[133]

(iii) Metals in Group VIIA

The 15 electron compound $Mn(Bu^t\text{-DAB})_2$ has been prepared from $Na(Bu^t\text{-DAB})$ and manganese acetylacetonate.[134] Mixed complexes have been made by the reaction of $MX(CO)_5$ with R-DAB{R',H} affording $MX(CO)_3(R\text{-DAB}\{R',H\})$ with M = Re, X = Cl, R = Pr^i, Cy, p-tol and M = Mn, X = Br, R = Pr^i, Bu^t (R' = H or Me), aryl. The X group is *cis* to the R-DAB ligand.[135] This anion may be substituted by metal carbonyl anions like $[M'(CO)_5]^-$ (M' = Mn, Re) and $[Co(CO)_4]^-$ leading to the metal–metal-bonded compounds $MM'(CO)_8(R\text{-DAB})$[136] and MnCo-$(CO)_7(R\text{-DAB})$.[137] The last compound is unstable and is converted to $MnCo(CO)_6(R\text{-DAB})$ containing a $\sigma\text{-}N,\mu^2\text{-}N',\eta^2\text{-}CN'$ (six electron) R-DAB group in which the $\eta^2\text{-}C{=}N$-bonded group has replaced one CO group by intramolecular substitution **(30)**.[137] The complex **(30)** shows an MnCo bond (2.639(3) Å) bridged by a $\sigma\text{-}N,\mu^2\text{-}N',\eta^2\text{-}CN'$ (six electron) donor unit and a semibridging carbonyl group. The $\eta^2\text{-}(H)C{=}N$-bonded hydrogen and carbon atoms show a ^1H signal at 4.85 p.p.m. and a ^{13}C signal at 170.5 p.p.m. respectively indicating π-backbonding from the Co atom into the π^* C—N bond.

(30)

(iv) Metals in Group VIII; Fe, Ru, Os

In the case of iron the paramagnetic 16 electron tetrahedral complex $Fe(R\text{-DAB})_2$ could be prepared from $FeCl_2$ and $Na(R\text{-DAB})$ (R = alkyl, aryl).[138,139] These complexes take up CO to yield $Fe(CO)(R\text{-DAB})_2$ which with excess CO irreversibly produces $Fe(CO)_3(R\text{-DAB})$ and R-DAB. $Fe(CO)_3(R\text{-DAB})$ may also be prepared from $Fe_2(CO)_9$ or $Fe(CO)_5$ and R-DAB.[112,119,138-140] It has been found that during the reaction 2-imidazolinone, $Bu^tNC(H){=}C(H)N(Bu^t)C{=}O$, is formed. Complexes $M(CO)_3(R\text{-DAB})$ could also be prepared for M = Ru and Os, but for Ru only with sterically bulky R groups.[141] The zerovalent $Ru(p\text{-MeOC}_6H_4\text{-DAB})_3$ (20 electron species) has been generated by the reaction of R-DAB with $RuH_2(PPh_3)_4$ or $RuH(C_6H_4PPh_2)(PPh_3)_2(C_2H_4)$.[142]

Complexes of divalent Fe^{II} and Ru^{II} have been well investigated. Examples are $[Fe(R\text{-DAB})_3]X_2$, $[Ru(R\text{-DAB})_3]X_2$ (R = alkyl, aryl; X = halide, BPh_4^-, PF_6^-), $RuHCl(PPh_3)_2(Pr^i\text{-DAB})$, $RuH_2(PPh_3)_2(Pr^i\text{-DAB})$ and $RuCl_2(Pr^i\text{-DAB})_2$.[112] The various properties have been discussed adequately in our review.[112]

(v) Metals in Group VIII; Co, Rh, Ir

Complexes $Co_2(CO)_6(R\text{-DAB})$ have been formed by reacting $Co_2(CO)_8$ with R-DAB (R = aryl). Further reaction with R-DAB afforded $Co_2(CO)_4(R\text{-DAB})_2$.[143] Heating of $Co_2(CO)_6(R\text{-DAB})$ with R-DAB in hexane resulted in the formation of $Co_4(CO)_8(R\text{-DAB})_2$. The R-DAB ligands in $Co_2(CO)_6(R\text{-DAB})$ and $Co_2(CO)_4(R\text{-DAB})_2$ are bonded as chelates, as shown by the characteristic low field imine ^1H NMR chemical shifts.[143]

In the case of rhodium, treatment of $[RhCl(cod)]_2$ with R-DAB (R = doubly or triply branched at C^α) afforded $[Rh(R\text{-DAB})(cod)]^+X^-$, which reacted with CO to give $[Rh(CO)_2(R\text{-DAB})]^+X^-$.[127,128,143,144]

Complexes $RhCl(CO)(\eta^2\text{-}C_2H_4)(R\text{-DAB})$ (R = Bu^t, $EtMe_2C$) containing five-coordinate Rh and chelating R-DAB have been reported[128] and extensively investigated by NMR. The R-DAB resides in the equatorial plane together with the alkene. Loss of ethylene provides $RhCl(CO)(R\text{-DAB})$. It has been established that in mixtures of $RhCl(CO)(\eta^2\text{-}C_2H_4)(R\text{-DAB})$ with R-DAB there is an intermolecular exchange of free and coordinated R-DAB.[128] The extensive investigations concerning Rh–R-DAB complexes and particularly the influence of the type of R group on the stability of the complexes have been covered elsewhere.[112,128]

(vi)   Metals in Group VIII; Ni, Pd, Pt

There is a rather extensive chemistry involving R-DAB complexes of d^{10} metals. In this section we mention some of the key compounds and refer to a review and the original literature.[112]

Well established are Ni(R-DAB)$_2$ and Ni(CO)$_2$(R-DAB) and the divalent metal complexes NiX$_2$(R-DAB) (X = halide, alkyl).[145,146] It has been claimed that the geometry may vary from tetrahedral to situations intermediate between tetrahedral and square-planar depending on the character of R.

An interesting compound is [NiBr(Pr$_2^i$CH-DAB)]$_2$ which formally contains Ni(I). Dimerization leads to a diamagnetic complex with a metal–metal bond. The compound may be prepared by various routes which may involve (i) reduction of NiBr$_2$(R-DAB) with Na, (ii) reaction of Ni(R-DAB)$_2$ with NiBr$_2$(R-DAB), (iii) treatment of Ni(R-DAB)$_2$ with R'Br and finally (iv) reaction of Ni(cod)(R-DAB) with aryl bromide.[147-149]

Zerovalent complexes of Pd and Pt are Pd(η^2-alkene)(Bu^t-DAB)[112] and Pt(cod)(R-DAB).[150,151] Divalent Pd^{II} and Pt^{II} compounds are abundant.[112] Key examples are five-coordinate PtCl$_2$(R-DAB)(η^2-alkene) and planar PtCl$_2$(R-DAB).[118,123] ^{1}H and ^{13}C NMR data showed that in the ground state the C=C skeleton of the alkene in PtCl$_2$(R-DAB)(η^2-alkene) is in the trigonal plane. When a phosphine is in the plane of the alkene the compound, *i.e.* PtCl$_2$(PR$_3$)(R-DAB) contains monodentate R-DAB. This molecule is fluxional and the five-coordinate geometry is then an intermediate in the σ-N $\rightleftarrows$ σ-N' fluxional process.

(vii)   Metals in Groups IB and IIB

Very little research has been carried out in this area. Zelewsky *et al.*[152] characterized complexes [CuX$_2$][Cu(R-DAB)$_2$] which were prepared from copper(I) halides and R-DAB[153] and which had earlier been wrongly formulated as CuX(R-DAB).

Complexes MX$_2$(R-DAB) with M = Zn, Hg and Cd have been isolated with R = Cy, *p*-MeOC$_6$H$_4$, Bu^t.[155] Organozinc complexes ZnR$_2$(Bu^t-DAB) have been prepared by reaction of ZnR$_2$ and Bu^t-DAB.[125] The reactivity of these complexes will be discussed later (Section 13.5.3.4.7).[125]

13.5.3.4.4   Bridging 1,4-diaza-1,3-butadiene ligands (σ-N,μ²-N',η²-C=N'; six electron)

The preparation, properties and chemistry of complexes containing σ-N,μ^2-N',η^2-C=N'-bonded (six-electron) R-DAB groups have been extensively reviewed before.[112] The structure of Fe$_2$(CO)$_6$(Bu^t-DAB) (26; Table 4)[119] has been mentioned previously and the six electron-donor mode has been discussed formally (Section 13.5.3.3).

This rather rare coordination mode has now also been demonstrated for Ru$_2$(CO)$_6$(R-DAB),[155-157] Os$_2$(CO)$_6$(R-DAB),[156] MnCo(CO)$_6$(R-DAB)[137] (30) and ReCo(CO)$_6$(Bu^t-DAB). A complex Ru$_2$-(CO)$_4$(R-DAB)$_2$ with two six electron donor R-DAB groups (32) has been obtained in reactions of Ru$_3$(CO)$_{12}$ and R-DAB, which will be discussed later (Section 13.5.3.4.6).

Characteristic for all complexes containing σ-N,μ^2-N',η^2-C=N (six electron) donor groups is the shift of both ^{1}H and ^{13}C NMR signals of the η^2-bonded ^{1}H^{13}C=NR group. The imino ^{1}H atom absorbs between 3.2 to about 5.5 p.p.m. while the amino ^{1}H signals absorb between 7.5 and 8.4 p.p.m. The imino ^{13}C signal lies between about 50 and 96 p.p.m., while the amino ^{13}C signal is found between 170 and 186 p.p.m. For a table of these NMR data and a table of carbonyl stretching frequencies the reader is referred to a review.[112]

An interesting point is that up till now it has been very difficult to prepare complexes containing the six electron donor mode with other substituents on the η^2-bonded C=N unit. The only exception is MnCo(CO)$_6$(Pr^c-DAB{Me,Me}) with a cyclopropyl substituent on the N atoms.[137] No doubt steric reasons are important but it should be noted that in the case of the eight electron donor mode for R-DAB{R',R"} stable complexes could also be obtained for R' = R" = Me.[158]

A final structural feature which we would like to stress is that the R-DAB group is in all cases virtually in a planar *E-s-cis-E* conformation. Clearly the R-DAB ligand strongly prefers a planar conformation in both the *E-s-cis-E* and *E-s-trans-E* conformations.

13.5.3.4.5   Bridging 1,4-diaza-1,3-butadiene ligands (σ-N,σ-N',η²-C=N,η²-C=N'; eight electron)

In addition to compounds containing six electron donor R-DAB ligands compounds with σ-N,σ-N',η^2-C=N,η^2-C=N' (eight electron) donor R-DAB bonding modes have also been formed.

Compounds containing such eight electron donor ligands are $Ru_2(CO)_4(\mu-C_2H_2)(Pr^i-DAB)$,[159,160] $Ru_4(CO)_8(Pr^i-DAB)_2$,[159,161] $Ru_3(CO)_8(R-DAB)$ **(27)**,[120,121,162] $Ru_3(CO)_9(Cy-DAB)$,[121] $Ru_2(CO)_5(Pr^i-DAB)$[120] and $Mn_2(CO)_6(Me-DAB\{Me,Me\})$,[158] the structures of which have all been crystallographically established.

In all cases the R-DAB is in a virtually planar *E-s-cis-s* conformation, as in the free ligand and in the two, four and six electron donor situations. The η^2-C=N-bonded groups have C=N bond lengths of about 1.40 Å (see Table II of our review[112]). The ^1H and ^{13}C NMR signals of the $^1H^{13}C=N$ groups have shifted upfield to a range of about 5.85 to about 6.5 p.p.m. and to about 100 p.p.m. respectively. Clearly ^1H and ^{13}C NMR spectroscopy is a good diagnostic tool to distinguish between the free ligand, the σ-N=C-bonded groups and the six electron and eight electron donating R-DAB groups.

The methods of preparation will be discussed in the course of the following subsections. Here we note that $Mn_2(CO)_6(Me-DAB\{Me,Me\})$ has been formed as a by-product in the reaction of $[Mn(CO)_4(CNMe)]^-$ and MeI.[158]

13.5.3.4.6 Synthesis and structural aspects in the formation of metal–R-DAB complexes

In this section we will mention as an example the very well investigated systems involving reactions between $Ru_3(CO)_{12}$ and R-DAB. This system throws some light on the possible course of the reactions and the influence of the R substituents on the formation and stability of the complexes.

In Figure 1 a reaction scheme is given which is a good representation of the reactions which may occur when $Ru_3(CO)_{12}$ is reacted with R-DAB. For all R-DAB ligands the first observable intermediate is mononuclear $Ru(CO)_3(R-DAB)$ **(31)**, which may be stabilized by branching of the carbon chain at both C^α and C^β (e.g. for R = Pr^i_2CH, 2,4,6-$Me_3C_6H_2$ and 2,6-$Me_2C_6H_3$) so that all π-bonds of the C=N groups are protected against metal coordination. Indeed for these R groups $Ru(CO)_3(R-DAB)$ is a stable, isolable end product. For all other studied R groups $Ru(CO)_3(R-DAB)$ is a transient species which further reacts with $Ru_3(CO)_{12}$ to give $Ru_2(CO)_6(R-DAB)$ **(33)**, which may be isolated for R = Bu^t (triply branched at C^α) and for R groups which are doubly branched at C^α and not or singly branched at C^β (R = Pr^i, Cy^c, Pr^c). In the case of R = Pr^i and Cy one may form $Ru_2(CO)_5(R-DAB)$ which contains an eight electron donor R-DAB group **(34)**. This compound may also be formed for R = $pent^n$ and Bu^i, which are singly branched at C^α and doubly branched at C^β. Clearly the steric bulk of these two R groups is such that they can move to one side of the planar N=C—C=N skeleton which is in the *E-s-cis-E* conformation, thereby facilitating eight electron donor bonding. The subsequent step in the reaction scheme is the formation of $Ru_4(CO)_8(R-DAB)_2$ which is produced on further heating of $Ru_2(CO)_5(Pr^i-DAB)$ (R = Pr^i and Cy) **(35)**. This tetranuclear compound is also stable for R = $pent^n$ and Bu^i. Reaction of $Ru_2(CO)_5(R-DAB)$ with $Ru_3(CO)_{12}$ affords $Ru_3(CO)_n(R-DAB)$ (R = Pr^i, Cy, $pent^n$ and Bu^i). $Ru_3(CO)_8(R-DAB)$ **(36)** and $Ru_3(CO)_9(R-DAB)$ **(37)**,* which both contain a symmetrically bonded eight electron R-DAB group, may be reversibly interconverted by CO addition/abstraction, which is a very facile process. Addition of one CO on the Ru atom in **(37)** which has sufficient room causes lengthening of all bonds around that Ru atom [Ru(2)] in **(36)**, since two electrons are donated to an electronically saturated trinuclear Ru_3 system.[121] In essence we have created a 20 electron configuration on Ru(2) in **(36)**.

It may be concluded that the steric bulk of the alkyl substituents (R) is an extremely important factor in stabilizing the various coordination modes and thus steers the course of the reactions. However, the tendency to form metal–metal bonds and the electronic influence of R may also be important. In this respect it is of interest that in the case of the analogous iron systems the number of products is much smaller. For example $Fe_2(CO)_9$ reacts with R-DAB according to equations (8) and (9).

Mononuclear $Fe(CO)_3(R-DAB)$ is again formed first, which may then react with iron carbonyls to give $M_2(CO)_6(R-DAB)$. This, however, is the end product[119,140] in the case of M = Fe, while for M = Ru further build-up of clusters may occur.

The electronic influence of R, *i.e.* the substitution of R = alkyl by R = aryl, on the product formation will be demonstrated in the following subsection.

---

* In ref. 112 asymmetric R-DAB bonding was reported for $Ru_3(CO)_9(R-DAB)$. However, further refinement of the structure showed symmetric eight electron R-DAB bonding.

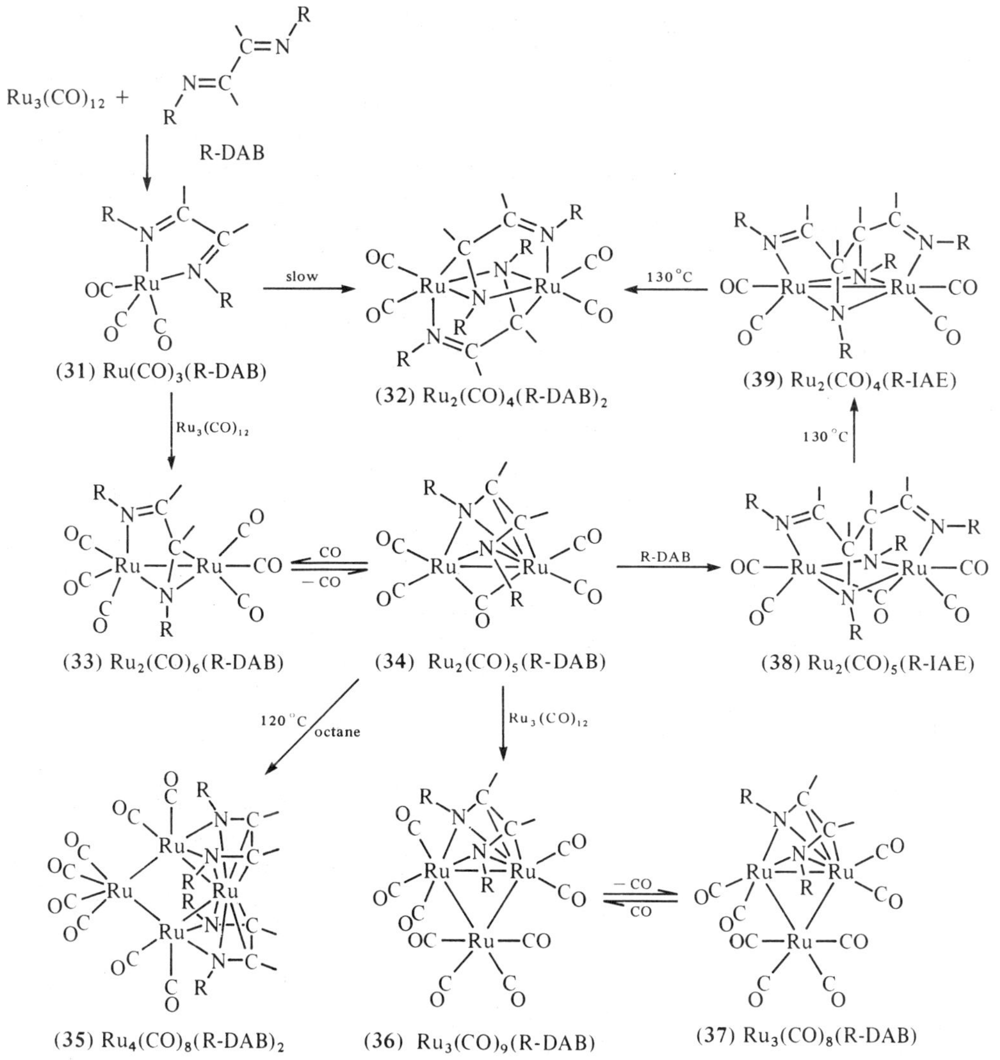

Figure 1 The known reaction steps and products in the reaction system $Ru_3(CO)_{12}$–R-DAB (see text)

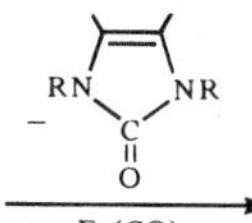

$$Fe_2(CO)_9 + 2R\text{-}DAB \xrightarrow[- Fe(CO)_5]{} Fe(CO)_3(R\text{-}DAB) \qquad (8)$$

$$Fe(CO)_3(R\text{-}DAB) \underset{+ CO, + R\text{-}DAB, \Delta T}{\overset{+ Fe_2(CO)_9, - CO, - Fe(CO)_5}{\rightleftharpoons}} Fe_2(CO)_6(R\text{-}DAB) \qquad (9)$$

13.5.3.4.7 *Activation of metal-coordinated R-DAB groups*

In Figure 1 it is shown that reaction of $Ru_2(CO)_6(R\text{-}DAB)$ (**34**) with excess R-DAB (R = Bu^t, Pr^i, Cy) produces an interesting dinuclear species $Ru_2(CO)_5(IAE)$[163] (IAE = bis[(alkylimino)(alkyl-amino)]ethane (**38**) containing two C—C-coupled R-DAB ligands). Since $Ru_2(CO)_5(R\text{-}DAB)$ may also be used as starting material it is very likely that this species is an intermediate when $Ru_2(CO)_6(R\text{-}DAB)$ (**34**) is used a reagent.

A completely similar reaction occurs when $Ru_2(CO)_6(R\text{-}Pyca\{R^1R^2\})$ (R-Pyca$\{R^1R^2\}$ = 6-$R^1NC_5H_3$-2-C(R^2) = NR with R = alkyl or aryl; R^1 = H, Me and R^2 = H, Me) is reacted with excess pyridinimine.[164] The facile reaction leads to the formation of $Ru_2(CO)_5(APM)$ (APM = bis(μ-t-butylamino)(2-pyridyl)methane-N) (ref. 162, footnote). (The crystal structure of this compound shows two C—C-coupled pyridin-2-imines which bridge a non-bonding Ru_2 metal pair. The

C—C bond is 1.47(1) Å, but not too much value should be placed on the bond length, since no satisfactory refinement could be obtained.) Further heating of $Ru_2(CO)_5(IAE)$ (38) leads to Ru_2-$(CO)_4(IAE)$ (39), which has lost a bridging CO group and gained a metal–metal bond. A similar reaction occurred for $Ru_2(CO)_5(APM)$ when R^1 was Me. However, when $R^1 = H$ the reaction proceeds *via* $Ru_2(CO)_4(APM)$ as an intermediate to $Ru_2(CO)_4(Pyca)_2$,[165] *i.e. via* a C—C cleavage reaction. $Ru_2(CO)_4(Pyca)_2$ has a structure completely analogous to that of $Ru_2(CO)_4(R\text{-}DAB)_2$ (32),[163] which is formed at 130°C from $Ru_2(CO)_4(IAE)$ (38). When R = aryl $Ru_2(CO)_4(R\text{-}DAB)_2$ (32) is very rapidly produced from $Ru_3(CO)_{12}$ and excess R-DAB probably likewise *via* (33), (34), (38) and (39) as intermediates.[163] There is also some evidence that (32) may also be formed directly from (31), but this is a slow process.

It is very interesting that in the case of the pyridinimine ligand complexes addition of carbon monoxide (1 atm) to solutions of $Ru_2(CO)_4(Pyca)_2$ reforms $Ru_2(CO)_5(APM)$,[164] while this is not the case for $Ru_2(CO)_4(R\text{-}DAB)_2$. It should be noted that C—C bond reformation by addition of CO to $Ru_2(CO)_4(Pyca)_2$ (32) leading to $Ru_2(CO)_5(APM)$ is interesting from the viewpoint of Fischer–Tropsch reactions, since we couple here two neighboring C atoms situated on an Ru_2 metal pair together to a C—C bond by addition of CO to that pair.

Carbon–carbon bond formation may also occur in reactions of $Ru_2(CO)_n(R\text{-}DAB)$ [$n = 6$ (33); $n = 5$ (34)] with carbodiimides ($R'N=C=NR'$),[165] sulfines ($R'_2C=S=O$)[165] and alkynes ($R'C\equiv CR'$),[160] leading to $Ru_2(CO)_5(R\text{-}DAB\text{-}R'NCNR')$ (40), $Ru_2(CO)_5(R\text{-}DAB\text{-}R'_2CS)$ with extrusion of CO_2 (41) and $Ru_2(CO)_5(AIB)$ (42). The crystal structure of $Ru_2(CO)_5$ (2-Ph-3-(t-butylamino)-4-(t-butylimino)-1-buten-1-yl) shows that the alkyne is coupled to the formally η^2-C=N-bonded unit *via* a C—C bond of 1.546(10) Å (see Figure 2).[160]

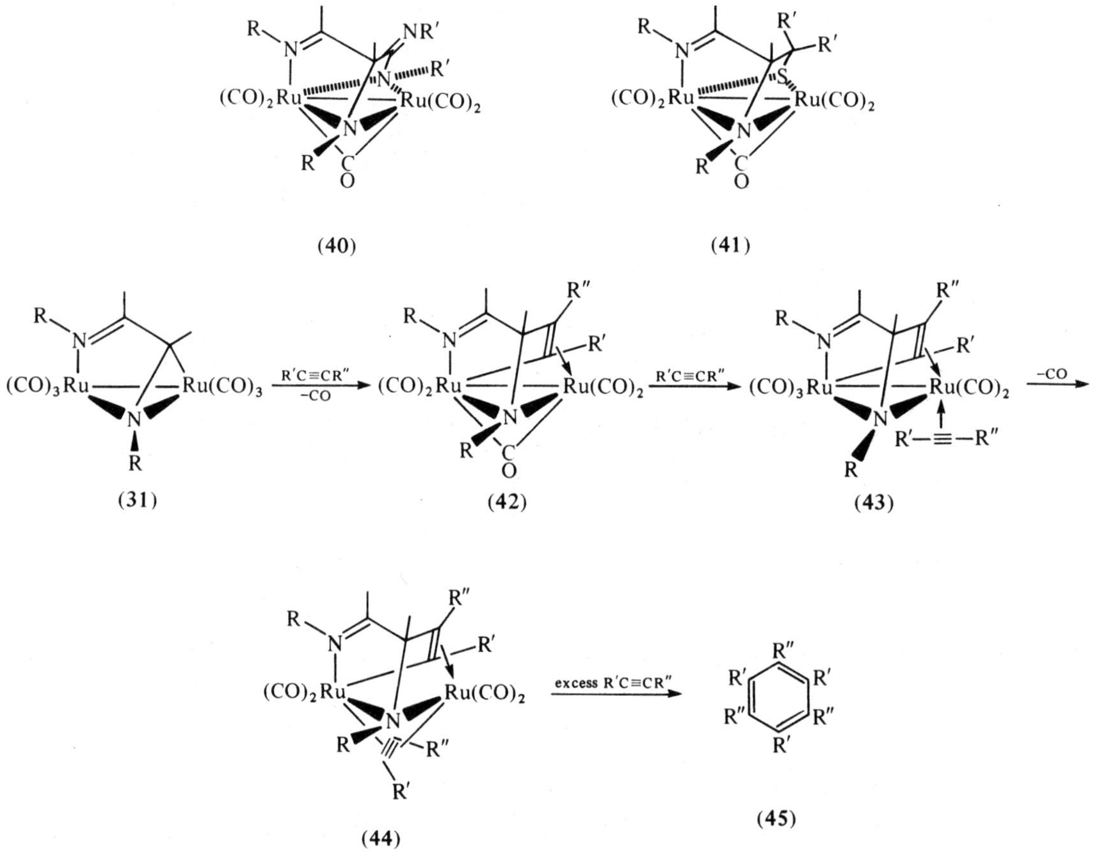

Figure 2 The observed intermediates in the catalytic reaction involving the production of substituted benzenes by trimerization of alkynes with $Ru_2(CO)_6(R\text{-}DAB)$ as starting complex

Further reaction of $Ru_2(CO)_5(AIB)$ (42) with alkynes afforded $Ru_2(CO)_5(AIB)(alkyne)$ (43) containing a monodentate two electron donating alkyne group. Heating of (43) causes extrusion of a carbonyl group and as a result $Ru_2(CO)_4(AIB)(alkyne)$ (44) is formed in which the alkyne now acts as a four electron bridging ligand. Finally addition of excess alkyne gives rise to a rapid catalytic reaction involving the regioselective formation of 1,3,5-trisubstituted benzenes (**45**) for example when RC≡CH is used as alkyne (see Figure 2).[160]

Very unusual reactions have been observed in the case of ketenes (*e.g.* $H_2C=C=O$).[166] In contrast to the C—C bond formation for $R'N=C=NR'$,[165] it has been found that reaction of $Ru_2(CO)_6(R-DAB)$ (R = Pr^i, Cy) with $H_2C=C=O$ produces $Ru_2(CO)_5[Pr^iN\cdots C(H)\cdots(H)CN(Pr^i)C(O)CH_2C(O)CH_2$ (**46**) in virtually quantitative yields. The crystal structure shows the presence of a head-to-tail-bonded diketene unit, which is bonded to one Ru atom and to one N atom of the R-DAB unit and not to a C atom. As a result the other part (NCC) of the NCCN skeleton is linked as an η^3-azaallyl unit to the other Ru atom of the Ru_2 pair.

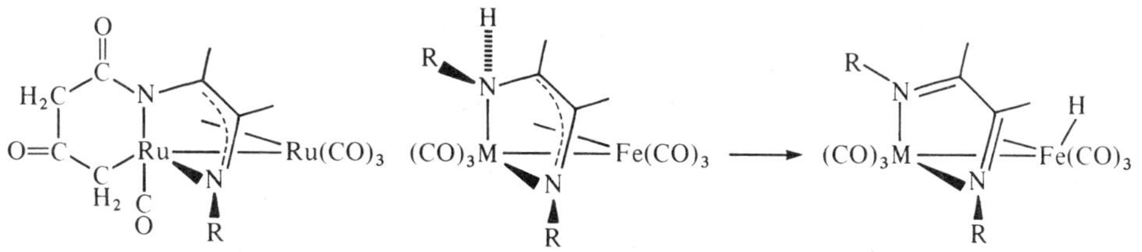

(46)   (47)   (48)

The η^3-azaallyl type of bonding can also be obtained by overall hydride transfer to one of the N atoms of the coordinated NCCN skeleton. For example reaction of $[FeH(CO)_4]^-$ with $MBr(CO)_3(R-DAB)$ (M = Mn, Re; R = Pr^i, Bu^t, Cy, *p*-tol) yielded the novel $FeM(RN\cdots C(H)\cdots C(H)N(H)R$. The crystal structure for M = Mn and R = Bu^t showed that the formerly neutral chelate-bonded Bu^t-DAB ligand is converted to an eight electron bonded monoanionic 3-*N*-alkylamino-1-*N*-alkylazaallyl moiety.[165] The amino substituent containing the new N—H bond coordinates to Mn, while the 1-azaallylic group is bonded as a four electron donor to the FeMn pair, *i.e.* η^3-bonded to Fe (two electrons) and σ-N bonded to Mn (two electrons) (**47**). It is interesting that heating of a toluene solution of $FeRe\{(Pr^iNC(H)C(H)N(H)(Pr^i)\}$ yielded the isomeric metal hydride $ReFeH(CO)_6(Pr^i-DAB)$ (**48**) in which the Pr^i-DAB has a six electron donor type of bond, *i.e.* the N—H bond is again broken and the hydride is now bridging the MnFe bond.

It is not easy to predict the reaction course of species which are similar in principle. For example, reaction of $MnBr(CO)_3(R-DAB)$ with $[Mn(CO)_5]^-$ does give, as predicted, the metal–metal-bonded $Mn_2(CO)_8(R-DAB)$.[136] However, reaction of $MnBr(CO)_3(R-DAB)$ with $[Mo(CO)_4(R-DAB)]^-$ gives first an unstable Mn—M-bonded species which then converts into $Mo_2(CO)_6(IAE)$ in which the Mo—Mo bond of 2.813(2) Å is bridged by the 10 electron IAE ligand. Therefore the structure of $Mo_2(CO)_6(IAE)$ is very similar to the one shown in Figure 1: (**39**) or (**38**).[167] $Mo_2(CO)_6(IAE)$ can also be formed by treating $Hg[Mo(CO)_3(R-DAB)]_2$ with acids such as acetic acid with the formation of Hg(II) species and H_2,[167] probably *via* the dimerization of the intermediate of composition $MoH(CO)_3(R-DAB)$. When comparing this reaction which leads to $Mo_2(CO)_6(IAE)$ with the process shown in Figure 1 involving the formation of $Ru_2(CO)_n(IAE)$ (**38** for n = 5 and **39** for n = 4) it is obvious that it is not clear at all which mechanism or mechanisms lead to this C—C coupling. In this respect it is of importance to discuss the reactions of ZnR_2 with R-DAB and with $Bu^tN=CH-2-C_5H_4N$ (= Bu^t-Pyca).[125,168] In Figure 3 the proposed reaction steps are shown for the reaction of $ZnEt_2$ with Bu^t-DAB and Bu^t-Pyca.[125] On the basis of chemical and spectroscopic measurements (ESR, NMR) it is clear that treatment of Et_2Zn with Bu^t-DAB gives first the chelated (four electron) $ZnEt_2(\alpha$-diimine) complex. Subsequently there are two major pathways then available for the reaction depending on whether we are dealing with Bu^t-DAB or Bu^t-Pyca. In the case of Bu^t-DAB the intermediate complex $ZnEt_2(R-DAB)$ (**49**) is converted to the *N*-alkylated product $ZnEt(Et)(Bu^t)NC(H)=C(H)N(Bu^t)$ (**51**), probably *via* transition state (**50**) and a fast intramolecular 1,2-shift of the ethyl group. A minor pathway involves the competitive escape of an ethyl radical from the solvent case giving a radical $ZnEt(R-DAB)\cdot$ (**52**), which is in equilibrium with the C—C coupled dimer $Zn_2Et_2(IAE)$ (**53**). The *C*-alkylated product (**54**) is not formed indicating that $k_N \gg k_{esc} \gg k_C$ which is compatible with the higher spin density on the N atoms than on the imine C atoms.[125]

Study of the reactivity of $Et_2Zn(Bu^t$-Pyca) shows that a different reaction course is chosen. *N*-Alkylation is impossible, since this would require loss of the conjugation in the ring. It is likely that the major pathway proceeds *via* transition state (**55**) or (**55'**), formed *e.g.* by a β-elimination route or a transition state (**50**) followed by elimination of ethylene from the ethyl radical and formation of the radical pair $[ZnEt(Bu^t$-Pyca)$\cdot$H$\cdot]$ (**55**) or (**55'**) (Figure 3). The formation of the

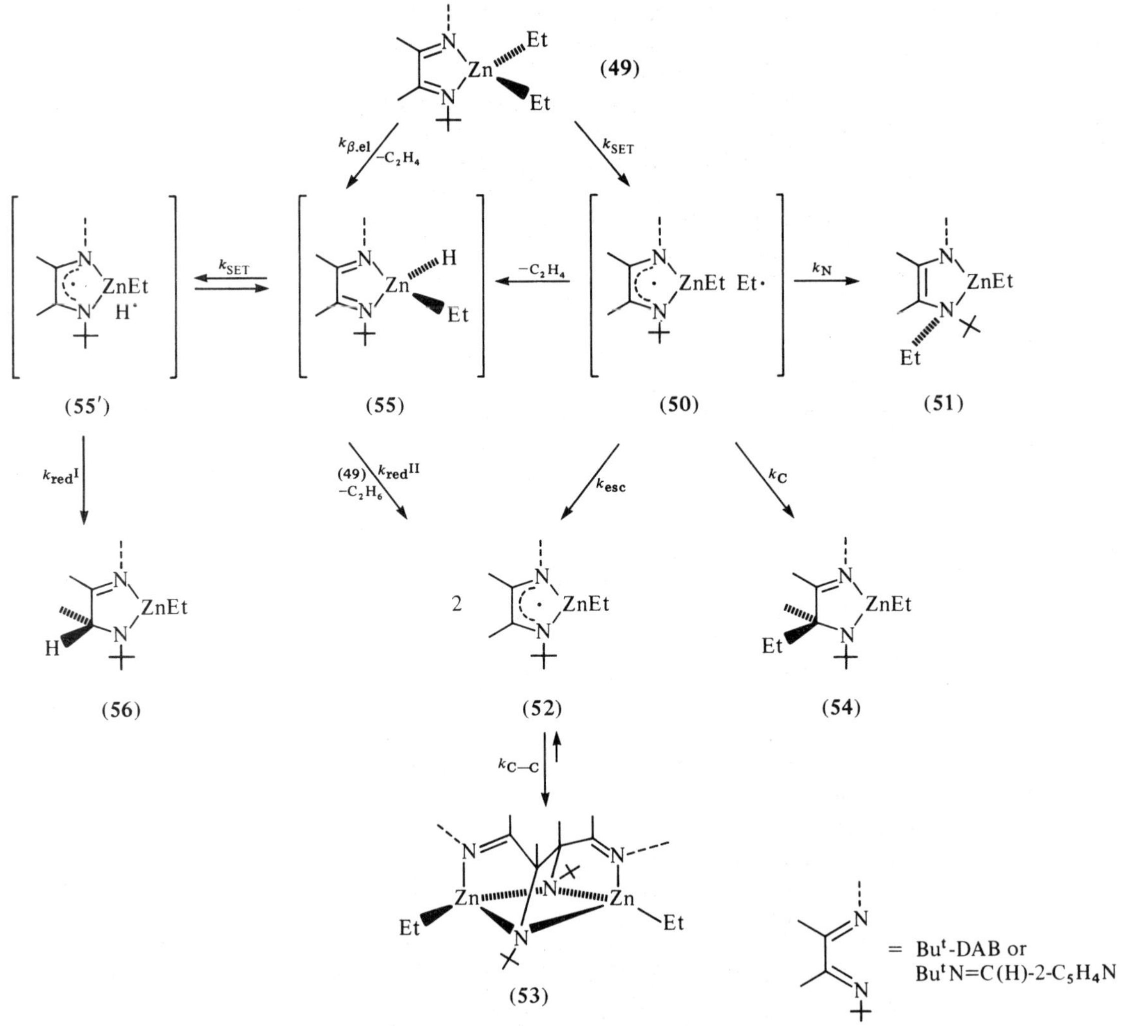

Figure 3 Proposed steps for the Et_2Zn/Bu^t-DAB and Et_2Zn/Bu^t-Pyca reactions[125]

reduction product $ZnEt(Bu^t$-N-CH$_2$-2-C$_5$H$_4$N) (**56**) is evidence for the formation of (**55**) or (**55′**). The participation of ethyl free radicals in the reaction scheme is unlikely, since 75% of one ethyl group is used to produce a 1:1 molar mixture of ethylene and ethane, while butane is absent.

Support for the above reaction scheme is further derived by the fact that the 1:1 reaction of ZnH(Et)(pyr) with Bu^t-Pyca gives (**56**) in quantitative yield, while this reaction in the presence of $ZnEt_2(Bu^t$-Pyca) afforded (**56**), (**54**) and (**53**) in the same yields, but ethane (one equivalent) is produced and not ethylene.[125] Similar types of reactions do occur when R-DAB is treated with AlR_3.[125] *N*-Alkylation, *C*-alkylation and the occurrence of persistent $AlR'_2(R$-DAB)· radicals were observed. The reactions have not yet been extensively investigated and will not be further discussed here.

Finally we wish to draw attention to a very recent publication in which the reaction of $Fe(CO)_3(R$-DAB) and of $Fe(CO)_3(R$-Pyca) with alkynes in the presence of CO was reported.[170] The reaction leads for R = Pr^i and alkyne = $MeCO_2C{\equiv}CCO_2Me$ *via* an unstable intermediate to the product $Fe(CO)_3[Pr^i{-}N{=}C(H)C(H)N(Pr^i)C(O)(MeO_2C)C{=}CCO_2Me]$ (**57**), which has been investigated by X-ray crystallography. The mechanism of the reaction is not clear, but it is of interest to see that there is probably an analogy to the reaction of $Fe_2(CO)_9$ and R-DAB which produces imidazolone in addition to metal complexes, *i.e.* a CO group is bonded to an R-DAB molecule in a 1,4-fashion.[140]

It might well be that the alkyne added is captured by an intermediate complex containing CO inserted between an N atom of R-DAB and the $Fe(CO)_3$ fragment.

Metal–R-DAB complexes have been used for catalytic experiments involving alkenes. The experimental details, however, do no lead us to believe that metal-R-DAB complexes are involved in the actual reactions.[171]

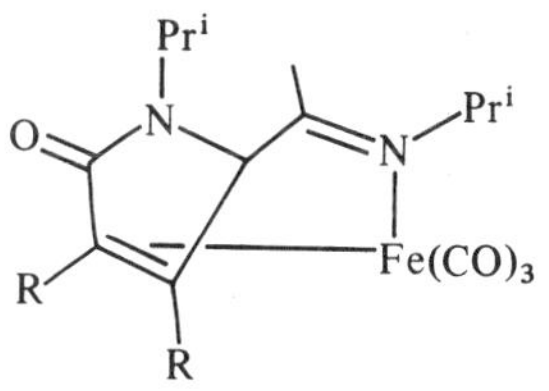

(57)

13.5.4 TETRAAZA-1,3-BUTADIENE METAL COMPLEXES

13.5.4.1 Syntheses of the Complexes M—R_2N_4

Closely related to the 1,4-diaza-1,3-butadiene ligand discussed in Section 13.5.3 is the tetraazadiene molecule RN=N—N=NR (R_2N_4, **58**) of which the coordination chemistry has been little explored. The first tetraazadiene metal complexes were reported only recently (1967).[172] This is due to the fact that the R_2N_4 ligand is unstable as a free molecule. Accordingly the R_2N_4 ligand has to be generated at the metal centre and subsequently stabilized by coordination.

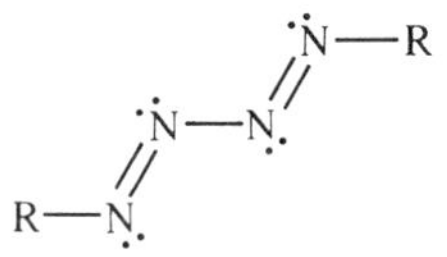

(58)

Table 5 contains the tetraazadiene–metal complexes reported to date. They all are derived from the Group VIIIB transition metals Fe, Co, Ni, Rh, Ir and Pt and have a 1/1 metal/R_2N_4 molar ratio (*cf.* **59–65** in Figure 4). Exceptions are the nickel complexes which in addition to various 1/1 complexes[173,174] also include 2/1 complexes $Ni(R_2N_4)_2$ (**66**).[175,176]

The greater part of the complexes have been synthesized by reacting a metal complex with excess of an organic azide (*cf.* equations 10 and 11). These reactions are often exothermic and the tetraazadiene–metal complexes are generally isolated from the reaction mixture in yields lower than 50%. The complexity of such mixtures is nicely illustrated by the products found in the reaction of $Fe_2(CO)_9$ with MeN_3 which produces (**67**) (20%) as well as three other identified products (equation 12).[172] The thermal reaction using an exact 1/2 $Fe_2(CO)_9/MeN_3$ molar ratio has been reported to produce N_2, (**67**), $Fe(CO)_5$, (**68**) and about 30–40% pyrophoric iron metal.[177]

$$Ni(cod)_2 \xrightarrow[-10\,^{\circ}C]{ArN_3} Ni(Ar_2N_4)_2 \quad \text{e.g. (66)} \tag{10a}$$

$$Ni(cod)_2 \xrightarrow[C_6F_5N_3]{-35\,^{\circ}C} Ni(cod)[(C_6F_5)_2N_4] \quad \text{(59)} \tag{10b}$$

$$M(\eta^5\text{-}Cp)_2 \xrightarrow[110\,^{\circ}C]{ArN_3} M(\eta^5\text{-}Cp)(Ar_2N_4) \tag{11}$$

M = Co, Ni $\qquad$ *e.g.* (61) and (62)

Equation (13) shows the use of a diazonium salt for the preparation of iridium–tetraazadiene complexes, *e.g.* (**65**), which are formed in addition to an aryldiazenido complex, *e.g.* (**69**).[187,194]

The mechanism of formation of a coordinated tetraazadiene ligand in equations (10) and (11) has been discussed extensively.[173,175,191,195] In Scheme 7 the proposed routes and intermediates have been summarized. One proposal (route a) involves initial formation of a monodentate, σ-N-coordinated R_2N_4 ligand (**71**) which then becomes bidentate σ,σ-N,N′-bonded (**73**) by attack of the non-coordinated N^{α} atom on the metal.[173] The second possibility (route b) comprises the intermediacy of metal–nitrene species (**70**) to which an organic azide coordinates (**72**), followed by a 3-N(azide)–1-

Table 5 Synthetic Routes and Some Structural Data of Metal 1,4-Disubstituted Tetraaza-1,3-butadiene Complexes

Compound	Route[a]	Isolated yield (%)	Bond distances (Å)[b]			Ref.[c]
			M—N	N^β—N^β	N^β—N^β	
Fe(CO)$_3$(R$_2$N$_4$), R = CH$_3$ (**67**), CD$_3$	RN$_3$/Fe$_2$(CO)$_9$ (r.t.)	5–20	1.83(3)	1.32(3)	1.32(3)	172, 177, (178)
Fe(CO)$_{3-n}$L$_n$(Me$_2$N$_4$), n = 1, L = PR′$_3$ with R′ = Me, Ph, OMe, OPh, Cy, MePh$_2$; n = 2, L = PPh$_3$ or L$_2$ = diphos	(**67**)/L (25–100°C)	d				177
n = 1 or 2, L = PR′$_3$ with R′ = Me, Ph, OMe, OPh	(**67**)/L (hν, r.t.)	d				177
n = 1–3, L = CNBu^t	(**67**)/nL (70°C; 3 d)	d				179
Co(η^5-Cp)(R$_2$N$_4$), R = Me, Ph, C$_6$F$_5$ (**61**), C$_6$H$_3$F$_2$-2,4, C$_6$H$_3$Me$_2$-2,6, C$_6$H$_4$Me-4	RN$_3$/Co(η^5-Cp)(CO)$_2$ (r.t.)	30–70	1.802(2) 1.819(2)	1.360(2) 1.355(2)	1.279(2)^e	180, 181 (182, 183)
	(RN$_3$/Co(η^5-Cp)$_2$ (110°C)	11				175
M(NO)(PPh$_3$)(R$_2$N$_4$), M = Rh, Ir with R = O$_2$SC$_6$H$_4$Me-4 (**62**)	RN$_3$/M(NO)(PPh$_3$)$_3$ (r.t.)	d				184
M(NO)(PPh$_3$)L(R$_2$N$_4$), M = Rh, Ir with R = O$_2$SC$_6$H$_4$Me-4, L = CO	(**62**)/L (r.t.)	d				184
[Ir(CO)(PPh$_3$)(R$_2$N$_4$)]BF$_4$, R = C$_6$H$_4$R′-4 with R′ = H, F (**65**), Cl, Br, CF$_3$ or OMe	(RN$_2$)BF$_4$/*trans* IrCl(CO)(PPh$_3$)$_2$ (r.t.)	7–45	1.941(13) 1.971(10)	1.400(16) 1.350(16)	1.270(16)^f	185, 186 (186, 187)
Ni(R$_2$N$_4$), R = C$_6$H$_4$R′-4 with R′ = Me, OMe, Cl and	RN$_3$/Ni(cod)$_2$ (−10°C)	20–30	1.851(2)	1.328(3)	1.309(3)^g	175, 176 (175, 176)
R = C$_6$H$_3$Me$_2$-3,5 (**66**)	RN$_3$/Ni(η^5-Cp)$_2$ (110°C)	15				176
R = C$_6$H$_4$Me-4	RN$_3$ (**62**) (110°C)	52				174, 175
Ni(η^5-Cp)(R$_2$N$_4$), R = C$_6$H$_4$Me-4 (**62**)	RN$_3$/Ni(η^5-Cp)$_2$ (110°C)	10	1.853(2) 1.843(2)	1.344(2) 1.346(2)	1.278(2)	174, 175 (174)
NiL$_2$(R$_2$N$_4$), R = C$_6$F$_5$, L$_2$ = cod (**59**)	RN$_3$/Ni(cod)$_2$ (−35°C)	79				173
R = C$_6$F$_5$, L = PR′$_3$ with R′ = Ph, Me, OMe and L$_2$ = bipy	(**59**)/L (0°C)	27–80				173
R = C$_6$H$_4$Me-4, L = CNBu^t	(**62**)/L (r.t.)	80				174, 175
	(**62**)/L (110°C)	11				176

Compound	Preparation[a]	Yield (%)				Ref.
R = C_6H_4Me-4, $C_6H_3Me_2$-3,5 $PtL_2(R_2N_4)$	(62) or (66)/$Ni(Bu^tNC)_2$	40–80				176, 188, 189
R = C_6H_4R'-4 with R′ = Me, Cl, NO_2 (60) and L = Bu^tNC, PEt_3 or L_2 = cod	$RN_3/Pt(cod)_2$ (r.t.)	30–40				190
	$RN_3/Pt_3(Bu^tNC)_6$	25				190
	$Ni(R_2N_4)_2/Pt(cod)_2/L$ (60°C)	40				188, 191
R = O_2SR' with R′ = C_6H_4Me-4 or Ph, L = PPh_3	$Pt(cod)(R_2N_4)/PEt_3$	30–40				190
	RN_3/PtL_4 or $PtL_2(C_2H_4)$ (r.t.)	d				192
$Pt(C_2Ph)_2(PEt_3)_2(R_2N_4)$ (64) R = $C_6H_4NO_2$-4	$RN_3/Pt(C_2Ph)_2(PEt_3)_2$ (r.t.)	15	2.11(1)	1.41(2) 1.40(1)	1.30(2)	193 (193)
$Pt(CHC(L)H(CH_2)_2CHCHCH_2CH_2)L(R_2N_4)$ R = $C_6H_4NO_2$-4 with L = PEt_3 (63)	$PEt_3/(60)$	62	2.169(5) 2.159(5)	1.385(7) 1.391(7)	1.268(8)	190 (190)

[a] Important reaction conditions are given in parentheses. [b] As obtained from X-ray crystal structure determinations which are taken from the references in parentheses. For N^α and N^β see (80) in text. [c] See note b. [d] Not specified. [e] For R = C_6F_5 (61). [f] For R′ = F (65). [g] For R = $C_6H_3Me_2$-3,5 (66).

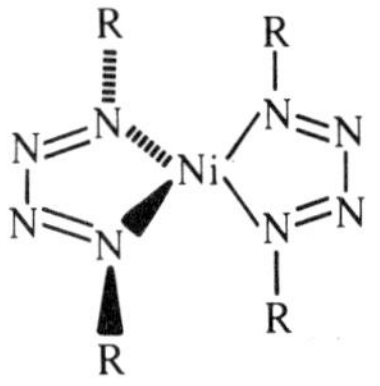

(66)[175,176] R = $C_6H_3Me_2$-3,5

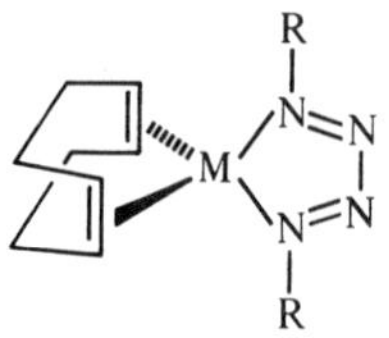

(59)[173] M = Ni, R = C_6F_5
(60)[188,190,191] M = Pt, R = $C_6H_4NO_2$-4

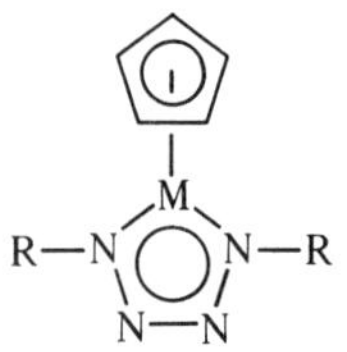

(61)[181,183] M = Co, R = C_6F_5
(62)[174] M = Ni, R = C_6H_4Me-4

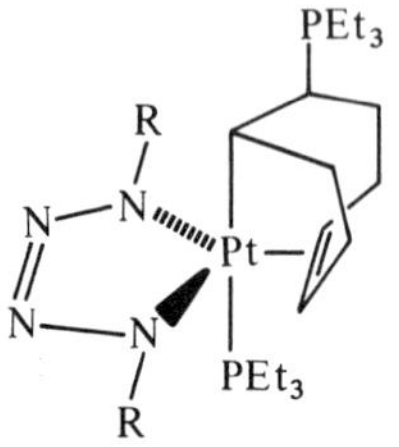

(63)[190] R = $C_6H_4NO_2$-4

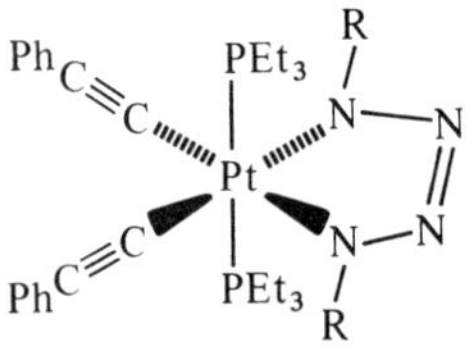

(64)[193] R = $C_6H_4NO_2$-4

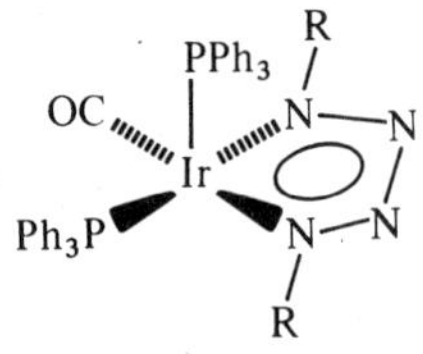

(65)[22] R = C_6H_4F-4

Figure 4 Schematic structures of some tetraazadiene–metal complexes (see also Table 5)

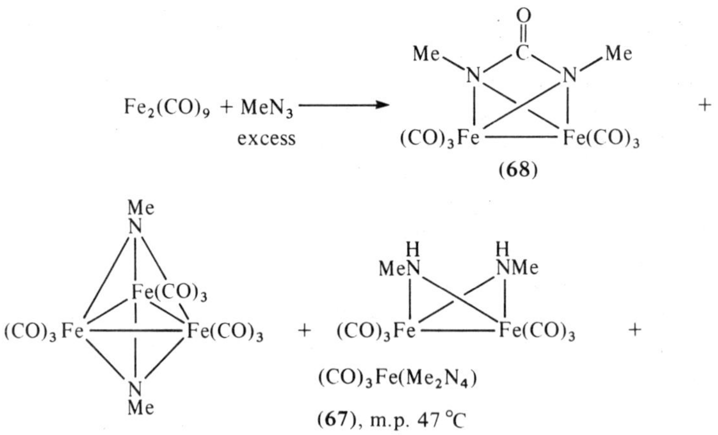
Fe2(CO)9 + MeN3
excess
(68)
+
(CO)3Fe(Me2N4)
(67), m.p. 47 °C

(12)

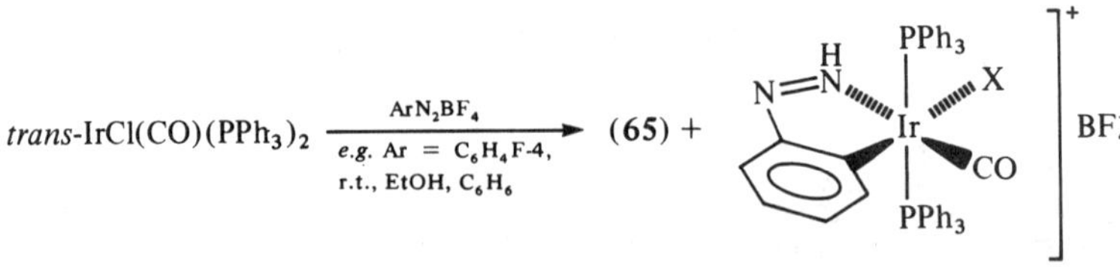
trans-IrCl(CO)(PPh3)2
ArN2BF4
e.g. Ar = C6H4F-4,
r.t., EtOH, C6H6
(65) +
BF4−

(69)

(13)

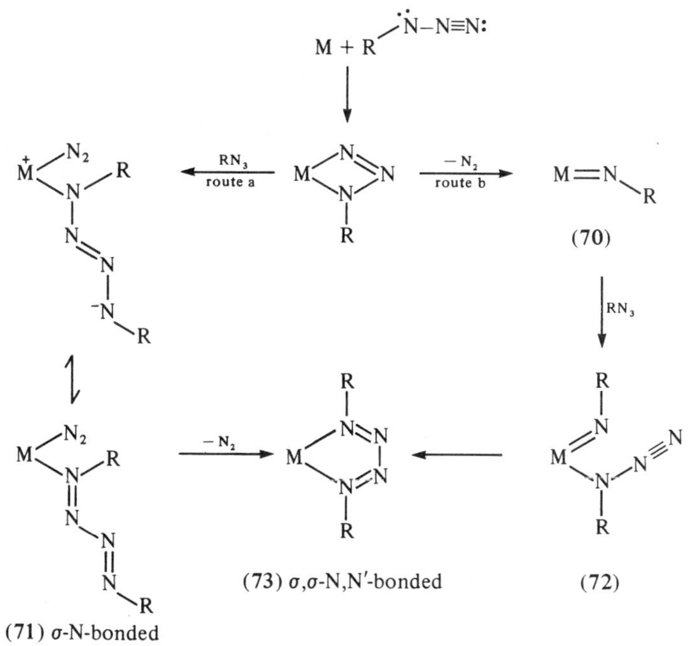

Scheme 7 Proposed mechanisms for the formation of the tetraaza-1,3-diene ligand at a metal center

N(nitrene) coupling leading to the metal–tetraazadiene species **(73)**.[191,195] Reactions which can be considered as models for the conversions of the intermediates **(70)**–**(72)** have been found.

The intermediate formation of nitrene species is suggested by the other products formed in the $Fe_2(CO)_9/MeN_3$ reaction (equation 12), *e.g.* **(68)** and $Fe_3(CO)_9(NMe)_2$.[172,195] Moreover, reactions of $Ni(\eta^5\text{-Cp})(R_2N_4)$ with $R'N_3$ lead not only to $Ni(R_2N_4)(R'_2N_4)$ but also to $Ni(\eta^5\text{-Cp})(RN_4R')$ containing a mixed diorganotetraazadiene ligand. The formation of the latter species supports the view that the tetraazadiene ligand can break up into an intermediate **(72)** containing a coordinated nitrene and organoazide ligand RN_3[184,191] which can exchange with $R'N_3$.[191] Furthermore the mixed $Ni(R_2N_4)(R'_2N_4)$ complex undergoes an intramolecular rearrangement of the coordinated R_2N_4 ligands upon heating, leading to $Ni(RN_4R')(R_2N_4)$, $Ni(RN_4R')_2$ and $Ni(RN_4R')(R'_2N_4)$ in a 2:1:2 molar ratio.[191] These products are also a strong support for the reversibility of the conversion **(72)** → **(73)**.

The reaction **(71)** → **(73)** is evidenced by the observation that tetraazadiene ligands can be transferred intact under mild conditions from $Ni(R_2N_4)_2$ to another nickel or platinum center, *e.g.* equations (14) and (15).[188] The heterodinuclear PtNi intermediate **(74)** could be isolated. Available spectroscopic data pointed to a structure with the R_2N_4 ligand bridging the Ni and Pt centers in a planar *s-trans* conformation *via* the N^α atoms. Various other structures were considered of which one comprises coordination of the lone pair of N^β of the chelate-bonded R_2N_4 ligand in $Ni(R_2N_4)_2$ to the PtL_2 unit. Also, this proposal may account for the structure of **(74)**[188] and is similar to the proposed interaction of BF_3 with $Fe(CO)_3(R_2N_4)$[179] (*vide infra*). Heating of **(74)** leads to transfer of one complete R_2N_4 ligand from Ni to Pt affording $Pt(R_2N_4)(Bu^tNC)_2$ **(76)**. Likewise the reaction of $Ni(R_2N_4)_2$ ($R = C_6H_3Me_2$-3,5) with $Ni(cod)_2$ in the presence of Bu^tNC leads to the formation of $Ni(R_2N_4)(Bu^tNC)_2$ in high yields (equation 15).[188]

The reactions of metal salts with organic azides RN_3 appear to be strongly dependent on the nature of R which can be accounted for by the view that both routes a and b in Scheme 7 may be

$$Ni(R_2N_4)_2 \xrightarrow[\text{excess } Bu^tNC]{Pt(cod)_2} \underset{\textbf{(74)}}{Ni(R_2N_4)_2Pt(Bu^tNC)_2} \xrightarrow{8\text{ h, }60\,^\circ C} \begin{cases} \underset{\textbf{(75)}}{Ni(R_2N_4)_2Pt(Bu^tNC)_2} \\ Pt(R_2N_4)(Bu^tNC)_2 \\ \textbf{(76)}\ 40\%,\ R = C_6H_4Me\text{-}4 \end{cases} \qquad (14)$$

$$Ni(R_2N_4)_2 \xrightarrow{Ni(cod)_2} [\text{brown complex}] \xrightarrow[\text{excess } Bu^tNC]{8\text{ h, }60\,^\circ C} \underset{\textbf{(77)}\ 80\%,\ R\ =\ C_6H_3Me_2\text{-}3,5}{Ni(R_2N_4)(Bu^tNC)_2} \qquad (15)$$

operative. When R is a strongly electron withdrawing group, *e.g.* C_6F_5, route a seems the dominating pathway[173] because route b would require nucleophilic attack by the nitrene nitrogen on the 1,3-dipolarophile RN_3 which would be expected to be rather slow. It must be noted that further reaction of $Ni(cod)[(C_6F_5)_2N_4]$ with $C_6F_5N_3$ to the bis-species $Ni[(C_6F_5)_2N_4]_2$ has not been observed. In contrast aryl groups with electron donating or mildly electron withdrawing substituents produce bis-products $Ni(R_2N_4)_2$ and for these reactions route b seems more likely, not only for the formation of the first coordinated R_2N_4 ligand, but also for the second one[190] (*cf.* the reactions of $Ni(R_2N_4)_2$ with $R'N_3$[191] discussed above). In various cases the formation of the R_2N_4 ligand can only be accomplished when the strong electron withdrawing NO_2 substituent is used, *e.g.* the formation of $Pt(C_2Ph)_2(PEt_3)_2(R_2N_4)$ **(64)**.[193]

13.5.4.2 Reactivity of the M—R₂N₄ Complexes

The greater part of the tetraazadiene–metal complexes are extremely stable to thermal decomposition as well as to further reaction with O_2 and H_2O. For example, $Co(\eta^5\text{-Cp})(Ph_2N_4)$ is stable in air and can be sublimed at 180 °C/1 mmHg.[181] The $Ni(R_2N_4)_2$ complexes can be refluxed in toluene for days in air without appreciable decomposition.[176] In particular this stability is striking when compared with the extreme sensitivity of the corresponding $Ni(RN=CHCH=NR)_2$ complexes (see Section 13.5.3.4.3.vi).

Only a few reactions involving conversion of the coordinated tetraazadiene ligand have been reported. $Pt(PPh_3)_2(R_2N_4)$ ($R = 4\text{-MeC}_6H_4SO_2$) reacts with HCl to give $PtCl_2(PPh_3)_2$, RN_3 and RNH_2.[196] A similar conversion of the R_2N_4 ligand was observed when reacting $M(NO)(PPh_3)(R_2N_4)$ ($R = 4\text{-MeC}_6H_4SO_2$; M = Ir or Rh) with HCl/Et_2O.[184]

The reaction of $Ni(R_2N_4)_2$ with Bu^tNC which results in the formation of $Ni(R_2N_4)(Bu^tNC)_2$ (equation 16) is the first example of a substitution of an R_2N_4 ligand by direct reaction with a neutral ligand. This reaction, however, requires drastic conditions.[176] Other examples are (i) the transfer of a complete R_2N_4 ligand from $Ni(R_2N_4)_2$ to another nickel or platinum center (see equations 14 and 15),[188] (ii) the extensive exchange and rearrangement processes taking place when a mixture of two different $Ni(R_2N_4)_2$ complexes is refluxed in toluene,[191] and (iii) the reaction of $Pt(cod)(R_2N_4)$ or $Ni(\eta^5\text{-Cp})(R_2N_4)$ with aryl azides $R'N_3$ which produces platinum and nickel complexes with non-symmetrical RN_4R' ligands.[191] The proposed mechanism for the latter type of reaction is shown in Scheme 8.

$$Ni(R_2N_4)_2 \xrightarrow[100\ ^\circ C]{2\,Bu^tNC(=\,L)} \quad \underset{R}{\overset{R}{\left[\begin{array}{c} L \\ L \end{array}\!Ni\!\begin{array}{c}N{=}N \\ N{=}N\end{array}\right]}} \quad \xleftarrow[25\ ^\circ C]{2\,Bu^tNC(=\,L)} \quad Ni(\eta^5\text{-Cp})(R_2N_4) \qquad (16)$$

By a novel photochemical process the intense colored cobalt *N*-phenyl-*O*-benzoquinone diimine complex **(78)** is formed which involves the irradiation of thermally stable $Co(\eta^5\text{-Cp})(Ph_2N_4)$ (equation 17). This unusual transformation most probably occurs *via* an intramolecular pathway.[180,197]

$$\xrightarrow[-\,N_2]{h\nu} \qquad (17)$$

(78)

Various reactions have been reported which comprise ligand substitution of L in complexes of the type $ML_n(R_2N_4)$; see Table 5. Thermal CO substitution in $Fe(CO)_3(Me_2N_4)$ proceeds readily to form monosubstituted products $Fe(CO)_2L(Me_2N_4)$ with various types of ligands (equation 18). Bis- and tris-substituted products are also observed in the case of Bu^tNC. The substitution proceeds by a second-order process with a rate law that is first order in $Fe(CO)_3(Me_2N_4)$ and in the entering ligand. It has been found that the rate is strongly dependent on the basicity as well as the size of L, the polarity of the solvents and to an even greater extent on solvents or reagents that can interact

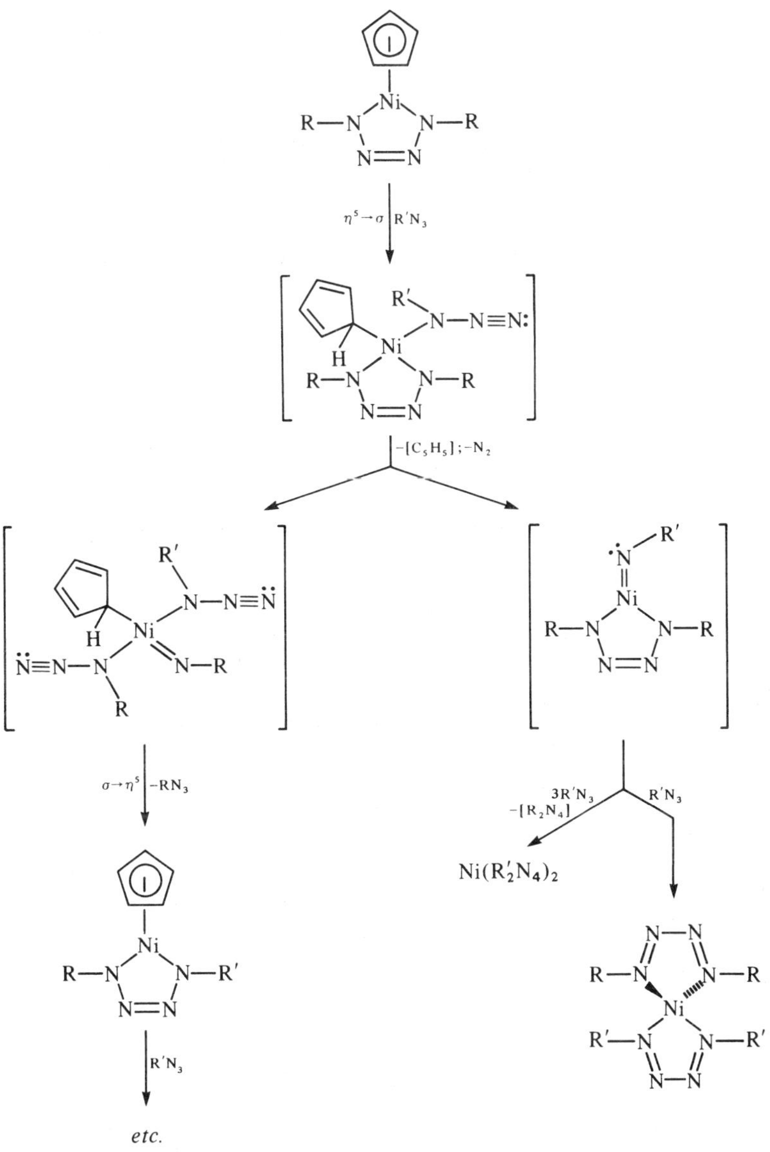

Scheme 8 Proposed pathways leading to all observed products formed in the reaction of Ni(η^5-Cp)(R$_2$N$_4$) with R'N$_3$[191]

with the lone pairs on the β-nitrogens (*cf.* **79**) of the Me$_2$N$_4$ ligand. On the basis of these observations an associative mechanism for these substitution reactions is proposed.[179]

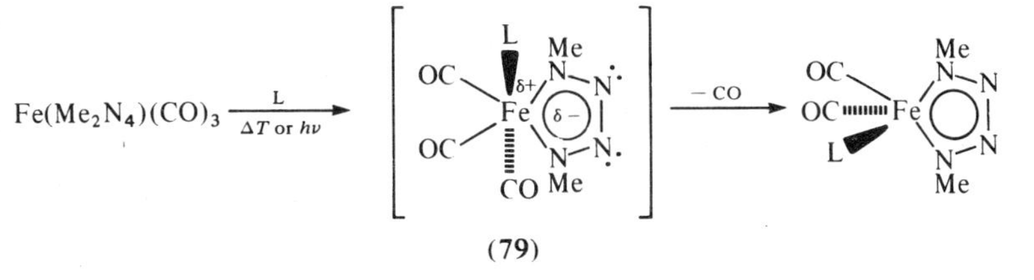

$$\text{Fe(Me}_2\text{N}_4)(\text{CO})_3 \xrightarrow[\Delta T \text{ or } h\nu]{L} \quad \text{(79)} \quad \xrightarrow{-\text{CO}} \quad (18)$$

(79)

L = phosphine, arsine, phosphite, pyridine or isocyanide

Photosubstitution of CO in Fe(CO)$_3$(Me$_2$N$_4$) and Fe(CO)$_2$L(Me$_2$N$_4$) complexes proceeds *via* a dissociative mechanism and allows the synthesis of a variety of mixed ligand iron tetraazadiene complexes (see Table 5).[177]

The cod ligand in Ni(cod)[(C$_6$F$_5$)$_2$N$_4$] can be readily displaced by various phosphine ligands to give NiL$_2$[(C$_6$F$_5$)$_2$N$_4$] complexes (see Table 5).[173] Likewise Ni(η^5-Cp)(R$_2$N$_4$) reacts with Bu^{t}NC to Ni(Bu^{t}NC)$_2$(R$_2$N$_4$) under mild conditions,[174] whereas the corresponding cobalt compound (R = C$_6$F$_5$ or C$_6$H$_4$Me-4, *cf.* **61** is unreactive.[175] Also Pt(cod)(R$_2$N$_4$) (R = C$_6$H$_4$R'-4 with R' = Me,

Cl, NO_2) undergoes cod substitutions with isocyanides and phosphines.[190] However, when R′ is the strongly electron withdrawing NO_2 group, reaction with PEt_3 leads to nucleophilic attack on the coordinated cod ligand affording (63) instead of cod substitution.

13.5.4.3 Structural and Bonding Aspects

With the report of the structural data of the first tetraazadiene complex, $Fe(CO)_3(Me_2N_4)$ (58),[172] the bonding in the unsaturated metallacycle became a major point of discussion.[173,174,176,177,179,182,183,185,187,190,196–199]

In principle the metallacycle may adopt either the 1,3-diene configuration (80), in which the R_2N_4 ligand acts as a neutral four electron donor, or the reduced 2-ene configuration (81), in which it is a formally double negative charged ligand and accordingly a two electron donor.

(80) M^{n+} (81) $M^{(n+2)+}$

In (81) the N^α atoms are σ-bonded to the metal. They may have either a tetrahedral configuration (sp^3-hybridization) or a configuration in which the N^α-N^β, N^α-M and N^αR bonds are in one plane (sp^2-hybridization). In the planar situation the N^α lone pairs resides in an orbital perpendicular to this plane and offers the possibility of π-delocalized bonding. The difference between these two extreme bonding situations is that in going from canonical structures (80) to (81) formally two electrons move from metal *d*-orbitals into the LUMO of structure (81). This orbital is bonding between N^β atoms and antibonding between N^α—N^β pairs. Accordingly, distinctly different sets of N—N distances are expected for tetraazabutadiene complexes belonging to either one of these extreme descriptions.[176] Indeed the N—N distances in the tetraazabutadiene complexes studied so far by X-ray structure determinations (see Table 5) seem to support these ideas.

The crystal structure of $(Me_3Si)_2N$—N=N—$N(SiMe_3)_2$ can be used as a model for obtaining reasonable bond length estimates for N—N (1.394(5)) and N=N (1.268(7) Å).[200] For example, while the $Ni(R_2N_4)_2$ complexes represent mainly form (80) (neutral ligand), the $M(\eta^5$-Cp)(R_2N_4) complexes (61) and (62), the Ir^{III} complex (65) and the Pt^{II} complexes (63) and (64) are clearly representatives of the form (81) (dianionic ligand, *cf.* the listed N—N distances in Table 5). The type of R in these complexes may have a further effect on the distribution of electron density in these complexes. In $Ni(R_2N_4)_2$ (66) the observation that the aryl rings in the R_2N_4 unit are coplanar with each other and with the NiN_4 ring points to the presence of a conjugated π-system in the $Ni(R_2N_4)$ units.[176] The alteration of the pattern of unsaturated bonds in the $M(R_2N_4)$ chelate by π-back-bonding is reflected by, for example, the short Co—N distances in (61).[182] Furthermore, the importance of additional mesomeric resonance structures in the bis-aryltetraazadiene–metal complexes is indicated by the structural features of (63). The involvement of the strongly electron withdrawing *p*-NO_2 substituents explains why the aryl rings are found to be coplanar with the PtN_4 system which makes (63) the most pronounced example of type (81) to date.[176] In the absence of such additional conjugation, *e.g.* if the substituent exerts only an inductive effect as in C_6H_4F-4, the aryl rings are no longer coplanar with the $M(R_2N_4)$ unit, *cf.* the structures of (62) and (65) with those of (63), (64) and (66) (Table 5).

The formation of (63) is also a nice example of the influence of the nature of R in the Pt(1,5-cod)–(R_2N_4) complexes on the sensitivity of the coordinated cod ligand for nucleophilic attack. Only if R contains the strongly electron withdrawing NO_2 group is reaction of the alkenic bonds with nucleophiles observed.[190]

Finally, extensive spectroscopic (UV, FES, ESCA, ESR, ^{13}C and ^{31}P NMR) studies as well as computational (X-α and Hückel) studies have been carried out to help the interpretation of the bonding patterns (extent of electron delocalization, formal oxidation state of the metal) in these tetraazadiene–metal complexes.

13.5.5 METAL COMPLEXES OF TRIATOMIC HETERO ANIONS: N_3^-, $(NCO)^-$, $(NCS)^-$, $(NCSe)^-$ AND $(NCTe)^-$*

13.5.5.1 Introduction

In this section some aspects of the coordination chemistry of the triatomic anions listed in Table 6 will be discussed. They all contain at least one N atom and are usually classified as pseudohalides. The latter term was first suggested by Birkenbach and Kellerman in 1925 to describe polyatomic groups whose chemical properties were very similar to those of the halides.[201]

There is considerable and widespread interest in the metal complexes of these anions and current research topics comprise for example: (i) the spectroscopic study of the binding in these anions (linkage isomerism) and their complexes, (ii) the synthesis of regular polymers of their transition metal complexes and study of the semiconducting properties of these polymers, (iii) the use of the pseudohalides in pharmacological (*e.g.* low toxicity of —SCN) and biochemical studies (easy complexation of SCN^- to metals), and (iv) the use of the activation of these triatomic anions by coordination to metals for their selective conversion in organic synthesis.

In the context of this review it is only possible to discuss a selected number of compounds. Based on these examples an idea of the scope and field of applications of these versatile triatomic anions can be provided. The fact that these and not other complexes are discussed is not because the latter lack importance. They can be found in reviews that are given in ref. 202–209. An account will be given of (i) the bonding aspects of the pseudohalides to transition metals based on reliable structural data and (ii) some synthetic aspects of the pseudohalides with special emphasis on their conversion in the coordination sphere of a metal.†

13.5.5.2 The Pseudohalides

Ionic azides with alkali metal (*e.g.* NaN_3)[210] or large organic (*e.g.* NR_4^+) counterions are chemically stable. Controlled decomposition occurs at higher temperatures, while they are also mechanically stable.[206]

The first elements of Group VI form stable complex ions with CN^- to make $(OCN)^-$ *via* KCN oxidation with PbO, and $(SCN)^-$ and $(SeCN)^-$ *via* melting KCN with either S or Se. The preparative procedures have been well documented.[211] In contrast the salts of tellurocyanate $(NCTe)^-$ with small polarizable cations (*e.g.* alkali metal cations) appear to be unstable. Only when large, weakly polarizable counterions are used can stable salts be isolated (*e.g.* $[NEt_4][NCTe]$).[212] The potassium pseudohalides readily undergo acid hydrolysis affording either COX (X = O for cyanate or S for thiocyanate) or give H_2X and free X (X = S, Se or Te) depending on the conditions as well as on the nature and concentration of the acid.

The molecular geometry of the anions has been subject of extensive spectroscopic and computational studies. The results have been reviewed.[203–205,213] Table 7 lists the data obtained for the ions in the solid. All pseudohalides appeared to be essentially linear.

13.5.5.3 Bonding Aspects

13.5.5.3.1 General remarks

As indicated in Table 6 the triatomic pseudohalides $(NCX)^-$ (X = O, S, Se or Te) are ambidentate

Table 6 Ambident Triatomic Anions: Pseudohalides[a]

Azide	NNN		
Cyanate[b]	OCN	Isocyanate	NCO
Thiocyanate	SCN	Isothiocyanate	NCS
Selenocyanate	SeCN	Isoselenocyanate	NCSe
Tellurocyanate	TeCN	Isotellurocyanate	NCTe

[a] When 'iso' is placed in parentheses, *e.g.* (iso)cyanate, the point of attachment of the pseudohalide to the metal(s) is not known. [b] The isoelectronic fulminate ion—ONC will be considered in Chapter 12.1.

---

* In this section, when 'iso' is placed in parentheses, *e.g.* (iso)cyanate, the point of attachment of the pseudohalide to the metal(s) is not known. Similarly, this is implied by the use of $(NCO)^-$ to represent NCO^- and OCN^-.

† Thanks are due to Mr H. L. Aalten for assistance in collecting literature data.

Table 7 Molecular Geometry of the N_3^- and $(NCX)^-$ Anions[a]

Anion $N\overset{a}{-}Y\overset{b}{-}X$ $\scriptstyle\alpha$	Bond distances (Å)		Angles (°)	
	a	b	α	Ref.
NNN[b]		1.167	180	210
NCO	1.17	1.23	180	c
NCS[d]	1.149 (14)	1.689 (13)	178.3 (1.2)	215
NCSe[d]	1.170 (26)	1.829 (25)	178.8 (2.5)	214
NCTe[e]	1.07 (1)	2.02 (1)	175 (1)	216

[a] X-Ray data unless otherwise stated. [b] α-NaN$_3$. [c] See ref. 204a, p. 236; calculated data.
[d] KCNX (X = S or Se). [e] [PPN][TeCN].

ligands. They can bind to a metal through both the N ('normal') or the X ('iso') end. Moreover, they can function as bridging ligands using either both ends, N and X, or only one end, *i.e.* N or X. Numerous spectroscopic investigations (*e.g.* IR and Raman, *cf.* refs. 203 and 204) have been carried out to elucidate the mode of attachments of the pseudohalide anions in new metal complexes. These studies have been particularly complicated by the fact that the actual bonding modes depend on a variety of factors, *e.g.* the type of the other ligands bonded to the metal and the nature of the solvent and anions, while a change of bonding mode (N–X switch, linkage isomerization[204b,217]) is often observed on going from the solid state to the solution. Accordingly, based on spectroscopic results, different, often conflicting, structural proposals have been put forward. It is through the recent studies of the structures in the solid by X-ray structure determinations that a more clear-cut picture of the bonding of these anions could be developed. In Table 8 the bonding modes known to date together with some examples are listed.

Some general conclusions can be drawn from this table. To date metal complexes containing the $(NCTe)^-$ anions, have not been reported. The most versatile in their bonding behavior seem to be the (iso)cyanate and azide anions while most of the bonding modes known for the last two pseudohalides in the table have also been established for (iso)cyanate. Notably missing are compounds containing an O-bonded isocyanate anion although the occurrence of this bonding type was suggested based on spectroscopic data (*cf.* ref. 204).

Examples of 'side on' bonding, *i.e.* involvement of the π-electrons, have not been established. This type of metal–ligand bonding is often encountered in metal complexes of the organic pseudohalides such as RNCS.

The azide ion is linear and symmetric, possessing equal N—N distances (see Table 7).[205] The covalent azides, *i.e.* organic azides, are linear, but possess unequal distances of which the (R—)N—N bond is longer than the —(N)—N—N one. This, together with the non-linear R—N—N angle, supports the view that the main contribution to the ground state geometry of the organic azides is provided by the two canonical structures (**82**) and (**83**). In contrast to this straightforward picture for the organic azides the valence bond structures for the coordinated N_3^- ligand is not as predictable (*vide infra*).

It has been calculated that the resonance energy of ionic N_3 is almost twice as high as that in the covalently bonded form. This explains why the latter complexes often are heat, light or shock sensitive whereas the ionic azides are reasonable stable.[208]

$$\diagup N\overset{+}{=}\overset{+}{N}=\overset{-}{N} \qquad \longrightarrow \qquad \diagup \overset{-}{N}-\overset{+}{N}\equiv N \qquad M-\overset{+}{N}\equiv C-\overset{-}{S} \qquad \overset{-}{N}=C=S$$
$$\qquad\qquad\qquad\qquad\qquad\qquad\qquad\qquad\qquad\qquad\qquad\qquad\qquad\qquad M$$

(82)   (83)   (84a)   (84b)

$$M\diagup^{S-C\equiv N} \qquad \longrightarrow \qquad M\diagup^{\overset{+}{S}=C=\overset{-}{N}}$$

(85a)   (85b)

The preference of the $(OCN)^-$ anions for N-bonding is difficult to explain, because INDO calculations predict an almost equal electron density on N and O.[218] In contrast N- or S-bonding of the $(SCN)^-$ anions can be rationalized in terms of the hardness (N-bonding) or softness (S-bonding) of the metal center.[219] However, many examples are available (see structures **102–104**) that point to a crucial contribution of steric[220] as well as electronic[221] factors. In a crowded complex steric interactions may cause N-coordination of the NCS ligand because the M—N—C arrangement (**84**; bond angle about 170°) is almost linear, whereas the alternative M—S—C bonding (**85**)

Table 8 Bonding Modes of the Ambident Pseudohalides

Ligand	Mode	Parameters[a]	Type	Example	Ref.
$(N_3)^-$	Terminal	$M \overset{a}{-} N \overset{b}{-} N \overset{d}{-} N$, α, β	I	$[Co(N_3)(trenen)][NO_3] \cdot H_2O$	218
	Bridging				
	$1,1$-μ	$M, M \overset{a}{-} N \overset{b}{-} N \overset{c}{-} N$, α, β	II	$[NEt_4][Mn_2(N_3)_3(CO)_6]$	219
	$1,1,1$-μ	$M, M, M \overset{a}{-} N \overset{b}{-} N \overset{c}{-} N$, α, β	III	$[Me_3Pt(N_3)]_4$	220
	$1,3$-μ	$M \overset{a}{-} N \overset{b}{-} N \overset{c}{-} N \overset{d}{-} M$, α, β, γ	IV	$[BPh_4][Cu_2(N_3)_2(Me_5dien)_2]$	221
$(NCX)^-$ $X = O, S, Se, Te$	Terminal *via* N	$M \overset{a}{-} N \overset{b}{-} C \overset{c}{-} X$, α, β	V	$Ti(\eta^5\text{-}Cp)(NCO)_2$ $Nb(OEt)_2(NCS)_2(dbm)$ $Ni(NCSe)_2(DMF)_4$	222 223 224
	via X^b	$M \overset{a}{-} X \overset{b}{-} C \overset{c}{-} N$, α, β	VI	$Pd(SCN)_2(dppm)$ $K[Co(SeCN)_2(DH)_2]$	225 226
	Bridging $1,1$-μ-N	$M, M \overset{a}{-} N \overset{b}{-} C \overset{c}{-} X$, α, β	VII	$Cu(NCO)_2(2,4\text{-lutidine})$ $[NBu_4^n]_3[Re_2(NCS)_{10}]$	227 228
	$1,1$-μ-X	$M, M \overset{a}{-} X \overset{b}{-} C \overset{c}{-} N$, α, β	VIII	$AuOs_3(NCS)(CO)_{10}(PPh_3)$	229
	$1,3$-μ	$M \overset{a}{-} X \overset{b}{-} C \overset{c}{-} N \overset{d}{-} M$, α, β, γ	IX	$[Ni_2(NCO)_2(tren)_2][BPh_4]_2$ $Mn(acac)_2(NCS)$	230 231
	$1,1,3$-μ	$M, M \overset{a}{-} X \overset{b}{-} C \overset{c}{-} N \overset{d}{-} M$, α, β, γ	X	$Ni(NCS)_2$	232

[a] Parameters a, b, c and d are the respective bonding distances (Å) and α, β and γ the bond angles (°) at the particular (hetero) atom. [b] 'Iso' form.

always has a bond angle close to $100°$. Based on the latter value it is assumed that the S atom is mainly sp^3-hybridized, *i.e.* canonical structure (**85a**) as a main contributor to the ground state.

Comparison of the canonical structures (**84**) and (**85a**) provides an assignment criterion of the $\nu(C\text{—}S)$ bond in the IR spectra for N- or S-bonded NCS.[204b,217] These structures suggest that the $\nu(C\text{—}S)$ should be higher in the N-bonded form (**84**) than in the S-bonded form (**85a**). This has generally been found, *i.e.* $860\text{–}780\,cm^{-1}$ for N-bonding and $720\text{–}690\,cm^{-1}$ for S-bonding, while the former bonds are usually more intense. However, care should be taken in using these assignment criteria in view of the easy occurrence of the N—S linkage switches in metal (iso)thiocyanate complexes.[217]

Electronic effects become apparent in the M—NCS/M—SCN linkage switches observed in a series of Pd^{II} (iso)thiocyanate complexes. Ligands positioned *trans* to the pseudohalide and that are suited to accept electron density from the metal into empty orbitals (backbonding) stabilize the Pd—NCS linkage isomer. However, this rationale is contradicted by the trend in Co—(NCS) bonding in a series of cobalt complexes (see ref. 204b for a review).

A coordination behavior similar to that of $(NCS)^-$ is observed for the $(NCSe)^-$ anion (for a review see ref. 202) although the influence of electronic effects of coligands on the type of bonding seems to be less important (*cf.* linkage isomers of $[Pd(NCSe)(Et_4dien)][BPh_4]$ with a linear PdNCSe linkage *vs.* $[Pd(SeCN)(Et_4dien)][BPh_4]$ containing bent PdSeCN[233]).

X-Ray structural determinations have shown a variety of bridging modes for the pseudohalides. A general rule seems to be that for NCS and NCSe, which show a tendency for linkage isomerization, end-to-end bridging is preferred. However, for the $(NCO)^-$ ions, which have a strong preference for terminal N-bonding bridging *via* only one atom is often found, while for the symmetrical N_3^- ion both end-to-end and bridging *via* a terminal N atom are observed.

Examples of complexes of the various pseudohalides with transition metals for which the structure has been confirmed by X-ray analysis will be discussed in more detail in the next section. For preparative methods see Section 13.5.5.4.

13.5.5.3.2 Transition metal–pseudohalide complexes

(i) Ti, Zr and Hf complexes (Table 9a)

Both terminal[234,235] (*e.g.* **86**) and 1,1-μ-$N^{236-238}$ (*e.g.* **87**) bonding is found for N_3^- in a series of titanium complexes. The N—N bond distances b and c in (**86**) are equal, which is unexpected for a covalently bonded azide (see Section 13.5.5.3.1). This has been explained with the increasing ion charge in the series $TiCl_3(N_3)$ polymer < $[TiCl_4(N_3)_2]^{2-}$ dimer (**87**) < $[TiCl_4(N_3)_2]^{2-}$ monomer (**86**) which favors canonical structure (**82**) of the azide. In (**82**) the negative charge is located on the 'outer' N atom, *i.e.* on the outside of the ion (**86**).[234]

(86)   (87)   (88)

(89)   (90)

Table 9a Complexes of Ti, Zr and Hf

Compound	Type[a]	Parameters[b]					Ref.
		a	b	c	α	β	
$[AsPh_4]_2[TiCl_4(N_3)_2]$	I	2.006	1.166	1.169	128.1	174.8	234
$Ti(\eta^5\text{-}Cp)_2(N_3)_2$	I	2.03	1.18	1.10	137		235
$[AsPh_4]_2[TiCl_4(N_3)]_2$	II	2.116	1.21	1.18	122.2	178	236
$TiCl_3(N_3)^c$	II	2.105	1.236	1.123	124.9	179.2	237
$[PPh_4]_2[ZrCl_4(N_3)]_2$	II	2.203	1.196	1.307	124[d]	179.9	238
$Ti(\eta^5\text{-}Cp)_2(NCO)_2$	V	2.012	1.154	1.188	173.7	179.5	222
$Zr(\eta^5\text{-}Cp)_2(NCO)_2$	V	2.110	1.148	1.175	175.0	179.2	222
$Ti(\eta^5\text{-}Cp)_2(NCS)_2$	V	2.021	1.17	1.611	177.5	179.2	240
$Zr(NCS)_4(bipy)_2$	V	2.182	1.160	1.590	163.1	179.3	239

[a] See Table 8. [b] For a, b, c (Å) and α, β (°), see Table 8. [c] Polymer. [d] Calculated from the Zr—N—Zr angle.

Comparison of the titanium–pseudohalide complexes (**86**) and (**88**) shows nicely the different bonding of the pseudohalide, *i.e.* bent Ti—N—N *vs.* linear Ti—N—C of the N-bonded isocyanate and isothiocyanate. Table 9a shows that in isostructural $M(\eta^5\text{-}Cp)_2(NCO)_2$ (M = Ti, Zr)[222] the N—C (b) bonds tend to be shorter than the C—O (c) bonds. This may point to a larger contribution of canonical structure (**89**) which has some C—O triple bond character in the bonding.

Finally, it is interesting to note the eight-coordination about Zr in the Zr–isothiocyanate complex which contains four linear N-bonded —NCO ligands.[239]

(ii)   V, Nb and Ta complexes (Table 9b)

Of this group of metals only a limited number of X-ray structure determinations have been reported.

Table 9b   Complexes of V, Nb and Ta

Compound	Type[a]	Parameters[b]					Ref.
		a	b	c	α	β	
[PPh₄][NbCl₅N₃]	I	1.92	1.47	1.05	138	173	241
[TaCl₄N₃]₂[c]	II	2.18	1.18	1.15	121.4	174.7	242
[NH₄]₂[VO(NCS)₄]·5H₂O	V	2.04	1.17	1.64	169.8	170.8	243
[AsPPh₄]₂[NbO(NCS)₅]	V	2.27[d]	1.16	1.67	172	171	244
		2.09[e]	1.18	1.57	168	175	
Nb(OEt)₂(NCS)₂(dbm)[f]		2.104	1.152	1.584	175.1	178.6	223

[a] See Table 8. [b] For *a*, *b*, *c* (Å) and *α*, *β* (°), see Table 8. [c] See text; only one set of structural data shown. [d] NCS *trans* to O. [e] NCS in equatorial sites.

The azide fragment in [PPh₄][NbCl₅N₃] is terminally bonded,[241] while the 1,1-μ-N-bonded form is observed in [TaCl₄N₃]₂.[242] The latter compound consists in the solid in two molecular geometries which differ with regard to the angle between the bridging linear N₃ fragment and the central Ta₂N₂ four-membered ring (17° and 2°, respectively).

The structures of the V[243] and Nb[223,239,244,245] thiocyanate complexes all contain terminal N-bonded isothiocyanate, *e.g.* square-pyramidal [NH₄]₂[VO(NCS)₄]·5H₂O[243] with the O ligand in axial position.

The NCS fragment *trans* to the oxo ligand in [AsPh₄][NbO(NCS)₅] is bonded *via* the longest Nb—N distance (2.27 Å) reported so far.[244] The other Nb—N distances are fully compatible with other known bond lengths[223,245] in, for example, Nb(OEt₂)(NCS)₂(dbm) (**91**).

(91)

(iii)   Cr, Mo and W complexes (Table 9c)

Crystal structures for the azides have been reported for Mo.[246–249]

[PPh₄]₂[Mo(NO)(H₂NO)(N₃)₄][246] contains a considerably elongated Mo—N bond as compared with the M—N bond lengths in other terminal N-bonded azide complexes. The only Mo–azide complex containing a 1,1-μ-N-bonded azide is dimeric [PPh₄][Mo₂O₂(μ-N₃){S(CH₂)₃S}₃].[248] The N₃ ligand has a slightly distorted linear conformation which has been ascribed to an intermolecular interaction of the N(3) atom with a methylene H atom.

Only a few cyanate complexes have been reported for Cr[250] and Mo.[251] Both in these complexes and in related thiocyanate complexes of Cr,[252,255] Mo[253,256] and W[254] the NCX ligands are terminal N-bonded.

Table 9c   Complexes of Cr, Mo and W

Compound	Type[a]	Parameters[b]					Ref.
		a	b	c	α	β	
[PPh$_4$]$_2$[Mo(NO)(H$_2$NO)(N$_3$)$_4$]	I	2.17	1.19	1.14	124.8	177.0	246
MoN(N$_3$)$_3$(pyridine)	I	2.043	1.16	1.15	129	175.1	247
[PPh$_4$]$_2$[Mo$_2$O$_2$(μ-N$_3$){S(CH$_2$)$_3$S}$_3$]	II	2.22	1.18	1.09	c	153.4	248
Cr(η^5-Cp)(NO)$_2$(NCO)	V	1.982	1.126	1.179	180.4	178.6	250
Mo$_2$(η^5-Cp)$_2$(NCO)(μ-Bu^{t}NC)(CO)$_3$	V	2.057	1.160	1.183	170.1	177.3	251
[Cr(NCS)$_6$]$^{3-}$	V	2.00	1.15	1.62	164	176	252
Mo$_2$(NCS)$_4$(μ-dppm)$_2$·2C$_3$H$_6$O	V	2.06	1.13	1.64	172	178	253
[Cs$_5$NH$_4$][W$_4$O$_8$(NCS)$_{12}$]·6H$_2$O	V	2.15	1.19	1.58	163	170	254

[a] See Table 8. [b] For a, b, c (Å) and α, β (°), see Table 8. [c] Not reported.

(iv)   Mn, Tc and Re complexes (Table 9d)

Azide complexes containing terminally,[257,262] 1,1-μ-[219] as well as 1,3-μ-bonded[258] coordination modes have been reported. In [NEt$_4$][Mn$_2$(N$_3$)$_3$(CO)$_6$][219] all three azide groups are 1,1-μ-bonded. Mn(acac)$_2$(N$_3$) is a coordination polymer as a result of 1,3-μ bridging of the azide groups between [Mn(acac)$_2$]$^+$ cations, cf. (**92**).

Table 9d   Complexes of Mn, Tc and Re

Compound	Type[a]	Parameters[b]							Ref.
		a	b	c	d	α	β	γ	
[Mn(TPP)(N$_3$)]·C$_6$H$_6$	I	2.045	1.168	1.168		127.0	178.3		257
[NEt$_4$][Mn$_2$(CO)$_6$(N$_3$)$_3$]	II	2.08	1.22	1.15		c	c		219
Mn(acac)$_2$(N$_3$)	IV	2.245	1.166			131.3	177.5		258
[BPh$_4$]$_2$[Mn$_2$(tren)$_2$(NCO)$_2$]	V	2.051	1.157	1.203		158.3	178.8		259
Mn(acac)$_2$(NCS)	VII	2.189	1.163	1.624	2.880	164.8	177.6	112	231
[NBu$_4^n$]$_3$[Re$_2$(NCS)$_{10}$]^c	V	2.026	1.16	1.60		171.7	178.8		228
	VIII	2.091	1.17	1.57		141.3	178.7		
Re(NCS)$_3$(PEt$_2$Ph)(dppe)	V	2.021	1.170	1.597		173.5	177.5		260
[NBu$_4^n$]$_3$[Tc(NCS)$_6$]	V	2.05	1.012	1.64		173	177		261

[a] See Table 8. [b] For a, b, c, d (Å) and α, β, γ (°), see Table 8. [c] See text.

Terminal N-bonded cyanate has been observed in [BPh$_4$][Mn$_2$(tren)$_2$(NCO)$_2$][259] which, however, becomes dimeric because of H bridge bonding between the O atoms and NH functions of the N(CH$_2$CH$_2$NH$_2$)$_3$ ligand.

A large number of thiocyanate complexes of Mn have been reported with either terminal N-[259,263] (see **93**) or 1,3-μ-bonded[231,264] NCS (see **94**). Comparison of polymeric (**92**)[219] with analogous (**94**)[231] reveals that in the thiocyanate complex the Mn atom in (**94**) is 0.12 Å out of the plane defined by the four O-donor sites of the acac ligand in the direction of the N (thiocyanate) atom.

The structure of [NBu$_4^n$]$_3$[Re$_2$(NCS)$_{10}$][228] is interesting because it contains both 1,1-μ-N-bonded and terminal N-bonded ligands. Surprisingly this difference in bonding is not reflected by the N—C and C—S distances. In the other Re[260,265] and Tc[261,266] thiocyanate complexes the terminally bonded form has been found.

(v)   Fe, Ru and Os complexes (Table 9e)

The N$_3$ ligands in the complexes listed in Table 9e are all terminally bonded.

The terminal N-bonding mode to Fe has also been found in the thiocyanate complexes of iron.[269-272] The Fe—N—C bond angle (α) of 160.6° in FeL(NCS)$_2$ (**95**) has been ascribed to the absence of d-electron density on the Fe center for π-bonding to the NCS fragment. This π-overlap would favor canonical structure (**84a**) and accordingly linear bonding.

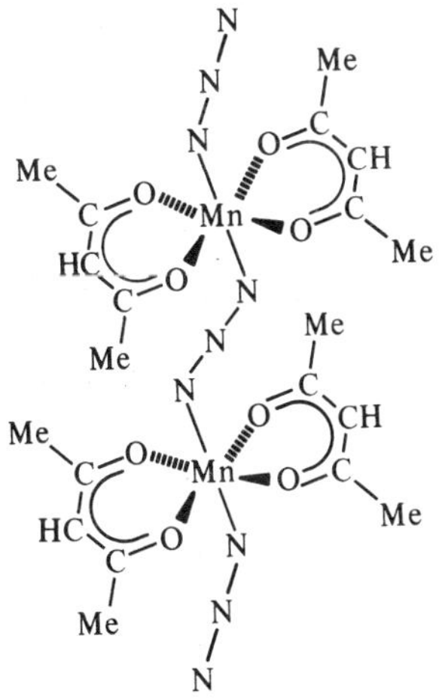

(92) 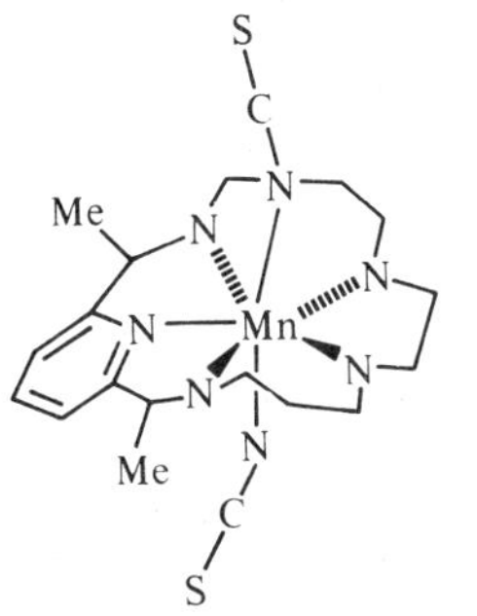 (93) 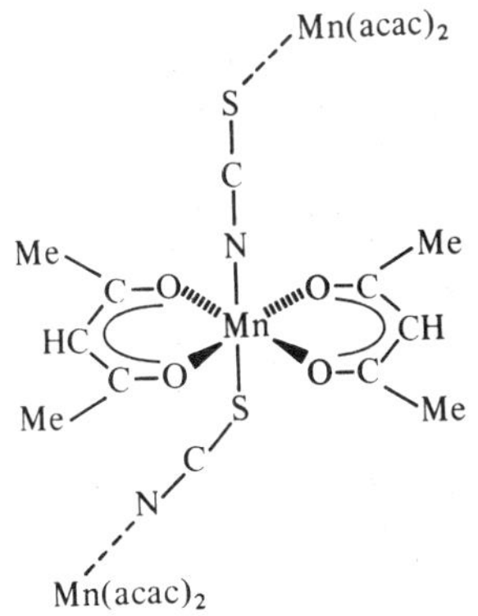 (94)

Table 9e Complexes of Fe, Ru and Os

Compound	Type[a]	Parameters[b]					Ref.
		a	b	c	α	β	
[AsPh₄]₂[Fe(N₃)₅]	I	2.01	1.16	1.16	125.0	178	267
[Ru(N₃)(N₂)(en)₂][PF₆]	I	2.121	1.179	1.146	116.7	180	268
Fe(η^5-Cp)(CO)₂(NCS)	V	1.89	1.10	1.73	169	168	269
Fe(L)(NCS)₂[c]	V	2.12	1.17	1.61	160.6	176.3	270
Fe(4-acpy)₂(H₂O)₂(NCS)₂	V	2.102	1.135	1.622	177.7	179.8	271
Os₃Au(CO)₁₀(PPh₃)(SCN)	VIII	2.43	1.73	1.12	d	175	218

[a] See Table 8. [b] For *a, b, c* (Å) and α, β (°), see Table 8. [c] See (**95**). [d] Not reported.

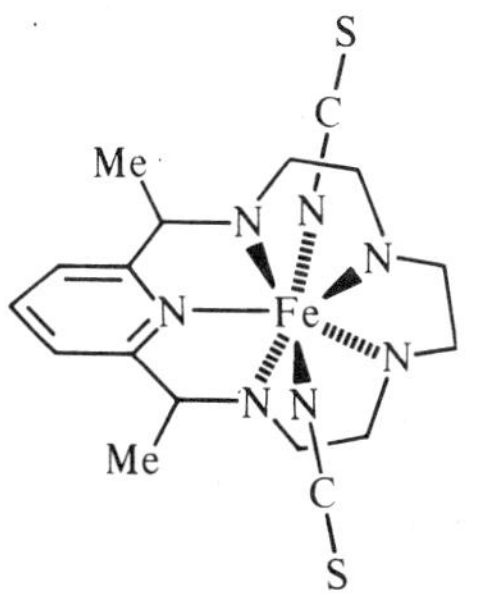

(95) 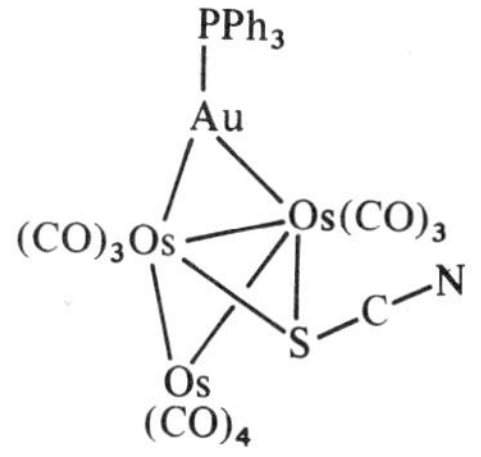 (96)

An example of the few 1,1-μ-S-bonded thiocyanate complexes is the recently reported Os₃Au(SCN)(CO)₁₀(PPh₃) (**96**).[229] The S atom bridges two Os atoms while the almost linear NCS ligand makes a considerable angle with the Os₂S plane (*cf.* canonical structure **85**).

(vi) Co, Rh and Ir complexes (Table 9f)

So far only terminal N-bonding of the azide ligand has been reported for Co azide complexes.[218,273,280] In the case of the Rh azide compounds both terminal[274,275] (see **97**) and 1,1-μ bonding[276] have been found.

Only one structure for a transition metal cyanate complex has been reported. In trigonal-bipyramidal Co(terpy)(NCO)₂ (**98**)[277] the cyanate ligand is terminal N-bonded. Based on the M—N—C bond angle (160.9°) an M—N(CO) bond distance of about 1.16 Å is expected whereas a much shorter bond is found.

Table 9f Complexes of Co, Rh and Ir

Compound	Type[a]	Parameters[b]							Ref.
		a	b	c	d	α	β	γ	
$[Co(trenen)(N_3)][NO_3]_2 \cdot H_2O$	I	1.957	1.209	1.152		119.0	176.4		218
$[Co(en)_2(N_3)_2][NO_3]$	I	1.96	1.12	1.18		120	179		273
$Rh(N_3)_3(Et_4dien)$	I	2.065	1.188	1.162		126.3	175.8		274
$Rh_2(\eta^5\text{-}Cp^*)_2(\mu\text{-}CH_2)_2(N_3)_2{}^c$	I	2.150	d	d		114	d		275
$Rh_2(\eta^5\text{-}Cp^*)_2(\mu\text{-}CH_2)_2(NCS)_2$	VI	2.393	d	d		104	d		275
$Rh_2(\eta^5\text{-}Cp^*)_2\{N_3C_2(CF_3)_2\}_3(N_3)$	II	2.12	1.22	1.14		118.0	179		276
$Co(terpy)(NCO)_2$	V	1.949	1.126	1.197		160.9	178.3		277
$[Co(PPQ)(NCS)_2]_2{}^e$	V	2.036	1.15	1.63		159.3	172.9		278
	IX	2.03	1.16	1.64	2.83	160.3	177	99.0	
$Hg(NCS)_4Co(pyridine)_2$	IX	2.118	1.144	1.654	2.539	170.5	178.9	94.4	279

[a] See Table 8. [b] For a, b, c, d (Å) and α, β, γ (°), see Table 8. [c] $Cp^* = C_5Me_5$. [d] Not reported. [e] Two types of NCS ligand bonding are present.

(97)   (98)   (99)

The NCS ligand in most Co complexes is terminal N-bonded,[281] although in $[Co(PPQ)(NCS)_2]_2$,[278] in addition to an N-bonded NCS ligand, a 1,3-μ-bonded one is also present. In $Hg(SCN)_4Co$-(pyridine)$_2$[279] and $Hg(SCN)_4Co(DMF)_2$[282] an NCS ligand is bridging between Co *via* N and Hg *via* S resulting in a polymeric structure (see **99**).

The two reported structures of Rh thiocyanate complexes both contain an S-bonded NCS unit.[275,283] The N-bonded isothiocyanate complex $IrCl(CN)(NCS)(CO)(PPh_3)_2$[284] is obtained by recrystallization of the S-bonded isomer (linkage isomerization). Its structure could not be refined and of the Ir—NCS fragment only the C—S distance (1.62 Å) has been given.

(vii) Ni, Pd and Pt complexes (Table 9g)

Nickel–azide complexes have been reported which contain the terminal-[285,295] or the 1,3-μ-bonded[286,295b] form. The N-coordination in $Ni(N_3)(NO)(PPh_3)_2$[285] is as expected, but not the N(1)—N(2) distance (b) which is unrealistically short (0.98 Å), most probably as a result of a contamination of Cl^- in the crystal.[285] In dimeric $[BPh_4]_2[Ni_2(tren)_2(N_3)_2]$[286] the azide is 1,3-$\mu$ bridging (see **100**).

Similar bonding patterns for the azide anion are found in the corresponding Pd complexes.[286–288,296] $[Pd(Et_4dien)(N_3)][NO_3]$[297] contains terminal-bonded azide, whereas in $[AsPh_4]_2[Pd_2(N_3)_6]$[288] both terminal and 1,3-μ bridging azides are found to be present in one molecule. A highly interesting structure is that of $PtMe_3(N_3)$ tetramer (**101**)[220] which like the corresponding halide complexes consists of a cubane-type Pt_4N_4 core. The azide ligand is therefore 1,1,1-μ bridging. Even in this case the N—N bond distances b and c are almost equal.

(100)   (101)

Table 9g   Complexes of Ni, Pd and Pt

Compound	Type^a	a	b	c	d	α	β	γ	Ref.
Ni(N$_3$)(NO)(PPh$_3$)$_2$	I	2.018	0.98	1.28		128.1	175.1		285
[BPh$_4$]$_2$[Ni$_2$(tren)$_2$(N$_3$)$_2$]	IV	2.069	1.173	1.174		135.3	177.1	123.3	286
[Pd(Et$_4$dien)(N$_3$)][NO$_3$]	I	2.077	1.13	1.15		123	172		287
[AsPh$_4$]$_2$[Pd$_2$(N$_3$)$_6$]^c	I	2.010	1.205	1.139		120.6	173.3		288
	IV	2.005	1.239	1.142		127.9	175.1		
[Pt(N$_3$)]$_4$	III	2.25	1.25	1.23		120	175		220
Ni(NCO)$_2$(phen)$_2$	V	2.06	1.11	1.20		168	174		263a
Ni(tren)$_2$(NCO)$_2$	IX	2.02	1.13	1.22	2.34	155	178.5	117.1	230
Ni(NCS)$_2$(PPh$_2$Me)$_2$	V	1.802	1.159	1.592		176.1	178.8		289
Ni(NCS)(NO)(PPh$_3$)$_2$	V	1.79	1.14	1.625		177.5	179.1		290
Ni(tam)$_2$(NCS)$_2$	IX	2.020	1.16	1.63	2.546	163.1	176.8	99.6	291
Ni(NCS)$_2$	X	2.511	1.653	1.151	1.995	100.9	178.2	161.7	232
Pd(SCN)$_2$(dppm)	VI	2.364	1.65	1.13		108.2	177.7		225
Pd(NCS)(SCN)(dppe)^c	V	2.062	1.078	1.645		163.5	178		225
	VI	2.364	1.653	1.160		115.5	175		
Pd(NCS)$_2$(dppp)	V	2.055	1.151	1.616		165.0	179.4		225
Pd(NCS)(SCN)(dpnp)^c	V	2.063	1.136	1.611		177.6	178.6		294
	VI	2.295	1.658	1.146		107.2	173.0		
cis-Pt(Ph$_2$PC≡CBu^t)$_2$(NCS)(SCN)^c	V	2.061	1.11	1.60		175	177		292
	VI	2.374	1.64	1.09		103	171		
α-Pt$_2$Cl$_2$(PPr$_3$)$_2$(NCS)$_2$	IX	2.078	1.124	1.164	2.327	164.9	179.3	103.6	293
β-Pt$_2$Cl$_2$(PPr$_3$)$_2$(NCS)$_2$	IX	1.965	1.168	1.641	2.408	167.3	178.6	102.9	293

^{a} See Table 8. ^{b} For a, b, c, d (Å) and α, β, γ (°), see Table 8. ^{c} Two types of pseudohalide bonding present; see Type.

The cyanate complexes of the Group VIII metals are less common. A terminal N-bonded ligand is present in Ni(NCO)$_2$(phen)$_2$[263a] and a 1,3-μ-bonded NCO in dimeric [BPh$_4$][Ni$_2$(tren)$_2$(NCO)$_2$] (*cf.* **100**).[230]

Thiocyanate complexes of nickel are likewise found to contain the NCS ligand either terminal N-[297,298] or 1,3-μ-bonded.[299] The NCS fragment in the polymeric structure of Ni(NCS)$_2$[232] is 1,1-μ-S bridge bonded over two Ni atoms and N-bonded to a third Ni atom.

Nice examples of the ambident character of the NCS ligand are provided by the various structures reported for Pd and Pt thiocyanate complexes. Structures are known containing terminal N-bonded NCS,[225,294,300] terminal S-bonded[225,292,294,300a,301] as well as 1,3-μ bridging NCS.[293,302]

The steric influence of the coligands on the actual bonding mode in these complexes is illustrated by the change from terminal S- (*i.e.* a bond angle M—S—C of about 110°) to terminal N-bonding (*i.e.* a bond angle M—N—C of about 180°) as the steric bulk of the coligands increases[225] (*cf.* structures **102–104**). However, electronic effects may also dominate the type of bonding as is apparent from the Pd (iso)thiocyanate bonding observed in Pd{P(Ph$_2$)(CH$_2$)$_3$NMe$_2$}(NCS)(SCN).[294] In this complex the NCS ligand is S-bonded in the position *trans* to the N atom of the bidentate coligand. A similar type of linkage isomer bonding is found in the platinum complex *cis*-Pt(P(Ph$_2$)C≡CBu^t)$_2$(NCS)(SCN).[292]

(102)   (103)   (104)   (105)

Examples of 1,3-μ bonding of NCS are found in the complexes Pd$_2$(NCS)$_2${P(O)Ph$_2$H}$_2$[302] and Pt$_2$Cl$_2$(PPr$_3$)(NCS)$_2$ **(105)**.[293]

(viii)   Cu, Ag and Au complexes (Table 9h)

A large amount of structural information is available for the pseudohalide complexes of Cu. Terminal,[303,304,317] 1,3-μ[304,318,319] and 1,1-μ[318,319b,320] bound azide ligands have been found. For example,

Table 9h   Complexes of Cu, Ag and Au

Compound	Type[a]	Parameters[b]							Ref.
		a	*b*	*c*	*d*	α	β	γ	
$CuBr(N_3)(Et_4dien)$	I	1.927	1.145	1.414		125.4	175.5		303
$Na_2Cu_2(N_3)_6(HMTA)\cdot 3H_2O^c$	I	1.979	1.188	1.150		128.2	176.3		304
	II	1.998	1.201	1.141		128.1			
$[BPh_4]_2[Cu_2(Me_5dien)_2(N_3)]$	IV	1.985	1.170	1.147	2.252	123.8	176.7	139.5	221
$[BPh_4]_2[Cu_2(tren)_2(NCO)_2]^d$	V	1.87	1.19	1.21		160	176		
$[BPh_4]_2[Cu_2(tren)_2(NCS)_2]^d$	V	1.946	1.15	1.622		162.7	179.3		
$Cu(NCO)_2(2,4\text{-lutidine})$	VII	2.15	1.178	1.178		e	174.4		227
$Ag(NCO)$	VII	2.115	1.195	1.180		128.2	178.2		305
$[NBu_4][Ag(NCO)_2]$	V	2.041	1.093	1.165		171.0	179.0		306
$[Cu(tn)_2(NCS)][ClO_4]$	V	2.11	1.17	1.61		173.8	178.3		307
$[Cu(trien)(SCN)][NCS]$	VI	2.607	1.646	1.164		89.5	178.0		308
$[Cu(C_6H_{14}NO)(SCN)]_2$	IX	1.941	1.15	1.63	2.851	167.0	177.9	99.1	309
$CuL(NCS)$	IX	1.930	1.152	1.633	2.709	169.0	178.2	98.3	310
$Cu_2L(NCS)_2$	VIII	2.38	1.70	1.13		108.0	e		311
$Cu_3L^1L^2(NCS)_4^c$	V	2.696	1.25	1.46		e	141.4		312
	VIII	2.85	1.48	1.25		e	147.0		
$Ag(NCS)(PPr_3^n)$	X	2.85	1.81	1.42	2.10	91.0	155.3	165.3	313
$AgL(NCS)$	VI	2.526	1.635	1.145		106.5	e		314
$Ag(tsc)(NCS)$	IX	2.24	e	e	2.99	160.5	e	104.7	315
$Au(PPh_3)_2(NCS)$	VI	2.468	1.647	1.157		107.6	178.2		316

[a] See Table 8. [b] For *a*, *b*, *c*, *d* (Å) and α, β, γ (°), see Table 8. [c] Two types of pseudohalide bonding present; see Type. [d] Dimeric structure as a result of H-bridge bonding. [e] Not reported.

the complex $CuBr(N_3)(Et_4dien)^{303}$ contains a terminally bonded N_3 while the corresponding ligands in dimeric $Na_2[Cu_2(N_3)_6(HMTA)]\cdot 3H_2O$ (**106**)[304] and $[BPh_4]_2[Cu_2(Me_5dien)_2(N_3)_2]^{221}$ are both $1,1\text{-}\mu$ and $1,3\text{-}\mu$ bridge bonded.

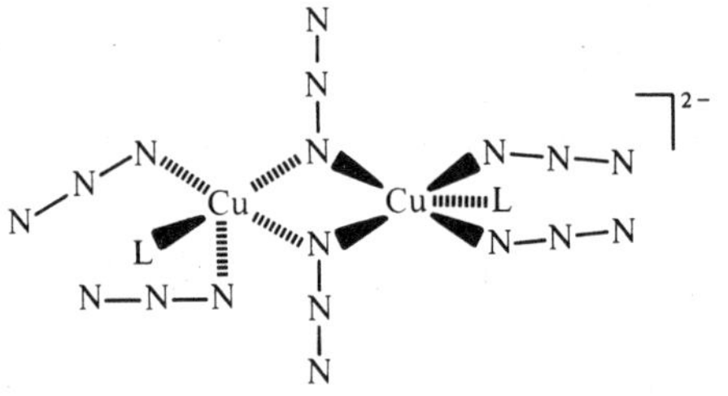

(106)

In the Cu–cyanate complexes the cyanate ligand can be bonded in the terminal N mode[321] as well as in the $1,1\text{-}\mu$ bridging form.[227] The structure of $Ag(NCO)^{305}$ consists of an infinite —N—Ag—N—Ag— chain as a result of $1,1\text{-}\mu$ N-bonding. In contrast $[NBu_4][Ag(NCO)_2]^{306}$ has a discrete molecular structure containing discrete OCN—Ag—NCO units.

Thiocyanate complexes of copper are a further illustration of the versatile bonding behavior of the NCS ligand: terminal N (*e.g.* **107**),[307,312,322] terminal S (**108**),[308] bridging $1,3\text{-}\mu$ (*e.g.* **109**)[309,310,323] and $1,1\text{-}\mu\text{-}S$ (*e.g.* **110**)[311,312] have been reported. In the recently reported structure of a copper NCS complex both terminal and $1,1\text{-}\mu$-bonded NCS fragments are present with as a special structural feature a considerable bending (140°) in the NCS ligands.[312]

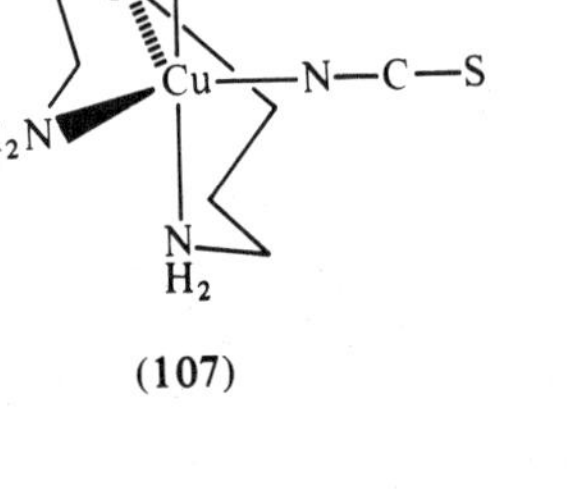

(107)

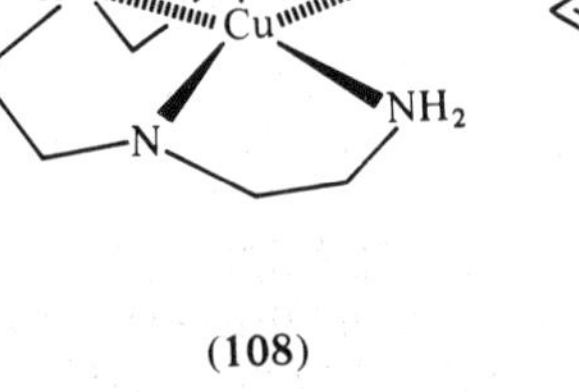

(108)

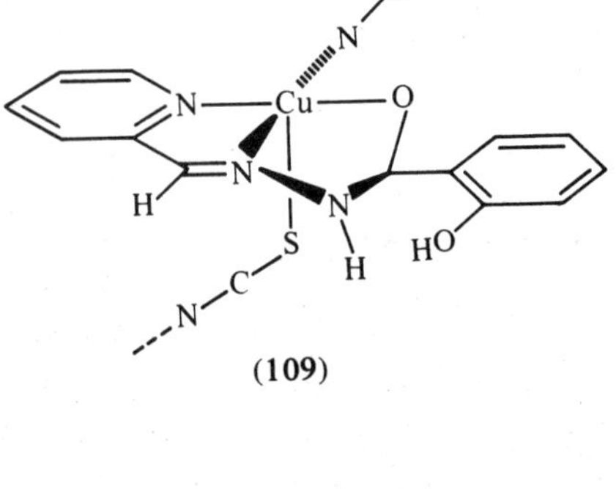

(109)

(110)   (111)

The polymeric structure of $[C_5H_6N][Cu_2(NCS)_3]$[324] contains a 1,1,3-μ-bonded NCS ligand. This can also be seen in the structure of $Ag(NCS)(PPr_3)$ (**111**).[313]

(ix) *Zn, Cd and Hg complexes (Table 9i)*

A few crystal structures are known for Zn, *i.e.* $Zn(N_3)_2(pyridine)_2$[325] and the corresponding NH_3 complex.[326] In analogous $Cd(N_3)_2(pyridine)_2$[327] the N_3 ligand is 1,3-μ bridging which leads to the polymeric structure observed. The Cd center has an octahedral coordination geometry, whereas that of Zn is tetrahedral.

Table 9i Complexes of Zn, Cd and Hg

Compound	Type[a]	Parameters[b]							Ref.
		a	*b*	*c*	*d*	α	β	γ	
$Zn(N_3)_2(C_5H_5N)_2$	I	1.94	1.16	1.13		129	174.5		325
$Zn(N_3)_2(NH_3)_2$	I	1.99	1.19	1.16		129	176		326
$Cd(N_3)_2(C_5H_5N)_2$	IV	2.35	1.14	1.17	2.34	129	179	139	327
CH_3HgN_3	I	2.22	1.18	1.23		123	170		328
$Hg(N_3)_2$	I	2.09	1.19	1.16		116	174		329
α-CF_3HgN_3[c]	II	2.02	1.20	1.11		126	176		330
		2.74							
CF_3HgNCO[c]	VII	2.03	1.13	1.18		137	176		330
		2.88				120			
$[Zn(tren)(NCS)][SCN]$	V	2.043	1.135	1.628		166.6	178.5		331
$Cd(en)_2(NCS)_2$	V	2.40	1.15	1.60		131.9	179.0		332
$Cd(ben)_2(NCS)_2$[c]	V	2.27	1.17	1.67		140.0	170.8		333
	VIII	2.36	1.12	1.67	2.93	164.4	177.0	97.9	
$[Cd_2(NCS)_4(butrz)_3]_\infty$[c]	V	2.24	d	d		d	d		334
	VII	2.39	d	d		130	d		
$Hg(NCS)_2\{P(C_6-H_{11})_3\}$[c]	VI	2.471	1.622	1.63		101.0	177.8		335
	VIII	2.516	1.113	1.663	2.553	169.9	177.5	102.8	

[a] See Table 8. [b] For *a, b, c, d* (Å) and α, β, γ (°), see Table 8. [c] Two types of pseudohalide bonding present; see Type. [d] Not reported.

Recent structures of $KCd(N_3)_3 \cdot H_2O$,[336] $K_2Cd(N_3)_4$[336] and $Tl_8Cd_3(N_3)_{14}$[337] are interesting networks all containing Cd atoms which are octahedrally surrounded by azide ligands. The latter are linear with mean N—N distances of 1.17 Å and all bonding modes for azide known to date are observed in these structures, *i.e.* terminal, 1,1-μ, 1,3-μ and 1,1,1-μ bridging.

The azide in $HgMe(N_3)$[328] is terminally bonded. The structure of $Hg(N_3)_2$[329] consists of N_3—Hg—N_3 units, but in the crystal structure each Hg atom interacts with another five N atoms of different N_3—Hg—N_3 units.

Asymmetric 1,1-μ bridging azides are present in $Hg(CF_3)N_3$[330] and the corresponding cyanate complex (**112**). In the latter complex the Hg—N bond distances differ considerably (2.03 *vs.* 2.88 Å).

The thiocyanate complexes of Zn all contain N-bonded NCS ligands[331,338] whereas for the Cd

Nitrogen Ligands

complexes not only the terminal N,[332–334,339] but also the 1,3-μ[298a,333,340] and the 1,1-μ[334] bonding modes have been established. The smallest M—N—C bond angle (131.9°) for transition metal–isocyanate complexes reported so far is observed in $Cd(NCS)_2(en)_2$[332] which is probably a result of a combination of electronic and crystal packing effects.

Both terminal N and 1,3-μ bridge bonded NCS ligands are present in the structure of $Cd(ben)(NCS)_2$ (113).[333] $[Cd_2(NCS)_4(butrz)_3]_n$ (114)[334] is the only known structure containing a 1,1-μ-N bridge bonded NCS unit.

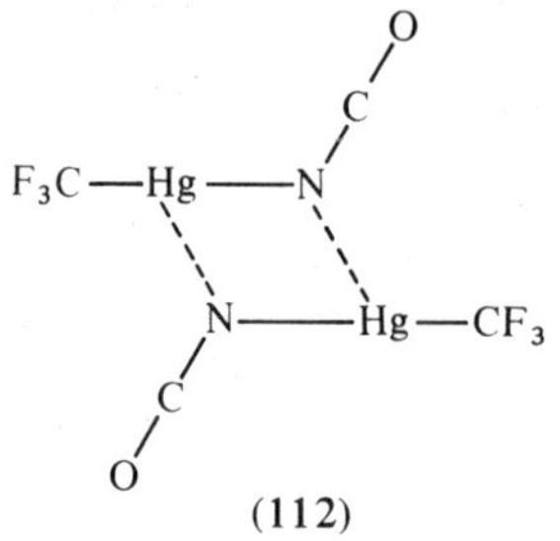

(112)

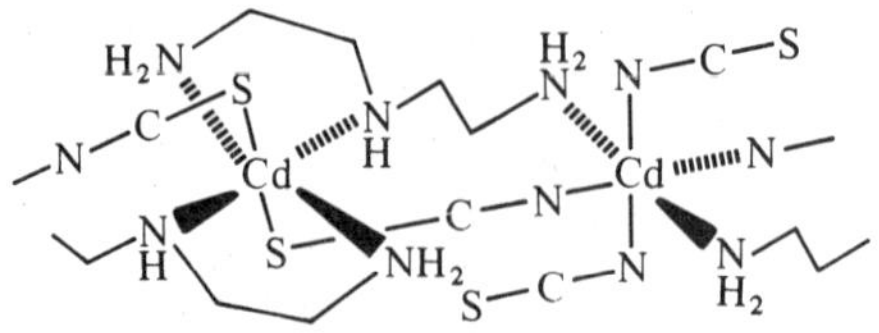

(113)

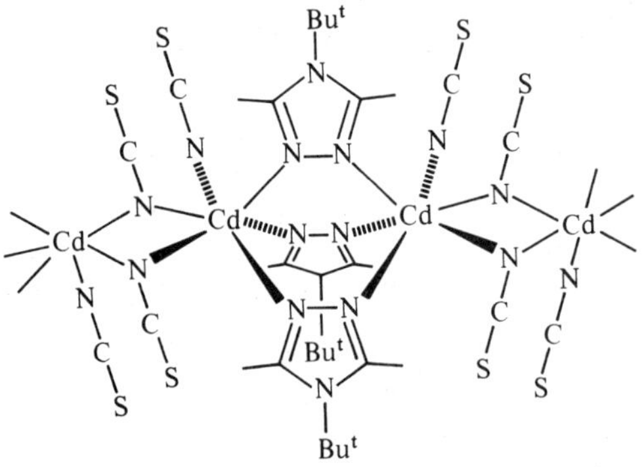

(114)

The thiocyanate ligands in the mercury complexes are either terminal S or 1,3-μ bridge bonded. An example containing both bonding modes is $Hg(NCS)_2(PCy_3)$.[335] Heterodinuclear complexes have been discussed in section (vi) above.

(x) Complexes with other metals

Only a few structures of thiocyanate complexes are known for the lanthanides and actinides, *i.e.* for Er,[341] Th[342] and U.[343] They all contain terminal N-bonded NCS and have as an interesting aspect a high coordination number, *e.g.* as in $[NEt_4]_4[Th(NCS)_8]$.[342]

13.5.5.4 Synthetic Aspects

13.5.5.4.1 General

A variety of preparative methods are available for the synthesis of metal pseudohalide complexes.[202–209] The use of a large counterion like NEt_4^+, PPh_4^+ or $(PPh_3)_2N^+$ is often important for the isolation of the water-free complexes. Furthermore, the presence of bulky counterions can have a stabilizing influence on the more or less explosive complexes containing azide anions.

A simple and straightforward approach to their synthesis is the reaction between simple metal salts and the alkali metal or complex anion pseudohalides. Mixed ligand complexes can easily be made *via* addition of a neutral ligand to the anionic complex species (*cf.* preparation of *trans*-$ML_2(NCS)_2$, *e.g.* 103). However, all these methods have been well reviewed[202–209] and representative examples can be found in the references of Section 13.5.5.3.

A new line of studying the metal–pseudohalide complexes comprises on the one hand the synthesis of the pseudohalide in the coordination sphere of a metal. On the other hand a coordinated pseudohalide is used as a building block for a new molecule which thus is constructed in the metal coordination sphere. A short outline of the first aspect will be given below, while for examples of the second aspect the reader is referred to refs. 344–347.

13.5.5.4.2 *Syntheses of cyanate and thiocyanate in the coordination sphere of a metal*

The synthetic route for the NCO and NCS ligand in the coordination sphere of a metal has a parallel in organic chemistry. This is illustrated by equations (19) and (20) in which the inorganic equivalent (equation 20)[348] is compared with its organic counterpart, the Curtius rearrangement of an acyl azide. In the latter reaction the azide anion attacks the electrophilic carbonyl carbon atom followed by a rearrangement with concomitant N_2 elimination. The general applicability of this reaction route is apparent from the successful synthesis of other transition metal–isocyanate complexes by reacting a metal carbonyl (*e.g.* $Re_2(CO)_{10}$, $Os_3(CO)_{12}$, $[Fe(\eta^5\text{-Cp})(CO)_3]^+$, $M(CO)_6$ (M = Cr, Mo) and $PtCl(PPh_3)_2(CO)^{[349]}$) with N_3^-. The reaction may be initiated by addition of $P(OPh)_3$ or a phosphine (PPh_3). The reaction proceeds most probably *via* initial substitution of the N_3^- ion by the phosphine followed by attack of the N_3^- ion on the coordinated CO to give the NCO anion, *e.g.* conversion of $Fe(\eta^5\text{-Cp})(CO)L_2(N_3)$ to $Fe(\eta^5\text{-Cp})(NCO)(CO)(phosphine)$.[350]

$$R\!-\!C\!\!\overset{O}{\underset{N=\overset{+}{N}=\overset{-}{N}}{\diagdown}} \xrightarrow{-N_2} \left[R\!-\!C\!\!\overset{O}{\underset{\ddot{N}:}{\diagdown}}\right] \longrightarrow RN{=}C{=}O \tag{19}$$

$$(CO)_5WCO + N_3^- \longrightarrow [(CO)_5W\!\!\overset{C=O}{\underset{N_3}{\diagdown}}]^- \xrightarrow{-N_2} [(CO)_5W\!-\!N{=}C{=}O]^- \tag{20}$$

The reversed reaction, *i.e.* attack of CO on the corresponding transition metal azido complex, sometimes also provides an attractive route for the synthesis of isocyanate complexes, *e.g.* the reaction of CO with, for example, $Co(N_3)(DH)_2(PPh_3)$,[344] $Rh(\eta^5\text{-}C_5Me_5)(N_3)_4^{276}$ or $Rh(N_3)(cod)$.[344]

Kinetic studies indicate that depending on the steric requirements of the coligands either prior CO coordination followed by the intramolecular rearrangement shown in equation (20)[351] or intramolecular attack of the coordinated azido group by CO is operative.[344]

The formation of isocyanates has also been observed in reactions outlined in equations (21)–(23). It has been shown that the reactions of equation (21) most probably proceed *via* coordination of the amine function followed by rearrangement with elimination of HX. The reaction with PPh_3 is interesting because of the initial formation of a nitrene species *via* N—O bond cleavage. Formation of a nitrene intermediate as a crucial step in the formation of the NCO ligand has also been proposed for the reaction with diazomethane.

Reactions with $H_2NX(X = Cl,^{202} OH,^{202} NH_2^{352})$:

$$[M\!-\!CO]^+ \xrightarrow{H_2NX} M\!-\!C\!\!\overset{O}{\underset{NH_2X^+}{\diagdown}} \xrightarrow{-H^+} M\!-\!\!\overset{C=O}{\underset{\underset{H\ \ \ X}{\diagdown N \diagup}}{|}} \xrightarrow{-HX} M\!-\!N{=}C{=}O \tag{21}$$

Reactions with PPh_3:[51]

$$\tag{22}$$

$$(CO)_2(\eta^5\text{-}Cp^*)Mo\equiv Mo(\eta^5\text{-}Cp^*)(CO)_2 + Me_2C{=}N_2 \longrightarrow (CO)_2(\eta^5\text{-}Cp^*)Mo \overset{\displaystyle \underset{\text{N}}{\overset{\text{Me}\diagdown_C\diagup\text{Me}}{\parallel}}\text{N}}{-\!\!\!-\!\!\!-} Mo(\eta^5\text{-}Cp^*)(CO)_2 \xrightarrow{\Delta T}$$

$$(CO)(\eta^5\text{-}Cp^*)Mo \overset{\overset{\text{Me}\diagdown_C\diagup\text{Me}}{\underset{\text{N}}{\parallel}}}{=\!\!=} Mo(\eta^5\text{-}Cp^*)(CO)_2 \underset{-\,CO(\Delta T)}{\overset{CO(0\,^\circ C)}{\rightleftharpoons}} (CO)_2(\eta^5\text{-}Cp^*)Mo \overset{\text{Me}\diagdown_C\diagup\text{Me}}{-\!\!\!-\!\!\!-} Mo(\eta^5\text{-}Cp^*)(CO)_2 \qquad (23)$$

$$\underset{\underset{O}{\overset{|}{C}}}{|} \qquad\qquad\qquad\qquad\qquad\qquad \underset{\underset{O}{\overset{|}{C}}}{|} \qquad Cp^* = C_5Me_5$$

$$\begin{array}{c} M{-}\overset{+}{N}{=}\overset{}{N}{=}\overset{-}{N} \\ + \\ (S)C{=}S \end{array} \longrightarrow \begin{array}{c} \text{(cyclic structure)} \end{array} \quad or \quad \begin{array}{c} \text{(cyclic structure)} \end{array} \qquad (24)$$

$$M{-}N{=}C{=}S$$

Isothiocyanate complexes are likewise accessible *via* a route analogous to equation (21) by reacting, for example, $Fe(\eta^5\text{-}Cp)(CO)_2(CS)$ with N_3^- to give $Fe(\eta^5\text{-}Cp)(NCS)(CO)_2$.[353]

Furthermore, isothiocyanate complexes can be prepared starting from the metal–azido complexes and reacting these with CS_2. These reactions most probably proceed *via* 1,3-dipolar cycloaddition reactions.[354] Syntheses of isothiocyanato complexes *via* this route are known for several metals (see, for example, ref. 355 for a synthesis involving Co). In these reactions formation of a cyclic intermediate (*cf.* equation 24) could be established.

13.5.6 REFERENCES

1. D. C. Moody, R. R. Ryan and A. C. Larson, *Inorg. Chem.*, 1979, **18**, 227.
2. D. M. P. Mingos, *Transition Met. Chem.*, 1978, **3**, 1.
3. G. J. Kubas, *Inorg. Chem.*, 1979, **18**, 182.
4. G. Leandri, V. Buzetti, G. Valle and M. Mammi, *Chem. Commun.*, 1970, 413.
5. A. Gieren and F. Pertlik, 2nd European Crystallographic Meeting, 1974.
6. J. Kuyper, P. H. Isselmann, F. C. Mijlhoff, A. Spelbos and G. Renes, *J. Mol. Struct.*, 1975, **29**, 247
7. K. Vrieze and G. van Koten, *Recl. Trav. Chim. Pays-Bas*, 1980, **99**, 145.
8. J. Kuyper and K. Vrieze, *J. Organomet. Chem.*, 1974, **74**, 289.
9. J. R. Grunwell and W. C. Danison, Jr., *Tetrahedron*, 1971, **27**, 5315.
10. R. Meij, A. Oskam and D. J. Stufkens, *J. Mol. Struct.*, 1979, **51**, 37.
11. J. N. Louwen and A. Oskam, personal communication.
12. R. T. Kops, E. van Aken and H. Schenk, *Acta Crystallogr., Sect. B*, 1973, **29**, 913.
13. J. Kuyper, P. I. van Vliet and K. Vrieze, *J. Organomet. Chem.*, 1976, **108**, 257.
14. J. Kuyper and K. Vrieze, *J. Organomet. Chem.*, 1975, **86**, 127.
15. J. Kuyper, L. G. Hubert-Pfalzgraf, P. C. Keyzer and K. Vrieze, *J. Organomet. Chem.*, 1976, **108**, 271.
16. R. Meij, J. Kuyper, D. J. Stufkens and K. Vrieze, *J. Organomet. Chem.*, 1976, **110**, 219.
17. R. Meij, T. A. M. Kaandorp, D. J. Stufkens and K. Vrieze, *J. Organomet. Chem.*, 1977, **128**, 203.
18. R. Meij and R. Olie, *Cryst. Struct. Commun.*, 1975, **4**, 515.
19. R. Meij, D. J. Stufkens, K. Vrieze, W. van Gerresheim and C. H. Stam, *J. Organomet. Chem.*, 1979, **164**, 353.
20. R. Meij, D. J. Stufkens, K. Vrieze, E. Roosendaal and H. Schenk, *J. Organomet. Chem.*, 1978, **155**, 323.
21. D. Walther and C. Pfützenreuter, *Z. Chem.*, 1977, **17**, 426.
22. G. la Monica, M. Pizzotti and S. Cenini, *Gazz. Chim. Ital.*, 1978, **108**, 611.
23. H. C. Ashton and A. R. Manning, *Inorg. Chem.*, 1983, **22**, 1440.
24. R. Meij, D. J. Stufkens and K. Vrieze, *J. Organomet. Chem.*, 1978, **144**, 239.
25. R. Meij, D. J. Stufkens, K. Vrieze, A. M. F. Brouwers and A. R. Overbeek, *J. Organomet. Chem.*, 1978, **155**, 123.
26. R. Meij, D. J. Stufkens, K. Vrieze, A. M. F. Brouwers, J. D. Schagen, J. J. Zwinselman, A. R. Overbeek and C. H. Stam, *J. Organomet. Chem.*, 1979, **170**, 337.
27. J. Kuyper, P. C. Keyzer and K. Vrieze, *J. Organomet. Chem.*, 1976, **116**, 1.
28. P. Hendricks, J. Kuyper and K. Vrieze, *J. Organomet. Chem.*, 1976, **120**, 285.
29. J. M. Klerks, R. van Vliet, G. van Koten and K. Vrieze, *J. Organomet. Chem.*, 1981, **214**, 1.
30. F. E. Brinckman and H. S. Haiss, *Chem. Ind. (London)*, 1963, 1124.

31. K. R. Laing, S. D. Robinson and M. F. Uttley, *J. Chem. Soc., Dalton Trans.*, 1974, 1205.
32. O. Dimroth, *Ber.*, 1906, 3906.
33. E. Pfeiffer, A. Oskam and K. Vrieze, *Transition Met. Chem.*, 1977, **2**, 240.
34. W. H. Knoth, *Inorg. Chem.*, 1973, **12**, 38.
35. (a) G. Bombieri, A. Immirzi and L. Toniolo, *Inorg. Chem.*, 1976, **15**, 2428.
35. (b) A. Immirzi, G. Bombieri and L. Toniolo, *J. Organomet. Chem.*, 1976, **118**, 355.
35. (c) G. Bombieri, A. Immirzi and L. Toniolo, *Transition Met. Chem.*, 1976, **1**, 130.
35. (d) L. D. Brown and J. A. Ibers, *Inorg. Chem.*, 1976, **11**, 2794.
35. (e) A. Immirzi, W. Porzio, G. Bombieri and L. Toniolo, *J. Chem. Soc., Dalton Trans.*, 1980, 1098.
36. F. E. Brinckman, H. S. Haiss and R. A. Robb, *Inorg. Chem.*, 1965, **4**, 936.
37. P. Peringer, *Z. Naturforsch., Teil B*, 1978, **33**, 1091.
38. A. N. Nesmeyanov, E. I. Fedin, A. S. Peregudov, L. A. Fedorov, D. N. Kravtsov, E. V. Borisov and F. Yu. Kiryazev, *J. Organomet. Chem.*, 1979, **169**, 1.
39. J. Kuyper, P. I. van Vliet and K. Vrieze, *J. Organomet. Chem.*, 1976, **105**, 379.
40. L. Toniolo, A. Immirzi, U. Croatto and G. Bombieri, *Inorg. Chim. Acta*, 1976, **19**, 209.
41. E. Pfeiffer and K. Vrieze, *Transition Met. Chem.*, 1979, **4**, 385.
42. L. Toniolo, G. Biscontin, M. Nicolini and R. Cipollini, *J. Organomet. Chem.*, 1977, **139**, 349.
43. L. Toniolo, *Inorg. Chim. Acta*, 1979, **35**, L367.
44. L. Toniolo and G. Cavinato, *Inorg. Chim. Acta*, 1979, **35**, L301.
45. D. M. Grove, G. van Koten, H. J. C. Ubbels, K. Vrieze, L. C. Niemann and C. H. Stam, *J. Chem. Soc., Dalton Trans.*, 1986, 717.
46. A. M. F. J. van der Ploeg, G. van Koten and K. Vrieze, *Inorg. Chem.*, 1982, **21**, 2026.
47. T. Boschi, U. Belluco, L. Toniolo, R. Favez and R. Roulet, *Inorg. Chim. Acta*, 1979, **23**, 37.
48. W. Bradley and I. Wright, *J. Chem. Soc.*, 1956, 640.
49. A. Bino, F. A. Cotton and W. Kaim, *Inorg. Chem.*, 1979, **18**, 3566.
50. F. A. Cotton, G. W. Rice and J. C. Sekutowski, *Inorg. Chem.*, 1979, **18**, 1143.
51. F. A. Cotton, T. Inglis, M. Kilner and T. R. Webb, *Inorg. Chem.*, 1975, **14**, 2023.
52. W. de Roode, K. Brieze, E. A. Koerner von Gustorf and A. Ritter, *J. Organomet. Chem.*, 1977, **135**, 183.
53. W. H. de Roode and K. Vrieze, *J. Organomet. Chem.*, 1978, **145**, 207.
54. M. H. Chisholm, D. A. Haitko, J. C. Huffman and K. Folting, *Inorg. Chem.*, 1981, **20**, 2211.
55. M. H. Chisholm, K. Folting, J. C. Huffman and I. P. Rothwell, *Inorg. Chem.*, 1981, **20**, 2215.
56. F. A. Cotton and L. W. Shive, *Inorg. Chem.*, 1975, **14**, 2027.
57. F. A. Cotton, W. H. Ilsley and W. Kaim, *Inorg. Chem.*, 1980, **19**, 2360.
58. N. G. Connelly, H. Daykin and Z. Demidowicz, *J. Chem. Soc., Dalton Trans.*, 1978, 1532.
59. E. W. Abel and S. J. Skittrall, *J. Organomet. Chem.*, 1980, **193**, 389.
60. P. Hendricks, K. Olie and K. Vrieze, *Cryst. Struct. Commun.*, 1975, **4**, 611.
61. P. Hendricks, J. Kuyper and K. Vrieze, *J. Organomet. Chem.*, 1976, **120**, 285.
62. F. P. Dwyer, *J. Am. Chem. Soc.*, 1941, **63**, 78.
63. F. P. Dwyer and D. P. Mellor, *J. Am. Chem. Soc.*, 1941, **63**, 81.
64. M. Corbett, B. F. Hoskins, N. J. McLeod and B. P. O'Day, *Aust. J. Chem.*, 1975, **28**, 2377.
65. L. Toniolo, G. Deganello, P. L. Sandrini and G. Bombieri, *Inorg. Chim. Acta*, 1975, **15**, 11.
66. S. Candeloro de Sanctis, N. V. Pavel and L. Toniolo, *J. Organomet. Chem.*, 1976, **108**, 409.
67. S. Candeloro de Sanctis, L. Toniolo, T. Boschi and G. Deganello, *Inorg. Chim. Acta*, 1975, **12**, 251.
68. T. Jack and J. Powell, *J. Organomet. Chem.*, 1971, **27**, 133.
69. C. M. Harris, B. F. Hoskins and R. L. Martin, *J. Chem. Soc.*, 1959, 3728.
70. I. D. Brown and J. D. Dunitz, *Acta Crystallogr.*, 1961, **14**, 480.
71. J. E. O'Connor, G. Janusonis and E. R. Corey, *Chem. Commun.*, 1968, 445.
72. P. I. van Vliet, G. van Koten and K. Vrieze, *J. Organomet. Chem.*, 1979, **179**, 89.
73. M. Corbett and B. F. Hoskins, *Inorg. Nucl. Chem. Lett.*, 1970, **6**, 261.
74. J. Kuyper, P. I. van Vliet and K. Vrieze, *J. Organomet. Chem.*, 1975, **96**, 289.
75. J. Kuyper, P. I. van Vliet and K. Vrieze, *J. Organomet. Chem.*, 1976, **107**, 129.
76. P. I. van Vliet, G. van Koten and K. Vrieze, *J. Organomet. Chem.*, 1979, **182**, 105.
77. R. T. Kops and H. Schenk, *Cryst. Struct. Commun.*, 1976, **5**, 193.
78. R. T. Kops, A. R. Overbeek and H. Schenk, *Cryst. Struct. Commun.*, 1976, **5**, 125.
79. A. M. F. J. van der Ploeg, G. van Koten and C. Brevard, *Inorg. Chem.*, 1982, **21**, 2878.
80. A. M. F. J. van der Ploeg, G. van Koten and K. Vrieze, *Inorg. Chem.*, 1982, **21**, 2026.
81. A. M. F. J. van der Ploeg, G. van Koten and K. Vrieze, *J. Organomet. Chem.*, 1982, **226**, 93.
82. A. M. F. J. van der Ploeg, G. van Koten, K. Vrieze, A. L. Spek and A. J. M. Duisenberg, *Organometallics*, 1982, **1**, 1066.
83. P. I. van Vliet, G. van Koten and K. Vrieze, *J. Organomet. Chem.*, 1980, **188**, 301.
84. P. I. van Vliet, M. Kokkes, G. van Koten and K. Vrieze, *J. Organomet. Chem.*, 1980, **187**, 413.
85. M. H. Chisholm, J. C. Huffman and R. L. Kelly, *Inorg. Chem.*, 1979, **18**, 3554.
86. R. B. King and K. C. Nainan, *Inorg. Chem.*, 1975, **14**, 271.
87. W. H. de Roode, D. G. Prins, A. Oskam and K. Vrieze, *J. Organomet. Chem.*, 1978, **154**, 273.
88. E. Pfeiffer, J. Kuyper and K. Vrieze, *J. Organomet. Chem.*, 1976, **105**, 371.
89. E. Pfeiffer, K. Vrieze and J. A. McCleverty, *J. Organomet. Chem.*, 1979, **174**, 183.
90. E. Pfeiffer and K. Olie, *Cryst. Struct. Commun.*, 1975, **4**, 605.
91. T. Inglis and M. Kilner, *J. Chem. Soc., Dalton Trans.*, 1975, 930.
92. W. H. de Roode, M. L. Beekes, A. Oskam and K. Vrieze, *J. Organomet. Chem.*, 1977, **142**, 337.
93. T. Inglis, M. Kilner, T. Reynoldson and E. E. Robertson, *J. Chem. Soc., Dalton Trans.*, 1975, 925.
94. E. W. Abel and S. J. Skittrall, *J. Organomet. Chem.*, 1980, **185**, 391.
95. R. Rossi, A. Duatti, L. Magon and L. Toniolo, *Inorg. Chim. Acta*, 1981, **48**, 243.
96. R. Graziani, L. Toniolo, U. Casellato, R. Rossi and L. Magon, *Inorg. Chim. Acta*, 1981, **52**, 119.
97. E. W. Abel and I. D. H. Towle, *J. Organomet. Chem.*, 1978, **155**, 299.

98. J. G. M. van der Linden, A. H. Dix and E. Pfeiffer, *Inorg. Chim. Acta*, 1980, **39**, 271.
99. A. Sahaipal and S. D. Robinson, *Inorg. Chem.*, 1979, **18**, 3572.
100. A. D. Harris, S. D. Robinson, A. Sahaipal and M. B. Hursthouse, *J. Chem. Soc., Dalton Trans.*, 1981, 1327.
101. L. D. Brown and J. A. Ibers, *Inorg. Chem.*, 1976, **15**, 2788.
102. C. J. Creswell, M. A. M. Queiros and S. D. Robinson, *Inorg. Chim. Acta*, 1982, **62**, 157.
103. W. R. Krigbaum and B. Rubin, *Acta Crystallogr., Sect. B*, 1973, **29**, 749.
104. M. Corbett and B. F. Hoskins, *Aust. J. Chem.*, 1974, **27**, 665.
105. E. Pfeiffer, M. W. Kokkes and K. Vrieze, *Transition Met. Chem.*, 1979, **4**, 393.
106. E. Pfeiffer, M. W. Kokkes and K. Vrieze, *Transition Met. Chem.*, 1979, **4**, 389.
107. L. Toniolo and G. Cavinato, *Inorg. Chim. Acta*, 1979, **35**, L301.
108. P. E. Jaitner, P. Peringer, G. Huttner and L. Zsolnai, *Transition Met. Chem.*, 1981, **6**, 86.
109. G. B. Deacon, *Inorg. Chim. Acta*, 1979, **37**, L528.
110. E. Wehman, personal communication.
111. G. van Koten and K. Vrieze, *Recl. Trav. Chim. Pays-Bas*, 1981, **100**, 129.
112. G. van Koten and K. Vrieze, in *Adv. Organomet. Chem.*, **21**, 151.
113. J. Keijsper, H. van der Poel, L. H. Polm, G. van Koten and K. Vrieze, *Polyhedron*, 1983, **2**, 1111.
114. O. Borgen, B. Mistvedt and I. Skauvik, *Acta Chem. Scand., Ser. A*, 1976, **30**, 43.
115. R. Benedix, P. Birner, F. Birnstock, H. Hennig and H.-J. Hoffmann, *J. Mol. Struct.*, 1979, **51**, 99.
116. J. Reinhold, R. Benedix, P. Birner and H. Hennig, *Inorg. Chim. Acta*, 1979, **33**, 209.
117. H. van der Poel, G. van Koten, K. Vrieze, M. Kokkes and C. H. Stam, *Inorg. Chim. Acta*, 1980, **39**, 197.
118. H. van der Poel, G. van Koten, M. Kokkes and C. H. Stam, *Inorg. Chem.*, 1981, **20**, 2491.
119. H. W. Frühauf, A. Landers, R. Goddard and C. Krüger, *Angew. Chem.*, 1978, **90**, 56.
120. J. Keijsper, L. H. Polm, G. van Koten, K. Vrieze, G. Abbel and C. H. Stam, *Organometallics*, 1983, **4**, 2142.
121. J. Keijsper, L. H. Polm, G. van Koten, K. Vrieze, F. A. B. Seignette and C. H. Stam, *Inorg. Chem.*, 1985, **24**, 518.
122. H. tom Dieck and H. Bock, *Chem. Commun.*, 1968, 678.
123. H. van der Poel, G. van Koten and K. Vrieze, *Inorg. Chem.*, 1980, **19**, 1145.
124. L. H. Staal, D. J. Stufkens and A. Oskam, *Inorg. Chim. Acta*, 1978, **26**, 255.
125. J. T. B. H. Jastrzebski, G. van Koten and K. Vrieze, *J. Organomet. Chem.*, 1983, **250**, 49.
126. K. J. Cavell, D. J. Stufkens and K. Vrieze, *Inorg. Chim. Acta*, 1980, **47**, 67.
127. H. van der Poel, G. van Koten and K. Vrieze, *J. Organomet. Chem.*, 1977, **135**, C63.
128. H. van der Poel, G. van Koten and K. Vrieze, *Inorg. Chim. Acta*, 1981, **51**, 241.
128a. H. tom Dieck and J. Klaus, *J. Organomet. Chem.*, 1983, **246**, 301.
129. A. Kinzel, Ph.D. Thesis, University of Hamburg (1979).
130. H. tom Dieck and A. Kinzel, *Angew. Chem.*, 1979, **91**, 344.
131. A. J. Graham, D. Abrigg and B. Sheldrick, *Cryst. Struct. Commun.*, 1976, **5**, 891; 1977, **6**, 253.
132. R. W. Balk, D. J. Stufkens and A. Oskam, *J. Mol. Struct.*, 1980, **60**, 387.
133. W. Majunke, D. Liebfritz, T. Mack and H. tom Dieck, *Chem. Ber.*, 1975, **108**, 3025.
134. P. Bruder, Ph.D. Thesis, University of Frankfurt, Frankfurt am Main (1979).
135. L. H. Staal, A. Oskam and K. Vrieze, *J. Organomet. Chem.*, 1979, **170**, 235.
136. L. H. Staal, G. van Koten and K. Vrieze, *J. Organomet. Chem.*, 1979, **175**, 73.
137. L. H. Staal, J. Keijsper, G. van Koten, K. Vrieze, J. A. Cras and W. P. Bosman, *Inorg. Chem.*, 1981, **20**, 555.
138. H. tom Dieck and H. Bruder, *J. Chem. Soc., Chem. Commun.*, 1977, 24.
139. D. Liebfritz and H. tom Dieck, *J. Organomet. Chem.*, 1976, **105**, 255.
140. L. H. Staal, L. H. Polm and K. Vrieze, *Inorg. Chim. Acta*, 1980, **40**, 165.
141. L. H. Staal, G. van Koten and K. Vrieze, *J. Organomet. Chem.*, 1981, **206**, 99.
142. B. Chaudret, H. Köster and R. Poilblanc, *J. Chem. Soc., Chem. Commun.*, 1981, 266 and B. Chaudret, C. Cayret, H. Köster and R. Poilblanc, *J. Chem. Soc., Dalton Trans.*, 1983, 941.
143. H. van der Poel, G. van Koten and K. Vrieze, *Inorg. Chim. Acta*, 1981, **51**, 253.
144. L. H. Staal, P. Bosma and K. Vrieze, *Inorg. Chim. Acta*, 1980, **43**, 125.
145. H. Bock and H. tom Dieck, *Angew. Chem.*, 1966, **18**, 159.
146. M. Svoboda and H. tom Dieck, *J. Organomet. Chem.*, 1980, **191**, 321.
147. H. tom Dieck, M. Svoboda and J. Kopf, *Z. Naturforsch., Teil B*, 1978, **33**, 1381.
148. A. L. Balch and R. H. Holm, *J. Am. Chem. Soc.*, 1966, **88**, 5201.
149. D. Walther, *Z. Anorg. Allg. Chem.*, 1977, **431**, 17.
150. P. Overbosch, G. van Koten, D. M. Grove, A. L. Spek and A. J. M. Duisenberg, *Inorg. Chem.*, 1982, **21**, 3253.
151. W. Beck and F. H. Holsboer, *Z. Naturforsch., Teil B*, 1973, **28**, 511.
152. H. Aryanci, C. Daul, M. Zobrist and A. von Zelewsky, *Helv. Chim. Acta*, 1975, **58**, 1732.
153. H. tom Dieck and I. W. Renk, *Chem. Ber.*, 1971, **104**, 92; H. tom Dieck and L. Stamp, *Acta Crystallogr., Sect C*, 1983, **39**, 841.
154. A. T. T. Hsiek and K. L. Ooi, *J. Inorg. Nucl. Chem.*, 1976, **38**, 604.
155. L. H. Staal, L. H. Polm, R. W. Balk, G. van Koten and A. M. F. Brouwers, *Inorg. Chem.*, 1980, **19**, 3343.
156. L. H. Staal, G. van Koten and K. Vrieze, *J. Organomet. Chem.*, 1981, **206**, 99.
157. L. H. Staal, J. Keijsper, L. H. Polm and K. Vrieze, *J. Organomet. Chem.*, 1981, **204**, 101.
158. R. D. Adams, *J. Am. Chem. Soc.*, 1980, **102**, 7476.
159. L. H. Staal, L. H. Polm, K. Vrieze, F. Ploeger and C. H. Stam, *J. Organomet. Chem.*, 1980, **199**, C13.
160. L. H. Staal, G. van Koten, K. Vrieze, B. van Santen and C. H. Stam, *Inorg. Chem.*, 1981, **20**, 3598.
161. L. H. Staal, L. H. Polm, K. Vrieze, F. Ploeger and C. H. Stam, *Inorg. Chem.*, 1981, **20**, 3590.
162. L. H. Polm, G. van Koten, C. J. Elsevier, K. Vrieze, B. F. K. van Santen and C. H. Stam, *J. Organomet. Chem.*, 1986, **304**, 353.
163. L. H. Staal, L. H. Polm, R. W. Balk, G. van Koten, K. Vrieze and A. M. F. Brouwers, *Inorg. Chem.*, 1980, **19**, 3343.
164. L. H. Polm, C. J. Elsevier, O. J. Stufliens, G. van Koten, K. Vrieze, R. R. Andréa and C. H. Stam, to be published.
165. J. Keijsper, L. Polm, G. van Koten and K. Vrieze, *Inorg. Chim. Acta*, 1985, **103**, 137.
166. L. H. Polm, G. van Koten, K. Vrieze, C. H. Stam and W. C. J. van Tunen, *J. Chem. Soc., Chem. Commun.*, 1983, 1177.

167. L. H. Staal, A. Oskam, K. Vrieze, E. Roosedaal and K. Vrieze, *Inorg. Chem.*, 1979, **18**, 1634.
168. G. van Koten, J. T. B. H. Jastrzebski and K. Vrieze, *J. Organomet. Chem.*, 1983, **250**, 49.
169. J. M. Klerks, D. J. Stufkens, G. van Koten and K. Vrieze, *J. Organomet. Chem.*, 1979, **181**, 271.
170. H. W. Frühauf, F. Seils, M. J. Romao and R. J. Goddard, *Angew. Chem. Suppl.*, 1983, 1435.
171. H. tom Dieck and R. Diercks, *Angew. Chem.*, 1983, **95**, 801; *Angew. Chem. Suppl.*, 1983, 1138.
172. M. Dekker and G. R. Knox, *Chem. Commun.*, 1967, 1243.
173. J. Ashley-Smith, M. Green and F. G. A. Stone, *J. Chem. Soc., Dalton Trans.*, 1972, 1805.
174. P. Overbosch, G. van Koten, A. L. Spek, G. Roelofsen and A. J. M. Duisenberg, *Inorg. Chem.*, 1982, **21**, 3908.
175. P. Overbosch, G. van Koten and O. K. Overbeek, *J. Am. Chem. Soc.*, 1980, **102**, 2091.
176. P. Overbosch, G. van Koten and O. K. Overbeek, *Inorg. Chem.*, 1982, **21**, 2373.
177. C. E. Johnson and W. C. Trogler, *J. Am. Chem. Soc.*, 1981, **103**, 6352.
178. R. J. Doedens, *Chem. Commun.*, 1968, 1271.
179. C. Chang, C. E. Johnson, T. G. Richmond, Y. Chen, W. C. Trogler and F. Basolo, *Inorg. Chem.*, 1981, **20**, 3167.
180. M. E. Gross and W. C. Trogler, *J. Organomet. Chem.*, 1981, **209**, 407.
181. S. Otsuka and A. Nakamura, *Inorg. Chem.*, 1968, **7**, 2542.
182. M. E. Gross, W. C. Trogler and J. A. Ibers, *Organometallics*, 1982, **1**, 732.
183. M. E. Gross, W. C. Trogler and J. A. Ibers, *J. Am. Chem. Soc.*, 1981, **103**, 192.
184. G. La Monica, S. Sandrini, F. Zingales and S. Cenini, *J. Organomet. Chem.*, 1973, **50**, 287.
185. A. B. Gilchrist and D. Sutton, *Can. J. Chem.*, 1974, **52**, 3387.
186. F. W. B. Einstein, A. B. Gilchrist, G. W. Rayner-Canham and D. Sutton, *J. Am. Chem. Soc.*, 1971, **93**, 1826.
187. F. W. B. Einstein and D. Sutton, *Inorg. Chem.*, 1972, **11**, 2827.
188. P. Overbosch and G. van Koten, *J. Organomet. Chem.*, 1982, **229**, 193.
189. P. Overbosch, G. van Koten and K. Vrieze, *J. Organomet. Chem.*, 1981, **208**, C21.
190. P. Overbosch, G. van Koten, D. M. Grove, A. L. Spek and A. J. M. Duisenberg, *Inorg. Chem.*, 1982, **21**, 3253.
191. P. Overbosch, G. van Koten and K. Vrieze, *J. Chem. Soc., Dalton Trans.*, 1982, 1541.
192. W. Beck, M. Bauder, G. La Monica, S. Cenini and R. Ugo, *J. Chem. Soc. (A)*, 1971, 113.
193. J. Geisenberger, U. Nagel, A. Sebald and W. Beck, *Chem. Ber.*, 1983, **116**, 911.
194. N. Farrell and D. Sutton, *J. Chem. Soc., Dalton Trans.*, 1977, 2124.
195. S. Cenini and G. La Monica, *Inorg. Chim. Rev.*, 1976, **18**, 279.
196. S. Cenini, P. Fantucci, M. Pizzotti and G. La Monica, *Inorg. Chim. Acta*, 1975, **13**, 243.
197. M. E. Gross, J. A. Ibers and W. C. Trogler, *Organometallics*, 1982, **1**, 530.
198. W. C. Trogler, C. E. Johnson and D. E. Ellis, *Inorg. Chem.*, 1981, **20**, 980.
199. C. E. Johnson and W. C. Trogler, *Inorg. Chem.*, 1982, **21**, 427.
200. M. Veith, *Acta Crystallogr., Sect. B*, 1975, **31**, 678.
201. L. Birkenbach and K. Kellerman, *Chem. Ber.*, 1925, **58**, 786, 2377.
202. W. Beck and W. P. Fehlhammer, in 'MTP International Review of Science, Inorganic Chemistry, Series One', Butterworth, London, 1972, vol. 2, p. 253.
203. R. A. Bailey, S. L. Kozak, T. W. Michelsen and W. N. Mills, *Coord. Chem. Rev.*, 1971, **6**, 407.
204. (a) A. H. Norbury, *Adv. Inorg. Chem. Radiochem.*, 1975, **17**, 231.
204. (b) A. H. Norbury and A. I. P. Shina, *Q. Rev., Chem. Soc.*, 1970, **24**, 69.
205. (a) Z. Dori and R. F. Ziolo, *Chem. Rev.*, 1973, **73**, 247.
205. (b) A. D. Yoffe, in 'Developments in Inorganic Nitrogen Chemistry, 1966–1973', ed. C. B. Colburn, Elsevier, Amsterdam, 1966, vol. 1, p. 72.
206. U. Müller, *Struct. Bonding (Berlin)*, 1973, **14**, 141.
207. S. Patai (ed.), 'The Chemistry of Cyanates and their Thio Derivatives', New York, 1977.
208. H. D. Fair and R. F. Walker (eds.), 'Energetic Materials I', Plenum, New York, 1977.
209. A. A. Newman (ed.), 'Chemistry and Biochemistry of Thiocyanic Acid and its Derivatives', Academic, London, 1975.
210. G. E. Pringle and D. E. Noakes, *Acta Crystallogr., Sect. B*, 1968, **24**, 262; U. Müller, *Z. Anorg. Allg. Chem.*, 1972, **392**, 159.
211. Gmelin's Handbuch der Anorganischen Chemie (Kohlenstoff, D1), Verlag Chemie, Weinheim, 1971, p. 316.
212. T. Austad, J. Songstad and K. Åse, *Acta Chem. Scand.*, 1971, **25**, 331; A. W. Downs, *Chem. Commun.*, 1968, 1290.
213. Z. Igbal, *Struct. Bonding (Berlin)*, 1972, **10**, 25; O. H. Ellestad, P. Klåboe and J. Songstad, *Acta Chem. Scand.*, 1972, **26**, 1724.
214. D. D. Schwank and R. D. Willet, *Inorg. Chem.*, 1965, **4**, 499.
215. C. Akers, S. W. Peterson and R. D. Willett, *Acta Crystallogr., Sect. B*, 1968, **24**, 1125.
216. A. S. Foust, *J. Chem. Soc., Chem. Commun.*, 1979, 414.
217. J. L. Burmeister, *Coord. Chem. Rev.*, 1968, **3**, 225.
218. I. E. Maxwell, *Inorg. Chem.*, 1971, **10**, 1782.
219. R. Mason and G. A. Rushholme, *Chem. Commun.*, 1971, 496.
220. M. Atam and U. Müller, *J. Organomet. Chem.*, 1974, **71**, 435.
221. T. R. Felthouse and D. N. Hendrickson, *Inorg. Chem.*, 1978, **17**, 444.
222. S. J. Anderson, D. S. Brown and K. J. Finney, *J. Chem. Soc., Dalton Trans.*, 1979, 152.
223. F. Dahan, R. Kergoat, M.-C. Tocquer and J. E. Geurchais, *Acta Crystallogr., Sect. B*, 1976, **32**, 1038.
224. G. V. Tsintsadze, M. A. Porai-Koshits and A. S. Antsyshkina, *Zh. Strukt. Khim.*, 1967, **8**, 296.
225. G. J. Palenik, M. Mathew, W. L. Steffen and G. Beran, *J. Am. Chem. Soc.*, 1975, **97**, 1059.
226. A. V. Ablov and I. D. Samus, *Dokl. Akad. Nauk SSSR*, 1962, **146**, 1071.
227. F. Valach, M. Dunaj-Jurco, J. Carai and M. Hvastijova, *Collect. Czech. Chem. Commun.*, 1974, **39**, 380.
228. F. A. Cotton, A. Davison, W. H. Ilsley and H. S. Trop, *Inorg. Chem.*, 1979, **18**, 2719.
229. B. F. G. Johnson, D. A. Kaner, J. Lewis and P. R. Raithby, *J. Organomet. Chem.*, 1981, **215**, C33.
230. D. M. Duggan and D. N. Hendrickson, *Inorg. Chem.*, 1974, **13**, 2056.
231. B. R. Stults, R. O. Day, R. S. Marianelli and V. W. Day, *Inorg. Chem.*, 1979, **18**, 1847.
232. E. Dubler, A. Peller and H. R. Oswald, *Z. Kristallogr.*, 1982, **161**, 265.

233. J. L. Burmeister and H. J. Gysling, *Chem. Commun.*, 1967, 543; J. L. Burmeister, H. J. Gysling and J. C. Lim, *J. Am. Chem. Soc.*, 1969, **91**, 44.
234. W. M. Dyck, K. Dehnicke, F. Weller and U. Müller, *Z. Anorg. Allg. Chem.*, 1980, **470**, 89.
235. E. R. de Gil, M. de Burguera, A. V. Rivera and P. Maxfield, *Acta Crystallogr., Sect. B*, 1977, **33**, 578.
236. U. Müller, W. M. Dyck and K. Dehnicke, *Z. Anorg. Allg. Chem.*, 1980, **468**, 172.
237. H. O. Wellern and U. Müller, *Chem. Ber.*, 1976, **109**, 3039.
238. W. M. Dyck, K. Dehnicke, G. Beyendorff-Gulba and J. Strahle, *Z. Anorg. Allg. Chem.*, 1981, **482**, 113.
239. E. J. Peterson, R. B. von Dreele and T. M. Brown, *Inorg. Chem.*, 1976, **15**, 309.
240. A. C. Villa, A. G. Manfredotti and C. Guastini, *Acta Crystallogr., Sect. B*, 1976, **32**, 909.
241. R. Dubgen, U. Müller, F. Weller and K. Dehnicke, *Z. Anorg. Allg. Chem.*, 1980, **471**, 89.
242. J. Strähle, *Z. Anorg. Allg. Chem.*, 1974, **405**, 139.
243. A. C. Hazell, *J. Chem. Soc.*, 1963, 5745.
244. B. Kamenar and C. K. Prout, *J. Chem. Soc. (A)*, 1970, 2379.
245. F. Dahan, R. Kergoat, M.-C. Benechal-Tocquer and J. E. Geurchais, *J. Chem. Soc., Dalton Trans.*, 1976, 2202.
246. K. Wieghardt, G. Backes-Dahman, W. Swiridoff and J. Weiss, *Inorg. Chem.*, 1983, **22**, 122.
247. E. Schweda and J. Strahle, *Z. Naturforsch., Teil B*, 1981, **36**, 662.
248. P. T. Bishop, J. R. Dilworth, J. Hutchinson and J. A. Zubieta, *J. Chem. Soc., Chem. Commun.*, 1982, 1052.
249. K. Dehnicke, J. Schmitte and D. Fenske, *Z. Naturforsch., Teil B*, 1980, **35**, 1070; E. Schweda and J. Strahle, *Z. Naturforsch., Teil B*, 1980, **35**, 1146; J. R. Dilworth, P. Dahlstrom, J. R. Hyde and J. Zubieta, *Inorg. Chim. Acta*, 1983, **71**, 21.
250. M. A. Bush and G. A. Sim, *J. Chem. Soc. (A)*, 1970, 605.
251. A. T. McPhail, G. R. Knox, C. G. Robertson and G. A. Sim, *J. Chem. Soc. (A)*, 1971, 205; W. A. Herrmann, L. K. Bell, M. L. Ziegler, H. Pfisterer and C. Pahl, *J. Organomet. Chem.*, 1983, **247**, 39.
252. J. Kay, J. W. Moore and M. D. Blick, *Inorg. Chem.*, 1972, **11**, 2818.
253. E. H. Abbott, K. S. Bose, F. A. Coton, W. T. Hall and J. C. Sekutowski, *Inorg. Chem.*, 1978, **17**, 3240.
254. J. P. Launy, Y. Jeannin and A. Nel, *Inorg. Chem.*, 1983, **22**, 277.
255. Y. Takeuchi and Y. Saito, *Bull. Chem. Soc. Jpn.*, 1957, **30**, 319; H. M. Colquhoun and K. Henrick, *Inorg. Chem.*, 1981, **20**, 4074.
256. A. Muller, U. Seyer and W. Eltzner, *Inorg. Chim. Acta*, 1979, **32**, L65; A. Bino and F. A. Cotton, *Inorg. Chem.*, 1979, **18**, 1381; A. Muller, N. Mohan, S. Sarka and W. Eltzner, *Inorg. Chim. Acta*, 1981, **55**, L33; A. Muller and N. Mohan, *Z. Anorg. Allg. Chem.*, 1981, **480**, 157; J. R. Knox and K. Eriks, *Inorg. Chem.*, 1968, **7**, 84; B. Kamenar, M. Penavic and B. Korpar- Colig, *Acta Crystallogr., Sect. A*, 1978, **34**, S136; M. Green, H. P. Kirsch, F. G. A. Stone and A. J. Welch, *Inorg. Chim. Acta*, 1978, **29**, 101; T. Glowiak, M. Sabat, H. Sabat and M. F. Rudolf, *J. Chem. Soc., Chem. Commun.*, 1975, 712.
257. V. W. Day, B. R. Stults, A. L. Tasset, R. S. Marianelli and L. J. Boucher, *Inorg. Nucl. Chem. Lett.*, 1975, **11**, 505.
258. B. R. Stults, R. S. Marianelli and V. W. Day, *Inorg. Chem.*, 1975, **14**, 722.
259. E. J. Laskowski and D. H. Hendrickson, *Inorg. Chem.*, 1978, **17**, 457.
260. J. E. Hahn, T. Nimry, W. R. Robinson, D. J. Salmon and R. A. Walton, *J. Chem. Soc., Dalton Trans.*, 1978, 1232.
261. H. S. Trop, A. Davison, A. G. Jones, M. A. Davis, D. J. Szalda and S. J. Lippard, *Inorg. Chem.*, 1980, **19**, 1105.
262. B. C. Schardt, F. J. Hollander and C. L. Hill, *J. Chem. Soc., Chem. Commun.*, 1981, 765; V. W. Day, B. R. Stults, E. L. Tasset, R. O. Day and R. S. Marianelli, *J. Am. Chem. Soc.*, 1974, **96**, 2650.
263a. G. V. Tsintvadze, T. I. Tsivtsivadze and F. V. Orbeladze, *Zh. Strukt. Khim.*, 1974, **15**, 306.
263b. M. V. Veidis, B. Dockum, F. F. Charron, Jr., W. M. Reiff and T. F. Brennan, *Inorg. Chim. Acta*, 1981, **53**, L197; K. Tomita, *Acta Crystallogr., Sect A*, 1981, **37**, C226; F. Bigoli, A. Braibanti, M. A. Pellinghelli and A. Tiripicchio, *Acta Crystallogr., Sect. B*, 1973, **29**, 39; M. C. Weiss and V. L. Goedken, *Inorg. Chem.*, 1979, **18**, 274; M. G. B. Drew, A. H. Bin Othman, S. G. McFall, P. D. A. McIlroy and S. M. Nelson, *J. Chem. Soc., Dalton Trans.*, 1977, 438; M. G. B. Drew, A. H. Bin Othman, S. G. McFall, P. D. A. McIlroy and S. M. Nelson, *J. Chem. Soc., Dalton Trans.*, 1977, 1173; G. V. Tsintvadze, T. I. Tsivsivadze and F. V. Orbeladze, *Zh. Strukt. Khim.*, 1975, **16**, 320; D. W. Engelfriet, G. C. Verschoor and W. J. Vermin, *Acta Crystallogr., Sect. B*, 1979, **35**, 2927; P. Lumme, J. Mutikainen and E. Lindell, *Inorg. Chim. Acta*, 1983, **71**, 217.
264. J. N. McElearney, L. Balagdt, J. A. Muir and R. D. Spence, *Phys. Rev. V*, 1979, **19**, 306.
265. M. A. A. F. De C. T. Carrondo, R. Shakir and A. C. Skapski, *J. Chem. Soc., Dalton Trans.*, 1978, 844.
266. J. Hauck, K. Schwochau and R. Bucksch, *Inorg. Nucl. Chem. Lett.*, 1973, **9**, 927.
267. J. Drummond and J. S. Wood, *J. Chem. Soc. (D)*, 1969, 1373.
268. B. R. Davis and J. A. Ibers, *Inorg. Chem.*, 1970, **9**, 2768.
269. A. F. Berndt and K. W. Barnett, *J. Organomet. Chem.*, 1980, **184**, 211.
270. M. G. B. Drew, A. H. Bin Othman and S. M. Nelson, *J. Chem. Soc., Dalton Trans.*, 1976, 1394.
271. G. J. Long, G. Galeazzi, U. Russo, G. Valle and S. Calogero, *Inorg. Chem.*, 1983, **22**, 507.
272. G. Dessy and V. Fares, *Cryst. Struct. Commun.*, 1981, **10**, 1025; C. Cairns, S. M. Nelson and M. G. B. Drew, *J. Chem. Soc., Dalton Trans.*, 1981, 1965; C. L. Raston, W. G. Sly and A. H. White, *Aust. J. Chem.*, 1980, **33**, 221; J. H. Enemark, R. D. Feltham, B. T. Huie, P. L. Johnson and K. B. Swede, *J. Am. Chem. Soc.*, 1977, **99**, 3285.
273. V. M. Padmanabham, R. Balasubramanian and K. V. Muralidharan, *Acta Crystallogr., Sect. B*, 1968, **24**, 1638.
274. R. F. Ziolo, R. M. Shelby, R. H. Stanford, Jr. and H. B. Gray, *Cryst. Struct. Commun.*, 1974, **3**, 469.
275. K. Isobe, P. M. Bailey, P. Schofield, J. T. Gauntlett, A. Nutton and P. M. Maitlis, *J. Chem. Soc., Chem. Commun.*, 1982, 425.
276. W. Rigby, P. M. Bailey, J. A. McCleverty and P. M. Maitlis, *J. Chem. Soc., Dalton Trans.*, 1979, 371.
277. D. L. Kepert, E. S. Kucharski and A. H. White, *J. Chem. Soc., Dalton Trans.*, 1980, 1932.
278. A. Mangia, M. Nardelli and G. Pelizzi, *Acta Crystallogr., Sect. B*, 1974, **30**, 487.
279. A. L. Beachamp, L. Pozdornik and R. Rivest, *Acta Crystallogr., Sect. B*, 1976, **32**, 650.
280. A. Clearfield, R. Gopal, R. J. Kline, M. Sipski and L. O. Urban, *J. Coord. Chem.*, 1978, **7**, 163; L. F. Druding and F. D. Sancilio, *Acta Crystallogr., Sect. B*, 1974, **30**, 2386; R. J. Restivo, G. Ferguson, R. W. Hay and D. P. Piplani, *J. Chem. Soc., Dalton Trans.*, 1978, 1131.
281. T. J. Westcott, A. H. White and A. C. Willis, *Aust. J. Chem.*, 1980, **33**, 1853; R. Faggiani, C. J. L. Lock, D. R. Eaton

and S. Ah-Yeung, *Cryst. Struct. Commun.*, 1981, **10**, 1091; C. A. Ghilardi and A. B. Orlandini, *J. Chem. Soc., Dalton Trans.*, 1972, 1698; I. Grenthe and E. Nordin, *Inorg. Chem.*, 1979, **18**, 1109; J. C. Stevens, P. J. Jackson, W. P. Schammel, G. G. Chrisoph and D. H. Busch, *J. Am. Chem. Soc.*, 1980, **102**, 3283; D. W. Engelfriet, G. C. Verschoor and W. den Brinker, *Acta Crystallogr., Sect. B*, 1980, **36**, 1554; G. Albertin, G. Pelizzi and E. Bordigno, *Inorg. Chem.*, 1983, **22**, 515.

282. M. R. Udupa and B. Krebs, *Inorg. Chim. Acta*, 1980, **42**, 37.
283. Yu. A. Simonov, A. A. Dvorkin and O. A. Bologa, *Izv. Akad. Nauk Mold. SSR*, 1980, 48.
284. J. A. Ibers, D. S. Hamilton and W. H. Baddley, *Inorg. Chem.*, 1973, **12**, 229.
285. J. H. Enemark, *Inorg. Chem.*, 1971, **10**, 1952.
286. C. G. Pierpont, D. N. Hendrickson, D. M. Duggan, F. Wagner and E. K. Barefield, *Inorg. Chem.*, 1975, **14**, 604.
287. N. B. Pahor, M. Calligaris and L. Randaccio, *J. Chem. Soc., Dalton Trans.*, 1976, 725.
288. W. P. Fehlhammer and L. F. Dahl, *J. Am. Chem. Soc.*, 1972, **94**, 3377.
289. J. J. MacDougall, J. H. Nelson, M. W. Babich, C. C. Fuller and R. A. Jacobson, *Inorg. Chim. Acta*, 1978, **27**, 201.
290. K. J. Haller and J. H. Enemark, *Inorg. Chem.*, 1978, **17**, 3552.
291. L. Capacchi, G. F. Gasparri, M. Nardelli and G. Pelizzi, *Acta Crystallogr., Sect. B*, 1968, **24**, 1199.
292. B. Wong, B. Jacobson, P. C. Chieh and A. J. Carty, *Inorg. Chem.*, 1974, **13**, 284.
293. U. A. Gregory, J. A. J. Jarvis, B. T. Kilbourn and P. G. Owston, *J. Chem. Soc. (A)*, 1970, 2770.
294. G. R. Clark and G. J. Palenik, *Inorg. Chem.*, 1970, **9**, 2754.
295. (a) M. J. d'Aniello, Jr., M. T. Mocella, F. Wagner, E. K. Barefield and I. C. Paul, *J. Am. Chem. Soc.*, 1975, **97**, 192.
295. (b) F. Wagner, M. T. Mocella, M. J. d'Aniello, Jr., A. H.-J. Wang and E. K. Barefield, *J. Am. Chem. Soc.*, 1974, **96**, 2625.
296. B. Hendiksen, W. C. Riley, M. W. Babich, J. H. Nelson and R. A. Jacobson, *Inorg. Chim. Acta*, 1982, **57**, 29.
297. A. E. Shvelashvili, E. B. Miminoshvili, B. M. Shchedrin, A. I. Kvitashvili, M. N. Kandelaki, T. N. Sakvarelidze and M. G. Tavberidze, *Koord. Khim.*, 1980, **6**, 1251.
298. G. V. Tsintvadze, T. I. Tsivtsivadze and F. V. Orbeladze, *Zh. Strukt. Khim.*, 1975, **16**, 319; A. R. Davis, F. W. B. Einstein and A. C. Willis, *Acta Crystallogr., Sect. B*, 1982, **38**, 437; G. Nieuwpoort and G. C. Verschoor, *Inorg. Chem.*, 1981, **20**, 4079; B. M. Foxman, P. L. Goldberg and H. Mazurek, *Inorg. Chem.*, 1981, **20**, 4368; K. R. Adam, G. Anderegg, K. Henrick, A. J. Leong, L. F. Lindoy, H. C. Lip, M. McParlin, R. J. Smith and P. A. Tasker, *Inorg. Chem.*, 1981, **20**, 4048; W. E. Hill, J. G. Taylor, C. A. McAuliffe, K. W. Muir and L. Manojlovic-Muir, *J. Chem. Soc., Dalton Trans.*, 1982, 833.
299. M. Nardelli, G. F. Gasparri, A. Musatti and A. Manfredotti, *Acta Crystallogr.*, 1966, **21**, 910; A. E. Shvelashvili, M. A. Porai-Koshits and A. S. Antsyshkina, *Zh. Strukt. Khim.*, 1969, **10**, 650; J. Lipowski and G. D. Andreetti, *Transition Met. Chem.*, 1978, **3**, 117.
300. (a) A. J. Carty, P. C. Chieh, N. J. Taylor and Y. S. Wong, *J. Chem. Soc., Dalton Trans.*, 1976, 572.
300. (b) U. Behrens and K. Hoffmann, *J. Organomet. Chem.*, 1977, **129**, 273; G. Ferguson and M. Parvez, *Acta Crystallogr., Sect. B*, 1979, **35**, 2207.
301. G. Beran, A. J. Carty, P. C. Chieh and H. A. Patel, *J. Chem. Soc., Dalton Trans.*, 1973, 488; H. J. Gysling, H. R. Luss and D. L. Smith, *Inorg. Chem.*, 1979, **18**, 2696; J. J. McDougall, A. M. Holt, P. de Meester, N. W. Alcock, F. Mathey and J. H. Nelson, *Inorg. Chem.*, 1980, **19**, 1439.
302. D. V. Naik, G. J. Palenik, B. Jacobson and A. J. Carty, *J. Am. Chem. Soc.*, 1974, **96**, 2286.
303. R. F. Ziolo, A. Allen, D. D. Titus, M. B. Bray and Z. Dori, *Inorg. Chem.*, 1972, **11**, 3044.
304. J. Pickardt, *Z. Naturforsch., Teil B*, 1982, **37**, 110.
305. D. Britton and J. D. Dunitz, *Acta Crystallogr.*, 1965, **18**, 424.
306. G. D. Andreetti, L. Coghi, M. Nardelli and P. Sgarabotto, *J. Cryst. Mol. Struct.*, 1971, **1**, 147.
307. M. Cannas, G. Carta and G. Marongiu, *J. Chem. Soc., Dalton Trans.*, 1974, 550.
308. G. Marongiu, E. C. Lingafelter and P. Paoletti, *Inorg. Chem.*, 1969, **8**, 2763.
309. A. Pajunen and K. Smolander, *Finn. Chem. Lett.*, 1974, 99.
310. P. Domiano, A. Musatti, M. Nardelli, C. Pelizzi and G. Predieri, *J. Chem. Soc., Dalton Trans.*, 1975, 2357.
311. S. M. Nelson, F. S. Esho and M. G. B. Drew, *J. Chem. Soc., Chem. Commun.*, 1981, 388.
312. R. Kivekas, A. Pajunen and K. Smolander, *Finn. Chem. Lett.*, 1977, 256.
313. C. Panattoni and E. Frasson, *Gazz. Chim. Ital.*, 1963, **93**, 601.
314. R. Louis, D. Pilessard and R. Weiss, *Acta Crystallogr., Sect. B*, 1976, **32**, 1480.
315. L. C. Capacchi, G. F. Gasparri, M. Ferrari and M. Nardelli, *Chem. Commun.*, 1968, 910.
316. J. A. Muir, M. N. Muir and S. Arias, *Acta Crystallogr., Sect. B*, 1982, **38**, 1318.
317. A. P. Gaughan, R. F. Ziolo and Z. Dori, *Inorg. Chem.*, 1971, **10**, 2776.
318. Y. Agnus, R. Louis and R. Weiss, *J. Am. Chem. Soc.*, 1979, **101**, 3381; R. Louis, Y. Agnus and R. Weiss, *Acta Crystallogr., Sect. A*, 1981, **37**, C230; J. Comarmond, P. Plumeré, J. M. Lehn, Y. Agnus, R. Louis, R. Weiss, O. Kahn and I. M. Baradau, *J. Am. Chem. Soc.*, 1982, **104**, 6330.
319. (a) R. F. Ziolo, A. P. Gaughan, Z. Dori, C. G. Pierpont and R. Eisenberg, *Inorg. Chem.*, 1971, **10**, 1289.
319. (b) I. Agrell, *Acta Chem. Scand.*, 1969, **23**, 1667.
320. K. D. Karlin, J. C. Hayes, J. P. Hutchinson and J. Zubieta, *J. Chem. Soc., Chem. Commun.*, 1983, 376.
321. D. P. Anderson and J. C. Marschall, *Inorg. Chem.*, 1978, **17**, 1258; L. Merz and W. Haase, *J. Chem. Soc., Dalton Trans.*, 1978, 1594.
322. M. G. B. Drew, C. Cairns, S. M. Nelson and J. Nelson, *J. Chem. Soc., Dalton Trans.*, 1981, 942; S. Tyagi and B. J. Hathaway, *J. Chem. Soc., Dalton Trans.*, 1981, 2029; J. Gazo, M. Kabesová and J. Soldánová, *Inorg. Chim. Acta*, 1983, **76**, 4203.
323. W. Haase, R. Mergehenn and W. Krell, *Z. Naturforsch., Teil B*, 1976, **31**, 85; M. Cannas, G. Carta and G. Marongiu, *J. Chem. Soc., Dalton Trans.*, 1974, 556; M. Cannas, G. Carta and G. Marongiu, *Gazz. Chim. Ital.*, 1974, **104**, 581; A. P. Gaughan, R. F. Ziolo and Z. Dori, *Inorg. Chim. Acta*, 1970, **4**, 640.
324. C. L. Raston, B. Walter and A. H. White, *Aust. J. Chem.*, 1979, **32**, 2757.
325. I. Agrell, *Acta Chem. Scand.*, 1970, **24**, 1247.
326. I. Agrell, N. G. Vanneberg, *Acta Chem. Scand.*, 1971, **25**, 1630.

327. I. Agrell, *Acta Chem. Scand.*, 1970, **24**, 3575.
328. U. Müller, *Z. Naturforsch., Teil B*, 1973, **28**, 426.
329. U. Müller, *Z. Anorg. Allg. Chem.*, 1974, **399**, 183.
330. D. J. Brauer, H. Burger, G. Pawelke, K. H. Flegler and A. Hass, *J. Organomet. Chem.*, 1978, **160**, 389.
331. G. D. Andreetti, P. C. Jain and E. C. Lingafelter, *J. Am. Chem. Soc.*, 1969, **91**, 4112.
332. A. E. Shvelashvili, M. A. Porai-Koshits, A. I. Kvitashvili, B. M. Shchedrin and L. P. Sarishvili, *Zh. Strukt. Khim.*, 1974, **15**, 315.
333. M. Cannas, G. Carta, A. Cristini and G. Marongiu, *Inorg. Chem.*, 1977, **16**, 228.
334. L. R. Groeneveld, G. Vos, G. C. Verschoor and J. Reedijk, *J. Chem. Soc., Chem. Commun.*, 1982, 620.
335. E. C. Alyea, G. Ferguson and R. J. Restivo, *J. Chem. Soc., Dalton Trans.*, 1977, 1845.
336. W. Clegg, H. Krischner, A. I. Saracoglu and G. M. Sheldrick, *Z. Kristallogr.*, 1982, **161**, 307.
337. H. Krischner, C. Kratky and H. E. Maier, *Z. Kristallogr.*, 1982, **161**, 225.
338. F. Bigoli, A. Braibanti, M. A. Pellinghelli and A. Tiripicchio, *Acta Crystallogr., Sect. B*, 1973, **29**, 2344; M. G. B. Drew and S. M. Nelson, *Acta Crystallogr., Sect. A*, 1975, **31**, S140; F. Bigoli, A. Braibanti, M. A. Pellinghelli and A. Tiripicchio, *Acta Crystallogr., Sect. B*, 1973, **29**, 2708; M. R. Caira and L. R. Nassimbeni, *Cryst. Struct. Commun.*, 1976, **5**, 309; A. E. Shvelashvili and M. A. Porai-Koshits, *Koord. Khim.*, 1975, **1**, 463.
339. I. Matsichek, V. K. Trunov and R. I. Machkhoshvili, *Zh. Neorg. Khim.*, 1981, **26**, 1690; M. Cannas, G. Cart, A. Cristini and G. Marongiu, *J. Chem. Soc., Dalton Trans.*, 1976, 210; R. E. Marsh and V. Schomaker, *Inorg. Chem.*, 1979, **18**, 2331.
340. F. Bigoli, A. Braibanti, M. A. Pellinghelli and A. Tiripicchio, *Acta Crystallogr., Sect. B*, 1972, **28**, 963; R. G. Goel, W. P. Henry, M. J. Olivier and A. L. Beauchamp, *Inorg. Chem.*, 1981, **20**, 3924.
341. J. L. Martin, L. C. Thompson, L. J. Radonovich and M. D. Glick, *J. Am. Chem. Soc.*, 1968, **90**, 4493; P. I. Lazarev, L. A. Aslanov and M. A. Porai-Koshits, *Koord. Khim.*, 1975, **1**, 706.
342. P. Charpin, M. Lance and A. Navara, *Acta Crystallogr., Sect. C*, 1983, **39**, 190.
343. G. Bombieri, F. Benotello, K. W. Bagnall, M. J. Plerus and D. Brown, *J. Chem. Soc., Dalton Trans.*, 1983, 45; D. L. Kepert, J. M. Patrick and A. H. White, *J. Chem. Soc., Dalton Trans.*, 1983, 559.
344. T. Kemmerich, J. H. Nelson, N. E. Takach, H. Boehme, B. Jablonski and W. Beck, *Inorg. Chem.*, 1982, **21**, 1226; L. Busetto, A. Pallazi and R. Ros, *Inorg. Chem.*, 1970, **9**, 2792; G. La Monica, C. Monti, M. Pizotti and S. Cenini, *J. Organomet. Chem.*, 1983, **241**, 241.
345. R. Goddard, S. D. Killops, S. A. R. Knox and P. Woodward, *J. Chem. Soc., Dalton Trans.*, 1978, 1255.
346. A. J. Deeming, I. Ghatak, D. W. Owen and R. Peeters, *J. Chem. Soc., Chem. Commun.*, 1982, 392.
347. K. Burgess, B. F. G. Johnson and J. Lewis, *J. Organomet. Chem.*, 1983, **247**, C42.
348. H. Werner, W. Beck and H. Engelman, *Inorg. Chim. Acta*, 1969, **3**, 331.
349. K. Dehnicke and R. Dübgen, *Z. Anorg. Allg. Chem.*, 1978, **444**, 61; R. J. Angelici and L. Busetto, *J. Am. Chem. Soc.*, 1969, **91**, 3197; M. Graziani, L. Busetto and A. Pallazi, *J. Organomet. Chem.*, 1971, **26**, 261; M. L. Blohm, D. E. Fjare and W. L. Gladfelter, *Inorg. Chem.*, 1983, **22**, 1004; W. Beck, H. Werner, H. Engelman and H. S. Smedal, *Angew. Chem.*, 1966, **78**, 267; *Angew. Chem., Int. Ed. Engl.*, 1966, **5**, 253; *Chem. Ber.*, 1968, **101**, 146; R. J. Angelici and G. C. Faber, *Inorg. Chem.*, 1971, **10**, 514; W. Beck and K. von Werner, *Chem. Ber.*, 1977, **104**, 2901.
350. D. A. Brown, F. M. Hussein and C. L. Arora, *Inorg. Chim. Acta*, 1978, **29**, L215.
351. W. Beck, M. Bauder, W. P. Fehlhammer, P. Pöllmann and H. Schachl, *Chem. Ber.*, 1969, **102**, 1976.
352. J. T. Moelwyn-Hughes, A. W. B. Garner and A. S. Howard, *J. Chem. Soc. (A)*, 1971, 2361; *J. Chem. Soc. (A)*, 1971, 2370.
353. L. Busetto, M. Graziani and U. Belluco, *Inorg. Chem.*, 1971, **10**, 78.
354. R. Huisgen, *J. Org. Chem.*, 1976, **41**, 403.
355. W. P. Fehlhammer and W. Beck, *Z. Naturforsch., Teil B*, 1983, **38**, 546.

13.6

Polypyrazolylborates and Related Ligands

ALAN SHAVER
McGill University, Montreal, Quebec, Canada

13.6.1 INTRODUCTION

In the late 1960s and early 1970s Swiatoslaw Trofimenko prepared the polypyrazolylborate class of ligand and investigated its coordination chemistry.[1,2] From the very beginning it was obvious that these ligands would have a widespread and varied chemistry. Complexes containing these ligands are known for almost every transition metal and many of the compounds display unusual structural or chemical features. The area has been reviewed[1-4] and the reader is directed to these articles to supplement the material presented here.

The ligands consist of two to four pyrazolyl groups attached to a boron atom, *i.e.* $H_2B(pz)_2^-$, $HB(pz)_3^-$ and $B(pz)_4^-$, where pz = 1-pyrazolyl. Thus, excellent bidentate (**1**) and tridentate (**2**)

(1)   (2)

ligands result. Since the single most important feature of these ligands is their ability to form chelate rings only those pyrazolylborate ligands with more than one pyrazolyl ring attached to the boron atom will be discussed here.

13.6.2 GENERAL FEATURES

13.6.2.1 Preparation of Ligands

The ligands are prepared by reaction of pyrazole with the tetrahydroborate ion. By regulating the temperature, the reaction can be stopped at any of the stages of substitution at the boron atom (Scheme 1). The resulting salts are stable, colourless materials with excellent shelf life. The potassium and sodium salts are most often used; however, quaternary ammonium salts and even the free acids can be prepared. The ligands $H_2B(3,5-Me_2pz)_2^-$ and $HB(3,5-Me_2pz)_3^-$, where $3,5-Me_2pz = 3,5$-dimethylpyrazolyl, are prepared in the same way but in this case substitution stops at three $3,5-Me_2pz$ rings. The free acid of the tetrasubstituted species, $B(3,5-Me_2pz)_3(3,5-Me_2pzH)$, was recently isolated from the reaction of $[TiCp_2Cl_2]$ with $HB(3,5-Me_2pz)_3^-$ ion.[5a] The unsymmetrical ligand $H_2B(pz)(3,5-Me_2pz)^-$ has been prepared.[5b] Ligands with alkyl or aryl substituents on the boron atom may be prepared from BR_4^- or H_3BR^- to give $R_2B(pz)_2^-$ and $RB(pz)_3^-$, respectively. Thus, a great variety of ligands is possible. The most commonly used are: $H_nB(pz)_{4-n}^-$, where $n = 0$, 1, 2; $H_nB(3,5-Me_2pz)_{4-n}^-$, where $n = 1$, 2; $Et_2B(pz)_2^-$; and $Ph_2B(pz)_2^-$. The incorporation of Et, Ph or $3,5-Me_2pz$ groups leads to bulky ligands which can sometimes have profound effects on the reactivity of their complexes.

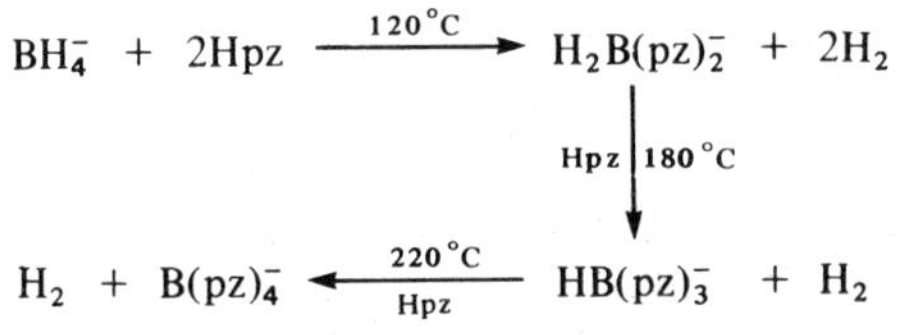

$$BH_4^- + 2Hpz \xrightarrow{120\,°C} H_2B(pz)_2^- + 2H_2$$

$$Hpz \downarrow 180\,°C$$

$$H_2 + B(pz)_4^- \xleftarrow[Hpz]{220\,°C} HB(pz)_3^- + H_2$$

Scheme 1

13.6.2.2 Coordination Properties

13.6.2.2.1 Bidentate ligands

The ligands of type (1) form stable complexes (3) with most of the first row transition metals with either tetrahedral geometry (M = Mn, Fe, Co, Zn) or square-planar geometry (M = Cr, Ni, Cu). An analogy between (1) and the acac ligand has been drawn[1,2] since both are uninegative four-electron ligands. Unlike many acac complexes those of type (3) are always monomeric because the nitrogen atoms bonded to the metal are sterically hindered from bridging to another metal. Another difference is that while tris acac complexes are common, tris $R_2B(pz)_2^-$ complexes are very rare.[6] The ring formed upon chelation of a bidentate pyrazolylborate ligand to a metal is not planar. The two five-membered pyrazolyl rings require that the metal and boron atoms be coplanar with each of them. Thus, the $B(pz)_2M$ ring has only two planes and the natural conformation is boat shaped as in (4) (the carbon atoms of the pz rings have been omitted for clarity). This leads to some very interesting structural and stereochemical consequences, the details of which will be discussed below. Generally speaking, stereochemical nonrigidity can occur as a result of ring inversion. In addition the boat conformation brings the R group attached to the boron atom quite close to the metal. This can lead to obstruction of the coordination site as in $[Ni(Et_2B(pz)_2)_2]$ which will not add pyridine in contrast to the $H_2B(pz)_2^-$ analogue. In the case of $[Mo(H_2B(3,5-Me_2pz)_2)(CO)_2(\eta^3\text{-allyl})]$ the hydrogen atom is actually attached to the molybdenum atom by a three-center two-electron bond (5). One methylene proton is similarly attached in the $Et_2B(pz)_2^-$ analogue.

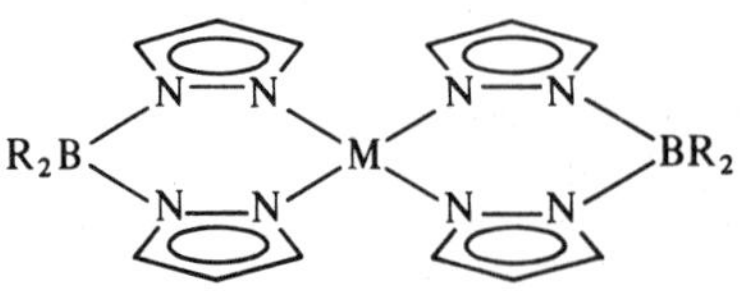

(3) 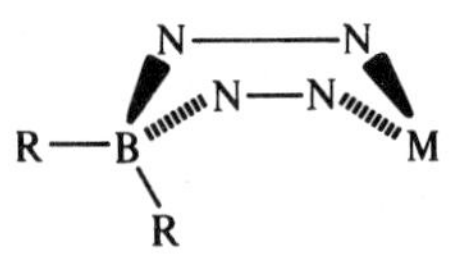   (4) 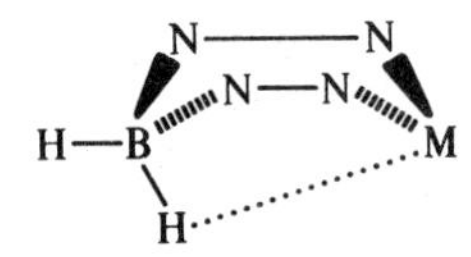   (5)

13.6.2.2.2 *Tridentate ligands*

Ligand (**2**) is an ideal tripod that also forms bis complexes with many transition metals (**6**). The complexes are octahedral and the ligands tridentate. The $RB(pz)_3^-$ ligand is formally analogous to the cyclopentadienyl (Cp) ligand since both are uninegative six-electron donors which occupy three coordination sites. This analogy has been explored extensively in the chemistry of complexes containing one $RB(pz)_3^-$ or $HB(3,5-Me_2pz)_3^-$ ligand. The bands in the ^1H NMR spectrum due to the proton attached to C-4, a triplet for the pz ring, appear at a different chemical shift when the ring is bonded to a metal than those for an unbonded ring. In some cases, however, all four pz rings are equivalent on the NMR timescale due to a dissociative exchange process referred[1,2] to as 'tumbling'. One pz ring dissociates and is replaced by the unattached ring after the bidentate chelate ring inverts. In complexes of the type (**7**), where A = B ≠ C, a type commonly encountered, three types of pz ring are usually detected in the ratio 2:1:1 in the ^1H NMR spectrum of $B(pz)_4^-$ complexes, corresponding to one unbound and three bound rings, the latter split into two sets by the C_s symmetry. However, sometimes the three bound rings are equivalent due to rotation of the triangular face formed by the three attached nitrogen atoms about the boron to metal axis. This form of fluxionality ('rotation') is nondissociative since the environment of the unbound pz ring is never exchanged.

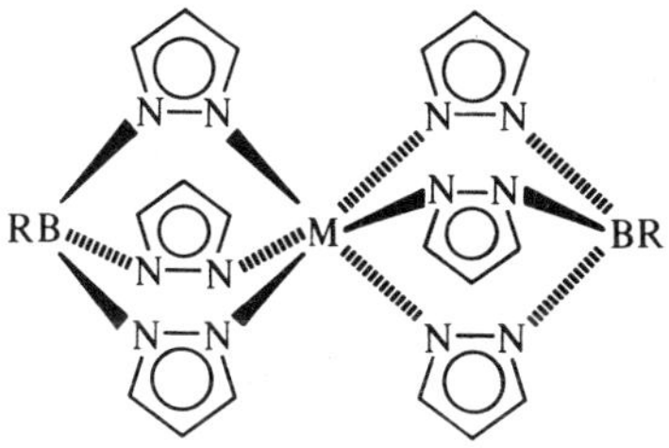

(6) 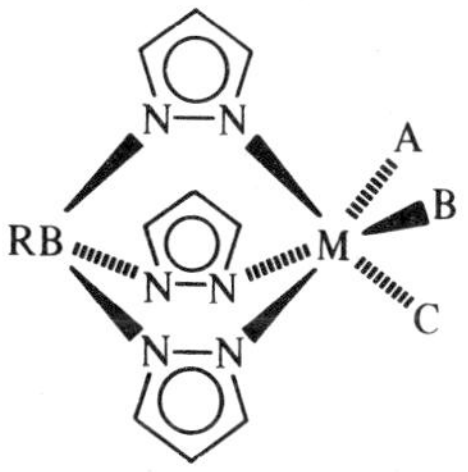   (7)

In the great majority of complexes, the $RB(pz)_3^-$ and $HB(3,5-Me_2pz)_3^-$ ligands are tridentate. However, these ligands can be bidentate and, like bispyrazolyl ligands, can exhibit conformational isomerism. As before the general case will be outlined here and detailed examples given below. For any system $XYB(pz)_2MAB$ two boat conformations exist. In the case where X = Y or A = B the two conformers (**8a**) and (**8b**) will be equivalent. However, if X ≠ Y and A ≠ B then (**8a**) and (**8b**) will be different *and* there will be another way to arrange the four groups to give (**9a**) and (**9b**). Thus, (**8**) and (**9**) are geometrical isomers. The conformers can be interconverted by rotations about single bonds, in this case ring inversion, but the geometrical isomers cannot. The ring inversion has been observed in a number of complexes and is usually rapid on the NMR time scale. This has led to variable temperature studies. The slow exchange limiting spectrum is usually obtained by $-100\,°C$. Interestingly, there is to date no confirmed example of the detection of geometrical isomers of types (**8**) and (**9**).

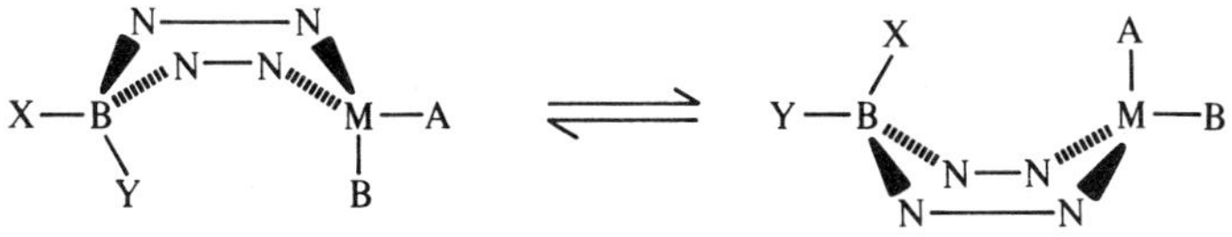

(8a)   (8b)

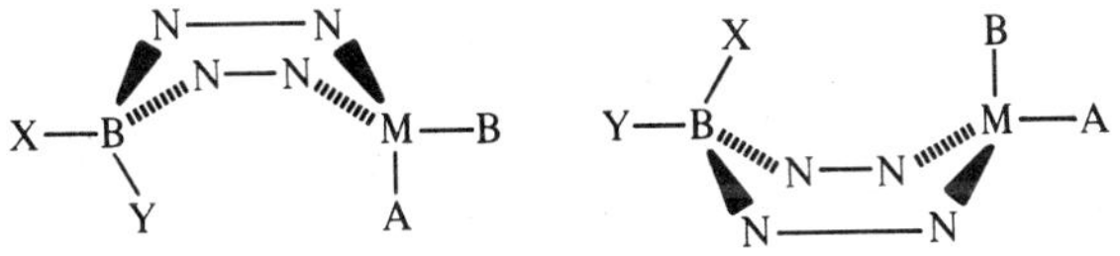

(9a)   (9b)

13.6.2.3 Preparation of Complexes

By far the most common synthesis technique is to introduce polypyrazolylborate ligands by replacing halide groups. As might be expected this leads to a great variety of complexes. In some cases neutral ligands can be displaced as with the reaction of $Mo(CO)_6$ to give anions of the type $[Mo(HB(pz)_3)(CO)_3]^-$. The general bonding and stereochemical possibilities having been outlined above, a group by group description of the complexes formed by these ligands follows. The purpose is to outline what structural types exist, how they bond and their spectroscopic properties. Accordingly some details considered to be of a general nature have been omitted. Some emphasis has been placed on work reported since the last reviews.[3,4]

13.6.3 COMPLEXES OF THE TRANSITION METALS

13.6.3.1 Titanium, Zirconium, Vanadium, Niobium, Tantalum

The complexes $[Ti(HB(pz)_3)CpCl_x]$, where $x = 1$, 2; $[Ti(H_yB(pz)_{4-y})Cp_2]$, where $y = 0$, 1, 2; $[M(HB(pz)_3)(Cl)_2(THF)]$ and $[M(HB(pz)_3)(Cl)_3]$, where M = Ti, V, have been isolated *via* halide replacement.[7] The polypyrazolylborate ligands are probably bidentate in the Cp_2 complexes. One of these, $[Ti(B(pz)_4)Cp_2]$, reacts with $[TiCp_2Cl]$ and $NaBPh_4$ to give $[Ti_2(B(pz)_4)Cp_4]BPh_4$ where the $B(pz)_4^-$ ligand bridges the two metal atoms (**10**). Complexes of zirconium have been prepared that illustrate: (1) the ability of the ligands to stabilize other functional groups on the metal and (2) the difference in steric requirements of the pz and $3,5\text{-Me}_2\text{pz}$ ligands. Thus, the stable alkyl complexes $[Zr(HB(3,5\text{-Me}_2\text{pz})_3)(OBu^t)_{3-x}R_x]$, where $x = 2$, R = Me, CH_2Ph; $x = 1$, R = Me, $C\equiv CR'$, were prepared from the appropriate chloro complexes.[8] The complex $[Zr(HB(3,5\text{-Me}_2\text{pz})_3)(OBu^t)Cl_2]$ shows two types of rings in its ^1H NMR and so is rigid on the NMR timescale while the complexes $[Zr(RB(pz)_3)(OBu^t)Cl_2]$, where R = Bu^n, Pr^i, H, pz, are nonrigid showing only one type of bound ring. The unbound pz ring never exchanges and a trigonal twist is proposed.[9]

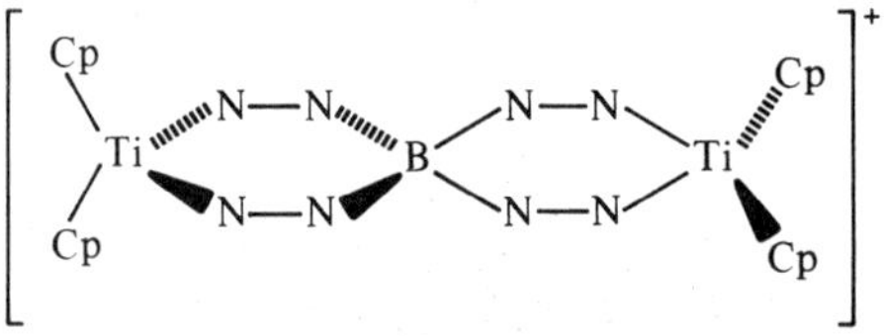

(10)

The only tris-$H_2B(pz)_2^-$ complex reported is of vanadium(II), namely $K[V(H_2B(pz)_2)_3]$. The X-ray structure confirmed the classic tris chelate octahedral arrangement but steric crowding resulted in very flat boat conformations for the BN_4V rings.[6] A limited number of niobium compounds have been reported[5a,10] but their structures have not been determined. Seven-coordinate tantalum complexes include $[Ta(H_2B(pz)_2)_2Me_3]$[11] and $[Ta(H_2B(3,5\text{-Me}_2\text{pz})_2)Me_3Cl]$.[12] In the latter the seventh coordination site is occupied by a boron hydrogen atom attached to the metal atom by a three-center two-electron bond (**11**). Resonances due to all three of the possible geometric isomers are observed in the low temperature ^1H NMR spectrum. At room temperature rapid rotation of the triangular faces of the molecule results in an averaged spectrum.

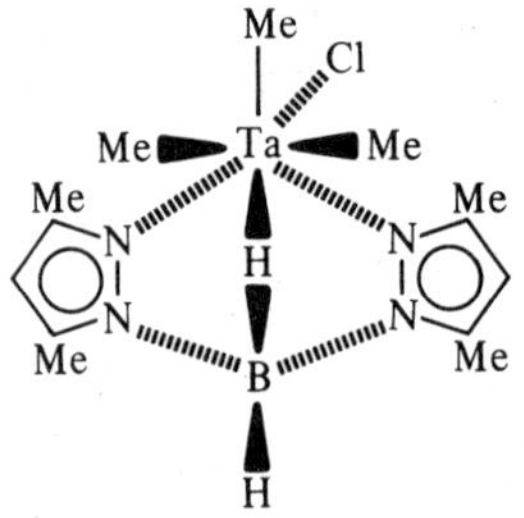

(11)

13.6.3.2 Molybdenum, Tungsten

Complexes of molybdenum form the largest set of any of the transition metals. Many of these compounds display spectroscopic properties, isomerism and structural features that illustrate the coordination chemistry of polypyrazolylborate complexes. In addition, this system illustrates that trispyrazolylborate ligands are more bulky than their formal analogue C_5H_5.

Complexes of the type [MoCp(CO)$_3$X] are formally seven-coordinate but only the complexes [Mo(HB(pz)$_3$)(CO)$_3$X] (12), where X = H,[13] Br,[14] I,[15] and [Mo(HB(pz)$_3$)(CO)$_2$(CS)I][15] are seven coordinate. Whereas the radical [MoCp(CO)$_3$] dimerizes to give [MoCp(CO)$_3$]$_2$, the analogous radical [Mo(HB(pz)$_3$)(CO)$_3$] is stable and must be heated to give the dimer [Mo(HB(pz)$_3$)(CO)$_2$]$_2$ with loss of two CO ligands.[16] The complex obtained by treatment of [Mo(HB(pz)$_3$)(CO)$_3$]$^-$ with MeI does not contain a σ-bonded Me group analogous to that in [MoCp(CO)$_3$Me]. Instead the steric bulk of the HB(pz)$_3^-$ ligand results in formation of the η^2-acyl [Mo(HB(pz)$_3$)(CO)$_2$(η^2-(C(O)Me)] (13).[14] An unusual carbyne complex, [Mo(HB(pz)$_3$)(CO)$_2$CX], where X is a halogen atom, has been reported.[17] Complexes containing the HB(3,5-Me$_2$pz)$_3^-$ ligand are even more crowded. The structure of Et$_4$N[Mo(HB(3,5-Me$_2$pz)$_3$)(CO)$_3$] showed that a very small pocket remained for other ligands.[18] This crowding probably accounts for the air stability of K[Mo(HB(3,5-Me$_2$pz)$_3$)(CO)$_3$] since Na[MoCp(CO)$_3$] is very air sensitive. The complex [MoCp(NO)I$_2$] is a dimer while [Mo(HB(3,5-Me$_2$pz)$_3$)(NO)I$_2$] is monomer.[19] The bulkiness of the HB(3,5-Me$_2$pz)$_3$ ligand has been exploited by McCleverty and others[20-23] who have made derivatives from the latter complex of the type [Mo(HB(3,5-Me$_2$pz)$_3$)(NO)XQ] (14), where X is a halogen ion and Q = OR, SR, NHR and NHNR^{1}R^2. These monomers are formally 16-electron complexes which cannot add additional ligands due to steric crowding. The Mo—O and Mo—N bond lengths are unusually short due to $p\pi \rightarrow d\pi$ bonding.[22c,24] The complexes containing NHR and NHNR^{1}R^2 ligands are examples of relatively rare species. The latter ligand is dihapto in the complex [MoCp(NO)(NHNR^{1}R^2)I]. The complexes [M(HB(3,5-Me$_2$pz)$_3$(CO)$_2$NS], where M = Mo and W, containing the thionitrosyl ligand are unknown for the analogous C$_5$H$_5$ systems.[25]

(12) (13) (14)

Upon treatment of the anion [Mo(HB(pz)$_3$)(CO$_3$)]$^-$ with allyl halides or treatment of [MoI(CO)$_2$(C$_7$H$_7$)] with HB(pz)$_3^-$, the complexes[2] [Mo(HB(pz)$_3$)(CO)$_2$(η^3-enyl)] (15) are isolated, where enyl = C$_3$H$_5$, CH$_2$CMeCH$_2$, C$_7$H$_7$. These are stable 18-electron complexes in which the trigonal face of the HB(pz)$_3$ ligand is rapidly rotating. A related class of complexes, [Mo{R$_2$B(pz)$_2$}(N$_2$C$_3$H$_4$)(CO)$_2$(η^3-enyl)], contains a bidentate R$_2$B(pz)$_2^-$ ligand (R = Et, Ph) and a single pyrazole group.[1,2,3,26] The X-ray structure of the complex where R = Et (16) revealed a chair conformation for the B(pz)$_2$Mo ring which is very unusual. Similar complexes containing Et$_2$B(pz)$_2^-$, or H$_2$B(3,5-Me$_2$pz)$_2$ but lacking the pyrazole ligand, are stabilized by the interesting three-center two-electron bonds from the Et or H groups attached to the boron atom. Thus, the complexes [Mo(H$_2$B(3,5-Me$_2$pz)$_2$)(CO)$_2$(η^3-enyl)] where enyl = C$_7$H$_7$ and CH$_2$CMeCH$_2$ contain a steeply puckered boat-shaped B(N$_2$)$_2$Mo ring (17), with the boron pulled towards the metal atom by a B—H $\cdots$ Mo interaction. This effectively locks the chelate ring and tends to satisfy the electronic requirements of the complex. Without this interaction the complex would be formally a 16-electron system. There was no indication of the B—H $\cdots$ Mo bond in the IR spectrum and the BH$_2$ protons are never observed in the ^{1}H NMR spectrum in any case due to broadening by the boron isotopes. In the case of complexes of the type [Mo(Et$_2$B(pz)$_2$)(CO)$_2$(η^3-enyl)], where

(15) (16) (17)

enyl = C_3H_5, C_7H_7, $CH_2CHPhCH_2$, Trofimenko[2] predicted a C—H $\cdots$ Mo interaction on the basis of low C—H stretching vibrations in the IR (in 2660 cm^{-1}) and a high chemical shift (-3 to -4 p.p.m.) for one methylene group in the ^{1}H NMR. The crystal structures of the complexes for which enyl = $CH_2CHPhCH_2$ and C_7H_7 confirm the interaction. The CH_2 group directs one C—H bond at the metal atom with the Mo—H distance being about 2.2 Å (**18**). Variable temperature NMR studies are consistent with rotation of the 'tridentate' ligand and exchange of the protons of the HC—H $\cdots$ Mo unit. Above 100 °C the C—H $\cdots$ Mo interaction dissociates and a ring flip occurs which brings the unbound ethyl group into interaction with the metal atom.

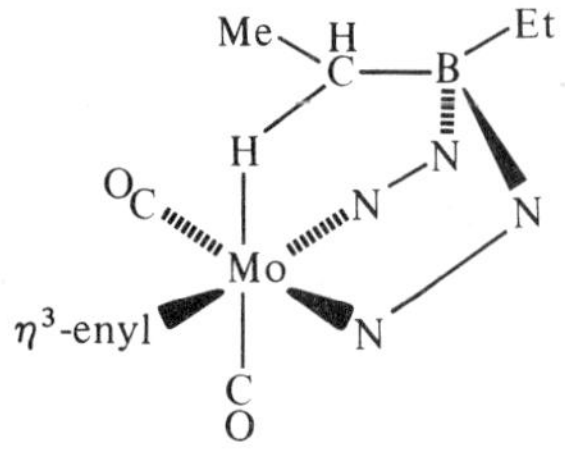

(18)

Replacement of the Et groups by Ph results in a 16-electron complex. Thus, the sixth coordination site in $[Mo(Ph_2B(pz)_2)(CO)_2(\eta^3\text{-}CH_2CPhCH_2]$ (**19**) is blocked by the flat side of one Ph ring. The two Ph rings are locked in 'edge-face' relationship by the two coordinated pz rings; thus, the expected interaction between the *ortho* hydrogen on the Ph ring and the metal atom cannot occur.

Conformational isomerism of the type described in (**8**) was observed for the complexes $[Mo(XYB(pz)_2)(CO)_2Cp]$ (**20**, **21**), where X = Y = pz, Et, and X = H, Y = pz. The IR spectrum in the carbonyl region displayed two sets of closely spaced bands (*i.e.* four bands in total) consistent with the presence of conformational isomers. The boat-shaped $B(pz)_2$Mo rings are flipping rapidly on the NMR timescale at room temperature and an averaged ^{1}H NMR spectrum is observed. The activation energy for interconversion is about 42 kJ mol^{-1}. The structure when X = Y = pz corresponds to (**20**) with a shallow boat conformation and the same type of edge–face relationship for the two pz rings as was observed for the Ph rings in (**19**). In $[Mo(H(pz)B(pz)_2)(CO)_2Cp]$ geometrical isomerism is possible but only one geometric isomer and its conformers were detected. It seems that only one isomer is produced during the preparation; its structure was not assigned.

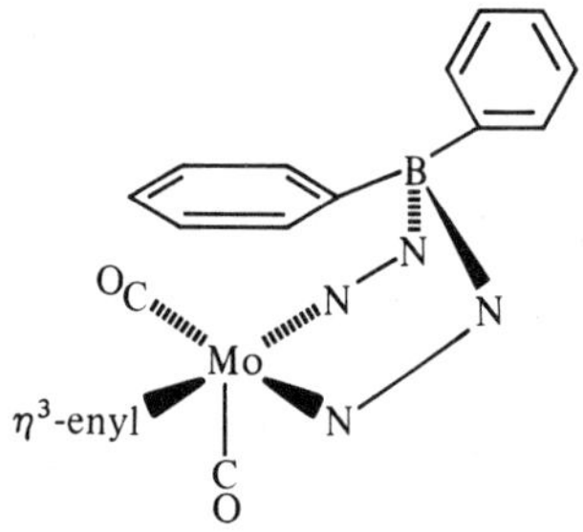

(19) 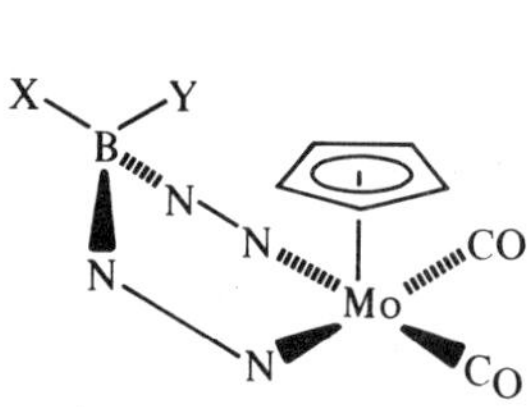   (20) 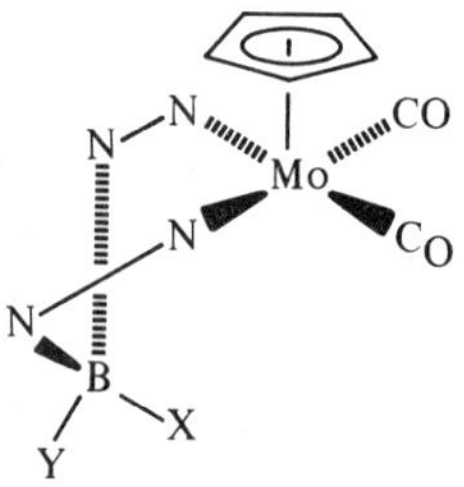   (21)

Treatment of $Mo_2(O_2CMe)_4$ with the appropriate ligand leads to the complexes $[Mo_2(Et_2B(pz)_2)_2(O_2CMe)_2]$, $[Mo_2(HB(pz)_3)_2(O_2CMe)_2]$ and $[Mo_2(Et_2B(pz)_2)_2(Et_2B(pz)OH)_2]$ which contain metal–metal quadruple bonds. The last of these is unusual, not only because it has an unsupported quadrupole bond but also because one of the pz rings has been lost and ligand (**22**) is coordinated to the metal *via* the pyrazole nitrogen and oxygen atoms.[27]

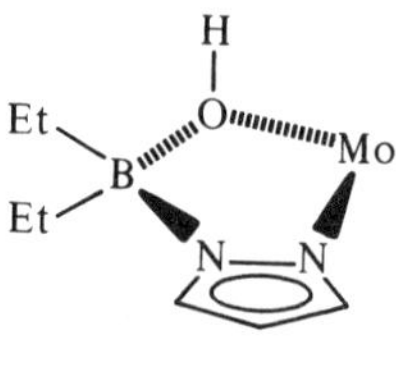

(22)

13.6.3.3 Manganese, Technetium, Rhenium

The analogues to $[M(CO)_3Cp]$ using tridentate pz and $3,5\text{-}Me_2pz$ ligands have been prepared for $M = Mn$ and Re, $(HB(pz)_3^-)$. The manganese complexes undergo carbonyl substitution reactions. The triangular face bridged by the $(pz)B(pz)_3^-$ ligand in (23) rotates about the Mn—B axis in $[Mn(B(pz)_4)(CO)_2L]$ while the analogous $HB(3,5\text{-}Me_2pz)_3$ complexes are rigid.

Recently the complexes $[M(HB(pz)_3)(O)Cl_2]$, where $M = Re$, Tc (24), and the complexes $[M(HB(pz)_3Cl_2L]$, where $M = Re$, $L = PPh_3$ and $M = Tc$, $L = PPh_3$, Ph_3PO and pz, were prepared as part of the development of radiopharmaceuticals of technetium.[28] These complexes are more lipophilic than other derivatives.

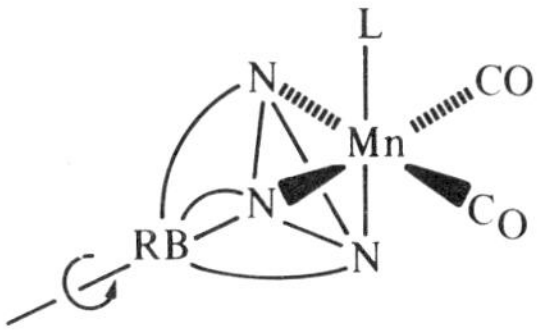

(23) 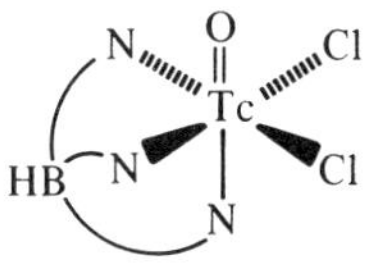   (24)

13.6.3.4 Iron, Ruthenium

The ferrocene analogues $[Fe(HB(pz)_3)_2]$ and $[Fe(HB(3,5\text{-}Me_2pz)_3)_2]$ are easily prepared, even from iron carbonyl compounds. The former is a low spin complex (Fe—N = 1.973 (7) Å) and the latter is high spin (Fe—N = 2.172 (22) Å).[29] If MeI is added to the reaction of $[Fe_2(CO)_9]$ and $HB(pz)_3^-$ or if $[Fe(I)(CO)_3(\eta^3\text{-}C_3H_5)]$ is treated with $HB(pz)_3^-$ the complexes $[Fe(HB(pz)_3)(CO)_2(C(O)R)]$ result, where $R = Me$ (25) or *cis*-CHCHMe (26). There are three possible orientations of the acyl groups (*i.e.* 27a–c). Compound (25) is in the form (27a) in the solid state while in solution at least two conformers are in rapid equilbrium since splitting occurs in the bands in the carbonyl region in the IR spectrum but only one peak is observed for the methyl group in the ^1H NMR spectrum. Complex (26) also shows splitting in the IR spectrum. Both (25) and (26) undergo facile thermal decarbonylation to give the complexes $[Fe(HB(pz)_3)(CO)_2R]$ where $R = Me$ and *trans*-CHCHMe. This sharply contrasts with the analogous Cp complex $[Fe(C(O)Me)(CO)_2Cp]$ which requires UV irradiation. Two minor side products (28, 29) from the preparation of (26) result from cleavage of pz rings from the $HB(pz)_3^-$ ligand.

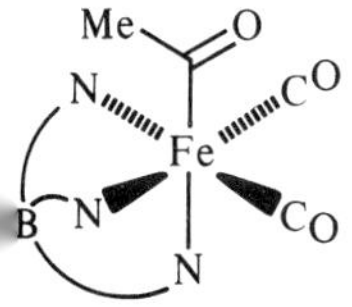

(25) 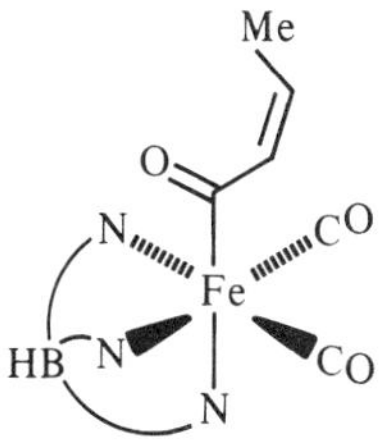   (26) 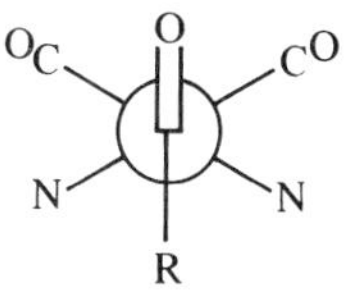   (27a) 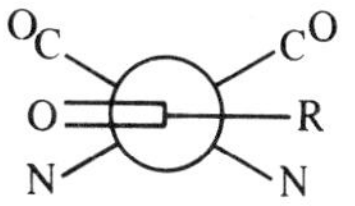   (27b) 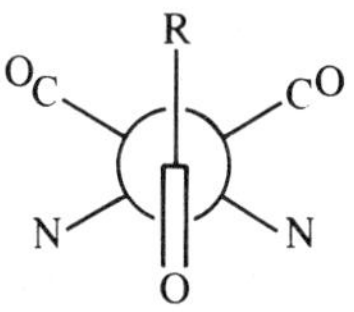   (27c)

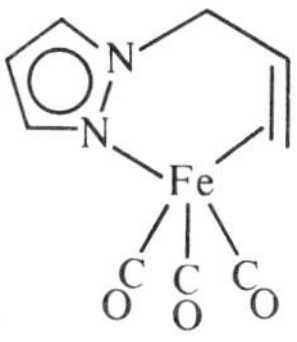

(28) 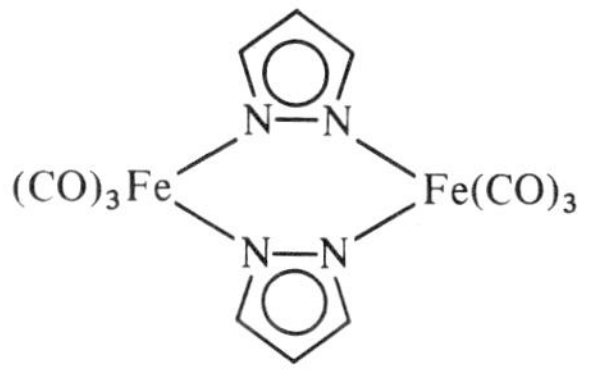   (29)

The ruthenium complexes $[Ru(HB(pz)_3)(CO)_2X]$, where $X = Cl$, Br, I, were prepared from $[Ru_3(CO)_{12}]$ and $[Ru(CO)_3Cl]_2$. Half-sandwich complexes of the type $[Ru(RB(pz)_3)(\eta^6\text{-}C_6H_6)]PF_6$ have been prepared.

13.6.3.5 Cobalt, Rhodium

Other half-sandwich complexes (**30**) have been prepared for cobalt. Since complete substitution often occurs with trispyrazolylborate ligands to give the metallocene analogue, complexes (**30**) are of note. In addition, (**30c**) contains the neutral ligand $HC(pz)_3$, a pyrazolylalkane analogue of the borates. There are large paramagnetic shifts associated with the Co^{II} atom in the octahedral complexes $[Co(p\text{-}RC_6H_4B(pz)_3)_2]$, where R = D, Bu^n, CO_2H, CO_2Me, which have been prepared to study their use as covalently bound paramagnetic shift reagents.[30]

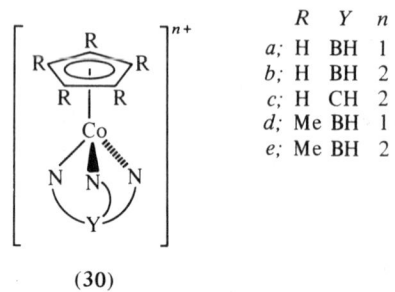

	R	Y	n
a;	H	BH	1
b;	H	BH	2
c;	H	CH	2
d;	Me	BH	1
e;	Me	BH	2

(30)

A large number of rhodium(I) complexes of the type $[Rh(XYB(pz)_2)L_2]$ (**31**) have been reported. Complexes containing three or more pyrazolyl rings undergo 'tumbling' type of exchange of all the rings. There does not appear to be any evidence for $B\text{—}H \cdots Rh$ or $HC\text{—}H \cdots Rh$ type interactions in (**31c, f** and **i**). The structures of (**31e**), (**31g**) and (**31h**), where dq = duroquinone, are of interest.[31] On the basis of low temperature ^1H NMR spectra complexes (**31e**) and (**31h**) and possibly (**31g**) were assigned five-coordinate geometry with tridentate ligands in solution. In the solid state, however, only the dq complex (**31h**) remains five-coordinate; the other two are square planar with bidentate ligands. A number of rhodium(III) complexes have been prepared from either oxidation of $[Rh_2(RB(pz)_3)_2(CO)_3]^{32}$ or substitution reactions of $[Rh_2(HB(pz)_3)_2Cl_4]$.[33] They include complexes of the type $[Rh(RB(pz)_3)(CO)XY]$, where R = H or pz and X = Y = Cl, Br or I; $[Rh(HB(pz)_3)ClY]$, where Y = acac, hfac, S_2CNEt_2; $[Rh(HB(pz)_3)Y_2L]$, where L = H_2O and Y = O_2CMe, O_2CCF_3; $[Rh(HBP_3)Cl_2L]$, where P = pz or $3,5\text{-}Me_2pz$, L = PR_3, AsR_3, py, NR_3, RCN, and are of structural type (**7**). Rhodium(I) complexes containing the $Et_2B(pz)_2$ ligand may also be oxidized[34] by I_2, MeI and $HgCl_2$ to give the compounds $[Rh(Et_2B(pz)_2)(RNC)_2XY]$, where X = Y = I; X = HgCl, Y = Cl; X = Me, Y = I.

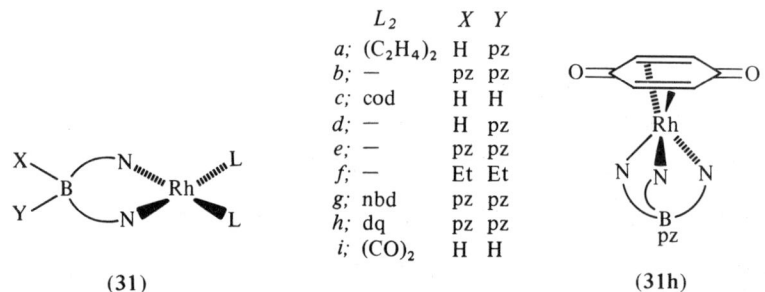

	L_2	X	Y
a;	$(C_2H_4)_2$	H	pz
b;	—	pz	pz
c;	cod	H	H
d;	—	H	pz
e;	—	pz	pz
f;	—	Et	Et
g;	nbd	pz	pz
h;	dq	pz	pz
i;	$(CO)_2$	H	H

(31)   (31h)

13.6.3.6 Nickel, Palladium, Platinum

The complexes $[Ni(R_2B(pz)_2)_2]$ (**32**), where R = H, Me, Et, Ph, have been examined for possible interactions between the R group and the nickel atom. The structure when R = H is not consistent with a $B\text{—}H \cdots Ni$ interaction while the structure of the Et complex was interpreted as not having an $HC\text{—}H \cdots Mo$ interaction. This is unusual since four methylene protons are perturbed in the ^1H NMR, consistent with a rapid interchange between bound and unbound protons of two of the methylene groups but with rigid boat-shaped rings. However, the boat conformers in the complexes $[Ni(Me_2M'(pz)_2)_2]$, where M' = B and Ga are rapidly flipping at room temperature. The relationship of the phenyl rings to the metal atom in $[Ni(Ph_2B(pz)_2)_2]$ is very similar to that in (**19**).

A series of papers have reported[35] palladium complexes of the type $[Pd(B(pz)_4)(A\text{—}B)]$, where A—B is a chelate with a carbon atom and a nitrogen or sulfur atom attached to the metal, (**33**). All four pz rings are equivalent in the ^1H NMR spectrum and the 'tumbling' mechanism is proposed.

Many platinum(II) and platinum(IV) complexes are known. Treatment of [PtMe$_3$I]$_4$ with RB(pz)$_3^-$ ligands, where R = H, pz, gave the platinum(IV) compounds [Pt(RB(pz)$_3$)Me$_3$] of structure type (**7**). Here too all four pz rings are equivalent in the ^{1}H NMR spectrum. The similarly obtained complex [Pt(H$_2$B(pz)$_2$)Me$_3$] (**34**) has a B—H $\cdots$ Pt interaction which completes the six-coordination. A band at 2039 cm^{-1} in the IR was attributed to the H $\cdots$ Pt 'bond'. Complex (**34**) reacts *via* displacement of this interaction to give [Pt(H$_2$B(pz)$_2$)Me$_3$L], where L = P(OMe)$_3$, CO, Hpz, H-3,5-Me$_2$pz. Complexes of platinum(IV) containing bispyrazolylalkane ligands have been reported: [Pt(R$_2$C(pz)$_2$)Me$_2$I$_2$], where R = H, Me and [Pt(H$_2$C(3,5-Me$_2$pz)$_2$)Me$_2$I$_2$].[36]

(32)   (33)   (34)

An insoluble polymer, possibly [Pt(HB(pz)$_3$)Me]$_n$, is produced by treatment of [Pt(Me)-(cod)(acetone)]PF$_6$ with HBpz$_3^-$. The structure is not known but it reacts with L to give five-coordinate complexes of the type [Pt(HB(pz)$_3$)MeL] (**35**), where L = alkynes, phosphines, alkenes and allenes. The HB(pz)$_3^-$ ligand does not rotate and the unbound pz ring in the B(pz)$_4$ analogue does not exchange with the bound rings. The complexes are stable with respect to insertion. The alkyne, alkene and allene ligands are attached such that the C—C bond lies in the equatorial plane (**36**). Geometrical isomers (**36a, b**), which interconverted rapidly at 120 °C, possibly by rotation, were detected for unsymmetrical alkenes such as methylacrylate. Cationic analogues of (**35**) are obtained when the HC(pz)$_3$ ligand was used.[37] Interestingly, the HC(pz)$_3$ ligand rotates in these complexes.

(35)   (36a)   (36b)

When L = CO and NCBu^t the complexes of type (**35**) are five-coordinate in solution; however, they are both square-planar four-coordinate in the solid (**37, 38**),[38] with boat-shaped rings. All of the above complexes (**34–38**) are characterized by extensive ^{195}Pt–^{1}H coupling in their NMR spectrum which usually permits structural assignment. When [Pt(Me)(cod)(acetone)]PF$_6$ was treated with R$_2$B(pz)$_2^-$, where R = Et, Ph, the square-planar complexes [Pt(R$_2$B(pz)$_2$)(cod)Me] which contain monodentate cod ligands were isolated. The latter are easily displaced by ligands to give [Pt(R$_2$B(pz)$_2$)MeL], where R = Et, L = alkynes, isocyanide, phosphines and CO; R = Ph, L = Ph-C≡CPh. The structure of [Pt(Et$_2$B(pz)$_2$)(Me)(PhC≡CPh)] (**39**) revealed no interaction between the metal and the methylene hydrogens nor was there any evidence for this in the IR or ^{1}H NMR spectra. The boat conformers are in rapid equilibrium. Insertion of the alkynes CF$_3$C≡CCF$_3$ and MeO$_2$CC≡CCO$_2$Me into the Pt—Me bond occurs to give (**40**) and (**41**).

(37)   (38)   (39)

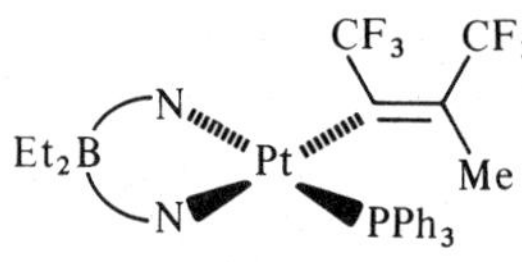

(40)

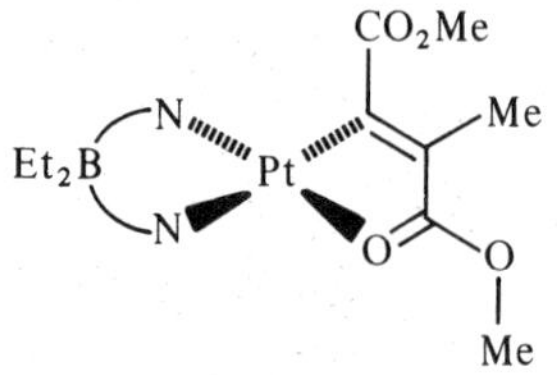

(41)

Treatment of [PtMe$_2$cod] with Y(pz)$_3$, where Y = HB or C, gave the expected products [PtMe$_2$(Y(pz)$_3$)]. However, these underwent a surprising reaction in refluxing pyridine to give square-planar products in which one of the pz rings was bonded to the metal *via* the C-5 carbon atom as shown in (42).[39]

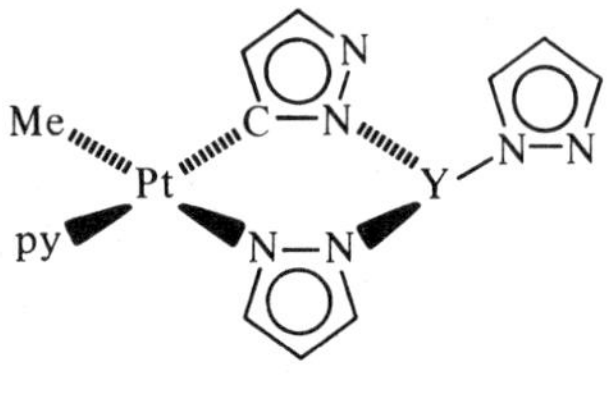

(42)

13.6.3.7 Copper, Silver, Gold

The reaction of [Cu$_2$Cl$_2$] with HB(pz)$_3^-$ in the absence of other ligands gave a dimer [Cu(HB(pz)$_3$)]$_2$ (43), which contains very unusual bridging HB(pz)$_3^-$ groups.[40] All the pz rings are equivalent in the ^{1}H NMR spectrum even at $-130\,^\circ$C. The copper(II) dimer {Cu[HB(pz)$_3$]Cl}$_2$ (44) contains bridging chloride groups.[41] If ligands were present [Cu$_2$Cl$_2$] gave complexes of the type [Cu(HB(pz)$_3$)L] (45), where L = CO, phosphines, phosphites, isocyanide, etc.[42] The CO complex is unusually stable, much more stable than the Cp analogue. The B(pz)$_4^-$ analogues show two types of pz rings in the ^{1}H NMR at $-90\,^\circ$C and one type at room temperature consistent with rapid 'tumbling'. If L = diars the B(pz)$_4^-$ ligand is bidentate (46), and all four pz rings are equivalent in the ^{1}H NMR spectrum at room temperature. At low temperature two types of rings in the ratio 1:1 are observed.[43]

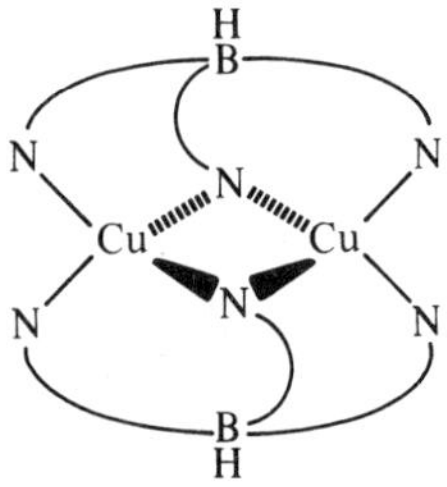

(43)

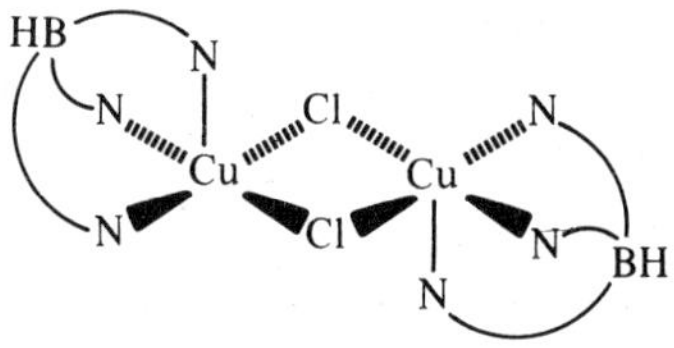

(44)

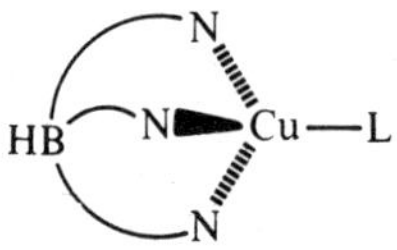

(45)

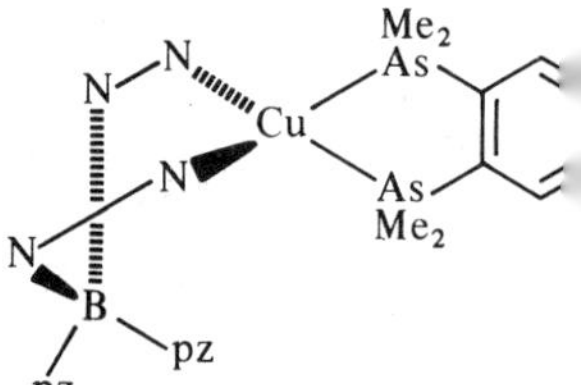

(46)

The steric requirements of the 3,5-Me$_2$pz ligands are reflected in the structures of the planar complexes [Cu(Me$_2$M'(pz)$_2$)$_2$], where M = B, Ga, and the pseudotetrahedral complex [Cu(Me$_2$Ga(3,5-Me$_2$pz)$_2$)$_2$]. Binary complexes of copper(I) and silver(I) of the type ML$_n$, where M = Cu or Ag and L = H$_2$B(pz)$_2^-$, Ph$_2$B(pz)$_2^-$, HB(pz)$_3^-$, HB(3,5-Me$_2$pz)$_3^-$ or B(pz)$_4^-$, include some of those mentioned above.[44] Other silver complexes include those of the type [Ag(L)Q], where L includes H$_2$B(pz)$_2^-$, Ph$_2$B(pz)$_2^-$, HB(pz)$_3^-$, HB(3,5-Me$_2$pz)$_3^-$ or B(pz)$_4^-$ and Q is a variety of phosphines or phosphite ligands.[45] Complexes of gold(III) include [Au(L)Cl$_2$], where L = Ph$_2$B(pz)$_2^-$, HB(pz)$_3^-$, HB(3,5-Me$_2$pz)$_3^-$ and B(pz)$_4^-$. The silver and gold complexes display the same type of fluxionality as the copper compounds. An unsymmetrical bidentate bonding mode was observed[46] for [Ag(Ph$_2$B(pz)$_2$)P(p-C$_6$H$_4$Me)$_3$] (47), where one Ag—N bond distance is 2.194 (4) Å and the other is 2.411 (4) Å. The complexes [Au(L)Me$_2$] and [Hg(L)Me] have been reported.[47]

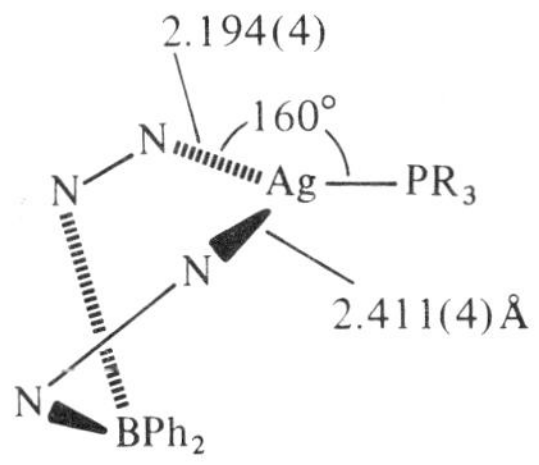

(47)

13.6.3.8 Ytterbium, Thorium, Uranium

A rare ytterbium(III) complex [Yb(HB(pz)$_3$)$_3$] has an eight-coordinate bicapped trigonal prismatic structure (48) with two tridentate ligands and one bidentate ligand. The complex is stereochemically rigid in solution.[48] A number of thorium and uranium complexes have been reported. These include [U(HB(pz)$_3$)$_4$], [U(H$_2$B(pz)$_2$)$_4$], [U(HB(pz)$_3$)$_2$Cl$_2$], and mono- and di-Cp metal–halogen complexes.[49]

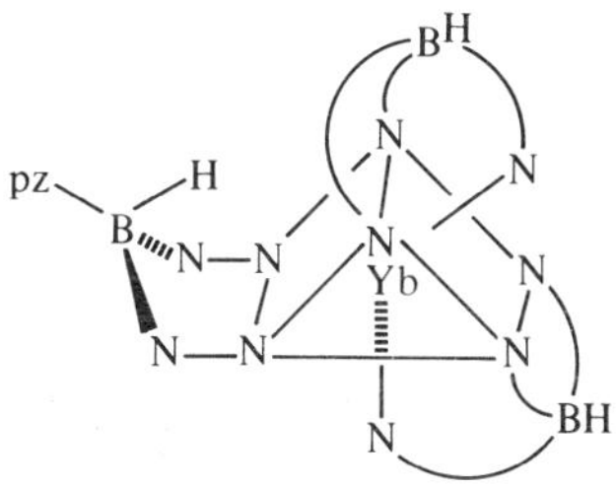

(48)

13.6.4 BIOLOGICAL APPLICATIONS

The unusual bonding mode observed in the copper complex (43) may be a model of the histidine-chelating sites in some copper proteins.[40] The complexes [Cu(HB(3,5-Me$_2$pz)$_3$)L]$^{n+}$, where $n = 0, 1$ and L = O-ethylcysteinate, approximate the N$_3$S donor set of blue copper electron transfer proteins.[50] The metal atom in copper(II) poplar plastocyanin is attached to two histidine nitrogen atoms, one cysteine sulfur atom and one methionine sulfur atom in a distorted tetrahedral arrangement. The mixed ligand HB(3,5-Me$_2$pz)$_2$(S-p-C$_6$H$_4$Me)$^-$ was prepared and its complex [Cu(HB(3,5-Me$_2$pz)$_2$(S-p-C$_6$H$_4$Me))L] (49), where L = O-ethylcysteinate, has spectral properties very similar to the natural material.[51] When [Fe(ClO$_4$)$_3$·10H$_2$O], NaO$_2$CMe·3H$_2$O and NaHB(pz)$_3$ were combined in aqueous medium, the dimer [Fe$_2$O(HB(pz)$_3$)$_2$(O$_2$CMe)$_2$] (50) was isolated.[52] This complex is an

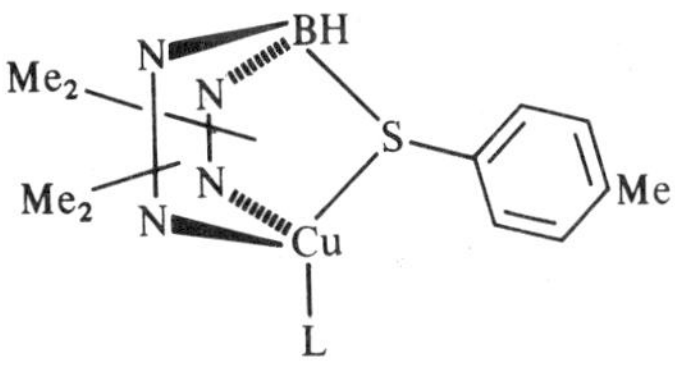

(49)

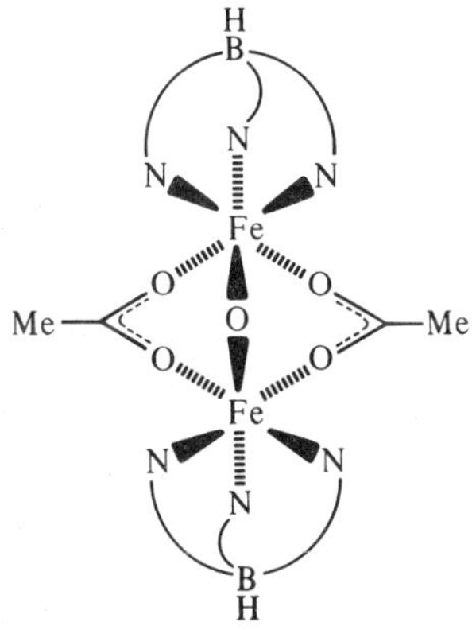

(50)

exact replica of the core of azidomet forms of hemerythrin and myohemerythrin and is a good example of 'spontaneous self-assembly'.[53] The stable complex ion [Mo(HB(3,5-Me$_2$pz)$_3$)Cl$_3$]$^-$, of structure type (7), has been prepared to support the proposal that molybdenum(III) intermediates stabilized by histidine ligands are a real possibility.[54] Thus, polypyrazolylborate ligands show promise as models for histidine ligands in biological systems. The stimulus for the preparation of the technetium complexes mentioned earlier is to develop new radiopharmaceuticals.

13.6.5 RELATED POLYPYRAZOLYL LIGANDS

Trofimenko also envisioned polypyrazolyl compounds of Be, Al, Ga, In, C, Si and P and prepared some alkane ligands and their complexes.[1,2] A few other applications of polypyrazolylalkane ligands have been mentioned in passing in the sections dealing with cobalt and platinum complexes. The reactions of dipyrazolyl ketone to produce [Co(OC(pz)$_2$)Cl$_2$] and [Co(Me$_2$C(pz)$_2$)Cl$_2$], the latter from acetone, are very unusual[55] but do not seem to have been exploited further (Scheme 2).

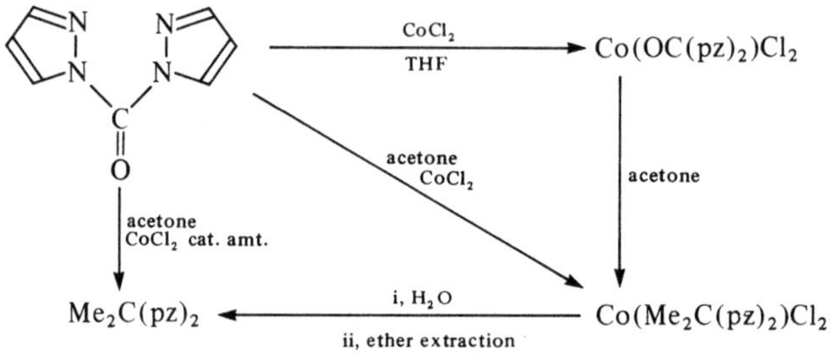

Scheme 2

Of the remainder only the gallium ligands have been shown to be successful. Storr and coworkers have prepared many polypyrazolylgallate complexes as part of a general study of organogallate chemistry. The ligand Me$_2$Ga(pz)$_2^-$ was mentioned in the section on nickel and copper complexes. The complexes [M(Me$_2$Ga(3,5-Me$_2$pz)$_2$)(η^3-C$_3$H$_5$)], where M = Ni and Pd, have been reported.[56] Rhodium and iridium compounds of the type [M(Me$_2$Ga(pz)$_2$)(LL)] have been prepared, where LL = cod, (CO)$_2$ and (PPh$_3$, CO).[57] The boat-shaped rings in these systems have been shown to be nonrigid by means of D NMR studies. The complexes with metal carbonyls, [M(Me$_2$Ga(pz)$_2$)-(N$_2$C$_3$H$_4$)(CO)$_3$]$^{n-}$, where n = 0, M = Mn and n = 1, M = Mo, W,[58] contain a free pyrazole group bonded to the metal (51). However, while the pyrazolyl and pyrazole rings are observed in the ^{1}H NMR in the ratio 2:1 for M = Mn, the Mo and W complexes show only one type of ring. This is very surprising since it implies that the pyrazolyl and pyrazole rings exchange roles rapidly. Similar complexes of rhenium have also been observed to display the same phenomenon.[59]

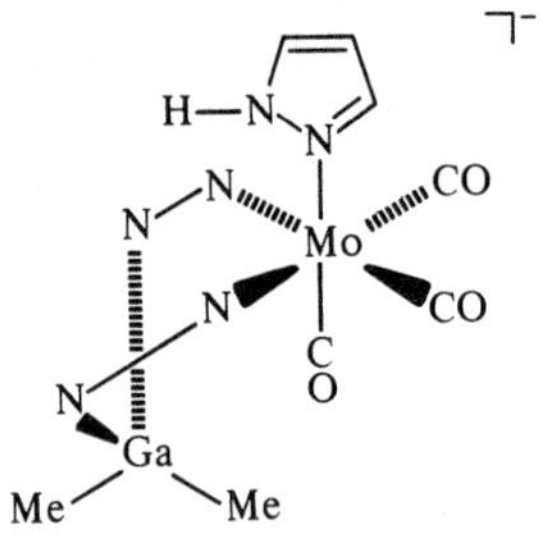

(51)

The tridentate ligand MeGa(pz)$_3^-$ forms octahedral complexes of the type [M(MeGa(pz)$_3$)$_2$], where M = Mn, Fe, Co, Ni, Cu, Zn, and of the type [M(MeGa(pz)$_3$)(CO)$_2$L], where M = Mn, L = CO; M = Mo, W; L = NO, η^3-C$_3$H$_5$, η^3-C$_4$H$_7$.[60] The (CO)$_3$ and (CO)$_2$NO complexes can also be prepared for Mo and W using the MeGa(3,5-Me$_2$pz)$_3^-$ ligand but attempts to prepare the η^3-enyl compounds led to the unexpected conversion[61] to [M(MeGa(3,5-Me$_2$pz)$_2$(OH)(CO)$_2$L] (52), where M = Mo and W; L = η^3-C$_3$H$_5$ and η^3-C$_4$H$_7$. It was felt that this conversion was promoted by the excessive steric requirements of the MeGa(3,5-Me$_2$pz)$_3^-$ ligand. Comparison of IR spectra of the above carbonyl complexes of the MeGa(pz)$_3^-$ ligand to those of the borate analogues indicated that

(52)

the gallate ligand increased the electron density on the metal. Comparison of the structures of $[Mo(MeGa(pz)_3)(CO_2)(\eta^3\text{-}C_3H_5)]$ and $[Mo(HB(pz)_3)(CO)_2(\eta^3\text{-}C_4H_7)]$ revealed that the Ga—N bond distance (1.926 Å) is longer than the B—N distance (1.545 Å) which accounts for much of the difference in the electronic properties and the increased steric requirements of the polypyrazolyl-gallate ligands.[61] Another complex containing the $MeGa(pz)_3^-$ ligand is the dimer $[Rh_2(MeGa(pz)_3)_2(CO)_3]$ which contains three bridging carbonyl ligands.[62] Storr and coworkers have also investigated monopyrazolylgallate ligands but these are beyond the scope of this chapter.[63]

13.6.6 MISCELLANEOUS PHYSICAL STUDIES

Low frequency ($650\text{--}150\,cm^{-1}$) IR spectral assignments for $[M(HB(pz)_3)_2]$, $[M(HB(3,5\text{-}Me_2pz)_3)_2]$ and $[M(B(pz)_4)_2]$, where M = Fe, Co, Ni, Cu, Zn, have been reported.[64] A low temperature X-ray analysis of the deformation of the electron density distribution in $[Ni(H_2B(pz)_2)]$ indicates delocalization over the five-membered rings.[65] Thermal analysis of the complexes $[M(H_2B(3,5\text{-}Me_2pz)_2)]$, where M = Co, Ni, Cu, Zn, has been reported.[66] The electrochemistry of $[Fe(HB(pz)_3)_2]$ has been compared to that of $[FeCp_2]$ and $[Fe(C_5Me_5)_2]$.[67] Finally the He-I and He-II excited photoelectron spectra of $HB(pz)_3^{-}$[68] and some of its complexes[69] with divalent transition metals have been reported.

13.6.7 CONCLUSIONS

The complexes formed by polypyrazolylborate ligands include most of the transition metals and there seems to be no reason why all the metals in the Periodic Table should not form complexes of some type. The bonding modes of this class of ligand are varied and flexible. High and low coordination numbers can be accommodated. The ligands have a stabilizing effect thus permitting normally unstable entities to be isolated. The development of these ligands from their preparation less than 20 years ago to their established well-known importance is a tribute to the genius of their invention.[1,2,4]

13.6.8 REFERENCES

1. S. Trofimenko, *Acc. Chem. Res.*, 1971, **4**, 17.
2. S. Trofimenko, *Chem. Rev.*, 1972, **72**, 497.
3. A. Shaver, *J. Organomet. Chem. Libr.*, 1976, **3**, 157.
4. S. Trofimenko, in 'Inorganic Compounds with Unusual Properties', ed. R. B. King, *Advances in Chemistry Series*, vol. 150, American Chemical Society, Washington, DC, 1976, 289.
5. (a) D. C. Bradley, M. B. Hursthouse, J. Newton and N. P. C. Walker, *J. Chem. Soc., Chem. Commun.*, 1984, 188. (b) E. Frauendorfer and G. Agrifoglio, *Inorg. Chem.*, 1982, **21**, 4122.
6. P. Dapporto, F. Mani and C. Mealli, *Inorg. Chem.*, 1978, **17**, 1323.
7. (a) L. E. Manzer, *J. Organomet. Chem.*, 1975, **102**, 167; (b) P. Burchill and M. G. H. Wallbridge, *Inorg. Nucl. Chem. Lett.*, 1976, **12**, 93; (c) J. K. Kouba and S. S. Wreford, *Inorg. Chem.*, 1976, **15**, 2313.
8. D. L. Reger, M. E. Tarquini and L. Lebioda, *Organometallics*, 1983, **2**, 1763.
9. D. L. Reger and M. E. Tarquini, *Inorg. Chem.*, 1983, **22**, 1064.
10. L. G. Hubert-Pfalzgraf and M. Tsunoda, *Polyhedron*, 1983, **2**, 203.
11. D. H. Williamson, C. Santini-Scampucci and G. Wilkinson, *J. Organomet. Chem.*, 1974, **77**, C25.
12. (a) D. L. Reger, C. A. Swift and L. Lebioda, *J. Am. Chem. Soc.*, 1983, **105**, 5343; (b) D. L. Reger, C. A. Swift and L. Lebioda, *Inorg. Chem.*, 1984, **23**, 349.

13. S. Trofimenko, *J. Am. Chem. Soc.*, 1969, **91**, 588.
14. M. D. Curtis, K.-B. Shiu and W. M. Butler, *Organometallics*, 1983, **2**, 1475.
15. W. W. Greaves and R. J. Angelici, *J. Organomet. Chem.*, 1980, **191**, 49.
16. K.-B. Shiu, M. D. Curtis and J. C. Huffman, *Organometallics*, 1983, **2**, 936.
17. T. Desmond, F. J. Lalor, G. Ferguson and M. Parvez, *J. Chem. Soc., Chem. Commun.*, 1983, 457.
18. C. P. Marabella and J. H. Enemark, *J. Organomet. Chem.*, 1982, **226**, 57.
19. J. A. McCleverty, D. Seddon, N. A. Bailey and N. W. Walker, *J. Chem. Soc., Dalton Trans.*, 1976, 898.
20. J. A. McCleverty, *Inorg. Chim. Acta*, 1982, **62**, 67 and references therein.
21. J. A. McCleverty, G. Denti, S. J. Reynolds, A. S. Drane, N. El Murr, A. E. Rae, N. A. Bailey, H. Adams and J. M. A. Smith, *J. Chem. Soc., Dalton Trans.*, 1983, 81.
22. (a) J. A. McCleverty, A. S. Drane, N. A. Bailey and J. M. A. Smith, *J. Chem. Soc., Dalton Trans.*, 1983, 91; (b) D. Condon, G. Ferguson, F. J. Lalor, M. Parvez and T. Spalding, *Inorg. Chem.*, 1982, **21**, 188; (c) T. Begley, D. Condon, G. Ferguson, F. J. Lalor and M. A. Khan, *Inorg. Chem.*, 1981, **20**, 3420.
23. J. A. McCleverty, A. E. Rae, I. Wolochowicz, N. A. Bailey and J. M. A. Smith, *J. Chem. Soc., Dalton Trans.*, 1982, 429.
24. J. A. McCleverty and N. El Murr, *J. Chem. Soc., Chem. Commun.*, 1981, 960.
25. D. L. Lichtenberger and J. L. Hubbard, *Inorg. Chem.*, 1984, **23**, 2718.
26. F. A. Cotton, C. A. Murrillo and B. R. Stults, *Inorg. Chim. Acta*, 1977, **22**, 75.
27. F. A. Cotton, B. W. S. Kolthammer and G. N. Mott, *Inorg. Chem.*, 1981, **20**, 3890.
28. (a) R. W. Thomas, G. W. Estes, R. C. Elder and E. Deutsch, *J. Am. Chem. Soc.*, 1979, **101**, 4581; (b) R. W. Thomas, A. Davison, H. S. Trop and E. Deutsch, *Inorg. Chem.*, 1980, **19**, 2840; (c) M. J. Abrams, A. Davison and A. G. Jones, *Inorg. Chim. Acta*, 1984, **82**, 125.
29. J. D. Oliver, D. F. Mullica, B. B. Hutchinson and W. O. Mulligan, *Inorg. Chem.*, 1980, **19**, 165.
30. D. L. White and J. W. Faller, *J. Am. Chem. Soc.*, 1982, **104**, 1548.
31. (a) M. Cocivera, G. Ferguson, B. Kaitner, F. J. Lalor, D. J. O'Sullivan, M. Parvez and B. Ruhl, *Organometallics*, 1982, **1**, 1132; (b) M. Cocivera, G. Ferguson, F. J. Lalor and P. Szczecinski, *Organometallics*, 1982, **1**, 1139.
32. M. Cocivera, T. J. Desmond, G. Ferguson, B. Kaitner, J. F. Lalor and D. J. O'Sullivan, *Organometallics*, 1982, **1**, 1125.
33. S. May, P. Reinsalu and J. Powell, *Inorg. Chem.*, 1980, **19**, 1582.
34. H. C. Clark and S. Goel, *J. Organomet. Chem.*, 1979, **165**, 383.
35. (a) M. Onishi, K. Sugimara and K. Hiraki, *Bull. Chem. Soc. Jpn.*, 1978, **51**, 3209; (b) M. Onishi, Y. Ohama, K. Sugimura and K. Hiraki, *Chem. Lett.*, 1976, 955; (c) M. Onishi, K. Hiraki, M. Shironita, Y. Yamaguchi and S. Nakagawa, *Bull. Chem. Soc. Jpn.*, 1980, **53**, 961; (d) O. Masayoshi, H. Yamamoto and K. Hiraki, *Bull. Chem. Soc. Jpn.*, 1980, **53**, 2540; (e) M. Onishi, K. Hiraki, A. Ueno, Y. Yamaguchi and Y. Ohama, *Inorg. Chim. Acta*, 1984, **82**, 121.
36. H. C. Clark, G. Ferguson, V. K. Jain and M. Parvez, *Organometallics*, 1983, **2**, 806.
37. H. C. Clark and M. A. Mesubi, *J. Organomet. Chem.*, 1981, **215**, 131.
38. J. D. Oliver and N. C. Rice, *Inorg. Chem.*, 1976, **15**, 2741.
39. (a) A. J. Canty and N. J. Minchim, *J. Organomet. Chem.*, 1982, **226**, C14; (b) A. J. Canty, N. J. Minchin, J. M. Patrick and A. H. White, *J. Chem. Soc., Dalton Trans.*, 1983, 1253.
40. C. S. Arcus, J. L. Wilkinson, C. Mealli, T. J. Marks and J. A. Ibers, *J. Am. Chem. Soc.*, 1974, **96**, 7564.
41. S. G. N. Roundhill, D. M. Roundhill, D. R. Bloomquist, C. Landee, R. D. Willet, D. M. Dooley and H. B. Gray, *Inorg. Chem.*, 1979, **18**, 831.
42. (a) O. M. Abu Salah, M. I. Bruce and J. D. Walsh, *Aust. J. Chem.*, 1979, **32**, 1209; (b) O. M. Abu Salah, M. I. Bruce and C. Hameister, *Inorg. Synth.*, 1982, **21**, 107.
43. O. M. Abu Salah, M. I. Bruce, P. J. Lohmeyer, C. R. Raston, B. W. Skelton and A. H. White, *J. Chem. Soc., Dalton Trans.*, 1981, 962.
44. M. I. Bruce and J. D: Walsh, *Aust. J. Chem.*, 1979, **32**, 2753.
45. O. M. Abu Salah, G. S. Ashby, M. I. Bruce, E. A. Pederzolli and J. D. Walsh, *Aust. J. Chem.*, 1979, **32**, 1613.
46. M. I. Bruce, J. D. Walsh, B. W. Skelton and A. H. White, *J. Chem. Soc., Dalton Trans.*, 1981, 956.
47. A. J. Canty, N. J. Minchin, J. M. Patrick and A. H. White, *Aust. J. Chem.*, 1983, **36**, 1107.
48. (a) M. V. R. Stainer and J. Takats, *Inorg. Chem.*, 1982, **21**, 4050; (b) M. V. R. Stainer and J. Takats, *J. Am. Chem. Soc.*, 1983, **105**, 410.
49. K. W. Bagnall, A. Beheshti, J. Edwards, F. Heatley and A. C. Tempest, *J. Chem. Soc., Dalton Trans.*, 1979, 1241.
50. J. S. Thompson, T. J. Marks and J. A. Ibers, *Proc. Natl. Acad. Sci. USA*, 1977, **74**, 3114.
51. J. S. Thompson, J. L. Zitzmann, T. J. Marks and J. A. Ibers, *Inorg. Chim. Acta*, 1980, **46**, L101.
52. (a) W. H. Armstrong and S. J. Lippard, *J. Am. Chem. Soc.*, 1983, **105**, 4837; (b) W. H. Armstrong, A. Spool, G. C. Papaefthymiou, R. B. Frankel and S. J. Lippard, *J. Am. Chem. Soc.*, 1984, **106**, 3653; (c) W. H. Armstrong and S. J. Lippard, *J. Am. Chem. Soc.*, 1984, **106**, 4632.
53. J. A. Ibers and R. H. Holm, *Science*, 1980, **209**, 223.
54. M. Millar, S. Lincoln and S. A. Koch, *J. Am. Chem. Soc.*, 1982, **104**, 288.
55. K. I. The and L. K. Peterson, *Can. J. Chem.*, 1973, **51**, 422.
56. K. S. Chong, S. J. Rettig, A. Storr and J. Trotter, *Can. J. Chem.*, 1981, **59**, 996.
57. (a) S. Nussbaum, S. J. Rettig, A. Storr and J. Trotter, *Can. J. Chem.*, 1985, **63**, 692; (b) B. M. Louie, S. J. Rettig, A. Storr and J. Trotter, *Can. J. Chem.*, 1985, **63**, 503; (c) B. M. Louie, S. J. Rettig, A. Storr and J. Trotter, *Can. J. Chem.*, 1984, **62**, 1057.
58. S. E. Anslow, K. S. Chong, S. J. Rettig, A. Storr and J. Trotter, *Can. J. Chem.*, 1981, **59**, 3123.
59. (a) B. M. Louie and A. Storr, *Can. J. Chem.*, 1984, **62**, 1344; (b) B. M. Louie, S. J. Rettig, A. Storr and J. Trotter, *Can. J. Chem.*, 1985, **63**, 703.
60. (a) K. R. Breakell, S. J. Rettig, D. L. Singbeil, A. Storr and J. Trotter, *Can. J. Chem.*, 1978, **56**, 2099; (b) S. J. Rettig, A. Storr and J. Trotter, *Can. J. Chem.*, 1979, **57**, 1823.
61. K. R. Breakell, S. J. Rettig, A. Storr and J. Trotter, *Can. J. Chem.*, 1979, **57**, 139.

62. B. M. Louie, S. J. Rettig, A. Storr and J. Trotter, *Can. J. Chem.*, 1984, **62**, 633.
63. S. J. Rettig, A. Storr and J. Trotter, *Can. J. Chem.*, 1981, **59**, 2391.
64. B. Hutchinson, M. Hoffbauer and J. Takemoto, *Spectrochim. Acta, Part A*, 1976, **32**, 1785.
65. D. A. Clemente, G. Bandoli and M. B. Cingi, *Congr. Naz. Chim. Inorg. (Atti), 12th*, 1979, 355.
66. U. B. Ceipidor, M. Tomassetti, F. Bonati, R. Curini and G. D'Ascenzo, *Thermochim. Acta*, 1983, **63**, 59.
67. P. R. Sharp and A. J. Bard, *Inorg. Chem.*, 1983, **22**, 2689.
68. G. Bruno, E. Ciliberto, I. Fragala and G. Granozzi, *Inorg. Chim. Acta*, 1981, **48**, 61.
69. I. Fragala, G. Condorelli, B. Bruno, E. Ciliberto and G. Granozzi, *Congr. Naz. Chim. Inorg. (Atti), 12th*, 1979, 242.

13.7
Nitriles

HELMUT ENDRES
University of Heidelberg, Federal Republic of Germany

13.7.1 INTRODUCTION

Metal complexes of organonitriles — reported for the first time in 1851[1] — are of interest because the $RC\equiv N$ group is isoelectronic with CO and N_2. They are widely used as precursors for other coordination and organometallic compounds. They are conveniently prepared by dissolving a metal halide, carbonyl or another appropriate metal complex in the nitrile and either boiling or irradiating. Many nitrile complexes are accidentally obtained during a reaction carried out in a nitrile as a solvent. The metal atoms in nitrile complexes have low or medium oxidation numbers. High oxidation states are prevented by the reducing properties of nitriles. Acetonitrile, MeCN, is mostly employed in ligand exchange reactions, but higher boiling homologues have been recommended, *e.g.* for the preparation of arene complexes of Group VI metals.[2] Reviews on structures, spectral properties and reactions of organonitrile complexes have appeared in 1965[3] and in 1977,[4] and a review on complexes of platinum metals with weak donor ligands[5] contains a chapter on nitriles.

Organonitriles are usually taken as weak σ-donors and π-acceptors (contrasting in that aspect the isoelectronic CO and explaining the lability of the complexes), but recent LCAO–MO calculations[6] indicate a potential π-donor character in MeCN. Relevant orbital energies for this ligand are shown in Figure 1.

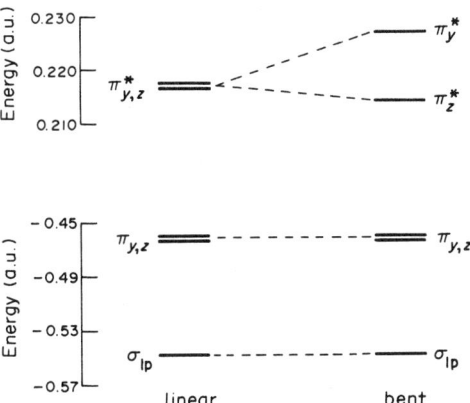

Figure 1 Molecular orbital diagram of valence orbitals of linear and bent (NCR = 165°) NCMe. Reproduced with permission from ref. 6

13.7.2 STRUCTURES

13.7.2.1 Mononitriles

Organonitriles usually coordinate in an end-on fashion (Figure 2a). The M—N≡C group is close to linear, but significant deviations from linearity are found in nearly all cases. They are discussed as an sp^2 contribution to the electronic state of N, and are not inconsistent with the π-donor character mentioned above. The M—N≡C angles scatter around $175 \pm 5°$, but values as low as $160°$ have been observed. Packing forces may well contribute to these deviations. The N≡C—R angles, on the other hand, deviate to a much lesser extent from linearity. The N≡C bond length usually shortens on coordination. Distances between 1.11 and 1.15 Å are most frequently found, compared to 1.16 Å in free MeCN.

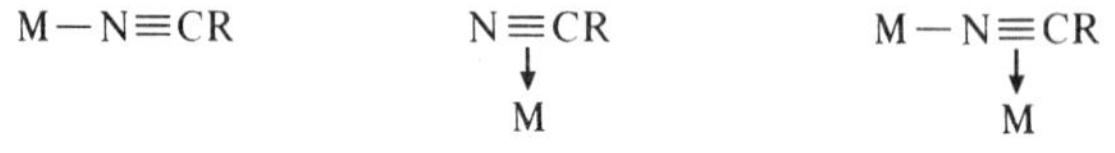

$$M—N≡CR \qquad\qquad N≡CR \qquad\qquad M—N≡CR$$
$$\downarrow \qquad\qquad\qquad \downarrow$$
$$M \qquad\qquad\qquad M$$

(a)   (b)   (c)

Figure 2 Possible bonding modes of nitriles to metals: (a) end-on (σ), (b) side-on (π) and (c) side-on in addition to end-on

The ability of the —C≡N group to coordinate in a side-on (or π-) fashion (Figure 2b) has been discussed.[4] Arguments are normally based on IR spectra (*vide infra*), but in most cases they proved to be erroneous. The only exception is the Pt[0] complex (1), the structure of which is said to be revealed by (unpublished) X-ray structure analysis.[7] Side-on, in addition to end-on, coordination (Figure 2c) has been detected by X-ray structure analysis[8,9] in a few polynuclear complexes (2, 3) as well as in some related complexes with N≡C—X ligands, with X other than alkyl or aryl. In these cases the C≡N distances are lengthened (1.26 Å, approaching the value for a C=N double bond, 1.28 Å), the N≡C—R angles are strongly reduced (to 131–135°), but the M—N≡C angles are much less influenced (160 and 167°).

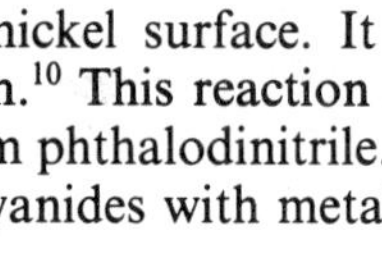

(1) 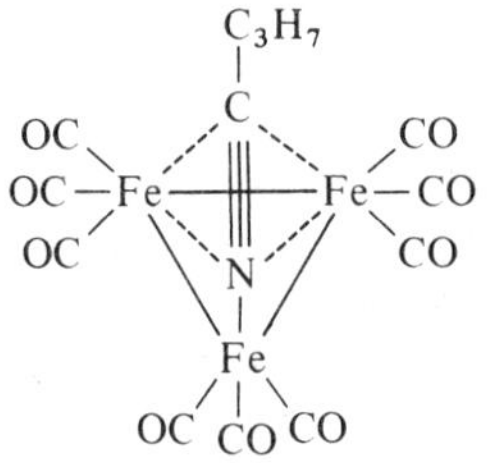   (2) 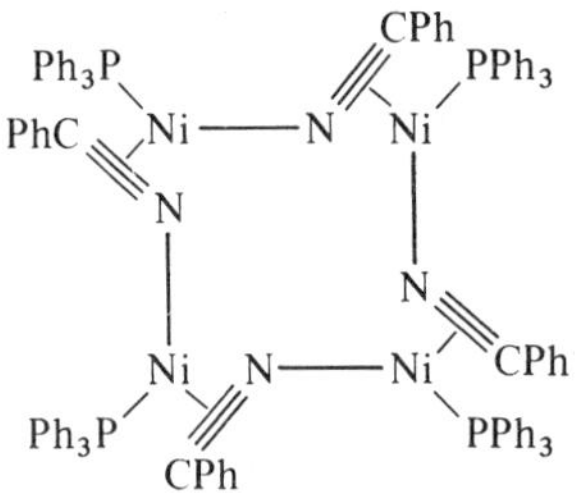   (3)

In this context it is interesting to note that benzonitrile, Ph—C≡N, trimerizes to a triazine on a Raney nickel surface. It was assumed that π-bonded nitriles were involved in the reaction mechanism.[10] This reaction resembles the well-known template synthesis of phthalocyanine complexes from phthalodinitrile. Formation of linear polymers $[—C(R)=N—]_n$ occurs on heating aryl or alkyl cyanides with metal halides.[11]

13.7.2.2 Polyfunctional Nitriles

Concerning dinitriles, N≡C—$(CH_2)_n$—C≡N, there has been some speculation about their ability to act as chelating ligands *via* π-coordination of both nitrile groups. This has never been observed and normal end-on coordination occurs, making chelate complexes impossible (at least for reasonably small values of n). Either only one nitrile group coordinates or, and more commonly, dinitrile-bridged dimers or polymers form. In the latter case chains, two-dimensional and three-dimensional networks have been detected.[12] The Cu[I] complexes of aliphatic dinitriles may have technological implications in connection with the absorption of Cu[I] ion on polyacrylonitrile, facilitating the fixation of dyes on the fibres.[13]

A topotactic solid state polymerization reaction has been reported for the Ni complex with tris(2-cyanoethyl)phosphine, $P(CH_2CH_2CN)_3$: the square planar $NiBr_2\{P(CH_2CH_2CN)_3\}_2$ complexes, with only uncoordinated cyano groups, polymerize in the crystals to give hexa-coordinated Ni^{II}. The reaction yielding (**4**) involves axial attack of one CN group of each phosphine ligand at the adjacent Ni complex.[14] The analogous chloro polymer is also known.

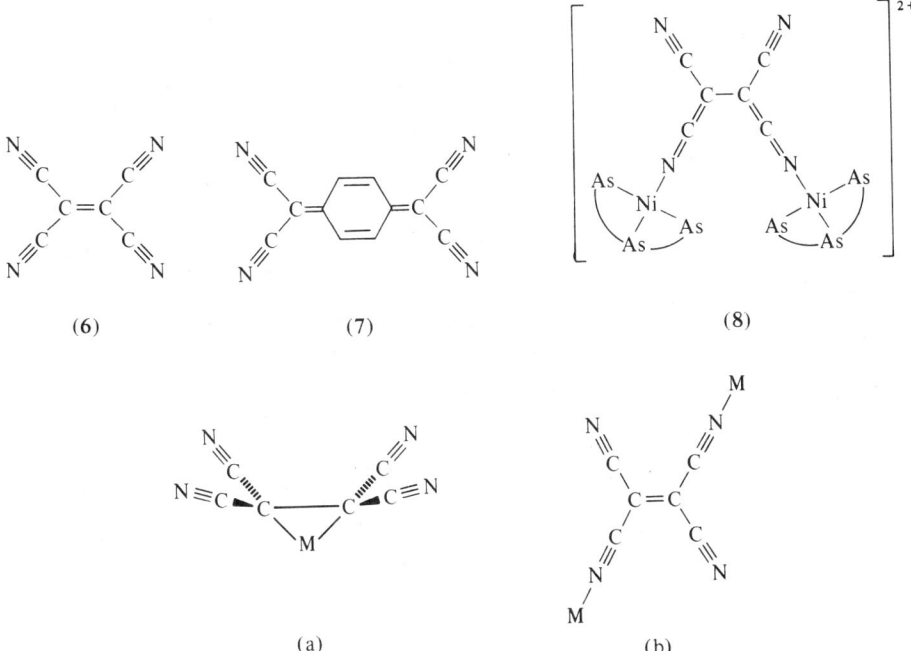

(4)   (5)

It is a general observation that nitriles containing another strongly donating group, like the above phosphine, preferentially coordinate *via* the stronger donor. Additional coordination through the CN group may, but need not, take place. An interesting class of complexes is formed by α,β-unsaturated nitriles like acrylonitrile: these molecules may link two metal atoms by σ-coordination of the CN group and π-coordination of the C=C group, *e.g.* in the $Mo_2(CO)_2(PBu_3)_4(\mu\text{-}CH_2\text{=}CH$ $-C\equiv N)_2$ complex (**5**).[15]

Dinuclear or polymeric complexes may result as well from the coordination of polynitriles like tricyanomethanide, $[C(CN)_3]^-$, dicyanonitrosomethanide, $[C(CN)_2NO]^-$, tetracyanoethylene and hexacyanobutene. Some of them have been reviewed by Baddley.[16] Tetracyanoethylene (TCNE) (**6**) and tetracyanoquinodimethane (TCNQ) (**7**) are of special interest, as different types of complexes may be formed: (1) sandwich type (charge transfer) complexes without close metal–nitrile interaction; (2) π-complexes involving the C=C bond; (3) metallacyclopropanes (Figure 3a); and (4) polynuclear species (Figure 3b) *via* coordination of the CN groups.[17] In some cases reduction of TCNE to coordinated $TCNE^{2-}$ has been proposed,[18] and bridged structures like in the cation (**8**) are assumed where the triarsine ligand is bis(3-dimethylarsinopropyl)phenylarsine. TCNQ hardly ever acts as a nitrile ligand[17,18] and normally forms complexes of the charge transfer type with metal ions.[19]

(6)   (7)   (8)

(a)   (b)

Figure 3 Two of the possible bonding modes of tetracyanoethylene, TCNE: (a) metallacyclopropane and (b) end-on bridging (*cis* bridging also possible)

13.7.3 VIBRATIONAL SPECTRA

Infrared spectra of nitrile complexes have been extensively discussed in previous reviews.[3,4] Free MeCN exhibits two absorptions in the C≡N stretching frequency region, a sharp intense band at $2255 \, cm^{-1}$ (interpreted as the C≡N stretching vibration) and a weak band at $2293 \, cm^{-1}$ (taken as a combination band of C—C stretching and CH_3 deformation). Consequently the latter one is missing in benzonitrile, where only the C≡N stretching vibration is observed at $2250 \, cm^{-1}$. These bands are usually shifted to *higher* frequencies on coordination, consistent with the described shortening of the C≡N bond. In acetonitrile complexes the two absorptions are frequently found in the range 2270–$2300 \, cm^{-1}$ and 2290–$2320 \, cm^{-1}$ depending, of course, on the nature of the metal and the other ligands in the complexes. Solvent dependence (by several wave numbers) is also observed. The bands may be weak or even not observable. A frequency shift to lower wave numbers has sometimes been taken as indication of side-on coordination of the C≡N group. Unless the reduction is very pronounced (to about $1750 \, cm^{-1}$ or below[7-9]) this interpretation has proved erroneous whenever X-ray structure determinations were performed. Even if nitriles are weak π-acceptors,[4] strong π-back-bonding complex fragments may reduce the C≡N stretching frequencies[4,20] by up to $100 \, cm^{-1}$.

The metal–nitrogen stretching vibrations are difficult to identify.[4] Their values may vary from 100 to $525 \, cm^{-1}$ and they are easily confused with other bands.

Infrared and Raman spectra are useful, of course, in determining the symmetry of poly(nitrile) complexes. An example is dichlorobis(benzonitrile)platinum, a versatile starting material in platinum chemistry. There is a dispute in the literature as to whether it has a *cis* or *trans* configuration. (There is a general agreement that the corresponding Pd complex is *trans*.) It has recently been shown by X-ray structure analysis that both *cis* and *trans* complexes exist, and their vibrational spectra have been discussed.[21]

13.7.4 REACTIONS

13.7.4.1 Formation of Nitrile Complexes

Nitrile complexes are obtained by dissolving a metal halide in the nitrile or by replacing CO from carbonyl complexes. A kinetic investigation of the reaction of acetonitrile on Group VI hexacarbonyls[22] showed that substitution by the first acetonitrile leads to increased replacement rates for the next two molecules. The second MeCN is built in *cis* to the first one, and replacement of the third CO gives *fac*-$Mo(CO)_3(MeCN)_3$. Further substitution does not occur. The rate law comprises two terms and is of the form[22] of equation (1).

$$\text{Rate} = k_1[Mo(CO)_6] + k_2[Mo(CO)_6][MeCN] \tag{1}$$

The second term, which outweighs the first one, is said to arise from an S_N2 contribution in the reaction mechanism. These observations have been taken as evidence in favour of the proposed π-donating properties of acetonitrile,[6] and it was stated that MeCN is as effective a nucleophile towards $Mo(CO)_6$ as triphenylphosphine. The *trans*-labilizing properties of nitriles, on the other hand, are less pronounced than those of, for example, carbonyl or dimethylphenylphosphine. This is consistent with the weak π-acceptor character of nitriles.

Nitrile complexes may even form from derivatives of nitriles. An interesting reaction has been observed[23] when nitrile oxide R—CNO reacts with Pt^0 complexes $[(Ph_3P)_2PtL]$ (L = Ph_3P, C_2H_4): nitrile complexes (9) result from migration of a phenyl group from P to Pt and of the oxygen atom from N to P.

$$
\begin{array}{c}
\text{PPh}_3 \\
|\\
\bigcirc\!\!-\text{Pt}-\text{N}\!\equiv\!\text{C}-\text{R} \\
|\\
\text{Ph}_2\text{P}\!=\!\text{O}
\end{array}
$$

(9)

13.7.4.2 Displacement Reactions

Displacement of nitriles, especially acetonitrile, by other ligands, either charged or neutral ones, belongs to the most important class of reactions of nitrile complexes.[4] Examples range from replacement of one MeCN by another monodentate ligand like isocyanide[24] or phosphine[25] to replacement of up to three MeCN by a polyhapto ligand like fulvene.[26] Nitrile exchange is also important in cluster chemistry, as it may take place without destroying the polymetal core.[27]

In the special case of displacement of MeCN by pyridine in the cation $[Mo(CO)_5(MeCN)]^+$, it has been assumed that the rate-determining loss of MeCN is preceded by attack of pyridine to a CO *cis* to the nitrile.[28]

13.7.4.3 Reactions of the Nitrile Group

Some oligomerization or polymerization reactions of nitriles in the coordination sphere of a metal have already been mentioned in Section 13.7.2.1. Dimerization reactions are also known: trifluoroacetonitrile, CF_3CN, was reported to dimerize[29] when reacted with $(Ph_3P)_4Pt^0$ to give the complex (10). The carbanion ligand in (11), generated from coordinated benzyl nitrile, is intramolecularly oxidized to the nitrile radical.[30] This dimerizes immediately to stilbene dinitrile. Reductive coupling of two acetonitrile ligands has also been observed.[31] The bridging ligand in the anion (12) is formed, which may be regarded as the tetraanion of diaminobutene. Reactions like this should be borne in mind when treating nitrile complexes with reducing agents.

$$[(NH_3)_5Co-N\equiv C-\overset{|}{\underset{H}{\bar{C}}}-Ph]^{2+}$$

(10)   (11)   (12)

(13)

Nitrile–nitrido ligand interconversion[32] leading to compounds like (13) takes place when halogenated nitriles (*e.g.* Cl_3CCN) react with hexachlorides of Mo and W. The formation of the nitrido ligand may be taken as an addition of two Cl atoms to the $C\equiv N$ group. Similar complexes with $M\equiv N$ triple bonds may occur as intermediates in the reduction of WCl_6 with alkyl cyanides giving $WCl_4(NCR)_2$. An infrared absorption around $1300\,cm^{-1}$ has been attributed to the $M\equiv N$ bond. Nucleophilic addition to coordinated nitriles and the catalytic effects of metal ions, which enhance reaction rates by several orders of magnitude, have been widely investigated.[4] Such reactions include attack of alcohols, thiols, water or amines yielding iminoether, iminothioether, amido or amidino complexes. The hydrolysis of coordinated acetonitrile to acetamide was observed as early as 1908, and was one of the first known reactions of a coordinated organic ligand. Acetamido and related complexes of Pt lead to the 'platinum blues' — oligomeric mixed valence Pt complexes which found renewed recent interest in connection with cancer chemotherapy.[33] Amino acids may be prepared by intra- and inter-molecular condensation of CN^- and NH_2^- with coordinated acetonitrile followed by saponification. This reaction may be important for the synthesis of amino acids with labelled C atoms.[34]

Kinetic investigations of reaction (2) reveal neighbouring-group participation of the substituent R in the rate determining step.[35] The reaction is especially fast with $R = CONH_2$. These studies complete earlier ones which state that strongly electron withdrawing groups at the coordinated nitrile accelerate hydrolysis reactions as they make nucleophilic attack at the nitrile C atom more easy. The question of whether the nucleophile attacks externally or whether coordination of the nucleophile to the metal atom precedes the reaction seems not to be definitely settled.[36] In the catalytic action of metal ions in the hydrolysis of nitriles[4] the mechanism may involve either coordinated OH^- attacking free nitrile, or coordinated nitrile being attacked by free or coordinated OH^-. There are arguments in favour of the second version, assuming activation of the nitrile group upon coordination. In dimeric nitrile complexes like (14) it is quite evident that nucleophilic attack starts with breaking of a metal–nitrile bond and coordination of the nucleophile to the metal.

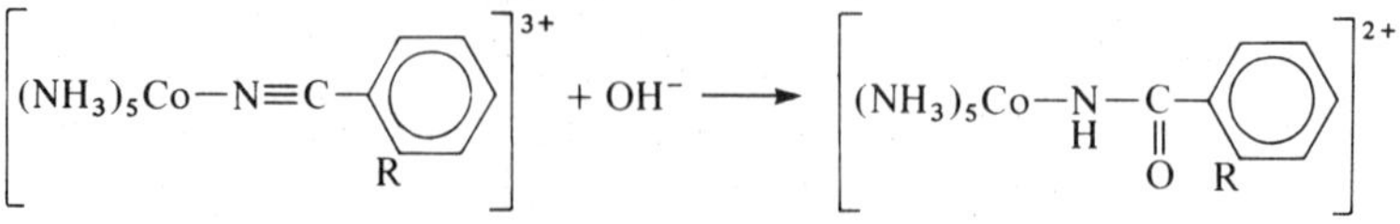

266   *Nitrogen Ligands*

(2)

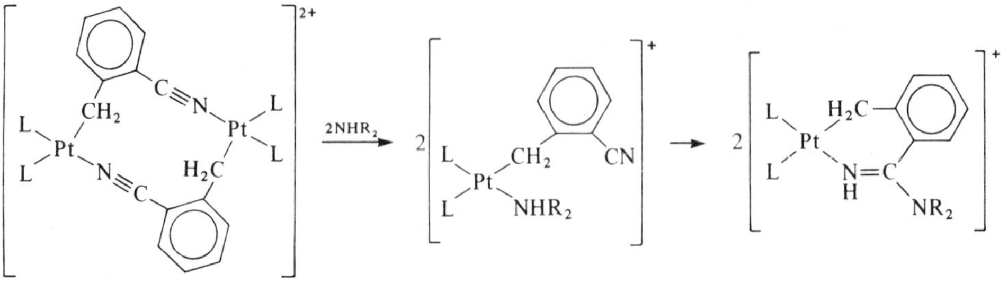

(3)

(14)   (15)   (16)

Investigations about the mechanism of the aminolysis reaction (3) have been carried out.[37] The dimeric complex (14) forms chelate amidino complexes of type (16) upon treatment with primary or secondary amines. The first stage of the reaction is thought to involve rapid formation of (15) through breaking of the nitrile bridges in (14) by the entering amine. In the analogous reaction with the azide ion N_3^- the intermediate (17) can be isolated.[38] It reacts through an intramolecular 1,3-cycloaddition to the tetrazolate complex (18). It has been proposed[37] that the nitrile group in the intermediate (15) is activated by side-on coordination to Pt, but there is no agreement about this.[39]

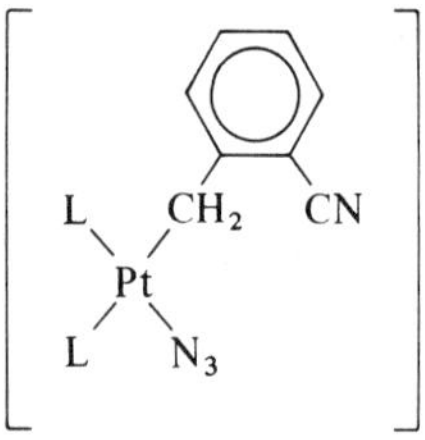

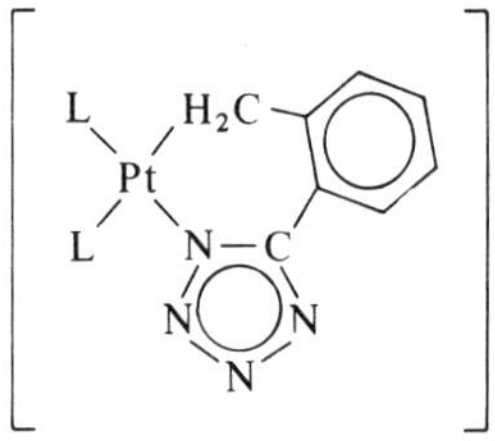

(17)   (18)

13.7.5 REFERENCES

1. W. Henke, *Justus Liebigs Ann. Chem.*, 1858, **106**, 280.
2. G. J. Kubas, *Inorg. Chem.*, 1983, **22**, 692.
3. R. A. Walton, *Q. Rev., Chem. Soc.*, 1965, **19**, 126.
4. B. N. Storhoff and H. C. Lewis, Jr., *Coord. Chem. Rev.*, 1977, **23**, 1.
5. J. A. Davies and F. R. Hartley, *Chem. Rev.*, 1981, **81**, 79.
6. J. A. S. Howell, J.-Y. Saillard, A. Le Beuze and G. Jaouen, *J. Chem. Soc., Dalton Trans.*, 1982, 2533.
7. W. J. Bland, R. D. W. Kemmitt and R. D. Moore, *J. Chem. Soc., Dalton Trans.*, 1973, 1292.
8. M. A. Andrews, C. B. Knobler and H. D. Kaesz, *J. Am. Chem. Soc.*, 1979, **101**, 7260.
9. I. W. Bassi, C. Benedicenti, M. Calcaterra, R. Intrito, G. Rucci and C. Santini, *J. Organomet. Chem.*, 1978, **144**, 225.
10. W. Z. Heldt, *J. Organomet. Chem.*, 1966, **6**, 292.
11. E. Oikawa and S. Kambara, *J. Polym. Sci. B, Polym. Lett.*, 1964, **2**, 649.
12. S. Gorter and G. C. Verschoor, *Acta Crystallogr., Sect B*, 1976, **32**, 1704.
13. Y. Kinoshita, I. Matsubara, T. Higuchi and Y. Saito, *Bull. Chem. Soc. Jpn.*, 1959, **32**, 1221.
14. K. Cheng and B. M. Foxman, *J. Am. Chem. Soc.*, 1977, **99**, 8102.
15. F. Hohmann, H. tom Dieck, C. Krüger and Y.-H. Tsay, *J. Organomet. Chem.*, 1979, **171**, 353.
16. W. H. Baddley, *Inorg. Chim. Acta Rev.*, 1968, **2**, 7.
17. A. B. Cornwell, C. A. Cornwell and P. G. Harrison, *J. Chem. Soc., Dalton Trans.*, 1976, 1612.
18. B. L. Booth, C. A. McAuliffe and G. L. Stanley, *J. Chem. Soc., Dalton Trans.*, 1982, 535.
19. H. Endres, in 'Extended Linear Chain Compounds', ed. J. S. Miller, Plenum, New York, 1983, vol. 3, p. 263.
20. I. W. Bassi, C. Benedicenti, M. Calcaterra and G. Rucci, *J. Organomet. Chem.*, 1976, **117**, 285.
21. H. H. Eysel, E. Guggolz, M. Kopp and M. L. Ziegler, *Z. Anorg. Allg. Chem.*, 1983, **499**, 31.
22. K. M. Al-Kathumi and L. A. P. Kane-Maguire, *J. Inorg. Nucl. Chem.*, 1972, **34**, 3759.
23. W. Beck, M. Keubler, E. Leidl, U. Nagel, M. Schaal, S. Cenini, P. del Buttero, E. Licandro, S. Maiorana and A. C. Villa, *J. Chem. Soc., Chem. Commun.*, 1981, 446.
24. P. M. Treichel and W. J. Knebel, *Inorg. Chem.*, 1972, **11**, 1289.
25. D. Drew, D. J. Darensbourg and M. Y. Darensbourg, *Inorg. Chem.*, 1975, **14**, 1579.
26. B. Lubke, F. Edelmann and U. Behrens, *Chem. Ber.*, 1983, **116**, 11.

27. K. Burgess, H. D. Holden, B. F. G. Johnson and J. Lewis, *J. Chem. Soc., Dalton Trans.*, 1983, 1199.
28. P. A. Bellus and T. L. Brown, *J. Am. Chem. Soc.*, 1980, **102**, 6020.
29. W. J. Bland, R. D. W. Kemmitt, I. W. Nowell and D. R. Russell, *Chem. Commun.*, 1968, 1065.
30. I. I. Creaser and A. M. Sargeson, *J. Chem. Soc., Chem. Commun.*, 1975, 324.
31. P. A. Finn, M. S. King, P. A. Kilty and R. E. McCarley, *J. Am. Chem. Soc.*, 1975, **97**, 220.
32. F. Weller, U. Müller, U. Weiher and K. Dehnicke, *Z. Anorg. Allg. Chem.*, 1980, **460**, 191.
33. S. J. Lippard, *Acc. Chem. Res.*, 1978, **11**, 211.
34. I. I. Creaser, S. F. Dyke, A. M. Sargeson and P. A. Tucker, *J. Chem. Soc., Chem. Commun.*, 1978, 289.
35. R. J. Balahura and W. L. Purcell, *Inorg. Chem.*, 1981, **20**, 4159.
36. C. R. Clark and R. W. Hay, *J. Chem. Soc., Dalton Trans.*, 1974, 2148.
37. L. Calligaro, R. A. Michelin and P. Uguagliati, *Inorg. Chim. Acta*, 1983, **76**, L83.
38. R. Ros, J. Renaud and R. Roulet, *J. Organomet. Chem.*, 1976, **104**, 393.
39. D. Schwarzenbach, A. Pinkerton, G. Chapuis, J. Wenger, R. Ros and R. Roulet, *Inorg. Chim. Acta*, 1977, **25**, 255.

13.8

Oximes, Guanidines and Related Species

RAM C. MEHROTRA
University of Rajasthan, Jaipur, India

13.8.1 OXIMES

The name oxime may be considered to be derived from oxy-imine, $>C=NOH$. The oxime group is amphiprotic with a slightly basic nitrogen and a mildly acidic hydroxyl group. In his doctoral research under Hantzsch, Werner[1] explained the observed isomerism of oximes on the basis of differences in 'spatial arrangement' of the groups attached to the $C=N$ moiety.

The most significant early discovery[2] in the chemistry of metal oximates was the reaction between nickel(II) salts and dimethylglyoxime, which is the best known example of a *vicinal* dioxime (abbreviated as *vic*-dioxime). The discoverer Tschugaeff[2] correctly identified the bidentate nature of *vic*-dioximes. However, the chelate ring size remained uncertain and went through the incorrect seven- and six-membered formulations before the correct five-membered ring was fully established. Interesting historical accounts of oximes are well documented.[3-6] Active interest has continued on various facets of metal oximate chemistry as evinced by publication of a number of review articles;[7-15] it may, however, be mentioned that the last two mainly deal with *O*-bonded oximate derivatives of organometallic moieties and metals.

13.8.1.1 Structure of the Oxime Group and Modes of Bonding in Complexes

Oximes are generally represented by structure (1), but the alternative nitrone structure (2) was debated actively as late as 1952.[16] Neutron diffraction work[17] on dimethylglyoxime definitely established the presence of O—H bonds (1.02 ± 0.04 Å) in it. Structure (1) is now usually accepted for the oximes, which are generally associated[17-21] in the solid state *via* O—H $\cdots$ N hydrogen bonds of length ~ 2.8 Å. The calculated $C=N$ and N—O distances based on covalent radii and electronegativity data are 1.27 and 1.44 Å respectively.[22] Actual experimental values for $C=N$ distances lie within ± 0.02 Å of the calculated figure, but the observed N—O distances are generally lower and lie in the range of 1.40 ± 0.02 Å.

$$>C=NOH \qquad >C=N\overset{\nearrow O}{\underset{H}{}} \qquad >C=N\overset{O(H)}{\underset{M}{}} \qquad >C=N\overset{O--H--O}{\underset{M}{}}N=C< \qquad >C=N\overset{O}{\underset{M}{}} \qquad >C=N-O\underset{M}{}$$

(1)   (2)   (3)   (4)   (5)   (6)

As a ligand, the oxime group is potentially ambidentate with possibilities of coordination through nitrogen and/or oxygen atom(s), as depicted in (3)–(6). Although a good number of oxygen-bonded complexes are known,[14,15] the coordination in actual practice generally occurs through nitrogen. In the latter cases, coordination can occur through the oxime or its conjugate base, as depicted by putting the hydrogen atom in parenthesis in (3); in (4), one oxime molecule is coordinating as such while the second one does so in the form of conjugate base, with the single hydrogen atom shared as $O \cdots H \cdots O$.

13.8.1.2 Simple Oximes

These are ligands which have merely one oxime group as the only coordination site; the simplest examples of these are acetaldoxime (Hado) and acetoxime (Hato). Hieber and Leutert[23] showed in 1927–29 that such oximes yield complexes of the type $M(oxime)_nX_2$ (where n is usually 2 or 4 and X is a halogen) with copper(II), nickel(II) and cobalt(II) salts. The three-dimensional crystal structure of $Ni(Hado)_4Cl_2$ has shown the molecule to be of *trans* octahedral NiN_4Cl_2 type, with the NiN_4 fragment being planar and the oxime protons being hydrogen-bonded with the coordinated chloride ions intramolecularly.[24] The work has been extended[25] to the synthesis of the corresponding $Ni(cyclohexanone\ oxime)_4Cl_2$ complex, which also indicated an octahedral structure, with the oxime being bonded through nitrogen only.

Complexes ML_4X_2 and $M'L_2X_2$ (M = Ni^{II}, Co^{II} and M' = Co^{II} and Cu^{II}; L = acetaldoxime or benzaldoxime and X = Cl or Br) have been prepared and characterized on the basis of conductivity, magnetic moment, IR and electron spectra data.[26]

A number of planar platinum(II) complexes, $[Pt(oxime)_2(NH_3)_2]Cl_2$ and $[Pt(oxime)_4]Cl_2$, have been isolated,[27,28] which have been shown to function as stronger protonic acids than the corresponding free oximes; in some cases neutralized species such as $[Pt(Hato)(ato)_2]$ can be readily obtained.[29]

Dimeric compounds with the formula $[M(ato)Me_2]_2$, where M = B, Al or Ga, with the 'ato' (acetoximato) group bridging have been reported.[30] Similarly, in the trimeric palladium(II) complexes $[Pd(acetato)(acetoximato)]_3$, both the acetato and acetoximato groups have been shown to bridge the three palladium atoms which constitute an equilateral triangle.[31]

Following the elucidation[32] of the crystal structure of an interesting complex, $Me_2CNOMo(CO)_2Cp$ (7), prepared by the reaction of excess of 2-bromo-2-nitrosopropane with $NaMo(CO)_3Cp$, King and Chen[33] have described the synthesis of a number of complexes of similar type, $CpM(CO)_2ONCRR'$ (where M = Mo or W; R = Me, R' = Me, $CH=CH_2$, $CH=CMe_2$ and $(CH_2)_2CH=CH_2$; and $R + R' = (CH_2)_5$), by simpler routes involving the reactions of $CpM(CO)_3Cl$ with the oxime in the presence of pyridine or with $RR'C=NOLi$ or $RR'C=NOSnMe_3$, indicating the requirement of UV irradiation for the formation of tungsten compounds only.

(7)   (8)   (9)

Khare and Doedens[34] determined the crystal structure (8) of an interesting product described by King and Douglas[35] by the reaction of $Na_2[Fe(CO)_4]$ with 2-bromo-2-nitrosopropane. Aime and coworkers[36] prepared the same derivative (8) by the reaction of Pr^iNO_2 with $[Fe_3(CO)_{12}]$ along with another product (9) in even greater yield.

13.8.1.3 *vic*-Dioximes

The best known example of a *vic*-dioxime is dimethylglyoxime (**10**) (H_2DMG), which gives the well-known red nickel complex (**11**) with Ni^{2+} in ammoniacal solution.[2] The detailed structures of a fairly large number of metal *vic*-dioximates are known from X-ray investigations. The important features, apart from N_4 planar binding, are the strong $O \cdots H \cdots O$ hydrogen bondings and the stacking of the planar units parallel to each other in the crystal, in the cases of Ni^{II} and Pd^{II} complexes in general. The nature of metal–metal interactions in such chains has been reviewed recently.[37,38]

A special mention[13] may be made of cobalt complexes; the $[Co(HDMG)_2]^{n\pm}$ ($n = 0, 1$) has been named 'cobaloxime' by analogy with 'cobalamin', which is another name of vitamin B_{12}.

Depending upon the metal and its oxidation state, structures of the type (**11**) may accommodate additional ligands on one or both[39] [*e.g.* $Rh(HDMG)_2Cl(PPh_3)$] the axial positions, perpendicular to the plane of (**11**). Ideally (**11**) should have the symmetry D_{2h}, but it is often found to be lower due to distortions arising out of the peculiarities in bonding or crystal packing. The $C{=}N$ distance ($\sim 1.30 \text{Å}$) in most of these *vic*-dioximates is almost the same as in parent dioximes, but the N—O distances are generally lowered to $\sim 1.34 \text{Å}$ from $\sim 1.40 \text{Å}$ in the free ligands. The structural details along with supporting spectral evidence have been well summarized in a recent review.[13] Selectivity and analytical applications of dimethylglyoxime and related ligands have also been extensively reviewed.[40]

A series of chiral bis-dioximato complexes of Fe^{II}, Co^{II}, Ni^{II} and Pd^{II} have been described from three geometrical isomers (α, β and δ) of D-camphorquinone dioxime (Hcqd) and two (β and δ) of L-nopinoquinone dioxime (Hnqd).[41] Opposite Cotton effects around $20\,000 \text{cm}^{-1}$ have been observed for $Co(\alpha\text{-cqd})_2 \cdot H_2O$ and $Co(\delta\text{-nqd})_2 \cdot H_2O$ indicating their 'quasi-enantiomeric' stereochemistries.

Although square-planar configuration is customarily considered classical for *vic*-dioximate of nickel(II), attempts have been made repeatedly over the years for preparing the above complexes in other configurations also. By employing weakly polar solvents and some other variations, success has been claimed in the preparation of mono(dioxime) complexes of nickel(II).[42,43] The dichlorobis(1,2-cyclohexanedione dioximato)nickel(II) has been shown to have an octahedral *vic* structure.[44] Examples of tris(dioxime) complexes of transition metals in general[45-48] and of bivalent atoms[46,47] in particular are rare and structural details of only a tris(dioxime) complex of cobalt(III) are known.[48] In a more recent publication,[49] the crystal structure of tris(1,2-cyclohexanedione dioximo)nickel(II) sulfate dihydrate has been elucidated.

13.8.1.3.1 *Oxidation of* **vic**-*dioximates*

$Ni(HDMG)_2$ dissolves in strongly alkaline media to give yellow solutions which are believed to contain species such as $[Ni(DMG)(OH)]^-$ and $Ni(DMG)_2^{2-}$.[50] The colour of the solution gradually changes to red on absorption of oxygen; the changes involved occur probably in two stages: (i) a rapid and reversible attachment of oxygen, followed by (ii) irreversible oxidation, leading to the probable formation of Ni^{IV} species.

Species like $M(Hdo)_2X$ and $M(Hdo)_2X_2$, obtained by the oxidation of $M(Hdo)_2$ (where M = Ni, Pd and Pt; and H_2do is *vic*-dioxime) with bromine and iodine, were initially believed to involve metal atoms in higher oxidation states. However, later work has shown that the halogen molecules may be held in these either by charge-transfer interactions or in the form of species like X_3^- ions.[51] On the other hand, there are various reports that species like $K_2[Ni(DMG)_3]$ can be synthesized by the oxidation of the bis species in alkaline medium with an excess of two-electron oxidant (hypoiodide).[52]

13.8.1.3.2 Clathrochelate or encapsulating ligands

Oxime complexes have been employed[13] for the construction of clathrochelate or encapsulating ligands in which three oxygen atoms of the chelating oximes are 'capped' by a suitable group[53] such as BF, $SnCl_3$ or SiO. For example, $[Co(DMG)_3(BF)_2]BF_4$ (12) was synthesized by the reaction of $K_3[Co(DMG)_3]$ with two equivalents of Lewis acid BF_3. The suitable geometry of the triangular arrays of oxygen atoms on the opposite faces of the $[Co(DMG)_3]^{3-}$ octahedron has been exploited[54] to synthesize 'three metal' species of the type $[Co(DMG)_3(ML_3)_2]^{n\pm}$ (where ML_3 = Cr(diethylene-triamine), Co(diethylenetriamine), etc.).

(12)

13.8.1.3.3 Substituted oximes

The *vic*-dioximes can be considered as monooximes with another oxime group in the molecule located such as to render a bidentate ligand; the same bi- or even poly-dentate nature can be achieved by the presence of other donor moieties in the oxime molecule. A few of these will be described below.

(i) Carbonyl oximes and 2-nitrosophenols

The *vic*-dioximes are really oximes of carbonyl oximes, the α- (13) and β-isomers (14) of which are represented below.

(13)   (14)   (15)

The simplest examples of (13) are diacetyl monooxime (R = R' = Me) and α-benzil monooxime (R = R' = Ph). While the β-isomer (14) of the latter is not generally capable of chelation, the α-isomer does exhibit chelating characteristics (15), although weaker than *vic*-dioximes. Their greyish complexes with nickel(II) are usually amorphous and unstable and correspond to hydroxy products like [Ni(L)(LH)(OH)], where LH represents (14). Attempts to prepare palladium(II) complexes of diacetyl monooxime[55] appear to have been unsuccessful. There is considerable confusion in the literature about the nature of these complexes.[56] A collection of magnetic data on different iron(II) complexes containing carbonyl oxime functions is available.[57] Phadke and Haldar[58] have studied the magnetic and spectral properties of complexes of nickel(II), palladium(II) and platinum(II) with α-benzil monooxime. A similar physicochemical study[59] has been carried out of the complexes MLX_2 and ML_2X_2 (where M = Mn, Co, Ni, Cu, Zn, Cd; X = Cl, NCS, NO_3 or ClO_4 and L = benzil monooxime).

Considerable work has been published[13] on 2-pyridine oximes (16) which can be considered as imine oximes in which the imine fragment is a part of a heteroatomic ring. For example, Cu(II) complexes of (17; R = Ph or Me) have shown[60] a subnormal magnetic moment of ~1.0 BM and their EPR spectra suggest a strong coupling between the copper atoms.

2-Nitrosophenols have a pronounced tendency to tautomerize to the corresponding quinone monooximes, the two resonance forms of which are represented below (17 and 18). In view of the above, it would be appropriate to give their account along with carbonyl oximes.[13]

Systematic physicochemical studies on copper(II) and nickel(II) complexes of (18; HYnP) have been reported.[61] The green iron(II) pigment, ferroverdin, produced by the species *Streptomyces*[62] has

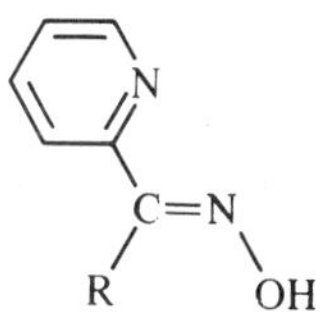

(16) 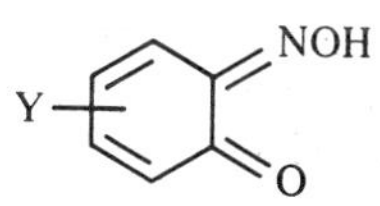   (17) 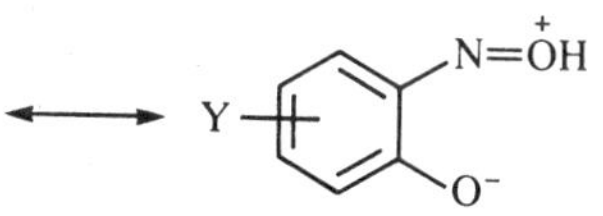   (18)

been identified as Na[Fe(Ynp)$_3$], where Y is H$_2$C=CH(p-C$_6$H$_4$)OC(O)—. A few polymeric species, *e.g.* metal complexes of 2,4-dinitrosoresorcinol, have also been described.[63] The products of the reactions between phenols and Na$_3$[Co(NO$_2$)$_6$] reported by Feigel[4] have been identified recently as tris(*o*-quinone monooximato)cobalt(III), which has been synthesized by two other alternative routes also.[64]

(ii) Imine oximes

The imine oximes (HYio; **19**) are Schiff bases of carbonyl oximes (**14**), and in their simplest forms these are also bidentate ligands,[13] if Y is an alkyl or aryl group.

Similarly, tridentate (**20, 21**), tetradentate (**22**) and hexadentate (**23**) ligands have been synthesized and their coordination chemistry has been investigated extensively.[13]

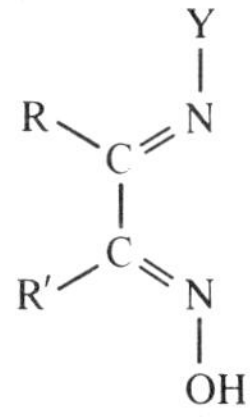

(19) 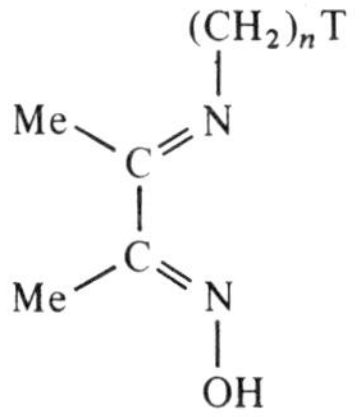   (20) 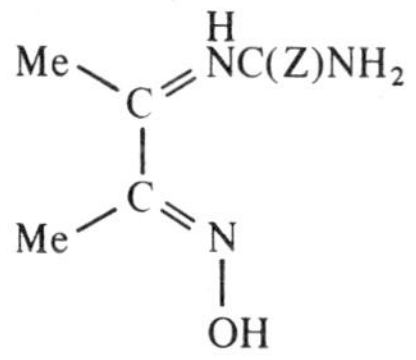   (21)

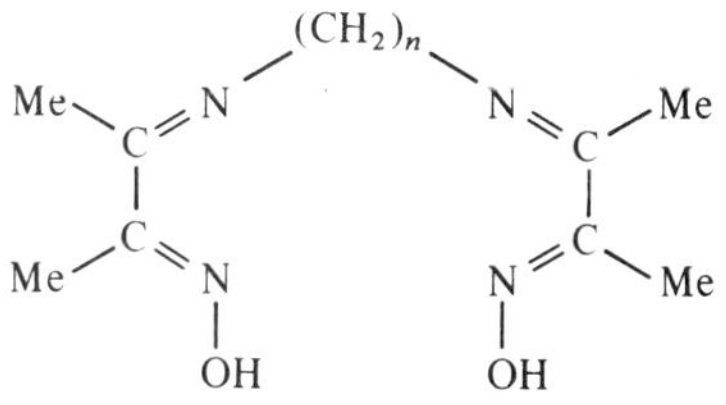

(22) 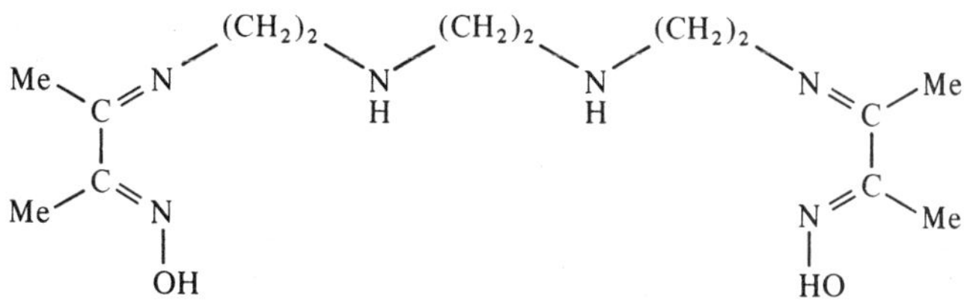   (23)

The ligands (**20**) contain a coordination site such as pyridyl or amino nitrogen, or a hydroxyl oxygen. Complexes of (**20**; T = NH$_2$ and $n = 2$) with iron(II) have been studied recently.[65] A copper(II) dimer (with bridging oxime groups) of (**20**; T = OH, $n = 2$ or 3) was described by Ablov and coworkers[66] in 1972 and its crystal structure has been elucidated[67] in 1974.

Upon exposure to air, the pseudooctahedral complex of (**23**) with Ni(IV) becomes paramagnetic and an EPR study of a single crystal of the material at 90 K indicates the formation of Ni^{III} centres from Ni^{IV} ions of the unexposed lattice.[68]

(iii) Hydroxy oximes

α-Acyloin oximes (**24**), the most common example of which is α-benzoin oxime, form green water-insoluble copper(II) complexes;[13,69] the monooxime of cyclohexane-1,2-dione produces polymeric copper(II), nickel(II) and cobalt(II) complexes[70] in which the ligand appears to have an enolate configuration (**25**).

The most common hydroxy oxime ligands are salicylaldoximes, complexes of which with copper(II), nickel(II) amd palladium(II), cobalt(II), iron(II), iron(III) and manganese(II) have been investigated extensively including crystal and molecular structures[13] of the first three. In an interesting study,[71] the reactions of cobalt bis chelates of this type have been studied with aluminum isopropoxide.

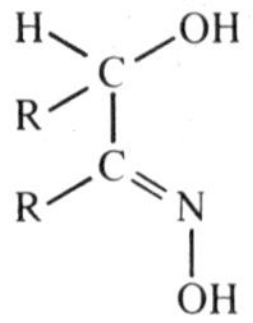

(24) 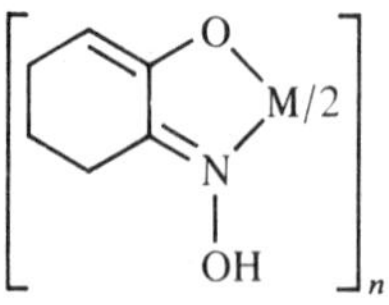   (25)

The P5000 Acorga extraction reagents contain as the major complexing agent 5-nonylsalicyl-aldoxime, the copper complexes of which were studied in 1983 by electronic and ESR spectroscopy.[72]

(iv) Amino oximes

Nickel complexes of amino oximes[13] (**26**; R = alkyl; HRabo) and (**27**; Haboen when $n = 2$, Habopn when $n = 3$) have been investigated at various pH values by Murmann[73] and their structures have been elucidated by X-ray[74] and neutron diffraction[75] techniques. The structural study of Cu(Haboen)Br has disclosed the presence of dimeric $[Cu_2(Haboen)_2]^{2+}$ cations.[76]

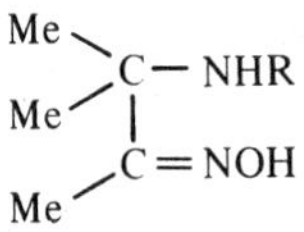

(26) 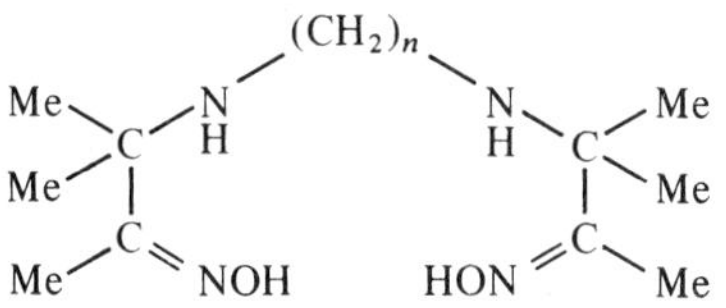   (27)

A series of papers have been published on amino oxime derivatives of later transition metals, including a study of the complexes of N,N'-bis[4'-benzo(15-crown-5)]diaminoglyoxime with Cu^{II}, Ni^{II}, Co^{II}, Co^{III}, Pd^{II}, Pt^{II} and $U^{II}O_2$.[77]

(v) Amide oximes

Amide oximes[13] can be represented by the general formula $RC(=NOH)NH_2$ and the ligand of this class which has received maximum attention is benzamide oxime (R = Ph). A good test of amide oxime function is the formation of a red-brown colour with iron(III) in neutral solution. The use of amide oximes as analytical reagents for the estimation of various metal ions has been reviewed.[78] However, as stated in a later review,[13] not much definitive structural information is available on these complexes. For example, although on the basis of IR data, benzamide oxime complexes of copper(II) and nickel(II) have been assigned the structure (**28**), yet even magnetic and spectral data are not available.

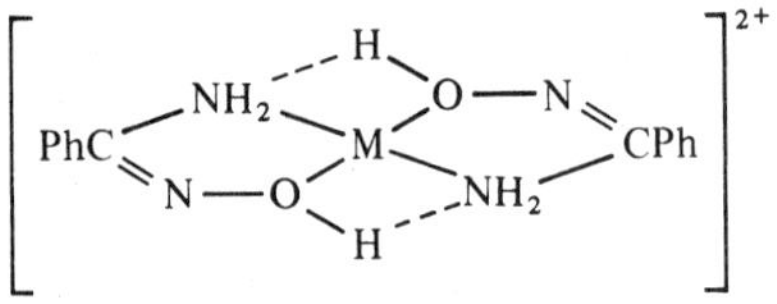

(28)

The oxidation of $Ni(NO_3)_2$ and benzamide oxime by H_2O_2 in ethanolic solution has been reported to yield red diamagnetic needles of $Ni(benzamidoximate)_4$, in which nickel and the ligand have been presumed to be tetra- and mono-valent respectively on the basis of IR data.[79]

(vi) Azo oximes

Azo oximes, in view of the unique combination of oxime and azo functions, make excellent ligands and have been investigated extensively[13] by Chakravorty and coworkers, who have recently published[80] an account of synthesis of bis(arylazooximato)palladium(II) and have tried to elucidate the palladium–nitrogen bond lability and redox activity in it.

13.8.2 AMIDINE COMPLEXES

Amidines have been known to coordinate with metals in a variety of monomeric,[81,82] dimeric[83] and cluster[84] systems, but it was only in 1983 that amidine clusters were structurally characterized.[85]

Following their observations on the stability of a copper(I) derivative of *N,N'*-di-2-anthraquin-onylformamidine,[86] Bradley and Wright[81] made a more detailed study of the derivatives of *N,N'*-diarylformamidines, ArNHCH=NAr, and some related ligands with copper(I), copper(II), silver(I) and nickel(II).

Patel and coworkers[87–89] and others[90] have been carrying out detailed investigations on the analytical applications of a number of amidines, *e.g.* *N*-hydroxy-*N*-phenyl-*N'*-benzylbenzamidine,[91] *N,N'*-diaryl substituted *N*-hydroxy-*p*-toluamidines[87] and hydroxydiarylbenzamidines.[88–90]

On the synthetic side, the reaction[92] of $Me_3SiNLiR$ with $F_3CC(Cl)=NR'$ gave 75–90% of seven products, $Me_3SiNRC(CF_3)=NR'$ (R = 2,4-, 2,6-$Me_2C_6H_3$, *m*-$F_3CC_6H_4$, *o*-FC_6H_4, Ph; R' = Ph, *m*-$F_3CC_6H_4$), which with XBR''_2 gave 80–95% of 16 products, $R''_2BNRC(CF_3)=NR'$ (R'' = Me, Me_2N, Et_2N). The reaction of $F_3CC(X)=NR$ (X = Cl, Br) with $Me_3SiNMeCH_2CH_2NMeLi$ gave 90% of six products, $Me_3SiNMeCH_2CH_2NMeC(CF_3)=NR$, which with R''_2BX (X = Cl, Br) gave 75–85% of eight products, $R''_2BNMeCH_2CH_2NMeC(CF_3)=NR$.

IR data have indicated[93] that (**29**; R = Ac, cyclohexyl, $MeSO_2$, Me, H; R' = Pr, Bu; R'' = Me, Ph) exist as tautomer (**29**) in the crystalline state and in proton acceptor solvents and as tautomer (**30**) in proton donating solvents.

(29)   (30)   (31)

Following the well-recognized role of carboxylato anions as bridging ligands in binuclear complexes with metal–metal bonds, Cotton and coworkers[94–97] have explored the use of the isostructural and isoelectronic amidinato (NR')RC(NR'') anions (**31**).

When $Mo(CO)_6$ is heated with *N,N'*-diphenylbenzamidine, $PhC(NPh)(NHPh)$, a red air-stable crystalline product $Mo_2[PhC(NPh)_2]_4$ (**32**) is obtained with a Mo≡Mo distance of 2.090 (1) Å.[94]

(32)   (33)

Two different types of crystalline products were obtained by fusing together $(Bu_4N)_2Re_2Cl_8$ with N,N'-diphenylbenzamidine under nitrogen and their crystal structures have been found[95] to correspond to $Re_2(N_2CPh_3)Cl_4$ and $Re_2(N_2CPh_3)Cl_4 \cdot THF$. The ReRe bond distance in the former, 2.177(2) Å, is the shortest yet observed, whereas it is only a shade longer, 2.209(1) Å, in the latter.

A yellow crystalline product, $Cr_2(MeNCPhNMe)_4$, has been isolated[96] by the reaction of $Cr_2(O_2CMe)_4$ with $Li(MeNCPhNMe)$ in hexane. The molecule has virtual D_{4h} symmetry and there is no axial coordination, since the four methyl groups on each end screen the axial positions very effectively (33). The Cr—Cr distance is 1.843(2) Å, which represents one in the shorter range of the 'super short' $Cr\equiv Cr$ quadrupole bonds.

The reaction of $W_2(dmhp)_4$ (dmhp = the anion of 2,4-dimethyl-6-hydroxypyrimidine) with the lithium salt of N,N'-diphenylacetamidine, $Li(PhN)_2CMe$, yields a red crystalline air sensitive but thermally stable product.[97] It crystallizes in space group $P2_{1/n}$ with $Z = 2$, requiring the molecules to lie on crystallographic inversion centres. The $W\equiv W$ distance, 2.174(1) Å, is very similar to that in $W_2(dmph)_4$.

With a view to examining whether bulky substituents at the amidine nitrogen atoms can prevent axial coordination in quadruply bonded dimetal compounds, Cotton and coworkers have studied the reactions of $(Bu_4N)_2Re_2Cl_8$ with $PhN(CMe)NHPh$ and $MeN(CPh)NHMe$.[83] The structures of the crystalline products can be represented as (34) and (35) respectively.

Structure (34) with $Re\equiv Re$ bond distance as 2.208(2) Å is quite similar to that found previously for $Re_2[(PhN)_2CPh]_2Cl_4$.[87] Structure (35) is of interest as the methyl groups keep the axial Cl^- ligands at a greater distance than for the $Re_2(O_2CR)_4Cl_2$ compounds.

(34)   (35)

N,N'-Dibenzyl- or -diisopropyl-formamidines react[84] with $Os_3(CO)_{12}$ or $Os_3(CO)_{10}(cyclooctene)_2$ to give different types of nonacarbonyl, $HOs_3(CO)_9(Pr^iNCHNPr^i)$ and $H_2Os_3(CO)_9$-$(PhCH_2NCHNCH_2C_6H_4)$, the difference being in the ability of the benzyl group to be *ortho*-metallated.

Reactions of $[Os_3(CO)_{10}(NCMe)_2]$ with amidines have been used to synthesize the complexes $[Os_3(\mu\text{-}H)(CO)_{10}\{NHC(Me)NH\}]$ and $[Os_3(\mu\text{-}H)(CO)_{10}\{NPhC(Ph)NH\}]$.[85]

These amidine clusters have been thermally decarbonylated to the corresponding nonacarbonyl compounds of which $[Os_3(\mu\text{-}H)(CO)_9(MeCN_2H_2)]$ alone exists as two tautomeric forms (36) and (37):

(36) .   (37)   (38)

The structure of $[Os_3(\mu\text{-}H)(CO)_9NPhC(Ph)NH]$ (38) was established by X-ray crystallography. There are two independent but structurally similar molecules of (38) per asymmetric unit. In each molecule, the Os atoms define an isosceles triangle, one edge of which is bridged by the protonated nitrogen atom of the $NPhC(Ph)NPh$ ligand, the other N atom being terminally bound to the third Os atom in an axial site. In this way the amidino ligand caps one triangular face of the metal framework and formally donates five electrons to the cluster.

13.8.3 IMIDATE ESTERS

Imidic acid (40) is the lactim tautomer of an acid amide (39; lactam form). Although no substantiated claim has been put forward for the isolation of free imidic acids, imidates (41) (also called imino ethers, imido esters, imidic esters and imidoates) can be easily obtained by alkylation of an amide in the presence of silver oxide.[98]

$$RC \overset{NH_2}{\underset{O}{\diagup}} \qquad RC \overset{NH}{\underset{OH}{\diagup}} \qquad RC \overset{NH}{\underset{OR'}{\diagup}}$$

$$(39) \qquad\qquad (40) \qquad\qquad (41)$$

Following the well-known Pinner synthesis of imidates by the reaction of nitriles with alcohols, there has been considerable interest in recent years in treating a nitrile bond within the coordination sphere of metal ions with water or alcohols.[99-107] It has been suggested that the initial step consists of nucleophilic attack of an external or coordinated hydroxide or alkoxide anion on the nitrile carbon atom. A number of imidate complexes have been isolated[108,109] from the reactions of 2-cyanopyridine with metal [Cu^{II}, Ni^{II}, Co^{II} and Fe^{II}] salts in alcohols, of *trans*-$PtMeClL_2$ with pentafluorobenzonitrile and silver hexafluorophosphate in alcohols,[110] and of *o*-cyanobenzylplatinum complexes.[111]

A series of imino ether complexes with the typical composition [$PtMeL_2$(NH=$C(OMe)C_6F_4$-$C(OMe)$=$NH)PtMeL_2$][(BF_4)_2] (where L = PMe_2Ph) have been synthesized[110] by the reactions of *trans*-$PtMeClL_2$ with pentafluorobenzonitrile and silver hexafluorophosphate in methanol (alcohol). The comparative role of PtMe and $PtCF_3$ in such reactions has been discussed[111] to elucidate the mechanism of such reactions.[112] The preparation of iridium imino ether complexes has been reported by a similar route.[113]

Abstraction of X, followed by nucleophilic attack by alcohols on *trans*-$PtX(CH_2C_6H_4CN)L_2$ (where L = PPh_3, $AsPh_3$; X = Cl, Br), obtained by oxidative addition of *o*-$XCH_2C_6H_4CN$ to PtL_4, yields *cis* imino ether complexes.[114]

Cationic pentachlorophenylnickel(II) complexes, *trans*-[$C_6Cl_5Ni(P)_2(R)$]$^+ClO_4^-$ (where P = $PPhMe_2$, PPh_2Me; R = Me, CH_2Ph, Ph), synthesized[112] by the reaction of *trans*-$C_6Cl_5Ni(P)_2Cl$ with silver perchlorate, and nitriles in benzene were treated with methanol or ethanol in the presence of triethylamine, to yield the corresponding imidate complexes, *trans*-[$C_6Cl_5Ni(P)_2NH$=$C(R)(OR')$]$^+ClO_4^-$ (42). The ^1H NMR spectra indicated that the products obtained, when P = $PPhMe_2$, were mixtures of complexes containing isomeric imidate groups (42; *E* and *Z*). The major isomer (when P = $PPhMe_2$, R = Ph, R' = Me) was shown by ^1H NMR spectroscopy to be the *cis* adduct (43).

$$(42E) \qquad\qquad (42Z) \qquad\qquad (43)$$

The oxidative addition of *o*-$XCH_2C_6H_4CN$ to PtL_4 gives *trans*-$PtX(CH_2C_6H_4CN)L_2$ (L = PPh_3, $AsPh_3$; X = Cl, Br). Abstraction of X yields *cis* cationic complexes in which the CN group is coordinated and is prone to nucleophilic attack by alcohols to give stable *cis* imino ether complexes (44).[111] The formation of an imino group is revealed by the decrease of ν(CN) by *ca.* 600 cm^{-1} down to the region normally assigned to C=N stretching frequencies and by the appearance of a ν(NH) around 3300 cm^{-1}. When $ReCl_4(MeCN)_2$ was briefly heated with methanol or ethanol, green complexes were obtained. The spectral studies indicate that these complexes are undoubtedly derivatives of imidate esters, $ReCl_4[MeC(=NH)OR]_2$ (45; R = Me, Et), formed by nucleophilic addition of alcohols to the nitrile group.[115]

$$(44)$$

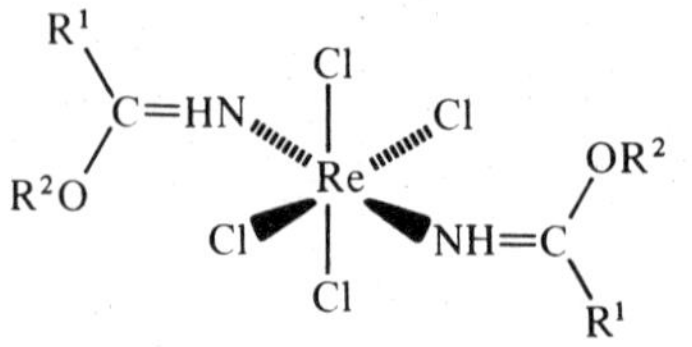

(45)

Cationic methylplatinum(II)–nitrile complexes of the type *trans*-PtMeL$_2$(NCR)$^+$X$^-$ have been isolated by the reaction of *trans*-PtMeClL$_2$, where L = dimethylphenylphosphine or trimethylarsine, with an aryl nitrile and AgX, where X$^-$ = BF$_4^-$, PF$_6^-$. Use of pentafluorobenzonitrile and 2,3,5,6-tetrafluorotetraphthalonitrile in alcohol has led to the synthesis of a series of imino ether complexes. A mechanism for imino ether formation, involving nucleophilic attack by an alcohol at a coordinated nitrile, is suggested and the course of the reaction is shown to be dependent not only on the alcohol but also on the size of the anion used.[110]

Perfluoroaryl nitriles react with methylplatinum cations in methanol to yield imino ether complexes.[110,113] These reactions also proceed *via* a π-bonded intermediate, in which the C≡N group is activated towards nucleophilic attack. This is a smooth reaction for Ir^{III} to give the imino ether complex (46) in good yield.[114] The imino ether complex [47; L = NH=C(OMe)(C$_6$F$_5$)] was synthesized from the reaction of PtMeCl(cod) (where cod = cyclo-octa-1,5-diene) and AgPF$_6$ with pentafluorobenzonitrile in methanol (Scheme 1).[116] These workers have previously examined a number of similar reactions resulting in the formation of Pt^{II},[117] Pt^{IV}[118] and Ir^{III}[113] imino ether complexes. A mechanism has been suggested[118] involving initial π-coordination of pentafluorobenzonitrile, delocalization of charge from platinum onto the —C≡N bond, and activation of the bond towards nucleophilic attack by the alcohol.

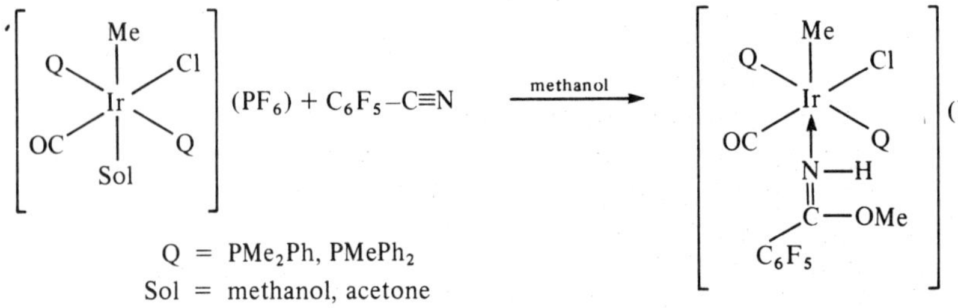

Q = PMe$_2$Ph, PMePh$_2$
Sol = methanol, acetone

(46)

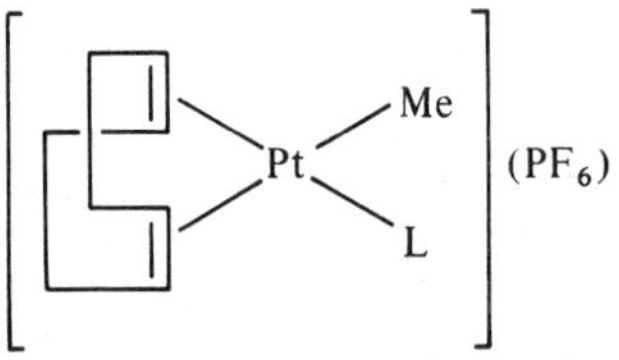

(47)

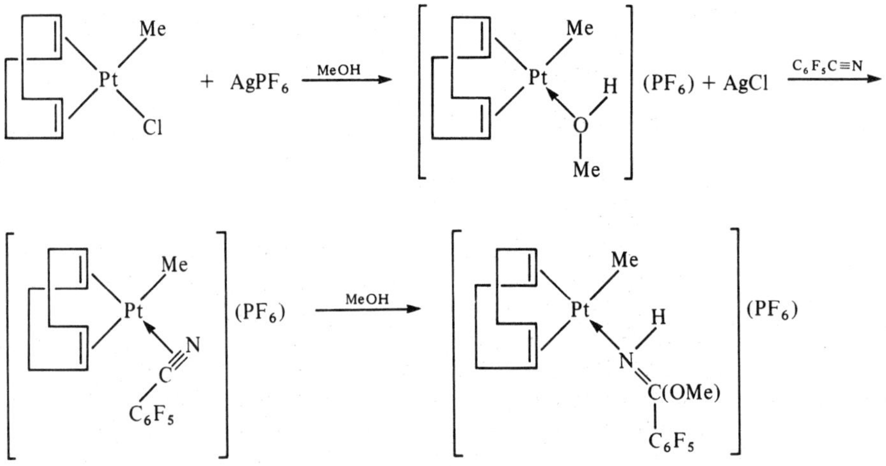

Scheme 1

The 2-cyanopyridine ligand is alcoholized with transition metal ions (*e.g.* Fe^{II}, Cu^{II}, Ni^{II} and Co^{II}) in alcohols to afford complexes of *O*-alkyl pyridine-2-carboximidate of type (48).[119] It has been suggested that the nitrile group of 2-cyanopyridine is activated by chelation with the metal ion and therefore the attack of the nucleophiles (*e.g.* OR⁻ or OH⁻) on it is promoted. Suzuki and coworkers[109] reported complexes of type (49) and type (50).

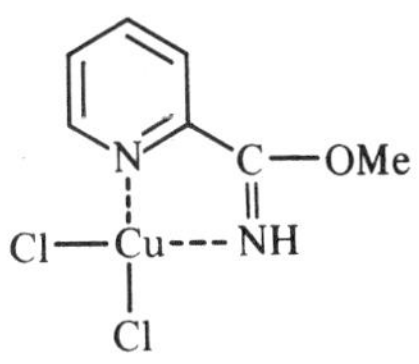

(48) 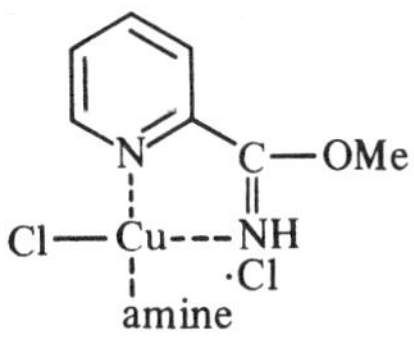   (49)

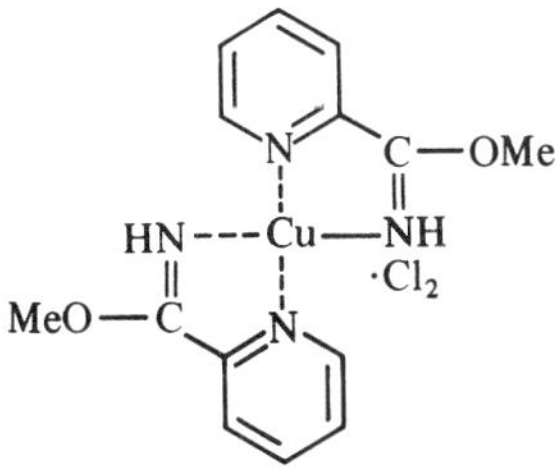

(50)

13.8.4 IMINE (RN=CR₂) COMPLEXES

Transition metal complexes containing alkylidenimino ligands, $R_2C=N-$, have been known for some time.[120] These are potentially strong ligands and may be considered to resemble nitrosyl (NO) ligands in functioning not only as one-electron but also as three-electron donors (giving rise to linear π-bonded groups). For example, in some silicon[121] and titanium[122] complexes, the M—N—C system is bent (51) with sp^2-hybridized nitrogen, while in some molybdenum[120] and boron[123-125] species, the M—N—C system is linear (52) with *sp*-hybridized nitrogen. These cases have been treated in terms of the alkylidenimino ligands behaving as one- and three-electron donors respectively. In addition to the above, these ligands have also a strong tendency to act as bridging group (53).

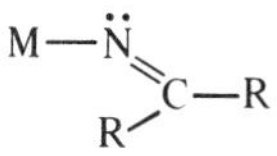

(51)   (52)   (53)
Bent   Linear   Bridged

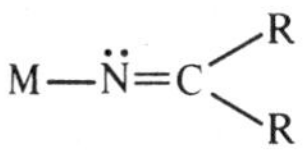

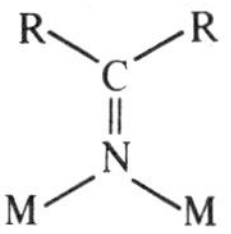

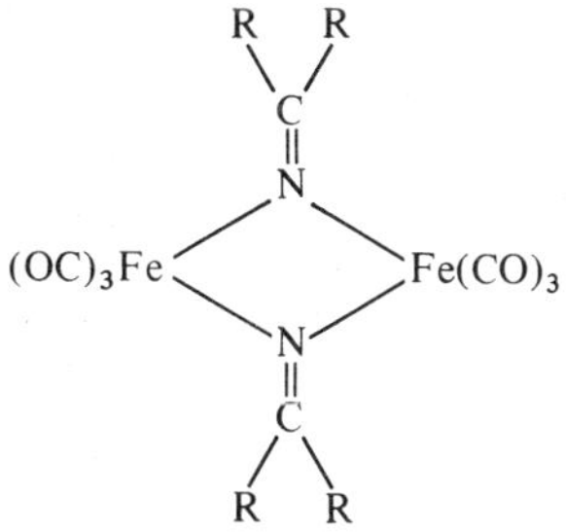

(54)

Fe—Fe	2.40 Å
Fe—N	1.94 Å
N—C	1.24 Å
∠NFeN	77.0°
∠FeNFe	77.1°
Dihedral angle between FeNFe planes	70.4°

The reaction of $(p\text{-tol})_2C=N-N=C(p\text{-tol})_2$ with $Fe(CO)_5$ yields the bridging alkylidenimino complex $[(p\text{-tol})_2C=N]_2Fe_2(CO)_6$, which has been shown[126] to have structure (54; R = p-tol).

A general method of synthesizing imine derivatives is the reaction of metal complex halides with $LiN=CR_2$ reagents, which are in turn obtained from compounds such as $Ph_2C=NH$ or $Bu_2^tC=NH$ by action of Bu^nLi.[127-131] These complexes may also be prepared by treating the ketimines with metal complex halides. Alternatively, trialkylstannyl derivatives, such as $Me_3SnN=C(CF_3)_2$, can also be used in metathetical reactions.[132] Treatment of $[CpM(CO)_3X]$ (X = Cl, Br, I) with $Me_3SiN=CRR'$ or $LiN=CRR'$ gives complexes of three types: $[CpM(CO)_2N=CRR']$ (55; R = R' = Ph, Bu^t, p-MeC$_6$H$_4$; R = Ph; R' = Bu^t), $[CpM(CO)N=CRR']_2$ (56; R = R' = Ph) and $[CpM(CO)_2(RR'CN)(RR')]$ (R = R' = Ph, p-MeC$_6$H$_4$).[133-138] In the complex $[CpMo(CO)_2N=C(Bu^t)_2]$ (55) the Mo—N—C skeleton has been shown to be almost linear and the molybdenum–nitrogen distance (1.87 Å) appears to be short compared to the Mo—N single-bond distance (2.32 Å) in $Mo(CO)_3(dien)$.[139]

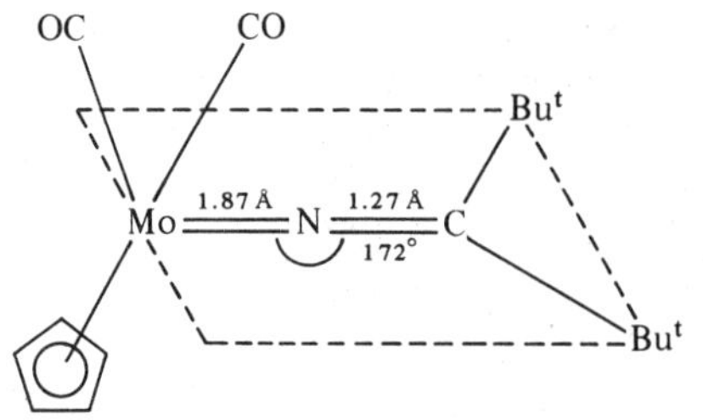

(55)

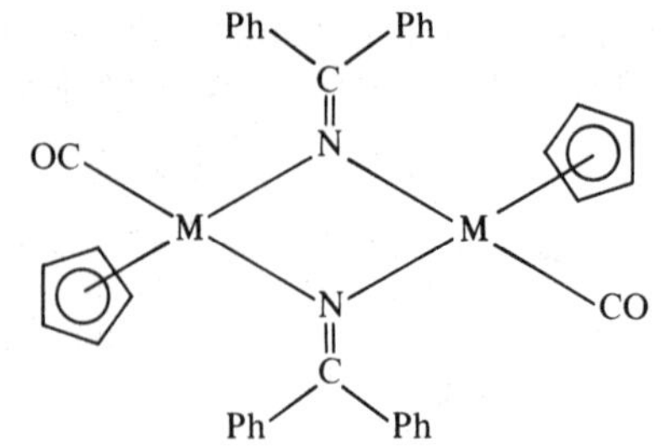

(56; M = Mo, W)

The plane of the ligand moiety $Bu^t_2C=N-$ adopts a staggered configuration similar to that reported for carbene derivatives for which π-bonding to the metal has also been postulated.[140] The spectral characteristics of the complex have been explained in terms of the configurational changes about the multiple metal–nitrogen bond.[136] The corresponding diphenyl derivatives,[134] unlike the di-*p*-tolyl,[120] and *tert*-butylphenyl complexes,[138] form the dinuclear complex (56) by the loss of CO. With M = Mo in the complex (56), the CO groups are in the *cis* position, and with M = W, these groups are *trans* to each other.

The complex (56) does not react with neutral ligands even under forcing conditions. This apparent coordinative saturation suggested strong bonding in the M_2N_2 ring and involvement of 10 electrons which is required if the effective atomic number rule is to be satisfied.

Reaction of the diphenyl derivative (56) with I_2 yielded products of the type $[Cp_3M_3I_3O_4]$ (M = Mo, W).[134] However, $[CpM(CO)_2N=CRR']$ (R = Ph, Bu^t; R' = Bu^t) with I_2 yielded the oxidation products of the type $[CpM(CO)I_2(N=CRR')]$.[137,138] The complex $[CpW(CO)(PPh_3)-N=Bu^tPh]$[138] has been found to be the only simple substitution product, which indicates the dependence of the properties of imine complexes on the nature of the substituent groups present.

The reaction of $Mn(CO)_5Br$ and $Me_3SnN=C(CF_3)_2$ yields a complex which was initially reported[141] to be $[(OC)_4Mn(\mu-NC(CF_3)_2)_2Mn(CO)_4]$ analogous to the phenyldiazo derivative, $[(OC)_4Mn(\mu-N_2Ph)_2Mn(CO)_4]$.[142,143] However, later on with the help of an X-ray diffraction study, it has been shown[144,145] to have a structure corresponding to $[(OC)_3Mn(\mu-CO)(\mu-NC(CF_3)_2)_2Mn(CO)_3]$ (57). The complex (57) consists of two $Mn(CO)_3$ groups which are held together by two bridging $(CF_3)_2C=N$ ligands, one bridging carbonyl ligand and a Mn—Mn bond 2.518 (2) Å in length. An unexpected feature of the complex (57) is the highly asymmetric ('semi-bridging') nature of the bridging carbonyl ligand, for which Mn(1)—CO = 1.944 (9) Å and Mn(2)—CO = 2.173 (9) Å. This asymmetry is apparently compensated by a small, but significant, contrary asymetry in Mn—N distances to the two bridging $(CF_3)_2C=N$ ligands.

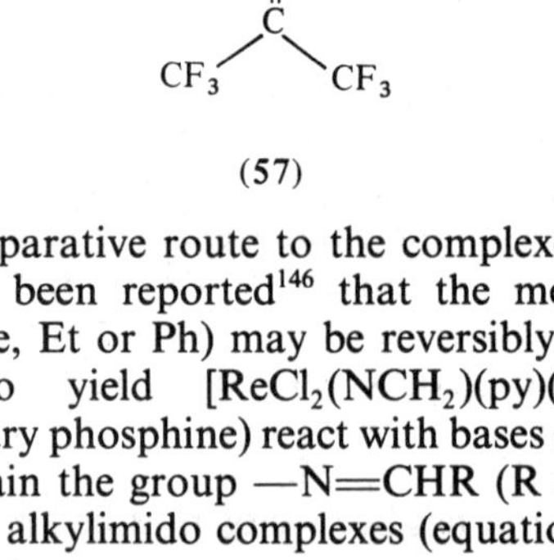

(57)

Little is known about the preparative route to the complexes of the unsubstituted methylene-amido ligand $N=CH_2$. It has been reported[146] that the methylimido ligand in the complex $[ReCl_3(NMe)(PPh_2R)_2]$ (R = Me, Et or Ph) may be reversibly deprotonated by, for example, an excess of pyridine (py) to yield $[ReCl_2(NCH_2)(py)(PPh_2R)_2]$. Similarly complexes $[ReCl_3(NR)(PR'_3)_2]$ (PR'_3 = tertiary phosphine) react with bases such as pyridine to yield methylene-diamido complexes, which contain the group $-N=CHR$ (R = H or alkyl). Acids reconvert the methyleneamido complexes into alkylimido complexes (equation 1).[147]

$$(PPh_2Me)_2Cl_3Re=NMe \underset{HCl}{\overset{py}{\rightleftharpoons}} (PPh_2Me)_2(py)Cl_2Re-N=CH_2 \qquad (1)$$

It has been observed that in the presence of added substrates such as Ph_2CO, Me_2CO, $PhCN$, $PhNCO$, $PhNH_2$ and *trans*-$PhCH\!=\!CHPh$, the reaction of $Ph_2C\!=\!NLi$ with $CpM(CO)_3Cl$ (M = Mo, W) yields $CpM(CO)_2(N\!=\!CPh_2)$ in contrast to the reaction without the added substrates which produces $CpM(CO)_2(Ph_2CNCPh_2)$.[148] The role of the added substrates and various types of environment has been reported.[128] Two forms of the complex $[CpMo(CO)_2(N\!=\!CPh_2)]$ are compared, one being assigned a monomeric and the other a dimeric formulation.

The complex $[CpFe(CO)N\!=\!CBu_2^t]$ (58) has been isolated from the reaction of $CpFe(CO)_2X$ (X = Cl, Br, I) with $Bu_2^tC\!=\!NLi$.[127] $Fe(CO)_4X_2$ (X = I, Br) reacts with $R'R''C\!=\!NLi$ (R' = R'' = Ph, *p*-tolyl; R' = Ph, R'' = Bu^t) to form $Fe_2(CO)_6I(N\!=\!CR'R'')$ (59) and $[Fe(CO)_3N\!:\!CR'R'']_2$ (60) for which structures with bridging methylenamino groups are suggested on the basis of IR and Mössbauer spectroscopic data.

(58)   (59)   (60)

Although all these complexes are formed in low yield, the complexes $Fe_2(CO)_6I(N\!=\!CR'R'')$ are of structural interest since they contain bridging groups of widely differing sizes and bridging characteristics.[127]

The crystal and molecular structures of isomeric cluster complexes, $HFe_3(MeC\!=\!NH)(CO)_9$ and $HFe_3(N\!=\!CHMe)(CO)_9$, derived from reduction of acetonitrile have been reported.[149,150] In addition to these, a wide variety of complexes with ligands containing azomethine groups ($>\!C\!=\!N\!-$) along with other functional groups (*e.g.* metal complexes of salicylaldiimine) are known and these will be discussed in Chapter 20.

13.8.5 GUANIDINES, BIGUANIDES AND RELATED LIGANDS

The interrelationships amongst the ligands of this section can be illustrated by their formulae (61–67).

$HOCO_2H$	H_2NCOH	H_2NCONH_2	$HN\!=\!C(OH)NH_2$
Carbonic acid	Carbamic acid	Urea (carbamide)	Pseudourea
(61)	(62)	(63a)	(63b)

$H_2NC(=NH)NH_2$	$H_2\overset{5}{N}C\overset{3}{N}HC\overset{1}{N}H_2$ $\underset{\overset{4}{NH}}{\parallel}\ \underset{\overset{2}{NH}}{\parallel}$	$H_2\overset{4}{N}C\overset{2}{N}HC\overset{1}{N}H_2$ $\underset{\overset{3}{NH}}{\parallel}\ \underset{}{\parallel}$ O	$H_2\overset{3}{N}C\overset{2}{N}HC\overset{1}{N}H_2$ $\underset{}{\parallel}\ \underset{}{\parallel}$ O O
Guanidine	Guanylguanidine or BIGUANIDE	Guanylurea	Biuret
(64)	(65)	(66)	(67)

The resemblance amongst the above ligands may be traced to their derivation from the same parent source, *i.e.* carbonic acid (61) or its amides, carbamic acid (62) and urea (63). Guanidine (64), an amidine of carbamic acid, can also be included in this group. Just as biuret (67) is formed during the thermal decomposition of urea nitrate, by condensation of two molecules of urea with the elimination of a molecule of ammonia, guanylurea and biguanide can be prepared by condensation of guanidine hydrochloride with urea and another molecule of guanidine respectively (equations 2–4).

$$H_2NCOH_2 + H_2NCONH_2 \longrightarrow H_2NCONHCONH_2 + NH_3 \qquad (2)$$
Biuret

$$H_2NC(=NH)NH_2 + H_2NCONH_2 \longrightarrow (H_2NC(=NH)NHCONH_2 + NH_3 \qquad (3)$$

Guanylurea

$$H_2NC(=NH)NH_2 + H_2NC(=NH)NH_2 \longrightarrow H_2NC(=NH)NHC(=NH)NH_2 + NH_3 \qquad (4)$$

Biguanide

In view of the presence of active functional groups at suitable sites in the above ligands and their substituted derivatives, these tend to form coordination compounds/chelates with a large number of metals. Excellent reviews have been published on the complex compounds of biguanides[151,152] and guanylureas,[151] and biuret.[153,154]

The close relationship between biguanide and guanylurea is evident from their formation from one and the same source material, dicyandiamide, by its union with a molecule of water and one of ammonia respectively (Scheme 2).

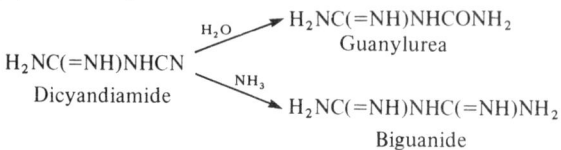

$H_2NC(=NH)NHCN$ $\overset{H_2O}{\nearrow}$ $H_2NC(=NH)NHCONH_2$
Guanylurea

Dicyandiamide $\overset{NH_3}{\searrow}$ $H_2NC(=NH)NHC(=NH)NH_2$
Biguanide

Scheme 2

Some substituted biguanides undergo facile hydrolysis in acid solution. A solution of paludrine, N^1-(p-chlorophenyl)-N^5-isopropylbiguanide, on being kept in moderately strong hydrochloric acid, deposited crystals of N^1-(p-chlorophenyl)-N^4-isopropylguanylurea,[155] but much more rapid hydrolysis occurs[156] in the case of N^1-(p-sulfamolylphenyl)biguanide, and some of its substituted derivatives. Obviously, in an aqueous solution of a biguanide, there tends to be an equilibrium between the biguanide molecule and the products of its hydrolysis, a guanylurea and ammonia (equation 5). The equilibrium tends to be shifted to the right in acidic solution, owing to removal of ammonia as ammonium salt.

$$RNHC(=NH)NHC(=NH)NH_2 + H_2O \rightleftharpoons RNHCONHC(=NH)NH_2 + NH_3 \qquad (5)$$

The metal complexes of biguanides are comparatively more stable than those of guanylureas.[157] The importance of the central N^3 atom of biguanide in the stability of transition metal complexes $[M(bigH)_2]^{2+}$ (M = Cu, Ni) is strikingly demonstrated by replacing the NH by CH_2.[158] The novel ligand thus obtained is malonic diamidine, $H_2NC(=NH)CH_2C(=NH)NH_2$, the corresponding Cu and Ni complexes of which are (by contrast to biguanide complexes) not planar and much less stable.

13.8.5.1 Guanidinium Complexes

Guanidine, $H_2N(C=NH)NH_2$, is the amidine of carbamic acid, $H_2N(CO)OH$. Guanidine forms three types of complexes with metals: cationic (in which the guanidinium cation is formed by taking up a proton), adducts with neutral molecules or coordination products with ionic salts, and substitution products. A brief account of each type is presented below.

Many cationic complexes of guanidine have been prepared and studied physicochemically.[159,160] The guanidinium rare earth (Ln = La to Lu) complexes[161] $[C(NH_2)_3]_x[Ln(CO_3)_y \cdot (H_2O)_z] \cdot wH_2O$ ($x = 1$, $y = 2$; $x = 3$, $y = 3$; and for Ln = Tb to Lu only: $x = 5$, $y = 4$; $z + w = 1-5$) were prepared and studied with the help of their IR spectra. The structure of the guanidinium cation in various thorium complexes has been studied by X-ray photoelectron techniques.[162]

Tetramethylguanidine adducts of Me_3Al, Et_3Al and $AlCl_3$ have been synthesized by Snaith and coworkers.[163] These adducts, with the formula $(Me_2N)_2C=NH \cdot AlX_3$ (X = Me or Et), are monomeric in dilute benzene solution and are believed to involve a coordination of the imino-N to Al. When heated, these lose XH, forming $(Me_2N)_2C=N \cdot AlX_2$, for which a four-membered ring structure has been proposed. Although the adduct $(Me_2N)_2C=NH \cdot AlCl_3$ does not similarly afford $(Me_2N)_2C=N \cdot AlCl_2$ when heated, the latter can be prepared from $(Me_2N)_2C=NLi$ and $AlCl_3$. Attempts have been made to elucidate the structures of these compounds by IR, ^1H NMR and mass spectra.[164,165]

A study of the complexing of triarylboranes with guanidine has been carried out by Yuzhakova and coworkers.[166] Complexes $(H_2N)_2C=NH \cdot BR_3$ (R = Ph; 2,3,4-MeC_6H_4; 2,5-, 3,4-xylyl; 2,4-$MeOC_6H_4$; 2- and 4-$EtOC_6H_4$) were prepared by treating ethereal solutions of BR_3 with the guanidine. The work has been extended to a study of thermochemical transformations of complexes of triarylboranes with symmetrical diphenylguanidine.

A detailed study of spectroscopic and magnetochemical properties of complexes of tetramethyl guanidine, $HNC[NMe_2]_2$, with transition metals Co^{II}, Cu^{II}, Zn^{II}, Pd^{II}, Ni^{II} and Cr^{III} has shown[167] the imine nitrogen as the donor site and with tetrahedral geometry of the cobalt complex which is confirmed[168] by X-ray powder diffraction patterns. The crystal structure of zinc guanidinum sulfate has been reported by Morimoto and coworkers.[169] Complex formation of Cu^{II} and Zn^{II} with diphenylguanidine has been studied by extraction with chloroform.[170]

N-Phenyl-*N'*-(2-thiazolyl)guanidine forms 2:1 complexes with Cu^{II} and Hg^{II} and 1:1 complexes with W^{VI} and Mo^{VI} as indicated by absorption spectra, conductometric and amperometric measurements.[171] Room temperature Mössbauer parameters have been tabulated for dehydro-*N,N',N"*-trialkylguanidinohexacarbonyldiiron(0) and related compounds.[172]

Complexes $[PdA_2L_2]X_2$ (L = 2-guanidiniumbenzimidazole; A = NH_3, C_5H_5N; X = Cl, NO_3, $0.5SO_4$) were prepared by treating $[PdL_2]X_2$ with gaseous NH_3 or C_5H_5N.[173] The adducts have tetragonal structure with NH_3 or C_5H_5N in axial positions.

Various complexes of Rh^{III}, Ir^{III} and Pt^{II} with 2-guanidiniumbenzimidazole and related ligands have been synthesized and characterized.[174] A tris-chelated chromium(III) complex with 2-guanidinobenzimidazole has been prepared by complete replacement of thiocyanate groups in $K_3[Cr(NCS)_6]$. The 10 Dq value (20 200 cm^{-1}), obtained from the $^4T_{2g} \leftarrow {}^4A_{2g}$ transition, is slightly lower than those of $Cr^{III} N_6$ chromophores in general.[175]

Vanadium(II) reacted with SCN^- and diphenylguanidine (L) to form a ternary complex with a V:SCN:L ratio of 1:2:2.[176] Other ternary complexes like phenylguanidine iron(II) cyanide are being exploited in titrimetric determination of cyanide ions, for example.

Reactions of Rh^{II} carboxylates with guanidines were studied[177] in which complexes of various compositions were formed. $Rh_2(O_2CR)_4 \cdot 2H_2O(I)$ (R = H, Me, Et) reacted with cyanoguanidine (L) in aqueous solution to give $Rh_2(O_2CH)_4 \cdot 2L$, $Rh_2(O_2CMe)_4 \cdot L \cdot H_2O$ and $Rh_2(O_2CEt)_4 \cdot L \cdot 2H_2O$. Alcoholic solutions of I reacted with aminoguanidine hydrochloride (L'HCl) to give $Rh_2(O_2CR)_4 \cdot L' \cdot HCl$, which, when treated with Na_2CO_3 solution, gave $Rh_2(O_2CR)_4 \cdot L' \cdot H_2O$. In addition to the above, guanidine (L") derivatives with the composition $(L"H)_2[(Rh_2(O_2CR)_4Cl_2]$, $Rh_2(O_2CH)_4 \cdot 2L" \cdot 2H_2O$ and $(L"H)_4Rh_2(CO_3)_4 \cdot 4H_2O$ were also prepared. The complexes were characterized by IR spectra and thermal analyses; the general conclusion was that whereas cyano-guanidine forms only coordinate bonds, aminoguanidine and guanidine form outer sphere protonated cations also.

The complex *trans*-$[CoX(dfg)(dpg)] \cdot nH_2O$ (Hdfg = diphenylglyoxime; Hdpg = diphenyl-guanidine; X = Cl, Br, I, NCO, NCS, NCSe, NO_2) were prepared and characterized by IR spectra, diffractometry and thermal stability curves.[178] The 'dpg' is coordinated through the nitrogen atom of the C=NH group and can replace H_2O, $PhNH_2$, C_5H_5N, guanidine and cyanoguanidine from the inner coordination sphere. ^1H NMR spectra of coordination compounds of cobalt(III) with dimethylglyoxime and diphenylguanidine have also been studied.[179] Substitution complexes of guanidines have received much less attention. Phenylmercury(II) replaces a proton of creatine, $H_2NC^+(NH_2)NMeCH_2CO_2^-$, in basic solution to form the zwitterionic complex, $PhHgNHC^+(NH_2)NMeCH_2CO_2^-$. Creatine and its metabolite creatimine ($C_4H_7N_3O$) react with $PhHg\{(OH)(NO_3)\}_{\frac{1}{2}}$ in aqueous ethanol to form a 2:1 complex $[(PhHg)_2C_4H_6N_3O]NO_3$, existing in two crystalline forms. X-Ray analysis of the crystalline 2:1 complex has been described.[180] Some novel mononuclear and binuclear phenylmercury(II) compounds with benzimidazole and 2-guanidinobenzimidazoles have been reported.[181]

Lithioguanidine, $LiN=C(NMe_2)_2$, as well as lithioketimine, $LiN=Bu^t$, have been shown[182] to have hexameric structures, based on slightly folded chain-shaped Li_6 rings held together by triply-bridging methylamino groups, $N=CR_2$, thus providing examples of electron-deficient bridging by the nitrogen atoms of organonitrogen ligands.

Novel substituted guanidines continue to be explored for their coordination abilities, *e.g.* nitro-aminoguanidine has been shown[183] to form bis complexes with cobalt(II), nickel(II) and copper(II) of the type $M\{H_2NNHC(=NH)NHNO_2\}_2X_2 \cdot nH_2O$ (when M = Co or Ni, X = Cl, Br, I, SCN, NO_3, $\frac{1}{2}SO_4$, ClO_4, BF_4, and when M = Cu, X = Cl, Br, NO_3, $\frac{1}{2}SO_4$, ClO_4, BF_4; n = 0.2).

13.8.5.2 Biguanide Complexes

The biguanide molecule, $H_2NC(=NH)NHC(=NH)NH_2$, has a strong proton affinity; its basic strength in aqueous solution is only slightly lower than that of hydroxyl (OH^-) ion.[184] Biguanide and its substitution products form highly coloured chelate complexes with a number of transition metals, *e.g.* Cu(II),[185-187] Ni(II),[188] Ni(III),[189] Co(I),[190] Co(II),[191] Co(III),[192,193] Cr(III),[194] Rh(III),[195] Rh(IV),[196] Ir(III),[197] Ir(IV),[198] Ru(II),[199] Ru(III),[200] Ag(I),[201] Ag(III),[202] Mo(III),[203] Mo(V),[204] Mo(VI),[205] VO(II),[206] Pt(II),[207] Pt(IV),[207] Pd(II),[208,209] Pd(IV),[203] Mn(II),[210] Mn(III),[211] Mn(IV),[212] Os(II),[213] Os(III),[213] Os(IV),[213] Os(VI),[214] Os(VIII),[215] Fe(II),[216] Fe(III),[217] Au(I),[218] Au(III),[218] Ln(III)[219] and U(IV).[220] A few reports on biguanide derivatives of main group elements [Be(II),[221] Al(III),[221] B(III),[222] Zn(II),[223] Hg(II)[224] and Si(IV)[225]] have also appeared in the literature.

Biguanides have a marked tendency of stabilizing unusually high oxidation states of the metals, *e.g.* Ag[III][203] and Ni[III].[189]

Many authors have classified the biguanide complexes into 'complex bases', $[M(bigH)_n](OH)_n$, and 'anhydro bases', $M(big)_n$. The former can be formulated as $[M(big)_n] \cdot mH_2O$ also and are generally converted into $[M(big)_n]$ on heating to 110°C. Detailed electronic and IR spectral studies have shown[226] that there is no spectral change observed in the transition from 'complex base' to 'anhydro base'. Biguanide complexes of metals could, therefore, be classified into two groups only, *i.e.* (i) uncharged metal biguanides, $[M(big)_n]$ or $[M(big)_n] \cdot mH_2O$, and (ii) charged biguanide. complexes of the type $[M(bigH)_n]X_n$ (where $X = Cl^-$, Br^-, $\frac{1}{2}SO_4^{2-}$, *etc.*). Sen[227] has suggested structures (68) ands (69) for the two types of complexes.

(68)   (69)

These structures with extensive electronic charge delocalization have been supported by LCAO Hückel calculations. The single crystal X-ray studies of tris(biguanide)cobalt(III)[228] and tris-(biguanide)chromium(III)[229] have further confirmed these structures. In the complex Cr(big)₃, there is no hydrogen atom attached to the nitrogen atom in the —C—N—C portion of the molecule, whereas in $[Co(bigH)_3]Cl_3$ the protonation occurs on the nitrogen atom of the —C—N—C fragment of biguanide. The photoelectron spectra data[230] indicate the presence of two types of nitrogen atoms in compounds with structure (69) in the ratio of 2:3 abundance. The data also indicate that the π-electron delocalization extends to all nitrogen atoms including those outside the chelate ring. The NMR data for $[Ni(bigH)_2]Cl_2$ also support the absence of quaternary nitrogen atoms.[231]

It may be appropriate here to list a few more X-ray structural studies of a few biguanide complexes, *e.g.* $[Ni(bigH)_2]Cl_2 \cdot 2H_2O$,[231] $[Cu(chlorobigH)_2(OH)_2]Cl_2$,[232] $[Cu(aminoethbigH)(NCS)]SCN[233]$ and $[Cu(aminoethbigH)(C_2H_4N_4)]SO_4 \cdot H_2O[234]$ as well as some biguanide complexes, *e.g.* ethylenedibiguanide derivatives of Cu[II],[235] Ni[II][235,236] and Ag[III].[202,237,238] Extensive work has been carried out on the resolution of optically active biguanide complexes and specific mention may be made in this connection of the contributions[239] on the resolution and crystal structure of tris(biguanide)cobalt(III) chloride.

In view of the strong colours of many of the biguanide complexes, these have been used as colorimetric reagents, *e.g.* the use of phenylbiguanide for spectrophotometric determination of Fe^{2+} and CN^- ions.[240] A polarographic study[241] of a bis(biguanide)copper(II) complex gives a kinetically controlled wave prior to a two-electron transfer, followed by a diffusion controlled second wave due to the reduction of bis species in solution.

Extensive magnetochemical studies of biguanide complexes of transition metals confirm[151,152] their inner level chelate character except in cases of manganese(IV) and rhenium(V) derivatives. In a thermochemical study of the effect of π-delocalization in metal biguanide complexes, it has been shown[242] that the reactions of Cu^{2+} and Ni^{2+} ions with biguanide are strongly exothermic and further, the largely negative enthalpy changes in these more than compensate for the remarkably unfavourable entropy contributions.

A number of interesting studies have been carried out on the kinetics and mechanism of biguanide complexes, *e.g.* bis(biguanide) derivatives of Ni[II] and Cu[II].[243] These kinetic studies have in some cases indicated the formation of transient species also; for example, a kinetic study indicates that the oxidation of an Ag^I complex of ethylenedibiguanide by peroxydisulfate occurs in one-electron steps, involving an Ag^{II} intermediate.[244]

To summarize, biguanide ligands have not only stabilized a number of unusual oxidation states of transition metals but have yielded coordination complexes of very large formation constants. These complexes generally exhibit aromatic character. The synthetic, structural, magnetic, spectroscopic, chromatographic, medicinal and various other studies of metal biguanides have thus contributed significantly to the richness of modern coordination chemistry.

13.8.6 UREA AND URYLENE COMPLEXES

While the adducts formed by urea with metal salts are generally oxygen bonded,[245] an interesting series of urylene (RNC(=O)NR) complexes with metal–nitrogen bonding have been isolated[246,247] by the reactons of metal carbonyls (iron, rhodium and iridium), with organic azides (RN_3; R = Me, Bu^n, Ph) or isocyanate (RNCO; R = Me, Bu^n, Ph).

Manuel[248] had reported that the reaction of $Fe_3(CO)_{12}$ with phenylisocyanate and azidobenzene yields hexacarbonylbis(phenylisocyanate)diiron(0), for which was suggested the formulation of the type $[Fe(CO)_3(RNCO)]_2$. However, later investigations[249,250] have shown these complexes to have the composition $[Fe_2(CO)_7(RN)_2]$ with the urylene group, PhN—CO—NPh, bridging the two iron atoms and each of the iron atoms having three carbonyl groups attached. It has now been established that the reaction of a metal carbonyl with organic azides in general results in the formation of derivatives containing urylene ligands. Dimeric[249-251] as well as monomeric[252,253] derivatives of the types shown in equations (6)–(10) have been isolated.

$$Fe_2(CO)_9 \xrightarrow{RN_3} (CO)_3Fe \text{------} Fe(CO)_3 \qquad R = Me, Ph \qquad (6)$$

$$Fe_3(CO)_{12} \xrightarrow{RN_3} (CO)_3Fe \text{------} Fe(CO)_3 \qquad R = Ph, p\text{-}MeOC_6H_4 \qquad (7)$$

$$Ru(CO)_3(PPh_3)_2 \xrightarrow[\text{r.t.}]{RN_3} (PPh_3)_2(CO)_2Ru \qquad C=O \qquad R = p\text{-}MeC_6H_4SO_2 \qquad (8)$$

$$M(NO)(CO)(PPh_3)_2 \xrightarrow{RN_3} (PPh_3)_2(NO)M \qquad C=O \qquad (9)$$
$$M = Rh, Ir, R = p\text{-}MeC_6H_4SO_2$$

$$Pd(PPh_3)_2(CO) \xrightarrow{RN_3} (PPh_3)_2Pd \qquad C=O \qquad R = p\text{-}MeC_6H_4SO_2 \qquad (10)$$

In the case of $Pt(PPh_3)_n(CO)_{4-n}$ ($n = 2,3$), the above reaction leads to the isolation of a more complex compound $[(PPh_3)_2Pt(R_2N_4CO)]$, which represents the precursor leading to the monomeric urylene complexes[253] by loss of nitrogen. However, urylene derivatives of iron seem to originate from the spontaneous decomposition of derivatives of the type $[Fe(CO)_3NR]_2$. Related intermediates were considered to be involved in the reaction of ruthenium carbonyls and $PhNO_2$ in the presence of H_2/CO mixtures.[254]

Urylene derivatives may also be synthesized by oxidative addition of toluene-*p*-sulfonylurea to the appropriate substrate (equations 11–13).[252,253]

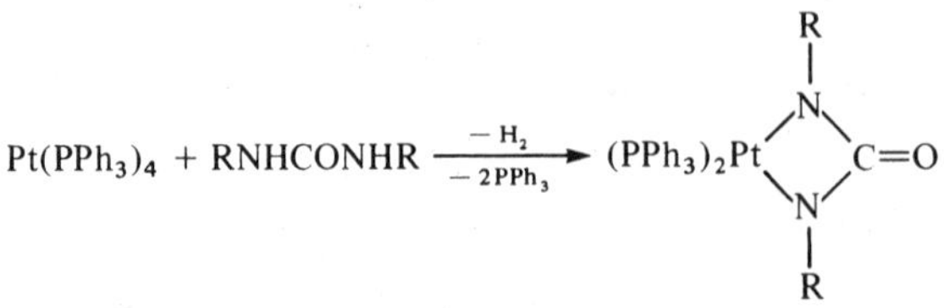

$$Pt(PPh_3)_4 + RNHCONHR \xrightarrow[-2PPh_3]{-H_2} (PPh_3)_2Pt \underset{\underset{R}{|}}{\overset{\overset{R}{|}}{\underset{N}{\overset{N}{\diagdown}}}}C=O \qquad (11)$$

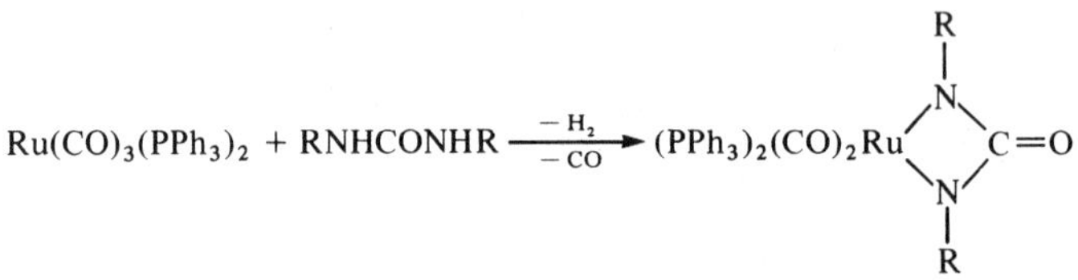

$$Ru(CO)_3(PPh_3)_2 + RNHCONHR \xrightarrow[-CO]{-H_2} (PPh_3)_2(CO)_2Ru \underset{\underset{R}{|}}{\overset{\overset{R}{|}}{\underset{N}{\overset{N}{\diagdown}}}}C=O \qquad (12)$$

$$Rh(PPh_3)_3Cl + RNHCONHR \xrightarrow[-PPh_3]{-H_2} (PPh_3)_2ClRh \underset{\underset{R}{|}}{\overset{\overset{R}{|}}{\underset{N}{\overset{N}{\diagdown}}}}C=O \qquad (13)$$

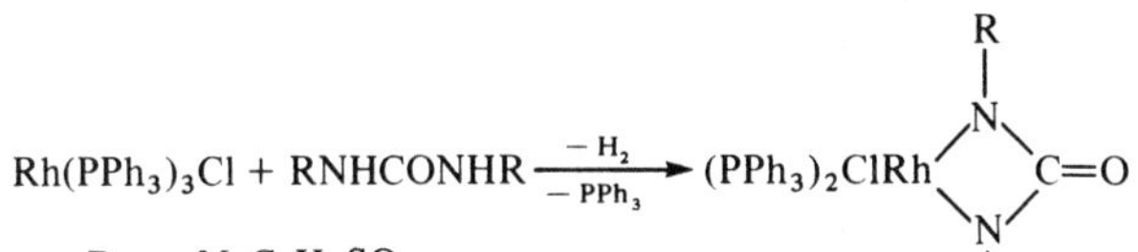

$$R = p\text{-}MeC_6H_4SO_2$$

The nature of bonding in the above monomeric urylene derivatives has been studied by spectral studies. The IR spectra exhibit a characteristic absorption due to the carbonyl group of the urea at about $1680-1690\,cm^{-1}$.

The X-ray structures of the μ-(dimethylurylene)bis(tricarbonyliron)[255] and μ-(diphenylurylene)bis(tricarbonyliron)[256,257] have been reported. The crystal structure of $[(PhNCONPh)Fe_2(CO)_6]$ (70) indicated a normal Fe—Fe bond length for a nitrogen-bridged structure, and a small dihedral angle (94.2°).[256,257]

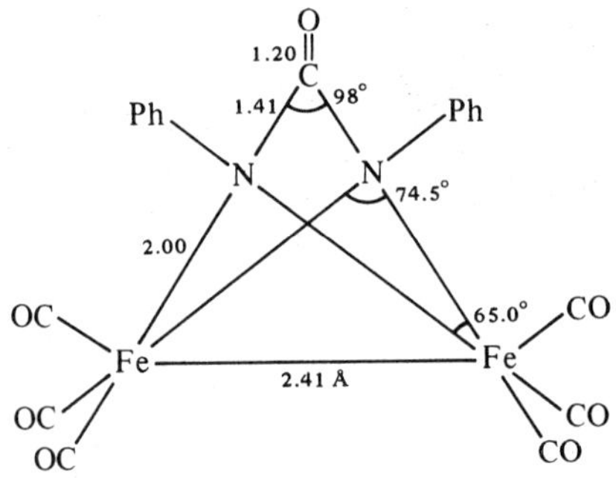

(70)

It is interesting to mention here that the NCN angle in the complex (70) is much less than expected for sp^2-hybridized carbon, and that the Fe—C bond length (1.79 Å) is also long. The replacement of the phenyl group by methyl group in (70) produced a shortening of the Fe—N distances by ~ 0.03 Å. The presence of a short Fe—Fe bond length (2.42 Å) in $[(MeNCONMe)Fe_2(CO)_6]$ is associated with a nitrogen-bridged structure. These Fe—Fe bond distances are in fact among the shortest known distances for nitrogen-bridged complexes.

A urylene derivative of cobalt, $[(CpCo)_2(Bu^tNCONBu^t)]$ (71) has been obtained by the reaction of N-*t*-butylsulfurimide, $(Bu^tN)_2S$, with $[CpCo(CO)_2]$. The reaction has been suggested to proceed *via* the formation of an intermediate nitrene complex, $[CpCo(CO)NBu^t]$.[257,258] The X-ray crystal structural determination has shown the presence of a short Co—Co distance (2.37 Å), which is 0.1 Å smaller than that reported in the related acetylene complex, $Co_2(CO)_6(C_2Ph_2)$.[258,260]

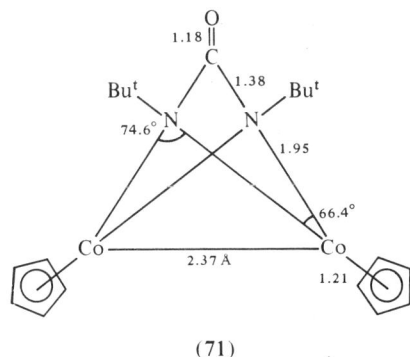

(71)

Another interesting series of ureadiide derivatives of manganese(I) and rhenium(I) carbonyls with the formulae $M'[M(CO)_n\{(RN)_2C{=}O\}]$ (where $M' = Li, K; M = Mn, Re; R = Et, Pr^i$, cyclohexyl, $PhCH_2$, Ph, *p*-biphenyl, 1-naphthyl; $n = 5, 4$), in which ureadiide ions act as mono- as well as bi-dentate ligands, were prepared by the reactions of $M'_2(RN)_2C{:}O$ with $M(CO)_5Br$.[261]

13.8.7 REFERENCES

1. A. Hantzsch and A. Werner, *Chem. Ber.*, 1890, **23**, 11.
2. L. Tschugaeff, *Chem. Ber.*, 1905, **38**, 2520; *J. Chem. Soc.*, 1914, **115**, 2187.
3. F. J. Welcher, 'Organic Analytical Reagents', Van Nostrand, New York, 1947, vol. 3, pp. 157–354.
4. F. Feigl, 'Chemistry of Specific, Selective and Sensitive Reactions', Academic, New York, 1949, pp. 251–280.
5. A. E. Martell and M. Calvin, 'Chemistry of Metal Chelate Compounds', Prentice-Hall, Englewood Cliffs, New Jersey, 1952, pp. 328–333.
6. J. C. Bailar and D. H. Busch, in 'The Chemistry of the Coordination Compounds', ed. J. C. Bailar, Jr., Reinhold, New York, 1956, pp. 76–78.
7. C. V. Banks, 'Analytical Chemistry', Elsevier, Amsterdam, 1963.
8. B. Egneus, *Talanta*, 1972, **19**, 1387.
9. P. N. Butler, F. L. Scott and T. A. F. O'Mahony, *Chem. Rev.*, 1973, **73**, 93.
10. B. C. Haldar, *J. Indian Chem. Soc.*, 1974, **51**, 224.
11. G. N. Schrauzer, *Angew. Chem.*, 1976, **88**, 465.
12. N. M. Samus and A. V. Ablov, *Coord. Chem. Rev.*, 1979, **28**, 177.
13. A. Chakravorty, *Coord. Chem. Rev.*, 1974, **13**, 1.
14. A. Singh, V. D. Gupta, G. Srivastava and R. C. Mehrotra, *J. Organomet. Chem.*, 1974, **64**, 145.
15. R. C. Mehrotra, A. K. Rai, A. K. Singh and R. Bohra, *Inorg. Chim. Acta*, 1975, **13**, 91.
16. J. Donohue, *J. Am. Chem. Soc.*, 1956, **78**, 4172.
17. D. Hall, *Acta Crystallogr.*, 1965, **18**, 955.
18. M. Callery, G. Ferraris and D. Viterbo, *Acta Crystallogr.*, 1966, **20**, 73.
19. P. Groth, *Acta Chem. Scand.*, 1968, **22**, 128.
20. A. N. Barrett and R. A. Palmer, *Acta Crystallogr., Sect. B*, 1969, **25**, 688.
21. F. Bachechi and L. Zambonelli, *Acta Crystallogr., Sect. B*, 1972, **28**, 2489.
22. I. N. Levine, *J. Chem. Phys.*, 1963, **38**, 2326.
23. W. Hieber and F. Leutert, *Chem. Ber.*, 1927, **60**, 2296; 1929, **62**, 1839.
24. M. E. Stone, B. E. Robertson and E. Stanley, *J. Chem. Soc. (A)*, 1971, 3632.
25. M. E. Stone and K. E. Johnson, *Can. J. Chem.*, 1973, **51**, 1260.
26. N. A. Patel, J. R. Shah and R. P. Patel, *Indian J. Chem., Sect. A*, 1980, **19**, 236.
27. A. A. Grinberg, A. I. Stetsenko and S. G. Strelin, *Russ. J. Inorg. Chem. (Engl. Transl.)*, 1968, **13**, 569.
28. Yu. N. Kukushkin, A. I. Stetsenko, S. G. Strelin and V. G. Duibanova, *Russ. J. Inorg. Chem. (Engl. Transl.)*, 1972, **17**, 561.
29. A. I. Stetsenko, S. G. Strelin and M. I. Gel'fman, *Russ. J. Inorg. Chem. (Engl. Transl.)*, 1970, **15**, 68.
30. J. R. Jennings and K. Wade, *J. Chem. Soc. (A)*, 1967, 1333.
31. A. Mawby and G. E. Pringle, *J. Inorg. Nucl. Chem.*, 1971, **33**, 1989.
32. G. P. Khare and R. J. Doedens, *Inorg. Chem.*, 1977, **16**, 907.
33. R. B. King and K. N. Chen, *Inorg. Chem.*, 1977, **16**, 1164.
34. G. P. Khare and R. J. Doedens, *Inorg. Chem.*, 1976, **15**, 86.
35. R. B. King and W. M. Douglas, *Inorg. Chem.*, 1974, **13**, 1339.
36. S. Aime, G. Gervasio, L. Milone, R. Rossetti and P. L. Stanghellini, *J. Chem. Soc., Dalton Trans.*, 1978, 534.
37. T. W. Thomas and A. E. Underhill, *Chem. Soc. Rev.*, 1972, **1**, 99.
38. H. Endres, H. J. Keller, R. Lehmann, A. Poveda, H. H. Rupp and H. Van de Sand, *Z. Naturforsch., Teil B*, 1977, **32**, 516.
39. F. A. Cotton and J. G. Norman, *J. Am. Chem. Soc.*, 1971, **93**, 80.
40. K. Burger, in 'Chelates in Analytical Chemistry', ed. H. A. Flaschka and A. J. Barnard, Jr., Dekker, New York, 1969, vol. 3, pp. 179–212.
41. A. Nakamura, A. Konishi and S. Otsuka, *J. Chem. Soc., Dalton Trans.*, 1979, 488.
42. F. Paneth and E. Thilo, *Z. Anorg. Allg. Chem.*, 1925, **147**, 196.
43. J. V. Dubsky, F. Brychta and M. Kuras, *Pub. Faculte Sci. Univ. Masaryk, No. 129*, 1931, 1 (*Chem. Abstr.*, 1932, **26**, 2943).

44. Yu. M. Simonov, M. M. Botoshanskii, T. I. Malinovskii, D. G. Batyr, L. D. Ozols and I. I. Bulgak, *Dokl. Akad. Nauk SSSR*, 1979, **246**, 609.
45. A. Nakahara and R. Tsuchida, *J. Am. Chem. Soc.*, 1954, **76**, 3103.
46. D. G. Batyr, I. I. Bulgak and L. D. Ozols, *Koord. Khim.*, 1978, **4**, 84.
47. K. I. Turta, R. A. Stukan, I. I. Bulgak, D. G. Batyr and L. D. Ozols, *Koord. Khim.*, 1978, **4**, 1391.
48. O. Bekaroglu, S. Sarisaban, A. R. Koray, B. Nuber, K. Weidenhammer, J. Weiss and M. L. Ziegler, *Acta Crystallogr., Sect B*, 1978, **34**, 3591.
49. Yu. A. Simonov, M. M. Botoshanskii, L. D. Ozols, I. I. Bulgak, D. G. Batyr and T. I. Malinovskii, *Sov. J. Coord. Chem. (Engl. Transl.)*, 1981, **7**, 302.
50. A. Okac and M. Simek, *Collect. Czech. Chem. Commun.*, 1951, **15**, 977; 1959, **24**, 2699.
51. H. J. Keller and K. Seibold, *J. Am. Chem.Soc.*, 1971, **93**, 1309.
52. R. K. Panda, S. Acharya and G. Neogi, *J. Chem. Soc., Dalton Trans.*, 1983, 1225, 1233 and 1239.
53. D. R. Boston and N. J. Rose, *J. Am. Chem. Soc.*, 1973, **95**, 4163.
54. R. S. Drago and J. H. Elias, *J. Am. Chem. Soc.*, 1977, **99**, 6570.
55. D. A. White, *J. Chem. Soc. (A)*, 1971, 233.
56. J. V. Quagliano and D. H. Wilkins, in 'Chemistry of the Coordination Compounds', ed. J. C. Bailar, Jr., Reinhold, New York, 1956, p. 672.
57. E. Konig, *Landolt-Börnstein*, Springer-Verlag, New York, 1970, vol. 2, pp. 2–105.
58. P. M. Dhadke and B. C. Haldar, *J. Indian Chem. Soc.*, 1979, **56**, 461.
59. B. K. Mohapatra, R. C. Mishra and D. Panda, *J. Indian Chem. Soc.*, 1981, **58**, 1154.
60. M. Mohan and B. D. Paramhans, *Indian J. Chem., Sect. A*, 1981, **20**, 1204.
61. J. Charalambous, M. J. Frazer and F. B. Taylor, *J. Chem. Soc. (A)*, 1971, 602.
62. J. B. Nielands, *Struct. Bonding (Berlin)*, 1966, **7**, 59.
63. P. W. W. Hunter and G. A. Webb, *J. Inorg. Nucl. Chem.*, 1970, **32**, 1386.
64. K. C. Kalia and Anil Kumar, *Indian J. Chem., Sect. A*, 1978, **16**, 49.
65. A. N. Singh, *Indian J. Chem., Sect. A*, 1981, **19**, 1215.
66. A. V. Ablov, N. I. Belichuk and M. S. Pereligina, *Russ. J. Inorg. Chem. (Engl. Transl.)*, 1972, **17**, 534.
67. J. A. Bertrand, J. H. Smith and P. G. Eller, *Inorg. Chem.*, 1974, **13**, 1649.
68. A. McAuley and K. F. Preston, *Inorg. Chem.*, 1983, **22**, 2111.
69. G. Rindorf, *Acta Chem. Scand.*, 1971, **25**, 774.
70. J. Bassett, J. Bensted and R. Grzeskowiak, *J. Chem. Soc. (A)*, 1969, 2873.
71. I. Rani, K. B. Pandeya and R. P. Singh, *J. Indian Chem. Soc.*, 1982, **59**, 235.
72. B. McCudden, P. O'Brien and J. R. Thornback, *J. Chem. Soc., Dalton Trans.*, 1983, 2043.
73. R. K. Murmann, *Inorg. Chem.*, 1963, **2**, 116.
74. E. O. Schlemper, *Inorg. Chem.*, 1968, **7**, 1130.
75. E. O. Schlemper, W. C. Hamilton and S. J. La Place, *J. Chem. Phys.*, 1971, **54**, 3990.
76. J. W. Fraser, G. R. Hedwig, H. K. J. Powell and W. T. Robinson, *Aust. J. Chem.*, 1972, **25**, 747.
77. A. Gul and O. Bekaroglu, *J. Chem. Soc., Dalton Trans.*, 1983, 2537.
78. V. Stuzka, J. Kuapil and M. Kuras, *Z. Anal. Chem.*, 1961, **179**, 401.
79. K. R. Manolov and B. M. Angelov, *Monatsh. Chem.*, 1971, **102**, 763.
80. P. Bandyopadhyay, P. K. Mascharak and A. Chakravorty, *J. Chem. Soc., Dalton Trans.*, 1981, 623.
81. W. Bradley and I. Wright, *J. Chem. Soc.*, 1956, 640.
82. G. H. Searle, *Aust. J. Chem.*, 1980, **33**, 2159.
83. F. A. Cotton, W. H. Ilsley and W. Kaim, *Inorg. Chem.*, 1980, **19**, 2360.
84. A. J. Deeming and R. Peters, *J. Organomet. Chem.*, 1980, **202**, C39.
85. K. Burgess, H. D. Holden, B. F. G. Johnson and J. Lewis, *J. Chem. Soc., Dalton Trans.*, 1983, 1199.
86. W. Bradley and E. Leete, *J. Chem. Soc.*, 1951, 2147.
87. K. S. Patel and R. K. Mishra, *J. Indian Chem. Soc.*, 1978, **55**, 773.
88. K. S. Patel and R. K. Mishra, *Bull. Chem. Soc. Jpn.*, 1979, **52**, 592.
89. K. S. Patel, K. K. Deb and R. K. Mishra, *Bull. Chem. Soc. Jpn.*, 1979, **52**, 595.
90. K. M. M. Rao, S. K. Satyanarayana and R. K. Mishra, *Anal. Chem.*, 1979, **295**, 47.
91. K. K. Deb and R. K. Mishra, *J. Indian Chem. Soc.*, 1978, **55**, 289.
92. W. Maringgele and A. Meller, *Z. Anorg. Allg. Chem.*, 1978, **443**, 148.
93. V. A. Dorokhov, M. N. Bochkareva, O. G. Boldyreva, B. V. Rassadin and B. M. Mikhailov, *Izv. Akad. Nauk SSSR, Ser. Khim.*, 1979, 411.
94. F. A. Cotton, T. Inglis, M. Kilner and T. R. Webb, *Inorg. Chem.*, 1975, **14**, 2023.
95. F. A. Cotton and L. W. Shive, *Inorg. Chem.*, 1975, **14**, 2027.
96. A. Bino, F. A. Cotton and W. Kaim, *Inorg. Chem.*, 1979, **18**, 3566.
97. F. A. Cotton, W. H. Ilsley and W. Kaim, *Inorg. Chem.*, 1979, **18**, 3569.
98. R. Roger and D. G. Nielson, *Chem. Rev.*, 1961, **61**, 179.
99. K. Sakai, T. Ito and K. Watanabe, *Bull. Chem. Soc. Jpn.*, 1967, **40**, 1660.
100. S. Komiya, S. Suzuki and K. Watanabe, *Bull. Chem. Soc. Jpn.*, 1971, **44**, 1440.
101. R. Breslow, R. Fairweather and J. Keana, *J. Am. Chem. Soc.*, 1967, **89**, 2135.
102. R. Breslow and M. Schmir, *J. Am. Chem. Soc.*, 1971, **93**, 4960.
103. D. Pinnell, G. B. Wright and R. B. Jordan, *J. Am. Chem. Soc.*, 1972, **94**, 6104.
104. D. A. Buckingham, F. R. Keene and A. M. Sargeson, *J. Am. Chem. Soc.*, 1973, **95**, 5649.
105. K. B. Nolan and R. W. Hay, *J. Chem. Soc., Dalton Trans.*, 1974, 914.
106. C. R. Clark and R. W. Hay, *J. Chem. Soc., Dalton Trans.*, 1974, 2149.
107. D. G. Butler, I. I. Creaser, S. F. Dyke and A. M. Sargeson, *Acta Chem. Scand., Ser. A*, 1978, **32**, 789.
108. P. F. B. Barnard, *J. Chem. Soc. (A)*, 1969, 2140.
109. S. Suzuki, M. Nakahara and K. Watanabe, *Bull. Chem. Soc. Jpn.*, 1971, **44**, 1441.
110. H. C. Clark and L. E. Manzer, *Inorg. Chem.*, 1971, **10**, 2699.
111. R. Ros, J. Renaud and R. Roulet, *J. Organomet. Chem.*, 1975, **87**, 379.

112. M. Wada and T. Shimohigashi, *Inorg. Chem.*, 1976, **15**, 954.
113. T. G. Appleton, M. H. Chisholm, H. C. Clark and L. E. Manzer, *Inorg. Chem.*, 1972, **11**, 1786.
114. H. C. Clark and L. E. Manzer, *J. Organomet. Chem.*, 1973, **47**, C17.
115. G. Rouschias and G. Wilkinson, *J. Chem. Soc. (A)*, 1968, 489.
116. L. E. Manzer, *J. Chem. Soc., Dalton Trans.*, 1974, 1535.
117. H. C. Clark and L. E. Manzer, *J. Chem. Soc., Chem. Commun.*, 1971, 387.
118. H. C. Clark and L. E. Manzer, *Inorg. Chem.*, 1972, **11**, 2749.
119. P. F. B. Barnard, *J. Chem. Soc. (A)*, 1969, 2140.
120. M. Kilner, *Adv. Organomet. Chem.*, 1972, **10**, 115.
121. L.-H. Chan and E. G. Rochow, *J. Organomet. Chem.*, 1967, **9**, 231.
122. M. R. Collier, M. F. Lappert and J. McMeeking, *Inorg. Nucl. Chem. Lett.*, 1971, **7**, 689.
123. V. A. Dorokhov and M. F. Lappert, *J. Chem. Soc. (A)*, 1969, 433.
124. G. J. Bullen and K. Wade, *Chem. Commun.*, 1971, 1122.
125. C. Summerford and K. Wade, *J. Chem. Soc. (A)*, 1970, 201.
126. D. Bright and O. S. Mills, *Chem. Commun.*, 1967, 245.
127. M. Kilner and C. Midcalf, *J. Chem. Soc., Dalton Trans.*, 1974, 1620.
128. H. R. Keable, M. Kilner and E. E. Robertson, *J. Chem. Soc., Dalton Trans.*, 1974, 639.
129. M. Kilner and J. N. Pinkney, *J. Chem. Soc. (A)*, 1971, 2887.
130. C. Kruger, E. G. Rochow and U. Wannagat, *Chem. Ber.*, 1963, **96**, 2132.
131. L.-H. Chan and E. G. Rochow, *J. Organomet. Chem.*, 1967, **9**, 231.
132. J. Keable, D. G. Othen and K. Wade, *J. Chem. Soc., Dalton Trans.*, 1976, 1.
133. K. Farmery and M. Kilner, *J. Organomet. Chem.*, 1969, **16**, 51.
134. K. Farmery, M. Kilner and C. Midcalf, *J. Chem. Soc. (A)*, 1970, 2279.
135. H. R. Keable and M. Kilner, *Chem. Commun.*, 1971, 349.
136. M. Kilner and C. Midcalf, *Chem. Commun.*, 1970, 552.
137. M. Kilner and C. Midcalf, *J. Chem. Soc. (A)*, 1971, 292.
138. M. Kilner and J. N. Pinkney, *J. Chem. Soc. (A)*, 1971, 2887.
139. F. A. Cotton and R. M. Wing, *Inorg. Chem.*, 1965, **4**, 314.
140. O. S. Mills and A. D. Redhouse, *J. Chem. Soc. (A)*, 1968, 642.
141. E. W. Abel, C. A. Burton and R. Rowley, Abstracts of papers, Sixth International Conference on Organometallic Chemistry, University of Massachusatts, Amherst, MA, August 13–17, 1973, paper 141.
142. M. R. Churchill and K.-K. G. Lin, *Inorg. Chem.*, 1975, **14**, 1133.
143. E. W. Abel, C. A. Burton, M. R. Churchill and K.-K. G. Lin, *J. Chem. Soc., Chem. Commun.*, 1974, 268.
144. E. W. Abel, C. A. Burton, M. R. Churchill and K.-K. G. Lin, *J. Chem. Soc., Chem. Commun.*, 1974, 917.
145. M. R. Churchill and K.-K. G. Lin, *Inorg. Chem.*, 1975, **14**, 1675.
146. J. Chatt, R. J. Dosser and G. J. Leigh, *J. Chem. Soc., Chem. Commun.*, 1972, 1243.
147. J. Chatt, R. J. Dosser, F. King and G. J. Leigh, *J. Chem. Soc., Dalton Trans.*, 1976, 2435.
148. H. R. Keable and M. Kilner, *J. Chem. Soc., Dalton Trans.*, 1972, 153.
149. M. A. Andrews and H. D. Kaesz, *J. Am. Chem. Soc.*, 1979, **101**, 7238.
150. M. A. Andrews, G. van Buskirk, C. B. Knobler and H. D. Kaesz, *J. Am. Chem. Soc.*, 1979, **101**, 7245.
151. P. Ray, *Chem. Rev.*, 1961, **61**, 313.
152. A. Syamal, *J. Sci. Ind. Res.*, 1978, **37**, 661.
153. F. Kurzer, *Chem. Rev.*, 1956, **56**, 95.
154. H. Sigel and R. B. Martin, *Chem. Rev.*, 1982, **82**, 385.
155. F. H. S. Curd, D. G. Davey and D. N. Richardson, *J. Chem. Soc.*, 1949, 1732.
156. N. Kundu and P. Ray, *J. Indian Chem. Soc.*, 1952, **29**, 811.
157. R. L. Dutta, *J. Indian Chem. Soc.*, 1960, **37**, 499.
158. G. Schwarzenbach and D. Schwarzenbach, *J. Indian Chem. Soc.*, 1977, **54**, 23.
159. W. Neumann, *Phys. Status Solidi*, 1969, **36**, K105.
160. M. Tamatani, T. Ban and I. Tsujikawa, *J. Phys. Soc. Jpn.*, 1971, **30**, 481.
161. F. Fromage, P. Delorme and V. Lorenzelli, *C.R. Hebd. Seances Acad. Sci., Ser. A*, 1973, **276**, 651.
162. V. I. Nefedov, *Zh. Neorg. Khim.*, 1974, **19**, 2628.
163. R. Snaith, K. Wade and B. K. Wyatt, *J. Chem. Soc. (A)*, 1970, 380.
164. C. M. Flynn, Jr. and M. T. Pope, *Inorg. Chem.*, 1971, **10**, 2524, 2745.
165. R. D. Peacock and T. J. R. Weakley, *J. Chem. Soc. (A)*, 1971, 1836.
166. G. A. Yuzhakova, R. P. Drovneva, M. I. Vakhrin and I. I. Lapkin, *Zh. Obshch. Khim.*, 1978, **48**, 811.
167. G. A. Yuzhakova, I. I. Lapkin, R. P. Drovneva and M. V. Mazilkina, *Zh. Obshch. Khim.*, 1981, **51**, 880.
168. R. Longhi and R. S. Drago, *Inorg. Chem.*, 1965, **4**, 11.
169. C. N. Morimoto and E. C. Lingafelter, *Acta Crystallogr., Sect. B*, 1970, **26**, 335.
170. M. M. Tananaiko and F. V. Mirzoyan, *Zh. Anal. Khim.*, 1971, **26**, 2333.
171. W. U. Malik, P. K. Srivastava and S. C. Mehra, *J. Indian Chem. Soc.*, 1973, **50**, 739.
172. E. Von Meerwall, J. Worth, W. Greenlae and M. F. Farona, *Spectrosc. Lett.*, 1974, **7**, 311.
173. A. K. Banerjee and S. P. Ghosh, *J. Indian Chem. Soc.*, 1974, **51**, 720.
174. S. P. Ghosh, P. Bhattacharjee, L. Dubey and L. K. Mishra, *J. Indian Chem. Soc.*, 1977, **54**, 230.
175. A. Mishra, *Indian J. Chem., Sect. A*, 1977, **15**, 919.
176. C. J. Martinez, R. F. Bosch, M. P. F. Caridad and M. C. Gracia Alvarez-Coque, *An. Quim. Ser. B.*, 1982, **78**, 217.
177. T. A. Vemeva and V. N. Shafranskii, *Zh. Obshch. Khim.*, 1979, **49**, 488.
178. I. V. Dranka, Yu. Ya Kharitonov and V. N. Shafranskii, *Zh. Neorg. Khim.*, 1981, **26**, 643.
179. V. N. Shafranskii, A. Stratulat and I. V. Dranka, *Zh. Neorg. Khim.*, 1981, **26**, 1807.
180. A. J. Canty, M. Fyfe and B. M. Gatehouse, *Inorg. Chem.*, 1978, **17**, 1467.
181. R. L. Dutta and S. K. Satapathi, *J. Inorg. Nucl. Chem.*, 1981, **43**, 1533.
182. W. Clegg, R. Snaith, H. M. M. Shearer, K. Wade and G. Whitehead, *J. Chem. Soc., Dalton Trans.*, 1983, 1309.
183. N. Saha and B. Mallick, *Indian J. Chem., Sect. A*, 1983, **22**, 39.

184. G. Schwarzenbach and G. Anderegg, *Pharm. Acta Helv.*, 1963, **38**, 547.
185. C. Gheorghiu, M. Constantinescu, O. Constantinescu and V. Pocol, *Rev. Roum. Chim.*, 1974, **19**, 1909.
186. C. W. Catron, D. R. Lorenz and J. R. Wasson, *Inorg. Nucl. Chem. Lett.*, 1976, **12**, 385.
187. A. Syamal and V. D. Ghanekar, *J. Inorg. Nucl. Chem.*, 1978, **40**, 1606.
188. J. Belgaumkar and B. K. Samantary, *Z. Kristallogr.*, 1971, **133**, 491.
189. D. Sen and C. Saha, *J. Chem. Soc., Dalton Trans.*, 1976, 776.
190. P. K. Das and A. K. De, *Proc. Convention of Chemists (India)*, Allahabad, 1972, p. 4.
191. S. S. Gupta and R. Kaushal, *J. Indian Chem. Soc.*, 1974, **51**, 649.
192. A. Syamal and P. K. Mandal, *Transition Met. Chem.*, 1978, **3**, 292.
193. K. Igi, J. Hidaka and B. E. Douglas, *J. Coord. Chem.*, 1978, **7**, 155.
194. A. Swinarski and H. Zawadzki, *Rocz. Chem.*, 1976, **50**, 2021.
195. S. P. Ghosh, P. Bhattacharjee and A. Mishra, *J. Less-Common Met.*, 1975, **40**, 97.
196. C. Gheorghiu, *Rev. Roum. Chim.*, 1968, **13**, 1181.
197. S. P. Ghosh and A. I. P. Sinha, *J. Inorg. Nucl. Chem.*, 1964, **26**, 1703.
198. P. Spacu and C. Gheorghiu, *Z. Anorg. Allg. Chem.*, 1969, **364**, 331.
199. P. Spacu and C. Gheorghiu, *Rev. Roum. Chim.*, 1967, **12**, 1459.
200. S. P. Ghosh and U. Prasad, *J. Less-Common Met.*, 1976, **44**, 345.
201. C. R. Saha, *J. Inorg. Nucl. Chem.*, 1976, **38**, 1635.
202. M. L. Simms, J. L. Atwood and D. A. Zatko, *J. Chem. Soc., Chem. Commun.*, 1973, 46.
203. S. P. Ghosh and K. M. Prasad, *J. Less-Common Met.*, 1974, **36**, 223.
204. P. Spacu, C. Gheorghiu, M. Constantinescu and L. Antonescu, *J. Less-Common Met.*, 1976, **44**, 161.
205. A. Maitra, *J. Indian Chem. Soc.*, 1973, **50**, 361.
206. M. M. Ray, K. De and S. N. Poddar, *Indian J. Chem.*, 1965, **3**, 228.
207. D. Sen, *J. Indian Chem. Soc.*, 1974, **51**, 183.
208. D. J. MacDonald, *J. Inorg. Nucl. Chem.*, 1968, **30**, 1971.
209. D. K. Nag and S. Guha, *Indian J. Phys.*, 1971, **45**, 139.
210. S. N. Poddar, *Sci. Cult.*, 1963, **29**, 212.
211. K. Dey, K. C. Ray, S. N. Poddar and N. G. Poddar, *Indian J. Chem., Sect. A*, 1976, **14**, 205.
212. J. Bera and D. Sen, *Indian J. Chem., Sect. A*, 1976, **14**, 880.
213. P. Spacu and C. Gheorghiu, *J. Less-Common Met.*, 1969, **18**, 117.
214. M. M. Ray, *Sci. Cult.*, 1964, **30**, 190.
215. N. C. Ta and D. Sen, *J. Inorg. Nucl. Chem.*, 1981, **43**, 209.
216. D. Sen, *Sci. Cult.*, 1961, **27**, 502.
217. S. Lahiry and V. K. Anand, *Inorg. Chem.*, 1981, **20**, 2789.
218. D. Sen, *Sci. Cult.*, 1961, **27**, 548.
219. I. Albescu, *Rev. Roum. Chim.*, 1973, **18**, 829.
220. M. C. Chakravorti and N. Bandyopadhyay, *J. Inorg. Nucl. Chem.*, 1971, **33**, 2565.
221. S. D. Nandi and P. Banerjee, *Z. Naturforsch., Teil B*, 1974, **29**, 347.
222. A. Maitra and D. Sen, *Indian J. Chem.*, 1974, **12**, 183.
223. P. V. Babykutty, P. Indrasenan, R. Anantaraman and C. G. Ramachandran Nair, *Thermochim. Acta*, 1974, **8**, 271.
224. R. Tsuchiya, A. Uehara, K. Otsuka and E. Kyuno, *Chem. Lett.*, 1974, 833.
225. A. Maitra and D. Sen, *Inorg. Nucl. Chem. Lett.*, 1972, **8**, 793.
226. R. H. Skabo and P. W. Smith, *Aust. J. Chem.*, 1969, **22**, 659.
227. D. Sen, *J. Chem. Soc. (A)*, 1969, 2900.
228. M. R. Snow, *Acta Crystallogr., Sect. B*, 1974, **30**, 1850.
229. L. Coghi, M. Nardelli and G. Pelizzi, *Acta Crystallogr., Sect. B*, 1976, **32**, 842.
230. W. E. Swartz and P. D. Alfonso, *J. Electron Spectrosc. Relat. Phenom.*, 1974, **4**, 351.
231. T. C. Creitz, R. Gsell and D. L. Wampler, *Chem. Commun.*, 1969, 1371.
232. C. R. Saha, D. Sen and S. Guha, *J. Chem. Soc., Dalton Trans.*, 1975, 1701.
233. G. D. Andreetti, L. Coghi, M. Nardelli and P. Sgarabotto, *J. Cryst. Mol. Struct.*, 1971, **1**, 147.
234. L. Coghi, A. Mangia, M. Nardelli, G. Pelizzi and L. Sozzi, *Chem. Commun.*, 1968, 1475.
235. M. Matthew and N. R. Kunchur, *Acta Crystallogr., Sect. B*, 1970, **26**, 2054.
236. B. L. Holian and R. E. Marsh, *Acta Crystallogr., Sect. B*, 1970, **26**, 1049.
237. N. R. Kunchur, *Nature (London)*, 1968, **217**, 539.
238. L. Coghi and G. Pelizzi, *Acta Crystallogr., Sect. B*, 1975, **31**, 131.
239. T. Tada, Y. Kushi and H. Yoneda, *Bull. Chem. Soc. Jpn.*, 1982, **55**, 1063.
240. J. M. Calatayud, F. B. Reig and M. C. G. Alvarez-Coque, *Talanta*, 1982, **29**, 139.
241. B. Datia and K. Singh, *Indian J. Chem., Sect. A*, 1981, **20**, 926.
242. L. Fabbrizzi, M. Micheloni and P. Paoletti, *Inorg. Chem.*, 1978, **17**, 494.
243. D. Banerjea, *Transition Met. Chem.*, 1982, **7**, 22.
244. J. D. Miller, *J. Chem. Soc. (A)*, 1968, 1778.
245. S. Calogero, U. Russo, P. W. C. Barnard and J. D. Donaldson, *Inorg. Chim. Acta*, 1982, **59**, 111.
246. M. Kilner, *Adv. Organomet. Chem.*, 1972, **10**, 115.
247. S. Cenini and G. La Monica, *Inorg. Chim. Acta Rev.*, 1976, **18**, 279.
248. T. A. Manuel, *Inorg. Chem.*, 1964, **3**, 1703.
249. M. Dekker and G. R. Knox, *Chem. Commun.*, 1967, 1243.
250. W. T. Flannigan, G. R. Knox and P. L. Pauson, *Chem. Ind. (London)*, 1967, 1094.
251. P. C. Ellgen and J. N. Gerlach, *Inorg. Chem.*, 1974, **13**, 1944.
252. S. Cenini, M. Pizzotti, F. Porta and G. La Monica, *J. Organomet. Chem.*, 1975, **88**, 237.
253. W. Beck, W. Rieber, S. Cenini, F. Porta and G. La Monica, *J. Chem. Soc., Dalton Trans.*, 1974, 298.
254. F. L'Eplattenier, P. Matthys and F. Calderazzo, *Inorg. Chem.*, 1970, **9**, 342.

255. R. J. Doedens, *Inorg. Chem.*, 1968, **7**, 2323.
256. J. A. J. Jarvis, B. E. Job, B. T. Kilbourn, R. H. B. Mais, P. G. Owston and P. F. Todd, *Chem. Commun.*, 1967, 1149.
257. J. Piron, P. Piret and M. Van Meerssche, *Bull. Soc. Chim. Belg.*, 1967, **76**, 505.
258. Y. Matsu-ura, N. Yasuoka, Y. Ueki, N. Kasai, M. Kakudo, T. Yoshida and S. Otsuka, *Chem. Commun.*, 1967, 1122.
259. S. Otsuka, A. Nakamura and T. Yoshida, *Inorg. Chem.*, 1968, **7**, 261.
260. Y. Matsu-ura, N. Yasuoka, T. Ueki, N. Kasai and M. Kakudo, *Bull. Chem. Soc. Jpn.*, 1969, **42**, 881.
261. W. Dennecker and H. W. Mueller, *Z. Naturforsch., Teil B*, 1982, **37**, 318.

14

Phosphorus, Arsenic, Antimony and Bismuth Ligands

CHARLES A. McAULIFFE
University of Manchester Institute of Science and Technology, UK

Because of factors beyond the editors' control, the submission of the manuscript for this chapter was delayed. In order to minimize any delay in publishing 'Comprehensive Coordination Chemistry' as a whole, the coverage of phosphorus, arsenic, antimony and bismuth ligands appears at the end of this volume, commencing on page 989.

15.1

Water, Hydroxide and Oxide

JOHN BURGESS
University of Leicester, UK

15.1.1 INTRODUCTION

Water, hydroxide and oxide are ligands which form complexes with one or more ions of almost every metal. Specific compounds are discussed according to element in Volumes 3 to 5. Here we deal with general features of water, hydroxide and oxide as ligands, and give some illustrations of how such chemistry, particularly of aquo ions, depends on position in the Periodic Table.

These ligands, with the recently established $H_3O_2^-$, have a very close relation with solvent water. Therefore throughout this section chemistry in solution and in the solid state will be juxtaposed and related in a way not possible for other ligands. Indeed the pH dependence of species present in solution has an important bearing on solid complexes, in that the latter are often prepared from solution.[1] The parent complexes, aquo ions $[M(OH_2)_x]^{n+}$, are important both in salt hydrates and in solution, and have even been studied to a limited extent in the gas phase.[2] Other complexes containing water, hydroxide and oxide in various proportions are all related to these aquo cations by simple and rapidly established proton transfer equilibria. Eventually deprotonation leads to oxo anions. Some of these, such as chromate or permanganate, can be considered as complexes; others, such as PbO_4^{4-} and GeO_4^{4-}, are on the borderline with mixed metal oxides.[3] Most such species have tetrahedral metal centres, but square pyramidal and octahedral complexes are also known, for example WO_5^{4-} and ReO_6^{5-}.[4] Intermediate between $[M(OH_2)_x]^{n+}$ and MO_x^{n-} are the hydroxo complexes such as $[Sb(OH)_6]^-$ and $[Pt(OH)_6]^{2-}$ (*cf.* Section 15.1.3.2). The aquo, hydroxo and oxo ligands may be combined in a variety of ways at a given metal ion; all three are simultaneously bound in the $[VO(OH)(OH_2)_4]^+$ cation and, in a much more complicated manner, in an octanuclear iron(III) complex in which the oxide and hydroxide ligands act as bridges. Simultaneous bonding of hydroxo and oxo ligands to a given metal ion is not common; other examples include $[Sn(O)(OH)]^-$,[6] $[OsO_2(OH)_4]^{2-}$[7] and $[OsO_4(OH)_2]^{2-}$.[8] Simultaneous attachment of oxo and aquo ligands is exemplified by *trans*-$[MoO(OH_2)(CN)_4]^{3-}$[9] and the familiar $VO(aq)^{2+}$. On electrostatic grounds one would expect to find hydroxo and oxo ligands more frequently associated with higher oxidation states. On redox grounds one would expect to find only the very difficultly reducible oxide ligand bonded to highest oxidation states, as indeed in FeO_4^{2-} and MnO_4^- (*cf.* OsO_4 and RuO_4). Only for the very large actinide elements does one find water ligands bonded to metals in oxidation states $5+$ and $6+$, albeit in the form of aquodioxo units $ActO_2(aq)^+$ and $ActO_2(aq)^{2+}$. Such species have long been known for Act = U, Np and Pu, and have been more recently established for Act = Cf, for example.[10] The UO_2^{2+} cation is known to have five water molecules coordinated to the uranium both in aqueous solution and in the solid state.[11] In all these oxo cations the O—M—O unit is linear, as it is in the ternary complex $[MoO_2(CN)_4]^{4-}$;[9] *cis*-MO_2 units are known in VO_2^+, and in ternary complexes ReO_2Me_3 and $ReO_2I(PPh_3)_2$.[12] The dependence of the aquo–hydroxo–oxo ligand balance on oxidation state is well illustrated by the aqueous solution chemistry of vanadium. In oxidation state $2+$ this occurs as $V(aq)^{2+}$, in oxidation state $3+$ as $V(aq)^{3+}$, which is weakly acidic, being in equilibrium with $V(OH)(aq)^{2+}$. In oxidation state $4+$ the normal species is $VO(aq)^{2+}$, while vanadium(V) exists in alkaline solution as VO_3^-, in neutral or acidic solution in the form of polyvanadates or $VO_2(aq)^+$, and in 'nitric oleum' as VO^{3+}.

Water is a poor bridging ligand, but hydroxide and oxide are good bridging ligands for metal ions, *e.g.* dichromate, decavanadate and polytungstates (though MnO_4^-, TcO_4^- and ReO_4^- have no condensed analogues). Polynuclear cations containing mainly or exclusively hydroxo bridges are known for up to, for example, 12 iron atoms ($[Fe_{12}(OH)_{34}]^{2+}$[13]) or 13 aluminum atoms (*vide infra*). Sometimes it is difficult to establish whether two metal ions are joined by two hydroxide bridges or one oxide bridge. Spectroscopic techniques are of limited value here, and even X-ray diffraction and EXAFS investigations may be ambiguous, as in the controversy over iron(III).[14] These strictures apply mainly to the solution state, where there is the additional complication for some ions [*e.g.* iron(III)[15]] of extremely slow approach to equilibrium. Polynuclear aquo cations, both homonuclear and heteronuclear, are of importance in a variety of areas of applied chemistry: iron in rusting and in geochemistry, aluminum in deodorants, titanium in paints, and titanium–iron and titanium–aluminum in metal extractions, to give but a very few examples.

It is possible to have aquo, hydroxo and oxo ligands in this type of polynuclear ion. A simple example is provided by the $[(UO_2)_2(OH)_2(aq)]^{2+}$ species in solution,[16] a complicated but important example by the Al_{13} species mentioned above. This consists of a central AlO_4 tetrahedron surrounded by 12 AlO_6 octahedra sharing edges, both in the solid state[17] and in solution.[18] Further complications are provided by the possibility of having ions from different metals or of different oxidation states in such polynuclear complexes,[19,20] for example the $GeAl_{12}$ species containing a central GeO_4 tetrahedral unit surrounded by 12 AlO_6 octahedra.[21] The chemistry of molybdenum illustrates many of the features mentioned.[22] Molybdenum(VI) exists as molybdate, MoO_4^{2-}, and polymolybdates, molybdenum(V) as $Mo_2O_4(aq)^{2+}$, and molybdenum(IV) as $Mo_3O_4(aq)^{4+}$ — in all these cases with various oxo bridges. Molybdenum(III) gives $Mo(aq)^{3+}$, $Mo(OH)(aq)^{2+}$, and a di-μ-hydroxo dimer, whereas molybdenum(II) introduces metal–metal bonding in the form of $Mo_2(aq)^{4+}$. All these species have been established in the solid and solution states by X-ray diffraction and EXAFS.[23]

The next stage of this general discussion recognizes that there are many complexes containing other ligands as well as aquo, hydroxo and oxo ligands singly or in combination. The presence of such ligands can have a profound effect in several ways. There may be a large effect on the pK_a for coordinated water (Section 15.1.4.4). The presence of the other ligands may stabilize an unusually high or low oxidation state (Table 1)[24–34] or assist a 'soft' metal ion to form a strong bond to water, hydroxide or oxide. Bridging hydroxide or oxide ligands may be supported by other bridging ligands (see Sections 15.1.2.2 and 15.1.2.4 below).

The topic of ternary aquo complexes is also relevant to salt hydrates, their structures and the products of their dissolution in water. In many cases the situation is simple, with, for instance, a salt $M(OH_2)_xX = MX \cdot xH_2O$ consisting of $[M(OH_2)_x]^{n+}$ and X^{n-} units in the solid, and dissolving in water to give these same species in solution. Such is the case for $Mg[ReCl_6] \cdot 6H_2O$ and $CoSO_4 \cdot 6H_2O$

Table 1 Examples of the Stabilization of High and Low Oxidation State Aquo and Hydroxo Complexes

		Low oxidation states				High oxidation states		
Ion	Ligand	Complex	Ref.	Ion	Ligand	Complex	Ref.
WATER							
Os^{II}	Benzene	$[Os(\eta^6\text{-}C_6H_6)(OH_2)_3]^{2+}$	24	Co^{III}	Ammonia	$[Co(NH_3)_5(OH_2)]^{3+}$	
Rh^{II}	Acetate	$[Rh_2(OAc)_4(OH_2)_2]$		Pt^{IV}	Methyl	$[PtMeCl_4(OH_2)]^{-a}$	30
	Me_5Cp	$[(C_5Me_5)Rh(aq)]^{2+}$		Sn^{IV}	Butyl	$[SnBu_3(OH_2)_2]^{+b}$	31
Tc^I	Nitrosyl	$[Tc(NO)(OH_2)(NH_3)_4]^{2+}$	25	Ti^{IV}	Cp	$[Ti(\eta^5\text{-}C_5H_5)_2(OH_2)_2]^{2+}$	32
Re^0	Carbonyl	$[Re_2(CO)_9(OH_2)]$	26				
HYDROXIDE							
Re^I	Carbonyl	$[Re(OH)(CO)_3]_4$	26				
Os^I	Carbonyl	$[OsH_3(\mu\text{-}OH)(CO)_{12}]$	27				
Mixed	Carbonyl	$[FeRu_2(\mu\text{-}OH)_2(PPh_3)_2(CO)_8]$	28				
Cr^{-I}	Carbonyl	$[Cr(OH)(CO)_5]^-$	29				

[a] But $[PtCl_2(OH)_2(NH_3)_2]$ and its 2-NH_2Pr analogue are very reluctant to take up protons (ref. 33).
[b] Again, $[Sn(OH)_2(uroporphyrin)]$ is reluctant to be protonated (ref. 34).

for example.[35] There are also many cases where the hydrate contains further waters of crystallization in its lattice, but again dissolves as $[M(OH_2)_x]^{n+}$ and $X(aq)^{n-}$. Such compounds as $FeSO_4 \cdot 7H_2O$ illustrate this. Basic hydrated aluminum chloride is more complicated, for this contains water coordinated to the aluminum, water immobilized in the lattice (presumably hydrogen-bonded to Al-coordinated water or to chloride), and absorbed water.[36] Prussian Blue contains a similar variety of three types of water molecule.[37]

If there is insufficient water of crystallization, then a compound may contain a ternary complex in the solid, but still give simple aquo-metal ions on dissolution. Examples include $CuCl_2 \cdot 2H_2O$ and $FeCl_2 \cdot 4H_2O$. The latter solid consists of octahedral *trans*-$[FeCl_2(OH_2)_4]$ units.[38] The metastable form of $MnCl_2 \cdot 4H_2O$ has an analogous structure; the stable form has *cis*-$MnCl_2$.[39] *trans*-$[MCl_2(OH_2)_4]$ units are also found in such compounds as $Cs_3VCl_6 \cdot 4H_2O$, $Cs_2CrCl_5 \cdot 4H_2O$, $Cs_2TiCl_5 \cdot 4H_2O$ and $Cs_2VBr_5 \cdot 4H_2O$.[40] Octahedral halide aquo complexes are also found with other stoichiometries, for example $[MCl_5(OH_2)]^{2-}$ for M = Fe, Ru, Sc, Tl, In[41] and $[Fe^{II}F_4(OH_2)_2]^{2-}$ (all F corner-shared with $Fe^{III}F_6$ units) in $Fe_3F_8 \cdot 2H_2O$.[42] In contrast to these octahedral species, trigonal prismatic $SrX_4(OH_2)_2$ entities are found in $SrCl_2 \cdot H_2O$ and $SrBr_2 \cdot H_2O$.[43] But $SrCl_2 \cdot 6H_2O$ has each Sr^{2+} coordinated by nine water molecules (Table 1).

Several hydrates contain sufficient water of crystallization to give the expected aquo ion, yet in practice consist of ternary complexes plus lattice water. $NiCl_2 \cdot 6H_2O$ and $CoCl_2 \cdot 6H_2O$ are examples of this; in the solid they contain *trans*-$[MCl_2(OH_2)_4]$ units analogous to those in $FeCl_2 \cdot 4H_2O$ (*cf.* above), plus two lattice waters.[44] Complementary examples exist amongst the lanthanide trichloride hexahydrates, which are $[LnCl_2(OH_2)_6]Cl$ for Ln = Nd to Lu;[45] $LaCl_3 \cdot 7H_2O$ and $PrCl_3 \cdot 7H_2O$ are di-μ-chloro dimers.[45]

Ternary aquo ligand complex units are also found for a variety of non-halide ligands, *e.g.* octahedral *trans*-$Mn(NCS)_2 \cdot 4H_2O$.[46] Sr^{2+} in $Sr[Hg(SCN)_4] \cdot 3H_2O$ has tricapped trigonal prismatic coordination: five oxygens and four nitrogens.[47] $Th(NO_3)_4 \cdot 5H_2O$ demonstrates a preference for non-aquo ligands on the part of the thorium reminiscent of $NiCl_2 \cdot 6H_2O$. The thorium has the unusual coordination number of 11, made up of eight oxygens from four bidentate nitrate ligands and three water molecules; the remaining two waters are elsewhere in the lattice.[48]

Nonetheless it is perfectly possible to have complexes such as $[FeCl_2(OH_2)_4]$ in solution, but it is necessary to add chloride to a solution of iron(II) chloride to generate this. Addition of chloride to a solution containing $Co(aq)^{2+}$ generates $[CoCl(OH_2)_5]^+$, characterized by X-ray diffraction and by EXAFS;[49] see Chapter 44 in Volume 4 for the important but controversial case of Fe^{3+}/Cl^- in aqueous media. The general area of stabilities of complexes in solution is dealt with in Chapter 9, Volume 1. To conclude this aspect, it should be mentioned that there are several important complexes in which one or two water ligands only are complemented by several other ligands, or one polydentate or macrocyclic ligand. Thus $[Fe(CN)_5(OH_2)]^{3-}$ is the precursor of a range of pentacyanoferrate(II) anions; similar species include $[W(CN)_7(OH_2)]^{3-}$ and its conjugate base $[W(CN)_7(OH)]^{4-}$.[50] Biochemical examples include iron porphyrins, cobalt in vitamin B_{12} derivatives, zinc in carbonic anhydrase and NADH coenzyme, and magnesium in several enzymes.

The solvation of metal ions in mixtures of solvating solvents, for example aqueous alcohols or aqueous dimethyl sulfoxide, provides examples of a special case of ternary complexes and, in preferential solvation, an indication of the relative affinities of the metal ion for water and for the organic cosolvent in question. The preferential hydration of Co^{2+} in aqueous methanol is shown by the ^{1}H NMR detection of $[Co(MeOH)(OH_2)_5]^{2+}$ and $[Co(MeOH)_4(OH_2)_2]^{2+}$ in methanol containing only small amounts of water. Such preferential solvation can be established by a range of approaches, including thermodynamic probes such as transfer chemical potentials[51] and enthalpies, as well as by several forms of spectroscopy.[52]

15.1.2 WATER, HYDROXIDE AND OXIDE AS BRIDGING LIGANDS

The role of these ligands in acting as bridges between metal ions in di- and poly-nuclear complexes has been mentioned in the previous section. Here are set out the various modes of bonding which have been established in such species.

15.1.2.1 Water

Bridging water molecules occur in several crystal hydrates,[53] and in a very few ternary complexes (Table 2).[53-55]

(1)   (2)   (3)   (4)

Table 2 Water as a Bridging Ligand Between Metal Ions

	Illustration	*Further examples*
Single	$BaCl_2 \cdot H_2O$ (1)[54]	Fe^{II}, Co^{II}, Ni^{II} hippurates[55] $LiOH \cdot H_2O$, $KF \cdot 2H_2O$[53]
Double	$Na_2B_4O_5(OH)_4 \cdot 8H_2O$ (2)[53]	
Triple	$LiClO_4 \cdot 3H_2O$ (3)[53]	$SrCl_2 \cdot 6H_2O$ (4)[53]

15.1.2.2 μ_2-Hydroxide

15.1.2.2.1 *Single hydroxide bridge*

The long-established examples of a single hydroxide bridge connecting two metal centres are $[(H_3N)_5CoOHCo(NH_3)_5]^{5+}$[56] and $[(H_3N)_5CrOHCr(NH_3)_4L]^{n+}$, where L = NH_3, OH_2, OH^- or Cl^-.[57] For L = NH_3, the CrOCr angle is 154° (contrast 180° in the μ-oxo analogue).[58] More recently the aquo analogue $[(H_2O)_5CrOHCr(OH_2)_5]^{5+}$ has been characterized (it often proves much harder to crystallize aquo-chromium(III) complexes than their ammine analogues). This cation is stable with respect to forming the di-μ-OH cation only at relatively high acidities.[59] An *sp*-block example is provided by $[HgOHHg]^{3+}$ (bent bridge).[60]

Often a single hydroxide bridge is supported by another bridging ligand, as in complexes (5),[56] (6)[61] and the peraquo ion (7).[62] Normally these complexes are, as (5)–(7), symmetrical, but unsymmetrical species, *e.g.* (8), are also known.[63] A much-supported hydroxide bridge (9) occurs in the organometallic butterfly compound $[Os_4(CO)_{12}H_3(\mu\text{-}OH)]$.[27] There are also many examples in which a large macrocyclic ligand encircles two metal ions which are themselves linked by a hydroxide bridge, *e.g.* (10).[64]

(5)   (6)   (7)

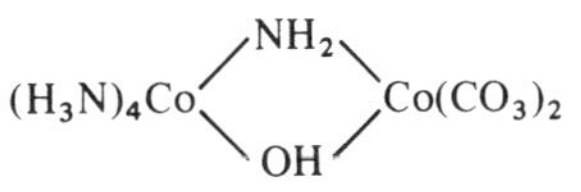

(8) 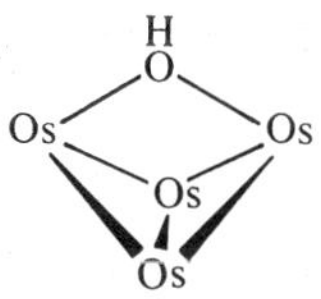   (9) 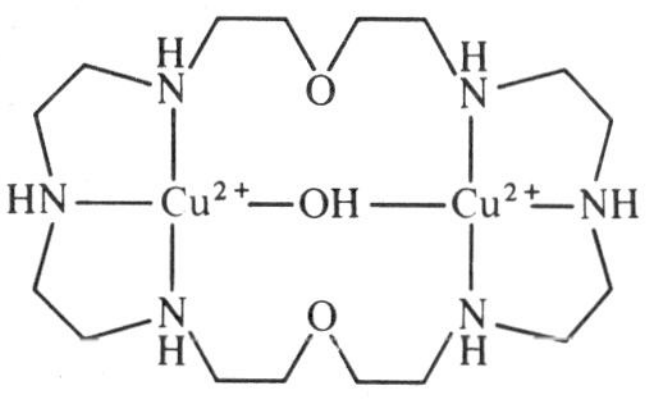   (10)

15.1.2.2.2 Double hydroxide bridges

Di-μ-hydroxide dicobalt(III) and dichromium(III) complexes have long been known;[56,59,65] analogous species are also known for many other ions, including tin,[66] copper,[67] rhenium,[68] platinum[69] and, in an organometallic environment, rhodium(I).[70]

Double hydroxo bridging may be assisted by a third bridge, (11),[56] with M = Cr, Mo or Rh, L = NH_2^-, CO_3^{2-}, OAc$^-$, *etc.*, or even edta bonded to chromium(III),[71] and $L_3' = (NH_3)_3$ or the tridentate ligand (12).[72] An iron atom acts as third bridge in $Ru_2Fe(\mu\text{-}OH)_2(CO)_8(PPh_3)_2$, in which X-ray diffraction has revealed the framework (13).[28]

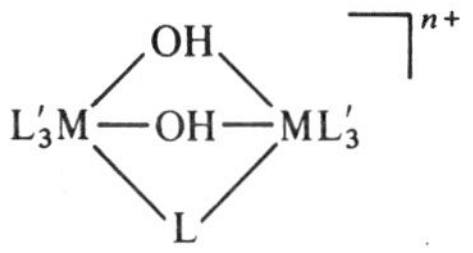

(11) 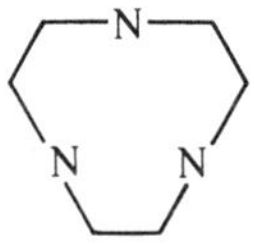   (12) 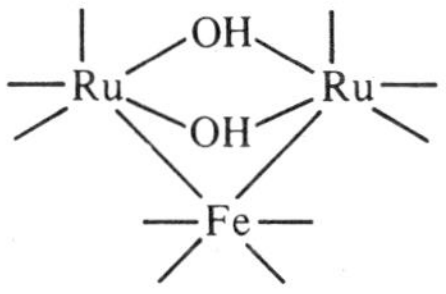   (13) 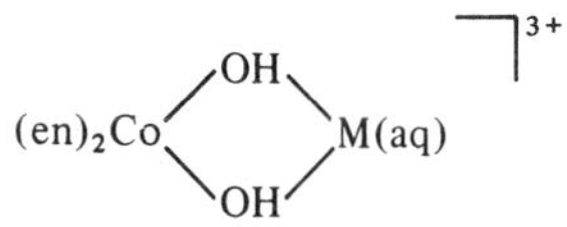   (14)

All the examples so far have been homonuclear. Complexes (14) provide examples of heteronuclear di-μ-OH cations, with M = Cu, Mn, Co, Ni or Zn, in solution.[73] Ni^{2+}/Al^{3+} dimers with di-μ-OH have been characterized in the solid state, and may well also exist in solution.[74] Although cocrystallization of cobalt and chromium hydroxo complexes normally gives the statistically expected mixture of Co_2, Cr_2 and CoCr di-μ-OH species (difficult to separate), if optically active starting materials are used a very high yield of the mixed dinuclear species $[(en)_2Cr(\mu\text{-}OH)_2Co(en)_2]^{4+}$ can be obtained.

15.1.2.2.3 Triple hydroxide bridges

The hydroxide ligand can form three bridges between a given pair of atoms, *e.g.* Cr, Co, Rh, though such linkages are rare and easily revert to a di-μ-OH linkage.[56] The triazacyclononane ligand (12) appears to stabilize tris-μ-OH complexes.[75] Triple hydroxide bridges are also known in a few organometallic derivatives, *e.g.* (15)[76] and (16).[77]

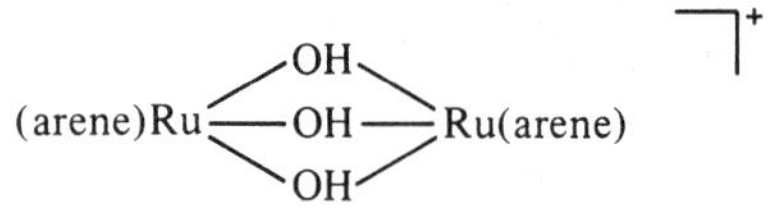

(15) 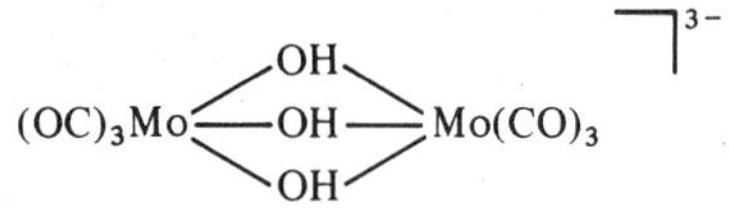   (16)

15.1.2.2.4 Polynuclear μ-hydroxo complexes

The $[Hg_3(OH)_2]^{4+}$ cation has a zigzag structure;[78] the $[Hg(OH)]^+$ cation, in its chlorate and bromate salts, is an infinite chain analogue (17).[79] A related cyclic μ-hydroxo trimer (18) is found in a platinum(II)–diaminocyclohexane complex.[69]

(17)   (18)

Di-μ-hydroxo bridges (19) are commonly encountered in polynuclear complexes of chromium(III) and of cobalt(III), and in the long established 'hexol' cation (20).[80,81] Similar polynuclear species are known for indium ([In$_4$(OH)$_6$]$^{6+}$),[82] bismuth ([Bi$_6$O$_4$(OH)$_4$]$^{6+}$),[83] zirconium [in the zirconyl halides 'ZrOX$_2$·8H$_2$O', which contain (21)],[84] titanium[85] and ruthenium. In the last case there are such species as [Ru$_4$(OH)$_{12}$]$^{4+}$, [Ru$_4$(OH)$_8$]$^{6+}$, [Ru$_4$(OH)$_6$]$^{7+}$ and [Ru$_4$(OH)$_4$]$^{8+}$, which contain ruthenium in formal oxidation states 4+, 3.5+, 3.25+ and 3+.[86]

(19)   (20)   (21)

(22)   (23)

Several polynuclear hydroxochromium(III) species[57] contain μ_2-OH, sometimes singly bridging, sometimes doubly bridging, sometimes in association with μ_3-OH (*e.g.* (22), Pfeiffer's cation).[87,88] Finally, polynuclear cobalt(III) species can have triple μ-OH bridges, as in (23).[56]

15.1.2.3 μ_3-Hydroxide

Hydroxide relatively rarely acts as a trifurcated bridge linking three metal ions. There are a few examples containing three copper atoms, (24), generally with bidentate carboxylato ligands also linking the coppers in pairs.[89] There is also a trinuclear chromium(III) complex containing μ_3-OH, (25).[88] More frequently μ_3-OH ligands are found at four corners of tetranuclear species with distorted cubane structures (26). Examples include [Cr$_4$(OH)$_4$]$^{8+}$, [Ni$_4$(OH)$_4$]$^{4+}$ and [Pb$_4$(OH)$_4$]$^{4+}$, which exist both in aqueous media and in the solid state,[90] and several organometallic derivatives, such as [Pt(OH)Me$_3$]$_4$, [Ru(OH)(arene)]$_4$,[91] [Re(OH)(CO)$_3$]$_4$[26,92] and [MH(OH)(CO)$_3$]$_4$ (M = Mo or W).[93] A more complicated polynuclear example is provided by the 6-methyl-2-oxopyridine complex [Co$_{12}$(OH)$_6$(OAc)$_6$(mop)$_{12}$], where all six hydroxides are μ_3 (pyramidal) and each is bonded to the central cobalt(II).[94]

(24)   (25)   (26)

15.1.2.4 μ_2-Oxide

15.1.2.4.1 Single oxide bridge

A single oxo bridge may subtend an angle between 140° and 180°, this angle being determined by steric or electronic factors (Table 3).[95–103] Almost all these examples refer to the solid state, but there are also several homo- and hetero-nuclear M—O—M and M—O—M—O—M species known in solution. Often these are intermediates in, or products of, electron transfer reactions with oxide-bridging inner-sphere mechanisms. Examples include V—O—V in V(aq)$^{2+}$ reduction of VO(aq)$^{2+}$, and Act—O—Cr in Cr(aq)$^{2+}$ reduction of UO$_2^{2+}$ or PuO$_2^{2+}$; a useful and extensive list of such species has been compiled.[104] The most recent examples are another V—O—V unit, this time from VO(aq)$^{2+}$ and VO$_2^+$,[105] and an all-actinide species containing neptunium(VI) and uranium-(VI).[106] An example of a trinuclear anion of this type, with the metal in two oxidation states, is provided by (31).[107]

Table 3 Single Oxide Bridges Between Metal Ions

BINUCLEAR

Linear $\quad$ [(H$_2$O)$_5$CrOCr(OH$_2$)$_5$]$^{4+}$ [95] $\qquad$ [X$_5$MOMX$_5$]$^{4-}$ [97a] $\qquad$ ZrOZr in (27) [99]

$\quad\quad\quad\quad$ [(H$_3$N)$_5$CrOCr(NH$_3$)$_5$]$^{4+}$ [58,96] $\qquad$ [Nb$_2$OCl$_9$]$^{-}$ [98] $\qquad$ [(pc)FeOFe(pc)] [100b]

Bent $\quad$ Ir—O—Ir in (28) [101] $\qquad\qquad$ [Cl$_3$Fe—O—FeCl$_3$]$^{2-}$ [102] $\qquad$ Cr$_2$O$_7^{2-}$, W$_2$O$_{11}^{2-}$

$\quad\quad\quad\quad$ Ru—O—Ru in (29) [101]

POLYNUCLEAR

Linear $\quad$ [(H$_3$N)$_5$RuORu(NH$_3$)$_4$ORu(NH$_3$)$_5$]$^{n+c}$

Bent $\quad$ [Al$_4$O$_{13}$]$^{14-}$ (30); [Al$_5$O$_{16}$]$^{17-}$ and sodalites [103]

[a] M = Ru, Os or Re; X = Cl or Br.
[b] pc = phthalocyanine.
[c] Ruthenium red when $n = 6$.

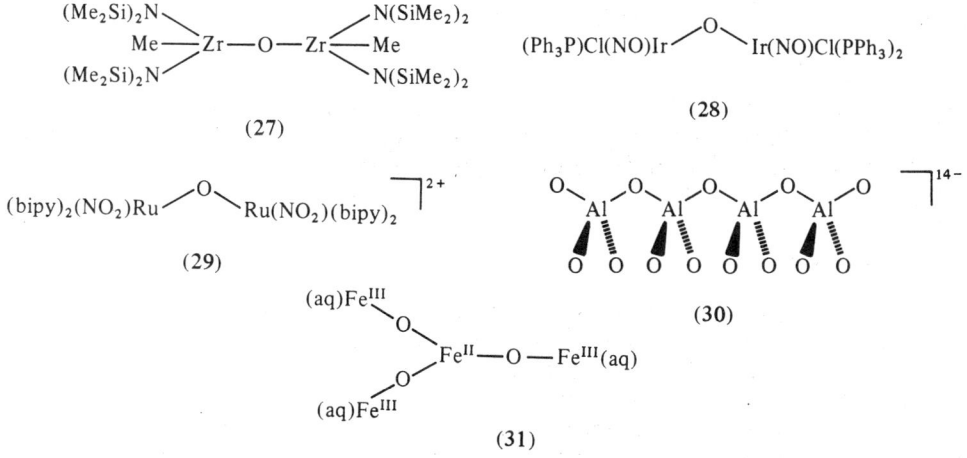

(27) $\qquad$ (28) $\qquad$ (29) $\qquad$ (30) $\qquad$ (31)

Oxo bridges can also be involved in cyclic entities, as in the tetranuclear frameworks (32) and (33), in [Ag$_4$O$_4$]$^{4-}$ (as in KAgO) and in [Au$_4$O$_4$]$^{4-}$,[108] in the mixed valence [W$_4$O$_8$(NCS)$_{12}$]$^{6-}$ anion[109] and in the adamantane skeleton (34) of the [Ta$_4$O$_6$F$_{12}$]$^{4-}$ anion.[110]* As in earlier sections, the majority of complexes are homopolynuclear, but (35) provides an instance of a heteropolynuclear version.[111] Also as earlier, μ-oxo bridges may be supported by other bridging ligands, as in (36),[112] (37)[113] (L = 12) and (38);[114] μ-O,μ-OH bridging represents a special case and a link with Section 15.1.2.2.[115]† A more complicated but in essence similar example is provided by haemerythrin, where μ-glutamate and μ-aspartate support μ-oxo in linking the two iron atoms.[116]

---

* The novel M$_3$O$_6$ unit in the [(η^5-C$_5$Me$_5$)Re$_3$O$_6$]$^{2+}$ cation, where three rheniums cap the square faces of an O$_6$ trigonal prism, also contains μ-oxo bridges (W. Herrmann, R. Serrano, M. L. Ziegler, H. Pfisterer and B. Nuber, *Angew. Chem., Int. Ed. Engl.*, 1985, **24**, 50).

† Bridging by μ-O and μ-OH$_2$ has recently been reported in the binuclear copper complex [LCu(O)(OH$_2$)CuL]$^{2+}$, where L = (12) (P. Chaudhuri, D. Ventur, K. Wieghardt, E.-M. Peters, K. Peters and A. Simon, *Angew. Chem., Int. Ed. Engl.*, 1985, **24**, 57).

(32)   (33)   (34)

(35)   (36)   (37)

(38)

15.1.2.4.2 Double and triple oxide bridges

These are most commonly encountered in polytungstates and related homo- and hetero-nuclear anions, in which double bridging is equivalent to sharing edges, and triple bridging to sharing faces, of the coordination tetrahedra. Simpler examples (39) are known for many elements, for example $M = Cr^{III},$[115] Tc^{IV}[117] and Mo^{IV}.[118] For lead(IV), di- and tri-nuclear (40) units have been characterized, in $K_2Li_6[Pb_2O_8]$ and $M_2Li_{14}[Pb_3O_{14}]$ (M = K, Rb or Cs), and a mixed lead(II)–lead(IV) analogue occurs in $Rb_2[Pb_2^{II}Pb_4^{IV}O_7]$.[119] The $[Al_3O_8]^{7-}$ anion consists of units of six AlO_4 tetrahedra, each unit being linked to the next by double oxide bridging to give an infinite chain.[120]

All these di-μ-oxo linkages are symmetrical, but there is one well-established example of an asymmetric form (41), in di-μ-oxo-bis[(cyclohexane-1,2-dithiolato)(oxo)(quinuclidine)osmium-(VI)].[121]

(39)   (40)   (41)

15.1.2.5 μ_3-Oxide

Three-connected oxide ligands are found in a variety of complexes. The M_3O unit is generally symmetrical, *i.e.* with essentially equal M—O distances, but may be flat or pyramidal, with the MOM angle as low as 90° when the oxide ligand forms a corner of a cube. One of the simplest examples of μ_3-oxide occurs in $Mo_3O_4(aq)^{4+}$, the normal form of molybdenum(IV) in aqueous solution.[122] This can be reduced to $Mo_2^{III}Mo^{IV}$ and then to Mo_3^{III}.[123] Note, here and later, that the metal atoms are also generally linked by μ_2-bridges.

Planar M_3O units occur in the basic carboxylates of such metals as iron, ruthenium, manganese, vanadium and chromium; the chromium compound has been known since 1919. The metal atoms are also linked by pairs of carboxylate ligands (often acetates), and have terminal ligands (generally pyridine or water). Mixed metal units, *e.g.* Fe_2CrO,[124] and mixed oxidation states, *e.g.* $Cr^{II}Cr_2^{III}O$,[125]

$Cu_2^{II}Cu^{III}O^{126}$ and $Mn^{II}Mn_2^{III}O$,[127] are also known, and there is a trinuclear Ru_3O-based unit whose charge can be varied from $+3$ to -2.[128]

Many other M_3O derivatives are related to these basic carboxylates. Thus replacement of acetate bridges by amino acids (LL) such as alanine or glycine[129] in $[Fe_3O(LL)_6L_3]$ gives analogues considered to be ferritin models. Vèzes's Red Salt (1893) contains a Pt_3O unit (**42**) held together by six nitrate instead of six carboxylate bridges,[130] while the product of prolonged exposure of $[Ir(NO_3)_6]^{2-}$ to concentrated nitric acid is thought to contain an analogous Ir_3O unit with six bridging nitrates (*cf.* Delépine's Salt).[131] The $[Ir_3O(SO_4)_9]^{10-}$ anion has three iridium atoms formally in oxidation state $3.33+$, linked by the μ_3-O.[132]

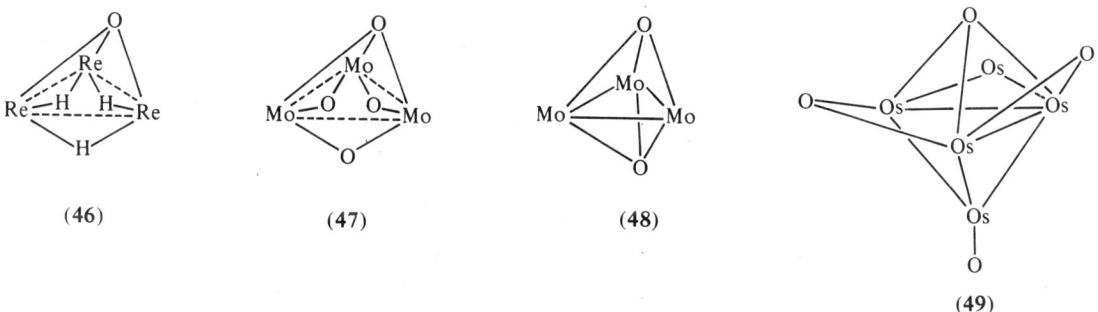

(42)   (43)   (44)

(45)

Examples of compounds containing two linked M_3O units can be found for such diverse elements as uranium (**43**),[133] aluminum (**44**)[134] and iron(III) in an octanuclear cation containing two μ_3-O moieties.[135] The first-named contains an almost planar U_4O_6 skeleton, with uranium in oxidation states $5+$ and $6+$. Another *sp*-block M_3O unit occurs in $[Hg_3O(OH_2)_3]^{4+}$ (in solution);[60,136] $[Hg_3O_2](SO_4)$ contains slightly undulating layers of μ_3-O linking linearly coordinated mercury (**45**).[137]

The M_3O unit is slightly pyramidal in basic iron(III) pivalate, with the μ_3-O 0.24 Å above the Fe_3 plane.[138] Pyramidal M_3O units are commonly found where the M atoms are also linked by other bridges. Examples include μ-OH in the $[Fe_8(O)_2(OH)_{12}(tacn)_6]^{8+}$ cation (tacn = **12**),[5] μ-OH and μ-SO_4 in $[Sn_3O(OH)_2(SO_4)]$,[139] μ-H (**46**) in the $[Re_3O(H)_3(CO)_9]^{2-}$ anion,[140] μ-O (**47**) in $Mo_3O_4(aq)^{4+}$ and numerous derivatives,[122] or another μ_3-O (**48**) as in $[Mo_3O_2(OAc)_6(OH_2)_3]^{2+}$ and related species (one of which contains μ-SO_4 *vice* μ-OAc).[141] There are also organometallic examples with μ_3-CO, *e.g.* (**49**) in $[Os_6(\mu_3-O)(\mu_3-CO)(CO)_{18}]$.[142]

(46)   (47)   (48)   (49)

A special case of pyramidal M_3O units involves the cubane framework (*cf.* **26**) consisting of a tetrahedron of M atoms linked in threes by the four μ_3-O ligands comprising the interpenetrating tetrahedron of oxygens. This framework is generally distorted, often having only D_2 symmetry. It tends to occur in low oxidation state organometallic compounds, for instance $[Os_4O_4(CO)_{12}]$[143] and $[Cr_4O_4(\eta^5\text{-}C_5H_5)_4]$,[144] and, probably, $[Mo_4O_8(\eta^5\text{-}C_5H_5)_4]$.[145] Pyramidal μ_3-O ligands of this type are also found capping triangular faces of octahedra, as in $[Ti_6O_8(\eta^5\text{-}C_5H_5)_6]$, and of trigonal bipyramids, as in $[V_5O_6(\eta^5\text{-}C_5H_5)_5]$.[144]*

15.1.2.6 μ_4-Oxide

This configuration, though rare, occurs in the well-known basic beryllium acetate (**50**)[146] and in a range of related compounds $[M_4O(O_2CR)_4]$ (M = Zn^{2+}, Fe^{3+}, Co^{4+}, *etc.*; R = a range of alkyl groups).[147] μ_4-Oxide also occurs in the analogous complex $[Cu_4OCl_6(Ph_3PO)_4]$ (**51**),[148] and in the $[Pb_6O(OH)_6]^{4+}$ cation.[149] Both in solution and in its perchlorate salt this latter cation consists of a tetrahedron of lead atoms with μ_4-O at its centre and μ-hydroxo-linked lead atoms above two faces (**52**). There is a Th_6O_7 unit of identical geometry.[150] In $TlPb_8O_4Br_9$ the four oxides are at the corners of a Pb_4O_4 pseudocubane unit, but are themselves all μ_4-O, each being bonded to a lead atom outside the cubane core (**53**).[151]

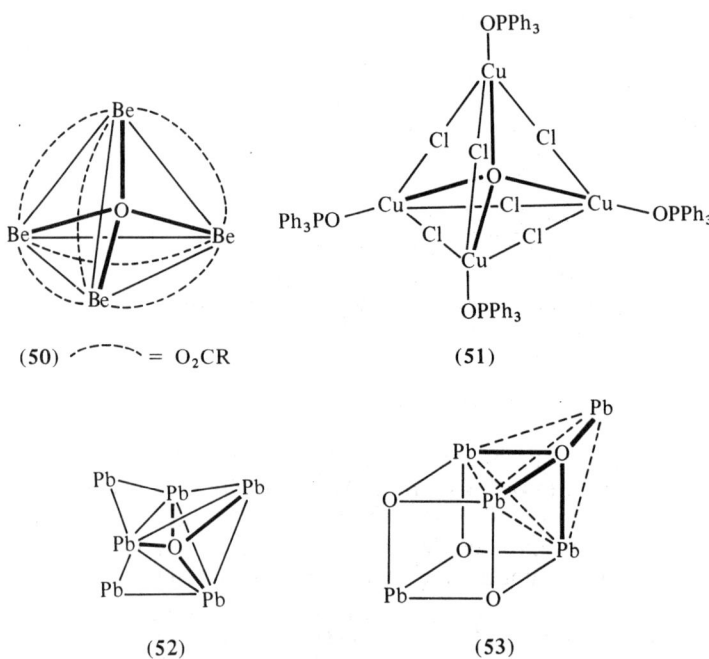

(50)   ⌐⌐⌐ = O_2CR (51)

(52)   (53)

15.1.2.7 μ_5-Oxide and μ_6-Oxide

μ_5-Oxide has been claimed, on the basis of infrared and mass spectrophotometric evidence, in $[Fe_5O(OAc)_{12}]^+$.[152] There seems as yet no established μ_6-oxide ligand; $[Pb_6O(OH)_6]^{4+}$ contains μ_4-O (*cf.* Section 15.1.2.6). As μ_6-C and μ_6-N are both known, the former for instance at the centre of an octahedron of ruthenium atoms, the latter at the centre of a trigonal prism of rhodium atoms, there seems a real possibility that μ_6-O may eventually be discovered in a polynuclear metal carbonyl oxide or related species. However, the greater electronegativity of oxygen than of carbon or nitrogen may make incorporation of oxide into an appropriate metal cluster difficult. Of course, μ_6-O is well known in the solid state in, for instance, the alkaline earth oxides CaO, SrO and BaO, and in such transition metal oxides as VO and MnO.

---

* A novel example of μ_3-O ligands in an M_nO_n cluster is provided by the Sn_6O_6 unit, drum-like in structure:

This occurs in $[PhSnO(O_2CC_6H_{11})]_6$ (V. Chandrasekhar, R. O. Day and R. R. Holmes, *Inorg. Chem.*, 1985, **24**, 1970).

15.1.2.8 μ-$H_3O_2^-$

Such bridges (**54**) have recently been established in polynuclear propionate complexes of molybdenum and of tungsten, and in the $[(\text{bipy})_2\text{Cr}(\mu\text{-}H_3O_2)_2\text{Cr}(\text{bipy})_2]^{4+}$ cation. This species is particularly important in relation to the formation of polynuclear metal ions in approximately neutral or mildly aqueous media. They may exist as such, or may be important intermediates in the formation of μ-OH or μ-O polynuclear cations from mononuclear $M(aq)^{n+}$ at appropriate pHs.[153] Very recent investigations suggest that hydrogen bonding between water and hydroxo ligands on adjacent metal atoms may give rise to an $H_3O_2^-$ bridge supporting the hydroxo bridge in compounds of the type $[(\text{en})_2(H_2O)\text{IrOHIr(OH)}(\text{en})_2]^{4+}$, and chromium(III) and ammine analogues.

(54)

15.1.3 PREPARATIVE ASPECTS

15.1.3.1 Aquo-metal Ions

Many of these have been known for a very long time, but others are difficult to generate — indeed several have only been characterized in recent years. Difficulties stem mainly from redox reactions with water or air, or through strong complexing tendencies. The problems and their solution are illustrated by such species as Mo(aq)^{3+}, Ir(aq)^{3+}, Ru(aq)^{2+}, $\text{Mo}_2\text{(aq)}^{4+}$ and Pt(aq)^{2+}, detailed under their respective elements elsewhere.

15.1.3.2 Hydroxo Salts

A number of elements form alkali metal or alkaline earth metal salts of anions $[M(OH)_x]^{n-}$ ($x = 3$–8; Table 4)[154–166] which are often only stable in solution under alkaline conditions. The reported formation of some species of this type, *e.g.* $[\text{Ln(OH)}_4]^-$ and $[\text{U(OH)}_5]^-$, seems doubtful, while thallium, for example, forms no such hydroxo anions.[167]

Table 4   Metal Ions Which Form Hydroxo Salts[a]

M^{2+}	Be, Mg, Sn, Pb	Mn, Fe,[b] Co, Ni, Cu,[155] Zn, Cd
M^{3+}	Al, Ga, In, Bi	Sc,[156] Yb,[157] Lu,[157] Cr, Mn,[158] Fe, Rh,[159] Ag[160]
M^{4+}	Ge,[c] Sn, Pb[161]	Ru,[162] Pd,[163] Pt,[164] Tc[165]
M^{5+}	Sb[166]	

[a] From ref. 154 unless otherwise indicated.
[b] Forms several species of this type, *e.g.* $[\text{Fe(OH)}_4]^-$, $[\text{Fe(OH)}_7]^{4-}$ and $[\text{Fe(OH)}_8]^{5-}$.
[c] Known only in stoettite, $\text{Fe}[\text{Ge(OH)}_6]$.

15.1.4 PROPERTIES

This section concentrates on physical and chemical properties, and deals mainly with aquo complexes. The intention is to focus on trends, particularly in relation to the Periodic Table, which will not be apparent in later volumes with their metal-oriented arrangement. Trends in stability constants and in kinetic properties, both substitution and redox, of aquo ions are covered in Chapters 7 and 9 in Volume 1.

15.1.4.1 Hydration Numbers

15.1.4.1.1 Crystal hydrates

Many crystal hydrates contain discrete $[M(OH_2)_x]^{n+}$ units, which are simply binary aquo complexes. The most common coordination number is, of course, six, with octahedral stereochemistry, as found in many hydrated salts with weakly or non-complexing anions such as SO_4^{2-}, ClO_4^-, $CF_3SO_3^-$, PF_6^- or BF_4^-. Many examples are established for the $2+$ and $3+$ ions of the first row of the transition metals, for *sp*-block ions such as Mg^{2+} and Al^{3+}, and even for such ions as Hg^{2+} [*e.g.* $Hg(ClO_4)_2\cdot 6H_2O^{168}$] and Tl^{3+} [*e.g.* $Tl(ClO_4)_3\cdot 6H_2O^{169}$]. Binary aquo complexes $[M(OH_2)_x]^{n+}$ with $x < 6$ are rarely found in crystalline materials. Tetrahedral $[Be(OH_2)_4]^{2+}$ is found in $BeSO_4\cdot 4H_2O$, for example, which has a slightly distorted cesium chloride structure with discrete aquo cations.[170] The elements which favour square planar coordination are generally reluctant to coordinate to water; ions such as Ag^+ also interact only weakly with water, but the mercury(I) ion has a total hydration number of two (linear O—Hg—Hg—O) in $Hg_2(ClO_4)_2 2H_2O$.[171]

A few metal ions which favour coordination numbers greater than six, particularly lanthanides and actinides, give aquo complexes $[M(OH_2)_x]^{n+}$ with $x = 8$ or 9 (as well as 6, *e.g.* in $Ln(ClO_4)_3\cdot 6H_2O$ with $Ln = La$, Tb or Er^{169}). Thus $Y[Tc_2Cl_8]\cdot 9H_2O$ contains slightly distorted square antiprismatic $[Y(OH_2)_8]^{3+}$ entities.[172] Yttrium also gives an enneahydrate complex $[Y(OH_2)_9]^{3+}$ in $Y(CF_3SO_3)_3\cdot 9H_2O$. Analogous tricapped trigonal prismatic (D_{3h}) units $[Ln(OH_2)_9]^{3+}$ occur in the ethyl sulfates $Ln(EtSO_4)_3\cdot 9H_2O$ ($Ln = Y$, Pr, Ho, Yb),[173] and in the trifluoromethylsulfonates of La, Gd and Lu,[174] as in the classic example of $Nd(BrO_3)_3\cdot 9H_2O^{175}$ and the more recently studied $Ln(BrO_3)_3\cdot 9H_2O$ ($Ln = Pr$, Sm and Yb).[173] Some ternary aquo complexes are also tricapped trigonal prisms, *e.g.* $Tl(NO_3)_3(OH_2)_3$,[176] whereas $[Th(tfac)_4(OH_2)]$ is a monocapped square antiprism[177] (a geometry unknown for $[M(OH_2)_9]^{n+}$).

Binary $[M(OH_2)_x]^{n+}$ entities are not known in crystal hydrates for $x > 9$ (indeed no ML_{10}^{n+} (L monodentate) has yet been characterized[178]), but ternary aquo complexes are known in which the coordination number of the metal ion is 10, 11 or 12 (Table 5).[178-183] The $[M(OH_2)_7]^{n+}$ unit does not appear to have been characterized crystallographically, but again ternary aquo complexes of this coordination number exist, *e.g.* $[UO_2(OH_2)_5]^{2+}$ in the salt $UO_2(ClO_4)_2\cdot 7H_2O$.[11]

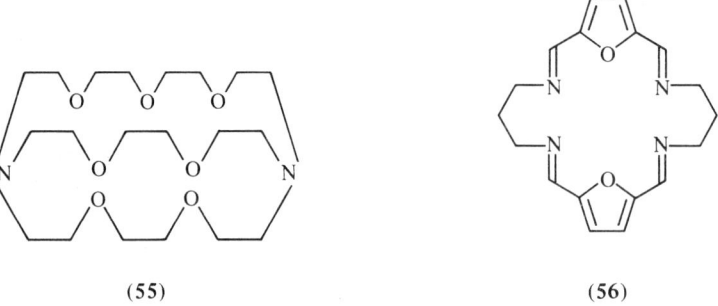

(55)   (56)

Table 5 Examples of Crystal Hydrates Containing Ternary Complexes of High Coordination Number Which Include Coordinated Water

Coordination number	Complex unit[a]	Salt	Ref.
10	$[U(ox)_4(OH_2)_2]^{4-}$	$K_4[(U(ox)_4]\cdot 4H_2O^b$	178
	$[Y(NO_3)_3(OH_2)_4]$	$Y(NO_3)_3\cdot 5H_2O$	179
11	$[La(NO_3)_3(OH_2)_5]$	$La(NO_3)_3\cdot 6H_2O$	180
	$[Th(NO_3)_3(OH_2)_5]$	$Th(NO_3)_4\cdot 5H_2O$	48
	$[Ba(crypt)(OH_2)_2]^{2+}$	$[Ba(crypt)](NCS)_2\cdot 2H_2O^c$	181
	$[BaL_2(OH_2)_2]^{2+}$	$[BaL_2][Co(NCS)_4]\cdot 2H_2O^d$	182
12	$[La(NO_3)_5(OH_2)_2]^{2-}$	$K_2[La(NO_3)_5]\cdot 2H_2O^e$	183

[a] Nitrates generally bidentate in these.
[b] Th salt analogous.
[c] crypt = ligand (55).
[d] L = ligand (56).
[e] Also $(NH_4)_2[La(NO_3)_5]\cdot 4H_2O$.

15.1.4.1.2 *In solution*

Ions in aqueous solution have primary and secondary hydration shells; the former can generally be related to the coordination shell about the metal ion in hydrates containing aquo cations (*cf.* previous section). NMR studies of aqueous solutions containing 'slow-exchange' cations, at low temperatures if need be, have established a primary solvation number of six for water as for many other donor solvents at many metal ions, indicating octahedral $[M(OH_2)_6]^{n+}$ complex ions in solution as in the solid state. The very small Be^{2+} ion has a primary hydration number of four, as have the d^8 Pd^{2+} and Pt^{2+} ions.[184] Recent EXAFS and ESEM studies have been interpreted in terms of a hydration number of four, presumably with tetrahedral geometry, for the Ag^+ cation.[185] Hg^{2+} is said to have a hydration number of two[171] (or six![168]) whereas the large Th^{4+} ion has, from ^1H NMR evidence, a hydration number of nine.[186]

In recent years, X-ray diffraction studies of aqueous solutions have established primary hydration numbers for several 'fast-exchange' cations;[45,187-190] the timescale of X-ray diffraction is very much shorter than that of NMR spectroscopy. Octahedral hydration shells have been indicated for Tl^{3+},[191] Cd^{2+}, Ca^{2+}, Na^+ and K^+, for example. For the lanthanides, $[Ln(OH_2)_9]^{3+}$ is indicated for La, Pr and Nd, but $[Ln(OH_2)_8]^{3+}$ for the smaller Tb to Lu.[192,193] Sometimes there are difficulties and uncertainties in extracting primary hydration numbers from X-ray data. Thus hydration numbers of eight and of six have been suggested for Na^+ and for K^+,[194] and for Ca^{2+},[195] and 8 and 9 for La^{3+}.[196] In some cases rates of water exchange between primary and secondary hydration shells are so fast as to raise philosophical questions in relation to specific definitions of hydration numbers.[197]

Both NMR and diffraction experiments suggest that for $[M(OH_2)_x]^{n+}$ in aqueous solution, x is generally unaffected by even large changes in temperature or pressure.[184,198] However, there are indications that $[Co(OH_2)_4]^{2+}$ may coexist alongside $[Co(OH_2)_6]^{2+}$ at elevated temperatures.[199] It should always be borne in mind that in the relatively strong solutions used for NMR and diffraction studies there is a real possibility of interaction between aquo cations and anions. Consequent ion-pairing may affect the perceived hydration number; complex formation will certainly lead to an apparent hydration number that is too low. This has been a problem in relation to Zn^{2+},[200] though apparently not for Ni^{2+}[187-190,201] for example (which is interesting in comparison with the crystal structure of $NiCl_2 \cdot 6H_2O$; *vide supra*).

Hydration numbers can also be determined for oxo cations. Thus UO_2^{2+} has five waters of primary hydraton in solution as in the solid state.[11,202]

15.1.4.2 Metal Ion–Water Distances

These are available from X-ray diffraction, neutron diffraction and related techniques such as EXAFS and SANS, both for crystalline hydrates[203] and for aqueous solutions. X-Ray diffraction studies of the latter started just before 1930,[204] but have become fairly common and precise only during the last two decades.[187-190] Neutron diffraction studies of ions in aqueous solution[205] commenced only recently,[206] and are making only slow progress because of the inherent difficulties of this technique and the great expense in both instrument time and isotopes required. However, neutron diffraction does offer the enormous advantage of locating hydrogen (deuterium) atoms and therefore of establishing the precise geometry of hydrating water, both in solution and in solids.[189,207] To some extent the solid and liquid phase results are complementary. Thus metal ion to water distances are available for many ternary species $[M(OH_2)_n L_m]^{x\pm}$ in crystal hydrates, but for few such species in solution. On the other hand, for cations such as K^+, Rb^+, Cs^+ and Ba^{2+} it is easier to establish a cation–water distance for the aquo ions in solution than from the rare and generally odd crystal hydrates.* Thus $K_4XeO_6 \cdot 9H_2O$ contains $[K(OH_2)_6]^+$ and $[K(OH_2)_7]^+$ units, with K—O distances ranging from 2.67 to 3.22 Å,[208] while Ba—O distances in $[Ba(OH_2)_n]^{2+}$ range from 2.96 to 3.18 Å.[203] Ca—O distances in Ca^{2+} salt hydrates generally range from 2.30 to 2.45 Å, with a few up to 2.50 Å.[209] The contrast between these variable ion–water distances and the clearly defined hexaaquo units of constant geometry in, for instance, hexahydrates of transition metal salts is discussed in ref. 210. The constancy of M—O distances in such hydrates in the solid state is well established, both from X-ray[54,203,211] and neutron[212] diffraction studies. Similar constancy in solution is shown in Table 6.[188] Slight differences in M—O distances in solutions of different salts of a given metal may arise from ion-pairing or even from complex formation, since diffraction studies are conducted on strong solutions. Both in solution and in the solid state, differences in the coordination number of the metal ion in different environments will lead to small but significant changes in M—O distances.[209,213]

---

* Li—O is difficult by any method; Li^+ is too small for good X-ray determinations, in solution or in the solid (and there are very few lithium salt hydrates), but neutron diffraction indicates Li—O = 1.90 to 1.95 Å (N. Ohtomo and K. Arakawa, *Bull. Chem. Soc. Jpn.*, 1980, **53**, 1789).

Oxygen Ligands

Table 6 M—O Distances (Å) in Aqueous Solutions of Transition Metal Salts

	ClO_4^-	SO_4^{2-}	NO_3^-	Cl^-	BF_4^-
Co^{2+}	2.08	2.15	2.09–2.11	2.1	
Ni^{2+}	2.04	2.15	2.07	2.05–2.1	*ca.* 2.15
Zn^{2+}	2.08	2.08, 2.15	2.09		

Table 7 compares M—O distances as determined in aqueous solution by diffraction and EXAFS with distances in the solid state, and with crystal radii for the respective ions. In general there is good agreement between the various solution results, and with M—O distances in hydrates. Thus for Ni^{2+} there have been at least 11 independent X-ray determinations of Ni—O, six in solution and five on crystal hydrates, giving results in the narrow range 2.04–2.07 Å.[213,214] The data in Table 7 show the variation of M—O distances down the Periodic Table and along the first row transition metal ions M^{2+} and M^{3+}, and show the expected correlation with ionic radii. It is interesting to see the Jahn–Teller distortion for d^9 copper(II), which is clearer in the X-ray diffraction results (though apparently not by neutron diffraction) than in solid state hydrates. In $Cu(ClO_4)_2 \cdot 6H_2O$ there are two water oxygens at 2.09 Å, two at 2.16 Å and two at 2.29 Å; at least two other salts have apical waters at 2.29–2.31 Å.[215] The $[Cu(OH_2)_6]^{2+}$ cation becomes a more elongated rhombic octahedron as the temperature decreases, with Cu—O = 1.96, 2.10 and 2.22 Å at 298 K, but 1.97, 2.01 and 2.28 Å at 123 K, in $(NH_4)_2Cu(SO_4)_2 \cdot 6H_2O$.[216]

Metal–water and metal–oxo ligand distances have been established for several aquo and dioxo actinide cations. Thus Th^{4+}, U^{4+} and U^{VI} to water distances all lie between 2.40 and 2.50 Å, in solution[202,217] and in the solid state,[11,48] while U^{VI}=O is between 1.70 and 1.80 Å.[218] For comparison, V^{IV}—OH_2 = 2.00 (axial), 2.03 (equatorial), and V^{IV}=O = 1.59 Å, in $VOSO_4 \cdot 6H_2O$.[219]

Metal–oxygen and metal–metal distances have been established for many oxo- and hydroxo-bridged dimers and polynuclear cations, for example (57)[220] and several of type (58).[217] Early work was on solids, but recently X-ray diffraction studies have established the geometries of a few such species in solution. EXAFS has proved valuable in establishing the geometry of the various di- and tri-nuclear forms of the various oxidation states of molybdenum in aqueous solution.[23]

Table 7 M—O Distances (Å) in Aqueous Solution and in Hydrates, from X-Ray and Neutron Experiments[a]

Cation	X-Ray diffraction	EXAFS[b]	Neutron diffraction	Crystal hydrates	Ionic radii[c]
Na^+	2.38–2.40		2.50		1.16
K^+	2.87–2.92		2.70	2.67–3.22	1.52
Ca^{2+}	2.40–2.49		2.40	2.30–2.45[d]	1.14
Mn^{2+}	2.20	2.18		2.00–2.22	0.96
Fe^{2+}	2.12	2.10[e]		2.08[e]	0.91
Ni^{2+}	2.04–2.07	2.05–2.07	2.07–2.10	2.02–2.11	0.84
Cu^{2+}	1.93–2.00[f]	1.93–1.97[g]	2.05	1.93–2.00	0.87
	2.33–2.43[f]	2.46[g]		h	
Zn^{2+}	2.08–2.10	1.94		2.08	0.89
Cd^{2+}	2.28–2.31			2.24–2.32	1.09
Hg^{2+}	2.40–2.41[i]			2.24–2.34[i,j]	1.16
Hg_2^{2+}	~2[k]			2.10–2.15	
Cr^{3+}	1.94–2.03	1.97–1.98		1.93–2.02	0.76
Fe^{3+}	2.04–2.05	1.99			0.79
Nd^{3+}	2.51[l]		2.48[m]	2.47	1.14
Ag^+		2.31–2.36[n]			1.29

[a] From refs. 188–190.
[b] T. K. Sham, J. B. Hastings and M. L. Perlman, *J. Am. Chem. Soc.*, 1980, **102**, 5904; D. R. Sandstrom, *J. Chem. Phys.*, 1979, **71**, 2381; P Lagarde, A. Fontaine, D. Raoux, A. Sadoc and P. Migliardo, *J. Chem. Phys.*, 1980, **72**, 3061.
[c] R. D. Shannon and C. T. Prewitt, *Acta Crystallogr., Sect. B*, 1970, **26**, 1046 (six-coordinate).
[d] From ref. 209.
[e] EXAFS on $Fe(OH_2)_6^{2+}$ in solid hydrates gives Fe—O = 1.99–2.00 Å.
[f] See ref. 215.
m s220[g] M. Sano, K. Taniguchi and H. Yamatera, *Chem. Lett.*, 1980, 1285.
[h] See text.
[i] Ref. 168.
[j] L. K. Templeton, D. H. Templeton and A. Zalkin, *Acta Crystallogr.*, 1964, **17**, 933.
[k] G. Johansson, *Acta Chem. Scand.*, 1966, **20**, 553.
[l] A set of Ln^{3+}—OH_2 distances is given in ref. 192.
[m] Ref. 193.
[n] Ref. 185.

(57)   (58)

Finally, the configuration of the water (D_2O) molecule with respect to the metal ion can be probed by neutron diffraction methods. These give metal–deuterium distances which, when compared with metal–oxygen distances, give the angle between the plane of the water (D_2O) and the M—O bond. This is 55° for Nd^{3+};[193] it is said to vary with concentration for Ni^{2+} in solution.[189]

15.1.4.3 Metal Ion–Water Bond Strengths

There are several ways of assessing strengths of interactions between metal ions and water molecules in $[M(OH_2)_x]^{n+}$ complexes. The most widely applicable is vibrational spectroscopy. This can be used for solid hydrates and for aquo cations in solution. One can also monitor metal ion–water vibrations directly or assess the strength of interaction by the effect of coordinating the metal ion on O—H frequencies. Unfortunately, frequencies[221] are determined by the reduced mass as well as the bond strength, so extraction of the latter is difficult (but see ref. 168 for an example of the derivation of force constants, in this case for $M(ClO_4)_2 \cdot 6H_2O$ with M = Zn, Cd or Hg). UV–visible spectroscopy and thermochemistry give more directly applicable measures of metal ion–water interactions, though the former approach is restricted to transition metal ions of d^1 to d^9 configuration.

15.1.4.3.1 Ultraviolet–visible spectroscopy

Values of the crystal field splitting parameters Δ or $10Dq$ are listed in Table 8.[222] These give a good idea of relative strengths of ion–water interactions; they are available from solution transmission spectra and from diffuse reflectance spectra for solids containing $[M(OH_2)_x]^{n+}$ complex cations. The greater interaction with M^{3+} than with M^{2+}, and with second-row M^{3+} than with first-row M^{3+}, as well as the dependence on d^n electron configuration and spin all accord satisfactorily with crystal field theory.

A few data are available relating to the oxide ligand in tetrahedral $[MO_4]^{n-}$ species, but assignments and Dq values are often equivocal.[223]

Table 8 Values of $10Dq$ (cm^{-1}) for Transition Metal Cations in Aqueous Solution[a]

d^1			Ti^{3+}	20 300		
d^2			V^{3+}	18 600		
d^3	V^{2+}	12 300	Cr^{3+}	17 000	Mo^{3+}	26 300
d^4	Cr^{2+}	14 100	Mn^{3+}	20 000		
d^5	Mn^{2+}	8500	Fe^{3+}	~14 000	Ru^{3+}	28 600[b]
d^6	Fe^{2+}	10 000	Co^{3+}	18 200	Rh^{3+}	~27 000
d^7	Co^{2+}	9300				
d^8	Ni^{2+}	8900				
d^9	Cu^{2+}	12 000				

[a] From ref. 223.
[b] Z. Harzion and G. Navon, *Inorg. Chem.*, 1980, **19**, 2236.

15.1.4.3.2 Hydration enthalpies

The overall strength of interaction between a metal ion and its hydration shell, which will parallel the strength of bonding within the $[M(OH_2)_x]^{n+}$, can be assessed from ion hydration enthalpies. The values listed in Table 9[224] are based on the single ion assumption of Halliwell and Nyburg (hydration enthalpy for $H^+ = -1091\,kJ\,mol^{-1}$).[225] For 'hard' metal ions there is the expected dependence on cation charge and radius, but enthalpies of hydration of other cations tend to be more negative (by up to about $400\,kJ\,mol^{-1}$) than expected from simple electrostatics. Crystal field effects are also apparent for d-block cations.

Table 9 Single-ion Hydration Enthalpies ($kJ\,mol^{-1}$) for Metal Cations (at 298.2 K)[a,b]

Li^+ -515	Be^{2+} -2487					
Na^+ -405	Mg^{2+} -1922		Al^{3+} -4660			
K^+ -321	Ca^{2+} -1592	Zn^{2+} -2044	Sc^{3+} -3960	Ga^{3+} -4658		
Rb^+ -296	Sr^{2+} -1445	Cd^{2+} -1806	Y^{3+} -3620	In^{3+} -4109	Sn^{2+} -1544	
Cs^+ -263	Ba^{2+} -1304		La^{3+} -3283^c	Tl^{3+} -4184^d	Pb^{2+} -1480	
	Ra^{2+} -1259					

					Cu^+ -594
Cr^{2+} -1850	Mn^{2+} -1845	Fe^{2+} -1920	Co^{2+} -2054	Ni^{2+} -2106	Cu^{2+} -2100
Cr^{3+} -4402		Fe^{3+} -4376			
					Ag^+ -475

[a] From Chapter 7 of ref. 20; estimated for zero ionic strength.
[b] See P. R. Tremaine and J. C. Leblanc, *J. Chem. Thermodyn.*, 1980, **12**, 521 for values at elevated temperatures.
[c] Enthalpies of hydration of lanthanide(3+) cations increase steadily until Lu^{3+}, -3758; the enthalpy of hydration of Ce^{4+} is -6489; that of Pu^{3+} is -3441 ($kJ\,mol^{-1}$).
[d] Tl^+ $-326\,kJ\,mol^{-1}$.

15.1.4.4 Other Properties of Aquo Ions

The acidic properties of coordinated water in aquo cations vary enormously with the cation. Table 10 contains p^*K_1 values,[19,226] where $^*K_1 = [M(OH)(OH_2)_{x-1}^{(n-1)+}][H^+(aq)]/M(OH_2)_x^{n+}][H_2O]$, with $[H_2O]$ taken as unity.[†] There is an approximate correlation with electrostatics (charges and ionic radii), but such properties as oxidizing power and 'softness' complicate the pattern.[227]

Redox potentials of aquo ions have been the subject of intensive study for many decades. In fact most values were established many years ago, and have been tabulated and discussed frequently.[228] Both in relation to redox potentials and to pK values it is important to bear in mind that ternary complexes $[M(OH_2)_yL_z]^{n+}$ may have properties very different from those of their binary parent ions $[M(OH_2)_x]^{n+}$.

Table 10 Selected p^*K_1 Values for Aquo-metal Ions (at 298.2 K and zero ionic strength)[a]

Li^+ 13.9								
Na^+ 14.7	Mg^{2+} 11.4		Al^{3+} **5.0**					
	Ca^{2+} **12.6**	Zn^{2+} **9.5**	Sc^{3+} 4.8	Ga^{3+} 2.6			
	Sr^{2+} **13.1**	Cd^{2+} 7.9^b	Y^{3+} (8)	In^{3+} 4.3	Sn^{2+} (4)^c		
Tl^+ **13.3**	Ba^{2+} **13.3**		La^{3+} (9)	Tl^{3+} (1)	Pb^2 7.2^d	Bi^{3+} 12.0^e

	Cr^{2+} 9	Mn^{2+} 10	Fe^{2+} (7)	Co^{2+} (9)	Ni^{2+} 10	Cu^{2+} 8.1^f	
Ti^{3+} (2)^9	Cr^{3+} **3.9**	Mn^{3+} (0)	Fe^{3+} **2.0**	Co^{3+} (1)			
			Ru^{3+} 2.4^h	Rh^{3+} 3	Pd^{2+} (2)	Ag^+ 12	
				Ir^{3+} 4.8	Pt^{2+} (4)	Ag^{2+} i	

$Ce^{3+} \rightarrow Lu^{3+}$ $9.3 \rightarrow 8.2$				Np^{3+} 7.4	Pu^{3+} 7.0
Ce^{4+} 0.3^j		Th^{4+} 3.0^k	U^{4+} 1	Np^{4+} 2.3	Pu^{4+} **1.5**

[a] Values in bold type are mean values with small standard deviations from several independent determinations; values in parentheses are approximate or doubtful (from ref. 227 unless stated otherwise). Table 9.3 of ref. 20 gives fuller details and outlines ionic strength effects.
[b] Or 7.3 (I. B. Mizetskaya, N. A. Oleinik and L. F. Prokopchuk, *Russ. J. Inorg. Chem. (Engl. Transl.)*, 1983, **28**, 956).
[c] M. Pettine, F. J. Millero and G. Macchi, *Anal Chem.*, 1981, **53**, 1039.
[d] See also R. N. Sylva and P. L. Brown, *J. Chem. Soc., Dalton Trans.*, 1980, 1577; K. Kogure, M. Okamoto, H. Kakihana and M. Maeda, *J. Inorg. Nucl. Chem.*, 1981, **43**, 1561.
[e] I. Hataye, H. Suganama, H. Ikegami and T. Kuchiki, *Bull. Chem. Soc. Jpn.*, 1982, **55**, 1475.
[f] A. J. Paulson and D. R. Kester, *J. Solution Chem.*, 1980, **9**, 269.
[g] B. S. Brunschwig and N. Sutin, *Inorg. Chem.*, 1979, **18**, 1731; Ya. I. Tur'yan and L. M. Maluka, *J. Gen. Chem. USSR (Engl. Transl.)*, 1983, **53**, 222.
[h] Z. Harzion and G. Navon, *Inorg. Chem.*, 1980, **19**, 2236.
[i] See C. Baiocchi, G. Bovio and E. Mentasti, *Int. J. Chem. Kinet.*, 1982, **14**, 1017.
[j] S. B. Hanna and J. T. Fenton, *Int. J. Chem. Kinet.*, 1983, **15**, 925.
[k] P. L. Brown, J. Ellis and R. N. Sylva, *J. Chem. Soc., Dalton Trans.*, 1983, **31**, 35 ($I = 0.1$ M).

[†] Formerly $[H_2O] = 1$ was by convention (though pre-1950 Bjerrum used the actual $[H_2O]$), but the current expression of equilibrium constants in dimensionless form by using ratios of concentrations (activities) to standard states results in the water term being unity since solvent water is, to a very close approximation, in its standard state.

15.1.5 REFERENCES

1. W. Feitknecht and P. Schindler, *Pure Appl. Chem.*, 1963, **6**, 130.
2. P. Kebarle, in 'Modern Aspects of Electrochemistry', ed. B. E. Conway and J. O'M. Bockris, Plenum, New York, 1974, vol. 9, chap. 1.
3. B. Nowitzki and R. Hoppe, *Z. Anorg. Allg. Chem.*, 1983, **505**, 105, 111.
4. Z. Zikmund, *Acta Crystallogr., Sect. B*, 1974, **30**, 2587; V. V. Fomichev, A. D. Savel'eva, O. I. Kondratov and K. I. Petrov, *Russ. J. Inorg. Chem. (Engl. Transl.)*, 1982, **27**, 22.
5. K. Wieghardt, K. Pohl, I. Jibril and G. Huttner, *Angew. Chem., Int. Ed. Engl.*, 1984, **23**, 77.
6. R. Nesper and H. G. van Schnering, *Z. Anorg. Allg. Chem.*, 1983, **499**, 109.
7. I. V, Lin'ko, A. K. Molodkin, B. E. Zaitsev, V. P. Dolganev and N. U. Venskovskii, *Russ. J. Inorg. Chem. (Engl. Transl.)*, 1983, **28**, 998.
8. G. Neogi, S. Acharya, R. K. Panda and D. Ramaswamy, *Int. J. Chem. Kinet.*, 1983, **15**, 881; B. N. Ivanov-Emin, N. A. Nevskaya, B. E. Zaitsev, N. N. Nevskii and Yu. N. Medvedev, *Russ. J. Inorg. Chem. (Engl. Transl.)*, 1983, **28**, 704; H. C. Jewiss, W. Levason, M. Tajik, M. Webster and N.P.C. Walker, *J. Chem. Soc., Dalton Trans.*, 1985, 199; B. N. Ivanov-Emin, N. A. Nevskaya, B. E. Zaitsev, N. V. Nevskii and A. S. Izmailovich, *Russ. J. Inorg. Chem. (Engl. Transl.)*, 1984, **29**, 710.
9. K. Wieghardt, G. Backes-Dahmann, W. Holzbach, W. J. Swiridoff and J. Weiss, *Z. Anorg. Allg. Chem.*, 1983, **499**, 44.
10. V. N. Kosyakov, E. A. Erin, V. M. Vityutnev, V. V. Kopytov and A. G. Rykov, *Radiokhimiya*, 1982, **24**, 551.
11. N. W. Alcock and S. Esperås, *J. Chem. Soc., Dalton Trans.*, 1977, 893.
12. G. F. Ciani, G. D'Alfonso, P. F. Romiti, A. Sironi and M. Freni, *Inorg. Chim. Acta*, 1983, **72**, 29.
13. L. Ciavatta and M. Grimaldi, *J. Inorg. Nucl. Chem.*, 1975, **37**, 163; A. J. Leffler, *Inorg. Chem.*, 1979, **18**, 2529.
14. See for example M. Magini, A. Saltelli and R. Caminiti, *Inorg. Chem.*, 1981, **20**, 3564; T. I. Morrison, G. K. Shenoy and L. Nielsen, *Inorg. Chem.*, 1981, **20**, 3565 and references therein.
15. W. Feitknecht, R. Giovanoli, W. Michaelis and M. Müller, *Z. Anorg. Allg. Chem.*, 1975, **417**, 114.
16. D. L. Perry, *Inorg. Chim. Acta*, 1982, **65**, L211.
17. G. Johansson, *Acta Chem. Scand.*, 1960, **14**, 771; *Ark. Kemi*, 1962, **20**, 321.
18. J. W. Akitt and A. Farthing, *J. Chem. Soc., Dalton Trans.*, 1981, 1233, 1606, 1617, 1624.
19. C. F. Baes and R. E. Mesmer, 'The Hydrolysis of Cations', Wiley, New York, 1976.
20. J. Burgess, 'Metal Ions in Solution', Ellis Horwood, Chichester, 1978.
21. S. Schönherr and H. Görz, *Z. Anorg. Allg. Chem.*, 1983, **503**, 37.
22. D. T. Richens and A. G. Sykes, *Comments Inorg. Chem.*, 1981, **1**, 141.
23. S. P. Cramer, P. K. Eidem, M. T. Paffett, J. R. Winkler, Z. Dori and H. B. Gray, *J. Am. Chem. Soc.*, 1983, **105**, 799.
24. Y. Hung, W.-J. Kung and H. Taube, *Inorg. Chem.*, 1981, **20**, 457.
25. R. A. Armstrong and H. Taube, *Inorg. Chem.*, 1976, **15**, 1904.
26. D. R. Gard and T. L. Brown, *J. Am. Chem. Soc.*, 1982, **104**, 6340.
27. B. F. G. Johnson, J. Lewis, W. J. H. Nelson, J. Puga, K. Henrick and M. McPartlin, *J. Chem. Soc., Dalton Trans.*, 1983, 1203.
28. D. F. Jones, P. H. Dixneuf, A. Benoit and J.-Y. Le Marouille, *Inorg. Chem.*, 1983, **22**, 29.
29. J. L. Cihonski and R. A. Levenson, *Inorg. Chem.*, 1975, **14**, 1717.
30. V. V. Zamashchikov and S. A. Mitchenko, *Kinet. Katal.*, 1983, **24**, 254.
31. A. G. Davies, J. P. Goddard, M. B. Hursthouse and N. P. C. Walker, *J. Chem. Soc., Chem. Commun.*, 1983, 597.
32. H.-P. Klein and U. Thewalt, *Z. Anorg. Allg. Chem.*, 1981, **476**, 62.
33. R. Kuroda, S. Neidle, I. M. Ismail and P. J. Sadler, *Inorg. Chem.*, 1983, **22**, 3620; C. F. J. Barnard, P. C. Hydes, W. P. Griffiths and O. S. Mills, *J. Chem. Res. (S)*, 1983, 302; *(M)*, 1983, 2801.
34. J. A. Shelnutt, *J. Am. Chem. Soc.*, 1983, **105**, 7178.
35. A. Zalkin, H. Ruben and D. H. Templeton, *Acta Crystallogr.*, 1962, **15**, 1219.
36. P. Brand, U. Seltmann, D. Müller and U. Büchner, *Z. Anorg. Allg. Chem.*, 1983, **502**, 132.
37. S. Ganguli and M. Bhattacharya, *J. Chem. Soc., Faraday Trans. 1*, 1983, **79**, 1513.
38. B. R. Penfold and J. A. Grigor, *Acta Crystallogr.*, 1959, **12**, 850.
39. A. Zalkin, J. D. Forrester and D. H. Templeton, *Inorg. Chem.*, 1964, **3**, 529.
40. P. J. McCarthy and M. F. Richardson, *Inorg. Chem.*, 1983, **22**, 2979; D. Michalska-Fong, P. J. McCarthy and K. Nakamoto, *Spectrochim. Acta, Part A*, 1983, **39**, 835.
41. G. Thiele and B. Grunwald, *Z. Anorg. Allg. Chem.*, 1983, **498**, 105.
42. E. Herdtweck, *Z. Anorg. Allg. Chem.*, 1983, **501**, 131.
43. B. Engelen, C. Freiburg and H. D. Lutz, *Z. Anorg. Allg. Chem.*, 1983, **497**, 151.
44. J. Mizuno, K. Ukei and T. Sugawara, *J. Phys. Soc. Jpn.*, 1959, **14**, 383; J. Mizuno, *J. Phys. Soc. Jpn.*, 1960, **15**, 1412; 1961, **16**, 1574.
45. G. W. Brady, *J. Chem. Phys.*, 1960, **33**, 1079; L. O. Morgan, *J. Chem. Phys.*, 1963, **38**, 2788; W. Urland and U. Schwanitz-Schüller, *Angew. Chem., Int. Ed. Engl.*, 1983, **22**, 1009 and references therein; D. L. Kepert, J. M. Patrick and A. H. White, *Aust. J. Chem.*, 1983, **36**, 477.
46. B. Beagley, C. A. McAuliffe, A. G. Mackie and R. G. Pritchard, *Inorg. Chim. Acta*, 1984, **89**, 163.
47. K. Brodersen and H.-U. Hummel, *Z. Anorg. Allg. Chem.*, 1983, **499**, 15.
48. T. Ueki, A. Zalkin and D. H. Templeton, *Acta Crystallogr.*, 1966, **20**, 836; J. C. Taylor, M. H. Mueller and R. L. Hitterman, *Acta Crystallogr.*, 1966, **20**, 842.
49. G. Paschina, G. Piccaluga, G. Pinna and M. Magini, *Chem. Phys. Lett.*, 1983, **98**, 157.
50. B. Sieklucka and A. Samotus, *J. Inorg. Nucl. Chem.*, 1980, **42**, 1003.
51. M. J. Blandamer and J. Burgess, *Pure Appl. Chem.*, 1982, **54**, 2285; 1983, **55**, 55.
52. See chapter 6 of ref. 20.
53. R. C. Evans, 'An Introduction to Crystal Chemistry', Cambridge University Press, London, 1966, chap. 12.
54. A. F. Wells, 'Structural Inorganic Chemistry', 5th edn., Oxford University Press, Oxford, 1984.

55. H. Eichelberger, R. Majeste, R. Surgi, L. Trefonas, M. L. Good and D. Karraker, *J. Am. Chem. Soc.*, 1977, **99**, 616; M. M. Morelock, M. L. Good, L. M. Trefonas, R. Majeste and D. G. Karraker, *Inorg. Chem.*, 1982, **21**, 3044.
56. A. G. Sykes and J. A. Weil, *Prog. Inorg. Chem.*, 1970, **13**, 1.
57. R. Duval and C. Duval, in 'Nouveau Traité de Chimie Minérale', ed. P. Pascal, Masson, Paris, 1959, vol. 14, p. 415.
58. A. Urushiyama, T. Nomura and M. Nakahara, *Bull. Chem. Soc. Jpn.*, 1970, **43**, 3971.
59. M. Thompson and R. E. Connick, *Inorg. Chem.*, 1981, **20**, 2279.
60. G. Johansson, *Acta Chem. Scand.*, 1971, **25**, 2799.
61. S. Fallab, M. Zehnder and U. Thewalt, *Helv. Chim. Acta*, 1980, **63**, 1491.
62. J. E. Finholt, K. Caulton, K. Kimball and E. Uhlenhopp, *Inorg. Chem.*, 1968, **7**, 610.
63. M. R. Churchill, G. M. Harris, R. A. Lashewycz, T. P. Dasgupta and K. Koshy, *Inorg. Chem.*, 1979, **18**, 2290.
64. R. J. Motekaitis, A. E. Martell, J.-P. Lecomte and J.-M. Lehn, *Inorg. Chem.*, 1983, **22**, 609.
65. I. Ostrich and A. J. Leffler, *Inorg. Chem.*, 1983, **22**, 921.
66. K. C. Molloy and J. J. Zuckerman, *Acc. Chem. Res.*, 1983, **16**, 386.
67. L. Banci, A. Bencini and D. Gatteschi, *J. Am. Chem. Soc.*, 1983, **105**, 761.
68. D. Fischer and B. Krebs, *Z. Anorg. Allg. Chem.*, 1982, **491**, 73.
69. G. S. Muraveiskaya, V. S. Orlova, I. F. Golovaneva and R. N. Shchelokov, *Russ. J. Inorg. Chem. (Engl. Transl.)*, 1981, **26**, 994; D. S. Gill and B. Rosenberg, *J. Am. Chem. Soc.*, 1982, **104**, 4598.
70. I. Tanaka, N. Jin-No, T. Kushida, N. Tsutsui, T. Ashida, H. Suzuki, H. Sakurai, Y. Moro-Oka and T. Ikawa, *Bull. Chem. Soc. Jpn.*, 1983, **56**, 657.
71. P. Leupin, A. G. Skyes and K. Weighardt, *Inorg. Chem.*, 1983, **22**, 1253; K. Wieghardt, M. Hahn, W. Swiridoff and J. Weiss, *Inorg. Chem.*, 1984, **23**, 94.
72. K. Wieghardt, W. Schmidt, R. van Eldik, B. Nuber and J. Weiss, *Inorg. Chem.*, 1980, **19**, 2922.
73. R. D. Cannon and S. Benjarvongkulchai, *J. Chem. Soc., Dalton Trans.*, 1981, 1924.
74. E. C. Kruissink, L. L. van Reijen and J. R. H. Ross, *J. Chem. Soc., Faraday Trans. 1*, 1981, **77**, 649; M. Holz, H. L. Friedman and B. L. Tembe, *J. Magn. Reson.*, 1982, **47**, 454.
75. K. Wieghardt, W. Schmidt, B. Nuber and J. Weiss, *Chem. Ber.*, 1979, **112**, 2220; K. Wieghardt, P. Chaudhari, B. Nuber and J. Weiss, *Inorg. Chem.*, 1982, **21**, 3086.
76. D. R. Robertson and T. A. Stephenson, *J. Organomet. Chem.*, 1976, **116**, C29.
77. W. Hieber, K. Englert and K. Rieger, *Z. Anorg. Allg. Chem.*, 1959, **300**, 295, 304.
78. G. Björnlund, *Acta Chem. Scand., Ser. A*, 1974, **28**, 169.
79. A. Weiss, S. Lyng and A. Weiss, *Z. Naturforsch., Teil B*, 1960, **15**, 678.
80. S. M. Jörgensen, *Z. Anorg. Allg. Chem.*, 1898, **16**, 184.
81. A. Werner, *Chem. Ber.*, 1907, **40**, 2103.
82. G. Biedermann and D. Ferri, *Acta Chem. Scand., Ser. A*, 1982, **36**, 611.
83. B. Sundvall, *Acta Chem. Scand., Ser. A*, 1980, **34**, 93; I. Grenthe and I. Toth, *Inorg. Chem.*, 1985, **24**, 2405.
84. A. Clearfield and P. A. Vaughan, *Acta Crystallogr., Sect. B*, 1956, **9**, 555; D. H. Devia and A. G. Sykes, *Inorg. Chem.*, 1981, **20**, 910; K. A. Burkov, G. V. Kozhevnikova, L. S. Lilich and L. A. Myund, *Russ. J. Inorg. Chem. (Engl. Transl.)*, 1982, **27**, 804.
85. H. Einaga, *J. Chem. Soc., Dalton Trans.*, 1979, 1917.
86. W. D'Olieslager, L. Heerman and M. Clarysse, *Polyhedron*, 1983, **2**, 1107.
87. M. T. Flood, R. E. Marsh and H. B. Gray, *J. Am. Chem. Soc.*, 1969, **91**, 193.
88. H. Stünzi and W. Marty, *Inorg. Chem.*, 1983, **22**, 2145; E. Bang, *Acta Chem. Scand., Ser. A*, 1984, **38**, 419.
89. K. Smolander, *Acta Chem. Scand., Ser. A*, 1983, **37**, 5; D. M. Ho and R. Bau, *Inorg. Chem.*, 1983, **22**, 4079.
90. See pp. 295 and 299 of ref. 20; L. Akhter, W. Clegg, D. Collison and C. D. Garner, *Inorg. Chem.*, 1985, **24**, 1725.
91. R. O. Gould, C. L. Jones, D. R. Robertson and T. A. Stephenson, *J. Chem. Soc., Chem. Commun.*, 1977, 222.
92. M. Herberhold, G. Süss, J. Ellermann and H. Gäbelein, *Chem. Ber.*, 1978, **111**, 2931.
93. U. Sartorelli, L. Garlaschelli, G. Ciani and G. Bonora, *Inorg. Chim. Acta*, 1971, **5**, 191; R. G. Teller and R. Bau, *Struct. Bonding (Berlin)*, 1981, **44**, 1.
94. W. Clegg, C. D. Garner and M. H. Al-Samman, *Inorg. Chem.*, 1983, **22**, 1534.
95. R. A. Holwerda and J. S. Petersen, *Inorg. Chem.*, 1980, **19**, 1775.
96. R. F. Johnston and R. A. Holwerda, *Inorg. Chem.*, 1983, **22**, 2942.
97. K.-F. Tebbe ad H. G. von Schnering, *Z. Anorg. Allg. Chem.*, 1973, **396**, 66.
98. E. Hey, F. Weller and K. Dehnicke, *Z. Anorg. Allg. Chem.*, 1983, **502**, 45.
99. R. P. Planalp and R. A. Andersen, *J. Am. Chem. Soc.*, 1983, **105**, 7774.
100. C. Ercolani, G. Rossi and F. Monacelli, *Inorg. Chim. Acta*, 1980, **44**, L215; C. Ercolani, M. Gardini, G. Pennesi and G. Rossi, *J. Chem. Soc., Chem. Commun.*, 1983, 549.
101. P.-T. Cheng and S. C. Nyburg, *Inorg. Chem.*, 1975, **14**, 327; D. W. Phelps, E. M. Kahn and D. J. Hodgson, *Inorg. Chem.*, 1975, **14**, 2486.
102. P. C. Healy, B. W. Skelton and A. H. White, *Aust. J. Chem.*, 1983, **36**, 2057; K. Dehnicke, H. Prinz, W. Massa, J. Pebler and R. Schmidt, *Z. Anorg. Allg. Chem.*, 1983, **499**, 20.
103. M. G. Barker, P. G. Gadd and S. C. Wallwork, *J. Chem. Soc., Chem. Commun.*, 1982, 516; F. Griesfeller, J. Köhler and R. Hoppe, *Z. Anorg. Allg. Chem.*, 1983, **507**, 155; I. Hassan and H. D. Grundy, *Acta Crystallogr., Sect. B*, 1984, **40**, 6; W. Depmeier, *Acta Crystallogr., Sect. C*, 1984, **40**, 226.
104. R. D. Cannon, 'Electron Transfer Reactions', Butterworths, London, 1980, pp. 350–351.
105. P. Blanc, C. Madic and J. P. Launay, *Inorg. Chem.*, 1982, **21**, 2923.
106. B. Guillaume, R. L. Hahn and A. H. Narten, *Inorg. Chem.*, 1983, **22**, 109.
107. T. Misawa, K. Hashimoto, W. Suëtaka and S. Shimodaira, *J. Inorg. Nucl. Chem.*, 1973, **35**, 4159; T. Misawa, K. Hashimoto and S. Shimodaira, *J. Inorg. Nucl. Chem.*, 1973, **35**, 4167.
108. H. Sabrowsky and R. Hoppe, *Z. Anorg. Allg. Chem.*, 1968, **358**, 241; K. Hestermann and R. Hoppe, *Z. Anorg. Allg. Chem.*, 1968, **360**, 113.
109. J. P. Launay, Y. Jeannin and A. Nel, *Inorg. Chem.*, 1983, **22**, 277.
110. J. Sala-Pala, J.-E. Guerchais and A. J. Edwards, *Angew. Chem., Int. Ed. Engl.*, 1982, **21**, 870.
111. M. T. Beck, 'Chemistry of Complex Equilibria', Van Nostrand Reinhold, New York, 1970, p. 227.

112. T. Shibahara, H. Kuroya, K. Matsumoto and S. Ooi, *Bull. Chem. Soc. Jpn.*, 1983, **56**, 2945.
113. K. Wieghardt, K. Pohl and W. Gebert, *Angew. Chem., Int. Ed. Engl.*, 1983, **22**, 727.
114. J. E. Armstrong, W. R. Robinson and R. A. Walton, *Inorg. Chem.*, 1983, **22**, 1301.
115. K. Michelsen, E. Pedersen, S. R. Wilson and D. J. Hodgson, *Inorg. Chim. Acta*, 1982, **63**, 141.
116. W. T. Elam, E. A. Stern, J. D. McCallum and J. Sanders-Loehr, *J. Am. Chem. Soc.*, 1982, **104**, 6369.
117. G. Anderegg, E. Müller, K. Zollinger and H.-B. Bürgi, *Helv. Chim. Acta*, 1983, **66**, 1593.
118. M. Ardon, A. Bino and G. Yahav, *J. Am. Chem. Soc.*, 1976, **98**, 2338.
119. B. Brazel and R. Hoppe, *Z. Anorg. Allg. Chem.*, 1982, **493**, 93; 1983, **497**, 176; 1983, **498**, 167; 1983, **499**, 153.
120. M. G. Barker, P. G. Gadd and M. J. Begley, *J. Chem. Soc., Chem. Commun.*, 1981, 379.
121. B. A. Cartwright, W. P. Griffith, M. Schröder and A. C. Skapski, *J. Chem. Soc., Chem. Commun.*, 1978, 853.
122. P. Souchay, M. Cadiot and M. Duhameaux, *C.R. Hebd. Séances Acad. Sci., Ser. C*, 1966, **262**, 1524; R. K. Murmann and M. E. Shelton, *J. Am. Chem. Soc.*, 1980, **102**, 3984.
123. D. T. Richens and A. G. Sykes, *Inorg. Chim. Acta*, 1981, **54**, L3; *Inorg. Chem.*, 1982, **21**, 418.
124. S. C. Chang and G. A. Jeffrey, *Acta Crystallogr., Sect. B*, 1970, **26**, 673.
125. F. A. Cotton and Wenning Wang, *Inorg. Chem.*, 1982, **21**, 2675.
126. D. Datta and A. Chakravorty, *Inorg. Chem.*, 1983, **22**, 1611.
127. A. R. E. Baikie, M. B. Hursthouse, D. B. New and P. Thornton, *J. Chem. Soc., Chem. Commun.*, 1978, 62.
128. J. A. Baumann, S. T. Wilson, D. J. Salmon, P. L. Hood and T. J. Meyer, *J. Am. Chem. Soc.*, 1979, **101**, 2916.
129. R. V. Thundathil, E. M. Holt, S. L. Holt and K. J. Watson, *J. Am. Chem. Soc.*, 1977, **99**, 1818.
130. A. E. Underhill and D. M. Watkins, *J. Chem. Soc., Dalton Trans.*, 1977, 5.
131. B. Harrison, N. Logan and A. D. Harris, *J. Chem. Soc., Dalton Trans.*, 1980, 2382.
132. A. G. Ryzhov and N. M. Sinitsyn, *Russ. J. Inorg. Chem.*, 1983, **28**, 383.
133. A. J. Zozulin, D. C. Moody and R. R. Ryan, *Inorg. Chem.*, 1982, **21**, 3083.
134. J. L. Atwood and M. J. Zaworotko, *J. Chem. Soc., Chem. Commun.*, 1983, 302.
135. K. Wieghardt, K. Pohl, I. Jibril and G. Huttner, *Angew. Chem., Int. Ed. Engl.*, 1984, **23**, 77.
136. I. Ahlberg, *Acta Chem. Scand.*, 1962, **16**, 887.
137. G. Nagorsen, S. Lyng, A. Weiss and A. Weiss, *Angew. Chem., Int. Ed. Engl.*, 1962, **1**, 115.
138. A. B. Blake and L. R. Fraser, *J. Chem. Soc., Dalton Trans.*, 1975, 193.
139. S. Grimvall, *Acta Chem. Scand.*, 1973, **27**, 1447; *Acta Chem. Scand., Ser. A*, 1975, **29**, 590.
140. G. Ciani, A. Sironi and V. G. Albano, *J. Chem. Soc., Dalton Trans.*, 1977, 1667.
141. A. Bino, F. A. Cotton and Z. Dori, *J. Am. Chem. Soc.*, 1981, **103**, 243; A. Bino and D. Gibson, *J. Am. Chem. Soc.*, 1982, **104**, 4383; F. A. Cotton, Z. Dori, D. O, Marler and W. Schwotzer, *Inorg. Chem.*, 1983, **22**, 3104.
142. R. J. Goudsmit, B. F. G. Johnson, J. Lewis, P. R. Raithby and K. H. Whitmire, *J. Chem. Soc., Chem. Commun.*, 1983, 246.
143. D. Bright, *J. Chem. Soc., Chem. Commun.*, 1970, 1169.
144. F. Bottomley, D. E. Paez and P. S. White, *J. Am. Chem. Soc.*, 1982, **104**, 5651; F. Bottomley and F. Grein, *Inorg. Chem.*, 1982, **21**, 4170.
145. M. Cousins and M. L. H. Green, *J. Chem. Soc. (A)*, 1969, 16.
146. W. H. Bragg, *Nature (London)*, 1923, **111**, 532.
147. V. Auger and I. Robin, *C.R. Hebd. Séances Acad. Sci.*, 1924, **178**, 1546; R. M. Gordon and H. B. Silver, *Can. J. Chem.*, 1983, **61**, 1218.
148. J. A. Bertrand, *Inorg. Chem.*, 1967, **6**, 495; Yu. A. Simonov, A. A. Dvorkin, M. A. Yampol'skaya and V. E. Zavodnik, *Russ. J. Inorg. Chem.*, 1982, **27**, 684.
149. Å. Olin and R. Söderquist, *Acta Chem. Scand.*, 1972, **26**, 3505.
150. R. Caminiti, G. Licheri, G. Piccaluga, G. Pinna and M. Magini, *Rev. Inorg. Chem.*, 1979, **1**, 333 (see p. 375).
151. H.-L. Keller, *Angew. Chem., Int. Ed. Engl.*, 1983, **22**, 324.
152. J. Catterick, P. Thornton and B. W. Fitzsimmons, *J. Chem. Soc., Dalton Trans.*, 1977, 1420.
153. M. Ardon and A. Bino, *J. Am. Chem. Soc.*, 1983, **105**, 7747.
154. R. Scholder, in 'Handbook of Preparative Inorganic Chemistry', 2nd edn., ed. G. Brauer, Academic, New York, 1965, vol. 2, p. 1677.
155. Fu-Tang Chen, Chung-Shin Lee and Chung-Sun Chung, *Polyhedron*, 1983, **2**, 1301; B. N. Ivanov-Emin, L. P. Petrishcheva, B. E. Zaitsev and A. S. Izmailovich, *Russ. J. Inorg. Chem. (Engl. Transl.)*, 1984, **29**, 860.
156. B. N. Ivanov-Emin, N. A. Nevskaya, B. E. Zaitsev and V. I. Tsirel'nikov, *Russ. J. Inorg. Chem. (Engl. Transl.)*, 1982, **27**, 1258.
157. N. N. Nevskii, B. N. Ivanov-Emin and N. A. Nevskaya, *Russ. J. Inorg. Chem. (Engl. Transl.)*, 1983, **28**, 1074; B. N. Ivanov-Emin, N. A. Nevskaya, B. E. Zaitsev and A. S. Izmailovich, *Russ. J. Inorg. Chem. (Engl. Transl.)*, 1984, **29**, 777.
158. H. Schwarz, *Z. Naturforsch., Teil B*, 1967, **22**, 554; B. N. Ivanov-Emin, N. A. Nevskaya, B. E. Zaitsev and T. M. Ivanova, *Russ. J. Inorg. Chem. (Engl. Transl.)*, 1982, **27**, 1755.
159. B. N. Ivanov-Emin, N. A. Nevskaya, B. E. Zaitsev and V. I. Tsirel'nikov, *Russ. J. Inorg. Chem.*, 1983, **28**, 557.
160. E. T. Borish and L. J. Kirschenbaum, *J. Chem. Soc., Dalton Trans.*, 1983, 749; L. J. Kirschenbaum and J. D. Rush, *Inorg. Chem.*, 1983, **22**, 3304.
161. C. Levy-Clément, *Ann. Chim.*, 1975, **10**, 105.
162. T. L. Popova, N. G. Kisel', V. I. Krivobok and V. P. Karlov, *Russ. J. Inorg. Chem. (Engl. Transl.)*, 1981, **26**, 1610.
163. N. U. Venskovskii, B. N. Ivanov-Emin and I. V. Lin'ko, *Russ. J. Inorg. Chem. (Engl. Transl.)*, 1983, **28**, 605; B. N. Ivanov-Emin, L. P. Petrishcheva, B. E. Zaitsev, V. I. Ivlieva, A. S. Izmailovich and V. P. Dolganev, *Russ. J. Inorg. Chem. (Engl. Transl.)*, 1984, **29**, 1169.
164. G. Bandel, C. Platte and M. Trömel, *Z. Anorg. Allg. Chem.*, 1981, **477**, 178.
165. R. Münze, *Isotopenpraxis*, 1978, **14**, 81.
166. M. J. Blandamer, J. Burgess and R. D. Peacock, *J. Chem. Soc., Dalton Trans.*, 1974, 1084.
167. B. N. Ivanov-Emin, V. I. Rybina and V. I. Kornev, *Russ. J. Inorg. Chem.*, 1965, **10**, 544.
168. M. Sandström, I. Persson and S. Ahrland, *Acta Chem. Scand., Ser. A*, 1978, **32**, 607.
169. J. Glaser and G. Johansson, *Acta Chem. Scand., Ser. A*, 1981, **35**, 639.

170. See pp. 286–287 of ref. 53.
171. G. Johansson, *Acta Chem. Scand.*, 1966, **20**, 553.
172. F. A. Cotton, A. Davison, V. W. Day, M. F. Fredrich, C. Orvig and R. Swanson, *Inorg. Chem.*, 1982, **21**, 1211.
173. J. Albertsson and I. Elding, *Acta Crystallogr., Sect. B*, 1977, **33**, 1460; R. W. Broach, J. M. Williams, G. P. Felcher and D. G. Hinks, *Acta Crystallogr., Sect. B*, 1979, **35**, 2317 and references therein.
174. J. MacB. Harrowfield, D. L. Kepert, J. M. Patrick and A. H. White, *Aust. J. Chem.*, 1983, **36**, 483.
175. L. Helmholz, *J. Am. Chem. Soc.*, 1939, **61**, 1544.
176. R. Faggiani and I. D. Brown, *Acta Crystallogr., Sect. B*, 1978, **34**, 1675.
177. T. W. Hambley, D. L. Kepert, C. L. Raston and A. H. White, *Aust. J. Chem.*, 1978, **31**, 2635.
178. M. C. Favas, D. L. Kepert, J. M. Patrick and A. H. White, *J. Chem. Soc., Dalton Trans.*, 1983, 571.
179. B. Eriksson, *Acta Chem. Scand., Ser. A*, 1982, **36**, 186.
180. B. Eriksson, L. O. Larsson and L. Niinistö, *J. Chem. Soc., Chem. Commun.*, 1978, 616; B. Eriksson, L. O. Larsson, L. Niinistö and J. Valkonen, *Inorg. Chem.*, 1980, **19**, 1207.
181. B. Metz, O. Moras and R. Weiss, *Acta Crystallogr., Sect. B*, 1973, **29**, 1388.
182. M. G. B. Drew, F. S. Esho and S. M. Nelson, *J. Chem. Soc., Dalton Trans.*, 1983, 1653.
183. B. Eriksson, L. O. Larsson, L. Niinistö and J. Valkonen, *Acta Chem. Scand., Ser. A*, 1980, **34**, 567; B. Eriksson, L. O. Larsson and L. Niinistö, *Acta Chem. Scand., Ser. A*, 1982, **36**, 465.
184. Ö. Gröning, T. Drakenberg and L. I. Elding, *Inorg. Chem.*, 1982, **21**, 1820.
185. T. Yamaguchi, O. Lindqvist, J. B. Boyce and T. Claeson, *Acta Chem. Scand., Ser. A*, 1984, **38**, 423; T. Yamaguchi, G. Johansson, B. Holmberg, M. Maeda and H. Ohtaki, *Acta Chem. Scand., Ser. A*, 1984, **38**, 437.
186. A. Fratiello, R. E. Lee and R. E. Schuster, *Inorg. Chem.*, 1970, **9**, 391.
187. G. W. Brady, *J. Chem. Phys.*, 1958, **28**, 464; 1959, **29**, 1371.
188. H. Ohtaki, *Rev. Inorg. Chem.*, 1982, **4**, 103.
189. G. W. Neilson and J. E. Enderby, *Annu. Rep. Prog. Chem., Sect. C*, 1979, **76**, 185; J. E. Enderby and G. W. Nielson, in 'Water — A Comprehensive Treatise', ed. F. Franks, Plenum, New York, 1979, vol. 6, chap. 1.
190. See chapter 4 of ref. 20.
191. J. Glaser and G. Johansson, *Acta Chem. Scand., Ser. A*, 1982, **36**, 125.
192. A. Habenschuss and F. H. Spedding, *J. Chem. Phys.*, 1979, **70**, 2797, 3758.
193. A. H. Narten and R. L. Hahn, *J. Phys. Chem.*, 1983, **87**, 3193.
194. N. Ohtomo and K. Arakawa, *Bull. Chem. Soc. Jpn.*, 1980, **53**, 1789.
195. G. Licheri, G. Piccaluga and G. Pinna, *J. Chem. Phys.*, 1975, **63**, 4412.
196. L. S. Smith and D. L. Wertz, *J. Am. Chem. Soc.*, 1975, **97**, 2365.
197. J. O'M. Bockris, *Q. Rev. Chem. Soc.*, 1949, **3**, 173; J. O'M. Bockris and P. P. S. Saluja, *J. Electrochem. Soc.*, 1972, **119**, 1060.
198. G. W. Nielson, *Chem. Phys. Lett.*, 1979, **68**, 247.
199. T. W. Swaddle and L. Fabes, *Can. J. Chem.*, 1980, **58**, 1418.
200. See p. 144 of ref. 20.
201. H. Weingärtner, C. Müller and H. G. Hertz, *J. Chem. Soc., Faraday Trans. 1*, 1979, **75**, 2712.
202. M. Åberg, D. Ferri, J. Glaser and I. Grenthe, *Inorg. Chem.*, 1983, **22**, 3986.
203. R. G. Wyckoff, 'Crystal Structures', 2nd edn., Interscience, New York, 1960, vol. 3, p. 529.
204. J. A. Prins, *Z. Phys.*, 1929, **56**, 617; *Nature (London)*, 1929, **123**, 84; *J. Chem. Phys.*, 1935, **3**, 72; J. A. Prins and R. Fonteyne, *Physica*, 1935, **2**, 570.
205. G. W. Nielson and J. E. Enderby, *Adv. Phys.*, 1980, **29**, 323; *Rep. Prog. Phys.*, 1981, **44**, 593.
206. A. H. Narten, F. Vaslow and H. A. Levy, *J. Chem. Phys.*, 1973, **58**, 5017; R. A. Howe, S. W. Howells and J. E. Enderby, *J. Phys. C*, 1974, **7**, L111.
207. J. V. Leyendekkers, *J. Chem. Soc., Faraday Trans. 1*, 1982, **78**, 3383.
208. A. Zalkin, J. D. Forrester, D. H. Templeton, S. M. Williamson and C. W. Koch, *J. Am. Chem. Soc.*, 1964, **86**, 3569.
209. H. Einspahr and C. E. Bugg, *Acta Crystallogr., Sect. B*, 1980, **36**, 264.
210. R. J. P. Williams, *Struct. Bonding (Berlin)*, 1982, **50**, 79.
211. A. F. Wells, *Q. Rev. Chem. Soc.*, 1954, **8**, 380.
212. G. Chiari and G. Ferraris, *Acta Crystallogr., Sect. B*, 1982, **38**, 2331.
213. K. E. Newmann and A. E. Merbach, *Inorg. Chem.*, 1980, **19**, 2481.
214. T. Fujita and H. Ohtaki, *Bull. Chem. Soc. Jpn.*, 1982, **55**, 455.
215. M. Magini, *Inorg. Chem.*, 1982, **21**, 1535; A. Pajunen and S. Pajunen, *Acta Crystallogr., Sect. B*, 1982, **38**, 3093; W. Henke, S. Kremer and D. Reinen, *Z. Anorg. Allg. Chem.*, 1982, **491**, 124.
216. N. W. Alcock, M. Duggan, A. Murray, S. Tyagi, B. J. Hathaway and A. Hewat, *J. Chem. Soc., Dalton Trans.*, 1984, 7.
217. S. Pocev and G. Johansson, *Acta Chem. Scand.*, 1973, **27**, 2146.
218. C. K. Jorgensen and R. Reisfeld, *Struct. Bonding (Berlin)*, 1982, **50**, 121.
219. M. Tachez and F. Théobald, *Acta Crystallogr., Sect. B*, 1980, **36**, 249.
220. S. J. Cline, R. P. Scaringe, W. E. Hatfield and D. J. Hodgson, *J. Chem. Soc., Dalton Trans.*, 1977, 1662.
221. K. Nakamoto, 'Infrared and Raman Spectra of Inorganic and Coordination Compounds', 3rd edn., Wiley, New York, 1978; D. M. Adams, 'Metal–Ligand and Related Vibrations', Arnold, London, 1967.
222. See chapter 3.1.1 of ref. 20.
223. S. L. Holt, in 'Inorganic Interactions', ed. S. Petrucci, Academic, New York, vol. II, chap. 8.
224. See chapter 7 of ref. 20.
225. H. F. Halliwell and S. C. Nyburg, *Trans. Faraday Soc.*, 1963, **59**, 1126; D. F. C. Morris, *Struct. Bonding (Berlin)*, 1968, **4**, 63.
226. See chapter 9 of ref. 20.
227. J. Burgess, 'Ions in Solution I', Royal Society of Chemistry/ETSG, London, 1984.
228. G. Milazzo and S. Caroli, 'Tables of Standard Electrode Potentials', Wiley, New York, 1978; and see chapter 8 of ref. 20; 'Standard Potentials in Aqueous Solution', ed. A. J. Bard, R. Parsons and J. Jordan, Dekker, New York, 1985.

15.2

Dioxygen, Superoxide and Peroxide

H. ALLEN O. HILL
University of Oxford, UK

and

DAVID G. TEW
University of California, San Francisco, CA, USA

15.2.1 INTRODUCTION

The coordination of dioxygen to transition metals is a subject of extreme interest due to its utilization by biological systems.[1] Hemoglobin forms, perhaps, the best-known metal–dioxygen complex, but it is not the only metalloprotein involved in dioxygen transport and storage. The latter role is performed by myoglobin, also a heme protein. Hemerythrin contains iron bound directly *to the protein*, having no prosthetic group, and is derived from the blood of some worms. Hemocyanin, a complex copper protein, is derived from the blood of arachnoids. Dioxygen is bound to heme iron and copper in cytochrome *c* oxidase where it acts as the terminal electron acceptor in the mitochondrial electron transport chain. Elsewhere in biological processes, activation of molecular oxygen occurs by coordination in metalloproteins which catalyze oxygen insertion reactions and oxidation.

The biological utilization of dioxygen has led to an additional interest in the investigation of many, much simpler, model systems, *e.g.* reaction of low-valent transition metal complexes, notably platinum metal complexes, with ligands such as phosphines, arsines and other π-acceptors. Given the metal and ligands they may be considered to be of little relevance to biological systems, though they do provide a basis for the understanding of systems of more chemical relevance. These involve first-row transition metal ions, in a variety of ligand environments and lead to dioxygen being bound 'end-on'.

15.2.2 DIOXYGEN

A simple molecular orbital diagram for dioxygen is shown in Figure 1, in which it can be seen that oxidation of O_2 would result in the loss of an antibonding electron, while reduction would lead to the gain of an antibonding electron. Thus it would be predicted that bond lengths should increase along the series $O_2^+ < O_2 < O_2^- < O_2^{2-}$ and X-ray studies confirm this (Table 1). The frequency of the O—O stretch also decreases through the series O_2^+ to O_2^{2-} corresponding to a weakening of the

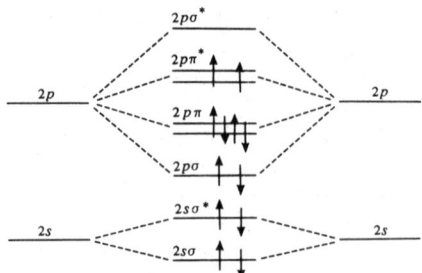

Figure 1   The molecular orbital diagram for the dioxygen molecule

Table 1   Physical Data for Dioxygen Species

Dioxygen species	Bond order	$O-O$ (Å)	ν_{O-O} (cm^{-1})	Ref.
O_2^+ (O_2AsF_6)	2.5	1.12	1858	2, 3
O_2	2	1.2074	1556, 1554.7	2, 4
O_2^-	1.5	1.32–1.35	1145 (KO$_2$)	5, 6
O_2^{2-}	1	1.48–1.49	842 (Na$_2$O$_2$·8H$_2$O)	7, 8
			738, 794 (Na$_2$O$_2$)	8
H_2O_2	1	1.475; 1.453 (7)	880	9, 10

$O-O$ bond. As will be seen later, these data provide a valuable insight into the nature of bound dioxygen.

Coordination of dioxygen can occur in four ways, two of which may be considered to correspond to one-electron reduction (superoxo complexes) and the others to two-electron reduction (peroxo complexes). The four possible coordination modes are illustrated in Figure 2. As can be seen, each type of complex may be further categorized as to whether it is a mono- or bi-nuclear complex. Whether a peroxo- or a superoxo-type complex is formed is dependent upon the metal involved. For mononuclear complexes, superoxo complexes would be expected with metals which readily undergo one-electron oxidation, *e.g.* Fe^{II} or Co^{II}, while peroxo complexes would be expected for metals with a preference for two-electron oxidation, *e.g.* Ir^I, Rh^I, Pt^0.

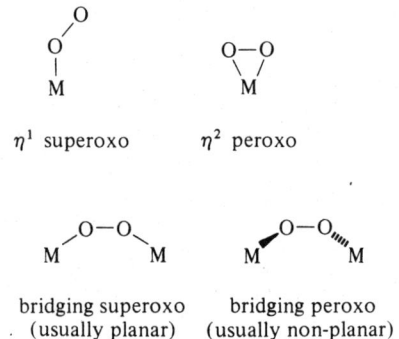

Figure 2

15.2.3   THEORETICAL CONSIDERATIONS

There have been many attempts to rationalize the bonding in metal–dioxygen complexes. In particular, there has been much debate over the description of the dioxygen ligand as 'superoxo' type or 'peroxo' type depending upon the extent of electron transfer from the metal centre to the dioxygen moiety. This problem may be rather neatly avoided by adopting a convention used by Enemark and Feltham.[11] For a metal complex L$_x$MXY, where XY is a diatomic ligand, the electronic configuration is specified as $\{MXY\}^n$ where n is the sum of the number of d-electrons on the metal fragment and the number of σ- and σ*-electrons on the XY ligand. Using this convention Hoffman *et al.*[12] have rationalized the bonding modes of a range of diatomic ligands. Beginning with a pyramidal ML$_4$ fragment, the following assumptions are made:

(a) Omit the core and M—L bonding orbitals of the ML_4 fragment, as well as the highest lying $d_{x^2-y^2}$ orbital.

(b) Omit the 's-' and 'p-' based orbitals of the ML_4 fragment though these may contribute significantly to L_4MXY bonding.

(c) Omit the $2\sigma_g$ and $2\sigma_u$ orbitals of the diatomic ligand. These orbitals have sizeable overlap with the d_{z^2} orbital of the ML_4 fragment but the energy gap separating them is large, consequently the net interaction is small.

(d) Omit the $3\sigma_u$ (or σ^*) orbital of the diatomic ligand. Except for the case where XY is the dihalogen ligand, this orbital is too high in energy to interact strongly with ML_4.

(e) Omit the diatomic $1\pi_u$ orbitals, the bonding orbitals of the diatomic. Even in the case of the dioxygen ligand, where these orbitals are of low energy, their influence is felt indirectly.

There are now seven molecular orbitals to be considered for the L_4MXY complex, d_{xy}, d_{z^2}, d_{xz} and d_{yz} from the L_4M fragment, and $3\sigma_g$ and $1\pi_u$ from the diatomic ligand. The $3\sigma_g$ and $1\pi_u$ orbitals will be referred to as the n and π^* orbitals. However, it is convenient to construct linear combinations of some of these orbitals for the two extremes of bonding geometry, linear η^1 coordination and η^2 coordination. This leads to a further simplification. The lowest lying molecular orbital for both the η^1 and η^2 geometries, $d_{z^2} + n$ and $n-\pi$ respectively, may be omitted because the orbital is not important in determining the geometry of the ligand but is in a diatomic donor orbital whose electrons are not counted in the $\{MXY\}^n$ notation. Six orbitals now remain and these are to be filled only by the n electrons counted on the $\{MXY\}^n$ notation. The Walsh diagram in Figure 3 may consequently be constructed.

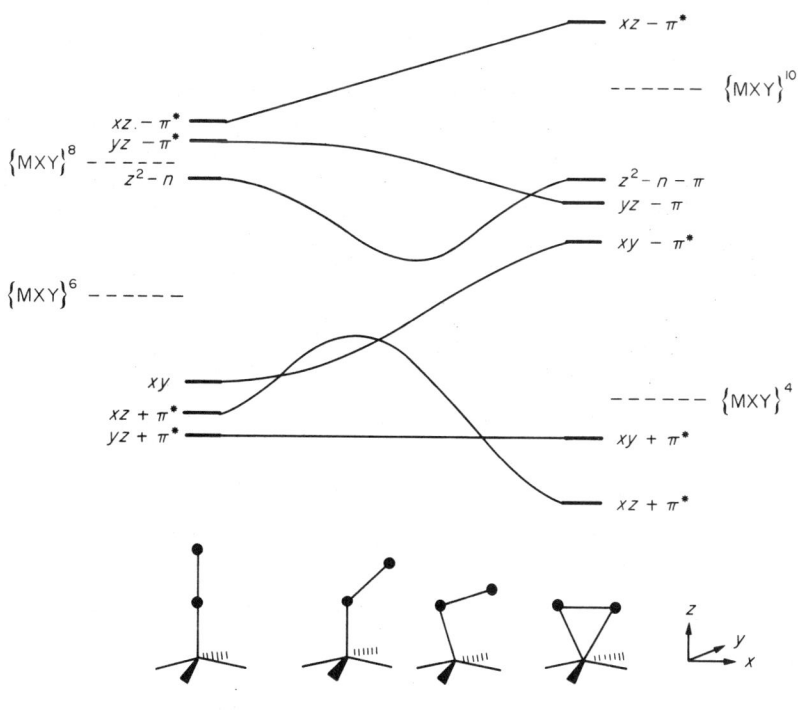

Figure 3

Several conclusions may be drawn from this diagram concerning the geometry of dioxygen complexes. Complexes of the type $\{MXY\}^4$ should exhibit the side-on η^2 geometry. Many structures of this type are known, e.g. $Ti(OEP)(O_2)$,[13] $MoOF_4(O_2)^{2-}$[14] and $VO(H_2O)(dipic)(O_2)^-$.[15] The next known dioxygen complex occurs at $\{MXY\}^7$. This is the compound $Mn(TPP)(O_2)$.[16] No structural data are available.

Dioxygen adducts of iron(II) porphyrins constitute the $\{MXY\}^8$ class of complexes. The Walsh diagram indicates that the dioxygen ligand will be either linear or bent. However, the Walsh diagram was constructed for a general diatomic ligand XY. The π^* levels of O_2 are by far the lowest lying of all the diatomics considered. It is likely that the O_2 π^* orbital lies below the metal d_{z^2} orbital. Consequently, the dioxygen ligand will bend. Structural data on $Fe(TpivPP)(O_2)$[17] *etc.* confirm this. $\{MXY\}^9$ complexes are represented by a multitude of cobalt(II) dioxygen adducts. These complexes are expected to contain a bent η^1 dioxygen ligand. Structural studies are in full agreement with this. The unpaired electron is expected to lie in the π^* orbital which is not in the plane of bonding and this is supported by ESR studies. On adding one more electron, complexes of type $\{MXY\}^{10}$ are reached. These complexes are expected to show η^2 coordination of dioxygen. Structural studies on compounds such as $IrCl(CO)(O_2)(PPh_3)_2$,[18] and many similar compounds show this to be the case.

The Walsh diagram implies that $\{MXY\}^{12}$ species should be unstable. However, the levels were derived for the ML_4 fragment (the ML_5 fragment is very similar). If fewer ligands are coordinated to the metal then $\{MXY\}^{12}$ complexes are stabilized. This is the case for dioxygen complexes such as $(Bu^tNC)_2Ni(O_2)$[19] and $(PPh_3)_2Pt(O_2)$[20] which contain η^2 dioxygen.

This strategy has deliberately avoided the superoxo/peroxo dichotomy. The extent of electron transfer to dioxygen in these adducts is difficult to assess and is a subject which remains controversial. However, it is convenient to divide the wealth of metal–dioxygen complexes into superoxo-type and peroxo-type upon the basis of structural data from the point of view of ease of discussion. We have done this purely on the basis of bond lengths and IR data and do not intend this to be interpreted as a measurement of the extent of electron transfer from the metal to the dioxygen ligand.

15.2.4 MONONUCLEAR PEROXO COMPLEXES

The metals which form this type of complex by reaction with dioxygen are mainly the later second and third row transition series metals, notably Ru^0, Os^0, Rh^I, Ir^I, Pt^0 and Pd^0. It is not surprising to see metals noted for their ability to undergo oxidative addition reactions amongst those which form this type of mononuclear dioxygen complex as a formal two-electron reduction of dioxygen is required for complex formation. The other metals known to form mononuclear peroxo-type complexes with dioxygen are Ni^0, Co^I and one example of Co^{II}.

Some of the early transition metals are known to form mononuclear peroxo complexes. These complexes are not formed by reacting a metal complex in a low oxidation state with dioxygen, as this usually results in the formation of metal oxo species, but rather by reaction of a metal complex in a high oxidation state, *e.g.* Ti^{IV}, Nb^V or Mo^{VI}, with the peroxide anion. This frequently leads to more than one peroxide ligand per metal centre.

The first mononuclear peroxo complex of the later transition metals to be characterized was $IrCl(CO)(O_2)(PPh_3)_2$ **(1)**[18] prepared by Vaska. The square planar Ir^I complex $IrCl(CO)(PPh_3)_2$ is converted into the square pyramidal Ir^{III} complex simply by bubbling oxygen through a solution of the Ir^I complex. In the structure shown in **(1)**, the O—O distance is anomalously short (1.30 Å) for a peroxo complex. More recently a ^{17}O nuclear quadrupole resonance (NQR) study was carried out on this dioxygen adduct and a comparison with hydrogen peroxide drawn.[21] The NQR spectrum of the iridium complex showed transitions at very similar frequencies to those for hydrogen peroxide. This led the authors to postulate that the O—O separation in the iridium complex was, in fact, nearer 1.5 Å than the 1.30 Å reported for the crystal structure. Some dioxygen adducts of the later transition metals lose oxygen on heating or placing under vacuum. Several workers attempted to correlate reversibility of dioxygen uptake with O—O bond length in the adduct and 'extent' of electron transfer from the metal to the dioxygen moiety. No such correlation exists and nearly all O—O distances for mononuclear peroxo complexes lie in the range 1.4–1.5 Å (Table 2).

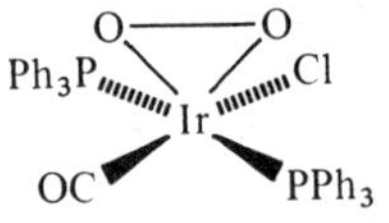

(1)

The peroxo nature of the dioxygen moiety in these complexes is further supported by the reactions of a cobalt diarsine complex. The Co^I complex takes up oxygen to give a dioxygen adduct, resulting in oxidation of the cobalt to Co^{III}. The same dioxygen adduct may be formed by reacting a cobalt(III) complex with hydrogen peroxide.[73]

IR studies on mononuclear peroxo complexes give a range for O—O stretching of 820–910 cm^{-1}. Hydrogen peroxide has $\nu_{O-O} = 880$ cm^{-1},[10] while $Na_2O_2 \cdot 8H_2O$ has $\nu_{O-O} = 842$ cm^{-1}.[18] This is consistent, therefore, with the peroxo assignment of the dioxygen ligand. Mixed isotopic substitution of these dioxygen complexes, *i.e.* $^{16}O_2$, $^{18}O_2$ and $^{16}O—^{18}O$, results in three bands due to O—O stretching being observed. This is consistent with the symmetrical side-bound coordination of dioxygen. A bent, end-on coordination geometry would give rise to four bands. Otsuka *et al.*[74]

Table 2 Mononuclear Peroxo Complexes

Complex	O—O (Å)	Ref.
$K_2Ti_2O(O_2)_2(dipic)_2 \cdot 5H_2O$	1.45(1)	22
$K_2TiF_2(O_2)(dipic) \cdot 2H_2O$	1.463(5)	23
$Ti(O_2)(H_2O)_2(dipic) \cdot 2H_2O$	1.464(2)	24
$Na_4Ti_2O(O_2)_3(NTA)_2 \cdot 11H_2O$	1.469(3)	25
$Ti(O_2)(pic)_2(HMPT)$	1.419(6)	26
$(OEP)Ti(O_2)$	1.445(5)	13
$(NH_4)_4[O\{VO(O_2)_2\}_2]$	1.444(2)	27
$(NH_4)[VO(O_2)(H_2O)(dipic)] \cdot 1.3H_2O$	1.441(3)	15
$(NH_4)VO(O_2)_2(NH_3)$	1.472(4)	28
$NbCl(Cp)(O_2)$	1.47(1)	29
$(NH_4)_3Nb(O_2)_2(C_2O_4)_2 \cdot H_2O$	1.483(6)	30
$KNb(O_2)_3(1,10\text{-phen}) \cdot 3H_2O$	1.488(1)	31
$KMgNb(O_2)_4 \cdot 7H_2O$	1.500(4)	32
$(8\text{-OHquinH})_2Nb(O_2)F_5$	1.17(9)	33
$\{1,10\text{-phen}(H)_2\}Nb(O_2)F_5$	1.462(1)	34
$[Ta(O_2)F_5]^{2-}$	1.39	35
$[Ta_2O(O_2)_2F_8]^{4-}$	1.64	35
$[NEt_4][Ta(O_2)F_4(2\text{-MepyO})]$	1.55(7)	36
$CrO(O_2)_2(py)$	1.404(2)	37
$Cr(O_2)_2(NH_3)_3$	1.429(2)	38
$K_3Cr(O_2)_2(CN)_3$	1.44(1)	39
$Cr(O_2)_2(H_2O)(en) \cdot H_2O$	1.45(1)	39
$CrO(1,10\text{-phen})(O_2)_2$	1.40(2)	39
$CrO(O_2)_2(2,2'\text{-bipy})$	1.40(2), 1.5(1)[a]	40
$K_2[MoO(H_2O)(O_2)_2]_2O \cdot 2H_2O$	1.483(16)	41
$K_6Mo_7O_{22}(O_2)_2 \cdot 8H_2O$	1.38(6)	42
$K_2MoO(O_2)_2(C_2O_4)$	1.457(2)	43
$MoO(O_2)\{OP(NMe_2)_3\}(H_2O)$	1.496(8)	44
$MoO(O_2)\{OP(NMe_2)_3\}(py)$	1.441(2)	44
$K_2MoOF_4(O_2) \cdot H_2O$	1.44(3)	14
$(pyH)_2O[MoO(O_2)_2(H_2O)]_2$	1.484(1)	45
$(pyH)_2[MoO(O_2)_2(H_2O)]_2$	1.470(1)	45
$MoO(O_2)(dipic)(H_2O)$	1.447(8)	46
$[MoO(O_2)_2(pic)_2]^-$	1.465(3)	46
$[Et_4N][F\{MoO(O_2)(dipic)\}_2]$	1.43(1)	47
$MoO(O_2)(N\text{-PhBz hydroxamato})_2$	1.21	48
$Mo(O_2)_2O\{(S)\text{-MeCH(OH)CONMe}_2\}$	1.455(7)	49
$[NH_4][MoO(O_2)F(dipic)]$	1.46(1)	50
$(TPP)Mo(O_2)_2$	1.40	51
$[WO(O_2)F_4]^{2-}$	1.20	33
$K_2W_2O_{11} \cdot 4H_2O$	1.52(3)	52
$[Co(R,R\text{-}C_{24}H_{38}As_4)(O_2)]^+$	1.424(1)	53
$Co(2\text{-phos})_2(O_2)BF_4 \cdot 2C_6H_6$	1.420(1)	54
$Co_2(O_2)(CN)_4(PMe_2Ph)_5 \cdot \frac{1}{2}C_6H_6$	1.441(1)	55
$RhCl(O_2)(PPh_3)_2 \cdot 2CH_2Cl_2$	1.414	56
$[Rh(PMe_2Ph)_4(O_2)][BPh_4]$	1.43(1)	57
$RhCl(O_2)(PPr^i_3)_2$	1.03(1)[b]	58
$[Rh(dppe)_2(O_2)][PF_6]$	1.418(1)	59
$[RhCl(O_2)(PPh_3)_2]_2 \cdot 2CH_2Cl_2$	1.44	60
$[Rh(O_2)(AsPhMe_2)_4][ClO_4]$	1.46(2)	61
$[Rh(O_2)(dppa)_2][PF_6]$	1.419(1)	62
$[RhO_2(2\text{-phos})_2]^+$	1.43	63
$[Ir(O_2)(dppe)_2][PF_6]$	1.52(1)	64
$[Ir(O_2)(dppm)_2][ClO_4]$	1.486(1)	65
$[Ir(O_2)(dppm)_2][PF_6]$	1.453(2)	65
$[Ir(O_2)(PPhMe_2)_4][BPh_4]$	1.485(2)	66
$IrCl(CO)(O_2)(PPh_3)_2$	1.30(3)	67
$IrCl(CO)(O_2)(PPh_2Et)_2$	1.469(1)	68
$IrI(CO)(O_2)(PPh_3)_2$	1.509(3)	69
$[Ir(O_2)(2\text{-phos})_2]^+$	1.38	63
$Ni(O_2)(CNBu^t)_2$	1.45(1)	19
$Pd(O_2)(PPhBu^t_2)_2 \cdot PhMe$	1.37(2)	70
$Pt(O_2)(PPh_3)_2 \cdot C_6H_6$	1.45(4)	71
$Pt(O_2)(PPh_3)_2 \cdot PhMe$	1.26	72
$Pt(O_2)(PPhBu^t_2)_2 \cdot PhMe$	1.43(2)	70
$Pt(O_2)(PPh_3)_2 \cdot 2CHCl_3$	1.505(2)	20

[a] Orthorhombic and triclinic forms respectively. [b] Severally disordered.

calculated the isotropic splitting and force constants for a number of $^{16}O_2$, $^{18}O_2$ and ^{16}O—^{18}O substituted dioxygen complexes of this type. They found that the frequency of the so-called O—O stretching mode (A_1 mode) was increased by an increase in the M—O stretching frequency (A_1 mode). However, an increase in O—O force constant will, in general, lead to a reduction in the M—O force constant and *vice versa*. Consequently the observed O—O stretch is expected to remain constant for peroxo complexes. Thus, IR data, whilst being consistent with the peroxo assignment of these complexes, are not expected to be a reliable guide to the extent of electron transfer. IR data for mononuclear peroxo complexes are summarized in Table 3.

A rather interesting mononuclear peroxo complex is formed by reaction of a Co^{II} complex with dioxygen. Normally Co^{II} complexes would be expected to form either superoxo complexes or peroxo-bridged dimers. However, on reacting $Co(CN)_2(PPh_2Me)_3$ with dioxygen, a compound having structure (2) is formed.[55] The O—O distance is 1.44 Å, consistent with the μ-peroxo assignment. The reaction presumably proceeds *via* an intermediate Co^{III} superoxo complex which reacts with a molecule of the Co^{II} complex, not, as is usually the case, to give a peroxo bridge but rather a CN-bridged species which undergoes an inner-sphere electron transfer.

An almost unique variation on the μ-peroxo coordination mode is exhibited by a rhodium dimer, $[RhCl(O_2)(PPh_3)_2]_2$,[60] which is composed of two identical subunits which have the dioxygen moiety coordinated in the μ-peroxo mode. These subunits are linked, not by a chlorine bridge as in other rhodium complexes such as $[RhCl(CO)_2]_2$, but *via* the coordinated dioxygen group as shown in (3).

Table 3 IR and Raman Data on Mononuclear Peroxo Complexes

Complex	ν_{O-O} (cm^{-1})	Ref.
(OEP)Ti(O$_2$)	895	13
(TPP)Ti(O$_2$)	895	13
(MPOEP)Ti(O$_2$)	855	13
CrO(O$_2$)$_2$py	880	37
K$_3$CrO$_8$	875	75
Cr(O$_2$)$_2$(NH$_3$)$_3$	865	38
[Co(2-phos)(O$_2$)][BF$_4$]·2C$_6$H$_6$	909	54
Co$_2$(O$_2$)(CN)$_4$(PMe$_2$Ph)$_5$·½C$_6$H$_6$	881	55
RuCl(NO)(PPh$_3$)$_2$(O$_2$)	875	76
Ru(CO)(O$_2$)(*p*-MePhCN)(PPh$_3$)$_2$	835	77
Ru(CO)$_2$(O$_2$)(PPh$_3$)$_2$	849	78
Os(CO)$_2$(O$_2$)(PPh$_3$)$_2$	820	78
[Co(2-phos)(O$_2$)][BF$_4$]·2C$_6$H$_6$	909	54
Co$_2$(O$_2$)(CN)$_4$(PMe$_2$Ph)$_5$·½C$_6$H$_6$	881	55
RhCl(O$_2$)(PPr$^i_3$)$_2$	990	58
RhCl(O$_2$)(PPh$_3$)$_2$·2CH$_2$Cl$_2$	845	56
RhCl(O$_2$)(PPh{(CH$_2$)$_3$PPh$_2$}$_2$)	862	80
RhCl(O$_2$)(PPh$_3$)$_2$(MeNC)	890	81
[Rh(CO)(PPh$_3$)$_2$(O$_2$)][CF$_3$CO$_2$]	833	82
[Rh(PPhMe$_2$)$_4$(O$_2$)]$^+$	841, 870	83
[Rh(AsPhMe$_2$)$_4$(O$_2$)]$^+$	862, 867	83
IrCl(CO)(O$_2$)(PPh$_3$)$_2$	858	69
IrBr(CO)(O$_2$)(PPh$_3$)$_2$	862	69
IrI(CO)(O$_2$)(PPh$_3$)$_2$	862	69
Ir(N$_3$)(CO)(O$_2$)(PPh$_3$)$_2$	855	69
Ir(NO$_3$)(CO)(O$_2$)(PPh$_3$)$_2$	862	84
Ir(C≡CEt)(CO)(O$_2$)(PPh$_3$)$_2$	832	85
Ir(OTs)(CO)(O$_2$)(PPh$_3$)$_2$	835	86
Ir(SC$_6$F$_5$)(CO)(O$_2$)(PPh$_3$)$_2$	860	87
IrCl(CO)(O$_2$)(AsPh$_3$)$_2$	860	88
IrI(CO)(O$_2$)(AsPh$_3$)$_2$	850	88
IrCl(CO)(O$_2$)(PPhMe$_2$)$_2$	835, 845	89
IrBr(CO)(O$_2$)(PPhEt$_2$)$_2$	837	89
IrCl(O$_2$)(C$_2$H$_4$)(PPh$_3$)$_2$	880	90
IrCl(O$_2$)(py)(PPh$_3$)$_2$	843	90
IrCl(O$_2$)(PPh$_3$)$_3$	853	91
[Ir(dppe)$_2$(O$_2$)][PF$_6$]	845	64
[Ir(CO)(PPh$_2$Me)$_3$(O$_2$)]$^+$	840	92
[Ir(CO)(PPh$_2$Et)$_3$(O$_2$)]$^+$	845	92
[Ir(AsPhMe$_2$)$_4$(O$_2$)]$^+$	838	83
Ni(O$_2$)(CNBu^t)$_2$	898	74
Ni(O$_2$)(CNCy)$_2$	904	74
Pd(O$_2$)(CNBu^t)$_2$	893	74
Pd(O$_2$)(PPh$_3$)$_2$	880	93
Pt(O$_2$)(PPh$_3$)$_2$	830	20

(2)

(3)

Each subunit forms a bond between one of its oxygen atoms and the rhodium atom of the other subunit. The O—O bond length of 1.44 Å is in the region expected for a peroxo-type ligand. The dioxygen moiety is bound asymmetrically, reflecting the bridging character of one of the oxygen atoms (Rh—O^1 = 1.980, Rh—O^1 = 2.198 and Rh'—O^2 = 2.069 Å). The only other fully characterized example of this coordination mode is the vanadium complex $(NH_4)_4[O\{VO(O_2)_2\}_2]$.[27]

15.2.5 MONONUCLEAR SUPEROXO COMPLEXES

These complexes are formed almost exclusively by metals of the first transition series due to the requirement of a formal one-electron oxidation of the metal. Cobalt is the only metal to form this type of complex without a ligating porphyrin ring. Simple complexes of other first row metals either do not react with dioxygen, or form oxo or similar species involving O—O bond cleavage. Nearly all Co^{II} species react with dioxygen to form mononuclear superoxo complexes. In many cases further reaction of the superoxo complex leads to the formation of a binuclear peroxo-bridged species. There is a fine balance between mononuclear superoxo species and binuclear peroxo-bridged species. To isolate the mononuclear superoxo adduct the reaction must be stopped before reaction with a second metal centre can occur. This may be achieved by working at low temperatures and high dilutions and also by isolating the dioxygen adduct rapidly!

The earliest known dioxygen complex contained $[(NH_3)_{10}Co_2O_2]^{4+}$, first described by Werner and Myelius in 1893.[94] A surge of interest in dioxygen complexes arose after the Co^{II} chelate of bis(salicylaldehyde)ethylenediimine was shown to take up oxygen reversibly by Tsumaki in 1938.[95] It is only recently, however, that the existence of mononuclear superoxo complexes has been established and structural studies performed upon them. A vast array of Co^{II} chelates based on Schiff bases' such as (4) and (5) have been synthesized.

(4)

(5)

A rather oversimplified view of the reaction between dioxygen and Co^{II} complexes is that the Co^{II} complex is oxidized to a Co^{III} complex (low-spin) while the dioxygen is reduced to give a superoxide ligand. This results in a complex with $S = \frac{1}{2}$. The ESR spectra of such complexes have been interpreted as supportive of a superoxo ligand, showing an ESR signal with $g > 2$ and weak coupling to the cobalt nucleus (^{59}Co, $I = \frac{7}{2}$). Many ESR data have been accumulated on cobalt–dioxygen 1:1 adducts and in general $> 90\%$ of the unpaired spin lies upon the dioxygen moiety though it does not mean that $> 90\%$ electron transfer from cobalt to dioxygen has occurred. Drago et al.[96] have shown that in order to make an estimate of the extent of electron transfer from cobalt to dioxygen a detailed analysis of the cobalt hyperfine interactions is necessary. Dori et al.[97] have measured the ^{17}O anisotropic hyperfine coupling constants for a Co—O$_2$ adduct. This was interpreted by Drago as showing that the unpaired electron resides in a molecular orbital consisting mainly of oxygen p orbitals. Consequently the molecular orbital scheme shown in Figure 4 was proposed. One π^* orbital of the dioxygen moiety may overlap with the cobalt d_{z^2} orbital to form a molecular orbital (ψ_1), which will be occupied by two electrons as indicated. The other dioxygen π^* orbital (ψ_2) is orthogonal to the first π^* orbital and contains the unpaired electron. This description means that the unpaired electron will always reside upon the dioxygen moiety whatever the extent of electron transfer. The two coefficients α and β determine the extent of electron transfer: when $\alpha = 0$ we have $Co^{III}O_2^-$; when $\beta = 0$ we have $Co^IO_2^+$ and when $\alpha = \beta$, we have $Co^{II}O_2$. It is worthy of note here that if $\alpha = 0$ then there would be no cobalt hyperfine anisotropy. The existence of cobalt hyperfine anisotropy means $\alpha \neq 0$ and so the limiting description of $Co^{III}O_2^-$ is not achieved. The results of this semiquantitative analysis are expressed as extent of electron transfer calculated as $2(1 - \alpha'^2) - 1$ (α' is used rather than α as an allowance has been made for a contribution to ψ_1 from the cobalt $4s$ orbital). The extent of electron transfer varies from 0.1 to 0.8 reflecting an increase in ligand field strength about cobalt. These values represent an upper limit for electron transfer.

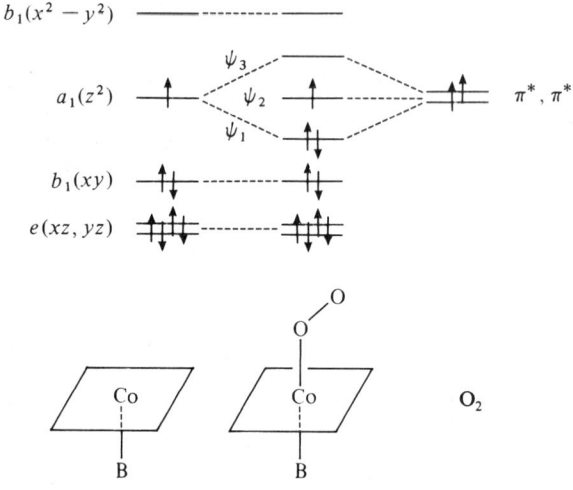

Figure 4

Although the assignment of the dioxygen moiety in these complexes as a superoxide ligand is not strictly correct, it does provide a useful basis for the interpretation of structural data. X-Ray studies reveal a range of O—O bond lengths from 1.24 to 1.35 Å and an MOO angle near to 120°. Alkali metal superoxides show an O—O bond length of 1.32–1.35 Å. Disorder of the dioxygen moiety may be a problem with a possibility of two-fold disorder occurring within each of two possible dioxygen orientations. This may lead to a large degree of uncertainty in both O—O bond length and bond angle. The relevant structural information for mononuclear superoxo complexes, including metal porphyrin systems, is summarized in Table 4.

IR studies on mononuclear superoxo complexes reveal a range of 1075–1220 cm^{-1} for O—O stretching (Table 5). This correlates well with ν_{O-O} for KO_2 and NaO_2 which are 1145 and 1142 cm^{-1} respectively.

Table 4 Mononuclear Superoxo Complexes

Complex	O—O (Å)	M—O—O (°)	Ref.
$Co(O_2)(3\text{-}Bu^t\text{-saltmen})$	1.257 (1)[a]	117.5 (6)	98
$(Et_4N)_3Co(O_2)(CN)_5 \cdot 5H_2O$	1.240 (2)	153.4 (2)	99
$Co(O_2)(saltmen)(1\text{-}BzIm) \cdot THF$	1.277 (3)	120.0	100
$Co(O_2)(2\text{-}(2'\text{-py})Etsalen) \cdot MeCN$	1.06 (3)[a]	136	101
$Co(O_2)(3\text{-}MeOsaltmen)(H_2O) \cdot (MeOCH_2)_2$	1.282 (2)[c]	117	102
$Co(O_2)(py)(3\text{-}Bu^t salen)$	1.350 (1)	116.4 (5)	103
$Co(O_2)(3\text{-}Prsalmen) \cdot C_6H_6$	1.06 (5)	137 (4), 133 (4)[c]	104
$Co(O_2)(bzacen)(py)$	1.26 (4)	126 (2)	105
$Co(O_2)(3\text{-}F\text{-saltmen})(1\text{-}MeIm) \cdot 2Me_2CO$	1.302 (3)	117.4	106
$Fe(O_2)(1\text{-}MeIm)(TpivPP)$	1.17 (4)	129 (2)	107
$Fe(O_2)(THT)(TpivPP) \cdot 2THT$	1.23 (8)	135 (4)[d]	108
	1.26 (8)	137 (4)	108
$Fe(O_2)(2\text{-}MeIm)(TpivPP) \cdot EtOH$	1.205 (2)	129.0 (1)[d]	109
	1.232 (2)	128.5 (2)	109
$Fe(O_2)(1\text{-}MeIm)(TpivPP)$	1.23 (8)	135 (4)[d]	110
	1.26 (8)	137 (4)	110
Oxymyoglobin	1.4	121[e]	111

[a] Disordered. [b] At -152 °C. [c] Two-fold disorder. [d] Two disordered sites. [e] $R = 20\%$.

Table 5 IR Data for Mononuclear Superoxo Complexes

Complex	ν_{O-O} (cm^{-1})	Ref.
$NbBr_2(TPP)(O_2)$	1220	117
$Cr(TPP)(py)(O_2)$	1142	118
$Fe(TpivPP)(O_2)(1\text{-MeIm})$	1159	116
$Co(bzacen)(O_2)(py)$	1128	112
$Co(acacen)L(O_2)$	1140	112
$Co(mesoalen)L(O_2)$	1128	112
$Co(3\text{-MeOsalen})(py)(O_2)$	1140	113
$Co(cyen)(py)(O_2)$	1137 (R)	114
	1078 ($^{18}O_2$) (R)	114
$Co(cyen)(MeIm)(O_2)$	1337 (R)	114
$[Co(CN)_5(O_2)]^{3-}$	1138	115
$Co(TpivPP)(NTrIm)(O_2)$	1153	116
$Co(TpivPP)(MeIm)(O_2)$	1150	116
$Co(O_2)(2\text{-}(2'\text{-py})Etsalen)\cdot MeCN$	1135	101

R = Raman.

15.2.6 BINUCLEAR PEROXO COMPLEXES

This type of complex is derived from the mononuclear superoxo species *via* a further one-electron reduction of the dioxygen moiety. Cobalt is the only metal to form these complexes by reaction with dioxygen in the absence of a ligating porphyrin ring. Molybdenum and zirconium form peroxo-bridged complexes on reaction with hydrogen peroxide. In most cases the mononuclear dioxygen adducts of cobalt will react further to form the binuclear species unless specific steps are taken to prevent this.

The first binuclear peroxo complex, reported by Werner and Myelius in 1893,[94] contained $[Co_2(NH_3)_{10}(O_2)]^{4+}$. It was not, however, until 1967 that the structure of this cation was determined[119] establishing a peroxo-bridged species. The O—O bond length was found to be 1.47 Å, very close to the 1.49 Å found for the O—O distance in Na_2O_2. The Co—OO—Co grouping would be expected to be non-planar in the case of a peroxo bridge and planar for a superoxo bridge. In this case a dihedral angle of 146° was found. Subsequent crystal structures, however, have shown that the non-planarity of the MOOM grouping is not diagnostic and some peroxo-bridged compounds are found to be planar in this respect.

There are two other types of peroxo-bridged complex which have been characterized. Firstly, a molybdenum complex containing $[Mo_4O_{12}(O_2)_2]^{4-}$[120] in which the peroxo groups bridge not two but four molybdenum atoms. The O—O bond length is 1.48 Å and so is consistent with the peroxo assignment. Secondly, both a molybdenum complex and a cobalt one exhibit a novel coordination mode (6) for a peroxo group bridging between two metal centres. In both of these complexes the peroxo group bridges through one oxygen atom only. The cobalt complex $[Co_2(HO_2)(NH_2)(en)_4]$-$(NO_3)_4\cdot2H_2O$[121] has an O—O bond length of 1.42 Å whilst the molybdenum complex $[MoO(O_2)_2(O_2H)](pyH)_2$[45] has an O—O bond length of 1.458(10) Å. The structural information on binuclear peroxo complexes is summarized in Table 6. Very few IR data have been obtained for binuclear peroxo complexes. However, what little data exist (Table 7) show a range of 790–830 cm^{-1} for ν_{O-O}. This is consistent with the peroxo formulation.

(6)

15.2.7 BINUCLEAR SUPEROXO COMPLEXES

Although there are no binuclear superoxo complexes formed simply by the interaction of a metal complex and dioxygen, they are worthy of discussion here if only to complement mononuclear

Oxygen Ligands

Table 6 Binuclear Peroxo Complexes

Complex	O—O (Å)	CoOOCo dihedral (°)	Ref.
$(pyH)_2\{MoO(O_2)_2OOH\}_2$	1.458(1)[a]		45
$[Mo_4O_{12}(O_2)_2]^{4-}$	1.48[b]		120
$K_8Co_2(O_2)(CN)_{10}(NO_3)_2 \cdot 4H_2O$	1.447(4)	180	122
$Co_2(O_2)(NH_3)_{10}(SCN)_4$	1.469(6)	180	123
$Co_2(O_2)(NH_3)_{10}(SO_4)_2 \cdot 4H_2O$	1.473(1)	146	119
$Co_2(O_2)(tren)_2(NH_3)_2(SCN)_4 \cdot 2H_2O$	1.511(9)	180	124
$Co_2(papd)_2(O_2)(S_2O_6)(NO_3)_2 \cdot 4H_2O$	1.486(7)	180	125
$Co_2(O_2)(en)_2(dien)_2(ClO_4)_4$	1.488	180	126
$Co_2(O_2)(NO_2)_2(en)_4(NO_3)_2 \cdot 4H_2O$	1.529(9)	180	127
$Co_2(O_2)(OH)(tren)_2(ClO_4)_3 \cdot 3H_2O$	1.462(3)	60.7	128
$Co_2(O_2)(OH)(en)_4(S_2O_6)(NO_3)_2 \cdot 2H_2O$	1.465	60.7	129
$Co_2(O_2)(tren)_3(ClO_4)_4 \cdot 2H_2O$	1.485(2)	19.8	130
$Co_2(O_2)(NH_2)(en)_4(SCN)_3 \cdot H_2O$	1.46	62.5	131
$Co_2(O_2)(OH)(en)_4(ClO_4)_3$	1.460(1)	64.5(5)	132
$Co_2(O_2)(NH_2)(en)_4(NO_3)_3(AgNO_3)_2 \cdot H_2O$	1.43(3)	61[c]	133
$Co_2(O_2)(salen)_2 \cdot DMF$	1.339(6)	110.1	134
$Co_2(O_2)(salen)_2(pip)_2 \cdot \frac{2}{3}Me_2CO\frac{1}{3}C_5H_{11}N$	1.383(7)	121.9(4)	135
$Co_2(O_2)(salptr)_2 \cdot PhMe$	1.45(2)	149	136
$Co_2(O_2)(1,11-py_2-2,6,10-triaz)I_4 \cdot 3H_2O$	1.456(9)	162	137
$Co_2(O_2)(1,9-py_2-2,5,8-triazanon)I_4$	1.489(8)		138
$Co_2(O_2)(OH)(4,7-Me_2-1,4,7,10-tetraz)(ClO_4)_3 \cdot 2H_2O$	1.429(2)	68	139
$Co_2(en)_4(HO_2)(NH_2)(NO_3)_4 \cdot 2H_2O$	1.42	a	121
$Pt_2(OH)(O_2)(PPh_3)_3^+$	1.547(2)	79	140

[a] Peroxo group bridges through one oxygen only.
[b] Peroxo group coordinated to four Mo atoms.
[c] Ag^+ Coordinated to peroxo bridge.

Table 7 IR Data on Binuclear Peroxo Complexes

Complex	ν_{O-O} (cm^{-1})	Ref.
$[Co_2(O_2)(NH_3)_{10}]^{4+}$ solid	800	141
solution	808	141
$Co_2(O_2)(His)_4$	805	141
$Co_2(O_2)(OH)(His)_4$	790	141
$[Co_2(O_2)(NH_2)(NH_3)_8]^{3+}$	793	141

superoxo complexes and binuclear peroxo complexes. The coordination geometry of binuclear superoxo-bridged complexes is similar to that for the peroxo analogus though a planar configuration for the MOOM grouping is expected. This may be rationalized, on a very simple level, in terms of π-bonding. The oxygen atoms of the dioxygen bridge may be considered to be sp^2 hybridized. A σ-bond may be formed between each cobalt–oxygen pair and between the two oxygen atoms. A lone pair is also localized on each oxygen atom. This accomodates a total of 10 electrons. For a peroxo bridge there will be a total of 14 electrons, one from each cobalt and six from each oxygen. A superoxo bridge will involve 13 electrons as it is the result of oxidation of the peroxo bridge. The remaining three electrons involved in the superoxo bridge may be accommodated in the two unhybridized oxygen p orbitals. However, these p orbitals have the correct symmetry to overlap with filled cobalt d_{xz} orbitals thus forming a π bond between the four atoms. This results in seven electrons, four of these arising from the two cobalt d_{xz} orbitals, filling both bonding and antibonding levels, giving a net bonding interaction. Consequently the Co—OO—Co bridge would be expected to be planar to maximize this bonding interaction. In the case of peroxo-bridged complexes there is one more electron to accommodate. This should go into a π* orbital, thus making the overall effect of the Co—OO—Co interaction non-bonding. The Co—OO—Co bridge would then be expected to lose its planarity to minimize non-bonding interactions, or, to return to the original assumption about oxygen orbital hybridization, the oxygen atoms in a peroxo bridge may be considered to be sp^3 hybridized.

X-Ray studies on superoxo-bridged complexes give an O—O bond length between 1.26 and 1.36 Å (Table 8). Although a planar configuration is predicted for the CoOOCo bridge, it would appear that the potential energy surface for the CoOOCo dihedral angle is quite 'flexible' and crystal packing effects may dominate. ESR studies on superoxo-bridged complexes show that the unpaired electron resides predominantly upon the oxygen atoms. However, weak coupling to two equivalent cobalt nuclei is also observed indicating that the bridge is symmetrical. IR and Raman studies on cobalt superoxo-bridged complexes show a range of O—O stretching from 1068 to 1122 cm^{-1} (Table 9).

Table 8 Binuclear Superoxo Complexes

Complex	O—O (Å)	CoOOCo dihedral (°)	Ref.
$Co_2(O_2)(NH_3)_{10}(NO_3)_5$	1.317(15)	180	142
$Co_2(O_2)(NH_2)(NH_3)_8(NO_3)_4$	1.320(5)	a.p.	143
$Co_2(O_2)(NH_2)(en)_4(NO_3)_4 \cdot H_2O$	1.353(11)	23.4	144, 145
$Co_2(O_2)(OH)(en)_4(NO_3)_4 \cdot H_2O$	1.339	22.0	146
$[Co_2(O_2)(3\text{-F-salen})_2(H_2O)]_2 \cdot 2CHCl_3 \cdot pip$	1.308(28)	122	147
$K_5Co_2(O_2)(CN)_{10} \cdot H_2O$	1.26	166	148
$Co_2(O_2)(NH_3)_{10}(SO_4)(HSO_4)_3 \cdot 3H_2O$	1.315(20)	a.p.	149
$Co_2(O_2)(NH_3)_{10}(NO_3)_2Cl_3 \cdot 2H_2O$	1.290(2)		150

a.p. = almost planar

Table 9 IR Data for Binuclear Peroxo Complexes

Complex	O—O (cm^{-1})	Ref.
$Co_2(O_2)(His)_4$	1120	141
$Co_2(O_2)(NH_3)_{10}Cl_5 \cdot 4H_2O$	1122 (R)	152
$[Co_2(O_2)(CN)_{10}]^{5-}$	1104	152
$Co_2(O_2)(NH_3)_{10}(NO_3)_5$	1122 (R)	152
$Co_2(O_2)(NH_3)_{10}(SO_4)_2(HSO_4) \cdot 3H_2O$	1110 (R)	152
$Co_2(O_2)(NH_2)(NH_3)_8(NO_3)_4$	1078, 1068 (R)	152

R = Raman

15.2.8 METAL–PORPHYRIN COMPLEXES

An iron porphyrin is the prosthetic group in the oxygen transport and storage proteins, hemoglobin and myoglobin. Consequently there has been much interest in porphyrin complexes, especially of first row transition metals, as model systems for oxygen transport and storage. Much interest has also been shown in metal porphyrins as models for oxidases, in particular cytochrome P-450.

Dioxygen binds to metal porphyrins in the three expected modes, *i.e.* μ-superoxo, peroxo and bridging peroxo. In contrast to the 'simple' complexes discussed previously, dioxygen coordination occurs with a wide range of transition metals from titanium and niobium through to the Group VIII metals.

Titanium(III) octaethylporphyrin (OEP) has been found to react with dioxygen at low temperatures to give a titanium(IV) μ-peroxo species. Characterization was based on a comparison with the reaction product from H_2O_2 oxidation of (OEP)TiO. This compound was shown to have a μ-peroxo group by a single crystal X-ray study, O—O = 1.445 Å, $\nu_{O-O} = 895$ cm^{-1}.[13]

No dioxygen adducts of vanadium porphyrins have been reported. However, a niobium(IV) porphyrin, $NbBr_2TPP$, has recently been reported to react with dioxygen. Characterization was based upon ESR data. A 10-line signal ($g = 2.002$), due to weak coupling to the ^{93}Nb nucleus ($I = \frac{9}{2}$), was seen. This implied a superoxo-type complex. IR data are consistent with the superoxo formalism showing $\nu_{O-O} = 1220$ cm^{-1}.[117]

A chromium(II) porphyrin $Cr(TPP)(py)$ was found to take up oxygen in the solid state to give a superoxo-type complex. Characterization was based upon IR evidence ($\nu_{O-O} = 1142$ cm^{-1}) and magnetic susceptibility measurements ($\mu = 2.7$ BM), which indicated two unpaired electrons. This suggested that the third t_{2g} electron is coupled with the unpaired electron on the dioxygen moiety in a manner similar to that proposed to account for the diamagnetism of oxyhemoglobin.[118] No dioxygen complexes of molybdenum formed by reaction with dioxygen have been reported. However, a bis-μ-peroxo molybdenum porphyrin has been formed by H_2O_2 oxidation of $Mo(TPP)(O)OH$[51] in a manner analogous to the titanium porphyrin described above. An X-ray study revealed two perpendicular μ-peroxo moieties, one each side of the porphyrin ring, O—O = 1.40 Å. The existence of this type of compound indicates the possibility of forming a molybdenum dioxygen adduct.

Manganese(II) porphyrins react with dioxygen giving Mn^{IV} μ-peroxo species as judged by an ESR study of the ^{17}O-substituted dioxygen adducts which showed three manganese-based unpaired electrons and, therefore, a diamagnetic oxygen moiety.[16] This is in disagreement with theoretical calculations which predict two unpaired electrons on manganese and one on the bound dioxygen. No X-ray data are available to clarify the mode of dioxygen binding.

The reaction of iron(II) porphyrins is an area of extreme interest and importance and will be discussed separately. The second row analogues of Fe^{II} porphyrins, Ru^{II} porphyrins, have been found to bind dioxygen reversibly[153] but only UV data are available on these dioxygen adducts.

Cobalt(II) porphyrins bind dioxygen as would be expected by analogy with the wealth of cobalt(II) complexes which display this property. The initial addition product is invariably a superoxo-type species, confirmed by ESR, IR and X-ray studies.[154] Subsequent reaction of the oxygenated complex with more Co^{II} porphyrin leads to a peroxo-bridged dimer. No X-ray data are available for cobalt porphyrin peroxo-bridged dimers but the formation of such dimers is well established in cobalt chemistry.

Rhodium, like cobalt, forms a superoxo-adduct with dioxygen. On reaction of $[Rh(OEP)]_2$ in toluene at $-80\,^{\circ}C$ with dioxygen, a complex with $\nu_{O-O} = 1075\,cm^{-1}$ was formed.[155] The ESR spectrum showed a signal due to the unpaired electron localized on the dioxygen moiety and weakly coupled to the ^{103}Rh nucleus ($I = \frac{1}{2}$). On warming, a second complex, formulated as a rhodium(III) peroxo-bridged dimer, appeared. The IR and NMR data are in accordance with this formulation. A complex, previously formulated as Rh(TPP), is now thought to be $Rh(TPP)(O_2)$. The ESR spectrum was consistent with a superoxo complex rather than a rhodium(II) complex. On heating to $150\,^{\circ}C$ under vacuum, the complex became diamagnetic. The diamagnetic product regenerates the starting material on reaction with dioxygen.[153] No X-ray data are available on any of these rhodium porphyrins.

Iron porphyrins, as previously discussed, lie at the centre of the work on oxygen binding by metalloporphyrins. If iron(II) tetraphenylporphyrin or protoporphyrin IX is allowed to react with dioxygen at room temperature, in the absence of a large excess of a strongly coordinating ligand, it is irreversibly oxidized to an iron(III) μ-oxo-bridged dimer.[156] This is obviously not the case for hemoglobin or myoglobin where 1:1 iron:dioxygen complexes are formed with each heme group.

There are three different approaches which can be employed to prevent irreversible oxidation of iron(II) porphyrins and promote dioxygen adduct formation: (1) low temperatures to reduce the rate of dimerization; (2) steric hindrance to prevent dimerization; and (3) immobilization of the heme to prevent dimerization.

(1) *Low temperatures.* The porphyrin in Figure 5 was synthesized by Chang and Traylor in 1973.[157] The imidazole side chain resulted in the iron being pentacoordinate. It bound O_2 reversibly in the solid state and, when dissolved in a polystyrene film, took up oxygen in CH_2Cl_2 solution at $-45\,^{\circ}C$. However, on warming to room temperature an irreversibly oxidized μ-oxo dimer was formed. The reversibility of O_2 binding was attributed to the low temperature employed and the covalently attached imidazole ligand.

Further work on porphyrins of the type $Fe(porphyrin)L_2$ showed that these complexes also bind dioxygen reversibly at low tempertures.[158,159,160] This was due to the kinetic stabilization of the dioxygen adduct at reduced temperatures. The presence of a covalently bound axial ligand was not important in stabilizing the dioxygen adduct relative to the μ-oxo dimer.

(2) *Steric hindrance.* An obvious way of preventing μ-oxo dimer formation, *i.e.* irreversible oxidation of iron(II) porphyrins, is to introduce bulky substituents at the porphyrin such that the resultant dioxygen adducts are unable to approach each other closely enough for dimer formation to occur.

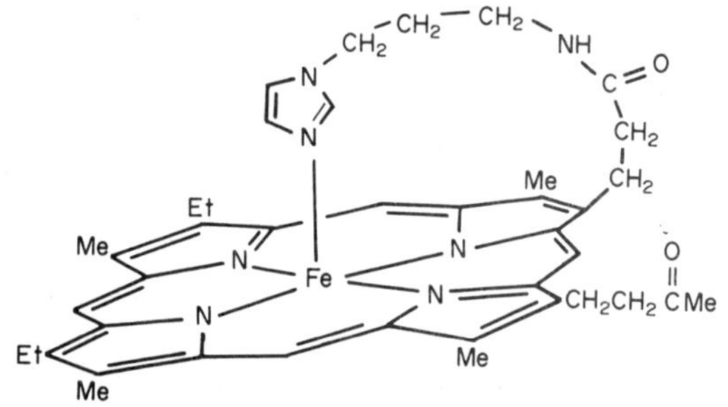

Figure 5

The first example of a sterically hindered porphyrin which bound oxygen reversibly at room temperature was Collman's 'picket fence' porphyrin, Fe(TpivPP),[161] illustrated in Figure 6. The dioxygen adduct formed in the presence of 1-methylimidazole was characterized by Mössbauer[162] and IR spectroscopy ($\nu_{O-O} = 1159\,cm^{-1}$),[116] which is consistent with an Fe^{3+}—O_2^- formalism. An X-ray study has also been performed on the dioxygen adduct but the crystal structure showed a high degree of disorder and a rather 'unrealistic' O—O distance of 1.15 Å was found.[107] The Mössbauer study revealed quadrupole splittings and an isomer shift similar to oxyhemoglobin. Magnetic susceptibility measurements are the subject of controversy. Measurements by Cerdonio show a thermal equilibrium between a singlet ground state and a triplet excited state with an energy separation J of $146\,cm^{-1}$.[163]

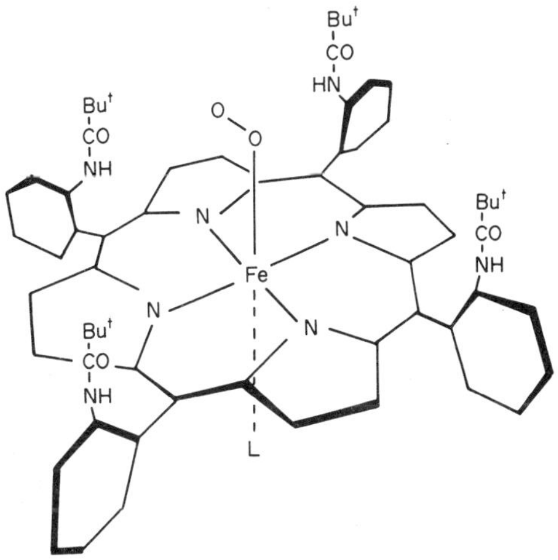

Figure 6

Baldwin has synthesized the 'capped' iron(II) porphyrin shown in Figure 7. An axial ligand can bind to the uncapped side of the porphyrin while dioxygen binds to the iron in the 'pocket' between the cap and the porphyrin. The cap prevents dimer formation.[164] It is interesting to note that, although this iron(II) porphyrin bound oxygen reversibly at 25°C when the solvent was pyridine or 5% 1-methylimidazole in benzene, irreversible autoxidation occurred when the solvent was changed to 5% pyridine in benzene. This was presumably due to oxygen coordinating to the open face of the porphyrin in the presence of pyridine which is a poorer ligand than 1-methylimidazole. Consequently irreversible oxidation to a μ-oxo dimer was able to take place.

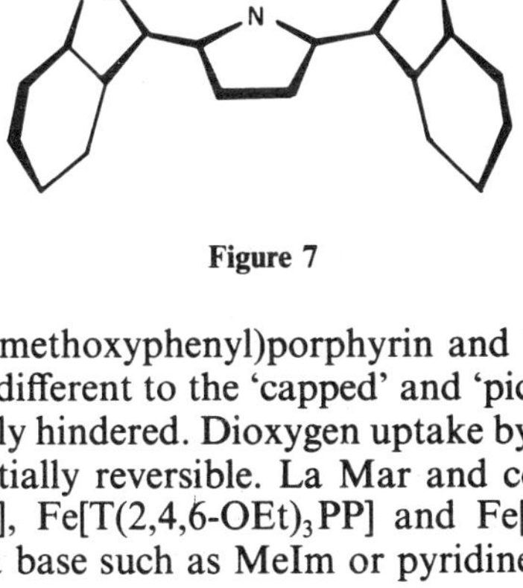

Figure 7

Iron(II) meso-tetrakis(2,4,6-trimethoxyphenyl)porphyrin and the triethoxy analogue have been synthesized by Vaska. These are different to the 'capped' and 'picket fence' porphyrins in that *both* sides of the porphyrin are sterically hindered. Dioxygen uptake by these iron(II) porphyrins at 25°C in solution was found to be partially reversible. La Mar and co-workers[166] have shown that the porphyrins Fe[T(2,4,6-OMe)₃PP], Fe[T(2,4,6-OEt)₃PP] and Fe[TpivPP] take up oxygen at low temperatures, in the absence of a base such as MeIm or pyridine, to give dioxygen adducts of the type Fe(porph)(O$_2$). On warming, these adducts can then be seen to form peroxo-bridged dimers, Fe(porph)O$_2$Fe(porph). Further warming results in the formation of the more familiar μ-oxo dimers [Fe(porph)]$_2$O. This is the first report of the postulated peroxo dimers being observed.

(3) *Immobilization.* The immobilization of iron(II) porphyrins is based upon the same principle as steric hindrance. If the porphyrin is attached to a rigid support at sufficiently high dilution then the dioxygen adduct will be unable to react with a second iron(II) porphyrin. Thus dimer formation and irreversible oxidation cannot take place. The first successful report of reversible oxygen binding by an immobilized iron(II) porphyrin was by Wang.[167] 1-(2-Phenylethyl)imidazole was incorporated into an amorphous polystyrene matrix. The diethyl ester of iron(II) protoporphyrin IX was then bound to the matrix, *via* the immobilized imidazole, to give a five-coordinate iron(II) porphyrin in a hydrophobic environment. The heme environment was considered to be rather similar to that for hemoglobin. A rather idealized structure is shown in Figure 8. This immobilized iron(II) porphyrin was found to bind oxygen reversibly. More recently it has been reported that cross-linked polystyrene containing imidazole is not rigid enough to prevent irreversible oxygenation of Fe(TPP) when treated with oxygenated benzene.[168] However, the possibility that the Fe(TPP) was bleached off the support prior to oxidation was not discussed. Immobilization of Fe(TPP) upon a modified silica gel support resulted in reversible dioxygen binding.[169]

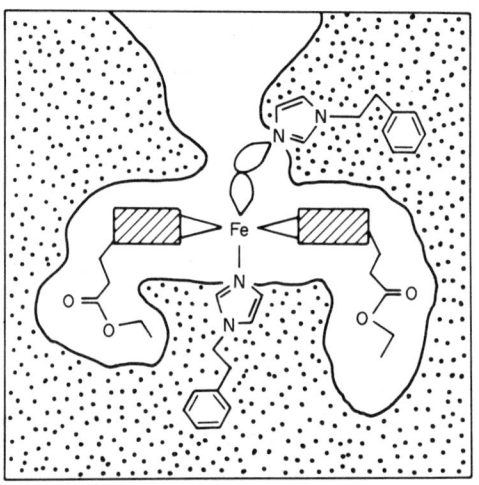

Figure 8

15.2.9 REACTIONS OF METAL–DIOXYGEN COMPLEXES

15.2.9.1 Mononuclear Peroxo Complexes

These complexes, more than any other type of dioxygen complex, have been studied with a view to their potential as oxidation catalysts, particularly in organic chemistry.[170] The dioxygen moiety in these complexes reacts readily with polar bonds.

Reaction of Pt(O$_2$)(PPh$_3$)$_2$ with the optically active substrate PhEtCHBr gives PtBr(O$_2$CHEtPh)(PPh$_3$)$_2$ with inversion of configuration.[171] This is supportive of a mechanism involving nucleophilic attack by the dioxygen moiety. The formation of an organic peroxide on reacting an active halide with a mononuclear peroxo complex is of general applicability. Nucleophilic attack on unsaturated polar bonds results in the initial formation of a coordinated cyclic peroxide. For example, the reaction of Pt(O$_2$)(PPh$_3$)$_2$ with acetone gave complex (7).[173] The crystal structure of this complex has been determined. It is interesting to note that when hexafluoroacetone was used instead of acetone, then a diadduct (8) was formed.[174] On reaction with PPh$_3$ this diadduct gave complex (9).[174] Iridium–dioxygen complexes appear to be less reactive than their platinum counterparts. For example, Pt(O$_2$)(PPh$_3$)$_2$ reacts with Bu^{n}Li to give Pt(PPh$_3$)$_2$(Bu^n)$_2$, while IrI(O$_2$)-(CO)(PPh$_3$)$_2$ does not react.[175]

Mononuclear peroxo complexes, such as MoO(O$_2$)$_2$(HMPT), show a more electrophilic character in their reactions. This particular complex may be used for the epoxidation of alkenes.[176] Much interest has been shown in the reaction of small molecules with dioxygen complexes of this type with a view to catalytic oxidation. These reactions and their products are sumarized in Table 10. The

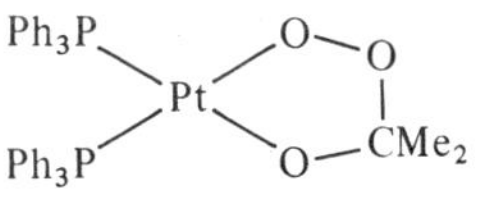

(7) 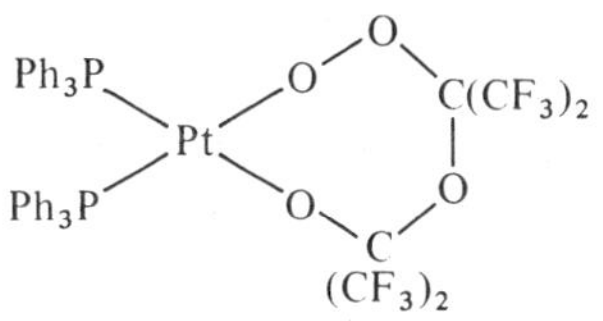   (8) 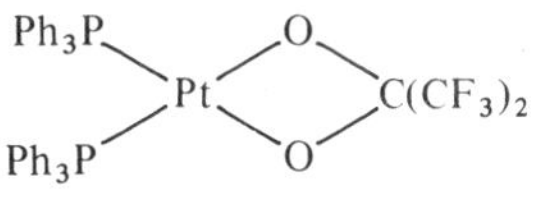   (9)

Table 10 The Reaction of Small Molecules with η^2 Metal Dioxygen Complexes[172]

Molecule	Product	Metal
SO_2	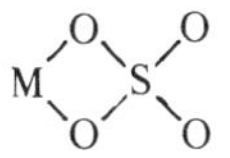	Ru, Rh, Ir Ni, Pd, Pt
CO_2	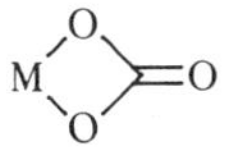	Ir, Ni, Pd, Pt
CO	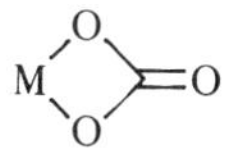	Rh, Ni, Pd, Pt
NO_2 NO	$M(NO_3)_2$ $M(NO_3)_2$	Rh, Ir, Ni, Pd, Pt Ni, Pd, Pt
NO^+	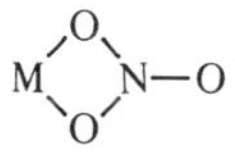	Pt
C_3O_2	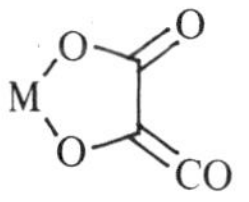	Pt

mechanism of some of these reactions has been elucidated by isotopic labelling. A five-membered cyclic peroxy intermediate has been proposed.[171]

A number of metal–dioxygen complexes, including $Ni(O_2)(RNC)_2$ and $Pt(O_2)(PPh_3)_2$, are known to undergo apparent metal-to-ligand oxygen atom transfer reactions. Halpern and Sen[177] showed that intramolecular oxygen atom transfer does not occur for $Pt(O_2)(PPh_3)_2$. The oxidation of PPh_3 proceeds *via* release of H_2O_2 after coordination of free phosphine. The peroxide thus formed oxidizes more free phosphine to the phosphine oxide. This mechanism is also likely for other apparent metal-to-ligand oxygen atom transfer reactions.

15.2.9.2 Mononuclear Superoxo Complexes

Mononuclear superoxo complexes may act as nucleophiles. The most frequently occurring reaction of this type is the formation of a binuclear complex by nucleophilic attack of the superoxo complex upon a metal cation. Nucleophilic attack can be considered to be followed by electron transfer to finally yield a peroxo bridge. Kinetic studies on the oxidation of cobalt(II) solutions show that the rate of the second step is faster for the more powerfully reducing cobalt(II) species.[178] The reduction of $[Co(cyclam)(O_2)(H_2O)]^{2+}$ by Fe^{2+} is slow enough for the intermediate CoOOFe complex to be observed using stopped flow methods.[179] A mixed metal dimer is formed on reaction of $[Co(CN)_5(O_2)]^{3-}$ with $MoCl_5$. This gives a CoOOMo species which decomposes to a mononuclear molybdenum peroxo complex. Isotopic labelling indicates that the coordinated dioxygen in the $Mo(O_2)$ complex does not arise from the CoOOMo bridge.[180] Presumably an Mo^{IV} complex is formed which reacts with dioxygen.

It has been shown that mononuclear cobalt superoxo complexes react with 2,4,6-tri-*t*-butyl-phenol. The anion $[Co(CN)_5(O_2)]^{3-}$ acts as a base in the oxidation of the phenol. The neutral complex $Co(salptr)(O_2)$ forms a peroxy adduct with 2,4,6-tri-*t*-butylphenol. An X-ray study on this adduct has revealed the structure shown in (10).[181]

A reaction specific to mononuclear cobalt superoxo complexes is free radical abstraction. Much

(10)

evidence has been produced in favour of these abstraction reactions. Both a cobalt Schiff base dioxygen complex and the dioxygen adduct of vitamin B_{12} have been shown to abstract a hydrogen atom from active hydrogen compounds such as N,N'-tetramethyl-*p*-phenylenediamine and hydroquinones.[182] Hydrogen atom abstraction is also involved in the catalytic autoxidation of phenols by cobalt Schiff base complexes bound to polymer supports.[183]

15.2.9.3 Binuclear Complexes

Few binuclear complexes are known for metals other than cobalt. Consequently the vast majority of reactions of binuclear complexes studies are those of cobalt complexes. The best established reaction of binuclear dioxygen complexes is the peroxo-superoxo interconversion. All known binuclear superoxo complexes are prepared by oxidation of the corresponding peroxo complex using strong oxidizing agents such as cerium(IV) or persulfate. The reduction of binuclear superoxo complexes is reasonably facile, proceeding via an outer-sphere mechanism, and may be effected using reducing agents such as Cr^{2+}, V^{2+}, Fe^{2+}, I^-, SO_3^{2-}, etc.[171] A notable exception is $[Co_2(O_2)(CN)_{10}]^{5-}$ which seems to be quite stable towards reduction. The reduction of superoxo complexes occurs via a simple electron-transfer mechanism. However, attempts to establish redox potentials for these species have not always taken into account effects due to protonation or dissociation.

As with other types of dioxygen complexes, the dioxygen moiety in binuclear complexes retains its nucleophilic character. The protonation of a peroxo bridge may lead to isomerization.[121] Protonation of the bridging dioxygen moiety is also thought to be involved in reactions where reduction is accompanied by O—O bond fission. For example, $[Co_2(NH_2)(O_2)(en)_4]^{3+}$ is reduced to $[Co_2(NH_2)(OH)(en)_4]^{4+}$ by iodide.[184] It has been shown that only the two protonated forms react and of these the isomerized form reacts much faster. Disproportion of this binuclear complex is catalysed by both bromide and chloride ions. The mechanism is thought to be similar to that for reduction by I^-. Di-bridged binuclear peroxo complexes give straightforward reactions with SO_2,[185] resulting in a bridging sulfato group. The reagents SeO_2[186] and NO react similarly,[187] forming a bridging selenato group and a bridging nitrito group respectively. Addition of small molecules to mono-bridged binuclear peroxo complexes may often be complicated by dissociation and the formation of mononuclear complexes.[187]

15.2.10 REFERENCES

1. M. Brunori, M. Coletta and B. Giardina, in 'Metalloproteins. Part II: Metal Proteins with Non-redox Roles', ed. P. Harrison, Macmillan, London, 1985, pp. 262–332.
2. J. C. Abrahams, *Q. Rev. Chem. Soc.,* 1956, **10**, 407.
3. J. Shamir, J. Beneboym and H. H. Classen, *J. Am. Chem. Soc.,* 1968, **90**, 6223.
4. G. Herzberg, in 'Molecular Spectra and Molecular Structure', 2nd edn., Van Nostrand, New York, 1950.
5. F. Halverston, *Phys. Chem. Solids,* 1962, **23**, 207.
6. J. A. Creighton and E. R. Lippincott, *J. Chem. Phys.,* 1964, **40**, 1779.
7. R. L. Tallman, J. L. Margrave and S. W. Bailey, *J. Am. Chem. Soc.,* 1957, **79**, 2979.
8. J. C. Evans, *Chem. Commun.,* 1969, 682.
9. R. L. Readington, W. B. Olson and P. C. Cross, *J. Chem. Phys.,* 1962, **36**, 1311.
10. F. J. Blunt, P. J. Hendra and J. R. McKenzie, *Chem. Commun.,* 1969, 278.

11. J. H. Enemark and R. D. Feltham, *Proc. Natl. Acad. Sci. USA*, 1972, **69**, 3534.
12. R. Hoffman, M. M.-L. Chen and D. L. Thorn, *Inorg. Chem.*, 1977, **16**, 503.
13. G. Guilard, M. Fontesse, P. Founari, C. Lecomte and J. Protas, *J. Chem. Soc., Chem. Commun.*, 1976, 161.
14. D. Grandjean and R. Weiss, *Bull. Soc. Chim. Fr.*, 1967, 3044.
15. R. E. Drew and F. W. B. Einstein, *Inorg. Chem.*, 1973, **12**, 829.
16. B. M. Hoffman, C. J. Weschler and F. Basolo, *J. Am. Chem. Soc.*, 1976, **98**, 5473.
17. J. P. Collman, R. R. Gagne, C. A. Reed, W. T. Robinson and G. A. Rodley, *Proc. Natl. Acad. Sci. USA*, 1974, **71**, 1326.
18. S. J. La Placa and J. A. Ibers, *J. Am. Chem. Soc.*, 1965, **87**, 2581.
19. M. Matsumoto and K. Nakatsu, *Acta Crystallogr., Sect. B*, 1975, **31**, 2711.
20. P.-T. Cheng, C. D. Cook, S. C. Nybang and K. Y. Wan, *Can. J. Chem.*, 1971, **49**, 3772.
21. O. Lumpkin, W. T. Dixon and J. Poser, *Inorg. Chem.*, 1979, **18**, 982.
22. D. Schwarzenbach, *Inorg. Chem.*, 1970, **9**, 2391.
23. D.Schwarzenbach, *Helv. Chim. Acta*, 1972, **55**, 2990.
24. D. Schwarzenbach, *Z. Kristallogr.*, 1976, **143**, 429.
25. D. Schwarzenbach and K. Gingis, *Helv. Chim. Acta*, 1975, **58**, 2391.
26. H. Mimoun, M. Postel, F. Casabianca, J. Fischerand and A. Mitschler, *Inorg. Chem.*, 1982, **21**, 1303.
27. I. B. Svensson and R. Stomberg, *Acta Chem. Scand.*, 1971, **25**, 898.
28. R. E. Drew and F. W. B. Einstein, *Inorg. Chem.*, 1972, **11**, 1079.
29. I. B. Konche-Waksman, C. Bois, J. Sala-Pala and J. E. Guerchais, *J. Organomet. Chem.*, 1980, **195**, 307.
30. G. Mathern and R. Weiss, *Acta Crystallogr., Sect. B*, 1971, **27**, 1572.
31. G. Mathern and R. Weiss, *Acta Crystallogr., Sect. B*, 1971, **27**, 1582.
32. G. Mathern and R. Weiss, *Acta Crystallogr., Sect. B*, 1971, **27**, 1598.
33. Z. Ruzic-Toros, B. Kojic-Prodic, F. Gabela and M. Syukic, *Acta Crystallogr., Sect. B*, 1977, **33**, 692.
34. R. Stomberg, *Acta Chem. Scand., Ser. A*, 1982, **36**, 101.
35. V. W. Massa and G. W. Pausewang, *Z. Anorg. Allg. Chem.*, 1979, **456**, 169.
36. J. C. Dewan, A. J. Edwards, J. Y. Calves and J. E. Guerchais, *J. Chem. Soc., Dalton Trans.*, 1977, 981.
37. R. Stomberg, *Ark. Kemi*, 1964, **22**, 29.
38. R. Stomberg, *Ark. Kemi*, 1964, **22**, 49.
39. R. Stomberg, *Ark. Kemi*, 1965, **24**, 283.
40. R. Stomberg and I. B. Ainalem, *Acta Chem. Scand.*, 1968, **22**, 1439.
41. R. Stomberg, *Acta Chem. Scand.*, 1968, **22**, 1076.
42. I. Larking and R. Stomberg, *Acta Chem. Scand.*, 1972, **26**, 3708.
43. R. Stomberg, *Acta Chem. Scand.*, 1970, **24**, 2024.
44. J.-M. Le Carpentier, R. Schlupp and R. Weiss, *Acta Crystallogr., Sect. B*, 1972, **28**, 1278.
45. J. M. Le Carpentier, A. Mitschler and R. Weiss, *Acta Crystallogr., Sect. B*, 1972, **28**, 1288.
46. S. E. Jacobson, R. Tang and F. Maves, *Inorg. Chem.*, 1978, **17**, 3055.
47. A. J. Edwards, D. R. Slim, J. E. Guerchais and J. R. Kengoat, *J. Chem. Soc., Dalton Trans.*, 1980, 289.
48. H. Tomioka, K. Takai, K. Oshima and H. Nozaki, *Tetrahedron Lett.*, 1980, **21**, 4843.
49. W. Winter, C. Mark and V. Schunig, *Inorg. Chem.*, 1980, **19**, 2045.
50. A. J. Edwards, D. R. Slim, J. E. Guerchais and J. E. Kengoat, *J. Chem. Soc., Dalton Trans.*, 1977, 1966.
51. B. Chevrier, T. Diebold and R. Weiss, *Inorg. Chim. Acta*, 1976, **19**, L57.
52. F. W. B. Einstein and B. R. Penfold, *Acta Crystallogr.*, 1964, **17**, 1127.
53. D. B. Cramp, R. F. Stepaniak and N. C. Payne, *Can. J. Chem.*, 1977, **55**, 438.
54. N. W. Terry, E. L. Amma and L. Vaska, *J. Am. Chem. Soc.*, 1972, **94**, 653.
55. J. Halpern, B. L. Goodall, G. P. Khare, H. S. Lim and J. J. Pluth, *J. Am. Chem. Soc.*, 1975, **97**, 2301.
56. M. J. Bennett and P. S. Donaldson, *Inorg. Chem.*, 1977, **16**, 1581.
57. M. Nolte and E. Singleton, *Acta Crystallogr., Sect. B*, 1976, **32**, 1410.
58. C. Busetto, A. D'Alfonso, F. Maspero, G. Perego and A. Zazzetta, *J. Chem. Soc., Dalton Trans.*, 1977, 1828.
59. J. A. McGinnety, N. C. Payne and J. A. Ibers, *J. Am. Chem. Soc.*, 1969, **91**, 6301.
60. M. J. Bennett and P. B. Donaldson, *Inorg. Chem.*, 1977, **16**, 1385.
61. M. J. Nolte and E. Singleton, *Acta Crystallogr., Sect. B*, 1975, **31**, 2223.
62. J. Ellerman, E. F. Hohenberger, W. Kehr, A. Pürzer and G. Thiele, *Z. Anorg. Allg. Chem.*, 1980, **464**, 45.
63. A. G. Gosh, N. W. Terry and E. L. Amma, *Trans. Am. Crystallogr. Assoc.*, 1973, 40 (Winter Meeting).
64. M. J. Nolte, E. Singleton and M. Laing, *J. Am. Chem. Soc.*, 1975, **97**, 6396.
65. M. Nolte, E. Singleton and M. Laing, *J. Chem. Soc., Dalton Trans.*, 1976, 1979.
66. M. Nolte and E. Singleton, *Acta Crystallogr., Sect. B*, 1976, **32**, 1838.
67. S. J. La Placa and J. A. Ibers, *J. Am. Chem. Soc.*, 1965, **87**, 2581.
68. M. S. Weininger, E. A. H. Griffith, C. T. Sears and E. L. Amma, *Inorg. Chim. Acta*, 1982, **66**, 67.
69. J. A. McGinnety, R. J. Doedens and J. A. Ibers, *Inorg. Chem.*, 1967, **6**, 2243.
70. T. Yoshida, K. Tatsumi, M. Matsumoto, K. Nakatsu, A. Nakamura, T. Fueno and S. Otsuka, *Nouv. J. Chim.*, 1979, **3**, 761.
71. T. Kashiwagi, N. Yasuoka, N. Kasai, M. Kakudo, S. Takahasi and N. Hagihara, *J. Chem. Soc. (A)*, 1969, 743.
72. C. D. Cook, P.-T. Cheng and S. C. Nyberg, *J. Am. Chem. Soc.*, 1969, **91**, 2123.
73. B. Bosnich, W. G. Jackson, S. T. D. Lo and J. W. McLaren, *Inorg. Chem.*, 1974, **13**, 2605.
74. S. Otsuka, A. Nakamura and Y. Tatsumo, *J. Am. Chem. Soc.*, 1969, **91**, 6994.
75. J. A. Connor and E. A. V. Ebsworth, *Adv. Inorg. Chem. Radiochem.*, 1964, **6**, 279.
76. K. R. Laing and W. R. Roper, *Chem. Commun.*, 1968, 1568.
77. D. F. Christian and W. R. Roper, *Chem. Commun.*, 1971, 1271.
78. B. E. Cavit, K. R. Grundy and W. R. Roper, *J. Chem. Soc., Chem. Commun.*, 1972, 60.
79. M. M. Taqui Khan, R. K. Andaland and P. T. Manoharan, *Chem. Commun.*, 1971, 561.
80. T. E. Nappier and D. W. Meek, *J. Am. Chem. Soc.*, 1972, **94**, 306.
81. A. L. Balch and J. Miller, *J. Organomet. chem.*, 1971, **32**, 263.

82. R. W. Mitchell, J. D. Ruddick and G. Wilkinson, *J. Chem. Soc. (A)*, 1971, 3224.
83. L. M. Haines and E. Singleton, *J. Organomet. Chem.*, 1971, **30**, C81.
84. D. N. Cash and R. O. Harris, *Can. J. Chem.*, 1971, **49**, 3821.
85. C. K. Brown and G. Wilkinson, *Chem. Commun.*, 1971, 70.
86. C. A. Reed and W. R. Roper, *Chem. Commun.*, 1971, 1556.
87. M. H. B. Stiddard and R. E. Townsend, *J. Chem. Soc. (A)*, 1970, 2719.
88. J. J. Levison and S. D. Robinson, *J. Chem. Soc. (A)*, 1971, 762.
89. A. J. Deeming and B. L. Shaw, *J. Chem. Soc. (A)*, 1969, 1128.
90. H. Van Gaal, H. G. A. M. Cuppers and A. Van der Ent, *Chem. Commun.*, 1970, 1694.
91. J. Valentine, D. Valentine, Jr. and J. P. Collman, *Inorg. Chem.*, 1971, **10**, 219.
92. G. R. Clark, C. A. Reed, W. R. Roper, B. W. Skelton and T. N. Waters, *Chem. Commun.*, 1971, 758.
93. G. Wilke, H. Schott and P. Heimbach, *Angew. Chem., Int. Ed. Engl.*, 1967, **6**, 92.
94. A. Werner and A. Myelius, *Z. Anorg. Chem.*, 1893, **16**, 252.
95. T. Tsumaki, *Bull. Chem. Soc. Jpn.*, 1938, **13**, 252.
96. B. T. Tourog, D. J. Kitko and R. S. Drago, *J. Am. Chem. Soc.*, 1976, **98**, 5144.
97. D. Getz, E. Melamund, B. L. Silver and Z. Dori, *J. Am. Chem. Soc.*, 1975, **97**, 3846.
98. R. S. Gall, J. R. Rogers, W. P. Schaefer and G. G. Christoph, *J. Am. Chem. Soc.*, 1976, **98**, 5135.
99. L. D. Brown and K. N. Raymond, *Inorg. Chem.*, 1975, **14**, 2595.
100. R. S. Gall and W. P. Schaefer, *Inorg. Chem.*, 1976, **15**, 2758.
101. G. B. Jameson, W. T. Robinson and G. A. Rodley, *J. Chem. Soc., Dalton Trans.*, 1978, 191.
102. B. T. Huie, R. M. Leyden and W. P. Schaefer, *Inorg. Chem.*, 1979, **18**, 125.
103. W. P. Schaefer, B. T. Huie, M. G. Murilla and S. E. Ealick, *Inorg. Chem.*, 1980, **19**, 340.
104. R. Cini and P. Orioli, *J. Chem. Soc., Chem. Commun.*, 1981, 196.
105. G. A. Rodley and W. T. Robinson, *Nature (London)*, 1972, **235**, 438.
106. A. Avdeef and W. P. Schaefer, *J. Am. Chem. Soc.*, 1976, **98**, 5153.
107. G. B. Jameson, G. A. Rodley, W. T. Robinson, R. R. Gagne, C. A. Reed and J. P. Collman, *Inorg. Chem.*, 1978, **17**, 850.
108. W. T. Robinson, G. A. Rodley and G. B. Jameson, *Acta Crystallogr., Sect. A*, 1975, **31**, S49.
109. G. B. Jameson, F. S. Molinaro, J. A. Ibers, J. P. Collman, J. I. Brauman, E. Rose and K. S. Suslick, *J. Am. Chem. Soc.*, 1980, **102**, 3224.
110. J. P. Collman, R. R. Gagne, C. A. Reed, W. T. Robinson and G. A. Rodley, *Proc. Natl. Acad. Sci. USA*, 1974, **71**, 247.
111. S. E. V. Phillips, *Nature (London)*, 1978, **273**, 247.
112. B. M. Hoffman, D. L. Diemente and F. Basolo, *J. Am. Chem. Soc.*, 1970, **92**, 61.
113. C. Floriani and F. Calderazzo, *J. Chem. Soc. (A)*, 1969, 946.
114. T. Szymanski, T. W. Cape, F. Basolo and R. P. Van Duyne, *J. Chem. Soc., Chem. Commun.*, 1979, 5.
115. D. A. White, A. J. Solodan and M. M. Baizer, *Inorg. Chem.*, 1977, **11**, 2160.
116. J. P. Collman, J. I. Brauman, T. R. Halbert and K. S. Suslick, *Proc. Natl. Acad. Sci. USA*, 1976, **73**, 3333.
117. P. Richard and R. Gillard, *J. Chem. Soc., Chem. Commun.*, 1983, 1454.
118. S. K. Cheung, C. J. Grimes, J. Wong and C. A. Reed, *J. Am. Chem. Soc.*, 1976, **98**, 5028.
119. W. P. Schaefer, *Inorg. Chem.*, 1968, **7**, 725.
120. R. Stomberg, L. Trysberg and I. Lanking, *Acta Chem. Scand.*, 1970, **24**, 2678.
121. U. Thewalt and R. E. Marsh, *J. Am. Chem. Soc.*, 1967, **89**, 6364.
122. F. R. Fronczek and W. P. Schaefer, *Inorg. Chim. Acta*, 1974, **9**, 143.
123. F. R. Fronczek, W. P. Schaefer and R. E. Marsh, *Acta Crystallogr., Sect. B*, 1974, **30**, 117.
124. U. Thewalt, M. Zehnder and S. Fallab, *Helv. Chim. Acta*, 1977, **60**, 867.
125. M. Zehnder and U. Thewalt, *Z. Anorg. Allg. Chem.*, 1980, **461**, 53.
126. J. R. Fritch, G. G. Christoph and W. P. Schaefer, *Inorg. Chem.*, 1973, **12**, 2170.
127. T. Shibahara, S. Koda and M. Mori, *Bull. Chem. Soc. Jpn.*, 1973, **46**, 2070.
128. M. Zehnder and U. Thewalt, *Helv. Chim. Acta*, 1976, **59**, 2290.
129. U. Thewalt and G. Struckmeier, *Z. Anorg. Allg. Chem.*, 1976, **419**, 163.
130. M. Zehnder, U. Thewalt and S. Fallab, *Helv. Chim. Acta*, 1979, **62**, 2099.
131. U. Thewalt, *Z. Anorg. Allg. Chem.*, 1972, **393**, 1.
132. S. Fallab, M. Zehnder and U. Thewalt, *Helv. Chim. Acta*, 1980, **63**, 1491.
133. T. Shibahara, M. Mori, K. Matsumoto and S. Ooi, *Bull. Chem. Soc. Jpn.*, 1981, **54**, 433.
134. M. Calligaris, G. Nardin, L. Randaccio and A. Ripamonti, *J. Chem. Soc. (A)*, 1970, 1069.
135. A. Avdeet and W. P. Schaefer, *Inorg. Chem.*, 1976, **15**, 1432.
136. L. A. Lindblom, W. P. Schaefer and R. E. Marsh, *Acta Crystallogr., Sect. B*, 1971, **27**, 1461.
137. J. H. Timmons, A. Clearfield, A. E. Martell and R. H. Niswander, *Inorg. Chem.*, 1979, **18**, 1042.
138. J. H. Timmons, R. H. Niswander, A. Clearfield and A. E. Martell, *Inorg. Chem.*, 1979, **18**, 2977.
139. W. Macke, M. Zehnder, U. Thewalt and S. Fallab, *Helv. Chim. Acta*, 1979, **62**, 1804.
140. S. Bhaduri, L. Casella, R. Ugo, P. R. Raithby, C. Zuccaro and M. B. Hursthouse, *J. Chem. Soc., Dalton Trans.*, 1979, 1624.
141. T. Freedman, G. Yoshida and T. M. Loehr, *J. Chem. Soc., Chem. Commun.*, 1974, 1016.
142. R. E. Marsh and W. P. Schaefer, *Acta Crystallogr., Sect. B*, 1968, **24**, 246.
143. G. C. Christoph, R. E. Marsh and W. P. Schaefer, *Inorg. Chem.*, 1969, **8**, 291.
144. U. Thewalt and R. Marsh, *J. Am. Chem. Soc.*, 1967, **89**, 6364.
145. U. Thewalt and R. Marsh, *Inorg. Chem.*, 1972, **11**, 351.
146. U. Thewalt and G. Struckmeier, *Z. Anorg. Allg. Chem.*, 1976, **419**, 163.
147. B.-C. Wangand and W. P. Schaefer, *Science*, 1969, **166**, 1404.
148. F. R. Fronczek, W. P. Schaefer and R. E. Marsh, *Inorg. Chem.*, 1975, **14**, 611.
149. W. P. Schaefer and R. E. Marsh, *J. Am. Chem. Soc.*, 1966, **88**, 178.
150. U. M. Miskowski, B. D. Santargiro, W. P. Schaefer, G. E. Ansok and H. B. Gray, *Inorg. Chem.*, 1984, **23**, 172.

151. R. D. Jones, D. A. Summerville and F. Basolo, *Chem. Rev.*, 1979, **79**, 139.
152. T. Shibahara, *J. Chem. Soc., Chem. Commun.*, 1973, 863.
153. N. Farrell, D. Dolphin and B. R. James, *J. Am. Chem. Soc.*, 1978, **100**, 324.
154. T. D. Smith and J. R. Pilbrow, *Coord. Chem. Rev.*, 1981, **39**, 295.
155. B. B. Wayland and A. R. Newman, *Inorg. Chem.*, 1981, **20**, 3093.
156. A. H. Corwin and Z. Reyes, *J. Am. Chem. Soc.*, 1956, **78**, 2437.
157. C. K. Chang and T. G. Traylor, *Proc. Natl. Acad. Sci. USA*, 1973, **70**, 2647.
158. G. C. Wagner and R. J. Kassner, *J. Am. Chem. Soc.*, 1974, **96**, 5593.
159. W. S. Brinigar, C. K. Chang, J. Geibel and T. G. Traylor, *J. Am. Chem. Soc.*, 1974, **96**, 5597.
160. J. Almog, J. E. Baldwin, R. C. Dyer, J. Huff and C. J. Wilkerson, *J. Am. Chem. Soc.*, 1974, **96**, 5600.
161. J. P. Collman, R. R. Gagne, T. R. Halbert, J.-C. Marckon and C. A. Reed, *J. Am. Chem. Soc.*, 1973, **95**, 7868.
162. K. Spartalian, G. Lang, J. P. Collman, R. R. Gagne and C. A. Reed, *J. Chem. Phys.*, 1975, **63**, 5375.
163. M. Cerdonio, A. Congiu-Castellano, F. Mogno, B. Pispisa, G. L. Romani and S. Vitale, *Proc. Natl. Acad. Sci. USA*, 1977, **74**, 398.
164. J. Almog, J. E. Baldwin and J. Huff, *J. Am. Chem. Soc.*, 1975, **97**, 227.
165. A. R. Amundsen and L. Vaska, *Inorg. Chim. Acta*, 1975, **14**, L49.
166. L. Latos-Grazynski, R.-J. Cheng, G. N. La Mar and A. N. Balch, *J. Am. Chem. Soc.*, 1982, **104**, 5992.
167. J. H. Wang, *Acc. Chem. Res.*, 1970, **3**, 90.
168. J. P. Collman and C. A. Reed, *J. Am. Chem. Soc.*, 1973, **95**, 2048.
169. O. Leal, D. L. Anderson, R. G. Bowman, F. Basolo and R. L. Burwell, Jr., *J. Am. Chem. Soc.*, 1975, **97**, 5125.
170. S. G. Davies, 'Organotransition Metal Chemistry: Applications to Organic Synthesis', Pergamon, Oxford, 1982, p. 304.
171. Y. Tatsuno and S. Otsuka, *J. Am. Chem. Soc.*, 1981, **103**, 5832.
172. M. H. Gubelmann and A. F. Williams, *Struct. Bonding (Berlin)*, 1983, **55**, 1.
173. R. Ugo, F. Conti, S. Cenini, R. Mason and G. B. Robertson, *Chem. Commun.*, 1968, 1498.
174. P. J. Hayward and C. J. Nyman, *J. Am. Chem. Soc.*, 1971, **93**, 617.
175. S. L. Regenand and G. M. Whitesides, *J. Organomet. Chem.*, 1973, **59**, 293.
176. H. Mimoun, I. Seree de Roch and L. Sajus, *Tetrahedron*, 1970, **26**, 37.
177. A. Sen and J. Halpern, *J. Am. Chem. Soc.*, 1977, **99**, 8337.
178. C. L. Wong, J. A. Switzer, J. F. Endicott and K. P. Balakrishnan, *J. Am. Chem. Soc.*, 1980, **102**, 5511.
179. J. F. Endicott and K. Kumar, *ACS Symp. Ser.*, 1982, **198**, 425.
180. H. Armouzanian, R. Lai, R. Lopez Alvarez, J. F. Petriogniani, J. Metzger and J. Furhop, *J. Am. Chem. Soc.*, 1980, **102**, 845.
181. A. Nishinaga, H. Tomita, K. Nishizawa, T. Masuura and K. Hirotsa, *J. Chem. Soc., Dalton Trans.*, 1981, 1504.
182. A. Nishinaga, T. Shimizu, T. Toyoda, T. Matsuura and K. Hirotsu, *J. Org. Chem.*, 1982, **47**, 2278, and references therein.
183. R. S. Drago, J. Gaul, A. Zombeck and D. K. Straub, *J. Am. Chem. Soc.*, 1980, **102**, 1033.
184. R. Davies and A. G. Sykes, *J. Chem. Soc. (A)*, 1968, 2237.
185. J. D. Edwards, C. H. Yang and A. G. Sykes, *J. Chem. Soc., Dalton Trans.*, 1974, 1561.
186. K. Garbett and R. D. Gillard, *J. Chem. Soc. (A)*, 1968, 1725.
187. C. H. Yang, D. P. Keeton and A. G. Sykes, *J. Chem. Soc., Dalton Trans.*, 1974, 1089.

15.3

Alkoxides and Aryloxides

MALCOLM H. CHISHOLM
Indiana University, Bloomington, IN, USA

and

IAN P. ROTHWELL
Purdue University, West Lafayette, IN, USA

15.3.1   FOREWORD

Compounds of empirical formula $M(OR)_n$ and $L_xM(OR)_y$, where R is an alkyl or aryl group, L_x represents a combination of other ligands, which may be either neutral or anionic, and n and y are integers $\geqslant 1$, are the subjects of this chapter. Excluded from consideration are catecholates and semiquinones, and chelated ligands containing alkoxy or phenoxy functionalities, *e.g.* compounds derived from the ligand formed by deprotonation of 2-hydroxypyridine.

15.3.2   SYNTHESIS OF METAL ALKOXIDES AND PHENOXIDES

A large number of synthetic routes to complexes containing these ligands have been developed. The preferred method typically depends on the electronegativity of the element for which the alkoxy derivative is needed and also the availability of suitable starting materials.[1-5]

15.3.2.1   From the Metal

15.3.2.1.1   By direct reactions of the bulk metal with alcohols and phenols

This method for the preparation of metal alkoxides and phenoxides is of only limited usefulness given the relatively low pK_a values of alcohols and even common phenols. Hence, the method is confined to the more electropositive elements where reaction can occur either directly with the alcohol or phenol or sometimes in the presence of a catalyst. For the Group I elements (Li, Na, K, *etc.*) the dissolution of the metal into the neat alcohol or phenol at close to reflux temperatures can lead to the pure alkoxides or phenoxides with the evolution of hydrogen (equation 1).[6,7]

$$M + ROH \longrightarrow MOR + \tfrac{1}{2}H_2 \tag{1}$$

The rate of the reaction is highly dependent on the nature of the alcohol with more sterically demanding (and also less acidic) alcohols such as Bu^tOH reacting very slowly.[1] For sodium and potassium the amalgamated metals can also be used.[8,9]

Magnesium and aluminum metal will react directly with phenol at its boiling point.[10,11] However, in both cases the inert oxide film must be mechanically removed or chemically dissolved.

For the less electropositive metals, direct reaction between the bulk metal and alcohols does not readily occur. However, for the Group II, III and lanthanide metals, reaction will occur in the presence of a catalyst.[12,15] Typical catalysts are iodine or a mercury(II) halide and their action is believed to be either in cleaning the metal surface or in forming intermediate halide derivatives which then undergo facile reaction with the alcohol (equation 2).[12]

$$Mg + 2ROH \xrightarrow[\text{catalyst}]{I_2} Mg(OR)_2 + H_2 \tag{2}$$

Alkoxides of aluminum,[16] calcium,[14] strontium[15] and barium[15] have also been synthesized using iodine as a catalyst.

Treatment of scandium, yttrium and the lanthanides with refluxing isopropyl alcohol in the presence of $HgCl_2$ (10^{-3}–10^{-4} equivalents) leads to the metal isopropoxide, which can then be crystallized from the excess isopropyl alcohol (equation 3).[17,18]

$$Ln + 3Pr^iOH \xrightarrow{HgCl_2} Ln(OPr^i)_3 + \tfrac{3}{2}H_2 \qquad (3)$$

Similarly, yttrium metal shavings were shown to react with phenol at its boiling point in the presence of $HgCl_2$ to form the phenoxide.[17] However, the amount of $HgCl_2$ catalyst added must be small if secondary reactions leading to impurities are to be avoided.

15.3.2.1.2 By electrochemical methods

The direct electrochemical synthesis of metal alkoxides by the anodic dissolution of metals into alcohols containing conducting electrolytes was initially demonstrated by Szilard in 1906 for the methoxides of copper and lead.[19] More recently the method has received some attention particularly in the patent literature.[20–25] The preparation of the ethoxides of silicon, titanium, germanium, zirconium and tantalum by electrolysis of ethanolic solutions of NH_4Cl has been patented, although the production of the ethoxides was found to cease after several hours.[24,25]

In a recent more thorough study the effect of supporting electrolyte was examined in detail.[26] Salts containing NH_4^+ or Cl^- ions were found to be inferior, while the use of $(Bu_4N)Br$ and $(Bu_4N)BF_4$ allowed the processes to be run indefinitely. Using these conductive admixtures the isopropoxides of Y, Sc, Ga and Zr and the ethoxides of Ti, Ge, Nb and Ta were obtained in current yields between 60–97%. Typically potentials of 30–100 V are used, giving currents of 0.05–0.25 A.[26] Although studies of this process have been limited so far, it does appear to be the most useful for the direct conversion of the less electropositive metals to their alkoxides.

15.3.2.1.3 By reactions of metal atom vapors with alcohols

The development of metal atom vapor technology over the last 15 years has made available to the chemist a new and useful synthetic technique. A number of novel transition metal compounds have been isolated by the method, particularly in the field of organometallic chemistry. However, the use of vaporized metal atoms for the synthesis of metal alkoxides and phenoxides by condensation into the neat alcohol has been only briefly mentioned in the literature.[27]

15.3.2.2 From Metal Halides

The halide derivatives of metals are by far the most useful and widely exploited starting materials for the synthesis of metal alkoxides and phenoxides.[1,2]

15.3.2.2.1 By direct reactions with alcohols and phenols

With the less electropositive elements such as boron,[28] silicon[29] and phosphorus,[30] alcoholysis of the chlorides goes to completion with the elimination of the corresponding amount of HCl (equation 4).

$$SiCl_4 + 4EtOH \longrightarrow Si(OEt)_4 + 4HCl \qquad (4)$$

With the more electropositive metals such as the lanthanides,[31] the dissolution of their chlorides into alcohols results only in the formation of solvates such as $LaCl_3 \cdot 3Pr^iOH$.[31] For most of the early d-block elements, only partial replacement of halide ligands occurs on alcoholysis (equations 5 and 6).[32–34]

$$TiCl_4 + 3EtOH \longrightarrow TiCl_2(OEt)_2 \cdot EtOH + 2HCl^{33} \qquad (5)$$

$$2MoCl_5 + ROH(solvent) \longrightarrow [MoCl_3(OR)_2]_2 + 4HCl^{34} \qquad (6)$$

However, a problem with this type of synthetic method is the potential for reaction of the liberated HCl with the alcohol (particularly tertiary ones) to generate water and a mixture of other organic products.[1]

Because of their increased acidity, it is possible for phenols to effect total substitution of halide ligands in cases where alcohols cause only partial substitution. Hence, although WCl_6 reacts with ethanol to give the W^V complex $[WCl_3(OEt)_2]_2$,[35] phenol in benzene gives $W(OPh)_6$ in excellent yields.[4,36] Similarly, the phenoxides of Ti,[37] Nb and Ta[38] have been obtained. In cases where only partial replacement by phenol occurs, the use of substituents that increase the phenolic acidity can sometimes aid the reaction, for example as shown in equation (7).[39,40]

$$VOCl_3 + 3ArOH \longrightarrow VO(OAr)_3 + 3HCl \ (Ar = p\text{-}ClC_6H_4) \tag{7}$$

15.3.2.2.2 By reactions with alcohols and phenols in the presence of base

The addition of a base, typically ammonia, to mixtures of transition metal halides and alcohols allows the synthesis of homoleptic alkoxides and phenoxides for a wide range of metals. Anhydrous ammonia was first used in the preparation of titanium alkoxides where the reaction is forced to completion by the precipitation of ammonium chloride.[41] Although useful for the synthesis of simple alkoxides and phenoxides of Si, Ge, Ti, Zr, Hf, V, Nb, Ta and Fe, as well as a number of lanthanides,[42-47] the method fails to produce pure t-butoxides of a number of metals.[58] Presumably, secondary reactions between HCl and Bu^tOH take place. However, mixing $MCl_4(M = Ti, Zr)$ with the Bu^tOH in the presence of pyridine followed by addition of ammonia proves successful, giving excellent yields of the $M(OBu^t)_4$ complexes.[59]

15.3.2.2.3 By reactions with alkali metal alkoxides and phenoxides

Metathetic exchange of alkoxide or phenoxide for halide is possible using either lithium or sodium salts. For instance, thorium tetraalkoxides[55] are best obtained from the reaction shown in equation (8).

$$ThCl_4 + 4NaOR \longrightarrow Th(OR)_4 + 4NaCl \tag{8}$$

The use of sodium salts has been successful in the synthesis of a large number of metal alkoxides, including nearly all of the lanthanides, typically in an alcohol–hydrocarbon solvent mixture.[60-62] However, one problem sometimes encountered in this method is the formation of double alkoxides with alkali metals. In particular, zirconium forms complexes of the type $M_2Zr(OR)_6$ from which removal of the parent alkoxide is difficult.[63,64] The use of sterically demanding alkoxides and phenoxides of the alkali metals can sometimes lead to only partial substitution. Hence, although lithium 2,6-dimethylphenoxide will totally substitute $TaCl_5$ to give the mononuclear penta-aryloxides, the much more sterically demanding 2,6-di-t-butylphenoxide (OAr′) will only substitute twice to yield $Ta(OAr′)_2Cl_3$.[65] Similarly, the extremely crowded tri-t-butylmethoxide (tritox) ion will only substitute the halide $ZrCl_4$ twice to yield initially a double salt containing the $Zr(tritox)_2Cl_3^-$ anion.[66]

The advantages of using lithium salts instead of sodium salts for the preparation of insoluble metal methoxides has been discussed.[67,68]

15.3.2.3 By Reactions of Metal Hydroxides and Oxides with Alcohols and Phenols

Alkali metal alkoxides and phenoxides may be obtained by reacting the metal hydroxide with either alcohol or phenol. With alcohols the generated water must be removed, typically as an azeotrope, as with sodium hydroxide dissolved in ethanol–benzene followed by reflux.[69] With the more acidic phenols the phenoxides of Li, Na, K, Rb and Cs are more readily formed by simply heating MOH in absolute ethanol with phenol followed by recrystallization.[70-72]

Hydroxides and oxides of the non-metals and less electropositive metals behave as oxyacids and react with alcohols to form esters (alkoxides) and water. Typically such reactions involve an equilibrium and removal of the generated water by fractionation is necessary to obtain the pure products. By this method the alkoxides of most of the metalloid main group elements have been obtained (equations 9 and 10).[1]

$$B(OH)_3 + 3ROH \longrightarrow B(OR)_3 + 3H_2O \tag{9}$$

$$B_2O_3 + 6ROH \longrightarrow 2B(OR)_3 + 3H_2O \tag{10}$$

Magnesium phenoxide is obtained in a commercial process by heating MgO with phenol,[10] while V_2O_5 has been reported to yield $OV(OPh)_3$ on heating with phenol.[73]

15.3.2.4 By Reactions of Metal Dialkylamides with Alcohols and Phenols

This method is particularly useful for the synthesis of the alkoxide and phenoxide derivatives of the earlier transition elements. The method is extremely convenient in view of the high volatility of the generated dialkylamines, which are readily removed in vacuum. One major drawback is the synthetic availability of the corresponding metal dialkylamide complex. In some cases the method represents not only the most convenient but also the only synthetic route to an alkoxide derivative. Hence, zirconium tetra-*t*-butoxide is formed in excellent yield from $Zr(NEt_2)_4$ and Bu^tOH, and the V^{IV} and Cr^{IV} *t*-butoxides are also readily obtained *via* this pathway (equation 11).[74]

$$M(NR_2)_4 + 4ROH \longrightarrow M(OR)_4 + 4HNR_2 \tag{11}$$

Similarly, alcoholysis of $Mo(NMe_2)_4$ allows a ready entry into Mo^{IV} alkoxide chemistry,[75] although reactions with silanols[76] and phenols[77] typically produce adducts in which the liberated dimethylamine remains coordinated to the metal center (equation 12).

$$Mo(NMe_2)_4 + 4Me_3SiOH \longrightarrow Mo(OSiMe_3)_4(HNMe_2)_2 + 2HNMe_2 \tag{12}$$

The dinuclear alkoxides and phenoxides of Mo^{III} and W^{III} can only be obtained in any reasonable yield *via* the amido intermediates.[78,79] With bulky phenols the reaction can be slow and in some cases lead only to partial substitution.[80]

The alcoholysis of a lower valent metal dialkylamide can also lead to oxidation of the metal, typically with evolution of hydrogen (equations 13 and 14).[74,81]

$$Nb(NEt_2)_4 + 5ROH \longrightarrow Nb(OR)_5 + 4HNMe_2 + \tfrac{1}{2}H_2 \tag{13}$$

$$2W_2(NMe_2)_6 + 16EtOH \longrightarrow W_4(OEt)_{16} + 12HNMe_2 + 2H_2 \tag{14}$$

15.3.2.5 From Exchange Reactions Involving Metal Alkoxides

15.3.2.5.1 By alcohol interchange

The use of alkoxides to synthesize new alkoxides by the process of alcohol interchange has been widely applied for a large number of elements (equation 15).

$$M(OR)_n + nR'OH \longrightarrow M(OR')_n + nROH \tag{15}$$

In general the facility of interchange of alkoxy groups by alcoholysis follows the order tertiary < secondary < primary.[82] Hence the *t*-butoxides of titanium and zirconium will undergo rapid exchange with methanol or ethanol.[83] An extra driving force here is the larger degree of oligomerization of methoxides or ethoxides in general over *t*-butoxides.[83] However, it is possible in some cases, by fractionating out the more volatile components, partly to reverse this order of reactivity (equations 16 and 17).[51,84]

$$Al(OPr^i)_3 + 2Bu^tOH \longrightarrow Al(OPr^i)(OBu^t)_2 + 2Pr^iOH \tag{16}$$

$$Ta(OMe)_5 + 4Bu^tOH \longrightarrow Ta(OMe)(OBu^t)_4 + 4MeOH \tag{17}$$

15.3.2.5.2 By phenol for alcohol exchange

This method is particularly useful for the synthesis of phenoxides and tends to work well due to both the increased acidity of phenols and their lower volatility relative to alcohols. Hence phenoxides of a large number of elements have been prepared by this route including main group,[85] *d*-block,[86] lanthanides[87] and actinides[88] (equations 18, 19 and 20).

$$Ge(OEt)_4 + 4PhOH \longrightarrow Ge(OPh)_4 + 4EtOH \tag{18}$$

$$Zr(OBu^t)_4 + 5PhOH \longrightarrow Zr(OPh)_4 \cdot PhOH + 4Bu^tOH \tag{19}$$

$$Ln(OPr^i)_3 + xPhOH \longrightarrow Ln(OPh)_x(OPr^i)_{3-x} + xPr^iOH \tag{20}$$

In some cases, control of the stoichiometry allows the isolation of mixed alkoxide–phenoxide derivatives.[87] In other cases only partial substitution can be achieved either for steric[89] or electronic reasons[88] (equations 21 and 22).

$$U(OEt)_5 + PhOH(xs) \longrightarrow U(OPh)_4(OEt) + 4EtOH \tag{21}$$

$$Mo_2(OPr^i)_6 + HOAr(xs) \longrightarrow Mo_2(OPr^i)_2(OAr)_4 + 4HOPr^i \tag{22}$$

$$(OAr = 2,6\text{-dimethylphenoxide})$$

15.3.2.5.3 By transesterification

The addition of an ester to a solution of an alkoxide sets up an equilibrium (equation 23).

$$M(OR)_n + xMeCO_2R' \longrightarrow M(OR)_{n-x}(OR')_x + xMeCO_2R \tag{23}$$

Hence a new alkoxide or phenoxide can be obtained if the ester produced is more volatile and is fractionated out of the mixture. This method has proved very useful for the preparation of tertiary alkoxides as it appears to be much less prone to steric factors than alcohol exchange. Thus the *t*-butoxides of the Group IV metals Ti, Zr and Hf are readily synthesized from their isopropoxides (equation 24).[90]

$$M(OPr^i)_4 + 4MeCO_2Bu^t \longrightarrow M(OBu^t)_4 + 4MeCO_2Pr^i \tag{24}$$

The phenoxides of niobium[91] and tantalum[90] have been obtained from their isopropoxides by a similar reaction carried out in cyclohexane (equation 25).

$$M(OPr^i)_5 + 5MeCO_2Ph \longrightarrow M(OPh)_5 + 5MeCO_2Pr^i \tag{25}$$

15.3.2.6 From Metal Alkyls and Hydrides

15.3.2.6.1 By reactions with alcohols or phenols

Carbanionic metal alkyls and hydridic metal hydrides will react with alcohols or phenols to give alkoxides and phenoxides, typically in excellent yields. The reaction is also important as it forms the basis for the calorimetric measurement of a large number of metal–alkyl bond dissociation energies.[93,94] This synthetic method tends to be very convenient due to the volatility of the generated alkane or hydrogen side products. Monoalkyl alkoxides of Be,[95] Mg[96] and Zn[97] can be obtained in this way (equation 26).

$$MR_2 + R'OH \longrightarrow RMOR' + RH \tag{26}$$

Similarly, alkylzinc phenoxides have been obtained.[97] The success of the method for the stoichiometric substitution of only one alkyl in this latter case has been attributed to the reaction shown in equation (27).[97]

$$Zn(OPh)_2 + ZnR_2 \longrightarrow 2Zn(OPh)R \tag{27}$$

In all of these cases substitution of the second alkyl can then occur to yield the dialkoxide or diphenoxide. This allowed the isolation of the monomeric beryllium phenoxide $Be(OAr')_2$ $(OAr' = 2,6$-di-t-butylphenoxide).[98] The alkyls of the Group IV metals, $MR_4(M = Ti, Zr, Hf)$, undergo rapid reactions with common alcohols and phenols yielding eventually the corresponding tetra-alkoxides or -phenoxides and four equivalents of alkane.[97,100] With very bulky substituted alcohols or phenols the reactivity can be very sluggish, in some cases leading to only partial substitution (equation 28).[66,100]

$$Zr(CH_2Ph)_4 + Ar'OH(xs) \longrightarrow Zr(OAr')_2(CH_2Ph)_2 + 2MePh \tag{28}$$

$$(OAr' = 2,6\text{-di-}t\text{-butylphenoxide})$$

Most of the alkyls of the other early transition metals demonstrate similar reactivity.[101,102] Hydridic metal hydrides will similarly react with the weakly acidic OH groups of alcohols and phenols, liberating hydrogen and forming alkoxides or phenoxides. Aluminum hydride will undergo stepwise substitution by alcohols (equation 29).[103]

$$AlH_3 + nROH \longrightarrow Al(OR)_nH_{3-n} + nH_2 \tag{29}$$

Hydrides of the early transition metals tend to react similarly. Hence, the molybdenum hydride $MoH_4(PMe_2Ph)_4$ reacts with p-cresol to yield the corresponding phenoxide (equation 30).[77]

$$MoH_4(PMe_2Ph)_4 + 4ArOH \longrightarrow Mo(OAr)_4(PMe_2Ph)_2 + 4H_2 + 2PMe_2Ph \tag{30}$$

Mixed hydrido–alkoxides and -phenoxides of thorium have been synthesized by the reaction shown (equation 31).[104]

$$Cp_2^*Th(H)_2 + HOX \longrightarrow Cp_2^*Th(H)OX + H_2 \tag{31}$$

$$(X = R, Ar)$$

15.3.2.6.2 *By reactions with aldehydes and ketones*

The great synthetic utility of the reaction of alkyllithium and Grignard reagents with ketonic functions has been well documented.[105] These reactions take place *via* the intermediacy of alkoxy derivatives formed by addition of the M—C bond across the C=O function. Hence ketones, aldehydes and formaldehyde will lead to tertiary, secondary and primary alkoxides, respectively. This type of reactivity is known for a number of other carbanionic metal alkyl derivatives, both main group and transition metals, although the synthetic utility of the reactivity has in most cases not been well documented.

However, in a comprehensive study, Seebach and coworkers have shown the usefulness of titanium and zirconium alkyls as selective nucleophilic reagents.[106,107] Typically compounds of the type $(RO)_3MR'$ $(M = Ti, Zr)$ are reacted with aldehydes or ketones to yield asymmetric alkoxides (equation 32).[107] The method has been extended to allow enantioselective addition using chiral organotitanium reagents where the chirality is imposed by the initial alkoxide substituents.[107,108]

$$(RO)_3MR' + R^1R^2CO \longrightarrow (RO)_3M(OCR^1R^2R') \tag{32}$$

The commonly used reducing agents AlH_4^- and BH_4^- will reduce aldehydes and ketones to primary and secondary alkoxides.[109] Similar reactivity is found for some other main group hydrides, *e.g.* gallium (equation 33).[110]

$$HGaCl_2 + R^1R^2CO \longrightarrow GaCl_2(OCHR^1R^2) \tag{33}$$

The hydridic nature of early transition metal hydrides is normally inferred chemically by their ability to reduce acetone to isopropoxide ligands.[111]

15.3.2.7 From Miscellaneous Reactions

15.3.2.7.1 *By insertion of O_2 into metal–alkyl bonds*

While some metal–alkyl complexes react violently with molecular oxygen and others are inert, there are a few well-documented examples of reactions which lead to the formation of alkoxide ligands. Since only one oxygen atom is required per metal–alkyl bond, the reactions require a multistep process.

The oxidation of R_2Zn has been shown to lead initially to the alkyl peroxide $Zn(OOR)_2$, which then yields the alkoxide as shown (equation 34).[112] Both boron and aluminum alkyls react similarly, yielding the corresponding alkoxide.[112,113]

$$Zn(OOR)_2 + ZnR_2 \longrightarrow 2Zn(OR)_2 \tag{34}$$

Schwartz and coworkers, in their development of hydrozirconation, studied the stereochemistry at carbon during the alkoxide forming reaction (equation 35).[114,115]

$$2Cp_2ZrCl(R^*) + O_2 \longrightarrow Cp_2ZrCl(OR) + Cp_2ZrCl(OR^*) \tag{35}$$

The formation of the alkoxide ligand occurred with 50% retention (R*) and 50% racemization (R). This finding was suggested to occur by a two-step reaction, the first step involving formation of a reactive alkyl peroxide intermediate by a radical process involving racemization, followed by a bimolecular reaction, proceeding with retention (equations 36 and 37).

$$Cp_2ZrCl(R^*) + O_2 \longrightarrow Cp_2ZrCl(OO\cdot) + R\cdot \longrightarrow Cp_2ZrCl(OOR) \tag{36}$$

$$Cp_2ZrCl(R^*) + Cp_2ZrCl(OOR) \longrightarrow Cp_2ZrCl(OR^*) + Cp_2ZrCl(OR) \tag{37}$$

Consistent with this proposal was the finding that reactions between $Cp_2ZrCl(R^*)$ and each of H_2O_2 and Bu^tOOH led to formation of R*OH.[115]

The reactions involving $(tritox)_2MMe_2$, where M = Ti and Zr,[66] and $(tritox)TiMe_3$[66] (tritox = Bu_3^tCO) with molecular oxygen have been studied.[116] The products $(tritox)_2M(OMe)_2$ and $(tritox)$-$Ti(OMe)_nMe_{3-n}$ are formed in near quantitative yield.[116] These reactions are also believed to proceed *via* M—OOMe containing intermediates which then react with M—Me bonds by either intra- or inter-molecular pathways.

The proposed alkyl peroxide ligands may well be η^2-bonded to these d^0 early transition metals as has been seen in the vanadium(V) complex $(dipic)VO(OOBu^t)(H_2O)$, where dipic = 2,6-pyridinedicarboxylate.[117] Support for the formation of an alkylperoxy ligand by the combining of an alkyl radical with either free or coordinated oxygen is seen in the oxidative addition of Pr^iI to Me_2Pt-(phen) in the presence of O_2, which yields the octahedral η^1-OOPr^i-containing Pt^{IV} compound $PtMe_2(Pr^iOO)(I)(phen)$, where phen is 1,10-phenanthroline.[118]

15.3.2.7.2 *By phenolysis of metal sulfides*

A mixture of phenol and aluminum sulfide is rapidly converted to the phenoxide on heating with the evolution of H_2S.[20] Similarly, titanium and silicon phenoxides can be prepared directly from their sulfides (equation 38).[119,120]

$$TiS_2 + 4PhOH \longrightarrow Ti(OPh)_4 + 2H_2S \tag{38}$$

15.3.2.7.3 *By hydrogenation of carbon monoxide*

The conversion of $CO + H_2$ (syn-gas) to hydrocarbons and oxygenates (Fischer–Tropsch chemistry)[119] is of considerable industrial importance and recently the activation and fixation of carbon monoxide in homogeneous systems has been an active area for research.[120,121] The early transition elements and the early actinide elements, in particular zirconium[124] and thorium,[125,126] supported by two pentamethylcyclopentadienyl ligands have provided a rich chemistry in the non-catalytic activation of CO. Reactions of alkyl and hydride ligands attached to the Cp_2^*M centers with CO lead to formation of reactive η^2-acyl or -formyl compounds.[125,126] These may be viewed in terms of the resonance forms (**1**) and (**2**) shown below.

(1) X = H or alkyl (2)

The carbene-like character of these η^2-acyl and -formyl derivatives is seen in their coupling to give enediolate complexes containing the functionality M—OCH=CHO—M.[125-129] In the presence of H_2, conversion to alkoxide ligands has also been noted in some instances: M—η^2-OCX + H_2 → M—OCH_2X, where X = H or alkyl. These may be taken as plausible steps for the hydrogenation of CO to methanol or ethylene glycol. However, the early transition elements and actinides are oxophilic, and formation of the alkoxide ligand or enediolate ligand represents attainment of a thermodynamic well. A further reaction with H_2 does not occur to regenerate metal hydride and alcohol. However, the hydrogenolysis of $[Cu(OBu^t)]_4$ under mild conditions (H_2, 1 atm, + PPh_3, THF, 24 °C) is known to give $(HCuPPh_3)_6$ and Bu^tOH.[130] Thus the likelihood of obtaining a homogeneous catalyst for the formation of methanol or ethylene glycol employing either a heterometallic complex or a homometallic complex, involving a transition element with properties intermediate between Zr/Th and Cu, seems quite high.

15.3.2.7.4 By reactions between metal halides and organoxysilanes

A convenient method for the stepwise substitution of halide groups by alkoxide ligands involves the use of alkoxysilanes. By this method, Handy and coworkers have synthesized and studied a large number of mixed alkoxy halides of molybdenum and tungsten (equation 39).[131-134] The method has also been used for the preparation of phenoxy and substituted phenoxy derivatives.[133]

$$WX_6 + nR_3Si(OR') \longrightarrow (R'O)_nWX_{6-n} + nR_3SiX \qquad (39)$$

15.3.2.7.5 By electrophilic attack on a μ_3-acylate ligand

A rare example of the formation of an alkoxide ligand in a metal cluster compound is seen in the reaction between methyl fluorosulfonate and the triiron carbonyl clusters anion $[Fe_3(CO)_9(\mu_3\text{-}MeCO)]^-$, which gives $Fe_3(CO)_9(\mu_3\text{-}CMe)(\mu_3\text{-}OMe)$ by C—O bond cleavage.[135,136] The C—O bond cleavage provides a possible model for a step in Fischer–Tropsch chemistry.[121]

15.3.2.8 Perfluoroalkoxides

Ionic fully fluorinated alkoxides of the heavier alkali metals may be prepared by the reaction between the anhydrous metal fluoride and various perfluorinated carbonyl compounds in donor solvents (equations 40 and 41).[137]

$$R_FCOF + MF \longrightarrow M^+R_FCF_2O^- \qquad (40)$$

$$(CF_3)_2CO + MF \longrightarrow M^+(CF_3)_2CFO^- \qquad (41)$$

$$R_F = F, CF_3 \text{ or } C_2F_5; \quad M = K, Rb \text{ and } Cs$$

In any of the fluorinated alkoxides made this way there must always be a fluorine attached to the α-carbon. There is in fact an equilibrium between the metal perfluoroalkoxide and the metal fluoride and carbonyl fluoride compound. Thus these Group IA metal alkoxides cannot be used in metathetic reactions to prepare other metal alkoxides. For example, R_3SiCl and $NaOCF(CF_3)_2$ react to yield R_3SiF, NaCl and $(CF_3)_2CO$ rather than $R_3SiOCF(CF_3)_2$.[138]

Perfluoro tertiary alcohols are known, *e.g.* $(CF_3)_3COH$,[139,140] $(C_6F_5)_3COH$[141] and $(C_2F_5)_3COH$,[142] and these will react by alcohol exchange to give metal tertiary perfluoroalkoxides, but these do not appear to have been systematically studied.

More is known of the chemistry of perfluoropinacol (H_2PFP), first prepared in the reaction between hexafluoroacetone and isopropyl alcohol.[143] An alternative and sometimes synthetically more useful procedure involves the reductive coupling of hexafluoroacetone with an alkali metal in a non-aqueous solvent such as THF (equation 42). By metathetic reactions involving Na_2PFP, a variety of perfluoropinacolate derivatives of the main group elements (B, Si, Ge, Sn and S) were prepared.[144] Since perfluoropinacol is acidic ($pK_a = 5.95$) and the anion quite stable to water, a variety of salts were prepared according to equation (43).[145]

$$2(CF_3)_2CO + 2Na \xrightarrow{\text{THF}} Na_2PFP \qquad (42)$$

$$nH_2PFP + M^{n+} + 2nKOH \longrightarrow K_nM(PFP)_n + 2nH_2O \qquad (43)$$

$$n = 2 \text{ or } 3; \quad M = Zn, Cu, Ni, Co, Al, Fe; \quad H_2PFP = \text{perfluoropinacol}$$

The perfluoropinacolate ligand has also been shown to yield a rare example of a d^1-Cr^V complex $K^+CrO(PFP)_2^-$, formed in the reaction between potassium chromate and H_2PFP (1:2 molar ratio) in an ethanol–water mixture.[146]

A variety of metal complexes containing fluorinated aminoalkoxy ligands have been prepared by template reactions involving amines and fluorinated 2,4-pentanediones.[147,148]

15.3.2.9 Mixed Metal Alkoxides and Aryloxides

Alkoxides and aryloxides demonstrate similar chemistry to that of hydroxides in that it is possible to prepare mixed metal or double metal derivatives similar to hydroxo salts such as $Na_2[Sn(OH)_6]$. The formation of mixed metal alkoxides, *e.g.* $Na_2[Zr(OEt)_6]$ and $Ln[Al(OPr^i)_4]_3$, is typically a result of the electron deficiency (Lewis acidity) of the metal centers in units of the type $M(OR)_x$ or $M(OAr)_x$ ($x < 5$). This then leads either to oligomerization *via* alkoxide bridges or, in the presence of other alkoxides, to the formation of mixed metal compounds.

Hence the most straightforward synthesis of mixed metal derivatives involves the direct reaction of two metal alkoxides with each other.[149-153] Some representative examples are given below (equations 44–47).

$$M(OR) + Al(OR)_3 \longrightarrow M[Al(OR)_4]$$

$$M(OBu^t) + Zr(OBu^t)_4 \longrightarrow M[Zr(OBu^t)_5] \qquad (44)$$

$$MOR + Ta(OR)_5 \longrightarrow M[Ta(OR)_6]$$

$$M = Li, Na, K, Rb, Cs$$

$$M'(OPr^i)_2 + 4Hf(OPr^i)_4 \longrightarrow M'[Hf_2(OPr^i)_9]_2$$

$$M'(OPr^i)_2 + 2Ta(OPr^i)_5 \longrightarrow M'[Ta(OPr^i)_6]_2 \qquad (45)$$

$$M' = Mg, Ca, Sr, Ba$$

$$Ln(OPr^i)_3 + 3Al(OPr^i)_3 \longrightarrow Ln[Al(OPr^i)_4]_3 \qquad (46)$$

$$Ln = La, Pr, Nd, Sm, Gd, Ho, Er, Y, Yb, Sc$$

$$Ta(OPr^i)_5 + Al(OPr^i)_3 \longrightarrow [TaAl(OPr^i)_8] \qquad (47)$$

The reduction of transition metal alkoxides and phenoxides with alkali metals can generate mixed metal derivatives which may or may not be stable, for example equation (48).[154]

$$Ti(OAr)_4 + Na/Hg \xrightarrow{THF} [(THF)_2Na(\mu\text{-}OAr)_2Ti(OAr)_2] \qquad (48)$$

$$OAr = \text{2,6-diisopropylphenoxide}$$

15.3.3 PHYSICAL AND SPECTROSCOPIC PROPERTIES OF METAL ALKOXIDES AND ARYLOXIDES, $M(OR)_x$

Although the high electronegativity of oxygen would be expected to cause significant ionic character in M—OR bonds, the alkoxides or aryloxides of most elements demonstrate significant volatility and solubility in organic solvents. This apparent decrease in the polarity of the metal–oxygen bond may be rationalized in terms of the electron releasing effect of the alkoxide substituents, the presence of oxygen-p to metal-d π-bonding for earlier transition elements and the formation of oligomeric species through alkoxide or aryloxide bridges. This latter behavior is a dominant characteristic of metal alkoxides and aryloxides that greatly affects a number of physical and spectroscopic properties. Results indicate that the degree of oligomerization for a given stoichiometry $M(OR)_x$ and $M(OAr)_x$ is dependent on the following considerations: (i) oligomerization increases as the metal center becomes more electron deficient; (ii) the larger the size of the metal atom, then the greater is its ability to expand its coordination number and form bridging alkoxide or aryloxide groups; and (iii) the steric effects of the alkoxides or aryloxide substituents, which with increasing bulk oppose oligomerization, have been found to be of more importance than the electronic nature of the substituents in determining the ultimate degree of oligomerization. The unusually low solubilities and volatilities of metal methoxides apparently arises from a combination of these factors and the fact that the very small external methyl groups result in a high lattice energy between oligomers.

Of the vast array of physical techniques available to the modern chemist, many have been applied to the study of metal alkoxides and aryloxides. We have decided to summarize the physical and spectroscopic properties of alkoxides and aryloxides according to stoichiometry and include techniques only where they highlight a particularly interesting characteristic of these compounds. The mass of more routine data that has been accumulated cannot be covered in detail here.

In order to highlight the major aspects of the physical and spectroscopic properties of metal alkoxides and phenoxides, representative examples of homoleptic compounds of stoichiometries $M(OR)_x$ or $M(OAr)_x$ will be considered.

15.3.3.1 Monoalkoxides ($x = 1$)

Most alkali metal alkoxides and phenoxides tend to be involatile and undergo thermal decomposition on heating, even *in vacuo*. This property reflects the highly ionic and polymeric nature of these compounds. However, lithium *t*-butoxide containing the smallest and least electropositive metal combined with a large alkyl substituent can be sublimed (110 °C/0.1 Torr) and is readily soluble in most organic solvents.[155] Molecular weight measurements of $LiOBu^t$ in benzene indicate a hexameric oligomer is present.[155,156] Mass spectrometric measurements also indicate that hexamers $[LiOBu^t]_6$ are initially evaporated from the solid before being ionized and fragmented in the probe of the spectrometer.[156] The ^1H NMR spectrum of $LiOBu^t$ in CCl_4 shows only one signal, indicative of either equivalent CMe_3 groups or fast exchange between different environments.[155] Cryoscopic molecular weight measurements of TlOPh in benzene have indicated a concentration dependence between a dimer and a tetramer.[157]

15.3.3.2 Dialkoxides ($x = 2$)

The dialkoxides of the metals Zn, Cu, Cr, Fe, Ni, Co and Mn are non-volatile, insoluble compounds presumably due to their polymeric nature. However, by using a combination of studies of their magnetic properties and reflectance spectral characteristics, some information concerning the stereochemistry about the M^{2+} ion was obtained.[158] The tertiary alkoxides of the Group II metals are more volatile and soluble in organic media.[95] In contrast to the sparingly soluble phenoxides $[Be(OPh)_2]_n$ and $[Mn(OPh)_2]_n$, the use of the sterically demanding 2,6-di-*t*-butylphenoxide (OAr′) leads to soluble compounds $M(OAr′)_2$ (M = Be,[98] Mn[159]) which have been shown to be monomeric in benzene solution.

The alkylmetal alkoxides and phenoxides of the alkaline earth metals and zinc are much more soluble and volatile and have been extensively studied. Cryoscopic molecular weight measurements on compounds RZn(OR′) and RZn(OAr) in benzene indicate tetrameric species in solution.[95–98] ^1H NMR studies of a number of such compounds have been carried out.[98]

15.3.3.3 Trialkoxides ($x = 3$)

The alkoxides and phenoxides of boron are all generally readily distillable or sublimable compounds and tend to be monomeric in solution.[160] However, the *t*-butoxide undergoes thermal degradation to produce an alkene, presumably *via* a carbonium ion intermediate.[160]

The alkoxide derivatives of aluminum and gallium are all thermally stable and volatile, although the methoxides require high temperatures and low pressures for sublimation.[162] The *normal* alkoxides of both metals have been shown to be tetrameric in solution, decreasing to a dimeric form for the *t*-butoxides.[84] In the case of the simple phenoxides a dimer–trimer equilibrium is indicated cryoscopically in benzene.[20] An important observation by Mehrotra that freshly distilled $Al(OPr^i)_3$ contained a trimeric form which slowly 'aged' to a tetrameric form apparently explained some initially conflicting evidence concerning the solution molecular weight of this compound.[1,84] Further strong support for the molecular complexity of aluminum and gallium alkoxides in solution comes from NMR methods. Hence, the presence of two signals of intensity 1:2 in the ^1H NMR spectra for $[M(OBu^t)_3]_2$ (M = Al, Ga) is consistent with a dimeric structure with two bridging and four terminal alkoxides which do not undergo facile exchange with each other.[163] In the case of the aluminum isopropoxide, the ^1H NMR spectrum indicates two types of isopropoxide ligands of equal intensity with one set containing diastereotopic OCHMe_2 methyl groups.[163] This result, coupled with the ^{27}Al NMR spectrum containing two resonances in the ratio 3:1,[164] is consistent with a tetrameric solution structure which allows one of the aluminum atoms to obtain an octahedral environment (see Figure 1 on p. 348). Although considerable study has been made of the IR spectrum of aluminum alkoxides, its use as a probe of structure has proved difficult. Bands in the range 500–700 cm^{-1} typically belong to $\bar{v}$(Al—O) stretching frequencies with the characteristic v(C—O) bands lying between 950 and 1100 cm^{-1}.[165]

The trialkoxides of iron have certain physical similarities to those of aluminum.[52] Thus all of the derivatives except the methoxide are volatile and solution molecular weight measurements show them to be trimeric for primary alkoxides decreasing to dimers for the *t*-butoxide in refluxing benzene.[166] The oligomeric nature of iron(III) alkoxides has been used to explain their anomalous magnetic behavior.[167] Thus, at higher temperatures, magnetic moments approach the high-spin d^5 limit, while at lower temperatures the drop in magnetic moments can be attributed to antiferromagnetic coupling of the $S = 5/2$ iron centers within trimeric clusters.[167]

In the case of the alkoxides and phenoxides of molybdenum(III) the interaction between metal atoms becomes dominant, leading to diamagnetic compounds of stoichiometry $Mo_2(OR)_6$ or $Mo_4(OR')_{12}$ (R' = the less bulky Me or Et).[78] These compounds are readily sublimable, (70–90 °C/10^{-3} Torr for $Mo_2(OPr^i)_6$) and retain either their dimeric or tetrameric nature both in solution and the gas phase, as determined by cryoscopic measurements and mass spectrometry, respectively.[78] An analogous tungsten(III) complex $W_2(OBu^t)_6$ can be obtained, but for other alkoxides the simple homoleptic compounds cannot be isolated.[79,168]

A large number of alkoxides of the lanthanides have been synthesized.[169] Although the methoxides and ethoxides are involatile and insoluble, the isopropoxides will sublime and dissolve in organic solvents. A combination of mass spectral and cryoscopic measurements indicate a tetrameric structure for these compounds.[170]

The monomeric three-coordinate aryloxides of scandium[171] and titanium,[172] $M(OAr')_3$ (OAr' = 2,6-di-*t*-butylphenoxide), have been reported. The electronic and ESR spectra of deep blue $Ti(OAr')_3$ are consistent with a trigonal planar geometry and a strong $Ti(OAr')_3^+$ ion was observed in the mass spectrum.[172]

15.3.3.4 Tetraalkoxides ($x = 4$)

Tetraalkoxides and phenoxides are by far the most numerous of this series of compounds. Their physical properties have been extensively studied and they are very illustrative of factors that can affect the degree of oligomerization and volatilities.

Although the alkoxides and phenoxides of silicon[173] and germanium[174] are all monomeric and volatile, those of tin range from monomers for the *t*-butoxides to the involatile, highly associated methoxide $[Sn(OMe)_4]_n$.[175]

The tetra-*t*-butoxides of Ti, V, Cr, Zr and Hf are all readily distillable liquids, soluble in common organic solvents. Solution molecular weight measurements indicate a monomeric nature while the ^1H NMR spectra of the diamagnetic Group IV metal derivatives show only one Bu^t resonance consistent with this.[176] On moving to smaller, less branched alkyl substituents the molecular complexities of the titanium and zirconium alkoxides increase, reaching a maximum of four for the methoxides in solution.[177] For a given substituent the molecular complexity and boiling point of the zirconium analogue is higher, consistent with its larger covalent radius.[177,178] Interestingly, when the effect of oligomerization is not present, as in the $M(OBu^t)_4$ compounds, then the volatility increases Hf > Zr > Ti. However, although the volatilities of alkoxides of these metals decrease as oligomerization increases, there is evidence that the vaporization process in a large part involves deoligomerization on going from the liquid to the gaseous state. This is indicated by anomalously high values of the entropies of vaporization for the more associated alkoxides, as well as vapor density measurements.[180]

On the basis of ^1H NMR data it has not been possible to make any definite conclusions concerning the structures of oligomeric titanium and zirconium alkoxides. At room temperature only one set of CH_2CH_3 signals was observed for $[Ti(OEt)_4]_4$ and although the presumably rapid bridge–terminal exchange was slowed down at lower temperatures, no simple assignment could be made.[176] In the case of the isopropoxides of Zr and Hf, freshly distilled samples indicate only one type of $OCHMe_2$ group, while aged samples indicate the presence of a number of environments for the Pr^i groups. This was accommodated into a picture in which trimer to tetramer condensation occurs on aging, consistent with separate molecular weight measurements.[176]

The electronic and magnetic properties of vanadium tetramethoxide are consistent with a trimeric structure involving six-coordinate vanadium atoms, while the electronic spectrum of deep blue $V(OBu^t)_4$ is consistent with a slightly distorted monomeric, tetrahedral structure.[181]

Consistent with their much larger covalent radii, the tetraalkoxides of cerium and thorium are much less volatile than those of the Group IV metals.[182] Even for the *t*-butoxides, solution measurements indicate considerable oligomerization, with a molecular complexity as high as 3.4 for $Th(OBu^t)_4$.[183]

The physical properties of the tetraalkoxides of molybdenum and tungsten are dominated by the tendency of these metals to form strong metal–metal interactions. Hence, although $Mo(OBu^t)_4$ has been shown to be essentially monomeric both in solution and the gas phase with a magnetic moment of 1.38 BM, the dimeric isopropoxide $Mo_2(OPr^i)_8$ is diamagnetic, indicative of strong metal–metal bonding.[76] Its low-temperature ^1H NMR spectrum is consistent with its dimeric structure, but at higher temperatures rapid bridge–terminal exchange of alkoxide groups occurs.[76] The slightly paramagnetic $Mo(OEt)_4$ was found to be tetrameric both in solution and the gas phase[76] and presumably is isostructural with the diamagnetic compounds $W_4(OR)_{16}$ (R = Me, Et).[168] ^1H NMR spectra of these latter complexes indicate eight non-equivalent alkoxide environments with no facile exchange among them on the NMR timescale.[168] This contrasts with the labile behavior of $[Ti(OEt)_4]_4$ noted above.

15.3.3.5 Pentaalkoxides ($x = 5$)

For the penta-alkoxides and -phenoxides of niobium and tantalum the molecular complexity appears to be limited to a maximum of two. Hence even the methoxides are distillable and give solution molecular weights corresponding to $M_2(OMe)_{10}$ in non-coordinating solvents.[47,148] In the low-temperature ^1H NMR spectra of these methoxides, three signals of intensity ratio 2:2:1 were observed, consistent with a dimeric bioctahedral structure with two bridging methoxides. Interestingly the temperature dependence of their spectra indicates that terminal–terminal exchange of alkoxides occurs faster than bridge–terminal exchange.[185] Although solution measurements indicate that $M(OPr^i)_5$ and $M(OBu^t)_5$ (M = Nb, Ta) exist primarily as monomers at room temperature, the ^1H NMR spectra of the isopropoxides were found to be both temperature and concentration dependent consistent with the equilibrium shown in equation (49).[186] As expected, the dimeric form was slightly more favorable for M = Ta than M = Nb under identical conditions.

$$M_2(OPr^i)_{10} \rightleftharpoons 2M(OPr^i)_5 \qquad (49)$$

Considerable effort has been made to explain the IR spectra of $M_2(OEt)_{10}$ compounds. By using ^{18}O labelling and by making careful comparisons of spectra, both the bridging and terminal $\bar{v}(M—O)$ vibrations were identified.[165]

The simple phenoxides $M_2(OPh)_{10}$ (M = Nb, Ta) similarly behave as dimers in solution,[187] while there is ^1H NMR and IR spectroscopic evidence for the terminal and bridging aryloxides.[188] With the more sterically demanding 2,6-dimethylphenoxides $M(OAr)_5$, physical and spectroscopic data are consistent with their being monomeric.[65]

15.3.3.6 Hexaalkoxides ($x = 6$)

Hexaalkoxides are only known for the two elements tungsten and uranium. In both cases, physical and spectroscopic data are consistent with monomeric, octahedral complexes. Tungsten hexaalkoxides are pale yellow or colorless substances that readily sublime and exhibit the expected ^1H NMR and mass spectra.[189] The totally symmetric $\bar{v}(W—O)$ stretch occurs between 530 and 500 cm^{-1} in the IR spectrum, depending on the alkyl substituents. In contrast, the homoleptic phenoxide $W(OPh)_6$ is intense red in color due to the presence of a strong oxygen-to-metal charge transfer band in this d^0 complex which absorbs in the visible part of the spectrum.[36]

The volatility of $U(OMe)_6$ has recently attracted attention for laser-induced uranium isotope separation with a CO_2 laser.[190] At 330 °C, $U(OMe)_6$ has a vapor pressure of 17 mTorr and $\Delta H^\circ_{sublimation} = 96 \pm 13\ kJ\ mol^{-1}$ and $\Delta S^\circ_{sublimation} = 318 \pm 17\ J\ K^{-1}\ mol^{-1}$. From IR and Raman spectroscopic studies of the ^{16}O and ^{18}O labelled methoxide, a good vibrational analysis has been performed, allowing the assignments of the $U-^{16}O$ stretching frequencies: 505.0 cm^{-1} (A_{1g}), 464.8 cm^{-1} (T_{1u}) and 414 cm^{-1} (E_g). The predominant photoproducts are $U(OMe)_5$, MeOH and CH_2O and the enrichment of unreacted $U(OMe)_6$ in ^{235}U is a maximum at *ca.* 927 cm^{-1}, near what is thought to be a U—O stretching overtone. This is suggestive of a multiphoton U—O bond homolysis to produce $U(OMe)_5$ and methoxy radicals.

15.3.4 SOLID STATE AND MOLECULAR STRUCTURES

15.3.4.1 Prototypal Structures for Homoleptic Alkoxides

The primary factors influencing the adoption of one structural type in preference to another are (i) the empirical formula, (ii) the size and formal oxidation state of the metal atoms, (iii) the bulk of the alkyl/aryl group and (iv) for the transition elements the d^n configuration. The latter may lead to pronounced ligand field effects or metal–metal bonding. When the large body of data collected from spectroscopy and other physicochemical measurements are combined with knowledge of structures obtained from X-ray studies, a general classification of structural types is possible. It is, however, still not possible to predict with certainty the structure for any given RO ligand and metal and sometimes more than one structural type is possible. Prototypal structures for discrete oligomers are summarized in Figure 1.

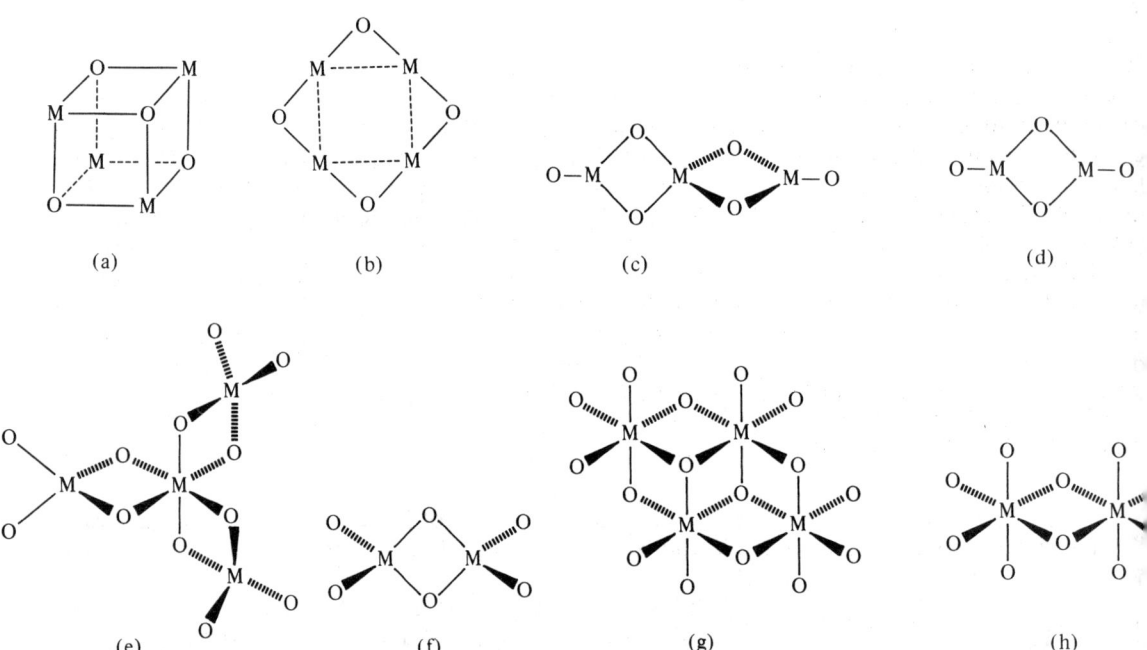

Figure 1 Prototypal structures for small oligomeric metal alkoxides: (a) the cubane-M_4O_4 unit in $[Tl(OMe)]_4$ and $[Na(OBu^t)]_4$; (b) the planar $[Cu(OBu^t)]_4$ structure; (c) and (d) fused trigonal and fused trigonal-tetrahedral units as seen in $[Be(OR)_2]_n$ compounds, where $n = 3$ and 2; (e) fused tetrahedral-octahedral units seen in $[Al(OPr^i)_3]_4$; (f) edge-shared tetrahedra as in $[Al(OBu^t)_3]_2$; (g) fused octahedra as in $[M(OEt)_4]_4$, where M = Ti, V, W; and (h) edge-shared octahedra as in $[Nb(OMe)_5]_2$

Polymeric methoxides of lithium[191] and sodium[192] are known to adopt a two-dimensional layer structure. Each metal atom is coordinated to four oxygen atoms. A different polymeric, double layer structure is found for KOMe.[193] Both potassium and oxygen atoms form two sheets and the methyl groups lie above and below these sheets. The potassium atoms are pentacoordinate and the oxygen atoms hexacoordinate (5K + 1C). Since the MeO$^-$ ligand has only three lone pairs a delocalized form of covalent bonding, characteristic of an electron deficient structure, must be envisaged.

A cubane structure is found for $[Tl(OMe)]_4$[194] and also for the *t*-butoxides of K, Rb and Cs

(Figure 1).[195] In this structure, each metal atom is coordinated to three oxygen atoms and the oxygen atoms are in a pseudo-tetrahedral environment.

An alternative to the cubane structure for a formula [M(OR)]$_4$ is seen in the planar structure of [Cu(OBu^t)]$_4$,[196] (see Figure 1). This structure is presumably favored because it allows Cu^I to have essentially linear O—Cu—O coordination. The Cu$\cdots$Cu distances fall in the range 2.65–2.77 Å, suggestive of some weak Cu—Cu bonding.

Polymeric structures are also proposed for many M(OR)$_2$ compounds, where R = Me or Et and M = a divalent metal such as Be, Zn and many of the first-row transition metals.[1] With bulkier alkyl or aryl groups lower oligomers are found, *e.g.* Be(OBu^t)$_2$[197] is trimeric and is believed to adopt a linear arrangement of Be atoms involving trigonal and tetrahedral coordination (Figure 1). The related perfluoro-*t*-butoxide is a dimer [Be(OC$_4$F$_9$-*t*)$_2$]$_2$[198] having trigonally coordinated Be atoms. Monomeric compounds M(OAr)$_2$ are seen for Ar = 2,6-di-*t*-butylphenyl and M = Ge, Sn and Pb.[199] These are found to adopt structures where the O—M—O angle is close to 90°.

Many M(OR)$_3$ compounds are also infinite polymers.[1] For example, the d^3 Cr^{3+} ion in Cr(OR)$_3$ compounds, where M = Me, Et, Pr^n or Pr^i, is almost certainly in a pseudo-octahedral environment since the electronic absorption spectra (diffuse reflectance) closely resemble that for the Cr(H$_2$O)$_6^{3+}$ ion. Al^{3+}, which is of a similar size to Cr^{3+} but lacks the ligand field stabilization of the d^3 configuration, forms lower oligomers. Al(OPr^i)$_3$, in one of its forms, adopts a tetrameric structure in which a central six-coordinate Al atom is surrounded by three four-coordinate Al atoms (Figure 1). With the more bulky Bu^{t}O group the dimer [Al(OBu^t)$_3$]$_2$ is obtained involving two AlO$_4$ units sharing an edge (Figure 1). The use of the extremely bulky aryloxide 2,6-di-*t*-butyl-4-methylphenoxide (OAr) has given rise to the isolation of a series of monomeric M(OAr)$_3$ compounds for Sc ($3d^0$), Y ($4d^0$) and the lanthanides ($4f^n$ where n = 0, 1, 2, 3, 9, 10, 11 and 13).[171] The structural characterization of the scandium complex reveals the expected trigonal planar coordination of the metal atom.[171]

Many monomeric M(OR)$_4$ compounds are known for bulky OR ligands,[1] *e.g.* M(OBu^t)$_4$ for M = Sn, Ti, V and Cr. These have essentially MO$_4$ tetrahedral units. With less bulky alkoxy ligands and/or larger metal ions, association to give five- or six-coordination is possible. The tetrameric structure involving four fused MO$_6$ octahedral units, M$_4$(μ_3-O)$_2$(μ_2-O)$_4$O$_{10}$ (Figure 1), is seen for [Ti(OR)]$_4$ compounds, where R = Me[200] and Et[201] and for W$_4$(OEt)$_{16}$.[168] A similar structural unit is seen in a wide variety of alkoxide and oxide structures of general formula M$_4$X$_a$Y$_b$Z$_c$, where X = OR, Y and Z are anionic ligands such as O^{2-}, halide or neutral ligands such as pyridine or ROH, and $a + b + c = 16$.[1,202]

M(OR)$_5$ compounds are known to be monomeric or dimeric. Examples of the latter include [Nb(OMe)$_5$]$_2$[203] and [U(OBu^t)$_5$]$_2$[204] which have edge-shared octahedral M$_2$O$_{10}$ units (Figure 1). An interesting example of a mixed valence compound of U^{4+} and U^{5+}, U$_2$(OBu^t)$_9$, has been found to have a confacial bioctahedral U$_2$O$_9$ moiety.[204] In both [U(OBu^t)$_5$]$_2$ and U$_2$(OBu^t)$_9$, the long U$\cdots$U distances, *ca.* 3.5 Å, preclude any direct metal–metal bonding.

Discrete MO$_6$ octahedra are seen in W(OR)$_6$ compounds[189] (R = Me, Et, Pr^i, CH$_2$Bu^t, Ph), (NEt$_4$)W(OPh)$_6$[205] and U(OR)$_6$ compounds.[190]

15.3.4.2 Metal–Metal Multiple Bonds and Clusters

The presence of strong M—M bonding dominates the structures of molybdenum and tungsten alkoxides.[5] The compounds Mo$_2$(OCH$_2$Bu^t)$_4$(PMe$_3$)$_4$,[206] Mo$_2$(OPr^i)$_4$[206] and Mo$_2$(OPr^i)$_4$(HOPr^i)$_4$[206] contain Mo≡Mo bonds with M—M distances in the range 2.10–2.20 Å. The compounds M$_2$(OR)$_6$ and M$_2$(OR)$_6$L$_2$, where M = Mo[78] or W[79], R = Bu^t, Pr^i or CH$_2$Bu^t, and L = a neutral donor ligand, have M≡M bonds with M—M distances in the range 2.22–2.35 Å. The triply and quadruply M—M bonded compounds are unbridged by alkoxide ligands. All known examples of M—M double and single bonded alkoxides of Mo and W have alkoxide bridges, however. There are M=M double bonds in Mo$_2$(OPr^i)$_8$[75] and W$_2$Cl$_4$(EtO)$_4$(EtOH)$_2$,[207] for example, with M—M distances of *ca.* 2.5 Å. The d^1–d^1 dimers M$_2$X$_4$(OR)$_6$, where M = Mo[208] and W[209], have distances in the range 2.70–2.75 Å with terminal M—X bonds (X = Cl, Br, I). The M-to-M distances are roughly 0.8 Å less than those seen in edge-shared octahedral d^0–d^0 dimers such as [Nb(OMe)$_5$]$_2$.[203]

The triangulo oxo-capped compounds of Mo^{IV}[210,211] and W^{IV}[212] of formula M$_3$(μ_3-O)(μ_3-OR)(μ_2-OR)$_3$(OR)$_6$ have M—M distances in the range 2.52–2.55 Å, consistent with M—M single bonds. Uranium forms a structural analogue U$_3$(μ_3-O)(μ_3-OBu^t)(μ_2-OBu^t)$_3$(OBu^t)$_6$[213] but the U-to-U distances of 3.57 Å are typical non-bonding M—M distances.[5] Related imido capped compounds W$_3$(μ_3-NH)(μ_3-OR)(μ_2-OR)$_3$(OR)$_6$[214] and alkylidyne capped compounds W$_3$(μ_3-CMe)(μ_2-OR)$_3$(OR)$_6$[215,216] are known, where R = Pr^i and CH$_2$Bu^t.

The tetranuclear compounds of formula $Mo_4X_4(OR)_8$ adopt structures dependent on the nature of X and R. When X = F and R = Bu^t there is a bisphenoid of Mo atoms having two short (2.24 Å) and four long (3.75 Å) Mo····Mo distances corresponding to Mo—Mo triple and non-bonding distances, respectively.[217,218] The molecule may be viewed as a dimer in which bridging fluoride ligands bring together two unbridged Mo≡Mo units: $[(\mu\text{-F})(Bu^tO)_2Mo\equiv Mo(OBu^t)_2(\mu\text{-F})]_2$. The compound $Mo_4Cl_4(OPr^i)_8$ has virtual D_{4h} symmetry with eight μ-OPr^i ligands and four terminal Mo—Cl bonds radiating from the square of Mo atoms.[219] The coordination geometry about each Mo is a square-based pyramid and the Mo—Mo distance is 2.37 Å. The compounds $Mo_4X_4(OPr^i)_8$, where X = Br^{219} and I,[220] have butterfly-Mo_4 units with five short Mo—Mo distances, *ca.* 2.5 Å, and one long distance, 3.1 Å. The structure has two μ_3-OR ligands, four μ_2-OR ligands and two terminal OR groups on the wing-tip Mo atoms. The local geometry for the MoO_4X moieties is again based on a square-based pyramid with the halide ligands in the axial position. Other tetranuclear Mo and W cluster alkoxides include $W_4(OEt)_{16}$,[168] which has the $[Ti(OEt)_4]_4$[201] type structure but with much shorter M—M distances, $W_4(H)_2(OPr^i)_{12}$,[221] $W_4(\mu\text{-CSiMe}_3)_2(OEt)_{14}$[216] and $W_4(C)(NMe)(OPr^i)_{12}$.[5,222]

The hexanuclear Mo^{II} methoxides $Na_2Mo_6(OMe)_{14}$ and $Na_2Mo_6Cl_8(OMe)_6$ contain central 24-electron octahedral Mo_6 clusters.[223] There are eight μ_3-X ligands (X = Cl or OMe) and six terminal OMe ligands which radiate from the center of the Mo_6 octahedra.

Though the occurrence of M—M bonds and clusters in alkoxide chemistry is presently only well documented for Mo and W, an extension to some neighboring elements seems likely. The d^1–d^1 dimer $Re_2O_3(OMe)_6$ shows a typical M—M single bond distance, 2.56 Å, for a confacial bioctahedral structure in $(O)(MeO)_2Re(\mu\text{-O})(\mu\text{-OMe})_2Re(O)(OMe)_2$.[224]

15.3.4.3 Oxoalkoxides

Of all the various classes of compounds which contain alkoxide ligands, brief mention is restricted to oxoalkoxides. The limiting situations are mononuclear species, such as phosphate esters and transition metal analogues like $MoO_2(OBu^t)_2$, $WO(OBu^t)_4$ and polymeric species. The preparation of the latter group has often involved partial hydrolysis of polymeric alkoxides. These neutral molecules are seen to have M_xO_y skeletons involving fused octahedra and they resemble subunits of the structures of isopoly- and heteropoly-oxoanions of Mo, W and V. The central Ti_7O_{24} unit of $Ti_7O_4(OEt)_{20}$[225] and the Nb_8O_{30} unit in $Nb_8O_{10}(OEt)_{20}$[226] are shown in Figure 2. Analogy with the polyoxoanion structures is even more striking when one recognizes that O-alkylation of $M_{12}PO_{40}^{3-}$ (M = Mo and W) with $Me_3O^+ BF_4^-$ yields $MeOM_{12}PO_{39}^{2-}$ salts.[227] A single-crystal X-ray study on the molybdenum complex revealed that the OMe ligand resides at a μ_2-site.[227] The molecular structure of $Zr_{13}O_8(OMe)_{36}$[228] is also of interest in this context. The $Zr_{13}O_{44}$ unit is shown in Figure 3. There is a central Zr atom surrounded by eight oxygen atoms which are themselves bonded to four Zr atoms. The central ZrO_8 unit is a cube. Surrounding this central Zr atom are 12 Zr atoms each of which is seven-coordinate, being bonded to two μ-oxo groups, four μ_2-OMe groups and one terminal OMe group. Again a resemblance to an oxide structure is seen, namely the baddeleyite form of ZrO_2. Compounds of this type form bridges between polymeric metal alkoxides and the extended arrays found in the solid state structures of metal oxides. They may well provide hydrocarbon soluble complexes capable of modeling the chemical reactivity of heterogeneous metal oxide catalysts.[202]

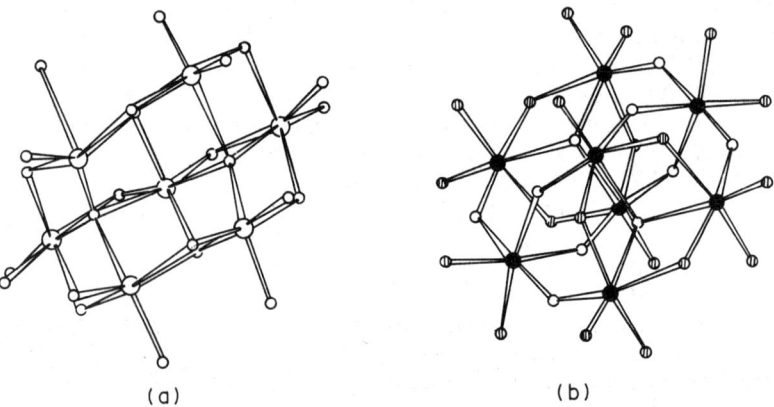

(a)   (b)

Figure 2   (a) The Ti_7O_{24} unit in $Ti_7O_4(OEt)_{20}$ and (b) the Nb_8O_{30} unit in the $Nb_8O_{10}(OEt)_{20}$ molecule

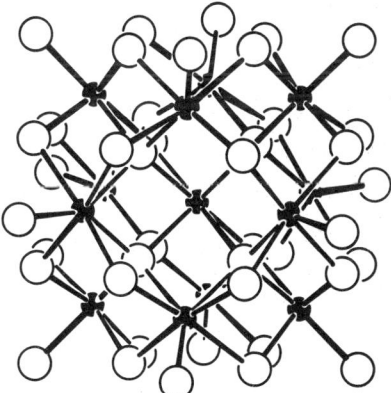

Figure 3 The $Zr_{13}O_{44}$ unit in $Zr_{13}O_8(OMe)_{36}$

15.3.4.4 Metal–Oxygen π-Bonding

M—O bond distances follow the order M—OR $(\mu_3) > (\mu_2) >$ terminal. Within the same molecule the distances can differ by up to 0.4 Å. Shorter terminal M—O distances are also characteristically associated with larger M—O—C angles which may be close to 180°. Conversely, relatively long terminal M—OR distances are associated with small M—O—C angles, typically in the range 120–130°. However, although one would expect aryloxide ligands to be poorer π-donors than their aliphatic counterparts, structural studies show that M—O—Ar angles are characteristically larger than M—O—R angles, even in the absence of steric effects. Angles of 160–180° are common for these ligands, although M—OAr distances are slightly longer than M—OR distances in related molecules.[89,100] The ability of aryloxides to readily increase the angle at oxygen can be accommodated into a resonance form of the type M≕O⇌Ar. A comparison of the bonding of alkoxide and aryloxide ligands to the same metal center has been reported.[89] These structural changes can, at least in most cases, be traced directly to metal–oxygen π-interactions. If the RO^- ligand is counted as a two-electron σ-donor, then this can be supplemented by π-interactions ranging from π^2 to π^4. If the metal is electron deficient, then extremely short M—O distances and large M—O—C angles are seen. For example, the $(tritox)_2ZrX_2$ compounds,[66] where tritox = Bu_3^tCO, may be viewed as analogues of Cp_2ZrX_2 compounds with the tritox ligand acting as a six-electron donor to the Zr^{4+} center. Similarly, the extremely short U—O distances and large U—O—C angles reported[213,214] for uranium alkoxides imply extensive RO-to-U π-bonding. A fine gradation of M—O and M—O—C angles has been noted[5] in the chemistry of molybdenum and tungsten alkoxides as a function of the number of alkoxide ligands competing for one or two vacant d-orbital sites. When there are no vacant d-orbitals, as in $(dppe)_2Pt(OMe)(Me)$,[215] or the d-orbitals are otherwise occupied in M—M bonding, *e.g.* in $Mo_2(OR)_4L_4(M≡M)$[206] compounds and $Na_2Mo_6(OMe)_{14}$,[223] then bonds approaching those anticipated for single M—O bond distances are seen with M—O—C angles close to 120°. The π-donating properties of the alkoxide ligand can have pronounced effects on the bonding and chemical reactivity of neighboring ligands. Also the extent of π-bonding influences the reactivity of the M—OR group toward electrophilic attack and insertion reactions. These effects are discussed in Sections 15.3.5.4 and 15.3.5.5.

15.3.5 REACTIONS OF METAL ALKOXIDES AND ARYLOXIDES

15.3.5.1 Ligand Coordination

This is an important reaction in metal alkoxide and aryloxide chemistry as it typically represents the initial step whereby reagents enter the metal coordination sphere. Reactions that appear to occur by direct interaction with M—OR bond are rare. The Lewis acidity of the metal centre in $M(OR)_x$ compounds is one of the main reasons for oligomerization. Clearly the coordination expansion of the metal can occur using lone pair electrons from added donor ligands just as easily as using the lone pairs of alkoxide ligands to form μ_2- or μ_3-OR linkages. However, the added donor ligand will have to compete thermodynamically with the oligomerization process (equation 50).

$$L_nM \overset{\overset{\displaystyle R}{\overset{\displaystyle |}{\underset{\displaystyle O}{}}}}{\underset{\underset{\displaystyle R}{\underset{\displaystyle |}{O}}}{}} ML_n + 2L \rightleftharpoons 2L_nM(OR)L \qquad (50)$$

For alkoxides of the Group I and II elements and the lanthanides, typical adducts involve coordinating O-donors such as Et_2O or THF. For example, the sparingly soluble, anhydrous aryloxide LiOAr′ (OAr′ = 2,6-di-*t*-butylphenoxide) will dissolve in ethers to give dimeric adducts of the type $[(OAr′)Li(L)]_2$ (L = Et_2O, THF, py), which have been structurally characterized.[229]

Aluminum alkoxides form few well-behaved adducts with nitrogen bases. However, species of the type $[Al_2(OR)_6(en)]$ (en = ethylenediamine) have been isolated.[230] The parent alcohol itself can in some cases act as the donor group. Hence, the isopropoxide of zirconium can be readily recrystallized from isopropyl alcohol as the adduct $Zr(OPr^i)_4 \cdot Pr^iOH$.

The dinuclear molecules $Mo_2(OR)_6$ will readily coordinate donor ligands in hydrocarbon solvents to set up equilibria such as equation (51).[231] In the case of the tungsten analogues the reaction is so favorable that it is not possible in some cases to obtain the base-free alkoxide. In these cases the act of synthesis, *via* the parent amide, generates the coordinating ligand (equation 52).[5,79,232] Similarly, the alcoholysis of $Mo(NMe_2)_4$ can lead to dimethylamine adducts that in some cases will sublime unchanged (equation 53).[76]

$$Mo_2(OR)_6 + 2L \rightleftharpoons Mo_2(OR)_6(L)_2 \qquad (51)$$

$$L = HNMe_2, py, PR_3$$

$$W_2(NMe_2)_6 + 6HOR \longrightarrow W_2(OR)_6(HNMe_2)_2 + 4HNMe_2 \qquad (52)$$

$$Mo(NMe_2)_4 + 4Me_3SiOH \longrightarrow Mo(OSiMe_3)_4(HNMe_2)_2 + 2HNMe_2 \qquad (53)$$

The dimeric complexes $M_2(OAr)_{10}$ (M = Nb, Ta) will form monomeric adducts with ethers or nitrogen bases (equation 54).[233]

$$Nb_2(OAr)_{10} + 2THF \longrightarrow 2Nb(OAr)_5(THF) \qquad (54)$$

The coordination of an alkoxide or aryloxide can lead to a mixed metal derivative.[149] In some cases, ligand exchange can take place to generate a neutral adduct (equation 55).[154]

$$Ti(OAr)_4^- + 2py \longrightarrow Ti(OAr)_3(py)_2 + OAr^- \qquad (55)$$

The replacement of alkoxide ligands by halide groups generally results in a more Lewis acidic metal center and more stable adducts result.[1-5]

15.3.5.2 Protonolysis Reactions

An important class of reactions of alkoxides and aryloxides involves the substitution of an alkoxide or aryloxide by protonation and elimination of alcohol or phenol (equation 56).

$$L_nM(OR) + HX \longrightarrow L_nM(X) + HOR \qquad (56)$$

A large number of molecules can react in this way and typically HX contains an H—S or H—O bond or else is a hydrohalic acid. There are both kinetic and thermodynamic considerations as to whether this type of reaction can take place. Firstly, the mechanism of the reaction rarely involves direct protonation of the M—OR bond. Instead, initial coordination of HX through lone pairs of electrons on X is necessary prior to transfer of the proton. Hence, the rate of the reaction will be dependent on the steric constraints of both HX and the metal coordination sphere as well as the electronic donor–acceptor properties of the two substrates. Thermodynamically the position of the equilibrium will depend on a number of variables, the relative strengths of the M—O and M—X bonds being important ones.

The hydrolysis of early transition metal alkoxides is an extremely facile reaction that makes the manipulation of these materials non-routine. The typical insolubility and high molecular complexity of the resulting metal oxides or mixed oxoalkoxides is an added driving force to these reactions (equations 57 and 58).

$$L_nMOR + H_2O \longrightarrow L_nMOH + ROH \tag{57}$$

$$L_nMOH + L_nM{-}OR \longrightarrow (L_nM)_2O + ROH \tag{58}$$

An in-depth study of the industrially important hydrolysis of titanium alkoxides has been carried out by Bradley.[234,235] A number of intermediate complexes were isolated and characterized. The alcohol exchange reaction has been discussed previously. The addition of hydrohalous acids to alkoxides is clearly related to the reverse reaction, the addition of alcohols to metal halides. In general, the products of these two reactions will be the same (equation 59). Hence, complete substitution will occur to give metal halides that are known to form only alcoholates with alcohols (equations 60 and 61).[31,236]

$$M(OR)_n + mHX \longrightarrow M(OR)_{n-m}(X)_m + mHOR \tag{59}$$

$$Fe(OPr^i)_3 + 3HCl \longrightarrow FeCl_3 \cdot 2Pr^iOH + Pr^iOH \tag{60}$$

$$Ln(OR)_3 + 3HCl \longrightarrow LnCl_3 \cdot 3ROH \tag{61}$$

$$Ln = \text{a lanthanide}$$

In general the organic carboxylic acids will react with complete removal of alkoxides to form metal carboxylates. An important thermodynamic driving force here is the formation of the bidentate O_2CR linkage (equation 62). This method represents a convenient route to carboxylates, and has been applied to a number of metals.[237]

$$M(OR)_n + nR'CO_2H \longrightarrow M(O_2CR')_n + nROH \tag{62}$$

Although formally involving the reaction with a C—H bond, metal alkoxides will react with β-diketones and β-keto esters to form six-membered chelates and alcohols. Hence the acetylacetonate (acac) derivatives of aluminum can be obtained (equation 63).[238]

$$Al(OR)_3 + 3acacH \longrightarrow Al(acac)_3 + 3ROH \tag{63}$$

In some cases the rate of substitution of alkoxide ligands drops off dramatically before total substitution has occurred. This clearly reflects both the electronic and steric saturation of the metal coordination sphere by the bidentate acac ligands. Hence for titanium it is difficult to substitute the last alkoxide,[239] while for Nb and Ta the reaction stops at the eight-coordinate tris-substitution products (equation 64).[240,241]

$$M(OEt)_5 + 3MeCOCH_2COR \longrightarrow (OEt)_2M(OCMeCHCOR)_3 + 3EtOH \tag{64}$$

There are also a vast number of reactions known between alkoxides and non-trivial alcoholic reagents such as glycol, catechols, alkanolamines and hydroxylamines. The fundamental reaction involves elimination of alcohol, but with the reaction products complicated by the nature of chemistry of the other substituents.[1]

15.3.5.3 The Meerwein–Ponndorf–Verley Reaction

It was noticed as early as 1925 that alkoxides of calcium, magnesium and particularly aluminum could catalyze the reduction of aldehydes by ethanol as shown in equation (65).[242,243] Removal of very volatile acetaldehyde is easily achieved to drive the reaction to the right. In 1926, Ponndorf devised a method in which both aldehydes and ketones could be reduced to alcohols by adding excess alcohol and aluminum isopropoxide.[244] Such reductions are today referred to as Meerwein–Ponndorf–Verley reactions. Although alkoxides of a number of metals, *e.g.* sodium, boron, tin, titanium and zirconium, have been used for these reactions, those of aluminum are by far the best.

$$RCHO + MeCH_2OH \longrightarrow RCH_2OH + MeCHO \tag{65}$$

A number of mechanistic studies of the reaction have been made and a cyclic transition state in which hydrogen transfer occurs from the β-C—H bond of an alkoxide to a coordinated ketone or aldehyde has been postulated.[245] However, kinetic measurements are complicated by the oligomeric nature of the aluminum reagent (see **3**).[246]

$$
\begin{array}{c}
RO \quad\quad RO \\
\diagdown \quad \diagup \\
Al \\
\diagup \quad\quad \diagdown \\
O \quad\quad\quad O \\
| \quad\quad\quad\quad | \\
R'-C \quad\quad\quad C-R'' \\
\diagup \quad\quad\quad\quad \diagdown \\
R' \quad\quad H \quad\quad R''
\end{array}
$$

(3)

Another synthetic use of the M–P–V reaction has been exploited in the synthesis of new alkoxide derivatives.[247,248] Hence, 2-hydroxybenzaldehyde (salicylaldehyde) will react with isoproxides of titanium and zirconium to initially displace isopropyl alcohol and give mixed derivatives which on heating undergo reduction of the aldehyde function with elimination of acetone. The stoichiometry of the overall reaction can be represented as shown in equation (66).[248]

$$ L_nM(OCHMe_2)_2 + HOC_6H_4CHO \longrightarrow L_nM(OC_6H_4CHO) + Me_2CHOH + Me_2CO \quad (66) $$

15.3.5.4 Insertion Reactions

15.3.5.4.1 Involving OCO, RNCO or RNCNR

Some metal alkoxides will react with some or all of the three isoelectronic molecules carbon dioxide, aryl or alkyl isocyanates, and carbodiimides to form the corresponding adducts as shown in equation (67). The resulting ligand can adopt a bidentate bonding mode and this can be a significant thermodynamic driving force for these reactions, some of which have been shown to be reversible. A good example is the insertion of CO_2 into the metal–alkoxide bonds of the compounds $Mo_2(OR)_6$. Only two moles of CO_2 are absorbed reversibly to give the bis(alkyl carbonates) shown to contain these new ligands bridging the dimetal center (equation 68).[249]

$$
\begin{array}{c}
\quad\quad\quad\quad\quad Y \\
\quad\quad\quad\quad\quad \| \\
MOR + X{=}C{=}Y \rightarrow MX-C-OR \leftrightarrow M\!\!\begin{array}{c}\diagup Y\\ \diagdown\end{array}\!\!COR \\
\quad (X, Y = O \text{ or } NR) \quad\quad\quad\quad X
\end{array}
\quad (67)
$$

$$ Mo_2(OR)_6 + 2CO_2 \rightleftharpoons Mo_2(OR)_4(O_2COR)_2 \quad (68) $$

Similar reactivity towards phenylisocyanate is found, while carbodiimide yields a 1:1 adduct in which insertion into the M—OR bond has not taken place.[250,251] Copper(II) methoxide undergoes the reversible insertion of CO_2 to yield the corresponding carbonate.[252]

The alkoxides of titanium and zirconium will undergo reaction with both phenyl isocyanate and carbodiimide.[253-255] Both types of reaction were shown to be reversible (equation 69).[253]

$$ M(OR)_4 + nR'NCO \longrightarrow (RO)_{4-n}M(OC(OR)NR')_n \quad (69) $$

15.3.5.4.2 Involving alkenes

The compound (dppe)PtMe(OMe),[256] which is prepared by a metathesis reaction involving NaOMe and (dppe)PtMe(Cl) in a mixed benzene/methanol solvent system (dppe = bis(1,2-diphenylphosphino)ethane), does not react with ethylene or pentene but does react with activated alkenes such as acrylonitrile, methylacrylate and fluoroalkenes. The reaction involving tetrafluoroethylene has been shown to give (dppe)PtMe(CF_2CF_2OMe), providing the first example of an alkene insertion into an M—OR bond.[256] Interestingly, no insertion into the Pt—Me bond was observed.

A mechanistic study of the insertion reaction in THF-d_8 showed[257] that the reaction was first order in (dppe)PtMe(OMe) and first order in C_2F_4 over a wide alkene concentration. Activation parameters were $\Delta H = 52.3\,\text{kJ mol}^{-1}$ and $\Delta S = -134\,\text{J mol}^{-1}\text{K}^{-1}$ and NMR studies showed that at low temperatures there was formation of a 1:1 alkene–Pt complex. Studies of the insertion reaction involving the protio-methoxy compound and perfluorocyclopentene in the presence of CD_3OD (10 equiv.) showed less than 8% incorporation of OCD_3 into the β-methoxyfluoroalkyl ligand. Collectively these results provide a picture of the alkene insertion into the Pt—OMe bond which directly parallels that of alkene insertion into Pt—alkyl bonds involving initial reversible π-complex formation followed by migratory insertion. Of particular note is that the reaction does not involve nucleophilic attack on a coordinated alkene by free methoxide ion. The replacement of an M—OR bond by an M—R bond is presumably thermodynamically favorable in this instance because a late transition metal–oxygen bond is replaced by a metal–carbon bond of a fluoroalkyl ligand.

15.3.5.4.3 *Involving carbon monoxide*

The compound (dppe)Pt(OMe)$_2$, which is formed in the reaction between NaOMe (2 equiv.) and (dppe)PtCl$_2$, reacts with CO to give (dppe)Pt(CO$_2$Me)$_2$.[258] Mechanistic studies reveal[259] that the reaction rate is first order in (dppe)Pt(OMe)$_2$ and first order in CO concentrations and, furthermore, when the reaction is carried out in CD_3OD less than 10% of CD_3O is incorporated into the Pt—CO$_2$Me ligand. This provides the first example of a carbonyl migratory insertion into an M—OR bond, rather than the more common attack by OR$^-$ on a carbonyl ligand in the formation of metallacarboxylates.

15.3.5.5 Ether Syntheses

The so-called Williamson synthesis of ethers is by far the most important ether synthesis because of its versatility: it can be used to make unsymmetrical ethers as well as symmetrical ethers, and aryl alkyl ethers as well as dialkyl ethers. These reactions involve the nucleophilic substitution of alkoxide ion or phenoxide ion for halide (equation 70).[260]

$$RX + NaOR \text{ or } NaOAr \longrightarrow ROR \text{ or } ROAr + NaX \qquad (70)$$

The reaction must involve an alkoxide or aryloxide of a highly electropositive metal, *i.e.* the M—O bond must have high ionic character. Alkoxides of molybdenum and tungsten of formula M$_2$(OR)$_6$, which have strong M—O σ- and π-bonds will not, for example, react with alkyl halides to yield ether.[261]

15.3.6 ORGANOMETALLIC CHEMISTRY SUPPORTED BY ALKOXIDE LIGANDS

15.3.6.1 Carbonyls of Molybdenum and Tungsten

Addition of carbon monoxide to hydrocarbon solutions of M$_2$(OR)$_6$ (M≡M) compounds leads to M(CO)$_6$ and alkoxy derivatives of Mo and W in oxidation states ranging from +4 to +6. The course of the reaction depends upon the metal and the alkoxy ligand. In the case of the reaction between Mo$_2$(OBu^t)$_6$ and CO, the stoichiometric reaction (71) has been established.[262]

$$2\text{Mo}_2(\text{OBu}^t)_6 + 6\text{CO} \longrightarrow \text{Mo(CO)}_6 + 3\text{Mo(OBu}^t)_4 \qquad (71)$$

The first step in these reactions involves the formation of monocarbonyl adducts containing the M$_2$(μ-CO) moiety. The compounds M$_2$(OBu^t)$_6$(μ-CO), where M = Mo[262] and W[263], have been fully characterized and shown to contain metal atoms in a pseudo-square based pyramidal geometry: (Bu^{t}O)$_2$M(μ-CO)(μ-OBu^t)$_2$. In the case of the less bulky alkoxides Pr^{i}O and Bu^{t}CH$_2$O, and in the presence of pyridine, compounds of formula M$_2$(OR)$_6$(py)$_2$(μ-CO), where M = Mo[264] and W[263], have been isolated. These adopt structures based on a confacial bioctahedral geometry (py)(RO)$_2$-M(μ-CO)(μ-OR)$_2$M(OR)$_2$(py) in which the pyridine ligands occupy positions *trans* to the bridging CO ligand. These compounds show unprecedentedly low μ-CO stretching frequencies for neutral compounds: $\bar{\nu}$(CO) *ca.* 1650 cm^{-1} for M = Mo and *ca.* 1570 cm^{-1} for M = W. The ^{13}C chemical shifts for the μ-CO carbon atoms are also downfield, below 300 p.p.m. relative to Me$_4$Si. Both of these observations have been reconciled with the oxycarbyne resonance forms shown below in (**4a**), (**4b**) and (**4c**).

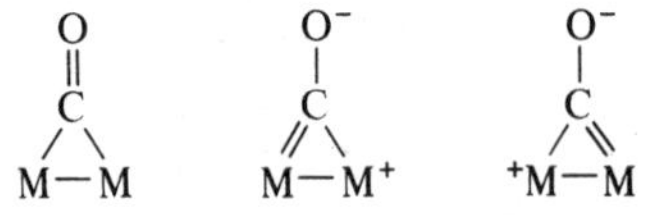

(**4a**)　　　　(**4b**)　　　　(**4c**)

This view leads to the expectation that the oxygen atom of the μ-CO ligand should be nucleophilic. Support for this view is seen in the fact that $W_2(OPr^i)_6(py)_2(\mu$-CO) reacts in solution to give $W_4(\mu$-CO$)_2(OPr^i)_{12}(py)_2$[263,265] and addition of Pr^iOH to $W_2(OBu^t)_6(\mu$-CO) gives $W_4(\mu$-CO$)_2(OPr^i)_{12}$.[263] The central $[W_2(\mu$-CO$)]_2$ moiety in these compounds can be viewed in terms of the valence bond descriptions (**5a**) and (**5b**) shown below. The W—W distances, 2.67 Å, and the C—O distances, 1.35 Å, may be taken as evidence for C—O and W—W single bond character, as can $v(^{12}C—^{16}O) = 1272 \, cm^{-1}$ (identified by ^{13}CO labelling studies).[263]

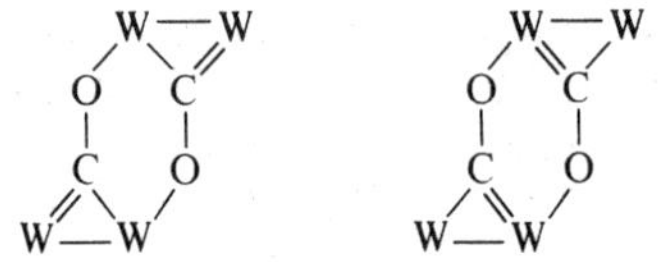

(**5a**)　　　　　　　(**5b**)

In the case of carbonylation of $Mo_2(OPr^i)_6$, the oxidized form of molybdenum is $Mo_2(OPr^i)_8$-$(CO)_2$, which is believed to have an edge-shared bioctahedral geometry, $(Pr^iO)_4Mo(\mu$-$OPr^i)_2Mo(OPr^i)_2(CO)_2$ with *cis* CO ligands.[217] Carbonylation of $W_2(OPr^i)_6(py)_2$ has been shown to give a shared bioctahedral structure $(Pr^iO)_4W(\mu$-$OPr^i)_2W(CO)_4$, with a pair of very asymmetric μ-OPr^i ligands and a long W—W distance, 3.1 Å, implying a $W(CO)_4$, t_{2g}^6 fragment, ligated to a $W(OPr^i)_6$ (W^{6+}) group by a pair of RO bridges.[266]

Carbonylation of $Mo_2(OBu^t)_6$ in the presence of pyridine led to the isolation of $Mo(OBu^t)_2(py)_2$-$(CO)_2$ containing a distorted octahedral Mo^{2+} center having *cis* CO, *cis* py and *trans* OBu^t ligands. The molecule shows unusually low CO stretching frequencies, 1906 and 1776 cm^{-1}, for a *cis* dicarbonyl Mo^{2+}-containing compound. This may reflect a combination of (i) the strong π-donor properties of the Bu^tO ligands and (ii) the small C—Mo—C angle, 72°.

15.3.6.2 Nitrosyl Derivatives

A series of nitrosyl derivatives is known for Cr, Mo and W. These all have linear M—NO groups and include $(Bu^tO)_3Cr(NO)$,[267] $(Pr^iO)_2(NO)M(\mu$-$OPr^i)_2$ (where $M = Cr$[268] and Mo),[269] $(W(OR)_3NO)_x$ (where $R = Bu^t$ and Pr^i),[270] and $W(OBu^t)_3(NO)py$.[270] These compounds are synthetically accessible by either addition of NO to $M_2(OR)_6(M≡M)$ compounds or by alcoholysis reactions involving $Cr(NPr^i)_3(NO)$.[268] Though the formulas of these compounds differ they all contain metal atoms in either trigonal (C_{3v}) or trigonal bipyramidal environments. If the linear M—N—O functions are viewed as $M^-(NO^+)$, then the formal oxidation state of the metal atoms is $+2$ and backbonding to the π-acid ligand NO^+ involves d^4 metal centers: $(d_{xz}, d_{yz})^4$ if the M—N—O axis is taken as the z axis. The values of $v(NO)$ fall in the range 1720 cm^{-1} ($M = Cr$), 1640 cm^{-1} ($M = Mo$) and 1550 cm^{-1} ($M = W$), This reflects the greater π-backbonding capabilities of the metal atoms, $W > Mo > Cr$, in their middle oxidation states.

An extensive series of compounds of formula $Mo(pz)(NO)(X)(Y)$ has been synthesized and characterized, where $pz = $ a substituted tridentate pyrazolylborate ligand and X, Y = various combinations of halides, OR, SR and NHR.[271,272] The values of $\bar{v}(NO)$ for the linear Mo—NO groups reflect the backbonding capabilities of the formally Mo^{2+} centers and fall within the range 1720 to 1630 cm^{-1}. From this series the relative electron donating properties of the ligands X and Y, both σ and π, are seen to follow the order $NHR > OR > SR > $ halide. The order $OR > SR$ is particularly interesting since the electronegativity of oxygen is greater than that of sulfur. It appears that this is offset by the greater π-donating ability of alkoxides relative to thiolates. Another comparison of the ligands can be made when $\bar{v}(NO) = 1640 \, cm^{-1}$ in $[Mo(OR)_3NO]_2$ compounds is compared to $\nu(NO) = 1675\,cm^{-1}$ in the anion $[Mo(SPh)_4NO]^-$.[273] In these compounds molybdenum is in a distorted trigonal bipyramidal geometry with a linear M—NO group occupying one of the axial positions.

15.3.6.3 Alkyne Complexes

The reactions between alkynes and $M_2(OR)_6(M\equiv M)$ compounds have been extensively studied and lead to a wide variety of products depending upon the metal (M = Mo or W), the alkoxide and the alkyne. For both molybdenum[274] and tungsten,[275,276] reactions involving acetylene give poly-acetylene and in the presence of pyridine 1:1 adducts, $M_2(OR)_6(\mu\text{-}C_2H_2)(py)_n$, where $n = 1$ or 2, and products of C—C coupling, $M_2(OR)_6(\mu\text{-}C_4H_4)(py)_n$, where $n = 1$ or 0. A variety of structural types is seen for the alkyne adducts with respect to the presence of one or two μ-OR ligands and the number and placement of the pyridine ligands at the dimetal center. However, common to all structures is the presence of a pseudo-tetrahedral M_2C_2 core. The W—W, W—C and C—C distances are approaching single bond distances which has led to the suggestion that these compounds be viewed as dimetallatetrahedranes. Also the ^{13}C–^{13}C coupling constants for the μ-C_2H_2 ligands are small, *ca.* 15 Hz for M = W and 24 Hz for M = Mo.

The $M_2(OR)_6(\mu\text{-}C_2H_2)(py)_n$ compounds are labile toward C–C coupling reactions with alkynes[274,276] and, for M = W, with nitriles.[277] Structurally characterized examples of the latter include $W_2(OBu^t)_6(CHCHCPhN)$, $W_2(OPr^i)_7(CH_2CHCPhN)$, $W_2(OCH_2Bu^t)_6(N(CMe)_4N)py$ and $W_2(OPr^i)_7(NHCMeCHCHCMeN)$, which provide examples of M—C, M=C, M=N, M—N and $M_2(\mu\text{-}N)$ bonds within heterocyclic ligands fused to the dinuclear center.[277] The $Mo_2(\mu\text{-}C_2H_2)$- and $Mo_2(\mu\text{-}C_4H_4)$-containing compounds have been shown to be intermediates in the catalytic trimerization of acetylene to benzene.[274] Tungsten forms $W_2(OPr^i)_6(\mu\text{-}C_4R_4)(\eta^2\text{-}C_2R_2)$ compounds[276], where R = H and Me, in which two alkyne units have been coupled to give a metal-lacyclopentadiene ligand π-bonded (η^4) to the neighboring tungsten atom which is η^2-bonded to an alkyne ligand. Though these compounds appear as likely models for intermediates in the cyclo-trimerization reactions observed for molybdenum, the tungsten compounds will not form benzenes catalytically.[276]

A number of reactions between $M_2(OR)_6$ compounds and alkynes have been found to give alkylidyne compounds by a metathesis of the $M\equiv M$ and $C\equiv C$ bonds (see Sections 15.3.6.4) and evidence has been presented[275] for an equilibrium between a ditungstatetrahedrane and a methyl-idyne–tungsten complex supported by Bu^tO ligands (equation 72).

$$W_2(OBu^t)_6(\mu\text{-}C_2H_2) \rightleftharpoons 2(Bu^tO)_3W\equiv CH \qquad (72)$$

15.3.6.4 Alkylidene and Alkylidyne Complexes

Alkoxide ligands play an important spectator role in the chemistry of metal–carbon multiple bonds.[278] Schrock and coworkers[278] have shown that niobium and tantalum alkylidene complexes are active toward the alkene metathesis reaction. One of the terminating steps involves a β-hydrogen abstraction from either the intermediate metallacycle or the alkylidene ligand. In each case the β-hydrogen elimination is followed by reductive elimination. The net effect is a [1,2] H-atom shift, as shown in equations (73) and (74), and a breakdown in the catalytic cycle. Replacing Cl by OR ligands suppresses these side reactions and improves the efficiency of the alkylidene catalysts.[279]

$$\qquad (73)$$

$$\qquad (74)$$

The compounds $(Bu^tO)_3W\equiv CR$ are alkyne metathesis catalysts.[280,281] Probable intermediates in these reactions are tungstacyclobutadienes and the latter are seen in the presence of Cl ligands.[282] Addition of alkynes to the latter leads to the formation of substituted η^5-cyclopentadiene derivatives of lower valent tungsten. Again the RO ligand is favored over Cl by stabilizing the alkylidyne ligand relative to the metallacyclobutadiene and by stabilizing the W^{6+} oxidation state, thus preventing reduction to lower valent complexes.

A fairly general route to alkylidyne complexes of the type $W(CX)(OBu^t)_3$ involves the metathesis reaction shown in equation (75).[283,284]

$$W_2(OBu^t)_6 + XC\equiv CX \text{ or } XC\equiv CR \longrightarrow (Bu^tO)_3W\equiv CX \qquad (75)$$

$$R = Me, Et; X = Me, Et, Pr, CMe_3, Ph, CH=CH_2, CH_2NR_2, CH_2OMe,$$

$$CH_2OSiMe_3, CH(OEt)_2, CO_2Me, CH_2CO_2Me, COMe, SCMe_3 \text{ or } H$$

A similar reaction involving nitriles and $W_2(OBu^t)_6$ leads to 1:1 mixtures of $(Bu^tO)_3W\equiv N$ and $(Bu^tO)W\equiv CR$.[283] The generality of these reactions has not been extended to reactions involving $Mo_2(OBu^t)_6$, though $PhC\equiv CH$ does react to give $(Bu^tO)_3Mo\equiv CPh$ and $(Bu^tO)_3Mo\equiv CH$.[285] These reactions occur rapidly at 25 °C in hydrocarbon solvents. Under somewhat different conditions and with other than 1:1 alkyne:$W_2(OBu^t)_6$ ratios, a variety of products have been obtained and structurally characterized: $(Bu^tO)_2W(\mu\text{-}CPh)_2W(OBu^t)_2$,[286] $(Bu^tO)_2W(\mu\text{-}C_2Ph_2)_2W(OBu^t)_2$,[286] $[W_3(OBu^t)_5(\mu\text{-}O)(\mu\text{-}CEt)O]_2$[287] and $W_3(\mu_3\text{-}CMe)(OPr^i)_9$.[215] A high yield preparation of the latter compound involves the conproportionation of $W_2(OPr^i)_6$ and $(Bu^tO)_3W\equiv CMe$ in the presence of Pr^iOH.[216]

15.3.6.5 Alkyl, Aryl and Benzyl Complexes

15.3.6.5.1 *Of Ti, Zr and Hf*

Seebach and coworkers have developed an extensive chemistry of mixed alkyl alkoxides of titanium and zirconium and their usefulness towards a number of organic synthetic problems has been demonstrated.[107,108]

Although alcoholysis or phenolysis of the alkyls of titanium and zirconium can eventually lead to total substitution, with bulky ligands partial substitution is possible.[66,100] The mixed benzyl aryloxides of zirconium, $Zr(OAr')_x(CH_2Ph)_{4-x}$ ($OAr' = 2,6$-di-t-butylphenoxide; $x = 1$ or 2) have been studied both structurally and spectroscopically.[100] For $Zr(OAr')(CH_2Ph)_3$ one of the benzyl groups interacts with the electron deficient metal in a π-fashion through part of the aromatic ring, while for $Zr(OAr')_2(CH_2Ph)_2$ only σ-bound benzyls are present.[100] This latter compound will insert alkyl isocyanides to generate the bis-iminoacyl complex $Zr(OAr')_2(RNCCH_2Ph)_2$ ($R = Bu^t$, 2,6-dimethylphenyl) in which both iminoacyl groups are bound in a dihapto fashion.[129] In contrast, insertion of isocyanide followed by CO leads to the complex $Zr(OAr')_2(OC(CH_2Ph)C(CH_2Ph)NR)$, formed by the coupling of an acyl and iminoacyl function.[129]

15.3.6.5.2 *Of Nb and Ta*

The alkylation of alkoxides and phenoxides of niobium and tantalum has been shown to initially yield complexes of the type $M(Z)_2(R)_3$ ($M = Nb$, Ta; $Z = OR$, OAr).[288] Structurally a trigonal bipyramidal geometry is adopted with *trans* axial oxygen atoms. Substitution of the remaining two alkoxides/aryloxides is slow. Using the sterically demanding 2,6-di-t-butylphenoxide (OAr'), alkylation of $Ta(OAr')_2Cl_3$ with $LiCH_2X$ ($X = SiMe_3$ or Ph) leads to the mixed alkyl alkylidene compounds $Ta(OAr')_2(=CHX)(CH_2X)$.[289] Although the trimethyl complex can be isolated ($Ta(OAr')_2Me_3$), it will undergo very efficient photochemical α-hydride abstraction to the analogous $Ta(OAr')_2(=CH_2)(Me)$.[290]

15.3.6.5.3 *Of Mo and W*

Alcoholysis reactions involving $M_2R_2(NMe_2)_4$ have led to a series of 1,2-$M_2R_2(OR')_4$ compounds with elimination of $HNMe_2$, where $M = Mo$ or W, $R = a$ β-hydrogen stabilized alkyl (Me, CH_2CMe_3, CH_2SiMe_3 and CH_2Ph), and $R' = Bu^t$, Pr^i and CH_2CMe_3.[291] With less bulky alcohols, a further reaction occurs involving elimination of alkane. When the alkyl groups bound to molybdenum contain β-hydrogen atoms, compounds of formula $Mo_2R(OR')_5$ are obtained with elimination of one equivalent of alkane. Mechanistic studies reveal[291] a rather intricate reaction pathway involving (i) reductive elimination of alkane and alkene; (ii) oxidative addition of R'OH to 'Mo$_2$(OR)$_4$' intermediates; and (iii) alkene insertion into the Mo$_2$—H moiety formed by (ii). Analogous reactions involving W$_2$ compounds proceed similarly but the initial β-hydrogen elimination process is reversible and the rate of reductive elimination is slower. Compounds of formula Mo$_2$(OR)$_4$(L)$_4$, where L = HOR, H$_2$NMe and PMe$_3$, have been isolated[206] but the tungsten compounds are unknown.

The compounds W$_2$(CH$_2$Ph)$_2$(OR)$_4$, where R = Pr^i and Bu^{t}CH$_2$, have been found to react with MeC≡CMe to give (RO)$_2$W(μ-C$_4$Me$_4$)(μ-CPh)W(OR)$_2$H compounds with release of 1 equivalent of toluene.[292]

15.3.6.6 Mixed Alkoxy and Aryloxy Hydrides

The anionic alkoxy hydrides of boron and aluminum, HM(OR)$_3^-$, have been used as organic reducing agents.[293] Compounds such as these are intermediates in reactions involving the tetrahydrides and carbonyl groups.

The tungsten complexes, W(PMe$_3$)$_4$(H)$_3$(X), where X = OMe, OPh, were first synthesized by Wilkinson and coworkers.[294] However, recent work by Green has shown that hydrogenolysis of the formaldehyde complex W(PMe$_3$)$_4$(H$_2$CO)(H)$_2$ initially gives the identical methoxy complex, which then undergoes relatively mild hydrogenolysis of the W—OMe bond, a step of considerable catalytic importance (equation 76).[295]

$$L_nW(CH_2O)(H)_2 + H_2 \longrightarrow L_nW(OMe)(H)_3 + H_2 \longrightarrow L_nW(H)_4 + HOMe \qquad (76)$$

W$_4$(H)$_2$(OPr^i)$_{14}$[296] provides the only example of a well-characterized transition metal hydride supported exclusively by alkoxide ligands. The hydride ligand is not acidic and does not exchange with Pr^{i}OD though alcohol/alkoxide exchange is rapid. There is a reversible reaction with ethylene to give a W—Et containing compound. Reactions with l-alkenes (C$_4$–C$_6$) have been shown to give slow but catalytic conversion to the *cis*-2-alkene at 40 °C.

Formation of NaW$_2$(H)(OPr^i)$_8$, formed in the reaction between W$_2$(OBu^t)$_6$ and NaOPr^i in Pr^{i}OH/hexane, has recently been reported[297] and structurally characterized as a diglyme adduct W$_2$(H)(OPr^i)$_8$Na(diglyme). No chemical reactivity was reported other than its conversion to W$_4$(H)$_2$(OPr^i)$_{14}$ by the addition of H$_2$NMe$_2^+$ salts in Pr^{i}OH.

15.3.6.7 The Activation of CH Bonds in Alkoxide and Aryloxide Ligands

A common decomposition pathway for alkoxide ligands, particularly those bonded to the later transition metals, involves β-hydrogen abstraction leading to metal hydride and ketones or aldehydes. Further reactions may occur to give alkene/alkane and carbon monoxide. These decomposition pathways are kinetically less favorable for alkoxides of the early transition elements, lanthanides and actinides, though some examples are well documented. For example, the compounds Cp*TaMe$_3$X, where X = alkyl, NR$_2$ or OR and Cp* = η^5-C$_5$Me$_5$, have been shown[298] to react *via* β-hydrogen abstraction from the X group with elimination of methane. The facility for β-H atom transfer followed the order R > NR$_2$ > OR, which is the inverse of the Ta—X bond strength (O > N > C).

However, the activation of the γ-C—H bonds has been shown to occur in some early transition metal alkoxides without decomposition. Hence Nugent has shown that some transition metal ethoxides can catalyze H/D exchange in MeCH$_2$OD,[299] deuterium being incorporated exclusively into the methyl group. The reversible formation of a four-membered metallacycle was proposed with a subsequent alcoholysis of the metal–carbon bond to regenerate the metal alkoxide. An interesting observation was that multiple exchange into the methyl group can take place.[299]

The ligand 2,6-di-*t*-butylphenoxide has been shown to undergo cyclometallation with a number of early transition metals.[300] The reaction involves the intramolecular activation of the CH bonds of the Bu^t groups to generate six-membered rings. Besides alkyl leaving groups,[301] alkylidene[290,302] and benzyne[303] functions have been shown to be potent for the activation of these normally inert C—H bonds.

15.4 REFERENCES

1. D. C. Bradley, R. C. Mehrotra and D. P. Gaur, 'Metal Alkoxides', Academic, New York, 1978.
2. K. C. Melhotra and R. L. Martin, *J. Organomet. Chem.*, 1982, **239**, 159.
3. D. C. Bradley, in 'Perspective Inorganic Reactions', ed. W. L. Jolly, Interscience, New York, 1965, vol. 2.
4. R. Masthoff, H. Kohler, B. Bohland and F. Schmeil, *Z. Chem.*, 1965, **5**, 122.
5. M. H. Chisholm, *Polyhedron*, 1983, **2**, 681.
6. Ethyl Corp., *Br. Pat.* 727 923 (1955) (*Chem. Abstr.*, 1956, **50**, 5018).
7. L. Lochmann, D. Lim and J. Coupek, *Ger. Pat.* 2 035 260 (1971) (*Chem. Abstr.*, 1971, **74**, 87 355).
8. Mathieson Alkali Works, *Fr. Pat.* 850 126 (1939) (*Chem. Abstr.*, 1942, **36**, 1619).
9. D. J. Loder and D. D. Lee, *U.S. Pat.* 2 278 550 (1942) (*Chem. Abstr.*, 1942, **36**, 4830).
10. R. Muller and O. Sussenguth, *Ger. Pat.* 695 280 (1940) (*Chem. Abstr.*, 1942, **35**, 29 065).
11. J. Lukasiak, L. A. May, I. Ya. Strauss and R. Piekos, *Rocz. Chem.*, 1970, **44**, 1675.
12. C. A. Cohen, *U.S. Pat.* 2 287 088 (1943) (*Chem. Abstr.*, 1943, **37**, 141).
13. A. G. Rheinpreussen, *Ger. Pat.* 1 230 004 (1966) (*Chem. Abstr.*, 1967, **66**, 37 426).
14. K. S. Mazdiyasni, R. T. Dolloff and J. S. Smith, II, *J. Am. Ceram. Soc.*, 1969, **523**, 523.
15. J. S. Smith, II, R. T. Dolloff and K. S. Mazdiyasni, *J. Am. Ceram. Soc.*, 1970, **53**, 91.
16. J. H. Gladstone and A. Tribe, *J. Chem. Soc.*, 1881, **39**, 4.
17. K. S. Mazdiyasni, C. T. Lynch and J. S. Smith, *Inorg. Chem.*, 1966, **5**, 342.
18. L. M. Brown and K. S. Mazdiyasni, *Inorg. Chem.*, 1970, **9**, 2783.
19. B. Szilard, *Z. Electrochem.*, 1906, **12**, 393.
20. *U.S. Pat.* 2 438 963 (1948) (*Chem. Abstr.*, 1948, **42**, 5045).
21. H. Lehmkuhl and M. Eisenbach, *Justus Liebigs Ann. Chem.*, 1975, 672.
22. *Ger. Pat.* 2 349 561 (1974) (*Chem. Abstr.*, 1974, **81**, 20 237).
23. *Can. Pat.* 1 024 466 (1978) (*Chem. Abstr.*, 1978, **88**, 160 737).
24. *Ger. Pat.* 2 121 732 (1972) (*Chem. Abstr.*, 1972, **77**, 4885).
25. *Fr. Pat.* 2 091 229 (1972) (*Chem. Abstr.*, 1972, **77**, 96 297).
26. V. A. Shreider, E. P. Turevskaya, N. I. Koslova and N. A. Tunova, *Inorg. Chim. Acta*, 1981, **53**, L73.
27. W. J. Power and G. A. Ozin, *Adv. Inorg. Chem. Radiochem.*, 1982, **23**, 140.
28. T. Colclough, W. Gerrard and M. F. Lappert, *J. Chem. Soc.*, 1966, 3006.
29. D. C. Bradley, R. C. Mehrotra and W. Wardlaw, *J. Chem. Soc.*, 1952, 5020.
30. G. O. Doak and L. D. Freeman, *Chem. Rev.*, 1961, **61**, 31.
31. S. N. Misra, T. N. Misra and R. C. Mehrotra, *J. Inorg. Nucl. Chem.*, 1965, **27**, 105.
32. D. C. Bradley, F. M. A. Halim and W. Wardlaw, *J. Chem. Soc.*, 1950, 3450.
33. D. C. Bradley, M. A. Saad and W. Wardlaw, *J. Chem. Soc.*, 1954, 2002.
34. D. C. Bradley, R. K. Multani and W. Wardlaw, *J. Chem. Soc.*, 1958, 4647.
35. O. Klejnot, *Inorg. Chem.*, 1965, **4**, 1668.
36. V. H. Funk and W. Baumann, *Z. Anorg. Chem.*, 1965, **231**, 264.
37. A. Rosenheim and C. Nernst, *Z. Anorg. Chem.*, 1933, **214**, 209.
38. V. S. Nayar and R. D. Peacock, *J. Chem. Soc.*, 1964, 2827.
39. V. H. Funk, G. Mohaupt and A. Paul, *Z. Anorg. Chem.*, 1959, **302**, 199.
40. V. H. Funk, W. Weiss and M. Zeising, *Z. Anorg. Chem.*, 1958, **296**, 41.
41. J. Nelles, *Br. Pat.* 512 452 (1939) (*Chem Abstr.*, 1940, **34**, 3764).
42. R. C. Mehrotra and B. C. Pant, *J. Indian Chem. Soc.*, 1962, **39**, 65.
43. S. Mathur, G. Chandra, A. K. Rai and R. C. Mehrotra, *J. Organomet. Chem.*, 1965, **4**, 294.
44. D. F. Herman, *U.S. Pat.* 2 654 770 (1953) (*Chem. Abstr.*, 1954, **48**, 13 710).
45. D. C. Bradley and W. Wardlaw, *J. Chem. Soc.*, 1951, 280.
46. D. C. Bradley, F. M. A. Halim, E. A. Sadek and W. Wardlaw, *J. Chem. Soc.*, 1952, 2032.
47. P. N. Kapoor, R. N. Kapoor and R. C. Mehrotra, *Chem. Ind. (London)*, 1968, 1314.
48. D. C. Bradley, R. C. Mehrotra and W. Wardlaw, *J. Chem. Soc.*, 1953, 1634.
49. D. C. Bradley, B. N. Chakravarti and W. Wardlaw, *J. Chem. Soc.*, 1956, 2381.
50. D. C. Bradley, B. N. Chakravarti and W. Wardlaw, *J. Chem. Soc.*, 1956, 4439.
51. D. C. Bradley, B. N. Chakravarti, A. K. Chatterjee, W. Wardlaw and A. Whitley, *J. Chem. Soc.*, 1958, 99.
52. D. C. Bradley, R. K. Multani and W. Wardlaw, *J. Chem. Soc.*, 1958, 126.
53. D. C. Bradley, B. N. Chakravarti and A. K. Chatterjee, *J. Inorg. Nucl. Chem.*, 1957, **3**, 367.
54. D. C. Bradley, B. N. Chakravarti and A. K. Chatterjee, *J. Inorg. Nucl. Chem.*, 1959, **12**, 71.
55. D. C. Bradley, M. A. Saad and W. Wardlaw, *J. Chem. Soc.*, 1954, 1091.
56. D. C. Bradley, B. Harder and F. Hudswell, *J. Chem. Soc.*, 1957, 3318.
57. V. H. Funk and R. Masthoff, *J. Prakt. Chem.*, 1956, **4**, 35.
58. N. M. Cullinane and S. J. Chard, *Nature (London)*, 1949, **164**, 710.
59. N. M. Cullinane, S. J. Chard, G. F. Price and B. B. Millward, *J. Soc. Chem. Ind., London*, 1950, **69**, 538 (*Chem. Abstr.*, 1951, **45**, 7950).
60. A. J. Bloodworth and A. G. Davies, 'Organotin Compounds', ed. A. K. Sawyer, Dekker, New York, 1971, vol. 1.
61. W. J. Reagan and C. H. Brubaker, Jr., *Inorg. Chem.*, 1970, **9**, 827.
62. D. C. Bradley and K. J. Fisher, 'M.T.P. International Review of Science', ed. H. J. Emeleus and D. W. A. Sharp, Butterworth, London, 1972, vol. 5, p. 65.
63. H. Meerwein, *US. Pat.*. 1 689 356 (1929) (*Chem. Abstr.*, 1929, **23**, 156).
64. D. C. Bradley and W. Wardlaw, *Nature (London)* 1950, **165**, 75.
65. L. Chamberlain, I. P. Rothwell and J. C. Huffman, *Inorg. Chem.*, 1984, **23**, 2575.
66. T. V. Lubben, P. T. Wolczanski, G. D. Van Duyne, *Organometallics*, 1984, **3**, 977.
67. H. Gilman, R. G. Jones, G. Karmas and G. A. Martin, Jr., *J. Am. Chem. Soc.*, 1956, **78**, 4285.
68. D. C. Bradley and M. M. Faktor, *Chem. Ind. (London)*, 1958, 1332.
69. E. F. Caldin and G. Long, *J. Chem. Soc.*, 1954, 3737.

70. G. E. Revzin and V. D. Zamedyanskaya, *Metody Poluch. Khim. React. Prep.*, 1967, **16**, 45.
71. G. E. Revzin and V. D. Zamedyanskaya, *Metody Poluch. Khim. React. Prep.*, 1967, **16**, 64.
72. K. Szymanski, H. Koesszny, *Pol. Pat.* 102 436 (1979) (*Chem. Abstr.*, 1979, **91**, 192 983).
73. G. Cannari and D. Cozzi, *Wien. Chem.-Ztg.*, 1943, **46**, 193.
74. I. M. Thomas, *Can. J. Chem.*, 1961, **39**, 1386.
75. M. H. Chisholm, F. A. Cotton, M. W. Extine and W. W. Reichert, *Inorg. Chem.*, 1978, **17**, 2944.
76. M. H. Chisholm, W. W. Reichert and P. Thornton, *J. Am. Chem. Soc.*, 1978, **100**, 2744.
77. S. Beshourie, T. W. Coffindaffer, S. Pirzad and I. P. Rothwell, *Inorg. Chim. Acta*, 1985, **103**, 111.
78. M. H. Chisholm, F. A. Cotton, C. A. Murillo and W. W. Reichert, *Inorg. Chem.*, 1977, **16**, 1801.
79. M. Akiyama, M. H. Chisholm, F. A. Cotton, M. W. Extine, D. A. Haitko, D. Little and P. E. Fanwick, *Inorg. Chem.*, 1979, **18**, 2266.
80. T. W. Coffindaffer, I. P. Rothwell and J. C. Huffman, *Inorg. Chem.*, 1984, **23**, 1433.
81. M. H. Chisholm, J. Leonelli and J. C. Huffman, *J. Chem. Soc., Chem. Commun.*, 1981, 270.
82. I. D. Verma and R. C. Mehrotra, *J. Chem. Soc.*, 1960, 2966.
83. R. C. Mehrotra, *J. Indian Chem. Soc.*, 1954, **31**, 904.
84. R. C. Mehrotra, *J. Indian Chem. Soc.*, 1953, **30**, 585.
85. R. C. Mehrotra and G. Chander, *J. Indian Chem. Soc.*, 1962, **39**, 235.
86. R. N. Kapoor and R. C. Mehrotra, *J. Am. Chem. Soc.*, 1960, **82**, 3495.
87. R. C. Mehrotra and J. M. Batwara, *Inorg. Chem.*, 1985, **9**, 1815.
88. K. W. Bagnall, A. K. Bhandari and D. Brown, *J. Inorg. Nucl. Chem.*, 1975, **37**, 1815.
89. T. W. Coffindaffer, I. P. Rothwell and J. C. Huffman, *Inorg. Chem.*, 1983, **22**, 2906.
90. R. C. Mehrotra, *J. Am. Chem. Soc.*, 1954, **76**, 2266.
91. R. C. Mehrotra and R. N. Kapoor, *J. Less-Common Met.*, 1964, **7**, 98.
92. R. N. Kapoor and R. C. Mehrotra, *J. Less-Common Met.*, 1966, **10**, 66.
93. J. A. Connor, *Top. Curr. Chem.*, 1977, **71**, 71.
94. J. W. Bruno, T. J. Marks and L. R. Morss, *J. Am. Chem. Soc.*, 1983, **105**, 6824.
95. G. E. Coates and A. H. Fishwick, *J. Chem. Soc. (A)*, 1968, 477.
96. R. A. Andersen and G. E. Coates, *J. Chem. Soc., Dalton Trans.*, 1972, 2153.
97. G. E. Coates and D. Ridley, *J. Chem. Soc.*, 1965, 1870.
98. E. C. Ashby, *Quart. Rev.*, 1967, **21**, 259.
99. L. Durfee, I. P. Rothwell and J. C. Huffman, *Inorg. Chem.*, 1985, **24**, 4569.
100. S. Latesky, A. K. McMullen, I. P. Rothwell and J. C. Huffman, *Organometallics*, 1985, **4**, 902.
101. R. R. Schrock and G. W. Parshall, *Chem. Rev.*, 1976, **76**, 243.
102. P. J. Davidson, M. F. Lappert and R. Pearce, *Chem. Rev.*, 1976, **76**, 219.
103. H. Noeth and H. Suchy, *Z. Anorg. Allg. Chem.*, 1968, **358**, 44.
104. P. J. Fagan, K. G. Moloy and T. J. Marks, *J. Am. Chem. Soc.*, 1981, **103**, 6959.
105. M. F. Kharesch and O. Reinmuth, in 'Grignard Reactions of Nonmetallic Substances', Prentice-Hall, New York, 1954.
106. B. Weidmann and D. Seebach, *Angew. Chem., Int. Ed. Engl.*, 1983, **22**, 31.
107. D. Seebach, A. K. Beck, M. Schiess, L. Wilder and A. Wonnacott, *Pure Appl. Chem.*, 1983, **55**, 1807.
108. D. Seebach and V. Prelog, *Angew. Chem. Int. Ed. Engl.*, 1982, **21**, 654.
109. E. R. H. Walker, *Chem. Soc. Rev.*, 1976, **5**, 23.
110. H. Schmidbaur and H. F. Klein, *Chem. Ber.*, 1967, **100**, 1129.
111. J. M. Mayer and J. E. Bercaw, *J. Am. Chem. Soc.*, 1982, **104**, 2157.
112. M. H. Abraham, *J. Chem. Soc.*, 1960, 4130.
113. M. H. Abraham and A. G. Davies, *J. Chem. Soc.*, 1959, 429.
114. J. Schwartz and J. A. Labiner, *Angew. Chem., Int. Ed. Engl.*, 1976, **15**, 333.
115. T. F. Blackburn, J. A. Labinger and J. Schwartz, *Tetrahedron Lett.*, 1976, 3041.
116. T. V. Lubben and P. T. Wolczanski, *J. Am. Chem. Soc.*, 1985, **107**, 701.
117. H. Mimoun, P. Chanmette, L. Saussine, J. Fischer and R. Weiss, *Nouv. J. Chim.*, 1983, **7**, 467.
118. G. Ferguson, M. Parvex, P. K. Monoghan and R. J. Puddephatt, *J. Chem. Soc., Chem. Commun.*, 1983, 267.
119. Monsanto Chemical Co., *US Pat.* 2 579 414 (1951) (*Chem. Abstr.*, 1951, **46**, 6139).
120. L. Malatesta, *Gazz. Chim. Ital.*, 1948, **78**, 753.
121. H. Fischer and H. Tropsch, *Brennst.-Chem.*, 1926, **7**, 97.
122. P. C. Ford, *ACS Symp. Ser.*, 1981, **152**.
123. C. Masters, *Adv. Organomet. Chem.*, 1979, **17**, 61.
124. J. E. Bercaw and P. J. Wolczanski, *Acc. Chem. Res.*, 1980, **13**, 121.
125. T. J. Marks, *Science*, 1982, **217**, 989.
126. P. J. Fagan, E. A. Maatta and T. J. Marks, *ACS Symp. Ser.*, 1981, **152**, 53.
127. D. A. Katahara, K. G. Moloy and T. J. Marks, *Organometallics*, 1982, **1**, 1723.
128. P. T. Barger, B. D. Santarsiero, J. Armentrout and J. E. Bercaw, *J. Am. Chem. Soc.*, 1984, **106**, 5178.
129. A. K. McMullen, I. P. Rothwell and J. C. Huffman, *J. Am. Chem. Soc.*, 1985, **107**, 1072.
130. G. V. Goeden and G. Caulton, *J. Am. Chem. Soc.*, 1981, **103**, 7354.
131. L. B. Handy, K. G. Sharp and F. E. Brinkman, *Inorg. Chem.*, 1972, **11**, 523.
132. L. B. Handy and F. E. Brinkman, *Chem. Commun.*, 1970, 214.
133. L. B. Handy, C. Benham, F. E. Brinkman and R. B. Johannesen, *J. Fluorine Chem.*, 1976, **8**, 55.
134. Am. Nobel and J. M. Winfield, *J. Chem. Soc. (A)*, 1970, 2574.
135. W.-K. Wong, G. Wilkinson, A. M. R. Galas, M. A. Thornton-Pett and M. B. Hursthouse, *Polyhedron*, 1982, **1**, 842.
136. W.-K. Wong, K. W. Chiu, G. Wilkinson, A. M. R. Galas, M. Thornton-Pett and M. B. Hursthouse, *J. Chem. Soc., Dalton Trans.*, 1983, 1557.
137. M. E. Redwood and C. J. Willis, *Can. J. Chem.*, 1965, **43**, 1893.
138. M. E. Redwood and C. J. Willis, *Can. J. Chem.*, 1967, **45**, 389.
139. I. L. Knunyants and B. L. Dyatkin, *Izv. Akad. Nauk SSSR, Ser. Khim*, 1964, 923.

140. R. Filler and R. M. Schure, *J. Org. Chem.*, 1967, **32**, 1217.
141. R. Filler, G.-S. Wang, M. A. McKinney and F. N. Miller, *J. Am. Chem. Soc.*, 1967, **89**, 1026.
142. D. P. Graham and V. Weinmayr, *J. Org. Chem.*, 1966, **31**, 957.
143. W. J. Middleton and R. V. Linsey, *J. Am. Chem. Soc.*, 1964, **86**, 4948.
144. M. Allan, A. F. Janzen and C. J. Willis, *Can. J. Chem.*, 1968, **46**, 3671.
145. M. Allan and C. J. Willis, *J. Am. Chem. Soc.*, 1968, **90**, 5343.
146. C. J. Willis, *J. Chem. Soc., Chem. Commun.*, 1972, 944.
147. J. W. L. Martin and C. J. Willis, *Can. J. Chem.*, 1977, **55**, 2459.
148. I. Chang and C. J. Willis, *Can. J. Chem.*, 1977, **55**, 2465.
149. R. C. Mehrotra, *Adv. Inorg. Chem. Radiochem.*, 1983, **26**, 269.
150. H. Meerwein and T. Bersin, *Justus Liebigs Ann. Chem.*, 1929, **476**, 113.
151. R. C. Mehrotra, M. M. Agrawal and P. N. Kapoor, *J. Chem. Soc. (A)*, 1968, 2673.
152. R. C. Mehrotra and A. Mehrotra, *J. Chem. Soc., Dalton Trans.*, 1972, 1203.
153. A. Mehrotra and R. C. Mehrotra, *Inorg. Chem.*, 1972, **11**, 2170.
154. L. Durfee and I. P. Rothwell, *J. Am. Chem. Soc.*, submitted for publication.
155. M. S. Bain, *Can. J. Chem.*, 1964, **42**, 945.
156. G. E. Hartwell and T. L. Brown, *Inorg. Chem.*, 1966, **5**, 1257.
157. N. V. Sidgwick and L. E. Sutton, *J. Chem. Soc.*, 1931, 1461.
158. R. W. Adams, E. Bishop, R. L. Martin and G. Winer, *Aust. J. Chem.*, 1966, **19**, 207.
159. B. Horvath, R. Mosler and E. G. Horvath, *Z. Anorg. Chem.*, 1979, **449**, 41.
160. H. Steinberg, 'Organoboron Chemistry', Interscience, New York, 1963, vol. 1.
161. C. H. DePuy and R. W. King, *Chem. Rev.*, 1960, **60**, 431.
162. D. C. Bradley and M. M. Faktor, *Nature (London)*, 1959, **184**, 55.
163. V. J. Shiner, D. Whittaker and V. P. Fernandez, *J. Am. Chem. Soc.*, 1963, **85**, 2318.
164. J. W. Akih and R. H. Duncan, *J. Magn. Reson.*, 1974, **15**, 162.
165. C.G. Barraclough, D. C. Bradley, J. Lewis and I. M. Thomas,. *J. Chem. Soc.*, 1961, 2601.
166. D. C. Bradley, R. K. Mutani and W. Wardlaw, *J, Chem. Soc.*, 1958, 4153.
167. R. W. Adams, C. G. Barraclough, R. L. Martin and G. Winter, *Inorg. Chem.*, 1966, **5**, 346.
168. M. H. Chisholm, J. C. Huffman, C. C. Kirkpatrick and J. Leonelli, *J. Am. Chem. Soc.*, 1981, **103**, 6093.
169. R. C. Mehrotra and J. M. Batwara, *Inorg. Chem.*, 1970, **9**, 2783.
170. K. S. Mazdiyasni, C. T. Lynch and J. S. Mith, *Inorg. Chem.*, 1966, **5**, 342.
171. P. B. Hitchcock, M. F. Lappert and A. Singh, *J. Chem. Soc., Chem. Commun.*, 1983, 1499.
172. S. Latesky, J. Keddington, A. K. McMullen, I. P. Rothwell and J. C. Huffman, *Inorg. Chem.*, 1985, **24**, 995.
173. F. Glockling, 'The Chemistry of Germanium', Academic, New York, 1962.
174. J. A. Zubieta and J. J. Zuckerman, *Prog. Inorg. Chem.*, 1978, **24**, 251.
175. R. C. Mehrotra and V. D. Gupta, *J. Indian Chem. Soc.*, 1964, **41**, 537.
176. D. C. Bradley and C. E. Holloway, *J. Chem. Soc. (A)*, 1968, 1316.
177. D. C. Bradley, R. C. Mehrotra, J. D. Swanwick and W. Wardlaw, *J. Chem. Soc.*, 1953, 202J.
178. R. L. Martin and G. Winter, *J. Chem. Soc.*, 1961, 2947.
179. D. C. Bradley and J. D. Swanwick, *J. Chem. Soc.*, 1959, 748.
180. D. C. Bradley, R. C. Mehrotra and W. Wardlaw, *J. Chem. Soc.*, 1952, 4204.
181. D. C. Bradley, R. H. Moss and K. D. Sales, *Chem. Commun.*, 1969, 1255.
182. D. C. Bradley, A. K. Chatterjee and W. Wardlaw, *J. Chem. Soc.*, 1957, 2260.
183. D. C. Bradley, A. K. Chatterjee and W. Wardlaw, *J. Chem. Soc.*, 1957, 2600.
184. D. C. Bradley, W. Wardlaw and A. Whitley, *J. Chem. Soc.*, 1955, 726.
185. C. E. Holloway, *J. Coord. Chem.*, 1972, **1**, 253.
186. D. C. Bradley and C. E. Holloway, *J. Chem. Soc. (A)*, 1968, 219.
187. K. C. Malhotra, U. K. Banerjee and S. C. Chaudhry, *J. Indian Chem. Soc.*, 1980, **57**, 868.
188. L. N. Lewis and M. F. Garbauskas, *Inorg. Chem.*, 1985, **24**, 363.
189. D. C. Bradley, M. H. Chisholm, M. W. Extine and M. E. Stager, *Inorg. Chem.*, 1977, **16**, 1794.
190. E. A. Cuellar, S. S. Miller, T. J. Marks and E. Weitz, *J. Am. Chem. Soc.*, 1983, **105**, 4580.
191. P. J. Wheatley, *J. Chem. Soc.*, 1961, 4270.
192. E. Weiss and H. Alsdorf, *Z. Anorg. Allg. Chem.*, 1970, **372**, 206.
193. E. Weiss, *Helv. Chim. Acta*, 1963, **46**, 2051.
194. L. F. Dahl, G. L. Davies, D. L. Wampler and K. West, *J. Inorg. Nucl. Chem.*, 1962, **24**, 357.
195. E. Weiss, H. Alsdorf and H. Kuhr, *Angew. Chem., Int. Ed. Engl.*, 1967, **6**, 801.
196. T. Greiser and E. Weiss, *Chem. Ber.*, 1976, **109**, 3142.
197. N. A. Bell and G. E. Coates, *J. Chem. Soc. (A)*, 1968, 823.
198. R. A. Andersen and G. E. Coates, *J. Chem. Soc., Dalton Trans.*, 1975, 1244.
199. J. L. Atwood, W. E. Hunter, R. D. Rogers, J. Holton, J. McMeeking, R. Pearce and M. F. Lappert, *J. Chem. Soc., Chem. Commun.*, 1978, 140.
200. D. A. Wright and D. A. Williams, *Acta Crystallogr., Sect. B*, **24**, 1107.
201. J. A. Ibers, *Nature (London)*, 1963, **197**, 686.
202. M. H. Chisholm, *ACS Symp. Ser.*, 1983, **211**, 243.
203. A. A. Pinkerton, D. Schwarzebach, L. G. Hubert and J. G. Reiss, *Inorg. Chem.*, 1976, **15**, 1196.
204. F. A. Cotton, D. O. Marler and W. Schwotzer, *Inorg. Chem.*, 1984, **23**, 4211.
205. J. I. Davies, J. F. Gibson, A. C. Skapski, G. Wilkinson and W.-K. Wong, *Polyhedron*, 1982, **1**, 641.
206. M. H. Chisholm, K. Folting, J. C. Huffman and R. J. Tatz, *J. Am. Chem. Soc.*, 1984, **106**, 1153.
207. L. B. Anderson, F. A. Cotton, D. DeMarco, A. Fang, W. H. Isley, B. W. S. Kolthammer and R. A. Walton, *J. Am. Chem. Soc.*, 1981, **103**, 5078.
208. M. H. Chisholm, J. C. Huffman and C. C. Kirkpatrick, *Inorg. Chem.*, 1981, **20**, 871.
209. F. A. Cotton, D. DeMarco, B. S. W. Kolthammer and R. A. Walton, *Inorg. Chem.*, 1981, **20**, 3048.
210. M. H. Chisholm, K. Folting, J. C. Huffman and C. C. Kirkpatrick, *Inorg. Chem.*, 1984, **23**, 1021.

211. M. H. Chisholm, F. A. Cotton, A. Fang and E. M. Kober, *Inorg. Chem.*, 1984, **23**, 749.
212. M. H. Chisholm, K. Folting, J. C. Huffman and E. M. Kober, *Inorg. Chem.*, 1985, **24**, 241.
213. F. A. Cotton, D. O. Marler and W. Schwotzer, *Inorg. Chim. Acta*, 1984, **95**, 207.
214. M. H. Chisholm, D. M. Hoffman and J. C. Huffman, *Inorg. Chem.*, 1985, **24**, 796.
215. M. H. Chisholm, D. M. Hoffman and J. C. Huffman, *Inorg. Chem.*, 1984, **23**, 3683.
216. M. H. Chisholm, K. Folting, J. A. Heppert, D. M. Hoffman and J. C. Huffman, *J. Am. Chem. Soc.*, 1985, **107**, 1234.
217. M. H. Chisholm, J. C. Huffman and R. L. Kelly, *J. Am. Chem. Soc.*, 1979, **101**, 7615.
218. M. H. Chisholm, D. L. Clark and J. C. Huffman, *Polyhedron*, 1985, **4**, 1203.
219. M. H. Chisholm, R. J. Errington, K. Folting and J. C. Huffman, *J. Am. Chem. Soc.*, 1982, **104**, 2025.
220. D. L. Clark, Ph.D. Thesis, Indiana University, 1986.
221. M. Akiyama, M. H. Chisholm, D. A. Haitko, D. Little, F. A. Cotton and M. W. Extine, *J. Am. Chem. Soc.*, 1979, **101**, 2504.
222. M. H. Chisholm, K. Folting, J. C. Huffman, N. S. Marchan, C. A. Smith and L. C. Taylor, *J. Am. Chem. Soc.*, 1985, **107**, 3722.
223. M. H. Chisholm, J. A. Heppert and J. C. Huffman, *Polyhedron*, 1984, **3**, 475.
224. P. G. Edwards, G. Wilkinson, M. B. Hursthouse and K. M. Abdul Malik, *J. Chem. Soc., Dalton Trans.*, 1980, 2467.
225. K. Watenpaugh and C. N. Caughlan, *Chem. Commun.*, 1967, 76.
226. D. C. Bradley, M. B. Hursthouse and P. F. Rodesiler, *Chem. Commun.*, 1968, 1112.
227. W. H. Knoth and R. L. Harlow, *J. Am. Chem. Soc.*, 1981, **103**, 4265.
228. B. Morisin, *Acta Crystallogr., Sect. B*, 1977, **33**, 303.
229. B. Cetinkaya, I. Gumükcü, M. F. Lappert, I. L. Atwood and R. Shakir, *J. Am. Chem. Soc.*, 1980, **102**, 2086.
230. M. S. Bains and D. C. Bradley, *Can. J. Chem.*, 1962, **40**, 2218.
231. M. H. Chisholm, F. A. Cotton, M. W. Extine and W. W. Reichert, *J. Am. Chem. Soc.*, 1978, **100**, 153.
232. M. J. Chetcuti, M. H. Chisholm, J. C. Huffman and J. Leonelli, *J. Am. Chem. Soc.*, 1983, **105**, 292.
233. T. W. Coffindaffer and I. P. Rothwell, unpublished results.
234. D. C. Bradley, *Coord. Chem. Rev.*, 1967, **2**, 299.
235. D. C. Bradley, 'Inorganic Polymers', ed. F. G. A. Stone and W. A. Graham, Academic, New York, 1962.
236. G. A. Kakos and G. Winter, *Aust. J. Chem.*, 1969, **22**, 97.
237. R. C. Mehrotra, *Nature (London)*, 1953, **172**, 74.
238. A. G. Charles, N. C. Peterson and G. H. Franke, *Inorg. Synth.*, 1967, **9**, 25.
239. D. M. Puri, K. C. Pandey and R. C. Mehrotra, *J. Less-Common Met.*, 1962, **4**, 393.
240. R. C. Mehrotra, R. N. Kapoor, S. Prakash and P. N. Kapoor, *Aust. J. Chem.*, 1966, **19**, 2079.
241. R. N. Kapoor, S. Prakash and P. N. Kapoor, *Bull. Chem. Soc. Jpn.*, 1967, **40**, 1384.
242. A. Verley, *Bull. Soc. Chim. Fr.*, 1925, **37**, 537.
243. H. Meerwein and R. Schmidt, *Justus Liebigs Ann. Chem.*, 1925, **444**, 221.
244. W. Penndorf, *Z. Angew. Chem.*, 1926, **39**, 138.
245. L. M. Jackman and A. K. J. MacBeth, *J. Chem. Soc.*, 1952, 3252.
246. M. S. Bains and D. C. Bradley, *Chem. Ind.*, 1961, 1032.
247. D. C. Bradley, R. P. N. Sinha and W. Wardlaw, *J. Chem. Soc.*, 1958, 4651.
248. R. C. Mehrotra and I. D. Verma, *J. Less-Common Met.*, 1961, **3**, 321; *J. Am. Chem. Soc.*, 1978, **100**, 1727.
249. M. H. Chisholm, F. A. Cotton, M. W. Extine and R. L. Kelly, *J. Am. Chem. Soc.*, 1978, **100**, 5764.
250. M. H. Chisholm, F. A. Cotton, K. Folting, J. C. Huffman, A. L. Ratermann and E. S. Shamshoum, *Inorg. Chem.*, 1984, **23**, 4423.
251. F. A. Cotton, W. Schwotzer and E. S. Shamshoum, *Organometallics*, submitted for publication.
252. T. Tsuda and T. Saegusa, *Inorg. Chem.*, 1972, **11**, 2561.
253. O. Metho-Cohn, D. Thorpe and H. J. Twitchett, *J. Chem. Soc. (C)*, 1970, 132.
254. P. C. Bahara, V. D. Gupta and R. C. Mehrotra, *Indian J. Chem.*, 1975, **13**, 156.
255. H. Burger, *Monatsh. Chem.*, 1964, **95**, 671.
256. H. E. Bryndza, J. C. Calabrese and S. S. Wreford, *Organometallics*, 1984, **3**, 1603.
257. H. E. Bryndza, *Organometallics*, 1985, **5**, 1686.
258. H. E. Bryndza and T. H. Tulip, *Organometallics*, 1985, **4**,
259. H. E. Bryndza, personal communication.
260. R. H. Morrison and R. N. Boyd, 'Organic Chemistry', 3rd edn., Allyn and Bacon, London, 1981, p. 556.
261. M. H. Chisholm, J. C. Huffman, A. L. Ratermann and C. A. Smith, *Inorg. Chem.*, 1984, **23**, 1596.
262. M. H. Chisholm, F. A. Cotton, M. W. Extine and R. L. Kelly, *J. Am. Chem. Soc.*, 1979, **101**, 7645.
263. M. H. Chisholm, D. M. Hoffman and J. C. Huffman, *Organometallics*, 1985, **4**, 986.
264. M. H. Chisholm, J. C. Huffman, J. Leonelli and I. P. Rothwell, *J. Am. Chem. Soc.*, 1982, **104**, 7030.
265. F. A. Cotton and W. Schwotzer, *J. Am. Chem. Soc.*, 1983, **105**, 4955.
266. F. A. Cotton and W. Schwotzer, *J. Am. Chem. Soc.*, 1983, **105**, 5639.
267. D. C. Bradley and C. W. Newing, *Chem. Commun.*, 1970, 219.
268. D. C. Bradley, C. W. Newing, M. H. Chisholm, R. L. Kelly, D. A. Haitko, D. Little, F. A. Cotton and P. E. Fanwick, *Inorg. Chem.*, 1980, **19**, 3010.
269. M. H. Chisholm, F. A. Cotton, M. W. Extine and R. L. Kelly, *J. Am. Chem. Soc.*, 1978, **100**, 3354.
270. M. H. Chisholm, F. A. Cotton, M. W. Extine and R. L. Kelly, *Inorg. Chem.*, 1979, **18**, 116.
271. A. S. Drane and J. A. McCleverty, *Polyhedron*, 1983, **2**, 53.
272. J. A. McCleverty, *Chem. Soc. Rev.*, 1983, **12**, 331.
273. P. T. Bishop, J. R. Dilworth, J. Hutchinson and J. A. Zubieta, *Inorg. Chim. Acta*, 1984, **84**, L15.
274. M. H. Chisholm, K. Folting, J. C. Huffman and I. P. Rothwell, *J. Am. Chem. Soc.*, 1982, **104**, 4389.
275. M. H. Chisholm, K. Folting, D. M. Hoffman and J. C. Huffman, *J. Am. Chem. Soc.*, 1984, **106**, 6794.
276. M. H. Chisholm, D. M. Hoffman and J. C. Huffman, *J. Am. Chem. Soc.*, 1984, **106**, 6806.
277. M. H. Chisholm, D. M. Hoffman and J. C. Huffman, *J. Am. Chem. Soc.*, 1984, **106**, 6815.
278. R. R. Schrock, *ACS Symp. Ser.*, 1983, **221**, 369.
279. R. R. Schrock, S. M. Rocklage, J. D. Fellman, G. A. Ruprecht and L. W. Messerle, *J. Am. Chem. Soc.*, 1981, **103**, 1440.

280. J. H. Wengrovius, J. Sancho and R. R. Schrock, *J. Am. Chem. Soc.*, 1981, **103**, 3932.
281. J. Sancho and R. R. Schrock, *J. Mol. Catal.*, 1982, **15**, 75.
282. R. R. Schrock, S. F. Pedersen, M. R. Churchill and H. J. Wasserman, *J. Am. Chem. Soc.*, 1982, **104**, 6808.
283. R. R. Schrock, M. L. Listemann and L. G. Sturgeoff, *J. Am. Chem. Soc.*, 1982, **104**, 4291.
284. M. L. Listemann and R. R. Schrock, *Organometallics*, 1985, **4**, 74.
285. H. Strutz and R. R. Schrock, *Organometallics*, 1984, **3**, 1600.
286. F. A. Cotton and W. Schwotzer, *Organometallics*, 1983, **2**, 1167.
287. F. A. Cotton, W. Schwotzer and E. S. Sahmshoum, *Organometallics*, 1983, **2**, 1340.
288. L. Chamberlain, J. Keddington and I. P. Rothwell, *Organometallics*, 1982, **1**, 1098.
289. L. Chamberlain, I. P. Rothwell and J. C. Huffman, *J. Am. Chem. Soc.* 1982, **104**, 7338.
290. L. Chamberlain, A. P. Rothwell and I. P. Rothwell, *J. Am. Chem. Soc.*, 1984, **106**, 1847.
291. M. H. Chisholm, J. C. Huffman and R. J. Tatz, *J. Am. Chem. Soc.*, 1983, **105**, 2075.
292. M. H. Chisholm, B. W. Eichhorn and J. C. Huffman, *J. Chem. Soc., Chem. Commun.*, submitted for publication.
293. H. C. Brown and S. Krishnamurthy, *Aldrichimica Acta*, 1979, **12**, 3.
294. K. W. Chiu, R. A. Jones, G. Wilkinson, A. M. R. Galas, M. B. Hursthouse and K. M. Abdul Malik, *J. Chem. Soc., Dalton Trans.*, 1981, 1204.
295. M. L. H. Green, G. Parkin, K. J. Moynihan and K. Prout, *J. Chem. Soc., Chem. Comm.*, 1984, 1540.
296. M. Akiyama, M. H. Chisholm, F. A. Cotton, M. W. Extine, D. A. Kaitko, J. Leonelli and D. Little, *J. Am. Chem. Soc.*, 1981, **103**, 779.
297. M. H. Chisholm, J. C. Huffman and C. A. Smith, *J. Am. Chem. Soc.*, 1986, **108**, 222.
298. J. M. Mayer, C. J. Curtis and J. E. Bercaw, *J. Am. Chem. Soc.*, 1983, **105**, 2651.
299. W. B. Nugent, D. W. Overall and S. J. Holmes, *Organometallics*, 1983, **2**, 161.
300. I. P. Rothwell, *Polyhedron*, 1985, **4**, 177.
301. L. Chamberlain, J. Keddington, J. C. Huffman and I. P. Rothwell, *Organometallics*, 1982, **1**, 1538.
302. L. Chamberlain, I. P. Rothwell and J. C. Huffman, *J. Am. Chem. Soc.*, 1984, **102**, 7338.
303. L. Chamberlain and I. P. Rothwell, *J. Am. Chem. Soc.*, 1983, **105**, 1665.

15.4

Diketones and Related Ligands

ALLEN R. SIEDLE
3M Central Research Laboratories, St. Paul, MN, USA

15.4.1 INTRODUCTION

1,3-Diketones exist in solution as mixtures of keto (**1a**) and enol (**1b**) forms, related by a 1,3 hydrogen shift. Keto–enol equilibria have been measured by NMR techniques[1] and the residence time for hydrogen atoms in the two-well potentials of unsymmetrical O—H ⋯ O hydrogen bonds in unsymmetrically substituted acetylacetone derivatives has been obtained from ^{13}C NMR spectra.[2] X-Ray studies of dibenzoylmethane[3] and ring-substituted derivatives[4,5] indicate that these diketones exist in the solid state in the enol form and have, excepting dibenzoylmethane, equidimensional C—O distances and symmetric intramolecular hydrogen bonds. Electron diffraction results[6,7] indicate that, in the gas phase, acetylacetone adopts the enol configuration with the keto:enol ratio decreasing with temperature. It is remarkable that until recently the solid state structure of this archetypal β-diketone was not known. An X-ray diffraction study has been performed on a crystal complex containing acetylacetone, diphenylhydantoin and 9-ethyladenine in which the dione does not appear to interact with the other components of the lattice. The structure of acetylacetone in this material is shown in Figure 1 and it discloses an enol form but with a quite unsymmetrical hydrogen bond.[8] While the structure of acetylacetone in this ternary matrix is secure, it remains uncertain how the molecular geometry changes on vaporization or, alternatively, precisely how redistribution of the C—O and O—H bond lengths affects the molecular energy.

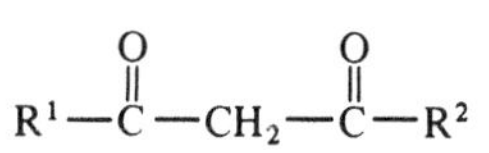

O   O
‖   ‖
R¹—C—CH₂—C—R²   R¹—C=CH—C—R²

$$R^1-\overset{\overset{\displaystyle O}{\|}}{C}-CH_2-\overset{\overset{\displaystyle O}{\|}}{C}-R^2 \qquad R^1-\overset{\overset{\displaystyle OH}{|}}{C}=CH-\overset{\overset{\displaystyle O}{\|}}{C}-R^2$$

(1a) 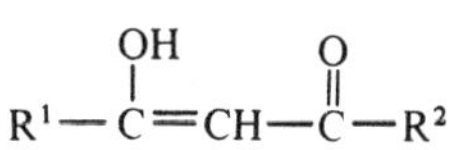   (1b)

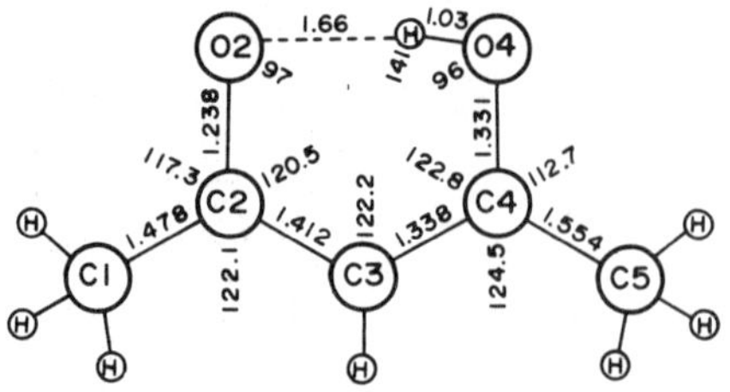

Figure 1 Bond distances and angles in acetylacetone (reproduced with permission from *J. Am. Chem. Soc.*, 1983, **105**, 1584)

The methine proton in the keto form and the hydroxyl proton in the enol form of β-diketones are acidic and their removal generates 1,3-diketonate anions (**2**), which are the source of an extremely broad class of coordination compounds referred to generically as diketonates or acetylacetonates. The synthesis, structure and properties of these compounds form the focus of this chapter. Diketonate anions are powerful chelating species and form complexes with virtually every transition and main group element. The scope of this chemistry is very large and it has been assessed earlier in several excellent reviews.[9-14]

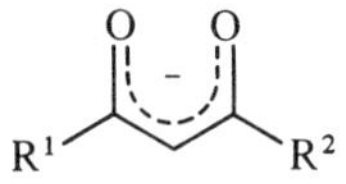

(2)

Complexes of β-diketonates are to be contrasted with those of simple ketones, in which bonding is appreciably less robust. Ketones in their transition metal complexes are generally labile[15,16] and often serve as sources of 'lightly stabilized' reactive cationic metal species, an example of which is $[(C_5H_5)Rh(acetone)_3]PF_6$.[17,18] The structural data base for metal ketone complexes is quite small. In $[(\eta_6\text{-mesitylene})Ru(4\text{-hydroxy-4-methylpentanone})(acetone)](BF_4)_2$, formed by isomerization of $[(\eta_6\text{-mesitylene})Ru(acetone)_3](BF_4)_2$, $d(\text{Ru—O})$ involving the coordinated acetone molecule is 2.11 Å.[19] The long (2.22–2.24 Å) Ir—O bonds in $[IrH_2(Me_2CO)_2(PPh_3)_2]BF_4$ are associated with a structural *trans* effect of the hydride ligands. Acetone in this iridium compound is so readily displaced that it is active in alkane and alkene dehydrogenation.[20]

So pronounced is the chelating tendency of the diketonate anion that even alkali metal complexes may be isolated, as illustrated by $Rb_2[(CF_3COCHCOCF_3)_3Na]$, in which sodium is surrounded by a trigonal prismatic array of donor oxygen atoms.[21] For complexes derived from dibenzoylmethane, stabilities in dioxane–water are in the order Li > Na > K > Cs.[22,23] For divalent metal ions[24] formation constants increase in the order Ba > Sr > Ca > Mg > Cd > Mn > Pb > Zn > Co, Ni, Fe > Cu, and for higher valent metal ions the first formation constants for chelates are in the order $Fe^{3+} > Ga^{3+} > Th^{4+} > In^{3+} > Sc^{3+} > Y^{3+} > Sm^{3+} > Nd^{3+} > La^{3+}$.[25-37] More recently, such studies have been extended to heptane-3,5-dionate complexes.[38] Detailed interpretation of these trends is complicated by the fact that many of the different complexes adopt different, *i.e.* aggregated, structures, at least in the solid state, by varying degrees of hydration and by conformational changes in the structure of the β-diketonate ion. Low-temperature ^{13}C NMR measurements on the sodium salt of 3-methylpentane-2,4-dione(1 −) reveal the presence of both the Z,Z and E,Z conformations, the former presumably coordinating to Na^+. Addition of Li^+, which forms a more stable complex, shifts the slow equilibrium in favor of the Z,Z form.[39]

The standard enthalpies of formation of the acetylacetonate complexes of Al, Ga, Cr^{III} and Mn^{III} have been determined by solution reaction calorimetry[40-43] and the enthalpies of combustion of the Al, Ga and In tris(chelates) reported.[44] Such data are relatively scarce in inorganic chemistry and quite valuable. For example, from solution calorimetric hydrolysis experiments on $(MeCOCHCOMe)_2Pt$, the Pt—O bond energy was found to be 183.1 kJ mol^{-1}.[45]

Different tautomers of β-diketones may react with metal ions at quite different rates. For example, the keto form of thenoyltrifluoroacetone does not react at all; the enol form reacts by two pathways, one independent and one inversely dependent on hydrogen ion concentration. Deprotonation of the ligand is evidently an important, rate determining factor.[46] Both tautomers of acetylacetone react with iron(III). The rate constants for reaction of the keto form with Fe^{3+} and $FeOH^{2+}$ are 0.29 and 5.4 $M^{-1}s^{-1}$ respectively and the rate constants for the reaction of undissociated enol form with Fe^{3+} and $FeOH^{2+}$ are 5.2 and $4.4 \times 10^3 M^{-1}s^{-1}$ respectively.[47] Rate constants for the reaction of the enol and keto tautomers of acetylacetone with Ti^{3+} are 2.4×10^3 and 0.7 $M^{-1}s^{-1}$ respectively.[48]

Metal β-diketonate coordination compounds may be divided on a structural basis into several broad classes, which are summarized below.

(1) η_2 Oxygen-bonded diketonates, O,O-RCOCHCOR^1, as found, for example, in Pd(CF$_3$COCHCOCF$_3$)$_2$ (3). In β-keto ester complexes, R corresponds to an alkoxy group. In a variant of this type of bonding the oxygen donor atom may be shared by two metal atoms, as in [Ni(MeCOCHCOMe)$_2$]$_3$ (Figure 2).[49]

R
$\overset{\text{R}}{\underset{\text{R}^1}{\bigg\langle}} \overset{=O}{\underset{=O}{\bigg\rangle} \text{M}}$

(3)

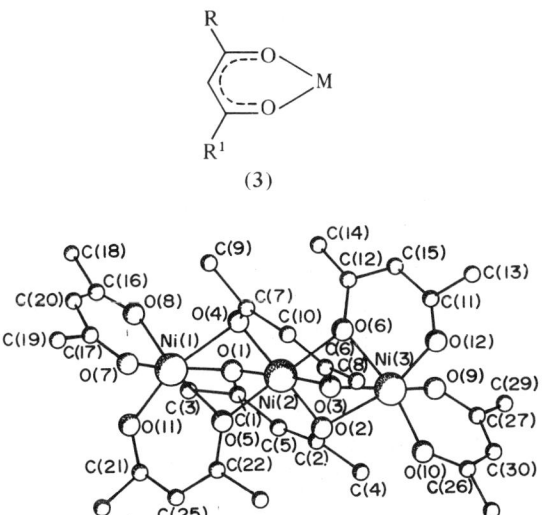

Figure 2 Molecular structure of [(MeCOCHCOMe)$_2$Ni]$_3$ (reproduced with permission from *J. Chem. Soc., Dalton Trans.*, 1982, 307)

(2) A small number of compounds exhibit semi-chelating β-diketonate bonding, so named by analogy to semi-bridging carbonyl ligands. The usual η nomenclature is inapplicable here. In this bonding type, exemplified by Pd{Ph$_2$PC$_2$H$_4$P(Ph)C$_2$H$_4$PPh$_2$}(CF$_3$COCHCOCF$_3$)$^+$ derivatives, the metal coordination plane and that of the semi-chelating hexafluoroacetylacetonate ligand are nearly perpendicular so that one diketonate oxygen is effectively in an axial position below the PdP$_3$O plane. The axial Pd—O bond is long, 2.653 Å, compared with the in-plane 2.110 Å Pd—O contact, and corresponds to much weaker metal–oxygen bonding.[50]

(3) Monodentate oxygen-bonded diketonate ligands, O-RCOCHCOR′, may occur in Pd(4-ClC$_5$H$_4$N)$_4$(CF$_3$COCHCOCF$_3$)$_2$ but in the solid state the Pd—O distances are very long (3.019 Å) and it is arguable that these are not chemically significant.[51] An unambiguous example of this type of bonding is found in Pd{(o-tolyl)$_3$P}(O-CF$_3$COCHCOMe)(O,O-CF$_3$COCHCOMe). Here the Pd—O distance is 2.011 Å and the CF$_3$CO and the Pd—O moieties are in a *trans* relationship relative to the double bond in the enolic η_1-ligand, as shown in Figure 3.[52]

(4) η_3-Acetylacetonates, in which the metal is σ-bonded to a terminal CH$_2$ group and π-bonded to the C=C double bond in the enolic form of the ligand, are known for a series of palladium complexes of the type Pd(CH$_2$COCHCHMe)(Cl)(L) (L = Ph$_3$P, Ph$_3$As), *e.g.* (4).[53]

$$\begin{array}{c}
\text{O—C} \overset{\text{CH}_2}{\diagup} \overset{\diagup}{\underset{\diagdown}{\text{Pd}}} \overset{\diagup}{\underset{\text{Cl}}{}}\text{L} \\
\text{H} \quad \text{CH} \\
\text{O=C} \\
\quad\quad \text{Me}
\end{array}$$

(4)

(5) In the above structure, if Ph$_3$P is replaced by a bidentate ligand such as bipyridyl, there results a C-η_1-acetylacetonate complex in which palladium is σ-bonded to a terminal CH$_2$ group. These presumably exist as tautomeric keto and enol forms (equation 1).[53]

(6) Carbon-bonded η_1-diketonate ligands, in which the metal is bonded to the central carbon, *i.e.* that flanked by two acetyl groups, are rather common in palladium and platinum chemistry and were first demonstrated in K[Pt(Cl)(C-MeCOCHCOMe)(O,O-MeCOCHCOMe)] (Figure 4).[54]

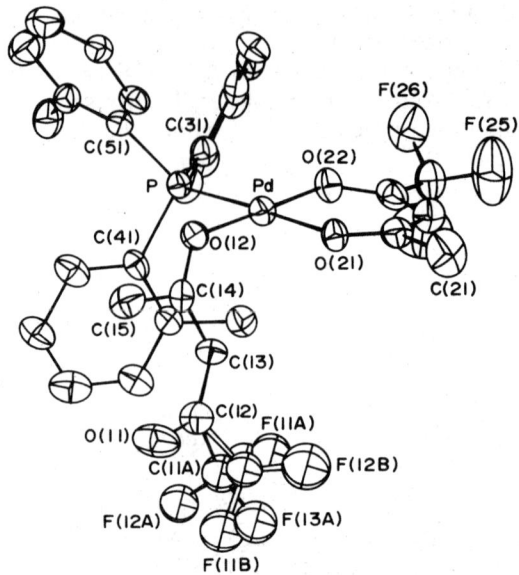

Figure 3 Molecular structure of Pd{(o-tolyl)$_3$P}(O-CF$_3$COCHCOMe)(O,O-CF$_3$COCHCOMe) (reproduced with permission from *Bull. Chem. Soc. Jpn.*, 1983, **56**, 3297)

$$\text{keto} \rightleftharpoons \text{enol} \tag{1}$$

keto   enol

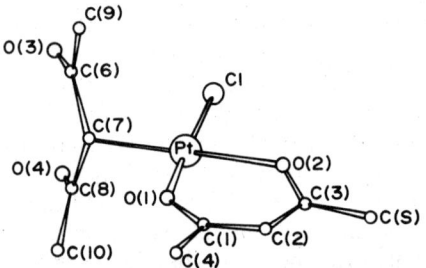

Figure 4 Molecular structure of the Pt(C-MeCOCHCOMe)(O,O-MeCOCHCOMe)(1−) ion (reproduced with permission from *J. Chem. Soc. (A)*, 1969, 485)

(7) Tridentate bonding involving both acetyl oxygen atoms and the methine carbon atom occurs in dimeric [Pt(C,O,O-C$_3$H$_7$COCHCOC$_3$H$_7$)Me$_3$]$_2$, whose structure, shown schematically in (**5**) is established by both X-ray[55] and neutron diffraction studies.[56]

(**5**)

(8) Several compounds have been reported which contain the dianion of acetylacetone (or of a β-keto ester), the extra negative charge attending removal of one of the methyl protons. In $Pd(CH_2COCHCOMe)(bipyridyl)$, the acetylacetonate(2−) ligand is considered to bind to palladium through C-1, C-2, and C-3, *e.g.* as in (6).[57] Isomeric with this are η_2 *C,O*-diketonate dianions. This mode of bonding is adopted by trifluoroacetylacetonate(2−) in $Pt(C,O$-$CH_2COCHCOCF_3)(PPh_3)_2$.[58]

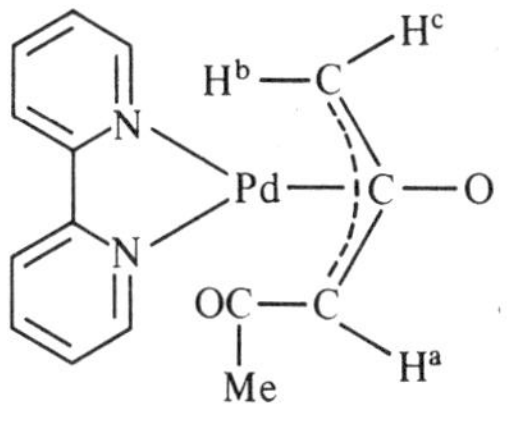

(6) 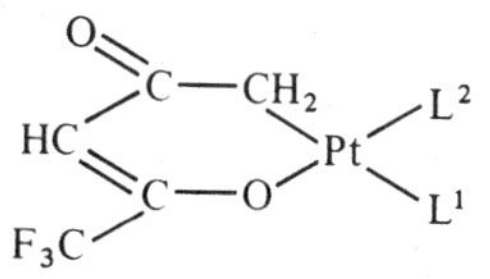   (7)

(9) The bis(chloromercury) compound $MeCOC(HgCl)_2COMe$ can formally be regarded as a derivative of the η_2 *C,C*-acetonylacetonate(2−) ion.[59]

(10) η_2 *O,O*-Acetylacetonate(2−) occurs as a dienediolate ligand in $Mn[(O,O$-CH_2=$C(O)$-CH=$C(O)Me]P(Ph)(CO)_2C_5H_5$ (Figure 5).[60,61]

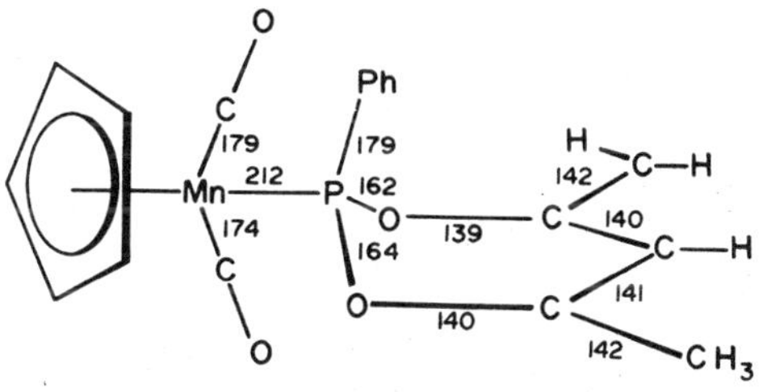

Figure 5 Structure of $Mn(O,O$-$CH_2COCHCOMe)P(Ph)(CO)_2C_5H_5$ (reproduced with permission from *Angew. Chem., Int. Ed. Engl.*, 1977, **16**, 858)

(11) An interesting bonding type, which eludes formal description, is found in Ag-$Ni(MeCOCHCOMe)_3 \cdot 2AgNO_3 \cdot H_2O$. In this compound, octahedral nickel is coordinated to three acetylacetonate ligands. Tetrahedral silver is surrounded by three oxygen atoms and is also 2.34 Å distant from the acetylacetonate methine carbon [cf.$Fe(MeCOCHCOMe)_2$] so that the diketonate ligand may be loosely considered to be tridentate (Figure 6).[62]

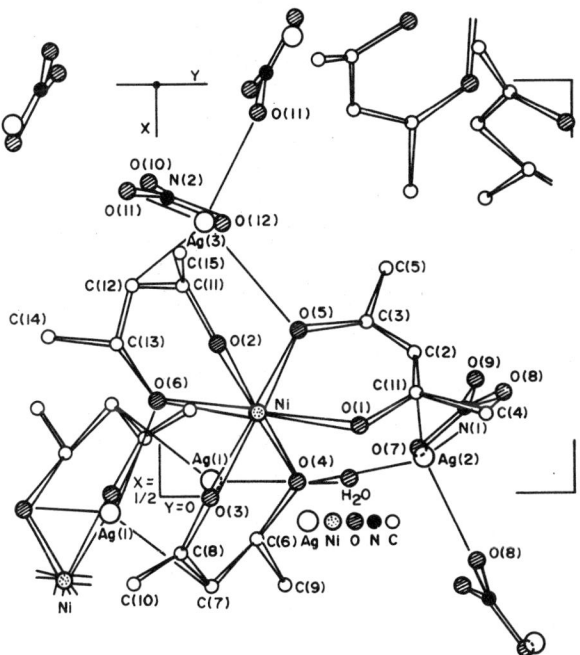

Figure 6 One molecular unit in $AgNi(MeCOCHCOMe)_3 \cdot 2AgNO_3 \cdot H_2O$ (reproduced with permission from *Inorg. Chem.*, 1966, **5**, 1074)

15.4.2 STRUCTURE OF METAL β-DIKETONATE COMPLEXES

Complete characterization of metal β-diketonate complexes relies heavily on X-ray crystallographic data, which will be a major focus of this section. The largest body of structural information pertains to η_2 oxygen-bonded diketonates. In these, the diketonate ligands are usually essentially planar with nearly equivalent C—C, C—O and M—O distances, consistent with delocalized bonding. Commonly encountered deviations include a folding along the O—O axis so that the metal is slightly displaced from the C_3O_2 plane; this may be due to packing forces. Variations in intramolecular metal–oxygen bond lengths, notable in Mn^{II} and Cu^{II} complexes, may be traceable to Jahn–Teller distortions or, in unsymmetrical complexes, to structural *trans* effects.

The dihydrate of Na(MeCOCHCOMe) is soluble in toluene and *may* contain four-coordinate sodium.[63] Similarly, Li($CF_3COCHCOCF_3$)($Me_2NC_2H_4NMe_2$) is monomeric at low concentrations in benzene.[64] Two classes of anionic complexes of the types $M(CF_3COCHCOCF_3)_2^-$ (M = Li, Na, K, Rb, Tl) and $M(CF_3COCHCOCF_3)_3^{2-}$ (M = K, Cu, Mg, Ni, Mn) have been prepared, and the complex $Rb_2[Na(CF_3COCHCOCF_3)_3]$, which contains sodium in a trigonal prismatic coordination geometry,[21] has been noted earlier. Sodium acetylacetonate shares with complexes of the first-row transition metals a propensity for polymorphism and an irreversible phase transition from the orthorhombic to the triclinic form occurs at about 82 °C.[65] Regular octahedral coordination of magnesium occurs in $[1,8-(Me_2N)_2C_{10}H_8 \cdot H]^+ Mg(CF_3COCHCOCF_3)_3^-$,[66] and monomeric beryllium acetylacetonate contains, as expected, four-coordinate, tetrahedral Be.[67]

The chemistry of 1,3-diketone complexes of nonmetallic elements (B, Si, Ge, Sb, Te) has been reviewed.[68] Condensation of acetylacetone with $TeCl_4$ yields $Te(CH_2COCH_2COCH_2)Cl_2$ (**8a**), which may be reduced to $Te(CH_2COCH_2COCH_2)$ (**8b**). The crystal structures of these compounds demonstrate an unusual coordination of tellurium to the terminal carbons of the diketonate.[69–72] In the dichloride complex, Te^{IV} exhibits a distorted trigonal bipyramidal geometry with a stereochemically active lone electron pair in the equatorial plane. In $S_2(MeCOCHCOMe)_2$ the acetylacetonate methine carbons are bonded to a disulfur unit.[73]

(8a)   (8b)

The molecular structure of the mixed acetylacetonate–acetate B(MeCOCHCOMe)(O_2CMe)_2 consists of discrete units having approximate C_{2v} symmetry. This compound is notable for its method of preparation, having been obtained from a bizarre reaction of vanadium diboride with acetic acid.[74]

β-Diketone complexes of the first-row transition metals are of interest as structural archetypes and because of the tendency of compounds containing elements on the right-hand side of the periodic table to adopt originally unanticipated oligomeric structures so that, by means of oxygen-bridging diketonate ligands, the central metal atom achieves a coordination number greater than four. Complexes of metals at each end of the first row are mononuclear. Scandium tris(acetylacetonate) contains a ScO_6 coordination core which is close to trigonal prismatic, while in both polymorphs of V(MeCOCHCOMe)_3 there is a slight trigonal distortion of the VO_6 octahedra.[76] In fact tris(acetylacetonates) are not expected to have precisely regular octahedral geometry since the chelate bite angle is less than 90°, as pointed out in a recent tabulation of shape parameters for these compounds.[77] Acetylacetonates of Cu^{II} and VO^{2+} are mononuclear. In the former, copper resides in a square planar array of oxygen atoms [d(Cu—O) = 1.92 Å] and the axial approach of the metal to the C_3O_2Cu ring of an adjacent molecule is *ca.* 3.1 Å.[78] Mononuclear structures have also been reported for Cu[MeCOC(Ph)COMe]_2,[79] Cu[MeCOC(Me)COMe]_2,[80] Cu(PhCOCHCOMe)_2,[81] bis(cyclodecane-1,3-dionate)copper and bis(cyclotridecane-1,3-dionate)copper.[82] The coordination geometry in vanadyl β-diketonates is square pyramidal, an arrangement apparently unique to VO^{2+} derivatives.[83,84]

Illustrative of an aggregated diketonate complex is nickel(II) acetylacetonate, which is trimeric in the solid state.[51] The bridging Ni—O—Ni bonds are cleaved by Lewis bases and thus Ni(MeCOCHCOMe)_2(H_2O)_2 is monomeric.[85] Replacement of the terminal methyl groups by larger alkyl groups tends to inhibit aggregation and leads to monomeric materials, examples of which are the 2,2,6,6-tetramethylheptane-3,5-dionate complexes of Ni^{II},[86] Co^{II} and Zn^{II}.[87] In these symmetrical chelates the C—O ring carbon–ring carbon and ring carbon–aliphatic carbon bond lengths correspond to bond orders of ~ 1.3, ~ 1.5 and ~ 1.0 respectively and are consistent with delocalized bonding in the C_3O_2M rings. Monomer–trimer equilibria for $Ni(MeCOCHCOMe)_2$ and related compounds have been observed at elevated temperatures by recording electronic spectra in diphenylmethane solvent.[88] Noteworthy in this regard is a series of alkane-soluble nickel(II) acetylacetonate derivatives bearing long hydrocarbon chains on C-1 and C-5. The transition between diamagnetic, purple monomers containing square planar nickel and the green, paramagnetic octahedral oxygen-bridged trimers is quite sensitive to the length of the hydrocarbon substituent in $Ni(RCH_2COCHCOCH_2R)_2$, occuring at 17 °C for $R = n\text{-}C_7H_{15}$ and 42 °C for $R = n\text{-}C_9H_{19}$.[89] Materials of this type may find application in thermochromic or liquid crystal devices.

Chromium(II) bis(acetylacetonate), which is isomorphous with the Cu^{II} and Pd^{II} analogues, crystallizes as an essentially planar monomer but the distance between ligand planes is 3.24 Å, somewhat less than the 3.35 Å interplanar spacing in graphite. There are, however, long (3.048(5) Å) interactions between chromium and the methine carbons of adjacent molecules.[90] Acetylacetonates of Zn^{II} and Co^{II} are trimeric and tetrameric respectively, aggregation again occuring *via* bridging diketonate ligands.[91,92]

Dissolution of metal acetylacetonates in organic solvents may lead to disruption of oligomers and so solution molecular weight measurements may not always provide reliable clues to the degree of aggregation in the solid phase. This point is underscored by a recent structural analysis of $Fe(MeCOCHCOMe)_2$. The compound is tetrameric but aggregation is achieved in a surprising way: $[Fe(MeCOCHCOMe)_2]$ units are connected by 2.785(9) Å Fe—C bonds involving the methine carbon atoms in adjacent dimers. This iron–carbon interaction is evidently sufficiently strong that polymerization by complete oxygen bridging does not occur (Figure 7).[93]

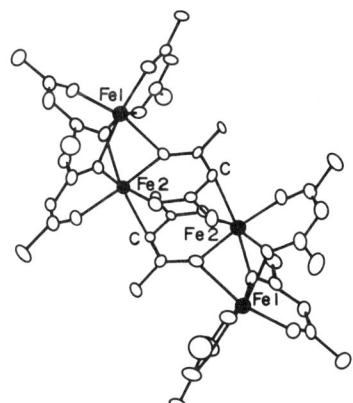

Figure 7 Schematic representation of tetrameric $Fe(MeCOCHCOMe)_2$ (reproduced with permission from *Nouv. J. Chim.*, 1977, **1**, 301)

An important series of papers has described the structures of $[Co(MeCOCHCOMe)_2]_4$, $[Co(MeCOCHCOMe)_2]_3H_2O$, $[Co(MeCOCHCOMe)_2]_2(H_2O)$ and *trans*-$Co(MeCOCHCOMe)_2(H_2O)_2$ and proposed a conceptual set of transformations by which the tetranuclear complex may be sequentially degraded upon hydration (Figure 8).[94,95]

In acetylacetonate complexes of trivalent metals, it appears that several phases or polymorphs of each may exist. Their confusing identification as α, β, γ, *etc.* is rooted in history. Chelates of Al, Co, Cr, V, Mn, Ga, Ru and Rh may crystallize in a so-called β-form, monoclinic, $P2_1/c$ for which $a \approx 14$, $b \approx 7.5$, $c \approx 16.5$ Å, $\beta \approx 99°$ and having $d(M\text{—}O)$ in the range 1.89–1.99 Å. A second class of chelates is orthorhombic, P_{bca}, for which $a \approx 15.5$, $b \approx 13.5$, $c \approx 16.7$ Å and having slightly longer M—O separations (~ 1.98–~ 2.07 Å). The γ-form of $Mn(MeCOCHCOMe)_3$ discussed below is distinctive, as is $Mo(MeCOCHCOMe)_3$, monoclinic, $P2_1/n$ with $a = 16.515$, $b = 13.052$, $c = 8.152$ Å and $\beta = 90.74°$.[96] The structures of the isomorphous acetylacetonates of Al^{III} and Co^{III}[97] and of monoclinic $Mo(MeCOCHCOMe)_3$[96] have been published. Crystal structures for both α- and β-$V(MeCOCHCOMe)_3$ have been determined. In the latter polymorph, the vanadium lies off the least-squares planes for all three chelate rings but off the plane of only one ring in the α-form.[98]

Manganese tris(acetylacetonate) exists in two crystalline forms, β and γ. High quality structures of both have been published.[99,100] Each exhibits definite evidence of Jahn–Teller distortion. In the β-form the average Mn—O distances are 1.95 and 2.00 Å in a compressed MO_6 octahedron; the

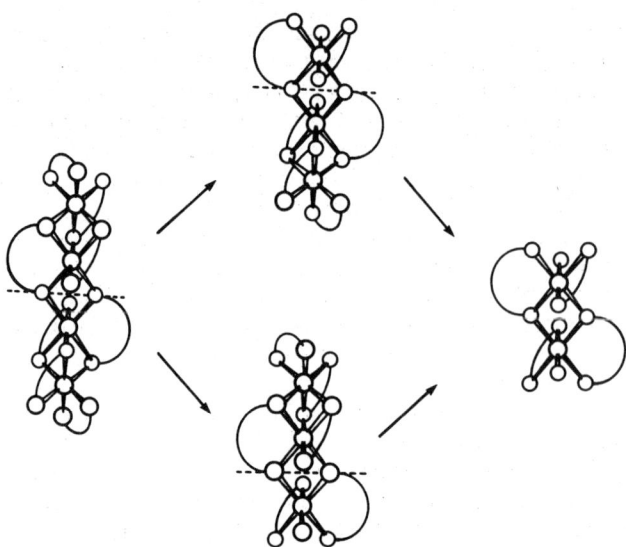

Figure 8 Scheme relating transformation of [Co(MeCOCHCOMe)₂]₄ to enantiomorphic [Co(MeCOCHCOMe)₂]₃(H₂O)₂ structures and *meso*[Co(MeCOCHCOMe)₂]₂(H₂O)₂. Large open circles represent oxygen atoms and curved lines represent the chelate rings (reproduced with permission from *J. Am. Chem. Soc.*, 1968, **90**, 38)

O—O bite distance is 2.79 Å. Detailed crystal field calculations have been performed on this molecule. The average short and long bonds in the γ-form, which is tetragonally elongated, are 1.935 and 2.111 Å respectively. Phenomenologically, the differences in the Mn—O bond lengths seem to propagate in two of the C₃O₂Mn chelate rings in that bond lengths and angles are not symmetrical within the same ligand but are related in pairs between the two ligands by a pseudo-C_2 molecular axis. The shorter C—O bonds are adjacent to the longer Mn—O bonds and the longer C—C bonds are adjacent to the shorter C—O bonds. Similar curious effects occur in dodecahedral, stereochemically nonrigid Zr(MeCOCHCOMe)₃NO₃[101] and Zr(MeCOCHCOMe)₂(NO₃)₂.[102] The gas phase structures of the hexafluoroacetylacetonates of Cu^{II} and Cr^{III} have been determined by electron diffraction but the data do not appear to have sufficient precision to establish a Jahn–Teller distortion in the d^9 system.[104]

Titanyl acetylacetonate is dinuclear, containing a planar di-μ-oxo-dititanium ring. The longer bond between the metal and the acetylacetonate oxygen atom is *trans* to the bridging oxide, thus representing a structural *trans* effect.[105]

In thallium(I) hexafluoroacetylacetonate[106] there are two structural units in the crystal. One is a monomeric Tl(CF₃COCHCOCF₃) unit; the other is a dimer in which four oxygen atoms from the hexafluoroacetylacetonate ligand and two *trans* thallium atoms form an octahedron. Two such units are linked by oxygen bridges to form a linear polymeric chain which must be easily ruptured since the compound is quite volatile. Displacement of Tl from the diketonate ligand plane in the dimer is quite large, 1.56 Å. The cause of such displacements is not precisely established and may represent a combination of lattice and electronic effects. In four-coordinate transition metal complexes the O—M—O plane is often slightly canted with respect to the mean plane of the ligand atoms and the dihedral angle between these planes in Co^{II}, Ni^{II} and Zn^{II} tetramethylheptanedionates averages 160°, considerably smaller than in five- [*cf.* VO(MeCOCHCOMe)₂, 168°], six- [*cf.* Co(MeCOCHCOMe)₂ (H₂O)₂, 164°] and eight-coordinate [*cf.* Zr(MeCOCHCOMe)₄, 158°] diketonates.

Unlike the dihydrates of Co(MeCOCHCOMe)₂ and Ni(MeCOCHCOMe)₂,[107] in which the water molecules are *trans* to one another, they are *cis* in [Ca(MeCOCHCOMe)₂(H₂O)₂]·H₂O.[108] However, Mg(MeCOCHCOMe)₂(H₂O)₂ is isomorphous with the nickel and cobalt analogues, indicating a *trans* arrangement of water ligands and implying that the tetragonal distortions observed in these transition metal derivatives may be associated with crystal packing effects.[109] Zinc acetylacetonate crystallizes as a monohydrate containing five-coordinate, distorted tetragonal pyramidal zinc.[110,111] In cadmium bis(acetylacetonate) there are infinite, parallel chains with octahedral cadmium being bridged by pairs of oxygen atoms from each of two diketonate ligands.[112] The tris(chelate) complex K[Cd(MeCOCHCOMe)₃] contains six-coordinate trigonal prismatic cadmium.[113] This material has been used as a host so that Mn^{II} can be studied in the same coordination environment.[114]

In Sn(PhCOCHCOMe)₂, divalent tin exhibits essentially a trigonal bipyramidal coordination geometry with the stereochemically active lone electron pair in an equatorial site.[115] For tetravalent tin complexes of the type $SnR_2(diketonate)_2$, simple diketonates which have large bite distances yield *trans* isomers, while tropolonates, which have smaller bites, tend to adopt a *cis* geometry.[116,117] In solution, such complexes may be structurally labile, a case in point being $SnMe_2(MeCOCHC-OMe)_2$[118,119] and *cis–trans* interconversion may be effected by rapid ligand exchange.[120,121] The methyl- and phenyl-tin substituted diketonates are considered, on the basis of NMR and vibrational spectra, to have *trans* configurations.[122]

The uranyl complex $[UO_2(CF_3COCHCOCF_3)_2]_3$ crystallizes as a trimer. Three uranium atoms form an equilateral triangle, the U—U edges of which are bridged by three of the six uranyl oxygen atoms rather than by the diketonate ligands. This is a most unusual structural feature for an uranyl derivative.[123] This compound and the dipivaloylmethane derivative exist in the gas phase as partially dissociated dimers and thermodynamic parameters for the vaporization processes have been obtained.[124] The work serves as a reminder that significant structural changes may accompany vaporization of metal acetylacetonates.

Fluid solution structures have been explicitly considered by workers who have used light scattering techniques to demonstrate a *trans* geometry for $SnR_2(tropolonate)_2$ and $SnR_2(PhCOCHCOPh)_2$ ($R = c\text{-}C_6H_{11}, n\text{-}C_8H_{17}$).[125] Large, positive Kerr constants and nonzero dipole moments are reported to be indicative of distorted *trans* geometries better described in $SnR_2(PhCOCHCOPh)_2$ ($R = Me$, Et) and $SnEt_2(tropolonate)_2$ in cyclohexane as skew-trapezoidal-bipyramidal.[126] Interesting geometrical variations are seen in $SnR_3(diketonate)$ compounds which contain trigonal bipyramidal Sn^{IV}. Mössbauer spectra indicate that when $R = Me$ a *mer* structure is adopted and that the phenyl analogues are all-*cis*. This is confirmed by an X-ray study of $SnPh_3(PhCOCHCOPh)$ in which the dibenzoylmethanate ligand spans an equatorial and an axial site with $d(Sn—O_{ax})$, 2.276 Å, being much longer than $d(Sn—O_{eq})$, 2.180 Å.[127]

The lanthanide elements, in their complexes with β-diketones, tend to adopt interesting, higher coordination geometries. These compounds frequently crystallize as hydrates from which water removal without decomposition of the compound is difficult. Some structural information is summarized in Table 1. 2,2,6,6-Tetramethylheptanedionate chelates of the lighter lanthanides (La to Dy) can be obtained in nonsolvated form, crystallize in the monoclinic system and contain dimer units whereas the heavier analogues (Ho to Lu) tend to be orthorhombic with isolated six-coordinate monomers.[128]

In the 1:1 adduct of $Yb(MeCOCHCOMe)_3$ with acetylacetonimine, the ytterbium coordination core comprises a capped trigonal YbO_7 prism with an acetylacetonate oxygen as the capping atom; the imine group participates in hydrogen bonding but is not bonded to the metal.[129] Eight coordinate $M(MeCOCHCOMe)_4$ chelates ($M = Ce, Th, Zr, Hf, U, Np$) exist in two crystalline modifications, the α and β forms. Proper description of these materials requires care since idealized eight-coordinate geometries are closely related by rather small deformations along an interconnecting reaction pathway. Analysis of the shape parameters for $\alpha\text{-}Ce(MeCOCHCOMe)_4$ and $\alpha\text{-}Th(MeCOCHC-OMe)_4$ indicate that the metal coordination polyhedra most closely approximate C_{2v} bicapped

Table 1 Stereochemistries of High Coordination Number β-Diketonates

Compound	Approximate coordination geometry	Ref.
$Zr(MeCOCHCOMe)_4$	Dodecahedron	619
$Ce(PhCH_2COCHCOCH_2Ph)_4$	Trigonal dodecahedron	620
$NH_4[Pr(thenoyltrifluoroacetate)_4]\cdot H_2O$	Dodecahedron	621
$Ho(MeCOCHCOMe)_3\cdot H_2O$	Capped octahedron	622
$Ho(Bu^tCOCHCOBu^t)_3(4\text{-}MeC_5H_4N)_2$	Distorted square antiprism	623
$Lu(Bu^tCOCHCOBu^t)_3\cdot(3\text{-}MeC_5H_4N)$	Distorted capped trigonal prism	624
$Lu[CF_3(CF_2)_2COCHCOBu^t]_3\cdot H_2O$	Capped trigonal prism	625
$Dy(Bu^tCOCHCOBu^t)_3\cdot H_2O$	Capped trigonal prism (H_2O at cap)	626
$Pr_2\{CF_3(CF_2)_2COCHCOBu^t\}_6(H_2O)_2$	Dodecahedron; bicapped trigonal prism	627
$Y(MeCOCHCOMe)_3(H_2O)_3$	Distorted square antiprism	628
$Y(MeCOCHCOMe)_3\cdot H_2O$	Capped trigonal prism	629
$La(MeCOCHCOMe)_3(H_2O)_2$	Distorted square antiprism	630
$Eu(thenoyltrifluoroacetonate)_3H_2O$	Square antiprism	631
$Pr_2(Bu^tCOCHCOBu^t)_6$	Distorted trigonal prism	632
$Y(PhCOCHCOMe)_3$	Capped octahedron	633
$Cs[Y(CF_3COCHCOCF_3)_4]$	Dodecahedron	634
$Er(Bu^tCOCHCOBu^t)_3$	Trigonal prism	635
$Zr(MeCOCHCOMe)_3Cl$	Distorted pentagonal bipyramid; axial Cl	636
$Nb(Bu^tCOCHCOBu^t)_4$	Square antiprism	637
$Th(CF_3COCHCOMe)_4\cdot H_2O$	Capped square antiprism	638
$Th(thenoyltrifluoroacetonate)_4$	Dodecahedron	639

trigonal prisms. The β-tetrakis(acetylacetonates) of Zr, Ce, U and Np adopt square antiprismatic coordination geometries.[130] Similar considerations apply, of course, to seven- and nine-coordination polyhedra.

The oxide cluster $Er_8O(Me_3CCOCHCOCMe_3)_{10}(OH)_{12}$, obtained as a minor byproduct in the synthesis of the tris(chelate) complex, is interesting in that two of the dipivaloylmethanate ligands are bridging; hydroxyl groups are triply bridging and erbium has approximately capped trigonal prismatic coordination geometry (Figure 9).[131]

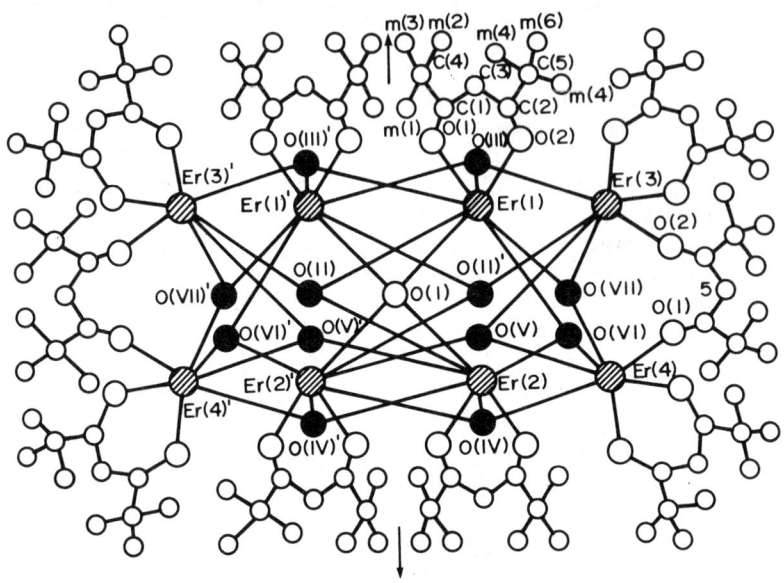

Figure 9 Schematic representation of the molecular structure of $Er_8O(OH)_{12}$(tetramethylheptanedionate)$_{10}$ (reproduced with permission from *J. Cryst. Mol. Struct.*, 1972, **2**, 197)

While complexes in which the metal is coordinatively unsaturated frequently oligomerize utilizing bridging β-diketonate ligands, a different mode of aggregation is believed to occur in $M(CO)_2$-(MeCOCHCOMe) (M = Rh, Ir). The near planarity of the acetylacetonate permits these flat molecules to stack so that short (3.20 Å for Ir) metal–metal contacts are formed. This leads to highly anisotropic DC electrical transport properties and these compounds are semiconductors.

Porphyrin-containing chelate complexes of the type $M(Ph_4porphyrin)(MeCOCHCOMe)$ (M = Y, Th) presumably contain bidentate oxygen-bonded acetylacetonate ligands.[133] These, and phthalocyanine complexes, represent novel classes of β-diketonate derivatives. Two types of rare earth materials, $M(PC)(PhCOCHCOPh)(PhCOCH_2COPh)$(acetone) (M = Sm, Eu, Gd, Tb, Dy, Ho, Er, Tm, Y, Yb; PC = phthalocyanine) and $M(PC)(PhCOCHCOPh)(MeOH)_2$ (M = Eu, Gd, Tb, Dy, Ho, Er, Tm, Yb, Y) are obtained from Li_2PC and the metal dibenzoylmethanate in acetone or methanol. The structure proposed for the latter class of complexes features an η_2 oxygen-bonded diketonate ligand attached to the metal, which is considerably displaced from the phthalocyanine plane. The first class of complexes is thought to be similar but to contain as well a coordinated, neutral molecule of dibenzoylmethane.[134] Similar syntheses, when carried out in THF and in the presence of oxygen, yield complexes of the type $M(PC)(Me_3CCOCHCOCMe_3)$, which are proposed to contain the phthalocyanine anion radical.[135] In the absence of oxygen, $M_2(\mu\text{-}PC)(Me_3CCOCHC\text{-}OCMe_3)_4$ is formed. The crystal structure of the samarium derivative (Figure 10) reveals a phthalocyanine ligand bridging two samarium atoms, one on either side of the phthalocyanine ring. Each is bonded to four nitrogen atoms and to two η_2-dipivaloylmethanate ligands, thus achieving a coordination number of eight.[136]

Atropisomeric cobalt(III) chelates in which optical activity derives from restricted rotation of an aromatic ring, such as napthyl or 2,4,6-trinitrophenyl, attached to the methine carbon atom have been resolved and characterized by X-ray crystallography.[137,138]

15.4.3 SYNTHESIS OF METAL β-DIKETONATE COMPLEXES

The very great majority of systematic preparative methods for metal diketonates employs the neutral β-diketone as a starting material. This may be reacted with various types of substrates, *e.g.*

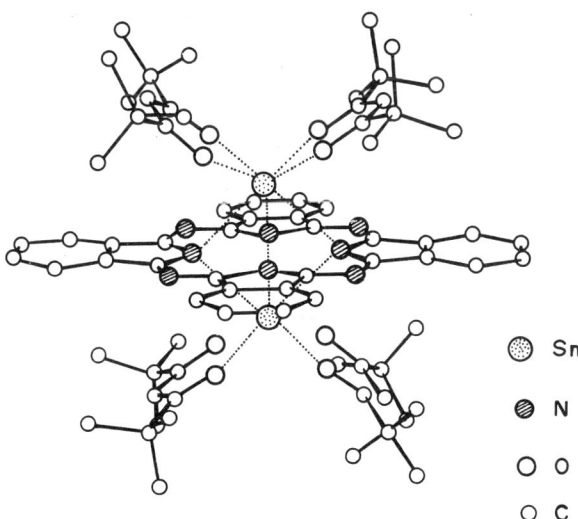

Figure 10 Molecular structure of Sm_2(tetramethylheptanedionate)$_4$(μ-phthalocyanine)$_2$ (reproduced with permission from *J. Chem. Soc., Chem. Commun.*, 1983, 1234)

Sm

N

O

C

metals, metal salts, oxides, *etc.*, as discussed below. Exemplary syntheses of $Fe(C_3F_7COCHCOCMe_3)_3$, $Ce(MeCOCHCOMe)_4$, $Ce(CF_3COCHCOMe)_4$, $Tl(CF_3COCHCOCF_3)$, $Mn(CF_3COCHCOCF_3)(CO)_4$, $Cr[MeCOCH(CN)COMe]_3$, $ZrX_{4-n}(MeCOCHCOMe)_n$ ($n = 2, 3$; $X = Cl, Br$) and $ZrI(MeCOCHCOMe)_3$ have been published.[139]

15.4.3.1 From Low- and Zero-valent Metal Species

Electropositive elements, the alkali and alkaline earth metals, react directly with hydrocarbyl substituted β-diketones to provide the corresponding metal chelates. The methine protons in perfluoroalkyl-substituted diketones are more acidic so that, in these cases, the method is applicable to other metals such as Fe, In, Ga, Mn, Pb, V and Sc.[140,141] Electrochemical syntheses of $M(MeCOCHCOMe)_2$ (M = Fe, Co, Ni) using the appropriate metal anode have been described[142] and the method should be widely applicable.

The carbonyls of Cr, Mo[143] and Co[144] react with acetylacetone and hexafluoroacetylacetone to form the metal chelates. Reactions of β-diketones with $Re(CO)_5Cl$ produce, *inter alia*, $Re_2(CO)_6$(diketonate)$_2$ and $Re_2Cl_2(CO)_6$(diketone)$_2$. The structure of $Re_2(CO)_6(PhCOCHCOPh)_2$ reveals two *fac*-$Re(CO)_3$ units joined by two bridging oxygen atoms, one from each dibenzoylmethanate ligand.[145] Thus, compounds of this type react with donor ligands to produce $Re(CO)_3L$(diketonate) (L = CO, C_5H_5N) or $Re(CO)_2(PPh_3)_2$(diketonate).[146]

β-Diketones react with platinum group metal hydrides and with $Ru(CO)_3(PPh_3)_2$ to give, for example, $MH(CO)(PPh_3)_2$(diketonate), $M(PPh_3)_2$(diketonate)$_2$ (M = Ru, Os), $RuCl(CO)(PPh_3)_2$(diketonate), $IrH_2(PPh_3)_2(MeCOCHCOMe)$ and $IrHCl(PPh_3)_2$(diketonate).[147] Replacement of hydride by hexafluoroacetylacetonate, with concomitant loss of Ph_3P, occurs when complexes of ruthenium or rhodium hydrides are treated with $Pd(CF_3COCHCOCF_3)_2$. Illustrative is $RuH_2(CO)(PPh_3)_3$, which forms successively $RuH(CO)(PPh_3)_2(CF_3COCHCOCF_3)$ and $Ru(CO)(PPh_3)(CF_3COCHCOCF_3)_2$.[148] Similarly, $Ru(CO)(NO_3)_2(PPh_3)_2$ reacts with acetylacetone and triethylamine to form $Ru(CO)(NO_3)(PPh_3)_2(MeCOCHCOMe)$, which is converted by methanolysis and β-elimination to *trans*-$RuH(CO)(PPh_3)_2(MeCOCHCOMe)$.[149]

Oxidative addition of acetylacetone to $Mo(C_2H_4)(Ph_2PC_2H_4PPh_2)_2$ produces $MoH(Ph_2PC_2H_4PPh_2)_2(MeCOCHCOMe)$; this material is also formed when $Mo(MeCOCHCOMe)_3$ is reduced with Et_3Al in the presence of $Ph_2PC_2H_4PPh_2$.[150] Reduction of $Fe(MeCOCHCOMe)_3$ in the presence of $Ph_2PC_2H_4PPh_2$ or addition of acetylacetone to $Fe(C_2H_4)(Ph_2PC_2H_4PPh_2)_2$ yields $Fe(Ph_2PC_2H_4PPh_2)(MeCOCHCOMe)_2$. In contrast, $Fe[(Me_3O)_3P]_5$ is protonated by hexafluoroacetylacetone to give $HFe[P(OMe)_5]^+CF_3COCHCOCF_3^-$.[151] Some mechanistic information about the behavior of metal acetylacetonates under highly reducing conditions derives from studies of their reactions with Grignard reagents. These produce, independent of the metal, the radical dianion of acetylacetone which is presumably complexed in some way to a magnesium-containing fragment.[152,153]

Formal reduction of rhenium occurs in the reaction of $ReH_4(PPh_3)_3$ with sodium acetylacetonate in ethanol, which provides $ReH_2(PPh_3)_3(MeCOCHCOMe)$. This can be used as the starting material for synthesis of a wide variety of rhenium acetylacetonate derivatives such as $Re(CO)_2(PPh_3)_2(MeCOCHCOMe)$ and $Re(CS_2)(PPh_3)_2(MeCOCHCOMe)$, formed with carbon monoxide and carbon disulfide respectively. Sequential reaction with hydrogen chloride yields $ReH(PPh_3)_3Cl(MeCOCHCOMe)$ and $Re(PPh_3)_2Cl_4$. Treatment with $CHCl_3$ or iodine leads to $ReH-(PPh_3)_2X_2(MeCOCHCOMe)$ (X = Cl, I), which are members of the comparatively rare class of paramagnetic transition metal hydrides.[154]

Treatment of $[Mo(\eta_7\text{-}C_7H_7)(\eta_6\text{-}toluene)]BF_4$ with acetylacetone leads to paramagnetic $Mo(\eta_7\text{-}C_7H_7)(MeCOCHCOMe)^+$, which in turn may be converted to $Mo(\eta_7\text{-}C_7H_7)(MeCOCHCOMe)X$ (X = Cl, Br, I, SCN, $P(OMe)_3$). The thiocyanate derivative has been structurally characterized.[155,156] The chlorine-bridged dimer $[Rh(C_2H_4)Cl]_2$ reacts with potassium acetylacetonate to provide monomeric $Rh(C_2H_4)_2(MeCOCHCOMe)$.[157] Dichroic $Ir(CO)_2(MeCOCHCOMe)$ has been prepared from acetylacetone and $cis\text{-}Ir(CO)_2Cl_2(Pr^iNH_2)$. Phenyl- and trifluoromethyl-substituted diketonate analogues are better synthesized from $Na_2Ir(CO)_4Cl_{4.8}$, a substance derived from high-pressure carbonylation of Na_3IrCl_6 in the presence of copper metal.[158]

15.4.3.2 Synthesis from Metal Salts

The simplest and most generally useful synthetic method for metal diketonates is from the diketone and a metal such as a halide, hydroxide, oxide, sulfate, carbonate, carboxylate, *etc.* in a variety of solvents such as water, alcohol, carbon tetrachloride or neat diketone. Since many β-diketones are poorly soluble in water, use of an organic solvent or cosolvent may be helpful. Optionally, a base such as sodium carbonate, triethylamine or urea may be added. Addition of a base early in the reaction converts the diketone to its conjugate base, which usually has greater solubility in aqueous media.[159] In some cases, metal halide complexes of the diketone form as intermediates, *e.g.* $SnCl_4(MeCOCH_2COMe)$, which has been formulated as $[SnCl_2(MeCOCH_2COMe)_2]SnCl_6$.[160] 3,3-Dimethylacetylacetone lacks acidic methine hydrogens and its complexes with such metal halides as TiX_4 (X = Cl, Br, I), $ZrCl_4$ and $SnCl_4$ serve as models for intermediates in preparative reactions.[161]

Titanium(IV) chelates may be obtained from the metal halide and a diketone. As a class, these materials have been carefully studied and exhibit interesting solution phase properties such as *cis–trans* isomerization, stereochemical nonrigidity and diketonate exchange. The tris(chelate) derivatives, *e.g.* $TiCl(MeCOCHCOMe)_3$, react with Lewis acids such as $FeCl_3$ or $SbCl_5$ to form $Ti(MeCOCHCOMe)_3^+$ salts.[162,163] Materials containing $Ti(MeCOCHCOMe)_3^+$ are alternatively available by extracting an acidified solution of titanium dioxide with acetylacetone in chloroform.[164] Similarly, neutral cyclopentadienyl chelates of the type $cis\text{-}M(C_5H_5)Cl(diketonate)$ are obtained from β-diketones and $M(C_5H_5)Cl_2$ (M = Ti, Zr) in the presence of triethylamine. Tropolone is distinctive and affords $Zr(C_5H_5)(tropolonate)_3$.[165] Ternary titanium complexes of the type $TiCl(OR)(MeCOCHCOMe)_2$ (R = lower alkyl) may be prepared from $TiCl_3(MeCOCHCOMe)$ and an alcohol in the presence of pyridine.[165a-167]

Rhenium tris(acetylacetonate) is prepared by heating solid sodium acetylacetonate with $ReCl_2(MeCOCHCOMe)_2$ or $ReCl_2(PPh_3)(MeCOCHCOMe)$; the α-form sublimes from the reaction mixture.[168] Monomeric *trans*-$ReCl_2(MeCOCHCOMe)_2$ in turn is derived from, *inter alia*, acetylacetone and $ReOCl_2PPh_3(MeCOCHCOMe)$, water rather than hydrogen chloride being eliminated.[169]

The tris(chelate) $Rh(CF_3COCHCOCF_3)_3$ is prepared from rhodium(III) nitrate, hexafluoroacetylacetone and aqueous base.[170] The course of the reaction of rhodium(III) chloride with ethanolic hexafluoroacetylacetone differs in that only two chloride ions are displaced. Fractional sublimation of the product affords two materials having empirical formulas corresponding to $Rh(CF_3COCHCOCF_3)_2Cl \cdot 3H_2O$. One of these compounds is dimeric; the other is poorly soluble and may be an isomer or higher oligomer.[171] Tetrachloroplatinate(2−) presents an extreme case of resistance to halide displacement by hexafluoroacetylacetonate and a successful synthesis of $Pt(CF_3COCHCOCF_3)_2$ requires use of $Pt(H_2O)_4^{2+}$.[172] The diamagnetic niobium(III) complexes $Nb_2(MeCOCHCOMe)_2Cl_4(PhPMe_2)_2 \cdot 2(MeCOCH_2COMe)$, $Nb_2(Me_3CCOCHCOCMe_3)_3Cl_5\text{-}(PhPMe_2)$ and $Nb_2(Me_3CCOCHCOCMe_3)_3Cl_3$ have been synthesized from the corresponding diketone and $Nb_2Cl_6(PhPMe_2)_4$ or $Nb_2Cl_6(SMe_2)_3$.[173] Eight-coordinate tetrakis(β-diketonate) and tropolonate complexes of niobium have been prepared by reacting the diketone with $NbCl_4$ in the presence of triethylamine or from $Tl(MeCOCHCOMe)$. The tantalum complex

Ta(PhCOCHCOPh)$_4$ was synthesized similarly; in the absence of base, Ta(PhCOCHCOPh)Cl$_4$ and Ta[(PhCOCHCOPh)Cl$_3$]$_2$O are formed.[174]

In aqueous systems or when water is a reaction byproduct, hydrated diketonate complexes may be obtained and the nature of the product obtained from water and organic solvents may be quite different. For example, Zr(MeCOCHCOMe)$_3$Cl is obtained from ZrCl$_4$ and acetylacetone in chloroform but in water the product is hydrated Zr(MeCOCHCOMe)$_4$, which may be freed of water by recrystallization from acetylacetone.[175]

The tetrahydrofuranates MCl$_3$(THF)$_3$ (M = Ti, V, Cr) react with one equivalent of diketones to form M(diketonate)Cl$_2$(THF)$_2$. The remaining chlorine atoms are susceptible to nucleophilic displacement by reagents such as LiC$_6$H$_4$-o-CH$_2$NMe$_2$, LiN(SiMe$_3$)$_2$, TlC$_5$H$_5$ and potassium hydrotris(pyrazolyl)borate, thus providing routes to a very extensive series of mixed diketonate complexes. The tetrahydrofuran ligands are also labile and may be displaced by, *inter alia*, Me$_3$P. The versatility of this chemistry contrasts with that of the less reactive Cr(MeCOCHCOMe)-Cl$_2$(MeCOCH$_2$COMe), obtained from CrCl$_3$(THF)$_3$ and acetylacetone.[176]

Ostensibly simple synthetic approaches occasionally yield unexpected products. For example, the reaction of acetylacetone with WOCl$_4$ yields WO$_2$(MeCOCHCOMe)Cl$_2^-$ as the 2,4,6-trimethyl-3-acetylpyrilium salt and the analogous reaction with MoOCl$_4$ produces 1,3,5,7-tetramethyl-2,4,6,8-tetraoxoadamantane.[177] During an attempt to synthesize copper-doped Zn(CF$_3$COCHCOCF$_3$)$_2$(C$_5$H$_5$N)$_2$, crystalline Cu(CF$_3$CO$_2$)$_2$(C$_5$H$_5$N)$_4$, probably a product of air oxidation, was isolated.[178] However, reliable syntheses of the bis(hexafluoroacetylacetonates) of Co^{II} and Ni^{II}, as well as of their bis(dimethylformamide) complexes, have been published.[179]

Rare earth acetylacetonates can be prepared from the metal oxide and acetylacetone; however, the degree of hydration of the products depends on the experimental conditions. Trihydrates precipitate from 60% aqueous ethanol, dihydrates from cold 95% ethanol and monohydrates from hot 95% ethanol.[180] An extensive series of hydrated lanthanide hexafluoroacetylacetonates has been prepared from the appropriate metal oxide and NH$_4^+$ CF$_3$COCHCOCF$_3^-$.[181–183] Trimeric [UO$_2$(CF$_3$-COCHCOCF$_3$)$_2$]$_3$ may be prepared from UO$_3$·2H$_2$O and hexafluoroacetylacetone in diethyl ether. An etherate is initially isolated which protects the product from hydrolysis but which can be readily desolvated by heating in vacuum.[184] Tested syntheses of the tris(tetramethylheptanedionates) of Sc, Y, La, Pr, Nd, Sm, Eu and Gd have been published.[185]

Reduction of MoO$_2$(MeCOCHCOMe)$_2$ is effected by tertiary phosphines which yield a coupled product, Mo$_2$O$_3$(MeCOCHCOMe)$_4$.[186] Mixed complexes, exemplified by Cu(MeCOCHCOMe)-(CF$_3$COCHCOMe), have been prepared by reaction in chloroform of CuCO$_3$·Cu(OH)$_2$ with equimolar amounts of acetylacetone and trifluoroacetylacetone.[187] An extensive series of rhenium diketonates has been prepared starting with ReOCl$_2$(OEt)(PPh$_3$)$_2$. The initial product of the reaction of this compound with acetylacetone is ReOCl$_2$(MeCOCHCOMe)(PPh$_3$)$_2$, which is reduced on heating with additional acetylacetone in benzene to yield ReCl$_2$(MeCOCHCOMe)-(PPh$_3$)$_2$. In neat acetylacetone, ReCl(MeCOCHCOMe)(PPh$_3$) and, subsequently, [ReCl$_2$(Me-COCHCOMe)$_2$]$_2$ are produced.[188]

Triarylantimony(V) diketonates have recently been described. Compounds of the types SbPh$_3$(diketonate)$_2$ and Sb(OMe)Ph$_3$(diketonate) were prepared from SbPh$_3$Br$_2$ and SbPh$_3$(OMe)$_2$ respectively.[189]

In some cases, alkoxide displacements at a metal center proceed sequentially and at different rates so that, if redistribution processes do not intervene, mixed diketonates can be obtained by adjusting the ratios of reactants.[190,191] Recently, bimetallic β-diketonates of the type M$_2$(diketonate)$_2$(OR)$_4$ (M = Mo, W; R = alkyl) have been prepared from diketones and M$_2$(OR)$_6$. X-Ray studies show that one diketonate is attached to each metal atom so that unbridged M≡M units are retained although rapid rotation about this multiple bond can occur.[192] Diketonate complexes with metal–metal multiple bonds are rather uncommon, one other example being C_{2v} *cis*-Mo$_2$(Me-COCHCOMe)$_2$(O$_2$CMe)$_2$.[193]

Metal alkoxides constitute a useful class of starting materials for the synthesis of the metal β-diketonates. The ethoxides of Nb^V, Ta^V and U^V react with diketones. Here, only partial substitution of the ethoxy groups occurs and materials of the type M(diketonate)$_3$(OEt)$_2$ are formed.[194,195] Similar reactions with lanthanide alkoxides, however, provide pure, unsolvated lanthanide tris(diketonates). The virtue of such syntheses lies in their ability to yield anhydrous diketonate complexes. Removal of water from the hydrates without decomposition is sometimes difficult.[196,197]

Cleavage by a β-diketone of hydrocarbyl groups from metal alkyls is a useful and potentially general synthetic route that has, thus far, been applied to prepare diketonates of boron[198,199] and thallium(III).[200] In both cases, only one alkyl group is cleaved and chelates of the type (diketonate) MR$_2$ (M = B, Tl) result. Cleavage of both methylcyclopentadienyl groups from Sn(MeC$_5$H$_4$)$_2$ by acetylacetone provides the interesting carbenoid $Sn(MeCOCHCOMe)_2$.[201] Alternatively, redistribution reactions, such as those between $Al(MeCOCHCOMe)$ and trialkylaluminum compounds may be used to produce mixed diketonate alkylaluminum complexes.[202]

triangulo-Trirhenium cluster alkyls react with β-diketones with elimination of alkane to provide novel rhenium diketonates. Thus, $Re_6Me_{12}(CF_3COCHCOMe)_6$ and $Re_3Me_6(MeCOCHCOMe)_3$ are obtained from Re_3Me_9 and trifluoroacetylacetone or acetylacetone respectively. Both classes of compounds, hexa- and tri-nuclear, retain the triangular Re_3 unit present in the starting material (Figure 11). Trinuclear $Re_3Cl_3(CH_2SiMe_3)(CF_3COCHCOMe)_3$ is similarly prepared from Re_3Cl_3-$(CH_2SiMe_3)_6$.[203]

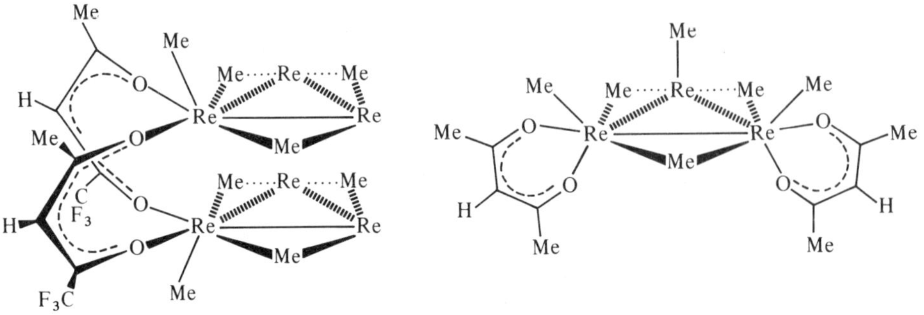

Figure 11 Proposed structures of $Re_6Me_{12}(CF_3COCHCOMe)_6$ and $Re_3Me_6(MeCOCHCOMe)_3$. The ligands around only two rhenium atoms are shown for clarity

15.4.4 REACTIONS OF METAL β-DIKETONATES

15.4.4.1 Substitution Reactions at Carbon

Electrophilic substitution at the methine carbon atom (C-3) of β-diketonates is, in many cases, a facile reaction. The process is of interest as a synthetic method for new diketonate complexes as well as from a mechanistic standpoint, for it is considered that such reactions imply, by their similarity to aromatic substitutions, significant bond delocalization in the C_3O_2M ring. Much exploratory work has been carried out with acetylacetonates; the PhCOCHCOMe and PhCOCHCOPh analogues, in general, react more sluggishly, an effect attributed to steric hindrance. Illustrative substitution reactions include halogenation with bromine,[204] N-halosuccinimides[205] and iodine monochloride.[206] Similarly, nitration has been effected with $Cu(NO_3)_2 \cdot 3H_2O$ and other metal nitrates[207] as well as with $N_2O_4 \cdot BF_3$.[208] Other substitution reactions which have been applied to metal diketonates are thiocyanation,[209,210] acylation,[211,212] formylation,[213] chlorosulfenylation[214] and the Mannich reaction.[215] Methine-substituted acetylacetonates may be further transformed by reactions commonly used in organic chemistry.[216]

Functionalization of $Ni(MeCOCHCOMe)_2$ occurs in reactions with isocyanates, diethyl azodicarboxylate and dimethyl acetylenedicarboxylate, which proceed by formal insertion of the methine C—H unit into the substrate multiple bonds to form respectively amides and ester-substituted hydrazines and alkenes. Similar additions of acetylacetone to these electrophiles is catalyzed by nickel acetylacetonate.[217,218]

15.4.4.2 Solvolysis and Displacement Reactions

Hydrolysis of metal β-diketone complexes is usually just the reverse of the preparative reaction but detailed study of such processes provides considerable insight into the mechanisms of inorganic substitution reactions.

Beryllium bis(acetylacetonate) undergoes rapid cleavage of both diketonate ligands under either acidic or basic conditions; the respective products are Be^{2+} and $MeCOCH_2COMe$ or $Be(OH)_4^{2-}$ and $MeCOCHCOMe^-$. Detailed kinetic studies show that the hydrolysis is first order in both substrate and hydrogen ion concentration, indicating that protonation of a 'dangling' or monodentate oxygen-bonded acetylacetonate group is involved. This appears to be a general effect of $[H^+]$ in chelate substitution reactions. Addition of nucleophilic ions which could participate in an S_N2 process, such as SCN^-, halide or $MeCO_2^-$, have little effect on the rate of the acid cleavage reaction, possibly because a five-coordinate intermediate would be too sterically congested. Vanadyl acetyl-acetonate, VO(MeCOCHCOMe)$_2$, hydrolyzes in aqueous acid in a stepwise fashion. Hydrolysis of the intermediate VO(MeCOCHCOMe)$^+$ is slower than that of the bis(chelate) by a factor of about 150 at 25 °C. Both steps are first order in hydrogen ion concentration in dilute solutions. As with the hydrolysis of Ni(MeCOCHCOMe)$_2$, the reaction rate is somewhat enhanced by addition of nucleophilic anions capable of coordinating to the metal.[219,220]

Hydrolysis of Pd(MeCOCHMe)$_2$ in aqueous methanol is considered to involve Pd(O,O-MeCOCHCOMe)(O-MeCOCHCOMe)(MeOH) as an intermediate from which the monodentate acetylacetonate ligand is then solvolyzed.[221] Subsequent studies on Lewis base complexes of palladium bis(diketonate) complexes provide ample support for the proposed intermediate. A pulse radiolysis study of the kinetics of aquation of M(MeCOCHCOMe)$_3^-$ (M = Cr, Co) indicates that an η_1–η_2 equilibrium involving one or more of the acetylacetonate ligands occurs, associated with an acid-catalyzed removal of the monodentate ligand.[222] Treatment of Cu(MeCOCHCOMe)$_2$ with picric acid in moist dichloromethane affords a partially hydrolyzed material, Cu(MeCOCHC-OMe)(H$_2$O)$_2$[C$_6$H$_2$(NO$_2$)$_3$O], proposed to contain square pyramidal five-coordinate copper with the oxygen atom from the picrate moiety at the apex.[223]

Hydrolysis of Si(MeCOCHCOMe)$_3^+$ salts in the presence of a variety of nucleophiles has been studied. The reaction is subject to general acid catalysis.[224] Isotopic labeling has been used to differentiate between attack at a ligand site and at the silicon center. When Si(PhCOCHCOPh)$_3^+$ was hydrolyzed with H$_2^{18}$O, one sixth of the oxygen atoms in the resulting dibenzoylmethane was enriched with ^{18}O. This implies that initial attack by OH$^-$ is at the diketonate carbon and not at silicon and, further, that subsequent hydrolysis of the remaining two dibenzoylmethanate ligands may proceed by a different mechanism. Similar results were obtained with B(tropolonate)$_2^+$, Si(tropolonate)$_3^+$ and P(tropolonate)$_3^{2+}$ but little or no enrichment occurred in the hydrolysis of the neutral tropolone complexes of Al, Ga, In, Fe^{III}, Sn^{IV} or Ce^{IV}. Ligand attack was the dominant pathway in hydrolysis of the eight-coordinate cation Nb(tropolonate)$_4^+$.[225]

Early studies of alcoholysis reactions of first-row metal acetylacetonates showed that mixed alkoxy metal acetylacetonates, many of which were oligomeric in solution, could be obtained.[226] In this way, the alkoxide bridged dimers [Cu(MeCOCHCOMe)OR]$_2$ were prepared from copper(II) acetylacetonate and ROK (R = PhCH$_2$, Me).[227,228] The chromium(III) analogues [Cr(MeCOCX-COMe)$_2$OMe]$_2$ (X = H, Cl, Br) and [Cr(MeCOCBrCOMe)$_2$OEt] have been crystallographically characterized and shown to contain bridging alkoxy ligands also. The magnetic properties of these materials have been studied in detail. Empirically, the energy of the triplet state is inversely related to the dihedral angle between the O—C vector in the μ-OMe groups and CrO$_2$ plane.[229–231]

Tetrameric clusters of the type [M(MeCOCHCOMe)OMe]$_4 \cdot$ 4MeOH (M = Co, Ni) are formed from M(MeCOCHCOMe)$_2$ and potassium hydroxide. An X-ray study showed that the cobalt complex adopts a cubane type structure with the metal and oxygen atoms of the methoxy ligands occupying alternate vertices of a cube. The magnetic properties of the isostructural nickel(II) analogue are of particular interest since it exhibits both inter- and intra-molecular ferromagnetic coupling.[232] Methanolysis of Pd(CF$_3$COCHCOCF$_3$)$_2$ in the absence of base yields [Pd(CF$_3$COCHC-OCF$_3$)OMe]$_2$, which has been structurally characterized. Elimination reactions involving the μ_2-methoxide ligands are much more facile than with the nickel acetylacetonate analogue and, on heating, this compound forms methanol, methyl formate, dimethoxyethane and palladium metal. The complex has been proposed as a model for surface methoxy groups on palladium. Both hexafluoroacetylacetonate groups are displaced by croconate(2−) to yield [Pd(C$_5$O$_5$)$_2$]$^{2-}$.[233]

In mixed complexes containing both β-diketonate and alkoxide ligands, the latter are more labile and undergo alkoxide interchange with alcohols rather than diketonate cleavage.[234–236] Similarly, reactions of M(MeCOCHCOMe)(o-Pr^i)$_2$ (M = Al, Gd, Er) with 8-hydroxyquinoline are reported to yield new complexes in which one or both isopropoxy groups are displaced,[237] but total displacement of acetylacetonate from the V, Mn, Fe, Co, Ni and Cu complexes is achieved with pyridine-2,6-dicarboxylic acid.[238,239] Protonolysis of *cis*-Ti(β-diketonate)$_2$(OR)$_2$ with phenol[240] or hydrogen chloride[241] also results on alkoxide cleavage. Retention of stereochemisty is indicated by an X-ray study of *cis*-Ti(MeCOCHCOMe)$_2$(2,6-Pr$^i_2$PhO)$_2$.[242]

High-spin manganese(III) complexes of the type Mn(MeCOCHCOMe)$_2$X (X = N$_3$, SCN, Br, Cl) were synthesized by displacement of one acetylacetonate ligand from Mn(MeCOCHCOMe)$_3$. A tetragonally elongated octahedral coordination geometry was demonstrated in the azide-bridged complex.[243] The manganese centers in these materials display antiferromagnetic intradimer coupling. The magnetic susceptibilities of the azide and thiocyanate derivatives undergo pronounced increases prior to phase transitions at 11.3 and 6.9 K respectively, possibly as a result of a small canting of the spins arising from zero-field splitting.[244]

Reactions of metal acetylacetonates with strong protic acids appear, compared with alcoholysis, to yield more complex products. Using trichloroacetic acid, $Cu(MeCOCHCOMe)(O_2CCCl_3)$, $Zn_2(MeCOCHCOMe)(O_2CCCl_3)_3$ and $Fe(MeCOCHCOMe)(O_2CCCl_3) \cdot H_2O$ were prepared.[245] Trinuclear materials of the type $Ni_3Br_{6-x}(MeCOCHCOMe)_x$ ($x = 2, 4, 5$) were isolated from the reaction of anhydrous nickel(II) acetylacetonate with controlled amounts of hydrogen bromide[246] and similar techniques have been used to prepare $M(MeCOCHCOMe)_2X$ (M = Mn, Fe; X = Cl, Br).[247]

Treatment of $K[Pt(O,O\text{-}MeCOCHCOMe)(C\text{-}MeCOCHCOMe)X]$ with mineral acids affords a material, formulated as $Pt(O,O\text{-}MeCOCHCOMe)Cl(MeCOCH_2COMe)$ in which neutral, enolic acetylacetone is π-bonded to platinum *via* the C=C bond. On standing, solutions of this material in chloroform yield red $[Pt(\mu\text{-}Cl)(MeCOCHCOMe)]_n$. Similar acidification of $K[Pt(MeCOCHCOMe)_3]$ with HX (X = Cl, Br) yields $Pt(1,3,5,7\text{-tetramethyl-}2,6,9\text{-oxabicyclo}[3.3.1]nona\text{-}3,7\text{-diene})_2X_2$,[248] whereas sulfuric, phosphoric or nitric acid leads to polymeric $[Pt(MeCOCHCOMe)_2]$. This compound, which is isomeric with the familiar $Pt(O,O\text{-}MeCOCHCOMe)_2$, may aggregate by means of tridentate $C,O,O\text{-}MeCOCHCOMe$ bridging ligands.[249]

Sequential displacement reactions of $Pd(CF_3COCHCOCF_3)_2$ with triarylformazans provides the highly colored chelates $Pd(CF_3COCHCOCF_3)[(aryl)_3CN_4]$ and $Pd[(aryl)_3CN_4]_2$.[250]

Exchange reactions of the ligands in metal diketonate complexes with additional β-diketone have been extensively studied by radiotracer and DNMR techniques, depending on the time scale of the process.[251] Analysis of the kinetics of exchange of $[2,4\text{-}^{14}C]$-labeled acetylacetone with $Cr(MeCOCHCOMe)_3$ in acetonitrile reveals the operation of two processes. When the acetylacetone concentration is > 0.1 M, attack by the diketone on the six-coordinate metal is proposed to generate an intermediate containing two monodentate acetylacetonate ligands, one of which is protonated. Proton transfer between these two groups followed by ejection of acetylacetone results in exchange. At acetylacetone concentrations $\leqslant 0.03$ M, attack at five-coordinate chromium in $Cr(O,O\text{-}MeCOCHCOMe)_2(O\text{-}MeCOCHCOMe)$ was proposed. Water seems to catalyze the intramolecular proton transfer step but similar exchange rate constants were observed in gas phase reactions.[252] Interestingly, $\Delta H^{\ddagger}$ for this second step is similar to that for racemization and isomerization of $Cr(MeCOCHCOMe)_3$ and $Co(PhCOCHCOMe)_3$, which reactions are also considered to proceed *via* bond rupture mechanisms.[253,254] The kinetics of exchange between $V(MeCOCHCOMe)_3$ and ^{14}C-labeled acetylacetone were interpreted as indicative of an associative process involving the enol tautomer. Reaction rates were solvent dependent, thus suggesting that an additional pathway, displacement of solvent from $V(O,O\text{-}MeCOCHCOMe)(O\text{-}MeCOCHCOMe)(solvent)$ by acetylacetone, may be significant. This system also exhibits acid catalysis.[255] Exchange in the complexes $M(MeCOCHCOMe)_3$ increases in the order Al < Ga < In. Coordinatively unsaturated intermediates are also implicated by radiotracer studies of exchange of free diketone with $Co(MeCOCHCOMe)_2$,[256] $Fe(MeCOCHCOMe)_3$,[257,258] $Fe(CF_3COCHCOMe)_3$,[259] $M(MeCOCHCOMe)_3$ (M = Cr, Co, Ru),[260] $Al(MeCOCHCOMe)_3$[261,262] and $Pd(MeCOCHCOMe)_2$.[263,264]

Exchange between $M(CF_3COCHCOMe)$ (M = Zr, Hf, Th) and free diketone is rather complicated. In chlorobenzene the kinetics are first order in each reactant, but in benzene the order in diketone varies with its concentration.[265] Rates of exchange of a series of Sn^{IV} and trivalent Au, Ga and In acetylacetonates with free acetylacetone have been measured. The rate of intramolecular exchange appears to be much greater than that for intermolecular exhange.[266] In indium(III) chelates of the type $In(CF_3COCHCOR)_3$, the rate-controlling process in exchange with free ligand was identified by ^{19}F DNMR studies as rotation of one monodentate diketonate ligand about a partial double bond prior to intramolecular proton transfer to a second monodentate ligand in a species containing two η_1 and two η_2 diketonate groups.[267] Kinetics of ligand substitution in bis(*N*-alkyl-salicylaldiminato)nickel(II) chelates by β-diketones has been reported. Substitution of the first ligand is rate determining. The rate law contains two terms, one dependent and one independent of diketone concentration, but Ni—O bond rupture appears to be involved in both of the separate pathways.[268]

Both displacement and condensation reactions occur when $M(MeCOCHCOMe)_4$ (M = Zr, Hf) is treated with (3)-1,2- or (3)-1,7-$B_9C_2H_{11}^{2-}$. The cation in the product $[M_4(OH)_{11}(MeCOCHCOMe)_4]^+ B_9C_2H_{12}^-$ is proposed to contain terminal acetylacetonate and bridging hydroxyl groups.[269]

15.4.4.3 Ligand Exchange Reactions

Exchange processes involving a metal diketonate and a neutral diketone have been discussed above in the context of solvolysis reactions. Many of these proceed slowly and radiotracer techniques are needed to examine the kinetics. Much faster are exchange reactions between two distinct metal complexes. These essentially represent disproportionation steps and the unsymmetrical products may be conveniently studied by NMR, although it should be pointed out that unambiguous assignment of spectral features in an equilibrium mixture to a particular component is sometimes difficult.

Rapid ligand exchange is particularly notable with Group IVB metals. Combination of $Zr(MeCOCHCOMe)_4$ and $Zr(CF_3COCHCOMe)_4$ in benzene yields, by ^1H NMR analysis, all of the mixed chelates $Zr(CF_3COCHCOMe)_x(MeCOCHCOMe)_{4-x}$.[270] Similar phenomena occur with the anionic, eight-coordinate yttrium chelates $Y(CF_3COCHCOCF_3)_4^-$ and $Y(CF_3COCHCOMe)_4^-$. These reactions presumably all involve metal–oxygen bond cleavage to produce (as intermediates or transition states) species containing one or more monodentate diketonate ligands.[272] In the $SnMe_2(MeCOCHCOMe)_2$–$SnPh_2(MeCOCHCOMe)_2$ system, DNMR studies indicate that the rate controlling step in acetylacetonate exchange is Sn—O bond rupture in the phenyltin complex.[273]

A thorough analysis of equilibria in the $M(MeCOCHCOMe)_4$–$M(CF_3COCHCOCF_3)_4$ system (M = Zr, Hf, Ce, Th) has been carried out. ^{19}F NMR was employed as the analytical technique and it is especially useful because of the large chemical shift dispersion. Equilibrium constants for reactions leading to $(MeCOCHCOMe)_x(CF_3COCHCOMe)_{4-x}Zr$ (x = 1, 2, 3) were obtained.[274] Ligand exchange for chromium and cobalt diketonates are slower than for the Group IVB analogues. Complexes of the type $M(MeCOCHCOMe)_x(CF_3COCHCOCF_3)_{3-x}$ may be obtained by heating solutions containing the symmetrical chelates. Since chromatographic mobility depends on the number of CF_3 groups in the complex and on its dipole moment, the mixed species can be isolated by column chromatography.[275]

Equilibrium constants for ligand exchange in aluminum chelates derived from acetylacetone, hexafluoroacetylacetone and tetramethyl-3,5-heptanedione have been evaluated by NMR methods. Mixed ligand species are particularly favored when one of the ligands is hexafluoroacetylacetonate. In the unsymmetrical chelates, *e.g.* $Al(CF_3COCHCOCF_3)(MeCOCHCOMe)_2$, the Me (and CF_3) groups are chemically nonequivalent and readily resolved in the 60 MHz ^1H NMR spectra. However, on heating the separate peaks coalesce due to optical inversion of the complex and thermodynamic parameters for these intramolecular averaging processes were obtained. Indeed, rates of racemization of unsymmetrical complexes can be estimated without prior resolution of the optical isomers.[276] Compounds of the type $Ti(diketonate)_2X_2$ exist as stereochemically nonrigid *cis* isomers, which rearrange by an intramolecular mechanism. In addition, they undergo rapid redistribution reactions, which exchange both monodentate ligands, X, such as halide or alkoxide, and the bidentate diketonate ligands.[277-279]

15.4.4.4 Rearrangement Reactions

Intramolecular rearrangements of β-diketone complexes, which are degenerate in the sense that compositional integrity is maintained, have received considerable attention in the literature. Rearrangements of materials containing an MO_6 coordination core are particularly important as they involve fundamental questions about reaction mechanisms of inorganic compounds.

Rearrangements in four-coordinate complexes are of interest because plausible pathways need not depend on the presence of low energy *d*-orbitals. Trialkylsilicon acetylacetonates possess an enol ether structure, $MeCOC(H)=C(Me)OSiR_3$ in which the acetyl and $OSiR_3$ groups may have either a *cis* or *trans* relationship with respect to the C=C double bond. Thermodynamic parameters for the *cis–trans* isomerization, which is thought to involve a five-coordinate silicon intermediate having an *O,O*-MeCOCHCOMe ligand, have been derived from DNMR studies. Interestingly, the *cis* isomer undergoes fast intramolecular rearrangement which interconverts chemically nonequivalent methyl groups while the *trans* isomer, which is not adapted to formation of a five-coordinate transition state, is static.[280-282]

Rearrangements of six-coordinate tris(diketonate) metal complexes have been the subject of numerous intense and careful investigations and the great body of literature has been incisively reviewed.[77,283] A principal conclusion is that in no case has a unique rearrangement been unambiguously established and indeed, one may not exist. The difficulty arises from the consideration that the various five-coordinate transition states (or intermediates) which result from bond rupture processes can have quite similar energies. Thus, combinations of mechanisms can obtain which lead to extremely complicated DNMR spectra or isomerization kinetics.

Rearrangement of diketonate derivatives of trivalent Co, Rh, Cr and Ru are sufficiently slow that kinetic parameters for isomerization or racemization may be obtained after chromatographic separation of geometric or optical isomers.[284,285] For the acetylacetonates of these four metals, $\Delta H^{\ddagger}$ and $\Delta S^{\ddagger}$ for rearrangement are quite similar, implying that a similar inversion mechanism operates in each case. This is considered to be a bond rupture process with a high percentage of trigonal bipyramidal character in the transition states. A similar analysis was applied to the four separated diastereoisomers of the $(+)$-3-acetylcamphor ruthenium chelate (9). For the $c\Delta$ and $c\Lambda$ isomers the data suggest a bond rupture mechanism with either square pyramidal–axial or 50% each trigonal bipyramidal–axial and –equatorial transition states.[286] An extremely detailed mechanistic analysis of rearrangements in $M(Pr^iCOCHCOCH_2Ph)_3$ (M = Al, Ga) has been carried out. The isopropyl group serves as a probe for both isomerization and inversion.[287]

(9)

Correlations of the reaction rate of chelates of trivalent d^0 or d^{10} metals with metal ionic radii (in parentheses) are available:[77]

$$In(0.79),\ Sc(0.73) \gg Ga(0.62) > Al(0.53)$$

A similar trend does not obtain for d^2–d^6 metals but rates decrease on descending a column in the Periodic Table; thus, $Fe^{III} > Ru^{III}$ and $Co^{III} > Rh^{III}$. Vanadium tris(trifluoroacetylacetonate) is stereochemically rigid up to $100\,°C$.[288]

Gas phase isomerization of $Cr(CF_3COCHCOMe)_3$ has been studied in an effort to eliminate solvent effects. For $cis \rightarrow trans$, $\Delta H^{\ddagger} = 117 \pm 4\,kJ\,mol^{-1}$ and $\Delta S^{\ddagger} = -23.0 \pm 12\,JK^{-1}\,mol^{-1}$. Because the Cr—O bond energy is estimated to be $\sim 230\,kJ\,mol^{-1}$, a twist mechanism which does not involve complete bond rupture was proposed.[289] More recently, the kinetics, steric course and mechanism of intramolecular stereoisomerization of the aluminum complexes $Al(Pr^iCOCHC\text{-}OPr^i)_3$ and $Al(CF_3COCHCOCF_3)_2(Pr^iCOCHCOPr^i)$ have been evaluated in detail. Principal mechanistic conclusions are that compounds containing alkyl or aryl substituents on the chelate ring probably stereoisomerize by a rhombic twist mechanism, and that those with fluoroalkyl substituents rearrange by a bond rupture mechanism involving a square pyramidal–apical five-coordinate intermediate.[290]

Transition metal tropolonates also contain an MO_6 coordination core and their rearrangements are amenable to analysis by DNMR. Particularly useful are the α-isopropyl- and α-isopropenyl-tropolonates. On the basis of lineshape changes for the methyl proton resonances, the temperatures at which the fast exchange limits are reached fall into four categories: $<0\,°C$ (V^{III}, Mn^{III}, Ga^{III}); $<100\,°C$ (Al^{III}, Co^{III}); $>100\,°C$ (Ge^{IV}); stereochemically rigid (Rh^{III}, Ru^{III}).[79] While it is generally true that tris(tropolonates) rearrange faster than the β-diketonate analogues, the occurrence of a nonrigid Co^{III} tropolonate is noteworthy.

DNMR spectra of isopropyl- and isopropenyl-tropolone complexes of Ga^{III} disclose two separate kinetic processes in partially overlapping temperature ranges. The pattern of averaging of the methyl proton resonances indicates that the low-temperature process, occurring between -83 and $-44\,°C$, is inversion of configuration by means of a trigonal twist. The high-temperature ($>-44\,°C$) process results in cis–$trans$ isomerization but the mechanism is not established. The rearrangement in $[Ge(\alpha\text{-isopropyltropolonate})_4]PF_6$ probably also involves a trigonal twist.[291,292]

Clues as to the influence of solid state geometry on rearrangement mechanisms derive from evaluation of molecular structures and shape parameters. Tris(β-diketonate) complexes have geometries near the octahedral limit while tropolonates exhibit a twist toward D_{3h} geometry as well as a slight flattening of the MO_6 coordination core. Thus the former tend to rearrange by bond rupture mechanism(s), at least in solution, and some tropolonates rearrange by a twist pathway. Further favoring trigonal twist processes are the rigid, planar nature of the tropolonate ligand, which inhibits M—O or M—O—C bond deformation, and its short ($ca.$ 2.5 Å) bite distance. The semi-chelating hexafluoroacetylacetonate ligand in $Pd\{Ph_2PC_2H_4P(Ph)C_2H_4PPh_2\}(CF_3COCHC\text{-}OCF_3)^+$ (Figure 12) provides a plausible and structurally characterized model for the 'dangling' or monodentate diketonate group thought to be involved in rearrangement and displacement processes which proceed via metal–oxygen bond rupture mechanisms.[50]

In $M(MeCOCHCOMe)_3^+ ClO_4^-$ (M = Si, Ge), racemization is intramolecular since it occurs in

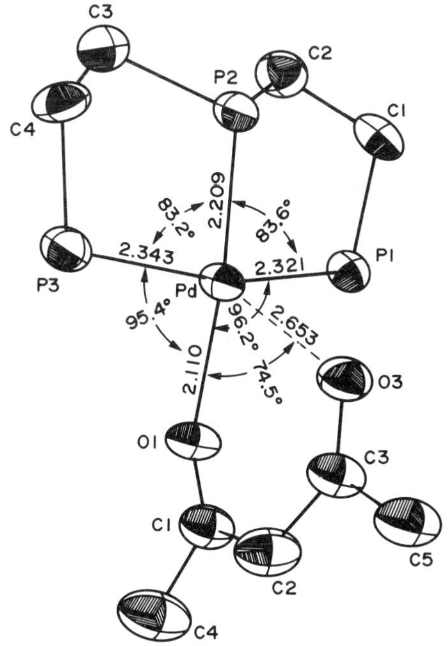

Figure 12 Coordination core of $Pd\{Ph_2PC_2H_4P(Ph)C_2H_4PPh_2\}(CF_3COCHCOCF_3)^+$

the presence of ^{14}C labeled acetylacetone without isotope incorporation. This reaction apparently proceeds by two mechanisms, one of which involves protonation or coordination of an intermediate as the rate law is of the form $k = k_1 + k_2$[acid or base].[251,293,294] Racemization of the germanium chelate, whose absolute configuration has been shown to be Δ in the $(-)_{589}$ isomer, has been studied in detail.[295,296] Investigation of the effect of pressure on reaction rates reveals a volume of activation larger than that expected for a twist mechanism. In donor solvents, a bond rupture process and participation of solvent in the transition are indicated.[297] Rearrangements in *cis*-Sn(MeCOCHC-OMe)$_2$X$_2$ have been studied by ^1H NMR spectroscopy. Activation energies for methyl group exchange are about the same for X = Cl, Br and I and only slightly smaller for X = F. These reactions are intramolecular because $^1H-^{117,119}Sn$ spin coupling is maintained and because halogen exchange with $^{36}Cl^-$ is much slower.[298,299] A permutational mechanistic analysis of rearrangements in SnRCl(MeCOCHCOMe)$_2$ indicates that, provided only one reaction pathway is operative, a trigonal prismatic transition state is involved.[300]

The titanium, zirconium and hafnium analogues M(MeCOCHCOMe)$_2$X$_2$ adopt a *cis* octahedral configuration in solution but are stereochemically nonrigid and, as in the tin system, rapid intra-molecular exchange of acetylacetonate methyl groups between the two nonequivalent sites occurs.[301,302] Barriers to rearrangement in *cis*-Ti(MeCOCHCOMe)$_2$(OR)$_2$ are *ca.* 42 kJ mol^{-1} when R = Me, Et and Bu but *ca.* 54 kJ mol^{-1} for R = Bu^t. Thus steric hindrance involving the alkoxy group tends to inhibit rearrangement. This argues in favor of a twist mechanism as in a bond rupture process; relief of steric strain should facilitate the reaction.[303] For GeCl$_2$(Me$_3$CCOCHCOCMe$_3$)$_2$, facile isomerization occurs in solution. The *cis*: *trans* ratio at 40 °C in benzene, 6.6, is greatly different from the statistical value. This may reflect stabilization of structures in which ligands of disparate π-bonding ability are *trans* to one another, a situation which also prevails for Sn^{IV} and Ti^{IV} chelates.[304] Dynamic behavior has also been observed in SbPh$_2$Cl$_2$(*O,O*-MeCOCHCOMe), which is considered to be in equilibrium with SbPh$_2$Cl$_2$(*C*-MeCOCHCOMe).[305]

Stereochemical nonrigidity is observed in higher coordinate, unsymmetrical metal diketonate complexes. Barriers to rearrangement in seven-coordinate ZrCl(MeCOCHCOMe)$_3$[306] and closely related M(MeCOCHCOMe)$_3$X compounds (M = Zr, Hf; X = Cl, Br, I) are quite low.[307,308] In Zr(C$_5$H$_5$)Cl(MeCOCHCOMe)$_2$, eight-coordinate zirconium has dodecahedral stereochemistry.[309] This compound rearranges in solution so that the nonequivalent acetylacetonate methyl groups interconvert in an intramolecular process. In the benzoylacetonate analogue, interconversion of all four geometric isomers present in solution accompanies interchange of the two nonequivalent diketonate ligands in each isomer.[310,311]

Nine-coordinate zirconium occurs in ZrC$_5$H$_5$(CF$_3$COCHCOCF$_3$)$_3$ using the formalism according to which an η^5-cyclopentadienyl ligand occupies three coordination sites but the stereochemistry is not readily described in terms of idealized nine-vertex polyhedra.[312,313] Compounds of the type

ZrC$_5$H$_5$(diketonate)$_3$ undergo two types of stereochemical rearrangements which have, notwithstanding the X-ray results, been analyzed in terms of a pentagonal bipyramidal geometry in which the C$_5$H$_5$ group is situated at an apex. The faster process interchanges nonequivalent terminal groups (Me or CF$_3$) on the equatorial diketonate ligands and the slower one exchanges these equatorial ligands with the unique diketonate group which spans an equatorial–axial edge. DNMR studies of ZrC$_5$H$_5$(CF$_3$COCHCOMe)$_3$ implicate a bond rupture or digonal twist mechanism.[314] Subsequently, further analysis of DNMR data for ZrCl$_2$(diketonate)$_2$, ZrCl(diketonate)$_3$, ZrC$_5$H$_5$(diketonate)$_3$, Zr(C$_5$H$_5$)Cl(diketonate)$_2$ and ZrC$_5$H$_5$(diketonate)$_2^+$ has been carried out. In this series it was considered that, in a structural sense, C$_5$H$_5$ is replaced by Cl. Such a substitution leads to a 10^{18} decrease in barrier(s) to rearrangement even though d(Zr—O)$_{av}$ is slightly longer in the cyclopentadienyl compounds [2.13 Å in ZrCl(MeCOCHCOMe)$_3$, 2.15 Å in Zr(C$_5$H$_5$)Cl(MeCOCHCOMe)$_2$]. The large differences in rates of ligand interchange thus appear to be inconsistent with a bond rupture mechanism. Similarly inconsistent is the greater lability of ZrC$_5$H$_5$(Me$_3$C-COCHCOCMe$_3$)$_2^+$ than of Zr(C$_5$H$_5$)Cl(diketonate)$_2$ and ZrC$_5$H$_5$(diketonate)$_3$ since Zr—O bond cleavage should be more facile in a neutral complex with a higher metal coordination number. It was proposed that the C$_5$H$_5$ group restricts the availability of empty metal d-orbitals for rehybridization and increases the energy needed to achieve ligand interchange by facial and/or digonal twists.[315]

The 1:1 adduct UO$_2$(CF$_3$COCHCOCF$_3$)$_2$·THF contains pentagonal bipyramidal, seven-coordinate uranium with the two uranyl oxygen atoms in apical positions.[316] Polymorphism, however, prevails for the 1:1 trimethyl phosphate complex and three phases have been detected by differential scanning calorimetry.[317] X-ray crystallographic studies of the α and β forms have characterized the grossly different packing patterns. These compounds, and other Lewis base complexes of UO$_2$(CF$_3$-COCHCOCF$_3$)$_2$, are fluxional and coalescence of the two ^{19}F NMR signals occurs above *ca.* $-70\,^\circ$C. The novel mechanism proposed for permutation of CF$_3$ groups involves migration of the Lewis base to an opposite equatorial site, possibly *via* a transition state having a capped octahedral geometry in which the base occupies the capping position.[320,321] Later work with UO(CF$_3$COCHC-OMe)$_2$·Me$_2$SO reveals that, in the ^{19}F DNMR spectra, peaks assigned to the *trans* isomer broaden and coalesce independently of those due to the *cis* isomer. This is consistent with base migration but not with diketonate rotation. Significantly, heats of reaction of UO$_2$(CF$_3$COCHCOCF$_3$)$_2$·THF with Lewis bases do not correlate with rates of rearrangement in UO$_2$(CF$_3$COCHCOCF$_3$)$_2$·base.[322]

15.4.4.5 Photochemical Reactions

Mechanistic photochemistry of metal β-diketonate complexes has received increasing attention. Near-ultraviolet irradiation of diketonate complexes of Mn^{III}, Fe^{III}, Co^{III}, No^{II} and Cu^{II} eventuates in one-electron reduction by the ligand of the metal center whereas Cr^{III} complexes photo-isomerize.[323] Irradiation of *trans*-Rh(CF$_3$COCHCOMe)$_3$ appears to lead to two photoactive excited states which may have significant radical character. Photoisomerization to the *cis* form occurs in inert solvents whereas decomposition takes place in the presence of potential hydrogen atom donors such as alcohols.[324-326]

Near-UV flash photolysis of *trans*-Co(CF$_3$COCHCOMe)$_3$ yields a long-lived transient which is proposed to be a Co^{II} complex containing a carbon-bonded diketonate ligand. This results from intramolecular rearrangement of the reduced metal–oxidized ligand radical formed in the primary photochemical process. Hydrogen atom donors are considered to intercept the radical pair, inhibiting transient formation.[327] Irradiation of *trans*-Cr(CF$_3$COCHCOMe)$_3$ at $\geqslant 366$ nm causes isomerization but at 254 nm decomposition occurs.[328] Irradiation of Cr(MeCOCHCOMe)$_3$ in aqueous ethanol leads to replacement of one acetylacetonate group by two molecules of water. The ligand-localized state was proposed to be more reactive than the lower-energy ligand field state.[329]

Metal acetylacetonates quench triplet species generated by flash photolysis of aromatic ketones and hydrocarbons.[330-333] More recently, these reactions have been studied from a synthetic standpoint. Triplet state benzophenone sensitizes photoreduction of Cu(MeCOCHCOMe)$_2$ by alcohols to give black, presumably polymeric, [Cu(MeCOCHCOMe)]$_n$. This reacts with Lewis bases to provide complexes of the type CuL$_2$(MeCOCHCOMe) (L = bipyridyl/2, ethylenediamine/2, carbon monoxide, Ph$_3$P). Disubstituted alkynes yield Cu(C$_2$R$_2$)(MeCOCHCOMe) but terminal alkynes form CuC$_2$R acetylides.[334] The bipyridyl complex of copper(I) acetylacetonate catalyzes the reduction of oxygen to water and the oxidation of primary and secondary alcohols to aldehydes and ketones.[335]

The tetrahydrofuranate $UO_2(CF_3COCHCOCF_3)_2 \cdot THF$ vaporizes without dissociation. Photochemistry of this compound has been extensively studied in part because of the prospect of using it in gas-phase uranium isotope separation schemes. Irradiation with the $10.6\,\mu m$ output from a CO_2 laser excites the asymmetric UO_2 stretching mode. Absorption of 5–10 photons, followed by energy migration on the picosecond or nanosecond timescale, leads to dissociation to THF and trimeric $UO_2(CF_3COCHCOCF_3)_2$. The $^{238,235}U$ and $^{16,18}O$ isotope effects in the excited mode are 0.7 and $17\,cm^{-1}$ respectively so that selective uranium or oxygen enrichment can be achieved.[336-340] Photolysis at 360–450 nm, where electronic transitions in the uranyl moiety occur, leads to a product formulated as $U(CF_3COCHCOCF_3)_4$, possibly by way of a U^{IV} peroxide species.[341]

15.4.4.6 Decomposition Reactions

In addition to the chemical reactions described above, which are respectable in that discrete, recognizable compounds result, metal β-diketone complexes may be decomposed to yield technologically useful deposits of elemental metals or of metal oxides. A key to the development of this technology has been the accessibility of metal chelates which are volatile and readily soluble in organic solvents.[342] In particular, complexes derived from hexafluoroacetylacetone and decafluoroheptane-4,6-dione[343] and dipivaloylmethane[344] exhibit both properties. Thermogravimetric studies show that replacement of hydrogen in metal acetylacetonates by fluorine significantly increases the vapor pressure of the compounds.[345] Also, complexes of the rare earths with smaller ionic radii are more volatile than those with larger ionic radii.[346] The bulky *t*-butyl groups inhibit coordination of water to the central metal atom as illustrated by trigonal prismatic $Er(Bu^tCOCHCOBu^t)_3$ **(10)**.[347] Thus, gas chromatographic separation of mixtures of these rare earth chelates is feasible, as is separation of mixtures of alkaline earth and transition metal hexafluoroacetylacetonates.[348]

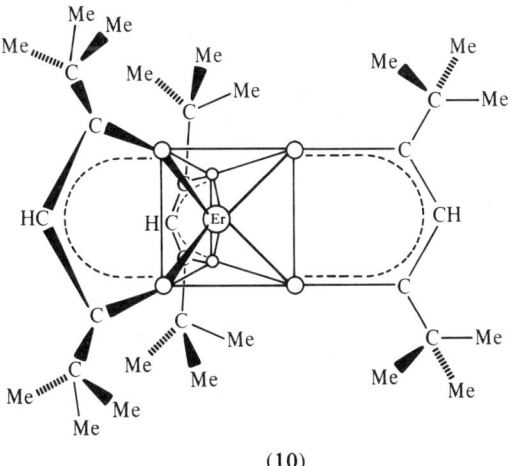

(10)

The $C_3F_7COCHCOBu^t$ ligand (often abbreviated as fod) has both a high fluorine content and bulky side chains, a combination of features which leads to quite high volatility in its complexes. Chelates of this ligand are often isolated as hydrates but these can be readily desolvated without significant decomposition.[349]

Gas phase reduction of the hydrated hexafluoroacetylacetonates of Cu^{II} and Ni^{II} and of the trifluoroacetylacetonates of Cu^{II} and Rh^{III} leads to deposition of thin films of the respective metals. Reduction can be carried out at as low as $250\,°C$ and the diketone byproduct can, in principle, be recycled.[350,351] The facile reduction by hydrogen or hydrocarbons of $Pd(CF_3COCHCOCF_3)_2$ and its Lewis base complexes provides thin palladium films useful as catalysts or primers for electroless plating.[352] In this case, reduction is facilitated by complex formation since, in the complexes, palladium has a formal positive charge and thus an increased electron affinity. Use instead of hydrogen sulfide allows chemical vapor deposition of metal sulfides such as CdS.[353]

Rare earth chelates derived from alkyl- and fluoroalkyl-substituted β-diketones are useful, even as mixtures, as anti-knock additives in gasoline.[354,355]

15.4.5 COMPLEX FORMATION AND β-DIKETONATE DISPLACEMENT

Many metal β-diketonates are coordinatively unsaturated and reactions with Lewis bases to form complexes are a pervasive feature of their chemistry.[14] In previous sections, base cleavage of M—O—M bridge bonds in oligomeric acetylacetonates and formation of hydrated lanthanide diketonates having high, odd coordination numbers have been noted. Mechanistically, acid–base complexes are quite likely to be involved in hydrolysis, displacement and *ortho*-metallation reactions, albeit that the interactions may be weak.

In the weak complex regime, proton NMR line broadening has been used to probe the dynamics of the interactions of Cr(MeCOCHCOMe)$_3$ with weak bases such as acetonitrile and methanol, which act as second sphere ligands.[356] Stoichiometric inclusion compounds of the types M(MeCOCHCOMe)$_3$·(urea)$_3$ and M(MeCOCHCOMe)$_3$·(thiourea)$_2$ (M = Al, Co, Cr, Fe, Rh) have been prepared,[357] as have clathrates containing halomethanes.[358] Hydrogen bonding between square planar and octahedral (but not tetrahedral) metal acetylacetonates and donors such as chloroform, methanol and water has been detected by IR spectroscopy.[359,360] Electronic spectra of solutions containing iodine and the acetylacetonates of Be, Al, Sc, Zr and Th show absorptions at *ca.* 360 nm which have been attributed to weak charge transfer complexes. These decompose or dissociate upon attempted isolation.[361] However, the crystalline material formed from platinum(II) acetylacetonate and iodine is the *trans* oxidative addition product PtI$_2$(MeCOCHCOMe)$_2$, as determined by X-ray crystallography. In carbon tetrachloride this compound partially dissociates into the starting materials, a process which may be photoassisted.[362]

Spectrophotometric studies of the reactions of [Co(MeCOCHCOMe)$_2$]$_4$ and [Ni(MeCOCHCOMe)$_2$]$_3$ with pyridine furnish equilibrium constants for the formation of intermediate complexes involved in oligomer cleavage. The 2:1 adducts [M(MeCOCHCOMe)$_2$]$_2$·pyridine (M = Co, Ni) have been isolated in pure form.[363,364] The crystal structures of these as well as of [Ni(MeCOCHCOMe)$_2$]·c-C$_5$H$_{10}$NH and [Ni(MeCOCHCOMe)$_2$]$_3$ (redetermination) have been published.[365] The [M(MeCOCHCOMe)$_2$]·pyridine adducts adopt structure (11a) with acetylacetonate groups bridging the metal centers in the two edge-shared LMO$_5$ octahedra. Bridging acetylacetonate ligands also occur in [Ni(MeCOCHCOMe)$_2$]$_2$·piperidine (11b) so that LMO$_5$ and LMO$_6$ octahedra share a common face. Both [Ni(MeCOCHCOMe)$_2$]$_3$ and [Ni(MeCOCHCOMe)$_2$]$_2$·pyridine exhibit ferromagnetic coupling while that in [Ni(MeCOCHCOMe)$_2$]$_2$·piperidine is antiferromagnetic. This alternation may be associated with the wider Ni—O—Ni angle in the latter compound, although the angular effect is difficult to quantify precisely.[366] The formally analogous [M(MeCOCHCOMe)$_2$]$_2$·Ph$_3$AsO (M = Co, Ni), which are obtained form M(MeCOCHCOMe)$_2$(H$_2$O)$_2$ and triphenylarsine oxide, adopt a similar structure with the important difference that one of the bridging oxygen atoms is furnished by the arsine oxide and one by a μ-acetylacetonate ligand.[367]

(11a)   (11b)

Critical to the utilization of lanthanide chelates, such as the heptafluorodimethyloctanedionate complexes of Eu and Pr, as NMR shift reagents is their participation in the formation of weak complexes in which chemical shifts are affected by dipolar or pseudocontact interactions between the lanthanide center and the nucleus of of interest.[368,369] In this context, numerous complexes of lanthanide β-diketonates with organic bases have been isolated.[370–372] In both Ho(Bu^{t}COCHCOBu^t)$_3$(4-MeC$_5$H$_4$N)$_2$[373] and Eu(Bu^{t}COCHCOBu^t)$_3$(pyridine)$_2$[374] the metals exhibit distorted square antiprismatic coordination geometries in which the pyridine ligands are *trans* to one another and on opposite square faces. Curiously, in the analogous diaqua lanthanum complex the water molecules are *cis* and on the same face.[375] Calculations of the magnitude of the dipolar shift utilize a relationship which assumes that the complex between the lanthanide diketonate and a substrate has real or effective axial symmetry.[376] Such high symmetry is not found in crystalline Eu(Bu^{t}COCHCOBu^t)$_3$(Me$_2$NCHO)$_2$ in which the two dimethylformamide ligands are *cis* to one another on the same square face of the typical distorted square antiprismatic O$_6$N$_2$ coordination core.[377] The actual geometries of these and other complexes of lanthanide diketonates in fluid solution phase may be very difficult to ascertain because of the close energetic proximity of other coordination poly- hedra, here dodecahedral. The situation is complicated by aggregation, ligand exchange and intramolecular rearrangement processes. Indeed, recent osmometric molecular weight measure- ments on lanthanide tris(acetylacetonate) dihydrates indicate dimer formation in benzene with the average number of monomers per solute molecule decreasing from Pr to Lu. The multiplicity of methyl resonances in the proton spectrum of $Lu(MeCOCHCOMe)_3 \cdot 2H_2O$ was interpreted in terms of a monomer–dimer equilibrium.[378] It appears that at least some of the molecular weight data remain controversial and that the precise role of solvent as a determinant of aggregation remains to be established.

An interesting, related 1:1 complex formed from N,N'-ethylenebis(salicylideniminato)cobalt(II) [Co(salen)] and $Cu(CF_3COCHCOCF_3)_2$ has been structurally characterized. The remarkable result is that, in this complex, copper and cobalt have effectively been interchanged so that the hexafluoroacetylacetonate ligands are coordinated to cobalt rather than to copper. The donor sites in the Cu(salen) portion of the molecule are the oxygen atoms.[379] The complex $Mg(CF_3COCHCOCF_3)_2 \cdot Cu(salen)$ has architecture similar to that in the cobalt diketonate analogue, no metal transposition having taken place. Because the Cu—Mg and Cu—Mn deriva- tives are isomorphous, it appears that Cu—Mn interchange has not occurred.[381] X-Ray crystal- lography also demonstrates, in the 1:1 complex between Cu(salen) and $Cu(CF_3COCHCOCF_3)_2$, coordination of the acceptor copper to the salen oxygen atoms. Detailed magnetic susceptibility and EPR studies indicate a weak $(-20.4\,cm^{-1})$ antiferromagnetic exchange interaction between Cu^{II} in four- and six-coordinate sites.[382]

An extensive series of unsymmetrical bimetallic complexes of the type M(2-hydroxypropioph- enoniminato) $\cdot M'(CF_3COCHCOCF_3)_2$ (M,M' = Cu,Cu, Ni,Cu, Cu,Co, Ni,Co, Cu,Mn and Ni,Mn) has been structurally characterized. Differences in J values can be accounted for by changes in geometric features, particularly the ligand bridging angles.[383–387]

Valuable thermodynamic data[388–390] and formation constants[391] have been obtained for the in- teraction of $Cu(CF_3COCHCOCF_3)_2$ and various Lewis bases. Calorimetric data indicate that $Cu(CF_3COCHCOCF_3)_2$ lies between iodine and trimethylborane in terms of covalent character of softness as an acceptor, but acceptor behavior is significantly modulated by solvent effects.[392] More recently, the thermodynamics of the reactions of (β-diketonate)$_2$Cu (diketone = hexafluoroacetyl- acetone, trifluoroacetylacetone, benzoyltrifluoroacetone, trifluorophenylbutanedione trifluoro-2- thenylbutanedione, acetylacetone or phenylbutanedione) with pyridine derivatives, bipyridyl and terpyridyl have been evaluated. As in previous studies, substitution of fluorine for hydrogen is seen to lead to a pronounced increase in acceptor strength.[393]

Octahedral coordination occurs in yellow-green $Cu(O,O\text{-}CF_3COCHCOCF_3)_2 \cdot 1,4$-diaza- bicyclo[2.2.2]nonane, each copper being coordinated to two *trans* nitrogen atoms from different donor molecules. The axial Cu—N bonds are rather long, 2.566 Å, possibly because of steric factors. Recrystallization of this compound from carbon tetrachloride affords the red 1:2 complex $Cu(CF_3\text{-}COCHCOCF_3)_2$(diazabicyclononane)$_2$.[394] Reaction of $Cu(CF_3COCHCOCF_3)_2 \cdot H_2O$ with biden- tate amines such as N,N'-dimethylethylenediamine affords a 2:1 complex whose crystal structure[395] discloses a square planar CuN_4 unit. The two hexafluoroacetylacetonate groups are each bonded to copper through one of the oxygen atoms. While these formally comprise $O\text{-}CF_3COCHCOCF_3$ ligands, the axial Cu—O bonds are so long, 2.793 Å, that they may not be chemically impor- tant.[396–399] Further indication of the propensity of copper(II) hexafluoroacetylacetonate derivatives to undergo Cu—O bond cleavage is found in the structures of $Cu(O,O\text{-}MeCOCHCOMe)(O\text{-}CF_3COCHCOCF_3)$(phenanthroline) and $[Cu(O,O\text{-}MeCOCHCOMe)(phenanthroline)(H_2O)]^+\text{-}CF_3COCHCOCF_3^- \cdot H_2O$. The former contains five-coordinate copper and an oxygen-bonded monodentate hexafluoroacetylacetonate ligand which, in the second compound, is replaced by water.[400] Other, related, ternary complexes of mixed copper(II) diketonates have also been repor- ted.[401–403]

In the bis(4-methylpyridine) complexes of (thienyltrifluorobutanedionate)$_2$M (M = Co, Cu, Zn), the central metal is six coordinate with the nitrogen donors *cis* to one another. In this series a linear relationship between force constants for stretching the M—O bond *cis* to nitrogen and $1/d$(M—O) was reported.[404] *Cis* complexes are also formed between pyridine and $M(CF_3COCHCOCF_3)_2$ (M = Cu, Zn). The zinc compound is interesting in that alternating short and long bond lengths occur in the ZnO_2C_3 ring but chemically significant alternations appear not to occur in the copper(II) analogue.[405] In these compounds and in $Cu(CF_3COCHCOCF_3)_2 \cdot$ bipyridyl, deviations from regular coordination geometries are attributable to Jahn–Teller effects.[406,407] The g values of $cis\text{-}Cu(CF_3COCHCOCF_3)_2(C_5H_5N)_2$ doped into the zinc analogue as host show an interesting temperature dependence. The CuO_4N_2 coordination core is tetragonally elongated while that of ZnO_4N_2 is compressed; thus ESR data may be interpreted on the basis of dynamic behavior associated with packing forces.[408] The complexes $Co(MeCOCHCOMe)_2(C_5H_5N)_2$ and $Ni(MeCOCHCOMe)_2(C_5H_5N)_2$, in contrast, crystallize as the *trans* isomers. In the Co^{II} derivative the pyridine rings are staggered with respect to one another but in the Ni^{II} material they are eclipsed, a dichotomy which has been explained on the basis of crystal packing forces.[410,411] A detailed analysis of the average and principle magnetic susceptibilities of $Ni(MeCOCHCOMe)_2(C_5H_5N)_2$ has been reported.[412]

The acceptor properties of the central metal atom may be important determinants of the stereochemistry of $M(diketonate)_2(ligand)_2$ compounds, as low-temperature NMR studies of the bis(4-methylpyridine) complexes of nickel(II) diketonates indicate formation of increasing amounts of the *cis* isomer as acetylacetonate is successively replaced by tri- and hexa-fluoroacetylacetonate.[413] Reaction of $Mg(MeCOCHCOMe)_2 \cdot 2H_2O$ and its tri- and hexa-fluoroacetylacetonate analogues with chelating bases such as ethylene glycol, phenanthroline and tetramethylethylenediamine yields 1:1 complexes.[414] As with complexes of Cu, Co and Ni, a six-coordinate metal center in these compounds is assumed and plausible, although displacement of one or more of the fluorinated diketonate ligands may have occurred.

Examination of a series of complexes of the type $(MeCOCHCOMe)_2VO \cdot RC_5H_4N$ indicates that *cis* or *trans* isomers may be obtained, depending on the nature of the pyridine substituent.[415-419]

High spin, monomeric 1:1 and 2:1 complexes of $Mn(MeCOCHCOMe)_2$ with pyridine[420] and 2:1 complexes with primary amines and ethylenediamine are obtained from either $Mn(MeCOCHCOMe)_3$ or $Mn(MeCOCHCOMe)_2(H_2O)_2$. In the synthesis employing the Mn^{III} diketonate, acetylacetone is liberated and oxidation of the amine occurs.[421] Free radical polymerization of vinyl monomers is effectively initiated by $Mn(MeCOCHCOMe)_3$, a process promoted by propylenediamine.[422] Tracer experiment show that each growing radical chain carries one acetylacetonate moiety, suggesting that, in the presence of amines, acetylacetonyl radicals are formed.[423] The allylamine complex $[Mn(MeCOCHCOMe)_2(CH_2{=}CHCH_2NH_2)]_2$ is dimeric and contains bridging acetylacetonate groups.[424]

Complexes formed with stable free radicals, exemplified by nitroxyls, comprise a significant aspect of metal β-diketonate coordination chemistry[425] and particular attention has focused on spin–spin interaction in systems in which the metal is also paramagnetic. Thus, electron–electron interaction in nitroxide complexes of $Cu(CF_3COCHCOCF_3)_2$ and $VO(CF_3COCHCOCF_3)_2$ can be probed by ESR spectroscopy and used to observe changes in spin delocalization associated with changes in metal ion,[426,427] in metal–nitroxide bonding and in nitroxide conformation.[428,429] The sign and magnitudes of exchange interactions in nitroxide–metal diketonate complexes can vary greatly. Strong antiferromagnetic coupling obtains in $Cu(CF_3COCHCOCF_3)_2$ adducts with 2,2,6,6-tetramethylpiperidinoxyl[430] and di-*t*-butylnitroxyl[431] and in $VO(CF_3COCHCOCF_3)_2 \cdot$ (tetramethylpiperidinoxyl), but in $Cu(CF_3COCHCOCF_3)_2 \cdot$ [4-hydroxy-2,2,6,6-tetramethylpiperidinoxyl], in which six-coordinate copper is bonded to both the nitroxyl and hydroxy oxygen atoms, coupling is quite small.[433] The crystal structure of an exemplary complex, $Cu(CF_3COCHCOCF_3)_2 \cdot$ (tetramethylpiperidinoxyl), demonstrates a distorted, square-pyramidal coordination geometry for copper.[434] Nitroxide complexes are presumably similar in a structural sense to closed-shell bis(heterocyclic amine *N*-oxide) complexes of nickel(II) and cobalt(II) acetylacetonates, which can participate in exchange reactions.[435,436] The pyridine *N*-oxide complex $Ni(MeCOCHCOMe)_2(C_5H_5NO)_2$ crystallizes as the *cis* isomer (*cf.* the pyridine and water analogues which are *trans* in the solid state).[437] Extensive studies have also been carried out on complexes containing nitroxyl-substituted pyridine ligands in which the donor atom is quite likely the pyridine nitrogen.[438]

A whole panoply of β-diketonate bonding modes is revealed by a systematic crystallographic and NMR investigation of the reactions of $Pd(CF_3COCHCOCF_3)_2$ with Lewis bases.[439] These are summarized in Figure 13, which shows relationships among the $Pd(CF_3COCHCOCF_3)_2(ligand)_n$ ($n = 1, 2, 3, 4$) complexes. Coordination of Lewis bases to palladium proceeds with sequential cleavage of the four Pd—O bonds leading, ultimately, to $[Pd(ligand)_4]^{2+}[CF_3COCHCOCF_3^-]_2$. The maximum value of n attainable for a particular base depends, *inter alia*, on its steric properties, being 4 for ammonia and methylamine, 1 or 2 for dimethylamine and 1 for trimethylamine; and, similarly, 4 for triphenylstibine, 2 for triphenylarsine and 1 for triphenylphosphine.

Two isomeric, but evidently not interconvertible, structures are observed in the $n = 1$ class of complexes, *i.e.* $Pd(O,O\text{-}CF_3COCHCOCF_3)(C\text{-}CF_3COCHCOCF_3)(ligand)$ and $Pd(O,O\text{-}CF_3COCHCOCF_3)(O\text{-}CF_3COCHCOCF_3)(ligand)$. Compounds containing the carbon bonded hexafluoroacetylacetonate ligand are formed by a rearrangement process associated with cleavage of the first Pd—O bond. This reflects the tendency of palladium, compared with elements further

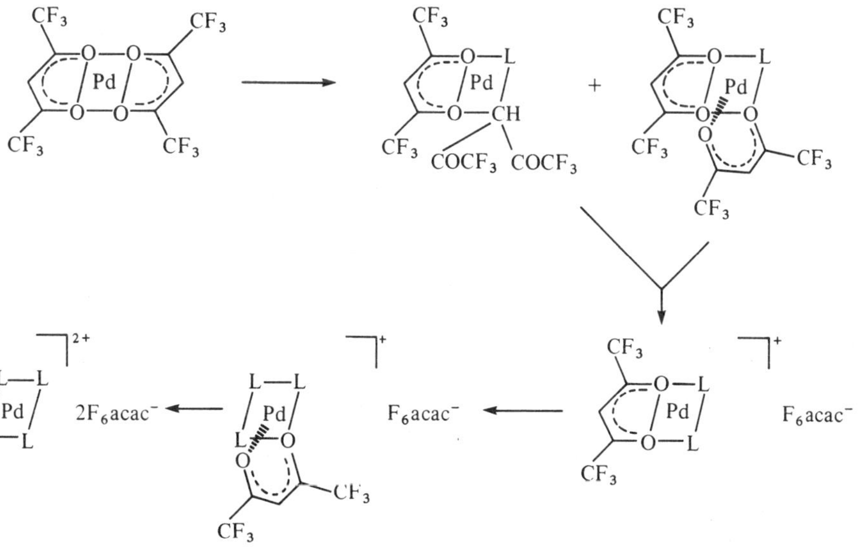

Figure 13 Sequential displacement reactions in Pd(CF$_3$COCHCOCF$_3$)$_2$

to the left in the Periodic Table, to retain a coordination number of four, which is not possible without rearrangement or ligand bond rupture.[440] Compounds of the second type, which contain a semichelating, oxygen bonded hexafluoroacetylacetonate ligand, are formed predominantly with bulky phosphine donors such as triphenyl- and tricyclohexyl-phosphine. Such complexes are fluxional in that, according to an analysis of ^{19}F DNMR spectra, concurrent permutation of all four nonequivalent CF$_3$ environments occurs.[441] The structure of the tri(o-tolyl)phosphine analogue is, after allowing for the greater bulk of the phosphine, somewhat similar. The barrier to rearrangement in these compounds is in fact a sensitive function of the steric properties of the phosphine.[441,442]

Complexes of the $n = 2$ class which have the general formula [Pd(O,O-CF$_3$COCHCOCF$_3$)-(ligand)$_2$]$^+$ CF$_3$COCHCOCF$_3^-$, result when two adjacent Pd—O bonds are cleaved and one hexa-fluoroacetylacetonate group is ejected as the anion. The bipyridyl complex has been structurally characterized and, in this material, exchange between coordinated and ionic hexafluoroacetyl-acetonate groups is rapid on the NMR timescale.[443] Such complexes are of particular interest as there is strong evidence that surface species of this structural type are formed when Pd(CF$_3$COCHC-OCF$_3$)$_2$ reacts with condensed phase donors such as metal oxides. The Pd(CF$_3$COCHCOC-F$_3$)$_2$–alumina surface complex is reduced with exceptional ease, for example on heating with benzene or propylene. As no reduction occurs with C$_6$F$_6$, it is thought that cracking of the hydrocarbon is involved in the reduction of Pd^{II} to palladium metal. In this process the hexafluoro-acetylacetonate ligands presumably emerge as tightly bound hexafluoroacetylacetone but X-ray photoelectron spectroscopy indicates that, subsequently, fluorine is transferred from carbon to aluminum.[444] Surface complexes formed on colloidal silica dispersed in 2-ethoxyethanol have been studied by narrow line ^{19}F NMR spectroscopy. In this system the ^{19}F spin–lattice relaxation time is reduced by rotational processes involving the 350 Å mean diameter silica particles and the CF$_3$ groups and, quite possibly, by ionic–covalent hexafluoroacetylacetonate exchange.[445]

Cleavage of three Pd—O bonds generates the $n = 3$ class of compounds, [Pd(O-CF$_3$COCHCOC-F$_3$)(ligand)$_3$]$^+$ CF$_3$COCHCOCF$_3^-$, which contain an oxygen-bonded, semi-chelating hexafluoro-acetylacetonate ligand.[50] Crystalline members of this class may be isolated when chelating tridentate ligands such as Ph$_2$PC$_2$H$_4$P(Ph)C$_2$H$_4$PPh$_2$ and terpyridyl are employed. Finally, rupture of the remaining Pd—O bond by an attacking ligand leads to formation of $n = 4$ complexes, [Pd(ligand)$_4$]$_2^{2+}$[CF$_3$COCHCOCF$_3^-$]$_2$. A representative member, the tetra(4-chloropyridine) deriva-tive, has been structurally characterized.[51] It is interesting that, notwithstanding the high reactivity of Pd(CF$_3$COCHCOCF$_3$)$_2$ towards nucleophiles, its Pd—O bond length, 1.967 Å,[439] is quite com-parable to that in the nonfluorinated analogue Pd(MeCOCHCOMe)$_2$, 1.979 Å.[446] A similar con-clusion, that susceptibility to nucleophilic attack does not correlate well with metal–oxygen bond distances, can be reached on consideration of the structure of Fe(CF$_3$COCHCOCF$_3$)$_3$, which is very similar indeed to that of the nonfluorinated analogue.[447]

Other workers have concurrently developed the extensive acid–base chemistry of palladium and platinum β-diketonates.[448–452] Sequential displacement reactions also occur with Pt(CF$_3$COCHC-OCF$_3$)$_2$, Pt(CF$_3$COCHCOMe)$_2$ and Pd(CF$_3$COCHCOMe)$_2$. Pdtetrakis(aniline)$^{2+}$ diketonates are exceptional in that, on heating or treatment with base, deprotonation occurs to give anilide-bridged dimers exemplified by [PdNHPh(CF$_3$COCHCOMe)]$_2$. These react with primary amines or pyridine to form [PdNHPh(amine)$_2$]$_2$(diketonate)$_2$.

From Pd(MeCOCHCOMe)$_2$ and diethylamine are obtained [Pd(Et$_2$NH)$_2$(*O,O*-MeCOCHCOMe)][MeCOCHCOMe] or Pd(Et$_2$NH)$_2$(*O,O*-MeCOCHCOMe)(*C*-MeCOCHCOMe) depending on reaction conditions. The kinetics of the reactions of Pd(MeCOCHCOMe)$_2$ with a wide variety of amines indicate participation by hydrogen bond donor solvents. Interestingly, the 1:1 complexes containing *C*-MeCOCHCOMe ligands were found to arise not directly but from the 2:1 complexes.[453] Differences have been noted between the structure of complexes formed from palladium and platinum bis(acetylacetonate). Thus, reaction of Pt(MeCOCHCOMe)$_2$ with excess piperidine produces both [Pt(C$_5$H$_{11}$N)$_2$(MeCOCHCOMe)][MeCOCHCOMe] and *trans*-Pt(C$_5$H$_{11}$N)$_2$(*O*-MeCOCHCOMe)$_2$.[454]

Acetylacetone is incompletely displaced from Pd(MeCOCHCOMe)$_2$ by triethylphosphine at $-10\,°C$. The product, Pd[MeCOC(H)=C(O)Me]$_2$(PEt$_3$)$_2$, contains two monodentate, oxygen bonded enolic acetylacetonate ligands.[455] Triphenylphosphine reacts at room temperature to form Pd(*O,O*-MeCOCHCOMe)(*C*-MeCOCHCOMe)PPh$_3$ but use instead of two equivalents of phosphine in refluxing THF leads to elimination of one equivalent of acetylacetone and to Pd(MeCOCHCOCH$_2$)(PPh$_3$)$_2$.[456] An extremely thorough study of the reactions of bis(β-diketonates) of palladium and platinum has been carried out and factors influencing the relative stability of the various types of products identified. *O*-Monodentate bonding of the diketonate, as in M(*O,O*-diketonate)(*O*-diketonate)(phosphine) and M(*O*-diketonate)$_2$(phosphine)$_2$, is more favorable for Pt than Pd, whereas carbon bonded ligands, as in M(*O,O*-diketonate)(*C*-diketonate)(phosphine), occurs more readily for Pd than Pt. Trifluoroacetylacetonate appears to stabilize the *O*-monodentate bonding mode relative to acetylacetonate and acetylacetonate stabilizes the *C*-bonding mode relative to tri- and hexa-fluoroacetylacetonate. In unsymmetrical chelates the fluorinated diketonates are cleaved or displaced more readily than acetylacetonate itself. The electronic properties of the phosphorus donor are also important: triarylphosphines tend to stabilize carbon bonding in mondentate diketonate ligands whereas trialkylphosphines, which are better σ-donors and poorer π-acceptors, tend to favor bonding to oxygen.[450] Tri-*o*-tolylphosphine is exceptional on account of its large steric bulk. The kinetics of the reaction of this donor with Pd(CF$_3$COCHCOMe)$_2$ have been studied spectrophotometrically,[457] and the crystal structure of Pd(*O,O*-CF$_3$COCHCOMe)(*O*-CF$_3$COCHCOMe)P(*o*-tolyl)$_3$ determined. The enolic *O*-monodentate ligand is *trans* to the CF$_3$CO portion of the *O,O*-bidentate trifluoroacetylacetonate group. The 'dangling' CF$_3$CO and ligating oxygen atom are *trans* with respect to the C=C double bond. Palladium has a square planar coordination geometry and so, for reasons yet to be deduced, this compound does not contain a semi-chelating trifluoroacetylacetonate ligand.[458] A study of the substitution reactions of Pt(*O,O*-MeCOCHCOMe)(*C*-MeCOCHCOMe)PPh$_3$ showed that neutral β-diketones existing predominantly in the enol form displace the η^2 oxygen-bonded acetylacetonate whereas those in the keto form tend to displace the carbon bonded diketonate.[459]

Numerous compounds which formally contain diketonate dianions have been prepared. Reaction of Pd(PhCN)$_2$Cl$_2$ with acetylacetone at $0\,°C$ yields [Pd(*O,O*-MeCOCHCOMe)Cl]$_2$, which slowly rearranges to a material in which palladium is considered to bond to C-1/C-3 (equation 2). Cleavage of the chloride bridge bonds in the former compound with monodentate ligands produces PdCl(*O,O*-MeCOCHCOMe)L (L = Ph$_3$P, Ph$_3$As) and, with bipyridyl, PdCl(bipyridyl)(*C*-MeCOCHCOMe). In contrast, the η^3-isomer reacts with bipyridyl to form PdCl(bipyridyl)(MeCOCH$_2$COCH$_2$).[460] The methine protons in this product are acidic and it behaves as a methyl-substituted acetylacetonate. Reaction with M(MeCOCHCOMe)$_2$ (M = Be, Pd) forms bimetallic materials (**12**) in which acetylacetonate (2−) is considered to be coordinated through the C-1 carbon and both oxygen atoms.[461] Further, reaction of PdCl(bipyridyl)(MeCOCH$_2$COCH$_2$) with thallium acetylacetonate, which behaves as a base in this system, results in elimination of hydrogen chloride and in formation of Pd(bipyridyl)(CH$_2$COCHCOMe) (**13**) in which the acetylacetonate dianion acts as a tridentate ligand with bonding to the C-1/C-2/C-3 carbon atoms.[462] Related materials having the proposed structures (**14**) and (**15**) have been prepared from Pd(bipyridyl)(MeCOCHCOCH$_2$) and Ph$_2$PC$_2$H$_4$PPh$_2$; subsequent condensation with [Pd(Ph$_2$PC$_2$H$_4$PPh$_2$)(H$_2$O)$_2$](ClO$_4$)$_2$ in acetone provides the cationic, dinuclear palladium chelate. Similar reactions with M(ClO$_4$)$_2$ (M = Mg, Ni) afford trinuclear complexes but these appear to be much less stable than the dinuclear complexes.[463,464] Coordination to oxygen and the terminal carbon atoms in Pd(2,6-Me$_2$C$_5$H$_3$N)(CF$_3$COCHCOCH$_2$)PPh$_3$·C$_6$H$_6$, which effectively contains the trifluoroacetylacetonate dianion (**16**), has been established. Compounds of this type are prepared by the reaction of Pd(CF$_3$COCHCOMe)$_2$ with triphenylphosphine then pyridine.[465]

(2)

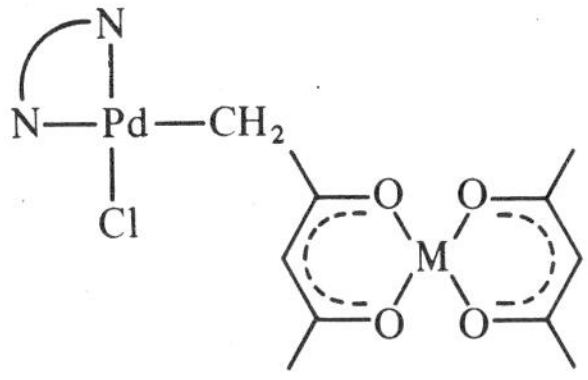

(12) 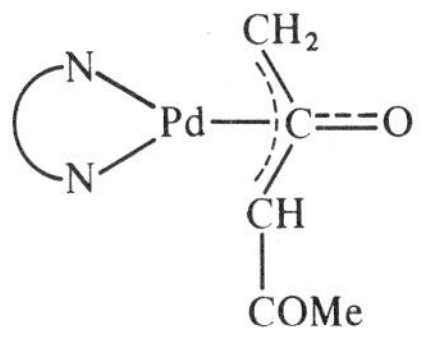 (13) 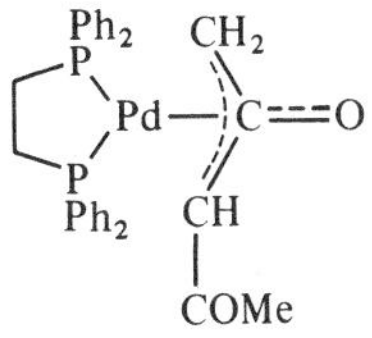 (14)

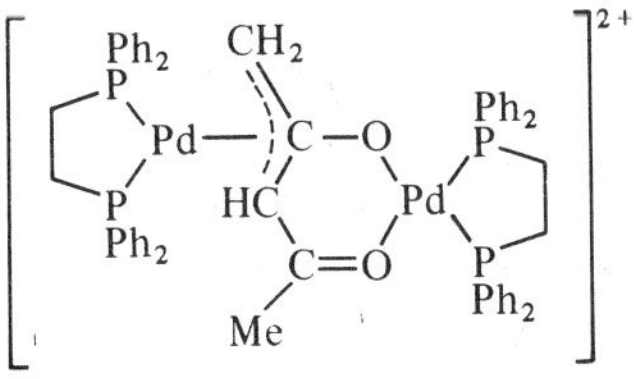

(15) 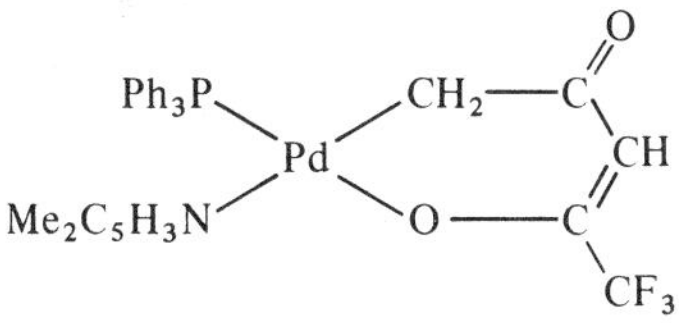 (16)

Platinum bis(acetylacetonate) reacts with $(p\text{-}ClC_6H_4)_3P$ to form $Pt(C_{1,2,3}\text{-}CH_2COCHCOMe)[P(C_6H_4Cl)_3]_2$; triphenylphosphine reacts with $Pt(CF_3COCHCOCF_3)_2$ to generate $Pt(C,O\text{-}CF_3COCHCOCH_2)(PPh_3)_2$. The former product is protonated by pyridinium perchlorate to yield $Pt(C_5H_5N)[(C\text{-}MeCOCHCOMe)\{P(C_6H_4Cl)_3\}_2]ClO_4$. It is remarkable that β-diketonate dianions are formed so readily in these systems in the absence of added base.[466] Apparently, with acetylacetonate(2−), η_3 carbon bonding is favored over the C,O-chelation exhibited by trifluoroacetylacetonate(2−).

For the formation from $Pt(MeCOCHCOMe)_2$ of $Pt(C_5H_5N)(O,O\text{-}MeCOCHCOMe)(C\text{-}MeCOCHCOMe)$, E_a, $\Delta H^{\ddagger}$ and $\Delta S^{\ddagger}$ have been determined to be 87.4 and 80.7 kJ mol^{-1} and -86.5 J K^{-1} mol^{-1} respectively. For the conversion of this compound to $Pt(C_5H_5N)_2(C\text{-}MeCOCHCOMe)_2$, the corresponding values are 68.1 and 65.2 kJ mol^{-1} and -136.7 J K^{-1} mol^{-1}.[467] The activation energy for exchange of ^{14}C labeled acetylacetone with $Pd(MeCOCHCOMe)_2$ is 96.1 kJ mol^{-1},[468] and the barrier for exchange in $Ni(C_2H_5)(MeCOCHCOMe)PPh_3$ is much lower, *ca.* 42 kJ mol^{-1}.[469]

Hydrogen–deuterium exchange involving the N—H protons in $\{Pd(RNH_2)_4\}(R'_3CCOCHCOMe)_2$ or $\{Pd(RNH_2)_2(O,O\text{-}R'_3CCOCHCOMe)\}(R'_3CCOCHCOMe)$ (R' = CF$_3$, H) and CDCl$_3$ occurs readily at room temperature. For solutions of primary or secondary amines in deuterochloroform, exchange is catalytic with respect to the palladium complex. Exchange with the methine protons also occurs and is faster in the cationic diketonate complex.[470]

Acetyl migration occurs in the reaction of $Tl(MeCOCHCOMe)$ with *trans*-$Pd(PhCN)_2Cl_2$, which yields an *N*-acetyl-β-ketoimine chelate (17).[471] Mössbauer data indicate that $Sn(MeCOCHCOMe)_2$ has a trigonal pyramidal geometry with one oxygen atom from each acetylacetonate ligand in an axial position.[472] The equatorial lone electron pair is both stereochemically and chemically active. Complexes in which the metal acts as the donor rather than the acceptor are formed by photolysis of $Sn(MeCOCHCOMe)_2$ and Group VI metal hexacarbonyls in THF. The products, $Sn(MeCOCHCOMe)_2M(CO)_5$ (M = Cr, Mo, W), evidently contain tin–metal bonds. The low

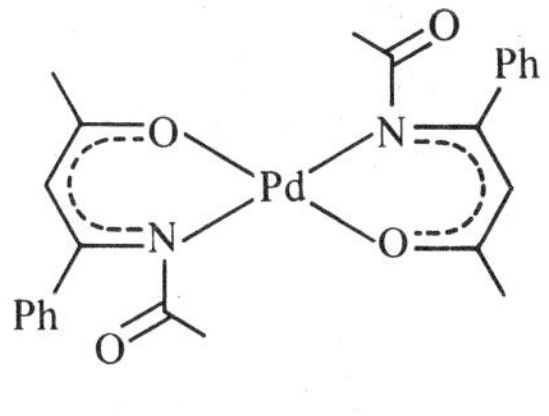

(17)

[119m]Sn isomer shifts are in the region characteristic of Sn[IV] but Mössbauer spectroscopy may not be useful for assessing tin valence states in compounds of this type.[473] Combination of $SnCl_2$, $Fe_2(CO)_9$ and a β-diketone yields $[Sn(diketonate)_2 Fe(CO)_4]_2$. Such materials contain dimeric units as the IR spectra show bridging carbonyl groups; the dimers may be cleaved with pyridine.[474] Formal oxidative addition of methyl iodide to $Sn(MeCOCHCOMe)_2$ leads to the Sn[IV] derivative $SnMeI(MeCOCHCOMe)_2$.[475] Insertions of $Sn(MeCOCHCOMe)_2$ into C—Hg and C—Tl bonds are also known and, more recently, this type of reaction has been extended to B-substituted carboranyl mercury compounds. Spontaneous loss of mercury from $\{9-(1,2-B_{10}C_2H_{11})\}Sn(MeCOCHCOMe)_2Hg\{9-(1,2-B_{10}C_2H_{11})\}$ produces $\{9-(1,2-B_{10}C_2H_{11})\}_2Sn(MeCOCHCOMe)_2$.[476] Fluorinated diketone complexes of Sn[II], such as $Sn(CF_3COCHCOCF_3)_2$, behave as Lewis acids, forming 1:1 complexes with bipyridyl.[477]

Apparently little studied since their first synthesis in 1931 are the interesting 1:1 complexes formed between sulfur dioxide and $Be(MeCOCHCOMe)_2$ and $Be(MeCOCHCOOEt)_2$.[478,479]

15.4.6 ORGANOMETALLIC CHEMISTRY AND CARBON-BONDED β-DIKETONATES

While the majority of metal β-diketonate complexes feature oxygen as the ligating atom(s), there is a discrete class of materials containing bonds between a metal and the methine carbon.[480] These are considered in this section along with other compounds and reactions which emphasize the organometallic aspects of metal diketonates.

Many carbon-bonded β-diketone compounds are acid–base complexes formed from starting materials originally containing bidentate, oxygen-bonded diketonate ligands. Empirically, in order that the coordination number of the metal does not increase, one or more of these η_2 ligands rearranges to an η_1. It is difficult to predict, except under rather circumscribed conditions, whether η_1 carbon or oxygen bonding will predominate. This has been discussed in Section 15.4.4 in the context of tertiary phosphine complexes of platinum and palladium diketonates. Binary carbon-bonded diketonate derivatives are uncommon. Examples include $(C\text{-MeCOCHCOMe})_2S_n$ $(n = 1, 2)$ discussed in Section 15.4.1. Although their solid state structures are established, IR data suggest that enolization may occur in solution.[481]

Mercury(II) derivatives of β-diketonates are usually insoluble, presumably polymeric materials which may contain Hg—C bonds. However, $Hg(C_3F_7COCHCOBu^t)_2$, in which the t-butyl and perfluoro-n-propyl groups tend to inhibit oligomerization, is monomeric in solution and amenable to study by ^{1}H DNMR spectroscopy. This technique reveals that in acetone two interconvertible isomers are present: $Hg(C\text{-}C_3F_7COCHCOBu^t)_2$ and $Hg(C\text{-}C_3F_7COCHCOBu^t)\{O\text{-}C_3F_7COCH{=}C(O)Bu^t\}$. In the latter isomer, mercury is considered to be bonded to one enolic oxygen atom, that adjacent to the Bu^t group.[482]

In phosphine and arsine complexes of the type LM(diketonate) (M = Ag, Au) and CuL_2(diketonate), η_2 oxygen bonded diketonate ligands occur in the copper(I) and silver(I) derivatives but the gold(I) analogues feature the η_1 carbon bonded form.[483]

Linkage isomerism in $AuMe_2(MeCOCHCOMe)$, which contains a bidentate, oxygen-bonded diketonate ligand, has been incisively probed by ^{1}H NMR spectroscopy. This compound reacts with phosphines to form $AuMe_2(C\text{-MeCOCHCOMe})(phosphine)$, which is isolable when $PhPMe_2$ is employed. Equilibrium mixtures are formed when triarylphosphines are used. The bulky $(c\text{-}C_6H_{11})_3P$ leads to $AuMe_2\{MeCOCH{=}C(O)Me\}P(c\text{-}C_6H_{11})_3$. The benzoylacetonate chelate $AuMe_2(O,O\text{-PhCOCHCOMe})$ is stereochemically rigid at ambient temperature but, upon addition of small amounts of Ph_3P, interconversion of the two Au—Me groups takes place, possibly *via* an intermediate containing a semi-chelating acetylacetonate ligand (Scheme 1). This process is even more facile for $AuMe_2(CF_3COCHCOMe)$ as its observation does not require addition of a Lewis base.[484]

Historically, the platinum group metals yielded the first examples of materials containing the C-MeCOCHCOMe ligand, although they were probably not recognized as such.[485] The X-ray diffraction investigation of $K[PtCl(O,O\text{-MeCOCHCOMe})(C\text{-MeCOCHCOMe})]$ provided seminal characterization of this type of bonding.[486] Additionally, it was found that $[PtMe_3(C,O,O\text{-MeCOCHCOMe})]_2$ reacts with bidentate Lewis bases such as bipyridyl[487] and ethylenediamine[488] to produce mononuclear $PtMe_3(C\text{-MeCOCHCOMe})(bipyridyl)$ and $\{PtMe_3(C\text{-MeCOCHCOMe})\}_2(\mu\text{-}NH_2C_2H_4NH_2)$ respectively.[489,490] A normal-coordinate analysis has been carried out on $Na_2[PtCl_2(C\text{-MeCOCHCOMe})_2]$.[491]

Reaction of $Pt(O,O\text{-MeCOCHCOMe})(C\text{-MeCOCHCOMe})X^-$ (X = Cl, Br) with a wide variety of divalent metal salts affords novel chelates of the type $\{Pt(MeCOCHCOMe)_2(X)\}_2M$ (M = Mn,

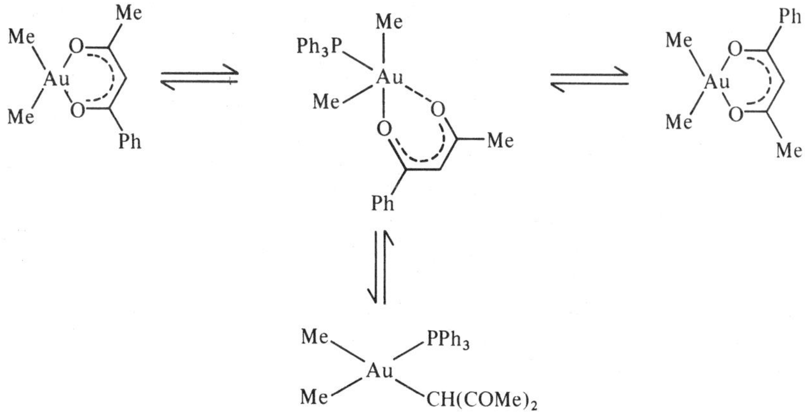

Scheme 1

Fe, Co, Ni, Cu, Zn, Cd, Pd, VO and UO_2) for which the structure (18) was proposed. Bonding to the oxygen atoms of the C-MeCOCHCOMe ligand was inferred from IR spectra. Since the magnetic behavior of the nickel(II) and cobalt(II) complexes is indicative of octahedral coordination, additional Pt—Cl—M bridge bonding, suggested by the dotted line, was proposed. A similar series of the type {Pt(MeCOCHCOMe)$_3$}$_2$M (M = Mn, Fe, Co, Cu, Cd and Pd) was prepared from Pt(O,O-MeCOCHCOMe)(C-MeCOCHCOMe)$_2^-$.[492] Detailed analyses of the IR spectra of the {PtCl(MeCOCHCOMe)}$_2$M compounds have been carried out using metal isotopic substitution to assign the M—O stretching bands. It was found that the donor C=O groups retain substantial double bond character upon coordination to M. This same coordination produces such a small ($\leqslant 15\,cm^{-1}$) effect on v_{Pt-Cl} that Pt—Cl—M bonding is not indicated and other explanations of the magnetic properties of these materials may be needed.[493]

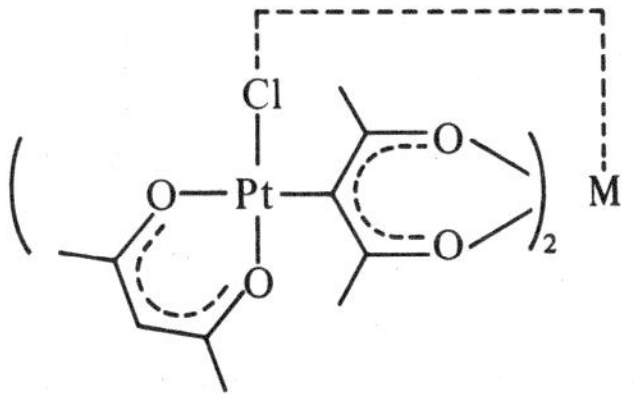

(18)

Reactions of Pt(O,O-MeCOCHCOMe)(MeCOCH$_2$COMe)X (X = Cl, Br) with a wide variety of donors, including phosphines, arsines, amines, alkenes, diamines, *etc.* have been examined. In all cases the coordinated neutral acetylacetone is displaced. When monodentate ligands, L, are employed, products are of the type Pt(O,O-MeCOCHCOMe)XL. Bidentate ligands, LL, yield Pt(C-MeCOCHCOMe)X(LL) and bridging alkenes form {PtCl(O,O-MeCOCHCOMe)}(dialkene).[494]

Organometallic compounds of the type M(β-diketonylcyclooctenyl)(diketonate) (M = Pd, Pt) are prepared by the reaction of the thallium(I) salt of a diketonate with M(1,5-cyclooctadiene)Cl$_2$. Formally, addition of the methine carbon of the diketonate anion to cyclooctadiene converts the diene to a C-MeCOCHCOMe-substituted allyl ligand.[495] Attack of nucleophiles, such as triphenylphosphine, on Pt{[(MeCO)$_2$CH]C$_8$H$_{12}$}(O,O-MeCOCHCOMe) leads to a rearrangement in which transfer of the acetylacetonate ligand from the metal to the cyclooctenyl ring occurs, thus forming Pt{1,5-[(MeCO)$_2$CH]$_2$C$_8$H$_{12}$}(PPh$_3$)$_2$.[496] Exemplary syntheses of [M(C$_8$H$_{12}$)(O,O-MeCOCHCOMe)]BF$_4$ (M = Pd, Pt) and Pt(AsPh$_3$){8-(1-acetylacetonyl)-4-cycloocten-1-yl} have been published.[497]

A very extensive series of pentamethylcyclopentadienylrhodium diketonate compounds has recently been prepared and characterized. From sodium acetylacetonate and Rh$_2$(Me$_5$C$_5$)$_2$Cl$_4$ was obtained Rh(Cl)(Me$_5$C$_5$)(O,O-MeCOCHCOMe). Displacement of the remaining chlorine from the iridium analogue affords Ir(Me$_5$C$_5$)(O,O-MeCOCHCOMe)(C-MeCOCHCOMe). Chlorine abstraction from Rh(Cl)(Me$_5$C$_5$)(O,O-MeCOCHCOMe) with silver fluoroborate yields [Rh(Me$_5$C$_5$)(C,O,O-MeCOCHCOMe)]$_2$ whose crystal structure reveals a dimeric cage structure in which rhodium is bonded to two oxygen atoms from one η_3-acetylacetonate ligand and to the central carbon atom of the other, as in [PtMe$_3$(C,O,O-MeCOCHCOMe)]$_2$.[498]

Addition of hexafluoro-2-butyne to Pd(CF$_3$COCHCOMe)$_2$ yields Pd{OC(Me)CH(COMe)-

$C(CF_3){=}C(CF_3)\}_2$ (19). An X-ray structure determination shows that the acetylene has inserted between palladium and the central, methine carbon atoms, displacing one acetyl group. An analogous material was prepared by addition of hexafluorobutyne to $Pd\{2\text{-}(Me_2NCH_2)C_6H_4\text{-}C^1,N\}$(MeCOCHCOMe).[499] In contrast, 1,4-addition of hexafluorobutyne to $Rh(C_8H_{12})$-(MeCOCHCOMe) occurs, along with cyclotrimerization of the alkyne, to provide an $\eta_4\text{-}(CF_3)_6C_6$ complex of an unusual bicyclic diketonate derivative.[500–502] Similar 1,4-additions have been reported for other Rh^I and Pd^{II} diketonates.[503,504] Reaction of hexafluorobutyne with $Rh(CO)_2(Bu^tCOCHCOBu^t)$ at elevated temperatures yields, *inter alia*, a product $Rh_2(CO)_2\{(Bu^tCOCHCOBu^t)(C_4F_6)_3\}$ (21) in which not only has 1,4-addition to two dipivaloylmethanate-Rh rings occurred but also addition of one alkyne unit across two $Rh(CO)_2$ units, which are joined by a metal–metal bond, to form a dirhodacyclobutene ring.[505]

(19)   (20)

(21)

Orthometallation reactions of $Pd(CF_3COCHCOCF_3)_2$ are quite facile. In three cases, 1:1 complexes with Lewis bases, $PdL(O,O\text{-}CF_3COCHCOCF_3)(C\text{-}CF_3COCHCOCF_3)$ (L = $PhCH_2SMe$ or $(p\text{-}MeOC_6H_4)_2CS$) and the semi-chelating hexafluoroacetylacetonate derivative $Pd(PhCH_2NMe_2)(O,O\text{-}CF_3COCHCOCF_3)(O\text{-}CF_3COCHCOCF_3)$, which are quite probably formed as intermediates in the orthometallation of the respective ligands, have been detected by ^1H and ^{19}F NMR.[506] The orthometallated azobenzene derivative phenylazophenylpalladium hexafluoroacetylacetonate undergoes a remarkable solid state rearrangement on heating to *ca.* 90 °C. This process is accompanied by a decrease in the *N*-phenyl torsional angle from 46 to 15°, a shift in absorption maximum from 415 to 480 nm and a sudden, discontinuous expansion of about 10% along the needle crystal axis with no change in width. The transformation has been interpreted in terms of a displacive or Martensitic-like transformation in which molecular stacks in the low-temperature yellow form interdigitate in the red rearranged form (Figure 14).[507] An unusual orthometallated compound is obtained from *trans*-$IrCl(N_2)(Ph_3P)_2$ and dibenzoyldiazomethane. In $IrCl\{P(C_6H_4)Ph_2\}(PhCOCHCOPh)PPh_3$, transfer of a hydrogen atom from the metallated ring to the central carbon position in $(PhCO)_2CN_2$ and loss of dinitrogen effectively creates a β-diketonate ligand which is η_2 oxygen bonded to iridium.[508]

The reaction of $H_2Os_3(CO)_{10}$ with $Rh(C_2H_4)_2$(MeCOCHCOMe) yields the mixed metal cluster $RhOs_3H_2$(MeCOCHCOMe)$(CO)_{10}$ whose structure is shown in Figure 15. One of the acetylacetonate oxygen atoms, O(3), unsymmetrically bridges Os(3) and Rh, while the other, O(2), is coordinated only to rhodium so that this ligand can be considered as a five-electron donor.[509]

Reaction of $[Pd_2(PhCH{=}CHCOCH{=}CHPh)_3]\cdot CHCl_3$ and the cyclopropenium salt $[C_3Ph(p\text{-}MeOPh)_2]Br$ produces a product which, when treated with thallium(I) acetylacetonate, yields $Pd_3\{C_3Ph(p\text{-}MeOPh)_2\}_2$(MeCOCHCOMe)$_2$. This material, shown in Figure 16, is composed

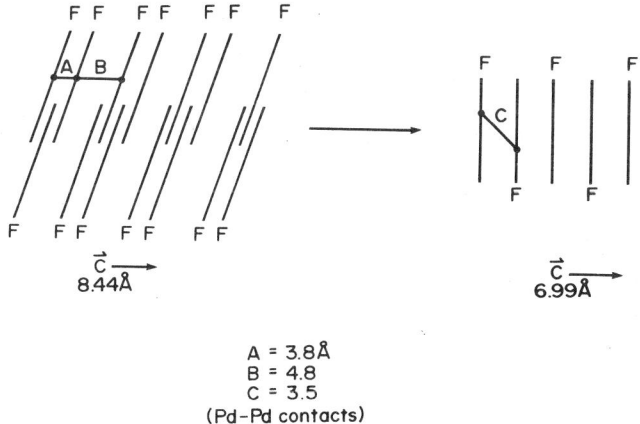

A = 3.8Å
B = 4.8
C = 3.5
(Pd–Pd contacts)

Figure 14 Schematic diagram of solid state rearrangement. Straight lines represent mean molecular planes and F indicates the Pd(CF₃COCHCOCF₃) terminus. The low temperature form is shown on the left

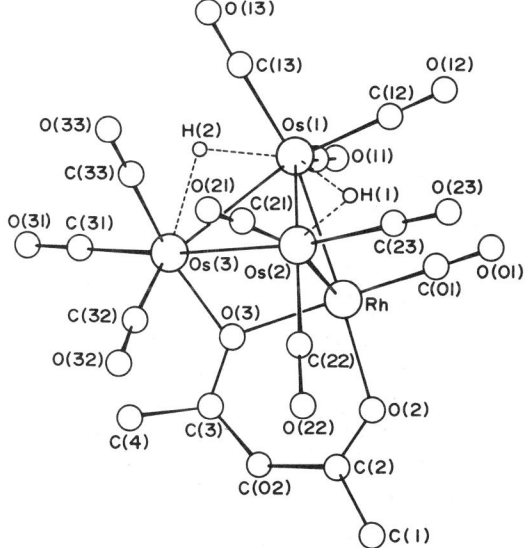

Figure 15 Structure of RhOs₃H₂(MeCOCHCOMe)(CO)₁₀ (reproduced with permission from *J. Chem. Soc., Dalton Trans.*, 1981, 171)

predominantly of an enantiomer which is considered to arise from a ring opening addition of palladium to the cyclopropenium ring, which metal is, in the final product, also bonded to an O,O-MeCOCHCOMe ligand. Two of these units are linked by an additional palladium atom so that the three metal atoms form a triangular array.[510] Compounds of this general class, Pd₃{C₃(aryl)₃}-(ligand)₄, exhibit dynamic behavior in solution. Detailed NMR studies show that this process involves racemization around the central metal atom or, alternatively, movement of the two PdC₃ metallocycle units relative to one another between two equivalent positions about the center palladium atom.[511] Triphenylmethylmetal complexes of the type M(CPh₃)(O,O-MeCOCHCOMe) (M = Pd, Pt) are produced by reaction of M(dibenzylidene)₂ with Ph₃CCl then Tl(MeCOCHC-OMe). For these, three independent fluxional processes have been observed. At low temperatures the M(MeCOCHCOMe) migrates from the C(α),C(1),C(2) to the C(α),C(1),C(6) ligand set (equation 3). In the second dynamic process the triphenylmethyl ligand effectively rotates with respect to the metal, somewhat in the manner of a three-bladed propeller (equation 4). Finally, at higher temperatures, the acetylacetonate ligand undergoes exchange.[512]

Only a small amount of research has been published dealing with the reactions of β-diketones with clean metal surfaces.[513,514] The interaction of acetylacetone with iron and nickel films under ultra high vacuum conditions has been investigated. X-Ray photoelectron spectroscopy is a particularly useful analytical probe as data on gas phase metal acetylacetonates are available for comparison.[515] On iron, dissociative adsorption giving acetylacetonate occurs at 90 K. This decomposes at about 290 K to form surface oxide, chemisorbed oxygen and a species considered to contain Fe—C bonds.

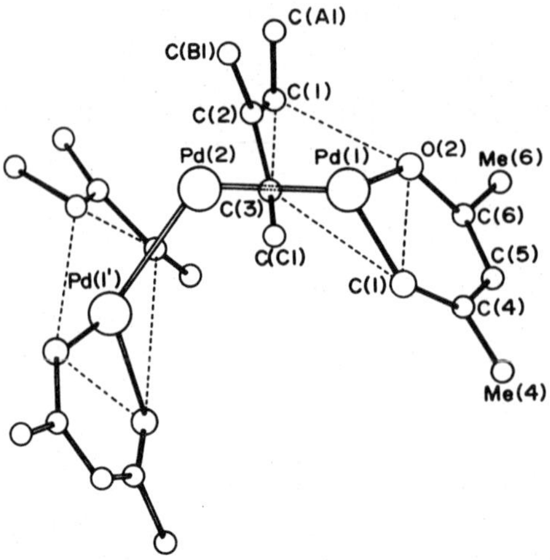

Figure 16 Projection of the molecular structure of Pd$_3${C$_3$Ph(*p*-MeOPh)$_2$(*O*,*O*-MeCOCHCOMe)} in the Pd plane (reproduced with permission from *J. Chem. Soc., Dalton Trans.*, 1978, 1825)

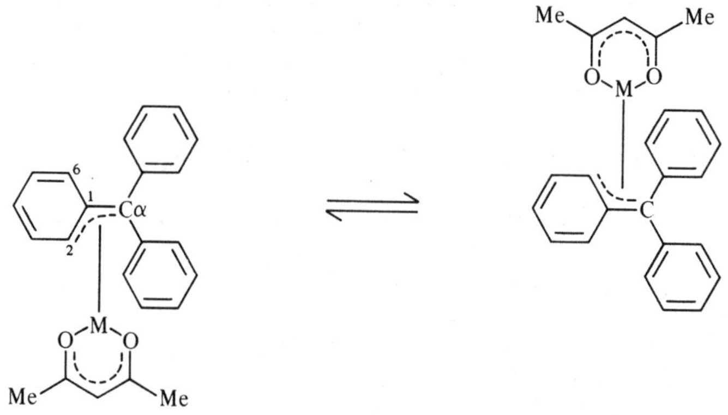

(3)

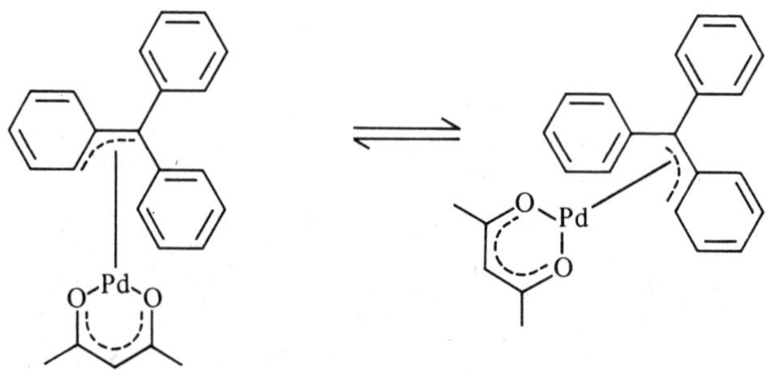

(4)

Pretreatment of the surface with oxygen to a coverage of $\theta \approx 0.5$ inhibits decomposition and acetylacetonate is formed by deprotonation of the diketone by O$_{ads}$. Pyridine adsorbed on iron inhibits absorption of acetylacetone but this base is displaced from the preoxidized surface by acetylacetone on which it would appear to occupy different binding sites.[516] Deprotonation of acetylacetone by dissociatively adsorbed oxygen has parallels in molecular rhodium–oxygen complexes. The μ-dioxygen compound {Rh(1,5-C$_8$H$_{12}$)}$_2$O$_2$ reacts with acetylacetone to form Rh(1,5-C$_8$H$_{12}$)(*O*,*O*-MeCOCHCOMe).[517,518]

Related to β-diketones are metalla-β-diketones, whose distinctive chemistry is a recent development. In these, the methine group of a conventional diketonate is replaced with an organometallic moiety such as *cis*-Mn(CO)$_4$, *cis*-Re(CO)$_4$, *fac*-Mn(CO)$_3$(RNC) or C$_5$H$_5$Fe(CO).[519] Exemplary is Al{Mn(MeCO)$_2$(CO)$_4$}$_3$, which is prepared in a two-step synthesis. Addition of methyllithium to Mn(CO)$_5$COMe forms Li[Mn(MeCO)$_2$(CO)$_4$] which, when treated with aluminum chloride, provides the tris(chelate) compound.[520] An X-ray study showed that the MnC$_2$O$_2$Al ring is essentially planar with Mn—C(acyl) bond distances indicative of a bond order of *ca.* 1.2. The O—O bite distance, 2.73 Å, is about the same as that of acetylacetonate in Al(MeCOCHCOMe)$_3$.[521] As in the carbocyclic chelate analogues, isomerization of the *cis* and *trans* forms of Al{Mn(CO)$_4$(MeCO)-(PhCH$_2$CO)}$_3$ occurs, most probably by an intramolecular, bond rupture pathway.[522] In Cu{Re(CO)$_4$(COMe)$_2$} each chelate ring has a nonplanar, boat conformation leading to a 'chaise longue' molecular shape. The ESR spectrum at $-166\,°C$ reveals coupling between the unpaired electron on copper and the two rhenium nuclei. Thus, whatever π-delocalization may obtain in metalladiketone complexes is not strictly dependent on a planar conformation.[523]

15.4.7 TROPOLONATES AND QUINONES

Tropolone is formally an α-diketone with two oxygen atoms exocyclic to the seven-membered ring. However, in many respects its coordination chemistry resembles that of the more familiar congener acetylacetone.[524,525] It was noted earlier that the planar, rigid nature and small bite angle of tropolonate favors twist rearrangement pathways in (tropolonate)$_3$M complexes. Such features, with which compactness is associated, make the tropolonate ligand well suited for synthesis of compounds in which the central metal atom has a high coordination number. This was first exemplified by ionic complexes such as B(tropolonate)$_2^+$, Si(tropolonate)$_3^+$, P(tropolonate)$_2^{2+}$ and In(tropolonate)$_4^-$.[526] Seven- and eight-coordinate complexes are formed by the rare earth elements. The tris(tropolonates) of lanthanum and the larger rare earth ions (Ce^{+3} through Ho^{3+}) are insoluble in common solvents and presumably polymeric. Smaller rare earth ions provide H$^+$[M(tropolonate)$_4$]$^-$ (M = Er, Tm, Yb, Lu). These lose tropolone on heating to form M(tropolonate)$_3$ compounds which, as they are easily soluble in dimethyl sulfoxide, are probably monomolecular or exist as low molecular weight aggregates in solution; coordination of the solvent is quite possible. Tropolone exchanges slowly with Ta(tropolonate)$_4^+$ and more rapidly with Pr(tropolonate)$_4^-$.[527] The pentakis(tropolonate) anions M(tropolonate)$_5^-$ (M = Th^{IV}, U^{IV}) are thought to contain ten-coordinate metal centers.[528]

Solution phase behavior of these tropolone chelates has been thoroughly studied. The tetrakis-(tropolonates) of Zr, Hf and U^{IV} are monomeric but that of Th^{IV} is associated. Exchange with free tropolone is slow and in the order Th > U > Hf $\approx$ Zr > Ta. The thorium complex also reacts with Lewis bases such as water and dimethyl sulfoxide to form nine-coordinate adducts, behavior not shared with the Zr, Hf and U analogues.[529] Uranyl bis(tropolonate) forms 1:1 adducts with a wide variety of donors (H$_2$O, Me$_2$SO, Ph$_3$PO, cyclohexanone, PhNH$_2$, C$_5$H$_5$N). The structure of UO$_2$(tropolonate)$_2$·C$_5$H$_5$N shows the usual pentagonal bipyramidal geometry.[530] X-Ray studies confirm the presence of pentagonal bipyramidal geometry in Sn(tropolonate)$_3$X (X = Cl, OH) where X occupies an axial position.[531] In Th(tropolonate)$_4$·DMF, the ThO$_9$ coordination core approximates C_s-m symmetry, having monocapped square antiprismatic form.[532] Careful analysis of the shape parameters for the tris(tropolonates) of Al^{III},[533] Sc^{III},[534] Fe^{III}[535] and Mn^{III} indicate a significant twisting towards a D_{3h} geometry.[536] The structure of Mn(tropolonate)$_3$ is unusual. There are two separate molecular geometries in the unit cell. One type of molecule is tetragonally elongated while another exhibits orthorhombic distortion (*cf.* axial compression in Mn(MeCOCHCOMe)$_3$).[537] Like acetylacetone, tropolone yields an anhydrous, presumably polymeric complex with nickel(II). However, [Ni(tropolonate)$_2$]$_n$ forms only a monohydrate. This is a dimer which contains two terminal and two bridging tropolonate ligands.[538]

o-Quinone, like tropolone, is formally an α-diketone and it has a significant coordination chemistry, important features of which are its ability, upon reduction, to provide stable paramagnetic ligands[539] and also potential relevance to bacterial photosynthesis.[540] The electronic structure of *o*-quinone is such that it is capable of undergoing successive one-electron reductions and is the terminal member in the series shown in equation (5). Likewise, tris(quinone) metal complexes form an electron transfer series of the type M(quinone)$_3^{n-}$ in which *n* may be 0, 1, 2 or 3. This is nicely illustrated by the tris(3,5-di-*t*-butylquinone)chromium series which has been characterized by, *inter alia*, X-ray, EPR and electrochemical techniques. The *n* = 0 and *n* = 3 members, prepared respectively from 3,5-di-*t*-butyl-*o*-benzoquinone and Cr(CO)$_6$ and from 3,5-di-*t*-butylcatechol and CrCl$_3$, may be isolated as pure compounds. Individual molecules of the neutral complex have trigonally distorted octahedral coordination cores and this material spontaneously resolves into the Λ-*cis* isomer on crystallization. In it, the ligands retain significant quinone character with alternating C—C bond lengths and d(C—O) = 1.285 Å. In contrast, the trianionic catechol analog has more regular C$_6$ rings and d(C—O) is 1.349 Å. The ground state of tris(3,5-di-*t*-butylsemiquinone)-chromium has $S = 0$ and the compound is nearly diamagnetic.[541-543] In M(benzoquinone)$_3$ (M = V, Cr, Fe) the Mössbauer spectra of the iron compound and the small ^{51}V hyperfine coupling in the ESR spectrum of the vanadium analogue indicate a localized electronic structure with the unpaired electron centered predominantly on the ligands. Magnetic susceptibility measurements suggest that there is an antiferromagnetic exchange among the coordinated semiquinone ligands.[544]

$$\tag{5}$$

Relatively weak antiferromagnetic interactions obtain in Fe(semiquinone)$_3$ (quinone = tetra-chloro-o-benzoquinone, phenanthroquinone), which exhibit temperature dependent magnetic susceptibility behavior. In contrast, (di-t-butylsemiquinone)$_3$Fe has μ_{eff} essentially independent of temperature. It behaves as an $S = 1$ complex with two unpaired electrons in the ground state, which is explicable in terms of an $S = 5/2$ metal interacting with three $S = 1/2$ ligands. The bulky t-butyl groups in the 3,5-positions prevent intermolecular interactions in the crystal lattice which, in the phenanthroquinone analogue, provide an additional spin coupling pathway. The Cr^{III} analogues behave similarly; indeed, Cr(3,5-di-t-butylsemiquinone)$_3$ is diamagnetic.[545,546] The one-electron reduction products of this compound show large ^{53}Cr hyperfine coupling constants and thus may be formulated as having one quinone or catecholate ligand respectively, leaving spin density concentrated on the chromium center.[547]

Solid Co(bipyridyl)(3,5-di-t-butylbenzoquinone)$_2$ contains trivalent cobalt; one of the ligands is coordinated as a semiquinone and the other as a catecholate. In solution, this material appears to be in equilibrium with the Co^{II} form which would contain two semiquinone ligands.[548] A somewhat similar dichotomy between solid and solution phase electronic structures has been found in high-spin Fe^{III} complexes such as Fe(bipyridyl)(9,10-phenanthroquinone)$_2$. Mixed valence electronic character is substantiated by the observation of an intervalence charge transfer band at 1100 nm in the solid compound and in solutions of low, but not high, dielectric constant.[549] The nickel(II) compound Ni(C$_5$H$_5$N)$_2$(9,10-phenanthroquinone)$_2$ contains two *cis* pyridine and two chelating o-semiquinone ligands. Dimerization in the solid state involves intermolecular π–π interactions between the planar phenanthroquinone rings. The magnetic susceptibility behavior of this material, as well as of the tetrameric [M(3,5-di-t-butylsemiquinone)$_2$]$_4$ (M = Ni, Co) aggregates[550] is indicative of weak, intramolecular antiferromagnetic exchange between the metal ion and the semiquinone ligands.

The spectroelectrochemistry of catecholate(2$-$) complexes of the types NiL$_2$(C$_6$H$_4$O$_2$) and NiL$_2$(C$_6$Cl$_4$O$_2$) (L = Ph$_3$P, Ph$_2$PC$_2$H$_4$PPh$_2$/2) has been carefully studied. The frozen-solution phase ESR spectra of the materials produced by quasi-reversible one-electron reduction show large, anisotropic ^{31}P hyperfine coupling to two equivalent phosphorus nuclei and anisotropic g values characteristic of d^9 Ni^I compounds. The ESR spectra of the one-electron oxidation products reveal diminished ^{31}P hyperfine splittings and almost isotropic g values consistent with Ni^{II} derivatives containing coordinated semiquinones.[551]

While reactions of o-quinones with first-row transition metal carbonyls are usually a straightforward route to M(semiquinone)$_n$ compounds, the chemistry of Mo(CO)$_6$ is rather different. It reacts with tetrachloro-o-benzoquinone to give the Mo^{VI} dimer [Mo(O$_2$C$_6$Cl$_4$)$_2$]$_2$, which has fully reduced ligands.[552] In contrast, 9,10-phenanthroquinone yields the *cis*-dioxomolybdenum complex Mo$_2$O$_5$(phenanthroquinone)$_2$.[553] A novel chelate in which Mo^{VI} has only one oxo ligand, [MoO{3,5-di-t-butylcatechol(2$-$)}$_2$]$_2$, is formed from the quinone and molybdenum hexacarbonyl.[554] The catecholate(2$-$) ion is able to stabilize high metal oxidation states, *e.g.* Os^{VI},[555] and the $+4$ states of Hf, U, Th and Ce. The crystal structures of Na$_4$[M(C$_6$H$_4$O$_2$)$_4$]·21H$_2$O (M = U, Th, Hf and Ce) have been determined. The metal coordination polyhedra are all quite close to idealized, trigonal-faced dodecahedra. In strongly basic solutions containing excess catechol, the Ce^{IV} complex undergoes a quasi-reversible one-electron reduction, presumably to a Ce^{III} product.[556,557] Compounds of this type are of interest as models for chelating agents which might be used to sequester Pu^{IV} under physiological conditions.[558] Recently, three catechol complexes of titanium have been structurally characterized. Nearly octahedral Ti^{IV} occurs in (Et$_3$NH)$_2$[Ti(O$_2$C$_6$H$_4$)$_3$]. The complex K$_4$[TiO(O$_2$C$_6$H$_4$)$_2$]$_2$·9H$_2$O contains oxo bridges in a Ti$_2$O$_2$ unit and bidentate catechol(2$-$) ligands. The remarkable compound (Et$_3$NH)$_2$[Ti(DTBC)$_2$(HDTBC)]$_2$ (DTBC is 3,5-di-t-butylcatechol) contains one catechol monoanion and two catechol dianions coordinated to titanium; one of the bidentate catechol dianion oxygen atoms is shared by two metal atoms.[559,560]

Facile oxidative addition of o-quinones to low-valent metal complexes containing Pd^0, Pt^0, Rh^I, Ir^I, Ru^0 and Ru^{II} have been reported[561–565] as well as the enthalpy changes associated with such reactions.[566]

15.4.8 POLYKETONE COMPLEXES

1,3,5-Triketones are homologues of 1,3-diketones and in these, too, keto–enol tautomerism has been probed by ^{1}H NMR spectroscopy.[567,568] Triketones and tetraketones may coordinate to one or more metal ions per molecule. In the latter case the metal centers are held in such close proximity that, in some cases, interesting magnetic effects may be observed. Structural and magnetic properties of polynuclear transition metal β-polyketonates have been thoroughly reviewed.[569,570]

Depending on experimental conditions, mono- (2:1) (22) and di-nuclear (1:1) (23) copper(II) complexes of diacetylacetone (2,4,6-heptanetrione) may be obtained, the former as a dihydrate.[571,572] Synthesis of Be^{II}, Co^{II}, Ni^{II} and Pd^{II} chelates proceeds analogously.[573,574] Introduction of methyl groups at the methine carbon positions tends to favor formation of the monometallic complexes,[575] possibly for steric reasons. While monometallic complexes are formed in higher yields when excess ligand is employed, Ni^{II} and Co^{II} induce condensation of diacetylacetone. Thus 2:1 complexes of these metals are better made by ligand exchange methods using the metal bis(acetylacetonate) and the triketone.[573]

(22)   (23)

Divalent cobalt, nickel and copper bimetallic triketone complexes readily react with four equivalents of Lewis bases to form octahedral adducts. Of these, $M_2(C_5H_5N)_4(PhCOCHCOCHCOPh)_2 \cdot 4C_5H_5N$ (M = Co, Ni), which also contain lattice pyridine, have been structurally characterized.[576,577] The metal atoms and the triketonate ligands are nearly coplanar in these compounds, in contrast to the β-diketonate complexes in which small displacements from the ligand planes occur. The copper(II) complex $Cu_2(C_5H_5N)_2$-$(MeCOCHCOCHCOMe)_2$ contains square pyramidal copper with one coordinated pyridine ligand on each side of the Cu_2O_6 plane from which the metals are slightly displaced.[578] An analogue, $Cu_2(C_5H_5N)_2\{p\text{-MeOPhCH}_2COCHCOCHCOCH_2(p\text{-MeOPh})\}_2$, has a similar structure and strong intramolecular antiferromagnetic coupling occurs in both compounds ($2J \approx -825\,\text{cm}^{-1}$). However, for the d^1–d^1 vanadyl system the exchange constant is only ca. $160\,\text{cm}^{-1}$, indicating the importance of orbital symmetry in superexchange mediated by oxygen bridges.[579–581]

X-Ray structure determinations have been carried out on $Cu_2(MeOH)_2(1,1,1,7,7,7\text{-hexafluoro-}2,4,6\text{-heptanetrionate})$ and $Ni_2(C_5H_5N)_2(1,1,1\text{-trifluoro-2,4,6-heptanetrionate})$. The metal–metal separations are 3.04 and 3.16 Å respectively and antiferromagnetic exchange prevails. Evaluation of exchange interactions in a series of structurally characterized dinuclear Cu^{II} complexes indicates that displacement of the metal ions from the ligand plane is associated with weaker coupling.[582,583] A detailed study of the magnetic properties of the trinuclear complex $Ni_3(OH)_2(H_2O)_6(\text{trifluoro-2,4,6-}$ heptanetrionate)$_2$ shows that, in the linear Ni_3 array, adjacent Ni^{II} centers are ferromagnetically coupled ($J = 10\,\text{cm}^{-1}$) but the terminal Ni^{II} ions are antiferromagnetically coupled ($J = -6\,\text{cm}^{-1}$),[584] a pattern also seen in $[Ni(MeCOCHCOMe)_2]_3$.[585] Complexes of dibenzoyl-acetylacetone with two or three divalent metal ions (Co, Ni, Cu) may be obtained depending on experimental conditions. Spin coupling in trimetallic cobalt complexes is antiferromagnetic but ferromagnetic for the nickel analogues.[586] Iron triketone complexes of the type $Fe_2(OH)_2(H_2O)_2$-(triketone)$_2$ (triketone = dibenzoylacetone, benzoylacetylacetone) have been prepared. Room temperature Mössbauer parameters are consistent with their formulation as high-spin Fe^{III} complexes and antiferromagnetic exchange between the metal centers prevails.[587]

Polyketone complexes of the UO_2^{2+} ion have been well studied and two such materials are structurally characterized. In $UO_2(MeOH)(1,5\text{-diphenyl-1,3,5-pentanetrionate})_2 \cdot MeOH$ the triketone is present at a 1− ligand. Uranium is bonded to four enolic oxygen atoms from the triketones, two uranyl oxygen atoms and one oxygen from the coordinated methanol. Interestingly, the $PhCOCH_2$ portion of one ligand appears to be ketonic ($d_{C-O} = 1.19\,\text{Å}$) while the other $PhCOCH_2$ unit is enolic ($d_{C-O} = 1.32\,\text{Å}$). A novel trimetallic uranium complex, $(Et_3NH)_2[(UO_2)_3O(2,2',8,8'\text{-tetramethyl-3,5,7-nonanetrionate})_3]$, has also been described. This contains three UO_2 units in a nearly equilateral triangle with a central μ_3 oxide ligand. One triketonate group spans each edge of the triangle with the central, enolic oxygen atom bridging two uranium atoms and the terminal oxygens bonded to one uranium atom (Figure 17).[588] In both these structures, and in copper–triketone complexes, the metal–oxygen bond lengths involving the central carbon atom of the 1,3,5-triketone are longer than that between the metal and oxygen atoms on the terminal portion of the ligand.[580] Additional systematic studies of lanthanide and actinide complexes of such polyketone ligands would be of great interest.

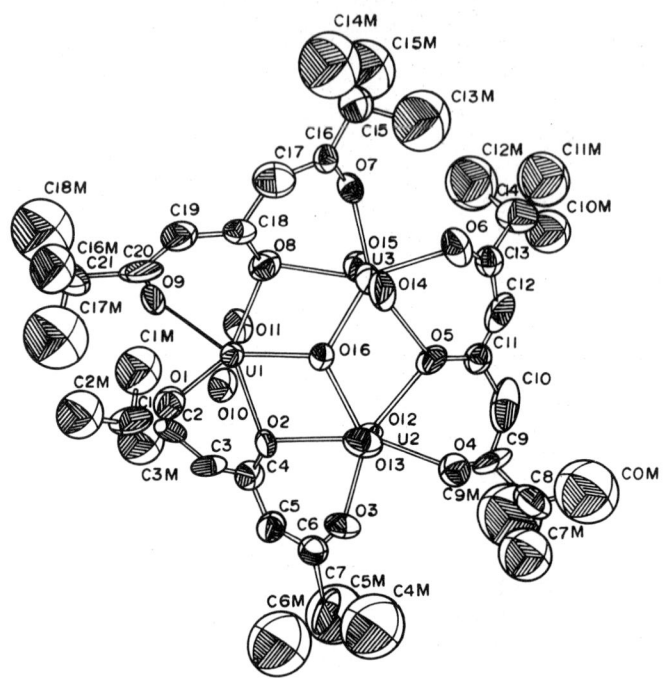

Figure 17   Structure of the [tris(2,2′,8,8′-tetramethyl-3,5,7-nonanetrionato)-μ_3-oxotris(dioxourante)](2−) ion (reproduced with permission from *Inorg. Chem.*, 1982, **21**, 2350)

Chelates containing both uranyl and palladium centers have been prepared using a novel, stepwise synthesis.[589] Because UO_2^{2+} requires five equatorial donor atoms to achieve the favored pentagonal bipyramidal coordination geometry, its reaction in methanol with benzoylacetylacetone (and related triketones) leads to a material with two η_2 triketonate ligands *cis* to one another. The two terminal, noncoordinated oxygen atoms are then favorably positioned to bind to a second metal ion, as shown in Scheme 2. This strategy has subsequently been extended to 1,3,5,7-tetraketonates and an entire series of trimetallic chelate complexes of the type $(UO_2)_2 M(solvent)_4(dibenzoylacetylacetonate)_2$ (M = Mn, Co, Ni, Cu, Zn) is thus available. The compound where M = Ni and solvent = pyridine has been structurally characterized.[590]

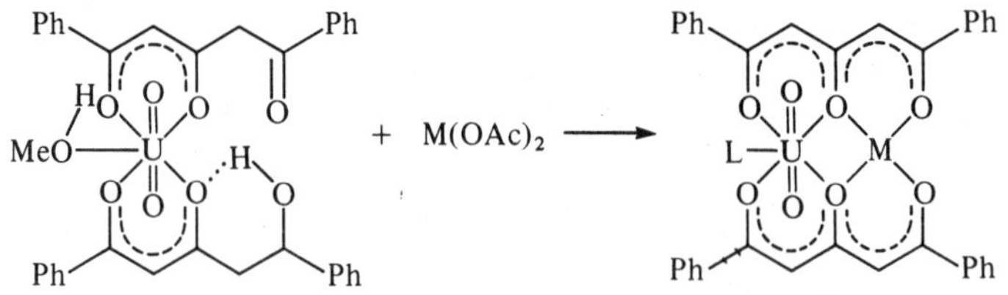

Ph ⟍ ⸺ ⟋ Ph   Ph ⟍ ⸺ ⟋ Ph

MeO—U⋯H   + M(OAc)₂ ⟶   L—U   M

Ph ⟋ ⸺ ⟍ Ph   Ph ⟋ ⸺ ⟍ Ph

Scheme 2

Detailed electrochemical studies have been carried out on bis(1,3,5-triketonate)dicopper(II) chelate complexes. Reversible two-electron reduction of the Cu^{II},Cu^{II} compounds to a Cu^{I},Cu^{I} product involves sequential electron transfer at the same electrode potential. Thus the diamagnetic Cu^{II},Cu^{II} pair behaves empirically as one electroactive center.[591–593]

A very large series of organotin complexes of the types $SnR'_3(RCOCHCOCH_2COR')$ and $(SnR'_3)_2(RCOCHCOCHCOR')$ (R, R′, R″ = alkyl, phenyl) has been prepared. Both *E* and *Z* enol forms of the former have been detected by ^{1}H NMR spectroscopy.[594] Analogous complexes of Sn^{II} and Sn^{IV} have also been synthesized and characterized by ^{119m}Sn Mössbauer spectroscopy.[595]

Mono- (**24**) and bis-(allyl) (**25**) dinickel complexes have been synthesized from the triketone and Ni(C$_3$H$_5$)$_2$; or from the dithallium salt of the triketone and [Ni(C$_3$H$_5$)Br]$_2$ respectively.[596] Other polyketone ligands may be constructed by using heteroatoms or aromatic groups to join two β-diketone units, *e.g.* (**26**)–(**28**).[597–598] Yet other variants of triketones are obtained from their conversion by α,ω-alkanediamines into Schiff bases, *e.g.* (**29**).[599] Compounds of this type may be regarded as compartmental ligands which contain N$_2$O$_2$ and O$_4$ donor sets. In the mixed metal complexes containing Cu—UO$_2$, Cu—VO, Cu—Zn, Ni—Cu, Ni—UO$_2$, Ni—VO and Ni—Zn, the metal listed first is contained in the N$_2$O$_2$ compartment.[600–604] Reversible positional isomerism occurs in the monocobalt complex. In the solvent-free material, cobalt resides in the N$_2$O$_2$ compartment as indicated by the appearance of IR bands due to the uncoordinated carbonyl groups. However, in derivatives containing coordinated or solvated pyridine or methanol, a shift to the O$_4$ binding site occurs.[605]

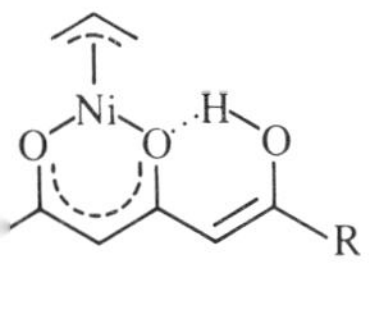

(24) R = CO$_2$Et   (25) R = Me 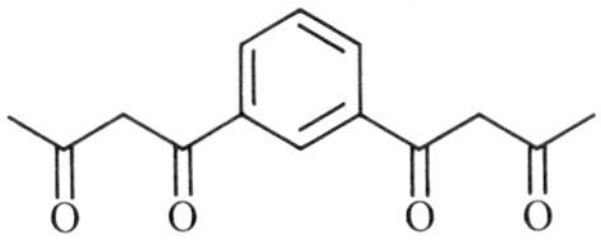   (26) 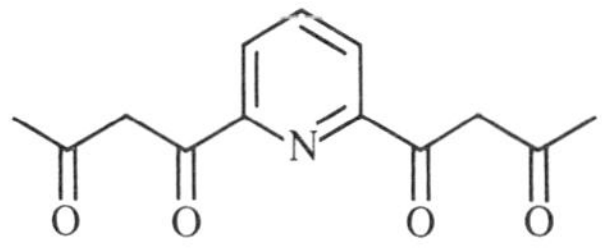   (27)

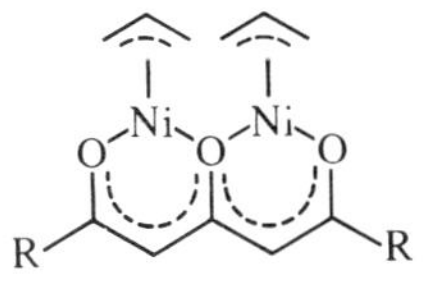

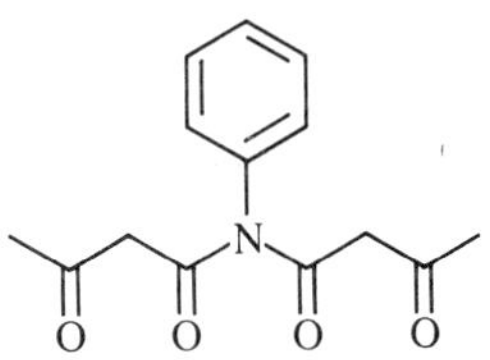

(28) 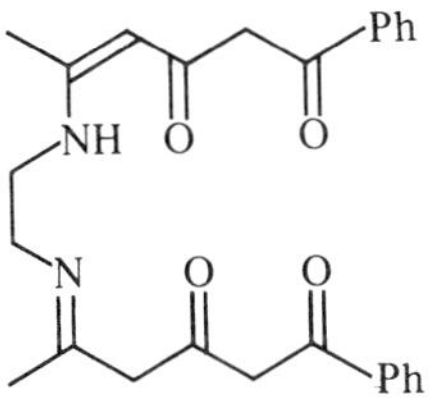   (29)

The phosphine-functionalized diketone [*o*-(diphenylphosphino)benzoyl]pinacolone has an extensive and interesting coordination chemistry. Its dimeric, head-to-tail complex with Cu^I contains trigonally coordinated copper atoms bonded to the O$_2$P ligand set.[606] Reaction with diaroyl peroxides produces a mixed valence Cu^I,Cu^{II} material. The phosphinodiketone ligands rearrange so that Cu^{II} is located in the 'hard' β-diketonate site and the Cu^I emerges in a P$_2$O$_2$ ligand set, as shown in Scheme 3.[607] In the synthesis of bimetallic Cu—Pt and Cu—Pd complexes the 'soft' platinum(II) is first added to the P$_2$ donor site and, subsequently, 'hard' Cu^{II} is introduced into the O$_4$ site. Chlorine abstraction leads to shift of the platinum from a P$_2$ to an O$_2$P$_2$ ligand set, as depicted in Scheme 4. Using similar synthetic methods, Cl$_2$Ru—Cu and Cl(CO)Ir—Cu analogues may be prepared. The electronic spectrum of the latter compound exhibits a band at 960 nm which may be due to Ir^I → Cu^{II} charge transfer.[608]

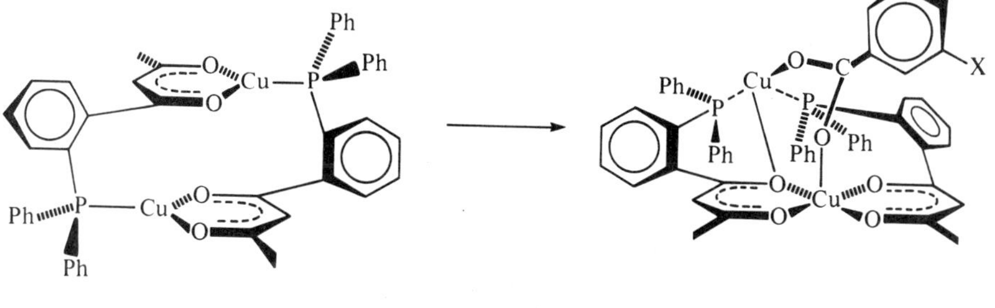

Scheme 3

15.4.9 COMPLEXES OF NEUTRAL β-DIKETONES

A small number of metal complexes containing a neutral β-diketone is known. These materials are generally prepared from a metal salt, often a perchlorate, and acetylacetone, or by controlled

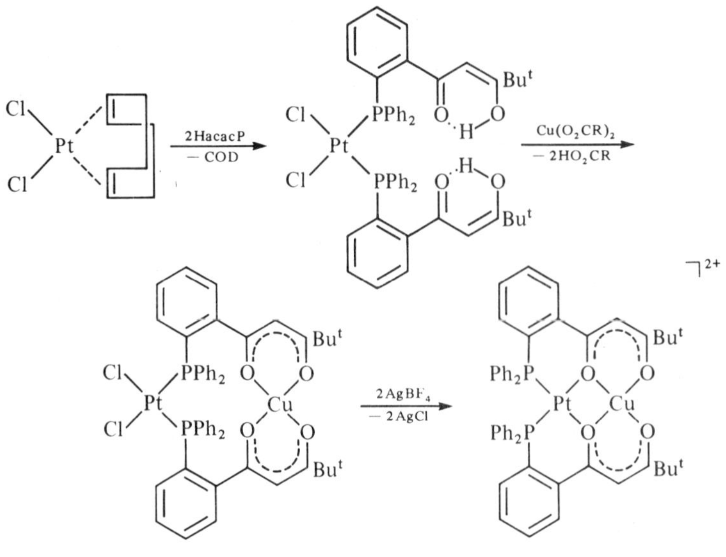

Scheme 4

hydrolysis of a metal acetylacetonate. Characterization has often proved difficult owing to ready conversion to diketonate complexes and, indeed, such compounds containing neutral diketones as ligands may be present in low concentrations as intermediates which, upon loss of a proton, form diketonates.

Direct combination of $CoCl_2$, $ZnCl_2$ or $MnBr_2$ with acetylacetone yields the bis(diketone) adducts. The crystal structure of $MnBr_2(MeCOCH_2COMe)_2$ reveals an infinite $MnBr_2$ chain with manganese coordinated to two axial, enolic oxygen-bonded molecules of acetylacetone. The Mn—O distance is 2.20 Å, rather longer than the 1.95—2.00 Å range found in β-$Mn(MeCOCHCOMe)_2$.[609] Reaction of $[Ni(MeCOCHCOMe)_2]_3$ with hydrogen bromide and excess acetylacetone yields *trans*-$NiBr_2(MeCOCH_2COMe)_2$. This compound contains octahedral nickel and two acetylacetone molecules in the keto form act as nearly symmetrical bidentate ligands which span equatorial sites. The average Ni—O separation is 2.05 Å, quite comparable with the 2.015 Å distance in $Ni(H_2O)_2(MeCOCHCOMe)_2$.[610] However, detailed analysis of the metal—oxygen contacts in these two acetylacetone adducts is not warranted since the *r* factors for both structures are quite large.

From $Ni(MeCO_2H)_6(ClO_4)_2$ and $Ni(H_2O)_6(ClO_4)_2$ were prepared $[Ni(MeCOCH_2COMe)_3]$-$(ClO_4)_2$, $[Ni(MeCOCH_2COMe)_2(MeCO_2H)_2](ClO_4)_2$ and $[Ni(MeCOCH_2COMe)_2(H_2O)_2]$-$(ClO_4)_2$[611] and similar methods were used to otbain $[Zn(MeCOCH_2COMe)_2(H_2O)](ClO_4)_2$ and $[Zn(MeCOCH_2COMe)_2](ClO_4)_2$.[612] Structural characterization of these compounds would be of interest but, like most perchlorates, the solids are potentially explosive. The reaction of $CrCl_3(C_4H_8O)_3$ with acetylacetone leads to displacement of the tetrahydrofuran and formation of the octahedral, mixed complex $CrCl_2(MeCOCHCOMe)(MeCOCH_2COMe)$.[613] Proton NMR studies of $[Mg(MeCOCH_2COMe)_2(H_2O)_2](ClO_4)_2$ indicate that it is the keto form of the diketone which is coordinated to magnesium.[614]

Other examples of materials containing neutral β-diketone ligands include $MoOCl_3(Me-COCH_2COMe)$,[615] $[ReCl(CO)_3(diketone)]_2$[616,617] and $UO_2(MeCOCHCOMe)_2(MeCOCH_2C-OMe)$.[618]

ACKNOWLEDGEMENTS

The author is grateful to Dr. T. E. Wood for commentary on the manuscript and to Pam Highstrom and Gail McCauley of the 3M Technical Library for their help in collecting the literature. Research in coordination chemistry in the 3M Central Research Laboratories has benefited greatly from collaboration with Prof. L. H. Pignolet and Dr. R. A. Newmark and from the support of Dr. E. Fatuzzo and Dr. H. G. Bryce.

15.4.10 REFERENCES

1. L. W. Reeves, *Can. J. Chem.*, 1957, **35**, 1351.
2. K. I. Lazar and S. H. Bauer, *J. Phys. Chem.*, 1983, **87**, 2411.
3. D. E. Williams, *Acta Crystallogr.*, 1966, **21**, 340.
4. D. E. Williams, W. L. Dumke and R. E. Rundle, *Acta Crystallogr.*, 1962 **15**, 627.
5. G. R. Engebretson and R. E. Rundle, *J. Am. Chem. Soc.*, 1964, **86**, 574.
6. A. H. Lowry, C. George, P. D'Antonio and J. Karle, *J. Am. Chem. Soc.*, 1971, **93**, 6399.
7. A. L. Andreassen and S. H. Bauer, *J. Mol. Struct.*, 1972, **12**, 381.
8. A. Camerman, D. Mastropaolo and N. Camerman, *J. Am. Chem. Soc.*, 1983, **105**, 1584.
9. R. C. Mehrotra, R. Bohra and D. P. Gaur, 'Metal β-Diketonates and Allied Derivatives', Academic, New York, 1978.
10. K. C. Joshi and V. N. Pathak, *Coord. Chem. Rev.*, 1977, **22**, 37.
11. D. W. Thompson, *Struct. Bonding (Berlin)*, 1971, **9**, 27.
12. J. P. Fackler, Jr., *Prog. Inorg. Chem.*, 1966, **7**, 362.
13. B. Bock, K. Flatau, H. Junge, M. Kuhr and H. Musso, *Angew. Chem., Int. Ed. Engl.*, 1971, **10**, 225.
14. D. P. Graddon, *Coord. Chem. Rev.*, 1969, **4**, 1.
15. J. A. Davies and F. R. Hartley, *Chem. Rev.*, 1981, **81**, 79.
16. O. W. Howarth, C. H. McAteer, P. Moore and G. E. Morris, *J. Chem. Soc., Dalton Trans.*, 1981, 1481.
17. S. J. Thompson, C. White and P. M. Maitlis, *J. Organomet. Chem.*, 1977, **136**, 87.
18. C. White, S. J. Thompson and P. M. Maitlis, *J. Chem. Soc., Dalton Trans.*, 1977, 1654.
19. M. A. Bennett, T. W. Matheson, G. B. Robertson, W. L. Steffen and T. W. Turney, *J. Chem. Soc., Chem. Commun,* 1979, 32.
20. R. H. Crabtree, G. G. Hlatky, C. P. Parnell, B. E. Segmuller and R. J. Uriarte, *Inorg. Chem.*, 1984, **23**, 354.
21. D. E. Fenton, C. Nave and M. R. Truter, *J. Chem. Soc., Dalton Trans.*, 1973, 2188.
22. W. C. Fernelius and L. G. VanUitert, *Acta Chem. Scand.*, 1954, **8**, 1726.
23. B. B. Martin and D. F. Martin, *J. Inorg. Nucl. Chem.*, 1975, **37**, 1079.
24. R. M. Izatt, C. G. Haas, B. P. Block and W. C. Fernelius, *J. Phys. Chem.*, 1954, **58**, 1133.
25. R. M. Izatt, W. C. Fernelius, C. G. Haas, Jr. and B. P. Block, *J. Phys. Chem.*, 1955, **59**, 80.
26. L. G. VanUitert, W. C. Fernelius and B. E. Douglas, *J. Am. Chem. Soc.,* 1953, **75**, 457.
27. L. G. VanUitert, W. C. Fernelius and B. E. Douglas, *J. Am. Chem. Soc.*, 1953, **75**, 2736.
28. L. G. VanUitert, W. C. Fernelius and B. E. Douglas, *J. Am. Chem. Soc.*, 1953, **75**, 2739.
29. L. G. VanUitert, W. C. Fernelius and B. E. Douglas, *J. Am. Chem. Soc.*, 1953, **75**, 3577.
30. L. G. VanUitert and W. C. Fernelius, *J. Am. Chem. Soc.*, 1953, **75**, 3862.
31. B. E. Bryant, W. C. Fernelius and B. E. Dougals, *J. Am. Chem. Soc.*, 1953, **75**, 3784.
32. R. M. Izatt, W. C. Fernelius and B. P. Block, *J. Phys. Chem.*, 1955, **59**, 235.
33. R. M. Izatt, W. C. Fernelius and C. G. Haas, Jr. and B. P. Block, *J. Phys. Chem.*, 1955, **59**, 170.
34. R. M. Izatt, C. G. Haas, Jr., B. P. Block and W. C. Fernelius, *J. Phys. Chem.*, 1954, **58**, 1133.
35. L. G. VanUitert and W. C. Fernelius, *J. Am. Chem. Soc.*, 1954, **76**, 5887.
36. B. E. Bryant and W. C. Fernelius, *J. Am. Chem. Soc.*, 1954, **76**, 4864.
37. B. E. Bryant and W. C. Fernelius, *J. Am. Chem. Soc.*, 1954, **76**, 1696.
38. J. L. Ault, H. J. Harries and J. Burgess, *J. Chem. Soc., Dalton Trans.*, 1973, 1095.
39. M. Raban and D. Haritos, *J. Chem. Soc., Chem. Commun.*, 1978, 965.
40. J. O. Hill and R. J. Irving, *J. Chem. Soc. (A)*, 1969, 2690.
41. J. O. Hill and R. J. Irving, *J. Chem. Soc. (A)*, 1966, 971.
42. J. O. Hill and R. J. Irving, *J. Chem. Soc. (A)*, 1968, 1052.
43. J. O. Hill and R. J. Irving, *J. Chem. Soc. (A)*, 1968, 3116.
44. K. J. Cavell and G. Pilcher, *J. Chem. Soc., Faraday Trans. 1*, 1977, 1590.
45. G. Al-Takhin, G. Pilcher, J. Bickerton and A. Zaki, *J. Chem. Soc., Dalton Trans.*, 1983, 2657.
46. M. R. Jaffe, D. P. Fay, M. Cefola and N. Sutin, *J. Am. Chem. Soc.*, 1971, **93**, 2878.
47. D. P. Fay, A. R. Nichols, Jr. and N. Sutin, *Inorg. Chem.*, 1971, **10**, 2096.
48. B. H. Berrie and J. E. Earley, *Inorg. Chem.*, 1984, **23**, 774.
49. G. J. Bullen, R. Mason and P. Pauling, *Inorg. Chem.*, 1965, **4**, 456.
50. A. R. Siedle, R. A. Newmark and L. H. Pignolet, *J. Am. Chem. Soc.*, 1981, **103**, 4947.
51. A. R. Siedle and L. H. Pignolet, *Inorg. Chem.*, 1982, **21**, 135.
52. S. Ooi, T. Matsushita, K. Nishimoto, S. Okeya, Y. Nakamura and S. Kawaguchi, *Bull. Chem. Soc. Jpn.*, 1983, **56**, 3297.
53. Z. Kanda, Y. Nakamura and S. Kawaguchi, *Inorg. Chem.*, 1978, **17**, 910.
54. R. Mason, G. B. Robertson and P. J. Pauling, *J. Chem. Soc. (A)*, 1969, 485.
55. A. G. Swallow and M. R. Truter, *Proc. R. Soc. London, Ser. A*, 1960, **252**, 205.
56. R. N. Hargreaves and M. R. Truter, *J. Chem. Soc. (A)*, 1969, 2282.
57. N. Yanase, Y. Nakamura and S. Kawaguchi, *Inorg. Chem.*, 1980, **19**, 1575.
58. S. Okeya, Y. Nakamura, S. Kawaguchi and T. Hinomoto, *Inorg. Chem.*, 1981, **20**, 1576.
59. L. E. McCandlish and J. W. Macklin, *J. Organomet. Chem.*, 1975, **99**, 31.
60. J. vonSeyerl, D. Neugebauer and G. Huttner, *Angew. Chem., Int. Ed. Engl.*, 1977, **16**, 858.
61. J. vonSeyerl, D. Neugebauer, G. Huttner, C. Kruger and Y. H. Tsay, *Chem. Ber.*, 1979, **112**, 637.
62. W. H. Watson, Jr. and C. T. Lin, *Inorg. Chem.*, 1966, **5**, 1074.
63. N. V. Sidgwick and F. M. Brewer, *J. Chem Soc.*, 1925, **127**, 2379.
64. K. Shobatake and K. Nakamoto, *J. Chem. Phys.*, 1968, **49**, 4792.
65. R. Kamel, M. Hilal, A. H. Eid and A. Sawaby, *Mol. Cryst. Liq. Cryst.*, 1975, **31**, 9.
66. M. R. Truter and B. L. Vickery, *J. Chem. Soc., Dalton Trans.*, 1972, 395.
67. J. M. Stewart and B. Morosin, *Acta Crystallogr., Part B*, 1975, **31**, 116.
68. R. M. Pike, *Coord. Chem. Rev.*, 1967, **2**, 163.
69. J. C. Dewan and J. Silver, *J. Chem. Soc., Dalton Trans.*, 1977, 644.

70. J. C. Dewan and J. Silver, *Inorg. Nucl. Chem. Lett.*, 1976, **12**, 647.
71. C. L. Raston, R. J. Secomb and A. H. White, *J. Chem. Soc., Dalton Trans.*, 1976, 2307.
72. J. C. Dewan and J. Silver, *J. Organomet. Chem.*, 1977, **125**, 125.
73. R. D. G. Jones and L. F. Power, *Acta Crystallogr., Part B*, 1976, **32**, 1801.
74. F. A. Cotton and W. H. Ilsley, *Inorg. Chem.*, 1982, **21**, 300.
75. T. J. Anderson, M. A. Neuman and G. A. Melson, *Inorg. Chem.*, 1973, **12**, 927.
76. B. Morosin and H. Montgomery, *Acta Crystallogr., Part B*, 1969, **25**, 1354.
77. L. H. Pignolet, *Top. Curr. Chem.*, 1975, **56**, 91.
78. T. S. Piper and R. L. Belford, *Mol. Phys.*, 1962, **5**, 169.
79. J. W. Carmichael, Jr., L. K. Steinkrauf and R. L. Belford, *J. Chem. Phys.*, 1965, **43**, 3959.
80. I. Robertson and M. R. Truter, *J. Chem. Soc. (A)*, 1967, 309.
81. P. K. Hon, C. E. Pfluger and R. L. Belford, *Inorg. Chem.*, 1966, **5**, 516.
82. C. L. Modenbach, F. R. Fronczek, E. W. Berg and T. C. Taylor, *Inorg. Chem.*, 1983, **22**, 4083.
83. R. P. Dodge, D. H. Templeton and A. Zalkin, *J. Chem. Phys.*, 1961, **35**, 55.
84. P.-K. Hon, R. L. Belford and C. E. Pfluger, *J. Chem. Phys.*, 1965, **43**, 1323.
85. H. Montgomery and E. C. Lingafelter, *Acta Crystallogr.*, 1964, **17**, 1481.
86. F. A. Cotton and J. L. Wise, *Inorg. Chem.*, 1966, **5**, 1200.
87. F. A. Cotton and J. S. Wood, *Inorg. Chem.*, 1964, **3**, 245.
88. J. P. Fackler, Jr. and F. A. Cotton, *J. Am. Chem. Soc.*, 1961, **83**, 3775.
89. C. S. Chamberlain and R. S. Drago, *Inorg. Chim. Acta*, 1979, **32**, 75.
90. F. A. Cotton, C. E. Rice and G. W. Rice, *Inorg. Chim. Act*, 1977, **24**, 231.
91. M. J. Bennett, F. A. Cotton and R. Eiss, *Acta Crystallogr., Part B*, 1968, **24**, 904.
92. F. A. Cotton and R. C. Elder, *Inorg. Chem.*, 1965, **4**, 1145.
93. F. A. Cotton and G. W. Rice, *Nouv. J. Chim.*, 1977, **4**, 301.
94. F. A. Cotton and R. C. Elder, *Inorg. Chem.*, 1966, **5**, 423.
95. F. A. Cotton and R. Eiss, *J. Am. Chem. Soc.*, 1968, **90**, 38.
96. C. L. Raston and A. H. White, *Aust. J. Chem.*, 1979, **32**, 507.
97. P. K. Hon and C. E. Pfluger, *J. Coord. Chem.*, 1973, **3**, 67.
98. B. Morosin and H. Montgomery, *Acta Crystallogr., Part B*, 1969, **25**, 1354.
99. J. P. Fackler, Jr. and A. Avdeef, *Inorg. Chem.*, 1974, **13**, 1864.
100. B. R. Stults, R. S. Marianelli and V. W. Day, *Inorg. Chem.*, 1979, **18**, 1853.
101. V. W. Day and R. C. Fay, *J. Am. Chem. Soc.*, 1975, **97**, 5136.
102. E. G. Muller, V. W. Day and R. C. Fay, *J. Am. Chem. Soc.*, 1976, **98**, 2165.
103. B. G. Thomas, M. L. Morris and R. L. Hildebrandt, *J. Mol. Struct.*, 1976, **35**, 241.
104. B. G. Thomas, M. L. Morris and R. L. Hildebrandt, *Inorg. Chem.*, 1978, **17**, 2901.
105. G. D. Smith, C. N. Caughlin and J. A. Campbell, *Inorg. Chem.*, 1972, **11**, 2989.
106. S. Tachiyashiki, N. Yakayama, R. Kurodo, S. Sato and Y. Saito, *Acta Crystallogr., Part B*, 1975, **31**, 1483.
107. H. Montgomery and E. C. Lingafelter, *Acta Crystallogr.*, 1964, **17**, 1481.
108. J. J. Sahbari and M. M. Olmstead, *Acta Crystallogr., Part C*, 1983, **39**, 208.
109. B. Morosin, *Acta Crystallogr.*, 1967, **22**, 316.
110. E. E. Lippert and M. R. Truter, *J. Chem. Soc. (A)*, 1970, 4996.
111. H. Montgomery and E. C. Lingafelter, *Acta Crystallogr.*, 1963, **16**, 748.
112. E. N. Maslen, T. M. Greaney, C. L. Raston and A. H. White, *J. Chem. Soc., Dalton Trans.*, 1975, 400.
113. T. M. Greaney, C. L. Raston, A. H. White and E. N. Maslen, *J. Chem. Soc., Dalton Trans.*, 1975, 876.
114. R. B. Birdy and M. Goodgame, *Inorg. Chem.*, 1979, **18**, 472.
115. P. F. R. Ewings, P. G. Harrison and T. J. King, *J. Chem. Soc., Dalton Trans.*, 1975, 1455.
116. E. O. Schlemper, *Inorg. Chem.*, 1967, **6** , 2012.
117. G. A. Miller and E. O. Schlemper, *Inorg. Chem.*, 1973, **12**, 677.
118. V. B. Ramos and R. S. Tobias, *Spectrochim. Acta, Part A*, 1974, **30**, 181.
119. R. LeBlanc and W. H. Nelson, *J. Organomet. Chem.*, 1976, **113**, 257.
120. N. Serpone and K. A. Hersh, *Inorg. Chem.*, 1974, **13**, 2901.
121. R. W. Jones and R. C. Fay, *Inorg. Chem.*, 1973, **12**, 2599.
122. M. M. Gray and R. S. Tobias, *J. Am. Chem. Soc.*, 1965, **87**, 1909.
123. J. C. Taylor, A. Ekstrom and C. H. Randall, *Inorg. Chem.*, 1978, **17**, 3285.
124. A. Ekstrom and C. H. Randall, *J. Phys. Chem.*, 1978, **82**, 2180.
125. W. F. Howard, Jr. and W. H. Nelson, *Inorg. Chem.*, 1982, **21**, 2283.
126. S. K. Brahma and W. H. Nelson, *Inorg. Chem.*, 1982, **21**, 4076.
127. G. M. Bancroft, B. W. Davies, N. C. Payne and T. K. Sham, *J. Chem. Soc., Dalton Trans.*, 1975, 973.
128. V. A. Mode and G. S. Smith, *J. Inorg. Nucl. Chem.*, 1969, **31**, 1857.
129. M. F. Richardson, P. W. R. Corfield, D. E. Sands and R. E. Sievers, *Inorg. Chem.*, 1970, **9**, 1632.
130. W. L. Steffen and R. C. Fay, *Inorg. Chem.*, 1978, **17**, 779.
131. J. C. A. Boyens and J. P. R. DeVilliers, *J. Cryst. Mol. Struct.*, 1972, **2**, 197.
132. C. G. Pitt, L. K. Monteith, L. F. Ballard, J. C. Collman, J. C. Morrow, W. R. Roper and D. Ulko, *J. Am. Chem. Soc.*, 1966, **88**, 4286.
133. C. P. Wong and W. D. Horrocks, Jr., *Tetrahedron Lett.*, 1975, 2637.
134. H. Sugimoto, T. Higashi and M. Mori, *Chem. Lett.*, 1982, 801.
135. H. Sugimoto, T. Higashi and M. Mori, *Chem. Lett.*, 1983, 1167.
136. H. Sugimoto, T. Higashi, A. Maeda, M. Mori, H. Masuda and T. Taga, *J. Chem. Soc., Chem. Commun.*, 1983, 1234.
137. Y. Nakano and S. Sato, *Inorg. Chem.*, 1980, **19**, 3391.
138. Y. Nakano and S. Sato, *Inorg. Chem.*, 1982, **21**, 1315.
139. J. P. Fackler, Jr. *et al.*, *Inorg. Synth.*, 1970, **12**, 70.
140. R. E. Sievers and J. W. Connoly, *Inorg. Synth.*, 1970, **12**, 72.
141. W. D. Ross and R. E. Sievers, *Anal. Chem.*, 1969, **41**, 1109.

142. H. Lehmkuhl and W. Eisenbach, *Ann. Chim.*, 1975, **691**, 672.
143. M. Dunne and F. A. Cotton, *Inorg. Chem.*, 1963, **2**, 263.
144. M. Kilner, F. A. Hartman and A. Wojcicki, *Inorg. Chem.*, 1967, **6**, 406.
145. M. C. Fredette and C. J. L. Lock, *Can. J. Chem.*, 1973, **51**, 317.
146. M. C. Fredette and C. J. L. Lock, *Can. J. Chem.*, 1975, **53**, 2481.
147. M. A. M. Queiros and S. D. Robinson, *Inorg. Chem.*, 1978, **17**, 310.
148. A. R. Siedle, *Inorg. Chem.*, 1981, **20**, 1318.
149. P. B. Critchlow and S. D. Robinson, *Inorg. Chem.*, 1978, **17**, 1902.
150. T. Ito, Kokubo, Y. Yamamoto, A. Yamamoto and S. Ikeda, *J. Chem. Soc., Dalton Trans.*, 1974, 1783.
151. S. D. Itel, *Inorg. Chem.*, 1977, **16**, 1245.
152. C. L. Kwan and J. K. Kochi, *J. Am Chem. Soc.*, 1976, **98**, 4903.
153. A. Stasko, A. Tkac, V. Laurinc and L. Malik, *Org. Magn. Reson.*, 1976, **8**, 237.
154. M. Freni, P. Romiti and D. Giusto, *J. Inorg. Nucl. Chem.*, 1970, **32**, 145.
155. M. Bockman, M. Cooke, M. Green, H. P. Kirsch, F. G. A. Stone and A. J. Welch, *J. Chem. Soc., Chem. Commun.*, 1976, 381.
156. M. Green, H. P. Kirsch, F. G. A. Stone and A. J. Welch, *Inorg. Chim. Acta*, 1978, **29**, 101.
157. R. Cramer, *Inorg. Synth.*, 1974, **15**, 14.
158. F. Bonati and R. Ugo, *J. Organomet. Chem.*, 1968, **11**, 341.
159. W. C. Fernelius, *Inorg. Synth.*, 1957, **5**, 113, 130, 188; 1960, **6**, 147.
160. R. C. Mehrotra and V. D. Gupta, *J. Indian Chem. Soc.*, 1963, **40**, 911.
161. A. L. Allred and D. W. Thompson, *Inorg. Chem.*, 1968, **7**, 1196.
162. R. C. Fay and R. N. Lowry, *Inorg. Chem.*, 1970, **9**, 2048.
163. D. W. Thompson, W. A. Somers and M. O. Workman, *Inorg. Chem.*, 1970, **9**, 1252.
164. M. L. Reynolds, *J. Inorg. Nucl. Chem.*, 1964, **26**, 667.
165. M. J. Frazer and W. E. Newton, *Inorg. Chem.*, 1971, **10**, 2137.
165. (a) D. W. Thompson, W. C. R. Munsey and T. V. Harris, *Inorg. Chem.*, 1973, **12**, 2190.
166. D. W. Thompson, *Inorg. Nucl. Chem. Lett.*, 1971, **7**, 931.
167. C. E. Holloway, *Can. J. Chem.*, 1971, **49**, 519.
168. W. D. Courrier, W. Forster, C. J. L. Lock and G. Turner, *Can. J. Chem.*, 1972, **50**, 8.
169. W. D. Courrier, C. J. L. Lock and G. Turner, *Can. J. Chem.*, 1971, **50**, 1797.
170. J. P. Collman, R. L. Marshall, W. L. Young and S. D. Goldby, *Inorg. Chem.*, 1962, **1**, 704.
171. S. C. Chattoraj and R. E. Sievers, *Inorg. Chem.*, 1967, **2**, 408.
172. S. Okeya and S. Kawaguchi, *Inorg. Synth.*, 1980, **20**, 65.
173. L. G. Hubert-Pfalzgraf, M. Tsunoda and D. Katochi, *Inorg. Chim. Acta*, 1981, **51**, 81.
174. R. L. Deutscher and D. L. Keppert, *Inorg. Chim. Acta*, 1970, **4**, 465.
175. G. T. Morgan and A. R. Brown, *J. Chem. Soc.*, 1924, **125**, 1259.
176. L. E. Manzer, *Inorg. Chem.*, 1978, **17**, 1552.
177. M. G. B. Drew, G. W. A. Fowles, D. A. Rice and K. J. Shanton, *J. Chem. Soc., Chem. Commun.*, 1974, 614.
178. J. Pradilla-Sorzano, H. W. Chen, F. W. Koknat and J. P. Fackler, Jr., *Inorg. Chem.*, 1979, **18**, 3519.
179. R. L. Pecsok, W. D. Reynolds, J. P. Fackler, Jr., I. Lin and J. Pradilla-Sorzano, *Inorg. Synth.*, 1974, **15**, 96.
180. M. F. Richardson, W. F. Wagner and D. E. Sands, *Inorg. Chem.*, 1968, **7**, 2495.
181. M. F. Richardson, W. F. Wagner and D. E. Sands, *J. Inorg. Nucl. Chem.*, 1968, **30**, 1275.
182. M. Ismail, S. J. Lyle and J. E. Newbury, *J. Inorg. Nucl. Chem.*, 1969, **31**, 1715.
183. M. Janghorbani, E. Wetz and K. Stark, *J. Inorg. Nucl. Chem.*, 1976, **38**, 41.
184. A. Ekstrom, H. Loeh, C. H. Randall, L. Szego and J. C. Taylor, *Inorg. Nucl. Chem. Lett.*, 1978, **14**, 301.
185. K. J. Eisentraut and R. E. Sievers, *Inorg. Synth.*, 1968, **11**, 94.
186. G. Chen. J. W. McDonald and W. E. Newton, *Inorg. Chem.*, 1976 , **15**, 2612.
187. M. F. Farona, D. C. Perry and H. A. Kuska, *Inorg. Chem.*, 1968, **7**, 2415.
188. D. E. Grove, N. P. Johnson, C. J. L. Lock and G. Wilkinson, *J. Chem. Soc.*, 1965, 490.
189. V. K. Jain, R. Bohra and R. C. Mehrotra, *J. Organomet. Chem.*, 1980, **184**, 57.
190. R. C. Mehrotra and S. R. Bindal, *Can. J. Chem.*, 1969, **47**, 2661.
191. C. M. Puri, K. C. Pande and R. C. Mehrotra, *J. Less-Common Met.*, 1962, **4**, 393.
192. M. H. Chisholm, K. Folting, J. C. Huffman and A. L. Ratterman, *Inorg. Chem.*, 1984, **23**, 613.
193. C. D. Garner, S. Parkes, I. B. Walton and W. Clegg, *Inorg. Chim. Acta*, 1978, **31**, L451.
194. R. C. Mehrotra, R. N. Kapoor, S. Prakash and P. N. Kapoor, *Aust. J. Chem.*, 1966, **19**, 2079.
195. S. Dubey, S. N. Misra and R. N. Kapoor, *Z. Naturforsch., Teil B*, 1970, **25**, 476.
196. M. Hasan, K. Kumar, S. Dubey and S. N. Misra, *Bull. Chem. Soc. Jpn.*, 1968, **41**, 2619.
197. M. Hasan, S. N. Misra and R. N. Kapoor, *Indian J. Chem.*, 1969, **7**, 519.
198. M. F. Hawthorne and M. Reintzes, *J. Am. Chem. Soc.*, 1964, **86**, 5016.
199. R. Koster and G. W. Rotermund, *Ann.*, 1965, **40**, 689.
200. J. R. Cook and D. F. Martin, *J. Inorg. Nucl. Chem.*, 1964, **26**, 1249.
201. P. F. R. Ewings, P. G. Harrison and D. E. Fenton, *J. Chem. Soc., Dalton Trans.*, 1975, 821.
202. W. R. Kroll and W. Naegele, *J. Organomet. Chem.*, 1969, **19**, 439.
203. P. G. Edwards, F. Felix, K. Mertis and G. Wilkinson, *J. Chem. Soc., Dalton Trans.*, 1979, 361.
204. J. P. Collman, R. A. Moss, H. Maltz and C. C. Heindel, *J. Am. Chem. Soc.*, 1961, **83**, 531.
205. P. R. Singh and R. Sahai, *Aust. J. Chem.*, 1967, **20**, 639.
206. P. R. Singh and R. Sahai, *Indian J. Chem.*, 1970, **8**, 178.
207. P. R. Singh and R. Sahai, *Aust. J. Chem.*, 1967, **20**, 649.
208. C. Djorjevic, J. Lewis and R. S. Nyholm, *Chem. Ind.*, 1959, 122.
209. K. Kuroda, S. Kajita and S. Fujioka, *Synth. React. Inorg. Metal-Org. Chem.*, 1984, **14**, 97.
210. P. R. Singh and R. Sahai, *Inorg. Chim. Acta*, 1968, **2**, 102.
211. J. P. Collman, R. L. Marshall, W. L. Young, III and C. T. Sears, Jr., *J. Org. Chem.*, 1963, **28**, 1449.
212. P. R. Singh and R. Sahai, *Indian J. Chem.*, 1970, **8**, 178.

213. J. P. Collman, R. L. Marshall, W. L. Young III and S. D. Goldby, *Inorg. Chem.*, 1962, **1**, 704.
214. R. W. Kluiber, *J. Am. Chem. Soc.*, 1961, **83**, 3030.
215. R. H. Barker, J. P. Collman and R. L. Marshall, *J. Org. Chem.*, 1964, **29**, 3216.
216. J. P. Collman and M. Yamada, *J. Org. Chem.*, 1963, **28**, 3017.
217. R. P. Eckberg, J. H. Nelson, J. W. Kenney, P. N. Howells and R. A. Henry, *Inorg. Chem.*, 1977, **16**, 3128.
218. K. Uehara, Y. Ohashi and M. Tanaka, *Bull. Chem. Soc. Jpn.*, 1976, **49**, 1447.
219. R. G. Pearson and J. W. Moore, *Inorg. Chem.*, 1966, **5**, 1523.
220. R. G. Pearson and J. W. Moore, *Inorg. Chem.*, 1966, **5**, 1528.
221. R. G. Pearson and D. A. Johnson, *J. Am. Chem. Soc.*, 1964, **86**, 3983.
222. D. Meisel, K. Schmidt and D. Meyerstein, *Inorg. Chem.*, 1979, **18**, 971.
223. R. D. Gillard and G. Wilkinson, *J. Chem. Soc.*, 1963, 5399.
224. R. G. Pearson, D. N. Edginton and F. Basolo, *J. Am. Chem. Soc.*, 1962, **84**, 3234.
225. E. L. Muetterties and C. M. Wright, *J. Am. Chem. Soc.*, 1965, **87**, 21.
226. J. A. Bertrand and D. Caine, *J. Am. Chem. Soc.*, 1964, **86**, 2298.
227. J. A. Bertrand and R. I. Kaplan, *Inorg. Chem.*, 1965, **4**, 1657.
228. J. E. Andrew and A. B. Blake, *J. Chem. Soc., Dalton Trans.*, 1973, 1102.
229. E. D. Estes, R. P. Scaringe, W. E. Hatfield and D. Hodgson, *Inorg. Chem.*, 1976, **15**, 1179.
230. E. D. Estes, R. P. Scaringe, W. E. Hatfield and D. Hodgson, *Inorg. Chem.*, 1977, **16**, 1605.
231. H. R. Fischer, J. Glerup, D. J. Hodgson and E. Peterson, *Inorg. Chem.*, 1982, **21**, 3063.
232. J. A. Bertrand, A. P. Ginsberg, R. I. Kaplan, C. E. Kirkwood, R. L. Martin and R. C. Sherwood, *Inorg. Chem.*, 1971, **10**, 240.
233. A. R. Siedle and L. H. Pignolet, *Inorg. Chem.*, 1982, **21**, 3090.
234. U. B. Saxena, A. K. Rai and R. C. Mehrotra, *Indian J. Chem.*, 1971, **9**, 709.
235. M. Hasan, K. Kumar, S. Dubey and S. N. Misra, *Bull. Chem. Soc. Jpn.*, 1968, **41**, 2619.
236. M. Hasan, S. N. Misra and R. N. Kapoor, *Indian J. Chem.*, 1969, **7**, 519.
237. I. Vevere, L. Maijs and D. Reikstina, *Chem. Abstr.*, 1971, **74**, 27067.
238. D. L. Hoof and R. A. Walton, *Inorg. Chim. Acta*, 1975, **12**, 71.
239. R. W. Mathews, A. D. Hammer, D. L. Hoof, D. G. Tisley and R. A. Walton, *J. Chem. Soc., Dalton Trans.*, 1973, 1035.
240. J. F. Harrod and K. R. Taylor, *Inorg. Chem.*, 1975, **14**, 1541.
241. D. M. Puri and R. C. Mehrotra, *J. Indian Chem. Soc.*, 1962, **39**, 499.
242. P. H. Bird, A. R. Fraser and C. F. Lau, *Inorg. Chem.*, 1973, **12**, 1322.
243. B. R. Stults, R. S. Marianelli and V. W. Day, *Inorg. Chem.*, 1975, **14**, 722.
244. A. K. Gregson and N. T. Moxon, *Inorg. Chem.*, 1982, **21**, 586.
245. M. M. Aly, *Inorg. Nucl. Chem. Lett.*, 1973, **9**, 253.
246. K. Isobe, K. Noda, Y. Nakamura and S. Kawaguchi, *Bull. Chem. Soc. Jpn.*, 1973, **46**, 1699.
247. K. Isobe, K. Takeda, Y. Nakamura and S. Kawaguchi, *Inorg. Nucl. Chem. Lett.*, 1973, **9**, 1283.
248. R. Mason and G. B. Robertson, *J. Chem. Soc. (A)*, 1969, 492.
249. D. Gibson, J. Lewis and C. Oldham, *J. Chem. Soc. (A)*, 1967, 72.
250. A. R. Siedle and L. H. Pignolet, *Inorg. Chem.*, 1980, **19**, 2052.
251. K. Saito, *Pure Appl. Chem.*, 1974, **38**, 325.
252. H. Kido and K. Saito, *Inorg. Chem.*, 1977, **16**, 397.
253. R. C. Fay, A. Y. Girgis and U. Klabunde, *J. Am. Chem. Soc.*, 1970, **92**, 7056.
254. A. Y. Girgis and R. C. Fay, *J. Am. Chem. Soc.*, 1970, **92**, 7061.
255. A. Wanatabe, H. Kido and K. Saito, *Inorg. Chem.*, 1981, **20**, 1107.
256. H. Kido and K. Saito, *Bull. Chem. Soc. Jpn.*, 1979, **52**, 3545.
257. H. Kido and K. Saito, *Bull. Chem. Soc. Jpn.*, 1980, **53**, 424.
258. T. Sekine, H. Honda, M. Kokiso and T. Tosaka, *Bull. Chem. Soc. Jpn.*, 1979, **52**, 1046.
259. T. Sekine and K. Inaba, *Chem. Lett.*, 1983, 1669.
260. H. Kido, *Bull. Chem. Soc. Jpn.*, 1980, **53**, 82.
261. K. Saito and K. Masuda, *Bull. Chem. Soc. Jpn.*, 1968, **41**, 384.
262. K. Saito and K. Masuda, *Bull. Chem. Soc. Jpn.*, 1970, **43**, 119.
263. K. Saito and M. Takahashi, *Bull. Chem. Soc. Jpn.*, 1969, **42**, 3462.
264. A. Barabas, *Inorg. Nucl. Chem. Lett.*, 1970, **6**, 774.
265. A. C. Adams and E. M. Larsen, *Inorg. Chem.*, 1966, **5**, 814.
266. G. E. Glass and R. S. Tobias, *J. Organomet. Chem.*, 1968, **15**, 481.
267. G. M. Tanner, D. G. Tuck and E. J. Wells, *Can. J. Chem.*, 1972, **50**, 3950.
268. M. Schumann, A. von Holtum, K. J. Wannowius and H. Elias, *Inorg. Chem.*, 1982, **21**, 616.
269. A. R. Siedle, *J. Inorg. Nucl. Chem.*, 1973, **35**, 3429.
270. A. C. Adams and E. M. Larsen, *J. Am. Chem. Soc.*, 1963, **85**, 3508.
271. A. C. Adams and E. M. Larsen, *Inorg. Chem.*, 1966, **5**, 228.
272. F. A. Cotton, P. Legzdins and S. J. Lippard, *J. Chem. Phys.*, 1966, **45**, 3461.
273. N. Serpone and R. Ishayek, *Inorg. Chem.*, 1974, **13**, 52.
274. T. J. Pinnavaia and R. C. Fay, *Inorg. Chem.*, 1966, **5**, 233.
275. R. A. Palmer, R. C. Fay and T. S. Piper, *Inorg. Chem.*, 1964, **3**, 875.
276. J. J. Fortman and R. E. Sievers, *Inorg. Chem.*, 1967, **6**, 2022.
277. R. C. Fay and R. N. Lowry, *Inorg. Nucl. Chem. Lett.*, 1967, **3**, 117.
278. R. C. Fay and R. N. Lowry, *Inorg. Chem.*, 1970, **9**, 2048.
279. R. C. Fay and R. N. Lowry, *Inorg. Chem.*, 1974, **13**, 1309.
280. T. J. Pinnavaia, W. T. Collins and J. J. Howe, *J. Am. Chem. Soc.*, 1970, **92**, 4544.
281. T. J. Pinnavaia and J. A. McClarin, *J. Am. Chem. Soc.*, 1974, **96**, 3016.
282. J. J. Howe, W. T. Collins and T. J. Pinnavaia, *J. Am. Chem. Soc.*, 1969, **91**, 5378.
283. F. Basolo and R. G. Pearson, 'Mechanisms of Inorganic Reaction', 2nd. edn., Wiley, New York, 1967.

284. A. Y. Girgis and R. C. Fay, *J. Am. Chem. Soc.*, 1972, **92**, 7061.
285. J. G. Gordon, II and R. H. Holm, *J. Am. Chem. Soc.*, 1972, **92**, 5319.
286. G. W. Everett, Jr. and R. R. Horn, *J. Am. Chem. Soc.*, 1974, **96**, 2087.
287. J. R. Hutchinson, J. G. Gordon, II and R. H. Holm *Inorg. Chem.*, 1971, **10**, 1004.
288. J. G. Gordon, II, M. J. O'Connor and R. H. Holm, *Inorg. Chim. Acta*, 1971, **5**, 381.
289. C. Kutal and R. E. Sievers, *Inorg. Chem.*, 1974, **13**, 897.
290. M. Pickering, B. Jurado and C. S. Springer, Jr., *J. Am. Chem. Soc.*, 1976, **98**, 4503.
291. S. S. Eaton, J. R. Hutchinson, R. H. Holm and E. L. Muetterties, *J. Am. Chem. Soc.*, 1972, **94**, 6411.
292. S. S. Eaton, G. R. Eaton, R. H. Holm and E. L. Muetterties, *J. Am. Chem. Soc.*, 1973, **95**, 1116.
293. T. Inoue and K. Saito, *Bull. Chem. Soc. Jpn.*, 1973, **46**, 2417.
294. T. Inoue, J. Fujita and K. Saito, *Bull. Chem. Soc. Jpn.*, 1975, **48**, 1228.
295. A. Nagasawa and K. Saito, *Bull. Chem. Soc. Jpn.*, 1978, **51**, 2015.
296. T. Ito, K. Toriumi, F. Ueno and K. Saito, *Acta Crystallogr., Part B*, 1980, **36**, 2998.
297. F. Ueno, A. Nagasawa and K. Saito, *Inorg. Chem.*, 1981, **20**, 3504.
298. D. G. Bickley and N. Serpone, *Inorg. Chem.*, 1974, **13**, 2908.
299. R. W. Jones, Jr. and R. C. Fay, *Inorg. Chem.*, 1973, **12**, 2599.
300. N. Serpone and K. A. Hersh, *Inorg. Chem.*, 1974, **13**, 2901.
301. N. Serpone and R. C. Fay, *Inorg. Chem.*, 1967, **6**, 1835.
302. R. C. Fay and R. N. Lowry, *Inorg. Chem.*, 1967, **6**, 1512.
303. D. C. Bradley and C. E. Holloway, *J. Chem. Soc. (A)*, 1969, 282.
304. T. J. Pinnavaia, L. J. Matienzo and Y. Λ. Peters, *Inorg. Chem.*, 1970, **9**, 993.
305. H. A. Meinema and J. G. Noltes, *J. Organomet. Chem.*, 1969, **16**, 257.
306. R. B. VonDreele, J. J. Stezowski and R. C. Fay, *J. Am. Chem. Soc.*, 1971, **93**, 2887.
307. T. J. Pinnavaia and R. C. Fay, *Inorg. Chem.*, 1968, **7**, 502.
308. R. C. Fay and T. J. Pinnavaia, *Inorg. Chem.*, 1968, **7**, 508.
309. J. J. Stezowski and H. A. Eick, *J. Am. Chem. Soc.*, 1969, **91**, 2890.
310. T. J. Pinnavaia and A. L. Lott, II, *Inorg. Chem.*, 1971, **10**, 1388.
311. M. J. Frazer and W. E. Newton, *Inorg. Chem.*, 1971, **10**, 2137.
312. M. Elder, J. G. Evans and W. A. G. Graham, *J. Am. Chem. Soc.*, 1969, **91**, 1245.
313. M. Elder, *Inorg. Chem.*, 1969, **8**, 2103.
314. J. J. Howe and T. J. Pinnavaia, *J. Am. Chem. Soc.*, 1970, **92**, 7342.
315. T. J. Pinnavaia, J. J. Howe and R. E. Teets, *Inorg. Chem.*, 1974, **13**, 1704.
316. G. M. Kramer, M. B. Dines, R. B. Hall, A. Kaldor, A. J. Jacobsen and J. C. Scanlon, *Inorg. Chem.*, 1980, **19**, 1340.
317. J. H. Levy and A. B. Waugh, *J. Chem. Soc., Dalton Trans*, 1977, 1628.
318. J. C. Taylor and A. B. Waugh, *J. Chem. Soc., Dalton Trans.*, 1977, 1630.
319. J. C. Taylor and A. B. Waugh, *J. Chem. Soc., Dalton Trans.*, 1977, 1636.
320. G. M. Kramer, M. B. Dines, R. Kastrup, M. T. Melchior and E. T. Maas, Jr., *Inorg. Chem.*, 1981, **20**, 3.
321. E. T. Maas, Jr., G. M. Kramer and R. G. Bray, *J. Inorg. Nucl. Chem.*, 1981, **43**, 2053.
322. G. M. Kramer and E. T. Maas, Jr., *Inorg. Chem.*, 1981, **20**, 3514.
323. R. L. Lintvedt, in 'Concepts of Inorganic Photochemistry', ed. A. Adamson and P. Fleischauer, Wiley, New York, 1975, chap. 7.
324. C. Kutal, P. A. Grutsch and G. Ferraudi, *J. Am. Chem. Soc.*, 1979, **101**, 6884.
325. X. Yang and C. Kutal, *J. Am. Chem. Soc.*, 1983, **105**, 6038.
326. N. Filipescu and H. Way, *Inorg. Chem.*, 1969, **8**, 1863.
327. G. Ferraudi, P. A. Grutsch and C. Kutal, *Inorg. Chim. Acta*, 1982, **59**, 249.
328. C. Kutal, D. B. Yang and G. Ferraudi, *Inorg. Chem.*, 1980, **19**, 2907.
329. E. Zinato, P. Riccieri and P. S. Sheridan, *Inorg. Chem.*, 1979, **18**, 720.
330. A. J. Fry, R. S. H. Liu and G. S. Hammond, *J. Am. Chem. Soc.*, 1961, **88**, 4781.
331. F. Wilkinson and A. Farmilo, *J. Chem. Soc., Faraday Trans. 2*, 1976, 604.
332. G. Porter and M. R. Wright, *Discuss. Faraday Soc.*, 1959, **27**, 18.
333. J. A. Bell and H. Linschitz, *J. Am. Chem. Soc.*, 1963, **85**, 528.
334. Y. L. Chow and G. E. Buono-Cere, *Can. J. Chem.*, 1983, **61**, 795.
335. M Munakata, S. Nishibayashi and H. Sakamoto, *J. Chem. Soc., Chem. Commun.*, 1980, 219.
336. A. Kaldor, R. B. Hall, D. M. Cox, J. A. Horsley, P. Rabinowitz and G. M. Kramer, *J. Am. Chem. Soc.*, 1979, **101**, 4465.
337. D. M. Cox, R. B. Hall, J. A. Horsley, G. M. Kramer, P. Rabinowitz and A. Kaldor, *Science*, 1979, **205**, 390.
338. E. T. Maas, Jr., *US Pat.* 4 321 116 (1982).
339. R. B. Hall, *US Pat.* 4 324 766 (1982).
340. T. G. Dietz, M. A. Duncan, R. E. Smalley, D. M. Cox, J. A. Horsley and A. Kaldor, *J. Chem. Phys.*, 1982, **77**, 4417.
341. G. M. Kramer, M. B. Dines, A. Kaldor, R. Hall and D. McClure, *Inorg. Chem.*, 1981, **20**, 1421.
342. R. E. Sievers and J. E. Sadlowski, *Science*, 1978, **201**, 217.
343. M. F. Richardson and R. E. Sievers, *Inorg. Chem.*, 1971, **10**, 498.
344. K. J. Eisentraut and R. E. Sievers, *J. Am. Chem. Soc.*, 1965, **87**, 5254.
345. R. E. Sievers and K. J. Eisentraut, *J. Inorg. Nucl. Chem.*, 1967, **29**, 1931.
346. J. E. Sicre, J. J. Dubois, K. J. Eisentraut and R. E. Sievers, *J. Am. Chem. Soc.*, 1969, **91**, 3476.
347. J. C. A. Boeyens and J. P. R. deVillier, *Acta Crystallogr., Part B*, 1971, **27**, 2335.
348. R. W. Mosher and R. E. Sievers, 'Gas Chromatography of Metal Chelates', Pergamon, New York, 1965.
349. C. S. Springer, Jr., D. W. Meek and R. E. Sievers, *Inorg. Chem.*, 1967, **6**, 1105.
350. R. L. VanHemert, L. B. Spendlove and R. E. Sievers, *J. Electrochem. Soc.*, 1965, **112**, 1123.
351. R. W. Mosher, R. E. Sievers and L. B. Spendlove, *US Pat.* 3 356 527 (1967).
352. A. R. Siedle, *US Pat.* 4 424 352 (1984).
353. S. C. Chattoraj, A. G. Cupta, Jr. and R. E. Sievers, *J. Inorg. Nucl. Chem.*, 1966, **28**, 1937.
354. K. J. Eisentraut, R. L. Tischer and R. E. Sievers, *US Pat.* 3 794 473 (1974).

355. R. E. Sievers and T. J. Wensel, *US Pat.* 4 251 233 (1981).
356. G. S. Vigee and C. L. Watkins, *Inorg. Chem.*, 1977, **16**, 709.
357. A. Merijanian and H. M. Neumann, *Inorg. Chem.*, 1967, **6**, 165.
358. F. R. Clark, J. R. Sternbach and W. F. Wagner, *J. Inorg. Nucl. Chem.*, 1964, **26**, 1311.
359. T. S. Davis and J. P. Fackler, Jr., *Inorg. Chem.*, 1966, **5**, 242.
360. J. P. Fackler, Jr., T. S. Davis and I. D. Chawla, *Inorg. Chem.*, 1965, **4**, 130.
361. P. R. Singh and R. Sahai, *Aust. J. Chem.*, 1970, **23**, 269.
362. P. M. Cook, L. F. Dahl, D. Hopgood and R. A. Jenkins, *J. Chem. Soc., Dalton Trans.*, 1973, 294.
363. J. P. Fackler, Jr., *J. Am. Chem. Soc.*, 1962, **84**, 24.
364. J. P. Fackler, Jr., *Inorg. Chem.*, 1963, **2**, 266.
365. M. B. Hursthouse, M. A. Laffey, P. T. Moore, D. B. New, P. R. Raithby and P. Thornton, *J. Chem. Soc., Dalton Trans.*, 1982, 307.
366. M. A. Laffey and P. Thornton, *J. Chem. Soc., Dalton Trans.*, 1982, 313.
367. J. H. Binks, G. D. Dorward and G. P. McQuillan, *Inorg. Chim. Acta*, 1981, **49**, 251.
368. R. E. Sievers (ed.) 'Nuclear Magnetic Shift Reagents', Academic, New York, 1973.
369. L. A. Burgett and P. Warner, *J. Magn. Reson.*, 1972, **8**, 87.
370. J. E. Schwarberg, D. P. Gere, R. E. Sievers and K. J. Eisentraut, *Inorg. Chem.*, 1967, **6**, 1933.
371. J. Selbin, N. Ahmad and N. S. Bhacca, *Inorg. Chem.*, 1971, **10**, 1385.
372. M. S. Shameem and N. Ahmad, *J. Inorg. Nucl. Chem.*, 1975, **37**, 2009.
373. W. D. Horrocks, Jr., J. P. Sipe and J. R. Luber, *J. Am. Chem. Soc.*, 1971, **93**, 5258.
374. R. E. Cramer and K. Seff, *J. Chem. Soc., Chem. Commun.*, 1972, 400.
375. T. Phillips, II, D. E. Sands and W. F. Wagner, *Inorg. Chem.*, 1968, **7**, 2295.
376. H. M. McConnell and R. E. Robertson, *J. Chem. Phys.,*, 1957, **29**, 1361.
377. J. A. Cunningham and R. E. Sievers, *Inorg. Chem.*, 1980, **19**, 595.
378. J. A. Kemlo, J. D. Nielsen and T. M Shepherd, *Inorg. Chem.*, 1977, **16**, 1111.
379. N. B. O'Bryan, T. O. Maier, I. C. Paul and R. S. Drago, *J. Am. Chem. Soc.*, 1973, **95**, 1640.
380. D. J. Kitko, K. E. Wiegers, S. G. Smith and R. S. Drago, *J. Am. Chem. Soc.*, 1977, **99**, 1410.
381. D. E. Fenton, N. Bresciani-Pahor, M. Calligaris, G. Nardin and L. Randaccio, *J. Chem. Soc., Chem. Commun.*, 1979, 39.
382. K. A. Leslie, R. S. Drago, G. D. Stucky, D. J. Kitko and J. A. Breese, *Inorg. Chem.*, 1979, **18**, 1885.
383. C. J. O'Connor, D. Freyberg and E. Sinn, *Inorg. Chem.*, 1979, **18**, 1077.
384. L. Banci, A. Bencini and D. Gatteschi, *Inorg. Chem.*, 1981, **20**, 2734.
385. L. Banci, C. Benelli and D. Gatteschi, *Inorg. Chem.*, 1981, **20**, 4397.
386. L. Banci, A. Bencini and D. Gatteschi, *Inorg. Chem.*, 1982, **21**, 1572.
387. M. Calligaris, G. Manzini, G. Nardin and L. Randaccio, *J. Chem. Soc., Dalton Trans.*, 1972, 543.
388. W. R. May and M. M. Jones, *J. Inorg. Nucl. Chem.*, 1963, **25**, 507.
389. A. F. Garito and B. B. Wayland, *J. Am. Chem. Soc.*, 1969, **91**, 866.
390. D. P. Graddon and K. B. Heng, *Aust. J. Chem.*, 1971, **24**, 1781.
391. W. R. Walker and N. C. Li, *J. Inorg. Nucl. Chem.*, 1965, **27**, 225.
392. W. Partenheimer and R. S. Drago, *Inorg. Chem.*, 1970, **9**, 47.
393. D. P. Graddon and W. K. Ong, *Aust. J. Chem.*, 1974, **24**, 741.
394. R. C. E. Belford, D. E. Fenton and M. R. Truter, *J. Chem. Soc., Dalton Trans.*, 1973, 2208.
395. D. E. Fenton, R. S. Nyholm and M. R. Truter, *J. Chem. Soc. (A)*, 1971, 1577.
396. W. L. Kwik and K. P. Ang, *Aust. J. Chem.*, 1978, **31**, 459.
397. Y. Fukuda, A. Shimura, M. Mukaida and K. Sone, *J. Inorg. Nucl. Chem.*, 1974, **36**, 1265.
398. D. F. Colton and W. J. Geary, *J. Chem. Soc. (A)*, 1971, 2457.
399. M. F. Richardson and R. E. Sievers, *J. Inorg. Nucl. Chem.*, 1970, **32**, 1895.
400. N. A. Bailey, D. E. Fenton, M. V. Franklin and M. Hall, *J. Chem. Soc., Dalton Trans.*, 1980, 984.
401. M. F. Farona, D. C. Perry and H. A. Kuska, *Inorg. Chim. Acta*, 1973, **7**, 144.
402. L. F. Nicholas and W. R. Walker, *Aust. J. Chem.*, 1970, **23**, 1135.
403. M. F. Farona, D. C. Perry and H. A. Kuska, *Inorg. Chem.*, 1968, **7**, 2415.
404. J. A. Pretorius and J. C. A. Boeyens, *J. Inorg. Nucl. Chem.*, 1978, **40**, 407.
405. J. Pradilla-Sorzano and J. P. Fackler, Jr., *Inorg. Chem.*, 1973, **12**, 1174.
406. M. V. Veidis, G. H. Schreiber, T. E. Gough and G. J. Palenik, *J. Am. Chem. Soc.*, 1969, **91**, 1859.
407. F. Izumi, R. Kurosawa, H. Kawamoto and H. Akaiwa, *Bull. Chem. Soc. Jpn.*, 1975, **48**, 3188.
408. J. Pradilla-Sorzano and J. P. Fackler, Jr., *Inorg. Chem.*, 1973, **12**, 1182.
409. R. C. Elder, *Inorg. Chem.*, 1968, **7**, 1117.
410. R. C. Elder, *Inorg. Chem.*, 1968, **7**, 2316.
411. J. T. Hashagen and J. P. Fackler, Jr., *J. Am. Chem. Soc.*, 1965, **87**, 2821.
412. M. Gerloch, R. F. McMeeking and A. M. White, *J. Chem. Soc., Dalton Trans.*, 1976, 655.
413. R. W. Kluiber, R. Kukla and W. D. Horrocks, Jr., *Inorg. Chem.*, 1970, **9**, 1319.
414. D. E. Fenton, *J. Chem. Soc. (A)*, 1971, 3481.
415. J. J. R. F. DaSilva and R. Wootton, *Chem. Commun.*, 1969, 421.
416. M. R. Caira, J. M. Haigh and L. R. Nassimbeni, *J. Inorg. Nucl. Chem.*, 1972, **34**, 3171.
417. J. Laugier and J. P. Mathieu, *Acta Crystallogr., Part B*, 1975, **31**, 631.
418. J. P. Mathieu and J. Chappert, *Phys. Status Solidi*, 1976, **75**, 163.
419. C. Nicolini, J. Chappert and J. P. Mathieu, *Inorg. Chem.*, 1977, **16**, 3112.
420. D. P. Graddon and G. M. Mockler, *Aust. J. Chem.*, 1964, **17**, 1119.
421. Y. Nishikawa, Y. Nakamura and S. Kawaguchi, *Bull. Chem. Soc. Jpn.*, 1972, **45**, 155.
422. C. H. Bamford and A. N. Ferrar, *Chem. Commun.*, 1970, 315.
423. C. H. Bamford and A. N. Ferrar, *Proc. R. Soc. London, Ser. A*, 1971, **321**, 425.
424. Y. Nishikawa, Y. Nakamura and S. Kawaguchi, *Inorg. Nucl. Chem. Lett.*, 1972, **8**, 89.
425. G. R. Eaton and S. S. Eaton, *Coord. Chem. Rev.*, 1978, **26**, 207.

426. K. M. More, S. S. Eaton and G. R. Eaton, *J. Am. Chem. Soc.*, 1981, **103**, 1087.
427. B. M. Sawant, A. L. W. Schroyer and G. R. Eaton, *Inorg. Chem.*, 1982, **21**, 1093.
428. K. M. More, S. S. Eaton and G. R. Eaton, *Inorg. Chem.*, 1981, **20**, 2641.
429. B. M. Sawant, G. A. Braden, R. E. Smith, G. R. Eaton and S. S. Eaton, *Inorg. Chem.*, 1981, **20**, 3349.
430. Y. Y. Lim and R. S. Drago, *Inorg. Chem.*, 1972, **11**, 1334.
431. R. A. Zelonka and M. C. Baird, *J. Am. Chem. Soc.*, 1971, **93**, 6066.
432. R. S. Drago, T. C. Kuechler and M. K. Kroger, *Inorg. Chem.*, 1979, **18**, 2337.
433. O. P. Anderson and T. C. Kuechler, *Inorg. Chem.*, 1980, **19**, 1417.
434. M. H. Dickman and R. J. Doedens, *Inorg. Chem.*, 1981, **20**, 2677.
435. R. W. Kluiber and W. D. Horrocks, Jr., *J. Am. Chem. Soc.*, 1965, **87**, 5350.
436. R. W. Kluiber and W. D. Horrocks, Jr., *J. Am. Chem. Soc.*, 1966, **88**, 1399.
437. W. D. Horrocks, Jr., D. H. Templeton and A. Zalkin, *Inorg. Chem.*, 1968, **7**, 1552.
438. J. K. More, K. M. More, G. R. Eaton and S. S. Eaton, *Inorg. Chem.*, 1982, **21**, 2455.
439. A. R. Siedle, R. A. Newmark and L. H. Pignolet, *Inorg. Chem.*, 1983, **22**, 2281.
440. A. R. Siedle and L. H. Pignolet, *Inorg. Chem.*, 1981, **20**, 1849.
441. A. R. Siedle, R. A. Newmark and L. H. Pignolet, *J. Am. Chem. Soc.*, 1982, **104**, 6584.
442. S. Okeya, T. Miyamoto, S. Ooi, Y. Nakamura and S. Kawaguchi, *Inorg. Chim. Acta*, 1980, **45**, L135.
443. A. R. Siedle, R. A. Newmark, A. A. Kruger and L. H. Pignolet, *Inorg. Chem.*, 1981, **20**, 3399.
444. A. R. Siedle, P. M. Sperl and T. W. Rusch, *Appl. Surf. Sci.*, 1980, **6**, 149.
445. A. R. Siedle and R. A. Newmark, *J. Am. Chem. Soc.*, 1981, **103**, 1240.
446. A. R. Siedle, T. J. Kistenmacher, R. M. Metzger, C. S. Kuo, R. P. VanDuyne and T. Cape, *Inorg. Chem.*, 1980, **19**, 2048.
447. C. E. Pfluger and P. S. Haradem, *Inorg. Chim. Acta*, 1983, **64**, 141.
448. S. Okeya, S. Ooi, K. Matsumoto, Y. Nakamura and S. Kawaguchi, *Bull. Chem. Soc. Jpn.*, 1981, **54**, 1085.
449. S. Okeya, H. Sazaki, M. Ogita, T. Takemota, Y. Onuki, Y. Nakamura, B. K. Mohapatra and S. Kawaguchi, *Bull. Chem. Soc. Jpn.*, 1982, **54**, 483.
450. S. Okeya, H. Yoshimatsu, Y. Nakamura and S. Kawaguchi, *Bull. Chem. Soc. Jpn.*, 1982, **55**, 483.
451. S. Okeya, Y. Nakamura and S. Kawaguchi, *Bull. Chem. Soc. Jpn.*, 1982, **55**, 1460.
452. S. Okeya, Y. Nakamura and S. Kawaguchi, *Bull. Chem. Soc. Jpn.*, 1981, **54**, 3396.
453. S. Matsumoto and S. Kawaguchi, *Bull. Chem. Soc. Jpn.*, 1980, **53**, 1577.
454. S. Okeya, F. Egawa, Y. Nakamura and S. Kawaguchi, *Inorg. Chim. Acta*, 1978, **30**, L319.
455. T. Ito, T. Kiriyama and A. Yamamoto, *Chem. Lett.*, 1976, 835.
456. T. Ito and A. Yamamoto, *J. Organomet. Chem.*, 1978, **161**, 61.
457. S. Matsumoto and S. Kawaguchi, *Bull. Chem. Soc. Jpn.*, 1981, **54**, 1704.
458. S. Ooi, T. Matsushita, K. Nishimoto, S. Okeya, Y. Nakamura and S. Kawaguchi, *Bull. Chem. Soc. Jpn.*, 1983, **56**, 3297.
459. T. Ito and A. Yamamoto, *J. Organomet. Chem.*, 1979, **174**, 237.
460. Z. Kanda, Y. Nakamura and S. Kawaguchi, *Inorg. Chem.*, 1978, **17**, 910.
461. S. Okeya, N. Yanase, Y. Nakamura and S. Kawaguchi, *Chem. Lett.*, 1978, 699.
462. N. Yanase, Y. Nakamura and S. Kawaguchi, *Inorg. Chem.*, 1980, **19**, 1575.
463. Y. Otani, Y. Nakamura, S. Kawaguchi, S. Okeya and T. Hinomoto, *Chem. Lett.*, 1981, 11.
464. Y. Otani, Y. Nakamura, S. Kawaguchi, S. Okeya and T. Hinomoto, *Bull. Chem. Soc. Jpn.*, 1982, **55**, 1467.
465. S. Okeya, Y. Kawakita, S. Matsumoto, Y. Nakamura, S. Kawaguchi, N. Kanehisa, K. Miki and N. Kasai, *Bull. Chem. Soc. Jpn.*, 1982, **55**, 2134.
466. S. Okeya, Y. Nakamura, S. Kawaguchi and T. Hinomoto, *Bull. Chem. Soc. Jpn.*, 1982, **55**, 477.
467. T. Ito, T. Kiriyama, N. Nakamura and A. Yamamoto, *Bull. Chem. Soc. Jpn.*, 1976, **49**, 3257.
468. K. Saito and M. Takahashi, *Bull. Chem. Soc. Jpn.*, 1969, **42**, 3462.
469. T. Yamamoto, T. Saruyama, Y. Nakamura and A. Yamamoto, *Bull. Chem. Soc. Jpn.*, 1976, **49**, 589.
470. S. Okeya, Y. Nakamura and S. Kawaguchi, *J. Chem. Soc., Chem. Commun.*, 1977, 914.
471. T. Uchiyama, K. Takagi, K. Matsumoto, S. Ooi, Y. Nakamura and S. Kawaguchi, *Bull. Chem. Soc. Jpn.*, 1981, **54**, 1077.
472. R. H. Herber and A. E. Smelkinson, *Inorg. Chem.*, 1977, **16**, 953.
473. A. B. Cornwell and P. G. Harrison, *J. Chem. Soc., Dalton Trans.*, 1975, 1486.
474. A. B. Cornwell and P. G. Harrison, *J. Chem. Soc., Dalton Trans.*, 1975, 2017.
475. K. D. Bos, H. A. Budding, E. J. Butler and J. G. Noltes, *Inorg. Nucl. Chem. Lett.*, 1973, **9**, 961.
476. V. I. Bregadze, G. Z. Suleimanov, V. T. Kampel, M. V. Petriashvill, S. G. Mamedova, P. V. Petrovski and N. N. Godovikov, *J. Organomet. Chem.*, 1984, **263**, 131.
477. P. F. R. Ewings, P. G. Harrison and D. E. Fenton, *J. Chem. Soc., Dalton Trans.*, 1975, 821.
478. H. S. Booth and V. D. Smiley, *J. Phys. Chem.*, 1933, **37**, 171.
479. H. S. Booth and G. G. Torrey, *J. Phys. Chem.*, 1931, **35**, 2476.
480. D. Gibson, *Coord. Chem. Rev.*, 1969, **4**, 225.
481. D. H. Dewan, J. E. Furgusson, P. R. Hentschel, C. J. Wilkins and P. P. Williams, *J. Chem. Soc.*, 1964, 688.
482. R. H. Fish, *J. Am. Chem. Soc.*, 1974, **96**, 6664.
483. D. Gibson, B. F. G. Johnson and J. Lewis, *J. Chem. Soc. (A)*, 1970, 367.
484. S. Komiya and J. K. Kochi, *J. Am. Chem. Soc.*, 1977, **99**, 3695.
485. A. Werner, *Chem. Ber.*, 1901, **34**, 2584.
486. R. Mason, G. B. Robertson and P. J. Pauling, *J. Chem. Soc. (A)*, 1969, 485.
487. A. G. Swallow and M. R. Truter, *Proc. R. Soc. London, Ser. A*, 1962, **266**, 527.
488. A. Robson and M. R. Truter, *J. Chem. Soc.*, 1965, 630.
489. J. Lewis, R. F. Long and C. Oldham, *J. Chem. Soc.*, 1965, 6740.
490. D. Gibson, J. Lewis and C. Oldham, *J. Chem. Soc. (A)*, 1966, 1453.
491. G. T. Behnke and K. Nakamoto, *Inorg. Chem.*, 1967, **6**, 440.
492. J. Lewis and C. Oldham, *J. Chem. Soc. (A)*, 1966, 1456.

493. Y. Nakamura and K. Nakamoto, *Inorg. Chem.*, 1975, **14**, 63.
494. G. Hulley, B. F. G. Johnson and J. Lewis, *J. Chem. Soc (A)*, 1970, 1732.
495. B. F. G. Johnson, J. Lewis and M. S. Subramanian, *J. Chem. Soc. (A)*, 1968, 1993.
496. B. F. G. Johnson, T. Keating, J. Lewis, M. S. Subramanian and D. A. White, *J. Chem. Soc., Dalton Trans.*, 1969, 1793.
497. D. A. White, *Inorg. Synth.*, 1972, **13**, 55.
498. W. Rigby, H.-B. Lee, P. M. Bailey, J. A. McCleverty and P. M. Maitlis, *J. Chem. Soc., Dalton Trans.*, 1979, 387.
499. D. R. Russell and P. A. Tucker, *J. Chem. Soc., Dalton Trans.*, 1975, 1743.
500. D. R. Russell and P. A. Tucker, *J. Chem. Soc., Dalton Trans.*, 1975, 1749.
501. A. C. Jarvis and R. D. W. Kemmitt, *J. Organomet. Chem.*, 1977, **136**, 121.
502. D. M. Barlex, J. A. Evans, R. D. W. Kemmitt and D. R. Russell, *Chem. Commun.*, 1971, 331.
503. D. M. Barlex, A. C. Jarvis, R. D. W. Kemmitt and B. Y. Kimura, *J. Chem. Soc., Dalton Trans.*, 1972, 2549.
504. J. A. Evans, R. D. W. Kemmitt, B. Y. Kimura and D. R. Russell, *J. Chem. Soc., Chem. Commun.*, 1972, 509.
505. A. C. Jarvis, R. D. W. Kemmitt, D. R. Russell and P. A. Tucker, *J. Organomet. Chem.*, 1978, **159**, 341.
506. A. R. Siedle, *J. Organomet. Chem.*, 1981, **208**, 115.
507. M. C. Etter and A. R. Siedle, *J. Am. Chem. Soc.*, 1983, **105**, 641.
508. M. Cowie, M. Gauthier, S. J. Loeb and I. R. McKeer, *Organometallics*, 1983, **2**, 1057.
509. L. J. Farrugia, J. A. K. Howard, P. Mitrprachachon, F. G. A. Stone and P. Woodward, *J. Chem. Soc., Dalton Trans.*, 1981, 171.
510. P. M. Bailey, A. Keasey and P. M. Maitlis, *J. Chem. Soc., Dalton Trans.*, 1978, 1825.
511. A. Keasey and P. M. Maitlis, *J. Chem. Soc., Dalton Trans.*, 1978, 1830.
512. A. Sonoda, B. E. Mann and P. M. Maitlis, *J. Chem. Soc., Chem. Commun.*, 1975, 108.
513. K. Kishi, S. Ikeda and K. Hirota, *J. Phys. Chem.*, 1967, **71**, 4384.
514. K. Kishi and S. Ikeda, *J. Phys. Chem.*, 1969, **73**, 15.
515. J. S. H. Perera, D. C. Frost and G. A. McDowell, *J. Chem. Phys.*, 1980, **72**, 5151.
516. K. Kishi and S. Ikeda, *Appl. Surf. Sci.*, 1980, **5**, 37.
517. F. Sakurai, H. Suzuki, Y. Moro-oka and T. Ikawa, *J. Am. Chem. Soc.*, 1980, **102**, 1749.
518. P. J. Chung, H. Suzuki, Y. Moro-oka and T. Ikawa, *Chem. Lett.*, 1980, 63.
519. C. M. Lukehart, *Acc. Chem. Res.*, 1981, **14**, 109.
520. C. M. Lukehart, G. P. Torrence and J. V. Zeile, *Inorg. Synth.*, 1978, **18**, 56.
521. C. M. Lukehart, G. P. Torrence and J. V. Zaile, *J. Am. Chem. Soc.*, 1975, **97**, 6903.
522. C. M. Lukehart and G. P. Torrence, *Inorg. Chem.*, 1978, **17**, 253.
523. P. G. Lenhert, C. M. Lukehart and L. T. Warfield, *Inorg. Chem.*, 1980, **19**, 311.
524. J. W. Cook, A. R. Gibb, R. A. Raphael and A. R. Somerville, *J. Chem. Soc.*, 1951, 503.
525. R. Shiono, *Acta Crystallogr.*, 1961, **14**, 42.
526. E. L. Muetterties and C. M. Wright, *J. Am. Chem. Soc.*, 1964, **86**, 5132.
527. E. L. Muetterties and C. M. Wright, *J. Am. Chem. Soc.*, 1965, **87**, 4706.
528. E. L. Muetterties and C. M. Wright, *J. Am. Chem. Soc.*, 1966, **88**, 305.
529. E. L. Muetterties and C. W. Alegranti, *J. Am. Chem. Soc.*, 1969, **91**, 4420.
530. S. Degetto, G. Maragoni, G. Bombieri, E. Forsellini, L. Baracco and R. Gaziani, *J. Chem. Soc., Dalton Trans.*, 1974, 1933.
531. J. J. Park, D. M. Collins and J. L. Hoard, *J. Am. Chem. Soc.*, 1970, **92**, 3636.
532. V. W. Day and J. L. Hoard, *J. Am. Chem. Soc.*, 1970, **92**, 3626.
533. E. L. Muetterties and L. J. Guggenberger, *J. Am. Chem. Soc.*, 1972, **94**, 8046.
534. T. J. Anderson, M. A. Newman and G. A. Nelson, *Inorg. Chem.*, 1974, **13**, 158.
535. T. A. Hamor and D. J. Watkin, *Chem. Commun.*, 1969, 440.
536. A. Avdeef and J. P. Fackler, Jr., *Inorg. Chem.*, 1975, **14**, 2002.
537. A. Avdeef, J. A. Costamagna and J. P. Fackler, Jr., *Inorg. Chem.*, 1974, **13**, 1854.
538. R. J. Irving, M. L. Post and D. C. Povey, *J. Chem. Soc., Dalton Trans.*, 1973, 697.
539. C. G. Pierpont and R. M. Buchanan, *Coord. Chem. Rev.*, 1981, **38**, 45.
540. C. A. Wraight, *Photochem. Photobiol.*, 1979, **30**, 767.
541. K. N. Raymond, S. S. Isied, L. D. Brown, F. R. Fronczek and J. H. Niebert, *J. Am. Chem. Soc.*, 1976, **98**, 1767.
542. S. R. Sofen, D. C. Ware, S. R. Cooper and K. N. Raymond, *Inorg. Chem.*, 1979, **18**, 234.
543. C. G. Pierpont and H. H. Downs, *J. Am. Chem. Soc.*, 1976, **98**, 4834.
544. R. M. Buchanan, H. H. Downs, W. B. Shorthill, C. G. Pierpont, S. L. Kessel and D. N. Hendrickson, *J. Am. Chem. Soc.*, 1978, **100**, 1318.
545. R. M. Buchanan, S. L. Kessel, H. H. Downs, C. G. Pierpont and D. N. Hendrickson, *J. Am. Chem. Soc.*, 1978, **100**, 7894.
546. R. M. Buchanan, H. H. Downs, W. B. Shorthill, C. G. Pierpont, S. L. Kessel and D. N. Hendrickson, *J. Am. Chem. Soc.*, 1978, **100**, 4318.
547. H. H. Downs, R. M. Buchanan and C. G. Pierpont, *Inorg. Chem.*, 1979, **18**, 1736.
548. R. M. Buchanan and C. G. Pierpont, *J. Am. Chem. Soc.*, 1980, **102**, 4951.
549. M. W. Lynch, M. Valentine and D. N. Hendrickson, *J. Am. Chem. Soc.*, 1982, **104**, 6982.
550. R. M. Buchanan, B. J. Fitzgerald and C. G. Pierpont, *Inorg. Chem.*, 1979, **12**, 3439.
551. G. A. Bowmaker, P. D. W. Boyd and G. K. Campbell, *Inorg. Chem.*, 1982, **21**, 2403.
552. C. G. Pierpont and H. H. Downs, *J. Am. Chem. Soc.*, 1975, **97**, 2123.
553. C. G. Pierpont and R. M. Buchanan, *J. Am. Chem. Soc.*, 1975, **97**, 6450.
554. R. M. Buchanan and C. G. Pierpont, *Inorg. Chem.*1979, **18**, 1616.
555. M. B. Hursthouse, T. Fram, L. New, W. P. Griffith and A. J. Nielson, *Transition Met. Chem.*, 1978, **3**, 255.
556. S. R. Sofen, K. Abu-Dari, D. P. Freyberg and K. N. Raymond, *J. Am. Chem. Soc.*, 1978, **100**, 7882.
557. S. R. Sofen, S. R. Cooper and K. N. Raymond, *Inorg. Chem.*, 1979, **18**, 1611.
558. K. N. Raymond, G. E. Freeman and M. J. Kappel, *Inorg. Chim. Acta*, 1984, **94**, 193.
559. S. R. Cooper, Y. B. Yoh and K. N. Raymond, *J. Am. Chem. Soc.*, 1982, **104**, 5092.

560. B. A. Borgias, S. R. Cooper, Y. B. Koh and K. N. Raymond, *Inorg. Chem.*, 1984, **23**, 1009.
561. C. G. Pierpont and H. H. Downs, *Inorg. Chem.*, 1975, **15**, 343.
562. M. Ghedini, G. Dolcetti, B. Giovanitti and G. Denti, *Inorg. Chem.*, 1977, **16**, 1725.
563. Y. S. Sohn and A. L. Balch, *J. Am. Chem. Soc.*, 1972, **94**, 1144.
564. O. Gandolfi, G. Dolcetti, M. Ghedini and M. Cais, *Inorg. Chem.*, 1980, **19**, 1785.
565. M. Ghedini, G. Denti and G. Dolcetti, *Isr. J. Chem.*, 1977, **15**, 271.
566. J. A. Mondal, R. Bulls and D. M. Blake, *Inorg. Chem.*, 1982, **21**, 1668.
567. F. Sagara, H. Kobayashi and K. Ueno, *Bull. Chem. Soc. Jpn.*, 1972, **45**, 900.
568. C. W. Dudley, T. N. Huckerby and C. Oldham, *J. Chem. Soc. (A)*, 1970, 2605.
569. M. D. Glick and R. L. Lintvedt, *Prog. Inorg. Chem.*, 1976, **21**, 223.
570. U. Casellato and P. A. Vigato, *Coord. Chem. Rev.*, 1977, **23**, 31.
571. F. Sagara, H. Kobayashi and K. Ueno, *Bull. Chem. Soc. Jpn.*, 1968, **41**, 266.
572. Y. Taguchi, F. Sagara, H. Kobayashi and K. Ueno, *Bull. Chem. Soc. Jpn.*, 1970, **43**, 2470.
573. F. Sagara, H. Kobayashi and K. Ueno, *Bull. Chem. Soc. Jpn.*, 1973, **46**, 484.
574. D. E. Fenton, S. E. Gayda and P. Holmes, *Inorg. Chim. Acta*, 1977, **21**, 187.
575. F. Sagara, H. Kobayashi and K. Ueno, *Bull. Chem. Soc. Jpn.*, 1972, **45**, 794.
576. M. Kuszaj, B. Tomlonovic, D. P. Murtha, R. L. Lintvedt and M. D. Glick, *Inorg. Chem.*, 1973, **12**, 1297.
577. R. L. Lintvedt, L. L. Borer, D. P. Murtha, J. M. Kuszaj and M. D. Glick, *Inorg. Chem.*, 1974, **13**, 18.
578. A. B. Blake and L. R. Fraser, *J. Chem. Soc., Dalton Trans.*, 1974, 2554.
579. D. Baker, C. W. Dudley and C. Oldham, *J. Chem. Soc. (A)*, 1970, 2608.
580. M. J. Heeg, J. L. Mack, M. D. Glick and R. L. Lintvedt, *Inorg. Chem.*, 1981, **20**, 833.
581. D. P. Murtha and R. L. Lintvedt, *Inorg. Chem.*, 1970, **9**, 1532.
582. J. W. Guthrie, R. L. Lintvedt and M. D. Glick, *Inorg. Chem.*, 1980, **19**, 2949.
583. R. L. Lintvedt, M. D. Glick, B. K. Tomlonovic, D. P. Gavel and J. M. Kuszaj, *Inorg. Chem.*, 1976, **15**, 1633.
584. G. L. Long, D. L. Lindner, R. L. Lintvedt and J. W. Guthrie, *Inorg. Chem.*, 1982, **21**, 1431.
585. A. P. Ginsberg, R. L. Martin and R. C. Sherwood, *Inorg. Chem.*, 1968, **7**, 932.
586. P. Andrelczyk and R. L. Lintvedt, *J. Am. Chem. Soc.*, 1972, **94**, 8633.
587. L. L. Borer and W. Vanderbout, *Inorg. Chem.*, 1979, **18**, 526.
588. R. L. Lintvedt, M. J. Heeg, N. Ahmad and M. D. Glick, *Inorg. Chem.*, 1982, **21**, 2350.
589. R. L. Lintvedt and N. Ahmad, *Inorg. Chem.*, 1982, **21**, 2356.
590. R. L. Lintvedt, B. A. Schoenfelner, C. Ceccarelli and M. D. Glick, *Inorg. Chem.*, 1982, **21**, 2113.
591. D. E. Fenton and R. L. Lintvedt, *J. Am. Chem. Soc.*, 1978, **100**, 6367.
592. R. L. Lintvedt and L. S. Kramer, *Inorg. Chem.*, 1983, **22**, 796.
593. R. L. Lintvedt, G. Ranger and B. A. Schoenfelner, *Inorg. Chem.*, 1984, **23**, 688.
594. A. Kumar, B. P. Bachlas and J. C. Maire, *Polyhedron*, 1983, **2**, 907.
595. B. P. Bachlas, H. Sharma, J. C. Maire and J. J. Zuckerman, *Inorg. Chim. Acta*, 1983, **71**, 227.
596. B. Bogdanovic and M. Yus, *Angew. Chem., Int. Ed. Engl.*, 1979, **18**, 681.
597. D. E. Fenton, C. M. Regan, U. Castellato, P. A. Vigato and M. Vidali, *Inorg. Chim. Acta*, 1982, **58**, 83.
598. D. E. Fenton, J. R. Tate, U. Casellato, S. Tambrini, P. A. Vigato and M. Vidali, *Inorg. Chim. Acta*, 1984, **83**, 23.
599. U. Casellato, P. A. Vigato, D. E. Fenton and M. Vidali, *Chem. Soc. Rev.*, 1977, **8**, 199.
600. D. E. Fenton and S. Gayda, *J. Chem. Soc., Dalton Trans.*, 1977, 2109.
601. H. Adams, N. A. Bailey, D. E. Fenton, M. S. L. Gonzalez and C. A. Phillips, *J. Chem. Soc., Dalton Trans.*, 1983, 371.
602. R. L. Lintvedt, M. D. Glick, B. K. Tomlonovic and D. P. Gavel, *Inorg. Chem.*, 1976, **15**, 1646.
603. M. D. Glick, R. L. Lintvedt, D. P. Gavel and B. Tomlonovic, *Inorg. Chem.*, 1976, **15**, 1654.
604. M. D. Glick, R. L. Lintvedt, T. J. Anderson and J. L. Mack, *Inorg. Chem.*, 1976, **15**, 2258.
605. R. P. Kreh, A. E. Rodriguez and S. P. Fox, *J. Chem. Soc., Chem. Commun.*, 1984, 130.
606. T. B. Rauchfuss, S. R. Wilson and D. A. Wrobleski, *J. Am. Chem. Soc.*, 1981, **103**, 6769.
607. D. A. Wroblewski, S. R. Wilson and T. B. Rauchfuss, *Inorg. Chem.*, 1982, **21**, 2114.
608. D. A. Wroblewski and T. B. Rauchfuss, *J. Am. Chem. Soc.*, 1982, **104**, 2314.
609. S. Koda, H. Kuroya, Y. Nakamura and S. Kawaguchi, *Chem. Commun.*, 1971, 280.
610. S. Koda, S. Ooi, H. Kuroya, Y. Nakamura and S. Kawaguchi, *Chem. Commun.*, 1971, 1321.
611. P. W. N. M. VanLeeuwen, *Recl. Trav. Chim. Pays-Bas*, 1968, **37**, 396.
612. R. E. Cramer and M. A. Chudyk, *Inorg. Chem.*, 1973, **12**, 1193.
613. Y. Nakamura, K. Isobe, H. Morita, S. Yamazaki and S. Kawaguchi, *Inorg. Chem.*, 1972, **11**, 1573.
614. P. W. N. M. VanLeeuwen and A. P. Praat, *Inorg. Chim. Acta*, 1970, **4**, 101.
615. M. L. Larsen and F. W. Moore, *Inorg. Chem.*, 1966, **5**, 801.
616. M. C. Fredette and C. J. Lock, *Can. J. Chem.*, 1973, **51**, 1116.
617. M. C. Fredette and C. J. Lock, *Can. J. Chem.*, 1975, **53**, 2481.
618. J. M. Haigh and D. A. Thornton, *Inorg. Nucl. Chem. Lett.*, 1970, **6**, 231.
619. J. V. Silverton and J. L. Hoard, *Inorg. Chem.*, 1963, **2**, 243.
620. V. L. Wolf and H. Barnighausen, *Acta Crystallogr.*, 1960, **13**, 778.
621. R. A. Lalancette, M. Cefola, W. C. Hamilton and S. J. LaPlaca, *Inorg. Chem.*, 1967, **6**, 2127.
622. A. Zalkin, D. H. Templeton and D. G. Karraker, *Inorg. Chem.*, 1969, **8**, 2680.
623. W. D. Horrocks, Jr., J. P. Sipe, III and J. R. Luber, *J. Am. Chem. Soc.*, 1971, **93**, 5258.
624. S. J. S. Wasson, D. E. Sands and W. F. Wagner, *Inorg. Chem.*, 1973, **12**, 187.
625. J. C. A. Boeyens and J. P. R. deVilliers, *J. Cryst. Mol. Struct.*, 1971, **1**, 297.
626. C. S. Erasmus and J. C. A. Boeyens, *J. Cryst. Mol. Struct.*, 1971, **1**, 83.
627. J. P. R. deVilliers and J. C. A. Boeyens, *Acta Crystallogr., Part B*, 1971, **27**, 692.
628. J. A. Cunningham, D. E. Sands and W. F. Wagner, *Inorg. Chem.*, 1967, **6**, 499.
629. J. A. Cunningham, D. E. Sands, W. F. Wagner and M. F. Richardson, *Inorg. Chem.*, 1969, **8**, 22.
630. T. Phillips, II, D. E. Sands and W. F. Wagner, *Inorg. Chem.*, 1968, **7**, 2295.
631. S. J. Lippard, *Prog. Inorg. Chem.*, 1967, **8**, 124.
632. C. S. Reasmus and J. C. A. Boeyens, *Acta Crystallogr., Part B*, 1970, **26**, 1843.

633. F. A. Cotton and P. Legzdins, *Inorg. Chem.*, 1968, **7**, 1777.
634. M. J. Bennett, F. A. Cotton, P. Legzdins and S. J. Lippard, *Inorg. Chem.*, 1968, **7**, 1770.
635. J. C. A. Boeyens and J. P. R. deVilliers, *Acta Crystallogr., Part B*, 1972, **28**, 2335.
636. R. B. vonDreele, J. J. Stezkowski and R. C. Fay, *J. Am. Chem. Soc.*, 1971, **93**, 2887.
637. T. J. Pinnavaia, B. L. Barnett, G. Podolsky and A. Tulinsky, *J. Am. Chem. Soc.*, 1975, **97**, 2712.
638. T. W. Hambley, D. L. Keppert, C. L. Raston and A. H. White, *Aust. J. Chem.*, 1978, **31**, 2635.
639. M. Lenner and O. Lindqvest, *Acta. Crystallogr., Part B*, 1979, **35**, 600.

15.5

Oxyanions

BRIAN J. HATHAWAY
University College, Cork, Republic of Ireland

15.5.1 INTRODUCTION

Classical coordination chemistry has been developed in terms of the traditional ligands[1,2] such as water, ammonia and the chloride anion, all of which contain at least one lone pair of electrons which may be used in the coordination of the ligand to a metal by the formation of a dative covalent bond.

$$n\text{L:} \ + \ M^{m+} \ \rightarrow \ M(:L)_n^{m+}$$

With ligands like the chloride anion with more than one lone pair of electrons available for donation, the same chloride anion may bond to more than one metal atom to form a bridging ligand. In anhydrous $FeCl_3$ and $FeCl_2$ each chloride ion bridges two and three iron atoms, respectively, to form an $FeCl_6$ chromophore in each structure.

By definition,[4] all oxyanions, $[XO_n]^{m-}$, contain a central element X (generally a non-metal, such as sulfur in a sulfate anion, $[SO_4]^{2-}$, but X may also be a metal,[5] such as chromium in the chromate anion, $[CrO_4]^{2-}$) and a number of terminal oxygen atoms (Table 1). Each terminal oxygen atom involves up to three lone pairs of electrons and consequently each oxyanion $[XO_n]^{m-}$ may be involved in coordination to n different metal atoms if each terminal oxygen atom coordinates to only one metal atom. If more than one lone pair on each terminal oxygen atom is used in coordination

Table 1 Polyatomic Oxyanion Species, their Shape, Point Group Symmetry and Representative Examples that are Known to Act as Ligands

Structural type	Shape	Symmetry	Examples
A. Mononuclear			
(i) XO_2	bent	C_{2v}	$[NO_2]^-$, $[ClO_2]^-$, $[H_2PO_2]^-$
(ii) XO_3	planar	D_{3h}	$[NO_3]^-$, $[BO_3]^{3-}$
	pyramidal	C_{3v}	$[HPO_3]^{2-}$, $[SO_3]^{2-}$, $[IO_3]^-$
(iii) XO_4	tetrahedral	T_d	$[SiO_4]^{4-}$, $[PO_4]^{3-}$, $[AsO_4]^{3-}$, $[SO_4]^{2-}$, $[SeO_4]^{2-}$, $[ClO_4]^-$, $[VO_4]^{3-}$, $[CrO]^{2-}$, $[MoO_4]^{2-}$, $[ReO_4]^-$
(iv) XO_6	octahedral	O_h	$[TeO_6]^{6-}$, $[IO_6]^{5-}$
B. Polynuclear			
X_2O_n	—	—	$[S_2O_3]^{2-}$, $[S_2O_8]^{2-}$, $[P_2O_7]^{4-}$, $[Cr_2O_7]^{2-}$

(maximum three), oxyanion coordination to $3n$ different metal atoms may occur up to a maximum of 18. This number is referred to as the *coordination number* (1–18) of the oxyanion. In general, the oxyanions available (Table 1)[5] are restricted, in mononuclear oxyanions, to two, three, four or six oxygen atoms and negative charges of 1–6, which together limit the coordination numbers of the oxyanions to values of less than three or four. In oxyanions $[XO_n]^{m-}$, where $n = 2$ or 3, such as in the $[NO_2]^{6-}$ or $[SO_3]^{2-7}$ anions, the X atoms may also have a lone pair of electrons and hence a potential ligand function and the ligand role of the oxygen atoms will depend upon the relative donor properties of the X and O atom lone pairs of electrons.

Three factors constrain the role of oxyanions as ligands: (a) their size (Table 2);[8] (b) their relatively rigid geometry (trigonal planar, pyramidal, tetrahedral or octahedral);[5] and (c) their weak donor properties[9–11] relative to nitrogen donors and especially relative to that of water.

Table 2 The Thermochemical Radii[8] (Å), and Mean X—O Distances (Å) in Parentheses, of Some Polyatomic Anions, XO_n^{m-}

XO_2		XO_3		XO_4		XO_4	
$[NO_2]^-$	1.55 (1.25)	$[NO_3]^-$	1.89 (1.25)	$[SiO_4]^{4-}$	2.40 (1.63)	$[VO_4]^{3-}$	2.46 (1.75)
$[ClO_2]^-$	— (1.58)	$[BO_3]^{3-}$	1.91 (1.38)	$[PO_4]^{3-}$	2.38 (1.52)	$[CrO_4]^{2-}$	2.40 (1.63)
$[H_2PO_2]^-$	— (1.49)	$[PO_3]^{3-}$	— (1.53)	$[AsO_4]^{3-}$	2.48 (1.76)	$[MoO_4]^-$	2.54 (1.74)
		$[SO_3]^{2-}$	— (1.51)	$[SO_4]^{2-}$	2.30 (1.48)	$[ReO_4]^-$	— (1.74)
		$[ClO_3]^-$	2.00 (—)	$[SeO_4]^{2-}$	2.43 (1.52)	$[MnO_4]^-$	2.40 (1.59)
		$[IO_3]^-$	1.82 (1.81)	$[ClO_4]^-$	2.36 (1.44)		

The size (a) of oxyanions is not too serious a factor as this is only slightly larger than that of the heavier halide ions (Table 2). Their rigid geometry restricts their function as bidentate chelate ligands in mononuclear oxyanions as this must lead to a rather strained four-membered chelate ring, which can only be relieved by one or two long bond distances (Figure 1).[12,13] Consequently, bidentate (and higher) functions, for mononuclear oxyanions, generally involve bridging to separate metal atoms to give coordination numbers of two and upwards. The relatively weak donor properties of the oxyanions towards metals relative to nitrogen ligands, and particularly water, make the preparation of oxyanion complexes very solvent dependent. In aqueous solution the perchlorate anion is generally considered[9,10] to be a non-associating or non-coordinating anion, while the use of non-aqueous solvents, such as $HClO_4$[14,15] or Cl_2O_6,[16] or a fused salt like $NOClO_4$[17,18] or NO_2ClO_4,[15] yields $[Cu(ClO_4)_2(OH_2)]$,[14] $[Fe(ClO_4)_3]$,[16] $[Co(ClO_4)_2]$,[19] $[Ni(ClO_4)_2]$[20] and $[Cu(ClO_4)_2]$,[17,18] and $[ClO_2][Cu(ClO_4)_3]$ complexes,[15] respectively, involving clear anion coordination to the metal. The nitrate ion is a slightly stronger ligand[9,12] in water than the perchlorate anion, and is known to form nitrato complexes even in aqueous solution.[12,13,21] In a non-aqueous solution such as anhydrous nitric acid,[11,14] N_2O_4[11,22,23] or N_2O_5,[24] metal nitrate dihydrates such as $[Fe(NO_3)_2(OH_2)_2]$[14] and anhydrous metal nitrates[22] (for example, $[Cu(NO_3)_2]$,[25] $[Co(NO_3)_3]$[26] and $[Ti(NO_3)_4]$[27]) are formed, respectively. As a number of these anhydrous oxyacid salts are volatile under reduced pressure,[16,17,28] there can be little doubt about the covalency of the metal–oxygen bond. As the electron diffraction structure of anhydrous $[Cu(NO_3)_2]$ in the gaseous state[29] involves a rhombic coplanar CuO_4 chromophore with a normal[30] Cu—O distance of ca. 2.0 Å, it is now clear that the oxyanions may function as ligands to transition metal ions, especially in the absence of traditional ligands such as water.

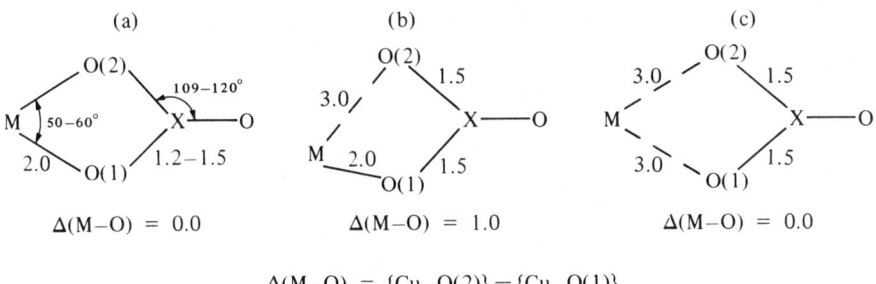

$$\Delta(M–O) = \{Cu–O(2)\} - \{Cu–O(1)\}$$

Figure 1 Bidentate chelate oxyanions: (a) symmetrical bidentate, (b) asymmetrical bidentate and (c) long bonded symmetrical

15.5.2 CLASSIFICATION OF OXYANIONS AS LIGANDS

The oxyanions as ligands may be classified according to: (a) the structural type of the oxyanion (XO_2, XO_3, XO_4 or XO_6); (b) the coordination number of the oxyanion (1–18), *i.e.* the number of metal atoms to which a single oxyanion may be coordinated; (c) the mode of coordination of the oxyanion, *i.e.* monodentate, bidentate, tridentate, *etc.*; and (d) the number of oxyanions per metal atom, the stoichiometry p, from one to six, *i.e.* [$M(XO_n)_p$]. Table 1 lists the oxyanions that will be considered in this section according to their structural types, with their approximate stereochemistry and point group symmetry. The carbon-containing oxyanions will be described in Chapter 15.6, and the cyanates in Chapter 13.5. For reasons of space this review will be primarily restricted to mononuclear oxyanions. Figures 2–5 illustrate the mode of coordination of the oxyanions as a function of their coordination number 1–18.

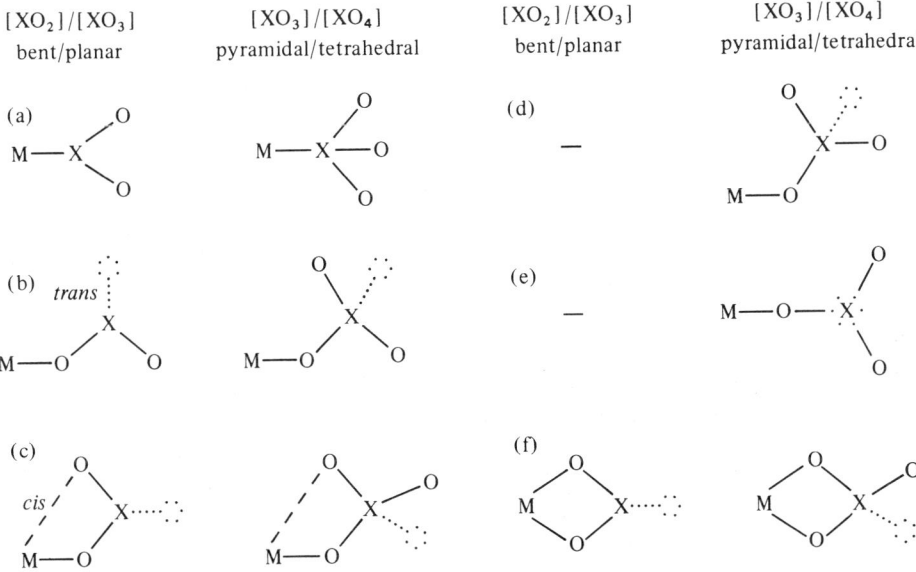

Figure 2 One metal atom. The notation $\cdots$ refers to a lone pair of electrons or to a normal —O unit

Within each coordination number the oxyanion may function as a monodentate, a bidentate or, very occasionally, a tridentate ligand to an individual metal cation. With the higher anion coordination numbers the M$\cdots$O distances to different metal atoms may not be identical. In practice, coordination numbers of 1 and 2 predominate and are independent of the three main structural types XO_2, XO_3 and XO_4. The coordination number 3 is essentially confined to tetrahedral XO_4 anions, and the much less numerous coordination numbers of 4–12 also involve mainly tetrahedral XO_4 type anions, particularly in their anhydrous oxyacid salts $M(XO_n)_q$ or double salts $M'_rM(XO_n)_{q+r}$. The stoichiometry number, p, is very often a funciton of the oxyanion/metal ratio of the preparative conditions. The higher p, the lower the coordination number of the oxyanion and the more the bonding is likely to involve a monodentate rather than a bidentate function. Nevertheless, the latter is very little influenced by the stoichiometry p. This is illustrated for the bidentate nitrate ion in the six structures (1)–(6),[29,31–34] in which p increases from one to six and the bonding role of the nitrato group is essentially unchanged.

(1) [$Cu(Ph_3P)_2(O_2NO)$][31]

(2) Gaseous [$Cu(O_2NO)_2$][29]

(3) [$Co(O_2NO)_3$][26]

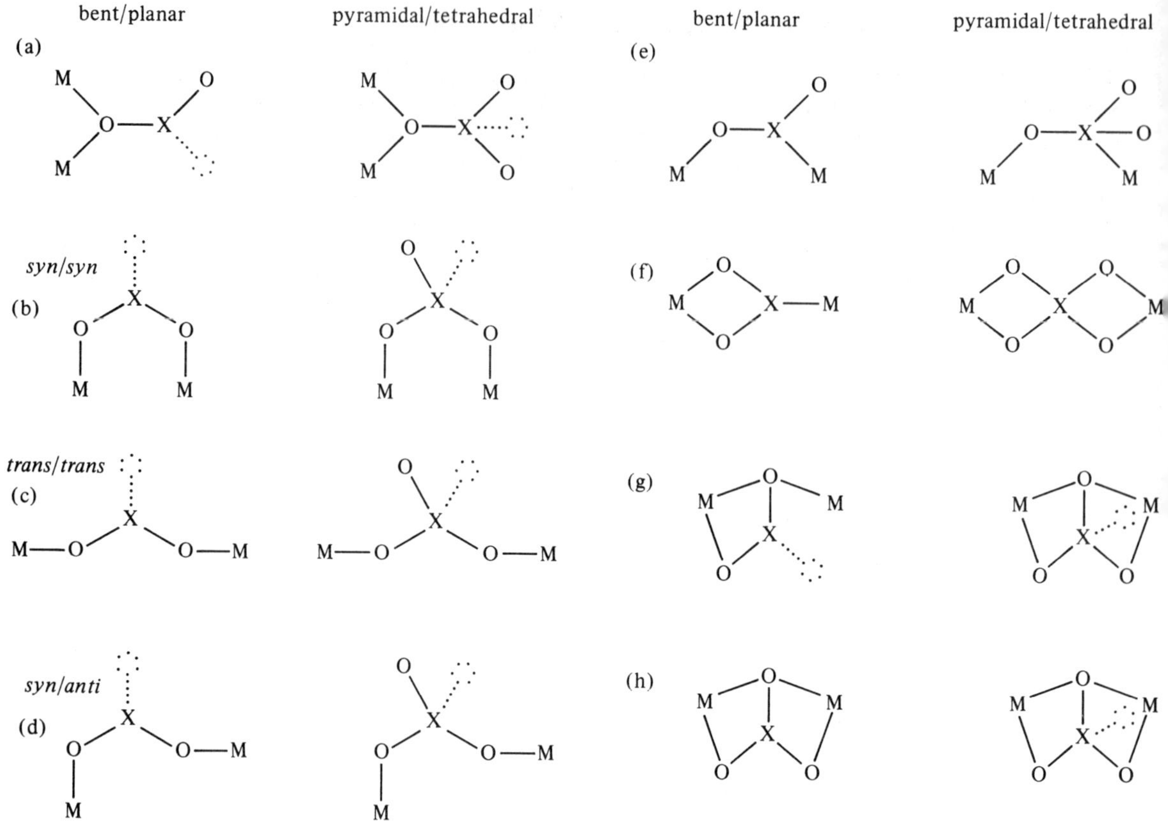

bent/planar   pyramidal/tetrahedral   bent/planar   pyramidal/tetrahedral

(a)   (e)

(b) *syn/syn*   (f)

(c) *trans/trans*   (g)

(d) *syn/anti*   (h)

Figure 3 Two metal atoms

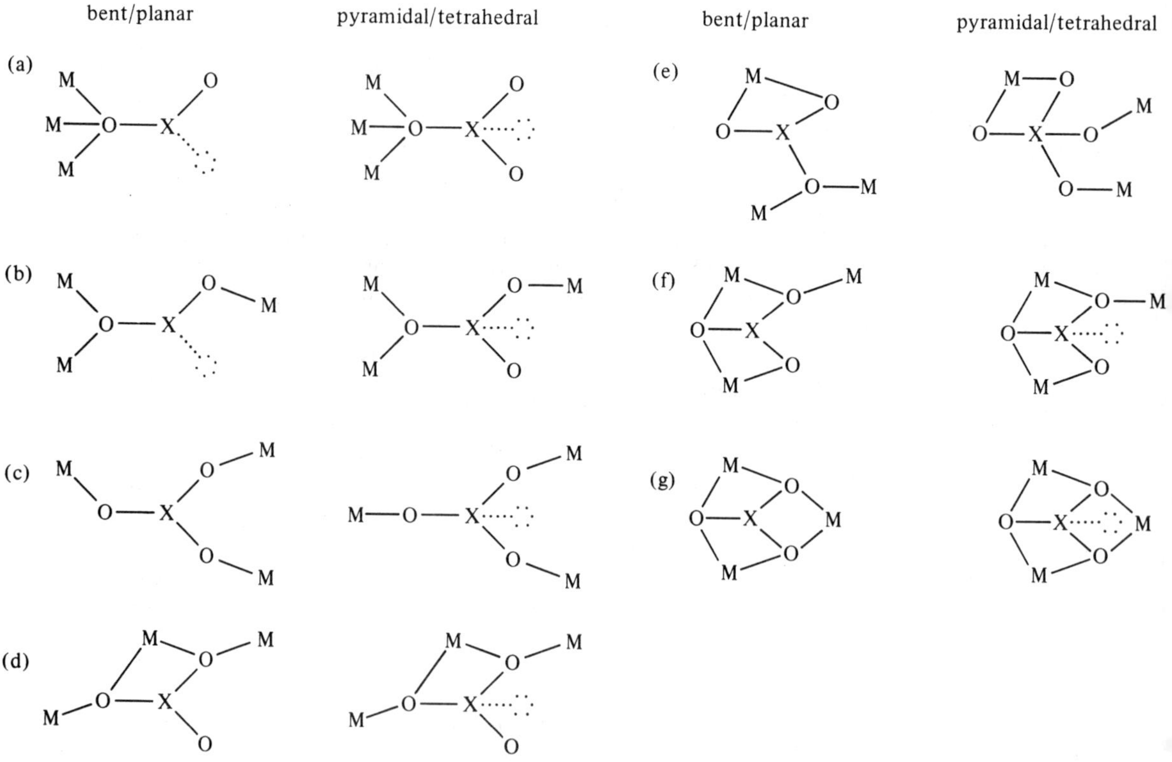

bent/planar   pyramidal/tetrahedral   bent/planar   pyramidal/tetrahedral

(a)   (e)

(b)   (f)

(c)   (g)

(d)

Figure 4 Three metal atoms

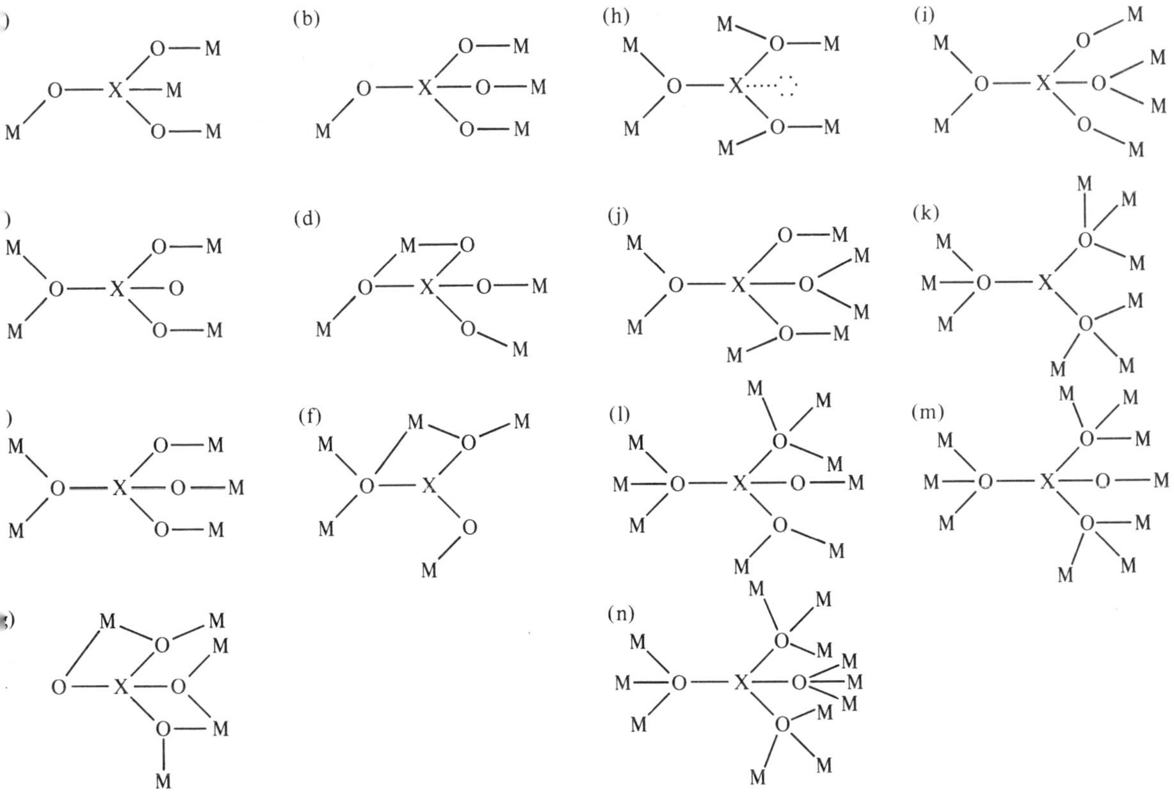

Figure 5 Four to twelve metal atoms

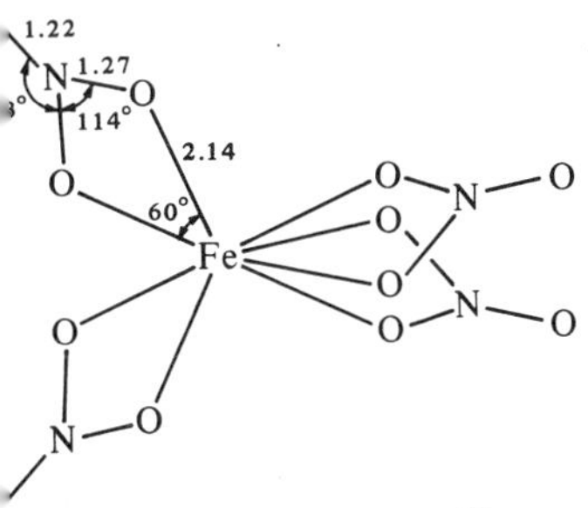

(4) [Ph₄As][Fe(O₂NO)₄] [32]

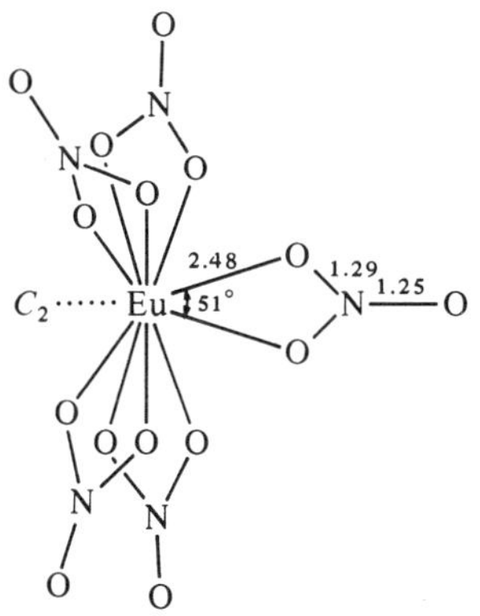

(5) [Ph₄As]₂[Eu(O₂NO)₅] [33]

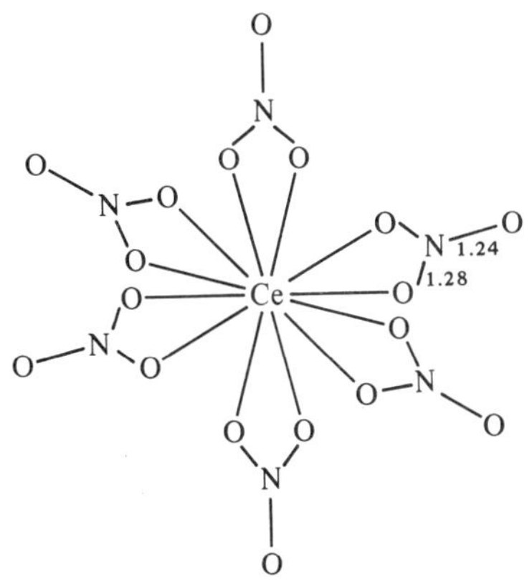

(6) [NH₄]₂[Ce(O₂NO)₆] [34]

Consequently, in describing the function of oxyanions as ligands, the stoichiometry factor p is less important than the coordination number 1–12, and for this reason the following pages classify oxyanions as ligands in terms of their coordination number. The number of oxyanions per metal atoms, p, shows a significant variation with structural type, $[XO_{2-4}]$ (Table 3). All four main structural types form stoichiometries of one to three, but while the XO_2 oxyanions form the additional stoichiometries of four, five and six with a pronounced maximum at six, these stoichiometries are less common for the XO_3 type oxyanions and are even less common for the XO_4 oxyanions. While the charge (m) on the anion influences p, in the sense that the higher the value of m the lower is the value of p, the size of the oxyanion must also be important (Table 2) and the larger bulk size of the XO_4 type oxyanions may also account for the absence of stoichiometries (p) greater than four. It is probably this relatively large three-dimensional size of XO_4 anions that is responsible for their tendency to form the high coordination numbers, 3, involving bridging oxyanions.

Only a brief mention will be made of the role of polynuclear oxyanions as ligands $[X_qO_n]^{m-}$ as these tend to involve infinite rigid lattices (one-, two- or three-dimensional) and thus lose their ability to act as independent ligands. In these rigid oxyanion structures the oxygens still act as ligands, as

Table 3 The Number, p, of Oxyanions (XO_n) per Metal Atom M as a Function of the Oxyanion Structural Type

XO_n	Shape/donor	$p=1$ $M(XO_n)_1$	$p=2$ $M(XO_n)_2$	$p=3$ $M(XO_n)_3$	$p=4$ $M(XO_n)_4$	$p=5$ $M(XO_n)_5$	$p=6$ $M(XO_n)_6$
XO_2	bent						
	X—M	$[Co(NH_3)_5(NO_2)]Br_2$[35]	$[Ni(NH_3)_4(NO_2)_2]$[36]	$[Co(NH_3)(en)(NO_2)_3]$[37]	$[Co(en)_2(NO_2)_2]$- $[Co(en)(NO_2)_4]\cdot H_2O$[38]	—	$K_2[PbNi(NO_2)_6]$[39]
	O—M	$[Co(NH_3)_5(ONO)]Cl_2$[40]	$[Ni(N,N'\text{-dimen})_2(ONO)_2]$[41]	—	$K_2[Cd(ONO)_4]$[42]	—	—
XO_3	planar						
	O—M	$[Cu(Ph_3P)_2(O_2NO)]$[31]	$[Cu(O_2NO_2(MeCN)_2]$[43]	$[Co(O_2NO)_3]$[26]	$[Sn(O_2NO)_4]$[44]	$[Ph_4As]_2[Eu(O_2NO)_5]$[33]	$[NH_4]_2[Ce(O_2NO)_6]$[34]
XO_3	pyramidal						
	X—M	$[Co(NH_3)_5(SO_3)]Cl$[45]	$Na[Co(en)_2(SO_3)_2]\cdot 3H_2O$[46]	—	$Na_6[Pd(SO_3)_4]\cdot 2H_2O$[47]	—	—
	O—M	α-$[Fe(O_2SO)(OH_2)_3]$[48]	—	—	$[Zr(O_2ClO_2)_4]$[51b]	—	$[NH_4]_9[Fe(OSO_2)_6]$[49]
XO_4	tetrahedral						
	O—M	$[Cu(O_4S)]$[50]	$[Zr(O_3SO)_2]$[51a]	$[NH_4]_3[In(O_2SO_2)_3]$[52]	$Na[Gd(OReO_3)_4(H_2O)_4]$[53]	—	—
XO_6	octahedral						
	O—M	$K[Ni(O_6I)]$[54]	$Na_3KH_3[Cu(O_2IO_4)_2]\cdot 14H_2O$[55]	—	—	—	—

in the ion exchange properties[56] of the $[Zr(SO_4)_2]$ structures,[51] which take up $[M(OH_2)_6]^{m+}$ cations and when dehydrated must involve only restricted coordination of the metal to the oxygens of the rigid $[Zr(SO_4)_2]$ lattice. For this reason this review will restrict discussion to q values of one, with only a brief mention of q values of two.

In listing the classification of structural types of XO_n anions (Figures 2–5), a crucial question is the length of a metal–oxygen distance that constitutes a normal metal–oxygen coordinate bond. For the first-row transition metals a normal short M—O distance[5] lies in the range 1.9 2.2 Å. For the heavier transition metals and main group metals slightly longer distances are appropriate, up to *ca.* 2.7 Å, and are listed in Table 4 based upon the metal–oxygen distance observed in covalent metal–oxygen systems.[5] In addition, for many oxyanions a second considerably longer M—O distance may be present, which arises from the conformation of an oxyanion with respect to a metal atom which is already involved in a normal short M—O distance (Figure 1b). It has already been suggested[12] for asymmetric bonding of the nitrate ion (Figure 1b) that if $\Delta(M–O) = \{M—O(2)\} - \{M—O(1)\}$ is less than 0.7 Å, the M—O(2) distance involves weak but definite covalent bonding. If this criterion is extended to all oxyanion ligands and all metals, the range of long M—O distances of Table 4 can be suggested. In the particular case of the copper(II) ion,[13,30] these long M—O distances in the range 2.3–2.8 Å arise from the non-spherical symmetry[57,30] of the d^9 configuration, and result in definite bonding of the oxyanion even at these long distances. The term 'semi-coordination' has been introduced[58] to describe this situation. In Figures 2–5 no attempt will be made to distinguish these short and long M—O distances, but in the M—O distances of the molecular structures (1)–(74) normal short distances are represented as a full line (—), and the longer distances as a dashed line (– – –). This notation is also used in the three structures of Figure 1. In an attempt to indicate the bonding role of the oxygen ligands, the oxyanions are written with the number of short bonded oxygen atoms in front of the central atom X; thus the monodentate, bidentate and tridentate perchlorate groups are represented $(OClO_3)$, (O_2ClO_2) and (O_3ClO), respectively.

Table 4 The Range of Normal Short and Long Metal–Oxygen Distances for the Transition Metal and Main Group Elements

Elements	Short	Long
Ti–Zn–Ge	1.90–2.2	2.5–3.0
Zr–Cd–Sb	2.0–2.4	2.7–3.2
La, Hf–Hg–Bi	2.2–2.6	2.8–3.5

15.5.3 STRUCTURAL DATA ON COORDINATED OXYANIONS

Information on the coordination of oxyanions to metals can be obtained from a number of sources:

(i) Single-crystal X-ray structure determination

Single-crystal X-ray structure determination is the most definitive source of information on the coordination of oxyanions in the solid state, and is rapidly accumulating such a substantial body of reliable information on the oxyanion shape, X—O distances, and O—X—O and M—O—X angles that it currently exceeds all other sources put together. The most important information is that of the oxygen–metal distance, which is usually in the range 2.0–3.0 Å (Table 4). The lower range (2.0–2.2 Å) represents strong covalent bonding and the higher range (2.7–3.0 Å) represents weak covalent bonding (or semi-coordination,[58] in the case of the copper(II) ion). For a particular metal it is not yet clear what distance within the range 2.0–3.0 Å represents the cut-off point for significant covalent bonding (Table 4). X-Ray crystallography provides information on the M—O—X angle that is not available from any other physical technique.

In general, no oxyanion undergoes a major change of internal shape upon coordination to a metal, but small differences in the X—O distances and O—X—O angles do occur.[6,12,59] As most oxyanions involve some degree of X—O double bonding, the X—O* distance to the coordinated oxygen atom should increase slightly upon coordination as the X—O distance reverts to a single bond. Correspondingly, the X—O distances of the non-coordinated oxygen atoms should decrease slightly as these X—O distances assume more double bond character. Unfortunately, the maximum change in X—O distance upon coordination is rarely as large as 0.1 Å for X—O distances of 1.2–1.8 Å (Table 2), and the differences observed in the X—O distances (MO_n^{m-}) are frequently less than 3σ (where σ is the estimated standard deviation of the X—O distance) and are therefore not significant. Equally the changes in the O—X—O bond angles upon coordination, $\pm 3°$, are only small and hardly significant, except in bidentate chelate coordination when changes of 10° may occur.

(ii) *Infrared and Raman spectroscopy*

The modes of vibration of polyatomic anions[60] are primarily determined by the number of atoms present and the symmetry of the free ions (Table 1), and the IR frequencies of these high-symmetry free ions are well characterized (Figure 6a).

Upon coordination of an oxyanion to a metal the symmetry of the anion may change,[6,12,18,60–64] which may result in the following: (a) degenerate modes of vibration of the free ion may be resolved

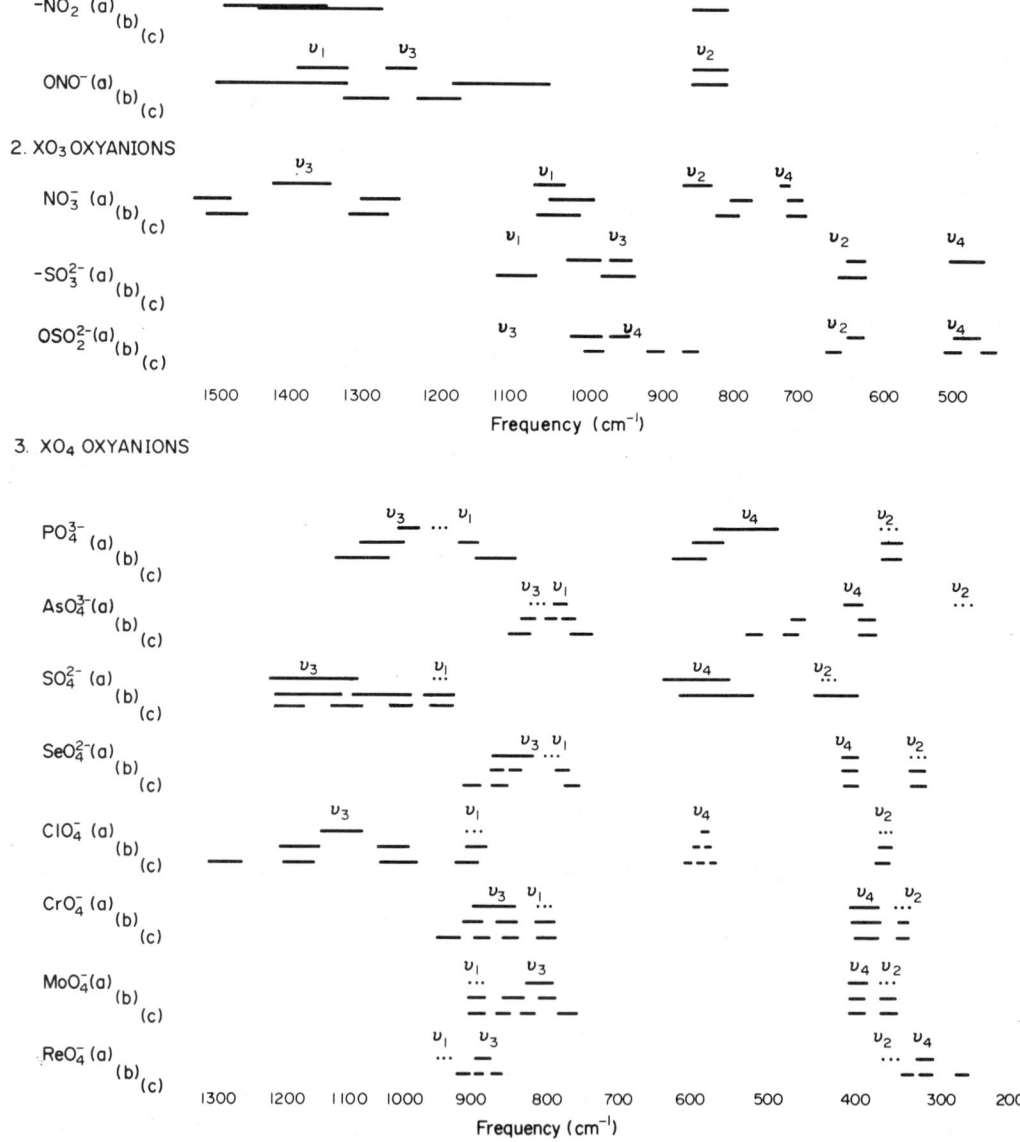

Figure 6 IR frequencies of XO_2, XO_3 and XO_4 type oxyanions: (a) free-ion, (b) monodentate and (c) bidentate; —, IR allowed; $\cdots$, IR forbidden

into nondegenerate modes; (b) the selection rules associated with the modes of vibration may change, and forbidden modes become allowed in both the IR and Raman spectra;[60,65] and (c) due to changes in the X—O bond orders, and hence in the force constants, a change in band frequency may occur. Taken together, these three effects may change the number of bands observed, their frequency and the relative intensities of the IR vibrations, and will also affect the polarization ratios of bands in the Raman spectra.[65] Table 5 sets out these changes in the selection rules for the XO_2^-, the XO_3^{n-} and the XO_4^{n-} anions, respectively; Figure 6 illustrates the range of frequencies of the free ions and of the coordinated oxyanions; and Figure 7 illustrates the changes in the IR spectrum of the perchlorate anion upon coordination. Historically, this 'infrared criterion' for the coordination of polyatomic oxyanions was first demonstrated for the tetrahedral sulfate ion,[61] the nitrate ion,[62] perchlorate ion,[18] sulfite[63] ion and nitrite ion.[64] These criteria were relatively successful (Figure 6) in distinguishing the free ion from the coordinated anion, but only partially successful in distinguishing the mono- and bi-dentate coordination or in distinguishing a bidentate chelate coordination from a bridging function. Consequently, IR spectroscopy is of little diagnostic value for identifying the different coordination numbers of coordinated oxyanions as revealed by X-ray crystallography. Two excellent reviews summarize this situation for the coordinated nitrate[12] and nitrite ion.[6] Both demonstrate that it must be X-ray crystallographic studies that give the most reliable information on the coordinating role of these polyanions as ligands, but both limit the discussion mainly to the coordination numbers of one and two.

Table 5 Correlation of the Modes of Vibration of the (a) XO_2, (b) XO_3 and (c) XO_4 oxyanions, their Infrared (I) and Raman (R) Selection Modes as a Function of the Idealized Symmetry from the Free Anion to the Coordinated Anion

		v_1	v_2	v_3	v_4
(a) XO_2 Bent-ionic					
XO_2	C_{2v}	A_1(I, R)	A_1(I, R)	B_1(I, R)	
*XO_2	C_{2v}	A_1(I, R)	A_1(I, R)	B_1(I, R)	
O*XO	C_s	A'(I, R)	A'(I, R)	A'(I, R)	
(b) XO_3 (i) Pyramidal-ionic					
XO_3	C_{3v}	A_1(I, R)	A_1(I, R)	E(I, R)	E(I, R)
O*XO_2	C_s	A'(I, R)	A'(I, R)	A'(I, R) + A''(I, R)	A''(I, R) + A''(I, R)
XO_3 (ii) Planar-ionic					
XO_3	D_{3h}	A_1'(R)	A_2''(I)	E'(I, R)	E'(I, R)
O*XO_2	C_{2v}	A_1(I, R)	B_1(I, R)	A_1(I, R) + B_2(I, R)	A_1(I, R) + B_2(I, R)
O$_2^*$XO	C_s	A(I, R)	A''(I, R)	A'(I, R) + A'(I, R)	A'(I, R) + A'(I, R)
(c) XO_4 Tetrahedral-ionic					
XO_4	T_d	A_1(R)	E(R)	T_2(I, R)	T_2(I, R)
O*XO_3	C_{3v}	A_1(I, R)	E(I, R)	A_1(I, R) + E(I, R)	A_1(I, R) + E(I, R)
O$_2^*$XO_2	C_{2v}	A_1(I, R)	A_1(I, R) + A_2(R)	A_1(I, R) + B_1(I, R) + B_2(I, R)	A_1(I, R) + B_1(I, R) + B_2(I, R)

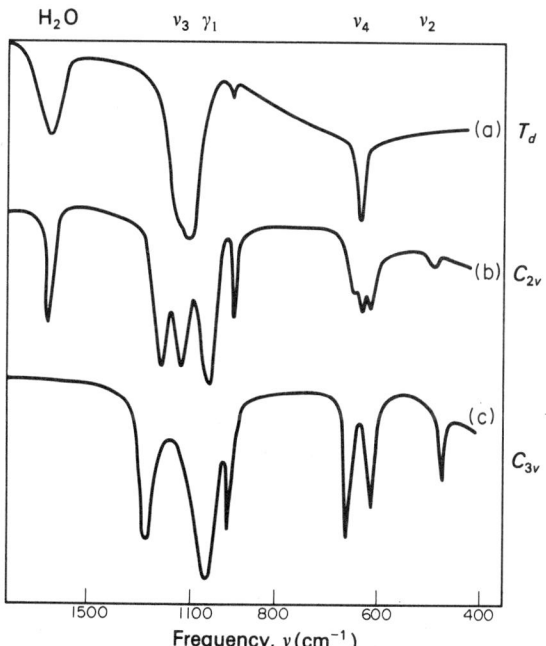

Figure 7 The IR spectra[19] of the perchlorate anion, $[ClO_4]^-$: (a) ionic $[ClO_4]^-$, T_d symmetry, $[Co(OH_2)_6][ClO_4]_2$; (b) bidentate $[ClO_4]^-$, C_{2v} symmetry, $[Co(OH_2)_2(OClO_3)_2]$; (c) tridentate $[ClO_4]^-$, C_{3v} symmetry, anhydrous $[Co(O_3ClO)_2]$

(iii)   EXAFS spectroscopy

In principle, the EXAFS spectra of coordination complexes[66] will give information on the number of metal–ligand distances and on the metal–ligand distances (but not the angles) themselves. In practice, although the number of short metal–ligand distances (1.9–2.2 Å) can be determined, information on any accompanying long metal–ligand distances can only be obtained with confidence if scattering from light atoms up to 4.0 Å distant from the central metal atom is included. The ability[67] to predict the short and long Cu—Cl distances of $CuCl_2(OH_2)_2$ (Table 6) lends confidence to the prediction[67] of the Cu—O distances of 1.94 and 2.68 Å of anhydrous $[Cu(ClO_4)_2]$ (Table 6), and to the proposed structure (Figure 8). With such data a promising future is predicted for the use of EXAFS spectroscopy in determining the M—O distances in oxyanion salts which cannot be obtained as single crystals suitable for X-ray crystallography. Equally useful data should soon be available from ^{17}O NMR spectroscopy,[68] as for oxyanion ligands this should yield information on the terminal oxygen atom environment, namely one X—O distance to an oxyanion of essentially fixed geometry, and up to three short M—O distances. This would be complicated by the presence of different terminal oxygen atom environments and especially by mixed short and long M—O distances but, combined with the EXAFS spectra, would yield unique information on the oxyanion coordination number.

Table 6   Cu—L Distances (Å) from EXAFS Spectra for (a) $[CuCl_2 \cdot 2H_2O]$ and (b) $[Cu(ClO_4)_2]$[67]

(a) [CuCl$_2 \cdot$2H$_2$O]			*(b) [Cu(ClO$_4$)$_2$]*	
Bond	*X-ray*	*EXAFS*	*Bond*	*EXAFS*
Cu—O	1.94	1.94	Cu—O	1.96
Cu—Cl	2.28	2.30	Cu—O	2.68
Cu—Cl	2.91	2.88	Cu—Cl	2.73
Cu—Cu	3.70	3.70	Cu—Cu	3.01

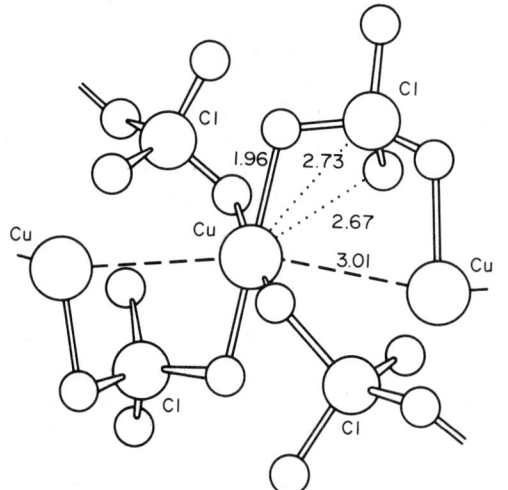

Figure 8   The proposed structure of anhydrous $[Cu(O_3ClO)_2]$ using EXAFS spectroscopic data

Finally, recent developments in profile analysis of X-ray powder data using synchrotron radiation[69] offer the hope that structure determination for polycrystalline samples will soon be possible. This will be especially relevant in the area of anhydrous metal oxyacid salts, such as the perchlorates, for which single crystals are difficult to obtain[17] and which samples are hygroscopic[15-20] but readily handled in a dry box. A further attraction is the small number of structure parameters required, as no organic ligands are present, and in view of the relatively rigid geometry of the oxyanions present the number of parameters could be further reduced by using rigid body analysis, as available[70] in SHELX-76.

15.5.4   ONE-COORDINATE OXYANIONS

All the oxyanions of Table 1 have been characterized as one-coordinate ligands, with both a monodentate and a bidentate coordination, and together represent the most prolific coordination number of the oxyanions. Monodentate coordination predominates (Figures 2a–e), but symmetrical bidentate coordination (Figure 2f) is less common, along with the asymmetric bidentate coordination (Figure 2c) of a *cis* monodentate function, especially for copper(II) which constrains a second oxygen atom to lie within bonding distance of the metal such that Δ(M—O) = M—O(2) − M—O(1) < 0.7 Å (Figure 1). Monodentate coordination through the X atom of XO_2 oxyanions occurs to yield nitro coordination, as in $[Co(NH_3)_5(NO_2)]Br_2$ (7),[35] and for the XO_3^{2-} oxyanion in the sulfito ligand as in $[Co(NH_3)_5(SO_3)]Cl \cdot H_2O$ (8).[45] Initially, nitro coordination was thought to dominate the structural chemistry of the NO_2^- oxyanion as a ligand but, increasingly,[6,71] cases of nitrito coordination have been characterized as in $[Co(NH_3)_5(ONO)]Cl_2$ (9)[40] (see also ref. 6). The various one-coordinate modes of bonding of the nitrite ion all arise in the two independent chromophores in $K_3[Cu(NO_2)_5]$ (10 and 11):[72,73] nitro (Figure 2a), monodentate (Figure 2b), symmetrical chelate (Figure 2f) and asymmetric chelate (Figure 2c).

(7) $[Co(NH_3)_5(NO_2)]Br_2^{35}$ (8) $[Co(NH_3)_5(SO_3)]Cl \cdot H_2O^{45}$ (9) $[Co(NH_3)_5(ONO)]Cl_2^{40}$

(10) $K_3[Cu(NO_2)_5]^{72}$ (11) $K_3[Cu(NO_2)_5]^{72}$ (12) $[Mn(O_2NO)_3(bipy)]^{74}$

While there is no crystallographic evidence for coordination of the nitrogen atom of the nitrate group (Figure 2a) (as initially suggested[25] in the sandwich structure of volatile copper nitrate), there is also little evidence for a clear *trans* (Figure 2b) monodentate nitrate coordination as the bidentate coordination role dominates the coordination chemistry of the nitrate ion as a ligand. Both symmetric bidentate coordination (Figure 2f; 1–6)[26,29,31–34] and asymmetric *cis* coordination of the nitrate anion (Figure 2c) with Δ(M—O) < 0.7 Å (see Figure 1b) predominate, with M—O—N angles less than 120° (108–120°). All three types occur in the same structure[74] as in $[Mn(O_2NO)_3(bipy)]$ (12), where Δ(M—O) = 0.0 and 0.284 Å. Even the monodentate nitrato group of (12) involves a 'laid back' *cis*-type coordination (Figure 2c) with a M—O—N angle of 131.2° and Δ(M—O) = 1.186 Å, rather than the *trans*-type structure (Figure 2b) with a M—O—N angle of 150° and Δ(M—O) < 0.7 Å. The *cis* asymmetric-type of coordination (Figure 2c) also occurs in $[Cu(tmen)(O_2NO)_2]$ (13),[75] where tmen = *N,N,N',N'*-tetramethylethylenediamine, and in $[Ph_4As]_2$-$[Co(O_2NO)_4]$ (14),[76] with Δ(M—O) = 0.33–0.43 Å. While the M—O distances for the symmetrical bidentate group (Figure 2f; 1–6) are generally less than 2.2 Å (except for heavy metals, as in 5 and 6) the nitrate group may bond reasonably symmetrically, with significantly longer M—O distances, in a fifth ligand position of a square pyramidal stereochemistry. This occurs in $[Fe(TPP)(O_2NO)]$ (15),[77] where TPP = tetraphenylporphyrin, and $[Cu(MeTAAB)(O_2NO)]NO_3$ (16),[78] where MeTAAB = tetrabenzo[*b,f,j,n*][1,5,9,13]tetraoxacyclohexadecine, in both of which it may form a bicapped square pyramidal stereochemistry.

In all of the XO_2 and XO_3 type anions (Figures 2a–c, f) the metal atom generally lies in the plane of the X,O(1),O(2) atoms and only in the *cis*-type coordination (Figure 2c) with Δ(M—O) > 0.7 Å does the M atom move out of the ligand plane, especially when long M—O distances are involved as in $[Cu(ONO_2)(OH_2)_3(bipy)]NO_3$ (17).[79]

The pattern of one-coordinate oxyanions for the pyramidal XO_3 and XO_4 type oxyanion species is somewhat different. X-atom coordination only occurs significantly *via* the sulfur atom of the

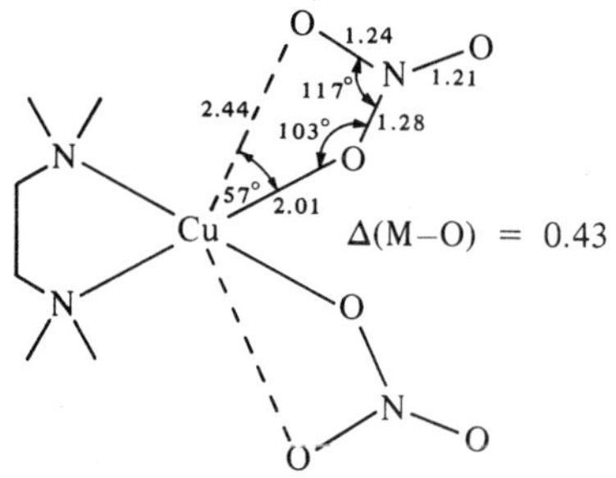

$\Delta(M-O) = 0.43$

$\Delta(M-O) = 0.43$

(13) [Cu(tmen)(O$_2$NO)$_2$] [75]

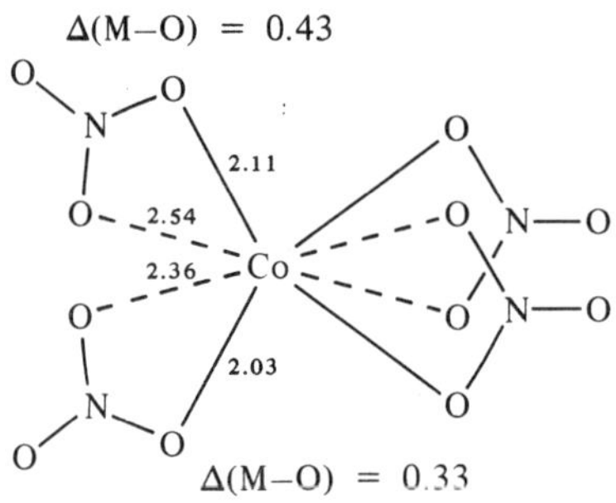

$\Delta(M-O) = 0.43$

$\Delta(M-O) = 0.33$

(14) [Ph$_4$As]$_2$[Co(O$_2$NO)$_4$] [76]

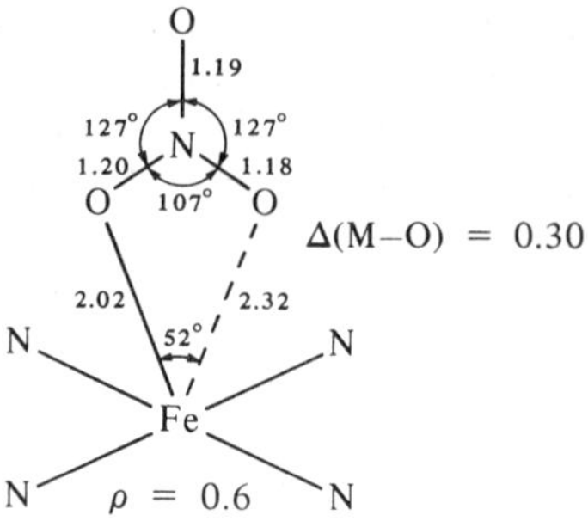

$\Delta(M-O) = 0.30$

$\rho = 0.6$

(15) [Fe(TPP)(O$_2$NO)] [77]

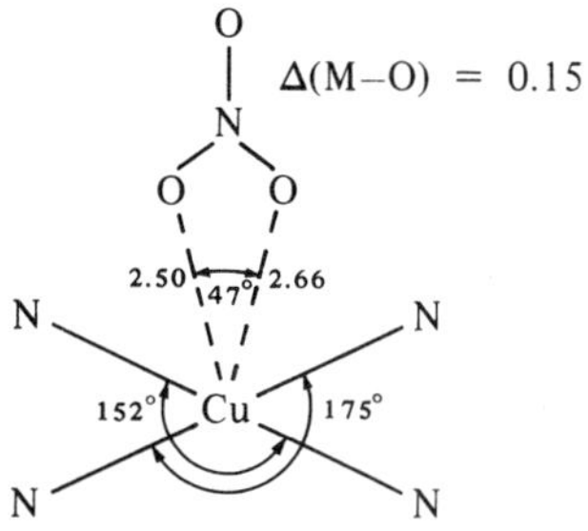

$\Delta(M-O) = 0.15$

(16) [Cu(MeTAAB)(O$_2$NO)]NO$_3$ [78]

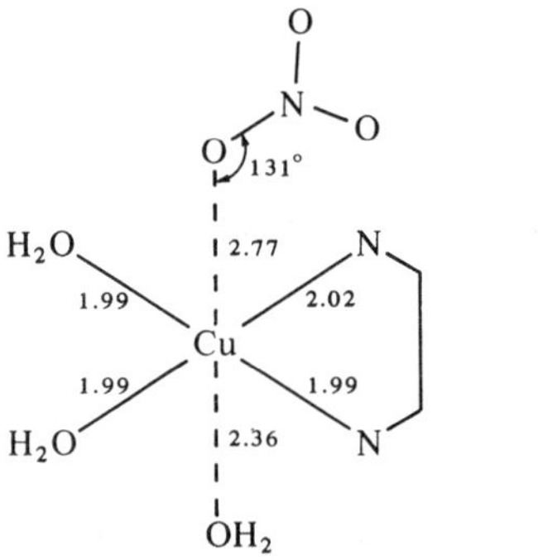

(17) [Cu(ONO$_2$)(OH$_2$)$_3$(bipy)]NO$_3$ [79]

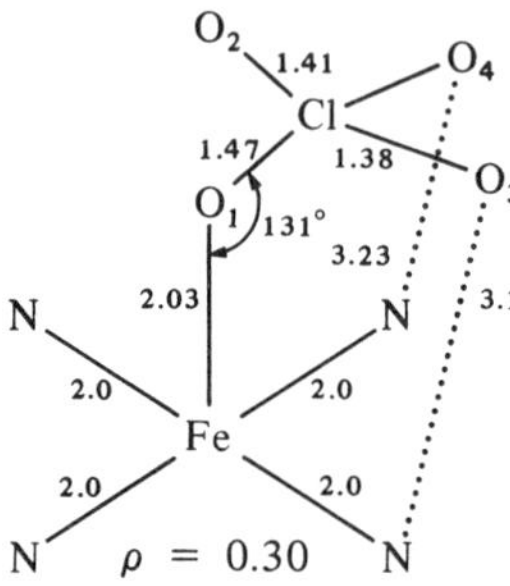

$\rho = 0.30$

(18) [Fe(TPP)(OClO$_3$)] [80]

sulfite anion, while P-atom coordination for the [HPO$_3$]$^{2-}$ anion or I-atom coordination for the [IO$_3$]$^-$ anion are unknown for short M—X distances, although long, secondary bonding distances are known (see ref. 8). There are four quite distinct modes of bonding of a tetrahedral XO$_3$ and XO$_4$ oxyanion as a monodentate ligand (Figures 2b–e), as well as the symmetrical bidentate function (Figure 2f); of the former four modes, Figures 2b and 2f predominate. The *trans* monodentate coordination occurs in [Fe(TPP)(OClO$_3$)] (18),[80] [NH$_4$]$_9$[Fe(OSO$_2$)$_6$] (19)[49] and for the less common [ReO$_4$]$^{2-}$ anion in Na[Gd(OReO$_3$)$_4$(OH$_2$)$_4$] (20).[53] In (18) the Fe,O(1),O(2) direction is virtually linear, the Fe,O(1),O(2),Cl atoms are coplanar and the O(3) and O(4) atoms lie well off this plane at long Fe$\cdots$O distances of 3.19 and 3.23 Å. For long M$\cdots$O(1) distances as in [Cu$_2$(bpim)(O$_2$NO)$_2$(OClO$_3$)(OH$_2$)]·H$_2$O (21),[81] where bpim = 4,5-bis(2-(2-pyridyl)ethylimino)methylimidazole, the Cu$\cdots$O(1),O(2) distances are all well over 3.0 Å, too long for even weak Cu—O bonding. This complex is the only complex of known crystal structure containing both coordinated nitrate and perchlorate oxyanions. In contrast to the common occurrence of Figure 2c with the NO$_2^-$ and NO$_3^-$ oxyanions, this conformation is most unusual for the tetrahedral XO$_3$ and XO$_4$ anions; it occurs in [Cu(tmen)(OSO$_3$)(OH$_2$)H$_2$O] (22)[82] with a short Cu—O(1) distance of 1.97 Å and a long Cu$\cdots$O(3) distance of 3.56 Å, which is enhanced by a Cu—O(1)—S angle of 128°.

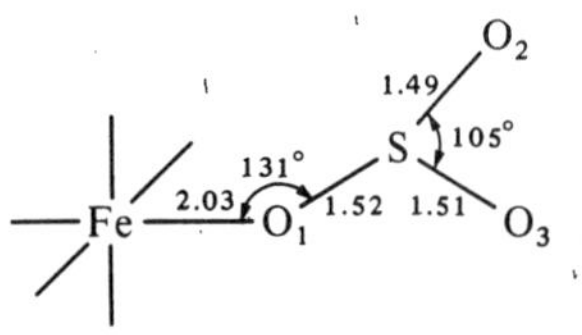

(19) [NH$_4$]$_9$[Fe(OSO$_2$)$_6$] [49]

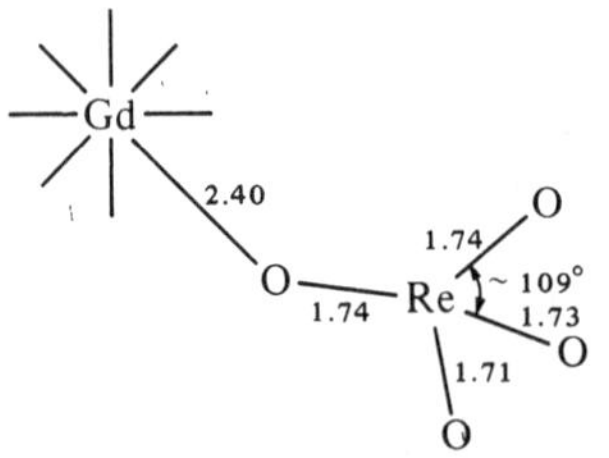

(20) Na[Gd(OReO$_3$)$_4$(OH$_2$)$_4$] [53]

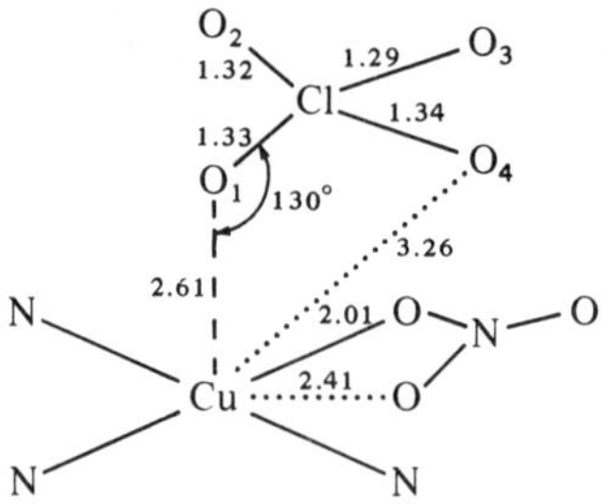

(21) [Cu$_2$(bpim)(O$_2$NO)(OClO$_3$)(OH$_2$)]·H$_2$O [81]

(22) [Cu(tmen)(OSO$_3$)(OH$_2$)]·H$_2$O[82]   (23) [Cu(pn)$_2$(OSO$_3$)]·H$_2$O[83]   (24) [Ni(p$_3$)(O$_2$SeO$_2$)][84]

Zr—O
2.13–2.23

(25) [Zr(O$_2$ClO$_2$)$_4$][85b]

In general, all tetrahedral XO$_3$ and XO$_4$ oxyanions bond with a 'laid back' M—O—X angle > 120° (125–160°). Consequently, Figure 2d is unknown as it involves a M—O—X angle ~109°. Figure 2e is equally uncommon, but does occur in [Cu(pn)$_2$(OSO$_3$)]·H$_2$O (23),[83] where pn = 1,3-propanediamine, with a Cu—O—S angle of *ca.* 95°. Surprisingly, the one-coordinate bidentate tetrahedral oxyanions are only moderately well established, as in [Nip$_3$)(O$_2$SeO$_2$)] (24)[84] and in the corresponding [Co(p$_3$)(O$_2$SO$_2$)],[84] where p$_3$ = 1,1,1-tris(diphenylphosphinomethyl)ethane. In all of these structures the bidentate chelate coordination results in a small bite angle at the metal of *ca.* 70° and the normal tetrahedal O—X—O angle is reduced to *ca.* 98°. The M,O(1),O(2),X atoms are coplanar and the M—O—X angles are *ca.* 95°. In all of these XO$_4$ oxyanions the one-coordinate symmetrical bidentate function only occurs with metal oxyanion stoichiometries, *p*, of 1:1; this contrasts with the occurrence of *p* values up to six for the bidentate nitrate ions (see Table 3 and structures 1–6). However, the *p* value in [Zr(O$_2$ClO$_2$)$_4$] (25) is an elegant exception.[85b]

In all of these one-coordinate XO$_{2-4}$ oxyanions the coordination is characterized as follows. (a) The basic geometry of the anions is retained, namely trigonal (120°) or tetrahedral (109°) (an exception is the symmetrical bidentate XO$_4$); (b) the X···O distances to the coordinated anion increase in length (<0.1 Å), while the uncoordinated X—O distances decrease (<0.1 Å); (c) the M—O—X angles are characteristically <120° (108–120°) for the trigonal oxyanions and >120° (125–160°) for the tetrahedral oxyanions; and (d) for short M—O distances with symmetric and asymmetric bidentate coordination the M,O(1),O(2),X atoms are generally coplanar.

While X-ray crystallography has been essential in establishing the bonding role of oxyanions to metal ions, it is in these cases of one-coordination (Figures 2a–f) that IR spectroscopy (see Section 15.5.3) is most informative in distinguishing the free ion from the monodentate and from the bidentate anion (see Section 15.5.3(ii) above).

15.5.5 TWO-COORDINATE OXYANIONS

The coordination number of two for oxyanions (Figure 3) is almost as common as a coordination number of one. Single oxygen atom bridging occurs with the XO$_2$ and planar XO$_3$ species, but is unknown for the pyramidal XO$_3$ and tetrahedral XO$_4$ oxyanions. Symmetrical bridging (Figure 3a) occurs[86] in anhydrous [Zn(O$_2$PH$_2$)$_2$]$_2$ (26), with short M—O distances and a pyramidal bridging oxygen atom. By far the most common bridging role is the symmetrical *syn/syn* coordination through separate oxygen atoms (Figure 3b). It occurs with almost all oxyanions, as in [Zn(O$_2$PH$_2$)$_2$]

(26) with short Zn—O distances, in [Nd(O₂MoO₂)(O₂ReO₂)] (27)[87] with slightly longer Nd—O distances, and in [Cu₃(O₂IO)₂(OH₂)₂] (28)[88] with long Cu···O distances of 2.4–2.5 Å. This type of short symmetrical bridging (Figure 3b) occurs[89] in K[Fe(O₂CrO₂)₂(OH₂)] (29), and for the novel bridging perchlorate group in [Sn₃O₂Cl₄(OClO₃)(O₂ClO₂)₃]₂[90] and [Sb₂Cl₂(OH)O(O₂ClO₂)] (30)[91] with short M—O distances. With the tetraphosphato and tetrasulfato species Cs₂[Mo₂(O₂PO₂-H)₄(OH₂)₂] (31)[92] and K₄[Mo₂(O₂SO₂)₄(OH₂)₂],[93] respectively, novel Mo—Mo bridged species are involved, with very short Mo—Mo distances consistent with multiple bonding.[94] In these Figure 3b systems the M—O—X angles are generally greater than 120°; consequently, in structures with short M—M distances (< 2.0 Å) the M—O and M'—O' directions are parallel and the XOO' plane is tilted away from the MOM'O' plane. In Figure 3b systems with the M···M distance > 3.0 Å, the M—O directions are not parallel and the M atoms are spread apart with the XO₂ plane generally lying in the approximate MOM'O' plane. The *trans/trans* bridging role (Figure 3c) is only slightly less common than *syn/syn* (Figure 3b). It occurs in anhydrous [Cu(O₂NO)₂] (32),[95] [Cu₃(O₂IO)₂(OH₂)] (28)[88] and in [Sb₂(O₂POH)₃] (33),[96] all with short M—O bond distances. In these *trans/trans* conformations the M—O/M—O distances are not linear but bent towards the X atom to reduce the M—O—X angle to only slightly greater than 120°. A near symmetrical *trans/trans* bridging ClO₄⁻ group occurs in [Cu(bipy)₂(O₂ClO₂)]ClO₄ (34)[97] involving long semi-coordinate Cu—O distances (2.51—2.75 Å). While the *syn/anti* conformation (Figure 3d) is the least common bridging structure, it does occur in [Zn(O₂Cl)(OH₂)₂] (35)[98] and in the linear chain structure of the tris-sulfato structure of [NH₄]₃[In(O₂SO₂)₃] (36).[52] Both involve short M—O distances and the symmetry of the bridging [O₂SO₂]²⁻ anion of (36) produces a near trigonal symmetry of the InO₆ chromophore.

The unsymmetrical bridging role of Figure 3e is not uncommon but limited to the nitrite ligand, with —O and N— coordination as in [Ni(3-Mepy)₂(O₂N)(ONO)]₃ (37)[99] with short Ni—O and Ni—N distances. Asymmetric bonding (Figure 3e) is unknown for tetrahedral XO₃ anions. One of the simplest bridging modes is that of the linear system (Figure 3f) which primarily occurs with the nitro ligand, as in K₂[PbNi(NO₂)₆] (38)[100] involving nitro coordination *via* nitrogen plus symmetric

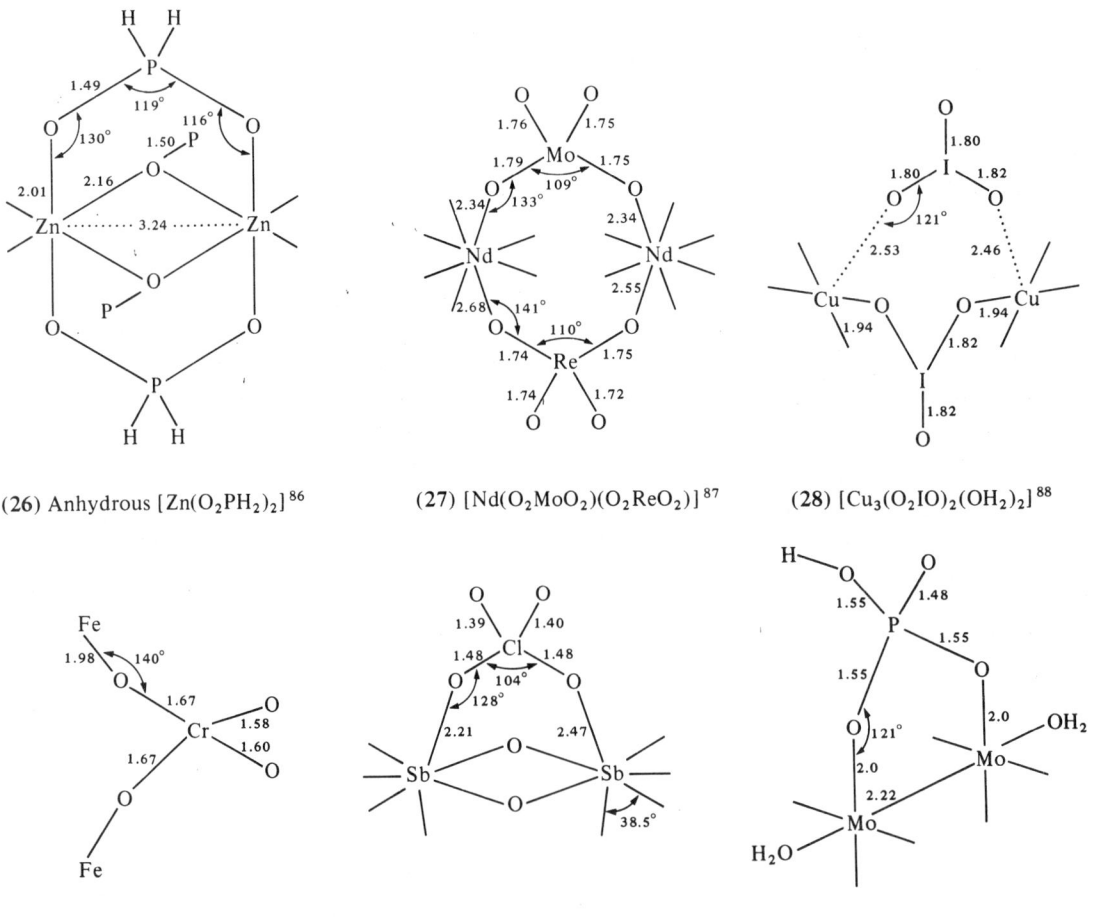

(26) Anhydrous [Zn(O₂PH₂)₂] [86]

(27) [Nd(O₂MoO₂)(O₂ReO₂)] [87]

(28) [Cu₃(O₂IO)₂(OH₂)₂] [88]

(29) K[Fe(O₂CrO₂)₂(OH₂)] [89]

(30) [Sb₂Cl₆(OH)(O)(O₂ClO₂)] [91]

(31) Cs₂[Mo₂(O₂PO₂H)₄(OH₂)₂] [92]

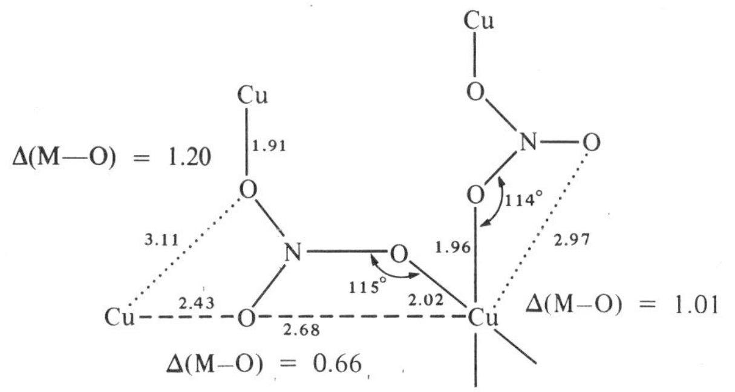

Cu

Δ(M—O) = 1.20   1.91

3.11

Cu

2.43   O   2.68

Δ(M—O) = 0.66

O   1.96

115°

N

114°

Cu

N—O

O

2.97

2.02 Cu Δ(M—O) = 1.01

(32) Anhydrous $[Cu(O_2NO)_2]$ [95]

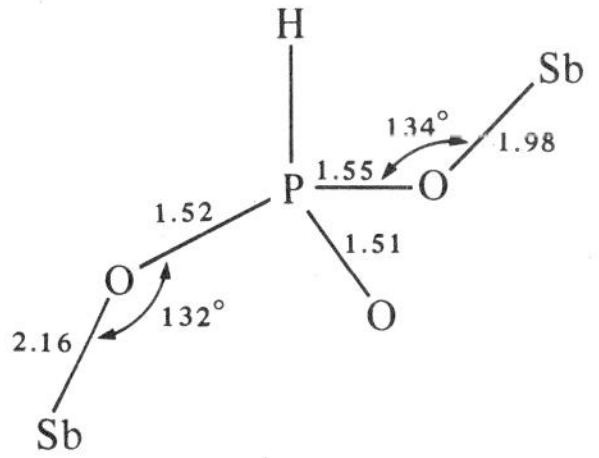

H

P   1.55   O   134°   1.98   Sb

1.52   1.51

O   132°

2.16

Sb

(33) $[Sb_2(O_2POH)_3]$ [96]

Cu

2.75

O

1.44   1.41   O

Cl

1.42   1.43

O

123°

2.51

Cu

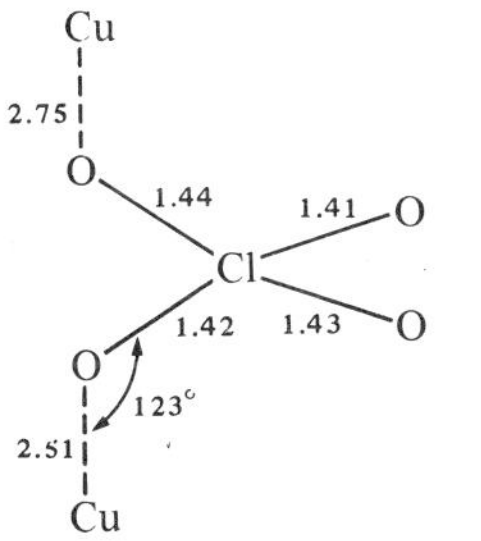

(34) $[Cu(bipy)_2(O_2ClO_2)]ClO_4$ [97]

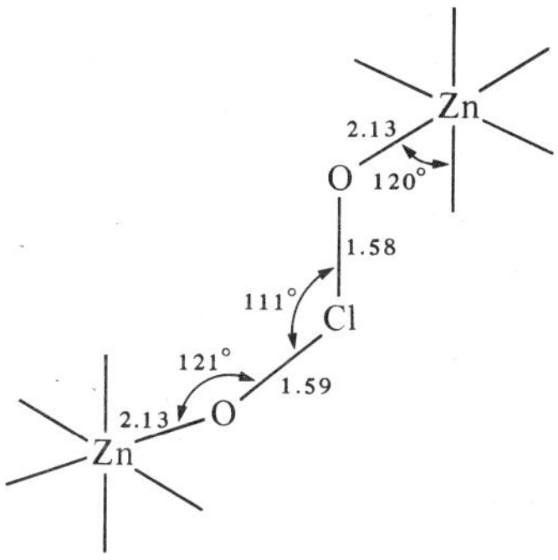

Zn

2.13

O   120°

1.58

Cl

111°

121°   1.59

2.13   O

Zn

(35) $[Zn(O_2Cl)_2(OH_2)_2]$ [98]

O   O

1.44   1.45

S

1.50   1.48

O   130°   O

2.15

2.14 O

In   133°   In

2.11

O   S   O

141°

O   O

S

O   O

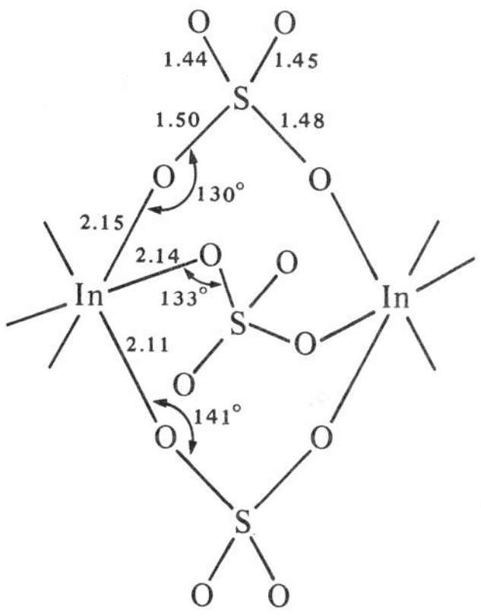

(36) $[NH_4]_3[In(O_2SO_2)_3]$ [52]

chelate nitrito coordination to a second metal. In general, the M—N distance is short (2.0 Å) and the M—O distances longer (2.77 Å), but sufficient covalent bonding must occur to form the high-symmetry face-centred cubic lattice that so frequently occurs in these $M_2^I[M^{II}M^{II}(NO_2)_6]$ complexes. Less common is the slightly asymmetrical bis-chelate structure of $Na[Th(O_4P)(P_2O_7)]_2$ (39),[101] with relatively short Th—O distances. Only the nitrite ion coordinates with a short M—O distance with a bidentate chelate plus monodentate coordination (Figure 3g) as in $[Ni(en)_2(O_2N)]_2[BPh_4]_2$ (40),[102] while a less symmetrical coordination, one short and two long Cu—O distances, frequently occurs with the nitrate ion as in $[Cu(ONO_2)_2(OH_2)_{2.5}]$ (41).[103] Near symmetrical bis-chelation (Figure 3h) occurs in $[Ag(Ph_3P)(O_2NO)_2]$ (42),[104] while the nearest approach yet known to a tridentate chelate function (asymmetric) of an oxyanion occurs with the $[PO_4]^{3-}$ oxyanion in $[Pb(O_3POH)_2]$ (43).[105]

15.5.6 THREE-COORDINATE OXYANIONS

This coordination number (Figure 4) is an order of magnitude less common than for one- and two-coordinate oxyanions (Figures 2 and 3). The coordination number does not occur for XO_2-type

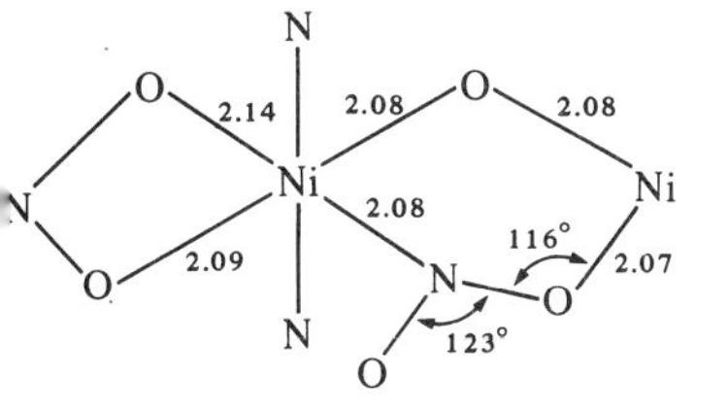

N

O   2.14   2.08   O   2.08

Ni   2.08   Ni

N   O

2.09

N

O   116°   2.07

N

123°

O

') $[Ni(3\text{-Mepy})_2(O_2N)(ONO)]_3 \cdot C_6H_6$ [99]

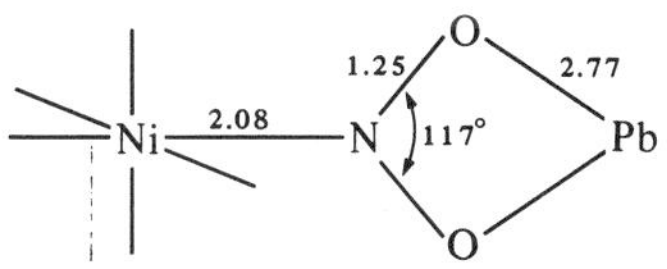

Ni   2.08   N

1.25

O   2.77

117°

Pb

O

(38) $K_2[PbNi(NO_2)_6]$ [100]

2.79   O   1.58

95°

Th   54°   102°   P   105°   1.48   O   2.48

1.48

56°   Th

109°

2.39   O   1.50   1.56   O   2.75

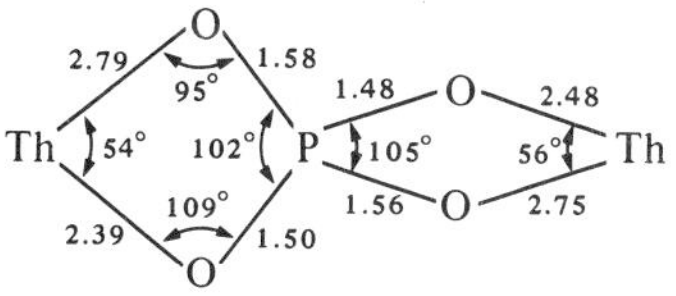

(39) $Na[Th(O_4P)(P_2O_7)_2]$ [101]

(40) [Ni(en)$_2$(O$_2$N)$_2$][BPh$_4$]$_2$ [102]   (41) [Cu(OH$_2$)$_{2.5}$(O$_2$NO)$_2$] [103]   (42) [Ag(PPh$_3$)(O$_2$NO)] [104]

oxyanions, but is well characterized with planar and pyramidal XO$_3$-type anions and most common with XO$_4$-type anions. Single oxygen bridging (Figure 4a) to three separate copper(II) ions occurs in [Cu$_2$(OH)$_3$(ONO$_2$)] (44)[106] with three comparable intermediate Cu—O distances, but has not been observed with the tetrahedral XO$_3$- and XO$_4$-type anions. Single oxygen bridging (Figure 4b) to two separate iron(II) atoms occurs in β-[Fe(O$_2$SO)(OH$_2$)$_3$] (45),[107] along with a monodentate oxygen coordination O(2) and a non-coordinated O(3) atom, with three short Fe—O distances. A simple tridentate coordination of the pyramidal XO$_3$-type (Figure 4c) occurs[108] in [Zn(O$_3$S)(OH$_2$)$_{2.5}$] (46), and of the XO$_4$-type occurs in [Zn(O$_3$POH)(OH$_2$)] (47)[109] and [Co$_3$(CF$_3$CO$_2$)$_3$Cl(O$_3$SO)-(dien)$_3$] (48),[110] all involving near symmetrical short M—O distances with the M—O—X angles > 120°; (47) involves the largest observed M—O—X angle of 160°.

A nearly symmetrical tridentate nitrate ion occurs in [Cu$_3$(OH)(pz)$_3$(Hpz)$_2$(O$_3$N)(ONO$_2$)]·H$_2$O (49),[111] where pz = pyrazine, with long Cu—O distances of 2.4–2.7 Å. More complex tridentate behaviour involving a bidentate chelate function is known (Figures 4d–f). A single bidentate chelate function (Figure 4d) is known for the nitrate ion in [Ag(bipy)(ONO$_2$)$_2$] (50),[112] but with rather irregular Ag—O distances. An asymmetrical bis-bidentate chelate coordination (Figure 4f) occurs in [Cu(O$_2$NO)$_2$] (32), with the asymmetric long and short Cu—O distances associated with the non-spherical symmetry of the copper(II) ion.[57]

(43) [Pb(O$_3$POH$_2$)$_2$] [105]   (44) [Cu$_2$(OH)$_3$(ONO$_2$)] [106]   (45) [β-Fe(O$_2$SO)(OH$_2$)$_3$] [107]

(46) [Zn(O$_3$S)(OH$_2$)$_{2.5}$] [108]   (47) [Zr(O$_3$POH)$_2$(OH$_2$)] [109]   (48) [Co$_3$(CF$_3$CO$_2$)$_3$Cl(O$_3$SO)(dien)$_3$]

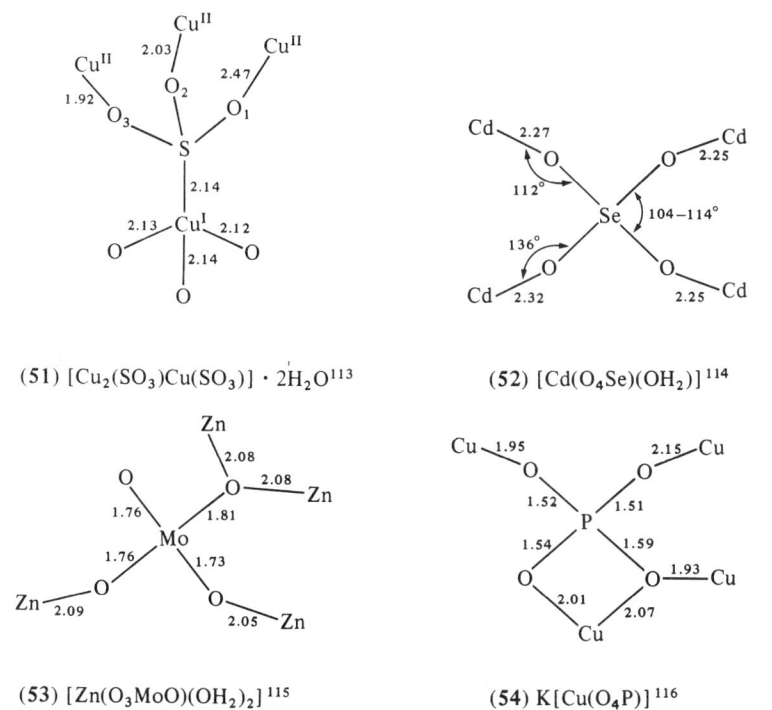

(49) [Cu₃(OH)(pz)₃(Hpz)₂(O₃N)(ONO₂)] · H₂O¹¹¹

(50) [Ag(bipy)(ONO₂)₂]¹¹²

15.5.7 FOUR- TO TWELVE-COORDINATE OXYANIONS

Coordination numbers for oxyanions greater than three and up to twelve occur (Figure 5), but with ever decreasing frequency. They are unknown for the XO_2-type anions, but do occur for the planar and pyramidal XO_3-type and especially XO_4-type anions. The simple four-coordinate, monodentate species (Figures 5a, 5b) occur in $[Cu_2(SO_3)Cu(SO_3)]\cdot 2H_2O$ (51)[113] and $[Cd(O_4Se)(OH_2)]$ (52)[114] respectively, and a single bridging oxygen occurs (Figure 5c) in $[Zn(O_3MoO)(OH_2)_2]$ (53),[115] along with an uncoordinated O atom and two monodentate O atoms. In all three complexes the M—O distances are short. A chelate (Figure 5d) coordination occurs in $K[Cu(O_4P)]$ (54),[116] with four short and one slightly longer Cu—O distances.

(51) [Cu₂(SO₃)Cu(SO₃)] · 2H₂O¹¹³

(52) [Cd(O₄Se)(OH₂)]¹¹⁴

(53) [Zn(O₃MoO)(OH₂)₂]¹¹⁵

(54) K[Cu(O₄P)]¹¹⁶

The most simple fivefold coordination (Figure 5e) of a tetrahedral oxyanion occurs[117] in α-Na[Cu(O₄P)] (55), with four short monodentate Cu—O distances plus one oxygen involved in an additional slightly longer Cu—O bridging role at 2.31 Å. A single chelate function occurs (Figure 5f) in [Pb₂Cu₅(SeO₃)₆(UO₂)₂(OH)₆]·2H₂O (56),[118] involving coordination of the selenite ligand to three different metals: Cu, Pb and U. An unusual bis-chelate fivefold coordination occurs with the silicate anion of Na[Gd(O₄Si)]·2NaOH (57),[119] with three of the O atoms involved in an additional oxygen bridging role to three further separate Gd atoms, all at relatively short Gd—O distances. A nearly regular sixfold coordination (Figure 5h) occurs in [Fe(O₃S)] (58),[120] with six short Fe—O distances. In anhydrous [Cu(O₄S)] (59)[50] the [SO₄]²⁻ group lies in a mirror plane with two O atoms symmetrically bridging, with one short Cu—O distance (2.00 Å) and one long Cu—O distance (2.37 Å), and with the remaining two O atoms symmetrically monodentate with a short Cu—O distance and yielding a Cu—Cu distance of 3.3 Å (Figure 5i).

(55) α-Na[Cu(O_4P)] [117]　　　(56) [Pb$_2$Cu$_5$(O$_3$Se)$_6$(UO$_2$)$_2$(OH)$_6$]·2H$_2$O [118]　　　(57) Na[Gd(O$_4$Si)]·0.2(NaOH) [119]

(58) [Fe(O$_3$S)] [120]　　　　　　　　(59) [Cu(O$_4$S)] [50]

Seven-coordinate oxyanions (Figure 5j) are very uncommon but do exist in [Cu$_3$(O$_4$P)$_2$] (60), [121] involving three bridging O atoms and one monodentate O atom of the tetrahedral [PO$_4$]$^{3-}$ oxyanion. No example of an eight-coordinate oxyanion has been observed, but an unusually regular nine-coordinate planar XO$_3$ anion exists (Figure 5k): not for the nitrate ion, as might be expected in view of its prolific role as a ligand, but in the [BO$_3$]$^{3-}$ anion of [Ni$_3$(O$_3$B)$_2$] (61). [122] The [BO$_3$]$^{3-}$ anion is essentially trigonal planar while the nine Ni—O distances all occur in the range 2.02–2.15 Å, with Ni—O—Ni angles *ca.* 96°, considerably lower than the tetrahedral angle of 109° 29′. A less regular nine-coordination occurs with the [VO$_4$]$^{3-}$ anion in [Cu$_5$(O$_4$V)$_2$O$_2$] (62) [123] involving two O atoms bonded to three Cu atoms, one O atom bridging two Cu atoms and one monodentate O atom. Seven short Cu—O distances are involved and two longer distances typical of the copper(II) ion. In [Cu$_5$(O$_4$V)$_2$O$_2$] there are two different [VO$_4$]$^{3-}$ anions; the second is ten-coordinate (63; Figure 5m) and together they generate five different copper(II) oxygen environments: two trigonal bipyramidal, two elongated rhombic octahedral and one *cis*-distorted octahedral. The highest observed coordination number for an XO$_4$ oxyanion is that of twelve (Figure 5n) in [Hg$_2$(O$_4$Ge)] (64); [124] each O atom of the tetrahedral germanate anion [GeO$_4$]$^{4-}$ involves three non-equivalent bond distances to the mercury(II) ion.

(60) [Cu$_3$(O$_4$P)$_2$] [121]　　　　　(61) [Ni$_3$(O$_3$B)$_2$] [122]　　　　　(62) [Cu$_5$(O$_4$V)$_2$O$_2$] [123]

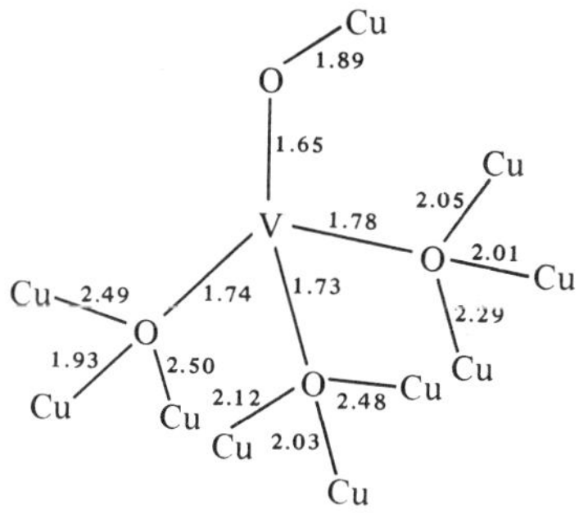

(63) $[Cu_5(O_4V)_2O_2]$ [123]

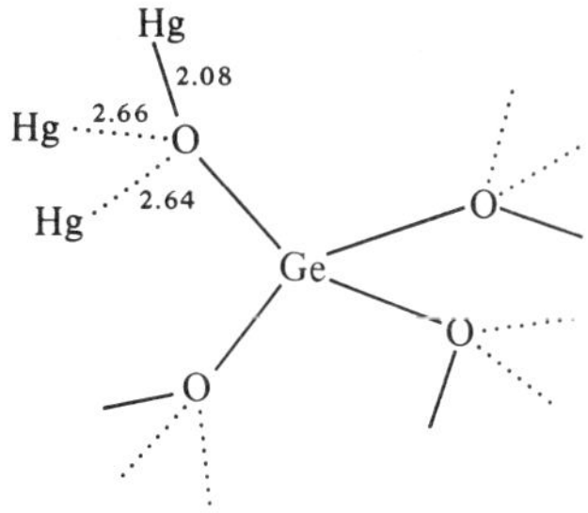

(64) $[Hg_2(O_4Ge)]$ [124]

15.5.8 XO$_5$ AND XO$_6$ OXYANIONS

The oxyanions as ligands are most prolific (Table 1) for the tetrahedral XO$_4$ species and pyramidal XO$_3$. The nitrite ion dominates the XO$_2$ species, and the nitrate ion the planar XO$_3$ species. In contrast, no XO$_5$ oxyanion species exist and the XO$_6$ species is very restricted to oxyanions such as $[TeO_6]^{6-}$ and $[IO_6]^{5-}$. Both involve regular octahedral structures, which have a regular coordination number of six in $K[Ni(O_6I)]$ (65)[54] with a linear I—O—Ni angle and an 18-fold coordination in $[Cu_3(O_6Te)]$ (66),[125] which is cubic with the CuO$_6$ chromophores involved in four short and two long Cu—O distances. However, the $[IO_6]^{5-}$ anion is one-coordinate in $Na_3KH_3[Cu(O_2IO_4)_2(OH_2)]\cdot 13H_2O$ (67),[55] generating a square pyramidal CuO$_4$O′ chromophore with a long Cu···OH$_2$ distance of 2.7 Å reminiscent of copper(II) rather than the reported copper(III) oxidation state.

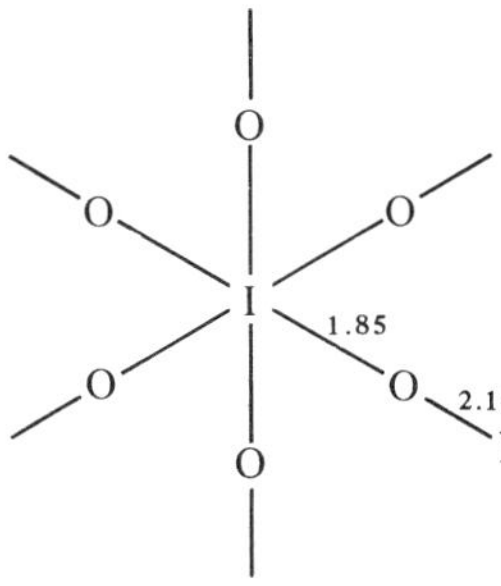

(65) $K[Ni(O_6I)]$ [54]

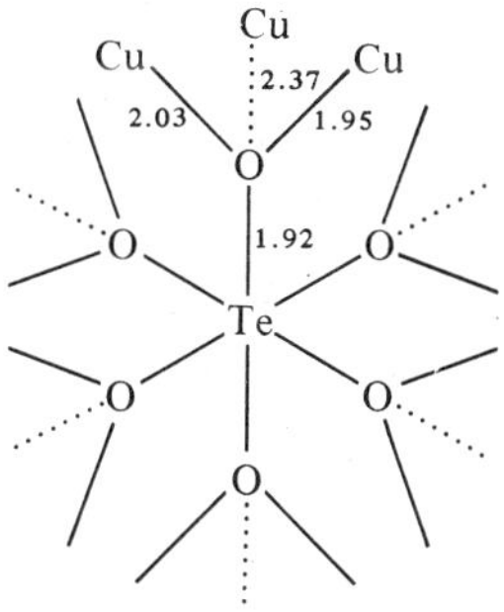

(66) $[Cu_3(O_6Te)]$ [125]

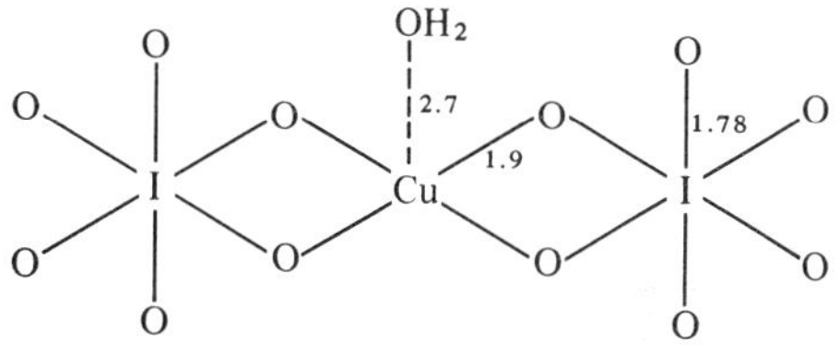

(67) $Na_3KH_3[Cu(O_2IO_4)_2(OH_2)]\cdot 3H_2O$ [55]

15.5.9 POLYNUCLEAR OXYANIONS X$_q$O$_n^{m-}$

For $q = 2$ this may involve X—X connections as in $S_2O_3^{2-}$ and $S_2O_4^{2-}$, or an X—O—X linkage as in $S_2O_7^{2-}$, $Se_2O_5^{2-}$ or $Cr_2O_7^{2-}$. The thiosulfate anion can coordinate through the terminal sulfur atom as a one-coordinate ligand (Figure 2a), as in $[Zn(tu)_3(S_2O_3)]\cdot H_2O$ (68),[126] or as a two-coordinate ligand (Figure 3a), as in $Na_4[Cu(NH_3)_4][Cu(S_2O_3)]_2\cdot H_2O$ (69).[127] Alternatively, it acts as a one-coordinate chelate ligand (Figure 2c) as in $[Ni(tu)_4(OS_2O_2)]\cdot H_2O$ (70).[128] IR spectroscopy has been used to suggest[60,129] that the $S_2O_3^{2-}$ anion of $[UO_2(S_2O_3)\cdot H_2O]$ is involved in oxygen-to-uranium coordination. Owing to their size and flexibility, polynuclear oxyanions readily act as bidentate ligands, chelating one-coordinate species, as in $[Te(O_2S_2O_5)_2]$ (71),[130] or bridging two-coordinate, as in $[Sn_2(O_4S_2)_2]$ (72),[131] both with short M—O distances. Alternatively, four-coordinate bridging oxyanions may occur, as in the $Cr_2O_7^{2-}$ anion of $[Cu(OH_2)_2(O_2Cr_2O_5)]$ (73),[132] with mixed short and long Cu—O distances, or as in the complex chain structure of $[Cu(O_4Se_2O)]$ (74).[133]

15.5.10 SUMMARY OF THE OXYANIONS AS LIGANDS

(a) They are good electron donor ligands when not in competition with traditional ligands such as water.

COC 2—O

(68) [Zn(tu)₃(S₂O₃)] · H₂O[126]   (69) Na₄[Cu(NH₃)₄][Cu(S₂O₃)₂]₂·H₂O[127]   (70) [Ni(tu)₄(OS₂O₂)]·H₂O[128]

(71) [Te(O₂S₂O₅)₂][130]   (72) [Sn₂(O₄S₂)₂][131]   (73) [Cu(O₂Cr₂O₅)(OH₂)₂][132]

(74) [Cu(O₄Se₂O)][133]

(b) They retain the geometric structure of the free anion on coordination.

(c) They generally form short M—O distances, *ca.* 2.0 Å for first-row metals and up to 2.7 Å for heavy metals.

(d) M—O—X angles are non-linear, $< 120°$ for XO_2 and XO_3 planar species and $> 120°$ for tetrahedral XO_3 and XO_4 species.

(e) The stoichiometry, p, is 1–6 for XO_2 and XO_3 anions, 1–4 for XO_4 anions but only 1–2 for XO_6 anions.

(f) The coordination numbers 1 and 2 predominant with $1 \approx 2 > 3 \gg 4$–18.

(g) With lower coordination numbers, monodentate coordination > bidentate predominates with tridentate virtually absent; with higher coordination numbers, anion bridging is involved.

(h) The metal atom generally involves normal coordination numbers, namely $6 \gg 5 \approx 4 > 7, 8,$ 9. Only very occasionally are unusual stereochemistries formed, such as *cis*-octahedral and bicapped square pyramidal.

(i) The copper(II) ion is unique in generating less regular five- and six-coordinate stereochemistries with short and long M—O distances, and hence much more varied copper oxyanion structures.

15.5.11 REFERENCES

1. F. A. Cotton and G. Wilkinson, 'Advanced Inorganic Chemistry', 4th edn., Wiley-Interscience, New York, 1980.
2. T. Moeller, 'Inorganic Chemistry — A Modern Introduction', Wiley-Interscience, New York, 1982, pp. 373–394.
3. R. W. G. Wyckoff, 'Crystal Structures', 2nd edn., 1964, vol. 1, p. 272; 1965, vol. 2, p. 45.
4. A. Muller and E. Diemann, 'MTP International Review of Science, Inorganic Chemistry Series Two', vol. 5, 'Transition Metals Part 1', Butterworths, London, 1975, pp. 71–111.
5. A. F. Wells, 'Structural Inorganic Chemistry', 5th edn., Clarendon, Oxford, 1984.
6. M. A. Hitchman and G. L. Rowbottom, *Coord. Chem. Rev.*, 1982, **42**, 55.
7. C. L. Raston, A. H. White and J. K. Yandell, *Aust. J. Chem.*, 1979, **32**, 291.
8. H. D. B. Jenkins and K. P. Thakur, *J. Chem. Educ.*, 1979, **56**, 576.
9. M. R. Rosenthal, *J. Chem. Educ.*, 1973, **50**, 331; N. M. N Gowda, S. B. Naikar and G. K. N. Reddy, *Adv. Inorg. Chem. Radiochem.*, 1984, **28**, 255.
10. L. Johansson, *Coord. Chem. Rev.*, 1974, **12**, 241.
11. C. C. Addison, *Chem. Rev.*, 1980, **80**, 21.
12. C. C. Addison, N. Logan, S. C. Wallwork and C. D. Garner, *Chem. Soc. Rev.*, 1971, **25**, 289.
13. B. J. Hathaway, *Struct. Bonding (Berlin)*, 1973, **14**, 49.
14. B. J. Hathaway and A. E. Underhill, *J. Chem. Soc.*, 1960, 648.
15. Z. K. Nikitina and V. Ya. Rosolovskii, *Russ. J. Inorg. Chem. (Engl. Transl.)*, 1980, **25**, 715.
16. M. Chaabouni, J.-L. Pascal, A. C. Pavia, J. Potier and A. Potier, *C. R. Hebd. Seances Acad. Sci., Ser. C*, 1978, **287**, 419.
17. B. J. Hathaway, *Proc. Chem. Soc.*, 1958, 344.
18. B. J. Hathaway and A. E. Underhill, *J. Chem. Soc.*, 1961, 3091.
19. S. V. Loginov, Z. K. Nikitina and V. Ya. Rosolovskii, *Russ. J. Inorg. Chem. (Engl. Transl.)*, 1980, **25**, 508.
20. S. V. Loginov, Z. K. Nikitina and V. Ya. Rosolovskii, *Russ. J. Inorg. Chem. (Engl. Transl.)*, 1980, **25**, 562.
21. R. D. Larsen and G. H. Brown, *J. Phys. Chem.*, 1964, **68**, 3060.
22. C. C. Addison and N. Logan, *Adv. Inorg. Chem. Radiochem.*, 1964, **6**, 71.
23. C. C. Addison and D. Sutton, *Prog. Inorg. Chem.*, 1967, **8**, 195.
24. B. O. Field and C. J. Hardy, *Chem. Soc. Rev.*, 1964, **18**, 361.
25. C. C. Addison and B. J. Hathaway, *Proc. Chem. Soc.*, 1957, 19.
26. J. Hilton and S. C. Wallwork, *Chem. Commun.*, 1968, 871.
27. C. D. Garner and S. C. Wallwork, *J. Chem. Soc. (A)*, 1966, 1496.
28. C. C. Addison and B. J. Hathaway, *J. Chem. Soc.*, 1958, 3099.
29. R. E. LaVilla and S. H. Bauer, *J. Am. Chem. Soc.*, 1963, **85**, 3597.
30. B. J. Hathaway and D. E. Billing, *Coord. Chem. Rev.*, 1970, **5**, 143.
31. G. G. Messmer and G. J. Palenik, *Inorg. Chem.*, 1969, **8**, 2750.
32. T. J. King, N. Logan, A. Morris and S. C. Wallwork, *Chem. Commun.*, 1971, 554.
33. J.-C. G. Bunzli, B. Klein, G. Chapuis and K. J. Schenk, *J. Inorg. Nucl. Chem.*, 1980, **42**, 1307.
34. T. A. Beineke and J. Delgaudio, *Inorg. Chem.*, 1968, **7**, 715.
35. F. A. Cotton and W. T. Edwards, *Acta Crystallogr., Sect. B*, 1968, **24**, 474.
36. B. N. Figgis, P. A. Reynolds and S. Wright, *J. Am. Chem. Soc.*, 1983, **105**, 434.
37. K. G. Jensen, H. Soling and N. Thorup, *Acta Chem. Scand.*, 1970, **24**, 908.
38. Y. Kushi, M. Kuramoto, S. Yamamoto and H. Yoneda, *Inorg. Nucl. Chem. Lett.*, 1976, **12**, 629.
39. S. Tagaki, M. Joesten and P. Galen-Lenhert, *Acta Crystallogr., Sect. B*, 1975, **31**, 1968.
40. I. Grenthe and E. Nordin, *Inorg. Chem.*, 1979, **18**, 1869.
41. A. F. Finney, M. A. Hitchman, C. L. Raston, G. L. Rowbottom and A. H. White, *Aust. J. Chem.*, 1981, **34**, 2047.
42. P. Phavanantha, Ph.D. Thesis, University of London, 1970.
43. B. Duffin, *Acta Crystallogr., Sect. B*, 1968, **24**, 396.
44. C. D. Garner, D. Sutton and S. C. Wallwork, *J. Chem. Soc. (A)*, 1967, 1949.
45. R. C. Elder, M. J. Heeg, M. D. Payne, M. Trkula and E. Deutsch, *Inorg. Chem.*, 1978, **17**, 431.
46. G. D. Fallon, C. L. Raston, A. H. White and J. K. Yandell, *Aust. J. Chem.*, 1980, **33**, 665.
47. D. Messer, D. K. Breitinger and W. Haegler, *Acta Crystallogr., Sect. B*, 1981, **37**, 19.
48. L.-G. Johansson and O. Lindqvist, *Acta Crystallogr., Sect. B*, 1979, **35**, 1017.
49. L. O. Larsson and L. Niinisto, *Acta Chem. Scand.*, 1973, **27**, 859.
50. B. R. Rao, *Acta Crystallogr.*, 1961, **14**, 321.
51. I. J. Bear and W. G. Mumme, *J. Solid State Chem.*, 1970, **1**, 497.
52. B. Jolibois, G. Laplace, F. Abraham and G. Nowogrocki, *Acta Crystallogr., Sect. B*, 1980, **36**, 2517.
53. Z. A. A. Slimane, J.-P. Silvestre, W. Freundlich and A. Rimsky, *Acta Crystallogr., Sect. B*, 1982, **38**, 1070.
54. L. P. Eddy and N. G. Vannerberg, *Acta Chem. Scand.*, 1966, **20**, 2886.
55. I. Hadinec, L. Jensovsky, A. Linek and V. Synecek, *Naturwissenschaften*, 1960, **47**, 377.
56. L. Alagna, A. A. G. Tomlinson, C. Ferragina and A. L. Ginestra, *J. Chem. Soc., Dalton Trans.*, 1981, 2376.
57. R. J. Gillespie, *J. Chem. Soc.*, 1963, 4672.
58. I. M. Procter, B. J. Hathaway and P. Nicholls, *J. Chem. Soc.*, 1968, 1678.
59. P. A. Tasker and L. Sklar, *J. Cryst. Mol. Struct.*, 1975, **5**, 329.
60. K. Nakamoto, 'Infrared and Raman Spectra of Inorganic and Coordination Compounds', 3rd edn., Wiley-Interscience, New York, 1978, pp. 220–278.
61. C. G. Barraclough and M. L. Tobe, *J. Chem. Soc.*, 1961, 1993; R. Coomber and W. P. Griffith, *J. Chem. Soc. (A)*, 1968, 1128.
62. B. M. Gatehouse and A. E. Comyns, *J. Chem. Soc.*, 1958, 3965.
63. G. Newman and D. B. Powell, *Spectrochim. Acta*, 1963, **19**, 213.
64. D. M. L. Goodgame and M. A. Hitchman, *Inorg. Chem.*, 1964, **3**, 1389.
65. R. E. Hester, *Coord. Chem. Rev.*, 1967, **2**, 319.
66. S. P. Cramer and K. O. Hodgson, *Prog. Inorg. Chem.*, 1979, **25**, 1.

67. J.-L. Pascal, J. Potier, D. J. Jones, J. Roziere and A. Michalowicz, *Inorg. Chem.*, 1984, **23**, 2068.
68. W. G. Klemperer, *Angew. Chem., Int. Ed. Engl.*, 1978, **17**, 246.
69. SERC Powder Diffraction User Group, Daresbury, 1983.
70. G. M. Sheldrick, 'SHELX-76, Program for Crystal Structure Determination', University of Cambridge, 1976.
71. M. A. Hitchman, R. Thomas, B. W. Skelton and A. H. White, *J. Chem. Soc., Dalton Trans.*, 1983, 2273.
72. K. A. Klanderman, W. C. Hamilton and I. Bernal, *Inorg. Chim. Acta*, 1977, **23**, 117.
73. J. Foley, W. Fitzgerald, S. Tyagi, B. J. Hathaway, D. E. Billing, R. Dudley, P. Nicholls and R. C. Slade, *J. Chem. Soc., Dalton Trans.*, 1982, 1439.
74. F. W. B. Einstein, D. W. Johnson and D. Sutton, *Can. J. Chem.*, 1972, **50**, 3332.
75. S. F. Pavkovic, D. Miller and J. N. Brown, *Acta Crystallogr., Sect. B*, 1977, **33**, 2894.
76. F. A. Cotton and J. G. Bergman, *J. Am. Chem. Soc.*, 1964, **86**, 2941.
77. M. A. Phillipi, N. Baenziger and H. M. Goff, *Inorg. Chem.*, 1981, **20**, 3904.
78. A. J. Jircitano, R. I. Sheldon and K. B. Mertes, *J. Am. Chem. Soc.*, 1983, **105**, 3022.
79. H. Nakai, S. Ooi and H. Kuroya, *Bull. Chem. Soc. Jpn.*, 1977, **50**, 531.
80. C. A. Reed, T. Mashiko, S. P. Bentley, M. E. Kastner, W. R. Scheidt, K. Spartalian and G. Lang, *J. Am. Chem. Soc.*, 1979, **101**, 2948.
81. J. C. Dewan and S. J. Lippard, *Inorg. Chem.*, 1980, **19**, 2079.
82. J. Balvich, K. P. Fivizzani, S. F. Pavkovic and J. N. Brown, *Inorg. Chem.*, 1976, **15**, 71.
83. B. Morosin and J. Howatson, *Acta Crystallogr., Sect. B*, 1970, **26**, 2062.
84. C. Benelli, M. Di Vaira, G. Noccioli and L. Sacconi, *Inorg. Chem.*, 1977, **16**, 182.
85. (a) J. Reed, S. L. Soled and R. Eisenberg, *Inorg. Chem.*, 1974, **13**, 3001; (b) E. A. Grunkina, V. P. Babaevia and V. Ya. Rosolovskii, *Koord. Khim.*, 1984, **10**, 1415.
86. T. J. R. Weakley, *Acta Crystallogr., Sect. B*, 1979, **35**, 42.
87. D. Argeles, J.-P. Silvestre and M. Quarton, *Acta Crystallogr., Sect. B*, 1982, **38**, 1690.
88. S. Ghose and C. Wan, *Acta Crystallogr., Sect. B*, 1974, **30**, 965.
89. V. Debelle, P. Gravereau and A. Hardy, *Acta Crystallogr., Sect. B*, 1974, **30**, 2185.
90. C. Belin, M. Chaabouni, J.-L. Pascal, J. Potier and J. Roziere, *J. Chem. Soc., Chem. Commun.*, 1980, 105.
91. C. H. Belin, M. Chaabouni, J.-L. Pascal and J. Potier, *Inorg. Chem.*, 1982, **21**, 3557.
92. A. Bino and F. A. Cotton, *Inorg. Chem.*, 1979, **18**, 3562.
93. F. A. Cotton, B. A. Frenz, E. Pederson and T. R. Webb, *Inorg. Chem.*, 1975, **14**, 391.
94. F. A. Cotton and R. A. Walton, 'Multiple Bonds Between Metal Atoms', Wiley-Interscience, New York, 1982.
95. S. C. Wallwork and W. E. Addison, *J. Chem. Soc.*, 1965, 2925.
96. J. Loub and H. Paulus, *Acta Crystallogr., Sect. B*, 1981, **37**, 1106.
97. J. Foley, D. Kennefick, D. Phelan, S. Tyagi and B. J. Hathaway, *J. Chem. Soc., Dalton Trans.*, 1983, 2333.
98. T. Pakkanen and T. Pakkanen, *Acta Crystallogr., Sect. B*, 1979, **35**, 2670.
99. A. J. Finney, M. A. Hitchman, C. L. Raston, G. L. Rowbottom and A. H. White, *Aust. J. Chem.*, 1981, **34**, 2125.
100. S. Tagaki, M. D. Joesten and P. Galen-Lenhert, *Acta Crystallogr., Sect. B*, 1975, **31**, 1968.
101. B. Kojic-Prodic, M. Sljukic and Z. Ruzic-Toros, *Acta Crystallogr., Sect. B*, 1982, **38**, 67.
102. A. Gleizes, A. Meyer, M. A. Hitchman and O. Kahn, *Inorg. Chem.*, 1982, **21**, 2257.
103. B. Morosin, *Acta Crystallogr., Sect. B*, 1970, **26**, 1203.
104. R. A. Stein and C. Knobler, *Inorg. Chem.*, 1977, **16**, 242.
105. P. Vasic, B. Prelesnick, R. Herak and M. Curic, *Acta Crystallogr., Sect. B*, 1981, **37**, 660.
106. B. Bovio and S. Locchi, *J. Crystallogr. Spectrosc. Res.*, 1982, **12**, 507.
107. L.-G. Johansson and E. Ljungstrom, *Acta Crystallogr., Sect. B*, 1979, **35**, 2683.
108. B. Nyberg, *Acta Chem. Scand.*, 1973, **27**, 1541.
109. A. Clearfield and G. D. Smith, *Inorg. Chem.*, 1969, **8**, 431.
110. J. E. Davis, *Acta Crystallogr., Sect. B*, 1982, **38**, 2541.
111. F. B. Hulsbergen, R. W. M. ten Hoedt, G. C. Verschoor, J. Reedijk and A. L. Spek, *J. Chem. Soc., Dalton Trans.*, 1983, 539.
112. G. W. Bushnell and M. A. Khan, *Can. J. Chem.*, 1972, **50**, 315.
113. P. Kierkegaard and B. Nyberg, *Acta Chem. Scand.*, 1965, **19**, 2189.
114. C. Stalhandske, *Acta Crystallogr., Sect. B*, 1981, **37**, 2055.
115. J. Y. Le Marouille, O. Bars and D. Grandjean, *Acta Crystallogr., Sect. B*, 1980, **36**, 2558.
116. G. L. Shoemaker, E. Kostiner and J. B. Anderson, *Z. Kristallogr.*, 1980, **152**, 317.
117. M. Quarton and A. W. Kolsi, *Acta Crystallogr., Sect. C*, 1983, **39**, 664.
118. D. Ginderow and F. Cesbron, *Acta Crystallogr., Sect. C*, 1983, **39**, 824.
119. G. A. Fallon and B. M. Gatehouse, *Acta Crystallogr., Sect. B*, 1982, **38**, 919.
120. G. Bugli and D. Carre, *Acta Crystallogr., Sect. B*, 1980, **36**, 1297.
121. G. L. Shoemaker, J. B. Anderson and E. Kostiner, *Acta Crystallogr., Sect. B*, 1977, **33**, 2969.
122. J. Pardo, M. Martinez-Ripoll and S. Garcia-Blanco, *Acta Crystallogr., Sect. B*, 1974, **30**, 37.
123. R. D. Shannon and C. Calvo, *Acta Crystallogr., Sect. B*, 1973, **29**, 1338.
124. K.-F. Hesse and W. Eysel, *Acta Crystallogr., Sect. B*, 1981, **37**, 429.
125. L. Falck, O. Lindqvist and J. Moret, *Acta Crystallogr., Sect. B*, 1978, **34**, 896.
126. G. D. Andreetti, L. Cavalca, P. Domiano and A. Musatti, *Ric. Sci.*, 1968, **38**, 1100.
127. B. Morosin and A. C. Larson, *Acta Crystallogr., Sect. B*, 1969, **25**, 1417.
128. G. F. Gasparri, A. Mangia, A. Musatti and M. Nardelli, *Acta Crystallogr., Sect. B*, 1969, **25**, 203.
129. A. N. Freedman and B. P. Straughan, *Spectrochim. Acta, Part A*, 1971, **27**, 1455.
130. F. W. Einstein and A. C. Willis, *Acta Crystallogr., Sect. B*, 1981, **37**, 218.
131. A. Magnusson and L.-G. Johansson, *Acta Chem. Scand., Ser. A*, 1982, **36**, 429.
132. D. Blum and J. C. Guitel, *Acta Crystallogr., Sect. B*, 1980, **36**, 667.
133. G. Meunier, C. Svensson and A. Carpy, *Acta Crystallogr., Sect. B*, 1976, **32**, 2664.

15.6

Carboxylates, Squarates and Related Species

COLIN OLDHAM
University of Lancaster, UK

15.6.1 INTRODUCTION

This chapter is principally concerned with ligand systems which contain a carbon atom linked to two oxygen atoms. In the vast majority of the complexes to be considered the donor atom is the relatively small, less polarizable oxygen atom, *i.e.* 'hard' under the Pearson classification. As a consequence these ligands are seldom found in conjunction with low oxidation state metals. Considerable advances have been made in this area in recent years. The most notable of these has been the work of Cotton, who has made extensive use of the ability of such ligands to bring metal atoms close together. The impact of diffraction methods, without which many of the advances would not have been possible, has been considerable over the whole area. These developments have provided the impetus for a continuing and expanding interest in the chemistry of carbonate and CO_2 complexes and it is anticipated that this will lead to systems of utility in both catalysis and biological inorganic systems.

This chapter is organized according to the nature of X in the grouping $X—CO_2^-$ (Table 1) and also includes oxocarbon systems as exemplified by squaric acid, $C_4H_2O_4$.

Table 1 Classification of $X—COO^-$ Species According to the Nature of X

X	*Structural formula*	*Name*
Hydrocarbon derived	$R—C\begin{smallmatrix}O\\O^-\end{smallmatrix}$	Monocarboxylates
Carboxylate	$\begin{smallmatrix}^-O\\O\end{smallmatrix}C—(CH_2)_n—C\begin{smallmatrix}O\\O^-\end{smallmatrix}$	Dicarboxylates
Oxygen	$^-O—C\begin{smallmatrix}O\\O^-\end{smallmatrix}$	Carbonates
	$^-O—O—C\begin{smallmatrix}O\\O^-\end{smallmatrix}$	Peroxocarbonates
Amino	$R_2N—C\begin{smallmatrix}O\\O^-\end{smallmatrix}$	Carbamates
Absent	$O=C=O$	Carbon dioxide

15.6.2 Monocarboxylates

Monocarboxylic acid complexes have occupied a central position in coordination chemistry from the 19th century through to the present day.[1,2] Ubiquitous is an adjective frequently and appropriately used in conjunction with this ligand in any overview of coordination chemistry. Reasons for this omnipresence of, particularly, acetic acid (ethanoic acid), an early member of this series, include its convenient properties (b.p. 117.7 °C, f.p. 16.6 °C), ready availability (millions of tonnes produced annually) and its versatile coordination behaviour. Despite this long and intensive exposure as a ligand, carboxylic acids still continue to provide coordination chemists with exciting new species as any survey of the current literature will confirm.

15.6.2.1 Modes of Coordination

Early members of the carboxylate series (< four carbon atoms) form many stable coordination complexes. Formic acid offers the advantage of minimal steric requirements, while thermodynamic and kinetic data reveal that frequently propionic acid forms stable complexes at a faster rate than acetic acid. Derivatives of higher acids (> four carbon atoms) show the same chemical characteristics except that, as the carbon chain of the acid grows, the tendency to coordinate decreases.

The usual ligand behaviour of carboxylic acids is as oxygen donor, uninegative entities, but there are a limited number of examples of organometallic behaviour for acetic acid. Although the divisions may not be sharply distinct, three structural types of carboxylate oxygen atom coordination have been identified, *i.e.* unidentate, chelating and bridging. The first two of these are shown in Table 2; however, the most extensively investigated is that of bridging oxygen coordination where no less than five types of carboxylate bridge have been identified (Table 3).

A large number of carboxylate structures have been unambiguously determined by diffraction methods. This must be the preferred physical technique for all carboxylates and is essential for those complexes which do not fall clearly within one of these seven modes of bonding. In addition to

Table 2 Unidentate and Chelate Carboxylate Oxygen Atom Coordination

Type	Description	Example	Ref.
$-C\underset{\diagdown O}{\overset{\diagup O}{\langle}}M$	Ionic (A)	$Na(HCOO)$	3
$-C\underset{\diagdown O}{\overset{\diagup O-M}{\langle}}$	Unidentate (B)	$B(O_2CMe)_2(acac)$	4
$-C\underset{\diagdown O}{\overset{\diagup O}{\langle}}M$	Chelate (C)	$Zn(O_2CMe)_2\cdot 2H_2O$	5

Table 3 Bridging Carboxylate Oxygen Atom Coordination

Type	Bridging description	Example	Ref.
$-C\underset{\diagdown O-M}{\overset{\diagup O-M}{\langle}}$	*Syn–syn* (D)	$[Cr(O_2CMe)_2H_2O]_2$ $[Cd(O_2CCF_3)_2PPh_3]_2$	6
$-C\underset{\diagdown O}{\overset{\diagup O}{\langle}}$ with M above and below	*Anti–anti* (E)	$Cu(O_2CH)_2\cdot 4H_2O$	7
$-C\underset{\diagdown O}{\overset{\diagup O-M}{\langle}}$ M	*Anti–syn* (F)	$Cu(O_2CH)_2$	8
$-C\underset{\diagdown O-M}{\overset{\diagup O-M}{\langle}}$ M	Tridentate (G)	$Cu(O_2CMe)$	9
$-C\underset{\diagdown O-M}{\overset{\diagup O}{\langle}}$ M	Monatomic (H)	$[Cu(O_2CMe)L]_2$ [a]	10

[a] L = salicylaldimine derivative.

coordinated derivatives, ionic metal carboxylates (A) are well established and have considerable importance as 'benchmarks' for much carboxylate work. Many structures have been deduced from other physical techniques, *e.g.* IR, NMR, but frequently these result in errors and a complication of the issues. Typical of these is vibrational spectroscopy where difficulties arise from the low symmetry of the carboxylate grouping (C_{2v}). As a consequence of this, carboxylate coordination to a metal does not result in appreciable lowering of symmetry together with splitting of degenerate modes, and this has necessitated less than a rigorous approach to the vibrational spectroscopic study of carboxylate systems. As the CO_2^- stretching frequencies are usually the most prominent feature of the vibrational spectrum, attention has naturally focused upon those. The usual approach has been to relate the positions and separations (Δ) of the antisymmetric and symmetric carboxylate frequencies in the complexes to those of an authentic ionic salt.[11]

There have been numerous magnetic susceptibility and NMR studies of carboxylates, particularly those systems containing bridging coordination. Such studies have revealed much information about the metal interactions, often with valuable support from ESR, electronic and mass spectrometry. Yet still these techniques cannot be said to yield, sufficiently frequently, definitive information about the mode of coordination without the support of diffraction studies.

15.6.2.2 Unidentate and Chelate Coordination

Unidentate coordination (B) is well established and indeed extensive particularly in circumstances where only one coordination site is available, *e.g.* cobalt(III) pentaammine series.[12] Within the various forms of coordination available to carboxylates the unidentate has been the 'poor' relation until recently, but over the last few years there has been an upsurge of interest triggered by its possible role as an intermediate in the formation of more complicated structures.[13]

As unidentate ligands, carboxylates are expected to (i) lose the equivalence of the two carbon–oxygen bonds found in the anion; and (ii) have one metal–oxygen distance considerably shorter than the next shortest M—O contact. Lithium acetate dihydrate exemplifies this[14] with C—O distances of 133 and 122 pm and Li—O distances of 227 and 257 pm. Most examples of unidentate carboxylate complexes have this classical configuration of M(O—C) and C=O respectively so certainly the presence of features (i) and (ii) unambiguously determine this mode of coordination.

Interaction of the free oxygen atom with other entities in the molecule always remains a possibility and this serves to modify these distinctive structural features. For example,[15] hydrogen bonding with this oxygen atom can reduce the carbon–oxygen difference to 5 pm. Recent interesting examples of approximate equality of the formally single and double bond lengths of unidentate acetate coordination occur in the chromium, molybdenum and tungsten complexes $M(CO)_5(O_2CMe)$. In these complexes the C—O bond length to the coordinated oxygen is very similar to that of the free oxygen atom. There is doubtless a relationship between these dimensions and the reported chemical reactivity of these acetate groups as *cis* labilizing groups, where metal interaction with the free carboxylic oxygen may well stabilize a five-coordinate intermediate. Significantly, replacement of one carbon monoxide in the $M(CO)_5$ unit by PEt_3 results in the more usual acetate dimensions of 126 pm for M(O—C) and 119 pm for C=O and this structural change is accompanied by a reduction in the *cis* labilizing effect.[16]

This inequivalence of the two carbon–oxygen bonds also forms the basis of the IR method of detecting this form of coordination.[17] Unidentate carboxyl ligation is generally associated with a shift of the *ca.* 1600 cm^{-1} band of the free ion to a higher energy. This should increase Δ (the separation between the so-called $v_{as}(CO_2^-)$ and $v_{sym}(CO_2^-)$ stretching frequencies) relative to those of the free ion and, indeed, high values of Δ have found extensive use as indicators of unidentate carboxylate coordination. For example, in the best charted area of acetate complexes one would typically observe a strong antisymmetric $v(CO_2^-)$ vibration ~ 1680 cm^{-1} and a somewhat weaker symmetric $v(CO_2^-)$ vibration ~ 1420 cm^{-1}, a Δ of 260 cm^{-1} (the Δ value for ionic acetate is 164 cm^{-1}). Generally $\Delta \geqslant 200$ cm^{-1} is found in systems where structural studies show unidentate acetate coordination.[11] Assignment of these bands is seldom a problem but ^{13}C labelling of the CO_2^- carbon can be (and has been) used as confirmation.[13] The $v_{as}(CO_2^-)$ vibration is particularly susceptible to interactions of the free oxygen and, as was noted in the structural information, a variety of effects can lower this.[18] In addition, the $v(CO_2^-)$ vibrational modes are not sufficiently sensitive to detect unidentate coordination when there is an approximate equality of the C—O bond lengths. Thus care must be exercised in the application of this approach to unidentate coordination, particularly when this is coupled with the observation[11] that sample preparation can also have an effect upon Δ.

Symmetrical chelate coordination (C) is less common than unidentate coordination. Both symmetrical chelation and bridging carboxylates maintain the equivalence of the two carbon–oxygen bonds found in the free ion,[19] although in chelating systems the OCO angle is usually smaller than that found for bridging systems. When this is combined with the 'bite' in chelates (223 pm)[20] then the symmetrical four-membered ring occurs far less than is found in nitrate systems (a frequently compared system). Relief of the steric factors induced by these constraints results in an extensive chemistry of appreciably unsymmetrical forms,[21] *e.g.* $Hg(O_2CMe)_2(phos)_3$. Symmetrical chelate formation is exemplified[22] by the complex $Ru(O_2CMe)H(PPh_3)_3$ with C—O distances of 125 and 126 pm and with the OCO angle of 115°.

The equivalence of the two C—O distances in both symmetrical chelation and bridging means that the technique of IR spectroscopy has a very limited use for differentiating between these modes. Δ separations significantly less than ionic values are certainly indicative of chelating and/or bridging carboxylates. Generally, smaller Δ values for chelating than bridging may be anticipated but this should not form the sole basis of structural conclusions.

15.6.2.3 Bridging Systems

There are numerous complexes where carboxylate ligands serve as a bridge between metal atoms (similar or different metals). Within this group there can be little doubt that the classic structural type with *syn–syn* bridging as shown in (D) is the most common. The published literature on such systems is approximately four times that of all other carboxylate systems together. Within this *syn–syn* bridging class the two most widely studied systems are the so-called 'paddlewheel' structure for dimetal complexes and the 'triangular' carboxylates for trimetal systems (Figure 1). This mode of bridging demands that the metal atoms are adjacent and superficially may be regarded as replacement of the enolic hydrogen atoms in the hydrogen-bonded form of the parent acid. The other bridging modes (E) and (F) are usually found in polymeric carboxylate complexes and it is noteworthy that the formate ligand is particularly versatile in its bridging ability. Carboxylate bridging in the *anti–syn* or *anti–anti* configurations results in large metal–metal separations, which in part accounts for the less extensive study of these complexes. Special mention should be made of the monatomic carboxylate bridge between two metals (H) often found in alkoxide chemistry. This one-atom bridge[9] is seldom the only link between the metal atoms; frequently the other carboxylate oxygen atom is also bound to a metal, making tridentate a better description of this mode of bonding (G).[23]

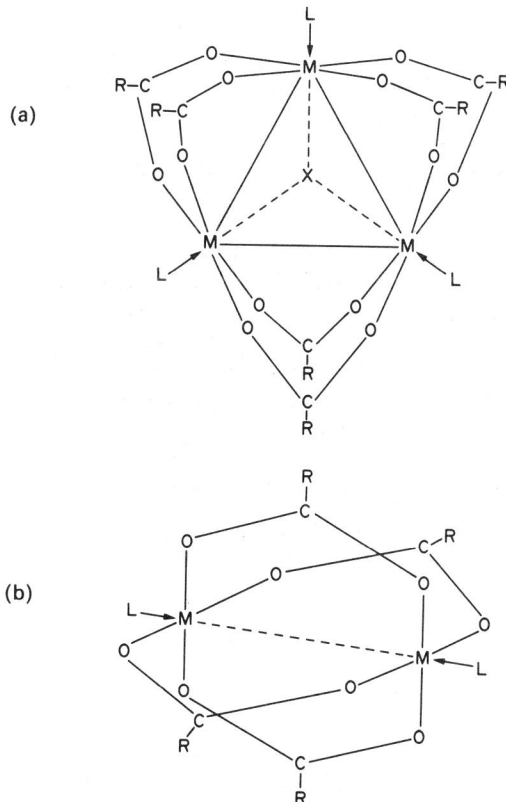

Figure 1 *Syn–syn* carboxylate bridges: (a) trinuclear 'triangular' carboxylate, $[M_3(O_2CR)_6L_3]^{n\pm}$; (b) dinuclear 'paddle-wheel' carboxylate, $[M_2(O_2CR)_4L_2]$

Bridging carboxylate groupings are best, and sometimes can only be recognized by structural determinations. Structural dimensions are singularly unexceptional; usually the equivalence of the carbon–oxygen bonds is retained. In the *syn–syn* bridging mode there is frequently a correlation between the C—O bond length and the OCO angle so as to retain the O—O 'bite'. C—O distances of 125 pm with OCO approximately 125° are typical but a dependence upon the acid strength has been reported.[24,25] In this mode the main structural interest is concerned more with the distance between the metal atoms than the dimensions of the ligand system.

This equivalence of the C—O bonds further underlines the difficulties vibrational spectroscopists have in differentiating bridging and chelating systems. A correlation of Δ with bridging or chelation of the ligand has been attempted for acetate systems, where a small Δ value ($< 105\,cm^{-1}$) may indicate chelation rather than bridging, but this is not a definitive method of distinction.[11] More recently a new criterion has been proposed for determination of the carboxylate bonding mode. Unidentate acetate shows three absorptions arising from CO_2^- deformations ($720–920\,cm^{-1}$) together with a strong band at $540\,cm^{-1}$ from the out-of-plane CO_2^- bending. These absorption bands are said to be absent for bridging and reduced in number for chelating acetate.[26] Magnetically the presence of a reduced magnetic moment may be used to detect *syn–syn* carboxylate bridging,[27] *e.g.* copper(II) complexes. However, such properties are not demanded by *syn–syn* bridging so the role of this technique is better as a supplement to other physical methods.

15.6.2.4 Paddlewheel Systems

Many carboxylic acids exist in the free state as hydrogen-bonded dimers with an oxygen–oxygen separation (between oxygens linked to the same carbon atom) close to 220 pm. Replacement of these hydrogens by two metal atoms results in the 'close' approach of the two metal atoms.[28] Much of the interest in these dinuclear paddlewheel systems has been generated by a need to understand the nature of these metal–metal interactions. The natural desire of chemists is to use formal bond orders as an index of this interaction and much has been published to this end. However, there are growing indications that such a formal concept is not entirely appropriate for such systems.[29]

The tetrakis(carboxylato) dimetal complexes, although first prepared in 1844 as chromium(II) acetate monohydrate, were not documented until the 1953 structural determination of copper(II) acetate monohydrate. This structure exhibits a spectrum of metal–metal interactions ranging from no interaction, through weak spin-pairing, to strong interactions. Associated with these interactions there is a wide range of metal–metal distances that can be accommodated by the paddlewheel, ranging from 370 pm in the vanadium trifluoroacetate dimer to approximately 200 pm for molybdenum complexes. However, the M—M distances for a given metal usually fall in a fairly narrow range characteristic of the metal. For example, bridging in this paddlewheel style leads to Cu—Cu distances typically about 260 pm; Cr—Cr, 230 pm; Re—Re, 225 pm; and Mo—Mo around 210 pm. Much effort has been directed towards an understanding of the relationship between the carboxylate ligands and the lengths of these M—M distances. Extra ligands, L, may occupy sites along the M—M direction (referred to as axial ligands) and these too have a role to play in determining these M—M distances.

The nature of the Cu—Cu interaction in dimeric copper(II) complexes has been the object of many experimental and theoretical investigations.[30,44] Temperature dependent magnetic susceptibility and ESR data are usually interpreted in terms of a model based upon two components: (i) direct Cu—Cu interaction in combination with (ii) superexchange *via* the carboxylate ligand. The direct Cu—Cu interaction can be represented by a singlet ground state separated from a triplet excited state by $-2J$ (typically $300\,cm^{-1}$ in acetate complexes). This small energy separation results in appreciable thermal population of the triplet state at ambient temperatures, also recently observed in Os^{III} carboxylates.[31] This direct interaction is dependent upon the M—M separation and also, in addition to an axial ligand dependency, has a small dependence upon the donor strength of the carboxylate ligand. The role of the carboxylate ligand as a vehicle for superexchange is not limited to the *syn–syn* bridging mode but is found in systems with large intermetal distances, such as the *syn–anti* and *anti–anti* bridging arrangements of copper(II) formate.[32] Possible pathways for this exchange are *via* the π-electron cloud and the σ-framework of the carboxylate ligand. Variation of the carboxyl carbon substituent will thus have a marked effect on the extent of the superexchange in these systems and the great strength of carboxylates in this role is that a wide variation of this substituent (for fine tuning in an electronic sense) can be achieved.

At the other end of the spectrum from these weakly interacting metal derivatives lie the strongly M—M bonded species such as $Mo_2(O_2CR)_4$, which contain a formal quadruple bond. Arguably the role of the carboxylate ligand is ancillary to the main thrust of the chemical interest in the metal–metal bonds. The main distinction of the carboxylates was to provide the first examples of strongly bonded metal systems combined with a capacity to provide a large number of closely related complexes. Again, simple variation of the carboxyl carbon substitutent does perturb the M—M bond length. These factors, when coupled with the favourable geometry of the *syn–syn* bridge, make carboxylates ideal ligands to bridge two metal atoms linked by a multiple bond. Many reviews are available which chart this extensive area[33] but special attention should be drawn to two aspects, tungsten(II) carboxylates and the class of substituents provided by amino acids and peptides. Repeated failures to obtain stable tungsten(II) carboxylate derivatives have hindered the general development of W—W multiply bonded species. These failures were not due to the inability of the carboxylate ligand to bridge two tungsten atoms, but rather to the susceptibility of the products to oxidation. This problem has recently been overcome and a route to these systems is now available.[34] Much of the amino acid carboxylate coordination chemistry remains to be elucidated but stable, well-characterized examples of *syn–syn* bridging have been isolated and further developments in the coordination chemistry of these biologically significant molecules are anticipated.[35]

Occupying a central position in the spectrum of metal interaction in *syn–syn* bridged carboxylate systems lie dirhodium compounds. The Rh—Rh bond lengths in such compounds are generally in the range 239–247 pm, shorter than anticipated for what now seems well established as a single bond. This area has been reviewed.[36]

15.6.2.5 Trinuclear Carboxylates

The essential components of these systems (Figure 1) are a triangle of metal atoms with six edge-bridging carboxylate groups and unidentate ligands (L) attached to each metal. Such systems may usefully be viewed as three octahedra sharing a common vertex at the centre of an equilateral triangle. A triply bridging oxygen is often the atom at this vertex and a large class of so-called 'basic' carboxylates have this arrangement. A number of cations of stoichiometry $[M_3(O)-(O_2CR)_6(H_2O)_3]^+$ have been known and structurally characterized for many years, *e.g.* M = Cr, V.[37] This stoichiometry confers on each of the three metal atoms of the triangle a formal oxidation state of three. Over the past few years an increasing number of systems have been reported with the same triangular shape but with metal atoms in non-integer oxidation states. The first example of this[38] was provided by $[Ru_3(O)(O_2CMe)_6(PPh_3)_3]$ (from the triphenylphosphine reduction of basic ruthenium(III) acetate) in which the metal atoms have a formal oxidation state of 8/3. Further reductions to the +2 state are also reported and in this oxidation state the central oxygen is lost from the triangle of ruthenium atoms, but can be readily reinserted by molecular oxygen. Thus it is not a requirement for such triangular complexes that the triply bridging atom be in the plane of the metal atoms; indeed, examples are known where the bridging atoms lie above and/or below this triangular plane, resulting in a mono- or bi-capped arrangement.[39]

As was observed for the dinuclear systems, the role of the carboxylate ligand in these trinuclear *syn–syn* bridged systems is very much that of a supporting role and structurally the RCO_2^- unit is little different in either system. A recent tabulation of structural data for these systems is available.[40] Other physical properties of these trinuclear transition metal carboxylates have attracted the attention of investigators for more than 20 years and, although much work remains to be done for a complete understanding to be achieved, a number of parallels with the dinuclear systems can be drawn. Vibrations of the $v(CO_2^-)$ unit are expected to be, and indeed are, very similar to those of the dinuclear complexes with the added difficulty of band assignment in these larger molecules. Many of the cations $[M_3(O)(O_2CR)_6L_3]^+$ have variable-temperature magnetic susceptibility properties reminiscent of the paddlewheel system.[41] Indeed, the molecule $[Cr_3(O)(O_2CMe)_6(H_2O)_3]Cl$ has played a similar role in the development of the understanding of these, as $Cu_2(O_2CMe)_4(H_2O)_2$ has for the dinuclear series. It is inconceivable that the interpretation for these triangular systems should differ qualitatively from the dinuclear unit but quantitatively these are one metal atom, two carboxylate ligands and a central vertex atom different. When this is combined with the less extensive experimental and theoretical work relative to the dinuclear system it is easy to understand why many problems still remain in this area. There is general agreement that metal–metal interaction occurs between the three magnetically coupled atoms. Whether this coupling is equivalent or non-equivalent amongst the three atoms seems to depend upon the mode of preparation, the metal atom itself and the associated ligands.[42] Mössbauer and electronic spectroscopy are just two techniques[43] which have been used in the search for a definitive result. It is often concluded for these trivalent species that metal atoms in 'slightly inequivalent sites' are present, yet it is ironic that in systems where inequivalent sites might be expected (*i.e.* valence trapping in non-integer metal oxidation state systems) there is little evidence for this.[40] Much work remains to be done on the physical characteristics of these systems. In both dinuclear and trinuclear systems the axial ligand, L, may also serve as a bridge to other units and it is appropriate to note that this ligand may be a carboxylate ligand.[45]

15.6.2.6 Other Polynuclear Carboxylates

Syn–syn carboxylate bridging of four metal atoms often results in a tetrahedron of metal atoms with an oxygen atom at the centre of the tetrahedron and a carboxylate bridge along each of the six edges, *e.g.* $[M_4(O)(O_2CR_6)]$, but this is by no means the only arrangement of four metal atoms.[43] As is anticipated for such *syn–syn* bridged systems, substantial cooperative effects in physical properties are observed, but full interpretation for these more complicated systems has yet to be reported.

One-atom carboxylate bridging (monatomic) results in the presence of an M—O—M unit in these complexes. This is a well established bridging mode for most ligands but, as yet, there are relatively few examples in carboxylate chemistry. Dimers with monatomic bridges between paramagnetic atoms have an extensive history of magnetic interactions between these centres and this is found in various copper(II) complexes.[46] In carboxylate systems this mode of bonding has structural characteristics resembling unidentate coordination in that the carboxyl oxygen coordinated to two metals is significantly longer (~ 129 pm) than the other C—O length[47] (~ 121 pm). However, the 'pendant' oxygen is available for further interaction (either hydrogen-bonding[48] or coordination to another metal), giving tridentate carboxylate coordination.[23]

15.6.2.7 Carboxylates as Organometallic Ligands

Acetic acid complexes are known which have various transition metals (*e.g.* Pd, Mo, Fe) σ-bonded to the methylene of the acid, giving rise to the grouping M—CH_2COOH. These linkage isomers of unidentate oxygen coordination result from COOH activation of the saturated methyl C—H unit.[49] The physical characteristics of the resulting metalated acids resemble those of acetic acid itself, frequently forming hydrogen-bonded dimers even though the acids themselves are much weaker than the corresponding hydrocarbon (pK_a acetic acid, 4.8; pK_a when M = Fe, 6.7). This mode of coordination, again best detected by diffraction methods, is also readily detected by ^{1}H NMR. There are also rare examples of chelate formation *via* both the carbon and oxygen atoms of acetic acid where formally the acid behaves as a dinegative ligand (reaction 1).

In this palladium complex the Pd—O bond appears to be more readily cleaved than the Pd—C unit, resulting in Pd—CH_2COOH coordination.[50]

$$(1)$$

15.6.2.8 Synthesis and Reactivity

The wide variety of carboxylic acids (particularly in acidic and steric properties) has led to an equally large range of preparative methods for these coordination complexes. Commonly used methods of synthesis include refluxing the appropriate acid (or anhydride) with metal carbonate, sulfate or oxide or treatment of the sodium or silver salt of the acid with the metal halide. There have been recent notable achievements in the synthesis of previously elusive tungsten carboxylates.[34] Carboxylate exchange reactions provide another synthetic route and have proved particularly successful for trifluoroacetate derivatives. Sacrificial anode electrochemical methods have proved successful as a route to some copper(II) complexes, while vapour synthesis appears capable of providing a very general route to carboxylate complexes.[51] The insertion of carbon dioxide into some, but not all, σ-bonded organotransition metal complexes also results in carboxylate complexes.[52] The ability of carboxylic acids to add oxidatively to lower oxidation state metal complexes has found wide use not only with platinum group metals but also with metal carbonyls.[53] One development of significance concerns the low yield of $Mo_2(O_2CMe)_4$ from the oxidation of molybdenum carbonyl by acetic acid/acetic anhydride mixtures. Although, by careful control of experimental conditions, yields of the dimer can be optimized,[54] in the original preparative work less than 25% of the molybdenum is used to form the dimer. In a search for the destination of the remaining metal, Cotton[55] has found that a triangular Mo_3 unit is also produced and there is evidence for a unidentate carboxylate $Mo(CO)_5(O_2CMe)$ precursor. Thus it appears possible that the formation of multinuclear metal carboxylates may be capable of control so as to give the desired product. Reactions of the paddlewheel structure have been extensively studied and are usually classified according to whether there is retention or cleavage of the metal interactions.[56]

Thallium(III) and mercury(II) carboxylates are frequently used reagents for the metalation of a variety of aromatic materials. The trifluoroacetate is often more effective than the corresponding hydrocarbon, which reflects the greater lability of the M—O bond in the trifluoro series. This lability of the trifluoroacetate group in $Mo(CO)_5(O_2CCF_3)$ also provides a rapid room temperature route to substituted $Mo(CO)_5L$ derivatives of this compound.[57] The bridged cationic ruthenium and rhodium carboxylates catalyze the hydrogenation of alkenes and alkynes[58] and various industrial polymerization processes make extensive use of coordination complexes of carboxylates.[59] The strong affinity of rhodium(II) carboxylates for axial substituents has led to a number of catalytic and anticancer studies involving this system.[60] Thermal decarboxylation can, in some instances, be a convenient route to the corresponding alkyl derivative.[61] Carbon monoxide reacts with the carboxylates of both platinum(II) and mercury(II), when alkoxycarbonyl derivatives result.[62] The hydrolysis reactions of cobalt(III) carboxylates have been reviewed,[63] while iron organometallic carboxylic acids have a role to play in the understanding of the water-gas shift reaction.[176]

15.6.3 DICARBOXYLATES AND RELATED SPECIES

Oxalic acid (ethanedioic acid) is one of the oldest and also the simplest of the dibasic acids. Complexes derived from this have been extensively investigated although, surprisingly, no review of the complexes has been undertaken since 1961.[64] Coordination complexes of oxalic acid occur naturally, *e.g.* as the iron complex in the mineral humboltine, and have wide application in the industrial field from chrome tanning through to the biochemical field, *e.g.* molybdenum. The extraordinary versatility of dicarboxylate ligands upon coordination is well known and this together with the many polymeric species makes their structural chemistry particularly interesting. One useful view when considering the coordination chemistry of dibasic acids is as two carboxylate groups linked by a carbon–carbon bond(s). So they show characteristic carboxylate coordination but, in addition, the close proximity of two CO_2^- groups enables oxalates to act as chelating ligands using one oxygen from each of the CO_2^- groups. From the number of C—O stretching vibrations it is, in principle, possible to make a distinction between the differing modes of coordination, although only a full X-ray structural analysis can solve the problem unambiguously. As before, their coordination chemistry will be considered in terms of the three structural types, *viz.* unidentate, chelating and bridging.

15.6.3.1 Unidentate and Chelate Coordination

Although there are examples of dibasic acids behaving as unidentate ligands it is more common to find them as chelating ligands. Chelation to metal can occur in two different ways and the literature nomenclature is frequently confusing. This review will adapt from organometallic chemistry the *hapto* notation, illustrated here for oxalate complexes. Chelation as in (1) will be referred to as η^3 chelation, denoting that three ligand atoms participate in the chelate ring. The more widespread chelation, as in (2), thus becomes η^4 chelation.

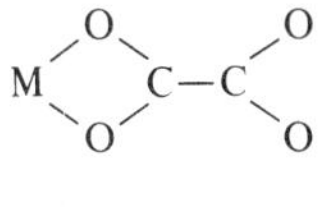

(1) 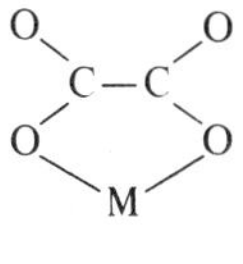   (2)

Complexes containing unidentate coordination have depended for their characterization upon physicochemical methods rather than structural determinations. The monomeric octahedral cobalt(III) complex $[Co(en)_2X(ox)]$, where X = halogen or OH, provides one of the few examples of unidentate oxalate coordination; the conformation was determined from IR data.[65] By contrast, there are numerous examples of structurally characterized chelating dicarboxylate systems. There are several structural determinations of different forms of oxalic acid and its ions but the most useful for comparative purposes are those of α-oxalic acid and anhydrous sodium oxalate. Using the planar (D_{2h}) oxalate ion and acid as benchmarks it can be seen (Table 4) that the η^4 oxalate ligand (bite = 265 pm)[66] has two long and two short C—O bonds. The coordinated C—O(M) bond resembles the C—O bond of the acid while the free C=O bonds are similar to those in the oxalate ion. The C—C distances show little variation over the complexes while the OCC angles reflect their involvement in the chelate ring. The characteristic chelate coordination of the CO_2^- group (η^3 chelation) occurs more widely with the higher members of the dicarboxylate series. The dimensions of the four-membered malonic acid chelate in the calcium complex $Ca(Mal)(H_2O)_2$ are typical of those found for monobasic acids and this is generally true of all η^3 chelating polycarboxylic acids.[74] The steric strain produced by η^3 chelation results in a strong preference for these acids to form chelate rings with more than four atoms. We have already seen this with the preference for five-atom ring chelation in the oxalate series and further evidence of this is provided by the six-membered ring (η^5 chelation) shown by complexes of malonic acid.[75] The conformational possibilities of these large ring systems confer additional interest to these complexes.[76] A normal coordinate analysis of divalent metal η^4 oxalate complexes suggests a linear relationship between the M—O, C—O and C—O(M) stretching frequencies[77] and, more generally, IR spectroscopy and ^{13}C or ^1H NMR can be useful probes to provide structural information.[78,81]

Table 4 Bond Distances and Angles for η^4 Oxalate Complexes

Compound	C—O(M) (pm)	C=O (pm)	C—C (pm)	OCC (°)	Ref.
$[Cr(en)_2(C_2O_4)]^+$	130	122	153	114	67
$[Co(C_2O_4)N_4]$	130	121	156	114	68
$[MoO(O_2)_2(C_2O_4)]^{2-}$	129	121	156	115	69
$[Cu(C_2O_4)_2(H_2O)_2]^{2-}$	128	123	154	115	70
$[Mo_2O_5(C_2O_4)_2(H_2O)_2]^{2-}$	126	118	161–156	112	71
α-$C_2O_4H_2$	129	119	156	—	72
$Na_2C_2O_4$	123	123	154	121	73

15.6.3.2 Bridging Systems

There are many possible bridging modes available to dibasic acid complexes and some of these are illustrated in Table 5.

All three of the illustrated modes (A), (B) and (C) are known for bridging systems which involve two of the four acid oxygen atoms. It is only for the organic molecule dimethyl oxalate, $Me_2C_2O_4$, that a *trans* structure has been confirmed by diffraction methods,[79] but this mode has also been proposed for the oxalate intermediate[80] $[(NH_3)_5Co(C_2O_4)Cr(H_2O)_5]^{4+}$. A *cis* bridging system is indicated[81] by IR measurements upon the cobalt(III) complex $[(NH_3)_4(H_2O)Co(C_2O_4)-Co(NH_3)_5]^{4+}$, while *syn–syn* bridging using just one of the two available acid groupings is found in cobalt(III) amine complexes.[80] The complexes $Sc(OH)(C_3H_2O_4)\cdot 2H_2O$ and $[\{(NH_3)_3Co\}_2(NH_3)_5Co(OH)(C_2O_4)]^{5+}$ provide examples of the tridentate bridges (D)[82] and (E)[83] respectively.

Tetradentate bridging of dicarboxylate groups is by far the most common form of bridging with the planar, doubly chelating system (F) the most prevalent. Two η^4 chelates in the oxalate system $\{Bu_3P(Cl)Pd\}_2(C_2O_4)$ were identified in 1938 but the first complete structural analysis[84] was of the iron oxalate mineral humboltine in 1957. Since that date, many crystal structure determinations have demonstrated the extent of this coordination mode. These are typified by the complexes $[(py_4Ru)_2(C_2O_4)]^{2+}$ and $[(triammineCu)_2(C_2O_4)]^{2-}$, from which it can be seen that the molecular dimensions are relatively invariant and closely resemble those of ionic oxalate systems in both C—O (126 pm) and C—C (154 pm) bond distances.[85] Two *syn–syn* bridges (G) often result in linked paddlewheel type dimers, as seen[86] in the succinate and glutarate complexes of copper(II). Such coordination is not favoured by either oxalates or malonates as insufficient CH_2 groups are present to allow linear bridge formation. Tetradentate ligand behaviour, producing one six- and two four-membered rings (H), has been structurally characterized for the malonate complexes of both europium, $Eu_2(mal)_3\cdot 8H_2O$, and cadmium, $Cd(mal)\cdot H_2O$. The bridging oxygen atoms are not equally shared between the metal atoms as the C—O bond is shorter (124 pm) on the six-membered chelate ring side.[87] Differences in ligand geometry are also reflected in the 'bite' of the two rings which are ~ 217 pm for the four-membered carboxylate ring as opposed to 300 pm in the larger, less constrained six-membered ring. Finally in this series, the four oxygen atoms of the two carboxylates are thought to bridge as in (I) and (J) in the oxalate complex[88] $[\{(NH_3)_3Co\}_4(\mu\text{-}OH)_4(\mu\text{-}C_2O_4)]^{6+}$ and the malonate complex[75] $Nd(C_3H_2O_4)_3\cdot 8H_2O$.

Many of the usual physical techniques are of limited use for the identification of these various coordination modes, but two reactions that have aided characterization are those of protonation and reduction. The lack of protonation of the η^4 oxalate chelate $[Co(NH_3)_4(C_2O_4)]^+$ contrasts with

Table 5 Dibasic Acid Complexes — Bridging Modes

Bridging via *two oxygen atoms (bidentate)*

(A) *cis*   (B) *trans*   (C) *syn–syn*

Bridging via *three oxygen atoms (tridentate)*

(D) unidentate + η^n chelation   (E) unidentate + *syn–syn*

Bridging via *four oxygen atoms (tetradentate)*

(F) two-(η^n chelation)   (G) two-(*syn–syn*)   (H) two-(η^3 chelation) + η^n chelation

(I) two-(η^3 chelation)   (J) two-(η^3 chelation + monatomic)

the ease of the protonation for the oxalate entity in the unidentate and bridging mode (C), both of which have uncoordinated CO_2^- groups.[89] Further distinctions have been possible, in the case of cobalt(III) complexes, by examining rates of chromium(II) reduction. This method is dependent upon individual oxygen atoms of the CO_2^- group having little tendency to bind to more than one metal. Thus, for bridging systems involving all four oxygen atoms, reduction is very slow, in contrast to those systems which contain non-coordinated carboxylic oxygen atoms where fast rates of reduction are observed.[90] By a combination of both reactions it is possible to gain a valuable insight into the nature of dicarboxylate coordination.[81]

As is to be expected from such extensively delocalized bridging systems, studies of the various metal interactions have provided a rich field for magnetochemists. As with the monocarboxylate series, it is again the copper complexes that have been most extensively studied. With the exception of copper malonate (which is magnetically 'normal'), all compounds give anomalous magnetic data. The three oxalate bridged[91] complexes $Cu(C_2O_4)\cdot\frac{1}{3}H_2O$, $Cu(C_2O_4)(NH_3)_2\cdot 2H_2O$ and $[Cu(C_2O_4)_2]^{2-}$ exemplify the various contributions of antiferromagnetic and ferromagnetic interactions to the magnetic properties. So variable are the magnetic properties associated with an oxalate bridge network that it has been suggested that they may well form the basis of tunable interaction systems.[92]

15.6.3.3 Synthesis and Reactivity

There are two principal synthetic routes to dicarboxylate complexes. One of these uses an aqueous solution of the alkali metal dicarboxylate and the corresponding metal halide,[93] while the other depends upon the dicarboxylic acid reduction of higher oxidation state metals. This reductive property of oxalic acid results in its ready dissolution of iron oxides and hence a cleaning utility in nuclear power plants.[94] Mention must also be made of the successful ligand exchange synthesis of molybdenum dicarboxylates, $Mo(dicarboxylate)_2 \cdot nH_2O$, from the corresponding acetate complex. Unfortunately the polymeric, amorphous and insoluble nature of these complexes has restricted the study of these systems, which may well provide examples of multiple M—M bonding in dicarboxylate coordination chemistry.[95]

Dicarboxylate complexes (particularly oxalato systems) have been extensively studied and a flavour of their varied chemistry is presented below. Bis(oxalato)platinum complexes have η^4 oxalate chelation with a square-planar stereochemistry about the platinum. Many of these are bronze coloured, mixed valence systems, typical of which is the complex $K_{1.6}Pt(C_2O_4)_2(H_2O)_x$, which has the planar units stacked above one another so that the Pt—Pt separation is $\sim 275\,pm$. Crystals of this belong to the class of highly conducting inorganic materials which depend for their properties upon this columnal stacking, with the high electrical conductivity along the Pt—Pt direction making them of interest as possible one-dimensional metals.[96] Werner[97] was the first to resolve the η^4 chelate oxalate complex $[Cr(C_2O_4)_3]^{3-}$, and since then numerous related studies have been made.[64] Ligand substitution reactions of chelated dicarboxylate complexes are further areas of intensive investigation, no doubt related to the considerable anticancer activity shown by these systems.[98] Kinetic data suggest a ring opening mechanism for many of these substitution reactions with intermediates involving unidentate coordination.[99]

Photolysis of oxalate complexes show that there is a strong tendency to undergo photoredox decompositions, resulting in oxides of carbon. Two useful applications of this are (i) the system based on the redox photolysis of aqueous $[Fe(C_2O_4)_3]^{3-}$ is widely used for chemical actinometry;[100] and (ii) UV irradiation of $(phos)_2M(C_2O_4)$ complexes (M = Pd, Pt) result in loss of two molecules of carbon dioxide and production of the synthetically useful, coordinatively unsaturated M^0 complex $(phos)_2M$.[19] One reaction which, if generally applicable to dicarboxylate complexes, may have considerable impact upon the validity of physical measurements upon these systems is the rather unusual, room temperature, solid-state reaction (2).[102]

$$2[Cu(dipy)(C_2O_4)] \longrightarrow [Cu(dipy)_2][Cu(C_2O_4)_2] \qquad (2)$$

15.6.4 CARBONATES

Carbonate is an important ligand not only in chemistry but also biochemistry and geochemistry. The reason for this is to be found in the wide occurrence in Nature of carbon dioxide and its ready hydration to give carbonate based systems. Thus there are many naturally occurring carbonates *e.g.* calcite, malachite and these have been reviewed.[103] The range of review literature prior to 1970 is extensive and the reader is encouraged to consult this.[104,105]

The carbonate ligand, isoelectronic with nitrate, is a versatile ligand, which is as expected in the context of the ligand behaviour of carboxy groups, for carbonate can be viewed as three carboxy groups back to back. This extensive chemistry has had one unfortunate consequence in that the nomenclature used to describe the various forms of coordination is confusing and lacks a system. This section will attempt to systematize this using as its basis the concept of three carboxy groupings bound together and the structural types as identified in the carboxylic acid section.

Structural determinations have allowed three classes of carbonate complexes to be recognized. The first two of these are illustrated in Table 6, with the ionic carbonate for reference, while the more complex bridging carbonate modes are in Table 7.

Vibrational spectroscopy might be anticipated to be of more definitive use for carbonate systems than those so far considered, as the D_{3h} symmetry of the carbonate ion is lowered upon coordination. For the non-bridging modes this is true and much use has been made of this[15] but, for bridging modes of coordination, it is of much more limited use. In the η^3 chelating mode the $v(C=O)$ of the uncomplexed carbon–oxygen bond has been used as a probe to study the effect of ligand changes at the metal.[104] ^{13}C NMR may be used to give some qualitative ideas as to the coordination of the carbonate in solution but, yet again, reference to structural studies underpins the usefulness of such an approach.[109] As is expected, magnetic coupling between paramagnetic centres bridged by the

Table 6 Unidentate and Chelate Carbonate Coordination

Type	Description	Example	Ref.
$O=C\begin{smallmatrix}O\\O\end{smallmatrix}$	Ionic (A)	$CaCO_3$ (calcite)	106
$O=C\begin{smallmatrix}O-M\\O\end{smallmatrix}$	Unidentate (B)	$[Co(NH_3)_5(CO_3)]^+$	107
$O=C\begin{smallmatrix}O\\O\end{smallmatrix}M$	η^3 Chelate (C)	$[Co(NH_3)_4(CO_3)]^+$	108

Table 7 Complex Bridging Carbonate Coordination

Type	Description	Example	Ref.
Bridging using two oxygen atoms			
$O=C\begin{smallmatrix}O-M\\O-M\end{smallmatrix}$	Syn–syn (D)	$[Rh_2(CO_3)_4(H_2O_2)]^{4-}$	114
$O=C\begin{smallmatrix}O-M\\O\end{smallmatrix}M$	Syn–anti (E)	$[Cu(CO_3)_2]^{2-}$	115
Bridging using three oxygen atoms			
$O\begin{smallmatrix}M-O\\O-C\\M-O\end{smallmatrix}$	Anti–anti + monatomic or two η^3 chelate rings (F)	$[(CuL)_2(CO_3)]^{2+}$	116
$M-O-C\begin{smallmatrix}O-M\\O\end{smallmatrix}M$	Anti–anti–anti (G)	$[\{CuL(H_2O)\}_3(CO_3)(NO_3)_4]$	117
$M-O-C\begin{smallmatrix}O\\O\end{smallmatrix}M$	Chelate + unidentate (H)	$(Ph_4Sb)_2CO_3$	118
$M\begin{smallmatrix}M\\O-C\\M\end{smallmatrix}\begin{smallmatrix}O-M\\O\end{smallmatrix}$	Anti–syn + Monatomic (I)	$Cu_2(OH)_2(CO_3)$	119

carbonate group is observed but, surprisingly, much less extensive investigations relative to the carboxylate series have been made.[110]

15.6.4.1 Unidentate and Chelate Coordination

When the carbonate group acts as a unidentate ligand the coordinated C—O(M) bond is longer (~ 131 pm) than that established for the free ion (129 pm), while the non-coordinated C—O bonds are appreciably shorter (~ 120 pm). These two oxygen atoms are also free to interact with other moieties in the molecule and this is often reflected in their dimensions.[111] Unidentate coordination also distorts the D_{3h} symmetry of the ion and this is reflected in the C—O vibrational modes. The carbonate ion exhibits different vibrational spectra as the site symmetry of the group may be in either a calcite or aragonite environment. Using the calcite structure as a benchmark the strong single absorption peak in the 1430–1490 cm^{-1} region is replaced by two intense absorptions, the splitting of which is smaller (typically 80 cm^{-1}) than in chelate complexes. There are a number of other absorption bands in the spectra of unidentate complexes which are absent in the spectra of ionic carbonates and detailed studies have been made on these.[112] However, it should always be borne in mind that the role played by the 'free' oxygen atoms can materially affect the usefulness of vibrational studies.

Symmetrical chelation of the carbonate group to a metal results in planar four-membered ring formation. The usual carbonate bonding features are two equal C—O bond lengths for the oxygen atoms coordinated to the metal (134 pm); the remaining C—O bond is shorter (124 pm), reflecting the greater double bond character. The strain involved in the four-membered ring is evident from the OCO angle which is usually close to 111°. Structural data have been tabulated.[113] The splitting of the $v(CO_2)$ for chelation is typically 300 cm^{-1} or greater, but as the chelation becomes asymmetrical the carbonate ligand increasingly resembles the unidentate form and this is reflected in the magnitude of this splitting.

15.6.4.2 Bridging Systems

For the identification of the various bridging modes, diffraction data are essential. The higher charge on the carbonate ligand, relative to monocarboxylates, favours anionic entities requiring balancing cations with the increased possibility of their involvement in a bridged structure. When the carbonate uses two of its three oxygen atoms to bridge two metal atoms, (D) and (F), there is a very close analogy with similar carboxylate systems. This is to be expected if the carbonate is considered to be a monocarboxylate with the organic group replaced by O or OH. The presence of dinuclear *syn–syn* bridging in a paddlewheel arrangement has been found in both the chromium(II) and rhodium(II) carbonates.[83] As was observed for the carboxylates, the main structural interest in these systems is the distance between the metal atoms. The three C—O distances of the carbonate show little (Cr) or no (Rh) differences, possibly reflecting the involvement of the 'free' oxygen with the water molecules in the crystal. There is certainly not the same range of *syn–syn* bridged complexes as is found in the monocarboxylate series, but this more reflects endeavours in this area rather than any inherent limitation within the carbonate ligand. *Anti–syn* metal bridging, as found in the anhydrous $Na_2Cu(CO_3)_2$, has the 'free' C—O bond appreciably shorter (122 pm) than the two coordinated C—O bonds (129 and 136 pm). Considerable distortion of the carbonate ligand is indicated by the variation of these latter two C—O distances and this is reflected in the IR spectrum, where the asymmetric CO vibrations are appreciably split. In bridging carbonate systems of this and subsequent types, planarity of the four ligand atoms and the metal is not always found. The p_π orbital of the coordinated oxygen may be involved in the metal–oxygen link and thus not be available for π-bonding within the carbonate ligand itself; this may well be revealed by different C—O(M) distances. Magnetic properties of *syn–anti* linked metal atoms are consistent with ferromagnetic coupling between the paramagnetic centres.[120]

The predominant bridging mode where the carbonate uses all three oxygen atoms is represented by (F). Despite the apparent symmetry of this mode of bonding, the OCO angles do not correspond to the idealized 120° value; the chelate portion of the ligand has smaller angles to accommodate the four-membered ring. Carbon–oxygen bond lengths range from 125 to 131 pm, reflecting the differing degrees of interactions with the various metals. The two bridging modes (G) and (H) are much less common, but the planar carbonate grouping of (H) again produces antiferromagnetic exchange when bridging paramagnetic centres.[121] Malachite $[Cu_2(OH)_2CO_3]$ provided the first example of the carbonate ligand bridging four metal atoms (I) and an example where the balancing cation is involved in this type of bridging is provided by the trihydrate of $Na_2Cu(CO_3)_2$.[122]

15.6.4.3 Synthesis and Reactivity

As many carbonate complexes are synthesized usually in aqueous solution under fairly alkaline conditions, the possibility of contamination by hydroxy species is often a problem. To circumvent this, the use of bicarbonate ion (*via* saturation of sodium carbonate solution with CO_2) rather than the carbonate ion can often avoid the precipitation of these contaminants. Many other synthetic methods use carbon dioxide as their starting point. Transition metal hydroxo complexes are, in general, capable of reacting with CO_2 to produce the corresponding carbonate complex.[123] The rate of CO_2 uptake, which depends upon the nucleophilicity of the OH entity, proceeds by a mechanism that can be regarded as hydroxide addition across the unsaturated CO_2.[124] There are few non-aqueous routes to carbonate complexes but one reaction (3), illustrative of a synthetic pathway of great potential, is that used to prepare platinum[125] and copper complexes.[126] Ruthenium and osmium carbonate complexes result from the oxidation of coordinated carbon monoxide by dioxygen insertion (4).[127]

$$(PPh_3)_2M + O_2 + CO_2 \xrightarrow{\text{ether}} (PPh_3)_2MCO_3 + OPPh_3 \tag{3}$$

$$M-CO \xrightarrow{O_2} M\underset{O}{\overset{O}{\diagdown}}C=O \tag{4}$$

Decarboxylation of carbonate complexes is usually effected by acid hydrolysis with the formation of a CO_2 free oxide or hydroxide complex.[128] All such reactions involve a protonated (bicarbonate) intermediate but there are some useful differences which, in many instances, may be reconciled with the three main structural types of carbonate complexes. Both unidentate and chelate carbonates readily yield CO_2 on acidification, while there is a greater resistance to CO_2 loss when the carbonate is a bridging ligand. Unidentate carbonate complexes decarboxylate with the initial formation of a bicarbonate intermediate and subsequent loss of CO_2 without rupture of the M—O bond, *viz.* structure (3). By contrast, in chelate carbonate complexes, cleavage of the M—O bond occurs (with ring opening) with the formation of a bicarbonate aqua ion before the loss of CO_2, *viz.* equation (5).[29]

$$
\begin{array}{c}
M-O-\!\!\!\!\mid\!\!\!\!-C=O \\
\vdots \qquad \mid \\
H\cdots\!\!-\!\!\cdots O
\end{array}
$$

(3)

$$M\underset{O}{\overset{O}{\diagdown}}C=O \longrightarrow M\underset{OH_2}{\overset{O(CO_2)H}{\diagdown}} \longrightarrow M\underset{OH}{\overset{OH}{\diagdown}} \tag{5}$$

Decarboxylation of bridged carbonate systems, while far less investigated, appears to be influenced by the stability of a doubly bridging bicarbonate species. Careful treatment of the paddlewheel bridged $[Rh_2(CO_3)_4]^{4-}$ with aqueous acid initially converts all four carbonate groups to bicarbonate bridges, but only two of these bridges slowly lose CO_2 to give a stable aqua species containing two bicarbonate bridging units per dimer.[130] However, complexes which contain just one carbonate bridge lose CO_2 very rapidly.[131] These interesting systems obviously warrant further investigation. Protonation of carbonate bridges in the dicopper complexes occurs when these complexes are used as initiators in the oxidative coupling of phenols.[132] Alkali metal carbonate complexes find industrial use both in fuel cells and the destruction of toxic chemical wastes.[133] Carbonate systems are often used in the purification and storage of the transuranic elements as in carbonate media the elements can exist in oxidation states unstable in acid solution.[134] Carbonate complexes also have considerable value in the synthesis of other inorganic complexes. This utility for ready replacement, arising because carbonate is the anion of a weak acid, is illustrated by reaction (6).[135]

$$[Cr_2(CO_3)_4]^{4-} + CF_3COOH \longrightarrow Cr_2(O_2CCF_3)_4 \tag{6}$$

15.6.4.4 Peroxocarbonates[136]

These contain a C—O—O linkage and may be synthesized either by the oxidation of carbonate complexes or by treatment of carbon dioxide with an alkali metal hydroxide and H_2O_2. Two classes of peroxocarbonates are found in the literature, *viz.* peroxomonocarbonate, $[CO_4]^{2-}$, and peroxodicarbonates, $[C_2O_6]^{2-}$. The $[CO_4]^{2-}$ dianion, which is probably planar, has a number of potential donor atoms and, while no complexes have been structurally characterized, chelation to give a five-membered ring containing the peroxo linkage is anticipated to be favoured over the typical four-membered ring of carbonate chelation. Vibrational spectra for the complex $(PPh_3)_2PtCO_4$ is consistent with the five-atom chelate ring illustrated in (4). The O—O stretch in the free dianion near to $900 \, cm^{-1}$ falls to $780 \, cm^{-1}$ in the complex, while the uncoordinated C=O observed above $1600 \, cm^{-1}$ is generally lower than in the carbonate systems. The splitting of the C—O stretching vibrations in the ion ($> 300 \, cm^{-1}$) is reduced by 50–70 cm^{-1} upon chelation.[137]

Potassium peroxodicarbonate ($K_2C_2O_6$) is readily prepared, for example by anodic oxidation of saturated potassium carbonate at $-20 \, °C$, but attempts at complex formation have been hampered by its ready hydrolysis by water and low solubility in organic solvents. Vibrational spectral data for the dianion have been interpreted[38] on the basis of the *trans* planar structure (5).

(4)   (5)

15.6.5 CARBAMATES

Carbamic acid, $HOOCNH_2$, the monoamide derivative of carbonic acid, is not known in the free state and usually occurs in the form of salts. Carbamate coordination chemistry is much less extensive than the ligands so far considered, which contain the ligating group O_2CX. This is principally for two reasons; firstly the synthetic routes to these complexes are often limited to carbon dioxide reactions with dialkylamide complexes, and secondly the unstable character of the complexes themselves due to reaction (7). Carbamate complexes are electronically and sterically similar to monocarboxylate systems, although the presence of two alkyl groups per ligand enhances their solubility.

$$H_2NCO_2^- + H_2O \longrightarrow NH_4^+ + CO_3^{2-} \qquad (7)$$

The three types of coordination (unidentate, chelating and bridging) expected for O_2CX groups are found in carbamate systems. In the bridging mode, not all the varieties anticipated by comparison with the RCO_2^- system have been reported. *Syn–syn* bridging is the most extensive to date, but examples of tridentate coordination *via* a monatomic bridge are known. As has been noted earlier, definitive structural information is heavily dependent upon diffraction methods. Again IR spectroscopy is a valuable, but not infallible, technique for these systems. Plausibility is enhanced by isotopic substitution, made easy from the commercial availability of labelled CO_2. Chisholm[139] has established that a strong absorption band in the 1550–1600 cm^{-1} region is characteristic of $v(O_2CN)$ in chelating carbamate systems, while an absorption above 1620 cm^{-1} is found in unidentate coordination. The two alkyl groups attached to the nitrogen atom not only provide extra solubility, but are useful probes for both ^1H and ^{13}C NMR studies. In comparison with carboxylate systems, surprisingly few detailed magnetic studies have been made on carbamate complexes.

15.6.5.1 Unidentate and Chelate Coordination

The complex $W(NMe_2)_3(O_2CNMe_2)_3$ provides a typical example of unidentate carbamate coordination[140] with the C—O(W) bond of 130 pm shortened relative to the C=O(free) of 124 pm and also a stretching frequency of 1632 cm^{-1}. Much greater equivalence of the carbon–oxygen bond lengths is found in the chelating (bite ~ 218 pm) carbamate complex $Cr_2(\mu\text{-}NR_2)_2(O_2CNR_2)_4$ where both C—O distances are close to 128 pm.[141] The extra strain in the four-membered ring is again reflected in the smaller OCO angle (116°) of the chelate when compared with that of the unidentate form (123°). There appears to be little variation in the C—N bond length (135 pm) in both of these coordination modes.

Changes arising from isotopic substitution of ^{18}O in $v(^{16}O_2CN)$ have been used to differentiate between unidentate and chelating ligands. A small shift (< 5 cm^{-1}) is indicative of chelation while much larger shifts (20 cm^{-1}) are associated with unidentate carbamate ligands, *e.g.* in the niobium complex $Nb(O_2CNMe_2)_5$.[142] X-Ray diffraction confirmed that eight coordination was achieved for the niobium *via* three chelate and two unidentate carbamate groups. NMR studies have been used to demonstrate that the O_2CNMe_2 groups show dynamic behaviour in solution.

15.6.5.2 Bridging Systems

Syn–syn bridging of the carbamate ligand is well established and gives rise to metal–metal multiply bonded systems. This is exemplified by the paddlewheel structure of the complexes[141,143]

$Cr_2(O_2CNR_2)_4 \cdot 2R_2NH$, $W_2(O_2CNR_2)_6$ and $W_2(O_2CNR_2)_4$ and is postulated from NMR data for the closely related $Mo_2(NR_2)_2(O_2CNR_2)_4$ complex. In both tungsten complexes the two bridging carbamate ligands have similar C—O(W) bond lengths (127 and 129 pm) together with an OCO angle of 125°, which reflects the lack of strain in the five-membered ring. These tungsten complexes also provide examples of typical chelating carbamate groups. In the $W_2(O_2CR_2)_6$ system, one of the methyl groups of $W_2Me_2(O_2CNR_2)_4$ is replaced by a unidentate carbamate ligand for which there is evidence of weak interaction between the free C=O and the axial position of the W≡W bond. ^{1}H and ^{13}C NMR show that at $-60\,°C$ a spectrum consistent with the crystal structure is observed but, as the temperature is raised, scrambling processes occur which at 77 °C result in a sharp, one-line resonance.[144]

Anti–anti bridging of the carbamate ligand is thought to occur in the tin(IV) complex $Me_3Sn(O_2CNR_2)$,[145] while the tetrameric uranium complex $U_4O_2(O_2CNR_2)_{12}$ provides an authentic example of a monatomic (tridentate) carbamate bridge between two metal atoms.[146]

15.6.5.3 Synthesis and Reactivity

Insertion of carbon dioxide into the metal–nitrogen bond of dialkylamide complexes is the principal route to complexes containing the carbamate anion as a ligand. In some cases, amides react to convert all the R_2N groups to the fully carbonated R_2NCO_2, while in others, *e.g.* W, only partially carbonated products result even in the presence of excess carbon dioxide. The reasons for this different behaviour appear to depend upon a subtle mixture of both electronic and steric factors. However, in the cases where fully carbonated products are obtained, the use of $< n$ equivalents of CO_2 usually allows the isolation of the mixed carbamate–amide complex.[147] In most cases these synthetic routes are aided by the presence of trace amounts of the appropriate amine in solution. There is not always need for a preliminary synthesis of the dialkylamido complex as in the synthesis of the chelating carbamate complex $U(O_2CNR_2)_4$ from UCl_4. This complex also illustrates the utility of carbamate systems as intermediates in the preparation of other complexes by their ability to react with proton active materials with replacement of the carbamate ligand,[148] *e.g.* reaction (8).

$$U(O_2CNR_2)_4 + HOOCR \longrightarrow 4CO_2 + 4R_2NH + U(O_2CR)_4 \qquad (8)$$

15.6.6 CARBON DIOXIDE COMPLEXES

In recent years there has been a growing interest in the chemistry of carbon dioxide, stimulated by current anxieties about alternative petrochemical feedstocks. One aspect under active exploration involves carbon dioxide activation *via* coordination to a transition metal, and indeed transition metal ions do form CO_2 complexes.[177] The number of simple and reasonably stable complexes is still relatively small and usually limited to low oxidation state metal ions. There are many systems where CO_2 is used as a 'reagent' giving rise to systems which, while not true CO_2 complexes, may simplistically be viewed as the products of insertion into metal–ligand bonds, *e.g.* reaction (9), where if L = H, formates are produced; if L = OH, carbonates or bicarbonates result; L = NR_2 yields dialkylcarbamates; and if L = R, carboxylate products result. Much of this area has recently been reviewed and will not be considered further.[149]

$$\text{Metal–L} + CO_2 \rightleftharpoons \text{Metal–O}_2C\text{–L} \qquad (9)$$

(6a) Side-on coordination   (6b) Carbon coordination   (6c) Oxygen coordination

X-Ray diffraction data for simple CO_2 systems is severely limited but discrete CO_2 complexes may be divided into three types, which are illustrated diagrammatically in (6a–c).

15.6.6.1 Side-on Coordination

This coordination (**6a**) requires electron density to be transferred from the filled $1\pi_u$ molecular orbital of the CO_2 into a vacant metal orbital with the simultaneous back-donation of electron density into an empty CO_2 orbital, in an analogous manner to the Dewar–Chatt–Duncanson model used in low oxidation state transition metal alkene complexes. Structural data[150] are provided for this bonding mode by the Ni^0 complex $Ni(PCy_3)_2(CO_2)$ ($Cy = C_6H_{11}$) and the main structural dimensions are shown in (**7**). It may be noted that (i) the CO_2 molecule is markedly bent on coordination, and (ii) the coordinated C—O bond length is much longer than that of the CO_2 molecule (116 pm), both of which are consistent with the proposed bonding scheme. IR data[151] for this molecule must, in the absence of more extensive structural work, serve as a reference point for this mode of coordination with the uncoordinated $v(C=O)$ in the region of $1740\,\mathrm{cm}^{-1}$ and the coordinated $v(C=O)$ much lower at $1150\,\mathrm{cm}^{-1}$. A similar platinum complex $(PPh_3)_2PtCO_2$ is also reported to show this mode of coordination.[151] Recent *ab initio* calculations[152] indicate that this side-on coordination may well be the preferred bonding mode for CO_2. Mention should be made of the little investigated observation that nickel vapour and carbon dioxide react according to equation (10).[153] Unfortunately the nature of the initial coordination of the CO_2 is not clear.

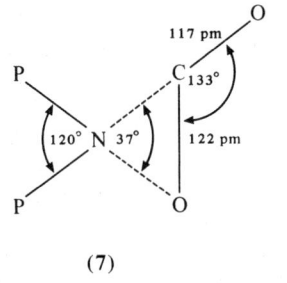

(7)

$$Ni \;+\; CO_2 \;\longrightarrow\; NiO \;+\; CO \;+\; Ni(CO)_4 \tag{10}$$

15.6.6.2 Carbon Atom Coordination

This mode of coordination, giving oxycarbonyl complexes, arises from the electrophilic character of CO_2. Transfer of electron density from the metal to the $2\pi_u$ antibonding orbital of the CO_2 can be aided by the association of the oxygen atoms with an electron acceptor, *e.g.* a second metal ion. This is exemplified by the N,N'-ethylenebis(salicylideniminato)cobalt(I) reaction (11). In this complex the coordination of the CO_2 carbon atom to the cobalt is aided by potassium coordination to the oxygen atoms, resulting in CO_2 interaction with three metal centres.[154] This is shown diagrammatically in (**8**), together with relevant dimensions from which it may be noted that (i) the CO_2 molecule is bent with approximately an sp^2 carbon atom, and (ii) the CO_2 bond lengths are equal and longer than in CO_2 itself.

$$[Co(Pr\text{—}salen)K] \;+\; CO_2 \;\xrightarrow{\;THF\;}\; [Co(Pr\text{—}salen)K(CO_2)THF] \tag{11}$$

(8)

This complex has four strong, sharp IR bands arising from the bound CO_2 at 1650, 1280, 1215 and $745\,\mathrm{cm}^{-1}$. A second structurally characterized example of this enhancement of the electrophilic CO_2 character is provided by the iridium bis(chelating phosphine) complex $(P_2)_2IrCO_2Me$, which shows similar features to the cobalt complex above. There are many other systems for which this mode of coordination is proposed and an extended discussion of these is available.[149]

15.6.6.3 Oxygen Atom Coordination

Electron donation from the $1\pi_g$ orbital of the CO_2 to a metal should result in classical Lewis acid/Lewis base complexes. The poor natural Lewis basicity of CO_2 usually requires the additional presence of an electron donating group (E) for stable complex formation, *e.g.* equation (12). In such molecules the integrity of the CO_2 is lost and such complexes are to be found elsewhere in this publication.

$$\text{M}-\text{E} \;\xrightleftharpoons{\;CO_2\;}\; \text{M}-\text{O}-\underset{\underset{\text{O}}{\|}}{\text{C}}-\text{E} \tag{12}$$

15.6.7 SQUARATES AND RELATED SPECIES

Commonly occurring cyclic oxocarbon acids include deltic ($C_3H_2O_3$), squaric ($C_4H_2O_4$), croconic ($C_5H_2O_5$) and rhodizonic ($C_6H_2O_6$) acids (see Table 8). The organic chemistry of these systems has been studied for over 150 years and is the subject of a recent review.[155] It is of passing interest to note that the croconate dianion shares with benzene the distinction of being amongst the first two aromatic compounds isolated in a pure form.[156] It is the dianions of these generally strong acids that aroused the interests of coordination chemists; no doubt this is related to their chemical similarity to the oxalate dianion with extensive π-electron delocalization and a planar stereochemistry. The coordination chemistry of the dianions squarate and croconate is the most extensive[157] and their planarity with cyclic conjugation led to an expectation that π-type coordination complexes may occur. To date no such π complexes have been characterized, but it is known that the same ligand features provide a relatively poor pathway for superexchange of electron spins between paramagnetic centres. Complexes of squaric acid will be considered initially to illustrate the chemistry of these ring systems and the coordination chemistry of the other oxocarbon acids will be related to this.

Table 8 Commonly Occurring Cyclic Oxocarbon Acids

Name	Formula	Acidity
Deltic acid	(structure: 3-membered ring, $O={}$, with two OH groups)	pK_1 2.6 pK_2 6.0
Squaric acid	(structure: 4-membered ring, two O, two OH)	pK_1 0.5 pK_2 3.5
Croconic acid	(structure: 5-membered ring, three O, two OH, HO)	pK_1 0.8 pK_2 2.2
Rhodizonic acid	(structure: 6-membered ring, four O, two OH)	pK_1 4.3 pK_2 4.7

15.6.7.1 Squaric Acid Complexes

X-Ray crystal structure analyses of both squaric acid[158] (3,4-dihydroxy-3-cyclobutene-1,2-dione) and the dianion[159] are the usual benchmarks used in consideration of structural data of metal complexes. The relevant dimensions are shown in Table 9, from which it can be seen that there is substantial conjugation and C_{2v} symmetry in the acid molecule while the dianion has D_{4h} symmetry with four equal C—C bonds, four equal C—O bonds and CCC angles close to 90°. Complexes have been characterized where two or four of the oxygen atoms may coordinate to the metal, but there are no well authenticated unidentate coordination complexes (Table 10).

Table 9   Molecular Dimensions in Squaric Acid and the Dianion

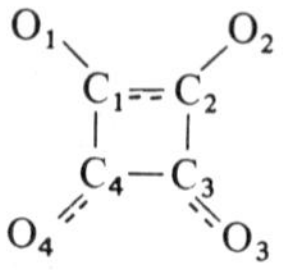

	C_1—C_2	C_2—C_3	C_3—C_4	C_4—C_1	C_1—O_1	C_2—O_2	C_3—O_3	C_4—O_4
				Bond lengths (pm)				
Acid	140.9	145.7	150.1	145.8	129.0	129.2	122.9	122.8
Dianion	146.9	149.9	144.4	144.4	126.0	126.0	125.8	125.8

	C_1	C_2	C_3	C_4
		Bond angles CC_xC (°)		
Acid	91.8	91.8	88.2	88.2
Dianion	90.1	90.1	89.9	89.9

Table 10   Coordination Modes of Squaric Acid

(a) *Chelation*

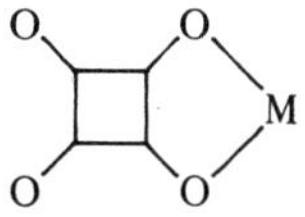

η^4 Chelation

(b) *Two-atom bridging*

(i)

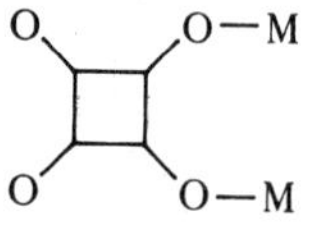

1,2-Bridging

(ii)

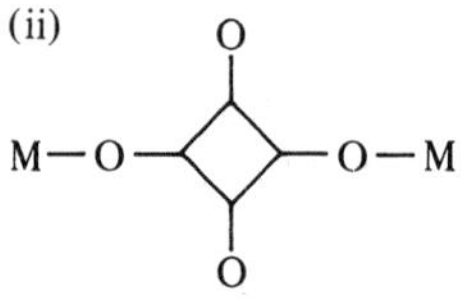

1,3-Bridging

(c) *Four-atom bridging*

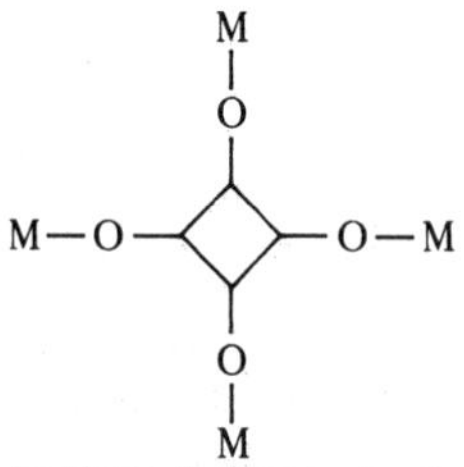

15.6.7.2   Two-atom Coordination of Squaric Acid

Much has been written about the close relation between oxalate and squarate ions, and this has led to frequent structural proposals of η^4 chelation for the complexes (see (a) in Table 10). The steric constraints consequent upon incorporation of this chelation with a four-carbon ring system are such that there are no structurally characterized examples of this mode of coordination with squaric acid itself. Relief from steric strain would be anticipated if the two adjacent oxygen atoms were to participate in bridging (see (b)(i) in Table 10) rather than chelation. This is the situation for trivalent metal complexes of stoichiometry [M(squarate)(OH)(H$_2$O)]. Initially these were predicted to be chelating squarate, hydroxy-bridged dimers, but a recent structural determination has shown the chromium(III) dimer to have two bridging squarate and two bridging hydroxy groups.[160] Predictably there is a lower C—O bond order for the oxygen atoms coordinated to the chromium (127.2 and 124.5 pm) and the average of 145 pm for the four C—C bond lengths is very similar to that in the dianion. Similar bridging is found in the complex Ba$_2$[Pt$_2$(squarate)$_4$], where the four squarate bridges adopt a paddlewheel arrangement about the two platinum atoms which are separated by 306 pm, consistent with interaction between the centres. The molecular dimensions of the squarate ligand indicate a similar electron distribution to that found in the chromium(III) complex.[161]

Two oxygen atoms of the squarate dianion are involved in a very different form of bridging, exemplified by the imidazole complex Ni(squarate)(imidazole)(H$_2$O)$_2$. Diffraction data[162] show chains of 1,3-bridging squarate groups (see (b)(ii) in Table 10). The dianion is almost square with the average of 146.7 pm for the four C—C bond lengths slightly larger than that of the dianion. The four C—O bond lengths of 125 pm reflect the involvement of the two uncoordinated oxygen atoms with the water molecules in the structure.

15.6.7.3 Four-atom Coordination of Squaric Acid

All four of the squarate oxygen atoms are used to bridge four different nickel atoms together in the Ni(squarate)·2H$_2$O molecules.[163] The octahedral nickel atoms are at the edges of a cube formed by the bridging squarate ligands and there is evidence for clathration of molecules within this cube. There are four equal C—C bond lengths and with four equal C—O bond lengths the squarate ligand resembles the dianion. This strain-free structure (see (c) in Table 10) appears to have the most extensive occurrence to date as the Mn^{II}, Fe^{II} and Co^{II} complexes are isostructural.

The croconate dianion,[164] with an average of 145 pm for the five C—C bonds and 126 pm for the five C—O bonds, has an overall D_{5h} symmetry. The relief of steric strain, resulting in an opening of the internal angles by $\sim 20°$ relative to the squarate dianion, would be expected to have considerable impact upon its coordination chemistry, particularly upon the η^4 chelation mode. Although to date no monomeric examples of η^4 chelation have been structurally characterized, such a chelate ring is commonly formed by the croconate ligand albeit as part of tridentate bridging behaviour. Three complexes of stoichiometry M(croconate)(H$_2$O)$_3$, where M = Mn^{II},[165] Cu^{II} and Zn^{II},[166] have been structurally characterized as polymeric chains formed by the bonding of each metal atom to two adjacent oxygen atoms of one croconate ring (η^4 chelation) and a single oxygen of a second croconate group. In this 1,2,4-bridging with the copper and manganese complexes the p-electron delocalized D_{5h} symmetry is retained with five equivalent C—C and C—O bonds (146 and 125 pm respectively). By contrast, in the zinc complex the croconate ring approaches the C_{2v} symmetry found in croconic acid since the C—C bond length in the η^4 chelate ring is shortened to 142 pm relative to the 149 pm of the remaining four C—C bonds, while the C—O(Zn) bonds average 126 pm.

IR spectra of squarate containing complexes appear characteristic of the mode of coordination. Complexes in which the squarate ligand is coordinated *via* all four oxygen atoms (see (c) in Table 10) exhibit a strong absorption band near 1500 cm^{-1}, which West[167] assigned to a mixture of C—C and C—O stretching vibrations. The 1,2-bridging mode (see (b)(i) in Table 10), with nominal C_{2v} symmetry, has a much richer IR spectrum[168] including two extra absorption bands above 1600 cm^{-1}, which have been assigned to C=C and C=O stretching modes. The nickel complex with 1,3-bridging (see (b)(ii) in Table 10) is reported to have the highest IR absorption[162] at 1480 cm^{-1}. The IR spectra of the croconate complexes show similar characteristics to the squarate systems. For example, tridentate croconate complexes have a strong absorption at 1725 cm^{-1} (non-coordinated C=O) and a strong band at 1300–1700 cm^{-1}. More extensive use of ^{13}C NMR is to be anticipated for these systems and encouraging results have been obtained for [Pd(croconate)$_2$]$^{2-}$.[169]

The transmission of magnetic effects through the ligand has been shown to be weak. Variable temperature magnetic susceptibility measurements reveal antiferromagnetic interactions in squarate bridged nickel(II) dimers which are only a fraction of those in the related oxalate dimers.[170] Polymeric Ni(squarate)(H$_2$O)$_2$, and the structurally isomorphic iron(II) complex, both have magnetic moments (3.20 and 5.50 BM respectively) consistent with their pseudo-octahedral oxygen environment. The nickel complex undergoes a transition to a ferromagnetically ordered state at very low temperatures while no such feature occurs in the iron(II) system. As the nickel complex contains clathrated solvent while the iron complex does not, this magnetic ordering may well be associated with the clathrated molecule rather than the squarate ligand.[171] The magnetic properties of the trivalent metal complexes with both squarate and hydroxy bridges are low, indicating antiferromagnetic spin pairing, but again it would appear that this may arise from the hydroxy rather than the squarate ligand.

15.6.7.4 Synthesis and Reactivity

The pioneering work of West[167] demonstrated that metal complexes may be readily prepared from an aqueous solution of the dipotassium salt of the acid and the appropriate metal ion. Some of the formulation problems experienced by the early workers are now much better understood in the light of the structural work mentioned earlier. Squaric acid itself has only a limited solubility in water (*ca.* 3% by weight at 25 °C), but the oxocarbon dianions appear to be kinetically stable in aqueous solution, although the rhodizonate dianion readily rearranges in base to the croconate dianion. Non-aqueous solvents have also been used to prepare complexes and this allows the use of either the acid or its alkali metal dianion.[172] The thermodynamics of complex formation for the squarate ligand with various metals have been investigated and indicate a preference of this ligand for higher oxidation state metals.[173]

The divalent metal squarate complexes, $M(squarate)(H_2O)_2$, decompose in a vacuum in two stages. First there is a reversible loss of two water molecules, followed at higher temperatures by decomposition to the metal. The nature of the anhydrous complexes is not fully understood but the metal atoms appear to be four-coordinate.[174]

15.6.8 REFERENCES

1. C. Oldham, *Prog. Inorg. Chem.*, 1968, **10**, 223.
2. U. Casellato, P. Vigato and M. Vidali, *Coord. Chem. Rev.*, 1978, **26**, 85; R. C. Mehrotra and R. Bohra, 'Metal Carboxylates', Academic, London, 1983.
3. W. H. Zachariasen, *J. Am. Chem. Soc.*, 1940, **62**, 1011.
4. F. A. Cotton and W. H. Ilsley, *Inorg. Chem.*, 1982, **21**, 300.
5. J. N. van Niekerk, F. R. L. Schoening and J. H. Talbot, *Acta Crystallogr.*, 1953, **6**, 720.
6. J. N. van Niekerk, F. R. L. Schoening and J. F. de Wet, *Acta Crystallogr.*, 1953, **6**, 501; T. Allman, R. C. Goel, N. K. Jha and A. L. Beauchamp, *Inorg. Chem.*, 1984, **23**, 914.
7. R. Kiriyama, H. Ibamoto and K. Matsuo, *Acta Crystallogr.*, 1954, **7**, 482.
8. G. A. Barclay and C. H. L. Kennard, *J. Chem. Soc.*, 1961, 3289.
9. R. D. Mounts, T. Ogura and Q. Fernando, *Inorg. Chem.*, 1974, **13**, 802.
10. A. M. Greenaway, C. J. O'Connor, J. W. Overman and E. Sinn, *Inorg. Chem.*, 1981, **20**, 1508.
11. G. B. Deacon and R. J. Phillips, *Coord. Chem. Rev.*, 1980, **33**, 227.
12. E. Zinato, C. Furlani, G. Lanna and P. Riccieri, *Inorg. Chem.*, 1972, **11**, 1746.
13. F. A. Cotton, D. J. Darensbourg and B. W. S. Kolthammer, *J. Am. Chem. Soc.*, 1981, **103**, 398.
14. V. M. Padmanabhan, *Acta Crystallogr.*, 1958, **11**, 896.
15. J. C. Speakman and H. H. Mills, *J. Chem. Soc.*, 1961, 1164; J. P. Costes, F. Dahan and J. P. Laurent, *Inorg. Chem.*, 1985, **24**, 1018.
16. F. A. Cotton, D. J. Darensbourg, B. W. S. Kolthammer and R. Kudaroski, *Inorg. Chem.*, 1982, **21**, 1656.
17. K. Nakamoto, 'Infrared and Raman Spectra of Inorganic and Coordination Compounds', 3rd edn., Wiley, New York, 1978.
18. E. Farkas, I. Sovago and A. Gergely, *J. Chem. Soc., Dalton Trans.*, 1983, 1545.
19. N. W. Alcock, V. M. Tracy and T. C. Waddington, *J. Chem. Soc., Dalton Trans.*, 1976, 2243.
20. N. W. Alcock and V. M. Tracy, *Acta Crystallogr., Sect. B*, 1979, **35**, 80.
21. P. J. Roberts, G. Ferguson, R. G. Goel, W. O. Ogini and R. J. Restivo, *J. Chem. Soc., Dalton Trans.*, 1978, 253.
22. A. C. Skapski and F. A. Stephens, *J. Chem. Soc., Dalton Trans.*, 1974, 390.
23. F. A. Cotton, G. E. Lewis and G. N. Mott, *Inorg. Chem.*, 1983, **22**, 1825.
24. Y. B. Koh and G. G. Christoph, *Inorg. Chem.*, 1979, **18**, 1122.
25. I. D. Brown, *J. Chem. Soc., Dalton Trans.*, 1980, 1118.
26. D. Stoilova, G. Nikolov and K. Balarev, *Izv. Khim.*, 1976, **9**, 371.
27. R. J. Doedens, *Prog. Inorg. Chem.*, 1976, **21**, 209.
28. F. A. Cotton, *Rev. Pure Appl. Chem.*, 1967, **17**, 25.
29. F. A. Cotton, *Chem. Soc. Rev.*, 1983, **12**, 35; J. C. A. Boeyens, F. A. Cotton and S. Han, *Inorg. Chem.*, 1985, **25**, 1750.
30. V. Mohan Rao, D. N. Sathyanarayana and H. Manohar, *J. Chem. Soc., Dalton Trans.*, 1983, 2167.
31. T. Behling, G. Wilkinson, T. A. Stephenson, D. A. Tocher and M. D. Walkinshaw, *J. Chem. Soc., Dalton Trans.*, 1983, 2109.
32. M. Inoue and M. Kubo, *Inorg. Chem.*, 1970, **9**, 2310.

33. F. A. Cotton and R. A. Walton, 'Multiple Bonds between Metals', Wiley-Interscience, New York, 1982.
34. D. J. Santure, K. W. McLaughlin, J. C. Huffman and A. P. Sattelberger, *Inorg. Chem.*, 1983, **22**, 1877; D. J. Santure, J. C. Huffman and A. P. Sattelberger, *Inorg. Chem.*, 1985, **24**, 371; F. A. Cotton, Z. Dori, D. O. Marler and W. Schwotzer, *Inorg. Chem.*, 1984, **23**, 4738.
35. A. Bino, F. A. Cotton and P. E. Fanwick, *Inorg. Chem.*, 1979, **18**, 1719.
36. T. R. Felthouse, *Prog. Inorg. Chem.*, 1982, **29**, 73.
37. B. J. Allin and P. Thornton, *Inorg. Nucl. Chem. Lett.*, 1973, **9**, 449; E. B. Boyer and S. D. Robinson, *Coord. Chem. Rev.*, 1983, **50**, 109.
38. A. Spencer and G. Wilkinson, *J. Chem. Soc., Dalton Trans.*, 1972, 1570.
39. A. Birnbaum, F. A. Cotton, Z. Dori, D. O. Marler, G. M. Reisner, W. Schwotzer and M. Shaia, *Inorg. Chem.*, 1983, **22**, 2723.
40. F. A. Cotton, G. E. Lewis and G. N. Mott, *Inorg. Chem.*, 1982, **21**, 3316.
41. J. T. Wrobleski, C. T. Dziobkowski and D. B. Brown, *Inorg. Chem.*, 1981, **20**, 684.
42. M. Takano, *J. Phys. Soc. Jpn.*, 1972, **33**, 1312.
43. J. Catterick and P. Thornton, *Adv. Inorg. Chem. Radiochem.*, 1977, **20**, 291; F. A. Cotton, S. A. Duraj and W. J. Roth, *Inorg. Chem.*, 1984, **23**, 4042.
44. M. Melnik, *Coord. Chem. Rev.*, 1981, **36**, 1; L. S. Erre, G. Micera, P. Piu, F. Cariati and G. Ciani, *Inorg. Chem.*, 1985, **24**, 2297.
45. L. W. Hessel and C. Romers, *Recl. Trav. Chim. Pays-Bas*, 1969, **88**, 545.
46. E. D. Estes, W. E. Estes, R. P. Scaringe, W. E. Hatfield and D. J. Hodgson, *Inorg. Chem.*, 1974, **14**, 2564; J. P. Costes, F. Dahan and J. P. Laurent, *Inorg. Chem.*, 1985, **24**, 1018.
47. J. N. Brown and L. M. Trefonas, *Inorg. Chem.*, 1973, **12**, 1730.
48. B. Chiari, W. E. Hatfield, D. Piovesana, T. Tarantelli, L. W. ter Haar and P. F. Zanazzi, *Inorg. Chem.*, 1983, **22**, 1468.
49. Y. Zenitani, K. Inoue, Y. Kai, N. Yasuoka and N. Kasai, *Bull. Chem. Soc. Jpn.*, 1976, **49**, 1531.
50. S. Baba, T. Ogura, S. Kawaguchi, H. Tokunan, Y. Kai and N. Kasai, *J. Chem. Soc., Chem. Commun.*, 1972, 910.
51. N. D. Cook and P. L. Timms, *J. Chem. Soc., Dalton Trans.*, 1983, 239.
52. M. E. Volpin and I. S. Kolomnikov, *Organomet. React.*, 1975, **5**, 313.
53. A. Dobson, S. D. Robinson and M. F. Uttley, *Inorg. Synth.*, 1977, **17**, 124.
54. A. Bino, F. A. Cotton, Z. Dori and B. W. S. Kolthammer, *J. Am. Chem. Soc.*, 1981, **103**, 5779.
55. A. Bino, F. A. Cotton and Z. Dori, *J. Am. Chem. Soc.*, 1981, **103**, 243.
56. F. A. Cotton and G. N. Mott, *Inorg. Chem.*, 1983, **22**, 1132; G. S. Girolami and R. A. Andersen, *Inorg. Chem.*, 1981, **20**, 2040.
57. C. D. Garner and B. Hughes, *J. Chem. Soc., Dalton Trans.*, 1974, 735.
58. P. Legzdins, G. L. Rempel and G. Wilkinson, *J. Chem. Soc., Chem. Commun.*, 1969, 825.
59. J. C. Marechal, F. Dawans and P. Teyssie, *J. Polym. Sci.*, 1970, **8**, 1993.
60. R. Pellicciari, R. Fringuelli, P. Ceccherelli and E. Sisani, *J. Chem. Soc., Chem. Commun.*, 1979, 959; M. S. Chinn, M. R. Colsman and M. P. Doyle, *Inorg. Chem.*, 1984, **23**, 3684.
61. J. E. Connett, A. G. Davies, G. B. Deacon and J. H. S. Green, *J. Chem. Soc. (C)*, 1966, 106.
62. D. M. Barlex and R. D. W. Kemmitt, *J. Chem. Soc., Dalton Trans.*, 1972, 1436.
63. M. E. Farago, M. A. R. Smith and I. M. Keefe, *Coord. Chem. Rev.*, 1972, **8**, 95.
64. K. V. Krishnamurty and G. M. Harris, *Chem. Rev.*, 1961, **61**, 213.
65. S. C. Chan and M. C. Choi, *J. Inorg. Nucl. Chem.*, 1976, **38**, 1949.
66. E. Hansson, *Acta Chem. Scand.*, 1973, **27**, 823.
67. J. W. Lethbridge, L. S. Dent-Glasser and H. F. W. Taylor, *J. Chem. Soc. (A)*, 1970, 1862.
68. S. Yano, S. Yaba, M. Ajioka and S. Yoshikawa, *Inorg. Chem.*, 1979, **18**, 2414.
69. R. Stomberg, *Acta Chem. Scand.*, 1970, **24**, 2024.
70. A. Gleizes, F. Maury and J. Galy, *Inorg. Chem.*, 1980, **19**, 2074.
71. F. A. Cotton, S. M. Morehouse and J. S. Wood, *Inorg. Chem.*, 1964, **3**, 1603.
72. E. G. Cox, M. W. Dougill and G. A. Jeffrey, *J. Chem. Soc.*, 1952, 4854.
73. G. A. Jeffrey and G. S. Parry, *J. Am. Chem. Soc.*, 1954, **76**, 5283.
74. A. Karipides, J. Ault and A. T. Reed, *Inorg. Chem.*, 1977, **16**, 3299.
75. E. Hansson, *Acta Chem. Scand.*, 1973, **27**, 2441.
76. K. R. Butler and M. R. Snow, *J. Chem. Soc., Dalton Trans.*, 1976, 259.
77. J. Fujita, A. E. Martell and K. Nakamoto, *J. Chem. Phys.*, 1962, **36**, 324, 331.
78. J. N. Cooper, C. A. Pennell and B. C. Johnson, *Inorg. Chem.*, 1983, **22**, 1956.
79. M. W. Dougill and G. A. Jeffrey, *Acta Crystallogr.*, 1953, **6**, 831.
80. K. L. Scott, M. Green and A. G. Sykes, *J. Chem. Soc. (A)*, 1971, 3651.
81. K. L. Scott, K. Wieghardt and A. G. Sykes, *Inorg. Chem.*, 1973, **12**, 655.
82. E. Hansson, *Acta Chem. Scand.*, 1973, **27**, 2841.
83. K. Wieghardt, *Z. Anorg. Allg. Chem.*, 1972, **391**, 142.
84. F. Mazzi and G. Garavelli, *Period. Mineral.*, 1957, **26**, 269.
85. P. T. Cheng, B. R. Loescher and S. C. Nyburg, *Inorg. Chem.*, 1971, **10**, 1275; T. R. Felthouse, E. J. Laskowski and D. N. Hendrickson, *Inorg. Chem.*, 1977, **16**, 1077.
86. B. H. O'Connor and E. N. Maslen, *Acta Crystallogr.*, 1966, **20**, 824.
87. E. Hansson, *Acta Chem. Scand.*, 1973, **27**, 2827; M. L. Post and J. Trotter, *J. Chem. Soc., Dalton Trans.*, 1974, 1922.
88. H. Siebert and G. Tremmel, *Z. Anorg. Allg. Chem.*, 1972, **390**, 292.
89. S. F. Ting, H. Kelm and G. M. Harris, *Inorg. Chem.*, 1966, **5**, 696; C. Andrade and H. Taube, *Inorg. Chem.*, 1966, **5**, 1087.
90. K. L. Scott and A. G. Sykes, *J. Chem. Soc., Dalton Trans.*, 1972, 1832.
91. J. J. Girerd, O. Kahn and M. Verdaguer, *Inorg. Chem.*, 1980, **19**, 274; A. Michalowicz, J. J. Girerd and J. Goulon, *Inorg. Chem.*, 1979, **18**, 3004.
92. M. Julve, M. Verdaguer, O. Kahn, A. Gleizes and M. Philoche-Levisalles, *Inorg. Chem.*, 1983, **22**, 368; M. Julve, M. Verdaguer, A. Gleizes, M. P. Levisalles and O. Kahn, *Inorg. Chem.*, 1984, **23**, 3808.

93. R. D. Gillard, J. P. DeJesus and P. S. Sheridan, *Inorg. Synth.*, 1980, **20**, 58, and references therein.
94. E. Baumgartner, M. A. Blesa, H. A. Marinovich and A. J. G. Maroto, *Inorg. Chem.*, 1983, **22**, 2224.
95. R. J. Mureinik, *J. Inorg. Nucl. Chem.*, 1976, **38**, 1275.
96. J. S. Miller and A. J. Epstein, *Prog. Inorg. Chem.*, 1976, **20**, 1.
97. A. Werner, *Ber. Dtsch. Chem. Ges.*, 1912, **45**, 3061.
98. M. J. Cleare, *Coord. Chem. Rev.*, 1974, **12**, 349.
99. A. Giacomelli and A. Indelli, *Inorg. Chem.*, 1972, **11**, 1033.
100. A. W. Adamson, W. L. Waltz, E. Zinato, D. W. Walts, P. L. Fleischauer and R. D. Lindholm, *Chem. Rev.*, 1968, **68**, 541.
101. D. M. Blake and C. J. Nyman, *J. Am. Chem. Soc.*, 1970, **92**, 5359.
102. P. O'Brien, *Transition Met. Chem.*, 1980, **5**, 314.
103. J. M. Hunt, M. P. Wisherd and L. C. Bonham, *Anal. Chem.*, 1950, **22**, 1478.
104. K. V. Krishnamurty, G. M. Harris and V. S. Sastri, *Chem. Rev.*, 1970, **70**, 171.
105. C. R. Piriz Mac-Coll, *Coord. Chem. Rev.*, 1969, **4**, 147.
106. R. L. Sass, R. Vidale and J. Donohue, *Acta Crystallogr.*, 1957, **10**, 567.
107. H. C. Freeman and G. Robinson, *J. Chem. Soc.*, 1965, 3194.
108. M. R. Snow, *Aust. J. Chem.*, 1972, **25**, 1307.
109. L. Ciavatta, D. Ferri, I. Grenthe and F. Salvatore, *Inorg. Chem.*, 1981, **20**, 463.
110. F. W. B. Einstein and A. C. Willis, *Inorg. Chem.*, 1981, **20**, 609.
111. R. L. Harlow and S. H. Simonsen, *Acta Crystallogr, Sect B*, 1976, **32**, 466.
112. E. Koglin, H. J. Schenk and K. Schwochau, *Spectrochim. Acta, Part A*, 1979, **35**, 641.
113. M. C. Favas, D. L. Kepert, J. M. Patrick and A. H. White, *J. Chem. Soc., Dalton Trans.*, 1983, 571.
114. F. A. Cotton and T. R. Felthouse, *Inorg. Chem.*, 1980, **19**, 320.
115. P. C. Healy and A. H. White, *J. Chem. Soc., Dalton Trans.*, 1972, 1913.
116. A. R. Davis and F. W. Einstein, *Inorg. Chem.*, 1980, **19**, 1203.
117. G. Kolks, S. J. Lippard and J. V. Waszczak, *J. Am. Chem. Soc.*, 1980, **102**, 4832.
118. G. Ferguson and D. M. Hawley, *Acta Crystallgr., Sect B*, 1974, **30**, 103.
119. V. P. Süsse, *Acta Crystallogr.*, 1967, **22**, 146.
120. A. K. Gregson and N. T. Moxon, *Inorg. Chem.*, 1981, **20**, 78.
121. M. H. Meyer, P. Singh, W. E. Hatfield and D. J. Hodgson, *Acta Crystallogr., Sect B*, 1972, **28**, 1607.
122. P. D. Brotherton and A. H. White, *J. Chem. Soc., Dalton Trans.*, 1973, 2338.
123. M. R. Churchill, G. M. Harris, R. A. Lashewycz, T. P. Dasgupta and K. Koshy, *Inorg. Chem.*, 1979, **18**, 2290.
124. U. Spitzer, R. van Eldik and H. Kelm, *Inorg. Chem.*, 1982, **21**, 2821.
125. C. J. Nyman, C. E. Wymore and G. Wilkinson, *Chem. Commun.*, 1967, 407.
126. M. R. Churchill, G. Davies, M. A. El-Sayed, M. R. El-Shazly, J. P. Hutchinson and M. W. Rupich, *Inorg. Chem.*, 1980, **19**, 201.
127. K. R. Laing and W. R. Roper, *Chem. Commun.*, 1968, 1568.
128. J. F. Glenister, K. E. Hyde and G. Davies, *Inorg. Chem.*, 1982, **21**, 2331.
129. R. van Eldik and U. Spitzer, *Transition Met. Chem.*, 1983, **6**, 351.
130. C. R. Wilson and H. Taube, *Inorg. Chem.*, 1975, **14**, 405.
131. M. R. Churchill, R. A. Lashewycz, K. Koshy and T. P. Dasgupta, *Inorg. Chem.*, 1981, **20**, 376.
132. G. Davies, M. F. El-Shazly and M. W. Rupich, *Inorg. Chem.*, 1981, **20**, 3757.
133. G. B. Dunks and D. Stelman, *Inorg. Chem.*, 1983, **22**, 2168.
134. J. Y. Bourges, B. Guillaume, G. Kochly, D. E. Hobart and J. R. Peterson, *Inorg. Chem.*, 1983, **22**, 1179; C. Madic, D. E. Hobart and G. M. Begun, *Inorg. Chem.*, 1983, **22**, 1494.
135. F. A. Cotton, M. W. Extine and G. W. Rice, *Inorg. Chem.*, 1978, **17**, 176.
136. D. P. Jones and W. P. Griffith, *J. Chem. Soc., Dalton Trans.*, 1980, 2526.
137. P. J. Hayward, D. M. Blake, G. Wilkinson and C. J. Nyman, *J. Am. Chem. Soc.*, 1970, **92**, 5873.
138. P. A. Giguere and D. Lemaire, *Can. J. Chem.*, 1972, **50**, 1472.
139. M. H. Chisholm and M. W. Extine, *J. Am. Chem. Soc.*, 1977, **99**, 782.
140. M. H. Chisholm and M. W. Extine, *J. Am. Chem. Soc.*, 1974, **96**, 6214.
141. M. H. Chisholm, F. A. Cotton, M. W. Extine and D. C. Rideout, *Inorg. Chem.*, 1978, **17**, 3536.
142. M. H. Chisholm and M. Extine, *J. Am. Chem. Soc.*, 1975, **97**, 1623.
143. M. H. Chisholm, F. A. Cotton, M. W. Extine and B. R. Stults, *Inorg. Chem.*, 1977, **16**, 603.
144. M. H. Chisholm and W. W. Reichert, *Inorg. Chem.*, 1978, **17**, 767.
145. R. F. Dalton and K. Jones, *J. Chem. Soc. (A)*, 1970, 590.
146. F. Calderazzo, G. dell'Amico, M. Pasquali and G. Perego, *Inorg. Chem.*, 1978, **17**, 474.
147. F. A. J. J. van Santvoort, H. Krabbendam, A. L. Spek and J. Boersma, *Inorg. Chem.*, 1978, **17**, 388.
148. F. Calderazzo, G. dell'Amico, R. Netti and M. Pasquali, *Inorg. Chem.*, 1978, **17**, 471.
149. R. P. A. Sneeden, in 'Comprehensive Organometallic Chemistry', ed. G. Wilkinson, F. G. A. Stone and E. W. Abel, Pergamon, Oxford, 1982, vol. 8, p. 225.
150. M. Aresta, C. F. Nobile, V. G. Albano, E. Forni and M. Manassero, *J. Chem. Soc., Chem. Commun.*, 1975, 636.
151. M. E. Vol'pin and I. S. Kolomnikov, *Pure Appl. Chem.*, 1973, **33**, 567.
152. S. Sakaki, K. Kitaura and K. Morokuma, *Inorg. Chem.*, 1982, **21**, 760.
153. P. L. Timms, *Adv. Inorg. Chem. Radiochem.*, 1972, **14**, 121.
154. G. Fachinetti, C. Floriani and P. F. Zanazzi, *J. Am. Chem. Soc.*, 1978, **100**, 7405.
155. R. West, 'Oxocarbons', Academic, London, 1980.
156. L. Gmelin, *Ann. Phys. Chem.*, 1825, **4**, 3.
157. R. West, *J. Am. Chem. Soc.*, 1963, **85**, 2580.
158. D. Semmingsen, *Acta Chem. Scand.*, 1973, **27**, 3961.
159. W. M. Macintyre and M. S. Werkema, *J. Chem. Phys.*, 1964, **40**, 3563.
160. J. P. Chesick and F. Doany, *Acta Crystallogr., Sect B*, 1981, **37**, 1076.
161. O. Simonsen and H. Toftlund, *Inorg. Chem.*, 1981, **20**, 4044.
162. J. A. C. van Ooijen, J. Reedijk and A. L. Spek, *Inorg. Chem.*, 1979, **18**, 1184.

163. M. Habenschuss and B. C. Gerstein, *J. Chem. Phys.*, 1974, **61**, 852.
164. N. C. Baenziger and J. J. Hegenbarth, *J. Am. Chem. Soc.*, 1964, **86**, 3250.
165. M. D. Glick and L. F. Dahl, *Inorg. Chem.*, 1966, **5**, 289.
166. M. D. Glick, G. L. Downs and L. F. Dahl, *Inorg. Chem.*, 1964, **3**, 1712.
167. R. West and H. Y. Niu, *J. Am. Chem. Soc.*, 1963, **85**, 2589.
168. J. T. Wrobleski and D. B. Brown, *Inorg. Chem.*, 1978, **17**, 2959.
169. A. R. Siedle and L. H. Pignolet, *Inorg. Chem.*, 1982, **21**, 3090.
170. D. M. Duggan, E. K. Barefield and D. N. Hendrickson, *Inorg. Chem.*, 1973, **12**, 985.
171. C. J. Long, *Inorg. Chem.*, 1978, **17**, 2702.
172. C. Santini-Scapucci and G. Wilkinson, *J. Chem. Soc., Dalton Trans.*, 1976, 807.
173. D. Alexandersson and N. Vannerberg, *Acta Chem. Scand., Ser. A.*, 1974, **28**, 423.
174. R. A. Bailey, W. N. Mills and W. J. Trangredi, *J. Inorg. Nucl. Chem.*, 1971, **33**, 2387.
175. F. A. Cotton, M. P. Diebold and I. Shim, *Inorg. Chem.*, 1985, **24**, 1510.
176. K. R. Lane, L. Sallans and R. R. Squires, *Inorg. Chem.*, 1984, **23**, 1999.
177. C. Mealli, R. C. Hoffman and A. Stockis, *Inorg. Chem.*, 1984, **23**, 56.

15.7

Hydroxy Acids

JULIO D. PEDROSA DE JESUS
University of Aveiro, Portugal

15.7.1 INTRODUCTION

The capacity of hydroxycarboxylic acids to act as ligands is an important property, relevant in several areas of chemistry and in other sciences. This is well illustrated by examples drawn from quite distinct fields of scientific and technological activity. For instance, the use of hydroxy acids in analytical chemistry is very common, their action as masking agents having been reviewed.[1] The Cotton effect, the basis of circular dichroism and other chiroptical techniques, was observed[2] for the first time in solutions of Cu^{II} tartrates. Fehling's solution and Benedict's solution, both widely used in organic chemistry, are solutions of complexes of Cu^{II} with tartrate and citrate respectively. It is well established[3] that the biological processes associated with the Krebs' cycle involve the coordination of hydroxy acids to metal centres as fundamental steps. In other areas of applied science, several good examples of the relevance of hydroxy acids as ligands can be found. For example, the pharmacological action of some hydroxy acids is associated with coordination of these compounds to metal ions.[4,5] There is increasing interest in using hydroxy acids in electroplating baths for improving surface quality and other aspects of the process.[6-8]

The aliphatic hydroxy acids which have been by far most studied as ligands are the low molecular weight 2-hydroxyalkanoic acids (glycolic, mandelic, lactic, malic, tartaric and citric). Salicylic acid and its derivatives are, on the other hand, the aromatic acids which have received most attention. We will concentrate on these two groups of ligands and will refer to other classes of hydroxy acids only when they play some relevant specific role as ligands.

The description and discussion of the properties considered most relevant in understanding and studying the coordination compounds of hydroxy acids will be our main aim in this review. Among other aspects, attention will be given to their acid–base behaviour, and to the stereochemistry and spectroscopic properties of these molecules as free ligands and coordinated to metals. The preparative chemistry and reactivity of hydroxy acids will not be discussed, but excellent reviews are available on this matter.[9,10] The list of references is by no means exhaustive; however it does cover the main points and provides an entry to the literature of the subject. In other chapters of this work dealing with the chemistry of specific elements the reader will find relevant information concerning the coordination chemistry of hydroxy acids which complements what is given in this section.

15.7.2 NOMENCLATURE

IUPAC rules[11] establish two methods for naming carboxylic acids. In the first, a suffix '-oic' is attached to the name of the parent aliphatic chain (with elision of terminal 'e' if present) and the word 'acid' is added thereafter. Thus, the change '-ane' to '-anoic acid' denotes the change of —Me into —CO_2H, *e.g.* butanoic acid, $MeCH_2CH_2CO_2H$. The second method consists of treating the —CO_2H group as a substituent, denoted by the ending 'carboxylic acid', which includes the carbon atom of the carboxyl group. This is the only method recommended when the characteristic group is attached to a ring, *e.g.* 1,2,3-propanetricarboxylic acid, $HO_2CCH_2CH(CO_2H)CH_2CO_2H$, or 1,3,5-naphthalenetricarboxylic acid (1). When carboxyl is the principal group, the numbering of an aliphatic acid is arranged to give the lowest number to this group irrespective of the numbering of the parent compound. Hydroxy acids are named considering the hydroxyl group as a substituent in a parent carboxylic acid. The numbering of aliphatic hydroxy acids is made so as to give the number 1 to a carboxyl group and the lowest possible locant to the hydroxyl group. In Table 1, systematic names for the more common hydroxy acids are given together with the accepted trivial names. Most of the hydroxy acids of interest here are usually known by their trivial names. These names will, therefore, be used in this work for most compounds. In trivial names the location of substituents in aliphatic acids is given by Greek letters. It should be noted that when Greek letters are used as locants, the lettering begins with a carbon atom adjacent to a CO_2H group, whereas when numbers are used the numbering begins with the carbon of the CO_2H group.

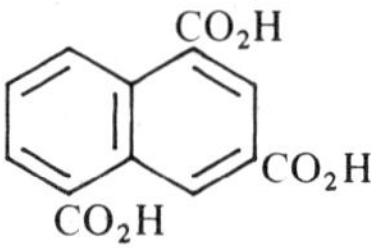

(1)

Several hydroxy acids relevant in coordination chemistry have one or more chiral carbon atoms. These compounds can, for that reason, appear as enantiomeric forms. This is the case, for example, for lactic, mandelic, glyceric, malic and tartaric acids (Figures 1 and 2). According to IUPAC recommendations,[11] enantiomers whose absolute configurations are not known may be differentiated as dextrorotatory (prefix +) or laevorotatory (prefix −) depending on the direction in which, under specified experimental conditions, they rotate the plane of polarized light. The methods adopted[11] to assign the labels (R) or (S) for chirality have been developed by Cahn, Ingold and Prelog.[12,13] The use of this nomenclature is strongly recommended and the old *l*- and *d*- symbols, quite often meaningless or confusing, should be abandoned. Enantiomers represented in Figures 1 and 2 are named there according to IUPAC rules. Tartaric acid can have a structural arrangement in which there is an equal number of enantiomeric (mirror images) groups identically linked (Figure 2c). This stereoisomer, usually called *meso*-tartaric acid, will be called here (2R,3S)-tartaric acid, according to these rules.

Figure 1 Fischer projections of (R)- and (S)-enantiomers of lactic acid (X = Me), mandelic acid (X = Ph), glyceric acid (X = CH_2OH) and malic acid (X = CH_2CO_2H)

Table 1 Trivial and Systematic (IUPAC) Names of Hydroxy Acids

Trivial	Systematic	Formula	$[\alpha]_D$[a]
Glycolic	Hydroxyethanoic	HOCH₂CO₂H	—
(+)-Mandelic[b]	(S)-(+)-2-Hydroxy-2-phenylethanoic[b]	HOCCH(Ph)CO₂H	+156.6[c]
Benzilic	2-Hydroxy-2,2-diphenylethanoic	HOC(Ph)₂CO₂H	—
(+)-Lactic[b]	(S)-(+)-2-Hydroxypropanoic[b]	MeCHOHCO₂H	+3.82[d]
(+)-Atrolactic[b]	(S)-(+)-2-Hydroxy-2-phenylpropanoic[b]	MeC(Ph)(OH)CO₂H	+37.7[e]
β-Hydroxypropionic	3-Hydroxypropanoic	CH₂(OH)CH₂CO₂H	—
Tartronic	Hydroxypropanedioic	HO₂CCH(OH)CO₂H	—
(−)-Glyceric[b]	(R)-(−)-2,3-Dihydroxypropanoic[b]	CH₂(OH)CH(OH)CO₂H	—
α-Hydroxybutyric[b]	2-Hydroxybutanoic[b]	MeCH₂CH(OH)CO₂H	—
(+)-Malic[b]	(R)-(+)-Hydroxybutanedioic[b]	HO₂CCH₂CH(OH)CO₂H	+2.92[c]
(+)-Tartaric[b]	(2R,3R)-(+)-2,3-Dihydroxybutanedioic[b]	HO₂CCH(OH)CH(OH)CO₂H	+11.98[e]
Citric	2-Hydroxypropane-1,2,3-tricarboxylic	HO₂CCH₂C(OH)(CO₂H)CH₂CO₂H	—
(+)-Isocitric[b]	(1R,2S)-(+)-1-Hydroxypropane-1,2,3-tricarboxylic[b]	HO₂CCH(OH)CH(CO₂H)CH₂CO₂H	+34.6
α-Hydroxyisobutyric	2-Hydroxy-2-methylpropanoic	MeC(Me)(OH)CO₂H	—
β-Hydroxybutyric[b]	(R)-(−)-3-Hydroxybutanoic[b]	MeCH(OH)CH₂CO₂H	−24.9
γ-Hydroxybutyric	4-Hydroxybutanoic	HOCH₂CH₂CH₂CO₂H	—
(−)-Gluconic[b]	(−)-2,3,4,5,6-Pentahydroxyhexanoic[b]	HOCH₂{CH(OH)}₄CO₂H	−6.7[c]
(+)-Saccharic[b]	(+)-2,3,4,5-Tetrahydroxyhexanedioic[b]	HO₂C{CH(OH)}₄CO₂H	+6.86 → +20.60[f]
Salicylic	2-Hydroxybenzoic	(benzene: CO₂H, OH) (2)	—
2-Pyrocatechuic	2,3-Dihydroxybenzoic	(benzene: CO₂H, OH, OH) (3)	—
γ-Resorcylic	2,6-Dihydroxybenzoic	(benzene: OH, CO₂H, OH) (4)	—
Gallic	3,4,5-Trihydroxybenzoic	(benzene: CO₂H, OH, HO, HO) (5)	—

Table 1 Trivial and Systematic (IUPAC) Names of Hydroxy Acids

Trivial	Systematic	Formula	$[\alpha]_D{}^a$
—	3-Hydroxy-2-naphthoic	(6)	—
—	1-Hydroxy-2-naphthoic	(7)	—
—	2-Hydroxy-1-naphthoic	(8)	—
Aluminon (aurintricarboxylic)	—	(9)	—

[a] All data from ref. 10(g), except for (−)-gluconic.[164] [b] Chiral hydroxy acids. [c] At 20 °C in H_2O. [d] At 15 °C in H_2O. [e] At 16.5 °C in ethanol. [f] At 19 °C in H_2O.

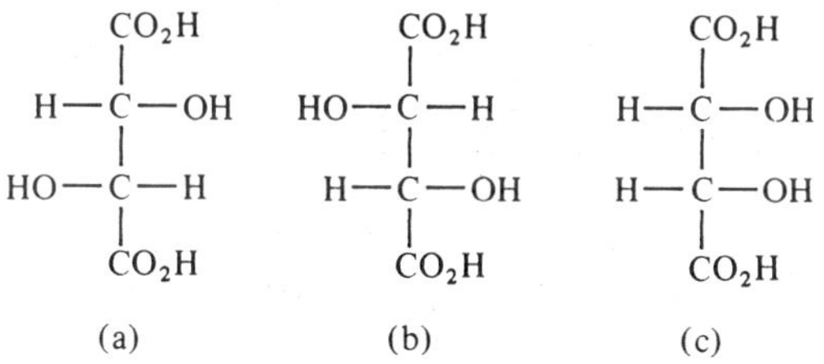

```
      CO₂H                CO₂H                CO₂H
       |                   |                   |
  H—C—OH            HO—C—H             H—C—OH
       |                   |                   |
 HO—C—H             H—C—OH             H—C—OH
       |                   |                   |
      CO₂H                CO₂H                CO₂H

       (a)                 (b)                 (c)
```

Figure 2 Fischer projections of (a) (2*R*,3*R*)-tartaric acid, (b) (2*S*,3*S*)-tartaric acid and (c) (2*R*,3*S*)-tartaric acid (*meso*-tartaric acid)

15.7.3 ACID–BASE PROPERTIES

The coordination of a metal ion, being a reaction competing with the addition of a proton to the ligand, always depends on the ligand protonation constant. This relationship between the formation constant of a metal complex and the basicity of the ligand can be very simple and has been used on several occasions to discuss relative stabilities of complexes of a metal ion with analogous ligands. It has also been used as the basis for explaining the differences in stabilities between complexes of a certain metal ion with monodentate carboxylates and complexes of the same metal ion with similar bidentate hydroxy acids.[14-16]

Although the situations are often complex, it is undoubtedly the case that the nature of the species formed in solutions of ligands and metal ions and their relative abundances are determined to a great extent by the ligand basicity. The enormous effort which has been devoted to the study of these equilibria is reflected in the voluminous data published on the formation constants of metal complexes. These results have been critically reviewed,[17] including the protonation constants of ligands.[18] These publications, together with some older works,[19,20] offer an excellent collection of data on protonation constants for hydroxy acids over a wide range of experimental conditions. A limited number of ionization constants are collected in Table 2. It should be emphasized, however, that those constants refer mainly to the ionization of carboxyl groups. As expected, there are few values available concerning the ionization of the hydroxyl groups of hydroxy acids. The results available show that the free ligand hydroxyl group is still protonated in most of the experimental conditions in which complexation with hydroxy acids is studied. It should be noted, however, that coordination of the hydroxyl group, even though not always resulting in release of the hydroxyl proton, always increases acidity of this group. For instance, a pK_a value of 3.51 has been obtained[24] for the coordinated hydroxyl group of (*S*)-lactic acid in the ion $[Co(S\text{-lac})(phen)_2]^{2+}$. The acidity of the hydroxyl group of tartaric acid in $[(en)_2Co(R,R\text{-tart})Co(en)_2]^{2+}$ is also much higher $(pK_a^{OH} = 2.45)^{25}$ than in free ligand $(pK_1^{OH} = 13.9; pK_2^{OH} = 15.5)$.[26] This effect of coordination on the acidity of hydroxyl groups of α-hydroxy acids can be confirmed by several other examples: $[Co(en)_2(glycolate)]^{2+}$, $pK_a^{OH} = 3.3 \pm 0.1$;[27] $[Co(en)_2(lactate)]^{2+}$, $pK_a^{OH} = 3.4 \pm 0.1$;[27] $[Co(en)_2(salicylate)]^{2+}$, $pK_a^{OH} = 1.1 \pm 0.3$[28] Even when the hydroxyl group is not directly coordinated to a metal, the coordination of a hydroxycarboxylate ligand results in higher acidity of that group. This is clearly shown, for instance, in complexes of (*R*,*R*)-tartaric acid in which it coordinates through one α-hydroxycarboxylate group and leaves one dangling α-hydroxycarboxylate end. The titration of the hydroxyl proton in the latter group gave pK^{OH} values of 3.85 and 3.72 for Δ- and Λ-[Co(*R*,*R*-tart)(bipy)_2]^+ and 4.08 and 3.81 for Δ- and Λ-[Co(*R*,*R*-tart)(phen)_2]^+, respectively.[24] Analogous results have been obtained for complexes of salicylates in which the ligand coordinates only through the carboxylate group. In the complexes of general formula $[(NH_3)_5Co\{O_2CC_6H_3(X)OH\}]^{n+}$, values of $pK_{OH} = 9.69, 9.83$ and 8.81 have been reported for X = 5-SO_3^-, 5-Br and 5-NO_2, respectively.[29]

The ionization of citric acid has been studied in detail by several authors using distinct techniques.[30-36] More recently, the protonation of citrate, tartrate and malate have been studied using potentiometric and calorimetric methods to obtain protonation constants and other thermodynamic data for a wide range of ionic strengths and two different temperatures (25 and 37 °C).[37] The complexation of Na^+ by those ligands[37] and of Li^+, Na^+, K^+, Rb^+, Cs^+ and NH_4^+ by citrate ions[38] has also been investigated. The latter results show how the cation of the ionic medium can influence the ionization equilibria and should be taken into account in corrections for this effect.

Table 2 Ionization Constants of Hydroxy Acids in Aqueous Solutions

Acid	Equilibrium constants	pK_a values (a)	(b)	(c)	(d)	(e)	Ref.
Glycolic	$[HL][H]/[H_2L]$	3.831	3.63	3.57	3.62	3.91	17
(R)-Mandelic	$[HL][H]/[H_2L]$	3.40	3.19	3.12	3.17	3.49	17
Benzilic	$[HL][H]/[H_2L]$	3.05[f]	2.87[f]	2.80[f]	2.80[f]	—	17
(S)-Lactic	$[HL][H]/[H_2L]$	3.860	3.66	3.61	3.64	3.81[g]	17
β-Hydroxypropionic	$[HL][H]/[H_2L]$	—	4.33[h]	—	4.32	4.56[g]	17
(±)-α-Hydroxybutyric	$[HL][H]/[H_2L]$	—	3.68	—	—	3.80[g]	17
(±)-β-Hydroxybutyric	$[HL][H]/[H_2L]$	—	4.28[j]	—	4.35[k]	4.53[g]	17
γ-Hydroxybutyric	$[HL][H]/[H_2L]$	—	—	—	4.57[k]	4.85[g]	17
(±)-Malic	$[H_2L][H]/[H_3L]$	3.459	3.24	—	3.11	—	12
	$[HL][H]/[H_2L]$	5.097	4.71	—	4.45	—	
(R,R)-Tartaric	$[H_3L][H]/[H_4L]$	3.036	2.82	2.62	2.69	2.98	17
	$[H_2L][H]/[H_3L]$	4.366	3.95	3.67	3.73	3.93	
meso-Tartaric	$[H_3L][H]/[H_4L]$	3.17	2.99	—	2.86		17
	$[H_2L][H]/[H_3L]$	4.91	4.44	—	4.10		
Citric	$[H_3L][H]/[H_4L]$	3.128	2.87	—	2.80	2.90[g]	17
	$[H_2L][H]/[H_3L]$	4.761	4.35	—	4.08	4.16[g]	
	$[HL][H]/[H_2L]$	6.396	5.69	—	5.33	5.18[g]	
(±)-Isocitric	$[H_3L][H]/[H_4L]$	—	3.02[l]	—	2.29	—	17, 21
	$[H_2L][H]/[H_3L]$	—	4.28[l]	—	3.73	—	
	$[HL][H]/[H_2L]$	—	5.75[l]	—	5.06	—	
	$[L][H]/[HL]$	—	—	—	12.19	—	
Salicylic	$[HL][H]/[H_2L]$	2.97	2.81	2.78	2.78	3.16	17
	$[L][H]/[HL]$	13.74	13.4	—	13.15	13.12	
3-Methylsalicylic	$[HL][H]/[H_2L]$	2.95	2.82	—	—	—	17
4-Methylsalicylic	$[HL][H]/[H_2L]$	3.04	2.97	—	—	—	17
	$[L][H]/[HL]$	14.26	—	—	—	—	17
5-Methylsalicylic	$[HL][H]/[H_2L]$	2.88	2.90	—	—	—	17
	$[L][H]/[HL]$	14.57	—	—	—	—	17
5-Chlorosalicylic	$[HL][H]/[H_2L]$	2.64	—	—	—	—	17
	$[L][H]/[HL]$	12.95	—	—	—	—	17
5-Bromosalicylic	$[HL][H]/[H_2L]$	2.64	—	—	—	—	17
	$[L][H]/[HL]$	12.84	—	—	—	—	
3-Nitrosalicylic	$[HL][H]/[H_2L]$	1.87	—	—	—	—	17
	$[L][H]/[HL]$	10.33	—	—	—	—	
5-Nitrosalicylic	$[HL][H]/[H_2L]$	2.12	2.20	—	—	—	17
	$[L][H]/[HL]$	10.34	10.11	—	—	—	
3,5-Dinitrosalicylic	$[HL][H]/[H_2L]$	2.14	2.14[i]	—	—		17
	$[L][H]/[HL]$	7.22	7.29[i]	—	—		
5-Sulfosalicylic	$[H_2L][H]/[H_3L]$	—	—	—	0.75	—	17
	$[HL][H]/[H_2L]$	2.84	2.49	2.35	2.32	2.84	
	$[L][H]/[HL]$	12.53	11.72	11.51	11.40	12.53	
3,5-Disulfosalicylic (H_3L)	$[HL][H]/[H_2L]$	2.69	2.03	1.70	1.71	—	17
	$[L][H]/[HL]$	12.50	11.55	11.04	10.95	—	
4-Aminosalicylic	$[H_2L][H]/[H_3L]$	—	1.78	—	1.95	2.08	17
	$[HL][H]/[H_2L]$	—	3.63	—	3.68	4.08	17
	$[L][H]/[HL]$	—	—	—	—	13.7	
4-Hydroxysalicylic	$[H_2L][H]/[H_3L]$	—	—	3.12	—	—	22
	$[HL][H]/[H_2L]$			8.56	—	—	
	$[L][H]/[HL]$	—	—	13.37	—	—	
5-Hydroxysalicylic	$[H_2L][H]/[H_3L]$	—	—	2.73	—	—	22
	$[HL][H]/[H_2L]$	—	—	10.00	—	—	
	$[L][H]/[HL]$	—	—	12.74	—	—	
6-Hydroxysalicylic	$[H_2L][H]/[H_3L]$	—	—	1.20	—	—	22
	$[HL][H]/[H_2L]$	—	—	12.57	—	—	
	$[L][H]/[HL]$	—	—	13.28	—	—	
3,5-Dihydroxybenzoic	$[H_2L][H]/[H_3L]$	—	—	3.81	—	—	22
	$[HL][H]/[H_2L]$	—	—	8.99	—	—	
	$[L][H]/[HL]$	—	—	10.74	—	—	
3-Hydroxy-2-naphthoic	$[HL][H]/[H_2L]$	2.75	2.54	2.41	2.41	—	17
	$[L][H]/[HL]$	12.84	12.48	12.37	12.31	—	
3-Hydroxy-5-,7-disulfo-2-naphthoic (H_4L)	$[HL][H]/[H_2L]$	2.18	2.37	2.14	2.18	—	17
	$[L][H]/[HL]$	12.03	11.28	10.91	10.81	—	
1-Hydroxy-4-sulfo-2-naphthoic (H_3L)	$[HL][H]/[H_2L]$	—	2.50	—	—	—	23
	$[L][H]/[HL]$	—	11.65	—	—	—	
1-Hydroxy-7-sulfo-2-naphthoic (H_3L)	$[HL][H]/[H_2L]$	—	2.73	—	—	—	23
	$[L][H]/[HL]$	—	12.37	—	—	—	

| Acid | Equilibrium constants | pK_a values | | | | | |
		(a)	(b)	(c)	(d)	(e)	Ref.
1-Hydroxy-4,7-disulfo-	$[HL][H]/[H_2L]$	—	2.208	—	—	—	23
2-naphthoic (H_4L)	$[L][H]/[HL]$	—	11.119	—	—	—	

[a] $25\,°C$, $I = 0\,mol\,dm^{-3}$. [b] $25\,°C$, $I = 0.1\,mol\,dm^{-3}$. [c] $25\,°C$, $I = 0.5\,mol\,dm^{-3}$. [d] $25\,°C$, $I = 1.0\,mol\,dm^{-3}$. [e] $25\,°C$, $I = 3.0\,mol\,dm^{-3}$. [f] $18\,°C$, I as above. [g] $25\,°C$, $I = 2.0\,mol\,dm^{-3}$. [h] $30\,°C$, $I = 0.1\,mol\,dm^{-3}$. [i] $20\,°C$, $I = 0.1\,mol\,dm^{-3}$. [j] $25\,°C$, $I = 0.2\,mol\,dm^{-3}$. [k] $20\,°C$, $I = 1.0\,mol\,dm^{-3}$. [l] Me_4NCl used as a background electrolyte and corrected for Cl^-.

15.7.4 SPECTROSCOPIC AND CHIROPTICAL PROPERTIES

15.7.4.1 NMR Spectra

NMR spectroscopy is certainly a powerful method for studying the structure of these ligands in solution. Both ^1H and ^{13}C NMR have been of great help in discussing metal–ligand bonding, in determining stoichiometries and in finding ligand conformations in a wide variety of situations where, for instance, malate, tartrate and citrate were involved. The use of multinuclear NMR in structure determinations of hydroxy acids in the presence of lanthanide ions in solutions has been reviewed.[39a]

The ^1H NMR spectra of a large number of aliphatic carboxylic acids and their anions have been studied in H_2O and D_2O solutions.[40] The ^1H chemical shifts for several hydroxy acids (glycolic, lactic, 2-hydroxybutyric, 2-hydroxyisobutyric, 3-hydroxypropionic, 3-hydroxybutyric, 4-hydroxy-butyric, tartaric, tartronic) and their respective anions are given. The factors affecting the chemical shift differences between the acids and the corresponding anions for the various proton positions are discussed. Evidence for intramolecular hydrogen bonding has been found in the 2-hydroxy acids (and possibly in 3-hydroxy acids). A detailed ^1H NMR study of citric acid in solution has been carried out by Loewenstein and Roberts.[31] The chemical shifts of the methylene hydrogens were measured as a function of pH for various concentrations of acid. The results were used to evaluate the chemical shifts associated with each ionization step. It was found that the first and second ionizations of citric acid take place predominantly at the terminal carboxyl groups. However, Martin studied the ionization of citric acid using a titration method and arrived at the opposite conclusion, *i.e.* that the ionization of the central carboxyl group is predominant.[32] The NMR result gave an incorrect answer due to an inappropriate choice of pK_1, to which the interpretation is sensitive. By selecting a value of pK_1 appropriate to the conditions of the experiment, the NMR results are consistent with the titration results.

Another area in which the use of NMR techniques has proved to be particularly useful is the study of the conformations of the hydroxy acids in solution. The main conformations of (R)-malic acid in relation to rotations around the C(2)—C(3) bond are represented in Figure 3 by their Newman projections. They correspond to two synclinal, or *gauche*, conformers ($+sc$ = a and $-sc$ = c) and to one antiperiplanar, or *trans* conformer (ap = b). The ^1H NMR spectrum of malic acid is of the ABX type and can be analyzed as if the observed vicinal H—H coupling constants were weighted averages of the corresponding values for the three staggered conformers. This analysis has enabled the determination of relative populations of the three conformations in solutions of the acid, the corresponding anions and their complexes. Gil and coworkers[41] have used ^1H NMR to show that (R)-malic acid in H_2O and D_2O solutions adopts the following conformations: ($+sc$) 42%, (ap) 7% and ($-sc$) 51%. In the corresponding malate ion the relative abundance of the ($+sc$) conformer increases to 60%, that of (ap) is 3% and the contribution of ($-sc$) decreases to 37%. The values for the anion have been redetermined.[42] The new values found for $0.1\,mol\,dm^{-3}$ solutions of (R)-malate anion are: ($+sc$) 71%, (ap) 9% and ($-sc$) 20%. It is expected that coordination of the hydroxy acid to a metal ion may alter the relative abundances of the conformations. Conformer ($-sc$) may be associated with the formation of tridentate complexes of (R)-malate. This has been observed for species formed in solutions of this ligand and Zn^{II},[41] Mo^{VI}[43] and W^{VI}.[44] On the other hand, it is expected that in the formation of bidentate complexes conformers ($+sc$) and ($-sc$) may be preferred. This is observed for complexes of Co^{III}[45] and in complexes of Mo^{VI},[43] W^{VI}[44] and UO_2^{2+}.[46]

Since the normal spectra of tartaric acids only have one line, the ^{13}C–^1H satellites had to be used for studying the conformations of these ligands in solution. Ascenso and Gil[47] studied the ^1H and ^{13}C NMR spectra of (R,R)- and (R,S)-tartaric acids [(R,R) = (2R,3S), (R,S) = (2R,3S), (S,S) = (2S,3S)] in D_2O at a wide range of pH values and found that conformer (a) in Figure 3A is favoured by (*R,R*)-tartaric acid in solution, as happens in the solid state. The relative population of that conformation increases with pH and this result is explained by the expected increasing repulsion between ionized carboxyl groups in (c). For *meso*-tartaric acid[(*R,S*)-tartaric acid] the conformations (b') and (c') in Figure 3B are preferred, as is the case in the solid state. Perfectly analogous results were obtained by Marcovich and Tapscott,[48] who have also discussed the conformations of methyl tartrates.

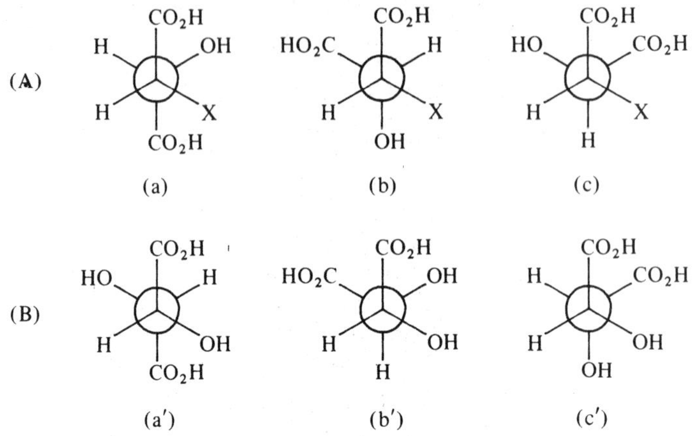

Figure 3 Newman projections of the staggered conformers of (A) (*R*)-malic acid (X = H) and (2*R*,3*R*)-tartaric acid
(X = OH); (B) (2*R*,3*S*)-tartaric acid (*meso*-tartaric acid)

Although the simplicity of the ^{1}H NMR spectra of tartaric acids may imply a lack of structural information, it has been proved that it is worth studying the ^{1}H NMR spectra of metal complexes of these ligands. The same could be said for the ^{13}C NMR spectra of these hydroxy acids, as is well illustrated by several of their complexes. Diastereoisomeric ternary complexes of Co^{III} with 1,10-phenanthroline and (*R,R*)-tartaric acid have been studied by several techniques.[24,45,49,50] The stereospecific formation (claimed by Haines and Bailey[49]) of Λ-[Co(phen)$_2$(*R,R*-tartH$_2$)]$^+$ by reacting Δ,Λ-[Co(phen)$_2$CO$_3$]$^+$ with (*R,R*)-tartaric acid in aqueous solutions has been questioned by Tatehata.[24,50] This author, although finding an excess of Λ diastereoisomer in the final product, showed that the formation ratio was Δ:Λ = 27.6:72.4. A very elegant ^{1}H NMR study[45] very convincingly established the proportion of Δ and Λ diastereoisomers in the product of that reaction and the ligand conformation in the complexes. The J_{AB} coupling constants observed for the diastereoisomers (1.82 ± 0.05 Hz for Λ and 1.53 ± 0.05 Hz for Δ) of the (*R,R*)-tartaric acid complex are typical of an essentially *gauche* arrangement of the C—H atoms on adjacent carbons. This result precludes any significant population of rotamer (b) in Figure 4. Both conformations (a) and (c) could explain the values obtained for J_{AB} and they are also the two conformations more suitable for intramolecular hydrogen bonding between the uncoordinated carboxyl group and the coordinated α-hydroxy oxygen atom. This situation is also favoured in the analogous *S*-(−)-malato complex. This coordination mode of the tartrate ligand has also been observed[51,52] by ^{1}H and ^{13}C NMR techniques in complexes formed between chiral tartaric acids and Mo^{VI} or W^{VI} in aqueous and DMSO solutions. These metal ions were shown to form complexes with metal:ligand ratios equal to 1:2 in which the ligand bonding modes have been identified as those represented by conformations (a) and (c) in Figure 4. The coordination of tartrate as a bridging ligand between two metal centres has been carefully discussed by Tapscott and coworkers.[53] NMR work[51] has suggested that this coordination mode is present in various species formed in solutions of Mo^{VI} and (*R,R*)-tartaric acid at various pH values and for several metal:ligand ratios. One species has been identified as the dimer represented in (**10**). This proposition has been strongly supported by the observation[52] of similar complexes of W^{VI}, and of mixed-metal dimers in solutions 1:1:2 in Mo^{VI}:W^{VI}:(*R,R*)-tartrate. Other complexes of Mo^{VI} in which bridging tartrate ligand is present have also been observed in these NMR studies.

15.7.4.2 Infrared Spectra

The IR spectra of hydroxy acids are not very different from the spectra of the corresponding carboxylic acids. The additional hydroxyl groups usually produce minor displacements of the carbonyl vibrational bands to higher energies and, of course, contribute additional characteristic

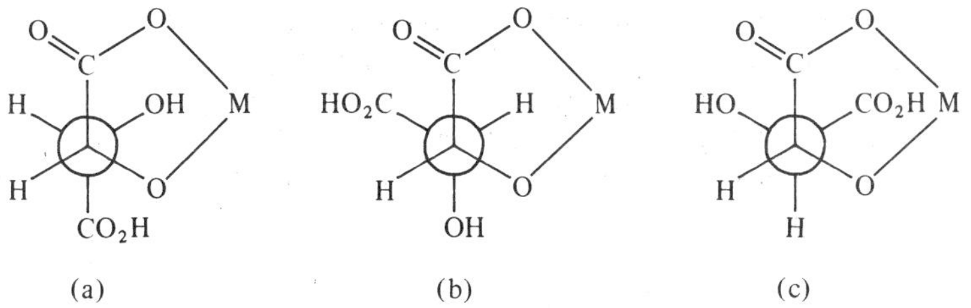

Figure 4 Representation of the staggered conformers of coordinated (2*R*,3*R*)-tartaric acid in the complex [Co(phen)$_2$-(H$_2$tart)]$^+$

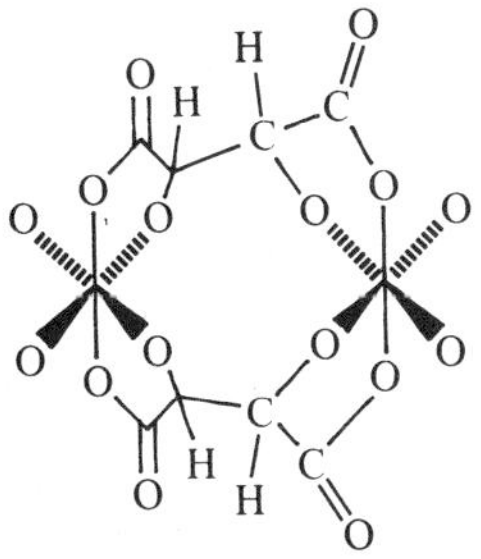

(10)

bands. Therefore the extensive information available[54] on the IR spectra of carboxylic acids and alcohols may be extremely valuable for the interpretation of the IR spectra of hydroxy acids and of their salts and coordination compounds.

The IR absorption spectra of these ligands have been studied in detail by several authors.[55-64] Part of this work has been reviewed by Shevchenko,[64] who shows how the absorption due to the alcoholic hydroxyl deformation can give valuable information about the location of that group in the acid skeleton and its involvement in coordination. The interpretation of the lactate spectra put forward by Goulden[62] is critically discussed, particularly for the bands around 1360 cm^{-1}. The spectra of salicylates are also reviewed in that work. Of particular interest is the observation that the C=O stretching vibration energy for salicylic[54,64] and 3-hydroxy-2-naphthoic[64] acids is lower (1690 cm^{-1}) than in acetic (1712 cm^{-1}) and benzoic acids (1740 cm^{-1}), while the hydroxyl stretching band is observed around 3200 cm^{-1}, owing to the presence of a strong intramolecular hydrogen bond. In methoxybenzoic acid, where such a hydrogen bond is impossible, the O—H stretching vibration occurs at 3400 cm^{-1}.

The effects of coordination on the IR bands of salicylate have been well illustrated very recently[65] for compounds M(Hsal)$_2$·8H$_2$O (M = Mn^{II}, Fe^{II}, Co^{II}, Ni^{II} and Zn^{II}, H$_2$sal = salicylic acid) and Cu(Hsal)$_2$(H$_2$O)$_2$. Deuteration studies showed that the in-plane bending vibration, δ(OH), comes at 1250 cm^{-1} for the first series of compounds and at 1210 for Cu(Hsal)$_2$(H$_2$O)$_2$. After deuteration, these bands shift to 1120 and 1095 cm^{-1}, respectively. X-Ray diffraction has shown that in the series of compounds M(Hsal)$_2$·8H$_2$O the salicylate ions are not directly bonded to the metal centres, but rather form second sphere complexes with the M(H$_2$O)$_6^{2+}$ species. On the other hand, in Cu(Hsal)$_2$-(H$_2$O)$_2$ the salicylate ligand is coordinated to two distinct Cu^{II} ions through the protonated alcoholic oxygen and the oxygen of the deprotonated carboxyl group, respectively. As far as the v(CO$_2$) absorptions are concerned, almost identical values have been obtained for all the complexes studied (v_{as} = 1593–1598 cm^{-1}; v_s = 1395–1398 cm^{-1}).

The IR spectra of coordinated ligands have been of great help in the characterization of the ligand coordination mode in situations where a deprotonated coordinated carboxylate and a protonated nonbonded carboxyl group are present in a complex. This has been the case, for example, for complexes of tartrate and malate ions chelated to Co^{III} [24,50] and for citrate[66] coordinated to Mo^{VI}. The protonated noncoordinated carboxyl group has an absorption in the region 1750–1700 cm^{-1}, which is typical of C=O stretching vibrations of CO$_2$H. The corresponding coordinated carboxylate shows an absorption at lower energy (1650–1590 cm^{-1}) and the expected band is observed in those complexes.

The interpretation of the IR spectra of hydroxy acids in the range 1500–1000 cm^{-1} is rather difficult.[54,59] In this region, absorptions corresponding to carbon–hydrogen deformations (CH$_3$, CH$_2$, CH groups) can be observed, as well as C—O stretching vibrations and in-plane deformations both of carboxyl C—O—H and hydroxyl C—O—H. In the region 1000–650 cm^{-1}, absorptions are observed due to out-of-plane deformation of acid C—OH, at around 930 cm^{-1}. Bolard[59] has studied the IR spectra of several α-hydroxy acids and some of their salts and metal complexes, namely glycolic, lactic, tartaric and malic acids. Goulden[63] has studied the spectra of M^{II} complexes of hydroxyisobutyrate and associated a band observed at 1370 cm^{-1} (stronger for Cu^{II} than for Ca^{II} and Sr^{II} complexes) with in-plane bending of coordinated alcoholic OH, δ(OH), and a band at 1330 cm^{-1} with δ(OH) of the free ligand. These assignments were made by comparison with the spectra of tertiary and secondary alcohols, and lactate ion and its complexes in the same region. The spectra of lactate, glycolate and their respective esters and metal complexes have been studied,[62] with particular emphasis on the effect of coordination on the in-plane deformation of the OH group. The displacement of δ(OH) from 1275 cm^{-1} in the free ion to frequencies around 1390 cm^{-1} in the complex, but variable with the cation, were used to infer the relative stability of the lactate complexes studied.

The presence of a hydroxyl group in the molecule of a carboxylic acid is readily established by the observation of intense bands due to OH stretching vibrations, in the region 3200–3650 cm^{-1}. These bands are quite sensitive to hydrogen bonding and have been used[55-58] to evaluate the formation of intramolecular hydrogen bonds in aliphatic and aromatic hydroxy acids in CCl$_4$ solutions. The absence of bands due to OH stretching vibrations can be used as evidence for deprotonation of coordinated hydroxyl groups. This situation has been reported[59] for potassium [bis{lactate(2−)}borate(III)], where the deprotonated hydroxyl and carboxyl groups chelate to B^{III} in a tetrahedral arrangement.

15.7.4.3 Electronic Spectra and Chiroptical Properties

Aliphatic hydroxy acids usually show an absorption band around 210 nm in their electronic spectra, which has been associated[67,68] with an $n \to \pi^*$ transition of the carboxyl group. In dissymmetric hydroxy acids this transition gives rise[67-70] to a Cotton effect (CE) for a wavelength close to that of the absorption band. Quite often a second, less intense, CE is observed[69,70] at longer wavelength (240 nm), the origin of which is the subject of dispute. Richardson *et al.*[68] have performed calculations on the electronic structure and spectra of the conformational isomers of (*S*)-lactic acid. They discussed the origin of the lowest energy electronic transitions in the acid conformers and suggest that both observed circular dichroism bands at 210 nm and 240 nm for (*S*)-lactic acid should be assigned to an $n \to \pi^*$ transition of two distinct conformers of the hydroxy acid. The results suggest that the sign of the $n \to \pi^*$ Cotton effect may be related to the three conformations represented in Figure 5 for (*S*)-lactic acid. Isomers (I) and (III) of Figure 5 should give a positive $n \to \pi^*$ Cotton effect for (*S*)-lactic acid, whereas a negative Cotton effect is expected for rotamer (II). The calculation also predicts that the relative values of λ_{max} for the $n \to \pi^*$ transitions should be in the order $\lambda_{max}(I) < \lambda_{max}(III) < \lambda_{max}(II)$. These results fully support the hypothesis put forward by Listowsky *et al.*,[69] that the positive band centred around 210 nm in the CD spectra of α-hydroxy acids is due to a conformational isomer of type (I), and that the negative band observed at longer wavelength (~240 nm) in particular solvents and pH conditions is associated with a rotamer of type (II). These authors considered that isomers of type (III) were the least stable and did not contribute significantly to the observed CD. The binding energies calculated by Richardson *et al.* supported the view that rotamer (I) should be energetically favoured over rotamers (II) and (III), which is consistent with the circular dichroism results obtained at varied temperatures and in different solvents. The CD band at lower wavelengths, associated with (I), is always much more intense. A decrease in temperature should decrease the population of rotamers (II) and (III) in favour of rotamer (I) (as is observed, with the negative Cotton effect at ~240 nm decreasing and the positive CE at 210–215 nm increasing).

The chiroptical properties of 2-hydroxyalkanoic acids have been discussed in relation to their stereochemical features in solution. The potential of circular dichroism techniques was emphasized

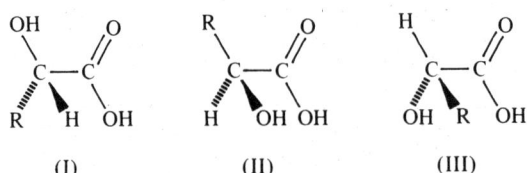

(I)   (II)   (III)

Figure 5 Conformations of α-hydroxy acids around the C(1)—C(2) bond

for conformational isomerism of hydroxy acids, although only empirical and semiempirical approaches to relate the information to electronic and conformational states of the molecule are available. The use of CD, ORD and polarimetry to study complexes of hydroxy acids is widely illustrated both in labile[71-76] systems and in inert complexes.[24,45,50,77,78] Another chiroptical technique, circularly polarized luminescence (CPL), has been extensively used[79-85] to study complexes of lanthanides(III) with optically active 2-hydroxy acids. Both simple and mixed ligand complexes of Mn^{III} with a wide range of 2-hydroxy acids have been studied. By comparing the changes in CPL bands of solutions of the complexes for different pH values and metal:ligand ratios, the authors infer the coordination mode of the ligand and the stoichiometry, and discuss several other properties of the complexes.

15.7.5 STRUCTURAL PROPERTIES OF COORDINATED HYDROXY ACIDS

15.7.5.1 General Features

The stereochemistry of 2-hydroxyalkanoate salts, of 2-hydroxyalkanoic acids, and of their respective coordination compounds has been the subject of a recent comprehensive review by Tapscott.[86] Geometric data available from X-ray and neutron diffraction studies up to 1980 have been critically compared, and the author has obtained average geometries for protonated, ionized and esterified 2-hydroxyalkanoates. These are represented in Figure 6. A characteristic feature[87] of these ligands is the coplanarity, or near coplanarity, of the 2-hydroxy and carboxyl groups observed in crystal structures. Experimental[69] and theoretical[68] results concerning the chiroptical properties of 2-hydroxyalkanoic acids have also led to the conclusion that the conformation represented in Figure 6 is favoured by these compounds in solution.

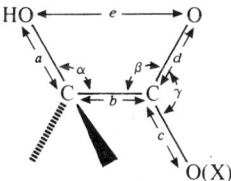

Figure 6 Average geometry found for the *ionized* ($\alpha = 111.1°$; $\beta = 118.0°$; $\gamma = 125.6°$; $a = 1.420$ Å, $b = 1.539$ Å, $c = 1.254$ Å, $d = 1.246$ Å, $e = 2.651$ Å), *protonated* ($\alpha = 109.7°$, $\beta = 123.1°$, $\gamma = 124.6°$; $a = 1.418$ Å, $b = 1.519$ Å, $c = 1.310$ Å, $d = 1.209$ Å, $e = 2.685$ Å) and *esterified* ($\alpha = 109.3°$, $\beta = 124.4°$, $\gamma = 124.0°$; $a = 1.409$ Å, $b = 1.534$ Å, $c = 1.342$ Å, $d = 1.197$ Å, $e = 2.717$ Å) α-hydroxycarboxylate groups (after ref. 86)

When coordinated to metal ions, hydroxy acids adopt bonding modes which will be described in detail later, when dealing with the stereochemistry of each ligand in metal complexes. It is, however, worth mentioning that the 2-hydroxyalkanoic acids have the common feature of forming very stable five-membered chelate rings of the type represented in (11), the hydroxyl group being ionized or not. This has been well documented in Tapscott's review and is corroborated by more recent results from solids and solution studies. For instance, the crystal structure of cesium *cis*-[bis{(*S*)-malato}dioxo-molybdate(VI)] monohydrate shows[88] that the two (*S*)-malato ligands are bonded to the metal centre in a way that forms two such chelate rings. (*R,R*)-Tartaric acid and (*S*)-malic acid both chelate Co^{III} in that way, each in ternary complexes with 1,10-phenanthroline as shown by ^1H NMR and chiroptical studies.[45] Both ^{13}C and ^1H NMR have also been used[51] to study the species in solutions of Mo^{VI} and (*R,R*)-tartrate. Here again the coordination involves the formation of the five-membered ring. A similar bonding mode has been observed[89] for citrate in the complex triethylenetetraaminecitratocobalt(III) hydrate, both in solution and in the solid state.

(H)O—M—O   R
C—C   CO₂H
O   OH

(11)   (12)

A feature of the structures of complexes of hydroxyalkanoates is that the metal–ionized-hydroxyl bond length is usually shorter than the corresponding metal–protonated-hydroxyl bond length.

The coordination chemistry of aromatic hydroxy acids has been mainly concerned with 2-hydroxybenzoic acid (salicylic acid) and related compounds. The structure of these ligands does not offer the same variety of molecular arrangements that is characteristic of most 2-hydroxyalkanoic acids. A wide range of substituted salicylates (**12**) has been studied, but particular attention has been given to compounds where R = OH, Cl, Br, I, NO_2 or $HOSO_2$. The formation of a five-membered chelate ring is not so common here as with aliphatic acids. On the other hand, coordination *via* the monodentate or bidentate carboxyl group only is much more usual for the aromatic carboxylates.

15.7.5.2 Aliphatic Hydroxy Acids

Those organic hydroxy acids which have been most extensively studied as coordinating entities are aliphatic. As can be seen from their formulae (Table 1), this group of carboxylic acids includes very simple ligands like glycolic acid or lactic acid which do not have many possible coordinating modes when forming complexes. However, it also includes more complex ligands, such as malic acid, tartaric acid or citric acid, that show an immense variety of possibilities for coordinating inorganic ions. This is well illustrated by the many proposed structures of complexes of these ligands in solution, very often mutually contradictory.

In this section the stereochemistry of the coordinated molecules will be reviewed, starting with the monohydroxymonocarboxylic acids (glycolic, lactic, mandelic), then turning to monohydroxypoly-carboxylic (malic, citric, isocritic) and polyhydroxypolycarboxylic (tartaric) and finally reviewing briefly a series of less common ligands (tartronic, saccharic, gluconic, glyceric and a few long-chain hydroxy acids). We will be mainly concerned with the ligand coordination modes which have been clearly identified for the different ligands. Detailed structural data will not be given, but they can easily be found in Tapscott's review[86] or in the original literature referred to in that review and in this chapter.

15.7.5.2.1 Glycolic acid

Glycolic acid is the simplest of these hydroxy acids. Metal complexes of this ligand with a wide variety of metal ions have been studied[14-16,90-96] and their stereochemistry reviewed.[86,87a]

The chelation of glycolate to metal centres through the hydroxyl oxygen and one of the carboxylate oxygen atoms is well established. This is the more common coordination mode adopted by this ligand, both in the solid state and in solution. The stability constants of 1:1 complexes with actinide(VI) oxo cations are much higher than the corresponding stability constants of complexes with analogous monocarboxylic acids. This additional stability has been taken[15] as meaning that chelation of the glycolate occurs. Cu^{II} complexes show a similar effect in solution. For solid bis(glycolato)copper(II) the ligand molecules are chelated[14] to the Cu^{II} centre with a *trans* arrangement. The Cu^{II} ion has a tetragonally distorted octahedral environment, with the longer axis positions occupied by carbinol oxygen atoms. A similar coordination mode has been proposed[92,93] for complexes of uranyl, although in these polymeric compounds the metal environment is more complex and ill defined.

The complex diaquabis(glycolato)zinc(II) is isomorphous with the Mn^{II} and Co^{II} complexes. The structures of these compounds, while still showing bidentate glycolato coordination, have the two ligands adopting a *cis* arrangement around the metal centre (the hydroxyl group of one glycolate is *trans* to the carboxyl group of the other, as shown in **13**). The crystal structure of the Zn^{II} complex shows the presence of equal amounts of Δ and Λ enantiomers in the unit cell. Molecular models suggested[86] that for the Λ absolute configuration the chelate rings with hydroxyl groups *trans* to water molecules should adopt λ conformations to reduce interaction between hydrogen atoms of water molecules and the methylene groups. On the other hand, rings with carboxyl groups *trans* to water molecules should adopt δ conformations. The crystal structure of the Zn^{II} complex shows that the predicted δ puckering in the Λ isomers is observed (although it is slight), but the ring predicted to be λ is planar. Aqua{glycolato(1−)}{glycolato(2−)}erbium(III) monohydrate[95] is a good example for illustrating some new ways of the glycolate behaving as a ligand. This compound is a network solid containing two chelated ligands, crystallographically nonequivalent. One of them is deprotonated only at the carboxyl group and the other is deprotonated both at the carboxyl and hydroxyl groups. Apart from the four positions occupied by chelated glycolato ligands, the coordination sphere of Er^{III} in this complex comprises one water molecule and three oxygens, two of them from carbonyl groups and one from a hydroxyl group of neighbouring glycolate. The metal to ionized hydroxyl oxygen bond lengths are shorter (about 0.2 Å in these cases) than the corresponding bond lengths involving protonated hydroxyl oxygen. This is a feature common to all 2-hydroxycarboxylate complexes. Polymeric networks involving glycolate bridging atoms are more common as the metal coordination number increases and are observed[86] in several structures of lanthanide complexes.

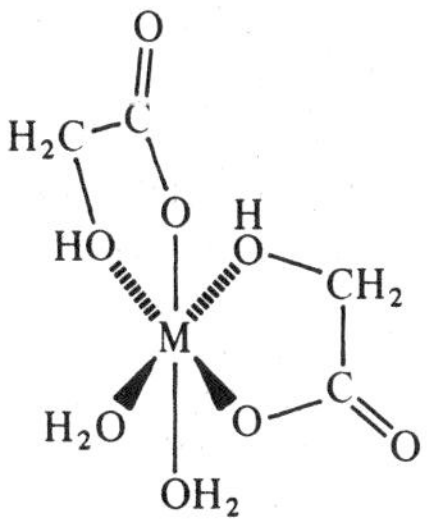

(13)

15.7.5.2.2 *Lactic acid*

2-Hydroxypropanoic acid, $MeCH(OH)CO_2H$, usually known as lactic acid, is a naturally occurring compound present in many foodstuffs. It is chiral and consequently can be found in two enantiomeric forms (Figure 1). The isomer usually called L-(+)-lactic acid has the (S) absolute configuration and D-(−)-lactic acid has the (R) absolute configuration.

Lactic acid has a preferred conformation in solution with the hydroxyl group synperiplanar with the carbonyl group. The conformational dependence of its chiroptical properties has also been examined in detail. Empirically based relationships[69,70] between CD spectra and conformations have been supported[68] by calculations.

Complexes of lactic acid with a wide variety of metal ions are known and their stability constants have been determined.[17,19] A variety of techniques has been used in the study of the species formed in the solutions of this hydroxy acid and inorganic ions. Electrochemical[15,96,97] and spectrophotometric[98–100] methods have been used with the main aim of determining stability constants of the complex ions. CD and optical rotatory dispersion (ORD) have also proved to be powerful methods in studying, for instance, Mo^{VI},[73] Mo^{II},[71] Cu^{II} and Co^{II},[101] as well as lanthanide complexes.[102] Very recently, Brittain *et al.*[79,81–83,103] used circularly polarized luminescence (CPL) techniques in the study of simple and mixed ligand complexes of lactic acid with lanthanide(III) ions. The general trend is for bidentate binding of the lactate ligand *via* both the hydroxyl and carboxyl groups. This coordinating mode is also extensively observed in complexes in the solid state. Crystal structures of lactato complexes have been reviewed by Tapscott.[86] Although four of the six known structures refer to Cu^{II} complexes, an interesting range of different stereochemistries has been observed. The complex aqua{(R)-lactato}{(S)-lactato}copper(II) (14) has two *trans*-chelated lactate ligands of opposite chirality bonded through the protonated hydroxyl oxygen and one deprotonated carboxyl oxygen.[14] The axial positons of the tetragonally distorted octahedron are occupied, one by the water molecule and the other by an oxygen atom of a neighbouring carboxyl group. Another bis-chelate is the *cis*-diaqua{(R)-lactato}{(S)-lactato}zinc(II) monohydrate complex, (15).[104] *cis*-(*N,N,N′,N′*-Tetramethylethylenediamine)bis{(S)-lactato}copper(II) and the analogous nickel(II) complex have,

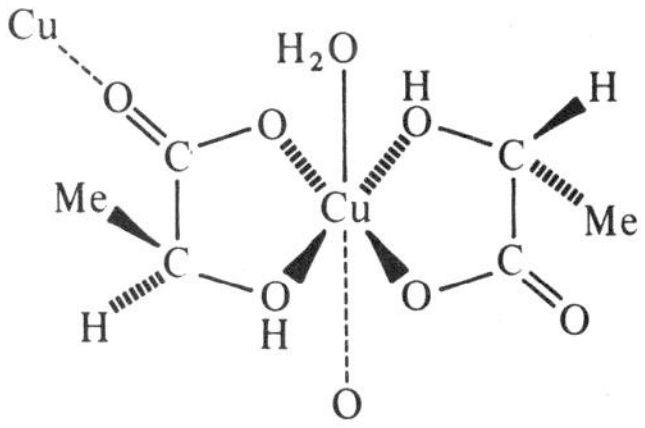

(14)

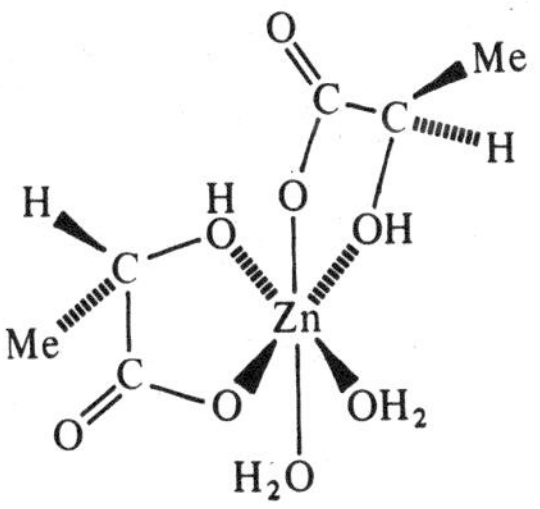

(15)

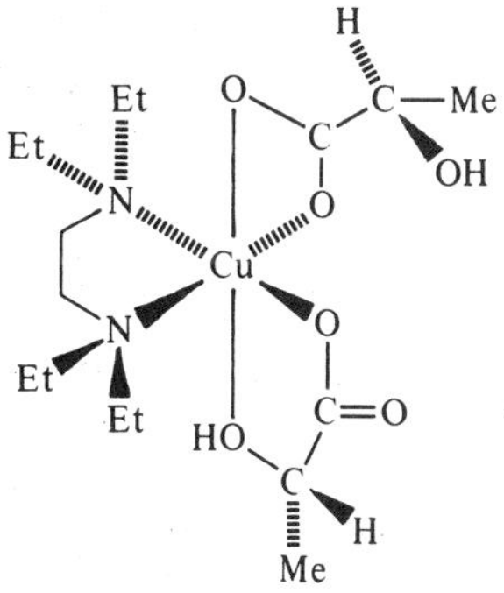

(16)

again, two chelated ligands. Here the ligands have the same chirality and the hydroxyl groups are in *trans* positions to each other. The complex packs stereoselectively in the Δ configuration, a situation rather different from the zinc(II) bis-chelate where units of Δ and Λ configuration are found in the lattice. A rather different binding mode is found in bis{(S)-lactato}(N,N,N',N'-tetraethylethylenediamine)copper(II) hemihydrate. Although a Δ absolute configuration is again adopted, one of the chelate rings results from the bonding of lactate through the two oxygen atoms of the carboxylate group (**16**), while the other binds in the usual way *via* carboxyl and hydroxyl oxygen atoms. In bis(N-isopropyl-2-methylpropane-1,2-diamine){(S)-lactato}copper(II) (S)-lactate monohydrate, and in the analogous compound of Zn(II),[186] the lactate ligand (one lactate is present as a noncoordinating entity) is monodentate *via* one of the carboxyl oxygens.

15.7.5.2.3 *Mandelic acid*

2-Hydroxy-2-phenylethanoic acid is usually called by its trivial name, mandelic acid; the name phenylhydroxyacetic acid is also used. It has a chiral carbon atom and can, therefore, be found in two enantiomeric forms (Figure 1). While the crystal structure of the racemic compound has been known for some time,[105] the structure of solid (R)-mandelic acid has been determined only recently.[106] The results for this enantiomer agree with the assignment of the (R) absolute configuration to what is usually called D-mandelic acid. The conformation of (R)-mandelic acid and substituted mandelic acids, in solution and in the solid state, have been recently discussed[107,108] in connection with the assignment of their CD spectra. As far as rotation around the C(2)—C(3) bond is concerned, two conformations predominate[107] for (R)-mandelic acid in the crystalline state (Figure 7a,b). Every molecule in one conformation is accompanied by another one in the other conformation. The preferred conformations in solution, with respect now to the C(1)—C(2) torsion angle, have been discussed by Håkansson and Gronowitz,[108] who also measured the absorption and CD spectra of (R)-mandelic acid in several solvents. The absorption spectrum in acetonitrile shows a band with vibrational structure centred at 258 nm, another at around 205 nm and a third absorption below 190 nm. The corresponding CD spectrum has a positive structured band centred at 265 nm, a strong negative band at 223 nm, a weaker one at 205 nm and a second positive absorption at 197 nm.

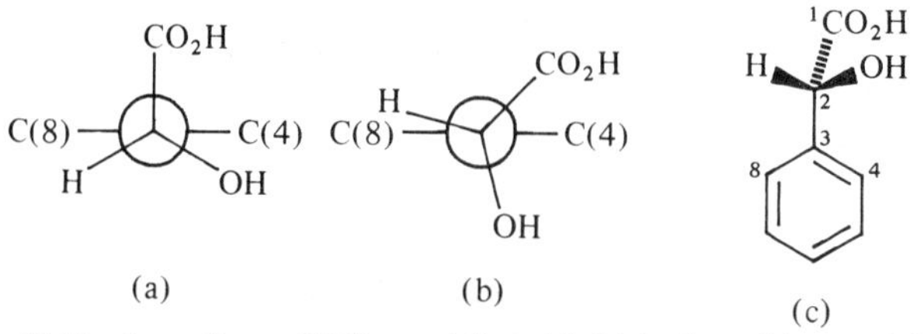

Figure 7 (a, b) Conformations of (R)-mandelic acid, (c) in the solid state (after ref. 107)

Although few complexes of mandelate have known crystal structures,[86] quite distinct coordination modes have been identified in those few. A bidentate ligand, coordinating *via* deprotonated hydroxyl and carboxyl oxygens, is found[109] in the compound *trans*-[diaquabis(mandelato)germanium(IV)] dihydrate (**17**). The crystal structure of the complex tetrakis{(R)-mandelato}dimolybdenum(II) bis(tetrahydrofuranate) has been recently solved.[106] Here each hydroxy acid anion is coordinated to two Mo centres through the two oxygen atoms of the carboxyl group, a coordinating mode reminiscent of that observed in dinuclear acetate complexes of Mo[II] and Cr[II]. Yet another binding mode is found[86] in a dinuclear complex of Ca[II] (**18**). The mandelato(2−) ligand is bidentate (*via* ionized hydroxyl and carboxylate) to one calcium centre and bridges to another calcium atom through the already bonded oxygen atom of the carboxyl group.

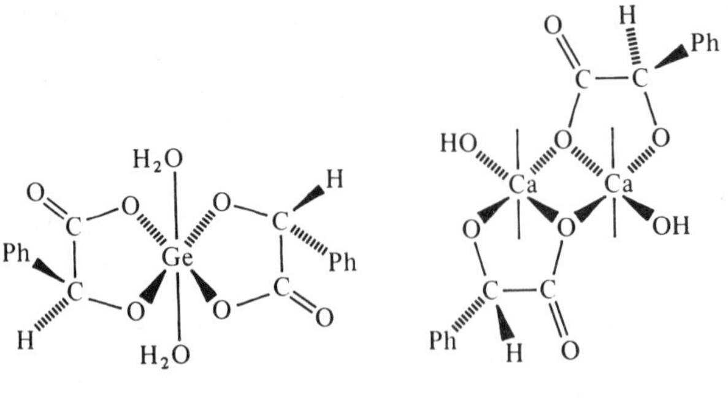

(17)   (18)

15.7.5.2.4 *Malic acid*

The systematic name of malic acid is 2-hydroxy-1,4-butanedioic acid, but it is usually called by the trivial name. It is a white crystalline substance. The molecule $HO_2CCH_2CH(OH)CO_2H$, having a chiral centre, has two enantiomers (Figure 1). The (*S*)-isomer is a natural constituent and common metabolite of plants and animals, being involved in the Krebs cycle and in the glyoxylic acid cycle.[110] It is the principal acid in apples, but is also present in many other fruits.[111] Some important uses of malic acid take advantage of its complexing ability, as in its use to protect vegetable oils from oxidation during processing, a reaction that is catalyzed by trace metals.

The stereochemistry of malic acid and its coordination compounds has been the subject of careful investigations.[36,188] The planarity of the five-atom α-hydroxycarboxylate group is again observed here. Conformational arrangements associated with rotation around C(2)—C(3) bond have been studied by NMR methods and have been discussed in Section 15.7.4.1.

The addition of metal ions to solutions of malic acid or malate ions may result in the formation of a wide variety of complex species. This is the expected consequence of the coordinating properties of this multidentate ligand, which is capable of binding inorganic ions in several ways. The most common coordinating mode is bidentate, through the hydroxyl and α-carboxyl groups to give a five-membered chelate ring. Tridentate chelation through the hydroxyl and the two carboxyl groups is observed very often, and involves the formation of both five-and six-membered chelate rings. Bridging *via* carboxyl oxygens is also common. Apart from recent suggestions[79] that in some mixed ligand complexes of lanthanide ions this may be the preferred coordination mode, there has not been enough evidence to support the bidentate binding *via* two carboxyl groups as a typical coordinating mode of malic acid. Tridentate chelation is adopted[41] in Zn^{II} complexes and also in some complexes of Mo^{VI}[43] and W^{VI},[44] in solution. This binding mode requires the (+ *sc*) conformation for (*S*)-malate, as has been shown by NMR studies. This technique has also been used to show[45] that bidentate binding *via* the hydroxyl and carboxyl group is favoured by conformer (− *sc*) and (+ *sc*).

The crystal structures of several complexes of malic acid have been determined and most of them have been reviewed.[86] We give only the general features of the ligand stereochemistry in these compounds. Bidentate malate(1 −) is found very frequently. The chelate ring involves one oxygen atom of a deprotonated carboxyl group and a protonated hydroxyl oxygen. It is also common to find the malate ion binding through the two carboxyl groups and the hydroxyl group. It may have the β-carboxyl group binding to the same metal centre as the α-hydroxycarboxylate group when the ligand is tridentate, or it can be bidentate on one metal in the usual way and use the β-carboxyl group to bind to another metal centre. The latter mode gives polymeric structures and is observed, for instance, in bis{(*S*)-malato}copper(II) dihydrate[112] and in diaqua{(*S*)-malato}manganese(II) monohydrate[113] in which the hydroxy acid is coordinated as shown in (**19**) and (**20**), respectively. The β-carboxyl groups in the Cu^{II} complex are protonated and one coordinates axially to a neighbouring Cu^{II} centre using a protonated hydroxyl oxygen, while the other coordinates through a carboxyl oxygen atom. The latter gives an M—O bond shorter than the corresponding *trans*-M—OH bond. The (*S*)-malato adopts (− *sc*) conformations in these two compounds. In tetraaqua{(*S*)-malato}magnesium(II) monohydrate (**21**),[114] the (*S*)-malato binds to the metal centre through the protonated hydroxyl group and one oxygen of the carboxyl group, giving a five-membered chelate ring. The deprotonated β-carboxyl group hydrogen bonds to one of the coordinated water molecules and the ligand adopts the (− *sc*) conformation. Bidentate malate ligands are also found in the following very different compounds: diaquabis{(*S*)-malato}nickel(II) dihydrate,[115] tetraaquabis{(*S*)-malato}calcium(II) dihydrate[116] and K[bis(malato)borate(III)]·H_2O.[117] In the first two the α-hydroxycarboxylate moiety has a protonated hydroxyl group while in the latter both the coordinated hydroxyl and the coordinated carboxyl groups are deprotonated. The ligand conformation

(19)　　　　　　　　　(20)　　　　　　　　　(21)

is $(-sc)$ and the noncoordinated β-carboxyl group is protonated in the three complexes. Another example of bidentate (S)-malato has been recently[88] observed in the crystal structure of dicesium *cis*-[bis{(S)-malato}dioxomolybdate(VI)] monohydrate.

Examples of tridentate malate are found in the isostructural diaqua{(S)-malato}zinc(II)[118,119] and cobalt(II)[120] monohydrate, where bridging through oxygen atoms of carboxyl groups is also observed. The $(+sc)$ conformation required for tridentate coordination is observed here, as has been found for the Zn^{II} complexes in solution. Stereochemical requirements, demonstrated by molecular models,[86] force the nonplanarity of the α-hydroxycarboxylato group in these two compounds. The binding mode of malate in these Co^{II} and Zn^{II} complexes is reminiscent of the aspartate(2−) coordination mode in the respective Zn^{II}, Co^{II} and Ni^{II} complexes,[121] where the ligand is tridentate to one metal centre and bridges to neighbouring metal ions through the β-carboxyl group. A more complex coordination mode has been found[122] in the anion bis{(malato(3−)}undecaoxotetramolybdate(VI). Here the two malato ligands are fully deprotonated, the α-hydroxycarboxylate groups being nearly planar with short metal to ionized hydroxyl bonds.

15.7.5.2.5 *Citric acid and related ligands*

The coordination chemistry of citric acid has attracted great attention for a very long time and significant progress has been made in understanding the structures of its complexes, both in solution and in the solid state. A great deal of this interest in the complexes of citric acid has to do with its biological activity. In particular, great effort has been devoted to coordination compounds related to the Krebs cycle[3] and to the transport of inorganic ions in biofluids[123,124] and natural waters.[125] In a series of recent articles, attention has been given to the study of complexes of citric acid with Cu^{II},[126] Ni^{II},[127] Fe^{II},[34,35,128] Mn^{II},[34,35] Ca^{II},[125g] Mg^{II},[125g] Al^{III},[36,129] Fe^{III}[130,131] and lanthanide(III)[39,132,133] in aqueous solutions. With few exceptions,[39,128,131,133] most of this work is dedicated to the determination of formation constants of the complexes, usually using potentiometric methods. Citric acid has a great tendency to form monomeric species in which it coordinates to the metal ion through one terminal carboxyl group, the hydroxyl group and the central carboxyl group as a tridentate ligand (see **22**). This binding mode has been suggested[129] for one of the predominant species in solutions of Al^{III} and citric acid in which the ligand can be protonated or deprotonated at the β-carboxyl group. A species with an analogous coordinating mode has been suggested[131] for Fe^{III} in the pH range 1–2 and also as an important compound present in solutions of this ligand and Fe^{II},[128] or lanthanides.[39] These complexes with Fe^{II} have been identified on the basis of ^{13}C NMR of paramagnetic solutions, molecular weight determinations and magnetic susceptibility measurements. These studies also support the proposition of other important species in cold alkaline solutions of citrate and Ni^{II} or Fe^{II} ions. Discrete tetrameric complexes with tetraionized citrate in which the M^{II}:citrate ratio is 1:1 are formed in these solutions. These complexes have a cubane type structure of S_4 symmetry.

(22)

A bridging citrate ligand has been suggested[66] for a dimeric complex of Mo^{VI} in which the ligand:metal ratio is 1:2 and for heterodinuclear complexes of Cu^{II}–Ni^{II}, Ni^{II}–Zn^{II} and Cu^{II}–Zn^{II}.[134] An elegant ESR study[135] of Cu^{II} complexation by citrate ions in aqueous solutions also shows the formation of dimeric citrate-bridged species in the pH range 7–11.

The possibility of having citrate binding as a bidentate ligand through the central carboxyl group and the hydroxyl group, forming a five-membered chelate ring, has also been suggested.[129] We are, therefore, dealing with a ligand of great versatility in its binding mode to metal ions and capable of forming very strong compounds in solution. Further work has to be done, using adequate techniques, to characterize in more detail the structure of the species formed in solution.

The crystal structures of metal complexes of citric acid have also received much attention. In particular, Glusker and coworkers[3] have made available crystal data for a series of complexes of citrate and related ligands. X-Ray determinations have been done on citric acid (monohydrated[136a] and anhydrous[136b]), on several alkali metal salts and ethylenediammonium citrate(2−).[137] Two main conformational arrangements have been identified in these structures. One extended arrangements (Figure 8a, c) has been observed for anhydrous and monohydrate citric acid, for potassium (+)-isocitrate[138] and for ethylenediammonium citrate salts. Conformers having a rotated terminal carboxyl group, (Figure 8b) have been observed in rubidium citrate(2−)[139] and in the isostructural citrate(2−) salts of lithium and sodium.[140]

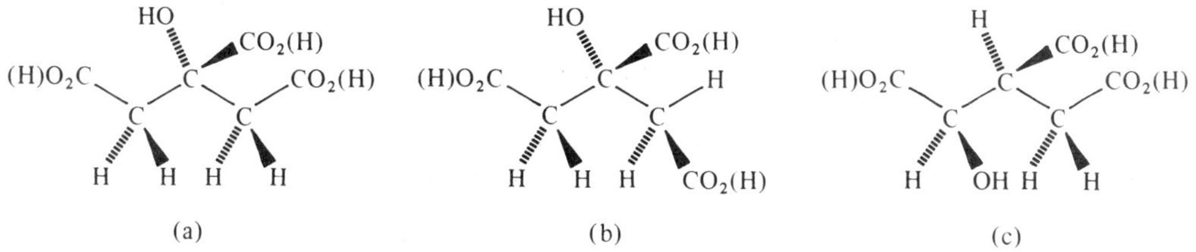

Figure 8 (a) Extended and (b) rotated conformations of citric acid and (c) extended conformation of (+)-isocitric acid

Crystal structure work on citrate compounds has been reviewed in relation to their enzymatic implications[3] and in a more general context.[86] An even greater variety of coordinating modes can be found in solid complexes than in solution species. Two bridging citrate ligands, tridentate to one metal centre and monodentate to the other one, have been observed[141] in the crystal structure of the dimer $K_2[Ni(C_6H_5O_7)(H_2O)_2]_2 \cdot 4H_2O$ (23). The coordination mode displayed in this compound is rather akin to that observed in the isomorphous series of complexes of Mn^{II},[142] Mn^{II}[143] and Fe^{II}[144] with triply ionized citrate, $[M(H_2O)_6][M\{O_2CCH_2C(OH)(CO_2)CH_2CO_2\}(H_2O)]_2 \cdot 2H_2O$ (24). In the latter complexes chain networks are formed, instead of discrete dimers, and both oxygens of the bridging carboxylate group are involved on bonding, one with each neighbouring metal centre. Tridentate citrate has been proposed by several authors for monomeric solution species of this group of cations. This is the expected result of the disruption of the carboxylate bonding to the neighbouring metal centres, a situation which is likely to occur in solution.

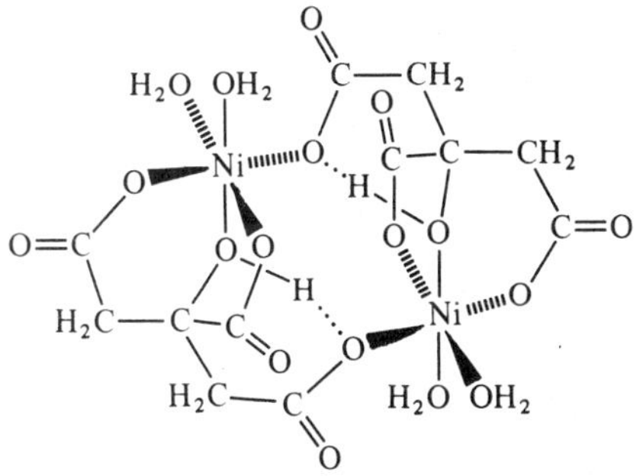

(23)

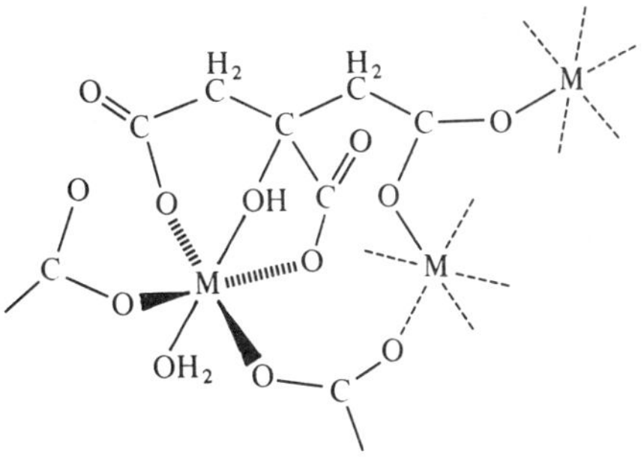

(24)

Bidentate citrate ligand has been found[89] in the crystal structure of β-[triethylenetetramine-citratoCo^{III}]·5H₂O. The citrate ion forms a five-membered chelate ring through the hydroxyl group and the central carboxyl group, a coordination mode which has also been proposed[120] for Al^{III} solution species.

Two metal complexes, one of Cu^{II} and one of Ni^{II}, with tetraionized citrate ligands have had their structures solved. In the compound $Cu_2(C_6H_4O_7) \cdot 2H_2O$ (25) there are no discrete molecules in the crystal and the three carboxylate groups and the ionized hydroxyl group bridge Cu^{II} centres in a complex array.[145] The crystal structure of $\{(NMe_4)_5[Ni_4(C_6H_4O_7)_3(OH)(H_2O)] \cdot 18H_2O\}_2$ has been determined[144] at −156 °C. This complex of Ni^{II} with tetraionized citrate contains the discrete centrosymmetric clusters $[Ni_8(C_6H_4O_7)_6(OH)_2(H_2O)_2]$ surrounded by a hydrogen bonded network of water molecules. Three crystallographically independent citrate ligands with similar configurations are present in these units. In two of them, the ionized hydroxyl groups are coordinated to three Ni^{II} centres (26a) and each of the three carboxylate groups is bonded to one of the three Ni^{II} ions which are coordinated to the ionized hydroxyl group of the same citrate ligand. The third independent citrate ligand (26b) bridges two Ni^{II} ions through the ionized hydroxyl group and has one terminal and the central carboxyl groups coordinated to one of these Ni^{II} centres. The other terminal carboxyl group has one oxygen atom bridging two Ni^{II} ions and the other binding a fourth Ni^{II} ion.

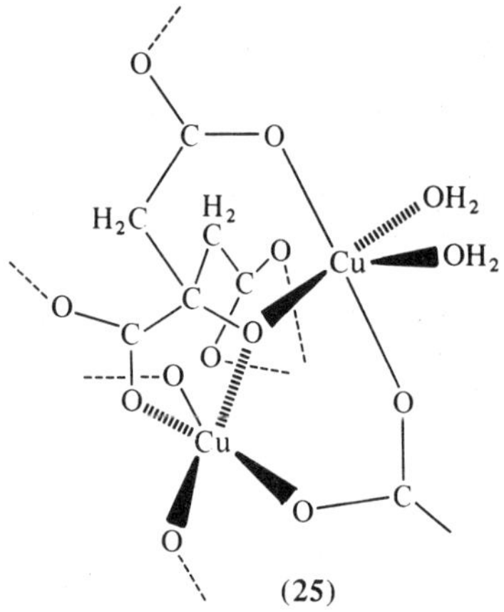

(25) 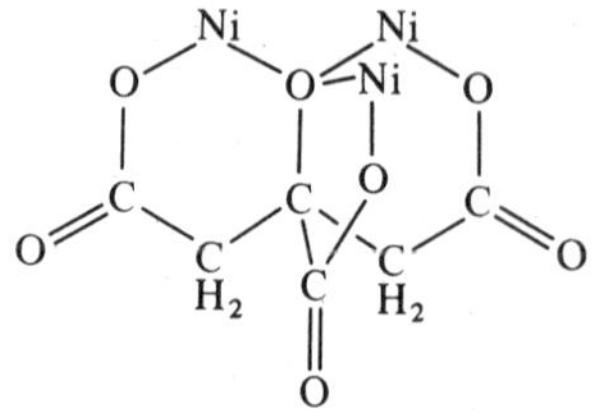   (26a) 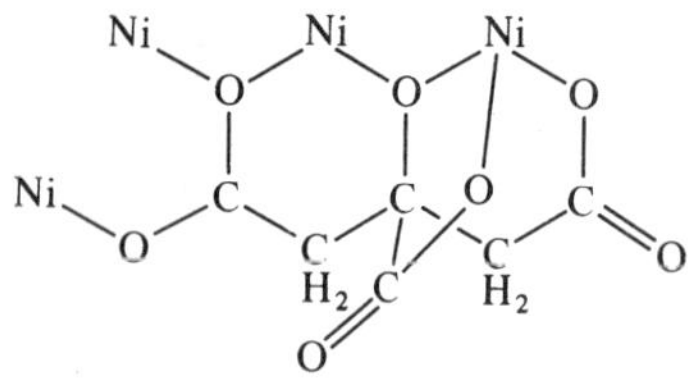   (26b)

Ligands closely related to aconitase have had their structures and the structures of some of their salts studied. These include the determination of the absolute configuration of naturally occurring (+)-isocitric acid[146] as (1R,2S)-1-hydroxy-1,2,3-propanetricarboxylic acid and the determination of the structure and absolute configuration of potassium dihydrogen isocitrate isolated from *Bryophyllum calycinum*.[138] In the latter the ligand adopts an extended conformation (Figure 8c) and coordinates as a bidentate chelate to one potassium ion and as a tridentate chelate to another. This study confirms the absolute configuration determined for the biologically active isomer of isocitric acid. The crystal structures of two diastereoisomeric hydroxycitrates have also been described by Glusker and coworkers.[147] These authors also discuss the several possible modes of coordination of hydroxycitrates to metal ions in connection with the biological activity of these ligands.

The absolute configuration of the isomer of fluorocitrate that inhibits aconitase has also been shown[148] to be (2R,3R) (Figure 9). It has been suggested[149] that a metal complex involving coordination through fluorine, in a mode similar to that observed in a metal fluorocitrate which has had the crystal structure determined, is formed in the enzyme. By coordinating in this way the fluorocitrate binds in a way that is not analogous to that of citrate, inhibiting enzyme activity.

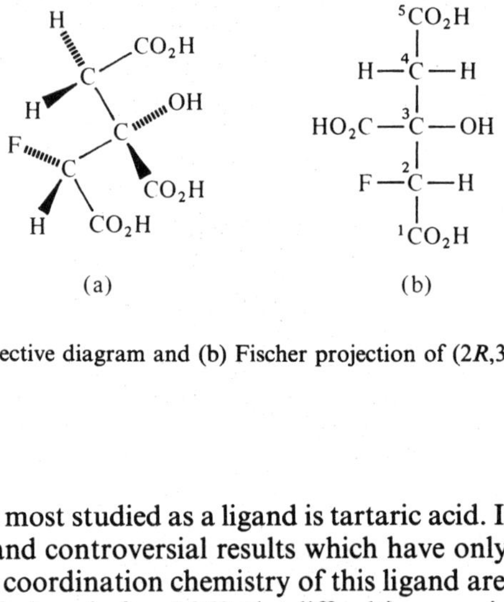

(a)   (b)

Figure 9   (a) Perspective diagram and (b) Fischer projection of (2R,3R)-fluorocitric acid

15.7.5.2.6   *Tartaric acid*

One of the α-hydroxy acids most studied as a ligand is tartaric acid. It is a good example of a great accumulation of qualitative and controversial results which have only recently begun to be understood. Several reviews on the coordination chemistry of this ligand are available. Pyatnitskii[150] gave a good description of the work done before 1963; the difficulties associated with the preparation and characterization of complexes of tartaric acid are well illustrated. Tapscott[86] reviews very thoroughly the stereochemistry of the complexes of this ligand, with particular emphasis on compounds which have had their crystal structures solved. One particular class of these compounds is that of the complexes containing bridging tartrate(4−) ligands, whose stereochemistry has also been reviewed in great detail.[53] With this series of reviews available, we shall summarize only the more important features and try to emphasize recent progress on understanding the coordinating modes of this particularly complex ligand.

As has been mentioned before in this chapter, tartaric acid has several stereoisomers (Figure 2). This is one reason for the complexity of its coordination chemistry, although in certain situations it may help in the understanding of the species formed. Unfortunately, it is very common to find reports which do not mention the chirality of the compounds used. This and the frequent occurrence of situations in which authors do not take advantage of this important property of common hydroxy acids show how little many coordination chemists appreciate the potential usefulness of this property.

It should be emphasized that the extensive coordination chemistry known for tartaric acids refers almost exclusively to the chiral ligands. *meso*-Tartaric acid is not a good complexing agent, most of the structural work done with this molecule being concerned with conformational studies of the acid[189] and its salts.

In tartrate bridged dinuclear complexes, the hydroxy acid acts as a tetradentate ligand, binding each metal centre through one carboxyl group and one hydroxyl group as shown in (**10**). Dinuclear species of this kind have been identified in several crystal structures of tartrate complexes which have been reviewed in great detail.[53] Since tartaric acid can exhibit three stereoisomers (*R,R*, *R,S* and *S,S*), several isomeric forms can be found in tartrate-bridged dinuclear complexes. Tapscott[151] has identified the possible isomers in these bridged species and discussed their relative stabilities. Bridging tartrate ligands have been identified in several coordination compounds in solution. This is the case in complexes of (*R,R*)-tartrate with Mo^{VI} or W^{VI} and also of their respective mixed-metal species studied by ^1H and ^{13}C NMR spectroscopy.[51,52] This last technique was also used to characterize analogous compounds of tartrate ligands with Sb^{III} or As^{III} and in the corresponding As^{III}–Sb^{III} dimers.[48] Interesting observations on the stereoselectivity in the formation, reactivity and other properties of dinuclear tartrate complexes of Cu^{II} [152] and V^{IV} [76,153] have been reported.

Bidentate chelation of tartrate ions using only one carboxyl group and the corresponding α-hydroxyl group in the bonding to the metal is also very common. This binding mode has been recognized[24,45,50,77,154] in a series of monomeric complexes of general formula $[M(NN)_2(H_2tart)]^+$, where $M = Co^{III}$ or Cr^{III}, NN = 2,2′-bipyridyl or 1,10-phenanthroline and $H_2tart = [HO_2CCH(OH)CH(O)CO_2]^{2-}$. The ligand conformations in some of these compounds have been discussed in Section 15.7.4.1. $[Cr(R,R\text{-tart})_3]^{6-}$ is another complex in which the tartrate ligand is bidentate through one deprotonated hydroxyl group and one deprotonated carboxyl group. The complex has had the Λ absolute configuration assigned.[155] The two more uncommon types of coordination in tartrates are the monodentate and tridentate binding modes. Two transition metal complexes where the ligand is monodentate are Cu^{II} complexes.[156] Tridentate chelation has been observed in only one crystal structure, that of sodium {(2*R*,3*R*)-tartrato}{(2*S*,3*S*)-tartrato}ferrate(III) 14-hydrate.[157] This complex has a centrosymmetric octahedral coordination polyhedron around Fe^{III} which is defined by two tetraionized tartrate ligands, each of them coordinated through one carboxylate and two deprotonated hydroxyl groups.

The crystal structures of the free acid and of several alkaline earth metal salts where the hydroxy acid is not coordinated to a metal ion have been determined.[86] They show that (2*R*,3*R*)-tartaric acid prefers a (+*sc*) conformation, as has been shown to be the case for the free ligand in aqueous solution over a wide range of pH values.[47] (2*S*,3*S*)-Tartaric acid is, of course, seen only as the (−*sc*) conformer, but, on the other hand, (2*R*,3*S*)-tartaric acid (*meso*-tartaric) appears in racemic mixtures of (+*sc*) and (−*sc*) conformations, in their respective crystalline structures. These, again, are the preferred conformations in solution. All the alkali metal tartrates which have had their structures determined exhibit polymeric structures with relatively rare chelation situations. The metal to hydroxyl oxygen distances tend to be longer than the metal to carboxyl oxygen distances among these compounds, as is the case with other α-hydroxycarboxylates. Strontium tartrate(2−) trihydrate and calcium tartrate(2−) tetrahydrate have coordination polyhedra which include two chelated α-hydroxycarboxylate groups and contacts with carboxyl oxygens from ligands chelated to neighbouring metal ions.

It is clear that a great deal of progress has been made on the understanding of the stereochemistry of complexes of tartrate ligands both in the solid state and in solution. However, we are a long way from a satisfactory picture of the structures of the enormous number of complexes which have been prepared as solids and of those which have been observed in solution.

15.7.5.2.7 *Other aliphatic hydroxy acids*

We now describe, very briefly, other hydroxy acids which are important ligands.

A few complexes of simple α-hydroxycarboxylates, which have known crystal structures, chelate to metal centres through ionized hydroxyl and carboxyl groups. This is the case for two complexes of oxo cations with analogous structures: potassium bis(2-hydroxy-2-methylbutyrate)oxochromate(V) monohydrate[158] and sodium tetraethylammonium bis(benzilato)oxovanadate(IV).[159] The ligands are fully deprotonated and the metal to ionized hydroxyl bonds are very short. Tartronic acid, $HO_2CCH(OH)CO_2H$, forms a complex with Cu^{II}, (tri-μ-tartronato)tricuprate(3 −), with a characteristic structure in which each ligand chelates two distinct Cu^{II} ions, forming two five-membered chelate rings.[160]

A series of hydroxy acids derived from or related to carbohydrates has been investigated as ligands.[86] They have several hydroxyl groups and one or more carboxyl groups in their respective molecules and may have several stereoisomers. Glyceric acid, $HO_2CCH(OH)CH_2OH$, the simplest acid in the group, exists in two enantiomeric forms: each enantiomer has three main staggered conformations in solution (two synclinal and one antiperiplanar). The coordination of glycerates and related ligands to metal ions has been studied by several techniques. Bidentate coordination to Dy^{III} ions has been suggested from ^{17}O NMR studies,[39a] although tridentate chelation to Eu^{III} ions is proposed from ^1H NMR studies.[161] The crystal structure of diaqua{($\pm$)-glycerato}calcium(II) has been reported.[162,163] The ligands chelate one metal centre through the α-hydroxycarboxylate moiety (containing a protonated hydroxyl group) and bind as monodentates through the protonated α-hydroxyl group to a second metal ion.

Gluconic acid and gluconate salts have been extensively used as sequestering agents. The coordination chemistry of this hydroxy acid has been reviewed by Sawyer[164] and a detailed NMR study of the ligand is also available.[165] The crystal structures of the potassium salt[166] and of a complex of Mn^{II}[167] have been determined. The latter shows the presence of gluconate ligands chelated by the carboxylate group and the protonated 2-hydroxy group, and of bridging ligands bonded by the carboxylate group to one metal centre and through the terminal hydroxyl group to another one. The crystal structure of the complex $[Pb(C_6H_{15}O_7)_2]$ has also been solved.[196] In this compound the coordination sphere of the metal centre includes two chelated gluconate ligands, bonded by one oxygen atom of the carboxyl group and by the α-hydroxyl group. The two remaining coordination sites are occupied by carboxylate oxygen atoms from two other gluconate ligands. A recent[129] potentiometric study of Al^{III} complexes suggests the presence of bidentate ligands coordinated through the deprotonated carboxylate group and the deprotonated 2-hydroxyl group. The authors also propose the formation of species containing a gluconate ligand tridentate to the same metal centre through deprotonated 2-, 4- and 6-hydroxyl groups. This study of Al^{III} complexes in solution includes results on saccharic acid species which are explained by the presence of a complex in which the hydroxy acid is tetradentate to the same Al^{III} ion through two fully deprotonated α-hydroxycarboxylate groups. González Velasco and coworkers[168] have studied the complexation of several metal ions by (+)-saccharic acid, but have been mainly concerned with the determination of the stoichiometries and stability constants of the complexes.

The polyether antibiotic lasalocid A or X-537A (27) is a hydroxy acid which has recently been studied as a complexing agent. The complexes of this and related ligands are very important in relation to the transport of metal ions through membranes and with other biological processes, and deserve more attention from coordination chemists. Garcia-Rosas and Schneider[169] refer to relevant literature and a series of other recent publications reports several coordination properties of this α-hydroxy acid and related molecules.[170] Another area of recent work on the coordination chemistry of long-chain α-hydroxy acids is related to their role as components of nervous system lipids.[171]

(27)

15.7.5.3 Aromatic Hydroxy Acids

Reported coordinating properties of aromatic hydroxy acids have been concerned mainly with salicylic acid (2) and substituted salicylic acids (12). Most published work refers to solution, where detailed characterization of the stereochemistry of the species formed is very rare. A series of recent publications by Lajunen *et al.*[172–174] on the coordination chemistry of substituted hydroxynaphthoic acids deals with the complexation of Fe^{III}, Al^{III}, Be^{II} and lanthanides(III) by these ligands. The structures of some of these hydroxy acids have been studied in solution by ^{13}C NMR,[175] and in the solid state[174] for two of them. In this section we review the literature related to the stereochemistry of these two groups and of a few other aromatic hydroxy acids.

15.7.5.3.1 Salicylic acid and substituted salicylic acids

The 2-hydroxybenzoic acid, normally called salicylic acid, is well known due to its medical applications, namely as an antipyretic and in the treatment of certain types of rheumatism. The biological action of this compound is connected[4,5] with its ability to bind metal ions. Not only is the efficacy of the treatment increased, but toxicity problems related to the injection of salicylic acid are diminished by administering salicylate metal complexes.

Although coordination compounds of salicylic acid with a wide variety of metal ions have been studied, only a few crystal structures of salicylato complexes have been determined. The usual bidentate mode of coordination, observed for other hydroxy acids, in which one α-hydroxyl oxygen and a carboxyl oxygen are involved, is very uncommon in these solids. Hanic and Michalov[176] have determined the crystal structure of $Cu\{C_6H_4(OH)CO_2\}_2 \cdot 4H_2O$. Molecules containing two water molecules and two salicylate ligands bonded to the same Cu^{II} ion, in a planar, centrosymmetric arrangement as in (28), are found in the crystal. Each salicylate is bonded to the metal centre through a single oxygen atom in *trans* positions to each other, with a Cu—O distance of 1.84 Å. The two water molecules bonded to the metal are also in *trans* positions at 1.92 Å from the Cu^{II} centre. Another compound of Cu^{II} which has had the crystal structure determined[177] is $Cu\{C_6H_4(OH)-CO_2\}_2(H_2O)_2$. The coordinating modes of the salicylate ligands in each molecule of this complex are of two kinds. One is analogous to that described for the previous compound, *i.e.* through one oxygen of the carboxyl group. The two other salicylate ions bonded to the square pyramidal Cu^{II} centre act as bridging ligands with neighbouring Cu^{II} ions as shown in (29). Another compound which has a known crystal structure is $Zn\{C_6H_4(OH)CO_2\}_2(H_2O)_2$.[178] Here, again, the salicylate coordinates to the metal through a single oxygen of the ionized carboxyl groups, as in (30). That molecule possesses a twofold axis and exists as a unit in the lattice. A more complex structure is found in the isostructural aquatris(salicylato)samarium(III) and aquatris(salicylato)americium(III).[179] There are no discrete molecules in the crystal, each metal ion being linked to six different salicylate ions through a variety of coordinating modes, using a total of eight oxygen atoms. The ninth coordination position is occupied by a water molecule. Figure 10 shows the coordination sphere of the M^{III} centres in these complexes and schematic representations of the three distinct coordination modes which the salicylate ligands use in binding to M^{III}. Salicylate ligands bidentate through the carboxylate groups have been reported for *cis* and *trans* isomers of bis(4-aminosalicylato)copper(II).[180]

(28)   (29)   (30)

Solution studies of complexes of salicylate and substituted salicylate ions have been carried out very extensively for a wide range of metal ions. A great deal of this work is concerned with the determination of stability constants of simple and mixed ligand complexes of salicylate and substituted salicylate ions.[17,19] A series of reports on Be^{II} complexes with hydroxysalicylic acids have been published by Lajunen *et al.*[181,182] They have studied the complexation of Be^{II} by 2,4-, 2,5-, 2,6- and 3,5-dihydroxybenzoic acids and have identified several species in solution. In the solutions of 2,4-dihydroxybenzoic acid (H_3L) and Be^{II} studied, two complexes, $BeHL$ and $BeH_2L_2^{2-}$, are suc-

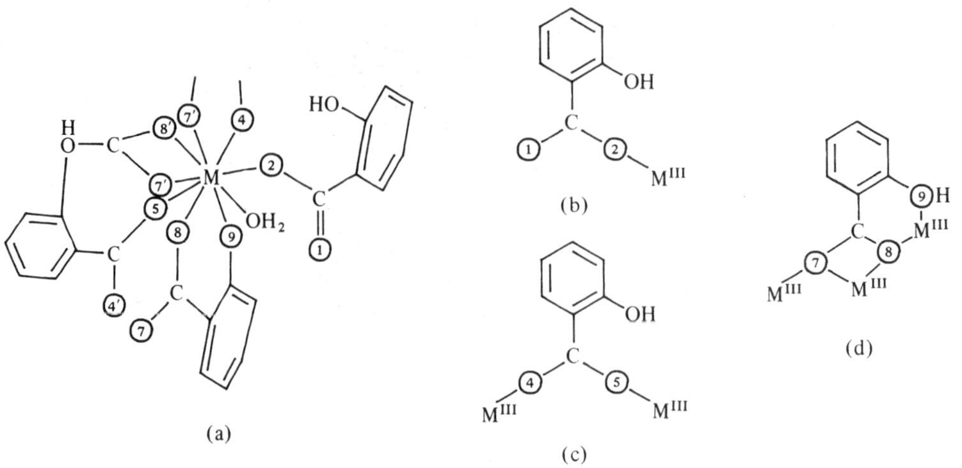

Figure 10 (a) M^{III} coordination sphere and (b–d) salicylato binding modes in the complexes aquatris(salicylato)M^{III}, where $M^{III} = Sm^{III}$ or Am^{III} (after ref. 179)

cessively formed in the acidic region. In alkaline solutions, when an excess of ligand is present, the species $BeHL_2^{3-}$ and BeL_2^{4-} are formed as a result of the ionization of $BeH_2L_2^{2-}$. Equilibria constants associated with the formation of these complexes have been determined. For 2,6-dihydroxybenzoate solutions a different pattern is observed. In the acidic region the species Be_2L^+ is formed besides BeHL and $BeH_2L_2^{2-}$, when an excess of metal ion is present. The authors suggest the structures represented in Figure 11 for these complexes. The study of Be^{II} complexes with the salicylate ions deserves some more attention in connection with the action of aurintricarboxylic acid (9) and salicylates as beryllium antidotes.[4]

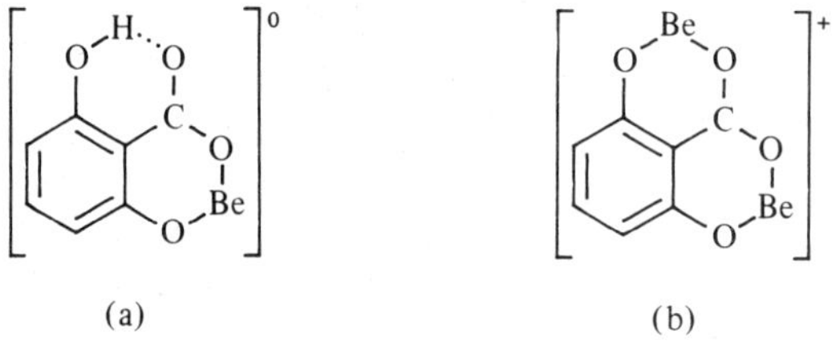

Figure 11 Structures proposed for the complexes $[Be(HL)]^0$ and $[Be_2L]^+$, where H_3L = 2,6-dihydroxybenzoic acid (after ref. 180)

A series of complexes of 2,6-dihydroxybenzoate has been prepared and studied as model compounds for the interactions of metal ions with humic acids.[65,183]

Bidentate coordination of salicylate through the α-hydroxycarboxylate group can be found in tetrahedral complexes of a non-metallic element.[59] The results with B^{III} and Be^{III} complexes seem to suggest that the increasing covalent bonding tends to favour chelation *via* deprotonated hydroxyl and carboxyl oxygen atoms. This coordinating mode has also been found in a number of complexes of Co^{III} studied by Gillard and coworkers.[28]

15.7.5.3.2 Hydroxynaphthoic acids and substituted hydroxynaphthoic acids

The coordination chemistry of hydroxynaphthoic acids and related ligands has been studied mainly with the aim of obtaining stability constants for the complexes formed. There is not, therefore, much work on the stereochemistry and coordination modes of these ligands.

The crystal and molecular structures of sodium 3-hydroxy-7-sulfonato-2-naphthoic acid trihydrate and of 1-hydroxy-4-sulfonato-2-naphthoic acid dihydrate have been determined, and their IR spectra and thermal behaviour studied.[174] Each 3-hydroxy-7-sulfonato-2-naphthoic acid coordinates to two sodium ions through a bidentate sulfonate group and one oxygen atom of the carboxyl group, respectively. The other three sites in the coordination octahedron are occupied by water molecules. The 1-hydroxy-4-sulfonato-2-naphthoic acid, as well as the bidentate sulfonate and monodentate carboxylate, also uses the oxygen atom of the hydroxyl group in coordinating the sodium atoms. Each ligand is bonded to three distinct metal ions, using one coordinating group to each of them. A series of coordination compounds of these ligands has been studied in solution. 1-Hydroxy-7-sulfo-2-naphthoic and 1-hydroxy-4,7-disulfo-2-naphthoic complexes of Fe^{III} have been studied using spectrophotometric and potentiometric methods.[23] Stability constants and species distribution diagrams have been determined for the complexes FeL, FeL_2^{3-} and FeL_3^{6-}. Details are not given for the structures of the complexes. The results are compared with those obtained for Fe^{III} complexes with 1-hydroxy-4-sulfo-2-naphthoic acid,[172] and for analogous studies of the complexes with Be^{II} and Cu^{II}. Similar studies for Al^{III} complexes have also been published.[173]

A series of complexes of Cu^{II} with 2-hydroxy-3-naphthoic acid has been recently prepared and the magnetic properties and ESR spectra of the complexes have been discussed. The complexes are supposed to be monomeric with two carboxylate groups coordinated to the Cu^{II} centre.[184] Some of these ligands had their ^{13}C NMR[175] and IR spectra[185] studied in detail very recently.

15.7.6 REFERENCES

1. D. D. Perrin, in 'Treatise of Analytical Chemistry', 2nd edn., ed. I. M. Kolthoff and P. J. Elving, Wiley, New York, 1979, part 1, vol. 2, p. 599.
2. A. Cotton, *C.R. Hebd. Seances Acad. Sci.*, 1895, **120**, 989, 1044.
3. J. P. Glusker, *Acc. Chem. Res.*, 1980, **13**, 345.
4. M. B. Chenoweth, *Pharmacol. Rev.*, 1956, **8**, 57.
5. J. R. J. Sorenson, *J. Med. Chem.*, 1976, **19**, 135.
6. G. Kadziene, J. Bubelis and J. Matulis, *Electrolit. Osazhdenie Splavov*, 1968, **1**, 24 (*Chem. Abstr.*, 1971, **74**, 8936).
7. T. Tamura, Y. Tsuru and K. Hosokawa, *Kyushu Kogyo Daigaku Kenkyu Hokoku, Hogaku*, 1976, **32**, 61(*Chem. Abstr.*, 1976, **85**, 69 981).
8. A. Tawada, *Jpn. Kokai*, 1978, 93132 (*Chem. Abstr.*, 1978, **89**, 223 201).
9. S. M. Roberts, in 'Comprehensive Organic Chemistry', ed. I. O. Sutherland, Pergamon, Oxford, 1979, vol. 2, p. 739.
10. (a) D. St. C. Black, G. M. Blackburn and G. A. R. Johnston, in 'Rodd's Chemistry of Carbon Compounds', 2nd edn., ed. S. Coffey, Elsevier, Amsterdam, 1965, vol. I, part D, p. 80; (b) B. J. Coffin, in 'Rodd's Chemistry of Carbon Compounds', 2nd edn., ed. M. F. Ansell, Elsevier, Amsterdam, 1983, vol. I, part E, p. 235; (c) R. A. Hill, in 'Rodd's Chemistry of Carbon Compounds', 2nd edn., ed. M. F. Ansell, Elsevier, Amsterdam, 1983, vol. I, parts F, G, p. 360; (d) J. E. G. Barnett and P. W. Kent, in 'Rodd's Chemistry of Carbon Compounds', 2nd edn., ed. S. Coffey, Elsevier, Amsterdam, 1976, vol. I, part G, p. 1; (e) J. Grimshaw, in 'Rodd's Chemistry of Carbon Compounds', 2nd edn., ed. S. Coffey, Elsevier, Amsterdam, 1976, vol. III, part D, p. 141; (f) A. B. Turner, in 'Rodd's Chemistry of Carbon Compounds', 2nd edn., ed. S. Coffey, Elsevier, Amsterdam, 1974, vol. III, part E, p. 71; (g) J. R. A. Pollock and R. Stevens (eds), 'Dictionary of Organic Compounds', 4th edn, Eyre and Spottiswoode, London, 1965.
11. J. Rigaudy and S. P. Klesney (eds.), 'Nomenclature of Organic Chemistry,' Pergamon, Oxford, 1979.
12. R. S. Cahn, C. Ingold and V. Prelog, *Angew. Chem., Int. Ed. Engl.*, 1966, **5**, 385.
13. R. S. Cahn, C. Ingold and V. Prelog, *Experientia*, 1956, **12**, 81.
14. C. K. Prout, R. A. Armstrong, J. R. Carruthers, J. G. Forrest, P. Murray-Rust and F. J. C. Rossotti, *J. Chem. Soc. (A)*, 1968, 2791.
15. L. Magon, G. Tomat, A. Bimondo, R. Portanova and U. Croatto, *Gazz. Chim. Ital.*, 1974, **104**, 967.
16. A. Lorenzotti, D. Leonesi, A. Cingolani and P. Di Bernardo, *Inorg. Chim. Acta*, 1981, **52**, 149.
17. A. E. Martell and R. M. Smith, 'Critical Stability Constants', Plenum, New York, 1977, vol. 3.
18. E. P. Serjeant and B. Dempsey, 'Ionisation Constants of Organic Acids in Aqueous Solutions', Pergamon, Oxford, 1979.
19. L. G. Sillén and A. E. Martell, 'Stability Constants of Metal-Ion Complexes', The Chemical Society, London, 1971.
20. G. Kortüm, W. Vogel and K. Andrussov, 'Dissociation Constants of Organic Acids in Aqueous Solution', Butterworths, London, 1961.
21. M. M. Petit-Ramel, G. Chottard and J. Bolard, *J. Chim Phys.*, 1976, **73**, 181.
22. L. H. J. Lajunen, J. Saarinen and S. Parhi, *Talanta*, 1980, **27**, 71.
23. L. H. J. Lajunen, E. Aitta and S. Parhi, *Talanta*, 1981, **28**, 277 and refs. therein.
24. A. Tatehata, *Inorg. Chem.*, 1976, **15**, 2086.
25. R. D. Gillard and M. G. Price, *J. Chem. Soc. (A)*, 1969, 1813.
26. M. T. Beck, B. Csiszár and P. Szarvas, *Nature (London)*, 1960, **188**, 846.
27. L. E. Bennett, R. H. Lane, M. Gilroy, F. A. Sedor and J. P. Bennett, Jr., *Inorg. Chem.*, 1973, **12**, 1200.
28. R. D. Gillard, J. R. Lyons and P. R. Mitchell, *J. Chem. Soc., Dalton Trans.*, 1973, 233.
29. A. C. Dash, R. K. Nanda and H. K. Patnaik, *Aust. J. Chem.*, 1975, **28**, 1613.
30. R. G. Bates and G. D. Pinching, *J. Am. Chem. Soc.*, 1949, **71**, 1274.
31. A. Loewenstein and J. D. Roberts, *J. Am. Chem. Soc.*, 1960, **82**, 2705.
32. R. B. Martin, *J. Phys. Chem.*, 1961, **65**, 2053.
33. E. Bottari and M. Vicedomini, *J. Inorg. Nucl. Chem.*, 1973, **35**, 1657.
34. J. T. H. Roos and D. R. Williams, *J. Inorg. Nucl. Chem.*, 1977, **39**, 367.
35. P. Amico, P. G. Daniele, V. Cucinotta, E. Rizzarelli and S. Sammartano, *Inorg. Chim. Acta*, 1979, **36**, 1.
36. L.-O. Öhman and S. Sjöberg, *J. Chem. Soc., Dalton Trans.*, 1983, 2513.
37. G. Arena, R. Cali, M. Grasso, S. Musumeci, S. Sammartano and C. Rigano, *Thermochim. Acta*, 1980, **36**, 329.
38. V. Cucinotta, P. G. Daniele, C. Rigano and S. Sammartano, *Inorg. Chim. Acta*, 1981, **56**, L45.
39. (a) J. A. Peters and A. P. G. Kieboom, *Recl. Trav. Chim. Pays-Bas*, 1983, **102**, 381; (b) C. A. M. Vijverberg, J. A. Peters, A. P. G. Kieboom and H. van Bekkum, *Recl. Trav. Chim. Pays-Bas*, 1980, **99**, 403; (c) A. P. G. Kieboom, C. A. M. Vijverberg, J. A. Peters and H. van Bekkum, *Recl. Trav. Chim. Pays-Bas*, 1977, **96**, 315.

40. K. B. Dillon, M. R. Harrison and F. J. C. Rossotti, *J. Magn. Reson.*, 1980, **39**, 499.
41. J. S. Mariano and V. M. S. Gil, *Mol. Phys.*, 1969, **17**, 313.
42. T. Taura and H. Yoneda, *Chem. Lett.*, 1977, 71.
43. M. M. Caldeira, M. E. T. L. Saraiva and V. M. S. Gil, *Inorg. Nucl. Chem. Lett.*, 1981, **17**, 295.
44. V. M. S. Gil, M. E. T. L. Saraiva, M. M. Caldeira and A. M. D. Pereira, *J. Inorg. Nucl. Chem.*, 1980, **42**, 389.
45. J. A. Chambers, R. D. Gillard, P. A. Williams and R. S. Vagg, *Inorg. Chim. Acta*, 1983, **72**, 263.
46. (a) J. D. Pedrosa and V. M. S. Gil, *J. Inorg. Nucl. Chem.*, 1974, **36**, 1803; (b) M. T. Nunes, V. M. S. Gil and A. V. Xavier, *Can. J. Chem.*, 1982, **60**, 1007.
47. J. Ascenso and V. M. S. Gil, *Can. J. Chem.*, 1980, **58**, 1376.
48. D. Marcovich and R. E. Tapscott, *J. Am. Chem. Soc.*, 1980, **102**, 5712.
49. R. A. Haines and D. W. Bailey, *Inorg. Chem.*, 1975, **14**, 1310.
50. A. Tatehata, *Inorg. Chem.*, 1978, **17**, 725.
51. A. M. V. S. V. Cavaleiro, V. M. S. Gil, J. D. Pedrosa de Jesus, R. D. Gillard and P. A. Williams, *Transition Met. Chem.*, 1984, **9**, 62.
52. A. M. V. S. V. Cavaleiro, J. D. Pedrosa de Jesus, R. D. Gillard and P. A. Williams, *Transition Met. Chem.*, 1984, **9**, 81.
53. R. E. Tapscott, R. L. Belford and I. C. Paul, *Coord. Chem. Rev.*, 1969, **4**, 323.
54. (a) L. J. Bellamy, 'The Infrared Spectra of Complex Molecules', 3rd edn., Chapman and Hall, London, 1975, vol. 1, pp. 107–128, 183–202; (b) L. J. Bellamy, 'The Infrared Spectra of Complex Molecules', 2nd edn., Chapman and Hall, London, 1980, vol. 2, pp. 95–105, 176–179, 240–268.
55. W. O. George, J. H. S. Green and D. Pailthorpe, *J. Mol. Struct.*, 1971, **10**, 297.
56. N. Mori, Y. Asano, T. Irie and Y. Tsuzuki, *Bull. Chem. Soc. Jpn.*, 1969, **42**, 482.
57. N. Mori, S. Omura, N. Kobayashi and Y. Tsuzuki, *Bull. Chem. Soc. Jpn.*, 1965, **38**, 2149.
58. N. Mori, S. Omura, O. Yamamoto, T. Suzuki and Y. Tsuzuki, *Bull. Chem. Soc. Jpn.*, 1963, **36**, 1401.
59. J. Bolard, *J. Chim Phys.*, 1965, 887, 894, 900, 908.
60. J. Bolard, *C.R. Hebd. Seances Acad. Sci.*, 1963, **256**, 4388.
61. J. Bolard, *C.R. Hebd. Seances Acad. Sci.*, 1963, **257**, 3145.
62. J. D. S. Goulden, *Spectrochim. Acta*, 1960, **16**, 715.
63. J. D. S. Goulden, *Chem. Ind. (London)*, 1960, 721.
64. L. L. Shevchenko, *Russ. Chem. Rev. (Engl. Transl.)*, 1963, **32**, 201.
65. F. Cariati, L. Erre, G. Micera, A. Panzanelli, G. Ciani and A. Sironi, *Inorg. Chim. Acta*, 1983, **80**, 57.
66. J. D. Pedrosa de Jesus, M. D. Farropas, P. O'Brien, R. D. Gillard and P. A. Williams, *Transition Met. Chem.*, 1983, **8**, 193.
67. L. I. Katzin and E. Gulyas, *J. Am. Chem. Soc.*, 1968, **90**, 247.
68. F. S. Richardson and R. W. Strickland, *Tetrahedron*, 1975, **31**, 2309.
69. I. Listowsky, G. Avigad and S. Englard, *J. Org. Chem.*, 1970, **35**, 1080.
70. J. C. Craig and W. E. Pereira, Jr., *Tetrahedron*, 1970, **26**, 3457.
71. G. Snatzke, U. Wagner and H. P. Wolff, *Tetrahedron*, 1981, **37**, 349.
72. J. Frelek, A. Perkowska, G. Snatzke, M. Tima, U. Wagner and H. P. Wolff, *Spectrosc. Int. J.*, 1983, **2**, 274.
73. W. Voelter, E. Bayer, G. Barth, E. Bunnenberg and C. Djerassi, *Chem. Ber.*, 1969, **102**, 2003.
74. A. Beltrán, A. Cervilla Avalos and J. Beltrán, *J. Inorg. Nucl. Chem.*, 1981, **43**, 1337.
75. A. Cervilla, A. Beltran and J. Beltran, *Can. J. Chem.*, 1979, **57**, 773.
76. R. D. Gillard and R. A. Wiggins, *J. Chem. Soc., Dalton Trans.*, 1973, 125.
77. A. Tatehata, *Inorg. Chem.*, 1977, **16**, 1247.
78. K. Garbett and R. D. Gillard, *J. Chem. Soc. (A)*, 1968, 979.
79. M. Ransom and H. G. Brittain, *Inorg. Chem.*, 1983, **22**, 2494.
80. R. A. Copeland and H. G. Brittain, *J. Lumin.*, 1982, **27**, 307.
81. R. A. Copeland and H. G. Brittain, *Polyhedron*, 1982, **1**, 693.
82. M. Ransom and H. G. Brittain, *Inorg. Chim. Acta*, 1982, **65**, L147.
83. H. G. Brittain, *Inorg. Chem.*, 1981, **20**, 4267.
84. H. G. Brittain, *Inorg. Chem.*, 1980, **19**, 2136.
85. Z. Konteatis and H. G. Brittain, *Inorg. Chim. Acta*, 1980, **40**, 51.
86. R. E. Tapscott, in 'Transition Metal Chemistry', ed. G. A. Melson and B. N. Figgs, Dekker, New York, 1982, vol. 8, p. 253.
87. (a) M. D. Newton and G. A. Jeffrey, *J. Am. Chem. Soc.*, 1977, **99**, 2413; (b) J. A. Kanters, J. Kroon, A. F. Peerdeman and J. C. Schoone, *Tetrahedron*, 1967, **23**, 4027.
88. C. B. Knobler, A. J. Wilson, R. N. Hider, I. W. Jensen, B. R. Penfold, W. T. Robinson and C. J. Wilkins, *J. Chem. Soc., Dalton Trans.*, 1983, 1299.
89. R. Job, P. J. Kelleher, W. C. Stallings, Jr., C. T. Monti and J. P. Glusker, *Inorg. Chem.*, 1982, **21**, 3760.
90. U. Casellato, P. A. Vigato and M. Vidali, *Coord. Chem. Rev.*, 1978, **26**, 85.
91. S. Harada, Y. Okuue, H. Kan and T. Yasunaga, *Bull. Chem. Soc. Jpn.*, 1974, **47**, 769.
92. G. Sbrignadello, G. Tomat, G. Battiston, P. A. Vigato and O. Traverso, *J. Inorg. Nucl. Chem.*, 1978, **40**, 1647.
93. B. F. Mentzen and H. Sauterau, *Inorg. Chim. Acta*, 1979, **35**, L347.
94. E. Mentasti, *Inorg. Chem.*, 1979, **18**, 1512.
95. I. Grenthe, *Acta Chem. Scand.*, 1969, **23**, 1253.
96. R. S. Ambulkar and K. N. Munshi, *Indian J. Chem., Sect. A*, 1976, **14**, 424.
97. I. Filipović, I. Piljac, A. Medved, A. Savić, A. Bujak, B. Bach-Bragutinović and B. Mayer, *Croat. Chem. Acta*, 1968, **40**, 131.
98. C. Miyaka and H. W. Nürnberg, *J. Inorg. Nucl. Chem.*, 1967, **29**, 2411.
99. A. Beltrán-Porter, A. Cervilla, F. Caturla and M. J. Vila, *Transition Met. Chem.*, 1983, **8**, 324.
100. K. M. Jones and E. Larsen, *Acta Chem. Scand.*, 1965, **19**, 1205.
101. J. Bolard and G. Chottard, *Inorg. Nucl. Chem. Lett.*, 1974, **10**, 991.
102. L. I. Katzin, *Inorg. Chem.*, 1968, **7**, 1183.

103. H. G. Brittain, *Inorg. Chem.*, 1981, **20**, 959.
104. K. D. Singh, S. C. Jain, T. D. Sakore and A. B. Biswas, *Acta Crystallogr., Sect. B*, 1975, **31**, 990.
105. (a) T. S. Cameron and M. Duffin, *Cryst. Struct. Commun.*, 1974, **3**, 539; (b) H. A. Rose, *Anal. Chem.*, 1952, **24**, 1680.
106. F. A. Cotton, L. R. Falvello and C. A. Murillo, *Inorg. Chem.*, 1983, **22**, 382.
107. O. Korver, S. De Jong and T. C. van Soest, *Tetrahedron*, 1976, **32**, 1225.
108. R. Håkansson and S. Gronowitz, *Tetrahedron*, 1976, **32**, 2973.
109. C. Sterling, *J. Inorg. Nucl. Chem.*, 1967, **29**, 211.
110. H. R. Mahler and E. H. Cordes, 'Biological Chemistry', Harper and Row, New York, 1966, p. 466, 526.
111. S. E. Berger, in 'Kirk-Othmer Encyclopedia of Chemical Technology', 3rd edn., Wiley, New York, 1981, vol. 13, p. 103.
112. W. Van Havere and A. T. H. Lenstra, *Bull. Soc. Chim. Belg.*, 1978, **87**, 419.
113. A. Karipides and A. T. Reed, *Inorg. Chem.*, 1976, **15**, 44.
114. A. Karipides, *Inorg. Chem.*, 1979, **18**, 3034.
115. W. Van Havere and A. T. H. Lenstre, *Bull. Soc. Chim. Belg.*, 1980, **89**, 427.
116. A. T. H. Lenstra and W. Van Havere, *Acta Crystallogr., Sect. B*, 1980, **36**, 156.
117. R. A. Mariezcurrena and S. S. Rasmussen, *Acta Crystallogr., Sect. B*, 1973, **29**, 1035.
118. A. T. H. Lenstra and W. Van de Mieroop, *Bull. Soc. Chim. Belg.*, 1976, **85**, 721.
119. A. T. Reed and A. Karipides, *Acta Crystallogr., Sect B*, 1976, **32**, 2085.
120. L. Kryger and S. E. Rasmussen, *Acta Chem. Scand.*, 1972, **26**, 2349.
121. T. Doyne and R. Pepinsky, *Acta Crystallogr.*, 1957, **10**, 438.
122. See ref. 88 and references therein.
123. M. T. Martin, K. F. Licklider, J. G. Brushmiller and F. A. Jacobs, *J. Inorg. Biochem.*, 1981, **15**, 55.
124. G. H. Godo and H. M. Reisenauer, *Soil Sci. Soc. Am. J.*, 1980, **44**, 993.
125. (a) N. F. Ng Kee Kwong and P. M. Huang, *Nature (London)*, 1978, **271**, 336; (b) N. F. Ng Kee Kwong and P. M. Huang, *Soil Sci. Soc. Am. J.*, 1979, **43**, 1107; (c) K. F. Ng Kee Kwong and P. M. Huang, *Soil Sci.*, 1979, **128**, 337; (d) R. M. Cornell and P. W. Schindler, *Colloid Polym. Sci.*, 1980, **258**, 1171; (e) R. M. Cornell and U. Schwertmann, *Clays Clay Miner.*, 1979, **27**, 402; (f) M. K. Razzaghe and M. Robert, *Ann. Agron.*, 1979, **30**, 493; (g) P. Amico, P. G. Daniele, C. Rigano and S. Sammartano, *Ann. Chim. (Rome)*, 1982, **72**, 1.
126. E. R. Still and P. Wikberg, *Inorg. Chim. Acta*, 1980, **46**, 147.
127. E. R. Still and P. Wikberg, *Inorg. Chim. Acta*, 1980, **46**, 153.
128. J. Strouse, *J. Am. Chem. Soc.*, 1977, **99**, 572.
129. R. J. Motekaitis and A. E. Martell, *Inorg. Chem.*, 1984, **23**, 18.
130. S. K. Dhar and S. Sichak, Jr., *J. Inorg. Nucl. Chem.*, 1979, **41**, 126.
131. E. Mentasti and C. Baiocchi, *J. Coord. Chem.*, 1980, **10**, 229.
132. D.-R. Svoronos, S. Boulhassa, R. Guillaumont and M. Quarton, *J. Inorg. Nucl. Chem.*, 1981, **43**, 1541.
133. L. Spaulding and H. G. Brittain, *J. Lumin.*, 1983, **28**, 385.
134. P. Amico, P. G. Daniele, G. Ostacoli, G. Arena, E. Rizzarelli and S. Sammartano, *Inorg. Chim. Acta*, 1980, **44**, L219.
135. R. H. Dunhill, J. R. Pilbrow and T. D. Smith, *J. Chem. Phys.*, 1966, **45**, 1474.
136. (a) G. Roelofson and J. A. Kanters, *Cryst. Struct. Commun.*, 1972, **1**, 23; (b) J. P. Glusker, J. A. Minkin and A. L. Patterson, *Acta Crystallogr., Sect B*, 1969, **25**, 1066.
137. N. Gavrushenko, H. L. Carrell, W. C. Stallings and J. P. Glusker, *Acta Crystallogr., Sect B*, 1977, **33**, 3936.
138. D. van der Helm, J. P. Glusker, C. K. Johnson, J. A. Minkin, N. E. Burow and A. L. Patterson, *Acta Crystallogr., Sect B*, 1968, **24**, 578.
139. C. E. Nordman, A. S. Weldon and A. L. Patterson, *Acta Crystallogr.*, 1960, **13**, 414.
140. J. P. Glusker, D. van der Helm, W. E. Love, M. L. Dornberg, J. A. Minkin, C. K. Johnson and A. L. Patterson, *Acta Crystallogr.*, 1965, **19**, 561.
141. E. N. Baker, H. M. Baker, B. F. Anderson and R. D. Reeves, *Inorg. Chim. Acta*, 1983, **78**, 281.
142. C. K. Johnson, *Acta Crystallogr.*, 1965, **18**, 1004.
143. H. L. Carrell and J. P. Glusker, *Acta Crystallogr., Sect B*, 1973, **29**, 638.
144. J. Strouse, S. W. Layten and C. E. Strouse, *J. Am. Chem. Soc.*, 1977, **99**, 562.
145. D. Mastropaolo, D. A. Powers, J. A. Potenza and H. J. Schugar, *Inorg. Chem.*, 1976, **15**, 1444.
146. A. L. Patterson, C. K. Johnson, D. van der Helm and J. A. Minkin, *J. Am. Chem. Soc.*, 1962, **84**, 309.
147. W. C. Stallings, J. F. Blount, P. A. Srere and J. P. Glusker, *Arch. Biochem. Biophys.*, 1979, **193**, 431.
148. W. C. Stallings, C. T. Monti, J. F. Belvedere, R. K. Preston and J. P. Glusker, *Arch. Biochem. Biophys.*, 1980, **203**, 65.
149. H. L. Carrell, J. P. Glusker, J. J. Villafranca, A. S. Mildvan, R. J. Dummel and E. Kun, *Science*, 1970, **170**, 1412.
150. I. V. Pyatnitskii, *Russ. Chem. Rev. (Engl. Transl.)*, 1963, **32**, 45.
151. (a) R. E. Tapscott, *Inorg. Chim. Acta*, 1974, **10**, 183; (b) R. E. Tapscott and D. Marcovich, *J. Am. Chem. Soc.*, 1978, **100**, 2050.
152. (a) L. Johansson, *Acta Chem. Scand., Ser. A*, 1980, **34**, 495; (b) J. H. Dunlop, D. F. Evans, R. D. Gillard and G. Wilkinson, *J. Chem. Soc. (A)*, 1966, 1260.
153. (a) R. M. Holland and R. E. Tapscott, *J. Coord. Chem.*, 1981, **11**, 17; (b) L. D. Pettit and J. L. M. Swash, *J. Chem. Soc., Dalton Trans.*, 1979, 286.
154. A. Tatehata, *Chem. Lett.*, 1972, 561.
155. L. Johansson and B. Norden, *Inorg. Chim. Acta*, 1978, **29**, 189.
156. See ref. 86, p. 308.
157. M. A. Ivanov and A. L. Kosoy, *Acta Crystallogr., Sect B*, 1975, **31**, 2843.
158. M. Krumpolc, B. G. De Boer and J. Roček, *J. Am. Chem. Soc.*, 1978, **100**, 145.
159. N. D. Chasteen, R. L. Belford and I. C. Paul, *Inorg. Chem.*, 1969, **8**, 408.
160. A. B. Ablov, G. A. Popovich, G. I. Dimitrova, G. A. Kiosse, I. F. Burshtein, T. I. Malinovoskii and B. M. Schedrin, *Dokl. Akad. Nauk SSSR*, 1976, **229**, 611.
161. T. Taga, Y. Kuroda and M. Ohashi, *Bull. Chem. Soc. Jpn.*, 1978, **51**, 2278.
162. E. J. Meehan, H. Einspahr and C. E. Bugg, *Acta Crystallogr., Sect. B*, 1979, **35**, 828.

163. T. Taga, M. Ohashi and K. Osaki, *Bull. Chem. Soc. Jpn.*, 1978, **51**, 1967.
164. D. T. Sawyer, *Chem. Rev.*, 1964, **64**, 633.
165. D. T. Sawyer and J. R. Brannan, *Anal. Chem.*, 1966, **38**, 192.
166. N. C. Panagiotopoulos, G. A. Jeffrey, S. J. La Placa and W. C. Hamilton, *Acta Crystallogr., Sect. B*, 1974, **30**, 1421.
167. T. Lis, *Acta Crystallogr., Sect. B*, 1979, **35**, 1699.
168. J. González Velasco, S. Ayllón and J. Sancho, *J. Inorg. Nucl. Chem.*, 1979, **41**, 1075, and references therein.
169. J. Garcia-Rosas and H. Schneider, *Inorg. Chim. Acta*, 1983, **70**, 183.
170. (a) J. Bolte, C. Demuynck, G. Jeminet, J. Juillard and C. Tissier, *Can. J. Chem.*, 1982, **60**, 981; (b) Y. Pointud, C. Tissier and J. Juillard, *J. Solution Chem.*, 1983, **12**, 473; (c) J. Juillard, Y. Pointud, C. Tissier and G. Jeminet, in 'Physical Chemistry of Transmembrane Ions Motions', ed. G. Spach, Elsevier, Amsterdam, 1983, p. 239; (d) Y. Pointud, J. Juillard, G. Jeminet and L. David, *J. Chim. Phys.*, 1982, **79**, 67.
171. (a) R. F. Boyer, M. E. Wernette, S. van Wylen, E. Fraustman and R. Titus, *J. Inorg. Biochem.*, 1979, **10**, 205; (b) R. F. Boyer, *J. Inorg. Nucl. Chem.*, 1980, **42**, 155.
172. L. H. J. Lajunen and E. Aitta, *Talanta*, 1981, **28**, 603.
173. L. H. J. Lajunen, R. Petrola, P. Schildt, O. Korppi-Tommola and O. Makitie, *Talanta*, 1980, **27**, 75.
174. L. H. J. Lajunen, M. Leskela and J. Valkonen, *Acta Chem. Scand., Ser. A.*, 1981, **35**, 551.
175. K. Räisänen and L. H. J. Lajunen, *Org. Magn. Reson.*, 1978, **11**, 12.
176. V. F. Hanic and J. Michalov, *Acta Crystallogr.*, 1960, **13**, 299.
177. S. Jagner, R. G. Hazell and K. P. Larsen, *Acta Crystallogr., Sect. B*, 1976, **32**, 548.
178. H. P. Klug, L. E. Alexander and G. G. Sumner, *Acta Crystallogr.*, 1958, **11**, 41.
179. J. H. Burns and W. H. Baldwin, *Inorg. Chem.*, 1977, **16**, 289.
180. K. Moore and G. S. Vigee, *Inorg. Chim. Acta*, 1984, **91**, 53.
181. L. H. J. Lajunen, A. Kostama and M. Karvo, *Acta Chem. Scand., Ser. A.*, 1979, **33**, 681.
182. L. H. J. Lajunen and M. Karvo, *Anal. Chim. Acta*, 1978, **97**, 423.
183. (a) F. Cariati, L. Erre, G. Micera, A. Panzanelli and P. Piu, *Thermochim. Acta*, 1983, **66**, 1; (b) F. Cariati, G. Deiana, L. Erre, G. Micera and P. Piu, *Inorg. Chim. Acta*, 1982, **64**, L213; (c) F. Cariati, L. Erre, G. Micera, D. A. Clemente and M. B. Cingi, *Inorg. Chim. Acta*, 1983, **79**, 205.
184. B. V. Agarwala, *Inorg. Chim. Acta*, 1979, **36**, 209.
185. J. Bachs and E. Melendez, *Ann. Quim.*, 1979, **75**, 327.
186. M. Ahlgrén, U. Turpeinen and R. Hämäläinen, *Acta Chem. Scand., Ser. A*, 1982, **36**, 841.
187. M. Ahlgrén, U. Turpeinen and R. Hämäläinen, *Acta Chem. Scand., Ser. A*, 1984, **38**, 169.
188. H. J. Geise, J. F. J. Van Loock and A. T. H. Lenstra, *Acta Crystallogr., Sect. C*, 1983, **39**, 69.
189. J. Kroon, in 'Molecular Structure and Biological Activity', ed. J. F. Griffin and W. L. Duax, Elsevier, New York, 1982, p. 151.
190. J. Kroon, A. J. M. Duisenberg and A. F. Peerdeman, *Acta Crystallogr., Sect C*, 1984, **40**, 645.
191. A. J. De Vries and J. Kroon, *Acta Crystallogr., Sect. C*, 1984, **40**, 1542.
192. A. J. A. R. Blankensteyn and J. Kroon, *Acta Crystallogr., Sect. C*, 1985, **41**, 182.
193. W. Moerman, M. Ouwerkerk and J. Kroon, *Acta Crystallogr., Sect. C*, 1985, **41**, 1205.
194. Von L. Bohatý, R. Fröhlich and K.-F. Tebbe, *Acta Crystallogr., Sect. C*, 1983, **39**, 59.
195. L. K. Templeton, D. H. Templeton, D. Zhang and A. Zalkin, *Acta Crystallogr., Sect. C*, 1985, **41**, 363.
196. T. Lis, *Acta Crystallogr., Sect. C*, 1984, **40**, 374.

15.8

Sulfoxides, Amides, Amine Oxides and Related Ligands

PETER. L. GOGGIN

University of Bristol, UK

15.8.1 SULFOXIDES AND SELENOXIDES

Although dimethyl sulfoxide (DMSO) was discovered in 1867, it attracted little attention for 90 years. In the late 1950s it became commercially available as a solvent and the growth of the study of its complexes was catalyzed by claims of its wide usefulness in medicine.[1] However, it can be irritating to the skin and be absorbed through it, carrying toxic solutes with it. In physiological tests, subjects have reported adverse reactions such as nausea, cramps and drowsiness. Violent reactions can occur with covalent chlorides such as $MeCOCl$, $SiCl_4$, $POCl_3$ or SCl_2, with oxidizing agents such as IF_5, $Mg(ClO_4)_2$, HIO_4 or SO_3, and with reducing agents such as NaH or P_4O_6. Ignition can occur with potassium powder, sodium isopropoxide or even $(CF_3CO)_2O$.[2]

General characteristics of DMSO are well covered in a monograph by Martin and Hauthal,[1] including its physical and chemical properties, its reactions and its applications in technology and medicine up to 1970. It includes a useful survey of solubilities:[3] DMSO dissolves three times its volume of CO_2 at room temperature, ethylene to a similar extent, and some 30 times its volume of acetylene. Extraordinarily high solubilities are shown by sulfur dioxide, methanethiol and ethylene oxide. Many organic compounds dissolve in it: phenanthrene, sucrose, urea and polyacrylonitriles all do so to a level greater than 10% by weight. Many metal salts also give solutions of high concentration and examples are given in Table 1.

Comprehensive literature reviews on coordination of sulfoxides have appeared. Reynolds[4] has surveyed the chemical and physical properties of DMSO and its interactions with metals up to 1968. Davies[5] has covered the coordination chemistry of transition elements with the full range of sulfoxides up to 1979, including a discussion of the physical methods applied.

Table 1 Solubility[a] of Metal Salts in DMSO[3]

	Cl	Br	I	NO₃	ClO₄
Li	10.2	31.4	41.1	10	31.5
Na	0.5	6.2	30	20	24.2
K	2.2	6.5	45.5	10	38
Ca	—	—	—	30[b]	—
Sr	10[b]	5[b]	—	—	—
Pb	10	—	—	20	—
Zn	30	—	—	550[b]	—
Cd	20	—	30	—	—
Hg	114	90	100	—	—
Ag	2.13×10^{-4}	1.81×10^{-4}	0.43×10^{-4}	130	—

[a] g 100 cm⁻³; 25 °C. [b] Hydrated salts.

The commentary here concentrates on DMSO as the most widely investigated example, but $(CH_2)_4SO$, Et_2SO, Pr_2SO, $p\text{-}Tol_2SO$, Ph_2SO^6 and a range of chelating sulfoxides[7] have been used. Long straight and branched chain sulfoxides have been explored in relation to solvent extraction.[8]

All metal ions undergo some form of coordination with sulfoxides. Alkali metal ions are solvated by DMSO in much the same way as they are by water and show many parallels with corresponding hydrates.

For studies of sulfoxide complexes of simple cations, perchlorate has been used in almost every case to provide an uncompetitive anion. There are many cautionary notes about explosive properties and it must be concluded that the latent potential for being oxidized to sulfones renders the combination of sulfoxides and perchlorates hazardous and unpredictable.

Preparative methods employed have been simple. Where coordinative saturation is desired the metal salt may be dissolved in DMSO and a non-polar solvent such as benzene added to induce crystallization of the product;[9] commonly, a hydrated metal salt may be used without drying and gives the same product. It is also equally effective in many cases to dissolve the metal salt in acetone or ethanol and add an excess of ligand in the same solvent, concentrating the solution if necessary to obtain crystals: this method is obviously more useful when it is undesirable or impracticable to use the neat ligand as a solvent, *e.g.* with Ph_2SO.[10]

In several cases it has proved advantageous to remove water from hydrated salts. In the preparation of Pr_2^nSO complexes of lanthanide nitrates[11] and Bu_2^nSO complexes of lanthanide perchlorates,[12] 2,2'-dimethoxypropane and triethyl orthoformate respectively were used as reaction media as 50% solutions in ethanol.

In the case of the platinum group metals, complexes with high numbers of DMSO ligands have been made by treating complexes containing coordinated halide with $AgClO_4$ or $AgBF_4$ to bring about halide removal by precipitation as silver halide, but this method does not always ensure complete abstraction. It is also a good method for preparing mixed ligand systems, *e.g.* [Pd(dien) DMSO][ClO_4].[13]

Hundreds of sulfoxide complexes are known but in many cases the only structural conclusion which can be drawn with confidence from the evidence presented is that linkage of the ligand to the metal is through O or S. However, where the coordination arrangement for a particular stoichiometry has been firmly established (*e.g.* from a crystal structure determination), it is often a good clue to the structure of related complexes. One source of confusion is the tendency of DMSO to take up both coordinated and lattice roles in crystals. Thus $Fe(ClO_4)_3 \cdot 6DMSO$, $Fe(ClO_4)_3 \cdot 7DMSO$, $Co(ClO_4)_2 \cdot 6DMSO$ and $Co(ClO_4)_2 \cdot 8DMSO$ have been isolated but there is no doubt that all contain $[M(DMSO)_6]^{n+}$ cations. The perchlorates of the lanthanides crystallize with different numbers of DMSO molecules according to cation size, 8 for La–Nd, 7 for Sm–Ho and 6 for Er–Lu, and these probably reflect coordination numbers.

Stoichiometries of metal halide complexes can depend on preparative conditions. According to the amount of ligand used, CoI_2 gives either $[Co(DMSO)_6]I_2$ or $CoI_2 \cdot 3DMSO$. The latter should be formulated as $[Co(DMSO)_6][CoI_4]$, illustrating that stoichiometry is not a good guide to coordination number or structure.[14] Other types of mixed anion–cation structures have been shown to account for $FeCl_3 \cdot 2DMSO$ as *trans*-$[FeCl_2(DMSO)_4][FeCl_4]$[15] and $Ga_2Cl_6 \cdot 3DMSO$ as [Ga-$(DMSO)_6][GaCl_4]_3$.[16] Of course, these structures are not necessarily a guide to the sole or even major species in the solutions from which they were obtained (*e.g.* see ref. 14), and in using solution techniques to draw inferences about solids which have been dissolved the potentiality of the solvent displacing the sulfoxide or other coordinated group must be considered.[10]

Sulfoxide complexes can be formed for a wide range of metal oxidation states, as shown by the chromium derivatives $CrO_2Cl_2 \cdot 3DMSO$, $CrCl_3 \cdot 3DMSO$, $CrCl_2 \cdot 2DMSO$ and $Cr(CO)_5 \cdot DMSO$.

This is not to say that DMSO plays an entirely passive role; when the Nb^V derivative $NbOCl_3 \cdot 2\text{-}DMSO$ is formed from $NbCl_5$, the initial action of DMSO is O–Cl exchange to form $NbOCl_3$, which then coordinates with more DMSO (Scheme 1).[17] Comparable O–F exchange does not occur and NbF_5 simply gives $NbF_5 \cdot 2DMSO$. A somewhat similar result occurs for boron halides: BF_3 gives the sublimable $[BF_3(DMSO)]$, whereas BCl_3 gives a mixture of products including $BOCl \cdot MeSCH_2Cl$ and $BCl_3 \cdot MeSCH_2Cl$.

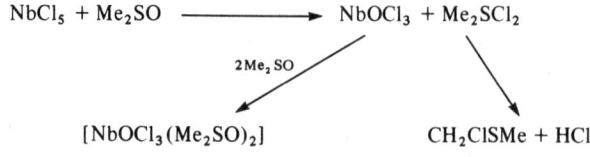

Scheme 1

Using a valence-bond description, DMSO can be represented by two structures, (1) and (2). The S—O bond length in the gas phase[18] is 1.477 Å, significantly less than the 1.66 Å expected for an S—O single bond. The bond angles about S are CSC = $96°\,23'$ and CSO = $106°\,43'$. There have been two reports of the crystal structure which agree that the bond angles are little different from those in the gas phase but they give markedly different S—O bond lengths: 1.471[19] and 1.531.[20] It is therefore unwise to use small differences between S—O bond lengths in complexes to draw inferences on S—O bond orders.

$$\text{Me}\backslash\overset{\text{Me}}{\underset{}{\diagup}}\!\!\!\overset{}{\underset{S=O}{}} \qquad \text{Me}\backslash\overset{\text{Me}}{\underset{}{\diagup}}\!\!\!\overset{}{\underset{^+S-O^-}{}}$$

$$(1) \qquad\qquad (2)$$

With electron pairs available for donation from either S or O, the ligand can display ambidentate character and examples of each are known. In O-bonded cases, values of S—O bond lengths between 1.45 and 1.57 Å have been reported.[5] The highest accuracies are claimed for *trans*-[FeCl$_2$(OSMe$_2$)$_4$][FeCl$_4$]* at 1.541(6)[15] and *trans*-[CuCl$_2$(OSMe$_2$)$_2$] at 1.531(4).[21] The S—O—metal angles vary from 118 to 138° but point to an essentially sp^2 hybridized O-atom as required if one p-orbital is not involved in hybridization but is available for π-bonding to S as in (1).

In S-bonded examples the reported S—O bond lengths span 1.45–1.51 Å but are mostly around 1.47 Å, *e.g.* 1.476(2) Å in *trans*-[PdCl$_2$(SMe$_2$O)$_2$];[22] MSC and MSO angles fall between 109 and 118° with the latter commonly the larger. On balance, S—O bonds seem to be shorter in the S-bonded cases, suggesting an enhanced contribution from (1) in these structures.

The O-atom can fulfil a bridging role as in [Hg(OSMe$_2$)$_4$(μ-OSMe$_2$)$_2$Hg(OSMe$_2$)$_2$][ClO$_4$]$_4$,[23] resulting in a mixture of octahedral and tetrahedral coordination in the complex cation. Such bridging may explain some of the curious formulations such as M(ClO$_4$)$_3\cdot$5.5Pr$_2^n$SO (M = Gd, Dy, Er).[24]

Crystallographic characterization has been applied only in a few cases and the majority of structural deductions are based on spectroscopic results, particularly IR measurements. The main criterion used has been the S—O stretching frequency, which is expected to be higher when structure (1) is more important, as in S-bonded sulfoxides.[25] Shifts attribtued to coordination are often quoted but the concept is arbitrary because S—O stretching of the ligand is sensitive to environment. For DMSO the accepted assignment is 1101 cm^{-1} in the gas phase, but varies between 1055 and 1085 cm^{-1} in a range of organic solvents. The IR band of the liquid at 1058 cm^{-1} is usually taken as the benchmark for coordination shifts, but it does not coincide with the strongest component of the Raman band in this range (1042 cm^{-1}) and associated species, probably dimers, are being measured.[26] In any case the vibrations in this region are not pure: S—O stretching, methyl deformation and methyl rocking coordinates are coupled; in O-bonded DMSO complexes, features relating to 'S—O stretching' and 'CH$_3$ rocking' fall in the same wavenumber range (*ca.* 900–1020 cm^{-1}) and there is ample scope for confusion. If it is desired to specify the wavenumber of the most S—O bond-dominated band, comparison betweeen DMSO and d^6-DMSO analogues clarifies the position because the 'S—O' vibration moves up some 10 cm^{-1} for the latter and rocking modes descend to around 800 cm^{-1}; however, the CD$_3$ deformations now come in the region 1000–1040 cm^{-1}.[25] Despite these reservations, with judicious application the IR method is useful. For S-bonded species the S—O stretches are strong and sharp and occur above 1080 cm^{-1} and their number can be a guide to stereochemisty: *trans*-[PdCl$_2$(SMe$_2$O)$_2$] displays a single feature at 1116 cm^{-1} whilst *cis*-[PtCl$_2$-(SMe$_2$O)$_2$] shows two bands at 1157 and 1134 cm^{-1}.[27]

In UCl$_4\cdot$7DMSO, bands at 1047 and 942 cm^{-1} indicate lattice and O-bonded DMSO respectively.[28] In [H(OSMe$_2$)$_2$][AuCl$_4$] the IR band of the cation is at 937 cm^{-1}.[29]

The far-IR has also been considered for identification of coordination type. Mixed metal–donor stretching and internal ligand deformation character makes unprofitable the application of precise mode descriptions of ligand-associated bands. However, an empirical observation, that the highest such bands below 600 cm^{-1} are in the range 470–550 cm^{-1} for O-bonded DMSO but in the range 400–450 cm^{-1} for the S-bonded ligand, can be a useful ancillary criterion for establishing coordination type.

^{1}H NMR spectroscopy has been quite widely applied to studying sulfoxide complexes, but as a solution-based technique the insight into precise structural details depends on the complex remaining intact on the NMR timescale. In diamagnetic complexes which meet this condition, δ(CH$_3$) of O-bonded DMSO ligands remains close to the free ligand value (2.53 p.p.m. in CH$_2$Cl$_2$, relative to

---

* When the form OSR$_2$ is used, bonding through O is implied; when SR$_2$O is used, S-bonding is implied.

tetramethylsilane), generally about 2.7 p.p.m., whereas in S-bonded cases there is much greater deshielding and the resonances fall between 3.2 and 3.8 p.p.m. With $I = \frac{1}{2}$, spin-active metals $^3J(MH)$ coupling has been observed (*e.g. ca.* 23 Hz when M = ^{195}Pt, *ca.* 0.5 Hz when M = ^{103}Rh) but $^4J(MH)$ has not been observed in O-bonded systems. NMR studies have uncovered the very complicated solution behaviour of some systems where both O- and S-bonding occur.[30] Four species (3)–(6) were identified when $RhCl_3 \cdot 3DMSO$ was dissolved in CH_2Cl_2. The crystal structure has been determined as (3) and this is the main structure present in solution, but there are two other linkage isomers (4) and (5), and a small amount of hydrolysis product (6) from water in the solvent replacing O-bonded DMSO from (3). In this work, ^{103}Rh shifts were very sensitive, being 4133, 3478, 4918 and 3926 p.p.m. with respect to ΞRh = 3.16 MHz as reference zero for (3), (4), (5) and (6), respectively.

As the example above shows, there is no sharp demarcation line between the occurrence of O- and S-bonding. It is probable that S-bonding does not occur for main group elements or for the lanthanides. Some IR data on actinides have been interpreted as indicating S-bonding, *e.g.* $Th(ClO_4)_4 \cdot 6DMSO$,[9] but as all contain perchlorate the appearance of bands between 1110 and 1150 cm^{-1} may equally well be due to the reduction of the symmetry of the anionic group through coordination.

The only clear evidence for S-bonding in complexes of the first transition series or other transition elements in groups III–VI is for very low oxidation states, *e.g.* for $[Fe(CO)_4(SMe_2O)]$, $v(S—O)$ = 1140 cm^{-1}, for $[W(CO)_5(SMe_2O)]$, $v(S—O)$ = 1115 cm^{-1}. However, in most (but not all) sulfoxide derivatives of the platinum group metals and for gold there are some S-bonded ligands. Where there are three or more sulfoxide ligands on Rh^{III} or Ru^{II}, a mixed coordination arrangement is preferred and structures with mutually *trans*-S-bonded sulfoxide ligands tend to be avoided. It appears that for S-bonding a highly polarizable metal centre is required but the balance between electronic and steric factors is a subtle one. Whereas there are two *cis*-S-bonded ligands opposite two O-bonded ligands in $[Pd(SMe_2O)_2(OSMe_2)_2][BF_4]_2$, all the ligands in the diisopentyl sulfoxide analogue are O-bonded. In $[Rh_2(\mu\text{-}O_2CR)_4(DMSO)_2]$ the bonding is through S when R is Me or Et but through O when R is CF_3.[31]

In contrast to sulfoxides, selenoxides have been relatively little studied as ligands. They can be prepared in high yields by oxidation of the corresponding dialkyl or diaryl selenides with $NaIO_4$ or $PhICl_2$.[32] Most of the reports of the preparation and characterization of their complexes emanate from Paetzold and his coworkers. These include complexes of metal perchlorates: $[M(OSeMe_2)_6]$-$[ClO_4]_n$ ($n = 2$, M = Mn, Co, Ni, Zn, Cd, Mg; $n = 3$, M = Cr, Fe), $[Cu(OSeMe_2)_4][ClO_4]_2$, $[Ag(OSeMe_2)_2][ClO_4]$ and $[UO_2(OSeMe_2)_5][ClO_4]_2$.[33] With the lanthanides, all the complexes were of the form $[M(OSeMe_2)_8][ClO_4]_3$.[34] Complexes of dimethyl selenoxide[35] and diphenyl selenoxide[36] with a range of metal halides have been reported and have compositions similar to known sulfoxide derivatives. In all these complexes the Se—O stretching frequency is lower than in the free ligand and coordination is deduced as being through oxygen, but systems which correspond to those where S-bonding has been found for sulfoxides have not been studied.

From calorimetric studies of enthalpy of formation of complexes with I_2 or iodine halides, the donor ability of some O-donors is ranked as $Ph_2CO < Ph_2SO < Ph_3PO < Ph_2SeO < Ph_3AsO$.[37]

15.8.2 FORMAMIDE, ACETAMIDE AND RELATED LIGANDS

This section deals with the coordination chemistry of amides of carboxylic acids, predominantly formamide, acetamide and their *N*-substituted derivatives. Lactams are also mentioned briefly.

Some of the stimulus for studying these systems arose from the availability and use of some of these ligands as solvents, a subject reviewed by Vaughan[38] who includes details of methods of purification and electrochemical parameters. The dielectric constants are much higher when these molecules contain an N—H bond than when they are *N,N'*-dimethyl substituted. Formamide, its *N*-methylated derivaties and dimethylacetamide (DMA) are liquids at room temperature; *N*-methyl-acetamide (NMA) melts at 29.8 °C and so can be used as a solvent at only slightly elevated temperatures.

These simple amide derivatives are miscible with water. Their vapours are irritating to the eyes and respiratory system. Skin contact should be avoided, and it is suspected that frequent and prolonged contact with the vapour can cause liver damage.[2] Many simple salts are appreciably soluble in these solvents and especially in formamide, the usefulness of which is limited by its poor thermal and photochemical stability. In DMF, KI, KNO_3, $LiClO_4$, NH_4SCN, $MnCl_2 \cdot 4H_2O$, $Co(NO_3)_2 \cdot 6H_2O$ and $FeCl_3$ are all soluble to the level of 1 mol dm^{-3} or so; however, NaCl and KCl are not soluble to a useful extent.

These amides have two possible donor sites, N or O. In the ligand there is some delocalization which can be represented by forms (7) and (8), taking DMF as an example. Force field studies of DMF calculate the bond-order of the CO bond as 1.75 and the CN bond as 1.5,[39] *i.e.* (8) makes a strong contribution to the structure. In all the coordination compounds that have been characterized by X-ray studies the bonding is through oxygen.

(7)   (8)

Carboxylic acid amides are 'hard' ligands. Consequently most attention has been given to their complexes with main group, first transition series and lanthanide elements. Water is a competitor with them for coordination, so in preparing their complexes it is advisable either to use strictly anhydrous reagents or to effect dehydration in the course of reactions by means of triethyl orthoformate. Table 2, adapted from ref. 40, shows the variety of steps used in preparing a range of NMF complexes of metal chlorides. If dry conditions have not been used, or if complexes are crystallized from water, products containing both amide and water are obtained (*e.g.* ref. 41). Many of the complexes are highly hygroscopic.

Formamide complexes have also been obtained by indirect means: $[FeCl_2(HCONH_2)_2]$ resulted from the reaction of almost dry HCN with $FeCl_2$,[42] whilst the reaction of uranyl formate with ammonia gave $UO_3 \cdot 2HCONH_2 \cdot 2H_2O$.[43] Perchlorate salts of di- or tri-valent ions of elements in the first transition series give complexes with six amide ligands,[44] or lactams.[45] Lanthanides yield eight-coordinate $[Ln(DMF)_8][ClO_4]_3$ with DMF but only six-coordinate complexes with diphenylformamide; presumably the coordination number is restricted by the bulk of the phenyl groups even though these are relatively remote for the coordination site.[46] A progressive difference is observed when DMA is the ligand, with coordination number eight for La–Nd, seven for Sm–Er and six for Tm–Lu.[47]

Anionic ligands with any significant coordinating power can compete with amides for coordination sites. Lanthanide nitrates lead to complexes $Ln(NO_3)_3 \cdot 4DMF$ (Ln = La, Pr, Nd or Sm)[48] and even perrhenate gives $Ln(ReO_4)_3 \cdot 4DMA$ (Ln = Ce, Pr, Nd, Sm, Eu, Dy, Ho, Er or Y) in which at least some of the perrhenate ions must be coordinated. With DMA, trichlorides of the lanthanides give compounds of stoichiometries $LaCl_3 \cdot 4DMA$, $LnCl_3 \cdot 3.5DMA$ (Ce–Dy) and $LnCl_3 \cdot 3DMA$ (Ho–Lu, and Y) and all are practically non-conductors in nitromethane.[49] The fractional nature of some of these formulations might be accounted for by one or more of the amide ligands acting as an O-bridging group; such a situation has been found in $(facam)_3Pr(\mu\text{-}DMF)_3Pr(facam)_3$ (facam = 3-trifluoroacetyl-*d*-camphor).

The products isolated from reactions of amides with transition metal halides usually contain coordinated halide (*e.g.* the formulations in Table 2). In some cases such as $[Co(NMF)_6][CoCl_4]$, halide and amide are coordinated to different metal atoms, but when such compounds are dissolved in the neat ligand, halide can be replaced and at high dilution all the metal ions may be fully coordinated by the amide alone. The electronic spectrum resulting when this cobalt complex is dissolved in nitromethane has been interpreted as relating solely to the tetrahedral complex $[CoCl_2(NMF)_2]$.

The platinum metals are not particularly good receptors for O-donors but a number of complexes with amides have been isolated. These include $[MCl_2L_2]$ where M = Pt or Pd and L = DMF or DMA. Interest in them stemmed from the possibility of N-coordination since urea is N-bonded in $[PtCl_2(urea)_2]$; however, IR shows them to be O-bonded. They are not very stable and allow the amide to detach itself even in as innocuous a solvent as dichloromethane.[50] In $CDCl_3$,

Table 2 Preparation of NMF Complexes Starting from 0.01 mol of Hydrated Metal Salts, using Triethyl Orthoformate (EOF) as Desiccant[40]

Step	$MgCl_2 \cdot 6H_2O$	$MnCl_2 \cdot 4H_2O$	$CoCl_2 \cdot 6H_2O$	$NiCl_2 \cdot 6H_2O$	$CuCl_2 \cdot 2H_2O$	$ZnCl_2$	$CdCl_2 \cdot 2.5H_2O$
1	20 cm³ NMF, stir until dissolved	6 cm³ NMF, dissolve	5 cm³ NMF, dissolve	20 cm³ EOF heat, stir 1 h	10 cm³ NMF, dissolve	5 cm³ NMF, dissolve	7 cm³ NMF, 1 cm³ H₂O, stir 0.5 h
2	10 cm³ EOF, stir 1 h	5 cm³ EOF, stir 1 h	5 cm³ EOF, stir 1 h	Slowly add 20 cm³ NMF, stir vigorously	5 cm³ EOF, stir 1 h	5 cm³ EOF, stir 1 h	Obtain thick paste, strong stirring needed
3	Triturate 4 times with 50 cm³ portions of Et₂O	50 cm³ Et₂O, stir 1 h, decant, repeat	50 cm³ Et₂O, stir 0.5 h, decant	Initial green gelatinous solid, stir	25 cm³ Et₂O, stir	15 cm³ Et₂O, stir overnight	10 cm³ EOF, stir 1.5 h
4	5 cm³ abs. EtOH, stir 10 min	5 cm³ abs. EtOH, stir 0.5 h, 10 cm³ Et₂O, stir	5 cm³ abs. EtOH, stir 0.25 h	After 1–2 h gives fine yellow powder	Gives green powder, continue stirring 1–2 h	Filter, wash with Et₂O, dry *in vacuo*	10 cm³ abs. EtOH, stir overnight
5	10 cm³ Et₂O, stir until fine powder	Scrape solid off walls, crush, stir until fine powder	25 cm³ Et₂O, stir overnight	Stir until no trace of green		Recrystallize from benzene	
6	Wash with 5 cm³ EtOH + 10 cm³ Et₂O, then with Et₂O	Wash with 5 cm³ EtOH + 10 cm³ Et₂O, then with Et₂O	Wash with Et₂O	Wash with Et₂O	Wash with Et₂O	Wash with benzene, then with Et₂O	Wash with EtOH, then with Et₂O
Product	[MgCl₂(NMF)₄]·2H₂O	{MnCl₂(NMF)₂}₂ₓ Hygroscopic	[Co(NMF)₆][CoCl₄]	[NiCl₂(NMF)₂] Hygroscopic	[CuCl₂(NMF)₂] Slightly hygroscopic	[ZnCl₂(NMF)₂]	[CdCl₂(NMF)₂]

[PtCl$_2$(py)DMF] is in equilibrium with [Pt$_2$Cl$_4$py$_2$] and free DMF.[51] The 'platinum blue' product from the reaction of K$_2$[PtCl$_4$] with acetamide now appears to contain some Pt^{III} and to have MeCONH$^-$ groups and not neutral acetamide.[52]

Complexes with halides of the main group elements have been widely investigated. Changes in ^{1}H shifts in NMR spectra of DMF as a result of coordination have been used as a basis for ranking Lewis acids. For 1:1 complexes the order was InCl$_3$ < BiCl$_3$ < BF$_3$ < SnCl$_4$ < PF$_5$ < AlCl$_3$ < SbF$_5$ < BCl$_3$ < SbCl$_5$ < BBr$_3$.[53] Addition compounds between amides and molecular I$_2$ have been widely studied.[54]

Coordination of O and H$^+$ gives cations such as [(acetamide)$_2$H]$^+$. A neutron diffraction study of its chloride salt implies two short, symmetrical H-bonds between O atoms,[55] but this has been questioned on spectroscopic grounds.[56]

Crystal structures have been determined for a variety of complexes. The structure of FeCl$_2$-(HCONH$_2$)$_2$ is polymeric using Cl-bridges to achieve a *trans*-octahedral arrangement about the metal.[42] Dimensions of metal-bonded DMF have been reported for [Fe(DMF)$_6$][ClO$_4$]$_3$,[57] acetamide for *trans*-[Ni(acetamide)$_4$(OH$_2$)$_2$]Cl$_2$[58] and DMA for tetrahedral [CoCl$_2$(DMA)$_2$].[59]

The structures of some NMA solvates of the alkali and alkaline earth metals have been reported by Chakrabarti *et al.*[60] The interest in these was in the relationship they might have to the bonding between such metal ions and peptides. The compounds studied were LiCl·4NMA, NaClO$_4$·2NMA, KSCN·NMA, MgCl$_2$·6NMA and CaCl$_2$·4NMA·2H$_2$O. Because the amide ligands remain planar or close to it in all the cases studied, NMA could display a *cis* or *trans* structure; both forms were evident in some cases but the arrangement with methyl groups mutually *trans* was the only one in the [Mg(NMA)$_6$]$^{2+}$ ion.

Some bond lengths in the examples selected here are shown in Table 3 and compared with those of the standard peptide group.[60] In all cases the CO bond is longer and its adjacent CN bond shorter in the coordinated ligands, *i.e.* form (8) has been enhanced by coordination.

Table 3 Bond Lengths (Å) and Angles at Oxygen in Complexed Carboxylic Acid Amides

	C'=O	C'—N	C'—C	N—C"	MOC'	Ref.
[Ni(acetamide)$_4$(OH$_2$)$_2$]Cl$_2$ [a]	1.249(3)	1.310(4)	1.496(5)	—	136.5(1)	58
	1.247(3)	1.311(4)	1.492(5)	—	139.1(1)	
[Mg(NMA)$_6$]Cl$_2$ [b]	1.245(6)	1.269(6)	1.623(9)	1.407(9)	140.28(39)	60
	1.232(4)	1.313(6)	1.538(7)	1.431(7)	146.63(25)	
	1.245(4)	1.327(5)	1.503(6)	1.442(6)	140.07(23)	
[CoCl$_2$(DMA)$_2$] [a]	1.258	1.325	1.509	1.480, 1.461	136.8	59
	1.249	1.319	1.509	1.487, 1.458	133.6	
[FeCl$_2$(formamide)$_2$]$_n$	1.23(1)	1.33(1)	—	—	132.02(36)	42
[Fe(DMF)$_6$][ClO$_4$]$_3$ [b]	1.25(1)	1.25(2)	—	—	128.9(5)	57
	1.26(1)	1.30(1)	—	—	124.8(4)	
	1.27(1)	1.30(1)	—	—	127.6(4)	
Standard peptide group [c]	1.24	1.325	1.51	1.455	—	
	1.229	1.335	1.522	1.449	—	

[a] Two independent ligand molecules. [b] Three independent ligand molecules. [c] Two different sources quoted in ref. 60.

The effect of coordination on structural parameters is paralleled by the changes in vibrational frequencies assigned as CO and CN stretching. The modes are actually of mixed character,[61] but the wavenumber of the band around 1700 cm^{-1}, traditionally called CO stretching, drops by *ca.* 30 cm^{-1} on coordination whilst the 'CN stretching' band moves to higher wavenumber,[62] as shown in Table 4.

Pisaniello and Lincoln have made detailed ^{1}H NMR studies of the amide complexes [ScL$_6$][ClO$_4$]$_3$ in CD$_3$CN and CD$_3$NO$_2$.[63] With NMF, DMF, *N,N*-diethylformamide and *N,N*-dibutylformamide, exchange between free and bound ligand occurred in the fast exchange limit at all accessible

Table 4 Effect of Coordination on Stretching Wavenumbers of Carboxylic Acid Amides[62]

Compound	v(C=O)	v(C—N)	Compound	v(C=O)	v(C—N)
HCONH$_2$	1690	1320	CH$_3$CONH$_2$	1680	1410
ZrOCl$_2$·4HCONH$_2$	1675	1375	ZrOCl$_2$·2CH$_3$CONH$_2$	1620	1440
NMF	1672	—	NMA	1670	1300
CuCl$_2$(NMF)$_2$	1642	—	ZrOCl$_2$·2NMA	1640	1335
DMF	1680	1255	DMA	1670	1220
ZrOCl$_2$·2DMF	1650	1300	ZrOCl$_2$·2DMA	1615	1255

temperatures. With NMA, DMA, *N,N*-diethylacetamide and *N*-phenylacetamide, intermediate exchange rates were observed; at low temperatures (typically 225 K) solutions of the complexes in the presence of excess free ligand show separate signals for free and bound ligands. From a study of coalescence behaviour, rate constants for the co-occurring associative and dissociative exchange processes were established. An interesting feature of the NMA system was the detection of both *cis* and *trans* disposition of the methyl groups across the CN bond. The proportion of *cis* isomer in the complex was *ca.* 9.5% at 350 K, three times its abundance in the free ligand, indicating almost equal $[Sc(\textit{trans}\text{-NMA})_6]^{3+}$ and $[Sc(\textit{trans}\text{-NMA})_5(\textit{cis}\text{-NMA})]^{3+}$ species. On the smaller Be^{2+} centre, exchange of DMF with $[Be(DMF)_4]^{2+}$ is slow enough to be measured.[64]

Many of the studies with amides have included evaluation of ligand-field parameters if appropriate. The differences between different amides are quite small. For $[NiL_6]^{2+}$, *Dq* varies from $749\,cm^{-1}$ when L is *N*-methylcaprolactam to $850\,cm^{-1}$ when L is DMF.[44,45] However, the spectrochemical order in complexes $[UO_2L_5]^{2+}$ is DMA < NMA $\approx$ DMF < NMF[65] compared with NMA < DMA $\ll$ NMF < DMF for Ni^{2+}; neither series follows the order of donor strengths reflected by the Gutmann donor numbers, which rank DMA marginally above DMF as a donor in terms of the enthalpies of adduct formation with $SbCl_5$.[66]

15.8.3 AMINE *N*-OXIDES

The majority of studies has involved aromatic N—O ligands, a selection of which is given by (9)–(13) together with abbreviations which will be used in referring to them. All of these offer scope for inclusion of substituents with a range of electronic characteristics.

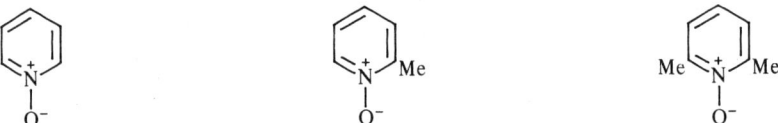

(9) Pyridine *N*-oxide (pyNO)   (10) 2-Picoline *N*-oxide (2-picNO)   (11) 2,6-Lutidine *N*-oxide (2,6-lutNO)

(12) Quinoline *N*-oxide (QuinNO)   (13) Bipyridyl bis-*N*-oxide (bipy2NO)

There have been comprehensive reviews on the coordination chemistry of aromatic *N*-oxides by Garvey *et al.* (up to 1968)[67] and Karayannis *et al.* (up to 1971),[68] the latter including a short section on aliphatic amine *N*-oxides and secondary amine nitroxide free radicals. A further review by Karayannis *et al.* (up to 1975)[69] covers mono- and di-oxides of bipyridyl, *o*-phenanthroline and some diazines.

Because of lack of available orbitals on nitrogen, trialkylamine N—O ligands can only possess σ-bonding between N and O. In aromatic N—O systems, negative charge is not confined to the oxygen atom and can be delocalized by π-bonding as depicted by (14)–(16). It has been suggested that forms (14) and (16) make nearly equal contributions to the structure of pyNO. Electron withdrawing groups at positions 2 and 4 should enhance the importance of form (15) and reduce the effective donor capacity of the ligand. The NO stretching frequency in pyNO has been assigned as $1243\,cm^{-1}$; in 4-NO_2pyNO it increases to $1279\,cm^{-1}$ whereas when the electron releasing methoxy group is in the 4-position it is reduced to $1212\,cm^{-1}$.[68] The N—O bond order in these systems is

(14)   (15)   (16)

estimated to be *ca.* 1.5. In Me_3NO the NO stretching vibration has been assigned as $948\,cm^{-1}$, consistent with the lower bond order necessarily present,[70] although it must be appreciated that such a vibrational description is only approximate.

Nelson *et al.*[71] have defined a pyridine *N*-oxide substituent constant, σ_{pyNO}, from the relationship $\sigma_{pyNO} = \Delta pK_{BH+}/2.09$ where ΔpK_{BH+} is the difference between the ionization constants of $pyNOH^+$ and substituted $pyNOH^+$ acids. The value of 2.09 has been chosen to give approximate numerical comparability with other sets of substituent constants in common use. Fairly consistent correlations between σ_{pyNO} and experimental parameters related to donor strength have been observed, *e.g.* the reduction of $v(V{=\!=}O)$ in $[VO(acac)_2]$ when 4-substituted pyNO ligands are added.

Many aromatic *N*-oxides are commercially available. They may be prepared in good yields by addition of 1.7 mol equivalents of 35% aqueous hydrogen peroxide to a solution of the amine in glacial acetic acid at 70–80 °C.[72] These ligands form complexes with nearly every element. They form 1:1 complexes with I_2.[73]

The majority of the complexes are decomposed by water through ligand displacement, although they are not necessarily sensitive to atmospheric moisture. They are generally prepared by direct addition of the amine *N*-oxide to solutions of metal salts in organic solvents. Triethyl orthoformate or 2,2′-dimethoxypropane have often been added to the reaction media as desiccants.

The pyNO ligand has relatively small spatial demands. With perchlorates of Mg^{II}, Ca^{II}, Sr^{II}, Mn^{II}, Fe^{II}, Co^{II}, Ni^{II}, Cu^{II}, Zn^{II}, Cd^{II} and Hg^{II}, the hexacoordinated $[M(pyNO)_6][ClO_4]_2$ salts are readily isolated. The crystal structures of several examples have been determined.[74,75] Because the bonding about O is angled, cubic T_h symmetry is not achieved; the structure about the metal is S_6 as shown in Figure 1. The environmental symmetry of the $[ClO_4]^-$ ion in the crystal lattice is C_{3v}, and the IR spectrum of the Hg complex shows doublets at frequencies associated with v_3 and v_4 of the free ion as well as allowing v_1 to be rendered active. This is a warning that observation of degraded symmetry is not a sure guide to anion coordination in solids.[76] The M—O bond lengths of these cations of first transition series metals are Fe = 2.112(2), Co = 2.089(2), Ni = 2.060(1), Cu = 2.086(2) and Zn = 2.102(1) Å. The N—O bond lengths are 1.33Å and do not appear to be sensitive to the metal atom; MON angles are close to 120°, consistent with bonding to sp^2 oxygen. The Cu^{II} complex is particularly interesting because it does not show any of the distortion expected for a $3d^9$ electron configuration. However, the analogous 4-picNO complex gives monoclinic crystals and exhibits a static Jahn–Teller distortion with two bonds of 2.385(11) Å being about 0.39 Å longer than the other four. The magnetic properties of these copper complexes depend on the structural form.[77]

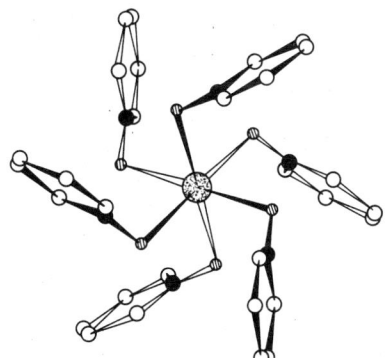

Figure 1 Projection of the $[Hg(pyNO)_6]^{2+}$ cation down the threefold axis (reprinted with permission of the Royal Society of Chemistry from *J. Chem. Soc., Dalton Trans.*, 1973, 760)[76]

If the $Cu(ClO_4)_2$ complex with pyNO is prepared with ethanol instead of methanol as solvent, $[Cu(pyNO)_4][ClO_4]_2$ is obtained; it has been shown to have a square-planar CuO_4 skeleton (approximately C_{4h} point symmetry) with, surprisingly, no significant contact between Cu and the anion, but at 1.92(1) Å the Cu—O bonds are significantly shorter than for the octahedral complex.[78]

The perchlorates of Y^{III} and the lanthanides(III) form $[M(pyNO)_8][ClO_4]_3$ and that of Th^{IV} forms $[Th(pyNO)_8][ClO_4]_4$. Two different arrangements of the eight ligands have been found crystallographically, as shown in Figure 2.[79]

Substitutents at the 4-position on pyNO do not affect the stoichiometry of the complex formed, nor does the single fused benzene ring of quinNO. However, substituents *ortho* to N can lead to complexes in which the coordination number of the metal is lower. Thus with 2-picNO, $[Co(2-picNO)_5][ClO_4]_2$ can be obtained; the cation has slightly distorted trigonal bipyramidal

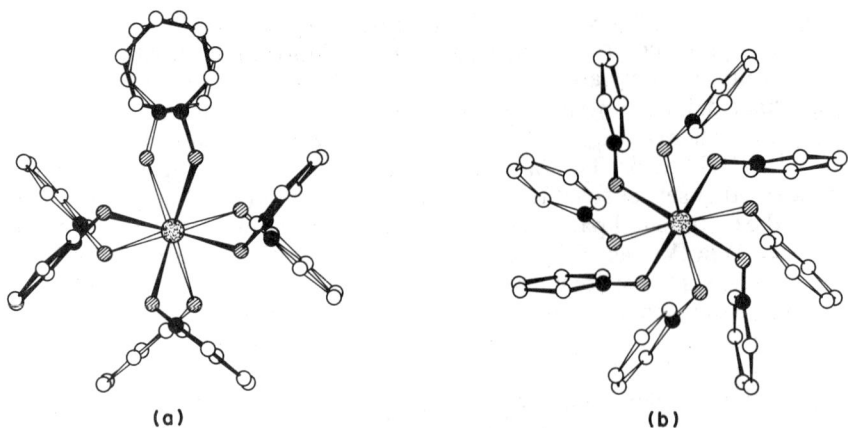

Figure 2 Structures of (a) $[La(pyNO)_8]^{3+}$ and (b) $[Nd(pyNO)_8]^{3+}$ viewed along the respective crystal c axes (reprinted with permission of the American Chemical Society from *Inorg. Chem.*, 1979, **18**, 1177)[79]

geometry with Co—O equatorial bonds marginally shorter than the axial ones.[80] Using 2,6-lutNO the predominant product for Co^{II} is $[Co(2,6-lutNO)_4][ClO_4]_2$ in which the cation has a tetrahedral structure, but one must be cautious about sweeping statements on steric requirements since small amounts of $[Co(2,6-lutNO)_6][ClO_4]_2$ were obtained from the mother liquor of the same reaction after it had been left for a day at $-30\,^{\circ}C$.

With bipy2NO, bidentate binding occurs readily[69] and the resulting coordination numbers are usually the same as for pyNO, *viz.* $[Co(bipy2NO)_3][ClO_4]_2 \cdot 2H_2O$.

With metal halides the products isolated commonly contain coordinated halide. In the salt $[Ni(pyNO)_6]I_2$ the iodide is not coordinated but pyNO is lost at $100\,^{\circ}C$ to give $[NiI_2(pyNO)_4]$. The halide is not necessarily coordinated to the same metal centre as the O-donor; $CoCl_2 \cdot 3pyNO$ is almost certainly $[Co(pyNO)_6][CoCl_4]$. The iron complexes $FeCl_3 \cdot 2L$, where L is pyNO, 2-picNO, 4-picNO and 2,4-lutNO, are all of the form $[FeCl_2L_4][FeCl_4]$.

The crystal structure determination of $[ZnI_2(pyNO)_2]$ demonstrates its molecular behaviour, with a tetrahedral environment about Zn, whereas the $CdI_2(pyNO)$ complex is polymeric with alternating O and I bridges between Cd atoms which are five-coordinate. The $HgCl_2 \cdot pyNO$ compound displays six-coordination about Hg but two *trans* HgCl bonds are much shorter than those to the other four ligands.[81]

The ability of the O atom of aromatic NO ligands to act as a bridging group has some interesting manifestations in Cu^{II} chemistry. Complexes such as $CuCl_2 \cdot 2L$ can be prepared in either green or yellow forms, sometimes both. The magnetic susceptibilities of the green complexes ($\mu_{eff} \approx 1.9\,BM$) are close to the expected value for a $3d^9$ system, whilst those of the yellow ones are often very much lower but not invariably so. The green $[CuCl_2(4-picNO)_2]$ has a *trans* square-planar structure[82] but the yellow-green $CuCl_2 \cdot 2pyNO$ ($\mu_{eff} \approx 0.5\,BM$) has the bridged structure (**17**).[83] Compounds in which the terminally bonded pyNO ligands have been replaced by other O-donors such as H_2O or DMSO have also been reported. Magnetic properties of these systems have been surveyed by Hodgson,[84] but the nature of the spin–spin interaction still arouses controversy.[85]

$$
\begin{array}{c}
py-O \qquad\qquad py \qquad\qquad Cl \\
Cl_{\prime\prime\prime\prime}\;\;\;\; O_{\prime\prime\prime\prime} \;\;\;\; \\
\quad Cu \quad\quad\quad Cu \\
Cl \qquad\qquad O \qquad\qquad Cl \\
\qquad\qquad py \qquad\quad O-py
\end{array}
$$

(17)

Main group Lewis acids readily form molecular complexes such as $[BCl_3(pyNO)]$ and $[SnI_4(pyNO)_2]$. The platinum group metals may not give particularly stable complexes with these relatively hard ligands; they do not appear to have been widely studied although $[RuCl_3(2,6-lutNO)_3]$, $[RhCl(2,6-lutNO)_5]Cl_2$, $[PdCl_2(4EtO-pyNO)_2]$ and $[PtCl_3(2,6-lutNO)_3]Cl$ have been reported. In contrast to the general paucity of information with these metals, the complexes *trans*-$[PtCl_2(pyNO)Y]$ have been studied in detail in cases where Y is CO, alkene or alkyne.[86]

Substituent effects on pyNO and on alkenes have been explored in depth. The stability of these compounds is probably due to the positioning of the O-donor opposite a Y ligand of low *trans*-influence. Palladium(II) analogues are known but are much less stable, although that is not necessarily because of the O-donor.

Many transition metal nitrates will form complexes which retain nitrate in the coordination sphere. $Ni(NO_3)_2$ gives $[Ni(O_2NO)_2(pyNO)_2]$ with both nitrate groups in bidentate coordinated form, but IR and electronic spectra suggest that the corresponding 2,6-lutNO complex is five-coordinate with one monodentate and one bidentate nitrate.

In contrast to the complexes with aromatic *N*-oxides, studies of aliphatic *N*-oxide complexes have been sparse and relatively superficial. It is established that trialkylamine *N*-oxide ligands form complexes with fewer such ligands attached than when pyNO is the donor. Although Cr^{III} and Sc^{III} give $[M(Me_3NO)_6][ClO_4]_3$ salts, Mn^{II}, Co^{II}, Ni^{II}, Cu^{II} and Zn^{II} give $M(ClO_4)_2 \cdot 4Me_3NO$. It has been suggested that the Mn, Co and Zn complexes contain tetrahedral $[M(Me_3NO)_4]^{2+}$ cations, that $[Cu(Me_3NO)_4]^{2+}$ is close to being square-planar, but that in the Ni derivative the coordination about the metal is distorted from tetrahedral. These stoichiometries are similar to those most readily obtained using 2,6-lutNO but the parallel does not hold in all cases: the uranyl group only takes on four Me_3NO ligands[87] to give $[UO_2(Me_3NO)_4][ClO_4]_2$ (as it does with Ph_3PO) whereas it forms $[UO_2(2,6\text{-lutNO})_5][ClO_4]_2$.

Addition of Et_3NO to a solution of $[Co(Et_3NO)_4][ClO_4]_2$ gave changes to the electronic spectrum which were interpreted as formation of $[Co(Et_3NO)_5]^{2+}$, but no such unit could be isolated in a solid salt.[88] This acts as a reminder that greater stereochemical diversity is sometimes possible in solution and that, in interpreting data from solutions in solvents which are potential donors, ligand displacement or solvent incorporation are possibilities to be borne in mind.

The structure of $CoCl_2 \cdot 2Me_3NO$ is still something of a mystery. Its electronic spectrum in solution is compatible with a four-coordinate tetrahedral structure but that of the solid is different. On the basis of negative criteria it has been postulated that, in the solid, Co is in a five-coordinate environment as a result of O-bridging.[89] Complexes with most main group Lewis acids (*e.g.* $[BF_3(Me_3NO)]$, $[SnCl_4(Me_3NO)_2]$) are readily rationalized but those of $SiCl_4$, $SiBr_4$ and $GeCl_4$ of stoichiometry $MX_4 \cdot 4Me_3NO$ have not been adequately accounted for. The $[BH_3(Me_3NO)]$ complex results from reaction of $Me_3NO \cdot 2H_2O$ and $Na(BH_4)$ in THF and is docile, whereas direct reaction between $Me_3NO \cdot 2H_2O$ and B_2H_6 in dichloromethane at $-78\,°C$ gives an insoluble addition compound which is shock sensitive and is postulated to be $[BH_2(Me_3NO)_2][BH_4]$.[90]

In contrast to the large number of structures of complexes with aromatic *N*-oxides which have been established by X-ray methods, Me_3NO complexes have been neglected. One compound has been studied which contains Me_3NO somewhat incidentally as a ligand, $[(\mu\text{-H})(\{Re_2(CO)_7\}(\mu\text{-}NC_5H_4)(Me_3NO)]$.[91] The N—O bond length is not significantly different from that in Me_3NO itself, but the ReON angle of 128° is larger than might have been expected about sp^3 hybridized O.

From IR studies it has been concluded that the N—O stretch of trialkylamine *N*-oxides is practically insensitive to coordination, unlike their aromatic counterparts, but that the predominantly M—O stretching frequencies are higher than in the aromatic cases.[92] This has been attributed, in part, to the lack of π-character in the N—O bond which makes these the more effective ligands.

The potentiality of NF_3O as a ligand has been explored.[93] However, although this molecule forms complexes with bases such as BF_3 and SbF_5, vibrational and ^{19}F NMR studies show conclusively that the derivatives are the salts $[NF_2O][BF_4]$ and $[NF_2O][SbF_6]$.

15.8.4 PHOSPHINE OXIDES, ARSINE OXIDES AND RELATED LIGANDS

There are four main catagories of OPX_3 neutral ligand: X = alkyl or aryl (R), X = OR, X = NR_2 and X = halide. There have also been studies in which these catagories are combined in ligands, *e.g.* $OP(OR)_2R$, the phosphonates. Arsine oxides, $OAsR_3$, although less extensively studied, are also included in this section.

The literature on coordination compounds of these ligands up to 1971 has been comprehensively reviewed by Karayannis *et al.*,[64] and whilst there has been much published since then, most of it is concerned with clarifying details on complexes of types reported earlier.

The introduction of OR or NR_2 groups on to phosphoryl ligands is readily achieved by reaction of alcohols or amines with $OPCl_3$. The phosphine oxide ligands are prepared by oxidation of the corresponding phosphine. For the most part, H_2O_2 has been used as the oxidizing agent.[95,96] In the case of $OPPh_3$ and $OAsPh_3$ the products were originally formulated as monohydrates; however,

Copley *et al.*[97] characterized 'OPPh$_3$·H$_2$O' as (OPPh$_3$)$_2$·H$_2$O$_2$ and it is not clear in subsequent work that this or its implications have been appreciated. A new mild and promising method using SO$_3$ or SO$_2$FCl as the oxidizing agent for the preparation of phosphine oxides has been reported by Olah *et al.*[98]

The common method used for preparing complexes of all these ligands has been the direct reaction between a metal salt and the ligand in a non-aqueous solvent such as an alcohol, acetone or halogenated hydrocarbon, selected according to the solubility of the metal salt. Triethyl orthoformate or 2,2′-dimethoxypropane have been added as an internal desiccant where necessary, because water is a competitive ligand in these systems, and where crystallization was not spontaneous it has been induced by addition of hydrocarbons or diethyl ether.

The most extensively studied phosphine oxide ligand has been OPPh$_3$, which gives well defined complexes with a wide range of metals. In nearly all cases it is safe to assume that all OPPh$_3$ molecules, represented by an empirical formula, are coordinated to the metal. Exceptions include LiI·5OPPh$_3$, which has four P—O bonds directed at Li$^+$ with a fifth OPPh$_3$ uncomplexed in the lattice,[99] and the gold system [H(OPPh$_3$)$_2$][AuCl$_4$] where the phosphine oxide is part of the hydrogen-bonded cation.[100]

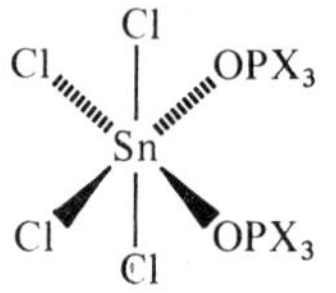

X = Cl; P—O = 1.41(5) Å;
SnOP = 148(6)°

(18)[101]

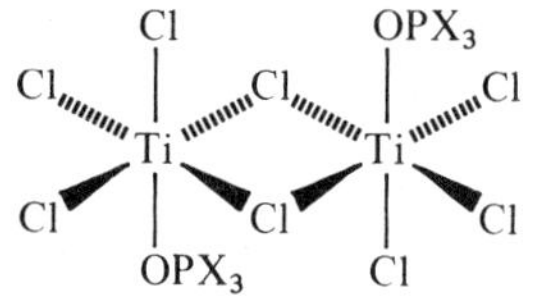

X = Cl; P—O = 1.44(2) Å;
TiOP = 152(1)°

(19)[102]

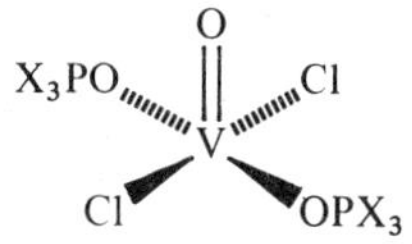

X = Ph; P—O = 1.495(12) Å;
VOP = 146°, 159°

(20)[103]

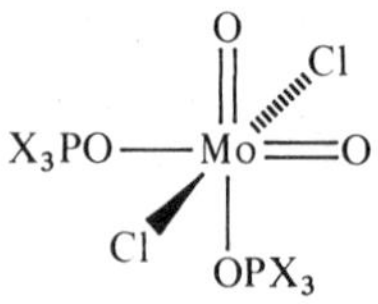

X = Ph; P—O = 1.50(1) Å;
MoOP = 162(4)°

(21)[104]

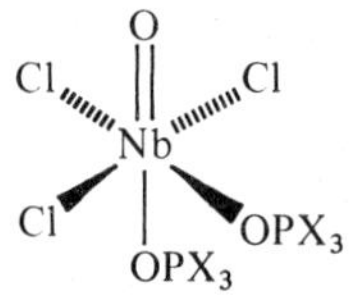

X = NMe$_2$; P—O = 1.50(1) Å;
NbOP = 154(4)°

(22)[105]

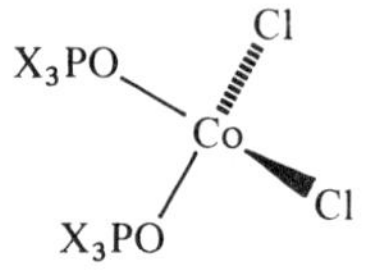

X = Bz; P—O = 1.50(1) Å;
CoOP = 153°

(23)[106]

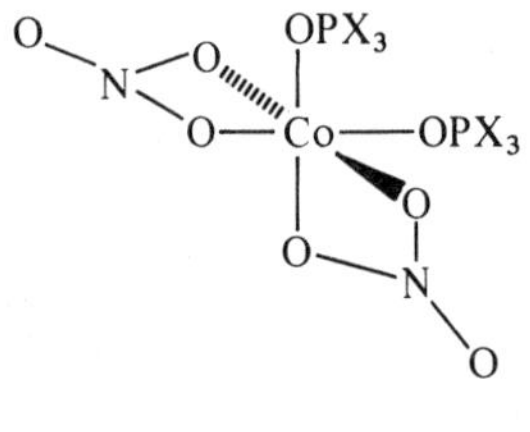

X = Me; P—O = 1.54(3) Å;
CoOP = 133°, 140°

(24)[107]

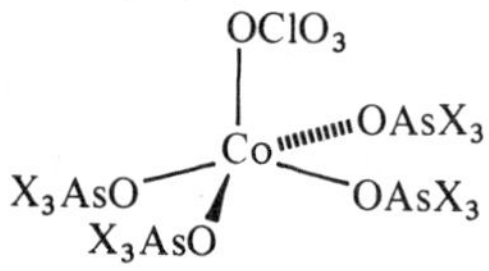

X$_3$ = Ph$_2$Me; As—O = 1.70 Å;
CoOAs = 129°

(25)[108]

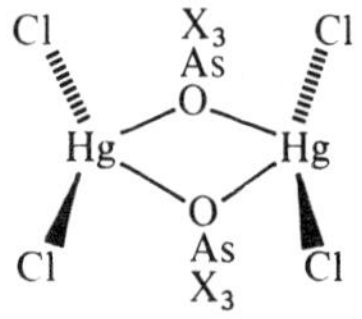

X = Ph; As—O = 1.66(2) Å;
HgOAs = 129°

(26)[109]

Structural formulae (18)–(26) represent some of the types which have been established crystallographically for OPX$_3$ or OAsR$_3$ ligands, all of which bond to the metal through oxygen. Details of P—O or As—O bond lengths and M—O—P or M—O—As bond angles are included, and are average values unless significant differences occur within the same structure. The dimensions for the free ligands are known for OPCl$_3$ [P—O = 1.45(3) Å],[110] OPPh$_3$ [P—O = 1.483(2) Å][111] and OAsPh$_3$ [As—O = 1.644(7) Å],[112] and P—O has been quoted as 1.48 Å for OPMe$_3$ without an indication of its accuracy.[113] Extreme electronic descriptions of the bonding in the ligand are shown in (27)–(30).[114] Bonding to a metal would tend to enhance (27) or (30). This would be expected to lead to longer P—O bond lengths in the complexes. On average there appear to be increases of *ca.* 0.02 Å on complexation but it is difficult to make a definitive judgement because the estimated standard deviations are comparable with such differences. Taking account of the implied arrangements of potential electron-donor pairs in (27)–(30), P—O—M angles of about 109° for contributions from (27) and (30), 120° from (28) and 180° from (29) might be anticipated. Although in a number of the OAsR$_3$ complexes such angles are between 125 and 130°, for the OPR$_3$ systems angles of about 150° are common. In [{Cu(OPEt$_3$)}$_4$OCl$_6$], one of the four CuOP angles is 180°.[115] Because there is no correlation between these angles and other properties, it is not possible to assign particular significance to them; of course, on a purely electrostatic bonding model, 180° angles would be anticipated.

(27)   (28)   (29)   (30)

One consequence of a bent MOP linkage is that the space required to accommodate the ligand around a metal centre is large, and the coordination numbers displayed by metals are often less than shown with other O-donors.

Perchlorate salts of the first transition series metals usually form octahedral complexes with O-donors but not with phosphine oxides. Cr^{III}, Mn^{II}, Fe^{II}, Fe^{III}, Co^{II} and Ni^{II} will only take on four OPPh$_3$. The complexes have been formulated in a variety of ways on the basis of electronic spectra, IR spectra and magnetic measurements. However, a crystal structure determination of the Co^{II} complex with OAsPh$_2$Me (25) showed that one of the perchlorate anions was bonded to the metal to give a square-based pyramidal structure with OClO$_3^-$ at the apex and the metal above the basal plane.[108] X-ray powder photographs established that the Mn^{II}, Fe^{II}, Ni^{II}, Cu^{II} and Zn^{II} analogues were isomorphous with the Co^{II} compound, and that there were UV spectral similarities between the Ni^{II} complex with this ligand and with OPPh$_3$.[116] Thus, it is likely that all these complexes with OPPh$_3$ are isostructural.

The ligands OPMe$_3$ and OAsMe$_3$ require somewhat less space than those with phenyl groups. Mn^{II}, Co^{II} and Ni^{II} react with an excess of these ligands to give [ML$_5$][ClO$_4$]$_2$ complexes,[117] for which the crystal structure of [Ni(OAsMe$_3$)$_5$][ClO$_4$]$_2$[118] shows only non-coordinated [ClO$_4$]$^-$. When a deficiency of OAsMe$_3$ was used, Mn^{II} and Ni^{II} gave [M(OClO$_3$)(OAsMe$_3$)$_4$][ClO$_4$], and this type of complex was the only one formed by Fe^{II} and Cu^{II}. Curiously, the Zn^{II} complex with OAsMe$_3$ was interpreted as having a tetrahedral [Zn(OAsMe$_3$)$_4$]$^{2+}$ cation, in contrast to its OAsPh$_2$Me analogue. That attempts to rationalize the position may be premature is indicated by the isolation of an octahedral [Co(OPMe$_3$)$_6$]$^{2+}$ complex with the [Co(CO)$_4$]$^-$ counterion.

In a comprehensive study of HMPA complexes, de Bolster and Groeneveld[119] obtained adducts of M(BF$_4$)$_2$ with 4HMPA for M = Mg, Ca, Mn, Fe, Co, Ni, Cu, Zn and Cd. Although the spectral data were interpreted on the basis of a tetrahedral structure, the authors do not preclude interaction between the metal and the anionic ligand, and since work on the corresponding perchlorates predates the recognition of coordinated perchlorates in such systems the true position is unclear. Certainly HMPA gives octahedral [M(HMPA)$_6$][ClO$_4$]$_3$ for M = Fe or Cr, in contrast to OPPh$_3$, so it would be surprising if HMPA were more restricting on coordination number on M^{II} centres but less so on M^{III} centres. The bidentate octamethylpyrophosphoramide ligand (a toxic anticholinesterase) achieves six-coordination on Mg^{II}, Co^{II} and Cu^{II}.[120]

With o-phenylene-bis(dimethylarsine oxide), Zn(ClO$_4$)$_2$ forms an adduct with two ligand molecules and achieves five-coordination with O-bridging by the neutral ligand[121] in preference to bonding to ClO$_4^-$.

Complexes of phosphine oxide and related ligands with metal nitrates often incorporate relatively few neutral ligand molecules but achieve higher coordination numbers by using NO$_3^-$ as a chelating bidentate ligand, e.g. six-coordinate [Co(O$_2$NO)$_2$(OPMe$_3$)$_2$] (24), eight-coordinate [UO$_2$(O$_2$NO)$_2$(OPBu$_3^n$)$_2$][122] and ten-coordinate [Ce(O$_2$NO)$_4$(OPPh$_3$)$_2$].[123]

Complexes with metal halides are commonly molecular. Those of the divalent cations of the first transition series are predominantly of the form [MX$_2$L$_2$] and tetrahedral, e.g. (23). FeCl$_3$ and InI$_3$ give complexes of the stoichiometry MX$_3 \cdot$2OPPh$_3$, but these have been shown to contain [MX$_4$]$^-$ ions and so should be formulated as [MX$_2$(OPPh$_3$)$_4$][MX$_4$]. It is surprising, particularly in the case of I$^-$, that the halide can be so readily accommodated. It is less surprising that higher coordination numbers can be achieved with the more slender NCS$^-$ ligand, as shown by [Co(NCS)$_2$(OPMe$_3$)$_4$],

[Ni(NCS)$_2$(OPPh$_3$)$_4$] and [U(NCS)$_4$(OPMe$_3$)$_4$][124] where they are two greater than shown by the corresponding chlorides.

With chlorides of the lanthanides, HMPA forms complexes [LnCl$_3$(HMPA)$_3$] and the structure determination of the praseodymium complex shows the *mer* configuration.[125] With OPPh$_3$ complexes of either type, [LnX$_3$(OPPh$_3$)$_3$] or [LnX$_3$(OPPh$_3$)$_4$] are formed for X = Cl or NCS, and there was no evidence to suggest that X had been displaced from the coordination sphere of the latter type.[126] Although rarely established, halide can be displaced by a P—O donor as shown by [UCl(OPMe$_3$)$_5$]Cl$_3$.[127]

Phosphine oxides are highly tolerant of metal oxidation state; thus Mo complexes *cis*-[Mo(OPPh$_3$)$_2$(CO)$_4$] and [MoO$_2$Cl$_2$(OPPh$_3$)$_2$] (21) are known. The high oxidation state metal halides may be reactive towards OPPh$_3$ through O—Cl exchange: WCl$_6$ furnishes [WO$_2$Cl$_2$(OPPh$_3$)$_2$].

Some of the earliest work on phosphoryl ligands was with OPCl$_3$, which is a poorer donor than ligands with P—C bonds because of the greater electronegativity of Cl. Complexes are formed with the higher oxidation state chlorides, *e.g.* [TiCl$_4$(OPCl$_3$)$_2$], [TiCl$_4$(OPCl$_3$)]$_2$ (19) and [SbCl$_5$(OPCl$_3$)]. Wherever crystallographic characterization has been undertaken the OPCl$_3$ has been shown to remain intact as a ligand. Ionization of many of the complexes with OPCl$_3$ was found to occur in solution and it was postulated that chloride transfer to the metal occurred by ionization of OPCl$_3$ to [OPCl$_2$]$^+$Cl$^-$. However, no conclusive evidence of the [OPCl$_2$]$^+$ ion was forthcoming and it appears that halide transfer as in Scheme 2 is responsible, especially since the same processes occur with triethyl phosphate.[128] For FeCl$_3$ in OPCl$_3$ the predominant species are probably [FeCl$_2$(OPCl$_3$)$_4$]$^+$ and [FeCl$_4$]$^-$.

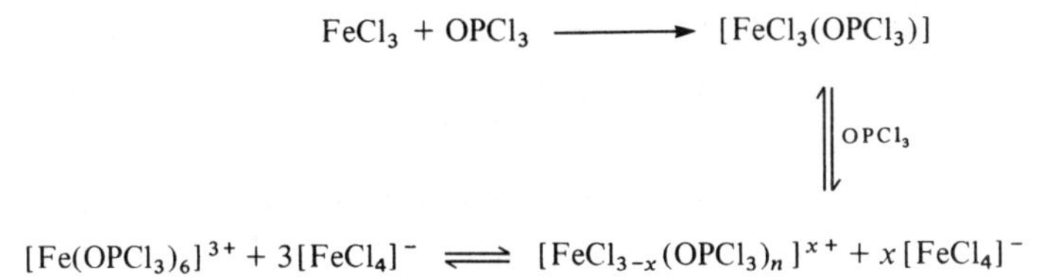

$$FeCl_3 + OPCl_3 \longrightarrow [FeCl_3(OPCl_3)]$$

$$\Big\Updownarrow OPCl_3$$

$$[Fe(OPCl_3)_6]^{3+} + 3[FeCl_4]^- \rightleftharpoons [FeCl_{3-x}(OPCl_3)_n]^{x+} + x[FeCl_4]^-$$

Scheme 2

Many complexes of trialkyl phosphates and the phosphonates OP(OR)$_2$R have been isolated, especially with shorter alkyl chains. The dependence of the type of complex formed on the steric demands of the ligand is illustrated by the formulations proposed for complexes of Co(ClO$_4$)$_2$: [Co(OClO$_3$)(OH$_2$){OP(OBu^n)$_3$}$_4$][ClO$_4$], [Co(OH$_2$)$_2${OPMe(OPr^i)$_2$}$_4$][ClO$_4$]$_2$, [Co{OP(OMe)$_3$}$_5$]-[ClO$_4$]$_2$ and [Co(OClO$_3$){OPMe(OPr^i)$_2$}$_4$][ClO$_4$].

One of the main methods of detecting interaction between these ligands and a metal has been the lowering of P—O or As—O stretching frequencies on complexation.[94] Examples are given in Table 5. The bands are usually of high intensity and rather broad, at least for solids, so that it is not generally possible to deduce structures from the multiplicity of these bands. For HMPA the lowering of wavenumber is rather small in most cases,[119] about a third of that seen for analogous OPPh$_3$ complexes, and there is no quantitative correlation between the trends for a series of metal centres with different ligands.

Table 5 P=O and As=O Stretching Wavenumbers in Ligands and their Complexes M(ClO$_4$)$_2$·4L

Ligand	OPPh$_3$	HMPA	OAsPh$_3$	OAsMe$_3$
Free ligand	1195	1208	880	870
M = Mn	1152	1190	873	868, 855
Fe	1147	1185	862	—
Co	1146	1190	860	861
Ni	1143	1193	863	866
Cu	1131	1191	844	854, 833
Zn	1156	1185	—	—
Ref.	129, 94	94	129	117

There is a sharp discontinuity between the extent of frequency lowering on lighter metal atom centres and those commonly observed for thorium and beyond where the P—O wavenumber is lowered by more than 100 cm^{-1}. The most spectacular reduction is seen for complexes [UX$_5$L]

(X = Cl or Br) where it exceeds $200 \, cm^{-1}$. For $[NbCl_5(OPPh_3)]$, $\nu(P\!-\!O)$ is $999 \, cm^{-1}$, but it is not just a result of the high oxidation state since in $[NbOCl_3(OPPh_3)_2]$ the one $P\!-\!O$ band observed is at $1145 \, cm^{-1}$.

Several works have assigned $M\!-\!O$ stretching frequencies with these ligands,[117,119] typically in the range $300\text{--}450 \, cm^{-1}$, but to what extent these are pure stretching coordinates has not been established.

On the basis of Co^{II} complexes assumed to be tetrahedral, de Bolster and Groeneveld[119] have related the position of a range of ligands in the spectrochemical series as

$$I^- < Br^- < Cl^- < HMPA \approx SP(NMe_2)_3 \approx OPPh_3 < OPMe_3 < SPMe_3 \approx N_3^- < NCO^- < NCS^-$$

and in the nephelauxetic series as

$$SPMe_3 \approx SP(NMe_2)_3 < N_3^- \approx I^- < NCS^- \approx Br^- \approx Cl^- < NCO^- < OPPh_3 < OPMe_3 \approx HMPA.$$

Although ^{31}P NMR might seem an attractive method of establishing the structures of the complexes, its usefulness is limited. Even with diamagnetic systems, fast ligand exchange usually gives time-averaged spectra at ambient temperatures.[130] In some instances the exchange can be halted at low temperatures. At $200 \, K$, CD_2Cl_2 solutions of $Mg(ClO_4)_2$ with 4–5 $OPPh_3$ show separate singlet signals for $[Mg(OPPh_3)_4]^{2+}$ and $[Mg(OPPh_3)_5]^{2+}$; the latter must still be internally dynamic as it must have more than one phosphorus environment.[131]

The rates of the exchange processes depend on the ligand. Systems containing $Cd(SbF_6)_2$ and phosphine oxides at $P\!:\!Cd > 6\!:\!1$ show that intermediate rates of exchange persist for $OPPh_3$ at $220 \, K$,[132] whereas the bulkier $OPCy_3$ gives $[Cd(OPCy_3)_4]^{2+}$, which displays $^2J(^{113}Cd\!-\!^{31}P)$ coupling even at room temperature.[133]

Undoubtedly a major cause of interest in ligands containing the phosphoryl group has been their usefulness as solvent extraction agents, and this has been reviewed by Peppard.[134] For separation of the actinides, $OP(OBu^n)_3$ has been widely used. Extraction by $OP(OBu^n)_3$ from aqueous phases occurs for a range of metals including Co^{II} and the lanthanides. Conditions have a crucial effect on extraction efficiency, and counter ions such as NO_3^-, Cl^- and SCN^- are often constituent ligands of the complex extracted.

The phosphorus compound may be used as a neat liquid or as a phase transfer agent between water and a hydrocarbon or other immiscible solvent. For a given alkyl group R, extracting efficiency follows the order: $OPR_3 > OP(OR)R_2 > OP(OR)_2R > OP(OR)_3$. Amongst the phoshine oxides, tri-n-octylphosphine oxide has been the most widely used and research into its use continues.[135]

15.8.5 REFERENCES

1. I. Beger and H. G. Hauthal, in 'Dimethyl Sulfoxid', ed. D. Martin and H. G. Hauthal, Akademie-Verlag, Berlin, 1971, chap. 12 (translated E. S. Halberstadt, Van Nostrand Reinhold, Wokingham, 1975).
2. L. Bretherick and G. D. Muir, 'Hazards in the Chemical Laboratory', 3rd edn., ed. L. Bretherick, The Royal Society of Chemistry, London, 1981, p. 307.
3. M. Steinbrecher, in ref. 1, chap. 5.
4. W. L. Reynolds, *Prog. Inorg. Chem.*, 1970, **12**, 1.
5. J. A. Davies, *Adv. Inorg. Chem. Radiochem.*, 1981, **24**, 115.
6. F. A. Cotton, S. J. Lippard and J. T. Mague, *Inorg. Chem.*, 1965, **4**, 508.
7. A. P. Zipp and S. K. Madan, *Inorg. Chim. Acta*, 1974, **22**, 49.
8. W. J. McDowell and H. D. Harmon, *J. Inorg. Nucl. Chem.*, 1971, **33**, 3107.
9. V. Krishnan and C. C. Patel, *J. Inorg. Nucl. Chem.*, 1964, **26**, 2201.
10. V. V. Savant and C. C. Patel, *J. Inorg. Nucl. Chem.*, 1969, **31**, 2319.
11. J. R. Behrendt and S. K. Madan, *J. Inorg. Nucl. Chem.*, 1977, **39**, 449.
12. V. K. L. Osario and A. M. P. Felicíssimo, *Inorg. Chim. Acta*, 1976, **19**, 245.
13. P.-K. F. Chin and F. R. Hartley, *Inorg. Chem.*, 1976, **15**, 982.
14. D. W. Meek, D. K. Straub and R. S. Drago, *J. Am. Chem. Soc.*, 1960, **82**, 6013.
15. M. J. Bennett, F. A. Cotton and D. L. Weaver, *Acta Crystallogr.*, 1967, **23**, 581.
16. A. J. Carty, H. A. Patel and P. M. Boorman, *Can. J. Chem.*, 1970, **48**, 492.
17. D. B. Copley, F. Fairbrother, K. H. Grundy and A. Thompson, *J. Less-Common Met.*, 1964, **6**, 407.
18. H. von Dreizler and G. Dendl, *Z. Naturforsch., Teil A*, 1964, **19**, 512.
19. M. A. Viswamitra and K. K. Kannan, *Nature (London)*, 1966, **209**, 1016.
20. R. C. Thomas, C. B. Shoemaker and K. Eriks, *Acta Crystallogr.*, 1966, **21**, 12.
21. R. D. Willet and K. Chang, *Inorg. Chim. Acta*, 1970, **4**, 447.
22. M. J. Bennett, F. A. Cotton, D. L. Weaver, R. J. Williams and W. H. Watson, *Acta Crystallogr.*, 1967, **23**, 788.
23. M. Sandström, *Acta Chem. Scand., Ser. A*, 1978, **32**, 527.
24. J. R. Behrendt and S. K. Madan, *J. Inorg. Nucl. Chem.*, 1976, **38**, 1827.

25. F. A. Cotton, R. Francis and W. D. Horrocks, Jr., *J. Phys. Chem.*, 1960, **64**, 1534.
26. M.-T. Forel and M. Tranquille, *Spectrochim. Acta, Part A*, 1970, **26**, 1023.
27. J. H. Price, A. N. Williamson, R. F. Schramm and B. B. Wayland, *Inorg. Chem.*, 1972, **11**, 1280.
28. K. W. Bagnall, D. Brown, D. H. Holah and F. Lux, *J. Chem. Soc. (A)*, 1968, 465.
29. R. A. Potts, *Inorg. Chem.*, 1970, **9**, 1284.
30. J. R. Barnes, P. L. Goggin and R. J. Goodfellow, *J. Chem. Res.*, 1979, S118, M1610.
31. F. A. Cotton and T. R. Felthouse, *Inorg. Chem.*, 1980, **19**, 2347.
32. M. Cinquini, S. Colonna and R. Giovini, *Chem. Ind. (London)*, 1969, 1737.
33. R. Paetzold and G. Bochmann, *Z. Anorg. Allg. Chem.*, 1969, **368**, 202.
34. R. Paetzold and G. Bochmann, *Z. Anorg. Allg. Chem.*, 1971, **385**, 256.
35. K. A. Jensen and V. Krishnan, *Acta Chem. Scand.*, 1967, **21**, 1988.
36. R. Paetzold and P. Vordank, *Z. Anorg. Allg. Chem.*, 1966, **347**, 294.
37. H.-P. Sieper and R. Paetzold, *Z. Phys. Chem. (Leipzig)*, 1974, **255**, 1125.
38. J. W. Vaughan, 'The Chemistry of Non-Aqueous Solvents', ed. J. J. Lagowski, Academic, New York, 1967, vol. 2, chap. 5.
39. G. Kaufmann and M. J. F. Leroy, *Bull. Soc. Chim. Fr.*, 1967, 402.
40. R. A. Mackay and E. J. Poziomek, *J. Chem. Eng. Data*, 1969, **14**, 271.
41. W. E. Bull, S. K. Madan and J. E. Willis, *Inorg. Chem.*, 1963, **2**, 303.
42. G. Constant, J. C. Daran and Y. Jeannin, *J. Inorg. Nucl. Chem.*, 1971, **33**, 4209.
43. B. Claudel, J. P. Puaux, D. Rehorek, and H. Sautereau, *J. Solid State Chem.*, 1981, **39**, 181.
44. R. S. Drago, D. W. Meek, M. D. Joesten and L. LaRoche, *Inorg. Chem.*, 1963, **2**, 124.
45. J. H. Bright, R. S. Drago, D. M. Hart and S. K. Madan, *Inorg. Chem.*, 1965, **4**, 18.
46. S. S. Krishnamurthy and S. Soundararajan, *Can. J. Chem.*, 1969, **47**, 995.
47. T. Moeller and G. Vicentini, *J. Inorg. Nucl. Chem.*, 1965, **27**, 1477.
48. S. S. Krishnamurthy and S. Soundararajan, *J. Inorg. Nucl. Chem.*, 1966, **28**, 1689.
49. G. Vicentini and R. Najjar, *J. Inorg. Nucl. Chem.*, 1968, **30**, 2771.
50. J. M. Gioria and B. P. Susz, *Helv. Chim. Acta*, 1971, **54**, 2251.
51. P.-C. Kong and F. D. Rochon, *Can. J. Chem.*, 1979, **57**, 682.
52. S. Durand, G. Jugie and J.-P. Laurent, *Transition Met. Chem.*, 1982, **7**, 310.
53. S. J. Kuhn and J. S. McIntyre, *Can. J. Chem.*, 1965, **43**, 375.
54. R. L. Middaugh, R. S. Drago and R. J. Neidzielski, *J. Am. Chem. Soc.*, 1964, **86**, 388.
55. J. C. Speakman, M. S. Lehmann, J. R. Allibon and D. Semmingsen, *Acta. Crystallogr., Sect. B*, 1981, **37**, 2098.
56. E. Spinner, *J. Chem. Soc., Perkin Trans. 2*, 1980, 395.
57. E. M. Holt, N. W. Alcock, R. H. Sumner and R. O. Asplund, *Cryst. Struct. Commun.*, 1979, **8**, 255.
58. M. E. Stone, B. E. Robertson and E. Stanley, *J. Chem. Soc. (A)*, 1971, 3632.
59. E. Lindner, B. Perdikatsis and A. Thasitis, *Z. Anorg. Allg. Chem.*, 1973, **402**, 67.
60. P. Chakrabarti, K. Venkatesan and C. N. R. Rao, *Proc. R. Soc. London, Ser. A*, 1981, **375**, 127.
61. R. Fussenegger, P. Peringer and B. M. Rode, *Monatsh. Chem.*, 1977, **108**, 265.
62. R. C. Paul, A. K. Moudgil, S. L. Chadha and S. K. Vasisht, *Indian J. Chem.*, 1970, **8**, 1017.
63. D. L. Pisaniello and S. F. Lincoln, *Inorg. Chem.*, 1981, **20**, 3689.
64. S. F. Lincoln and M. N. Tkaczuk, *Ber. Bunsenges. Phys. Chem.*, 1982, **86**, 221.
65. S. F. Lincoln, A. Ekstrom, and G. J. Honan, *Aust. J. Chem.*, 1982, **35**, 2385.
66. U. Meyer and V. Gutmann, *Struct. Bonding (Berlin)*, 1972, **12**, 113.
67. R. G. Garvey, J. H. Nelson and R. O. Ragsdale, *Coord. Chem. Rev.*, 1968, **3**, 375.
68. N. M. Karayannis, L. L. Pytlewski and C. M. Mikulski, *Coord. Chem. Rev.*, 1973, **11**, 93.
69. N. M. Karayannis, A. N. Speca, D. E. Chasan and L. L. Pytlewski, *Coord. Chem. Rev.*, 1976, **20**, 37.
70. S. Kida, *Bull. Chem. Soc. Jpn.*, 1963, **36**, 712.
71. J. H. Nelson, R. G. Garvey and R. O. Ragsdale, *J. Heterocycl. Chem.*, 1967, **4**, 591.
72. E. Ochiai, *J. Org. Chem.*, 1953, **18**, 534.
73. R. O. Gardner and R. O. Ragsdale, *Inorg. Chim. Acta*, 1968, **2**, 139.
74. C. J. O'Connor, E. Sinn and R. L. Carlin, *Inorg. Chem.*, 1977, **16**, 3314.
75. D. Taylor, *Aust. J. Chem.*, 1978, **31**, 713.
76. D. L. Kepert, D. Taylor and A. H. White, *J. Chem. Soc., Dalton Trans.*, 1973, 670.
77. J. S. Wood, R. O. Day, C. P. Keijzers, E de Boer, A. E. Yildirim and A. A. K. Klaassen, *Inorg. Chem.*, 1981, **20**, 1982.
78. J. D. Lee, D. S. Brown and B. G. A. Melsom, *Acta Crystallogr., Sect. B*, 1969, **25**, 1378.
79. A. R. Al-Karaghouli and J. S. Wood, *Inorg. Chem.*, 1979, **18**, 1177.
80. I. Bertini, P. Dapporto, G. Gatteschi and A. Scozzafava, *Inorg. Chem.*, 1975, **14**, 1639.
81. G. Sawitzki and H. G. von Schnering, *Chem. Ber.*, 1974, **107**, 3266.
82. D. R. Johnson and W. H. Watson, *Inorg. Chem.*, 1971, **10**, 1068.
83. J. C. Morrow, *J. Cryst. Mol. Struct.*, 1974, **4**, 243.
84. D. J. Hodgson, *Prog. Inorg. Chem.*, 1975, **19**, 173.
85. R. L. Carlin, R. Burriel, R. M. Cornelisse and A. J. van Duyneveldt, *Inorg. Chem.*, 1983, **22**, 831.
86. M. Orchin and P. J. Schmidt, *Coord. Chem. Rev.*, 1968, **3**, 345.
87. P. A. Vigato, U. Casellato and M. Vidali, *Gazz. Chim. Ital.*, 1977, **107**, 61.
88. D. W. Cunningham and M. O. Workman, *J. Inorg. Nucl. Chem.*, 1971, **33**, 3861.
89. D. W. Herlocker, *Inorg. Chem.*, 1969, **8**, 2037.
90. R. A. Geanagel, *J. Inorg. Nucl. Chem.*, 1974, **36**, 1397.
91. P. O. Nubel, S. R. Wilson and T. L. Brown, *Organometallics*, 1983, **2**, 515.
92. S. H. Hunter, V. M. Langford, G. A. Rodley and C. J. Wilkins, *J. Chem. Soc. (A)*, 1968, 305.
93. C. A. Wamser, W. B. Fox, B. Sakornik, J. R. Holmes, B. B. Stewart, R. Juurnik, N. Vanderkool and D. Gould, *Inorg. Chem.*, 1969, **8**, 1249.
94. N. M. Karayannis, C. M. Mikulski and L. L. Pytlewski, *Inorg. Chim. Acta Rev.*, 1971, **5**, 69.
95. R. L. Shriner and C. N. Wolf, *Org. Synth.*, 1950, **30**, 97.

96. M. L. Denniston and D. R. Martin, *Inorg. Synth.*, 1977, **17**, 183.
97. D. B. Copley, F. Fairbrother, J. R. Miller, and A. Thompson, *Proc. Chem. Soc.*, 1964, 300.
98. G. A. Olah, B. G. B. Gupta, A. Garcia-Luna, and S. C. Narang, *J. Org. Chem.*, 1983, **48**, 1760.
99. A. R. Hands and A. J. H. Mercer, *J. Chem. Soc. (A)*, 1968, 449.
100. R. A. Potts, *Inorg. Chem.*, 1970, **9**, 1284.
101. C.-I. Brändén, *Acta Chem. Scand.*, 1963, **17**, 759.
102. C.-I. Brändén and I. Lindqvist, *Acta. Chem. Scand.*, 1960, **14**, 726.
103. M. R. Caira and B. J. Gellatly, *Acta Crystallogr., Sect. B*, 1980, **36**, 1198.
104. R. J. Butcher, B. R. Penfold and E. Sinn, *J. Chem. Soc., Dalton Trans.*, 1979, 668.
105. L. G. Hubert-Pfalzgraf, and A. A. Pinkerton, *Inorg. Chem.*, 1977, **16**, 1895.
106. R. H. de Almeida Santos and Y. Mascarenhas, *J. Coord. Chem.*, 1979, **9**, 59.
107. F. A. Cotton and R. H. Soderberg, *J. Am. Chem. Soc.*, 1963, **85**, 2402.
108. P. Pauling, G. B. Robertson and G. A. Rodley, *Nature (London)*, 1965, **207**, 73.
109. C.-I. Brändén, *Ark. Kemi*, 1964, **22**, 485.
110. Q. Williams, J. Sheridan, and W. Gordy, *J. Chem. Phys.*, 1952, **20**, 164.
111. G. Ruban and V. Zabel, *Cryst. Struct. Commun.*, 1976, **5**, 671.
112. G. Fergusson and E. W. Macauley, *J. Chem. Soc. (A)*, 1969, 1.
113. C.-I. Brändén and I. Lindquist, *Acta Chem. Scand.*, 1961, **15**, 167.
114. J. R. Cox and O. B. Ramsay, *Chem. Rev.*, 1964, **64**, 317.
115. M. R. Churchill, B. G. de Boer and S. J. Mendak, *Inorg. Chem.*, 1975, **14**, 2496.
116. J. S. Lewis, R. S. Nyholm and G. A. Rodley, *Nature (London)*, 1965, **207**, 72.
117. A. M. Brodie, S. H. Hunter, G. A. Rodley and C. J. Wilkins, *Inorg. Chim. Acta*, 1968, **2**, 195.
118. Y. S. Ng, G. A. Rodley and W. T. Robinson, *Inorg. Chem.*, 1976, **15**, 303.
119. M. W. G. de Bolster and W. L. Groeneveld, *Recl. Trav. Chim. Pays-Bas*, 1971, **90**, 477.
120. M. D. Joesten, M. S. Hussain and P. G. Lenhert, *Inorg. Chem.*, 1970, **9**, 151.
121. S. H. Hunter, G. A. Rodley and K. Emerson, *Inorg. Nucl. Chem. Lett.*, 1976, **12**, 113.
122. J. H. Burns, *Inorg. Chem.*, 1981, **20**, 3868.
123. Mazhar-Ul-Haque, C. N. Caughlan, F. A. Hart and R. van Nice, *Inorg. Chem.*, 1971, **10**, 115.
124. C. E. F. Rickard and D. C. Woollard, *Aust. J. Chem.*, 1979, **32**, 2181.
125. L. J. Radonovich and M. D. Glick, *J. Inorg. Nucl. Chem.*, 1973, **35**, 2745.
126. D. R. Cousins and F. A. Hart, *J. Inorg. Nucl. Chem.*, 1968, **30**, 3009.
127. G. Bombieri, E. Forsellini, D. Brown and B. Whittaker, *J. Chem. Soc., Dalton Trans.*, 1976, 735.
128. D. W. Meek, in 'The Chemistry of Non-Aqueous Solvents' ed. J. J. Lagowski, Academic, New York, 1966, vol. 1, chap. 1.
129. S. H. Hunter, R. S. Nyholm and G. A. Rodley, *Inorg. Chim. Acta*, 1969, **3**, 631.
130. S. O. Grim and L. C. Satek, *J. Coord. Chem.*, 1976, **6**, 39.
131. S. F. Lincoln, D. L. Pisaniello, T. M. Spotswood and M. N. Tkaczuk, *Aust. J. Chem.*, 1981, **34**, 283.
132. P. A. W. Dean and M. K. Hughes, *Can. J. Chem.*, 1980, **58**, 180.
133. P. A. W. Dean, *Can. J. Chem.*, 1981, **59**, 3221.
134. D. F. Peppard, *Adv. Inorg. Chem. Radiochem.*, 1966, **9**, 1.
135. S. Kusakabe and T. Sekine, *Bull. Chem. Soc. Jpn.*, 1981, **54**, 2930.

15.9
Hydroxamates, Cupferron and Related ligands

RAM C. MEHROTRA
University of Rajasthan, Jaipur, India

15.9.1 HYDROXAMATES

Hydroxamic acids* have been widely used as colorimetric[1-5] and gravimetric[6-8] reagents. Attempts have been made more recently to isolate and actually characterize hydroxamates of transition[9-14] and main group[15-24] elements.

Hydroxamic acid exists in two tautomeric forms, (1) and (2), and such keto–enol tautomerism provides a number of sites for coordination and chelation. The keto form (1) predominates in acid media and the enol form (2) in alkaline media;[25] this has been corroborated by the extraction of vanadium benzohydroxamic acid complexes in organic solvents.[26]

(1)   (2)   (3)   (4)

The preference of hydroxamic acids to form metal complexes through the hydroxamide functional group (1) and not through the hydroxyoxime structure (2) is confirmed by IR,[27,28] UV,[29] ESR[30] and NMR[28] spectral studies. Hydroxamic acids bind metal ions to form complexes with structure (3) rather than structure (4), for which evidence has been claimed[31] in one case only.

The only X-ray diffraction study[32] reported is that of iron(III) benzohydroxamate dihydrate, in which normal octahedral coordination of the iron atom by the oxygen atoms of the ligand was observed. Interestingly, the past two decades have witnessed a marked increase in our knowledge of naturally occurring hydroxamic acids (*cf.* Chapter 22), where they function variously as growth factors, antibiotics, antibiotic antagonists, tumour inhibitors and cell division factors. Their biological activity seems to be especially related to their ability to chelate iron specifically.[33]

The overall stability constants of several metals in solution have been determined by many workers[34] employing Bjerrum's technique, pH titration and distribution and potentiometric methods. It has been shown by Irving and Williams,[35] as well as by Mellor and Maley,[36] that the stability of chelates of bivalent $3d$ metals increases regularly from Mn^{2+} to Cu^{2+} and decreases from

---

*It may be mentioned that an alternative name for hydroxamic acids, treating them as derivatives of hydroxylamine, continues to be used. For example *N*-phenylbenzohydroxamic acid [PhCON(Ph)OH] has also been named alternatively as *N*-benzoyl-*N*-phenylhydroxylamine even in recent literature.

Cu^{2+} to Zn^{2+}, irrespective of the nature of the ligand used. The stability constants of a few divalent metals with N-aryl o-substituted phenylhydroxamic acids are tabulated[37] in Table 1.

Table 1 Stability Constants of Metal Complexes of N-Aryl o-Substituted Phenylhydroxamine Acids[a]

o-Substituted group	Cu^{II}		Zn^{II}		Ni^{II}		Mn^{II}	
	$\log K_1$	$\log K_2$	$\log K_1$	$\log K_2$	$\log K_1$	$\log K_2$	$\log K_1$	$\log K_2$
Me	10.25	8.56	7.86	6.21	7.82	7.77	6.46	4.77
OMe	10.35	8.68	8.63	6.65	5.86	6.66	6.80	5.67
F	10.03	8.08	7.40	5.69	7.36	5.54	6.70	4.88
Cl	9.96	8.04	7.52	5.92	7.23	5.45	6.40	4.93
Br	9.90	7.97	7.49	5.87	7.21	5.33	6.07	4.76
I	9.96	8.12	7.40	6.64	7.23	5.30	6.02	4.68
NO_2	9.45	7.49	7.12	5.44	7.07	5.04	5.74	4.38
H	10.29	8.73	7.71	6.63	7.20	—	—	—

[a] At 35 °C in 50% (v/v) dioxane–water.

The extensive use of hydroxamic acids as metal precipitants can be illustrated by the work of Shome and coworkers on N-phenylbenzohydroxamic acid for precipitation of beryllium,[38] gallium,[39,40] indium,[39,40] lanthanum,[41] molybdenum(V),[42] thorium,[43] titanium,[44] uranium(VI)[45] and tungsten.[46] Generating the reagent by slow hydrolysis of O-acetyl-N-benzoyl-N-phenylhydroxylamine, PhCON(Ph)OCOMe, a procedure for homogeneous precipitation from solution has also been suggested.[47] The use of other hydroxamic acids can also be illustrated by the use of N-(o-tolyl)benzohydroxamic acid[48] for precipitation of niobium in the presence of tantalum,[49] salicylhydroxamic[49] acid for estimation of aluminum, gallium and indium, N-phenylacetohydroxamic acid[50] for separation of titanium and zirconium form niobium, 2-thiophenecarbohydroxamic acid[51] for vanadium and quinaldohydroxamic acid[52] and benzohydroxamic acid[53] for estimation of niobium and tantalum.

15.9.1.1 Metal hydroxamates

As the literature published up to 1976 on hydroxamates, with special reference to transition metals, has been well reviewed,[37] this account will mainly concentrate on adding a few subsequent findings in this direction along with a brief account of the hydroxamic acid derivatives of main group elements.

N-Phenylbenzohydroxamic acid (LH) derivatives of the types $TiL_2(OR)_2$ and TiL_2Cl_2 have been synthesized[54] by the reactions of the ligand with titanium alkoxides and chlorides, respectively.

Formation constants of $3d$ metal ions[55] with N-m-tolyl-p-substituted benzohydroxamic acids and of rare earths[56] with thenoylhydroxamic acid have been determined. Formation constants of proton and metal complexes of N-phenyl-2-thenoyl- and N-p-tolyl-2-thenoyl-hydroxamic acids[57] have also been determined. In addition, study has been made of the mixed ligand complexes involving nicotino- and isonicotino-hydroxamic acids.[58] A method of extraction and spectrophotometric determination of vanadium with chlorophenylmethylbenzohydroxamic acid has also been published.[59] It may be mentioned that hydroxamic acids (in particular, the N-phenylbenzohydroxamic acid) have been widely used[60,61] as analytical reagents for metal ions. Solvent extraction of titanium by benzo- or salicyl-hydroxamic acid in the presence of trioctylamine in the form of coloured complexes has been reported.[62] N-m-Tolyl-p-methoxybenzohydroxamic acid has been used[63] for extraction and spectrophotometric determination of Mo^{VI} and W^{VI} from hydrochloric acid media containing thiocyanate.

A large volume of work[64] has been published on the determination of stability constants for complexes of hydroxamic acids, *e.g.* acetohydroxamic acid.[65] The stability of $3d$ transition metal ions (Mn^{2+} to Zn^{2+}) with salicylhydroxamic and 5-methyl-, 5-chloro-, 5-bromo-, 5-nitro-, 4-chloro-, 4-bromo- and 3-chloro-salicylhydroxamic acids,[66] as well as with methyltolylbenzohydroxamic acid,[67] has been studied potentiometrically. Stability constants of iron(III) with a number of hydroxamic acids have been determined by redox potential studies.[68]

Following their earlier work[69] on iron(III) acetohydroxamate as a suitable iron source in iron-deficient anaemia, Brown *et al.*[70] have carried out a detailed comparative structural and spectroscopic study of the complexes of Fe^{III}, Co^{II}, Ni^{II} and Cu^{II} with aceto-, propiono-, benzo-, N-methyl-benzo- and *N*-phenylbenzo-hydroxamic acids. Spectral and magnetic properties of the complexes of Fe^{III}, Co^{II} and Ni^{II} indicate octahedral coordination, with the latter two metal ions forming polymeric species. These monohydroxamic acid complexes show slightly larger values of 10Dq compared with the corresponding aqua and acetylacetonato complexes; Cu^{II} has been shown to form a square planar complex, probably with a d_{xy} ground state.

In view of the earlier assignments[71] for the IR spectra in these systems being purely on an empirical basis, Brown and coworkers[70] have carried out a normal coordinate treatment of *N*-methyl-hydroxamates of copper, iron and nickel employing a 1:1 ligand model and a Urey–Bradley force field, treating the methyl group as a point mass. Very satisfactory agreements of observed and calculated IR frequencies were obtained for copper(II) and iron(III) complexes; the metal–oxygen bonds are equivalent, having force constants of approximately $1.08 \times 10^{-8}\,\text{N\,Å}$. Delocalization does occur over the chelate systems, resulting in significant double bond character of the C—N bond and to a lesser extent of the N—O bond. The spectrum of the nickel complex showed significant differences, probably due to the complex being an octahedrally coordinated oxygen bridged tetramer.

Bhattacharya and Dhar[72] have recently described adducts of bases like pyridine, 2-, 3- and 4-methylpyridine, 2,4- and 2,6-butadienes, quinoline, aniline and piperidine with hydroxamic acids (R'CONROH, where R = Ph, *o*-MeC$_6$H$_4$, *o*-NO$_2$C$_6$H$_4$, *o*-ClC$_6$H$_4$; R' = Ph, *o*-MeC$_6$H$_4$), investigating the effect of steric factors, electronic effects, base strength and concentration in the stabilization of the adducts.

Following their earlier work[13] on the synthesis of oxalohydroxamates of rare earths, Brahma and Chakraburtty[73] have also reported the synthesis of some lanthanon/salicylhydroxamate complexes of the composition Ln(O-C$_6$H$_4$CONHOH)$_3$·2H$_2$O (where Ln = La, Ce, Nd, Sm or Eu). La, Ce and Sm complexes tend to lose both molecules of water in the temperature range 90–160 °C with endothermic peaks at 120, 110 and 115 °C, respectively.

Kundu and Bhattacharya[14] have isolated dioxouranium complexes of benzohydroxamic acid with the compositions M[UO$_2$(C$_7$H$_6$O$_2$N)$_3$] [where M = Li, Na, K, Cs, Tl, N$_4$, pyH$^+$ (pyridinium) or agH$^+$ (aminoguanidinium)] and M'[UO$_2$(C$_7$H$_6$O$_2$N)$_3$]$_2$ [where M' = enH$_2^{2+}$ (ethylenediammonium)]. All the complexes, with the exception of the sodium compound, are insoluble in common organic solvents but are soluble in DMSO and DMF. The complexes have been characterized on the basis of electronic, IR and molar conductance data in DMF. Their fairly stable character is indicated by thermogravimetric analysis and the stability order is: NH$_4^+$ < Tl$^+$ < Cs$^+$ < Li$^+$ ≈ Na$^+$ ≈ agH$^+$ < K$^+$ ≈ pyH$^+$ < enH$_2^{2+}$.

Pal and Kapoor[74] have studied the reactions of isopropoxides of aluminum, titanium and zirconium with benzo- and phenylaceto-hydroxamic acids in anhydrous benzene. Solid products of the types Al(OPr^i)$_{3-n}$L$_n$ and M(OPr^i)$_{4-n}$L$_n$ (where M = Ti or Zr and L is the hydroxamic acid) have been isolated; all the aluminum and zirconium products are white in colour whereas titanium ones are yellow. The mixed isopropoxide hydroxamates interchange their isopropoxy group with *t*-butoxy groups, yielding *t*-butoxide products.

Carboxylic acids (acetic, halo substituted acetic and benzoic acids, HA) have been shown[75] to interact with *N*-phenylbenzohydroxamates of copper(II), nickel(II) and cobalt(II) with the formation of adducts with the formula M(LL)$_2$(HA)$_2$ (where LL is the anion of hydroxamic acid).

A few complexes of silicon of the type (SiL$_3$)$^+$HCl$_2^-$ (LH = RCONR'OH,[15,16] where R = Ph, R' = *o*-MeC$_6$H$_4$, *o*-EtOC$_6$H$_4$, *p*-MeOC$_6$H$_4$; R = C$_4$H$_3$O, R' = Ph; R = C$_4$H$_3$OCH=CH, R' = Ph; R = *p*-MeOC$_6$H$_4$CH=CH, R' = *o*-MeOC$_6$H$_4$, *p*-MeOC$_6$H$_4$) have been described by the reaction of silicon tetrachloride with the respective hydroxamic acid (*N*-substituted hydroxylamines). These have been characterized on the basis of UV and IR spectra as well as conductivity measurements.

Following the synthesis[17] of a neutral complex, PhCON(Ph)OSiMe$_3$, Narula and Gupta[18] have described the synthesis of a variety of hydroxamic acid derivatives, *e.g.* mono(triorganosilyl)-, mono(diorganochlorosilyl)- and bis(trimethylsilyl)-hydroxamates. These products have been characterized on the basis of IR and NMR spectra.

A few *O*-trialkyltin hydroxylamines (hydroxamates), R$_3$SnONHCOPh (R = Me, Pr^n) and R$_3$SnON(Ph)COPh (R = Me, Pr^n, Ph), and organosilicon and organolead analogues, Me$_3$SiON(Ph)COPh and Ph$_3$PbON(Ph)COPh, have been synthesized.[19] Dissolution of Me$_3$SnONHCOPh in triethylamine leads to an ionic product, [NEt$_3$H]$^+$[Me$_3\overline{\text{SnONCPhO}}]^-$, the triphenyl analogue of which is obtained by the reaction of Ph$_3$SnCl with the hydroxamic acid in the presence of excess triethylamine. The structures of the derivatives both in solution as well as in solid state have been discussed on the basis of their IR, NMR, ^{119}Sn Mössbauer and mass spectra.

King and Harrison[20] have determined the crystal and molecular structure of triphenyltin-*N*-

benzoyl-*N*-phenylhydroxamate, which has been shown to possess a trigonal bipyramidal arrangement of groups about tin, to which the carbonyl oxygen is intramolecularly coordinated. This has been followed by the elucidation[20] of the structure of bis(*N*-methyl-*N*-acetylhydroxamato)dimethyltin. The geometry of tin has been shown to be distorted octahedral, the overall symmetry approximating to C_{2v}. The two *N*-acetylhydroxylamine residues function as bidentate ligands, forming one short covalent and one long coordinate bond to tin, whilst the Me—Sn—Me group is not linear, the C—Sn—C bond angle being 145.8°. Bond distances within the two ligand residues indicate significant multiple bond character for the C—N bonds and single bond zwitterionic charater for the C=O bonds.

In the X-ray crystal structure of bis(*N*-benzoyl-*N*-phenylhydroxylaminato)dichlorotin(IV), $[Cl_2Sn\{ONPh(COPh)\}_2]$, the tin atom has been shown[21] to be coordinated in a distorted octahedral fashion, with two chlorine atoms occupying *cis* positions [Sn—Cl, 2.361(5) Å; Cl—Sn—Cl, 99.0(2)°] and two *N*-benzoyl-*N*-phenylhydroxylaminato residues chelating tin almost symmetrically [Sn—O, 2.04(2), 2.11(2) Å].

The crystal structure of $Me_2Sn(ONHCOMe)_2$ (I) and its monohydrate (II) have also been determined.[22] Crystals of (I) are monoclinic, space group $C2/c$, with $a = 13.7918(7)$, $b = 8.6803(5)$, $c = 13.4680(7)$ Å and $\beta = 139.53(5)°$. Crystals of (II) are also monoclinic, space group $P2_1/c$, with $a = 14.1107(7)$, $b = 20.0146(11)$, $c = 8.8636(5)$ Å and $\beta = 106.95(4)°$. In the crystal lattice of (I), neighbouring octahedral molecules are held together by two NH···O=C hydrogen bonds connecting adjacent *N*-hydroxylamino groups, giving rise to infinite linear stacks with no significant bonding interactions between neighbouring stacks. The crystal lattice of (II) is made up of alternate layers, each composed of two crystallographically distinct $Me_2Sn(ONHOCMe)_2$ molecules and separated by water molecules, which connect the layers by a network of hydrogen bonds. $Me_2Sn(ONHOCMe)_2$ molecules within each layer are also held together by hydrogen bonds, although these are formed in this case between NH and an oxygen atom bonded directly to the tin atom of an adjacent molecule.

A number of dichlorobis(hydroxamato)tin(IV) complexes of the general formula L_2SnCl_2 have been prepared[22] from $SnCl_4$ and hydroxamic acids (LH, where LH = RCONR'OH, with R = Ph, *o*-HOC_6H_4, *o*-IC_6H_4, 3,5-$(NO_2)_2C_6H_3$; R' = H, Ph, *o*-MeC_6H_4, *m*-MeC_6H_4, *p*-ClC_6H_4). The complexes have been found, on the basis of various physicochemical data, to be octahedral with two chlorine atoms *cis* to each other. The work has been extended[23] to some bromide and iodide analogues, with similar conclusions.

Earlier (1977), Pradhan and Ghosh[24] had reported the preparation of some interesting derivatives which can be represented by (a) Ph_2SnL_2, (b) Ph_2SnLX, (c) $PhSnL_2X$ and (d) $PhSnLX(OMe)$ (where LH = *N*-phenylbenzohydroxamic acid and X = Cl, Br, I, SCN) by interconversions of the type shown in Scheme 1.

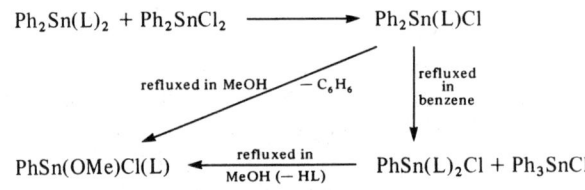

Scheme 1

Narula and Gupta[76] have measured the thermodynamic stepwise stability constants for the formation of complexes of dialkyltin(IV) ions, R_2Sn^{2+} (R = Me or Bu^n), with hydroxamic acids, RCONHOH (R = Ph, Me_2CH, Me) potentiometrically in 75% dioxane–water media. This has been followed by determination of formation constant of *N*-arylhydroxamic acids–dialkyltin(IV) ions.[77]

Das *et al.*[78–80] have described the synthesis of a number of diorganobis(hydroxamato)tin(IV) complexes of the type R_2SnL_2 (R = Bu^n or Ph, HL = monohydroxamic acid) and *O*-(triphenylstannyl)-*N*-acyl-*N*-arylhydroxylamines. Attempts have been made to elucidate their structural features by a variety of physicochemical techniques like IR, NMR and Mössbauer spectroscopy.

Dutta and Satapathi[81] have described the synthesis of a variety of phenylmercury(II) complexes with benzo-, *p*-chlorobenzo-, picolino-, quinolino- and quinoline-8-hydroxamic acids and characterized them by their IR spectra and conductivity measurements.

15.9.2 CUPFERRON, NEOCUPFERRON AND RELATED LIGANDS

Soon after the introduction of dimethylglyoxime as a specific reagent for nickel by Tschugaeff–Kraut–Brunck (1905–1907), Baudisch discovered a compound which precipitates copper and iron quantitatively from acid solutions.[82] He appropriately named this reagent as 'cupferron'. It is the water soluble ammonium salt of nitrosophenylhydroxylamine (5). When dissolved in chloroform, the whitish-grey copper compound gives a bright yellow solution and the brown yellow iron(III) compound a deep red solution. This behaviour reveals the 'inner complex' character of these derivatives (6).

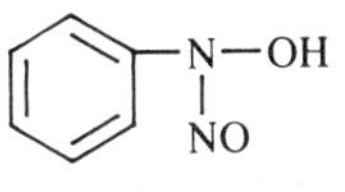

(5) 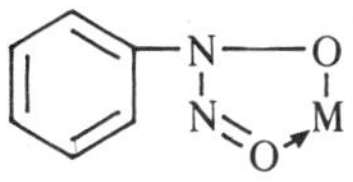 (6) (where M = Cu/2 or Fe/3)

The earliest expectations of 'cupferron' being a selective and specific reagent for copper and iron only were not justified as the reagent has been found to precipitate, with the exception of aluminum and chromium,[83] almost all the tri- to penta-valent metals which form insoluble hydrous oxides. However, the reagent has acquired permanent value as an aid in separating certain metals, *e.g.* titanium from aluminum and chromium.[84]

During precipitation of metal cupferronates from mineral acid solutions, some nitrosophenyl-hydroxylamine is always coprecipitated. If attempts are made to wash out this water insoluble 'free acid' with an organic solvent or ammonia, some metal is also lost and the remaining precipitate is often hydrolyzed. Consequently, cupferronates cannot be used as final weighing forms, and they serve only for purposes of precipitation and separation. These precipitates often have distinct colours which can form the basis of their estimation after extraction with suitable organic solvents.[85]

The following tautomeric forms have been suggested for cupferron:[86]

$$Ph—N(NO)—OH \rightleftharpoons Ph—N(=O)NOH$$

The main evidence adduced for the latter is the formation of nitrosobenzene on decomposition of the ligand. However, the common form on the left appears to be involved almost uniformly in the complexes which the ligand forms with a large variety of metals.

The technique of homogeneous precipitation, through production *in situ* by the reaction of phenylhydroxylamine and sodium nitrate, has been applied for quantitative precipitation of copper and titanium. The active inner-complex salt-forming group in cupferron is —N(OH)NO. The activity of the OH group and its enhanced acidity appear to be due to the inductive and electronic effects of the Ph and NO groups, which appears to be corroborated by the observation that, on replacement of the phenyl group by an alkyl group, the ligand loses its characteristic power to precipitate metal ions in acidic solutions. It is also interesting that the quantitative precipitability of iron and copper is retained when the active group —N(OH)NO is bound to aromatic groups other than Ph. A few illustrative examples of such ammonium salts are cited in (7)–(12), mainly from the findings of Baudisch's school. The colours of copper salts of (7)–(12) in the solid state and in chloroform solution are given in Table 2.

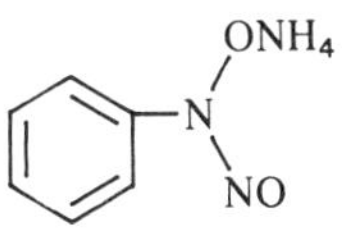

(7) 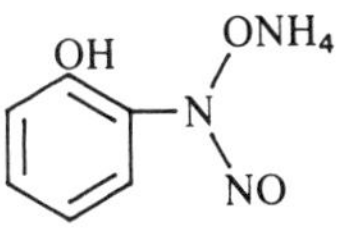 (8) 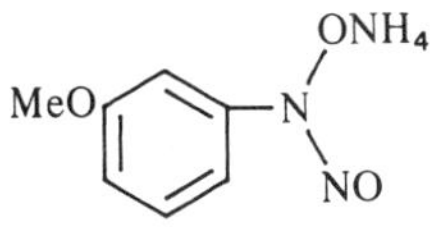 (9) 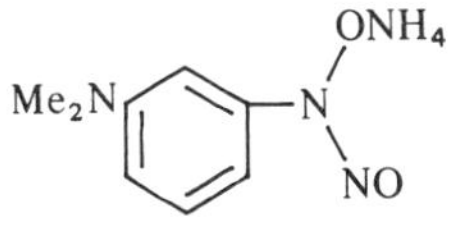 (10)

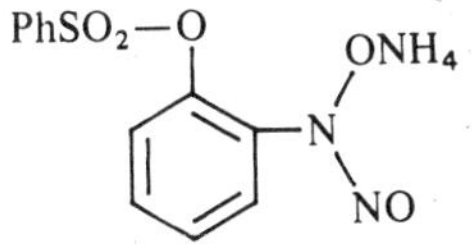

(11) 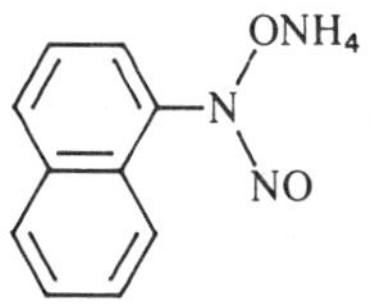 (12)

Table 2 The Colours of Copper Salts of Ligands (7)–(12) in the Solid State and in Chloroform Solution

Ligand	Colour of copper(II) salt	
	Solid state	CHCl₃ solution
(7)	Whitish-grey	Bright yellow
(8)	Grey	Blue-green
(9)	Light grey	Dark blue
(10)	Yellowish-green	Dark blue (slightly soluble)
(11)	Grey-blue	Blue
(12)	Grey	Yellow

Of the above derivatives, the ammonium salt of 1-nitrosonaphthylhydroxylamine (12, named as neocupferron) appears to have attracted maximum attention.[87] Its uranium, neodymium and cadmium salts (in contrast to the corresponding cupferron precipitates) are soluble in organic liquids. Neocupferron also appears to give a more voluminous (discernable quantity) precipitate with small amounts of iron.

The literature on cupferron and neocupferron ligands, especially as analytical reagents, is well documented up to the middle of this century.[88–92] The literature on cupferron since 1950 is rather diffuse and deals with analytical applications (quantitative precipitation and separations followed by solvent extraction) coupled with a few publications on the composition and structures of some metal cupferronates. As a survey of the whole literature, particularly dealing with analytical aspects, would not be directly relevant here a few illustrative examples of these will be given, but the main emphasis will be on investigations which have thrown some significant light on the composition and structural feactures of the derivatives.

Cupferron, being a sensitive reagent, has been widely used in the separation and precipitation of lanthanide and actinide elements; for example, composition and conditions for complete precipitation and extraction of cupferronates of trivalent lanthanide metals and scandium have been described.[93] If precipitated at pH 6.5 and above, the cupferronates of lighter rare earths up to gadolinium are insoluble and depict normal 1:3 composition. These can be dried at 100 °C, with the exception of cerium(III) cupferronate which is stable up to 80 °C only. The cupferronates of scandium, yttrium and rare earths heavier than gadolinium are appreciably soluble in water but these can be quantitatively precipitated in the presence of a large excess of ammonium salts. For example, under these conditions the composition of the scandium precipitate corresponds to $NH_4Sc(PhN_2O_2)_4$.

Protactinium can be separated from natural ore concentrates by cycles consisting of adsorption on MnO_2 precipitates followed by solvent extraction of the cupferron complex with pentyl acetate.[94]

The formation of complex compounds of neptunium with cupferron and a number of other complexing agents has been studied.[95] More recently, the extraction of neptunium(IV and VI) from HCl, HBr and HNO₃ media has been investigated.[96]

Plutonium tetracupferronate, $Pu(PhN_2O_2)_4$, with plutonium in oxidation state IV only, has been reported to be formed in the reactions of cupferron with Pu^{III}, Pu^{IV} and Pu^{VI} solutions.[97]

Partition behaviour of americium(III) chelates with cupferron and other bidentate reagents was studied spectrophotometrically between a number of inert solvents and dilute $HClO_4$ solutions.[98] Of special interest may be the data on their extractability and colours of chloroform extracts, collected in a tabular form for cupferronate derivatives of 58 metals. The pH ranges for the formation of cupferronates of 39 metal ions have been shown graphically in this publication.[99] Solvent extraction and polarographic techniques were employed to study the possible adducts between technetium and cupferron.[100] Evidence indicates a Tc^{III} cupferronate and possibly a pertechnitate adduct, but no indication of a technetium(IV) complex was obtained.

The composition of iron(III) cupferronate in chloroform was determined photometrically at 325 and 385 nm by the molar ratio methods.[101] At both wavelengths the composition of the complex was found to be $Fe(PhN_2O_2)_3$. The crystal structure of the complex was determined by three-dimensional Fourier and least-square methods.[102] The crystals are monoclinic with $a = 12.50$, $b = 17.45$, $c = 11.15$ Å, $\beta = 122° 19'$ and $z = 4$; the space group is $P2_1/a$. The cupferron groups are unsymmetrically attached to iron and are, therefore, crystallographically independent. The Fe—O bond seems to be ionic.

An interesting study of interactions of cupferron with metal ions (2×10^{-3} to 5×10^{-5} M Th^{4+}, Mo^{6+} and U^{6+}) and mixtures of metal ions has been carried out by the techniques of high-frequency and potentiometric titrations.[103] It has been shown that Th^{4+} (6×10^{-5} to 2.4×10^{-4} M) forms a 1:4 complex with cupferron at pH 3.1–3.6; Mo^{6+} forms 1:1 and 1:2 complexes at pH 2.4 and only the

1:1 complex at pH 4.6 buffer; the U^{6+} forms a 1:2 complex at pH 4.6–4.8. In the case of U^{6+} the formation of a 1:2 complex is confirmed by spectrophotometry at 470 nm and it has been shown that Beer's law is applicable for less than 6×10^{-4} M U^{6+} ions.

Iron in the concentration range of the order of 3.5×10^{-4} to 7×10^{-4} M can be determined by high-frequency titration in the presence of other metal ions in $Fe^{3+}/Ti^{4+} = 1$ and $Fe^{3+}/Zr^{4+} = 1$. Total concentration of the two cations can be determined by high-frequency titration with cupferron, followed by a redetermination after masking the interfering ions.

A variety of spectroscopic studies of metal cupferronates have been carried out by a number of investigators. The IR spectra of Cu^{2+}, Hg^{2+}, Al^{3+}, Fe^{3+}, Ga^{3+}, Bi^{3+}, Ti^{4+}, Zr^{4+}, V^{4+}, Th^{4+}, U^{4+}, V^{5+} and Nb^{4+} cupferronates indicate a close parallelism in the spectra of different metals in the same oxidation state, indicating similarity of structural features. The absorption maxima in the 930–400 cm^{-1} region move to the lower frequency side, as expected, with increasing atomic weight of the cations and these, therefore, apparently correspond to the vibrations of the metal–oxygen bonds.[104]

The IR and UV absorption spectra of cupferron (HL) and ML_2 (M = Mn, Co, Ni, Cu and Zn) have been studied.[105] The IR absorption bands appearing in the ranges 560 ± 15, 507 ± 17 and 402 ± 15 cm^{-1} were assigned to M—O vibrations. These bands showed a high dependence on the number of electrons in the *d*-orbitals and the order of change was Mn < Co < Ni < Cu > Zn. In the UV spectra (200–340 nm) a blue shift for the observed two bands was found in the metal complexes compared with the ammonium salts (cupferron) and this was explained on the basis of an electron shift from the benzene ring through the O ligand caused by the replacement of the NH_4 group by a stronger electron acceptor metal cation.

IR and far-IR spectra of cupferronates of Mn^{2+}, Fe^{3+}, Co^{2+}, Ni^{2+}, Cu^{2+}, Zn^{2+}, Cd^{2+}, Sn^{2+}, Pb^{2+} and Bi^{3+} have been studied in the range 4000–32 cm^{-1}.[106] The bands in the region 1700–500 cm^{-1} and the C—H valence vibrations can be assigned to the vibrations of the pure ligand; differentiation of individual metal chelates in this range is rather difficult. By contrast, the far-IR region (below 500 cm^{-1}) contains many absorption bands (the metal chelate frequency region) which differ in position as well as in intensity with the nature of the metal cation. The highest M—O character which can be recognized on the basis of the large metal-dependent frequency shifts depends linearly on the second ionization potential of the metal, except for the Mn and Sn chelates. The relative absorptions, E_{M-O}/E, of the far-IR bands of these metal cupferronates allow classification of the chelates into groups corresponding to the metal–ligand bonding (tetrahedral, octahedral or planar) involved. The characteristic value used for this classification (the quotient of the absorption at the M—O frequency and the absorption of the strongest remaining band in the far-IR) lies in the range 0.92–2.17 for tetrahedral chelates and 5.30–6.20 for planar chelates, with the octahedrally coordinated chelates having intermediate values.

Calculations of the vibration frequencies expected for a chelate ring structure of copper cupferronate, calculated on the basis of an assumed geometry and set of force constants, have been found[107] to be in much better agreement with experimentally observed values than those calculated on the basis of a single ring structure. They also confirm that the observed changes at 935, 680, 580, 523 and 439 cm^{-1} in the spectrum of the metal complex are due to the mass of the metal.

A method has been described[108] for the vanadium(V) complexes with MeOH, utilizing the shift of the frequencies for the *sym* and *asym* vibrations of Me from 2835 and 2950 cm^{-1} in free MeOH to 2810 and 2910 cm^{-1}, respectively, in VO(OMe)L_2 complexes, where HL = cupferron, 8-quinolinol, *etc.* The IR spectra of VO(OH)L_2 complexes of the above two ligands have been recorded in this publication.

In an ESR study of the interaction of oxovanadium(IV) cupferronate with basic organic solvents, it has been shown that the coordination of basic solvent molecules leads to the destruction of the dimeric nature of the complex molecules.[109] Parameters of the spin Hamiltonian of the ESR spectra of the adducts thus formed in 15 cases have been correlated with the basicity of the organic solvents. By contrast, VO(cupferronate)$_2$ retains its diamagnetic dimeric form in the frozen solution of non-coordinating solvents such as chloroform, carbon tetrachloride, cyclohexane or benzene.

An interesting study of the ESR spectra of uranium(IV) chelates with cupferron and neocupferron has been carried out;[110] in particular, the uranium(IV) ion can be considered to be sufficiently diluted in these systems to avoid dipole–dipole interactions between the neighbouring ions. ESR spectra of tetrakis(cupferronato)- and tetrakis(neocupferronato)-uranium(IV) at room temperature and 77 K showed *g*-values of 4.27 and 4.50 respectively. In the solution of degassed THF the same signals were observed at 77 K only. Since these signals did not appear in the corresponding diamagnetic complex of Th^{IV}, Ce^{IV}, Hf^{IV}, Zr^{IV} and Ti^{IV}, the above signals were discussed on the basis of the crystal field effect with D_{4d} symmetry on the $^5H_4$ term of U^{IV} with a $5f^2$ electronic configuration. On the other

512   *Oxygen Ligands*

hand, ESR signals having g-values of 2.003–2.005 were also observed in spectra of U^{IV} as well as in the spectra of all the other metal derivatives above. The appearance of these signals has been explained on the basis of availability of an unpaired electron produced by a N—N bond fission on complex formation.

Magnetic susceptibilities of eight-coordinated U^{IV} cupferronates and a few other chelates were measured in the range of room to liquid nitrogen and helium temperatures.[111] The Curie–Weiss law holds above the temperature of liquid nitrogen with a magnetic moment of 3.5–2.9 μB, and a temperature independent paramagnetism was observed below liquid nitrogen temperatures. An attempt has been made to interpret the magnetic behaviour on the basis of the crystal field in dodecahedron/square antiprism models.

It has been reported[112] on the basis of a spectrometric study, using the method of continuous variations, that UO_2^{2+} ions form a soluble 1:2 complex $[UO_2(PhN_2O_2)_2]$ initially, but with excess of cupferron an insoluble product, $NH_4[UO_2(PhN_2O_2)_3]$, is obtained which has a solubility product of $5.8 \pm 2.5 \times 10^{-10}$.

Kundu *et al.*[113] have isolated mixed chelate complexes of identical composition of the type $(NH_4)_2[UO_2(L)_2B]$ and $(NH_4)_2[UO_2(L)_2X_2]$, where L = anion of cupferron, B = CO_3^{2-}, $C_2O_4^{2-}$ and X = F^-, Cl^-. Two different forms (one orange-yellow and one yellow) of both fluoro and chloro complexes were obtained, indicating the possibility of their existence in geometrical isomeric forms. The orange-red complex *cis*-$[UO_2L_2(H_2O)_2]$ becomes yellow after refluxing in acetone. Various physicochemical studies (conductance, DTA, NMR, IR and UV spectra) indicate it to be a hydroxy-bridged dinuclear complex, $H[(UO_2)_2L_4OH(H_2O)_2]$. In a later study, Kundu and Roy[114] have evaluated, by the thermogravimetric technique, parameters for the first stage of the dissociation of the orange-yellow and yellow forms of $(NH_4)_2[UO_2L_2F_2]$.

This has been followed by a similar study of the heat and kinetics of thermal decomposition of the complexes $(NH_4)_2[UO_2L_2X]$, where HL = cupferron and X = CO_3^{2-} or $C_2O_4^{2-}$.[115] The ligand dissociation reactions were found to be of zero order in these latter cases, compared with first in the former cases with monovalent chloro or fluoro substituents.

In continuation of the earlier work[116] on dioxouranium(VI) complexes of neocupferron, new complexes with the composition $M[UO_2(C_{10}H_7N_2O_2)_3]\cdot nH_2O$ (where M = Tl^+, Ag^+, $N_2H_5^+$, NH_3OH^+, $C_5H_6N^+$, α-MeC$_5$H$_5$N$^+$, $C_9H_8N^+$, $PhNH_3^+$, Ph_4As^+ and *cis*- and *trans*-[Co(en)$_2$Cl$_2$]$^+$), $M'[UO_2(C_{10}H_7N_2O_2)_3]_2$ (where M' = Zn^{2+} and enH$_2^{2+}$), $M''[UO_2(C_{10}H_7N_2O_2)_2L]$ (where M'' = NH_4^+ or H^+; L = $C_2O_4^{2-}$, 2MeCO$_2^-$, $C_9H_6NO^-$) and $[UO_2(C_{10}H_7N_2O_2)L']$ (where L' = pyridine or 2,2'-bipyridyl) have been synthesized and characterized on the basis of conductance, electronic and IR spectral and TG data.[117]

Complexes of Th^{4+} with cupferron (as well as oxime) were investigated by determining the distribution of tracer amount of ^{234}Th between the organic (chloroform or isobutyl methyl ketone) and aqueous phases.[118] The distribution curves show the presence of complexes other than the tetra ones, but formation constants for the neutral tetrakis products were calculated on the basis of the two-parameter equation method of Dryssen and Sillén.

The crystal structure of zirconium cupferronate has been shown to contain discrete $Zr(PhN_2O_2)_4$ molecules, the ligands being related by a twofold axis through the zirconium atom.[119] Zirconium has eightfold coordination by oxygen atoms situated at the vertices of a somewhat irregular dodecahedron with triangular faces. The Zr—O distances range from 2.168 to 2.222 Å.

Hyperfine quadrupole couplings and the extent of π-bonding in d^0 dodecahedral hafnium(IV) tetracupferronates and some other chelates have been estimated by time-differential perturbed angular correlation experiments using the 133–482 keV γ–γ cascade of the ^{181}Ta nucleus.[120] On analyzing the hyperfine data within the frame of both the point-charge (on the basis of the actual geometry[119]) and valence bond models, it has been deduced that π-bonding makes a significant contribution to the stability of the complex. The overall charge donated by the ligands into the valence orbitals of the metal lies between -0.7 and $1.2\,e$, which signifies the important ionic character of the hafnium–oxygen bonds in these eight-coordinate chelates.

An interesting series of dicarbonyl complexes of rhodium(I) containing singly charged bidentate ligands like cupferron has been synthesized by three different routes,[121] *i.e.* (i) from tetracarbonyl-μ,μ'-dichlorodirhodium(I), (ii) from solutions of rhodium chloride in DMF under reflux and (iii) from a carbonylated solution of rhodium chloride in boiling absolute alcohol. These dicarbonyls react with PPh_3, $AsPh_3$ and $SbPh_3$ to yield monocarbonyl derivatives.

Very little work has been carried out on cupferron derivatives of main group elements. Some aspects of cupferronates of tin and antimony have been studied[122] with a view to developing a suitable method of extraction of the metalloids from a mixture of Sn, Sb, As and Bi.

Diaryltin dicupferronates have been isolated by the reactions of acetonic solutions of diaryltin dichloride with a solution of cupferron in the same solvent.[123] On the basis of IR spectra, a hexacoordinate geometry is suggested for tin in these complexes.

15.9.3 REFERENCES

1. A. S. Bhaduri and P. Ray, *Fresenius' Z. Anal. Chem.*, 1957, **154**, 103.
2. R. L. Dutta, *J. Indian Chem. Soc.*, 1958, **35**, 243.
3. V. C. Bass and J. H. Yoe, *Talanta*, 1966, **13**, 735.
4. I. P. Alimarin, F. P. Sudakov and B. G. Golovkin, *Russ. Chem. Rev. (Engl. Transl.)*, 1962, **31**, 466.
5. F. Baroneelli and G. Grossi, *J. Inorg. Nucl. Chem.*, 1965, **25**, 1085.
6. A. K. Majumdar, 'N-Benzoyl Phenylhydroxylamine and Its Analogs', Pergamon, Oxford, 1971.
7. C. Musante, *Gazz. Chim. Ital.*, 1948, **78**, 536.
8. Y. K. Agarwal and J. P. Shukla, *J. Indian Chem. Soc.*, 1974, **51**, 373.
9. R. L. Dutta and S. Ghosh, *J. Indian Chem. Soc.*, 1967, **44**, 311.
10. N. N. Ghosh and G. Siddhanta, *J. Indian Chem. Soc.*, 1968, **45**, 1049.
11. R. S. Misra, *J. Indian Chem. Soc.*, 1969, **46**, 1074.
12. B. Chatterjee, *J. Indian Chem. Soc.*, 1973, **50**, 758.
13. S. K. Brahma and A. K. Chakraburtty, *J. Inorg. Nucl. Chem.*, 1977, **39**, 1723.
14. P. C. Kundu and S. Bhattacharya, *Indian J. Chem., Sect. A*, 1980, **19**, 233.
15. G. Schott and K. Golz, *Z. Anorg. Allg. Chem.*, 1973, **399**, 7.
16. T. Seshashdri, *J. Inorg. Nucl. Chem.*, 1974, **36**, 519.
17. P. G. Harrison, *J. Organomet. Chem.*, 1972, **38**, C5.
18. C. K. Narula and V. D. Gupta, *Indian J. Chem., Sect. A*, 1980, **19**, 1095.
19. P. G. Harrison, *Inorg. Chem.*, 1973, **12**, 1545.
20. T. J. King and P. G. Harrison, *J. Chem. Soc., Chem. Commun.*, 1972, 815.
21. P. G. Harrison, T. J. King and R. A. Richards, *J. Chem. Soc., Dalton Trans.*, 1975, 826; 1976, 1414.
22. P. G. Harrison, T. J. King and R. C. Phillips, *J. Chem. Soc., Dalton Trans.*, 1976, 2317.
23. M. K. Das and M. R. Ghosh, *J. Indian Chem. Soc.*, 1980, **57**, 678; *Indian J. Chem., Sect. A*, 1981, **20**, 738.
24. B. Pradhan and A. K. Ghosh, *J. Organomet. Chem.*, 1977, **131**, 23.
25. A. E. Harvey and D. L. Manning, *J. Am. Chem. Soc.*, 1950, **72**, 4498.
26. D. O. Meller and Y. H. Yoe, *Talanta*, 1960, **7**, 107.
27. R. L. Dutta and B. Chatterjee, *J. Indian Chem. Soc.*, 1967, **44**, 781.
28. L. Bauer and O. Exner, *Angew. Chem., Int. Ed. Engl.*, 1975, **13**, 376.
29. O. Exner and J. Holubeck, *Collect. Czech. Chem. Commun.*, 1965, **30**, 940.
30. M. Shiotani, D. Moruichi and J. Sohma, *Hokkaido Daigaku Kogakubu Kenyu Hokoku*, 1968, **49**, 455 (*Chem. Abstr.*, 1969, **70**, 82 866u).
31. E. P. Dubini, C. Ricca and G. Bargigia, *Chem. Ind. (London)*, 1965, **47**, 517.
32. H. J. Lindner and S. Goettlicher, *Acta Crystallogr., Sect. B*, 1969, **25**, 832.
33. B. G. Malstrom, 'Biochemical Function of Iron in Iron Deficiency', Academic, New York, 1970.
34. S. Mizukami and K. Nigata, *Coord. Chem. Rev.*, 1968, **3**, 267.
35. H. Irving and R. J. P. Williams, *Nature (London)*, 1948, **161**, 740.
36. L. E. Mellor and D. P. Malley, *Aust. J. Sci. Res.*, 1949, **A2**, 579.
37. B. Chatterjee, *Coord. Chem. Rev.*, 1978, **26**, 281, and refs. therein.
38. J. Das and S. C. Shome, *Anal. Chim. Acta*, 1961, **24**, 37.
39. H. R. Das and S. C. Shome, *Anal. Chim. Acta*, 1962, **27**, 545.
40. I. P. Alimarin and S. A. Hamid, *Zh. Anal. Khim.*, 1963, **18**, 1332 (*Chem. Abstr.*, 1964, **60**, 6204h).
41. B. Das and S. C. Shome, *Anal. Chim. Acta*, 1965, **32**, 52.
42. S. K. Sinha and S. C. Shome, *Anal. Chim. Acta*, 1961, **24**, 33.
43. B. Das and S. C. Shome, *Anal. Chim. Acta*, 1965, **33**, 462.
44. V. R. M. Kaimal and S. C. Shome, *Anal. Chim. Acta*, 1963, **29**, 286.
45. J. Das and S. C. Shome, *Anal. Chim. Acta*, 1962, **27**, 58.
46. V. R. M. Kaimal and S. C. Shome, *Anal. Chim. Acta*, 1964, **31**, 268.
47. P. R. Ellefsen, L. Gorden, R. Belcher and W. G. Jackson, *Talanta*, 1963, **10**, 701.
48. A. K. Majumdar and B. K. Pal, *J. Indian Chem. Soc.*, 1956, **42**, 43.
49. S. M. Qaim, *Anal. Chim. Acta*, 1964, **31**, 447.
50. A. K. Mazumdar and B. K. Pal, *Anal. Chim. Acta*, 1963, **29**, 168.
51. J. Minczewski and Z. Skorko-Trybula, *Talanta*, 1963, **10**, 1063.
52. A. K. Mazumdar and B. K. Pal, *Fresenius' Z. Anal. Chem.*, 1961, **184**, 115.
53. A. K. Mazumdar and B. K. Pal. *Anal. Chim. Acta*, 1962, **27**, 356.
54. S. K. Pandit and C. Gopinathan, *Indian J. Chem.*, 1976, **14**, 132.
55. Y. K. Agarwal, *Bull. Soc. Chim. Belg.*, 1977, **86**, 565; *Monatsh. Chem.*, 1977, **108**, 713.
56. Y. K. Agarwal and H. L. Kapoor, *J. Inorg. Nucl. Chem.*, 1977, **39**, 479.
57. S. A. Abbasi, *Rocz. Chem.*, 1977, **51**, 821 (*Chem. Abstr.*, 1977, **87**, 91 552).
58. S. A. Abbasi, R. S. Singh and M. C. Chattopadhyaya, *Rocz. Chem.*, 1977, **57**, 1821 (*Chem. Abstr.*, 1978, **88**, 28 495).
59. Y. K. Agarwal, P. C. Verma and P. V. Khadikar, *Ann. Soc. Sci. Bruxelles*, 1977, **91**, 264 (*Chem. Abstr.*, 1978, **89**, 122 495y).
60. R. L. Dutta, *J. Indian Chem. Soc.*, 1960, **37**, 167.
61. T. P. Sharma and Y. K. Agarwal, *J. Inorg. Nucl. Chem.*, 1975, **37**, 1830.

62. O. R. Zmievokaya and V. I. Fedeeva, *Zh. Anal. Khim.*, 1979, **34**, 908 (*Chem. Abstr.*, 1979, **91**, 116 682g).
63. S. B. Ghosh and R. B. Kharat, *Indian J. Chem., Sect. A*, 1981, **20**, 423.
64. S. Mizukami and K. Nigata, *Coord. Chem. Rev.*, 1968, **3**, 267.
65. G. Anderegg, F. L. Epplattenier and G. Schwarzenbach, *Helv. Chim. Acta*, 1963, **46**, 11 400.
66. G. D. Ratnakar and D. V. Jahagirder, *J. Inorg. Nucl. Chem.*, 1977, **39**, 1385.
67. Y. K. Agarwal, *Bull. Soc. Chim. Belg.*, 1978, **87**, 89 (*Chem. Abstr.*, 1978, **88**, 198 676).
68. V. A. Shenderovich, E. F. Strizhev and V. I. Ryaboi, *Zh. Neorg. Khim.*, 1978, **23**, 2681 (*Chem. Abstr.*, 1978, **89**, 221 827q).
69. D. A. Brown, M. V. Chidambaran, J. J. Clarke and D. M. McAlleese, *Bioinorg. Chem.*, 1978, **9**, 3.
70. D. A. Brown, D. McKeith (*née* Byrne) and W. K. Glass, *Inorg. Chim. Acta*, 1979, **35**, 5, 57.
71. D. R. Agarwal and S. G. Tandon, *J. Indian Chem. Soc.*, 1971, **48**, 571.
72. B. C. Bhattacharyya and M. Dhar, *Indian J. Chem., Sect. A*, 1982, **21**, 420.
73. S. K. Brahma and A. K. Chakraburtty, *J. Indian Chem. Soc.*, 1981, **58**, 615.
74. M. Pal and R. N. Kapoor, *Indian J. Chem., Sect. A*, 1980, **19**, 912.
75. B. C. Bhattacharyya, S. Bandyopadhyay and M. Dhar, *Indian J. Chem., Sect. A*, 1983, **22**, 301.
76. C. K. Narula and V. D. Gupta, *Indian J. Chem., Sect. A*, 1980, **19**, 491.
77. G. Singh, C. K. Narula and V. D. Gupta, *Indian J. Chem., Sect. A*, 1982, **21**, 738.
78. M. K. Das and M. Nath, *Indian J. Chem., Sect. A*, 1981, **20**, 1224.
79. M. K. Das and M. Nath, *Z. Naturforsch., Teil B*, 1982, **37**, 889.
80. M. K. Das, M. Nath and J. J. Zuckerman, *Inorg. Chim. Acta*, 1983, **71**, 49.
81. R. L. Dutta and S. K. Sạtapathi, *Indian J. Chem., Sect. A*, 1981, **20**, 136.
82. O. Baudisch, *Chem. -Ztg.*, 1909, **33**, 1298; 1911, **35**, 122, 913; *Ber. Dtsch. Chem. Ges.*, 1912, **45**, 1164, 3426; 1915, **48**, 1665; 1916, **49**, 172, 180, 191, 203; 1916, **50**, 325.
83. L. Lehrman, E. A. Kabat and H. Weisberg, *J. Am. Chem. Soc.*, 1933, **55**, 359.
84. W. W. Scott, 'Standard Methods of Chemical Analysis', 4th edn., Van Nostrand, New York, 1927, p. 547.
85. W. R. Bennett, *J. Am. Chem. Soc.*, 1934, **56**, 277.
86. A. H. A. Heyn and N. G. Dave, *Talanta*, 1966, **13**, 27, 33.
87. O. Baudisch and S. Holmes, *Fresenius' Z. Anal. Chem.*, 1940, **119**, 16.
88. G. F. Smith, 'Cupferron and Neocupferron', C. F. Smith Chemical Co., Columbus, 1938.
89. W. Prondinger, 'Organic Reagents Used in Quantitative Inorganic Analysis' (translated by S. Holmes), Elsevier, New York, 1940, pp. 72–81.
90. J. H. Yoe and L. A. Sarver, 'Organic Analytical Reagents', Wiley, New York, 1941, p. 220.
91. F. J. Welcher, 'Organic Analytical Reagents', Van Nostrand, New York, 1947, vol. III, pp. 355–404.
92. F. Feigel, 'Chemistry of Specific, Selective and Sensitive Reagents', Academic, New York, 1949, pp. 262–266.
93. A. K. Das and S. Banerjee, *Indian J. Chem., Sect. A*, 1970, **8**, 284.
94. A. G. Maddock and G. L. Miles, *J. Chem. Soc.*, 1949, S253.
95. A. Zolotov and I. P. Alimarin, *J. Inorg. Nucl. Chem.*, 1963, **25**, 691.
96. R. Shabana, *Radiochim. Acta*, 1978, **25**, 691.
97. I. V. Moiseev, N. N. Borodina and V. T. Tsvetkova, *Zh. Neorg. Khim.*, 1961, **6**, 543 (*Chem. Abstr.*, 1962, **56**, 5456g).
98. M. Wada and T. Kanno, *Nippon Kagaku Kaishi*, 1979, **10**, 1332 (*Chem. Abstr.*, 1979, **91**, 217 658j).
99. A. T. Pilipenko, A. V. Shpak and E. A. Shpak, *Ukr. Khim. Zh.*, 1975, **41**, 78 (*Chem. Abstr.*, 1975, **82**, 164 454v).
100. G. B. Salaria, C. L. Rulfs and P. J. Elving, *Anal. Chem.*, 1964, **36**, 146.
101. Si-Joong Kim and Doo-Soonshin, *Daehan Hwahak Hwoejee*, 1963, **7**, 280 (*Chem. Abstr.*, 1964, **61**, 6384).
102. D. Vander Helm, L. L. Merritt, Jr., R. Degeilt and C. H. MacGillavry, *Acta Crystallogr.*, 1965, **18**, 355.
103. C. B. Riolo, T. F. Soldi and J. Spini, *Anal. Chim. Acta*, 1968, **41**, 388.
104. A. T. Pilipenko, L. L. Shevchenko and V. N. Strokan, *Zh. Neorg. Khim.*, 1971, **16**, 2399 (*Chem. Abstr.*, 1971, **75**, 145 718y).
105. A. H. Abou-el-Ela, F. M. Abdel Keum, H. H. Afifi and H. F. Aly, *Z. Naturforsch., Teil B*, 1973, **28**, 610 (*Chem. Abstr.*, 1974, **80**, 101 809c).
106. R. Kellner and P. Prokopowski, *Anal. Chim. Acta*, 1976, **86**, 175.
107. L. L. Shevchenko, *Ukr. Khim. Zh.*, 1977, **43**, 430 (*Chem. Abstr.*, 1977, **87**, 31 359z)
108. A. T. Pilipenko, L. L. Shevchenko and L. A. Zyuzya, *Zh. Neorg. Khim.*, 1973, **18**, 130 (*Chem. Abstr.*, 1973, **78**, 91 956r).
109. V. V. Zhukov, I. N. Marov, N. B. Kalinichenko, O. M. Petrukin and A. N. Ermakov, *Zh. Neorg. Khim.*, 1974, **19**, 3279 (*Chem. Abstr.*, 1975, **82**, 147 821i).
110. T. Yoshimura, C. Miyake and S. Imoto, *J. Inorg. Nucl. Chem.*, 1975, **37**, 739.
111. T. Yoshimura, C. Miyake and S. Imoto, *Bull. Chem. Soc. Jpn.*, 1975, **47**, 515 (*Chem. Abstr.*, 1974, **80**, 138 521d).
112. A. E. Klygin and N. S. Kolyada, *Zh. Neorg. Khim.*, 1961, **6**, 216 (*Chem. Abstr.*, 1962, **56**, 8289).
113. P. C. Kundu, P. S. Roy and R. K. Banerjee, *J. Inorg. Nucl. Chem.*, 1980, **42**, 851.
114. P. C. Kundu and P. S. Roy, *J. Inorg. Nucl. Chem.*, 1980, **42**, 1773.
115. P. C. Kundu and P. S. Roy, *J. Inorg. Nucl. Chem.*, 1981, **43**, 43.
116. P. C. Kundu and A. K. Bera, *Indian J. Chem., Sect. A*, 1978, **16**, 865; 1979, **18**, 62.
117. P. C. Kundu and A. K. Bera, *Indian J. Chem., Sect A*, 1982, **21**, 1132.
118. D. Dryssen, *Svensk Kem. Tid.*, 1953, **65**, 43 (*Chem. Abstr.*, 1953, **47**, 10 968i).
119. W. Mark, *Acta Chem. Scand.*, 1970, **24**, 1398.
120. A. Baudry, P. Boyer, A. Tissier and P. Vulliet, *Inorg. Chem.*, 1979, **18**, 3427.
121. K. Goswami and M. Singh, *Transition Met. Chem.*, 1980, **5**, 83 (*Chem. Abstr.*, 1980, **92**, 225 836b).
122. S. J. Lyle and A. D. Shendrikar, *Anal. Chim. Acta*, 1966, **36**, 286.
123. T. N. Srivastava, *J. Inorg. Nucl. Chem.*, 1975, **37**, 1546.

16.1

Sulfides

ACHIM MÜLLER and EKKEHARD DIEMANN
University of Bielefeld, FRG

16.1.1 GENERAL CHARACTERISTICS OF METAL–SULFUR BONDS

The high affinity of metal for sulfur is clearly demonstrated by the enormous variety of metal sulfide minerals in nature. Qualitatively this might be understood using Pearson's concept of hard and soft acids and bases. It is better described, however, by the covalency of the metal–sulfur bond, particularly with the 'soft' metals.

In contrast to oxygen and nitrogen as donor atoms, sulfur has low lying, unoccupied $3d$-orbitals available. MO calculations reported in the literature, however, indicate that this influence might be overestimated. It has to be assumed that the high polarizability of the electrons on the sulfur atoms is responsible for the variety of structures and the reactivity of metal complexes with sulfur-containing ligands.

All degrees of covalency of M—S bonds are known, from almost completely ionic bonds (in alkali metal sulfides) to those with bond orders of about 3. If no other ligands are present, a higher covalency is usually obtained when the metal's d valence orbital ionization potential (VOIP) (of the open shell transition metals, but also of Cu^I, Ag^I and Au^I) matches the sulfur $3p$ VOIP. Thus, relatively high oxidation states are found for the metals in the most stable thiometalates of the early transition elements (*e.g.* MoS_4^{2-}), and lower oxidation states occur in complexes of the electron rich metals (Hg^{II}, Cu^I, *etc.*; see Section 16.1.4.3). Stabilization may, however, also be achieved by extension of the coordination number of the metal center, or, in multinuclear complexes, through the formation of metal–metal bonds. Additionally, π-accepting ligands will decrease the electron density on the metal and can thus allow for an increased electron density in the M—S bond.

The (remaining) electron density on the coordinated sulfur atom mainly determines its reactivity. This can be reduced by the formation of stronger M—S bonds, and also by the substituents H and R, as in SH_2, SH^-, SR^- and SR_2. However, complexes with the hydrogen-containing ligands may easily undergo deprotonation reactions and are, therefore, less stable. This is the main reason for there being relatively few complexes with H_2S and SH^- as ligands but many examples of compounds with SR^- and SR_2 ligands.

The coordinating ability (*i.e.* the stability constants of the complexes) of sulfur-containing ligands increases with the dipole moment in the order $H_2S < RSH < R_2S$. The stability of the complexes decreases in the order $S^{2-} > RS^- > R_2S$. Here the polarizability and the number of lone pairs are the dominant factors.

In the spectrochemical series, R_2S is placed between H_2O and NH_3, but RS^- and HS^- come between F^- and Cl^-.[1]

The reactivity of metal–sulfur bonds has been discussed in a review by Kuehn and Isied,[2] but other reviews also include chapters on this topic.[3–7]

16.1.2 H₂S AND RSH AS LIGANDS

In contrast to the large number of complexes with H_2O as a ligand, very few examples have been reported containing H_2S ligands. This is probably mainly due to the fact that the primary products can easily react to yield sulfhydryl or multinuclear complexes with bridging sulfur ligands.[7] For example, during the reaction of $[Pt(PPh_3)_n]$ ($n = 2, 3$) with H_2S in solution an intermediate is formed which, from its ^1H NMR spectrum, is thought to be $[(PPh_3)_2Pt(SH_2)]$. The final product, however, is $[(PPh_3)_2Pt(H)(SH)]$.[8]

'Thiohydrates' were regarded for a long time as the only 'H₂S complexes' which could be isolated.[9] They are formed by addition of H_2S to Lewis acids in liquid hydrogen sulfide (for example $AlCl_3 \cdot H_2S$ and $TiCl_4 \cdot nH_2S$, $n = 1, 2$) and decompose thermally below room temperature. To stabilize complexes with the neutral ligands H_2S and RSH, metal ions in low oxidation states, preferably with inert and bulky units, are required.

The first stable H_2S complexes were reported in 1976.[10,11] By direct displacement of H_2O by H_2S in aqueous solution the two Ru^{II} complexes (1) and (2) can be obtained as BF_4^- salts.[10] In another example the bulky organometallic unit $W(CO)_5$ is used for coordination. When H_2S is bubbled through a solution of tungsten hexacarbonyl in pentane, while photolyzing the reaction mixture, $[W(CO)_5H_2S]$ is formed.[11,12] The synthesis is also possible from $[W(CO)_5THF]$ (by substitution of THF) or, more conveniently, by hydrolysis of $[W(CO)_5S(EMe_3)_2]$ (E = Si, Sn).[13] By replacing an aqua ligand by hydrogen sulfide the compound $[Mn(CO)_4(PPh_3)H_2S]BF_4$ can be prepared.[14]

Both $[W(CO)_5H_2S]$ and the pentaammine complex (1) decompose in the solid state even under vacuum. When isonicotinamide (isn) is *trans* to H_2S in the Ru complex (2) it becomes less stable to the replacement of H_2S by H_2O but more stable towards oxidation. This has been attributed to

$$[(H_3N)_5Ru(H_2S)]^{2+}$$

(1)

(2)

S → Ru back-donation, as there is an increase in acidity of the coordinated H_2S when isn replaces NH_3 in the coordination sphere of Ru^{II}.[10]

Some reactions of $[W(CO)_5H_2S]$ have also been reported.[12] Surprisingly, this complex is quite stable thermally, it decomposes at 90 °C in a nitrogen atmosphere. With diazomethane or diazoethane, alkylation of the coordinated H_2S occurs, yielding the corresponding thioether complexes. The increased acidity of the H_2S ligand allows deprotonation (equation 1).

$$[W(CO)_5SH_2] + (Et_4N)Br \xrightarrow{CH_2Cl_2} (Et_4N)^+[W(CO)_5SH]^- + HBr \quad (1)$$

Although one might expect thiol (RSH) complexes to be more stable than H_2S complexes, only a few well characterized examples have been reported. The analogs of the Ru^{II} compounds (1) and (2) with EtSH have been prepared.[10,15] By oxidation with silver trifluoroacetate the corresponding pentaammine–Ru^{III} complex can be obtained.[10] Spectral evidence for the formation of thiol complexes has also been given, *e.g.* for the systems Fe^{II}/PrSH and PhSH,[16] Zn^{II}/BuSH[17] and $[Pt(pyz)Cl_2]$/PhSH;[18] however, no further descriptions of the complexes were given.

Iron(III) (tetraphenylporphyrinato)(benzenethiolato)benzenethiol, Fe(TPP)(PhS)(PhSH), which serves as a model compound for cytochrome P-450, and for which ESR, Mössbauer and magnetic measurements indicate the presence of both low-spin and high-spin iron(III), has been studied crystallographically.[19] Although attempts to locate the thiol hydrogen were unsuccessful, it can be argued that in the high-spin, five-coordinate system (Fe—$S_{thiolate}$ = 2.32 Å) the thiol remains uncoordinated, while in the low-spin complex it is clearly coordinated (Fe—S_{thiol} = 2.43 Å, Fe—$S_{thiolate}$ = 2.27 Å).

16.1.3 HYDROSULFIDO (SH⁻) COMPLEXES

The complex chemistry of the hydrosulfido ligand has emerged during the last decade. The bonding versatility of this ligand is immediately obvious from Figure 1, and for each of the types A–C examples have been reported in the literature. This flexibility is mainly due to the fact that the SH group can act as a one-, three- or five-electron donor. However, polynuclear systems with and

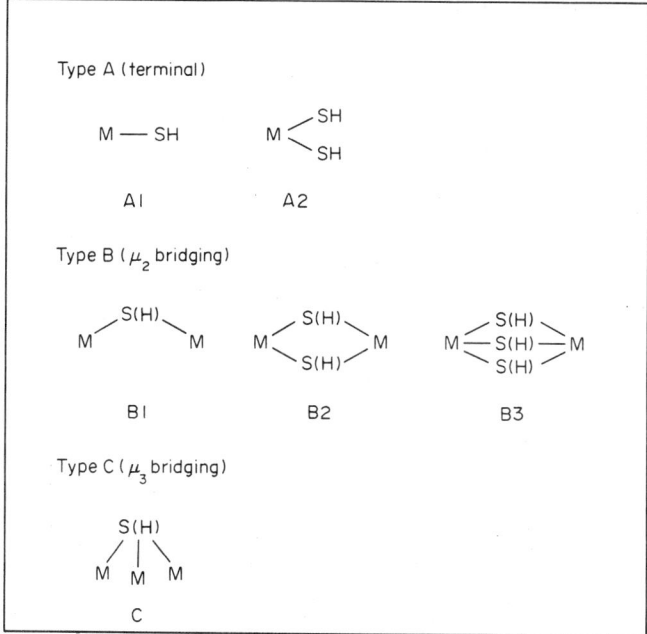

Figure 1 Structural types of central groups of SH complexes

without metal–metal bonds can be formed and, depending on the ligand environment, monomer–polymer equilibria can exist. In principle the SH$^-$ ligand possesses the same coordinating abilities as the SR$^-$ ligand (*vide infra*), although there are some additional possibilities for side reactions, *i.e.* deprotonation followed by destructive processes, derivatization and polymerization.

16.1.3.1 Synthesis of SH$^-$ Complexes

Routes to prepare SH$^-$ complexes have, in principle, to overcome two major problems; the formation of thermodynamically preferred metal sulfides and, *via* partial ligand elimination, the formation of polymeric species. This requires both a rigid stereochemistry brought about by bulky ligands and sufficient electron density on the metal centers.

Possible reagents include sulfur-containing species such as PbS, Ag$_2$S, S$_2$O$_3^{2-}$ and polysulfide.[20] The SH$^-$ ion can, for instance, substitute ligands such as chloride in the initial substrate.[21] Other reagents are a mixture of Et$_3$N and H$_2$S in Et$_2$O,[22] gaseous H$_2$S (substituting an azide ion[23]) or propylene sulfide (insertion into a metal–hydride bond[24]). Another well developed route is the hydrolytic cleavage of S—E bonds (E = Sn, Ge) in trimethylstannylthio[25] or triethylgermylthio complexes.[26]

16.1.3.2 Survey of Compound Types

The most common type of hydrosulfido complex is type A1 (Figure 1) with one terminal SH$^-$ ligand. Ardon and Taube[20] described as the first example the preparation of [(H$_2$O)$_5$CrSH]$^{2+}$ from Cr^{2+} and polysulfide. Other examples are a series of compounds with the poly(tertiary phosphine) ligands ppp, pnp, n$_2$p$_2$H, pp$_3$ and np$_3$ (where np$_3$ denotes one N and three P donors in the ligand),[27] *e.g.* [Fe(SH)pp$_3$]BPh$_4$ or [Ni(SH)n$_2$p$_2$H](BF$_4$)$_2$. De Simone and coworkers[28] succeeded in preparing a completely sulfur-coordinated sulfur-bridged dinuclear Mo^{II} species [Mo$_2$(SH)$_2$(16-aneS$_4$)$_2$] (CF$_3$SO)$_2$·2H$_2$O with a terminal SH$^-$ group at each metal (16-ane S$_4$ is a cyclic thioether, see Chapter 16.2). The stereochemically demanding ligand L = difluoro-3,3'-(trimethylenedinitrilo)bis-(2-pentanone oximato)borate offers a square planar N$_4$ donor system in which Rh^{III} can react directly with H$_2$S to give [Rh^{III}L(H)(SH)].[26] An example of an A1 type SH$^-$ complex where the metal is in a high oxidation state is the W^{VI}S$_3$SH$^-$ ion.[29] Here the SH$^-$ ligand is, as expected, much more acidic and the complex can only be stabilized through the use of large inert cations like Ph$_4$P$^+$ or Ph$_4$As$^+$.

For complexes of structural type A2 the two terminal SH$^-$ groups can be in a *cis* or *trans* position. The first example of the former was [Cp$_2$Ti(SH)$_2$][22] and some others have been prepared since then, for example [(PPh$_3$)$_2$Pt(SH)$_2$], [(diphos)Ni(SH)$_2$] and [(diphos)Pd(SH)$_2$].[23,30] The reaction of HCl and [(NH$_4$)$_2$WS$_4$] suspended in Me$_2$O at dry ice temperature is suspected to yield [S$_2$W(SH)$_2$]. This material is very unstable, probably due to the high acidity of the two SH protons, and has not been completely characterized. Data for this and related products (which probably also contain terminal SH groups) are compiled in ref. 31.

The *trans* A2 type can be obtained when the equatorial positions in the complex are already occupied by a rigid system as, for instance, with the above mentioned borate-N$_4$ ligand L. [Rh^{III}LCl$_2$] reacts first with Et$_3$GeS$^-$ to give *trans*-[Rh^{III}L(SGeEt$_3$)$_2$], which can be hydrolyzed directly to *trans*-[Rh^{III}L(SH)$_2$]. Another example is *trans*-[Pt(SH)$_2$(PEt$_3$)$_2$], which has been prepared by direct exchange of chloride ligands of the corresponding complex.[32]

To stabilize SH$^-$ in a bridging situation is much more difficult and consequently fewer examples are known. Type B1 is found in [M$_2$(CO)$_{10}$(μ-SH)]$^-$ (M = Cr, Mo, W)[33] and type B2 occurs in [Mn$_2$(CO)$_8$(μ-SH)$_2$].[24] A stable trihydrosulfido dinuclear system of the type B3 is [Fe$_2$(μ-SH)$_3$(ppp)$_2$]$^+$ (3; ppp = bis(2-diphenylphosphinoethyl)phenylphosphine).[35] Type C is claimed to be represented by (Me$_3$PtSH)$_4$,[34] which has, however, not been fully characterized.

(3)

16.1.3.3 Reactivity, Structures and Bonding

SH^- complexes may undergo deprotonation reactions; however, due to their generally low acidity, quite strong bases are required and these, in principle, are always capable of attacking the metal centers. Therefore a deprotonation is often paralleled by destructive processes. Different preparative routes to the deprotonated complexes have been reported.[39,40] Another type of reaction utilizing the acidity of the proton is alkylation with diazoalkanes. Derivatives of hydrosulfido complexes can also be obtained by inserting, for example, acetylene[37] or phenylacetylene[36] into the S—H bond. If the system is sufficiently electron rich, as in A2 type $[Cp_2Mo(SH)_2]$, a tetrasulfido complex is formed with S_8, accompanied by evolution of H_2S.[38]

Structural data for SH^- complexes have only been reported in a few cases. In $[Ni(SH)n_2p_2H]BF_4$ one N donor, two P donors and the SH^- ligand form a distorted square-planar environment around the metal.[27] The Ni—S distance, $d(Ni—S) = 2.144$ Å, is rather short when compared to other systems. In $[M(SH)pp_3]BPh_4$ (M = Fe, Ni; trigonal bipyramidal coordination with the sulfur apical), M—S distances of $d(Ni—S) = 2.256$ and $d(Fe—S) = 2.247$ Å have been determined, respectively.[27] The Mo—SH distances, which have been obtained for the above mentioned complexes, $[Mo_2^{II}(SH)_2(16\text{-}aneS_4)_2](CF_3SO_3)_2 \cdot 2H_2O$ ($d(Mo—SH) = 2.47$ Å, $d(Mo—S_{ring}) = 2.46\text{–}2.54$ Å) and $[Mo^{IV}O(SH)(16\text{-}ane\text{-}S_4)](CF_3SO_3)$ ($d(Mo—SH) = 2.49$, $d(Mo—S_{ring})_{av} = 2.47$ Å), correspond to values expected for M—S single bonds.[28]

No X-ray structure determinations are available for the above mentioned W^{VI}–SH complexes. For S_3WSH^- the approximate bond orders $N(WS_{term}) = 2.0$ and $N(WSH) = 1.2$ (as determined from the vibrational spectra) indicate that there is still a π contribution in the W—SH bond.[29] Correspondingly, this group shows a high acidity which increases even further in $[S_2W(SH)_2]$. Here the acid strength is comparable to that of sulfuric acid.[31]

16.1.4 SULFIDE (S^{2-}) AS A LIGAND

Our knowledge concerning soluble metal complexes with sulfide ions as ligands has increased considerably during the last two decades and this kind of compound is still of topical interest. Some of the reasons for this are the development of a very flexible and fascinating structural chemistry of multinuclear metal–sulfur complexes, the fact that the active sites of some electron transfer proteins contain metal ions and labile sulfur,[41,42] and also the relation of metal–sulfur cluster compounds to some heterogeneous catalysts. In addition, apart from the numerous binary and ternary sulfides which occur in nature, we have at our disposal a rich solid state chemistry of metal sulfides, which has been reviewed elsewhere and will be excluded here.[43-47]

A general procedure for the synthesis of metal–sulfur complexes does not exist. However, the affinity between sulfur and the transition metals can be easily observed considering the fact that such very different reagents as, for instance, H_2S, S_8, S_x^{2-}, SCN^-, $S_2O_3^{2-}$, SO_3^{2-}, RNSNR or RSH may be the precursors of sulfide ligands, apart from S^{2-} itself. Very little is known, however, about the reaction pathways.

16.1.4.1 Survey of Compound Types

The high structural flexibility of the sulfide ligand is easily seen from the structural types shown in Figure 2. For each of these types examples have been reported in the literature, and for some particular types the number of compounds has increased so much in recent years that we have to restrict our survey here to some arbitrarily chosen, but typical, examples.

Type A structures with a terminal metal–sulfur bond are quite common with d^0 metal ions of the early transition elements and with several main group elements. With the latter, *e.g.* in PS_4^{3-}, AsS_4^{3-} and SbS_4^{3-}, M—S single bonds are anticipated (type A1),[48] while some d^0 metal ions show strong π contributions to the M—S bond (as in MS_4^{n-} species with bond orders $N \approx 1.5\text{–}2$). A list of ions containing sulfide and selenide ligands of type A2 is given in Table 1 together with some spectral and structural data to facilitate their identification.[31,49] These chalcogenometalates can be prepared either by solid state reactions (K_3VS_4, Tl_3MX_4; M = V, Nb, Ta; X = S, Se) or by the reaction of H_2S (or H_2Se) with an aqueous solution of the corresponding oxometalate. Since all ions have strong and characteristic absorption bands in the UV/VIS region, their reactions can be readily followed by spectrophotometric methods (Figure 3). All species $MO_{4-n}S_n^{m-}$ are formed in succession and from the existence of isobestic points it follows that only two species can coexist in the solution

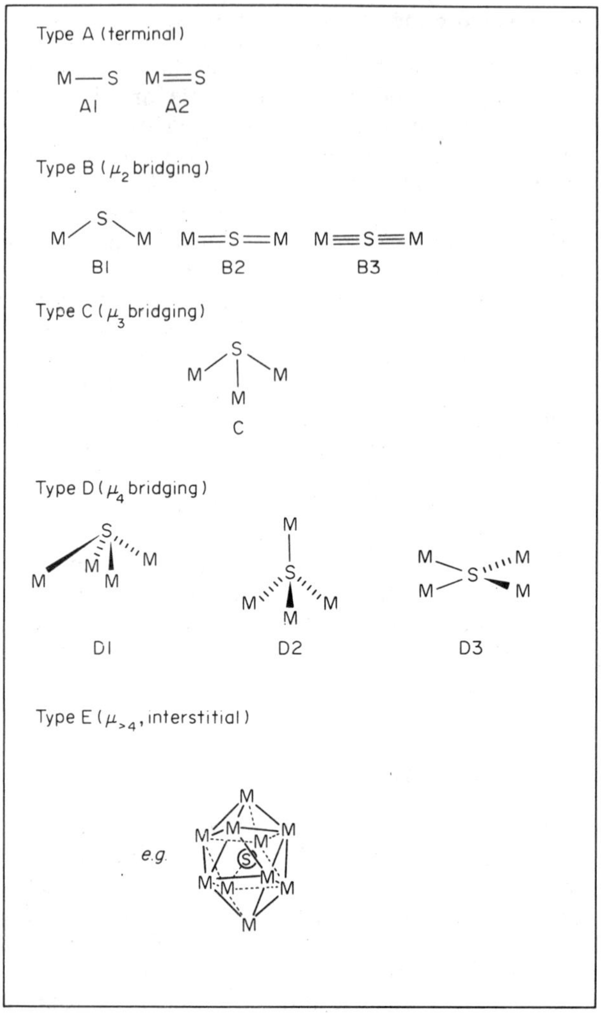

Type A (terminal)

M—S   M≡S
AI   A2

Type B (μ_2 bridging)

M⟍S⟋M   M≡S≡M   M≡S≡M
  BI   B2   B3

Type C (μ_3 bridging)

M⟍ | ⟋M
   M
   C

Type D (μ_4 bridging)

DI   D2   D3

Type E ($\mu_{>4}$, interstitial)

e.g.

Figure 2 Structural types of central groups of sulfido complexes

at the same time.[50] Thioanions can also be used as ligands. Their properties and bonding are described in some detail in Chapter 16.3. Type A2 bonds are also observed in various thiohalides, *e.g.* MSX_3 (M = Nb, Ta; X = Cl, Br), MSX_4 (M = Mo, W; X = F, Cl) and ReS_3Cl,[46] and ions like $W_3S_9^{2-}$ (**4**)[3] and $W_4S_{12}^{2-}$ (**5**).[49]

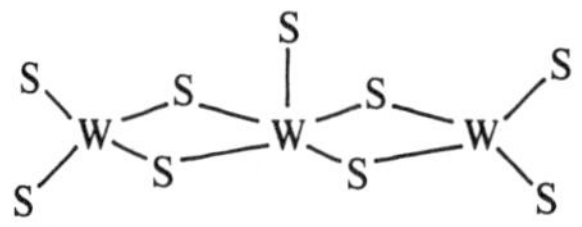

(4) 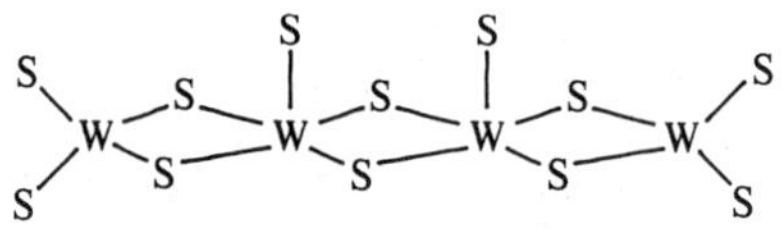   (5)

Structures of type B include three subtypes of a μ_2-bridging sulfide ligand. Among these, B1 with two μ_2 ligands is clearly the most common type. Examples for the alternatives, *i.e.* B2 and B3, demonstrate the bonding flexibility of the sulfide in that it can operate as a two-, four- or six-electron donor in very similar compounds. A single μ_2 bridge has been found in the Pt complex (**6**).[62] Two μ_2-bridging sulfides in the anion of the red Roussin's salt $[Fe_2S_2(NO)_4]^{2-}$ (**7**), one of the oldest metal–sulfur complexes known, were claimed by Seel[63] as early as 1942 and this was confirmed later by the structural determination of the corresponding ethyl ester.[64] Other representative compounds of type B1 are the synthetic 2Fe analogs of non-heme iron–sulfur proteins (**8**) and (**9**),[41,51,52] and also some Mo, W and Re complexes of general structure (**10**).[53–56] The anions (**3**) and (**4**) may also be quoted as examples, although a different description (complexes with WS_4^{2-} ligands) seems possible.[3]

In cases where only one sulfide ion acts as a bridge without the formation of M—M bonds, higher

Table 1 Discrete Thio and Seleno Anions of V, Nb, Ta, Mo, W and Re[3,31]

Compound	Color in solution	Vis/UV v $(10^3 \text{cm}^{-1})^{c,4}$	IR/Raman v $(\text{cm}^{-1})^{b,5}$	Typical $M\!-\!S(Se)$ bond length (average) (Å)
$^{3-}$	Yellow-orange	21.8, 27.8, 32.8	—	—
	Red	19.2, 21.8, 30.8, 33.9	—	—
	Red-violet	18.6 (3.76), 25.4, 28.5, 37.5	404.5, 193.5, 470, 193.5	2.15 (NH_4 salt)
bS_4	Red-brown	24.0^a	408, 163, 421, 163^d	—
aS_4	Brown	25.5^a	424, 170, 399, 170^d	—
e_2^{3-}	Red	19.2 (?)	—	—
$_3^{3-}$	Red-violet	15.6, 19.2	—	—
	Violet	15.6, 21.45	(232), 121, 365, 121	—
bSe_4	Deep violet	19.3^a	239, 100, 316, 100^d	—
aSe_4	Yellow-green	21.2^a	249, 103, 277, 103^d	—
$_3S^{2-}$	Yellow	25.4, 34.7, 44.7, 52.0	—	—
$_2S_2^{2-}$	Orange	25.4, 31.4, 34.7	—	—
S_3^{2-}	Orange-red	21.5, 25.5, 32.0, ~38.5, 44.1	—	2.178 (Cs salt)
	Red	21.4 (4.11), 31.5, 41.3, 48.3	458, 184, 472, 184	2.178 (NH_4 salt)
$_3Se^{2-}$	Orange	22.2, 31.75	—	—
$_2Se_2^{2-}$	Red	22.2, 28.5, 32.0, 40.8	—	—
Se_3^{2-}	Red-violet	17.9, 22.0, 28.45, 35.5, 40	—	—
e_4^{2-}	Blue-violet	18.0, 27.8, 37.2	255, 120, 340, 120	—
Se_3^{2-}	Red-violet	18.5, 28.4, 38.0	—	—
$_2Se_2^{2-}$	Red-violet	19.65, 29.5, 39.45	—	—
$_3Se_2^{2-}$	Red	20.6, 30.6, 40.35	—	—
S^{2-}	Pale yellow	30.6, 41.0	—	—
S_2^{2-}	Yellow	30.6, 36.6, 41.0	—	2.193 (NH_4 salt)
$_3^{2-}$	Yellow	26.7, 29.9, 37.0, 41.1	—	2.20 ($K_3(WOS_3)Cl$)
	Yellow	25.5 (4.27), 36.1, 46.2	479, 182, 455, 182	2.177 (NH_4 salt)
Se^{2-}	Yellow	27.0, 38.0	—	—
Se_2^{2-}	Orange	27.0, 34.0, 38.0	—	—
$_3^{2-}$	Red	21.9, 32.5, 42.6	—	—
e_3^{2-}	Orange	22.1, 26.0, 34.1, 38.2	—	—
	Red	21.6 (4.20), 31.6, 41.3	281, 107, 309, 107	2.344 (Cs salt), 2.317 (NH_4 salt)
S_2Se^{2-}	Red	24.3, 31.0, 42.5	—	—
$_2Se^{2-}$	Yellow	29.1, 36.0, 41, 48.4	—	—
SSe_2^{2-}	Red	18.5, 23.15, 30.75	—	—
Se_2^{2-}	Orange	23.25, 26.75, 33.5	—	—
S^-	Yellow	28.6, 33.6, 46.5	—	2.14 (Rb salt)
$_2S_2^-$	Orange	25 (?), 32 (?)	—	—
S_3^-	Red	19.8, 25.5, 32.3	—	—
	Red-violet	19.8 (3.98), 32.0, 44.0	501, 200, 488, 200	2.155 (PPh_4 salt)

...ectance in $MgSO_4$. ^{b} Band maxima sequence for the ions with T_d symmetry: $v_1(A_1)$, $v_2(E)$, $v_3(F_2)$, $v_4(F_2)$ (in some solid state spectra v_2 and v_4 do not coincide). ...es of log ε for v_1 in brackets. ^{d} Tl salts with strong Te—S interactions.

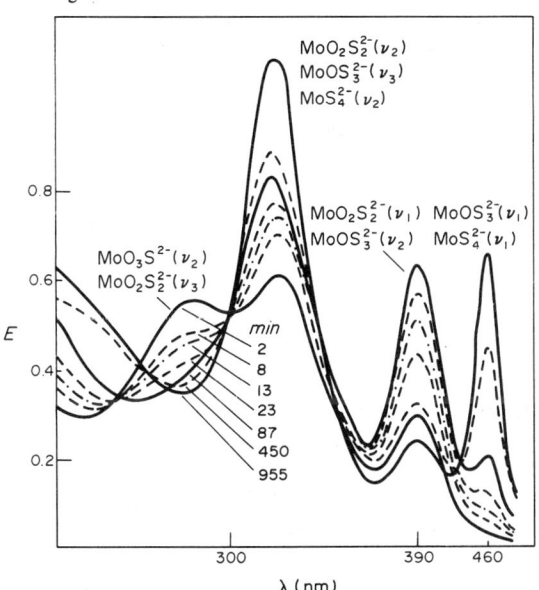

Figure 3 Electronic absorption spectra of the reaction products from MoO_4^{2-} and H_2S in aqueous solution, as a function of time

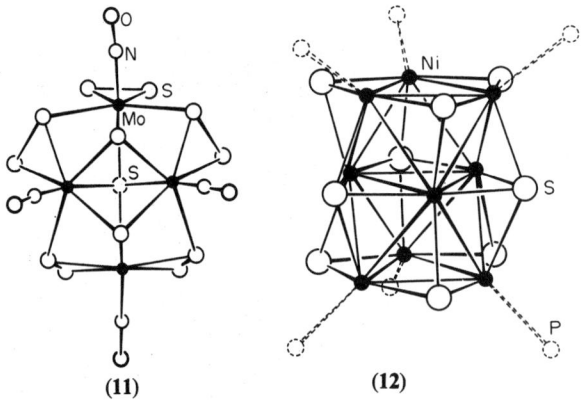

M—S bond orders are expected and structural types B2 and B3 are encountered. There is no clear-cut borderline between these two types. An assignment can be made using experimental M—S bond lengths; however, not much material for comparison is available. Sacconi and coworkers[57] have reported complexes of Ni^{II} and Co^{I} with poly(tertiary phosphines), $[\{(ppp)Ni\}_2S](BPh_4)_2 \cdot 1.6\ DMF$ and $[\{(np_3)Co\}_2S]$. Both contain linear M—S—M units, with $d(Co—S) = 2.128$ Å (indicative of type B2) and $d(Ni—S) = 2.034$ Å. The latter is the shortest distance ever found for such a linkage and should, therefore, be assigned to type B3. The linear units $Cr≡S≡Cr$ ($d(Cr—S) = 2.074$ Å)[58] and very recently $V≡S≡V$ ($d(V—S) = 2.172$ Å)[59] (with CO, Cp and diphos ligands additionally) have been reported. The $Mo=S=Mo$ moiety in $[(CN)_6MoSMo(CN)_6]^{6-}$ (type B2), however, shows a slight deviation from linearity in one salt[60] but has been found to be linear in another one.[61]

Type C coordination (μ_3) of the sulfide ligand is a very common one and occurs for all other types of ligands, with all degrees of metal–metal interactions, and has also been detected in solid state structures. An arbitrary selection of examples for the μ_3-bridging ligand is displayed in Figure 4 together with some appropriate references. There may be one or two ('bicapped') μ_3-sulfide ligands apical to the M_3 plane. Another possibility is that the μ_3-sulfide forms part of a cage (e.g. a cubane one). An interesting example of asymmetric coordination of a μ_3-sulfide ligand of the latter subtype has been found in $[Mo_4S_4(NO)_4(CN)_8]^{8-}$ which has a distorted cubane-like structure with two Mo—Mo bonds.[72a]

For structures of type D (μ_4 bridging), three alternatives are possible, all rather rare. Type D1 was first detected in the compound $[Co_4S_2(CO)_{10}]$.[73] More recent examples are $[Mo_4(NO)_4S_{13}]^{4-}$ (11)[3] and $[Ni_9(\mu_4\text{-}S)_3(\mu_3\text{-}S)_6(PEt_3)_6]^{2-}$ (12).[74] Type D2 has been found in $[SZn_4(S_2AsMe_2)_6]$[75] and type D3 in $[Fe_4S(SR)_2(CO)_{12}]$.[76] Here the coordination corresponds to that of the zincblende or wurtzite structure.

Very few examples are known, apart from solid state structures, where the sulfide acts as a bridge to more than four metal atoms. Two Rh clusters $[Rh_{17}(S)_2(CO)_{33}]^{3-}$ and $[Rh_{10}S(CO)_{22}]^{2-}$ have been reported in the literature[77,78] where the sulfide is found interstitially in the center of the metal cluster and has contacts to nine and eight Rh atoms, respectively. In the former example an almost linear S—Rh—S unit (with $d(Rh—S)$ as short as 2.16 Å) is encapsulated in an Rh_{16} cluster, with four Rh—S contacts at about 2.33 Å and four more at about 2.8 Å.[77] Other related systems are, for example, the $[M_6S_{17}]^{4-}$ ions (13; M = Nb, Ta) which have been recently prepared by Holm and co-workers.[80] Here, among other types of sulfide coordination, a μ_6-S has been found in the base of the bell-shaped ions.

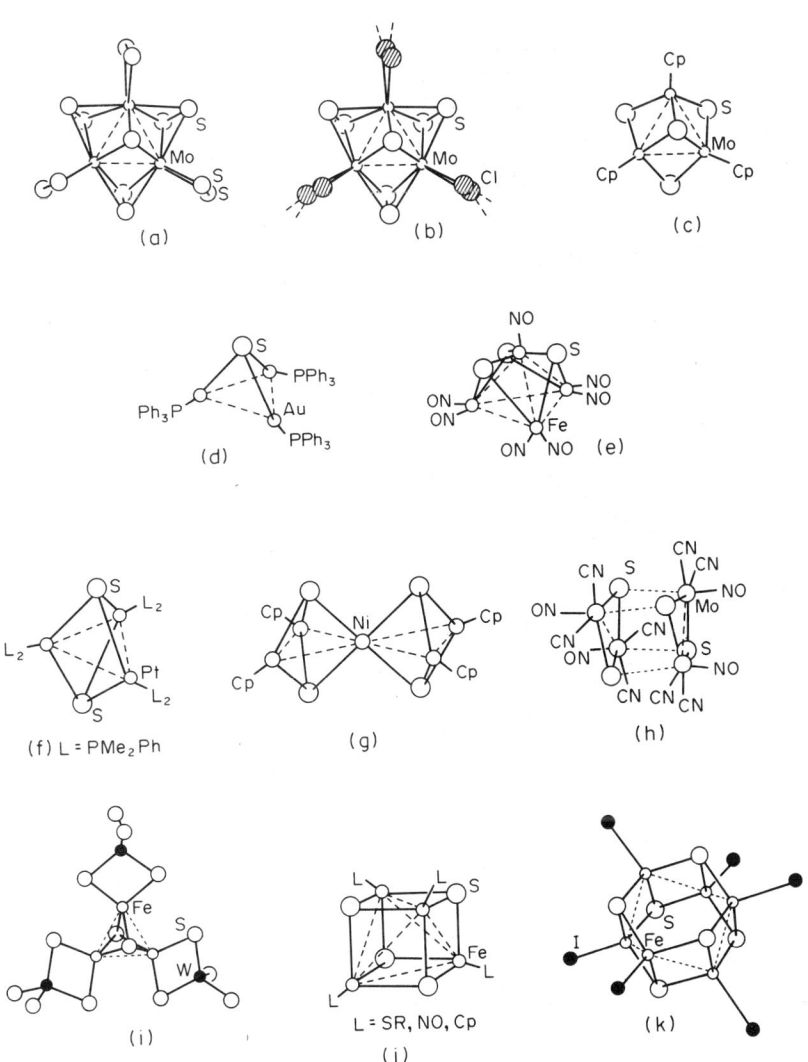

Figure 4 A selection of examples of complexes with an η_3-S ligand: (a) $[Mo_3S_{13}]^{2-}$ (ref. 3); (b) $[Mo_3S_7Cl_4]$ (ref. 66); (c) $[Mo_3S_4Cp_3]^+$ (ref. 65); (d) $[\{(PPh_3)Au\}_3S]^+$ (ref. 67); (e) $[Fe_4S_3(NO)_7]^-$ (ref. 68); (f) $[Pt_3S_2(PMe_2Ph)_6]^{2+}$ (ref. 62); (g) $[Ni_5Cp_4S_4]$ (ref. 69); (h) $[Mo_4S_4(NO)_4(CN)_8]^{8-}$ (ref. 72a); (i) $[Fe_3S_2(WS_4)_3]^{4-}$ (ref. 70); (j) $[Fe_4S_4L_4]^{n-}$ (ref. 41, 71); and (k) $[Fe_6S_6I_6]^{2-}$ (ref. 72)

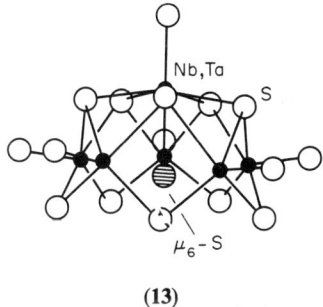

(13)

16.1.4.2 Some Aspects of the Nature of the Metal Sulfide Bond

The examples described in the previous paragraph clearly show that the sulfide ligand can donate a variable number of electrons to the metal. This ability is responsible for the extremely flexible and fascinating structure and electrochemistry of systems with metal ions and sulfide ligands.

Some typical M—S distances for thioanions are given in Table 1. For d(Nb—S) and d(Ta—S) in sulfido complexes, examples have only recently been reported. d(Nb—S$_{term}$) = 2.09–2.13 Å and $d(\text{Nb–S}_{\text{bridge}})_{\text{av}} = 2.3\,\text{Å}$ have been determined for some tetrahydrothiophene adducts of Nb^{V} thiohalides;[81] $d(\text{Ta–S}_{\text{term}}) = 2.181\,\text{Å}$ has been found in $[\text{Ta}^{\text{V}}\text{S}(\text{S}_2\text{CNEt}_2)_3]$[82] and $d(\text{Nb–S}) = 2.24\text{–}2.28$ and $2.24\text{–}2.30\,\text{Å}$ have been found in solid state structures containing NbS_4^{3-} and NbOS_3^{3-}, respectively.[83]

Holm's thioniobate (**13**) contains four different types of sulfide bonding: terminal ($d(\text{Nb–S})_{\text{av}} = 2.18\,\text{Å}$), μ_2-bridging ($d(\text{Nb–S})_{\text{av}} = 2.41\,\text{Å}$), μ_3-bridging ($d(\text{Nb–S})_{\text{av}} = 2.59\,\text{Å}$) and μ_6-bridging ($d(\text{Nb–S}) = 2.636, 2.94\,\text{Å}$).[80] As expected (also for related systems), the bond length increases the series $d(\text{M–S}_{\text{term}}) < d(\text{M–}\mu_n\text{-S}) < d(\text{M–}\mu_{n+1}\text{-S})$.

An important feature of M–S bonds in general is the occurrence of π-bonding over a wide range of bond lengths (*ca.* 2.1–2.4 Å), if S $3p$ lone pairs are available. The overlap integral $S_\pi(\text{M-}nd, \text{S-}3p)$ varies only to a relatively small extent with the M–S distance. Besides, if the M-nd and S-$3p$ VOIPs are nearly equal, depending on geometry, electron population and additional ligands, very different electron density distributions within the M–S bonds are possible.

The negative charge on the sulfide ligand decrease in the series

$$\text{M—S} > \text{M}\diagdown\overset{\text{S}}{\diagup}\text{M} > \text{M}\diagdown\overset{\overset{\text{S}}{|}}{\underset{\text{M}}{\diagup}}\text{M} \approx \text{M}{=}\text{S}{=}\text{M}$$
(bent)

There may be a relatively large charge separation, *e.g.* in MoS_4^{2-} and almost equal net charges on the metal and sulfur atoms, *e.g.* in $[\text{Mo}_2\text{S}(\text{CN})_{12}]^{6-}$.[78a] MO calculations on the latter ion show that there are π MOs delocalized over three centers. Resonance Raman studies further indicate that the delocalization extends over the whole linear N—C—Mo—S—Mo—C—N system. The $\pi(\text{S}) \to d(\text{Mo})$ donation induces a decrease of electron density on the sulfur and is, therefore, responsible for an unusual charge transfer transition Mo $\to$ S (band at $27\,100\,\text{cm}^{-1}$), quite the reverse assignment to that in examples with terminal sulfide where a considerable p contribution also has to be anticipated (see Table 1).[3,4]

Many species containing M_xS_y units can also undergo reversible electrochemical redox reactions. Those with the cubane-like Fe_4S_4 system with several different ligands have been studied in much detail due to their being models for biological systems.[41] All known cyanothiomolybdates (with Mo_xS_y units; Figure 5) also show reversible redox reactions[78a] and exist with different electron populations. It has been proposed that cyanothiometalates (and particularly molybdates) could have been precursors of metal enzymes.[3,79]

Electrons taken away from or added to complexes with M_xS_y units will originate *mainly* from or

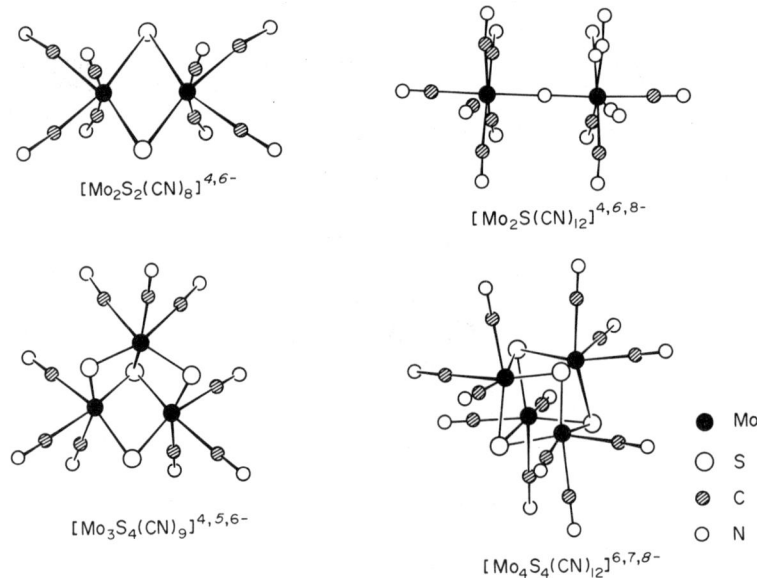

Figure 5 Thio(cyano)molybdates incorporating sulfide ligands. The species with italicized ionic charges have been isolated, the others have been detected in CV experiments

will be going to the sulfide ligands, although the MOs which donate or receive these electrons might be composed largely of metal AOs in the initial species.[84] In this context, results of investigations of the multimetal aggregates $[S_2WS_2CoS_2WS_2]^{n-}$ should be considered. Although upon reduction from $n = 2$ to $n = 3$ the number of electrons is increased in a level with predominantly Co-$3d$ character, the destabilization of this level causes a change of the relative populations, particularly of the σ(Co—S) MOs. As a result the charge changes mainly at the sulfur atoms and remains almost unaltered at the three metal centers.[85]

It turns out that small M_xS_y units with delocalized ground state MOs have low-lying LUMOs which are related to the conducting and semiconducting properties of metal sulfides. We will discuss some general properties of these, from a different point of view, in the following section.

16.1.4.3 A Classification of Metal Sulfides and Their Types of Metal—Sulfur Bonds

Metal sulfides belong to the most important classes of compounds because they are of general significance for geochemistry (they are the most important ores for many metals), analytical and structural chemistry, and biochemistry (metal sulfide systems act as electron transfer systems) as well as catalysis (a high percentage of industrially used heterogeneous catalysts are sulfides) and materials science.

A useful classification can be made by grouping the metal sulfides (practically ionic compounds like the alkali metal sulfides are not considered) according to their solubility in acids and bases, *i.e.* whether they themselves can be regarded as acidic, amphoteric or basic, which corresponds to the difference in electron density distribution within the metal–sulfur bonds or different electron density on the sulfur atoms (which determines the reactivity).[86] *Group B* includes the sulfides of *Mn, Fe, Co, Ni, (Zn and Cd)*, which dissolve in acids (but not in ammonium or alkali metal sulfide solutions). *Group S* includes sulfides which are not soluble in diluted mineral acids, but dissolve in bases (*e.g.* S^{2-}) and form thioanions. We distinguish here between *group S(a)*, which includes the sulfides of *Sn, As, Sb, V, Mo, W and Re* (all with high oxidation states), and *group S(b)*, which consists of the sulfides of the 'soft metals' *Cu, Ag, Au, Hg, Pb, Bi, Pd and Pt* (mainly in low oxidation states).

The solubility of the group S(a) sulfides in an excess of sulfide ions is well established and used for analytical separation procedures. For the group S(b) sulfides, however, there are few reports in the older literature regarding their solubility in alkali metal sulfide solutions.

It is evident that the sulfides with the lowest solubility in water show the highest tendency to form discrete sulfido and polysulfido complexes and clusters corresponding to their solubility in bases, a fact which is interesting for future studies in synthetic and structural chemistry.

A straightforward explanation for the different reactivities or the solubility properties is found in the character of the particular metal–sulfur bond, *i.e.* whether it is covalent or ionic. This determines the charge on the sulfur and metal atoms and, implicitly, which kind of attack, by H^+ or S^{2-}, will take place. Decreased electron density on the sulfur atoms of group S(a) and S(b) metal–sulfur bonds caused by strong covalent contributions stabilizes discrete thioanions like MoS_4^{2-} but also HgS_2^{2-}.

Our SCCC–EH calculations for species of the type MX_4^{n-}, *e.g.* MoS_4^{2-}, CoS_4^{6-} and HgS_4^{6-}, yield considerable covalency for the d^0 complexes of group S(a) (including strong π-bonding for the transition metal species), rather low M—S bond orders for the group B complexes, and high covalency for group S(b) thioanions. The covalency of the M—S bonds thus is mainly determined by the energy values of the low-lying unoccupied metal orbitals (s and p for the main group elements and the elements with d^{10} configuration, $d_{x^2-y^2}$ for Ni, Pd and Pt, and d in general for all other transition metals) relative to the S $2p$ levels and whether metal orbitals (antibonding $\pi^*(2e)$ in the case of Mn, Fe and Co) are occupied.

The above given classification is important in many respects.

(i) In the nature of ore forming solutions: there is no consistent explanation for the transport of 'metals and sulfur together' in the ore forming solutions (the existence of pure soluble metal sulfide clusters has only been reported quite recently). It turns out that only the sulfides with the lowest solubility contants (in water) show increasing solubility in alkaline S^{2-} solutions.

(ii) In the understanding of metal sulfide clusters as active sites in electron transfer proteins and enzymes, their reactivity and formation of their precursors from minerals: the known hetero metal clusters (*e.g.* several Mo enzymes) contain metals from groups B *and* S. Obviously, on the precambrian Earth an aqueous solution of biometals such as Cu could also be obtained in the absence of CN^- (see for instance ref. 3 and references cited therein).

(iii) To heterogeneous catalysis: sulfides of group S(a) often show catalytic activity which may be strongly enhanced in the presence of metal centers of group B elements (*cf.* the important hydrodesulfurization reactions utilizing Co—Mo—S catalysts). The catalytic activity may be discussed in terms of the different electron densities at the metal centers and the sulfur atoms attached to them.

(iv) To analytical separations: here the separation tables introduced by C.R. Fresenius in 1841 are extended and the significance of the well known 'groups' can be understood (especially the solubility of highly insoluble sulfides (in water) in alkaline S^{2-} solutions).

(v) To heavy metal toxicology.

16.1.5 THIOLATES (SR⁻)

Although metal thiolates have been known for a long time, only in the last decade have some fascinating aspects of their chemistry become evident as thiolate (and sulfide) coordination occurs for many metal ions in metalloenzymes such as nitrogenase, molybdooxidases, metallothioneins, ferredoxins and hemocyanins. On the other hand, the structural chemistry of thiolato species, mainly that of the polynuclear type, is very attractive. The monothiolate group RS⁻ is a fundamental monodentate ligand type. The electronic structure and reactivity are comparable to some extent to those of halide ligands as, for instance, a terminal monodentate SR⁻ ligand can often replace or be replaced by a halide one.

16.1.5.1 General Survey of Compound Types

16.1.5.1.1 *Mononuclear complexes with monodentate ligands*

Comparatively few examples of this type of complex are known, as thiolates have a strong tendency to act as bridging ligands. Some representative examples of complexes containing only, or predominantly, thiolate ligands are $[M(SPh)_4]^{2-}$ (M = Mn, Fe, Co, Ni, Zn, Cd),[87,88] $[Mo(SBu^t)_4]$[89] and $[OM(SPh)_4]^-$ (M = Mo, W, Re)[90,91] with either distorted tetrahedral or square pyramidal configurations. Mainly four- and five-coordinate species have been reported. It is difficult to get complexes of higher coordination number because of the steric interactions between the thiolate alkyl or aryl groups and the easy elimination of disulfide (causing reduction of the metal atom).*

Mononuclear Mo^V complexes with sterically hindered monodentate thiolato ligands are of interest as models for special molybdenum enzymes (the Mo oxidases). Here 'dimer' formation *via* thiolato bridges and sulfide formation by C—S bond cleavage do not occur. Such a ligand is TIPTH. The complex $[Mo(CO)_2(TIPT)_3]^-$ (14), with essentially trigonal prismatic coordination about Mo and *trans* CO groups in the axial sites, can be obtained.[91]

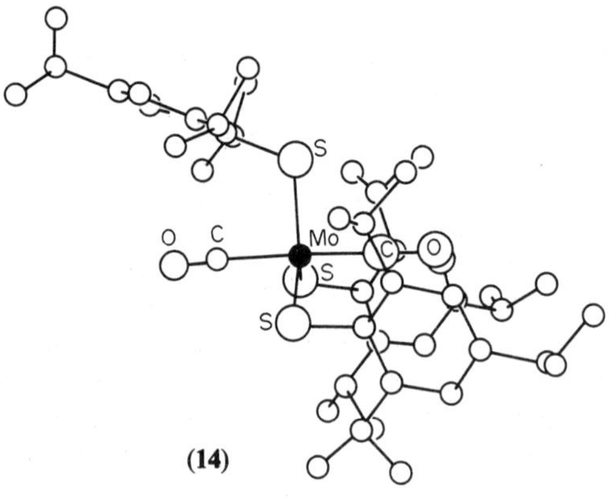

(14)

Complexes of the type $[Fe(SR)_4]^{n-}$ can be regarded as models for rubredoxin.

---

* There is an analogy to some extent to the pseudohalide ligand.

16.1.5.1.2 *Metal sulfide aggregates stabilized by terminal ligands*

An interesting aspect of the chemistry of thiolates is their ability to stabilize metal sulfide aggregates. Polynuclear species like $[S_4M_{10}(SPh)_{16}]^{4-}$ (**15**; M = Zn, Cd)[92] with four nonbridging ligands (molecular structure: *tetrahedro-(μ_3-S)$_4$-octahedro-M$_6$-truncated tetrahedro-(μ-SPh)$_{12}$-tetrahedro-(MSPh)$_4$*) and $[Co_8S_6(SPh)_8]^{4-}$ (**16**)[93] have been obtained, in which metal–sulfur arrays have the 'structure of minerals', in these cases that of sphalerite and pentlandite, respectively. The $Cd_{10}S_{20}$ array for instance is congruent with a fragment of the lattice of sphalerite.

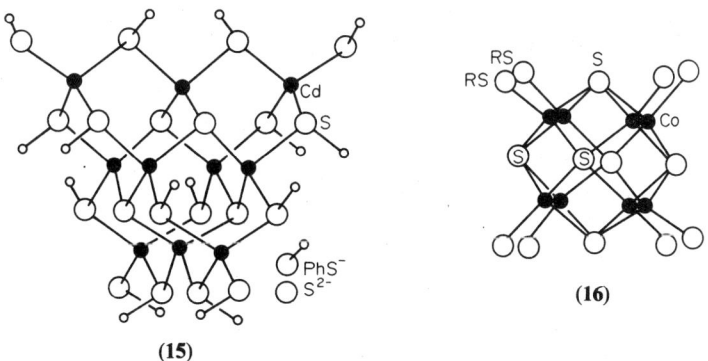

(**15**)   (**16**)

Other examples are $[Fe_2S_2(SR)_4]^{2-}$ (**8**), $[Fe_3S_4(SR)_4]^{3-}$ (**17**), $[Fe_4S_4(SR)_4]^{2-/3-}$ (R = Et; Figure 4) and $[Fe_6S_9(SR)]^{4-}$.[94,41] The spectroscopic and electronic properties of these species are very close to those of the ferredoxins, the external ligation being provided by cysteinyl sulfurs. The thiolato ligands can readily be exchanged by treatment with an excess of another thiol. This provides an excellent method for the extrusion of Fe_4S_4 cores from 4Fe4S ferredoxins by addition of an excess of thiophenol. (The tetranuclear cluster is 'extracted' in nearly quantitative yield as $[Fe_4S_4(SPh)_4]^{2-}$).

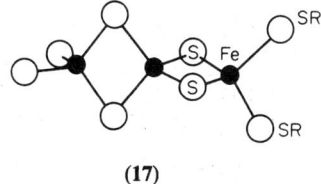

(**17**)

16.1.5.1.3 *Di- and poly-nuclear complexes with μ_2-bridging ligands*

Thiolate ligands show a high tendency to utilize an additional sulfur lone-pair of electrons to form dinuclear complexes with two bridging ligands. Representative examples of this type of complex are $[Fe_2(SR)_6]^{2-}$ (**18**)[87] and $[Ni_2(SCH_2Ph)_2L_2]$ (**19**; L = S$_2$CSCH$_2$Ph).[95] Five different geometric isomers (see Figure 6) are in principle possible for di-μ_2-bridged thiolato complexes depending on whether the M(μ_2-SR)$_2$M unit is planar or folded and on the disposition of the substituents. Whereas examples of the *syn-exo* form are rarely found, several complexes of the other types have been reported.[95]

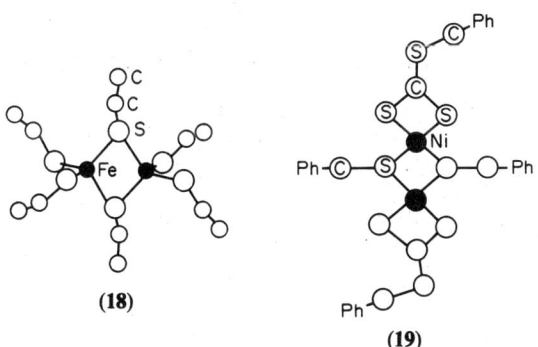

(**18**)   (**19**)

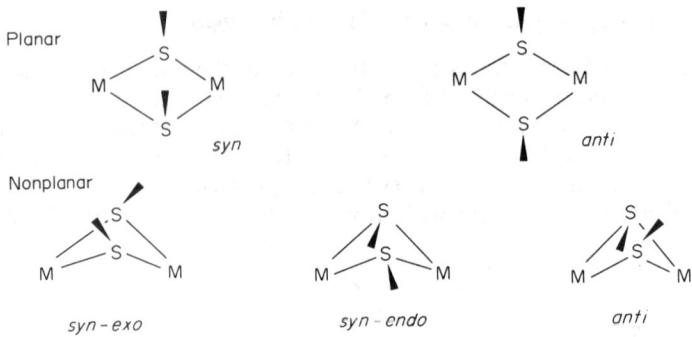

Figure 6 Possible geometries for M(SR)₂M units

Doubly bridging thiolate ligands can also 'glue' more than two metals together, forming $M_x(\mu\text{-}SR)_y$ rings (as in $[Cu_3(SPh)_3(PPh_3)_4]$; **20**[100]) or cage-type compounds. In tetranuclear aggregates of the type $[M_4(SPh)_{10}]^{2-}$ (M = Fe (**21**), Co, Zn, Cd; with adamantane structure) there are six μ_2-bridging and four terminal ligands.[101,102]

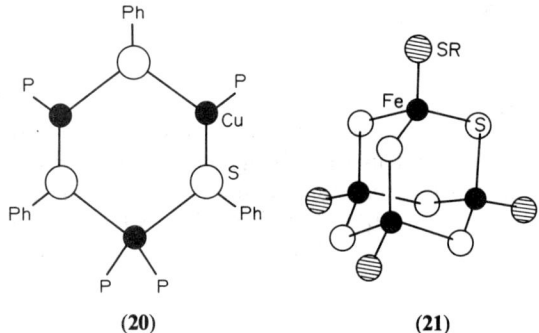

(**20**)   (**21**)

Clusters with interesting structural features are known for lower thiolato: metal ratios. Structures (**22a**)–(**22c**) illustrate how the structures of $[M_4(SPh)_6]^{2-}$, $[M_5(SPh)_7]^{2-}$ and $[M_6(SPh)_8]^{2-}$ (M = Ag) are related.[96] The pentanuclear species can formally be obtained by replacing one S vertex of the distorted tetrahedral $[M_4(SPh)_6]^{2-}$ species (with six μ_2-bridging ligands) by a linear RS—M—SR group. Repetition of this process at the opposite vertex generates the $[M_6(SR)_8]^{2-}$ structure. The cluster $[Ag_{12}(SR)_{16}]^{4-}$ can be derived from linkage of two $[Ag_6(SPh)_8]^{2-}$ units at two Ag vertices.[96]

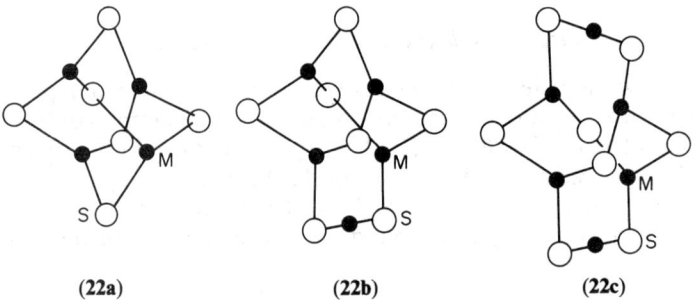

(**22a**)   (**22b**)   (**22c**)

A fascinating compound with 14 bridging thiolato ligands is $[(Ph_3P)_4(AgSBu^t)_{14}]$ (**23**)[97] containing a 28-membered cycle of alternating sulfur and silver atoms. Also $[Cu_5(SBu^t)_6]^-$ (**24**; a cluster with a metal–metal bonded trigonal bipyramid of Cu^I atoms) contains only μ_2-bridging ligands.[98]

The hexanuclear cluster $[Ni_6(SEt)_{12}]$ (**25**)[99] has a crown-like structure with a ring of six square-planar coordinated nickel ions and six μ_2-bridging ligands.

Fewer complexes with three μ_2-bridging thiolates are known. An example is $[W_2Cl_4(\mu_2\text{-}SEt)_3(SMe_2)_2]$, a mixed oxidation state dinuclear tungsten species with a W—W distance of 2.505 Å.[103] Though triple μ_2-thiolato bridges are usually accompanied by a strong metal–metal bond, this is not the case for the clusters $[Mo_2Fe_6S_8(SPh)_9]^{3-/5-}$ (**26**) with two cubane-like MoFe₃S₄ units linked at the molybdenum atoms by a triple thiolato-bridge.[41] (These types of clusters have been claimed to be spectroscopic models for the nitrogenase iron–molybdenum protein and iron–

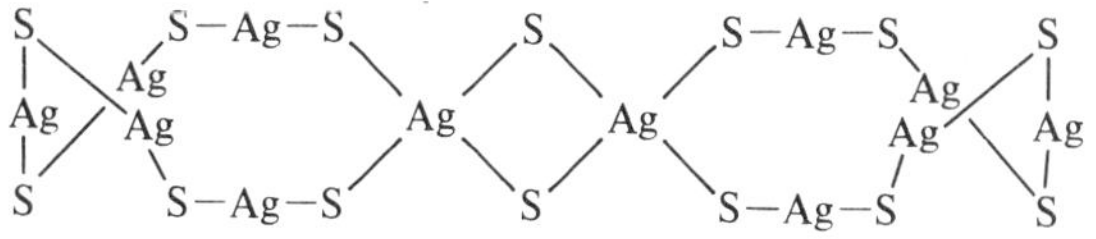

(23)

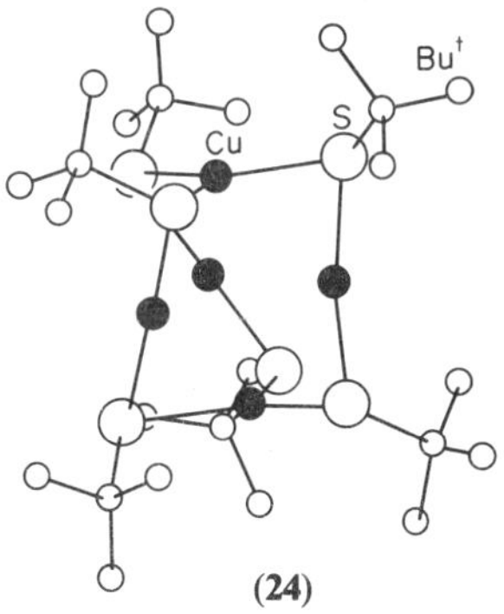

(24)

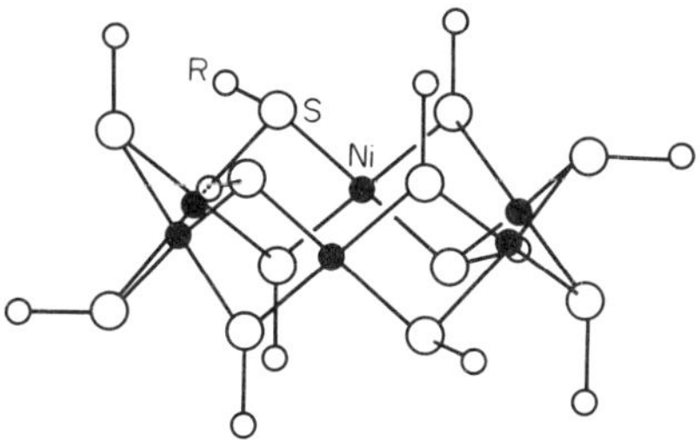

(25)

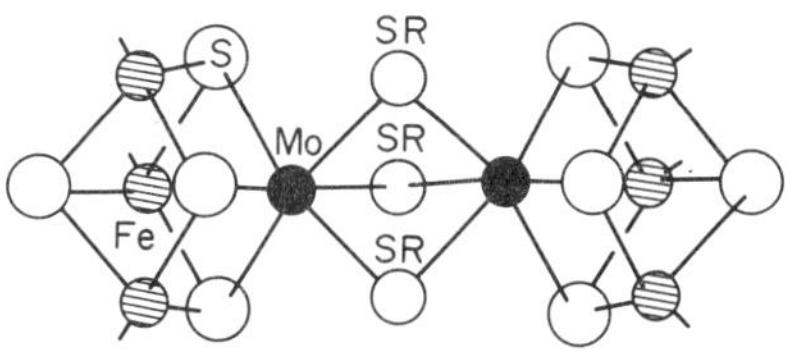

(26)

molybdenum cofactor.) It is possible to cleave the triple-bridge by reactions with sterically hindered catechols to give reactive monocubane clusters such as $[MoFe_3S_4(SEt)_3(cat)_3]^{3-}$.[41]

16.1.5.1.4 Polynuclear complexes with μ_3-bridging ligands

The sulfur atom of a μ_2-bridging thiolato ligand has still one lone pair of electrons available for donation to a third metal atom, thereby creating a μ_3-bridging ligand. Some cluster systems such as $[Ag_{12}(SPh)_{16}]^{4-}$ contain μ_3- as well as μ_2-bridging ligands.[96] The bridging system may be symmetric as in $[\{Re(CO)_3(SMe)\}_4]^{104}$ or asymmetric as in $[\{MeZn(SPr^i)\}_8]$ (with rather different Zn —S distances).[105]

16.1.5.1.5 Complexes with polydentate ligands

The simplest ligands belonging to this class are the bidentate thiols $HS(CH_2)_nSH$. The corresponding complexes are, in general, more stable than those with monodentate ligands. Examples of complexes with different types of bidentate ligands (chelating and bridging) are $[SRe(SCH_2CH_2S)_2]^-$ (27),[107] $[Ni_2(SCH_2CH_2SCH_2CH_2S)_2]^{108}$ and $[Fe_2(S_2$-o-xyl)_3]^{2-}$ (28).[109]

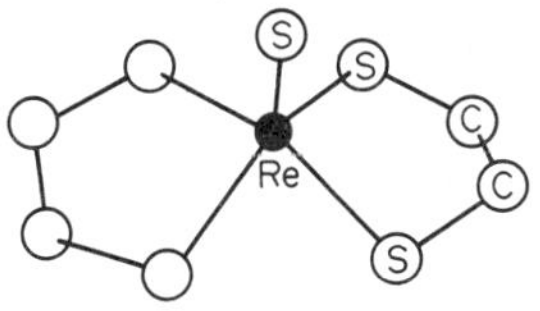

(27)

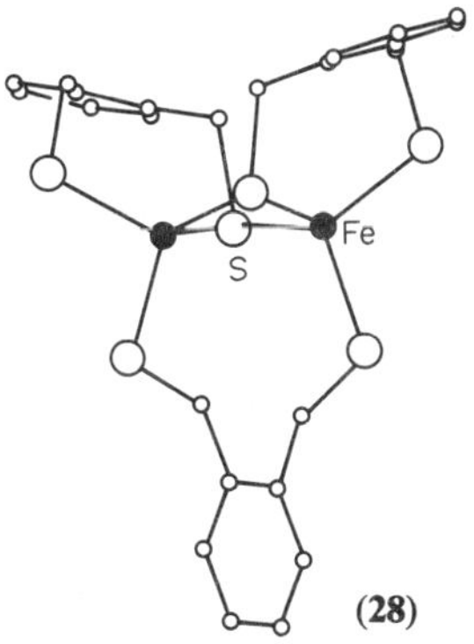

(28)

Not much is known about trithiolato complexes. In the homoleptic complex $[Re\{(SCH_2)_3CMe\}_2]^-$ (29)[106] the central atom has trigonal prismatic coordination.

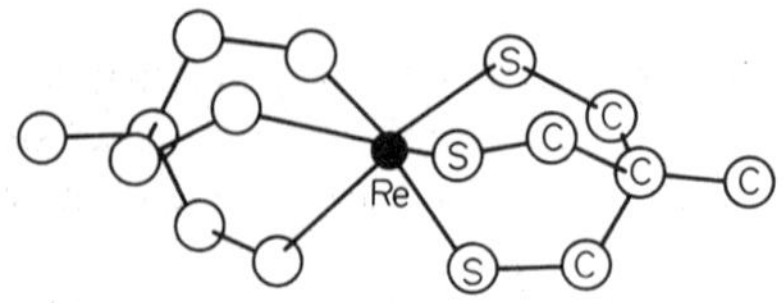

(29)

Polydentate ligands with thiolato and other coordination sites have also been studied but cannot be discussed here in detail (including other S donor atoms from thioether groups as in $[Ni_2(SCH_2CH_2SCH_2CH_2S)_2]^{108}$ or other N and P donors).[91]

16.1.5.2 Sulfur–Sulfur Interactions in Complexes with Monodentate and Polydentate Ligands

An interesting aspect of the coordination chemistry is the presence of S–S interactions. Electron transfer in $Mo(SR)_2$ groups with partial disulfide formation is probably significant in biochemistry. Stiefel *et al.* have proposed a scheme with the reacting units (**30a–c**).[110,111]

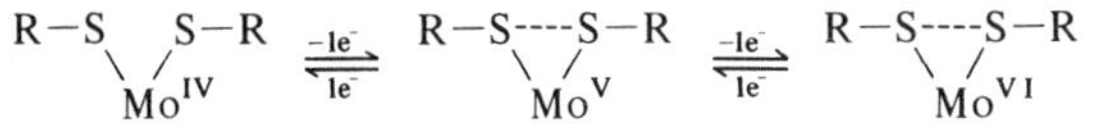

(30a)   (30b)   (30c)

These interactions are less common with monodentate ligands than with polydentate ones. One example of the first case is the complex $[Mo_2O_2(\mu_2\text{-}O)_2(SPh)_4]^{2-}$ ($d(\text{S–S}) = 2.94\,\text{Å}$),[112] and the compound $[MoO_2(MeNHCH_2CMe_2CS)_2]$ ($d(\text{S–S}) = 2.76\,\text{Å}$)[110] is an example of the second case. The limiting factor for the interaction of the monodentate ligands is disulfide formation accompanied by reduction of the metal atom.

16.1.5.3 Synthesis and Reactions

Thiolato complexes are, in general, prepared by reaction of halogeno complexes with thiolate anions. Some typical reactions will be summarized here. It will also be shown that the thiolate often acts as a reducing agent and that the type of species obtained is often highly dependent on the relative amounts of the reagents used.

The mononuclear complex $[Fe(SR)_4]^{2-}$ (R = aryl or alkyl) can be obtained by the reaction of Fe^{II} salts with an excess of thiolate anion SR^- in methanol. Using smaller amounts of thiolate anion SR^- in methanol the species $[Fe_2(SR)_6]^{2-}$, $[Fe_4(SR)_{10}]^{2-}$ and $[Fe_3Cl_6(SR)_3]^{3-}$ are formed (with 3, 2.5 and 1 equivalents, respectively).[87] In general, Fe^{III} is reduced to Fe^{II} (but see ref. 113).[87]

Mononuclear tetrahedral complexes of Co^{II}, Zn^{II} and Cd^{II} are also formed with high molar ratios of arylthiolate whereas at lower thiolate concentrations polynuclear species can be obtained. The equilibria existing in acetonitrile solution containing Co^{II}, PhS^- and halide have been postulated as shown in Scheme 1.[114]

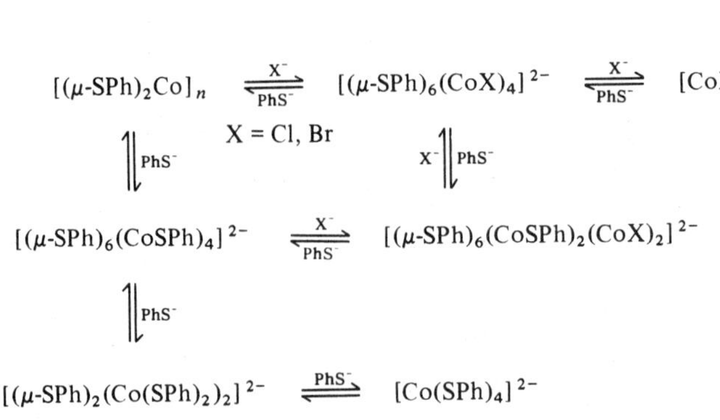

Scheme 1

The $[M_4(SPh)_{10}]^{2-}$ type of complex is stereochemically nonrigid (owing to the occurrence of terminal/bridge ligand scrambling) and can undergo rapid metal ion exchange.[101] The iron complex of this type undergoes an interesting quantitative reaction with elemental sulfur to give $[Fe_4S_4(SPh)_4]^{2-}$.

Aliphatic thiolates often show C—S bond cleavage as shown in equation (2).[107] The C—S fission reaction often tends to generate rather unreactive sulfido-bridged species, but sometimes, as in the given example, a mononuclear sulfido complex is produced.

Another type of redox reaction is the formation of disulfide with a two-electron reduction of the metal center M^n (equation 3).[115] The disulfide formation may be regarded as a limiting case of the sulfur–sulfur interactions which occur in thiolato complexes.

$$[Re^{II}Cl_6]^{2-} \; + \; HSCH_2CH_2SH \; \longrightarrow \; [Re^VS(SCH_2CH_2S)_2]^- \; + \; CH_2{=}CH_2 \tag{2}$$

$$M^n\!\!\begin{array}{c}\diagup SPh \\ \diagdown SPh\end{array} \to M^{n-2} \; + \; PhSSPh \tag{3}$$

$[Mo(SBu^t)_4]$, which can be prepared by reaction of $MoCl_4$ with $LiSBu^t_4$ in 1,2-dimethoxyethane, undergoes a combined substitution and redox reaction with Bu^tNC to give an Mo^{II} complex (equation 4).[117]

$$[Mo(SBu^t)_4] \; + \; 4Bu^tNC \; \longrightarrow \; \textit{cis}\text{-}[Mo(SBu^t)_2(CNBu^t)_4] \; + \; Bu^tSSBu^t \tag{4}$$

Though thiolato complexes often show electrochemically reversible redox processes, the electrochemistry of complexes with monodentate thiolato ligands is very often complicated by irreversible reactions (*e.g.* liberating a thiol, which is itself subject to an oxidation process), such as during the oxidation of $[MoO(SPh)_4]^-$. The complex shows a quasireversible one-electron reduction process corresponding to the generation of $[MoO(SPh)_4]^{2-}$. The overall oxidation process, however is as shown in equation (5).[117,118] By contrast, the complex $[MoO(SC_6H_2Pr^i_3)_4]^-$ undergoes a *reversible* one-electron oxidation process, reflecting the reluctance of the sterically demanding thiol to form the dinuclear thiolato-bridged species.[91]

$$2[MoO(SPh)_4]^- \; \xrightarrow{\;-2e-\;} \; [Mo_2O_2(SPh)_6(MeCN)] \; + \; PhSSPh \tag{5}$$

The $[Fe_4S_4(SR)_4]^{n-}$ and $[Mo_2Fe_6S_8(SR)_9]^{n-}$ clusters undergo extensive redox processes predominantly localized in the M_4S_4 clusters. These types of reaction are also of bioinorganic interest.[41,119,120]

16.1.6 THE DISULFIDO LIGANDS[6]

Complexes of stable diatomic ions and molecules (*e.g.* O_2, N_2, CO, CN^-) are well known and have been extensively studied. In recent years a number of metal complexes containing one or more coordinated S_2^{2-} units have been prepared and characterized by X-ray structure determination. Complexes with S_2^{2-} units can be formed for many metals under a variety of conditions and coordinated S_2^{2-} ligands exhibit a rich chemistry.

16.1.6.1 Survey of Compound Types

The complexes show a remarkable variety of structures (see Table 2). This variety results from extension of the fundamental structures Ia (side-on coordination) and IIa and IIb (*cis* and *trans* end-on (doubly) bridging coordination) which are known for dioxygen complexes by using the remaining lone pairs of electrons on sulfur to coordinate additional metal atoms. Such coordination of additional metal atoms occurs for all three fundamental structural types (Ia, IIa, IIb). Representative examples (including the definition of the different types) are shown in Table 2. In contrast to dioxygen, end-on coordination of a disulfido ligand to a *single* metal atom to give a terminal MS_2 unit is unknown.

Another type of structural unit found in many cluster compounds is type III, in which both sulfur atoms of the ligand are bonded to each of two metal atoms. The S—S bond is orientated approximately normal to the metal–metal vector.

Table 2 Typical Geometries of S_2^{2-} Complexes

Structural type		Example
Ia	(η^2) M$\diagdown$S–S (S$\cdots$S bridged)	$[Ir(dppe)_2(S_2)]^+$
Ib	M$\diagdown$S–S–M	$[Mo_4(NO)_4(S_2)_5S_3]^{4-}$
Ic	M$\diagdown$S(–M)–S	$[Mn_4(S_2)_2(CO)_{15}]$
Id	M$\diagdown$S(–M)–S(–M)(–M)	$[Mn_4(S_2)_2(CO)_{15}]$
IIa	$(\eta^1{-}\eta^1)$ M–S–S–M	$[(NH_3)_5Ru(S_2)Ru(NH_3)_5]^{4+}$
IIb	M–S–S–M	$[Cp_2Fe_2(S_2)(SEt)_2]$
IIc	M–S–S(–M)(–M)	$[Cp_4Co_4(S_2)_2S_2]$
IId	(M)(M)–S–S(–M)(–M)	$[(SCo_3(CO)_7)_2(S_2)]$
III	$(\eta^2{-}\eta^2)$ S–S bridging two M	$[Mo_2(S_2)_6]^{2-}$
IV	S(–M)(–M)	$[Cp_2'Cr_2(S_2)_2S]^a$

^{a} Cp′ = C$_5$Me$_5$

A very remarkable type of bridging ligand has been discovered in which two metal atoms are attached to the same sulfur atom of the ligand (structure type IV).

16.1.6.1.1 Complexes with type I structures (η^2 basic unit)

Complexes with type I structures are listed in Table 3 along with their S—S distances and S—S vibrational frequencies. The most common mode of coordination is type Ia in which the ligand occupies two coordination sites of the metal atom. One example of this type of binding is the complex [Ir(dppe)$_2$(S$_2$)]Cl (**31**) whose dioxygen analog exhibits a similar structure. The Ir atom is in a distorted octahedral environment, as would be expected for a six-coordinate metal with a d^6 electronic configuration. Higher coordination numbers are found for complexes of the early transition elements. In Cs$_2$[MoO(S$_2$)$_2$(C$_2$O$_3$S)] (**32**) the Mo atom adopts pentagonal bipyramidal coordination geometry with the S atoms of S_2^{2-} and of the thiooxalate ligand in the pentagonal plane. It is interesting to note that with NMe$_4^+$ as cation both of the S_2^{2-} ligands are asymmetrically bonded to the Mo atoms (Mo—S = 2.42, 2.43, 2.38, 2.39 Å).

Several dinuclear Mo^V complexes contain type Ia ligands. The central [MoO(η^2-S)$_2$MoO]$^{2+}$ structural unit in (NMe$_4$)$_2$[Mo$_2$O$_2$S$_2$(S$_2$)$_2$] (**33**) has the Mo=O groups in *syn* stereochemistry.

Table 3 Complexes with Side-on Coordinated S_2^{2-} Ligands (Type I)

	Type	$d(S-S)^a$ (Å)	$v(S-S)^a$ (cm^{-1})	Color
$[Ir(dppe)_2(S_2)]^+$	Ia	2.07	528	Orange
$[Rh(dmpe)_2(S_2)]^+$	Ia	—	525	Orange
$[Rh(L1)(S_2)Cl]^b$	Ia	—	546	Orange
$[Os(CO)_2(PPh_3)_2(S_2)]$	Ia	—	—	Orange
$[Rh(vdiars)_2(S_2)]^{+c}$	Ia	—	554	Red-brown
$[Rh(L2)_2(S_2)]^{+d}$	Ia	—	—	Brown
$[MoO(S_2)_2(mtox)]^{2-e}$	Ia	2.01	530	Dark red
$[MoO(S_2)(dtc)_2]^f$	Ia	2.02	558	Blue
$[Cp_2Mo(S_2)]$	Ia	—	536	Red
$[Cp_2Nb(S_2)Cl]$	Ia	—	540	Red
$[Cp_2Nb(S_2)Me]$	Ia	2.01	540	Orange
$[Mo_2O_2S_2(S_2)_2]^{2-}$	Ia	2.08	510	Red-orange
$[Mo_2S_4(S_2)(S_4)]^{2-}$	Ia	2.07	—	Red-violet
$[Mo_4(NO)_4(S_2)_5S_3]^{4-}$	Ia	2.04	536	Red
$[Mo_4(NO)_4(S_2)_5S_3]^{4-}$	Ib	2.05	550	Red
$[Cp_2Fe_2(S_2)_2CO]$	Ib	1.99	—	Green
$[Mn_4(S_2)_2(CO)_{15}]$	Ic	2.07	—	Red
$[Mn_4(S_2)_2(CO)_{15}]$	Id	2.09	—	Red

^{a} Mean value. ^{b} L1 = PPh(CH$_2$CH$_2$CH$_2$PPh$_2$)$_2$. ^{c} vdiars = Ph$_2$AsCHCHAsPh$_2$. ^{d} L2 = R$_2$PNHPR$_2$. ^{e} mtox = O$_2$CCOS^{2-}. ^{f} dtc = S$_2$CNPr$_2^-$.

(31)   (32)   (33)

Disulfido ligands of structure type Ib also exist in the interesting compound $(NH_4)_4[Mo_4(NO)_4(S_2)_5S_3]\cdot 2H_2O$ (11), which contains sulfur atoms in five different bonding situations. Four of the five S_2^{2-} ligands unsymmetrically bridge pairs of Mo atoms (structure type Ib). One S_2^{2-} ligand is side-on coordinated to a single Mo atom (structure type Ia).

Structural types Ic and Id (Table 2), in which the ligand is coordinated to three and four metal atoms, respectively, both occur in $[Mn_4(S_2)_2(CO)_{15}]$ (34).

(34)

16.1.6.1.2 Complexes with type II structures ($\eta^1-\eta^1$ basic unit)

Many diatomic ligands favor end-on coordination to a single metal atom. This mode of coordination is also known for several O$_2$ complexes, although doubly bridging end-on coordination (structural types IIa and IIb) is more common. For the S_2^{2-} ligand only type IIa and type IIb structures are known.

Complexes with type II structures are listed in Table 4 along with their S—S distances and S—S vibrational frequencies.

There are several examples of complexes with type IIa structures (planar *trans* end-on bridging coordination). Vibrational spectroscopy indicates that this structure occurs in $[(CN)_5Co(S_2)Co(CN)_5]^{6-}$ (35). *Trans* arrangement of the metal atoms has been proved by X-ray structure determination for $[(NH_3)_5Ru(S_2)Ru(NH_3)_5]Cl_4$.

Structure type IIb, *cis* end-on bridging coordination, is found in $[Cp_2Fe_2(S_2)(SEt)_2]$ (36) where

Sulfur Ligands

Table 4 Complexes with Type II S_2^{2-} Ligands

	Type	$d(S—S)^a$ (Å)	$\nu(S—S)^a$ (cm^{-1})	Color
$[(CN)_5Co(S_2)Co(CN)_5]^{6-}$	IIa	—	490	Red-brown
$[(NH_3)_5Ru(S_2)Ru(NH_3)_5]^{4+}$	IIa	2.01	514	Green
$[Cp(CO)_2Mn(S_2)MnCp(CO)_2]$	IIa	2.01	—	Dark green
$[(Re_6S_8)S_{4/2}(S_2)_{2/2}^{4-}]_n$	IIa	2.09	—	Dark red
$[Cp_2Fe_2(S_2)(SEt)_2]$	IIb	2.02	507	Dark green
$[Mo_4(NO)_4(S_2)_6O]^{2-}$	IIb	2.08	480	Violet
$[Cp'_2Mo_2(S_2)_5]^c$	IIb^b	2.04	—	Black
$[Cp_4Co_4(S_2)_2S_2]$	IIc	2.01	—	Black
$[Cp_4Fe_4(S_2)_2S_2]$	IIc	2.04	503	Black
$[(SCo_3(CO)_7)_2(S_2)]$	IId	2.04	—	Black
$[CpMn(NO)(S_2)]_n$	IId	—	491 (?)	Red-brown
$[(Cp_4Fe_4(S_2)_2S_2)_2Ag]^+$	IId	2.05	478	Black

^{a} Mean value. ^{b} Mo—S—S—Mo torsion angle: $\sim 59.7°$. ^{c} Cp' = Me$_5$P.

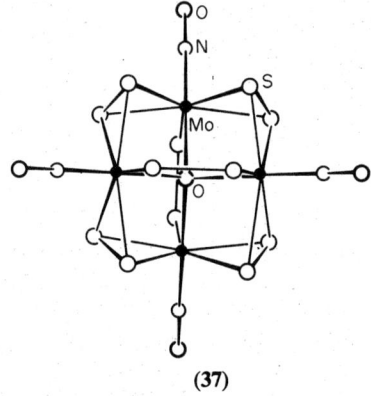

(35)   (36)

it is dictated by the two bridging SEt groups. The Fe(S$_2$)Fe group is planar in (**36**), just as with type IIa structures.

The same kind of disulfur bridging occurs in $[Mo_4(NO)_4(S_2)_6O]^{2-}$ (**37**). This complex represents a distorted tetrahedral cage of Mo atoms with an interstitial four-coordinated oxygen atom. The two non-adjacent edges of the tetragonal bisphenoid are bridged by type IIb S_2^{2-} ligands. The remaining four edges are bridged by ligands of type III.

(37)

Expansion of structures IIa and IIb through coordination of additional metal atoms to the ligand destroys the planarity of the central [M(S$_2$)M] unit.

The complex $[Cp_4Co_4(S_2)_2S_2]$ has a cage structure with two S_2^{2-} ligands, each of which bridges three Co atoms (type IIc).

A few complexes have been structurally characterized in which an S_2^{2-} ligand bridges four metal atoms (structural type IId, Table 2). In the cluster compound $[\{SCo_3(CO)_7\}_2(S_2)]$ (**38**) the ligand is bonded to one edge of each of two Co$_3$ triangles. The resulting structure consists of two Co$_3$S$_2$ planes which have a common S$_2$ edge.

(38)

16.1.6.1.3 Complexes with type III structures (η^2–η^2 basic unit)

Another class of complexes containing bridging S_2^{2-} ligands are those with type III structures in which the S—S bond is oriented approximately perpendicular to the metal–metal vector so that each sulfur atom is bonded to both metal atoms. These complexes are listed in Table 5. Of particular interest are the cluster anions which contain only molybdenum and sulfur. The unusual dinuclear compound $(NH_4)_2[Mo_2(S_2)_6]\cdot 2H_2O$ (39a) contains only S_2^{2-} ligands. Two of the S_2^{2-} ligands bridge the two Mo atoms (type III) and four of the S_2^{2-} ligands are bonded to a single Mo atom (type Ia geometry). An unusual feature of the crystal structure is the orientation of the two independent $[Mo_2(S_2)_6]^{2-}$ anions relative to the crystallographic C_2 axes. One anion has a C_2 axis along the Mo—Mo vector, the other a C_2 axis perpendicular to the Mo—Mo vector. The idealized symmetry for each anion is D_2. The coordination geometry about each Mo atom is a distorted dodecahedron. An interesting feature of the type III bridging S_2^{2-} ligands is their asymmetric bonding to the Mo atoms. Another cluster anion with type III ligands and which contains only molybdenum and sulfur atoms occurs in $(NH_4)_2[Mo_3S(S_2)_6]\cdot nH_2O$ (39b; $n = 0$–2; with variable nonstoichiometric amounts of water, which are disordered in the crystal lattice).

Table 5 Complexes with Type III S_2^{2-} Ligands

	$d(S-S)^a$ (Å)	$v(S-S)^a$ (cm^{-1})	Color
$[Mo_2(S_2)_6]^{2-}$	2.04	550	Green-black
$[Mo_2(S_2)_2Cl_4Cl_{4/2}]_n$	1.98	561	Dark brown
$[Nb_2(S_2)_2Cl_4]$	2.03	588	Brown
$[Mo_3S(S_2)_6]^{2-}$	2.02	545	Red
$[Mo_3S(S_2)_3Cl_4]$	2.03	562	Red
$[Fe_2(S_2)(CO)_6]$	2.01	555	Red
$[Ta_2(S_2)_2(PS_4)_2]$	2.05	—	Grey
$[Mo_2(n\text{-BuCp})_2(S_2)Cl_4]$	2.02	—	Black
$[Mo_2(S_2)(S_2C_2Ph_2)_4]$	2.04	518	Green-black
$[Nb_2S(S_2)Br_4(tht)_4]^b$	2.01	—	Green
$[Mo_2(S_2)(SO_2)(CN)_8]^{4-}$	2.00	520	Violet

^{a} Mean value. ^{b} tht = tetrahydrothiophene.

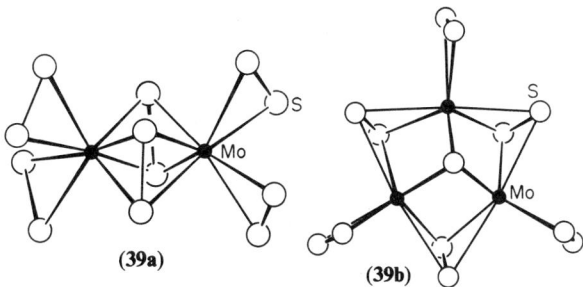

(39a) (39b)

In $[Cp_2Cr_2S(S_2)_2]$ (40) two Cr atoms are bridged by a type III disulfido ligand, by a sulfido group and by a type IV ligand. This compound is the first example of type IV ligand. Both the type III and type IV S—S distances are unusually long (2.15 and 2.10 Å, respectively).

(40)

16.1.6.2 Structural Features of Disulfido Complexes

Several common features of the structures can be noted.

Structural diversity is exhibited through the utilization of the nonbonded pairs of electrons on the S_2^{2-} ligand of both type I and type II complexes to coordinate additional metal atoms.

S—S distances of known complexes range from 1.98 to 2.15 Å, with standard deviations of 0.002 Å or larger for individual determinations. Most S—S distances are intermediate between the distance of 1.89 Å for S_2 $(^3\Sigma_g^-)$ and 2.13 Å for S_2^{2-} $(^1\Sigma_g^+)$ in Na_2S_2. The mean S—S distances show no clear systematic trend with structural type: type Ia, 2.04(3) Å (seven compounds, Table 2); types IIa and IIb, 2.04(4) Å (six compounds, Table 3); type III, 2.02(3) Å (11 compounds, Table 4).

The average type Ia S—S distances for $[Mo_2(S_2)_6]^{2-}$ and $[Mo_3S(S_2)_6]^{2-}$ are slightly longer than the average type III S—S distances (0.006–0.04 Å, 1–3 σ). The data for $[Mo_4(NO)_4(S_2)_6O]^{2-}$ indicate that type IIb S—S distances may be slightly longer than type III S—S distances (0.018 Å, 3–6 σ). Caution must be exercised in interpreting these trends because of the known tendency for type III ligands to be slightly asymmetrically bound to the metal (*vide infra*).

Partial disorder of an asymmetrically bound type III disulfur ligand between structures (**41a**) and (**41b**) will be manifested as a slightly shorter S—S distance, but a slightly shorter S—S distance for type III ligands is also consistent with vibrational spectra.

(41a)   (41b)

High coordination numbers of the metal atom are favored for type I and type III structures by the small coordination angle of bidentate disulfido ligand. High coordination numbers also protect the metal center from nucleophilic attack, an important factor in stabilizing metal–metal bonds.

M—S distances are similar to those for metal complexes of other simple sulfur-containing ligands. Ligands of types Ia and III exhibit slightly asymmetric M—S distances. This inequality is usually 0.03–0.04 Å; however, it can be as large as 0.06–0.08 Å in complexes which contain structual units such as (**42**) in which the sulfur atoms from two different disulfido ligands are approximately *trans* to one another.

(42)

Other coordination types for disulfido complexes can be envisioned in addition to those of Table 2. In principle, six electron pairs are available to coordinate to metals

16.1.6.3 Syntheses of S_2^{2-} Complexes

The complexes have been prepared by various methods. The preparative routes can be grouped into five classes.

16.1.6.3.1 *Reactions of low valence metal complexes with elemental sulfur*

Oxidative addition of elemental sulfur to an electron-rich metal, which is coordinatively unsaturated, is a convenient method for preparing S_2^{2-} complexes (equation 6). The reaction has a high yield for complexes having metal atoms with d^8 electronic configurations, such as Ir^I, Rh^I, Ru^0 and Os^0. Many reactions of complexes with low-valence metal atoms produce mixtures of various metal–sulfur complexes, often in low yield.

$$[ML_4] + S_8 \longrightarrow [ML_4(S_2)] \tag{6}$$

16.1.6.3.2 *High-temperature reactions of metals or metal halides with S_8 and/or S_2Cl_2*

Several compounds with network structures which contain S_2^{2-} units can be prepared by high-temperature reactions (250–700 °C). An example of a reaction of a metal halide with sulfur and S_2Cl_2 is shown in equation (7). The conditions under which these S_2^{2-}-containing cluster compounds are formed illustrate their high stability.

$$MoCl_3 + S_8/S_2Cl_2 \longrightarrow [Mo_3S(S_2)_3Cl_4] + [Mo_2(S_2)_2Cl_6] \tag{7}$$

16.1.6.3.3 *Reactions of metal complexes with S_x^{2-}*

One method for directly introducing the ligand is by substitution of S_2^{2-} for other ligands. For this purpose, Na_2S_2 (equation 8) or a polysulfide solution (equations 9 and 10) can be used. Aqueous polysulfide solutions are especially useful reagents for the synthesis of S_2^{2-} complexes. The solutions are prepared by saturating an aqueous solution of ammonia, to which sulfur is added, with H_2S. In the case of the reaction of MoO_4^{2-} with this S_x^{2-} solution, Mo^{VI} is reduced and the products obtained depend upon the sulfur content of the polysulfide solutions. From polysulfide solutions containing 6 g S/500 ml H_2O the Mo^{IV} cluster $(NH_4)_2[Mo_3S(S_2)_6]\cdot nH_2O$ ($n = 0$–2) (**39b**) is precipitated at 90 °C; the corresponding Mo^V complex $(NH_4)_2[Mo_2(S_2)_6]\cdot 2H_2O$ (**39a**) is obtained from the filtrate at room temperature. If a more concentrated polysulfide solution is used (90 g S, 500 ml H_2O), then a nearly quantitative yield of $(NH_4)_2[Mo_3S(S_2)_6]\cdot nH_2O$ is obtained after heating for a few hours.

$$[Cp_2MoCl_2] + Na_2S_2 \longrightarrow [Cp_2Mo(S_2)] + 2NaCl \tag{8}$$

$$[Mo_3S_4(CN)_9]^{5-} + S_x^{2-} \longrightarrow [Mo_3S(S_2)_6]^{2-} \tag{9}$$

$$Mo_3S(S_2)_3Cl_4 + S_x^{2-} \longrightarrow [Mo_3S(S_2)_6]^{2-} \tag{10}$$

16.1.6.3.4 *Oxidation of sulfur-containing ligands*

An interesting way of forming S_2^{2-} ligands is by the oxidation of two SR ligands. One example is reaction (11). Such a reaction also proceeds when R is a good leaving group, for example $SnMe_3$. Intramolecular redox reactions within metal–sulfido moieties are another important route to S_2^{2-} complexes. An example is the formation of $[Mo_2O_2S_2(S_2)_2]^{2-}$ from $[MoO_2S_2]^{2-}$. The redox processes (12)–(14) might be involved.

$$[Fe(CO)_4(SR)_2] \xrightarrow{ox} [Fe_2(S_2)(CO)_6] \tag{11}$$

$$Mo^{VI} \xrightarrow{red} Mo^{IV} \tag{12}$$

$$(Mo^{IV} + Mo^{VI} \longrightarrow 2Mo^V) \tag{13}$$

$$2S^{2-} \xrightarrow{ox} S_2^{2-} \tag{14}$$

16.1.6.3.5 *Reactions with other sulfur-containing reagents*

The formation of S_2^{2-} complexes by other sulfur-containing reagents such as H_2S, P_4S_{10}, COS, Na_2S_4 or Na_2S, R_2S_x and Na_2SO_3 has also been reported.

16.1.6.4 Reactions of the Coordinated S_2^{2-} ligands

16.1.6.4.1 *Electron transfer and intramolecular redox reactions*

The redox behavior of S_2^{2-} complexes is of particular interest because it can probably provide a foundation for understanding the course of reactions in relevant enzymes and catalysts (especially hydrodesulfurization catalysts). Intramolecular redox reactions related to type Ia S_2^{2-} ligands can be summarized by Scheme 2. Examples of reaction (i) in Scheme 2 are the oxidations of $=$S and —SR groups. Reaction (ii) involves reduction of the S—S bond to form two sulfido groups. An example of reaction (iii) is the thermal decomposition of $Cs_2[Mo_2(S_2)_6]\cdot nH_2O$ to give S_2 as the main gaseous product. Examples of reaction (iv) are the syntheses of S_2^{2-} complexes from reactions employing elemental sulfur.

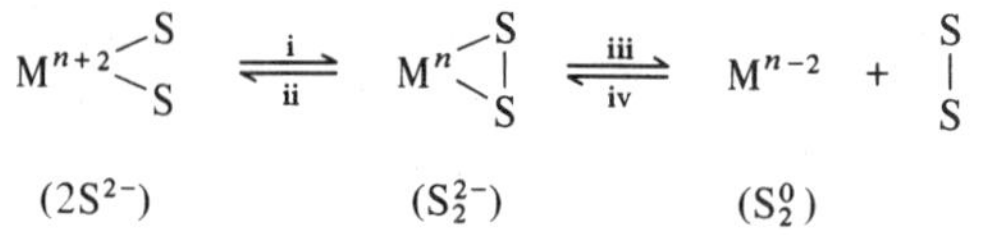

$$M^{n+2}\diagdown_{S}^{S} \underset{ii}{\overset{i}{\rightleftharpoons}} M^{n}\diagdown_{S}^{S}| \underset{iv}{\overset{iii}{\rightleftharpoons}} M^{n-2} + {S \atop |\atop S}$$

$$(2S^{2-}) \qquad\qquad (S_2^{2-}) \qquad\qquad (S_2^0)$$

Scheme 2

16.1.6.4.2 Reactions with nucleophiles (abstraction of S^0)

A characteristic reaction of S_2^{2-} ligands is the abstraction of a sulfur atom by nucleophiles (N) such as PPh_3, SO_3^{2-}, SR^-, CN^- and OH^-, *e.g.* reaction 15 for type III ligands. The reaction involves transfer of a neutral sulfur atom from the complex with no change in the oxidation state of the metal atoms. In this regard the reaction of $[Mo_3S(S_2)_6]^{2-}$ with CN^- (giving $[Mo_3S_4(CN)_9]^{5-}$) is particularly interesting. The bridging S_2^{2-} ligands are converted to bridging sulfido ligands and the terminal S_2 ligands are replaced by CN^- groups.

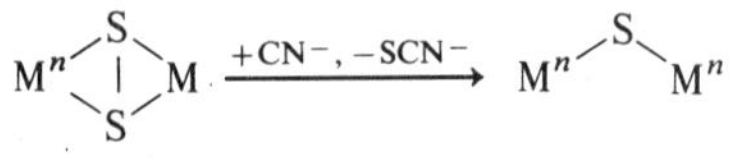

$$M^n\diagdown_{S}^{S}| \diagup M \xrightarrow{+CN^-,\ -SCN^-} M^n\diagdown_{S}\diagup M^n \qquad\qquad (15)$$

In $[Mo_4(NO)_4(S_2)_4(S_2)'_2O]^{2-}$ both end-on bridging S_2^{2-} ligands of type IIb and the type III bridging S_2^{2-} ligands react with CN^- to produce a strongly distorted $[Mo_4S_4]$ cube (equation 16). The cube results from abstraction of a sulfur atom from each of the four type III ligands by reaction (15). The main product of the reaction is $[Mo_2S_2(NO)_2(CN)_6]^{6-}$. A reasonable mechanism for the reaction of the two type IIb ligands is the stepwise sequence shown in equation (17), where a pair of two-electron reductions occurs forming two metal–metal bonds within the $[Mo_4S_4]$ cube.

$$[Mo_4(NO)_4(S_2)_4(S_2)'_2O]^{2-} + CN^- \longrightarrow [Mo_4S_4(NO)_4(CN)_8]^{8-} + SCN^- \qquad (16)$$

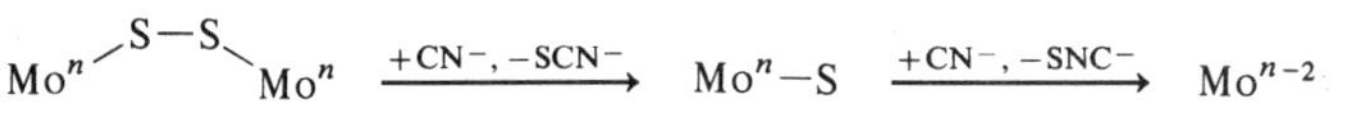

$$Mo^n\diagup^{S-S}\diagdown_{Mo^n} \xrightarrow{+CN^-,\ -SCN^-} Mo^n-S \xrightarrow{+CN^-,\ -SNC^-} Mo^{n-2} \qquad (17)$$

Type III bridging ligands are more susceptible to nucleophilic attack with extrusion of a neutral sulfur atom than are type Ia ligands. This is consistent with a generally lower electron density on the S atoms of type III ligands (bonded to two metal atoms) than of the type Ia ligands (bonded to one metal atom).

16.1.6.4.3 Oxidation of the ligand by external agents

In $[Ir(dppe)_2(S_2)]Cl$, for instance, the S_2^{2-} ligand can be oxidized stepwise on the metal to form S_2O and S_2O_2. Complexes with bridging S_2O ligands are also known.

16.1.6.4.4 Thermal decomposition (generation of S_2)

The main gaseous product of thermal decomposition at rather low temperatures (100–200 °C) of $Cs[Mo_2(S_2)_6]\cdot nH_2O$ is the S_2 molecule, which results from a reductive elimination process. This has been proven by mass spectroscopy and matrix isolation Raman, UV/VIS and IR spectroscopy.

16.1.6.5 Spectroscopic Properties of Disulfido Complexes

The $v(S-S)$ frequencies range from *ca.* 480 to 600 cm^{-1}. Comparison of the $v(S-S)$ values for the discrete diatomic sulfur species S_2 ($^3\Sigma_g^-$, 725 cm^{-1}), S_2^- ($^2\Pi_g$, 589 cm$^-$) and S_2^{2-} ($^1\Sigma_g^+$, 446 cm^{-1}) leads to the conclusion that the approximate charge distribution in disulfido complexes is somewhere between that for S_2^- and that for S_2^{2-}. However, in this comparison the strong coupling of the $v(S-S)$ vibration with the $v(M-S)$ vibrations, which leads to higher $v(S-S)$ values, must also be considered. This coupling is proven by the shift of 1–2 cm^{-1} in $v(S-S)$ upon substitution of ^{92}Mo by ^{100}Mo in $[Mo_2(S_2)_6]^{2-}$ and $[Mo_3S(S_2)_6]^{2-}$.

The frequencies of the type III bridging disulfido ligands are generally higher than those for type I structures.

For $[Mo_3S(S_2)_6]^{2-}$ with both types of S_2^{2-} ligands the vibrational band at $544\,cm^{-1}$ can be assigned to the type III ligands and the bands at 504 and $510\,cm^{-1}$ to the type I ligands. The higher frequencies found for type III ligands are consistent with the slightly shorter S—S distances for this structure type.

The intensities of the $\nu(S—S)$ bands of type III ligands are normally high in the IR as well as in the Raman spectra. The type Ia ligands show intense $\nu(S—S)$ bands in the IR, but in the Raman spectrum these bands are usually weak.

For structure type IIa ligands both the $\nu(S—S)$ and the totally symmetric $\nu(M—S)$ vibrations are practically forbidden in the IR spectrum, but the corresponding bands (very intense and strongly polarized) can easily be observed in the Raman spectrum.

The X-ray photoelectron spectra of disulfido complexes have been measured in order to obtain additional information about the effective charge on the sulfur atoms in these ligands.

The sulfur 2*p* binding energies lie between 162.9 and 164.4 eV, indicating that the sulfur atoms are generally more negatively charged than in neutral sulfur ($E_B(2p)$ for S_8 is 164.2 eV). The corresponding binding energy for sulfur in complexes with reduced sulfur-containing ligands such as S^{2-}, $\equiv C—S^-$, $\equiv C—S—C\equiv$ or $\equiv C=S$ are in the range 161.5–163.5 eV; Na_2S has a binding energy of 162.0 cV, and thiometalates have binding energies of 162.2–163.4 eV. Thus, the S 2*p* binding energies for disulfido complexes are consistent with the conclusion drawn from S—S bond distances and from vibrational spectroscopy, *i.e.* the *effective charge* on the sulfur atoms in the disulfido ligand in metal complexes is between 0 and −2.

Additional evidence about the oxidation state of the sulfur can be obtained by comparison of the metal binding energies in disulfido complexes to the metal binding energies in complexes with known oxidation states. In particular it is clear that the Mo binding energies of $[Mo_2(S_2)_6]^{2-}$ and $[Mo_3S(S_2)_6]^{2-}$ can be understood if the ligands are formulated as $(S_2)^{2-}$ units rather than neutral S_2 fragments. Similar results are found for the metal binding energies of Ir and Os complexes.

Both the sulfur and metal XPS results are consistent with the disulfur ligand having an effective negative charge of about −2. However, neither the sulfur nor the metal XPS data are sufficiently sensitive to distinguish among the various modes of coordination of the S_2^{2-} ligand.

For bonding type Ia the π^* orbital of S_2^{2-} splits into a strongly interacting π_h^* orbital in the MS_2 plane and a weakly interacting π_v^* orbital perpendicular to the MS_2 plane. The longest wavelength band in the complexes $[Cp_2Nb(S_2)X]$ (X = Cl, Br, I) and $[MoO(S_2)(dtc)_2]$ occurs at $\sim 20\,000\,cm^{-1}$ and is assigned to an LMCT (ligand to metal charge transfer) of the type $\pi_v^*(S) \rightarrow d(M)$. This assignment is probably also correct for other complexes of structural type Ia which contain metal atoms in a high oxidation state, *e.g.* $[Mo_2O_2S_2(S_2)_2]^{2-}$.

The position of this first band is influenced by the oxidation state of the metal, by the metal–metal bonding and by the nature of the other ligands, which determine the energy of the LUMO and its metal character and thereby influence the optical electronegativety of the metal.

For complexes of structural type III the corresponding absorption band is expected to occur at higher energy because both π^* orbitals of the ligand interact strongly with the metals.

Especially interesting is the very intense band at $14\,200\,cm^{-1}$ in $[(NH_3)_5Ru(S_2)Ru(NH_3)_5]Cl_4$. A comparable band is not found in the related compounds with type IIa structures, $[(CN)_5Co(S_2)-Co(CN)_5]^{6-}$ and aqueous $[Cr(S_2)Cr]^{4+}$. It has been proposed that the central unit in $[(NH_3)_5Ru(S_2)-Ru(NH_3)_5]^{4+}$ is best formulated as $Ru^{II}—(S_2)^-—Ru^{III}$. Such a mixed valence complex could exhibit the intense band observed.

16.1.6.6 Chemical Bonding

One framework for discussing the bonding of metal complexes with diatomic ligands is to partition the complex conceptually into the units M^{m+} and Y_2^{n-}. The values for *m* and *n* are determined by comparing the physical properties of the complex (Y—Y distance, $\nu(Y—Y)$ (band position and intensity), electronic spectra, XP spectra, magnetic properties) with those of the isolated Y_2^{n-} species.

The physical data of dioxygen complexes have been discussed in detail elsewhere. Here it is sufficient to note that the complexes have been divided into superoxide (O_2^-) and peroxide (O_2^{2-}) complexes, primarily on the basis of O—O distance and $\nu(O—O)$ (Table 6). The superoxide ligand has $d(O—O) \sim 1.30\,Å$ and $\nu(O—O) \sim 1125\,cm^{-1}$; the peroxide ligand has $d(O—O) \sim 1.45\,Å$ and $\nu(O—O) \sim 860\,cm^{-1}$. The electronic spectra of the complexes have also been used to classify dioxygen complexes. The S—S distances range continuously from 1.98 to 2.15 Å, and do not correlate strongly with structural type. Correspondingly the $\nu(S—S)$ frequencies range continuously from 480 to $600\,cm^{-1}$. Moreover, the $v(S—S)$ vibration is strongly coupled to the $v(M—S)$ vibrations so that the $v(S—S)$ frequencies are not a simple indicator of the formal charge on the ligand. However, comparison of the S—S distances *and* frequencies in Table 6 with those in Tables 2–4 indicates that an effective charge on the ligand in the complexes between -1 and -2 seems reasonable. A similar conclusion can be drawn from the sulfur and metal binding energies of the ESCA spectra. All the physical data are consistent with a charge delocalization from the polarizable (soft) S_2^{2-} ligand to the metal.

Table 6 Comparison of the Distances and Frequencies for the Species X_2^{n-} (X = O, S; n = 0, 1, 2)

	X_2 ($^3\Sigma_g^-$)	X_2^- ($^2\pi_g$)	X_2^{2-} ($^1\Sigma_g^+$)
$d(O—O)$ (Å)	1.21	1.33	1.49
$d(S—S)$ (Å)	1.89	2.00	2.13
$v(O—O)$ (cm^{-1})	1580	1097	802
$v(S—S)$ (cm^{-1})	725	589	446

The Y_2^- ligand seems to be less important compared to dioxygen complexes (almost all complexes are diamagnetic, consequently EPR spectroscopy cannot be used to probe the electron distribution in the HOMO in these complexes). The S_2^- classification has been advocated for the ligand in $[(NH_3)_5Ru(S_2)Ru(NH_3)_5]^{4+}$ from analysis of the electronic spectra and for $[Cp_2Fe_2(S_2)(SR)_2]$ from similarities of the ligand geometry to the geometry of the O_2 ligand in superoxide complexes.

An alternative approach to describing the bonding in MY_2 complexes is to consider the triatomic MY_2 unit as a covalent inorganic functional group. Enemark and Feltham have shown that the diverse properties of metal nitrosyl complexes can be conveniently interpreted by describing the complexes as derivatives of the $[MNO]^n$ group, where n is the total number of electrons in the d-orbitals of the metal and the π^*-orbitals of the nitrosyl ligand.

Relevant to our case is the analysis of the bonding in η^2 side-on MO_2 complexes which show (see Figure 7) that the primary bonding interaction involves the metal d_{xz} orbital and the $\pi_z^*(\pi_h^*)$ orbital of the O_2 ligand (π-bonding) and the d_{z^2} orbital and the π_z (π_h) orbital (σ-bonding) according to the qualitative Dewar–Chatt–Duncanson bonding scheme (the interaction of the π_y (π_v) and the $\pi_y^*(\pi_v^*)$ orbitals are considered to be negligible). The major contribution is the π-bonding as shown by more quantitative molecular orbital calculations of side-on MO_2 complexes.

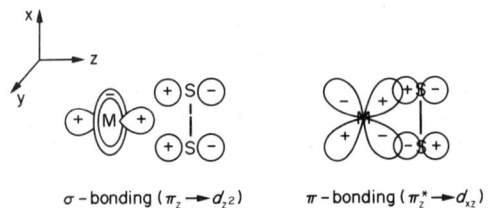

σ – bonding ($\pi_z \rightarrow d_{z^2}$) π – bonding ($\pi_z^* \rightarrow d_{xz}$)

Figure 7 σ- and π-bonding in MS_2 complexes (η^2 coordination)

For type Ia complexes the π-bonding should also be the major bonding interaction. For type III $[M(1)—(S_2)—M(2)]$ complexes this bonding interaction would occur twice, once between π_z^* of S_2 and M(1) and once between π_y^* of S_2 and M(2). If M(2) is a positively charged metal ion with few d-electrons, then there should be a lower electron density on the disulfur unit in a type III complex relative to a type Ia complex. Such depopulation of the π^*-orbitals of the S_2^{2-} ligand is consistent with the greater susceptibility of type III complexes to nucleophilic attack and with the shorter S—S distances and higher $v(S—S)$ values for such complexes. These properties are consistent with the π-donor character of the S_2^{2-} ligand.

16.1.7 POLYSULFIDO LIGANDS

The increased solubility of heavy metal sulfides in alkaline polysulfide-containing solutions is not only of general chemical but also of analytical and mineralogical interest.

Sulfide minerals are mostly formed hydrothermally from post-magnetic fluids; it appears remarkable that the formation of many ore deposits cannot be conclusively explained because of the very low solubility of the corresponding sulfide. Though it was postulated earlier[121] that polysulfido complexes might have been responsible for 'the transport metals and sulfur together',[122] it was only very recently proven that discrete polysulfido clusters of metals exist in polysulfido solutions.[123,124]

The formation of soluble polysulfido clusters or complexes is also responsible for the fact that some classical analytical separation procedures based on polysulfide solutions (distinguishing between the classical thioanion forming elements like As, Sb, Sn and others like Cu) sometimes fail.

It is now clear that the S_x^{2-} ions are fascinating and versatile ligands from the structural point of view and that metal aggregates can be 'glued' by these ligands due to their high and variable number of coordination sites, which keeps the charge of the complex and of the chain low.[125] In $[Cu_6(S_4)_3(S_5)]^{2-}$ only four ligands are capable of stabilizing a cluster with six metal atoms.[125] It is remarkable that practically all possible ions S_x^{2-} ($x = 2, 3, 4, 5, 6, 8, 9$) occur in complexes, although S_9^{2-} has not been reported until now as isolated ion.

16.1.7.1 Polysulfide Ions and Polysulfide Solutions

Upon digesting aqueous sulfide with sulfur, solutions containing *mainly* S_4^{2-} and S_5^{2-} are obtained. Solutions in methanol and ethanol are prepared correspondingly. Salts containing the S_x^{2-} ($x = 2-7$) ions can be obtained not only from these solutions but also in dry reactions and from liquid NH_3. All anions have a chain structure whereby the average S—S bond length is smaller in S_x^{2-} ions ($x > 2$) than in S_2^{2-} and the length of the S—S terminal bond decreases from S_3^{2-} (2.15 Å) to S_7^{2-} (1.992 Å; Table 7).[126] These facts indicate that the negative charge (filling a π^*-antibonding molecular orbital) is delocalized over the whole chain but that the delocalization along the chain is less in higher polysulfides. These considerations are important for the comparison of the S—S bond length of the free ions with those of the metal complexes.

Table 7 Comparison of the Terminal S—S Distances and S—S—S Angles in the S_n^{2-} Polysulfides[126]

Polysulfide	S_t—S (Å)	S_t—S—S (°)
S_2^{2-}	2.13	—
S_3^{2-}	2.15	103
S_4^{2-}	2.074(1)[a]	109.76(2)[a]
S_5^{2-}	2.043(4)[a]	109.2(1)[a]
S_6^{2-}	2.01(3)[a]	109.6(20)[a]
S_7^{2-}	1.992(a)[a]	111.3(1)[a]

[a] These are average values of the two S_t—S and S_t—S—S terminal distances and angles, respectively.

16.1.7.2 Survey of Compound Types

16.1.7.2.1 *Mononuclear complexes with M(μ_2-S)$_x$ ring systems*

Only a few compounds containing the S_3^{2-} ion as a ligand are known. $[Cp_2'TiS_3]$ (43) contains a nonplanar four-membered TiS_3 ring with a dihedral angle of 49° between $TiS(1)S(2)$ and $S(1)S(2)S(3)$.[127] On the other hand, quite a few complexes with S_4^{2-} ions as ligands have been prepared and structurally characterized.

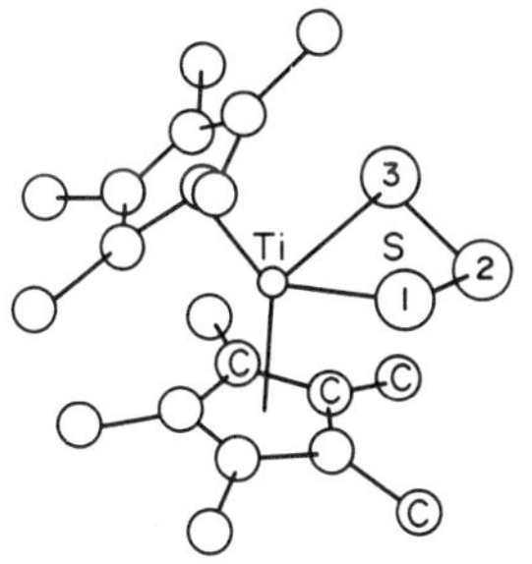

(43)

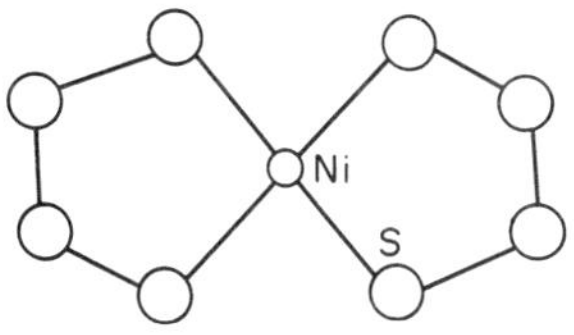

(44)

In compounds where the ion acts as a bidentate ligand (that means forming a five-membered MS_4 ring system), an 'envelope' as well as a 'half-chair' conformation is possible. In $[Cp_2M(S_4)]$ (M = Mo, W) (complexes with tetrahedral coordination of the metal ion)[128] and in $[M(S_4)_2]^{2-}$ (M = Ni (**44**), Pd) the 'half-chair' conformation is found,[129] whereas several complexes with a square-pyramidal coordination of Mo and W show the 'envelope' conformation (for details see Section 16.1.7.4).[130,131]

A few complexes with the bidentate pentasulfido ligand are known. The first one reported was the homoleptic and optically active complex $[Pt(S_5)_3]^{2-}$ (**45**).[133,147] The MS_5 moiety has a chair conformation in all known mononuclear complexes. It is worth noting that the 'bite' varies strongly and that the largest one is found in $[(S_5)Fe(MoS_4)]^{2-}$.[132]

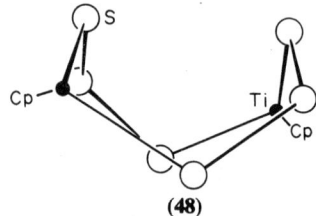

(45)

Some mononuclear complexes containing bidentate S_6^{2-} ligands could be isolated, *e.g.* the homoleptic $[M(S_6)_2]^{2-}$ [M = Zn, Cd, Hg (**46**)][134] and one containing the S_9^{2-} ligand, namely $[AuS_9]^-$ (**47**) with a ring structure.[135]

(46)   (47)

The following homoleptic mononuclear complexes are known at present: $[M(S_4)_2]^{2-}$ (M = Ni,[129] Pd,[138] Hg[138]), $[Pt(S_5)_2]^{2-}$,[136] $[M(S_5)_3]^{n-}$ (Pt with n = 2;[133,147] Rh with n = 3),[137] $[M(S_6)_2]^{2-}$ (M = Zn, Cd,[138] Hg[134]), but also some polynuclear ones exist (see following chapter): $[Cu_3(S_4)_3]^{3-}$, $[Cu_3(S_6)_3]^{3-}$, $[Cu_4(S_4)_3]^{2-}$, $[Cu_4(S_5)_3]^{2-}$ and $[Ag_2(S_6)_2]^{2-}$.

16.1.7.2.2 Dinuclear complexes with polysulfido ions as bridging ligands

Very few dinuclear complexes are known where the S_x^{2-} ions (x = 3, 4, 5, 8) act as bridging ligands.

The structure of $[(MeCp)_4Ti_2S_6]$ (**48**) consists of an eight-membered ring containing alternating Ti atoms and S_3 fragments. In contrast to cyclo-S_8 the ring adopts a 'cradle' conformation with the Ti atoms positioned at the sites adjacent to the apical sulfur atoms. In the complex the S—S—S angles are expanded relative to those known for other cyclic sulfur ring species.[139]

(48)

The $S_5Mo_2(S_2)$ moiety in $[Mo_2(NO)_2(S_2)_3(S_5)(OH)]^{3-}$ (**49**), however, has practically the same geometry as that of cyclooctasulfur (if one ring member is assumed to be at the center of the bridging S_2^{2-} group).[140]

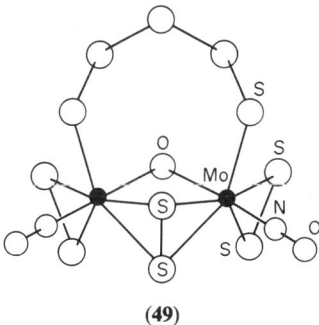

(49)

$[Os_2-\mu-(S_5)-\mu-(S_3CNR_2)(S_2CNR_2)_3]$ (50) contains the S_5^{2-} ion as a 'half-bridging' ligand.[141] The six-membered OsS_5 ring has a chair conformation.

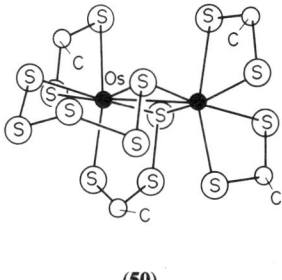

(50)

The remarkable ten-membered highly symmetrical $[Au_2S_8]^{2-}$ ring (51) consists of two Au^I atoms and two bridging S_4^{2-} ligands. The ring has approximately D_2 symmetry.[123] The dinuclear complexes $[(S_6)M(S_8)M(S_6)]^{4-}$ (52; M = Cu, Ag) have two bidentate S_6^{2-} ligands (forming rings) and the novel doubly bridging S_8^{2-} ligand.[142,143]

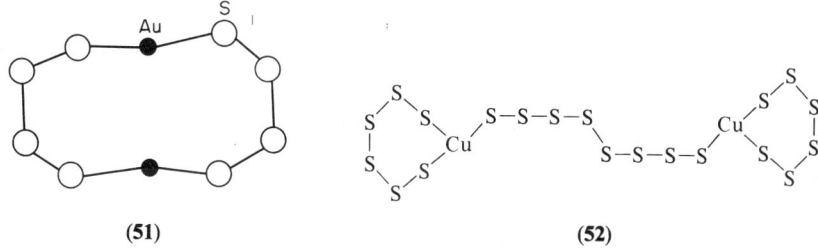

(51)   (52)

16.1.7.2.3 Polynuclear species with μ_n-S atoms (n > 2) — condensed ring systems and clusters

One would expect that, because of the large number of coordination sites of the polysulfido ligands, 'soft metal aggregates' could be 'glued' by S_x^{2-} ions.

A very remarkable species with very different types of coordination is the hexanuclear complex $[Cu_6S_{17}]^{2-}$ (53) which contains 10 μ_3-S atoms and a novel aggregate of six Cu^I ions which can be (only approximately) described as two Cu_4 tetrahedra sharing one edge.[125,144]

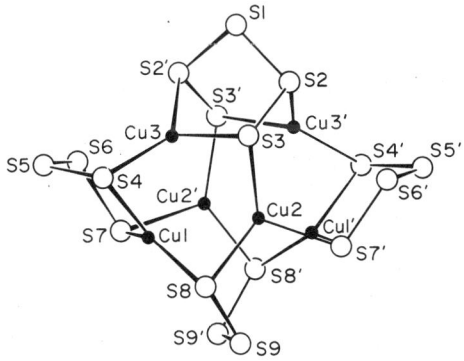

53)

Whereas only the two terminal S atoms of the three S_4 ligands in (53) are bonded to copper atoms, coordination of four S atoms of the S_5^{2-} ligand [S(3)—S(2)—S(1)—S(2′)—S(3′)] is found. Each of the two equivalent S_4^{2-} ions [S(4)—S(5)—S(6)—S(7) and S(4′)—S(5′)—S(6′)—S(7′)] acts as bridging and chelating ligand for the two 'Cu tetrahedra' with coordination to three different Cu atoms. S(4) and S(7) [correspondingly S(4′) and S(7′)] are bonded not only to the same atom Cu(1) [Cu(1′)] but also to Cu(3) and Cu(2′) [Cu(3′) and Cu(2)], respectively. The other S_4^{2-} [S(8)—S(9)—S(9′)—S(8′)] and the S_5^{2-} ligands are responsible for the connection of the 'tetrahedra'. Both ligands show coordination to four different Cu atoms.

The tetranuclear clusters $[Cu_4(S_n)_3]^{2-}$ (**54**; $n = 4, 5$), $[Cu_4(S_5)_2(S_4)]^{2-}$ and $[Cu_4(S_5)(S_4)_2]^{2-}$ have six μ_3-S atoms from three S_x^{2-} ligands ($x = 4, 5$) as bridges for the six edges of the metal tetrahedra and the trinuclear species $[Cu_3(S_6)_3]^{3-}$ and $[Cu_3(S_4)_3]^{3-}$ (see Figure 8b) have three μ_3-S atoms forming the central Cu_3S_3 ring with a chair conformation.[123]

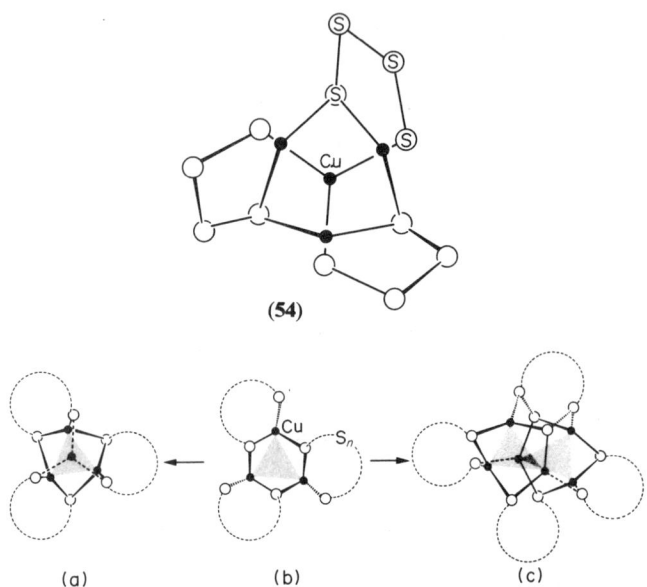

(54)

(a)   (b)   (c)

Figure 8 Correlation of the structures of $[Cu_4(S_n)_3]^{2-}$ ($n = 4, 5$) (a) to those of $[Cu_3(S_n)_3]^{3-}$ ($n = 4, 6$) (b) and $[Cu_6(S_4)_3(S_5)]^{2-}$ (c)

The basic structure of the species $[Cu_4(S_x)_3]^{2-}$ can formally be obtained from $[Cu_3(S_x)_3]^{3-}$ (novel polycyclic inorganic species with different puckered copper–sulfur heterocycles, a central Cu_3S_3 ring with a chair conformation and three outer CuS_6 or CuS_4 rings[143,145]) by coordination of each of the three 'end-on bonded sulfur atoms' of the three polysulfide ligands to an additional fourth Cu^I as indicated in Figure 8a.[123] An alternative description of the structure is that the Cu and S of the Cu_4S_6 core are inequivalent, as there are three Cu atoms coordinated to two different S_x^{2-} ligands (those of the marked $Cu_3(\mu$-S$)_3$ ring in Figure 8) and one Cu coordinated to all three ligands.

The structure of $[Cu_6S_{17}]^{2-}$ can formally be derived by connecting two $Cu_3(\mu$-S$)_3$ units by the four polysulfido ligands (see Figure 8c).[123]

It turns out that the six-membered $Cu_3(\mu$-S$)_3$ rings are paradigmatic units. This type of ring system has been incorporated into current models of metallothioneins, low molecular weight proteins which are believed to play a key role in metal metabolism (see ref. 123). The structural chemistry of Ag species seems to be different. Monocyclic $[Ag(S_x)]^-$ rings can be linked *via* bridging ligands as in $[(S_6)Ag(S_8)Ag(S_6)]^{4-}$ or condensed as in $[Ag_2(S_6)_2]^{2-}$ (**55**).

(55)

16.1.7.2.4 *Compounds with solid state structures and S_x^{2-} ligands containing μ-S_n type atoms ($n > 2$)*

Regarding the ability of S_x^{2-} species to 'glue' metal aggregates together, compounds such as $(NH_4)_2PdS_{11}\cdot 2H_2O^{146}$ (with Pd atoms linked *via* S_6^{2-} chains in a three dimensional array)* and $(NH_4)CuS_4$ [148] (with 'solid state' $(CuS_4)^-$ chelate rings linked *via* additional Cu—S bonds to form one-dimensional polymeric anions) are similar to those discussed in the last section.

16.1.7.3 Synthesis and Reactions

Polysulfido complexes have been prepared by various methods. The main preparative routes and some typical reactions of the complexes will be summarized here.

Oxidative addition of elemental sulfur to an electron-rich metal which is coordinatively unsaturated is a useful method for preparing polysulfido complexes. Examples are the reaction of $[\pi\text{-}CpRh(PPh_3)_2]$ yielding $[\pi\text{-}CpRh(PPh_3)(S_5)]^{149}$ or of $[ML_4]$ (M = Pd, Pt; L = PPh_3) to give $[L_2MS_4]$.[150] Other interesting reactions of complexes with sulfur without a change in the oxidation state of the central metal are shown in equations (18) and (19).[151]

$$[Cp_2MoS_2] + \tfrac{1}{4}S_8 \longrightarrow [Cp_2Mo(S_4)] \tag{18}$$

$$[Cp_2Mo(SH)_2] + \tfrac{3}{8}S_8 \longrightarrow [Cp_2Mo(S_4)] + H_2S \tag{19}$$

By reaction of thioanions of Mo or W as, for instance, MoS_4^{2-}, $MoOS_3^{2-}$ and WS_4^{2-} with sulfur, complexes containing MS_4 ring systems, such as $[SMo(S_4)]^{2-}$, $[OMo(S_4)_2]^{2-}$, $[Mo_2O_2S_2(S_4)_2]^{2-}$ and $[W_2S_4(S_4)_2]^{2-}$ (56), have been obtained.[130,131] $[OMo^{IV}(S_4)_2]^{2-}$ could be formed from $[OMo^{VI}S_3]^{2-}$ according to equation (20). The basic type of reactions should be as shown in equation (21) and (22). The redox reaction shown in equation (23) and (24) might be responsible for the formation of $[W_2S_4(S_4)_2]^{2-}$.

$$[(O)Mo^{VI}(S^{2-})_3]^{2-} \longrightarrow \left[OMo^{IV} \begin{smallmatrix} S^{2-} \\ \diagup \\ -S^{(-)} \\ \diagdown \diagup \\ S^{(-)} \end{smallmatrix} \right]^{2-} \xrightarrow{S_8} [OMo^{IV}(S_4)_2]^{2-} \tag{20}$$

$$Mo^{VI}(S^{2-}) + S_8^0 \longrightarrow Mo^{IV}(S_4^{2-}) \tag{21}$$

$$Mo^{IV}(S_2^{2-}) + S_8^0 \longrightarrow Mo^{IV}(S_4^{2-}) \tag{22}$$

$$2WS_4^{2-} + S_n \longrightarrow [W_2^V S_4(S_2)_2]^{2-} + S_n^{2-} \tag{23}$$

$$[W_2^V S_4(S_2)_2]^{2-} + \tfrac{1}{2}S_8 \longrightarrow [W_2^V S_4(S_4)_2]^{2-} \tag{24}$$

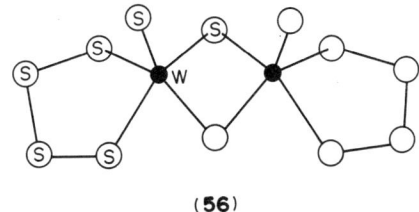

(56)

Most of the known complexes have been prepared by reaction of metal salts or metal complexes with polysulfide solutions. In general, the S_x^{2-} ligand is introduced by substitution of other ligands. All homoleptic complexes have been obtained in this way. Solvents used in most cases were H_2O, MeOH, EtOH, MeCN or DMF. It seems that the type of S_x^{2-} ligand which occurs in the complex need not necessarily have a high abundance in the polysulfide solution: $[Cp_2TiCl_2]$ reacts with Na_2S_n ($n = 2$–7) to yield $[Cp_2Ti(S_5)]$.[152]

The formation of polysulfido complexes with other sulfur-containing reagents such as Cl_2S_x, R_2S_x has also been reported.[153,154] The former undergoes a particularly interesting reaction, shown in equation (25).[153]

---

*The crystal structure is a composite of different possible linkages, in which some chains are shortened (principally at the middle sulfurs of the chains), whereby sulfur absences are responsible for the PdS_{11} composition.

$$M \overset{}{\underset{S \diagdown S}{\diagup}} + Cl_2S_2 \rightarrow S \overset{M}{\underset{S-S}{\diagup \diagdown}} S + 2Cl^- \qquad (25)$$

It has been pointed out[154] that the synthesis of $[Fe_2S_2(S_5)_2]^{2-}$ from $[Fe(SPh)_4]^{2-}$ with dibenzyl trisulfide would have implications regarding the enzymatic biosynthesis of the metal clusters in the 2Fe ferredoxins since trisulfides seem to be present in biological systems.

Some reactions of polysulfido complexes with sulfur-abstracting reagents have been reported. $[Pt(S_5)_3]^{2-}$ (45) reacts with CN^- yielding $[Pt(S_5)_2]^{2-}$.[136] During the course of this reaction an interesting two-electron transfer occurs reducing Pt^{IV} to Pt^{II}. The same complex reacts with PPh_3 (equation 26).[155,156] The perthiocarbonate ligand is formed by the electrophilic attack of CS_2 on the coordinated S_4^{2-} ligand in $[SMo(S_4)_2]^{2-}$ yielding $[SMo(CS_4)_2]^{2-}$.[157] The same substrate reacts with bis(carboxymethyl)acetylene giving $[MoS_2C_2(CO_2Me)_2]_3^{2-}$.[158] In both reactions the unit (A) present in $[(S_4)MoS(S_2)]^{2-}$, formed as a possible intermediate, may react.

$$[PtS_5)_3]^{2-} + 12PPh_3 \longrightarrow S^{2-} + [(PPh_3)_2Pt(S_4)] + 10PSPh_3 \qquad (26)$$

An interesting reaction is the oxidation of polysulfide ligands. $[(S_2)Mo(S)(S)_2Mo(S)(S_4)]^{2-}$ can be oxidized giving $[(S_2)Mo(O)(S)_2Mo(O)(S_3O_2)]^{2-}$ (57)[161] and $[(S_2)Mo(S)(S)_2Mo(S)(S_3O)]^{2-}$ (58).[160]

(A)   (57)   (58)

16.1.7.4 A Few Remarks on Chemical Bonding and the Geometry of S_x^{2-} Ligands

Not much is known about this topic as most of the interesting species have only recently been reported. Whereas upon coordination of the S_2^{2-} ligand to metal atoms a significant shortening of the S—S bond is always observed, the corresponding situation is much more complicated in the case of the polysulfido complexes. In general, the shortening of the S—S bond is less in polysulfido complexes (the S—S bond shortening of S_2^{2-} complexes is caused by sulfur-to-metal π-backbonding).

In Table 8 bond lengths for S—S bonds in compounds with S_4 structural units have been compared (uncoordinated and coordinated). It turns out that in several S_4^{2-} complexes (in addition to a shortening of the S—S bonds compared to the length in S_4^{2-} units in $BaS_4 \cdot H_2O$) variations of the S—S bond length in the S_4^{2-} chelate are observed. It has been postulated that the high oxidation state of Mo and W in the corresponding complexes contributes to sulfur-to-metal π-backbonding and a shortening of the central S(2)–S(3) bond of the $[S(1)–S(2)–S(3)–S(4)]^{2-}$ ligand.

Table 8 Bond Lengths for S—S Bonds in Complexes with Bidentate S_4^{2-} Ligands[131,138,155]

	S(1)—S(2) (Å)	S(2)—S(3) (Å)	S(3)—S(4) (Å)
$[(PPh_3)_2Pt(S_4)]$	2.024(8)	2.022(10)	2.081(10)
$(PPh_4)_2[Hg(S_4)_2]$	2.050	2.043	2.048
$(Et_4N)_2[Ni(S_4)_2]$	2.073(2)	2.037(4)	2.073(2)
$(Et_4N)_2[Pd(S_4)_2]$	2.062(8)	2.054(6)	2.065(8)
$(AsPh_4)_2[Mo_2S_2(S_4)_2]$ (a)	2.019(5)	1.970(6)	2.115(5)
$(AsPh_4)_2[Mo_2S_2(S_4)_2]$ (b)	2.096(16)	1.936(19)	2.169(14)
$(PPh_4)_2[Mo_2O_2S_2(S_4)_2]$	2.066	2.024	2.084
$(PPh_4)_2[W_2S_4(S_4)_2] \cdot 0.5 DMF$	2.044(11)	2.013(14)	2.112(12)
$(PPh_4)_2[W_2O_2S_2(S_4)_2] \cdot 0.5 DMF$	2.062	2.011	2.100
$(PPh_4)[W_2S_4(S_2)(S_4)] \cdot 0.5 DMF$	2.039	2.016	2.067
$(Et_4N)_2[MoS(S_4)_2]$	2.107(1)	2.012(1)	2.166(1)
$[Cp_2Mo(S_4)]$	2.081(8)	2.018(9)	2.085(7)
$[Cp_2W(S_4)]$	2.105(7)	2.016(8)	2.116(9)
$BaS_4 \cdot H_2O$	2.079(3)	2.062(4)	2.079(3)
$[C_6H_2(OEt)_2(S_4)_2]$ (a)	2.028(5)	2.068(5)	2.027(5)
$[C_6H_2(OEt)_2(S_4)_2]$ (b)	2.034(5)	2.067(5)	2.024(5)

(a) and (b) denote two independent S_4 chains.

The different conformations occurring in MS_4 ring systems can be explained by varying degree of interligand interactions. In the $[Cp_2M(S_4)]$ complexes with a tetrahedral coordination of the metal atoms (Table 8) and half-chair conformation of the MS_4 rings these interactions are not likely to be significant.[130]

In Mo and W complexes of the type $[M_2S_2X_2(S_4)(S_x)]^{2-}$ and $[XMo(S_4)_2]^{2-}$ (X = O, S; x = 2, 4) the envelope conformation for the five-membered rings and the obligatory orientation of the lone pairs on the Mo-bound sulfur atoms result in a structure with S—S interligand interactions within the XMS_4 pyramidal units being minimized. In these units the half-chair conformation would bring the lone pairs of the sulfur atoms bound to Mo in a position with closer interligand lone pair contacts.[130]

For the six-membered MS_5 ring system the preferred conformation is that of a half-chair. The molecular parameters of heteroatomic ring systems are comparable to those of S_6 in rhombohedral cyclohexasulfur. The S_6 molecules also have a chair-like structure with an S—S bond length of 2.057(18) Å, an S—S—S bond angle of 102.2(1.6)° and a torsional angle of 74.5(2.5)°. A formal replacement of a sulfur atom by $TiCp_2$ or VCp_2 fragments, for instance, leads to longer M—S distances (2.40–2.46 Å) (compared to the S—S bond length), these being compensated by smaller S—M—S bond angles (89–95°). Therefore, the geometry of the resultant S_5 ligand is similar to that of the S_6 molecule.[159]

16.1.8 ADDENDUM

The S_7^{2-} ion acts as a tridentate ligand in $(Me_3P)_3OsS_7$, where the bicyclic OsS_7 shows an *exo–endo* conformation similar to S_8^{2+} ion in $S_8(AsF_6)_2$.[162] While complexes with the S_8^{2-} ion as a bridging ligand have been known for some time (see above), the preparation of tetraphenylphosphonium octasulfide has been reported only recently.[163] Another example of the S_9^{2-} ion as a ligand has been found in the highly symmetric $[Ag(S_9)]^-$ (59).[164]

A series of new polysulfido complexes have been reported: $[Sn^{IV}(S_4)_3]^{2-}$, $[Sn^{IV}(S_4)_2(S_6)]^{2-}$, $[Cr^{III}(NH_3)_2(S_5)_2]^{2-}$ and, as an example of bonding of polysulfide ligands to 'hard' metal cations, $[Nb_2^V(OMe)_2(S_2)_3(S_5)O]^{2-}$.[164] Coucouvanis *et al.* have prepared the complexes $[Zn(S_4)_2]^{2-}$ and $[(S_5)Mn(S_6)]^{2-}$.[165] A seven-membered 1,4-dirhenia cycloheptasulfur ring has been found by Herberhold and co-workers in $Cp_2Re_2(CO)_2(\mu\text{-}S_2)(\mu\text{-}S_3)$.[166]

(59)   (60)   (61)

A second structural isomer (61) of Cp_2TiS_5 (60)[167] is obtained by heating the latter at reflux for one day.[168] This transformation provides an interesting example of the migration of a π-complexed organic unit to an inorganic sulfur ligand. Brick-red $(NH_4)_2[Pt(S_5)_3]\cdot 2H_2O$ crystallizes from the reaction of $K_2[PtCl_6]$ with aqueous $(NH_4)S_x$ solution. Addition of concentrated HCl results in the separation of maroon $(NH_4)_2[PtS_{17}]\cdot 2H_2O$.[169] The $[Pt(S_5)_3]^{2-}$ ion crystallizes from the solution as a racemate which could be resolved by forming diastereoisomers. On crystallization $[PtS_{17}]^{2-}$ undergoes a second-order asymmetric transformation so that the solid contains an excess of the $(-)$ enantiomer.[169] (For a recent review on polysulfido complexes see ref. 170).

16.1.9 REFERENCES

1. C. K. Jørgensen, *Inorg. Chim. Acta Rev.*, 1968, **2**, 65.
2. C. G. Kuehn and S. S. Isied, *Prog. Inorg. Chem.*, 1980, **27**, 153.
3. A. Müller, E. Diemann, R. Jostes and H. Bögge, *Angew. Chem.*, 1981, **93**, 957; *Angew. Chem., Int. Ed. Engl.*, 1981, **20**, 934.

4. A. Müller, E. Diemann and C. K. Jørgensen, *Struct. Bonding (Berlin)*, 1973, **14**, 23.
5. K. H. Schmidt and A. Müller, *Coord. Chem. Rev.*, 1974, **14**, 115.
6. A. Müller, W. Jaegermann and J. H. Enemark, *Coord. Chem. Rev.*, 1982, **46**, 245.
7. H. Vahrenkamp, *Angew. Chem.*, 1975, **87**, 363; *Angew. Chem., Int. Ed. Engl.*, 1975, **14**, 322.
8. D. Morelli, A. Segre, R. Ugo, G. La Monica, S. Cenini, F. Copti and F. Bonati, *Chem. Commun.*, 1967, 524; R. Ugo, G. La Monica, S. Cenini, A. Segre and F. Conti, *J. Chem. Soc. (A)*, 1971, 522.
9. W. Biltz and E. Keunecke, *Z. Anorg. Allg. Chem.*, 1925, **147**, 171.
10. C. G. Kuehn and H. Taube, *J. Am. Chem. Soc.*, 1976, **98**, 689.
11. M. Herberhold and G. Süss, *Angew. Chem.*, 1976, **88**, 375; *Angew. Chem., Int. Ed. Engl.*, 1976, **15**, 366.
12. M. Herberhold and G. Süss, *J. Chem. Res. (S)*, 1977, 246; *J. Chem. Res. (M)*, 1977, 2720.
13. H. Vahrenkamp, in 'Sulfur — Its Significance for Chemistry, for the Geo-, Bio- and Cosmosphere and Technology', ed. A. Müller and B. Krebs, Elsevier, Amsterdam, 1984.
14. P. J. Harris, S. A. R. Knox, R. J. McKinney and F. G. A. Stone, *J. Chem. Soc., Dalton Trans.*, 1978, 1009.
15. C. G. Kuehn, Ph.D. Thesis, Stanford University, 1976.
16. J. P. Collman and T. N. Sorrell, *J. Am. Chem. Soc.*, 1975, **97**, 4133; J. P. Collman, T. N. Sorrell and B. M. Hoffman, *J. Am. Chem. Soc.*, 1975, **97**, 913.
17. M. Nappa and J. S. Valentine, *J. Am. Chem. Soc.*, 1978, **100**, 5075.
18. U. Belluco, L. Cattalini, F. Basolo, R. G. Pearson and A. Turco, *J. Am. Chem. Soc.*, 1965, **87**, 241.
19. J. P. Collman, T. N. Sorrell, K. O. Hodgson, A. K. Kulshrestha and C. E. Strouse, *J. Am. Chem. Soc.*, 1977, **99**, 5180.
20. M. Ardon and H. Taube, *J. Am. Chem. Soc.*, 1967, **89**, 3661.
21. T. Miyamoto, *J. Organomet. Chem.*, 1977, **134**, 335.
22. H. Köpf and M. Schmidt, *Angew. Chem.*, 1965, **77**, 965; *Angew. Chem., Int. Ed. Engl.*, 1965, **4**, 953.
23. B. Kreutzer, P. Kreutzer and W. Beck, *Z. Naturforsch., Teil B*, 1972, **27**, 461.
24. W. Beck, W. Danzer and R. Höfer, *Angew. Chem.*, 1973, **85**, 87; *Angew. Chem., Int. Ed. Engl.*, 1973, **12**, 77.
25. V. Küllmer and H. Vahrenkamp, *Chem. Ber.*, 1976, **109**, 1560, 1569; 1977, **110**, 3799, 3811.
26. J. P. Collman, R. K. Rothrock and R. A. Starke, *Inorg. Chem.*, 1977, **16**, 437.
27. M. DiVaira, S. Midollini and L. Sacconi, *Inorg. Chem.*, 1977, **16**, 1518; 1978, **17**, 816.
28. J. Cragel, Jr., V. B. Pett, M. D. Glick and R. E. De Simone, *Inorg. Chem.*, 1978, **17**, 2885.
29. E. Königer-Ahlborn, H. Schulze and A. Müller, *Z. Anorg. Allg. Chem.*, 1977, **428**, 5.
30. M. Schimdt, G. G. Hoffmann and R. Höller, *Inorg. Chim. Acta*, 1979, **32**, L19.
31. E. Diemann and A. Müller, *Coord. Chem. Rev.* 1973, **10**, 79, and references cited therein.
32. I. M. Blacklaws, E. A. V. Ebsworth, D. W. H. Rankin and H. E. Robertson, *J. Chem. Soc., Dalton Trans.*, 1978, 753.
33. M. K. Cooper, P. A. Duckworth, K. Henrick and M. McParflin, *J. Chem. Soc., Dalton Trans.*, 1981, 2357.
34. R. Graves, J. M. Homan and G. L. Morgan, *Inorg. Chem.*, 1970, **9**, 1592.
35. M. DiVaira, S. Midollini and L. Sacconi, *Inorg. Chem.*, 1979, **18**, 3466.
36. M. Sato, F. Sato, N. Takemoto and K. Iida, *J. Organomet. Chem.*, 1972, **34**, 205.
37. M. Rakowski Du Bois, M. C. Van Dervcer, D. L. Du Bois, R. C. Haltiwanger and W. K. Miller, *J. Am. Chem. Soc.*, 1980, **102**, 7456
38. H. Köpf, S. K. S. Hazari and M. Leitner, *Z. Naturforsch., Teil B*, 1978, **33**, 1398.
39. R. G. W. Gingerich and R. J. Angelici, *J. Am. Chem. Soc.*, 1979, **101**, 5604.
40. R. Kury and H. Vahrenkamp, *J. Chem. Res. (M)*, 1982, 0401, 0417.
41. R. H. Holm, *Chem. Soc. Rev.*, 1981, **10**, 455.
42. M. N. Hughes, 'The Inorganic Chemistry of Biological Processes', 2nd edn., Wiley, New York, 1981, and references cited therein.
43. F. Jellinek, in 'Inorganic Sulphur Chemistry', ed. G. Nickless, Elsevier, Amsterdam, 1968, p. 669.
44. F. Hulliger, *Struct. Bonding (Berlin)*, 1968, **4**, 83.
45. F. Jellinek, *MTP Int. Rev. Sci., Inorg. Chem. Ser. 1*, 1973, **5**, 339.
46. M. J. Atherton and J. H. Holloway, *Adv. Inorg. Chem. Radiochem.*, 1979, **22**, 171.
47. J. Fenner, A. Rabenau and G. Trageser, *Adv. Inorg. Chem. Radiochem.*, 1979, **23**, 330.
48. A. Müller, E. Diemann and M. J. F. Leroy, *Z. Anorg. Allg. Chem.*, 1970, **372**, 113.
49. F. Sécheresse, J. Lefebre, J. C. Daran and Y. Jeannin, *Inorg. Chim. Acta*, 1980, **45**, L45.
50. P. J. Aymonino, A. C. Ranade and A. Müller, *Z. Anorg. Allg. Chem.*, 1969, **371**, 295; P. J. Aymonino, A. C. Ranade, E. Diemann and A. Müller, *Z. Anorg. Allg. Chem.*, 1969, **371**, 300.
51. J. Cambray, R. W. Lane, R. W. Johnson and R. H. Holm, *Inorg. Chem.*, 1977, **16**, 2565.
52. T. Herskowitz, B. V. De Pamphilis, W. Gillum and R. H. Holm, *Inorg. Chem.*, 1975, **14**, 1427.
53. M. G. B. Drew, P. C. H. Mitchell and C. G. Pygall, *Angew. Chem.*, 1976, **88**, 855; *Angew. Chem., Int. Ed. Engl.*, 1976, **15**, 784.
54. D. L. Stevenson and L. F. Dahl, *J. Am. Chem. Soc.*, 1967, **89**, 3721.
55. G. Bunzey and J. H. Enemark, *Inorg. Chem.*, 1978, **17**, 682.
56. B. Spivack and Z. Dori, *J. Chem. Soc., Chem. Commun.*, 1973, 909.
57. C. Mealli, S. Midollini and L. Sacconi, *Inorg. Chem.*, 1978, **17**, 632.
58. T. C. Greenhough, B. W. S. Kolthammer, P. Legzdins and J. Trotter, *Inorg. Chem.*, 1979, **18**, 3543.
59. J. Schiemann, P. Hübner and E. Weiss, *Angew. Chem.*, 1983, **95**, 1021; *Angew. Chem., Int. Ed. Engl.*, 1983, **22**, 980.
60. A. Müller and P. Christophliemk, *Angew. Chem.*, 1969, **81**, 752; *Angew. Chem., Int. Ed. Engl.*, 1969, **8**, 753; M. G. B. Drew, P. C. H. Mitchell and C. F. Pygall, *J. Chem. Soc., Dalton Trans.*, 1979, 1213.
61. C. Potvin, J. M. Manoli, J. M. Brégault and G. Chottard, *Inorg. Chim. Acta*, 1983, **72**, 103.
62. J. Chatt and D. M. P. Mingos, *J. Chem. Soc. (A)*, 1970, 1243.
63. F. Seel, *Z. Anorg. Allg. Chem.*, 1942, **249**, 308.
64. J. T. Thomas, J. H. Robertson and E. G. Cox, *Acta Crystallogr.*, 1958, **11**, 599.
65. W. Beck, W. Danzer and G. Thiel, *Angew. Chem.*, 1973, **85**, 625; *Angew. Chem., Int. Ed. Engl.*, 1973, **12**, 582; P. J. Vergamini, H. Vahrenkamp and L. F. Dahl, *J. Am. Chem. Soc.*, 1971, **93**, 6327.
66. J. Marcoll, A. Rabenau, D. Mootz and H. Wunderlich, *Rev. Chim. Miner.*, 1974, **11**, 607.
67. C. Kowala and J. M. Swan, *Aust. J. Chem.*, 1966, **19**, 547.

68. G. Johansson and W. N. Lipscomt, *Acta Crystallogr.*, 1958, **11**, 594.
69. H. Vahrenkamp and L. F. Dahl, *Angew. Chem.*, 1969, **81**, 152; *Angew. Chem., Int. Ed. Engl.*, 1969, **8**, 144.
70. A. Müller, W. Hellmann, H. Bögge, R. Jostes, M. Römer and U. Schimanski, *Angew. Chem.*, 1982, **94**, 863; *Angew. Chem., Int. Ed. Engl.*, 1982, **21**, 860.
71. R. S. Gall, C. T. W. Chu and L. F. Dahl, *J. Am. Chem. Soc.*, 1974, **96**, 4019.
72. W. Saak, G. Henkel and S. Pohl, *Angew. Chem.*, 1984, **96**, 153; *Angew. Chem., Int. Ed. Engl.*, 1984, **23**, 150.
72a. A. Müller, W. Eltzner, W. Clegg and G. M. Sheldrick, *Angew. Chem.*, 1982, **94**, 555; *Angew. Chem., Int. Ed. Engl.* 1982, **21**, 536.
73. C. H. Wei and L. F. Dahl, as cited in ref. 7.
74. C. A. Ghilardi, S. Midollini and L. Sacconi, *J. Chem. Soc., Chem. Commun.*, 1981, 47.
75. D. Johnstone, J. E. Fergusson and W. T. Robinson, *Bull. Chem. Soc. Jpn.*, 1972, **45**, 3721.
76. J. A. de Beer and R. J. Haines, *J. Organomet. Chem.*, 1970, **24**, 757.
77. J. L. Vidal, R. A. Fiato, L. A. Cosby and R. L. Pruett, *Inorg. Chem.*, 1978, **17**, 2574; J. L. Vidal, R. C. Schoening, R. L. Pruett and R. A. Fiato, *Inorg. Chem.*, 1979, **18**, 1821.
78. G. Giani, L. Gardaschelli, A. Sironi and S. Marinengo, *J. Chem. Soc., Chem. Commun.*, 1981, 563.
78a. A. Müller, R. Jostes, W. Eltzner, Chong-shi-Nie, E. Diemann, H. Bögge, M. Zimmermann, M. Dartmann, U. Reinsch-Vogell, Shun Che, S. J. Cyvin and B. N. Cyvin, *Inorg. Chem.*, 1985, **24**, 2872.
79. M. T. Beck, 'Prebiotic Coordination Chemistry, The Possible Role of Transition Metal Complexes in the Chemical Evolution', in 'Metal Ions in Biological Systems', ed. H. Sigel, Dekker, New York, 1978, vol. 7.
80. J. Sola, Y. Do, J. M. Berg and R. H. Holm, *J. Am. Chem. Soc.*, 1983, **105**, 7784.
81. M. G. B. Drew and R. J. Hobson, *Inorg. Chim. Acta*, 1983, **72**, 233; M. G. B. Drew, D. A. Rice and D. M. Williams, *J. Chem. Soc., Dalton Trans.*, 1983, 2251.
82. E. J. Peterson, R. B. v. Dreele and T. M. Brown, *Inorg. Chem.*, 1978, **17**, 1410.
83. L. E. Rendon-Diazmiron, C. F. Campona and H. Steinfink, *J. Solid State Chem.*, 1983, **47**, 322.
84. P. J. M. Geurts, J. W. Gosselink, A. v. d. Avoird, E. J. Baerends and J. C. Snijders, *Chem. Phys.*, 1980, **46**, 133.
85. A. Müller, W. Hellmann, U. Schimanski, R. Jostes and W. E. Newton, *Z. Naturforsch., Teil B*, 1983, **38** 528.
86. A. Müller, E. Diemann and R. Jostes, *Naturwissenschaften*, in press.
87. K. S. Hagen and R. H. Holm, *J. Am. Chem. Soc.*, 1982, **104**, 5496; *Inorg. Chem.*, 1984, **23**, 418.
88. D. Swenson, N. C. Baenziger and C. Coucouvanis, *J. Am. Chem. Soc.*, 1978, **100**, 1932; S. A. Koch, L. E. Maelia and M. Millar, *J. Am. Chem. Soc.*, 1983, **5**, 5944.
89. S. Otsuka, M. Kamata, K. Hirotsu and T. Higuchi, *J. Am. Chem. Soc.*, 1981, **103**, 3011.
90. J. R. Bradbury, M. F. Mackay and A. G. Wedd, *Aust. J. Chem.*, 1978, **31**, 2423.
91. J. R. Dilworth, in 'Sulfur — Its Significance for Chemistry, for the Geo-, Bio- and Cosmosphere and Technology', ed. A. Müller and B. Krebs, Elsevier, Amsterdam, 1984.
92. A. Choy, D. Craig, I. Dance and M. Scudder, *J. Chem. Soc., Chem. Commun.*, 1982, 1246.
93. G. Christou, K. S. Hagen and R. H. Holm, *J. Am. Chem. Soc.*, 1982, **104**, 1744.
94. K. S. Hagen, A. D. Watson and R. H. Holm, *J. Am. Chem. Soc.*, 1983, **105**, 3905.
95. J. P. Fackler, Jr. and W. J. Zegarski, *J. Am. Chem. Soc.*, 1973, **95**, 8566.
96. I. G. Dance, *Inorg. Chem.*, 1981, **20**, 1487.
97. I. G. Dance, L. Fitzpatrick, M. Scudder and D. Craig, *J. Chem. Soc., Chem. Commun.*, 1984, 17.
98. I. G. Dance, *J. Chem. Soc., Chem. Commun.* 1976, 68.
99. P. Woodward, L. F. Dahl, E. W. Abel and B. C. Crosse, *J. Am. Chem. Soc.*, 1965, **87**, 5251.
100. I. G. Dance, L. J. Fitzpatrick and M. L. Scudder, *J. Chem. Soc. Chem. Commun.*, 1983, 546.
101. K. S. Hagen, D. W. Stephan and R. H. Holm, *Inorg. Chem.*, 1982, **21**, 3928.
102. I. G. Dance, *Inorg. Chem.*, 1981, **20**, 2155, and references cited therein.
103. P. M. Boorman, V. D. Patel, K. A. Kerr, P. W. Codding and P. V. Roey, *Inorg. Chem.*, 1980, **19**, 3508.
104. E. W. Abel, W. Harrison, R. A. N. McLean, W. C. Marsh and J. Trotter, *Chem. Commun.*, 1970, 1531.
105. G. W. Adamson and H. M. M. Shearer, *Chem. Commun.*, 1969, 897.
106. P. J. Blower, J. R. Dilworth, J. P. Hutchinson, and J. A. Zubieta, *Transition Met. Chem.* 1982, **7**, 353.
107. P. J. Blower, J. R. Dilworth, J. P. Hutchinson and J. A. Zubieta, *Inorg. Chim. Acta*, 1982, **65**, L225.
108. G. A. Barclay, E. M. McPartlin and N. C. Stephenson, *Inorg. Nucl. Chem. Lett.*, 1967, **3**, 397.
109. G. Henkel, W. Tremel and B. Krebs, *Angew. Chem.*, 1983, **95**, 317.
110. E. I. Stiefel, K. F. Miller, A. E. Bruce, J. L. Corbin, J. M. Berg and K. O. Hodgson, *J. Am. Chem. Soc.*, 1980, **102**, 3624.
111. E. I. Stiefel and R. R. Chianelli, in 'Nitrogen Fixation, The Chemical–Biochemical–Genetic Interface', ed. A. Müller and W. E. Newton, Elsevier, Amsterdam, 1983, p. 341.
112. I. G. Dance, A. G. Wedd and I. Boyd, *Aust. J. Chem.*, 1978, **31**, 519.
113. M. Millar, J. F. Lee, S. A. Koch and R. Fikar, *Inorg. Chem.*, 1982, **21**, 4105.
114. I. G. Dance, *J. Am. Chem. Soc.*, 1979, **101**, 6264.
115. P. M. Boorman, T. Chivers, K. N. Mahadev and B. D. O'Dell, *Inorg. Chim. Acta*, 1976, **19**, L35.
116. M. Kamata, T. Yoshida, S. Otsuka, K. Hirotsu and T. Higuchi, *J. Am. Chem. Soc.*, 1981, **103**, 3572.
117. J. R. Bradbury, A. G. Wedd and A. M. Bond, *J. Chem. Soc., Chem. Commun.*, 1979, 1022.
118. I. W. Boyd, I. G. Dance, A. E. Landers and A. G. Wedd, *Inorg. Chem.*, 1979, **18**, 1875.
119. W. E. Newton, in 'Sulfur — Its Significance for Chemistry, for the Geo-, Bio- and Cosmosphere and Technology', ed. A. Müller and B. Krebs, Elsevier, Amsterdam, 1984.
120. C. D. Garner, S. R. Acott, G. Christou, D. Collison, F. E. Mabbs, V. Petrouleas and C. J. Pickett in 'Nitrogen Fixation, The Chemical–Biochemical–Genetic Interface', ed. A. Müller and W. E. Newton, Elsevier, Amsterdam, 1983, p. 245.
121. P. L. Cloke, *Geochim. Cosmochim. Acta*, 1963, **27**, 1265; P. L. Cloke, *Geochim. Cosmochim. Acta*, 1963, **27**, 1299; S. Matthes, 'Mineralogie, Eine Einführung in die spezielle Mineralogie, Petrologie und Lagerstättenkunde', Springer, Berlin, 1983, p. 237.
122. K. B. Krauskopf, 'Introduction to Geochemistry', McGraw Hill, New York, 1979, p. 393; H. S. Wunderlich, 'Einführung in die Geologie', 2nd ed., BI Hochschultaschenbücher, 1970, Mannheim, vol. 2, p. 141.

123. A. Müller, M. Römer, H. Bögge, E. Krickemeyer and K. Schmitz, *Inorg. Chim. Acta*, 1984, **85**, L39.
124. A. Müller, M. Römer, E. Krickemeyer and H. Bögger, *Naturwissenschaften*, 1984, **71**, 43.
125. A. Müller, M. Römer, H. Bögge, E. Krickemeyer and D. Bergmann, *J. Chem. Soc., Chem. Commun.*, 1984, 348.
126. M. G. Kanatzidis, N. C. Baenziger and D. Coucouvanis, *Inorg. Chem.*, 1983, **22**, 290.
127. P. H. Bird, J. M. McCall, A. Shaver and U. Siriwardane, *Angew. Chem.*, 1982, **94**, 375; *Angew. Chem. Int. Ed. Engl.*, 1982, **21**, 384.
128. H. Köpf, *Angew. Chem.*, 1969, **81**, 332; *Angew. Chem. Int. Ed. Engl.*, 1969, **8**, 375; B. R. Davis, I. Bernal and H. Köpf, *Angew. Chem.*, 1971, **83**, 1018; *Angew. Chem., Int. Ed. Engl.*, 1971, **10**, 921.
129. A. Müller, E. Krickemeyer, H. Bögge, W. Clegg and G. M. Sheldrick, *Angew. Chem.*, 1983, **95**, 1030; *Angew. Chem., Int. Ed. Engl.*, 1983, **22**, 1006.
130. M. Draganjac, E. Simhon, L. T. Chan, M. Kanatzidis, N. C. Baenziger and D. Coucouvanis, *Inorg. Chem.*, 1982, **21**, 3321; W. Clegg, G. Christou, C. D. Garner and G. M. Sheldrick, *Inorg. Chem.*, 1981, **20**, 1562.
131. A. Müller, M. Römer, C. Römer, U. Reinsch-Vogell, H. Bögge and U. Schimanski, *Monatsh. Chem.* 1985, **116**, 711.
132. D. Coucouvanis, N. C. Baenziger, E. D. Simhon, P. Stremple, D. Swenson, A. Kostikas, A. Simopoulos, V. Petrouleas and V. Papaefthymiou, *J. Am. Chem. Soc.*, 1980, **102**, 1730; D. Coucouvanis, P. Stremple, E. D. Simhon, D. Swenson, N. C. Baenziger, M. Draganjac, L. T. Chan, A. Simopoulos, V. Papaefthymiou, A. Kostikas and V. Petrouleas, *Inorg. Chem.*, 1983, **22**, 293.
133. K. A. Hofmann and F. Höchtlen, *Ber. Dtsch. Chem. Ges.*, 1904, **37**, 245; K. A. Hofmann and F. Höchtlen, *Ber. Dtsch. Chem. Ges.*, 1903, **36**, 3090; P. E. Jones and L. Katz, *Acta Crystallogr., Sect. B*, 1969, **25**, 745; R. D. Gillard and F. L. Wimmer, *J. Chem. Soc., Chem. Commun.*, 1978, 936.
134. A. Müller, J. Schimanski and U. Schimanski, *Angew. Chem.*, 1984, **96**, 158; *Angew. Chem., Int. Ed. Engl.*, 1984, **23**, 159.
135. G. Marbach and J. Strähle, *Angew. Chem.*, 1984, **96**, 229; *Angew. Chem., Int. Ed. Engl.*, 1984, **23**, 246.
136. A. E. Wickenden and R. A. Krause, *Inorg. Chem.*, 1969, **8**, 779; M. Schmidt and G. G. Hoffmann, *Z. Anorg. Allg. Chem.*, 1979, **452**, 112.
137. R. A. Krause, *Inorg. Nucl. Chem. Lett.*, 1971, **7**, 973.
138. A. Müller, J. Schimanski and H. Bögge, *Z. Naturforsch., Teil B*, 1985, **40**, 1277.
139. C. M. Bolinger, T. B. Rauchfuss and S. R. Wilson, *J. Am. Chem. Soc.*, 1981, **103**, 5620.
140. A. Müller, W. Eltzner, H. Bögge and E. Krickemeyer, *Angew. Chem.*, 1983, **95**, 905; *Angew. Chem., Int. Ed. Engl.*, 1983, **22**, 884.
141. L. F. Maheu and L. H. Pignolet, *J. Am. Chem. Soc.*, 1980, **102**, 6346.
142. A. Müller, F. W. Baumann, H. Bögge, M. Römer, E. Krickemeyer and K. Schmitz, *Angew. Chem.*, 1984, **96**, 607; *Angew. Chem., Int. Ed. Engl.*, 1984, **23**, 632.
143. A. Müller, M. Römer, H. Bögge, E. Krickemeyer, F. W. Baumann and K. Schmitz, *Inorg. Chim. Acta*, 1984, **89**, L7.
144. A. Müller, M. Römer, H. Bögge, E. Krickemeyer and D. Bergmann, *Z. Anorg. Allg. Chem.*, 1984, **511**, 84.
145. A. Müller and U. Schimanski, *Inorg. Chim. Acta*, 1983, **77**, L187.
146. P. S. Haradem, J. L. Cronin, R. A. Krause and L. Katz, *Inorg. Chim. Acta*, 1977, **25**, 173.
147. M. Spangenberg and W. Bronger, *Z. Naturforsch., Teil B*, 1978, **33**, 482; P. E. Jones and L. Katz, *Chem. Commun.*, 1967, 842.
148. C. Burschka, *Z. Naturforsch., Teil B*, 1980, **35**, 1511.
149. Y. Wakatsuki and H. Yamazaki, *J. Organomet. Chem.*, 1974, **64**, 393.
150. J. Chatt, D. M. P. Mingos, *J. Chem. Soc. (A)*, 1970, 1243.
151. H. Köpf, S. K. S. Hazari and M. Leitner, *Z. Naturforsch., Teil B*, 1978, **33**, 1398.
152. H. Köpf and B. Block, *Chem. Ber.* 1969, **102**, 1504.
153. N. B. Egen and R. A. Krause, *Inorg. Nucl. Chem.*, 1969, **31**, 127; H. Köpf, *Chem. Ber.*, 1969, **102**, 1509.
154. D. Coucouvanis, D. Swenson, P. Stremple and N. C. Baenziger, *J. Am. Chem. Soc.*, 1979, **101**, 3392.
155. D. Dudis and J. P. Fackler, Jr., *Inorg. Chem.*, 1982, **21**, 3577.
156. B. Kreutzer, P. Kreutzer and W. Beck, *Z. Naturforsch., Teil B*, 1972, **27**, 461.
157. D. Coucouvanis and M. Draganjac, *J. Am. Chem. Soc.*, 1982, **104**, 6820.
158. M. Draganjac and D. Coucouvanis, *J. Am. Chem. Soc.*, 1983, **105**, 139.
159. E. G. Muller, J. L. Petersen and L. F. Dahl, *J. Organomet. Chem.*, 1976, **111**, 91.
160. A. Müller and U. Reinsch-Vogell, unpublished; U. Reinsch-Vogell, Dissertation, University of Bielefeld, 1984.
161. A. Müller, U. Reinsch-Vogell, E. Krickemeyer and H. Bögge, *Angew. Chem.*, 1982, **94**, 784; *Angew. Chem., Int. Ed. Engl.*, 1982, **21**, 796.
162. J. Gotzig, A. L. Rheingold and H. Werner, *Angew. Chem.*, 1984, **96**, 813.
163. B. Czeska and K. Dehnicke, *Z. Naturforsch., Teil B*, 1985, **40**, 120.
164. A. Müller, J. Schimanski, M. Römer, H. Bögge, F. W. Baumann, W. Eltzner, E. Krickemeyer and U. Billerbeck, *Chimia*, 1985, **39**, 25.
165. D. Coucouvanis, P. R. Patil, M. G. Kanatzidis, B. Detering and N. C. Baenziger, *Inorg. Chem.*, 1985, **24**, 24.
166. M. Herberhold, D. Reiner, K. Ackermann, U. Thewalt and T. Debaerdemaeker, *Z. Naturforsch., Teil B*, 1984, **39**, 1199.
167. F. N. Tebbe, E. Wasserman, W. G. Peet, A. Vatvars and A. C. Hayman, *J. Am. Chem. Soc.*, 1982, **104**, 4971; R. Steudel and R. Straus, *J. Chem. Soc., Dalton Trans.*, 1984, 1775.
168. D. M. Giolando and T. B. Rauchfuss, *J. Am. Chem. Soc.*, 1984, **106**, 6455.
169. R. D. Gillard, F. L. Wimmer and J. P. G. Richards, *J. Chem. Soc., Dalton Trans.*, 1985, 253.
170. A. Müller and E. Diemann, *Adv. Inorg. Chem.*, 1987, **31**, 89.

16.2
Thioethers

ACHIM MÜLLER and EKKEHARD DIEMANN
University of Bielefeld, FRG

Thioethers, R_2S (R = alkyl, aryl), have two lone pairs of electrons on the sulfur, and even though the formation tendency of complexes with these ligands is relatively low, they possess, mainly due to the absence of ionizable protons, fewer pathways for decomposition and side reactions than H_2S and SH^- ligands, for example. Consequently, much more stable complexes with thioether ligands, monodentate or bridging, have been described in the literature, some of them quite early.[1,2] Since, by variation of R, more than 100 different R_2S ligands have been used in coordination compounds, the survey has to be restricted here to a description of only the most significant types of ligands and therefore excludes a comprehensive compilation of examples. One such can be found in a recent review by Murray and Hartley.[3] The report by Kuehn and Isied[4] on the reactivity of metal–sulfur bonds also contains much information about thioether complexes.

16.2.1 SURVEY OF COMPOUND TYPES WITH THIOETHER LIGANDS

16.2.1.1 Ligands of the Types R_2S, $RS(CH_2)_nSR$ and $RS(CH_2)_nS(CH_2)_nSR$

In this section we describe not only metal complexes of simple dialkyl sulfides but also those with ligands which contain more than one thioether group. Both types were studied in the early 1900s by the famous Russian coordination chemist Tschugajeff (Chugaev),[5] who mainly contributed to our knowledge about platinum and other d^8 metal complexes.[6] The isomers (**1**) and (**2**) were among the first compounds reported by his group. Another example of dialkylsulfido complexes is the series

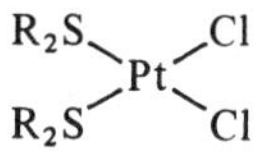

(1) 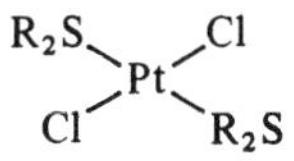   (2)

551

of compounds $Cl_3M(SEt_2)_3$ with M = Rh, Ir, Ru.[7,8] These ligands are not strongly bound and can easily be displaced, for example by pyridine and aniline.

It was again Tschugajeff who introduced multifunctional organic sulfides as ligands into coordination chemistry. He realized that the formation tendency of five-membered rings dominates. In the case of Pt^{II} as central ion 1,2-dithioethers form the most stable complexes. The stability of 1,3- and 1,5-dithioether complexes is lower (and even more so with the 1,1-ligand).[11]

As typical examples of complexes with bidentate $RS(CH_2)_nSR$ and tridentate $[RS(CH_2)_n]_2S$ we may quote here a series of mononuclear Ru compounds which have been prepared and characterized by Chatt *et al.*[7] Another example of a tridentate ligand is trithiane, which forms complexes with Ag^I and Hg^{II}.[9,10] More complexes are given in a review concerned with chelating dithioether complexes of d^8 metal ions.[11].

Because there are two nonbonding electron pairs available, thioethers may also act as bridging ligands, as for instance in (3), where an interesting competition with halide bridging (*e.g.* with SPh_2 as ligand) may occur.[3] Other examples can be found in a series of complexes reported by Stein and Taube,[12] such as (4) and (5), which may exist with mixed valencies. Dinuclear complexes with a bridging spirocyclic ligand have also been reported, *e.g.* (6), with M = Hg, X = Cl and R = Bu^t.[13]

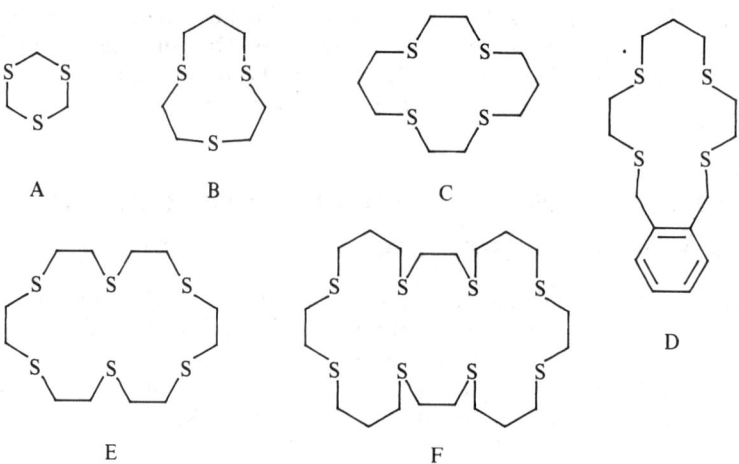

(3)   (4)   (5)   (6)

16.2.1.2 Cyclic and Macrocyclic Thioethers

Both macrocyclic and open-chain cyclic thioethers have been used as ligands. Figure 1 shows a collection of typical examples of macrocyclic species. They offer very flexible coordination behavior and their metal complexes sometimes exist with different electron populations.

The Cu^{II} complex of ligand C in Figure 1, 14-ane-S_4, is mononuclear as expected, with copper in the square planar environment of the four sulfur atoms.[14] However, the Cu^I compound is polymeric with only three S atoms of one ligand molecule coordinating to one copper, its tetrahedral environment being completed with one sulfur atom of a neighboring ligand molecule.[15] Surprisingly, the same ligand forms the dinuclear complex Cl_2HgS_2-14-ane-S_2HgCl_2 with Hg^{II}.[16] The macrocyclic ligand F in Figure 1 is also capable of fixing two metal ions, *e.g.* Ni^{2+}, in a square planar coordination.[3]

Another type of cyclic thioether is 1,3,5-tris(alkylthia)cyclohexane (7), which is related to the open-chain multifunctional ligands mentioned above. With R = Me, the formation of weak Cu^{II} complexes with this ligand has been observed.[17]

Figure 1 Some examples of macrocyclic thioether ligands

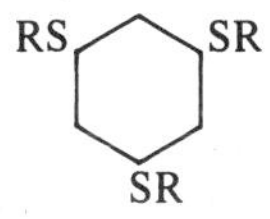

(7)

16.2.1.3 RSSR Ligands

Organic disulfides can act as monodentate ligands. The simplest compound of this type, dimethyl disulfide, acts as a bridge in $[Re_2Br_2(CO)_6(S_2Me_2)]$, where each sulfur coordinates to one Re.[18] Other acyclic disulfides have been shown to form complexes of Cu^I and Cu^{II}.[18-20] An example incorporating a monodentate cyclic disulfide ligand, the Ru^{II} complex ion (8), has been reported by Stein and Taube.[12]

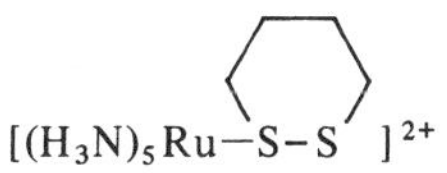

$$[(H_3N)_5Ru-S-S]^{2+}$$

(8)

16.2.1.4 Chelating Ligands with Thioether Groups and Other Donor Atoms

Finally, we comment here on a type of thioether ligand which contains, in addition to the sulfur, one or more other typical donor atoms, *e.g.* nitrogen or oxygen. Representative examples of this kind of ligand have been reported by Schugar *et al.*[21]

The sulfur and nitrogen atoms of $H_2NCH_2CH_2SMe$ (L) both coordinate to Cu^{II} in $[CuL_2]$-$(ClO_4)_2$. Two ligands form a square planar environment with two weakly coordinating perchlorate anions in the apical positions.

Tridentate $S(CH_2CH_2NH_2)_2$ forms five-coordinate $CuLCl_2$[22] and tetradentate $N(CH_2CH_2SMe)_3$ also forms five coordinate trigonal planar $[CuLBr]Br$.[23] An example of a dinuclear system employing three different kinds of donor is the complex (9).[24] If the donor atoms are bound to an aromatic system, or a part of it, as for instance in the ligands (10) and (11), it is expected that the π-delocalization from the ring will enhance the stability of the corresponding complexes.[25] In analogy to the above-mentioned macrocyclic thioethers, chelating ligands of this type may also be employed; complex (12) is an example.[26]

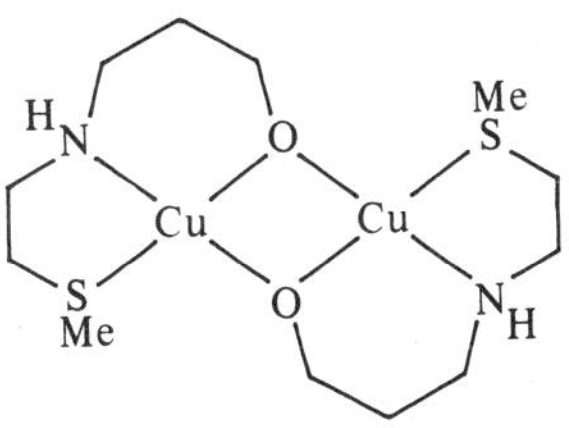

(9) 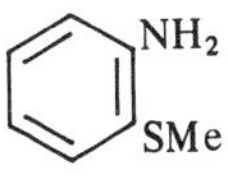   (10) 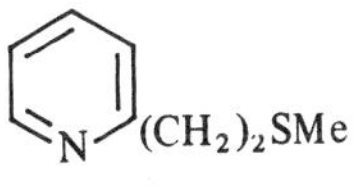   (11) 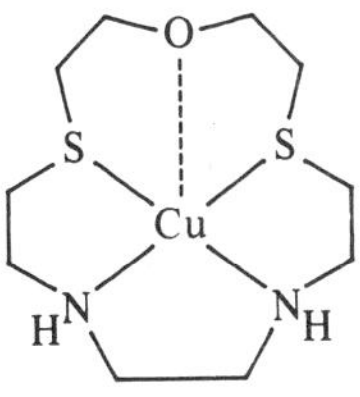   (12)

16.2.2 FORMATION AND DISPLACEMENT REACTIONS

16.2.2.1 Formation Reactions of Thioether Complexes

In many cases thioether complexes can be obtained in a direct reaction of aqua or halogeno complexes with the thioether in aqueous or alcoholic solutions by simple addition or substitution. The actual stoichiometry, however, may also depend strongly on the concentrations used. Diethyl sulfide reacts with VCl_3 to give trigonal bipyramidal *trans*-$VX_3(SEt_2)_2$; however, in Et_2S solution the six-coordinate $VX_3(SEt_2)_3$ is formed.[27] A series of Nb^{IV} compounds NbX_4L and NbX_4L_2 have also been reported,[28] where for L = Me_2S in benzene an equilibrium between the five- and the six-coordinate species was detected. The stability of the bis-adduct with respect to the mono-adduct decreases in the order I > Br > Cl for X.

Another related procedure has been employed by Flint and Goodgame,[29] who reacted finely ground $Ni(ClO_4)_2 \cdot aq$ directly with 2,5-dithiahexane for several weeks. The blue solid formed by this reaction was repeatedly ground and then treated further with the ligand until the $\nu(OH)$ absorption of water could no longer be detected in the IR spectrum of the product.

A different route is the S-alkylation of thiol ligands, as for example in the reaction of (diethyl-2-mercaptoethylamine)gold(III) with bromoethane to give the corresponding thioether complex.[30] The polymeric complexes of Pd^{II} with 1,2-ethanedithiol and 1,3-propanedithiol also react easily with alkyl halides to form mononuclear and dinuclear complexes containing thioether ligands.[31] We might also mention here the S-alkylation of a mercaptoamine complex of Ni^{II} (13) with alkyl halides in DMF to give the thioether complex (14).[32]

(13)   (14)

16.2.2.2 Equilibrium and Kinetic Studies

The relative stabilities of the thioether complexes vary depending on both the metal and the other ligands present. Cobalt, nickel and copper in their +2 oxidation state form more stable complexes with thioethers than with ether ligands, whereas Zn^{II} shows the opposite behavior. For $RSCH_2CO_2H$ the equilibrium constants decrease in the order $Ag^I \gg Cu^{II} \gg Cd^{II} \approx Co^{II} > Mn^{II} \approx Zn^{II} > Ni^{II}$. For the ligand $S(CH_2CO_2H)_2$ a similar series has been found, with Ni^{II} (more typical) between Cu^{II} and Co^{II}.[33]

Equilibrium data are also available for some Ag^I complexes with alkyl-, alkenyl- and phenyl-thioacetic acids. An important conclusion here is that the strength of the Ag—S bond is mainly determined by its σ component.[34] In addition, the fact that the *cis* isomer (1) isomerizes to the *trans* isomer (2) upon moderate heating suggests that there is no strong metal-to-sulfur π-interaction.[35] Affinity data for other systems have been compiled by Kuehn and Isied.[4]

There are also some studies concerning the kinetics of the displacement of amine and halide ligands by the R_2S ligands of Pd^{II}, Pt^{II} and Au^{III} complexes.[3] The results again suggest that the coordination properties of the thioether are dominated by its σ-donor ability.

16.2.2.3 S-Dealkylation and S-Alkylation at Coordinated Sulfur

If, for example, complexes of Pt^{II} and Pd^{II} with the ligand (15; N-SMe) are heated in DMF for some hours S-demethylation occurs as in equation (1).[3] Such a reaction was described in the last century,[1] when dimethyl sulfide was demethylated by $PtCl_2$ [to give $Pt(SMe)_2$], and many other examples have been reported since.[3] This occurs mainly with complexes of d^8 metal centers and Au^{III}, which both form relatively strong bonds to these ligands. This may facilitate the cleavage of the C—S bond *via* polarization, but the fact that the S-dealkylated ligand is more strongly bound to the central atom than the thioether also makes the reaction thermodynamically viable.

(15) N-SMe

$$M(N\text{-}SMe)Br_2 \longrightarrow M_2(N\text{-}S)_2Br_2 \tag{1}$$

An example of a thermal dealkylation is the thermolysis of $[(Bu_2^tS)HgCl_2]_2$, which yields $(Bu^tS)HgCl$ and some other products. This reaction is of first order and a mechanism involving an interaction between the ligand side-chain and the chloride has been discussed.[36]

S-Alkylation has already been mentioned as a general pathway for the preparation of new thioether complexes. It appears likely that S-alkylation is facilitated by a high nucleophilicity of the coordinated mercaptide ligand. If this nucleophilicity is reduced, either by the presence of electron-withdrawing groups at the sulfur or by a bridging coordination of the mercaptide between two metal centers, this reaction does not usually occur. In the case where polymeric Pd^{II} complexes with μ-1,2-ethanedithiol could be alkylated, *e.g.* with methyl iodide, to form a mononuclear system containing a thioether donor,[31] no other ligands are present, which probably allows sufficient electron density to remain on the sulfur. If the complex is stabilized by an organometallic unit, coordinated thioethers can be further alkylated, as for example in equation (2) where a ligand of the general type $R_2R'S^+$ is formed.[37]

$$\eta\text{-}CpMn(CO)_2(SR_2) + MeSO_3F \rightarrow [\eta\text{-}CpMn(CO)_2(SMeR_2)]SO_3F \qquad (2)$$

16.2.2.4 Inversion

Inversion at a coordinated thioether was first recognized by the low temperature magnetic nonequivalence of the methyl protons in the NMR spectrum of $PtCl_2$(2,5-dithiohexane) (due to the presence of two isomers, **16**), the signals of which coalesce to a single triplet at about 95 °C.[38] The low activation energy for this process and the retention of the $^{195}Pt-^1H$ coupling at higher temperature have led to a mechanism for the inversion in which complex (**17**) reacts *via* the intermediate (**18**) (with both lone pairs bonded to the metal) to give the isomer (**19**),[39,40] and consequently rule out a dissociative procedure. When bidentate thioether ligands coordinate, the coalescence temperatures are significantly higher than those for *trans* unidentate ligands. The *trans* influence of halide ions on the inversion barriers here is in the order $Cl^- > Br^- > I^-$.[3]

meso-(16)   ±-(16)   (17)   (18)   (19)

Inversion at the sulfur is also observed with a series of dinuclear Pt complexes with the ligand MeS—CHR—SMe (R = Me, H). However, at higher temperatures a fluxional behavior is encountered where the sulfur atoms obviously swap between the Pt centers.[41]

16.2.3 STRUCTURE AND BONDING

16.2.3.1 X-Ray Diffraction Studies

Murray and Hartley have reviewed more than 60 X-ray structure determinations of thioether complexes.[3] Thus, we may restrict our report here to the principal features.

In many cases the M—S bond has been found to be 'normal' in length, *i.e.* the bond distances agree fairly well with the sum of the covalent radii (Pauling's scale, data appropriate to coordination number and oxidation state). However, in some complexes the M—S distances are significantly shorter than the calculated ones, for example in thioether compounds with Cr^0, Pd^{II}, Pt^{II}, Cu^I and Au^I. This observation has often led to the suggestion that the shortening is due to some π-back-donation from the metal to the sulfur.

When the thioether is a bridging ligand the M—S bond lengths are usually shorter compared with those of terminal ligands, probably due to the fact that there is no repulsion by the lone pair of electrons. This situation has been found in complexes of Pd^{II}, Pt^{II}, Ta^{III}, Nb^{III} and Mo^{II}. However, in the Cu^I compound $(Et_2S)_3Cu_4I_4$ (**20**) the bridging M—S bonds are longer than the terminal ones by almost 0.04 Å.[3]

In general, the geometry about the sulfur in thioether complexes is approximately tetrahedral (considering the lone pairs of electrons too). Exceptions are examples with a rigid stereochemistry, *e.g.* when the metal or the sulfur forms part of a ring.

(20)

16.2.3.2 Spectroscopic Measurements

In the vibrational spectra the ν(M—S) bands are mostly weak in intensity and are observed, in general, between 290 and 350 cm^{-1}.[3] In cases where the X-ray structural results indicate shorter bond lengths the M—S stretching frequency may shift to wavenumbers above 400 cm^{-1}. This applies for instance to the thioether-bridged Pt^{II} complexes $(R_2S)_2Pt_2X_4$, where the ν(M—S) frequency around 400 cm^{-1} demonstrates a stronger M—S bond in the bridging situation (in accordance with the structure determination).[42]

Vibrational data of other ligands may also be used as a source of information regarding the nature of the M—S bond or the influence of a ligand in a position *trans* to the thioether. For example, the variation of the ν(CO) frequency in complexes $[ML(CO)_5]$ (M = Cr, Mo, W) describes the π-acceptor ability of L (dien $\approx$ pyridine < R_2S < PPh_3 < PCl_3).[43] An analysis of the UV-vis spectra of such compounds suggests that the wavenumber of the longest wavelength band could also be characteristic for the strength of the M—S interaction.[44] The chain length of the alkyl group of thioether ligands has a significant effect on the ligand field splitting, as for example in *trans*-$[PdCl_2(R_2S)_2]$ (R = Pr^n, Bu^n, Bu^i, Bu^s), where Δ_1 varies from 23.2 to 24.3 $\times$ 10^3 cm^{-1}.[45] The R_2S ligands were placed between the halides and the tertiary arsines and phosphines in the spectrochemical series.

16.2.3.3 Electrochemistry of Thioether Complexes

Extensive data for this section are only available for Ru complexes, which have almost exclusively been studied by Taube's group.[4] By measurement of the Ru$^{II/III}$ redox potential it could be demonstrated that two SMe$_2$ or the bidentate MeS(CH)$_2$SMe ligand attached to the $(NH_3)_4$Ru moiety stabilize the Ru^{II} complex over the ligation with a single thioether. From the difference in potentials for the pentaammine ruthenium(II) complexes of 1,2-dithiane, 1,3-dithiane and 1,4-dithiane, when compared to the corresponding complex with dimethyl sulfide, it could further be shown that a sulfur atom which is not directly coordinating may have a substantial influence on the stability. There is also a pronounced ring size effect of the macrocyclic ligands on the redox potentials (which vary over a range of 200 mV for the above-mentioned Ru thioether complexes). The stabilizing influence decreases with an increasing sulfur–sulfur distance in the ligand, as for example in (8) and related systems.

16.2.3.4 Bonding

In the preceding chapters we have described several features pertinent to bonding in thioether complexes. Simplified, they all contain sp^3-hybridized sulfur (initially with two lone pairs of electrons). If only one lone pair is involved in the bonding to metal ions then the other may have a pronounced steric effect (Gillespie–Nyholm approach). In principle, however, this second lone pair may also take part in bonding *via* π-donation. Where the metal ions have unoccupied d-levels with suitable symmetry they may act as acceptors and π-donation from the sulfur may be anticipated. In addition, the sulfur atom has empty d-orbitals which can principally take part in a metal-to-sulfur π-back-donation if the symmetry matches. Although some of the features reported above could be explained straightforwardly by involvement of such kinds of π-bonding, it turns out to be difficult to attribute these effects unequivocally to this kind of interaction only. The variation of the formation enthalpies of $MCl_4 \cdot 2L$ (L = pyridine, tetrahydrofuran and tetrahydrothiophene) from MCl_4 and 2L for M = Zr^{IV} and Mo^{IV} has been ascribed to π-donation.[52] In any case, such a donation is small and definitely far less important than for phosphine complexes, for example.

The absence of strong π-interactions and the only 'normal' σ-donor abilities of the thioethers make this class of compounds relatively 'poor' ligands.

16.2.4 METAL–THIOETHER INTERACTIONS OF BIOLOGICAL INTEREST

Metal–thioether interactions of biological or medical interest concentrate on complexes of copper. D-Biotin (vitamin H) involves tetrahydrothiophene,[46] and the 'blue' copper proteins are believed to contain interactions between different thioether units of methionine (as well as cysteine and histidine)[47] (see also ref. 48).

In general, the affinity of Cu^{II} for thioethers is much less than that of Cu^I and also the structure of the complexes changes considerably when the oxidation state is altered. For example, py(CH$_2$)$_2$-S(CH$_2$)$_2$S(CH$_2$)$_2$py (py = pyridine) forms a Cu^I complex with two N- and two S-donors producing a tetrahedral environment. The Cu^{II} complex, however, has a square planar coordination of two N and two S atoms (with one perchlorate anion apical to the basal plane).[49,50] Two nitrogen and two sulfur donors have also been found for the copper-binding site in plastocyanin[51] and in azurin from *Pseudomonas aeruginosa* (*cf.* refs. 3 and 4). The low affinity of Cu^{II} towards thioether ligands, however, mandates that its environment in the protein must provide protection against substitution by other ligands, particularly by H_2O. However, there is still the possibility that other ligands around Cu^{II} may enhance the stability as well as sulfur–sulfur interactions between different ligands. The latter have also been suspected of producing the intense blue color, as for example in macrocyclic thioether Cu^{II} complexes.[4]

Amino acids with coordinating side-chains such as methionine (and the binding sites of some metals in metalloproteins) are described in more detail in Chapter 20.2.

6.2.5 REFERENCES

1. C. W. Blomstrand, *J. Prakt. Chem.*, 1883, **27**, 161, 196.
2. A. Loir, *Liebigs Ann. Chem.*, 1889, **107**, 234.
3. S. G. Murray and F. R. Hartley, *Chem. Rev.*, 1981, **81**, 365.
4. C. G. Kuehn and S. S. Isied, *Prog. Inorg. Chem.*, 1980, **27**, 153.
5. E. Fritzmann, *Z. Anorg. Allg. Chem.*, 1924, **134**, 277.
6. L. Tschugajeff and W. Subbotin, *Chem. Ber.*, 1910, **43**, 1200; L. Tschugajeff and J. Benewolensky, *Z. Anorg. Allg. Chem.*, 1903, **82**, 420.
7. J. Chatt, G. J. Leigh and A. P. Storace, *J. Chem. Soc. (A)*, 1971, 1380.
8. B. R. James and F. T. T. Ng, *J. Chem. Soc., Dalton Trans.*, 1972, 1321.
9. A. Domenicano, L. Scaramuzza, A. Vaciago, R. S. Ashworth and C. K. Prout, *J. Chem. Soc. (A)*, 1968, 866.
10. W. R. Costello, A. T. McPhail and G. A. Sim, *J. Chem. Soc. (A)*, 1966, 1190.
11. M. Schmidt and G. G. Hoffmann, *Phosphorus Sulfur*, 1978, **4**, 239.
12. C. Stein and H. Taube, *Inorg. Chem.*, 1979, **18**, 2212.
13. D. C. Goodall, *J. Chem. Soc. (A)*, 1967, 1387.
14. M. D. Glick, O. P. Gavel, L. Diaddario and D. B. Rorabacher, *Inorg. Chem.*, 1976 **15**, 1190.
15. E. R. Dockal, L. L. Diaddario, M. D. Glick and D. B. Rorabacher, *J. Am. Chem. Soc.*, 1977, **99**, 4530.
16. N. W. Alcock, N. Herron and P. Moore, *J. Chem. Soc., Chem. Commun.*, 1976, 886.
17. A. R. Amundsen, J. Whelan and B. Bosnich, *J. Am. Chem. Soc.*, 1977, **99**, 6730.
18. I. Bernal, J. L. Atwood, F. Calderazzo and D. Vitali, *Isr. J. Chem.*, 1976, **15**, 153.
19. J. A. Thich, D. Mastropaolo, J. Potenza and H. Schugar, *J. Am. Chem. Soc.*, 1974, **96**, 726; C. I. Branden, *Acta Chem. Scand.*, 1967, **21**, 1000.
20. M. Micheloni, P. M. May and D. R. Willliams, *J. Inorg. Nucl. Chem.*, 1978, **40**, 1209.
21. C. C. Ou, V. M. Miskowski, R. A. Lalancette, J. A. Potenza and H. J. Schugar, *Inorg. Chem.*, 1976, **15**, 3157.
22. L. T. Taylor and E. K. Barefield, *J. Inorg. Nucl. Chem.*, 1969, **31**, 3831.
23. M. Ciampolini, J. Gelsomini and N. Nardi, *Inorg. Chim. Acta*, 1968, **2**, 343.
24. M. Mikuriya, H. Okawa and S. Kida, *Inorg. Chim. Acta*, 1979, **34**, 13.
25. A. C. Braithwaite, C. E. F. Rickard and T. N. Waters, *J. Chem. Soc., Dalton Trans.*, 1975, 1817.
26. F. Arnaud-Neu, M. J. Schwing-Weill, J. Juillard, R. Louis and R. Weiss, *Inorg. Nucl. Chem. Lett.*, 1978, **14**, 367.
27. R. J. H. Clark and G. Natile, *Inorg. Chim. Acta*, 1970, **4**, 533.
28. J. B. Hamilton and R. E. McCarley, *Inorg. Chem.*, 1970, **9**, 1333.
29. C. D. Flint and M. Goodgame, *J. Chem. Soc. (A)*, 1968, 2178.
30. R. V. G. Ewens and C. S. Gibson, *J. Chem. Soc.*, 1949, 431.
31. L. Cattalini, J. S. Coe, S. Degetto, A. Dondoni and A. Vigato, *Inorg. Chem.*, 1972, **11**, 1519.
32. D. H. Busch, J. A. Burke, D. C. Jicha, M. C. Thompson and M. L. Morris, *Adv. Chem. Ser.*, 1963, **37**, 125; D. H. Busch, D. C. Jicha, M. C. Thompson, J. W. Wrathall and E. Blinn, *J. Am. Chem. Soc.*, 1964, **86**, 3642.

33. R. M. Tichane and W. E. Bennett, *J. Am. Chem. Soc.*, 1957, **79**, 1293; G. J. Ford, P. Gans, L. D. Pettit and C. Sherrington, *J. Chem. Soc., Dalton Trans.*, 1972, 1763.
34. L. D. Pettit, A. Royston and R. J. Whewell, *J. Chem. Soc. (A)*, 1968, 2009; L. D. Pettit and C. Sherrington, *J. Chem. Soc. (A)*, 1968, 3078.
35. E. A. Andronov, Yu. N. Kukushkin, T. M. Lukicheva, L. V. Konovalov, S. I. Bakhireva and E. S. Postnikova, *Russ. J. Inorg. Chem. (Engl. Transl.)*, 1976, **21**, 1343.
36. P. Biscarini, L. Fusina and G. D. Nivellini, *J. Chem. Soc., Dalton Trans.*, 1972, 1921.
37. R. D. Adams and D. F. Chodosh, *J. Organomet. Chem.*, 1976, **120**, C39.
38. E. W. Abel, R. P. Bush, F. J. Hopton and C. R. Jenkins, *Chem. Commun.*, 1966, 58.
39. P. Haake and P. C. Turley, *J. Am. Chem. Soc.*, 1967, **89**, 4611, 4617.
40. J. H. Eekhof, H. Hogeveen, R. M. Kellogg and E. Klei, *J. Organomet. Chem.*, 1978, **161**, 183.
41. E. W. Abel, A. R. Khan, K. Kite, K. G. Orrell and V. Sik, *J. Chem. Soc., Chem. Commun.*, 1979, 126.
42. P. L. Goggin, R. J. Goodfellow and F. J. S. Reed, *J. Chem. Soc., Dalton Trans.*, 1974, 576.
43. E. W. Ainscough, E. J. Birch and A. M. Brodie, *Inorg. Chim. Acta*, 1976, **20**, 187.
44. *cf.* P. S. Braterman and A. P. Walker, *Discuss. Faraday Soc.*, 1969, **47**, 121; F. A. Cotton, W. T. Edwards, F. C. Rauch, M. A. Graham, R. N. Perutz and J. J. Turner, *J. Coord. Chem.*, 1973, **2**, 247.
45. B. E. Aires, J. E. Fergusson, D. T. Howarth and I. M. Miller, *J. Chem. Soc. (A)*, 1971, 1144.
46. D. B. McCormick, R. Griesser and H. Sigel, *Met. Ions Biol. Syst.*, 1973, **1**, 213; E. I. Solomon, K. W. Penfield and D. E. Wilcox, *Struct. Bonding (Berlin)*, 1983, **53**, 1.
47. N. S. Ferris, W. H. Woodruff, D. B. Rorabacher, T. E. Jones and L. A. Ochrymowycz, *J. Am. Chem. Soc.*, 1978, **100**, 5939.
48. K. D. Karlin and J. Zubieta, *Inorg. Perspect. Biol. Med.*, 1979, **2**, 127.
49. H. A. Goodwin and F. Lions, *J. Am. Chem. Soc.*, 1960, **82**, 5013.
50. G. R. Brubaker, J. N. Brown, M. K. Yoo, R. A. Kinsey, T. M. Kutchan and E. A. Mottel, *Inorg. Chem.*, 1979, **18**, 299.
51. P. M. Colman, H. C. Freeman, J. M. Guss, M. Murata, V. A. Norris, J. A. M. Ramshaw and M. P. Venkatappa, *Nature (London)*, 1978, **272**, 319.
52. F. M. Chung and A. D. Westland, *Can. J. Chem.*, 1969, **47**, 195; A. D. Westland and V. Uzelac, *Can. J. Chem.*, 1970, **48**, 2871.

16.3

Metallothio Anions

ACHIM MÜLLER and EKKEHARD DIEMANN
University of Bielefeld, FRG

Owing to their range of colors, the thioanions of the early transition metals have long been of interest to chemists. For example, at the beginning of the 19th century Berzelius investigated their formation by passing H_2S into aqueous solutions of MoO_4^{2-} or WO_4^{2-}. However, their true composition was not established until the turn of the century. The thiometalate ions formed from V, Nb, Ta, Mo, W and Re in their highest oxidation states exhibit interesting properties.[1]

The reactivities of the thiometalates and their application in complex chemistry are especially interesting, since they display unique ligand properties. These anions can be used to produce multimetal complexes which were described for the first time some 15 years ago.[2] The complexes are either interesting from the structural point of view and/or have unusual electronic properties. Some of the complexes comprise four-membered metal–sulfur ring systems $M'S_2M$ with very different formal valencies of the M and M' atoms and metal–metal interactions. Moreover, poly(thiometalates), homonuclear thiometalato complexes with mixed valencies, can be produced from the thiometalates by new types of condensation–redox reaction. In addition, the thiomolybdate and thiotungstate anions are currently of interest since they play a part in certain problems of bioinorganic chemistry, nutrition physiology and veterinary medicine.

16.3.1 DISCRETE THIOMETALATE IONS—PROPERTIES, ELECTRONIC STRUCTURE AND REACTIVITY[1]

Table 1 of Chapter 16.1 shows the known transition metal thioanions with a d^0 configuration of the central atom together with some characteristic data.

The chalcogenometalates are prepared either by reactions in the solid state from the constituent elements themselves or by the reaction of H_2S with an aqueous solution of the corresponding oxometalate. Several of the known thiometalates have only been isolated in pure form or detected in solution within the last 20 years. The preparation of selenometalates, and particularly that of the mixed thioselenometalates, is more difficult (*inter alia* because of their high sensitivity towards oxidation of the selenide).

The salts with organic cations are important, since they can be used for the synthesis of heteronuclear multimetal complexes in organic solvents. Most of them can be obtained simply from aqueous solutions of their alkali metal salts.

Since all of the chalcogenometalates have strong and characteristic absorption bands in the UV-vis region, the reactions in which they are formed and decomposed can be readily followed by spectrophotometric methods. When hydrogen sulfide is passed into an aqueous solution of an oxometalate the electronic spectrum changes, the bands of all the species $MO_{4-n}S_n^{m-}$ ($n = 1$–4; $M = V$, Mo, W, Re) appearing in succession (*cf.* Figure 3 of Chapter 16.1).

The alkali metal and ammonium salts are soluble in water (the solubility decreasing from Na^+ to Cs^+), and the XR_4^+ salts ($X = N$, P, As; $R = $ aryl or alkyl) are soluble in various organic solvents. The ammonium salts are relatively unstable even at room temperature.

For aspects of complex formation it is important to realize that thiometalates are not very stable in aqueous solution, especially at low pH. Their decomposition may be caused by hydrolysis to oxometalates, by intramolecular redox processes, or by their marked tendency to form sulfides. With molybdates and tungstates the stability decreases with increasing oxygen content and increasing electron density on the ligands. When they decompose in acid media, binary sulfides (or with selenometalates, selenides) such as MoS_3 are formed.

Chalcogenometalate ions are distinguished by characteristic high-intensity absorption bands in the UV-vis region but they can also be traced by IR and Raman spectroscopy. Total assignments and normal coordinate analyses are available for many of the species (also with recourse to isotope shifts, as in the case of $^{92/100}MoO_2S_2^{2-}$, $^{92/100}MoOS_3^{2-}$ or $^{92/100}MoS_4^{2-}$).

The first indication of the type of bonding, which is important for the understanding of the bonding in the complexes, is given by the metal–sulfur bond lengths (Table 1 of Chapter 16.1), which are significantly shorter than the sum of the ionic or covalent radii, and thus suggest a bond order greater than one, *i.e.* the involvement of π-bonds. The MoS bond length in MoS_4^{2-} or $MoOS_3^{2-}$ ranges between those for a single and a double bond. The involvement of π-bonding MOs is also demonstrated by MO calculations and various physical measurements [for the MS_4^{m-} ions with T_d symmetry from the stretching force constants, from the Raman intensities of the $v_1 (A_1)$ band and from the oscillator strengths of the $v_1 (t_1 \rightarrow 2e)$ electronic transition]. Both calculations and empirical findings, *e.g.* the linear relationship between the energy of the longest-wavelength electronic transition v_1 and the stretching force constants, demonstrate that there is a stabilization of the strongly π-bonding MOs (see Figure 1).

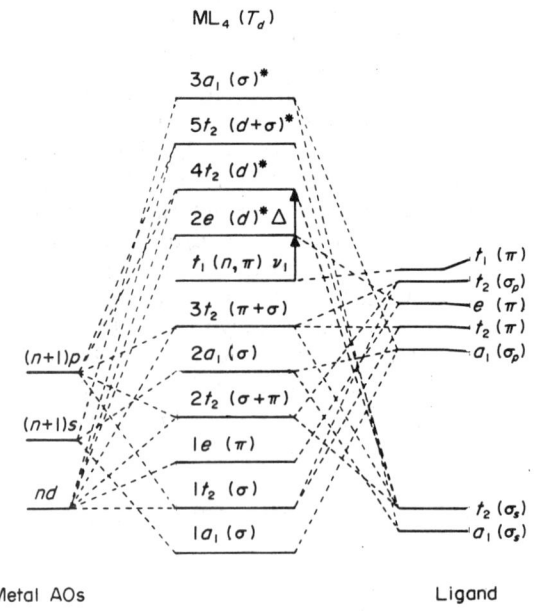

Figure 1 Simplified MO scheme for MS_4^{m-} ions with T_d symmetry

Unambiguous assignments of the band maxima of the allowed electronic transitions are only possible for the longest-wavelength bands of ions with T_d and C_{3v} symmetry. However, all bands of Table 1 of Chapter 16.1 can be assigned qualitatively to metal reduction transitions (from MOs with predominantly ligand character to ones essentially localized on the central atom). Empirically, this follows, for example, from the practically linear relationship between the transition energy and the optical electronegativity of the central atom (for a given ligand) or of the ligand (for a given central atom) (Figure 2). For ions with T_d symmetry the assignment $t_1 \rightarrow 2e$ (Figure 1) for the longest-wavelength transition is, furthermore, clearly demonstrable by means of MCD measurements.

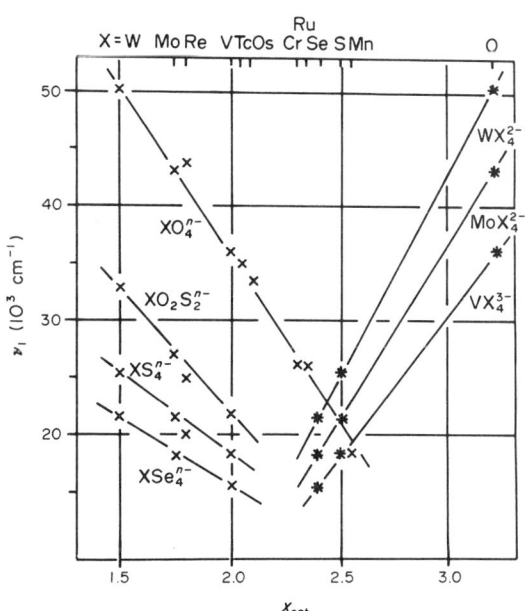

Figure 2 $v_1 (t_1 \rightarrow 2e)$ transition energies of species with T_d symmetry (together with corresponding data for the dithiometalates), as a function of the optical electronegativity of the central atom or the ligands O, S and Se

On the basis of empirical considerations and the influence of various coordination centers (see Section 16.3.3.2) on the UV-vis bands of MoS_4^{2-}, the assignments $v_2 \triangleq 3t_2 \rightarrow 2e$ and $v_3 \triangleq t_1 \rightarrow 4t_2$ are probable. The positions of the metal reduction bands can also be related to the redox behavior of the thioanions.

The longest-wavelength bands of the MOX_3^{2-} species $MoOS_3^{2-}$ and $WOSe_3^{2-}$ can be unambiguously assigned to $\pi(X) \rightarrow d(M)$ (or $a_2 \rightarrow e$) transitions, since in the resonance Raman spectra (excitation line within the band in question) only the intensities of the lines pertaining to vibrations in corresponding MX_3 chromophores are enhanced.

Particularly informative resonance Raman spectra[3] are obtained from PPh_4^+ salts (or those with other bulky organic cations with low anion–anion interaction), since the number of observed overtones and combination tones is particularly large, *e.g.* in the spectra of the corresponding compounds with ReS_4^- and MoS_4^{2-}, where the series of combination tones $v_2 + nv_1$ could be observed up to $n = 7$ and 4, respectively. These are interesting because of the relevance to the corresponding spectra of the complexes, which give information concerning the type of metal–ligand interaction in the complex (Figure 3).

The consideration of the reactivity is necessary to understand the formation of homonuclear thiometalato complexes formed from MS_4^{2-} species. M===S groups with π-bond contributions (*e.g.* in the thiometalates) can react with either nucleophiles (Nu; equation 1) or electrophiles (E; equation 2).

$$Mo'\!\!=\!\!=\!\!=\!\!S + Nu \longrightarrow Mo'^{-2} + NuS \tag{1}$$

(abstraction of S/two-electron reduction)

$$Mo'\!\!=\!\!=\!\!=\!\!S + E \longrightarrow Mo' + ES^{2-} \tag{2}$$

(abstraction of S^{2-})

A further important type of reaction (equations 3 and 4) corresponds to the intramolecular redox process. The formation of $(S_2)^{2-}$ can formally occur *via* the intramolecular reaction of (S^0) (produced according to equation 3) with S^{2-}. This type of reaction is of decisive importance for molybdenum–sulfur chemistry (and probably in various catalytic and biochemical problems as well).

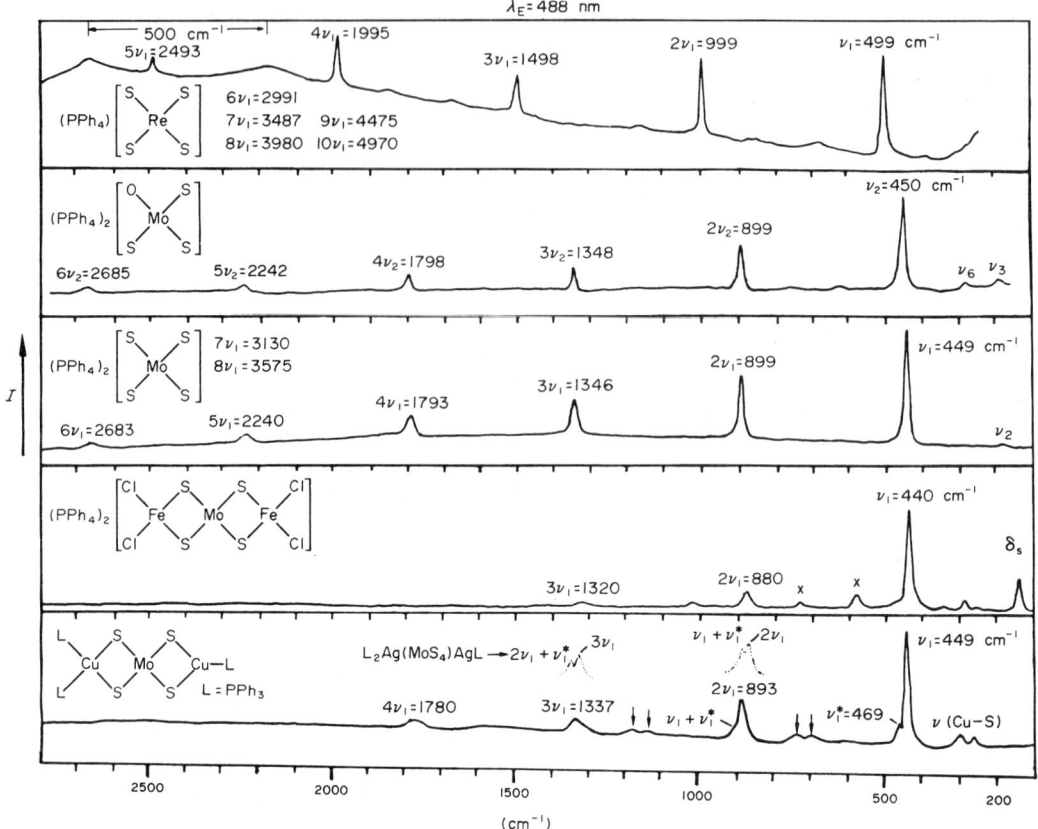

Figure 3 Resonance Raman spectra of MoS_4^{2-}, $MoOS_3^{2-}$ and ReS_4^- and of some thiomolybdato complexes with doubly bridging MoS_4^{2-} ligands; $\downarrow$: combination bands of $n\nu_s(MoS)$ with $\nu(CuS)$; x: further combination bands of stretching and bending vibrations

$$Mo^r(S^{2-}) \longrightarrow Mo^{r-2}(S^0) \tag{3}$$

$$Mo^r(S^{2-})_2 \longrightarrow Mo^{r-2}(S_2^{2-}) \tag{4}$$

16.3.2 GENERAL SURVEY OF COMPOUND TYPES

16.3.2.1 Homonuclear Complexes with Thiometalato Ligands (Chalcogenometalates with Mixed Valencies)

While a great deal of research has been done on condensations of oxometalates in aqueous solutions, there has been little such work in the field of thiometalates. The condensation reactions subsequent to protonation of thiometalate ions take place at lower pH than those of the oxoanions, since the proton affinity of S is appreciably lower than that of O. Investigation of condensation behavior of thiometalate ions is more difficult, since various complex decomposition processes are also involved. For example, no definite condensation product has yet been isolated from an aqueous solution containing MoS_4^{2-}.

The behavior of WS_4^{2-} is again different from that of MoS_4^{2-} and of the oxothiometalates (the proton affinity of WS_4^{2-} is lower than that of MoS_4^{2-}, since the electron density on the sulfur atoms in MoS_4^{2-} is larger). When an aqueous WS_4^{2-} solution is acidified, the condensed species $[WO(WS_4)_2\text{-}(H_2O)]^{2-}$ (1) (an ion with mixed valencies) can be precipitated from the solution by various cations, such as Cs^+.[4,5] The corresponding $[MoO(MoS_4)]^{2-}$ without an H_2O ligand could also be obtained.[6]

The ion (1), which can also be formulated as $W_3OS_8^{2-}$ since the very weakly bound H_2O can be regarded as water of crystallization, formally contains two W^{VI} and one W^{IV}. It is possibly formed as shown in Scheme 1.

The $W_3S_9^{2-}$ ion (2), the first reported complex of this type of species only containing sulfur,[7] could not be obtained directly from an aqueous solution but was prepared quite simply, for example by

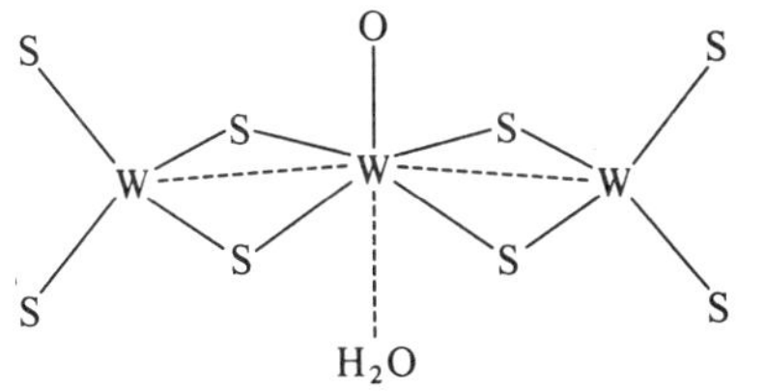

(1)

$$WS_4^{2-} + 2H_2O \longrightarrow WO_2S_2^{2-} + 2H_2S$$

$$WO_2S_2^{2-} + H^+ \xrightarrow{aq} OW^{VI}(OH)S_2(aq)^-$$

$$OW^{VI}(OH)S_2(aq)^- \longrightarrow OW^{IV}(OH)(S_2)(aq)^-$$

$$OW^{IV}(OH)(S_2)(aq) + 2WS_4^{2-} \longrightarrow [OW(WS_4)_2(H_2O)]^{2-} + OH^- + S_2^{2-}$$

Scheme 1

heating a solution of $(NH_4)_2WS_4$ in an organic solvent. $Mo_3S_9^{2-}$ is prepared in the same way.[8] The homologous species $W_4S_{12}^{2-}$ (3) was also isolated; this formally contains the central unit $W_2^V S_4^{2+}$ and two $W^{VI}S_4^{2-}$ ligands.[9] The WS_4^{2-} ligands can be 'abstracted' from $W_3S_9^{2-}$ and $W_4S_{12}^{2-}$ with $FeCl_2$ to give $[Cl_2Fe(WS_4)]^{2-}$.

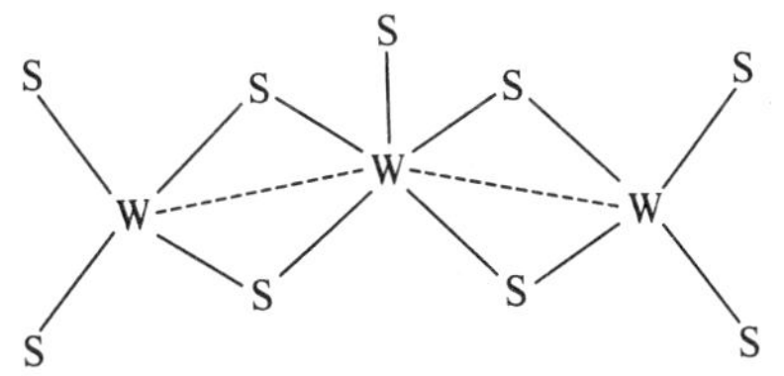

(2) 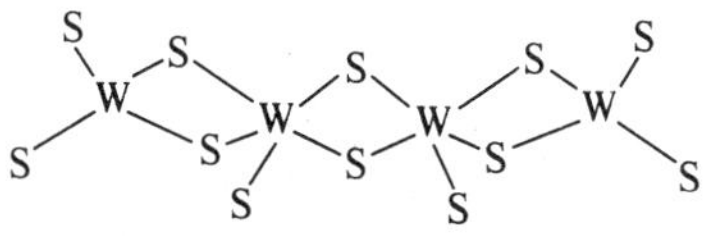   (3)

$W_3OS_8^{2-}$ can be obtained easily from $W_3S_9^{2-}$ in organic solvents containing a small amount of water. (This must be kept in mind in preparing the latter by heating WS_4^{2-} in organic solvents.)

EH-SCCC-MO calculations on the diamagnetic polythiometalates $M_3XS_8^{2-}$ (X = O, S) also show that these can be described as coordination compounds of $M^{IV}X^{2+}$ with MS_4^{2-} ligands. For instance, an occupied MO is about 50% localized on the central W atom, but owing to the appreciable (*ca.* 20%) participation of the outer orbitals of W atoms there is a clear metal–metal interaction comparable to the M′–TM interaction in heteronuclear systems (TM = thiometalate).]

There is a metal–metal single bond in the central $W_2^V S_4^{2+}$ unit of $W_4S_{12}^{2-}$ (3).

The $[Mo_2O_2S_2(S_2)_2]^{2-}$ anion has been isolated from a solution of $MoO_2S_2^{2-}$, and in this reaction an S_2^{2-} ligand is thus formally generated by an oxidation of two S^{2-} ligands (with reduction of Mo). The corresponding complex $[Mo_2O_2S_2(S_2)(MoS_4)]^{2-}$ (4), where an S_2^{2-} is replaced by the MoS_4^{2-} ligand, is obtained by a redox condensation process of MoS_4^{2-}.[10]

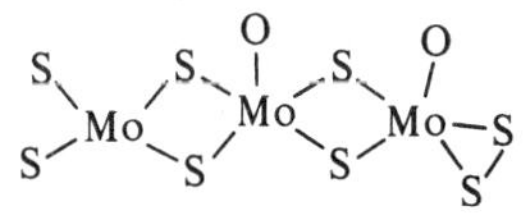

(4)

16.3.2.2 Heteronuclear Complexes with Thiometalato Ligands[1]

Thioanions have remarkable ligand properties, *e.g.* versatile coordination behavior and very low-lying unoccupied orbitals. Thus, in general, complexes show strong metal–ligand interactions. Multimetal complexes of this type exist with a variety of electron populations and some with coordinatively unsaturated metal centers (*e.g.* $[Fe(WS_4)_2]^{2-}$). Some complexes catalyze the photoreduction of water.

16.3.2.2.1 Trinuclear bis(thiometalato) complexes

These were the first thiometalato complexes to be isolated.

The reaction of various divalent transition metal cations with thiometalates in solution (e.g., in H_2O) takes place as shown in equation (5), leading to the formation of bis(thiometalato) complexes, which can be isolated, *e.g.* as quaternary phosphonium salts (exception: the combination M′ = Fe, M = Mo). Spectroscopic investigations show that in solution other species may be present prior to the precipitation: for example, solutions of Co^{II} and $WO_{4-n}S_n^{2-}$ (n = 4, 3) in water do not show the characteristic electronic absorption spectra of $[Co(WO_{4-n}S_n)_2]^{2-}$ complexes (although they do so in some organic solvents); from an orange solution of Fe^{II} and WS_4^{2-} in H_2O a green precipitate of $(PPh_4)_2[Fe(WS_4)_2]$ is obtained.[1,6,12] The structures of $[Ni(WS_4)_2]^{2-}$ (5), $[Co(WS_4)_2]^{2-}$ (6) and $[Fe(H_2O)_2(WS_4)_2]^{2-}$ (7)[6] are given as examples.

$$2XR_4^+ + 2MO_{4-n}S_n^{2-} + M'^{2+} \longrightarrow (XR_4)_2[M'(MO_{4-n}S_n)_2] \tag{5}$$

M′ = Fe, Co, Ni, Pd, Pt, Zn, Cd, Hg; M = Mo, W; n = 2, 3, 4; X = P, As

(5) M′ = Ni (square planar)
(6) M′ = Co (tetrahedral)
X = O, S; M = Mo, W

(7)

Species in which the ligand acts as a doubly bridging entity can also be formed, particularly with tetrathiometalate ions due to the high electron density on the S atoms (see next section). In some cases inhomogeneous and X-ray amorphous products with polymeric ions are formed correspondingly. For example, in the case of Ni^{II}, the reaction in H_2O produces not only the species $(XR_4)_2[Ni(MoS_4)_2]$ but also amorphous products which, in contrast to the complex, are insoluble in nitromethane and thus can be separated off. (However, using $MeCN/H_2O$ as a solvent only the complex without by-products is formed.) If dioxothiometalates are used, as expected, practically no by-products are produced in water. However, it is necessary to work quickly and with cooled solutions because of the high decomposition rate of the dithio anions.[11]

In all known complexes the $MO_{4-n}S_n^{2-}$ ions coordinate *via* sulfur.

There are some bis(thiometalato) complexes with different electron populations, for example $[Co(WS_4)_2]^{n-}$ (n = 2, 3) and $[Fe(WS_4)_2]^{n-}$ (n = 2, 3).[6,12,13,14]

The reactions in the system $Fe^{2+}/MS_4^{2-}/H_2O$ (M = Mo, W),[6,12] which is of bioinorganic interest, are complicated. Whereas $[Fe(WS_4)_2]^{n-}$ (n = 2,3) and $[Fe(H_2O)_2(WS_4)_2]^{2-}$[6] (as also, for example $[Co(WS_4)_2]^{2-}$, as well as other complexes with a double negative charge*) can be isolated from aqueous media, the corresponding reaction between Fe^{II} and MoS_4^{2-} according to equation (5) generally gives various amorphous products, the composition being dependent on the nature of the precipitation conditions (approximate composition $(PPh_4)_2\{Fe(MoS_4)_2\}$); these samples *do not* contain the discrete $[Fe(MoS_4)_2]^{2-}$ ion. However, they could not be clearly characterized because of their amorphous state. The Mössbauer spectrum of the frozen solution shows that in the $Fe^{2+}/MoS_4^{2-}/H_2O$ system, the anion $[Fe(MoS_4)_2]^{3-}$ with a triple negative charge is present. This indicates the high stability of $[Fe(MoS_4)_2]^{3-}$, and demonstrates that additional sulfur-containing organic ligands are not necessary for the reduction. It is, therefore, possible to isolate the $[Fe(MoS_4)_2]^{3-}$ ion in high yield as the NMe_4^+ salt from a solution of $(NH_4)_2MoS_4$ and $FeSO_4 \cdot 7H_2O$ in $MeCN/H_2O$ (1:1).

While the greater stability of the triply negatively charged compared to the doubly negatively charged complex can be understood on the basis of the MO scheme, the mechanism of reduction has not been clarified. (During the reaction of Cu^{II} with MoS_4^{2-} or with WS_4^{2-}, Cu^I complexes are also formed.)

---

* A compound with the stoichiometry $(PPh_4)Fe(WS_4)$, probably having a chain structure of the anion, is also known[12] (for other Fe/WS_4^{2-} species see Section 16.3.2.2.6).

16.3.2.2.2 *Trinuclear and chain-type complexes with doubly bridging thiometalato ligands*

The tendency of the tetrathiometalate ligands to form bridges is illustrated by equation (6). In the trinuclear complex (**8**), MoS_4^{2-} acts as a doubly bridging ligand.[15] The corresponding complexes $[(Cl_2Fe)_2(VS_4)]^{3-}$ [34] and $[(RSCu)_2(MoS_4)]^{2-}$ [16] have also been reported (*cf.* Section 16.3.4).

$$[Cl_2Fe(MoS_4)]^{2-} + \{FeCl_2\} \longrightarrow [Cl_2Fe(MoS_4)FeCl_2]^{2-}c \qquad (6)$$

(8)

$(PPh_4)Cu(MoS_4)$ (red-black), $(PPh_4)Cu(WS_4)$ (orange), $(PPh_4)Ag(MoS_4)$ (dark red) and $(PPh_4)Ag(WS_4)$ (yellow) are obtained by reaction of $CuSO_4 \cdot 5H_2O$ or $AgNO_3$ with $(PPh_4)_2MS_4$ in MeCN, and green $(PPh_4)Fe(MoS_4)$ (^{57}Fe Mössbauer (290 K): IS = 0.41, QS = 0.79 mm s^{-1}) is obtained by dissolving the corresponding amorphous product of reaction (5) in DMF (O_2 free and precipitation with diethyl ether; the thiomolybdato complexes of Cu and Ag cannot yet be obtained in analytically pure form) (for $(PPh_4)Fe(WS_4)$, see Section 13.3.2.2.1).

The compounds, according to their well-defined and nearly identical powder diffraction patterns, are isostructural and have practically the same unit cell dimensions. According to the vibrational spectra and especially the resonance Raman spectrum of $(PPh_4)M'(MoS_4)$ (M' = Cu, Ag) they contain doubly bridging MS_4^{2-} ligands and chains of $M'S_4$ and MS_4 tetrahedra (**9**) connected *via* edges [as in $(NH_4)Cu(MS_4)$].

(8)   (9)

16.3.2.2.3 *Binuclear mixed ligand complexes*

Whereas the discrete bis(tetrathiomolybdato) complex of Fe^{II} (formally) is unstable, mixed ligand complexes of the type $[X_2Fe(MS_4)]^{2-}$ (M = Mo, W; see Table 1) could be isolated. They contain Fe^{II} and one thiometalate ligand (the group of orbitals with predominantly Fe$3d$ character at lower energy is destabilized relative to the corresponding orbitals in pure bis(thiometalato) complexes). Binuclear species of other metals (see Table 1 and *cf.* Section 16.3.4) have also been obtained, like $[NCCu(MoS_4)]^{2-}$ (see the following chapter).

Table 1 Multimetal Complexes with Thiometalate Ligands (Averaged M—S and M$\cdots$M' Distances in Å)^a

Compound^a	Color	M—S$_{br}$	M—S$_{term}$	M$\cdots$M'
$(PPh_4)_2[Mo_3S_9]^8$	Black	2.241	2.141	2.948^h
$(PPh_4)_2[Mo_3OS_8]^6$	Black	2.245	2.137	2.984^h
$(PPh_4)_2[W_3S_9]^5$	Dark red	2.254	2.142	2.965^h
$Cs_2[W_3OS_8(H_2O)]^5$	Red	2.253	2.164	2.990^h
$(PPh_4)_2[W_3OS_8(DMF)]^6$	Red	2.230	2.144	2.978^h
$(PPh_4)_2[W_4S_{12}]^9$	Dark red	2.266	2.131	3.131^h
$(Ph_3PNPPh_3)_2(NEt_4)[Fe(MoS_4)_2]$	Violet	2.256	2.171	2.740
$(NEt_4)_2[(PhS)_2Fe(MoS_4)]$	Dark red	2.246	2.154	2 750
$(NEt_4)_2[(PhS)_2Fe(MoS_4)]$	Dark red	2.255	2.153	2.756
$(PPh_4)_2[(PhS)_2Fe(MoS_4)]^{26}$	Red-brown	2.262	2.149	2.740
$(NEt_4)_2[Cl_2Fe(MoS_4)]$	Dark brown	e	e	2.786
$[Fe(DMSO)_6][Cl_2Fe(MoS_4)]$	Dark red	e	e	2.775
$(PPh_4)[N(C_7H_7)Me_3][Cl_2Fe(MoS_4)]$	Dark red	e	e	2.775
$(NEt_4)_2[(PhS)_2Fe(WS_4)]^{26}$	Red	2.246	2.157	2.775
$[Fe(DMSO)_6][Cl_2Fe(WS_4)]$	Brown	e	e	2.77
$[Fe(DMF)_6][Cl_2FeWS_4]^{27}$	Black (dark brown)	e	e	2.796
$(PPh_4)[N(C_7H_7)Me_3][Cl_2Fe(WS_4)]$	Red	e	e	2.808
$[Fe(DMSO)_6][Cl_2Fe(MoOS_3)]$	Dark brown	e	e	2.78
$(PPh_4)_2[Cl_2Fe(MoO_2S_2)]^{27}$	Black (violet)	e	e	2.716
$(PPh_4)_2[Cl(N_3)Fe(WS_4)]^{27}$	Red	e	e	2.789
$(PPh_4)_2[(S_5)Fe(MoS_4)] \cdot 0.5DMF^{26}$	Dark brown	2.261	2.148	2.737
$(PPh_4)_2[(S_5)Fe(WS_4)] \cdot 0.5DMF^{26}$	Brown	2.260	2.157	2.752
$(PPh_4)_2[(Cl_2Fe)_2(MoS_4)]$	Brown	2.204	—	2.775
$(PPh_4)_2[(Cl_2Fe)_2(WS_4)]^{28}$	Orange-red	2.209	—	2.801
$(NMe_4)_3[(Cl_2Fe)_2(VS_4)] \cdot DMF^{34}$	Red-black	2.178	—	2.730
$(NEt_4)_2[(PhO)_2Fe_2(MoS_4)]^{35}$	d	2.273	2.126	2.797

Table 1 *(continued)*

Compound[a]	Color	$M—S_{br}$	$M—S_{term}$	$M \cdots M'$
$(Ph_3PNPPh_3)_2[(NO)_2Fe(MoS_4)]$[28]	d	2.259	2.182	2.835
$(PPh_4)_2[(DMF)_2Fe(WS_4)_2]$[28]	d	2.213	2.164	3.044
$(NEt_4)_2[(H_2O)_2Fe(WS_4)_2]$[6]	Black	2.219	2.174	3.010
$(Ph_3PNPPh_3)_2(NEt_4)[Fe(MoS_4)_2]\cdot 2DMF$[28]	Violet	2.255	2.171	2.740
$(Ph_3PNPPh_3)_2(NEt_4)[Fe(WS_4)_2]\cdot 2MeCN$[14]	Red-brown	2.239	2.166	2.749
$(Ph_3PNPPh_3)_2(NEt_4)_2[Fe_2S_2(WS_4)_2]\cdot 3MeCN$[6]	Violet	2.234	2.173	2.816
$(PPh_4)(NEt_4)_3[Fe_3S_2(WS_4)_2]$[29]	Black	2.238	2.152	2.774
$(PPh_4)_2[Cl_2FeMoOS_4]\cdot DMF$[30]	Black-green	2.303	1.660[c]	2.752
$(PPh_4)_2[Br_2FeWOS_4]\cdot DMF$[31]	Black	2.259	1.713[c]	2.749
$(NEt_4)_3[(p\text{-}MePhS)_2FeS_2Fe(MoS_4)]$[35]	d	2.244	2.163	2.778
$(PPh_4)_2[Co(WS_4)_2]$	Olive-green	2.219	2.139	2.798
$(Ph_3PNPPh_3)_2(NEt_4)[Co(MoS_4)_2]\cdot 2MeCN$[6]	Black	2.23	2.15	2.75
$(Ph_3PNPPh_3)_2(NEt_4)[Co(WS_4)_2]\cdot 2MeCN$[13]	Black-violet	2.230	2.175	2.727
$(PPh_4)_2[Ni(MoS_4)_2]$	Dark brown	2.227	2.151	2.798
$(AsPh_4)_2[Ni(WS_4)_2]$[5]	Red brown	2.232	2.151	2.817
$(PPh_4)_2[Zn(WS_4)_2]$	Orange	2.233	2.156	2.927
$(PPh_4)_2[Hg(WS_4)_2]$[27]	Orange	2.246	2.173	3.088
$NH_4Cu(MoS_4)$	Red	2.19	—	2.70
$NH_4Cu(WS_4)$	Bright red	d	—	2.72[b]
$(Cu_3MoS_3Cl)(PPh_3)_3S$	Red	2.254	2.118	2.700
$(Cu_3MoS_3Cl)(PPh_3)_3O$	Dark red	2.259	1.769[c]	2.718
$(Cu_3WS_3Cl)(PPh_3)_3S$	Yellow	2.251	2.131	2.717
$(Cu_3WS_3Cl)(PPh_3)_3O$	Yellow	2.241	1.754[c]	2.738
$(Cu_3MoS_3Br)(PPh_3)_3O\cdot 0.5Me_2CO$[27]	Dark red	2.262	1.853[c]	2.718
$(Cu_2MoS_3)(PPh_3)_3O\cdot 0.8CH_2Cl_2$[18]	Red	2.238	1.700[c]	2.719
$(Cu_2WS_3)(PPh_3)_3O\cdot 0.8CH_2Cl_2$[18]	Yellow	2.235	1.676[c]	2.738
$(Cu_4W_2S_6)(PPh_3)_4O_2$	Orange-red	2.251	1.696[c]	2.780
$(Cu_4W_2S_6)[P(C_7H_7)_3]_4O_2$	Orange	2.248	1.70[c]	2.784
$(Ag_4Mo_2S_6)(PPh_3)_4S_2$	Violet	2.227	2.108	2.975
$(Ag_4W_2S_6)(PPh_3)_4S_2$	Orange-red	2.234	2.121	2.997
$(Ag_4W_2S_6)(PMePh_2)_4S_2$	Orange	2.225	2.131	3.002
$(PMePh_2)_2Au_2(WS_4)$	d	2.219	—	2.840
$(PPh_3)_3Ag_2(MoS_4)\cdot 0.8CH_2Cl_2$	Red	2.205	—	2.945
$(PPh_3)_3Ag_2(WS_4)\cdot 0.8CH_2Cl_2$	Yellow	2.207	—	2.971
$(PPh_3)_3Cu_2(MoS_4)\cdot 0.8CH_2Cl_2$	Red	2.208	—	2.709
$(PPh_3)_3Cu_2(WS_4)\cdot 0.8CH_2Cl_2$	Yellow	2.214	—	2.740
$(PPh_3)_3Ag_2(WSe_4)\cdot 0.8CH_2Cl_2$[18]	Red	2.328[i]	—	2.990
$(Ph_3PNPPh_3)_2(NEt_4)[Cu(WS_4)_2]\cdot 2MeCN$[27]	Orange-yellow	2.198	2.164	2.750
$(Ph_3PNPPh_3)(NEt_4)[Ag(MoS_4)]\cdot MeCN$[36]	Dark red	2.203	2.159	2.935
$(NPr_4^n)_2[(PhSCu)(MoS_4)]$[16]	Red	2.221	2.162	2.636
$(NPr_4^n)_2[(PhSCu)_2(MoS_4)]$[16]	Dark red	2.206	—	2.632
$(PPh_4)_2[(CuCl)_3(MoS_4)]\cdot MeCN$[32]	Dark red	2.227	—	2.621
$(PPh_4)_2[(CuCl)_3(WS_4)]$[33]	Orange	2.227	—	2.636–2.645
$(PPh_4)_2[(CuBr)_3(MoS_4)]\cdot MeCN$[27]	Dark red	2.224	—	2.621
$(PPh_4)_2[(CuCl)_3(MoOS_3)]$[18]	Red	2.263	1.693[c]	2.638
$(PPh_4)_2[NCCu(MoS_4)]$	Red	2.245	2.163	2.630
$(NPr_4^n)_2[NCCu(MoS_4)]$[20]	Bright red	2.226	2.150	2.624
$(AsPh_4)_2[(NCCu)_2(MoS_4)]\cdot H_2O$[20]	Orange-red	2.210	—	2.635
$(PPh_4)_2[NCAg(MoS_4)]$[18]	Red	2.218	2.160	2.868
$(PPh_4)_2[NCAg(WS_4)]$[18]	Yellow	2.223	2.166	2.894
$(NPr_4^n)_2[NCAg(WS_4)]$[20]	d	2.227	2.156	2.890
$(NMe_4)_2NCCu(MoS_4)(CuCN)$	Dark brown	2.218	—	—
$(PPh_4)_2[Au_2(WS_4)_2]$	Dark red	e	2.136	3.169
$(PPh_4)_2[Au_2(WOS_3)_2]$	Orange	e	1.745[c]	3.221
$(PPh_4)_4[Sn_2(WS_4)_4]$	Orange	2.231[f]	2.154[f]	—
		2.199[g]	2.168[g]	

[a] M = Mo, W; M′ = Mo, W, Fe, Co, Ni, Zn, Hg, Cu, Ag, Au, Sn. For all compounds see ref. 1 unless otherwise stated. [b] Corresponding to half the length of the c-axis. [c] M—O distance. [d] Not given. [e] Not determined because of statistical disorder. [f] Values for the bidentate ligands. [g] Values for the tridentate ligands. [h] M is the central atom of the terminal 'MS$_4^{2-}$' ligands; M′ here is the reduced M atom. [i] M—Se distance.

16.3.2.2.4 Versatility of coordination in polynuclear complexes with the soft metal ions Cu^I, Ag^I and Au^I

In the reaction system[17,18] $M'^+/MO_{4-n}S_n^{2-}/PR_3$ (M′ = Cu, Ag, Au; M = Mo, W; n = 4, 3; $R_3 = Ph_3, (C_7H_7)_3, MePh_2$) the following types of compounds are formed, which are interesting from the structural point of view and contain doubly, triply and quasiquadruply bridging thiometalate ligands (see Figure 4).

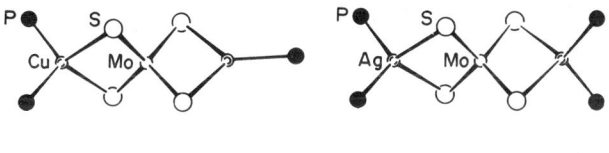

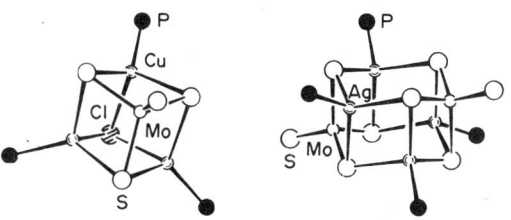

Figure 4 Examples of multimetal complexes in the system $M'^+/MO_{4-n}S_n^{2-}/PPh_3$ ($n = 3, 4$; $M' = Cu, Ag$; $M = Mo, W$): $(PPh_3)_3Cu_2(MoS_4)$, $(PPh_3)_4Ag_2(MoS_4)$, $(Cu_3MoS_3Cl)(PPh_3)_3S$ and $(Ag_4Mo_2S_6)(PPh_3)_4S_2$

(a) Compounds with a central cubane-type unit $(Cu_3MS_3Cl)(PPh_3)_3X$ ($M = Mo, W$; $X = S, O$)

These have a strongly distorted cube as their central unit (see Figure 4). The MXS_3^{2-} ligand here is triply bridging. The coordination polyhedra of the three Cu atoms are not equivalent: while one Cu atom has a distorted tetrahedral coordination, in the case of the other two atoms there is a transition to trigonal planar coordination.

(b) Compounds with a novel prismatic type of cage structure of the composition $(M'_4M_2S_6)(PR_3)_4X_2$ ($M' = Cu, Ag$; $M = Mo, W$; $X = S, O$; $R_3 = Ph_3, (C_7H_7)_3, MePh_2$)

The central unit represents a new type of cage, formally arising from a fusion of two $MS_2M'_2S$ rings (see Figure 4). Here, the MXS_3^{2-} ligand acts as a quasi-quadruple bridge (with bonds to all four M' atoms).

(c) Trinuclear complexes of the composition $(PR_3)_nM'_2(MS_4)$ ($M' = Cu, Ag, Au$; $M = Mo, W$; $R_3 = Ph_3, MePh_2$; n = 2, 3, 4)

In these compounds each pair of M' atoms is connected *via* a doubly bridging MS_4^{2-} ligand. The M' atoms each carry one or two PR_3 ligands, and hence for $n = 4$ both atoms are surrounded by distorted tetrahedra, for $n = 2$ both have almost trigonal planar environments, and for $n = 3$ one atom is coordinated tetrahedrally while the other has a trigonal planar environment (see Figure 4 and Table 1).

It is worth noting that one thiometalate ligand provides six bonds to Cu and Ag atoms in the cage structure compounds.

Since the compound types (a) to (c) above have a comparable formation tendency, different compounds can easily appear side by side in a given reaction; slight variations of the concentration of the reactants can shift the equilibrium strongly in favor of one of the possible products.

From the $Cu^+/MoS_4^{2-}/CN^-/XR_4^+$ system[19,20] different species could also be isolated: $[NCCu(MoS_4)]^{2-}$ (**10**), $[NCCu(MoS_4)CuCN]^{2-}$ (**11**) and $(NMe_4)_2NCCu(MoS_4)\{CuCN\}$ (**12**). This third compound is worth noting, since here the complex $[NCCuMoS_4]^{2-}$ is coordinated to the Cu atoms of an infinite zigzag CuCN chain. (It is also interesting that MoS_4^{2-} can be coordinated to a CuCN molecule as well as to a CuCN chain.)

Thiometalate ligands are coordinated *via* sulfur to soft cations such as Cu^+, Ag^+ and Au^+. Thus, it has not yet been possible to prepare any compound of the structural type (c) with MOS_3^{2-} ligands. Following this principle, using oxotrithiometalates, compounds with a cubane-like structure were deliberately prepared.

The only non-polymeric thiometalato complexes with elements of group Ib without additional

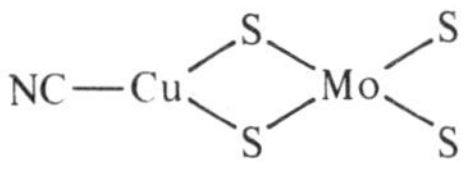

(10) 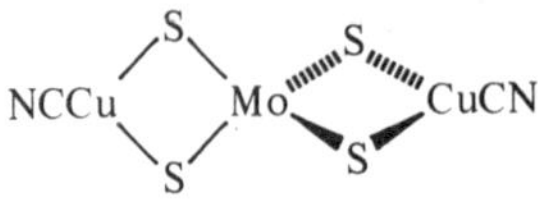   (11) 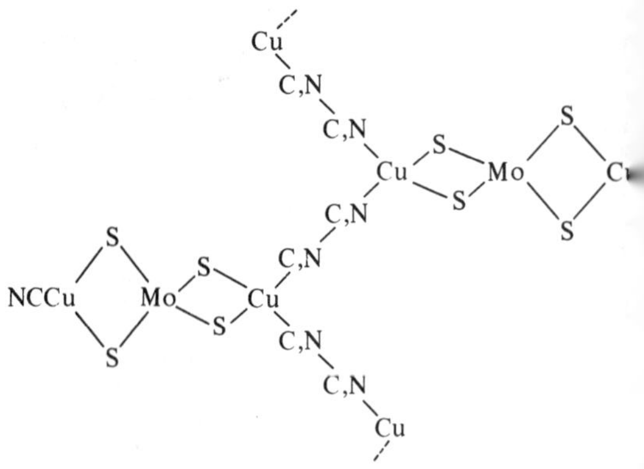   (12)

ligands that have been isolated are $[Cu(WS_4)_2]^{3-}$, $[Ag(MoS_4)]^{3-}$, $[Au_2(WS_4)_2]^{2-}$ (13) and $[Au_2(WOS_3)_2]^{2-}$ (pure metal–sulfur ring systems without organic ligands). In each case the anions consist of two Au atoms bridged by two thiotungstate ligands (see Table 1 and ref. 1).

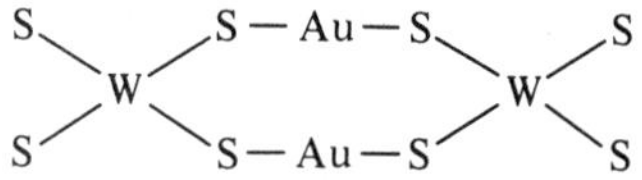

(13)

16.3.2.2.5 Heterometallic clusters from thiometalate ligands by 'unit construction' and the possibility of 'adding' these ligands to coordinatively unsaturated species

According to the high formation tendency of M—S bonds (M = soft cations like Cu^+ or Ag^+), coordinatively unsaturated species like CuCN, CuCl or $Cu(PPh_3)_n$ ($n = 1$, 2) can be 'added' to thioanions and stabilized. Following this principle polynuclear heterometallic compounds can be obtained with the intact thiometalate unit (see also last chapter). By 'addition' of CuCl to $MoOS_3^{2-}$ the novel heterometallic species $[(CuCl)_3(MoOS_3)]^{2-}$ (14) (having roughly C_{3v} symmetry) is obtained.[18]

The central unit corresponds to a distorted cube with one corner missing (the Cu atoms are trigonally planar coordinated). Remarkably, the $\{Cu_3MoS_3\}$ unit can be completed by 'addition' of another atom, for example chlorine (see the schematic structure of $[(CuPPh_3)_3(MoOS_3)Cl]$ (15) with the $\{Cu_3MoS_3Cl\}$ cube[18] and Figure 5).

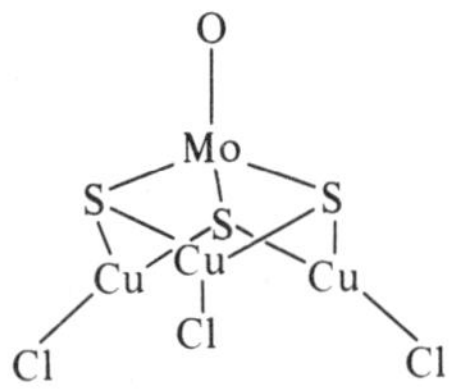

(14) 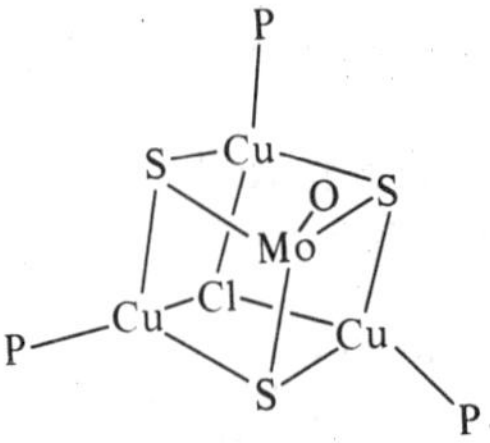   (15) 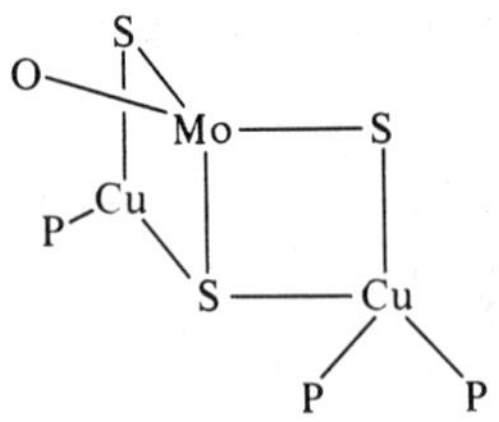   (16) 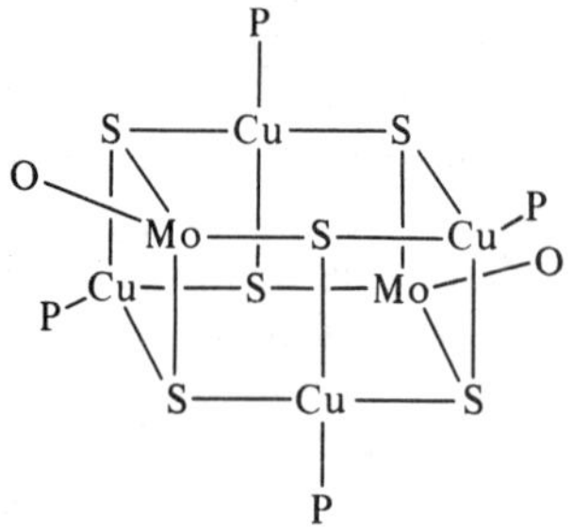   (17)

The compound $[(CuPPh_3)(Cu(PPh_3)_2)(MoOS_3)]\cdot 0.8CH_2Cl_2$ (16) with two missing corners having the central $\{Cu_2MoS_3\}$ moiety could also be isolated.[18] If we compare the structure with that of $[(CuPPh_3)_4(MOS_3)_2]$ (M = Mo) (17) we realize that it represents almost one half of it (it has only one more PPh_3 ligand at one of the two nonequivalent Cu atoms). Owing to the preferred formation of Cu—S bonds in solution, the reaction shown in equation (7) takes place, which is an impressive example of the construction of a cage from its two constituent halves (see Figure 5).

$$2[(CuPPh_3)(CuPPh_3)_2)(MoOS_3)] \xrightarrow{-2PPh_3} [(CuPPh_3)_4(MoOS_3)_2] \qquad (7)$$

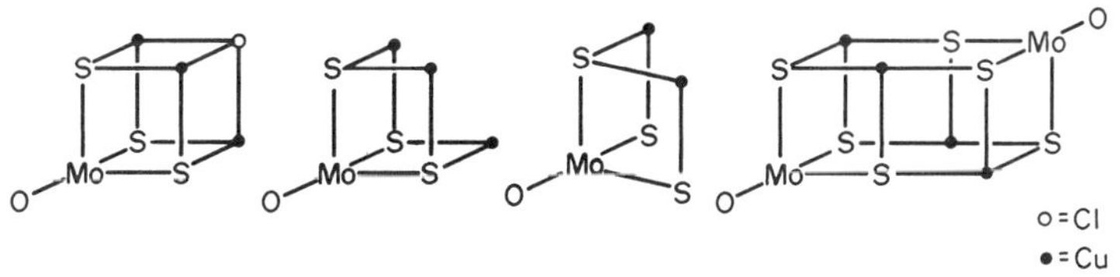

Figure 5 Structural comparison of the central units of [(CuPPh$_3$)$_3$(MoOS$_3$)Cl], [(CuCl)$_3$(MoOS$_3$)]$^{2-}$ and [(CuPPh$_3$)(Cu(PPh$_3$)$_2$)(MoOS$_3$)] and the relation to the prismatic structure of [(CuPPh$_3$)$_4$(MoOS$_3$)$_2$]

Heteronuclear clusters, where CuCl is added to MS_4^{2-} (M = Mo, W), such as [(CuCl)$_3$(MoS$_4$)]$^{2-}$ have also been obtained.

16.3.2.2.6 *Thiometalato ligands coordinated to metal-sulfide aggregates*

These kinds of species are of considerable bioinorganic and catalytic interest.[25] Complexes with central Fe$_x$S$_y$ units could be obtained, such as the remarkable linear [Fe$_2$S$_2$(WS$_4$)$_2$]$^{4-}$ (**18**)[17] and the hexanuclear [Fe$_3$S$_2$(WS$_4$)$_3$]$^{4-}$ (**19**)[21] with an almost planar array of six metal atoms. It is also worth noting that four other complexes could be isolated from the system Fe^{II}/WS$_4^{2-}$/solvent (see Section 16.3.2.2.1).

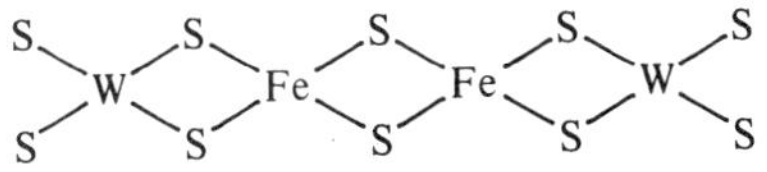

(18)

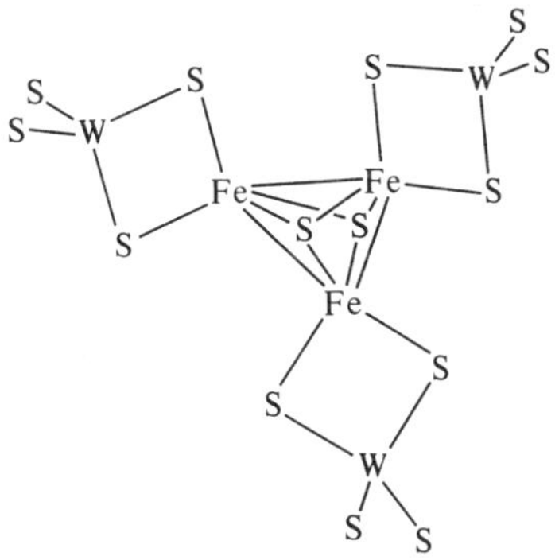

(19)

16.3.2.2.7 *Complexes containing metal ions with lone pair electrons*

Only compounds with thiometalate ligands involving elements of the 4th main group in the valence state 2 + are known so far. In these species a type of coordination occurs which has not been found in transition metal complexes. In the novel polynuclear complexes of the type [M$_2'$(MS$_4$)$_4$]$^{4-}$ (M' = Sn, Pb; M = Mo, W) (**20**) there are nonequivalent bridging and terminal bidentate ligands. Corresponding anions are very probably present in the compounds (PPh$_4$)$_4$[Pb$_2$(MoS$_3$)$_4$].

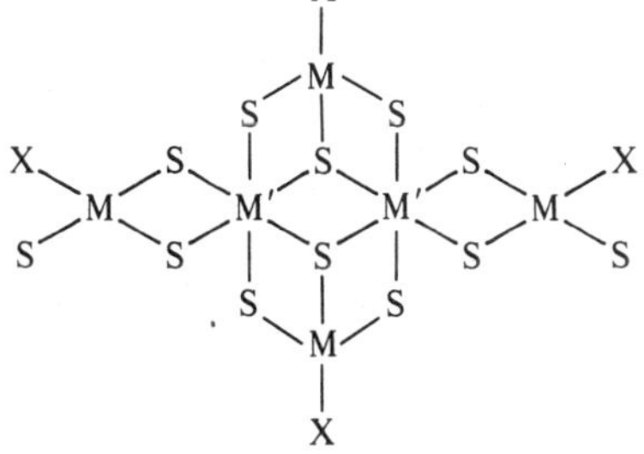

(20) X = O, S;
M' = Sn, Pb;
M = Mo, W

16.3.2.2.8   *Substituted thiometalato ligands*

Only two complexes are known which contain a substituted thiometalate as a ligand. In $[Cl_2FeS_2-MoO(S_2)]^{2-}$ [22] the $[S_2MoO(S_2)]^{2-}$ group can formally be regarded as a bidentate perthiomolybdate ligand. However, the electron-withdrawing power is so large that the electronic structure is better described in terms of the valence structures (21) and (22). The situation in $[Br_2FeS_2WO(S_2)]^{2-}$ is similar.

Fe^{II}⟨S⟩Mo^{VI} Fe^{III}⟨S⟩Mo^{V}

(21)   (22)

16.3.3   SPECTROSCOPIC PROPERTIES AND CHEMICAL BONDING

Spectroscopic studies were carried out, both to determine the molecular structure and to investigate the interesting electronic properties. Vibrational spectra are very useful in determining the type of coordination.

16.3.3.1   Vibrational Spectra[1,23]

The structures of the first reported complexes (type $[M'(MO_{4-n}S_n)_2]^{2-}$) have been determined by IR spectroscopic investigations. The isotope substitution technique ($^{54}Fe/^{56}Fe$, $^{58}Ni/^{62}Ni$, $^{63}Cu/^{65}Cu$, $^{64}Zn/^{68}Zn$, $^{92}Mo/^{100}Mo$) has made it possible to produce detailed information concerning the spectra of the complexes $[Ni(MoS_4)_2]^{2-}$, $[Ni(WS_4)_2]^{2-}$, $[Ni(WO_2S_2)_2]^{2-}$, $[Zn(MoS_4)_2]^{2-}$, $[Zn(WS_4)_2]^{2-}$, $[NCCu(MoS_4)]^{2-}$ and $[Cl_2Fe(MoS_4)]^{2-}$, and to show that the $\nu(MS_{br})$ and especially the $\nu(MS_{term})$ vibrations are relatively characteristic ($\nu_{as,term}$ is the most, and $\nu_{s,br}$ the least characteristic vibration). Thus, the presence of corresponding bonds can easily be deduced from the vibrational spectra (see Figure 6).

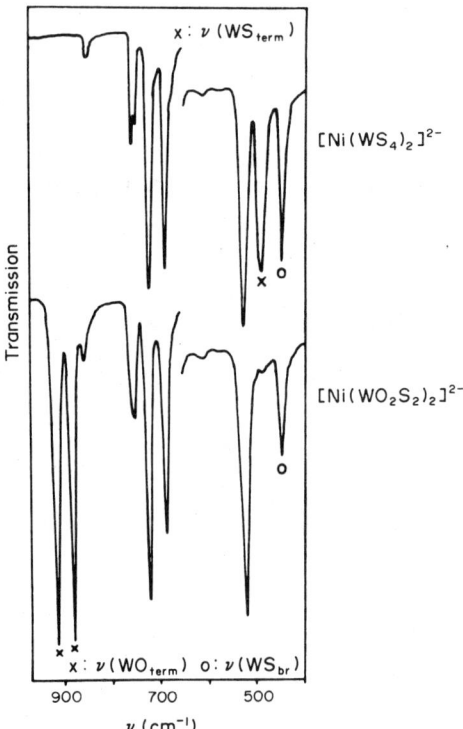

Figure 6   IR spectra of $(PPh_4)_2[Ni(WS_4)_2]$ and $(PPh_4)_2[Ni(WO_2S_2)_2]$ in the region of the WO and WS stretching vibrations (demonstrating the bidentate nature of the ligands and the S coordination)

Since the ligand internal transitions in thiomolybdato complexes correspond roughly to those of free MoS_4^{2-}, one can obtain resonance Raman spectra (using the 488 nm line of an Ar^+ laser) from almost all the complexes in Table 1. In these spectra the lines due to the totally symmetric vibrations of the MoS_4 chromophore (usually stretching vibrations) are significantly intensified. The resonance raman effect is a sensitive probe for the detection of MoS_4^{2-} ligands (in principle also at Cu or Fe centers in biologically relevant systems). Among other things, significant distinctions can be made between bridging and bidentate ligands. This method offers especially a unique and sensitive probe for doubly bridging ligands (Figure 3).

16.3.3.2 Spectra and Electronic Structure[1,24,39,40]

Various physical measurements have demonstrated that in thiometalate complexes with central atoms possessing open d-shells, there are strong metal–ligand interactions, *i.e.* delocalized ground state molecular orbitals. Thus, the assignment of the electronic transition bands for such systems to chromophores can only be approximate. The electronic interactions in FeS_2M systems are interesting due to their relevance in the bioinorganic field.

The known complexes of the type $[M'(MO_{4-n}S_n)_2]^{n-}$ (M' = Fe, Co, Ni, Pd, Pt) show characteristic absorption bands whose positions are only roughly comparable to those of the free thiometalates (*i.e.* L $\rightarrow$ L* assignment). The reduction of the symmetry due to complex formation often leads to a splitting of the bands, whereas the total intensity remains virtually unchanged. However, because of the strong M'—L interactions, the spectra in the region of the ligand internal transitions depend markedly on the nature of the central atom (*i.e.* the influence is quite noticeable). For example, there is a characteristic splitting of the ν_1 band of WS_4^{2-} in the Ni complex. In contrast, the corresponding longest-wavelength band in $[Ni(MoS_4)_2]^{2-}$ is strongly shifted in the direction of longer wavelengths. This demonstrates different interactions, which should be particularly large in the MoS_4^{2-} complex.

According to EH-SCCC-MO calculations, strong $3d_{xz}$(Ni)– and $3d_{yz}$(Ni)–π interactions with the ligand orbitals occur in the $[Ni(MS_4)_2]^{2-}$ complexes (M = Mo, W). In the trimetallic complexes with a square planar coordination of the central atom, the coordination geometry gives rise to a considerable perturbation of the electronic structure of the ligands. Since the opposite S bridge atoms of the two thiometalate ligands are separated by only *ca.* 2.8 Å, the energies of the occupied nonbonding $3p$ (S) ligand orbitals are, in part, markedly increased. Particularly in the MoS_4^{2-} complex, the Ni–ligand interaction is so strong that it can no longer be accurately formulated as an Ni^{II}–Mo^{VI} complex, and one must think in terms of a formal reduction of the Ni center and a partial disulfide bond between the bridging S atoms.

The spectra of complexes with d^{10} central atoms show essentially ligand internal transitions corresponding to those of the free thiometalate ions. It is worth noting, for example, that in $[NCCu(MoS_4)]^{2-}$ and $[Zn(MoS_4)_2]^{2-}$, of the three longest-wavelength bands the first (ν_1) and the third (ν_3) are clearly changed in their structure by complex formation, while the second (ν_2) is practically unaltered in its position, intensity and half-width (see Figure 7). (The fact that this is not true for MoS_4^{2-} complexes of central atoms with open d-shells, for example those of Fe and Ni, can be explained in terms of MO theory.) MO calculations show that on complex formation the t_1 orbitals of the MS_4^{2-} ligands, as expected, are much more strongly disturbed than the $3t_2$, $2e$ and $4t_2$ MOs. It follows that in the transition due to the ν_2 band in MoS_4^{2-}, the t_1 orbital is not involved; this lends support to the assignments $\nu_2 \cong 3t_2 \rightarrow 2e$ and $\nu_3 \cong t_1 \rightarrow 4t_2$. This probably constitutes an essential contribution to the old assignment problem of the electron transfer transitions of tetrahedral transition metal chalcogenometalates. It is also interesting to note that the type of perturbation of the ν_2 band of complexes with open-shell metal ions reflects the type of M'—L interaction.

The Fe and Co complexes of the above type show remarkable bands of high intensity ($\varepsilon \approx 10^3$–$10^4\,M^{-1}\,cm^{-1}$) in the near-IR–vis region (Figure 8), which are not present in 'classical complexes' without strong metal–ligand interactions. (In the complexes $[M'(TM)_2]^{3-}$ (M' = Fe, Co) two charge transfer transitions of the type L $\rightarrow$ d(M') are found.) An assignment can be achieved on the grounds of empirical relations and from MO calculations.

The following trends concerning the energy of this type of corresponding absorption bond are observed for analogous complexes:
    (1) Fe > Co complexes;
    (2) M'^I > M'^{II} complexes (formal consideration)*;

---

*Differences in the electronic spectra of the oxidized and reduced species (especially in the region of ligand internal transitions) show that the reaction $[M'(TM)_2]^{2-} \rightarrow [M'(TM)_2]^{3-}$ cannot simply be explained by a reduction $M'^{II} \rightarrow M'^I$, but also imply a change in the electronic structure of the TM ligands.

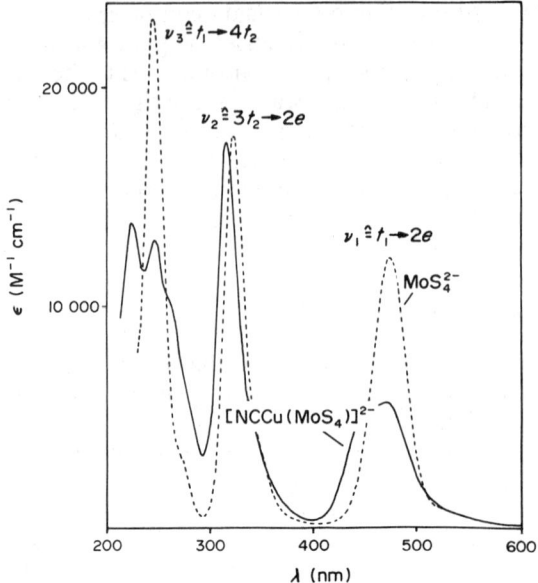

Figure 7 Influence of a coordination center on the longest-wavelength bands of MoS_4^{2-}; electronic spectra of $(NMe_4)_2[NCCu(MoS_4)]$ and $(NMe_4)_2MoS_4$ (in acetonitrile)

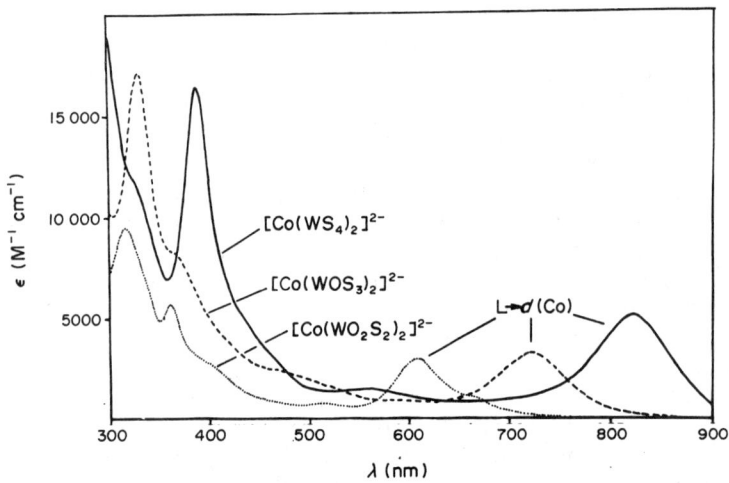

Figure 8 Electronic absorption spectra of the complexes $[Co(WO_{4-n}S_n)_2]^{2-}$ in CH_2Cl_2 (PPh_4^+ salts)

(3) tungstato > molybdato complexes;

(4) dithiometalato > trithiometalato > tetrathiometalato complexes.

All four series can be rationalized *approximately* by the concept of optical electronegativity which states that the energy of a ligand → metal charge transfer transition is a function of the optical electronegativities of the ligand ($x_{opt}(TM)$) and the metal ($x_{opt}(M')$) as well as the ligand field splitting energy Δ and the difference in the spin pairing energies of the ground and the excited state D_{ESP} according to equation (8). The first and the second series are explained by the relations $x_{opt}(Fe) < x_{opt}(Co)$ and $x_{opt}(M'^I) < x_{opt}(M'^{II})$. The third and fourth series also fit into this concept as increasing delocalization reduces the spin pairing energy by reduction of the Racah parameters B and C; additionally, $x_{opt}(TM)$ is expected to decrease in the series dithiometalato > trithiometalato > tetrathiometalato.

$$\nu_{CT} = (x_{opt}(TM) - x_{opt}(M')) \times 30 \times 10^3\,cm^{-1} + \Delta + \Delta_{ESP} \qquad (8)$$

Furthermore, there is a linear relationship between the transition energies of the $[Co(MoO_{4-n}S_n)_2]^{2-}$ complexes and those of the corresponding $[Co(WO_{4-n}S_n)_2]^{2-}$ species. All these findings indicate that the above-mentioned intense bands should be assigned to a charge transfer transition of the type $L \rightarrow d(M')$.

This assignment is supported by EH-SCCC-MO calculations on corresponding Co and Fe complexes, which lead to the qualitative MO scheme displayed in Figure 9. (The problem of the sequence of the MOs in these complicated systems is evident; however, the MO scheme given here can provide answers to certain interesting questions of stability.) In this scheme the following classes of MOs with approximately similar energies are included: (1) M—S σ-bonding MOs originating from positive overlap of $3d(M')$ AOs and ligand t_1 levels; (2) nonbonding occupied L-MOs with a predominantly $3p$ (S) character; (3) closed or open shell MOs of predominantly $3d(M')$ character (corresponding to the e and t_2 orbitals in a crystal field theory description); and (4) unoccupied L* MOs of predominantly $4d$(Mo) or $5d$(W) character (corresponding to the $2e$-MOs of the free thiometalate ions).

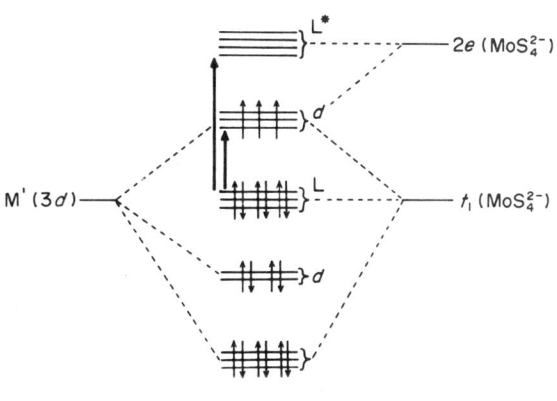

$$[\text{Fe(MoS}_4)_2]^{3-} / [\text{Co(MoS}_4)_2]^{2-}$$

Figure 9 Qualitative MO scheme of the isoelectronic complexes $[\text{Fe(MoS}_4)_2]^{3-}$ and $[\text{Co(MoS}_4)_2]^{2-}$, and analogous bis(thiometalato) complexes of Fe and Co with half-occupied and completely occupied 'd-levels', respectively. The braces link together MOs of approximately the same energy (see also ref. 39).

Charge transfer transitions of the type $d(M') \rightarrow$ L* are to be expected at higher energies, possibly in the region of the first ligand internal transitions. (However, as indicated in Table 2 the band assignment is only approximate, because of the strong interaction between central atom and ligand; this also applies to the assignments of $d \rightarrow d$ transitions given in Table 2).

As two transitions with predominant $d \rightarrow d$ character can be observed for the complexes $[\text{Co(MO}_2\text{S}_2)_2]^{2-}$ (M = Mo, W) (here the M'—TM interaction is smaller than in the complexes having higher sulfur content), it is possible to calculate approximate ligand field parameters for the $\text{MO}_2\text{S}_2^{2-}$ ligands (under the assumption of a pseudo-tetrahedral Co^{II}S_4: $\Delta = 5.4 \times 10^3 \text{ cm}^{-1}$, $\beta = 0.35$ (M = Mo) and $\Delta = 4.8 \times 10^3 \text{ cm}^-$, $\beta = 0.54$ (M = W); thiometalates thus are 'pseudostrong' ligands, in both the spectrochemical and nephelauxetic series, which can be seen by comparison of these values with those of other complexes containing Co^{II}S_4 chromophores). The given assignment of the low energy and low intensity absorptions is strongly supported by the spectrum of $[\text{Co(WO}_2\text{S}_2)_2]^{2-}$, as the band around $8 \times 10^3 \text{ cm}^{-1}$ shows the typical spin–orbit splitting pattern of the $\nu_2(^4A_2 \rightarrow {}^4T_1(F))$ ligand field transition.[24]

The remarkable properties of thiometalate ligands can also be demonstrated by calculations: the MO coefficients of the $3d$-type (M') MOs show a high electron delocalization (which increases with an increasing sulfur content of the ligand), and hence a strong electron acceptor ability (which can be explained by the low-lying unoccupied L* orbitals). Since the $3d(M')$ level has an energy comparable to that of the nonbonding MOs of the ligands (t_1 of the free MS_4^{2-} ions), it also follows that there is a remarkable M'—L σ-interaction.

The overlap integrals between the $3d(M')$ and $4d$(M) or $5d$(M) functions show that at an M' —M separation of *ca.* 2.8 Å direct metal–metal interactions should also be taken into consideration. The four-membered ring systems $\text{M}'(\text{S}_{br})_2\text{M}$ can be described (in terms of VB theory) with the resonance structures (**23**) and (**24**).

$$\text{M}'^n \overset{S}{\underset{S}{\diamond}} \text{M}^{VI} \quad\longleftrightarrow\quad \text{M}'^{n+1} \overset{S}{\underset{S}{\diamond}} \text{M}^{V}$$

(23)　　　　　　(24)

The calculations also show that due to complex formation the electron density on the M′ (M′ = Fe, Co) center is diminished, while that on the M center is increased, relative to the free MS_4^{2-}. The electron density on the terminal S atoms is hardly altered by electron delocalization compared to that of the free MS_4^{2-}, especially in Fe—Mo complexes.

MOs delocalized (occupied) over the metal centers M′ and M occur particularly clearly in complexes with $M'S_2MS_2M'$ units (*e.g.* with M′ = Fe, and notably also when M′ = Cu^I with a closed *d*-shell; *cf.* the results of the resonance Raman spectra mentioned below).

As expected, the electronic spectra of complexes of the type $[X_2Fe(MoS_4)]^{2-}$ (X = Cl, PhS, *p*-MeC$_6$H$_4$S, $\frac{1}{2}$S$_5$), where the M′—L interaction is smaller than in the trimetallic species, are very similar. This is likewise true for the $[X_2Fe(WS_4)]^{2-}$ complexes. For example, the Fe → WS_4^{2-} electron delocalization is greater in $[Fe(WS_4)_2]^{2-}$ than in $[Cl_2Fe(WS_4)]^{2-}$. In the first of these complexes, this is responsible for a lower energy, both of the longest-wavelength ligand internal transition and of the L → *d*(Fe) charge transfer transition (Table 2).

The stability of the complexes $[Fe(MS_4)_2]^{n-}$ and $[Co(MS_4)_2]^{n-}$ (M = Mo, W), and the special behavior of the Fe species (see above) can be explained as follows: MOs with appreciable $3d$(M′) participation are lower in energy for M′ = Co compared to M′ = Fe, but also for (formal) M′^{II} compared to M′^{I}; furthermore MOs with appreciable $2e$(MS_4^{2-}) contribution are lower in energy for M = Mo than for M = W. Thus the low-energy MOs of the $3d$(M′) type in the Co complexes (these are occupied here, in contrast to those of the Fe complexes with $n = 2$) and in $[Fe(MoS_4)_2]^{2-}$ are situated at a lower energy than the nonbonding L orbitals, and in the case of $[Fe(WS_4)_2]^{2-}$ they are at about the same energy. Thus, the unstable, discrete $[Fe(MoS_4)_2]^{2-}$ is stabilized by taking up one electron to form $[Fe(MoS_4)_2]^{3-}$, and $[Fe(WS_4)_2]^{n-}$ exists with $n = 2$ and 3. Strong reducing reagents are necessary to obtain the trianions of Co as the difference between Mo and W is not decisive in this case, since the 'redox MO' is definitely higher in energy than the nonbonding ligand levels.

The ^{57}Fe Mössbauer spectra of complexes of the type $[X_2Fe(MS_4)]^{2-}$ show unusually low values for the isomer shift (IS) of the order of $0.5\,\mathrm{mm\,s^{-1}}$ (all values relative to α-Fe at room temperature). This also demonstrates the strong electron delocalization Fe → MS_4^{2-}. As expected, this delocalization is greater in the Mo complexes than in the analogous W complexes. The effect of a thiometalate ligand is clearly illustrated by the decrease of the IS values in the series $[Cl_2FeCl_2]^{2-}$ (IS = $0.76\,\mathrm{mm\,s^{-1}}$), $[Cl_2Fe(WS_4)]^{2-}$ (IS = $0.52\,\mathrm{mm\,s^{-1}}$), $[(WS_4)Fe(WS_4)]^{2-}$ (IS = $0.43\,\mathrm{mm\,s^{-1}}$). The Mössbauer spectrum of $[Fe(MoS_4)_2]^{3-}$, with an isomer shift of $0.38\,\mathrm{mm\,s^{-1}}$, shows that the electron density on the Fe center is much lower than would be expected for a formal oxidation state of $+1$.

Various complexes of the type $[M'(MO_{4-n}S_n)_2]^{2-}$ (M′ = Fe, Co, Ni, Pd) and also $[(FeCl_2)_2\text{-}MoS_4]^{2-}$ can be reduced reversibly in a one-electron process (equation 9) and some also in a second one-electron process (equation 10) according to cyclic voltammetric investigations.

$$[M'(MO_{4-n}S_n)_2]^{2-} \xrightarrow[-e]{e} [M'(MO_{4-n}S_n)_2]^{3-} \tag{9}$$

$$[M'(MO_{4-n}S_n)_2]^{2-} \xrightarrow[-e]{e} [M'(MO_{4-n}S_n)_2]^{4-} \tag{10}$$

This redox behavior is typical of complexes of first row transition metal ions chelated by two thiometalates and Table 2 shows data for many of these species. The following conclusions can be drawn on the basis of these data.

It is evident that the order for the difficulty of reduction of complexes with first row metals ligated by WS_4^{2-} is Fe < Co < Ni. This trend doubtless reflects the increasing electron density on the central metal (d^6 for Fe^{II}, d^7 for Co^{II}, and d^8 for Ni^{II}).

Secondly, it is possible to follow systematically the effect of varying only M in $[M'(MS_4)_2]^{2-}$ (M′ = Fe, Ni; M = Mo, W) and in $[M'(MO_2S_2)_2]^{2-}$ (M′ = Co; M = Mo, W). For the MS_4–Ni system, only a minor effect on the $2-/3-$ couple was observed on substitution of W for Mo, while the Co–$MO_{4-n}S_n$ systems showed a significant shift to negative potentials on changing M from Mo to W. The Fe–MS_4 system shows the largest shift negative potential on substitution of W for Mo. This trend seems to indicate that the reduction is increasingly centered in the central MS_4 chromophore as one moves from Fe through Co to Ni.

Finally, the electrochemical data for the cobalt system enable evaluation of the effect on redox properties of the thiometalate ligands (TM) varying from $MO_2S_2^{2-}$ to MS_4^{2-}. It is evident that an increasing oxygen content of the ligand makes the reduction more difficult.

The different electron delocalization capabilities of the different TM species can be rationalized in terms of the differences in the nd(M) $\leftrightarrow$ mp(X) interaction (Mo $4d$/W $5d$ $\leftrightarrow$ O $2p$/S $3p$). Our MO

Table 2 Physical Properties of Thiometalato Complexes[1,13,14,24]

Anion	Color	Near IR/vis/UV (10^3 cm^{-1})			IR (cm^{-1})			$-E^{1}_{1/2}$ a (V)
		$L \rightarrow d(M')$	$L \rightarrow L^*$	$d \rightarrow d$	$\nu(MO_{term})$	$\nu(MS_{term})$	$\nu(MS_{br})$	
$[Fe(MoS_4)_2]^{3-}$	Violet	15.9, 17.3	19.7, 24.4			495/472	438	-0.07 (E^{ox}_{irr})
$[Fe(WS_4)_2]^{2-}$	Green	16.2	23.4, 26.7	8.7		493/484	438	0.27
$[Fe(WS_4)_2]^{3-}$	Orange	18.0, 20.3	22.9, 27.6			470	431	0.36
$[Co(MoS_4)_2]^{2-}$		10.7	22.3, 32.5	11 (?)		482/466	444	0.45
$[Co(MoS_4)_2]^{3-}$	Black	12.3, 15.9	18.6, 25.9	8.7				0.78
$[Co(MoOS_3)_2]^{2-}$		12.3	21.7, 26.5, 30.8, 35.5	8.3, 13.0	903/884		453/447	
$[Co(MoO_2S_2)_2]^{2-}$	Green	14.7	22.6, 29.1, 33.1		865/840		445	
$[Co(MoO_2S_2)_2]^{3-}$		17.6						
$[Co(WS_4)_2]^{2-}$	Olive green	12.2	25.8	8.6		500/491	450/442	0.55
$[Co(WS_4)_2]^{3-}$	Violet	15.4, 18.0	28.3			467	438	
$[Co(WOS_3)_2]^{2-}$	Olive green	13.8	27.2, 30.3	8.3	917/907	490/485	445	0.62
$[Co(WOS_3)_2]^{3-}$		18.8, 21.4			890	470	435	
$[Co(WO_2S_2)_2]^{2-}$	Blue-green	16.4	27.9, 31.7	8.0, 14.3	927/892		440	0.84
$[Co(WO_2S_2)_2]^{3-}$		20.5, 22.0			885/850		435	
$[^{58}Ni(^{92}MoS_4)_2]^{2-}$	Dark brown		19.5, 25.3, 30.0			513/494.0	455.5/442.5	0.52
$[Ni(MoOS_3)_2]^{2-}$	Brown		21.8, 27.6, 31.0	14.3	896/884	500/492	458/446	0.58
$[Ni(MoO_2S_2)_2]^{2-}$	Brown		27.0, (31.0), 34.7	14.3	896/880	496/490	463/453	0.62
$[^{58}Ni(WS_4)_2]^{2-}$	Red-brown		23.8, 26.3	24.5		496/486	449	0.54
$[Ni(WOS_3)_2]^{2-}$	Brown		26.3, 30.1		921/908	513/497	450	0.58
$[^{58}Ni(WO_2S_2)_2]^{2-}$	Brown-yellow		31.5, 34.7	15.0	916.1/882.4	499/491	451.1	0.68
$[Pd(MoS_4)_2]^{2-}$	Red		21.2, 26.1, 28.6, 31.6			510/496	450/435	1.09
$[Pd(WS_4)_2]^{2-}$	Brown-red		25.2, 27.2, 31.9			498/489	440	1.05
$[Pt(MoS_4)_2]^{2-}$	Red		19.5, 24.3, 27.5, 32.5			516/499	454/433	
$[Pt(WS_4)_2]^{2-}$	Brown-orange		24, 32.0			502	446/438	
$[^{64}Zn(^{92}MoS_4)_2]^{2-}$	Brown		17.5, 21.4, 31.9, 40.5, 42.9			492.4/487.2	456/434.5	
$[Zn(MoOS_3)_2]^{2-}$	Orange		21.7, 25.1, 33.0		898	495/485	456/437	
$[^{68}Zn(WS_4)_2]^{2-}$	Orange		21.7, 25.3			493/486	445.5/435	
$[Zn(WOS_3)_2]^{2-}$	Yellow		25.3, 29.5		918/904	493/486	440	
$[Cd(WS_4)_2]^{2-}$	Orange		21.6, 25.3			503/488	435	
$[Hg(WS_4)_2]^{2-}$	Orange		22.3, 25.3			496/483.5	431	
$[Cl_2Fe(^{92}MoS_4)]^{2-}$	Dark red	17.1, 19.2	21.2, 23.0, 31.9, 35.2	9.5		492/482	452/(427)	
$[Cl_2Fe(^{100}MoS_4)]^{2-}$	v. sup.	v. sup.	v.sup.	v. sup.		487	447.5/(427)	
$[Cl_2Fe(WS_4)]^{2-}$	Red	19.4, 22.6	24.1, 26.8	8.7		484/472	443/438	
$[(NO)Fe(WS_4)_2]^{2-}$	Dark red	22.5	25.6	(15.6)		500/486	450/443	
$[(DMF)_2Fe(WS_4)_2]^{2-}$	Red		21.2, 22.2, 31.5, 40.5, 44.6			493/481	462/446	
$[NCCu(^{92}MoS_4)_2]^{2-}$	Red						447/416	
$[NCCu(^{100}MoS_4)_2]^{2-}$	v. sup.		v. sup.				443/415	

a CV data of all Co, Fe and Pd complexes relative to SCE in DMF; those of Ni complexes relative to Ag/AgCl/LiCl(sat.)/EtOH in DMF.

calculations[33] show that an electron delocalization is mainly achieved by the low-lying unoccupied L* levels of the TM ligands ($2e$ for MS_4^{2-} species, *cf.* ref. 6 and literature cited therein). As π-bonding is stronger for M—O compared to M—S bond systems, the corresponding π*-levels are lower in energy (better accessible acceptor orbitals) for the latter. Additionally, the unoccupied lowest energy π*-levels of the species with the ligands MOS_3^{2-} and $MO_2S_2^{2-}$ are mainly located in the M—S bond systems (relatively low contribution of oxygen AOs); thus in ligands having higher sulfur content the acceptor capability is also increased due to a 'spatially larger' electron reservoir. Differences between Mo and W species can also be explained by the results of MO calculations, namely by the fact that the $2e$ level (with predominant nd(M) character) is more delocalized (higher $3p$ S participation) for Mo species compared to W species.

These findings including the spectroscopic data also illustrate the marked electron delocalization $M' \rightarrow MO_{4-n}S_n^{2-}$ (M' = Fe, Co), since the stabilization of the formal 0 and +1 oxidation states of the metals M' can only be explained by the pronounced electron acceptor properties of the thiometalato ligands.

MO calculations give further insight into the electronic implications of the reductions $[M'(TM)_2]^{2-} \rightarrow [M'(TM)_2]^{3-}$: although the number of electrons in an MO with predominant $3d$(M') character is raised by the reduction, a destabilization of these levels results in a change of the relative populations, especially of the Co—S σ-MOs. Thus, the net charge of three metal centers remains essentially unchanged while the charge density on all sulfur atoms is increased. This charge delocalization implies a significant decrease of the M—S_{term} bond order but not of the other bonds (as inferred from overlap populations). This is in accordance with X-ray studies on salts of both $[M'(TM)_2]^{2-}$ and $[M'(TM)_2]^{3-}$ species, which show a 'contraction' of the $M'(S_{br})_2M$ moieties and an increase of the M—S_{term} bond length upon reduction.

Resonance Raman spectroscopy also allows an investigation of the kind of bonding in complexes in general. Whereas in the Raman spectra of tetrathiometalato complexes ($[M'(WS_4)_2]^{2-}$ with M' = Pt, Zn), the intensities I_R of the $\nu(MS)$ bands are basically given by $I_R[\nu(MS_{term})] > I_R[\nu(MS_{br})]$ (reason: higher π bond order in MS_{term}), the preresonance and resonance Raman spectra can exhibit clear deviations from this rule. (In the interpretation of the spectra it should be kept in mind that $\nu(MS_{term})$ vibrations are more characteristic than $\nu(MS_{br})$, *e.g.* $\nu_s(MS_{br})$ also contains a clear $\nu_s(MS_{term})$ component.[23])

Magnetic measurements have mainly been used to clarify the structures of the Ni and Co complexes. The isolated Ni complexes are diamagnetic, as would be expected for pseudo-square planar coordinated Ni^{II}. The magnetic moments of the Co complex are essentially closer to spin-only values than to values of 'typical' compounds with a tetrahedral CoS_4 chromophore, because of the strong electron delocalization $M' \rightarrow MO_{4-n}S_n^{2-}$. This behavior is even more clearly expressed in $[Fe(MoS_4)_2]^{3-}$.

16.3.4 ADDENDUM

Interesting complexes with the ReS_4^- ion, which can be reversibly reduced, could be obtained, *e.g.* $[(Cl_2Fe)_2ReS_4]^{2-}$ (containing one electron more than the Mo analog and which can be quasireversibly oxidized)[36] and $[Cl_7Cu_5(ReS_4)]^{3-}$ (with a novel cubane structure.[37] A new type of poly(thiometalate), $[(S_2WS_2)WS\{WS_3(S_2)\}]^{2-}$, has also been prepared.[10]

The electronic structure of nearly all types of metallothio anions has been discussed in the light of EH–SCCC MO calculations.[38] Resonance Raman spectra of several species have been measured to get information about their electronic structure.[39]

16.3.5 REFERENCES

1. A. Müller, E. Diemann, R. Jostes and H. Bögge, *Angew. Chem.*, 1981, **93**, 957; *Angew. Chem., Int. Ed. Engl.*, 1981, **20**, 934; see also E. Diemann and A. Müller, *Coord. Chem. Rev.*, 1973, **10**, 79.
2. A. Müller and E. Diemann, *Chem. Commun.*, 1971, 65.
3. A. Müller, A.-M. Dommröse, W. Jaegermann, E. Krickemeyer and S. Sarkar, *Angew. Chem.*, 1981, **93**, 1119; *Angew. Chem., Int. Ed. Engl.*, 1981, **20**, 1061.
4. A. Müller, R. G. Bhattacharyya, E. Königer-Ahlborn and R. C. Sharma, *Inorg. Chim. Acta*, 1979, **37**, L493.
5. A. Müller, H. Bögge, E. Krickemeyer, G. Henkel and B. Krebs, *Z. Naturforsch., Teil B*, 1982, **37**, 1014.
6. A. Müller, W. Hellmann, C. Römer, M. Römer, H. Bögge, R. Jostes and U. Schimanski, *Inorg. Chim. Acta*, 1984, **83**, L75.
7. E. Königer-Ahlborn and A. Müller, *Angew. Chem.*, 1975, **87**, 598; *Angew. Chem., Int. Ed. Engl.*, 1975, **14**, 573.
8. W.-H. Pan, M. E. Leonowicz and E. Stiefel, *Inorg. Chem.*, 1983, **22**, 672.

9. F. Secheresse, J. Lefebvre, J. C. Daran and Y. Jeannin, *Inorg. Chim. Acta*, 1980, **45**, L45.
10. A. Müller, unpublished results.
11. A. Müller, E. Königer-Ahlborn, E. Krickemeyer and R. Jostes, *Z. Anorg. Allg. Chem.*, 1981, **483**, 69.
12. A. Müller, W. Hellmann, J. Schneider, U. Schimanski, U. Demmer, A. Trautwein and U. Bender, *Inorg. Chim. Acta*, 1982, **65**, L41.
13. A. Müller, W. Hellmann, U. Schimanski, R. Jostes and W. E. Newton, *Z. Naturforsch., Teil B*, 1983, **38**, 528.
14. J. W. McDonald, G. D. Friesen, W. E. Newton, A. Müller, W. Hellmann, U. Schimanski, A. Trautwein and U. Bender, *Inorg. Chim. Acta*, 1983, **76**, L297.
15. D. Coucouvanis, *Acc. Chem. Res.*, 1981, **14**, 201.
16. S. R. Acott, C. D. Garner, J. R. Nicholson and W. Clegg, *J. Chem. Soc., Dalton Trans.*, 1983, 713.
17. A. Müller, H. Bögge and U. Schimanski, *Inorg. Chim. Acta*, 1983, **69**, 5.
18. A. Müller, U. Schimanski and J. Schimanski, *Inorg. Chim. Acta*, 1983, **76**, L245.
19. A. Müller, M. Dartmann, C. Römer, W. Clegg and G. M. Sheldrick, *Angew. Chem.*, 1981, **93**, 1118; *Angew. Chem., Int. Ed. Engl.*, 1981, **20**, 1060.
20. S. F. Gheller, T. W. Hambley, J. R. Rodgers, R. T. C. Brownlee, M. J. O'Connor, M. R. Snow and A. G. Wedd, *Inorg. Chem.*, 1984, **23**, 2519.
21. A. Müller, W. Hellmann, H. Bögge, R. Jostes, M. Römer and U. Schimanski, *Angew. Chem.*, 1982, **94**, 863; *Angew. Chem., Int. Ed. Engl.*, 1982, **21**, 860.
22. A. Müller, S. Sarkar, H. Bögge, R. Jostes, A. Trautwein and U. Lauer, *Angew. Chem.*, 1983, **95**, 574; *Angew. Chem., Int. Ed. Engl.*, 1983, **22**, 561.
23. A. Müller, W. Jaegermann and W. Hellmann, *J. Mol. Struct.*, 1983, **100**, 559.
24. A. Müller, R. Jostes, W. Hellmann, C. Römer, H. Bögge, U. Schimanski, B. Zuhang, L. D. Rosenheim, J. W. McDonald and W. E. Newton, *Z. Anorg. Allg. Chem.*, 1986, **533**, 125.
25. B. A. Averill, *Struct. Bonding (Berlin)*, 1983, **53**, 59.
26. D. Coucouvanis, P. Stremple, E. D. Simhon, D. Swenson, N. C. Baenziger, M. Draganjac, L. T. Chan, A. Simopoulos, V. Papaefthymiou, A. Kostikas and V. Petrouleas, *Inorg. Chem.*, 1983, **22**, 293.
27. U. Schimanski, Ph.D. Thesis, University of Bielefeld, 1984.
28. D. Coucouvanis, in 'Nitrogen Fixation, The Chemical–Biochemical–Genetic Interface', ed. A. Müller and W. E. Newton, Plenum, New York, 1983, p. 211.
29. A. Müller, W. Hellman, H. Bögge, R. Jostes, M. Römer and U. Schimanski, *Angew. Chem.*, 1982, **94**, 863; *Angew. Chem., Int. Ed. Engl.*, 1982, **21**, 860.
30. A. Müller, S. Sarkar, H. Bögge, R. Jostes, A Trautwein and U. Lauer, *Angew. Chem.*, 1983, **95**, 574; *Angew. Chem., Int. Ed. Engl.*, 1983, **22**, 561.
31. M. Römer, Ph.D. Thesis, University of Bielefeld, 1984.
32. W. Clegg, C. D. Garner and J. R. Nicholson, *Acta Crystallogr., Sect. C*, 1983, **39**, 552.
33. J. M. Manoli, C. Potvin and F. Secheresse, *J. Chem. Soc., Chem. Commun.*, 1982, 1159.
34. Y. Do, E. D. Simhon and R. H. Holm, *J. Am. Chem. Soc.*, 1983, **105**, 6731.
35. B. K. Teo, M. R. Antonio, R. W. Tieckelmann, H. Craig Silvis and B. A. Averill, *J. Am. Chem. Soc.*, 1982, **104**, 6126.
36. A. Müller, E. Krickemeyer, F.W. Baumann, R. Jostes and H. Bögge, *Chimia*, 1986, **40**, 310.
37. A. Müller, E. Krickemeyer and H. Bögge, *Angew. Chem.*, 1986, **98**, 98; *Angew Chem., Int. Ed. Engl.*, 1986, **25**, 990.
38. A. Müller and R. Jostes, *J. Mol. Struct. (Theochem.)*, in press.
39. A. Müller and W. Hellmann, *Spectrochim. Acta, Part A*, 1985, **41**, 359.

16.4

Dithiocarbamates and Related Ligands

JOANNUS A. CRAS and JACOBUS WILLEMSE
University of Nijmegen, The Netherlands

16.4.1 INTRODUCTION

On substitution of the two CR$_2$ groups in the allene molecule by isoelectronic hetero groups or atoms X and Y (X, Y = O, S, NR), molecules X=C=Y are formed, which are called pseudo- or hetero-allenes. This chapter deals with the heteroallenes CS$_2$ and COS and the ligands based on these two molecules, the heteroallyls. The heteroallenes themselves have been reviewed by Gattow and Behrendt[1] and Reid[2] respectively. Their coordinating properties have been compared recently by Ibers.[3] Assuming X to be preferred over Y for coordination, both end-on coordination by X and side-on (η^2) coordination by the C=X double bond occurs, whereas for X=Y=S bridging coordination between two metals is also found.

The reactivity of free and coordinated heteroallenes towards nucleophilic and electrophilic agents is determined by the cumulation of the unsaturated bonds and the polarity within the molecule. Although some reactions can be explained as an initial electrophilic attack at the sulfur hetero atom (*cf.* the reaction of a Grignard reagent with carbon disulfide[4]), the majority of reactions are nucleophilic additions to the central electrophilic carbon atom, leading to the formation of a heteroallylic anion (**1**). This anion, although it has a four-electron three-centre π-system analogous to the allylic ion, tends to coordinate through X and Y with the metal situated in the plane determined by X, Y and Z, whereas a common coordination for allyls is η^3 with the metal out of the plane formed by the allylic ion. The reasons for these preferences are not yet understood. Perhaps the extension of the π delocalization towards both the metal and the group Z, or the bulkiness of the atoms X and Y in comparison with carbon, plays a role. Only two examples of an η^3 out-of-plane coordinated heteroallyl are known, surprisingly both upon coordination towards molybdenum (see sections 16.4.3.5 and 16.4.3.8) and both in combination with in-plane coordination of the same ligand.

In metal heteroallyl complexes the carbon atom is susceptible to attack by nucleophiles, resulting in replacement of Z or addition to the carbon atom.[5] We do not deal with the ligands resulting from the latter reaction.

In coordination compounds with heteroallyls ZC(X)S (X = S, O) the resonance structure (2) is of great importance. Ions with the fragment Z'C(X)Y can also be dinegative, *e.g.* (3). These ligands can be thought to stem from the heteroallyls when Z = Z'H *via* the loss of an H atom and a shift of the lone pair left on Z' into the Z'—C bond.

$$Z\!\!=\!\!C\underset{Y}{\overset{X}{\big<}}\;(-) \qquad \overset{(+)}{Z}\!\!=\!\!C\underset{S\,(-)}{\overset{X\,(-)}{\big<}} \qquad Z'\!\!=\!\!C\underset{Y\,(-)}{\overset{X\,(-)}{\big<}} \qquad Os_3H(CO)_9(\mu\text{-}\eta^2\text{-}HC\!\!=\!\!NPh)$$

$$(1) \qquad\qquad (2) \qquad\qquad (3) \qquad\qquad\qquad (4)$$

The CS_2^- and COS-based ligands are all dealt with in Section 16.4.3. Dimeric oxidation products of the heteroallyls and other ligands in which a CS_2 or COS fragment is found are treated briefly in Section 16.4.4. No further attention is paid to ligands like S—S—Me and methanedithiolate or dithioacetal, which occurs for example in the reaction of two moles of (4) with one mole of CS_2 in a two-step hydrogen migration.[6]

16.4.2 CARBON DISULFIDE AND CARBON MONOSULFIDE AS LIGANDS

16.4.2.1 Introduction

Carbon disulfide forms complexes in which the metal has a low oxidation state with almost every transition metal. The complexes have been reviewed extensively.[1,7,8] Three bonding modes are found: end-on *via* S, η^2 bonded and bridging between two metal atoms. The evidence for these three bonding types is largely spectroscopic and therefore limited. CS_2 shows a variety of insertion and disproportionation reactions.

16.4.2.2 Complexes with CS_2 End-on Bonded *via* S

As a characteristic IR frequency for complexes containing the rather weakly bonded, linear M—S—C—S, the ν(S—C—S) at about $1500\,\text{cm}^{-1}$ is considered. On this evidence, Cr, Fe, Ru, Os, Rh and Ir complexes are claimed.[8]

16.4.2.3 Complexes with CS_2 η^2 Bonded

The complexes with the coordination mode (5) have a characteristic IR frequency in the region $1000\text{--}1200\,\text{cm}^{-1}$ (ν(C—S, exocyclic)). They are generally prepared by ligand substitution. A few X-ray structures confirm the coordination mode of these complexes,[9] *e.g.* compound (6): (C—S' 147.8, C—S 172, Pt—S 232.8, Pt—C 206.3 pm, S—C—S 136.2° C—Pt—S 45.5°.[10] Extended Hückel calculations show that a reaction path starting from an end-on approach of the CS_2 molecule is energetically most accessible.[10b]

$$M-\!\!\!\overset{\displaystyle C\!\!=\!\!S'}{\underset{\displaystyle S}{|}} \qquad\qquad \underset{Ph_3P}{\overset{Ph_3P}{\big>}}Pt-\!\!\!\overset{\displaystyle C\!\!=\!\!S'}{\underset{\displaystyle S}{|}}$$

$$(5) \qquad\qquad\qquad (6)$$

The complexes are reactive towards nucleophilic as well as electrophilic attack. Attack by alkyl halides yields thioalkylated compounds. Mostly an alkylation at one sulfur atom takes place, but examples of alkylation at both sulfur atoms are known (reactions 1 and 2). The facile sulfur abstraction with PPh_3 yielding a thiocarbonyl complex is a nucleophilic reaction (reaction 3).

$$Ru(CO)_2(PPh_3)_2(\eta^2\text{-}CS_2) + MeI \longrightarrow [Ru(CO)_2\{C(S)SMe\}(PPh_3)_2]I \qquad (1)$$

$$Pt(PPh_3)_2(\eta^2\text{-}CS_2) + MeI \xrightarrow{\text{reflux}} [Pt\{C(SMe)_2\}(PPh_3)_2(I)]I \longrightarrow Pt\{C(SMe)_2\}(PPh_3)(I)_2 \qquad (2)$$

$$Mn(CO)_2(\eta^2\text{-}CS_2)(\eta^5\text{-}C_5H_5) + PPh_3 \longrightarrow Mn(CO)_2(CS)(\eta^5\text{-}C_5H_5) + PPh_3S \qquad (3)$$

η^2-Coordinated CS_2 groups can react *via* dimerization and abstraction of CS_2, as can be seen in a survey of reactions of a rhodium(I) phosphine complex with CS_2 (Scheme 1).[11] The dimerization can be throught to proceed through a nucleophilic attack on the carbon atom of CS_2 *via* an end-on intermediate of the heteroallene fragment.[10b]

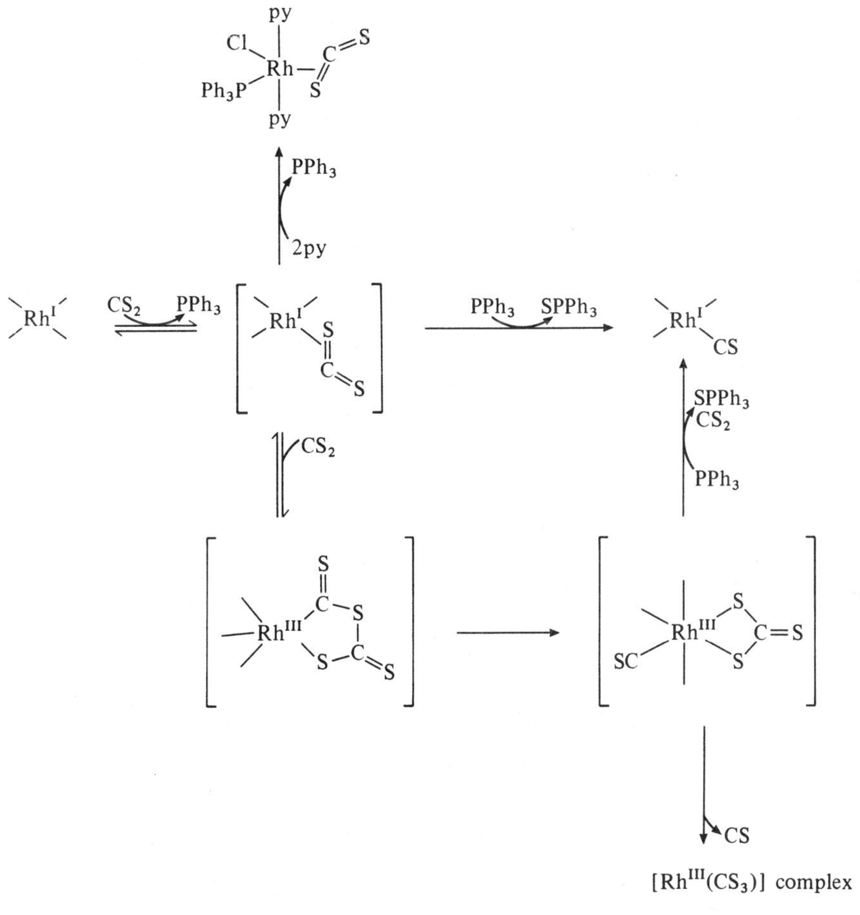

Scheme 1

The reaction of CS_2 with (**7**) in benzene gives (**8**), (**9**), (**10**) and (**11**). Compound (**8**) appears to be the principal source of (**9**), (**10**) and (**11**). The slow thermal decomposition of (**8**) in chloroform gives only (**9**) and (**11**), due to the principal heteroallene reaction (reaction 4) proceeding perhaps *via* (**12**).[12]

Apart from compounds with a $C_2S_4^{2-}$ ring, the existence of which is confirmed by X-ray analysis of (**13**),[13] complexes with CS_3^{2-} (—C(S)S—S—), $C_2S_5^{2-}$ and $C_3S_5^{2-}$ rings were also postulated.[14-16] Whereas reaction (4) yields complex (**14**),[17] the product from reaction (5) is (**15**).[9]

$$Co(CO)L(\eta^5\text{-}C_5H_5) \qquad\qquad CoL(\eta^2\text{-}CS_2)(\eta^5\text{-}C_5H_5) \qquad\qquad Co(CS)L(\eta^5\text{-}C_5H_5)$$

(**7**) L = tert. phosphine, RNCS   (**8**)   (**9**)

$Co_3(\eta^5\text{-}C_5H_5)(\mu_3\text{-}CS)(\mu_3\text{-}S)$ $Co_4(CS_3)L(\eta^5\text{-}C_5H_5)$ $Co(C_2S_4)L_n(\eta^5\text{-}C_5H_5)$

(10)   (11)   (12)

$Rh(C_2S_4)(PMe_3)(\eta^5\text{-}C_5H_5)$

(13)   (14)   (15)

$$Ni(cod)_2 + CS_2 + PR_3 \longrightarrow Ni(CS)_2(PR_3) \qquad (4)$$

$$(R = Ph, \textit{p}\text{-tolyl, cyclohexyl})$$

$$Ni(cod)_2 + 2CS_2 + 2PR_3 \longrightarrow \text{'}Ni(CS_2)_2(PR_3)_2\text{'} + 2cod \qquad (5)$$

16.4.2.4 Complexes with CS_2 Bridging Between Two Metals

Since the synthesis of the first complex (16) containing the CS_2 group as a bridge between two metal atoms, from (17) and CS_2 by Wilkinson,[18] several CS_2-bridging derivatives have been synthesized. The structure of this dinuclear Co complex has been described in two different ways. Wilkinson suggested that the compound contained a unidentate metallodithiocarboxylate group —C(S)S—, whereas Mizuta *et al.*[19] formulated the compound as having a symmetrical linear —SCS— group.

$$K_6[(CN)_5CoCS_2Co(CN)_5] \qquad\qquad K_3Co(CN)_5$$

(16)   (17)

X-Ray structural determination has revealed the existence of two classes of compounds with a carbon disulfide bridge between two metals. In the first class one metal is η^2 bonded to CS_2 (Table 1). The second class can be described as metallodithiocarboxylato complexes in which the ligand has a metal bonded to the carbon atom of the CS_2 group (Table 2). In general the transition metal groups are less effective in the stabilization of an E—CS_2 bond than the first-row main-group donor atoms.[20] Dithiocarbene complexes of the type (18) were also synthesized[21,22] and a crystal structure confirmed the bonding mode.[21]

The reaction of (19) with CS_2 yields, apart from products that do not contain CS_2 or COS, several products that have been characterized by X-ray structure determination: (20) (two isomers), (21) and (22).[23-26]

Table 1 Complexes with CS_2 Bridge Bonding

M	M'	Type, remarks[a]	Ref.
$Fe(CO)_2\{P(OMe)_3\}$	$Mn(CO)_3(\eta^5\text{-}C_5H_5)$	A, X-ray structure	130, 131
$Fe(CO)_2(PMePh_3)$	$Mn(CO)_3(\eta^5\text{-}C_5H_5)$	A	130, 131
$Fe(CO)_2(PMe_3)$	$Mn(CO)_5$	A	130, 131
$Co(\eta^5\text{-}C_5H_5)(PMe_3)$	$Cr(CO)_5$	A	132
$Co(\eta^5\text{-}C_5H_5)(PMe_3)$	$Mn(CO)_2(\eta^5\text{-}C_5H_5)$	A	132
Co(triphos)	$Cr(CO)_5$	A	133
Co(triphos)	$Mn(CO)_2(\eta^5\text{-}C_5H_5)$	A	133
Co(triphos)	Co(triphos)	B, X-ray structure	133
$Mn(CO)_2(\eta^5\text{-}C_5H_5)$	$Mn(CO)_2(\eta^5\text{-}C_5H_5)$	A	134

[a] Type A: M with S; type B: M—M' with S.

Table 2 Metallodithiocarboxylato Complexes

M	M'	Type, remarks[a]	Ref.
Pt(PPh$_3$)$_2$	PtCl(PPh$_3$)$_2$	C, cation, X-ray structure	135, 136
PtCl(PPh$_3$)	PtCl(PPh$_3$)$_2$	C	136
NiCl(PPh$_3$)	PtCl(PPh$_3$)$_2$	C	136
PdCl(PPh$_3$)	PtCl(PPh$_3$)$_2$	C	136
Fe(CO)(η^5-C$_5$H$_5$)	Fe(CO)$_2$(η^5-C$_5$H$_5$)	C	136, 137
Fe(CO)$_2$(η^5-C$_5$H$_5$)	Fe(CO)$_2$(η^5-C$_5$H$_5$)	D	139
Mn(η^5-C$_5$H$_5$)(NO)	Fe(CO)$_2$(η^5-C$_5$H$_5$)	C	137
Re(CO)$_4$	Fe(CO)$_2$(η^5-C$_5$H$_5$)	C	138
Pt(PPh$_3$)$_2$	Fe(CO)$_2$(η^5-C$_5$H$_5$)	C	136
Pt	Fe(CO)$_2$(η^5-C$_5$H$_5$)	C, PtL$_2$	136
Cr	Fe(CO)$_2$(η^5-C$_5$H$_5$)	C, CrL$_3$, molecular weight not appropriate	139
SnPh$_3$	Fe(CO)$_2$(η^5-C$_5$H$_5$)	D	139
Mn(CO)$_4$	SnPh$_3$	C	140
Re(CO)$_4$	SnPh$_3$	C	140
Re(CO)$_5$	SnPh$_3$	D	140
Pt$_2$Cl$_2$(μ-CS$_2$)(dppm)$_2$		D, X-ray structure	141

[a] Type C:

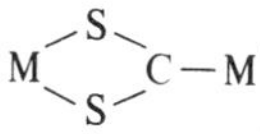

type D:

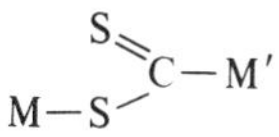

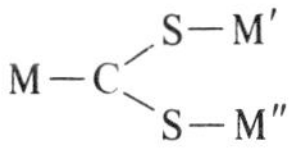

(18) (19) (20)

[Co$_3$(CO)$_7$(CO)(CS$_2$)Co$_3$(CO)$_7$S] [{Co$_3$(CO)$_9$C}$_2$(SCO)]

(21) (22)

16.4.2.5 Carbonyl Sulfide

Complexation of COS with metals is an exception rather than a rule.[3] To stabilize the M—COS bond formation, a basic metal centre is needed. This, however, weakens the C—S bond, promoting disproportionation reactions or sulfur abstraction.[27] Complexes of Ni,[28] Pd, Pt,[29] Ru,[30] Ir[20] and Rh[31] were prepared. COS is η^2CS bonded in these complexes. Decomposition reactions, in which the M—COS complexes play a role, give either dithiocarbonates[30] or carbonyls (reactions 6 and 7).

$$M(CO)_2(COS)(L)_2 \xrightarrow{COS} M(CO)_2(L)_2(S_2CO) + CO \quad (M = Fe, Ru) \quad (6)$$

$$M(CO)_2(COS)(L)_2 \xrightarrow{L} M(CO)_3(L)_2 + SL \quad (L = PPh_3) \quad (7)$$

In contrast to a head-to-tail bis(heteroallene) dimer intermediate, as is found for the metal-promoted disproportionation of several heteroallenes (see CS$_2$), a mechanism *via* an η^2 intermediate and **(23)** is suggested for reaction (8). Reaction (9) is also thought to proceed *via* an η^2-COS intermediate.[31] Apparently, although C—S bond cleavage occurs in the transition metal chemistry of CS$_2$, it is more important in the transition metal chemistry of COS.[3]

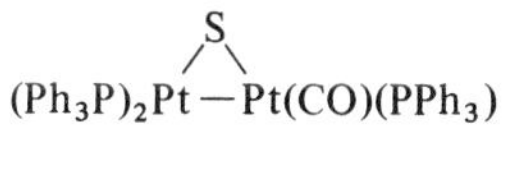

(23)

$$Pt(PPh_3)_4 + \text{excess COS} \longrightarrow Pt(PPh_3)_2(S_2CO) \quad (8)$$

$$Rh(H)(PPh_3)_4 + COS \longrightarrow Rh(CO)(PPh_3)_2(SH) \quad (9)$$

CO_2 is rather unreactive[32] and only three examples of $M—CO_2$ complexes have been characterized[33-35] (see Chapter 15.6).

16.4.3 CS_2- AND COS-BASED LIGANDS

16.4.3.1 Ligands with the Fragment HC(S)S or HC(O)S

Alkali dithioformates can be prepared according to reaction (10) in a methanolic medium. The unstable free monomeric acid, which trimerizes within a few minutes, is formed when an aqueous solution of these compounds is treated with HCl. The trimeric acid turns into the polymeric product (24) in DMF or DMSO. The alkali metal compounds are light and oxygen sensitive. Its complexes with Cu^{II}, Tl^I, In^{III} and Pb^{II} are stable, whereas the Ag^I, Fe^{II}, Co^{II} and Ni^{II} compounds decompose and give the metal sulfides.[1]

$$2M_2S + CHCl_3 \longrightarrow M(S_2CH) + 3MCl \qquad (10)$$

$$[HCS(SH)]_x \qquad\qquad Co(B_9C_2H_{10})_2$$

$$(24) \qquad\qquad\qquad (25)$$

Insertion of CS_2 in metal–hydride bonds gives metal dithioformates. Bonding of the ligands to the metal can take place in several ways: (a) ionic in alkali and alkaline earth metal compounds; (b) symmetric bidentate in most of the transition metal complexes; and (c) as a positive symmetrically bonded ligand to (25), in which the $HC(S)S^-$ has donated two electrons to the two electron-deficient dicarbanonaborane groups.[36] Little is known about monothioformic acid and its complexes.

16.4.3.2 Ligands with the Fragment BC(S)S

Reactions of R_2BH with CS_2 or COS are not described in the literature and no compounds with the fragment BC(S)S or BC(O)S are known. Noteworthy, however, is the fact that BH_3 is isoelectronic with oxygen and sulfur and H_3BCO can be compared with CO_2 and CS_2. The product (26) of the reaction of H_3BCO with NH_3 can be compared with a carbamate. The many derivatives and complexes of H_3BCO that are known are not within the scope of this chapter.[37]

$$NH_4^+[H_3BC(O)NH_2]$$

$$(26)$$

16.4.3.3 Ligands with the Fragment CC(S)S or CC(O)S

16.4.3.3.1 Dithiocarboxylic and monothiocarboxylic acids and their complexes

The preparation and properties of the dithiocarboxylic acids and their metal complexes have been reviewed several times.[38-41] The formation of C—C bonds in the direct reaction of CS_2 requires sufficiently nucleophilic carbon bases, directly or potentially accessible in the form of ambifunctional phenoxides, organometallic compounds, CH acidic compounds, enamines or ketimines. Carbanions react with CS_2 to give dithiocarboxylates. The preparation and purification of the acids is performed *via* their salts. Metal complexes are in general readily available. The bonding in these complexes is mostly of the type (27) but a bonding mode (28) is also found. Action of elemental sulfur upon heavy metal complexes of (29) aromatic dithiocarboxylic acids yields the perthio complexes (29) of these compounds.

$$(27) \qquad\qquad (28) \qquad\qquad (29)$$

Several cyanodithioformates are mentioned in ref. 1. The coordination chemistry of monothiocarboxylates deals almost exclusively with thioacetates and thiobenzoates. Structures of several monothiobenzoates are based on IR data.[42] Those, however, of the copper(I) and silver(I) complexes are doubtful. They are probably polymeric and of the same structural class as the mono- and di-thiocarbamates[43-45]

16.4.3.3.2 Di-, tri- and tetra-thiooxalic acids and their salts

The potassium salt of monothiooxalate can be prepared.[46] Two isomers of the dithiooxalic dianion (1,1 and 1,2) and the trithiooxalato anion can also be prepared.[46] Transition metal complexes of these ligands are unknown until now. The tetrathiooxalato anion was synthesized and characterized in in the form of the nickel complex. The ligand coordinates as a 1,2-dithiolate.[47]

16.4.3.3.3 1,1-Dithiolates

A variety of 1,1-dithiolate ligands (38) have been investigated. A good survey of their complexes is given by Coucouvanis.[39] Compounds with the ligands (31) and (32) are especially interesting. Ligand (31) makes only a minor contribution to the resonance form (33) and the major influence is that of (34).[48] Ligand (32) is quite effective in the stabilization of high formal oxidation states (*e.g.* Fe^{IV}, Mn^{IV}, Cu^{III}). Worthy of mention are the Cu^{I} complexes of (32), which have a Cu_8S_{12} 'cubane' core. Proton addition leads to a Cu_{10} complex, which consists of two copper tetrahedra bridged by two copper atoms. Upon addition of sulfur a Cu_5 cluster results, which can be described as a mixed valence Cu^{I}–Cu^{III} compound.[39] The crystal structure reveals a rectangular pyramidal copper core with the formally Cu^{III} atom in the axial position.

$$[RR'C{=}C(S)S]^{2-} \qquad [(CH_2)_5CS_2]^{2-} \qquad [\{RO(O)C\}_2CCS_2]^{2-}$$

| (30) | (31) | (32) | (33) | (34) |

M_8S_{12} clusters of isomaleonitrile (35; M = Cu, Ag) have also been characterized.[49-51] A recent study of the addition of alkynyllithium reagents to the S—S bond of (μ-dithio)bis(tricarbonyliron) revealed the formation of the ligand (36) bridging the two iron atoms of a hexacarbonyldiiron group. Several further reaction products of this compound were synthesized.[52]

$$[(CN)_2CCS_2]^{2-} \qquad\qquad [Li(R)C{=}CS_2]^{2-}$$

| (35) | (36) |

16.4.3.4 Ligands with the Fragment NC(S)S or NC(O)S

16.4.3.4.1 Dithiocarbamates

Ammonium dithiocarbamates can be prepared through reaction (11). With HCl or H_2SO_4 at lower temperatures the free acids are formed. Only one transition metal complex (37) is known, with a pseudo-octahedral coordination of Ni: four sulfur atoms of the dithiocarbamate in the equatorial plane and two pyridine solvent molecules at the apical positions. Sulfides can be eliminated from monosubstituted dithiocarbamates to give isothiocyanates.[53]

$$2NH_3 + CS_2 \longrightarrow NH_4(S_2CNH_2) \qquad\qquad (11)$$

$$Ni(H_2dtc)_2 \cdot 2py$$

(37)

A hexameric Cu compound, in which each Cu atom is coordinated through the non-esterified sulfur and the nitrogen atom, was prepared from the ester of monophenyl dithiocarbamate after deprotonation of the nitrogen atom.[54]

Unsubstituted and mono substituted ammonium dithiocarbamates react with sodium chloroacetate to produce different rhodanines, from which numerous metal complexes have been made.[55]

The most investigated group of dithiocarbamates is that derived from the secondary amines because of their inertness, stability and interesting redox properties.[39–41,56–59]

Solutions of dialkylammonium N,N-dialkyldithiocarbamates can be prepared from an excess of the amine with CS_2 in water. Almost all main group and transition element dithiocarbamates can be obtained from these solutions in a quantitative yield. In complexes of transition elements the bonding to the central atom is through the two sulfur atoms of the ligand and both M—S distances are equal. In the planar MS_2CN system an extensive π delocalization exists with a high contribution to the resonance structure (38),[60] and a relatively high electron density on the metal can be expected. Extended redox series are known and oxidation in general is easier than reduction.[59,61] Trends in the redox properties were described by van Gaal.[62] Several complexes with a high formal oxidation state (*e.g.* Fe^{IV}, Cu^{III}, Ni^{IV}) could be isolated, for example by oxidation with halogens.[56]

The strongly donating properties of the ligands are lost when the nitrogen is bonded to aryl groups or in an aromatic system (diphenyl dithiocarbamate and pyrrole dithiocarboxylate respectively[63]). The preparation of the alkali salts can only be performed by means of the corresponding amides. The C—N stretching frequency, which is located around $1500 \, cm^{-1}$ in the aliphatic dithiocarbamates, is about $300 \, cm^{-1}$ lower in the aromatic dithiocarbamates and the oxidation potentials are high. The decreased electron donor ability of the pyrrole dithiocarboxylate ligand relative to that of the alkyl analogues is chemically evident in the lability of the carbonyl group in (39), which greatly enhances the rate of formation of the bis(alkyne) products.[64]

$$\underset{\text{(38)}}{\overset{(+)}{R_2N}=C\overset{S\,(-)}{\underset{S\,(-)}{\diagdown}}} \qquad \underset{\text{(39)}}{Mo(CO)(RC_2R)(S_2CNC_4H_4)} \qquad \underset{\text{(40)}}{Zn(R_2dtc)_3^-}$$

NMR studies of the fluxionality of octahedral transition element dithiocarbamates were performed by Pignolet and coworkers.[65] The synthesis and X-ray structure of the pseudo-octahedral complex ion (40) were described by McCleverty *et al.* in their investigations of the inorganic aspects of rubber vulcanization.[66] Our investigations point to the existence of a barium analogue.[67]

In many complexes, mixed oxidation states for the transition metals are found. Complex (41) is a diamagnetic compound with the actual composition (42).[68] Copper dithiocarbamates are synthesized not only with copper in the oxidation states I, II or III but also with the metal in mixed oxidation states III and I (43, 44, 45), II and I (46, 47) and even III and I [48, 49].[56] Complex (50), which has two isomeric structures α and β, has Ru either in a trapped oxidation state II,III or in a delocalized oxidation state of 2.5. The β-isomer is the most stable.[69] Slow, controlled potential electrochemical oxidation of (51) yields (52), in which the metal has a trapped oxidation state II,III.[70]

$$\underset{\text{(41)}}{AuBr(R_2dtc)} \qquad \underset{\text{(42)}}{[Au^{III}(R_2dtc)_2][Au^IBr_2]} \qquad \underset{\text{(43)}}{[Cu_3(R_2dtc)_6][Cd_2Br_6]} \qquad \underset{\text{(44)}}{[Cu_2(R_2dtc)_4][I_3]} \qquad \underset{\text{(45)}}{[Cu_2(R_2dtc)_3][Br_2]}$$

$$\underset{\text{(46)}}{[Cu(R_2dtc)_2]\cdot nCuBr} \qquad \underset{\text{(47)}}{CuCl(R_2dtc)_2} \qquad \underset{\text{(48)}}{[Cu(R_2dtc)_2][Cu_{n-1}Br_n]} \qquad \underset{\text{(49)}}{[Cu(R_2dtc)_2]_2[Cu_2Br_4]} \qquad \underset{\text{(50)}}{Ru_2(R_2dtc)_5}$$

$$\underset{\text{(51)}}{[Ru(CO)(Et_2dtc)_2]_2} \qquad \underset{\text{(52)}}{[Ru(CO)_2(Et_2dtc)_4]^+}$$

A special feature of compounds (53) is their magnetism. The magnetic properties can be interpreted by assuming a mixture of different spin states or by a single mixed-spin state.[71] In general the first interpretation is followed.[71,72] Ligand substituents R influence the position of this spin state equilibrium through inductive, mesomeric and steric effects. Shifts of the spin state equilibrium can be effected by changes in temperature, pressure and incorporation of solvent molecules in the lattices. Owing to the presence of different types of solvents in the lattices, some of the magnetic data in the literature may well be misleading.

$$\underset{\text{(53)}}{Fe(R_2dtc)_3} \qquad \underset{\text{(54)}}{Me_3Sb(R_2dtc)_2} \qquad \underset{\text{(55)}}{Me_2ISb(R_2dtc)_2}$$

In contrast to the transition elements, the main group element dithiocarbamates often have asymmetrical metal–sulfur bonds due to the lack of $p\pi$–$d\pi$ MOs. In theses compounds the σ-bonds are responsible for metal–sulfur interaction. High oxidation states for these dithiocarbamates are only found when high electron density is brought upon the metal by σ-donating groups. For instance, (54) exists whereas (55) does not.[73]

16.4.3.4.2 Monothiocarbamates

Reaction of COS with primary and secondary amines in alcoholic or ethereal media yields the unstable monothiocarbamic acid. A substantial excess of COS promotes the decomposition.[74] The ligand is stabilized in alkaline medium, so either an excess of amine is maintained or the preparation is carried out in aqueous NaOH or KOH. Monothiocarbamato complexes, recently reviewed,[75] are mostly prepared from a metal salt and (56). The ligand field strength of monothiocarbamate is somewhat less than that of the dithio analogue.[76]

$M(R_2mtc)$ (+)R_2N=$C$$\diagdown$^{O}(−) ... R_2N—$C$$\diagup^{O}\diagdown$$_{S}$(−) $[Mn(CO)_3(Me_2mtc)]_2$

(56) M = Na, K, NR_4 (57)   (58)   (59)

A normal vibration analysis of the dimethylmonothiocarbamato ligand shows that in metal complexes the most important resonance structures are (57) and (58).[75,77] In the complexes, a broad band—usually split up—in the region 1510–1590 cm^{-1} is ascribed to the C=O and C=N stretching frequencies. Bidentate coordination (S and O) is found for hard or borderline metals, with a CO/CN frequency in the region 1510–1550 cm^{-1}, whereas monodentate S coordination occurs with soft metals, correlated with a v(CO/CN) from 1550 to 1590 cm^{-1}.[77] Exceptions exist, however, e.g. Tl^{I} is a soft acid but bidentate coordinated[77] and a crystal structure determination of (59) with v(CO/CN) = 1585 cm^{-1} also reveals a bidentate monothiocarbamate.[78] For the class of monothiocarbamates derived from aromatic amines, the resonance structure (57) is of minor importance, which has its consequences for the CO and CN stretching frequencies.[75,79] Apart from the coordination modes mentioned the following types are found: (60), e.g. (61),[43,44] and (62), e.g. (63),[80] whereas (64) is proposed for (65), and (66)[81] for (67).[82] The fact that in all of the reported structures the sulfur atoms are located *cis* is striking.[75]

R_2N—$C$$\diagup$$^{O—M}$$\diagdown$$_{S}$$\diagdown$$_{M}$ (with M above) $[M(Pr_2mtc)]$ R_2N—$C$$\diagup^{O}\diagdown$$_{S}$ ...M $Ni(Bu_2mtc)_2$ R_2N—$C$$\diagup^{O}\diagdown$$_{S}$$\diagup^{M}\diagdown$$_{M}$

(60)   (61) M = Cu, Ag   (62)   (63)   (64)

$Pd(R_2dtc)(R_2mtc)$ R_2N—$C$$\diagup$$^{O—M}$$\diagdown$$_{S—M}$ $Mo_2(R_2mtc)_4$

(65)   (66)   (67)

Electrochemical studies of (68) show an irreversible oxidation at +1 V, which is higher than that of the dithiocarbamato complex, and a quasi-reversible reduction at almost the same potential as the dithiocarbamate (−0.3 to −0.4 V).[83] The oxidation potentials of (69) in acetonitrile and dichloromethane increase in the direction of X(Y) = Se(Se) < S(Se) < S(S) ≈ Se(O) < S(O).[84]

$Fe(R_2mtc)_3$ $Fe\{XC(Y)NMe_2\}(CO)_2(\eta^5\text{-}C_5H_5)$

(68)   (69)

The magnetic properties of (68) are still under discussion. Dependent upon the method of preparation the complex with R = Me is high spin or has a mixture of different spin states, $S = \frac{1}{2}$ ↔ induced by lattice effects. For R = Et only the spin state mixture is found, whereas the compounds with $R_2 = (CH_2)_4$ and $(CH_2)_5$ are high spin.[85]

Complexes with only monothiocarbamate as ligand easily react with amines and phosphines.[75] A noticeable reaction is that of (70) with (71),[86] which yields first (72) (monodentate, v(CO/CN) = 1603 cm^{-1}); heating gives complex (59) (bidentate, v(CO/CN) = 1575 cm^{-1}; for crystal structure see ref. 78), *via* (73) (bidentate, v(CO/CN) = 1540 cm^{-1}; product identified in solution).

$MnBr(CO)_5$ $[Me_2NH_2][Me_2mtc]$ $Mn(Me_2mtc)(CO)_5$ $Mn(Me_2mtc)(CO)_3$

(70)   (71)   (72)   (73)

16.4.3.4.3 Other compounds with NC(S)S ligands

Numerous other systems with an NC(S)S fragment are known; see for instance refs. 87 and 88. The most interesting complexes among them are those that have azidodithioformates together with phosphines, π-allyls or dienes.[1] The clear yellow solution which results from addition of CS_2 to (74) in chloroform evolves N_2, colloidal sulfur and a solution of (75) upon photolysis. Azido-bridged Pt^{II} and Pd^{II} complexes of the type (76) undergo 1,3-cycloaddition with CS_2 to form complexes of the type (77), in which the metals are sulfur bridged by thiatriazolethiolate groups (78).[89] The monomeric (79), however, seems to have a nitrogen-bonded thiatriazolethiolate group.[90]

$Cu(N_3)(PPh_3)_2$ $Cu(NCS)(PPh_3)_2$ $[M(diene-OMe)(N_3)]$

(74)   (75)   (76)

$[M(diene-OMe)(S_2CN_3)]_2$ (78) $Ni(PBu_3)(\eta^5\text{-}C_5H_5)(S_2CN_3)$

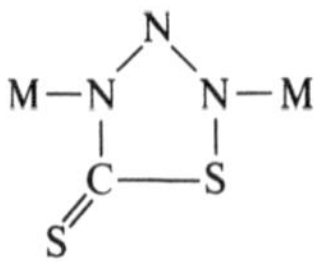

(77)   (79)

16.4.3.4.4 Dithiocarbimates

The coordination chemistry of the dithiocarbimates (80), with R = Me or Ph, has been described for complexes of Ni and Pr.[1,40] Dubois *et al.* reported the synthesis of (81):[91] complex (82), in which each Mo atom has one double bonded and two bridged sulfur atoms, was reacted with two moles of RNC. The isocyanide molecules bridge two pairs of sulfur atoms equally.

$[RN=CS_2]^{2-}$ $[Mo(\eta^5\text{-}C_5H_5)(S_2C=NR(S)SH]_2$ $[Mo(\eta^5\text{-}C_5H_5)]_2$

(80)   (81)   (82)

The cyanodithiocarbimato complexes (R = CN) have been prepared with the metals Ni, Co, V and Sn.[1,40] Upon sulfur addition to the bis(dithiocarbimato) nickel ion a perthiocarbimato compound is formed.[39]

16.4.3.5. Ligands with the Fragment OC(S)S or OC(O)S

Free xanthic acids (83), also referred to as *O*-alkyl- and *O*-aryl-dithiocarbonic acids, are not well-characterized compounds. They are strong acids and stable only in an extremely pure state. The isomeric acids (84) are unknown.

H[SC(S)OR]   H[SC(O)SR]

(83)   (84)

Alkyl xanthates (Rxan) of most elements have been known for a long time.[1,40] Aryl xanthates have been prepared *via* potassium xanthate (of 2-substituted phenols only) and thallium aryl xanthates.[92]

The ligand field strength of the ligands is between that of the dithiocarbamates and water.[49] IR studies show characteristic bands near 1250, 1100, 1020 and $550 \, cm^{-1}$.[93] The contribution of the resonance form (85) in transition metal complexes is less than that of the analogous structure in the dithiocarbamates.[60] The electron density on the metal is not very high, which accounts for the fact that abnormal high oxidation states are exceptional and a strong interaction of bases with the square planar nickel (and some other metal) xanthates is found.[94]

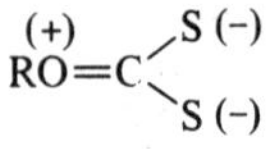

(85) $Ni(Etxan)_2$

(86)

Electrochemical measurements show that the oxidation and reduction of (86) are irreversible, but that a new, reversible couple is present after addition of K(Etxan). This couple (at $+0.265 \, V$) is suggested to be as shown in reaction (12).[95]

$Ni(Etxan)_3^- \longrightarrow Ni(Etxan)_3 + e^-$ (12)

As very few examples of metal haloxanthates are known, the preparation of copper(II) halo-xanthates is worthy of mention.[96] Pyrolytic decomposition of metal xanthates has been found to produce alkenes.[97]

Upon reaction of PR_3 with Pd and Pt bisxanthato complexes, complexes (87) are formed. The same reaction is found for other dithio acids. With excess of phosphine a dithiocarbonato complex is formed according to the pathway suggested in reaction (13).[98]

$$M(PR_3)(R'xan)_2 \qquad Mo_2(Rxan)_4I_2$$
$$(87) \qquad\qquad (88) \qquad\qquad\qquad (89)$$

$$M(R'xan)_2 \xrightarrow{PR_3} M(PR_3)(R'xan)_2 \xrightarrow{PR_3} [M(PR_3)_2(R'xan)](R'xan)$$

$$\longrightarrow M(PR_3)_2(S_2CO) + R'S_2COR'$$

$$[M(PR_3)(R'xan)](R'xan) \xrightarrow{Et_2O} M(PR_3)(R'xan)_2 \qquad (13)$$

The structure of (88) is remarkable; two of the xanthato ligands are bonded in a normal, terminal way whereas the other two ligands are terminally bonded to one molybdenum and η^3 bonded to the other other molybdenum atom as in (89).[99] The latter bonding type shows that a xanthate can act like an allyl ion.

16.4.3.6 Ligands with the Fragment FC(S)S

The only known complex with a fluorodithioformato ligand, compound (90), can be prepared by insertion of carbon disulfide into the Pt—F bond of (91).[100]

$$[Pt(PPh_3)_2(S_2CF)][HF_2] \qquad [Pt(PPh_3)_2F][HF_2]$$
$$(90) \qquad\qquad\qquad (91)$$

16.4.3.7 Ligands with the Fragment PC(S)S, AsC(S)S or PC(O)S

Unsubstituted phosphino- and arsino-dithioformic acids and their dithioformates have not yet been described. In addition, the monosubstituted primary alkyl and aryl compounds have not been investigated very well. Disubstituted phosphinodithioformates can be prepared easily. Because of the difference in basicity of Ph_2PH and Et_2PH they react with CS_2 in different ways (reactions 14 and 15). By addition of metal ions to the resulting ethyl compound, one CS_2 is split off.[1] Only the potassium complexes have a second CS_2 molecule bonded to the phosphorus atom.

$$PPh_2H + CS_2 \longrightarrow H[S_2CPPh_2] \qquad (14)$$

$$PEt_2H + 2CS_2 \longrightarrow H[(S_2C)_2PEt_2] \qquad (15)$$

Reactions of secondary arsines with CS_2 proceed in the same way as those of the corresponding phosphines. The phosphino- and arsino-dithioformates coordinate *via* one sulfur atom and the phosphorus or arsenic atom. ^{1}H NMR, IR and UV–visible spectra show that the second sulfur atom of the CS_2 fragment is exocyclic.[101,102] Reaction of (92) with (93) gives hexacoordinated (94) with the evolution of 1 mol of CO.[103]

Transition metal thiophosphinyl- and thioarsinyl-dithioformates are formed when sulfur is added to a dioxane solution of the corresponding phosphino- or arsino-dithioformates.[104] The transition metals coordinate to one sulfur of the CS_2 unit, with the sulfur atom bonded to the phosphorus or

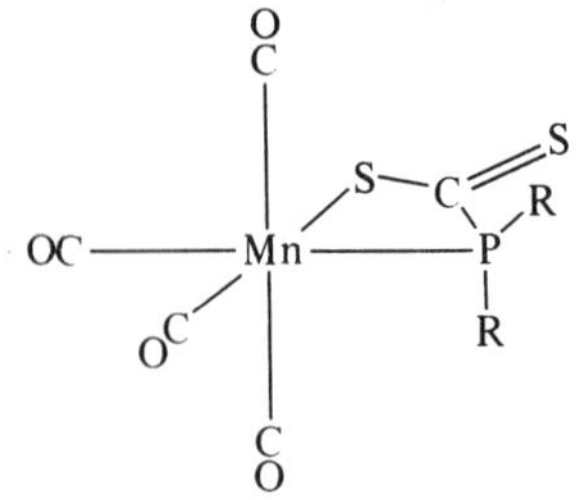

MnBr(CO)$_5$ K[S$_2$CPR$_2$]

(92)   (93) R = Ph, cyclohexyl

(94)

arsenic atom (forming a five-membered ring) and the second sulfur atom of the CS$_2$ unit being exocyclic to that ring.

Tertiary phosphines react with CS$_2$, giving the compounds (95). X-Ray structure analysis of the ethyl compound[105] reveals the phosphorus atom to be bonded to four carbon atoms in a nearly tetrahedral coordination. The carbon atom of the CS$_2$ fragment is coplanar with the phosphorus and the two sulfur atoms. The compounds react with MeI yielding (96).[1] Well-characterized complexes of this zwitterion are known with the metals Fe, Ir, Ru, Cr, Mo, Pd and Ag.[17,106,107] An interesting compound of Mn^I is given in ref. 108.

R$_3$PCS$_2$ [R$_3$PC(S)SMe]I   SnPh$_3$(PPh$_2$)

(95)   (96)   (97)

Neither stibino- nor bismuthino-dithioformic acid, nor their dithioformates, are known. A triphenyltin complex of diphenylphosphinomonothioformate has been prepared by Schmidt *et al.* from COS and (97).[109]

16.4.3.8 Ligands with the Fragment SC(S)S or SC(O)S

Thioxanthato (alkyl- or aryl-trithiocarbonato) (RSxan) complexes are known for a few metals only. The methods of preparation are mentioned in ref. 1. The ligand field strength does not differ much from that of the xanthates.[110] IR spectra give strong carbon–sulfur stretching vibrations in the regions 990–980 and 950–940 cm^{-1}.[1] The contribution of the resonance form (98) is low.[60] Electrochemical studies show extended redox series, *e.g.* (99)[111] [compare with dithiocarbamates (100)].[112]

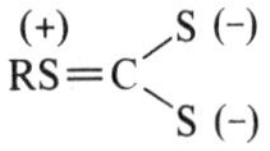

$$\underset{(98)}{\overset{(+)}{RS}=C\overset{S\,(-)}{\underset{S\,(-)}{\diagdown}}}$$

Mo(RSxan)$_4^n$ ($n = -2$ to $+1$)   Mo(R$_2$dtc)$_4^n$ ($n = -1$ to $+2$)   MoO(Pr^{i}Sxan)$_2$

(99)   (100)   (101)

Characteristic of the heavy metal complexes is their carbon disulfide elimination, yielding mixed thiol–thioxanthato complexes with trimeric structures and thiol bridges.[1,39] The facile nucleophilic substitution of RS$^-$ by R$_2$N$^-$ is interesting.[113,114]

Although the normal bonding type of thioxanthates is bidentate through two sulfur atoms, the structure of (101) reveals η^3 coordination of the CS$_2$ fragment of one of the ligands.[115] Trithio-(CS$_3^{2-}$) and dithio-carbonato (CS$_2$O^{2-}) complexes have been reviewed by Gattow[1] and Coucouvanis.[40]

Noteworthy is the structure of complex (102), which has a square planar arrangement of copper atoms with the dppm ligands bridging the edges. Each trithiocarbonate anions is bonded to the four copper atoms through two sulfur atoms.[116] Other types of bridging trithiocarbonate anions that are found are (103),[14,117] (104)[14] and (105).[118] A combination of thioxanthate and trithiocarbonate ligands is found in complex (106).[119] Insertion of sulfur in the C—S bond of a trithiocarbonate leads to the formation of a perthiocarbonate (CS$_4^{2-}$).[120]

Cu$_4$(dppm)$_4$(CS$_3$)$_2$

(102)   (103) 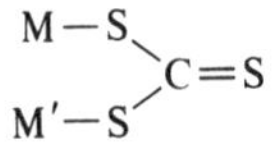   (104)    (105)

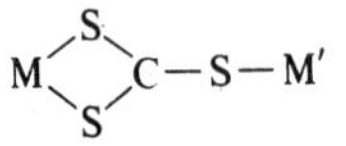

[Et$_4$N]$_2$[Fe(CS$_3$)(SEt)(S$_2$CSEt)]$_2$

(106)

16.4.3.9 Ligands with the Fragment ClC(S)S

Apart from one fluorodithioformato complex (see Section 16.4.3.6), the only halogenodithioformates known are the compounds (107) in which M = Na, K, Rb or Cs. They are formed from the corresponding alkali metal chlorides and CS_2 in acetonitrile, in the presence of solid NaOH which functions, according to the authors, as a catalyst. The yellow, unstable compounds are well characterized.[121]

$$M[S_2CCl]$$

(107)

16.4.4 COUPLING PRODUCTS OF COMPOUNDS WITH CSS AND COS FRAGMENTS AS LIGANDS

A general feature of thiols is their oxidizability to disulfides. This holds equally well for dithio ligands. Products containing the fragment (108), in which x can be 1, 2 or greater than 2, are obtained by suitable reactions. As the two doubly bonded sulfur atoms provide two coordination places, they can act as a ligand. The compounds with $x = 2$ form seven-membered rings in which the metal is included, for instance with the Zn group metals.

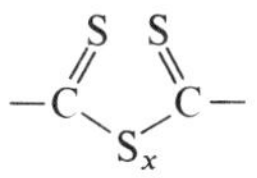

(108)	(109)	(110)	(111)	(112)

with structure (108) shown as $-C(=S)-S_x-C(=S)-$ and:
$CuBr_2$ (109), $Cu(R_2dtc)Br_2$ (110), $NiBr_2$ (111), $[Ni(R_2dtc)_3]Br$ (112)

Reverse reactions, in which the disulfides are reduced and S—S bond breaking takes place, are also possible. So thiuram disulfide reacts with elemental Cu or Ag yielding the metal dithiocarbamates, and (109) can be oxidized with thiuram disulfide to (110) and (111) to (112).

The complexes of thiuram disulfide can be oxidized by an excess of halogens to products containing the bis(iminio)trithiolane dication (113).[122,123] One of the sulfur atoms is removed as free sulfur or as sulfur halide. According to some reports the bis(iminio)tetrathiolane dication (114) also exists.[124] In our investigations, however, we could not obtain any proof of this ion. The electrochemical literature about the oxidation of thiuram disulfide is also contradictory.[125,126] Apart from the trithiolane ion a dithietane ion (115) is prepared by oxidation of dimethylformamide with bromine.[127] As far as we know, no other disulfide can be oxidized to an analogous dication.

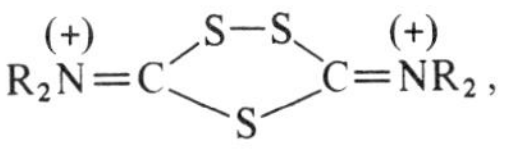

(113)   (114)

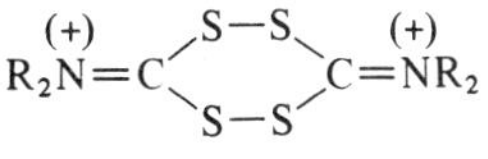

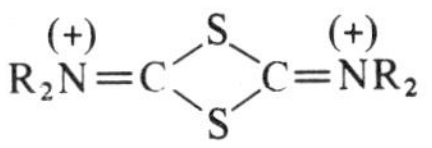

(115)

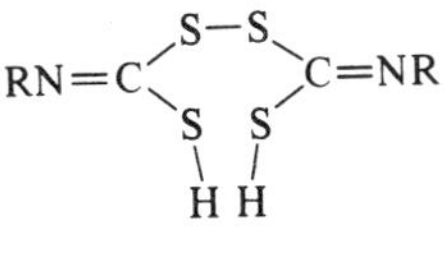

(116)

The ion (113) is an oxidant in several reactions.[128] In addition to the normal thiuram disulfides, isothiuram disulfide (116) also exists.[129] However, no complexes are known in either a neutral or (for R′ = H) a deprotonated form. Numerous other ring systems exist in which a CS_2 fragment is incorporated, *e.g.* rhodanines and thiazoles.[1] They can act as neutral ligands in coordination compounds.

16.4.5 REFERENCES

1. G. Gattow and W. Behrendt, 'Carbon Sulfides and Their Inorganic and Complex Chemistry' (Topics in Sulfur Chemistry, ed. A. Senning, vol. 2), Thieme, Stuttgart, 1977.
2. E. E. Reid, *Org. Chem. Bivalent Sulfur*, 1964, **4**, 386.
3. J. A. Ibers, *Chem. Soc. Rev.*, 1982, **11**, 57.
4. D. Paquer, *Bull. Soc. Chim. Fr.*, 1975, 1439.

5. C. Bianchini and A. Meli, *J. Chem. Soc., Dalton Trans.*, 1983, 2419.
6. R. D. Adams, *Acc. Chem. Res.*, 1983, **16**, 67 .
7. I. S. Butler and A. E. Fenster, *J. Organomet. Chem.*, 1974, **66**, 161.
8. P. V. Yanev, *Coord. Chem. Rev.*, 1977, **23**, 183; H. Werner, *Coord. Chem. Rev.*, 1982, **49**, 165.
9. M. G. Mason, P. N. Swepton and J. A. Ibers, *Inorg. Chem.*, 1983, **22**, 411.
10. (a) R. Mason and A. I. M. Rae, *J. Chem. Soc. (A)*, 1970, 1767; (b) C. Meally, R. Hoffmann and A. Stockis, *Inorg. Chem.*, 1984, **23**, 56.
11. D. H. M. W. Thewissen, *J. Organomet. Chem.*, 1980, **188**, 211.
12. J. Fortune and A. R. Manning, *Organometallics*, 1983, **2**, 1719.
13. H. Werner, O. Kolb, R. Feser and U. Schubert, *J. Organomet. Chem.*, 1980, **191**, 283.
14. M. Kubota and C. R. Carey, *J. Organomet. Chem.*, 1970, **24**, 491.
15. I. B. Benson, J. Hunt, S. A. R. Knox and V. Oliphant, *J. Chem. Soc., Dalton Trans.*, 1978, 1240.
16. G. Steimecke, R. Kirmse and E. Hoyer, *Z. Chem.*, 1957, **15**, 28.
17. C. Bianchini, C. A. Ghilardi, A. Meli, S. Midolini and A. Orlandini, *J. Chem. Soc., Chem. Commun.*, 1983, 753.
18. M. C. Baird, G. Hartwell and G. Wilkinson, *J. Chem. Soc. (A)*, 1967, 2037.
19. T. Mizuta, T. Suzuki and T. Kwan, *Nippon Kagaku Zasshi*, 1967, **88**, 573 (*Chem. Abstr.*, 1967, **67**, 39 679 m).
20. J. E. Ellis, R. W. Fennel and E. A. Flom, *Inorg. Chem.*, 1976, **15**, 2031.
21. L. Busetto, M. Monari, A. Palazzi, V. Albano and F. Demartin, *J. Chem. Soc., Dalton Trans.*, 1983, 1849.
22. H. Stolzenberg and W. P. Fehlhammer, *J. Organomet. Chem.*, 1982, **235**, C7.
23. G. Gervasio, R. Rosetti, P. L. Stanghellini and G. Bor, *Inorg. Chem.*, 1982, **21**, 3781.
24. G. Gervasio, R. Rosetti, P. L. Stanghellini and G. Bor, *J. Chem. Soc., Dalton Trans.*, 1983, 1613.
25. P. L. Stanghellini, G. Gervasio, R. Rosetti and G. Bor, *J. Organomet. Chem.*, 1980, **187**, C37.
26. G. Bor, G. Gervasio, R. Rosetti and P. L. Stanghellini, *J. Chem. Soc., Chem. Commun.*, 1978, 841.
27. T. R. Gaffney and J. A. Ibers, *Inorg. Chem.*, 1982, **21**, 2854.
28. E. Uhlig and W. Poppitz, *Z. Chem.*, 1979, **19**, 191.
29. T. R. Gaffney and J. A. Ibers, *Inorg. Chem.*, 1982, **21**, 2860.
30. T. R. Gaffney and J. A. Ibers, *Inorg. Chem.*, 1982, **21**, 2851.
31. T. R. Gaffney and J. A. Ibers, *Inorg. Chem.*, 1982, **21**, 2857.
32. R. Eisenberg and D. E. Hendriksen, *Adv. Catal.*, 1979, **28**, 119.
33. M. Aresta, C. F. Nobile, V. G. Albano, E. Forni and M. Manassero, *J. Chem. Soc., Chem. Commun.*, 1975, 636.
34. G. S. Bristow, P. B. Hitchcock and M. F. Lappert, *J. Chem. Soc., Chem. Commun.*, 1981, 1145.
35. G. Fachinetti, C. Floriani and P. F. Zanazzi, *J. Am. Chem. Soc.*, 1978, **100**, 7405.
36. J. N. Francis and M. F. Hawthorne, *Inorg. Chem.*, 1971, **10**, 594.
37. L. H. Long, *Adv. Inorg. Chem. Radiochem.*, 1974, **16**, 201.
38. S. Scheithauer and R. Mayer, 'Thio- and Dithiocarboxylic Acids and Their Derivatives' (Topics in Sulfur Chemistry, ed. A. Senning, vol. 4), Thieme, Stuttgart, 1979.
39. D. Coucouvanis, *Prog. Inorg. Chem.*, 1970, **11**, 233.
40. D. Coucouvanis, *Prog. Inorg. Chem.*, 1979, **26**, 301.
41. R. Eisenberg, *Prog. Inorg. Chem.*, 1970, **12**, 295.
42. V. V. Savant, J. Gopalakrishnan and C. C. Patel, *Inorg. Chem.*, 1970, **9**, 748.
43. R. Hesse and U. Aava, *Acta Chem. Scand.*, 1970, **24**, 1355.
44. P. Jennische and R. Hesse, *Acta Chem. Scand.*, 1971, **25**, 423.
45. R. Hesse, *Ark. Kemi*, 1963, **20**, 481.
46. W. Stork and R. Mattes, *Angew. Chem.*, 1975, **87**, 452.
47. J. J. Maj, A. D. Rae and L. F. Dahl, *J. Am. Chem. Soc.*, 1982, **104**, 4278.
48. R. D. Bereman, M. L. Good, J. Buttone and P. Savino, *Polyhedron*, 1982, **1**, 187.
49. L. E. McCandlish, E. C. Bissell, D. Coucouvanis, J. P. Fackler, Jr. and K. Knox, *J. Am. Chem. Soc.*, 1968, **90**, 7357.
50. H. Dietrich, *Acta Crystallogr., Sect. A*, 1978, **34**, S126.
51. P. J. M. W. L. Birker and G. C. Verschoor, *J. Chem. Soc., Chem. Commun.*, 1981, 322.
52. D. Seyfert, G. B. Womack and L. C. Song, *Organometallics*, 1983, **2**, 776.
53. J. March, 'Advanced Organic Chemistry: Reactions, Mechanisms, and Structure', 2nd edn., McGraw-Hill, New York, 1977, p. 824.
54. J. Willemse, W. P. Bosman, J. H. Noordik and J. A. Cras, *Recl. Trav. Chim. Pays-Bas*, 1983, **102**, 477.
55. F. G. Moers, J. W. M. Goossens and J. P. M. Langhout, *J. Inorg. Nucl. Chem.*, 1973, **35**, 855.
56. J. Willemse, J. A. Cras, J. J. Steggerda and C. P. Keyzers, *Struct. Bonding (Berlin)*, 1976, **28**, 83.
57. J. J. Steggerda, J. A. Cras and J. Willemse, *Recl. Trav. Chim. Pays-Bas*, 1981, **100**, 41.
58. J. P. Fackler, Jr., *Prog. Inorg. Chem.*, 1976, **21**, 55.
59. A. M. Bond and R. L. Martin, *Coord. Chem. Rev.*, 1984, **54**, 23.
60. D. F. Lewis, S. J. Lippard and J. A. Zubieta, *Inorg. Chem.*, 1972, **11**, 823.
61. A. M. Bond, A. R. Hendrickson, R. L. Martin, J. E. Moir and D. R. Page, *Inorg. Chem.*, 1983, **22**, 3440.
62. H. L. M. van Gaal and J. G. M. van der Linden, *Coord. Chem. Rev.*, 1982, **47**, 41.
63. R. D. Bereman, M. R. Churchill and D. Nalewajek, *Inorg. Chem.*, 1979, **18**, 3112.
64. R. S. Herrick, S. J. Nieter Burgmayer and J. L. Templeton, *Inorg. Chem.*, 1983, **22**, 3275.
65. L. H. Pignolet, *Top. Curr. Chem.*, 1975, **56**, 91.
66. J. A. McCleverty, N. J. Morrison, N. Spencer, C. C. Ashworth, N. A. Bailey, M. R. Johnson, J. M. A. Smith, B. A. Tabbiner and C. R. Taylor, *J. Chem. Soc., Dalton Trans.*, 1980, 1945.
67. A. Nieuwpoort, A. H. Dix, P. A. T. W. Porskamp and J. G. M. van der Linden, *Inorg. Chim. Acta*, 1979, **35**, 221.
68. P. T. Beurskens, H. J. A. Blaauw, J. A. Cras and J. J. Steggerda, *Inorg. Chem.*, 1968, **7**, 805.
69. A. R. Hendrickson, J. M. Hope and R. L. Martin, *J. Chem. Soc., Dalton Trans.*, 1976, 2032.
70. L. H. Pignolet and S. H. Wheeler, *Inorg. Chem.*, 1980, **19**, 935.
71. R. J. Butcher, J. R. Ferraro and E. Sinn, *Inorg. Chem.*, 1976, **15**, 2077.
72. B. Hutchinson, P. Neill, A. Finkelstein and J. Takemoto, *Inorg. Chem.*, 1981, **20**, 2000.
73. J. A. Cras and J. Willemse, *Recl. Trav. Chim. Pays-Bas*, 1978, **97**, 28.

74. Monsanto Co., *Br. Pat.* 956460 (1964) (*Chem. Abstr.*, 1964, **61**, 8196g).
75. B. J. McCormick, R. Bereman and D. Baird, *Coord. Chem. Rev.*, 1984, **54**, 99.
76. K. R. Kunze, D. L. Perry and L. J. Wilson, *Inorg. Chem.*, 1977, **16**, 594.
77. A. R. Crosby, R. J. Magee, K. N. Tantry and C. N. R. Rao, *Proc. Indian Acad. Sci., Sect. A*, 1979, **88**, 393.
78. E. W. Abel and M. O. Dunster, *J. Chem. Soc., Dalton Trans.*, 1973, 98.
79. R. D. Bereman, D. M. Baird, J. Bordner and J. R. Dorfman, *Polyhedron*, 1983, **2**, 25.
80. B. F. Hoskins and C. D. Pannan, *Inorg. Nucl. Chem. Lett.*, 1974, **10**, 229.
81. J. G. M. van der Linden, W. Blommerde, A. H. Dix and F. W. Pijpers, *Inorg. Chim. Acta*, 1977, **24**, 261.
82. R. D. Bereman, D. M. Baird and C. G. Moreland, *Polyhedron*, 1983, **2**, 59.
83. D. L. Perry and S. R. Cooper, *J. Inorg. Nucl. Chem.*, 1980, **42**, 1356.
84. G. Nagao, K. Tanaka and T. Tanaka, *Inorg. Chim. Acta*, 1980, **42**, 43.
85. D. L. Perry, L. J. Wilson, K. R. Kunze, L. Maleki, P. Deplano and E. F. Trogue, *J. Chem. Soc., Dalton Trans.*, 1981, 1294.
86. K. Tanaka, Y. Miya-Uchi and T. Tanaka, *Inorg. Chem.*, 1975, **14**, 1545.
87. H. Hlawatschek and G. Gattow, *Z. Anorg. Allg. Chem.*, 1983, **502**, 11.
88. M. K. Cooper, J. M. Downes, K. Henrick, H. J. Goodwin and M. McPartlin, *Inorg. Chim. Acta*, 1983, **76**, L159.
89. Z. Dori and R. F. Ziolo, *Chem. Rev.*, 1973, **73**, 247.
90. F. Sato, M. Etoh and M. Sato, *J. Organomet. Chem.*, 1972, **37**, C51.
91. D. L. Dubois, W. K. Miller and M. Rakowsi-Dubois, *J. Am. Chem. Soc.*, 1981, **103**, 3429.
92. J. P. Fackler, Jr., D. P. Schussler and H. W. Chen, *Synth. React. Inorg. Metal-Org. Chem.*, 1978, **8**, 27.
93. V. L. Agarwala and P. B. Rao, *Inorg. Chim. Acta*, 1968, **2**, 337.
94. D. Coucouvanis and J. P. Fackler, Jr., *Inorg. Chem.*, 1967, **6**, 2047.
95. A. R. Hendrickson, R. L. Martin and N. M. Rohde, *Inorg. Chem.*, 1975, **14**, 2980.
96. R. W. Gable and G. Winter, *Inorg. Nucl. Chem. Lett.*, 1980, **16**, 9.
97. J. P. Fackler, Jr., W. C. Seidel and Sister M. Myron, *Chem. Commun.*, 1969, 1133.
98. J. M. C. Alison and T. A. Stephenson, *J. Chem. Soc., Dalton Trans.*, 1973, 254.
99. F. A. Cotton, M. W. Extine and R. H. Niswander, *Inorg. Chem.*, 1978, **17**, 692.
100. J. A. Evans, M. J. Hacker, R. D. W. Kemmit, D. R. Russell and J. Stocks, *J. Chem. Soc., Chem. Commun.*, 1972, 72.
101. J. Kopf, R. Lenck, S. N. Olafsson and R. Kramolowski, *Angew. Chem., Int. Ed. Engl.*, 1976, **15**, 768.
102. A. W. Gal, J. W. Gosselink and F. A. Vollenbroek, *J. Organomet. Chem.*, 1977, **142**, 357.
103. K. G. Steinhauser, W. Klein and R. Kramolowski, *J. Organomet. Chem.*, 1981, **209**, 355.
104. P. August, Thesis, Hamburg, 1973; H. Burmeister, Thesis, Hamburg, 1973.
105. T. N. Margulis and D. H. Templeton, *J. Am. Chem. Soc.*, 1961, **83**, 995.
106. C. Bianchini, C. A. Ghilardi, A. Meli, S. Midollini and A. Orlandini, *Organometallics*, 1982, **1**, 778.
107. C. Bianchini, C. A. Ghilardi, A. Meli, A. Orlandini and G. Scapacci, *J. Chem. Soc., Dalton Trans.*, 1983, 1969.
108. F. W. Einstein, E. Enwall, N. Flitcroft and J. M. Leach, *J. Inorg. Nucl. Chem.*, 1972, **34**, 885.
109. H. Schumann, P. Jutzi and M. Schmidt, *Angew. Chem.*, 1965, **77**, 812.
110. D. F. Lewis, S. J. Lippard and J. A. Zubieta, *J. Am. Chem. Soc.*, 1972, **94**, 1563.
111. J. Hyde and J. A. Zubieta, *J. Inorg. Nucl. Chem.*, 1977, **39**, 289.
112. A. Nieuwpoort and J. J. Steggerda, *Recl. Trav. Chim. Pays-Bas*, 1976, **95**, 250.
113. J. P. Fackler, Jr. and W. C. Seidel, *Inorg. Chem.*, 1969, **8**, 1631.
114. R. A. Winograd, D. L. Lewis and S. J. Lippard, *Inorg. Chem.*, 1975, **14**, 2601.
115. J. Hyde, K. Venkatasubramanian and J. A. Zubieta, *Inorg. Chem.*, 1978, **17**, 414.
116. A. M. M. Lanfredi, A. Tiripicchio, A. Camus and N. Marsich, *J. Chem. Soc., Chem. Commun.*, 1983, 1126.
117. W. P. Fehlhammer, A. Mayr and H. Stolzenberg, *Angew. Chem.*, 1979, **91**, 661.
118. G. Thiele, G. Liehr and E. Lindner, *J. Organomet. Chem.*, 1974, **70**, 427.
119. G. W. Henkel, W. Simon, H. Strasdeit and B. Krebs, *Inorg. Chim. Acta*, 1983, **70**, 29.
120. D. Coucouvanis and M. Draganjac, *J. Am. Chem. Soc.*, 1982, **104**, 6820.
121. B. Sturm and G. Gattow, *Z. Anorg. Allg. Chem.*, 1983, **502**, 7.
122. J. Willemse, J. A. Cras and P. J. H. A. M. van de Leemput, *Inorg. Nucl. Chem. Lett.*, 1976, **12**, 255.
123. P. T. Beurskens, W. P. J. H. Bosman and J. A. Cras, *J. Cryst. Mol. Struct.*, 1972, **2**, 183.
124. E. W. Ainscough and A. M. Brodie, *J. Chem. Soc., Dalton Trans.*, 1977, 565.
125. G. Cauquis and A. Deronzier, *J. Chem. Soc., Chem. Commun.*, 1978, 809.
126. C. Scrimager and L. J. Dehayes, *Inorg. Nucl. Chem. Lett.*, 1978, **14**, 125.
127. W. Walter and R. F. Becker, *Liebigs Ann. Chem.*, 1972, **755**, 145.
128. J. Willemse and J. A. Cras, *Recl. Trav. Chim. Pays-Bas*, 1972, **91**, 1309.
129. J. von Braun, *Ber. Bunsenges. Phys. Chem.*, 1902, **35**, 827.
130. T. G. Southern, U. Oehmichen, J. Y. Le Marouille, H. Le Bozec, D. Grandjean and P. H. Dixneuf, *Inorg. Chem.*, 1980, **19**, 2976.
131. U. Oehmichen, T. G. Southern, H. Le Bozec and P. Dixneuf, *J. Organomet. Chem.*, 1978, **156**, C29.
132. H. Werner, K. Leonard and C. Burschka, *J. Organomet. Chem.*, 1978, **160**, 291.
133. C. Bianchini, C. Mealli, A. Meli, A. Orlandini and L. Sacconi, *Inorg. Chem.*, 1980, **19**, 2968.
134. M. Herberhold, M. Suesz-Fink and C. J. Kreiter, *Angew. Chem.*, 1977, **89**, 191.
135. J. M. Lisy, E. D. Dobrzynski, R. J. Angelici and J. Clardi, *J. Am. Chem. Soc.*, 1975, **97**, 656.
136. W. P. Fehlhammer, A. Mayr and H. Stolzenberg, *Angew. Chem.*, 1979, **91**, 661.
137. L. Busetto, M. Monari, A. Palazzi, V. Albano and F. Demartin, *J. Chem. Soc., Dalton Trans.*, 1983, 1849.
138. L. Busetto, A. Palazzi and M. Monari, *J. Chem. Soc., Dalton Trans.*, 1982, 1631.
139. J. E. Ellis, R. W. Fennel and E. A. Flom, *Inorg. Chem.*, 1976, **15**, 2031.
140. T. Hattig and U. Kunze, *Angew. Chem.*, 1982, **94**, 374.
141. T. S. Cameron, P. A. Gardner and K. R. Grundy, *J. Organomet. Chem.*, 1981, **212**, C19.

16.5

Dithiolenes and Related Species

ULRICH T. MUELLER-WESTERHOFF and BLAKE VANCE
University of Connecticut, Storrs, CT, USA

16.5.1 INTRODUCTION

The chemistry of the dithiolenes developed in a mode in which discoveries frequently appear to be made: although early work existed, it was largely ignored until a sudden swell in activity involving parallel efforts by several groups led the field into rapid development. Transition metal complexes with sulfur-containing ligands (thiols, for example) were not uncommon in the last century and even the usefulness of benzenedithiolate complexes for analytical purposes was known early on, but it was not until 1962 that three groups independently came to realize the unique nature of the dithiolene compounds and thereby to lead this field into active and rapidly expanding development.

The uniqueness of the dithiolenes stems from their abilty to exist in several clearly defined oxidation states, which are connected through fully reversible redox steps. Although the neutral and dianionic extremes of redox states in planar dithiolenes were discovered independently by different groups, it did not take very long until the connection was made between the results of the groups of Schrauzer,[1] Gray[2] and Davison,[3] which then provided the framework for the broader exploration of this field. The thermal and photochemical stability and the highly reversible redox chemistry of dithiolenes were recognized as indications of a type of bonding different from that which was previously known for transition metal complexes. The reason was soon found to be a high degree of electron delocalization in these systems, which also manifests itself in intense electronic transitions at unusually low energies, not unlike the absorptions seen in extended organic π-systems.

A few years after these events, a review article by McCleverty[4] in 1968 provided (in over 170 pages) the first thorough overview of this field. This review contains a multitude of details which still merits its mention as a valuable source of information on this class of compounds. Several other reviews dealing with different aspects of dithiolene chemistry have appeared in the meantime.[5-11] The field has expanded in different directions, spurred by the synthesis of several structurally unique compounds and of materials of particular use in quite diverse areas such as highly conducting molecular crystals and as infrared dyes for various applications. Dithiolenes thus represent a class of materials which are not only of significance on their own, but which also find some applications in which their unique electronic properties can be employed.

A qualitative description of the bonding in dithiolenes, as exemplified by the neutral d^8 complex $Ni(S_2C_2H_2)_2$, involves the structures (1)–(3) in which the metal assumes formal oxidation states of 0, 2+ and 4+, while the ligands are either neutral 'dithiodiketones' or dinegative 'dithiolates'.

(1)   (2)   (3)

Dithiolenes are best considered to be a resonance hybrid of the limiting structures (1)–(3). In both bis- and tris-dithiolenes the electron delocalization is not limited to the ligand, but includes the metals to give rise to cyclic delocalization ('aromaticity'). To symbolize this electron delocalization in dithiolenes, they can be represented, in a manner similar to that used for benzene, by formulas containing a ring inside the framework given by the metal, sulfur and carbon atoms. We will use this notation, shown in (4), throughout this chapter.

(4)

The name dithiolenes was chosen[4] to describe these compounds without prejudice towards one of the limiting structures. The equally descriptive name 'dithienes' has been coined[6] for the same reason, but it is now rarely used. The less fortunate description of dithiolenes as dithiolato complexes is found occasionally, but it does have a much more restricted meaning (see Section 16.5.2.4) and should be avoided for the neutral species. Nevertheless, *Chemical Abstracts* refers to dithiolenes as 'bis[1,2-ethenedithiolato(2 −)]' complexes of the respective central metal, for example the parent nickel complex (4) is listed as 'nickel, bis[1,2-ethenedithiolato(2 −)-S,S']-'; however, depending on the date of the *CA* issue, its tetraphenyl derivative will be found either under bis[α,α'-stilbenedithiolato(2 −)]-nickel or as bis[1,2-diphenyl-1,2-ethenedithiolato(2 −)-S,S']-nickel. Even less appropriate are the *CA* names for the radical anions and dianions of the dithiolenes, which are referred to as metallates(−) and metallates(2 −) of the respective ligands; the dianion of the parent nickel dithiolene thus is found as bis[1,2-ethenedithiolato(2 −)]-nickelate(2 −), a name which has little to do with the electronic structure of the compound.

To make it useful both to experts and novices in this field, this description of the dithiolenes attempts to present a compilation of the most important *recent* results, while covering the field with reasonable completeness. For this reason, only occasional mention is made — because of their factual or historical significance — of results which can be found in the earlier review literature.

It appears fair to say that the basic properties of dithiolenes have now been established with considerable thoroughness and that the field has matured. That is not to imply that new classes or new synthetic routes may not be waiting to be explored. However, the major efforts seem to be directed towards exploitation of the unique properties of dithiolenes.

In the present overview, we will use some common abbreviations for certain ligands, regardless of their oxidation state. For some less commonly encountered dithiolenes, abbreviations will be explained in the text. In comparing dithiolenes to other sulfur-containing ligand systems, we will also use the abbreviations listed below.

edt	$H_2C_2S_2$	the parent *ethylenedithiolene*
mnt	$(NC)_2C_2S_2$	*maleonitriledithiolene*
tfd	$(CF_3)_2C_2S_2$	*trifluoromethyldithiolene*
cmt	$(MeOCO)_2C_2S_2$	*carbomethoxydithiolene*
dmit	$S=CS_2C_2S_2$	trithionedithiolene (*dimercaptoisotrithione*)
bdt	$C_6H_4S_6$	*o-benzenedithiolate*
tdt	$MeC_6H_3S_2$	*toluenedithiolate*
ptt	$C_3H_3S_2$	dithiomalonaldehyde (*propenethionethiolate*)
SacSac	$CH(MeCS)_2$	dithioacetylacetonate (*Sulfur acac*)
dtc	R_2NCS_2	*dithiocarbamate*

16.5.2 SYNTHESES AND TYPES OF DITHIOLENES

The dithiolene ligands offer a variety of synthetic and structural possibilities and choices. The following description introduces in an intentionally general fashion (which some might consider to be superficial) the various classes and describes the methods used in their synthesis, while a more detailed discussion of their properties will be given in the subsequent parts of this section.

16.5.2.1 Bis- and Tris-dithiolenes

In most known bis-dithiolenes, the ligands form the strictly square planar arrangement indicated in (5), with all S—M—S bond angles very close to 90°. The planar geometry is also found in the majority of mono- and di-anions of these species, but here the dihedral angle between the two ligands can vary up to the tetrahedral extreme (see, for example, Section 16.5.2.5). Depending on the central metal, the neutral bis-dithiolenes may exist as planar monomers or form dimers with either metal–metal bonds, as shown in (6), or with metal–sulfur bonds as, for example, the dithiolene (7).

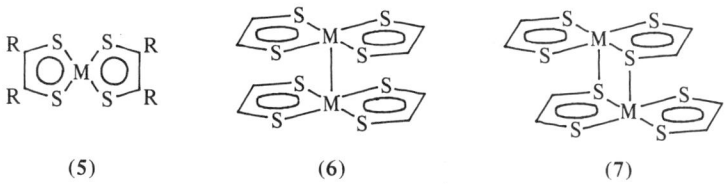

(5)   (6)   (7)

Even the simplest examples of bis-dithiolenes, the diamagnetic d^8 complexes of Ni, Pd and Pt, show different solid state properties: the parent Ni dithiolene is a planar monomer[12] of type (5), while its Pd and Pt analogs form[13] metal–metal-bonded dimers of type (6), which appear to undergo a monomer–dimer equilibrium in solution. This difference in structure may relate to their different degree of electron delocalization (see Section 16.5.3.9) and the size of the valence orbitals available for metal–metal bonding. In the dimers, the metal–metal distances are very short (2.79 and 2.75 Å, respectively); the intermolecular S—S distances are 3.03 Å and thus much shorter than the sum of sulfur van der Waals radii and considerably shorter than the intraligand S—S distances of 3.25 Å; the C_2H_2 groups bend away slightly from the central core to achieve more normal intermolecular distances.

One of the unusual features of tris-dithiolenes is their preference for trigonal prismatic rather than octahedral coordination geometries.[7] Examples of this structure are the dithiolenes (8) of Cr, Mo and W. Distortions towards octahedral structures are known to occur especially in anionic tris-dithiolenes, but the preference for trigonal prismatic arrangements is clearly established for the neutral species. The main reason is seen in favorable interligand interactions between the sulfur atoms.[7] In Mo(edt)$_3$ the intra- and inter-ligand S—S distances of 3.10 and 3.11 Å are virtually identical.[14] In the tris-dithiolene anions, the increased charge density on the S atoms leads to repulsive interactions for which the octahedral coordination is preferred.

(8)

16.5.2.2 General Synthetic Methods

Two main synthetic routes are available for the preparation of dithiolenes: in the first and most frequently used method, either the free ethylenedithiol or an appropriate salt of the ethylene-dithiolato ligand dianion is reacted with a metal salt to produce anions of the dithiolenes, which may or may not be subsequently oxidized to the neutral species; the second one, applied so far only to transition metal dithiolenes, converts vicinal diketones into dithiodiketones and reacts these either with zerovalent metals to form dithiolenes directly or uses metal salts to arrive at cationic species, which are reduced to the neutral dithiolenes either during the reaction or in a subsequent step.

The substituents R in (5) and (8) can be alkyl or aryl groups (which may be substituted by a variety of functional groups); they can be cyclic structures or they may be electron-accepting or -donating functional groups. Most of the known dithiolenes are symmetrically substituted, but there exist (see Section 16.5.2.3) a number of dithiolenes with different substituents. This flexibility in the structural choices has allowed the synthesis of a large variety of dithiolenes with tailored properties. Di-thiolenes of almost all transition metals and of many main group elements are accessible and the preparative potential seems to be limited only by the individual's ability to synthesize the desired organic ligand precursors. Once these are in hand, their conversion to dithiolenes generally follows well-established procedures. A minor limitation is the availabilty of transition metal compounds with solubility properties appropriate for reaction in nonaqueous media, but it applies only to reactions of ligands which might hydrolyze or decompose under standard aqueous reaction conditions.

16.5.2.2.1 *Syntheses using ethylenedithiolates*

In the synthesis of the parent dithiolenes, the reaction of the disodium salt of ethylene-*cis*-dithiolate with metal salts is the only useful method.[15] For the preparation of substituted dithiolenes the use of substituted ethylenedithiolates is the most efficient choice. In the synthesis of 1,2-dicyanoethylene-1,2-dithiolato ('mnt') complexes, it is the only available method.[16]

To obtain ethylenedithiols, the main options described in Scheme 1 are available. Vicinal dithio-ethers, which can be prepared from *cis*-dichloro- or dibromo-alkenes or from alkenes through a stepwise synthesis,[15,17,18] can be cleaved to produce dithiols (Route A). Ethylenedithiocarbonates can be prepared by several well-established procedures[19] and by some less general ones[20] as well. The ethylenedithiocarbonates are central intermediates for many dithiolene syntheses and for syntheses of derivatives of the electron donor tetrathiafulvalene (TTF), which has been intensively investigated in the past decade. This has widened the synthetic possibilities for ethylenedithio-carbonates.[21] Their cleavage by alkali metal hydroxides or alkoxides produces salts of ethylene-dithiolates (Route B). As an alternate route, their photolysis with loss of CO and generation of a dithioketone has been explored.[22]

Aromatic dithiols are stable and can be stored as such. 9,10-Phenanthrenedithiol, for example, is a crystalline (m.p. 134 °C), colorless and odorless solid.[23] Other dithiols (such as H_2bdt and H_2tdt) are commercially available. Dithiols of aliphatic alkenes are prone to decomposition and poly-merization.[17] They are best converted into alkali metal salts or salts of organic cations such as the tetrabutylammonium ion. These salts usually have a considerable shelf life.

16.5.2.2.2 *Syntheses using dithiodiketones*

Several 1,2-dithiones can be obtained from 1,2-diones by direct sulfurization with P_4S_{10} or B_2S_3. The usually heterogeneous reaction can be accelerated by the use of ultrasound.[24] Although most dithiones are prone to hydrolysis and decomposition (some cannot be prepared at all under the above conditions), they are most useful intermediates in the synthesis of more complex dithiolenes.

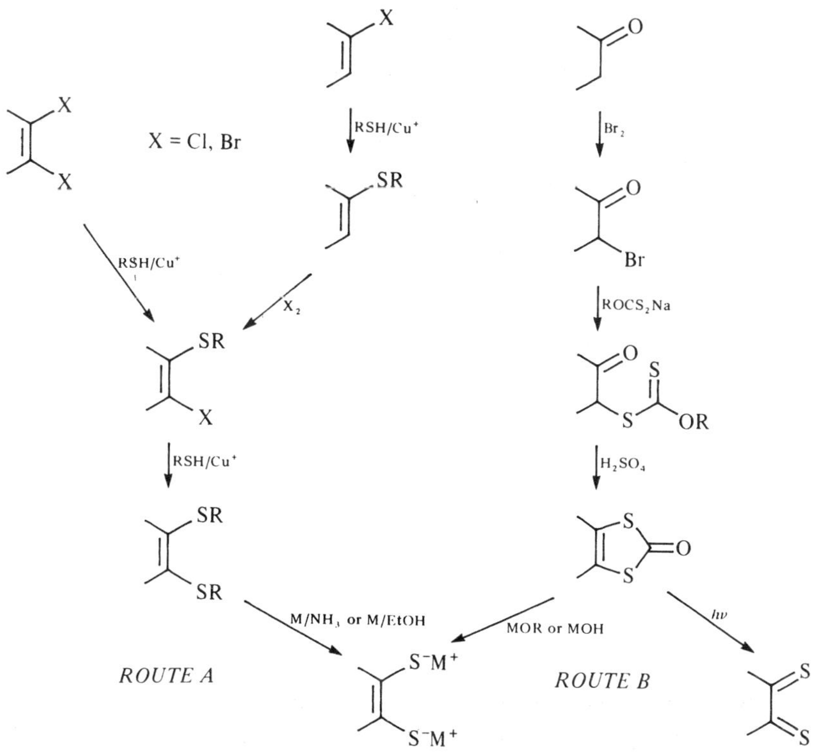

Scheme 1 Synthetic routes to ethylenedithiolato ligands

For example, the julolidine-substituted dithiolene (**9**) can be obtained[18] in low yield by standard procedures; the synthesis of the red, crystalline dithione and its conversion to (**9**) under reducing conditions proceeds in high yield.[24] Dithiones are only stable if they are substituted by strongly electron-donating or -withdrawing substituents, and their use is limited to these cases.

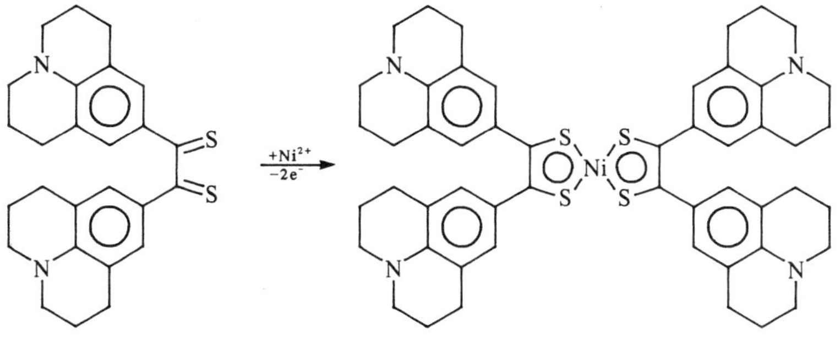

(9)

Unless steric and electronic constraints prevail, vicinal dithiodiketones (**10**) can exist in an equilibrium mixture with their dithiacyclobutene (dithiete or dithietene) isomers (**11**). Strongly electron-withdrawing substituents, such as in (**12**), appear to favor the latter form, but the donor-substituted dithione (**13**) is also known[22] to show such an equilibrium. As a *cis*-fixed dithiooxamide, the ligand (**14**) is particularly stable and presumably exists only in the dithione form.

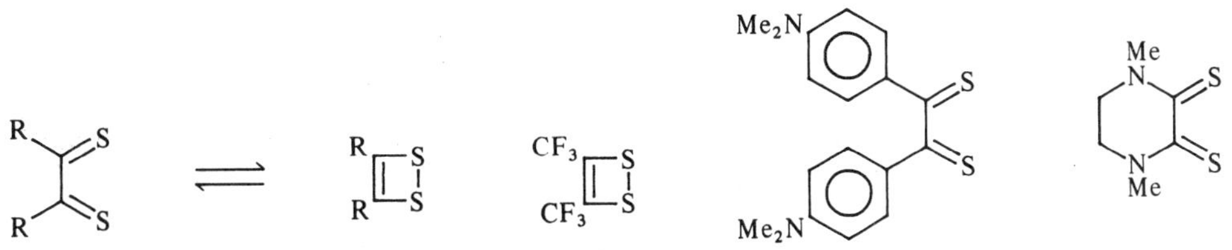

(10)   (11)   (12)   (13)   (14)

Three main options, described in Scheme 2, exist for the conversion of dithiodiketones to dithiolenes: (i) reaction with a transition metal carbonyl or other reactive zerovalent material; (ii) reduction to an ethylenedithiolate and reaction as described above for these species; and (iii) reaction with a metal salt to form a cationic dithiolene and reduction of the reaction product to the neutral compounds, whereby a suitable solvent (such as methanol) may serve the function of the reducing agent for *in situ* generation of the neutral species.

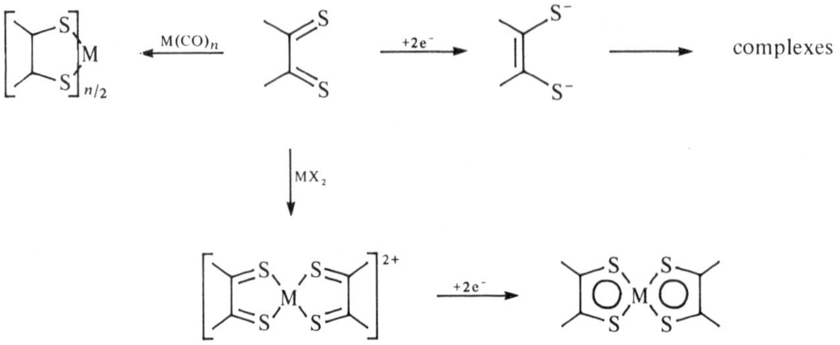

Scheme 2   Reactions of 1,2-dithiones to form dithiolenes

Dithiolene synthesis according to (i) was the first entry to the class of tris-dithiolenes: King[25] reacted the dithiete (12) with $Mo(CO)_6$ to obtain $Mo(tfd)_3$. The reduction of the julolidinedithione by $NaBH_4$ in THF–MeOH, followed by acidification and reaction with aqueous $NiCl_2$, led[24] to (9) in excellent yield. As is evident from the MO diagram of dithiolenes (see Section 16.5.3.7) and their established redox properties, method (iii) applies only to dithiolenes with strongly electron-donating substituents, which may stabilize the cationic intermediates.

16.5.2.2.3 *Other synthetic methods*

The original synthesis of alkyl- and aryl-substituted dithiolenes[26-28] by reacting acyloins and benzoins with P_4S_{10} in dioxane or toluene and treatment of this mixture with a metal salt has now been superseded by better and more reproducible methods. However, several dithiolenes have only been prepared by this method and, although alternate routes should clearly be advantageous, they apparently have not been explored. In some cases, this method produces 'sulfur rich' dithiolenes such as the compound $Mo_2S_2(S_2C_2Ph_2)_4$ (15).[28]

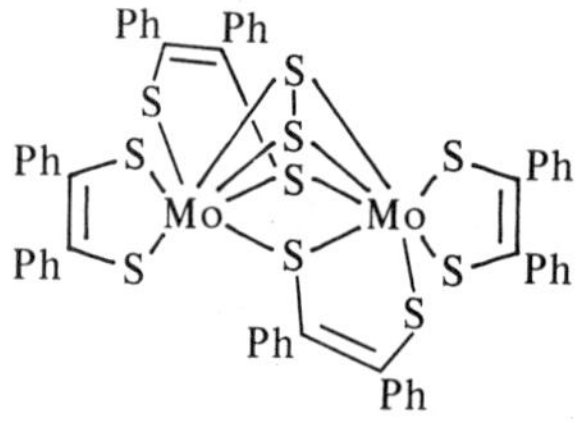

(15)

Some other syntheses of dithiolenes have also been reported,[29-33] although they generally would not qualify as preparative routes. There also exist several surprising ways of producing dithiolate complexes. Since these clearly are limited to the particular compound in question, these methods will be mentioned where such compounds are discussed.

16.5.2.3 **Unsymmetrically Substituted Bis-dithiolenes**

While the majority of dithiolene complexes are symmetric for reasons of synthetic simplicity (16; $R^1 = R^2 = R^3 = R^4$), there exist two groups of unsymmetrically substituted dithiolenes. In the first of these, exemplified by compounds (17) and (18), the ligand itself is unsymmetrically substituted, while the second group, shown in structure (19), contains two different ligands.

Dithiolenes with unsymmetrically substituted ligands are not uncommon, because such ligands can be obtained by the standard routes outlined above. In principle, they can exist either in a *cis*

$$(16) \quad R^1 = R^2 = R^3 = R^4$$
$$(17) \quad R^1 = R^4; \; R^2 = R^3 \; (cis)$$
$$(18) \quad R^1 = R^3; \; R^2 = R^4 \; (trans)$$
$$(19) \quad R^1 = R^2; \; R^3 = R^4$$

(**17**) or *trans* (**18**) arrangement, but the *trans* form seems to be generally preferred. Crystal structure determinations of these compounds are rare, but the mesomorphic nickel bis(4-*n*-octyl-phenyl)dithiolene (**20**; $R^1 = R^3 = H$, $R^2 = R^4 = 4\text{-}C_8H_{17}C_6H_4$) was shown[34] to exist only as the *trans* isomer.

The possibility of *cis–trans* equilibria exists, but has not been explored in depth. One indication for such equilibria comes from the NMR investigation of the parent nickel dithiolene in a nematic solvent: only a rapidly equilibrating compound could have produced the observed spectral pattern.[35] In the preparation of some mesomorphic dithiolenes related to the one cited above, two products were initially obtained, which during further work-up converted into the *trans* isomer only.[36] In Pt complexes of this type, the existence of two ^{195}Pt–^{1}H couplings was taken as an indication[36] that the *cis* and *trans* isomers coexisted.

Equally important in their relevance to our basic understanding of the dithiolenes are systems of type (**19**). Several examples of such compounds have been known for some time,[18,31] but this particular type of dithiolene has only recently been investigated with some degree of completeness.[38] Their novelty lies in the coexistence in one molecule of two different dithiolene ligands. The earlier work[37] had focused on the voltammetric study of ligand exchange rates in anionic Ni dithiolenes according to the equation

$$(R_2'C_2S_2)_2Ni^{2-} + (R_2''C_2S_2)_2Ni \rightarrow 2(R_2'C_2S_2)(R_2''C_2S_2)Ni^-$$

while the recent work has investigated the spectral properties.

Depending upon the nature of their substituents, these compounds may be dithiolene-like[38] (**21**) or approach a class of compounds, exemplified by (**22**), in which we see the existence of the dinegative dithiolato ligand together with the other extreme of a dithiodiketone ligand.[18] Such dithiolato complexes form a class by themselves.

Unsymmetrically substituted tris-dithiolenes appear not to have been investigated in depth, although the earliest known[39] tris-dithiolene, Mo(tdt)$_3$, falls into this category. In principle, interesting steric effects should manifest themselves in this class of dithiolene.

16.5.2.4 Dithiolato Complexes

In contrast to the dithiolenes, we refer to dithiolato complexes as those in which the typical dithiolene resonance (and the properties which are associated with it) are absent because the ligand exists entirely in its dinegative form. In the neutral compound (**22**) above, the 2− charge of the ethylenedithiolato ligand is balanced by the 2+ charge of the nickel atom and approximately square planar coordination is provided by the neutral dithiooxamide-type ligand.

Dianions of the planar dithiolenes also qualify as dithiolato complexes because they have the electron distribution shown approximately in (23), where the metal maintains the same 2+ charge as in the neutral compound, while the ligands become dianionic.

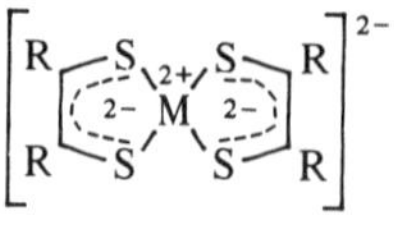

(23)

16.5.2.4.1 Dithiolato complexes with other sulfur ligands

In addition to dithiodiketones such as the one in (22), thioethers, sulfur heterocycles and dtc or SacSac ligands can be part of mono- or bi-nuclear dithiolato complexes. Examples of this type of structure are compounds (24) and (25).

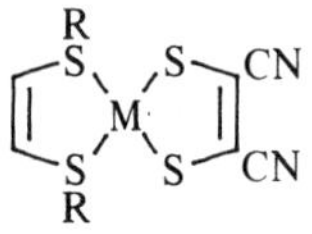

(24) 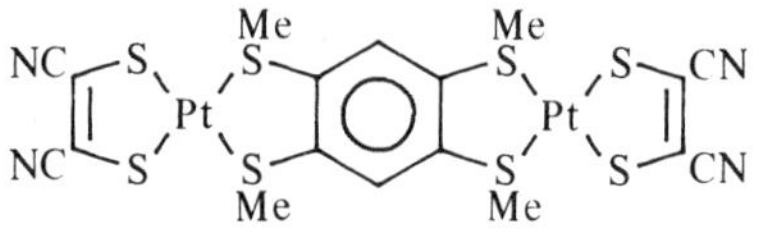   (25)

Thioether-containing dithiolato complexes can be obtained from dithiolenes. In the alkylation of dithiolene dianions, reaction occurs on the sulfur atoms, which are considered to be the centers of highest nucleophilicity. Both sulfur atoms of one ligand can be alkylated[40] to produce species of the type (24), but reaction of mnt complex dianions of Co, Ni, Cu and Rh with MeI leads to alkylation of all four S atoms, which is followed by dissociation of the complex.[41] The corresponding mono-anions appear to be much less reactive or even unreactive. Alkylation of only one sulfur atom of each ligand to give compounds of the type (26) apparently does not occur. However, an independent synthesis starting from the monoalkylated ligand anion has provided[42] both cis-(26) and trans-(26) (M = Ni, Pd, Pt) and the neutral tris-chelates of Co, Rh and Ir.

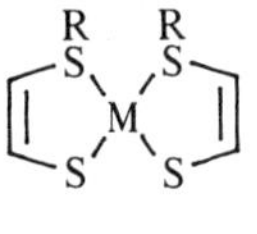

(26) *cis* 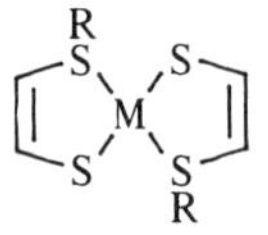   (26) *trans*

Mixed dithiocarbamato–dithiolene complexes show a number of interesting phenomena. For Au^{III} complexes containing N,N-di-n-butyl-dtc and mnt ligands, both the salt $(dtc)_2Au^+ Au(mnt)_2^-$ (27) and the neutral, square planar (dtc)Au(mnt) complex (28) have been isolated and structurally characterized.[43]

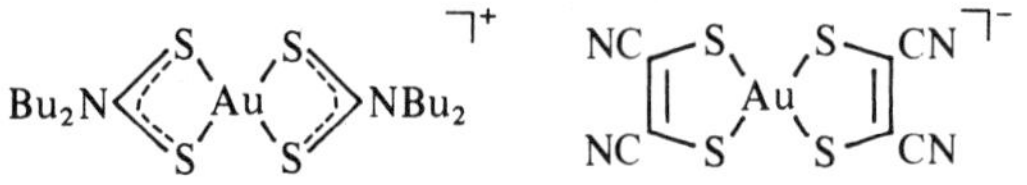

(27) 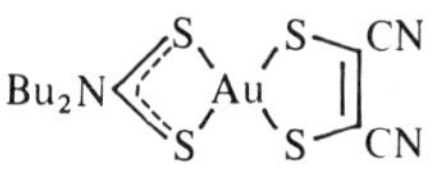   (28)

Related ligand exchange reactions have also been reported.[44] Mixed Cu and Ni species with S- and Se-containing ligands were characterized. For example, $Cu(mnt)_2^{2-}$ was reacted with neutral $Cu(dsc)_2$ (dsc = diselenocarbamate) to produce the anion of the mixed Cu(mnt)(dsc) complex (29).

In $(dtc)_2Fe(mnt)$ and $(dtc)_2Fe(tfd)$ (30) and their Ru analogs several novel observations were made:[45] the complexes are stereochemically nonrigid, they show a temperature dependent, singlet–triplet equilibrium and they can be subjected to oxidation and reduction over a series of four steps (2− to 1+), which, in the case of the Ru complexes, are fully reversible.

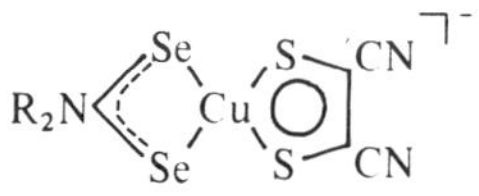

(29)

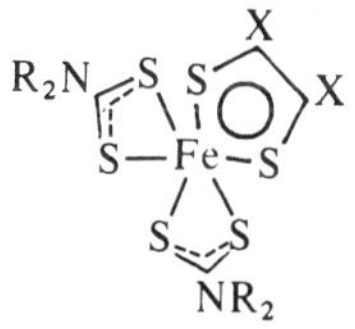

(30) X = CN, CF₃; R = alkyl, aryl

16.5.2.4.2 Dithiolato complexes with nitrogen and phosphorus ligands

Reaction of *o*-phenanthroline with the anions $Co(mnt)_2^-$ and $Co(tfd)_2^-$ leads[46] to addition and formation of octahedral complexes of type (31).

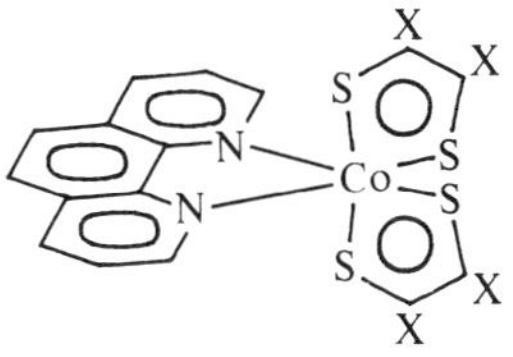

(31) X = CN, CF₃

On the other hand, the reaction of several dithiolenes with bipyridyl and related ligands was shown[47-49] to lead to dithiolato complexes of the type (32) and (33), in which the nitrogen-containing ligands have displaced one of the dithiolene ligands. The related bis-ammine complex $(NH_3)_2Pd(mnt)$ was the first dithiolato complex to be isolated.[16]

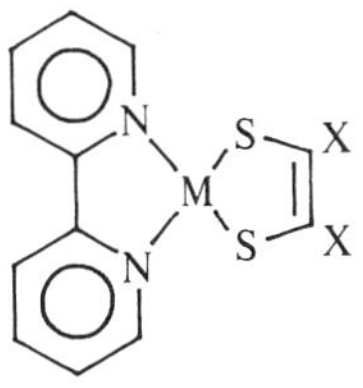

(32) 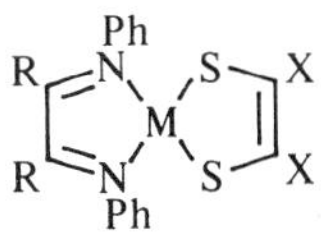   (33)

Addition of NO to $Fe(mnt)_2$ anions led to the square pyramidal adducts[50] (34). Addition of NO and displacement of a dithiolene ligand by bipyridyl affords square pyramidal Fe and Co complexes (35) of the type M(mnt)(bipy)(NO). An ionic complex [Fe(mnt)NO][Fe(bipy)NO] is also known.[51]

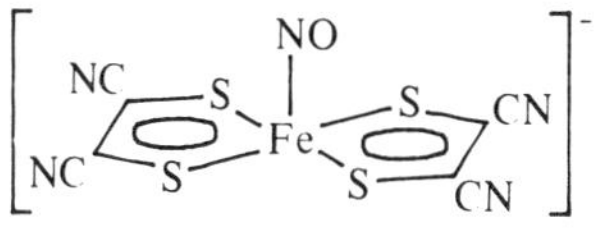

(34) 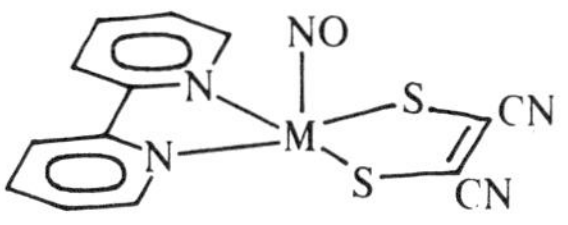   (35)

Phosphines also are able to (a) add to a vacant coordination site in planar dithiolenes or split dimers such as $[Fe(mnt)_2]_2$ to form[52] square pyramidal adducts of structure (36); and (b) displace certain dithiolene ligands to form dithiolato complexes of the general structure (37). These reactions also are briefly discussed in Section 16.5.3.8.

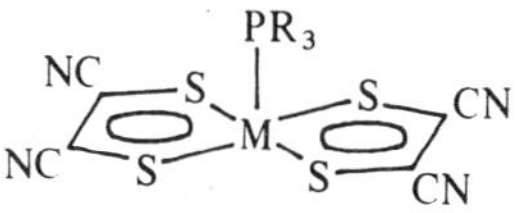

(36) 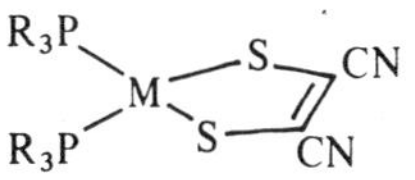   (37)

16.5.2.4.3 *Dithiolato complexes of metallocenes*

Titanocene dichloride was the first metallocene derivative to be converted[53] into dithiolato complexes, for example compound (38). Some other examples of this structural type exist, and their structural features are described in Section 16.5.3.1.

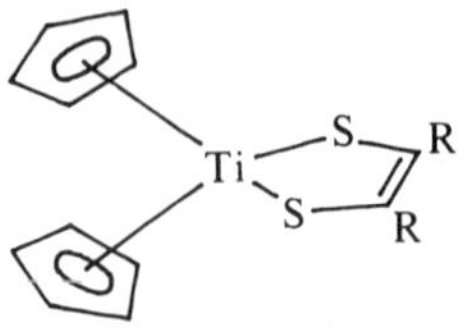

(38)

Related to this group of complexes are the transition metal half-sandwich dithiolates, almost exclusively with the mnt or tfd ligands, of the type $CpML_m(mnt)_n$, where $m = 0$–2 and $n = 1$ or 2. Examples[7] are the linear CpCo(mnt) and CpCo(tfd) complexes (39) and CpMo(tfd)$_2$ (40). The structure of the tetramethylcyclobutadieneNi(mnt) complex has been reported.[54] The synthesis and structure of Rh(cod)(Me-mnt) has also been published.[55]

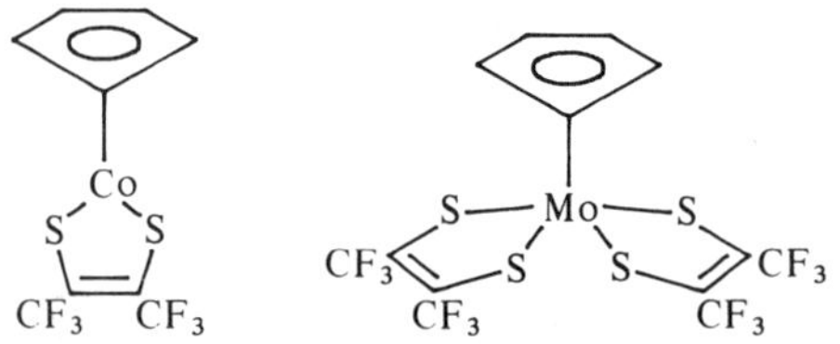

(39)   (40)

A very interesting linear dimer (41) containing two Me$_5$CpNi units bridged by a tetrathiooxalate ligand is known.[56] It was obtained by reaction of [Me$_5$CpNi(CO)]$_2$ in neat CS$_2$ through reductive head-to-head dimerization of CS$_2$.

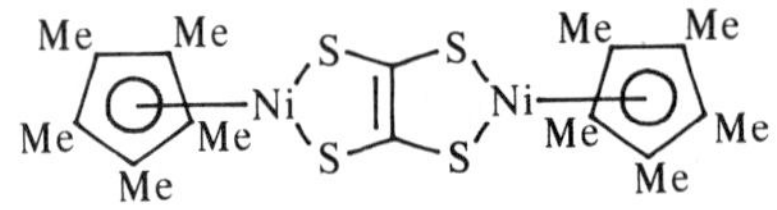

(41)

Several substituted cyclopentadienylcobalt dithiolenes were obtained[57] by a rather surprising route from CpCo carbyne clusters and elemental sulfur (Scheme 3). An astonishing solid state phenomenon was discovered[58] in the related CpCo benzenedithiolate: a dimerization reaction was observed at room temperature, which could be reversed at 150 °C (Scheme 4, see also Section 16.5.3.1).

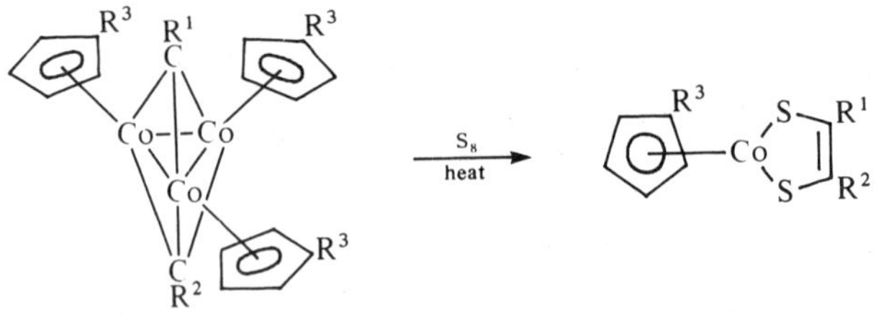

Scheme 3 CpCo dithiolenes from carbyne clusters

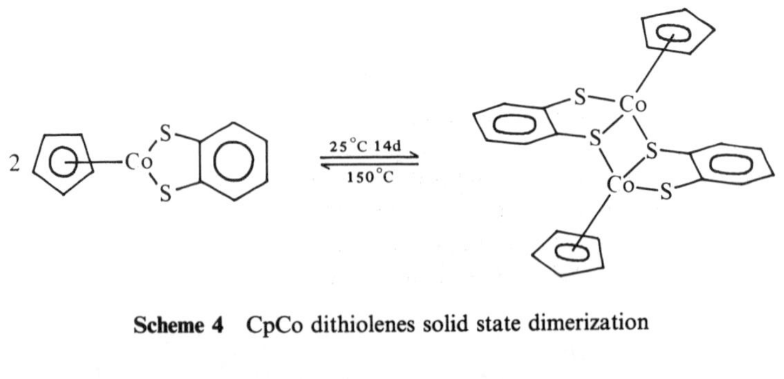

Scheme 4 CpCo dithiolenes solid state dimerization

The reaction of [CpFeS$_2$]$_2$ (**42**) with hexafluorobutyne has produced the bis-dithiolato bridged dimer (**43**), together with the mono-dithiolato and S$_2$-bridged half-reacted product of unknown structure.[59]

(42)   (43)

16.5.2.4.4 *Dithiolato complexes of the main group elements*

The dithiolene chemistry of the main group elements differs from that of the transition metals in several ways: (a) the typical dithiolene redox chemistry and IR absorptions are absent because *d*-orbitals are not involved and the ligands function as dithiolates only; (b) pure bis- or tris-dithiolato complexes are rare, since many of the known complexes also contain other ligands such as halides or nitrogen heterocycles; (c) dithiolene ligands frequently also act as bridging ligands to form extended chains; and (d) most of the known chemistry concerns the two ligands mnt and tdt.

In 1936, Mills and Clark[60] published their findings on 'new thio-salts' of Hg, Cd and Zn. They proposed a tetrahedral conformation and discovered a specific and reversible isomerization between two crystal forms. They noted that optical activity could possibly exist in these complexes and suggested it as the most probable explanation for the existence of the two modifications. The next, larger step in main group dithiolate chemistry came only in 1960, when Gilbert and Sandell[61] prepared Mo(tdt)$_3$ as the first neutral tris-dithiolate. While main groups dithiolates are still of broad interest, most of the recent work in this area has focused on structural questions.

AsCl(tdt) was found to be a molecular solid with indications of extended interactions through As—C bonds to the benzene rings of the ligand.[62] In the Ph$_4$As salt of Bi(mnt)$_2$, Hunter and Weakley[63] found an infinite linear polymer chain in which the Bi atoms are bridged by two mnt ligands, of which each provides one bridging S atom. The coordination geometry deviates from octahedral symmetry and a pentagonal bipyramid with a lone pair in one of the equatorial positions was proposed. One mnt-S forms an unsymmetrical bridge between two Bi (Bi—S distances = 2.791 and 3.210 Å); the other mnt-S forms a stronger bond with only one Bi (2.678 Å). There are four such mnt ligands in the coordination sphere of each Bi. Nonplanar chelate rings (31.6°), long intramolecular S—S distances (3.357 Å) and small chelate SBiS angles (75.7°) were other structural anomalies.

Ge and Sn tris-mnt dianions and the bis-mnt Pb dianion are known.[64] For the SnCl(tdt)$_2$ anion a rectangular pyramidal pentacoordination was found.[56] The anion of Sn(tdt)$_3$ is octahedral and shows no interligand S—S interactions.[53]

In the Ga, In and Tl group, the reaction of the trihalides with mnt^{2-} leads[66,67] to Ga(mnt)$_2^-$, a mixture of In(mnt)$_2^-$ and In(mnt)$_3^{3-}$ and to Tl(mnt)$_3^{3-}$, respectively. The reaction of In(mnt)$_2^-$ with bipyridyl led to In(bipy)(mnt)$_2^-$.[67] Indium(I) halides react with the neutral tfd ligand to form In(tfd) halides, which were reacted further with bipyridyl under aerobic conditions to form the salt [In(tfd)$_2$(bipy)][In(tfd)(bipy)$_2$] containing two In^{III} species.[68]

Almost 50 years after its first appearance[60] in the literature, the dianion Cd(tdt)$_2^{2-}$ has been examined by X-ray crystallography and found to indeed be tetrahedral.[69] The very poorly soluble neutral complex was assumed to be polymeric.

It also deserves mention that the complex Me$_2$Sn(mnt) has been used as a convenient source of the mnt ligand in the synthesis of other complexes.[70,71]

16.5.2.5 Benzenedithiolates

Because the benzene resonance energy is much higher than that of an isolated double bond, the oxidation of an ethylenedithiolate to the butadiene-like dithione is easier to accomplish than the oxidation of a benzenedithiolate to the corresponding dithio-*ortho*-benzoquinone: there is a difference in stability between the dianionic ligands (**44**) and (**46**) relative to the dithioketone forms (**45**) and (**47**) derived from them by a two-electron oxidation.

This stability difference means that the resonance forms (**1**)–(**3**) (Section 16.5.1), which are taken to be the basis of the dithiolene stability, will contribute with different weights in benzenedithiolates

(44)   (45)   (46)   (47)

than they would in ethylenedithiolenes. Therefore, benzenedithiolates of metals which can exist as neutral dithiolenes with the usual ligands should prefer to exist as monoanionic or dianionic 'dithiolato-like' species. It was observed early on by Gray and coworkers[72] that in the series $Ni(tdt)_2$, $Ni(tdt)_2^-$ and $Ni(tdt)_2^{2-}$ the paramagnetic monoanionic species was by far the most air stable.

The $Fe(bdt)_2$ dianion (48) is planar[73] and reacts[74] with CO to form the octahedral $Fe(bdt)_2(CO)_2$ dianion (49). This dianion can be converted[75] into a variety of bridged neutral complexes of the general structure (50).

(48)   (49)   (50)

A very recent study of anionic $Mn(tdt)_2$ complexes[76] has shown that the dianion adopts a distorted tetrahedral configuration (dihedral angle 89°), while the monoanion is strictly planar. A five-coordinate MeOH adduct of the monoanion with a square pyramidal geometry was also observed.

In the $Au(tdt)_2^-$ ion[77] a strictly square planar AuS_4 core exists but the toluene planes are twisted, possibly through packing effects.

Tris-dithiolenes of tdt-like ligands are, of course, stable as neutral species if a 6+ oxidation state is favorable, and many such examples exist. The first examples of structures relating to the dithiolates in fact came from this group of compounds: the tris-tdt-dithiolene.

In the main group dithiolenes, where the question of dithiolene resonance is without relevance, many examples of benzenedithiolato complexes are known (see Section 16.5.2.4.4).

16.5.2.6 Multinuclear Dithiolenes

Multifunctional dithiolene ligands have been used to prepare binuclear as well as polymeric complexes. Some examples of this type of complex were already mentioned above (25 and 41) and involve the ligands benzenetetrathiol (51) or its alkyl ethers,[78] the tetrathiosquaric acid (52), ethylenetetrathiol (53) and tetrathiooxalic acid (54) and its Se analog. The two-electron difference in oxidation state between (53) and (54) is much the same as that discussed for the parent dithiolene ligand. Crystallographic results (see Section 16.5.3.1) point out the equivalence of the latter two ligands.

(51)   (52)   (53)   (54)

Dithiolenes can serve quite well as bridging ligands (see compound 43). A regular dithiolene ligand has also been shown to coexist in a binuclear complex with a bridging one: the 'sulfur rich' dithiolene[28] $Mo_2(S_2)(S_2C_2Ph_2)_4$, cited earlier in a different context, is an example of this type of coexistence.

trans-Butadienetetrathiolate has been used[79] to generate a polymeric Ni dithiolene, for which the structure (55) was proposed. The complex showed near-IR transitions at 900 and 1800 nm but was not characterized further.

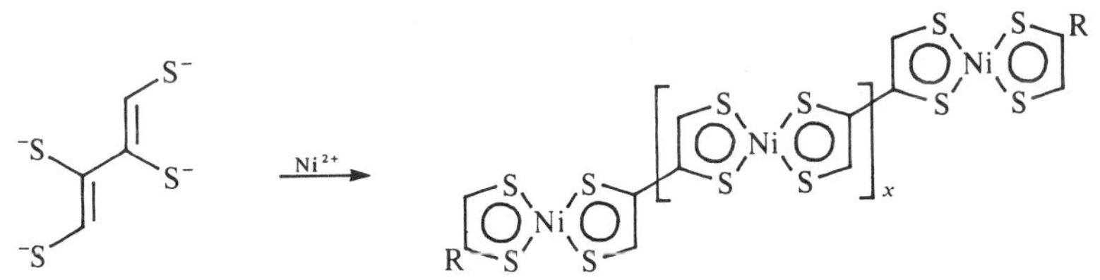

(55)

16.5.2.7 Selenium- and Tellurium-containing Dithiolene Analogs

Only relatively few examples exist for complexes of transition metals with Se-containing dithiolene-like ligands and those containing Te are unknown.Their synthesis generally follows the routes established for the sulfur complexes.

The early work by Davison *et al.*[80] on the selenium analog of the tfd complexes established their close similarity in electronic structure, as judged by their redox potentials.

Pierpont and Eisenberg[81] established a perfect trigonal prismatic D_{3h} geometry for the $MoSe_6$ complex core in $Mo[Se_2C_2(CF_3)_2]_3$. The Se_2C_2 planes form a dihedral angle of 18.6° with their respective $MoSe_2$ planes to give the molecule overall C_{3h} symmetry.

The reaction mentioned above to produce several CpCo dithiolenes (Scheme 3) has also been extended to the selenium analog (**56**).

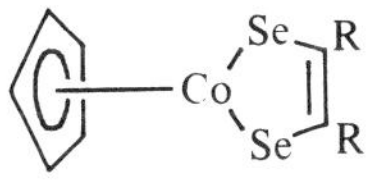

(56)

Bianchini *et al.*[82] have obtained triphos-capped tetrathio- and tetraseleno-oxalate bridged Rh dimer dications of the structure (triphos)Rh(C₂E₄)Rh(triphos)₃, where triphos = 1,1,1-tris-(diphenylphosphinomethyl)ethane and E = S, Se.

An interesting new entry into metallocene-containing dithiolato and diselenolato complexes (see also Section 16.5.2.4.3) are the reactions of titanocene pentasulfide (**57**) or pentaselenide (**58**) with activated alkynes.[83] Titanocene dithiolates (**59**) and diselenolates (**60**) with X = CF₃ and CO₂Me were obtained. The authors also performed a number of exchange reactions with the Se compound (**60**; X = CO₂Me).

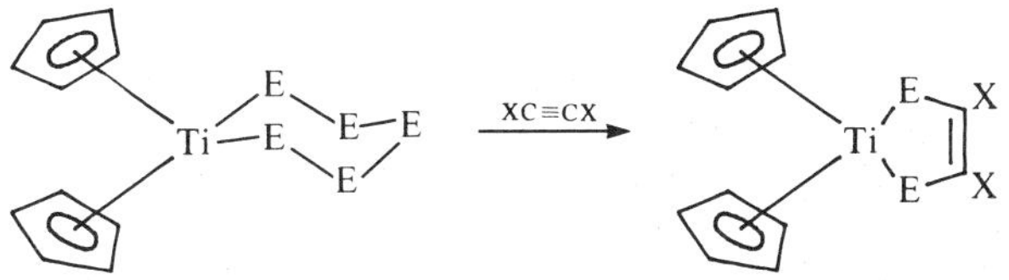

(57) E = S   (59) E = S, X = CF₃, CO₂Me
(58) E = Se   (60) E = Se, X = CF₃, CO₂Me

In the context of synthesizing highly conductive molecular solids, Sandman *et al.*[84] have prepared the Bu₄N salt of the planar benzenediselenole Ni complex anion (**61**).

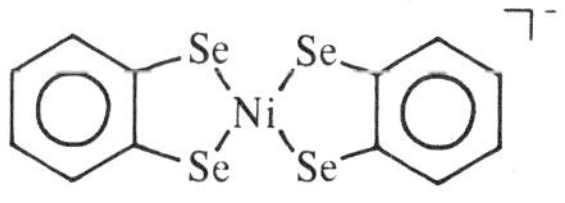

(61)

16.5.2.8 Complexes Related to Dithiolenes

In the transition metal complexes of ligands like *o*-phenylenediamine, we find a distinct analogy to the chemistry of dithiolenes. The full equivalence of their redox chemistry with that of dithiolenes suggests a comparably delocalized structure. The outward structural similarity between this ligand and the benzenedithiols is not reflected in the redox properties of their transition metal complexes

(62), which show a fully reversible five-membered redox series. Mainly through the work of Holm,[85] and others, this class of compounds has been well explored.

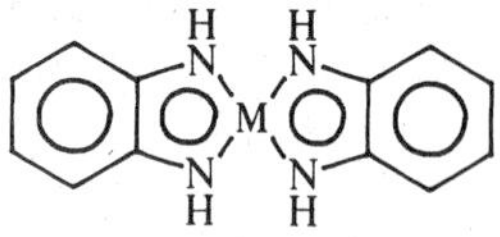

(62)

A recent report[86] by Peng *et al.* on the nitrogen-based analog **(63)** of the mnt complexes has pointed out the similarity and the difference of these two classes: while the basic structure of the Co complex must be very similar to the mnt complex, the crystal structure shows a different type of dimerization using a direct metal–metal bond.

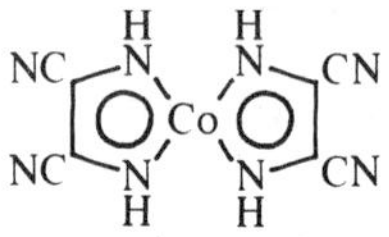

(63)

Although they bear considerable similarity to dithiolenes in their core structure and planarity, the bis-dithiotropolone complexes **(64)**, explored by Forbes and Holm,[87] are not true dithiolenes and instead belong to the 'odd' ligand complexes discussed in Section 16.5.3.9.

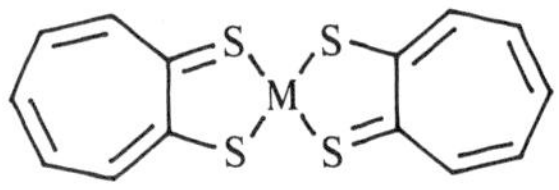

(64)

16.5.3 CHARACTERIZATION AND PROPERTIES OF DITHIOLENES

16.5.3.1 Crystal and Molecular Structures

Three main aspects are of importance in the structural characterization of transition metal dithiolenes: (a) the molecular structures, with particular relevance given to the variations in bonding as a function of the nature of substituents and of the different oxidation states; (b) molecular structures which show the existence of short range interactions such as dimer formation through different types of bridge bonding; and (c) crystal structures as a probe for the existence and nature of extended interactions and of stacking preferences in donor–acceptor complexes involving dithiolenes. In this section, we will consider only the first two aspects, while the third one will be discussed with the conductive and magnetic properties of dithiolenes in Section 16.5.4.1.

Over 20 1,2-dithiolene complexes were included in a structural review by Eisenberg[7] which covered the literature up to 1970. In his comparisons to the 1,1-dithiolato compounds, the major differences cited were the more extensive metal–ligand π-bonding and the less strained five-membered chelate rings of the 1,2-dithiolenes. He also stressed the planar geometry of bis-dithiolenes and focused on interligand S—S interactions as the reason for the preferred trigonal prismatic geometry in tris-dithiolenes.

Most of the recent work has focused on dithiolenes with organometallic ligands or addressed structural questions relating to their conductive properties (see Section 16.5.4.1). Nevertheless, a few reports on other dithiolenes require mention as well.

Complementing the early work by Sartain and Truter[88] on the first known neutral dithiolene $Ni(S_2C_2Ph_2)_2$, the structures of the Pt analog[89] and of the Bu_4N salt of the Ni complex radical anion[90] have recently been published. The Pt complex is strictly square planar (Pt—S bond length 2.245 Å) and the phenyl groups are twisted unequally by 36 and 67° with respect to the plane of the complex core. In the nickel dithiolene anion, Ni—S bond lengths of 2.137 Å are longer than those in the neutral complex (2.101 Å) and all phenyl groups are twisted by 48 to 54° out of the core plane in a close to uniform way.

In the acetone solvate of the methylene blue salt of $Cu(mnt)_2^{2-}$, Snaathorst *et al.*[91] found a rare case of a nonplanar dithiolene anion in which the dihedral angle between the ligands is 47.4°. This abnormality was attributed to the influence of the large cations.

Lindner *et al.*[92] have characterized the product of the reductive dimerization of thioaroyl chlorides with $NaMn(CO)_5$ in THF: a metal–metal bonded $Mn_2(CO)_6$ unit is bridged by a dithiolate. There exists significant Mn—C interaction between the dithiolate double bond and one of the Mn atoms.

Carbon disulfide and carbon diselenide react with several transition metal compounds to undergo head-to-head dimerizations leading to tetrathio- and tetraseleno-oxalate ligands of dithiolene character. The work of Maj *et al.*[56] (see Section 16.5.2.4.3) has found analogies in the work of Bianchini *et al.*[82,93] The structure of $(triphos)Rh(\mu-C_2S_4)Rh(triphos)^{2+}$ shows bond lengths in the bridging ligand (C—C = 1.36 Å and C—S = 1.73 Å) which make it appear more like an ethylenetetrathiolate rather than a tetrathiooxalate. The structure of the related Se complex is not known at this time.

Several structural studies involving metal dithiolenes and π-bonded carbocyclic ligands have been reported. Hemmer *et al.*[54] found that $(Me_4C_4)Ni(mnt)$ (65) crystallizes as long black needles in which the cyclobutadiene ring was 0.7° from being perpendicular to the five-membered chelate ring. The C—C bonds in the four-membered ring ranged from 1.45(1) to 1.50(1) Å indicating an approximately square formation; all methyl carbons were displaced 0.15 Å from the plane of the ring carbons, away from the metal center. The closest intermolecular approaches were between methyl carbons and nitrile nitrogens at 3.28 and 3.33 Å.

Work by Kutoglu[94] on $Cp_2Ti(edt)$ (66) and Debaerdemaeker and Kutoglu[95] on the related $Cp_2W(bdt)$ has established an unsymmetrical structure for both compounds. The Cp rings form dihedral angles of 51 and 43°, respectively, the S atoms lie in the symmetry plane defined by this arrangement, but the rest of the ligands are bent out of this plane by 46 and 48°, respectively.

(65)

(66) X = H
(67) X = CO₂Me

This structural type has now found another member in $(MeCp)_2Ti(cmt)$ (67), described by Bolinger and Rauchfuss.[83] The distances from Ti to the center of gravity of the cyclopentadienyl rings (Cg_a and Cg_b) were 2.052 and 2.072 Å, the S—S bite was 3.189 Å. The Cg_aTiCg_b angle equalled 132.43°, S_aTiS_b equalled 82.23(2)° and the STiCg angles ranged from 105.99 to 109.09°. The TiS_2 plane was bent 44° with respect to the S_2C_4 dithiolene plane. The deviation from planarity was attributed to π-donation from the formally dianionic dithiolene to an empty a_1 orbital of the Ti^{IV} center. $Cp_2Ti(cmt)$, $Cp_2TiSe_2C_2(CO_2Me)_2$ and $Cp_2TiSe_2C_2(CF_3)_2$ were predicted to have very similar structures based on spectroscopic similarities.

Several dinuclear complexes containing cyclopentadienyl rings and dithiolene ligands have been obtained by reacting sulfur bridged CpM dimers with activated alkynes. One example is the product of the reaction of reacting $Cp_2V_2S_4$ with hexafluorobutyne in toluene at 60 °C. The structural elucidation by Bolinger *et al.*[96] revealed a V_2 core separated by 2.574(3) Å, capped by π-bonded Cp rings, and bridged by a $\mu-\eta^2-S_2$ and a $\mu-\eta^2$-tfd ligand. The four sulfur atoms are coplanar.

Rakowski DuBois and coworkers[97] determined the structures of $(MeCp)_2Mo_2(S_2CH_2)(edt)$ (68) and $(Cp)_2W_2S_2(edt)_2$ (69).

(68)

(69)

An element of inaccuracy in the bond distances and angles was incurred for (68) as it crystallized about an inversion center disordering the two dithiolate bridges. A Mo—Mo separation of 2.601(1) Å and M—S bonds of 2.464(1) and 2.456(1) Å were found. Tungsten in (69) can be described as being in a four-legged piano stool environment. There are two independent molecules in the structure; the one with an inversion center forces the W_2S_2 core into strict planarity; the other molecule's W_2S_2 ring is folded 159.2° along the S—S vector. W—W and S—S bond lengths are shorter in the centrosymmetric molecule, possibly as a result of packing forces, though no unusually short intermolecular contacts were found.

Miller *et al.*[58] determined the crystal structure of CpCo(bdt) (**70**). The crystal contained two crystallographically different though chemically similar molecules packed head-to-tail. The Cp ring is 90.1° with respect to the CoS_2C_2 plane for one molecule and 85.9° for the other, yielding a two-legged piano stool coordination about Co. The four Co—S distances average 2.113 Å and S—C bonds of 1.737 Å; the Co—Cp centroid distance is 1.646 Å. The closest intermolecular approaches are 4.325(1) and 4.702(1) Å and these occur between symmetry-related molecules. What makes this compound so unique is its dimerization at room temperature with retention of crystallinity. The process constitutes a completely reversible solid phase chemical reaction, represented already in Scheme 4. In the dimer, Co*—S equals 2.246(1), 2.230(1) and 2.27(1); Co—Co equals 3.2893(4) Å. The analogy to the homoleptic bis-dithiolene dimers linked by M—S bonds is evident. However, in this case there is an equilibrium between the monomer and dimer which is shifted by subtle crystal packing and molecular deformation forces.

(70)

Several recent reports have dealt with work related to dithiolenes. Stach *et al.*[44] investigated a single crystal of the mixed-ligand Ni complexes with dtc and mnt ligands and the salt $(Bu_4N)[Ni(mnt)(Et_2dsc)]$. The crystal of the latter decayed severely during X-ray data collection. The bond between Ni and S (2.12 Å) was shorter and that between Ni and Se (2.34 Å) was longer than in the corresponding symmetrical parents. A C—C≡N bond angle of 146.7° is probably more indicative of the poor quality data set ($R = 0.101$) than reality.

Relating to the work of Heber and Hoyer[42] (see Section 16.5.2.4.1) on methyl ethers of the dithiolene ligand, Richter *et al.*[98] determined the structure of tris(*S*-methylethylenedithiolato)Rh and found an octahedral structure, a *cis* arrangement of the methyl groups and no evidence for interligand interactions of the S atoms.

16.5.3.2 Redox Properties

Davison *et al.*[99,100] made the astute observation that Schrauzer's neutral complex, $Ni(S_2C_2Ph_2)_2$, and Gray's anionic one, $[Ni(mnt)_2]^{2-}$, were electronically related. The number of electrons associated with the central metal and the two chelate rings, ignoring the substituents on the ligands, differ by two for the complexes. Based on the stability, they reasoned that complexes like Schrauzer's may undergo one- or two-electron reductions while one- or two-electron oxidations of complexes like Gray's may be possible. They proceeded to demonstrate the facile, reversible, two-step, one-electron redox interconversion of $[M(S_2C_2R_2)_2]$ and $[M(S_2C_2R_2)_2]^{2-}$, revealing the rich redox chemistry of the dithiolenes. A number of empirical observations from the polarographic half-wave potentials of 16 square planar, bis-1,2-dithiolenes follow: (i) in couples less positive than $\sim 0.00\,V$, the reduced species are readily air-oxidized in solution, whereas in couples more positive than $\sim 0.00\,V$, the reduced species are air-stable; (ii) in couples more positive than $\sim 0.20\,V$, the oxidized species are reduced by weakly basic solvents (*e.g.* ketones or alcohols), while for couples within the range -0.12 to $+0.20\,V$, the oxidized forms are reduced by stronger bases (*e.g.* aromatic amines); (iii) for couples more negative than $\sim -0.12\,V$, the oxidized species require strong reducing agents (*e.g.* hydrazine or sodium amalgam); and (iv) in couples less positive than $\sim 0.40\,V$, the reduced forms are oxidized by iodine.

For a given M and z, the ease of oxidation of $[M(S_2C_2R_2)_2]^z$ decreases in the order R = H, alkyl > Ph > CF_3 > CN and R = Ph > CF_3 > CN for the analogous tris-chelated complexes. The series parallels the electron-donating tendency of the substituent R. For the dianion, another stability trend was noted; the oxidative stability increases across the first row transition metals: Fe ≪ Co < Ni < Cu.

The redox properties of $[Co(mnt)_2]^{2-}$ were examined chemically and electrochemically by Vlcek and Vlcek.[101] Reduction of the dianion in THF at a DME was achieved by the application of -1.83 V (SCE) to yield $[Co(mnt)_2]^{3-}$. Lithium aluminum hydride in THF is also capable of carrying out the same reduction. The green solution is extremely air-sensitive and becomes an air-stable, brown-yellow solution upon the addition of excess $LiAlH_4$. The conclusion was that the reduction was metal-localized to yield Co^I. The oxidation of $[Co(mnt)_2]^{2-}$ was a two-step process; the dianion was first oxidized to the monomeric monoanion, $[Co(mnt)_2]^-$ in THF at a DME or RPE at -0.02 V (SCE). Chemically, this step could be accomplished with iodine, oxygen, tetracyanoethylene, tetrachloro-*p*-quinone, trityl chloride, benzoic acid and all stronger protic acids. The monomer then slowly dimerized to $[Co(mnt)_2]_2^{2-}$.

Geiger *et al.*[102] were able to extend the redox series of the mnt complexes of Ni, Pd and Pt to the $M(mnt)_2^{3-}$ trianions. These are believed to be M^{1+} complexes in which the metal has a d^9 configuration.

A substantial effort has been directed toward the synthesis of cation-deficient, anionic complexes to improve electrical conductivity (see Section 16.5.4.1). Dithiolenes with ligands such as mnt, tfd and dmit, usually stablest as the mono- or di-anion, have been partially oxidized by a battery of methods to achieve this goal.

16.5.3.3 NMR Studies

Although the majority of NMR applications would be considered routine characterizations, a few studies require elaboration.

Among the first to apply the NMR technique to dithiolene chemistry were Davison *et al.*[100] In 1964, the ^{19}F NMR spectra were reported for the series $M(cmt)_2^n$ (M = Ni, Fe; n = 0, -1, -2).

Schrauzer and Mayweg[103] observed a single peak at $\delta = 9.2$ in the ^1H NMR spectrum of $[Ni(edt)_2]$ in CS_2. The downfield shift relative to the single 6.8 p.p.m. resonance in Na_2edt was interpreted as being indicative of more aromatic character in the Ni complex. The diamagnetism necessary to allow the observation of an NMR spectrum appears to exclude the existence of an equilibrium containing a significant amount of tetrahedral species or the involvement of such a species in an exchange reaction between the planar configurations. While symmetrically substituted dithiolenes would not provide any insight into any such exchange reaction, unsymmetrically substituted dithiolenes should allow the observation of *cis* and *trans* isomers and/or an equilibrium between them. Indeed, a minor amount of the *cisoid* form of $[Pt(S_2C_2R^1R^2)_2]$ (R^1 = H, R^2 = *p*-$C_nH_{2n+1}C_6H_4$; n = 4–10) has been suggested[36] to exist in the NMR spectrum of the *trans* isomer from the presence of two slightly different $J(^{195}Pt-^1H)$ values of satellites of unequal intensity. Other unsymmetrical Pt dithiolenes have not shown this phenomenon. These data cannot tell whether the *cis* isomer is present as a minor component or as part of a *cis–trans* equilibrium.

Evidence for the existence of such equilibria comes from the eight-line spectrum obtained for the parent compound in 4-cyano-4'-*n*-pentylbiphenyl, an inert liquid crystal. Bailey and Yesinowski[35] found the only tenable rationalization for the fewer than expected 12-line spectrum was a rapid intramolecular rotation of one chelate ring with respect to the other, amounting to a degenerate *cis–trans* isomerization.

A sealed NMR tube containing $[(MeCp)Co(S_2C_2Ph_2)]$, $[CpCo(Se_2C_2Bu_2^n)]$ and C_6D_6 was heated to 90 °C. NMR data for the ensuing exchange revealed that scission and reattachment do not occur at the chalcogen–C bond ('alkyne crossover'). To decide between the other two exchange pathways, (a) Cp transfer and (b) CpCo transfer, a similar reaction between $CpRh(S_2C_2Ph_2)$ and $(MeCp)Co(S_2C_2Bu_2^n)$ was performed. The reaction resulted in formation of only CpRh and $[(MeCp)Co(S_2C_2R_2)]$ products. The authors concluded that the latter transfer mechanism was operative.[57]

Solid state ^{13}C NMR data at 120 °C supported the nematic nature of $[Ni(S_2C_2R^1R^2)_2]$ (R^1 = H; R^2 = *p*-$C_4H_9C_6H_4$). The transition to the nematic phase occurred at 117 °C and the complex decomposed at approximately 200 °C before becoming isotropic.[104]

Ion-pairing between the paramagnetic $[Co(tdt)_2]$ monoanion and nine different cations was examined by Tsao and Lim.[105] The cations belong to either of two classes, quaternary ammonium or substituted *N*-octylpyridinium ions. By recording the ^1H NMR spectra as a function of concentration (nitrobenzene, 307 K), the concentration association constants (K_{as}) were obtained. Substituent effects were found to influence the ion-pair geometry, as deduced from the isotropic shifts of the cationic protons and their shift ratios. In low dielectric constant solvents, speculation consistent with the magnetic anisotropy and the relation between the cationic proton shifts and concentration was tendered for cylindrically shaped aggregates.

Variable temperature ^{1}H NMR has been employed by several groups to determine inversions and rotation barriers. Bolinger and Rauchfuss[83] examined the inversion (Scheme 5) for the series $(Cp)_2Ti(E_2R)$ (E = S, Se; R = $C_2(CO_2Me)_2$, $C_2(CF_3)_2$, C_6H_4Me).

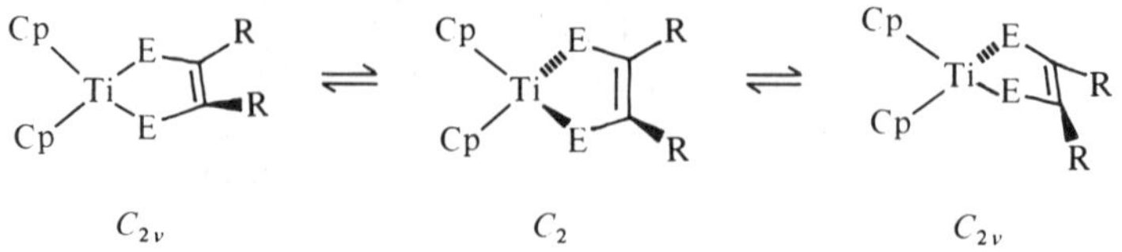

Scheme 5 Inversion of Cp_2Ti dithiolenes

The activation energy for the inversion at S is related to the energy difference between the C_2 and C_{2v} conformers. The difference in ΔG is only ~2 kJ mol^{-1} between S and Se complexes (R = $C_2(CO_2Me)_2$) and ~4 between $C_2(CO_2Me)_2$ and tdt (E = S). This appears to be a surprising insensitivity of the π-bonding to these two perturbations. The authors also noted a single ^{77}Se (19 MHz) resonance occurred at 1001 p.p.m. downfield of neat Me_2Se, one of the lowest field chemical shifts of any organic derivative of divalent Se. The π-acidity of Cp_2Ti^{IV} and the delocalization in the diselenene chelate are major factors.

The ligand $(MeS)_2C=C(SMe)_2$ was shown[78] to form a bidentate Pt^{II} complex (71) and a bridged, binuclear Pt^{IV} complex (72). Compound (71) shows four distinct ^{1}H NMR signals (1:1 v/v CS_2–CD_2Cl_2) for the chelate methyls. The two Pt-bound S—Me groups appear at the lowest field, δ 2.77 and 2.85 p.p.m. A large difference in their $^3J(^{195}Pt–^1H)$ values corresponds to the large difference in the ability of the directly metal-bound Cl and Me to exert a *trans* influence. A $^3J(^{195}Pt–^1H)$ value of ~17 Hz was observed in $Me_2Pt\{(SMe)_2C=C(SMe)_2\}$, supporting the assignment of the 16.8 Hz coupling constant to the SMe *trans* to Me and 70.3 Hz to the SMe *trans* to Cl. $^5J(^{195}Pt–^1H)$ values were not observed for the noncoordinated SMe groups. At $-114\,°C$, some broadening was evident due to the decreased inversion rate at the S *trans* to PtMe. A barrier of approximately 50 kJ mol^{-1} was estimated by comparison to related systems. The extremely low barrier was attributed to the large *trans* influence of PtMe and alkenic back-bonding stabilizing the transition state by (3p–2p) π-conjugation.

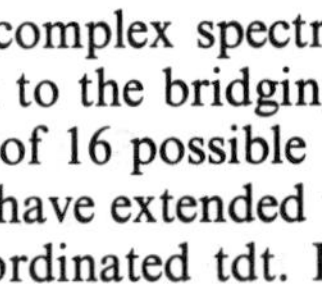

(71) 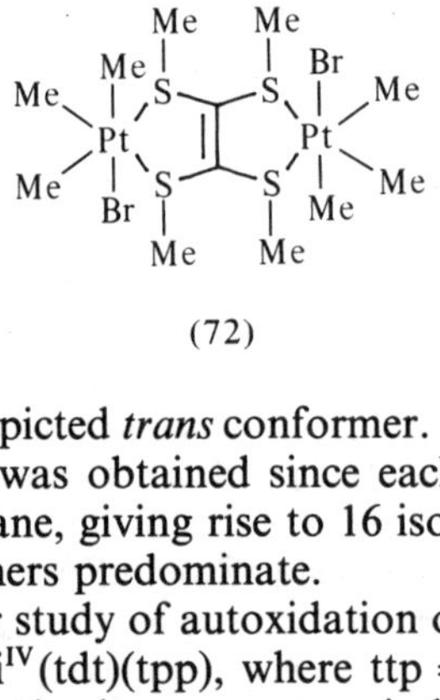   (72)

Complex (72) exists either as the *cis* or the depicted *trans* conformer. By lowering the temperature to $-70\,°C$, a much more complex spectrum was obtained since each S can occupy one of two conformations with respect to the bridging plane, giving rise to 16 isomers. A 400 MHz spectrum suggests only two or three of 16 possible isomers predominate.

Deronzier and Latour[106] have extended their study of autoxidation of Ti^{III} porphyrins to include the influence of axially coordinated tdt. In Ti^{IV}(tdt)(tpp), where ttp = tetraphenylporphyrin, the dithiolene ligand experiences the strong porphyrin ring currents as indicated by the upfield shift of tdt *ortho* and *meta* protons to 5.14 and 5.83 p.p.m., respectively. Lowering the temperature below 200 K was sufficient in a peroxo complex, Ti^{IV}(O$_2$)(tpp), to block rotation of the axial ligand. However, rotation of tdt around the C_2 axis (ignoring the Me on tdt) was still too rapid on the NMR timescale even at 180 K. Although the Ti d_{xy} orbital can interact with the tdt S atoms in a manner similar to the peroxo complex, the orbital overlap is much weaker and the resultant barrier to rotation is lower.

16.5.3.4 ESR Studies

The reviews of McCleverty,[4] and Burns and McAuliffe[11] should be consulted for coverage of early ESR findings.

Kirmse and coworkers[107–111] have shown the advantages of applying the ESR technique to transition metal dithiolenes: (i) a large delocalization of the unpaired spin density giving rise to the high covalency of the M—S bond; and (ii) the various electronic configurations and the corresponding different metal oxidation states. ESR was particularly useful in studying ligand-exchange behavior and one-dimensional D—A complexes. Since the ligand hyperfine data can yield direct information about the nature of the electronic ground state and the extent of electron delocalization of the ligand, double and triple resonance techniques and isotopic doping have been used. Mnt ligands, often paired with non-dithiolene ligands were central in these studies.

The $[Co(mnt)_2]^{n-}$ ($n = 1, 2, 3$) redox series was reexamined with the aid of ESR by Vlcek and Vlcek.[101] Upon chemical or electrochemical oxidation of $[Co(mnt)_2]^{2-}$, an initial spin-triplet monomer $[Co(mnt)_2]^-$ formed. Slow dimerization led to the final oxidation product $[Co(mnt)_2]_2^{2-}$, which in the solid state or in noncoordinating solvents was diamagnetic. Solution spectra in DMSO showed an ESR signal due to $[Co(mnt)_2]^-$; THF solution spectra were identical but indicated three times less monomer concentration. A THF solution of $[Co(mnt)_2]^{2-}$ underwent a reversible one-electron reduction process localized on the metal to yield a green solution of a Co^I complex, $[Co(mnt)_2]^{3-}$, with no ESR signal.

Snaathorst *et al.*[91,112] have investigated the acetone solvate of the salt $[Cu(mnt)_2]^{2-}[MB^+]_2$ (where MB = methylene blue) by single-crystal ESR techniques. They found that the complex consisted of dimerized anions each with a singlet ground state and a thermally accessible triplet state. At 4.2 K, triplet–triplet interactions between dimers along the a axis were responsible for satellite lines in the ESR spectrum.

Bowmaker *et al.*[113] have investigated the redox chemistry of $[M(S_2C_2Ph_2)_2]$ (M = Ni, Pd, Pt), the mixed-ligand complexes $[M(S_2C_2Ph_2)(dpe)]$ (M = Ni, Pd, Pt and dpe = 1,2-bis(diphenylphosphino)ethane) and $[M(S_2C_2Ph_2)(PPh_3)_2]$ (M = Pd). The heretofore unknown dithioketyl radical $(S_2C_2Ph_2)^-$ was characterized by ESR upon reversible one-electron oxidation of $[Ni(S_2C_2Ph_2)(dpe)]$. The alternative explanation, oxidation of Ni^{II} to Ni^{III}, seemed less reasonable based on the magnitude of g values. Further support was noted when $[M(S_2C_2Ph_2)(dpe)]$ and $[M(S_2C_2Ph_2)(PPh_3)_2]$ (M = Pd and Pt) underwent reversible oxidation at potentials similar to $[Ni(S_2C_2Ph_2)(dpe)]$, supporting the argument that the oxidation is based on the S ligand and therefore relatively independent of the metal ion or the phosphine ligand.

A typical single-crystal ESR study was part of the investigation by Brown and coworkers,[20] which also involved the crystal and molecular structure determination of $(Ph_4As)[Ni(cmt)_2]$. Just one ESR signal was observed at X-band (9.5 GHz) and Q-band (35 GHz) frequencies for the two magnetically and crystallographically nonequivalent complex anions in the unit cell due to spin–spin exchange between the nearly orthogonal ions even though they are separated by 12.14 Å. Data were consistent with a $^2B_{2g}$ ground state.

The nuclear quadrupole coupling constant for low-spin Co^{II} in a square planar environment was determined[114] by ESR of Co-doped $(Et_4N)_2Ni(mnt)_2$. A qualitative discussion was given to explain the finding that QD and QE were unexpectedly small.

Using FeTPP·$[Cu(mnt)_2]$·FeTPP (where TPP = *meso*-tetraphenylporphyrin) as a model compound for cytochrome *c* oxidase, Elliot and Akabori[115] have provided an alternative explanation for the enzyme's 'ESR silent' iron–copper pair. Their solid state ESR results supported a combination of dipolar coupling and a small amount of exchange coupling between the copper and the rapidly relaxing $S = \frac{3}{2}$ irons. This line of reasoning avoids the existence of an exceptionally large Fe–Cu coupling constant ($-J > 200\,cm^{-1}$) required by an older explanation.

16.5.3.5 Electronic Spectra and Photochemistry

One of the most obvious properties of the dithiolenes is their low energy absorption. Spectra of bis- and tris-dithiolenes were listed in the review by McCleverty.[4] New results exist for bis-dithiolenes, but not much can be added to the data on tris-dithiolenes.

The fairly broad but intense ($\varepsilon = 15\,000$ to $40\,000$) long wavelength ($\lambda_{max} > 700$ nm) electronic transitions of the dithiolenes are characteristic of this class of compound. In particular, the intensity of the transition is unmatched in any other transition metal compound, where low energy transitions usually are of *d–d* character and thus considerably weaker. Early theoretical treatments have been confirmed in their essential results by more refined calculations (see Section 16.5.3.7) in assigning this electronic absorption band to a π–π^* transition between the relatively high-lying highest filled MO and the lowest vacant MO, which is of unusually low energy. This latter fact is, of course, substantiated by the electron acceptor properties of the appropriately substituted dithiolenes, where this LUMO is filled by one or two electrons.

In the square planar dithiolenes of Ni, Pd and Pt this transition can be shifted far into the near-IR region by suitable substitution. In Table 1, the electronic absorption maxima λ_{max} are listed for nickel dithiolenes with a variety of substituents. Their λ_{max} values span the range from 720 nm for the parent dithiolene (R = H) to the present maximum of $\lambda_{max} = 1440$ nm. With the parent nickel complex already absorbing at 720 nm, it seems that this class is well suited to study how far into the IR region one can shift electronic transitions. Since no electronic transition has ever been observed in the standard IR region beyond 3 μm, the question may be asked in which energy range they might end. Dithiolenes might be able to provide this answer.

Table 1 Influence of Ligand Structure on the Lowest Energy Transitions for Neutral Nickel Dithiolenes

Substituents	Solvent	λ_{max} (nm)	ε^a	Ref.
Symmetrical ligands ($R^1 = R^2 = R^3 = R^4$)				
H	Hexane	720	14 300	136
Me	$CHCl_3$	774	28 200	4
CF_3	Pentane	715	12 300	4
SC_4H_9	$CHCl_3$	1004	—	200
Ph	$CHCl_3$	866	30 900	4
2-Naphthyl-	$CHCl_3$	905	—	200
3-Pyrenyl-	$CHCl_3$	910	—	200
2-Thienyl-	$CHCl_3$	982	—	200
$4\text{-}Me_2NC_6H_4$	CH_2Cl_2	1120	28 000	18
Julolidine	CH_2Cl_2	1270	45 000	18
$DETHQ^b$	CH_2Cl_2	1370	(42 000)	18
$DETHQ^b$	DMSO	1440	(42 000)	24
Cyclic ligands				
2,3-Indenedithiole	CH_2Cl_2	915	(25 000)	18
9,10-Phenanthrenedithiole	CH_2Cl_2	966	(30 000)	18
1,2-Acenaphthylenedithiole	CH_2Cl_2	1140	(30 000)	18
Unsymmetrical ligands ($R^1 = R^3$, $R^2 = R^4$)				
H, Ph	CH_2Cl_2	805	(25 000)	18
H, $4\text{-alkyl}C_6H_4$	Hexane	850	(25 000)	36
H, julolidine	CH_2Cl_2	1180	(35 000)	18
Ph, $4\text{-}Me_2NC_6H_4$	CH_2Cl_2	1060	28 000	196
Dithiolenes with two different ligands				
L^1 L^2				
mnt^c $Ph_2C_2S_2$	—	($\sim$900)	(20 000)	38
mnt $(CH_2)_2(NMe)_2C_2$	CH_2Cl_2	775	(20 000)	18

^{a} Extinction coefficients in parentheses are estimates.
^{b} 3-*N*,*N*'-Diethyltetrahydroquinazolyl.
^{c} Only the spectrum of the anion was reported. The data for the neutral species are estimated.

The simplified diagram in Scheme 6 of the frontier MOs of substituted dithiolenes, based on EHT results,[18] helps to explain this trend. Acceptor substituents lower the energy of the highest occupied and the lowest vacant orbital by roughly the same amount and no drastic shift in the absorption maximum is observed. Donor substituents also affect both MOs, but they raise the energy of the highest occupied MO more than that of the lowest vacant one. Consequently, they cause a bathochromic shift of the lowest energy transition. Very strong donors cause the HOMO to become antibonding, and such dithiolenes can be expected to be strong electron donors and to exist preferentially as dications.

Dithiolenes do not fluoresce, nor do they show any other type of emission when irradiated in solution at the lowest energy transition. All absorbed energy must therefore be transmitted to the environment as heat. As long as the surrounding solution remains stable, this irradiation does not lead to decomposition of the dithiolene.[116] This extreme photochemical stability and the position of the low energy transition in the near IR has led to several important uses of dithiolenes in laser applications (see Section 16.5.4.2).

Towards higher energy, the intense first band is followed by a weak *d–d* type transition near 600 to 700 nm and then by a series of very intense UV absorptions. Dithiolenes are relatively sensitive

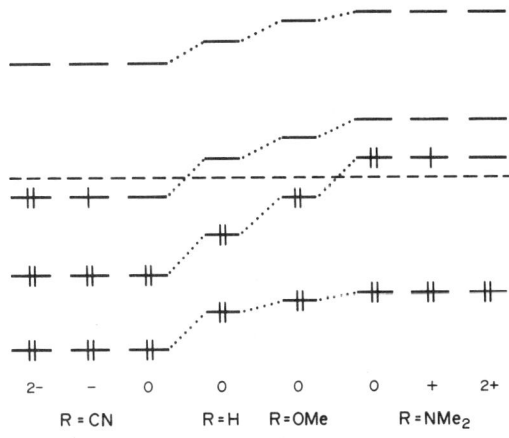

Scheme 6 Frontier MOs of donor- and acceptor-substituted dithiolenes

to irradiation into these latter bands and can undergo photoreactions, but their decomposition by exposure to UV light is not uncommon.

Miller and Marsh[31] studied the photochemical transformation of $[W(S_2C_2Me_2)_2(CO)_2]$ into $[W(S_2C_2Me_2)_3]$ as a function of wavelength ($\lambda_1 = 365$, $\lambda_2 = 500$ nm), solvent (chloroform, benzene or acetone) and concentration. From their results, a mechanism was suggested: (i) loss of a CO to form $[W(S_2C_2Me_2)_2(CO)]$; (ii) dimerization of the monocarbonyl intermediate upon nucleophilic attack by a sulfur lone pair of electrons; and (iii) rearrangement to yield the tris-dithiolene, CO, W and the free ligand. This heretofore uncharacterized 'photoinduced bimolecular bidentate ligand migration reaction' may also apply to $[Fe\{P(OMe)_3\}_3(tfd)]$, $[Fe(tfd)_2]$ or $[Ni(S_2C_4H_6)_2]^z$ ($z = 1-$, $2-$) on reacting with $W(CO)_6$.

A number of dithiolato–metal–dye complexes were synthesized[117] by adding thionine (TH), tolusafranine (SAF) or methylene blue (MB) to a mixture of a metal salt and $Na_2(mnt)$ or $H_2(tdt)$. The composition of the products was determined from elemental analyses and valence state considerations. Peculiar stoichiometries were found, *e.g.* $Mn_2(mnt)_5(SAF)_6(OH)_2 \cdot 11H_2O$, necessitating the assumption that equally unusual structures existed. Nonetheless, the molar absorption coefficients of the maximum absorption were larger for the complexes than the free dyes. In hexamethylphosphoramide, the metal dithiolene accelerated the rate of photochemical reduction *versus* the free ligand with TH but retarded the rate for the SAF and MB complexes.

A strong emission at 654 nm was observed when $[Pt(tdt)(bipy)]$ was irradiated at 366, 436 or 546 nm in an ethanol glass at 77 K. Vogler and Kunkely[118] assigned this emission to the relaxation of the ligand-to-ligand charge-transfer (LLCT) excited state to the ground state. An ethanol solution irradiated at room temperature proved stable, but failed to emit any observable light. Photolysis of chloroform solutions resulted in decomposition. In the latter case, the authors proposed that excitation of the complex to the LLCT excited state was followed by an electron transfer from the complex to $CHCl_3$:

$$[Pt(tdt)(bipy)]^* + CHCl_3 \longrightarrow [Pt(tdt)(bipy)]^+ + Cl^- + \cdot CHCl_2$$

In a related experiment by the same authors,[119] ESR data were found consistent with a ligand-centered radical, probably the easily oxidized tdt ligand. $[Pt(tdt)(bipy)]^+$ decomposed into unidentified products. Photolysis of $[Ni(S_2C_2Ph_2)(phen)]$ in $CHCl_3$ at room temperature promoted an analogous photooxidation. Here, subsequent reaction of the unstable cationic complex resulted in the formation of the symmetric complexes $[Ni(S_2C_2Ph_2)_2]$ and $[Ni(phen)_2]^{2+}$.

In another study by Vogler and Kunkely,[120] it was found that symmetric dithiolenes $[M(S_2C_2R_2)_2]^z$ (M = Ni, Pd, Pt, R = CN, $z = 2-$; M = Ni, R = Ph, $z = 1-$) can be oxidized electrochemically between $E_{1/2} = 0.1$ and 0.5 V *versus* SCE or photochemically ($300 < \lambda < 350$ nm) in $CHCl_3$. Closely related complexes in higher oxidation states, $[M(S_2C_2R_2)_2]^z$ (M = Ni, Pd, Pt, R = CN, $z = 1-$; M = Ni, Pt, R = Ph, $z = 0$), were stable to irradiation in $CHCl_3$ and require higher potentials for oxidation. A charge-transfer to solvent (CTTS) mechanism was suggested as being involved. The redox potential of the solvent, in addition to that of the complex, determines the energy of the CTTS state. As the oxidizing strength of the solvent increases, the photoactive wavelength region undergoes a bathochromic shift. Experimental support was obtained by conducting the photolysis in CCl_4 or CH_2Cl_2 where $E_{1/2}$ values *versus* SCE are -0.78 and $-2.33\,V$, respectively. Also characteristic of a CTTS mechanism was the increase in the photooxidation quantum yield Φ with decreasing wavelength. For example, $[Ni(mnt)_2]_2^{2-}$ in $CHCl_3$ showed the following trend:

λ (nm)	405	366	333	313 nm
Φ ($\pm 5\%$)	0.001	0.013	0.10	$0.25 \pm 5\%$

Dooley and Patterson[121] also noted the strong wavelength dependence of the quantum yield for the photooxidation of $[M(mnt)_2]_2^{2-}$ (M = Co, Ni, Pd, Pt or Cu). Furthermore, they observed (i) the lack of correlation between $E_{1/2}$ values and Φ, and (ii) the significantly lower quantum yields in 24:1 (v/v) $CHCl_3$/MeCN *versus* $CHCl_3$, solvents with almost identical redox potentials. From these observations, they suggested that additional excited states were reactive, in particular those with decreased bonding electron density on the ligands relative to the ground state.

Frank and Rau[122] investigated the reduction of $[Ni(S_2C_2Ph_2)_2]$ by 2-aminonaphthalene using $Ru(bipy)_3^{2+}$ as photocatalyst in homogeneous and heterogeneous media. Electron transfer from the excited Ru complex to the dithiolene generated the radical anion (and subsequently the dianion); irreversible electron transfer from the amine to the $Ru(bipy)_3^{3+}$ ion followed. Without the catalyst no reaction occurred between the amine and the dithiolene, although other amines, including diphenylamine and triethanolamine, reacted with dithiolenes in the dark and without catalyst.

Highly structured emissions were observed by Johnson *et al.*[123] from a number of mixed-ligand mnt complexes containing Ir, Rh or Pt. The phenomena were seen both in dilute, frozen glass matrices and in solids at ambient temperature; no luminescence was noted from solutions. These observations are consistent with an intramolecular process. Intraligand (IL) $\pi-\pi^*$ transitions were rejected since $Na_2(mnt)$ failed to emit at 77 K. Of the two remaining possible mechanisms, metal-to-ligand charge transfer (MLCT) and ligand-to-metal charge transfer (LMCT), the authors favored the former. Their conclusion was supported by the absorption–emission energy following trends in the metals and in the donor–acceptor influence of the non-dithiolene ligands. An interesting observation was that the Stokes shift was often less than $400\,cm^{-1}$ indicating a minimal distortion of the excited state relative to the ground state.

16.5.3.6 Other Characterization Techniques

Dithiolenes have been the subject of investigations by many physical methods besides the ones mentioned explicitly above. Unfortunately, these other methods have been applied only sporadically, and few generalizations can be made from them. We list briefly such investigations as an aid to locate recent literature. The low volatility of dithiolenes and their tendency to decompose near the melting point make them poor candidates for characterization techniques such as mass spectroscopy or vapor phase electron diffraction. Consequently, most investigations have concerned dithiolenes either in the solid phase or in solution.

Mass spectra can only be observed for simple Ni dithiolenes and few others. The fragmentation patterns generally are far too complex to allow a sensible analysis.

Mössbauer spectroscopy has been applied[124-127] to dithiolenes in several instances. In the early phases of dithiolene chemistry, it allowed the assignment of a dimer structure to the Fe dithiolenes before crystallographic data were available.[124] A more recent application of Mössbauer spectroscopy[125] concerned monomeric and polymeric Fe dithiolenes. Through [119]Sn Mössbauer spectroscopy, a large number of formally four-, five- and six-coordinate Sn^{IV} complexes with mnt ligands[126,127] and with other sulfur ligands[127] were investigated.

The 1975 publication by Schläpfer and Nakamoto[128] appears to be the most recent example of the application to dithiolenes of IR spectroscopy, in this case basing assignments on Ni isotope shifts, coupled with a normal coordinate analysis. An overview of standard IR spectroscopic data on bis- and tris-dithiolenes was provided in the review by McCleverty.[4]

X-Ray photoelectron spectroscopy (XPS) has been used[129-131] to investigate several dithiolene ligands and complexes. Sano *et al.*[130] correlated their results with $X\alpha$ calculations, while Blomquist *et al.*[131] related theirs to EHT results.

16.5.3.7 Theoretical Treatments

Ab initio and semiempirical methods have been applied to the interpretation of many aspects of dithiolene chemistry: electronic spectra, ESR, Mössbauer, XPS, charge distributions, redox properties, reaction mechanisms, metal binding in biological systems and ligand-exchange behavior. We shall focus our attention on the theoretical deductions of some representative research groups. For computational details, the reader is referred to the original papers and references therein.

Schrauzer and Mayweg[103] found that semiempirical Hückel-type approximations provided satisfactory agreement with observed electronic transitions for $[Ni(edt)_2]^{0,1-,2-}$. Less favorable results were obtained from the Wolfsberg–Helmholtz method, due to excessively large off-diagonal, H_{ij} values. A ground state configuration, $\ldots (2b_{1u})_2(2b_{2g})_2(3a_g)_2(2b_{3g})_2(4_{gg})_2(3b_{2g})^*(3b_{1g})^* \ldots$, favored the valence bond resonance form (73).

(73)

These results were severely criticized by McCleverty,[4] who considered those from Gray's group[132] to be more realistic.

Later, Schrauzer and Rabinowitz[133] determined the charge distribution for the same Ni series using the ω–β self-consistent HMO approach. The Ni charge decreased from -0.075 to -0.292 to -0.593 as the net charge became progressively more negative. Whereas, in the neutral complex, 82% L-π and 18% metal $3d_{xz}$ character had been calculated for the lowest unoccupied molecular orbital (LUMO), a one-electron reduction equalized these contribution. Taking into account ESR and polarographic results, a 'physically realistic conclusion' placed the L-π character in the $3b_{2g}$ LUMO between 50 and 85% for the monoanion. The calculated energy of the near-IR transition ($2b_{1u}$ to $3b_{2g}$) observed in $[Ni(edt)_2]^{0,1-}$ was found to be in good agreement with experiment. Furthermore, the bathochromic shift of this transition for the monoanion was correctly predicted.

A more recent and much more complete calculation at the MC–SCF–CI level by Blomberg and Wahlgren[134] on the dianion of $Ni(edt)_2$ produced the wrong ground state configuration (tetrahedral), and an estimate of d–d excitation energies, while made, lacked significance.

Zalis and Vlcek[135] studied the electronic structure of $[M(mnt)_2]^{n-}$ (M = Co, Ni or Cu; n = 1, 2 or 3) by semiempirical methods. Iterative extended Hückel methods (IEHT) and CNDO results were compared. Except for $[Co(mnt)_2]^{3-}$, both methods yielded $4b_{2g}$ as the HOMO for all d^8 complexes; for the exception, CNDO found $10A_{1g}$ as the HOMO. IEHT and CNDO revealed a greatly reduced d_{xz} metal contribution to the $4b_{2g}$ orbital along the isoelectronic series $[Co(mnt)_2]^{3-}$, $[Ni(mnt)_2]_2^{2-}$ and $[Cu(mnt)_2]^-$. Changes in the overall charge of the complex shifted orbital energies and altered the atomic character of specific orbitals, but IEHT showed no MO level crossing due to the redox process. On the other hand, the CNDO-derived MO schemes did not retain the same sequence of orbital levels. The cobalt series experienced the largest changes as a function of charge, whereas Cu complexes were almost charge insensitive. The authors attributed this artifact to the inability of IEHT to adequately account for the considerable orbital relaxation.

INDO calculations on $Ni(edt)_2$ were performed[136] together with calculations at the same level for the parent compound $Ni(S_2C_3H_3)_2$ of the SacSac complexes. The results agreed well with spectroscopic data and allowed a comparison to be made between the 'even' and 'odd' ligand systems. The correlations between these two classes will be discussed in Section 16.5.3.9.

To interpret and predict ligand exchange equilibria, Dietzsch *et al.*[137] performed EHT calculations on a large set of dithiolene-containing complexes. For all 24 paired combinations of four bis-1,2-dithiolenes (edt, mnt, tfd or dmit) with six bis-1,1-disubstituted ethylene-2,2-dichalcogenolates, EHT predicted the mixed-ligand product to be more stable than the symmetrical reactants. Based on ΔE, the tfd ligand gave the most stable mixed-ligand products. For the remaining combinations, neither the π-acceptor behavior nor the inductive effect of the dichalcogenolate substituents led to a predictable ordering of product stability. A few of the reactions proved experimentally feasible and were found to 'correlate pretty well' with the theoretical predictions.

Miller and Dance[47] used a qualitative MO description to compare the electrochemical, redox and absorption spectra of $(\alpha\text{-diimine})_n Ni(S_2C_2R_2)_{2-n}$ (R = CN, CF_3 and Ph, α-diimine = *o*-phenylenediimine, biacetylbisanil, 1,10-phenanthroline and derivatives, n = 0, 1 or 2). The diimine π-

orbitals were placed at a higher energy than those of the corresponding dithiolene. Some of their conclusions concerning the mixed-ligand complexes ($n = 1$) follow: (i) diimine character predominates in the LUMO; (ii) metal $3d_{yz}$ character prevails in the HOMO; (iii) the intense, long-wavelength absorption is mainly dithiolene to diimine; and (iv) a dipolar valence bond type resonance form (74) predominates in the ground state.

(74)

In an attempt to gain insight into metal binding in biological systems, Fischer-Hjalmars and Henriksson-Enflo[138] undertook a formidable project: an *ab initio*, MO–LCAO–SCF approach was used to examine trends in binding energies, electron affinities, orbital energies, the 'metal bridge effect', charge distributions, bond polarities, sulfur d-orbital effects and transition metal d-orbital effects. Their model system consisted of $[M(X^1X^2C_2H_2)_2]^n$ where the variables were M = Li, Be, Mg, Sc, Ti, Cr, Ni, Cu or Zn; $X^1 = X^2 =$ NH, O or S; $X^1 =$ O, $X^2 =$ S; and $n = 0, 1\pm, 2\pm$. Beryllium complexes were found to be more stable than other metal complexes suggesting an explanation for this metal's high toxicity. The electron affinity of the complexes was almost independent of the central metal. The highest occupied molecular orbital (HOMO) and LUMO were predominantly of ligand character. The large energy gap ($\sim 10\,eV$) between the HOMO and LUMO for a dication was appreciably narrowed (3–5 eV) on reduction to $n = 0$ or ± 1. Corresponding electronic shifts to longer wavelengths have been experimentally noted. The variation of this energy difference could be simulated by reduction of a *metal-free*, dimeric ligand. The 10–13 eV gap between the HOMO and the LUMO was almost unaffected by reducing a monomer of the ligand. Thus, although the orbital energy gap was primarily a property of the ligand, it demanded a ligand dimer. The authors surmised that the metal served two functions: (i) it was essential for dimerization of the ligand; and (ii) it increased the electronegativity of the system and therefore its stability to reduction. This 'metal bridge effect' was even operative with only a 1+ point charge. Sulfur d-orbital populations amounted to merely 0.01 to 0.1 in a test case, $[M(SH)_4]_n$ (M = Ni or Zn). From these results, the authors felt the sulfur d-orbitals made an insignificant contribution to metal–sulfur bonding and they were therefore justified in excluding these d-orbitals from the calculations.

Several groups have observed the generation of hydrogen from water in the presence of transition metal dithiolenes. Analogous to S-alkylation of the dithiolenes by alkyl halides, Alvarez and Hoffmann[139] considered the possibility of protonation at the dithiolene 1,1′-S atoms, followed by concerted elimination of H_2. With the aid of a correlation diagram, this process was dismissed on symmetry grounds for d^8 complexes (see the original paper for an explanation of their electron counting scheme). However, they noted that the process was accessible to an excited state for d^n systems with $n \leqslant 8$. The other combinations of diprotonation, at the 1,2-S or the 1,2′-S sites, were not mentioned. As a more reasonable mechanism, they suggested the metal could donate two electrons through its d_{z^2} orbital to an incoming proton, whereupon the proton is reduced to a hydride ligand and the metal is formally oxidized. A second electron reduces the metal, which then allows the complex to undergo a second protonation at either the hydride ligand, sulfur or the metal. The authors argued in favor of the hydride attack which would lead to a facile dissociation of H_2 by weakening the M—H bond and strengthening the H—H bond. The experimental aspects of the work on dithiolenes as catalysts have since been subject to revision[216] and the theoretical work is left without a direct basis for comparison.

A mostly qualitative MO approach was used by Alvarez, Vicente and Hoffmann[140] to interpret the regularities and irregularities in the dimerization and stacking of transition metal dithiolenes. While firm predictions could not be made, the balance between sulfur lone-pair repulsion, size and occupation of the metal d-orbitals and substitutent steric demand were pointed out. The coordination preference of ethylenetetrathiolate was determined to be of the 1,2-dithiolene type rather than the 1,1-dithiolene. This result was suggested to apply both to binuclear and polymeric species.

16.5.3.8 Substitution and Alkylation Reactions

Although dithiolenes are stable compounds, many of which do not react with strong acids or bases, they are nevertheless reactive enough to undergo ligand displacement reactions as well as reactions on the ligands, especially when electrophilic attack is facilitated by reduction of the dithiolenes to their mono- or di-anionic forms or when strong electron-withdrawing substituents induce a higher charge density in the ligands and weaken the metal–sulfur bond.

In the absence of a nucleophile, most dithiolenes decompose slightly above their melting points to give low yields of organic sulfur-containing materials. For example, Schrauzer *et al.*[26] showed that tetraphenylnickeldithiolene produces 2-phenylbenzothiophene (2-phenylthionaphthene; **75**), tetraphenylthiophene (**76**) and tetraphenyldithiane (**77**). Nitrogen bases such as pyridine or hydrazine reduce dithiolenes to their radical mono- or di-anions.

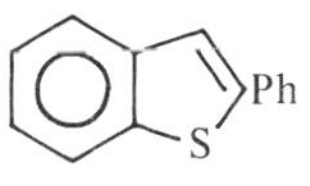

(75) 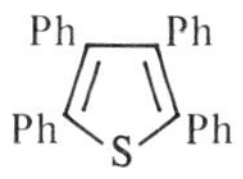   (76) 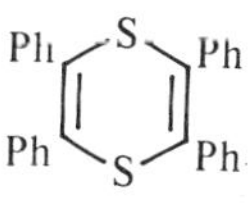   (77)

Schrauzer[6] also explored cycloaddition reactions between dithiolenes and alkenes or alkynes, which might have structures such as (**78**) as intermediates. Because of the dithiolene resonance, such reactions require energetic conditions and the intermediates were not observed.

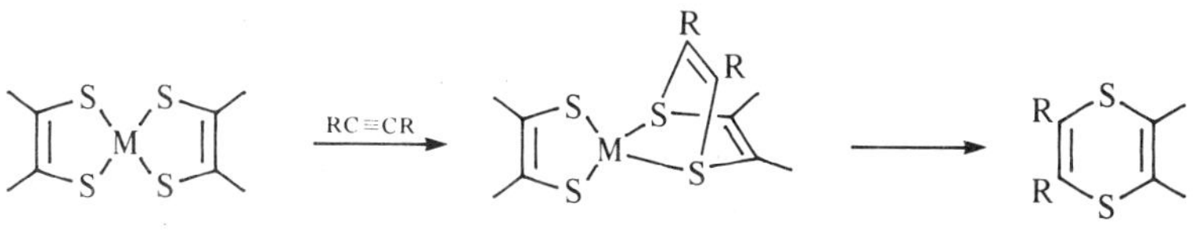

(78)

The only other known[141] direct addition reaction of neutral dithiolenes occurs with NO, which forms square pyramidal adducts in which NO occupies the axial position.

Reduction of dithiolenes to their dianions increases the nucleophilicity of the S atoms and allows reactions with alkyl halides to occur. Schrauzer *et al.*[142] described the formation of neutral complexes (**79**) with one dithioether and one dithiolato ligand when the tetraphenylnickeldithiolene dianion was reacted with methyl iodide or other alkyl halides.

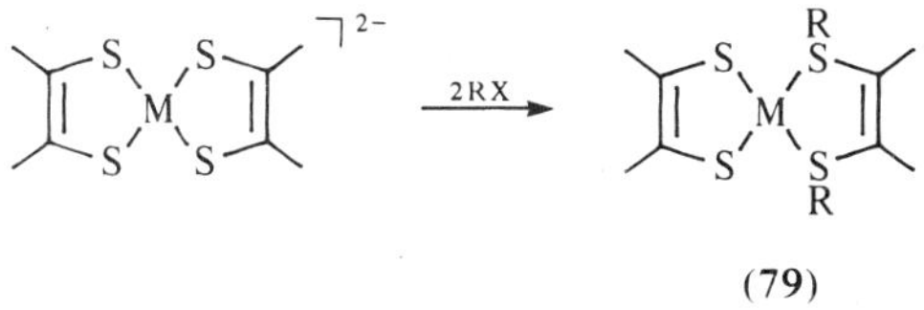

(79)

Allen *et al.*[143] showed that the reaction of a main group dithiolato complex dianion is rapid. When the tetrabutylammonium salt of $Pb(mnt)_2^{2-}$ was reacted with methyl iodide, $(R_4N)Pb(mnt)I$ precipitated and dimethyl-mnt was isolated from the supernatant solution in 81% yield. A rate comparison of the reaction of $Na_2(mnt)$ and $Pb(mnt)_2^{2-}$ showed clearly that complexation to Pb led to a dramatic rate increase.

One might thus wonder why the dianion $Ni(mnt)_2^{2-}$ was claimed[133,142] to be unreactive towards methyl iodide. A recent reinvestigation by Vlcek[41] of this reaction, motivated by a prediction that the dianionic nature in these mnt ligands should allow this reaction to occur, showed that complete reaction and dissociation of the complex occurs. The reaction can be described by the equation

$$M(mnt)_2^{2-} + 4MeI \rightarrow 2Me_2mnt + MI_4^{2-}$$

The reaction of $Pt(mnt)_2^{2-}$ with cyanide produces[144] the new dianion $Pt(mnt)(CN)_2^{2-}$, in which one ligand has been displaced.

There also exists a report[145] that Pd dithiolenes can be oxidized by bromine to give an undefined ESR active 'Pd^V' species. Whether this is a displacement or cleavage product instead remains to be seen.

Mayweg and Schrauzer[146] investigated the reactions of phosphines with the Ni, Pd and Pt tetraphenyldithiolenes. The nickel compound was heated with triphenylphosphine to 250 °C without apparent reaction. In benzene solution, an adduct $NiS_4C_4Ph_4 \cdot 2PPh_3$ was formed, which decomposed into dithiolene and phosphine above 120 °C. The Pd complex was converted into the planar bis-phosphinedithiolato complex and the Pt complex formed an isolable adduct as the only product. Davison and Howe[147] described the same reaction as producing only the substitution products (**80**).

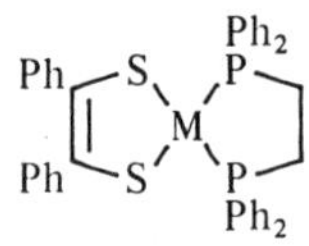

$$
\begin{array}{c}
\text{Ph} \\ \quad \text{S} \quad \text{P} \quad \text{Ph}_2 \\
\text{M} \\
\text{Ph} \quad \text{S} \quad \text{P} \quad \text{Ph}_2
\end{array}
$$

(80)

By contrast, diphos (1,2-bis(diphenylphosphino)ethane) was found to react[146] with the Ni and Pd complexes in solution to form planar complexes of type (**80**), while the Pt complex formed an isolable adduct of unknown structure, which could be thermally or photochemically converted into the planar product.

McCleverty and Ratcliff[148] investigated the redox chemistry of the complexes and found a three- or even four-step electron transfer series to exist in these compounds as well.

Vlcek and Vlcek[149] have summarized the reactivity of dithiolenes in a general way. The π-delocalization in the dithiolenes favors an extensive redox series but restricts axial interactions. Reaction in this position can only occur if the incoming ligand can incorporate its own π-orbitals into that of the planar complex. The dianions are nucleophilic at sulfur; oxidation to the anion or neutral species lets them become electrophilic and Diels–Alder-type additions may occur.

16.5.3.9 Comparison Between Dithiolenes and Other Complexes Containing Thioorganic Ligands

With the discovery of the extraordinary redox behavior of the dithiolenes and the establishment of a redox series between the neutral dithiolenes and their mono- and di-anionic analogs arose the question why these compounds are able to exist in several *stable* oxidation states while related species such as their SacSac analogs or even the isomeric 1,1-dithiolenes show only poorly defined and generally irreversible oxidation and reduction behavior.

An attempt to bring this into perspective by considering the ligand class topology was made by Schrauzer,[6] who defined dithiolenes as members of an 'even' class of ligands with an even number of conjugated atoms (four) in the central ligand unit. Diazabutadiene complexes, including the glyoximes and complexes of the S_2N_2 ligand, also belong in this group. On the other hand, the SacSac-type ligands (with five atoms) and the 1,1-dithiolates and dithiocarbamates (with three atoms) were assigned to the 'odd' ligand class with an odd number of ligand atoms. While this 'even/odd' classification has considerable merit in principle, it has often been abused to interpret data in a too simplistic way.

The basic idea is illustrated in Figure 1 as presented by Schrauzer.[6] A more detailed MO description by Herman *et al.*[136] of the dithiolene and SacSac complexes of Ni shows good correlation with the observed electronic spectral data as well as the electrochemical results and thus substantiates the merits of the even/odd classification. Considering only the frontier orbitals (Figure 2), we see that the LUMO of the dithiolene correlates with the HOMO of the SacSac series and the scheme easily explains the acceptor properties found in most of the dithiolenes as well as their absence in the SacSac compounds.

The synthesis by Mueller-Westerhoff and Alscher[150] of the Ni, Pd and Pt dithiomalonaldehyde ('ptt') complexes allowed the comparison of NMR chemical shift data between the parent even and odd systems with sulfur-containing conjugated ligands. The results are given in Table 2.

All proton signals show significant downfield shifts, but they clearly cannot be related to

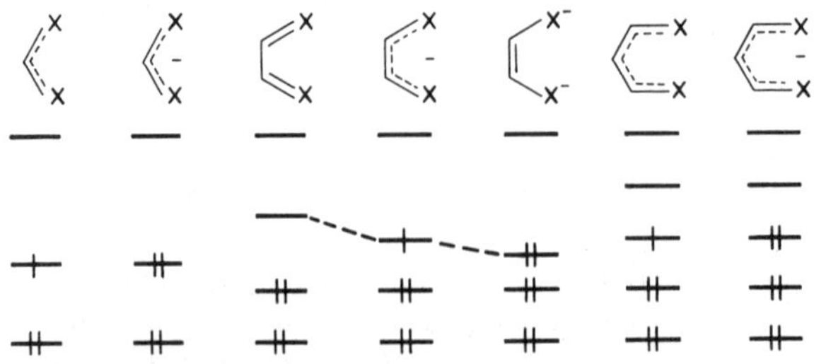

Figure 1 Simplified MO level diagrams for even and odd systems

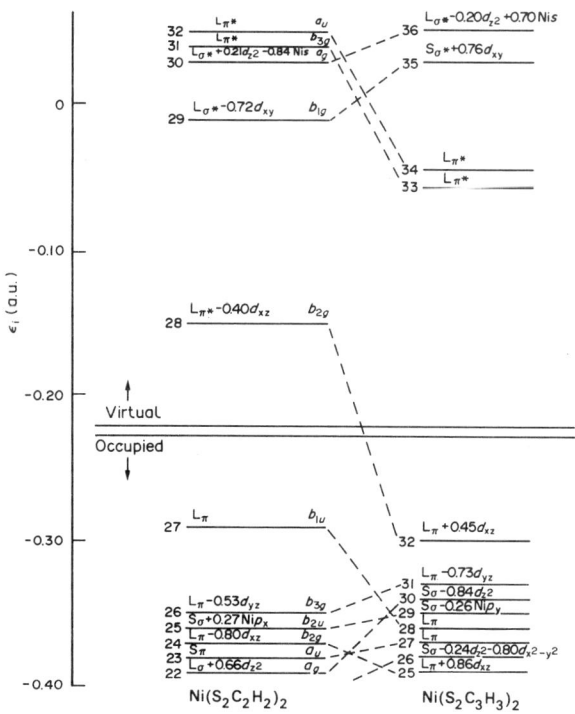

Figure 2 MO energies for $Ni(S_2C_2H_2)_2$ and $Ni(S_2C_3H_3)_2$ at the CNDO level

Table 2 ^1H Chemical Shifts in the Planar Parent Compounds of the Dithiolene and ptt Series[a]

Metal	$M(S_2C_2H_2)_2$[b]	$M(S_2C_3H_3)_2$ α-H	β-H
Ni	9.14	8.73	7.40
Pd	9.01	9.02	7.48
Pt	8.77[c]	9.35[d]	7.40[e]

[a] In p.p.m. *vs.* TMS (from ref. 150).
[b] Owing to the very poor solubility of the dimerized Pd and Pt dithiolenes the *t*-butyl derivatives were used for all three compounds.
[c] Pt–H coupling 116 Hz.
[d] Pt–H coupling 104 Hz. [e] Pt–H coupling 8 Hz.

delocalization alone: in the ptt series, the shifts of the α-protons increase from Ni to Pt, but those of the β-protons remain the same; if delocalization were the cause of the strong downfield shift, all protons should experience it. Apparently, differences in charge distribution are the major factor in determining these chemical shifts. Notable is the magnitude of the Pt–H coupling in both classes: the dithiolene ligand protons and the α-protons of the ppt complex show the same large coupling. The β-protons in the $Pt(ptt)_2$ complex show only an 8 Hz coupling with ^{195}Pt.

A semiquantitative study by Nazzal and Mueller-Westerhoff[151] of the extent of cyclic electron delocalization ('aromaticity') within each ligand ring in both classes was based on the NMR chemical shift differences observable in mesityl derivatives of these compounds. If a delocalized cyclic structure carries a mesityl substituent, the methyl groups in the *ortho* position will experience an upfield shift, while those in the *para* position do not . Their chemical shift difference Δδ can be used as a measure of delocalization. Compounds with Δδ > 0.30 Hz are considered to be 'aromatic', *i.e.* highly delocalized.

These experiments showed that the Ni dithiolene is highly delocalized while Pd and Pt dithiolenes show less delocalization, a trend which parallels the ^1H chemical shifts and the observation that the latter two form dimers while the Ni complex does not. In the ptt series, all three complexes show a lower as well as a completely undifferentiated degree of delocalization. These data are given in Table 3. For obvious reasons, similar measurements cannot be made for 1,1-dithiolenes.

Sulfur Ligands

Table 3 ¹H Chemical Shift Difference
($\Delta\delta$) Between *o*- and *p*-Methyl Groups in
Mesityl Derivatives of Dithiolenes and
SacSacs[a]

Metal	$M(S_2C_2H_2)_2$	$M(S_2C_3H_3)_2$
Ni	0.40	0.32
Pd	0.31	0.32
Pt	0.34	0.32

[a] The larger $\Delta\delta$, the more delocalized the system.

Although the ptt complexes are deeply colored,[150] they lack the intense absorption at low energy found in the class of dithiolenes and which is assigned to a $\pi-\pi^*$ transition of the delocalized system. The absence of such a transition in the ptt complexes is not a contradiction of the above results on the cyclic electron delocalization. It rather points out that the dithiolenes show delocalization throughout the entire complex unit as shown in (81), but the delocalization in the ptt complexes apparently involves the two ring systems independently, as shown in (82): the cyclic electron delocalization does involve the metal, but there is no crossover of delocalization from one half of the molecule to the other.

(81)   (82)

Bowden *et al.*[152] studied the electrochemistry of the Ni, Pd and Pt ptt complexes and of their SacSac derivatives and compared it with that of the dithiolenes, pointing out the dangers of the even–odd distinction. Both the dithiolenes and the ptt complexes undergo two ligand-based one-electron reductions. However, the reduction of the ptt complexes requires much more (approximately 1 V) negative potentials; the reductions are close to irreversible. This means that the anionic species derived from ptt and SacSac complexes are much less stable than the dithiolene mono- and di-anions, in accord with the MO results of Figure 2.

16.5.4 USEFUL PROPERTIES OF DITHIOLENES

16.5.4.1 Conductivity and Magnetism

Dithiolenes have been the subject of numerous studies relating to their ability to conduct electric currents and to show unusual magnetic properties. Both are consequences of extended inter-molecular interactions in molecular crystals of such order that the overlap between adjacent molecules is strong and uniform throughout the entire crystal.

16.5.4.1.1 General remarks on one-dimensional stacked dithiolenes

Much attention is focused on partially reduced dithiolenes, their stacking properties and the existence and nature of extended interactions in such stacks. Since both electrical conductivity and any form of magnetism necessarily require the presence of open shell systems, the exceptional stability of radical anions of the dithiolenes has made them a particularly interesting subject for the study of extended interactions in molecular solids. It is important to realize that nonintegral oxidation states are essential in many, but not all, cases to avoid a temperature dependent phase transition between a conductive and an insulating or semiconducting state. Integral oxidation states in radical anions or cations lead to half-filled bands which may reach a more stable phase for one-dimensional conductors by dimerization, creating an insulating phase; a partially filled band may avoid this so-called Peierls instability.

A second prerequisite for extended interactions is the arrangement of molecules in the solid in a way that intermolecular contact may be made. Beyond the formation of dimers or oligomers, long range interactions require the formation of radical ion stacks in which the intermolecular distances are as small as possible. The d^8 dithiolenes are particularly well suited for this, because their radical anions are planar and allow intermolecular interactions either through metal–metal bonds akin to those seen in the neutral dimers or through metal–sulfur bonds.

The magnetic properties and the magnitude of the electrical conductivity thus largely depend on the stacking arrangements in dithiolenes in nonintegral oxidation states. Many structural investigations concern these questions.

A particular subgroup of these low-dimensional conductors are the charge transfer complexes between dithiolenes and organic donor species such as TTF and related compounds. Here, nonintegral charge transfer is often seen as the reason for high conductivity down to low temperatures.

To deviate from strict one-dimensionality seems to be of great importance for the stability of the metallic phase. Consequently, the design of materials with highly anisotropic but not strictly one-dimensional interactions appears to be a promising research area of much current interest.

Several books, conference proceedings and reviews have summarized this active field with reasonable completeness.[21,153–159] The initial phases of the exploration of dithiolenes as conductive solids mostly concerned the partially reduced nickel and platinum dithiolenes of three types: the parent dithiolenes (edt), the cyano derivative (mnt) and the trifluoromethyl analog (tfd). Historically, however, a charge transfer complex was the first conductor involving a dithiolene. The most recent work involves the charge transfer complexes of the so-called isotrithione dithiolenes, because of the realization that many of these stacked complexes have significant three-dimensional interactions. The trithione units of the ligand provide such interstack exchange pathways and stabilize the metallic phase. These three areas will be discussed in this order.

The terminology 'partially reduced' and 'partially oxidized' dithiolenes used in this physics-dominated field may be confusing, and a few words to clarify this situation may be needed. (i) The Ni atom in an $Ni(mnt)_2^{n-}$ complex, in which $n = 0, 1, 2$ or 3, will have a formal oxidation state of $4+, 3+, 2+$ or $1+$, respectively. (ii) An $Ni(mnt)_2^{n-}$ complex with a nonintegral nickel oxidation state, for example $3.5+$ in a complex with $n = 0.5$, can be formally and experimentally derived either from $Ni(mnt)_2^0$ by partial reduction or from $Ni(nmt)_2^-$ by partial oxidation. (iii) A look at the formula of the product will convince the confused reader that the terms 'partially oxidized' and 'partially reduced' are used interchangeably and that they may indeed describe the same complex.

16.5.4.1.2 *Partially reduced ditholenes*

The earliest report[160] on one-dimensional dithiolene salts with innocent cations concerned the structure of the integral valence systems $Ni(mnt)_2^-$ and $Ni(mnt)_2^{2-}$. In the dianion, the nickel atoms were separated by 9.84 Å, while the interplanar distance shrank to 3.5 Å in the monoanion. The conductivities in some salts of $Ni(mnt)_2$ and $Pd(mnt)_2$ monoanions with closed shell cations were found[161] to be surprisingly high and an increase in conductivity in one such case as a consequence of partial hydration was noted but not recognized as an indication of the importance of nonintegral stoichiometries. This came shortly afterwards[162] when cation properties and anion disproportionation were shown to be all-important. Methylene blue as a redox active cation was shown to give the highest conductivities. This example has recently been reinvestigated.[163]

The groups of Kobayashi and Underhill have focused their attention on partially oxidized Ni and Pt mnt complex anions with nonstoichiometric proportions of different cations. Such nonintegral valence salts have the general structure $(cation)_n M(mnt)_2^{n-}(H_2O)_x$ ($x = 0–2$), in which n can be anywhere between 0.5 and 0.82.

In several separate or joint recent (1981–85) communications from these groups,[164–173] the structural, conductive and magnetic aspects of these complexes, including the nature of the Peierls instability,[168,173] were described.

There are many contributions to this field from other laboratories as well. The reader is referred to the cited literature for details on other results. To summarize the results, four conclusions can be drawn: (i) there still exists uncertainty about the exact compositions of the complexes involved; the formal oxidation states are not always known, but it is certain that nonintegral ones are essential for high conductivity; (ii) the formation of linear stacks of parallel ions and close stacking create the necessary intermolecular exchange interactions which result in band formation and conductivity; (iii) electron transport occurs through a hopping process with a very small activation barrier; and (iv) one-dimensional stacks are inherently unstable with respect to the formation of dimerized (Peierls) arrangements through which a gap opens at the Fermi level and an insulating phase results.

Electron transport is impossible in closed shell systems (the moving electron would have to occupy an antibonding orbital) and in half-filled open shell systems (spin pairing would prohibit the transport of electrons). Only the existence of hole states in partially oxidized dithiolenes allows electrons to flow freely.

16.5.4.1.3 Charge transfer salts of dithiolenes

The first reports in 1969 and 1971 from the group of Maki[174,175] (expanded later[176,177]) on charge transfer complexes between large, delocalized organic molecules and a dithiolene as acceptor were followed in 1973 by the first example of a highly conducting complex involving the strong electron donor tetrathiafulvalene (TTF). This report by Wudl *et al.*[178] was the beginning of a rapid and often extremely confusing development in this area. The confusion arose from widely diverging results on the magnitude of the highly anisotropic conductivity, which is strongly dependent on sample purity: in a one-dimensional stack each impurity interrupts the conduction path along an entire stack, so that conductivity differences of several orders of magnitude can be seen in the same compound and even in different crystals from the same batch.

In 1982, Ibers *et al.*[179] compiled crystallographic data on stacked metal complexes consisting of an electron acceptor, A and an electron donor, D. A large number of these D—A complexes contained a 1,2-dithiolene as the A component (ML_2; M = Fe, Co, Ni, Pt, Cu or Au; L = edt, mnt, tfd or bdt). Crystallographic data existed for their coupling to a variety of donors. D—A complexes composed of a transition metal moiety inherently allow for the tuning of the metal and ligand energy levels, thereby permitting chemical modifications unavailable to organic D—A systems. The authors forecast greater research activity in this field to maximize the already substantial conductivity and unusual magnetic properties exhibited by some members in this class of compounds. Alcacer and Novais limited their similar review[157] to 1,2-dithiolenes but their coverage was extended to the early transition metal and heavier main group metals; additional dithiolene substituents were also included ($S_2C_2R_2$; R = Me, Et or Ph). The complexes were categorized according to their stacking arrangement within the crystal. Magnetic and electrical properties were then discussed in relation to their findings.

The more recent work includes two detailed studies on tetramethylphenylenediamine (TMPD) complexes of Ni- and Pt-mnt[180,181] and an example of a complex in which both the donor and the acceptor are nickel complexes.[182] However, most recent activity has been devoted to materials in which the strict one-dimensionality of the dithiolene salts and complexes has given way to interstack coupled systems with lesser degrees of anisotropy.

16.5.4.1.4 Dmit complexes and their salts

The most frequently used designation 'dmit' (dimercaptoisotrithione) for the ligand (83) may not be the best choice (the name ethylenetrithiocarbonatedithiolate or trithionedithiolate would seem more appropriate and 1,3-dithiol-2-thione-4,5-dithiolate would seem accurate), but it has established itself throughout this corner of the chemical literature.

Steimecke *et al.*[183] prepared the ligand as a dianion, purified it *via* a Zn salt and prepared its nickel complex (84) as a dianion, monoanion and finally as the neutral complex, although the latter was isolated only as an insoluble powder. The anionic species was characterized[184,185] by X-ray crystallography. Ribas and Cassoux[186] also succeeded in preparing a highly conductive crystalline version of the neutral compound. Since then, several research groups have investigated a number of cation deficient $M_x[Ni(dmit)_2]$ complexes,[187–191] charge transfer complexes with various donors[186,187,189,191–195] and Se analogs of the dmit ligand and its complexes.[188]

(83)   (84) *n* = 0, 1, 2

An anionic Fe(dmit)₂ was prepared and its Bu_4N salt was found to be highly conductive.[194] Otherwise, Ni complexes have been the ones to receive all the attention. Only very recently have dmit complexes of Pd and Pt been investigated.[195]

The partially oxidized mixed-valence complexes were generated by electrochemical methods. In the case of the Bu_4N salt ($x = 0.29$), the extent of oxidation was determined by elemental analysis, electron microprobe analysis, mass spectra and the X-ray structure analysis; only elemental analysis was cited for the Et_4N ($x = -0.5$) derivative. Upon reduction, the essentially planar anions exhibit the same Ni—S and S—C bond lengthening (see Table 4) as mentioned in Section 16.5.3.1. Based on the relative changes in the last two entries in the above table (then the only available data), Lindqvist *et al.*[184,185] suggested the redox process was centered on the metal.

The earlier reports emphasized the intramolecular bond distances and angles to establish the basic geometry of [Ni(dmit)₂]. The report on the monoanion made no mention of interionic contacts; that

Table 4 Ni—S and S—C Bond Lengthening in M_x[Ni-(dmit)$_2$] Complexes on Reduction

M	x	Ni—S′(Å)	S′—C(Å)	Ref.
Et$_4$N	0.5	2.152	1.69	190
Bu$_4$N	1	2.156	1.72	184
Bu$_4$N	2	2.216	1.75	185

on the dianion noted a 12.07 Å Ni$\cdots$Ni separation and only normal van der Waals interactions for all other interionic contacts. Subsequent reports on the mixed-valence complexes concentrated on intermolecular associations. For the Bu$_4$N ($x = 0.29$) complex, two types of S$\cdots$S contacts were found between the anions stacked parallel to *b*. The π-type bridged the anions within a stack and the σ-type (3.47(2) Å) extended the interaction from stack to stack, creating a two-dimensional network. The electrical conductivity reflects the crystal's dual dimensionality: $\sigma_a = 10$, $\sigma_b = 5$, $\sigma_c = 10^{-3} \Omega^{-1} cm^{-1}$.

Note that the conductivity is not greatest along the stack. The Et$_4$N ($x = 0.5$) salt contains weakly coupled dimers of eclipsed [Ni(dmit)$_2$] stacked along *b*. The interplanar distance between dimers is 3.759 Å, whereas it is only 3.437 Å within a dimer, fostering S$\cdots$S contacts of 3.610 and 3.691 Å. Again, numerous S$\cdots$S contacts in the 3.499 to 3.687 Å range create a two-dimensional interstack network. Lastly, a 3.903 Å separation was noted between terminal thionyl sulfurs along *a*. The anisotropy of the electrical conductivity was markedly decreased: $\sigma_a:\sigma_b:\sigma_c = 1:50:8$.

In the series of TTF[M(dmit)$_2$]$_x$ complexes with M = Ni, Pd and Pt, the Ni and Pd systems are isostructural and have 1:2 stoichiometry ($x = 2$) while the Pt compound crystallizes in a 1:3 stoichiometry.[195]

In TTF[Ni(dmit)$_2$]$_2$, both components are stacked separately and uniformly along *b* (3.73 Å). The 3.55 and 3.65 Å interplanar spacing between TTF donors and between [Ni(dmit)$_2$] acceptors are slightly less than *b* because the molecules are tilted 18 and 12° respectively out of the *ac* plane. Nonetheless, interatomic distances are still greater than the sum of the appropriate van der Waals radii. The [Ni(dmit)$_2$] stacks are loosely zig-zag stitched down the length of the molecules by S$\cdots$S contacts ranging from 3.45 to 3.71 Å. Extending the network to three dimensions are S$\cdots$S interactions of only 3.39 and 3.59 Å between sulfur atoms of TTF and the thionyl sulfurs of [Ni(dmit)$_2$].

Although the anisotropy of the complexes' electrical conductivity was not mentioned, the three-dimensional nature of the complex facilitates a metal-like temperature dependence of its conductivity down to 4 K where $\sigma = 10^5 \Omega^{-1} cm^{-1}$ (room-temperature σ is a respectable 300 $\Omega^{-1} cm^{-1}$). The absence of a metal–insulator transition down to 4 K shows that the Peierls instability has been successfully avoided by increasing the interstack coupling.

The isostructural Pd complex has a metallic regime above 220 K with σ near 700 $\Omega^{-1} cm^{-1}$. However, it exhibits a semiconductive behavior below that temperature. In the Pt complex the conductivity is clearly thermally activated in the entire temperature region. The crystal structure established the reason for this behavior: alternating monomer and dimer molecules of the dmit complex form an irregular stack, while the uniformly stacked TTF molecules show large (6.31 Å) stacking distances.

16.5.4.2 Infrared Dyes and Their Uses

Several specialized applications of dithiolenes utilize in different ways their intense IR absorption. The high thermal stability of dithiolenes and their photostability towards long wavelength radiation and ambient light makes them useful as substrates in passive Q-switch and mode-locking applications for different IR lasers. In these processes, the dye molecules serve as reversibly bleachable absorbers of the right relaxation time to allow bleaching of a dye solution at a desirable level of photon flux. Placing such a dye solution in the laser cavity has the important effect of converting low power CW radiation into very short and extremely intense pulses. The first example of this application, discovered by Drexhage and Mueller-Westerhoff,[196] was Q-switching of the Nd glass and Nd YAG lasers (operating at 1.06 μm) by bis(*p*-dimethylaminostilbenedithiolato)nickel (1-phenyl-2-(4-dimethylaminophenyl)nickeldithiolene). While most organic dyes which absorb in this region show low photostability and decompose almost instantly, dithiolenes have been shown to withstand long term exposure to IR radiation. It has even been possible to use solid state Q-switch materials made from solutions of dithiolenes in polymers.[197,198] Dithiolenes are commercially available for this purpose.

The absorption maxima of dithiolenes are solvent dependent and can be shifted within limits to correspond to the laser wavelength. Several studies have been published which address this point.[199-201]

The bis(julolidinylethylenedithiolato) complexes (**9**) of Ni and Pt absorb in the 1.3 μm region and are thus suited to Q-switch the technically important iodine laser. This laser is one candidate of several thought to be able to initiate nuclear fusion reactions. At present, the julolidine-substituted Ni dithiolene is the best available Q-switch for the iodine laser. In appropriate solvents, the dialkylamino-substituted complexes can be shifted into the 1.3 μm region and also be used for this purpose.[202,203]

Although this process for using dithiolenes has been patented[204] and published,[196] a number of more recent Japanese[205] and East German[206] patents also deal with this application.

16.5.4.3 Mesomorphic dithiolenes

Planar dithiolenes are ideally suited to mimic the shape of many types of organic liquid crystals, because the shape of the dithiolene nucleus is not too different from that of a benzene ring.[207] Substituting a dithiolene for a benzene ring in structures known to exhibit mesomorphic properties leads to liquid crystals containing transition metals and thus incorporating some or all of the properties associated with these classes of compounds. A typical example is the replacement of the central benzene ring in the nematic (**85**) by a dithiolene–nickel complex ring to produce the liquid crystalline material (**86**).

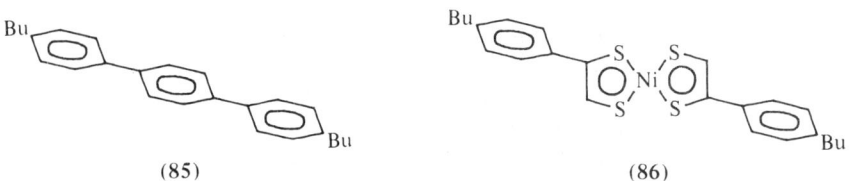

(85)   (86)

While the typical organic liquid crystals of type (**85**) are colorless, the dithiolene-containing ones have a strong absorption band at 850 nm and the redox properties typical for this class of dithiolenes. Polarized transmission microscopy as well as ^{13}C NMR spectroscopy in the mesophase temperature region show that both (**85**) and (**86**) have a nematic phase, while longer hydrocarbon chains in both classes (C_6 to C_{10}) cause the liquid crystalline phases to be smectic.[104] The transition temperature to the mesophases is considerably higher in the dithiolene liquid crystal than in its organic analog, possibly indicating stronger interactions in the crystal lattice. The crystal structure of the diphenylnickeldithiolene (**20**), substituted by *n*-octyl chains, shows a high degree of order in the crystalline phase.[34]

Platinum analogs in this series of C_4 to C_{10} show fully parallel behavior in forming nematic or smectic mesophases. By contrast, none of the analogous Pd complexes show any tendency to form a liquid crystalline phase. This is thought to be due to stronger intermolecular interactions and the formation of Pd—Pd-bonded dimers.[36]

The precursor dithiocarbonates were also used to prepare mesomorphic derivatives of TTF (tetrathiafulvalene), a strong electron donor compound used in the formation of highly conductive charge transfer complexes. Attempts to prepare CT complexes between the mesomorphic dithiolenes as acceptors and these mesomorphic TTF donors showed that the electron acceptor strength of this type of dithiolene does not suffice to form strong donor–acceptor complexes: mixtures of the two components form mixed crystals without any visible degree of charge transfer.[208]

Mixing mesomorphic dithiolenes with standard liquid crystals of high dielectric constant produced mesomorphic materials which combine the useful properties of both classes. A cell containing such a mixture has been used as an IR light valve: in the thermodynamically preferred parallel alignment of the rod-like molecules with the cell walls, the cell is opaque to IR light; application of a low voltage realigns the molecule to a vertical position with respect to the cell walls and the cell becomes transparent to IR light.[209]

Since it is well established that, with one exception,[123] dithiolenes show neither fluorescence nor phosphorescence, any absorbed energy must be released through nonradiative relaxation processes, implying that their local environment will be heated. Dithiolenes thus can serve as sensitizers in processes where IR radiation would otherwise not be absorbed. One example is the local melting by GaAs laser radiation (850 nm) of polystyrene, which itself is transparent in this spectral region, when appropriately substituted dithiolenes are added. Films cast from polystyrene containing 5 to 20% of a dithiolene can be used as recording materials for videodisk applications: localized irradiation causes small holes to be formed by local melting. A complementary process uses the same type of film in a reverse way: because dithiolenes are unstable toward UV radiation, the green films can be locally bleached by exposure to UV light, leaving a green film with transparent 'marks' which can be read by a low power GaAs laser. While the basic processes have been understood, there are no known commercial applications for these effects. The long alkyl chains of the mesomorphic dithiolenes provide these materials with high solubility in organic solvents, including polymers. They are thus particularly well suited to be used as polymer additives in the hole-burning applications described above. The 850 nm absorption of mixtures containing the mesomorphic dithiolenes makes them particularly useful for applications involving the GaAs laser. These processes are covered by invention disclosures[209] and patents.[210]

16.5.4.4 Dithiolenes as Catalysts

Dithiolenes with a variety of central metals and ligands were investigated by Kisch *et al.*[211-218] to determine their activity as catalysts in the photoproduction of hydrogen from water by irradiating H_2O/THF solutions of the complex for extended periods and measuring the amount of hydrogen produced. Only hydrogen evolution was observed, and any oxygen resulting from the photo-dissociation of water must have reacted with the medium. The most active complexes by far were those containing Zn and ligands such as mnt.

The initial results were interpreted as evidence of catalytic activity by dithiolenes themselves.[211,212] However, Kisch *et al.*[213] were soon able to demonstrate that the $Zn(mnt)_2$ complex dianion decomposed to ZnS under these conditions and it is now believed that the dithiolene is only the catalyst precursor and that ZnS is the actual catalyst. The sulfide must be extremely finely dispersed, approaching near homogeneous conditions, because filtration of the solution through a cellulose acetate filter of $0.2\,\mu m$ pore size did not reduce the rate of reaction.

After the initial enthusiasm created by several publications on this subject, the catalytic properties of dithiolenes in photochemical water-splitting processes are now somewhat less exciting. Nevertheless, this area continues to receive attention, both from the experimental[214-218] and the theoretical[140] side.

A few other applications of dithiolenes make use of their redox properties. Kumar *et al.*[219] proposed the use of dithiolenes as photosensitizers. Umezawa *et al.*[220] coated a Pt cathode with $(Et_4N)Ni(mnt)_2$ and saw a modest degree $(1.4 \times 10^{-4}\%)$ of light conversion upon irradiation. On the opposite side, Bradley *et al.*[221] used dithiolenes to stabilize *n*-type Si anodes against photoanodic decomposition.

A final example of the use of dithiolenes comes from the work of Grimaldi and Lehn,[222] and Ohki, Tagaki and Ueno.[223] These groups used tetraphenylnickeldithiolene as a redox potential driven electron carrier and cation carrier through artificial membranes. Lipophilic cocarriers were employed to generate a multicomponent carrier system, in which charge equalization occurs. Applications to biomembranes, ion separation and related processes were suggested.

16.5.5 REFERENCES

1. G. N. Schrauzer and V. P. Mayweg, *J. Am. Chem. Soc.*, 1962, **84**, 3221.
2. H. B. Gray and E. Billig, *J. Am. Chem. Soc.*, 1963, **85**, 2019.
3. A. Davison, N. Edelstein, R. H. Holm and A. H. Maki, *J. Am. Chem. Soc.*, 1963, **85**, 2029.
4. J. A. McCleverty, *Prog. Inorg. Chem.*, 1968, **10**, 49.
5. G. N. Schrauzer, *Transition Met. Chem. (N.Y.)*, 1968, **4**, 299.
6. G. N. Schrauzer, *Acc. Chem. Res.*, 1969, **2**, 72.
7. R. E. Eisenberg, *Prog. Inorg. Chem.*, 1970, **12**, 295.
8. R. H. Holm and M. J. O'Connor, *Prog. Inorg. Chem.*, 1971, **14**, 241.
9. E. Hoyer, W. Dietzsch and W. Schroth, *Z. Chem.*, 1971, **11**, 41.
10. J. A. McCleverty, *MTP Int. Rev. Sci.: Inorg. Chem. Ser. One*, 1972, **2**, 301.
11. R. P. Burns and C. A. McAuliffe, *Adv. Inorg. Chem. Radiochem.*, 1979, **22**, 303.
12. K. W. Browall and L. V. Interrante, *J. Coord. Chem.*, 1973, **3**, 27.
13. K. W. Browall, L. V. Interrante and J. S. Kasper, *J. Am. Chem. Soc.*, 1971, **93**, 6289; K. W. Browall, T. Bursh, L. V. Interrante and J. S. Kasper, *Inorg. Chem.*, 1972, **11**, 1800.
14. A. E. Smith, G. N. Schrauzer, V. Mayweg and W. Heinrich, *J. Am. Chem. Soc.*, 1966, **87**, 5798.

15. E. Hoyer, W. Dietzsch, H. Hennig and W. Schroth, *Chem. Ber.*, 1969, **102**, 603; E. Hoyer, W. Dietzsch and H. Müller, *Z. Chem.*, 1967, **7**, 354.
16. G. Bähr and G. Schleitzer, *Chem. Ber.*, 1955, **88**, 1771; *Chem. Ber.*, 1957, **90**, 438; A. Davison and R. H. Holm, *Inorg. Synth.*, 1967, **10**, 8.
17. W. Schroth and J. Peschel, *Chimia*, 1964, **18**, 171; E. Fromm, H. Benzinger and F. Schäfer, *Justus Liebigs Ann. Chem.*, 1912, **394**, 325.
18. A. Nazzal, A. Jaffe and U. T. Mueller-Westerhoff, unpublished; A. Nazzal, R. W. Lane, J. J. Mayerle and U. T. Mueller-Westerhoff, 1977, *Final Report USARO, United States NTIS 1978*, **78**, 137.
19. E. Campaigne, R. D. Hamilton and N. W. Jacobsen, *J. Org. Chem.*, 1964, **29**, 1708; A. K. Bhattacharya and A. G. Hortmann, *J. Org. Chem.*, 1974, **39**, 95.
20. R. K. Brown, T. J. Bergendahl, J. S. Wood and J. H. Waters, *Inorg. Chim. Acta*, 1983, **68**, 79.
21. P. J. Nigrey, in 'Extended Linear Chain Compounds', ed. J. S. Miller, Plenum, New York, 1983, vol. 3.
22. W. Kusters and P. deMayo, *J. Am. Chem. Soc.* 1974, **96**, 3502.
23. U. T. Mueller-Westerhoff, R. L. Lane and A. Nazzal, unpublished.
24. U. T. Mueller-Westerhoff, B. Vance and D. I. Yoon, to be published.
25. R. B. King, *Inorg. Chem.*, 1963, **2**, 641.
26. G. N. Schrauzer, V. P. Mayweg, H. W. Finck, U. T. Mueller-Westerhoff and W. Heinrich, *Angew. Chem.*, 1964, **76**, 345.
27. G. N. Schrauzer, V. P. Mayweg and W. Heinrich, *Inorg. Chem.*, 1965, **11**, 1615.
28. D. C. Bravard, W. E. Newton, J. T. Huneke, K. Yamanouchi and J. H. Enemark, *Inorg. Chem.*, 1982, **21**, 3795.
29. J. Jones and J. Douek, *J. Inorg. Nucl. Chem.*, 1981, **43**, 406.
30. M. Draganjac and D. Coucouvanis, *J. Am. Chem. Soc.*, 1983, **105**, 139.
31. J. S. Miller and D. G. Marsh, *Inorg. Chem.*, 1982, **21**, 2891.
32. G. Steimecke, R. Kirmse and E. Hoyer, *Z. Chem.*, 1975, **15**, 28.
33. E. Uhlemann and B. Zöllner, *Z. Chem.*, 1974, **14**, 245.
34. M. Cotrait, J. Gaultier, C. Polycarpe, A. M. Giroud and U. T. Mueller-Westerhoff, *Acta Crystallogr., Sect. C*, 1983, **39**, 833.
35. D. Bailey and J. P. Yesinowski, *J. Chem. Soc., Dalton Trans.*, 1975, 498.
36. U. T. Mueller-Westerhoff and A. Nazzal, unpublished observations; U. T. Mueller-Westerhoff, A. Nazzal, R. J. Cox and A. M. Giroud, *Mol. Cryst. Liq. Cryst. Lett.*, 1980, **56**, 249.
37. A. Davison, J. A. McCleverty, E. T. Shawl and E. J. Wharton, *J. Am. Chem. Soc.*, 1967, **89**, 830.
38. A. Vogler and H. Kunkely, *Angew. Chem., Int. Ed. Engl.*, 1982, **21**, 77.
39. T. W. Gilbert, Jr. and E. B. Sandell, *J. Am. Chem. Soc.*, 1960, **82**, 1087.
40. G. N. Schrauzer and H. N. Rabinowitz, *J. Am. Chem. Soc.*, 1968, **90**, 4297; 1969, **91**, 6522.
41. A. Vlcek, Jr., *Inorg. Chim. Acta*, 1980, **43**, 35.
42. R. Heber and E. Hoyer, *J. Prakt. Chem.*, 1973, **315**, 106.
43. J. H. Noordik and P. T. Beurskens, *J. Cryst. Mol. Struct.*, 1971, **1**, 339; J. H. Noordik, T. W. Hummelink and J. G. M. van der Linden, *J. Coord. Chem.*, 1973, **2**, 185.
44. J. Stach, R. Kirmse, U. Abram, W. Dietzsch, J. H. Noordik, K. Spee and K. P. Keijzers, *Polyhedron*, 1984, **3**, 433.
45. L. H. Pignolet, R. A. Lewis and R. H. Holm, *Inorg. Chem.*, 1972, **11**, 99; L. H. Pignolet, G. S. Patterson, J. F. Weiher and R. H. Holm, *Inorg. Chem.*, 1974, **13**, 1263.
46. J. A. McCleverty, N. M. Atherton, N. G. Connelly and C. J. Winscom, *J. Chem. Soc. (A)*, 1969, 2242.
47. T. R. Miller and I. G. Dance, *J. Am. Chem. Soc.*, 1973, **95**, 6970.
48. G. A. Bowmaker, P. D. W. Boyd and G. K. Campbell, *J. Chem. Soc., Dalton Trans.*, 1983, 1019.
49. A. Vogler, H. Kunkely, J. Hlavatsch and A. Merz, *Inorg. Chem.*, 1984, **23**, 506.
50. J. A. McCleverty, N. M. Atherton, J. Locke, E. J. Wharton and C. J. Winson, *J. Am. Chem. Soc.*, 1967, **89**, 6082.
51. P. Thomas, M. Lippmann, D. Rehorek and H. Hennig, *Z. Chem.*, 1980, **20**, 155.
52. G. N. Schrauzer, V. P. Mayweg, H. W. Fink and W. Heinrich, *J. Am. Chem. Soc.*, 1966, **88**, 4604; A. L. Balch, *Inorg. Chem.*, 1967, **6**, 2158.
53. J. Locke and J. A. McCleverty, *Inorg. Chem.*, 1966, **5**, 1157.
54. R. Hemmer, H. A. Brune and U. Thewalt, *Z. Naturforsch., Teil B*, 1981, **36**, 78.
55. D. G. VanDerveer and R. Eisenberg, *J. Am. Chem. Soc.*, 1974, **96**, 4994.
56. J. J. Maj, A. D. Rae and L. F. Dahl, *J. Am. Chem. Soc.*, 1982, **104**, 4278.
57. K. P. C. Vollhardt and E. C. Walborsky, *J. Am. Chem. Soc.*, 1983, **105**, 5507.
58. E. J. Miller, T. B. Brill, A. L. Rheingold and W. C. Fultz, *J. Am. Chem. Soc.*, 1983, **105**, 7580.
59. R. Weberg, R. C. Haltiwanger and M. Rakowski DuBois, *Organometallics*, 1985, **4**, 1315.
60. W. H. Mills and E. D. Clark, *J. Chem. Soc.*, 1936, 175.
61. T. W. Gilbert, Jr. and E. B. Sandell, *J. Am. Chem. Soc.*, 1960, **82**, 1087.
62. J. M. Kisenyi, G. R. Willey, M. G. B. Drew and S. O. Wandiga, *J. Chem. Soc., Dalton Trans.*, 1985, 69.
63. G. Hunter and T. J. R. Weakley, *J. Chem. Soc., Dalton Trans.*, 1983, 1067.
64. E. S. Bretschneider, C. W. Allen and J. H. Waters, *J. Chem. Soc. (A)*, 1971, 500.
65. A. C. Sau, R. O. Day and R. R. Holmes, *Inorg. Chem.*, 1981, **20**, 3076.
66. R. O. Fields, J. H. Waters and T. J. Bergendahl, *Inorg. Chem.*, 1971, **10**, 2808.
67. F. W. B. Einstein, G. Hunter, D. G. Tuck and M. K. Yang, *Chem. Commun.*, 1968, 423.
68. A. F. Berniaz, G. Hunter and D. G. Tuck, *J. Chem. Soc. (A)*, 1971, 3254.
69. L. Bustos, M. A. Khan and D. G. Tuck, *Can. J. Chem.*, 1983, **61**, 1146.
70. E. W. Abel and C. R. Jenkins, *J. Chem. Soc. (A)*, 1967, 1344.
71. R. Uson, J. Vicente and J. Oro, *Inorg. Chim. Acta*, 1981, **52**, 29.
72. M. J. Baker-Hawkes, E. Billig and H. B. Gray, *J. Am. Chem. Soc.*, 1966, **88**, 4870.
73. D. Sellmann, U. Kleine-Kleffmann, L. Zapf, G. Huttner and L. Zsolnai, *J. Organomet. Chem.*, 1984, **263**, 321.
74. D. Sellmann and E. Unger, *Z. Naturforsch., Teil B*, 1979, **34**, 1096.
75. D. Sellmann, H. E. Jonk, H. R. Pfeil, G. Huttner and J. v. Seyerl, *J. Organomet. Chem.*, 1980, **191**, 171; D. Sellmann and W. Ludwig, *J. Organomet. Chem.*, 1984, **269**, 171; D. Sellmann and W. Reisser, *Z. Naturforsch., Teil B*, 1984, **39**, 1268.

76. G. Henkel, K. Greiwe and B. Krebs, *Angew. Chem., Int. Ed. Engl.*, 1985, **24**, 117.
77. M. A. Mazid, M. T. Razi and P. J. Sadler, *Inorg. Chem.*, 1981, **20**, 2872.
78. E. W. Abel, K. Kite and B. L. Williams, *J. Chem. Soc., Dalton Trans.*, 1983, 1017.
79. J. R. Andersen, V. V. Patel and E. M. Engler, *Tetrahedron Lett.*, 1978, 239.
80. A. Davison and E. T. Shawl, *Inorg. Chem.*, 1970, **9**, 1820.
81. C. G. Pierpont and R. Eisenberg, *J. Chem. Soc. (A)*, 1971, 2285.
82. C. Bianchini, C. Mealli, A. Meli and M. Sabat, *J. Chem. Soc., Chem. Commun.*, 1984, 1647.
83. C. M. Bolinger and T. B. Rauchfuss, *Inorg. Chem.*, 1982, **21**, 3947.
84. D. J. Sandman, J. C. Stark, L. A. Acampora, L. A. Samuelson, G. W. Allen, S. Jansen, M. T. Jones and B. M. Foxman, *Mol. Cryst. Liq. Cryst.*, 1984, **107**, 1.
85. A. Balch and R. H. Holm, *J. Am. Chem. Soc.*, 1966, **88**, 5201; G. S. Hall and R. H. Soderberg, *Inorg. Chem.*, 1968, **7**, 2300.
86. S. M. Peng, D. S. Liaw, Y. Wang and A. Simon, *Angew. Chem., Int. Ed. Engl.* 1985, **24**, 210.
87. C. E. Forbes and R. H. Holm, *J. Am. Chem. Soc.*, 1970, **92**, 2297.
88. D. Sartain and M. R. Truter, *Chem. Commun.*, 1966, 382; *J. Chem. Soc. (A)*, 1967, 1264.
89. G. Dessy, V. Fares, C. Bellitto and A. Flamini, *Cryst. Struct. Commun.*, 1982, **11**, 1743.
90. C. Mahadevan, M. Seshasayee, P. Kuppusamy and P. T. Manoharan, *J. Crystallogr. Spectrosc. Res.*, 1984, **14**, 179.
91. D. Snaathorst, H. M. Doesburg, J. A. A. J. Perenboom and C. P. Keijzers, *Inorg. Chem.*, 1981, **20**, 2526.
92. E. Lindner, I. P. Butz, S. Hoehne, W. Hiller and R. Fawzi, *J. Organomet. Chem.*, 1983, **259**, 99.
93. C. Bianchini, C. Mealli, A. Meli and M. Sabat, *Inorg. Chem.*, 1984, **23**, 4125.
94. A. Kutoglu, *Acta Crystallogr., Sect. B*, 1973, **29**, 2891.
95. T. Debaerdemaeker and A. Kutoglu, *Acta Crystallogr., Sect. B*, 1973, **29**, 2664.
96. C. M. Bolinger, T. B. Rauchfuss and A. L. Rheingold, *J. Am. Chem. Soc.*, 1983, **105**, 6321.
97. M. McKenna, L. L. Wright, D. J. Miller, L. Tanner, R. C. Haltiwanger and M. Rakowski DuBois, *J. Am. Chem. Soc.*, 1983, **105**, 5329; O. A. Rajan, M. McKenna, J. Noordik, R. C. Haltiwanger and M. Rakowski DuBois, *Organometallics*, 1984, **3**, 831.
98. R. Richter, J. Kaiser, J. Sieler and L. Kutschabsky, *Acta Crystallogr., Sect. B*, 1975, **31**, 1642.
99. A. Davison, N. Edelstein, R. H. Holm and A. H. Maki, *Inorg. Chem.* 1963, **2**, 1227.
100. A. Davison, N. Edelstein, R. H. Holm and A. H. Maki, *Inorg. Chem.*, 1964, **3**, 814.
101. A. Vlcek, Jr. and A. A. Vlcek, *Inorg. Chim. Acta*, 1982, **64**, L273; *J. Electroanal. Chem. Interfacial Electrochem.*, 1981, **125**, 481.
102. W. E. Geiger, Jr., C. S. Allen, T. E. Mines and F. C. Sentleber, *Inorg. Chem.*, 1977, **16**, 2003.
103. G. N. Schrauzer and V. P. Mayweg, *J. Am. Chem. Soc.*, 1965, **87**, 3585.
104. A. M. Giroud, A. Nazzal and U. T. Mueller-Westerhoff, *Mol. Cryst. Liq. Cryst. Lett.*, 1980, **56**, 225.
105. N. Y. Tsao and Y. Y. Lim, *Aust. J. Chem.*, 1981, **34**, 2321.
106. A. Deronzier and J. M. Latour, *Nouv. J. Chim.*, 1984, **8**, 393.
107. R. Kirmse, J. Stach, W. Dietzsch, G. Steimecke and E. Hoyer, *Inorg. Chem.*, 1980, **19**, 2679.
108. J. Stach, R. Kirmse, W. Dietzsch, I. N. Marov and V. K. Belyaeva, *Z. Anorg. Allg. Chem.*, 1980, **466**, 36.
109. J. Stach, R. Kirmse, W. Dietzsch and P. Thomas, *Z. Anorg. Allg. Chem.*, 1981, **480**, 60.
110. R. Kirmse, R. Böttcher and C. P. Keijzers, *Chem. Phys. Lett.*, 1982, **87**, 467.
111. R. Kirmse, J. Stach, U. Abram, W. Dietzsch, R. Böttcher, M. C. M. Gribnau and C. P. Keijzers, *Inorg. Chem.*, 1984, **23**, 3333.
112. D. Snaathorst and C. P. Keijzers, *Mol. Phys.*, 1984, **51**, 509.
113. G. A. Bowmaker, P. D. W. Boyd and G. K. Campbell, *Inorg. Chem.*, 1983, **22**, 1208.
114. E. P. Duliba, E. G. Seebauer and R. L. Belford, *J. Magn. Reson.*, 1982, **49**, 507.
115. C. M. Elliott and K. Akabori, *J. Am. Chem. Soc.*, 1982, **104**, 2671.
116. U. T. Mueller-Westerhoff, unpublished experiments.
117. M. Kaneko and A. Yamada, *Makromol. Chem.*, 1981, **182**, 105.
118. A. Vogler and H. Kunkely, *J. Am. Chem. Soc.*, 1981, **103**, 1559.
119. A. Vogler and H. Kunkely, *Angew. Chem., Int. Ed. Engl.*, 1981, **20**, 386.
120. A. Vogler and H. Kunkely, *Inorg. Chem.*, 1982, **21**, 1172.
121. D. M. Dooley and B. M. Patterson, *Inorg. Chem.*, 1982, **21**, 4330.
122. R. Frank and H. Rau, *Z. Naturforsch., Teil A*, 1982, **37**, 1253.
123. C. E. Johnson, R. Eisenberg, T. R. Evans and M. S. Burberry, *J. Am. Chem. Soc.*, 1983, **105**, 1795.
124. T. Birchall, N. N. Greenwood and J. A. McCleverty, *Nature (London)*, 1967, **215**, 625.
125. H. Poleschner, E. Fanghaenel and H. Mehner, *J. Prakt. Chem.*, 1981, **323**, 919.
126. C. W. Allen and D. B. Brown, *Inorg. Chem.*, 1974, **13**, 2020.
127. D. Petridis and B. W. Fitzsimmons, *Inorg. Chim. Acta*, 1974, **11**, 105.
128. C. W. Schläpfer and K. Nakamoto, *Inorg. Chem.*, 1975, **14**, 1338.
129. G. Leonhardt, W. Dietzsch, R. Heber, E. Hoyer, J. Hedman, A. Berntsson, M. Klasson, R. Nilsson and C. Nordling, *Z. Chem.*, 1973, **13**, 24.
130. M. Sano, H. Adachi and H. Yamatera, *Bull. Chem. Soc. Jpn.*, 1981, **54**, 2636.
131. J. Blomquist, U. Helgeson, B. Folkesson and R. Larsson, *Chem. Phys.*, 1983, **76**, 71.
132. E. I. Stiefel, R. Eisenberg, R. C. Rosenberg and H. B. Gray, *J. Am. Chem. Soc.*, 1966, **88**, 2956.
133. G. N. Schrauzer and H. N. Rabinowitz, *J. Am. Chem. Soc.*, 1968, **90**, 4297.
134. M. R. A. Blomberg and U. Wahlgren, *Chem. Phys.*, 1980, **49**, 117.
135. S. Zalis and A. A. Vlcek, *Inorg. Chim. Acta*, 1982, **58**, 89.
136. Z. S. Herman, R. F. Kirchner, G. H. Loew, U. T. Mueller-Westerhoff, A. Nazzal and M. C. Zerner, *Inorg. Chem.*, 1982, **21**, 46.
137. W. Dietzsch, J. Lerchner, J. Reinhold, J. Stach, R. Kirmse, G. Steimecke and E. Hoyer, *J. Inorg. Nucl. Chem.*, 1980, **42**, 509.
138. I. Fischer-Hjalmars and A. Henriksson-Enflo, *Adv. Quantum Chem.*, 1982, **16**, 1.
139. S. Alvarez and R. Hoffmann, *An. Quim, Ser. B*, 1986, **82**, 52.

140. S. Alvarez, R. Vicente and R. Hoffmann, *J. Am. Chem. Soc.*, 1985, **107**, 6253.
141. J. A. McCleverty and B. Ratcliff, *J. Chem. Soc. (A)*, 1970, 1627.
142. G. N. Schrauzer, R. K. Y. Ho and R. P. Murillo, *J. Am. Chem. Soc.*, 1970, **92**, 3508.
143. C. W. Allen, D. E. Lutes, E. J. Durhan and E. S. Bretschneider, *Inorg. Chim. Acta*, 1977, **21**, 277.
144. M. J. Hynes and A. J. Moran, *J. Chem. Soc., Dalton Trans.*, 1973, 2280.
145. J. Stach, R. Kirmse, W. Dietzsch and E. Hoyer, *Z. Chem.*, 1980, **20**, 221.
146. V. P. Mayweg and G. N. Schrauzer, *Chem. Commun.*, 1966, 640.
147. A. Davison and D. V. Howe, *Chem. Commun.*, 1965, 290.
148. J. A. McCleverty and B. Ratcliff, *J. Chem. Soc. (A)*, 1970, 1631.
149. A. Vlcek, Jr. and A. A. Vlcek, *Proc. Conf. Coord. Chem.*, 1980, **8**, 445 (*Chem. Abstr.* 1981, **94**, 113 557).
150. U. T. Mueller-Westerhoff and A. Alscher, *Angew. Chem.*, 1980, **92**, 654.; *Angew. Chem., Int. Ed. Engl.*, 1980, **19**, 638.
151. A. Nazzal and U. T. Mueller-Westerhoff, *Transition Met. Chem. (Weinheim, Ger.)*, 1980, **5**, 318.
152. W. L. Bowden, J. D. L. Holloway and W. E. Geiger, Jr., *Inorg. Chem.*, 1978, **17**, 256.
153. J. S. Miller and A. J. Epstein, *Prog. Inorg. Chem.*, 1976, **20**, 106.
154. J.-J. André, A. Bieber and F. Gauthier, *Ann. Phys.*, 1976, **1**, 145.
155. L. V. Interrante, *Adv. Chem. Ser.*, 1976, **150**, 1.
156. L. V. Interrante, *Coord. Chem.*, 1981, **21**, 87 (*Chem. Abstr.*, 1981, **95**, 89 409).
157. L. Alcacer and H. Novais, *Ext. Linear Chain Compd.*, 1983, **3**, 319.
158. J. W. Bray, L. V. Interrante, I. S. Jacobs and J. C. Bonner, *Ext. Linear Chain Compd.*, 1983, **3**, 353.
159. P. Kuppusamy, B. L. Ramakrishna and P. T. Manoharan, *Proc. Indian Acad. Sci., Sect. A*, 1984, **93**, 977.
160. A. Kobayashi and Y. Sasaki, *Bull. Chem. Soc. Jpn.*, 1977, **50**, 2650.
161. E. A. Perez-Albuerne, L. C. Isett and R. K. Haller, *J. Chem. Soc., Chem. Commun.*, 1977, 417.
162. D. R. Rosseinsky and R. E. Malpas, *J. Chem. Soc., Dalton Trans.*, 1979, 749.
163. J. S. Tonga and A. E. Underhill, *J. Chem. Soc., Dalton Trans.*, 1984, 2333.
164. M. M. Ahmad and A. E. Underhill, *J. Chem. Soc., Dalton Trans.*, 1983, 165; 1982, 1065.
165. A. E. Underhill and M. M. Ahmad, *J. Chem. Soc., Chem. Commun.*, 1981, 67.
166. A. E. Underhill and M. M. Ahmad, *Mol. Cryst. Liq. Cryst.*, 1982, **81**, 223.
167. A. E. Underhill and M. M. Ahmad, *J. Chem. Soc., Chem. Commun.*, 1981, 67.
168. M. M. Ahmad, D. J. Turner, A. E. Underhill, C. S. Jacobsen, K. Mortensen and K. Carneiro, *Phys. Rev. B: Condens. Matter*, 1984, **29**, 4796.
169. A. E. Underhill, M. M. Ahmad, D. J. Turner, P. I. Clemenson, K. Carneiro, S. Yuequiran and K. Mortensen, *Mol. Cryst. Liq. Cryst.*, 1985, **120**, 369.
170. M. M. Ahmad, D. J. Turner, A. E. Underhill, A. Kobayashi, Y. Sasaki and H. Kobayashi, *J. Chem. Soc., Dalton Trans.*, 1984, 1759.
171. A. Kobayashi, Y. Sasaki, H. Kobayashi, A. E. Underhill and M. M. Ahmad, *J. Chem. Soc., Chem. Commun.*, 1982, 390.
172. A. Kobayashi, Y. Sasaki, H. Kobayashi, A. E. Underhill and M. M. Ahmad, *Chem. Lett.*, 1984, 305.
173. A. Kobayashi, T. Mori, Y. Sasaki, H. Kobayashi, M. M. Ahmad and A. E. Underhill, *Bull. Chem. Soc. Jpn.*, 1984, **57**, 3262.
174. R. D. Schmitt, R. M. Wing and A. H. Maki, *J. Am. Chem. Soc.*, 1969, **91**, 4394.
175. W. E. Geiger, Jr. and A. H. Maki, *J. Phys. Chem.*, 1971, **75**, 2387.
176. L. Alcacer and A. H. Maki, *J. Phys. Chem.*, 1974, **78**, 215.
177. A. Singhabhandhu, P. D. Robinson, J. H. Fang and W. F. Geiger, Jr., *Inorg. Chem.*, 1975, **14**, 318.
178. F. Wudl, C. H. Ho and A. Nagel, *J. Chem. Soc., Chem. Commun.*, 1973, 923.
179. J. A. Ibers, L. J. Pace, J. Martinsen and B. M. Hoffman, *Struct. Bonding (Berlin)*, 1982, **50**, 1.
180. T. Setoi, M. Inoue and D. Nakamura, *Bull. Chem. Soc. Jpn.*, 1982, **55**, 1691.
181. B. L. Ramakrishna and P. T. Manoharan, *Inorg. Chem.*, 1983, **22**, 2113.
182. P. Cassoux, L. Interrante and J. Kasper, *Mol. Cryst. Liq. Cryst.*, 1982, **81**, 293; *C. R. Hebd. Seances Acad. Sci., Ser. C*, 1980, **291**, 25.
183. G. Steimecke, H.-J. Sieler, R. Kirmse and E. Hoyer, *Phosphorus Sulfur*, 1979, **7**, 49.
184. O. Lindqvist, L. Andersen, J. Sieler, G. Steimecke and E. Hoyer, *Acta Chem. Scand., Ser. A*, 1982, **36**, 855.
185. O. Lindqvist, L. Sjolin, J. Sieler, G. Steimecke and E. Hoyer, *Acta Chem. Scand., Ser. A*, 1979, **33**, 445.
186. J. Ribas and P. Cassoux, *C. R. Hebd. Seances Acad. Sci., Ser. B*, 1981, **293**, 287.
187. G. C. Papavassiliou, *Z. Naturforsch., Teil B*, 1981, **36**, 1200.
188. G. C. Papavassiliou, *Z. Naturforsch., Teil B*, 1982, **37**, 825.
189. G. C. Papavassiliou, *Mol. Cryst. Liq. Cryst.*, 1982, **86**, 159.
190. R. Kato, T. Mori, A. Kobayashi, Y. Sasaki and H. Kobayashi, *Chem. Lett.*, 1984, 1.
191. L. Valade, M. Bousseau, A. Gleizes and P. Cassoux, *J. Chem. Soc., Chem. Commun.*, 1983, 110.
192. M. Bousseau, L. Valade, M.-F. Bruniquel, P. Cassoux, M. Garbauskas, L. Interrante and J. Kasper, *Nouv. J. Chim.*, 1984, **8**, 3.
193. L. Valade, J.-P. Legros, M. Bousseau, P. Cassoux, M. Garbauskas and L. V. Interrante, *J. Chem. Soc., Dalton Trans.*, 1985, 783.
194. F. Kubel, L. Valade, J. Straehle and P. Cassoux, *C. R. Hebd. Seances Acad. Sci., Ser. B*, 1982, **295**, 179.
195. P. Cassoux. L. Valade, M. Bousseau, J.-P. Legros, M. Garbauskas and L. V. Interrante, *Mol. Cryst. Liq. Cryst.*, 1985, **120**, 377.
196. K. H. Drexhage and U. T. Mueller-Westerhoff, *IEEE J. Quantum Electron.*, 1972, **QE-8**, 759.
197. R. W. Lane, J. J. Mayerle, U. T. Mueller-Westerhoff and A. Nazzal, *IBM Tech. Discl. Bull.*, 1979, **21**, 4175; J. J. Mayerle, U. T. Mueller-Westerhoff and A. Nazzal, *IBM Tech. Discl. Bull.*, 1979, **21**, 4176.
198. D. E. Johnson, G. L. Kellermeyer, H. M. Lo and U. T. Mueller-Westerhoff, *IBM Tech. Discl. Bull.*, 1973, **15**, 2619.
199. A. Graczyk, E. Bialkowska and A. Konarzewski, *Tetrahedron*, 1982, **38**, 2715.
200. W. Freyer, *Z. Chem.*, 1984, **24**, 32.
201. W. Freyer, *Z. Chem.*, 1985, **25**, 104.
202. D. Beaupere and J. C. Farcy, *Opt. Commun.*, 1978, **27**, 410.

203. V. A. Katulin, A. L. Petrov and V. Freier, *Kvantovaya Elektron. (Moscow)*, 1984, **11**, 115 (*Chem. Abstr.*, 1984, **100**, 182 772).
204. K. H. Drexhage and U. T. Mueller-Westerhoff, *US Pat.* 3 743 964 (1973).
205. Mitsubishi Electric Corp. *Jpn. Pat.* 80 24450 (1980) (*Chem. Abstr.*, 1980, **93**, 213 151); *Jpn. Pat.* 80 58 587 (1980) (*Chem. Abstr.*, 1980, **93**, 228 441); *Jpn. Pat.* 80 58 586 (1980) (*Chem. Abstr.*, 1980, **93**, 228 442); *Jpn. Pat.* 80 58 588 (1980) (*Chem. Abstr.*, 1980, **93**, 228 443).
206. W. Freyer, F. Fink, H. Poleschner and E. Fanghaenel, *Ger. (East) Pat.* 206 282 (1984) (*Chem. Abstr.*, 1984, **101**, 81 495); W. Freyer, A. L. Petrov and V. A. Katulin, *Ger. (East) Pat.* 210 416 (1984) (*Chem. Abstr.*, 1984, **101**, 201 174); W. Freyer, *Ger. (East) Pat.* 203 054 (1983) (*Chem. Abstr.*, 1984, **100**, 177 259).
207. A. M. Giroud and U. T. Mueller-Westerhoff, *Mol. Cryst. Liq. Cryst. Lett.*, 1977, **41**, 11.
208. U. T. Mueller-Westerhoff, A. Nazzal, R. J. Cox and A. M. Giroud, *J. Chem. Soc., Chem. Commun.*, 1980, 497.
209. U. T. Mueller-Westerhoff and A. Nazzal, *IBM Tech. Discl. Bull.*, 1979, **22**, 2442, 2858, 5052; 1980, **23**, 1188, 2510, 2511; 1982, **24**, 6186.
210. Anonymous (UK), *Res. Discl.*, 1982, **216**, 117 (*Chem. Abstr.*, 1982, **97**, 47 075); K. Enmanji, K. Takahashi and T. Kitagawa (Mitsubishi Electric Corp.), *Jpn. Pat.* 80 39 340 (1980) (*Chem. Abstr.*, 1980, **93**, 85 201); Mitsui Toatsu Chemicals, Inc., *Jpn. Pat.* 82 11 090 (1982) (*Chem. Abstr.*, 1982, **97**, 101 746).
211. R. Henning, W. Schlamann and H. Kisch, *Angew. Chem., Int. Ed. Engl.*, 1980, **19**, 645.
212. R. Battaglia, R. Henning and H. Kisch, *Z. Naturforsch., Teil B*, 1981, **36**, 396.
213. J. Buecheler, N. Zeug and H. Kisch, *Angew. Chem., Int. Ed. Engl.*, 1982, **21**, 783.
214. H. Kisch, *Ger. Pat.* 2 908 633 (1980) (*Chem. Abstr.*, 1981, **95**, 65 230); *US Pat.* 4 325 793 (1982) (*Chem. Abstr.*, 1982, **97**, 136 592).
215. R. Battaglia, R. Henning, B. Dinh-Ngoo, W. Schlamann and H. Kisch, *J. Mol. Catal.*, 1983, **21**, 239.
216. A. Fernandez and H. Kisch, *Chem. Ber.*, 1984, **117**, 3102.
217. H. Kisch, A. Fernandez, Y. Wakatsuki and H. Yamazaki, *Z. Naturforsch., Teil B*, 1985, **40**, 292.
218. A. Fernandez, H. Görner and H. Kisch, *Chem. Ber.*, 1985, **118**, 1936.
219. L. Kumar, K. H. Puthraya and T. S. Srivastava, *Inorg. Chim. Acta*, 1984, **86**, 173.
220. Y. Umezawa, T. Yamamura and A. Kobayashi, *J. Electrochem. Soc.*, 1982, **129**, 2378.
221. M. G. Bradley, T. Tysak and N. Vlachopoulos, *J. Electroanal. Chem. Interfacial Electrochem.*, 1984, **160**, 345.
222. J. J. Grimaldi and J. M. Lehn, *J. Am. Chem. Soc.*, 1979, **101**, 1333.
223. A. Ohki, M. Takagi and K. Ueno, *Chem. Lett.*, 1980, 1591.

16.6
Other Sulfur-containing Ligands

STANLEY E. LIVINGSTONE

University of New South Wales, Kensington, NSW, Australia

16.6.1   OXIDES OF SULFUR

16.6.1.1   Sulfur Dioxide

The complex $[RuCl(NH_3)_4(SO_2)]Cl$ was reported in 1936;[1] it was prepared by the reaction of HCl upon $[Ru(HSO_3)_2(NH_3)_4]$. However, another 30 years had elapsed before an extensive series of sulfur dioxide complexes was synthesized and investigated by spectroscopic and X-ray crystallographic studies.[2-4] The compound $[RuCl(NH_3)_4(SO_2)]Cl$ was shown by X-ray crystal analysis to be octahedral with the SO_2 and Cl ligands mutually *trans* and the Ru—S distance 2.07 Å.[5] The related pentaammine complexes $[Ru(NH_3)_5(SO_2)]X_2$ (X = Cl, Br, I) are known and the iodide has been prepared by the reaction of $[Ru(NH_3)_5(H_2O)]I_2$ with SO_2.[6]

The coordination chemistry of SO_2 is of especial interest because of the large quantity of this pollutant which is currently released into the atmosphere from industrial sources, including the smelting of sulfide ores and the combustion of coal. Sulfur dioxide has recently been recognized as a versatile ligand for transition metals, exhibiting a variety of modes of bonding. There is a close relationship between NO and SO_2 in their ligating behaviour but there are also important differences, particularly in regard to oxidative and oxidative addition reactions. Sulfur dioxide complexes are known with the transition metals Cr, Mo, W, Mn, Re, Fe, Ru, Os, Co, Rh, Ir, Ni, Pd and Pt. Usually only one SO_2 ligand is coordinated but several complexes are known with two; however, two seems to be the maximum number.

Electron diffraction,[7] X-ray diffraction[8] and microwave[9] studies have established that SO_2 has C_{2v} symmetry with an O—S—O bond angle of 119° and an S—O bond length of 1.43 Å. The dipole moment is 1.61 D[10] (*cf.* H_2O, 1.9 D) and the first ionization potential is 12.34 eV,[11] the latter value being comparable with that for H_2O (12.61 eV) and less than those for N_2 and CO. The SO_2 molecule has an empty orbital of b_1 symmetry capable of accepting electrons from the filled d-orbitals of a transition metal ion. The SO_2 ligand, like N_2 and CO, has π-acceptor capability when coordinated to a transition metal in a low oxidation state. The bonding of SO_2 in transition metal complexes has been discussed in detail in two recent reviews.[12,13]

The M—SO_2 bonding, excluding that occurring in SO_2-bridged complexes and insertion compounds, can be classified into three types: η^1-planar (1), η^1-pyramidal (2) and η^2 (3).[12] Structures (1) and (2) are similar to the linear and bent geometries known for NO.[14-16] The η^2 coordination mode has not been identified for NO but has been established for CS_2.[17] Enemark and Feltham[18] from a consideration of the molecular orbital diagrams for the nitrosyl ligand suggested a classification of the MNO moiety according to the number of d-type or π^* electrons present in the MNO moiety; thus an MNO moiety with the electronic configuration $(3a_1)^2(2e)^4$ would be written {MNO}^{4} and one with $(3a_1)^2(2e)^4(1b_2)^2$ would be {MNO}^{6} and so forth. Mingos[12] applied this nomenclature to MSO_2 complexes and proposed the following set of rules to predict the geometries of {MB} complexes (B = NO or SO_2).

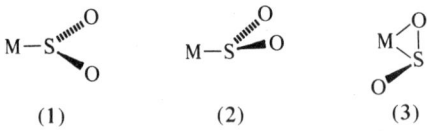

(1) In {MB}^{6} complexes based on the octahedron or square pyramid with B occupying an apical site, the linear and η^1-planar geometries are energetically preferred. For {MB}^{8} complexes the bent or η^1-pyramidal geometries are more stable.

(2) In square-pyramidal {MB}^{8} complexes of the type ML_2DAB, where D is a π-donor *trans* to an A π-acceptor, the NO or SO_2 ligand will bend in the DMA plane towards the π-acceptor.

(3) In related {MB}^{8} trigonal-bipyramidal complexes the ligand is less likely to bend than when occupying the apical site of the corresponding square pyramid.

(4) Nitrosyl and SO_2 ligands in axial positions in trigonal-bipyramidal and basal sites in a square-pyramidal {MB}^{8} complex prefer to be linearly (or η^1-planar) coordinated.

(5) Good π-acceptor ligands L in $\{L_4MB\}^8$ and $\{L_3MB\}^{10}$ complexes in general encourage linear (or η^1-planar) geometries, whereas good π-donors encourage the bent (or η^1-pyramidal) geometries.

Some examples of confirmation of these rules by X-ray structural evidence are as follows. In accordance with Rule 1 the $\{MB\}^6$ complex $[RuCl(NH_3)_4(SO_2)]Cl$ is η^1-planar,[5] while the $\{MB\}^8$ complex $[IrCl(CO)(SO_2)(PPh_3)_2]$ is η^1-pyramidal and, in accordance with Rule 2, in the latter complex the SO_2 ligand bends towards the π-acceptor CO group.[19]

The first η^2 complex to be established was $[Rh(NO)(SO_2)(PPh_3)_2]$ in which the SO_2 ligand is coordinated through the sulfur atom and one of the oxygen atoms; the complex is tetrahedral if it is accepted that the sulfur and oxygen atoms occupy only one coordination site.[20] Other η^2 complexes are $[RuCl(NO)(SO_2)(PPh_3)_2]$[21] and the Mo^0 complexes $[Mo(CO)_3(phen)(SO_2)]$ and $[Mo(CO)_2(bipy)(SO_2)_2]$.[22] IR evidence indicates that a series of Mo^0 and W^0 SO_2 complexes containing CO and phen, bipy or py, for which structural data are not available, also have η^2 geometry in regard to the SO_2 ligand.[13]

IR data of $\{MSO_2\}^6$ complexes are diagnostic for both η^1-planar and η^2 geometries: for η^1 complexes the SO_2 asymmetric and symmetric stretches, respectively, are in the ranges 1300–1225 and 1140–1065 cm^{-1}, while for η^2 complexes they occur at 1160–1100 and 950–850 cm^{-1}.[23] The η^1-pyramidal geometry is not expected for $\{MSO_2\}^6$ complexes and it appears that no examples are known.

All three types of geometry occur for $\{MSO_2\}^8$ complexes. If it is accepted that η^2-SO_2 effectively occupies only one coordination site, the majority of $\{MSO_2\}^8$ complexes are five-coordinate, either square-pyramidal or trigonal-bipyramidal. The $v(SO)$ frequencies are higher for η^1-planar than for η^1-pyramidal structures; the η^2 complexes can be safely assigned, since one of the $v(SO)$ values is much lower than for either η^1-planar or η^1-pyramidal geometries.[13]

All the structurally characterized $\{MSO_2\}^{10}$ complexes are tetrahedral except $[Ni(N-P-P-P)(SO_2)]$ (N—P—P—P = tris(2-diphenylphosphinoethyl)amine) which is trigonal-bipyramidal with an axial η^1-pyramidal SO_2.[24] All three types of geometry occur. The dominance of metal basicity is shown by the fact that first-row metal complexes have η^1-planar geometry, while second- and third-row metal complexes have η^1-pyramidal and η^2 geometries in regard to SO_2 coordination. The $v(SO)$ frequencies in $\{MSO_2\}^{10}$ complexes cannot, in general, be used for the assignment of SO_2 geometry, since the frequencies are more sensitive to the effects of the other ligands than to the geometry of the SO_2 group.[13]

The square-planar $[RhClL]$ $[L = PhAs(CH_2CH_2CH_2AsMe_2)_2]$ reacts with SO_2 to give $[RhClL(SO_2)]$.[25] The complexes $[M(SO_2)_2][AsF_6]_2$ (M = Mg, Mn, Fe, Co, Ni, Cu and Zn) have been prepared in liquid SO_2 by (a) oxidation of the metals with AsF_5, (b) reaction of MBr_2 with $AgAsF_6$, and (c) reaction of MF_2 with AsF_5.[26] The compound $[Zn(SO_2)_6][AsF_6]_2$ has also been prepared in liquid SO_2; the SO_2 is O-bonded.[27] The complexes cis-$[M(CO)_4L(SO_2)]$ (L = CO, PMe$_3$, PPh$_2$Me, P(OMe)$_3$) and mer-$[M(CO)_3(Ph_2PCH_2CH_2PPh_2)SO_2]$ (M = Cr, Mo, W) contain η^1-coordinated SO_2, whereas η^2-SO_2 occurs in fac-$[Mo(CO)_3(Ph_2PCH_2CH_2PPh_2)(SO_2)]$. The proton-catalyzed fac–mer isomerization of this compound is the first example of linkage isomerism of SO_2 caused by a change in coordination geometry.[28]

Complexes are known which contain an SO_2 group bound through sulfur bridging to two metal atoms.[13] These complexes may contain (i) an unsupported SO_2 bridge as in $\{CpFe(CO)_2\}_2SO_2$,[29] (ii) a bridging SO_2 and an M—M bond as in $\{Fe(CO)_4\}_2SO_2$,[30] (iii) an SO_2 bridge supported by another bridge and an M—M bond as in $[CpFe(CO)(SO_2)]$,[31] or (iv) an SO_2 bridge involving dative bonds from M to S and O to the second M atom as in $[Rh_2(PPh_3)_4(\mu\text{-}Cl)(\mu\text{-}SO_2)_2]SO_4$.[32] The complex $[Pt(SO_2)(PPh_3)_3]_3$ contains bridging SO_2 ligands and three Pt—Pt bonds.[33] All the bridged SO_2 complexes have considerable thermal and chemical stability.

There are some complexes in which SO_2 behaves as a Lewis acid, binding to a ligand rather than to the metal. Ligand-bound SO_2 was first established in $[Pt(Me)(I\text{—}SO_2)(PPh_3)_2]$; the I—$SO_2$ geometry is pyramidal and the I—S distance (3.39 Å) is greater than the sum of the covalent radii (2.36 Å).[34] Several other I—SO_2 complexes are known, including the iodo-bridged complex $[Cu_2I_2(PPh_2Me)_4(SO_2)]$.[35] Sulfur dioxide can also bind reversibly to some coordinated thiolo ligands, e.g. $[Cu(SPh)(PPh_2Me)_3]$ forms a 1:1 adduct with SO_2; the S—SO_2 distance is 2.53 Å.[36] A remarkable example of ligand-bound SO_2 occurs in $[Ru(CO)_2(\eta^2\text{-}SO_2\text{—}SO_2)(PPh_3)_2]$, in which SO_2 is weakly attached by sulfur to one oxygen of the η^2-coordinated SO_2 ligand.[37]

Sulfur dioxide is capable of reacting with metal alkyl, aryl or σ-allyl complexes in an insertion-type reaction to yield S-sulfinate (4), O-sulfinate (5) or O,O'-sulfinate complexes (6).[13] It can also insert into the metal–metal bond in the cobalt complex (7) to give the SO_2-bridged complex (8).[38]

The structures (4)-(8) appear here.

(4) 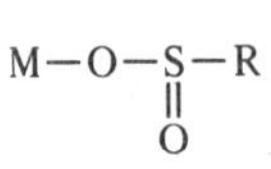   (5) 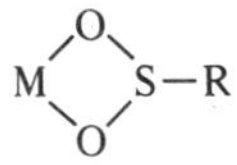   (6) 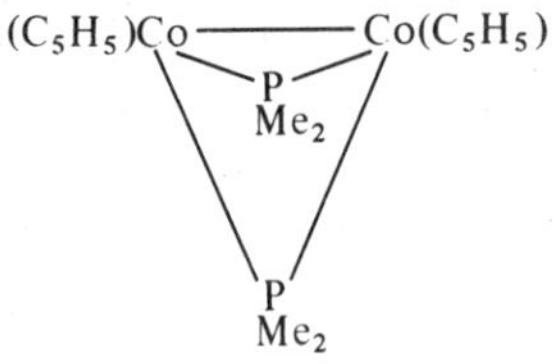   (7) 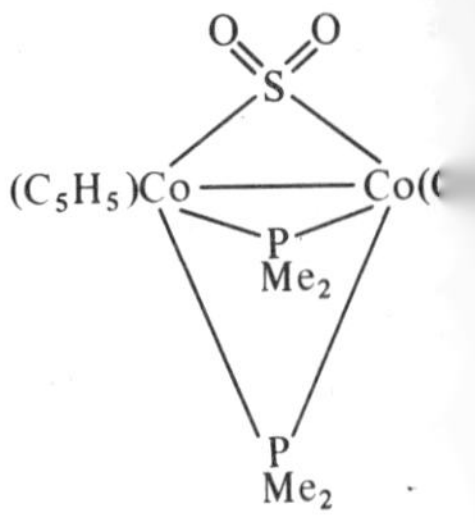   (8)

16.6.1.2 Sulfur Monoxide

The existence of sulfur monoxide SO was confirmed in 1932. It is formed when SO_2 is reduced by S vapour in a glow discharge at 0.5 Torr and 120 °C; it is unstable at all temperatures.[39] However, models for the gaseous composition of dense interstellar clouds and red giant stars predict a high abundance of SO.[40] Despite its instability SO can be trapped as a ligand. Oxidation of $[(Ph_2PCH_2CH_2PPH_2)_2Ir(S_2)]Cl$ with $NaIO_4$ yields $[(Ph_2PCH_2CH_2PPh_2)_2Ir(SO)_2]Cl$.[41] The complex $[Rh(SO)X(PPh_3)]_2$ was obtained by the reaction of $RhX(PPh_3)_2$ with stilbene episulfoxide in refluxing CH_2Cl_2.[42]

16.6.1.3 Disulfur Monoxide

Disulfur monoxide, S_2O, can be prepared, along with SO, by the reduction of SO_2 in a glow discharge or from the reaction of metal sulfides with $SOCl_2$. In the vapour phase at low pressure it slowly decomposes to S and SO_2 but in condensed phases it polymerizes. Its structure is similar to that of SO_2 with an SSO bond angle of 118°.[39,43] The complex $[(Ph_2PCH_2CH_2PPh_2)_2Ir(S_2O)]Cl$ (9) has been prepared by the reaction shown in equation (1).[41] The complex (9) is air stable and the IR spectrum shows only one $\nu(S—O)$ at 1043 cm^{-1}.[41,44] The complex $[Mo(S_2O)(S_2CNEt_2)_2]$ has also been isolated in small yield.[45]

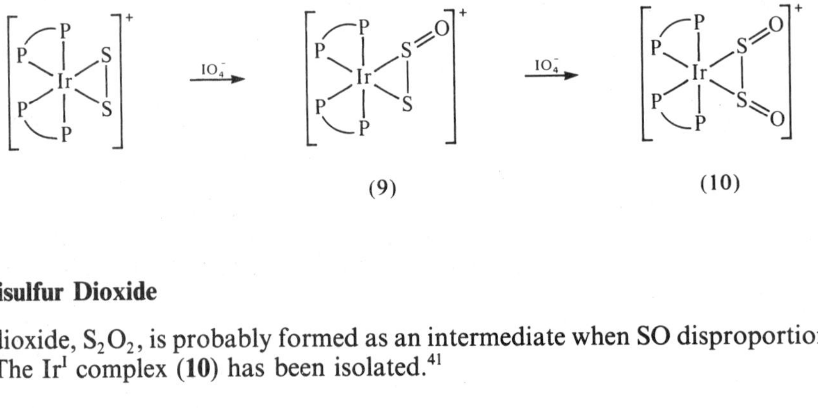

$$\left[\begin{smallmatrix}P&&&\\P&&|&&S\\&&Ir&&\\P&&|&&S\\&&P&&\end{smallmatrix}\right]^+ \xrightarrow{IO_4^-} \left[\begin{smallmatrix}P&&&\\P&&|&&S=O\\&&Ir&&\\P&&|&&S\\&&P&&\end{smallmatrix}\right]^+ \xrightarrow{IO_4^-} \left[\begin{smallmatrix}P&&&\\P&&|&&S=O\\&&Ir&&\\P&&|&&S=O\\&&P&&\end{smallmatrix}\right]^+ \quad (1)$$

(9)   (10)

16.6.1.4 Disulfur Dioxide

Disulfur dioxide, S_2O_2, is probably formed as an intermediate when SO disproportionates to S_2O and SO_2.[39] The Ir^I complex (10) has been isolated.[41]

16.6.1.5 Polysulfur Oxides

Polysulfur oxides are formed by the polymerization and partial disproportionation of S_2O[39] or by the oxidation of S_8.[46] The complexes $[(S_8O)SbCl_5]$ and $S_{12}O_2 \cdot 2SbCl_5 \cdot 3CS_2$ are known; the ligand is O-bonded.[46]

16.6.2 THIOSULFATE ION

Strong complexes are formed with (b) class metal ions, the stability constants for Hg^{II} (log β_2, 29.9), Pd^{II} and Pt^{II} (log β_4, 35.0 and 43.7, respectively) being quite high.[47,48] Alkaline thiosulfate solution will dissolve many insoluble salts of Cu^I, Ag^I, Hg^{II} and Pb^{II}.[47] It is unlikely that thiosulfate ion binds to hard acids such as UO_2^{2+}, ZrO^{2+} or Ln^{3+} (Ln = lanthanide).[49]

The thiosulfate ion is usually unidentate and S-bonded as in $[Hg(S_2O_3)_2]^{2-}$ and $[(NH_3)_5Co(S_2O_3)]Cl$; however, reduction of the latter complex by Cr^{II} proceeds by attack at an uncoordinated oxygen atom with the formation of an O-bonded thiosulfate complex of Cr^{III}, Co^{2+} and NH_4^+.[50] Other unidentate complexes include $[Cu(S_2O_3)_2]^{3-}$, $[Ag(S_2O_3)_2]^{3-}$, $[Au(S_2O_3)_2]^{3-}$, $[Pt(S_2O_3)_4]^{6-}$ and $[ML_n(S_2O_3)]^{x+}$ (M = Co^{II}, Co^{III}, Ni^{II}; L = N-donor ligand).[47,51] The copper complex has been shown to be an anti-inflammatory agent in rats.[52] The silver complex has long been used in photography and the gold complex $Na_3[Au(S_2O_3)_2] \cdot 2H_2O$ has been employed in the treatment of arthritis, tuberculosis and leprosy. In this compound the gold atom is linearly bound by two sulfur atoms as shown in (11); the Au—S distance is 2.28 Å.[53]

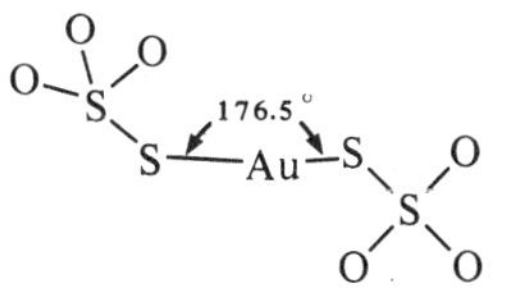

(11) 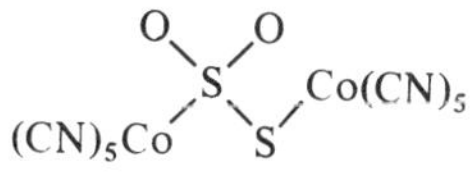 (12)

In $[Bi(S_2O_3)_3]^{3-}$, $[Pt(S_2O_3)Cl_2]^{2-}$ and $[Pt(S_2O_3)_2]^{2-}$ the ligand is chelated, being bound through one S atom and one O atom; *cis* and *trans* isomers of $[Pr(S_2O_3)_2]^{2-}$ have been prepared.[47] Spectral data indicate $CdS_2O_3 \cdot 2H_2O$ contains bridging groups being through both S atoms.[54] In Pt^{II} complexes the group *trans* to $S_2O_3^{2-}$ is labilized in the same way as in thiourea complexes.[47]

An interesting compound recently reported is one containing a bridging thiosulfito group bound by both sulfur atoms. Exposure to air of a saturated solution of $[(CN)_5CoSSCo(CN)_5]^{6-}$ at $-15\,°C$ leads to the formation of the dinuclear anion $[(CN)_5CoSSO_2Co(CN)_5]^{6-}$ in which the thiosulfito group bridges two cobalt atoms *via* the sulfur atoms as in (12).[55]

16.6.3 SULFITE ION

16.6.3.1 Methods of Bonding

The sulfito group can coordinate in a variety of ways: (i) S-bonded (13), (ii) O-bonded (14), (iii) O,O-bidentate (15), (iv) S,O-bidentate (16), (v) O,O-bridging (17), (vi) S,O-bridging (18), (vii) O,O,O-bridging (19), (viii) S,O,O,O-bridging (20).

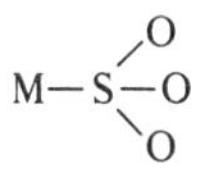

(13) 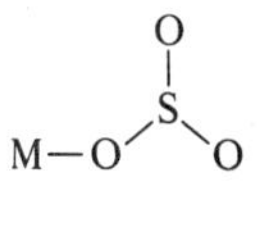 (14) 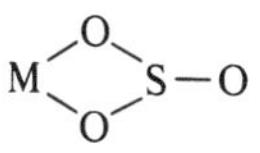 (15) 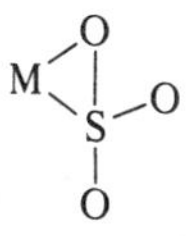 (16) 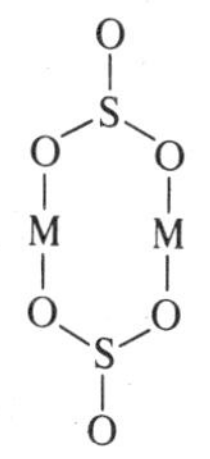 (17) 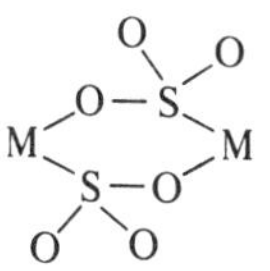 (18)

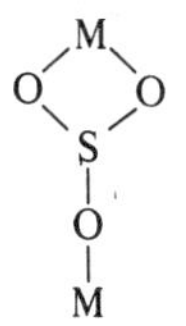

(19) 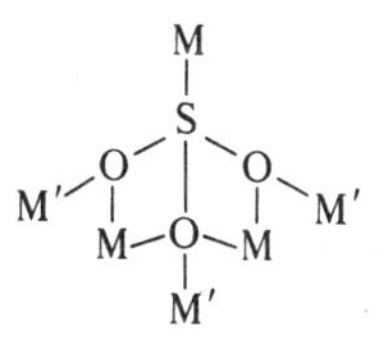 (20)

Sulfur Ligands

The SO_3^{2-} ion has C_{3v} symmetry, which is effectively unchanged if the ligand is S-bonded, giving rise to four IR and Raman active modes: v_1 (sym. stretch), v_2 (sym. bend), v_3 (asym. str.) and v_4 (asym. bend). However, in the crystalline state, site symmetry can lower the effective symmetry to C_3, C_s or C_1, causing splitting of the degenerate v_3 and v_4 modes and six vibrations would be expected. For Na_2SO_3 the following frequencies occur: v_1 (1010), v_3 (961), v_2 (633) and v_4 (496 cm^{-1}). For S-coordination (13), v_1 and v_3 shift to higher frequencies. For O-coordination (14) the symmetry is lowered to C_s and six fundamentals are expected, while v_1 and v_3 shift to lower frequencies. A band of high intensity above 975 cm^{-1} indicates S-bonding (13), while one below 960 cm^{-1} is indicative of O-bonding (14). Bands between 975 and 960 cm^{-1} indicate either ionic bonding or some mode of bonding involving both S- and O-coordination.[47,56,57] However, the IR spectra often display multiple splitting, making it difficult to designate the type of bonding but the Raman spectra are usually simple and more diagnostic.[58]

16.6.3.2 Unidentate S-Bonded Complexes

Complexes of (b) class metal ions are usually S-bonded. The values of the stability constants are quite high: log β_2 for Hg^{II}, 22.9,[47] log β_4 for Tl^{III}, 34,[47] for Pd^{II}, 29.1 and for Pt^{II}, 37.9.[59] X-Ray crystal structural determinations of [Pd(SO$_3$)(NH$_3$)$_3$] and *trans*-[Rh(SO$_3$)CN(NH$_3$)$_4$]·2H$_2$O show that the sulfito ligand is S-bonded as in (13).[60,61] Spectral data indicate that [Hg(SO$_3$)$_2$]$^{2-}$, [Pt(SO$_3$)$_4$]$^{6-}$, [Co(NH$_3$)$_5$(SO$_3$)]Cl, *trans*-[Co(en)$_2$(SO$_3$)$_2$]$^-$, [Co(en)$_2$(SO$_3$)Cl], *trans*-[Ir(SO$_3$)$_2$Cl$_4$]$^{5-}$, [Ir(SO$_3$)$_3$-(NH$_3$)$_3$]$^{3-}$, [Os(SO$_3$)$_3$(NH$_3$)$_3$]$^{4-}$, [Ir(SO$_3$)$_3$(NH$_3$)$_3$]$^{3-}$, [Ir(SO$_3$)$_4$Cl$_2$]$^{7-}$, [Pd(SO$_3$)$_4$]$^{6-}$, [Pt(SO$_3$)$_6$]$^{6-}$, [OsO$_2$(SO$_3$)$_4$]$^{6-}$, [Os(SO$_3$)$_3$(H$_2$O)$_3$]$^{4-}$ and [Ir(SO$_3$)$_4$Cl$_2$]$^{7-}$ are S-bonded as in (13).[56,58] Various sulfito complexes of gold are used in electroplating.

The S-bonded sulfito group has a marked *trans* effect. In *trans*-[Co(en)$_2$Cl(SO$_3$)]·H$_2$O the Co—S distance (2.20 Å) is short, while the Co—Cl distance (2.34 Å) is the longest six-coordinate Co—Cl distance reported, and in *trans*-[Co(en)$_2$NH$_3$(SO$_3$)]ClO$_4$ the Co—S distance is 2.23 Å, while the Co—N distance (2.07 Å) is remarkably long.[62] In Na$_6$[Pd(SO$_3$)$_4$]·2H$_2$O the rather long Pd—S distances (2.316 and 2.341 Å) are attributed to the strong *trans* influence of the sulfito ligand.[63] In K$_3$[Pt{(SO$_3$)$_2$H}Cl$_2$] the S-bonded sulfito groups are *cis* with rather short (2.246 Å) Pt—S bonds, while the Pt—Cl bonds are quite long (2.388 Å) between the two sulfito groups.[64] The S-bonded sulfito group has a high position in the spectrochemical series, comparable with that of NH$_3$.[47]

16.6.3.3 Unidentate O-Bonded Complexes

The complex (NH$_4$)$_9$[Fe(SO$_3$)$_6$] has been shown by an X-ray study to contain O-bonding as in (14);[65] the Raman and IR spectra differ from those of S-bonded complexes, v_1 and v_3 occurring at frequencies *ca.* 150 cm^{-1} lower.[59] Sulfur dioxide reacts with [Co(NH$_3$)$_5$(OH)]$^{2+}$ to give O-bonded [Co(NH$_3$)$_5$(SO$_3$)]$^+$ which slowly isomerizes to the S-bonded analogue.[66]

16.6.3.4 Bidentate Complexes

The complexes [Co(en)$_2$(SO$_3$)]$^+$ and [UO$_2$(C$_2$O$_4$)(SO$_3$)]$^{2-}$ probably contain O,O-bonding as in (15).[47,67] The complexes [Pd(SO$_3$)$_2$]$^{2-}$ and [Pt(SO$_3$)$_2$]$^{2-}$ may contain O,S-bonding as in (16) but a bridged structure is also possible.[47]

16.6.3.5 Bridged Complexes

The complex [Pd(SO$_3$)(phen)] is dimeric with two bidentate sulfito bridging ligands as in (18).[68] The compound [Rh(SO$_3$)$_3$]$^{3-}$ may contain S,O-bidentate sulfito groups but the complexity of its spectrum and its low solubility suggest that it is bridged as in (18) or has some more complicated structure.[58] The complex Tl$_2$[Cu(SO$_3$)$_2$] has the structure (20) (M = Tl, M' = Cu).[57] In the dimeric uranyl compound [(UO$_2$)$_2$(SO$_3$)$_2$C$_2$O$_4$]$^{2-}$ the sulfito group is bridging as in (19).[67] Other more complicated bridging arrangements have been reported.[57]

16.6.4 THIOUREA AND ITS DERIVATIVES

16.6.4.1 Thiourea

Thiourea, $(H_2N)_2C{=}S$ (tu), acts as a unidentate ligand forming strong complexes with (b) class metal ions, in particular Cu^I, Ag^I, Au^I and Hg^{II}; it reduces Cu^{II} to Cu^I, Au^{III} to Au^I, Pt^{IV} to Pt^{II} and Te^{IV} to Te^{II}, forming complexes with the metal in the lower oxidation state.[47] The only metal reported to be N-bonded is Ti^{IV}, all others being S-bonded; $v(M{-}S)$ occurs at 300–200 cm^{-1}.[69]

The complex $[Pd(tu)_4]Cl_2$ is distorted slightly from a square-planar to a tetrahedral arrangement with mean Pd—S bond lengths of 2.33 Å.[70] The complex $[Ni(tu)_4Cl_2]$ is *trans*-octahedral with the four S atoms at 2.54 Å, while $[Ni(tu)_2(NCS)_2]$ is octahedral and polymeric: each Ni atom is bound to four thiourea S atoms at 2.54 Å and to two thiocyanate N atoms at 1.99 Å; the S—C distance (1.77 Å) is much longer than in thiourea itself (1.64 Å). The compounds $[M(tu)_2(NCS)_2]$ (M = Mn, Co, Cd) are isostructural with $[Ni(tu)_2(NCS)_2]$ but $[Zn(tu)_2(NCS)_2]$ is not. The complex $[Cd(tu)_2Cl_2]$ is tetrahedral but $[Pb(tu)_2Cl_2]$ is polymeric with the Pb atom seven-coordinate and surrounded by four bridging S atoms at 2.9–3.1 Å and two bridging Cl atoms at 3.2 Å, while the terminal Cl atom is at a distance of 2.75 Å. In $[Te(tu)_4]Cl_2$ each Te atom has a square-planar arrangement of four S atoms.[47] The magnetic moment of $[Mo(tu)_3Cl_3]$ is 3.71 BM, while that of $[Mo_2(tu)_3Cl_6]$ is 0.59 BM; in the latter complex the two Mo centres are bridged by either three S or three Cl atoms and the low magnetic moment is ascribed to spin coupling due to Mo–Mo interaction, either directly or *via* the bridging atoms.[47] Complexes $Ln_4(tu)_5Br_{12}\cdot 20H_2O$ are known for all the lanthanides.[71] In $Ln(tu)_2(ClO_4)_3\cdot 10H_2O$ the tu is S-bonded.[72]

Thiourea, like other S-donor ligands, has a high *trans* effect,[73,74] which is the basis of the Kurnakov test for distinguishing *cis* and *trans* isomers of dihalogenodiammineplatinum(II).[75] This test relies on the *trans* effect of the ligands decreasing in the order: tu > Cl > NH_3. Thus reaction of thiourea with *cis*-$[Pt(NH_3)_2Cl_2]$ (**21**) yields $[Pt(tu)_4]^{2+}$ (**22**; Scheme 1), while with *trans*-$[Pt(NH_3)_2Cl_2]$ (**23**) thiourea reacts to give $[Pt(NH_3)_2(tu)_2]^{2+}$ (**24**; Scheme 2).

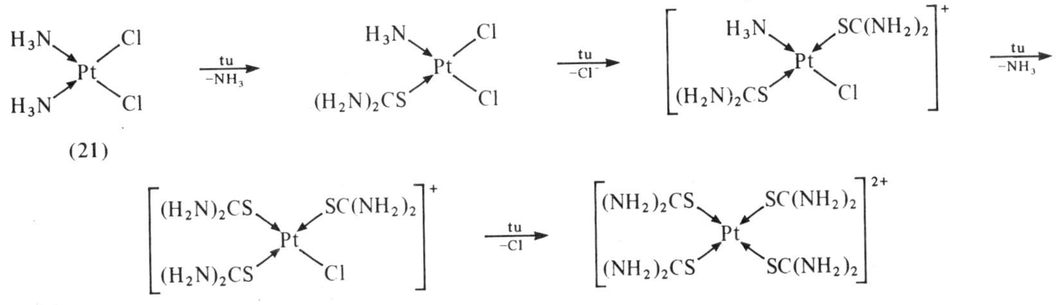

Scheme 1

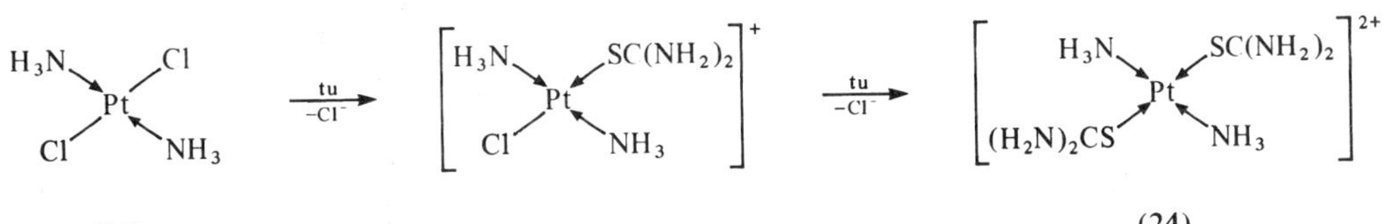

Scheme 2

16.6.4.2 Substituted Thioureas

Complexes of *N,N'*-substituted thioureas $(RR'N)_2C{=}S$ have been extensively studied, especially those of Cu^I and Ag^I. Ethylenethiourea (**25**; etu) forms complexes with Cu^I and Ag^I containing one, two, three and four etu molecules, whereas the only Au^I complexes known contain one or two etu molecules. Spectral data obtained for Co^{II} complexes indicate that etu is higher in the spectrochemical series than R_2S, lying between N_3^- and Ph_3PO; the Racah parameter B' is 0.66 that of the free ion, placing etu at the low end of the nephelauxetic series.[76] The paramagnetic compounds $[Ni(etu)_4X_2]$ (X = Cl, Br) are octahedral and have been obtained in *cis* and *trans* forms, but the iodo complex $[Ni(etu)_4I_2]$ is a rare example of a diamagnetic tetragonal complex, while $[Ni(etu)_4](ClO_4)_2$

is square-planar; on the other hand, [Ni(ntu)$_2$X$_2$] [X = Cl, Br; ntu = 1-(1-naphthyl)-2-thiourea] is tetrahedral.[76]

(25)   (26)   (27)

It has been reported that *N*-methylthiourea forms S-bonds with Zn^{II} and Cd^{II} but N-bonds with Pd^{II}, Pt^{II}, Pt^{IV} and Cu^{I};[77] this seems unlikely and needs confirmation. However, complexes of *N,N'*-dimethylthiourea and *N,N,N',N'*-tetramethylthiourea with a range of metals are all S-bonded.[78,79] Bis-chelated complexes [ML$_2$] (M = Co, Ni, Cu) are formed by a number of *N,N'*-diaryl-*N*-hydroxythioureas (26; LH) with the ligand bound through sulfur and oxygen.[47]

Diphenylphosphinothioyl derivatives of thiourea form complexes such as (27) in which the ligand acts as an S,S-donor.[80]

16.6.4.3 Dithiobiuret

Dithiobiuret (28; dtbH) by loss of a proton from its iminothiol tautomer (29) forms neutral complexes [M(dtb)$_2$] (M = Ni, Pd, Pt, Cu and Cd); the nickel complex is diamagnetic.[81] In the complex [Pd(dtb)$_2$] the four sulfur atoms are arranged in approximately square-planar arrangement with a mean Pd—S distance of 2.295 Å.[82]

(28)   (29)

16.6.5 SULFOXIDES

Dimethyl sulfoxide, Me$_2$S=O, can act as a ligand towards both transition and non-transition metal ions forming complexes such as [M(Me$_2$SO)$_6$](ClO$_4$)$_2$, [MX$_2$(Me$_2$SO)$_2$], [MX$_2$(Me$_2$SO)$_4$] and [MX$_3$(Me$_2$SO)$_3$] (X = Cl, Br, I, NCS). Early work based on IR spectral data indicated that in most complexes the metal is O-bonded but in Pd^{II} and Pt^{II} complexes it is S-bonded.[83-85] The structure of sulfoxides may be considered as a resonance hybrid of the structures (30) and (31). If coordination occurs through oxygen, the contribution of (31) will decrease and ν(S—O) will decrease. Conversely, if coordination is through sulfur, the contribution of (30) will decrease and this should result in an increase in ν(S—O). In dimethyl sulfoxide, ν(S—O) occurs at 1053 cm^{-1}; for O-bonded complexes it occurs in the range 985–1025 cm^{-1}, while for S-bonded complexes ν(S—O) occurs at 1116–1157 cm^{-1}.[83-86]

(30)   (31)   (32)

Complexes of diphenyl sulfoxide exhibit S—O stretching frequencies ranging from 931 cm^{-1} for Fe^{III} to 1012 cm^{-1} for Mg^{II}.[87,88] The IR spectra of the metal complexes of tetrahydrothiophene oxide (32) show that the complexes of Co^{II}, Ni^{II} and Cu^{II} are O-bonded, while those of the (b) class metal ions Pd^{II} and Pt^{II} are S-bonded.[89]

An X-ray crystal structure analysis of [PdCl$_2$(Me$_2$SO)$_2$] showed that the complex has a *trans* configuration with the sulfoxide ligands S-bonded; the interatomic distances were found to be as follows: Pd—S, 2.298; Pd—Cl, 2.287; S—O, 1.475 Å.[90] The IR and NMR spectra of [MCl$_2$(Me$_2$SO)$_2$] (M = Pd, Pt) and the corresponding complexes of dibenzyl sulfoxide, methyl benzyl sulfoxide and methyl isopropyl sulfoxide confirm S-bonding; the palladium complexes are *trans* but the platinum complexes have the *cis* configuration.[91]

The thiocyanato complexes $[Cr(NCS)_3(Me_2SO)_3]$ and $[M(NCS)_2(Me_2SO)_4]$ (M = Mn, Ni, Zn, Cd) contain N-bonded thiocyanate and S-bonded dimethyl sulfoxide.[92,93] The complex $[Ni(Me_2SO)Cl_2]$ displays antiferromagnetic coupling.[94] The frequency of the S—O stretching vibration and the occurrence of $v(Hg—O)$ in the IR spectra of a series of adducts of $HgCl_2$ with dialkyl and diaryl sulfoxides show that the sulfoxide ligands are O-bonded.[95] This is surprising in view of the pronounced (b) class character of Hg^{II}.

S-Bonding is not universal among dialkyl sulfoxide complexes of the platinum metals. The Ru^{III} complex $[RuCl_3(Me_2SO)_3]$ exhibits $v(S—O)$ at $1013\,cm^{-1}$, indicating O-bonding.[16] The Rh^I complexes trans-$[RhX(Me_2SO)(PPh_3)_2]$ (X = Cl, Br, I) display $v(S—O)$ at $1114–1120\,cm^{-1}$, indicating S-bonding.[97] The Rh^{III} complex $[Rh(Me_2SO)_5Cl]ClO_4$ contains two dimethyl sulfoxide ligands which are S-bonded and three which are O-bonded. The O-bonded ligands are labile and the complex can be readily converted to $Na[Rh(Me_2SO)_2Cl_4]$.[97] The Os^{II} complex $[Os(Me_2SO)_4\{\sigma\text{-}CH_2S(O)Me\}_2]\cdot 2Me_2SO$ has been prepared; the IR and ^1H and ^{13}C NMR spectra show that the two C-bonded $CH_2S(O)Me$ groups occupy trans positions, while the four equatorial positions are occupied by S-bonded dimethyl sulfoxide ligands.[98]

The complexes $[Pd(Me_2SO)_4]X_2$ (X = ClO_4, BF_4) contain two O-bonded and two S-bonded dimethyl sulfoxide groups in a cis configuration.[99] However, the complex $[Pd(Pe_2^iSO)_4](PF_6)_2$ (Pe^i = isopentyl) has four O-bonded sulfoxide groups, $v(S—O)$ occurring at $914\,cm^{-1}$, whereas $[PdCl_2(R_2SO)_2]$ (R = Me, Pr^n, Bu^n, Pe^i) displays $v(S—O)$ at $1119–1136\,cm^{-1}$, indicating that S-bonding is preferred by Pd(II) when steric influences are not predominant. The occurrence of both S- and O-bonding in the cationic complex $[Pd(Me_2SO)_4](BF_4)_2$ suggests that steric influences prohibit the formation of a complex with four S-bonded ligands, while electronic effects related to sulfur's π-acceptor properties dictate a cis configuration. In the complexes $[Pd(R_2SO)_4]^{2+}$ (R = Pr^n, Bu^n) increased steric influence brings about a trans arrangement. Finally in $[Pd(Pe_2^iSO)_4]^{2+}$ the steric effects become predominant leading exclusively to O-bonding.[100]

The complexes $[PdCl_2(Me_2SO)_2]$ and $[RhCl_3(Me_2SO)_3]$ can be reduced by $NaBH_4$ to give catalysts for the isomerization of alkenes and the hydrogenation of alkenes, dienes and alkynes.[101]

16.6.6 ORGANOPHOSPHORUS SULFIDES AND RELATED LIGANDS

16.6.6.1 The Ligands

The ligands include phosphine sulfides, $R_3P{=}S$, amino-substituted phosphine sulfides, e.g. $(Me_2N)_3P{=}S$, trialkylthiophosphates, $(RO)_3P{=}S$, and dithiophosphinates, $R_2P(=S)S^-$. These ligands, together with the oxygen and selenium analogues and their metal complexes, have been extensively reviewed.[102] The complex $Pt(Et_3PS)_2Cl_2$ was reported in 1857[103] but nearly 100 years were to elapse before further complexes were reported, viz. $[Hg(Et_3PS)_2Cl_2]$ by Malatests[104] and $[Pd(Ph_3PS)_2Cl_2]$ and $[Sn(Ph_3PS)_2Cl_4]$ by Bannister and Cotton.[105]

Phosphine sulfides, $R_3P{=}S$ (R = Me, Et, Pr^n, Bu^n, Ph), and amino-substituted phosphine sulfides, $am_3P{=}S$ (am = MeNH, Me_2N, cyclohexamino, piperidino, pyrrolidino or morpholino), are known to form metal complexes by coordination via the sulfur atom. The bond dissociation energy of P—S ($293–385\,kJ\,mol^{-1}$) in these ligands is less than that of P=O ($502–627\,kJ\,mol^{-1}$) in the oxygen analogues, indicating that the P=S bond has less π-character than P=O. The length of the P=S bond is $1.87–1.96\,\text{Å}$, which is indicative of a bond order of ca. 1.7; the bond order increases with increasing electronegativity of the substituent R. The bond dipole moments of phosphine sulfides (3.1 D) are greater than those of phosphine oxides (2.7 D); nevertheless these low values (ca. 35% expected) suggest that π-bonding produces a displacement of charge towards the phosphorus atom.[102] The $v(P{=}S)$ frequency of $Ph_3P{=}S$ is $637\,cm^{-1}$ but falls to ca. $590\,cm^{-1}$ on coordination.[106]

Trialkylthiophosphates, $(RO)_3P{=}S$ (R = Et, Pr^n), act as unidentate sulfur ligands.[102] Dialkylthiophosphinates (33) react with some metal ions forming complexes (34), being concomitantly oxidized to bis(dialkylthiophosphoryl)disulfanes (35) as shown in equation (2).[107] Dialkyldithiophosphates, $(RO)_2P(S)SH$, form complexes similar to (34).

$$3 \quad (33) \quad + \quad M^{n+} \quad \longrightarrow \quad (34) \quad + \quad (35) \tag{2}$$

(33)   (34)   (35)

16.6.6.2 Phosphine Sulfides

Phosphine sulfides, $R_3P\!=\!S$ (R = Me, Et or Ph), form complexes with both (a) and (b) class metals, although the latter are better suited for complex formation with these soft S-donor ligands. Complexes are known with the (a) class metals Al^{III}, Ti^{IV}, Nb^V, Ta^V, Sn^{IV} and Sb^V, and with the (b) class and borderline metals Fe^{II}, Co^{II}, Pd^{II}, Pt^{II}, Pt^{IV}, Cu^I, Ag^I, Au^I, Au^{III}, Zn^{II}, Cd^{II}, Hg^{II} and Sn^{II}.[102,106] The phosphine sulfide $Ph_3P\!=\!S$ forms 1:1 adducts with $AlBr_3$, $TiBr_4$, $SbCl_5$ and NbX_5 (X = Cl, Br), while $AlCl_3$ and $TiCl_4$ form both 1:1 and 1:2 adducts and SnX_4 forms the 1:2 adducts $[Sn(R_3PS)_2X_2]$ (R = Me, Et, Pr^n, Bu^n, Ph).[102] The relative donor strength of a series of related ligands towards the hard Lewis acid Sn^{IV} are: $(RO)_3P\!=\!O > R_3P\!=\!S > (RO)_3P\!=\!S > R_3'P\!=\!S$ (R = alkyl; R' = aryl).[108]

Copper(II) and Fe^{III} are reduced by $R_3P\!=\!S$ with the formation of Cu^I and Fe^{II} complexes, respectively. The complexes $[Co(Me_3PS)_2X_2]$ (X = Cl, Br, I) and $[Co(Me_3PS)_4](ClO_4)_2$ are tetrahedral. The 1:2 complexes $[M(R_3PS)_2X_2]$ (M = Pd, Pt, Zn, Cd, Hg, Sn; R = Me, Et or Ph) are known but Hg^{II} also forms the 1:1 complexes $[Hg(R_3PS)X_2]$ (R = Me, Ph; X = Cl, Br, I), which are dimeric and sulfur-bridged, and the 1:4 complex $[Hg(Ph_3PS)_4](ClO_4)_2$.[102,106] The complex $[Cu(Me_3PS)Cl]$ has a trimeric structure consisting of a six-membered ring of alternate copper and sulfur atoms; the trigonally coordinated copper atoms are bonded to two bridging sulfur atoms and a terminal chlorine atom.[109]

16.6.6.3 Amino-substituted Phosphine Sulfides

Tris(dimethylamino)phosphine sulfide is known to form the complexes $[M\{(Me_2N)_3PS\}_4](ClO_4)_2$ (M = Co, Ni), $[M\{(Me_2N)_3PS\}_2X_2]$ (M = Co, Pd, Hg; X = Cl, Br or I) and $[M\{(Me_2N)_3PS\}_2]ClO_4$ (M = Cu, Ag). The Co^{II} complexes are tetrahedral but surprisingly so is the Ni^{II} complex (μ, 3.5 BM), possibly because of steric hindrance among the four bulky ligand moieties. The $P\!=\!S$ stretching frequency is lowered from 565 to 545 cm^{-1} upon coordination.[102]

Tripiperidino-, tripyrrolidino-, trimorpholino-, tri(methylamino)- and tri(cyclohexylamino)-phosphine sulfides forms adducts with Sn^{IV} halides with a lowering of $\nu(P\!=\!S)$ of 45 cm^{-1}. The first two ligands form 1:2 adducts with $PdCl_2$ and both 1:1 and 1:2 adducts with CdX_2 (X = Br, I).[102]

16.6.6.4 Thiophosphoryl Halides

Thiophosphoryl halides, $X_3P\!=\!S$ (X = Cl, Br), are weaker donors than $Cl_3P\!=\!O$ and $Ph_3P\!=\!S$. Only a few complexes are known: $[Cl_3PS\cdot AlCl_3]$, $[Br_3RS\cdot AlBr_3]$, $[Cl_3PS\cdot SbCl_5]$ and $[Br_3PS\cdot SbCl_5]$. The ligands are S-bonded, as $\nu(P\!=\!S)$ decreases from 748 to *ca.* 650 cm^{-1} and $\nu(P–X)$ increases with adduct formation.[110]

16.6.6.5 Neutral Thiophosphoryl Esters

Trialkylthiophosphates, $(RO)_2(RS)P\!=\!O$ (L), form the complexes $AlCl_3\cdot L$, $TiCl_4\cdot 2L$, $2FeCl_3\cdot 2L$, $PtCl_4\cdot 2L$ and $HgCl_2\cdot 2L$, while trialkyldithiophosphates, $(RO)(RS)_2P\!=\!O$, form bis-adducts with SnX_4 (X = Cl, Br). In both types of complex the ligands are O-bonded.[111]

16.6.6.6 Thiophosphinato Complexes

Dithiophosphinic acids, $R_2P(S)SH$, have been important in the characterization of the hitherto unknown S—H$\cdots$S bridge bond. The bond enthalpies of S—H$\cdots$S bridges are *ca.* 0.1 times as great as those of hydrogen bridges involving oxygen. Hydrogen bridges can be inferred from $\nu(S—H)$; the greatest shifts in $\nu(S—H)$ (*ca.* 220 cm^{-1}) occur with dithiophosphinic acids, in which the hydrogen bridges have enthalpies of *ca.* 12 kJ mol^{-1}.[112]

Many dithiophosphinato complexes $[M\{R_2P(S)S\}_n]$ have been reported (M = Cr^{III}, Fe^{III}, Ru^{III}, Co^{II}, Co^{III}, Rh^{III}, Ir^{III}, Ni^{II}, Pd^{II}, Pt^{II}, Cu^I, Ag^I, Au^I, Zn^{II}, Cd^{II}, Hg^{II}, Al^{III}, Tl^I, Pb^{II}, As^{III}, Sb^{III} and Bi^{III}; R = alkyl, Ph).[113] The ligand in most cases is bidentate forming a four-membered PSSM ring; however, with Cu^I, Ag^I and Au^I oligomeric complexes $[M\{R_2P(S)S\}]_n$ (n = 2 or 4) are formed. The complexes, except those of Fe^{III} and Co^{III}, are very stable and many are deeply coloured. Those of dialkyldithiophosphinic acids are readily soluble but the diaryldithiophosphinato complexes are less so. The Ni^{II} and Ru^{III} complexes are low-spin, as expected for a complex with soft donor atoms. Many complexes form adducts with Lewis bases, *e.g.* $[Co\{Et_2P(S)S\}_2py_2]$.[113] Dipole moment measurements indicate *cis* and *facial (cis)* configurations for the square-planar and octahedral complexes, respectively.[114] This is in keeping with the *cis* and *facial* configurations found for complexes of other sulfur ligands.[115]

Because of their solubility in organic solvents some of the complexes can be extracted quantitatively from neutral and even acidic solutions. There are several patents for various industrial uses of some complexes.[113]

As stated above, dithiophosphinates reduce some metal ions, while being concomitantly oxidized to bis(diorganothiophosphoryl)disulfanes, $R_2P(S)SS(S)PR_2$. The latter react with the carbonyls $Cr(CO)_6$, $Fe(CO)_5$ and $Ni(CO)_4$ to give $[Cr(R_2PS_2)_3]$, $[Fe(R_2PS_2)_3]$ and $[Ni(R_2PS)_2]$, respectively, while $SnCl_2$ is oxidized to $[(R_2PS_2)_2SnCl_2]$.[107]

Only a few complexes of monothiophosphinic acids, $R_2P(S)OH$, are known. Diphenyl-thiophosphinic acid, $Ph_2P(S)OH$, forms polymeric complexes with Ni^{II} and Al^{III} but the Cd^{II} complex is monomeric. The Ni^{II} complex is paramagnetic (μ, 3.86 BM) with a typically tetrahedral spectrum.[116] Complexes of diethyldithiophosphinic acid, $Et_2P(S)OH$, with Co^{II}, Zn^{II}, In^{III} and Pb^{II} have been reported; their degree of association in organic solvents, involving ligand bridging, is concentration dependent.[113]

16.6.6.7 Imidodithiodiphosphinato Complexes

Imidodithiodiphosphinato complexes (**36**; R = Me, Ph; M = Fe, Co, Ni, Zn, Pd and Pt) are known.[117] The iron complex is readily oxidized in solution but the other complexes are stable. The iron, cobalt, nickel and zinc complexes are tetrahedral. The nickel complexes are paramagnetic (μ, 3.40 BM) and their solution electronic spectra are consistent with a tetrahedral configuration, which has been confirmed by an X-ray determination on the complex with R = Me.[118] These are rare examples of a nickel complex having a tetrahedral configuration both in the solid state and in solution. The tetrahedral stereochemistry of the nickel complexes is surprising, since there are no steric factors associated with the ligands which might be considered to destabilize the square-planar configuraiton expected for the NiS_4 chromophore.

(36)

16.6.6.8 Monoalkyl Dithiophosphonates

A few complexes of monoalkyl dithiophosphonates, $R(R'O)P(S)SH$, are known. The deprotonated ligand forms a four-membered PSSM chelate ring. However, in the complex $[Ni\{Et-(Pr^iO)PS_2\}_2]$ the two P—S bond lengths (2.03 and 1.90 Å) are not equivalent, suggesting one single and one double bond; moreover, the Ni—S bond lengths (2.16 and 2.26 Å) are not identical, showing the anisobidentate (unsymmetrical) nature of the ligand.[119]

16.6.6.9 Dialkyl and Diaryl Dithiophosphates

The O,O-diesters of dithiophosphoric acids, $(RO)_2P(S)SH$, were first reported in 1908.[120] Chemical interest in these compounds is threefold: (a) their coordination with transition metal ions; (b) their use as analytical reagents; (c) the biological activity of both the ligands and their metal complexes. They have been used for the vulcanization of rubber[121,122] and as insecticides; the latter activity is related to their enzyme-inhibition ability.[123]

The metal complexes of diethyl dithiophosphate were first investigated in 1931,[124] when it was shown that the transition metal complexes $[M\{(EtO)_2PS_2\}_n]$ ($n = 2$ or 3) were similar to those of xanthates and dialkyl dithiocarbamates, containing a four-membered XSMS (X = C or P) ring. A range of metal complexes of dialkyl, diaryl and mixed alkyl aryl dithiophosphates was investigated by Malatesta and Pizzotti.[125] These complexes are brightly coloured and very soluble in organic solvents. The purple Ni^{II} complexes are diamagnetic and square-planar but the black Fe^{III} complexes are high-spin. Surprisingly, diethyl dithiophosphate forms complexes with the lanthanides, which, being strong (a) class Lewis acids, would be expected to have a low affinity for soft S,S donors. Complexes of the type $[NEt_4][Ln\{(EtO)_2PS_2\}_4]$ (Ln = La, Ce, Pr, Nd, Sm, Eu, Tb, Ho, Yb) have been reported and their electronic spectra have been discussed.[126] Diethyl dithiophosphate has a low position in the electrochemical series and a large nephelauxetic effect.[127] The complexes $[M\{EtO)_2PS_2\}_2]$ (M = Ni, Pd) showed significant anticancer activity in the Walker[256] carcinosarcoma test system in mice and $[Ni\{(ClCH_2CH_2O)_2PS_2\}_2]$ showed larvicidal activity against *Tineola bisselliella*.[128] The complexes $[Ni\{(RO)_2PS_2\}_2]$ react with nitrogen heterocycles to form green, paramagnetic adducts $[Ni\{(RO)_2PS_2\}\cdot 2B]$ (B = py, γ-picoline, $\frac{1}{2}$bipy or $\frac{1}{2}$phen), containing the S_4N_2 chromophore.[128]

Quite a few structural determinations have been made on dialkyl dithiophosphate complexes; they have been recently reviewed by Haiduc.[119] In the square-planar complexes $[Ni\{(RO)_2PS_2\}_2]$ (R = Me, Et, Pr^i) the coordination is symmetrical (isobidentate) with virtually equivalent P—S distances at 1.95–1.98 Å and the Ni—S bonds are also identical at 2.21–2.22 Å. In the *trans* octahedral complex $[Ni\{(EtO)_2PS_2\}_2(py)_2]$ the Ni—S distance is much longer (2.49 Å). In the *cis* octahedral complexes $[Ni\{(RO)_2PS_2\}_2(phen)]$ (R = Me, Et) and $[Ni\{(MeO)_2PS_2\}_2(bipy)]$ the Ni—S bonds *trans* to the sulfur atoms are some 0.04 Å longer than those *trans* to the nitrogen atoms.[119,128]

In $[In\{(EtO)_2PS_2\}_3]$ the ligand is anisobidentate with long (2.137 Å) and short (1.90 Å) P—S bonds and unequal In—S bonds (2.578 and 2.607 Å). The octahedral complexes $[Sb\{(RO)_2PS_2\}_3]$ (R = Me, Pr^i) also contain non-equivalent P—S bonds (2.04 and 1.94 Å).[119] In the distorted tetrahedral complex $[NMe_4][Zn\{(MeC_6H_4O)_2PS_2\}_3]$ one ligand moiety is bidentate (mean Zn—S distance, 2.44 Å), while the other two are unidentate (mean Zn—S, 2.31 Å).[121] In $[Te\{(MeO)_2PS_2\}_2]$ and $[Ph_3Sn\{(EtO)_2PS_2\}]$ the ligand is unidentate.[119]

There are several compounds known in which the dialkyl dithiophosphate ligand bridges two metal atoms. In the dimeric complexes $[M\{(Pr^iO)_2PS_2\}_2]_2$ (M = Zn, Cd) two of the ligands are terminal and bidentate, while the other two bridge the metal atoms forming an eight-membered MSPSMSPS ring. The complex $[Zn\{(EtO)_2PS_2\}_2]$ is polymeric in the solid state with long chains of alternating zinc atoms and bridging diethyl dithiophosphato ligands with an additional bidentate ligand attached to each zinc atom. The mercury complex $[Hg\{(Pr^iO)_2PS_2\}_2]$ contains unsymmetrical bridges. One ligand is terminal and anisobidentate with one short (2.39 Å) and one long (2.89 Å) Hg—S bond. The second, bridging ligand forms a short intramolecular Hg—S bond (2.39 Å) and a long (2.75 Å) intermolecular Hg—S bond.[119]

16.6.7 ORGANO-ARSENIC AND -ANTIMONY SULFIDES

Triphenylarsine and triphenylstibine sulfides form complexes similar to those of Ph_3PS. The stabilities of the complexes of related ligands decrease in the order $Ph_3PS > Ph_3PSe \approx Ph_3AsS > Ph_3SbS$. The following complexes of Ph_3AsS have been prepared: $[Pd(Ph_3AsS)_2Cl_2]$, $[Au(Ph_3AsS)Cl]$ and $[Hg(Ph_3AsS)X_2]$ (X = Cl, Br, I). The only complexes of Ph_3SbS which could be isolated were $[Cu(Ph_3SbS)_2I]$ and $[Hg(Ph_3SbS)I_2]$.[106]

16.6.8 1,5-DITHIACYCLOOCTANE

1,5-Dithiacyclooctane (37) can be oxidized in two steps by nitrosyl fluoroborate $(NO)(BF_4)$ to give a long-lived doubly charged cation (38), which can be isolated as the colourless crystalline fluoroborate $(DTCO)(BF_4)_2$ (DTCO = 1,5-dithiacyclooctane dication). This dication can act as an electrophile and as a two-electron oxidizing agent.[129]

The Ru^{II} complex of (37), $[(NH_3)_5Ru(C_4H_8S_2)](PF_6)_2$, undergoes reversible two-electron oxidation.[130] Treatment of $(DTCO)(BF_4)_2$ with $[(C_2H_4)Pt(Ph_3P)_2]$ gives the Pt^0 complex $[(DTCO)Pt(Ph_3P)_2](BF_4)_2$ (39) and not an oxidative-addition product, as might be expected. The structure (39) has been confirmed by NMR spectral data obtained from solution and the solid state.[131]

(37)   (38)   (39)

16.6.9 THIOCARBOXYLIC ACIDS

16.6.9.1 Monothiocarboxylic Acids

Thioacetic acid, MeCOSH, by the loss of a proton can act as an S-bonded, S,O-bidentate or bridging ligand. Antimony tris(monothioacetate), $[Sb(MeCOS)_3]$, can be prepared from Sb_2O_3 and thioacetic acid in hexane at 50 °C; unlike the analogous triacetate it is unaffected by water. An X-ray structure determination showed that the ligand is bound primarily through sulfur (mean Sb—S distance, 2.47 Å) with weak secondary bonding through oxygen (Sb—O distances, 2.75–2.92 Å). One of the thiacetate groups bridges adjacent molecules (Sb–O distance, 3.04 Å), leading to polymeric chains. The overall stereochemistry of the antimony atom, inclusive of the lone pair, is distorted dodecahedral.[132]

The dimeric Rh^{II} complex $[Rh_2(MeCOS)_4(MeCOSH)_2]$ contains four bridging thioacetate groups; the two thioacetic acid ligands are in the thione form and are each bound *via* the sulfur atom to one of the rhodium atoms.[133] The Ru^{II} complex $[Ru(MeCOS)_2(CO)(Ph_3P)_2]$ is known in two isomeric forms.[134]

Monothiobenzoic acid, PhCOSH, acts as a bidentate ligand in the complex $[Cr(PhCOS)_3]$, although there is considerable strain in the chelate ring (40).[135] In the complex $[Ni(PhCOS)_2]$ the ligand may be chelated but it is more likely that this complex contains the NiS_4 chromophore with bridging sulfur atoms. In the complex $[Ni(PhCOS)_4]^{2-}$ the ligand is unidentate and S-bonded.[135,136] The complexes $[Ni(RCOS)_2]$ (R = Me, Ph) react with nitrogen heterocycles to form green, paramagnetic, octahedral complexes $[Ni(RCOS)_2L_2]$ (L = py, $\frac{1}{2}$bipy) but with phosphines to give red, diamagnetic, square-planar complexes $[Ni(RCOS)_2(R_3'P)_2]$ ($R_3'P$ = Ph_3P, Ph_2MeP, $PhMe_2P$), in which the monothiocarboxylate group is unidentate and S-bonded.[137]

(40)

The complex $[R_2MOCl_2]$ (R = η^5-cyclopentadienyl, η^5-indenyl; M = Mo, W) react with NaSCOR' (R' = H, Me, Et, Ph) in THF to give $[R_2MO(R'COS)_2]$.[138–140] Reaction of $[RTiCl_2]$ (R = Cp, indenyl) with NaSCOR' (R' = H, Me, Et, Ph, p-tolyl) gives $[RTi(R'COS)_2]$.[141]

An interesting reaction is that of the mononuclear $[W(CO)_5(SH)]^-$ and the SH^--bridged dinuclear μ-HS$\{W(CO)_5\}_2^-$ complexes with acetic anhydride to give the thioacetate complex $[(MeCOS)W(CO)_5]^-$. The SH^- complexes react with aliphatic ketones and aromatic aldehydes to yield the complexes $[(R_2C{=}S)W(CO)_5]$ and $[(RCHS)W(CO)_5]$ of these otherwise unstable thioketones and thioaldehydes.[142]

16.6.9.2 Dithiooxalic Acid

Dithiooxalic acid ($dtoH_2$)forms complexes $K_2[M(dto)_2]$ (M = Ni, Pd, Pt) which are isomorphous and square-planar (41); the M—S distances are: Ni—S, 2.30; Pd—S, 2.44; Pt—S, 2.44 Å.[143] The IR spectra of the Ni^{II}, Pd^{II} and Pt^{II} complexes and of the Co^{III} complex $[Co(dto)_3]^{3-}$ in the region 4000–280 cm^{-1} have been reported; a normal coordinate analysis for a 1:1 model of the platinum complex confirms that the Pt—S bands are at 430 and 320 cm^{-1}.[144]

(41)

A detailed study of the electronic spectra of the d^8 complexes $[M(dto)_2]^{2-}$ (M = Ni, Pd, Pt) and $[Au(dto)_2]^-$ has been reported. The spectra are virtually identical in a variety of solvents, indicating that axial perturbations due to the solvent are minimal. Dithiooxalate has a high position among square-planar NiS_4 chromophores in the spectrochemical series; for dithio ligands the series is: maleonitriledithiolate < $(CF_3)_2C_2S_2^{2-}$ < diethyl dithiophosphate < ethyl xanthate < diethyl dithiocarbamate < 2,3-dimercaptopropanol anion < dithiomalonate $\approx$ dithiooxalate.[145]

16.6.9.3 Dithiocarboxylic Acids

Dithiobenzoic acid, PhCSSH, and dithiophenylacetic acid, PhCH$_2$CSSH, form low-spin monomeric square-planar [M^{II}L$_2$] (M^{II} = Ni, Pd, Pt) and octahedral [M^{III}L$_3$] (M^{III} = Cr, Co, Rh, In) complexes. They are brightly coloured (Table 1); the colours of the indium and platinum complexes are quite unusual. These two ligands have a higher place in the spectrochemical series than other dithio anions, lying between H$_2$O and F$^-$; they also have a large nephelauxetic effect.[146]

Table 1 Colours of Metal Complexes of Dithiobenzoic Acid (dtbH) and Dithiophenylacetic Acid (dtpaH)

Complex	L = dtb	L = dtpa
CrL$_3$	Reddish brown	Greenish brown
CoL$_3$	Reddish brown	Dark green
RhL$_3$	Reddish brown	Orange
IrL$_3$	Brownish red	Reddish brown
InL$_3$	Brownish orange	Golden yellow
NiL$_2$	Deep blue	Deep blue
PdL$_2$	Reddish violet	Brownish yellow
PtL$_2$	Dark green	Greenish blue

A crystal structure determination of [Pd(dtb)$_2$] showed a distorted tetragonal arrangement of sulfur atoms about the palladium atom. The Pd—S distances in the square plane are 2.32–2.34 Å, while the Pd—S (apical) distance is 3.46 Å, suggesting some interaction.[147] The nickel complex [Ni(dtb)$_2$] can be readily oxidized to [Ni$_2$(dtb)$_4$S$_2$], which was postulated to be a Ni^{IV} complex,[136] but it is more likely that the complex contains Ni^{II} with the structure (42).[148]

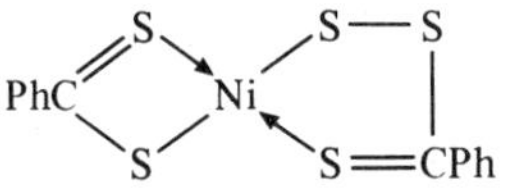

(42) 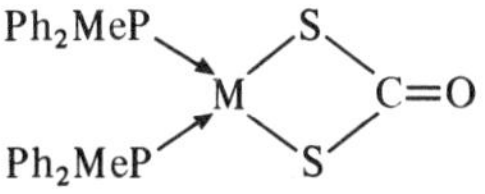   (43)

Reaction of methyldiphenylphosphine with the benzyl xanthate complexes [M(S$_2$COCH$_2$Ph)$_2$] (M = Pd, Pt) in chloroform yields the dithiocarbonato complexes (43). The platinum complex (43; M = Pt) undergoes a variety of reactions in chloroform solution, e.g. treatment with HCl yields cis-[(Ph$_2$MeP)$_2$PtCl$_2$].[149]

16.6.9.4 Trithiocarbonic Acid

An X-ray crystal structure determination of H$_2$CS$_3$ at −100 °C showed the CS$_3$ group to be planar.[150] The stability constants (log β_2) for the Co^{II} and Ni^{II} complexes of trithiocarbonate, CS$_3^{2-}$, were reported to be 8.1 and 9.0, respectively.[151] The complexes (Ph$_4$As)$_2$[M(CS$_3$)$_2$] (M = Ni, Pd, Pt) are isomorphous; in these complexes ν(C=S) occurs at ca. 1000 cm^{-1}. The red nickel complex is diamagnetic and its IR spectrum displays ν(Ni—S) at 380 cm^{-1}.[152]

Trithiocarbonato complexes resemble in their reactions and properties other 1,1-dithiolo complexes such as those of xanthates and dithiocarbamates (see Chapter 16.4). Oxidation of [Ni(CS$_3$)$_2$]$^{2-}$ by iodine or elemental sulfur yields [Ni(CS$_4$)$_2$]$^{2-}$, which probably has the structure (44). The frequency ν(S—S) occurs at 480 cm^{-1} in the Raman spectrum, while ν(C—S) occurs at 1035 cm^{-1}, i.e. at slightly higher energy than in the spectrum of [Ni(CS$_3$)$_2$]$^{2-}$. The complex (Ph$_4$As)$_2$[Ni(CS$_4$)$_2$] can also be prepared from KCS$_4$.[148]

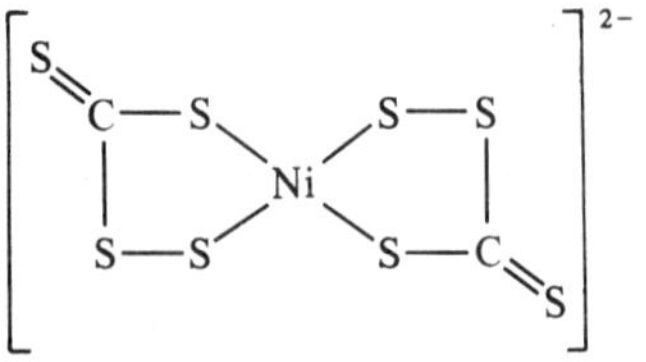

(44)

16.6.10 THIOCARBOXAMIDES

16.6.10.1 Thioacetamide, Thiobenzamide and Related Amides

Thioacetamide, $MeC(=S)NH_2$ (taa), is isoelectronic with thiourea which it resembles by acting as a unidentate S-donor ligand. Thiobenzamide, $PhC(=S)NH_2$ (tba), behaves similarly. Both thioamides form stable complexes with (b) class and borderline metals. Thioacetamide forms the tetrahedral complexes $[MX_2(taa)_2]$ (M = Fe, Co, Zn; X = Cl, Br, I, NCS).[47,153–157] The $v(M—S)$ frequency is rather low, ranging from 255 to 230 cm^{-1}.[145] The Mössbauer spectra of $FeX_2(taa)_2$ (X = Cl, Br) are consistent with a tetrahedral configuration.[158]

The square-planar complexes $[M(taa)_4]X_2$ (M = Cl, Br or I), $[Pt(taa)_2Cl_2]$, $[Ptpy(taa)Cl_2]$, $[Pt(tba)_4]Cl_2$ and $[Pt(tba)_2Cl_2]$ have been reported.[155,159–161] The bluish purple complexes $[Ni(taa)_4]X_2$ (X = Br, ClO_4, NO_3) are diamagnetic and square-planar[155] but an X-ray structure determination on the chloro complex $[Ni(taa)_4Cl_2]$ showed it to have a *trans* octahedral configuration with Ni—S distances of 2.46 Å.[162] The thiocyanato complex $[Ni(taa)_2(NCS)_2]$ is also octahedral, each nickel atom being surrounded by two sulfur atoms from the thioacetamide groups at 2.45 Å, two sulfur atoms from the thiocyanato ligands at 2.55 Å and two thiocyanato nitrogen atoms at 2.02 Å; the nitrogen atoms are *trans* with respect to the plane of the sulfur atoms.[163]

The 1:1 Ni^{II} complex $[Ni(taa)Cl_2]$ has a rather low room-temperature magnetic moment (2.62 BM); it may be antiferromagnetic, resembling the DMSO complex $[NiCl_2(Me_2SO)]$. The electronic spectrum has been interpreted in terms of a distorted octahedral configuration, achieved by chloro bridging in the plane and sulfur bridging in the apical positions.[154] In the chloro-bridged complex $[Cd(taa)Cl_2]$ each cadmium atom is surrounded octahedrally by one sulfur and five chlorine atoms and the coordination polyhedra are linked by chloro bridges into double chains.[165] The complex $[Hg(taa)_2Cl_2]$, surprisingly, is octahedral, each mercury atom being coordinated by two sulfur and four chlorine atoms.[166]

The complexes $[M(taa)_4]^+$ (M = Cu, Ag) are tetrahedral;[47] the single-crystal IR and Raman spectra of $[Cu(taa)_4]Cl$ have been fully discussed.[167] The 1:1 complex Cu(taa)Cl is also known.[164] The structure consists of hexagonal rings which lie in layers; each ring is composed of three copper and three sulfur atoms which alternate in a zig-zag pattern. Each copper atom is surrounded by one chlorine and two sulfur atoms and the thioacetamide moiety retains its planarity.[168] The complexes of Au^I are two-covalent and linear. The complex $[Au(taa)_2]Br$ has long been known.[47] But $[Au(tba)_2]X$ and $[AuLX]$ (L = N,N-dimethylthioformamide and N,N-dimethylthioacetamide; X = Cl, Br or I) have been recently reported; the $v(Au—S)$ frequency occurs in the range 320–260 cm^{-1}.[169]

Mixed ligand complexes of Co^{III} are known, *viz.* $[Co(DH)_2(taa)Cl]$, $[Co(DH)_2(taa)_2]^+$ (DH_2 = dioxime), $Na[Co(acac)_2(taa)]$ and $Na[Co(acac)_2(NO_2)(tba)]$.[47,170] The complexes $[MCl_3(taa)_3]$ (M = Rh, Ir) and $[RuCl_2(taa)_4]$ are also known.[159,171] The Rh^{II} complexes $[Rh_2(Me-CO_2)_4(taa)_2]$, $[Rh_2(MeCO_2)_4(tba)_2]$ and the analogous thiourea compound all contain the dimeric $Rh_2(MeCO_2)_4$ unit.[172] The Os^{II} complex $[Os(taa)_4(SnCl_3)_2]$ has been described.[173] The Mössbauer and ESR spectra of the nitrosyl iron complexes $[Fe(taa)(NO)_2X]$ and $[Fe(tba)(NO)_2X]$ (X = Cl, Br) have been reported.[174] The carbonyl complexes $[Mo(phen)(CO)_3(taa)]$ and $[W(taa)(CO)_5]$ have been prepared.[175,176]

16.6.10.2 Thioacetamidothiazole

2-Thioacetamidothiazole (**45**; tatz) forms the complexes $[MX_2(tatz)_2]$ (M = Zn, Cd, Hg; X = Cl, Br) and $[CuCl(tatz)_2]$. The ligand is coordinated through the thiazole nitrogen in the Zn^{II} and Cd^{II} complexes but *via* the thioamide sulfur in the complexes of the strongly (b) class metals Cu^I and Hg^{II}.[177]

(45)

16.6.11 MONOTHIO-β-DIKETONES

16.6.11.1 The Ligands

Although metal complexes of acetylacetone were first reported in 1887 by Coombes,[178] it was not until 1964 that the first complexes of monothio-β-diketones were reported by Chaston and Livingstone.[179] The first monothio-β-keto ester was synthesized by Mitra[180] in 1931 but the synthesis of monothioacetylacetone was first reported by Mayer *et al.* in 1963.[181] In 1964 three groups of workers, Chaston and Livingstone,[169] Uhlemann *et al.*[182] and Yokoyama *et al.*,[183] independently initiated work on the synthesis and properties of monothio-β-diketones and their metal complexes.

There are two main methods for the synthesis of monothio-β-diketones, both affording yields in excess of 50%. The first method involves the passage of H_2S through a dilute solution of the β-diketone in ethanol, followed by the rapid passage of HCl at $-70\,^{\circ}C$; the conditions are critical.[184,185] The second method involves the Claisen-type condensation of ketones with thionic esters, RC(S)OMe, or dithionic esters, RC(S)SMe, in ether in the presence of sodamide.[186-188] This method has the advantage that both isomers (**46**) and (**47**) can be prepared, whereas the first method yields only isomer (**46**) when R' has a greater electron-withdrawing power than R.

(**46**)   (**47**)

Evidence from IR,[187,189-192] visible-UV[193] and NMR spectra[189,192,194,195] and X-ray diffraction[196] shows that monothio-β-diketones exist either entirely or predominantly in the thienol form (**46**) and the broad IR band at 2450 cm^{-1} has been assigned to the intramolecularly bound thiol group —S—H$\cdots$O=.[192] The MS fragmentation patterns of a number of monothio-β-diketones have been reported;[190,187] the replacement of one oxygen by sulfur in a β-diketone brings about greater complexity in the mass spectrum.[197] Determinations by Livingstone[198,199] and Uhlemann[200,201] and their co-workers of the dissociation constants pK_D of a number of monothio-β-diketones in dioxane–water solution indicate that the thio derivatives are stronger acids than β-diketones, since they have pK_D values 2.0–2.7 log units lower than their oxygen analogues. The hydrolysis of the monothio-β-diketones (**46**; R = R' = Me: R = Me, R' = Ph; R = Me, m-BrC$_6$H$_4$, p-BrC$_6$H$_4$, R' = CF$_3$) have been studied and the rate constants have been determined.[202]

16.6.11.2 Metal Complexes

Monothio-β-diketones and their metal complexes have been reviewed by Cox and Darken,[203] Livingstone,[184] Uhlemann,[204] and, more recently, by Mehrotra *et al.*[205] Monothio-β-diketones (LH) readily form stable complexes [ML$_n$] with (b) class metal ions (M = Rh^{III}, Ni^{II}, Pd^{II}, Pt^{II}, Hg^{II} and Pb^{II}). Although complexes of most borderline metals, *viz.* V^{III}, Fe^{III}, Ru^{III}, Os^{III}, Co^{III}, Cu^{II}, Zn^{II}, Cd^{II}, Sn^{II} and Sn^{IV}, are known, reported complexes of (a) class metals are relatively few. Nevertheless, complexes of Cr^{III}, Mn^{III}, Ga^{III} and In^{III} are known. Unstable complexes [ML$_2$py$_2$] (M = Mn^{II} and Fe^{II}) have been prepared,[206] while Co^{II} complexes CoL$_2$ have been prepared under anaerobic conditions at low temperatures.[207,208] The hexacarbonyls [M(CO)$_6$] (M = Mo, W) react with LH in the presence of pyridine to give [ML$_4$].[209] Vanadyl(IV) complexes have also been reported.[206,210,211] Although α-alkyl-substituted monothio-β-diketones form complexes with Ni^{II} and Co^{III}, attempts to prepare complexes of Pd^{II}, Pt^{II}, Cu^{II}, Zn^{II} and Hg^{II} resulted in oxidation of the ligand to the disulfide.[212,213] Complexes of monothio-β-diketones are readily soluble in organic solvents, whereas β-diketonato complexes with few exceptions have negligible solubility in these solvents.

16.6.11.3 Physical Measurements

Stability constant determinations have been made on a range of monothio-β-diketonato complexes in dioxane–water solution. The results indicate that (b) class metal ions form more stable complexes with monothio-β-diketones than with β-diketones. The stability sequence for monothio-β-diketonato complexes is: Cu > Ni > Zn > Cd > Pb.[198-201]

With few exceptions, Ni^{II} complexes of β-diketones are green, octahedral and paramagnetic ($\mu \approx 3.2$ BM), whereas Ni^{II} complexes of monothio-β-diketones are brown, square-planar and diamagnetic.[190] However, they form green, octahedral and paramagnetic adducts [NiL$_2 \cdot$2B] (B = py; 2B = phen, bipy).[214]

Iron(III) complexes of β-diketones are high-spin (μ, 5.92–6.00 BM), whereas those of monothio-β-diketones display anomalous magnetic behaviour in that the moments vary between 2.3 and 5.7 BM at 298 K. Ho and Livingstone[215] found temperature dependence of the moments of seven Fe^{III} complexes: at 373 K the moments were in the range 3.65–5.77 BM, while at 83 K they ranged from 1.86 to 4.13 BM. This was attributed to a thermal equilibrium between the nearly equienergetic spin-paired $^6A_1$ (t_{2g}^5) and spin-free $^2T_2$ ($t_{2g}^3 e_g^2$) states of the iron atom, resulting from the approximately equal magnitudes of the ligand field and the spin-pairing energy. Cox *et al.*[216] reported that the Mössbauer spectra of four Fe^{III} complexes showed the presence of both high- and low-spin isomers at 298 and 80 K. Das *et al.*[217] measured the moments of nine Fe^{III} complexes of (46; R′ = CF$_3$) over the range 83–273 K and the Mössbauer spectra at 77, 193 and 293 K. The data were interpreted as the iron atom being in an intermediate crystal field of symmetry lower than octahedral where the ground state is always low-spin. The effective crystal field depends on the temperature and the substituent R, increasing in the order R = C$_4$H$_3$S < β-C$_{10}$H$_7$ < *p*-XC$_6$H$_4$ < Ph < *m*-XC$_6$H$_4$ (X = Me, F, Cl or Br). Anomalous magnetic behaviour has been reported for thiolo-bridged Mo^{III} complexes[218] and for Os$^{IV\ 206}$ and Ru$^{III\ 219}$ complexes.

The IR spectra of the metal chelates of monothio-β-diketones display ν(M—O) and ν(M—S) in the ranges 499–437 cm^{-1} and 399–376 cm^{-1}, respectively.[174] Proton NMR spectra of some monothio-β-diketonates have been measured and the chemical shifts have been interpreted in terms of π-delocalization in the metal chelate rings.[201,220–225]

Livingstone and co-workers[226–229] reported the mass spectra of a large number of metal chelates of fluorinated monothio-β-diketones (46; R′ = CF$_3$). The fragmentation pattern is determined by the metal ion, *i.e.* chelates of the same metal with different monothio-β-diketones have essentially the same fragmentation pattern; however, the fragmentation pattern differs markedly for each metal. Valence changes occur in the mass spectrometer with the complexes of Cr^{III}, Fe^{III}, Ru^{III}, Co^{III}, Rh^{III} and Ni^{II} but not for those of Pd^{II} and Pt^{II}. The complexes of Co^{II} and Fe^{III} are thermally degraded in the mass spectrometer, as shown by equation (3) (Met = Fe, Co; LH = monothio-β-diketone). Complexes of Cr^{III}, Ru^{III} and Rh^{III} react as given in equation (4) (M = molecular ion). Complexes of Cr^{III}, Co^{III} and Rh^{III} give a peak for $M - 3L$ (Met$^+$), denoting reduction in the mass spectrometer to the univalent state. Surprisingly, Ni^{II} complexes display a peak for $M - 2L$ (Ni$^+$).

$$\text{Met}^{III}\text{L}_3 \longrightarrow \text{Met}^{II}\text{L}_2 + \cdot\text{L} \qquad (3)$$

$$[\text{Met}^{III}\text{L}_3]^{\ddagger} \xrightarrow{-\cdot\text{L}} [\text{Met}^{II}\text{L}_2]^+ \xrightarrow{-\cdot\text{L}} [\text{Met}^{II}\text{L}]^+ \qquad (4)$$
$$\quad (M) \qquad\qquad\quad (M-\text{L}) \qquad\qquad (M-2\text{L})$$

Fluorine migration from the CF$_3$ group to the metal occurs with Cr^{III}, Fe^{III}, Co^{III}, Pt^{II} and Zn^{II} but not with Ru^{III}, Rh^{III}, Ni^{II} and Pd^{II}. The reaction pathway for the zinc complexes is shown in Scheme 3. The molecular ion (48) loses a $\cdot$CF$_3$ radical from one monothio-β-diketone moiety to give the ion (49), which then loses carbon monoxide to yield the fragment (50). Sulfur is ejected from (50) with the concomitant formation of a zinc–carbon bond in the species (51), which can subsequently lose the alkyne RC≡CH to give the mono-ligand ion [ZnL]$^+$ (52). The second ligand moiety loses :CF$_2$ with the formation of a zinc–fluorine bond in the ion (53).

Das and Livingstone[229–233] measured the dipole moments of over 200 monothio-β-diketonato complexes; the moments range from 1.96 D for [Zn(*p*-FC$_6$H$_4$CS=CHCOCF$_3$)$_2$] to 8.71 D for [Co(*p*-EtOC$_6$H$_4$CS=CHCOCF$_3$)$_3$]. The data show that the Ni^{II}, Pd^{II}, Pt^{II} and Cu^{II} complexes are *cis* square-planar, the Zn^{II} complexes are tetrahedral, and the Cr^{III}, Fe^{III}, Ru^{III}, Co^{III} and Rh^{III} complexes have the *facial (cis)* octahedral configuration (54) and not the alternative *meridional* octahedral configuration (55). Measurement of the moments of some of the complexes by both the static polarization and dielectric relaxation methods showed that atomic polarization is *ca.* 0.3 D in square-planar, 0.5 D in tetrahedral and 0.9 D in octahedral complexes (see Tables 2–4).

Proton NMR spectra of some V^{III} and Cr^{III} complexes indicated a *facial* octahedral configuration with all three sulfur atoms *cis*.[234] More recent ^{13}C and ^{19}F NMR data on a wide range of transition metal complexes of fluorinated monothio-β-diketones support the assignment of *cis* square-planar and *facial* octahedral geometries.[235,236] X-Ray structural data have established the *cis* square-planar configuration for a Pd^{II} and a Pt^{II} complex[237] and four Ni^{II} complexes,[238,239] the tetrahedral configura-

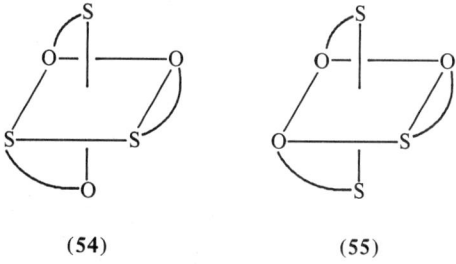

Scheme 3 Reactions of zinc monothio-β-diketonates in the mass spectrometer

tion for two Zn^{II} complexes[240] and the *facial* octahedral configuration for an Fe^{III} complex[240] and a Co^{III} complex.[241]

The preference for the *cis* and *facial* structures has been explained[230] as arising from strong d_π–d_π bonding between the metal and sulfur atoms; transition metal ions can form two π-bonds at 90° and, consequently, of the isomeric *cis* and *trans* structures, wherein only the two sulfur atoms of the four donor atoms can form d_π–d_π bonds, the *cis* isomer is the more stable. An alternative explanation[239] is that the *cis* configuration is due to weak non-bonded S—S interaction. Recent dipole moment and ^{13}C and ^{19}F NMR data on Ga^{III} and In^{III} complexes indicate a *facial–meridional (cis–trans)* equilibrium in solution.[242] For these non-transition elements d_π–d_π bonding cannot arise; consequently, the explanation of d_π–d_π stabilization of the configurations for transition metal complexes appears to be the correct one.

The dipole moments (μ_s) of square-planar, tetrahedral and octahedral complexes of a large number of fluorinated monothio-μ-diketones are listed in Tables 5 and 6.

For the square-planar complexes and the cobalt complexes — the only octahedral complexes for which data are available for a large number of ligands — the dipole moments decrease if the R groups are arranged in the order: $p\text{-EtOC}_6H_4 > 5\text{-MeC}_4H_2S > p\text{-MeOC}_6H_4 > m\text{-F},p\text{-MeOC}_6H_3 > m,p\text{-Me}_2C_6H_3 \approx p\text{-EtC}_6H_4 \approx m,m\text{-(MeO)}_2C_6H_3 > p\text{-MeC}_6H_4 \approx p\text{-Bu}^tC_6H_4 > p\text{-Pr}^iC_6H_4 >$

Table 2 Dipole Moments (μ)[a] and Relaxation Times (τ) of Square-planar Metal Complexes [M(RCS=CHCOR')$_2$]

R	R'		Ni	Pd	Pt	Cu
Me	Ph	μ_s (D)	2.89	2.58	b	3.43
		μ_d (D)	2.69	2.45		3.19
		τ (ps)	112	125		120
Ph	Ph	μ_s (D)	2.45	2.48	2.97	3.46
		μ_d (D)	2.28	2.16		2.64
		τ (ps)	108	169	80	189
Me	CF$_3$	μ_s (D)	4.14	4.39	b	
		μ_d (D)	4.08	4.29		
		τ (ps)	66	77		
C$_4$H$_3$S[c]	CF$_3$	μ_s (D)	5.74	5.88	6.09	4.96
		μ_d (D)	5.44	5.56	5.67	4.09
		τ (ps)	135	115	137	121
Ph	CF$_3$	μ_s (D)	4.92	5.02	5.35	4.39
		μ_d (D)	4.65	4.88	4.90	3.93
		τ (ps)	137	155	133	137
p-MeC$_6$H$_4$	CF$_3$	μ_s (D)	5.84	5.91	6.16	5.12
		μ_d (D)	5.35	5.54	5.62	4.63
		τ (ps)	168	173	170	175
p-BrC$_6$H$_4$	CF$_3$	μ_s (D)	2.85	2.97	3.21	2.45
		μ_d (D)	2.56	2.92	3.01	2.18
		τ (ps)	162	210	185	148

[a] μ_s = dipole moment determined by static polarization method. μ_d = dipole moment determined by dielectric relaxation method. [b] The Pt complex could not be isolated. [c] C$_4$H$_3$S = 2-thienyl.

Table 3 Dipole Moments (μ) and Relaxation Times (τ) of Octahedral Complexes [M(RCS=CHCOR')$_3$]

R	R'		Fe	Co
Me	Ph	μ_s (D)	4.81	3.27
		μ_s (D)	4.29	2.68
		τ (ps)	71	192
Ph	Ph	μ_s (D)	4.29	3.84
		μ_d (D)	3.30	2.05
		τ (ps)	105	171
Me	CF$_3$	μ_s (D)		5.40
		μ_d (D)		5.05
		τ (ps)		116
C$_4$H$_3$S	CF$_3$	μ_s (D)	7.00	7.14
		μ_d (D)	5.82	6.55
		τ (ps)	177	215
Ph	CF$_3$	μ_s (D)	6.03	6.54
		μ_d (D)	5.51	5.41
		τ (ps)	236	190
p-MeC$_6$H$_4$	CF$_3$	μ_s (D)	6.85	7.36
		μ_d (D)	5.74	6.15
		τ (ps)	233	237
p-BrC$_6$H$_4$	CF$_3$	μ_s (D)	3.59	3.62
		μ_d (D)	2.46	2.88
		τ (ps)	164	196

Table 4 Dipole Moments (μ) and Relaxation Times (τ) of Tetrahedral Complexes [Zn(RCS=CHCOR')$_2$]

R	R'	μ_s (D)	μ_d (D)	τ (ps)
Me	Ph	3.37	3.02	120
Ph	Ph	3.29	2.72	189
Me	CF$_3$	3.37	2.41	74
C$_4$H$_3$S	CF$_3$	3.57	3.45	120
Ph	CF$_3$	3.07	2.75	130
p-MeC$_6$H$_4$	CF$_3$	3.65	3.21	145
p-BrC$_6$H$_4$	CF$_3$	2.46	1.77	175

Table 5 Dipole Moments μ_s (D) of Square-planar and Tetrahedral Metal Complexes [M(RCS=CHCOCF$_3$)$_2$]

R	Square-planar				Tetrahedral
	Ni	Pd	Pt	Cu	Zn
Me	4.14	4.39			3.37
Pr^i	4.48	4.75	4.90	3.85	2.82
Bu^i	4.38	4.64	4.89	3.64	2.70
β-C$_{10}$H$_7$ [a]	5.47	5.55	5.68		3.48
Ph	4.92	5.02	5.35	4.39	3.07
m-ClC$_6$H$_4$	4.07	4.09	4.38	3.59	3.00
m-BrC$_6$H$_4$	3.93	4.01	4.32	3.55	2.90
p-FC$_6$H$_4$	2.92	3.28	3.54		1.96
p-ClC$_6$H$_4$	2.89	3.20	3.42		2.10
p-BrC$_6$H$_4$	2.85	2.97	3.21	2.45	2.46
m,p-Cl$_2$C$_6$H$_3$	2.80	2.97	3.01	2.45	2.16
m-MeC$_6$H$_4$	5.33	5.55	5.68	4.66	3.37
p-MeC$_6$H$_4$	5.84	5.91	6.16	5.12	3.65
p-EtC$_6$H$_4$	5.94	6.07	6.34	5.26	3.85
p-Pr^{i}C$_6$H$_4$	5.80	5.98	6.38		3.77
p-Bu^{t}C$_6$H$_4$	5.84	5.91	6.31		
m,p-Me$_2$C$_6$H$_3$	6.05	6.32	6.20	5.46	3.76
m-MeOC$_6$H$_4$	5.43	5.88	5.64	4.91	3.94
p-MeOC$_6$H$_4$	6.70	7.16	6.73	6.02	4.80
p-EtOC$_6$H$_4$	6.95	7.23	7.36	6.36	4.81
m,m-(MeO)$_2$C$_6$H$_3$	5.87	5.96		5.50	4.13
m,p-(MeO)$_2$C$_6$H$_3$	5.62	6.45	6.07	5.24	3.96
m,m,p-(MeO)$_3$C$_6$H$_2$	4.32	4.48		3.98	3.38
m-F,p-MeOC$_6$H$_3$	6.24	6.48	6.62	5.70	4.89
C$_4$H$_3$S [b]	5.74	5.88	6.09	4.96	3.57
5-ClC$_4$H$_2$S	4.12	4.25	4.32		
5-MeC$_4$H$_2$S	6.90	7.20	7.31	6.21	4.63

[a] β-naphthyl. [b] 2-thienyl.

Table 6 Dipole Moments μ_s (D) of Octahedral Metal Complexes [M(RCS=CHCOCF$_3$)$_3$]

R	Cr	Fe	Ru	Co	Rh
Me				5.40	
β-C$_{10}$H$_7$	6.19	6.09	6.69	6.90	7.16
Ph	6.27	6.03	6.11	6.54	6.52
m-ClC$_6$H$_4$		4.91	4.83	5.11	
m-BrC$_6$H$_4$		4.70	4.55	5.08	
p-FC$_6$H$_4$		3.68		3.79	4.13
p-ClC$_6$H$_4$		3.52		3.72	
p-BrC$_6$H$_4$		3.59		3.62	
m,p-Cl$_2$C$_6$H$_3$				3.29	
m-MeC$_6$H$_4$		6.62	6.20	6.79	
p-MeC$_6$H$_4$		6.85	6.91	7.36	7.16
p-EtC$_6$H$_4$				7.43	
p-Pr^{i}C$_6$H$_4$				7.29	
p-Bu^{t}C$_6$H$_4$				7.36	
m,p-Me$_2$C$_6$H$_3$				7.75	
m-MeOC$_6$H$_4$				6.91	
p-MeOCH$_6$H$_4$				8.57	8.14
p-EtOC$_6$H$_4$				8.71	
m,m-(MeO)$_2$C$_6$H$_3$				7.08	
m,p-(MeO)$_2$C$_6$H$_3$				6.89	7.04
m,m,p-(MeO)$_3$C$_6$H$_2$				5.15	
m-F,p-MeOC$_6$H$_3$				7.62	
C$_4$H$_3$S	7.07	7.00	6.91	7.14	7.19
5-ClC$_4$H$_2$S				5.20	
5-MeC$_4$H$_2$S				8.67	

$C_4H_3S > \beta\text{-}C_{10}H_7 \approx m\text{-}MeOC_6H_4 \approx m\text{-}MeC_6H_4 > Ph > Pr^i > Bu^i > Me > 5\text{-}ClC_4H_2S > m\text{-}ClC_6H_4 > m\text{-}BrC_6H_4 > p\text{-}FC_6H_4 > p\text{-}ClC_6H_4 > p\text{-}BrC_6H_4 > m,p\text{-}Cl_2C_6H_3$. The positions of the R groups $m,p\text{-}(MeO)_2C_6H_3$ and $m,m,p\text{-}(MeO)_3C_6H_2$ vary slightly with different metals.

This series can be better understood if we consider only the ligands having R = phenyl or substituted phenyl; the series reduces to: $p\text{-}EtOC_6H_4 > p\text{-}MeOC_6H_4 > m\text{-}F,p\text{-}MeOC_6H_3 > m,p\text{-}Me_2C_6H_3 \approx p\text{-}EtC_6H_4 \approx m,m\text{-}(MeO)_2C_6H_3 > p\text{-}MeC_6H_4 \approx p\text{-}Bu^tC_6H_4 > p\text{-}Pr^iC_6H_4 > m\text{-}MeOC_6H_4 \approx m\text{-}MeC_6H_4 > Ph > m,m,p\text{-}(MeO)_3C_6H_2 > m\text{-}ClC_6H_4 > m\text{-}BrC_6H_4 > p\text{-}FC_6H_4 > p\text{-}ClC_6H_4 > p\text{-}BrC_6H_4 > m,p\text{-}Cl_2C_6H_3$.

The differences in the value of the dipole moments of the complexes $[M(RCS=CHCOCF_3)_n]$ (R = Ph or X-substituted phenyl) depend upon: (a) the magnitude and vector direction of the Ph—X bond moments; (b) the inductive effect arising from the difference in electron density of the C-1 and C-5 atoms of the monothio-β-diketone (this is affected by the nature and position of the substituent X on the phenyl ring); (c) the change in moment brought about by the mesomeric effect (**56**) of X.[230,231]

(56)

Table 7 lists the dipole moments μ_s of some complexes of monothio-β-diketones, $RCSH=CHCOR'$ ($R' = C_3F_7$, C_2F_5, and CHF_2). With increasing fluorine content of the R' group the dipole moments increase for analogous complexes.[233] Dipole moments of monothio-β-diketonato complexes have been recently reviewed.[115]

Table 7 Dipole Moments μ_s (D) of Metal Complexes of the Fluorinated Monothio-β-diketones $RC(SH)=CHCOR'$ (LH)

R	Complex	C_3F_7	C_2F_5	CF_3	CHF_2
Ph	NiL_2	5.38	5.35	4.92	3.27
	PdL_2	5.63	5.55	5.02	
	CoL_3	6.68	6.54	6.54	
$\beta\text{-}C_{10}H_7$	NiL_2	6.07	5.86	5.47	
	PdL_2	6.24	6.00	5.55	
	CoL_3	7.36	7.20	6.90	
C_4H_3S	NiL_2	6.31	5.94	5.74	3.87
	PdL_2	6.38	6.18	5.88	
	CoL_3	7.60	7.48	7.14	4.95

16.6.12 DITHIO-β-DIKETONES

The first recorded attempt to prepare a thio derivative of a β-diketone resulted in the isolation of the colourless dimer (**57**; R = R' = Me) from the reaction of acetylacetone and hydrogen sulfide in hydrochloric acid.[243] Similar dimers with various R and R' groups have been reported[244] and the structure (**57**) has been confirmed from NMR and MS data.[245] Other attempts to prepare dithio-β-diketones (**58**) yielded 1,2-dithiolium salts (**59**).[246]

(57)   (58)   (59)   (60)

Acetylacetone reacts with H_2S and HCl in the presence of a metal ion to give the complex (60) with the unstable dithio-β-diketone before the latter dimerizes.[247] The complex [ML_2] (LH = dithio-β-diketone; M = Co, Ni, Pd, Pt, Zn, Cd and Hg) and ML_3 (M = Cr, Fe, Ru, Os, Co, Rh, Ir) are known; they are deeply coloured, stable in air, and, like complexes of monothio-β-diketones, insoluble in water but readily soluble in organic solvents.[248]

The nickel complexes, like those of monothio-β-diketones, are diamagnetic. The Co^{II} complex of dithioacetylacetone is low-spin (μ, 2.35 BM) and square-planar; the paramagnetic anisotropy and ESR spectrum indicate the ground-state configuration $(d_{xy})^2(d_{xz})^2(d_{x^2-y^2})^2(d_{z^2})^1$.[249,250] The complexes [ML_3] (M = Fe, Ru and Os) are low-spin.[248]

The MS fragmentation patterns of dithio-β-diketonato complexes differ from those of both β-diketonato and monothio-β-diketonato complexes. The mass spectra of bivalent dithioacetylacetonato complexes are characterized by a molecular ion peak $M^{\ddagger}$ and a much stronger peak for L^+ due to the formation of the 3,5-dimethyl-1,2-dithiolium ion; another peak at m/e 96 is due to the loss of HS from the dithiolium cation. The mass spectra of nickel complexes of dithio-β-diketones are dominated by fragmentation products of the uncomplexed oxidized ligand, which is consistent with the low stabiltiy of dithio-β-diketones and the greater stability of the 1,2-dithiolium ions.[251]

Like metal chelates of some other dithio ligands, *e.g.* maleonitrile dithiolate, dithio-β-diketonato complexes undergo reversible redox reactions in non-aqueous solvents. The square-planar complexes [ML_2] (M = Co, Ni, Pd and Pt) exhibit two consecutive one-electron reductions as shown in equation (5).[252,253]

$$[ML_2] + e^- \rightleftharpoons [ML_2]^- + e^- \rightleftharpoons [ML_2]^{2-} \qquad (5)$$

The reduction potentials of dithio-β-diketonato complexes are greatly influenced by the substituents at the 3-, 4- and 5-positions, irrespective of whether the electron transfer is ligand- or metal-based. The effect arises from the degree to which different substituents stabilize the reduced, compared with the oxidized, form of the ligand. The nature of R and R' in a series of Ni^{II} complexes can produce a variation in E_0 of as much as 1 volt.[248]

Although less work has been done on metal complexes of dithio-β-diketones compared with those of β-diketones and monothio-β-diketones, the available data allow interesting comparisons to be made among the three closely related classes of compounds, *e.g.* Ni^{II} complexes of β-diketones are green, paramagnetic and octahedral, those of monothio-β-diketones are brown, diamagnetic and square-planar, while those of dithio-β-diketones are dark red, diamagnetic and square-planar. The Ni^{II} complexes of monothio-β-diketones yield green, paramagnetic base adducts [$NiL_2 \cdot 2B$] (B = py,; 2B = phen, bipy), whereas the Ni^{II} complex of dithioacetylacetone does not form base adducts, being quite insensitive to axial perturbation and its electronic spectrum in donor solvents such as pyridine is unchanged.

Metal complexes of dithio-β-diketones have been reviewed by Lockyer and Martin[248] who discuss their IR, visibile-UV, NMR and ESR spectra.

16.6.13 β-MERCAPTO KETONES, ESTERS, THIOESTERS AND AMIDES

16.6.13.1 β-Mercapto Ketones

A number of β-mercapto ketones (61; R = 2-thienyl, 2-furyl; R' = Ph, 2-thienyl, 2-furyl) have been reported; they were prepared, in somewhat low yield, by the reaction shown in equation (6). Reaction of the *o*-mercapto ketone (61) with Cu^{II} gives the yellow Cu^I complex with concomitant oxidation of the mercapto ketone to the disulfide.[254] *o*-Mercaptoacetophenone (62) and β-mercapto-benzophenone (63) have been prepared but are unstable;[255] their chelating ability is greater than that of the ketones of type (61) because of the conjugated double bonds in the former.[256]

$$\text{RCH}{=}\text{CHCR'} + H_2S \longrightarrow \text{RCHCH}_2\text{CR'} \qquad (6)$$
$$\qquad\quad \overset{\|}{O} \qquad\qquad\qquad\quad \overset{|}{SH} \ \ \overset{\|}{O}$$

(61)

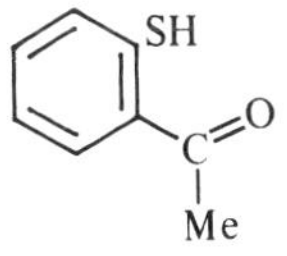

(62) 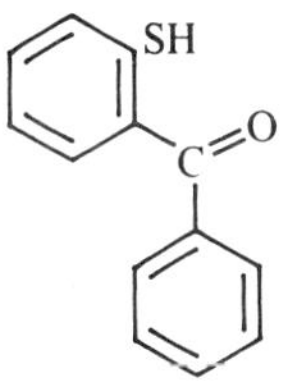   (63)

16.6.13.2 β-Mercapto Esters

Alkyl and aryl β-mercaptohydrocinnamates (64) have been prepared by the reaction shown in Scheme 4. The yellow Cu^I and red Ni^{II} complexes were characterized.[257] The α,β-unsaturated β-mercapto ester (65) forms complexes with Cu^I, Ni^{II}, Co^{III} and Fe^{III}.[190,258] The Ni^{II} complex of (66) has also been reported.[190] These complexes are stabilized by electron delocalization in the chelate ring, which is identical with that of monothio-β-diketones.

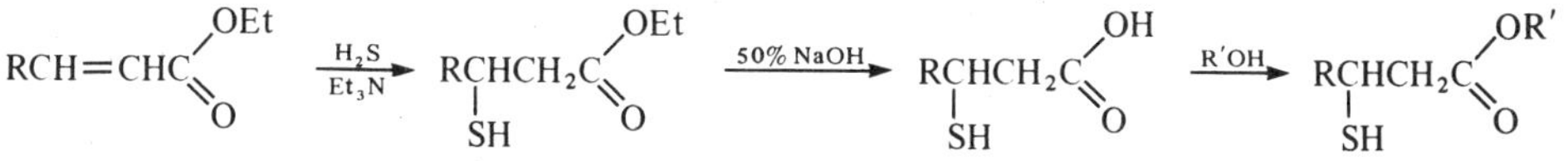

Scheme 4

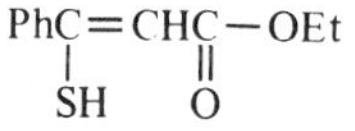

(65) 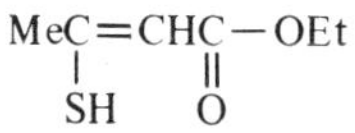   (66) 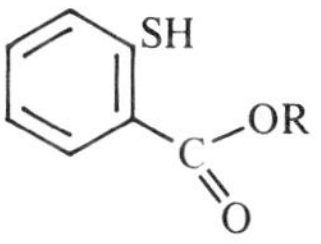   (67) 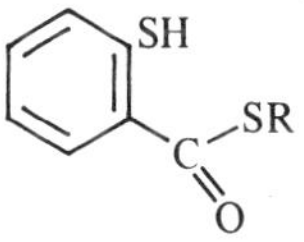   (68)

The o-mercaptobenzoic acid esters (67; R = Et, Pr^i) form complexes with Cu^I, Ni^{II} and Co^{III}.[255] These interesting ligands and their sulfur analogues (68) should be further investigated.

16.6.13.3 β-Mercapto Thioesters

The β-mercapto thioesters (69; R = Et, Pr^i) and the o-mercaptobenzoic acid thioesters (68; R = Et, Pr^i, Pe^i) form complexes with Cu^I, Ni^{II} and Co^{III}.[255,258] Since several (b) class and borderline metals form more stable complexes with the S-esters (68) than with the O-esters (67), it is likely that the S-esters behave as S,S-bidentates.

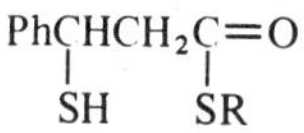

(69)

16.6.13.4 β-Mercapto Amides

The β-mercapto amides (70) and the o-mercaptobenzoic acid amides (71) behave as SO-bidentates giving complexes with Cu^I and Ni^{II}.[256,259] The stabilities of the complexes are lower than those of the corresponding thioesters, esters and ketones.[256]

It is noteworthy that the stabilities of (b) class and borderline metal complexes decrease if the ligands are arranged in the order: (65) > (68) > (67) > (63) > (69) > (71) > (64) > (70) > (61).[256]

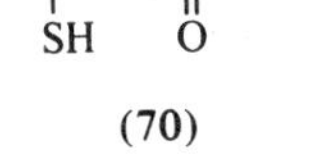

(70) 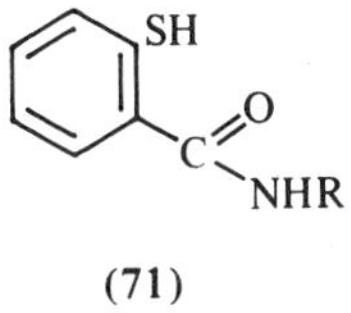   (71)

16.6.14 REFERENCES

1. K. Gleu and K. Rehm, *Z. Anorg. Allg. Chem.*, 1936, **227**, 237; K. Gleu, W. Breuel and K. Rehm, *Z. Anorg. Allg. Chem.*, 1938, **235**, 201, 211.
2. L. Vaska and S. S. Bath, *J. Am. Chem. Soc.*, 1966, **88**, 1333.
3. J. J. Levison and S. D. Robinson, *Chem. Commun.*, 1967, 198.
4. L. Vaska and D. L. Catone, *J. Am. Chem. Soc.*, 1966, **88**, 5324.
5. L. H. Vogt, J. L. Katz and S. E. Wiberly, *Inorg. Chem.*, 1965, **4**, 1157.
6. J. Chatt, G. J. Leigh and N. Thankarajan, *J. Chem. Soc. (A)*, 1971, 3168.
7. J. Hasse and M. Winnewisser, *Z. Naturforsch., Teil A*, 1968, **23**, 61.
8. B. Post, R. S. Schwartz and L. Fankuchen, *Acta Crystallogr.*, 1952, **5**, 372.
9. Y. Morino, Y. Kikuchi, S. Saito and E. Hiroto, *J. Mol. Spectrosc.*, 1964, **13**, 95.
10. J. E. Boggs, C. M. Crain and J. E. Whiteford, *J. Phys. Chem.*, 1957, **61**, 482.
11. S. Rothenbergy and J. F. Schaefer, *J. Chem. Phys.*, 1970, **53**, 3014.
12. D. M. P. Mingos, *Transition Met. Chem.*, 1978, **3**, 1.
13. R. R. Ryan, G. J. Kubas, D. C. Moody and P. G. Eller, *Struct. Bonding (Berlin)*, 1981, **46**, 47.
14. P. D. Dacre and M. Elder, *Theor. Chim. Acta*, 1972, **25**, 254.
15. R. R. Ryan and P. G. Eller, *Inorg. Chem.*, 1976, **15**, 495.
16. D. M. P. Mingos, *Nature (London), Phys. Sci.*, 1971, **229**, 193.
17. M. C. Baird, G. Hartwell, R. Mason, A. J. M. Rae and G. Wilkinson, *Chem. Commun.*, 1967, 92.
18. J. H. Enemark and R. D. Feltham, *Coord. Chem. Rev.*, 1974, **13**, 339.
19. S. J. La Placa and J. A. Ibers, *Inorg. Chem.*, 1966, **5**, 405; K. W. Muir and J. A. Ibers, *Inorg. Chem.*, 1969, **8**, 1921.
20. D. C. Moody and R. R. Ryan, *Inorg. Chem.*, 1977, **16**, 2473.
21. R. D. Wilson and J. A. Ibers, *Inorg. Chem.*, 1978, **27**, 2134.
22. G. J. Kubas, R. R. Ryan and V. McCarty, *Inorg. Chem.*, 1980, **19**, 3003.
23. G. J. Kubas, *Inorg. Chem.*, 1979, **18**, 182.
24. C. Mealli, A. Orlandini, L Sacconi and P. Stoppioni, *Inorg. Chem.*, 1978, **17**, 3020.
25. B. L. Booth, C. A. McAuliffe and G. L. Stanley, *J. Organomet. Chem.*, 1982, **226**, 191.
26. R. Hoppenheit, W. Isenberg and R. Mews, *Z. Naturforsch., Teil B*, 1982, **37**, 116.
27. H. W. Roesky, M. Thomas, M. Noltemeyer and G. M. Sheldrick, *Angew. Chem.*, 1982, **94**, 780.
28. W. A. Shenk and F. E. Baumann, *Chem. Ber.*, 1982, **115**, 2615.
29. M. R. Churchill, B. G. De Boer and K. L. Kalra, *Inorg. Chem.*, 1973, **12**, 1646.
30. J. Meunier-Piret, P. Piret and M. Van Meerssche, *Bull. Soc. Chim. Belg.*, 1967, **76**, 374.
31. M. R. Churchill and K. L. Kalra, *Inorg. Chem.*, 1973, **12**, 1650.
32. C. E. Briant, G. R. Hughes, P. C. Minshall and D. M. P. Mingos, *J. Organomet. Chem.*, 1982, **224**, C21.
33. D. C. Moody and R. R. Ryan, *Inorg. Chem.*, 1977, **16**, 1052.
34. M. R. Snow and J. A. Ibers, *Inorg. Chem.*, 1973, **12**, 224.
35. P. G. Eller, G. J. Kubas and R. R. Ryan, *Inorg. Chem.*, 1977, **16**, 2454.
36. P. G. Eller and G. J. Kubas, *J. Am. Chem. Soc.*, 1977, **99**, 4346.
37. D. C. Moody and R. R. Ryan, *J. Chem. Soc., Chem. Commun.*, 1980, 1230.
38. W. Hofmann and H. Werner, *Angew. Chem., Int. Ed. Engl.*, 1981, **20**, 1014.
39. P. W. Schenk and R. Steudel, in 'Inorganic Sulfur Chemistry', ed. G. Nickless, Elsevier, Amsterdam, 1968, p. 367.
40. S. S. Prasad and W. T. Huntress, *Astrophys. J.*, 1982, **260**, 590; H. R. Johnson, *Astrophys. J.*, 1982, **260**, 254.
41. G. Schmid and G. Ritter, *Chem. Ber.*, 1975, **108**, 3008; *Angew. Chem., Int. Ed. Engl.*, 1973, **14**, 645.
42. S. K. Arulaamy, K. K. Pandey and U. C. Agarwala, *Inorg. Chim. Acta*, 1981, **54**, L51.
43. B. Bock, B. Solouki, P. Rosmus and R. Stuedel, *Angew. Chem., Int. Ed. Engl.*, 1973, **12**, 933.
44. J. Knecht and G. Schmid, *Z. Naturforsch., Teil B*, 1977, **32**, 653.
45. M. Schappacher, L. Ricard and R. Weiss, *J. Less-Common Met.*, 1977, **54**, 91.
46. R. Steudel, J. Steidel and J. Pickardt, *Angew. Chem.*, 1977, **92**, 313; R. Steudel, T. Sandow and J. Steidel, *J. Chem. Soc., Chem. Commun.*, 1980, 180.
47. S. E. Livingstone, *Q. Rev., Chem. Soc.*, 1965, **19**, 386.
48. R. D. Hancock, N. P. Finkelstein and A. Evers, *J. Inorg. Nucl. Chem.*, 1977, **39**, 1031.
49. Z. Gabelica, *J. Less-Common. Met.*, 1977, **51**, 343.
50. R. J. Restivo, G. Ferguson and R. J. Balahura, *Inorg. Chem.*, 1977, **16**, 167.
51. J. R. Lusty, *J. Coord. Chem.*, 1980, **10**, 243.
52. M. W. Whitehouse and W. R. Walker, *Agents Actions*, 1978, **8**, 85.
53. H. Ruben, A. Zalkin, M. Faltens and D. H. Templeton, *Inorg. Chem.*, 1974, **13**, 1836.
54. Z. Gabelica, *Chem. Lett.*, 1979, 1419.
55. F. R. Fronczek, R. E. Marsh and W. P. Shaefer, *J. Am. Chem. Soc.*, 1982, **104**, 3382.
56. G. Newman and D. B. Powell, *Spectrochim. Acta*, 1963, **19**, 213.
57. B. Nyberg and R. Larsson, *Acta Chem. Scand.*, 1973, **27**, 63.
58. J. P. Hall and W. P. Griffith, *Inorg. Chim. Acta*, 1981, **48**, 65.
59. R. D. Hancock, N. P. Finkelstein and A. Evers, *J. Inorg. Nucl. Chem.*, 1977, **39**, 1031.
60. M. A. Spinnler and L. N. Becka, *J. Chem. Soc. (A)*, 1967, 1194.
61. L. M. Dikareva, J. B. Baranovskii and Z. G. Melzhtiev, *Russ. J. Inorg. Chem. (Engl. Transl.)*, 1972, **17**, 1772.
62. C. L. Raston, A. H. White and J. K. Yandell, *Aust. J. Chem.*, 1980, **33**, 419, 1123.
63. D. Messner, D. K. Breitinger and W. Haegler, *Acta Crystallogr., Sect. B*, 1981, **37**, 19.
64. W. G. Kehr, D. K. Breitinger and G. Bauer, *Acta Crystallogr., Sect B*, 1980, **36**, 2545.
65. L. O. Larsson and L. Niinsto, *Acta Chem. Scand.*, 1973, **27**, 859.
66. A. C. Dash, A. A. El-Awady and G. M. Harris, *Inorg. Chem.*, 1981, **20**, 3160.
67. R. N. Shchelokov, G. T. Bolotova and N. Golubkova, *Zh. Neorg. Khim.*, 1978, **23**, 2858.
68. R. Eskenazi, J. Raskovan and R. Levitus, *J. Inorg. Nucl. Chem.*, 1965, **27**, 371.
69. M. Nakamoto, 'Infrared Spectra of Inorganic and Coordination Compounds', 2nd edn., Wiley-Interscience, New York, 1970, p. 210.

70. D. A. Berta, W. A. Spofford, P. Boldrini and E. L. Amma, *Inorg. Chem.*, 1970, **9**, 136.
71. Y. G. Sakharova and V. N. Perov, *Zh. Neorg. Khim.*, 1978, **23**, 2844.
72. T. Y. Ashikhmina and A. S. Karnaukov, *Resp. Sb. Nauchn. Tr-Yarosl. Gos. Pedagog. Inst.*, 1978, **169**, 34 (*Chem. Abstr.* 1980, **92**, 51 196a).
73. J. V. Quagliano and L. Schubert, *Chem. Rev.*, 1952, **50**, 201.
74. F. Basolo and R. G. Pearson, 'Mechanisms of Inorganic Reactions', Wiley, New York, 1958, p. 172.
75. N. S. Kurnakov, *J. Prakt. Chem.*, 1894, **50**, 483.
76. R. L. Carlin and S. J. Holt, *Inorg. Chem.*, 1963, **2**, 849; *J. Am. Chem. Soc.*, 1964, **86**, 3017.
77. T. J. Lane, A. Yamaguchi, J. V. Quagliano, J. A. Ryan and S. Nizushima, *J. Am. Chem. Soc.*, 1959, **81**, 3824.
78. M. Schafer and C. Curran, *Inorg. Chem.*, 1966, **5**, 265.
79. R. K. Gosavi and C. N. R. Rao, *J. Inorg. Nucl. Chem.*, 1967, **29**, 1937.
80. I. Ojima, T. Iwamoto, T. Onishi, N. Inamoto and K. Tamaru, *Chem. Commun.*, 1969, 1301.
81. C. M. Harris and S. E. Livingstone, in 'Chelating Agents and Metal Chelates', ed. F. P. Dwyer and D. P. Mellor, Academic, New York, 1964, p. 95.
82. R. L. Girling and E. L. Amma, *Chem. Commun.*, 1968, 1487.
83. F. A. Cotton, R. Francis and W. D. Horrocks, *J. Phys. Chem.*, 1960, **64**, 1534.
84. D. W. Meek, D. K. Straub and R. S. Drago, *J. Am. Chem. Soc.*, 1960, **82**, 6013.
85. J. Selbin, W. E. Bull and L. H. Holmes, *J. Inorg. Nucl. Chem.*, 1961, **16**, 219.
86. R. S. Drago and D. W. Meek, *J. Phys. Chem.*, 1961, **65**, 1446.
87. P. W. N. M. van Leeuwen, *Recl. Trav. Chim. Pays-Bas*, 1967, **86**, 201.
88. P. W. N. M. van Leeuwen and W. L. Groenveld, *Recl. Trav. Chim. Pays-Bas*, 1967, **86**, 721.
89. R. Francis and F. A. Cotton, *J. Chem. Soc.*, 1961, 2078.
90. M. J. Bennett, F. A. Cotton, D. L. Weaver, R. J. Williams and W. H. Watson, *Acta Crystallogr.*, 1967, **23**, 788.
91. W. Kitching and C. Moore, *Inorg. Nucl. Chem. Lett.*, 1968, **4**, 691.
92. S. G. Patel, *Bol. Soc. Chil. Quim.*, 1970, **16**, 18.
93. G. V. Tsiutsadze, *Zh. Neorg. Khim.*, 1971, **16**, 1160.
94. C. R. Kanekar, S. V. Nipankar and C. C. Patel, *Indian J. Chem., Sect. A.*, 1970, **8**, 451.
95. P. Biscarini, L. Fusina and G. Nivellini, *J. Chem. Soc., Dalton Trans.*, 1972, 1003.
96. J. D. Gilbert, D. Rose and G. Wilkinson, *J. Chem. Soc. (A)*, 1970, 2765.
97. Yu. N. Kukushkin, N. D. Rubtsova and M. M. Singh, *Zh. Neorg. Khim.*, 1970, **15**, 1879, 2001.
98. M. M. Khan, M. Ahmed and A. Kumar, *Inorg. Chim. Acta*, 1980, **43**, 137.
99. B. B. Wayland and R. F. Schramm, *Inorg. Chem.*, 1969, **8**, 971.
100. J. H. Price, R. F. Schramm and B. B Wayland, *Chem. Commun.*, 1970, 1377.
101. Yu. A. Kopyttsev, L. Kh. Freidlin, N. M. Mazarova and I. P. Yakovlev, *Izv. Akad. Nauk SSSR, Ser. Khim.*, 1975, 977.
102. N. M. Karayannis, C. M. Mikulski and L. L. Pytiewski, *Inorg. Chim. Acta Rev.*, 1971, **5**, 69.
103. A. Cahours and A. W. Hoffman, *Ann. Chem. Pharm.*, 1857, **104**, 1.
104. L. Malatesta, *Gazz. Chim. Ital.*, 1947, **77**, 518.
105. E. Bannister and F. A. Cotton, *J. Chem. Soc.*, 1960, 1959.
106. M. G. King and G. P. McQuillan, *J. Chem. Soc. (A)*, 1967, 898.
107. H. Keck, W. Kuchen, J. Mathow, B. Meyer, D. Mootz and H. Wunderlich, *Angew. Chem., Int. Ed. Engl.*, 1981, **20**, 975.
108. H. Teichmann and G. Hilgetag, *Angew. Chem.*, 1967, **79**, 1029.
109. J. A. Teithof, J. H. Stalick, P. W. R. Corfield and D. W. Meek, *J. Chem. Soc., Chem. Commun.*, 1982, 1141.
110. W. Van der Veer and F. Jellinek, *Recl. Trav. Chim. Pays-Bas.*, 1968, **87**, 365; 1970, **89**, 933.
111. G. Hilgetag, H. Teichmann and M. Krüger, *Chem. Ber.*, 1965, **98**, 864; H. Teichmann and G. Hilgetag, *Chem. Ber.*, 1965, **98**, 856.
112. B. Krebs, *Angew. Chem., Int. Ed. Engl.*, 1983, **22**, 113.
113. W. Kuchen and H. Hertel, *Angew. Chem., Int. Ed. Engl.*, 1969, **8**, 89.
114. W. Kuchen, A. Judat and J. Metten, *Chem. Ber.*, 1965, **98**, 3981.
115. S. E. Livingstone, *J. Organomet. Chem.*, 1982, **239**, 143.
116. J. J. Pitts, M. A. Robinson and S. I. Trotz, *J. Inorg. Nucl. Chem.*, 1968, **30**, 1299.
117. A. Davison and E. S. Switkes, *Inorg. Chem.*, 1971, **10**, 837.
118. M. R. Churchill, J. Cooke, J. Wormald, A. Davison and E. S. Switkes, *J. Am. Chem. Soc.*, 1969, **91**, 6218.
119. I. Haiduc, *Rev. Inorg. Chem.*, 1981, **3**, 353.
120. P. S. Pitschimuka, *Chem. Ber.*, 1908, **41**, 3854.
121. R. S. Z. Kowalski, N. A. Bailey, R. Mulvaney, H. Adams, D. A. O'Cleirigh and J. A. McCleverty, *Transition Met. Chem.*, 1981, **6**, 64.
122. J. A. McCleverty, S. Gill, R. S. J. Kowalski, N. A. Bailey, H. Adams, K. W. Lumbard and M. A. Murphy, *J. Chem. Soc., Dalton Trans.*, 1982, 493.
123. A. K. S. Gupta, R. C. Srivastava and S. S. Parma, *Can. J. Chem.*, 1967, **45**, 2293; K. Rufenacht, *Helv. Chim. Acta*, 1968, **51**, 518.
124. L. Cambi and L. Szegö, *Chem. Ber.*, 1931, **64**, 2591; 1933, **11**, 656.
125. L. Malatesta and R. Pizzotti, *Chim. Ind. (Milano)*, 1945, **27**, 6.
126. M. Ciampolini and N. Nardi, *J. Chem. Soc., Dalton Trans.*, 1977, 2121.
127. C. K. Jørgensen, 'Inorganic Complexes', Academic, London, 1963, p. 131.
128. S. E. Livingstone and A. E. Mihkelson, *Inorg. Chem.*, 1970, **9**, 2545.
129. W. K. Musker, T. L. Wolford and P. B. Roush, *J. Am. Chem. Soc.*, 1978, **100**, 6416.
130. C. A. Stein and H. Taube, *J. Am. Chem. Soc.*, 1978, **100**, 1635.
131. A. S. Hirshon and W. K. Musker, *Transition Met. Chem.*, 1980, **5**, 191.
132. M. Hall and D. B. Sowerby, *J. Chem. Soc., Dalton Trans.*, 1980, 1292.
133. L. M. Dikareva, G. G. Sadikov, M. A. Porai-Koshits, M. A. Golubnichaya, J. B. Baronovskii and R. Shchelokov, *Zh. Neorg. Khim.*, 1977, **22**, 2013; 1978, **23**, 1044.
134. P. B. Critchlow and S. D. Robinson, *Inorg. Chem.*, 1978, **17**, 1902.

135. C. Furlani, M. L. Luciani and R. Candori, *J. Inorg. Nucl. Chem.*, 1968, **30**, 2121.
136. W. Hieber and R. Brück, *Z. Anorg. Allg. Chem.*, 1952, **269**, 13.
137. J. A. Goodfellow and T. A. Stephenson, *Inorg. Chim. Acta*, 1980, **41**, 19.
138. S. C. Hari, M. S. Bhalla and R. K. Multani, *Indian J. Chem., Sect. A*, 1978, **16**, 168.
139. R. S. Arora, S. C. Hari and R. K. Multani, *J. Inst. Chem., Calcutta*, 1980, **52**, 166.
140. K. C. Goyal and B. D. Khosla, *Curr. Sci.*, 1981, **50**, 128.
141. R. S. Arora, S. C. Hari, M. S. Bhalla and R. K. Multani, *J. Chin. Chem. Soc. (Taipei)*, 1980, **27**, 65.
142. R. G. W. Gingerich and R. J. Angelici, *J. Am. Chem. Soc.*, 1979, **101**, 5604.
143. E. G. Cox, W. Wardlaw, and K. C. Webster, *J. Chem. Soc.*, 1935, 1475.
144. J. Fujita and K. Nakomoto, *Bull. Chem. Soc. Jpn.*, 1964, **37**, 528.
145. A. R. Latham, V. C. Hascall and H. B. Gray, *Inorg. Chem.*, 1965, **4**, 788.
146. C. Furlani and M. L. Luciani, *Inorg. Chem.*, 1968, **7**, 1586.
147. M. Bonamico and G. Dessy, *Chem. Commun.*, 1968, 483.
148. D. Coucouvanis and J. P. Fackler, *J. Am. Chem. Soc.*, 1967, **89**, 1346, 1745.
149. J. P. Fackler and W. C. Seidel, *Inorg. Chem.*, 1969, **8**, 1631.
150. B. Krebs and G. Gattow, *Z. Anorg. Allg. Chem.*, 1965, **340**, 294.
151. G. Gattow and B. Krebs, *Angew. Chem.*, 1962, **74**, 29; *Z. Anorg. Allg. Chem.*, 1962, **323**, 13.
152. J. P. Fackler and D. Coucouvanis, *J. Am. Chem. Soc.*, 1966, **88**, 3913.
153. M. Nardelli and I. Chierici, *Gazz. Chim. Ital.*, 1957, **87**, 1478; 1958, **88**, 359.
154. M. Nardelli, G. F. Gasparri, A. Musatti and A. Manfredotti, *Acta Crystallogr.*, 1966, **21**, 910.
155. C. D. Flint and M. Goodgame, *J. Chem. Soc. (A)*, 1968, 750.
156. M. Rolies and C. J. De Ranter, *Cryst. Struct. Commun.*, 1977, **6**, 275.
157. K. K. Chatterjee, *Inorg. Chim. Acta*, 1972, **6**, 8.
158. T. Birchall and M. F. Morris, *Can. J. Chem.*, 1972, **50**, 211.
159. Yu. N. Kukushkin, S. A. Simanova, N. N. Knyazeva, V. P. Alashkevich, S. I. Bakireva and E. P. Leonenko, *Zh. Neorg. Khim.*, 1971, **16**, 2488.
160. V. Y. Sibirskaya, Yu. N. Kukushkin, V. N. Samuseva, Yu. N. Martynov, V. Strukov and O. V. Stefanova, *Zh. Obshch. Khim.*, 1976, **46**, 1883.
161. A. J. Aarts, H. O. Desseyn and M. A. Herman, *Transition Met. Chem.*, 1978, **3**, 144.
162. R. L. Girling, J. E. O'Connor and E. L. Amma, *Acta Crystallogr., Sect. B*, 1972, **28**, 2640.
163. L. Capacchi, G. F. Gasparri, M. Nardelli and G. Pelizzi, *Acta Crystallogr., Sect. B*, 1968, **24**, 1199.
164. R. R. Iyengar, D. N. Sathyanarayana and C. C. Patel, *J. Inorg. Nucl. Chem.*, 1972, **34**, 1088.
165. M. Rolies and C. J. De Ranter, *Acta Crystallog., Sect. B*, 1978, **34**, 3216.
166. M. Rolies and C. J. De Ranter, *Cryst. Struct. Commun.*, 1977, **6**, 157.
167. D. M. Adams and R. R. Smardzewski, *J. Chem. Soc. (A)*, 1971, 8.
168. C. J. De Ranter and M. Rolies, *Cryst. Struct. Commun.*, 1977, **6**, 399.
169. A. J. Aarts, H. O. Desseyn and M. A. Herman, *Transition Met. Chem.*, 1980, **5**, 10.
170. L. K. Mishra and H. Bhushan, *Indian J. Chem., Sect. A*, 1981, **20**, 415.
171. B. Singh and M. Chandra, *Indian J. Chem., Sect. A*, 1976, **14**, 676.
172. T. A. Mal'kova and V. N. Shafranskü, *Zh. Fiz. Khim.*, 1975, **49**, 2805.
173. P. G. Antonov, Yu. N. Kukushkin, V. I. Konnov, V. A. Varnek and G. B. Avetikyan, *Koord. Khim.*, 1978, **4**, 1889.
174. T. Birchall and K. M. Tun, *J. Chem. Soc., Dalton Trans.*, 1973, 2521.
175. L. W. Houk and G. R. Dobson, *Inorg. Chem.*, 1966, **5**, 2119.
176. E. M. Garrido and A. Granifo, *Contrib. Cient. Tecnol. (Univ. Tec. Estado, Santiago)*, 1979, **39**, 45 (*Chem. Abstr.*, 1980, **93**, 196837).
177. S. Burman and D. N. Sakhyanarayana, *J. Inorg. Nucl. Chem.*, 1981, **43**, 1940.
178. A. Coombes, *Ann. Chim.*, 1887, **12**, 199; *C.R. Hebd. Seances Acad. Sci.*, 1887, **105**, 869.
179. S. H. H. Chaston and S. E. Livingstone, *Proc. Chem. Soc.*, 1964, 111.
180. S. K. Mitra, *J. Indian Chem. Soc.*, 1931, **8**, 471.
181. R. Mayer, G. Hiller, M. Nitzchke and J. Jentzsch, *Angew. Chem., Int. Ed. Engl.*, 1963, **2**, 370.
182. E. Uhlemann, G. Klose and H. Müller, *Z. Naturforsch., Teil B*, 1964, **19**, 962.
183. A. Yokoyama, K. Ashida and H. Tanaka, *Chem. Pharm. Bull.*, 1964, **12**, 690.
184. S. E. Livingstone, *Coord. Chem. Rev.*, 1971, **7**, 59.
185. M. Das and S. E. Livingstone, *Inorg. Synth.*, 1976, **16**, 206.
186. E. Uhlemann and H. Müller, *Angew. Chem., Int. Ed. Engl.*, 1965, **4**, 154.
187. E. Uhlemann and P. Thomas, *J. Prakt. Chem.*, 1966, **34**, 180.
188. E. Uhlemann, H. Müller and P. Thomas, *Z. Chem.*, 1971, **11**, 401.
189. Z. Reyes and R. M. Silverstein, *J. Am. Chem. Soc.*, 1958, **80**, 6367.
190. S. H. H. Chaston, S. E. Livingstone, T. N. Lockyer, V. A. Pickles and J. S. Shannon, *Aust. J. Chem.*, 1965, **18**, 673.
191. E. Uhlemann and P. Thomas, *Z. Chem.*, 1967, **7**, 430.
192. A. Yokoyama, S. Kawanishi, M. Chikuma and H. Tanaka, *Chem. Pharm. Bull.*, 1967, **15**, 540.
193. S. H. H. Chaston and S. E. Livingstone, *Aust. J. Chem.*, 1967, **20**, 1979.
194. G. Klose, P. Thomas, E. Uhlemann and J. Marki, *Tetrahedron*, 1966, **22**, 2695.
195. K. Arnold, G. Klose, P. H. Thomas and E. Uhlemann, *Tetrahedron*, 1969, **25**, 2957.
196. L. F. Power, K. E. Turner and F. H. Moore, *Tetrahedron Lett.*, 1974, 875.
197. M. Das and S. E. Livingstone, *Org. Mass Spectrom.*, 1974, **9**, 781.
198. S. H. H. Chaston and S. E. Livingstone, *Aust. J. Chem.*, 1966, **19**, 2035.
199. S. E. Livingstone and E. A. Sullivan, *Aust. J. Chem.*, 1969, **22**, 1363.
200. E. Uhlemann and W. W. Suchan, *Z. Anorg. Allg. Chem.*, 1966, **342**, 41.
201. E. Uhlemann, P. Thomas, G. Klose and K. Arnold, *Z. Anorg. Allg. Chem.*, 1969, **364**, 153.
202. M. Leban, J. Fresco and S. E. Livingstone, *Aust. J. Chem.*, 1974, **27**, 2353; M. Leban, J. Fresco, M. Das and S. E. Livingstone, *Aust. J. Chem.*, 1974, **27**, 2357.
203. M. Cox and J. Darken, *Coord. Chem. Rev.*, 1971, **7**, 29.

204. E. Uhlemann, *Z. Chem.*, 1971, **11**, 401.
205. R. C. Mehrotra, R. Bohra and D. P. Gaur, 'Metal β-Diketonates and Allied Derivatives', Academic, New York, 1978.
206. E. Uhlemann and P. Thomas, *Z. Naturforsch., Teil B*, 1968, **23**, 275.
207. D. H. Gerlach and R. H. Holm, *Inorg. Chem.*, 1969, **8**, 2292.
208. E. Uhlemann and H. Müller, *Anal. Chim. Acta*, 1969, **48**, 115.
209. E. Uhlemann and U. Eckelmann, *Z. Chem.*, 1972, **12**, 298; 1974, **14**, 66.
210. R. K. Y. Ho, S. E. Livingstone and T. N. Lockyer, *Aust. J. Chem.*, 1966, **19**, 1179.
211. R. A. Bozis and B. J. McCormick, *Chem. Commun.*, 1968, 1592; *Inorg. Chem.*, 1970, **9**, 1541.
212. R. K. Y. Ho, S. E. Livingstone and T. N. Lockyer, *Aust. J. Chem.*, 1965, **18**, 1927.
213. E. Uhlemann and U. Eckelmann, *Z. Anorg. Allg. Chem.*, 1971, **383**, 321.
214. S. H. H. Chaston, S. E. Livingstone and T. N. Lockyer, *Aust. J. Chem.*, 1966, **19**, 1401.
215. R. K. Y. Ho and S. E. Livingstone, *Aust. J. Chem.*, 1968, **21**, 1987.
216. M. Cox, J. Darken, B. W. Fitzsimmons, A. W. Smith, L. F. Larkworthy and K. A. Rogers, *Chem. Commun.*, 1970, 105.
217. M. Das, R. M. Golding and S. E Livingstone, *Transition Met. Chem.*, 1978, **3**, 112.
218. L. F. Lindoy, S. E. Livingstone and T. N. Lockyer, *Aust. J. Chem.*, 1965, **18**, 1549.
219. S. E. Livingstone, J. H. Mayfield and D. S. Moore, *Aust. J. Chem.*, 1975, **28**, 2531.
220. S. G. Lippard and S. M. Morehouse, *J. Am. Chem. Soc.*, 1972, **94**, 6949.
221. J. A. Sadownick and S. J. Lippard, *Inorg. Chem.*, 1973, **12**, 2659; H. J. Heltner and S. J. Lippard, *Inorg. Chem.*, 1972, **11**, 1447.
222. G. Dorange and J. E. Guerchais, *Bull. Soc. Chim.*, 1971, 43.
223. A. Yokoyama, S. Kawanishi and H. Tanaka, *Chem. Pharm. Bull.*, 1970, **18**, 356, 363; 1972, **20**, 262; 1973, **21**, 2653.
224. G. N. La Mar, *Inorg. Chem.*, 1969, **8**, 581.
225. M. D. Glick and R. L. Lintvedt, *Prog. Inorg. Chem.*, 1976, **21**, 233.
226. M. Das and S. E. Livingstone, *Aust. J. Chem.*, 1974, **27**, 53, 749, 1177, 2115; 1975, **28**, 513.
227. S. E. Livingstone and N. Saha, *Aust. J. Chem.*, 1975, **28**, 1249.
228. S. E. Livingstone and D. S. Moore, *Aust. J. Chem.*, 1976, **29**, 283.
229. S. E. Livingstone and J. E. Oluka, *Aust. J. Chem.*, 1976, **29**, 1913.
230. M. Das, S. E. Livingstone, S. W. Fillipczuk, J. W. Hayes and D. V. Radford, *J. Chem. Soc., Dalton Trans.*, 1974, 1409; 1975, 886.
231. M. Das and S. E. Livingstone, *J. Chem. Soc., Dalton Trans.*, 1975, 452; 1977 663.
232. M. Das, S. E. Livingstone, J. H. Mayfield, D. S. Moore and N. Saha, *Aust. J. Chem.*, 1976, **29**, 767.
233. M. Das, *Inorg. Chim. Acta*, 1979, **36**, 79; 1981, **48**, 33; *Transition Met. Chem.*, 1980, **5**, 17.
234. R. H. Holm, D. H. Gerlach, J. G. Gordon and M. G. McNamee, *J. Am. Chem. Soc.*, 1968, **90**, 4184.
235. M. Das and D. T. Haworth, *J. Inorg. Nucl. Chem.*, 1981, **43**, 515; D. T.Haworth, D. L. Maas and M. Das, *J. Inorg. Nucl. Chem.*, 1981, **43**, 1807.
236. D. T. Haworth and M. Das, *J. Fluorine Chem.*, 1982, **20**, 487.
237. E. A. Shugam, L. M. Shkol'nikova and S. E. Livingstone, *Zh. Strukt. Khim.*, 1967, **8**, 550.
238. L. Kutchabsky and L. Beyer, *Z. Chem.*, 1971, **11**, 30; J. Sieler, P. Thomas, E. Uhlemann and E. Hohne, *Z. Anorg. Allg. Chem.*, 1971, **380**, 160; D. C. Craig, M. Das, S. E. Livingstone and N. C. Stephenson, *Cryst. Struct. Commun.*, 1974, **3**, 283.
239. O. Siiman, D. D. Titus, C. D. Cowman, J. Fresco and H. B. Gray, *J. Am. Chem. Soc.*, 1974, **96**, 2353.
240. B. F. Hoskins and C. D. Pannan, *Inorg. Nucl. Chem. Lett.*, 1975, **11**, 405, 409.
241. J. Ollis, M. Das, V. J. James, S. E. Livingstone and K. Nimgirawath, *Cryst. Struct. Commun.*, 1976, **5**, 679.
242. D. T. Haworth and M. Das, *Synth. React. Inorg. Metal-Org. Chem.*, 1982, **12**, 721.
243. E. Fromm and P. Ziersch, *Chem. Ber.*, 1906, **39**, 3599.
244. A. Fredja and A. Brändström, *Ark. Kemi Mineral. Geol.*, 1949, **26B**, no. 4; *Ark. Kemi*, 1950, **1**, 197.
245. K. Ollson, *Ark. Kemi*, 1967, **26**, 465.
246. H. Prinzbach and E. Futterer, *Adv. Heterocycl. Chem.*, 1967, **7**, 39.
247. R. L. Martin and I. M. Stewart, *Nature (London)*, 1966, **210**, 522; C. G. Barraclough, R. L. Martin and I. M. Stewart, *Aust. J. Chem.*, 1969, **22**, 891; G. A. Heath and R. L. Martin, *Aust. J. Chem.*, 1970, **23**, 1721.
248. T. N. Lockyer and R. L. Martin, *Prog. Inorg. Chem.*, 1980, **27**, 223.
249. A. K. Gregson, R. L. Martin and S. Mitra, *Chem. Phys. Lett.*, 1970, **5**, 310.
250. A. K. Gregson, R. L. Martin and S. Mitra, *J. Chem. Soc., Dalton Trans.*, 1976, 1458.
251. C. G. MacDonald, R. L. Martin and A. F. Masters, *Aust. J. Chem.*, 1976, **29**, 257.
252. A. M. Bond, G. A. Heath and R. L. Martin, *Inorg. Chem.*, 1971, **10**, 2026.
253. W. L. Bowden, J. D. L. Holloway and W. E. Geiger, *Inorg. Chem.*, 1978, **17**, 256.
254. H. Tanaka and A. Yokoyama, *Chem. Pharm. Bull.*, 1960, **8**, 260, 275, 1008, 1012.
255. H. Tanaka and A. Yokoyama, *Chem. Pharm. Bull.*, 1962, **10**, 25.
256. H. Tanaka and A. Yokoyama, *Chem. Pharm. Bull.*, 1962, **10**, 1129, 1133.
257. H. Tanaka and A. Yokoyama, *Chem. Pharm. Bull.*, 1961, **9**, 66, 110.
258. H. Tanaka and A. Yokoyama, *Chem. Pharm. Bull.*, 1962, **10**, 13, 19.
259. H. Tanaka and A. Yokoyama, *Chem. Pharm. Bull.*, 1962, **10**, 556.

17

Selenium and Tellurium Ligands

FRANK J. BERRY
University of Birmingham, UK

17.1 INTRODUCTION

The use of selenium and tellurium compounds as ligands has received relatively little attention until comparatively recently. This is well illustrated by a review article[1] in 1965 which cites less than 30 references to metal complexes with selenium and tellurium ligands. Although the situation has subsequently improved in as much as a variety of selenium ligands have been described during the last 20 years, the analogous chemistry of tellurium has taken longer to develop and is only now emerging as an area of significant growth. This recent development of activity in the ligand chemistry of selenium and tellurium has been excellently reviewed by Gysling within the last few years in two articles[2,3] which have outlined various aspects of the synthetic chemistry and techniques which may be used for the characterization of these compounds. Gysling's reviews must be considered as the most comprehensive surveys of the subject which have yet appeared.

17.2 SELENIUM LIGANDS

The steady growth of interest in selenium chemistry since the mid 1960s, which has led to the development of a variety of selenium ligands, has been progressively monitored in the literature[4-18] and latterly summarized[3] in Gysling's comprehensive appraisal of both the early work and the recent developments. Much of the ligand chemistry of selenium has involved monodentate ligands, although the development of bi-, tri- and tetra-dentate ligands represents an area of considerable scope for future investigations.

17.2.1 Monodentate Selenium Ligands

17.2.1.1 Selenocyanates

The selenocyante ligand has, like the other pseudohalide ligands OCN^- and SCN^- which involve group VI donor atoms, been used extensively for the formation of metal complexes. Indeed, the coordination chemistry of the selenocyanate ligand $SeCN^-$ has received more attention than any other selenium- or tellurium-containing ligand. The magnitude of the literature on this ligand has been reviewed in the recent past[10,14,18] and attention has been drawn to the bonding characteristics of simple selenocyanate complexes which appear to follow patterns predicted by hard–soft acid–base theory. It is interesting to note the limited occurrence of linkage isomeric pairs in selenocyanate ligand chemistry as compared to the thiocyanate ligand. It is also relevant to record the recent report[19] of an oxidative addition reaction in which a gold(I) substrate gave an insoluble complex containing gold in the unusual oxidation state of $+2$ (Scheme 1).

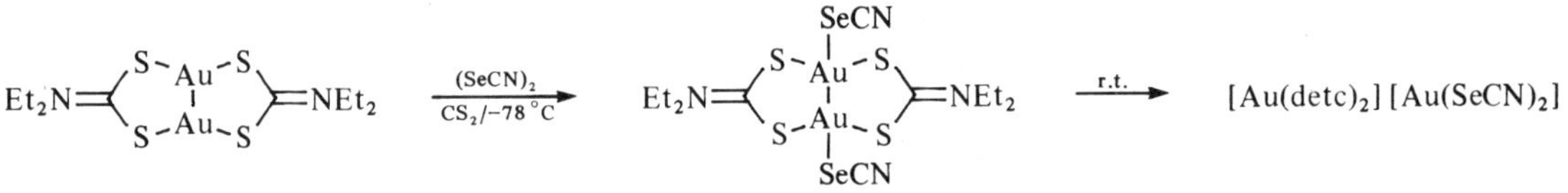

detc = *N, N*-diethyldithiocarbamato ($Et_2NCS_2^-$)

Scheme 1

17.2.1.2 Diorgano selenides

Complexes of the monodentate diorgano selenide ligands R_2Se (R = alkyl or aryl) have been known since 1911 and now form a substantial body of data which has been recently reviewed by Murray and Hartley.[16] The ligands were used in the classic studies of square planar complexes by Chatt *et al.* in the 1950s and 1960s and have found more recent application in the ^1H NMR studies by Abel *et al.* and Cross *et al.* of inversion at selenium atoms in metal complexes. Although the synthesis of the ligands and their complexes is well established,[16] it is interesting to note the recently reported[20] preparation of coordinated diorgano selenides by the alkylation of coordinated SeH^- (equation 1).

$$PPN[(OC)_5CrSeH] \xrightarrow{[Et_3O][BF_4]} (OC)_5CrSeEt_2 \tag{1}$$

The solution phase geometry of complexes containing diorgano selenide ligands has been found to be amenable to examination by IR and ^1H NMR spectroscopy, and dipole moment measurements.

17.2.1.3 Diorgano diselenides

The diorgano diselenides RSeSeR (R = alkyl or aryl) have been successfully used as reagents for the introduction of terminal and bridging RSe ligands by oxidative addition reactions. However, they may also function as bridging ligands in which the Se—Se bond remains intact, this being well illustrated by the recently reported rhenium(I) complexes with bridging diorgano diselenide ligands (Scheme 2).[21,22] These complexes were shown by X-ray crystallography to be dimeric involving two

$$\text{BrRe(CO)}_5 \xrightarrow[h\nu]{\text{THF}} \text{(OC)}_3\text{(THF)Re}(\mu_2\text{-Br})_2\text{Re(CO)}_3\text{(THF)} \xrightarrow[R = Ph, CH_2Ph]{\text{RSeSeR/PhMe}} \text{(OC)}_3\text{Re}(\mu\text{-Br})_2(\mu_2\text{-RSeSeR)Re(CO)}_3$$

Scheme 2

bridging bromide ions and a bridging RSeSeR ligand. Similar dimeric complexes of platinum(IV) have also been described.

17.2.1.4 Selenolates

Recent developments in the ligand chemistry of selenolates RSe^- (R = alkyl or aryl) have been well reviewed by Gysling[3] who has also discussed the synthetic routes to the different types of selenolate ligands and their complexes, the use of techniques such as column chromatography to separate isomers, and some of the structural properties of the complexes. It is pertinent to note the recent work of Ziegler *et al.* who have prepared a number of complexes with terminal and bridging SeR^- ligands and who have performed single-crystal X-ray diffraction studies of the monomeric $(\eta^7\text{-}C_7H_7)W(CO)_2SePh$,[23] the dimeric complexes[24] $Cp(NO)Cr(\mu\text{-SePh})_2Cr(NO)Cp$ and $Cp(NO)Cr(\mu\text{-Se-}n\text{-Bu})(\mu\text{-OH})Cr(NO)Cp$, and the first triply selenato bridged dimeric complex[23] $(\eta^7\text{-}C_7H_7)Mo(\mu_2\text{-SePh})_3Mo(CO)_3$ the structure of which is depicted in Figure 1.

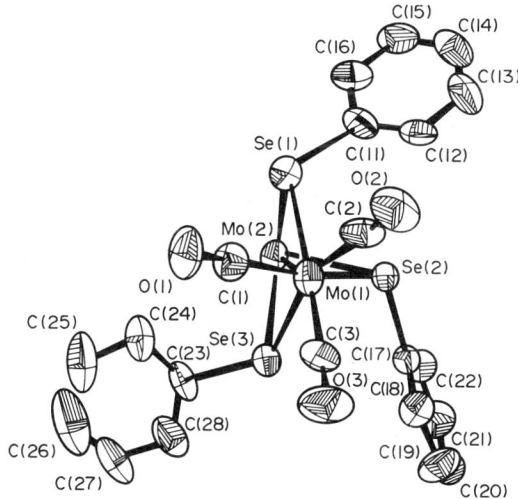

Figure 1 Structure of the triply selenato-bridged complex $(\eta^7\text{-}C_7H_7)Mo(\mu_2\text{-SePh})_3Mo(CO)_3$ (reproduced with permission from ref. 23)

17.2.1.5 Other monodentate selenium ligands

(i) Hydrogen selenide

Although the hydrogen selenide ligand SeH^- has not received extensive usage, it is pertinent to note its appraisal in Gysling's review[3] and its successful introduction into metal complexes by oxidative addition reactions with low-valent species or by metathetical reactions. Cleavage reactions of coordinated selenium ligands have also been used to generate the SeH^- ligand and dimeric complexes with bridging SeH^- ligands have been prepared.

(ii) Selenourea

Although the first reported[25] metal complex with a selenium ligand, $AgCl(su)_2$, involved selenourea (su), the low stability of this class of ligand has militated against its greater use.[11] However, complexes with the monodentate selenourea and ethyleneselenourea ligands have been prepared and successfully examined by single-crystal X-ray diffraction techniques.

(iii) Heterocycles

Several metal complexes of heterocyclic selenium compounds have been described[3] with compounds of the type *trans*-PtBr$_2$(SeCH$_2$CH$_2$OCH$_2$CH$_2$)$_2$ having been structurally characterized by X-ray crystallography.[26]

(iv) Organophosphine selenides

The first complexes involving ligands of this type, *e.g.* PdCl$_2$(SePPh$_3$)$_2$, were reported[27] in 1960 and have recently been subjected to single-crystal structural analysis.[28]

17.2.2 Bidentate Selenium Ligands

17.2.2.1 Organophosphine selenides

A number of metal complexes with bidentate phosphine selenide ligands, *e.g.* Ph$_2$P(Se)(CH$_2$)$_n$-P(Se)Ph$_2$ (n = 1 or 2), have been reported but, unlike the monodentate phosphine selenides, have yet to be successfully characterized by X-ray crystallography. The coordination of selenium in these types of ligands is amenable to examination by inspection of the ν_{PSe} vibration in the IR spectra.[29,30]

It is pertinent to note the chelation reaction which was reported[31] in 1981 for a diphenylphosphido selenide (Scheme 3).

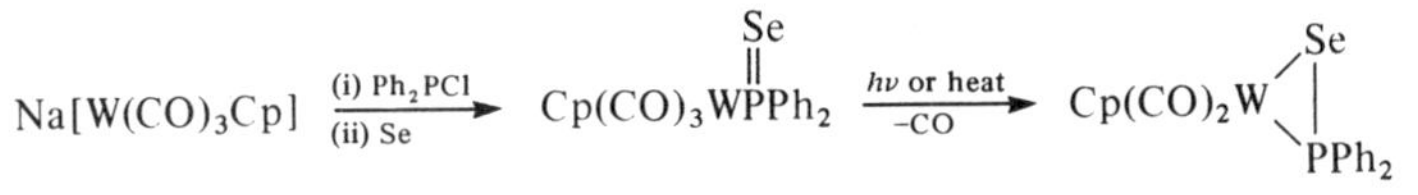

$$\text{Na}[\text{W(CO)}_3\text{Cp}] \xrightarrow[\text{(ii) Se}]{\text{(i) Ph}_2\text{PCl}} \text{Cp(CO)}_3\text{W}\overset{\overset{\text{Se}}{\|}}{\text{P}}\text{Ph}_2 \xrightarrow[-\text{CO}]{hv \text{ or heat}} \text{Cp(CO)}_2\text{W}\overset{\text{Se}}{\underset{\text{PPh}_2}{|}}$$

Scheme 3

It is also pertinent to mention the occurrence of bridging phosphido and arsenido selenide ligands (Scheme 4).[32] The phosphido selenide complexes (**1**; Scheme 4) were found to be very reactive and readily gave the six-membered heterocyclic complexes (**3**). In contrast the arsenido selenide systems gave both the four-membered ring intermediates (**1**; M = Mn, Re) and the six-membered heterocycles (**3**).

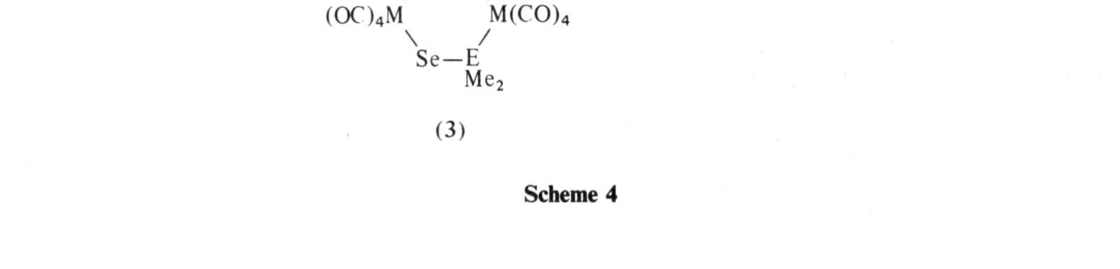

Scheme 4

17.2.2.2 Se, C ligands

The first complexes involving coordinated carbon diselenide, CSe$_2$, were reported[33] in 1973 from reactions with coordinatively unsaturated low-valent metals such as platinum, nickel and rhodium and were characterized by IR spectroscopy. Since then several η^2-CSe$_2$ complexes have been described, for example oxidative addition of carbon diselenide to *trans*-IrCl(CO)(PPh$_3$)$_2$ and Pt(PPh$_3$)$_4$ has given IrCl(CO)(PPh$_3$)$_2$(CSe$_2$) and Pt(PPh$_3$)$_2$(CSe$_2$) respectively.[34] Other reactions of the CSe$_2$ and CSSe ligands have recently been reviewed.[3]

17.2.2.3 Se, N and Se, P ligands

A few examples of complexes involving ligands with Se, N and Se, P donor sets have been described[3] but the use of these ligands, and the structural properties of their complexes, have not received substantial attention.

17.2.2.4 Se, O; Se, S and Se, Se ligands

The activity in this area of coordination chemistry involving bidentate ligands containing selenium and another group VI element is best considered according to the neutral or anionic character of the ligands.

(i) Neutral ligands

A variety of bidentate ligands of the type $RSe(CH_2)_nER$ (E = S or Se; $n = 1$–3) have been described with the complex chemistry of these types of ligands being developed in the 1960s and early 1970s by Hunter *et al.*[35-39] More recent work by Abel *et al.*[40-46] has involved variable temperature ^1H NMR studies of various intramolecular dynamic processes that occur in such complexes including pyramidal inversion at sulfur and selenium atoms, ligand ring reversal, and sulfur or selenium atom switching in bridged dimeric complexes.

(ii) Anionic ligands

A wide variety of Se, O; Se, S and diseleno ligands have been reported. The early work of Jensen *et al.* and the subsequent investigations by Hoyer and Dietzsch and co-workers have been well reviewed.[3] The complexes are generally less stable than the dithio analogues, although aspects of their electrochemistry, ESR and NMR (^{77}Se and ^{31}P) spectra, and X-ray crystal structural properties have been reported.

It is pertinent to note that the chemistry of the 1,2-diseleno ligands is less extensive than that of the analogous sulfur chelates. However, it is interesting to note that chemical or electrochemical reductive coupling of CSe_2 has provided 1,2-diseleno anions which can react with a variety of metal salts to give the corresponding square planar bis chelates (Scheme 5).[47,48]

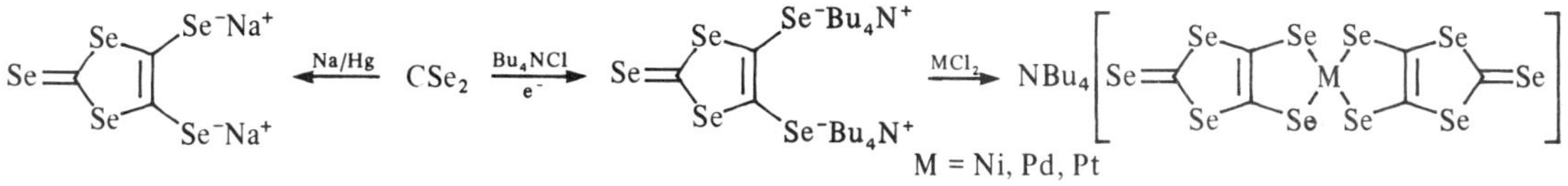

Scheme 5

The tetrabutylammonium salts of the bis chelates are semiconductors but exhibit enhanced conducting properties after doping with iodine. The corresponding salts which are formed when Bu_4N^+ is replaced by TTF^+ or $TSeF^+$ have higher conductivities than the analogous dithiolene complexes with magnitudes similar to that of TTF–TCNQ (TTF = tetrathiafulvalene, TSeF = tetraselenofulvalene, TCNQ = tetracyano-*p*-quinodimethane).

17.2.2.5 Selenoformaldehyde

Although polymeric modifications of selenoformaldehyde have been synthesized and characterized both in the free state, for example as the cyclic trimer 1,3,5-triselane,[49] and when coordinated to a metal, for example in $[\{(CH_2Se)_3\}_2Ag]AsF_6$,[50] monomeric selenoformaldehyde has only been stabilized by coordination to transition metals within the last year.[51-53] The structure of one of these, the selenoformaldehyde-bridged dimer $\{Cp(CO)_2Mn\}_2(CH_2Se)$, has been determined by single-crystal X-ray diffraction.[53]

17.2.3 Tridentate and Tetradentate Selenium Ligands

The synthetic routes to the limited number of tri- and tetra-dentate ligands containing selenium, which, with the exception of an open chain tetraselenoether,[54] are of the hybrid type incorporating nitrogen[55] or phosphorus[56] donors, have been recently reviewed.[3] The tetradentate selenoether, bsep, has been shown to react[54] with Na_2PdX_4 (X = Cl, Br or I) to give $Pd_2(bsep)X_4$ which, on the basis of IR spectroscopy, have been formulated as selenium-bridged dimers (**4**).

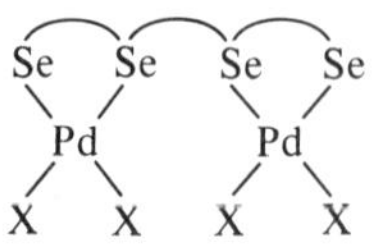

$$
\begin{array}{c}
\overset{\frown}{Se}\;\;\overset{\frown}{Se}\;\;\;\overset{\frown}{Se}\;\;\overset{\frown}{Se} \\
\diagdown\diagup\qquad\diagdown\diagup \\
Pd\qquad\quad Pd \\
\diagup\diagdown\qquad\diagup\diagdown \\
X\quad X\quad\; X\quad X
\end{array}
$$

(4)

17.2.4 Se_n Ligands

Although the synthesis of metal complexes with terminal Se^{2-} ligands has been known for many years and comprehensively discussed in the recent review literature,[3,7,13] significant activity has recently developed in transition metal complexes containing Se_n ligands (n = 2, 4, 5). The types of ligands involved are illustrated in Figure 2. The use of these materials in the synthesis of 1,2-diselenolate complexes (Scheme 6)[57] and mixed-metal clusters[58-62] indicates a potentially powerful applicability of the ligands for synthetic organometallic chemistry.

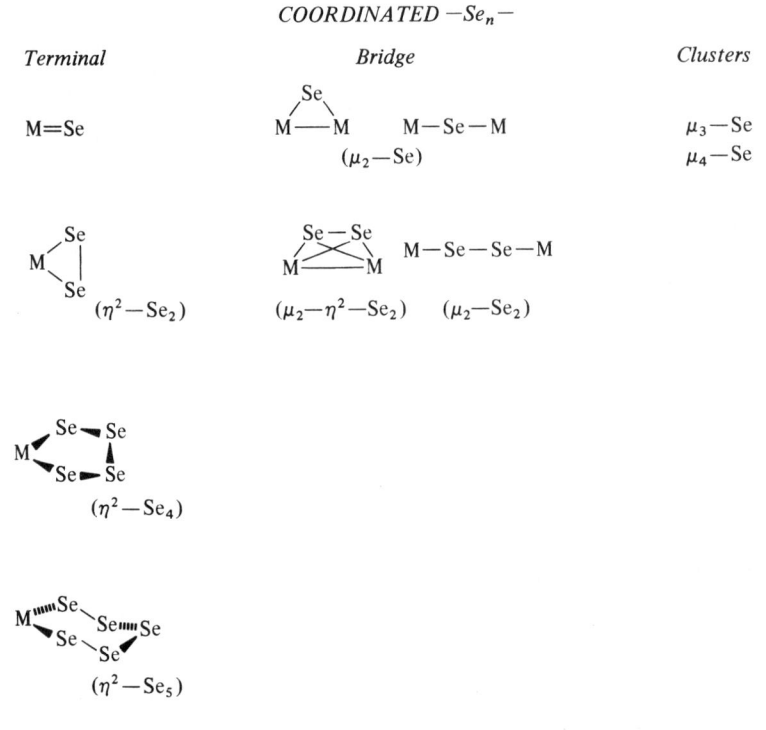

Figure 2 Coordinated Se_n ligands (reproduced with permission from ref. 3)

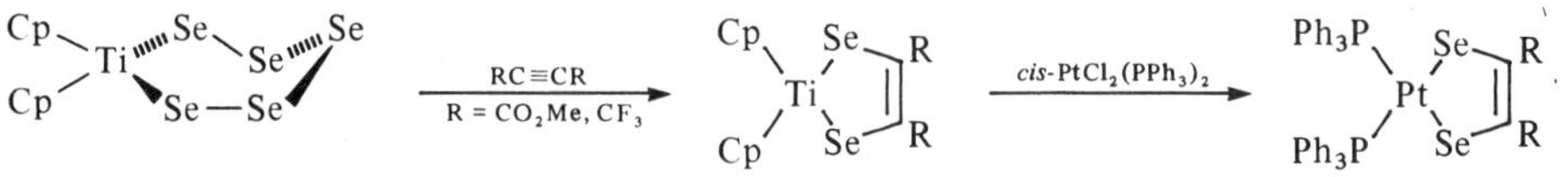

Scheme 6

17.2.4.1 Complexes with terminal Se_n ligands

The crystal structures[63,64] of $[Ir(dppe)_2Se_2]Cl$ (dppe $= Ph_2PCH_2CH_2PPh_2$) and $Os(Se_2)$-$(CO)_2(PPh_3)_2$ have shown the presence of side-on-bonded Se_2 (η^2-Se_2). The Se—Se bonds in these complexes have been found to be reactive to electrophilic alkylating reagents with retention of the Se—Se bond[64] whereas low-valent transition metal species undergo oxidative insertion into the Se—Se bond.[63] The reactions of these types of complexes, which involve oxidative addition and transition metal displacement, have recently been examined in detail.[63]

Three complexes involving the chelating η^2-Se_4 ligand have been reported[65,66] with the X-ray structure of $[Ir(Me_2PCH_2CH_2PMe_2)_2(Se_4)]Cl$ showing[66] *cis*-octahedral geometry with the Se_4 ligand being symmetrically chelated to the iridium at equatorial positions and the $IrSe_4$ ring adopting a half-chair conformation. SCF–X_α–SW calculations[66] were consistent with the description of the Se_4 ligand as an excited Se_4 molecule.

The reaction of Cp_2TiCl_2 with Na_2Se_5 has been reported[67] to give $CpTiSe_5$, which contains the η^2-Se_5 ligand with the $TiSe_5$ ring adopting a fixed conformation at temperatures below 298 K.

17.2.4.2 Complexes with bridging Se_n ligands

The preparation and structural features of complexes with bridging selenium (μ_2-Se) involving the M—Se—M and M—Se—M units,[68,69] and which have sometimes involved the insertion of selenium into M—M bonds, have recently been described in detail.[70–72]

Only one complex, $Cp(OC)_2Mn(SeSe)Mn(CO)_2Cp$, which contains a μ_2-Se_2 bridging M—Se—Se—M linkage, has been described,[69] although the related ligand η^1-SeSeMe has been reported.[64] The complex $(OC)_3Fe(\mu_2,\eta^2$-$Se_2)Fe(CO)_3$ has been synthesized and has been shown by X-ray crystallography[74] to contain Se—Se and Fe—Fe bonds. Recent work by Seyferth[61] and Ranchfuss[60,62] has demonstrated an extensive chemistry of the dimer which is based on the reactivity of the Se—Se bond (Figure 3). The dimer may be used as a substrate in either oxidative addition reactions with low-valent metal complexes to give a wide variety of mixed-metal clusters, or in substitution reactions of the terminal carbon monoxide ligands by phosphines.[58,59,62]

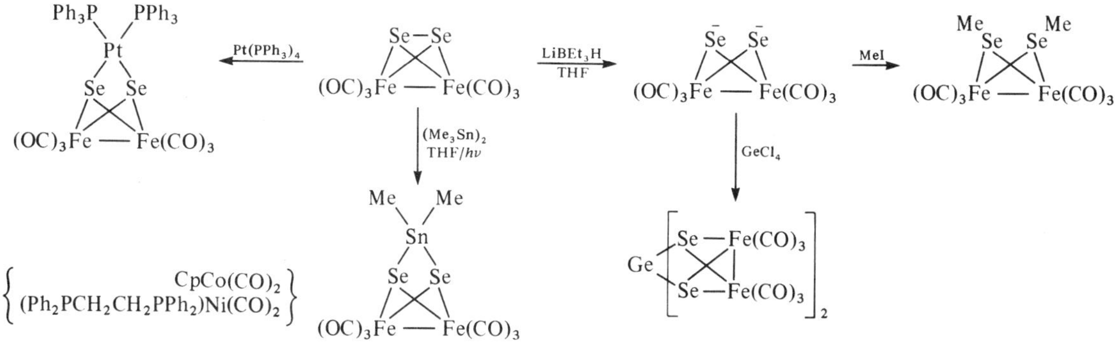

Figure 3 Chemistry of coordinated Se—Se bonds (reproduced with permission from ref. 3)

The μ_2,η^2-Se_2 ligand has also been found in the dimeric $Cl_4W(\mu_2$-Se$)(\mu_2,\eta^2$-$Se_2)WCl_4^{2-}$ complex[75] and the $Nb_4(MeCN)_4Br_{10}Se_3$ cluster.[76]

17.2.4.3 Metal clusters containing incorporated selenium

Several cluster compounds containing transition metals and selenium within the cluster framework have been reported. Most of the clusters involve the selenium atom acting as a μ_3-Se ligand being bonded to three transition metal atoms in the cluster framework. Most of the clusters have been structurally characterized by single-crystal X-ray crystallography, that of $Fe_3(\mu$-$Se_3)_2(CO)_9$ being illustrative[77] of the chemistry and structural types which may be involved. The structure of $Os_4(CO)_{12}H_2Se_2$ has a novel trigonal prismatic core[78] and $Nb_4(MeCN)_4Br_6(\mu_2$-Br$)_4(\mu_2$-$Se_2)(\mu_4$-Se$)$ contains[76] both a diselenide and a μ_4-Se ligand (Figure 4).

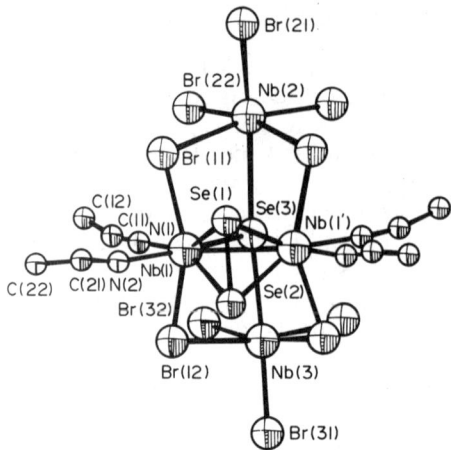

Figure 4 Crystal structure of the (μ-diselenido-μ-selenido)tetraniobium complex $Nb_4Br_{10}Se_3(NCMe)_4$ (reproduced with permission from ref. 76)

17.3 TELLURIUM LIGANDS

The ligand chemistry of tellurium has been the subject of only insignificant attention until relatively recently. The situation is well illustrated by reviews in 1965[1] and 1981[16] which reveal a distinct sparsity of activity in this field and which may be compared with Gysling's reviews in 1982[2] and 1983[3] which describe the developments to date and illustrate the potential scope in this area of chemistry for the future. The preparation of the more common organotellurium ligands has been summarized in detailed reviews of synthetic organotellurium chemistry by Irgolic[79,80] and has been updated by Gysling.[2,3]

Although the preparations of transition metal complexes with tellurium ligands often involve similar methods to those used for the synthesis of analogous selenium compounds, the ligand chemistry of tellurium has by no means achieved the scope of that now known for selenium.

17.3.1 Monodentate Tellurium Ligands

The monodentate tellurium ligands constitute the major category of the limited number of tellurium ligands now known. However, even these ligands show marked differences from their selenium counterparts. For example, although the ligand chemistry of selenocyanates and, to an even larger extent, thiocyanates is well established there appear to be no reports of transition metal tellurocyanates. Similarly, the extensive literature on triorganophosphine–sulfide and –selenide complexes may be compared with the absence of triorganophosphine tellurides.

17.3.1.1 Diorgano tellurides

A large number of the transition metal complexes with tellurium ligands involve the dialkyl and diaryl tellurides. Much of the chemistry is simple, for example diorgano telluride complexes of mercury halides, which were amongst the earliest tellurium complexes to be formed, have often been prepared by straightforward reaction between the telluride and mercury(II) salt in water, ethanol or acetone. The range and variety of complexes involving dialkyl, diaryl and alkyl aryl tellurides have been comprehensively surveyed in the recent past.[2] The structural properties of many of the complexes of the mercury and copper halides,[81,82] as well as those of palladium and platinum,[83–86] have been elucidated by techniques such as IR, Raman, ^{1}H NMR and [125]Te Mössbauer spectroscopy,[81,87] with the latter technique demonstrating that the ligands act primarily as σ donors by using lone pairs of p electrons. The order of Lewis acidity towards (p-EtOC$_6$H$_4$)$_2$Te has been shown to follow the trend Hg^{II} > Pt^{II} > Pd^{II} > Cu^{II}. It should also be noted that studies of the diorgano telluride complexes of silver iodides[88] have recently been extended[89,90] to complexes with other silver halides with light sensitive properties and potential application as photothermographic imaging materials.

The first examples of nickel complexes with a diorgano telluride ligand have recently been reported[91] from reactions involving the substitution of cyclopentadiene (Scheme 7).

$$[CpNi(C_5H_6)]BF_4 + 2Me_2Te \xrightarrow{Et_2O} [CpNi(Me_2Te)_2]BF_4 \longrightarrow [CpNi(Me_2Te)I] + Me_2Te$$

Scheme 7

Recently, a nickel[II] complex of formula RNiTeR′ has been reported as an intermediate in nickel(II)–phosphine-catalyzed carbon–carbon bond forming reactions between diorgano telluride and Grignard reagents.[92]

The reactions of iron carbonyls with diorgano tellurides[93] deserve mention, for example the reaction of $Fe_3(CO)_{12}$ with Ph_2Te gives $Ph_2TeFe(CO)_4$, whilst several ruthenium–carbonyl complexes have been prepared from reactions between diphenyl telluride and alcoholic carbon monoxide-saturated solutions of ruthenium trichloride hydrate.[94] Various other ruthenium–carbonyl complexes of diorgano tellurides, including di- and tri-substituted species, have also been described. The utility of diphenyl telluride in transition metal carbonyl chemistry has also been well illustrated during studies of manganese[93] and rhenium[95] compounds.

17.3.1.2 Diorgano ditellurides

Diorgano ditelluride complexes of many metal halides have been prepared, characterized and classified[2] and no extensive review is required here. It is pertinent however to note the recent preparation and characterization of mercury(II) halide complexes of diaryl ditellurides[81] and other studies[96,97] of complexes of copper halides which illustrate the applicability of modern spectroscopic techniques to the structural examination of these types of materials. The tellurium–tellurium bond in the ditellurides is generally amenable to cleavage during reactions with transition metal compounds to give bridging or terminal TeR^- ligands but is also capable of coordinating transition metals by retaining the tellurium–tellurium bond intact. Bridging diphenyl ditelluride ligands are found in $(OC)_3Re(\mu\text{-}Br)_2(\mu\text{-}Ph_2Te_2)Re(CO)_3$, the crystal structure of which (Figure 5) shows two pseudo-octahedral rhenium(I) centres joined by bromide and diphenyl ditelluride bridges.

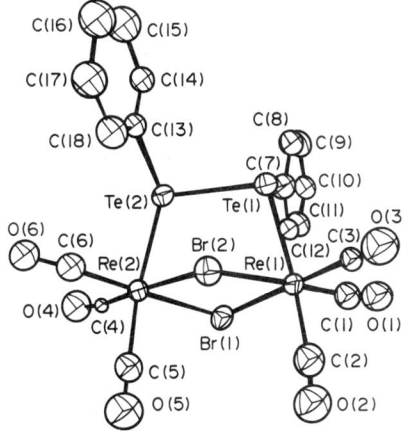

Figure 5 Crystal structure of $[Re_2Br_2(CO)_6(Te_2Ph_2)]$ (reproduced with permission from ref. 98)

17.3.1.3 Tellurols

Complexes involving tellurols have been known for many years. The preparative and reaction chemistry of these ligands with numerous metals, together with illustrations of the use of modern spectroscopic techniques in elucidating their structural and bonding characteristics, has been excellently summarized in the recent past.[2] A range of species is formed, for example the aryltellurol complexes of palladium (5)[81] which contain both terminal and bridging ligands and which were prepared by oxidative addition reactions between $Pd(PPh_3)_4$ and the corresponding ditelluride are dimeric and nonelectrolytes in solution, whilst polymeric species of composition $[Pd(TePh)_2]_n$ have been prepared by reaction of $PdCl_2(NCPh)_2$ with $PhTeCOPh^{99}$ or $PhTeGePh_3$.[100]

(5)

It is also relevant to record that several iron–carbonyl complexes with bridging, and in one case terminal, aryltellurol ligands have been prepared by reaction of $Fe(CO)_5$, $Fe(CO)_{12}$ or $[\pi\text{-}CpFe(CO)_2]_2$ with diaryl ditellurides and which, together with complexes containing other transition metal carbonyls, *e.g.* ruthenium, osmium and manganese, provide a substantial number of interesting compounds.[2]

17.3.1.4 Other monodentate tellurium ligands

(i) Hydrogen telluride

The first complexes containing the TeH^- ligand have recently been prepared by photochemical and thermal substitution reactions and have been characterized by IR and ^1H NMR spectroscopy.[101] The TeH^- ligand has been found to act as a bridging group in dinuclear complexes of the type $C^+[(OC)_5M(\mu\text{-}TeH)M(CO)_5]^-$, where M = Cr or W and C^+ = PPN^+ or $AsPh_4^+$.

(ii) Tellurourea

The first complexes with a tellurourea-type ligand, $M(CO)_5L$ (L = $Te{=}\overline{CNEtCH_2NEt}$), involving chromium, molybdenum and tungsten, were reported in 1980.[102] The solid complexes are indefinitely stable when kept under an inert atmosphere in the absence of light. A more recent study[103] has reported an iron complex involving a paramagnetic iron(I) species which was not isolated but which gave an IR spectrum in solution.

(iii) Heterocycles

The ligand chemistry of heterocyclic tellurium compounds has been examined over several years.[85,86,92,104–106] The study[105] of the monomeric (6) and dimeric (7) complexes, which were obtained from the reaction of Na_2PdCl_4 and tellurophene and separated by their solubility differences in acetone and chloroform, is illustrative of the preparative chemistry and characterization techniques used in this area of chemistry.

(6)   (7)

(iv) Ligands with Te—M (M = Ge, Sn, Pb, P and As) bonds

Several transition metal complexes of ligands containing tellurium bonded to germanium,[107] tin,[107–109] lead,[107] phosphorus[110,111] and arsenic[110] have been prepared and characterized. For example, the complexes $Cr(CO)_5Te(MMe_3)_2$ (M = Ge,[107] Sn,[107,109] Pb[107]) were prepared at 0 °C by substitution reactions with the photochemically generated complex $Cr(CO)_5THF$. The bonding in these complexes, which are sensitive to air and moisture and slowly decompose even at 0 °C, has been examined by IR and Raman spectroscopy and may be compared with the sulfur and selenium analogues in which the pure ligands themselves are more stable to light than the tellurides.

(v)   Tellurocarbonyl complexes

The first tellurocarbonyl complex, which is also the only osmium complex of an organotellurium ligand, is the recently reported[112] $OsCl_2(PPh_3)_2(CO)CTe$ species.

(vi)   Tellurium as a six-electron donor

The complex μ_3-Te[Mn(CO)$_2$Cp]$_3$ has, according to an X-ray crystal structure determination (Figure 6),[113] an almost planar μ_3-TeM$_3$ skeleton in which the central tellurium atom acts as a six-electron donor. The complex was formed from the reaction between CpMn(CO)$_2$(THF) and H$_2$Te and was obtained in the presence of a by-product formulated as Te$_2$[Mn(CO)$_2$Cp]$_3$.

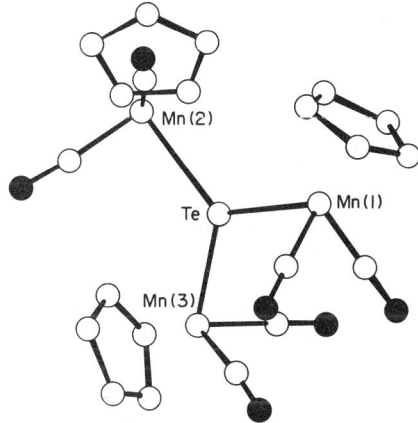

Figure 6   Structure of μ_3-Te[Mn(CO)$_2$Cp]$_3$ (reproduced with permission from ref. 113)

17.3.2   Bidentate Tellurium Ligands

The bidentate ligands are the only polydentate ligands reported for tellurium. Indeed, no complexes of 1,1- or 1,2-Te,Te or -Te,X (X = O, S, or Se) and several other analogues of common bidentate selenium ligands have been reported.

Complexes containing terminal chelating telluroformaldehyde (Cp(PMe$_3$)Rh(η^2-CH$_2$Te)) and bridging telluroformaldehyde[53] (Cp(CO)$_2$Mn(μ_2-CH$_2$Te)Mn(CO)$_2$Cp) have recently been described.[52] These complexes were prepared by routes similar to those used for the synthesis of the analogous selenoformaldehyde complexes.

17.3.3   Metal Clusters Incorporating Tellurium

Several cluster complexes containing both transition metals and tellurium within the cluster framework have been reported and recently reviewed.[2] A good example of the chemistry and structural characterization involved in this area of chemistry is provided by the mixed cluster compound FeCo$_2$(CO)$_9$Te which was obtained[114] from the stoichiometric reaction between Co$_2$(CO)$_8$ and Fe$_3$(CO)$_{12}$ with diethyl telluride. X-Ray crystallography revealed a tetrahedral FeCo$_2$Te cluster system formed by the symmetrical coordination of an apical tellurium atom to a basal FeCo$_2$(CO)$_9$ fragment containing three M(CO)$_3$ groups at the corners of an equilateral triangle and linked to one another by metal–metal bonds. Several other cluster compounds containing iron and tellurium have been reported and successfully examined by ^{13}C NMR and X-ray crystallography.[115–126] The kinetics of carbon monoxide isotopic exchange and substitution reactions have also been investigated in several cases.

Ru$_3$(CO)$_{12}$ reacts with elemental tellurium in *n*-octane under a CO/H$_2$ pressure of 35 atm to give a mixture of Ru$_3$(CO)$_9$Te$_2$ and Ru$_3$(CO)$_9$H$_2$Te, which may be separated by thin layer chromatography.[127] When performed under pure carbon monoxide the yield of the hydrido cluster was decreased.

17.4 CONCLUSION

Although there has been a substantial growth in the synthesis and characterization of coordination complexes containing selenium and tellurium, the ligand chemistry of tellurium remains less developed and there is a notable sparsity of polydentate tellurium ligands for which the selenium analogues are known. It is also clear that little success has been found so far for commmercial applications of these compounds. For example, although some tellurium complexes of silver and copper have been shown to possess light sensitive properties which may render them suitable for use as photothermographic imaging materials,[89,90,128,129] and whilst some metal selenide-containing electroless plating solutions have been reported,[130,132] no large scale applications of these materials can currently be identified. It is worth noting, however, that $PtCl_2(SePh_2)_2/SnCl_2$ may be an example of other compounds which are catalytically active for the homogeneous hydrogenation of non-aromatic alkenes[133] and that the conducting metal complexes with 1,2-diseleno ligands[47,48] may have applications which are amenable to development in the future.

17.5 REFERENCES

1. S. E. Livingstone, *Q. Rev. Chem. Soc.*, 1965, **19**, 386.
2. H. J. Gysling, *Coord. Chem. Rev.*, 1982, **42**, 133.
3. H. J. Gysling, in 'Proceedings of the Fourth International Conference on the Organic Chemistry of Selenium and Tellurium', eds. F. J. Berry and W. R. McWhinnie, The University of Aston in Birmingham, 1983, p. 32.
4. E. W. Abel and B. C. Crosse, *Organomet. Chem. Rev., Sect. A*, 1967, **2**, 443.
5. C. K. Jørgensen, *Inorg. Chim. Acta Rev.*, 1968, **2**, 65.
6. W. Kuchen and H. Hertal, *Angew. Chem., Int. Ed. Engl.*, 1969, **8**, 89.
7. E. Dieman and A. Muller, *Coord. Chem. Rev.*, 1973, **10**, 79.
8. K. A. Jensen and C. K. Jorgensen, in 'Organic Selenium Compounds: Their Chemistry and Biology', ed. D. L. Klayman and W. H. H. Gunther, Wiley, New York, 1973, p. 1017.
9. V. Krishnan and R. A. Zingaro, in 'Selenium', ed. R. A. Zingaro and W. C. Cooper, Van Nostrand Reinhold, New York, 1974, p. 337.
10. A. H. Norbury, *Adv. Inorg. Chem. Radiochem.*, 1975, **17**, 231.
11. G. B. Aitken and G. P. McQuillan, *Inorg. Synth.*, 1976, **16**, 83.
12. I. S. Butler, *Acc. Chem. Res.*, 1977, **10**, 359.
13. D. A. Rice, *Coord. Chem. Rev.*, 1978, **25**, 119.
14. P. P. Singh, *Coord. Chem. Rev.*, 1980, **32**, 33.
15. I. S. Butler, A. M. English and K. P. Plowman, *Inorg. Synth.*, 1981, **21**, 1.
16. S. G. Murray and F. R. Hartley, *Chem. Rev.*, 1981, **81**, 365.
17. W. H. Pan and J. P. Fackler, *Inorg. Synth.*, 1981, **21**, 6.
18. V. V. Skopenko, G. V. Tsintsadze and E. I. Ivanova, *Russ. Chem. Rev. (Engl. Transl.)*, 1982, **51**, 21.
19. D. C. Calabro, B. A. Harrison, G. T. Palmer, M. K. Moguel, R. L. Rebbert and J. L. Burmeister, *Inorg. Chem.*, 1981, **20**, 4311.
20. H. Hausmann, M. Hofler, T. Kruck and H. W. Zimmermann, *Chem. Ber.*, 1981, **114**, 975.
21. J. Korp, I. Bernal, J. L. Atwood, F. Calderazzo and D. Vitali, *J. Chem. Soc., Dalton Trans.*, 1979, 1492.
22. E. W. Abel, S. K. Bhargava, M. M. Bhatti, M. A. Mazid, K. G. Orrell, V. Sik, M. B. Hursthouse and K. M. A. Malik, *J. Organomet. Chem.*, 1983, **250**, 373.
23. A. Rettenmeier, K. Weidenhammer and M. L. Ziegler, *Z. Anorg. Allg. Chem.*, 1981, **473**, 91.
24. J. Rott, E. Guggolz, A. Rettenmeier and M. L. Ziegler, *Z. Naturforsch., Teil B*, 1982, **37**, 13.
25. M. A. Verneuil, *Ann. Chim. Phys.*, 1886, Ser. 6, **9**, 294.
26. J. C. Barnes, G. Hunter and M. W. Lown, *J. Chem. Soc., Dalton Trans.*, 1977, 458.
27. E. Bannister and F. A. Cotton, *J. Chem. Soc.*, 1960, 1959.
28. L. S. D. Glasser, L. Ingram, M. G. King and G. P. McQuillan, *J. Chem. Soc. (A)*, 1969, 2501.
29. J. A. Tiethof, A. T. Hetey and D. W. Meek, *Inorg. Chem.*, 1974, **13**, 2505.
30. T. S. Lobana and K. Sharma, *Transition Met. Chem.*, 1982, **7**, 333.
31. W. Malisch, R. Maisch, I. J. Colquhoun and W. McFarlane, *J. Organomet. Chem.*, 1981, **220**, C1.
32. V. Kullmer and H. Vahrenkamp, *Chem. Ber.*, 1977, **110**, 237.
33. K. A. Jenson and E. Huge-Jensen, *Acta Chem. Scand.*, 1973, **27**, 3605.
34. K. Kawakami, Y. Ozaki and T. Tanaka, *J. Organomet. Chem.*, 1974, **69**, 151.
35. N. N. Greenwood and G. Hunter, *J. Chem. Soc. (A)*, 1960, 1520.
36. R. Donaldson, G. Hunter and R. C. Massey, *J. Chem. Soc., Dalton Trans.*, 1974, 2889.
37. G. Hunter and R. C. Massey, *J. Chem. Soc., Dalton Trans.*, 1975, 2009.
38. G. Hunter and R. C. Massey, *J. Chem. Soc., Dalton Trans.*, 1976, 2007.
39. N. N. Greenwood and G. Hunter, *J. Chem. Soc. (A)*, 1969, 929.
40. E. W. Abel and G. V. Hutson, *J. Inorg. Nucl. Chem.*, 1969, **31**, 3333.
41. E. W. Abel, A. R. Khan, K. Kite, K. G. Orrell and V. Sik, *J. Chem. Soc., Dalton Trans.*, 1980, 1175.
42. E. W. Abel, A. R. Khan, K. Kite, K. G. Orrell and V. Sik, *J. Chem. Soc., Dalton Trans.*, 1980, 2208.
43. E. W. Abel, K. Kite, K. G. Orrell, V. Sik and B. L. Williams, *J. Chem. Soc., Dalton Trans.*, 1981, 2439.
44. E. W. Abel, S. K. Bhargava, T. E. Mackenzie, P. K. Mittal, K. G. Orrell and V. Sik, *J. Chem. Soc., Chem. Commun.*, 1982, 983.
45. E. W. Abel, S. K. Bhargava, M. M. Bhatti, K. Kite, M. A. Mazid, K. G. Orrell, V. Sik, B. L. Williams, M. B. Hursthouse and K. M. Abdul Malik, *J. Chem. Soc., Dalton Trans.*, 1982, 2065.

46. E. W. Abel, A. R. Khan, K. Kite, K. G. Orrell and V. Sik, *J. Chem. Soc., Dalton Trans.*, 1980, 1169.
47. G. C. Papavassiliou, *Mol. Cryst. Liq. Cryst.*, 1982, **86**, 159.
48. G. C. Papavassiliou, *Z. Naturforsch., Teil B*, 1982, **37**, 825.
49. L. Mortillaro, L. Credali, M. Russo and C. De. Checchi, *J. Polym. Sci., Polym. Lett. Ed.*, 1965, **3**, 581.
50. H. Hofmann, P. G. Jones, N. Noltemeyer, E. Peymann, W. Pinkert, H. W. Roesky and G. M. Sheldrick, *J. Organomet. Chem.*, 1983, **249**, 97.
51. C. E. L. Headford and W. A. Roper, *J. Organomet. Chem.*, 1983, **244**, C53.
52. W. Paul and H. Werner, *Angew. Chem., Int. Ed. Engl.*, 1983, **22**, 316.
53. W. A. Herrmann, J. Weichmann, R. Serrano, K. Belchschmitt, H. Pfisterer and M. L. Ziegler, *Angew. Chem., Int. Ed. Engl.*, 1983, **22**, 314.
54. W. Levason, C. A. McAuliffe and S. G. Murray, *J. Chem. Soc., Dalton Trans.*, 1976, 269.
55. G. Labauze and J. B. Raynor, *J. Chem. Soc., Dalton Trans.*, 1980, 2388.
56. G. Dyer and D. W. Meek, *Inorg. Chem.*, 1967, **6**, 149.
57. C. M. Bolinger and T. B. Rauchfuss, *Inorg. Chem.*, 1982, **21**, 3947.
58. R. Rossetti, G. Gervasio and P. L. Stanghellini, *Inorg. Chim. Acta*, 1979, **35**, 73.
59. S. Aime, G. Gervasio, R. Rossetti and P. L. Stanghellini, *Inorg. Chim. Acta*, 1980, **40**, 131.
60. D. A. Lesch and T. B. Rauchfuss, *J. Organomet. Chem.*, 1980, **199**, C6.
61. P. Seyferth and R. S. Henderson, *J. Organomet. Chem.*, 1981, **204**, 333.
62. V. W. Day, D. A. Lesch and T. B. Rauchfuss, *J. Am. Chem. Soc.*, 1982, **104**, 1290.
63. A. P. Ginsberg, W. E. Lindsell, C. R. Sprinkle, K. W. West and R. L. Cohen, *Inorg. Chem.*, 1982, **21**, 3666.
64. D. H. Farrar, K. R. Grundy, N. C. Payne, W. R. Roper and A. Wallker, *J. Am. Chem. Soc.*, 1979, **101**, 6577.
65. H. Kopf, W. Kahl and W. Wirl, *Angew Chem., Int. Ed. Engl.*, 1970, **9**, 801.
66. A. P. Ginsberg, J. H. Osborne and C. R. Sprinkle, *Inorg. Chem.*, 1983, **22**, 1781.
67. H. Kopf, B. Block and M. Schmidt, *Chem. Rev.*, 1968, **101**, 272.
68. E. Rottinger, V. Kullmer and H. Vahrenkamp, *J. Organomet. Chem.*, 1978, **150**, C6.
69. M. Heberhold, D. Reiner, B. Zimmer-Gasser and U. Schubert, *Z. Naturforsch., Teil B*, 1980, **35**, 1281.
70. W. Hofmann and H. Werner, *Angew. Chem., Int. Ed. Engl.*, 1981, **20**, 1014.
71. H. Brunner, J. Wachter and H. Wintergerst, *J. Organomet. Chem.*, 1982, **235**, 77.
72. W. A. Herrmann, C. Bauer and J. Weichmann, *J. Organomet. Chem.*, 1983, **243**, C21.
73. W. Hieber and J. Gruber, *Z. Anorg. Allg. Chem.*, 1958, **296**, 91.
74. C. F. Campana, F. Y. K. Lo and L. F. Dahl, *Inorg. Chem.*, 1979, **18**, 3060.
75. M. G. B. Drew, G. W. A. Fowles, E. M. Page and D. A. Rice, *J. Am. Chem. Soc.*, 1979, **101**, 5827.
76. A. J. Benton, M. G. B. Drew and D. A. Rice, *J. Chem. Soc., Chem. Commun.*, 1981, 1241.
77. L. F. Dahl and P. W. Sutton, *Inorg. Chem.*, 1963, **2**, 1067.
78. B. F. G. Johnson, J. Lewis, P. G. Lodge, P. R. Raithby, K. Henrick and M. McPartlin, *J. Chem. Soc., Chem. Commun.*, 1979, 719.
79. K. J. Irgolic and R. A. Zingaro, in 'Organometallic Reactions', ed. E. I. Becker and M. Tsutsui, Wiley, New York, 1971, vol. 2, p. 137.
80. K. J. Irgolic, 'The Organometallic Chemistry of Tellurium', Gordon and Breach, London, 1974, p. 257.
81. N. S. Dance and C. H. W. Jones, *J. Organomet. Chem.*, 1978, **152**, 175.
82. W. R. McWhinnie and V. Rattanphani, *Inorg. Chim. Acta*, 1974, **9**, 153.
83. H. J. Gysling, H. R. Luss and D. L. Smith, *Inorg. Chem.*, 1959, **18**, 2696.
84. H. J. Gysling, N. Zumbulyadis and J. A. Robertson, *J. Organomet. Chem.*, 1981, **209**, C41.
85. S. Sergi, F. Faraone, L. Silvestro and R. Peitropaolo, *J. Organomet. Chem.*, 1971, **33**, 403.
86. L. Y. Chia and W. R. McWhinnie, *J. Organomet. Chem.*, 1978, **148**, 165.
87. I. Davies, W. R. McWhinnie, N. S. Dance and C. H. W. Jones, *Inorg. Chim. Acta*, 1978, **29**, L203.
88. G. E. Coates, *J. Chem. Soc.*, 1951, 2003.
89. Asahi Chemical Industry, *Jpn. Pat.* 53 65 827 (1978) (*Chem. Abstr.*, 1978, **89**, 146 588).
90. Asahi Chemical Industry, *Jpn. Pat.* 53 143 216 (1978) (*Chem. Abstr.*, 1979, **90**, 195 607).
91. N. Kuhn and M. Winter, *J. Organomet. Chem.*, 1983, **249**, C28.
92. S. Uemura and S. Fukuzawa, *Tetrahedron Lett.*, 1982, **23**, 1181.
93. W. Hieber and T. Kruck, *Chem. Ber.*, 1962, **95**, 2027.
94. W. Hieber and P. John, *Chem. Ber.*, 1970, **103**, 2161.
95. W. Hieber, W. Opausky and W. Romm, *Chem. Ber.*, 1968, **101**, 2244.
96. I. Davies and W. R. McWhinnie, *Inorg. Nucl. Lett.*, 1976, **12**, 763.
97. I. Davies, W. R. McWhinnie, N. S. Dance and C. H. W. Jones, *Inorg. Chim. Acta*, 1978, **29**, L217.
98. F. Calderazzo, D. Vitali, R. Poli, J. L. Atwood, R. D. Rogers, J. M. Cummings and I. Bernal, *J. Chem. Soc., Dalton Trans.*, 1981, 1004.
99. S. A. Gardner and H. J. Gysling, *J. Organomet. Chem.*, 1980, **197**, 111.
100. S. A. Gardner, P. J. Trotter and H. J. Gysling, *J. Organomet. Chem.*, 1981, **212**, 35.
101. H. Hausmann, M. Hofler, T. Kruck and H. W. Zimmermann, *Chem. Ber.*, 1981, **114**, 975.
102. M. F. Lappert, T. R. Martin and G. M. McLaughlin, *J. Chem. Soc., Chem. Commun.*, 1980, 635.
103. M. F. Lappert, J. J. MacQuitty and P. L. Pye, *J. Chem. Soc., Dalton Trans.*, 1981, 1583.
104. E. H. Braye, W. Hubel and I. Caplier, *J. Am. Chem. Soc.*, 1961, **83**, 4406.
105. K. Ofele and E. Dotzauer, *J. Organomet. Chem.*, 1972, **42**, C87.
106. G. T. Morgan and F. H. Burstall, *J. Chem. Soc.*, 1931, 180.
107. H. Schumann, R. Mohtachemi, H. J. Kroth and U. Frank, *Chem. Ber.*, 1973, **105**, 2049.
108. V. Kullmer and H. Vahrenkamp, *Chem. Ber.*, 1977, **110**, 228.
109. H. Schumann and R. Weiss, *Angew. Chem., Int. Ed. Engl.*, 1970, **9**, 246.
110. J. Grobe and D. Le Van, *Z. Naturforsch., Teil B*, 1979, **34**, 1653.
111. J. Grobe and D. Le Van, *Z. Naturforsch., Teil B*, 1980, **35**, 694.
112. G. R. Clark, K. Marsden, W. R. Roper and L. J. Wright, *J. Am. Chem. Soc.*, 1980, **102**, 1206.
113. M. Herberhold, D. Reiner and D. Neugebauer, *Angew Chem., Int. Ed. Engl.*, 1983, **22**, 59.

114. C. E. Strouse and L. F. Dahl, *J. Am. Chem. Soc.*, 1971, **93**, 6032.
115. W. Hieber and J. Gruber, *Z. Anorg. Allg. Chem.*, 1958, **196**, 91.
116. D. A. Lesch and T. B. Rauchfuss, *J. Organomet. Chem.*, 1980, **199**, C6.
117. G. Cetini, P. L. Stanghellini, R. Rossetti and O. Gambino, *J. Organomet. Chem.*, 1968, **15**, 373.
118. M. K. Chaudhuri, A. Hass, M. Rosenberg, M. Velicescu and N. Welcman, *J. Organomet. Chem.*, 1977, **124**, 37.
119. S. Aime, L. Milone, R. Rossetti and P. L. Stanghellini, *J. Chem. Soc., Dalton Trans.*, 1980, 46.
120. R. Rossetti, P. L. Stanghellini, O. Gambino and G. Cetini, *Inorg. Chim. Acta*, 1972, **6**, 205.
121. G. Cetini, P. L. Stanghellini, R. Rossetti and O. Gambino, *Inorg. Chim Acta*, 1968, **2**, 433.
122. M. Schumann, M. Magerstadt and J. Pickardt, *J. Organomet. Chem.*, 1982, **240**, 407.
123. L. E. Brogen, D. A. Lesch and T. B. Rauchfuss, *J. Organomet. Chem.*, 1983, **250**, 429.
124. D. A. Lesch and T. B. Rauchfuss, *Inorg. Chem.*, 1983, **22**, 1854.
125. V. W. Day, D. A. Lesch and T. B. Rauchfuss, *J. Am. Chem. Soc.*, 1982, **104**, 1290.
126. T. B. Rauchfuss and T. D. Weatherill, *Inorg. Chem.*, 1982, **21**, 827.
127. B. F. G. Johnson, J. Lewis, P. G. Lodge, P. R. Raithby, K. Henrick and M. McPartlin, *J. Chem. Soc., Chem. Commun.*, 1979, 719.
128. H. J. Gysling. (Eastman Kodak Company), *US Pat.* 4 394 318 (1983).
129. H. J. Gysling. (Eastman Kodak Company), *US Pat.* 4 287 354 (1981).
130. R. A. Zingaro and D. O. Skovlin, *J. Electrochem. Soc.*, 1964, **111**, 42.
131. P. Pramanik and R. N. Bhattacharya, *J. Solid State Chem.*, 1982, **44**, 425.
132. R. N. Bhattacharya and P. Pramanik, *Bull. Mater. Sci.*, 1981, **3**, 403.
133. H. A. Tayim and J. C. Bailar, Jr., *J. Am. Chem. Soc.*, 1967, **89**, 4330.

18

Halogens as Ligands

ANTHONY J. EDWARDS
University of Birmingham, UK

18.1 INTRODUCTION

Although there are authoritative and extensive reviews of the chemistry of the halogens available, these usually cover the complete chemistry of the elements, and are not restricted to coordination chemistry.[1-4] Halogens are ubiquitous as ligands, in complexes with a wide variety of oxidation states for the central atom involved. The difficulty in this particular area of coordination chemistry is to draw a line between those compounds considered as coordination compounds, and those more appropriately treated as collections of ions.

In the majority of examples of halogens acting as ligands, the halogen is monodentate, and no problem arises, but for bridging halogen atoms there is a gradation from isolated, multinuclear species through to three-dimensional, totally bridged species, which can be more appropriately regarded as a collection of ions.

This review will attempt to deal with the chemistry of the halogens appropriate to those compounds considered as coordination compounds. It will not include reference to astatine, which is covered by the general reviews listed above and for which little chemistry is known at all, and almost none as a coordinated ligand. Since the chemistry of the halides of individual elements is treated in the chapters for those elements, this section will be concerned with an overview of the general properties of the group.

18.2 PHYSICAL PROPERTIES OF THE HALOGENS

The halogens are usually regarded as a cohesive chemical group, largely due to the electron configuration in the ground state, ns^2np^5, one electron short of the noble gas configuration, leading to formation of a single covalent bond (or of a uninegative ion). However, there are considerable differences in the behaviour of the individual halogens as ligands, particularly that of fluorine, which can be related to fundamental properties. They are all strongly electron accepting, with high electron affinities, but these do not vary regularly as might be expected. The electron affinity of fluorine is anomalously lower than that of chlorine, with bromine and iodine following a normal progression.

This low value of electron affinity for fluorine follows inexorably from the dissociation energy of the difluorine molecule,[5] which is very much lower than might be expected from an extrapolation up the halogen series of the values given in Table 1. From this extrapolation a value of some $260\,kJ\,mol^{-1}$ rather than the experimental value[5] of $158.8\,kJ\,mol^{-1}$ would have been expected. This low value of the dissociaton energy of fluorine compared with the values for the other halogens is of considerable importance in relation to the comparative chemical reactivities of the halogens.

Table 1 Properties of the Halogens

	F	Cl	Br	I	At
Electron affinity (kJ mol^{-1}) at 298 K	328	349	325	295	($\sim$270)
Dissociation energy of X_2 (kJ mol^{-1}) at 298 K	159	243	193	151	($\sim$116)
First ionization energy (kJ mol^{-1})	1681	1251	1140	1008	($\sim$926)
Electronegativity coefficient (Pauling)	4.0	3.0	2.8	2.5	($\sim$2.4)
Covalent radius (Å)	0.71	0.99	1.14	1.33	($\sim$1.41)
Anionic radius (Å) for six-coordination	1.33	1.82	1.98	2.20	($\sim$2.3)

Various explanations for the low strength of the F—F bond have been proposed. The possibility of partial multiple bonding for the higher halogens,[6] involving higher empty d orbitals, or the larger electron–electron repulsions and less penetration of charge clouds for fluorine have been suggested.[7] Another favoured proposal is based on the very small size of the fluorine atom, and the large increase in electron–electron repulsions for the seven electrons around each atom, when an extra electron is introduced to form the covalent bond in the diatomic molecule or the fluoride ion.[8]

Some recent calculations,[9] based on ionization potentials derived from photoelectron spectroscopy, favour the high lone-pair to lone-pair repulsions in the molecule as being very significant in the weakness of the bond.

The variations in ionization potentials and in electronegativity coefficients follow the expected trends as shown in Table 1. The covalent radii are derived from the interatomic distances in the halogen molecules and a smaller value for fluorine has been suggested, on the grounds that the F—F distance in the molecule represents an 'artificially' long bond, corresponding to the low dissociation energy. A value as low as 0.57 Å (derived from the C—F distance) has been suggested, but a working value of 0.64 Å appears more useful.[1a]

Although the low dissociation energy of fluorine compared to those of the other halogens leads to easier formation of fluorine atoms, the greater chemical reactivity of fluorine must also depend on the strength of the bonds formed between fluorine and other atoms, as compared with those formed by chlorine, bromine and iodine. Table 2 shows a representative set of average bond energies. In all cases there is an increase of values from iodine through to fluorine. In most cases the bonds to fluorine have significantly higher average energies than would be extrapolated up the series. However, in the case of Hg^{II} there are comparatively small changes over the halogen series and fluorine does not lie out of line. This behaviour is followed by the other elements which can be considered as class b acceptors, or 'soft' acids, and in all these cases the bond energies are comparatively low.

Table 2 Some Representative Values for Average Bond Energies for Bonds to Halogens (kJ mol^{-1}, 298 K)

	F	Cl	Br	I
Be^{II}	636	460	402	318
Al^{III}	590	425	364	285
Si^{IV}	597	400	330	249
P^{III}	497	329	270	184
Ti^{IV}	594	435	373	306
Nb^{V}	569	410	343	285
Mn^{II}	465	402	339	276
Ni^{II}	469	377	318	260
Ag^{I}	356	327	306	255
Hg^{II}	276	230	193	155

Thus, fluorides are often formed more readily than the other halides and there is also a gradation of chemical reactivity through the series, chlorine, bromine and iodine. This applies particularly to high oxidation states, for example the ready formation of SF_6 and SF_4 and the instability of SCl_4 and formation only of S_2Br_2 and no corresponding iodides. Quantitatively, these differences can be related to the energies involved in vaporization and atomization of sulfur, and the dissociation of the halogen molecules and formation of sulfur–halogen bonds.

The increase in size for the halogen atoms, from fluorine through to iodine, leads to a general decrease in coordination number as illustrated above for sulfur, although it is difficult to distinguish the size effect from the effect of change in oxidizing power of the halogen atom.

18.3 BOND FORMATION

Bonds formed by the halogens to other atoms have been shown above to vary in strength, increasing from iodine through to fluorine. This can be correlated with the decreasing size of the halogen atom, allowing increasingly efficient overlap of the half-filled p orbital with an appropriate orbital on the other atom. Although this σ bond formation accounds for the attainment of the formal s^2p^6 configuration for the bonded halide, there has been considerable speculation on the possibility of π bond formation involving the halogens.

For fluorine, the only π interaction possible involves the filled p orbitals on fluorine overlapping with suitable empty orbitals on the combining atom. The other halogens can π bond in this way but there is also the possibility of π bonding in the opposite sense, with filled orbitals on the central atom overlapping with empty d orbitals on the halogen atom.

Although π bonding has been assumed to occur, its importance relative to the σ-bonding interaction is very difficult to assess. Variations in the values of spectroscopically defined parameters have been interpreted to give some measure of the effect. A recent review[10] of calculations involving the angular overlap method has shown that good agreement between calculated and observed spectra, for transition metal complexes including halogen ligands in the coordination arrangement, can be obtained with considerable contributions from halogen–metal π bonding. Table 3 contains parameters derived for some chromium(III) complexes and shows a decrease in both σ and π parameter values as the halogen ligand changes from fluorine through to iodine, but relatively little change in the parameters for fluorine when the other ligands are changed.

Table 3 Angular Overlap Parameters (cm^{-1}) for some Chromium(III) Complexes

Complex	e_σ	e_π	Ref.
$[Cr(en)_2F_2]^+$	8033	2000	11
$[Cr(en)_2Cl_2]^+$	5857	1040	12
$[Cr(en)_2Br_2]^+$	5120	750	12
$[Cr(en)_2I_2]^+$	4292	594	13
$[Cr(pd)_2F_2]^+$	9093	2450	13
$[Cr(pd)_2Cl_2]^+$	5832	1052	13
$[Cr(pd)_2Br_2]^+$	5417	903	13
$[Cr(pd)_2I_2]^+$	3986	739	13
$[Cr(NH_3)_4F_2]^+$	7453	1753	14
$[Cr(H_2NPr)_4F_2]^+$	7617	1630	14

The participation in bonding of d orbitals on the central, non-transition element atom of a complex has also been the subject of some speculation. These orbitals are considered to be rather diffuse and of high energy, for example in sulfur, and a net positive charge on the central atom would contract them.[15] Thus, fluorine in particular would favour such participation, with a greater build-up of charge giving a larger contraction and hence a more suitable match, both energetically and spatially, for overlap of sulfur d orbitals with fluorine orbitals. However, it has been suggested from the results of calculations using *ab initio* generalized valence bond wave functions,[16] that the stability of SF_6 is mainly due to the incorporation of charge-transfer configurations, with $3d$ functions on sulfur being less important.

18.4 BRIDGE BOND FORMATION

The linking of two metal atoms by bridging halogen atoms occurs in compounds containing only halogen ligands and also in those with mixed ligand systems. A monodentate halogen ligand will have free electron pairs, which can be used for dative bonding with an acceptor orbital on another metal atom. Bridge formation will be favoured if the terminal sites on the metal atom are occupied by acceptor ligands.

Examples are known with single, double, triple and quadruple halogen bridges between two centres. For fluorine, single bridges are most common, with double bridges rare and only a single example of a triple bridge. This can be related to the smaller size of the fluorine atom, compared with those of the other halogens, where multiple bridges would bring the two metal atoms very close together.

For chlorine, bromine and iodine the double bridge system is common, with fewer triple bridges, single bridges comparatively rare, and a single example of a quadruple chlorine bridge.

There are also a few cases where the halogen atom links more than two other atoms and has a coordination number of three or more. These cases are often associated with a cluster of central atoms.

Halogen bridge systems can be symmetric or asymmetric. Asymmetry in the bridges can be associated with three effects. Firstly, the link can be between dissimilar elements; secondly, different ligands *trans* to the bridge system can affect the bond to the bridge atom in different ways; and thirdly, the inherent geometry of the coordination arrangement of the metal involved can produce this result.

Some representative examples of the effect of dissimilar elements are shown in Table 4. The asymmetry in the bond leads to a closer association of the bridge atom with one of its neighbours, and the compounds are usually formally represented as ions. Thus the compound[17] formed between niobium and antimony pentafluorides has single fluorine bridges between the metal atoms, with the fluorine atom much closer to antimony than to niobium, and can therefore be formally represented $[NbF_4]^+[SbF_6]^-$. The closer association of the bridging atom with one centre depends on the relative acceptor strengths of the elements involved, with Sb^V in a fluoride environment being a particularly strong acceptor.

Table 4 Asymmetry in Halogen Bridges Between Different Elements

	M—X (Å) (M in parentheses)	'Ionic' formulation	Ref.
$NbF_5 \cdot SbF_5$	2.17 (Nb) 1.95 (Sb)	$[NbF_4]^+ [SbF_6]^-$	17
$BrF_3 \cdot SbF_5$	2.29 (Br) 1.91 (Sb)	$[BrF_2]^+ [SbF_6]^-$	18
$ICl_3 \cdot SbCl_5$	2.93 (I) 2.43 (Sb)	$[ICl_2]^+ [SbCl_6]^-$	19
$SeCl_4 \cdot AlCl_3$	3.04 (Se) 2.14 (Al)	$[SeCl_3]^+ [AlCl_4]^-$	20
$(Cp_2 TiCl)_2 ZnCl_2 \cdot 2C_6H_6$	2.58 (Ti) 2.25 (Zn)	Not applicable	21
$(Cp_2 TiCl)_2 MnCl_2 \cdot 2OC_4H_8$	2.575 (Ti) 2.532 (Mn)	Not applicable	22

The two examples in Table 4 of chlorine bridges of this type for main-group elements are highly asymmetric,[19,20] with the chlorine atoms strongly associated with the anions. However, the two examples involving transition elements[21,22] are more symmetric, and an ionic formulation is inappropriate.

The best examples of the effect of ligands *trans* to the bridge system are provided by oxide halides, with a terminal oxygen atom weakening, and lengthening, the bond *trans* to it in the coordination arrangement. Thus, in the infinite zigzag chain arrangement[23] found in molybdenum tetrafluoride oxide, where octahedra are *cis*-linked through single bridges, the Mo—F(bridge) distances are 2.29 Å *trans* to oxygen and 1.94 Å *trans* to fluorine. The symmetric bridge in molybdenum pentafluoride[24] has an Mo—F distance of 2.06 Å, which is close to the average (2.11 Å) of the two asymmetric distances, and illustrates the 'redistribution' of the bonding under the influence of the *trans* oxygen ligand.

The same effect can be seen for molybdenum trichloride oxide,[25] where bridge Mo—Cl distances of 2.80 and 2.36 Å, *trans* to oxygen and chlorine respectively, average 2.58 Å, and the symmetric bridge distance in molybdenum pentachloride[26] is 2.53 Å.

Table 5 lists parameters for a selection of halogen-bridged species with this type of asymmetry. The effect of multiply bonded oxygen is larger than the other ligands listed, and the comparison of oxide as part of the ligand, in the pentane-2,4-dionate complex,[27] where the asymmetric Mo—Cl distances only differ by 0.091 Å, with the oxide trichloride, where the difference is 0.44 Å, is particularly striking.

Asymmetry of halogen bridges arising from the geometry of the coordination arrangement of the metal is illustrated by compounds containing five-coordinated metals, with an idealized arrange-

Table 5 Asymmetry in Halogen Bridges Between the Same Elements

	Bridge system	*M—X* (Å)	*Ligand* trans *to X*	*Ref.*
MoF$_4$O	Mo—F—Mo	1.94	F	23
		2.29	O	
MoCl$_3$O	Mo—Cl—Mo	2.36	Cl	25
		2.80	O	
MoOCl$_2$(hfacac)	Mo—Cl—Mo	2.369	Cl	27
		2.460	O in hfacac	
Rh$_2$Cl$_2$(cod)[P(OPh)$_3$]$_2$	Rh—Cl—Rh	2.387	C=C in cod	28
		2.404	P(OPh)$_3$	
[RhCl$_3$(PBu$^n_3$)$_2$]$_2$	Rh—Cl—Rh	2.394	Cl	29
		2.523	PBu$^n_3$	
[PdCl$_2$(PhCH=CH$_2$)]$_2$	Pd—Cl—Pd	2.32	Cl	30
		2.41	C=C	
[Rh$_2$I$_6$(COMe)$_2$(CO)$_2$]$^{2-}$	Rh—I—Rh	2.679	I	31
		3.001	C in COMe	
[PdCl$_2$(cdpdm)]$_2$ [a]	Pd—Cl—Pd	2.33	Cl	32
		2.41	P in cdpdm	

[a] cdpdm = 2-chloro-3-diphenylphosphinodimethylmaleate.

ment based on either a trigonal bipyramid or a tetragonal pyramid (Figure 1). In both cases there is a distinction between bond distances to the axial and equatorial ligands, with the axial distances in the bridge being longer. Some examples of such bridges are given in Table 6, derived mainly from complexes containing divalent copper.

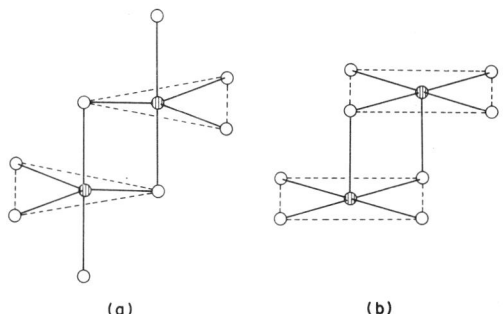

(a)   (b)

Figure 1 Asymmetric bridges arising from the coordination geometry of the central atom: (a) trigonal bipyramidal arrangement, (b) tetragonal pyramidal arrangement

Table 6 Asymmetry in Halogen Bridges Between the Same Elements Resulting from the Coordination Geometry

	Geometry[a]	*M—X* (Å)	Position	Ref.
[Cu$_2$Cl$_8$]$^{4-}$	tb	2.325	Equatorial	33
		2.703	Axial	
[CuCl$_2$(H$_2$DMG)$_2$]$_2$	tp	2.246	Equatorial	34
		2.715	Axial	
[CuBr$_2$(H$_2$DMG)$_2$]$_2$	tp	2.379	Equatorial	35
		2.884	Axial	
[CuBr$_2$(meox)$_2$]$_2$ [b]	tp	2.556	Equatorial	36
		2.710	Axial	
[CuBr$_2$(meapy)$_2$]$_2$ [c]	tb	2.468	Equatorial	37
		2.802	Axial	
[Co$_2$F$_2$(dmpz)$_6$]$^{2+}$ [d]	tb	1.924	Equatorial	38
		2.146	Axial	
[Cu$_2$F$_2$(tmpz)$_6$]$^{2+}$ [e]	tp	1.911	Equatorial	39
		2.183	Axial	
[CdCl$_2$(PCy$_3$)]$_4$ [f]	tb	2.465	Equatorial	40
		2.832	Axial	
[NiCl$_2$(dmphen)]$_2$ [g]	tb	2.378	Equatorial	41
		2.414	Axial	
[NiBr$_2$(dmphen)]$_2$	tb	2.468	Equatorial	41
		2.649	Axial	

[a] tb = trigonal bipyramid, tp = tetragonal pyramid. [b] meox = 4-methyloxazole. [c] meapy = 2-(2-methylaminoethyl)pyridine. [d] dmpz = 3,5-dimethylpyrazole. [e] tmpz = 3,4,5-trimethylpyrazole. [f] Cy = cyclohexyl. [g] dmphen = 2,9-dimethyl-1,10-phenanthroline.

The difference in the bridge bond lengths shows no correlation with geometry and varies from a rather small asymmetry, for example 0.154 Å in the methyloxazole bromide, to a very large extension, which must represent an extremely weak interaction, for example 0.505 Å in the dimethylglyoxime bromide. For the two examples[38,39] with fluorine bridges the asymmetry is more marked for copper than for cobalt. The asymmetry of the bridging in the two nickel complexes, and the increase on changing from chlorine to bromine, have been attributed to the steric effects of the organic ligand.[41]

There are also some examples of asymmetry in bridges which are not as large as most of those in the previous cases, nor as simply rationalized. Thus the $[Ni_2Cl_8]^{4-}$ anion[42] has a structure based on the linking of two tetragonal pyramids through an equatorial edge, which should lead to a symmetric bridge, but in which the Ni—Cl(bridge) distances are 2.365 and 2.449 Å. $V_2Cl_6(NCl)_2$ has a similar arrangement,[43] with the chloroimide ligands in the axial position and V—Cl(bridge) distances of 2.383 and 2.463 Å. One structure determination[21] of the $[Zn_2Cl_6]^{2-}$ anion, which has an arrangement based on the linking of two tetrahedra through an edge and therefore should be symmetrically bridged, gives Zn—Cl(bridge) distances of 2.319 and 2.455 Å, but another [44] has found symmetric bridges with a Zn—Cl(bridge) distance of 2.355 Å.

Although attempts have been made to explain the asymmetry, based on Cl—Cl interactions and M—M repulsions, it appears that other factors must be involved.

18.4.1 Single Bridges

A single bridge M—X—M is usually classified as linear or bent and examples are known with bridging angles from 180° to around 115°. Since this is the most common bridging arrangement for fluorine, and there are sets of structurally characterized compounds extending across the transition series, attempts have been made to explain the variation in bridge angle with position in the series in terms of bonding.

Most of these attempts have been based on possible π bonding between the bridging fluorine atom and its neighbouring metals.[45,46] A linear system has been correlated with the presence of π bonds, whereas for the angular arrangement, localized pairs of electrons on the bridge atom, with an approximately tetrahedral arrangement of two bonds and the two electron pairs, were assumed. This model correlates with transition metal pentafluoride structures particularly, since the structures are known[47] for all compounds prepared, except for that of gold.

These pentafluoride structures fall into three types. The first type, found for the pentafluorides of niobium, tantalum, molybdenum and tungsten, consists of tetramers with bridge angles close to 180°. The second type, found for the pentafluorides of vanadium, chromium, technetium and rhenium, consists of endless, zigzag chains, with angles close to 150°. The remaining pentafluorides, of ruthenium, osmium, rhodium, iridium and platinum (and also probably of gold), form tetramers with bridge angles close to 135°.

Thus the approximately linear bridge is associated with a d^0 or d^1 configuration and the bent bridge of 135° with a d^3, d^4, d^5 or d^6 configuration. The intermediate angle of 150° is associated with the d^2 configuration except for vanadium and chromium in the first transition series, where the smaller size of the transition metal atom is invoked.

The ability to form π bonds depends on the back-donation from filled fluorine p orbitals to empty orbitals on the metal, and this would obviously decrease as the number of d electrons increases.

It is interesting that the tetramer structures, in the solid state, correspond to close-packed arrangements of the fluorine atoms, cubic close-packed for the niobium pentafluoride type and hexagonal close-packed for the rhodium pentafluoride type. It has been convincingly argued[48] that the close-packed arrays are a consequence of the packing of the tetramer units, rather than the bridge angles being a consequence of the overall close-packing. This is supported by the almost linear bridge found in the $[Nb_2F_{11}]^-$ anion, in the structure[49] of the adduct $SeF_4 \cdot 2NbF_5$, and the bridge angle Ru—F—Ru of 145° in the structure[50] of $Ru_4(CO)_{12}F_8$. In both cases there is no overall close-packing of the light atoms involved.

The correlation of bridge angle with d electron configuration is not easy to reconcile with the structure of SbF_5.[51] This also consists of tetramers, but two bridge angles are close to 180° and two close to 135°, and the overall solid state structure can be described as a packing of fluorine atoms in alternating pairs of cubic and hexagonal close-packed layers.

Single bridges for chlorine, bromine and iodine are much less common than those for fluorine, with the majority of systems involving double or triple bridges. Thus, unlike the tetrameric transition metal pentafluorides discussed above, the solid pentachlorides are dimeric with a double bridge arrangement.

Linear single bridges have been found in the structures of μ-iodo-bis[tris(2-diphenylarsino-ethyl)aminenickel(I)] tetraphenylborate[52] and μ-iodo-bis[tetrakis(phenyl isocyanide)iodocobalt(II)] iodide.[53] In both cases the metal to bridging iodide distance is long (Ni—I 2.994 Å, Co—I 2.891 Å) and this is rationalized in terms of the steric requirements of the phenyl rings on the ligands.

In contrast the bridge found in the heptachlorodialuminate(III) anion has an angle close to 115°, with two examples with an eclipsed configuration and angles of 112.1°[54] and 115.6°,[55] and two examples with a staggered configuration and angles of 116.1°[54] and 110.8°.[56] Although from steric considerations the staggered configuration would be expected to have the smaller bridge angle, this is seen not to be the case.

The difficulty of attempting to correlate angles at the bridging halogen atom is illustrated by the structures of a series of diruthenium tetracarboxylate compounds.[57] In $Ru_2(O_2CMe)_4Cl \cdot 2H_2O$ and $Ru_2(O_2CEt)_4Cl$ the carboxylate-bridged diruthenium units are linked by linear chlorine bridges, whereas in $Ru_2(O_2CPr)_4Cl$ the chlorine bridges are angular with an Ru—Cl—Ru angle of 125.4°.

Thus, halogen bridges appear to be flexible rather than rigid, easily responding to the geometric demands imposed by the rest of the system.

18.4.2 Double Bridges

The double bridge system MX_2M can be characterized by the M—X—M angle. There has been considerable interest in the formation of metal–metal bonds in general bridge systems of this type and a smaller than theoretical bridge angle, with consequent closer approach of the metal atoms to each other, is often a good indication of such interactions for halides.

For fluorine such double bridges are rare. The solid state structures[58] of VF_3O and CrF_2O_2 (and presumably also VF_4, which has an almost isodimensional crystallographic unit cell) show pairs of metal atoms bridged by two fluorine atoms, with these units linked together through single fluorine bridges, and an overall octahedral coordination for the metal atom. The double bridges are symmetric in VF_3O and asymmetric in CrF_2O_2, due to the effect of the terminal oxygen atom *trans* to the bridge fluorine atom in the latter case. The bridge angles are 105.5° and 107° for the vanadium and chromium compounds respectively with metal–metal distances of 3.09 Å for vanadium and 3.18 Å for chromium. The increase in bridge angle, from the theoretical 90° for octahedra sharing an edge, would indicate a repulsive interaction, if any, in line with the d^0 configurations of the metal atoms. Unfortunately the detailed structure of VF_4 has not been established, since the magnetic properties of the compound are consistent with interaction between the metal atoms,[59] and a V—V distance and V—F—V angle would have given a useful comparison with the oxide trifluoride.

A similar system, with an isolated unit, has been reported[60] for the $[Mo_2F_6O_4]^{2-}$ anion, where the two molybdenum atoms are symmetrically bridged by two fluorine atoms with an Mo—F—Mo angle of 110.7° and an Mo—Mo distance of 3.535 Å. Again there is an increase in bridge angle for a metal with d^0 configuration.

The other examples of double fluorine bridges have been mentioned previously as examples containing asymmetric bridges and are found in some binuclear complexes of first transition series metals, formed from thermal reactions of mononuclear tetrafluoroborate complexes.[38] The compounds $M_2F_2(mpz)_6(BF_4)_2$ (M = Co[38] or Cu[39] and mpz = methyl-substituted pyrazole) have been characterized crystallographically. For both metals the double bridge system is asymmetric, with Co—F distances of 1.924 and 2.146 Å and Cu—F distances of 1.904 and 2.258 Å, which can be correlated with the equatorial and axial bonds respectively in trigonal bipyramidal (for Co) and square pyramidal (for Cu) coordination arrangements. The M—F—M angles of 98.8° and 97.2° for Co and Cu respectively show an increase over the theoretical 90° for linking of the coordination polyhedra and the metal–metal distances are 3.092 Å for Co and 3.131 Å for Cu. Magnetic measurements have been interpreted in terms of small interactions between the metal atoms, in agreement with the structural results. There is also one example of such a copper compound $Cu_2F_2(mppz)_4(BF_4)_2$ (mppz = 3-methyl-5-phenylpyrazole) with a symmetric double bridge,[61] with average Cu—F distance 1.922 Å, Cu—F—Cu angle 98.9° and Cu—Cu distance 2.922 Å. The coordination of the copper atom is distorted square planar, with long contacts to fluorine atoms of the $[BF_4]^-$ anions completing a six-coordinate arrangement. This compound shows a strong magnetic exchange coupling, which must result from the symmetric bridge, as the Cu—Cu distance is little shorter than before.

For the other halogens the double bridge system is very common and no attempt will be made to deal systematically with all examples. Hoffmann and his co-workers have made detailed theoretical studies of double bridged systems in general, with halogens as only one possible type of bridging ligand. They have analyzed the systems where the coordination number of the metals was either four[62] or six.[63] For four-coordination,[62] the combination of two tetrahedra sharing an edge gives a theoretical bridge angle of $70.5°$, and of two squares sharing an edge an angle of $90°$. The combination of squares is complicated by the possibility of a resulting hinged system, with a dihedral angle, as well as the planar arrangement. They considered three factors influencing the structures of the dimer molecules. These were the geometry of the monomer from which the dimer could be considered to be derived, indirect coupling through bridging group orbitals and direct metal–metal bonding.

For double halogen bridges the results were of particular significance for some tetrahedral complexes involving d^{10} and d^9 electron configurations for metals. Thus, as illustrated in Table 7, although the $[Zn_2Cl_6]^{2-}$ anion[21,44] and molecular Ga_2Cl_6[64] show the expected increase in bridge angle from the theoretical value, the copper(I) iodide complex,[65] also with d^{10} electron configuration, shows a definite decrease. That this was not due to metal–metal interaction was verified by considering a related monomer structure, where the X—M—X angle was equivalently greater than the theoretical tetrahedral angle.

Table 7 Structural Parameters for some Doubly Bridged Species with Four-coordinate Metals

(a) Tetrahedral coordination

Compound	M—X—M (°)	M—M (Å)	Ref.
$[Zn_2Cl_6]^{2-}$	87.5	3.30	21, 44
Ga_2Cl_6	86	3.12	64
$Cu_2I_2(nas)_2$ [a]	63.5	2.73	65
$Co_2Cl_2(NO)_4$	92	3.20	66
$Fe_2I_2(NO)_4$	75	3.05	67

[a] nas = (O-dimethylaminophenyl)dimethylarsine-As,N

(b) Square planar coordination

Compound	Dihedral angle (°)	M—M (Å)	Ref.
Pd_2Cl_2(2-methylallyl)$_2$	180	3.48	68
Pd_2Cl_2(1,3-dimethylallyl)$_2$	148	3.31	69
$Rh_2Cl_2(cod)_2$	180	3.50	70
Rh_2Cl_2(4-methylpenta-1,3-diene)	115.8	3.09	71

The geometry of the related monomer was also considered for $Fe_2I_2(NO)_4$, with a d^9 configuration for iron,[67] where the bridge angle has decreased as compared with the equivalent cobalt chloride complex[66] with a d^{10} configuration (Table 7). Although metal–metal bonding was postulated originally,[67] the calculated geometry of the equivalent monomer was very close to that of the dimer. It was not clear whether metal–metal bonding or the distortion associated with the d^9 configuration was responsible for the lower angle.[62]

Again, as with single bridges, the double halogen bridge system is seen to be flexible and to respond to the demands of the rest of the system. This was also evident in the inability to predict whether a dimer formed from square planar complexes containing metals with a d^8 electron configuration would be hinged or planar. Representative compounds are given in Table 7. The energy difference for the two arrangements was shown to be small, and steric and crystal packing forces are therefore more important. Such d^8 complexes with halogen bridges were predicted to have planar rather than tetrahedral geometry because of the π-donor properties of the halogen.

The double bridge system for six-coordinate metals is more common than for those with four-coordination. In their theoretical study, Hoffmann and his co-workers[63] were concerned with M_2L_{10} complexes in general, and particularly with the factors influencing formation of bridged or non-bridged species. Halogen-bridged systems form part of this group, with the halogen being a donor bridge ligand, and it was predicted[63] that with donor terminal ligands there should be no metal–metal bonding for d^1–d^1 to d^5–d^5 systems. This is illustrated in Table 8. The three chlorides[72-74] have very similar bridge parameters, all with bridge angle greater than $90°$ as does the d^4 complex

$Re_2Cl_6(Ph_2PCH_2CH_2PPh_2)_2$.[75] The d^6 rhenium complex $Re_2Br_2(CO)_6(P_2Ph_4)$,[76] included for comparison, also shows similar bridge parameters.

Table 8 Structural Parameters for some Doubly Bridged Species with Six-coordinate Metals

	Configuration	M—X—M (°)	M—M (Å)	Ref.
$[Ti_2Cl_{10}]^{2-}$	d^0	101.2	3.855	72
Nb_2Cl_{10}	d^0	101.3	3.95	73
Mo_2Cl_{10}	d^1	98.6	3.84	26
Re_2Cl_{10}	d^2	98.1	3.74	74
$Re_2Cl_6(Ph_2PCH_2CH_2PPh_2)_2$	d^4	99.5	3.809	75
$Re_2Br_2(CO)_6(P_2Ph_4)$	d^6	94.5	3.890	76
$Ta_2Cl_6(PMe_3)_4$	d^2	67.4	2.721	77
$W_2Cl_6(py)_4$	d^3	69.8	2.737	78
$Mo_2Cl_6(EtSCH_2CH_2SEt)_2$	d^3	69.8	2.735	79

However, metal–metal bonding has been shown to be present in some complexes to which the prediction should apply. The examples for tungsten,[78] with pyridine ligands, and tantalum,[77] with trimethylphosphine ligands, given in Table 8, clearly show a much decreased bridge angle. It has been suggested that the decrease in M—M distance and in M—X—M angle in changing from the tungsten (d^3–d^3) to tantalum (d^2–d^2) complex corresponds to a change from a formal single to a formal double metal–metal bond.[77] The molybdenum complex,[79] with 3,6-dithiaoctane as ligand, has an almost identical bridge system and has also been considered to show metal–metal interaction.

18.4.3 Triple Bridges

The triple bridge system MX_3M can also be characterized by the M—X—M angle. As for the double bridge systems, there has been much interest in metal–metal bonding in this arrangement also, with bridge angles less than the theoretical value again often being the indication of such interaction.

There is only one report as yet of fluorine forming triple bridges, other than in structures considered as more ionic. In the cationic complex $[\{(PMePh_2)_3H_2Mo\}_2(\mu\text{-}F)_3]^+$, crystallized as the fluoroborate,[80] the molybdenum atoms have an eight-coordinate arrangement, with dodecahedral geometry, and the two dodecahedra joined through a triangular face. With Mo—F—Mo angles averaging 97.7° the Mo—Mo distance of 3.256 Å is not unusually short and no metal–metal bond is formed.

For the other halogens triple bridge formation is not uncommon and usually results from joining two octahedra through a face. This geometrical arrangement has been studied by several workers. Cotton and Ucko[81] discussed the structural results for species $[M_2X_9]$ (X = halogen) and pointed out the interrelationships of the geometric parameters in the system and the importance of the bridge angle in any discussion of metal–metal interactions. Summerville and Hoffmann,[82] using qualitative molecular orbital arguments and extended Hückel calculations in a similar manner to their study of the doubly bridged dimer molecules described above,[62] have looked in detail at the general, triple-bridged system. Halogen bridges form part of this general set and are characterized by their π-donor properties.

The $[M_2X_9]$ unit has been particularly studied, as there are examples with many different d electron configurations of the metals, which allow useful comparisons to be made (Table 9). The series of ions $[M_2Cl_9]^{3-}$, for M = Cr,[83] Mo[84] and W,[85] shows a structural change involving a decrease in the M—Cl—M angle down the series as shown in Table 9, which is paralleled by the bromides of Cr and Mo.[84] This has been correlated with an increasing contribution from metal–metal bonding. Thus the value for Cr—Cl—Cr of 76.5° is greater than the theoretical value of 70.5°, suggesting a repulsive interaction, if any, whereas the value for W—Cl—W of 58.1° suggests a considerable attractive interaction, by comparison. The value of 64.6° for Mo—Cl—Mo is intermediate, but less than the theoretical value and closer to the value for the tungsten rather than chromium species. The greater interaction for tungsten has been attributed to the increase in size, and, whereas a single bond has been suggested for the Mo—Mo interaction, a triple bond has been assigned to that for tungsten. The $[W_2Br_9]^{2-}$ anion,[86] with one fewer electron, has rather larger values for M—M and M—X—M, as expected for the lower bond order.

Substitution of different ligands in the terminal positions of the confacial bioctahedral arrangement leads to little change in the bridge system, apart from asymmetry if different terminal ligands are involved. Some examples are given in Table 9.

Table 9 Structural Parameters for some Triply Bridged Species

	Average M—X—M (°)	M—M (Å)	Ref.
[Ti₂Cl₉]⁻	86.7	3.43	72
[Cr₂Cl₉]³⁻	76.5	3.12	83
[Mo₂Cl₉]³⁻	64.6	2.655	84
[W₂Cl₉]³⁻	58.1	2.41	85
[Cr₂Br₉]³⁻	80.1	3.317	84
[Mo₂Br₉]³⁻	64.9	2.816	84
[W₂Br₉]²⁻	60.0	2.601	86
[Rh₂Cl₉]³⁻	81.3	3.121	81
[Sb₂Br₉]³⁻	79.6	3.89	87
Ru₂Cl₃(CO)₅(SnCl₃)	80.9	3.157	88
[Ru₂Cl₃(PEtPh)₆]⁺	87.9	3.443	89
Rh₂Cl₆(PBuⁿ₃)₃	81.6	3.187	90
Mo₂Cl₆(Me₂S)₃	59.7	2.462	91
[Mo₂Cl₃(CO)₄(η³-C₃H₅)₂]⁻	86.5	3.53	92

18.4.4. Quadruple Bridges

Although confacial linking of octahedra leads to triply bridging systems, as described in the previous section, the six-coordinate arrangement cannot be involved in systems with two metal atoms linked by four bridging halogen atoms. However, higher coordination numbers for the metal can give such an arrangement, especially with a delocalized ligand which can be assumed to fill a number of positions opposite to the square bridge system.

Only one example appears to have been characterized structurally as yet. The structure of the monoisopropylcyclopentadienyl molybdenum dichloride dimer has been determined crystallographically[93] and shows an MoCl₄Mo bridge system with a square of chlorine atoms, with average Mo—Cl—Mo angle of 63.25° and Mo—Mo distance of 2.607 Å. This short distance suggests a direct Mo—Mo interaction. The cyclopentadienyl ring is symmetrically placed on the opposite side of the molybdenum atom to the square plane and can be assumed to fill three coordination positions in a seven-coordinate arrangement.

This structure can be usefully compared with that of the compound Ta₂Cl₆(PMe₃)₄H₂,[94] which contains a quadruple bridge system of two adjacent chlorine and two adjacent hydrogen atoms, assumed to be in an approximately square arrangement. Although the hydrogen atom positions were not revealed by the crystallographic analysis, the positions were inferred from the geometry of the system. The Ta—Cl—Ta angle of 61.8° and Ta—Ta distance of 2.621 Å are very similar to those in the molybdenum complex. The arrangement of the phosphorus atoms gives an overall, approximately square antiprismatic geometry around the tantalum atom.

18.5 HIGHER COORDINATION NUMBERS

In the previous sections halogens acting as bridging ligands have a coordination number of two. Examples with higher coordination numbers are less frequently encountered in compounds regarded as coordination complexes, and are usually associated with 'clusters' of atoms of the other element.

For a coordination number of three for the halogen atom, the simplest geometrical arrangement is that of a planar triangle of atoms of the other element, surrounding the halogen symmetrically. This has been observed for fluorine in the anion of NaSb₃F₁₀, where three [SbF₃] units are symmetrically linked by a three-coordinate fluorine atom, with Sb—F(bridge) bonds considerably longer than the terminal Sb—F distances, to give a planar [Sb₃F] unit.[95] These units are then linked into sheets through even longer Sb—F bonds. A similar arrangement has recently been found[96] in the ion [F(XeOF₄)₃]⁻. The [Xe₃F] unit is not planar, with the atom displaced 0.49 Å from the Xe₃ plane, giving Xe—F—Xe angles of 116.5°. In both these compounds the three-coordinate fluorine appears to have considerable ionic character.

The other arrangement, based on a triangle of metal atoms, consists of two halogen atoms on either side of the triangle, giving a mixed cluster of atoms with overall trigonal bipyramidal geometry. This has been found for complexes of copper(I)[97,98] and silver(I)[99] halides with bidentate chelate phosphine ligands. Examples are given in Table 10 together with a related arrangement,[100] where a triangle of nickel atoms, linked by singly bridging chlorine atoms, has chlorine and hydroxide ligands in the capping positions. The coordination of the halogen atom is pyramidal, with average M—X—M angles between 60° and 80°, although the coordination is not always regular.

Table 10 Species Containing Three-coordinate Halogen Atoms

	Geometry	Average M—X—M (°)	Average M—X (Å)	Ref.
$[Sb_3F_{10}]^-$	Planar triangle	120	2.38	95
$[F(XeOF_4)_3]^-$	Non-planar triangle	116.5	2.62	96
$[Cu_3Cl_2(dppm)_3]^+$	Trigonal bipyramid	78.9	2.53	97
$[Cu_3I_3(dppm)_2]$	Trigonal bipyramid	61.6	2.80	98
$[Ag_3Br_2(dppa)_3]^{+\ a}$	Trigonal bipyramid	71.4	2.83	99
$[Ni_3Cl_4(OH)(tmen)_3]^{+\ b}$	Trigonal bipyramid	73.1	2.55	100
$[CoF(etim)_3]_4^{4+\ c}$	'Cubane'	100.5	2.14	102
$[TiCl(cot)]_4$	'Cubane'	99.6	2.58	103
$[Ph_3PAgCl]_4$	'Cubane'	86.5	2.65	105
$[Et_3AsCuI]_4$	'Cubane'	62.6	2.68	105
$[Ph_3PAgI]_4$	Chair or step	78.8	2.88	106
$[Cu_2Cl_2dppm]_2$	Chair or step	77.5	2.47	107
$[Cu_2I_2dppm]_2$	Chair or step	70.4	2.75	108
$[TeCl_4]_4$	'Cubane'	94.8	2.929	109
$[Te_3Cl_{13}]^-$	Part 'cubane'	91.1	2.95	112
$[TeI_4]_4$	Slipped cube	93.0	3.232	110
$[BiBr_4]_4^{4-}$	Slipped cube	93.1	3.13	111
$[Mo_6Cl_8]^{4+}$	Octahedral cluster	63.5	2.485	113
$[Mo_5Cl_8]^{3+}$	Part octahedraon	62.6	2.473	114
$[Mo_4I_7]^{2-}$	Part octahedron	54.5	2.826	115
$[XeF_6]_6$	Octahedron	118.8	2.56	116

[a] dppa = methylbis(diphenylphosphino)amine. [b] tmen = *N,N,N',N'*-tetramethylethylenediamine. [c] etim = *N*-ethylimidazole.

An unusual arrangement for three-coordinate iodine is found in the structure[101] of the cation $[Rh_3(\mu\text{-}dpmp)_2(\mu\text{-}I)_2(\mu\text{-}CO)(CO)_2]^+$ (where dpmp = bis[[(diphenylphosphino)methyl]phenyl]phosphine). The iodine atom and three rhodium atoms lie in a plane and form a distorted rectangle with Rh—I—Rh angles of 58.3° and 60.3°. Although one Rh—I distance of 3.149 Å is long compared with the others of 2.801 and 2.865 Å, it does appear to represent a significant interaction.

The other examples of three-coordinate halogens involve coordination to the triangular faces of more complex geometric figures. Thus, in the most common arrangement, four halogen atoms coordinated to the four faces of a tetrahedron of atoms of the other element result in the 'cubane' type, with a distorted cube geometry for the E_4X_4 unit, as shown in Figure 2. The coordination of E is sometimes completed by a single ligand, for approximately tetrahedral four-coordination, or by three ligands, for approximately octahedral six-coordination. Examples for all the halogens are given in Table 10. The compound $Ti_4Cl_4(cot)_4$ illustrates[103] the completion of coordination by the delocalized ring ligand and also distortion from the cube arrangement to give Ti—Ti distances greater than Cl—Cl distances, presumably due to the steric effects of closer association of the metal with the ring ligand. The opposite type of distortion with X—X greater than M—M distances is illustrated by the compounds $[R_3EMX]_4$ (R = phenyl or ethyl, E = P or As, M = Cu or Ag and X = Cl, Br or I) for which variations in the structure with changes of the constituent atoms have been attributed to the different interatomic repulsions.[104,105]

In these compounds, an alternative structure to the 'cubane' cluster is found, for example in $[Ph_3PAgI]_4$.[106] This is the 'chair' or 'step' type of arrangement, illustrated in Figure 2, in which two of the four halogen atoms retain a coordination number of three. This arrangement has also been found in some copper(I) halide complexes with bis(diphenylphosphino)methane.[107,108] The pyramidal coordination of the halogen atom is similar to that in the cubane structure, although often more distorted.

The tetrametric tetrahalides of selenium and tellurium mainly adopt the 'cubane' arrangement,[109] but a similar alternative, with a slipped cube arrangement, as shown in Figure 2, is found for $[TeI_4]_4$[110] and the $[BiBr_4]_4^{4-}$ anion.[111] The ion $[Te_3Cl_{13}]^-$ has a derived structure,[112] now retaining only one three-coordinate chlorine atom (Figure 2).

Although two-coordinate halogen ligands are often found bridging two of the metal atoms in metal clusters, particularly in the M_6 octahedral arrangement, three-coordinate halogen ligands are found only in $[M_6X_8]$ units, with the eight X atoms at the corners and the six M atoms at the face centres of a cube. The idealized arrangement has M—X—M angles of 60°, compared with 90° in

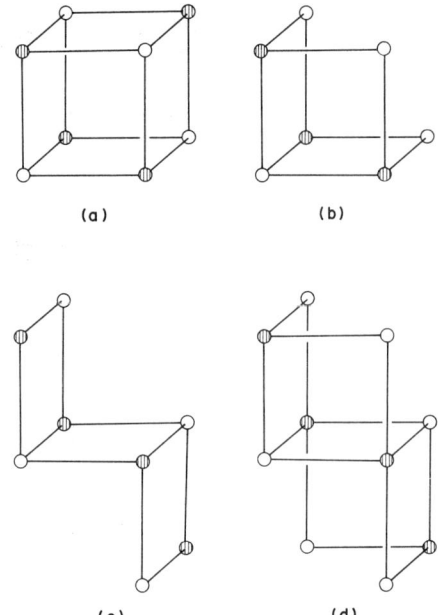

Figure 2 The idealized 'cubane' and related arrangements: (a) 'cubane', (b) part 'cubane', (c) 'chair' or 'step', (d) slipped cube

the idealized 'cubane' arrangement. The $[Mo_6Cl_8]$ unit[113] has angles close to this, as does the $[Mo_5Cl_8]$ unit,[114] which is derived by the loss of one molybdenum atom (Figure 3).

The $[Mo_4I_7]$ unit[115] can also be related to the cube arrangement, with two adjacent molybdenum atoms removed, and the two related chlorine atoms at corners of the cube replaced by one at the centre of the edge linking these corners as shown in Figure 3. In this unit the two three-coordinate iodine atoms have much smaller Mo—I—Mo angles, presumably due to the larger size of the iodine atom.

The hexamer in solid xenon hexafluoride[116] also contains three-coordinate fluorine ligands, but the Xe_6 unit is very different from the octahedral clusters described above. The six xenon atoms, in $[XeF_5]$ units, are weakly linked by six three-coordinate fluorine ligands, with Xe—F(bridge) distances of 2.56 Å and Xe—F—Xe angles of 118.8°. The bridging fluorine atoms lie above six faces of the Xe_6 octahedron, with two opposite faces uncapped. The Xe—F distance is similar to that in the $[F(XeOF_4)_3]^-$ anion[96] and the ligand again appears to have considerable ionic character.

A coordination number of four for a halogen ligand is rare, but has been observed in a complex of silver(I) bromide with methylbis(diphenylphosphino)amine.[99] Four silver atoms in a square have two bromine atoms above and below the square, completing an octahedral 'cluster' of atoms. Thus the coordination of the bromine is square pyramidal, with Ag—Br—Ag angles averaging 63.2°. The arrangement is related to the 'chair' cluster described above, with the coordination number of the halogen atom increased from three to four by a movement of the metal atoms to form an almost regular square.

The other example of a quadruply bridging halogen is found in the anion $[Sb_4Cl_{12}O]^{2-}$.[117] $[Sb_2Cl_7]$ and $[Sb_2OCl_4]$ groups are linked by a chlorine ligand with distorted, square pyramidal coordination by the four antimony atoms. The Sb—Cl distances average 3.10 Å and Sb—Cl—Sb angles 79.3°. The stereochemistry may well be determined by crystal packing, since the metal atoms are less rigidly held than in the silver bromide complex described above.

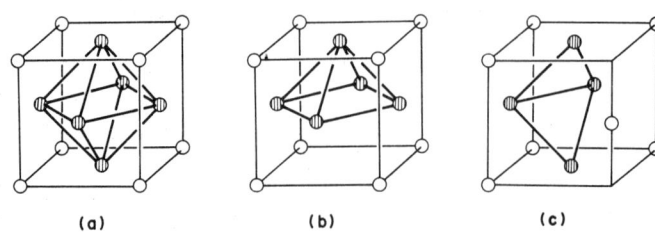

Figure 3 The idealized octahedral cluster and related arrangements: (a) the Mo_6Cl_8 unit, (b) the Mo_5Cl_8 unit, (c) the Mo_4I_7 unit

There is an example of a chlorine ligand with coordination number six, in the compound $Nd_6(OPr_2^i)_{17}Cl$.[118] The chlorine atom lies at the centre of a trigonal prism, formed by the six neodymium atoms which are held together by bridging isopropoxide groups. The Nd—Cl distance of 3.05 Å is compatible with an ionic formulation, and the stereochemistry does not appear significant.

18.6 REFERENCES

1. J. C. Bailar, Jr., H. J. Eméleus, R. Nyholm and A. F. Trotman-Dickenson (eds.), 'Comprehensive Inorganic Chemistry', Pergamon, Oxford, 1973; see especially (a) Fluorine, T. A. O'Donnell, chap. 25, pp. 1009–1106; and (b) Chlorine, Bromine, Iodine and Astatine, A. J. Downs and C. J. Adams, chap. 26, pp. 1107–1594.
2. 'Gmelin Handbuch der Anorganischen Chemie', 8th cdn. and Supplement Volumes, Springer-Verlag, Berlin, especially System-Nos. 5 (Fluorine), 6 (Chlorine), 7 (Bromine) and 8 (Iodine).
3. J. W. Mellor, 'A Comprehensive Treatise on Inorganic and Theoretical Chemistry', Supplement II, Part I, F, Cl, Br, I, Longmans, London, 1956.
4. (a) F. A. Cotton and G. Wilkinson, 'Advanced Inorganic Chemistry', 4th edn., Wiley-Interscience, New York, 1980; (b) K. F. Purcell and J. C. Kotz, 'Inorganic Chemistry', Holt-Saunders, Philadelphia, 1977; (c) N. N. Greenwood and A. Earnshaw, 'Chemistry of the Elements', Pergamon, Oxford, 1984.
5. A. A. Woolf, *Adv. Inorg. Chem. Radiochem.*, 1981, **24**, 1.
6. R. S. Mulliken, *J. Am. Chem. Soc.*, 1955, **77**, 884.
7. G. L. Caldow and C. A. Coulson, *Trans. Faraday Soc.*, 1962, **58**, 633.
8. P. Politzer, *J. Am. Chem. Soc.*, 1969, **91**, 6335.
9. W. L. Jolly and C. J. Eyermann, *Inorg. Chem.*, 1983, **22**, 1566.
10. M. Gerloch and R. G. Woolley, *Prog. Inorg. Chem.*, 1984, **31**, 371.
11. L. Dubicki, M. A. Hitchman and P. Day, *Inorg. Chem.*, 1970, **9**, 188.
12. L. Dubicki and P. Day, *Inorg. Chem.*, 1971, **10**, 2043.
13. T. J. Barton and R. C. Slade, *J. Chem. Soc., Dalton Trans.*, 1975, 650.
14. J. Glerup, J. Josephson, K. Michelsen, E. Pedersen and C. E. Schäffer, *Acta Chem. Scand.*, 1970, **24**, 247.
15. K. A. R. Mitchell, *J. Chem. Soc. (A)*, 1968, 2676.
16. P. J. Hay, *J. Am. Chem. Soc.*, 1977, **99**, 1003.
17. A. J. Edwards, *J. Chem. Soc. (A)*, 1972, 2325.
18. A. J. Edwards and G. R. Jones, *J. Chem. Soc. (A)*, 1969, 1467.
19. C. G. Vonk and E. H. Wiebenga, *Acta Crystallogr.*, 1959, **12**, 859.
20. B. A. Stork-Blaisse and C. Romers, *Acta Crystallogr., Sect. B*, 1971, **27**, 386.
21. D. G. Sekutowski and G. D. Stucky, *Inorg. Chem.*, 1975, **14**, 2192.
22. D. Sekutowski, R. Jungst and G. D. Stucky, *Inorg. Chem.*, 1978, **17**, 1848.
23. A. J. Edwards and B. R. Steventon, *J. Chem. Soc. (A)*, 1968, 2503.
24. A. J. Edwards, R. D. Peacock and R. W. H. Small, *J. Chem. Soc.*, 1962, 4486; A. J. Edwards and K. I. Khallow, unpublished redetermination of the structure, find the same value (2.063 Å) within experimental error.
25. G. Ferguson, M. Mercer and D. W. A. Sharp, *J. Chem. Soc. (A)*, 1969, 2415; M. G. B. Drew and I. B. Tomkins, *J. Chem. Soc. (A)*, 1970, 22.
26. D. E. Sands and A. Zalkin, *Acta Crystallogr.*, 1959, **12**, 723.
27. M. G. B. Drew and K. J. Shanton, *Acta Crystallogr., Sect. B*, 1978, **34**, 276.
28. J. Coetzer and G. Gafner, *Acta Crystallogr., Sect. B*, 1970, **26**, 985.
29. J. A. Muir, M. A. Muir and A. J. Rivera, *Acta Crystallogr., Sect. B*, 1974, **30**, 2062.
30. J. R. Holden and N. C. Baenziger, *J. Am. Chem. Soc.*, 1955, **77**, 4987.
31. G. W. Adamson, J. J. Daly and D. Forster, *J. Organomet. Chem.*, 1974, **71**, C17.
32. D. Fenske, H. Prokscha, P. Stock and H. J. Becker, *Z. Naturforsch., Teil B*, 1980, **35**, 1075.
33. D. J. Hodgson, P. K. Hale and W. E. Hatfield, *Inorg. Chem.*, 1971, **10**, 1061.
34. D. H. Svedung, *Acta Chem. Scand.*, 1969, **23**, 2865.
35. C. L. Raston, B. W. Skelton and A. H. White, *Aust. J. Chem.*, 1980, **33**, 1519.
36. W. E. Marsh, T. L. Bowman, W. E. Hatfield and D. J. Hodgson, *Inorg. Chim. Acta*, 1982, **59**, 19.
37. R. B. Wilson, W. E. Hatfield and D. J. Hodgson, *Inorg. Chem.*, 1976, **15**, 1712.
38. J. Reedijk, J. C. Jansen, H. van Koningsveld and C. G. van Kralingen, *Inorg. Chem.*, 1978, **17**, 1990.
39. F. J. Rietmeijer, R. A. G. De Graaff and J. Reedijk, *Inorg. Chem.*, 1984, **23**, 151.
40. N. A. Bell, T. D. Dee, M. Goldstein and I. W. Nowell, *Inorg. Chim. Acta*, 1982, **65**, L87.
41. R. J. Butcher and E. Sinn, *Inorg. Chem.*, 1977, **16**, 2334.
42. F. K. Ross and G. D. Stucky, *J. Am. Chem. Soc.*, 1970, **92**, 4538.
43. J. Strähle and H. Barnighausen, *Z. Anorg. Allg. Chem.*, 1968, **357**, 325.
44. F. A. Cotton, S. A. Duraj, M. W. Extine, G. E. Lewis, W. J. Roth, C. D. Schmulbach and W. Schwotzer, *J. Chem. Soc., Chem. Commun.*, 1983, 1377.
45. J. H. Canterford and R. Colton, 'Halides of the Second and Third Row Transition Metals', Wiley-Interscience, New York, 1968.
46. O. Glemser, *J. Fluorine Chem.*, 1984, **24**, 319.
47. A. J. Edwards, *Adv. Inorg. Chem. Radiochem.*, 1983, **27**, 83.
48. D. M. Adams, 'Inorganic Solids', Wiley, New York, 1974.
49. A. J. Edwards and G. R. Jones *J. Chem. Soc. (A)*, 1970, 1491.
50. C. J. Marshall, R. D. Peacock, D. R. Russell and I. L. Wilson, *Chem. Commun.*, 1970, 1643.
51. A. J. Edwards and P. Taylor, *Chem. Commun.*, 1971, 1376.
52. L. Sacconi, P. Dapporto and P. Stoppioni, *Inorg. Chem.*, 1977, **16**, 224.

53. D. Baumann, H. Endres, H. J. Keller, B. Nuber and J. Weiss, *Acta Crystallogr., Sect. B*, 1975, **31**, 40.
54. F. Stollmaier and U. Thewalt, *J. Organomet. Chem.*, 1981, **208**, 327.
55. G. Allegra, G. Tettamanti Casagrande, A. Imirzi, L. Porri and G. Vituli, *J. Am. Chem. Soc.*, 1970, **92**, 289.
56. T. W. Couch, D. A. Lokken and J. D. Corbett, *Inorg. Chem.*, 1972, **11**, 357.
57. A. Bino, F. A. Cotton and T. R. Felthouse, *Inorg. Chem.*, 1979, **18**, 2599.
58. A. J. Edwards and P. Taylor, *Chem. Commun.*, 1970, 1474, and unpublished results.
59. A. C. Gossard, F. J. Disalvo, W. E. Falconer, T. M. Rice, J. M. Voorhoeve and H. Yasuoka, *Solid State Commun.*, 1974, **14**, 1207.
60. J. Dirand, L. Ricard and R. Weiss, *Transition Met. Chem. (Weinheim, Ger.)*, 1975, **1**, 2.
61. W. C. Velthuizen, J. G. Haasnoot, A. J. Kinneging, F. J. Rietmeijer and J. Reedijk, *J. Chem. Soc., Chem. Commun.*, 1983, 1366.
62. R. H. Summerville and R. Hoffmann, *J. Am. Chem. Soc.*, 1976, **98**, 7240.
63. S. Shaik, R. Hoffman, C. R. Fisel and R. H. Summerville, *J. Am. Chem. Soc.*, 1980, **102**, 4555.
64. S. C. Wallwork and I. J. Worrall, *J. Chem. Soc.*, 1965, 1816.
65. R. Graziani, G. Bombieri and E. Forsellini, *J. Chem. Soc. (A)*, 1971, 2331.
66. S. Jagner and N.-G. Vannerberg, *Acta Chem. Scand.*, 1967, **21**, 1183.
67. L. F. Dahl, E. Rodulfo de Gil and R. D. Feltham, *J. Am. Chem. Soc.*, 1969, **91**, 1653.
68. A. E. Smith, *Acta Crystallogr.*, 1965, **18**, 331.
69. J. Lukas, J. E. Ramakers-Blom, T. G. Hewitt and J. J. de Boer, *J. Organomet. Chem.*, 1972, **46**, 167.
70. J. A. Ibers and R. G. Snyder, *J. Am. Chem. Soc.*, 1962, **84**, 495.
71. M. G. B. Drew, S. M. Nelson and M. Sloan, *J. Chem. Soc., Dalton Trans.*, 1973, 1484.
72. T. J. Kistenmacher and G. D. Stucky, *Inorg. Chem.*, 1971, **10**, 122.
73. A. Zalkin and D. E. Sands, *Acta Crystallogr.*, 1958, **11**, 615.
74. K. Mucker, G. S. Smith and Q. Johnson, *Acta Crystallogr., Sect. B*, 1968, **24**, 874.
75. J. A. Jaecker, W. R. Robinson and R. A. Walton, *J. Chem. Soc., Dalton Trans.*, 1975, 698.
76. J. L. Atwood, J. K. Newell, W. E. Hunter, I. Bernal, F. Calderazzo, I. P. Mavani and D. Vitali, *J. Chem. Soc., Dalton Trans.*, 1978, 1189.
77. A. P. Sattelberger, R. B. Wilson, Jr. and J. C. Huffman, *Inorg. Chem.*, 1982, **21**, 2392.
78. R. B. Jackson and W. E. Streib, *Inorg. Chem.*, 1971, **10**, 1760.
79. F. A. Cotton, P. E. Fanwick and J. W. Fitch, III, *Inorg. Chem.*, 1978, **17**, 3254.
80. R. H. Crabtree, G. G. Hlatky and E. M. Holt, *J. Am. Chem. Soc.*, 1983, **105**, 7302.
81. F. A. Cotton and D. A Ucko, *Inorg. Chim. Acta*, 1972, **6**, 161.
82. R. H. Summerville and R. Hoffmann, *J. Am. Chem. Soc.*, 1979, **101**, 3821.
83. G. J. Wessel and D. J. W. Ijdo, *Acta Crystallogr.*, 1957, **10**, 466.
84. R. Saillant, R. B. Jackson, W. E. Streib, K. Folting and R. A. D. Wentworth, *Inorg. Chem.*, 1971, **10**, 1453.
85. W. H. Watson and J. Waser, *Acta Crystallogr.*, 1958, **11**, 689.
86. J. L. Templeton, R. A. Jacobson and R. E. McCarley, *Inorg. Chem.*, 1977, **16**, 3320.
87. C. R. Hubbard and R. A. Jacobson, *Inorg. Chem.*, 1972, **11**, 2247.
88. M. Elder and D. Hall, *J. Chem. Soc. (A)*, 1970, 245.
89. K. A. Raspin, *J. Chem. Soc. (A)*, 1969, 461.
90. J. A. Muir, R. Baretty and M. M. Muir, *Acta Crystallogr., Sect. B*, 1976, **32**, 315.
91. P. M. Boorman, K. J. Moynihan and R. T. Oakley, *J. Chem. Soc., Chem. Commun.*, 1982, 899.
92. M. G. B. Drew, B. J. Brisdon and M. Cartwright, *Inorg. Chim. Acta*, 1979, **36**, 127.
93. M. L. H. Green, A. Izquierdo, J. J. Martin-Polo, V. S. B. Mtetwa and K. Prout, *J. Chem. Soc., Chem. Commun.*, 1983, 538.
94. A. P. Sattelberger, R. B. Wilson, Jr. and J. C. Huffman, *Inorg. Chem.*, 1982, **21**, 4179.
95. F. Fourcade, G. Mascherpa and E. Philippot, *Acta Crystallogr., Sect. B*, 1975, **31**, 2322.
96. J. H. Holloway, V. Kaučič, D. Martin-Rovet, D. R. Russell, G. J. Schrobilgen and H. Selig, *Inorg. Chem.*, 1985, **24**, 678.
97. N. Bresciani, N. Marsich and G. Nardin, *Inorg. Chim. Acta*, 1974, **10**, L5.
98. G. Nardin, L. Randaccio and E. Zangrando, *J. Chem. Soc., Dalton Trans.*, 1975, 2566.
99. U. Schubert, D. Neugebauer and A. A. M. Aly, *Z. Anorg. Allg. Chem.*, 1980, **464**, 217.
100. U. Turpeinen, *Ann. Acad. Sci. Fenn., Sect. AII, Chem.*, 1977, **182**, 5.
101. M. M. Olmstead, R. R. Guimerans and A. L. Balch, *Inorg. Chem.*, 1983, **22**, 2473.
102. J. C. Jansen, J. van Koningsveld and J. Reedijk, *Nature (London)*, 1977, **269**, 318.
103. H. R. van der Wal, F. Overset, H. O. van Oven, J. L. de Boer, H. J. de Liefde-Meijer and F. Jellinek, *J. Organomet. Chem.*, 1975, **92**, 329.
104. M. R. Churchill, B. G. De Boer and S. J. Mendak, *Inorg. Chem.*, 1975, **14**, 2041.
105. B.-K. Teo and J. C. Calabrese, *Inorg. Chem.*, 1976, **15**, 2467.
106. B.-K. Teo and J. C. Calabrese, *Inorg. Chem.*, 1976, **15**, 2474.
107. G. Nardin and L. Randaccio, *Acta Crystallogr., Sect. B*, 1974, **30**, 1377.
108. N. Marsich, G. Nardin and L. Randaccio, *J. Am. Chem. Soc.*, 1973, **95**, 4053.
109. B. Buss and B. Krebs, *Inorg. Chem.*, 1971, **10**, 2795.
110. B. Krebs and V. Paulat, *Acta Crystallogr., Sect. B*, 1976, **32**, 1470.
111. A. L. Rheingold, A. D. Uhler and A. G. Landers, *Inorg. Chem.*, 1983, **22**, 3255.
112. B. Krebs and V. Paulat, *Z. Naturforsch., Teil B*, 1979, **34**, 900.
113. P. C. Healy, D. L. Kepert, D. Taylor and A. H. White, *J. Chem. Soc., Dalton Trans.*, 1973, 646.
114. K. Jödden, H. G. von Schnering and H. Schäfer, *Angew. Chem., Int. Ed. Engl.*, 1975, **14**, 570.
115. S. Stensvad, B. J. Helland, M. W. Babich, R. A. Jacobson and R. E. McCarley, *J. Am. Chem. Soc.*, 1978, **100**, 6257.
116. R. D. Burbank and G. R. Jones, *J. Am. Chem. Soc.*, 1974, **96**, 43.
117. A. L. Rheingold, A. G. Landers, P. Dahlstrom and J. Zubieta, *J. Chem. Soc., Chem. Commun.*, 1979, 143.
118. R. A. Andersen, D. H. Templeton and A. Zalkin *Inorg. Chem.*, 1978, **17**, 1962.

19

Hydrogen and Hydrides as Ligands

ROBERT H. CRABTREE
Yale University, New Haven, CT, USA

19.1 INTRODUCTION

The hydrogen atom is the simplest of ligands, and its coordination chemistry has a number of unique and fascinating features. Apart from its intrinsic interest, the metal–hydrogen bond is also important because of its widespread occurrence in intermediates in both homogeneous and heterogeneous catalytic processes and in organometallic reactions. It is probably also present in certain biological systems.

The first complex metal hydrides were made by Walter Hieber in the 1930s during his work on metal carbonyl complexes. For example, he characterized $H_2Fe(CO)_4$[1] from the reaction of $Fe(CO)_5$ with OH^-. For 30 years, however, his compounds tended to be regarded somewhat as chemical curiosities. Electron diffraction studies showed that the molecule was tetrahedral[2] (in fact it is somewhat distorted from the pure tetrahedral geometry) and this appeared to suggest (but see Section 19.4.2) that the hydrogens were bound elsewhere than at the metal, probably at the CO oxygen atom. In his classic text[3] Sidgwick formulated the molecule as $Fe(CO)_2(COH)_2$. Edgell[4] used IR evidence and molecular orbital calculations to support a structure for $HCo(CO)_4$ in which the

Co—H bond was unusually long (*ca.* 2.0 Å) and the H showed greater bonding interaction with the CO groups than with the metal. It was only with Ginsberg's neutron diffraction data on K_2ReH_9 that the modern picture of the M—H bond as a normal covalency was unambiguously established.[5]

During the late 1930s, the first homogeneous catalytic processes were discovered. In 1938, Calvin[6] showed that copper(I) acetate hydrogenated benzoquinone at 100 °C and proposed a copper hydride as a possible intermediate. At that time Roelen[7] was introducing the cobalt-catalyzed hydroformylation of alkenes. Although he used a heterogeneous catalyst, he recognized that the reaction was probably due to soluble cobalt carbonyls of the type that Hieber had studied. This process is of great commercial importance, and modern versions are in very wide use. By the mid 1950s the idea that non-carbonyl compounds with M—H bonds might be isolable began to find new supporters, and efforts were made in several laboratories to synthesize examples. Three important complexes from this period (1955–65) are Wilkinson's Cp_2ReH (1), the first example of a non-carbonyl hydride,[8] Chatt's $[PtHCl(PEt_3)_2]$ (2),[9] and most surprising of all, Ginsberg's K_2ReH_9 (3).[5] Another notable contemporary example is Fischer's $CpM(CO)_3H$ (M = Mo and W). The new structural and spectroscopic methods that were then available made it possible to prove conclusively that all these stable compounds did indeed contain an M—H bond. Reactivity studies, particularly of (2), began to show some of the rich chemistry of the M—H group. Since that time, the hydride ligand has come to occupy a central position in transition metal chemistry.

(1)   (2)   (3)

The distinction between coordination and organometallic chemistry, implicit in the title of this series, is not clear cut for hydride complexes because so many coordination complexes of the hydride ligand either react with unsaturated organic compounds to give, or are formed from, organometallic species. We shall therefore cover all aspects of hydrides, while emphasizing coordination chemistry.

The purpose of this chapter is to survey very briefly the general features of the occurrence, preparations and properties of hydrides. Specific compounds will be covered under each element in Volumes 2–4. We have concentrated on transition metal hydrides but also briefly mention some features of main group hydride chemistry. Useful reviews have appeared on various aspects of metal hydride complexes. One by M. L. H. and J. C. Green[10] in 'Comprehensive Inorganic Chemistry' has very useful lists of compounds but dates from 1973. Teller and Bau[11] have covered the structural data on metal hydrides and give extensive tabulations of structural data. Humphries and Kaesz[12] have considered cluster hydrides, especially in terms of their reactivity. Hlatky and Crabtree[13] have reviewed polyhydrides. In each of these reviews, the authors have extensively tabulated the relevant data. We shall try to avoid duplication by emphasizing areas not previously covered.

19.2   TYPES OF HYDRIDE COMPLEX

19.2.1   With Molecular Hydrogen

This topic constitutes the newest and one of the most active areas in the field of hydride chemistry. It has recently been found[14] that the H_2 molecule can be bound in a side-on fashion as an undissociated two-electron donor ligand. The full implications are not yet clear but it may mean that a large number of di- and poly-hydrides will have to be reformulated. As early as 1976, Ashworth and Singleton[14a] suggested that $RuH_4(PR_3)_3$ might really be an Ru^{II} dihydrogen complex. There is now good evidence that it is $RuH_2(H_2)(PR_3)_3$.[14e] Kubas *et al.*[14b] discovered the first unequivocal case of a dihydrogen complex in $[W(H_2)(CO)_3(PCy_3)_2]$ (4), reported in 1984. Apart from the neutron diffraction, a second key piece of evidence was the observation of a $^1J(H,D)$ coupling of 33.5 Hz in the ^1H NMR of the HD complex (*cf.* 43.2 Hz for free HD). $[Cr(CO)_5(H_2)]$ has also been reported but is very unstable under ambient conditions.[14c] Crabtree and Lavin[14d] reported $[IrH(H_2)-(PR_3)_2(7,8-benzoquinolinate)]^+$, the HD analog of which gave a $^1J(H,D)$ of 29.5 Hz. Deprotonation of the coordinated H_2 molecule gave the neutral dihydride. Proton exchange between the coordinated H_2 and Ir—H group was invoked to account for the fluxionality. The same authors found that T_1 measurements could be used to identify H_2 complexes.[14d] The dipole–dipole contribution of nucleus X to the relaxation of nucleus Y depends on γ^2 and on r^{-6} (where r is the XY distance). This means that when r is small (*ca.* 0.8 Å in the case of H_2 complexes[14b]) then this contribution is very large. This method assisted in the identification of the non-classical polyhydride $[IrH_2(H_2)_2$-$(PCy_3)_2]^+$ ($T_1 = 60$ ms, 250 MHz, $-80\,°C$) formed by protonation of the classical pentahydride $IrH_5(PCy_3)_2$ ($T_1 = 870$ ms); partial deuteration of the non-classical polyhydride led to a large increase of T_1, due to the effect of the lower γ for the deuterium directly bound to H in the HD complex. The reason these complexes are not classical dihydrides is presumably that the two M—H bonds that would be formed are not sufficiently strong to compensate for breaking the H—H bond. There is therefore an analogy with C—H$\cdots$M bridged systems, in which a C—H bond of a ligand binds in the same way. It is notable that in all the examples of H_2 complexes so far described, a high *trans* effect ligand, CO or H, occupies a position *trans* to the H_2 ligand. This may help weaken the incipient M—H bond and favor non-classical binding. Many of the examples are also octahedral; the seven-coordinate geometry of the corresponding classical hydride might be an additional factor in disfavoring this form.

$$
\begin{array}{c}
CO \\
\big| \quad \diagup P \\
OC\!—\!W\!\!\leftarrow\!\! \begin{array}{c} H \\ \diagup \\ H \end{array} \\
\diagup \big| \\
P \quad CO
\end{array}
$$

(4)

The bonding of H_2 to the metal appears to be largely H_2 (σ) to M (d_σ) donation, which leaves the coordinated H_2 with a partial positive charge (Figure 1). This is perhaps the reason that the $(H_2)M$ system, like the CH$\cdots$M can often be easily deprotonated.

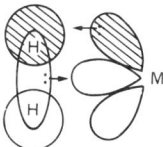

Figure 1 The bonding of H_2 to a metal

Dihydrogen complexes may be intermediates in the oxidative addition of H_2 to a metal[14b] (equation 1) and in the hydrogenolysis of d^0 alkyls. They will also need to be considered as possible species present in hydrogenases and nitrogenase.[14f]

$$
M + H_2 \longrightarrow M\!\!\leftarrow\!\!\begin{array}{c} H \\ \big| \\ H \end{array} \longrightarrow M\!\!\begin{array}{c} \diagup H \\ \diagdown H \end{array}
\tag{1}
$$

19.2.2 With Terminal Hydrides

Terminal hydride complexes contain M—H single bonds. Pure homoleptic hydrides, such as FeH_6^{4-} and ReH_9^{2-}, exist but are rare.[15,5] More usually, stabilizing ligands are present. The most common are tertiary phosphines, CO and cyclopentadienyl groups, but a wide variety of other coligands are known, *e.g.* $[IrH_2L_2(PPh_3)_2]^+$ (L = MeI, MeOH, H_2O, $PhNO_2$), $CoH(N_2)(PPh_3)_3$, $[Rh(NH_3)_5H]^{2+}$, $[ReCl(Et_2NCS_2)H_2(PMe_2Ph)_3]^{2+}$ and $[Co(CN)_5H]^{2+}$. In the vast majority of cases, the complexes conform to the 18-electron rule. Formulations not conforming to the rule have often later been shown to be erroneous. Hydride complexes can have any number form one to nine hydrogens per metal and can be cationic, anionic or, most often, neutral. The 'hydride' ligand can be hydridic or protonic in character, according to circumstances. All the transition metals form hydride complexes, but examples are especially numerous among the later third row metals (*e.g.* Re and Ir).

It is notable that hydrogen, as a very small ligand, encourages the attainment of high coordination numbers. Many eight- and nine-coordinate transition metal complexes are hydrides. The formal oxidation state of the resulting complexes is high. It is often stated that oxygen and fluorine are the elements most capable of stabilizing high oxidation states. It is now clear that hydrogen also has this property, *e.g.* $MoH_6(PCy_3)_3$, $TaH_5(dmpe)_2$, ReH_9^{2-} and $OsH_6(PMe_2Ph)_3$. Complexes like these, having both a coordination number greater than six, and four or more hydrogens bound to the metal, are referred to as 'polyhydrides'.[13] Most polyhydrides have only phosphines as coligands, but a few organometallic examples have recently been discovered, *e.g.* $(C_5Me_5)Ta(PMe_3)_2H_4$[16] and $[(C_5Me_4Bu^t)WH_4]_2$.[17]

Among the chemical forms adopted by the vitamin B_{12} coenzyme, one is a terminal hydride of the type $CoHL_5$ where L represents a nitrogen ligand of the corrin ring or axial base. Hydrides may be important *in vivo* both in this case and in that of hydrogenases.

The great majority of hydrides are diamagnetic, although a few paramagnetic ones are now known (*e.g.* the 17-electron $Ta(dmpe)_2H_2Cl_2$[18]). Lanthanide and actinide hydrides have recently been obtained, *e.g.* $[Th(C_5Me_5)_2H]$[19] and $[Th_2(C_5Me_5)_4H_4]$;[20] these will probably prove to have strongly hydridic hydrides because of the low electronegativity of the metals involved.

Main group hydrides are of particular interest in themselves but are relevant to coordination chemistry only to the extent that they are used to prepare transition metal hydrides or act as ligands.

19.2.3 With Bridging Hydrides

In polynuclear metal complexes, hydride often acts as a bridging ligand.[11,12] Two metals can be bridged by one to four hydrides (5)–(8), and cases are known where both three (9) and six (10) metal atoms are bridged by a single hydrogen. In the latter case (*e.g.* $[Co_6(CO)_{15}(\mu\text{-}H)]^{-}$[21]) the ligand is completely surrounded by an octahedron of metal atoms in which case the term 'interstitial' hydride is often used because the hydrogen is located in the interstices of the cluster. The same term is used for solid state structures in which absorbed hydrogen atoms are located in the interstices of the lattice, *e.g.* PdH_x, $FeTiH_2$. Carbon monoxide is by far the commonest coligand in these complexes, *e.g.* $[H_3Re_3(CO)_{10}]^{2-}$ (11), $[HCr_2(CO)_{10}]^{-}$ (12), $[H_6Re_4(CO)_{12}]^{2-}$ and $[HOs_3(CO)_{10}Cl]$, although other types are known, *e.g.* $[(codRhH)_4]$ (13),[22] $[(IrH_2L_2)_3(\mu_3\text{-}H)]^{2+}$ (14),[23] $[HNb_6I_{11}]$, $[Mo_2Cl_8H]^{3-}$,[24] $[Ir_4H_4(cod)_4(MeC=CH)]$,[25] $[Et_2ZnHZnEt_2]^{-}$ and $[Cp_2W(\mu\text{-}H)_2Rh(PPh_3)_2]^{+}$.

(5)   (6)   (7)   (8)   (9)   (10)

(11)   (12)   (13)

(14)   (15)

The chemistry of cluster hydrides has been reviewed very thoroughly by Kaesz.[12] These have very rich chemistry and constitute a most important class of hydrides. $Os_3(CO)_{10}(\mu_2\text{-}H)_2$ (15) is perhaps the most widely used starting material for cluster hydride chemistry.[27]

Bridging hydrogens between dissimilar elements, *i.e.* M—H—M′, are relatively common. In the case where both M and M′ are transition metals, there are several examples where a metal hydride M—H displaces a labile ligand L from a second metal M′—L to give the complex (equation 2).[26] This is likely to be a general reaction.[28]

$$M—H + L′—M \longrightarrow M—H—M + L′ \qquad (2)$$

In the case where one of the elements M is a transition metal, then M—H—M′ systems are known where M′ can be B, Al, Si or Zn as well as carbon. BH_4^- complexes are an extensive class; the ligand can bind *via* one, two or three H bridges (16)–(18). Compounds containing C—H—M bridges are of particular importance and have been reviewed by Brookhart and Green.[29] No pure cluster hydride has yet been observed, but there is no reason to think that these should not be stable.

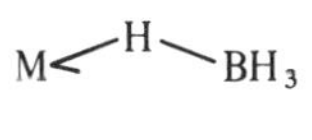

$$M{<}^H{-}BH_3 \qquad M{<}^H_H{>}BH_2 \qquad M{⫷}^H_H{⫸}BH$$

$$(16) \qquad\qquad (17) \qquad\qquad (18)$$

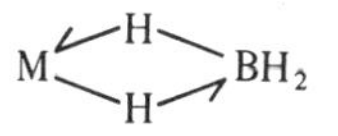

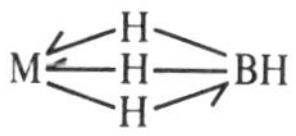

19.2.4 Other Types of Hydride

Some hydrides only exist in the solid state.[30] One class is formed on the absorption of H_2 into a solid material. Several of these are ternary hydrides, such as $FeTiH_2$. In general they contain hydrogen in interstitial, bridging positions. Other compounds contain a discrete anionic fragment such as FeH_6^{4-} or IrH_5^-.[31] They have been suggested for use as hydrogen storage material in fuel engineering applications, but may find other applications in the future. Oligoatomic hydrides are also known for many metals (*e.g.* AgH, AlH, CuH, MnH_2, CoH),[32] but often only in the gas phase. We shall not discuss these types of hydrides in detail.

19.2.5 Stability of Hydrides

Early ideas in this area were hampered by the lack of a clear notion of the possible decomposition pathways for hydride complexes. It has now become clear that the chief modes of decomposition (Scheme 1) are (a) by loss of H_2 (which can be bimolecular), (b) by reductive elimination with another ligand and (c) by loss of H^+. Loss of H_2 may occur from *cis* dihydrides rather easily. The presence of strong M—H bonds, such as occur in the second and third rows, tends to militate against H_2 loss, and it is for this reason that almost all polyhydrides come from these heavier elements. A stable complex would also result if the decomposition would give a very unstable metal fragment. Even monohydrides can decompose by H_2 loss if a stabilizing ligand is first lost from one metal center; an M—H—M-bridged species can then form and H_2 loss can occur (equation 3). Direct bimolecular H_2 loss *via* a four-center transition state may also occur in some cases. Strongly binding soft ligands are useful to stabilize hydrides. This accounts for the common occurrence of cyclopentadienyl, CO and PR_3 as coligands in hydrides. NH_3 and H_2O on the other hand are rare, and those cases known, *e.g.* $[Rh(NH_3)_5H]^+$, contain substitution-inert ions. Oxo, nitrido, fluoro[33] and hydroxo[34] are unknown or very rare as coligands. This may be because these ligands render the metal harder and less able to form strong covalencies to H. Ligands having good π-acceptor character seem to be more suitable for sites *trans* to hydrides[35] so competition for metal π-bonding orbitals is limited.

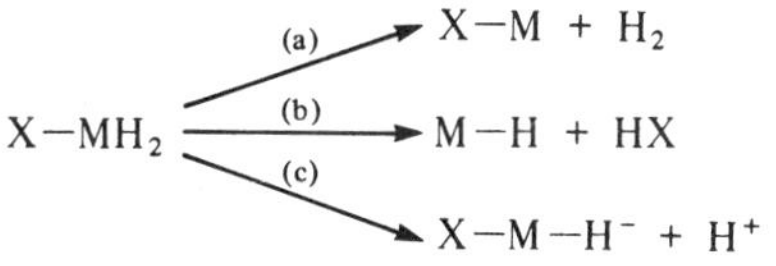

$$X—MH_2 \begin{array}{l} \xrightarrow{(a)} X—M + H_2 \\ \xrightarrow{(b)} M—H + HX \\ \xrightarrow{(c)} X—M—H^- + H^+ \end{array}$$

Scheme 1 Decomposition pathways of hydride complexes

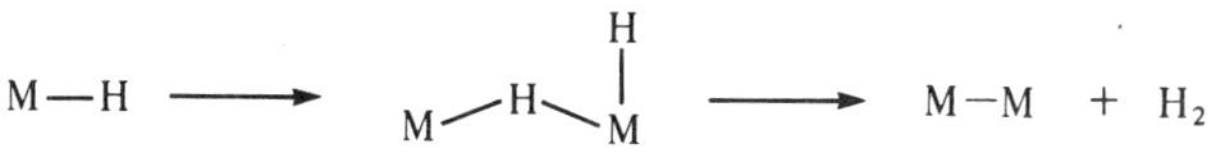

$$M—H \longrightarrow M{-}^H{-}M^{|}_H \longrightarrow M—M + H_2 \qquad (3)$$

Certain ligands appear to be generally unsuitable coligands. Alkenes, alkynes and alkyl groups tend to react to give alkyl and vinyl groups and loss of alkane, respectively (equations 4–6). If the (C=C) and M—H groups are mutually *trans*, or else *cis* but orthogonal to one another[36] or prevented from reacting in some other way, then an alkene hydride may be isolated (*e.g.* **19, 20**). Carbenes hydrides are also likely to be in dynamic equilibrium with the corresponding alkyl (equation 7).[37]

$$\parallel-M-H \rightleftharpoons \overset{\diagup}{M}\diagdown H \tag{4}$$

$$\parallel\parallel-M-H \rightleftharpoons \overset{\diagup}{M}\diagdown H \tag{5}$$

$$R-M-H \longrightarrow M + RH \tag{6}$$

$$H-M=CR_2 \rightleftharpoons M-CR_2-H \tag{7}$$

(19)   (20)

Loss of a proton is relatively rare but may take place in cases where there is a positive charge on the complex (*e.g.* $IrH_2(diene)L_2^+$ in equation 8) or the anionic conjugate base is stabilized (*e.g.* $HCo(CO)_4$ in equation 9).

$$(indenyl)IrHL_2^+ \underset{H^+}{\overset{-H^+}{\rightleftharpoons}} (indenyl)IrL_2 \tag{8}$$

$$HCo(CO)_4 \rightleftharpoons H^+ + Co(CO)_4^- \tag{9}$$

Metal hydrides are often stable to hydroxylic solvents and water but less often stable to air. Acid may decompose them completely, or give a stable protonated form (*e.g.* $WH_5(PR_3)_4^+$ from $WH_4(PR_3)_4$ and HBF_4; equation 10)[38] or give a reaction product (*e.g.* $WH_2(MeCN)_3(PR_3)_3^{2+}$ from $WH_6(PR_3)_3$ and HBF_4 in MeCN;[39] equation 11).

$$WH_4(PR_3)_4 \overset{H^+}{\longrightarrow} WH_5(PR_3)_4^+ \tag{10}$$

$$WH_6(PR_3)_3 \overset{H^+, MeCN}{\longrightarrow} WH_2(MeCN)_3(PR_3)_3^{2+} \tag{11}$$

Some hydrides are light-sensitive. M. L. H. Green developed the photolytic extrusion of H_2 from Cp_2WH_2 as a synthetic route in tungstenocene chemistry (Scheme 2).[40]

The stability of cluster hydrides seems to follow a similar pattern. Bridging seems to be the predominant structural form, but fluxional behavior, presumably involving intermediate terminal hydride species, is common. Norton has discussed binuclear elimination[41] as an important cluster-forming decomposition pathway for metal hydrides (see Section 19.5.3). Structural rules due to Wade,[42] Mingoes,[43] Hoffmann,[44] Lauher[45] and Teo[46] have been proposed to account for the structures of the higher nuclearity clusters. It is quite possible that clusters of several different electron counts will be accessible for any given polyhedron of metals, especially for the larger clusters. The relative usefulness of the different systems is still under discussion.

19.3 PREPARATION OF HYDRIDE COMPLEXES

A review by Kaesz and Saillant[47] covers preparative methods particularly fully, and the 'Inorganic Syntheses' series gives detailed, tested preparations of a large number of specific hydride complexes.

19.3.1 From Molecular Hydrogen

This is very common especially for d^8 species. A 16-electron species may add H_2 directly, but an 18-electron species must lose a ligand first. A concerted H_2 homolysis is often invoked, perhaps from

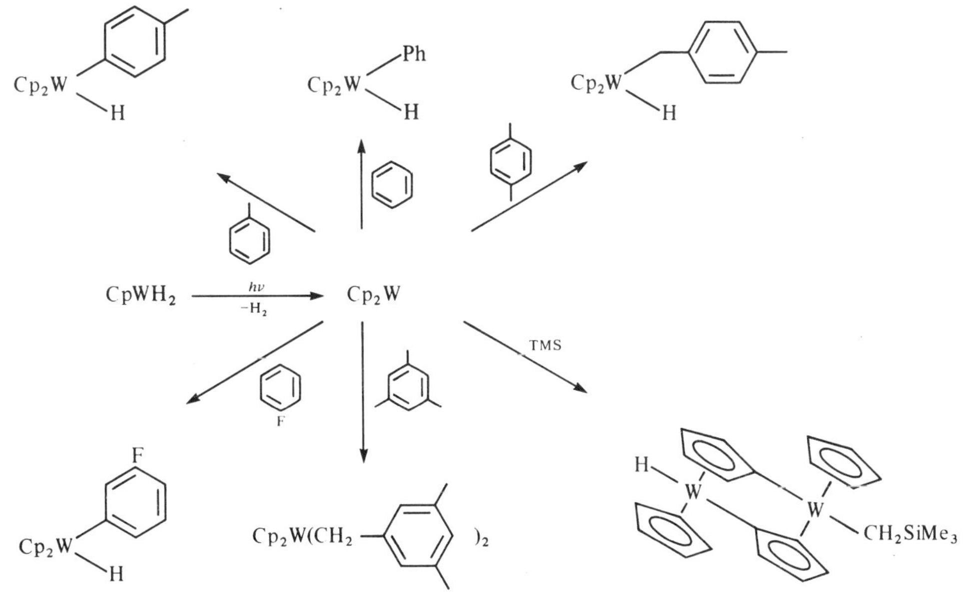

Scheme 2 contains structures:

Cp$_2$W with H and p-tolyl group (top left)

Cp$_2$W with Ph and H (top center)

Cp$_2$W with CH$_2$-p-tolyl and H (top right)

CpWH$_2$ $\xrightarrow[-H_2]{h\nu}$ Cp$_2$W (center)

Cp$_2$W with F-phenyl and H (bottom left)

Cp$_2$W(CH$_2$—(3,5-dimethylphenyl))$_2$ (bottom center)

Dinuclear W structure with H and CH$_2$SiMe$_3$ (bottom right, TMS)

Reagents around center: toluene, benzene, p-xylene (toluene), fluorobenzene, 3,5-dimethylbenzene, TMS

Scheme 2 Some reactions of tungstenocene

an η^2-H$_2$ species (equation 1). In all cases the two hydrogens initially seem to occupy *cis* sites in the product, although subsequent rearrangements or deprotonation may be fast and lead to rearrangement (*e.g.* equation 12). Several groups have considered theoretical aspects of the problem.[48a] The reaction is usually termed 'oxidative addition', although in the case of H$_2$, a reagent better noted for its reducing properties, the oxidation is often purely formal.[48b] The reaction seems to show definite regiochemistry (equations 13 and 14), although there is not yet full agreement on what factors are responsible.[48c]

$$(\text{diene})IrL_2^+ \xrightarrow{H_2} \textit{cis-}(\text{diene})IrH_2L_2^+ \xrightarrow{-H+} (\text{diene})IrHL_2 \xrightarrow{H+}$$

kinetic product

$$\textit{cis,trans-}(\text{diene})IrH_2L_2^+ \tag{12}$$

thermodynamic product

(diene = *sym*-dibenzocyclooctatetraene, L = PPh$_3$)

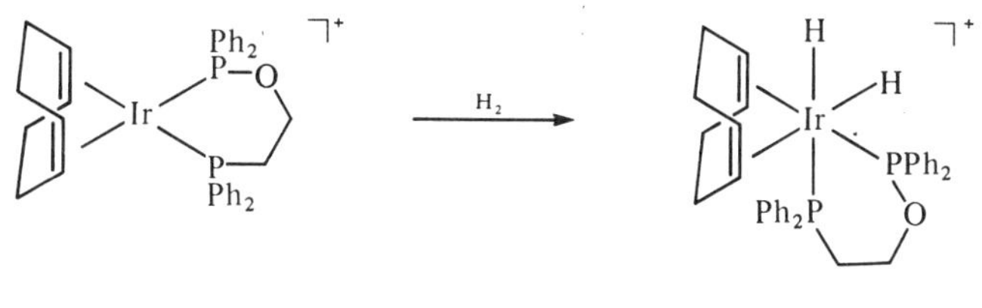

$$\text{(structure)} \xrightarrow{H_2} \text{(structure, exclusive product)} \tag{13}$$

exclusive product

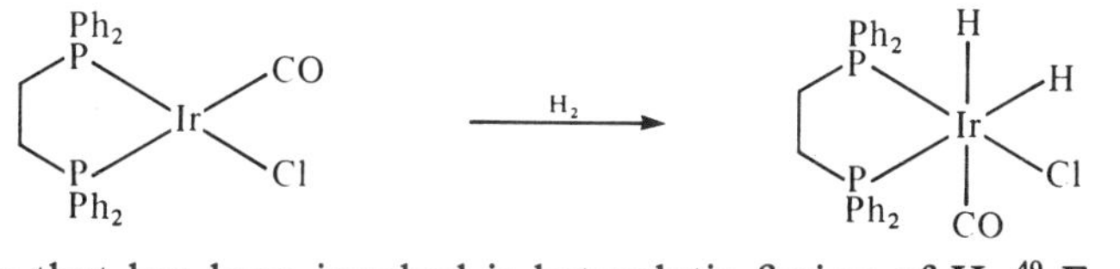

$$\text{(structure)} \xrightarrow{H_2} \text{(structure)} \tag{14}$$

A second process that has been invoked is heterolytic fission of H$_2$.[49] Examples are shown in equations (15) and (16). Although the overall process certainly involves heterolysis, some or all of these reactions may well go *via* initial oxidative addition followed by fast deprotonation or reductive elimination (equation 17). A third variant closely related to heterolytic fission is the addition of hydrogen across a metal–ligand or metal–metal bond. Examples are shown in equations (18) and (19). Here too, homolytic addition *via* equation (1) followed by fast reductive elimination may be taking place in some cases. In d^0 systems, the homolytic fission reaction of equation (1) is unlikely as the metal is already at its maximum valency, so examples of heterolytic fission and addition to

M—L bonds in d^0 systems may be authentic (*e.g.* equation 19). Elimination could still take place from an η^2-H_2 complex because H_2 can add in this way without requiring metal d electrons to be available. Molecular hydrogen can also give metal hydrides by more complicated processes, *e.g.* equation (20), involving ligand hydrogenation and multiple H_2 additions. The formation of hydrides from H_2 is involved in the catalysis of hydrogenation, hydroformylation and isomerization by metal complexes (see Section 19.5.5).

$$RuCl_6^{3-} + H_2 \longrightarrow [HRuCl_5]^{3-} + HCl \tag{15}$$

$$RuCl_2(PPh_3)_4 + H_2 \xrightarrow{NEt_3} RuHCl(PPh_3)_4 + NEt_3HCl \tag{16}$$

$$X—M + H_2 \longrightarrow X—MH_2 \longrightarrow M—H + HX \tag{17}$$

$$Co(CO)_4(COR) + H_2 \longrightarrow HCo(CO)_4 + RCHO \tag{18}$$

$$WMe_6 + PR_3 + H_2 \longrightarrow WH_6(PR_3)_3 + 6CH_4 \tag{19}$$

$$Ir(cod)L_2 + H_2 \xrightarrow{NEt_3} IrH_5L_2 + cyclooctane \tag{20}$$

Also synthetically useful is the combination of a metal halide, an alkali metal and hydrogen. For example, $MoCl_4L_2$, L, Na and H_2 (1 atm) give MoH_4L_4 at 25 °C (equation 21).[50] More severe conditions are required in other cases (*e.g.* 100 atm/115 °C for TaH_5dmpe_2 from $TaCl_5$). $NaReO_4$ is reduced by sodium metal in ethanol to $NaReH_9$ (equation 22). Cluster hydrides are also often synthesized from H_2 in reactions similar in principle but more complex in detail (equation 23).

$$MoCl_4(PR_3)_2 \xrightarrow{Na/Hg, H_2, L} MoH_4(PR_3)_4 \tag{21}$$

$$ReO_4^- \xrightarrow{Na, EtOH} Na_2[ReH_9] \tag{22}$$

$$Ir(cod)L_2^+ \xrightarrow{H_2} [(IrH_2L_2)_3(\mu_3\text{-}H)]^{2+} \tag{23}$$

19.3.2 From Acids

Many metal complexes protonate to give hydride complexes (equation 24). Often, the reaction of an acid having a potentially coordinating anion (Cl^-, $MeCO_2^-$) leads to the formation of a product in which the anion is also bound to the metal (equation 25). Use of HPF_6, CF_3SO_3H or $HSbF_6$ can often prevent this, but even such anions as ClO_4^-, BF_4^- and BPh_4^- can coordinate in certain cases (equation 26); BR_4^- loses F^- to the metal.[51] For anionic metal complexes, weaker acids suffice and the tendency to bind the counteranion is less because a neutral complex is generated (equation 27).

$$Cp_2ReH + H^+ \rightleftharpoons Cp_2ReH_2^+ \tag{24}$$

$$Pt(PPh_3)_4 + HCl \longrightarrow Pt(PPh_3)_2HCl + 2PPh_3 \tag{25}$$

$$IrCl(CO)(PR_3)_2 + HBF_4 \longrightarrow \tag{26}$$

$$Co(CO)_4^- + H_2O \rightleftharpoons HCo(CO)_4 + OH^- \tag{27}$$

Simple protonation is not expected in the case of formally d^0 metal complexes, as the metal has no lone pairs, but the proton may attack an M—H bonding electron pair and lead to products (equation 11). The protonation of many complexes of alkenes and other unsaturated hydrocarbons often occurs at carbon, not at the metal (*e.g.* equation 28[52]).

$$(C_5Me_5)Ir \quad \underset{-H^+}{\overset{H^+}{\rightleftharpoons}} \quad (C_5Me_5)Ir^+ \tag{28}$$

Lewis acids, such as BF_3 and $AlMe_3$, can also bind to metals directly (equation 29)[53] although they often attack ligands such as CO and N_2 (equation 30.)[54] Attack at the metal raises $v(CO)$ of attached carbonyls by reducing the available metal electron density, but attack at CO or N_2 lowers $v(CO)$ or $v(N_2)$ because the lone pair donated to the Lewis acid has C—O and N—N bonding character.

$$Cp_2MoH_2 + AlMe_3 \longrightarrow Cp_2Mo \overset{H}{\underset{H}{\lessgtr}} AlMe_3 \tag{29}$$

$$CpMo(CO)_3H + AlMe_3 \longrightarrow CpMo(CO)_2(COAlMe_3)H \tag{30}$$

Carbonyl hydrides, especially clusters, are commonly synthesized by reaction of a carbonyl complex with KOH to give CO_2 or CO_3^{2-} and an anionic complex. Quenching with acid then generates the hydride[12] (equation 31). The initial reduction can also be carried out with Na/Hg or a similar reagent. Trace water can be the proton source in some cases. Chini *et al.* have used 'redox condensation' to build up cluster hydrides. Typically a metal carbonyl anion is condensed with a metal carbonyl and the product protonated (equation 32).[55] Cationic metal carbonyls can hydrolyze to give CO_2 and a hydride.[56] This reaction (*e.g.* equation 33) is promoted by the net positive charge on the metal, which makes the carbonyl carbon much more sensitive to nucleophilic attack. The proposed intermediate metalacarboxylic acid eliminates to give CO_2 and the metal hydride. Deeming and Shaw[57] were the first to isolate a stable metalacarboxylic acid; they showed that it did indeed decompose as mentioned above.

$$Ir_4(CO)_{12} \xrightarrow{K_2CO_3/MeOH} HIr_4(CO)_{11}^- \xrightarrow{H^+} H_2Ir_4(CO)_{11} \tag{31}$$

$$Rh_6(CO)_{15}^{2-} + Rh_6(CO)_{12} \xrightarrow{25^\circ C} Rh_{12}(CO)_{30}^{2-} + CO \tag{32}$$

$$Mn(CO)_6^+ \xrightarrow{OH^-} (CO)_5Mn \overset{H \diagdown O}{\underset{}{-C=O}} \longrightarrow HMn(CO)_5 + CO_2 \tag{33}$$

Lewis and co-workers[58] have studied the reaction of $M_3(CO)_{12}$ (M = Ru, Os) with water, which gives a wide range of unusual polynuclear hydrides such as $[HOs_3(CO)_{10}OH]$ and $[H_2Os_7(CO)_{19}C]$.

19.3.3 From Other Hydrides

This is perhaps the most common laboratory method. Normally a borohydride or alumino-hydride salt is used to reduce a metal halide complex (equation 34), or a metal halide in the presence of stabilizing ligands (equation 35).[59] Often, the use of the complex hydrides $Na[H_2Al(OCH_2CH_2OMe)_2]$ or $Li[BEt_3H]$ can improve yields. There can be few reducing agents that have not been used to prepare metal hydrides, *e.g.* Zn (equation 36), hydrazine, sodium dithionite, sodium hypophosphite, formic acid and alcohols. Formic acid may act *via* the decarboxylation of hydrido formate complex (equation 37) and is a convenient H_2-equivalent reagent. Alcohols, especially in the presence of base are known to give hydrides. This usually occurs by substitution of a halo ligand by alkoxide, followed by abstraction of a hydrogen atom (equation 38) from the alkoxide. In the case of primary alcohols, the aldehyde product can subsequently decarbonylate to give unexpected products (equation 39), in which case a secondary alcohol (*e.g.* Pr^iOH) must be used (equation 40). Equation (40) also illustrates the phosphine disproportionation sometimes encountered. Even though the initial reagent has an L/Ru ratio of four, some of the reagent decomposed to give the extra mole of L, and indeed yields are improved by addition of excess L.

$$Cp_2WCl_2 \xrightarrow{NaBH_4} Cp_2WH_2 \tag{34}$$

$$WCl_6 + 4PR_3 \xrightarrow{LiBHEt_3} WH_4(PR_3)_4 \tag{35}$$

$$RhCl_3 + NH_3 \xrightarrow{Zn} [Rh(NH_3)_5H]^+ \tag{36}$$

$$M + HCOOH \longrightarrow H-M \overset{H \diagdown O}{\underset{}{-C=O}} \xrightarrow{-CO_2} MH_2 \tag{37}$$

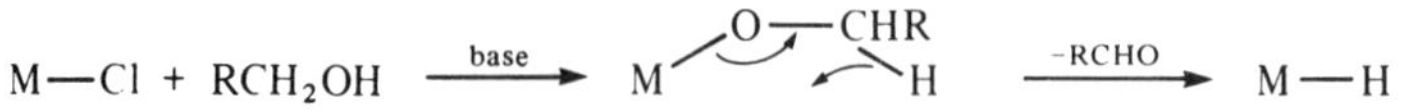

$$M\text{—Cl} + RCH_2OH \xrightarrow{\text{base}} M\overset{O\text{—CHR}}{\underset{H}{\diagdown}} \xrightarrow{-RCHO} M\text{—H} \tag{38}$$

$$[Ru_2Cl_3(PR_3)_6]^+ + EtOH \xrightarrow{OH^-} RuH_2(CO)(PR_3)_3 + CH_4 \tag{39}$$

$$[Ru_2Cl_3(PR_3)_6] + Pr^iOH \xrightarrow{OH^-} RuH_2(PR_3)_4 + Me_2CO \tag{40}$$

The use of Li[BHEt$_3$] as hydride donor avoids the formation of BH$_4$ complexes (as can happen with NaBH$_4$) because BEt$_3$ dissociates readily.[60a] An initially formed hydride may also react with H$_2$ to give a polyhydride. These reactions are, in general, rather sensitive to the exact conditions and reagents used. Low temperatures can be helpful. For example, MoH$_6$L$_3$ can only be obtained if the appropriate reduction (equation 41) is carried out at $-78\,^\circ$C with Na[AlH$_2$(OR)$_2$].[60b]

$$MoCl_4(THF)_2 \xrightarrow[PR_3, THF, -80^\circ C]{NaH_2Al(OCH_2CH_2OMe)_2} MoH_6(PR_3)_3 \tag{41}$$

Alkyl ligands containing β hydrogens can also decompose in a similar way to yield hydrides (equations 42–43). Normally this reaction is so rapid when an empty coordination site is available (equation 43) that to obtain such an alkyl complex requires that all the ligands be firmly bound (*e.g.* CpFe(dpe)Et in equation 42). The empty site is required to abstract the β hydrogen (equation 44).

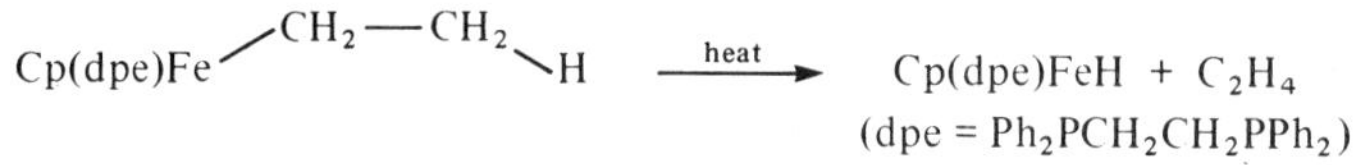

$$Cp(dpe)Fe\overset{CH_2\text{—}CH_2}{\diagdown}{}_H \xrightarrow{\text{heat}} Cp(dpe)FeH + C_2H_4 \tag{42}$$
$$(dpe = Ph_2PCH_2CH_2PPh_2)$$

$$Ir(cod)Cl + 2L + Pr^iMgBr \longrightarrow IrH(cod)L_2 + MeCH=CH_2 \tag{43}$$

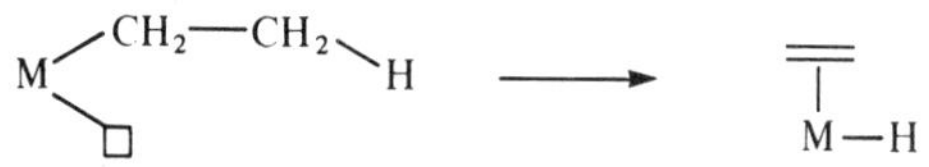

$$M\overset{CH_2\text{—}CH_2}{\underset{\square}{\diagdown}}{}_H \longrightarrow \overset{=}{\underset{M\text{—H}}{|}} \tag{44}$$

In cases where an alkyl has no β hydrogens (or no accessible β hydrogen), an important alternative process, α elimination, can occur. M. L. H. Green[40] has proposed the process shown in Scheme 3 to explain the formation of the ylide complex shown. Another interesting example (Scheme 4) is due to Shaw.[6] It is not known whether Schrock's[62] remarkable chemistry (equation 45), which led to the first examples of carbenes not stabilized by heteroatoms, also goes *via* α elimination or, perhaps more likely, by deprotonation of an alkyl at the α position by an organolithium reagent or other base.

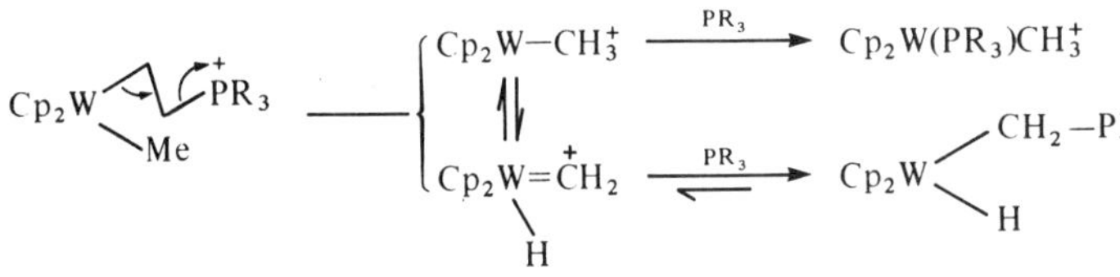

Scheme 3 An example of the equilibrium between a 16-electron methyl complex and an 18-electron methylene hydride

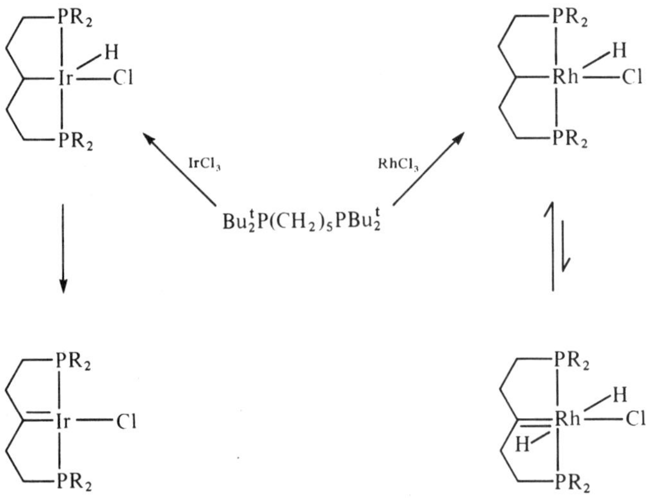

Scheme 4

$$Np_3TaCl_2 \xrightarrow{\text{LiNp}} Np_3Ta{=}CHBu^t \tag{45}$$

$$(Np = neopentyl)$$

In this connection observations of great importance have been made by M. L. H. Green and co-workers.[63] They have shown that in the 12-electron complexes $Ti(dmpe)Cl_3R$ [R = Me (**21**) and Et (**22**)] a hydrogen atom of the R ligand is bonded to the metal in a bridging arrangement. The β hydrogen is bridging in the ethyl case and naturally the α hydrogens bridge in the methyl case. These are termed agostic interactions and are only to be expected for metals having an electron count lower than 18. These agostic interactions are probably quite general and may represent intermediates on the pathway for α and β elimination.

(21) (22)

Metal complexes capable of reacting with molecular hydrogen (see Section 19.3.1) often also react with silanes (equation 46), such as R_3SiH, in the same manner as with H_2. This is particularly important for the catalysis of hydrosilation (equation 47) and related reactions by metal complexes.

$$IrH(CO)L_3 \xrightarrow{R_3SiH} IrH(CO)L_2(SiR_3)H \tag{46}$$

$$RCH{=}CH_2 + R_3SiH \xrightarrow{\text{catalysis}} RCH_2CH_2SiR_3 \tag{47}$$

Hydrides are also formed in a cyclometalation reaction[64] in which the metal breaks a C—H bond in one of its ligands (equation 48), but the hydride may only be an intermediate and subsequently evolve to give other products (equation 49). Any ligand containing CH bonds can in principle undergo this reaction. Particularly susceptible seem to be $P(OPh)_3$, R_2PPh (where R is a bulky ligand), propene (which can give a π allyl hydride product) and cyclopentadiene (which can give a cyclopentadienyl hydride).

$$Ru(Me_2PCH_2CH_2PMe_2)_2 \longrightarrow$$

$$\tag{48}$$

$$RhMe(PPh_3)_3 \longrightarrow$$

$$+ \quad CH_4 \tag{49}$$

There are also examples of intermolecular C—H bond-breaking reactions leading to stable hydrides. The first example of this happening in the case of alkanes was observed by Crabtree *et al.*,[65a] and several other cases are now known (equations 50 and 51).[65b-c] Mechanistically these reactions probably resemble H_2 addition and may go *via* C—H—M intermediates.[48a]

$$IrH_2(Me_2CO)_2(PPh_3)_2^+ \xrightarrow[\text{Bu}^t\text{CH}{=}\text{CH}_2]{\text{cyclopentane}} CpIrH(PPh_3)_2^+ \tag{50}$$

$$(C_5Me_5)IrH_2(PMe_3) \xrightarrow{h\nu, \text{ cyclohexane}} (C_5Me_5)Ir(Cy)H(PMe_3) \tag{51}$$

19.4 PHYSICAL PROPERTIES

19.4.1 The Strength of the M—H Bond

The M—H bond dissociation energies of the small number of simple complexes for which this value is known are all close to $250 \, kJ \, mol^{-1}$.[66] It is reasonable to suppose that this high value compared to the M—C bond strength (125–170 kJ mol^{-1}) in alkyl complexes[66] is related to the fact that H carries no electrons that are not involved in the M—H bond. Halpern[66b] has ascribed the difference to steric effects. In the vapor phase, fragments such as [M—X]$^+$ tend to have similar bond strengths for X = H and X = Me,[67] possibly because Me is more polarizable than H or because of C—H—M bonding.

19.4.2 Structure and Bonding

The review by Teller and Bau[11] is the key reference in the important area of structural studies. This covers both X-ray and neutron methods, illustrates all the main structural types, and tabulates all the relevant structural data. In view of the availability of this review, we will only sketch the main features of the topic and refer the reader to it for the detailed data.

The structures of hydride complexes are often relatively difficult to determine with precision because location of hydrides by X-ray crystallography in the neighborhood of heavy atoms is not always possible because H scatters X-rays only weakly. Neutron diffraction has been used in suitable cases and can provide completely reliable data because the H nucleus scatters neutrons very efficiently. The neutron experiment detects scattering from the hydrogen nucleus but the X-ray experiment uses scattering from the electrons in the M—H bond. Terminal M—H distances derived from X-ray studies are therefore usually *ca.* 0.1 Å shorter than the corresponding neutron-derived values. Similarly, in (bent) bridging systems an X-ray hydrogen position may be closer to the M—M axis than the neutron position.

While a neutron study is preferred, the large crystal size required and the limited number of groups involved in these studies have meant that in most cases structural determinations have been carried out by X-ray methods. Unfortunately, the hydrogens often escape detection or, even when detected, cannot often be reliably located by X-ray methods, especially in the neighborhood of one of the heavier transition metals, or when disorder is present. Collection of data at low temperature, and use of low angle data, seem to improve one's chances. From X-ray data collected at 200 K, for example, Howard and co-workers[68] were able to locate all six hydrogens in WH$_6$(PPr$_2^i$Ph)$_3$.

The hydrogen positions can often be estimated using the heavy atom positions alone. The coligands often reveal the presence of the unseen hydrogens by distortions from the geometry expected in the absence of these hydrogens. In most cases, spectroscopic evidence for the number and positions of hydrogens is also necessary (see Sections 19.4.3 and 19.4.4). It is prudent to obtain this evidence even if the hydrides have been crystallographically located, because errors can be made by relying exclusively on X-ray data.

The first X-ray structure of a hydride complex, PtHBr(PEt$_3$)$_2$,[69] showed a square planar arrangement with one vertex unoccupied by a heavy atom. [Re$_3$H$_2$(CO)$_{12}$]$^-$ provides an example in which the carbonyl ligands flanking the Re—H—Re bridges are distorted by *ca.* 10° away from the hydride positions.[70] Similarly, in [{MoH$_2$L$_3$}$_2$(μ-F)$_3$]$^{2+}$ the gaps where the hydrogens must be located turn out to complete a dodecahedral arrangement around each metal.[33] In many cases hydrogen atoms have been directly detected by X-ray methods.[71,72] As mentioned above, it is important to have other spectroscopic evidence to indicate or confirm the number and type of hydrogens present. This is not always simple to obtain, but a number of NMR methods are now available (see section 19.4.3). Chemical evidence, such as hydrogen evolution studies on protonation or thermolysis, is highly suspect since ligand rearrangements can occur to generate extra hydrogen, or some hydrogen may remain coordinated to the metal or attached to the organic thermolysis products. In cluster hydrides of the M$_3$(μ_2-H) type (*e.g.* Os$_3$H$_2$(CO)$_{12}$) the bridged M—M bond is often, but not always,[73] *ca.* 0.1 Å long than the other two. Combined neutron/X-ray refinement has been used to advantage in certain cases.[74]

In the majority of complexes the hydrides seem to be fully stereochemically active in the sense that the hydride occupies a coordination position and the other ligand positions are relatively undistorted from their expected positions in the new coordination polyhedron. However, in some cases, such as the protonation of Fe(CO)$_4^{2-}$, a hydride (*e.g.* H$_2$Fe(CO)$_4$) is formed which has a structure (Figure 2[75]) approximately half-way between the starting polyhedron, a tetrahedron in this case, and the ideal final polyhedron, which would be an octahedron for H$_2$Fe(CO)$_4$. The fact that this class of complex was the first to be studied structurally[2] led to some contention in the early literature as to whether hydrogen was stereochemically active or not. A complicating factor was the failure, in some of the early work, to detect the distortion from T_d symmetry now known to be present. Lauher[76] has suggested that the stereochemically less active hydrides tend to occur when the protonic character of the hydrogen ligand is the most pronounced, because the corresponding metal anion does not need to rehybridize so much on protonation. For complexes of type MHL_n, cases are known where the coordination geometry is very little distorted from that of an *n*-vertex polyhedron (*e.g.* RhH(PPh$_3$)$_4$[77]) to cases where the distortion is such as to produce an almost perfect $(n + 1)$-vertex polyhedron (*e.g.* MnH(CO)$_5$,[78] Figure 2).

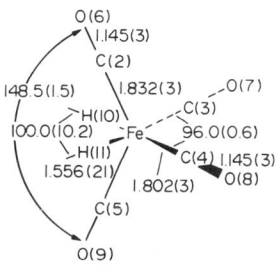

Figure 2 The structure of H$_2$Fe(CO)$_4$ (reproduced from ref. 75 by kind permission of the American Chemical Society)

Generally speaking, the usual M—H distances are 1.55 Å (M—H terminal) and 1.70 Å (M—H bridging) for the first row; and 1.65 Å (M—H terminal) and 1.80 Å (M—H bridging) for the second and third rows. The terminal M—H distances are approximately the sum of the covalent radii, *e.g.* for Ir this sum is 1.55 Å and for Fe, 1.47 Å. [Th$_2$(C$_5$Me$_5$)$_2$H$_4$], the only crystallographically characterized actinide hydride, has Th—H 2.03(1) Å.

High coordination number hydrides are not uncommon; seven-, eight- and nine-coordinate hydrides so far seem to have been found only in pentagonal biprismatic, dodecahedral and tricapped trigonal prismatic geometries, respectively.

Bridging M—H—M systems are of considerable interest. [HCr$_2$(CO)$_{10}$]$^-$ (12), an early example, was originally considered to have a linear M—H—M arrangement. The (OC)Cr···Cr(CO) system is very close to linearity. Careful neutron work[79] showed that four-fold disorder in the hydrogen position gave the appearance of M—H—M linearity, but that in fact the M—H—M angle is 160°. In each case so far studied in detail the M—H—M system has been found to be bent (M—H—M = 85–160°). Bau[11] has suggested that this arrangement may be general for all types of electron-deficient bonds.

The bonding scheme generally involved for this system involves three orbitals, one of σ type on each of the metals and the hydrogen 1s. The in-phase combination of these three has bonding character and is filled with two electrons (Figure 3). This arrangement helps explain the M—H—M bending. The metals can each successfully bond with the H 1s for any M—H—M angle, but as this angle falls below 180°, then M—M overlap increases. The angle cannot fall too greatly, because of L···L repulsions. The bond can therefore be regarded as having partial M—M bonding character. The representation (a) shown in (Figure 4), the simple three-center two-electron bridge, is to be avoided since this would imply the presence of a bridge and an M—M bond with four electrons (this arrangement does also occur but is distinct from the simple case). Bau[11] has used representation (b) and M. L. H. Green[80] representation (c); both of these are unobjectionable, but we shall use (c) as this makes it more convenient to count electrons.

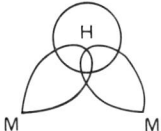

Figure 3 M—H—M bridge bonding. Two metal hybrids and the H 1s overlap in the same region of space. Filling the resulting combination with two electrons gives the two-electron three-center bond proposed for these systems.

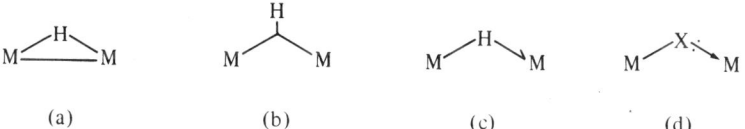

(a)	(b)	(c)	(d)

Figure 4 Common representations of M—H—M bridges. The very different M—X—M bridge is shown in (d)

Electron counting in these electron-deficient systems is slightly different than in electron-precise hydrides. As noted by Green,[80] canonical forms can be written in which the hydrogen is treated normally *vis-à-vis* one metal and the M—H bond is treated as a two-electron ligand to the second metal (*e.g.* **11**, **18**). Where a hydride bridges three metals one M—H bond is considered as a two-electron donor to *each* of the remaining two metals. The same general counting scheme is also effective for boron hydrides. Bridging groups carrying lone pairs, such as RS, Cl, Br or R_2P, bond quite differently from H, since a lone pair can now form the bond to the second metal, and M—M bonding is not intrinsic to the M—X—M bridge (Figure 4d).

A number of $M(\mu\text{-}H)_nM$ systems are known. These necessarily have both a bent M—H—M arrangement and, in view of the small covalent radius of H, a short M—M distance. Such systems also occur in metal clusters, *e.g.* the intensively studied $Os_3H_2(CO)_{10}$. A strongly unsymmetrical bridge has been observed in $[(dpe)Pt(\mu\text{-}H)_2PtH(dpe)]^+$ by neutron diffraction.[8]

$M(\mu\text{-}H)_nM$ systems where $n \geqslant 3$ are rare, perhaps because they seem only to occur in non-carbonyl cluster complexes. Examples are $[Fe_2(\mu\text{-}H)_3\{(Ph_2PCH_2)_3CMe\}_2]^+$, $[(IrHL_2)_2(\mu\text{-}H)_3]^+$, $[Re_2H_4(PEt_2Ph)_4(\mu\text{-}H)_4]$ and $[Ru_2H_4(PR_3)_4(\mu\text{-}H)_4]^+$. No $M(\mu\text{-}H)_5M$ systems are as yet known. Additional structural types arise in cluster compounds.[12] A hydrogen can bridge three metals. In this case M—M bonds are usually present in addition to the two-electron four-center bridge bond, and the hydrogen is out of the plane of the metals by *ca.* 0.8 Å. $M_3(\mu_3\text{-}H)$ (**9**) arrangements are proposed in several trinuclear clusters: $(CpRh)_3(\mu_3\text{-}H)(\mu_3\text{-}Cp)$ and $[(IrH_2LL')_3(\mu_3\text{-}H)]^{2+}$; as well as poly-nuclear ones: $(CpCo)_4(\mu_3\text{-}H)$ and $(CpNi)_4(\mu_3\text{-}H)_3$. An interesting structural analogy appears to exist between a hydride and its $Au(PPh_3)$ analog in that the same deformations from the ideal polyhedron are often seen in each case.[82]

19.4.3 Vibrational Spectroscopy

M—H stretching frequencies lie in the range 1500–2300 cm^{-1} but have variable, often low intensities in the IR and can be very broad and solvent-sensitive.[83] Some complexes show no IR absorption at all.[84] This is probably connected with the weak polarity of the M—H bond and the correspondingly small values of $d\mu/dr$. The generally rather small M—H bond moment is confirmed by dipole moment studies.[35] Cationic hydrides seem generally to have lower IR intensities than neutral ones, perhaps as a result of delocalization of the positive charge on to the hydrides (see Table 1).

Table 1 Typical Spectral Data for Hydride Complexes

Compound	IR $v(M\text{—}H)$	^1H NMR δ	Ref.
$TaH_5(dmpe)_2$ [a]	1544s	−0.70	165
$\{TaCl_2(PMe_3)_2\}_2(\mu\text{-}H)_4$	1225m	+8.79	166
$MoH_4(dppe)_2$ [a]	1725s	−4.06	167
$WH_4(dppe)_2$ [a]	1830, 1780s	−10.7	168
K_2ReH_9	1814, 1846 1931s	−9.1	5
$RuH_4(PPh_3)_2$	1950s	−7.1	169
$[RuH_5(PPh_3)_2]^-$	1750s	−7.64	170
$IrH_5(PPh_3)_2$	1948s	—	171
$WH_5(dppe)_2^+$ [a]	1812, 1895w	−1.7	38, 172

[a] dmpe = $Me_2PCH_2CH_2PMe_2$, dppe = $Ph_2PCH_2CH_2PPh_2$.

When observed at all, hydride vibrations follow similar coupling rules to those for carbonyl complexes. A *cis* dihydride will generally give two absorptions, for example.[87] Metal–hydride vibrations may also couple with CN, CO and N_2 vibrations. Deuterium labeling is useful in these cases, by shifting the M—H vibrations to lower energy ($v_H/v_D \approx 1.4$).

The position of the M—H stretching frequency has been found to correlate with the bond strength. There is a large frequency increase on going from the early to the late transition elements. There is also a smaller increase on going from the first to the third rows. The frequency falls when ligands of greater *trans* effect are in the *trans* position and as the acidity of the H increases.[88] In the same series of complexes $v(M\text{—}H)$ correlates with δMH, the chemical shift (see Section 19.4.4), on changing the *trans* ligand.

In an extensive study of iridium(III) hydrides of the type $[IrCl_{3-x}H_x(PR_3)_3]$, Chatt[89] found empirical ranges of $v(Ir\text{—}H)$ appropriate for different *trans* ligand types. For example, H *trans* to I gave $v(Ir\text{—}H)$ of 2190–2270 cm^{-1}, and H *trans* to H gave 1740–1790 cm^{-1}.

Vaska has shown that substantial $v(CO)$ shifts on deuteration correlate with the presence of a *trans* HM(CO) arrangement, which he ascribes to a dynamic coupling.[90] Stereochemistries can also be assigned by classical IR methods common to non-hydride complexes. For these, a review by Jesson is a good source.[91] Bending modes around 700–900 cm^{-1} can be assigned in the IR spectrum in some cases, *e.g.* IrH$_5$(PPr$_3^i$)$_2$, 875; RuH$_4$(PPh$_3$)$_2$, 805; ReH$_9^{2-}$, 735; and WH$_4$(PEt$_2$Ph)$_4$, 843 cm^{-1}.[87,92]

Incoherent inelastic neutron scattering is efficient in detecting M—H vibrations because the intensity of scattering is proportional to the square of the atomic vibrational amplitude ($\propto 1$/atomic mass) and the scattering cross section (moderately high for H). HCo(CO)$_4$ and H$_3$M$_3$(CO)$_{12}$ (M = Mn and Re) have been studied, for example.[93]

The bridging systems M—H—X also show IR absorptions. When X is a second metal, the asymmetric stretching frequency v_{as} often comes in the range 1200–1700 cm^{-1} and the symmetric frequency v_s in the range 800–1200 cm^{-1}. The ratio v_{as}/v_s has been correlated with the M—H—M angle.[94] This method is a useful complement to X-ray crystallographic studies.

Where X in M—H—X is a main group element, different behavior is observed. The case of carbon is particularly important,[29] and here a weak or medium weak asymmetric stretch is observed at a frequency slightly lower than $v(C—H)$ for the uncomplexed ligand. For example, $v(MHC)$ as appears at 2510 cm^{-1} for TaCp($CHMe_3$)Cl.[95] In metal borohydride complexes (X = B), the $v(BHM)_{as}$ vibrations are *ca.* 200 cm^{-1} to lower wavenumber than the $v(B—H)$ terminal vibrations.[96]

Raman spectroscopy has not been used routinely on metal hydrides, but useful conclusions can be drawn from this technique.[97]

19.4.4 Nuclear Magnetic Resonance Spectroscopy

^{1}H NMR spectroscopy is the most widely used spectroscopic method in studies on hydride complexes. This is because the resonances for diamagnetic complexes almost always appear in the otherwise vacant region to high field of TMS. The unusual shift appears to be due to a combination of factors[98] but the large diamagnetic contribution due to an incomplete d shell on the metal is probably important. In general, d^0 and d^{10} systems (*e.g.* ZrCp$_2$H(BH$_4$), $\delta + 4.3$ p.p.m.) tend to have lower field shifts compared to other d^n configurations (*e.g.* NiHCl(PPr$_3$)$_2$, -24.3 p.p.m.). The usual range to be expected runs from 0 to -40, with exceptional examples from $+30$ to -60 p.p.m. One interesting case is that of the hydride in [Co$_6$(CO)$_{15}$(μ_6-H)]$^-$, for which δMH lies are $+23.2$ p.p.m. Faller[99] has interpreted this as arising from the symmetrical chemical shift tensor in these molecules as contrasted with the unsymmetrical tensors in the case of terminal or bridging hydrides. Indeed (μ_6-H) ligands in unsymmetrical clusters have unexceptional shifts. Chatt[89] has found empirical characteristic ranges of chemical shifts for iridium(III) hydrides depending on the *trans* ligand (*e.g.* CO -7 to -10; H, -11 to -12; SnCl$_3$, -12 to -15). Such correlations are likely to be of use only in comparing closely related compounds.

The intensity of MH resonances, compared to ligand proton resonances, is often different to that expected only on the basis of the relative numbers of protons present. This means that it is often not protons present. This means that it is often not possible to rely on integration of the NMR spectrum for determining the number or hydrides present. This can be particularly misleading for polyhydride complexes: the IrH$_5$(PR$_3$)$_2$ series was originally considered to be IrH$_3$(PR$_3$)$_2$, for example. The probable reason is the very different relaxation times for MH and other ligand protons.[33a] Fortunately, several methods are available to solve this problem. One can add as a standard a known weight of a hydride having a similar relaxation time and of known formulation, and integrate against the hydride resonances of the standard; [ReH$_5$(PMe$_2$Ph)$_3$] has been used for this purpose.[33b] Alternatively, in a phosphine complex one can observe proton coupling to ^{31}P. The fully undecoupled ^{31}P NMR spectrum is unsuitable for such a study because coupling to ligand protons also occurs. Selective decoupling of the ligand protons is required.[100] If the ligand has both aliphatic and aromatic protons then decoupling the aromatic protons can lead to sufficient off-resonance decoupling of the aliphatic region to allow determination of the number of hydrides present.[33b] As a final check, it is important to show that broad band proton decoupling removes all coupling. In principle, an analogous experiment should be possible using Rh and Pt NMR in hydrides of these metals.[101]

Perhaps the most important information about the stereochemistry of metal hydride complexes comes from the coupling of the MH with other nuclei. The characteristic range of ^2J(P–H) for PR$_3$ *trans* to H (90–180 Hz) is very different from the range for PR$_3$ *cis* to H (10–30 Hz) and unambiguous assignment is often possible. Exceptions to this rule have been noted, however,[102] (μ_2-H) hydrides tend to show about a half and (μ_3-H) about a third of the expected coupling, respectively, presumably due to the lower bond order in these cases.[103]

Johnson and co-workers[104] have observed coupling of H to two different types of Os ([187]Os, 1.3%, $I = \frac{1}{2}$) in a 3:1 ratio in [HOs$_{10}$C(CO)$_{24}$]$^-$, and deduced that the H probably lies in a tetrahedral cavity. This may not be a secure method, however, as a coupling constant cannot distinguish between an H bridging to n atoms and a terminal H fluxionating over n atoms. Coupling to other $I = \frac{1}{2}$ nuclei such as [195]Pt (33.7%), [103]Rh (100%) and [183]W (14.3%) is always present in hydride complexes of these metals.[105] Faller[106] has observed a ^1J(Mo—H) coupling of 300 Hz in [(TPM)Mo(CO)$_3$H]$^+$ by [95]Mo NMR (TPM = tris-2-pyridylmethane). Coupling between (inequivalent) classical hydrides is small (^2J(H–H$'$) = 0–10 Hz) — often too small to observe.

The lack of an M—H resonance should not be taken as firm evidence for the absence of an H ligand as it can occasionally be broad, especially when bridging. Molecular hydrogen complexes also give broad resonances at least in part as a result of dipole–dipole interactions.[14] Paramagnetic metal hydrides, although rare, also occur. Broadening due to the presence of paramagnetic impurities is much more common. Finally, resonances can be broad due to interaction of the hydrogen with a quadrupolar nucleus, *e.g.* Co,[107] although this is rare.

The temperature dependence of the ^{1}H NMR of metal hydrides can give useful information. It can also lead to the unexpected 'absence' of hydride resonances in a hydride complex mentioned above. [Ir$_2$H$_5$(PPh$_3$)$_4$]$^+$, for example, has a hydride resonance pattern that is so broadened at room temperature as to be indistinguishable from the base line. Cooling by 30–60 °C gives the static spectrum.[103a]

The origin of these effects is the high susceptibility of hydrides to rearrangement within the coordination sphere of the metal, by which they readily exchange positons. This is known as

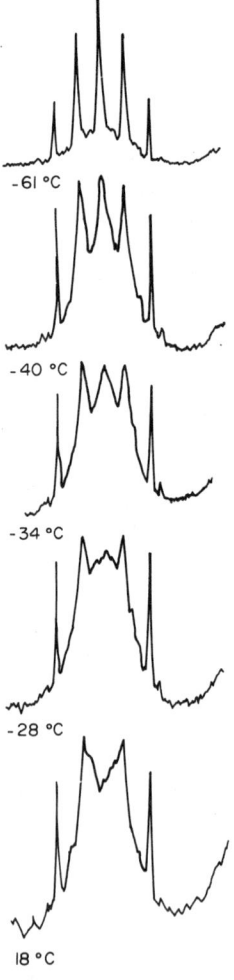

-61 °C

-40 °C

-34 °C

-28 °C

18 °C

Figure 5 The spectral changes found in the ^{1}H NMR spectra of WH$_4$(PMe$_2$Ph)$_4$ with changing temperature (reproduced from ref. 108 by kind permission of the American Chemical Society)

'fluxionality' and is particularly common in five-, seven-, eight- and nine-coordination. Rigid hydrides exhibiting the expected spectrum at room temperature can be found among four- and especially six-coordinate hydrides. Some (*e.g.* ReH_9^{2-}) retain the fluxional spectrum to all accessible temperatures. Rigid structures may become fluxional on warming; others (*e.g.* $IrH_3(PR_3)_3$) are rigid at all accessible temperatures. Figure 5 shows the reversible series of changes that occurs in the NMR of $WH_4(PMe_2Ph)_4$.[108]

A powerful method of detecting X—H—M bridges in cases where fluxionality leads to fast exchange of hydrogens is isotopic perturbation of resonance.[109] If a methyl group RCH_3 takes part in a bridging RH_2C—H—M system, but a very fast process exchanges terminal (H_t) and bridging (H_b) H atoms, and the system is now perturbed by using a CH_2D group, then the D will prefer the terminal site (equation 52). This is because the zero point energy difference for C—H and C—D favors D being in the position (D_t) having the higher force constant. The resonance for CH_2D shows a temperature-dependent shift from the resonance for CH_3, commonly by several tenths of a p.p.m. The appearance of such a resonance only shows that the system is unsymmetrical, not that it is necessarily bridged. Rapid CH bond breaking and remaking in an unbridged RCH_3—M system to give substantial amounts of RH_2C—M—H would also give an isotopic perturbation. The observation of a C—H_b stretching frequency by IR and of a C—H_b coupling constant by ^{13}C NMR help to establish the true nature of the interaction.[106]

$$ M\begin{matrix} H \\ \diagdown \end{matrix} C \begin{matrix} H \\ \diagup \\ \diagdown D \end{matrix} \quad \rightleftharpoons \quad M \begin{matrix} D \\ \diagdown \end{matrix} C \begin{matrix} H \\ \diagup \\ \diagdown H \end{matrix} \tag{52} $$

In cluster hydrides, the resonance for bridging hydrides tends to lie further upfield (-15 to -25 p.p.m.) than those for terminal M—H groups (0 to -15 p.p.m.), but exceptions exist.[103] Nematic phase NMR has been used to determine M—H distances in cluster hydrides.[110]

19.4.5 Electron Paramagnetic Resonance

Few paramagnetic hydrides are known, but EPR has been applied to complexes such as $TiHCp_2(THF)$, $TiHCp_2(PPh_3)$ and $[TiH_2Cp_2]^-$, and coupling to $TiH(10\,G)$, $CH(0.5\,G)$ and P ($24\,G$) is observed.[111] The 17-electron hydrides $[M(dmpe)_2H_2Cl_2]$ (M = Ta, Nb)[18] have also been studied in this way ($g = 1.96$, $a(Nb) = 109\,G$, $a(Ta) = 126\,G$, $a(P) = 25\,G$ and $a(H) = 11\,G$).

19.4.6 Mössbauer Spectroscopy

This method has essentially been applied only to iron hydrides. A correlation between Mössbauer and ^1H NMR chemical shifts has been noted.[112] The method has also been applied in cluster hydrides, and the non-equivalence of the metals in $[Fe_4H(CO)_{13}]$ was shown in this way, for example.

19.4.7 Dipole Moment

Dipole moment studies[85] substantiate the suggestion that the M—H bond tends not to be strongly polar, at least for neutral complexes of the platinum metals.

19.4.8 Mass Spectrometry

The mass spectra of a variety of volatile carbonyl hydrides have been measured.[113] Ionization without loss of H is often possible.

19.4.9 *Trans* Influence and *Trans* Effect of Hydrogen

The weakening effect of a hydride on metal–ligand bonds in the *trans* position (the *trans* effect[114]) is moderately strong compared with the effect of other ligands. Two sorts of effects have been distinguished. The term '*trans* influence' is often reserved for effects on ground state properties, usually detected spectroscopically or in structural studies. For example, $v(MCl)$ falls (IR) and the M—Cl bond lengthens (X-ray crystallography) *trans* to H when compared to an M—Cl bond *trans* to Cl.[114] On the other hand, kinetic labilization is usually considered as a '*trans* effect', although the latter term is sometimes used to cover both manifestations. An example of the *trans* effect would be the more rapid substitution of a ligand *trans* to H than *trans* to PR_3.[103a]

The high *trans* effect of the H ligand is probably connected with its high ligand field strength. The origin of these effects is not completely understood,[115] although they are well documented, dating from the pioneering investigations of Chernyaev.[116] Hydrogen is a particularly interesting case because its bond has pure σ character and so provides an obstacle to theories that consider π bonding as being of most significance in the origin of the *trans* effect. Probably, rehybridization of the metal σ framework takes place so as to give most advantageous overlap for the M—H bond, and this leaves the *trans* ligand in a less advantageous situation.

In the kinetic *trans* effect, the departure of the *trans* ligand is probably aided by a stabilization of the transition state *via* the same mechanisms operative for the *trans* influence.[114] Both associative and dissociative ligand substitution processes seem to be facilitated in this way.[117]

Vaska[118] found a shift of $v(CO)$ in *trans*- but not *cis*-X—M—CO arrangements of about $20-30\,\mathrm{cm}^{-1}$ to higher frequency when X = D. He ascribed this result to coupling effects. Halpern[119] showed that k_H/k_D for the replacement of Cl^- by pyridine in *trans*-$PtHClL_2$ is 1.44, unexpectedly large for a secondary isotope effect. He ascribed this to Pt—H bond weakening in the transition state.

19.5 CHEMICAL PROPERTIES

19.5.1 Acidity and the Polarity of the M—H Bond

Although hydrogen complexes are covalent compounds, the coordinated hydrogen can exhibit H^-, H and even H^+ character in its reactions. In considering this, we have to distinguish carefully between the ground state properties of a complex which may be revealed by spectroscopy or suggested by calculation or structural parameters, and the pattern appearing in its reactions which derive from the properties of the transition state and are strongly dependent on the nature of the reagent with which it reacts.

An overall positive charge or the presence of electron-withdrawing ligands in the coordination sphere may predispose an H ligand to depart as H^+, *e.g.* equation (12). Some hydrides (*e.g.* H—$Co(CO)_4$, $K_a(H_2O) = 2$) can even be moderately strong acids in suitable solvent (*e.g.* 8–9). This often leads to rapid isotopic exchange between the hydride and D_2O. $D_2Os_3(CO)_{10}$ and $D_2Re_2(CO)_8$ have been observed to exchanged with water protons on chromatography.

Conversely, the presence of an overall negative charge or of good σ-donor ligands can predispose an H ligand to depart as H^-. An example is $[Ru(PPh_3)(o\text{-}C_6H_4PPh_2)H_2]^-$, which appears to be a good H^- donor reduces ketones and esters with H_2.[120] Reactions in which H transfer occurs have been described especially for first row metals.[66a,121] Sometimes, however, reactions that appear to be H transfer can take place by an initial one-electron redox step followed by H^+ transfer.[122]

The effect of bond polarity on the regiochemistry of insertion may be involved in the case of the Rh complex of equation (53) which adds to acrylonitrile as if it were Rh^+—H^-, in contrast with the Ir complex of equation (54) which appears to add as Ir^-—H^+.[123] Relatively little is known about the pK_a values of metal hydrides.[124] In addition to the reactivity properties mentioned above, the M—H bond can also tend towards M^+—H^- or M^-—H^+ character in its ground state structure. This seems to have an influence on spectroscopic (Section 19.4.3), physical (Section 19.4.7) and structural (Section 19.4.2) properties.

$$RhHCl_2L_3 \xrightarrow{\quad \overset{\displaystyle CN}{=\!\!/} \quad} L_3Cl_2Rh-\!\!\!\overset{\displaystyle CN}{\underset{\displaystyle CN}{\big\langle}} \tag{53}$$

$$IrH(CO)L_3 \xrightarrow{\quad \overset{\displaystyle CN}{=\!\!/} \quad} (CO)L_2(CH_2=CHCN)Ir \diagdown\!\!\diagup\!\!\diagdown^{CN} \tag{54}$$

19.5.2 Reactions with Halogens and Halides

Reaction with halogens and halocarbons commonly leads to replacement of the hydride with halide (equation 55) and often to subsequent oxidation as well. I_2 is useful in giving a more selective oxidation in many cases (equation 56).

$$Cp_2WH_2 + 2CCl_4 \longrightarrow Cp_2WCl_2 + 2CHCl_3 \tag{55}$$

$$ReH(CO)_3L_2 \xrightarrow{I_2} Re(CO)_3L_2I + \tfrac{1}{2}H_2 \tag{56}$$

19.5.3 Loss of H₂ or HX

Donor molecules can displace H_2 leading to substitution. Loss of one hydride ligand and one other anionic ligand (*e.g.* Cl, OAc, Me) is also possible (equation 57). Norton[125] has discussed binuclear elimination as an important route for reductive elimination in metal hydrides. An interesting development has been the use of a base such as NEt_3 to dehydrochlorinate a metal complex. Richards *et al.*[126] have shown the versatility of this technique (equation 58).

$$RuH(OAc)L_3 + CO \longrightarrow Ru(CO)_3L_2 + AcOH \tag{57}$$

$$WH_2Cl_2L_4 \xrightarrow{NEt_3,\ L'} WL_3L'_3 + 2NEt_3HCl + L \tag{58}$$

$$(L = PMe_2Ph,\ L' = CO,\ C_2H_4,\ MeNC)$$

Transfer of a hydride to another ligand on substitution is also possible (equation 59). Alternatively, substitution reactions can occur without loss of the H ligand. This is common where only one H is present (equation 60). The presence of the coordinated hydrogen can even encourage such reactions by its high *trans* effect and perhaps even by transfer to a CO (or related ligand, where one is present) to give a transient formyl and an open site (equation 61).

$$Cp_2ReH + CO \longrightarrow CpRe(CO)(\eta^4\text{-}C_5H_6) \tag{59}$$

$$HMn(CO)_5 \xrightarrow{L} HMn(CO)_{5-x}L_x \tag{60}$$

$$HMn(CO)_5 \rightleftharpoons \square\text{-}Mn(CHO)(CO)_4 \tag{61}$$

$$(\square = \text{empty site})$$

19.5.4 Reactions with Unsaturated Organic Compounds

The great ease of insertion of M—H bonds into unsaturated organic compounds is perhaps their most important chemical property, because through the hydride one has access to a vast range of organometallic species. Not only are such species accessible, but they can subsequently react by stoichiometric or catalytic pathways to give organic compounds. Chatt[127] studied the decomposition of $PtEtBr(PEt_3)_2$ and showed by isotopic labelling that the equilibrium of equation (62) was involved. More recently, Osborn[128] observed the same process in a molybdenum ethylene hydride complex (equation 63). Two isomeric cyclooctadiene hydrides (**23**) and (**24**) were recently described.[36] Isomer (**24**), containing the coplanar M(C=C)H grouping, underwent insertion at least 40 times faster than isomer (**23**), which contains a *cis* but orthogonal M(C=C)H arrangement. The requirement for this reaction therefore seems to be the presence of a *cis* and coplanar M(C=C)H group. In general the metal hydride side of the equilibrium seems to be favored thermodynamically, so that an alkyl containing β hydrogens will decompose to a metal hydride and alkenes, often free alkene, if an open site is available in the *cis* position to accept the β hydrogen, and the alkyl can adopt a suitable conformation to allow the reaction to occur (equation 44). Specially stable 'alkyls' therefore either lack β hydrogens (Ph, PhCH₂, Me, Bu^{t}CH₂, MeOCH₂) or cannot easily adopt a conformation that permits reaction (1-adamantyl, 1-norbornyl and, to some extent, cyclohexyl).

$$Pt\text{---}H \underset{}{\overset{C_2H_4}{\rightleftharpoons}} Pt\text{---}Et \tag{62}$$

$$Mo(N_2)_2dpe_2 \xrightarrow{C_2H_4} Mo(C_2H_4)_2dpe_2 \xrightarrow{H^+} [(C_2H_4)dpe_2Mo\text{---}Et]^+ \rightleftharpoons \left[(C_2H_4)dpe_2Mo\diagdown_H \right]^+ \tag{63}$$

Insertions into substituted alkenes[129] involve the problem of selectivity (equation 64). For catalytic and organic synthetic applications, this raises a fundamental question, because reliable, predictable and selective methods of functionalizing alkenes are of great interest. For aliphatic alkenes, the bulk

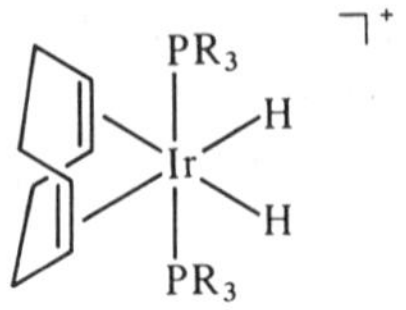

(23) 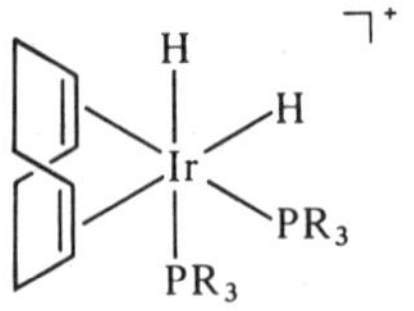   (24)

of the metal fragments in general lead to the linear product (26). This is used to good effect in the hydroformylation of linear alkenes to linear aldehydes and alcohols (equation 65), a transformation of great commercial importance. In contrast, when R is aromatic the insertion tends to give branched products of type (25). This has been ascribed[130] to the formation of an η^3-benzyl intermediate (27; equation 66); radical pathways can also give the same product.

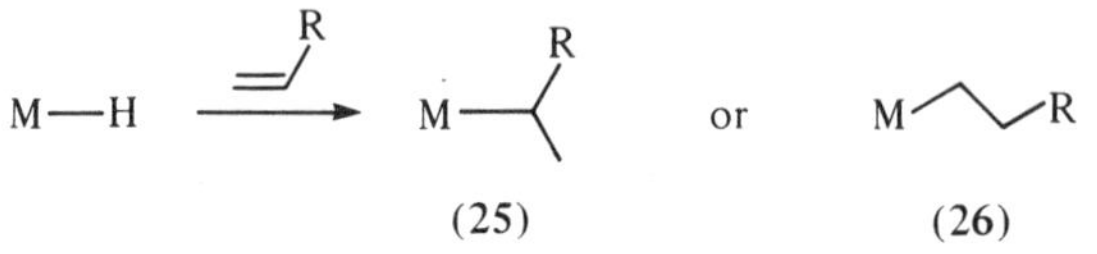

(64)

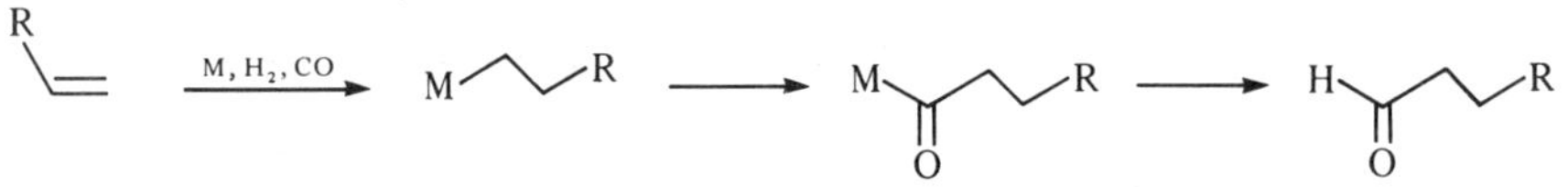

(25)   (26)

(65)

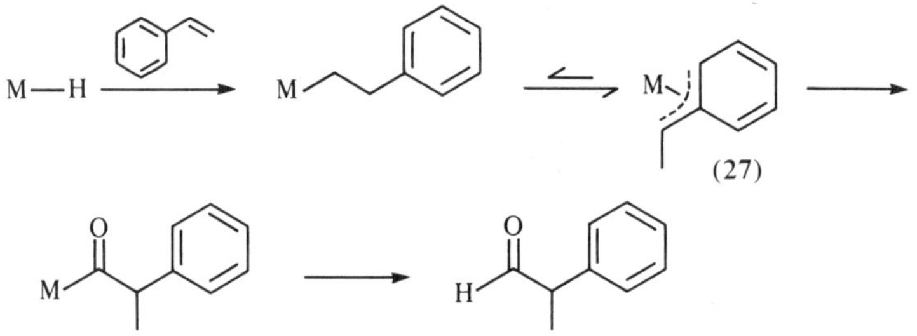

(66)

(27)

The reaction of an alkene (in excess) with a polyhydride can strip hydrogen from the metal. The open sites formed in this process can now react, whether by cyclometalation or by coordinating the alkenes. If the alkene has accessible C—H bonds (*e.g.* propene) these may be broken by the metal. Indeed dehydrogenation may proceed until a stable 16-electron compound has been formed. In most cases this hydrogen stripped from the ligand is passed to the excess of alkene. Examples of these types of reaction have been described by Wilkinson (Scheme 5).[131]

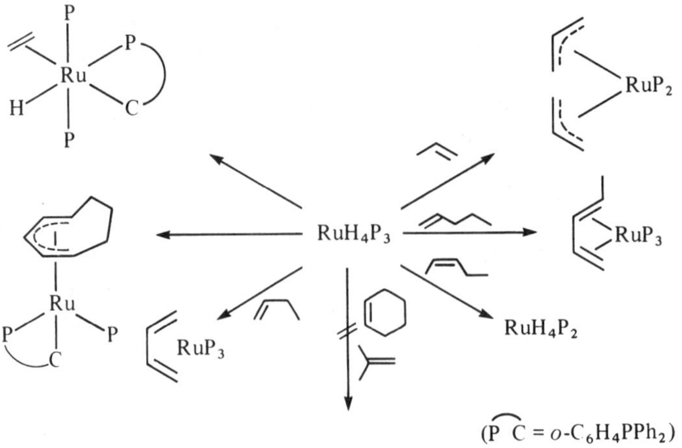

$(\overset{\frown}{P\ C} = o\text{-}C_6H_4PPh_2)$

Scheme 5 Some reactions of RuH$_4$(PPh$_3$)$_3$

An alkene which is reactive enough to strip hydrogen from a metal, yet does not seem to bind strongly to the open sites it creates, is *t*-butylethylene. This reagent was first used by Crabtree *et al.* in their studies of alkane C—H and C—C bond activation by metal complexes (equation 50).[65a,132,133]

Alkynes also insert into M—H bonds. These are of interest in that the stereochemistry of insertion is often apparent in the metal vinyl complex formed. First row metals have been observed to give *trans* products in which case a radical mechanism may be operative[134], but second and third row metals tend to give *cis* products as expected from a concerted insertion.

Other compounds have also been found to insert into MH bonds. For example, CO_2 gives metal formates;[135] CO gives formyls;[136] isocyanides give iminoacyl;[137] isocyanates give formamides;[137] carbodiimides give amidinates;[137] and Adams *et al.* have even found that CS_2 can react with two MH units to give $MSCH_2SM$.[137]

19.5.5 Homogeneous Catalysis

Homogeneous catalysis involving transition metal hydride complexes is such a large subject that several monographs have been devoted to it.[138-140] Table 2 shows some typical homogeneous catalysts, and lists the main reactions they catalyze. Coordination complexes are of direct relevance only to homogeneous catalysis, in which they are used in solution. Catalysts in which metal crystallites are supported on high surface area carbons or aluminas (heterogeneous catalystes) will not be covered.[141,142]

Table 2 Typical Homogeneous Catalysts Involving Hydride Intermediates and the Reactions they Catalyze

Catalyst	Reaction	Ref.
$RhCl(PPh_3)_3$	Alkene isomerization	148
	Alkene hydrogenation	148
$[Rh(nbd)(PPh_3)_2]$	Alkene hydrogenation	150
	Alkyne hydrogenation	150
	Directed hydrogenation	155a
$[Ir(cod)(PCy_3)(py)]^+$	Alkene hydrogenation	154
	Directed hydrogenation	155b, c
$[Co(CN)_5]^{3-}$	Hydrogenation	173, 156
$[Cr(CO)_6]$	*cis* 1,4-hydrogenation of dienes	174
$[Rh(nbd)(diop)]^+$	Asymmetric hydrogenation	152
H_2PtCl_6	Alkene hydrosilation	175
$HCO(CO)_4$	Alkene hydroformylation	7
$Ni\{P(OR)_3\}_4$	Alkene hydrocynation	130

A large proportion of homogeneous catalyst systems involve hydride complexes and we will discuss some of the major types. In many cases, the key features of hydride chemistry that lead to successful catalysis are (a) their ready formation from molecular hydrogen or other hydrogen-containing molecules, (b) the facile insertion of the resulting M—H bond into an unsaturated group, (c) insertions and other transformations of the group produced in step (b) and (d) the ready dissociation or reductive elimination of the product.

Often the coordination complex charged into the reactor, loosely called the 'catalyst' for the reaction, undergoes a change prior to entering the catalytic cycle. $[Rh(nbd)(PR_3)_2]^+$ irreversibly loses norbornane under H_2, for example. 'Catalyst precursor' or 'precatalyst' are therefore more precise terms for the complex charged. In the catalytic solutions themselves, essentially only one chemical species can often be present. This may or may not be on the direct catalytic cycle. It is sometimes convenient to refer to this species specifically as the 'majority species'. Finally, a deactivation process, often irreversible, can end the life of the catalyst either before or after the consumption of the reagents. This gives the 'deactivation product' or 'dead catalyst'. Kinetic studies can give valuable mechanistic information on catalytic cycles, but the complexity of many systems means that the analysis can be very difficult.[143] Spectroscopic studies are an important part of mechanistic work but, as mentioned above, species observed in catalytic reaction mixtures do not necessarily lie directly on the catalytic cycle. Authentic catalytic species are observed often enough to make the effort worth while, however.[117,118] Similar considerations apply as have been discussed in the case of enzyme kinetics.[145]

The catalytic cycle can often split up conceptually into several phases: the activation steps in which the reagents are bound at the metal; the rearrangement steps in which the product is assembled; and the product liberation step in which the catalyst is regenerated. An unsuccessful catalyst can fail at any point, and closer study can often show where this point is in the desired cycle. Steps can sometimes be taken to improve the system in the light of this knowledge. We will not discuss high oxidation state catalysts that do not involve hydride intermediates.[146]

Among the first homogeneous catalytic reactions to be investigated was hydrogenation. Calvin[6] described the reduction of quinone by copper(I) acetate in 1938. It was not until the development of the Wilkinson catalyst, however, that a practical system was discovered.[147] A detailed review by Jardine[148] covers the important properties of the complex. The catalyst now has a very great number of applications to organic synthesis. It has also been used for isomerization, decarbonylation and dehydrogenation, as well as its more traditional use in the selective *syn* addition of H_2 to C=C and C≡C groups. It picks up H_2 to form the dihydride (Figure 6) ($K = 18$ atm) which loses PPh_3 to allow access to the substrate. Insertion followed by reductive elimination complete the cycle. The mechanism has in fact proved to be rather more complicated than this simple picture suggests.[148,149] Catalysts of a closely related type can be formed from $[Rh(cod)Cl]_2$ and $2n$ moles of PPh_3. Optimum activity has been demonstrated for $n = 2$. In the case of PEt_3, the catalyst is active at $n = 2$ but quite inactive at $n = 3$ because PEt_3 dissociates less readily than PPh_3 from $[RhH_2Cl(PR_3)_3]$.

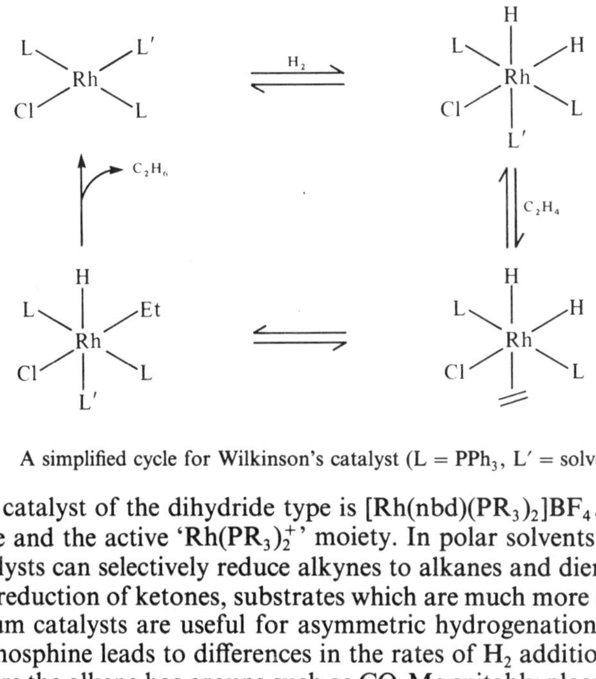

Figure 6   A simplified cycle for Wilkinson's catalyst (L = PPh_3, L′ = solvent or L)

Another important catalyst of the dihydride type is $[Rh(nbd)(PR_3)_2]BF_4$.[150] This reacts with H_2 to liberate norbornane and the active '$Rh(PR_3)_2^+$' moiety. In polar solvents $[RhH_2(solv)_2(PR_3)_2]^+$ is obtained. Such catalysts can selectively reduce alkynes to alkanes and dienes to monoenes. They are also active for the reduction of ketones, substrates which are much more difficult to reduce.[150,151]

The cationic rhodium catalysts are useful for asymmetric hydrogenation.[152] In this variant, the presence of a chiral phosphine leads to differences in the rates of H_2 addition to the two faces of a prochiral alkene. Where the alkene has groups such as CO_2Me suitably placed to bind to the metal, the selectivity can become very great; enantiomeric excesses of the product over its enantiomer can reach 95–98% (equation 67). The mechanism has recently been elucidated by Halpern.[153]

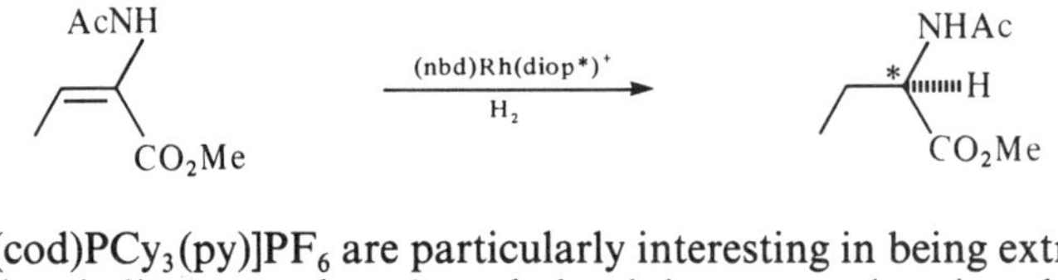

(67)

The catalysts $[Ir(cod)PCy_3(py)]PF_6$ are particularly interesting in being extremely active and able to reduce even hindered alkenes, such as 1-methylcyclohexene, and various keto-steroids.[154] Potential poisons such as SR or Hal groups or even air do not affect the catalyst. The cationic rhodium and iridium complexes are significant in binding hard ligands such as keto and alcohol groups. This leads to the phenomenon of directivity (equation 68),[155] in which a C=O, OH or similar group on the substrate alkene binds the catalyst in such a way that H_2 is added preferentially to one face of the substrate or to one C=C bond rather than to one another. This will certainly find many applications in organic synthesis,[155c] where control of stereochemistry is so important. The cationic

Rh catalysts are also especially effective in asymmetric hydrogenation[152] for the same reason: because the metal can bind to ligating groups in the substrate to give a rigid chelated metal–substrate complex. The reason that hard groups, such as OH, are bound seems to be that the net positive charge makes the metal 'harder'.

$$\text{Ir(cod)PCy}_3\text{(py)}^+ \quad (68)$$

A different class of catalyst is exemplified by $HCo(CN)_5^{3-}$, formed by homolytic activation of H_2 by $Co(CN)_5^{3-}$ ion. These catalysts are particularly effective for reducing activated alkenes, for example $XCH{=}CH_2$ to XEt (where X = CO_2H, CO_2R, Ph, CN) and 1,3-dienes to monoenes. Radical intermediates have been implicated in these reactions.[156]

Ruthenium compounds are also active catalysts.[157] For example $[(C_6Me_6)RuCl_2]_2$ efficiently reduces arenes. Transfer hydrogenation, in which the hydrogen comes from the solvent (*e.g.* Pr^iOH) is, also commonly observed.[159] Carbonylation reactions such as hydroformylation[7] are of great commercial importance.[138]

A very recent discovery is stoichiometric and catalytic alkane dehydrogenation.[65a,65b,132,160] In this process hydrogen is abstracted from an alkane, to give the corresponding alkene and passed to a hydrogen acceptor: $Bu^tCH{=}CH_2$ seems to be the reagent of widest applicability.[65a] This is particularly important in view of the long-standing interest in alkane activation in general.[160] Careful work was required to show that these systems really are homogeneous. This is a general problem in the field of homogeneous catalysis and experimental methods for making the distinction between homogeneous and heterogeneous systems are being developed.[161,162]

19.5.6 Photochemistry

The photochemistry of transition metal hydrides was first studied by M. L. H. Green *et al.*[40] who showed that reactions due to tungstenocene were observed on photoextrusion of H_2 from Cp_2WH_2. Reaction with benzene generated Cp_2WPhH, for example. More recently, it has become clear that this is a general reaction. For example, $IrH_3(PR_3)_3$ gives $IrH(PR_3)_3$ [163] and $CpIr(PR_3)H_2$ gives alkane C—H activation, chemistry believed to be derived from '$CpIr(PR_3)$'.[65c,48a] Concerted H_2 loss seems most probable but the details of the reaction are unclear. It is not known whether the photoproduct fragments react in an electronically excited state or decay to the ground state first. Rothwell *et al.*[164] have observed a photoinduced α hydrogen abstraction in $(RO)_2TaMe_3$ to give $(RO)_2Ta(CH_2)Me$ and methane, which presumably goes *via* a hydride intermediate.

ACKNOWLEDGEMENTS

I thank Dr Gregory Hlatky and Dr Mark Burk for helpful comments.

19.6 REFERENCES

1. (a) W. Hieber and F. Leutert, *Naturwissenschaften*, 1931, **19**, 360; W. Hieber and F. Leutert, *Z. Anorg. Allg. Chem.*, 1932, **204**, 145.
2. R. V. G. Ewens and M. W. Lister, *Trans. Faraday Soc.*, 1939, **35**, 681.
3. N. V. Sidgwick, 'The Chemical Elements and their Compounds', 1st edn., Oxford University Press, Oxford, 1950.
4. W. F. Edgell, C. Magee and G. Gallup, *J. Am. Chem. Soc.*, 1956, **78**, 4185, 4188.
5. S. C. Abrahams, A. P. Ginsberg and K. Knox, *Inorg. Chem.*, 1964, **3**, 558.
6. M. Calvin, *Trans. Faraday Soc.*, 1938, **34**, 1181.
7. O. Roelen, *Ger. Pat.*, 103 362 (1938).
8. J. M. Birmingham and G. W. Wilkinson, *J. Am. Chem. Soc.*, 1955, **77**, 3421.
9. J. Chatt, R. Duncanson and B. L. Shaw, *Proc. Chem. Soc.*, 1957, 343.
10. J. C. Green and M. L. H. Green, in 'Comprehensive Inorganic Chemistry', Pergamon, Oxford, 1973, vol. 4, p. 355.
11. R. G. Teller and R. Bau, *Struct. Bonding. (Berlin)*, 1981, **44**, 1.
12. A. P. Humphries and H. D. Kaesz, *Prog. Inorg. Chem.*, 1979, **25**, 145.
13. G. G. Hlatky and R. H. Crabtree, *Coord. Chem. Rev.*, 1985, **65**, 1.

="712"

14. (a) T. V. Ashworth and E. Singleton, *J. Chem. Soc., Chem. Commun.*, 1976, 705; (b) G. J. Kubas, R. R. Ryan, P. J. Vergamini and H. Wasserman, *J. Am. Chem. Soc.*, 1984, **106**, 451; G.J. Kubas, R.R. Ryan and D.A. Wrobleski, *J. Am. Chem. Soc.*, 1986, **108**, 1339; H.J. Wasserman, G.J. Kubas and R.R. Ryan, *J. Am. Chem. Soc.*, 1986, **108**, 2294; (c) R. K. Upmakis, G. E. Gadd, M. Poliakoff, M. B. Simpson, J. J. Turner, R. Whyman and A. F. Simpson, *J. Chem. Soc., Chem. Commun.*, 1985, 27; S. P. Church, F.-W. Grevels, H. Herman and K. Schaffner, *J. Chem. Soc., Chem. Commun.*, 1985, 30; (d) R. H. Crabtree and M. Lavin, *J. Chem. Soc., Chem. Commun.*, 1985, 794 and 1161; *J. Am. Chem. Soc.*, 1986, **108**, 4032; (e) R. H. Crabtree and D. Hamilton, *J. Am. Chem. Soc.*, 1986, **108**, 3321; (f) R. H. Crabtree, *Inorg. Chim. Acta*, 1986, **125**, L7.
15. R. Bau, D. M. Ho and S. G. Gibbins, *J. Am. Chem. Soc.*, 1981, **103**, 4960.
16. J. M. Mayer and J. E. Bercaw, *J. Am. Chem. Soc.*, 1982, **104**, 2157.
17. S. J. Holmes and R. R. Schrock, *Organometallics*, 1983, **2**, 1463.
18. M. L. Luetkens, Jr., W. L. Elcesser, J. C. Huffman and A. P. Sattelberger, *J. Chem. Soc., Chem. Commun.*, 1983, 1072.
19. R. D. Wilson, T. F. Koetzle, D. W. Hart, A. Kvick, D. L. Tipton and R. Bau, *J. Am. Chem. Soc.*, 1977, **99**, 1775.
20. J. Manriquez, P. J. Fagan, T. J. Marks, V. W. Day and C. S. Day, *J. Am. Chem. Soc.*, 1978, **100**, 7112; J. Manriquez, P. J. Fagan and T. J. Marks, *J. Am. Chem. Soc.*, 1978, **100**, 3939.
21. D. W. Hart, R. G. Teller, C. Y. Wei, R. Bau, G. Longoni, S. Campanella, P. Chini and T. F. Koetzle, *J. Am. Chem. Soc.*, 1981, **103**, 1458.
22. M. Kulzick, R. T. Price, E. L. Muetterties and V. W. Day, *Organometallics*, 1982, **1**, 1256.
23. D. F. Chodosh, R. H. Crabtree, H. Felkin, S. Morehouse and G. E. Morris, *Inorg. Chem.*, 1982, **21**, 1307.
24. A. Bino and F. A. Cotton, *J. Am. Chem. Soc.*, 1979, **101**, 4150.
25. J. Muller, H. Menig and J. Pickardt, *Angew. Chem., Int. Ed. Engl.*, 1982, **20**, 401.
26. N. W. Alcock, O. W. Howarth, P. Moore and G. E. Morris, *J. Chem. Soc., Chem. Commun.*, 1979, 1160.
27. J. B. Kiester and J. R. Shapley, *Inorg. Chem.*, 1982, **21**, 3304; R. D. Adams, D. A. Katahira and J. P. Selegue, *J. Organomet. Chem.*, 1981, **213**, 259.
28. L. M. Venanzi, *Coord. Chem. Rev.*, 1982, **43**, 251.
29. M. Brookhart and M. L. H. Green, *J. Organomet. Chem.*, 1983, **250**, 395.
30. E. L. Muetterties, 'Transition Metal Hydrides', Dekker, New York, 1971.
31. J. Zhuang, J. M. Hastings, L. M. Corliss, R. Bau, C. Y. Wei and R. O. Moyer, Jr., *J. Solid State Chem.*, 1981, **40**, 352.
32. J. Halpern, *Pure Appl. Chem.*, 1979, **51**, 2171.
33. (a) R. H. Crabtree, A. E. Segmuller and R. J. Uriarte, *Inorg. Chem.*, 1985, **24**, 1949; (b) R. H. Crabtree, G. G. Hlatky and E. M. Holt, *J. Am. Chem. Soc.*, 1983, **105**, 7302.
34. B. N. Chaudret, D. J. Cole-Hamilton, R. S. Nohr and G. Wilkinson, *J. Chem. Soc., Dalton Trans.*, 1977, 1546; T. Yoshida, T. Matsuda, T. Okano, T. Kitani and S. Otsuka, *J. Am. Chem. Soc.*, 1979, **101**, 2027.
35. J. Chatt and B. T. Heaton, *J. Chem. Soc. (A)*, 1968, 2745.
36. R. H. Crabtree, H. Felkin, T. Fillebeen-Khan and G. E. Morris, *J. Organomet. Chem.*, 1979, **168**, 183.
37. M. L. H. Green, *Pure Appl. Chem.*, 1978, **50**, 27; R. B. Calvert and J. R. Shapley, *J. Am. Chem. Soc.*, 1978, **100**, 7726.
38. E. Carmona-Guzman and G. W. Wilkinson, *J. Chem. Soc., Dalton Trans.*, 1977, 1716.
39. L. F. Rhodes, J. D. Zubkowski, K. Folting, J. C. Huffman and K. G. Caulton, *Inorg. Chem.*, 1982, **21**, 4185.
40. M. L. H. Green, *Pure Appl. Chem.*, 1978, **50**, 27.
41. J. R. Norton, *Acc. Chem. Res.*, 1979, **12**, 139.
42. K. Wade, *Adv. Inorg. Chem. Radiochem.*, 1976, **18**, 1.
43. R. Mason, K. M. Thomas and D. M. P. Mingos, *J. Am. Chem. Soc.*, 1973, **95**, 3802.
44. R. Hoffman, *Science*, 1981, **211**, 995.
45. J. W. Lauher, *J. Am. Chem. Soc.*, 1978, **100**, 5305.
46. B. K. Teo, *J. Chem. Soc., Chem. Commun.*, 1983, 1362.
47. H. D. Kaesz and R. B. Saillant, *Chem. Rev.*, 1972, **72**, 231.
48. (a) J. Y. Saillard and R. Hoffmann, *J. Am. Chem. Soc.*, 1984, **106**, 2006 and refs. therein; (b) R. H. Crabtree and G. G. Hlatky, *Inorg. Chem.*, 1980, **19**, 571; (c) C. E. Johnson, B. J. Fisher and R. Eisenberg, *J. Am. Chem. Soc.*, 1983, **105**, 7772.
49. P. J. Brothers, *Prog. Inorg. Chem.*, 1981, **28**, 1.
50. B. Bell, J. Chatt and G. J. Leigh, *Chem. Commun.*, 1970, 842.
51. J. Reedijk, *Comments Inorg, Chem.*, 1982, **1**, 379.
52. C. White, S. J. Thompson and P. M. Maitlis, *J. Chem. Soc., Dalton Trans.*, 1977, 1654.
53. D. F. Shriver, *Acc. Chem. Res.*, 1970, **3**, 231.
54. J. C. Kotz and D. G. Pedrotty, *Organomet. Chem. Rev. (A)*, 1969, **4**, 479.
55. P. Chini, G. Longoni and V. G. Albano, *Adv. Organomet. Chem.*, 1976, **14**, 285.
56. E. O. Fischer, K. Fichtel and K. Oefele, *Chem. Ber.*, 1962, **95**, 249.
57. A. J. Deeming and B. L. Shaw, *J. Chem. Soc. (A)*, 1969, 443.
58. C. R. Eady, B. F. G. Johnson and J. Lewis, *J. Chem. Soc., Dalton Trans.*, 1977, 838.
59. J. Chatt, R. S. Coffey and B. L. Shaw, *J. Chem. Soc.*, 1965, 7391.
60. (a) R. H. Crabtree and G. G. Hlatky, *Inorg. Chem.*, 1982, **21**, 1273; (b) R. H. Crabtree and G. G. Hlatky, *J. Organomet. Chem.*, 1982, **238**, C21.
61. C. Crocker, R. J. Errington, W. S. McDonald, K. J. Odell, B. L. Shaw and R. J. Goodfellow, *J. Chem. Soc., Chem. Commun.*, 1979, 498 and refs. therein.
62. R. R. Schrock, *Acc. Chem. Res.*, 1979, **12**, 98.
63. Z. Dawoodi, M. L. H. Green, V. S. B. Mtetwa and K. Prout, *J. Chem. Soc., Chem. Commun.*, 1982, 802, 1410.
64. M. I. Bruce, *Angew. Chem., Int. Ed. Engl.*, 1977, **16**, 73; I. Omae, *Coord. Chem. Rev.*, 1980, **32**, 235.
65. (a) R. H. Crabtree, J. M. Mihelcic, J. M. Quirk and M. F. Mellea, *J. Am. Chem. Soc.*, 1979, **101**, 7738; 1982, **104**, 107; M.J. Burk and R.H. Crabtree, *J. Chem. Soc., Chem. Commun.*, 1985, 1829; R.H. Crabtree, *Chem. Rev.*, 1985, **85**, 245; (b) D. Baudry, M. Ephritikhine and H. Felkin, *J. Chem. Soc., Chem. Commun.*, 1980, 1243; 1982, 606; (c) A. H. Janowicz and R. G. Bergman, *J. Am. Chem. Soc.*, 1982, **104**, 352; (d) J. K. Hoyano and W. A. G. Graham, *J. Am. Chem. Soc.*, 1982, **104**, 3722; (e) W. D. Jones and F. J. Feher, *J. Am. Chem. Soc.*, 1982, 104, 4240.

66. (a) J. Halpern, *Acc. Chem. Res.*, 1982, **15**, 238; *Pure Appl. Chem.*, 1979, **51**, 2171; J. A. Connor, *Top. Curr. Chem.*, 1977, **71**, 71; (b) J. Halpern, personal communication, 1983.
67. L. F. Halle, P. B. Armentrout and J. L. Beauchamp, *Organometallics*, 1982, **1**, 963.
68. D. Gregson, J. A. K. Howard, J. N. Nicholls, J. L. Spencer and D. G. Turner, *J. Chem. Soc., Chem. Commun.*, 1980, 572.
69. P. G. Owston, J. M. Partridge and J. M. Rowe, *Acta Crystallogr.*, 1960, **13**, 246.
70. M. R. Churchill, P. H. Bird, H. D. Kaesz, R. Bau and R. Fontal, *J. Am. Chem. Soc.*, 1968, **90**, 7135.
71. S. J. LaPlaca and J. A. Ibers, *J. Am. Chem. Soc.*, 1963, **85**, 3501; *Acta Crystallogr.*, 1965, **18**, 511.
72. R. D. Wilson and R. Bau, *J. Am. Chem. Soc.*, 1976, **98**, 4687.
73. M. R. Churchill, B. R. deBoer, F. J. Rotella, E. W. Abel and R. J. Rowley, *J. Am. Chem. Soc.*, 1975, **97**, 7158.
74. A. G. Orpen, A. V. Rivera, E. G. Bryan, D. Pippard and G. M. Sheldrick, *J. Chem. Soc., Chem. Commun.*, 1978, 723.
75. E. A. McNeill and F. R. Scholer, *J. Am. Chem. Soc.*, 1977, **99**, 6243.
76. J. W. Lauher and K. Wald, *J. Am. Chem. Soc.*, 1981, **103**, 7648.
77. R. W. Baker and P. Pauling, *Chem. Commun.*, 1969, 1495.
78. S. J. LaPlaca, W. C. Hamilton, J. A. Ibers and A. Davison, *Inorg. Chem.*, 1969, **8**, 1928.
79. J. Roziere, J. M. Williams, R. P. Stewart, Jr., J. L. Petersen and L. F. Dahl, *J. Am. Chem. Soc.*, 1977, **99**, 4497.
80. M. Berry, N. J. Cooper, M. L. H. Green and S. J. Simpson, *J. Chem. Soc., Dalton Trans.*, 1980, 29.
81. M. Y. Chiang, R. Bau, G. Minghetti, A. L. Bandini, G. Banditelli and T. F. Koetzle, *Inorg. Chem.*, 1984, **23**, 122.
82. J. W. Lauher and K. Wald, *J. Am. Chem. Soc.*, 1981, **103**, 7648; J. Silvestre and T. A. Albright, *Isr. J. Chem.*, 1983, **23**, 139.
83. F. Farmery and M. Kilner, *J. Chem. Soc. (A)*, 1970, 634.
84. M. Freni and P. Romiti, *Inorg. Nucl. Chem. Lett.*, 1970, **6**, 167.
85. J. Chatt and G. J. Leigh, *Angew. Chem., Int. Ed. Engl.*, 1978, **17**, 400.
86. P. Rigo, M. Bressan and A. Morvillo, *J. Organomet. Chem.*, 1975, **92**, C15; 1975, **93**, C34.
87. V. B. Polyakova, A. P. Borisov, V. D. Makhaev and K. N. Semenenko, *Koord. Khim.*, 1980, **6**, 743.
88. B. L. Shaw, *J. Organomet. Chem.*, 1975, **94**, 251.
89. J. Chatt, R. S. Coffey and B. L. Shaw, *J. Chem. Soc.*, 1965, 7391.
90. L. Vaska, *J. Am. Chem. Soc.*, 1966, **88**, 4100.
91. J. P. Jesson, in 'Transition Metal Hydrides', Dekker, New York, 1971, chap. 4.
92. P. G. Douglas and B. L. Shaw, *J. Chem. Soc. (A)*, 1970, 334; W. F. Edgell, G. Asato, W. Wilson and C. Angell, *J. Am. Chem. Soc.*, 1959, **81**, 2022.
93. J. W. White and C. J. Wright, *Chem. Commun.*, 1970, 970 and 971.
94. M. W. Howard, V. A. Jayasooriya, S. F. A. Kettle, D. B. Powell and N. Sheppard, *J. Chem. Soc., Chem. Commun.*, 1979, 18.
95. S. J. McLain, C. D. Wood and R. R. Schrock, *J. Am. Chem. Soc.*, 1977, **99**, 3519.
96. M. Ehenann and H. Noth, *Z. Anorg. Allg. Chem.*, 1971, **386**, 87.
97. A. Davison and J. W. Faller, *Inorg. Chem.*, 1967, **6**, 845.
98. L. L. Lohr, Jr. and W. N. Lipscomb, *Inorg. Chem.*, 1964, **3**, 22; A. D. Buckingham and P. J. Stephens, *J. Chem. Soc.*, 1964, 2747.
99. J. W. Faller, personal communication, 1983.
100. B. E. Mann, C. Masters and B. L. Shaw, *J. Inorg. Nucl. Chem.*, 1971, **33**, 2195; L. J. Archer and T. A. George, *Inorg. Chem.*, 1979, **18**, 2079.
101. M. J. Fernandez and P. M. Maitlis, *J. Chem. Soc., Chem. Commun.*, 1982, 310.
102. F. N. Tebbe, P. Meakin, J. P. Jesson and E. L. Muetterties, *J. Am. Chem. Soc.*, 1970, **92**, 1068.
103. (a) R. H. Crabtree, H. Felkin and G. E. Morris, *J. Organomet. Chem.*, 1977, **141**, 205; (b) D. F. Chodosh, R. H. Crabtree, H. Felkin, S. Morehouse and G. E. Morris, *Inorg. Chem.*, 1982, **21**, 1307.
104. E. C. Constable, B. F. G. Johnson, J. Lewis, G. N. Pain and M. J. Taylor, *J. Chem. Soc., Chem. Commun.*, 1982, 754.
105. R. K. Harris and B. E. Mann, 'NMR and the Periodic Table', Academic, London, 1978.
106. J. W. Faller, Personal Communication, 1984.
107. P. Chini, L. Colli and M. Peraldo, *Gazz. Chim. Ital.*, 1960, **90**, 1005.
108. P. Meakin, L. J. Guggenberger, W. G. Peet, E. L. Muetterties and J. P. Jesson, *J. Am. Chem. Soc.*, 1973, **95**, 1467.
109. M. Saunders, M. H. Jaffe and P. Vogel, *J. Am. Chem. Soc.*, 1971, **93**, 2558; R. B. Calvert, J. R. Shapley, A. J. Schulz, J. M. Williams, S. L. Suib and G. D. Stucky, *J. Am. Chem. Soc.*, 1978, **100**, 6240.
110. A. D. Buckingham, J. P. Yesinowski, A. J. Canty and A. J. Rest, *J. Am. Chem. Soc.*, 1973, **95**, 2732.
111. H. H. Brintzinger, *J. Am. Chem. Soc.*, 1967, **89**, 6871.
112. G. M. Bancroft, M. J. Mays, B. E. Prater and F. P. Stefanini, *J. Chem. Soc. (A)*, 1970, 2146.
113. B. F. G. Johnson, J. Lewis and P. W. Robinson, *J. Chem. Soc. (A)*, 1970, 1684.
114. T. G. Appleton, H. C. Clark and L. E. Manzer, *Coord. Chem. Rev.*, 1973, **10**, 335; E. M. Shustorovich, M. A. Porai-Koshits and Yu. A. Buslaev, *Coord. Chem. Rev.*, 1975, **17**, 1.
115. C. H. Lang Ford and H. B. Gray, 'Ligand Substitution Processes', Benjamin, New York, 1965, chap. 2.
116. I. I. Chernayev, *Izv. Inst. Platiny Akad. Nauk SSR*, 1926, **4**, 261.
117. P. Meakin, J. P. Jesson and C. A. Tolman, *J. Am. Chem. Soc.*, 1972, **94**, 3240.
118. L. Vaska, *J. Am. Chem. Soc.*, 1966, **88**, 4100.
119. C. D. Falk and J. Halpern, *J. Am. Chem. Soc.*, 1965, **87**, 3003.
120. R. Grey, G. P. Pez, A. Wallo and J. Corsi, *J. Am. Chem. Soc.*, 1981, **103**, 7528 and 7536.
121. M. Lappert and P. W. Lednor, *Adv. Organomet. Chem.*, 1976, **14**, 345.
122. J. Hanzlik and A. A. Vlcek, *Inorg. Chem.*, 1969, **8**, 669.
123. K. C. Dewhirst, *Inorg. Chem.*, 1966, **5**, 319; W. H. Baddeley and M. S. Fraser, *J. Am. Chem. Soc.*, 1969, **91**, 3661.
124. R. F. Jordan and J. R. Norton, *J. Am. Chem. Soc.*, 1982, **104**, 1255.
125. J. Evans, S. J. Okrasinski, A. J. Pribula and J. R. Norton, *J. Am. Chem. Soc.*, 1976, **98**, 4000.
126. H. Dadkhah, N. Kashef, R. L. Richards, D. L. Hughes and A. J. L. Pombeiro, *J. Organomet. Chem.*, 1983, **255**, C1.
127. J. Chatt, R. S. Coffey, A. Gough, D. T. Thompson, *J. Chem. Soc. (A)*, 1968, 190.
128. J. W. Byrne, H. U. Blaser and J. A. Osborn, *J. Am. Chem. Soc.*, 1975, **97**, 3871.
129. D. L. Thorn and R. Hoffman, *J. Am. Chem. Soc.*, 1978, **100**, 2079.

130. (a) R. H. Crabtree, M. F. Mellea and J. M Quirk, *J. Am. Chem. Soc.*, 1984, **106**, 2913; see also (b) C. A. Tolman, W. C. Seidel, J. D. Druliner and P. J. Domaille, *Organometallics*, 1984, **3**, 33.
131. D. J. Cole-Hamilton and G. Wilkinson, *Nouv. J. Chim.*, 1977, **1**, 141.
132. M. J. Burk, R. H. Crabtree, C. P. Parnell and R. J. Uriarte, *Organometallics*, 1984, **3**, 816.
133. R.H. Crabtree, R.P. Dion, D.J. Gibboni, D.V. McGrath and E.M. Holt, *J. Am. Chem. Soc.*, 1986, **108**, 7222.
134. B. R. Booth and R. G. Hargreaves, *J. Chem. Soc. (A)*, 1969, 2766.
135. D. J. Darensbourg and R. A. Kudaroski, *Adv. Organomet. Chem.*, 1983, **22**, 129.
136. B. A. Woods, B. B. Wayland, V. M. Minda, *Abstr. Papers*, 186th ACS Meeting, Washington DC, INOR 252.
137. R. D. Adams, *Acc. Chem. Res.*, 1983, **16**, 57.
138. G. W. Parshall, 'Homogeneous Catalysis', Wiley, New York, 1980.
139. C. Masters, 'Homogeneous Transition Metal Catalysis', Chapman and Hall, London, 1981.
140. B. R. James, 'Homogeneous Hydrogenation', Wiley, New York, 1976.
141. G. A. Somorjai, 'Chemistry in Two Dimensions', Cornell University Press, Ithaca, NY, 1981.
142. E. L. Muetterties, *Angew. Chem., Int. Ed. Engl.*, 1978, **17**, 545.
143. C. A. Tolman and J. W. Faller, in 'Homogeneous Catalysis with Metal Phosphine Complexes', ed. L. H. Pignolet, Plenum, New York, 1983, p. 13.
144. R. H. Crabtree, P. C. Demou, D. Eden, J. M. Mihelcic, C. A. Parnell, J. M. Quirk and G. E. Morris, *J. Am. Chem. Soc.*, 1982, **104**, 6994.
145. J. W. Cornforth, *Chem. Soc. Rev.*, 1973, **2**, 1.
146. R. A. Sheldon and J. K. Kochi, 'Metal Catalysed Oxidation of Organic Compounds', Academic, New York, 1981.
147. J. A. Osborn, F. H. Jardine, J. F. Young and G. Wilkinson, *J. Chem. Soc. (A)*, 1966, 1711.
148. F. H. Jardine, *Prog. Inorg. Chem.*, 1984, **31**, 265.
149. J. Halpern, *J. Mol. Catal.*, 1976, **2**, 65.
150. R. R. Schrock and J. A. Osborn, *J. Am. Chem. Soc.*, 1976, **98**, 2134 and 4450.
151. R. H. Crabtree, A. Gautier, G. Giordano and T. Khan, *J. Organomet. Chem.*, 1977, **141**, 113.
152. H. B. Kagan and T. P. Dang, *J. Am. Chem. Soc.*, 1972, **94**, 6429; B. R. James, *Adv. Organomet. Chem.*, 1979, **17**, 319.
153. A. S. C. Chan, J. J. Pluth and J. Halpern, *J. Am. Chem. Soc.*, 1982, **102**, 5952.
154. R. H. Crabtree and G. E. Morris, *J. Organomet. Chem.*, 1977, **135**, 395; R. H. Crabtree, H. Felkin and G. E. Morris, *J. Organomet. Chem.*, 1977, **141**, 205; J. W. Suggs, S. D. Cox, R. H. Crabtree and J. M. Quirk, *Tetrahedron Lett.*, 1981, **22**, 303.
155. (a) J. M. Brown and R. G. Naik, *J. Chem. Soc., Chem. Commun.*, 1982, 348; (b) R. H. Crabtree and M. W. Davis, *Organometallics*, 1983, **2**, 681; (c) G. Stork and D. E. Kahne, *J. Am. Chem. Soc.*, 1983, **105**, 1072.
156. J. Kwiatek, I. L. Mador and S. K. Seyler, *J. Am. Chem. Soc.*, 1962, **84**, 304.
157. D. Evans, J. A. Osborn, F. H. Jardine and G. Wilkinson, *Nature (London)*, 1965, **208**, 1203.
158. M. A. Bennett, T. N. Huang and T. W. Turney, *J. Chem. Soc., Chem. Commun.*, 1979, 312.
159. I. S. Kolomnikov, V. P. Kukolev and M. E. Volpin, *Russ. Chem. Rev. (Engl. Transl.)*, 1974, **43**, 399; G. Brieger and T. J. Nestrick, *Chem. Rev.*, 1974, **74**, 567.
160. E. L. Muetterties, *Chem. Soc. Rev.*, 1982, **11**, 283.
161. J. E. Hamlin, K. Hirai, A. Millan and P. M. Maitlis, *J. Mol. Catal.*, 1980, **7**, 543.
162. D. R. Anton and R. H. Crabtree, *Organometallics*, 1983, **2**, 855.
163. G. L. Geoffroy and R. Pierantozzi, *J. Am. Chem. Soc.*, 1976, **98**, 8054.
164. L. R. Chamberlain, A. P. Rothwell and I. P. Rothwell, *J. Am. Chem. Soc.*, 1984, **106**, 1847.
165. F. N. Tebbe, *J. Am. Chem. Soc.*, 1973, **95**, 5823.
166. R. B. Wilson, A. P. Sattleberger and J. C. Huffman, *J. Am. Chem. Soc.*, 1982, **104**, 858.
167. F. Penella, *Chem. Commun.*, 1971, 158.
168. B. Bell, J. Chatt and G. J. Leigh, *J. Chem. Soc., Chem. Commun.*, 1972, 34.
169. J. J. Levison and S. D. Robinson, *J. Chem. Soc. (A)*, 1970, 2947.
170. R. Wilczynski, W. A. Fordyce and J. Halpern, *J. Am. Chem. Soc.*, 1983, **105**, 2066.
171. J. Chatt, R. S. Coffey and B. L. Shaw, *J. Chem. Soc.*, 1965, 7391.
172. V. D. Makhaev, A. P. Borisov, G. N. Boiko and K. N. Semenenko, *Koord. Khim.*, 1982, **8**, 963.
173. J. Kwatiek, *Catal. Rev.*, 1967, **1**, 37.
174. A. Rejoan and M. Cais, 'Progress in Coordination Chemistry', Elsevier, 1968, p. 32.
175. J. L. Speier, *Adv. Organomet. Chem.*, 1979, **17**, 404.

20.1

Schiff Bases as Acyclic Polydentate Ligands

MARIO CALLIGARIS and LUCIO RANDACCIO
Università di Trieste, Italy

20.1.1 INTRODUCTION

20.1.1.1 General Considerations

The purpose of this introductory section is to indicate the salient ground state properties of Schiff bases by using illustrative examples viewed from the point of view of their coordination chemistry, rather than by a comprehensive review of this class of compounds based on their organic chemistry, intended for an organic chemistry textbook. This implies that the classification we have used is mainly based on their potential use as ligands at a metal centre. In fact we believe that this classification is more pertinent to inorganic chemists, and should therefore be more useful to the reader of this chapter.

In the following subsections, after having briefly described the general properties of the most common Schiff bases, we will classify and describe them following the number of their potential donor atoms. Since the N donor set ligands are treated in another chapter, the majority of the ligands which we will discuss are those with an O,N donor set, these having been the most commonly used as ligands to transition metals.

20.1.1.2 Synthesis

The condensation of primary amines with aldehydes and ketones gives products known as imines which contain a C=N double bond. These compounds rapidly decompose or polymerize unless there is at least an aryl group bonded to the nitrogen or to the carbon atom. The latter imines are called Schiff bases, since their synthesis was first reported by Schiff.[1] The most common method of obtaining a Schiff base (4) is straightforward, as indicated in the condensation reaction (1) between (1) and (2) with the formation of an intermediate hemiaminal (3).

$$
\begin{array}{ccccccc}
\underset{\text{O}}{\overset{\text{R}\ \ \text{R}'}{\underset{\|}{\text{C}}}} & + \ \text{R}''\text{NH}_2 & \longrightarrow & \underset{\text{OH}}{\overset{\text{NHR}''}{\text{R}-\text{C}-\text{R}'}} & \longrightarrow & \text{H}_2\text{O} + & \underset{\text{R}\ \ \text{R}'}{\overset{\overset{\text{R}''}{\text{N}}}{\underset{\|}{\text{C}}}} \\
(1) & (2) & & (3) & & & (4)
\end{array}
\tag{1}
$$

Other methods of synthesis are widely reviewed by Dayagi and Degani.[2] However, it must be observed that few Schiff bases commonly used as ligands have been prepared and characterized in the uncomplexed state, since the corresponding metal complexes have been directly obtained by other procedures. For example, many metal complexes containing salenH$_2$ may be obtained directly by reaction between metal ions, salicylaldehydes and ethylenediamine.[3]

20.1.1.3 Spectroscopic Properties

The C=N stretching frequencies of the ligands occur in the region between 1680 and 1603 cm^{-1} when H, alkyl or Ar groups are bonded to C and N atoms. The nature of the different substituents on these atoms determines the position of the stretching frequency in the above range, *e.g.* Ar groups on the C and N atoms cause a shift of the frequency towards the lower part of the range. Thus, for compounds such as ArCH=NR a frequency range of 1657–1631 cm^{-1} has been reported. Typical frequencies are given in Table 1. Compounds of the type ArCH=NAr, with variously substituted aryl groups, exhibit a range of 1631–1613 cm^{-1} for ν(C=N) (Table 1).

The presence of an OH group at the 2-position of the phenyl ring of the BA residue effects a bathochromic shift as shown in Table 1. This has been attributed to intramolecular hydrogen bond formation, with the benzenoid form (5) preferred over the quinoid one (6). In these compounds, the phenolic C—O stretching vibration occurs between 1288 and 1265 cm^{-1}. Upon coordination to

Table 1 Typical C=N Stretching Frequencies for some R'C$_6$H$_4$CH=NCH$_2$R" (a) and R'C$_6$H$_4$CH=NC$_6$H$_4$R" (b) Derivatives in CHCl$_3$ Solutions[4]

	R'	R"	ν (cm^{-1})
(a)	H	2-OHC$_6$H$_4$	1657
	2-OH	2-OHC$_6$H$_4$	1634
	H	CH$_2$N=CHPh	1646
	2-OH	CH$_2$N=CHC$_6$H$_4$-2-OH	1634
(b)	H	H	1630
	4-OH	H	1629
	4-MeCONH	H	1629
	H	4-OH	1630
	H	2-OH	1626
	2-OH	4-NH$_2$	1624
	2-OH	H	1622
	2-OH	2-OH	1621

metal ions through both O and N, a decrease of the C—N frequency is generally observed, as will be shown in the following subsections. IR spectra and properties, covering the literature until 1970, are reported by Curran and Siggia,[4] and Sandorfy.[5]

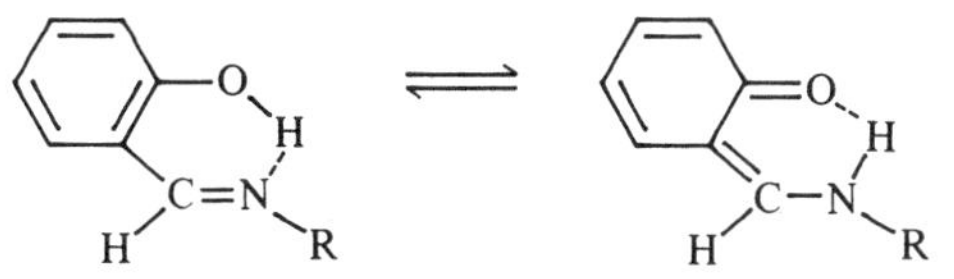

(5)   (6)

Scheme 1

NMR studies have been mainly applied to elucidating the structural features of Schiff bases in solution. These studies are mainly concerned with Schiff bases derived from benzaldehyde and its substituents, β-diketones, *o*-hydroxyacetophenones and *o*-hydroxyacetonaphthones. They were devoted to obtaining an insight into the keto–enol equilibrium (Scheme 1), *syn* and *anti* isomerism and steric distortions in different kinds of solvents. Relevant data and results are described in the following sections.

The UV spectra of compounds containing an unconjugated chromophore are characterized by bands due to $n \to \pi^*$ transitions in the range 235–272 nm. However, conjugation with alkenic or aryl groups causes large changes in the spectrum, since strong bands due to $\pi \to \pi^*$ transitions cover the rather weak $n \to \pi^*$ absorptions. Electronic spectra of many azomethines, including several Schiff bases, are reported by Bonnett.[6] Schiff bases derived from substituted SAL compounds and alkylamines generally exhibit four bands in the near UV that are solvent dependent. The band at about 400 nm, intense in ethanol, is almost absent in hexane. The presence of this band has been attributed to the tautomeric equilibrium of the kind shown in Scheme 1, the keto–amine tautomer being responsible for that absorption. This is confirmed by subsequent NMR studies (see Section 20.1.3).

Such tautomerization in the solid state has been suggested to be responsible for the thermochromism of many crystalline anils.[7] The band in the region of 240 nm, attributed to the $n \to \pi^*$ transition of the C=N chromophore, is optically active if the chromophore is dissymmetrically perturbed by appropriate substituents at the C and/or N atoms. Many studies of the ORD and CD spectra of the ligands have been made and those until 1970 have been widely reviewed by Bonnett.[6] Furthermore, the Cotton effect at 315 nm appears to be a good way of assigning the absolute configuration to imines of the kind $R^1R^2CHN=CHC_6H_4OH$, since the derivatives of (*S*) configuration exhibit a positive Cotton effect, while those of (*R*) configuration show a negative one.[8] The influence of the solvent appears to be important in determining the spectrum shape. However, examples are reported which show the difficulty of interpretation of the Cotton effects for open-chain compounds in simple conformational terms.[6]

20.1.1.4 Structural Properties

Despite the considerable number of papers which have appeared on the structure of Schiff base metal complexes, a relatively small number of free ligands have been structurally characterized. In addition, only two SALAN derivatives are reported in the reviews collecting structure determinations published up to 1965.[9] Therefore, no previous review on structural properties of free Schiff bases has been so far reported. Thus, these aspects will be fully described in the next subsections. However, it may be useful to report here the values quoted in ref. 9 of average C—N bond lengths for single, partial double and triple bonds (Table 2).

Table 2 Average Carbon–Nitrogen Bond Lengths

Bond	Bond length (Å)
R—N[a]	1.472(5)–1.479(5)
Ar—N[a]	1.426(12)
C—N (heterocyclic systems)	1.352(5)
N—C=O	1.322(2)
C≡N	1.158(2)

[a] R = alkyl, Ar = aryl.

20.1.1.5 Photochromism and Thermochromism

An interesting property of SALAN and its substituted derivatives is their ability to colour markedly under UV irradiation.[10] This photochromism has been attributed to the possible formation of a geometric isomer of the quinoid tautomer. On the other hand, these compounds also exhibit thermochromism, which consists of a thermally induced change in colour which increases with an increase of the temperature. This phenomenon has been attributed to intramolecular hydrogen transfer, such as that depicted in Scheme 1. Both processes are reversible and mutually exclusive for the same compound in a given crystalline form. However, since the same anil may exhibit polymorphism, it may be thermochromic in one crystalline form and photochromic in the other. Therefore, these processes should be correlated with differences in the crystal structure rather than with inherent properties of the molecule.[10] All the crystallographic results, which will be reported in Section 20.1.2.3, confirm the earlier hypothesis[11] that molecules which exhibit thermochromism are planar, while the others are non-planar, as shown in Figure 1.

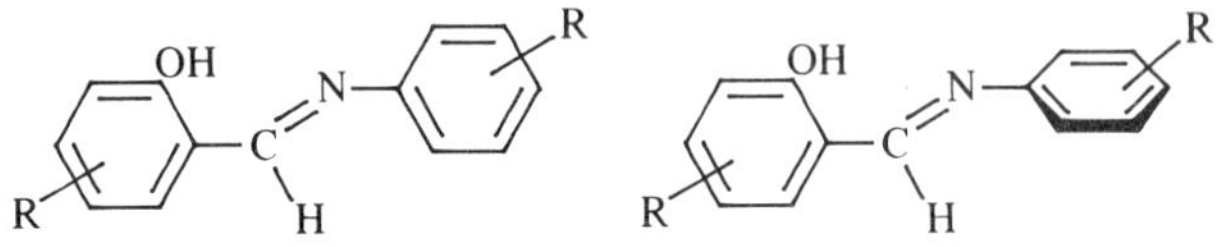

Molecular geometry	Planar	Not planar
Crystal packing	Stacking	Open
UV light	No change	Change
Heat	Change	No change

Figure 1 Relationship of thermochromism and photochromism with geometry and crystal structure of *N*-salicylaldehydoanilines

20.1.2 MONODENTATE SCHIFF BASES

20.1.2.1 Bonding to Metals

The monodentate Schiff bases should not be included in a section concerned with open polydentate ligands, but it is of interest to examine some of their most important properties which could be useful for comparison with polydentate analogues. It has been claimed that the basic strength of the C=N group is not sufficient to obtain stable complexes by coordination of the imino nitrogen atom to a metal ion.[12] Hence only the presence of at least one other donor atom, suitably near the nitrogen atom, should stabilize the metal–nitrogen bond through formation of chelate rings. On the other hand, it has been recently reported[13] that the Schiff base PhCH=NMe is able to act as a ligand in the Pd complex shown in Figure 2, although the suggested weak interaction Pd··· H may stabilize the coordination to the metal.

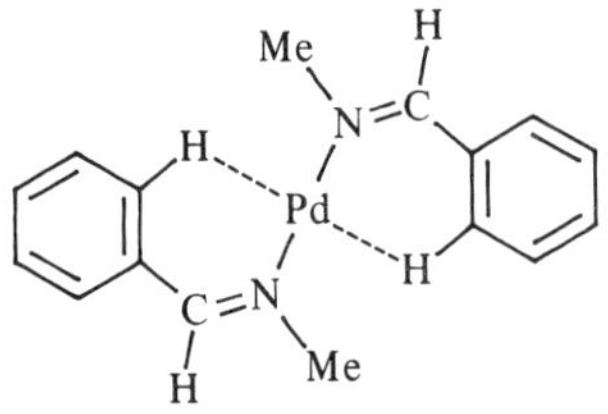

Figure 2 The structure of $PdCl_2(PhC=NMe)_2$, showing the coordination of the monodentate imine

20.1.2.2 Spectroscopic Properties

^1H and ^{13}C NMR spectra of monoimines have been reported and are shown in Table 3. These data fit the Hammett relationship and good correlations are found between various σ constants and ^{13}C chemical shifts at positions remote from the substituent.[14] ^{13}C chemical shifts of imines derived from BA and variously substituted benzylamines are affected by substituents through up to 11 bonds.[15] The influence of Z/E isomerism (Figure 3) and the effect of substituents on NMR spectra of some unsaturated imines have been studied.[16] The photochemically induced isomerization of several substituted Schiff bases derived from BA and arylamines and their chemical shifts, as well

Table 3 ¹H and ¹³C NMR Data for some Monoimines[a]

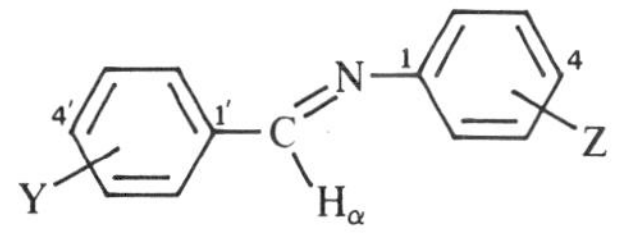

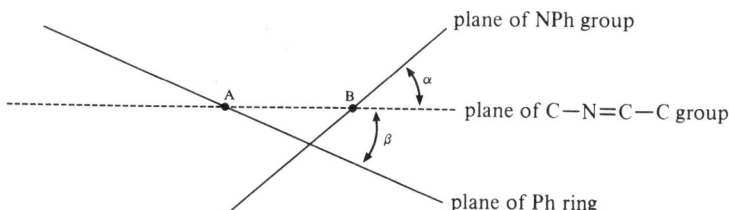

Y	Z	$\delta(H_a)$	$J(^{13}C–H_a)$	$\delta(C\text{-}4')$	$\delta(C\text{-}1')$	$\delta(C\text{-}7)$	$\delta(C\text{-}1)$	$\delta(C\text{-}4)$
p-NO₂	H	8.56	161.0	149.40	141.78	157.38	151.06	127.26
p-Cl		8.42	158.1	137.20	134.88	158.28	151.65	126.14
H		8.47	157.3	131.24	136.48	160.01	152.18	125.94
p-Me		8.41	156.5	141.58	134.02	159.87	152.45	125.73
p-OMe		8.38	156.1	162.39	129.52	129.52	152.58	125.60
p-NMe₂		8.28	154.1	153.24	124.74	124.74	152.65	125.07
H	*p*-NO₂	8.43	158.6	132.56	135.68	162.79	158.08	145.69
	p-Cl	8.43	157.5	131.57	136.21	160.40	150.59	131.57
	H	8.47	157.3	131.24	136.48	160.01	152.18	125.94
	p-Me	8.45	156.5	130.91	136.61	158.81	149.53	135.41
	p-OMe	8.48	156.6	131.04	136.81	157.95	144.96	158.55

[a] Chemical shifts are in p.p.m. from internal TMS; *J* values are in Hz. All data are for CDCl₃/CHCl₃ solutions (5:7 ratio).

as their possible photoisomers, have been observed in deuterated acetone solution by ¹H NMR spectroscopy. Differences up to about 0.8 p.p.m. between the two isomers of each Schiff base are observed.[17] Recently, it has also been reported[18] that lanthanide shift reagents cause isomerization of the stable form of this kind of Schiff base to the less stable *Z* form (Figure 3).

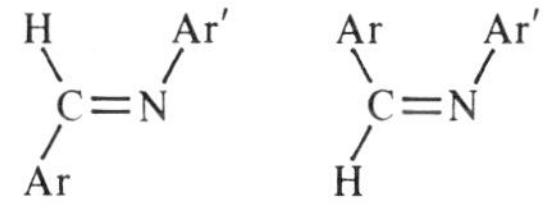

$$
\begin{array}{cc}
\overset{\displaystyle H}{\underset{\displaystyle Ar}{\diagdown \diagup}} C = N \overset{\displaystyle Ar'}{\diagup} & \overset{\displaystyle Ar}{\underset{\displaystyle H}{\diagdown \diagup}} C = N \overset{\displaystyle Ar'}{\diagup} \\
\textit{E} \text{ isomer} & \textit{Z} \text{ isomer}
\end{array}
$$

Figure 3 The *E* and *Z* isomers of arylimines

Photoelectron spectroscopic studies have shown that, in the gas phase, the angle α (Figure 4) is even smaller than that observed in the crystal (55°).[19]

Figure 4 Definition of α and β interplanar angles of Table 4. The positive sign refers to a counterclockwise rotation

20.1.2.3 Structural Properties

The ligand BAAN, variously substituted in the phenyl rings, has been the most extensively studied by X-ray analyses. Initial interest came from their anomalous solution spectral properties. In fact BAAN exhibits spectral differences with respect to the isoelectronic *trans*-azobenzene and *trans*-stilbene. These differences have been shown to be due to its non-planarity.[19] On the other hand, UV and ESR studies showed that other BAAN derivatives should be nearly planar or, alternatively, rather twisted in solution.[20] Since the spectral properties are a sensitive function of the molecular conformation, systematic studies on the solid state structure have been carried out. Furthermore, since the original work of Burgi and Dunitz,[21] the BAAN derivatives were thought to be a good model for development of the principles of 'crystal engineering',[22] which utilizes the solid state as a way of obtaining, among other objectives, desired molecular conformations, which determine some physical properties such as thermochromism and photochromism. The geometrical parameters defining the overall conformation of these Schiff bases and the geometry of the CN=CC moiety are collected in Table 4. The most interesting results may be summarized as follows.

Table 4 Relevant Structural Parameters of some BAAN Derivatives, $p\text{-}RC_6H_4\text{---}N\text{=}CH\text{---}C_6H_4\text{-}p\text{-}R'$

R	R'	Form[a]	Ph—N (Å)	N=C (Å)	C—Ph (Å)	α (°)[b]	β (°)[b]	Ref.
H	H		1.460(3)	1.237(3)	1.496(3)	55.2	−10.3	21
CO$_2$H	H		1.431(7)	1.281(12)	1.461(9)	41.1	−13.7	21
NO$_2$	Me		1.400(3)	1.269(5)	1.474(3)	50.2	−8.1	21
Cl	Cl	(I)	1.437(4)	1.262(4)	[1.437(4)]^c	0.4	0.4	29
Cl	Cl	(II)	1.429(5)	1.250(5)	[1.429(5)]^c	24.8	−24.8	30
Br	Br		1.45(1)	1.24(1)	[1.45(1)]^c	1.9	1.9	31
Br	H		1.41(1)^d	1.28(1)	1.48(1)^d	39.0; 46.2	−11.1; −9.4	32
H	Cl, o-Cl		1.41(1)	1.27(1)	1.48(1)	50.2	−8.1	33
NMe$_2$	NO$_2$		1.416(4)	1.258(4)	1.460(4)	−9.2	−3.5	34
NO$_2$	NMe$_2$	(I)	1.399(3)^d	1.279(4)^d	1.452(3)^d	41.5; 49.0	−11.4; −7.7	34
NO$_2$	NMe$_2$	(II)	1.394(3)^d	1.281(3)^d	1.440(3)^d	43.1; 45.8	−9.7; −8.7	35
NO$_2$	NMe$_2$	(III)	1.398(5)	1.282(5)	1.446(5)	43.7	−7.6	35
NO$_2$	NMe$_2$	(IV)	1.407(9)^d	1.272(9)^d	1.442(9)^d	15.1–50.4	−8.4–11.1	35
MeO	OH		1.425(3)	1.287(3)	1.454(3)	44.4	0.3	26
Me	Me	(I)	1.459(7)	1.253(7)	1.493(7)	2.3; 10.2	2.8; 9.9	36
Me	Me	(II)	1.411(4)	1.265(4)	1.456(4)	41.7	−3.0	37
Me	Me	(III)		e		4.9^f	4.9^f	38
Cl	Me		1.49(1)	1.25(2)	[1.49(1)]^c			39
Me	Cl		1.49(1)	1.29(2)	[1.49(1)]^c			39
PrO$_2$CC(Me)=CH	OMe		1.422(4)	1.274(4)	1.455(4)	36.1	−3.0	40

^{a} The roman numeral indicates the different crystalline forms.
^{b} α and β are the torsion angles of the aniline and benzylaldehydo groups, respectively, with respect to the C—N=C—C plane. The positive sign indicates a counterclockwise rotation.
^{c} Owing to the crystallographic disorder, the two Ph—N and C—Ph distances are equal.
^{d} Mean values of the two or more crystallographically independent molecules.
^{e} The disordered structure did not allow reliable measure of bond lengths.
^{f} The two angles are equal because of the statistically imposed equivalence of the two halves of the molecule.

(a) The BAAN derivatives variously substituted in their phenyl rings exhibit a wide range of conformations defined by the interplanar angles α and β, shown in Figure 4. The former varies from 0° to 55° whereas β ranges from −25° to 11°.

(b) Variations in the bond lengths of the nearly planar CN=CC group accompany the conformational changes, but no clear relationship between the two may be detected. On the other hand, no simple relationship is found between the above structural parameters and the electronic nature of the phenyl group substituents.

(c) Noticeable conformational changes occur in the polymorphs of the same compound with implications for their thermo- and photo-chromism.

(d) Smaller values of the C=N bond lengths generally correspond to larger values of the C—Ph and Ph—N distances, suggesting that the formal C=N double bond in these species is *ca.* 1.24 Å.

(e) Theoretical studies,[23] including *ab initio* calculations,[24] indicate that the conformation with α = 45° and β = 0° is favoured over the planar one by 6.57 kJ mol^{-1}. Hence the lattice and, possibly, disorder are expected to supply a similar amount of energy to stabilize the unfavoured planar conformation.

(f) The thermal mesomorphism of *para*-substituted BAAN is suppressed when the phenyl side groups are able to take part in the intermolecular hydrogen bonds.[25] It has been suggested that the loss of mesomorphism is due to the disruption of the nearly parallel arrangement of molecules in the crystal caused by the formation of intermolecular hydrogen bonds. The structure of a series of other monoimines derived from asymmetric amines has been recently determined and the absolute configuration has been assigned,[27] as well as the structure at −160 °C of the monoimine derived from diphenyl ketone and aniline.[28]

20.1.3 BIDENTATE SCHIFF BASES

20.1.3.1 Introduction

Bidentate Schiff bases have been among those most used in preparing metal complexes. In the following subsections we shall describe the potentially bidentate ligands according to their donor atom set.

20.1.3.2 N,N Donor Atom Set

Schiff bases having two nitrogen atoms as donors may be derived either from condensation of dialdehydes and diketones with two molecules of an amine, or from reaction of diamines with aldehydes or ketones. In Section 20.1.2.1, it has been pointed out that coordination through the N atom may occur only under particular circumstances. However, in the case of diimines the formation of chelate rings stabilizes the metal–nitrogen bond. Thus, they can form both mono-[41] and bis-chelate[42] complexes.

The structure of racemic crystals of BADPH has been reported[43] and some structural data are given in Table 5 (a). This ligand has been shown, by NMR investigations, to have a fluctuating structure, *i.e.* at 159–164 °C it exists as a mixture of equal amounts of D, L and *meso* isomers, while the stable DL form is the only one detected at temperatures below 115 °C.[44] Among the Schiff bases from (b) to (d) of Table 5, which are potentially bidentate ligands and whose structures have been reported,[45] only that in (b) could act as a chelate ligand. They are not planar in the solid state and some important geometrical parameters are given in Table 5. Coordination should not occur in the case of the diimine (e), since chelation will give large unstable chelate rings. Some of its geometrical parameters[46] at room temperature are given in Table 5. This compound exhibits mesomorphism with temperature, having nine phases in the range − 33 to + 236 °C, some of which are nematic and smectic.[47] Other diimines exhibit mesomorphic behaviour, as reported in Table 5 (f) together with some structural parameters.[48] As pointed out in Sections 20.1.1.5 and 20.1.2.2, a knowledge of the molecular structure and, mainly, of crystal packing is important in defining the specific properties of mesomorphism of these Schiff bases.

The Schiff bases with two benzylidenimine residues bridged by polysulfide chains of different lengths can act as potentially bidentate ligands. These compounds have been obtained by reaction of sulfur with benzylamine in the presence of PbO. They have been characterized by X-ray, NMR, IR and chromatographic techniques.[49] The structural analysis of three of them, having $n = 2$, 3 and 4 and R = H, showed that the overall geometry is dominated by the conformation of the polysulfide chain.[50]

Bidentate Schiff bases with N,N donor sets which can be deprotonated affording monoanionic ligands can also chelate metal atoms, as shown in Figure 5. They form $M^{II}(chel)_2$ complexes with M = Cu, R^1 = Ph, R^2 = Me and R^3 = H,[51] and with M = Ni, R^1 = p-ClC$_6$H$_4$, R^2 = H and R^3 = Ph.[52] The free ligands, with different R^1 substituents and R^2 = R^3 = H, have been studied in solution by ^{1}H NMR spectroscopy. At room temperature there is an equilibrium between the 'all-*cis*' (Figure 5a) and 'all-*trans*' (Figure 5b) isomers, the latter being predominant in non-polar solvents, whereas the former is stabilized in solvents forming hydrogen bonds.[53] The Schiff base derived from 2-pyrrolecarbaldehyde and NH$_3$ acts as a monoanionic ligand as shown in Figure 5c.[54]

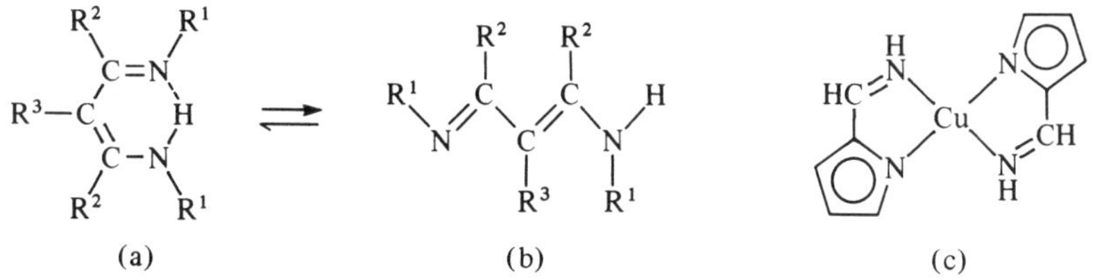

(a)   (b)   (c)

Figure 5 Bidentate Schiff bases with N,N donor set which act as monoanionic ligands: (a) all *cis*, $R_2 = R_3 = H$; (b) all *trans*, $R^2 = R^3 = H$; (d) a copper(II) complex with a bidentate monoanionic aldimine

Recently a series of condensation products of the oximes (7; R = H, Me; R' = Me, CH$_2$Ph, Ph, p-MeC$_6$H$_4$) and (8) with amines have been prepared by reactions (2) and (3) respectively, and the species formed investigated by X-ray, NMR and IR studies.[55] IR spectroscopic data obtained from the product of reaction (2) suggest structure (9) instead of (9') for the oxime (7), while ^{1}H NMR spectral data from the product of reaction (3) are in agreement with an equilibrium between (10) and (10'). A crystal structure analysis indicates that the tautomers (9; R = Me, R' = p-MeC$_6$H$_4$) and (10) are those preferred in the solid state.[56] Furthermore, compound (9) is the *syn*-Me,*syn*-Me isomer, while (10) is the *anti*-Me,*anti*-acetyl isomer.

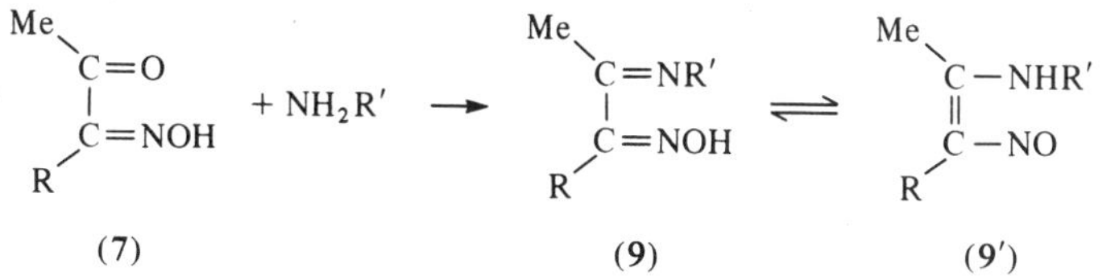

(7)   (9)   (9')

(2)

Table 5 Some Geometrical Parameters of Potentially Bidentate Schiff Bases Having an N,N Donor Atom Set[a]

Compound	C—N (Å)	C=N (Å)	C—C (Å)	Interplanar angles (°)	Ref.
(a) PhCH=N(CHPh)$_2$N=CHPh	1.437(3)	1.267(4)	1.465(4)		44
(b) [structure]	1.445(7)	1.229(7)	(1.445)[b]	α = 17.9; 8.9[b] β = −17.9; 8.9[b]	45
(c) [structure]	1.427(5)	1.274(5)	1.470(5)	α = 46.2; 51.3 β = 13.4; 2.5	45
(d) [structure]	1.427(10)	1.277(10)	1.449(10)	α = 19.2 β = −9.8	45
(e) [structure]	1.415(8) 1.402(9)	1.261(9) 1.272(8)	1.466(9) 1.449(8)	(1,2) 16.1 (1,3) 66.7 (1,4) 21.6 (2,3) 52.9 (2,4) 9.6 (2,5) 16.5 (3,5) 66.7	46
(f) [structure]	1.423(6) 1.421(6)	1.269(6) 1.273(6)	1.478(7) 1.470(7)	(1,2) 81.0 (1,5) −28.4 (1,3) 16.4 (2,4) 39.1 (2,4) 9.0 (3,5) 14.9 (4,6) 9.0	48

[a] α and β are defined in Figure 4.
[b] C—N and C—C are equal, α = −α and β = −β because of the statistical disorder in the crystal.
[c] The values refer to the two crystallographically independent molcules.

Me\
    C=O
   |
   C
Me—C N—OH
    O

  + *p*-NH₂C₆H₄Me →

Me\
    C=NC₆H₄Me
   |
   C=NOH
Me—C
    O

⇌

Me\
    C—NHC₆H₄Me
   ‖
   C—NO
Me—C
    O

(3)

(8)   (10)   (10′)

These potential ligands are of interest since they contain both oximinato and imino functions.

20.1.3.3 N,O Donor Atom Set

20.1.3.3.1 Bonding to metals

A large group of bidentate Schiff bases utilized as metal ligands is characterized by having an N,O donor set. Since the oxygen is often present as an OH group, these ligands generally act as chelating monoanions. In this section we will treat these Schiff bases as bidentate ligands. However, it must be stressed that since the earlier work[57] of Sacconi and co-workers it has been known that the hydroxyl oxygen atom, under particular circumstances, may bridge two metal atoms. In this case, the Schiff bases should be considered as tridentate ligands which favour the formation of binuclear complexes.[58]

On the other hand, few examples of coordinated neutral N,O bidentate Schiff bases have been reported. For example the potentially bidentate ligands derived from β-diketones and NH₂Prⁱ coordinate to zinc(II) only through their oxygen atom, preserving the intramolecular hydrogen bond[59] (Figure 6a) as suggested by IR spectral evidence.[60] The neutral acetylacetonimine in Figure 6b coordinates to the Ybᴵᴵᴵ ion through its oxygen atom in the complex Yb(acac)₃-(MeCOCH=C(Me)NH₂).[61] In this case, however, no intramolecular hydrogen bond is found, since the NH₂ group is involved in hydrogen bonds with the acac oxygen atoms of another complex molecule in the crystal. As a consequence, its N···O bite increases to 2.77 Å, as compared with that of 2.54 Å of the Schiff base of Figure 6a. In both cases the ketoamine form appears to be preferred over the enolimine one.

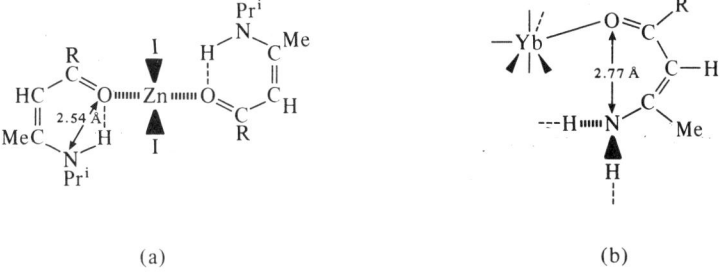

Figure 6 Coordination of neutral N,O Schiff bases to metals: (a) R = Me, Prⁱ; (b) R = Me, Ph

20.1.3.3.2 Spectroscopic properties

Bidentate Schiff bases derived from β-diketones have been shown to exist almost entirely as the tautomer (b) (Figure 7) in solution in common solvents. Proof of this structure has been obtained from the observed spin–spin splitting of the N—H protons by the proton of R′.[62]

Similarly the condensation products of amines with 1-hydroxy-2-acetonaphthone and 2-hydroxy-1-naphthaldehyde, in 1:1 ratio, have been shown to be ketoamines, the quinoid forms (d) and (f) being preferred over the forms (e) and (g), respectively (Figure 7).[63] The ¹⁵N NMR spectra of the above Schiff bases confirmed these assignments and the temperature dependence of the spectra gives information on the tautomeric equilibria.[64] However, the Schiff bases derived from hydroxymethylene ketones (Figure 8) exhibit tautomers (b) and (c) but not (a) in a variety of solvents. The ratio of the two isomers is influenced by the substituent R.[65] Similarly, a recent NMR study of the formylcamphor derivatives (d)–(f) in Figure 8 showed that the tautomer distribution was very strongly solvent dependent, and also that spectral changes occurred with time. Thus, form (e) of Figure 8 is the most abundant in CHCl₃ while form (f) is the most stable in acetone. In benzene

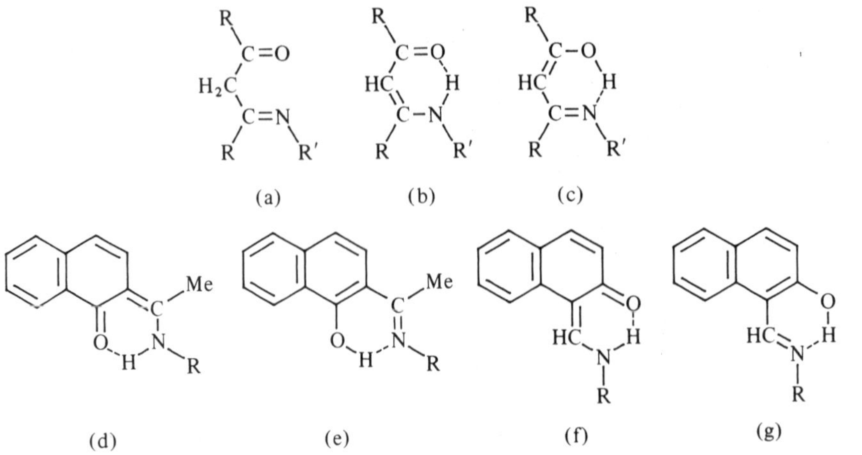

Figure 7 Tautomers of Schiff bases derived from β-diketone (a–c), from 1-hydroxy-2-acetonaphthone (d, e) and from 2-hydroxy-1-naphthaldehyde (f, g)

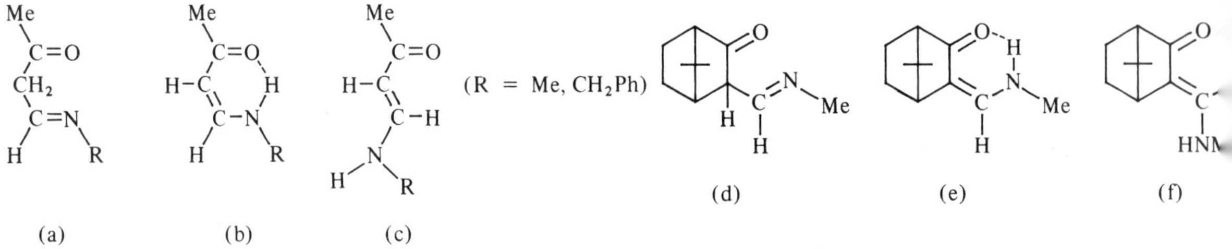

Figure 8 Isomers of Schiff bases derived from hydroxymethylene ketones (a–c) and formylcamphor (d–f)

solution the ratio (e)/(f) varied from 1/10 to 10/1 during 24 h. No evidence of the tautomer (d) was obtained.[66] The relevant data are given in Table 6.

Table 6 ^{1}H NMR Data for some Protons of the Schiff Bases Derived from Formylcamphor in Different Solvents[a]

Proton	Tautomer[b]	CDCl$_3$	CD$_3$OD	Solvent C$_6$D$_6$	CD$_3$COCD$_3$
N—H···O	e	7.39	—	7.55	—
NH	f	4.25	—	3.71	5.54
=CHN	e	6.29	6.51	5.85	6.41
	f	7.00	7.09	7.00	6.88
MeN	e	2.91	2.93	2.34	—
	f	2.94	2.96	2.11	2.93
CH	e	2.30	2.33	—	—
	f	2.48	2.63	—	2.65
Ratio e/f (after 24 h)		6:1	1.3:1	1:10 to 10:1 (during 24 h)	1:20

[a] Chemical shifts are in p.p.m. and are relative to TMS.
[b] The designation e/f refers to the tautomers shown in Figure 11.

The NMR and IR spectra of many Schiff bases derived from substituted SAL and RNH$_2$, 9 with R = alkyl and 28 with R = aryl, have been reported.[67] NMR results show that all these compounds exist in solution as the enolimine tautomer. In addition, the electron-withdrawing R substituent reduces the magnetic shielding of the hydroxyl proton. In contrast, the C=N stretching frequencies do not appear to be influenced by the nature of the R group in a systematic way. As expected, upon coordination to transition metals the C=N stretching frequencies at about 1620 cm^{-1} decrease, whereas those at about 1575 cm^{-1} increase. On the other hand, the two bands assigned to the C—O stretching frequency increase by about 30 and 40 cm^{-1}.

Many studies of the electronic spectra of bidentate N,O Schiff bases derived from chiral primary amines have been reported[68] in order to establish the absolute configuration of the amine. Three absorption bands, designated[69] I, II and III at about 315, 255 and 215 nm, respectively, are present in the EA spectra of hexane solutions of the Schiff bases derived from SAL and RNH_2 where R is a chiral group. These bands have been attributed to the chromophore C=N. As already pointed out, two additional bands, at about 400 and 280 nm, become apparent in polar solvents and have been attributed to the quinoid tautomer. The CD spectra usually show Cotton effects of the same sign in correspondence with bands I and II, which can be correlated with the absolute configuration of the amine by using the 'salicylidenimino chirality rule'.[70] Further CD studies have shown an additional Cotton effect opposite in sign to those of the bands I and II at about 275 nm.[71] Investigation of the Schiff bases derived from benzaldehyde, *o*-methoxybenzaldehyde and salicylaldehyde with chiral amines such as (*S*)-α-phenylethylamine, (*S*)-α-benzylethylamine and 17-β-aminoandrostan-3α-ol indicates that the lowest energy $n \rightarrow \pi^*$ transition of the azomethyne group occurs at about 275 nm. The corresponding Cotton effect of the CD spectrum of salicylaldehyde derivatives, whose EA spectra do not show any absorption for this transition, makes the application of the chirality rule less ambiguous for establishing the absolute configuration of the amine.

20.1.3.3.3 Structural properties

The Schiff bases (a) shown in Figure 9 have been prepared by reaction of 1,3-di-*t*-butylazidinone with Grignard reagents, and the structure of the derivative with R = Ph (Figure 9b) has been reported.[72] In this case, only the enolimine form is possible because of the lack of conjugation of the C=N bond with suitable substituents at both C and N atoms. Therefore the values of the C—N bond lengths of 1.250(3) and 1.479(3) Å may be assumed to be a good measure of the double and single C—N bonds respectively. The $C(sp^3)$—O single bond length is 1.430(3) Å. The intramolecular hydrogen bond between the N and O atoms determines the overall geometry of the molecule.

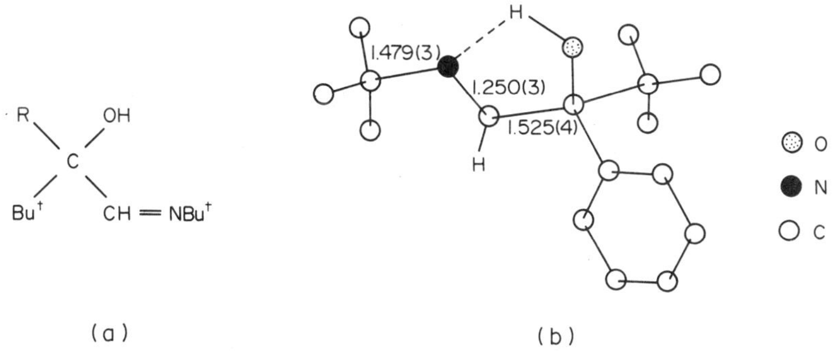

Figure 9 Schiff bases obtained by reaction of 1,3-*t*-butylazidinone with Grignard reagents (R = Me, Ph) (a), and the X-ray structure of *N*-*t*-butyl-2-hydroxy-3,3-dimethyl-2-phenylbutanimine (b)

The most studied bidentate Schiff bases containing an N,O donor set are those derived from substituted salicylaldehyde derivatives, since they have been widely used as ligands for many transition metal complexes.[73] They have also been investigated extensively because of their meso-morphism in the solid state. Those which have been structurally characterized are collected in Table 7 together with some structural parameters of interest.

Table 7 Bidentate Schiff Bases Having an N,O Donor Set Which Have Been Structurally Characterized

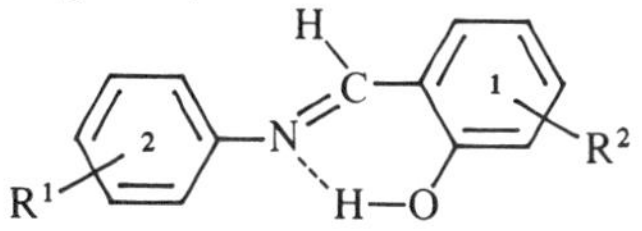

R^1	R^2	C—N (Å)	N=C (Å)	C—C (Å)	C—O (Å)	1,2 (°)	1,3 (°)	2,3 (°)	Ref.
H	H	1.466(5)	1.262(8)	1.529(5)	1.320(7)	49	2	49	77
H	5-Cl[a]	1.419(6)	1.270(6)	1.444(6)	1.351(6)	1	—	—	78
H	5-Cl[b]	1.414(7)	1.292(7)	1.438(7)	1.364(7)	1	—	—	78
2-Cl	H	1.421(4)	1.288(4)	1.452(4)	1.365(4)	51.5	—	—	11
H	5-SO₃[c]	1.419(10)	1.310(9)	1.409(10)	1.297(8)	0	0	0	79

[a] At 300 K.
[b] At 90 K.
[c] The Schiff base is in the ketoamine form.

As already observed for the corresponding benzaldehyde derivatives, these data confirm that, while bond lengths and angles are similar throughout the series, the interplanar angles vary to a large extent, determining the solid state behaviour of this class of compounds. In fact a planar geometry is responsible for thermochromism, while a bent geometry is associated with photochromism.[36] If the last compound of Table 7 is excluded, comparison of the C=N and C—N bond lengths, which vary from 1.262(8) to 1.288(4) Å and from 1.419(10) to 1.466(3) Å respectively, with that of 1.250(3) Å reported for (b) of Figure 9 shows that some degree of conjugation must be present in the CN=CN moiety. Significant changes in the bond lengths of the latter group take place when the Schiff bases, under particular circumstances, exhibit the ketoamine form, as happens with the last compound of Table 7. In fact, the data in Table 7 show that for the latter form the N=C distance increases and the C—C distance decreases with respect to those of the enolimine form, while the C—N bond length is essentially the same.

The keto tautomer is rather unusual, since it has been previously shown that Schiff bases derived from salicylaldehyde prefer form (a). When such Schiff bases coordinate to the metal as mono-anionic ligands the C—N and C—O bond lengths undergo variations which are in agreement with the IR spectral results reported in the previous section. For a series of metal complexes of Schiff bases derived from salicylaldehyde, the C—N and C—O bond lengths have mean values of 1.295 and 1.312 Å respectively.[14] The corresponding values for other metal complexes of Schiff bases derived from *o*-hydroxyacetophenone are 1.293 and 1.316 Å respectively.[75]

As expected, comparison of these data with those relative to the free ligands shows that a shortening of the C—O distance corresponds to a small, but significant, lengthening of the C=N distance. This is in agreement with the amount of variation of the corresponding stretching frequencies of C—N and C—O bonds upon coordination to a metal.

Recently an interesting result has been reported[76] which may be useful for understanding the formation of Schiff bases from aldehydes and amines. On crystallization from ethanol and chlorobenzene of the Schiff base (a) derived from 4-aminopyridine and salicylaldehyde (Figure 10), crystals containing the two hydrolysis products (b) and (c) (Figure 10) were obtained. These crystals, after heating above 45 °C, yielded a powder which exhibits photochromic behaviour, suggesting the formation of (a) in the solid state.

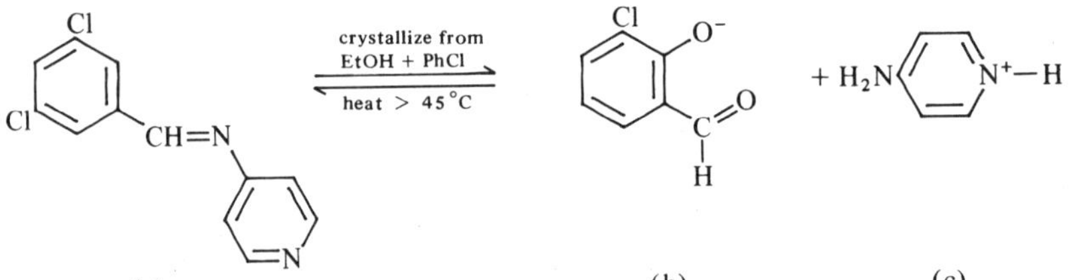

Figure 10 Solid state reaction giving a Schiff base

20.1.4 TRIDENTATE SCHIFF BASES

Many tridentate Schiff bases have been utilized as anionic ligands having N_2O, N_2S, NO_2 and NSO donor sets.[80-89] Some typical ligands forming metal complexes are shown in Table 8. These may be generally considered as derived from the bidentate analogues by addition of another donor group. However, only a few of them have been characterized as free ligands. It must be recalled that the oxygen donor atoms of ligands such as those of Table 8 may often act as a bridge between two metal centres, giving polynuclear complexes. Thus, they may give metal complexes with different stoichiometries even with the same metal.

NMR studies have been carried out on Schiff bases derived from pyridoxal phosphate and amino acids, since they have been proposed as intermediates in many important biological reactions such as transamination, decarboxylation, *etc.*[90] The pK_a values of a series of Schiff bases derived from pyridoxal phosphate and α-amino acids, most of which are fluorinated (Figure 11), have been derived from ^1H and ^{19}F titration curves.[91] The imine N atom was found to be more basic and more sensitive to the electron-withdrawing effect of fluorine than the pyridine N atom. Pyridoxal and its phosphate derivative are shown in Figure 12a. The Schiff base formation by condensation of both with octopamine (Figure 12b) in water or methanol solution was studied by ^{13}C NMR. The enolimine form is favoured in methanol, while the ketoamine form predominates in water.[92]

Table 8 Typical Tridentate Schiff Bases with Different Donor Atom Sets

Compound	X	R	Ref.
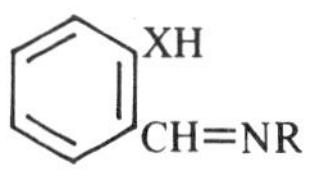	O	CH_2CO_2H	80
	O	$(CH_2)_nC_6H_4OH$-o ($n = 1, 2$)	81
	O	2-C_6H_4N	82
	O	$(CH_2)_nX$ ($n = 2, 3$; $X = OH, NMe_2, NEt_2, NH_2$)	83, 84
	S	C_6H_4OH	85
	S	$CH_2CH_2NH_2$	86
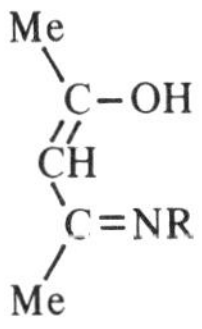	—	$(CH_2)_nOH$ ($n = 2, 3$)	87

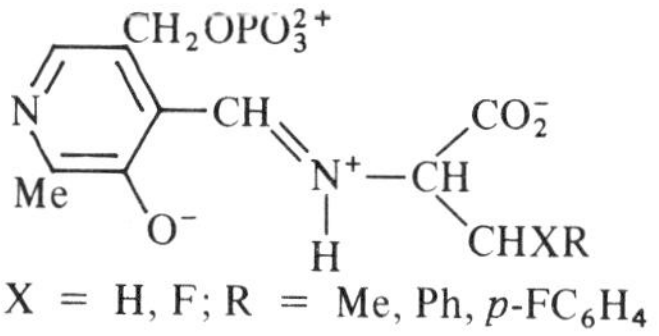

$X = H, F; R = Me, Ph, p$-FC_6H_4

Figure 11 Schiff base derived from pyridoxal phosphate and α-amino acids

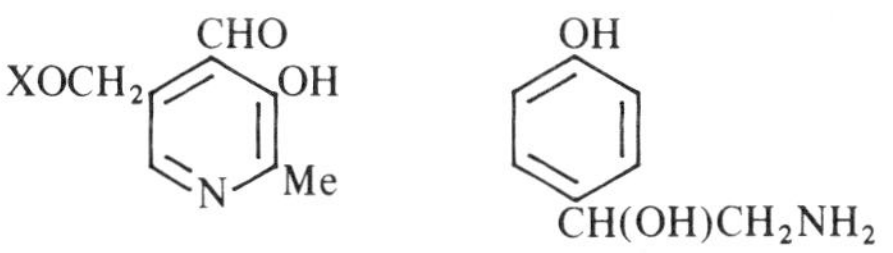

(a) (b)

Figure 12 (a) Pyridoxal ($X = H$) and pyridoxal phosphate ($X = PO_3H_2$); (b) octopamine

These Schiff bases have been shown to act as tridentate ligands through the imine nitrogen, the phenolic oxygen and one of the carboxylate oxygen atoms.[89,93] However, other donor atoms are present, and it has been shown that further coordination to a metal centre, through the pyridine nitrogen and/or the oxygen atom of the CH_2OH side group, may occur with formation of polynuclear complexes.[94]

The potentially tridentate Schiff bases derived from ring-substituted SAL and aminopyridine so far structurally characterized are reported in ref. 95. The molecules are not strictly planar. Since the relatively small deviations from planarity correspond to a weak thermochromism, it has been suggested[95] that thermochromism should be a monotonic function of the angles which define the relative orientation of the six-membered rings with respect to each other.

Some of these ligands have been shown to coordinate to three different copper atoms in polynuclear species, acting both as a chelate and as a bridging ligand.[96]

20.1.5 TETRADENTATE SCHIFF BASES

Tetradentate Schiff bases with an N_2O_2 donor set have been widely studied for their ability to coordinate metal ions. The properties of these complexes are determined by the electronic nature of the ligand as well as by its conformational behaviour.

Free ligands have been studied in order to obtain an insight into their structure, both in solution and in the solid state, and for comparison with their metal complexes. ^1H NMR spectroscopy has been used to investigate the keto–enol equilibrium and the nature of the hydrogen bonds. In the case of optically active Schiff bases UV and CD spectra provided information about structure in solution. The Schiff bases that have been most widely examined are derivatives of acetylacetone, salicylaldehyde and hydroxymethylenecamphor, whose prototypes with en are shown in Figure 13.

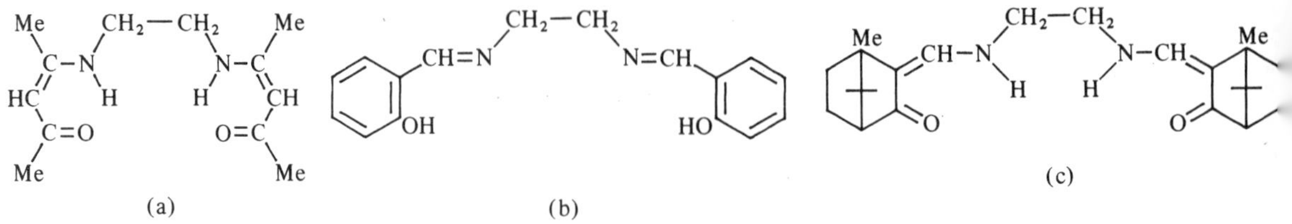

Figure 13 Tetradentate Schiff bases: (a) acacenH$_2$, (b) salenH$_2$ and (c) fmcenH$_2$

20.1.5.1 Derivatives of β-Diketones

Selecting acacenH$_2$ as the prototype of these compounds, forms (a)–(c) may be present in the tautomeric equilibrium (Figure 14). ^{1}H NMR spectra of Schiff bases derived from acacenH$_2$ and various diamines in CHCl$_3$ are consistent with the ketamine structure of Figure 14c.[65] Band positions and assignments for these Schiff bases in different solvents are listed in Table 9. Except for solvent shifts, most marked in aromatic solvents, the spectra are similar to that in CCl$_4$. In acetone a new signal appears at 3.07 p.p.m., which may be assigned to the methylene group of the tautomer of Figure 14a. From the ratio of the intensities of the methylene and vinyl signals it has been shown that the compound is about 22% in the ketimine form as shown in Figure 14a.[65] In other solvents the contribution of the ketimine form to the equilibrium is less than *ca.* 5%.

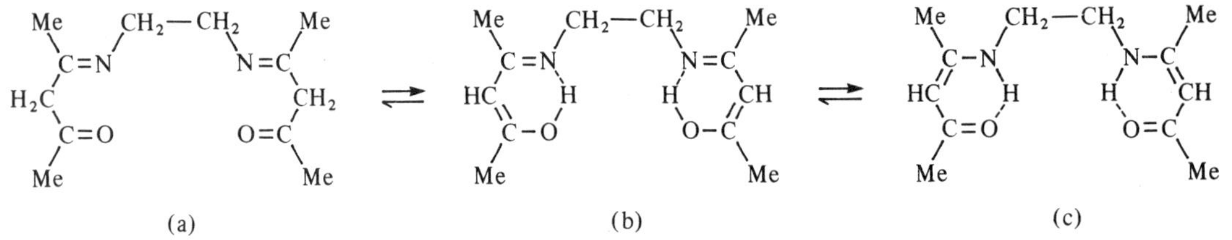

Figure 14 Tautomers of acacenH$_2$

It has been shown that neither NMR nor electronic spectra of these Schiff bases are concentration dependent, so that only intramolecular hydrogen bonding must be taken into account. Furthermore neither electronic nor CD spectra significantly change over a long period of time upon addition of 0.1 M sodium hydroxide, indicating a very strong kinetically protected and rigid hydrogen-bonded structure.[97]

Protons *a* appear as sharp singlets between 1.87 and 1.92 p.p.m., and while protons *b* are also singlets, their resonances are not so sharp. At high resolution they appear as doublets ($J < 1$ Hz), probably because of a very small long range coupling with protons *d* or *e*. Protons *c* generally give rise to a sharp singlet, but in the case of (−)-pn, the singlet is changed in two signals separated by 0.03 p.p.m. owing to the asymmetry of the diamines. The low field peaks due to protons of the bridge are rather complex, as expected from coupling with protons *d*. NMR spectral data for Schiff bases of various diamines with differently substituted β-diketones are reported in ref. 102.

The IR spectra in solution are very complex. However, the absorption above 3100 cm^{-1} is virtually identical for all compounds of this class. They have a rather broad band centred at 3350 cm^{-1}, which has been assigned to the N—H group involved in intramolecular hydrogen bonds, and two very weak bands at *ca.* 3210 and 3140 cm^{-1}, which are supposed to arise from overtones of the very strong bands at 1618 and 1580 cm^{-1}. Secondly, in the region from 2700 and 1500 cm^{-1}, there are always only three strong bands at about 1618, 1580 and 1515 cm^{-1}.[97]

The electronic spectra of this class of compounds are similar, and concentration independent over a large concentration range. The electronic spectra are characterized by two rather strong absorption bands, centred at about 325 and 300 nm.[97,98] The doublet may be explained as arising from coupling of two interacting chromophores[99] (*e.g.* the hydrogen-bonded C=C—C=O chromophore which should absorb in the region of 300 nm[100]), determined by their mutual arrangement.

Simulation of spectra for (*R*)-acacpnH$_2$ and its trifluoro derivative indicates that the *syn* conformation, predominant in solution, is that having the pn methyl group *anti* to the N atom, as found in the solid state.[101] A similar conformer distribution is expected also for acacenH$_2$ in view of the similarity of absorption spectra. In conclusion, it appears that these compounds exist in solution essentially in a unique conformation, maintained by strong intramolecular hydrogen bonds.

Table 9 ¹H NMR Data for Schiff Bases Derived from Acetylacetone and Diamines[a,b]

Scheme (structure):

a Me—C(B)=N ... b H ... c R ... with atoms labelled a Me, b H, c R, d H(H_d H), e O—C, and ring

B	Solvent	a	b	c	d	e	f	g	Ref.
CH₂CH₂ (e)	CCl₄	1.90	1.85	4.83	10.9	3.39, 3.49			65
	CCl₄	1.94(6)	1.87(6)	4.88(2)	10.9(2)	3.54(4)			97
	CHCl₃	2.00	1.91	4.99	10.9	3.37, 3.48			65
	CDCl₃	2.00	1.92	5.00	10.84	3.42			102
	(CD₃)₂CO	1.91	1.87	4.95	10.8	3.50			65
	C₆H₆	1.98	1.41	4.83	11.0	2.42, 2.53			65
	CS₂	1.82		4.81	10.9	3.29, 3.39			65
	py	2.03	1.80	5.02	11.2	3.18, 3.28			65
	Pyrrole	1.83	1.39	4.81	10.7	2.38, 2.48			65
	PhNO₂	1.97	1.91	4.95	11.1	3.37, 3.48			65
	PhBr	1.94	1.58	4.82	10.9	2.79, 2.89			65
CH—CH₂ (−) (e,f) | CH₃ (g)	CCl₄	1.84	1.89	4.79	11.1	3.37		1.24, 1.35	65
	CCl₄	1.90(6)	1.85(6)	4.78, 4.80(2)	10.8(2)	3.27, 3.65(2)		1.27(3)	97
(g)	CDCl₃	2.00	1.88, 1.91	4.96, 4.98	10.89	3.28		1.27	102
	(CD₃)₂CO	1.87	1.93	4.95	10.9	3.40		1.18, 1.29	65
	C₆H₆	1.97	1.47, 1.51	4.81	11.1	2.53		0.59, 0.70	65
CH—CH (+) (e) | Me Me (meso)	CCl₄	1.87(6)	1.80(6)	4.73(2)	10.95(2)	3.62(2)		1.30(6)	97
	CCl₄	1.88(6)	1.83(6)	4.80(2)	11.80(2)	3.50(2)		1.20(6)	97
CH₂ CH₂—CH₂—CH₂ (meso) (g) CH—CH (+) (e)	CCl₄	2.00 6	1.90(6)	5.00(2)	11.0(2)	3.68(2)		1.65(8)	97
	CCl₄	1.99(6)	1.86(6)	4.96(2)	11.0(2)	3.28(2)		1.5(8)	97
CH—CH (−) (e) | Ph Ph (meso) (g)	CCl₄	1.92(6)	1.78(6)	4.92(2)	11.85(2)	4.68(2)		7.15(10)	97
	CCl₄	1.93(6)	1.63(6)	4.82(2)	11.4(2)	4.70(2)		7.10(10)	97
CH₂CH₂CH₂ (e,f)	CCl₄	1.90	1.90	4.83	10.9	3.33, 3.43			65
	CDCl₃	2.00	1.93	4.99	10.85	3.36	1.87		102
	C₆H₆	1.42	2.02	4.86	11.1	2.68, 2.59	1.05		65
CH₂CH₂CH₂CH₂ (e,f)	CCl₄	1.42	1.87	4.82	10.9	3.32, 3.25	1.71		65
	C₆H₆		2.04	4.88	11.1	2.56, 2.46	1.01		65

[a] Chemical shifts in p.p.m. relative to TMS; relative intensities in parentheses.
[b] In the scheme R = Me.

So far, X-ray analyses of free tetradentate Schiff bases have been carried out only in a very few cases. For acacenH$_2$ derivatives only two structures have been reported[101,103] besides the thio analogue of acachxnH$_2$.[104] Some structural parameters of interest are listed in Table 10. The X-ray analysis results show that crystals consist, just as in solution, of discrete molecules characterized by an enaminoketone structure with intramolecular hydrogen bonding. However, the bond lengths (Table 10) and the planarity of the two chromophores show that there is a certain amount of conjugation within the enamino ketone groups. The structure found in the solid state, with an approximate *gauche* conformation of the diamines, is consistent with that proposed as predominant in solution.

Table 10 Bond Lengths (Å) and Torsion Angles (°) of Schiff Bases Derived from β-Diketones and Diamines

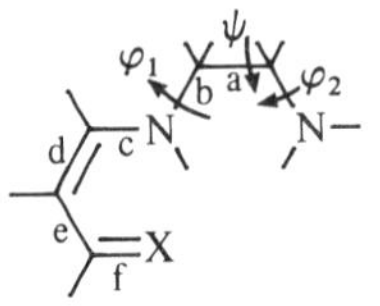

Diamine	X	a	b	c	d	e	f	ψ	φ$_1$	φ$_2$	Ref.
en[a]	O	1.521(3)	1.465(3)	1.331(3)	1.393(5)	1.441(4)	1.244(4)	−67.2	94.1	100.0	103
pn	O	1.525(7)	1.458(4)	1.332(4)	1.376(4)	1.434(4)	1.245(4)	−65.8	99.5	99.5	101
chxn[a]	S	1.542(16)	1.475(15)	1.319(15)	1.415(16)	1.401(16)	1.721(12)	−62.3	137.3	155.2	104
en[b]	O	1.503(12)	1.462(5)	1.308(4)	1.423(6)	1.366(5)	1.294(3)	—	—	—	110

[a] Mean values for the two equivalent halves.
[b] Weighted average values with their standard deviations for eight metal–acacen complexes.

20.1.5.2 Derivatives of Salicylaldehyde

The two possible tautomeric structures of these compounds are shown in Figure 15 where the main resonance structures are represented for the quinoid tautomer. ¹H NMR and IR spectra show that the predominant form is (a) in Figure 15.

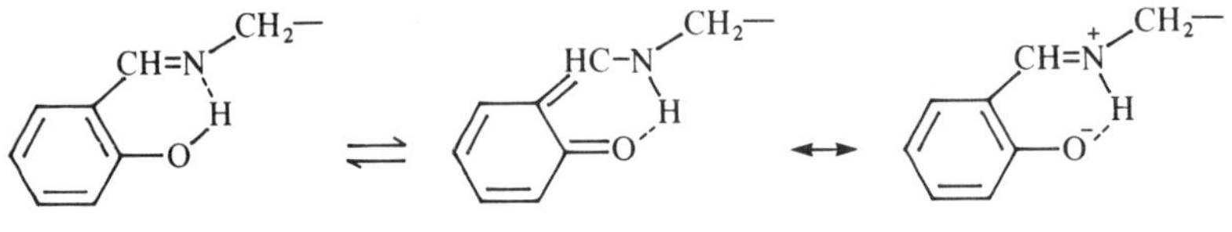

Figure 15 The aldimine tautomer (a), and the ketoamine tautomer in two resonance forms, (b) and (b′), of salenH$_2$. Only half a molecule is represented

The IR spectra in CHCl$_3$ are very similar in the region above 2500 cm^{-1}. They all display a rather broad peak at *ca.* 3500 cm^{-1}, which may be assigned to free or weakly hydrogen-bonded hydroxyl groups. From 3500 to 2500 cm^{-1} there is a very broad and strong absorption due to strongly hydrogen-bonded hydroxyl groups. The ratio of the intensities of these two broad absorptions varies with concentration, showing that intermolecular hydrogen bonds may also occur.

This appears also from NMR spectra (Table 11). The signal due to the azomethine group CH=N shows two or more sharp peaks, the more intense probably due to intramolecular hydrogen bonds, the less intense to intermolecular hydrogen bonds. Furthermore, the rather broad signal due to protons *c* in CDCl$_3$ has a chemical shift which slightly depends on concentration.[97]

The electronic spectra show two absorptions probably corresponding to $n \rightarrow \pi^*$ (415 nm) and $\pi \rightarrow \pi^*$ (320 nm) transitions. The $\pi \rightarrow \pi^*$ transition involves molecular orbitals essentially localized on the azomethine group.[111] The positive–negative couplet in the CD spectra centred at about 260 nm could arise from exciton splitting of a transition mainly involving the orbitals of the benzene rings of the salicylaldehyde group.[112] The similarities of the spectra suggest the presence of the same predominant conformation in solution throughout the series. The strong Cotton effects arise not only from the vicinal induction of asymmetric carbon atoms of the diamines, but also from the

Table 11 ^{1}H NMR Data of Schiff Bases Derived From SAL and Diamines[a]

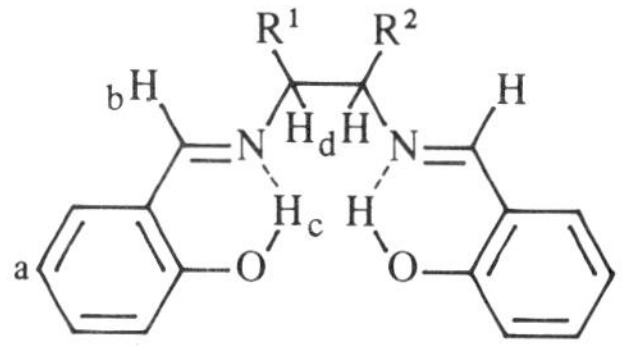

Diamine	R^1	R^2	a	b	c	d	e
en	H	H	7.1(8)	8.33(2)	3.81(4)	13.1(2)	—
(+)-pn	Me	H	7.1(8)	8.27(2)	3.80(3)	13.1(2)	1.24(3)
(−)-butn	Me	Me	7.1(8)	8.35(2)	3.45(2)	13.2(2)	1.25(6)
(*meso*)-butn	Me	Me	7.2(8)	8.40(2)	3.47(2)	13.2(2)	1.29(6)
(+)-chxn	$\frac{1}{2}(CH_2)_4$	$\frac{1}{2}(CH_2)_4$	7.1(8)	8.17(2)	3.21(2)	13.2(2)	1.75(8)
(*meso*)-chxn	$\frac{1}{2}(CH_2)_4$	$\frac{1}{2}(CH_2)_4$	7.1(8)	8.35(2)	3.54(2)	13.2(2)	1.85(8)
(−)-stien	Ph	Ph	7.1(18)	8.36(2)	4.77(2)	13.3(2)	7.30(18)
(*meso*)-stien	Ph	Ph	7.1(18)	8.10(2)	4.77(2)	13.1(2)	7.30(18)
(+)-phenen	H	Ph	7.1(13)	8.53(2)	4.08(1)	13.3(2)	7.10(13)

[a] Chemical shifts in p.p.m. relative to TMS in CCl_4, ref. 97. Relative intensites in parentheses.

intrinsic asymmetry of the conformers. Thus, of the three possible conformations, only (a) and (b) in Figure 16 are consistent with the CD spectra. From the comparison with the spectrum of the (+)-chxn derivative, whose conformation is necessarily (a) in Figure 16, it was concluded that the most stable conformation in solution is also (a) in Figure 16.[113] The optical activity of the (+)-pn and (−)-stien SAL derivatives suggests a larger amount of the conformers (b) and (c) in solution.

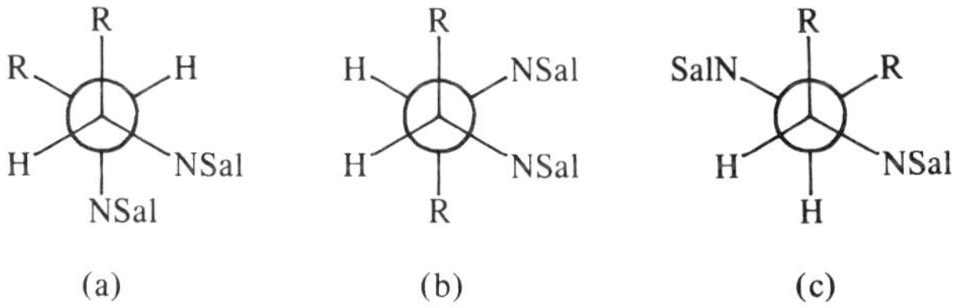

Figure 16   Newman projections of (*S*) conformers of methylene bridge-substituted salenH$_2$ derivatives

The X-ray structures so far reported (first three compounds of Table 12) confirm the enolimine structure found in solution, with intramolecular hydrogen bonds. In fact the N—CH and O—C distances are consistent with double N=C and single O—C bonds. Furthermore the C—C distances are close to those expected for $C(sp^2)$—$C(sp^2)$ single bonds.[105] In this class of compounds, however, the conformation found in the solid state is different from that present in solution (see Table 12). In fact, when there is a possibility of free rotation around the ethylene bridge, the N atoms are *trans* to each other, as in (c) of Figure 16.

Data for the fourth compound in Table 12 refer to a complex between $Ca(NO_3)_2$ and saltrienH$_2$. The X-ray structure results show that upon coordination of Ca^{2+} through the oxygen atoms,

Table 12   Bond Lengths (Å) and Torsion Angles (°) for Schiff Bases Derived from SAL and Diamines

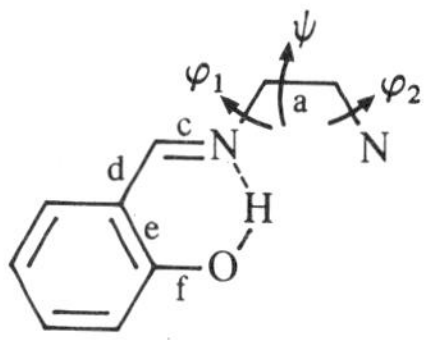

Diamine	a	b	c	d	e	f	ψ	ϕ_1	ϕ_2	Ref.
en[a]	1.499	1.461	1.270(3)	1.457(4)	1.412	1.345(3)	179.0	−65.1	65.1	105
stien	1.579(3)	1.469(3)	1.277(3)	1.455(4)	1.405(4)	1.349(3)	180.0	−94.6	94.6	106
phn[a,b]	1.408(7)	1.421(7)	1.288(7)	1.449(7)	1.421(7)	1.345(7)	8.2	−132.2	178.0	107
trien	—	1.466	1.300	1.422	1.416	1.306	—	—	—	109
en[c]	1.510(11)	1.481(3)	1.291(3)	1.434(3)	1.416(3)	1.321(3)	—	—	—	110

[a] Mean values for the two equivalent halves.
[b] Values from ref. 108 are similar, but are not reported because of their low accuracy.
[c] Weighted average values, with their standard deviation, for 17 metal–salen complexes.

protons are transferred to the nitrogen atoms forming intramolecular hydrogen bonds.[109] Bond lengths indicate contributions to the resonance of both (b) and (b') structures of Figure 15.

In the last row of Table 12 the average values for some metal–salen complexes are reported. It is interesting to observe that upon chelation of the doubly ionized ligand, there is a shortening of the C—O and C—C bond lengths accompanied by a lengthening of the N—CH distances.[105] In the saltrienH$_2$ complex bond lengths are remarkably close to those of the metal–salen complexes.[109]

20.1.5.3 Derivatives of Formylcamphor

The complexity of the stereochemistry of condensation products between formylcamphor and primary amines has been discussed in Section 20.1.3.3. When the condensation products of diamines with two molecules of formylcamphor are considered, the number of possibilities increases. It has been found that the tautomer distribution varies with solvent and temperature and, for CHCl$_3$ solutions, with time also.[66]

From ^{1}H NMR spectra of Schiff bases derived from fmc and various diamines in CDCl$_3$ solution, it was concluded that the *syn/anti* ratio was *ca.* 10:1 in aged CDCl$_3$ solutions. The absorption and CD spectra in methanol are very different from the spectra in aged chloroform solution. It was concluded that in MeOH the *anti* structure of ketoenamines dominates. The variation of dichroism with temperature of the en derivative in aged CHCl$_3$ solutions has been explained as a displacement of the equilibrium among the three rotamers (two *gauche* and one *trans*) towards the most stable one. This should be the one that has the opposite absolute configuration of the (*R*)-chxn Schiff base which exhibits no variation of its CD spectrum with temperature, as expected by the non-existence of rotamers. The temperature variation of CD spectra for derivatives of en, pn and bn diamines suggests the existence of rotamer equilibrium reflecting the increasing energy barriers towards rotation.[66]

The ^{13}C NMR spectrum of acetylcamphor shows that only one tautomer is present in CDCl$_3$ solution. The signals at 205.3 and 152.9 p.p.m., relative to TMS, are assigned to the camphor carbonyl carbon atom and to an enamine group, respectively.[115] ^{1}H NMR spectra of acetylcamphor derivatives show no variation with time and/or solvent. The chemical shift of the amine protons, *e.g.* acetylcamphor, is 9.0 p.p.m. relative to TMS, and is typical of hydrogen bonding. Its intensity corresponds to two protons, indicating a *syn* structure of the chromophore as the only species present in solution.[115]

To our knowledge, the only X-ray structure analysis reported for this kind of ligand is that of the en derivative.[116] The arrangement of the N atoms of the ethylenediamine bridge is *gauche*, similar to the arrangement found in the analogous acetylcamphor derivatives.[101,103] The conformation of the ketonenamine form of the formylcamphor groups can be described as an *anti* arrangement, making intramolecular hydrogen bonding impossible. The stereochemical aspects of the structure are in agreement with the predictions of Jensen and Larsen.[66,114]

20.1.5.4 Ligand Properties and Conformational Aspects

In most of the metal complexes with this kind of ligand the tetradentate Schiff bases are in their ionized form with an essentially planar arrangement, *i.e.* with the four donor atoms nearly coplanar (Figure 17a). In fact, the relative flexibility of these ligands allows distortions of the Schiff base according to the metal atom nature and/or the bulkiness of apical or axial ligands determining 'umbrella' or 'stepped' geometries.[110]

The deformations induced by steric interactions of the ethylene bridge substituents with apical ligands can produce distortions from a square pyramidal geometry towards a trigonal bipyramid.

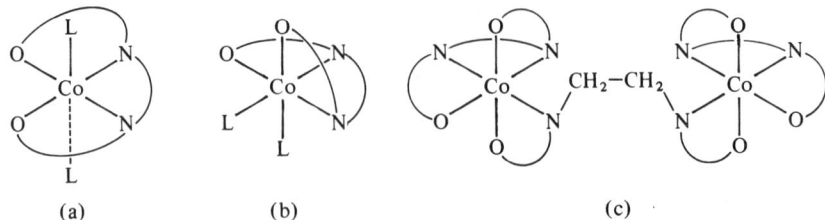

(a)   (b)   (c)

Figure 17 Possible arrangements of tetradentate Schiff bases in metal complexes

This appears relevant to the understanding of the role played by steric effects in the reversible oxygenation of Co compounds.[117] However, it has been suggested that the salen ligand exhibits a non-planar arrangement (Figure 17b) in a series of Cosalen(β-diketonate) complexes.[118] This has been confirmed by the X-ray structure analysis of Cosalen(acac)[119] and Co_2(3-MeOsalen)$_3$.[120] In the last compound it is also shown that salen ligands can act as bis-bidentate ligands bridging two metal atoms (Figure 17c).

It has been shown that besides this group of compounds, a second class is formed where the ligand is apparently un-ionized or partly ionized. Typical examples are $TiCl_3$(salenH$_2$) and $LaCl_3$-(salenH)$_2$. Although many examples of this second type are known,[121] to our knowledge only two X-ray structures have been reported.[109,122] In both these compounds the ligands act as bis-mono-dentate through their oxygen atoms. Unfortunately the X-ray analysis of the polymeric complex $SnCl_2Me_2 \cdot$ salenH$_2$ is of low accuracy and does not permit determination of the tautomeric form of the ligand.[122] On the contrary, the structure of $Ca(NO_3)_2$ saltrienH$_2$ (see Table 12) reveals that the charge-separated form (b') of Figure 15 predominates in this and related compounds. In fact, IR spectra of various calcium complexes containing salenH$_2$, salpnH$_2$ and salchxnH$_2$ are closely related to the saltrienH$_2$ compound.[109]

20.1.6 MULTIDENTATE SCHIFF BASES

As observed above, many Schiff bases have not been characterized in the free, *i.e.* uncomplexed, state. This is particularly so for Schiff bases with more than four donor atoms. Without entering into the details of the structures or spectroscopic properties of the metal complexes of such multidentate ligands, a brief survey of compounds will be presented here in order to exemplify the ligand ability of such Schiff bases.

20.1.6.1 Pentadentate Schiff Bases

Metal complexes of the potentially pentadentate ligands shown in Figure 18a have been widely investigated. Single crystal X-ray studies have shown that the ligand with X = N acts as penta-dentate for Ni^{II}[124–126] and Zn^{II},[127] while an unusually long bond is found in a Cu^{II} species,[126] so that the existence of a real Cu—N bond is questionable. In the five-coordinate complexes the coordination polyhedron may be described as a distorted trigonal bipyramid with the amine N atom and the two O atoms in the equatorial plane. It has been shown that the ligand when X = S is able to penta-coordinate Co^{II} and Ni^{II}.[123] However, in this case Lewis bases like pyridine complete the coordination polyhedron giving hexacoordinate complexes, with the N(py) atom *trans* to S. Therefore, changes in the coordination geometry may occur depending on the nature of the ligand itself and on the metal bound to it. Thus, in the case of X = N, Fe^{III} and Co^{III} form octahedral complexes, the sixth position being occupied by a chloride ion[128] and 1-Me-imidazole,[129] respectively. The flexibility of the ligand is shown by the different arrangements assumed by the pentadentate Schiff base.

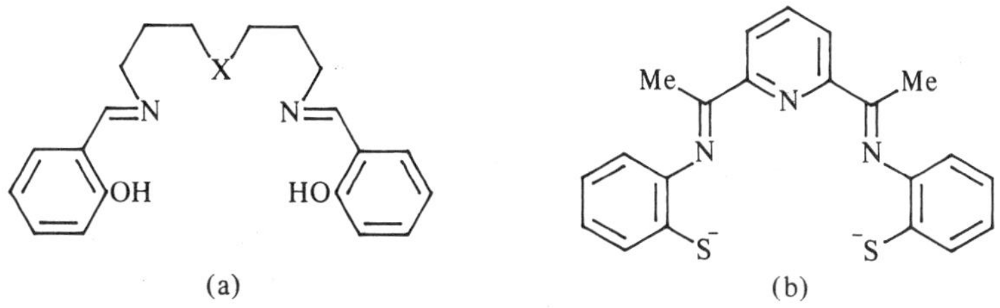

Figure 18 Some potentially pentadentate Schiff bases

Ligands like that depicted in Figure 18b have been shown to be able to give planar pentagonal coordination.[130] However the five-coordinate Zn^{II} complex has a coordination geometry that is approximately midway between a trigonal bipyramid and a square pyramid.[129] A distorted pen-tagonal geometry has been observed in *N,N'*(3-aza-1,5-pentanediyl)bis(salicylideniminato)dioxo-uranium(IV), where the five coordinated atoms form a rather puckered pentagon.[131]

The potentially pentadentate Schiff bases of Figure 19a actually act as bis-tridentate ligands with respect to a single metal centre, giving dinuclear Cu^{II} complexes as in Figure 19b.[132,133] Furthermore,

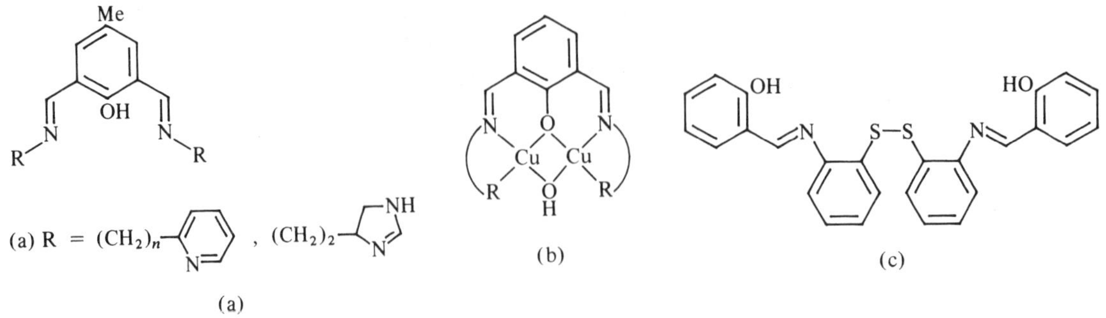

Figure 19 Some pentadentate Schiff bases (a) which act as bis-tridentate ligands in copper(II) complexes (b); pentadentate
Schiff base with a disulfide bridge (c)

in the case of the ethylpyridine derivative a dinuclear Cu^I complex has been isolated where the Schiff base does not coordinate through its pyridine N atom.[134]

Schiff bases containing disulfide groups (Figure 19c) may act, in the dianionic form, as pentadentate ligands coordinating through the two phenolic oxygens, the two imine nitrogens and one sulfur atom of the disulfide group. An octahedral Fe^{III} complex has been structurally characterized, in which the sixth position is occupied by a chloride ion.[135]

Pentadentate Schiff bases can give Fe^{II} and Co^{II} complexes which are relevant to the investigation of the stereochemical changes accompanying oxygenation. In fact, the N,N'-[2-(2'-pyridyl)ethyl]-ethylenebis(salicylideniminato) ligand forms five-coordinate Fe^{II} and Co^{II} complexes with a trigonal bypyramidal geometry.[136] While the iron complex is easily oxidized, the cobalt complex forms a dioxygen adduct where Co has an approximately octahedral coordination and the Schiff base moiety is nearly planar.[137]

20.1.6.2 Hexadentate Schiff Bases

The dianion of the Schiff base shown in Figure 20a, with R = H, is known to act as a hexadentate ligand in a variety of metal complexes, thereby forming a distorted octahedral arrangement about the metal atom.[138,139] The two oxygen atoms occupy the *cis* positions while the four nitrogen atoms (two *cis* amine and two *trans* imine) complete the coordination sphere. However, it has been shown that the ligand with R = Ph gives dinuclear Cu^{II} complexes. Each copper atom is bonded to two *cis* nitrogens and one phenolic oxygen, the fourth square planar coordination ligand being provided by an acetate ion.[140]

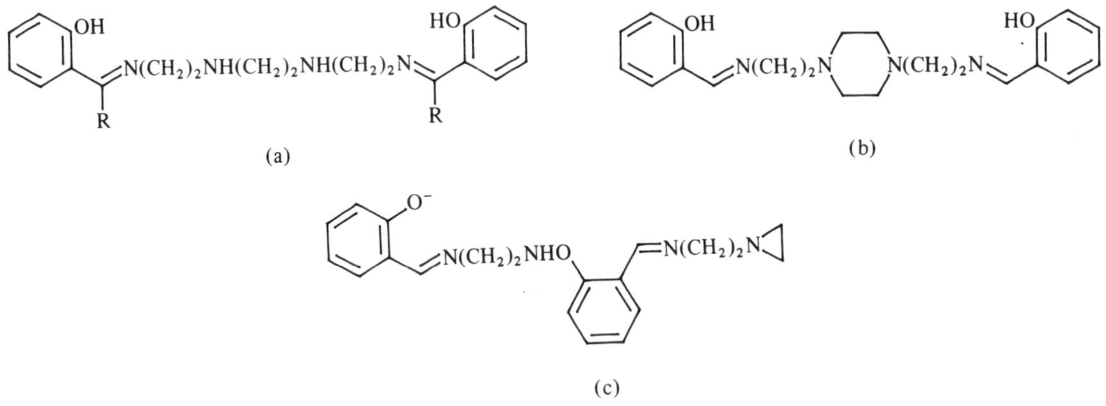

Figure 20 Some potentially hexadentate Schiff bases

Substitution of the central ethylenediamine bridge with a piperazine ring (Figure 20b) imposes steric constraints on the ligand in such a way that it gives dinuclear complexes, as shown by the X-ray structure of di-μ-methoxy-[dichloro-1,4-piperazine]bis(N-ethylenesalicylaldiminato)diiron(III).[141] The structure of the free Schiff base has been determined,[142] showing that upon coordination the ligand barely changes its conformation. The change mainly involves piperazine, which is 'chair'-shaped in the free ligand and 'boat'-shaped in the complex.

Reaction of bis[*N*-[2-(1-aziridinyl)ethyl]salicylaldiminate]nickel(II) with HBr yields Ni(salen).[143] During the reaction an intermediate has been isolated which appeared to be an octahedral Ni^{II} complex containing the hexadentate Schiff base anion depicted in Figure 20c. In the complex, the Schiff base nitrogen atoms are mutually *trans*, the aziridine nitrogen is *trans* to the ether oxygen, while the phenolate oxygen and the secondary amine nitrogen are *trans* to each other.

20.1.6.3 Heptadentate Schiff Bases

The potentially heptadentate Schiff base, obtained from tris(2-aminoethyl)amine and three moles of SAL and shown in Figure 21a, reacts with trivalent metal ions to form neutral 1:1 compounds, and in so doing acts as a hexadentate ligand. An X-ray structural analysis of an Fe^{III} complex has shown that the tertiary nitrogen atom of the amine is not bonded to Fe.[144]

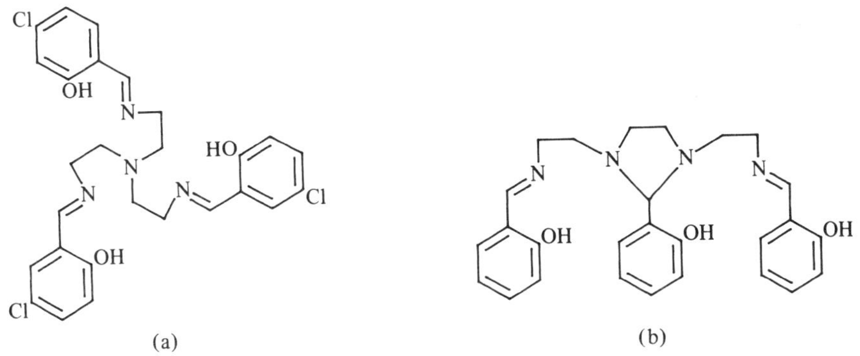

Figure 21 Some potentially heptadentate Schiff Bases

As previously observed with hexadentate Schiff bases, potentially heptadentate ligands easily form dinuclear metal complexes. This is the case with trisalicylidenetriethylenetetramine, shown in Figure 21b, which gives methoxo[145] and hydroxo[146] dinuclear Fe^{III} complexes.

20.1.7 APPENDIX

acacenH$_2$	*N,N'*-(Acetylacetonato)ethylenediamine
acachxnH$_2$	*N,N'*-(Acetylacetonato)-1,2-cyclohexylenediamine
acacpnH$_2$	*N,N'*-(Acetylacetonato)-1,2-propylenediamine
BA	Benzaldehyde
BAAN	*N*-(Benzaldehydo)aniline
BADPH	*N,N'*-(Benzaldehydo)-1,2-diphenylethylenediamine
butn	2,3-Butanediamine
chxn	1,2-Cyclohexylenediamine
fmc	Formylcamphor
fmcenH$_2$	3,3'-Ethylenebis(aminomethylene)di-2-bornanone
phenen	1,2-Diamino-1-phenylethane
phn	1,2-Phenylenediamine
SAL	Salicylaldehyde
SALAN	*N*-(Salicylaldehydo)aniline
salchxnH$_2$	*N,N'*-(Salicylaldehydo)-1,2-cyclohexylenediamine
salpn	*N,N'*-(Salicylaldehydo)-1,2-propylenediamine
saltrienH$_2$	*N,N'*-(Salicylaldehydo)-1,3-propylenediamine
stien	1,2-Diphenylethane-1,2-diamine

20.1.8 REFERENCES

1. H. Schiff, *Ann. Chim. (Paris)*, 1864, **131**, 118.
2. S. Dayagi and Y. Degani, in 'The Chemistry of the Carbon–Nitrogen Double Bond', ed. S. Patai, Wiley-Interscience, New York, 1970, p. 71.

3. F. P. Dwyer and D. P. Mellor, 'Chelating Agents and Metal Chelates', Academic, London, 1964.
4. D. J. Curran and S. Siggia, in 'The Chemistry of the Carbon–Nitrogen Double Bond', ed. S. Patai, Wiley-Interscience, New York, 1970, p. 149.
5. C. Sandorfy, in 'The Chemistry of the Carbon–Nitrogen Double Bond', ed. S. Patai, Wiley-Interscience, New York, 1970, p. 1.
6. R. Bonnett, in 'The Chemistry of the Carbon–Nitrogen Double Bond', ed. S. Patai, Wiley-Interscience, New York, 1970, p. 181.
7. M. D. Cohen and S. Flavian, *J. Chem. Soc. (B)*, 1967, 321.
8. H. E. Smith and R. Records, *Tetrahedron*, 1966, **22**, 813.
9. O. Kennard (ed.), 'Molecular Structures and Dimensions', N. V. A. Oosthoek's Uitgevers Mij, Utrecht, 1972, vol. A1.
10. M. D. Cohen, G. M. J. Schmidt and S. Flavian, *J. Chem. Soc.*, 1964, 2041; M. D. Cohen and G. M. J. Schmidt, *J. Phys. Chem.*, 1962, **66**, 2442.
11. J. Bregman, L. Leiserowitz and K. Osaki, *J. Chem. Soc.*, 1964, 2086.
12. J. W. Smith, in 'The Chemistry of the Carbon–Nitrogen Double Bond', ed. S. Patai, Wiley-Interscience, New York, 1970, p. 239.
13. L. G. Kuzmina and Yu. T. Struchkov, *Cryst. Struct. Commun.*, 1979, **8**, 715.
14. N. Inamoto, K. Kushida, S. Masuda, H. Ohta, S. Satoh, Y. Tamura, K. Tokumaru, K. Tori and M. Yoshida, *Tetrahedron Lett.*, 1974, 3617.
15. J. E. Arrowsmith, M. J. Cook and D. J. Hardstone, *Org. Magn. Reson.*, 1978, **11**, 160.
16. N. Naulet, M. L. Filleux, G. J. Martin and J. Pornet, *Org. Magn. Reson.*, 1975, **7**, 326.
17. M. Kobayashi, M. Yoshida and H. Minato, *J. Org. Chem.*, 1976, **41**, 3322.
18. L. M. N. Saleem, *Org. Magn. Reson.*, 1982, **19**, 176.
19. T. Bally, E. Haselbach, S. Lanyiova, R. Marschner and M. Rossi, *Helv. Chim. Acta*, 1976, **59**, 486.
20. P. Skrabal, J. Steiger and H. Zollinger, *Helv. Chim. Acta*, 1975, **58**, 800 and refs. therein.
21. H. B. Burgi and J. D. Dunitz, *Helv. Chim. Acta*, 1970, **53**, 1747.
22. G. M. J. Schmidt, *Pure Appl. Chem.*, 1971, **27**, 647.
23. H. B. Burgi and J. D. Dunitz, *Helv. Chim. Acta*, 1971, **54**, 1255.
24. J. Bernstein, M. Engel and A. T. Hagler, *J. Chem. Phys.*, 1981, **75**, 2346.
25. G. W. Gray, 'Molecular Structure and Properties of the Liquid Crystals', Academic, London, 1962, p. 162.
26. R. F. Bryan, P. Forcier and R. W. Miller, *J. Chem. Soc., Perkin Trans. 2*, 1978, 368.
27. I. Fonseca, S. Martinez-Carrera and S. Garcia-Blanco, *Acta Crystallogr., Sect. B*, 1982, **38**, 3121, 2735.
28. P. A. Tucker, A. Hoekstra, J. M. Teulate and A. Vos, *Acta Crystallogr., Sect. B*, 1975, **31**, 733.
29. J. Bernstein and G. M. J. Schmidt, *J. Chem. Soc., Perkin Trans. 2*, 1972, 951.
30. J. Bernstein and I. Izak, *J. Chem. Soc., Perkin Trans. 2*, 1976, 429.
31. J. Bernstein and I. Izak, *J. Cryst. Mol. Struct.*, 1975, **5**, 257.
32. B. T. Baylock and R. F. Bryan, *Trans. Am. Crystallogr. Assoc.*, winter meeting, 1972, abstract C2.
33. J. Bernstein, *J. Chem. Soc., Perkin Trans. 2*, 1972, 946.
34. H. Nakai, M. Shiro, K. Ezumi, S. Sakata and T. Kubota, *Acta Crystallogr., Sect. B*, 1976, **32**, 1827.
35. H. Nakai, K. Ezumi and M. Shiro, *Acta Crystallogr., Sect. B*, 1981, **37**, 193.
36. I. Bar and J. Bernstein, *Acta Crystallogr., Sect. B*, 1982, **38**, 121.
37. I. Bar and J. Bernstein, *Acta Crystallog., Sect. B*, 1977, **33**, 1738.
38. J. Bernstein, I. Bar and A. Christensen, *Acta Crystallogr., Sect. B*, 1976, **32**, 1609.
39. I. Bar and J. Bernstein, *Acta Crystallogr., Sect. B*, 1983, **39**, 266.
40. D. P. Lesser, A. de Vries, J. W. Reed and G. H. Brown, *Acta Crystallogr., Sect. B*, 1975, **31**, 653.
41. D. L. Johnston, W. L. Rohrbaugh and W. de W. Horrocks, Jr., *Inorg. Chem.*, 1971, **10**, 547.
42. A. J. Graham, D. Akrigg and B. Sheldrick, *Acta Crystallogr., Sect. C*, 1983, **39**, 192 and refs. therein.
43. B. Preslenik and W. Nowacky, *Z. Kristallogr.*, 1975, **142**, 239.
44. F. Vogtle and E. Goldshmitt, *Angew. Chem., Int. Ed. Engl.*, 1974, **14**, 520.
45. M. Wiebaka and D. Mootz, *Acta Crystallogr., Sect. B*, 1982, **38**, 2008.
46. J. Doucet, J. P. Mornon, R. Chevalier and A. Lifchitz, *Acta Crystallogr., Sect. B*, 1977, **33**, 1701.
47. J. Doucet, M. Levalut and M. Lambert, *Phys. Rev. Lett.*, 1974, **32**, 301 and refs. therein.
48. M. Cotrait, D. Sy and M. Ptak, *Acta Crystallogr., Sect. B*, 1975, **31**, 1869.
49. Y. Sasaki and F. P. Olsen, *Can. J. Chem.*, 1971, **49**, 283.
50. J. C. Barrick, C. Calvo and F. P. Olsen, *Can. J. Chem.*, 1973, **51**, 3691 and 3697; 1974, **52**, 2985.
51. D. Attanasio, A. A. G. Tomlinson and L. Alagna, *J. Chem. Soc., Chem. Commun.*, 1977, 618.
52. W. S. Sheldrick, R. Knorr and H. Polzer, *Acta Crystallogr., Sect. B*, 1979, **35**, 739.
53. C. P. Richards and G. A. Webb, *Org. Magn. Reson.*, 1976, **8**, 202.
54. R. Tewari and R. C. Srivastava, *Acta Crystallogr., Sect. B*, 1971, **27**, 1644.
55. A. C. Veronese, C. Cavicchioli, F. D'Angeli and V. Bertolasi, *Gazz. Chim. Ital.*, 1981, **111**, 153.
56. V. Bertolasi, G. Gilli and A. C. Veronese, *Acta Crystallogr., Sect. B*, 1982, **38**, 502.
57. P. L. Orioli, M. Di Vaira and L. Sacconi, *Inorg. Chem.*, 1966, **5**, 400.
58. R. M. Countryman, W. T. Robinson and E. Sinn, *Inorg. Chem.*, 1974, **13**, 2013 and refs. therein.
59. N. Bresciani-Pahor, L. Randaccio and E. Libertini, *Inorg. Chim. Acta*, 1980, **45**, L11.
60. F. A. Bottino, E. Libertini, O. Puglisi and A. Recca, *J. Inorg. Nucl. Chem.*, 1979, **41**, 1725.
61. M. F. Richardson, P. W. R. Corfield, D. E. Sands and R. E. Sievers, *Inorg. Chem.*, 1970, **9**, 1632.
62. G. O. Dudek and R. H. Holm, *J. Am. Chem. Soc.*, 1961, **83**, 2099; 1962, **84**, 2691.
63. G. O. Dudek, *J. Am. Chem. Soc.*, 1963, **85**, 694.
64. G. O. Dudek and E. Pitcher, *J. Am. Chem. Soc.*, 1964, **86**, 4283.
65. G. O. Dudek and G. P. Volpp, *J. Am. Chem. Soc.*, 1964, **85**, 2697.
66. H. P. Jensen and E. Larsen, *Acta Chem. Scand., Ser. A*, 1975, **29**, 157.
67. G. C. Percy and D. A. Thornton, *J. Inorg. Nucl. Chem.*, 1972, **34**, 3357 and 3369.
68. H. E. Smith, B. G. Padilla, J. R. Neergaard and F. M. Chen, *J. Org. Chem.*, 1979, **44**, 1690 and refs. therein.
69. H. E. Smith, E. P. Burrows, E. H. Massey and F. M. Chen, *J. Org. Chem.*, 1975, **40**, 2897.

70. H. E. Smith, E. P. Burrows, M. J. Marks, R. D. Lynch and F. M. Chen, *J. Am. Chem. Soc.*, 1977, **99**, 707.
71. H. E. Smith, J. R. Neergaard, E. P. Burrow and F. M. Chen, *J. Am. Chem. Soc.*, 1974, **96**, 2908.
72. H. E. Baumgarten, D. G. McMahan, V. J. Elia, B. I. Gold, V. W. Day and R. O. Day, *J. Org. Chem.*, 1976, **41**, 3798.
73. J. W. Smith, in 'The Chemistry of the Carbon–Nitrogen Double Bond', ed. S. Patai, Wiley-Interscience, New York, 1970, p. 235.
74. E. C. Lingafelter and R. L. Braun, *J. Am. Chem. Soc.*, 1966, **88**, 2951.
75. R. M. Kirchner, G. D. Andreetti, D. Barnhart, F. D. Thomas, II, D. Welsh and E. C. Lingafelter, *Inorg. Chim. Acta*, 1973, **7**, 17.
76. I. Moustakali-Mavridis and E. Hadjoudis, *Acta Crystallogr., Sect. C*, 1983, **39**, 365.
77. R. Destro, A. Gavezzotti and M. Simonetta, *Acta Crytallogr., Sect. B*, 1978, **34**, 2867.
78. J. Bregman, L. Leiserowitz and G. M. J. Schmidt, *J. Chem. Soc.*, 1964, 2068.
79. B. M. Gatehouse, *Cryst. Struct. Commun.*, 1982, **11**, 1793.
80. T. Ueki, T. Ashida, Y. Sasada and M. Kakudo, *Acta Crystallogr., Sect. B*, 1969, **25**, 328.
81. A. Syamal and K. S. Kale, *Inorg. Chem.*, 1979, **18**, 992.
82. J. Drummond and J. S. Wood, *J. Chem. Soc., Dalton Trans.*, 1972, 365.
83. A. P. Summerton, A. A. Diamantis and M. R. Snow, *Inorg. Chim. Acta*, 1978, **27**, 123.
84. P. C. Chieh and G. J. Palenik, *Inorg. Chem.*, 1972, **11**, 816.
85. Von H. Prent, H. J. Haupt, F. Huber, R. Cefalu and R. Barbieri, *Z. Anorg. Allg. Chem.*, 1974, **410**, 88.
86. G. D. Fallon and B. M. Gatehouse, *J. Chem. Soc., Dalton Trans.*, 1975, 1344.
87. J. A. Bertrand, J. A. Kelley and J. L. Breece, *Inorg. Chim. Acta*, 1970, **4**, 247.
88. S. Capasso, F. Giordano, C. Mattia, L. Mazzarella and A. Ripamonti, *J. Chem. Soc., Dalton Trans.*, 1974, 2228 and refs. therein.
89. R. R. Gagné, R. P. Kreh and J. A. Dodge, *J. Am. Chem. Soc.*, 1979, **101**, 6917.
90. A. E. Braunstein, 'Enzymes', 3rd edn., 1973, vol. 9, p. 379.
91. C. Beguin and S. Hamman, *Org. Magn. Reson.*, 1981, **16**, 129.
92. R. Haran, J.-P. Laurent, M. Massol and F. Nepveu-Juras, *Org. Magn. Reson.*, 1980, **14**, 45 and refs. therein.
93. R. H. Holm, in 'Inorganic Biochemistry', ed. G. L. Eichorn, Elsevier, New York, 1973, vol. 3, p. 1137. This chapter reviewed the literature covering vitamin B6 complexes.
94. J. F. Cutfield, D. Hall and T. N. Waters, *Chem. Commun.*, 1967, 785.
95. I. Moustakali-Mavridis, E. Hadjoudis and A. Mavridis, *Acta Crystallogr., Sect. B*, 1978, **34**, 3709; 1980, **36**, 1126.
96. J. Drummond and J. S. Wood, *J. Chem. Soc., Dalton Trans.*, 1972, 365 and refs. therein.
97. M. Gullotti, A. Pasini, P. Fantucci, R. Ugo and R. D. Gillard, *Gazz. Chim. Ital.*, 1972, **102**, 855.
98. E. Larsen, *Acta Chem. Scand., Ser. A*, 1969, **23**, 2158.
99. J. N. Murrel, 'The Theory of the Electronic Spectra of Organic Molecules', Methaveen, London, 1963, p. 133.
100. K. Ueno and A. E. Martell, *J. Phys. Chem.*, 1957, **61**, 257.
101. N. Bernth, E. Larsen and S. Larsen, *Tetrahedron*, 1981, **37**, 2477.
102. P. J. McCarthy and A. E. Martell, *Inorg. Chem.*, 1967, **6**, 781.
103. N. Bresciani-Pahor, M. Calligaris, G. Nardin, L. Randaccio and D. Viterbo, *Acta Crystallogr., Sect. B*, 1979, **35**, 2776.
104. H. P. Jensen, B. S. Kristensen, H. Mosbark and I. Sotofte, *Acta Chem. Scand., Ser. A*, 1978, **32**, 141.
105. N. Bresciani-Pahor, M. Calligaris, G. Nardin and L. Randaccio, *Acta Crystallogr., Sect. B*, 1978, **34**, 1360.
106. Von R. Senn and W. Novacky, *Z. Kristallogr.*, 1977, **145**, 16.
107. N. Bresciani-Pahor, M. Calligaris, P. Delise, G. Dodić, G. Nardin and L. Randaccio, *J. Chem. Soc., Dalton Trans.*, 1976, 2478.
108. C. Subrahmanian, M. Seshasayee and G. Aravarnudan, *Cryst. Struct. Commun.*, 1982, **11**, 1719.
109. J. I. Bullock, M. F. C. Ladd, D. C. Povey and H. A. Tajmir-Riahi, *Acta Crystallogr., Sect. B*, 1979, **35**, 2013.
110. M. Calligaris, G. Nardin and L. Randaccio, *Coord. Chem. Rev.*, 1972, **7**, 385.
111. B. Bosnich, *J. Am. Chem. Soc.*, 1968, **90**, 627.
112. R. S. Downing and F. L. Urbach, *J. Am. Chem. Soc.*, 1970, **92**, 5861.
113. A. Pasini, M. Gullotti and R. Ugo, *J. Chem. Soc., Dalton Trans.*, 1977, 346.
114. H. P. Jensen and E. Larsen, *Gazz. Chim. Ital.*, 1977, **107**, 143.
115. H. P. Jensen, *Acta Chem. Scand., Ser. A*, 1978, **32**, 149.
116. S. Larsen, *Acta Crystallogr., Sect. B*, 1981, **37**, 742.
117. N. Bresciani, M. Calligaris, G. Nardin and L. Randaccio, *J. Chem. Soc., Dalton Trans.*, 1974, 498.
118. S. N. Poddar and D. K. Biswas, *J. Inorg. Nucl. Chem.*, 1969, **31**, 565.
119. M. Calligaris, G. Manzini, G. Nardin and L. Randaccio, *J. Chem. Soc., Dalton Trans.*, 1972, 543.
120. M. Calligaris, G. Nardin and L. Randaccio, *Chem. Commun.*, 1970, 1079.
121. J. I. Bullock and H. A. Tajmir-Riahi, *J. Chem. Soc., Dalton Trans.*, 1977, 36.
122. L. Randaccio, *J. Organomet. Chem.*, 1973, **55**, C58.
123. E. M. Boge, G. M. Mockler and E. Sinn, *Inorg. Chem.*, 1977, **16**, 467.
124. M. Di Vaira, P. L. Orioli and L. Sacconi, *Inorg. Chem.*, 1971, **10**, 553.
125. M. Seleborg, S. L. Holt and B. Post, *Inorg. Chem.*, 1971, **10**, 1501.
126. P. C. Healy, G. M. Mockler, D. P. Freyberg and E. Sinn, *J. Chem. Soc., Dalton Trans.*, 1975, 691.
127. D. P. Freyber, G. M. Mockler and E. Sinn, *J. Chem. Soc., Dalton Trans.*, 1976, 447.
128. E. M. Holt, S. L. Holt and M. Vlasse, *Cryst. Struct. Commun.*, 1979, **8**, 645.
129. T. J. Kistenmacker, L. G. Marzilli and P. A. Marzilli, *Inorg. Chem.*, 1974, **13**, 2089.
130. V. L. Goedken and G. G. Christoph, *Inorg. Chem.*, 1973, **12**, 2316.
131. F. Benetollo, G. Bombieri and A. J. Smith, *Acta Crytallogr., Sect. B*, 1979, **35**, 3091.
132. R. R. Gagné, M. W. McCool and R. E. Marsh, *Acta Crystallogr., Sect. B*, 1980, **36**, 2420.
133. R. J. Majeste, C. L. Klein and E. D. Stevens, *Acta Crystallogr., Sect. C*, 1983, **39**, 52.
134. R. R. Gagné, R. P. Kreh and J. A. Dodge, *J. Am. Chem. Soc.*, 1979, **101**, 6917.
135. J. A. Bertrand and J. L. Breece, *Inorg. Chim. Acta*, 1974, **8**, 267.
136. G. B. Jameson, F. C. March, W. T. Robinson and S. S. Koon, *J. Chem. Soc., Dalton Trans.*, 1978, 185.
137. G. B. Jameson, W. T. Robinson and G. A. Rodley, *J. Chem. Soc., Dalton. Trans.*, 1978, 191.

138. P. D. Cradwick, M. E. Cradwick, G. G. Dodson, D. Hall and T. N. Waters, *Acta Crystallogr., Sect. B*, 1972, **28**, 45.
139. E. Sinn, G. Sim, E. V. Dose, M. F. Tweedle and L. J. Wilson, *J. Am. Chem. Soc.*, 1978, **100**, 3375.
140. B. Chiari, W. E. Hatfield, O. Piovesana, T. Tarantelli, L. W. Haar and and P. F. Zanazzi, *Inorg. Chem.*, 1983, **22**, 1468.
141. B. Chiari, O. Piovesana, T. Tarantelli and P. F. Zanazzi, *Inorg. Chem.*, 1982, **21**, 1396.
142. B. Chiari, O. Piovesana, T. Tarantelli and P. F. Zanazzi, *Acta Crystallogr., Sect. B*, 1982, **38**, 331.
143. K. R. Levan, C. E. Strouse and C. A. Root, *Inorg. Chem.*, 1983, **22**, 853.
144. D. F. Cook, D. Cummins and E. D. McKenzie, *J. Chem. Soc., Dalton Trans.*, 1976, 1369.
145. B. Chiari, O. Piovesana, T. Tarantelli and P. F. Zanazzi, *Inorg. Chem.*, 1982, **21**, 2444.
146. B. Chiari, O. Piovesana, T. Tarantelli and P. F. Zanazzi, *Inorg. Chem.*, 1984, **23**, 2542, 3398.

20.2
Amino Acids, Peptides and Proteins

STUART H. LAURIE
Leicester Polytechnic, UK

20.2.1 INTRODUCTION

The attention in this section will be mainly focused on the naturally occurring amino acids and the peptides and proteins derived from them. Synthetic aminopolycarboxylic acids will be considered in the next Chapter (20.3). Increasing awareness of the biological importance of the metal complexes of the ligands considered in this section has resulted in a tremendous growth in reported research over the past 20 years. Consequently, only those features which are deemed to be of general or of major importance are highlighted. Those aspects of particular biological significance will be featured in Chapter 22 of this volume and in Chapters 62.1 and 62.2 of Volume 6.

Relevant original papers are to be found in a wide range of journals covering many disciplines. Consequently, considerable use has been made of the excellent review series edited by Sigel[1] and of the annual review[2] 'Amino Acids, Peptides and Proteins'; interested readers are referred to these for up-to-date information.

20.2.2 AMINO ACIDS

These compounds, essential to life, have the general formula (1); the more important of them are listed in Table 1. The 'trivial' rather than systematic names are given since these are still in general use and the codes derived from these, as given in Table 1, are used to defined peptide sequences and genetic codes.

$$H_3\overset{+}{N}CH(R)CO_2H$$

(1)

Ignoring any ionization in the side chain R it can be seen from the data in Table 1 that all amino acids undergo two reversible proton ionizations steps (equation 1). Consequently, depending upon the solution pH, the amino acids can coordinate through either or both of the amino (NH_2) or carboxyl (CO_2^-) groups in aqueous media. In addition those with polar R groups, *e.g.* Asp, Cys, His, Glu, Pen, offer additional coordinating sites.

$$H_3\overset{+}{N}CH(R)CO_2H \underset{}{\overset{pH\,2-3}{\rightleftharpoons}} H_3\overset{+}{N}CH(R)CO_2^- \underset{}{\overset{pH\sim9}{\rightleftharpoons}} H_2NCH(R)CO_2^-$$ (1)

20.2.2.1 Amino Acids with Non-coordinating Side Chains

Monodentate coordination in solution is more readily proved with the kinetically inert metal ions. Thus, coordination through the amino N donor atom has been established with Cr^{III}, Co^{III}, Ir^{III}, Pt^{II} and Rh^{III}. Monodentate coordination through the weaker field carboxyl O donor atom is not so common but has been observed with Co^{III} and more recently with Pt^{II}. With the kinetically labile metal ions, monodentate coordination is sometimes observed at pH 3–5 as a minor species or can be obtained by means of blocking either the N or O donor atoms by substitution.

However, the more common mode of coordination is as a bidentate chelate through the N and O atoms (2), which gives rise to a thermodynamically stable five-membered ring for the α-amino acids (six-membered for β-alanine). The formation constants of some of the major species formed are shown in Table 2. The constants refer to the formation reactions (equation 2). Because of the wide range of conditions and methods employed, comparisons of the data in Table 2 should be treated as semiquantitative at best and the original papers should be consulted. It is important also to realize that since the species concentrations will be pH dependent a large numerical value for the formation constant does not necessarily mean that the species will be of major significance. Fortunately, over the past 10 years or so, improvements in potentiometric titration techniques and the availability of sophisticated computer programs (*e.g.* MINIQUAD) have resulted in greater confidence in results and the ability to determine the concentrations of major and minor species over a wide pH range. Examples of this type of treatment involving amino acid complexation can be

(2)

Table 1 The More Common Amino Acids, $H_3\overset{+}{N}CH(R)CO_2H$, and their Ionization Constants (at 25 °C, 0.1 mol dm^{-3})

Name	Symbol	R	pK_a *(ref. 7)*
Alanine (α)	Ala	Me	2.31, 9.70
Arginine	Arg	$CH_2(CH_2)_2NHC\overset{NH_3^+}{\underset{NH}{\diagdown}}$	2.03, 9.02, (12.1)
Asparagine	Asn	CH_2CONH_2	2.15, 8.72
Aspartic acid	Asp	CH_2CO_2H	1.94, 3.70, 9.62
Cysteine	Cys	CH_2SH	1.91, 8.16, 10.29
Dihydroxyphenylalanine	Dopa	CH_2- (dihydroxyphenyl, OH, OH)	2.20, 8.72, 9.78, (13.4)
Glutamine	Gln	$CH_2CH_2CONH_2$	2.16, 9.96
Glutamic acid	Glu	$CH_2CH_2CO_2H$	2.18, 4.20, 9.59
Glycine	Gly	H	2.36, 9.56
Histidine	His	CH_2- (imidazole)	1.7, 6.02, 9.09
Isoleucine	Ile	CHMeEt	2.21, 9.56
Leucine	Leu	CH_2CHMe_2	2.27, 9.28
Lysine	Lys	$CH_2(CH_2)_3\overset{+}{N}H_3$	2.19, 9.12, 10.68
Methionine	Met	CH_2CH_2SMe	2.10, 9.06
Penicillamine	Pen	CMe_2SH	1.90, 7.92, 10.6
Phenylalanine	Phe	CH_2Ph	2.17, 9.09
Serine	Ser	CH_2OH	2.13, 9.05
Threonine	Thr	CH(OH)Me	2.20, 8.96
Tryptophan	Trp	CH_2- (indole, NH)	2.34, 9.32
Tyrosine	Tyr	CH_2- (phenyl, OH)	2.17, 9.04, 10.11
Valine	Val	$CHMe_2$	2.26, 9.49
β-Alanine	β-Ala	$H_3\overset{+}{N}CH_2CH_2CO_2H$	3.53, 10.10
Cystine	Cys.Cys	$\overset{H_3N^+}{\underset{HO_2C}{\diagup}}CHCH_2SSCH_2CH\overset{NH_3^+}{\underset{CO_2H}{\diagdown}}$	(<2), (*ca.* 2) 8.03, 8.80
Proline	Pro	(pyrrolidine)$\overset{+}{N}H_2$–CHCO$_2$H	1.9, 10.41

Table 2 Log Formation Constants of Some Metal Amino Acid Complexes in Aqueous Media[a]

Metal ion	Gly $[H_2L]$[b]	α-Ala $[H_2L]$	Val $[H_2L]$	Leu $[H_2L]$	Ile $[H_2L]$	Phe $[H_2L]$	Tyr $[H_3L]$	β-Ala $[H_2L]$
Mg^{2+}	ML 1.34 (b)[c]	ML 1.96 (k)	—	—	—	—	—	—
Ca^{2+}	ML 1.4 (k)	ML 1.24 (k)	—	—	—	—	MHL 1.48 (k)	—
Mn^{2+}	ML 2.60 (b); ML_2 4.5 (b); ML_3 5.3 (b)	ML 2.39 (d); ML_2 4.29 (d); ML_3 5.7 (d)	ML 2.34 (d); ML_2 4.0 (d); ML_3 5.2 (d)	ML 2.4 (a)	—	ML 2.4 (a); ML_2 4.7 (a)	MHL 11.64 (a); $M(HL)_2$ 25.28 (a)	ML ~7 (k)
Co^{2+}	ML 4.67 (a); ML_2 8.46 (a); ML_3 10.8 (a)	ML 4.31 (a); ML_2 7.8 (a); ML_3 9.5 (a)	—	ML 4.32 (j); ML_2 3.6 (j)	—	ML 4.45 (i); ML_2 8.44 (i)	MHL 14.0 (a); $M(HL)_2$ 27.8 (a)	—
Ni^{2+}	ML 5.78 (a); ML_2 10.58 (a); ML_3 14.0 (a)	ML 5.40 (a); ML_2 9.9 (a); ML_3 12.9 (a)	ML 5.42 (a); ML_2 9.72 (a); ML_3 12.2 (a)	ML 5.53 (a); ML_2 9.46 (a); ML_3 14.38 (a)	ML 5.4 (a); ML_2 9.7 (a); ML_3 12.7 (a)	ML 5.35 (i); ML_2 10.49 (i)	MHL 16.1 (a); $M(HL)_2$ 29.6 (a); $M(HL)_3$ 42.4 (a)	ML 4.54 (a); ML_2 7.87 (a); ML_3 9.7 (a)
Cu^{2+}	ML 8.13 (a); ML_2 15.0 (a)	ML 8.15 (a); ML_2 14.9 (a)	ML 8.09 (a); ML_2 14.9 (a)	ML 8.2 (a); ML_2 15.4 (a)	ML 8.45 (a); ML_2 15.4 (a)	ML 7.90 (a); ML_2 14.8 (a)	MHL 17.9 (a); $M(HL)_2$ 34.9 (a)	ML 7.04 (a); ML_2 12.54 (a)
Zn^{2+}	ML 4.96 (a); ML_2 9.19 (a); ML_3 11.6 (a)	ML 4.56 (a); ML_2 8.55 (a); ML_3 10.6 (a)	ML 4.44 (d); ML_2 8.24 (d); ML_3 10.62 (d)	ML 4.51 (d); ML_2 8.56 (d)	ML 4.49 (b); ML_2 8.49 (b); ML_3 10.9 (b)	—	MHL 14.30 (a); $M(HL)_2$ 28.5 (a)	ML 3.90 (b); ML_2 7.20 (b); ML_3 10.40 (b)
Cd^{2+}	ML 4.24 (a); ML_2 7.71 (a); ML_3 9.76 (a)	ML 3.80 (f); ML_2 7.10 (f); ML_3 9.09 (f)	ML 3.46 (b); ML_2 6.46 (b)	ML 3.84 (f); ML_2 6.54 (f); ML_3 8.60 (f)	ML 7.0 (f); ML_2 8.8 (f)	ML 3.7 (a); ML_2 6.9 (a)	MHL 13.71 (a); $M(HL)_2$ 26.2 (g)	ML_2 5.70 (n); ML_3 6.78 (n)
Pb^{2+}	ML_2 7.7 (a)	ML_2 4.15 (e)	ML 4.02 (n); ML_2 5.89 (n)	—	—	ML 4.01 (g); ML_2 8.84 (g)	MHL 14.3 (g); $M(HL)_2$ 29 (g)	—
$MeHg^+$	ML 7.88 (c)	ML 7.56 (o)	ML 7.27 (f)	—	—	ML 8.29 (o)	—	ML 7.25 (c); MHL 12.5 (c)
Fe^{2+}	ML 4.13 (a); ML_2 7.65 (a)	ML 3.54 (l)	—	ML 3.42 (l)	—	ML 3.74 (i); ML_2 7.19 (i); ML_3 10.7 (i)	—	ML ~4 (k)
Hg^{2+}	ML 10.3 (f); ML_2 19.2 (f)	—	—	—	—	—	—	—

Metal ion	Dopa $[H_4L]$ (h)	Ser $[H_2L]$	Thr $[H_2L]$	Asp $[H_3L]$	Glu $[H_3L]$	Asn $[H_2L]$	Gln $[H_2L]$	Arg $[H_3L]$
Mg^{2+}	MHL 4.71 (f)	ML 1.43 (k)	—	ML 2.43 (a)	ML 1.9 (a)	—	—	MHL 11.22 (a)
Ca^{2+}	—	—	—	ML 1.60 (a)	ML 1.43 (a)	—	—	MHL 11.22 (a)
Mn^{2+}	—	ML 2.50 (a); ML_2 3.98 (a)	ML 2.58 (a); ML_2 3.96 (a)	ML 3.7 (a)	ML 3.3 (m)	ML 3.10 (i); ML_2 5.22 (a)	ML 2.86 (i); ML_2 4.6 (i)	MHL 11.56 (a)
Co^{2+}	MH_2L 3.75 (a); $M(H_2L)_2$ 7.25 (a)	ML 4.20 (d); ML_2 7.56 (d); ML_3 9.81 (d)	ML 4.16 (d); ML_2 7.45 (d); ML_3 8.82 (d)	ML 5.95 (a); ML_2 10.23 (a)	ML 4.56 (a); ML_2 7.86 (a)	ML 4.51 (a); ML_2 8.01 (a); ML_3 9.96 (a)	ML 4.04 (a); ML_2 7.32 (a); ML_3 11.41 (i)	MHL 16.0 (a); $M(HL)_2$ 31.2 (a); $M(HL)_3$ 45.5 (a)
Ni^{2+}	MH_2L 4.88 (a); $M(H_2L)$ 8.9 (a)	ML 5.40 (a); ML_2 9.9 (a); ML_3 13.1 (a)	ML 5.46 (a); ML_2 10.01 (a); ML_3 13.3 (a)	ML 7.15 (a); ML_2 12.39 (a); MHL 11.2 (a)	ML 5.60 (a); ML_2 9.76 (a)	ML 6.15 (i); ML_2 11.16 (i); ML_3 14.55 (i)	ML 5.56 (i); ML_2 10.28 (i); ML_3 13.82 (i)	MHL 17.1 (a); $M(HL)_2$ 33.3 (a); $M(HL)_3$ 37.8 (a)
Cu^{2+}	MH_2L 7.55 (a); $M(H_2L)_2$ 14.1 (a); $M(HL)$ 12.99 (f)	ML 7.89 (a); ML_2 14.5 (a)	ML 8.0 (a); ML_2 14.69 (a)	ML 8.89 (a); ML_2 15.93 (a)	ML 8.33 (a); ML_2 14.84 (a); MHL 12.48 (a)	ML 7.83 (a); ML_2 14.36 (a)	ML 7.76 (a); ML_2 14.23 (a)	MHL 19.6 (a); $M(HL)_2$ 38.0 (a)

Metal ion	Lys [H₃L]	His [H₃L]	Trp [H₂L]	Met [H₃L]	Cys [H₃L]	Pen [H₃L]	Cystine [H₄L]	Pro [H₂L]
Mg^{2+}								
Ca^{2+}								
Mn^{2+}	MHL 2.18 (m)	ML 3.24 (d); ML₂ 6.16 (d)	ML 2.84 (i); ML₂ 5.15 (i); ML₃ 8.0 (i)	ML 2.77 (a); ML₂ 4.57 (a)	ML 4.7 (a)			ML 2.84 (d); ML₂ 5.53 (d); ML₃ 6.7 (d)
Co^{2+}	MHL 14.5 (a); M(HL)₂ 28.4 (a); M(HL)₃ 41.5 (a)	ML 6.87 (a); ML₂ 12.38 (a); ML(HL) 18.37 (a)	ML 4.58 (i); ML₂ 8.90 (i); ML₃ 12.25 (i)	ML 4.14 (a); ML₂ 7.58 (a)	ML 9.3 (m); ML₂ 16.6 (m)			
Ni^{2+}	MHL 15.7 (a); M(HL)₂ 30.5 (a); M(HL)₃ 44.0 (a)	ML 8.66 (a); ML₂ 15.52 (a); ML(HL) 20.55 (a)	ML 5.76 (i); ML₂ 10.98 (i); ML₃ 15.46 (i)	ML 5.33 (a); ML₂ 9.89 (a); ML₃ 11.6 (a)	ML 9.8 (a); ML₂ 20.07 (a); M₂L₃ 33.0 (j)	ML 10.63 (a); ML₂ 22.97 (a)		ML 5.95 (a); ML₂ 10.90 (a)
Cu^{2+}	MHL 18.3 (a); M(HL)₂ 35.4 (a)	ML 10.16 (a); ML₂ 18.10 (a); ML(HL) 23.8 (a)	ML 8.25 (a); ML₂ 15.4 (a)	ML 7.86 (a); ML₂ 14.6 (a)			MHL 16.1 (d); M₂L₂ 28.07 (d)	ML 8.84 (a); ML₂ 16.36 (a)
Zn^{2+}	MH₂L 3.77 (a); MHL 9.94 (f); M(HL)₂ 18.06 (f); M(HL)₂ 7.6 (m)	ML 6.51 (a); ML₂ 12.04 (a); ML(HL) 17.84 (a); [4.47 (d); 8.31 (d); 10.6 (d)]	ML 5.01 (i); ML₂ 9.8 (i); ML₃ 13.5 (i); [ML 4.43 (d); ML₂ 8.14 (d); ML₃ 10.1 (d)]	MH₂L 5.58 (b); MHL 10.1 (b); M(HL)₂ 11.9 (d); ML 4.35 (a); ML₂ 7.55 (a); ML 4.38 (a); ML₂ 8.33 (a)	ML 9.04 (a); ML₂ 18.12 (a); M(HL) 24.5 (a); ML 12.9 (i); ML₂ 19.6 (i)	ML 9.5 (a); ML₂ 19.40 (a); ML(HL) 25.4 (a); ML(HL) 12.68 (i); ML₂ 20.68 (i)	MHL 12.2 (d); M(HL)₂ 24.3 (d)	ML 5.13 (d); ML₂ 9.69 (d); ML₃ 11.9 (d); M(HL)₂ 20.2 (a)
Cd^{2+}	MH₂L 3.61 (g)	ML 6.48 (i); ML₂ 11.11 (i); [ML 4.15 (i); ML₂ 7.86 (i); ML₃ 10.22 (i)]	ML 4.48 (i); ML₂ 8.6 (i); ML₃ 12.0 (i)	ML 3.69 (a); ML₂ 7.00 (a)	ML 12.21 (i); ML₂ 18.57 (i); MHL 15.7 (o); 25.48 (o)	ML 19.05 (i)		
Pb^{2+}	MH₂L 5.56 (g)	ML 6.90 (i); ML₂ 9.81 (i); [ML 5.05 (i); ML₂ 8.27 (i); ML₃ 10.0 (i)]	ML 4.9 (i); ML₂ 10.3 (i); [ML 4.43 (d); ML₂ 7.2 (d)]	ML 4.38 (a); ML₂ 8.62 (a); [ML 6.67 (i); ML₂ 9.4 (i)]				
$MeHg^{+}$		ML 3.43 (l)	ML 3.30 (l)	ML 4.34 (l)		ML 16.60 (p); MHL 25.34 (p)		MHL 3.20 (m)
Fe^{2+}	MHL 4.5 (m)	ML 5.88 (i); ML₂ 10.43 (i)	ML 3.92 (i); ML₂ 7.39 (i)	ML 3.24 (l)	ML 6.2 (m); ML₂ 11.77 (k)	ML 16.1 (a)		ML 4.07 (l)
Hg^{2+}				ML 6.52 (a); ML₂ 11.45 (a)	ML 43.57 (f)			MHL 14.35 (a); M(HL)₂ 28.23 (a)

[a] Data compiled from ref. 7.

[b] Formula indicates *fully* protonated form.

[c] Conditions of temperature and ionic strength (mol dm⁻³) are indicated by the letter in parentheses, as follows: (a) 25 °C, 0.1 mol dm⁻³ ionic strength; (b) 25 °C, 0.5; (c) 25 °C, 0.25; (d) 37 °C, 0.15; (e) 20 °C, 0.4; (f) 25 °C, 1.0; (g) 20 °C, 0.37; (h) data for equilibria $M + H_xL \rightleftharpoons M(H_xL)$,; (i) 25 °C, 3.0; (j) 20 °C, 1.0; (k) 25 °C, 0.1; (l) 20 °C, 0.01; (m) 20 °C, 0.01; (n) 30 °C, 1.0; (o) from data compiled by D. L. Rabenstein, *Acc. Chem. Res.*, 1978, **11**, 100: conditions not stated; (p) data from A. P. Arnold and A. J. Canty, *Can. J. Chem.*, 1983, **61**, 1428: 25 °C, 0.3.

$$pM^{a+} + qH^+ + r(aa)^{b-} \rightleftharpoons [M_pH_q(aa)_r]^{(pa + q - rb)} \qquad (2)$$

aa = amino acidate

found in refs. 3–6. Stability constants, other than those in Table 2, for other metal ion complexes are also available,[7] *e.g.* for Ag^+, Cu^+, Ga^{3+}, In^{3+}, lanthanides, Th^{4+}, UO_2^{2+} and VO_2^+. A compilation of related thermodynamic data has been prepared by Christensen *et al.*[8]

Not mentioned in Table 2 (and often not in the original papers!) is the optical form (chirality) of the amino acids used. All the amino acids, except for glycine (R = H), contain an asymmetric carbon atom (the C_α atom). In the majority of cases the optical form used, whether L, D or racemic DL, makes little difference to the stability constants, but there are some notable exceptions (*vide infra*). Examination of the data in Table 2 reveals: (i) that the order of stability constants for the divalent transition metal ions follows the Irving–Williams series; (ii) that for the divalent transition metal ions, with excess amino acid present at neutral pH, the predominant species is the neutral chelated $M(aa)_2$ complex; (iii) that the species formed reflect the stereochemical preferences of the metal ions, *e.g.* for Cu^{II} a 2:1 complex readily forms but not a 3:1 ligand:metal complex (see Volume 5, Chapter 53). Confirmation of the species proposed from analysis of potentiometric data and information on the mode of bonding in solution has involved the use of an impressive array of spectroscopic techniques, *e.g.* UV/visible, IR, ESR, NMR, CD and MCD (magnetic circular dichroism).

Chelate formation is also well established with the kinetically inert metal ions, *e.g.* Co^{III}, Cr^{III}, Rh^{III}, Ru^{III} and Pt^{II}, to give complexes of the type $M^{II}(aa)_2$, $M^{III}(aa)_3$ and $M^{III}(L)(aa)$, *etc.*, where L = amine ligand(s). In these cases, because of their inertness, there is generally agreement between the structure of the major species in solution and those found in the solid state from diffraction studies. The same does not necessarily hold true of the kinetically labile systems and correlation between the two phases must be treated carefully. Despite this caveat a large number of the complexes have been crystallized and their structures solved by X-ray diffraction measurements; some of these are given in Table 3. Not included in this table are the many studies of complexes of substituted amino acids. Structures reported prior to 1967 have been reviewed by Freeman[9] whose detailed analysis of the geometries adopted is still pertinent today. One important observation from all these studies is that the five-membered chelate ring is near planar.

Table 3 Structures of some Amino Acid Coordination Compounds from X-ray Crystallography

Compound	Geometry; ligating atoms[a]	Ref.
[Ag(GlyO)]	Linear chains; O (car), N (am')[b]	1
[Cd(L-GluO)(H₂O)]H₂O	Octa; N (am), Oˣ (car), 2 Oʸ (car'), O (H₂O)	2
[Cd(L-HisO)₂]·2H₂O	Octa; *cis* 2 N³ (imid), *trans* 2N (am), *cis* 2 O (car) (neutron diffraction)	3
[Coᴵᴵ(L-HisO)(D-HisO)]2H₂O	Dist octa; *cis* 2 O (car), *cis* 2 N (am), *cis* 2 N³ (imid)	4
[Coᴵᴵ(L-HisO)₂]	Octa; *cis* 2 N (am), *cis* 2 O (car), *trans* 2 N³ (imid)	5
[Coᴵᴵᴵ(L-AlaO)₃]H₂O	Octa; *mer* 3 N (am), 3 O (car)	6
[Coᴵᴵᴵ(β-AlaO)₃]4H₂O	Octa; *mer* 3 N (am), 3 O (car)	7
[Coᴵᴵᴵ(L-HisO)(D-PenO)]H₂O	Octa; *cis* 2 O (car), *cis* 2 N (am), *cis* N³ (imid) and S	8
K[Coᴵᴵᴵ(D-PenO)(L-PenO)]2H₂O	Octa; *cis* 2 O (car), *cis* 2 N (am), *cis* 2 S	9
[Coᴵᴵᴵ(L-HisO)(D-HisO)]Br	Octa; *cis* 2 O (car), *cis* 2 N (am), *cis* 2 N³ (imid)	10
[Coᴵᴵᴵ(L-HisO)₂](ClO₄)·2H₂O	Octa; *cis* 2 O (car), *trans* 2 N (am), *cis* 2 N³ (imid)	11
[Crᴵᴵᴵ(L-HisO)(D-PenO)]H₂O	Isostructural with Coᴵᴵᴵ analogue	12
[Crᴵᴵᴵ(GlyO)₃]H₂O	Octa; facial 3 N (am), 3 O (car)	13
Na[Crᴵᴵᴵ(L-CysO)₂]2H₂O	Octa; *cis* 2 O (car), *cis* 2 N (am), *trans* 2 S	14
[Cuᴵᴵ(L-HisO)(D-HisO)(H₂O)₂]4H₂O	Tetra octa; *trans* 2 N (am), *trans* 2 N³ (imid), axial 2 O (H₂O)	15
[Cuᴵᴵ(L-SerO)₂]	Sq pyr; *cis* 2 O (car), *cis* 2 N (am), axial O (car')	16
[Cuᴵᴵ(L-TyrO)₂]	Sq pyr; *trans* 2 O (car), *trans* 2 N (am), axial O (car'), Cu–π interaction	17
[Cuᴵᴵ(L-HisO)(L-AsnO)]	Sq pyr; *cis* 2 N (am), N³ (imid), O (car), axial O (car) from His	18
[Cuᴵᴵ(L-HisO)(L-ThrO)]	Sq pyr; *cis* 2 N (am), N (imid), O (car), axial O (car)	19
[Cuᴵᴵ(L-HisO)(L-SerO)(H₂O)]3H₂O	Sq pyr; *cis* 2 N (am), N (imid), O (car), axial O (H₂O)	20
Cuᴵᴵ(L-His)₂(H₂O)₂](NO₃)₂	Tetra octa; *trans* 2 O (car), 2 N (am), axial 2 O (H₂O), (imidH⁺)	21
[Cuᴵᴵ(L-MetO)₂]	Tetra octa; *trans* 2 N (am), 2 O (car), axial 2 O (car')	22
Tl⅝[Cu⅛ᴵCu₆ᴵᴵᴵ(D-PenO)₁₂Cl]·nH₂O	S-bridging Mⁿ⁺ cluster with CuᴵS₃ and CuᴵᴵN₂S₂ chromophores	23
[MeHg(L-TyrO)]H₂O	Linear?; N (am), weak O (car)	24
[MeHg(DL-MetO)]	Linear?; N (am), weak O (car)	25
[MeHg(L-Cys)]	Linear; S⁻, Me	26
[MeHg(DL-Pen)]H₂O	Linear; S⁻, Me	27
[HgᴵᴵCl₂(DL-Pen)₂]H₂O	Polymeric; Cl⁻, S, O (car) bridging	28

Table 3 (*continued*)

Compound	Geometry; ligating atoms[a]	Ref.
[Mn^{II}(DL-Ala)$_2$(H$_2$O)$_2$]Br$_2$·2H$_2$	Octa dimeric; *cis* O (car), O (car′), 4 O (H$_2$O)	29
Mn^{II}(DL-Pro)$_2$Br$_2$(H$_2$O)$_2$]	Octa; 2 O (car), 2 Br$^-$, 2 O (H$_2$O)	30
[Na$_2$Mo$_2^V$O$_4$(L-CysO)$_2$(H$_2$O)$_2$]3H$_2$O	Tridentate CysO^{2-}	31
[Mo$_2^V$O$_2$S$_2$(L-HisO)$_2$]H$_2$O	di-μ-S, *trans* 2 N (am), *trans* 2 N (imid), *cis* 2 O (car)	32
[Mo$_2^V$O$_4$(L-HisO)$_2$]3H$_2$O	di-μ-O, *trans* 2 N (am), *trans* 2 N (imid), *cis* 2 O (car)	33
[Mo$_2^{IV}$(π-C$_5$H$_5$)(L-ProO)](PF$_6$)	Bidentate ProO$^-$	34
[Ni^{II}(L-HisO)$_2$]H$_2$O	Octa; *cis* 2 O (car), *cis* 2 N (am), *trans* 2 N^3 (imid)	35
[Ni^{II}(L-TyrO)$_2$(H$_2$O)$_2$]H$_2$O	Octa; *trans* 2 N (am), *trans* 2 O (car), *trans* 2 O (H$_2$O)	36
[Pb^{II}(D-PenO)]	Tridentate PenO^{2-}	37
[Pd^{II}(L-SerO)$_2$]	Sq planar; *cis* 2 N (am), *cis* 2 O (car)	38
[Pd^{II}(DL-Met)Cl$_2$]	Sq planar; *cis* 2 Cl$^-$, *cis* N (am), S (CO$_2$H)	39
[Pd^{II}(L-TyrO)$_2$]	Sq planar; 2 O (car), 2 N (am), Pd–π interaction	40
[Pt^{II}(L-Met)Cl$_2$]	Sq planar; *cis* 2 Cl$^-$, *cis* N (am), S (CO$_2$H)	39
[Pt^{II}(GlyO)$_2$]	Sq planar; *trans* 2 O (car), *trans* 2 N (am)	41
[Pt^{II}(Gly)$_2$Cl$_2$]H$_2$O	Sq planar; *cis* 2 Cl$^-$, *cis* 2 N (am)	42
[MeSn^{IV}(GlyO)]	Trig bipyr polymeric; Me, N (am), N (am′), O (car), O (car′)	43
[Me$_2$Tl^{III}(L-PheO)]	Octa polymeric; 2Me, N (am), O (car), 2 O (car′)	44
[Zn^{II}(L-ThrO)$_2$(H$_2$O)$_2$]	Octa; *cis* 2 O (car), *cis* 2 N (am), *trans* 2 O (H$_2$O)	45
[Zn^{II}(β-Ala)$_2$(H$_2$O)$_4$](NO$_3$)$_2$	Octa; *trans* 2 O (car), 4 O (H$_2$O)	46
[Zn^{II}(L-AsnO)$_2$]	Dist octa; *trans* 2 N (am), *trans* 2 O (car), *trans* 2 O (car′)	47
[Zn^{II}(L-SerO)$_2$]	Sq pyr; N (am), O (car), O (car, bidentate), axial N (am)	48

[a] am, amine; car, carboxyl; imid, imidazole; octa, octahedral; dist, distorted; *mer*, meridional; sq pyr, square pyramidal; tetra, tetragonally distorted; trig bipyr, trigonal bipyramid; μ, bridging.
[b] Primes indicate ligands from adjacent molecules.

1. C. B. Acland and H. C. Freeman, *Chem. Commun.*, 1971, 1016.
2. R. J. Flook, H. C. Freeman and M. L. Scudder, *Acta Crystallogr., Sect. B*, 1977, **33**, 801.
3. H. Fuess and H. Bartunik, *Acta Crystallogr., Sect. B*, 1976, **32**, 2803.
4. R. Candlin and M. M. Harding, *J. Chem. Soc. (A)*, 1970, 384.
5. M. M. Harding and H. A. Long, *J. Chem. Soc. (A)*, 1968, 2554.
6. R. Herak, B. Prelesink and I. Krstanovic, *Acta Crystallogr., Sect. B*, 1978, **34**, 91.
7. H. Soling, *Acta Chem. Scand., Ser. A*, 1978, **32**, 361.
8. P. de Meester and D. J. Hodgson, *J. Am. Chem. Soc.*, 1977, **99**, 101.
9. H. M. Helis, P. de Meester and D. J. Hodgson, *J. Am. Chem. Soc.*, 1977, **99**, 3309.
10. N. Thorup, *Acta Chem. Scand., Ser. A*, 1977, **31**, 203.
11. N. Thorup, *Acta Chem. Scand., Ser. A*, 1979, **33**, 759.
12. P. de Meester and D. J. Hodgson, *J. Chem. Soc., Dalton Trans.*, 1977, 1604.
13. R. F. Bryan, P. T. Greene, P. F. Stokely and E. W. Wilson, *Inorg. Chem.*, 1971, **10**, 1468.
14. P. de Meester, D. J. Hodgson, H. C. Freeman and C. J. Moore, *Inorg. Chem.*, 1977, **16**, 1494.
15. N. Camerman, J. K. Fawcett, T. P. A. Kruck, B. Sarkar and A. Camerman, *J. Am. Chem. Soc.*, 1978, **100**, 2690.
16. D. W. Van der Helm and W. A. Franks, *Acta Crystallogr., Sect. B*, 1969, **25**, 451.
17. D. Van der Helm and C. E. Tatsch, *Acta Crystallogr., Sect. B*, 1972, **28**, 2307.
18. T. Ono, H. Shimanouchi, Y. Sasada, T. Sakurai, O. Yamauchi and A. Nakahara, *Bull. Chem. Soc. Jpn.*, 1979, **52**, 2229.
19. H. C. Freeman, J. M. Guss, M. J. Healey, R. P. Martin, C. E. Nockolds and B. Sarkar, *Chem. Commun.*, 1969, 225.
20. Y. Sasada, A. Takenaka and T. Furuya, *Bull. Chem. Soc. Jpn.*, 1983, **56**, 1745.
21. B. Evertsson, *Acta Crystallogr., Sect. B*, 1969, **25**, 1203.
22. C. C. Ou, D. A. Powers, J. A. Thich, T. R. Felthouse, D. N. Hendrickson, J. A. Potenza and H. J. Schugar, *Inorg. Chem.*, 1978, **17**, 34.
23. P. J. M. W. L. Birker and H. C. Freeman, *J. Am. Chem. Soc.*, 1977, **99**, 6890.
24. N. W. Alcock, P. A. Lampe and P. Moore, *J. Chem. Soc., Dalton Trans.*, 1978, 1324.
25. Y. S. Wong, A. J. Carty and P. C. Chieh, *J. Chem. Soc., Dalton Trans.*, 1977, 1157.
26. Y. S. Wong, N. J. Taylor, P. C. Chieh and A. J. Carty, *J. Chem. Soc., Chem. Commun.*, 1974, 625.
27. Y. S. Wong, A. J. Carty and P. C. Chieh, *J. Chem. Soc., Dalton Trans.*, 1977, 1801.
28. A. J. Carty and N. J. Taylor, *J. Chem. Soc., Chem. Commun.*, 1976, 214.
29. Z. Ciunik and T. Glowiak, *Inorg. Chim. Acta*, 1980, **19**, 1215.
30. T. Glowiak and Z. Ciunik, *Acta Crystallogr., Sect. B*, 1977, **33**, 3237.
31. J. R. Knox and C. K. Prout, *Acta Crystallogr., Sect. B*, 1969, **25**, 1857.
32. B. Spivack, A. P. Gaughan and Z. Dori, *J. Am. Chem. Soc.*, 1971, **93**, 5265.
33. L. J. T. Delbaere and C. K. Prout, *Chem. Commun.*, 1971, 162.
34. C. K. Prout, S. R. Critchley, E. Cannillo and V. Tazzoli, *Acta Crystallogr., Sect. B*, 1977, **33**, 456.
35. T. Sakurai, H. Iwasaki, T. Katano and Y. Nakahashi, *Acta Crystallogr., Sect. B*, 1978, **34**, 660.
36. R. Hamalainen, M. Ahlgren, U. Turpeinen and T. Raikas, *Cryst. Struct. Commun.*, 1978, **7**, 379.
37. H. C. Freeman, G. N. Stevens and I. F. Taylor, *J. Chem. Soc., Chem. Commun.*, 1974, 366.
38. R. S. Vagg, *Acta Crystallogr., Sect. B*, 1979, **35**, 341.
39. H. C. Freeman and M. L. Golomb, *Chem. Commun.*, 1970, 1523.
40. M. Sabat, M. Jezowska and H. Kozlowski, *Inorg. Chim. Acta*, 1979, **37**, L511.
41. H. C. Freeman and M. L. Golomb, *Acta Crystallogr., Sect. B*, 1969, **25**, 1203.
42. I. A. Baidina, N. V. Podberezskaya, S. V. Borisov, N. A. Shestakova, V. F. Kukhna and G. D. Mal'chikov, *Zh. Strukt. Khim.*, 1979, **20**, 544.
43. B. Y. K. Ho, K. C. Molloy, J. J. Zuckerman, F. Reidinger and J. A. Zubieta, *J. Organomet. Chem.*, 1980, **187**, 213.
44. K. Henrick, R. W. Matthews and P. A. Tasker, *Acta Crystallogr., Sect. B*, 1978, **34**, 1347.
45. R. Hamalainen, *Finn. Chem. Lett.*, 1977, 113.
46. I. Dejehet, R. Debuyst, B. Ledieu, J. P. Declereq, G. Germain and M. Van Meerssche, *Inorg. Chim. Acta*, 1978, **30**, 197.
47. F. S. Stephens, R. S. Vagg and P. A. Williams, *Acta Crystallogr., Sect. B*, 1977, **33**, 433.
48. D. Van der Helm, A. F. Nicholas and C. G. Fisher, *Acta Crystallogr., Sect. B*, 1970, **26**, 1172.

20.2.2.2 Amino Acids with Coordinating Side Chains

A number of the amino acids in Table 1 have polar side groups and hence the potential of becoming tridentate. As will be seen later it is these third sites which become important in protein binding of metal ions.

The most obvious sign of increased chelation is an increase in stability constant relative to that of a complex of a known bidentate amino acid (*e.g.* alanine). For example, bidentate Ni(aa)$_2$ complexes have log stability constants of the order 9.7–10.1 (note that glycine generally gives slightly higher values). However, with His, Cys and Pen the corresponding values are 15.52, 20.07 and 22.97 respectively, reflecting the substantial gain in thermodynamic stability with the extra donor atom. A similar gain in stability is observed for Cu(HisO)$_2$ as compared to Cu(AlaO)$_2$ (*cf.* data in Table 2); this cannot be attributed to tridentate behaviour because of the tetragonal distortion in Cu^{II} complexes, but does imply an involvement of the histidine side chain.

Simple comparisons of stability constants in this way have been criticized because no allowance is made for the differences in ligand basicities. Increased basicity as a rule parallels increased metal binding. Also, in the majority of cases competition between hydrogen ion and metal ion occurs (equation 3). The equilibrium constant (K) for reaction (3) is then given by equation (4).

$$M^{a+} + HL \rightleftharpoons ML^{(a-1)+} + H^+ \tag{3}$$

$$\log K = \log \beta_{ML} - pK_{a(HL)} \tag{4}$$

$$\log K = \log \beta_{ML} - 0.7\,pK_{a(HL)} \tag{5}$$

Martin advocates the use of equation (5) in order not to overemphasize the role of ligand basicity;[10] the more positive log K in equations (4) or (5) the greater the thermodynamic stability of the complex. Examination of the data in Table 2 in this way then reveals that Arg, Lys and Trp (with basic side chain) are bidentate in the normal way, Glu^{2-} and Gln$^-$ display weak tridentate behaviour towards Co^{II} and Ni^{II} in their 1:1 complexes, and His$^-$, Cys^{2-} and Pen^{2-} are most obviously tridentate in behaviour, as can be judged from the unadjusted stability constants themselves.

Other evidence is required to confirm side-chain involvement, particularly for the kinetically inert metal ions where equilibrium conditions are difficult to obtain. For aqueous media UV/visible electronic absorbance spectroscopy is particularly useful for Co^{II}, Co^{III} and Ni^{II}, *e.g.* for the basic amino acids ornithine and 2,4-diaminobutyric acid it has been used to establish that with Ni^{II} the amino acids act as bidentate ligands at neutral pH but become tridentate in basic media.[11] For optically active ligands circular dichroism is of particular value because of its sensitivity to the configuration at the metal centre and to the ligand configuration.[12,13] In many cases tridentate coordination results in a dramatic sign inversion of the *d–d* Cotton effects.[14] NMR shifts in diamagnetic systems and line broadening in paramagnetic systems of side-chain atoms are diagnostic of side-chain coordination, a classic example being the study of the Co^{II}–L–His complexation in D$_2$O by McDonald and Phillips.[15]

More recently an oil continuous microemulsion technique has been described,[16] which allows the study of specific interactions between amino acid side chains and metal ions. Both the metal ion and amino acid are microencapsulated as aqueous droplets in a dispersed phase. The technique is of particular relevance to metalloprotein and metal–membrane interactions where the local dielectric constant can be considerably less than that of bulk water.

Two recent reviews have appeared on the tridentate amino ligands.[10,17] These amino acids will now be considered individually.

20.2.2.2.1 Histidine[10,18]

The imidazole ring (**3**), as part of histidine and as its 5,6-dimethyl derivative, is an important ligand for transition metal ions in a number of biological systems. The ring is aromatic and hence planar. The pyrrole nitrogen N-1 (or N$^\delta$ in molecular biology) has a pK_a of 14.4, whilst the pyridine nitrogen (N-3 or N$^\varepsilon$) has a much lower pK_a of 6.7 and is thus mainly deprotonated at neutral pH. Chelation involving the imidazole ring of histidine invariably involves the N-3 atom (**4**). This creates a six-membered N-3 N and a seven-membered N-3 O chelate ring. Enhances thermodynamic stability must then predominantly arise from the N-3 N chelate ring formation. However, at higher pH N-1 can undergo deprotonation to become an alternative donor atom; it is also the favoured donor atom in many metalloproteins (*vide infra*). This latter observation may be related to hydrogen tautomerism between the two N atoms in the imidazole ring.

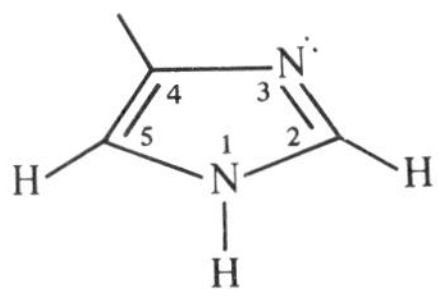

(3) 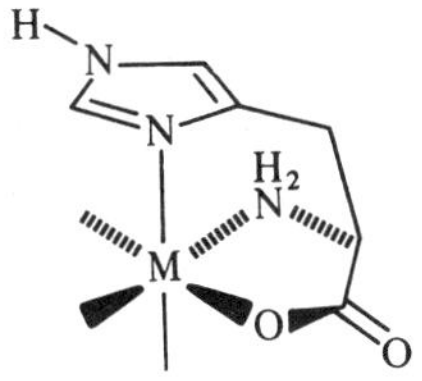   (4)

Tridentate chelation has been established in aqueous solution (see data in Table 2) and in the solid state (see Table 3) for $[Co^{II}(L\text{-}HisO)_2]$ (5) and $[Co^{II}(L\text{-}HisO)(D\text{-}HisO)]$ (6), in both phases for the bis complexes of Ni^{II}, Zn^{II} and Cd^{II}, for numerous complexes of Co^{III} and Cr^{III}, and is present in the Mo^V complex $[Mo_2O_4(L\text{-}HisO)_2]$ (7).

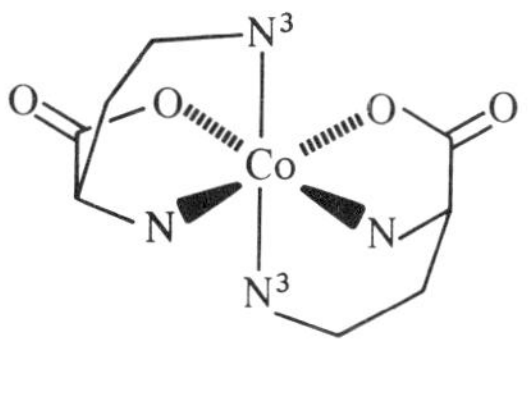

(5) 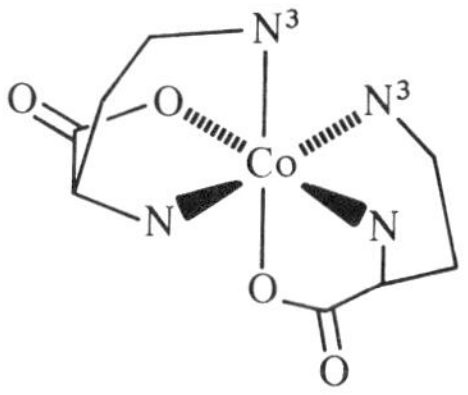   (6) 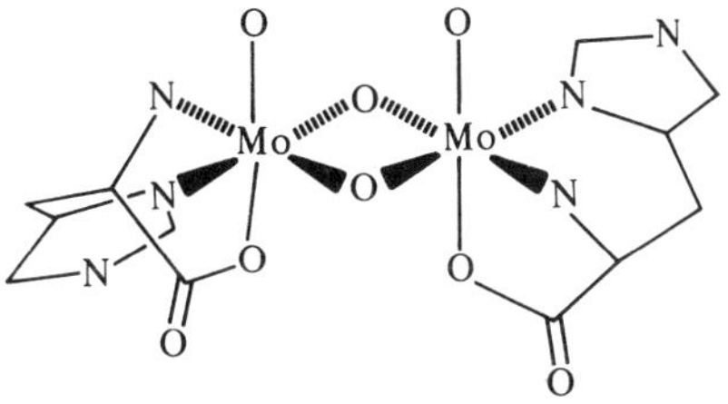   (7)

Histidine complexation, however, is not always so straightforward. Thus, at low pH, reaction with $HgCl_2$ and VO_2^{2+} gives complexation *via* the carboxyl O atom only and with Cu^{II} gives glycine-like chelation. More recently, Hg—C bond formation has been observed through the C-4 atom of the imidazole ring. In acid solution $[Ru(H_2O)(NH_3)_5]^{2+}$ binds initially to the the N-1 atom and then undergoes linkage isomerism to form an Ru—C bond to the C-2 atom of the imidazole moiety. A binuclear Ru^{III} complex in which the imidazole ring bridges the two metal centres as in (8) has also been reported[19] and may be of relevance to the spectroscopic changes observed for many metal–histidine solutions in the pH region in which the N-1 atom deprotonates.

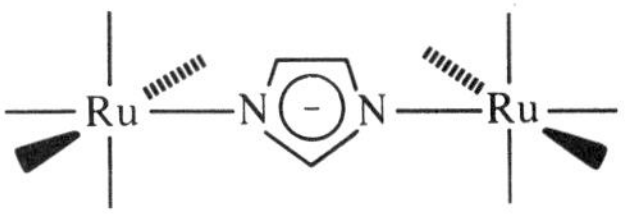

(8)

The aqueous copper(II)–L-histidine system has been the subject of a large number of investigations. Potentiometric studies show a number of pH-dependent species, but there is general agreement that at neutral pH the major species is the neutral $[Cu(L\text{-}HisO)_2]$ with minor amounts of the monoprotonated form $[Cu(L\text{-}His)(L\text{-}HisO)]^+$ present (see refs. 20 and 21 and refs. therein). As commented earlier, the enhanced stability constant for the neutral species cannot be attributed to tridentate chelation because of the Cu^{II} geometry. In this case it is the four donor atoms in the square plane around the Cu^{II} which primarily determine the ligand field strength. For $[Cu(L\text{-}HisO)_2]$, the *d–d* absorption maximum at 640 nm suggests a CuN_2O_2 chromophoric group.

It is the nature of the histidine binding to Cu^{II} that has intrigued so many chemists and involved so many spectroscopic techniques: CD, ESR, IR, ^1H and ^{13}C NMR and UV/visible. The general consensus of these studies is that the protonated form has the structure (9) and that the neutral species is a mixture of linkage isomers in facile equilibrium with each other (Scheme 1). The enigma is whether or not axial coordination (pseudo-tridentate behaviour) occurs in aqueous solution. If it does occur then, for L-His, *trans*-CuN_4 and *cis*-CuN_3O would be less likely to form. The solid-state structures are of little help for when His is tridentate, as in $[Cu(L\text{-}HisO)(L\text{-}ThrO)]$ and $[Cu(L\text{-}$

HisO)(L-AsnO)] (see Table 3), considerable distortion from octahedral symmetry results. On the other hand, solid [Cu(L-HisO)(D-HisO)(H$_2$O)$_2$]·4H$_2$O has a *trans*-CuN$_4$ structure with *no* axial carboxyl bonding, in contrast to solution studies which, from IR spectroscopy, support a metal–carboxyl bond. The structural studies also show that with *cis* imidazole groups distortion from coplanarity around the Cu^{II} would be required, with a consequent lowering of ligand field strength.

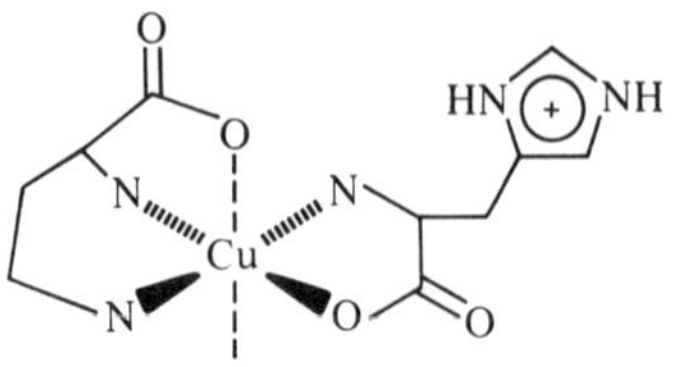

(9)

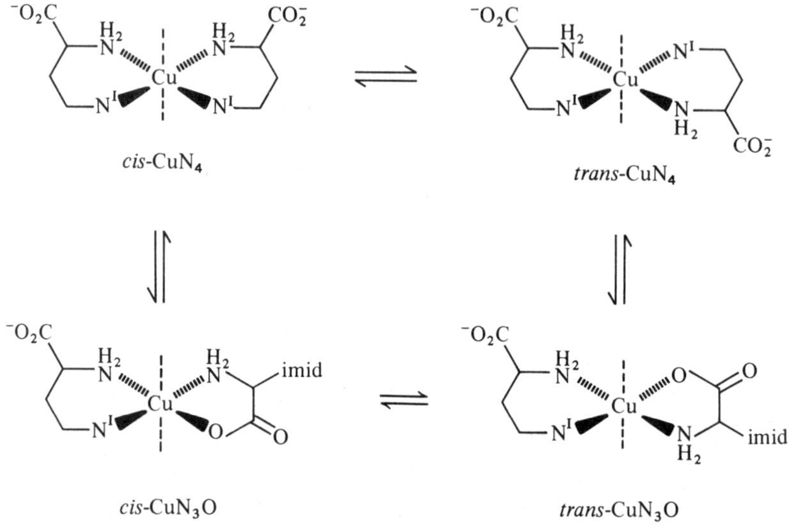

Scheme 1

20.2.2.2.2 Cysteine and penicillamine[22,23]

As with histidine, cysteine is a major metal binding site in proteins. Penicillamine (β,β-dimethylcysteine), a degradation product of penicillin, finds use as a therapeutic chelating agent (see Volume 6, Chapter 62.2). Both Cys and Pen have three dissociable protons (pK_a values approximately 2, 8 and 10) with Pen being the slightly more basic. Both amino acids undergo oxidation to the disulfide form (equation 6), a reaction which is catalyzed by metal ions, notably Cu^{II} and Fe^{III}.

$$2RSH \longrightarrow RSSR + 2H^+ + 2e^- \tag{6}$$

As ligands they are ambidentate in that they can be tridentate or bidentate through combinations of (S,N), (N,O) or (S,O) donor atoms (see Table 4). The strong affinity of 'soft' (or group B) metal ions for ionized thiol groups also gives rise to monodentate behaviour. Organomercury(II), for example, forms very stable 1:1 compounds *via* an Hg—S bond, even at low pH. The thiol anion is also a very effective metal bridging site.

It should be pointed out that most studies with cysteine have involved the naturally occurring L isomer. However, L-Pen is toxic so it is the enantiomeric D form that is therapeutically used. It is D-Pen that is used to remove the accumulated copper in sufferers of Wilson's disease; consequently the interaction between copper and Pen has been intensively studied. At neutral pH a deep violet colour develops, the extent of formation of the colour being dependent on a number of factors: solution pH, Cu^{II}:Pen ratio, O$_2$ content, chloride ion concentration. A crystalline Tl^I salt isolated from neutral solution was shown to be a mixed valence complex of composition Tl$_5^I$[Cu$_8^I$-Cu$_6^{II}$(PenO)$_{12}$Cl] (see Table 3). The same anion was shown to be the major species at neutral pH. A number of other mixed valence species involving Pen have since been isolated.[24]

Table 4 Observed Modes of Binding of Cysteine Showing
its Ambidentate Nature

Donor atoms	Metal ions
Tridentate (S,N,O)	Cd^{II}, Co^{II}, Co^{III}, Cr^{III}, In^{III}, Mo^{IV}, Mo^{V}, Pb^{II}, Sn^{II}
Bidentate (S,N)	Co^{II}, Co^{III}, Fe^{III}, Ni^{II}, Pd^{II}, Zn^{II}
Bidentate (S,O)	Cd^{II}, Co^{III}, Fe^{III}, In^{III}, Zn^{II}
Bidentate (N,O)	Co^{III}
Monodentate (S)	Ag^{I}, Cu^{I}, Hg^{II}, $MeHg^{+}$, Pt^{II}

In contrast, addition of cysteine to Cu^{II} solutions gives only a fleeting appearance of a violet colour before complete and rapid reduction to Cu^{I} occurs, with the latter stabilized by complex formation, giving the stoichiometry shown in equation (7).

$$Cu^{II}_{aq} + 2CysO \longrightarrow \frac{1}{n}\{Cu^{I}CysO\}_n + \frac{1}{2}cystine + 2H^+ \qquad (7)$$

The disulfides of Cys and Pen, formed by oxidation, chelate through the (N,O) donor groups at each end of the molecule, with weak or no interaction of the disulfide moiety with first row transition elements. However, with 'softer' metal centres, *e.g.* Pd^{II}, there is evidence from CD and ^1H NMR studies that a strong metal–sulfur bond can form.[25,26]

20.2.2.2.3 Methionine[27]

Compared to cysteine, the extra carbon atom in the side chain together with the thioether group would suggest considerably weaker metal–sulfur interactions for this ligand. Indeed, the data in Table 2 are compatible with bidentate (N,O) coordination. Any binding of the thioether would be expected with the larger 'soft' metal ions; this appears to be generally true, *e.g.* (S,N) coordination with protonation of the carboxyl group occurs in the complexes $[Pt^{II}(DL-Met)Cl_2]$ and $[Pt^{II}(L-Met)Cl_2]$. However, there are some notable exceptions, thus in solution Zn^{II}, Cd^{II} and Pb^{II} appear not to form a bond to the thioether whilst Ag^{I} does.[28] Sulfur bridging occurs in $[Hg^{II}(Met)_2(ClO_4)_2]$, but in $[MeHg(DL-Met)]$, surprisingly, the bond is to the N atom with no Hg—S contact (see Table 3). With Cu^{II} there is no evidence of any Cu—S interaction in aqueous media or in the solid state, but it appears to be likely in non-aqueous solution,[29] an observation reflected in some of the blue Cu^{II} proteins (*vide infra*).

20.2.2.2.4 Aspartic and glutamic acids[30]

Tridentate coordination of aspartic acid, involving the ionized side-chain carboxyl group (β), gives rise to six-membered (N,O_β) and seven-membered (O_α,O_β) chelate rings, apart from the normal five-membered (N,O_α) chelate rings. Not surprisingly, although there is a wealth of evidence for tridentate behaviour by $AspO^{2-}$, $GluO^{2-}$ appears to chelate as a glycine-like bidentate ligand. This appears evident from a comparison of the stability constant data in Table 3, but as Evans *et al.*[30] point out conclusions from titration data only for these ligands must be regarded as tentative because of the strong possibility of polynuclear complex formation. Thus, evidence of tridentate behaviour from stability constants for $GluO^{2-}$ with Fe^{III}, Ga^{III} and Mo^{VI} needs verification by other techniques. With the larger and highly charged lanthanide ions there is evidence of chelation through the two carboxyl groups occurring with no amino group involvement.

20.2.2.2.5 Asparagine and glutamine

From an examination of basicity-adjusted stability constants Martin[10] concludes that at neutral pH the γ-amide oxygen of $GlnO^-$ interacts only weakly with Co^{II} and Ni^{II} but not at all with Cu^{II}, whilst with $AsnO^-$ the interactions are much stronger, particularly with Cu^{II} and Ni^{II}. Crystal structure determinations of the neutral bis($AsnO$) complexes with Cd^{II}, Cu^{II} and Ni^{II}, however, reveal only bidentate glycine-like coordination.

Under more alkaline conditions, pH > 9.5, Cu^{II} and Asn solutions undergo deprotonation and a concomitant change in CD and visible spectra. These changes observed with Asn but not Gln are interpreted in terms of tridentate $AsnO^{2-}$ behaviour.[31] With $Pd(en)^{2+}$ as the coordinating centre deprotonations occur at considerably lower pH (apparent pK_a 6.4 for $AsnO^-$ and 9.0 for $GlnO^-$) to give bidentate N(amino),N^-(amide) chelation.[32]

20.2.2.2.6 Arginine and lysine

Although the pK_a of the ω-amine group for Lys is within the titratable range and that for Arg would probably become so on coordination, the relatively long methylene chains render any tridentate chelation unlikely, and this appears to be the case.[10,14] The shorter methylene side chains in the synthetic diamine-monocarboxylates ornithine, 2,4-diaminobutanoic acid and 2,3-diamino-propionic acid lead to increased tridentate behaviour in that order.

20.2.2.2.7 Serine and threonine

From the basicity-adjusted stability constants (equation 5), it is concluded[10] that Ser and Thr form additional weak bonds to Co^{II}, Ni^{II} and Zn^{II} *via* their side-chain hydroxyl groups in aqueous media. However, the same metal ions readily form tris(bidentate) complexes with these amino acids, with seemingly normal K_2 and K_3 values (Table 2). Much work has centred on Cu^{II} complexation: at neutral pH the stability constants and spectroscopic properties are consistent with the usual bidentate glycine-like behaviour. This is supported by crystal structure analyses of the complexes [Cu(L-HisO)(L-ThrO)] and [Cu(L-SerO)$_2$] (Table 3), both crystallizing at neutral pH, and both showing no hydroxyl group coordination. However, at lower pH, Pettit and Swash[33] have produced evidence of weak coordination *via* O(carboxyl) and O(hydroxyl) with the amino group protonated. Numerous workers (see ref. 34 and references therein) have reported substantial spectroscopic changes as the pH of the Cu^{II}–Ser or –Thr solutions is raised above 9.5. Theses changes coincide with ionizations of the hydroxyl groups and so are generally attributed to tridentate behaviour involving the ionized hydroxyl groups. This conclusion is by no means definite as is evident by very recent examinations of these same systems and by the observations of Gillard *et al.*[35] that high pH spectroscopic changes also occur with the [$Cu^{II}(aa)_2$] complexes of the strictly bidentate amino acids Ala, Pro and Val.

20.2.2.2.8 Dihydroxyphenylalanine (Dopa)[36]

Because of its biological and therapeutic importance L-Dopa and its coordination properties have been extensively studied. The weight of potentiometric and spectroscopic evidence points to the ligand chelating either as a normal bidentate amino acid (N,O) or through the two phenolate oxygens (O,O); the ligand is therefore classed as ambidentate (a similar situation occurs for tyrosine in which it is sterically impossible to achieve tridentate behaviour).

Depending upon the metal ion and the pH either (N,O) or (O,O) or a mixture of species of the two types occurs. The tendency to rearrange from (N,O) or mixed species to entirely (O,O) follows the order $Cu^{II} \sim Zn^{II} > Co^{II} > Mn^{II}$, Ni^{II}.[37] In general, in keeping with the pK_a values (see Table 1), (N,O) chelation is favoured at a lower pH range than is (O,O), while at intermediate pH values the mixed binding modes occur. As is predictable polymeric complexes are also known, particularly at low Dopa:metal ratios. With the 'hard' metal ions, *e.g.* Mg^{II}, Ca^{II}, Co^{III} and Fe^{III}, (O,O) coordination is the preferred mode provided the ligands are in excess.

20.2.2.3 Ternary Complex Formation[38]

In more recent years attention has turned from studying the equilibria of binary metal–amino acid complexes to that of ternary complex formation in aqueous media, particularly to complexes of the type (aa)—M^{II}—L, where L is some other ligand or a different amino acid to (aa), and M^{II} is a kinetically labile metal ion. Ternary complexes involving kinetically inert metal ions, *e.g.* Co^{III} and Pt^{II}, are more well known since they can be separated from mixtures and studied in isolation. Such is not the case with the labile systems. Because of the facile nature of their equilibria they must be studied *in situ* (claims regarding the separation of labile species by chromatographic procedures must therefore be treated with caution). The interest in such ternary systems arises because of their biological significance (see Chapter 22) and because their concentrations in solutions are often greater than predicted. Thus, considering the equilibrium reactions (8) and (9), the equilibrium is found to lie further to the right than is statistically expected. An excellent review of much of the earlier work, most of it done by his own group, is given by Sigel.[39] The general conclusions he reached hold good. These are, that the factors enhancing the formation of ternary species are: (1) statistical reasons; (2) neutralization of charge; (3) lower steric hindrance relative to the binary species; and (4) electronic effects such as favourable interactions between π systems, between hydrophobic systems or between oppositely charged side groups. The presence of an aromatic moiety in the amino acid, *e.g.* imidazole (His) or indole (Trp), frequently enhances formation of ternary complexes involving these amino acids. Some recent examples have been given by Sigel and his co-workers.[40,41] Other recent studies of ternary complexes involving amino acids are given in Table 5.

$$M(aa) + ML \rightleftharpoons [M(aa)L] + M_{aq} \tag{8}$$

$$M(aa)_2 + ML_2 \rightleftharpoons 2[M(aa)L] \tag{9}$$

Table 5 Examples of Recently Investigated Ternary Complex Systems

Amino acids	Metal ions	Other ligands	Ref.
Gly, L-Ala, L-Val, L-Phe	Cd[II]	Pyridoxamine	1
L-His	Cu[II]	Diaminocarboxylates	2
L-Ala, L-Trp	Cu[II], Zn[II]	Adenosine-5'-triphosphate	3
L- or D-His	Cu[II]	Amino acids	4
L-His	Cu[II]	Gly-Gly-L-His peptide	5
L-His	Cu[II]	Histamine, imidazole	6
L-His	Cu[II]	Thiodicarboxylates, pyridinedicarboxylates	7
L-His	Cu[II]	Nucleoside monophosphates	8
L-, D-Phe, D-Trp, D-Tyr	Cu[II]	Acetylacetonate	9
L-His	Cu[II]	2,2'-Bipyridyl, 1,10-phenanthroline	10
L-His	Co[II], Ni[II], Zn[II]	Gly, ethylenediamine, 2,2'-bipyridyl	11
L-Cys	Ni[II], Zn[II]	D-Pen	12
L-Phe, L-Trp, L-Tyr	Zn[II]	Adenosine-5'-triphosphate	13

1. B. A. Abd-el-Nabey and M. S. El-Ezaby, *J. Inorg. Nucl. Chem.*, 1978, **40**, 739.
2. M. S. Nair, K. Venkatachalapathi, M. Santappa and P. K. Murugan, *J. Chem. Soc., Dalton Trans.*, 1982, 55.
3. G. Arena, R. Cali, V. Cucinotta, S. Musumeci, E. Rizzarelli and S. Sammartano, *J. Chem. Soc., Dalton Trans.*, 1983, 1271.
4. O. Yamauchi, T. Sakurai and A. Nakahara, *J. Am. Chem. Soc.*, 1979, **101**, 4164.
5. T. Sakurai and A. Nakahara, *Inorg. Chim. Acta*, 1979, **34**, L245.
6. M. S. Nair, M. Santappa and P. Natarajan, *J. Chem. Soc., Dalton Trans.*, 1980, 1312.
7. D. N. Schelke, *Inorg. Chim. Acta*, 1979, **32**, L45.
8. J. B. Orenberg, B. E. Fischer and H. Sigel, *J. Inorg. Nucl. Chem.*, 1980, **42**, 785.
9. V. A. Pavolv, E. I. Klabunowski, S. R. Piloyan, Y. S. Airapetov and E. G. Rukhadzi, *Izv. Akad. Nauk SSSR, Ser. Khim.*, 1978, **5**, 1052.
10. B. E. Fischer and H. Sigel, *J. Am. Chem. Soc.*, 1980, **102**, 2998.
11. I. Sovago, T. Kiss and A. Gergely, *J. Chem. Soc., Dalton Trans.*, 1978 964.
12. I. Sovago, A. Gergely, B. Harman and T. Kiss, *J. Inorg. Nucl. Chem.*, 1979, **41**, 1649.
13. J. J. Toulme, *Bioinorg. Chem.*, 1978, **8**, 319.

20.2.2.4 Schiff Base Complexes[42]

It is well over 40 years since Pfeiffer discovered that certain reactions of α-amino acid esters, in particular, ester exchange, racemization and oxygenation, are effected very readily when their Schiff bases with salicylaldehyde are complexed to a transition metal ion (most notably Cu[II]). The Schiff bases result from a condensation reaction between a reactive carbonyl group and the amino group of the amino acids. Snell and his co-workers[43] were also one of the first to point out that similar reactions also occurred if pyridoxal was used instead of salicylaldehyde, and that there is a close analogy with pyridoxal phosphate-promoted enzymic reactions of α-amino acid metabolism. Since then much work has been due on these and other similar systems and their reactivities.

The principal structural types are formed as shown for salicylaldehyde in reaction (10) giving the Schiff base complex (10). Analogous reactions give the products (11), (12) and (13) from pyridoxal, pyruvic acid and (+)-(hydroxymethylene)camphor respectively. In all cases the ligands formed are planar tridentate (O,N,O) chelates, with M = Al^{III}, Cu^{II}, Fe^{II}, Fe^{III}, Ni^{II}, V^{IV}, V^{V} or Zn^{II}. The crystal structures of a number of these complexes have been reported and are listed in Table 6. Most work has been on the Cu and Zn complexes because of their biological relevance. The high thermodynamic stability associated with these complexes is attributed to their greater chelation over the bidentate nature of the parent compounds. An excellent review of the available thermodynamic data is given by Leussing.[44] While the complexes are particularly stable towards hydrolysis and dissociation in aqueous solution, they are, as mentioned, particularly reactive, a feature which is attributed in part to their ability to undergo ketoimine $\rightleftharpoons$ aldimine tautomerism (equation 11).

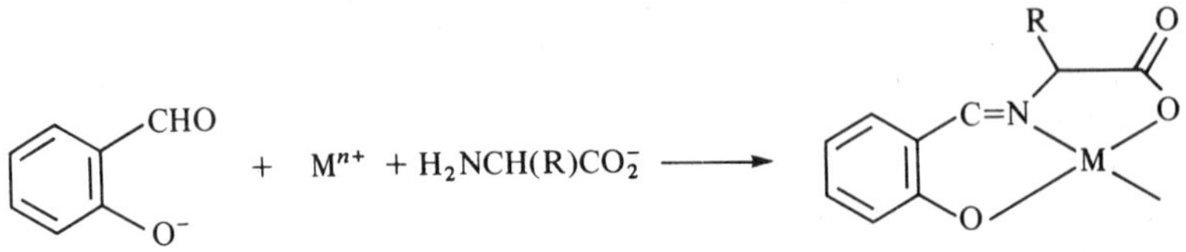

(10)

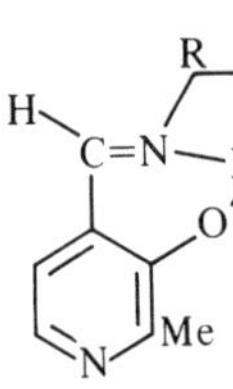

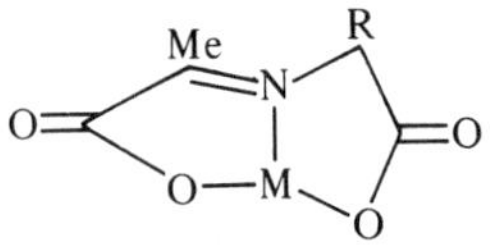

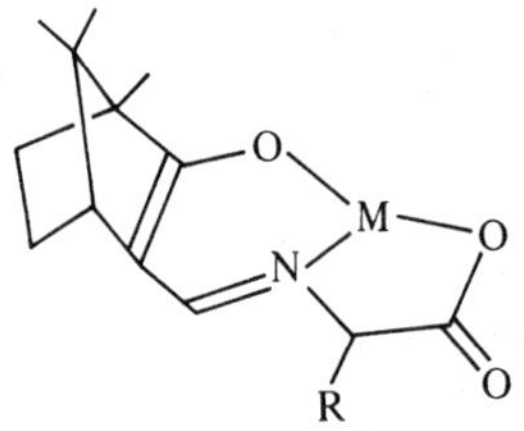

(11)   (12)   (13)

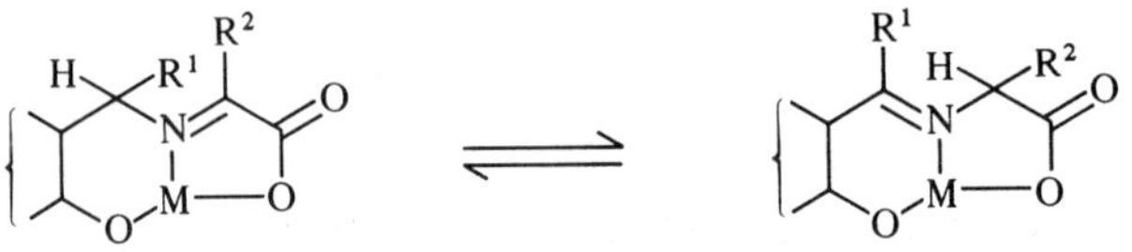

(11)

Table 6   Schiff Base Amino Acid Coordination Compounds whose Structures are Known from X-ray Crystallographic Studies

Compound	Ref.
μ-(*N*-Salicylidene-L-valinato-*O*)-*N*-(salicylidene-L-valinato)diaquadicopper(II)	1
(Pyruvylidene-β-alaninato)aquacopper(II) dihydrate	2
(Pyruvylideneglycinato)aquacopper(II) dihydrate	3
DL-Phenylalaninato(pyridoxylidine-5-phosphate)copper(II)	4
(*N*-Salicylideneglycinato)aquacopper(II) tetrahydrate	5
catena-μ-(*N*-Salicylidene-L-tyrosinato-*O*,*O*′)copper(II)	6
Bis(*N*-salicylidene-L-phenylalaninato)diaquadicopper(II)	7
(*R*,*S*)-[*N*,*N*′-Ethylenebis(serinato)]aquacopper(II)	8
(*N*-Salicylidene-L-threoninato)aquacopper(II) hydrate	9
(1*R*)-3-Hydroxymethylidenecamphorato-*N*-L-phenylalaninatocopper(II)	10
Sodium bis(3-methylsalicylidene-L-threoninato)cobalt(II) solvate	11

1. K. Korhonen and R. Hamalainen, *Acta Chem. Scand., Ser. A*, 1979, **33**, 569.
2. T. Ueki, T. Ashida, Y. Sasada and M. Kakudo, *Acta Crystallogr., Sect. B*, 1968, **24**, 1361.
3. A. Torii, H. Tamura-Kogayashi, K. Ogawa and T. Watanabe, *Z. Kristallogr.*, 1971, **133**, 179.
4. G. A. Boutley, J. M. Waters and T. N. Waters, *Chem. Commun.*, 1968, 988.
5. T. Ueki, T. Ashida, Y. Sasada and M. Kakudo, *Acta Crystallogr., Sect. B*, 1969, **25**, 328.
6. R. Hamalainen, M. Ahlgren, U. Turpeinen and M. Rantala, *Acta Chem. Scand., Ser. A.*, 1978, **32**, 235.
7. R. Hamalainen, U. Turpeinen, M. Ahlgren and M. Rantala, *Acta Chem. Scand., Ser. A*, 1978, **32**, 549.
8. F. Pavelak and J. Majer, *Acta Crystallogr., Sect. B*, 1980, **36**, 1645.
9. K. Korhonen and R. Hamalainen, *Acta Crystallogr., Sect. B*, 1981, **37**, 829.
10. L. Casella, M. Gullotti, A. Pasini, G. Ciani, M. Manassero and A. Sironi, *Inorg. Chim. Acta*, 1978, **26**, L1.
11. Y. N. Belokon, V. M. Belikov, S. V. Vitt, T. F. Saveleva, V. M. Burbelo, V. I. Bakhmutov, G. G. Aleksandrov and Yu. T. Struchkov, *Tetrahedron*, 1977, **33**, 2551.

Schiff base complexes formed from the tridentate amino acids are ambidentate in nature. Thus, with histidine, only pyruvate of those mentioned above leads to the His binding through the N(amino) and N-3(imidazole); with the others glycine-like chelation occurs. These different modes can be readily distinguished by their CD spectra. There appears to have been only a limited number of studies on the S-containing amino acids.[45-47]

20.2.2.5 Stereochemistry

20.2.2.5.1 Geometrical and optical isomerism

For the tris(amino acidato)metal(III) complexes four geometrical and chiral isomers are possible, as shown in Scheme 2, where N O is the chelated amino acid anion, *mer* and *fac* refer to the meridional and facial geometrical isomers and Δ and Λ refer to the configuration at the metal centre. For the glycine complexes Δ and Λ are an enantiomeric pair, while for the optically pure forms of the other amino acids they form a diastereomeric pair and hence are easier to separate. For most of the simple bidentate amino acids of the kinetically inert metal ions Cr^{III}, Co^{III} and Rh^{III}, the four isomers have been obtained.[48] The isomers are distinguishable by their UV/visible, CD and NMR spectra. Not all the isomers can be found in certain cases, for example, with L-proline the λ-*fac* isomer could not be prepared, a fact which was predicted on steric grounds.[49]

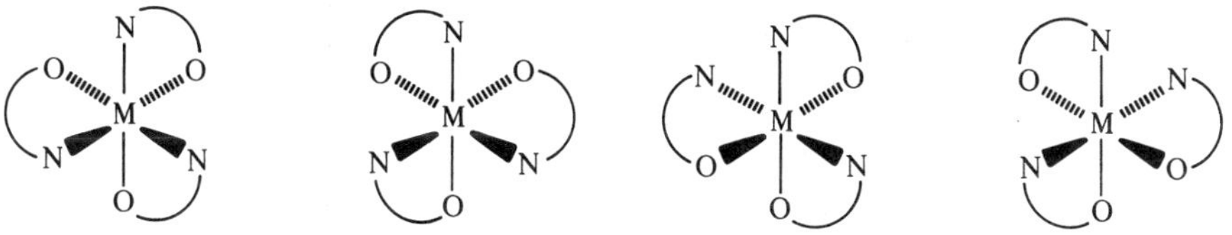

Δ-fac Λ-fac Δ-mer Λ-mer

Scheme 2

With mixed amino acid complexes the number of such isomers increases. In the case of $[Co(L-AspO^{2-})(L-HisO)]$ or $[Co(D-AspO^{2-})(L-HisO)]$, for example, all six isomers have been obtained and their isomerization reactions studies.[50]

For $M^{II}(aa)_2$ complexes with square planar geometry only geometric *cis* (14) and *trans* (15) isomers are possible. These isomers have been known for many years for the inert Pd^{II} and Pt^{II} ions. For the labile systems, *e.g.* Cu^{II}, deductions as to the major species in solution and their geometrical form has always been a problem because of their rapid interconversions. However, it has proved possible to isolate both geometrical isomers as crystalline solids in a number of cases, *e.g.* for $[Cu(GlyO)_2]$, $[Cu(L-AlaO)_2]$, and such isomerism probably accounts for the two solid forms for $[Cu(DL-PheO)_2]$ and $[Cu(DL-TyrO)_2]$.

(14) 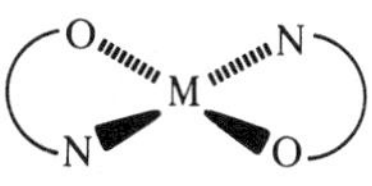 (15)

Intramolecular interactions between two coordinated amino acids can influence the position of *cis* $\rightleftharpoons$ *trans* equilibrium. These interactions can take the form of hydrophobic stacking interactions, as observed between the side chains of tyrosines in $[Cu(L-TyrO)_2]$ and $[Pd(L-TyrO)_2]$, or Coulombic attraction between oppositely charged side chains, as in $[Cu(HisO)L]$ (L = Arg, Lys or ornithine).[51] The optical configuration of the amino acids is of particular importance for these interactions. For amino acids of the same configuration *trans* geometry (15) is required, but for amino acids of opposite configuration *cis* geometry (14) is necessary around the metal centre.

20.2.2.5.2 Stereoselectivity

A metal ion coordinated to an optically active amino acid can, in principle, discriminate between incoming substrates. Using the system outlined in Scheme 3, and ignoring any differences that arise between *cis* and *trans* isomers, then if R^1 and R^2 are chemically different and $k_1 \neq k_2$, this merely reflects the differences in thermodynamic stabilities between the two ternary complexes. If, however, R^1 and R^2 are chemically identical but belong to an enantiomeric pair and $k_1 \neq k_2$, then this reflects stereoselectivity on the part of the 1:1 complex. (In this latter situation if *either* k_1 or k_2 is zero, then the reaction is said to be stereospecific.)

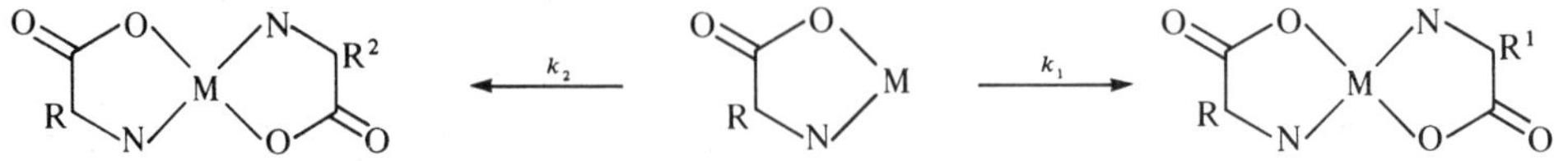

Scheme 3

With labile divalent metal ions absence of stereoselectivity has been definitely established for the bidentate amino acids but stereoselectivity is evident for the tridentate amino acids. Most notable among these is histidine. Thus, the equilibrium constant for the distribution reaction $[Co^{II}(L-His)_2] + [Co^{II}(D-His)_2] \rightleftarrows 2[Co^{II}(D-His)(L-His)]$ has a value of 11.5 as compared[15] to the statistically expected value of 4. This stereoselective favouring of the racemic complex is also found for the analogous histidine complexes of Ni^{II} and Zn^{II}, while the reverse is found for the Cu^{II} analogue. The finding of stereoselectivity favouring the racemic $[Ni(D-MetO)(L-MetO)]$ isomer over the $[Ni(L-MetO)_2]$ form is taken as an indication of tridentate coordination of Met with weak Ni—S bonding. Stereoselectivity in these complexes has been the subject of a number of reviews.[48,52,53] The finding of stereoselectivity in the above systems has generally relied upon the observation of non-superimposable titration curves between the racemic and optically active systems carried out under identical conditions. For the inert metal ions, such as Co^{III}, establishing the thermodynamic origin of any observed stereoselectivity is more difficult and the origin may even be kinetic.[48]

Coordination of an N-substituted amino acid introduces an extra asymmetric centre and further enhances stereoselectivity.[53] Two classical examples are the coordination of sarcosine (*N*-methylglycine) and proline. Thus, the complex ion $[Co(NH_3)_4(SarO)]^{2+}$ can be resolved into two chiral isomers because of the presence of the asymmetric coordinated N atom[54] and the chirality of $[Cu(L-ProO)_2]$ can be resolved into contributions from the two asymmetric centres.[55] Considerably enhanced stereoselectivity is also found in bis(*N*-benzylprolinato)copper(II).[56] *O*-Phosphate-substituted serine stereoselectively coordinates to Zn^{II}, but not with Co^{II}, Ni^{II} or Cu^{II}, in which cases the ligand is bidentate. S-substituted cysteine also leads to a new asymmetric centre with interesting stereoselective possibilities.[57] Further examples are to be found in the review by Pettit and Hefford.[53] It should be noted that in the above examples the observation of stereoselectivity in aqueous solution is very much pH dependent, for under conditions where the ligand atom can be protonated or further deprotonated a symmetric intermediate can form which leads to racemization.

20.2.2.5.3 *Stereoselectivity and optical resolution*

The resolution of racemic amino acid mixtures *via* coordination to a metal ion has been a popular field of study. $[Cu(L-aa)_2]$ complexes can be used to resolve DL-Asp, DL-Glu and DL-His.[58,59] $(-)-[Co(EDTA)]^-$ has been used to resolve DL-His having first resolved the racemic $[Co(EDTA)]^-$ ion using the L-histidinium cation.[60] Schiff base complexes of both Co^{III} and Ni^{II} have also been used to resolve amino acids.[61,62] A more esoteric finding is that the bacterium *Enterobacter cloacae* prefers to metabolize the $\Delta-(-)$ isomer of *fac*-$[Co(GlyO)_3]$ rather than the $\Lambda-(+)$ form,[63] an observation reminiscent of that made by Bailar using tris(ethylenediamine)cobalt(III) salts.

The above chemical separations work on the basis of forming diastereomeric pairs as intermediates; the same principle applies to the column chromatography techniques that have been developed over the past few years, in which a chiral metal complex is attached to the stationary phase. A number of complexes based on Cu^{II}, Ni^{II}, Zn^{II} and Co^{III} have been shown to be capable of resolving amino acids. A reverse-phase technique has also been described in which Cu^{II} or Zn^{II} peptide complexes are used in the mobile rather than the stationary phase.[64]

20.2.2.5.4 *Stereoselectivity and asymmetric synthesis*

From the properties outlined in the previous section it might well be expected that synthesis of an amino acid *via* reaction at a ligand coordinated to a chiral metal centre could lead to a chiral product or at least some stereoselectivity. This has certainly been achieved in a number of elegant syntheses, usually employing an unsaturated ligand. One of the first using classical coordination compounds was described by Asperger and Liu in which partially resolved alanine was obtained from a chiral Co^{III} complex of α-amino-α-methylmalonate. A more recent example from the same school is shown in Scheme 4.[65] Unfortunately, the final decomposition stage (formation of metal sulfides) causes substantial racemization.

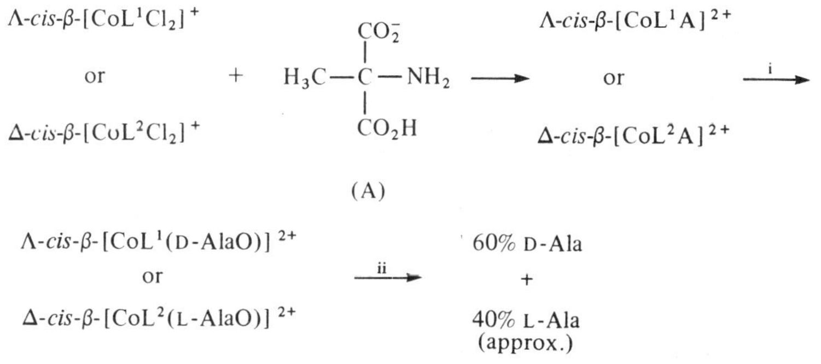

$$\Lambda\text{-}cis\text{-}\beta\text{-}[CoL^1Cl_2]^+$$

or

$$\Delta\text{-}cis\text{-}\beta\text{-}[CoL^2Cl_2]^+$$

$$+ \quad H_3C-\underset{\underset{CO_2H}{|}}{\overset{\overset{CO_2^-}{|}}{C}}-NH_2 \quad \longrightarrow$$

$$\Lambda\text{-}cis\text{-}\beta\text{-}[CoL^1A]^{2+}$$

or

$$\Delta\text{-}cis\text{-}\beta\text{-}[CoL^2A]^{2+}$$

$$\overset{i}{\longrightarrow}$$

(A)

$$\Lambda\text{-}cis\text{-}\beta\text{-}[CoL^1(D\text{-AlaO})]^{2+}$$

or

$$\Delta\text{-}cis\text{-}\beta\text{-}[CoL^2(L\text{-AlaO})]^{2+}$$

$$\overset{ii}{\longrightarrow}$$

60% D-Ala

+

40% L-Ala
(approx.)

i, decarboxylation; ii, decomposition.
L^1 = 3R isomer of 3-methyl-1,6-bis[(2S)-pyrrolidin-2-yl]-2,5-diazahexane,
L^2 = corresponding 3S isomer.

Scheme 4

The alternative approach is to make use of the well-known asymmetric hydrogenation reactions catalyzed by organometallic compounds of the platinum group metals. Among the most elegant of these syntheses are those developed by Bosnich and his group[66,67] based on chiral diphosphine ligands coordinated to Rh^I and α-*N*-acylaminoacrylic acids as the amino acid precursors. The catalyzed hydrogenation of these systems leads to highly optically pure products of Ala, Leu, Phe, Tyr and Dopa. The optical purity obtained can be related to the ring conformations of the diphosphine ligands. High optical purity Dopa and Phe have been obtained by analogous reactions from β-acylaminoacrylic acids.[68] Also related to these methods is the system described in Scheme 5, the D-(+)-imine giving mainly D-Phe, and the L-(−)-imine the L enantiomer. The optical purities are of the order of 77%.

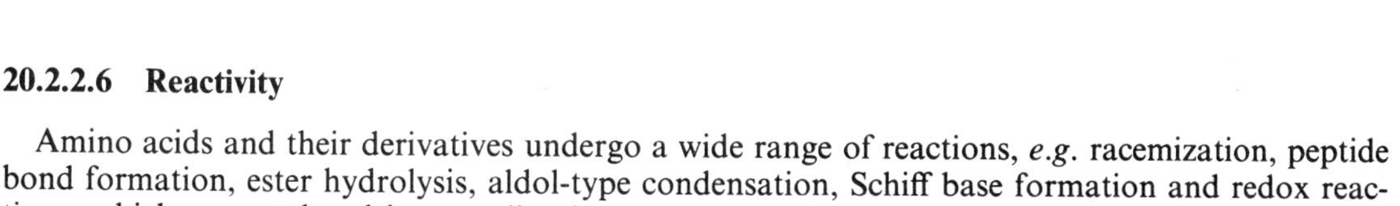

$$H-\underset{\underset{Ph}{|}}{\overset{\overset{Me}{|}}{C^*}}-N{=}C\underset{\underset{(CO)_4}{Fe}}{\overset{\overset{CO_2Et}{\diagup}}{\diagdown_H}} \quad + \quad RBr \quad \longrightarrow \quad H-\underset{\underset{Ph}{|}}{\overset{\overset{Me}{|}}{C^*}}-\underset{\underset{(CO)_3Br}{Fe}}{\overset{}{N}}-\underset{\underset{H}{|}}{\overset{\overset{R}{|}}{C^*}}-CO_2Et \quad \xrightarrow[ii,\ OH^-]{i,\ H_2/Pd} \quad H_2N-\underset{\underset{R}{|}}{\overset{\overset{CO_2^-}{|}}{C^*}}-H \quad + \quad PhEt$$

Scheme 5

20.2.2.6 Reactivity

Amino acids and their derivatives undergo a wide range of reactions, *e.g.* racemization, peptide bond formation, ester hydrolysis, aldol-type condensation, Schiff base formation and redox reactions, which are catalyzed by coordination to a metal centre. A number of reviews are available which cover some of these reactions.[48,69,70]

20.2.2.6.1 Kinetics and mechanisms of formation

The rate constants for the formation of divalent metal ion–amino acid complex formation have been determined by T-jump measurements; representative data are given in Table 7. As is generally found for bidentate complex formation, the mechanism involves a two-step process (equations 12 and 13), in which the initial outer-sphere association (equation 12) is followed by ring closure (equation 13). For the α-amino acid anions reaction (12), the displacement of a water molecule, is the rate-determining step, but for the β-amino acid anions, which form the less favourable six-membered chelate rings, it is the ring closure step which is the slower process and hence rate-determining.

Table 7 Rate Constants for the Formation Reactions of Metal Ion–Amino Acid Complexes[a]

Metal	Ligand	K_1 (M^{-1} s^{-1})	K_{-1} (s^{-1})	K_2 (M^{-1} s^{-1})	K_{-2} (s^{-1})	Ref.
Co^{2+}	L-AlaO	6.0×10^5	32	8.0×10^5	280	1
	β-AlaO	7.5×10^4	7.5	8.6×10^4	86	1
Ni^{2+}	L-AlaO	2.0×10^4	0.022	4.0×10^4	0.80	1
	β-AlaO	1.0×10^4	0.23	6.9×10^3	2.7	1
Ni^{2+}	GlyO	2.2×10^4				2
	L-Lys	4.4×10^3				2
Cu^{2+}	L-AlaO	1.3×10^9	12	1.5×10^8	33	3
	β-AlaO	2.0×10^8	11	8×10^6	19	3
	L-His	1.3×10^7	115	3.0×10^6	111	3
	GlyO	4.0×10^9	34	4.0×10^8	50	4
Cr^{2+}	GlyO	3.1×10^9		2.8×10^8		5
Zn^{2+}	Gly	5.9×10^7	5.1×10^7			6

[a] For conditions used the original references should be consulted.
1. K. Kustin, R. F. Pasternack and E. M. Weinstock, *J. Am. Chem. Soc.*, 1966, **88**, 4610.
2. J. C. Cassatt and R. G. Wilkins, *J. Am. Chem. Soc.*, 1968, **90**, 6045.
3. W. B. Makinen, A. F. Pearlmutter and J. E. Stuehr, *J. Am. Chem. Soc.*, 1969, **91**, 4083.
4. A. F. Pearlmutter and J. E. Stuehr, *J. Am. Chem. Soc.*, 1968, **90**, 858.
5. I. Nagypal, K. Micskei and F. Debreczeni, *Inorg. Chim. Acta*, 1983, **77**, L161.
6. S. Harada, Y. Uchida, M. Hiraishi, H. L. Kuo and T. Yasunga, *Inorg. Chem.*, 1978, **17**, 3371.

$$(\text{H}_2\text{O})_n\text{M} + \text{A} \frown \text{B} \underset{k_{-1}}{\overset{k_1}{\rightleftharpoons}} (\text{H}_2\text{O})_{n-1}\text{M:A} \frown \text{B} + \text{H}_2\text{O} \qquad (12)$$

$$(\text{H}_2\text{O})_{n-1}\text{M:A} \frown \text{B} \underset{k_{-2}}{\overset{k_2}{\rightleftharpoons}} (\text{H}_2\text{O})_{n-2}\text{M} \left(\genfrac{}{}{0pt}{}{\text{A}}{\text{B}}\right) + \text{H}_2\text{O} \qquad (13)$$

For Cu^{II} and Ni^{II} it has been demonstrated that the zwitterionic forms of the amino acids, H$_3$N$^+$CH(R)COO$^-$, are extremely unreactive, and this is taken as evidence that the initial donor atom (A in equation 12) must be N(amino). In apparent accord with this is the relatively slow rate of complex formation between Cu^{II} and the histidine zwitterion. This is attributed to the requirement of an initial proton-transfer step, from the amino to the imidazole group, to give glycine-like coordination. However, consideration of the micro- rather than macro-stability constants for histidine indicates that a facile proton tautomeric equilibrium between the protonated amino and imidazole forms should be taken into consideration. Furthermore, measurements made with the thiol-containing amino acids are consistent with initial coordination occurring *via* O or S donor atoms.[71] Measurements on Zn^{II}–glycine solutions at low pH (see data and references in Table 7) show that the unfavourable reaction with the zwitterion is attributable to a highly favourable dissociation step (k_{-1}).

20.2.2.6.2 *Activation of the α C—H bond*

The electron-withdrawing effect of the metal ion causes considerable enhancement of the acidity of the α C—H bond of coordinated amino acids. The most obvious illustration of this effect is that most (but not all, *vide infra*) metal ions increase the rates of racemization of amino acids. With Co^{III} complexes it has been clearly demonstrated that the rate of racemization is similar to the rate of H–D exchange at the α C atom (conveniently followed by ^{1}H NMR), results which are interpreted in terms of a symmetrical carbanion intermediate (equation 14).

One of the first reactions reported on the activation of the α C—H bond was the aldol condensation reaction of glycine, coordinated to Cu^{II}, with acetaldehyde to yield threonine.[72] The reaction, which is base catalyzed, proceeds under far milder conditions than for free glycine. Similar reactions have been reported with other metal ions and aldehydes; again the postulated intermediate is a carbanion.[69,70] By using resolved Co^{III} complexes, *e.g.* Λ-(+)-[Co(en)₂(GlyO)]²⁺, some stereoselectivity can be obtained in the threonine product.[73]

Co^{III}, Cr^{III}, Cu^{II}, Pd^{II} and Pt^{II} accelerate the racemization rates of amino acids. The mechanisms are not always as simple as that outlined above for Co^{III}. Thus, with [Cu(L-AlaO)₂] it has been known for some time that the rate of racemization was faster than that for H–D exchange under the same conditions. It was subsequently shown that at high pH, with O_2 present, pyruvate was formed from an oxidative deamination reaction which further catalyzed the racemization rate through formation of a Schiff base intermediate (16).[74] This combination of carbanion and Schiff base intermediates was further confirmed by studies on the Zn^{II} complex. Contrary to earlier reports it has now been shown that Ni^{II} actually retards the racemization rates of amino acids.[75]

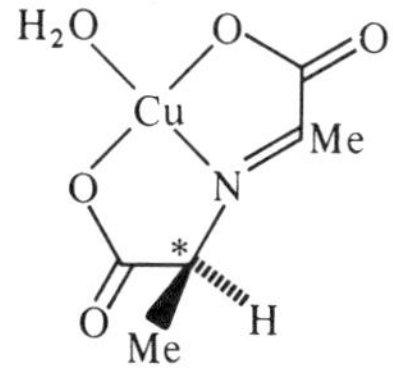

(16)

In invoking the Schiff base intermediate described above, Gillard *et al.*[74] were making use of the known extra activation of the α C—H bond in Schiff base complexes as compared to the simple amino acid complexes. Rates of racemization are enhanced several-fold, as are aldol condensation reactions.

20.2.2.6.3 *Hydrolytic reactions*[76]

The ability of metal ions to catalyze the hydrolysis of peptide bonds has been known for 50 years, while the catalytic effect on the hydrolysis of amino acid esters was highlighted in the 1950s. As Hay and Morris point out in their review,[76] the major problem with the kinetically labile systems is determining the nature of the reactive complex in solution. Such problems generally do not arise in the more inert systems and consequently reactions involving Co^{III} have been the more popular for study.

Amino acid esters may coordinate as a monodentate (amino N) or as a bidentate (N,O) ligand. In the latter case (17) significant polarization of the ester bond results which is of likely importance in the reaction mechanism (Scheme 6). Alternatively, hydroxide attack at the metal centre may occur (equation 15). Hydroxide ion rather than water is the predominant nucleophile, even at pH 5. Direct evidence for metal–ester bond formation comes from isolated Co^{III} complexes which contain such a bond. Rate enhancements, compared to the free amino acids, are often of the order of 10^5–10^6. For methyl histidinate, however, rate enhancement by Cu^{II} is only of the order of 265; this is ascribed to there being no Cu–ester bond formation. The order of decreasing reactivity in this study was established as being $CuE_2^{2+} > NiE_2^{2+} > CuE^{2+} > NiE^{2+} > EH^+ > CuEA^+ > CuEOH^+ > NiEA^+ > E$, where E is L-HisOMe and A is L-HisO⁻. The relative effectiveness of protons and metal ions as catalysts has been discussed by Martin.[77]

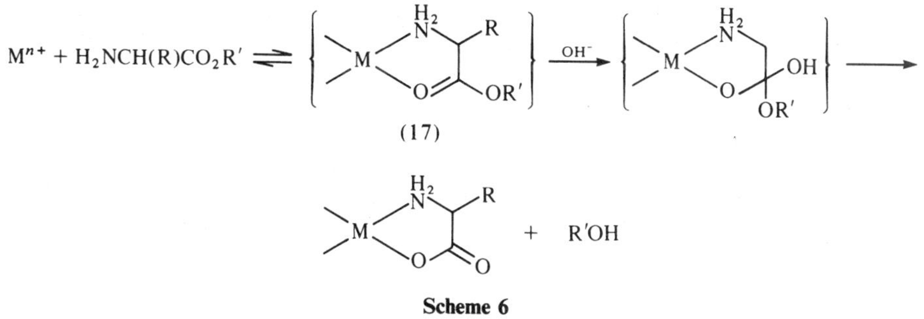

$$M^{n+} + H_2NCH(R)CO_2R' \rightleftharpoons$$

(17)

Scheme 6

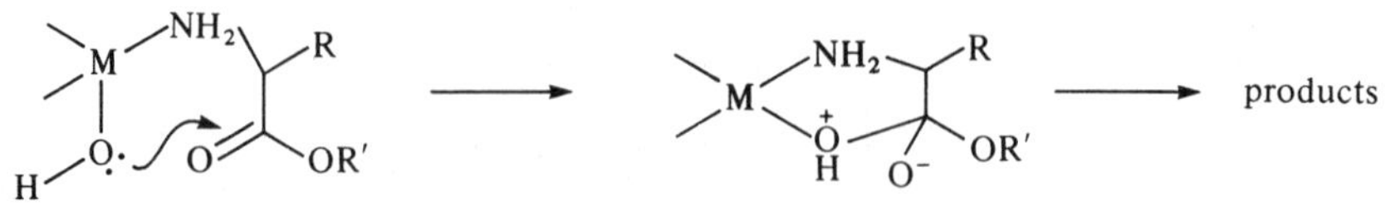

$$\text{(15)}$$

When the nucleophile is also coordinated to the metal ion, then intramolecular nucleophilic attack occurs to give rate enhancements of the order of 10^7–10^{11}. These, mainly Co^{III}, systems then become good models for the enzyme-catalyzed reactions.[76]

20.2.2.6.4 *Peptide bond formation*

Reaction of Co^{III} amine complexes with glycine esters gives products which are dependent on the solvent used (Scheme 7). In non-aqueous solvents peptide bond formation is observed.[78–80] It is concluded that the reaction proceeds *via* the initial formation of a bidentate glycine ester complex in which the coordinated carbonyl group is activated (by the metal ion) towards nucleophilic attack by a further molecule of the glycine ester. Similar reactions have been observed with other Co^{III} complexes.

$$[CoL(Gly\text{-}GlyOR)]^{3+} \xleftarrow{\text{DMSO or}}_{\text{DMF}} [CoLCl_2]Cl + GlyOR \xrightarrow{H_2O} [CoL(GlyOR)Cl]^{2+}$$

L = bis(ethylenediamine) or triethylenetetramine

Scheme 7

20.2.2.6.5 *Oxidation and reduction reactions*

Although such reactions have been known for a long time, it appears a somewhat neglected area of study. Most attention has been on cysteine and its oxidation to the disulfide which is catalyzed by metal ions, in particular Cu^{II} (see Section 20.2.2.2.2) and Fe^{III}.[81] The likely intermediates in these reactions are metal–cysteine complexes which undergo internal electron transfer. As noted earlier (Section 20.2.2.2.2), penicillamine differs from cysteine in its reactivity and gives rise to mixed valence species. More recently Mn^{II} has also been found to catalyze the oxidations of Cys and Pen.

Early reports of an Mo^{VI}–Cys complex have been shown to be erroneous because redox occurs to give the Mo^V complex $[Mo_2O_3(CysO)_4]^{4-}$ and cystine; further electrochemical reduction to Mo^{III} is also possible. Oxidation of Cys coordinated to Co^{III} can give a sulfinic acid or a sulfenamide, according to the reagents and conditions used.

Dopa can also undergo autocatalytic oxygenation with Fe^{III}, but Cu^{II} inhibits the reaction. This difference in reactivity reflects different coordination modes (see Section 20.2.2.2.8).[82] The enzyme superoxide dismutase (*vide infra*) catalyzes disproportionation of the superoxide ion. The same catalysis is exhibited by Cu^{II}–L-His solutions, the reactive species, from the pH dependence, appearing to be $[Cu(L\text{-}HisO)(L\text{-}His)]^+$.

The ability of many Co^{II} complexes to reversibly absorb molecular O_2 has interested many groups. One requirement is for three or more N donor atoms. This is satisfied by $\{Co(L\text{-}HisO)_2\}$ which, in solution, rapidly absorbs a half mole of O_2 to give a peroxo-bridged intermediate ($Co^{III}O_2^{2-}Co^{III}$); further reaction gives $[Co^{III}(L\text{-}HisO)_2]^+$. (For references to this reaction see the review by Chow and McAuliffe.[17])

20.2.3 PEPTIDES

The condensation reaction between two amino acid molecules, usually in the form of active esters, produces a dipeptide molecule (equation 16). The resulting amide bond (—CONH—) is often referred to as a peptide bond. This condensation reaction can be repeated to give a polypeptide chain (18). In naming the sequence of amino acid residues in the chain the customary habit of starting from the N-terminal (free amino) end will be adopted. Synthetic polypeptides made from a single amino acid type, *i.e.* $R^1 = R^2 = R^3 = \cdots = R^n$ in (18), are known as homopolypeptides. Cyclic peptide molecules are also known and can occur naturally.

$$H_3\overset{+}{N}CH(R^1)CO_2^- + H_3NCH(R^2)CO_2^- \longrightarrow H_3\overset{+}{N}CH(R^1)CONHCH(R^2)CO_2^- + H_2O \qquad (16)$$

$$H_3\overset{+}{N}CH(R^1)CONHCH(R^2)CONHCH(R^3)' \cdots CH(R^n)CO_2^-$$

$$(18)$$

The configuration adopted by a polypeptide molecule is determined by several factors, among which are the nature of the R groups and the need to accommodate the planar amide group. The geometry around the amide bond is *trans* and delocalization gives some double bond character in both the C—O and C—N bonds. Molecular orbital calculations indicate comparable amounts of the two canonical forms (equation 17). Deprotonation of the amide hydrogen does not occur for the non-coordinated group within the normal pH range; indirect measurements give $pK_a \approx 15$. In the following, general aspects of metal–peptide interactions will be considered. More specific details on naturally occurring peptides will be covered in Chapter 22.

$$(17)$$

20.2.3.1 Binding Sites and Stereochemistry

Both these aspects need to be considered together since they are intricately connected and more dependent on the nature of the R group (both polar and non-polar) than is the case with the amino acids. From the preceding sections the amino, carboxyl and polar R groups may be anticipated as being likely metal-binding sites. In addition to these the peptide bonds are also very important binding sites. An excellent review of the amide bond and its coordinating properties has recently appeared.[83]

The multiplicity of protonation and binding sites means that stability constants are less meaningful than with the amino acids; a selection of such values for some major sepcies formed is given in Table 8. Further information is necessary to delineate the binding sites. In this respect UV/visible, CD, ESR and NMR spectroscopic techniques have proved most informative, particularly for Cu^{II} which has been the most studied of the metal ions.[84] It is particularly gratifying that the correlation between solid state and solution structures is excellent for the kinetically labile systems. A review and detailed analysis of the structures known before 1967 has been given by Freeman.[85] Some more recent structural data are given in Table 9.

The conclusions that Freeman[85] made are still valid, namely: (1) the average dimensions of free and complexed peptides are very similar, except for the peptide C—O bond, which lengthens upon coordination, and the C—N bond, which shortens; (2) the peptide group remains very close to planar; (3) protonated N(peptide) never coordinates to a metal ion (presumably the tetrahedral geometry required would be energetically and geometrically unfavourable); (4) when a metal ion is bonded to three donor groups of a peptide molecule, the central one of which is a peptide N, then the three donor atoms and the metal must be coplanar.

Taking into account factor (3) then factor (1) can be explained in terms of the canonical forms in equation (18). The displacement of the peptide hydrogen by a metal ion requires a considerable lowering of the peptide pK_a. Values compiled by Sigel and Martin,[38] relating to the equilibrium reactions (19)–(21), are given in Table 10. These are in keeping with the general observation that the N-chelated Ni^{II} peptides require a higher pH for their formation than do the Cu^{II} analogues. For the divalent metal ions most commonly studied the ability to displace a peptide proton follows the order Pd (2) > Cu (4) > Ni (8) > Co (10), with the approximate pK_a values given in parentheses. Zn^{II} is unable to promote peptide ionization in the measurable pH range. The kinetically inert Co^{III} and Pt^{II} appear to be as effective as Pd^{II}. As an illustration of the pH dependence Cu^{II} and triglycine (Gly-Gly-Gly) solutions at low pH are green-blue, consistent with bidentate N(amino),O(peptide) chelation. However, on raising the pH the solution turns violet as the Cu displaces the peptide protons to give a tridentate N(amino),2N(peptide) chelate. This UV shift of the *d–d* absorption band reflects the greater ligand field associated with N(peptide) donation. The inert Co^{III} complexes also allow the study of the reverse protonation process. Thus, for $[Co(GlyGlyO)_2]^-$ the successive protonation constants (log values) are 1.46 and 0.10, the protonations occurring at the peptide oxygen atoms.[86]

Table 8 Logarithmic Equilibrium Constants of Some Metal–Peptide Complexes in Aqueous Media[a,b]

Ion	Equilibrium	Gly-Gly (H_3L^+)[c]	Gly-Ala (H_3L^+)	Gly-β-Ala (H_3L^+)	Gly-His (H_4L^{2+})	His-Gly (H_4L^{2+})	β-Ala-His (H_4L^{2+})
					Dipeptides		
H^+	$H_2L/HL.H$	8.08[d]	8.11	8.09	8.11	7.60	9.36
	$H_3L/H_2L.H$	3.13	3.07	3.91	6.68	5.81	6.76
	$H_4L/H_3L.H$				2.45	2.71	2.49
							3.69
Co^{2+}	$MHL/M.HL$	3.01	3.10			5.19[f]	
	$M(HL)_2/M.(HL)^2$	5.35	5.68			8.16[f]	
	$MHL/ML.H$					7.15	
Ni^{2+}	$MHL/M.HL$	4.05	4.23	4.19	3.9	6.84	5.42
	$M(HL)_2/M.(HL)^2$	7.22	7.60	7.53	8.82	12.39	
	$M(HL)_3/M.(HL)^3$	9.4	9.7	9.7	11.57		
	$MHL/ML.H$		8.79	9.24	5.5		9.14
	$M(HL)_2/ML_2.H^2$		19.84	18.9			
Cu^{2+}	$MHL/M.HL$	5.50	5.77	5.70	9.14	8.83	8.52
	$MHL/ML.H$	4.07	4.09	4.60	4.26	8.07	5.60
	$ML(HL)/ML.HL$	3.14	3.15	2.87	3.56		2.11
	$M(HL)_2/M.(HL)^2$				16.53		
Zn^{2+}	$MHL/M.HL$	3.44			3.65[f]	14.15[f]	14.39[f]
	$M(HL)_2/M.(HL)^2$	6.31			6.89[f]	4.25[f]	3.86[f]
	$M_2(HL)_2/(MHL)^2$				3.30[f]	8.46[f]	
	$M_2(HL)_2/M(HL)L.H$				6.19[f]		

Ion	Equilibrium	Tri-Gly (H_4L^+)	Gly-Gly-His (H_5L^{2+})	Gly-His-Gly (H_5L^{2+})	Asp-Ala-His-NHCH$_3$ (H_5L^{2+})
				Tripeptides	
H^+	$H_3L/H_2L.H$	7.89	8.02	7.56	7.73[g]
	$H_4L/H_3L.H$	3.21	6.77	6.24	6.47[g]
	$H_5L/H_4L.H$		2.72	2.91	2.95[g]
Co^{2+}	$M(H_3L)/M.H_2L$	3.05[g]			
	$M(HL)(H_2L)/M(HL)_2.H$	5.45[g]			9.03
	$M(H_2L)_2/M.(H_2L)^2$				
Ni^{2+}	$M(H_2L)/M.H_2L$	3.70	6.08		
	$M(H_2L)_2/M.(H_2L)^2$	6.73	12.26		
	$M(H_2L)_2/M.(H_2L)^2$	8.83			
	$M(H_2L)_2/M(HL)_2.H$	16.6			
	$M(H_2L)/ML.H^2$				
Cu^{2+}	$MH_2L/M.H_2L$	5.08	7.04[f]	8.52[f]	8.39[g]
	$MH_2L/MHL.H$	5.13	4.64	3.20	
	$MHL/ML.H$	6.72		9.01	
	$M(HL)_2/M.(HL)^2$	9.6		15.78[f]	
	$MH_2L/ML.H^2$				
Zn^{2+}	$MH_2L/M.H_2L$	3.18[g]	9.41		7.64[g]
	$M_2(H_2L)_2/(MHL)^2$		3.31[f]		4.75[g]
	$M_2(H_2L)_2/M(HL)(H_2L).H$		3.15		
	$M(H_2L)_2/M.(H_2L)^2$		6.46		8.15[g]

Ion	Equilibrium	TetraGly (H_5L^+) $n=3$	TriGly—His (H_6L^{2+}) $n=3$	PentaGly (H_6L^+) $n=4$
H^+	$H_{(n+1)}L/H_nL.H$	7.87	7.99	7.86ᵍ
	$H_{(n+2)}L/H_{(n+1)}.H$	3.18	6.73ᵍ	3.39ᵍ
	$H_{(n+3)}L/H_{(n+2)}.L.H$		2.90ᵍ	
Co^{2+}	$MH_nL/M.H_nL$	3.01ᵍ		
	$M(H_nL)_2/M.(H_nL)^2$	5.50ᵍ		
Ni^{2+}	$MH_nL/M.H_nL$	3.64		
	$MH_nL/MH_{n-1}.L.H$	⎱15.8	8.38ᵍ	
	$MH_{n-1}L/MH_{n-2}.L.H^2$	⎰	8.63ᵍ	
	$MH_{n-2}L/MH_{n-3}.L.H$	8.6	8.98ᵍ	
Cu^{2+}	$MH_nL/M.H_nL$	5.10		5.32ᵍ
	$MH_nL/MH_{n-1}.L.H$	5.40	6.23	6.00ᵍ
	$MH_{n-1}L/MH_{n-2}.L.H$	6.80	6.98	6.90ᵍ
	$MH_{n-2}L/MH_{n-3}.L.H$	9.14	8.58	8.04ᵍ
Zn^{2+}	$MH_nL/M.H_nL$	3.14		

[a] Data compiled from ref. 7.
[b] Amino acid residues assumed to have the L configuration unless stated otherwise.
[c] Formula indicates the maximum number of replaceable protons including the peptide hydrogens.
[d] Data at 25 °C and 0.1 mol dm⁻³ ionic strength unless indicated otherwise.
[e] 25 °C, 3.0.
[f] 37 °C, 0.15.
[g] 25 °C, 0.15.

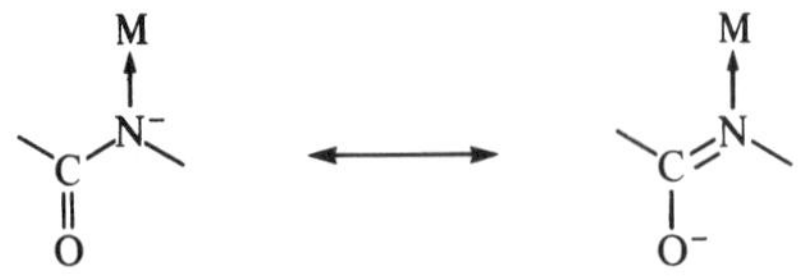

(18)

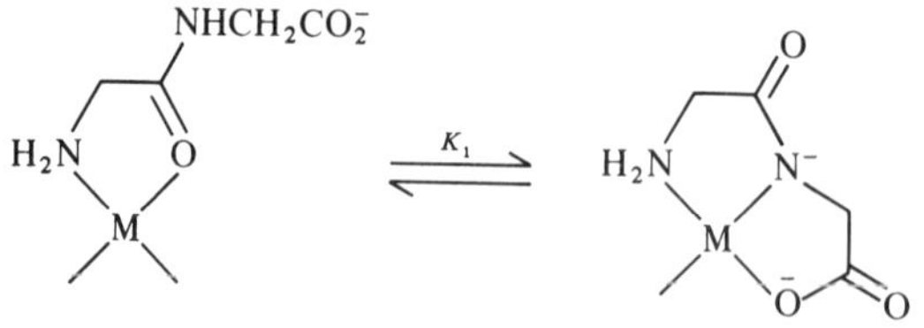

(19)

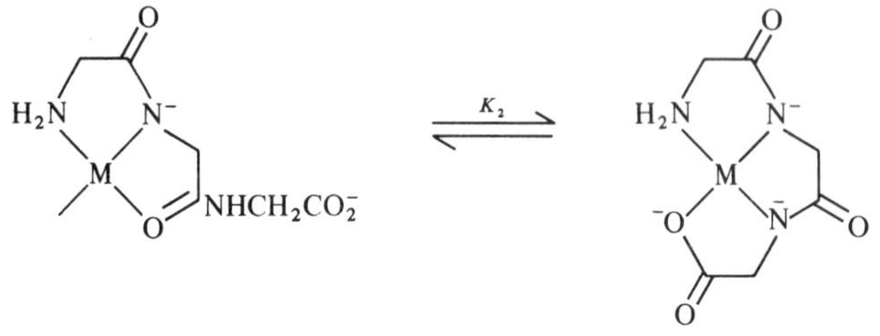

(20)

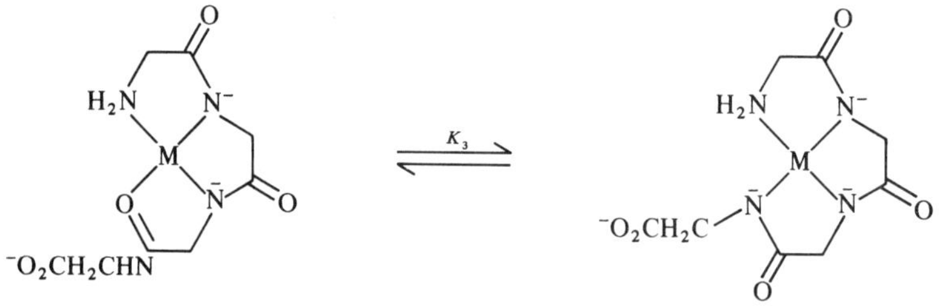

(21)

20.2.3.1.1 Dipeptides

Stability constants for Cu^{II} complexation of Gly—X and X—Gly dipeptides show that even relatively small alkyl side chains can influence coordination.[83] At low pH, for example, the bidentate chelates N(amino),O(peptide) of L-Leu-Gly and L-Ile-Gly are the thermodynamically less stable, presumably due to steric hindrance from the branched side chains. Similar coordination by Gly—X is virtually independent of the nature of X. This latter result is not unexpected since only the Gly moiety is coordinating. However, on raising the solution pH (> 4) the peptide deprotonation that occurs to give tridentate N(amino),N(peptide),O(carboxyl) chelation is slightly inhibited by the Leu and Ile residues (by 0.7–0.8 log units).

Metal–aromatic ring interactions are also evident in those dipeptides where Phe, Tyr or Trp are at the carboxyl end but not when they are at the amino end. At high pH (> 12) further ionizations and coordination changes occur, often accompanied by decarboxylation of the peptide.[87]

Earlier studies[85] established that imidazole coordination as part of a chelate structure requires the His residue to be near the amino end. This important positional feature of His is also apparent in the dipeptides Gly-His and His-Gly. As can be seen from Table 8 the His-Gly stability constants are invariably larger and the range of complexes formed also differs. Sundberg and Martin[18] have reviewed the solution and solid state properties of His-containing peptides. More recently Gergely and his group[88] have reexamined the coordinating properties of Gly-L-His, L-His-Gly and

Table 9 Structures of some Metal–Peptide Complexes from X-Ray Crystallography

Compound[a]	Geometry; Ligating Atoms[b]	Ref.
[Cd^{II}(HGly-Gly)$_2$Cl$_2$]	Octa polymeric; 2 Cl, O (pep), O (car'), 2 O (car")	1
[Cd^{II}(GlyO)(H$_2$O)]Cl	Octa polymeric; N (am), O (pep), Cl, O (H$_2$O), O (car'), O (car")	1

Table 9 (*continued*)

Compound[a]	Geometry; Ligating Atoms[b]	Ref.
$[Cd^{II}(Gly\text{-}GluO)(H_2O)]H_2O$	7 coord polymeric; N (am), O (pep), Cl, 2 O (car'), 2 O (car")	1
$[Co^{II}(H_2O)_6][Co^{III}(Gly\text{-}GlyO)_2]_2$ 6 or $12H_2O$	Λ, Δ 2 N (am), 2 N (pep), 2 O (car)	2
$[Co^{III}(Gly\text{-}GlyO)_2](ClO_4)$	Same as Co^{III} anion above, protonation at the O (pep) atoms	2
$[Cu^{II}(Gly\text{-}MetO)]$	Sq pyr; N (am), N (pep), O (car), O (car'), axial O (pep")	3
$[Cu^{II}(Met\text{-}GlyO)]$	Sq pyr; as Gly-Met complex	4
$[Cu^{II}(Gly\text{-}TyrO)(H_2O)_2]H_2O$	Sq pyr; N (am), N (pep) O (car), O (H$_2$O), axial O (H$_2$O)	5
$[Cu^{II}_2(\beta\text{-}Ala\text{-}HisO)_2]$	Sq planar; N (am), N (pep), O (car), N (pep) from 2nd peptide	6
$[Cu^{II}(Gly\text{-}His\text{-}GlyO)]2.5H_2O$	Sq planar; N (am), N (pep), N^3 (imid), O (car')	7
$[Cu^{II}(Gly\text{-}Gly\text{-}His\text{-}NHCH_3)]$	Sq planar; N (am), 2 N (pep), N^3 (imid)	8
$[Cu^{II}(Gly\text{-}Leu\text{-}TyrO)]8H_2O\cdot(C_2H_5)_2O$	Sq planar dimeric; N (am), O (pep), N (pep'), O (car'), Cu–π interaction	9
$Na_2[Cu^{II}(pentaGly\text{-}O)]4.5H_2O$	Sq. planar; N (am), 3 N (pep)	10
$[Cu^{II}_2\{cyclo(His\text{-}His)\}_2(H_2O)_2]ClO_4\cdot3.5H_2O$	Cu (1): 2 N^1 (imid), 2 O (pep), O (H$_2$O); Cu (2): 2 N^1 (imid), 2 N (pep)	11
$[Cu^{II}\{cyclo(His\text{-}His)\}_2](ClO_4)_2\cdot4H_2O$	Dist tetra; 4 N^1 (imid), 13-membered chelate rings	12
$[Cu^{III}(tri\text{-}\alpha\text{-}aminoisobutyrate)]2H_2O\cdot1.5NaClO_4$	Sq planar, N (am), 2 N (pep), O (car)	13
$Na_2[Ni^{II}(Gly\text{-}GlyO)_2]8$ or $9H_2O$	Octa; 2 N (am), 2 N (pep), 2 O (car)	14
$Na_2[Ni^{II}(tetraGly\text{-}O)]8$ or $10H_2O$	Sq planar; N (am), 3 N (pep)	15
$[Rb^I\{cyclo(Pro\text{-}Gly)_4\}]NCS\cdot3H_2O$	4 O (pep)	16
$[Pt^{II}(Gly\text{-}Met)Cl]H_2O$	Sq planar; N (am), N (pep), S, Cl	17
$[Ph_2Sn^{IV}(Gly\text{-}GlyO)]$	N (am), N (pep), O (car)	18
$[Zn^{II}(Gly\text{-}Gly\text{-}GlyO)(H_2O)_2]0.5SO_4\cdot2H_2O$	Trig bipyr; N (am), O (pep), O (car'), 2 O (H$_2$O)	19

[a] Amino acid residues have the L configuration unless stated otherwise.
[b] pep, peptide, other abbreviations as in Table 3.

1. R. J. Flook, H. C. Freeman, C. J. Moore and M. L. Scudder, *J. Chem. Soc., Chem. Commun.*, 1973, 753.
2. M. T. Barnet, H. C. Freeman, D. A. Buckingham, I.-N. Hsu and D. Van der Helm, *Chem. Commun.*, 1970, 367.
3. C. A. Bear and H. C. Freeman, *Acta Crystallogr., Sect. B*, 1976, **32**, 2534.
4. J. Dehard, J. Jordanov, F. Keck, A. Mosset, J. J. Bonnet and J. Galy, *Inorg. Chem.*, 1979, **18**, 1543.
5. P. A. Mosset and J. J. Bonnet, *Acta Crystallogr., Sect. B*, 1977, **33**, 2807.
6. H. C. Freeman and J. T. Szymanski, *Acta Crystallogr.*, 1967, **22**, 406.
7. P. de Meester and D. J. Hodgson, *Acta Crystallogr., Sect. B*, 1978, **33**, 3505.
8. N. Camerman, A. Camerman and B. Sarkar, *Can. J. Chem.*, 1976, **54**, 1309.
9. W. A. Franks and D. Van der Helm, *Acta Crystallogr., Sect. B*, 1971, **27**, 1299.
10. J. F. Blount, H. C. Freeman, R. V. Holland and G. H. Milburn, *J. Biol. Chem.*, 1970, **245**, 5177.
11. Y. Kojima, K. Hirotsu and K. Matsumoto, *Bull. Chem. Soc. Jpn.*, 1977, **50**, 3222.
12. F. Hori, Y. Kojima, K. Matsumoto, S. Ooi and H. Kuroya, *Bull. Chem. Soc. Jpn.*, 1979, **52**, 1076.
13. L. L. Diaddario, W. R. Robinson and D. W. Margerum, *Inorg. Chem.*, 1983, **22**, 1021.
14. H. C. Freeman and J. M. Guss, *Acta Crystallogr., Sect. B*, 1978, **34**, 2451.
15. H. C. Freeman, J. M. Guss and R. L. Sinclair, *Acta Crystallogr., Sect. B*, 1978, **34**, 2459.
16. Y. H. Chiu, L. D. Brown and W. N. Lipscomb, *J. Am. Chem. Soc.*, 1977, **99**, 4799.
17. H. C. Freeman and M. L. Golomb, *Chem. Commun.*, 1970, 1523.
18. F. Huber, H. J. Haupt, H. Preut, R. Barbieri and M. T. Lo Giudice, *Z. Anorg. Allg. Chem.*, 1977, **432**, 51.
19. D. Van der Helm and H. B. Nicholas, *Acta Crystallogr., Sect. B*, 1970, **26**, 367.

Table 10 Ionization Constants of Some Coordinated Glycine Peptides, Showing Metal-ion-promoted Peptide Deprotonations[a]

	Gly-Gly		Gly-Gly-Gly		TetraGly	
	Cu^{II}	Ni^{II}	Cu^{II}	Ni^{II}	Cu^{II}	Ni^{II}
pK_1	3.99	9.35	5.22	8.8	5.50	9.1
pK_2			6.60	7.7	6.89	7.1
pK_3					9.29	8.2

[a] 25 °C and 0.1–0.16 mol dm^{-3} ionic strength. Ref. 83.

β-Ala-L-His (carnosine). With Cu^{II} a range of species forms depending on the solution pH and the metal:ligand ratio. Interestingly they conclude that the imidazole residues in carnosine and His-Gly do not participate in chelate binding in the 1:1 species but can act as bridges between metal centres to form dimers. These authors also find, in contrast to earlier claims, that Zn^{II}, as well as Co^{II} and Ni^{II}, can promote peptide deprotonation in Gly-His. With His-Gly however, none of these metal ions caused peptide deprotonation up to pH 9, and the peptide coordinates through the N(amino) and N(imidazole) atoms, *i.e.* 'histidine-like'.

Gergely and Farkas[89] have also studied the influence on the coordination to Cu^{II} of the alcoholic, amide and carboxyl side groups in the dipeptides Gly-L-Ser, Gly-L-Asn, Gly-L-Asp and Gly-L-Glu. Whilst Kowalik *et al.*[90] have studied the involvement of the thioether group of Met in dipeptides coordinated to Pd^{II}, here again the stoichiometry of the species formed depends on the relative position of the Met residue.

20.2.3.1.2 Larger peptides

The data in Table 8 show that with triglycine and tetraglycine peptides further metal ion-promoted peptide ionizations can occur to ultimately give, with the tetrapeptide, four-N-donor chelation comprising the N(amino) and three N(peptide) atoms. Further extension of the peptide chain sees no further promotion of peptide ionization. The same data also reveal that these ionizations can proceed in a cooperative fashion, in that after the first peptide ionization the subsequent ones proceed relatively easily; in Ni^{2+}–triglycine, for example, $pK_{a2} < pK_{a1}$. A histidine residue at the third position is especially effective in the Cu^{II}- and Ni^{II}-promoted peptide ionizations. Of the known peptides Gly-Gly-L-His and L-Asp-L-His provide the strongest chelation modes for these metal ions, the predominant species in aqueous solution being the planar four-N-atom 1:1 chelate (**19**), as confirmed by spectroscopic and solid state studies (see Table 9). The latter of these two tripeptides mimics the Cu^{II} binding site of the protein albumin.[91] For a large number of long peptides that have been examined the preferred binding site for Cu^{II} and Ni^{II} is the N(amino) terminus with neighbouring ionized peptide nitrogens.

(19)

Cysteine-containing peptides are difficult to synthesize because of their easy oxidation to give cross-linked disulfide products (also catalyzed by metal ions). Polynuclear formation *via* S bridging also occurs. N-MercaptoacetylGly and N-mercapropropionylGly peptides have been studied as models for the blue Cu proteins, with particular emphasis being placed on the detection of Cu^{II}—S bonds.[92–94] The naturally occurring cysteine-containing peptide glutathione and its important metal-binding properties will be discussed in Chapter 22.

The influence of proline residues on coordination has been examined by Kozlowski and co-workers. Where Pro occurs in a peptide chain there is no peptide hydrogen (**20**), and from a variety of spectroscopic studies these workers have established that a Pro residue causes disruption of metal ion-promoted peptide ionizations, *i.e.* Pro serves as a 'breaking point' in metal ion coordination. On

Cu^{II} coordination to Pro-substituted (P) tetra-L-alanine (A_4) peptides only A_4 and PA_3 bind through four N atoms; APA_2, A_2PA and A_3P do not. The authors conclude that, although the proline peptide N is a potential coordinating site, the considerable stereochemical change required for its coordination (tetrahedral N) precludes its metal binding, nor is there any evidence for the Pro peptide O being able to coordinate.[95] With both Pro and Tyr in tetrapeptides the coordination mode depends very much on their relative positions, phenolate oxygen binding to Cu^{II} being evident, with consequent severe distortion from planarity around the metal ion. Spectroscopic examination of the Cu^{II} coordination to the octapeptide (all L isomers) Thr-Lys-Pro-Lys-Thr-Lys-Pro-Lys at neutral pH suggests three N donor atoms, and the unusual structure (21) that the authors propose results from the Pro breaking points.[96]

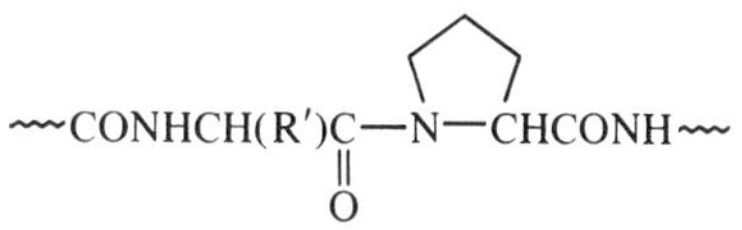

(20)

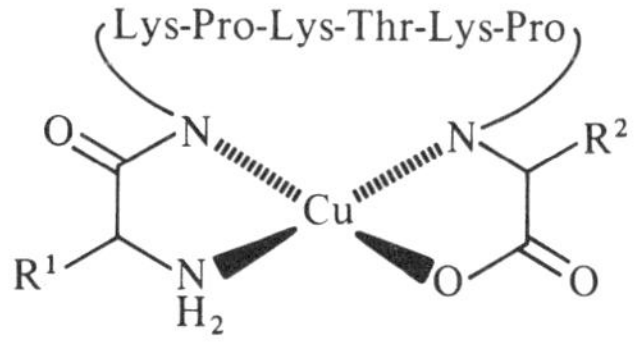

(21) R^1 = Thr residue, R^2 = Lys residue

20.2.3.1.3 *Homopolypeptides*

Polypeptide chains from a single amino acid source are easier to synthesize and their regularity of structure eases the interpretation of their coordinating properties. They also have interesting coordinating properties in their own right. Thus, for those homopolypeptides with potential coordinating side chains the amino terminal becomes less significant.

In their review Sigel and Martin distinguish three kinds of complexes that form in solution.[83] At low pH interaction is exclusively with the side chains (type S); a higher pH form involves both side chain and deprotonated peptide nitrogens (type SP); at pH 10–12 the binding is solely through the deprotonated nitrogens (biuret type B). The three forms show distinctive CD spectra. Type SP does not appear to form with those polypeptides containing non-chelating side groups, *e.g.* poly(L-Lys), poly(L-ornithine) and poly(L-Glu).

20.2.3.1.4 *Cyclic peptides*

These may be regarded as linear peptides in which the amino and carboxyl ends undergo intramolecular condensation giving an extra peptide bond. The simplest of these are the cyclic dipeptides (22).

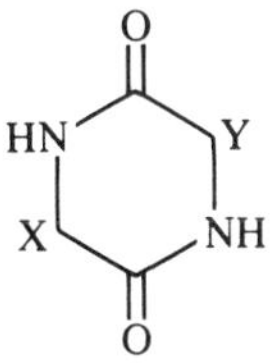

(22)

With the smaller cyclic peptides the lack of flexibility combined with the lack of an amino 'anchor point' means that deprotonation of the peptide groups requires more strongly alkaline conditions than for their linear counterparts. However, once formed the cyclic complexes can be thermodynamically and kinetically more stable.[97] This also applies to Cu in oxidation state III.

Two crystal structures have been reported for Cu^{II} complexes of cyclo(L-His-L-His) (22; X = Y = imidazole; see Table 9). The most interesting of these is the bis-ligand complex in which each ligand is bidentate through the N-1 imidazole atoms, thus forming 13-membered chelate rings. Presumably chelation through N-1 rather than the more usual N-3 atoms is preferred because the latter would create smaller 11-membered rings with increased strain. NMR, CD and stability constant measurements suggest the same structure forms in aqueous solutions.[98] The large chelate ring system forms a valuable precedent for the proposed Pro-containing linear peptide complexes (21) mentioned earlier.[96]

A number of larger cyclic peptides with important chelating properties occur naturally and will be mentioned in Chapter 22. With the synthetic compounds the greater flexibility enables the O atoms to bind non-transition metal ions as macrocylic chelates. Blout and his co-workers have studied a number of Gly- and Pro-containing cyclic peptides and their binding to Groups IA and IIA metal ions, their main interest being the study of the peptide conformations by means of CD and NMR techniques.[99-101] Bicyclic decapeptides in which two peptide rings are connected by a disulfide link also show strong affinity for the Groups IA and IIA metal ions, again by chelation through the peptide O atoms.[102] The absence of peptide N bonding also occurs in Zn[II] complexes of cyclic heptapeptides, the binding sites being polar side groups.[103]

20.2.3.1.5 *Schiff base complexes*

As with amino acids the terminal amino group can participate in Schiff base formation. Most interest has centred on compounds formed from salicyladehyde and dipeptides (23), and their ability to act as tetradentate chelates. The stability of Cu[II] complexes of Gly- and/or β-Ala-containing compounds appears to be a function of ring sizes, the inclusion of a six-membered ring in the fused-ring system enhancing stability.[104] Similar conclusions have also been made with tripeptide-containing Schiff base complexes.[105] The same Japanese school has also found that Cu[II] complexes of dipeptide Schiff bases formed from 2-acetylpyridine or pyridine-2-carbaldehyde undergo trans-amination reactions.

$$\text{OH}$$
$$\text{C=NCH(R}^1\text{)CONHCH(R}^2\text{)CO}_2\text{H}$$

(23)

20.2.3.2 Stereoselectivity

Since stereoselectivity in the amino acid complexes requires three-point attachment to the metal centre then the observation of such effects in peptides should not be unexpected. However, with dipeptides there is the additional requirement that both the amino acid residues must be chiral.[106] With non-glycyl dipeptides considerable differences in thermodynamic stabilities arise between LL and LD diastereomeric pairs on complexing to metal ions. This stereoselectivity can be related to differences in basicities[83] and to steric differences arising from the different dispositions of the side chain (which also includes favourable hydrocarbon interactions between non-polar side chains located on the same side of the metal centre). These effects can work in opposition to each other. Thus, for the [Cu(dipep)]$^+$ complexes that form at low pH the *meso* or DL peptide complexes have slightly higher stability constants than their chiral LL (or equally DD) counterparts, where dipep is Ala-Ala or Leu-Leu. These differences are ascribed to the greater basicities of the amino groups of the DL forms. On raising the pH sufficiently to cause peptide ionization, and hence coordination to the peptide N, steric factors favour the LL over the DL isomer, the differences being greater for the larger alkyl side groups of Leu as compared to Ala. The limited data available for Co[II], Ni[II] and Zn[II] complexes suggest that similar trends operate.

Tridentate coordination of Co[III] by dipeptide anions generates an asymmetric metal centre, and hence the mirror-image-related Λ- and Δ-[Co(dipep)$_2$]$^-$ isomers. With non-Gly-containing dipeptides stereoselective formation occurs, the stereoselectivity being pH dependent.[107] Stereoselectivity has also been reported in complexes of the type [CoL(dipep)],[24] where L is an amine ligand or ligands. The stereoselectivity primarily depends on the carboxyl-terminal amino acid, although enhanced stereoselectivity is obtained if both residues are of the same chirality.[48] ^{1}H NMR is particularly useful in assigning the configurations of individual isomers.[108]

20.2.3.3 Reactivity

20.2.3.3.1 *Kinetics of proton and ligand transfer reactions*

The predominant interest in the reactivity of the metal–N(peptide) bond has produced some interesting observations. Thus the reaction of this coordination group with H$_3$O$^+$ is several orders of magnitude slower than expected for a proton transfer reaction in aqueous solution, even with the kinetically labile metal ion complexes. Margerum and Duke in their review[109] give a partial explanation in terms of the sluggish making and breaking of the metal–N(peptide) bond. The same reaction exhibits acid catalysis with Cu^{II} complexes, occasionally so with Ni^{II}, but not at all with Pd^{II}. A proposed mechanism involves two steps, equations (22) and (23), where H_{-n} corresponds to n deprotonated peptide groups coordinated to the metal ion M. The species $M(H_{-n}Pep)H$ is a proposed moderately strong acid intermediate with a still intact M—N(peptide) bond. A precedent for this was mentioned earlier in the case of O(peptide)-protonated $[Co^{III}(Gly\text{-}GlyO)_2]^+$ ion.

$$M(H_{-n}Pep) + HX \rightleftharpoons M(H_{-n}Pep)H + X \qquad (22)$$

$$M(H_{-n}Pep)H \longrightarrow M(H_{-n+1}Pep) \qquad (23)$$

Nucleophilic reactions, in particular ligand replacement by EDTA and polyamine ligands, readily occur with Ni^{II} and Cu^{II} peptide complexes, the most reactive nucleophiles being those that contain an amine group to give an associative-type intermediate. The presence of histidine as the third residue in a peptide appears to reduce the susceptibility of Cu^{II} complexes to nucleophilic attack.

20.2.3.3.2 Peptide bond formation and hydrolysis

Whilst metal–N(peptide) bond formation inhibits hydrolysis of the peptide bond, coordination to O(peptide) has the opposite effect. These differences in reactivity can be readily demonstrated and put to practical use with the inert Co^{III} complexes. One of the first examples was the reaction of $[Co(trien)(H_2O)(OH)]^{2+}$ with peptides to give hydrolysis of the peptide bond at the N-terminal end. The proposed mechanism involving nucleophilic attack by hydroxide at the peptide carbon is shown in Scheme 7.[110] Similar selective hydrolyses of N-terminal peptide bonds have since been demonstrated with other Co^{III} amine complexes and the reaction has been examined as a method for determining the N-terminal amino acid residue in peptides and proteins.[111,112]

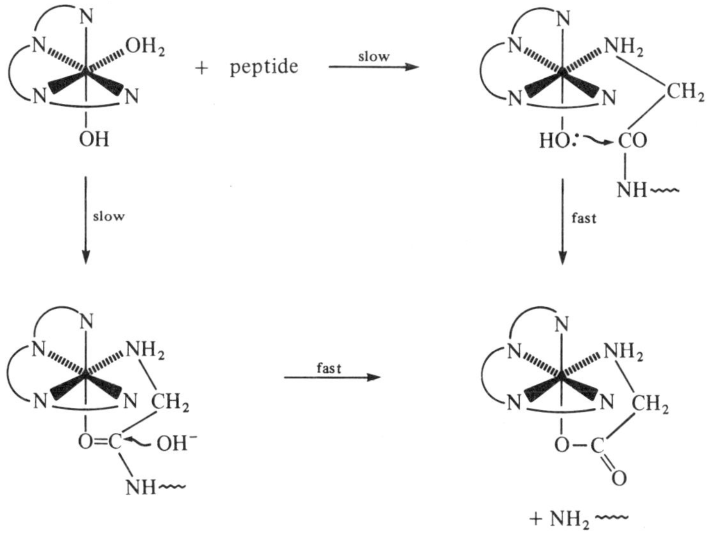

Scheme 8

Similarly, Cu^{II} coordination at the O(peptide) atom (pH 3.6–6) results in catalysis of peptide bond hydrolysis, whilst at higher pH the Cu^{II} becomes catalytically inactive as coordination to the peptide N atom takes over. Two studies worthy of further mention are, first, attempts to prepare crystals of a Cu^{II} hexaglycine complex in alkaline solution gave instead the pentaglycine complex and free glycine, the hydrolysis occurring at the carboxyl terminal.[113] Second, an attempt to prepare a Schiff base complex from salicylaldehyde and triglycine with Cu^{II} yielded the Schiff base complex of diglycine and free glycine, again from the carboxyl terminal.[114] Both these reactions reflect the stabilization of the peptide by ionization and coordination to a metal ion. Further examples of these and other hydrolytic reactions are given in the review by Hay and Morris.[76]

The amino acid ester complex ion [Co(en)$_2$(GlyOR)Cl]^{24} can be prepared in aqueous solution by reacting the glycine ester with [Co(en)$_2$Cl$_2$]$^+$. However, if non-aqueous solvents are used then reaction proceeds to give peptide bond formation,[115] the product being [Co(en)$_2$(Gly-GlyOR)]$^{3+}$. Peptide bond formation also occurs with other amine complexes. The proposed mechanism is that initial formation of the coordinated glycine ester complex occurs, activating the ester group towards nucleophilic attack by the nitrogen lone pair electrons of a second amino acid ester complex.[76]

20.2.3.3.3 Activation of α C—H bonds

Gillard[48] reported the differential reactivity of α C—H bonds in coordinated peptides. For example, under alkaline conditions H–D exchange occurs at a measurable rate in [Co(Gly-GlyO)$_2$]$^-$ at the carboxyl-terminal methylene group but not at the amino-terminal methylene group. A more recent result reminiscent of this finding is that racemization of L-Ala occurs faster in Gly-L-Ala than in L-Ala-Gly when they are coordinated to Co^{III}. In [Co(Gly-L-AlaO)$_2$]$^-$ the methine carbon of Ala is sandwiched between two negatively charged groups (24) and it is suggested that this would retard formation of a carbanion intermediate.[75]

(24)

Differential reactivity of α C—H bonds has also been reported for Ni^{II} complexes of peptides and peptide Schiff bases[116] in which it is the carboxyl-terminal bond that is the more reactive. However, *N*-salicylideneglycyl-L-valinatocopper is found to react with aqueous formaldehyde to give serine-L-valine containing highly optically pure serine. In this case the reaction obviously occurs stereoselectively at the amino-terminal residue.[116]

20.2.3.3.4 Redox reactions

Many laboratories have reported on the reaction of Co^{II} dipeptide complexes with molecular O$_2$. The initial reaction is reversible with formation of a brown dimeric Co^{II} complex. Further changes then occur resulting in the irreversible formation of a mononuclear Co^{III} complex. One important observation is that at least three N donor atoms coordinated to the initial Co^{II} reactant are necessary for the oxygenation to be reversible. Gillard and Spencer[107] have given a detailed scheme relating some of the intermediates formed. The reactions are complex, being dependent on the nature of the dipeptide and the solution pH. These dependences reflect that equilibria involving a number of species are present and, indeed, much of the work has gone into elucidating the reactive species.

Discordant results have been obtained particularly with histidine-containing dipeptides. Harris and Martell,[117] for example, claim the active species at pH 8 is [CoL$_2$O$_2$(OH)]$^-$, where L is the deprotonated anion of Gly-L-His or L-His-Gly, although an earlier study found no evidence for a hydroxyl species. A reexamination of the species equilibria by Gergely and co-workers[118] also suggests no hydroxyl species involvement; their conclusions are that for Gly-L-His the active complex is most likely [Co^{II}(Gly-L-HisO)], whilst for L-His-Gly and β-Ala-L-His (carnosine) a 2:1 ligand:Co^{II} complex is the active species. These complexes possess the minimum three coordinated N atoms required for O$_2$ uptake. Differences also exist over the nature of the active species responsible for the reversible oxygenation of ternary Co^{II}—carnosine—L-His solutions. One study[119] claims that the L-His stabilizes the Co^{II}–carnosine complex against stabilization, whilst Gergely and co-workers[118] conclude that the active species is [Co^{II}(HisO)$_2$] (see Section 20.2.2.6.5).

The most thoroughly studied peptide complexes are those of Cu^{II} and Ni^{II}; even so they are capable of giving unexpected results, *e.g.* the aforementioned Cu^{II}-catalyzed hydrolysis of hexaglycine.[113] Attempts to prepare a crystalline complex by reacting Gly-Gly-L-His with Cu(OH)$_2$ resulted in crystals of the decarboxylated product (25).[120] A similar reaction has been observed with Ni^{II}. These are examples of oxidative decarboxylation reactions. Just as unexpectedly Margerum and his co-workers found that neutral aqueous solutions of Ni^{II} or Cu^{II} tetra- and penta-peptide complexes absorbed O$_2$ to give a number of products. Resulting from these observations has come the establishment of peptide complexes of these metals in the unusual oxidation state III. The requirement for their stabilization appears to be three or more deprotonated peptide groups. The thermodynamic, kinetic and electron-transfer studies of these new Cu^{III} complexes have been reviewed.[121]

(25)

The first crystal structure of a Cu^{III} peptide, $[Cu^{III}(tri-\alpha-aminoisobutyrate)]\cdot 2H_2O\cdot 1.5NaClO_4$ (see Table 9), has recently been reported. As expected for a d^8 ion the Cu^{III} is strictly four-coordinate and coplanar with the donor atoms. The metal–ligand bond lengths are 12–17 pm shorter than in their Cu^{II} counterparts, reflecting the increased ligand field strength. The shorter bond lengths also mean that the conformation of the peptide molecule is significantly different to the Cu^{II} analogue.

20.2.4 PROTEINS

Metal–protein complexation is vital to all living systems.[122] The metal ions can be irreversibly bound in the protein molecule and vital to the protein's catalytic function. Such proteins are known as *metalloenzymes*; several hundred of them occur naturally. The wide range of functions in which they are involved will be considered in Chapter 22 and in Volume 6. In the metalloenzymes the metal ion can be directly bonded to side chains of the amino acid residues or to other ligands which are covalently linked to the protein. These ligands are referred to as prosthetic groups. Other enzymes require the presence of metal ions for their catalysis — these are referred to as *metal-activated enzymes* and mostly involve K^+, Mg^{2+} or Ca^{2+} ions. Reversible binding of metals to proteins also occurs in the transport and storage forms of the metals in living systems.[123] Finally, the specific binding of heavy metal atoms to proteins is of extreme importance in X-ray diffraction studies of protein structures.

The unusual coordination observed in many metalloenzymes, unusual in the sense that they differ thermodynamically and kinetically from the more familiar low-molecular-weight complexes, has proved a stimulating challenge to coordination chemists. In order to delineate the nature of the coordination site and its relation to the functional property of the enzyme an enormous array of physical and spectroscopic techniques have been commissioned. These include UV/visible, near-IR, CD, MCD,[124] ESR, ENDOR,[125] resonance Raman,[126] Mössbauer[127] and multinuclear NMR,[128] including such nuclei as ^7Li, ^{31}P, ^{35}Cl, ^{63}Cu, ^{113}Cd, ^{199}Hg and others. More recently EXAFS[129] is proving invaluable in giving accurate geometrical information of metal binding sites. Other techniques are still in their infancy in this field, *e.g.* optical fluorescence[130] and laser-induced luminescence.[131]

Despite their large size and complexity a surprising number of crystal structures have been solved and have proved invaluable in the understanding of the protein's role, as well as the metal ion's, in the catalytic activity; a number of these structural analyses are presented in Table 11. For the remainder of this section the emphasis will be on the nature of the coordination sites found in proteins with consideration of specific metal ions.

20.2.4.1 Binding Sites

A long-established method of determining protein concentrations is the biuret method which depends on the formation of a blue chromophore of Cu^{II} bound to ionized peptide N atoms. The high pH required for the development of this chromophore is indicative of the difficulty of achieving this type of coordination with proteins. From the peptide studies it is safe to conclude that such coordination also requires the availability of the amino terminal group. This is often unavailable in the proteins. To date there is only one genuine example of peptide N binding under more normal conditions, and that is the Cu^{II} and Ni^{II} transport site of serum albumin.[132] This specific binding site occurs at the amino-terminal end, and from comparison of spectroscopic properties with those of peptides which mimic this site the coordination is most likely as in (**19**).[91] In albumins the third amino acid residue is histidine. One exception is dog albumin which lacks this His residue and also lacks the ability to bind Cu^{II} and Ni^{II} in the same manner. Other claims to peptide N involvement in metal binding have subsequently been proved wrong.[83]

It could be hypothesized that in the case of Cu^{II}–albumin it is the metal ion that is dictating the requirements and, hence, achieving its more normal properties of microscopic reversibility and kinetic lability, properties which are also essential for the proposed biological transport role of this complex. On the other hand in the metalloenzymes it is the protein that is dictating the nature of the coordination site and matching this as nearly as possible to the optimum requirements of the catalytic reaction. This optimization is reflected in the unusual coordination geometries that are frequently observed (examples in Table 11) and in the lack of reversibility of metal coordination. Examination of the data in Table 11 reveals the lack of peptide N involvement but the preponderance of cysteine and histidine residues as coordination sites. In the latter case frequently the pyrrole N (N-1) rather than the pyridine N (N-3) of the imidazole moiety (**3**) is the ligating atom. This can be attributed to the lack of restraint that is imposed by chelation involving neighbouring peptide groups, and to hydrogen tautomerism (see Section 20.2.2.2.1). The high thermodynamic stability associated with the metalloenzyme coordination sites cannot be attributed to their being multidentate chelate sites because the donor atoms come from different parts of the polypeptide chains so that 'ring systems' formed are far larger than the five- or six-membered optimal chelate systems. Nor are the bond lengths or bond angles sufficiently different to cause the enhanced stability. On the contrary, they are frequently of the wrong order for maximum covalent bond formation with the valence orbitals of the metal ions (a good example of this is the Cu^{II} site in plastocyanin; see Table 11). The second feature evident from the crystal structure analyses in Table 11, and from solution properties, is that these coordination sites are located in cavities within the protein structures and, most importantly, the mouths of these cavities are generally constricted and lined with hydrophobic residues (from the amino acids with non-polar side chains). This hydrophobicity has the important effect of limiting both the access and the activity of water molecules within the cavities and results in the enhanced thermodynamic stabilities. Without this property water, being an effective ligand and being present at a very high concentration ($55.5 \, mol \, l^{-1}$), would swamp the coordination sites available for the incoming substrates. The types of protein coordination sites will now be considered for some of the more important biological metals.

20.2.4.1.1 Calcium[133]

Ca^{2+} binding to proteins is generally considered to be weak relative to that of the transition metal ions. As is predictable for a 'hard' metal ion the coordination appears to be entirely through O donor atoms, from carboxyl and peptide O atoms (see Table 11). Unfortunately Ca^{2+} is spectroscopically silent and much of the work on the elucidation of its binding sites relies upon the specific

Table 11 Metal Binding Sites in Proteins Established by X-ray Crystallography

Protein and metal ion(s)	Source	Resolution of crystallographic data (nm)	Metal binding sites[a]	Ref.
Azurin, Cu^{II}	*Pseudomonas aeruginosa*	0.30	S (Cys), 2 N (His), S (Met)?, N (Gln)?	1
Carbonic anhydrase C, Zn^{II}	Human erythrocytes	0.22	2 N^1 (His), N^3 (His), O (H_2O)	2
Carboxypeptidase, Zn^{II}	Bovine	0.20	2 N (His), O (Glu), O (H_2O)	3
Concanavalin A, Ca^{II}, Mn^{II}	Jack bean	0.20	Mn: 2 O (Asp), O (Glu), N (His), 2 O (H_2O) Ca: 2 O (Asp), O (Asn*), O (Tyr*), 2 O (H_2O) (two Asp ligands shared)	4
Cytochrome b_5, Fe^{II}	Calf liver	0.20	Haem, 2 N (His)	5
Cytochrome b_{562}, Fe^{II}	*Escherichia coli*	0.25	Haem, N (His), S (Met)	6

Table 11 (*continued*)

Protein and metal ion(s)	Source	Resolution of crystallographic data (nm)	Metal binding sites[a]	Ref.
Cytochrome c, Fe^{II} or Fe^{III}	Tuna	0.20	Haem, N^1 (His), S (Met)	7
Cytochrome c_3, $4Fe^{II}$	*Desulfovibrio desulfuricans*	0.25	4 {Haem, 2 N^1 (His)}	8
Erythrocruorin(oxy), Fe^{II}	Vertebrate	0.14	Haem, N^1 (His), O_2 ($< FeOO = 170°$)	9
Ferredoxin, $8Fe^{II}$	*Peptococcus aerogenes*	0.28	2 $Fe_4S_4\{S(Cys)\}_4$ clusters	10
Haemerythrin(met), $2Fe^{III}$	*Themiste dyscritum*	0.28	Fe (1): 3 N (His) Fe (2): 2 N (His), O (Tyr), 2 Fe linked by O (Asp), O (Glu), O (H_2O)	11
Haemoglobin, $4Fe^{II}$	Horse	0.28	4 {Haem, N^1 (His), 6th site vacant}	12
High-potential iron–sulfur protein (HIPIP), 4 Fe^{II}	*Chromatium vinosum*	0.225	$Fe_4S_4\{S(Cys)\}_4$ cluster	13
Liver alcohol dehydrogenase, $2Zn^{II}$	Horse	0.24	Zn(1): 2 S (Cys), N (His), O (H_2O or OH^-) Zn(2): 4 S (Cys) (Zn(1) is the catalytic atom)	14
Myoglobin (oxy) Fe^{II}	Sperm whale	0.20	Haem, N^1 (His), O_2 ($\langle FeOO = 121°$)	15
Parvalbumin, $2Ca^{II}$	Carp muscle	0.185	2 {2 O (pep), 4 O (car)}	16
Plastocyanin, Cu^{II}	Poplar	0.27	S (Cys), S (Met), S N^3 (His)	17
Rubredoxin, Fe^{II}	*Desulfovibrio vulgaris*	0.20	4 S (Cys)	18
Superoxide dismutase, Cu^{II}, Zn^{II}	Bovine	0.30	Cu: 4 N (His) Zn: 3 N (His), O (Asp*) (two metals bridged by one His)	19
Thermolysin, $4Ca^{II}$, Zn^{II}	*Bacillus thermoproteolyticus*	0.23	Zn: 2 N (His), O (Glu), O (H_2O) Ca(1): 3 O (Glu), 2 O (Asp), O (H_2O) Ca(2): 2 O (Glu), O (Asp), O (Asn*), 2 O (H_2O) Ca(3): 2 O (Asp), O (Glu), 3 O (H_2O) Ca(4): O (Asp), 2 O (Thr), O (Tyr), O (Ile*), O (H_2O)	20
Trypsin, Ca^{II}	Bovine	0.18	2 O (Glu), O (Asn*), O (Val*), 2 O (H_2O)	21

[a] Asterisk indicates O donor atom from the peptide group.

1. E. T. Adman, R. E. Stenkamp, L. C. Sieker and L. H. Jensen, *J. Mol. Biol.*, 1978, **123**, 35.
2. K. K. Kannan, B. Notstrand, K. Fridborg, S. Lovgren, A. Ohlsson and M. Petef, *Proc. Natl. Acad. Sci. USA*, 1975, **72**, 51.
3. W. N. Lipscomb, G. N. Reeke, Jr., J. A. Hartsuck, F. A. Quicho and P. N. Bethge, *Philos. Trans. R. Soc. London, Ser. A*, 1970, **257**, 117.
4. G. N. Reeke, Jr., J. W. Becker and G. M. Edelman, *J. Biol. Chem.*, 1975, **250**, 1513.
5. P. Argos and F. S. Mathews, *J. Biol. Chem.*, 1975, **250**, 747.
6. F. S. Mathews, P. H. Bethge and E. W. Czerwinski, *J. Biol. Chem.*, 1979 **254**, 1699.
7. R. Swanson, B. L. Trus, N. Mandel, G. Mandel, O. B. Kallai and R. E. Dickerson, *J. Biol. Chem.*, 1977, **252**, 759 and 776.
8. R. Haser, M. Pierrot, M. Frey, F. Payan, J. P. Astier, M. Bruschi and J. LeGall, *Nature (London)*, 1979, **282**, 806.
9. E. Weber, W. Steigemann, T. A. Jones and R. Huber, *J. Mol. Biol.*, 1978, **120**, 327.
10. E. T. Adman, L. C. Sieker and L. H. Jensen, *J. Biol. Chem.*, 1973, **248**, 3987.
11. R. E. Stenkamp, L. C. Sieker and L. H. Jensen, *Proc. Natl. Acad. Sci. USA*, 1976, **73**, 349.
12. M. F. Perutz, H. Muirhead, J. M. Cox, L. C. G. Goaman, F. S. Mathews, E. L. McGandy and L. E. Webb, *Nature (London)*, 1968, **219**, 29.
13. S. T. Freer, R. A. Alden, C. W. Carter and J. Kraut, *J. Biol. Chem.*, 1975, **250**, 46.
14. H. Eklund, B. Nordstrom, E. Zeppezauer, G. Soderlund, I. Ohlsson, T. Boiwe, B. O. Soderberg, O. Tapia, C. I. Branden and A. Akeson, *J. Mol. Biol.*, 1976, **102**, 27.
15. S. E. V. Phillips, *Nature (London)*, 1978, **273**, 247.
16. R. H. Kretsinger and C. E. Nockolds, *J. Biol. Chem.*, 1973, **248**, 3313.
17. P. M. Colman, H. C. Freeman, J. M. Guss, M. Murata, V. A. Norris, J. A. M. Ramshaw and M. P. Venkatappa, *Nature (London)*, 1978, **272**, 319.
18. E. T. Adman, L. C. Sieker, L. H. Jensen, M. Bruschi and J. Legall, *J. Mol. Biol.*, 1977, **112**, 113.
19. K. M. Beem, D. C. Richardson and K. V. Rajagopalan, *Biochemistry*, 1977, **16**, 1930.
20. B. W. Matthews, L. H. Weaver and W. R. Kester, *J. Biol. Chem.*, 1974, **249**, 8030.
21. W. Bode and P. Schwager, *J. Mol. Biol.*, 1975, **98**, 693.

substitution of Ca^{2+} by paramagnetic metal ions with essentially filled shell configurations, *i.e.* the lanthanides.[134,135] For the smaller Mg^{2+} ion Mn^{2+} is found to be a better substitute than the larger lanthanide ions.

20.2.4.1.2 Copper[136–138]

Because of its amenability to a range of spectroscopic techniques the binding of copper to proteins has received considerable attention over the past 20 years. Two sets of published conference proceedings summarize the earlier[136] and more recent[137] findings. Two sets of the volumes edited by Sigel[1] have also been devoted to copper in its biological complexes.[138]

In the copper-containing enzymes three classes of Cu sites are recognized. *Type 1*, Cu^{II}, is characterized by an intense absorption band at *ca.* 600 nm and an unusually small hyperfine coupling constant in its ESR spectrum. This large visible absorption gives rise to the intense blue colour of many of the animal and plant Cu enzymes. The unusual spectroscopic features are attributed to a combination of (Cys)S → Cu^{II} charge transfer transition and a tetrahedral coordination geometry. A near tetrahedral geometry and the S(Cys) ligands have been definitely confirmed for azurin and plastocyanin (see Table 11). The nature of the coordination site ensures a minimal activation energy for the electron transfer ($Cu^{II} \leftrightarrow Cu^{I}$) reactions in which these enzymes are involved. Many low-molecular-weight 'model' compounds have been synthesized to mimic the spectroscopic features of the type 1 Cu sites (*e.g.* ref. 139).

Type 2, Cu^{II}, present in multicopper oxidases, refers to those Cu^{II} ions which give rise to more usual visible and ESR spectra. This means that their visible absorption is generally too weak to be observed, although their ESR spectra are just observable. The type 2 Cu ions work in conjunction with the type 1 in the overall electron transfer process.

Type 3, Cu, also present in multicopper oxidases, is the 'ESR-silent' Cu ions, either Cu^{I} or binuclear coupled Cu^{II} ions. Both may be present in some cases, *e.g.* ceruloplasmin.[140] The binuclear centres give rise to intense absorptions around 330 nm and may be dioxygen coordination sites.

In all these classes His residues are involved in the Cu binding. A His residue also bridges Cu and Zn ions in superoxide dismutase (see Table 11) and is implicated between Cu^{II} and Fe haem in mitochondrial cytochrome *c* oxidase.

20.2.4.1.3 Iron[141,142]

Iron exhibits the widest range of types of coordination in proteins of any metal, involving both prosthetic groups and amino acid side chains. The coordination and hydrolytic properties of its two main oxidation states, II and III, are substantially different and this is reflected in their modes of binding and usage in proteins. Iron(III) occurs in the transport protein, transferrin, and in the storage protein, ferritin. In the former protein there are two specific Fe^{III} sites; the donor atoms are not known with certainty although there is strong evidence for tyrosine involvement, as there is in other proteins involving Fe^{III}.[143] Ferritin consists of a polymeric Fe^{III} hydrated oxide-phosphate core, containing upto 4500 Fe atoms, encased in a protein shell.

Fe^{II} in proteins mostly occurs complexed to haem prosthetic group (26), as in haemoglobin, myoglobin, the cytochrome enzymes and others. The haem moiety is covalently linked to the protein. The Fe^{II} is thus coordinated to the four coplanar N(haem) atoms, and the two remaining axial coordination sites can be either singly or both occupied by protein His residues. The extensive delocalization over the porphyrin ring aids electron transfer to and from the iron centre (as in the cytochromes) and stabilizes Fe^{II} against oxidation when reversible coordination to dioxygen occurs

(26)

(as in haemoglobin and myoglobin). A number of haemoglobin and myoglobin structures are now known from X-ray structural analyses (see Table 11).

The other major group of, formally, Fe^{II} proteins comprises the *iron–sulfur* group,[144,145] a ubiquitous range of relatively small proteins. A group of larger-sized proteins, the hydrogenases, also comes within this group. The commonality of these proteins is that their active sites comprise a near-tetrahedral FeS$_4$ chromophoric group, where the S donor atoms come from Cys residues and sulfide ions. The proteins can be classified according to the type of FeS$_4$ core: (27), (28) and (29). Rubredoxin, the simplest of these proteins, contains core (27), which contains no labile sulfur, a description referring to coordinated S^{2-} ion which can be liberated as H$_2$S by reacting with acid. Plant ferredoxins contain core (28), whilst other ferredoxins contain one or more of the cubane clusters (29). The function of these proteins is to act as electron transfer reagents (*e.g.* in photosynthetic bacteria), odd electrons being delocalized over the FeS$_4$ groups. Crystal structures of some of these proteins have been reported (Table 11). Fe$_4$S$_4$ clusters linked to Mo atoms by S bridging are at the active site of the enzyme nitrogenase, which catalyzes the conversion of atmospheric dinitrogen to ammonia.[146]

(27)   (28)   (29)

20.2.4.1.4 Manganese[147]

Mn is an essential element and occurs in a number of enzymes and is also known to stimulate enzymic activity by reversible association. ESR and NMR line broadening are particularly useful techniques for studying Mn^{II} environments in proteins.[148] These measurements and crystal structure studies (concanavalin, Table 11) suggest near regular octahedral coordination by O and N donors.

Surprisingly, too, there are claims of higher oxidation states of Mn in some systems, *e.g.* Mn^{IV} in photosynthetic(II) chloroplast systems and Mn^{III} in acid phosphatase. In the latter enzyme Tyr and Cys residues appear to form part of the metal-binding site. The metal is also involved in the phosphate binding. While superoxide dismutase (SOD) is more generally found with Cu and Zn as the active metals, an Mn–SOD form is found in certain bacteria. The Mn oscillates between different oxidation states in its catalytic activity.[149]

20.2.4.1.5 Molybdenum[150,151]

A number of Mo-containing enzymes are now known, their principal roles being redox reactions in which Mo acts as a multielectron donor or acceptor. Studies of the Mo sites are difficult because other prosthetic groups are present (*e.g.* Fe-heme, or FeS clusters) which hinder spectroscopic examination and it is not possible to replace the tightly bound Mo atoms with other transition metal ions. Hence, only limited knowledge is so far available. EXAFS appears to be a promising technique; for example, it suggests that in sulfide oxidase the Mo is in an octahedral environment which includes two oxo ligands and two or three thiolate groups.[152] Thiolate ligands are also implicated in other Mo enzymes. In the reduced forms of these enzymes, Mo^V or M^{IV}, these ligands appear to be protonated.

20.2.4.1.6 Zinc[153,154]

More Zn-containing enzymes are known, some 200, than for any other metal. Only a handful of these have been reasonably characterized, and some X-ray structural studies are given in Table 11.

These enzymes, which mainly catalyze hydrolytic reactions, have the zinc ions at their active sites. However, Zn ions also appear necessary in some cases for stabilization of the protein structure, *e.g.* in Cu/Zn SOD, insulin, liver alcohol dehydrogenase and alkaline phosphatases.

Zn^{II} is spectroscopically silent and therefore much of the solution work has utilized the specific replacement of the Zn^{II} by transition metal ions, most noticeably Co^{II}, which is of similar size and coordination properties. Vallee and his co-workers have been particularly successful in this approach.[155,156]

Finally, mention must be made of a group of proteins known as metallothioneins.[157] These are low-molecular-weight proteins with a large number of the amino acid residues being cysteine (30–35%). Histidine and the 'aromatic' amino acids are absent. These proteins strongly bind up to 6–10% by weight of metal ions (approximately one metal ion per three Cys residues). The absorption spectra of the metallothioneins are dominated by S → M charge transfer bands. Spectroscopic studies on the Co^{II} and other metal ion derivatives suggest tetrahedral MS_4 coordination sites, which is supported by an EXAFS study on the Zn protein.[158] The tetrahedral symmetry and 1:3 M:Cys ratio strongly suggest S bridging of the metal ions with probable metal ion–sulfur cluster formation. Support for this comes from evidence of metal–metal interactions in ^{113}Cd NMR and Co^{II} ESR studies.

20.2.5 ADDENDUM

Some recent texts and reviews of general relevance are the following: 'Complex formation between palladium(II) and amino acids, peptides and related ligands', by L. D. Pettit and M. Bezer, *Coord. Chem. Rev.*, 1985, **61**, 97; 'Metalloproteins' (Part 1 — those with a redox role, Part 2 — non-redox systems), by P. M. Harrison, McMillan, London, 1985; 'Copper Proteins and Copper Enzymes', ed. R. Lontie, CRC Press, Boca Raton, Florida, 1984, vols. 1–3; 'Zinc Enzymes', ed. T. G. Spiro, Wiley, New York, 1983; 'Long range electron transfer in peptides and proteins', by S. S. Isied, *Prog. Inorg. Chem.*, 1984, **32**, 443; 'Bioinorganic applications of MCD spectroscopy' (metal-protein studies), by D. M. Dooley and J. H. Dawson, *Coord. Chem. Rev.*, 1984, **60**, 1. The proceedings of the second international conference on bioinorganic chemistry, held in Portugal in 1985, have also been published (*Rev. Port. Quim.*, 1985, **27**).

20.2.6 REFERENCES

1. 'Metal Ions in Biological Systems', ed. H. Sigel, Dekker, New York, 1974, vol. 1; 1982, vol. 14.
2. 'Amino Acids, Peptides and Proteins', ed. R. C. Sheppard, Royal Society of Chemistry Specialist Periodical Reports, London, 1969, vol. 1; 1983, vol. 14.
3. C. W. Childs and D. D. Perrin, *J. Chem. Soc. (A)*, 1969, 1039.
4. T. P. A. Kruck and B. Sarkar, *Can. J. Chem.*, 1973, **51**, 3549.
5. A. M. Corrie, M. L. D. Touche and D. R. Williams, *J. Chem. Soc., Dalton Trans.*, 1973, 2561.
6. A. Gergely and T. Kiss, *Inorg. Chim. Acta*, 1976, **16**, 51.
7. A. E. Martell and R. M. Smith, 'Critical Stability Constants', Plenum, New York, 1974, vol. 1; 1982, vol. 5.
8. J. J. Christensen, L. D. Hansen and R. M. Izatt, 'Handbook of Proton Ionization Heats and Related Thermodynamic Quantities', Wiley, New York, 1978.
9. H. C. Freeman, *Adv. Protein Chem.*, 1967, **22**, 257.
10. R. B. Martin, *Met. Ions Biol. Syst.*, 1979, **9**, 1.
11. G. R. Brubaker and D. H. Busch, *Inorg. Chem.*, 1966, **5**, 2110.
12. R. D. Gillard, in 'Physical Methods in Advanced Inorganic Chemistry', ed. H. A. O. Hill and P. Day, Interscience, London, 1968, p. 167.
13. R. B. Martin, *Met. Ions Biol. Syst.*, 1974, **1**, 129.
14. E. W. Wilson, M. H. Kasperian and R. B. Martin, *J. Am. Chem. Soc.*, 1970, **92**, 5365.
15. C. C. McDonald and W. D. Phillips, *J. Am. Chem. Soc.*, 1963, **85**, 3736.
16. G. D. Smith, R. E. Barden and S. L. Holt, *J. Coord. Chem.*, 1978, **8**, 157.
17. S. T. Chow and C. A. McAuliffe, *Prog. Inorg. Chem.*, 1975, **19**, 51.
18. R. J. Sundberg and R. B. Martin, *Chem. Rev.*, 1974, **74**, 471.
19. R. Gulka and S. S. Isied, *Inorg. Chem.*, 1980, **19**, 2842.
20. A. Kayali and G. Berthon, *Polyhedron*, 1982, **1**, 371.
21. L. Casella and M. Gullotti, *J. Inorg. Biochem.*, 1983, **18**, 19.
22. C. A. McAuliffe and S. G. Murray, *Inorg. Chim. Acta Rev.*, 1972, **6**, 103.
23. A. Gergely and I. Sovage, *Met. Ions Biol. Syst.*, 1979, **9**, 77.
24. P. J. M. W. L. Birker, *J. Chem. Soc., Chem. Commun.*, 1980, 946.
25. S. Bunel, C. Ibarra and A. Urbana, *Inorg. Nucl. Chem. Lett.*, 1977, **13**, 259.
26. H. Kozlowski, G. Formicka-Kozlowski and B. Jezowska-Trzebiatowska, *Bull. Acad. Pol. Sci., Ser. Sci. Chim.*, 1978, **26**, 153.

27. D. B. McCormick and H. Sigel, *Met. Ions Biol. Syst.*, 1974, **1**, 214.
28. L. D. Pettit, K. F. Siddique, H. Kozlowski and T. Kowalik, *Inorg. Chim. Acta*, 1981, **55**, 87.
29. H. Kozlowski and T. Kowalik, *Inorg. Chim. Acta*, 1979, **34**, L231.
30. C. A. Evans, R. Guevremont and D. L. Rabenstein, *Met. Ions Biol. Syst.*, 1979, **9**, 41.
31. A. Gergely, I. Nagypal and E. Farkas, *J. Inorg. Nucl. Chem.*, 1975, **37**, 551.
32. M.-C. Lim, *J. Chem. Soc., Dalton Trans.*, 1977, 1398.
33. L. D. Pettit and J. L. M. Swash, *J. Chem. Soc., Dalton Trans.*, 1976, 2416.
34. P. Grenouillet, R.-P. Martin, A. Rossi and M. Ptak, *Biochim. Biophys. Acta*, 1975, **322**, 185.
35. R. D. Gillard, R. J. Lancashire and P. O'Brien, *Transition Met. Chem. (Weinheim, Ger.)*, 1980, **5**, 340.
36. A. Gergely and T. Kiss, *Met. Ions Biol. Syst.*, 1979, **9**, 143.
37. T. Kiss and A. Gergelyl, *Inorg. Chim. Acta*, 1983, **78**, 247.
38. R.-P. Martin, M. M. Petit-Ramel and J. P. Scharff, *Met. Ions Biol. Syst.*, 1973, **2**, 1.
39. H. Sigel, *Met. Ions Biol. Syst.*, 1973, **2**, 64.
40. H. Sigel, K. H. Scheller, U. K. Haring and R. Malini-Balakrishnan, *Inorg. Chim. Acta*, 1983, **79**, 277.
41. R. Tribolet, H. Sigel and K. Trefzer, *Inorg. Chim. Acta*, 1983, **79**, 278.
42. H. S. Maslen and T. N. Waters, *Coord. Chem. Rev.*, 1975, **17**, 137.
43. D. E. Metzler, M. Ikawa and E. E. Small, *J. Am. Chem. Soc.*, 1954, **76**, 648.
44. D. L. Leussing, *Met. Ions Biol. Syst.*, 1976, **5**, 1.
45. F. Jursik and B. Hajek, *Inorg. Chim. Acta*, 1975, **13**, 169.
46. L. G. McDonald, D. H. Brown and W. E. Smith, *Inorg. Chim. Acta*, 1979, **33**, L183.
47. L. G. McDonald, D. H. Brown and W. E. Smith, *Inorg. Chim. Acta*, 1982, **63**, 213.
48. R. D. Gillard, *Inorg. Chim. Acta Rev.*, 1967, **1**, 69.
49. R. G. Denning and T. S. Piper, *Inorg. Chem.*, 1966, **5**, 1056.
50. M. Watabe, H. Yano and S. Yoshikawa, *Bull. Chem. Soc. Jpn.*, 1979, **52**, 61.
51. G. Brookes and L. D. Pettit, *J. Chem. Soc., Dalton Trans.*, 1977, 1918.
52. K. Bernauer, *Top. Curr. Chem.*, 1976, **65**, 1.
53. L. D. Pettit and R. J. W. Hefford, *Met. Ions Biol. Syst.*, 1979, **9**, 173.
54. J. F. Blount, H. C. Freeman, A. M. Sargeson and K. R. Turnbull, *Chem. Commun.*, 1967, 324.
55. A. A. Kurganov, L. Y. Zhuchkova and V. A. Davenkov, *Izv. Akad. Nauk SSSR, Ser. Khim.*, 1977, **11**, 2540.
56. V. A. Davenkov, *J. Chem. Soc., Chem. Commun.*, 1972, 1328.
57. H. Kozlowski, B. Decock-Le-Reverend, J.-L. Delaruelle and C. Loucheux, *Inorg. Chim. Acta*, 1983, **78**, 31.
58. K. Harada and W.-W. Tso, *Bull. Chem. Soc. Jpn.*, 1972, **45**, 2859.
59. T. Sakurai, O. Yamauchi and A. Nakahara, *J. Chem. Soc., Chem. Commun.*, 1977, 718.
60. R. D. Gillard, P. R. Mitchell and H. L. Roberts, *Nature (London)*, 1968, **217**, 949.
61. Y. Fujii, M. Sano and Y. Nakano, *Bull. Chem. Soc. Jpn.*, 1977, **50**, 26.
62. P. Amico, P. G. Daniele, G. Arena, G. Ostacoli, E. Rizzarelli and S. Sammartano, *Inorg. Chim. Acta*, 1979, **35**, L383.
63. R. D. Gillard and C. Thorpe, *Chem. Commun.*, 1970, 997.
64. C. Gilon, R. Lesham, Y. Tapuhi and E. Grushka, *J. Am. Chem. Soc.*, 1979, **101**, 7612.
65. M.-J. Jun, N. M. Yoon and C. F. Liu, *J. Chem. Soc., Dalton Trans.*, 1983, 999.
66. M. D. Fryzuk and B. Bosnich, *J. Am. Chem. Soc.*, 1979, **101**, 3043.
67. P. A. MacNeil, N. K. Roberts and B. Bosnich, *J. Am. Chem. Soc.*, 1981, **103**, 2273.
68. K. Achiwa and T. Soga, *Tetrahedron Lett.*, 1978, 1119.
69. A. Pasini and L. Casella, *J. Inorg. Nucl. Chem.*, 1974, **36**, 2133.
70. D. A. Phipps, *J. Mol. Catal.*, 1979, **5**, 81.
71. J. E. Letter and R. B. Jordan, *J. Am. Chem. Soc.*, 1975, **97**, 2381.
72. M. Sato, K. Okawa and S. Akabori, *Bull. Chem. Soc. Jpn.*, 1957, **30**, 937.
73. J. C. Dabrowiak and D. W. Cooke, *Inorg. Chem.*, 1975, **14**, 1305.
74. R. D. Gillard, P. O'Brien, P. R. Norman and D. A. Phipps, *J. Chem. Soc., Dalton Trans.*, 1977, 1988.
75. G. G. Smith, A. Khatib and G. S. Reddy, *J. Am. Chem. Soc.*, 1983, **105**, 293.
76. R. W. Hay and P. J. Morris, *Met. Ions Biol. Syst.*, 1976, **5**, 174.
77. R. B. Martin, *J. Am. Chem. Soc.*, 1967, **89**, 2501.
78. M. D. Alexander and D. H. Burch, *Inorg. Chem.*, 1966, **5**, 602.
79. J. P. Collman and E. Kimura, *J. Am. Chem. Soc.*, 1967, **89**, 6096.
80. D. A. Buckingham, L. G. Marzilli and A. M. Sargeson, *J. Am. Chem. Soc.*, 1967, **89**, 2772.
81. A. D. Gilmour and A. McAuley, *J. Chem. Soc. (A)*, 1970, 1006.
82. J. E. Gorton and R. F. Jameson, *J. Chem. Soc. (A)*, 1968, 2615.
83. H. Sigel and R. B. Martin, *Chem. Rev.*, 1982, **82**, 385.
84. R. B. Martin, *Met. Ions Biol. Syst.*, 1974, **1**, 129.
85. H. C. Freeman, *Adv. Protein Chem.*, 1967, **22**, 257.
86. D. L. Rabenstein, *Can. J. Chem.*, 1971, **49**, 3767.
87. H. Kozlowski and Z. Siatecki, *Chem. Phys. Lett.*, 1978, **54**, 498.
88. I. Sovago, E. Farkas and A. Gergely, *J. Chem. Soc., Dalton Trans.*, 1982, 2159.
89. A. Gergely and E. Farkas, *J. Chem. Soc., Dalton Trans.*, 1982, 381.
90. T. Kowalik, H. Kozlowski and B. Decock-Le-Reverend, *Inorg. Chim. Acta*, 1982, **67**, L39.
91. K. S. N. Iyer, S.-J. Lau, S. H. Laurie and B. Sarkar, *Biochem. J.*, 1978, **169**, 61.
92. Y. Sugiura and Y. Hirayama, *J. Am. Chem. Soc.*, 1977, **99**, 1581.
93. Y. Sugiura, *Inorg. Chem.*, 1978, **17**, 2176.
94. B. Decock-Le-Reverend, C. Loucheux, T. Kowalik and H. Kozlowski, *Inorg. Chim. Acta*, 1982, **66**, 205.
95. G. Formicka-Kozlowska, H. Kozlowski, I. Z. Siemion, K. Sobczyk and E. Nawrocka, *J. Inorg. Biochem.*, 1981, **15**, 201.
96. G. Formicka-Kozlowska, D. Konopinska and H. Kozlowski, *Inorg. Chim. Acta*, 1983, **78**, L47.
97. D. W. Margerum and G. D. Owens, *Met. Ions Biol. Syst.*, 1981, **12**, 75.
98. Y. Kojima, *Chem. Lett.*, 1981, 61.

99. B. Bartman, C. M. Deber and E. R. Blout, *J. Am. Chem. Soc.*, 1977, **99**, 1028.
100. V. Madison, C. M. Deber and E. R. Blout, *J. Am. Chem. Soc.*, 1977, **99**, 4788.
101. D. Baron, L. G. Pease and E. R. Blout, *J. Am. Chem. Soc.*, 1977, **99**, 8299.
102. H. J. Moeschler, D. F. Sargent, A. Tun-Kyi and R. Schwyzer, *Helv. Chim. Acta*, 1979, **62**, 2442.
103. K. S. Iyer, J. P. Laussac, S.-J. Lau and B. Sarkar, *Int. J. Pept. Protein Res.*, 1981, **17**, 549.
104. Y. Nakao and A. Nakahara, *Bull. Chem. Soc. Jpn.*, 1973, **46**, 187.
105. O. Yamauchi, J. Nakao and A. Nakahara, *Bull. Chem. Soc. Jpn.*, 1973, **46**, 2119.
106. G. Brookes and L. D. Pettit, *J. Chem. Soc., Dalton Trans.*, 1975, 2302.
107. R. D. Gillard and A. Spencer, *Discuss. Faraday Soc.*, 1968, **46**, 213.
108. L. V. Boas, C. A. Evans, R. D. Gillard, P. R. Mitchell and D. A. Phipps, *J. Chem. Soc., Dalton Trans.*, 1979, 582.
109. D. W. Margerum and G. R. Dukes, *Met. Ions Biol. Syst.*, 1974, **1**, 158.
110. D. A. Buckingham, C. E. Davis, D. M. Foster and A. M. Sargeson, *J. Am. Chem. Soc.*, 1970, **92**, 5571.
111. S. K. Oh and C. B. Storm, *Bioinorg. Chem.*, 1973, **3**, 89.
112. M. D. Fenn and J. H. Bradbury, *Anal. Biochem.*, 1972, **49**, 498.
113. R. H. Andreatta, H. C. Freeman, A. V. Robertson and R. L. Sinclair, *Chem. Commun.*, 1967, 203.
114. A. Nakahara, K. Hamada, I. Miyachi and K. Sakurai, *Bull. Chem. Soc. Jpn.*, 1967, **40**, 2826.
115. T. Sakurai, Y. Nakao and A Nakahara, *J. Inorg. Nucl. Chem.*, 1980, **42**, 1673.
116. S. Suzuki, H. Narita and K. Harada, *J. Chem. Soc., Chem. Commun.*, 1979, 29.
117. W. R. Harris and A. E. Martell, *J. Am. Chem. Soc.*, 1977, **99**, 6746.
118. E. Farkas, I. Sovago and A. Gergely, *J. Chem. Soc., Dalton Trans.*, 1983, 1545.
119. C. E. Brown, D. W. Vidrine, R. Czernuszewicz and K. Nakamoto, *J. Inorg. Biochem.*, 1982, **17**, 247.
120. P. De Meester and D. J. Hodgson, *J. Am. Chem. Soc.*, 1976, **98**, 7086.
121. D. W. Margerum and G. D. Owens, *Met. Ions Biol. Syst.*, 1981, **12**, 75.
122. R. J. P. Williams, *Chem. Br.*, 1983, **19**, 1009.
123. S. H. Laurie, *J. Inherited Metab. Dis.*, 1983, **6**, Suppl. 1, 9.
124. T. A. Kaden, *Met. Ions Biol. Syst.*, 1974, **4**, 2.
125. J. E. Roberts, T. G. Brown, B. M. Hoffman and J. Peisach, *J. Am. Chem. Soc.*, 1980, **102**, 825.
126. J. A. Larrabee and T. G. Spiro, *J. Am. Chem. Soc.*, 1980, **102**, 4210.
127. E. Munck and P. M. Champion, *Ann. N.Y. Acad. Sci.*, 1975, **244**, 142.
128. R. A. Dwek, R. J. P. Williams and A. V. Xavier, *Met. Ions Biol. Syst.*, 1974, **4**, 62.
129. S. P. Cramer and K. O. Hodgson, *Prog. Inorg. Chem.*, 1979, **25**, 1.
130. M. Epstein, J. Reuben and A. Levitzki, *Biochemistry*, 1977, **16**, 2449.
131. W. de W. Horrocks, G. F. Schmidt, D. R. Sudnick, C. Kittrell and R. A. Bernheim, *J. Am. Chem. Soc.*, 1977, **99**, 2378.
132. B. Sarkar, *Met. Ions Biol. Syst.*, 1981, **12**, 233.
133. R. H. Wasserman, R. A. Corradino, E. Carafoli, R. H. Kretsinger, D. H. Maclennan and F. L. Seigel (eds.), 'Calcium Binding Proteins and Calcium Function', Elsevier, New York, 1977.
134. R. B. Martin and D. Richardson, *Q. Rev. Biophys.*, 1979, **12**, 181.
135. J. Reuben, in ref. 133, p. 21.
136. J. Peisach, P. Aisen and W. E. Blumberg (eds.), 'Biochemistry of Copper', Academic, New York, 1966.
137. 'Biological Roles of Copper', CIBA Foundation Symposium No. 79, Excerpta Medica, Amsterdam, 1980.
138. Ref. 1, 1981, vol. 12; 1981, vol. 13.
139. U. Sakaguchi and A. W. Addison, *J. Am. Chem. Soc.*, 1977, **99**, 5189.
140. S. H. Laurie and E. S. Mohammed, *Coord. Chem. Rev.*, 1980, **33**, 279.
141. A. Jacobs and M. Worwood (eds.), 'Iron in Biochemistry and Medicine', Academic, New York, 1974, vol. 1; 1980, vol. 2.
142. Ref. 1, 1978, vol. 7.
143. L. Que, *Coord. Chem. Rev.*, 1983, **50**, 73.
144. W. Lovenberg (ed.), 'Iron–Sulfur Proteins', Academic, New York, 1973, vol. 1; 1977, vol. 2; 1979, vol. 3.
145. R. Malkin and A. J. Bearden, *Coord. Chem. Rev.*, 1979, **28**, 1.
146. R. W. F. Hardy (ed.), 'Dinitrogen Fixation', Wiley-Interscience, New York, 1978.
147. A. R. McEuen, in 'Inorganic Biochemistry', ed. H. A. O. Hill, Royal Society of Chemistry Specialist Periodical Reports, 1981, vol. 2, p. 249.
148. N. Niccolai, E. Tiezzi and G. Valensin, *Chem. Rev.*, 1982, **82**, 359.
149. G. D. Lawrence and D. T. Sawyer, *Biochemistry*, 1979, **18**, 3045.
150. M. P. Coughlan (ed.), 'Molybdenum and Molybdenum Containing Enzymes', Pergamon, Oxford, 1980.
151. E. I. Stiefel, *Prog. Inorg. Chem.*, 1977, **22**, 1.
152. S. P. Cramer, R. Wahl and K. V. Rajagopalan, *J. Am. Chem. Soc.*, 1981, **103**, 7721.
153. G. L. Fisher, *Sci. Total Environ.*, 1975, **4**, 373.
154. J. E. Chlebowski and J. E. Coleman, *Met. Ions Biol. Syst.*, 1976, **6**, 1.
155. T. A. Kaden, B. Holmquist and B. L. Vallee, *Biochem. Biophys. Res. Commun.*, 1972, **46**, 1654.
156. K. F. Geoghegan, B. Holmquist, C. A. Spilburg and B. L. Vallee, *Biochemistry*, 1983, **22**, 1847.
157. M. Vasak, J. H. R. Kagi and H. A. O. Hill, *Biochemistry*, 1981, **20**, 2852.
158. C. D. Garner, S. S. Hasnain, I. Bremner and J . Bordas, *J. Inorg. Biochem.*, 1982, **16**, 253.

20.3

Complexones

GIORGIO ANDEREGG
ETH, Zurich, Switzerland

20.3.1 INTRODUCTION

The name complexone was introduced in 1945 by G. Schwarzenbach[1] for a series of organic ligands, containing normally at least one iminodiacetic group [—N(CH₂CO₂H)₂] or two amino-acetic groups (—NHCH₂CO₂H), which form stable complexes with almost all cations. Two of these substances were already known at that time under the names of Trilon A and B and were used to eliminate hardness of water due to calcium and magnesium ions without their separation. They had also found other applications in textile and photographic processing as well as in the paper industry. Schwarzenbach showed that in solution the aminopolycarboxylate anions are able to bind calcium and other cations strongly, sometimes so strongly that they cannot be detected by the usual classical precipitation or colorimetric reagents.

The high values for the stability constants of the complexes formed by these ligands are due to the presence of the basic secondary or tertiary amino groups and to the large negative charge of the ligand anions, as well as to the formation of stable five-membered chelate rings with the metal ions.

This chapter firstly gives some typical synthetic routes for the preparation of the organic ligands. These, classified on the basis of their structural features, are described in three subsections. For all ligands at least one reference is given which allows one to find the original literature and the synthesis used in the specific case. The extent of the description of each ligand and of its chelating properties is in direct relation on its actual importance in practice and in research. For complexones not included in this account, the easiest access to the relevant literature is *via* the use of the publications on stability constants.[2,3]

20.3.2 SYNTHESES

Nitrilotriacetic acid $N(CH_2CO_2H)_3$ (nta) was first obtained by Heintz in 1862 from an ammoniacal solution of monochloroacetic acid. Other products, *i.e.* glycolic, aminoacetic and iminodiacetic acids, were also formed. The next paper concerned with the preparation of aminopolycarboxylic acids appeared after 1930. A procedure similar to that given above, based on the treatment of ethylenediamine with monochloroacetic acid in presence of sodium hydroxide, was reported in 1935 by I.G.-Farbenindustrie.[5] In a more recent version of the chloroacetic acid process, reaction (1) is carried out at 40–90 °C. For the preparation of edta, good yields of the tetrasodium salt can be obtained using appropriate conditions. The conversion to the insoluble free acid is achieved by acidification with a mineral acid.

$$8NaOH + H_2CH_2CH_2NH_2 + 4ClCH_2CO_2H \rightarrow Na_4 \begin{bmatrix} {}^-O_2CCH_2 & CH_2CO_2^- \\ & NCH_2CH_2N \\ {}^-O_2CCH_2 & CH_2CO_2^- \end{bmatrix} + 4NaCl + 4H_2O \quad (1)$$

In 1937 I.G.-Farbenindustrie patented a method for the preparation of 'polyaminoacetonitriles, corresponding acids and their derivatives'.[6] The process, known as the hydrogen cyanide process, utilizes sodium cyanide in acid solution. The cyanomethylation takes place by treatment of the amine with formaldehyde and hydrogen cyanide. In the case of ethylenediamine reaction (2) takes place: ethylenediaminetetraacetonitrile (1) is isolated to ensure that the resulting tetrasodium salt, after hydrolysis, is not contaminated by by-products. This synthesis can be performed as a continuous process and is also used to obtain other complexones. The use of hydrogen cyanide and of an acid medium gives rise to corrosion and safety problems.

$$H_2NCH_2CH_2NH_2 + 4HCN + 4HCHO \rightarrow \quad \begin{matrix} NCCH_2 & CH_2CN \\ & NCH_2CH_2N \\ NCCH_2 & CH_2CN \end{matrix} + 4H_2O \quad (2)$$

$$(1)$$

A further method often used is the carboxymethylation developed by Bersworth.[7] It converts ethylenediamine to the tetrasodium salt of edta by simultaneous addition of sodium cyanide and formaldehyde to a sodium hydroxide solution of the diamine. The success of reaction (3) depends on the concurrent and complete removal of the ammonia liberated during the reaction. The eventual by-products, containing less than four acetic groups, can be separated from Na$_4$edta and returned to the reaction mixture. The process can be performed continuously. In the case of nta, the salt $Na_3nta \cdot H_2O$ is obtained after purification of the reaction product.

$$\begin{matrix} NCCH_2 & CH_2CN \\ & NCH_2CH_2N \\ NCCH_2 & CH_2CN \end{matrix} + 4H_2O + 4NaOH \rightarrow Na_4 \begin{bmatrix} {}^-O_2CCH_2 & CH_2CO_2^- \\ & NCH_2CH_2N \\ {}^-O_2CCH_2 & CH_2CO_2^- \end{bmatrix} + 4NH_3 \quad (3)$$

$$H_2NCH_2CH_2NH_2 + 4NaCN + 4HCHO \rightarrow Na_4 \begin{bmatrix} {}^-O_2CCH_2 & CH_2CO_2^- \\ & NCH_2CH_2N \\ {}^-O_2CCH_2 & CH_2CO_2^- \end{bmatrix} + 4NH_3 \quad (4)$$

Edta can also be obtained by catalytic oxidation of tetra(hydroxyethyl) ethylenediamine (2) by a process developed by the Carbide and Carbon Chemical Corporation (equation 5).[8,9] The amino alcohol is heated at 220–230 °C for several hours with sodium or potassium hydroxide and cadmium oxide as a catalyst. Hydrogen is liberated and Na$_4$edta is obtained.

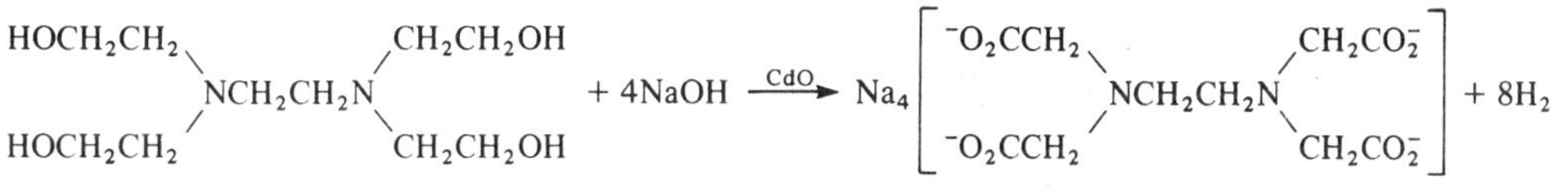

$$\underset{\text{HOCH}_2\text{CH}_2}{\overset{\text{HOCH}_2\text{CH}_2}{>}}\text{NCH}_2\text{CH}_2\text{N}\underset{\text{CH}_2\text{CH}_2\text{OH}}{\overset{\text{CH}_2\text{CH}_2\text{OH}}{<}} + 4\text{NaOH} \xrightarrow{\text{CdO}} \text{Na}_4\left[\underset{^-\text{O}_2\text{CCH}_2}{\overset{^-\text{O}_2\text{CCH}_2}{>}}\text{NCH}_2\text{CH}_2\text{N}\underset{\text{CH}_2\text{CO}_2^-}{\overset{\text{CH}_2\text{CO}_2^-}{<}}\right] + 8\text{H}_2 \quad (5)$$

(2)

Instead of monochloroacetic acid, the condensation with a primary amine can be done using methyl bromoacetate. This procedure was used for the syntheses of β-methoxyethyliminodiacetic acid: β-methoxyethylamine, methyl bromoacetate and potassium carbonate were reacted without solvent with formation of the corresponding ester (equation 6). This was separated, purified by distillation and hydrolyzed with a Ba(OH)$_2$ solution. The separation of the acid from the solution was made after quantitative removal of barium ions with sulfuric acid.[10]

$$\text{MeOCH}_2\text{CH}_2\text{NH}_2 + 2\text{BrCH}_2\text{CO}_2\text{Me} \xrightarrow{\text{K}_2\text{CO}_3} \text{MeOCH}_2\text{CH}_2\text{N}\underset{\text{CH}_2\text{CO}_2\text{Me}}{\overset{\text{CH}_2\text{CO}_2\text{Me}}{<}} + 2\text{KBr} + \text{CO}_2 + \text{H}_2\text{O} \quad (6)$$

Finally, *N*-(cyanomethyl)iminodiacetic acid was obtained from the ester of iminodiacetic acid and chloroacetonitrile as illustrated in Scheme 1.[10] The purified ester is hydrolyzed selectively at the methylester groups with zinc perchlorate in a 70% alcoholic solution and formation of the 1:1 complex (3) occurs. The substituted iminodiacetic acid is extracted with ether after zinc separation as ZnS by treatment with H$_2$S, and crystallized in a chloroform ethanol solution.

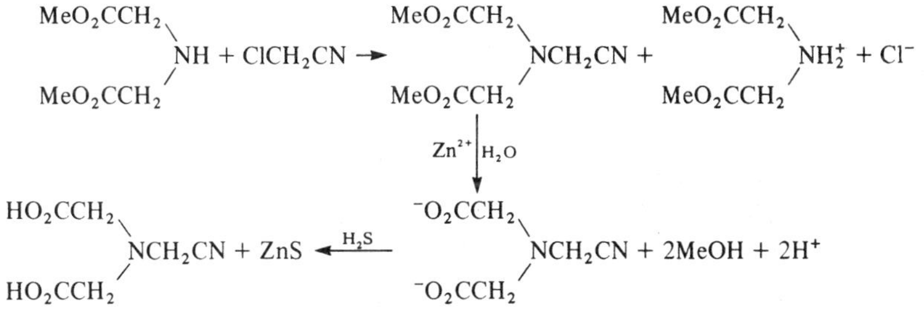

$$\underset{\text{MeO}_2\text{CCH}_2}{\overset{\text{MeO}_2\text{CCH}_2}{>}}\text{NH} + \text{ClCH}_2\text{CN} \rightarrow \underset{\text{MeO}_2\text{CCH}_2}{\overset{\text{MeO}_2\text{CCH}_2}{>}}\text{NCH}_2\text{CN} + \underset{\text{MeO}_2\text{CCH}_2}{\overset{\text{MeO}_2\text{CCH}_2}{>}}\text{NH}_2^+ + \text{Cl}^-$$

$$\Big\downarrow \text{Zn}^{2+}\Big|\text{H}_2\text{O}$$

$$\underset{\text{HO}_2\text{CCH}_2}{\overset{\text{HO}_2\text{CCH}_2}{>}}\text{NCH}_2\text{CN} + \text{ZnS} \xleftarrow{\text{H}_2\text{S}} \underset{^-\text{O}_2\text{CCH}_2}{\overset{^-\text{O}_2\text{CCH}_2}{>}}\text{NCH}_2\text{CN} + 2\text{MeOH} + 2\text{H}^+$$

(3)

Scheme 1

20.3.3 EDTA AND ITS HOMOLOGUES

These compounds are neutral acids and have the structure (4). The list of the substances already investigated comprises the terms for $n = 2$ (ethylenediaminetetraacetic acid, edta),[11] $n = 3$ (trimethylenediaminetetraacetic acid, tmta),[12] $n = 4$ (tetramethylenediaminetetraacetic acid, teta),[13] $n = 5$ (pentamethylenediaminetetraacetic acid, peta),[14] $n = 6$ (hexamethylenediaminetetraacetic acid, hdta) and $n = 8$ (octamethylenediaminetetraacetic acid, odta).[14]

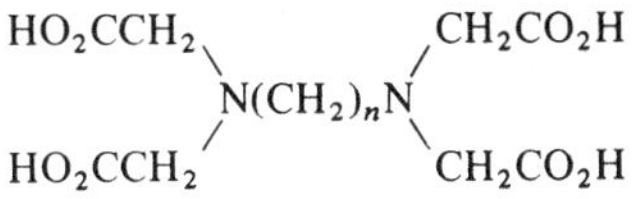

$$\underset{\text{HO}_2\text{CCH}_2}{\overset{\text{HO}_2\text{CCH}_2}{>}}\text{N(CH}_2)_n\text{N}\underset{\text{CH}_2\text{CO}_2\text{H}}{\overset{\text{CH}_2\text{CO}_2\text{H}}{<}}$$

(4)

20.3.3.1 Chelating Properties of edta

20.3.3.1.1 Acid–base equilibria of edta

The complexones are weak acids and therefore dissociate one or more protons in aqueous solution, depending on the pH of the solution. For them the general symbol H_pL is used, where p represents the number of dissociable protons which is identical with the number of carboxylic groups of the acid. It can be obtained experimentally from the titration of a solution of the neutral acid H_pL with a solution of strong base and measuring the pH value. The curve obtained in the case of edta is shown in Figure 1. It can be interpreted with four dissociation steps, as illustrated by Scheme 2. The given values for pK_1, pK_2, pK_3 and pK_4 were obtained in aqueous solution with an ionic strength I of 0.1 using KCl as inert salt and at a temperature of 20 °C.[16] They show that the first two protons are liberated between pH 1 and 4 and thus should be bound to two carboxylic groups. The other two values are much higher in magnitude and comparable to those of the ethylenediammonium cation, showing that in H_2L^{2-} and HL^{3-} the protons are bound to the N atoms as corroborated by IR spectra[17] in solution. An increase in the solubility of edta is found from pH 2 to pH 0, due to the formation of H_5L^+ and H_6L^{2+} (pK of $H_5L^+ = 1.4$ and that of $H_6L^{2+} = 0.2^{18}$).

$$H_4L \longrightarrow H^+ + H_3L^- \qquad K_1 = ([H][H_3L])/[H_4L] \qquad pK_1 = 1.99$$

$$H_3L^- \longrightarrow H^+ + H_2L^{2-} \qquad K_2 = ([H][H_2L])/[H_3L] \qquad pK_2 = 2.67$$

$$H_2L^{2-} \longrightarrow H^+ + HL^{3-} \qquad K_3 = ([H][HL])/[H_2L] \qquad pK_3 = 6.16$$

$$HL^{3-} \longrightarrow H^+ + L^{4-} \qquad K_4 = ([H][L])/[HL] \qquad pK_4 = 10.26$$

Scheme 2

20.3.3.1.2 *Formation of edta complexes*

As can be seen in Figure 1, the shape of the titration curve of a solution of the acid H_4L differs from that obtained in presence of an equal quantity of a metal ion, indicating that some reactions take place between edta and the cation. Let us consider the case of three different metal ions: Li^+, Mg^{2+} and Cu^{2+}, for which the stability constant $K_1 = [ML]/([M][L])$ is equal to $10^{2.8}$, $10^{8.7}$ and $10^{18.9}$ respectively ($I = 0.1\,M$ (KCl) and 20 °C).

For solutions with the same quantity of added strong base the formation of the complex is generally accompanied by a decrease of the pH with respect to the solution without metal ion. This can be explained assuming rections of types (7) and (8), as can be shown by a mathematical analysis of the experimental results. The curves begin to deviate from curve (1) of the acid H_4L when complex formation begins to take place. For Li^+ this is the case at pH $\approx$ 9. As can be seen in Figure 2, at this pH edta is present as HL^{3-} and it is found that only a small quantity of LiL^{3-} is formed. As magnesium(II) gives a complex which shows a stability constant six orders of magnitude larger with respect to that of Li^+, complex formation starts at pH $\approx$ 4, where the solutions contain mainly H_2L^{2-}. This situation can be followed using Figures 1 and 2. Copper(II) forms a complex with K_1 *ca.* 10^{10} times larger than that of magnesium(II); the complex is already formed mainly in the solution of the components before any base addition. The equimolar solution in the two components contains a large percentage of CuL^{2-} and $CuHL^-$, whereas the other ligand species and free

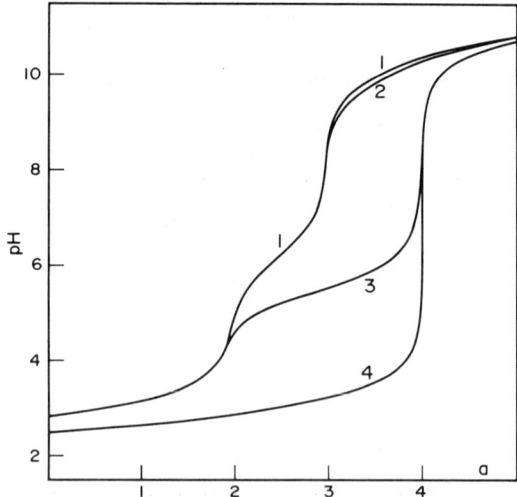

Figure 1 Titration curves for H_4edta. Curve 1: H_4edta; curve 2: H_4edta + lithium; curve 3: H_4edta + magnesium; curve 4: H_4edta + copper. The quantity a is in moles of strong base per mole H_4edta. Total ligand and metal concentration: 0.001 mol dm^{-3}

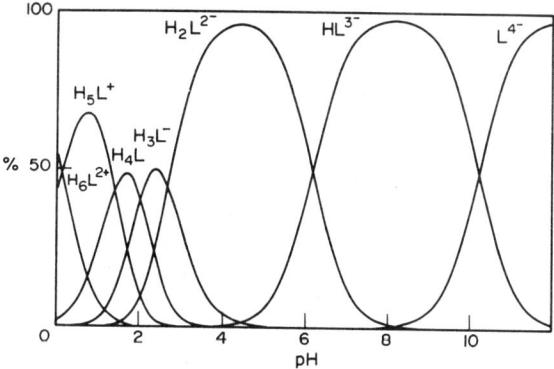

Figure 2 Distribution of species $H_pL^{(4-p)-}$ as a function of pH for curve 1 of Figure 1. The percentage of a species at a chosen pH value is obtained as the ordinate of the intersection of the vertical straight line with its curve

copper(II) are present in negligibly low concentrations. In the first two cases the experimental results of Figure 1 allow the determination of the equilibrium constants of the species formed and therefore the stability constant K_1 can be evaluated. This is not possible for copper(II) because the concentrations of Cu^{2+} and L^{4-} cannot be obtained exactly. In this case only the pK value of $CuHL^-$ can be obtained. This can also be understood by considering the graphical representation of the different edta species present *vs.* pH for Mg^{2+} and Cu^{2+} in the solutions used for the titrations shown in Figure 1. The percentage of the protonated species MHL^- reaches only 11% for magnesium(II) while it is 89.3% for copper (Figures 3 and 4).

$$HL^{3-} + M^{z+} \longrightarrow ML^{(4-z)-} \tag{7}$$

$$H_2L^{2-} + M^{z+} \longrightarrow MHL^{(3-z)-} \tag{8}$$

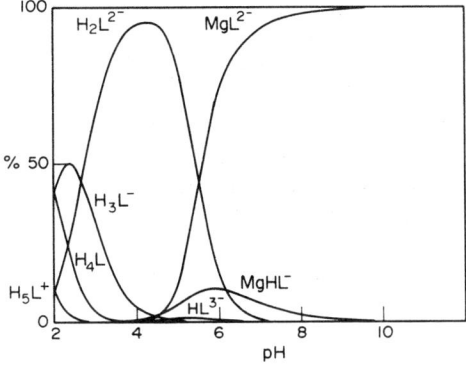

Figure 3 Distribution of species $H_pL^{(4-p)-}$ and $MgH_pL^{(2-p)-}$ for curve 3 of Figure 1 *vs.* pH. The percentage of a species at a chosen pH value is obtained as the ordinate of the intersection of the vertical straight line with its curve

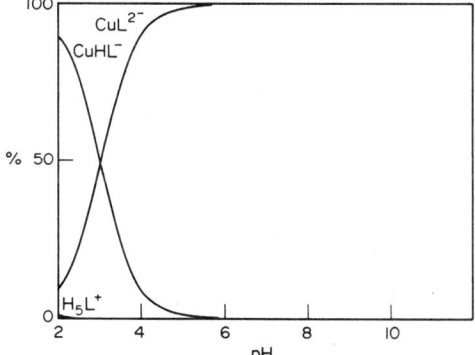

Figure 4 Distribution of species $H_pL^{(4-p)-}$ and $CuH_pL^{(2-p)-}$ for curve 4 of Figure 1 *vs.* pH. The percentage of a species at a chosen pH value is obtained as the ordinate of the intersection of the vertical straight line with its curve

A nearly complete list of the known stability constant K_1 of the edta complexes[11] is given in Table 1.

Table 1 Stability Constants K_1 of edta Complexes

Metal ion	T (°C)	I (M) and medium	$\log K_1$	Metal ion	T (°C)	I (M) and medium	$\log K_1$
Ag^+	20	0.1 (KNO_3)	7.22	MoO_3	25	0.1 ($NaClO_4$)	10
Al^{3+}	20	0.1 (KNO_3)	16.7	Na^+	20	0.1 (KCl)	1.66
Am^{3+}	20	0.1 (KNO_3)	16.9	Nd^{3+}	20	0.1(KNO_3)	16.6
Ba^{2+}	20	0.1 (KNO_3)	7.73	Ni^{2+}	20	0.1 (KNO_3)	18.67
Be^{2+}	20	0.1 ($KClO_4$)	9.27	NpO_2^+	25	0.1 (ClO_4^-)	7.33
Bi^{3+}	20	1 ($NaClO_4$)	25.7	Np^{4+}	25	1 ($HClO_4$)	24.55
Ca^{2+}	20	0.1 (KNO_3)	10.7	Pb^{2+}	20	0.1 (KNO_3)	18.3
Cd^{2+}	20	0.1 (KNO_3)	16.62	Pd^{2+}	20	1 (KBr)	25.5
Ce^{3+}	20	0.1 (KNO_3)	16.07	Pm^{3+}	20	0.1 (NH_4Cl)	16.96
Cf^{3+}	20	0.1 (NH_4NO_3)	17.9	Pr^{3+}	20	0.1 (KNO_3)	16.31
Cm^{3+}	20	0.1 (NH_4ClO_4)	17.1	Pu^{3+}	20	0.1 (KCl)	18.12
Co^{2+}	20	0.1 (KNO_3)	16.49	Pu^{4+}	20	1 (HNO_3)	25.75
Co^{3+}	25	0.1 (KNO_3)	41.5	Ra^{2+}	20	0.1 (NaCl)	7.07
Cr^{2+}	20	0.1 (KCl)	13.6	Rb^+	25	0.32 (CsCl)	0.59
Cr^{3+}	20	0.1 (KCl)	23.4	Sc^{3+}	20	0.1 (KNO_3)	23
Cs^+	25	0.32 (CsCl)	0.15	Sm^{3+}	20	0.1 (K^+)	17.0
Cu^{2+}	20	0.1 (KNO_3)	18.86	Sn^{2+}	20	1 ($NaClO_4$)	18.3
Dy^{3+}	20	0.1 (KNO_3)	18.20	Sr^{2+}	20	0.1 (KNO_3)	8.60
Er^{3+}	20	0.1 (KNO_3)	18.98	Tb^{3+}	20	0.1 (KNO_3)	17.83
Eu^{2+}	25	0.1 (KNO_3)	10	Th^{4+}	20	0.1 (KNO_3)	23.25
Eu^{3+}	20	0.1 (KNO_3)	17.22	Ti^{3+}	20	0.1 (ClO_4^-)	21.5
Fe^{2+}	20	0.1 (KCl)	14.3	TiO^{2+}	20	0.1 ($KClO_4$)	17.5
Fe^{3+}	20	0.1 (KCl)	25.1	Tl^+	20	0.1 (KNO_3)	6.54
Ga^{3+}	20	0.1 ($NaClO_4$)	20.18	Tl^{3+}	20	1 ($NaClO_4$)	37.8
Gd^{3+}	20	0.1 (KNO_3)	17.27	Tm^{3+}	20	0.1 (KNO_3)	19.6
Hg^{2+}	20	0.1 (KNO_3)	21.8	V^{4+}	25	0.1 (KNO_3)	25.8
Ho^{3+}	20	0.1 (KNO_3)	18.42	V^{2+}	20	0.1 (KCl)	12.7
In^{3+}	20	0.1 (KNO_3)	24.95	V^{3+}	20	0.1 (KCl)	25.9
K^+	25	0.1 (Me_4NCl)	0.55	VO^{2+}	20	0.1 (KCl)	18.76
La^{3+}	20	0.1 (KNO_3)	15.5	VO_2^+	20	0.1 (KCl)	15.55
Li^+	20	0.1 (KCl)	2.79	Y^{3+}	20	0.1 (KNO_3)	18.11
Lu^{3+}	20	0.1 (KNO_3)	20.03	Yb^{3+}	20	0.1 (KNO_3)	19.73
Mg^{2+}	20	0.1 (KNO_3)	8.65	Zn^{2+}	20	0.1 (KNO_3)	16.68
Mn^{2+}	20	0.1 (KNO_3)	13.95	Zr^{4+}	20	1 ($HClO_4$)	28.1

20.3.3.1.3 *Complexometric titrations*

The equilibrium between Mg^{2+} and the edta anion L^{4-} can be compared to that between NH_3 and H^+, because the corresponding equilibrium constants are very close in magnitude. In the latter case, it is possible to titrate NH_3 with a solution of a strong acid in order to determine quantitatively its total concentration. It is therefore quite evident that, based on the values of the equilibrium constants of Scheme 3, the quantitative determination of the Mg^{2+} using edta should be possible. Because the other cations form more stable complexes than Mg^{2+}, the complexometric titration should be of wide application. Some caution is necessary concerning the pH value at which the determination is done, because the ligand can be protonated, with consequent decrease of its chelating power. However, in the case of copper(II), its edta complex is already completely formed at pH 3 and therefore a titration is possible under these conditions.

$$Mg^{2+} + L^{4-} \longrightarrow MgL^{2-} \quad K = 10^{8.7}$$

$$NH_3 + H^+ \longrightarrow NH_4^+ \quad K = 10^{9.3}$$

Scheme 3

In the last 30 years more than 150 metallochromic indicators have been developed for complexometric titrations.[19] In some of them one notes the presence of the iminodiacetic group $RN(CH_2CO_2H)_2$ of high complexing capacity. Of particular interest are indicators with a good color change in neutral or acidic solutions. The possibility of selective complexometric determinations is strongly increased by change of the pH[20] of the solution as well as by use of masking agents.[21] In practice, the complexometric titration allows the determination of almost all metal ions (see Chapter 10 of this work, Volume 1).

20.3.3.1.4 *Composition of edta complexes*

Equilibrium studies between edta and metal ions have revealed that although the 1:1 complex $ML^{(4-z)-}$ is normally present over a wide pH range, some other species can also be formed (Table 2). In acidic solution at pH < 4 protonated species $MHL^{(4-p-z)-}$ are found. The more important $MHL^{(3-z)-}$ shows pK values which depend strongly on the charge of the metal ion.[22] In the specific cases of platinum(II)[23] and palladium(II)[24] it was possible to detect other species with more protons, *i.e.* MH_2L and MH_3L^+. Depending on the stability of the complex, these can also be formed with other cations.

Table 2 Hydrolytic M^{z+}–edta Species[11]

Species	Equilibrium	z	pK	Metal ions
$L^{(3-z)-}$	$MHL^{(3-z)-} \rightleftharpoons ML^{(4-z)-} + H^+$	1	6.5	Ag^+
	$MHL^{(3-z)-} \rightleftharpoons ML^{(4-z)-} + H^+$	2	3	Ca^{2+}, Mn^{2+}, Co^{2+}, Fe^{2+}, Cu^{2+}, Ni^{2+}, Hg^{2+}
	$MHL^{(3-z)-} \rightleftharpoons ML^{(4-z)-} + H^+$	3	1–2.5	Ln^{3+}, Al^{3+}, Ti^{3+}
$_2L^{(2-z)-}$	$MH_2L^{(2-z)-} \rightleftharpoons MHL^{(3-z)-} + H^+$	2	2.3	Pd^{2+}, Pt^{2+}
$_3L^{(1-z)-}$	$MH_3L^{(1-z)-} \rightleftharpoons MH_2L^{(2-z)-} + H^+$	2	0.9	Pd^{2+}, Pt^{2+}
$OH)L^{(5-z)-}$	$ML^{(4-z)-} \rightleftharpoons M(OH)L^{(5-z)-} + H^+$	2	9	Hg^{2+}
	$ML^{(4-z)-} \rightleftharpoons M(OH)L^{(5-z)-} + H^+$	3	4.4–10.6	Al^{3+}, Ga^{3+}, Tl^{3+}, Fe^{3+}, In^{3+}, Sc^{3+}, V^{3+}
	$ML^{(4-z)-} \rightleftharpoons M(OH)L^{(5-z)-} + H^+$	4	4.7–6.1	Zr^{4+}, Th^{4+}
$OH)_2L_2^{(10-z)-}$	$2ML^{(4-z)-} \rightleftharpoons M_2(OH)_2L_2^{(10-z)-} + 2H^+$	3	12.4	Fe^{3+}
		4	9.8	Th^{4+}

In basic solution the 1:1 complex can undergo partial hydrolysis forming $M(OH)L^{(5-z)-}$ or its dimer $M_2(OH)_2L_2^{(10-2z)-}$.[27] This is normally observed only for metal ions with the charge +3 or higher. As edta is only a hexadentate ligand, it cannot occupy all coordination sites of many metal ions. The formation of 1:2 complexes $ML_2^{(8-z)-}$ in solution, however, is uncertain.

20.3.3.2 Structure of edta Complexes in the Solid State

Structural information about edta complexes obtained by X-ray investigations in the last 20 years is summarized in Table 3, showing that different coordination numbers are observed. L represents the hexadentate ligand, HL and H_2L the respective mono- and di-protonated species in which the ligand is acting as penta- and ter-dentate, although a carboxylate group can sometimes be bound to a metal ion and to a proton at the same time. Because of this, the denticity of the ligand in the complex ions of Table 3 can be obtained by substracting the number of coordinated H_2O molecules from the number listed in the table. It appears that the coordination numbers of the metal ions are generally higher than six by further coordination of H_2O. The results of X-ray structure analyses of solid complexes have been successfully used to interpret the thermodynamic functions of complex formation in solution.[53] For this reason, it also seems appropriate to use the structures of Table 3 as representative for the complex ions in solution, although it has been general practice to formulate these 1:1 complexes as having an octahedral structure with hexadentate edta.

Table 3 The Coordination Numbers of Metal Ions in edta Complexes from X-Ray Structure Analyses

Coordination number	Complexes	Ref.
6	AlL^-, $Ga(H_2O)HL$, SbL^-, $Cr(H_2O)HL$, MnL^-, $Fe(H_2O)HL$	28–33
	CoL^{2-}, CoL^-, $Cu(H_2O)H_2L$, CuL, $Ni(H_2O)H_2L$, NiL^{2-}	34–39
7	$Sn(H_2O)L^{2-}$, $Mn(H_2O)L^{2-}$, $Fe(H_2O)L^-$, $Os(H_2O)L$, $Mg(H_2O)L^{2-}$	40–44
	$Cd(H_2O)L^{2-}$, $Pb(H_2O)L^{2-}$, $NaL(H_2O)^{3-}$	45–47
8	$Zr(H_2O)_2L$, $Yb(H_2O)_2L^-$, $Er(H_2O)_2L^-$, $Ca(H_2O)_2L^{2-}$	48–51
9	$La(H_2O)_3L^-$	52
10[a]	$La(H_2O)_4HL$	53

[a] The solid compound $Ba_2L \cdot 2.5H_2O$ shows a first Ba^{2+} bound to six donor atoms of edta, to three O atoms of a second edta and one O of an H_2O, and a second one bound to seven O atoms of four different edta anions and to one O of an H_2O, with the coordination numbers 10 and 8 respectively.

20.3.3.3. Complexes of edta Homologues

20.3.3.3.1 1:1 ML complexes

The stability constants of 1:1 complexes with a given metal ion show a decrease with the number n of methylene groups in (4) because of increase in size of the chelate ring between the two N atoms (Table 4). The stability constants of ML are given for the ligands with $n \leqslant 5$, because for the ligand with the highest n value the formation of polynuclear species sometimes occurs. As they have not been considered in the interpretation of the experimental measurements, some errors in the constants are possible. For the trend of the values of log K_1 vs. n, one notes that the cations can be divided into two groups, I and II. The first contains Fe^{2+}, Co^{2+}, Ni^{2+}, Cu^{2+} and Zn^{2+}; log K_2 decreases very slowly with an increase of n from 2 to 4, by an amount between 0.65 (Co^{2+}) and 1.47 (Cu^{2+}), followed by a sudden drop of 1 (Cu^{2+}), 2.2 (Fe^{2+}, Co^{2+}, Zn^{2+}) or 3.4 (Ni^{2+}) from $n = 4$ to $n = 5$. The second group comprises Mg^{2+}, Ca^{2+}, Ba^{2+}, Sr^{2+}, Mn^{2+}, Cd^{2+}, Pb^{2+}, La^{3+} and Fe^{3+} for which the largest decrease of log K_1 occurs on going from edta to tmta ($n = 3$). Its magnitude varies between 2.47 (Mg^{2+}) and 4.27 (La^{3+}). The values of log K_1 for the complexes of teta are generally lower than those of tmta: the difference lies between 0 (Mg^{2+}) and 3.25 (Pb^{2+}). A further increase of the ring size of one methylene group is accompanied by a decrease of log K_1 from 1 (Mg^{2+}) to 0.42 (Cd^{2+}). The values of K_1 for La^{3+} and Pb^{2+} with peta are not known.

Table 4 Equilibrium Constants at $I = 0.1$ and 20 °C

Metal ion	log K_1 of ML^{2-}				pK of MHL^-			
	edta	tmta	teta	peta	edta	tmta	teta	peta
Mg^{2+}	8.69	6.21	6.22	5.2	3.9	7.5	7.93	9.0
Ca^{2+}	10.7	7.28	5.66	5.2	3.1	6.34	8.68	9.2
Sr^{2+}	8.6	5.28	4.42	—	3.9	7.76	9.06	—
Ba^{2+}	7.76	3.95	3.77	—	4.57	8.72	9.47	—
Mn^{2+}	14.04	10.0	9.53	8.7	3.1	5.3	6.6	7.6
Fe^{2+}	14.2	13.42	13.27	10.8	2.8	3.34	< 3	5.26
Co^{2+}	16.11	15.55	15.66	13.38	3.0	2.4	< 3	5.8
Ni^{2+}	18.62	18.15	17.36	13.9	3.2	2.2	2.35	5.81
Cu^{2+}	18.8	18.92	17.33	16.24	3.0	2.2	4.19	5.85
Zn^{2+}	16.26	15.26	15.01	16.67	3.0	2.5	3.1	6.0
Cd^{2+}	16.46	13.90	12.02	11.6	2.9	3.06	5.5	8.46
Pb^{2+}	18.04	13.78	10.53	—	2.8	3.86	7.63	—
La^{3+}	15.5	11.23	9.13	—	2	4.67	7.7	—

These results can be interpreted by considering the thermodynamic data of complex formation and the structures of the edta complexes in the solid state. It appears that the complexes of the cations of Group II are stabilized not only by chelate ring formation, but also to a significant extent by retention of water molecules in the first coordination sphere. Only the complexes of the cations of Group I show, within a homologous series, a coordination number of six by coordination of six donor atoms of the ligand. The trend of log K_1 vs. n is also in accordance with that observed for series of other homologous ligands, in which the number of atoms between two donors is changed by insertion of methylene groups.[55]

20.3.3.3.2 Protonated MHL and binuclear M₂L complexes

The protonated species MHL^- has been detected for edta as well as for the other homologues. Comparing the pK values of these acids (Table 4), lower values were found for the complexes of the cations Co^{2+}, Ni^{2+}, Cu^{2+} and Zn^{2+} with tmta than for those with edta. This step corresponds to the deprotonation of an uncoordinated CO_2H group and its coordination to M. This reaction is favored with tmta because of the lower steric strain in the complex, as it forms an intermediate N—N six-membered chelate ring instead of a five-membered one.[54]

In the case of the other cations (Mg^{2+}, Ca^{2+} and Mn^{2+}) the protonated complex of tmta shows much higher pK values; the proton seems to be bound to a nitrogen atom as expected for species (5) in which only one iminodiacetate group is bound to M^{2+}. In the case of peta all given pK values of MHL^- are higher than 5.2 and this should correspond to a species of this type. The greater tendency for its formation is a consequence of the decrease in stability for a large chelate ring. For $n = 5$ the gain in stability for the simultaneous coordination of both iminodiacetate groups is so low that the two groups are coordinated in two steps. This is also in agreement with the fact that by increasing *n* a marked tendency to the formation of bimetallic species is observed, *i.e.* the 1:1 complex in the presence of M^{z+} forms $M_2L^{(4-2z)-}$ (6), with a cleavage of more chelate rings.[14]

$$
\begin{array}{cc}
\underset{\text{O}_2\text{CCH}_2}{\overset{\text{O}_2\text{CCH}_2}{\diagdown}}\!\!\!\!\overset{}{\underset{}{\text{M}}}\text{------------}\text{N(CH}_2)_n\overset{+}{\text{N}}\!\!-\!\!\text{H}\overset{\text{CH}_2\text{CO}_2^-}{\underset{\text{CH}_2\text{CO}_2^-}{\diagup}} &
\underset{\text{O}_2\text{CCH}_2}{\overset{\text{O}_2\text{CCH}_2}{\diagdown}}\!\!\!\!\overset{}{\underset{}{\text{M}}}\text{------------}\text{N(CH}_2)_n\text{N}\text{------------}\text{M}\overset{\text{CH}_2\text{CO}_2}{\underset{\text{CH}_2\text{CO}_2}{\diagup}} \\
(5) & (6)
\end{array}
$$

20.3.4 TETRACARBOXYLIC ACIDS WITH SUBSTITUENTS

20.3.4.1 In the N,N′ Chain

20.3.4.1.1 With two C atoms between the two N atoms

(*i*) *With methyl and phenyl as substituent*

A dozen such substances are known[23] with methyl or phenyl as substituents (7). The acids of type (7) show different pK values with respect to edta: aliphatic substituted acids have larger pK values of HL^{3-} from 0.5 (R^1 = Me, $R^2 = R^3 = R^4$ = H) to 1.4 [($\pm$), $R^1 = R^3$ = Me, $R^2 = R^4$ = H]. Phenyl derivatives show lower values for the pK of HL^{3-} and H_2L^{2-}. For $R^1 = R^3$ = Me and $R^2 = R^4$ = H the *meso* and the ($\pm$) acid have been prepared and investigated. The complexes of the *meso* form are invariably less stable than those of the corresponding ($\pm$) isomer. This is explained by the fact that the interference between the two methyl groups is less in the latter than in the former ligand; the conformation imposed by the methyl group in the *meso* isomer would make ring closure of the acetato group difficult. The lower stabilities[55] of the complexes of the *meso* isomer could be explained on the assumption that one chelate ring less is formed. In general an increase in the pK values of the ligand is accompanied by an increase in the stability constants of its complexes. Therefore, it is not astonishing that the ligands with aliphatic substituents show higher stability constants by a factor between 10 and 100 with respect to edta. This again is only effective in alkaline media for pH > pK_4 and therefore without practical interest. 1-Phenylethylenedinitrilotetraacetic acid and ($\pm$)-1,2-diphenylethylenedinitrilo-*N,N,N′,N′*-tetraacetic acid show remarkably high values for the stability constants in comparison to the lower basicity of the ligand L^{4-}; the increased stability constants as compared with edta have been attributed to the steric effect of the phenyl groups.[56]

$$
\underset{\text{HO}_2\text{CCH}_2}{\overset{\text{HO}_2\text{CCH}_2}{\diagdown}}\!\!\!\!\text{NCR}^1\text{R}^2\text{CR}^3\text{R}^4\text{N}\overset{\text{CH}_2\text{CO}_2\text{H}}{\underset{\text{CH}_2\text{CO}_2\text{H}}{\diagup}}
$$

(7)

(*ii*) *With the two C atoms forming part of a cyclic structure*

$$
\underset{\text{HO}_2\text{CCH}_2}{\overset{\text{HO}_2\text{CCH}_2}{\diagdown}}\!\!\!\!\text{N}-\underset{\underset{\text{(CH}_2)_n}{\rule{1.5em}{0.5pt}}}{\text{CH}\rule{1em}{0.5pt}\text{CH}}-\text{N}\overset{\text{CH}_2\text{CO}_2\text{H}}{\underset{\text{CH}_2\text{CO}_2\text{H}}{\diagup}} = \text{H}_4\text{L}
$$

(8) a: *n* = 2
    b: *n* = 3
    c: *n* = 4

(8d)

In these tetraacetic acids the two C atoms of the N,N′ chain are connected by a polymethylene chain. The compounds for *n* = 2 (8a), 3 (8b) and 4 (8c) are known.[2,3] In the last case, both *cis* and *trans* isomers are possible, but the data for the former are only given in an annual report and need confirmation.[57] A further acid (8d) was obtained in which the polymethylene chain for *n* = 4 is condensed through the two middle C atoms with a benzene ring. The effect of the basicity of the ligand anion on the stability of the calcium complex due to the presence of a ring system is shown in Table 5. Of the four ligands, *trans*-cdta (**8c**) has found particular interest and the stability constants of its complexes have been investigated by different authors.[2,3] The complexes with this ligand are more stable relative to edta, but in this case the effect is mainly due to the increased basicity of the ligand anion and is therefore evident in strong alkaline solution because of the extremely high value of pK_4.

Table 5 pK Values and Stability Constants of Calcium Complexes with Ligands of Type (**8**) at $I = 0.1$ and 20 °C

	(8a)	(8b)	(8c)	(8d)
pK of HL^{3-}	9.75	10.09	12.35	10.26
pK of H_2L^{2-}	5.8	7.48	6.12	5.96
log K_{CaL}	8	9.45	13.15	11.63

20.3.4.2 With Substituents in the CH_2CO_2H Chains

$$\begin{array}{c} HO_2CCHR^1 \\ \diagdown \\ HO_2CCH_2 \diagup \end{array} NCH_2CH_2N \begin{array}{c} \diagup CHR^2CO_2H \\ \diagdown CH_2CO_2H \end{array} = H_4L$$

(9)   a: $R^1 = R^2 = Me$
       b: $R^1 = R^2 = Et$
       c: $R^1 = R^2 = Pr^n$
       d: $R^1 = R^2 = Pr^i$

Ligands of type (**9**) are known for $R^1 = R^2 = Me$ (**9a**), Et (**9b**), Pr^n (**9c**) and Pr^i (**9d**).[2,3] The weakest complexes are formed by the ligand containing two isopropyl groups (Table 6). The effect of the methyl substituent is very small for the cations of Group I but more evident for those of Group II (see Section 20.3.3.3.1).

Table 6 pK Values and Stability Constants K_1 of Ligands of Type (**9**)

Ligand	pK of HL^{3-}	pK of H_2L^{2-}	Mg^{2+}	log K_1 Ca^{2+}	Zn^{2+}	Cu^{2+}
(9a)	10.42	6.65	8.58	10.01	16.02	18.69
(9b)	10.42	6.09	7.83	9.02	16.04	18.76
(9c)	10.47	6.15	7.96	9.24	16.10	18.89
(9d)	10.68	5.60	5.20	6.05	13.07	16.07

20.3.4.3 With More than Two C Atoms and Heteroatoms Between the Two N Atoms

The tetraacetic acids obtained from (**10**; $n = 1$) and (**11**; $X = O$, S and NMe) have been synthesized and studied as chelators.[14,15] Furthermore, the acids of type (**10**; $n = 3$ and 4) with O as heteroatom have also been investigated. The substance which has acquired more importance is egta, *i.e.* (**11**; $X = O$). In comparison with the tetraacetic acid with methylene groups instead of the atoms X, a large gain in stability is observed for Ca^{2+}, Mn^{2+} and Cd^{2+} and a much lower one for Co^{2+}, Ni^{2+}, Cu^{2+} and Zn^{2+}.[14] The difference between the constants for calcium (log $K_1 = 10.97$) and magnesium (log $K_1 = 5.2$) is remarkable; this allows the selective complexometric titration of Ca^{2+} in presence of Mg^{2+}.[19]

$$\begin{array}{c} HO_2CCH_2 \\ \diagdown \\ HO_2CCH_2 \diagup \end{array} N(CH_2)_nX(CH_2)_nN \begin{array}{c} \diagup CH_2CO_2H \\ \diagdown CH_2CO_2H \end{array}$$

(10)

$$\begin{array}{c} HO_2CCH_2 \\ \diagdown \\ HO_2CCH_2 \diagup \end{array} N(CH_2)_2X(CH_2)_2X(CH_2)_2N \begin{array}{c} \diagup CH_2CO_2H \\ \diagdown CH_2CO_2H \end{array}$$

(11)

In the case of (**10**; $n = 2$, X = O), the effects on the stability are similar to those for edta. However, much lower K_1 values are obtained for the homologous ligand with X = S. These values are nevertheless larger than those for peta (see Section 20.3.3.3.1) by a remarkable amount for Cd^{2+} and much less for Zn^{2+}, Co^{2+}, Mn^{2+}, Cu^{2+} and Ca^{2+}.[14]

20.3.4.4. Di- and Tetra-propionic Acid Derivatives

Chelate ring size can also be changed by substitution of acetic by propionic residues.[2,3] The ligands (**12**) and (**13**) of Scheme 4 form weaker complexes in comparison with edta (see Table 7). In the case of the tetrapropionic acid, the lowest decrease of K_1 is observed for Cu^{2+}. For the other cations the dipropionate derivative shows values of $\log K_1$ which are not more than two units lower than those for edta. The data for the two propionic acids have to be further controlled by new measurements, because the method used is not always the most appropriate to obtain reliable values of K_1.

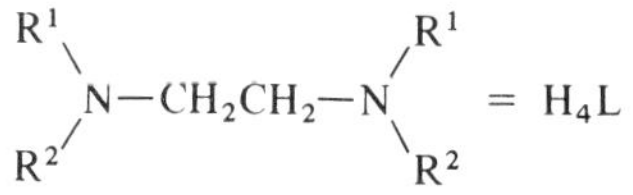

$$R^1 \diagdown \atop R^2 \diagup N - CH_2CH_2 - N {\diagup R^1 \atop \diagdown R^2} \quad = H_4L$$

(12) $R^1 = CH_2CO_2H$, $R^2 = CH_2CH_2CO_2H$

(13) $R^1 = R^2 = CH_2CH_2CO_2H$

Scheme 4

Table 7 Log K_1 of 1:1 Complexes with edta (20 °C, $I = 0.1$) and of Two Propionate Derivatives (**12**) and (**13**) (30 °C, $I = 0.1$)

	Mg^{2+}	Co^{2+}	Ni^{2+}	Cu^{2+}	Zn^{2+}	Cd^{2+}	Pb^{2+}	Ref.
edta	8.7	16.5	18.7	18.8	16.7	16.6	18.3	11
(12)	6.9	14.9	15.5	16.3	14.5	14.9	13.2	58
(13)	1.8	7.6	9.7	15.4	7.8	6.0	—	59

20.3.5 OTHER COMPLEXONES

20.3.5.1 With More than Four Acetic Acid Residues

Polyamines in which the nitrogen atoms are separated by two C atoms can be used as a basis for obtaining complexones. The more important ligands of this type are the derivatives of the linear polyamines $H_2N(CH_2CH_2NH)_nCH_2CH_2NH_2$ ($n = 1$, 2 and 3) given in Scheme 5. The ligands contain $6 + 2n$ donor atoms available for complex formation and can therefore be of interest for cations which show a coordination number larger than six. This is the case for the trivalent cations of the rare earths and for tetravalent thorium and zirconium, for instance. In these cases the constant K_1 shows a large increase on going from edta to dtpa[60,62] and to ttha[2,3] (see Table 8). The pK values of the corresponding protonated complexes are low; thus, all available donors seem to be bound in the complex $ML^{(4+n-z)-}$. In contrast to this, for cations showing coordination number six, an increase in K_1 is only observed for the dtpa complexes with respect to those of the edta complexes. This could be due to the coordination of three instead of two N atoms. In the case of ttha: (i) the K_1 values obtained are normally lower than those for dtpa; (ii) the protonated complex $MHL^{(5-z)-}$ dissociates to $ML^{(6-z)-}$ only in basic solution; and (iii) quite stable dinuclear complexes $M_2L^{(6-2z)-}$

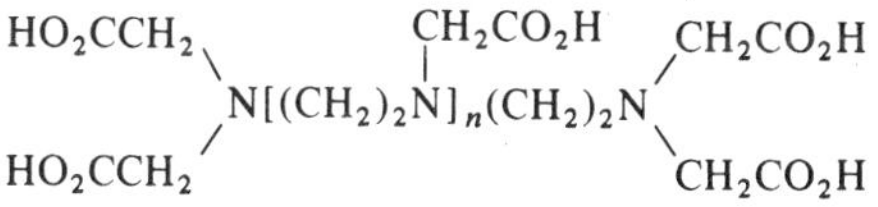

$$\begin{array}{c} HO_2CCH_2 \diagdown \qquad \qquad CH_2CO_2H \qquad CH_2CO_2H \\ \qquad\qquad N[(CH_2)_2N]_n(CH_2)_2N \\ HO_2CCH_2 \diagup \qquad\qquad\qquad\qquad\qquad \diagdown CH_2CO_2H \end{array}$$

$n = 1$: dtpa
$n = 2$: ttha
$n = 3$: tpha

Scheme 5

are formed. In addition, Fe^{3+} profits from the large donor number of dtpa, forming a complex without coordinated H_2O molecules and reaches heptacoordination involivng four carboxylate groups and the three N atoms in $Ba[Fe(dtpa)] \cdot 3H_2O$.[61] This explains the difference in hydrolytic behaviour between $[Fe(edta)(H_2O)]^-$ ($pK = 7.5$) and $[Fe(dtpa)]^{2-}$ ($pK = 9.9^{26}$).

Table 8 Equilibrium Data for dtpa and ttha Complexes[2,3]

		Ca^{2+}	Cu^{2+}	Zn^{2+}	Zr^{4+}	Th^{4+}
dtpa	pK of MHL	6.1	4.74	5.43	<1.5	2.16
	$\log K_1$	10.89	21.53	18.55	36.9	29
ttha	pK of MHL	8.86	8.03	8.05	—	3.05
	$\log K_1$	10.52	21.8	18.1	—	31.9

20.3.5.2 With Hydroxylic Groups as Substituents

Many complexones have been synthesized in the search for stronger ligands for the binding of Fe^{III} ions, since the stability of the Fe^{III} edta complex is not high enough to keep it from decomposing with precipitation of iron(III) hydroxide above pH 8–9. Some examples of complexones with hydroxylic groups are given in Scheme 6.[2,3] In the search for more effective chelating agents, it was found that the Fe^{III} complexes with dtpa and Hedta are stable at higher pH values but that they also decompose to iron(III) hydroxide in moderately alkaline solutions. A compound which is able to bind Fe^{III} ions more selectively is ehpg. Its Fe^{III} chelate does not decompose even in the most strongly alkaline solutions. In other situations Hedta and dtpa are used instead of edta for sequestering Fe^{III} ions, in the pH ranges 1–10 and 1–8 respectively. Another complexone of this type, N,N-di(hydroxyethyl)glycine, $(HOCH_2CH_2)_2NCH_2CO_2H$, has been prepared and used for Fe^{III} chelation in the pH range 6–12.5, although its equilibria have not been fully investigated.

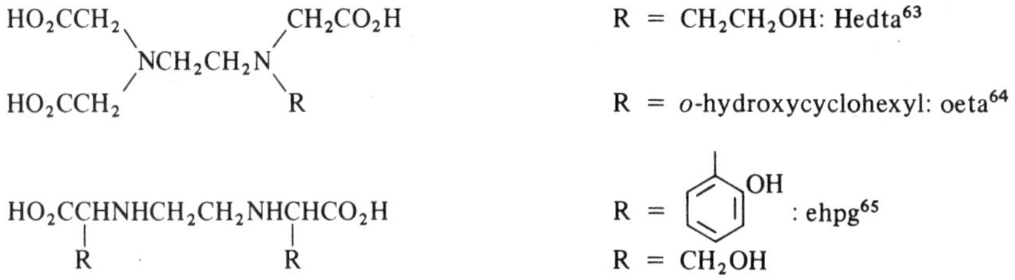

Scheme 6

20.3.5.3 Iminodiacetic Acid and Derivatives

Iminodiacetic acid and its derivatives have the formula (14) with an iminodiacetic group and a substituent R ($= Me$, Pr^n, $Bu^tCH_2CH_2$, Cy, Ph, $CH_2{=}CHCH_2$, NH_2, NO, NO_2, $CNCH_2$, $HOCH_2CH_2$, HO_2CCH_2, $HO_2CCH_2CH_2$, $MeOCH_2CH_2$, $HO_2CCH_2CH_2CH_2$, HO_2CCMe_2, $HSCH_2CH_2$, $MeSCH_2CH_2$, $H_2NCH_2CH_2$, $PhCOCH_2$, $MeCOMe$, p-nitrotolyl, m-nitrotolyl, uramil).

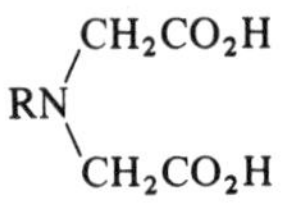

(14)

These acids have been investigated in relation to the particular influence of the substituent R on the stability of the metal complexes formed. This is because the tridentate iminodiacetate group ensures a generally quite elevated stability of the complexes formed. Of these ligands nitrilotriacetic acid (nta; $R = CH_2CO_2H$) is the most important because it has found various applications.

20.3.5.3.1 *Nitrilotriacetic acid (nta)*

The pK values of the triprotonic acid (H_3L) are 1.9, 2.5 and 9.73 at the ionic strength 0.1 with KCl or KNO_3 as inert salts and 20 °C. The anion L^{3-} forms stable water-soluble complexes with polyvalent metal ions M^{z+} of types $ML^{(3-z)-}$ and $ML_2^{(6-z)-}$, depending on the ratio between the total concentrations of the metal ion and of the ligand, on the pH value of the solution and on the values of the stability constants K_1 and K_2.[66] The 1:1 complexes $ML^{(3-z)-}$ with Ca^{2+} (log $K_1 = 6.4$) and Mg^{2+} (log $K_1 = 5.4$) are sufficiently stable to allow the elimination of hardness in water. In the 1:1 complexes, four coordination sites of the metal ions are already occupied and the coordination number of the cation is often six; the value of K_2 is therefore normally much lower than K_1. As the trivalent cations of the rare earth metals show a coordination number larger than eight, the difference log K_1 − log K_2 is low since there is no difficulty in binding eight donor atoms of two nta anions with the same central ion.

Table 9 Stability Constants K_1 and K_2 of nta Complexes at 20 °C and $I = 0.1$[66]

	Ca^{2+}	Mn^{2+}	Co^{2+}	Ni^{2+}	Cu^{2+}	Zn^{2+}	Pb^{2+}	La^{3+}	Lu^{3+}
log K_1	6.4	7.4	10.4	11.5	13.0	10.7	11.4	10.5	11.4
log K_2	2.5	3.6	4.0	4.9	4.4	3.6	—	7.4	9.0

5-Amino-2,4,6-trioxo-1,3-perhydrodiazine-*N*,*N*-diacetic acid (uramil-*N*,*N*-diacetic acid) also forms quite stable complexes with monovalent cations such as lithium (log $K_1 \approx 5$).[2,3]

20.3.5.4 Miscellaneous Ligands

20.3.5.4.1 *Macrocyclic complexones*

Macrocyclic complexones contain more than one aminoacetic group, which is bound over the N atom to a cyclic unit as in the case of the tetraaza macrocycles (**15**). Different selection effects are found in the case of these ligands (see Chapters 21.1–21.3 of this volume).[67] The pK values of the acids are much higher than those of edta: they almost explain the high values of the stability constants with alkaline earth and transition metal cations. The differences in log K_1 between Ca^{2+} and Mg^{2+} are very high and similar to those observed for egta. For the transition metals the cyclic complexones have no advantage over the non-cyclic ones such as edta and cdta.

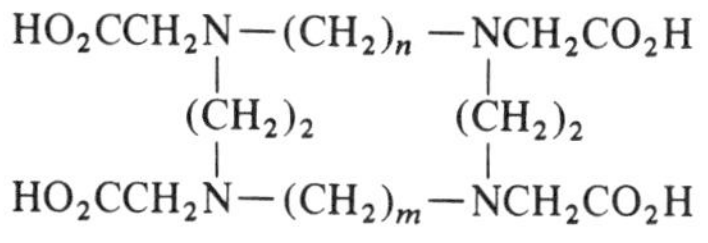

$$HO_2CCH_2N-(CH_2)_n-NCH_2CO_2H \qquad n = m = 2$$
$$(CH_2)_2 \qquad (CH_2)_2 \qquad n = 2, m = 3$$
$$HO_2CCH_2N-(CH_2)_m-NCH_2CO_2H \qquad n = m = 3$$

(15)

20.3.5.4.2 *Metallochromic complexones*

In the search for indicators for metal ions with good chelating properties (see Chapter 10 of Volume 1 of this work) known pH indicators, such as phthaleins and alizarin, have been used as substrates to which one or more iminodiacetate groups have been added as substituents.[19] Two examples of substances of this type are shown below. In acidic media xylenol orange (**16**) shows a yellow color and forms red complexes with Cd^{2+}, Hg^{2+}, Pb^{2+}, Bi^{3+} and Th^{4+}. In the case of the last two cations, the complexes are stable in acidic media and their determination is possible for solutions in the pH range 1–2. Zinc is titrated directly in the presence of an acetic acid/acetate buffer, Cd^{2+} and Hg^{2+} in the presence of hexamethylenetetramine. Back-titration of the edta excess in moderately acidic media is preferred for the determination of Ni^{2+}, Co^{2+}, Cu^{2+}, U^{4+}, V^{4+}, Cr^{3+} and Ti^{4+}, using a thorium or a thallium(III) standard solution (see Chapter 10 of Volume 1 of this work). Xylenol orange is an excellent indicator for about 50 metal ions and among the metallochromic indicators shows the sharpest color change at the endpoint. Xylenol orange[68] is an acid of type H_6L and forms mononuclear $MH_pL^{(6-p-z)-}$ and dinuclear $M_2H_pL^{(6-p-2z)-}$ complexes. The investigation of its equilibria is complicated because of the existence of protonated species. Errors are also caused by inadequate purity of the substances used. The only method employed in the determination of the equilibrium constants was spectrophotometry, which does not allow the exact determination of the values of p. For Th^{4+} the measurements in the pH range 1.8–3.5 have been interpreted assuming the formation of Th_2H_2L and $Th_2H_4L_2$.[69] Alizarin complexone[19] (17) is used for the determination of Co^{2+}, Pb^{2+} and Zn^{2+} as well as for the organic microelemental analysis of fluorine.

(16)   (17)

20.3.6 APPLICATIONS

20.3.6.1 Metal Buffering

Chemical and biological reactions often depend critically on the presence of low concentrations of certain free metal ions. Concentrations less than 10^{-4} M are difficult to maintain in presence of adventitious complexing agents, hydrolytic equilibria, adsorption and possibly contamination. Many problems can be overcome by the use of metal ion buffers which provide a controlled source of free metal ions in a manner similar to the regulation of hydrogen ion concentration by pH buffers. Because the quantity to be controlled is the concentration of the free metal ion M^{z+}, by analogy to pH, the concept of pM is introduced, *i.e.* the negative logarithm of $[M^{z+}]$: $pM = -\log[M^{z+}]$. For buffer systems in which only 1:1 complexes $ML^{(p-z)-}$ between M^{z+} and L^{p-} are formed, equimolar concentrations of the complex and of the ligand in excess are used in order to obtain $pM = \log K_1$, from $K_1 = [ML^{(p-z)-}]/([M^{z+}][L^{p-}])$ for $[ML^{(p-z)-}] = [L^{p-}]$. This is only true as long as the three different species in question are not involved in reaction with H_2O.[20] Because of the availability of a very large number of ligands giving complexes with different K_1 values,[2,3] it is possible to find a ligand which gives, under given conditions, the required pM value. Some caution is necessary in order to avoid the formation of ternary species from the complex $ML^{(p-z)-}$ in the final system of interest.

20.3.6.2 Other Applications

The many applications of complexones have in common the regulation of metal concentration in widely differing systems. In the examples which follow, grouped according to the overall methodology involved, examples of the agents used are given, together with cross-references to other chapters of this work where appropriate.

(1) *Conversion of the metal ion present in solution to a soluble complex with suppression of certain effects due to the presence of such metals.* The effects can be: (i) formation of insoluble products, for instance with soaps and detergents ($K_3H(edta)$, Na_3nta; see Chapter 66); (ii) oxidative degradation giving, for instance, rancidity (Na_2H_2edta); (c) catalytic decomposition such as decoloration of leather ($(NH_4)_2Hnta$; see Chapter 66) and dyes (see Chapter 58). Complexones are further used in this context in textile and paper processing solutions, ($Na_3H(edta)$, Na_3nta), cosmetic and pharmaceutical products (Na_2H_2edta), photographic developing solutions ($NH_4Fe(edta)$, $NaFe(edta)$, $Na_3H(edta)$, dtpa, $(NH_4)_2H_2edta$; see Chapter 59), chemical process water, beverages and foodstuffs [$Na_2Ca(edta)$].

(2) *Solubilization of a solid phase by complex formation.* This chelation solubilization is used for deposits and scale in boilers, processing tanks, and oxide films of metals [$Na_2Mg(edta)$, NH_4Fe (edta), $NaFe(edta)$, $(NH_4)_3H(edta)$].

(3) *Control of a given pM value.* (i) Electroplating or electroless plating (see Chapter 57); (ii) control of the activity of metal-dependent polymerization catalysts [for styrene and butadiene: $NaFe(edta)$]; (iii) biological growth systems supplying micronutrient metal ions (Fe^{3+}: $NaFe(edta)$,

Fe(Hedta), $Na_2Fe(edta)$; Zn^{2+}, Mn^{2+}, Mg^{2+}: edta complexes; see Chapter 62.1); (iv) removal of poisoning or radioactive metal ions from the body (Cd^{2+}: NaCa(edta); Pb^{2+}: $Na_3Ca(dtpa)$; see Chapter 62.2). Complexones have been used in the form of ^{99m}Tc complexes to obtain scintigraphic imaging of human organs in diagnostic nuclear medicine (see Chapter 65).

20.3.7 REFERENCES

1. G. Schwarzenbach, E. Kampitsch and R. Steiner, *Helv. Chim. Acta*, 1945, **28**, 1133.
2. L. G. Sillén and A. E. Martell, 'Stability Constants of Metal-Ion Complexes', Special Publication No. 17, Chemical Society, London, 1964: Supplement No. 1, Special Publication No. 25, Chemical Society, London, 1971; D. D. Perrin, 'Stability Constants of Metal-Ion Complexes. Part B, Organic Ligands', Pergamon, Oxford, 1979.
3. R. M. Smith and A. E. Martell, 'Critical Stability Constants', Plenum, New York, 1975, vols. 1–5.
4. W. Heintz, *Liebigs Ann. Chem.*, 1862, **122**, 260.
5. I. G. Farbenindustrie A. G., *Ger. Pat.*, 718 981 (1935).
6. I. G. Farbenindustrie A. G., *Ger. Pat.*, 694 790 (1937).
7. F. C. Bersworth (Martin Dennis Co.), *US Pat.* 2 387 735 (1945) (*Chem. Abstr.*, 1946, **40**, 1171).
8. H. C. Chitwood (Carbide and Carbon Chemical Corp.), *US Pat.* 2 384 817 (*Chem. Abstr.*, 1946, **40**, 354).
9. G. O. Curme, Jr., H. C. Chitwood and J. W. Clark (Carbide and Carbon Chem. Corp.), *US Pat.* 2 384 816 (*Chem. Abstr.*, 1946, **40**, 353).
10. G. Schwarzenbach, G. Anderegg, W. Schneider and H. Senn, *Helv. Chim. Acta*, 1955, **38**, 1147.
11. G. Anderegg, 'Critical Survey of Stability Constants of EDTA complexes', Pergamon, Oxford, 1977.
12. G. Schwarzenbach and H. Ackerman, *Helv. Chim. Acta*, 1948, **31**, 1029.
13. F. L'Eplattenier and G. Anderegg, *Helv. Chim. Acta*, 1964, **47**, 1792.
14. G. Anderegg, *Helv. Chim. Acta*, 1964, **47**, 1801.
15. G. Schwarzenbach, H. Senn and G. Anderegg, *Helv. Chim. Acta*, 1957, **40**, 1886.
16. G. Schwarzenbach and H. Ackermann, *Helv. Chim. Acta*, 1947, **30**, 1798.
17. K. Nakamoto, Y. Morimoto and A. E. Martell, *J. Am. Chem. Soc.*, 1962, **84**, 2081; 1963, **85**, 309.
18. G. Anderegg, *Helv. Chim. Acta*, 1967, **50**, 2333.
19. G. Schwarzenbach and M. Flaschka, 'Die Komplexometrische Titration', Enke, Stuttgart, 1965 (see also the English translation: Methuen, London, 1969).
20. A. Ringbom, 'Complexation in Analytical Chemistry', Interscience, New York, 1963.
21. D. D. Perrin, 'Masking and Demasking in Chemical Reactions', Wiley-Interscience, New York, 1970.
22. G. Schwarzenbach, R. Gut and G. Anderegg, *Helv. Chim. Acta*, 1954, **37**, 937.
23. H. Stünzi and G. Anderegg, *Helv. Chim. Acta*, 1973, **56**, 1698.
24. G. Anderegg and S. C. Malik, *Helv. Chim. Acta*, 1976, **59**, 1498.
25. G. Schwarzenbach and J. Heller, *Helv. Chim. Acta*, 1951, **34**, 576.
26. E. Bottari and G. Anderegg, *Helv. Chim. Acta*, 1967, **50**, 2349.
27. R. L. Gustafson and A. E. Martell, *J. Phys. Chem.*, 1963, **67**, 576.
28. T. N. Polynova, N. P. Bel'skaya, D. Tyurk de Garcia Banols, M. A. Porai-Koshits and L. I. Martynenko, *Zh. Strukt. Khim.*, 1970, **11**, 164.
29. C. H. L. Kennard, *Inorg. Chim. Acta*, 1967, **1**, 347.
30. M. Shimoi, Y. Orita, T. Uehiro, I. Kita, I. Iwamoto, A. Ouchi and Y. Yoshino, *Bull. Chem. Soc. Jpn.*, 1980, **53**, 3189.
31. L. Hoard, C. H. L. Kennard and G. S. Smith, *Inorg. Chem.*, 1963, **2**, 1316.
32. Ya. N. Nesterova, T. N. Polynova and M. A. Porai-Koshits, *Koord. Khim.*, 1975, **1**, 966.
33. C. H. L. Kennard, *Inorg. Chim. Acta*, 1967, **1**, 347.
34. E. F. K. McCandlish, T. K. Michael, J. A. Neal, E. C. Lingafelter and N. J. Rose, *Inorg. Chem.*, 1978, **17**, 1383.
35. H. A. Weakliem and J. L. Hoard, *J. Am. Chem. Soc.*, 1959, **81**, 549.
36. F. S. Stephens, *J. Chem. Soc. (A)*, 1969, 1723.
37. Ya. N. Neterova, M. A. Porai-Koshits and V. A. Logvinenko, *Zh. Neorg. Khim.*, 1979, **24**, 2273.
38. G. S. Smith and J. L. Hoard, *J. Am. Chem. Soc.*, 1959, **81**, 556.
39. Ya. N. Nesterova, M. A. Porai-Koshits and V. A. Logvinenko, *Zh. Strukt. Khim.*, 1980, **21**, 171.
40. F. P. Van Remoortere, J. J. Flynn, F. P. Boer and P. P. North, *Inorg. Chem.*, 1971, **10**, 1511.
41. S. Richards, B. Pedersen, J. V. Silverton and J. L. Hoard, *Inorg. Chem.*, 1964, **3**, 27.
42. M. D. Lind, M. J. Hamor and J. L. Hoard, *Inorg. Chem.*, 1964, **3**, 34.
43. M. Saito, T. Uehiro, F. Ebina, T. Iwamoto, A. Ouchi and Y. Yoshino, *Chem. Lett.*, 1979, 997.
44. E. Passer, J. G. White and K. L. Cheng, *Inorg. Chim. Acta*, 1977, **24**, 13.
45. A. I. Pozhidaev, T. N. Polynova and M. A. Porai-Koshits, *Acta Crystallogr., Sect. A*, 1972, **28**, 576.
46. P. G. Harrison, M. A. Healy and A. T. Steel, *Inorg. Chim. Acta*, 1982, **67**, L15.
47. E. O. Schlemper, *J. Cryst. Mol. Struct.*, 1977, **7**, 81.
48. A. I. Pozhidaev, M. A. Porai-Koshits and T. N. Polynova, *Zh. Strukt. Khim.*, 1974, **15**, 644.
49. L. R. Nassimbeni, M. R. W. Wright, J. C. van Niekerk and P. A. McCallum, *Acta Crystallogr., Sect. B*, 1979, **35**, 1341.
50. T. V. Filippova, T. N. Polynova, A. L. Il'inskii, M. A. Porai-Koshits and L. I. Martynenko, *Zh. Strukt. Khim.*, 1977, **18**, 1127.
51. B. L. Barnett and V. A. Uchtman, *Inorg. Chem.*, 1979, **18**, 2674.
52. J. L. Hoard, B. Lee and M. D. Lind, *J. Am. Chem. Soc.*, 1965, **87**, 1612; M. D. Lind, B. Lee and J. L. Hoard, *J. Am. Chem. Soc.*, 1965, **87**, 1611.
53. G. Anderegg, *Adv. Mol. Relax. Interact. Proc.*, 1980, **18**, 79.
54. P. Paoletti, L. Fabbrizzi and R. Barbucci, *Inorg. Chem.*, 1973, **7**, 1861.
55. H. M. N. H. Irving and K. Sharpe, *J. Inorg. Nucl. Chem.*, 1971, **33**, 203.

56. N. Okaku, K. Toyoda, Y. Moriguchi and K. Ueno, *Bull. Chem. Soc. Jpn.*, 1967, **40**, 2326.
57. H. Kroll, A. E. C. Contract (30-1) 2096, Annual Report, 1960.
58. S. Chaberek, Jr. and A. E. Martell, *J. Am. Chem. Soc.*, 1952, **74**, 6228.
59. R. C. Courtney, S. Chaberek, Jr. and A. E. Martell, *J. Am. Chem. Soc.*, 1953, **75**, 4814.
60. G. Anderegg, P. Nägeli, F. Müller and G. Schwarzenbach, *Helv. Chim. Acta*, 1959, **42**, 827.
61. E. B. Chuklanova, T. N. Polynova, A. L. Poznyak, L. M. Dikareva and M. A. Porai-Koshits, *Koord. Khim.*, 1981, **7**, 1729.
62. W. van der Linden and G. Anderegg, *Helv. Chim. Acta*, 1970, **53**, 569.
63. R. Skochdopole and S. Chaberek, *J. Inorg. Nucl. Chem.*, 1959, **11**, 222.
64. J. Schubert, G. Anderegg and G. Schwarzenbach, *Helv. Chim. Acta*, 1960, **43**, 410.
65. G. Anderegg and F. L. L'Eplattenier, *Helv. Chim. Acta*, 1964, **47**, 1067.
66. G. Anderegg, *Pure Appl. Chem.*, 1982, **54**, 2693.
67. H. Stetter and W. Frank, *Angew. Chem.*, 1976, **88**, 760; R. Delgado and J. J. R. Frausto da Silva, *Talanta*, 1982, **29**, 815.
68. J. Körbl, R. Přibil and A. Emr, *Collect. Czech. Chem. Commun.*, 1957, **22**, 961.
69. B. W. Budesinsky and A. Svec, *Anal. Chim. Acta*, 1972, **61**, 465.
70. D. D. Perrin and B. Dempsey, 'Buffers for pH and Metal Ions Control', Chapman and Hall, London, 1974.
71. H. Kroll and M. Knell, in 'Kirk-Othmer Encyclopedia of Chemical Technology', 1st edn., Wiley, New York, 1954, vol. 12, p. 164; A. E. Martell, in 'Kirk-Othmer Encyclopedia of Chemical Technology', 2nd edn., Wiley, New York, 1965, vol. 6, p. 1; A. L. McCrary and W. L. Howard, in 'Kirk-Othmer Encyclopedia of Chemical Technology', 3rd edn., Wiley, New York, 1979, vol. 5, p. 357.

20.4
Bidentate Ligands

ROBERT S. VAGG
Macquarie University, New South Wales, Australia

20.4.1 INTRODUCTION AND SCOPE

The scope of this chapter is defined broadly as the coordination chemistry of complexes containing the atomic grouping shown in (1). The normal use of the term bidentate applies, with X and Y being different non-metallic donor atoms, there being significant σ character in each M—X (or Y) bond. Here X and Y are linked through covalent bonds to form a chelate ring of three or more atomic members, but not containing another metal atom. Many dinuclear and bridged dimeric and polymeric species are thereby excluded. The chelate ring need not contain a carbon atom.

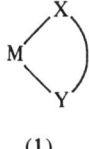

(1)

Some complexes which satisfy this definition are dealt with elsewhere, however. Those with X or Y as the Group V donors P, As or Sb are dealt with in Chapter 14 of this volume. Those with X or Y = Hg are described in Chapter 11 and the Group IV donors Si, Ge, Sn and Pb are included in Chapter 12.2. With very few exceptions this means that the relevant X,Y combinations are pairs derived from nitrogen, oxygen, sulfur or selenium donor groups. However, some specific examples of these combinations also occur elsewhere in this work. Amino acids (generally N—O bidentates) are covered in Chapter 20.2, Schiff base ligands (N—O or N—S) in Chapter 20.1, and several S—O bidentate ligands, including monothio-β-diketones and monothiocarboxylic acids, have been reviewed by Livingstone in Chapter 16.6. Hence a detailed discussion of these ligands is not included here.

It would seem that no review devoted solely to the subject of mixed atom bidentates has appeared in the literature, probably as a result of the very vast scope provided by this title. Certainly there is an almost infinite number of possible combinations of different N, O, S or Se donor groups defining (1), allowing also for different ring sizes, various substituent groups and side chains and various metal ions. In this chapter an attempt has been made to provide for the reader initial access to recent research literature on as wide a range of these mixed donor bidentates as possible. In doing this some emphasis has been placed either on relevant reviews, on individual articles with broad references or on more recent structural papers. The latter not only demonstrate an individual ligand's necessary bidentate behaviour but usually reference or provide details of their syntheses and physical properties, and of those of related complexes.

The coordination chemistry of individual N, O, S or Se donor groups is covered in other chapters in this volume and will not be reiterated here. In general the properties of mixed donor bidentates reflect their hybrid nature in comparison with the corresponding single donor ligands. This is more obvious with ligands combining both a 'hard' and a 'soft' atom donor. Often monodentate behaviour only results with such ligands in combination with metal ions which are not 'borderline'. Alternatively, the electronic and structural properties of a metal ion may be regulated by a deft choice of each individual donor atom and the overall ligand geometry. The bidentate nature of a ligand also may serve to allow a study of the chemical reactivity of one of its constituent coordinated donor groups.

The chapter has been sectioned according to the paired combinations of N—O, N—S, S—O or Se containing bidentates. Another Group VI donor element, tellurium, has not been included since in only one example has this element formed part of a mixed donor bidentate. This is the reported interaction of di(*o*-aminophenyl)ditelluride (2) with Cu^I and Cu^{II}, where N—Te bidentate behaviour was concluded for the tetrahedral Cu^I(N—Te)$_2$ chelate.[1]

(2)

20.4.2 BIDENTATE NITROGEN–OXYGEN LIGANDS

20.4.2.1 N—O Bidentates with Amine Group Donors

A broad review by Bertrand and Eller[2] on polynuclear complexes formed with amino alcohols and imino alcohols as ligands summarizes work reported to 1976 on the syntheses, structures and properties of such complexes. The neutral ROH group is usually a poorer donor than an amine group. However, the former is generally more acidic, and the amino alkoxides that result from deprotonation show an increased tendency both to chelate and to form additional bonds and therefore to bridge. The polynuclear complexes which result may be classified either as oxygen-bridged (3) or hydrogen-bonded (4). Often magnetic properties indicative of antiferromagnetic behaviour have been noted as a result of the allowed M$\cdots$M interactions. The mechanism of this magnetic coupling is now the subject of considerable investigation and discussion.

(3)   (4)

Cotton and Mott[3] reported the reaction of chromium(II) acetate with Me$_2$NCH$_2$CH$_2$ONa to form an unusual tetrameric product in which each *N,N*-dimethylaminoethanolate group is N—O bidentate but bridges through O to other Cr atoms in the tetramer. A recent protonation and aquation study on Cr^{III}/ethanolamine complexes[4] has shown a sequence of reactions involving initial protonation of the coordinated O atoms leading to ring opening at the alcohol function. A number of *N*-methylated 3-amino-1-propanols have been investigated as monoanionic N—O bidentates with Cu[II].[5] Again O-bridged dimeric structures are indicated. A short review of the several related structures of Cu[II] chelates of amino alcohols is provided by Muhonen *et al.*[6] The Cu[II] chelates of 2-aminoethanol (EtaH) and 2-amino-2-methylpropanol (MepH) in mixed protonated–deprotonated forms are hydrogen-bridged dimers showing antiferromagnetic coupling, which does not correlate either with the observed O···O or Cu···Cu distances in the compounds.[239] The fully deprotonated complex [Cu(Mep)$_2$], however, is *trans* planar and monomeric, although extensive hydrogen bonding is evident in the crystal structure.[7] The complexes [Co$_2$(Eta)$_3$(EtaH)$_3$]$^{3+}$ and [Ni$_2$(Eta)$_3$(EtaH)$_3$]$^{2+}$ each contain dinuclear cations made up of octahedral tris chelates joined face-to-face by O···H···O bonds. In the Ni dimer the six central O atoms form an octahedron, whereas a trigonal prism is formed in the Co complex with significantly shorter O···O distances.[8] The structures of some Ni[II] and Cu[II] chelates of 1-amino-2-propanol also have been reported.[9,10]

Richardson *et al.*, as part of a study of the reactions of metal ions with vitamins, have determined the structures of the bis chelates of pyridoxamine (PM) with Cu[II][11] and Zn[II].[12] In each case the pyridoxamine molecule behaves as an N—O bidentate in bonding through the 4-(aminomethyl) and phenolate groups. The PM molecules are zwitterionic, with the heterocyclic N atoms protonated.

The ligand 2-(methylaminomethyl)pyridine 1-oxide (5) acts either as an N—O bidentate or as an O monodentate to Cu[II]. In the former case the structure is *trans* octahedral involving the formation of the expected six-membered chelate rings,[13] whereas the second complex, which has unusual magnetic properties, is dimeric with each Cu coordination sphere being a distorted square pyramid.[14] The Cu[II] chelate of the *N,N*-dimethyl analogue is *trans* square planar, with long perchlorate–O interactions in tetragonal positions.[15]

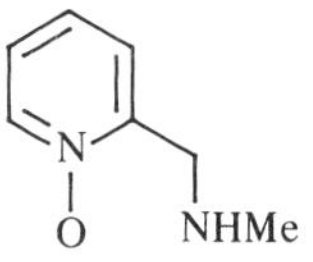

(5)

Anthranilic acid (2-aminobenzoic acid) is an N—O bidentate which commonly forms 2:1 complexes with a variety of bivalent metal ions. d'Avignon and Brown[16] provided details of some of the structures and applications of these chelates. The synthesis of ternary complexes of 2-aminobenzoic acid and 1-nitroso-2-naphthol with a variety of transition metal ions has been reported.[17] An X-ray structure shows an unusual N monodentate coordination mode to rhodium for the ligand.[18]

20.4.2.2 N—O Bidentates with N-Heterocyclic Donors

8-Hydroxyquinoline (oxine, 6), one of the earliest analytical reagents, also is one of the most widely studied N—O bidentates. Its early coordination chemistry was reviewed by Phillips.[19] Its use as an *in vivo* agent in microbiological systems has been reviewed by Schulman and Dwyer.[20] The extensive use of oxine and substituted forms, and closely related bidentates, for the analytical solvent extraction and colorimeteric determination of metal ions has been comprehensively reviewed.[21] An unusual bridged bonding mode for oxine has been reported in which N monodentate and O monodentate coordination to two different ruthenium atoms occurs.[240] Recently the Co[II] bis chelate of 2-methyl-8-hydroxyquinoline was shown to catalyze the disproportionation of nitric oxide.[22] The preparation and properties of some ternary Co[III] complexes of this ligand have also been reported.[23]

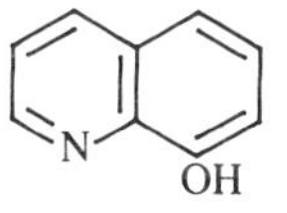

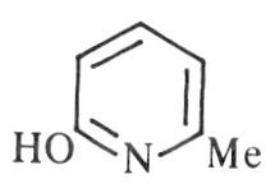

(6)   (7)

2-Hydroxypyridine does not normally act as a bidentate ligand but it commonly serves to bridge in dinuclear or polymeric species. The 6-methyl derivative (7; mhpH), however, has been shown to form four-membered chelate rings in the complex *trans*-[Ru(mhp)$_2$(PPh$_3$)$_2$].[24] The geometry of the bidentate mhp ligand differs considerably from that generally observed when bridging, suggesting a predominance of the hydroxy tautomeric form with little C—O double bond character. Clegg *et al.*[25] have reviewed the coordination chemistry of mhpH and reported the reactions of this ligand with Co^{II} to give the monomeric complexes [Co(mhp)$_2$] and [Co(mhp)$_2$(bipy)], as well as the novel species [Co$_{12}$(OH)$_6$(O$_2$CMe)$_6$(mhp)$_{12}$]. An X-ray structure of the latter complex shows an interior network of 12 Co atoms linked by the bridging O donors. The mhp molecules are again bidentate.

Complexes of pyridine-2-carboxylic acid (picolinic acid, picH) and its substituted derivatives commonly exhibit the N—O bidentate nature of this aromatic amino acid. From aqueous solutions chelates are obtained with the coordinated carboxylic group deprotonated, or neutral ligand forms may be isolated from non-aqueous media.[26] Bis chelates are common in either case with bivalent metal ions. The tris chelates of trivalent cobalt[27] and manganese[28] have been structurally characterized recently. The latter is tetragonally distorted in a structure similar to Mn^{III}(oxine)$_3$.

The closely related ligand pyrazine-2-carboxylic acid (**8**; 2-pzCO$_2$H) forms complexes analogous to those of picH. The crystal structures of some copper chelates of this ligand were recently reported.[29,30] The more versatile ligand 2,3-dicarboxypyrazine (**9**; 2,3-pz(CO$_2$H)$_2$) is potentially an N—O bis bidentate. It has been shown to perform this function in a 1:1 copper(II) chelate.[31] The 1:1 chelates of Zn^{II}[241] and Co^{II}[32] are bridged polymeric also, although in each complex the ligand acts singularly as an N—O bidentate with the second carboxylic group rotated perpendicular to the pyrazine ring to bridge adjacent molecules through both O atoms. The structural groupings and physicochemical properties of the 1:1 chelates of bivalent Mn, Fe, Co, Ni, Cu and Zn have been reported.[33] A report of the synthesis, magnetic properties and crystal structures of some Co^{II} bis chelates of both 2-pzCO$_2$H and 2,3-pz(CO$_2$H)$_2$ has shown all complexes to be structurally similar,[34] the N—O ligands adopting planar *trans* arrangements with octahedral coordination completed by water molecules.

(8)   (9)

The donor properties of the derivatives pyridine-2-carboxamide (**10**; picolinamide) and pyrazine-2,3-dicarboxamide (**11**) are both metal and pH dependent, with N—O bidentate behaviour involving the carbonyl O atom being observed for the neutral ligands. At high pH coordination of the deprotonated amide N atom occurs, with the ligands changing to N—N bidentates. The factors which influence this choice of amide coordination mode have been well reviewed.[35,36] An electrochemical study of the intense green complexes formed between Cr^{II} and pyrazine, pyrazine-2-carboxamide and pyrazine-2-carboxylic acid has shown both latter N—O bidentate chelates to undergo a reversible one-electron oxidation reaction which is not observed with the unsubstituted pyrazine complex.[37] The sulfonate analogue of picH, pyridine-2-sulfonic acid, has been shown to coordinate to Ag^I also as an N—O bidentate in a polymeric structure in which the sulfonate O atoms are involved in extensive bridging.[38]

(10)   (11)

A discussion of probable metal coordination sites involving phosphate, base and ribose moieties of nucleotides and nucleic acids has been provided by Izatt *et al.*[39] Coordination involving a base N atom and an O atom of either a phosphate or ribose group is indicated for several of the more commonly studied transition elements. Later reviews by Hodgson[40] and Marzilli[41] conclude from structural evidence that purines (**12**) do not act as chelating ligands to metals through N-7 and the exocyclic X atom attached to C-6 except in the case of 6-mercaptopurine (X = SH). Stereochemical reasons for this lack of N—O bidentate behaviour have been advanced by Sletten.[42] The nucleic acids cytosine and cytidine, however, have been shown to coordinate to Cu^{II} as N—O bidentates, although weak bonds to the O atoms only are indicated.[40,41]

(12)   (13)   (14)

Ainscough *et al.*[43] synthesized a variety of N—O bidentates containing the phenolate group and an N heterocycle in an attempt to model the specific iron-binding sites of lactoferrin and transferrin. From spectroscopic and structural studies it was concluded that in both proteins three tyrosyl O atoms, two N atoms of histidyl residues and one water molecule constitute the metal's coordination sphere. Orotic acid (13) (6-uracilcarboxylic acid) has been shown to act as a dianionic N—O bidentate with geometry similar to picolinic acid. The structures of the diammine–orotato complexes of both bivalent copper[44] and platinum[45] have been reported. The Pt chelate is planar whereas in the Cu complex weak intermolecular bridging of ring oxygen atoms results in a 4 + 2 coordination.

Several examples are known of N—O coordination of *N*-hydroxypiperidines resulting in three-membered chelate rings on deprotonation (14). The crystal structures and ^{95}Mo NMR studies of some complexes of Mo^{VI} containing this grouping have appeared.[46-48] A similar coordination mode was observed in the crystal structure of a Pd^{II} complex of the 2,2,4,4-tetramethylpiperidine derivative.[49]

20.4.2.3 N—O Bidentates with Oxime or Hydroxylamine Group Donors

The oxime group ($>$C=N—OH) is ambident with the possibility of coordination through either oxygen or nitrogen, the latter being far more common. The structural chemistry of transition metal complexes of oxime-containing ligands has been extensively reviewed.[242] Examples of N—O bidentates containing an oxime group are carbonyl oximes (15), 2-nitrosophenols (16), hydroxy oximes (17) and amide oximes (18). Generally, carbonyl oximes are poor chelating agents for bivalent nickel and copper. Cobalt(II) is rapidly oxidized to readily form octahedral complexes of the type [CoL$_3$]. The crystal structure of such a complex with R′ = Ph and R = Me has been reported,[50] as has that of the [Co(en)$_2$L]$^{2+}$ cation with this same ligand.[51] Each confirms the ligand's N—O bidentate behaviour.

(15)   (16)   (17)   (18)   (19)

The reaction of nitric oxide or nitrite ion with [M(acac)$_n$] chelates has been observed to result in a rearrangement of the coordination spheres to yield complexes of 3-hydroxyiminopentane-2,4-dione (inaaH) (19). In these products the resulting oxime groups demonstrate preferred coordination over the dissociated carbonyl O atoms. The Ni^{II}, Pd^{II} and Pt^{II} solid products are obtained in different isomeric forms, thought to be *cis* and *trans* species.[52] The structure of [Co(inaa)$_2$py$_2$] demonstrates a *cis* geometry,[53] in contrast with the *trans* analogue of acetylacetone. The octahedral tris complexes [Co(inaa)$_3$] and [Fe(inaa)$_3$]$^-$ are diamagnetic. The structure of the cobalt complex is *fac* octahedral,[54] similar to the tris(oxyiminomalonamido)ferrate(III) anion.[55] The compound [Fe(inaa)$_2$(3-mepy)$_2$] adopts a *cis* octahedral structure with the N atoms of inaa *trans* to the pyridyl nitrogen atoms.[56] An imine derivative of the Ni^{II} complex demonstrates an unusual N—O bidentate bonding mode involving oxime-O coordination,[57] which has been verified by an X-ray study.[58]

2-Nitrosophenol tautomerizes to a carbonyl oxime form (16), and thus coordinates as an N—O bidentate with a variety of transition metals. The green naturally occuring pigment ferroverdin, an iron(II) chelate of substituted nitrosophenol, is *cis* octahedral[59] and its neutral cobalt(III) analogue is known.[60] Amidoximes (RC(=NOH)NH$_2$) are amphoteric substances which have received wide use as analytical reagents for metal ions. Although many such complexes have been described in the literature, few have been characterized unambiguously. The reactions of α-acyloin oximes with Cu^{II}

have been widely studied. Typically these result in water-insoluble 1:1 neutral chelates which show antiferromagnetic behaviour and are thought to be polymeric in nature.[61] α-Benzoin oxime (17; R = R' = Ph; 'cupron') has received wide use as an analytical reagent for copper(II), as has salicylaldehyde oxime (20). Each has been found to chelate with most transition metals.

An unusual example has appeared of N—O bidentate behaviour by a bis-dioxime ligand. The ligand δ-camphorquinone dioxime (21; δ-H_2CQD) forms a mixed chelate with Ag^I and either Pd^{II} or Ni^{II}.[62] In these complexes each M(δ-HCQD)$_2$ unit is *cis* square planar with a hydrogen bond between one pair of oxime O atoms. The other pair of oxime O atoms bridge to Ag ions, thus forming part of a hexanuclear cluster. These cluster molecules do not form with Ni(HDMG)$_2$, suggesting that the oxime-O bonding mode is a necessary requirement for their formation. The reaction of barbituric acid with *trans*-Na$_2$[Ru(OH)(NO)(NO$_2$)$_4$] in aqueous solution results in the tris(dihydrogenviolurato)ruthenium(II) anion which was structurally characterized as its Ba^{2+} salt.[63] This compound was shown to contain the bidentate grouping (22). The heterocyclic rings are almost coplanar and the *fac* octahedral geometry observed is thought to represent an intrinsic stability of such ruthenium–ketoxime complexes, associated with the large *trans* effects of the oxime groups.

(20)   (21)   (22)

Some simple oximes have been shown to act as N—O bidentates by 'side-on' coordination of the oxime groups alone. The reaction of acetone with a hydroxylamido complex of molybdenum results in this bonding form for 2-propanone oxime.[64,243] The N—O and C—N bond lengths indicate single and double bond character respectively. The reaction was related to the prebiotic formation of HCN from formaldehyde and hydroxylamido–Mo chelates. A similar form of bonding for this ligand was concluded with some fluorotungstate(IV) complexes.[65]

Weighardt *et al.* have studied the interaction of substituted hydroxylamines with some molybdenum(VI) species. N-Methylhydroxylamine (MeNHOH) commonly acts as a bidentate through both the N and O atoms to form three-membered chelate rings (23; R = H, R' = Me), which has been demonstrated by both X-ray and ^1H NMR methods.[66,67] In such compounds N—O and C—O show normal single bond lengths. The ligand N-methyl-N-hydroxythiourea (24) also is shown to act as both an N—O and an S—O bidentate to Mo,[66] the former involving deprotonation and coordination of both the OH and NH$_2$ groups. An N,N-diethylhydroxylamine (23; R = R' = Et) complex of Mo^{VI} recently has been characterized.[68] The heavily substituted N,N-di-t-butyl-hydroxylamine (23; R = R' = Bu^t) also has been shown to bond to palladium(II) in this manner.[69] General methods for the synthesis of d^0 metallooxaziridines by the reaction of hydroxylamines with metal-oxo compounds have been reported.[70,71] The products also contain N and O in a three-membered chelate ring (25). A compound with R = Ph was produced using either phenylhydroxylamine or nitrosobenzene as reactants, and its crystal structure confirmed this bonding mode.[70]

(23)   (24)   (25)

20.4.2.4 Carbazide Derivatives as N—O Bidentates

Ligands of general form (26) commonly act as N—O bidentates by coordination through the carbonyl-O and hydrazide-N^3 atoms. Several structural studies on complexes of such ligands have demonstrated this coordination mode. These include the bivalent cadmium and cobalt chelates of (26) with R = Et,[71] the Co^{II} chelates with R = Pr^n[72] or m-HOC$_6$H$_4$,[73] the Zn^{II} chelate with R = Me[74] and a Pt^{II} chelate of (26) with R = p-NO$_2$C$_6$H$_4$.[75] The latter complex was prepared by the reaction of [Pt(PPh$_3$)$_2$(O$_2$)] with the aroylhydrazine and has both N^2 and N^3 atoms deprotonated. The short N—N distance suggests that the five-membered chelate ring has aromatic character. An N^3,N^3,N^2-trimethyl-substituted form of (26) with R = Ph also behaves as an N—O bidentate to Cu^{II}.[244]

(26)

Isoniazid (isonicotinoylhydrazine, inH) (**26**; R = 4-C$_5$H$_5$N) is a commonly used tuberculostatic drug with an activity increased by the presence of Cu^{2+} ions. The structure of the mono chelate formed with Cu^{II}Cl$_2$ shows the compound to coordinate through the O and N^3 atoms.[76] This same coordination mode is present in an Mn^{II} chelate.[77] A report of the preparation and properties of several 1:1 and 2:1 complexes formed with inH and Mn^{II}, Fe^{II}, Co^{II}, Ni^{II}, Cu^{II} and Zn^{II} has appeared recently.[78] Bridged polymeric structures were concluded for compounds of general form [M(inH)$_2$X$_2$] (X = Cl or Br), but N^2 deprotonation resulted in the more common N—O bidentate behaviour in the compounds [Cu^{II}(in)Cl]H$_2$O and [Cu^{II}(in)Br].

20.4.3 BIDENTATE NITROGEN–SULFUR LIGANDS

The coordination chemistry of S—N chelating agents was extensively reviewed in 1974 by Akbar Ali and Livingstone.[79] Here factors influencing the chemical and physical properties of metal chelates containing sulfur donor atoms were summarized, as were properties peculiar to S—N ligands. This includes a large number of N—S bidentates. Carcinostatic and antiviral activity of such ligands and their complexes were linked to these properties. Kuehn and Isied,[80] in an exhaustive review of the reactivity of metal ion–sulfur bonds, have given details of studies which deal with biological, spectral, thermodynamic and kinetic aspects of M–S interactions, including a large number of N—S and O—S chelates. Oxidation and alkylation reactions at coordinated thiols, and of N—S or O—S bidentate complexes in particular, are discussed in detail.

20.4.3.1 N—S Bidentates with Amine Donor Groups

Tetrasulfur tetranitride (S$_4$N$_4$) reacts with metal ions in hydrogen-containing solvents to yield thionitrosyl complexes of general formula M(HN$_2$S$_2$).[81] Using nickel chloride in ethanol three products were obtained, of varying S:N ratios, which were separated chromatographically. The complex [Ni(HN$_2$S$_2$)$_2$] and its Pd and Pt analogues (**27**) have been shown to have a *cis* planar geometry.[82] Reaction of the Ni^{II} chelate of (**27**) with methyl iodide generates a dimethyl derivative with a *trans* configuration.[83] The NH groups of these complexes are quite reactive, leading to a number of tricyclic derivatives.[84] Reaction of [Ni(HN$_2$S$_2$)$_2$] with formaldehyde and methanol results in the substitution of one amine proton by a CH$_2$OMe group in a planar structure showing parallel stacking in the solid state.[85] S$_4$N$_4$ reacts with [Co(Cp)(CO)$_4$] resulting in displacement of CO to yield the compound [Co(Cp)(S$_2$N$_2$)],[86] which again shows N—S bidentate coordination of the disulfur dinitride ligand.

(27)

One of the simplest N—S chelating agents is 2-aminoethanethiol (H$_2$NCH$_2$CH$_2$SH), which was first studied by Jensen[87] and later received detailed investigation by Jicha and Busch.[88] Deprotonated complexes of the form M(H$_2$NCH$_2$CH$_2$S)$_2$ (M = Ni or Pd) and [M$_3$(H$_2$NCH$_2$CH$_2$S)$_4$]Cl$_2$ (M = Co, Ni or Pd) were obtained, the latter complexes containing thiolo bridges.[89] The monomeric square planar nickel and palladium complexes react with alkyl halides to form the corresponding thioether complexes. The sulfur atom remains coordinated during these S-alkylation reactions.[90]

The closely related aromatic N—S bidentate 2-aminobenzenethiol (H$_2$NC$_6$H$_4$SH) forms complexes with VO^{IV}, Cr^{III}, Mn^{II}, Fe^{II}, Co^{II}, Ni^{II}, Ni^{IV}, Cu^I and Zn^{II},[91] and Co^{II}, Co^{III} and Pd^{II}.[92] Again, reaction of [Ni(SC$_6$H$_4$NH$_2$)$_2$] with methyl iodide leads to S-methylation of the coordinated ligands,[93] whereas oxidation in strongly alkaline solution yields a dark blue Ni^{IV} complex of formula

[Ni(SC$_6$H$_4$NH)$_2$]. The crystal structure of the tris chelate of molybdenum(VI) with the ligand in this deprotonated form shows the coordination geometry to be an irregular polyhedron midway between octahedral and trigonal prismatic.[94] A closely related bis chelate of Mo^V is trigonal prismatic with N$_3$ and S$_3$ faces.[95] In this structure the protons of the N—H group lie in the ligand planes, and short S—C and N—C bonds indicate considerable double-bond character.

Deutsch *et al.* have investigated in detail the reactivity of coordinated thiols using these simple N—S bidentates.[96] The cation [Co(en)$_2$(SCH$_2$CH$_2$NH$_2$)]$^{2+}$ has served as a prototype to demonstrate several oxidation reactions at coordinated sulfur. Reaction of this ion with hydrogen peroxide results in a coordinated sulfenic acid derivative, whereas with *N*-thiophthalimides coordinated disulfides result.[97] The latter compounds may be identified by a characteristic absorption band at *ca.* 340 nm not present in the spectra of other Co^{III}—S complexes. The thiolato precursors may be regenerated by reaction of the disulfides with alkali. Reaction of the cation with hydroxylamine-*O*-sulfonic acid yields an S-bonded sulfenamide complex,[245] whereas reaction with molecular iodine results in a dimeric sulfenyl iodide.[98] In general these reactions are facile, are thought to proceed *via* nucleophilic attack of the coordinated sulfur atom with concomitant atom or group transfer from the oxidant to the sulfur, and result in an oxidized ligand that is often considerably more stable than the corresponding non-coordinated species.[245] An X-ray analysis of the 2-aminobenzenethiol analogue of this cationic precursor shows all six bonds to Co to be equivalent to those in the NH$_2$CH$_2$CH$_2$S$^-$ complex.[99] The complex is shown to undergo oxidation, alkylation and adduct formation reactions which are characteristic of the ethyl analogue also. Thus there appears to be little distinction between the aromatic and aliphatic ligands in their oxidative reactivity or structural chemistry. A related compound, tris(2-aminoethanesulfenato)cobalt(III), has been resolved using (+)-tartaric acid to form a novel diastereoisomeric molecular compound.[100] The crystal structure shows that hydrogen bonds involving the acid and amino groups and sulfenato oxygen atoms of the bidentate allow the resolution of the neutral chelate.

The widely used analytical reagent dithiooxamide (28; 'rubeanic acid') forms insoluble 1:1 chelates with copper(II) and nickel(II) that are thought to be polymeric and involve N—S bis bidentate bridging behaviour by the ligand.[101] The ligand provides a very sensitive spot test for the detection of Co, Ni or Cu.[102] In solutions containing strong mineral acid it reacts only with certain platinum group elements, including Pt, Pd and Ru. It may be used as a reagent for the detection of Ru in the presence of Os, Pt or Pd.[102]

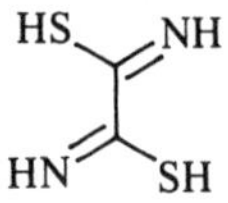

(28)

Nag *et al.* have investigated the coordination chemistry of ligands related to or derived from 2-aminocyclopentene-1-dithiocarboxylic acid, concluding N—S bidentate behaviour for the complexes in general.[103] With Mo^{VI}, however, this ligand and *N*-alkyl derivatives are observed to function rather as S—S bidentates.[104]

20.4.3.2 N—S Bidentates with N-Heterocyclic Donors

It has been suggested that π delocalization arising from an aromatic ring significantly enhances the stability of N—S chelates of 'soft' metal ions. For example, 8-mercaptoquinoline (thioxine) (29; R = R′ = H) in general forms stable, strongly coloured chelates with metals that form insoluble sulfides. These complexes are readily soluble in organic solvents, allowing their wide analytical use. Complexes of 3- and 5-halogenated derivatives show even higher solubility. Analytical extraction procedures have been reviewed.[105] The metal-exchange reaction of [Ni(thioxine)$_2$] with Cu^{II} ions has been investigated recently.[106] So also have the kinetics of the alkylation of the thiolato S atom of thioxine coordinated to Ru^{II}.[107]

A series of X-ray structural analyses of complexes of substituted forms of thioxine have appeared recently. The Cu^{II} and Zn^{II} bis chelates of the 5-methylthio-substituted form (29; R = H, R′ = SMe),[108] the Cd$^{II\,109}$ and Ni$^{II\,110}$ complexes of 2-methyl-8-mercaptoquinoline (29; R = Me, R′ = H) and the Ni$^{II\,111}$ and Co$^{II\,112}$ bis chelates of 2-isopropyl-8-mercaptoquinoline (29; R = Pr^i, R′ = H) all demonstrate the expected N—S bidentate behaviour of these ligands. In contrast the structures of the mono complexes of thioxine with triphenyl-lead and -tin show the ligand to be functioning solely as an S monodentate.[246]

(29)

The thioether 8-methylthioquinoline (**30**; R = H) forms complexes of the type $[M(N—SMe)_2X_2]$ with Pd^{II} and Pt^{II}, but with Na_3AuCl_6 S-demethylation occurs to yield the non-electrolyte $[Au(N—S)Cl_2]$,[79] which is identical to the compound formed with thioxine. Heating the Pd^{II} or Pt^{II} complexes in N,N'-dimethylformamide also results in dimeric S-demethylated compounds. Attempted S-demethylation of 2-(methylthio)methylpyridine (**31**; R = Me) complexes, however, results only in decomposition; it is thought that dealkylation reactions require an aromatic S substituent. The Au^{III} and Pd^{II} complexes of 2-methyl-8-methylthioquinoline do not demethylate, a result attributed to weakening of the M—S bond through steric hindrance effects.[113] In contrast bis{2-(2-mercaptoethyl)pyridine}-nickel(II) and -palladium(II) react with benzyl chloride leading to binuclear μ-thiolo complexes with the displaced ligand S-alkylated to form the corresponding benzyl thioether.

(30)   (31)

Octahedral complexes of the form ML_2X_2 (M = Co^{II} or Ni^{II}) are known with *trans*-(2-ethylthio)cyclopentylamine,[114] with 2-(methylthiomethylpyridine (**31**; R = Me)[115] and with 2-methyl-8-mercaptoquinoline (**29**; R = Me, R' = H).[116] With the latter sterically hindered ligand only the thiocyanate complexes are octahedral, the halide complexes being tetrahedral. Complexes of this ligand with Cu^{II} and Cu^{I} also were investigated. A number of complexes of 1-thiocarbamylpyrazole (**32**), and substituted forms, are known. These were discussed by Trofimenko in a review of pyrazole-derived ligands.[117]

Greenway *et al.*[118] investigated the blue and green crystalline products obtained from the reaction of Cu^{II} with the anticancer drug cimetidine (**33**). The crystal structure of the green complex shows it to be polymeric with the drug bonding through imidazole-N and thioether-S atoms. The blue form was determined as a Cu^{I} amide derivative resulting from the copper-catalyzed hydrolysis of the drug, and it was postulated that such metal ion interactions may be necessary for its biological activity.

(32)   (33)

As part of a study of Cu–thioether interactions, Ainscough *et al.* have investigated the coordination of 2-(3,3-dimethyl-2-thiabutyl)pyridine (**31**; R = Bu^t) with Cu^{I} and Cu^{II}.[119] The dimeric complexes of general formula $[(Cu^{II}L_2X_2)_2]$ (where X = Cl or Br) were prepared and characterized by electronic, ESR and IR spectroscopy. An X-ray analysis of the dibromo complex shows the Cu ion to be distorted five-coordinate, with the *t*-butyl groups blocking the sixth coordination site. A number of Cu^{I} complexes also were prepared and characterized.[120] The $[Cu^{I}L_2Br]$ complex, which was structurally characterized, shows one ligand molecule to be functioning as an N—S bidentate and the other as an S monodentate. In the monodentate form the pyridyl N atom does not participate in any inter- or intra-molecular bonding interactions. The crystal structure of bis(α-thiopicolinanildato)copper(II)[121] shows this ligand to function as a bidentate involving thiolate-S rather than imine-N. Jardine has reviewed the coordination chemistry of copper(I), including complexes of 2-pyridyl thioethers and 8-alkylthioquinolines.[122]

Several complexes of bivalent cobalt and zinc with imidazoline-($1H,3H$)-2-thione (**34**; R = H) of the forms ML_2X_2 (X = halide ion) and ML_4X_2 (X = NO_3^- or ClO_4^-) have been isolated and characterized by physicochemical techniques.[247] In contrast with the S monodentate behaviour concluded for the earlier-reported Ni^{II} complexes, the ligands are N—S bidentate in these metal salts to form four-membered chelate rings. Similar coordination behaviour has been observed for the ligand 2-thiobenzothiazole (**35**; X = C). The synthesis, properties and structures of several Cd^{II} complexes of this ligand have been reported.[123] They are thought to be important compounds for the activation of sulfur during metal-accelerated vulcanization of 'diene' rubbers. X-Ray analysis shows the tris-chelate geometry to be intermediate between octahedral and trigonal prismatic, with bidentate function through the N and exocyclic S atoms. The unusual bond angles observed at the donor N atom are thought to derive from the requirement of the Cd atom to achieve coordinative saturation. In the previously reported Zn^{II} chelate the ligand is N monodentate.

(34)   (35)

The formation of four-membered chelate rings involving N and S donors is not unusual and has been observed in several complexes of 2-thiol- or 2-thione-substituted N heterocycles. Foye and Lo[124] reported this form of coordination for a variety of metal chelates of antibacterial heterocyclic thiones. 2-Mercaptobenzimidazole (**35**; X = NH) was reported to form such chelates with Cu^{II}, Al^{II} and Fe^{III}. Complexes of 2-thiouracil (**36**), and substituted forms, also have been reported with this bonding mode to these metal ions.[125] The tetrakis chelate of W^{IV} with 2-mercaptopyrimidine has been shown to adopt a symmetrical dodecahedral geometry.[126] The mercapto S atom is thought to act as the better π-electron donor with the corresponding aromatic N atom acting as a π acceptor.

Goodgame *et al.*[127] reported on the versatile coordination behaviour of the closely related ligands 1-methylpyrimidine-2-thione (**37**; R = H) and 1,4,6-trimethylpyrimidine-2-thione (R = R′ = Me)[128] towards the metal ions of the first row transition series, concluding N—S coordination with Mn^{II}, Cu^{II}, Ni^{II} and Co^{II} in ligand:metal ratios of 1:1, 2:1 or 3:1. This coordination mode was confirmed for a Co^{III} tris chelate of 4,6-dimethylpyrimidine-2-thione by X-ray diffraction.[129] The Cu^{II} chelate of 1-methyl-2-mercaptoimidazole (**34**; R = Me) has received study as a model for metal–sulfur coordination in copper proteins. Again N—S coordination is concluded.[124,130,131]

(36)   (37)

Dance and Guerney[248] isolated a number of water-soluble complexes of (**38**) and X-ray analysis of a complex of Pb^{II} demonstrated its N—S coordination. The cyclic molecule 5-methyl-1-thia-5-azacyclooctane (**39**) forms complexes with Pd^{II} chloride and iodide in which it adopts a boat–chair configuration. These provide the first examples of bidentate chelates in which both thioether and tertiary amine groups are coordinated.[132] ^{1}H NMR analysis shows the inflexible chloride complex to remain bidentate in solution, whereas in the iodide chelate the ligand is S monodentate in dichloromethane.

(38)   (39)

20.4.3.3 Thiosemicarbazide and Related Ligands

Thiosemicarbazide (tscH) (**40a**) exists in the two tautomeric forms shown in (**41**). It may coordinate either as a neutral or (by loss of a proton) anionic N—S bidentate. The nickel(II) complexes in particular of thiosemicarbazide and related ligands (**40**) have been studied in considerable detail.

A very comprehensive review by Campbell[133] on the coordination chemistry of thiosemicarbazide and thiosemicarbazones provides details of the thermodynamic, structural, magnetic and vibration-

$$XNNHC(=S)Y$$

(40)

(41)

a; thiosemicarbazide; $X = H_2$, $Y = NH_2$
b; thiosemicarbazone; $X = R_2C$, $Y = NH_2$
c; S-methyldithiocarbazate; $X = H_2$, $Y = SMe$
d; dithiocarbazic acid; $X = H_2$, $Y = SH$
e; thiocarbohydrazide; $X = H_2$, $Y = NHNH_2$
f; N-substituted thiocarbohydrazide; $X = RH$, $Y = NHNH_2$

al and electronic spectral properties of their metal complexes. Both ligands usually act as S—N bidentate chelating agents through the sulfur and hydrazinic nitrogen atoms, although some examples of monodentate S bonding are known. The broad pharmacological activity of these species is thought to be dependent on metal chelate formation and this has resulted in a high degree of research activity in this area. Campbell concludes that the nature of the sulfur donor atom is the most important single factor affecting the behaviour of these ligands. This leads to a high degree of thermodynamic stability for complexes formed with 'soft' or 'marginally soft' metal ions. The stereochemistries of N^3-substituted complexes are largely influenced by steric effects involving the sulfur atoms.

The complexes $[Ni(tscH)_2X_2]$ with $X = Cl$, Br or I are diamagnetic, whereas with $X = ClO_4$ or SCN the complexes are spin-free. The general complexes $[Cu(tscH)_2X_2]$ are distorted octahedral, $[Zn(tscH)X_2]$ is tetrahedral and $[Ag(tscH)Cl]$ is polymeric with the ligand bound through sulfur only. Complexes of the deprotonated ligand of formula $[M(tsc)_2]$ (M = Ni, Pd, Pt) and the octahedral complexes $[Co(tscH)_3]Cl_3$ and $[Co(tsc)_3]$ are each thought to exist in both *cis* and *trans* isomeric forms.[79] *cis*-$[Co(tscH)_3]Cl_3$ is more stable than the *trans* isomer. Both cationic forms have been resolved and their observed inversion and isomerization reactions studied in aqueous solution.[144] The crystal structures of $[Co(tsc)_3]$ and $[Rh(tsc)_3]$ have been reported.[135,136] Several copper(II) mono and bis chelates have been isolated with tscH. The sulfate bis complex is square-based pyramidal with the two N—S ligands *cis* in the base and a sulfate O coordinated in the apex. *Cis* planar coordination of tscH in the 2:1 compounds appears to be a common structural feature.

Bidentate thiosemicarbazones, such as that derived from acetone (**40b**; $X = Pr^i$, $Y = NH_2$), form complexes similar to those of neutral tscH. An X-ray analysis shows the nickel chloride complex to have a trigonal bipyramidal structure. An isovaleryl-substituted form, however, was concluded to behave as an N—O bidentate only to Co^{II}, Ni^{II}, Cu^{II}, Cd^{II} and Zn^{II}.[137] The analytical applications of thiosemicarbazones have been reviewed.[138,139] Thiocarbohydrazide (tch, **40e**) acts as a neutral bidentate coordinating through the thione sulfur and terminal N atoms. Complexes of general form $[M(tch)_2X_2]$ (M = Zn, Cd, Hg, Pd, Fe or Co) and $M(tch)_3Cl_2$ (M = Ni or Cu) have been reported. N-Substituted derivatives (**40f**) coordinate in both neutral and ionic forms, often resulting in polymeric complex structures.[79]

Dithiocarbazic acid (**40d**) forms 2:1 complexes with bivalent nickel, platinum, zinc, cadmium and lead, with trivalent rhodium, and a 3:1 complex with chromium(III). Electronic spectral details are consistent with N—S rather than S—S coordination. Similar donor behaviour is exhibited by α-N-substituted derivatives, although β-N substituents may change the mode of coordination to S—S. S-Methyl derivatives are known to form stable complexes in both neutral and deprotonated forms, with bidentate N—S coordination being indicated.[140] Recent X-ray analyses of Pd^{II}, Ni^{II}, Co^{II} and Pt^{II} S-methyl derivatives of dithiocarbazic acid verify this coordination mode. The Pd^{II} chelate is *cis* planar[141] with a smaller tetrahedral distortion than was observed in a previously reported Ni^{II} analogue. Three closely related tris chelates of Ni^{II} were all reported to be distorted octahedral.[142] It was concluded that in two of these compounds one of the three ligands is deprotonated at the α nitrogen resulting in some structural disorder. The cobalt complex of the α-N-methyl-substituted form, $[Co(MeL)_2Cl_2]$, is *cis* octahedral, with evidence suggesting thermal disproportionation to a $[Co(MeL)_3][CoCl_4]$ species.[143] The structure of the Ni^{II} tris chelate of this ligand has been reported also.[144] A Pt^{II} bis chelate of a β-N-phenyl derivative shows the metal to have a *trans* planar geometry.[145] The Pt—N and Pt—S bond lengths are considerably shorter than in the corresponding thiosemicarbazide chelate, and the short N—N distance indicates significant double bond character.

3-Mercapto-1,5-diphenylformazan (dithizone, dzH) (**42**) was first prepared by Emil Fischer over a century ago.[146] In 1925 Hellmut Fischer[147] introduced dithizone as a versatile analytical reagent and subsequently explored its use for the solvent extraction and quantitative determination of a number of metals of industrial and toxicological interest. The conditions for the isolation and determination of a wide variety of metals as their dithizonates or substituted dithizonates have been well reviewed.[102,105] The ligand forms neutral complexes with 'soft' metal ions and with several 'borderline' metals.[148]

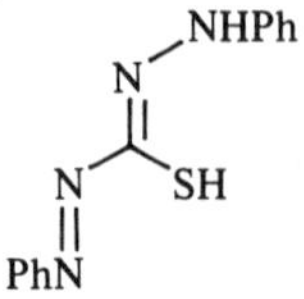

(42)

The crystal structures of methyl- and phenyl-mercury(II) dithizonates have been determined as part of a study on the photochroism of such compounds.[149] In both structures the chelate is planar, with an irregular three-coordination at the metal including N and S of the ligand. The article provides a short structural review of dithizone complexes.

20.4.3.4 N—S Bidentates with Other N Donor Groups

Some examples of three-membered chelate rings containing N and S are known. *N*-Sulfinyl-aniline, and substituted forms, has been shown to bond in this manner to give chelates of the type (43) with a number of metal ions in low oxidation states. Such compounds have been reported with Ni[0],[150] with Rh[I] and Ir[I],[151] and with Fe[0].[152] A similar bonding mode has been demonstrated for 4-(methyl)thionitrosobenzene in a bridged adduct of general form $[Fe_2(CO)_6L]$.[153]

N,N-Dimethyl-*N'*-phenylthiourea has been shown to coordinate to Rh[III] as an N—S bidentate involving four-membered chelate ring formation.[154] N-Substituted thioamides also may bond in this manner.[155,156] 1-Amidino-2-thioureas (44) may behave either as N—S or as N—N bidentates, with this donor choice being dependent mainly on pH and the nature of the metal ion.[157] As N—S donors they are known to stabilize lower oxidation states.[158] As part of a study on Mo—S-containing complexes as models for redox-active molybdoenzymes, Dilworth *et al.* have shown that some *p*-(substituted)phenylhydrazines may coordinate as N—S bidentates in three different ways to one metal atom.[159] The three diazenido, diazene and hydrazonido forms vary in their degree of deprotonation and therefore their anionic nature.

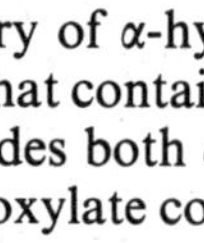

(43) 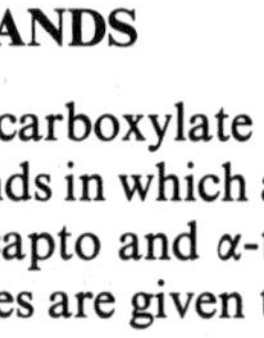   (44)

20.4.4. BIDENTATE SULFUR–OXYGEN LIGANDS

In a review of the stereochemistry of α-hydroxycarboxylate complexes, Tapscott[249] includes the structural chemistry of complexes that contain ligands in which a sulfur atom is attached to a carbon α to a carboxylic group. This includes both α-mercapto and α-thioether groups. Lists of structural bonding parameters for α-thiocarboxylate complexes are given together with parameters for chelate rings.

20.4.4.1 S—O Bidentates with Thiol Group Donors

Like their amine analogues (Section 20.4.2.1) thiol-substituted alcohols may act as bidentates with chelation often leading to polymeric products. The stability constants of the Cd[II] complexes of both 2-mercaptoethanol (mel) and 3-mercapto-1,2-propanediol (mpd) have been determined,[160] with polymeric structures being concluded similar to those of the previously determined Ni[II] and Zn[II] compounds. Similar data for the Pb[II] complexes[161] showed polymer formation with mpd and gave evidence for Pb–mel species with at least five different metal:ligand ratios. Reaction of the pertechnetate(VII) ion $[TcO_4]^-$ with sodium dithionite and 2-mercaptoethanol results in the chelated ligand having both O and S atoms deprotonated. In the square pyramidal $[Tc(O)(SCH_2CH_2O)_2]^-$ product the S—O bidentates adopt a *cis* configuration in the base.[162]

The potentially bridging ligand (45) was prepared in an attempt to form binuclear complexes involving thiolate bridging.[163] Mononuclear bis chelates only were isolated with Cu^{II} and Ni^{II} in either anhydrous or hydrated forms. The diamagnetic NiL_2 chelate is *cis* planar with weak bonding to the metal atom from O atoms of adjacent molecules. The structure of the Ni^{II} chelate of a closely related ligand was reported also.[164]

2-Mercaptopyridine 1-oxide (46) acts as an S—O bidentate on deprotonation of the thiol group. Formation constants for the 1:1 chelates of this ligand with a variety of metal ions were compiled by Garvey *et al.*[165] in a review of the coordination chemistry of heterocyclic *N*-oxides. Synthetic, structural and bonding aspects of these mixed donor ligands were covered in later reviews by Karayannis *et al.*[166,167]

MeO
=O
—SH
=O
MeO

(45)

N SH
O

(46)

The anionic ligands 2-mercaptoethanoate ($^-SCH_2CO_2^-$), 3-mercaptopropanoate ($^-SCH_2CH_2CO_2^-$) and *o*-mercaptobenzoate ($^-SC_6H_4CO_2^-$) have received detailed study. Their early coordination chemistry was reviewed by Harris and Livingstone.[168] The reactivity of the complex cation $[Co^{III}(en)_2(SCH_2CO_2)]^+$ has been investigated in detail by Deutsch and co-workers. Oxidation of this species results in the S—O-chelated monothiooxalate ion (mtox).[169] This complex ion also was prepared by direct reaction of $mtoxH_2$ with $[Co(en)_2Cl_2]^+$. The same species was prepared by Gainsford *et al.*[170] and its crystal structure reported. A detailed mechanism was proposed to account for the observed oxidation of the α carbon of the S—O bound 2-mercapto-ethanoate species. An earlier structural study of the $[Mo(O)(S_2)_2(mtox)]^{2-}$ ion demonstrated this S—O bonding mode.[250]

20.4.4.2 S—O Bidentates with Thioether Donor Groups

Complexes of substituted thioacetic acids of the form $RSCH_2CO_2H$ show decreasing stability constants in the order $Ag^I \gg Cu^{II} \gg Cd^{II} = Co^{II} > Mn^{II} = Zn^{II} > Ni^{II}$.[171] Comparative data for the silver complexes suggest that the strength of the M—S bond is largely due to its σ component, although π back-donation may vary linearly with inductive effects.[172] The early coordination chemistry of *S*-methyl- and *S*-ethyl-alkanoic acids was reviewed in 1964.[168] A more recent short review of the structural chemistry of alkylthioacetato complexes has been provided by Ouchi *et al.*[173] These authors reported the structure of $[Co^{II}(MeSCH_2CO_2)_2(H_2O)_2]$ to be a distorted *trans* octahedral monomer. Dehydration leads to a hexameric form similar to the *S*-propyl and *S*-isopropyl analogues.[174] The complex $[Cu^{II}(EtSCH_2CO_2)_2(H_2O)_2]$ also is a distorted *trans* octahedral monomer. The EPR spectra of this and related chelates have been reported.[175] Many thioether–carboxylate ligands are water soluble and show high selectivity for mercury(II) in an aqueous environment.[176]

Studies on Co^{III} complexes of the type $[Co(en)_2\{RS(CH_2)_nCO_2\}]^{2+}$ have been performed by several workers. The formation of these thioether complexes by alkylation of the corresponding coordinated thiolates has been reported by Elder *et al.*[177] A separate study of complexes of this type (with $n = 1$ or 2) has shown the thioether donors to coordinate stereospecifically.[178] Analogous complex cations of the form $(\pm)$-$[Co^{III}(tren)L]^{n+}$ (were $L = MeSCH(Me)CO_2^-$, $^-SCH(Me)CO_2^-$ or $^-O_2SCH(Me)CO_2^-$) have been isolated and their racemization and methine-H deuteration reactions investigated.[179] The *S*-methyl and sulfinato chelates racemize at pH 6–7, whereas the thiolato optical form was stable. Reactivity differences are explained in terms of the different net charges on the complex ion and the different electron-withdrawing effects of the S^-, Me and $S(O)_2^-$ donor groups. The structure of dichlorobis(2,2'-thiodiethanol)cobalt(II) shows the potentially O—S—O terdentate ligand to behave as a bidentate through thioether S and one OH group only in a *trans* octahedral geometry.[180]

20.4.4.3 S—O Bidentates with Other S Donor Groups

Sulfur dioxide is known to behave as an S—O bidentate to metal ions in low oxidation states, resulting in the formation of three-membered chelate rings. The structures of three such complexes of molybdenum have been described by Kubas and Ryan,[181,182] including an unusual dinuclear complex in which the SO_2 ligands bridge two Mo atoms.[182] A closely related complex of Ru also has been characterized.[183] This η^2-SO_2 bonding form was discussed in relation to these structures. Monothiocarboxylate ions (47) may coordinate in several ways; as simple S monodentates, as bridging ligands similar to carboxylate ions or as S—O bidentates. Further details of the coordination chemistry of these species are provided in Chapter 16.6 of this volume.

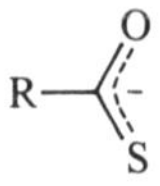

(47)

The reaction of carbonyl sulfide with $[M(O_2)(PPh_3)_2]$ (where M = Pd or Pt) has resulted in the first reported examples of transition metal complexes of the monothiocarbonate anion (47; R = O^-). Bidentate S—O coordination was concluded from $^{31}P\{^1H\}$ NMR analyses of these compounds.[184] A short structural review of metal complexes of monothiocarbamate ions (47; R = N(R′)R″) demonstrates their varied coordination chemistry.[185] In complexes of the dialkyl forms the sulfur atom is seen to have considerable 'mercaptide character', whereas aromatic amine derivatives demonstrate C—S and M—S partial multiple bonding.[185,251] A review on the coordination chemistry of these ligands has appeared.[186] Additional detail is provided in Chapter 16.4 of this volume.

The coordination chemistry of complexes of thiohydroxamic acids (48) was reviewed in 1968.[187] These species have been shown to behave as S—O bidentates to a wide variety of metal ions. The Cu^{II} chelates are known to play an important role as antibiotics.[188] Freyberb *et al.*[189] reported the preparation and structural characterization of the tris chelates of Co^{III}, Cr^{III}, Fe^{III} and Mn^{III} with *N*-methylthiobenzohydroxamic acid (48; R = Ph, R′ = Me). The complexes are isostructural, with the metal ions in trigonally distorted *cis* octahedral geometries, the Mn^{III} chelate showing some tetragonal distortion. Bond distances indicate a considerable contribution from tautomer (48b). A related tetrakis chelate of hafnium(IV) was prepared as part of a study on the design and synthesis of specific sequestering agents for actinide(IV) ions.[190] The crystal structure shows a symmetrical bicapped trigonal prismatic coordination. Becher *et al.*[191] reported the synthesis, spectral and electrochemical properties of a range of Cu^{II} thiohydroxamates. Detailed ESR analysis showed the chelates to have *trans* planar $\{S_2O_2\}$ chromophores. The closely related ligand 1,1-diethyl-3-benzoylthiourea (49) forms a bis chelate of Cu^{II} which exists in three crystal modifications.[192] X-Ray analysis of one form shows a distorted tetrahedral Cu geometry.

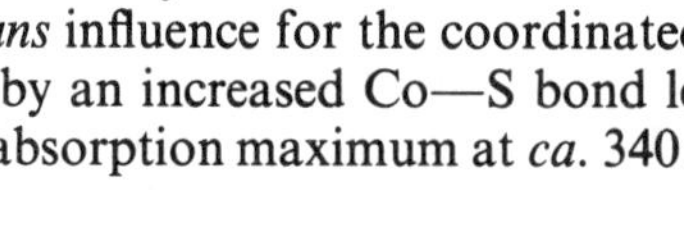

(48a)   (48b) 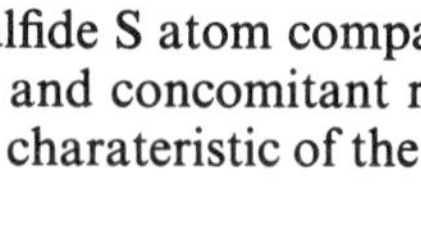   (49)

Lydon *et al.* have reported the synthesis of a series of S—O bidentate complexes of Co^{III} with disulfide S donors.[193] The crystal structure of the complex formed with $HO_2CCH_2SSCH_2CO_2H$ shows a decreased *trans* influence for the coordinated disulfide S atom compared with the thiolato precursor, evidenced by an increased Co—S bond length and concomitant reduction of the *trans* Co—N distance. An absorption maximum at *ca.* 340 nm is charateristic of the coordinated disulfide group.

20.4.5 MIXED DONOR BIDENTATES CONTAINING SELENIUM

20.4.5.1 Se—N Bidentates

The formation constants of the Cu^{II}, Zn^{II}, Ni^{II} and Pd^{II} chelates of selenophene-2-aldoxime (50) were determined by Bark and Griffen[194] as part of a study on O, S and Se analogous ligands. The unusual order Se ≫ S > O which was found was attributed to the aromatic nature of the chelates.

The aromatic ligand 2-(2-pyridyl)benzoselenophene (**51**) was investigated as a selective reagent for Pd^{II}.[195] The selenium analogue of thioxine, 8-selenoquinoline (**52**), has been studied in some detail. A report of the thin-layer chromatographic behaviour of a large number of mono, bis, tris and tetrakis chelates of this ligand has appeared.[196] An IR spectral study on chelates of the ligand gave evidence for two different structural types based on their pH-controlled precipitation. Diselenide formation on aerial oxidation results in dimeric impurities.[197]

The Pt^{II} bis chelates of some S- and Se-containing nucleosides or nucleotides have been synthesized and evaluated for antitumour activity in both *in vitro* and *in vivo* systems. The increased effectiveness of the *trans* chelate of selenoguanine is believed to be due to its slow dissociation and consequent release of the ligand into the blood stream.[198] The delayed cytotoxicity observed for the Se compound was not obtained with the S analogue.[199] The *cis*-diammineplatinum(II) 1:1 chelates of thio- and seleno-purines also have been prepared and evaluated against L1210 cells in mice. The Se chelate is active, with its stability affected by mouse serum, although the reason for the latter is unclear.[200]

The coordination chemistry of the Se—N bidentate selenosemicarbazide (**53**) has received wide study, generally in comparison with its thio analogue (Section 20.4.3.3). An IR study has concluded bidentate behaviour in the bivalent Ni, Cu and Zn chelates, but with monodentate Se bonding occurring with Hg^{II}, Ag^{I}, Cd^{II} and Cu^{I}.[201] The stability constants and heats of formation of the copper(II) bis chelates of selenosemicarbazide and its O and S analogues were determined by Goddard and Ajayi.[202] The results demonstrated the affinity order of the donor atoms to be O $\ll$ Se > S, whereas the enthalpy changes follow the order O $\ll$ S > Se. The tris chelates of bivalent nickel demonstrated the order Se < O < S for both of these parameters.[203] Detailed vibrational spectra analyses of several bis chelates of this ligand have appeared.[204,205]

The synthesis and characterization of a variety of Zn^{II}, Cd^{II} and Ni^{II} chelates of acetone- and cyclohexanone-derived selenosemicarbazones have been reported.[206,207] A benzaldehyde derivative is shown to be an Se—N bidentate in the Co^{II} and Ni^{II} chelates of general form [ML$_2$(NCX)$_2$] (where X = S or Se), but acts as a monodentate only in the [ML$_4$(NCX)$_2$] chelates.[252] The analytical applications of the selenium analogue of dithizone (**42**) have been investigated.[208]

The synthesis of a variety of metal complexes of selenocysteamine (NH$_2$CH$_2$CH$_2$SeH) was reported by Sakurai *et al.*[209] The structures proposed were similar to known complexes of cysteamine. Chelates of selenocysteamine-based ligands have demonstrated catalytic activity towards the reduction of acetylene, with this activity varying with change in metal ion and with O or S substitution for Se in the order S > Se $\gg$ O. This subject was surveyed systematically by Sugiura *et al.*[210] The crystal structure of the [Co(en)$_2$(SeCH$_2$CH$_2$NH$_2$)]$^{2+}$ anion shows a significant *trans* effect by the Se donor, leading to a lengthening of the opposite Co—N bond.[211] This cation and its closely related selenonyl and seleninyl derivatives have been resolved into their optical forms, which were characterized by their electronic and circular dichroism spectra.[212] The synthesis of several seleno-ether derivatives by reaction of this selenocysteamine chelate with alcohols, unsaturated species or alkyl halides has been reported.[213] A methyl selenoether analogue containing *N*,*N*-bis(2-amino-ethyl)-1,2-ethanediamine was prepared and optically resolved using chromatographic methods.[214] Inversion at Se was observed in the two isomers.

The formation of a large variety of bis chelates with selenosalicylaldimines containing bulky N-substituents has been reported.[215] The bivalent Co and Zn complexes are believed to be tetrahedral, whereas the Ni chelates are *cis* planar. A kinetic study of the enantiomerization of the tetrahedral Zn^{II}, Cd^{II} and Pb^{II} bis chelates of an isopropyl-substituted form, and its O and S analogues, showed the E_a values for the reactions to be affected by both choice of metal and O, S or Se donor atom.[216]

20.4.5.2 Se—O Bidentates

Preti and Tosi[253] reported a unique example of Se—O bidentate function of substituted benzeneselenic acids to form three-membered chelate rings in monomeric complexes of Ir^{III}. A dimeric dimethylcarbamoselenoato chelate of tricarbonylrhenium has been prepared by thermolysis of a monomeric pentacarbonyl precursor.[217] This product was identified by IR and mass spectrometry and shown to involve both Se—O bidentate and Se bridging behaviour in the dimer.

Cobalt(III) chelates of the α-selenoacetate dianion $SeCH_2CO_2^{2-}$ have been investigated by several workers. The crystal structure of the $[Co(en)_2(SeCH_2CO_2)]^+$ ion was reported by Elder *et al.*[218] to demonstrate a lengthening of the Co—N bond *trans* to Se, an effect similar to thiolato analogues. The photochemistry of this cation, and its thiolato analogue, has been investigated by Houlding *et al.*[219,220] The circular dichroism and magnetic CD of the ions were reported. The Se chelates showed photoreduction at all wavelengths, unlike the S analogues. The results obtained were rationalized in terms of the presence or absence of anisotropic π-donor interactions between the metal ion and the S- or Se-containing ligands.

Metal chelates of substituted seleno-β-dicarbonyl ligands of general form (54) have been investigated in some detail. The syntheses of complexes of (54) with X = CH, R = R' = Ph have been reported, of the form ML with Tl^I, ML$_2$ with Ni^{II}, Pd^{II}, Hg^{II}, Zn^{II}, Pb^{II} and Cd^{II}, ML$_3$ with Co^{III} and In^{III} and Ni^{II}L$_2$L' (where L' = *o*-phen or bipy).[221] A similar nickel(II) chelate with X = CPh, R = Ph and R' = H also was isolated.[222] A general method of synthesis of NiL$_2$ and CoL$_3$ chelates of (54) with X = CH and R = Ph, prepared from the reaction of the corresponding isoselenuronium salts with β-diketones, has been reported.[223] Diamagnetic chelates of 1,1-diethyl-3-benzoyl-selenourea (54; X = N, R = NEt$_2$ and R' = Ph) also have been reported. The 1:1 chelate with Tl^I, 2:1 complexes of Pd^{II} and Ni^{II} and the compound Co^{III}L$_3$ were studied by ^{1}H NMR methods.[224,225] Coordination of the ligand to these metals was observed to lead to an increase of the rotational barrier about the terminal N—C(Et) bond. These energy barriers are found to vary both with changes in the metal ion and with substitution of Se for S. An analysis of the ESR spectrum of the 2:1 Cu^{II} chelate of this ligand demonstrated the presence both of a rhombic distortion and of $4s$–$4p$ hybridization in the Se atom.[226]

(54)

An ESCA investigation of some closely related N-substituted forms of (54) with X = N, R = NMePh, NPr$_2$ or NBu$_2^i$ and R' = Ph, using the chemical shifts of N $1s$, Se $3p$ and metal nl ESCA lines, demonstrated pronounced σ character in the M—O bond and a decrease in electron density on Se on coordination.[227] A mass spectral study of chelates of Ni^{II} and Cu^{II} with the NBu$_2^i$-substituted form of the ligand, and analogues with Se substituted by O or S, demonstrated fragmentation of the M—O bonds to be easier than that of M—S or M—Se.[228]

20.4.5.3 Se—S Bidentates

Carbamoselenothioic acids have been shown in several examples to act as S—Se bidentates leading to the formation of four-membered chelate rings. In a study of the coordination chemistry and photochemical properties of tetravalent-metal tris chelates of dichalcogenocarbamates, for example, Eckstein and Hoyer[229] reported the formation of such complexes with Ni^{IV} and diethyl-carbamoselenothioic acid. A ligand exchange study involving Cu^{II} and Ni^{II} chelates of this same ligand also has been reported.[230] As part of an electrochemical study on mixed donor carbamate ligands the dimethylcarbamatoselenothioate analogue has been shown to bind in this way to iron(II).[231] A crystal structure analysis of [Ni(SeC(S)=C(CN)$_2$)$_2$] shows this rigid ligand to coordinate through both S and Se.[232]

A number of chelates of Pt^{II} and Pt^{IV} with the simple bidentates MeSCH$_2$CH$_2$SeMe (55) and *o*-MeSC$_6$H$_4$SeMe (56) have been studied by NMR band shape analysis in an investigation of the barrier to inversion at coordinated S and Se.[233,254] Evidence is provided for fluxional movement effecting a 180° rotation of the aromatic bidentate ligand. It was concluded that $3p$–$2p$ π conjugation between the inverting centre and the ligand backbone causes a 10–12 kJ mol^{-1} decrease in this inversion barrier on going from aliphatic to aromatic alkenic ligand backbones. The effects on these energetic barriers of differences in ligand ring size, coordinated halogen and metal oxidation state are discussed. The methylation and demethylation reactions of the Pd^{II} chelate of (56) have been the subject of a kinetic study using ^{13}C NMR methods.[234] The alkylation reaction was shown to be approximately 50 times faster than dealkylation, with an increase in the carbonium ion character of the methyl group indicated on coordination.

The ligand o-HSC_6H_4SeH (**57**), in deprotonated form, has been shown to stabilize the trivalent oxidation state of Ni, Cu and Co,[235] with rapid oxidation of the $[M^{II}L_2^{2-}]$ species to $[M^{III}L_2^-]$ being observed. The anionic chelates were characterized by magnetic and polarographic methods, and reduction back to the bivalent forms was achieved using hydrazine. A nickel bis chelate (**58**) of the anionic ligand α-(selenylmethylene)benzenemethanethiol has been reported as one product of the photolysis reaction of $[Ni(CO)_4]$ with a number of heterofulvalenes.[236]

Several mono, bis or tris chelates of Ni^{II}, Cu^I, Zn^{II} and Co^{III} have been isolated with ligands derived from thioseleno-β-dicarbonyl derivatives of the general form (**59**) with X = CH or N, R = NHPh, R' = Ph or R = Ph, R' = NHPh. Uhlemann *et al.* have reported the isolation of such compounds and their characterization using ESCA techniques.[237] A related bis chelate of Ni^{II} (**59**; with X = N, R = Ph and R' = NEt_2), which was characterized by electronic and mass spectroscopy, has been reported recently.[238]

20.4.6 REFERENCES

1. R. E. Cobbledick, F. W. B. Einstein, W. R. McWhinnie and F. H. Musa, *J. Chem. Res. (S)*, 1979, 145; *(M)*, 1901.
2. J. A. Bertrand and P. G. Eller, *Prog. Inorg. Chem.*, 1976, **21**, 29.
3. F. A. Cotton and G. N. Mott, *Inorg. Chem.*, 1983, **22**, 1136.
4. L. Kolosci and K. Schug, *Inorg. Chem.*, 1983, **22**, 3053.
5. T. Lindgren, R. Sillanpaa, T. Nortia and K. Pihlaja, *Inorg. Chim. Acta*, 1984, **82**, 1.
6. H. Muhonen, A. Pajunen and R. Hamalainen, *Acta Crystallogr., Sect. B*, 1980, **36**, 2790.
7. H. Muhonen, *Acta Crystallogr., Sect. B*, 1981, **37**, 951.
8. J. A. Bertrand, P. G. Eller, E. Fujita, M. O. Lively and D. G. VanDerveer, *Inorg. Chem.*, 1979, **18**, 2419.
9. G. Nieuwpoort and G. C. Verschoor, *Inorg. Chem.*, 1981, **20**, 4079.
10. G. Nieuwpoort, A. J. deKok and C. Romers, *Recl. Trav. Chim. Pays-Bas*, 1981, **100**, 177.
11. K. J. Franklin and M. F. Richardson, *Inorg. Chem.*, 1980, **19**, 2107.
12. D. M. Thompson, W. Balenovich, L. H. Hornich and M. F. Richardson, *Inorg. Chim. Acta*, 1980, **46**, 199.
13. S. F. Pavkovic and J. N. Brown, *Acta Crystallogr., Sect. B*, 1982, **38**, 274.
14. S. F. Pavkovic and S. L. Willie, *Acta Crystallogr., Sect. B*, 1982, **38**, 1605.
15. D. X. West, S. F. Pavkovic and J. N. Brown, *Acta Crystallogr., Sect. B*, 1980, **36**, 143.
16. D. A. d'Avignon and T. L. Brown, *Inorg. Chem.*, 1982, **21**, 3041.
17. R. C. Aggarwal, R. Bala and R. L. Prasad, *Indian J. Chem., Sect. A*, 1983, **22**, 955.
18. S. Yizhen, Q. Jusheng, F. Haifu, Q. Jin-Zi, X. Zhangbao, F. Chuangui and L. Dehong, *Sci. Sinica*, 1981, **24**, 11111.
19. J. P. Phillips, *Chem. Rev.*, 1956, **56**, 271.
20. A. Shulman and F. P. Dwyer, in 'Chelating Agents and Metal Chelates', ed. F. P. Dwyer and D. P. Mellor, Academic New York, 1964, p. 383.
21. J. Stary, 'The Solvent Extraction of Metal Chelates', Pergamon, Oxford, 1964, p. 80.
22. E. Miki, K. Saito, K. Mizumachi and T. Ishimori, *Bull. Chem. Soc. Jpn.*, 1983, **56**, 3515.
23. Y. Yamamoto and E. Toyota, *Bull. Chem. Soc. Jpn.*, 1984, **57**, 47.
24. W. Clegg, M. Berry and C. D. Garner, *Acta Crystallogr., Sect. B*, 1980, **36**, 3110.
25. W. Clegg, C. D. Garner and M. H. Al-Samman, *Inorg. Chem.*, 1983, **22**, 1534.
26. V. M. Ellis, R. S. Vagg and E. C. Watton, *J. Inorg. Nucl. Chem.*, 1974, **36**, 1031; *Aust. J. Chem.*, 1974, **27**, 1191.
27. C. Pelizzi and G. Pelizzi, *Transition Met. Chem. (Weinheim, Ger.)*, 1981, **6**, 315.
28. B. N. Figgis, C. L. Raston, R. P. Sharma and A. H. White, *Aust. J. Chem.*, 1978, **31**, 2545.
29. C. L. Klein, R. J. Majeste, L. M. Trefonas and C. J. O'Connor, *Inorg. Chem.*, 1982, **21**, 1891.
30. C. L. Klein, L. M. Trefonas, C. J. O'Connor and R. J. Majeste, *Cryst. Struct. Commun.*, 1981, **10**, 891.
31. C. J. O'Connor, C. L. Klein, R. J. Majeste and L. M. Trefonas, *Inorg. Chem.*, 1982, **21**, 64.
32. P. P. Richard, D. T. Qui and E. F. Bertant, *Acta Crystallogr., Sect. B*, 1974, **30**, 628.
33. R. L. Chapman, F. S. Stephens and R. S. Vagg, *Inorg. Chim. Acta*, 1978, **26**, 247.
34. C. J. O'Connor and E. Sinn, *Inorg. Chem.*, 1981, **20**, 545.
35. H. Sigel and R. B. Martin, *Chem. Rev.*, 1982, **82**, 385.
36. M. Nonoyama and K. Nonoyama, *Kagaku no Ryoiki*, 1977, **31**, 46.
37. J. Swartz and F. C. Anson, *Inorg. Chem.*, 1981, **20**, 2250.
38. F. Charbonnier, R. Fanre and H. Loiseleur, *Cryst. Struct. Commun.*, 1981, **10**, 1129.
39. R. M. Izatt, J. J. Christensen and J. H. Rytting, *Chem. Rev.*, 1971, **71**, 439.
40. D. J. Hodgson, *Prog. Inorg. Chem.*, 1977, **23**, 211.
41. L. G. Marzilli, *Prog. Inorg. Chem.*, 1977, **23**, 255.
42. E. Sletten, *Chem. Commun.*, 1971, 558.
43. E. W. Ainscough, A. M. Brodie, J. E. Plowman, K. L. Brown, A. W. Addison and A. R. Gainsford, *Inorg. Chem.*, 1980, **19**, 3655.
44. I. Mutikainen and P. Lumme, *Acta Crystallogr., Sect. B*, 1980, **36**, 2233.
45. T. Solin, K. Matsumoto and K. Fuwa, *Bull. Chem. Soc. Jpn.*, 1981, **54**, 3731.

46. S. Bristow, D. Collison, C. D. Garner and W. Clegg, *J. Chem. Soc., Dalton Trans.*, 1983, 2495.
47. K. Wieghardt, M. Hahn, J. Weiss and W. Swiridoff, *Z. Anorg. Allg. Chem.*, 1982, **492**, 164.
48. M. Minelli, J. H. Enemark, K. Wieghardt and M. Hahn, *Inorg. Chem.*, 1983, **22**, 3952.
49. M. H. Dickman and R. J. Doedens, *Inorg. Chem.*, 1982, **21**, 682.
50. H. Saarinen, J. Korvenranta and E. Nasakkala, *Acta Chem. Scand., Ser. A*, 1978, **32**, 303.
51. H. Saarinen, J. Korvenranta and E. Nasakkala, *Acta Crystallogr., Sect. B*, 1979, **35**, 963.
52. D. A. White, *J. Chem. Soc. (A)*, 1971, 233.
53. H. Tamura, K. Ogawa, R. Ryu, M. Tanaka, T. Shono and I. Masuda, *Inorg. Chim. Acta*, 1981, **50**, 101.
54. B. N. Figgis, C. L. Raston, R. P. Sharma and A. H. White, *Aust. J. Chem.*, 1978, **31**, 2437.
55. C. L. Raston, A. H. White and R. M. Golding, *J. Chem. Soc., Dalton Trans.*, 1977, 329.
56. B. N. Figgis, C. L. Raston, R. P. Sharma and A. H. White, *Aust. J. Chem.*, 1978, **31**, 2431.
57. M. J. Lacey, C. G. MacDonald, J. F. McConnell and J. S. Shannon, *Chem. Commun.*, 1971, 1206.
58. A. Sreekantan, C. C. Patel and H. Manohar, *Transition Met. Chem. (Weinheim, Ger.)*, 1981, **6**, 214.
59. S. Candelero, D. Gardenic, N. Taylor, B. Thompson, M. Viswamitra and D. C. Hodgkin, *Nature (London)*, 1969, **224**, 589.
60. A. Ballio, S. Barcellona, E. B. Chain, A. Tonolo and L. Verro-Barcellona, *Proc. R. Soc. London, Ser. B*, 1964, **161**, 384.
61. G. Rindorf, *Acta Chem. Scand.*, 1971, **25**, 774.
62. M. S. Ma, R. J. Angelici, D. Powell and R. A. Jacobson, *Inorg. Chem.*, 1980, **19**, 3121.
63. F. Abraham, G. Nowogrocki, S. Seuer and C. Bremard, *Acta Crystallogr., Sect. B*, 1980, **36**, 799.
64. A. Muller, N. Mohan, S. Sarka and W. Eltzner, *Inorg. Chim. Acta*, 1981, **55**, L33.
65. Yu. V. Kokunov, S. G. Sakharov, I. I. Moiseev and Yu. A. Buslaev, *Koord. Khim.*, 1979, **5**, 207.
66. K. Wieghardt, E. Hofer, W. Holzbach, B. Nuber and J. Weiss, *Inorg. Chem.*, 1980, **19**, 2927.
67. W. Holzbach, K. Wieghardt and J. Weiss, *Z. Naturforsch., Teil B*, 1981, **36**, 289.
68. S. F. Gheller, T. W. Hambley, P. R. Traill, R. T. C. Brownlee, M. J. O'Connor, M. R. Snow and A. G. Wedd, *Aust. J. Chem.*, 1982, **35**, 2183.
69. M. Okunaka, G. Matsubayashi and T. Tanaka, *Inorg. Nucl. Chem. Lett.*, 1976, **12**, 813.
70. L. S. Liebeskind, K. B. Sharpless, R. D. Wilson and J. A. Ibers, *J. Am. Chem. Soc.*, 1978, **100**, 7061.
71. D. A. Muccigrosso, S. E. Jacobson, P. A. Aggar and F. Mares, *J. Am. Chem. Soc.*, 1978, **100**, 7063.
72. J. Macicek, V. K. Trunov and R. I. Machkhoshvili, *Zh. Neorg. Khim.*, 1981, **26**, 1690.
73. J. Macicek and V. K. Trunov, *Koord. Khim.*, 1981, **7**, 1585.
74. J. Macicek, R. I. Machkhoshvili and V. K. Trunov, *Zh. Neorg. Khim.*, 1981, **26**, 1963.
75. I. A. Krol, V. M. Agre, M. S. Kvernadze, N. I. Pirtskhalava and A. G. Nyudochkin, *Koord. Khim.*, 1981, **7**, 800.
76. J. C. Hanson, N. Camerman and A. Camerman, *J. Med. Chem.*, 1981, **24**, 1369.
77. G. V. Tsintsadze, Z. O. Dzhavakhishvili, G. G. Aleksandrov, Yu. T. Struchkov and A. P. Narimanidze, *Koord. Khim.*, 1980, **6**, 785.
78. J. R. Allan, G. M. Baillie and N. D. Baird, *J. Coord. Chem.*, 1984, **13**, 83.
79. M. Akbar ALi and S. E. Livingstone, *Coord. Chem. Rev.*, 1974, **13**, 101.
80. C. A. Kuehn and S. S. Isied, *Prog. Inorg. Chem.*, 1980, **27**, 153.
81. H. W. Roesky, *Adv. Inorg. Chem. Radiochem.*, 1979, **22**, 239.
82. I. Lindquist and J. Weiss, *J. Inorg. Nucl. Chem.*, 1958, **6**, 184; J. Weiss and U. Thewalt, *Z. Anorg. Allg. Chem.*, 1968, **363**, 159.
83. J. Weiss and M. Ziegler, *Z. Anorg. Allg. Chem.*, 1963, **322**, 184.
84. I. Haiduc, 'The Chemistry of Inorganic Ring Systems', Wiley-Interscience, New York, 1970.
85. U. Thewalt, *Z. Anorg. Allg. Chem.*, 1979, **451**, 123.
86. F. Edelmann, *J. Organomet. Chem.*, 1982, **228**, C47.
87. K. A. Jensen, *Z. Anorg. Allg. Chem.*, 1936, **229**, 265.
88. D. C. Jicha and D. H. Busch, *Inorg. Chem.*, 1962, **1**, 872, 878, 884.
89. C. H. Wei and L. F. Dahl, *Inorg. Chem.*, 1970, **9**, 1878.
90. D. H. Busch, J. A. Bourke, D. C. Jicha, M. C. Thompson and M. L. Morris, *Adv. Chem. Ser.*, 1963, **37**, 125.
91. L. F. Larkworthy, J. M. Murphy and D. J. Philllips, *Inorg. Chem.*, 1968, **7**, 1436.
92. S. E. Livingstone, *J. Chem. Soc.*, 1956, 437, 1042.
93. L. F. Lindoy and S. E. Livingstone, *Inorg. Chem.*, 1968, **7**, 1149.
94. K. Yamanouchi and J. H. Enemark, *Inorg. Chem.*, 1978, **17**, 2911.
95. K. Yamanouchi and J. H. Enemark, *Inorg. Chem.*, 1978, **17**, 1981.
96. E. Deutsch, M. J. Root and D. L. Nosco, in 'Advances in Inorganic and Bioinorganic Reaction Mechanisms', ed. A. G. Sykes, Academic, London, 1982, pp. 269–389.
97. D. L. Nosco, R. C. Elder and E. Deutsch, *Inorg. Chem.*, 1980, **19**, 2545.
98. D. L. Nosco, M. J. Heeg, M. D. Glick, R. C. Elder and E. Deutsch, *J. Am. Chem. Soc.*, 1980, **102**, 7784.
99. M. H. Dickman, R. J. Doedens and E. Deutsch, *Inorg. Chem.*, 1980, **19**, 945.
100. M. Kita, K. Yamanari, K. Kitahama and Y. Shimura, *Bull. Chem. Soc. Jpn.*, 1981, **54**, 2995.
101. R. N. Hurd, G. De La Mater, G. C. McElheny and J. P. McDermott, in 'Advances in the Chemistry of Coordination Compounds', ed. S. Kirschner, Macmillan, New York, 1961, p. 350.
102. F. Feigl and V. Anger, 'Spot Tests in Inorganic Analysis', 6th edn., Elsevier, London, 1972.
103. S. K. Mondal, R. Roy, S. K. Mondal and K. Nag, *J. Indian Chem. Soc., Sect A*, 1981, **20**, 982.
104. M. Chaudhury, *J. Chem. Soc., Dalton Trans.*, 1984, 115.
105. J. Stary, 'The Solvent Extraction of Metal Chelates', Pergamon, Oxford, 1964, 132, 135.
106. K. Haraguchi and H. Freiser, *Inorg. Chem.*, 1983, **22**, 653.
107. M. J. Root and E. Deutsch, *Inorg. Chem.*, 1984, **23**, 622.
108. O. G. Matyukhina, J. Ozols, B. T. Ibragimov, J. Lejejs, L. E. Terenteva and N. V. Belov, *Zh. Strukt. Khim.*, 1981, **22**, 144.
109. L. Pecs, J. Ozols, A. Kemme, J. Bleidelis and A. Sturis, *Latv. PSR Zinat. Vestis Kim. Ser.*, 1979, 259.
110. L. Pecs, J. Ozols, A. Kemme, J. Bleidelis and A. Sturis, *Latv. PSR Zinat. Vestis Kim. Ser.*, 1980, 549.
111. L. Pecs, J. Ozols and A. Sturis, *Latv. PSR Zinat. Vestis Kim. Ser.*, 1979, 623.
112. L. Pecs, J. Ozols, I. Tetere and A. Sturis, *Latv. PSR Zinat. Vestis Kim. Ser.*, 1981, 161.

113. L. F. Lindoy, S. E. Livingstone and T. N. Lockyer, *Nature (London)*, 1966, **211,** 519; *Aust. J. Chem.*, 1966, **19,** 1391.
114. E. Wenschuch, B. Wendelberger and H. Hartung, *J. Inorg. Nucl. Chem.*, 1969, **31,** 2073.
115. P. S. K. Chia, S. E. Livingstone and T. N. Lockyer, *Aust. J. Chem.*, 1967, **20,** 239.
116. P. S. K. Chia, S. E. Livingstone and T. N. Lockyer, *Aust. J. Chem.*, 1968, **21,** 339.
117. S. Trofimenko, *Chem. Rev.*, 1972, **72,** 497.
118. F. T. Greenway, L. M. Brown, J. C. Dabrowiak, M. R. Thompson and V. M. Day, *J. Am. Chem. Soc.*, 1980, **102,** 7782.
119. E. W. Ainscough, E. N. Baker, A. M. Brodie and N. G. Larsen, *J. Chem. Soc., Dalton Trans.*, 1981, 2054.
120. E. W. Ainscough, E. N. Baker, A. M. Brodie, N. G. Larsen and K. L. Brown, *J. Chem. Soc., Dalton Trans.*, 1981, 1746.
121. V. A. Neverov, V. N. Biyushkin and M. D. Mazus, *Zh. Strukt. Khim.*, 1981, **22,** 188.
122. F. H. Jardine, *Adv. Inorg. Chem. Radiochem.*, 1975, **17,** 115.
123. J. A. McCleverty, S. Gill, R. S. Z. Kowalski, N. A. Bailey, H. Adams, K. W. Lumbard and M. A. Murrary, *J. Chem. Soc., Dalton Trans.*, 1982, 493.
124. W. O. Foye and J. R. Lo, *J. Pharm. Sci.*, 1972, **61,** 1209.
125. J. F. Villa and H. C. Nelson, *J. Indian Chem. Soc*, 1978, **55,** 631.
126. F. A. Cotton and W. H. Ilsley, *Inorg. Chem.*, 1981, **20,** 614.
127. D. M. L. Goodgame and G. A. Leach, *J. Chem. Soc., Dalton Trans.*, 1978, 1705.
128. D. M. L. Goodgame and G. A. Leach, *Inorg. Chim. Acta*, 1979, **32,** 69.
129. B. A. Cartwright, P. O. Langguth, Jr. and A. C. Skapski, *Acta Crystallogr., Sect. B*, 1979, **35,** 63.
130. M. Younes, W. Pilz and U. Weser, *J. Inorg. Biochem.*, 1979, **10,** 29.
131. Y. Henry and A. Dobry-Duclaux, *J. Chim. Phys. Phys.-Chim. Biol.*, 1976, **73,** 1068.
132. A. S. Hirshon, W. K. Musker, M. M. Olmstesd and J. L. Dallas, *Inorg. Chem.*, 1981, **20,** 1702.
133. M. J. M. Campbell, *Coord. Chem. Rev.*, 1975, **15,** 279.
134. K. K. W. Sun and R. A. Haines, *Can. J. Chem.*, 1970, **48,** 2327.
135. M. E. Rusanovskii, I. D. Samus, N. M. Samus, O. A. Bologa and N. V. Belov, *Dokl. Akad. Nauk SSSR*, 1981, **260,** 98.
136. I. D. Samus, M. E. Rusanovskii, O. A. Bologa and N. M. Samus, *Koord. Khim.*, 1981, **7,** 120.
137. M. M. Mostafa and A. A. El-Asmy, *J. Coord. Chem.*, 1983, **12,** 197.
138. R. B. Singh, B. S. Garg and R. P. Singh, *Talanta*, 1978, **25,** 619.
139. A. G. Asuero and M. Gonzalez-Balairon, *Microchem. J.*, 1980, **25,** 14.
140. M. Akbar Ali, S. E. Livingstone and D. J. Philips, *Inorg. Chim. Acta*, 1971, **5,** 119, 493.
141. T. Glowiak and T. Ciszewska, *Acta Crystallogr., Sect. B*, 1982, **38,** 1735.
142. T. I. Malinovsky, V. A. Neverov, C. F. Belyaeva and V. N. Biyushkin, *Acta Crystallog., Sect. A*, 1981, **37,** C235.
143. M. Lanfranchi, A. M. M. Lanfredi, A. Tiripicchio, A. Monaci and F. Tarli, *J. Chem. Soc., Dalton Trans.*, 1980, 1893.
144. G. Dessy and V. Fares, *Acta Crystallogr., Sect. B*, 1980, **36,** 944.
145. G. Dessy and V. Fares, *Acta Crystallogr., Sect. B*, 1980, **36,** 2266.
146. E. Fischer and E. Besthorn, *Ann. Chim.*, 1881, **212,** 316.
147. H. Fischer, *Wiss. Veroff. Siemens-Werken*, 1925, **4,** 158.
148. H. Irving and C. F. Bell, *J. Chem. Soc.*, 1954, 4253; H. Irving and J. J. Cox, *J. Chem. Soc.*, 1961, 1470.
149. A. T. Hutton, H. M. N. H. Irving, L. R. Nassimbeni and G. Gafner, *Acta Crystallogr., Sect. B*, 1980, **36,** 2064.
150. D. Walther and C. Pfuetzenreuter, *Z. Chem.*, 1977, **17,** 426.
151. R. Meij, D. J. Stufkens, K. Vrieze, W. Van Gerresheim and C. H. Stam, *J. Organomet. Chem.*, 1979, **164,** 353.
152. H. C. Ashton and A. R. Manning, *Inorg. Chem.*, 1983, **22,** 1440.
153. R. Meij, D. J. Stufkens, K. Vrieze, A. M. F. Brouwers, J. D. Schagen, J. J. Zwinselman, A. R. Overbeek and C. H. Stam, *J. Organomet. Chem.*, 1979, **170,** 337.
154. D. H. M. W. Thewissen, J. G. Noltes, J. Willemse and J. W. Diesveld, *Inorg. Chim. Acta*, 1982, **59,** 181.
155. M. Creswick and I. Bernal, *Inorg. Chim. Acta*, 1982, **57,** 171.
156. G. M. Reisner and I. Bernal, *J. Organomet. Chem.*, 1981, **220,** 55.
157. C. R. Saha and N. K. Roy, *J. Coord. Chem.*, 1983, **12,** 163.
158. N. K. Roy and C. R. Saha, *Indian J. Chem., Sect. A*, 1980, **19,** 889.
159. J. R. Dilworth, P. L. Dahlstrom, J. R. Hyde, M. Kustyn, P. A. Vella and J. Zubieta, *Inorg. Chem.*, 1980, **19,** 3562.
160. H. F. De Brabander and L. C. Van Poucke, *J. Coord. Chem.*, 1974, **3,** 301.
161. H. F. De Brabander, J. J. Tombeux and L. C. Van Poucke, *J. Coord. Chem.*, 1974, **4,** 87,
162. A. G. Jones, B. V. De Phamphilis and A. Davidson, *Inorg. Chem.*, 1981, **20,** 1617.
163. U. Casellato, P. A. Vigato, R. Graziani, M. Vidali and B. Acone, *Inorg. Chim. Acta*, 1980, **45,** L79.
164. R. Graziani, M. Vidali, U. Casellato and P. A. Vigato, *Transition Met. Chem. (Weinheim, Ger.)*, 1981, **6,** 166.
165. R. G. Garvey, J. H. Nelson and R. O. Ragsdale, *Coord. Chem. Rev.*, 1968, **3,** 375.
166. N. M. Karayannis, L. L. Pytlewski and C. M. Mikulski, *Coord. Chem. Rev.*, 1973, **11,** 93.
167. N. M. Karayannis, A. N. Speca, D. E. Chasen and L. L. Pytlewski, *Coord. Chem. Rev.*, 1976, **20,** 37.
168. C. M. Harris and S. E. Livingstone, in 'Chelating Agents and Metal Chelates', ed. F. P. Dwyer and D. P. Mellor, Academic, New York, 1964, p. 124.
169. J. D. Lydon, K. J. Mulligan, R. C. Elder and E. Deutsch, *Inorg. Chem.*, 1980, **19,** 2083.
170. G. J. Gainsford, W. G. Jackson and A. M. Sargeson, *Aust. J. Chem.*, 1980, **33,** 707.
171. G. J. Ford, P. Gans, L. D. Pettit and C. Sherrington, *J. Chem. Soc., Dalton Trans.*, 1972, 1763.
172. L. D. Pettit, *Q. Rev., Chem. Soc.*, 1971, **25,** 1.
173. A. Ouchi, M. Shimoi, T. Takeuchi and H. Saito, *Bull. Chem. Soc. Jpn.*, 1981, **54,** 2290.
174. T. Ogawa, M. Shimoi and A. Ouchi, *Acta Crystallogr., Sect. B*, 1980, **36,** 3114.
175. T. Ogawa, M. Shimoi and A. Ouchi, *Bull. Chem. Soc. Jpn.*, 1982, **55,** 126.
176. S. G. Murray and F. R. Hartley, *Chem. Rev.*, 1981, **81,** 365.
177. R. C. Elder, G. J. Kennard, M. D. Payne and E. Deutsch, *Inorg. Chem.*, 1978, **17,** 1296.
178. K. Yamanari, J. Hidaka and Y. Shimura, *Bull. Chem. Soc. Jpn.*, 1977, **50,** 2299.
179. M. Kojima and J. Fujita, *Bull. Chem. Soc. Jpn.*, 1983, **56,** 2958.
180. M. R. Udupa, B. Krebs and U. Seyer, *Inorg. Chim. Acta*, 1980, **41,** 31.
181. G. J. Kubas, R. R. Ryan and V. McCarty, *Inorg. Chem.*, 1980, **19,** 3003.

182. G. D. Jarvinen, G. J. Kubas and R. R. Ryan, *J. Chem. Soc., Chem. Commun.*, 1981, 305.
183. D. C. Moody and R. R. Ryan, *J. Chem. Soc., Chem. Commun.*, 1980, 1230.
184. T. R. Gaffney and J. A. Ibers, *Inorg. Chem.*, 1982, **21**, 2860.
185. R. D. Bereman, D. M. Baird, J. R. Dorfman and J. Bordner, *Inorg. Chem.*, 1982, **21**, 2365.
186. J. B. McCormick, R. D. Pereman and D. M. Baird, *Coord. Chem. Rev.*, 1984, **54**, 99.
187. S. Mizukami and K. Nagata, *Coord. Chem. Rev.*, 1968, 3, 267.
188. Y. Egawa, K. Umino, Y. Ito and T. J. Okuda, *J. Antibiot.*, 1971, **24**, 124, 140.
189. D. P. Freyberg, K. Abu-Dari and K. N. Raymond, *Inorg. Chem.*, 1979, **18**, 3037.
190. K. Abu-Dari and K. N. Raymond, *Inorg. Chem.*, 1982, **21**, 1676.
191. J. Becher, D. J. Brockway, K. S. Murray and P. J. Newman, *Inorg. Chem.*, 1982, **21**, 1791.
192. R. Richter, L. Beyer and J. Kaiser, *Z. Anorg. Allg. Chem.*, 1980, **461**, 67.
193. J. D. Lydon, R. C. Elder and E. Deutsch, *Inorg. Chem.*, 1982, **21**, 3186.
194. L. S. Bark and D. Griffin, *J. Inorg. Nucl. Chem.*, 1971, **33**, 3811.
195. L. S. Bark and D. Brandon, *Talanta*, 1967, **14**, 759.
196. G. Schneeweiss and K. H. Koenig, *Fresenius' Z. Anal. Chem.*, 1983, **316**, 16.
197. Y. Mido, I. Fujiwara and E. Sekido, *J. Inorg. Nucl. Chem.*, 1974, **36**, 1003.
198. M. Maeda, N. Abiko and T. Sasaki, *J. Med. Chem.*, 1981, **24**, 167.
199. F. Kanzawa, M. Maeda, T. Sasaki, A. Hoshi and K. Kuretani, *J. Natl. Cancer Inst.*, 1982, **68**, 287.
200. M. Maeda, N. Abiko and T. Sasaki, *J. Pharmacobio-Dyn.*, 1982, **5**, 81.
201. A. J. Adejumobi and D. R. Goddard, *J. Inorg. Nucl. Chem.*, 1977, **39**, 910.
202. D. R. Goddard and S. O. Ajayi, *J. Chem. Soc. (A)*, 1971, 2673.
203. J. A. Adejumobi and D. R. Goddard, *J. Inorg. Nucl. Chem.*, 1977, **39**, 912.
204. A. Zabokrzycka, B. B. Kedzia and P. M. Drozdzewski, *Bull. Acad. Pol. Sci., Ser. Sci. Chim.*, 1979, **27**, 291.
205. N. V. Mel'nikova, A. T. Pilipenko and V. P. Badekha, *Ukr. Khim. Zh.*, 1981, **47**, 42.
206. A. V. Ablov, N. V. Gerbeleu, A. M. Romanov and V. M. Vlad, *Zh. Neorg. Khim.*, 1971, **16**, 1357.
207. D. Negoiu and I. Ghelase, *Rev. Roum. Chim.*, 1983, **28**, 225.
208. R. S. Ramakrishna and H. M. N. H. Irving, *Anal. Chim. Acta*, 1969, **48**, 251.
209. H. Sakurai, A. Yokoyama and H. Tanaka, *Chem. Pharm. Bull.*, 1971, **19**, 1270.
210. Y. Sugiura, T. Kikuchi and H. Tanaka, *Adv. Chem. Ser.*, 1980, **191**, 393.
211. C. A. Stein, P. E. Ellis, Jr., R. C. Elder and E. Deutsch, *Inorg. Chem.*, 1976, **15**, 1618.
212. T. Konno, K. Okamoto and J. Hidaka, *Chem. Lett.*, 1982, 535.
213. L. Roecker, M. H. Dickman, D. L. Nosco, R. J. Doedens and E. Deutsch, *Inorg. Chem.*, 1983, **22**, 2022.
214. K. Nakajima, M. Kojima and J. Fujita, *Chem. Lett.*, 1982, 925.
215. V. P. Kurbatov, O. A. Osipov, V. O. Minkin, L. E. Nivorozhkin and L. S. Minkina, *Zh. Neorg. Khim.*, 1974, **19**, 2191.
216. V. I. Minkin, L. E. Nivorozhkin, L. E. Konstantinovskii and M. S. Korobov, *Koord. Khim.*, 1977, **3**, 174.
217. M. Nakamoto, K. Tanaka and T. Tanaka, *J. Chem. Soc., Dalton Trans.*, 1979, 87.
218. R. C. Elder, K. Burkett and E. Deutsch, *Acta Crystallogr., Sect. B*, 1979, **35**, 164.
219. V. H. Houlding, H. Maecke and A. W. Adamson, *Inorg. Chim. Acta*, 1979, **33**, L175.
220. V. H. Houlding, H. Maecke and A. W. Adamson, *Inorg. Chem.*, 1981, **20**, 4279.
221. G. Wilke and E. Uhlemann, *Z. Chem.*, 1975, **15**, 66.
222. G. Wilke and E. Uhlemann, *Z. Chem.*, 1974, **14**, 288.
223. G. Wilke and E. Uhlemann, *Z. Chem.*, 1975, **15**, 453.
224. L. Beyer, S. Behrendt, E. Kleinpeter, R. Borsdorf and E. Hoyer, *Z. Anorg. Allg. Chem.*, 1977, **437**, 282.
225. E. Kleinpeter, S. Behrendt and L. Beyer, *Z. Anorg. Allg. Chem.*, 1982, **495**, 105.
226. R. Kirmse, L. Beyer and E. Hoyer, *Z. Chem.*, 1975, **15**, 454.
227. Von J. V. Salyn, E. K. Zumadilov, V. I. Nefedov, R. Scheibe, G. Leonhardt, L. Beyer and E. Hoyer, *Z. Anorg. Allg. Chem.*, 1977, **432**, 275.
228. R. Herzschuh, B. Birner, L. Beyer, F. Dietze and E. Hoyer, *Z. Anorg. Allg. Chem.*, 1980, **464**, 159.
229. P. Eckstein and E. Hoyer, *Z. Anorg. Allg. Chem.*, 1982, **487**, 33.
230. W. Dietzsch, J. Reinhold, R. Kirmse, E. Hoyer, I. N. Marov and V. K. Belyaeva, *J. Inorg. Nucl. Chem.*, 1977, **39**, 1377.
231. G. Nagao, K. Tanaka and T. Tanaka, *Inorg. Chim. Acta*, 1980, **42**, 43.
232. W. Dietzsch, J. Kaiser, R. Richter, L. Golic, J. Siftar and R. Heber, *Z. Anorg. Allg. Chem.*, 1981, **477**, 71.
233. E. W. Abel, S. K. Bhargava, K. Kite, K. G. Orrell, V. Sik and B. L. Williams, *Polyhedron*, 1982, **1**, 289.
234. D. M. Roundhill, G. N. S. Roundhill, W. B. Beaulieu and U. Bagchi, *Inorg. Chem.*, 1980, **19**, 3365.
235. C. G. Pierpont, B. J. Corden and R. Eisenberg, *J. Chem. Soc. (D)*, 1969, 401.
236. E. Fanghaenel and H. Poleschner, *Z. Chem.*, 1979, **19**, 192.
237. E. Uhlemann, E. Ludwig, W. Huebner and R. Szargan, *Z. Chem.*, 1983, **23**, 32.
238. L. Beyer and A. Hautschmann, *Z. Chem.*, 1983, **23**, 230.
239. J. A. Bertrand, E. Fujita and D. G. Van Derveer, *Inorg. Chem.*, 1980, **19**, 2022.
240. J. A. van Doorn and P. W. N. M. van Leewen, *J. Organomet. Chem.*, 1981, **222**, 299.
241. P. P. Richard, D. T. Qui and E. F. Bertant, *Acta Crystallogr., Sect. B*, 1973, **29**, 1111.
242. A. Chakravorty, *Coord. Chem. Rev.*, 1974, **13**, 1.
243. A. Mueller and N. Mohan, *Z. Anorg. Allg. Chem.*, 1981, **480**, 157.
244. P. L. Bellon, S. Cenini, F. Demartin, M. Manassero, M. Pizotti and F. Porta, *J. Chem. Soc., Dalton Trans.*, 1980, 2060.
245. M. S. Reynolds, J. D. Oliver, J. M. Cassel, D. L. Nosco, M. Noon and E. Deutsch, *Inorg. Chem.*, 1983, **22**, 3632.
246. N. G. Furmanova, Yu. T. Struchkov, E. M. Rokhlina and D. N. Kraytsov, *Zh. Strukt. Khim.*, 1980, **21**, 87.
247. E. S. Raper and P. H. Crackett, *Inorg. Chim. Acta*, 1981, **47**, 159.
248. I. G. Dance and P. J. Guerney, *Aust. J. Chem.*, 1981, **34**, 57.
249. R. E. Tapscott, *Transition Met. Chem. (Weinheim, Ger.)*, 1982, **8**, 253.
250. K. Mennemann and R. Mattes, *Angew. Chem., Int. Ed. Engl.*, 1977, **16**, 260.
251. R. D. Bereman, D. M. Baird, J. Bordner and J. R. Dorfman, *Polyhedron*, 1983, **2**, 25.
252. A. N. Vedyanu and N. V. Gerbeleu, *Zh. Neorg. Khim.*, 1976, **21**, 3319.
253. C. Preti and G. Tosi, *Z. Anorg. Allg. Chem.*, 1977, **432**, 259.
254. E. W. Abel, K. S. Bhargava, K. Kite, K. G. Orrell, V. Sik and B. L. Williams, *J. Chem. Soc., Dalton Trans.*, 1982, 583.

21.1

Porphyrins, Hydroporphyrins, Azaporphyrins, Phthalocyanines, Corroles, Corrins and Related Macrocycles

TOSHIO MASHIKO and DAVID DOLPHIN
University of British Columbia, Vancouver, BC, Canada

21.1.1   PORPHYRINS AND HYDROPORPHYRINS

Porphyrin (**1**) consists of four pyrrolic units linked by four methine bridges. It contains an 18-electron π system and exhibits aromaticity. The resonance energy is estimated to be 1670–2500 kJ mol^{-1},[1] which, as expected, renders meso-hydrogenated derivatives such as porphodimethanes (**2**) and porphyrinogens (**3**) thermodynamically unfavourable, although their preparation under anaerobic conditions is possible. Addition of two and four hydrogen atoms to the β-positions of porphyrin gives chlorin (**4**) and bacteriochlorin (**5**) or isobacteriochlorin (**6**) respectively, which are also 18-π-electron aromatic macrocycles (Figure 1).

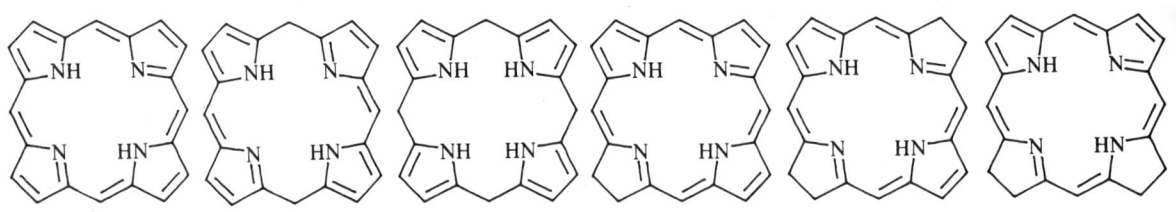

(1) Porphyrin   (2) Porphodimethane   (3) Porphyrinogen   (4) Chlorin   (5) Bacteriochlorin   (6) Isobacteriochlorin

Figure 1   Porphyrins and hydroporphyrins

21.1.1.1   Porphyrins

The abbreviations shown in Figure 2 are used in this chapter.

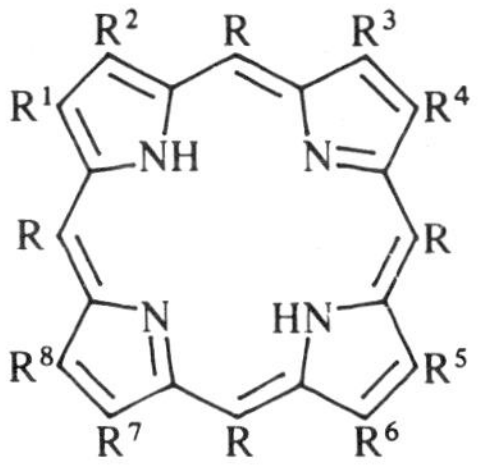

Figure 2 Porphyrin abbreviations

Por: unspecified
TPP: R = Ph, R^1–R^8 = H; TTP: R = $C_6H_4CH_3$, R^1–R^8 = H; TpyP: R = C_5H_4N, R^1–R^8 = H
$T_{p-X}PP$: R – p-XC_6H_4, R^1–R^8 = H
OMP: R = H, R^1–R^8 = Me; OEP: R = H, R^1–R^8 = Et
Etio I: R = H, $R^1 = R^3 = R^5 = R^7$ = Me, $R^2 = R^4 = R^6 = R^8$ = Et
PPIX (DME): R = H, $R^1 = R^3 = R^5 = R^8$ = Me, $R^2 = R^4 = CHCH_2$,
    $R^6 = R^7 = CH_2CH_2CO_2H$ (dimethyl ester)
DPIX (DME): R = H, $R^1 = R^3 = R^5 = R^8$ = Me, $R^2 = R^4$ =H,
    $R^6 = R^7 = CH_2CH_2CO_2H$ (dimethyl ester)
HPIX (DME): R = H, $R^1 = R^3 = R^5 = R^8$ = Me, $R^2 = R^4$ = Et,
    $R^6 = R^7CH_2CH_2CO_2Me$ (dimethyl ester)

21.1.1.1.1 General properties

Porphyrins are intensely coloured, with a strong absorption band at ~ 400 nm ($\varepsilon \approx 10^5$), the Soret band. The large ring current effect in the ^{1}H NMR ($\delta_{NH} \approx -5$, $\delta_{meso-H} \approx 10$ p.p.m.), the stable parent mass peaks, electrophilic substitutions on the ring peripheries, the planarity and intermediate bond lengths in their X-ray structures and the large heat of combustion all reflect the aromaticity of porphyrins.[2]

The location of the inner protons is of interest. It was shown by NMR that they are placed on two diagonal nitrogens and shift stepwise to the other pair of nitrogens with an activation energy $\Delta G^{\neq}$ of ~ 54 kJ mol^{-1}.[3]

21.1.1.1.2 Spectral properties

The absorption spectra of metalloporphyrins are classified into three types, regular, hypso and hyper, and summarized as follows.[4]

Regular-type spectra are observed for a majority of M(Por) where the relevant porphyrin π orbitals do not significantly interact with metal π (p_z or d_{xz}, d_{yz}) orbitals. The relaxation process is fluorescent. The absorption bands are as follows:

Q(0,0), Q(1,0) [500–600 nm, $a_{2u}(\pi) \rightarrow e_g(\pi^*)$, $\varepsilon \approx 10^4$ M^{-1} cm^{-1}], a quasi-allowed transition (α) and its vibronic overtone (β) which appear in the interval of ~ 1250 cm^{-1}, each splitting to two (~ 3000 cm^{-1}) in the metal-free porphyrins (H$_2$Por);

B(0,0) [Soret, γ; 380–420 nm, $a_{1u}(\pi) \rightarrow e_g(\pi^*)$, $\varepsilon \approx 10^5$ M^{-1} cm^{-1}], an allowed transition sometimes accompanied by a vibrational band on the ~ 1250 cm^{-1} blue side;

N band (~ 325 nm), M band (~ 215 nm) and L band (between N and M bands), which are all weak.

There are several M(Por) whose absorption spectra are normal but whose decay processes are radiationless or luminescent. These pseudo-normal-type metals have partly filled d or f levels too high to cause ligand to metal charge transfer transitions [V^{IV}=O, $S = \frac{1}{2}$; Cr^{II}, $S = 1, 2$; Mo^{IV}, $S = 0$; Mn^{II}, $S = \frac{5}{2}$ (very weak phosphorescence)]. The lanthanides and actinides ($S \neq 0$) belong to this type.

Hypso-type spectra follow the regular absorption pattern but are blue shifted ($\alpha < 570$ nm) due to filled metal d (π) to Por e_g (π^*) back donation. Metal ions of this type are limited to those of Groups VIII and IB in low-spin d^{5-10} states. The relaxation is radiationless (Fe, Co, Ni, Ag, some Ru and Os), phosphorescent (Rh, Pd, Ir, Pt, Au, some Ru and Os) or luminescent (Cu).

Hyper-type spectra contain charge transfer bands in addition to significant shifts of the Por ($\pi \rightarrow \pi^*$) bands. This type is further subdivided into three classes by the charge transfer types.

p-Type spectral patterns are caused by metal $a_{2u}(np_z) \rightarrow e_g(\pi^*)$ Por interactions. The metal ions are those of the main group in low oxidation states (IIIB M^I, IVB M^{II}, VB M^{III}). The spectra consist of three characteristic bands in the visible region, *i.e.* ~ 370, ~ 460 (B) and 535–650 nm (Q). They are radiationless or very weakly phosphorescent.

d-Type spectral patterns result from Por a_{1u} (π), a_{2u} (π) $\rightarrow e_g$ (d_{xz}, d_{yz}) metal interactions. Group VIA, VIIA and VIII (Fe, $Os^{IV,VI}$) metals in high oxidation states show this type of spectra. Four or more bands are observed in the visible region, among which band VI (~ 350 nm), band V (~ 440 nm, B) and bands IV and III (550–600 nm, Q) are relatively consistent. Because of the proximity of metal *d* and porphyrin HOMO and LUMO levels, the Fe(Por) spectra are variable and complicated. The decay process is radiationless (Mn^{III}, $S = 2$; Fe^{III}, $S = \frac{1}{2}, \frac{3}{2}, \frac{5}{2}$; Fe^{II}, $S = 1, 2$; Mo^V, $S = \frac{1}{2}$; W^V, $S = \frac{1}{2}$; Re^V, $S = 0$; Os^{IV}, $S = 0$), phosphorescent (Os^{VI}, $S = 0$) or luminescent (Cr^{III}, $S = \frac{3}{2}$).

Unusual spectral types include those with forbidden metal e_g (d_{xz}, d_{yz}) $\rightarrow e_g$ (π^*) Por charge transfer bands, observed for Group VIIA and VIII metals in low oxidation states (M^I or M^{II}).

M(Por) shows medium to strong IR bands due to ring deformation at ~ 1150, ~ 1060, ~ 990, ~ 760 and ~ 700 cm^{-1} (in-plane E_u mode) and ~ 850, ~ 840 and ~ 800 cm^{-1} (out-of-plane A_u mode) (M = Ni^{II}), and far IR bands at ~ 400, ~ 340 and ~ 280 cm^{-1} (N—M) and ~ 220 and ~ 160 cm^{-1} (ring).[2] The IR stretching bands of some ligands may be used for structural assignments, for example the metal-bound O_2 shows $\nu(O_2)$ at 1100–1160 cm^{-1} (end-on) or 800–1000 cm^{-1} (side-on), and the oxo metal stretch at $\nu(M{=}O) = 800$–1030 cm^{-1}.

The NMR spectra of diamagnetic porphyrins are characterized by the large diamagnetic shifts of inner ($\delta \approx -4$ p.p.m.) and peripheral ($\delta_{meso-H} \approx 10$ p.p.m., $\delta_{\beta-H} \approx 9$ p.p.m.) proton signals.[2] Some theoretical approaches have been made to estimate ring current effects.[5] ^{13}C NMR signals are observed at 145–150 p.p.m. (C_α), 130–140 p.p.m. (C_β) and 95–125 p.p.m. (C_{meso}).[2] NMR analyses of paramagnetic porphyrins are also the subject of numerous studies.[2] The Evans method[6] is often used to estimate the magnetic susceptibility of paramagnetic metalloporphyrins.

Reflecting their aromaticity, porphyrins usually give strong $\{M(Por)\}^+$ and $\{M(Por)\}^{2+}$ mass peaks, though some fragmentations of the side chains and loss of the metal from labile metalloporphyrins are often observed.[2] The ionization energy of M(TPP) (M = H_2, Cu) was reported to be 8.0 eV by mass spectrometry. MCD, CD, resonance Raman, Zeeman, ENDOR, ESR and Mössbauer spectroscopies have been reviewed in detail.[2]

21.1.1.1.3 *General synthesis*

D_{4h} symmetrical porphyrins such as porphin itself,[7] H_2TPP[8] and H_2OEP[9] are conveniently prepared from pyrrole and aldehyde by cyclocondensations (Scheme 1). Addition of Zn^{II} ion often improves the yield.[10] The product is often contaminated with chlorin by-product, which can be converted to porphyrin by oxidants such as DDQ,[10] or separated by chromatography.[11] Bulky substituents tend to lower the yield, and meso *o*-substituted phenyl groups give rise to atrope isomers.

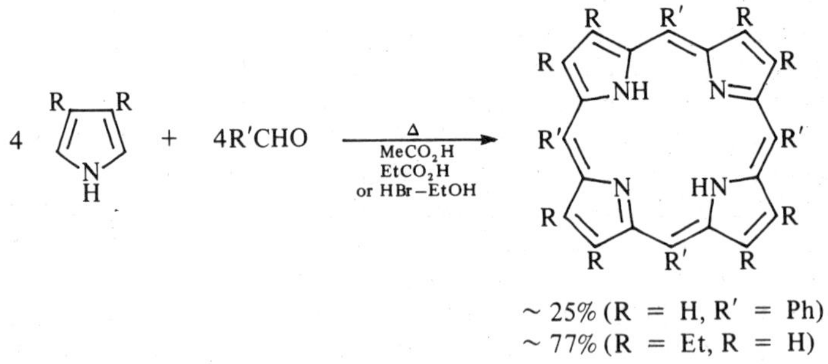

$$4 \quad \text{(pyrrole)} + 4R'CHO \xrightarrow[\substack{MeCO_2H \\ EtCO_2H \\ \text{or } HBr{-}EtOH}]{\Delta} \text{(porphyrin)}$$

$\sim 25\%$ (R = H, R' = Ph)
$\sim 77\%$ (R = Et, R' = H)

Scheme 1

For porphyrins of C_4 symmetry, *e.g.* coproporphyrin I[12] and etioporphyrin I,[13] cyclotetramerization of α-(functionalized methyl)pyrrole is suitable, though the pyrrole synthesis is not always facile (Scheme 2). Drastic conditions or prolonged reaction occasionally cause isomeric by-product formation.[2]

A porphyrin with a C_2 axis perpendicular to its plane may be similarly synthesized by dimerization of α-functionalized dipyrro-methanes or -methenes. If only a half of the molecule is C_2 symmetric, the condensation of dipyrrolic intermediates is useful. Unsymmetrical porphyrins are usually synthesized in a stepwise fashion *via* linear tetrapyrrolic intermediates.

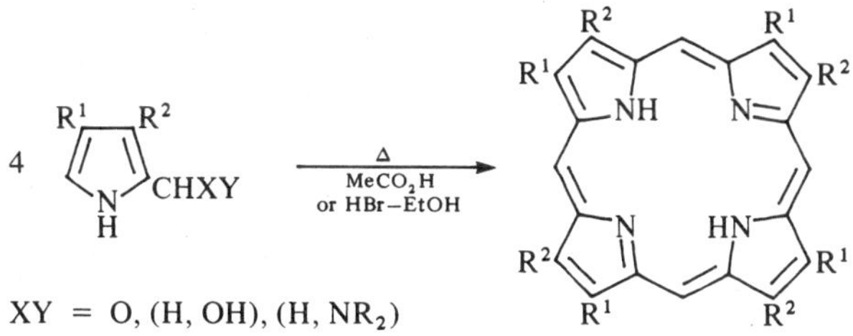

XY = O, (H, OH), (H, NR$_2$)

Scheme 2

21.1.1.1.4 Reactions at the porphyrin periphery

Porphyrins have four inner nitrogens and show four stepwise acid–base reactions (equation 1).[14] The pK_1 and pK_2 values are estimated to be ~16, pK_3 is ~5, and pK_4 is ~2 for β-alkylated porphyrins. Because of N protonation, electrophilic meso deuteration of metal-free porphyrins is slower in strong acids (H$_2$OEP in TFA-d, $\tau_{1/2}$ = 275 h at 90 °C) than in weak acids ($\tau_{1/2}$ = 140 ± 14 min in refluxing AcOD).[2] Metalloporphyrins are more easily deuterated in strongly acidic media, *e.g.* M(OEP) (M = Fe, Cu, Pt) are completely meso-deuterated in TFA-d_1:CDCl$_3$ (1:1) at room temperature within 20 min. Meso deuteration of M(OEP) is possible under basic conditions (MgI$_2$–D$_2$O or MeOD–hot pyridine).[2]

$$(\text{Por})^{2-} \underset{\text{H+}}{\overset{K_1}{\rightleftharpoons}} (\text{HPor})^- \underset{\text{H+}}{\overset{K_2}{\rightleftharpoons}} (\text{H}_2\text{Por}) \underset{\text{H+}}{\overset{K_3}{\rightleftharpoons}} (\text{H}_3\text{Por})^+ \underset{\text{H+}}{\overset{K_4}{\rightleftharpoons}} (\text{H}_4\text{Por})^{2+} \tag{1}$$

The frontier orbital theory predicts the reactivity order, $C_{meso} > C_\alpha > N \geqslant C_\beta$, for electrophilic attack and $C_{meso} > C_\beta \approx C_\alpha \geqslant N$ for nucleophilic attack.[2,19] In general, these predictions are observed experimentally, for example, Mg(porphin) is brominated to give Mg($\alpha,\beta,\gamma,\delta$-tetrabromoporphin). However, the site of reaction is highly dependent on steric and electronic factors, and may be summarized as follows.[2]

Nitrogen protonation or incorporation of high oxidation state metal ions facilitates nucleophilic substitution (especially at C_{meso}) whereas N-deprotonation or coordination to low valent metal ions enhances the electrophilic reaction (especially at C_{meso}). The effects of electron-donating substituents (alkyl) are less than those of electron-withdrawing groups (especially π acceptors, NO$_2$, COR).

In metal-free porphyrins, the inner nitrogen atoms are the strongest nucleophiles and undergo S_N2 reactions with electrophiles (R—X) under neutral to weakly basic conditions. For metalloporphyrins (M^{II} or M^{III}) small electrophiles (MeOSO$_2$F, Me$_2\overset{+}{N}$=CHCl, $\overset{+}{N}$O$_2$, Cl$_2$, Br$_2$, HSCN) select C_{meso}, but the less hindered C_β is attacked by bulky reagents (Friedel–Crafts type). A σ- or π-donor group (alkyl, Cl) at C_{meso} promotes electrophilic substitution at $C_{meso-\gamma}$, while $C_{meso-\beta}$ is preferentially attacked in porphyrins with an electron-withdrawing meso substituent (NO$_2$, CH=$\overset{+}{N}$Me$_2$).[2] Nitration and formylation of M(Por) occur selectively at $C_{meso-\alpha}$ and C_γ.[15]

The course of carbene reaction is metal dependent (see Scheme 37, Section 21.1.1.1.12).[16] Cu(Por) reacts with carbenes at C_β=C_β, and Co and Zn(Por) at M—N. On heating, Co(Por)–carbene adducts rearrange to Co(N,N'-methylene-Por), and Zn(Por)–carbene adducts to Zn(meso-alkyl-OEP) or Zn(β-alkyl-TPP). N-Ethoxycarbonylmethyl-TPP is converted to a homoporphyrin when heated with Ni(acac)$_2$. Homoazaporphyrin is also formed by nitrene addition to H$_2$OEP, though the same reaction gives meso-amino-OEP if Cu(OEP) is used. The N-ethoxycarbonylhomoazaporphyrin rearranges to meso-ethoxycarbonylamino-OEP when metallated with Cu^{II} or Zn^{II}.

Cyclo additions, such as Diels–Alder reactions of H$_2$(β-vinyl-Por), and OsO$_4$ oxidation occur at C_β=C_β, while radical reactions, Tl(OCOCF$_3$)$_3$ oxidation and catalytic hydrogenations result in addition to the meso positions.[2,19]

21.1.1.1.5 Metallation, demetallation and transmetallation

Porphyrins are generally metallated with metal salts in a solvent of high boiling point (100–250 °C).[17] The type of salt used depends on the metal and mainly on availability and stability. Since the metallation proceeds *via* ligand exchange, metal carriers without tight metal–counterion bonds, usually low oxidation state metal salts, are preferred (Table 1).

Table 1   Summary of Porphyrin Metallation Conditions[17]

MX_nL_m/AcOH, 100 °C	V, Mn, Zn, Rh, Ag, Au, In, Sn, Ir, Pd, Pt, Tl
MX_n/py, 115–185 °C	Groups IIA, IIIA, lanthanides and actinides, VIB, VB, IB, IIB, Tl, Sc
MX_n/PhOH, 180–240 °C	Mo, Ta–Os (phenoxy complex will be obtained)
MCl_n/PhCN, 190 °C	Ti, Group VIA, Zr–Nb, Pd, In, Pt
MCl_n/DMF, 153 °C	Groups IIA, VB, IIB, V–Zn, Rh–Sb
$M(acac)_n$/solvent, 180–240 °C	Groups IIIA, lanthanides and actinides, IVA, VIA, IIIB, Sc–Zn [Group IIIA, Ln, An will give *cis*-$M(acac)_2$(Por)]
$MX_n(CO)_m$/solvent, 80–200 °C	Groups VIA, VIIA, VIII [Tc, Re give $\{M(CO)_3\}_2$(Por), Rh gives $\{M(CO)_2\}_2$(Por)]
M(OR)/solvent, 35–80 °C	Groups IA, IIA
MR_m/solvent, 80–200 °C	Mg, B, Al, Ti

Several metal insertion mechanisms have been proposed, but none of them is conclusive.[18] The rate of metallation varies from square root to second order in metal salt from one system to another, and apparently there exists more than one pathway. Where the rate law is second order in metal salt, a so called sitting-atop metal ion–porphyrin complex intermediate or metal ion-deformed porphyrin intermediate, which then incorporates another metal ion into the porphyrin centre, has been postulated (Figure 3).[19] For the reactions with the square root dependence on the metal salt concentration, the aggregation of metal salts is suggested.[18] Of course, there are many examples which follow simple kinetics, *i.e.* $d[M(Por)]/dt = k[M \text{ salt}][H_2Por]$.

Figure 3   Sitting-atop and deformed intermediates in porphyrins metallation

The stability of a metalloporphyrin can be roughly estimated by the charge-to-radius ratio (Z/γ_i). Buchler *et al.*[17] found that the stability index, $SI = 100(ZX/\gamma_i)$, where X is the Pauling electronegativity, predicts the stability of a metalloporphyrin to acid demetallation, and correlated the SI to the Falk–Phillips stability criteria (Table 2).

Table 2   Correlation between Falk–Phillips Stability Criteria[a] and Stability Index[17]

	Classification (at r.t., 2 h)	$Z = 1$	2	3	4	5
			Range of SI (γ_i in Å) for			
I	100% H_2SO_4, incomplete demetallation	—	7	9	11	15
II	100% H_2SO_4, complete demetallation	—	6–7	6.8–9	7–11	—
III	HCl/H_2O/CH_2Cl_2, complete demetallation	—	4–5	5.9–6.8	5	11.5–12.5
IV	100% AcOH, complete demetallation	—	2–4	5.6	—	—
V	H_2O/CH_2Cl_2, complete demetallation	2	2	—	—	—

[a] J. N. Phillips, *Rev. Pure Appl. Chem.*, 1960, **310**, 35; J. E. Falk, 'Porphyrins and Metalloporphyrins', Elsevier, Amsterdam, 1964.

If a metal oxidation state change occurs under the demetallation conditions, the charge (Z) and the radius (γ_i) should be those of the actually demetallated species. As is obvious from the definition of SI, a metalloporphyrin with the metal ion of a high charge, a high electronegativity and a small radius is stable. The SI underestimates the stability of metalloporphyrins where there are strong metal–porphyrin π interactions.

The acid demetallation mechanism is simpler than the metallation mechanism.[18] In general, the rate law is expressed as $d[H_4Por^{2+}]/dt = k[M(Por)][H^+]^n$ [$n = 1$–3 depending on the stability of M(Por)] where aggregation is negligible, and N protonation induces the metal dissociation.

The relative stabilities of metalloporphyrins can be compared by transmetallation reactions.[18] As expected from the Z/γ_i ratio, monovalent metals are easily displaced by other metal ions. The order of stability is indicated as follows: Cu > Zn > Cd > Hg > Pb (> Ba) > Li > Na > K. In most cases the transmetallation rate is first order both in metal ion and in metalloporphyrin, and the intermediate is considered to be the dinuclear complex [M(Por)M′]. For Cu/Zn(2,4-disubstituted DPIX DME) in boiling pyridine, the substitution rate increases as the porphyrin basicity decreases, *i.e.* substituent = Br > CH=CH$_2$ > CH$_2$CH$_3$ (> Etio).

21.1.1.1.6 *Structure and spin states of transition metalloporphyrins*

Among the various structural types (Figure 4), square four-coordinate, square pyramidal five-coordinate and octahedral six-coordinate complexes are most commonly found in metalloporphyrins, especially for the first row transition metal series.

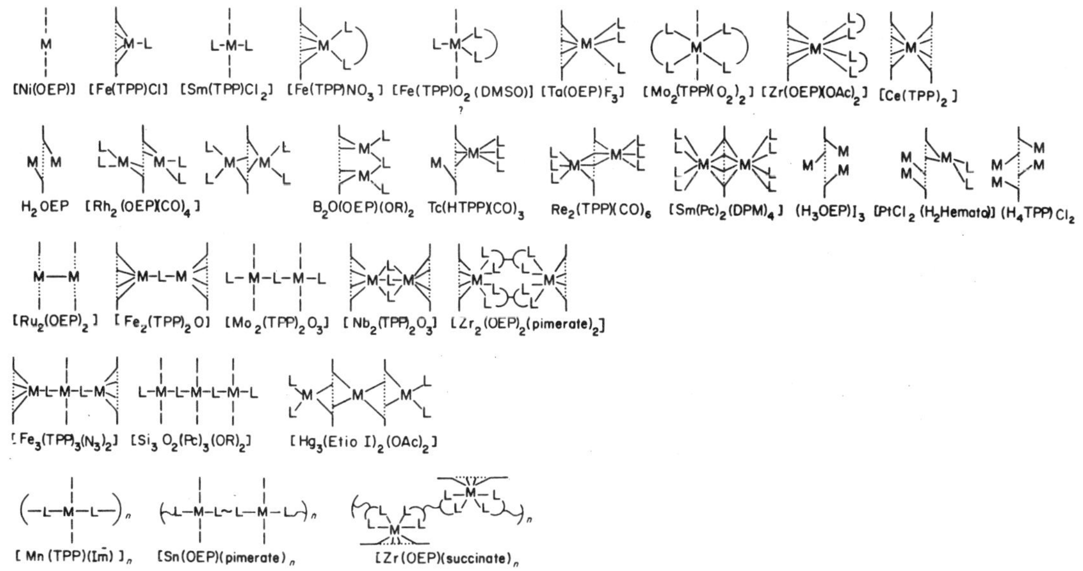

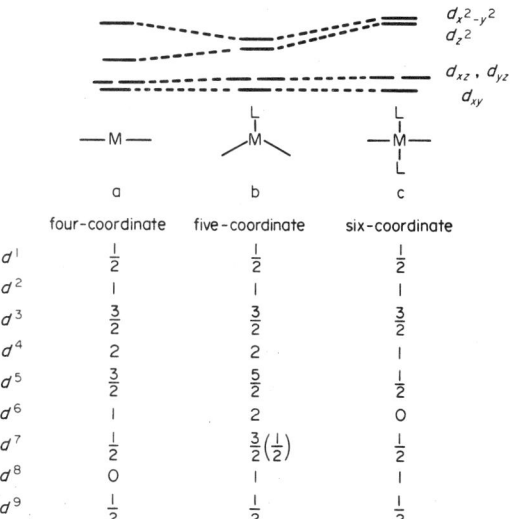

Figure 4 Various types of metalloporphyrins

In the square planar M(Por), the metal $d_{x^2-y^2}$ orbital is highly destabilized by the strong σ-type interactions with the porphyrin N σ orbitals. The d_{z^2} orbital also has some orbital overlaps with the N σ orbitals and is moderately destabilized. On the other hand, the energy levels of d_{xz} and d_{yz} depend on the extent of $d_\pi - p_\pi$ interactions. For the first row transition metals, the porphyrin π (e_g) to $d\pi$ (d_{xz}, d_{yz}) interactions exceed the d (π) to porphyrin e_g (π^*) interactions and locate the d_{xz} and d_{yz} levels slightly above the d_{xy} (Figure 5a). Addition of the fifth ligand raises the d_{z^2} energy level due to the axial σ interactions and, at the same time, lowers the $d_{x^2-y^2}$ level by displacing the metal ion out of the N_4 plane (Figure 5b). Both $d_{x^2-y^2}$ and d_{z^2} levels are raised in the six-coordinate complexes because of the strong σ interactions with the ligands (Figure 5c).

	a	b	c
	four-coordinate	five-coordinate	six-coordinate
d^1	$\frac{1}{2}$	$\frac{1}{2}$	$\frac{1}{2}$
d^2	1	1	1
d^3	$\frac{3}{2}$	$\frac{3}{2}$	$\frac{3}{2}$
d^4	2	2	1
d^5	$\frac{3}{2}$	$\frac{5}{2}$	$\frac{1}{2}$
d^6	1	2	0
d^7	$\frac{1}{2}$	$\frac{3}{2}\left(\frac{1}{2}\right)$	$\frac{1}{2}$
d^8	0	1	1
d^9	$\frac{1}{2}$	$\frac{1}{2}$	$\frac{1}{2}$

Figure 5 Qualitative *d*-orbital diagrams for the first row transition metalloporphyrins

From these qualitative pictures, the spin states of the first transition series are predicted as in Figure 5. Mn(Por) and Co(Por) show an unusually strong tendency to take high and low spin states respectively, and do not follow the general trends. Exceptions also arise for strong π donors (oxo and nitrido ligands) or π acceptors (:CO, :CN—R, :CR$_2$, :N—R, and ·NO) which significantly raise or lower the $d\pi$ levels, and for weak field ligands (neutral oxygenous ligands, and weak anionic ligands such as RSO$_3^-$, ClO$_4^-$, BF$_4^-$, PF$_6^-$, AsF$_6^-$, SbF$_6^-$) which only moderately destabilize the d_{z^2} level.

The situation is best exemplified by Fe(Por) whose spin state is delicately controlled by the axial ligand fields.[20] Fe^{II}(TPP), {Fe^{II}(TPP)(SPh)}$^-$ and Fe^{II}TPP(1-MeIm)$_2$ (1-MeIm = 1-methyl-imidazole) are four-coordinate intermediate-spin ($S = 1$), five-coordinate high-spin ($S = 2$) and six-coordinate diamagnetic ($S = 0$) complexes respectively, but five-coordinate diamagnetic Fe^{II}(TPP)(CO) ($S = 0$) and six-coordinate high-spin Fe^{II}(TPP)(THF)$_2$ ($S = 2$) also exist. While Fe^{III}(TPP)Cl shows high-spin character ($S = \frac{5}{2}$), the nature of Fe^{III}(TPP)ClO$_4$ is best explained by a quantum mechanically admixed intermediate-spin ($S = \frac{5}{2}/\frac{3}{2}$) state. [Fe^{III}(TPP)(py)$_2$]$^+$ ClO$_4^-$ and [Fe^{III}(TPP)(DMSO)$_2$]$^+$ ClO$_4^-$ take low-spin ($S = \frac{1}{2}$) and high-spin ($S = \frac{5}{2}$) states respectively. Scheidt and Reed[20] pointed out the Fe—N bond extension along the electron population in the antibonding $d_{x^2-y^2}$ orbital (Figure 6 and Table 3).

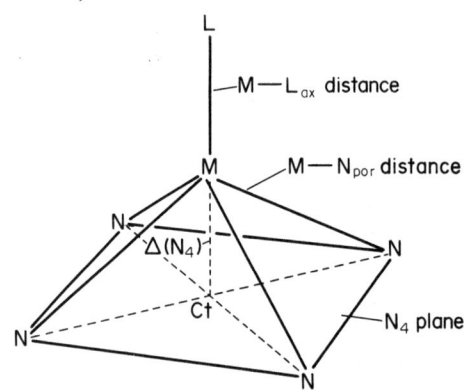

Figure 6 Abbreviations for structural parameters

Table 3 Fe—N$_{por}$ Bond Distances in Fe(Por)[20]

| | Fe—N$_{por}$ distance (Å) | | | |
| | Fe^{III}(Por) | | Fe^{II}(Por) | |
	HS	LS/IS	HS	LS/IS
Six-coordinate	2.05	1.99	2.06	1.99
Five-coordinate	2.07	2.00	2.08	1.99
Four-coordinate				1.97

The second and the third row transition metalloporphyrins, in general, take the lower spin states due to the larger energy gap between the metal d orbitals.

The size of the central porphyrin hole is such that it just accommodates metal ions of $\gamma_i \approx 0.60$–0.65 Å.[21] In M(Por) with a small metal size, the porphyrin is often buckled (S_4 distortion) to reduce the hole size. A core expansion or a doming occurs if the central metal is too large.

21.1.1.1.7 Coordination chemistry of metalloporphyrins

(i) Group IA (Li^I, Na^I, K^I, Rb^I, Cs^I)

Alkali metals form porphyrin complexes of the type M$_2$(Por) when their alkoxides are heated with porphyrin in dry pyridine, but are readily hydrolyzed.[17] The transmetallation experiments indicate the stability of M$_2$(Por) in the order K < Na < Li (< Zn, Hg).[18] Using radioactive sodium, Na$_2$(TPP) is shown to undergo rapid metal exchange with NaI in pyridine (equation 2).

$$M_2(\text{Por}) + 2M'I \underset{\text{pyridine}}{\rightleftharpoons} M_2'(\text{Por}) + 2MI \qquad (2)$$

(ii) Group IIA (Be^{II}, Mg^{II}, Ca^{II}, Sr^{II}, Ba^{II})

Alkaline earth metalloporphyrins, M(Por), are obtained by porphyrin metallation with metal perchlorates in pyridine or metal hydroxides in methanol.[17a] They dissociate under mild conditions, even in acetic acid.

Nitrogenous ligands coordinate to Mg(TPP) in two steps (equation 3).[18] With the increase of ligand pK_a, log K_1 increases linearly while log K_2 decreases linearly. K_1 (10^{2-5}) is much larger than K_2 (10^{1-0}).[22] The ionic radius of Mg^{II} (0.72 Å) is somewhat large for the central porphyrin hole and thus may contribute to the stable five-coordination with the metal out of the N_4-plane structure [$\Delta(N_4)$ 0.40 Å, Mg—N_{por} 2.072(10) Å in Mg(TPP)(OH$_2$)].[21] Oxygen-containing ligands such as water, alcohol and cyclic ethers also coordinate to Mg(Por), but soft sulfides are very weak ligands for the hard magnesium ion.[17a,21,23]

$$Mg(TPP) + 2L \underset{}{\overset{K_1}{\rightleftharpoons}} Mg(TPP)L + L \underset{}{\overset{K_2}{\rightleftharpoons}} Mg(TPP)L_2 \qquad (3)$$

(iii)   Group IIIB (B^{III}, Al^{III}, Ga^{III}, In^{III}, $Tl^{I,III}$)

Reaction of H$_2$(OEP) and diborane in THF and subsequent alkolysis gave {(ROB)$_2$O}(OEP) (Scheme 3), while {(HO)$_2$B}$_2$(TPP) was obtained after chromatographic purification of the unstable reaction product of H$_2$(TPP) and boron trichloride (Scheme 4).[24] Their structures have not been confirmed. They are slowly hydrolyzed in aqueous ethanol and rapidly in acid.

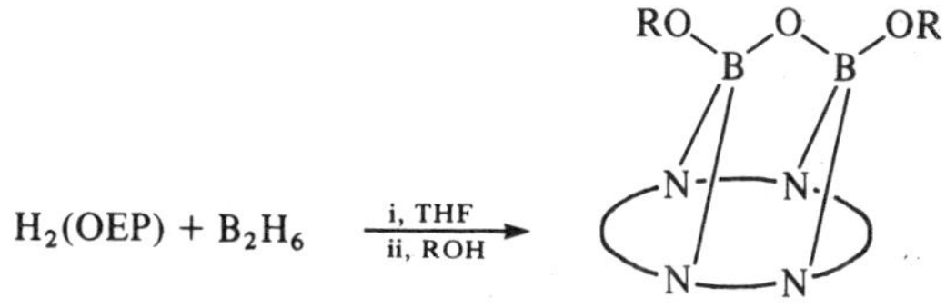

Scheme 3

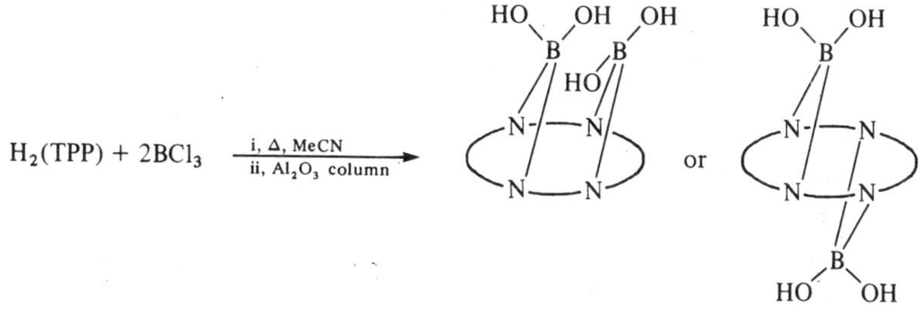

Scheme 4

Al^{III} and Ga^{III} porphyrins are prepared by metallation with metal acetylacetonates.[17a] InCl$_3$ or Tl(OCOCF$_3$)$_3$ is used for complex formation.[17a] Except for boron compounds, Group IIIB metalloporphyrins show strong tendencies to take a square pyramidal geometry with a fifth ligand such as halide, alkoxide or hydroxide.[17a,25] Al(OEP)(OH) can be condensed to a μ-oxo dimer, [Al(OEP)]$_2$O, at high temperature under vacuum, but the μ-oxide is easily hydrolyzed by water. A corresponding [Ga(OEP)]$_2$O has not been obtained. {In^{III}(TPP)(OSO$_2$R)} is unusual in this group since the indium atom exists in the expanded porphyrin plane and is connected to the neighbouring indium atoms by sulfonate bridges.[26] In Tl^{III}(OEP)(OCOMe), the acetate is considered to coordinate to the thallium atom as a bidentate ligand (ν(COO) = 1560 cm^{-1}).[17a]

Halide complexes of this group react with carbanions to form stable alkylmetalloporphyrins.[27] The thallium atom is significantly displaced (0.98 Å) from the N_4 plane with a large Tl—N_{por} distance (2.29 Å) in Tl(TPP)Me, formed by the reaction of H$_2$TPP and MeTl(OCOMe)$_2$ (Scheme 5). Monovalent thallium porphyrin complexes are also known.[17a] They can be prepared by treating a porphyrin with thallium(I) ethoxide in THF.[29] Unlike the other Group IIIB metalloporphyrins, [Tl^I]$_2$(Por) is easily hydrolyzed.

Scheme 5

(iv)   Group IVB, (Si^{IV}, Ge^{IV}, Sn^{II,IV}, Pb^{II,IV})

Silicon and germanium require rather drastic reaction conditions for complexation to porphyrins.[17a] SiCl$_4$ or GeCl$_4$ must be heated with the porphyrin at $\sim 170\,^\circ$C for several hours in pyridine in a sealed tube, and the complex is obtained as the dimethoxy form, M^{IV}(Por)(OMe)$_2$, by crystallization in the presence of excess methanol after hydrolysis and chromatographic isolation. Ge(Por)Cl$_2$ is also obtained by the reaction of GeCl$_4$ and porphyrin in quinoline at 230 $^\circ$C. Sn^{II} and Pb^{II} porphyrins are easily prepared from the divalent metal salts in pyridine.[17a] Sn^{II}(Por) is air sensitive and oxidized to the Sn^{IV} complex, Sn^{IV}(Por)X$_2$. In Sn(TPP)Cl$_2$, the porphyrin ring is expanded to accommodate the large Sn^{IV} atom (Sn—N$_{por}$ = 2.098(2) Å).[21] The Pb atom in Pb^{II} meso-tetrapropylporphyrin (Δ(N$_4$) 1.174 Å, Pb—N$_{por}$ $\sim$ 2.37 Å) is easily dissociated from it.[30] Pb^{II}(Por) can be oxidized to Pb^{IV}(Por)Cl$_2$ by chlorine.[17a]

(v)   Group VB (P^{III,V}, As^{III,V}, Sb^{III,V}, Bi^{III})

Reaction of Group VB halides, M^{III}X$_3$, and porphyrins in pyridine gives M^{III}(Por)X under anaerobic conditions.[17a,31] P^{III} and As^{III} porphyrins are very sensitive to oxygen and quickly autoxidize to P^V and As^V species, [M^V(Por)(OH)$_2$]X. Sb^{III} complexes are more stable and Sb^{III}(Por)Cl and [Sb^V(Por)(OH)$_2$]Cl are easily prepared. Reflecting the general order of stability, P^{III}, As^{III} < Sb^{III} < Bi^{III}, only Bi^{III}(Por) have been fully characterized.

The X-ray crystal structure of [P^{V}TPP(OH)$_2$]OH·2H$_2$O has been determined.[32] The porphyrin skeleton is distorted S_4 symmetrically to attain short P—N$_{por}$ distance ($\sim$ 1.90 Å). The hydroxy groups are strongly intermolecularly hydrogen bonded, and the phosphorus atom is displaced from the N$_4$ plane by 0.096 Å. The antimony(V) ion is centred in the porphyrin plane in [Sb(OEP)(OH)$_2$]$^+$ with a somewhat large Sb—N$_{por}$ distance of 2.065(6) Å.[31] The structure of [Bi(OEP)]NO$_3$ is of interest, since the four-coordinate Bi^{III} ion protrudes from the porphyrin N$_4$ plane by 1.09 Å with a very long Bi—N$_{por}$ distance of 2.32 Å.[31]

(vi)   Group VIB (Se, Te)

Spectral changes (Soret maxima $\sim$ 450 nm) during the reaction of SeCl$_4$ (or TeCl$_4$) and porphyrin in hot pyridine imply formation of the Se (or Te) porphyrin, but these complexes have not been characterized.[31]

(vii)   Group IIIA (Sc^{III}, Y^{III}, La^{III} and lanthanides^{III} except for Pm, Th^{IV}, U^{IV})

Trivalent yttrium and lanthanide metals, except for promethium, have been complexed to octaethylporphyrin by heating at 210 $^\circ$C in an imidazole melt.[17] The complexes obtained as hydroxides, M^{III}(OEP)(OH), are unstable in acidic media. As the charge:radius ratio rule predicts, the early lanthanide metalloporphyrins, M^{III}(OEP)(OH) (M = La, Ce, PR, Nd), are demetallated during purification, and the middle series (M = Sm, Eu, Gd, Tb, Dy) in 1% acetic acid in methanol, while the last five (M = Ho, Er, Tm, Yb, Lu) survive in 2% acetic acid in methanol but are dissociated in dilute hydrochloric acid. The M^{III}(OEP)(OH) appears to coordinate more than one equivalent of pyridine and piperidine, and dimerizes in noncoordinating solvents such as benzene and dichloromethane at $\sim 10^{-4}$ M concentration. The dimer is considered to be a di-μ-hydroxo-bridged species, different from the μ-oxo dimer, {Sc^{III}(OEP)}$_2$O (Scheme 6).

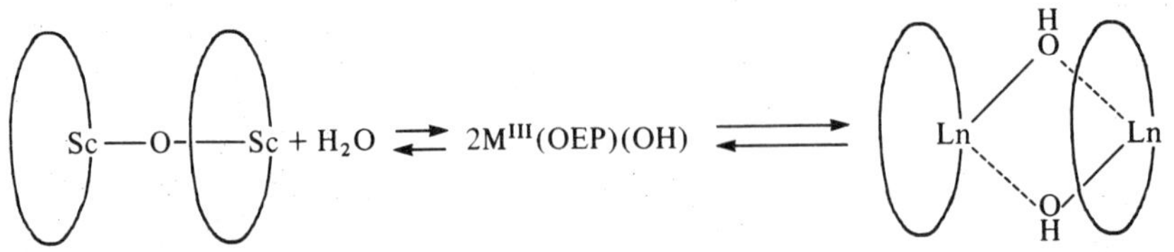

Scheme 6

If an M^{III} acetylacetonate is used for the porphyrin metallation, $M^{III}(Por)(acac)$ is formed, in which the acetylacetonato coordinates to the metal as a bidentate ligand (Scheme 7).[17a,33] The paramagnetic NMR shift of the phenyl proton in $Eu^{III}(TPP)(acac)$ suggests a large displacement of the europium atom from the centre of the porphyrin (~ 2.0 Å) with long $Eu-N_{por}$ bonds (~ 2.83 Å). Sandwich complexes $Ce(TPP)_2$ and $PrH(TTP)_2$ have been reported.[34]

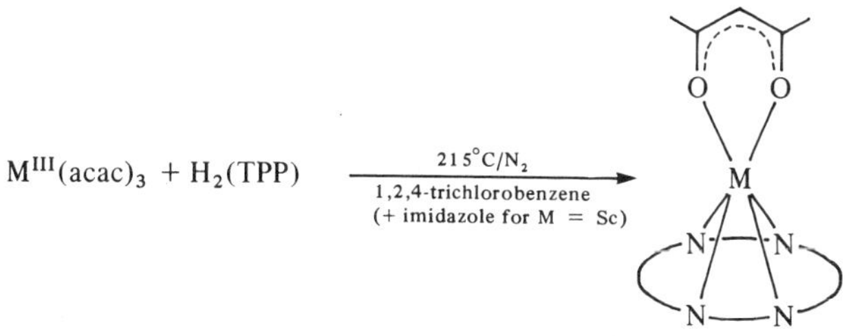

$$M^{III}(acac)_3 + H_2(TPP) \xrightarrow[\substack{1,2,4\text{-trichlorobenzene} \\ (+ \text{ imidazole for } M = Sc)}]{215\,°C/N_2}$$

Scheme 7

The actinide element, thorium, also forms a porphyrin complex, $Th^{IV}(Por)(acac)_2$, in the same way as the lanthanide derivatives (Scheme 8).[17a] The acetylacetonato groups most likely coordinate to the thorium atom on the same side of the porphyrin (Scheme 8). Th^{IV} and $U^{IV}(OEP)Cl_2L_2$ ($L = PhCN$, THF, py) are prepared from $H_2(OEP)$ and MCl_4.[35]

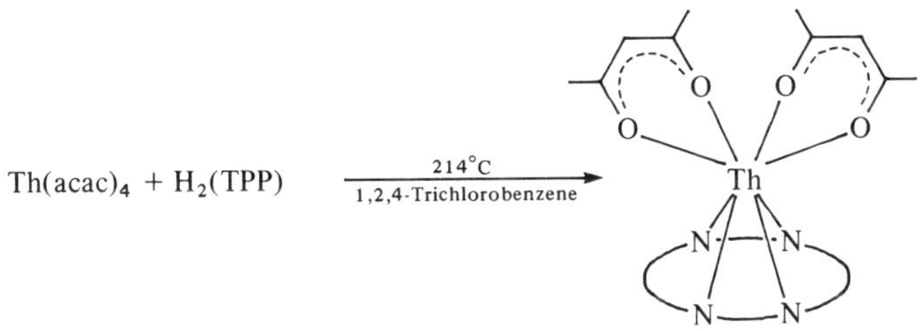

$$Th(acac)_4 + H_2(TPP) \xrightarrow[\text{1,2,4-Trichlorobenzene}]{214\,°C}$$

Scheme 8

(viii) Group IVA ($Ti^{III,IV}$, Zr^{IV}, Hf^{IV})

Titanium can be inserted into a porphyrin by heating $TiPh_2Cl_2$, $TiO(acac)_2$ or $TiCl_4$ in high boiling solvents.[17a] The complex is isolated as $TiO(Por)$, which serves as a precursor of halide derivatives.[36] The reaction of an oxotitanium porphyrin with benzoyl peroxide or hydrogen peroxide gives a peroxy compound ($\nu(O-O)$ 895 cm^{-1}). X-Ray structure analyses confirmed the triangular side-on structure with the O—O bond (~ 1.45 Å) parallel to one of the $N_{por}-Ti-N_{por}$ directions. In $Ti(OEP)(O_2)$, the O—O orientation is fixed at low temperatures, but starts oscillating between the two $N_{por}-Ti-N_{por}$ axes on the NMR timescale at $\sim -50\,°C$ in solution. Out-of-plane displacement of the titanium atom is somewhat larger in the peroxo complex ($\Delta(N_4)$ 0.620(6) Å) than in the oxo species ($\Delta(N_4)$ 0.555(6) Å) while the average $Ti-N_{por}$ distance (~ 2.11 Å) is identical in both complexes. $TiO(Por)$ is converted to halide complexes, $Ti(Por)X_2$, by the corresponding acid, HX (X = F, Cl, Br) and $Ti^{IV}(Por)F_2$ is reduced ($E_{1/2} - 0.45$ V vs. SCE in CH_2Cl_2, Por = TPP) to $Ti^{III}(Por)F$ by zinc amalgam. The paramagnetic Ti^{III} complex coordinates donor ligands such as THF, amide, imidazole, pyridine and phosphine, and shows ESR signals at $g = 1.96-1.97$ with coupling constant $a_F \approx 10$ G, $a_{Npor} \approx 2.3$ G and $a_{Nax} \approx 1$ G. $[Ti^{III}(OEP)F_2]^-$ also gives ESR signals in the same region with $a_F \approx 6.3$ G and $a_N \approx 2.25$ G in dichloromethane at 25 °C, consistent with the unpaired electron being in the metal d_{xy} orbital. $Ti^{III}(Por)F$ is alkylated to air sensitive organotitanium(III) complexes which, in contrast to the starting material, do not coordinate a sixth ligand, suggesting increased electron density at the alkylated titanium atom (Scheme 9).

$$TiO(TPP) \xrightarrow{HX} Ti(TPP)X_2 \xrightarrow{Zn-Hg} Ti(TPP)F \xrightarrow{RMgX} Ti(TPP)R \xrightarrow{O_2} TiO(TPP)$$

$$X = F, Cl, Br$$

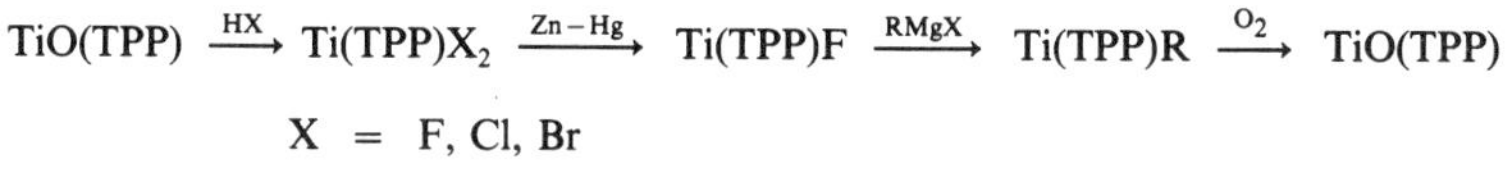

Scheme 9

Zr^{IV} and Hf^{IV} porphyrins are prepared from metal acetylacetonates and porphyrin, but the product depends on the solvent and the work-up procedure.[17a] $M^{IV}(Por)(acac)_2$, $M^{IV}(Por)(acac)(OPh)$ and $M^{IV}(Por)(O_2CMe)_2$ are all interconvertible, and the additional ligands are on the same side of the porphyrin (Scheme 10). The acetato groups serve as bidentate ligands and the metal atom is centred in a distorted square antiprism.[21] $Zr(OEP)(O_2CMe)_2$ and $Hf(OEP)(O_2CMe)_2$ are isodimensional with $M—N_{por} \approx 2.26\,\text{Å}$, $M—O \approx 2.28\,\text{Å}$ and $\Delta(N_4) \approx 1\,\text{Å}$.

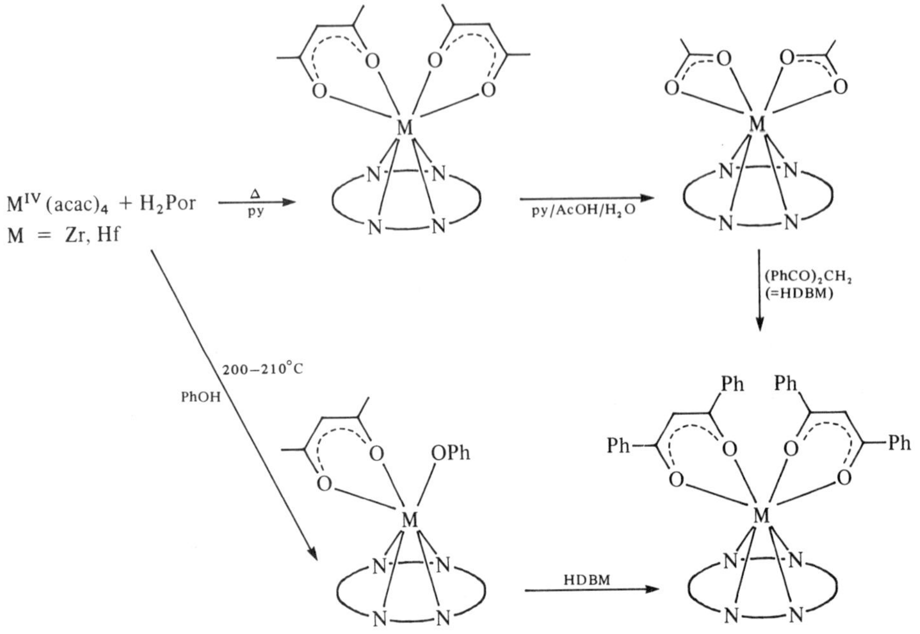

Scheme 10

There are marked differences between Group IVA and IVB complexes. For example, $M^{IV}(Por)LL'$ is invariably octahedral for IVB (Si, Ge, Sn), but can be six-, seven- or eight-coordinate for IVA (Ti, Zr, Hf). $M(Por)F_2$ (M = Ti, Zr, Hf) are easily hydrolyzed while Si, Ge and Sn analogues are rather stable to water.

(ix)   Group VA ($V^{II,IV}$, $Nb^{IV,V}$, Ta^V)

The reaction of vanadium salts (VCl_4, $V_2O_2(SO_4)_2$, $V(OCOMe)_4$, $VO(acac)_2$) and porphyrin gives a complex which, after work-up, is isolated as a very stable oxovanadium(IV) species, VO(Por) (for VO(OEP), $V—O = 1.620(2)\,\text{Å}$, $V—N_{por} = 2.102\,\text{Å}$, $\Delta(N_4) = 0.543\,\text{Å}$).[17a,21] In the presence of a large excess of nitrogenous ligand such as pyridine and piperidine, it forms a six-coordinate complex with a small equilibrium constant ($K = 10^{-1}$–$10^{-2}\,1\,M^{-1}$).[18] The effect of β substituents on the association constant (K_X) is expressed by the Hammett equation (equation 4).

$$VO(2,4\text{-}X_2DPIX\ DME) + py \underset{}{\overset{K_X}{\rightleftharpoons}} VO(2,4\text{-}X_2DPIX\ DME)(py) \qquad (4)$$

$$\log(K_X/K_H) = \varrho(2\sigma_X),\ \varrho \approx 0.65,\ \sigma = \text{Hammett substituent constant}$$

For the association of piperidine and VO(Tp-substituted PP), the remote substituent effect was shown to be very small ($\log(K_x/K_H) = \varrho(4\sigma_x)$, $\varrho = 0.113 \pm 0.003$, $K \approx 10^{-1}\,M^{-1}$). The negative $\Delta H\,(-23.4 \pm 1.7\,\text{kJ mol}^{-1})$ and somewhat more negative $\Delta S\,(-88 \pm 4\,\text{J K}^{-1}\,\text{mol}^{-1})$ values for this type of coordination suggest rigid and rotationless bonding between the vanadium and piperidine. VO porphyrins show ESR signals at $g_\| \approx 1.96$ and $g_\perp \approx 1.98$ with $A_\| \approx 160 \times 10^{-4}\,\text{cm}^{-1}$ and $A_\perp \approx 55 \times 10^{-4}\,\text{cm}^{-1}$.[2]

VO(Por) can be converted to $V(Por)Cl_2$, by oxalyl chloride or thionyl chloride in toluene.[37] The chloride has a magnetic moment of $\mu_{eff} = 1.62\,\text{BM}$ and the unpaired electron (d^1) is suggested to be in e_g (d_{xz}, d_{yz}) orbitals by CNDO/S MO calculations. The chloride ligands are substituted with bromide by treating $V(Por)Cl_2$ with HBr. EXAFS study on the $V(Por)Br_2$ revealed the V—Br bond length to be about 2.41 Å. $V^{IV}(Por)$ is reduced by zinc amalgam in the presence of phosphine ligands to $V^{II}(Por)(PR_3)_2$ which is autoxidized to $V^{IV}O(Por)$. Similarly sulfur oxidizes $V^{II}(Por)(THF)_2$ to $V^{IV}S(Por)$.

$NbCl_5$ and $TaCl_5$ metallate porphyrins in benzonitrile at 210 °C and in phenol at 295 °C respectively.[17a,38] The trichloride is hydrolyzed to $\{M(Por)\}_2(O)_3$ in which three oxygen atoms are bridging two metal ions.[21] The tri-μ-oxo dimer dissociates to oxoacetates $MO(Por)(OCOMe)$ in hot acetic acid and to trihalides $M(Por)X_3$ in hydrohalic acid (HF, HCl, HBr).[17a,38] $NbO(OEP)X$ (X = I_3, F, acac) complexes are also known (Scheme 11).[39]

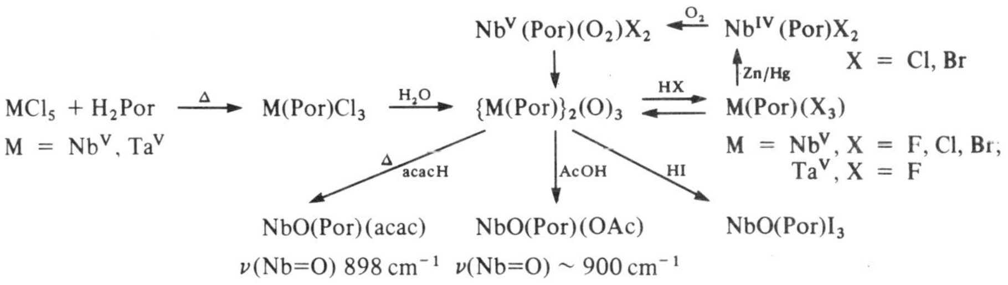

$$NbCl_5 + H_2Por \xrightarrow{\Delta} M(Por)Cl_3 \xrightarrow{H_2O} \{M(Por)\}_2(O)_3$$

Scheme involving:

$$Nb^V(Por)(O_2)X_2 \xleftarrow{O_2} Nb^{IV}(Por)X_2 \qquad X = Cl, Br$$

$$\{M(Por)\}_2(O)_3 \underset{\xrightarrow{HX}}{\overset{}{\rightleftharpoons}} M(Por)(X_3) \qquad M = Nb^V, X = F, Cl, Br, \qquad Ta^V, X = F$$

$$M = Nb^V, Ta^V$$

$$NbO(Por)(acac) \qquad NbO(Por)(OAc) \qquad NbO(Por)I_3$$

$\nu(Nb{=}O)$ 898 cm^{-1} $\qquad$ $\nu(Nb{=}O) \sim 900$ cm^{-1}

Scheme 11

In $NbO(TPP)(O_2CMe)$, the acetate coordinates to the niobium atom as a bidentate ligand on the same side of the porphyrin ring as the oxo group. The heptacoordinate niobium atom is displaced out of the N_4 plane by 0.99 Å, similar to that in $\{Nb(TPP)\}_2(O)_3$ $(\Delta(N_4) = 1.00$ Å).[38] The Nb=O distance, 1.720(6) Å, is consistent with its double bond character. The average length of the unsymmetric $Nb{-}N_{por}$ bonds is 2.23 Å.

Zn amalgam reduces $Nb(Por)X_3$ to $Nb^{IV}(Por)X_2$ (X = Cl, Br), whose ESR parameters (X = Br, $g_\| = 1.942$, $g_\perp = 1.965$, $A_\| = 224 \times 10^{-4}$ cm^{-1}, $A_\perp = 112 \times 10^{-4}$ cm^{-1} in THF at 77 K) indicate an octahedral structure with an unpaired electron in d_{xy}.[40] $Nb(Por)X_2$ irreversibly binds O_2 (X = Br, $\nu(O_2) = 1220$ cm^{-1}, $g = 2.002$, $A_{s.h.f} = 10.8 \times 10^4$ cm^{-1} in THF at r.t.).

(x) *Group VIA ($Cr^{II,III,IV,V}$, $Mo^{II,IV,V,VI}$, W^V)*

Chromium complexes can be prepared from $CrCl_2$, $Cr(acac)_3$ or $Cr(CO)_6$ and porphyrins.[17a] The initially isolated compound, $Cr^{III}(Por)OR$, is converted to $Cr^{III}(Por)Cl$ by hydrogen chloride.[17a,41] Electrochemical (for $(TPP)Cr^{II}/Cr^{III}$, $E_{1/2} = -0.86$ V vs. SCE in DMSO) or $Cr(acac)_2$ reduction of the chloride gives the square planar $Cr^{II}(Por)$.[42] The high-spin Cr(II) complex (d^4, $S = 2$, $\mu_{eff} \approx 4.8$ BM) coordinates one or two pyridine molecules to form five-coordinate intermediate-spin $Cr^{II}(Por)(py)$ ($S = 1$, $\mu_{eff} = 2.8$–2.9 BM), or six-coordinate diamagnetic $Cr^{II}(Por)(py)_2$. The $Cr{-}N_{por}$ distance is the same for the square planar $Cr(TPP)$ (2.033(1) Å) and octahedral $Cr(TPP)(py)_2$ (2.027(8) Å) complexes because of the empty $d_{x^2-y^2}(Cr)$ in both compounds.[21] In the solid state, $Cr^{II}(TPP)(py)$ irreversibly adds molecular oxygen, and the O_2 adduct shows $\nu(O_2) = 1142$ cm^{-1}, indicating an end-on superoxide structure.[42] $Cr^{II}TPP$ is oxidized by O_2 to the diamagnetic $CrO(TPP)$ ($\nu(Cr{=}O)$ 1020 cm^{-1}) in which the Cr(IV) atom is displaced out of the N_4 plane by 0.47 Å with $Cr{-}N_{por}$ and $Cr{=}O$ bond lengths of 2.032(8) Å and 1.572(6) Å respectively.[43] $Cr^{III}(Por)Cl$ is converted to $Cr^{III}(Por)Br$ by metathesis with Bu^n_4NBr.[41] Both chloride and bromide coordinate a sixth ligand such as O donors (acetone, THF, DMF), N donors (aliphatic and aromatic amines, nitriles) and S donors (thioethers). Unlike ordinary Cr^{III} complexes, $Cr^{III}(Por)XL$ exhibits rapid ligand (L) exchange, and the association constant (K) can be estimated from the equilibrium constant (K_{obs}) (equation 5). Log K is linearly correlated to pK_{BH+} of the sixth ligand (substituted pyridines and aliphatic amines) except for 1-MeIm. The anomalous behaviour of imidazole coordination is a common phenomenon for Cr, Mn and Fe porphyrins and may be ascribed to its π-donor character.

$$Cr(TPP)Cl + L \underset{}{\overset{K}{\rightleftharpoons}} Cr(TPP)ClL \qquad K = \frac{[Cr(TPP)ClL]}{[Cr(TPP)Cl][L]} \approx K_{obs} \text{ (L}' = \text{acetone)} \qquad (5)$$

$$Cr(TPP)ClL' + L \underset{\text{toluene}}{\overset{K_{obs}}{\rightleftharpoons}} Cr(TPP)ClL + L'$$

$$\log K \approx 4.28 \text{ (L = py)}, \; 6.71 \text{ (L = 1-MeIm)}$$

Cr(Por)Cl is solvolyzed to Cr(Por)OR by alcohols and hydrolyzed to Cr(Por)OH which has little tendency to μ-oxide formation.[17a,41] All the five-coordinate Cr^{III}(Por)X complexes are high-spin compounds (d^3, $S = \frac{3}{2}$, $\mu = 3.6$–3.8 BM), while Cr^{III}(Por)(OPh)(HOPh) and Cr^{III}(Por)Br(py) are high-spin (d^3, $S = \frac{3}{2}$, $\mu = 3.2$ BM) and low-spin (d^3, $S = \frac{1}{2}$, $\mu = 2.8$ BM) respectively.[41,42]

Oxidation of Cr(Por)(OH) with sodium hypochlorite in the presence of ammonia gives a stable nitridochromium(V) complex, CrN(Por) (d_{xy}^1, $S = \frac{1}{2}$, $g \approx 1.98$, $v(Cr\equiv N) = 1017 \, cm^{-1}$) in which the chromium atom is displaced from the N_4 plane by 0.42 Å with Cr—N_{por} and Cr≡N distances of 2.04 and 1.565 Å respectively.[44] $Cr^V O(TPP)Cl$ was reported to form when Cr(TPP)Cl was treated with iodosobenzene, sodium hypochlorite or *m*-chloroperbenzoic aicd.[45] The oxochromium species (λ_{max} 418 nm, $v(Cr—O)$ 1026 cm^{-1}, $\mu = 2.05$ BM) is stable in dichloromethane solution at room temperature for several hours, may be reduced back to the starting material by oxygen acceptors, and appears to undergo ligand exchange with F^-, OH^- or Bu^tNH_2 (Scheme 12).

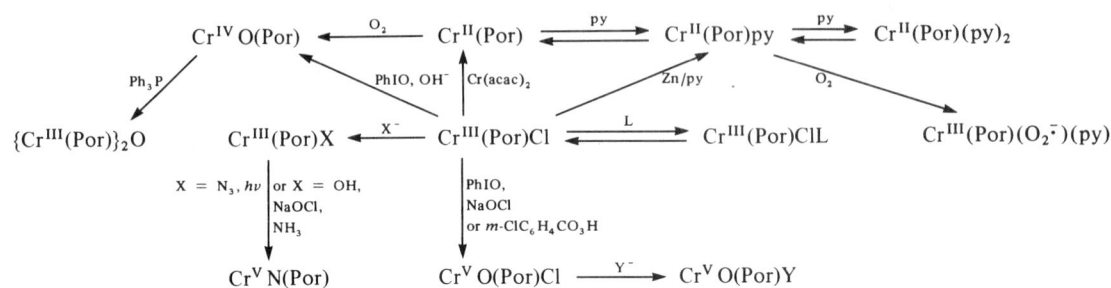

Scheme 12

Dimeric $\{M^V O(Por)\}_2 O$ (M = Mo, W) complexes are obtained by the metallation of porphyrins with metal oxide or chloride.[46] Molybdenum hexacarbonyl and acetylacetonate also serve as the metallating reagent.[17a] Unlike the neighbouring element Nb, Mo adopts octahedral geometry with a linear O=Mo—O—Mo=O array in $Mo(Por)_2O_3$. As expected, the Mo=O bond (1.707(3) Å) is shorter than the Mo—O bond (1.936(4) Å).[21] The Mo atoms are displaced from the N_4 plane towards the terminal oxygen by 0.09 Å and the average Mo—N_{por} distance is 2.094(3) Å. The μ-oxo dimer is cleaved by acids to MO(Por)X (M = Mo, W), and the halides are substituted by alkoxide or pseudohalogens.[17a] There is a relationship between $v(M=O)$ and polarizability of the *trans* ligand, *i.e.* $v(W=O)$ for X = F (930 cm^{-1}) < Cl (932 cm^{-1}) < Br (938 cm^{-1}) < I (960 cm^{-1}) for WO(OEP)X. MoO(TPP)X is paramagnetic (d^1, $S = \frac{1}{2}$, $\mu = 1.75$ BM) and exhibits ESR signals around $g = 1.96$ with $A_{Mo} \approx 50$ G.

$\{MoO(Por)\}_2 O$ can be reduced to $Mo^{IV} O(Por)$ ($v(Mo=O) \approx 900 \, cm^{-1}$) by such systems as NH_2NH_2/py, H_2S/CH_2Cl_2–MeOH, O_2^-/DMSO–CH_2Cl_2, or by sublimation at 250 °C *in vacuo*.[17a,47] The Mo^{IV} species is also obtained directly from $Mo(acac)_3$ and porphyrin.[48] MoO(TPP) is converted to the octahedral $Mo(TPP)Cl_2$ (Mo—N_{por} = 2.07(2) Å) with hydrogen chloride.

An interesting Mo^{VI} compound, $Mo(Por)(O_2)_2$, is formed by the reaction of MoO(Por)(OR) and hydrogen peroxide.[49] The molybdenum is octacoordinate with Mo—N_{por} and Mo—O bond lengths of 2.096(4) and 1.958(4) Å. The O—O bonds are rather short (1.399(6) Å) and parallel to one of the N—Mo—N axes while orthogonal to each other. $Mo(Por)(O_2)_2$ is photochemically converted to the *cis*-dioxo complex, $Mo^{VI}(O)_2(Por)$. A nitrido complex, $Mo^{VI}(N)(Por)(Br_3)$, is also known.[49]

Zinc amalgam reduction of $Mo^{IV}(TPP)Cl_2$ under NO leads to *cis*-$Mo^{II}(TPP)(NO)_2$ which is converted to *trans*-$Mo^{II}(TPP)(NO)(HOMe)$ on column chromatography (Scheme 13).[49]

(xi) Group VIIA ($Mn^{II,III,IV,V}$, Tc^I, $Re^{I,V}$)

Mn^{II} salts and porphyrin react smoothly in hot DMF or acetic acid in the presence of air.[17a] The Mn^{III} complex thus obtained is reduced to Mn^{II}(Por) by aqueous $Na_2S_2O_4$ or $NaBH_4$.[50] The manganese atom in Mn(TPP)(toluene) is high-spin (d^5, $s = \frac{5}{2}$, $\mu_{eff} = 6.2$ BM) and deviates from the centre of the porphyrin ($\Delta(N_4) = 0.19$ Å) with a little elongation of the Mn—N_{por} bonds (2.084(2) Å) due to half occupation of the antibonding $d_{x^2-y^2}$ orbital.[21] The out-of-plane conformation of the Mn^{II} atom implies preferential five-coordination. Indeed, square pyramidal complexes, Mn^{II}(Por)L, are isolated in the presence of various donor ligands.[51] The five-coordinate complexes also take a high-spin state, for example, Mn^{II}(Por)(1-MeIm) shows $\mu_{eff} = 6.2$–6.5 BM, typical of high-spin d^5. The large Mn—N_{por} (2.128(2) Å) and Mn—N_4 plane (0.515 Å) distances reflect the population of

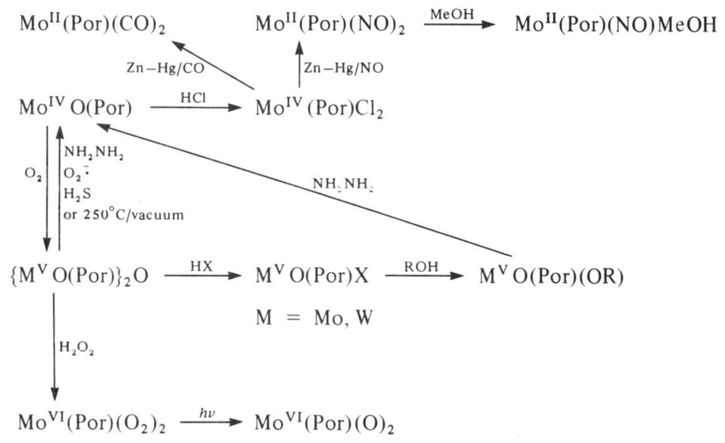

$$Mo^{II}(Por)(CO)_2 \qquad Mo^{II}(Por)(NO)_2 \xrightarrow{MeOH} Mo^{II}(Por)(NO)MeOH$$

$$\text{Zn–Hg/CO} \qquad \uparrow \text{Zn–Hg/NO}$$

$$Mo^{IV}O(Por) \xrightarrow{HCl} Mo^{IV}(Por)Cl_2$$

$$O_2 \left|\begin{array}{l} NH_2NH_2 \\ O_2^- \\ H_2S \\ \text{or } 250^\circ C/vacuum \end{array}\right. \qquad NH.NH_2$$

$$\{M^V O(Por)\}_2O \xrightarrow{HX} M^V O(Por)X \xrightarrow{ROH} M^V O(Por)(OR)$$

$$M = Mo, W$$

$$\left| H_2O_2 \right.$$

$$Mo^{VI}(Por)(O_2)_2 \xrightarrow{h\nu} Mo^{VI}(Por)(O)_2$$

Scheme 13

the $d_{x^2-y^2}$ orbital as well. The axial ligand association constant for Mn(TPP)L is not high ($K \approx 10^4$ for L = Im, py, DMSO), and increases moderately as the ligand pK_a increases [log (K/K_{Cl}) = 1.14 for L = 4-CNpy with pK_a = 1.86, log (K/K_{Cl}) = 3.24 for L = 4-Mepy with pK_a = 6.03].[52] The thermodynamic parameters for L = 4-CNpy were determined as shown in equation 6.

$$Mn(TPP) + 4\text{-CNpy} \underset{toluene}{\overset{K}{\rightleftharpoons}} Mn(TPP)(4\text{-CN-py}) \tag{6}$$

$$K \approx 3.61 M^{-1} \text{ at } 23\,^\circ C$$
$$\Delta H = -44.8 \pm 16 \, kJ\,mol^{-1}$$
$$\Delta S = -82.8 \pm 6.3 \, J\,K^{-1}\,mol^{-1}$$

Mn[II] porphyrins are easily oxidized in air at ambient temperature, but reversibly bind molecular oxygen at low temperature.[53] The ESR spectra of Mn(TPP)(O$_2$) and the IR spectra of Mn(TPP)O$_2$ ($\nu(O_2) = 983\,cm^{-1}$) and Mn(OEP)O$_2$ ($\nu(O_2) = 991\,cm^{-1}$) support the side-on Mn^{IV}(O$_2^{2-}$) formulation with the manganese ion in a high-spin (d^3, $S = \frac{3}{2}$) state.[54] The equilibrium constant for the oxygenation decreases as the donor power of the competitor ligand increases since ligand substitution is initiated by dissociation of the axial ligand which coordinates more tightly as its basicity increases (equation 7).[46]

$$Mn(TTP)L + O_2 \underset{toluene, -78\,^\circ C}{\overset{K_{obs}}{\rightleftharpoons}} Mn(TPP)O_2 + L \tag{7}$$

log $K_{obs} \approx -6.0$ (L = py), -8.0 (L = 1-MeIm), -8.5 (L = Bu$_3^n$P), -7.4 (L = PhSMe)

Nitroso complexes, Mn(Por)NO and Mn(TTP)(NO)(4-MePip) (4-MePip = 4-methylpiperidine), are also known.[55] In the latter, M—N=O is almost linear (176.2(5)°) in contrast to the bent Fe—N=O (143.7(6)°). The d_{yz}–p_y π interaction between Mn $(d_{xy})^2(d_{xz})^2(d_{yz})^1$ and N $(s)^2(p_x)^1(p_y)^1(p_z)^1$ is fully bonding.

Mn[III](Por)X exchanges its axial ligand when treated with excess of anion in aqueous methanol or the acid form in organic solvent.[17a,51,56] Very weak ligands such as perchlorate can be coordinated by heating Mn(Por)Cl and their Ag salts in THF.[52] All the five-coordinate complexes show high-spin character (d^4, $S = 1$, $\mu \approx 5\,BM$), but the Mn—N$_{por}$ bond length (~ 2.01 Å) and the metal displacement ($\Delta(N_4) \approx 0.27$ Å) are normal, consistent with the $d_{x^2-y^2}$ vacancy.[21] In the presence of an appropriate ligand, six-coordinate complexes are formed.[17a,51,57] The equilibrium constants are highly dependent on the counteranion and to some extent on the neutral ligand (equations 8 and 9).[51,52]

$$[Mn(TPP)X] + L \overset{K_1}{\rightleftharpoons} [Mn(TPP)LX] \tag{8}$$

$$[Mn(TPP)LX] + L \overset{K_2}{\rightleftharpoons} [Mn(TPP)L_2]^+ + X^- \tag{9}$$

L	log K_1 (X = Cl, in CH$_2$Cl$_2$)	log K_1	log K_2 (X = ClO$_4$, in CH$_2$ClCH$_2$Cl)
Im		4.35 ± 0.04	7.45 ± 0.04
py	1.10	4.08 ± 0.04	6.99 ± 0.07
DMSO		3.49 ± 0.01	5.74 ± 0.02
4-X-py	0.43 (X = CN) to 1.97 (X = Me)		

Most of the six-coordinate Mn^{III} porphyrins are also high-spin compounds, and form long axial bonds (*e.g.* $Mn-N_{ax} \approx 2.3$ Å) while the equatorial bond lengths are normal (~ 2.01 Å) due to the $(d_{xy})^1(d_{xz})^1(d_{yz})^1(d_{z^2})^1(d_{x^2-y^2})^0$ configuration.[21,57] This may explain the low association constant of $Mn(TPP)Cl(py)$ ($\log K_1 \approx 1.10$) compared with that of $Cr(TTP)Cl(py)$ ($\log K_1 \approx 4.28$).

Oxidations of Mn^{III} porphyrins take different courses depending on the nature of the axial ligands. If the axial ligands are weak donors, the electron will be removed from the porphyrin ring, *i.e.* $[Mn^{III}(TPP^{\ddagger})Cl]X^-$ ($X^- = ClO_4^-$, AsF_6^-).[58] Good donor ligands will stabilize the Mn^{IV} state, *i.e.* $Mn^{IV}(TPP)X_2$ ($X = {}^-OMe$, ${}^-N_3$, ${}^-NCO$, ${}^-CN$, ${}^-OI(OAc)Ph$) and $[Mn^{IV}(TPP)X]_2O$ ($X = {}^-N_3$, ${}^-NCO$, ${}^-OI(Cl)Ph$, ${}^-OI(Br)Ph$).[59] The manganese ion can be further oxidized to the pentavalent state, *i.e.* $Mn^V N(Por)$, in the presence of particularly strong donor ligands.[44] Mn^{IV} species are all high-spin six-coordinate compounds (d^3, $S = \frac{3}{2}$, $\mu \approx 3.8$ BM for monomeric compounds) but the bond parameters are normal ($N-N_{por} \approx 2.015$, $M-N_{ax} \approx 1.9$, $M-O_{ax} \approx 1.8$ Å) due to empty d_{z^2} and $d_{x^2-y^2}$ orbitals.[59] Both N_3 ($\nu(N_3) = 1997\,cm^{-1}$) and NCO ($\nu(NCO) = 2127\,cm^{-1}$) coordinate to the manganese atom with a bent $Mn-N-XY$ geometry ($125.0(5)°$ for N_3 and $\sim 135°$ for NCO). While Mn^{IV} porphyrins exhibit high spin states, nitrido Mn^V complexes are diamagnetic (d^2, $S = O$) because of the strong p_π to d_π interactions.[44] The X-ray structure of MnN(TPP) is consistent with the $N{\equiv}Mn^V$ formulation, *i.e.* $Mn-N_{por} \approx 2.02$, $Mn{\equiv}N = 1.515(3)$ Å. The $Mn{\equiv}N$ vibration gives an IR signal at $1036\,cm^{-1}$ (Scheme 14).

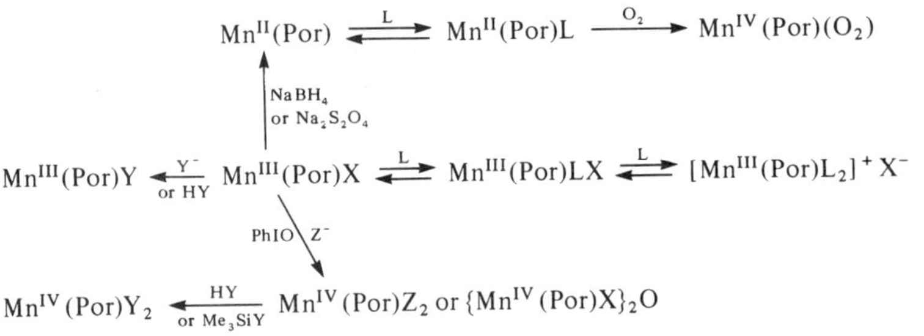

Scheme 14

Metallation of porphyrins with $M_2(CO)_{10}$ ($M = Tc$, Re) gives $M^I(CO)_3$ porphyrins.[19a,60] The reaction is stepwise and both mono- and bi-metallic forms can be obtained (Scheme 15). Each of the metal ions coordinates to three nitrogen atoms and is surrounded by six donor atoms in a quasi-octahedral arrangement. Theoretical calculations predict a high barrier to $Re(CO)_3$ rotation around the porphyrin normal through the Re atom (> 2 eV).[61] The structural parameters are $M-N_{por}$ (unshared) 2.16, $M-N_{por}$ (shared) 2.40, $M{\cdots}N_{por}$ (remote) 3.2 and $\Delta(N_4)$ 1.42; Re deviation from the S_2 axis (porphyrin normal) is 0.7–0.8 Å.[60] The Tc and Re complexes are virtually isodimensional. The Re^V complex, $ReO(Por)(OPh)$, is prepared by heating Re_2O_7 and porphyrins in phenol.[17a] Like Mo and W porphyrins, the oxorhenium(V) complex is converted to a μ-oxo dimer by aqueous alkali, and regenerated by acids (Scheme 16).

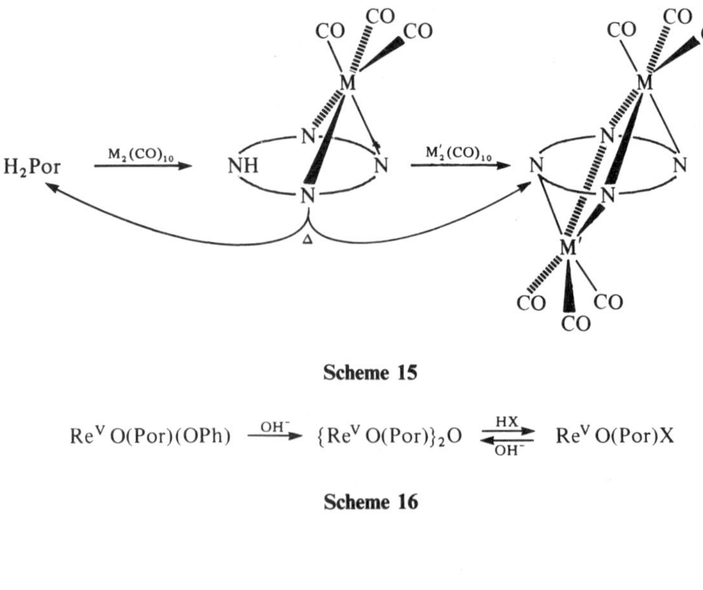

Scheme 15

$$Re^V O(Por)(OPh) \xrightarrow{OH^-} \{Re^V O(Por)\}_2O \underset{OH^-}{\overset{HX}{\rightleftharpoons}} Re^V O(Por)X$$

Scheme 16

(xii) Group VIII — 1 ($Fe^{I,II,III,IV}$, $Ru^{I,II,III,IV}$, $Os^{II,III,IV,Vi}$)

Iron is usually inserted into porphyrins by heating them with an Fe^{II} salt in a solvent.[14,17a] The systems most commonly used are $FeCl_2$/DMF, $FeSO_4$/py–AcOH, $Fe(OAc)_2$/HOAc and $FeBr_2$/base/C_6H_6–THF. $Fe(acac)_2$ may also serve as a metal carrier. $Fe(CO)_5$ is used to metallate sterically hindered porphyrins.[62]

If the preparations is carried out under strictly anaerobic conditions, divalent iron porphyrins are obtained. The iron(II) complexes can be prepared from iron(III) porphyrins by reduction with $Na_2S_2O_4$, $NaBH_4$ or $Cr(acac)_2$.[14,17a,63–65]

Four-coordinate Fe^{II}(Por) are intermediate-spin compounds (d^6, $S = 1$, $\mu_{eff} \approx 4.5$ BM) with the iron atom at the centre or porphyrin.[21,64] Addition of one axial ligand leads to square pyramidal complexes, Fe^{II}(Por)L, which exhibit a high-spin (d^6, $S = 2$, $\mu_{eff} \approx 5.0$ BM) state unless the ligand is a particularly good π acceptor such as C=X (X = O, S, Se, NR) and NO.[17a,66] If weak field ligands are coordinated, the six-coordinate complex stays in the high spin state, whereas strong field ligands cause a spin state change from $S = 2$ to $S = 0$.[20] Since the high-spin d^6 state populates odd electrons in antibonding $d_{x^2-y^2}$ and d_{z^2} orbitals, expansion of the coordination sphere and weakening of the axial bonds relative to the low spin state are expected. Indeed, the structural parameters of high-spin $Fe(TPP)(THF)_2$, *i.e.* Fe—N_{por} 2.057(3) and Fe—O_{ax} 2.351(3) Å, are larger than those of low-spin $Fe(TPP)(Pip)_2$ (Pip = piperidine), *i.e.* Fe—N_{por} 2.004(4) and Fe—N_{ax} 2.127(3) Å. Weak field ligands such as water, alcohols, ethers, ketones, amides and other neutral oxygen-containing ligands add to Fe(Por) in a stepwise fashion and five-coordinate complexes can be characterized spectroscopically.[67] On the other hand, six-coordinate complexes form predominantly in the presence of strong field ligands, *i.e.* aliphatic and aromatic nitrogenous ligands (equation 10).[17a,68]

$$Fe(Por) \underset{-L}{\overset{K_1 \ L}{\rightleftharpoons}} Fe(Por)L \underset{-L}{\overset{K_2 \ L}{\rightleftharpoons}} Fe(Por)L_2 \tag{10}$$

IS	HS	HS	L = weak field ligand, K_1 (0.1–10) > K_2
		LS	L = strong field ligand, K_1 (10^3–10^4) < K_2 (10^4–10^5)

		Por	TPP (p$K_3 \approx 4.4$)		DPIX DME (p$K_3 \approx 5.5$) in C_6H_6 at 25 °C	
L	pK_a		K_1 (M^{-1})	K_2 (M^{-1})	K_1(M^{-1})	K_2(M^{-1})
py	5.27		1.5×10^3	1.9×10^4		
Im	6.95		8.8×10^3	7.9×10^4	4.5×10^3	6.8×10^4
2-MeIm	7.86		2.4×10^4		1.25×10^4	
THF			5.8 ± 0.3		5.2 ± 0.4	

In general, K_1 and K_2 increase as the ligand pK_a increases, but the effects of π interactions are significant, especially for K_1. For example, a good π acceptor, Bu_3^nP (d_π–d_π), binds 10–10^2 times more tightly to $Fe(C_2Cap)$ (10) than aliphatic amines.[69] Fe(TPP) shows higher ligand affinity than Fe(DPIX DME) as expected from the *cis* effect.[67] The steric repulsion between the α substituent of axial ligands and the porphyrin (F strain) does not appear to be significant in the five-coordinate Fe(Por)(2-MeIm) (2-MeIm = 2-methylimidazole), but is strong enough to reduce K_2 markedly. The steric effect is operative for aliphatic amines, and the order of K_1 is $Pr^nNH_2 > Bu^sNH_2 > Bu^tNH_2$, opposite to the basicity order.[69] It is interesting that the observed entropy change for Fe(DPIX DME)(py)$_2$ formation agrees well with the calculated value of -293 J K^{-1} mol^{-1} for the loss of translational and rotational freedom for the coordination of two pyridine molecules to the bare heme in a solvent of low solvating power (equation 11).[67]

$$Fe(DPIX\ DME) + 2py \underset{in\ C_6H_6,\ 25\,°C}{\overset{K = 1.3 \pm 0.2 \times 10^8\,M^{-2}}{\rightleftharpoons}} Fe(DPIX\ DME)py_2 \tag{11}$$

$$\Delta H = -144.3 \pm 1.3\,kJ\,mol^{-1}$$
$$\Delta S = -326 \pm 41\,J\,K^{-1}\,mol^{-1}$$

Fe^{II} porphyrins (hemes) are air sensitive and easily oxidized to the μ-oxo dimer $[Fe^{III}(Por)]_2O$ (Scheme 17).[70] However, the irreversible oxidation can be prevented by a large excess of the axial ligand (retards process A), at a low temperature ($\leqslant -50$ °C, retards process D), or by protection of the O_2 binding site (prevents process C). The third approach has been intensively studied in relation to the biological oxygen carriers, myoglobin (Mb) and hemoglobin (Hb).

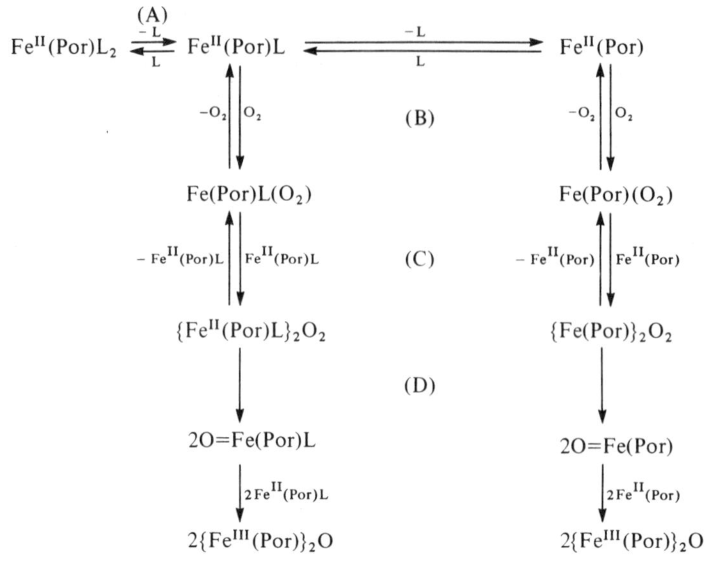

$$(A)$$

$$Fe^{II}(Por)L_2 \underset{L}{\overset{-L}{\rightleftharpoons}} Fe^{II}(Por)L \underset{L}{\overset{-L}{\rightleftharpoons}} Fe^{II}(Por)$$

$$\downarrow\uparrow \quad -O_2 \mid O_2 \qquad (B) \qquad \downarrow\uparrow \quad -O_2 \mid O_2$$

$$Fe(Por)L(O_2) \qquad\qquad Fe(Por)(O_2)$$

$$\uparrow \quad -Fe^{II}(Por)L \mid Fe^{II}(Por)L \qquad (C) \qquad \uparrow \quad -Fe^{II}(Por) \mid Fe^{II}(Por)$$

$$\{Fe^{II}(Por)L\}_2O_2 \qquad\qquad \{Fe(Por)\}_2O_2$$

$$\downarrow \qquad\qquad (D) \qquad\qquad \downarrow$$

$$2O{=}Fe(Por)L \qquad\qquad 2O{=}Fe(Por)$$

$$\downarrow 2Fe^{II}(Por)L \qquad\qquad \downarrow 2Fe^{II}(Por)$$

$$2\{Fe^{III}(Por)\}_2O \qquad\qquad 2\{Fe^{III}(Por)\}_2O$$

Scheme 17

The first reversible O_2 binding was observed with a heme–Im mixture (Im = imidazole) trapped in a polymer matrix where the μ-peroxy dimer formation (C) was avoided by spatial separation and immobilization of each heme.[71] Several synthetic models were also shown to coordinate O_2 reversibly (Figure 7 and equation 12). They belong to either the 'picket fence type' (7–9), in which bulky o-substituents of the TPP phenyl groups surround the heme-bound O_2,[63,70,72] or the 'cyclophane type' (10–15, 19) where the O_2 is covered by an appended bulky group.[70,73] The 'strapped type' (16–18) can bind O_2 reversibly, but the stability to oxidation appears to be lower than for the other two.[74] In addition to the reversible O_2 binding, Mb and deoxy Hb show substrate selectivity. Their CO affinity is much lower than that of most model hemes, and the steric hindrance at the coordination site, which compels a bent and/or tilted geometry to the otherwise linear Fe—C≡O group, has been suggested.[75a] This so-called distal-side steric hindrance, enforced by the imidazole (E7 His) and isopropyl (E11 Val) groups of the globin chain, is simulated by the cyclophane-type models with a short interplane distance. For example, Fe([6.6]-cyclophane)(1,5-DCIm) (11; $n = 1$, B = 1,5-di-cyclohexylimidazole) shows similar CO affinity ($P^{CO}_{1/2} = 8.4 \times 10^{-2}$ Torr) to that of Mb ($P^{CO}_{1/2} \approx 2 \times 10^{-2}$), while a high affinity ($P^{CO}_{1/2} = 9.1 \times 10^{-4}$ Torr) is observed for Fe[7.7]cyclophane)(1,5-DCIm) (11; $n = 2$) with a larger interplane distance.[75a] Dioxygen coordination is not hampered by the distal groups, because the Fe—O—O group is intrinsically bent ($\sim 130°$).[77] Indeed, the distal imidazole may even stabilize the O_2 adduct by intramolecular hydrogen bonding or polar effects. The amide-strapped heme (18) shows ~ 10 times higher O_2 affinity than the ether-strapped analogue (17) due presumably to the polar effects of the amide groups which stabilize the charges on the iron(III) superoxide contribution to the oxygenated species (Table 4).[74]

$$Fe(Por)L + L' \underset{k^{L'}_{off}}{\overset{k^{L'}_{on}}{\rightleftharpoons}} Fe(Por)LL' \qquad \begin{array}{l} P^{L'}_{1/2} = \text{half saturation pressure} \\ M = P^{O_2}_{1/2}/P^{CO}_{1/2} \end{array} \qquad (12)$$

Ligand affinities for the natural system are in the order, deoxy Hb (T state) < Mb < oxy Hb (R state). According to the Perutz theory of cooperativity[75] the proximal imidazole (F8 His) coordinating to the heme is moved by the globin chain, in the T state, so as to generate the five-coordinate Fe-out-of-plane structure, whereas in the R state the imidazole can approach the heme without strain and the six-coordinate Fe-in-plane geometry is easily attained. 2-Methylimidazoles are used to simulate the T state since they cannot approach the heme centre closely enough to realize the ideal N_{ax}—porphyrin centre distance in the six-coordinate state due to the F strain. Actually, in $Fe(T_{piv}PP)(2\text{-MeIm})(O_2)$ (7), the Fe atom is displaced out of the N_4 plane towards the imidazole ($\Delta(N_4) \approx 0.086$ Å) by extending the Fe—O_2 distance (1.898(7) Å) without significantly altering the Fe—N distance (2.107(4) Å) relative to the unstrained $Fe(T_{piv}PP)(1\text{-MeIm})(O_2)$ ($\Delta(N_4) \approx 0.03$ Å towards O_2, Fe—O_{ax} 1.75(2) Å, Fe—N_{ax} 2.07(2) Å).[77a] Changing the axial base from 1-MeIm to 1,2-Me_2Im reduces the O_2 affinity by a factor of 10^2 (Table 5).

Table 4 Ligand Binding Site Steric Effect on CO and O_2 Coordination to $Fe^{II}(Por)$

		k_{on}^{CO} (M^{-1} s^{-1})	k_{off}^{CO} (s^{-1})	$P_{1/2}^{CO}$ (Torr)	$k_{on}^{O_2}$ (M^{-1} s^{-1})	$k_{off}^{O_2}$ (s^{-1})	$P_{1/2}^{O_2}$ (Torr)	M	Ref.
Mb	aq. pH 7–7.4, 20 °C	3–5 × 10^5	0.0015–0.04	1.4–2.5 × 10^{-3}	1–2 × 10^7	10–30	0.37–1	20–40	a
HbA (R-state)	aq. pH 7–7.4, 20 °C	~4.6 × 10^6	0.009	1.4 × 10^{-3}	3.3 × 10^7	13.1	0.22	150	b
Fe(DPIX DME)(Im)	C$_6$H$_6$, 20 °C	1.2 × 10^7	0.28	2.4 × 10^{-4}					c
Chelated protoheme	C$_6$H$_6$, 20 °C	1.1 × 10^7	0.25	2.3 × 10^{-4}	6.2 × 10^7	4200	5.6	24000	a, d, e
Chelated mesoheme	C$_6$H$_6$, 20 °C	1.04 × 10^7	~0.05	4.9 × 10^{-4}	5.3 × 10^7	1700	2.8	5700	a, d, e
7,7-Cyclophane(1,5-DCI)	C$_6$H$_6$, 20 °C	6 × 10^6	0.05	9.2 × 10^{-4}	6.5 × 10^7	1000	1.4	1500	e
6,6-Cyclophane(1,5-DCI)	C$_6$H$_6$, 20 °C	3 × 10^4	0.05	1.69 × 10^{-1}	1 × 10^5	800	696	4100	e
FeCu5(THPIm)[g]	C$_6$H$_6$, 20 °C	9 × 10^4	0.02	0.02	1.8 × 10^5	91	5	250	f
FeCu4(THPIm)[g]	C$_6$H$_6$, 20 °C	2 × 10^4	0.02	0.1	5.2 × 10^5	160	31	310	f

a T. G. Traylor, *Acc. Chem. Res.*, 1981, **14**, 102.
b Q. H. Gibson, *J. Biol. Chem.*, 1970, **245**, 3285; J. S. Olson, M. E. Andersen and Q. H. Gibson, *J. Biol. Chem.*, 1971, **246**, 5919; V. S. Sharma, M. R. Schmidt and H. M. Ranney, *J. Biol. Chem.*, 1976, **251**, 4267; R. C. Steinmeier and L. Parkhurst, *Biochemistry*, 1975, **14**, 1564.
c Ref. 67; D. K. White, J. B. Cannon and T. G. Traylor, *J. Am. Chem. Soc.*, 1979, **101**, 2443.
d T. G. Traylor, D. K. White, D. H. Campbell and A. P. Berzines, *J. Am. Chem. Soc.*, 1981, **103**, 4932; T. G. Traylor, C. K. Chang, J. Geibel, A. P. Berzines, T. Mincey and J. Cannon, *J. Am. Chem. Soc.*, 1979, **101**, 6716.
e Ref. 75b.
f Ref. 73.
g THPIm = 4,5,6,7-tetrahydrobenz[e]imidazole. For other abbreviations, see Figure 7.

Table 5 Effect of *Trans* Axial Steric Hindrance on CO and O_2 Coordination to $Fe^{II}(Por)$[a]

		k_{on}^{CO} (M^{-1} s^{-1})	k_{off}^{CO} (s^{-1})	$P_{1/2}^{CO}$ (Torr)	$k_{on}^{O_2}$ (M^{-1} s^{-1})	$k_{off}^{O_2}$ (s^{-1})	$P_{1/2}^{O_2}$ (Torr)	M	Ref.
Hb (R-state)	aq. pH 7.0–7.4, 20 °C	4.6 × 10^6	0.009	1.4 × 10^{-3}	3.3 × 10^7	13.1	0.22	150	b
Mb	aq. pH 7.0–7.4, 20 °C	3–5 × 10^5	0.0015–0.04	1.4–2.5 × 10^{-2}	1–2 × 10^7	10–30	0.37–1	20–40	c
Hb (T-state)	aq. pH 7.0–7.4, 20 °C	2.2 × 10^5	0.09	3.0 × 10^{-1}	2.9 × 10^6	180	40	135	b
Fe(TpivPP)(1-MeIm)	PhMe, 25 °C						0.31		d
Fe(TpivPP)(1,2-Me$_2$Im)	PhMe, 25 °C	1.4 × 10^6	0.14	8.9 × 10^{-3}	1.6 × 10^8	4600	38	4280	d
Fe(MedPoc)(1-MeIm)	PhMe, 25 °C	1.5 × 10^6	0.0094	6.5 × 10^{-4}	1.7 × 10^7	71	0.36	550	e
Fe(MedPoc)(1,2-Me$_2$Im)	PhMe, 25 °C	2.1 × 10^5	0.053	2.6 × 10^{-2}	5.2 × 10^6	800	12.4	480	e
Fe(PocPiv)(1-MeIm)	PhMe, 25 °C	5.8 × 10^5	0.0086	1.5 × 10^{-3}	2.2 × 10^6	9	0.36	270	e
Fe(PocPiv)(1,2-Me$_2$Im)	PhMe, 25 °C	9.8 × 10^4	0.055	6.7 × 10^{-2}	1.9 × 10^6	280	12.1	216	d, e
Fe(C$_2$-cap)(1-MeIm)	PhMe, 25 °C	9.5 × 10^5	0.05	5.4 × 10^{-3}			23	4300	f
Fe(C$_2$-cap)(1,2-Me$_2$Im)	PhMe, 25 °C			2.0 × 10^{-1}			4000	20000	f

a For abbreviations, see Figure 7.
b Q. H. Gibson, *J. Biol. Chem.*, 1970, **245**, 3285; J. S. Olson, M. E. Andersen and Q. H. Gibson, *J. Biol. Chem.*, 1971, **246**, 5919; V. S. Sharma, M. R. Schmidt and H. M. Ranney, *J. Biol. Chem.*, 1976, **251**, 4267; R. C. Steinmeier and L. Parkhurst, *Biochemistry*, 1975, **14**, 1564.
c T. G. Traylor, *Acc. Chem. Res.*, 1981, **14**, 102.
d J. P. Collman, J. I. Brauman and K. M. Doxsee, *Proc. Natl. Acad. Sci. USA*, 1979, **76**, 6035.
e J. P. Collman, J. I. Brauman, B. L. Iverson, J. L. Sessler, R. M. Morris and Q. H. Gibson, *J. Am. Chem. Soc.*, 1983, **105**, 3052.
f Ref. 69; J. E. Linard, P. E. Ellis, Jr., J. R. Budge, R. D. Jones and F. Basolo, *J. Am. Chem. Soc.*, 1980, **102**, 1896; E. I. Rose, P. N. Venkatasubramanian, J. C. Swartz, R. D. Jones, F. Basolo and B. M. Hoffmann, *Proc. Natl. Acad. Sci. USA*, 1982, **79**, 5472.

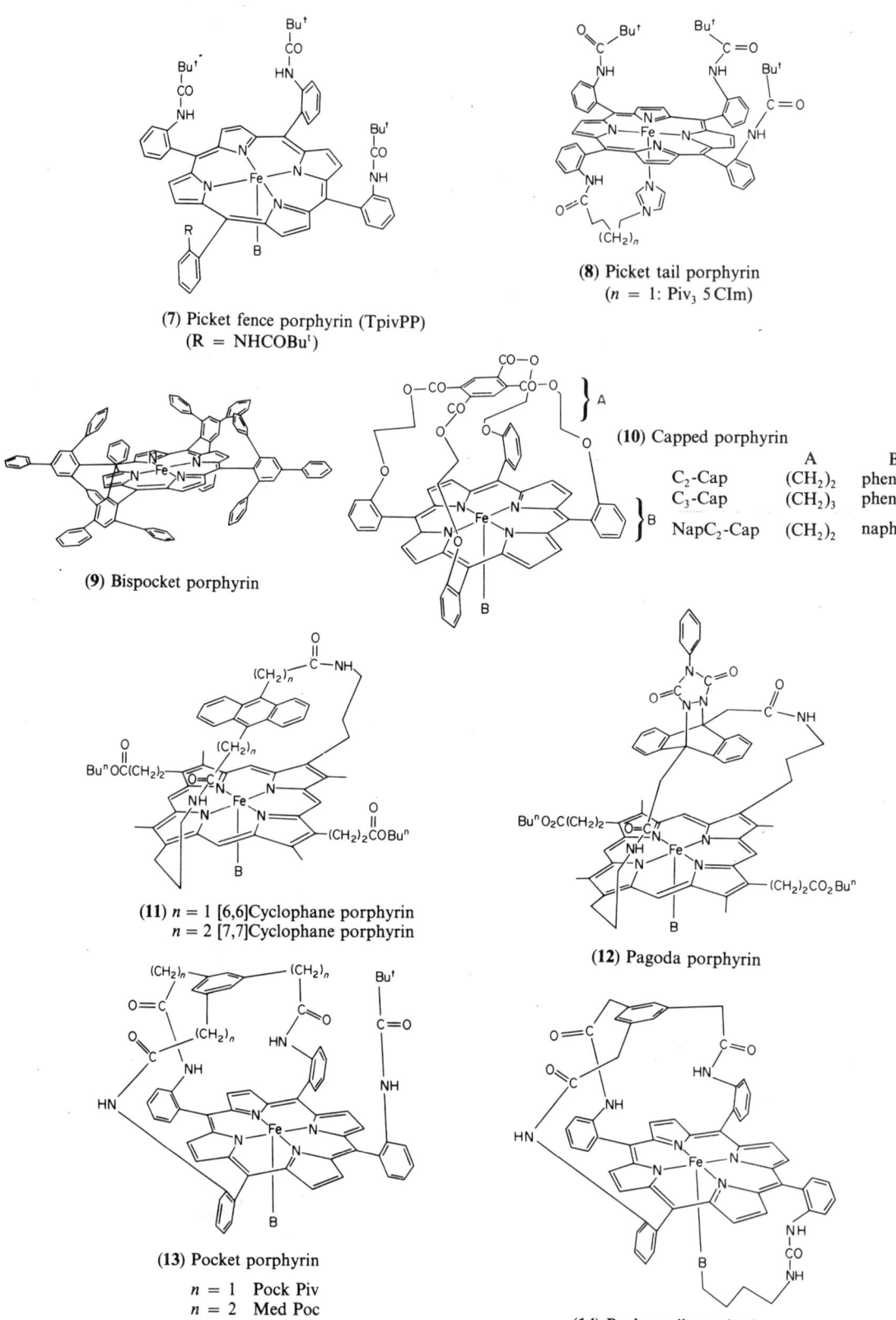

Figure 7 Iron porphyrins exhibiting reversible O_2 binding

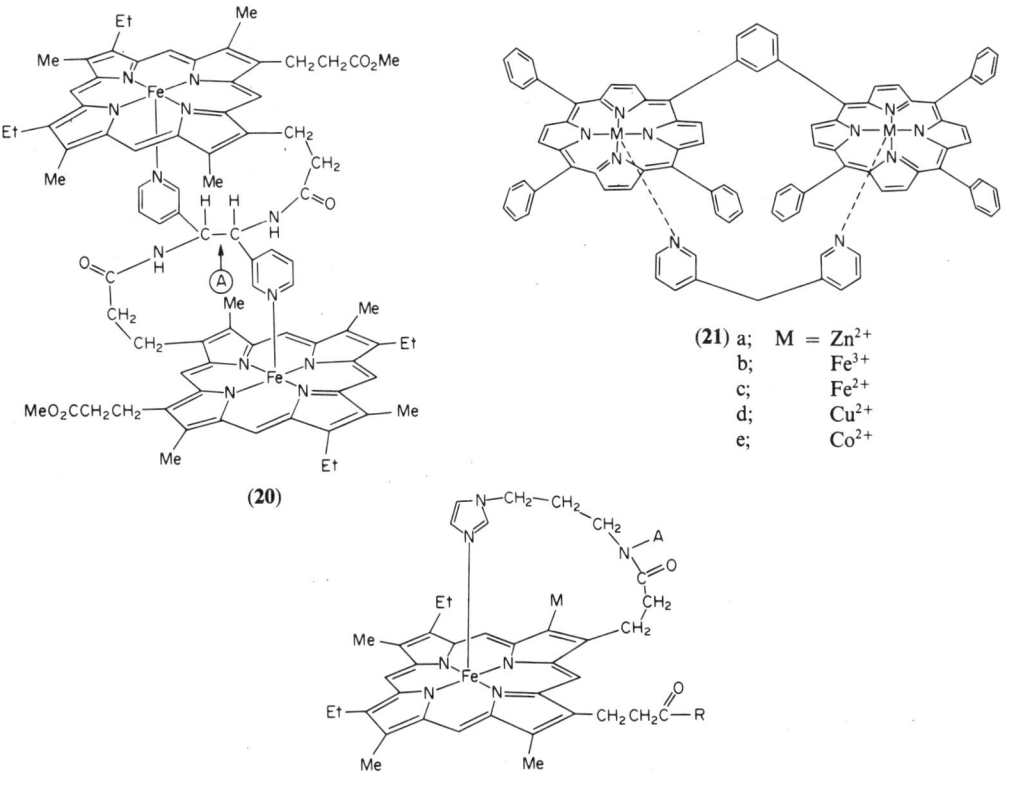

(**15**) Cofacial diporphyrin

R = *n*-pentyl
Fe-Cu-4   X = CH$_2$
Fe-Cu-5   X = (CH$_2$)$_2$

(**16**) Strapped porphyrin

n = 5   FeSP-13
n = 6   FeSP-14
n = 7   FeSP-15

(**17**) Ether-linked hanging-base porphyrin   (**18**) Amide-linked hanging-base porphyrin   (**19**) Double bridged porphyrin

(**20**)

(**21**) a;   M = Zn^{2+}
       b;   Fe^{3+}
       c;   Fe^{2+}
       d;   Cu^{2+}
       e;   Co^{2+}

(**22**) Chelated mesoporphyrin

Because of the cooperative interactions among its four subunits ($\alpha_2\beta_2$), Hb reduces its O_2 affinity at low oxygen pressure and increases it at high oxygen tension. A similar biphasic ligand binding behaviour was observed on a polylysine-bound heme.[76] The picket fence model, $Fe(T_{piv}PP)(2\text{-}MeIm)(EtOH)$, shows solid state cooperativity,[77] and the pyridine-chelated diheme (**20**) binds CO with two rate constants.[78] The tail base heme dimer (**22**, with a shorter methylene linkage) and the gabled Co porphyrin (**21**) also exhibit dual ligand affinity.[79]

Fe^{II} porphyrins also bind NO. Both $Fe(TPP)NO$ and $Fe(TPP)(NO)L$ and (L = 1-MeIm, 4-MePip) have been examined by X-ray crystallography, showing a bent Fe—NO geometry (149.2(6)° ~ 140°) and the typical low-spin Fe^{II}—N_{por} distance (2.001(3) and ~2.005 Å; $\Delta(N_4) \sim 0.21$ and ~0.08 Å).[21] The six-coordinate complexes readily lose the axial ligand *trans* to NO due to the odd election in d_{z^2}, and the weakness of the axial base ligation is reflected in the long Fe—N_{ax} distances, *i.e.* 2.180(4) Å for L = 1-MeIm, 2.33–2.46 Å for L = 4-MePip compared to 2.014(5) Å in $FeTPP(1\text{-}MeIm)_2$ and 2.127(3) Å in $Fe(TPP)(Pip)_2$.[21]

Fe^{III} porphyrins show strong ligand affinity and four-coordinate species have not been isolated. They form stable five-coordinate complexes (hemins) with a variety of anionic ligands, such as halides (F^-, Cl^-, Br^-, I^-), pseudohalides (OCN^-, NCS^-, $NCSe^-$, N_3^-), carboxylate, alkoxide (OR^-, OPh^-) mercaptides (SPh^-, SCH_2Ph^-) and imidazolate ($4\text{-}MeIm^-$).[17a,80,81] Cyanide forms six-coordinate complexes with $Fe^{III}(Por)$, while aliphatic mercaptides without electron-withdrawing groups reduce $Fe^{III}(Por)$ to heme. Some five-coordinate complexes with weak anionic ligands, *e.g.* RSO_3^-, ClO_4^-, BF_4^-, PF_6^-, AsF_6^-, SbF_6^-, are also known.[82] Nitrate has been shown to coordinate to $Fe^{III}TPP$ as a bidentate ligand, the only example of the six-coordinate iron porphyrin with the two extra ligands on the same side of the macrocycle.[83] Hydroxide complexes rapidly dimerize to μ-oxides (hematins) unless sterically protected.[84]

The five-coordinate iron(III) complexes are prepared from the μ-oxo dimer and the acid form of anionic ligands, or by metathesis of the chloride in dry solvents. Silver salts are used to convert $Fe(Por)Cl$ to the complexes of weak ligands.[82] All the five-coordinate complexes are high-spin compounds (d^5, $S = \frac{5}{2}$, $\mu_{eff} \approx 5.9$ BM) with structural parameters of Fe—$N_{por} \approx 2.07$ Å and $\Delta(N_4) \approx 0.5$ Å, and show ESR signals at $g \approx 6.0$ and 2.0, except for those of weak ligands (RSO_3^-, ClO_4^-, BF_4^-, PF_6^-, AsF_6^- SbF_6^-) which take a quantum mechanically admixed intermediate-spin (d^5, $S = \frac{5}{2}/\frac{3}{2}$, $\mu_{eff} \approx 5$ BM) state with Fe—$N_{por} \approx 2.00$ Å, $\Delta(N_4) \approx 0.27$ Å and $g \approx 4.7$ and 2.0.[20] μ-Oxo dimers show low magnetic susceptibility ($\mu_{eff} \approx 1.8$ BM) due to strong antiferromagnetic coupling between the two high-spin Fe atoms.[17a] Organo σ-alkyl and σ-aryl Fe(Por) are also reported to be low-spin compounds.[85]

Though the five-coordinate complexes, except for μ-oxo dimers, coordinate sixth ligands, the association constants are low ($K < 10^3 M^{-1}$) because of the strong *trans* effect of the anionic ligand (equation 13).[18] In the presence of excess ligand, the anionic ligand may be displaced (equation 14).[13]

$$Fe(Por)X + L \underset{}{\overset{K}{\rightleftharpoons}} Fe(Por)XL \qquad (13)$$

$$Fe(Por)X + 2L \underset{}{\overset{\beta_2}{\rightleftharpoons}} [Fe(Por)L_2]^+ X^- \qquad (14)$$

The equilibrium constants K and β_2 increase as the ligand pK_a increases. The increases in porphyrin basicity and solvent polarity also increase β_2, indicating the importance of the charge neutralization factor in the iron(III) porphyrin coordination chemistry (Table 6).[86] For preparative purposes, five-coordinate complexes of the weak ligands are conveniently used to avoid contamination of the mixed ligand species $Fe(Por)XL$.

The spin state of six-coordinate $Fe^{III}(Por)LL'$ varies from low to high and depends on the ligand fields of L and L'.[20,87]

Low spin: d^5, $S = \frac{1}{2}$, $\mu_{eff} \approx 2.2$ BM, $g = 3.0–2.4, 2.3–2.1, 1.4–1.9$, L, L' = strong field ligands, *i.e.* N, S donors, CN^-; L, L' = Cl^-, unhindered Im; L, L' = RO^-, N donors.

Intermediate spin: d^5, $S = \frac{3}{2}$, $\mu_{eff} \approx 5.4$ BM, $[Fe(TPP)C(CN)_3]_n$.

Admixed spin: d^5, $S = \frac{3}{2}/\frac{5}{2}$, $\mu_{eff} \approx 4.5$ BM, probably $[Fe(OEP)ClO_4(py)]$.

High spin: d^5, $S = \frac{5}{2}$, $\mu_{eff} \approx 5.8$ BM, $g \approx 6.0–2.0$, L, L' = weak field ligand, *i.e.* neutral O donors, F^-.

The sensitivity to structure of hemochrome spin states is represented by the two examples reported by Scheidt *et al.*,[88] *i.e.* $[Fe(OEP)(3\text{-}Clpy)_2]ClO_4$ exists in the low-spin state at 98 K but equilibrates to a 1:1 mixture of the low- and high-spin species at 293 K in the solid state, and $Fe(TPP)(NCS)(py)$ crystallizes as a low-spin compound with a bent Fe—N—CS geometry, while a high-spin species with a linear FeNCS structure is obtained for $Fe(OEP)(NCS)(py)$. Slightly longer Fe—N_{por} dis-

Table 6 Effect of Axial Ligand Basicity and *Cis* and *Trans* Effects on the Equilibrium Constants for Fe^{III}(Por)

Por	X	L	Solvent	t (°C)	Ligand pK_a	K (M⁻¹)	β_2 (M⁻²)		Ref.
TPP	N_3^-	py	CH_2Cl_2	23	5.17	1.75			a
		Im	CH_2Cl_2	23	6.65	79			
		1-MeIm	CH_2Cl_2	23	7.33	145			
TPP	Cl^-	py	$CHCl_3$	25	5.17	0.20 ± 0.2	~ 0.5 (?)		b
	Cl^-	3,4-Me$_2$py	$CHCl_3$	25	6.46	4 ± 4	$\sim 25 \pm 15$ (?)		b
	Cl^-	1-MeIm	$CHCl_3$	25	7.33	9 ± 2	$(1.50 \pm 0.24) \times 10^3$		
	Cl^-	1,2-Me$_2$Im	$CHCl_3$	25	7.85	3.4 ± 1.3	9.3 ± 1.1	Steric hindrance	
	Cl^-	4-Me$_2$Npy	$CHCl_3$	25	9.70	51 ± 12	$(2.63 \pm 0.12) \times 10^3$		
	Cl^-	X···H-Im	$CHCl_3$	25	(6.65)	630	$(1.00 \pm 0.10) \times 10^6$	Hydrogen bonding	
TPP	Cl^-	1-MeIm	$CHCl_3$	25		10 ± 2	$(1.50 \pm 0.25) \times 10^3$	$K_2' = 25 \, M^{-1}$	b
	Br^-	1-MeIm	$CHCl_3$	25		$> 1 \times 10^6$	$> 2.5 \times 10^7$	$K_2' = 100 \, M^{-1}$	
	I^-	1-MeIm	$CHCl_3$	25		$> 1 \times 10^5$	$> 1 \times 10^7$		
					Porphyrin pK_3				
TPP	Cl^-	1-MeIm	$CHCl_3$	25	4.4	9 ± 2	$(1.50 \pm 0.24) \times 10^3$		b,c
DPIX DME	Cl^-	1-MeIm	$CHCl_3$	25	5.5	36 ± 4	$(5.89 \pm 0.72) \times 10^3$		
OEP	Cl^-	1-MeIm	$CHCl_3$	25			$(6.75 \pm 1.05) \times 10^3$		
					Hammett σ for X				
$T_{p-X}PP$									b
X = OMe	Cl^-	1-MeIm	$CHCl_3$	25	−0.268	38 ± 23	$(3.6 \pm 0.6) \times 10^3$		
Me	Cl^-	1-MeIm	$CHCl_3$	25	−0.170	20 ± 5	$(2.70 \pm 0.19) \times 10^3$		
H	Cl^-	1-MeIm	$CHCl_3$	25	0	9 ± 2	$(1.50 \pm 0.24) \times 10^3$	$\log (\beta^X/\beta^H) = -0.39$ (4σ)	
F	Cl^-	1-MeIm	$CHCl_3$	25	0.062	14 ± 7	$(1.00 \pm 0.10) \times 10^3$		
Cl	Cl^-	1-MeIm	$CHCl_3$	25	0.227	22 ± 10	$(0.71 \pm 0.08) \times 10^3$		
					Solvent ε_{20}				
TPP	Cl^-	1-MeIm	C_6H_6	25	2.3	1.3 ± 0.2	5.8 ± 0.8		b
	Cl^-	1-MeIm	$CHCl_3$	25	4.8	10 ± 2	$(1.5 \pm 0.25) \times 10^3$		
	Cl^-	1-MeIm	CH_2Cl_2	25	9	88 ± 12	$(1.0 \pm 0.2) \times 10^4$		
	Cl^-	1-MeIm	DMF	25	37	126	$(6.11 \pm 0.54) \times 10^4$		

[a] K. M. Adams, P. G. Rasmussen, W. R. Scheidt and K. Hatano, *Inorg. Chem.*, 1979, **18**, 1892.
[b] F. A. Walker, M.-W. Lo and M. T. Ree, *J. Am. Chem. Soc.*, 1976, **98**, 5552.
[c] T. Yoshimura and T. Ozaki, *Bull. Chem. Soc. Jpn.*, 1979, **52**, 2268.

tances ($\sim 2.05\,\text{Å}$) in the high-spin compounds relative to the other six-coordinate complexes ($\sim 1.99\,\text{Å}$) reflect the population of an electron in the σ-antibonding $d_{x^2-y^2}$ orbital.

[Fe^I(Por)]$^-$ is formed when Fe^{III}(Por) or Fe^{II}(Por) is electrochemically reduced or treated with a stoichiometric amount of sodium anthracenide in THF.[89] [Fe^{I}TTP]$^-$ (Na-18-crown-6)$^+$ (THF)$_2$ is a low-spin square planar complex ($\mu_{eff} \approx 2.5\,\text{BM}$, $g \approx 2.1, 1.9$, Fe—N$_{por} \approx 2.023(5)\,\text{Å}$), and shows a similar visible spectrum to that of Fe^{II}(TPP) but with a much lower intensity. This species is isoelectronic to Co^{II}(Por), the anionic charge, in addition to an odd electron to the d_{z^2} orbital, diminishes the ligand affinity of the iron atom. Further reduction generates an Fe^{II}–porphyrin dianion.

An Fe^{IV} oxidation state, apart from those of catalase and peroxidase compounds I and II, is attained only in the carbene complexes, RR$'$C=Fe^{IV}(Por), nitrene complexes, R$'$N—N=Fe^{IV}(Por), or the dimeric compounds, Fe^{IV}(Por)=C=Fe^{IV}(Por), Fe^{IV}(Por)=N—Fe^{III}(Por) and [Fe^{IV}(Por)—O—Fe^{III}(Por)]$^+$.[90] Both carbene and nitrene complexes are diamagnetic, and the former appear to coordinate RNH$_2$, py, Im, ROH, Cl$^-$ and RS$^-$. They lose the axial ligand in the presence of an excess of pyridine, and form Fe^{II}(Por)(py)$_2$. Though the iron oxidation state in these formally Fe^{IV}(Por) complexes is still ambiguous, their reactivity implies an iron(III) oxidation state. Attempted synthesis of Fe^{IV} porphyrins by a one-electron oxidation of Fe^{III}(TPP)Cl resulted in formation of the corresponding porphyrin π cation radical Fe^{III}(TPP)$^{+\ddagger}$.[91] The high oxidation state iron porphyrins are of particular interest in relation to the cytochromes P-450, peroxidases and catalases, and Fe^{IV}O(Por)L (L = 1-MeIm, py, Pip) have been spectroscopically characterized (Scheme 18).[70,91]

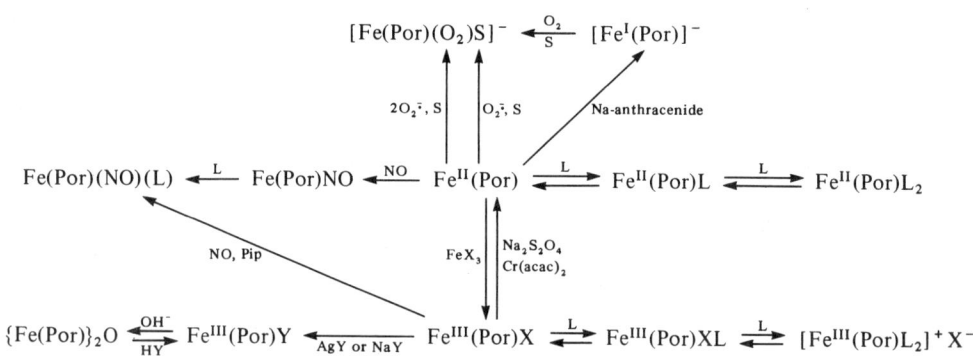

Scheme 18

Metallation of porphyrins with Ru$_3$(CO)$_{12}$ gives Ru(Por)(CO)L.[17a] The CO is bound tightly to the Ru, and ordinary σ donors (O, S, N donor) only substitute the *trans* ligand.[17a,92] Strong π-acceptor ligands such as phosphine [R$_3$P, (RO)$_3$P], isocyanide (RNC) and nitroxyl (NO), however, can replace the CO.[17a,93,94] The ligand substitution reactions occur by a dissociative mechanism, and are much slower than those of the Fe analogues, in general by a factor of 10^8, probably because the Ru (d^6) ion is in a low-spin state throughout the reaction.[94] The activation energy for ligand exchange is 79.5–96.2 kJ mol^{-1}, and is decreased by electron-donating substituents on the porphyrin ring.[92] The activation entropy of 21–42 J K^{-1} mol^{-1} is a reasonable value for a dissociative mechanism without developing charges (equation 15).

$$\text{Ru(T}_{p\text{-X}}\text{PP)(4-Bu}^t\text{py)} + \text{4-Bu}^t\text{py*} \rightleftharpoons \text{Ru(T}_{p\text{-X}}\text{PP)(CO)(4-Bu}^t\text{py*)} + \text{4-Bu}^t\text{py} \qquad (15)$$

$$\varrho = -0.17\ (25\,°\text{C}), -0.13\ (70\,°\text{C})\ \text{for}\ 4\sigma$$

Ru^{II}(Por)(L)$_2$ can be reversibly reduced to Ru^I ($E_{1/2} \approx -1.3\,\text{V}$ *vs.* SCE) and oxidized to the Ru^{III} state ($E_{1/2} \approx 0.8\,\text{V}$ *vs.* SCE) electrochemically. The one-electron oxidation of Ru^{II}(Por)(CO)L (ν(CO) $\approx 1945\,\text{cm}^{-1}$) gives Ru^{II}(Por)$^{+\ddagger}$(CO)L ($\nu$(CO) $\approx 1965\,\text{cm}^{-1}$), which forms [Ru^{III}(Por)L$_2$]$^+$ in the presence of donor ligands (N, P, As or CN$^-$) by an intramolecular electron shift.[95] Oxidation of Ru(Por)(CO)L with hydroperoxide affords the μ-oxo dimer {Ru^{IV}(Por)(OR)}$_2$O where the anionic ligand can be displaced by conjugated bases of acids HX (X = Cl, Br, OCOMe).[96] The Ru^{II}—Ru^{II} bond in [Ru^{II}(OEP)]$_2$, prepared by thermal removal of the axial ligands of Ru(OEP)(py)$_2$, is also oxidized by *t*-butyl hydroperoxide to the {RO—Ru^{IV}}$_2$O moiety.[97] The addition of a diazonium salt to Ru^{II}(Por)(CO)(HOEt) is reported to give Ru(Por)(N$_2$Ph)(EtOH). A dinitrogen adduct, Ru(Por)(N$_2$)(THF) (ν(N≡N) $\approx 2110\,\text{cm}^{-1}$), is also known.[98] Ru(Por)(L)$_2$ (L = MeCN) reversibly binds O$_2$ in DMF but not in toluene (Scheme 19).[99]

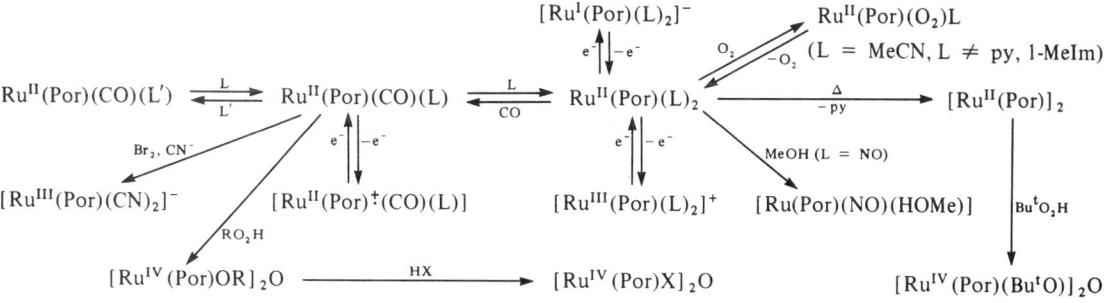

Scheme 19

Os^{II}(Por)(CO)(L) can be prepared from OsO$_4$ or K$_2$OsCl$_6$ and a porphyrin under CO.[17a] The CO
is more tightly bound to Os than to Ru, and P(OR)$_3$ or NO is required to dissociate it. Os^{II} porphyrin
complexes are rather inert to ligand substitution, but many mixed ligand complexes, *i.e.* Os(Por)-
(CO)L (L = neutral O, S, N, P or As donor, RNC, CN$^-$), Os(Por)(CS)py, Os(Por)(N$_2$)(THF),
Os(Por)(THT)(L) (L = py, 1-MeIm) as well as ordinary octahedral complexes Os(Por)(L)$_2$ (L = O,
S, N or P donor, RNC, CO, NO) have been prepared.[100] Os(Por)(NO)$_2$ is converted to Os(Por)-
(NO)(X) (X = OMe, F, OClO$_3$) with HX.[17a] A one-electron oxidation of Os(Por)(CO)(L)
(v(CO) ≈ 1900 cm^{-1}) gives [Os^{III}(Por)(CO)(L)]$^+$ (v(CO) ≈ 2010 cm^{-1}), and Os^{II}(Por)(L)$_2$ is revers-
ibly oxidized to [Os^{III}(Por)(L)$_2$]$^+$.[101] Air oxidation of Os^{II}(Por)(CO)(L) in the presence of 2,3-
dimethylindole leads to the diamagnetic {Os^{IV}(Por)(OR)}$_2$O,[96] while the highest oxidation state
complex, *trans*-Os^{VI}(O)$_2$(Por) (v(Os=O) 825 cm^{-1}), is obtained in the absence of an oxygen
acceptor.[17a] Os^{VI}(O)$_2$(Por) is reduced to Os^{IV}(Por)(OMe)$_2$ with SnCl$_2$ or Na–Hg in methanol, to
Os^{II}(Por)(L)$_2$ with diisobutylaluminium hydride or hydrazine in the presence of appropriate ligands,
or to a dinitrogen complex, Os^{II}(Por)(N$_2$)(THF) (v(N≡N) 2050 cm^{-1}), with hydrazine under
nitrogen.[17a] Since both axial ligands are easily substituted with other donors such as tetrahydrothiophene (THT) or ammonia, the dinitrogen complexes are useful intermediates.[101,102] Air oxidation
of the diammine complex Os^{II}(Por)(NH$_3$)$_2$ in methanol gives a nitrido complex Os^{VI}(N)(Por)(OMe)
[v(Os≡N) 2032 cm^{-1} (Por = OEP), 2050 cm^{-1} (Por = TTP)] (Scheme 20).[17a,100,101]

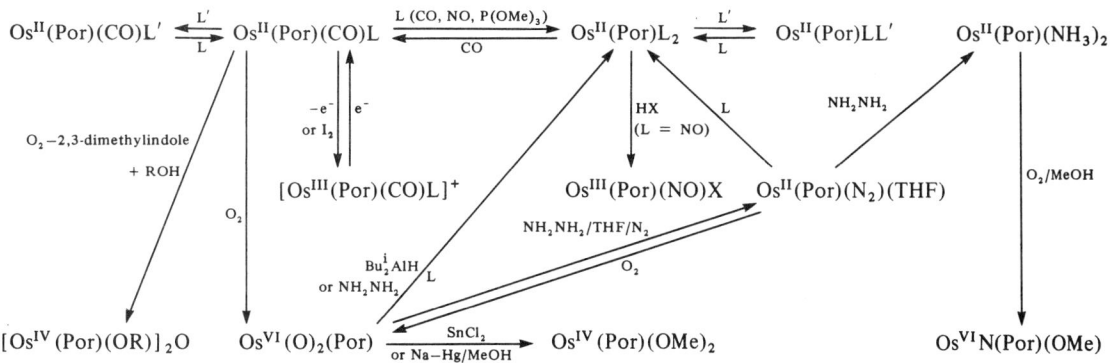

Scheme 20

Buchler *et al.*[101] pointed out that both metal d_π to the macrocyclic π^* and to the axial ligand π^*
back donation increase from the first to the third row element, *i.e.* for λ_{max}(Q): Fe(Por)(L)$_2$ >
Ru(Por)(L)$_2$ > Os(Por)(L)$_2$, and for v(CO): Fe(Por)(CO)(L) > Ru(Por)(CO)(L) > Os(Por)-
(CO)(L), but it does not necessarily result in an increase in charge density on the metal; *i.e.*
$E_{1/2}$ = −0.15 V (Fe(OEP)(py)$_2$), −0.02 V (Ru(OEP)(py)$_2$), −0.37 V (Os(OEP)(py)$_2$). The π-accep-
tor strength has been measured for Os(Por)(L)$_2$ by λ_{max}(Q) and for Os(Por)(CO)(L) by v(CO) as
follows: P(OMe)$_3$ > THT > MeCN > py > γ-Pic > 1-MeIm > NH$_3$ > Pip > NMe$_3$ or CO >
MeNC > Ph$_3$P > Ph$_3$As > THT > py > MeOH > THF > 1-MeIm > Et$_3$N > DMF (> CN). The
position of 1-MeIm in the series is anomalous because of its π-donor and -acceptor character.

(xiii) Group VIII—2 (Co^{I,II,III}, Rh^{I,II,III,IV}, Ir^{I,III})

Cobalt insertion into porphyrins with CoOAc, CoCl$_2$, Co(acac)$_2$ or Co$_2$(CO)$_8$ gives Co(Por),
which can be reduced to [Co^I(Por)]$^-$ electrochemically ($E_{1/2}$ ≈ −1 V *vs.* SCE) or with NaBH$_4$.[17a,103]

The Co^I species are strong nucleophiles and undergo S_N2 reactions with alkyl halides. The Co—C bond in R—Co(Por) is homolytically photolyzed to alkyl radical and Co^{II}(Por) (see Section 21.1.3.2.7).

Co^{II}(Por) complexes form five- and six-coordinate complexes with ordinary N-, P- and As-donor ligands; all the Co^{II}(Por) are low-spin d^7, $S = \frac{1}{2}$, with ESR signals at $g_\parallel \approx 1.9$, $g_\perp \approx 3.3$ ($A_\parallel^{Co} \approx 200 \times 10^{-4}\,cm^{-1}$, $A_\perp^{Co} \approx 300 \times 10^{-4}\,cm^{-1}$) for Co(Por), $g_\parallel \approx 2.025$, $g_\perp \approx 2.3$ ($A_\parallel^{Co} \approx 75 \times 10^{-4}\,cm^{-1}$, $A_\perp^{Co} \approx 20 \times 10^{-4}\,cm^{-1}$) for Co(Por)L and $g_\parallel \approx 2.08$, $g_\perp \approx 2.00$ ($A_\parallel^{Co} \approx 20 \times 10^{-4}\,cm^{-1}$, $A_\perp^{Co} \approx 10 \times 10^{-4}\,cm^{-1}$) for Co(Por)(O$_2$)L.[2,17a,21] The odd electron in a σ-antibonding d_{z^2} orbital in Co^{II}(Por) complexes weakens the Co^{II}—L$_{ax}$ bond. For instance, the M—N$_{ax}$ bond length of 2.436 Å in Co(TPP)(Pip)$_2$ is significantly longer than 2.127 Å in Fe(TPP)(Pip)$_2$,[21] and K_2 values for Co(Por)L$_2$ (1.0–0.1 M^{-1} for nitrogenous ligands) are considerably lower than those of Fe(Por)L$_2$ ($\sim 10^{4-5}$ M^{-1} for nitrogenous ligands) while K_1 values are in the same order (10^3 M^{-1}) for both Co(Por)L and high-spin Fe(Por)(L).[18] The K_1 values for Co(Por)L are related to the axial and equatorial ligand basicities by the Hammett equation (equation 16).[18,104]

$$Co^{II}(Por) \underset{-L}{\overset{\overset{K_1}{L}}{\rightleftharpoons}} Co^{II}(Por)L \underset{-L}{\overset{\overset{K_2}{L}}{\rightleftharpoons}} Co^{II}(Por)(L)_2 \qquad (16)$$

$\Delta H° \approx -29.3$ to $-46\,kJ\,mol^{-1}$ $\Delta H° \approx -8.4\,kJ\,mol^{-1}$
$\Delta S° \approx -41.8$ to $-83.7\,J\,K^{-1}\,mol^{-1}$ $\Delta S° \approx -41.8\,J\,K^{-1}\,mol^{-1}$

$\log K_1^X/K_1^H = \varrho\sigma$ for 4-substituted pyridines, $\varrho = -0.6$ (23 °C)
$\log K_1^X/K_1^H = \varrho(4\sigma)$ for T$_{p-X}$PP, L = piperidine, $\varrho = 0.145 \pm 0.017$ (25 °C)
$\qquad\qquad$ L = pyridine, $\varrho = 0.168 \pm 0.013$ (25 °C)

Co^{II}(Por) complexes also bind O$_2$, CO and NO.[17a,18] Relatively small ESR hyperfine coupling constants for the O$_2$ adducts suggest a Co^{III}—O$_2^-$ formulation is appropriate rather than Co^{II}—O$_2$ though it is not a discrete Co^{III}—O$_2^-$ species (ν(O—O) $\approx 1150\,cm^{-1}$). Oxygenation is enhanced in polar media, *i.e.* $P_{1/2} \approx 12.6$ mmHg in DMF compared with 417 mmHg in toluene for Co(PPIX DME)(1-MeIm). In general, Co(Por) binds O$_2$ by a factor of 10^{2-3} less tightly than do the Fe analogues (equation 17).

$$Co(Por)L + O_2 \overset{K_{O_2}}{\rightleftharpoons} Co(Por)(O_2)L \qquad (17)$$

$\Delta H° \approx -33.5$ to $-50.2\,kJ\,mol^{-1}$
$\Delta S° \approx 200–247\,J\,K^{-1}\,mol^{-1}$

$K_{O_2} = 10^3–10^4\,mm^{-1}$ ($P_{1/2}^{O_2} = 10^1–10^3$ mmHg at -23 to -65 °C)
$\varrho = -0.056$ (extrapolated to 20 °C) for 4-substituted pyridines

Co(TPP)(CO) and Co(TPP)(CO)(O$_2$) have only been characterized spectroscopically at low temperature (-150 °C).

Because the Co^{II} is in a $(d_{xy})^2(d_{xz}d_{yz})^4(d_{z^2})^1$ configuration, the Co—NO bond in Co(TPP)(NO) (ν(NO) = 1689 cm^{-1}) is weaker than the corresponding Fe and Mn analogues. The M—N$_{ax}$ distance, Δ(N$_4$) and MNO angle of M(TPP)(NO) reflect the strength of M—NO interactions, *e.g.* 1.637(2) Å, 0.34 Å and 177.8(5)° for M = Mn, 1.717(7) Å, 0.21 Å and 149.2(6)° for M = Fe, and 1.833(53) Å, 0.094 Å and 128.5° for M = Co.[21] Co(Por)L complexes, where L = SO$_2$ or RNC, are also known.[17a]

Autoxidation of Co^{II}(Por) with protic acids gives Co^{III}(Por)X (X = Cl, Br) and, in the presence of appropriate ligands, Co^{III}(Por)LX (X = Cl, Br, L = H$_2$O, py, NH$_3$) or [Co^{III}(Por)L$_2$]X (X = ClO$_4$, BF$_4$, L = H$_2$O, py, 1-MeIm).[103] [Co^{III}(Por)(OH$_2$)$_2$]ClO$_4$ is converted to [Co^{III}(Por)(OH$_2$)X] (X = I, N$_3$, SCN) with HX. Recrystallization of Co^{III}(Por)X(py) or [Co^{III}(Por)(OH$_2$)$_2$]X in the presence of excess pyridine gives [Co^{III}(Por)(py)$_2$]X (X = Cl, ClO$_4$). [Co^{III}(Por)(CN)$_2$]$^-$ is formed from Co^{III}(Por)X and cyanide ion. Oxidation of Co(Por)(NO)(3,5-lutidine) gives a nitro compound, Co(Por)(NO$_2$)(3,5-lutidine), the structure of which has been analyzed by X-ray crystallography.[21] Alkylation of Co^{III}(Por)X with carbanions gives organocobalt-porphyrins, RCo(Por)(OH$_2$) (Scheme 21).[17a]

Ligand substitution kinetics on water-soluble porphyrins have been measured.[18,105,106] The Co^{III} ion of the porphyrin exchanges axial ligands several orders of magnitude more readily than do simple amine or aqua complexes. Relatively low ligand discrimination ratios and positive activation entropies support a dissociative mechanism (equation 18, Table 7).

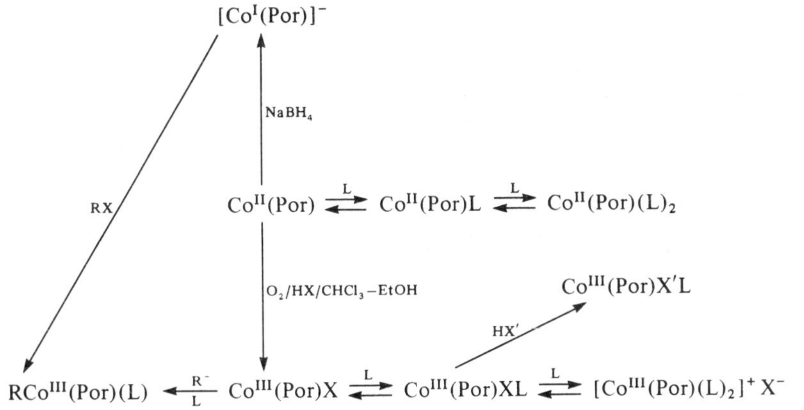

$$[Co^I(Por)]^-$$

Scheme 21

$$Co^{III}(Por)(OH_2)_2 \underset{-X,k_{-1}}{\overset{K_1 \atop X,k_1}{\rightleftharpoons}} Co^{III}(Por)(OH_2)X \underset{-X,k_{-2}}{\overset{K_2 \atop X,k_2}{\rightleftharpoons}} Co^{III}(Por)(X)_2 \text{ at } 25\,°C \qquad (18)$$

Table 7 Thermodynamic and Kinetic Parameters for the Coordination of Axial Ligands to $Co^{III}(Por)$ in Aqueous Solution

Por	X	K_1 (M^{-1})	k_1 (M^{-1}s^{-1})	k_{-1} (s^{-1})	$\Delta H_1^{\neq}$ (kJ mol^{-1})	$\Delta S_1^{\neq}$ (J K^{-1} mol^{-1})	K_2 (M^{-1})	k_2 (M^{-1}s^{-1})	k_{-2} (s^{-1})	Acid dissociation constants pK_1 and pK_2 of $Co(Por)(OH_2)$
$T_{N\text{-Me}}pyP$	py	10^6	0.7	6×10^{-7}			4.8×10^4	2.8	5.8×10^{-5}	5.46, 10.7
	SCN	6.4×10^3	2.9	3.1×10^{-4}	79.5	31.0	1.3×10	2.8×10^4	3.0×10^{-3}	6.0, 10.0
	I	3.4×10	1.6		83.7	43.5				
	Br	1.4×10^{-1}	0.25							
	Cl	8.0×10^{-2}	0.092							
$T_{p\text{-}CO_2H}PP$	py	$>10^6$	1.4×10^3	$\leqslant 10^{-3}$			9.5×10^4	2.1×10^3	2.2×10^{-2}	7.4, >9
	SCN	2.9×10^3	4.5×10^2	1.5×10^{-1}			6.2	1.8×10^6	2.9×10^5	
$T_{p\text{-}SO_3^-}PP$	SCN	2.6×10^3	3.2×10^2		77.0	60.2	2.8			5.72
	I	1.1×10	1.2×10^2		87.0	86.6				

[a] K. R. Ashley, M. Berggren and M. Cheng, *J. Am. Chem. Soc.*, 1975, **97**, 1422.
[b] Ref. 105.
[c] Ref. 106.

Metallation of porphyrin with $[Rh(CO)_2Cl]_2$ gives a binuclear complex, *trans*-$\{Rh^I(CO)_2\}_2(Por)$ *via* a sitting-atop intermediate, $\{Rh(CO)_2\}_2ClH(Por)$.[17a] On heating in $CHCl_3$, the bis-rhodium complex pyrolyzes to $Rh^{III}(Por)Cl$, which can be directly formed if the metallation is carried out in the presence of air. $Rh^{III}(Por)Cl$ complexes are reduced to $[Rh^I(Por)]^-$ with $NaBH_4$ and, like the Co analogues, the Rh^I complexes are alkylated with alkyl halides.[107] The organorhodium complexes are also obtained by the reaction of $Rh^{III}(Por)Cl$ with alkyllithium, vinyl ether or alkynes. Both $Rh(Por)X$ and $Rh(Por)R$ bind sixth ligands such as H_2O, py, R_3P, CO and carbanions.[17a,21,107,108] Six-coordinate complexes of the type $[Rh^{III}(Por)(L)_2]X$ (L = N or O donor) are also known. Halides and pseudohalides substitute the anionic ligand or both axial ligands. For the water-soluble complex, $[Rh^{III}(T_{p-SO_3^-}PP)(OH_2)_2]^{3-}$, thermodynamic and kinetic parameters have been reported and, as is the case for the Co analogues, a significant labilization of the axial ligand by the porphyrin (by a factor of 10^{6-7}) is noted (equation 19).[108]

Metal–metal-bonded dimers, $[Rh^{II}(Por)]_2$, are formed by thermolysis of the hydride complexes, $Rh^{III}(Por)H$, obtained by protonation of $[Rh^I(Por)]^-$ with acetic acid, or from $Rh^{III}(Por)CO_2Et$, generated from $Rh^{III}(Por)Cl(CO)$ ($\nu(CO) = 2100\,cm^{-1}$) and ethoxide.[109] Air oxidation of $[Rh^I(Por)]^-$ also gives $[Rh^{II}(Por)]_2$.[110] The Rh^{II} porphyrin binds O_2 at $-80\,°C$ in toluene in a end-on fashion ($g \approx 2.09, 2.01, 2.00$) and releases it when the solid is heated under vacuum, but decomposes to an Rh^{III} species at $20\,°C$ in solution.[109] In the presence of ligands, $Rh(Por)(O_2)L$ (L = py, Pip, $(RO)_3P$) are formed. NO converts the dimer to monomeric $Rh(Por)NO$. Reaction of $[Rh(Por)]_2$ and vinyl ether gives β-ketoalkyl $Rh^{III}(Por)$ (Scheme 22).

$$Rh(T_{p\text{-}SO_3^-}PP)(OH_2)_2 \underset{-X, k_{-1}}{\overset{K_1 \atop X, k_1}{\rightleftharpoons}} Rh(T_{p\text{-}SO_3^-}PP)(OH_2)X \underset{-X, k_{-2}}{\overset{K_2 \atop X, k_2}{\rightleftharpoons}} Rh(T_{p\text{-}SO_3^-}PP)(X)_2 \qquad (19)$$

$pK_{a_1} = 7.0$, $pK_{a_2} = 9.8$ $25\,°C$, $\mu = 1.00\,M$ (NaClO$_4$)

X	K_1 (M^{-1})	k_1 (M^{-1}s^{-1})	k_{-1} (s^{-1})
Cl	0.74 ± 0.11	$(2.64 \pm 0.11) \times 10^{-3}$	$(2.33 \pm 0.05) \times 10^{-3}$
Br	1.82 ± 0.10	$(5.71 \pm 0.14) \times 10^{-3}$	$(2.60 \pm 0.05) \times 10^{-3}$
I	$(4.12 \pm 0.18) \times 10$	$(2.01 \pm 0.03) \times 10^{-2}$	$(0.72 \pm 0.05) \times 10^{-3}$
NCS	$(1.22 \pm 0.04) \times 10^4$	$(2.73 \pm 0.06) \times 10^{-2}$	

X	$\Delta H^{\neq}$ (kJ mol^{-1})	$\Delta S^{\neq}$ (J K^{-1} mol^{-1})
Cl	56.1 ± 2.5	-106 ± 7.9
Br	57.7 ± 2.5	-94.6 ± 4.6
I	53.6 ± 3.3	-97.9 ± 10.5
NCS	69.0 ± 0.8	-43.1 ± 2.5

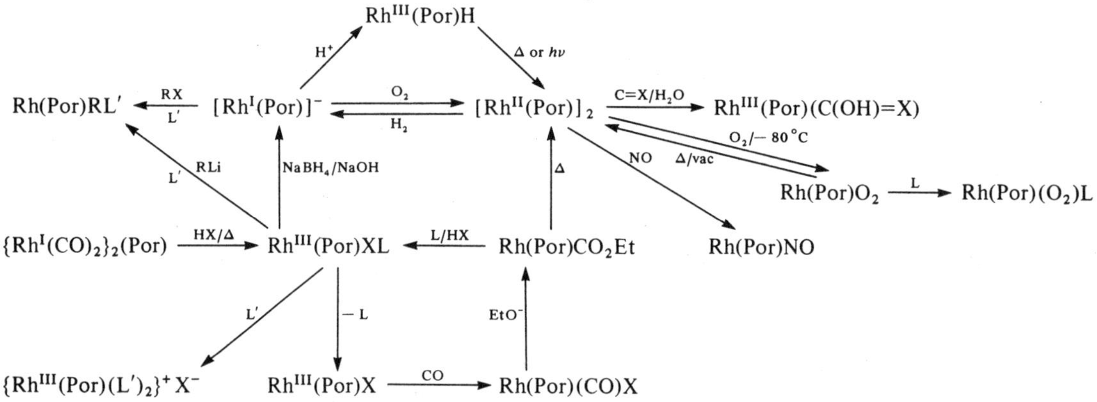

Scheme 22

Prolonged metallation of H$_2$TPP with {Rh(CO)$_2$Cl}$_2$ in benzene is reported to result in Rh^{IV}(TPP)(Ph)Cl formation.[17a]

Unlike the Rh systems, the reaction of Ir(CO)$_3$Cl and H$_2$Por gives {Ir^I(CO)$_3$}$_2$(Por) and/or Ir^{III}(Por)(CO)Cl.[17a] Ir^{III}(Por)(CO)Cl is reduced to [Ir^I(Por)]$^-$ with NaBH$_4$, and the anionic Ir^I complexes are alkylated with alkyl halides.[111] The organoiridium complexes can be directly obtained from Ir^{III}(Por)(CO)Cl and organolithium compounds, or BH$_4^-$ reduction of Ir^{III}(N-methyl-OEP)Cl$_2$. Both methyl and carbonyl Ir^{III}(Por) are stable and form six-coordinate complexes, Ir^{III}(Por)(R)L (L = N donor or CN$^-$) in the presence of appropriate ligands.[17a,112] Ir^{III}(OEP)-(Me)(CO) is also known. The stability of the Ir—CO moiety (v(CO) $\approx$ 2060 cm^{-1} cm^{-1}) is remarkable. The chloride in Ir(Por)(CO)Cl can be substituted with Br$^-$, HOSO$_3^-$ or MeCO$_2^-$ by treating with the corresponding acid, HX, and with OH$^-$, though it is hydrolyzed to Ir^{III}(Por)(OH)(OH$_2$) in hot aqueous alkali. Ir^{III}(Por)(CO)(OClO$_3$) can also be prepared by heating Ir(Por)(CO)Cl and AgClO$_4$ in benzene, and converted to Ir^{III}(Por)(CO)(BR$_4^-$) with HBF$_4$ (Scheme 23). The marked increase of the M—CO bond stability (Ir > Rh > Co) reflects the increasing d_π to p_π back-donation in this series. The IR stretching frequency, v(CO), decreases as the *trans* ligand (X) becomes a better donor, *i.e.* for v(CO), X = BF$_4$ > ClO$_4$ > CN > Br $\geqslant$ Cl > Me, and the NMR signal of the meso protons shifts to lower field along the series Ir^{III}(OEP)LX[(L, X) = (CO, BF$_4$) > (CO, ClO$_4$) > (CO, Cl) > (CO, Me) > (−, Me) > (1-MeIm, Me) > (CN$^-$, Me)], in agreement with the order of $E_{1/2}$ for the Ir$^{I/III}$ redox couple.

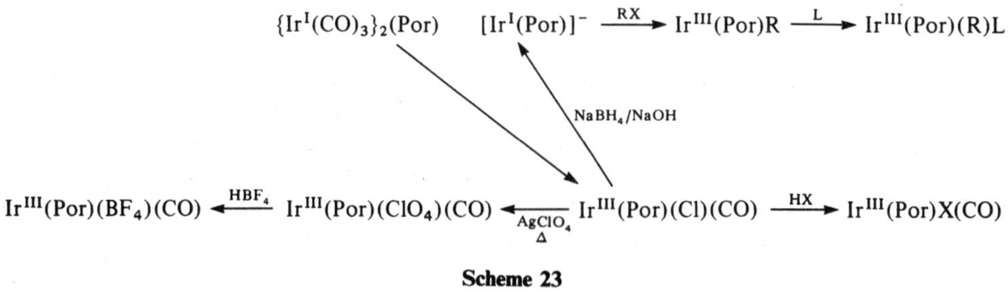

Scheme 23

(xiv) Group VIII—3 ($Ni^{II,III}$, $Pd^{II,IV}$, $Pt^{II,IV}$)

$Ni^{II}(Por)$ complexes, obtained by metallation of porphyrin with Ni salts,[17a] form five- or six-coordinate complexes only in the presence of ligands in high concentration.[18] The low ligand affinity of $Ni(Por)$ is attributed to the stable diamagnetic $(d_{xy})^2(d_{xz}d_{yz})^4(d_{z^2})^2$ configuration which converts to the high-spin $(d_{xy})^2(d_{xz}d_{yz})^4(d_{z^2})^1(d_{x^2-y^2})^1$ configuration on axial ligation. The two unpaired electrons in each σ-antibonding orbital expand the coordination sphere of the Ni ion, *i.e.* $Ni—N_{por} = 1.958$ Å in $Ni(OEP)$ *vs.* 2.038 Å in $[Ni(T_{N-Me}pyP)(Im)_2]^{4+}$.[21] The thermodynamics of $Ni(Por)$ are characterized by small enthalpy change on axial ligation ($\Delta H \approx 25.1$ kJ mol^{-1}) and rather large ϱ values in the Hammett equation (equation 20).[18,113]

$$Ni(Por) \underset{-L}{\overset{K_1, L}{\rightleftharpoons}} Ni(Por)L \underset{-L}{\overset{K_2, L}{\rightleftharpoons}} Ni(Por)(L)_2 \qquad \beta_2 = K_1 K_2 \qquad (20)$$

$Ni(2,4-X_2DPIX\ DME)(Pip)_2$	$\log (\beta_2^X/\beta_2^H) = 1.4\ (2\sigma)$ $(CHCl_3, 25\,°C)$
for X = Et	$\log \beta_2 = 1.92$, $\Delta H = -20.25 \pm 0.21$ kJ mol^{-1},
	$\Delta S = -105.4 \pm 1.3$ J K^{-1} mol^{-1} (THF, 7 °C)
$Ni(2,4-X_2DPIX\ DME)(py)$	$\log (K_1^X/K_1^H) = 0.8\ (2\sigma)$ (THF)
for X = CH(OH)Me	$\log K_1 = 1.5$ $(CHCl_3, 25\,°C)$
$Ni(T_{p-X}PP)(Pip)_n$	$\log (\beta_2^X/\beta_2^H) = 0.331 \pm 0.005\ (4\sigma)$ (toluene, 22 °C)
$Ni(T_{m-X}PP)(Pip)_n$	$\log (\beta_2^X/\beta_2^H) = 0.413 \pm 0.006\ (4\sigma)$ (toluene, 22 °C)
for X = H	$\log \beta_2 = -0.37 \pm 0.03$, $\Delta H = -23.4 \pm 2.1$ kJ mol^{-1},
	$\Delta S = -88 \pm 8.5$ J K^{-1} mol^{-1}, $\log K_1 = 2.3 \pm 0.3$
$Ni(TPP)(py)_2$	$\log \beta_2 = 2.7$ $(CHCl_3, 25\,°C)$, $\Delta G = -18 \pm 4.2$ kJ mol^{-1},
	$\Delta H = -26.4 \pm 1.3$ kJ mol^{-1}, $\Delta S = -29 \pm 17$ J K^{-1} mol^{-1} $(C_6H_6, 25\,°C)$
$Ni(T_{N-Me}pyP)(L)_2$, L = H_2O	$\log \beta_2 = 0.09$, $\Delta H° = -39$ kJ mol^{-1}, $\Delta S° = -130.1$ J K^{-1} mol^{-1}
	(acetone–H_2O, 25 °C), $\mu_{eff} = 2.1$ BM in D_2O
L = py	$\log K_1' = 0.54$ (substitution of H_2O) $(H_2O, 25\,°C)$,
	$\mu_{eff} = 3.1$ BM in py–D_2O
L = Im	$\log K_1' = 0.92$, $\log K_2' = -0.72$, $\Delta H° = -44.4$ kJ mol^{-1},
	$\Delta S° = -131.4$ J K^{-1} mol^{-1} (substitutions of H_2O) $(H_2O, 25\,°C)$,
	$\mu_{eff} = 3.3$ BM in Im–D_2O

Electrochemical oxidation of $Ni^{II}TPP$ in CH_2Cl_2 gives a red coloured radical, $Ni^{II}(TPP)^{\ddagger}$ ($g = 2.0041$), which changes, reversibly *via* an intramolecular electron transfer, to a green coloured $[Ni^{III}(TPP)OClO_3]$ on cooling to 77 K ($g_\perp = 2.286$, $g_\parallel = 2.086$).[17a]

Metallation of porphyrins with $PdCl_2$ or $PtCl_2$ in hot benzonitrile furnishes the square planar complexes, $M(Por)$ ($M = Pd^{II}$ or Pt^{II}), which show no tendency for axial coordination.[17a] Pd^{II} and $Pt^{II}(Por)$ are oxidized in acidic media with H_2O_2 (or HOCl), presumably to $M^{IV}(Por)(OAc)_2$, which are converted to $M^{IV}(Por)Cl_2$ with HCl (Scheme 24).[117]

$$M^{II}(Por) \xrightarrow[\text{AcOH}]{H_2O_2\,(\text{or HOCl})} M^{IV}(Por)(OAc)_2 \xrightarrow{\text{HCl}} M^{IV}(Por)Cl_2$$

M = Pd, Pt

Scheme 24

(xv) Group IB (Cu^{II}, $Ag^{I,II,III}$, Au^{III})

$Cu^{II}(Por)$ are obtained by metallation of porphyrins with Cu^{II} salts.[17a] The square planar complexes show weak ligand affinity due to the filled d_{z^2} orbital (equation 21).[18] ESR signals are observed at $g_\perp \approx 2.05$ and $g_\parallel \approx 2.19$ with hyperfine coupling constants $A_\perp^{Cu} \approx 35 \times 10^{-4}$ cm^{-1}, $A_\parallel^{Cu} \approx 200 \times 10^{-4}$ cm^{-1} and $A_\perp^N \approx 14 \times 10^{-4}$ cm^{-1}, $A_\parallel^N \approx 17 \times 10^{-4}$ cm^{-1}.[2,115]

Porphyrins react with AgOAc in warm pyridine to give $Ag_2^I(Por)$.[17a] A green solution of the disilver complex turns reddish brown on boiling due to disproportionation of the $Ag_2(Por)$ to $Ag^{II}(Por)$ and metallic Ag. $Ag^{II}Por$ forms directly if the metallation is carried out in acetic acid. The square planar $Ag^{II}(Por)$ ($g_\perp \approx 2.03$, $g_\parallel \approx 2.11$, $A_\perp^{CO} \approx 30 \times 10^{-4}$ cm^{-1}, $A_\parallel^{CO} \approx 60 \times 10^{-4}$ cm^{-1}, $A_\perp^N \approx 20 \times .10^{-4}$ cm^{-1}, $A_\parallel^N \approx 23 \times 10^{-4}$ cm^{-1})[2,115] is readily oxidized chemically, *e.g.* $Fe(ClO_4)_3$, or electrochemically ($E_{1/2} \approx 0.33$ V in $CHCl_3$–MeOH, 0.44 V in DMSO, 0.46 V in $CHCl_3$–MeCN) to $[Ag^{III}(Por)]^+$.[17a,116] The trivalent Ag complex, $Ag^{III}(OEP)(OClO_3)$, is diamagnetic and X-ray photoelectron spectroscopy suggests a five-coordinate square pyramidal complex.

$$Cu(Por) + L \underset{}{\overset{K}{\rightleftharpoons}} Cu(Por)L \qquad (21)$$

Cu(MPIX DME)(py)	$\log K = -0.676$, $\Delta H = -9.62 \pm 0.21 \text{ kJ mol}^{-1}$,
	$\Delta S = -48.5 \pm 1.3 \text{ J K}^{-1} \text{mol}^{-1}$ (THF, 7 °C)
Cu(HPIX DME)(py)	$\log K = -2.1$ (CHCl$_3$, 25 °C)
Cu(TPP)(py)	$\log K = -1.1$ (CHCl$_3$, 25 °C)
Cu(3,8-X$_2$DPIX DME)(py)	$\log (K^X/K^H) = 0.33$ (2σ) (C$_6$H$_6$)

Metallation of porphyrins with KAuCl$_4$–NaOAc in boiling acetic acid gives Au^{III}(Por)X; the counteranion may be (AuCl$_4$)$^-$, Cl$^-$ or OAc$^-$ depending on the work-up.[117] The X-ray structure of Au(TPP)Cl(CHCl$_3$)$_{0.7}$ shows the gold atom in the N$_4$ plane (Δ(N$_4$) $\leqslant$ 0.05 Å) with short Au—N$_{por}$ distances of ~2.00 Å and a long Au—Cl distance of 3.01 Å close to the sum of ionic radii, 3.18 Å, suggesting that the Au—Cl bond is essentially ionic.[21] The Au^{III} ion in porphyrins is electrochemically inert and NaBH$_4$ reduction gives an Au^{III} phlorin.[118]

(xvi) *Group IIB (Zn^{II}, Cd^{II}, Hg^{I,II})*

Zn^{II}(Por) complexes are easily prepared by briefly heating Zn(OAc)$_2$ and the porphyrin in acetic acid.[17a] Because of its large ionic radius (0.74 Å), the Zn atom in Zn(Por) preferentially takes a five-coordinate square pyramidal structure which allows the Zn—N$_{por}$ distance to increase from 2.04 Å in Zn(TPP) to 2.07 Å in Zn(TpyP)(py) by displacing the Zn atom from the N$_4$ plane by 0.3 Å.[21,119] The association constants for substituted pyridines are correlated to their pK_{BH+} and Hammett substituent constants (equation 22).[18]

$$Zn(TPP) + \text{substituted py} \underset{}{\overset{K}{\rightleftharpoons}} Zn(TPP) \text{ (substituted py)} \qquad (22)$$

Zn(TPP)py $\quad \Delta G = -20.5 \pm 2.5 \text{ kJ mol}^{-1}$, $\Delta H = -36.8 \pm 5.0 \text{ kJ mol}^{-1}$,

$\Delta S = -54 \pm 21 \text{ J K}^{-1} \text{mol}^{-1}$ (C$_6$H$_6$, 25 °C)

$\log K = 3.78 \pm 0.02$ (for X = H), $\log K = 0.236\text{p}K + 2.47$, $\log (K^X/K^H) = -1.50\sigma$
(C$_6$H$_6$, 25 °C)

$\log K = 3.82$ (for X = H), $\log K = 0.250\text{p}K + 2.47$ (CH$_2$Cl$_2$, 25 °C)

Log $K = 5.05 \pm 0.5$ for piperidine as a ligand in benzene fits equation (22), suggesting the Zn—N$_{ax}$ bond is predominantly a σ type, though imidazoles exhibit ~10 times higher K than expected from their pK_{BH+}.[120] As expected, Zn^{II}(Por)‡ and Zn^{II}(Por)$^-$ show both ligand affinity and ligand basicity dependence, higher and lower respectively than Zn^{II}(Por).

[Zn(TPP)py]‡ $\log K = 3.98$, $\log K = 0.492\text{p}K + 1.40$
Zn(TPP)py $\log K = 3.62$, $\quad \log K = 0.233\text{p}K + 2.40 \quad$ 0.1 M Bu$_4^n$NClO$_4$ in CH$_2$Cl$_2$, at 25°C
[Zn(TPP)py]$^-$ $\log K = 2.06$, $\log K = 0.095\text{p}K + 1.50$

The steric effect of an α substituent, $\log (K^{4\text{-Me}}/K^{2\text{-Me}})$, also increases as the net charge increases, *e.g.* Zn(TPP)‡ (2.31) > Zn(TPP) (1.57) > Zn(TPP)$^-$ (1.31).

Cd^{II}(Por) and Hg^{II}(Por), obtained by metallation of porphyrins with metal acetate in pyridine,[17a] resemble the Zn analogues in coordination behaviour. They are easily dissociated from the macrocyclic ligand, and form five-coordinate complexes with N-donor ligands.[18] In this group, the metal—ligand bonds, both *cis* and axial, become weaker as the metal becomes heavier (Zn > Cd > Hg) in contrast to other transition metal series but similar to other main group series (equation 23).

$$M^{II}(TPP) + \text{substituted py} \underset{C_6H_6, 25°C}{\overset{K}{\rightleftharpoons}} M^{II}(TPP)(\text{substituted py}) \qquad (23)$$

M	$\log K$ for X = H	ϱ for $\log (K^X/K^H) = \varrho\sigma$	a, b for $\log K^X = a\text{p}K + b$
Zn	3.78	−1.50	0.25, 2.45
Cd	3.51	−2.2	0.34, 1.65
Hg	1.21	−3.1	0.44, −1.10

Cd(TPP)(dioxane)$_2$ is interesting since its X-ray structure indicates a square planar complex (Δ(N$_4$) 0.03(1) Å towards O-1) with short Cd—N$_{por}$ bonds of 2.14(4) Å and long Cd—O distances of 2.64 Å (O-1) and 2.80 Å (O-2).[121]

Mercury(II) porphyrins seem to be rather irregular, since various types of Hg^{II} complexes have been reported (Scheme 25).[17a]

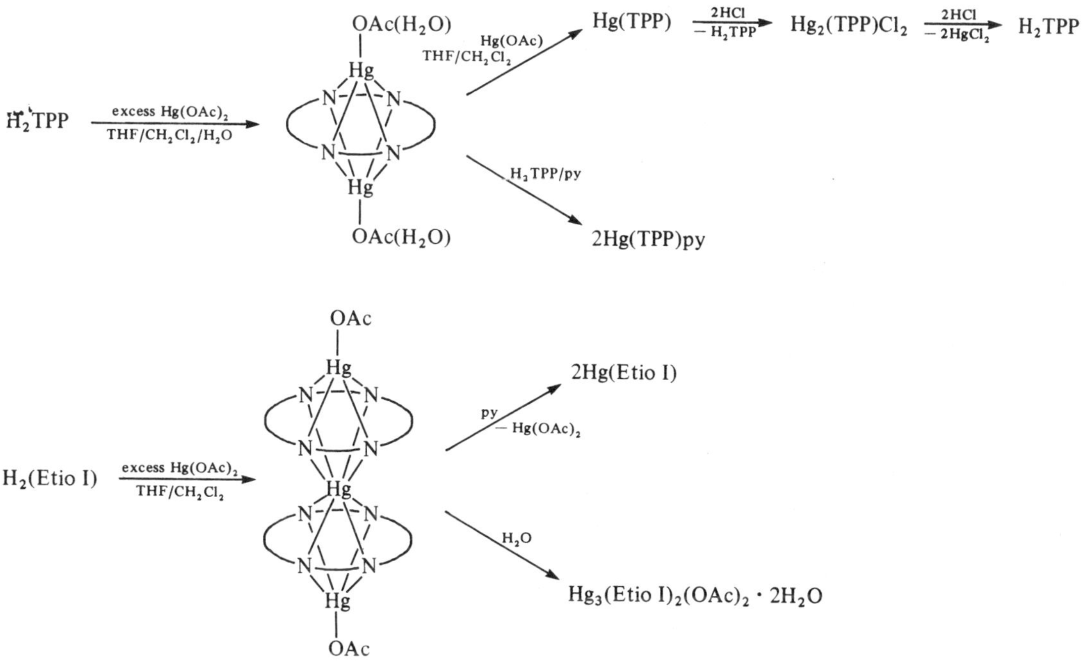

Scheme 25

21.1.1.1.8 Metal–metal bonding

Cr^{II}(MPIX DME) is a high-spin complex ($\mu = 5.19$ BM) in solution. The low magnetic susceptibility ($\mu = 2.84$ BM) in the solid state may be the result of Cr—Cr interactions.[19a] Oxidation of Rh^{I}(Por), thermolysis of HRh^{III}(Por) or EtO$_2$CRh^{III}(Por), and photolysis of HRh^{III}(Por) furnish diamagnetic dimers, [Rh^{II}(Por)]$_2$.[109,110] These dimers react with NO at room temperature or O$_2$ at $-80\,°$C in toluene to form monomeric Rh(Por)NO or Rh(Por)O$_2$(L). The Ru analogue, [Ru^{II}(Por)]$_2$, is prepared photochemically from Ru(Por)(CO)py, or thermolytically from solid Ru(Por)(py)$_2$ under vacuum.[97] Vacuum thermolysis of Mo(Por)(PhC≡CPh) and Os(Por)(py)$_2$ gives [Mo(Por)]$_2$ and [Os(Por)]$_2$.[122] According to valence bond theory, these dimeric compounds contain single (Rh—Rh), double (Ru=Ru, Os=Os) and quadrupole (Mo≡Mo) bonds respectively.

Metal–metal bonding is also known between different metals. Me$_3$Sn—Fe(TPP) made from [Fe^{I}(TPP)]$^-$ is diamagnetic and probably contains an Sn^{IV}—Fe^{II} bond.[19a,123] Reaction of anionic metal carbonyls and In(Por)Cl gives metal-bonded complexes which show characteristic hyper-type spectra ($\lambda_{max} \approx 440$ nm) (Scheme 26).[124] The precedent of this type of metathesis is the unexpected formation of (TPP)Sn—Mn(CO)$_4$—Hg—Mn(CO)$_5$ from Sn(TPP)Cl$_2$ and sodium amalgam reduced [Mn(CO)$_5$]$^-$.[125] In this compound, the linear Mn—Hg—Mn moiety is bonding to the Sn atom through the tetracarbonylated Mn and parallel to the macrocycle (Scheme 27).

In(OEP)Cl + [ML(CO)$_n$]$^-$ $\longrightarrow$ (OEP)In—ML(CO)$_n$

[ML(CO)$_n$]$^-$ = [Mn(CO)$_5$]$^-$, [Co(CO)$_4$]$^-$, [Mo(C$_5$H$_5$)(CO)$_3$]$^-$, [W(C$_5$H$_5$)(CO)$_3$]$^-$

In(OEP)Cl + [Fe(CO)$_4$]$^{2-}$ $\longrightarrow$ [(OEP)In]$_2$Fe(CO)$_4$

Scheme 26

$$Sn(TPP)Cl_2 \xrightarrow{Mn_2(CO)_{10}-NaHg} (TPP)Sn-Mn(CO)_4HgMn(CO)_5$$

Scheme 27

$M(Por)\{Re(CO)_3\}_2$ complexes (M = Mg, Zn, Co, Sn) are the first examples of this type of compound and are made from M(Por) or $Sn(Por)Cl_2$ and $Re_2(CO)_{10}$.[126] The absorption spectra and IR CO stretching frequencies of $M(Por)\{Re(CO)_3\}_2$ are similar to those of M(Por) and $\{Re(CO)_3\}_2(Por)$ respectively (Scheme 28).

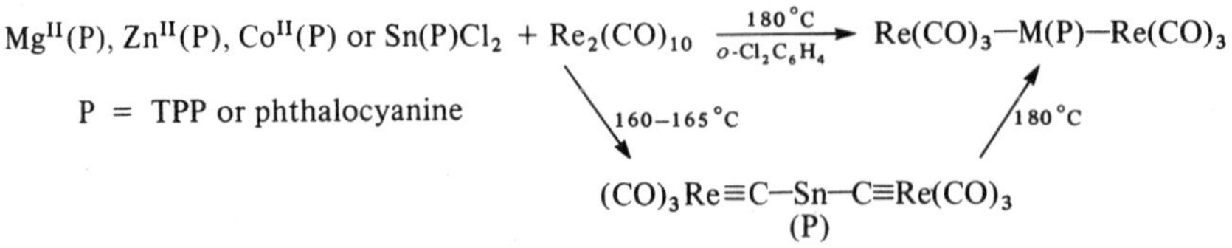
Mg^II(P), Zn^II(P), Co^II(P) or Sn(P)Cl2 + Re2(CO)10
180 °C
o-Cl2C6H4
Re(CO)3—M(P)—Re(CO)3
P = TPP or phthalocyanine
160–165 °C
180 °C
(CO)3Re≡C—Sn—C≡Re(CO)3
(P)

Scheme 28

21.1.1.1.9 Reactions at the metal

Biologically, hydrocarbons are oxidized to alcohols or epoxides by molecular oxygen in the presence of cytochromes P-450, which contain Fe(PPIX) as a prosthetic group coordinated by a cysteinyl mercaptide (Scheme 29).

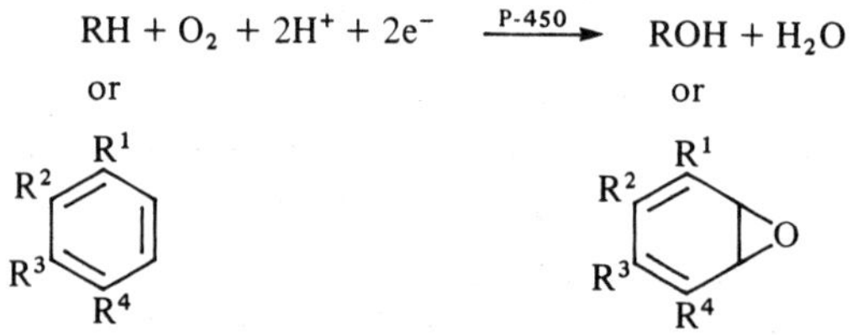
RH + O2 + 2H+ + 2e−
P-450
ROH + H2O
or
or
R1
R2
R3
R4
O

Scheme 29

Groves *et al.* found that a simple heme–iodosobenzene system mimics the enzymic reactions.[127] Cyclohexane and cyclohexene are oxidized to cyclohexanol and a mixture of cyclohexene oxide and cyclohexenol respectively by this system. Using meso-tetrakis-$\alpha,\beta,\alpha,\beta$-(*o*-acylamidophenyl)porphinatoiron(III) chloride where the acyl group is (*R*)-2-phenylpropionyl or (*S*)-2′-methoxycarbonyl-1,1′-binaphthyl-2-carbonyl, optically active styrene oxides are obtained in ~51% e.e. The Fe(TPP)Cl–PhIO system can also oxygenate arenes to arene oxides.[128] Based on the following observations, mechanisms involving O—$Fe^{IV}(Por)^{\ddagger}$ as the active species have been proposed (Scheme 30).[127]

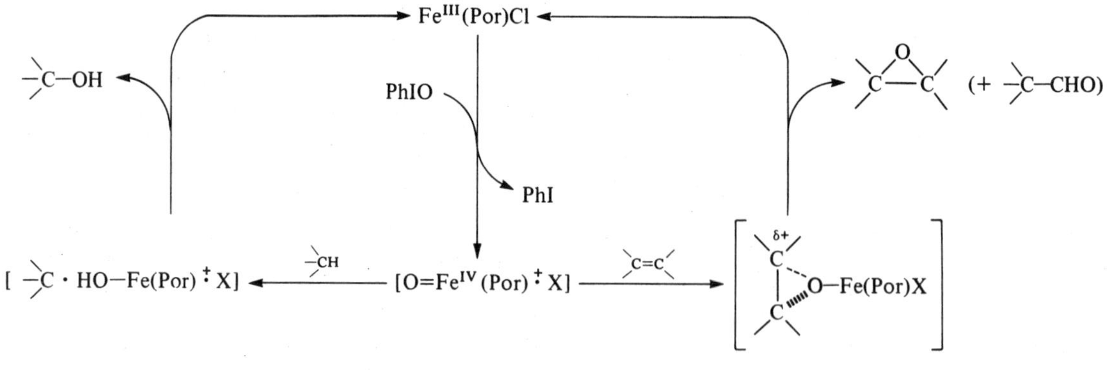
Fe^III(Por)Cl
PhIO
PhI
C—OH
C—C
(+ C—CHO)
[C · HO—Fe(Por) ‡ X]
CH
[O=Fe^IV(Por) ‡ X]
C=C
δ+
O—Fe(Por)X

Scheme 30

The hydroxylation occurs preferentially at tertiary position (adamantane C-1:C-2 = 25–48:1) unless steric hindrances exist. Some configuration inversion (~10% in *cis*-decaline hydroxylation at C-9) and bromination in the presence of $BrCCl_3$ have been observed. The double bond scrambling in allylic hydroxylations also supports a stepwise mechanism. The large isotope effect ($k_H/k_D = 12.9 \pm 1.0$ for cyclohexane hydroxylation) is compatible with a radical mechanism.

For epoxidations, electron-rich alkenes react more readily, in case of *p*-substituted styrenes, with a concomitant increase in the amount of aldehyde by-product, suggesting formation of an unsymmetric intermediate. However, *cis*-stilbene is converted to *cis*-stilbene oxide in 82% yield, and the intermediate, if any, must be short-lived. Under the same conditions, the reaction of *trans*-stilbene is slow. A close parallel approach of the double bond to the active heme site seems to be essential for epoxidation.

Isolation of the active species was reported and the spectral and physical properties support the $O{=}Fe^{IV}$ (Por)‡ formulation.[127] The Fe(IV)$^{Por\,\ddagger}$ species are considered to be intermediates of peroxidase and catalase enzymic reactions.

$Mn^{III}(Por)X$ also catalyze oxygen transfer from iodosobenzene to hydrocarbons.[129] The reactions are apparently stepwise, for example, cyclohexane hydroxylation accompanies halogenation by the counterion X, and *cis*-stilbene gives a mixture of *cis*- and *trans*-stilbene oxide (1.6:1). The active species is proposed to be $O{=}Mn^{V}(Por)X$.

$Mn^{III}(TPP)Cl{-}NaBH_4$ or $H_2/Pt{-}O_2$ systems are interesting, since the components are analogous to the natural P-450–NADPH–O_2 system (Scheme 31).[130]

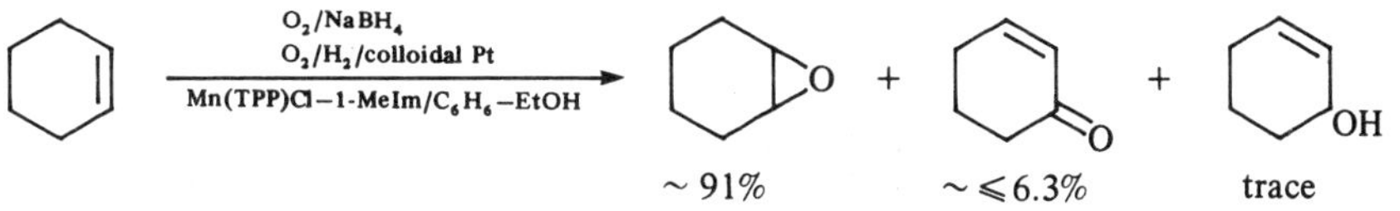

Scheme 31

Epoxidation of cyclohexene by $Mo^{V}O(TPP)X{-}Bu^{t}O_2H$ has been reported.[131] Demethylation of *N,N*-dimethylaniline by $Fe^{III}(TPP)Cl{-}PhIO$ offers a model for P-450-catalyzed *N*-dealkylation reactions.[132] The aziridine formation from acylimino Mn(Por) and alkene may be taken for the nitrogen analogue of the oxygen transfer reaction (Scheme 32).[133]

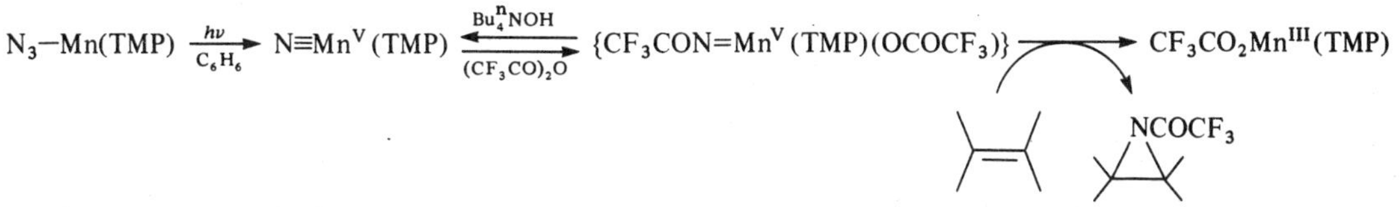

Scheme 32

$\{Rh(CO)_2\}_2(Por)$ complexes undergo addition reactions with carbonyl compounds and with alkyl halides to give acyl or alkyl $Rh^{III}(Por)$ similar to $Rh^{I}(CO)(phosphine)_3$ (Scheme 33).[134]

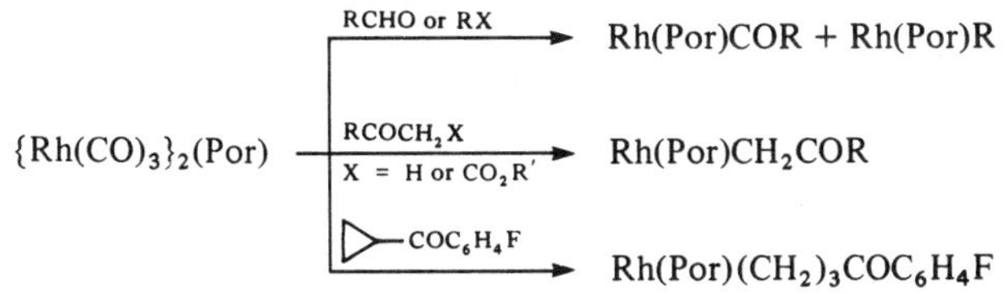

Scheme 33

In relation to B$_{12}$, the chemistry of organo-Co-, -Rh- and -Ir-(Por) is of interest (Scheme 34). Though there are some subtle differences in reactivity among the metals as well as between the macrocycles, the main reactions are common to both systems except that unlike the porphyrin macrocycle the corrin ring system has no biologically important redox chemistry.[17a,107,109,111]

$[Fe^{I}(Por)]^-$ are also good nucleophiles and react with alkyl halides to give organo-Fe(Por).[135]

$Ru(por)L$ (L = Pip, Bu_3^nP) are reported to serve as a decarbonylation catalyst for various aldehydes.[136] Interestingly, Rh(TPP)I catalyzes the alkylation of unactivated alkanes upon reaction with a diazoester.[137]

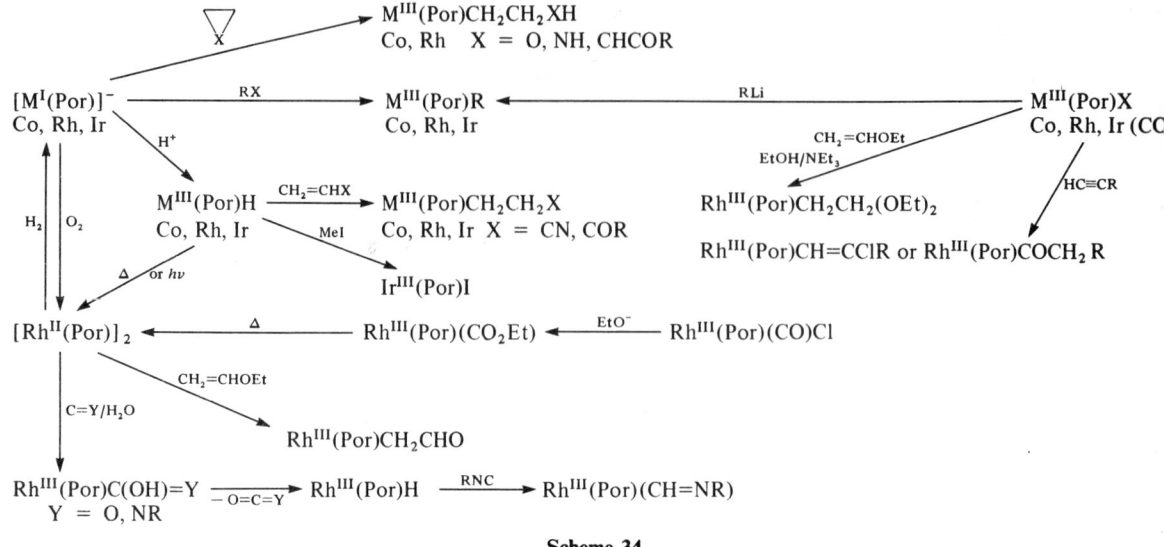

Scheme 34

21.1.1.1.10 Photochemistry

Porphyrins and certain metalloporphyrins are excellent photosensitizers because of similar energies of the singlet and triplet excited levels, high intersystem crossing yield and long triplet lifetimes.[138] The triplet state energies of M(Por) are estimated to be 159–193 kJ mol^{-1} from their phosphorescence spectra.[4] The acceptor can be a variety of other species (*e.g.* $^3O_2 \to {}^1O_2{}^*$), porphyrins ($^1Por \to {}^3Por^*$) or ligands such as 4-stilbazole and 1-(α-naphthyl)-2-(γ-pyridyl)ethylene($^1L \to {}^3L^*$).[138]

The excited state porphyrins can be both donors and acceptors and form exciplexes [*Por‡D^{-}] or [*Por‡A^{-}], which then relax to the ground state or undergo chemical reactions.

In acidic media, or in the presence of reductants such as ascorbic acid, tertiary amines or thiols, porphyrins are photoreduced to dihydro, tetrahydro and hexahydro derivatives. The hydrogen additions are stepwise and the intermediate radicals have been detected by ESR. The first reduction product of H_2Por is a phlorin (5,22-dihydroporphyrin) which isomerizes to chlorin or is further reduced to a porphyrinogen *via* a porphomethene (5,10,15,22-tetrahydroporphyrin). Chlorins are also converted to bacteriochlorins by photoreduction and isomerization. If the reductant can act as an oxidant after the electron donation, the processes may be reversible. Zn(Por) is similarly hydrogenated with visible light to Zn(Chl), but the intermediate is a Zn(porphodimethane), *i.e.* a 5,15-dihydroporphyrin complex. Further reduction gives tetra- and hexa-hydroporphyrins. Photochemical reduction of Sn^{IV}(Por) or Ge^{IV}(Por) with $SnCl_2/H_2O$/py gives M^{IV}Chl and then M^{IV}(iBchl). It is interesting that the photosynthesized chlorins have the 2,3-*cis* configuration. For the photoreduction of 5,15-dihydro or tetrahydro species, the remaining porphyrin molecules serve as the sensitizer.

If an appropriate electron acceptor is available, the photoexcited porphyrin may transfer an electron to it. The acceptors may be organic molecules such as *N*-alkylnicotinamide, metal complexes such as $[Fe(CN)_6]^{3-}$ or another porphyrin molecule.[139] The second order reaction rate for the Zn(uroporphyrin)–acceptor system at infinite ionic strength is estimated to be $\sim 10^8 M^{-1} s^{-1}$. The electron travels from Zn(Por) to *N*-benzylnicotiamide or to $[Fe(CN)_6]^{3-}$ over a distance of ~ 15 Å or ~ 30 Å respectively. Since the reaction creates charged species, polar media accelerate the process. The back reaction or other photosensitization processes such as $^1O_2$ generation become favourable in nonpolar solvents.

Photodissociation of axial ligands, especially of CO from Fe, Ru and Os, is an important process in the coordination chemistry of metalloporphyrins.[70,140] Insertion of CO_2 into RAl(Por)[141] and of O_2 into RCo(Por)[142] and into R_2M(Por) (M = Sn, Ge)[143] under light have also been reported.

21.1.1.1.11 Redox chemistry

Electrochemical redox reactions of metalloporphyrins occur either at the metal or at the ring.[144] The first and second ring reductions occur at -1.3 ± 0.3 V and -1.7 ± 0.3 V ($E_{1/2}$ *vs.* SCE) with an interval of 0.42 ± 0.05 V. The ring oxidation potentials are observed at $+0.7 \pm 0.3$ V and $+1.1 \pm 0.3$ V and the separation is 0.29 ± 0.05 V. The difference between the first ring reduction and oxidation potentials of 2.25 ± 0.15 V is in good agreement with an HOMO–LUMO energy gap of 2.18 V or 2.0 V after correction for solvent effects.

For some transition metal complexes, the metal redox processes intervene between the ring redox. For instance, $Mn^{III/II}$, $Fe^{III/II/I}$, $Co^{III/II/I}$ and $Ni^{III/II}$ couples are usually observed between the first ring redox potentials.

Relationships between the nature of porphyrin and the redox potentials have been found. For example, The reduction potentials of 3,8-disubstituted DPIX DME and their pK_1 values are correlated by equation (24).[145] Kadish *et al.*[146] have summarized porphyrin substituent effects by applying the Hammett–Taft equation (equation 25).

$$pK_3 = -5.9(E_{1/2}) - 5.2 \text{ in DMF} \tag{24}$$

$$E_{1/2}(X) - E_{1/2}(H) = \sigma\varrho_{EMP} \tag{25}$$

$E_{1/2}(H)$, $E_{1/2}(X)$: redox potentials for the parent and substituted porphyrins
σ: Hammett substituent constant
ϱ_{EMP}: sensitivity parameter

ϱ_{EMP} (mV) for the first row transition metalloporphyrins and metal free porphyrins:

M(T_{p-X}PP)	Ring Oxidation 2nd	Ring Oxidation 1st	Ring reduction 1st	Ring reduction 2nd	
M(T_{p-X}PP)	70 ± 10	70 ± 10	60 ± 10	70 ± 10	(M = VO, Mn, Fe, Co, Ni, Cu, Zn)
H$_2$(T_{p-X}PP)			73 (CH$_2$Cl$_2$)53 (DMSO)		
H$_2$(β-X-TPP)		170	280		
H$_2$(*meso*-X-OEP)		510			

The effect of substitution is in the order meso > β > phenyl, as expected from the shapes of HOMO and LUMO and the distances from the porphyrin π system. The ring substituent effects on the metal redox potential are smaller than those on the ring redox: for comparison, $\varrho_{EMP} = 18$ mV ($Ni^{III/II}$), 38 mV ($Fe^{III/II}$), 54 mV ($Mn^{III/II}$) for M(T_{p-x}PP) in CH$_2$Cl$_2$.

In general, ligands which bind strongly to the metal shift the metal redox potentials to the negative side, *i.e.* $E_{1/2}(Fe^{III/II})$ for Fe(TPP)X is in the order F > N$_3$ > Cl > Br > I > ClO$_4$, and the effects are larger for the metal ion at higher oxidation states. The ring oxidation potentials are not greatly influenced by the axial ligands (< 0.3 V).

Electron transfer rates for metalloporphyrins are of interest in relation to the biological electron carriers, the cytochromes. The ring redox rates are reported to be $\sim 4 \times 10^{-2}$ cm s^{-1} by cyclic voltammetry except for the ring oxidation of Mg(OEP) (~ 0.1 cm s^{-1}).[144] The metal reduction rate for Fe^{III}(Por) is spin-state dependent and increases in the order $Fe^{III}(S = \frac{5}{2}) \to Fe^{II}(S = 0)$ [0.02 cm s^{-1}] < $Fe^{III}(S = \frac{5}{2}) \to Fe^{II}(S = 2)$ [0.16 cm s^{-1}] < $Fe^{III}(S = \frac{1}{2}) \to Fe^{II}(S = 0)$ [4 cm s^{-1}]. The rate is also proportional to the basicity of substituted pyridine ligands.

There have been several model systems made to mimic cytochrome *c* where the heme group is linked to the peptide chain through two thioether linkages and coordinated with histidinyl imidazole on one side and the methionyl sulfide on the other. The model studies indicate that changing the axial ligand from bis-imidazole to mono-imidazole monothioether shifts the $Fe^{III/II}$ couple to 150–167 mV more positive, with a further 55 mV positive shift by ligation of two thioether groups.[147] The positive shift of the $Fe^{III/II}$ couple with thioether ligation is in the same direction as with the natural system, though the potentials themselves are 0.2–0.4 V more positive in the natural system probably due to the nonpolar protein environment. The more important finding may be that the bond lengths do not change significantly as the metal oxidation state changes in the thioether complexes (Fe^{II}—S ~ 2.34 Å *vs.* Fe^{III}—S ~ 2.33 and 2.35 Å, Fe^{II}—N$_{por}$ 1.996(6) Å *vs.* Fe^{III}—N$_{por}$ 1.982(6) Å). The rapidity of cytochrome *c* electron transfer may partly be attributed to the lack of substantial nuclear motion upon redox (*cf.* Franck–Condon principle) together with the fast low-spin Fe^{III} to low-spin Fe^{II} reduction rate.

21.1.1.1.12 N-*Substituted porphyrins*

Recently *N*-substituted porphyrins have increased in their biological importance in relation to the active species in P-450, peroxidase, catalase and deactivation mechanisms of P-450 and Hb.[148]

There have been several procedures reported for the *N*-alkylated porphyrin syntheses.[149] *N* alkylation with alkylating reagents tends to give a mixture of mono- and di-*N*-alkylporphyrins because the second alkylation is faster than the first. Since the basicity of (*N*-R)H(Por) is higher than that of $H_2(Por)$, N protonation may prevent the further alkylations (Scheme 35). Reaction of heme and phenylhydrazine is used to convert Hb to the green pigment from which an isomeric mixture of $(N\text{-}Ph)H(PPIX)$ was isolated.[148] The reaction is of a radical type but applicable to simple hemes. However, the demetallative N alkylation may be the most convenient method of $(N\text{-}R)H(Por)$ synthesis (Scheme 36).[150] Reactions of carbenes and porphyrins are metal dependent (Scheme 37).[19]

$$H_2(Por) + MeI \xrightarrow{\Delta} (N\text{-}Me)H(Por) + trans\text{-}N^{21},N^{22}\text{-}Me_2(Por) \xrightarrow{EtI} trans,trans\text{-}N^{21},N^{22}\text{-}Me_2\text{-}N^{23}\text{-}Et(Por)$$

$$H_2(Por) + MeOSO_2F \xrightarrow{CHCl_3} trans,trans\text{-}N^{21},N^{22},N^{23}\text{-}Me_3(Por) \longrightarrow cis\text{-}N^{21},N^{23}\text{-}Me_2(Por)$$

Scheme 35

$$R\text{-}M(Por) \xrightarrow[\text{(O)}]{H^+} (N\text{-}R)H(Por)$$

$$M = Co, Fe$$

Scheme 36

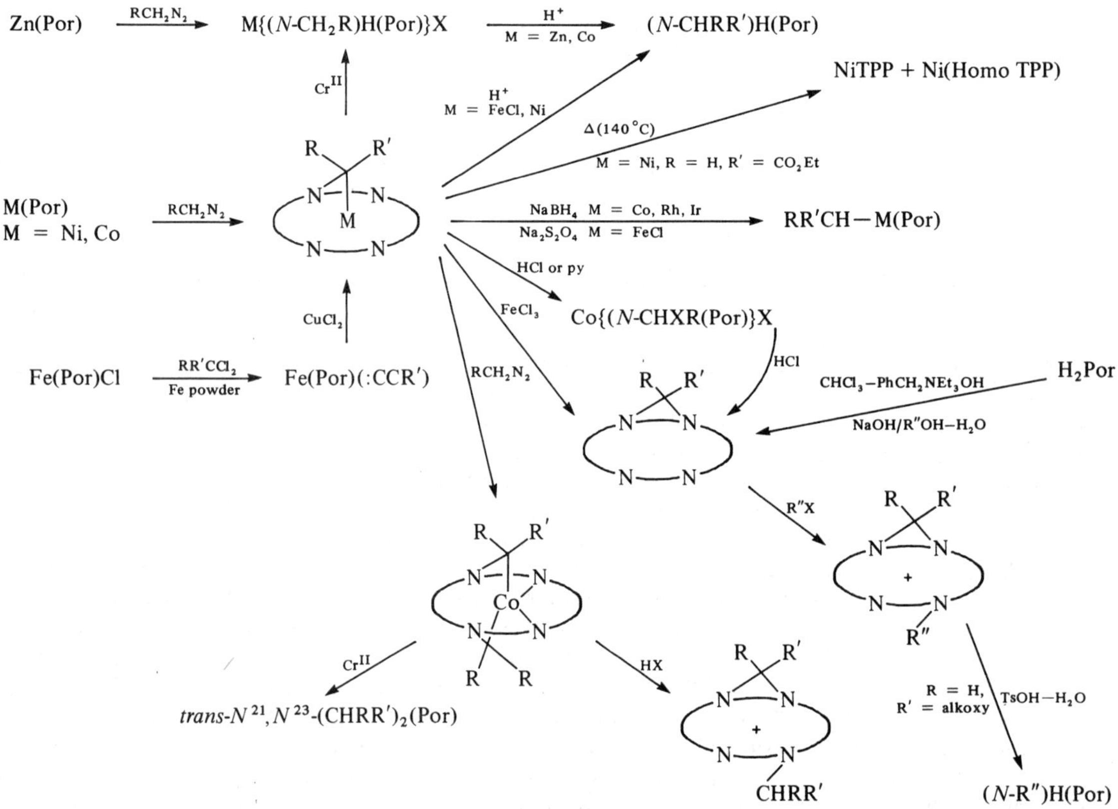

Scheme 37

Since the central hole of porphyrin is not large enough ($\leqslant 2.01$ Å) to accommodate even a methyl group, N-alkylporphyrins are distorted. In $[(N\text{-}CO_2Et)H_2(Por)]^+ I^-$, the N-substituted pyrrole ring is tilted up by $19.1°$ and the other three are declined by $4.8°$, $11.7°$ and $2.2°$ to reduce the Van der Waals contacts between the ester group and the pyrrolic nitrogens.[151] The deviation from planarity is much smaller in $[H_3(OEP)]^+[Re_2(CO)_6Cl_3]^-$ ($9.4°$ and -2.2 to $-2.8°$ respectively).[152] The strain is larger at C-5 and C-20 than in other positions and electrophilic deuterium exchange ($CF_3CO_2D\text{-}D_2SO_4$) or nitration occurs preferentially on those two meso carbons.[153]

The lack of ring deformation on the first N protonation in the predeformed N-alkylporphyrins renders them more basic [$pK_3 \approx 5.64$ for $(N\text{-}Me)H(TPP)$] than the planar porphyrins [$pK_3 \approx 4.38$ for $H_2(TPP)$], whereas their pK_4 values (3.85) are identical because their monoprotonated forms have already been distorted.[149]

The effects of ring distortion are also observed in the metallation and demetallation reactions (Scheme 38).[154] N-Alkylporphyrins are ~ 100 times more readily metallated than the unsubstituted

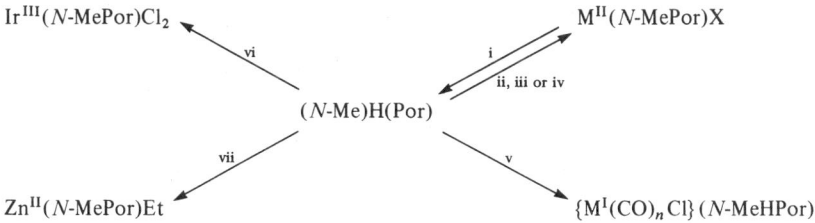

i, HCl or CF_3CO_2H; ii, MCl_2/$CHCl_3$, CH_2Cl_2, THF, acetone, alcohol/2,6-lutidine or 2,6-Bu_2^tpy (M = Mn, Fe, Co, Ni, Cu, Zn, Cd); iii, $M(ClO_4)_2$/DMF (M = Mn, Co, Ni, Zn, Cu); iv, $M(OAc)_2CHCl_3$–MeOH, Δ(M = Co, Ni, Zn, Hg); v, $M(CO)_nCl_2$/C_6H_6 [M = Rh(n = 2), Ir(n = 3)]; vi, Ir(cyclooctadien)$_2Cl_2$/CH_2Cl_2, r.t.; vii, $ZnEt_2$

Scheme 38

porphyrins. The metallation rate $(= k[M^{II}][(N\text{-Me})H(TPP)]$ increases in the order Cu > Zn, Co > Mn $\gg$ Ni. The initial outer sphere association of the solvated metal ion and porphyrin, and succeeding stepwise increases of the inner sphere interactions accompanied by the loss of a solvent ligand for every step are considered to be kinetically important. The predistortion of the N-substituted porphyrin ring facilitates the sitting-atop intermediate formation.

On the other hand, the alkylated N atom must take an sp^3 configuration to coordinate to the metal ion, and thus increase the ring distortion. Indeed, the N-substituted pyrrole ring in Co(N-CO$_2$Et-OEP)Cl is tipped by 44° from the mean plane of the porphyrin.[155] The formal charge on the N-alkylporphinate (-1) is also half that of the porphinate (-2). Therefore, the demetallation of N-alkylmetalloporphyrin is thermodynamically more favoured than that of the porphyrin analogue.

The ligand field of the macrocycle is also modified by ring distortion.[154] The metal to alkylated nitrogen distance is 2.33 Å (Fe) to 2.53 Å (Zn) which is considerably longer than the normal value of 2.0–2.1 Å for the first row transition metalloporphyrins in a high spin state, while the extension is small for the other three M—N_{por} bonds. The reduced and distorted ligand field of the macrocycle makes it possible for even the Co complexes as well as other metals to take a high spin state.

Partly because of the N-alkyl group being somewhat sensitive to nucleophilic attack, and partly because of the mechanistic interest, kinetic and thermodynamic studies on the N-alkylporphyrins have been directed to the metallation and demetallation reactions. Since the N-alkyl group is blocking one side of the porphyrin from ligand addition, only five-coordinate high-spin complexes of type $M^{II}(N$-R-Por)X or $[M^{II}(N$-R-Por)L]X have been reported for the first row transition series. The ligand substitution equilibrium constant $(K' \approx 1.66\,M^{-1})$ for Fe(N-Me-OEP)Cl with pyridine is the only data available at this time.[156] Reflecting the reduced metal–macrocycle interactions, the divalent state is more stable than the trivalent for M(N-Me-TPP) with M = Mn–Co.[154] The redox couple, $M^{II/III}$, is positively shifted by N methylation for M = Mn $(\Delta E_{1/2} = 0.99\,V)$, Fe($\Delta E_{1/2} = 0.78\,V$) and Co $(\Delta E_{1/2} = 0.28\,V)$ respectively, while the ring redox potentials ($E_{1/2} \approx 1.5$, 1.17, -0.78, $-1.24\,V$ vs. SCE) are less affected by N methylation.

The alkyl shifts between metal and nitrogen during redox reactions are mechanistically interesting (equation 26). The question as to whether the alkyl migration is intramolecular or intermolecular, or a dual process, does not seem to have a clear-cut answer as yet.[150] Reactions of the carbene adducts are summarized in Scheme 37.

$$R\text{—}M(Por) \underset{e^-}{\overset{-e^-}{\rightleftharpoons}} [M(N\text{—}RPor)]^+ \qquad M = Co, Fe \qquad (26)$$

Photocleavage of the ethyl–Zn(N-Me-TPP) bond in CH_2Cl_2 is reported to give ethane and Zn(N-Me-TPP)Cl via a radical mechanism.[157]

$X_2M(N^{21}\text{-}N^{22}$-alkoxymethylene-TPP) are the only examples of N,N'-dialkylporphyrin complexes (Scheme 39).[16] N-Aminoporphyrins are stable enough to be metallated only as their amide forms (Scheme 40).[158] The X-ray structure analysis of Ni(N-NTs-TPP) revealed that the Ni ion is coordinated with the macrocycle through three pyrrolic nitrogens and an amide nitrogen. A porphyrin N-oxide, $(N$-O)H$_2$(OEP), and its Cu complex have been reported, but the complex appears to be unstable in solution (Scheme 41).[159]

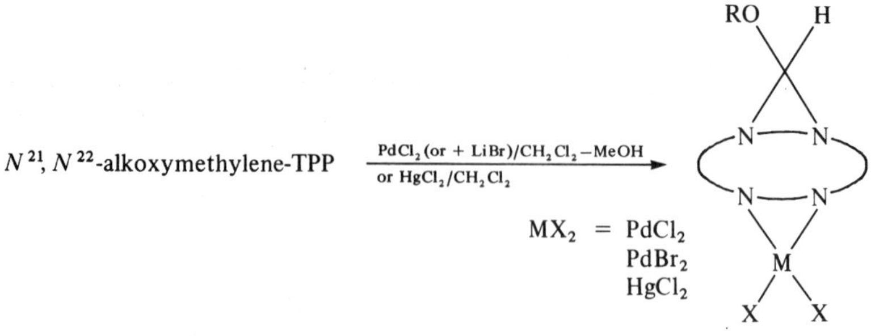

N^{21}, N^{22}-alkoxymethylene-TPP $\xrightarrow[\text{or } HgCl_2/CH_2Cl_2]{PdCl_2 \text{ (or } + LiBr)/CH_2Cl_2-MeOH}$

$$MX_2 = PdCl_2$$
$$PdBr_2$$
$$HgCl_2$$

Scheme 39

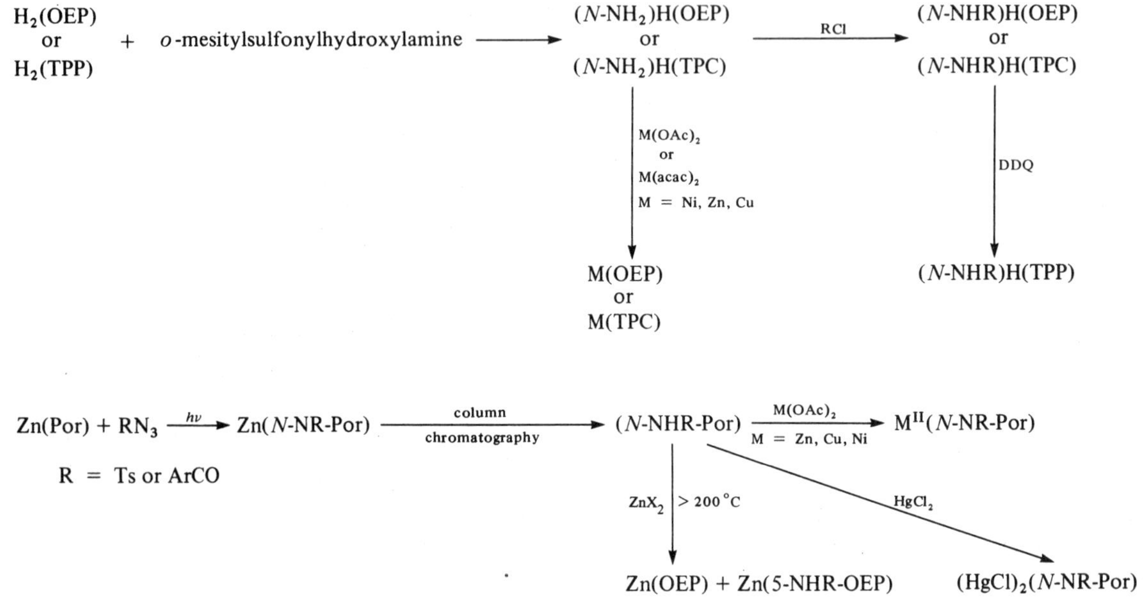

$\begin{matrix} H_2(OEP) \\ \text{or} \\ H_2(TPP) \end{matrix}$ + o-mesitylsulfonylhydroxylamine $\longrightarrow$ $\begin{matrix} (N\text{-}NH_2)H(OEP) \\ \text{or} \\ (N\text{-}NH_2)H(TPC) \end{matrix}$ $\xrightarrow{RCl}$ $\begin{matrix} (N\text{-}NHR)H(OEP) \\ \text{or} \\ (N\text{-}NHR)H(TPC) \end{matrix}$

$\downarrow$ $\begin{matrix} M(OAc)_2 \\ \text{or} \\ M(acac)_2 \\ M = Ni, Zn, Cu \end{matrix}$

$\downarrow$ DDQ

$\begin{matrix} M(OEP) \\ \text{or} \\ M(TPC) \end{matrix}$

$(N\text{-}NHR)H(TPP)$

$Zn(Por) + RN_3 \xrightarrow{h\nu} Zn(N\text{-}NR\text{-}Por) \xrightarrow[\text{chromatography}]{\text{column}} (N\text{-}NHR\text{-}Por) \xrightarrow[M = Zn, Cu, Ni]{M(OAc)_2} M^{II}(N\text{-}NR\text{-}Por)$

$R = Ts$ or $ArCO$

$ZnX_2 \bigg| > 200\,^\circ C$ $HgCl_2$

$Zn(OEP) + Zn(5\text{-}NHR\text{-}OEP)$ $(HgCl)_2(N\text{-}NR\text{-}Por)$

Scheme 40

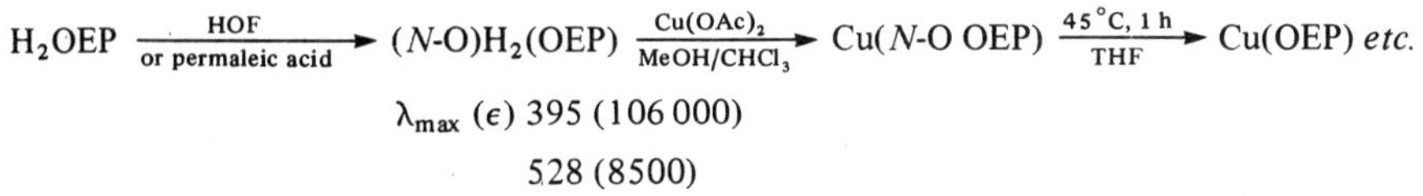

$H_2OEP \xrightarrow[\text{or permaleic acid}]{HOF} (N\text{-}O)H_2(OEP) \xrightarrow[MeOH/CHCl_3]{Cu(OAc)_2} Cu(N\text{-}O\ OEP) \xrightarrow[\text{THF}]{45\,^\circ C, 1\ h} Cu(OEP)\ etc.$

$\lambda_{max} (\epsilon)\ 395\ (106\ 000)$
$528\ (8500)$

Scheme 41

21.1.1.1.13 Heteroporphyrins

Several porphyrin analogues with the pyrrolic nitrogens substituted by heteroatoms have been synthesized, but only the oxa analogues (**23**; X = O, Y = NH) are reported to form stable metal complexes (Figure 8).[160] Formation of a Zn complex of the thia analogue (**23**; X = S, Y = NH) requires the presence of a large excess of Zn^{II} ion. An iron complex of dithiaporphyrin (**23**; X = Y = S) is also known.

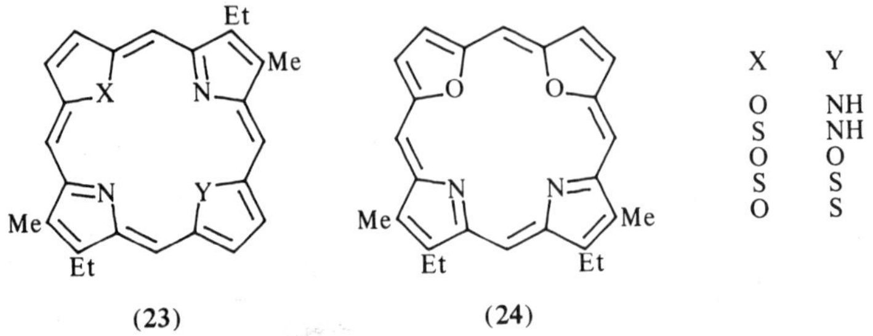

X	Y
O	NH
S	NH
O	O
S	S
O	S

(23)   (24)

Figure 8 Heteroporphyrins

21.1.1.1.14 Homoporphyrins

The homoporphyrin (**A**) is shown to be nonaromatic and easily oxidized to an aromatic cation (**B**) on metallation (Scheme 42).[161]

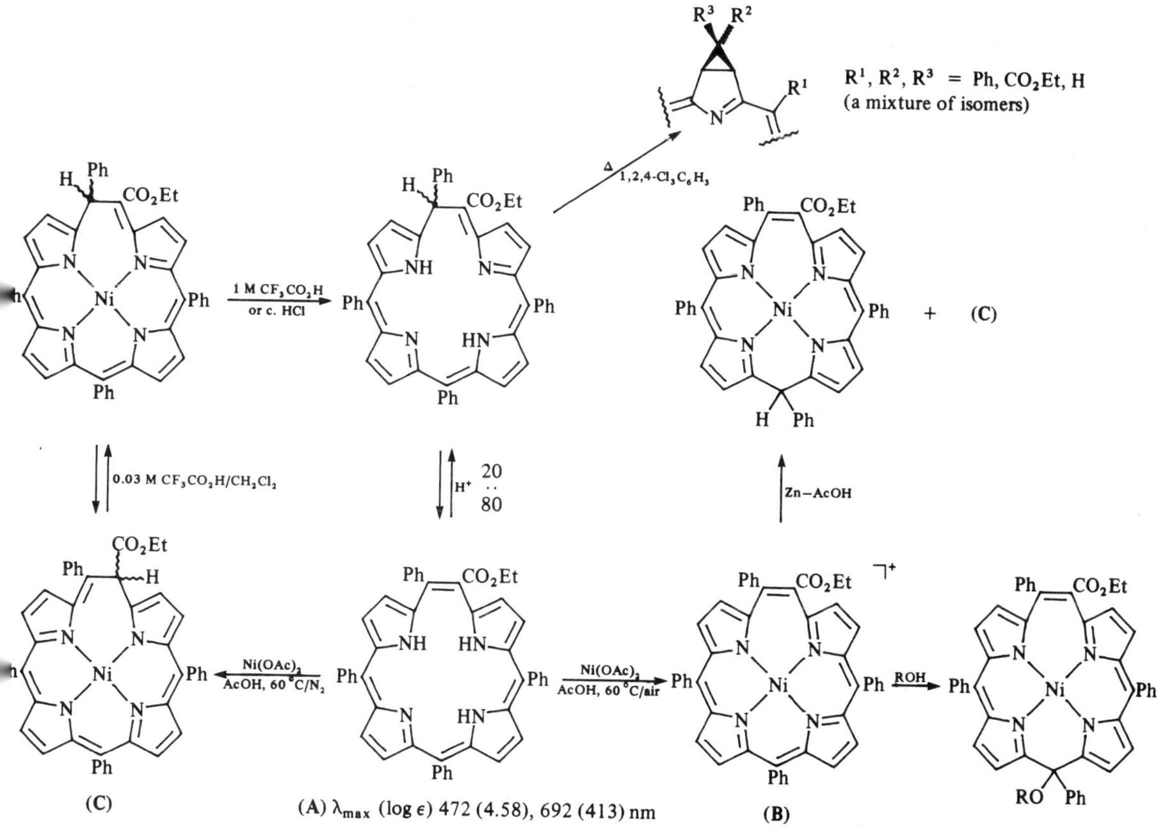

(A) λ_{max} (log ϵ) 472 (4.58), 692 (413) nm
δ (CDCl$_3$/TMS) 6.2, 7.5 (β-H) p.p.m.

Scheme 42

21.1.1.2 Chlorins, Bacteriochlorins and Isobacteriochlorins

21.1.1.2.1 General properties

Porphyrin can add up to six hydrogen atoms on the pyrrolic β-carbons without loss of aromaticity. Except for 2,3,12,13-tetrahydro derivatives (Bchl) (**27**), the porphyrin normal is no more than a C_s axis and the four nitrogens are nonequivalent in the other hydroporphyrins (Chl (**25**), iBchl (**26**) and hexahydroporphyrin). The X-ray structures of some metallochlorins [M(Chl)] show the asymmetry of the macrocycles (Figure 9 and Table 8).[21,162,163]

21.1.1.2.2 Spectroscopic properties

Chlorins and metallochlorins exhibit two intense absorptions corresponding to the porphyrin Soret (γ) and Q (α) bands, at ~ 400 nm ($\varepsilon \approx 10^5$, B) and ~ 660 nm [$\varepsilon \approx 10^5$, Qy(0,0)].[2] The former is usually accompanied by two shoulders (η_1 and η_2 bands) on the blue side, and the latter by three weak bands [Qy(1,0), Qx(0,0), Qx(1,0)]. Medium intensity double bands at ~ 500 nm ($\varepsilon \approx 10^4$) in the metal-free chlorin spectra disappear on metallation. In the Bchl spectra, the Qy(0,0) band bathochromically shifts to ~ 750 nm ($\varepsilon \approx 10^5$) accompanied by a weak band (~ 690 nm), and the Soret band splits in two (~ 360, ~ 400 nm; $\varepsilon \approx 10^5$). Between those two main bands, a medium intensity

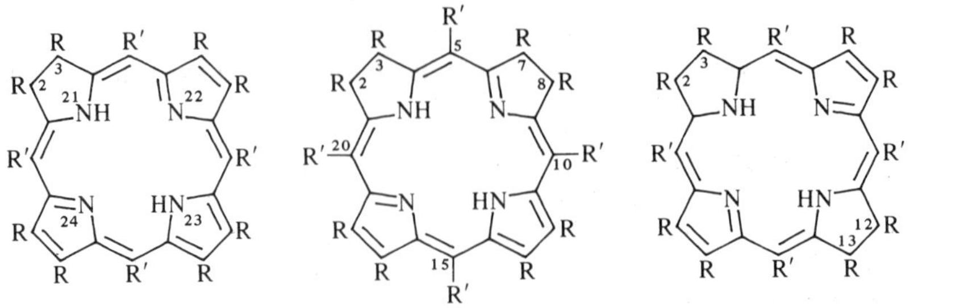

(25) Chlorin (H$_2$Chl)   (26) Isobacteriochlorin (H$_2$iBchl)   (27) Bacteriochlorin (H$_2$Bchl)

R = Et, R′ = H: H$_2$(OEC)   H$_2$(OEiBC)   H$_2$(OEBC)

R = H, R′ = Ph: H$_2$(TPC)   H$_2$(TPiBC)   H$_2$(TPBC)

Figure 9 Hydroporphyrins

Table 8 Structural Parameters of some Metallochlorins

	Fe^{II}(OEC)[a]	Ni^{II}(*meso*-tetramethylchlorin)[b]	Zn^{II}(TPC)py[c]
M—N(21)	2.002(4)	1.935(3)	2.130(1)
M—N(22)	1.963(3)	1.923(3)	2.060(1)
M—N(23)	1.986(4)	1.927(3)	2.072(1)
M—N(24)	1.963(3)	1.917(3)	2.063(1)
			Zn—N$_{ax}$ = 2.171(2), Δ(N$_4$) = 0.33
	Planar	S_4 ruffled	Square pyramidal
	Fe^{II}(TPP)[d]	Ni^{II}(*meso*-tetramethylporphyrin)[b]	Zn^{II}(TpyP)py[e]
M—N$_{por}$	1.972(4)	1.966(10)	2.073(6)
		1.952(10)	Zn—N$_{ax}$ = 2.143(4), Δ(N$_4$) = 0.33

[a] Ref. 162.
[b] Ref. 163.
[c] L. D. Spaulding, L. C. Andrews and G. J. B. Williams, *J. Am. Chem. Soc.*, 1977, **99**, 6918.
[d] Ref. 64.
[e] D. M. Collins and J. L. Hoard, *J. Am. Chem. Soc.*, 1970, **92**, 3761.

band intervenes at ~ 560 nm (ε ≈ 10^4). The two Soret bands are red-shifted by 10–20 nm and the Q band by 200–250 nm to the blue in the iBchl spectra relative to those of Bchl.

An IR band at ~ 1615 cm^{-1}, assigned to C=N stretching, is characteristic of chlorins and distinguishes them from porphyrins and bacteriochlorins.[2]

The ring current anisotropy is reduced in the hydroporphyrins compared with porphyrins, and the peripheral and inner proton signals are observed at δ = 9.9–9.2 (H-10 and H-15), 8.9–8.4 (H-5 and H-20), 9.0–8.0 (unsaturated β-H), − 1.0 to − 3.0 (NH) for (Chl), ~ 8.8 (meso-H), ~ 7.9 (unsaturated β-H), ~ − 2 (NH) for (Bch), and δ = 9.0–8.5 (H-15), 8.0–7.4 (H-10 and H-20), 7.5–6.8 (H-5), 7.5–6.8 (unsaturated β-H), ~ 2.0 (NH) for (iBchl).[2,14] The signal of a proton on the meso carbon attached to a pyrroline ring is located at ~ 1 p.p.m. higher frequency than the others partly because of the increased electron density on the position and partly because of the loss of pyrrolic ring current effects.

^{13}C and ^{15}N NMR signals are observed at ~ 93 (unsubstituted C-5, C-20), ~ 100 (unsubstituted C-10, C-15), ~ 102 (alkylated C-5, C-20), ~ 104 (alkylated C-10, C-15), ~ 224 (N-21), ~ 184 (N-23) and 164–166 (N-22, N-24) for chlorins.[2]

The mass spectral features of hydroporphyrins are dominated by the benzylic cleavage of β substituents on the saturated β carbons.[2,14]

21.1.1.2.3 *Syntheses*

Pyrrole condensation usually gives a mixture of porphyrin and chlorin which are separated by column chromatography.[11] The use of Cu^{II} as a template is reported to increase the (Chl/Por) ratio while Co^{II} reduces it.[164] The product ratio depends on the reactants and conditions in the Ni^{II} template condensation (Scheme 43).

More conveniently, hydroporphyrins are obtained by reduction of porphyrins.[2,14,165] For example, photochemical reduction of M(Por) (M = H$_2$, Zn, SnCl$_2$, GeCl$_2$) with reductants such as R$_3$N,

$$4 \text{ pyrrole} + 4RCHO \xrightarrow[\substack{\Delta \\ \text{solvent}}]{M^{II}} M(Por) + M(Chl)$$

Scheme 43

RSH, $SnCl_2$ or ascorbic acid gives M(Chl) and M(iBchl) (M $\neq$ Zn) (Scheme 44). The hydrogens are introduced with a *cis* configuration. Chemically, M(Por) (M = Mn, Fe) can be reduced to *trans*-Chl with Na in refluxing amyl alcohol, while excessive reduction gives a mixture of Chl, Bchl and iBchl from which iBchl may be isolated in $\sim 40\%$ yield (Scheme 45).[166] Reduction of Zn(TPP) to Zn(TPC) with Na anthracenide has been reported.[2]

$$M(Por) \xrightarrow[2e^-, \ 2H^+]{h\nu} M(cis\text{-Chl}) \xrightarrow[2e^-, \ 2H^+]{h\nu} M(iBch)$$

Scheme 44

$$M(Por) \xrightarrow[\substack{\text{amyl alcohol} \\ \Delta}]{Na} M(trans\text{-Chl}) \xrightarrow[\substack{\text{amyl alcohol} \\ \Delta}]{Na} M(iBch)$$

Scheme 45

Diimide reduction of H_2(TPP) proceeds in a stepwise fashion to give H_2(TPC) and H_2(TPBC) whereas, under similar conditions, Zn(TPP) is reduced to Zn(TPC) and then to Zn(TPiBC).[167] H_2(OEP) is also reduced to H_2(OEC) by diimide, but further reduction occurs at the meso positions (Scheme 46). H_2(TPC) is converted to H_2(TPBC) by catalytic hydrogenation (Raney Ni/H_2, 1 atm/ether).[2,14,165]

$$H_2(TPP) \xrightarrow[\text{py, 100 °C}]{TsNHNH_2 - K_2CO_3} H_2(TPC) \xrightarrow[\text{py, 100 °C}]{TsNHNH_2 - K_2CO_3} H_2(TPBC)$$

$$Zn(TPP) \xrightarrow[\text{py, reflux}]{TsNHNH_2 - K_2CO_3} Zn(TPC) \xrightarrow[\text{py, reflux}]{TsNHNH_2 - K_2CO_3} Zn(TPiBC)$$

$$H_2(OEP) \xrightarrow[\text{3-Mepy, reflux}]{TsNHNH_2 - K_2CO_3} H_2(OEC) \xrightarrow[\text{3-Mepy, reflux}]{TsNHNH_2 - K_2CO_3} \text{octaethylporphyrinogen}$$

Scheme 46

The ease of oxidation is Bchl, iBchl > Chl > Por,[166] and selective dehydrogenation is possible (Scheme 47).[2]

$$M(TPiBC) \xrightarrow{PhNO_2} M(TPC) \xrightarrow{PhNO_2} M(TPP)$$
$$M = SnCl_2, \ GeCl_2$$

$$M(OEiBC) \xrightarrow[\text{in py}]{h\nu/\text{air}} M(OEC) \xrightarrow[\text{in py}]{h\nu/\text{air}} M(OEP)$$
$$M = SnCl_2$$

$$H_2(Bchl) \xrightarrow[\substack{\text{or quinone } (h\nu) \\ \text{or } I_2 \\ \text{or } (p\text{-}BrC_6H_4)_3N^{\ddagger} SbCl_6^-}]{o\text{-chloranil}} H_2(Chl) \xrightarrow[\substack{\text{or quinone } (h\nu) \\ \text{or } I_2}]{o\text{-chloranil}} H_2(Por)$$

Scheme 47

21.1.1.2.4 Reactions of the macrocycle

The basicity of the macrocycle decreases along with hydrogenation and the order of pK_3 is H_2(Por) > H_2(Chl) > H_2(Bchl), H_2(iBch).[2,14] The pK_3 difference is often utilized to separate each macrocycle from a mixture.

The β hydrogens of the pyrroline rings are substituted with deuterium under basic conditions ($Bu^tOK–Bu^tOD/THF$, r.t) without isomerization to other hydroporphyrins. The acidic proton–

deuterium exchange occurs at the meso positions connected to pyrroline rings ($MeCO_2D/80\,°C$, 4 h for rhodochlorin dimethyl ester) faster than at those between two pyrrole rings.

The higher reactivity of chlorin C-5 and C-20 is generally observed in electrophilic substitution reactions such as nitration, halogenation, formylation and acetoxylation. However, N-methyl- and N,N',N''-trimethylchlorin are formed by the S_N2 attack of the inner nitrogens on methyl fluorosulfonate and methyl iodide (heating in a sealed tube) respectively.

Nucleophilic meso nitration of diprotonated chlorins with nitrous acid has been reported. Hydrogenations and dehydrogenations have been discussed in Section 21.1.1.2.3.

21.1.1.2.5   Metallation and demetallation

The metallation and demetallation procedures for porphyrins are applicable for hydroporphyrins, though M^{III} salts may cause some dehydrogenation. Perchlorate or acetate salts are usually used as the metal carrier for both metallations and transmetallations (Scheme 48).[17,167,168]

$$\text{hydroporphyrins} \xrightarrow[\text{MOMe/py/N}_2\,(M\,=\,Na)]{} \text{metallohydroporphyrins}$$

$$M(ClO_4)/py/N_2, \Delta\ (M = Mg,\ Ag,\ Co,\ Zn)$$

$$M(OAc)_2/DMSO,\ \text{dioxane, benzene or } MeOH-CHCl_3/N_2,\ \Delta\ (M = Mg,\ Cd,\ Zn,\ Cu,\ Pd)$$

$$M(OAc)_2/NaOAc-AcOH/N_2,\ \Delta\ (M = Fe,\ Ni,\ Pd,\ Sn)$$

$$M(OCOCF_3)_3/CH_2Cl_2-THF/N_2\ (M = Tl)$$

$$\text{Cd(Chl)} \xrightarrow[M\,=\,Mg,\,Zn,\,Mn,\,Fe,\,Co,\,Ni,\,Pb]{M(OAc)_2/MeOH/N_2,\,\Delta} M(\text{Chl})$$

Scheme 48

Mg^{II} insertion into labile hydroporphyrins is effected by the use of $IMg(2,6$-di-t-butyl-4-methylphenoxide)$-Li(2,2,6,6$-tetramethylpiperidine).[169]

Reflecting the lower basicity of $H_2(Chl)$, $M(Chl)$ are more readily demetallated than the corresponding $M(Por)$.[17b] The dissociation rate in AcOH–EtOH is described by equation (27). The ease of dissociation is in the order $M = Fe^{III}OAc$, $Pd < Cu < Ni < Co < Zn < Mg < Cd < Hg$ (Chl = pheophytin). Fe^{III}[17a] and Mn^{III}[170] are easily removed from the macrocycle under reductive conditions, and Cu^{II} by H_2S–TFA.[171]

$$-\frac{d[M(\text{Chl})]}{dt} = k_{\text{dissoc}}[M(\text{Chl})][H^+\text{-solvent}]^n - k_{\text{assoc}}[H_2(\text{Chl})][M(OAc)_2] \qquad (n \approx 2) \qquad (27)$$

21.1.1.2.6   Coordination chemistry of metallochlorins

The axial coordination chemistry of metallohydroporphyrins parallels that of their porphyrin analogues.

$Mg(TPC)(py)_2$ and $Zn(TPC)(py)$ are more stable to dissociation than the corresponding TPP derivatives probably because of the higher cationic charge on the central metals in the former.[172] Similarly, the association constants of Fe^{II} complexes are in the order $Fe(OEP)(THF)$ $(K = 0.7 \pm 0.3\,M^{-1}) < Fe(OEC)(THF)$ $(K = 15 \pm 3\,M^{-1}) < Fe(OEiBC)(THF)$ $(K = 54 \pm 5\,M^{-1})$.[162] The increase in stability with increasing hydrogenation may partly be attributed to the increase in flexibility of the macrocycle, because CO binding constants are not much different among them (~ 3.3, ~ 4.5 and $\sim 7.8 \times 10^4\,M^{-1}$ respectively) where the spin state change ($S = 1 \to 0$) does not require $Fe-N_{Chl}$ bond distance changes.

21.1.1.2.7   Photochemistry

The photochemistry of hydroporphyrins is similar to that of porphyrins (Scheme 49).[138] The photoexcited molecule crosses from a singlet to a triplet state and serves as an electron acceptor or donor (see Section 21.1.1.1.10). Since the oxidation potentials shift negatively with increasing hydrogenation,[166] the more hydrogenated macrocycles are expected to be the better electron donors. Therefore, the use of hydroporphyrins, *i.e.* chlorophylls, bacterio- and isobacterio-chlorophylls, may be beneficial to the biological photolytic energy fixing systems, coupled with the intense absorption bands in the low energy region of the hydroporphyrins.

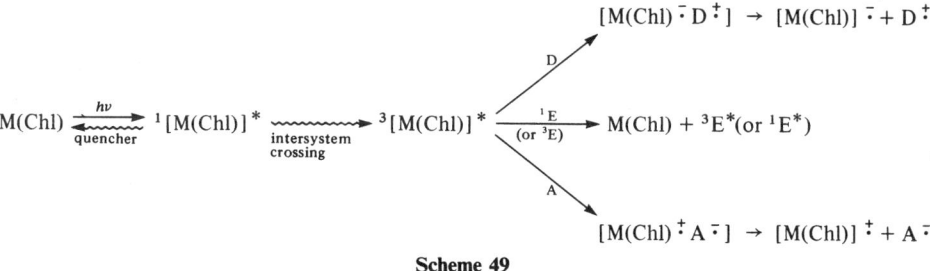

Scheme 49

21.1.1.2.8 Redox chemistry

As the peripheral double bonds are reduced, the first and second ring oxidation potentials shift negatively by ~ 0.3 and $\sim 0.2\,V$ respectively, while the ring reduction potentials are not much affected.[144,166] Thus the approximate ranges of redox potentials ($E_{1/2}$) are $+1.0 \pm 0.3$, $+0.5 \pm 0.5$, -1.0 ± 0.3 and $-1.5 \pm 0.3\,V$ for M(Chl), and $+0.8 \pm 0.3$, $+0.3 \pm 0.5$, -0.9 ± 0.3 and $-1.7 \pm 0.3\,V$ for M(Bchl) and M(iBchl) *vs.* SCE. The separations between the first and the second steps of both ring reduction and oxidation are $\sim 0.5\,V$.

21.1.1.2.9 Phlorins, porphodimethanes, porphyrinogens and corphins

There are several other nonaromatic hydroporphyrins known.[2,14] Phlorins (5,22-dihydroporphyrins) are stable only in acidic media ($E_{1/2} \approx -0.04\,V$ *vs.* SCE, pH 2/HCl) but easily oxidized to porphyrins under neutral to basic conditions ($E_{1/2} = -0.5$ to $-0.7\,V$ *vs.* SCE) unless the methylene bridge is blocked with substituents. Their spectra are characterized by two bands at $\sim 430\,nm$ ($\varepsilon \approx 10^5$) and $\sim 650\,nm$ (broad). The former is accompanied by a band at $\sim 370\,nm$ ($\varepsilon \approx 10^4$), and both red-shift by $\sim 20\,nm$ on metallation. The latter shifts to $\sim 750\,nm$ on protonation and to more than 800 nm on metallation. Phlorins are at the same oxidation level as chlorins, and are the intermediates of photochemical conversions of porphyrins to chlorins.

Porphodimethanes (5,15-dihydroporphyrin) are also at the same oxidation level as chlorins and form as intermediates during photoreductions of M(Por) to M(Chl). They show absorption spectra similar to the corresponding dipyrromethenes with λ_{max} at $\sim 420\,nm$ which shifts to $\sim 480\,nm$ on metallation.[2,14,173] If the meso methylene bridges carry alkyl substituents, the porphodimethanes are reasonably stable.[173] They are bent at the methylene bridges and take *syn*-diaxial conformations (Figure 10). A fifth ligand coordinates preferentially on the same side as the meso alkyl groups except for sterically encumbered *t*-butyl derivatives.[174] In general, coordination chemistry parallels that of the porphyrin analogues with some modifications imposed by the steric effects of the meso alkyl substituents, for example di(*t*-butyl) derivatives do not appear to form stable six-coordinate complexes, while both five- and six-coordinate complexes are obtained for the dimethyl analogues (Scheme 50).[173]

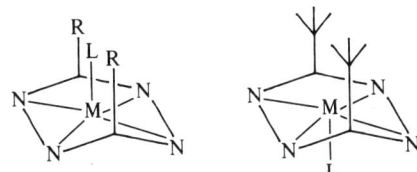

Figure 10 General structures of five-coordinate metallo-α,γ-*meso*-dialkylporphyrins M(PorR$_2$)

Metal oxidation is easier for M(PorR$_2$) than for M(Por), probably because of the smaller central hole in the former ($\Delta E_{1/2} = 0.10$–$0.17\,V$ for Fe$^{III/II}$ or Co$^{III/II}$).

Porphyrinogens (5,10,15,20,22,24-hexahydroporphyrins) are formed as intermediates in some porphyrin syntheses or by catalytic hydrogenation of porphyrins.[2] They are easily oxidized to porphyrins unless the methylene bridges are blocked by substituents.

Corphins (2,3,7,8,12,13-hexahydro-17*H*,21*H*-porphyrins) themselves are nonaromatic due to cross conjugation [$\lambda_{max} \sim 350$, $\sim 370\,nm$ ($\varepsilon \approx 10^4$) and $\sim 440\,nm$ ($\varepsilon \approx 10^4$ broad)], but become aromatic on protonation (probably at C-18).[2,175] Some metal complexes (Pd^{II}, Ni^{II}) are also known.

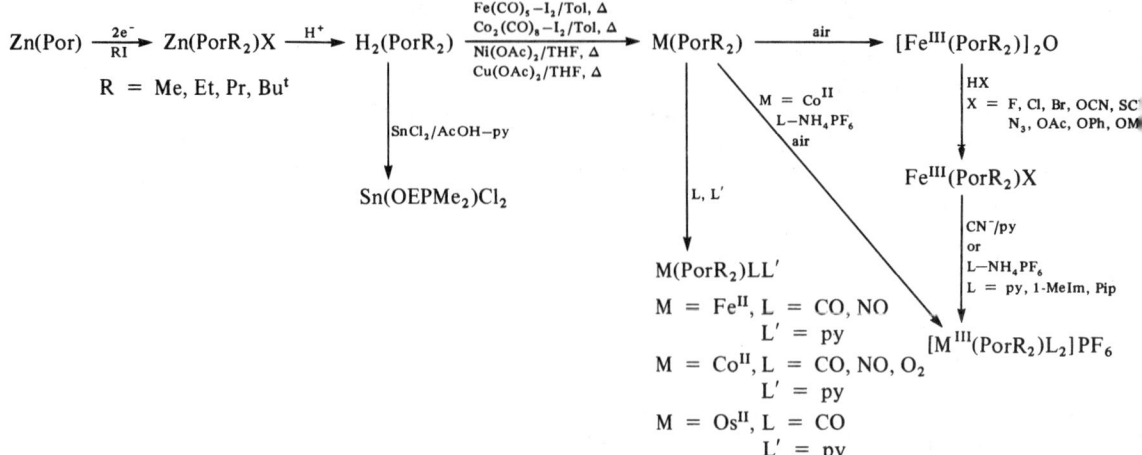

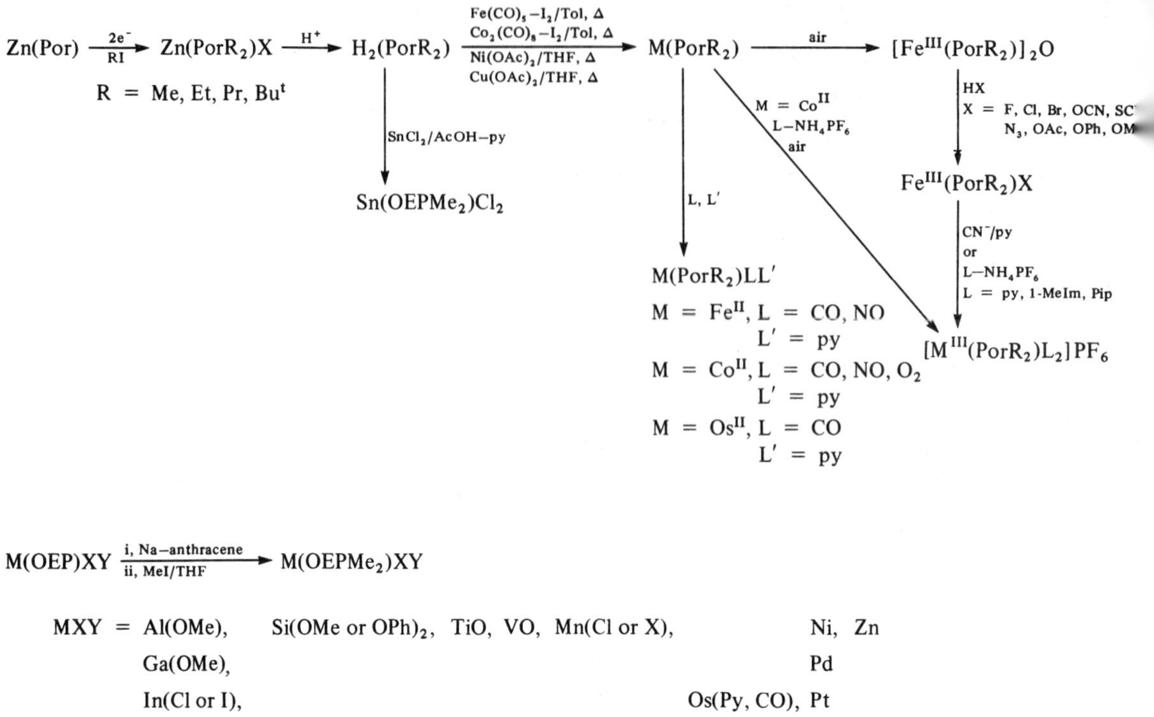

Scheme 50

21.1.2 AZAPORPHYRINS

Azaporphyrins with one to four meso aza bridges are divided into two classes: azaporphyrins with or without fused benzene rings on the β edges (Figure 11).

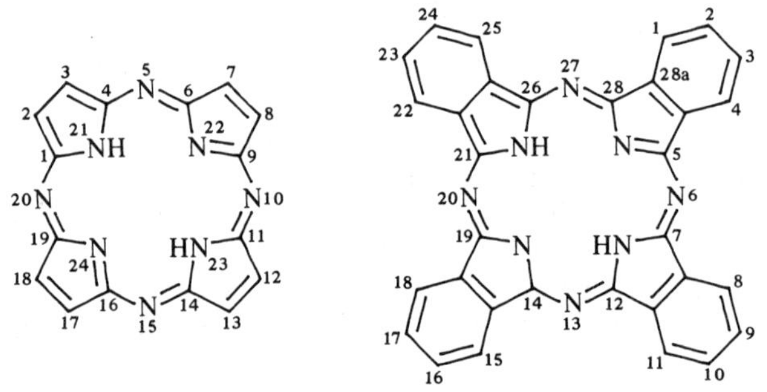

(28) Porphyrazine (H_2Tap)   (29) Phthalocyanine (H_2Pc)

Figure 11 Azaporphyrins

21.1.2.1 Azaporphyrins without Fused Benzene Rings

21.1.2.1.1 General properties

In general, azaporphyrins with more nitrogen bridges are thermally more stable but less resistant to chemical oxidations.[2]

The two inner protons are attached on the two diagonal nitrogens but rapidly migrate between the two possible positions. The thermodynamic parameters, $\Delta G^{\neq} = 41.4 \pm 0.8\,\mathrm{kJ\,mol^{-1}}$, $\Delta H^{\neq} = 42.3 \pm 3.3\,\mathrm{kJ\,mol^{-1}}$, $\Delta S^{\neq} = 0 \pm 1.3\,\mathrm{J\,K^{-1}\,mol^{-1}}$, suggest a more rapid motion of the pyrrolic protons ($1\,\mathrm{s^{-1}}$ at $-99\,^{\circ}\mathrm{C}$) in these rigid macrocycles relative to those in porphyrin ($\Delta G^{\neq} = 46$–$50\,\mathrm{kJ\,mol^{-1}}$).[3,176]

21.1.2.1.2 Spectral properties

Monoazaporphyrins, and their metal complexes, show absorption maxima at 380–410 nm (Soret, $\varepsilon \approx 10^5$), 500–550 nm (β, $\varepsilon \approx 10^4$) and 560–590 nm (α, $\varepsilon \approx 10^4$).[177] Both α and β bands of the metal-free bases split into two with an interval of ~ 35 nm and ~ 50 nm respectively. Similarly, λ_{max} of 320–350 nm (Soret, $\varepsilon \approx 10^4$), 530–550 nm (β, $\varepsilon \approx 10^4$) and 570–600 nm (α, $\varepsilon \approx 10^4$) are observed for the tetraaza analogues.[178] Detailed energy level calculations were reported on metal-free and metallo porphyrazines.[179]

^{1}H NMR signals of *meso*-H and β-H are observed at 8.5–10 p.p.m. due to the diamagnetic ring current of the macrocyclic π system.[177,180]

21.1.2.1.3 Synthesis

Monoazaporphyrins were first prepared using [2 + 2] dipyrromethane condensation methods by Fischer *et al.*[2] Later, Johnson *et al.* improved the yield to a moderate level (~ 58%) by cyclization of 1,19-dibromobiladiene-a,c with NaN_3.[177] Syntheses of metallomonoazaporphyrins *via* oxophlorin intermediates have also been reported (Scheme 51).[2,181]

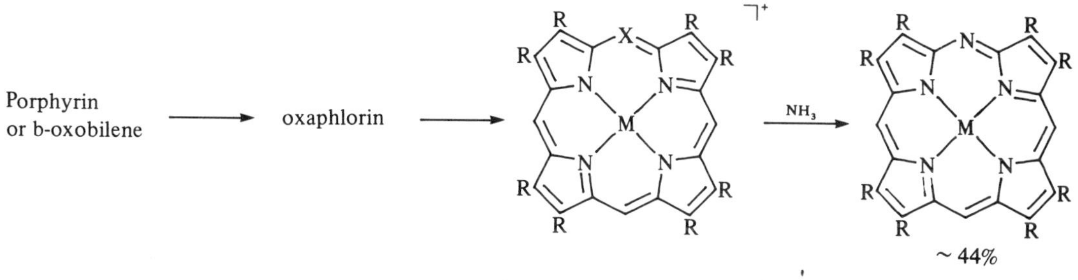

M = VO, MnCl, FeCl, Cu, Zn, Pd

Scheme 51

Fischer's [2 + 2] methods are also applicable to 5,15-diazaporphyrin synthesis and often give metal complexes in good yield in the presence of an appropriate metal template (Scheme 52).[2,182]

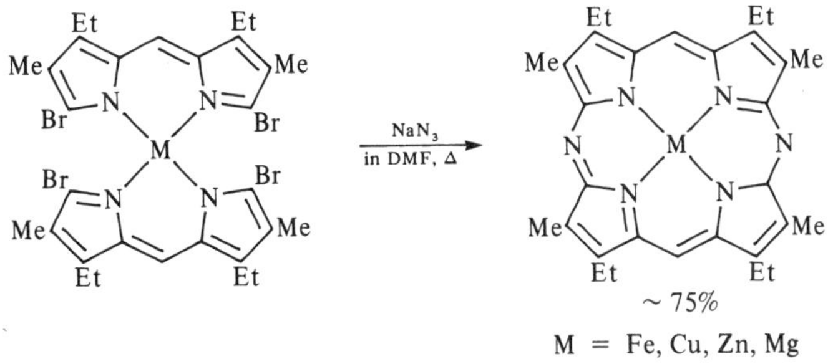

M = Fe, Cu, Zn, Mg

Scheme 52

Tetraazaporphyrins are easily prepared by cyclotetramerization of maleonitrile or its analogues, in the presence of metal salts (Scheme 53).[178,183]

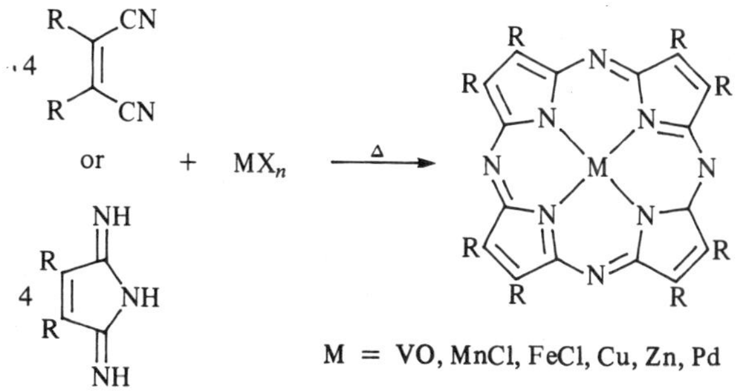

M = VO, MnCl, FeCl, Cu, Zn, Pd

Scheme 53

21.1.2.1.4 Reactions of the macrocycles

Azaporphyrins are monoprotonated in acetic acid, and addition of sulfuric acid (3.75×10^{-4} M) causes diprotonation. The third protonation (11.2 M H_2SO_4–AcOH) presumably occurs on the meso nitrogen atom.[184]

Metalloazaporphyrins often form salts with acids, probably by protonation of the meso nitrogen.[185]

Methyl iodide and dimethyl sulfate methylate $H_2(Map)$ in buffered chloroform to mono-, di- and tri-*N*-methyl-Map.[149]

Catalytic hydrogenation of porphyrazines or their Mg complexes with palladium black gives tetrahydrogenated compounds which are oxidized back to the starting materials by a stoichiometric amount of 2,3-dichloro-5,6-dicyanobenzoquinone (DDQ).[2] The royal blue products are degraded to *cis*-succinimide and maleimide, and considered to be the tetraaza analogues of either bacteriochlorins or isobacteriochlorins. Photochemical reduction of monoazaporphyrins is reported to occur at the meso positions.

21.1.2.1.5 Metallation and demetallation

Generally, the same procedure for porphyrin metallation seems to be applicable for azaporphyrins, but the examples are not abundant (Scheme 54).[183,186-188]

$$H_2(Map) + \{Rh(CO)_2Cl\}_2 \xrightarrow[\text{CHCl}_3, \, \Delta]{\text{K}_2\text{CO}_3} \{Rh(CO)_2\}_2(Map)$$

$$H_2(Tap) \xrightarrow[\substack{\text{NiCl}_2/o\text{-Cl}_2\text{C}_6\text{H}_4, \, \Delta \\ \text{Cu bronze}/o\text{-Cl}_2\text{C}_6\text{H}_4, \, \Delta}]{\text{M(OAc)}_2/\text{py or } o\text{-Cl}_2\text{C}_6\text{H}_4, \, \Delta} M(Tap)$$

$$M = VO, Co, Ni, Cu, Zn, Cd, Pd$$

Scheme 54

The ease of metallation of octaphenyltetraazaporphyrin (H_2Ph_8Tap) with metal acetates in pyridine is reported to be Cu > Zn > Cd > Ni.[189] The rate is pseudo-zero and first order in $M(OAc)_2$ and H_2Ph_8Tap respectively in the concentration range 10^{-5} to 10^{-3} M^{-1} for the former and 10^{-5} to 10^{-6} M^{-1} for the latter. An initial rapid N_{aza} coordination followed by rate-determining metal insertion into Ph_8Tap is suggested.

Mg and Cd are easily dissociated from their porphyrazine complexes in acetic acid.[186,190] Demetallation of Mg- and Cd-Ph_8Tap in pyridine–acetic acid is first order in the metal complexes and second order in the solvated proton. Similar results are reported on the Zn- and Ni-Ph_8Tap–DMSO–$H_2SO_4 \cdot H_2O$ system.

21.1.2.1.6 Axial coordination chemistry

The axial coordination chemistry of metalloazaporphyrins follows the same trend as that of porphyrin complexes, though these areas are little investigated.

21.1.2.1.7 Reactions at the metal

Like its porphyrin analogue, $[Rh(CO)_2]_2(Tap)$ reacts with alkyl halides, aldehydes, alkoxy formates and ketones to give R—$Rh^{III}(Tap)$ or RCO—$Rh^{III}(Tap)$ (Scheme 55).[134]

21.1.2.1.8 Photochemistry

Photochemical reduction of Mg- and Zn-(Ph_8Tap) with ascorbic acid is reported.[191] The anion radical product is kinetically stable at high pH, because the reduction is faster and the back reaction slower at higher pH (equation 28).

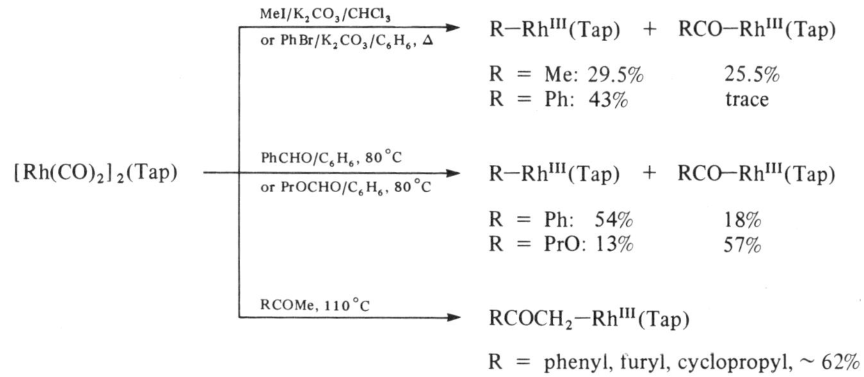

$$[Rh(CO)_2]_2(Tap) \quad \xrightarrow[\text{or PhBr/K}_2CO_3/C_6H_6,\ \Delta]{\text{MeI/K}_2CO_3/CHCl_3} \quad R-Rh^{III}(Tap) \quad + \quad RCO-Rh^{III}(Tap)$$

R = Me: 29.5%   25.5%
R = Ph: 43%   trace

$$\xrightarrow[\text{or PrOCHO/C}_6H_6,\ 80\,^\circ C]{\text{PhCHO/C}_6H_6,\ 80\,^\circ C} \quad R-Rh^{III}(Tap) \quad + \quad RCO-Rh^{III}(Tap)$$

R = Ph: 54%   18%
R = PrO: 13%   57%

$$\xrightarrow{\text{RCOMe, } 110\,^\circ C} \quad RCOCH_2-Rh^{III}(Tap)$$

R = phenyl, furyl, cyclopropyl, ∼ 62%

Scheme 55

$$M(Ph_8Tap) \quad \underset{ii}{\overset{i}{\rightleftharpoons}} \quad M(Ph_8Tap)^{\overline{\cdot}} \tag{28}$$

i, *hv*, ascorbic acid in py/Triton X-100; *k*
ii, dark; − *k*

$$k\ (\text{pH } 8.4) \ > \ k\ (\text{pH } 9.4)$$
$$-k\ (\text{pH } 8.4) \ < \ -k\ (\text{pH } 9.4)$$

21.1.2.1.9 Redox chemistry

Mg(Ph$_8$Tap) gives four successive ring reduction waves at $E_{1/2} \approx -0.68$, -1.11, -1.81 and -2.18 V *vs.* a saturated aqueous calomel electrode in DMF–Pr$_4^n$NClO$_4$ at 20 °C.[192] These half-wave reduction potentials are 0.7–0.9 V more positive than those of etioporphyrin I and its Zn^{II} complex. Colour changes during the reduction of Mg(Ph$_8$Tap), emerald → grey → brown, and for Ni(Me$_8$Tap), purple → indigo → brown, with Na metal in 2-methyltetrahydrofuran are attributed to the electron insertion into the ring π* orbitals. Ring oxidation potentials of H$_2$(Tap) and Mg(Tap) are reported to be $E_{1/2}^{ox} = 0.51$ and 0.73 V respectively (Hg electrode in 0.1 M Me$_4$NClO$_4$/DMF).[193]

5-Azamesoporphyrin dimethyl ester and its metal complexes are oxidized at 1.03 (metal-free), 0.63 (Zn), 0.79 (Cu), 0.93 (Pd) and 1.12 V (VO) *vs.* SCE in CH$_2$Cl$_2$–Bu$_4^n$NClO$_4$.[177] The Zn(MAP) cation radical may be generated chemically with DDQ, I$_2$ or Br$_2$ in CH$_2$Cl$_2$.

21.1.2.10 Octahydro-5,15-diazaporphyrins

Metal complexes of hexadecamethyloctahydro-5,15-diazaporphyrin (OHDap) are reported.[194] The dicationic complexes are good electrophiles and react with alcohols, amines and Grignard reagents on the C-1 and C-11 positions (Scheme 56). The second addition is highly selective and the *syn* isomer is predominantly formed. Asymmetric reactions using 1-alkyl derivatives are also reported. Thus, meso-2,3-butanediol is selectively *O*-monomethylated *via* the optically active Ni complex (X = 3-hdyroxy-2-butoxy, Y = Me) to (+)-(2R,3S)-3-methyoxy-2-butanol, and 2-ethylallyl alcohol and 4-hydroxy-2-butanone are reduced to (+)-(2R)-2-methyl-1-butanol with diimine and to (+)-(3S)-1,3-butanediol by hydride reduction respectively *via* the chiral complex (X = RO, Y = Me). The Co^I complex, obtained by Zn–NH$_4$Cl or electrochemical (− 1.2 V) reduction, undergoes nucleophilic reactions with alkyl halides. The alkyl Co^{III}(1-hydroxy-OHDap) loses the alkyl group oxidatively, thermally or reductively.

21.1.2.2 Azaporphyrins with Fused Benzene Rings

21.1.2.2.1 General properties

In a series of tetrabenzoazaporphyrins there are tendencies coupled to the increase in the number of meso nitrogens, such as the decrease in stability, the increase in sublimability, the decrease in basicity, the increase in oxidizability, the red shift of main absorption bands and the increase in metal complex stability.[195]

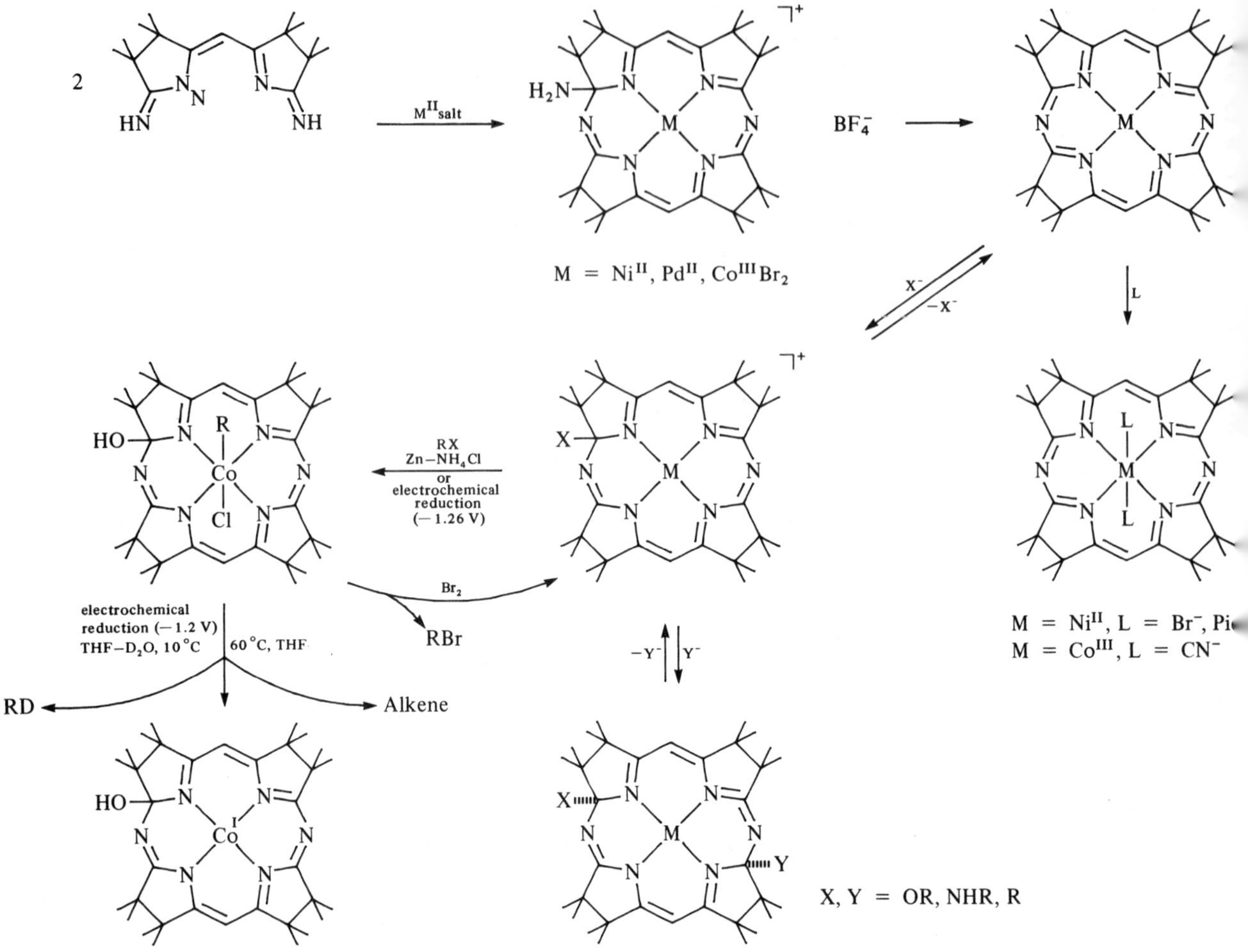

Scheme 56

Phthalocyanines have been most intensively studied among the azaporphyrins (Figure 12).[196–199]

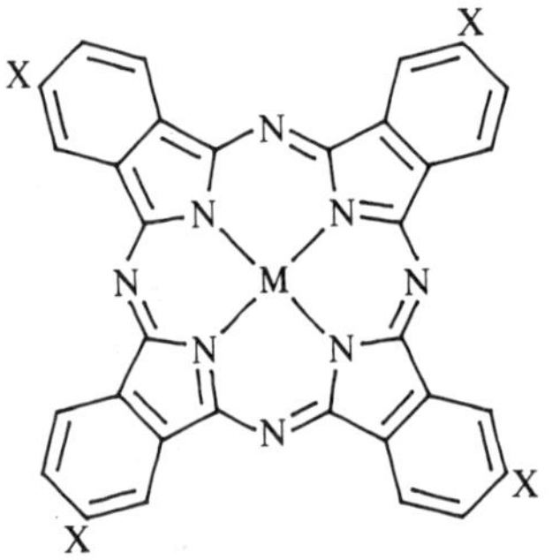

X = H: MPc
X = SO$_3^-$: M(PCTS)$^{4-}$
X = SO$_2$NHC$_{18}$H$_{37}$: M(PCTO)

Figure 12 Abbreviations for substituted phthalocyanines

H$_2$(Pc) and M(Pc) take one of the three forms, α, β and γ, in the solid state depending on their method of preparation.[197] These forms are crystallographically different and spectroscopically distinguishable. In solution, they associate to dimers through Pc–Pc (π) interactions with $K_{assoc} \approx 10^2$–10^6 M^{-1}.[198] A metal–metal distance of ~ 4.5 Å has been estimated by ESR analysis.

21.1.2.2.2 Spectroscopic properties

H$_2$(Pc) and M(Pc) show characteristic absorption spectra consisting of four intense π–π^* bands in the vis-UV region:[197,198]

Q (710–650 nm), a_{1u} $(\pi) \to e_g$ (π^*), an intense band ($\varepsilon \approx 10^5$), which splits into two in metal-free H_2Pc spectra, and is usually accompanied by one or two vibrational bands ($\varepsilon \approx 10^4$) on the blue side;

B (360–330 nm), a_{2u} $(\pi) \to e_g$ (π^*), the Soret band ($\varepsilon \approx 10^4$), broadened presumably by overlapping of Np $(\sigma) \leftarrow e_g$ (π^*) Pc transitions;

N ($\sim$ 270 nm), a'_{2u} $(\pi) \to e_g$ (π^*) ($\varepsilon \approx 10^5$);

L ($\sim$ 240 nm), a_{1u} $(\pi) \to e'_g$ (π^*) ($\varepsilon \approx 10^5$)

The assignment of these main bands to $\pi \to \pi^*$ transitions was confirmed by MO calculations and MCD studies.[179,198] For the first row transition metals with vacancies in d_{xz} and d_{yz} orbitals, a ligand to metal charge transfer band is observed at 380–500 nm.[200] A weaker metal to ligand charge transfer band is also expected to occur in this region.

The near IR bands in the 850–2000 nm region may be attributed to charge transfer, a_{1u} $(\pi) \to e_g$ (d_{xz}, d_{yz}),[200] or n (aza) $\to \pi^*$ (Pc) transitions[198] except for a band near 1660 nm (v(C—H) overtone) and another around 1000 nm (irradiation-damaged Pc). The metal dependency of these bands suggests the existence of intramolecular metal–N (aza) interactions.

Fluorescence from Mg, Zn, Cd and Pd(Pc) and phosphorescence from Mg, Zn, Cu, Cd, Pd and VO(Pc) have been observed.[198]

MPc has D_{4h} symmetry and shows IR absorption bands due to E_u mode vibrations around 645, 575 and 515 cm^{-1}.[198] These bands shift to the low frequency side and split into two (B_{2u} and B_{3u} mode) in the spectra of D_{2h} symmetric H_2(Pc).

In the far IR region three markedly metal independent bands are observed, *i.e.* out-of-plane vibration $\sim$ 340 cm^{-1}, E_u mode vibration $\sim$ 300 cm^{-1} and metal–ligand vibration 150–200 cm^{-1}.[198] The first two bands also exist in the H_2(Pc) spectra with the second split into two.

The Pc peripheral protons give NMR signals at δ = 9.7–9.0 p.p.m. (α-H) and 8.1–7.7 p.p.m. (β-H).[197,201] The diamagnetic shifts of the Pc π system are smaller than those of the porphyrin π system due to perturbations of ring current by the *meso* nitrogens and benzo π systems.[202]

The location of the two inner protons is of interest. The X-ray photoelectron spectrum of H_2Pc indicates the localization of these protons on two diagonal nitrogens.[198,199] This method (ESCA) also revealed the higher binding energy of metal $2P_{3/2}$, hence higher positive charge on the metal in M(Pc) than in metalloporphyrin because of the enhanced M–Pc π back-donation in the former.

In the first row M(Pc), the order of d orbital energy levels is estimated to be b_{1g} $(d_{x^2-y^2}) > a_{1g}$ $(d_{z^2}) > e_g$ $(d_{xz}d_{yz}) > b_{2g}$ (d_{xy}) by ESR spectroscopy and paramagnetic anisotropy measurements.[203]

Reflecting their aromaticity, M(Pc) are, in general, thermally stable and characterized by strong molecular ion mass peaks.[198] Exceptions are the extensive fragmentation of $Mn^{II}Pc$ and the lack of $[M(Pc)]^+$ signals in the spectra of Sc^{III} and $Yb^{III}(Pc)Cl$. $[Ta^V(Pc)Br_3]$ loses the metal and decomposes significantly under mass spectral conditions, while $[Nb^V(Pc)Br_3]$ gives $[Nb(Pc)Br_2]^+$ and $[Nb(Pc)Br]^+$. The ionization potentials of M(Pc) are reported to be in the range of 7.22–7.46 eV, in the gas phase, by electron impact mass spectroscopy. The electron is removed probably from Pc a_{1u} (π) or e_u $(N\sigma)$ orbitals.

21.1.2.2.3 Synthesis

Helberger *et al.* synthesized Cu^{II} complexes of tetrabenzo-mono-, -di- and -tri-azaporphyrins from CuCN, *o*-bromo- or cyano-benzophenone and various amounts of phthalonitrile in boiling quinoline.[204] Dent also obtained Cu^{II} tetrabenzotriazaporphyrin by heating a mixture of CuCl, phthalonitrile and 1-imino-3-methyleneisoindoline or its analogues.[205] But tetrabenzotriazaporphyrin was more easily prepared from phthalonitrile and methylmagnesium iodide in a moderate yield by Linstead *et al.*[195] If a large excess of Grignard reagent is used, Mg^{II} tetrabenzomonoazaporphyrin is formed (Scheme 57).

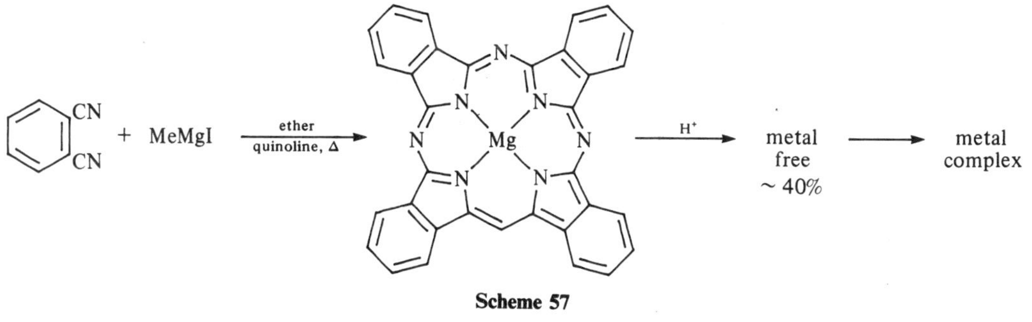

Scheme 57

Cyclotetramerization of phthalonitrile, or its analogues, is the traditional and facile method of phthalocyanine preparation.[196–199] In the presence of an appropriate metal template, a variety of metallophthalocyanines have been synthesized (Scheme 58).

phthalonitrile equivalent $\xrightarrow[\text{with or without solvent, } \Delta]{\text{catalyst [Pt, (NH}_4)_2\text{MoO}_4 \text{ or ZrCl}_4]}$ H$_2$Pc or MPc

template: metal, hydride, oxide, alkoxide, carboxylate, halide, acetylacetonate

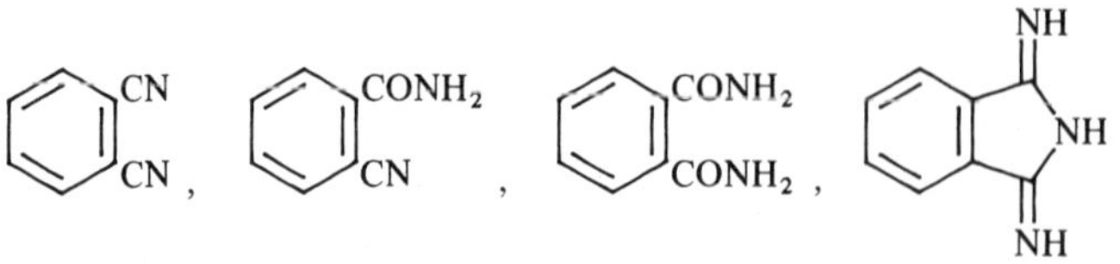

Phthalic acid equivalent $+$ (H$_2$N)$_2$CO $\xrightarrow[\Delta]{\text{catalyst [(NH}_4)_2\text{MoO}_4 \text{ or ZrCl}_4]}$ H$_2$Pc or MPc

template: metal, halide, acetate

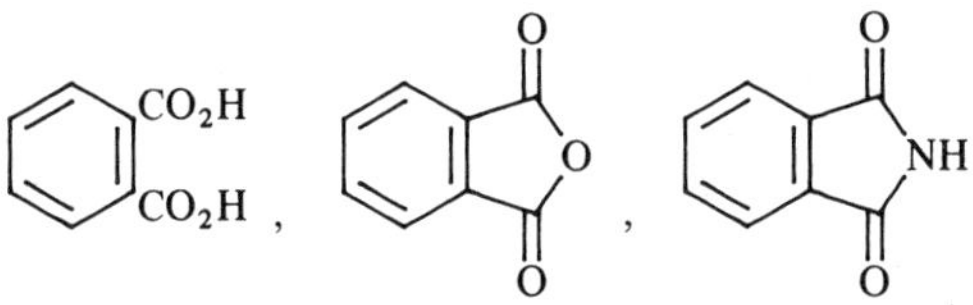

Scheme 58

These condensation procedures are generally applicable to phthalonitrile or its analogues with various substituents.[199,201,206] The substituted phthalocyanines are useful because of their higher solubility in organic or aqueous solvents than phthalocyanine itself (Figure 13).[201] Phthalocyanine analogues with various heterocyclic fused rings are also known (Figure 14).[207] It is possible to substitute two of the isoindoline units with other aromatics by condensing their diamino derivative with phthalonitrile (Scheme 59).[208]

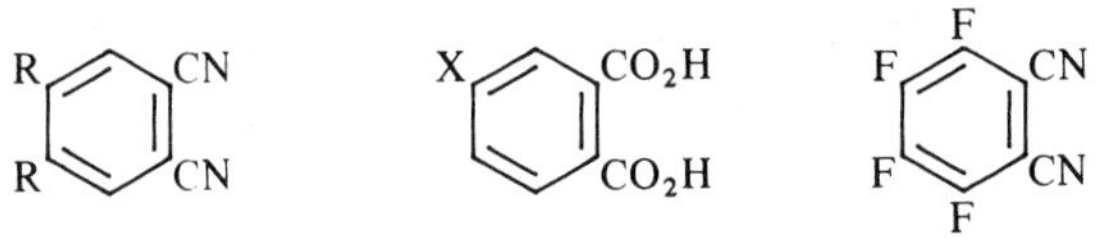

X = SO$_3$Na, NO$_2$, Cl, NH$_2$, OH

Figure 13 Substituted phtalocyanine precursors

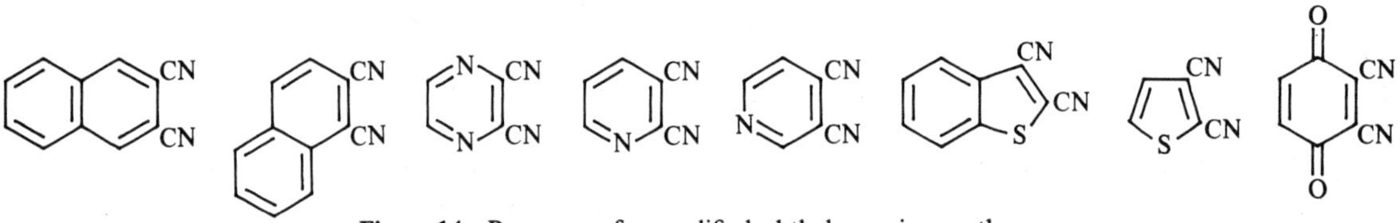

Figure 14 Precursors for modified phthalocyanine syntheses

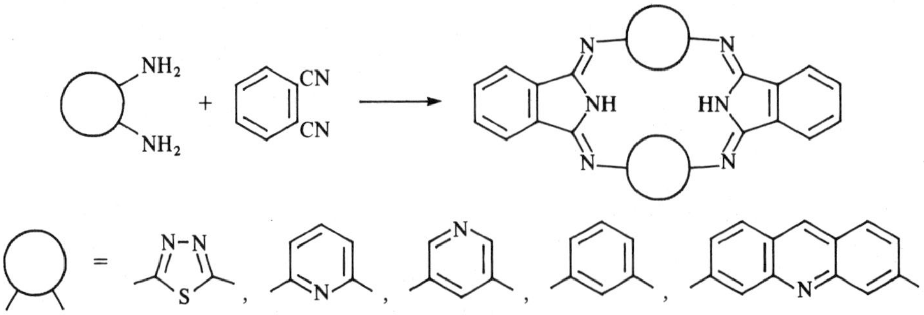

Scheme 59

21.1.2.2.4   Reactions of the macrocycle

Phthalonitrile condensation with some metal chlorides (Zn, Ir, Pd) gives M(Pc) with a chlorine substituent on the benzo ring.[197] The chlorine is removed with $AgNO_3$ in fuming HNO_3 at high temperature and pressure.[209]

M(Pc) (M = Fe, Co, Ni) can be nitrated with NO_2BF_4 in sulfurane at 20 °C under nitrogen, and the nitro substituent is reduced to an amino group with H_2S (M = Ni).[210] Sulfonation of $M^{III}(Pc)(HPc)$ with SO_3 vapour has been reported.[211]

21.1.2.2.5   Metallation and demetallation

Though direct synthesis of M(Pc) is possible for most metals, transmetallation of $Li_2(Pc)$ with a metal salt is often preferred for the M(Pc) preparation because $Li_2(Pc)$ is readily purified by recrystallization.[197] Usually, metal halides MCl_2, MCl_3 and MBr_5 are used as the metallating reagent (Scheme 60). Unstable Ag(Pc) is prepared from $Li_2(Pc)$ and $AgNO_3$ or Pb(Pc) and Ag_2SO_4. Reaction of $M(DBM)_3$ (M = Group IIIB and lanthanides, DBM = dibenzoylmethanate) and $Li_2(Pc)$ gives *cis*-$M^{III}(Pc)(DBM)(DBMH)$, while dinuclear $[M^{III}(DPM)_2]_2(Pc)$ is obtained from $M(DPM)_3$ (M = Y and lanthanide, DPM = dipivalomethanate) and $Li_2(Pc)$.[212] $Sn(Pc)Cl_2$ and $Na_2(Pc)$ form a sandwich complex, $Sn(Pc)_2$.[199]

$$Li_2(Pc) + MX_2 \rightarrow M^{II}(Pc)$$

$$Li_2(Pc) + MX_3 \rightarrow M^{III}(Pc)X$$

$$Li_2(Pc) + MX_5 \rightarrow M^V(Pc)X_3$$

Scheme 60

$H_2(Pc)$ is also metallated by metal halides such as $AlCl_3$ and $GeCl_4$ in boiling quinoline.[197] Metallation of $H_2(Pc)$ by $SnCl_2$ and $SnCl_4$ gives $Sn(Pc)Cl_2$ while $Sn^{II}(Pc)$ is obtained from $H_2(Pc)$ and metallic Sn. $(MeHg)_2(Pc)$ is formed when $H_2(Pc)$ is reacted with $MeHgN(SiMe_3)_2$.[198]

Among the metallophthalocyanines, $Na_2(Pc)$ and $K_2(Pc)$ are the most acid labile and dissociated in alcohols.[197] Demetallation of Group IA and IIA metals is rapid in acidic solution as expected. Group IIB, except for Zn^{II}, divalent Group IVB, trivalent Group VB metals and Fe^{II} and Mn^{II} are also dissociated from Pc in conc. H_2SO_4. Ag(Pc) is acid labile as well as thermally unstable.

In general, M(Pc) with metal ions of large charge–size ratio are stable, *i.e.* resistant to acid demetallation, and sublime without significant loss of the metal ion.

21.1.2.2.6   Coordination chemistry of metallophthalocyanines

(i)   Group IA (Li^I, Na^I, K^I)

Group IA metallophthalocyanines are prepared directly from phthalonitrile and a metal alkoxide, and easily demetallated in weak acids [$Li_2(Pc)$] or even in alcohols [$Na_2(Pc)$, $K_2(Pc)$].[197,199]

(ii)   Group IIA (Be^{II}, Mg^{II}, Ca^{II}, Ba^{II})

Be(Pc) and Mg(Pc) are obtained by the condensation reaction of phthalonitrile in the presence of metallic Be and Mg respectively.[197,198] They form very stable complexes with water. $Mg(Pc)(OH_2)$ is square pyramidal and the magnesium atom is displaced out of the N_4 plane by 0.5 Å with an Mg—N_{por} distance of ~2.04 Å. They lose the water on sublimation. Amines, alcohols, ketones, thiols and sulfides also coordinate to Mg(Pc). Ca(Pc) and Ba(Pc) are obtained from phthalonitrile and the metal oxide.[197,199]

(iii)   Group IIIB (Al^{III}, Ga^{III}, In^{III}, $Tl^{I,III}$)

Group IIIB metallophthalocyanines are synthesized by the reaction of *o*-cyanobenzamide and metal halide, and are obtained as $M^{III}(Pc)(OH)(OH_2)$ by reprecipitation from conc. H_2SO_4 and aqueous NH_4OH.[197,199] $Al(Pc)(OH)(OH_2)$ is deprotonated in aqueous alkali. It loses the water on heating at $110\,°C$ *in vacuo*, and is dehydrated to a μ-oxo dimer at $400\,°C$. $[Al(Pc)]_2O$ is stable to aqueous alkali and to acid, but hydrolyzed in conc. H_2SO_4 or hot $6\,N\,HCl$. Phenols and silanols react with $Al(Pc)(OH)(OH_2)$ to form phenoxo and siloxo complexes. $Al(Pc)(OPh)$ resists hydrolysis in hot aqueous NH_4OH, but reacts in hot $6\,N\,H_2SO_4$. $Al(Pc)(OSiPh_3)$ is more stable and requires conc. H_2SO_4 for hydrolysis. It is readily sublimed. Pyridinium bromide reacts with $Al(Pc)(O-H)(OH_2)$ to form $[Al(Pc)Br(py)](py)_2$ from which the pyridine molecules are easily lost.

(iv) Group IVB (Si^{IV}, $Ge^{II,IV}$, $Sn^{II,IV}$, Pb^{II})

Group IVB metallophthalocyanines can be prepared by phthalonitrile condensation in the presence of metal chloride (Si, Ge, Sn) or oxide (Pb).[197,198]

$Si(Pc)Cl_2$ is converted to $Si(Pc)(OH)_2$ in a refluxing mixture of pyridine and aqueous NH_4OH, but $Si(Pc)F_2$ resists hydrolysis. $Si(Pc)Cl_2$ and $Si(Pc)(OH)_2$ react with hydroxide at high temperature to form various oxides of the type $Si(Pc)(OR)_2$. For example, polymers of $HO[Si(Pc)O]_{10-100}H$ and $HO[Si(Pc)ORO]_nH$ are obtained by self condensation at $400\,°C$ and copolymerization of $Si(Pc)(OH)_2$ with diols respectively. The symmetrical compounds $Al(Pc)O[Si(Pc)O]_nAl(Pc)$, prepared from $Si(Pc)(OH)_2$ and $Al(Pc)(OH)(OH_2)$), may be useful since they are converted to $HO[Si(Pc)O]_n$ by HBr–pyridine. Alkoxides and mercaptides are also made from $Si(Pc)Cl_2$ or $Si(Pc)(OH)_2$ and alcohols or thiols. The reaction of $Si(Pc)Cl_2$ and $Ag(XCN)$ ($X = O, S, Se$) gives the N-bound pseudohalide, $Si(Pc)(NCX)$.

Interesting alkylsilyl complexes, $RSi(Pc)Cl$, are directly prepared from $RSiCl_3$ and phthalonitrile. The Si—C bond is chemically very stable and unaffected by conc. H_2SO_4 or HF but it is photolabile.

$Si(Pc)(OSiMe_3)_2$ was analyzed by NMR and the calculated SiOSi angle of $162°$ in solution is in close agreement with the solid state angle $157°$ determined by X-ray analysis. The short Si—O distance ($1.68\,Å$) reflects the strength of this bond.

The chemistry of Ge^{IV} phthalocyanines parallels that of Si analogues, though $Ge^{IV}(Pc)X_2$ are, in general, more stable to ligand substitution than the Si analogues.[197,198,202,213] Unlike $Si^{IV}(Pc)Cl_2$, $Ge^{IV}(Pc)Cl_2$ is reduced to $Ge^{II}Pc$ by BH_4^-. $Ge^{II}(Pc)$ is light sensitive, but unusually stable to oxidation. Strong oxidants affect the macrocycle prior to metal oxidation. Alkynyl complexes, $M(Pc)(C{\equiv}CR)_2$ ($M = Ge^{IV}, Sn^{IV}$), are formed from $M(Pc)Cl_2$ and ethynylmagnesium chlorides.[214]

Hydrolysis of $Sn(Pc)X_2$ in hot aqueous NH_4OH is halide dependent and the reaction rate decreases ($I > Cl \gg F$) as the covalency of the Sn—X bond increases.[197,198] $Sn(Pc)(OH)_2$ is deprotonated in aqueous alkali and converted to the halide, $Sn(Pc)X_2$, in aqueous HX ($X = F, Cl$). $Sn(Pc)(OR)_2$ complexes ($R = $ alkyl, silyl, germanyl) are also known. $Sn(Pc)_2$, obtained by the reaction of $Sn(Pc)Cl_2$ and $Na_2(Pc)$, is the first example of the sandwich-type complex[199] (see Group IIIA).

Reduction of $Sn(Pc)Cl_2$ with $SnCl_2$ or H_2 in quinoline gives $Sn^{II}(Pc)$, which is inert to weak acid (AcOH) and alkali ($3\,N\,NaOH$) but is oxidized to $Sn^{IV}(Pc)X_2$ by Br_2 or I_2.[197,198] Cl_2 halogenates the Pc ring as well.

Mössbauer spectra of a series of $Sn(Pc)X_2$ have been studied. The linear relationship between the isomer shift and the axial ligand electronegativity is interpreted in terms of an electron density increase on the Sn atom and a decrease in π donation from Pc to Sn with an increase in axial σ donation.[198] This is supported by X-ray photoelectron spectroscopy.

Both the Sn and Pb atoms in $M^{II}(Pc)$ are displaced out of the N_4 plane, by 1.11 and $0.92\,Å$ respectively.[198,215] $Pb(Pc)$ is unstable.[197]

(v) Group VB (P^{III}, As^{III}, Sb^{III}, Bi^{III})

Group VB $M^{III}(Pc)X$ are prepared by metallation of $H_2(Pc)$ ($M = P$), transmetallation of $Li_2(Pc)$ ($M = As, Sb$) or by template reactions ($M = Sb, Bi$) with metal halides (Cl or Br).[31,197,216,217]

(vi) Group IIIA (Sc^{III}, Y^{III}, La^{III}, lanthanide, actinide)

The phthalonitrile condensation with Group IIIA metal salts often gives a mixture of $M^{III}(Pc)X$ and $M^{III}(Pc)(PcH)$, or $M^{IV}(Pc)_2$[199,218] (see below). If UO_2X_2 ($X = Cl, OAc$) is used as a template, UO_2(superphthalocyanine) is formed (see Section 21.1.4).[201] Transmetallation of $Li_2(Pc)$

with $M^{III}(DBM)_3$ (M = Group IIIA and lanthanide, DBM = $(PhCO)_2CH^-$) gives $M^{III}(Pc)(DBM)(DBMH)$ in which DBM and DBMH coordinate to the metal on the same side of the Pc ring.[212] A series of interesting dinuclear complexes, $[M^{III}(DPM)_2]_2(Pc)$, are obtained from Li_2Pc and $M^{III}(DPM)_3$ (M = Y and the lanthanides Sm–Yb). The Nd^{III} complex is unstable and Lu^{III} forms the 1:1 complex, $Lu^{III}(Pc)(DPM)$. The X-ray structure of $[Sm^{III}(DPM)_2]_2(Pc)$ shows the metals in a twisted square antiprism structure, equivalently coordinated by the four Pc nitrogens and the four DMP oxygens, and the molecule in D_{2h} symmetry. The dinuclear complexes are stable in nonpolar solvents (CH_2Cl_2, C_6H_6, λ_{max} 700 nm) but dissociate to mononuclear complexes in polar solvents (DMF, DMSO, λ_{max} 670 nm).

A characteristic of the Group IIIA, lanthanide and actinide metals is to form stable sandwich complexes, $M^{III}(Pc)(PcH)$ (Group IIIA, lanthanide) or $M^{IV}(Pc)_2$ (actinide).[199,218] According to Moskalev and Kirin,[219] the metals in the trivalent or tetravalent state with covalent radius > 1.35 Å, which prohibits metal-in-plane structures, are possible candidates for this type of complex. Their list includes In, Zr, Sn, Hf, Tl, Pb, Bi and Po in addition to Group IIIA, lanthanide and actinide metals. However, $Sn^{IV}(Pc)_2$ is the only example so far characterized other than this group.[199,218] The trivalent metal sandwich complexes are deprotonated in aqueous NH_4OH. The $[M^{III}(Pc)_2]^-$ and $[M^{IV}(Pc)_2]$ complexes show blue to violet colour in solution and are characterized by two Q bands at ~ 630 and ~ 680 nm with an intensity ratio of 2:1. The 630 nm band is linearly correlated to the atomic number or the ionic radius of the metal, and blue-shifted from La (57; $\lambda_{max} = 639$ nm) to Lu (71; 618 nm). These complexes appear to be stable in the solid state and sublime without decomposition, though $Sn(Pc)_2$ was reported to lose one molecule of H_2Pc at 560 °C *in vacuo*. In strong acids they are likely to be protonated but not dissociated to monophthalocyanine complexes. The X-ray structures of $Sn(Pc)_2$, $U(Pc)_2$ and $Nd(Pc)(PcH)$ revealed a staggered conformation of the two Pc groups (42°, 37°, 45° respectively) with interplane distances of 2.70, 2.81 and 2.96 Å respectively. The Pc rings are convex in the first two and planar in the third.

Interesting electrochroism is observed; on electrochemical reduction of $M^{III}(Pc)(PcH)$, a full spectral range of colour is observed between $[M^{III}(Pc)(PcH)]^-$ (violet) and $[M^{III}(Pc)(PcH)]^{2+}$ (yellow-red).

(vii) Group IVA ($Ti^{III,IV}$, Zr^{IV}, Hf^{IV})

The phthalonitrile condensation with $TiCl_4(py)_2$ gives $Ti^{IV}O(Pc)$.[197] If $TiCl_3$ is the template, $Ti(Pc)Cl_2$ is obtained in a quantitative yield.[198] $Ti(Pc)Cl_2$ is hydrolyzed to $Ti(Pc)(OH)_2$ which dehydrates to a μ-oxo polymer, $HO[Ti(Pc)O]_nH$ on heating.[197,198] Copolymers, *e.g.* $HO[Ti(Pc)OPO(Ph)O]_nH$, are also known.

Transmetallation of $Li_2(Pc)$ with $TiCl_3$ in boiling quinoline under anaerobic conditions gives paramagnetic $Ti^{III}(Pc)Cl$ ($\mu = 1.79$ BM) which is stable in the solid state but oxidizes to $TiO(Pc)$ in solution.[197,220]

Reactions of $ZrCl_4$ and $HfCl_4$ with phthalonitrile at 170 °C give ring-chlorinated products $M^{IV}(PcCl)Cl_2$ which are hydrolyzed on reprecipitation from conc. H_2SO_4–H_2O.[197]

(viii) Group VA (V^{IV}, $Nb^{IV,V}$, Ta^V)

Paramagnetic $V^{IV}O(Pc)$ ($\mu = 1.71$ μB) is prepared from phthalonitrile and V_2O_5.[197,199] The V atom is displaced from the N_4 plane by 0.575 Å and the V—O and V—N distances are 1.580 and 2.026 Å respectively.[221]

Nb^V and $Ta^V(Pc)Br_3$ are obtained by transmetallation of $Li_2(Pc)$ with MBr_5.[198] In $Nb^{IV}(Pc)Cl_2$, the metal atom is displaced out of the macrocyclic ring by 0.98 Å and the two chlorines coordinate on the same side of the Pc.[222]

(ix) Group VIA ($Cr^{I,II,III,IV}$, $Mo^{V,VI}$, $W^{IV,VI}$)

$Cr(CO)_6$ react with phthalonitrile in hot 1-chloronaphthalene to precipitate the β form of $Cr^{II}Pc$ which is readily oxidized to $Cr^{III}(Pc)OH$ in air.[197,198] The hydroxide coordinates other ligands to form six-coordinate complexes, $Cr^{III}(Pc)(OH)X$ (X = ROH, H_2O, CN, SCN, SeCN). The hydroxy group is substituted by CN^- in conc. KCN solution and $\{Cr^{III}(Pc)(CN)_2\}^-$ is formed. Acetic anhydride acetylates $Cr(Pc)(OH)(OH_2)$ in a stepwise fashion to $Cr(Pc)(OAc)(OH_2)$ and $Cr(Pc)(OAc)(HOAc)$.

$Cr^{III}(Pc)(OH)$ is converted to the chloride in methanolic HCl, while a dianion, $[Cr(Pc)(OH)O]^{2-}$, is formed in aqueous alkali.

The α form of $Cr^{II}(Pc)$, obtained by sublimation of $Cr^{II}(Pc)(py)_2$, is inert, while the β form reacts with O_2 and NO.[198,199] The difference in reactivity may be attributed to the metal–metal distance change from 3.40 Å in the α to 4.80 Å in the β form. The NO complex ($v(NO) = 1690\,cm^{-1}$) binds a pyridine ligand to form $Cr(Pc)(NO)(py)$ ($v(NO) = 1680\,cm^{-1}$) and is oxidized by air to a compound with the composition of $[Cr(Pc)(NO)(OH)]$ ($v(NO) = 1690\,cm^{-1}$). The NO group dissociates from $Cr(Pc)(NO)$ at 280 °C *in vacuo*.

$Cr(Pc)$ is reduced with Na metal to a Cr^I complex whose electronic configuraton is assigned as $(e_g)^4(b_{2g})^1$ in THF and as $(b_{2g})^2(e_g)^3$ in HMPA.[198]

$Mo(Pc)$ and $W(Pc)$ complexes, $Mo^V(Pc)Cl_3$,[223] $Mo^{VI}(Pc)(O_2)$,[224] $W^{IV}(Pc)\{OC_6H_3(NO_2)_3\}_2$[225] and $W^{VI}(Pc)(O_2)$,[225,226] have been reported.

(x) Group VIIA ($Mn^{I,II,III}$, Tc^{II}, $Re^{II,IV}$)

Phthalonitrile condensation in the presence of metallic manganese or its acetate salt gives $Mn^{II}(Pc)$ ($S = \frac{3}{2}$) which is not sensitive by itself to oxygen but is oxidized in the presence of donor ligands to $Mn^{III}(Pc)(OH)(L)$.[197-199] The hydroxide condenses to a μ-oxo dimer $[Mn^{III}(Pc)(L)]_2O$ or exchanges its axial ligands to $Mn^{III}(Pc)XL$ ($X = Cl^-$, CN^-, RCO_2^-, OH^-, L = nitrogenous base, alcohol, OH^-, CN^-). The μ-oxo dimers lose their axial ligands at high temperature. The five-coordinate complex, $Mn^{III}(Pc)(OCOH)$, is also formed from $Mn^{III}(Pc)(OCOH)(HCO_2H)$ at 110 °C. $Mn^{III}(Pc)X$ ($X = Cl$, Br) are obtained by the oxidation of $Mn^{II}(Pc)$ with thionyl halides. Pseudohalogen (CN^-, NCO^-, NCS^-, $NCSe^-$) complexes are also known. All the $Mn^{III}(Pc)$ complexes are high-spin (d^4, $S = 2$) compounds.

$Mn^{II}(Pc)(py)_2$ (d^5, $S = \frac{1}{2}$), generated by photoreduction of $Mn^{III}(Pc)(OH)(py)$,[227] reversibly binds molecular oxygen in pyridine.[227] The mode of O_2 coordination is suggested to be end-on in the solid $[Mn^{II}(Pc)(py)(O_2)]$ ($v(O_2) = 1154\,cm^{-1}$) but side-on in matrix-trapped $Mn(Pc)(O_2)$ ($v(O_2) = 991\,cm^{-1}$ in Kr and $992\,cm^{-1}$ in Ar).[54] Nitroxide is also bound to $Mn^{II}(Pc)$.[198] Like its Cr analogue, $Mn(Pc)(NO)$ ($v(NO) = 1760\,cm^{-1}$) coordinates pyridine ($Mn(Pc)(NO)(py)$, $v(NO) = 1737\,cm^{-1}$) and is oxidized to $Mn(Pc)(NO)(OH)$ ($v(NO) = 1760\,cm^{-1}$).

$Mn(Pc)$ is reduced with Na metal in THF to $[Mn^I(Pc)]^-$ (d^6, $S = 1$).

Generation of $Tc^{II}Pc$ and $Re^{II}Pc$ by the nuclear reaction of $Cu(Pc)$ and high energy deuterium in the presence of Mo and W respectively was reported.[199]

Condensation of *o*-cyanobenzamide with $K_2[Re^{IV}Cl_6]$ or $K_4[Re_2^{IV}OCl_{10}]$ gives a μ-oxo dimer, $[Re^{IV}(Pc)Cl]_2O\cdot(H_2Pc)_2$, which is diamagnetic, probably due to antiferromagnetic coupling between the two Re atoms.[199] The crystalline solvent-free $[Re^{IV}(Pc)Cl]_2O$ showed a temperature independent, magnetic field dependent magnetic moment. The μ-oxide is hydrolyed to $[Re^{IV}(Pc)(OH)Cl]\cdot H_2O$ by reprecipitation from conc. H_2SO_4–H_2O.

(xi) Group VIIIA—1 ($Fe^{I,II,III}$, $Ru^{II,III}$, $Os^{II,III,IV}$)

Phthalonitrile reacts with $Fe(CO)_5$ in refluxing 1-chloronaphthalene to precipitate out $Fe^{II}(Pc)$.[197,198,206]

$Fe^{II}(Pc)$ (d^6, $S = 1$) coordinates various ligands.[228] Usually the second association constant is much larger than the first, and six-coordinate complexes are isolated (equation 29). The relative stability of the $Fe-L_{ax}$ bond is determined by the ligand exchange reaction (equation 30). The order of stability, Im > Pip > py, implies that the axial coordination is σ-character predominant (Pip > py) with some metal to ligand π back-donation (Im > Pip). For substituted pyridines, the order of complex stability is the same as that of the ligand σ basicity (pK_{BH^+}) except for the sterically hindered 2-Me derivative. The $Fe-L_{ax}$ π back-bonding in $Fe^{II}Pc$ is weaker than in Fe^{II}porphyrins. The strong π acceptor but poor σ donor ligand CO is several orders of magnitude more readily displaced from the former than the latter by amine ligands (equation 31).

$$Fe(Pc) \underset{L}{\overset{L}{\rightleftharpoons}} Fe(Pc)L \underset{L}{\overset{L}{\rightleftharpoons}} Fe(Pc)L_2 \qquad (29)$$

$$L = CO, RNC, Py, Im, Pip, NO, PR_3, P(OR)_3, DMF, H_2O$$

$$Fe(Pc)LL' + L'' \overset{K_{obs}}{\rightleftharpoons} Fe(Pc)LL'' + L' \qquad K_{obs} = \frac{[Fe(Pc)LL''][L']}{[Fe(Pc)LL'][L'']} \qquad (30)$$

$$\text{Fe(Pc)LL}' \underset{k_{-1}}{\overset{k_1}{\rightleftharpoons}} \text{Fe(Pc)L} \underset{k_{-2}}{\overset{k_2}{\rightleftharpoons}} \text{Fe(Pc)LL}'' \qquad (31)$$

The ligand exchange reaction is described by a dissociative mechanism in which the activation parameters are independent of the entering group (k_1, $k_{-2} \ll k_{-1}$, k_2). The rate of dissociation (k_1) is in the order 2-MeIm > MeC$_6$H$_4$NO > Pip > py > CO > Im > PhCH$_2$NC, and varies over a wide range (1–10^4). The reaction is much slower (10^3 or more) in the Pc complexes than in the porphyrin analogues. The *trans* labilizing effect increases in the order 2-MeIm > Pip > py > Im for L$'$ = CO and PhCH$_2$NC > Pip > py > 1-MeIm for L$'$ = RNC, reflecting the strong σ- and weak π-axial interactions in Fe(Pc)LL$'$. The sterically hindered 2-MeIm labilizes the *trans* ligand by pulling the iron atom out of the N$_4$ plane towards itself. The five-coordinate intermediates show little ligand selectivity ($k_{-1}/k_2 \approx 1$), but some variations in the ratio k_{-1}/k_2 indicate a short finite lifetime for this species.

The weak axial π interactions and high tendency to six-coordination of the Fe^{II}Pc complexes are attributed to the strong metal e_g ($d_{xz}d_{yz}$) to Pc e_g (π^*) back-donation which decreases the electron density on the Fe atom. At the same time, the strong metal–Pc σ and π interactions give rise to large energy gaps between $d_{x^2-y^2}$, d_{z^2} and d_{xz}, d_{yz}, d_{xy} levels in the Fe^{II} atom and destabilize the five-coordinate high- (or intermediate-) spin d^6 state relative to the six-coordinate low spin state. These arguments are supported by Mössbauer, MCD and X-ray photoelectron spectroscopic studies.[198]

Fe^{II}Pc is oxidized by molecular oxygen in basic media but reversibly binds O$_2$ at lower pH values.[198,199,229] The isolation of {Fe(Pc)}$_2$O$_2$ has been reported. Theoretical calculations predict the side-on structure for O$_2$ binding in Fe(Pc)O$_2$,[230] but the experimental data are not conclusive.

Solid Fe^{II}(Pc) absorbs NO.[198,199] However, Fe(Pc)(NO) (ν(NO) = 1737 cm^{-1}) is not oxidized by atmospheric oxygen, and loses the NO on exposure to pyridine vapour.

Li–benzophenone reduces Fe(Pc) to [Fe^I(Pc)]$^-$.

Ru^{II}(Pc) shows axial coordination chemistry similar to the Fe analogue, but the dissociation rate for Ru(Pc)LL$'$ is much smaller than for the Fe(Pc)LL$'$.[231] For example, k_1(Fe)/k_1(Ru) $\approx$ 20 000 (L$'$ = P(OBu)$_3$), 890 (LK$'$ = 1-MeIm) and 260 (L$'$ = py). The rate decrease is more marked for the better π acceptor, and the axial π interaction appears to be stronger in Ru(Pc) complexes than in the Fe analogues.

Ru^{II}(Pc) complexes bind CO more tightly than do the Fe^{II} analogues, but do not show O$_2$ binding probably because of their high oxidation potentials. For instance, ring oxidation ($E_{1/2}$ = 0.81 V *vs.* SCE) precedes metal oxidation for Ru^{II}(Pc)(py)$_2$.

Condensation reactions of phthalonitrile with Ru$_3$(CO)$_{12}$ or OsO$_4$ in the presence of CO give M^{II}(Pc)(CO) complexes which are isolated as pyridine or THF adducts, Ru^{II}(Pc)(py)(CO), Os^{II}(Pc)(L)(CO) (L = py or THF).[199] Ru^{III}(Pc)Cl, [Os^{VI}(O)$_2$(Pc)]{C$_6$H$_4$(CN)$_2$}, Os^{III}(Pc)Cl and Os^{IV}(Pc)(SO$_4$) have also been reported.[198,232]

(xii) Group VIIIA — 2 (Co^{I,II,III}, Rh^{II,III}, Ir^{II,III})

Co^{II}(Pc) forms five- and six-coordinate complexes in the presence of ligands.[198] Since the d_{z^2} orbital is half filled in d^7 Co^{II}(Pc), independent of its spin state, the axial coordination is weaker in this system than, for example, the low-spin Fe^{II} analogue. The existence of an electron in the antibonding d_{z^2} orbital is reflected in the metal–axial ligand bond length. The Co—N$_{ax}$ distance (2.30 Å) in Co(Pc)(γ-Pic)$_2$ (γ-Pic = 3-methylpyridine) is $\sim$ 0.30 Å larger than in Fe(Pc)(γ-Pic)$_2$ (2.00 Å) while the M—N$_{Pc}$ distances are similar to each other.

The weakness of the Co—L$_{ax}$ bonds is also observed in thermal dissociation experiments. Co(Pc)(py) and Co(Pc)(γ-Pic) in the solid state lose their axial ligands at 130–160 and 150–180 °C respectively, while the Zn analogues dissociate at a 40–80 °C higher temperature. Co(Pc)(py) can be crystallized from pyridine solution at 100 °C, while Co(Pc)(py)$_2$ forms at room temperature.

Co^{II}Pc complexes do not bind CO and O$_2$ at ambient temperature, but reversibly bind O$_2$ at low temperature. In the presence of axial ligands, 1:1 adducts [Co(Pc)L(O$_2$)] (L = HMPA, γ-Pic, py) are formed at liquid N$_2$ or even lower temperature. The ESR data on the O$_2$ complexes support a Co^{III}–O$_2^-$ formulation.

[Co^{II}(PCTS)]$^{4-}$ also forms a 1:1 O$_2$ adduct at -84 °C. However, above pH 12 at room temperature, it binds O$_2$ in 2:1 stoichiometry. If the axial ligand is strong electron donor, [Co^{II}(PCTS)(L)$_2$]$^{4-}$ is slowly oxidized by air to [Co^{III}(PCTS)(L)$_2$]$^{4-}$ (L = Im, CN$^-$).

Solid Co(Pc) coordinates NO (ν(NO) = 1705 cm^{-1}) which is substituted by pyridine on exposure to pyridine vapour.

$[Co^I(Pc)]^-$ is generated by Li–benzophenone reduction of $Co^{II}(Pc)$.

The condensation product of phthalonitrile or its analogues with $RhCl_3$ was reported to be $Rh^{III}(Pc)Cl$, which precipitates out from conc. H_2SO_4–H_2O as the bisulfate, $[Rh(Pc)(SO_4H)]$.[197,198,232] $Rh^{III}(Pc)Cl$ is reduced to $Rh^{II}(Pc)(H_2NPh)_2$ with aniline.

Reaction of $IrCl_3$ and phthalonitrile gives a ring-chlorinated compound, $[Ir^{III}(Pc\text{-}Cl)Cl]$, while use of *o*-cyanobenzamide leads to $Ir^{III}(Pc)Cl$ which is reduced to $Ir^{II}(Pc)(H_2NPh)_2$ upon extraction with aniline.[197,198,232]

(xiii)   Group VIIIA—3 (Ni^{II}, Pd^{II}, Pt^{II})

$Ni^{II}(Pc)$, $Pd^{II}(Pc)$ and $Pt^{II}(Pc)$ are prepared by the condensation method with metallic Ni or $NiCl_2 \cdot 6H_2O$, Pd black and $PtCl_2$ respectively.[197] $PdCl_2$ causes Pc ring chlorination and metallic Pt catalyzes $H_2(Pc)$ formation.

(xiv)   Group IB (Cu^{II}, Ag^{I,II}, Au^{II})

$Cu^{II}(Pc)$ and $Au^{II}(Pc)$ are obtained by the condensation method, while the labile $Ag(Pc)$ was prepared by transmetallion of $Li_2(Pc)$ with $AgNO_3$ or $Pb(Pc)$ with Ag_2SO_4.[197,233]

(xv)   Group IIB (Zn^{II}, Cd^{II}, Hg^{II})

The phthalocyanine condensations with Zn dust and Cd filings give $Zn^{II}(Pc)$ and $Cd^{II}(Pc)$ respectively.[197] If $ZnCl_2$ is used as the template, the ring-chlorinated product, $Zn^{II}(Pc\text{-}Cl)$ is formed. The square pyramidal structure of $Zn(Pc)(H_2NC_6H_{11})$ has been determined by X-ray analysis.[198] The metal displacement from the N_4 plane is 0.48 Å with a slightly long $M—N_{Pc}$ distance of 2.06 Å. The Hg^{II} ion is inserted into Pc by transmetallation of Li_2Pc with $HgCl_2$.[197]

21.1.2.2.7   Phthalocyanine complexes containing two metals

Mixed metallophthalocyanines, $[M^ILi(Pc)]$, where Pc is coordinated to two different metals, Li^I and M^I (M = Cu, Ag), are prepared from $\{(C_{12}H_{25})_3N(Bu)\}Li(Pc)$ and M^I salts.[234]

Reaction of $M(Pc)$ (M = Mg, Zn, $SnCl_2$) and $Re_2(CO)_{10}$ gives metal–metal-bonded complexes, $\{Re(CO)_3\}_2M(Pc)$.[126]

21.1.2.2.8   Reactions at the metal

Electrochemical or Li–benzophenone reduction of $Fe(Pc)$ and $Co(Pc)$ gives $[M^IPc]^-$ and $[M^I(Pc)]^{2-}$.[198,199] These monovalent metal complexes react with alkyl halides to give organometallic compounds which transfer the alkyl group to alkenes in the presence of Pd^{II} salts (Scheme 61). $[Co^IPCTS]^{5-}$ and $[Co^IPCTS]^{6-}$ reductively bind one and two molecules of oxygen respectively in DMF.[198] The second O_2 addition is reversible (Scheme 62).

21.1.2.2.9   Photochemistry

The $n(N_{aza}) \to \pi^*(Pc)$ excitation of $M^{II}Pc$ by UV light generates a chemically active species, which abstracts a hydrogen atom from a solvent molecule (Scheme 63).[235]

If the dimeric form is photoactivated, it splits into an ion pair *via* an electron transfer between the two associated molecules. Either a ring or a metal redox pair is formed depending on the central metal ion (Scheme 64).

Irradiation of the Q band causes $a_{1u}(\pi) \to e_g(\pi^*)$ electron excitation in $M(Pc)$, and the excited species catalyzes an electron transfer reaction between a donor–acceptor pair such as triethanolamine–methylviologen. There are two possible pathways and the course of electron transfer is determined by the redox potentials of the components. These processes are quenched by intersystem

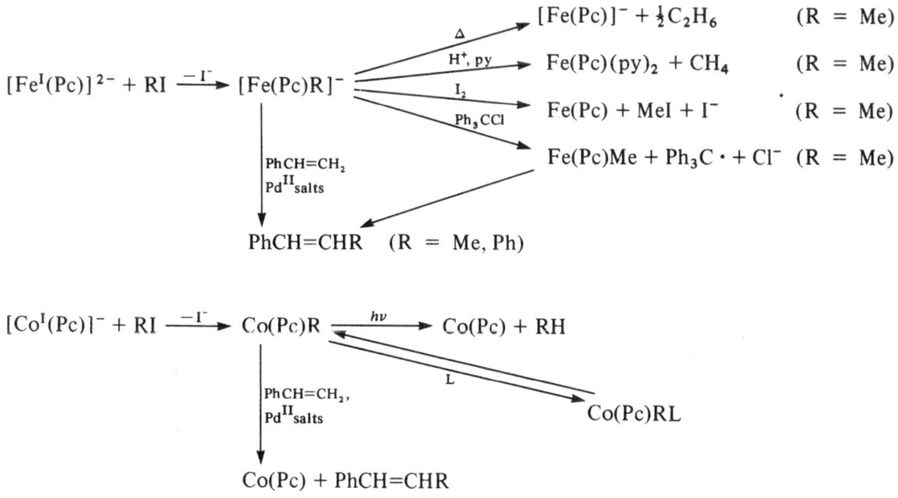

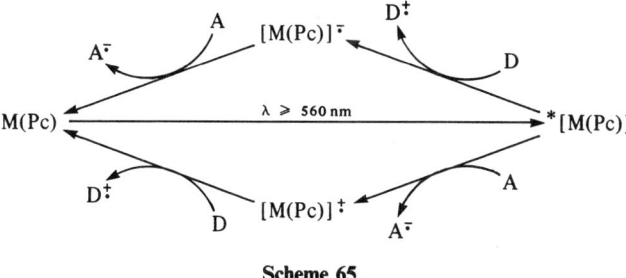

$$[Fe^I(Pc)]^{2-} + RI \xrightarrow{-I^-} [Fe(Pc)R]^-$$

$$\xrightarrow{\Delta} [Fe(Pc)]^- + \tfrac{1}{2}C_2H_6 \qquad (R = Me)$$

$$\xrightarrow{H^+, py} Fe(Pc)(py)_2 + CH_4 \qquad (R = Me)$$

$$\xrightarrow{I_2} Fe(Pc) + MeI + I^- \qquad (R = Me)$$

$$\xrightarrow{Ph_3CCl} Fe(Pc)Me + Ph_3C \cdot + Cl^- \quad (R = Me)$$

$$\xrightarrow[Pd^{II}salts]{PhCH=CH_2} PhCH=CHR \qquad (R = Me, Ph)$$

$$[Co^I(Pc)]^- + RI \xrightarrow{-I^-} Co(Pc)R \xrightarrow{h\nu} Co(Pc) + RH$$

$$Co(Pc)R \xrightarrow{L} Co(Pc)RL$$

$$Co(Pc)R \xrightarrow[Pd^{II}salts]{PhCH=CH_2,} Co(Pc) + PhCH=CHR$$

Scheme 61

$$[Co^I(PCTS)]^{5-} + O_2 \xrightarrow{DMF} [Co^{II}(PCTS)(O_2^{\bar{}})]^{5-}$$

$$[Co^I(PCTS)]^{6-} + O_2 \xrightarrow{DMF} [Co^I(PCTS)(O_2^{\bar{}})]^{6-} \xrightleftharpoons{O_2/DMF} [Co^{II}(PCTS)(O_2^{\bar{}})_2]^{6-}$$

Scheme 62

$$M(Pc) \xrightarrow{h\nu\,(UV)} {}^*M(Pc) \xrightarrow{SH} {}^*M(P\dot{c}-H) + S\cdot$$

$$M = Cu^{II}, Co^{II}$$

Scheme 63

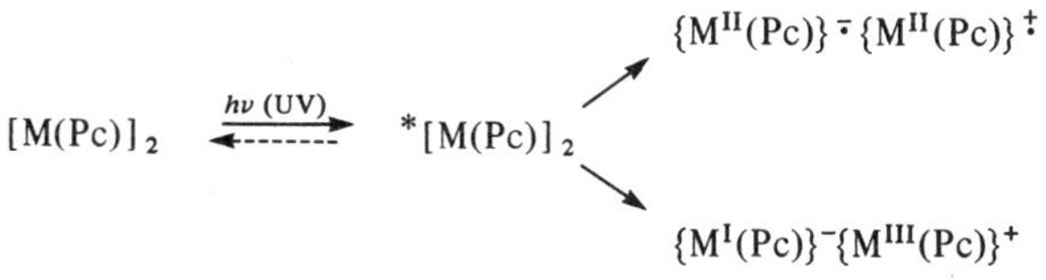

$$[M(Pc)]_2 \xrightleftharpoons[\text{------}]{h\nu\,(UV)} {}^*[M(Pc)]_2$$

$$\nearrow \{M^{II}(Pc)\}^{\bar{}}\{M^{II}(Pc)\}^{\ddagger}$$

$$\searrow \{M^I(Pc)\}^-\{M^{III}(Pc)\}^+$$

Scheme 64

crossing from the Q state to the Pc → metal charge transfer state, if the latter is available (Scheme 65).

$$M(Pc) \xrightarrow{\lambda \geqslant 560\,nm} {}^*[M(Pc)]$$

Scheme 65

Photodissociation is a useful method to convert a stable complex to other complexes (Scheme 66).[236]

$$M(Pc)(L)(RNC) \xrightarrow[L']{h\nu} M(Pc)(L)(L') + RNC$$

$$M = Fe, Ru$$

$$Ru(Pc)(L)(CO) \xrightarrow[L']{h\nu} Ru(Pc)(L)(L') + CO$$

$$Co(Pc)R(L) \xrightarrow[L']{h\nu} Co(Pc)(L)(L') + R\cdot$$

Scheme 66

21.1.2.2.10 Redox chemistry

The electrochemistry of phthalocyanines has been systematically studied by Lever et al.[237] The redox centre is either the metal or the macrocyclic π system, and usually two successive oxidations and four reductions are observed for the latter. The ring reduction sequence is generally reversible, but the oxidation steps are often irreversible.

For main group metallophthalocyanines, the ring centred redox is the only process to occur. The separation between the first oxidation and reduction potentials corresponds to the energy difference of the HOMO and LUMO, hence to the Q(0,0) absorption band at ~ 670 nm, and is about 1.56 V. Deviation from the mean value becomes large when the size of the metal significantly exceeds the cavity of the Pc ring. The first reduction and oxidation potentials themselves (E° vs. NHE) depend on the polarizing power of the metal ion (Ze/γ) and are approximated by equation (32).

$$\text{first oxidation: } (Ze/\gamma)(E^\circ - 1.410) = -0.012 \tag{32}$$
$$\text{first reduction: } (Ze/\gamma)(E^\circ + 0.145) = -0.012$$

The second reduction occurs at a potential some 0.42 V more negative, but more scattered than the first. The third and fourth reduction waves appear around 1.7 and 2.0 V respectively.

The metal redox intervenes the ring process in some transition metal phthalocyanines. Fe and Co change their oxidation state from M^{III} to M^{I}, and Mn and Cr from M^{III} to M^{II} between the first ring redox steps, while OTi^{IV}, OV^{IV}, Ni^{II}, Cu^{II} and Zn^{II} are unaffected.

The ring redox potentials for the transition metal complexes are similar to those of the main group with the metal atom at the same oxidation state and of analogous size. The effect of Pc peripheral substitution is small and usually less than 0.1 V unless the substituent is charged.

While solvent effects are small for the ring redox ($\leqslant 0.2$ V), the effect of donor solvents which stabilize the higher oxidation state is marked for the metal redox, especially so for those of low-spin d^6/d^7 and d^7/d^8. There is a linear relationship between E° for the Fe$^{II/I}$ couple and the Gutmann donicity number. The Co$^{III/II}$ and Co$^{II/I}$ couples shift positively as the pK_a of substituted pyridine, as solvent, decreases; and the ring oxidation precedes the Co$^{III/II}$ couple in noncoordinating solvents such as CH_2Cl_2.

The effect is reversed for the Fe$^{III/II}$ (low-spin d^5/d^6) and Cr$^{III/II}$ (low-spin d^3/d^4) pairs, where the electron is removed from (added to) the metal e_g (d_{xz}, d_{yz}) orbital. The σ donation from the axial ligands to the metal d_{z^2} stabilizes the e_g level by enhancing the metal to Pc π-bonding interactions.

Reduction of MPc by Li–benzophenone has been reported.[198] The assignments of species are not necessarily consistent with those of the electrochemistry (Scheme 67). Co^{II}(Pc) is reduced to Co^I(Pc) by NaBH$_4$ in DMA, and NH$_2$NH$_2$ or H$_2$S in H$_2$O. Conversion of Rh^{III}(Pc)Cl and Ir^{III}(Pc)Cl to M^{II}(Pc)(H$_2$NPh)$_2$ with aniline has been reported.

$[M(Pc)] \rightarrow [M(Pc)]^- \rightarrow [M(Pc)]^{2-} \rightarrow [M(Pc)]^{3-} \rightarrow [M(Pc)]^{4-}$

  Li–Ph$_2$CO/THF, M $=$ Mg^{II}, Al^{III}Cl, Zn^{II}, Ni^{II}

$[Cr^{II}(Pc)] \rightarrow [Cr^I(Pc)]^- \rightarrow [Cr^I(Pc)]^{2-}$

  Na or Li–Ph$_2$CO/THF

$Mn^{II}(Pc)] \rightarrow [Mn^I(Pc)]^- \rightarrow [Mn^{II}(Pc)]^{2-} \rightarrow [Mn(Pc)]^{3-} \rightarrow [Mn^I(Pc)]^{4-}$

  Na, or Li–Ph$_2$CO/THF

$[Fe^{II}(Pc) \rightarrow [Fe^I(Pc)]^- \rightarrow [Fe^I(Pc)]^{2-} \rightarrow [Fe^I(Pc)]^{3-} \rightarrow [Fe^I(Pc)]^{4-}$

  Na, or Li–Ph$_2$CO/THF

$[Co^{II}(Pc)] \rightarrow [Co^I(Pc)]^- \rightarrow [Co^I(Pc)]^{2-} \rightarrow [Co^{II}(Pc)]^{3-} \rightarrow [Co^I(Pc)]^{4-} \rightarrow [Co^I(Pc)]^{5-}$

  Li–Ph$_2$CO/THF

$[Cu^{II}(Pc)] \rightarrow [Cu^{II}(Pc)]^- \rightarrow [Cu^I(Pc)]^{2-} \rightarrow$ demetallation?

  Li–Ph$_2$CO/THF

Scheme 67

Thionyl halides (SOCl$_2$, SOBr$_2$) oxidize Mn^{II}(Pc) to Mn^{III}(Pc)X, and Cr^{II}, Fe^{II} and Co^{II}(Pc) to $[M^{III}(Pc)^{+\ddagger}$ Cl]Cl, but not Ni^{II}, Cu^{II}, OV^{IV} or H$_2$(Pc).

21.1.3 CORROLES AND CORRINS

Substitution of one of the four methine bridges of the porphyrin skeleton by direct linkage leads to a structure named corrole (**30**; Figure 15). In close relation to it is corrin (**31**) which is the central chromophore of coenzyme and vitamin B_{12}.

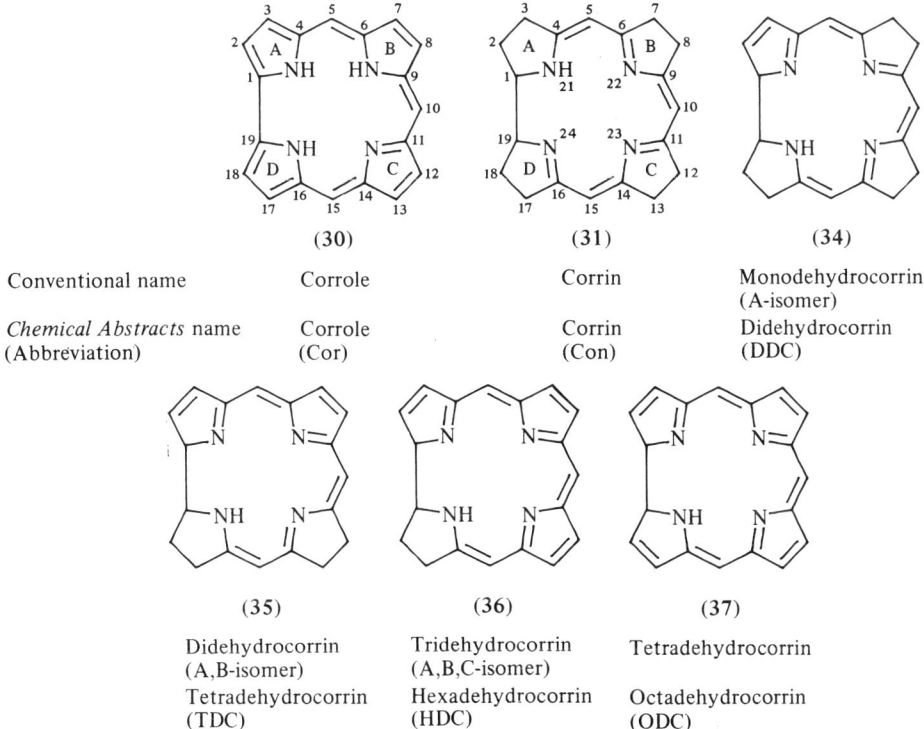

	Conventional name	Corrole	Corrin	Monodehydrocorrin (A-isomer)
	Chemical Abstracts name (Abbreviation)	Corrole (Cor)	Corrin (Con)	Didehydrocorrin (DDC)

	Didehydrocorrin (A,B-isomer)	Tridehydrocorrin (A,B,C-isomer)	Tetradehydrocorrin
	Tetradehydrocorrin (TDC)	Hexadehydrocorrin (HDC)	Octadehydrocorrin (ODC)

Figure 15 Corrole, corrin and dehydrocorrins

21.1.3.1 Corroles

21.1.3.1.1 General properties

Corrole contains an 18-π-electron system and shows aromaticity similar to the porphyrin chromophore, with strong ring current effects on the peripheral and inner-proton NMR, relatively stable parent mass ion and intermediary C—C bond length between single and double bonds.

The corrole ring is not strictly planar in the metal-free form.[238a] Because of the steric repulsion between inner protons, pyrrole ring A is tilted out of the mean plane. According to π-electron distribution calculations the three inner protons are attached to N-21, N-22 and N-24.[238b]

21.1.3.1.2 Synthesis

The corrole synthesis was established by Johnson's group.[239] Their first attempt to cyclize tetrapyrrole with various bridging elements did not lead to corroles but to their 10-hetero analogues (Scheme 68).

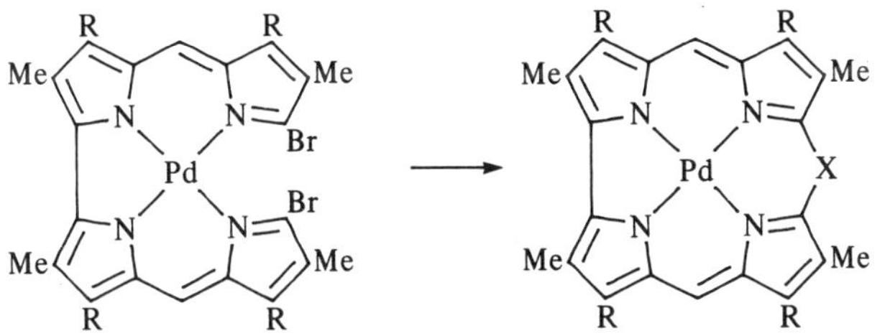

X = O, NH, NMe, S

Scheme 68

The most successful approach is to complete the macrocycle by A–D linkage formation (Scheme 69).[239,240] The ring closure is effected by light or oxidants, and the presence of an appropriate metal template facilitates metallocorrole synthesis.[239-242]

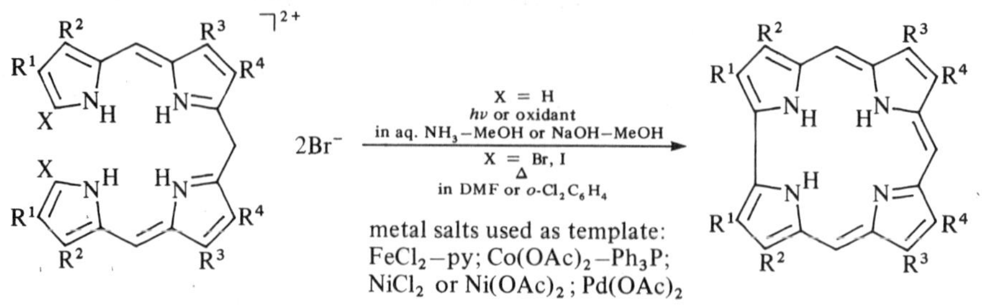

X = H
hv or oxidant
in aq. NH₃–MeOH or NaOH–MeOH
X = Br, I
Δ
in DMF or o-Cl₂C₆H₄

metal salts used as template:
FeCl₂–py; Co(OAc)₂–Ph₃P;
NiCl₂ or Ni(OAc)₂; Pd(OAc)₂

Scheme 69

A [2 + 2] route *via* a thiaphlorin intermediate (Scheme 70),[239] and a cobalt ion-assisted [2 + 2] method (Scheme 71) have also been reported.[239]

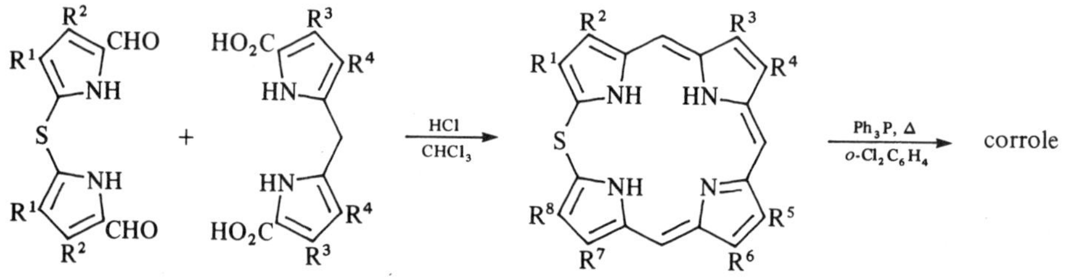

$\xrightarrow[\text{CHCl}_3]{\text{HCl}}$

$\xrightarrow[o\text{-Cl}_2\text{C}_6\text{H}_4]{\text{Ph}_3\text{P, }\Delta}$ corrole

Scheme 70

$\xrightarrow[\text{MeOH reflux}]{\text{Ph}_3\text{PCo(OAc)}_2}$ Ph₃PCo^{III} corrole

X = CHO, Y = CO₂H, or
X = CO₂H, Y = CHO

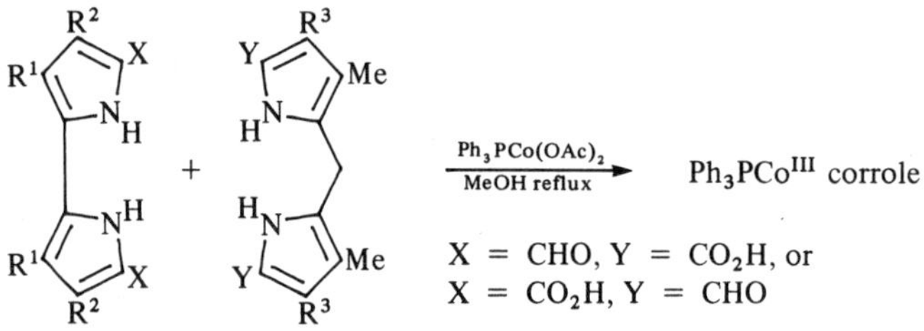

Scheme 71

21.1.3.1.3 Reactions of the corrole macrocycle

Corroles are stronger acids and weaker bases than porphyrins.[239] They are deprotonated in dilute alkali to form aromatic anions. In acidic media, the first protonation occurs on the pyrrole nitrogen, but the second proton adds to the meso carbon causing loss of the aromaticity (Scheme 72).

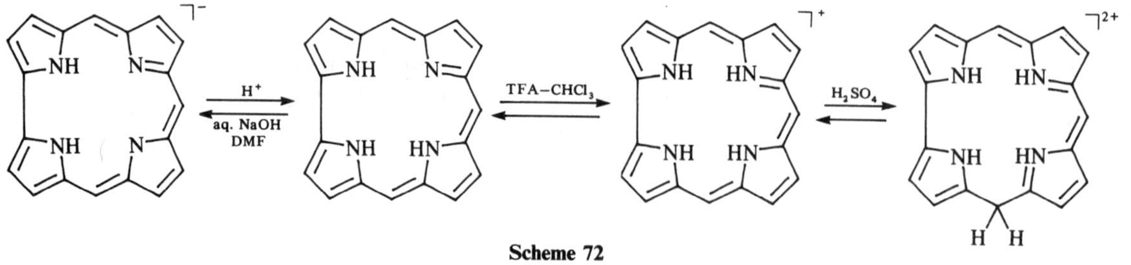

$\xrightleftharpoons[\substack{\text{aq. NaOH} \\ \text{DMF}}]{\text{H}^+}$

$\xrightleftharpoons{\text{TFA–CHCl}_3}$

$\xleftarrow{\text{H}_2\text{SO}_4}$

Scheme 72

Proton exchange at the meso position of corrole is facile, and the deuteration is complete in 15 min in trifluoroacetic acid at room temperature. Under similar condition, porphyrin does not exchange the meso protons.

Platinum-catalyzed hydrogenation occurs on the meso carbons and corrogen is obtained.[240] On the other hand, electrophiles attack the nitrogens.[239] The N^{21}-acetylcorrole is easily hydrolyzed back to the starting material. An N^{22}-alkyl group is thermally labile and either migrates to N-21 or dissociates (Scheme 73).

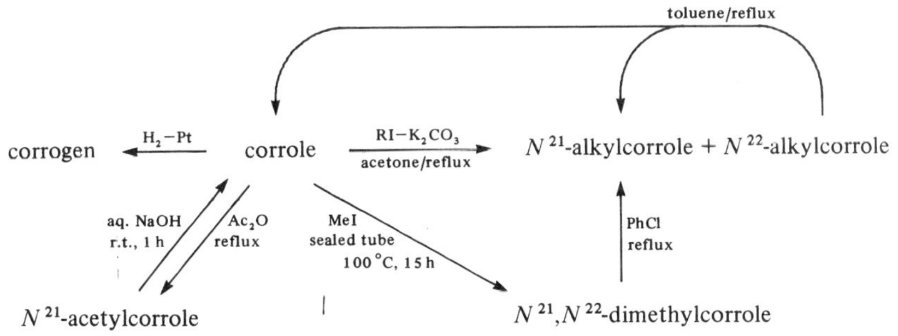

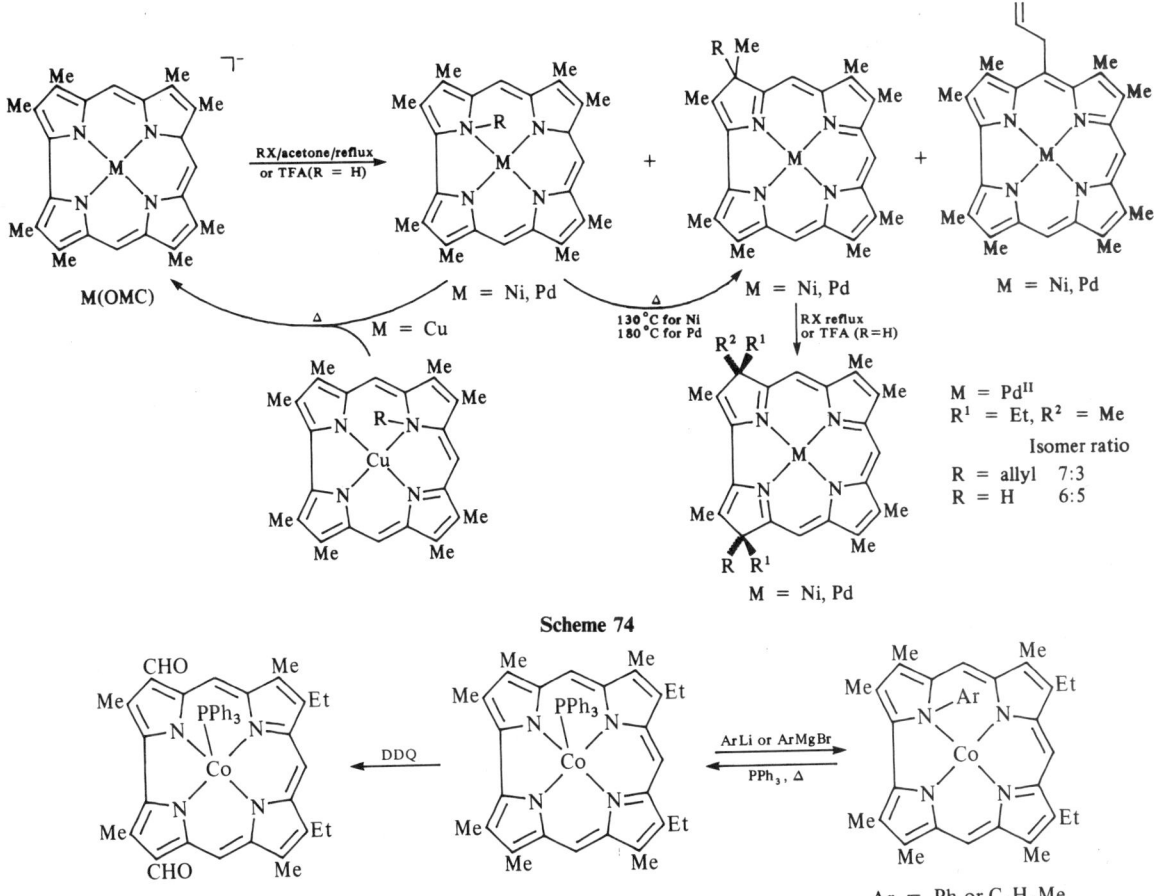

Scheme 73

Ni^{II}, Pd^{II} and Cu^{II} corroles have an extra hydrogen which is easily removed by base such as pyridine. The Ni^{II} and Pd^{II} corrole anions react with alkyl halide to give a mixture of N^{21}- and C^3-alkyl adducts. The C alkylation increases as the alkyl group becomes large. If allyl bromide is used, C^5-allylcorrole is formed together with the C^3-allyl adduct. The N-21 alkyl group migrates to C-3 thermally, probably by a [1,5]-sigmatropic mechanism. Further alkylation of the C-3 adduct occurs at C-17 with some stereoselectivity. Anionic Cu^{II} corrole also gives an N^{21}-alkyl derivative by alkylation, but both Cu N^{21}- and N^{22}-alkylcorroles lose the N-alkyl substituent at high temperature (Scheme 74). Unlike Co corrins, (Ph$_3$P)Co^{III}(Cor) is reductively N-alkylated by aryllithium or magnesium bromide. N-Aryl derivatives are dearylated back to the starting material by heating in the presence of triphenylphosphine (Scheme 75). DDQ oxidizes the 13- and 17-methyl groups to aldehydes.

Scheme 74

Scheme 75

21.1.3.1.4 Metallation and demetallation

Usually corroles are metallated with metal salts or metal carbonyls (Scheme 76).[239,243,244]

$$\text{Corrole} \xrightarrow{\text{reagents i–viii}} \text{metallocorrole}$$

i, $CrCl_2$–NaOAc/DMF, Δ; ii, $FeCl_2$; iii, $Co(OAc)_2$/py; iv, $NiCl_2$–aq.
NH_3/$CHCl_3$–MeOH; v, $Cu(OAc)_2$/py; vi, $MoCl_5$–NaOAc
or $Mo(CO)_6$/decalin, Δ; vii, $Pd(OAc)_2$/AcOH; viii, $Pd(OAc)_2$/AcOH

Scheme 76

For metallation of *N*-alkylcorroles, milder reaction conditions are required to avoid loss or migration of the *N* substituent (Scheme 77). Ni, Pd, Cu and Rh are complexed with *N*-alkylcorroles, but Cu insertion into the *N*-allyl derivative causes loss of the substituent.[239,244] *N*-Alkyl bond cleavage also occurs upon metallation with $Co(ClO_4)_2(EtOH)_2$ in acetonitrile.

$$\text{N-alkylcorrole} \xrightarrow{\text{reagents i–iv}} M^{II}\text{N-alkylcorrole}$$

i, $Ni(MeCN)_6(ClO_4)_2$; ii, $Pd(OAc)_2$/AcOH;
iii, $Cu(OAc)_2$/$CHCl_3$–MeOH, Δ;
iv, $\{Rh(CO)_2Cl\}_2$–NaOAc/CH_2Cl_2, r.t.

Scheme 77

21.1.3.1.5 Coordination chemistry of metallocorroles

(i) Cr^V, Mo^V

The high anionic charge (3−) and small central hole of the corrole ligand stabilize the high oxidation states of metals. For this reason, stable Cr^V and Mo^V oxo complexes are formed directly by the metallation reactions.[243] ESR spectra of these complexes suggest a distorted square pyramidal structure with an unpaired electron in the d_{xy} orbital due to strong $d_{xz}d_{yz}$–p_π interactions (Table 9). The smaller hyperfine and larger super hyperfine splitting for the Cr complex indicate stronger σ_N–$d_{x^2-y^2}$ bonding in it than in the Mo analogue.

Table 9 ESR Parameters for $O{=}Cr(MEC)$ and $O{=}Mo(MEC)$

	$v(M{=}O)$ (cm^{-1})	g	A_M $(10^4\,cm^{-1})$	A_M $(10^4\,cm^{-1})$	Ref.
CrO(MEC)	964	1.987	19.3	3.3 (CH_2Cl_2, r.t.)	a
MoO(MEC)	948	1.967	41.8	2.3 (CH_2Cl_2, r.t.)	b
		1.967 $(g_\parallel)$, 1.970 $(g_\perp)$	67.9 $(A_{Mo_\parallel})$, 26.7 $(A_{Mo_\perp})$	(CH_2Cl_2, 77 K)	

[a] Ref. 243.
[b] Y. Murakami, Y. Matsuda and S. Yamada, *Chem. Lett.*, 1977, 689.

(ii) Fe^{III}

A five-coordinate pyridine complex is known.[241]

(iii) $Co^{II,III}$, Rh^I

Co^{III} corroles form five- and six-coordinate complexes in the presence of ligands but the second association constant is much lower than the first. A five-coordinate triphenylphosphine complex was isolated and shown to have a distorted square pyramidal structure with the Co atom displaced out of the N_4 plane by 0.28 Å.[245] In pyridine, it is converted to a diamagnetic bis-pyridine complex.[239] Isocyanides displace pyridine ligands to form five-coordinate complexes, from which amine complexes may be obtained (Scheme 78).

$$Co^{III}(Cor)PPh_3 \xrightarrow[\text{in py}]{} Co(Cor)(py)_2 \xrightarrow[\text{in } CHCl_3]{p\text{-Tol-NC}} Co(Cor)(p\text{-Tol-NC}) \xrightarrow[\text{in MeOH, }\Delta]{p\text{-Tol-NH}_2} Co(Cor)(p\text{-Tol-NH}_2)$$

Scheme 78

Bis-pyridine complexes exist in equilibrium with five- and four-coordinate species in solution.[242] The four-coordinate complexes form polymers, which become predominant above a concentration of 5×10^{-3} M in chloroform. The association constants K_1 and K_2 for 3- or 4- substituted pyridine complexes are well correlated to their pK_a (BH^+) values and Hammett's σ values by equation (33) though deviations are rather large for 3-acetyl- and 2-methylpyridines. In general, K_2 is smaller than K_1 by a factor of 10^2–10^3.[87]

$$\{Co^{III}(OMC)\}_n \;\rightleftharpoons\; Co(OMC) \;\underset{-L}{\overset{K_1 \atop L}{\rightleftharpoons}}\; Co(OMC)L \;\underset{-L}{\overset{K_2 \atop L}{\rightleftharpoons}}\; Co(OMC)L_2 \qquad (33)$$

$$\log K = a.pK_a + b, \qquad \log K_i^X/K_i^H = \varrho\sigma$$

$a = 0.37$, $b = 2.5$, $\varrho = 2.0$, for K_1
$a = 0.26$, $b = 0.32$, $\varrho = 1.3$, for K_2 $\qquad$ $Co^{III}(OMC)$ in acetone at 25 °C

$Co^{III}(Cor)$ complexes are reduced to paramagnetic $Co^{II}(HCor)$ by $NaBH_4$. The Co^{II} complexes bind only one molecule of pyridine, and both four- and five-coordinate species are in the low spin state.[242] The $Co^{II}(HCor)$ contain an extra hydrogen, and the deprotonated $[Co^{II}(Cor)]^-$ do not coordinate pyridine (equation 34).[246]

$$Co^{II}(Cor)^- \;\underset{-H^+}{\overset{H+}{\rightleftharpoons}}\; Co^{II}(HCor) \;\underset{-py}{\overset{py}{\rightleftharpoons}}\; Co^{II}(HCor)py \qquad (34)$$

The metallation of corrole with $\{Rh(CO)_2Cl\}_2$ gives an isomeric mixture of $Rh^I(H_2Cor)(CO)_2$.[244] According to an X-ray structure analysis of the N-23,N-24 complex (N^{21}-Me,N^{22}-H), the two CO molecules coordinate to the approximately square planar Rh atom on the same side of the corrole which is considerably distorted and serves as a bidentate ligand. The N-23,N-24 isomer may be thermally equilibrated with the N-21,N-23 isomer (Scheme 79).

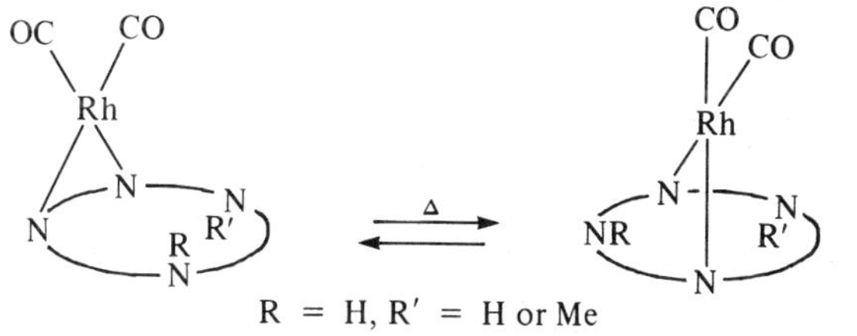

R = H, R' = H or Me

Scheme 79

(iv) Ni^{II}, Pd^{II}

Only four-coordinate species are known for Ni^{II} and Pd^{II} corroles.[239]

(v) Cu^{II}

$Cu^{II}(HCor)$ show similar absorption spectra to those of the *N*-alkyl analogue, and the extra hydrogen atom is probably attached on the pyrrolic nitrogen.[239] The X-ray structure of $Cu(N^{21}$-MeCor) indicates the N-21 atom to be in an sp^3 configuration and the pyrrole ring twisted out of the mean plane by 23°.[240]

21.1.3.1.6 *Reactions at the metal*

Fe^{III} and Co^{III} corroles catalyze the reaction between alkene and hydroxide. The product is either an alcohol or a ketone depending on the substrate.[241] The kinetics are first order in both alkene and hydroxide ion, and the rate increases in the order ethoxyethylene ≫ styrene ≫ 1-octene. No intermediates such as alkylmetal complexes have been detected spectroscopically. These observations suggest a mechanism involving an initial metal–alkene π complexation followed by rate-determining hydroxylation (Scheme 80).

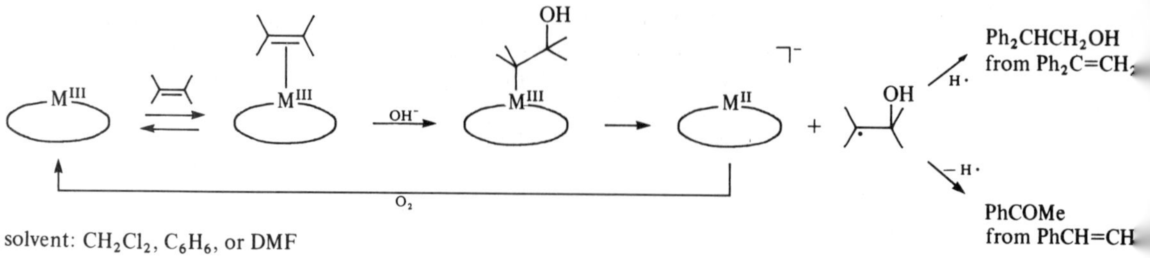

solvent: CH_2Cl_2, C_6H_6, or DMF

Scheme 80

21.1.3.1.7 Redox chemistry

Electrochemical reductions of $Ni^{II}(HCor)$ and $Cu^{II}(HCor)$ are both ligand-centred processes, and the first step accompanies hydrogen elimination (Scheme 81).[247] The *N*-alkyl analogues are reduced at 30–50 mV less negative potentials without loss of the substituent.

$$M^{II}(H-L) \xrightarrow{e^-, -H^\cdot} [M^{II}L]^- \xrightarrow{e^-} [M^{II}L]^{2-} \xrightarrow{e^-} [M^{II}L]^{3-}$$

M = Ni	−2.06 V	−2.34 V	−2.64 V
Cu	−1.96 V	−2.34 V	−2.58 V

L = MEC, dropping Hg electrode in DMF ($E_{1/2}$ *vs.* SCE)

Scheme 81

For $Fe^{III}(Cor)$[241] and $Co^{III}(Cor)$,[239,246] the first reduction wave corresponds to the $M^{III/II}$ couple. Co^{III} complexes lose an electron from the macrocycle on oxidation (Table 10).

Table 10 Redox Potentials of Fe(corrole) and Co(corrole)

	(L/L^+) $E_{1/2}$ *vs.* SCE (V)	$M(III/II)$ $E_{1/2}$ *vs.* SCE (V)		*Ref.*
Fe(MEC)py		−0.92	in DMF containing py	a
Co(Cor)	+0.52	−0.26		b
Co(Cor)py	+0.25	−0.35		b
Co(Cor)py$_2$	+0.7	−0.54		b
Co(Cor)PPh$_3$	+0.20	−0.65		b
Co(N-PhCor)	+0.58	−0.21		b

[a] Ref. 241.
[b] M. Conlon, A. W. Johnson, W. R. Overend, D. Rajapaksa and C. M. Elson, *J. Chem. Soc., Perkin Trans. 1*, 1973, 2281.

The metal redox precedes the ring process for MoO(Cor) (Scheme 82).[248]

$$[Mo^{VI}O(MEC)]^{2+} \xleftarrow{-e^-} [Mo^{VI}O(MEC)]^+ \xleftarrow{-e^-} [Mo^V O(MEC)] \xrightarrow{+e^-} [Mo^{IV}O(MEC)]^-$$

$E_{1/2}$ *vs.* SCE in CH_2Cl_2	+1.3 v	+0.70 V	−0.72 V

Scheme 82

21.1.3.1.8 Dioxacorroles

Corrole analogues with two inner nitrogen atoms substituted by oxygen may be synthesized by a [2 + 2] condensation (Scheme 83).[239]

Dioxacorroles are 18-π-electron aromatic systems like corroles. They exhibit basicity intermediate between that of porphyrin and corrole, and require 1 h at 100 °C in TFA for complete deuteration of the meso positions. The furan β protons are also substituted by deuterium under the same conditions after 100 h. Friedel–Crafts acylation occurs at C-5 while alkyl halides attack on the pyrrolic nitrogens to give a mixture of mono- and di-alkyl derivatives.

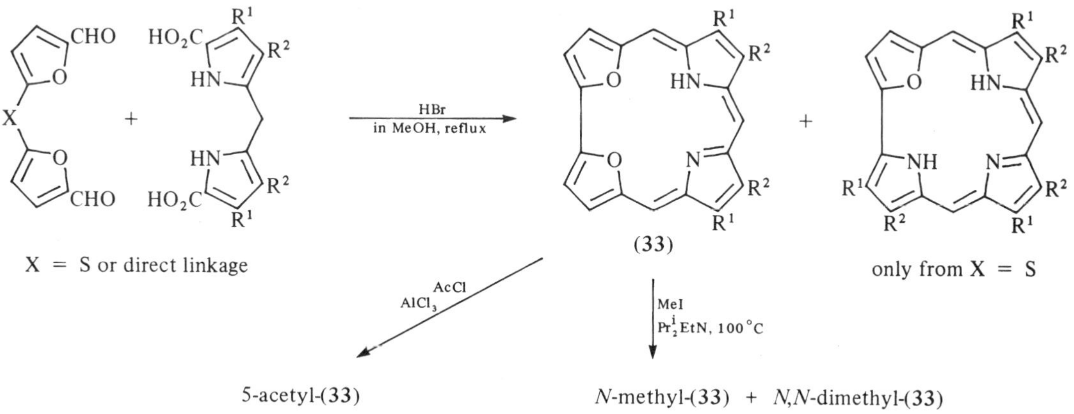

Scheme 83

Metallation of dioxacorrole with $\{Rh(CO)_2Cl\}_2$ gives a mono-Rh complex in which the monovalent rhodium atom is coordinated by two corrole nitrogens (Scheme 84).[244]

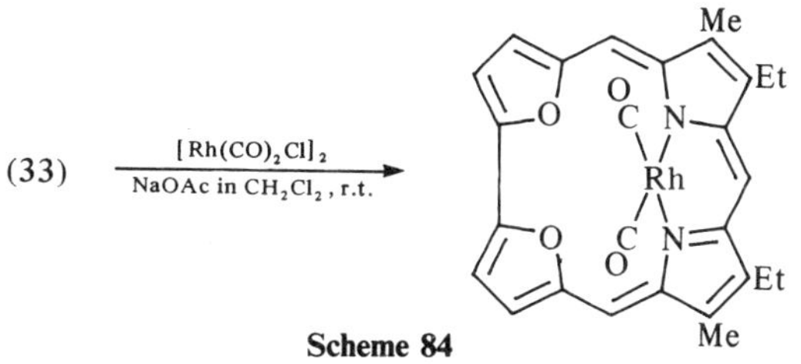

Scheme 84

21.1.3.2 Corrin and Dehydrocorrins

Corrin and a series of dehydrocorrins are known (Figure 15). Though conventional names are widely used for the dehydrocorrins, some of them are incorrect, and the names cited in *Chemical Abstracts* are used through this chapter.

21.1.3.2.1 General properties

In the corrin structure, the π system is interrupted between C-1 and C-19 and the macrocycle is not aromatic. C-1 (and C-19) is chiral and resolution of the ($\pm$) isomers is possible.[239,249] The naturally occurring corrin has the ($1R$) configuration, *i.e.* the Me group is on the α side of the ring.[250] A crystal structure of a cationic corrin salt has been analyzed.[251] The macrocycle is significantly ruffled and the pyrrole ring A is tipped out of the mean plane. The two inner protons are placed on N-21 and N-23. The deviation from the mean plane is small in the metal complexes.[252]

21.1.3.2.2 Spectral properties

The standard absorption spectra of corrins and metallocorrins are characterized as follows:[253]

α,β bands (600–420 nm), two to four bands ($\varepsilon \approx 10^4$) attributed to the forbidden $\pi_7 \rightarrow \pi_8{}^*$ transitions [$\alpha(0,0)$, $\beta(0,1)$];

DE bands (420–390 nm), two weak bands probably corresponding to an antisymmetric forbidden combination of the $\pi_6 \rightarrow \pi_8{}^*$ and $\pi_7 \rightarrow \pi_9{}^*$ excitations;

γ band (390–330 nm), the most intense band ($\varepsilon \approx 10^4$) (usually accompanied by a shoulder and a band on the shorter wavelength side), which arises by a symmetry-allowed combination of the $\pi_6 \rightarrow \pi_8{}^*$ and $\pi_7 \rightarrow \pi_9{}^*$ excitations;

δ bands (330–300 nm), a series of four low to medium intensity bands explained as γ-vibronic bands.

Charge transfer and metal *d–d* bands are probably buried in the $\pi \to \pi^*$ bands. The $\pi \to \pi^*$ absorption bands shift bathochromically with electron density increase or expansion of the π system. The effect of 7,8-dehydrogenation is larger than that of 2,3-unsaturation.[254]

NMR signals of the meso protons appear at $\delta = 5$–8 p.p.m. due to lack of aromaticity.

21.1.3.2.3 Synthesis

There are two main approaches to the corrin system. One of them completes the macrocycle by ring closure between C-1 and C-19. This is the main route and quite versatile. The first example was reported for the octadehydrocorrin synthesis.[239] Biladienes-a,c were cyclized in the presence of metal salts and air to 1,19-*trans*-ODC complexes, and radical 17π and cationic 16π conrotatory mechanisms have been suggested. The cyclization is also induced by light (Scheme 85). Similarly 1-ethoxy-2,3-dihydro-19-methyl-b-bilene is cyclized to 1-ethoxy-19-methyl-7,8,12,13,17,18-HDC (M = Ni).[239,249]

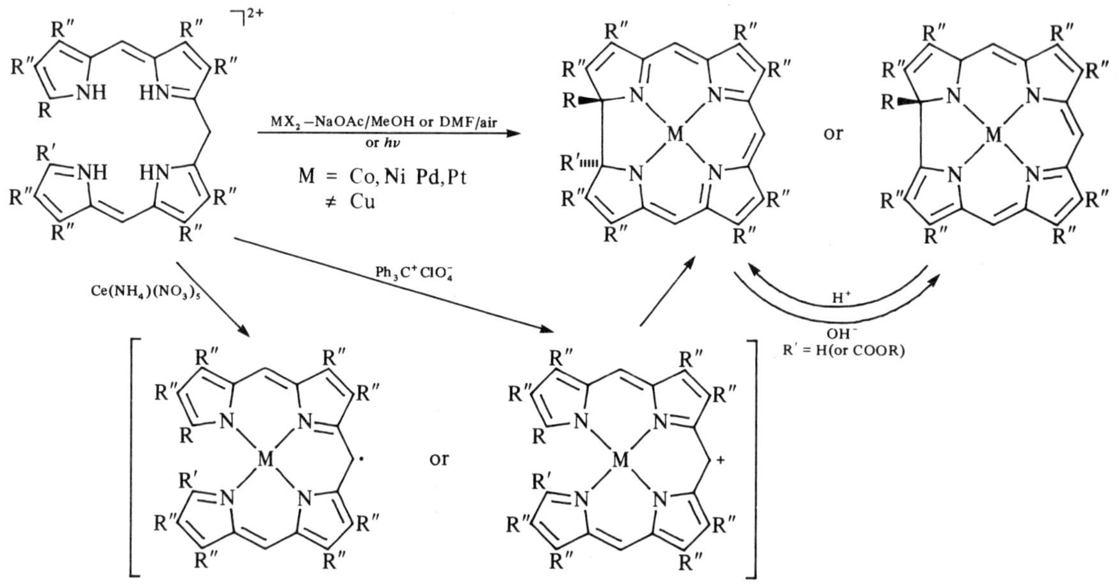

R, R′ = alkyl, carboxyalkyl, hydroxymethyl, alkoxycarbonyl or alkoxy
R′ = Br, I (lost after cyclization)
R″ = unspecified substituent

Scheme 85

1-Methylenebilatriene intermediates are used for the syntheses of corrins and various dehydrocorrins (Scheme 86).[250,255,256] The only one example of a non metal-assisted cyclization was achieved by this method.

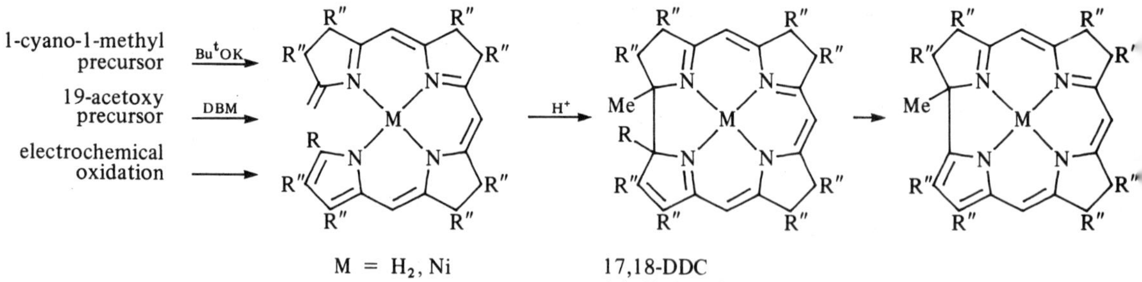

M = H₂, Ni

17,18-DDC
(similarly, 12,13,17,18-TDC and
7,8,12,13,17,18-HDC are obtained)

Scheme 86

The photocyclization (Scheme 87) is metal ion dependent, and those with low-lying metal *d* states are not suitable because of quenching of the excited state.[250,256,257] This reaction may be explained as a photoinduced 1,16-sigmatropic H shift $[_\sigma 2_s + _\pi 16_a]$ followed by an 18π conrotatory cyclization

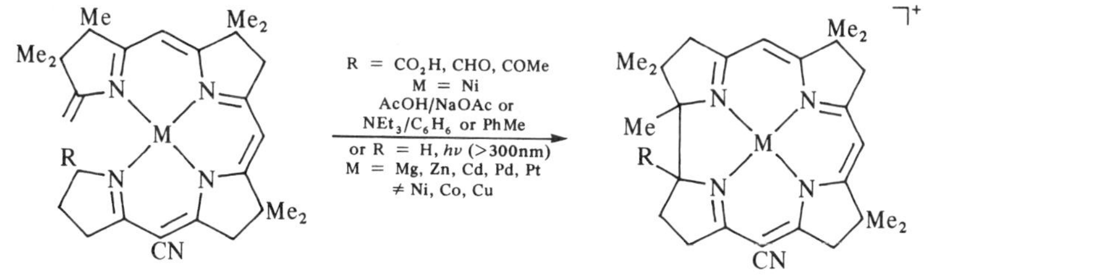

Scheme 87

$[_\pi 18_a]$. 1-Oxo analogues react similarly but under UV light. Electroreductive cyclizations and biomimetic transformations are also reported.

Another route to corrin *via* 4,5-seco intermediates has been established (Scheme 88).[250,255] It was shown by X-ray structure analyses of the Ni-4,5-secocorrin, and its cyclization product, that the pyrrolic A and B rings are properly oriented for the reaction.[250] A sulfur extrusion method was employed in the B$_{12}$ synthesis of Woodward and Eschenmoser.[253]

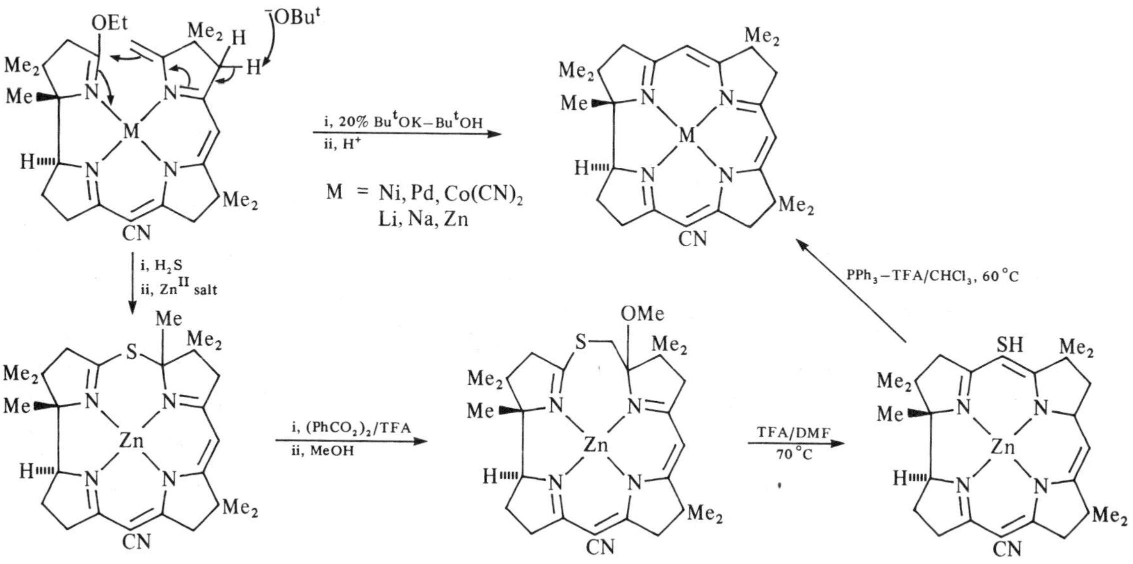

Scheme 88

2,3,17,18- and 7,8,12,13-TDC complexes are derived from ODC by catalytic hydrogenations.[239]

21.1.3.2.4 *Reactions of the corrin macrocycle*

Metal-free corrin is stable only in the protonated form and reversibly deprotonates at C-8 and C-13 in neutral or basic media (pK^*mcs 8.6 in dimethylcellosolve–water, 1:1).[250] Therefore, deuteration under basic conditions (NaOD–KCN/Bu^{t}OD–D$_2$O (1:1), r.t., M = H$_2$, Co^{III}(CN)$_2$, Ni^{2+}) occurs at those positions. The electrophilic proton–deuterium exchange (TFA/Bu^{t}OD–D$_2$O (1:1), r.t., M = H$_2$, Co^{III}(CN)$_2$, Ni^{2+}) is more facile at C-5 and C-15 than at C-10 as predicted from MO calculations.[239,250]

In general, [M(Con)]$^+$ [250,253] and [M(ODC)]$^+$ [239,258] react with electrophiles at C-5 (or C-15) under mild or stoichiometric conditions, and at C-5 and C-15 under more forcing conditions. Nitration appears to be less selective. Meso substitution is more or less reversible, and the substituents may be removed under acidic conditions if they are reasonably good leaving groups (Schemes 89 and 90).[253,255,259,260] For [Ni(ODC)]$^+$, nucleophiles also react at the meso positions (Scheme 91).[239,258]

Thermal rearrangement of Co^{III}(ODC)(CN)$_2$ accompanies metal reduction and Co^I(5-CN-ODC) and Co^I(5,15-(CN)$_2$ODC) are obtained, probably *via* radical reaction.[239,253]

Soft nucleophiles (X = CN$^-$, SCN$^-$, I$^-$) attack Ni(5-ClCH$_2$- or 5-HOCH$_2$-ODC) at C-10 to give Ni(5-Me-10-X-ODC), while S_N2 reactions occur with hard nucleophiles (HO$^-$, CF$_3$CO$_2^-$).[258]

Oxidation of natural cobalt(III) corrins by KMnO$_4$–pyridine gives 5- or 15-carboxy and 5,15-dicarboxy derivatives.[253,259] Acetoxylation of the meso methyl groups has also been reported.[260]

$M = Ni^{II+}, Co^{III}(CN)_2$	$ClSO_2NC/DMF$	$X = CN$	$Y = H$
$Co^{III}(CN)_2$	$Me_2\overset{+}{N}{=}CH_2 I^-$	CH_2NMe_2	H
	$PhSCH_2Cl, AgBF_4, Pr^i_2NEt/MeCN$	CH_2SPh	$H + CH_2SPh$
	$PhCH_2OCH_2Cl/sulfone, 75-80\,°C$	CH_2OCH_2Ph	CH_2OCH_2Ph

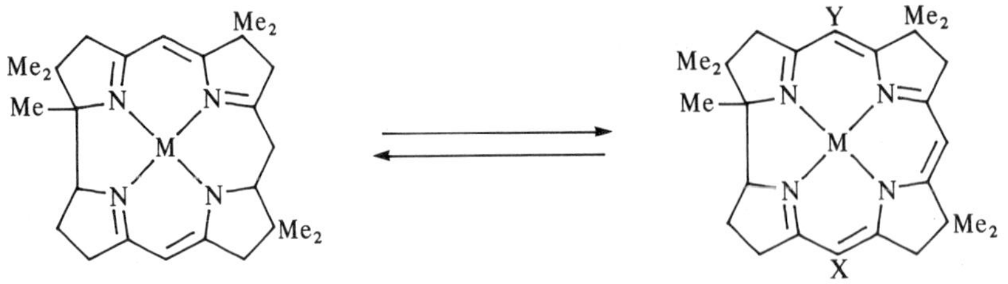

$M = Co^{III}(CN)_2$	$HS(CH_2)_3SH, HCl/CCl_4{-}CHCl_3$	$X = Me$	$Y = Me$
	$AcOH{-}H_2O$, Dowex 5W	CH_2OH	CH_2OH
	$AcOH{-}Ac_2O$, reflux/N_2	CO_2H	CO_2H
	0.1 M HCl, 230 °C, 40 h/N_2	CN	H
Zn^{II}	PPh_3, TFA/$CHCl_3$, 60 °C	SH	H

Scheme 89

$M = Ni^{II+}$	$MeI{-}SnCl_4$	$X = Me$	$Y = H$	$Z = H$
	chloromethylation	CH_2Cl	H	H
	hydroxymethylation, 50 °C	CH_2OH	H	H
	r.t.	H	H	CH_2OH
	$Me_2NH/HCHO/alcohol$	CH_2NMe_2	H	H
	$DMF{-}POCl_3/CHCl_3$	CHO	H	H
	$AcCl{-}SnCl_4$	Cl	Cl	H
	$Cl_2(Br_2)/CHCl_3$ or CH_2Cl_2	Cl (Br)	Cl (Br)	H
	$Br_2/AcOH$	Br	H	H
	$H_2SO_4{-}HNO_3$, 0 °C	NO_2	$NO_2 + H$	$H + NO_2$
	HNO_3 oleum	NO_2	NO_2	NO_2
$M = Ni^{II+}, Co^{III}(CN)_2$	$Cu(NO_3)_2, Ac_2O$	NO_2	$NO_2 + H$	$NO_2 + H$

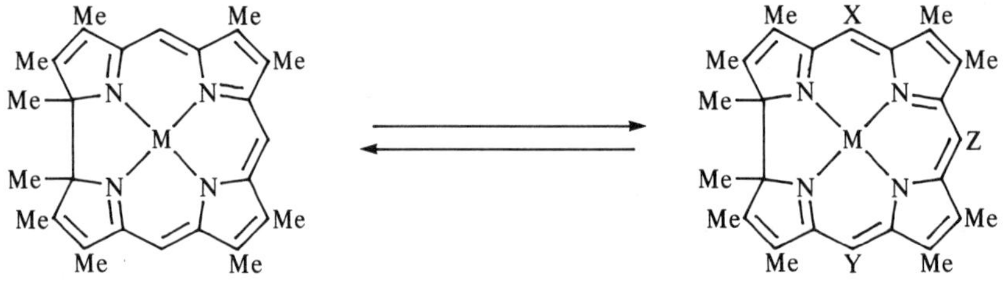

$M = Ni^{II+}$	$H^+/CHCl_3, \Delta$	$X = Br$	$Y = Br$	$Z = H$

Scheme 90

Interesting oxidative lactonizations and lactamizations of the natural corrins are known.[253] Under Udenfriend conditions, the acetate group of ring B cyclizes to C-6 accompanied by hydroxylation at C-5 (Scheme 92,i). If the acetate group forms a lactone ring at C-8, its migration to C-6 is compensated for by spiro lactone formation of the propionic side chain (Scheme 92,ii). In the absence of the acyl participation, the 5-methyl group shifts to C-6 giving a 5-oxo complex (Scheme 92,iii). The reaction is regio- and stereo-specific, and a Co ion-assisted mechanism has been suggested.

On the other hand, lactamization occurs at C-8 in aqueous alkali. The 7-hydroxyethyl derivative reacts similarly to give a cyclic ether. The reaction does not necessarily require O_2 and is inhibited by excess CN^-. For B_{12}, some ligand dependence was observed, and the order of reactivity is cyano > aqua > alkyl, sulfonato cobalamin (no reaction). In addition, colour changes during the reaction imply participation of the cobalt atom, but the mechanism is not clear (Scheme 93).

X = H	Y = H	OH⁻/CH₂Cl, MeOH or THF	Z = H

The following reagents and products are listed:

X	Y	Reagent	Z
H	H	OH^-/CH_2Cl, MeOH or THF	H
H	H	KCN/MeOH	CN
H	H	aq. NaOH/CHCl₃	CHO
NO₂	H	MeLi/moist THF	NO₂
Br	H	30% H₂O₂–NaOH	Br
Br	Br	MeLi/moist THF or NaHCO₃–H₂O₂	Br

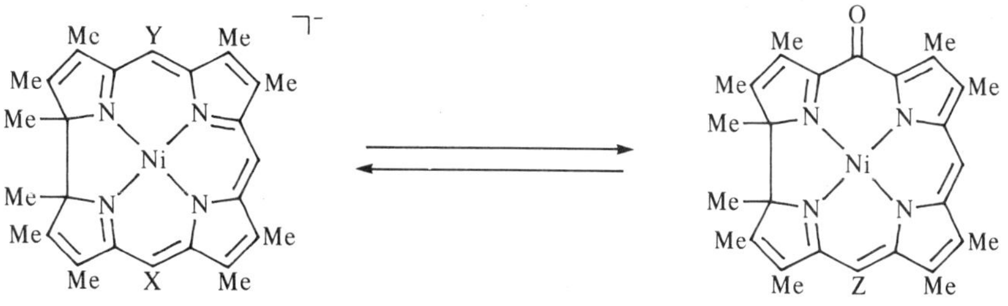

X = H   Y = OH   SnCl₂–HCl   Z = NO₂

Scheme 91

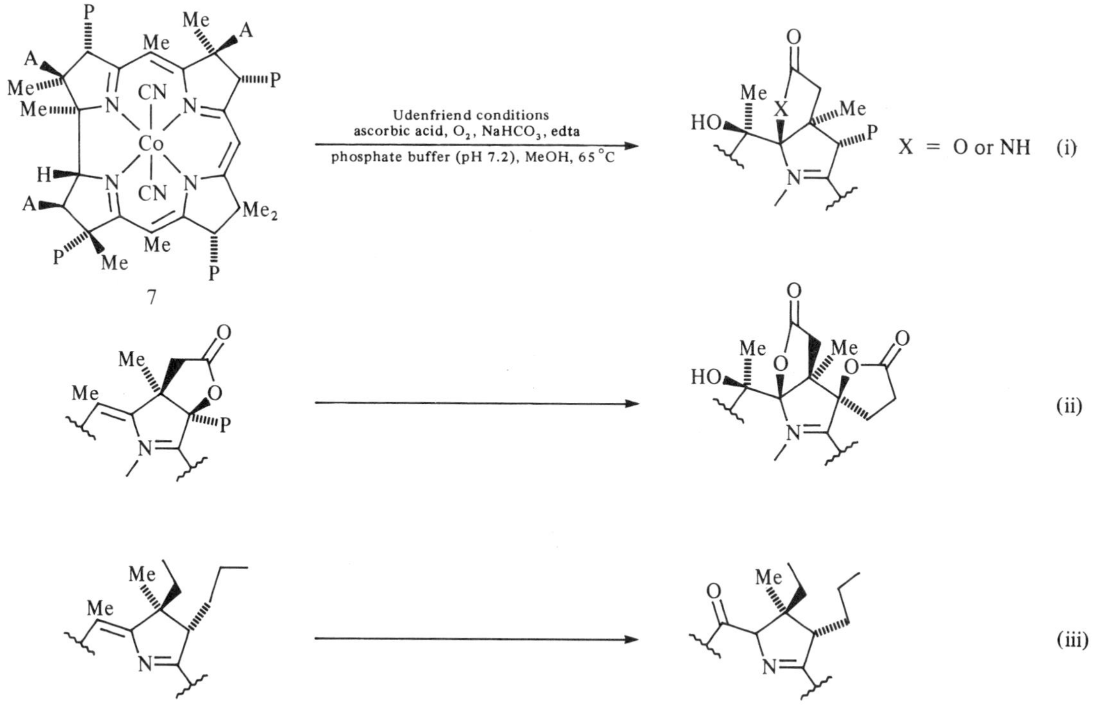

Udenfriend conditions
ascorbic acid, O₂, NaHCO₃, edta
phosphate buffer (pH 7.2), MeOH, 65 °C

X = O or NH   (i)

(ii)

(iii)

Scheme 92

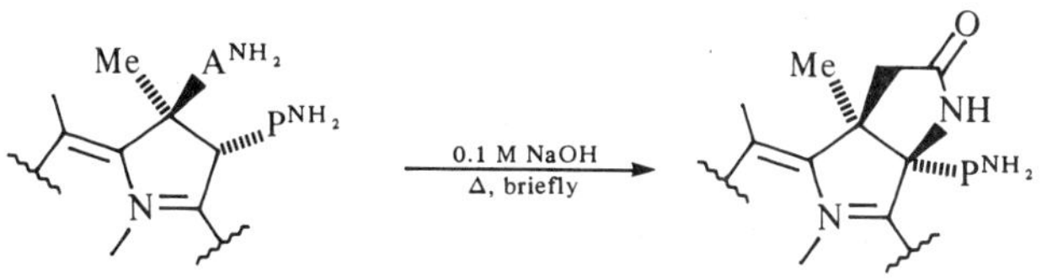

0.1 M NaOH
Δ, briefly

Scheme 93

Oxidation under acidic conditions gives a c-lactone instead of a c-lactam. The 7-acetamide reacts readily but the 7-acetic ester resists the cyclization. If an excess of oxidant is used, C-10 halogenation follows the lactonization (Scheme 94). The two reactions are competitive for cobalamins, and the course of the reaction depends on the β ligand and the reaction conditions.

Ni corrins lacking a quaternary carbon in rings B and C are dehydrogenated to the corresponding [Ni(7,8,12,13-TDC)]⁺ by aerial oxidation in ethanolic alkali.[239] [Ni(2,3,17,18-TDC)]⁺ is oxidized similarly to [Ni(ODC)]⁺, but if C-7 is quaternized by alkylation, only ring C is dehydrogenated.

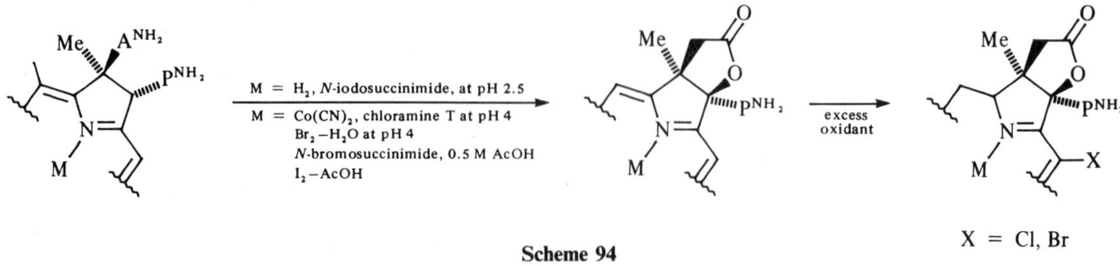

Scheme 94

$$X = Cl, Br$$

OsO$_4$ oxidation of [Ni(ODC)]$^+$ proceeds in a stepwise manner (rings B, D, A and C) and the 7,8-diol, 7,8,12,13-tetraol, 2,3,7,8,12,13-hexaol and 2,3,7,8,12,13,17,18-octaol are obtained.[239,261] The tetraol is converted to [Ni(2,3,17,18-TDC)]$^+$ by pinacol–pinacolone rearrangement followed by catalytic reduction (Scheme 95).

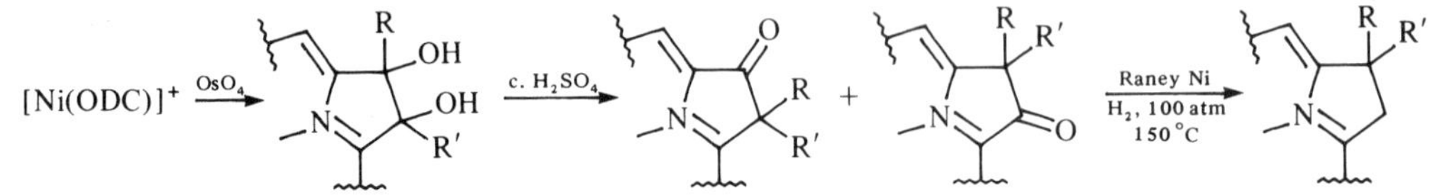

Scheme 95

Dehydrocorrins are catalytically hydrogenated to the less dehydrogenated analogues, but the course of the reaction depends on the steric constraints of the macrocycle (Scheme 96).[239] Zn–AcOH reduction of Co^{III}(17,18-DDC)(CN)$_2$ to Co^{III}(Con)(CN)$_2$ has also been reported.[256b]

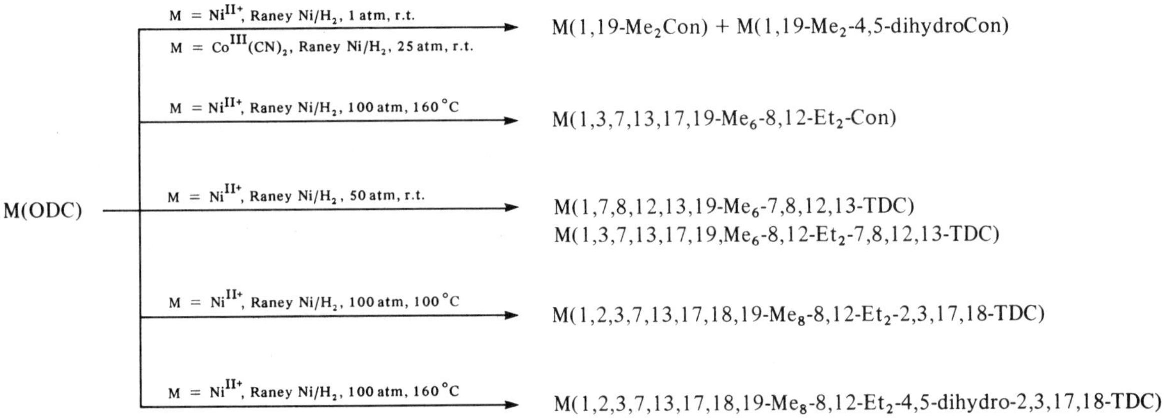

Scheme 96

The C-1—C-19 junction of ODC is labile and cleaved by Na-anthracenide. Optically active [Ni(*trans*-1,19-Me$_2$ODC)]$^+$ racemized at 80–190 °C depending on the counteranion and a Hoffmann-type ring opening was proposed.[239] Thermal rearrangement of 1,19-dialkyl-ODC complexes to porphyrin is considered to proceed *via* the same intermediate (Scheme 97). In the case of a 19-unsubstituted ODC, the 19-H is easily lost, and rearrangement of the C-1 substituent to the C-3 position occurs instead of a ring opening.[239,249] In some cases, the C-1 substituent migrates to C-2 and the C-2 substituent shifts to C-3. The reaction may be explained as a double [1,5]- or [1,17]-sigmatropic shift. The ease of migration is in the order allyl > ethoxycarbonyl > ethyl > methyl, and the 3,3-dimethylallyl group shifts without changing the position of its double bond.

19-Unsubstituted ODC are alkylated to 1,19-dialkyl-ODC with alkyl iodide or allyl bromide, while 1-ethoxycarbonyl- and 1,19-bis(ethoxycarbonyl)-ODC are decarboxylated to 19-unsubstituted ODC and corrole respectively by alcoholic alkali. 19-Acyl and -carboxyl groups are also eliminated by alcoholic alkali (2 N KOH, EtOH, 60 °C) and NEt$_3$–AcOH respectively.[262]

21.1.3.2.5 Metallation and demetallation

Metal ions are easily inserted into corrins, but their removal tends to cause ring opening of the macrocycle.[250,256a] So far, Li, Mg, Zn, Cd, Mn and Co have been removed from corrin and the ease

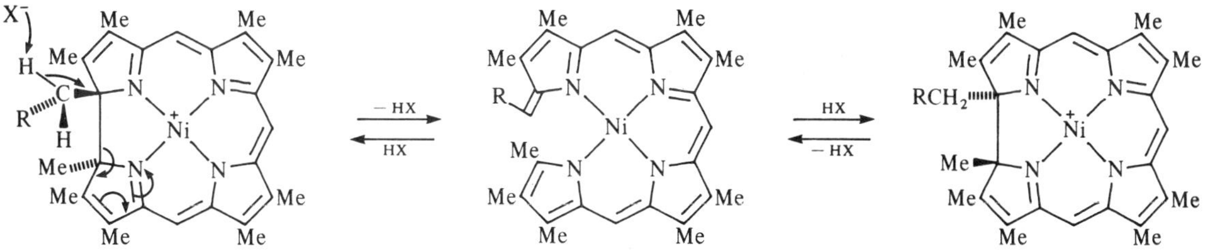

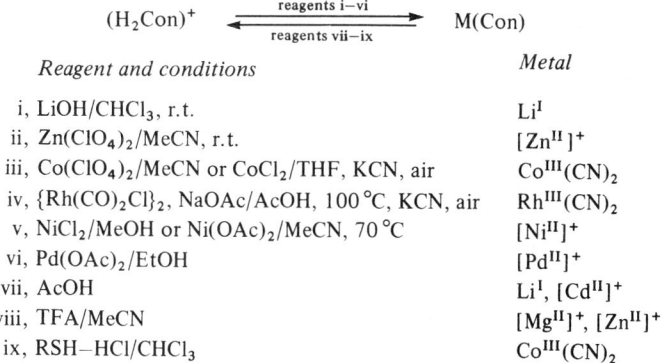

R = H, $\tau_{1/2}$ = 1–2 h, X = Br, I, 80 °C
              = camphorsulfonate, 110 °C
              = ClO$_4$, 189 °C

Δ in o-Cl$_2$C$_6$H$_4$

Ni(Por)(X = F, Cl, Br, I, NO$_3$, ClO$_4$) + Ni(5-R-Por) (X = ClO$_4$)
+ Ni(5-R-2,3-epoxy-Por)

Scheme 97

of dissociation is in the order Li > Cd > Zn > Mg > Mn > Co (Scheme 98).[250a,256,262b,263] For metallation of hydrogenobalamin, a metal salt–aqueous solvent system is frequently used (Scheme 99).[264] The insertion of Mn ion gives manganobalamin, but it is easily oxidized to the c-lactam (*cf.* Schemes 93 and 94).

$$(H_2Con)^+ \xrightleftharpoons[\text{reagents vii–ix}]{\text{reagents i–vi}} M(Con)$$

Reagent and conditions	*Metal*
i, LiOH/CHCl$_3$, r.t.	Li^I
ii, Zn(ClO$_4$)$_2$/MeCN, r.t.	[Zn^{II}]$^+$
iii, Co(ClO$_4$)$_2$/MeCN or CoCl$_2$/THF, KCN, air	Co^{III}(CN)$_2$
iv, {Rh(CO)$_2$Cl}$_2$, NaOAc/AcOH, 100 °C, KCN, air	Rh^{III}(CN)$_2$
v, NiCl$_2$/MeOH or Ni(OAc)$_2$/MeCN, 70 °C	[Ni^{II}]$^+$
vi, Pd(OAc)$_2$/EtOH	[Pd^{II}]$^+$
vii, AcOH	Li^I, [Cd^{II}]$^+$
viii, TFA/MeCN	[Mg^{II}]$^+$, [Zn^{II}]$^+$
ix, RSH–HCl/CHCl$_3$	Co^{III}(CN)$_2$

Scheme 98

$$\text{hydrogenobalamin} \xrightleftharpoons[\text{reagent vii}]{\text{reagents i–vi}} \text{metallobalamin}$$

i, Mn(OAc)$_2$/EtOH, Δ; ii, Fe(OAc)$_3$–Fe powder–NaOAc/AcOH, 80 °C/dark;
iii, CoCl$_2$/H$_2$O (alkaline pH) or EtOH (neutral), r.t. or 100 °C;
iv, {Rh(CO)$_2$Cl}$_2$, NaOAc/EtOH–AcOH, 100 °C or r.t.;
v, Cu salts/H$_2$O, 100 °C; vi, Zn(OAc)$_2$/H$_2$O, 100 °C; vii, Fe/MeCO$_2$H, or
NaBH$_4$ or Na$_2$S$_2$O$_4$/H$_2$O (for manganobalamin c-lactone or -lactam)

Scheme 99

21.1.3.2.6 *Coordination chemistry of the metallocorrins*

(i) Mn^{II,III}

Manganobalamin appears to exist in a six-coordinate state, *i.e.* as a diaqua and monoaqua Mn^{III} form ($\mu \approx 3.7$–3.9 BM) at pH 2.0 and neutral pH respectively, and as a cyanoaquo Mn^{II} complex ($\mu \approx 2$–2.2 BM) in 0.1 N KCN.[264]

(ii) Fe^{III}

Aqua-, cyano- and dicyano-ferribalamin have been reported.[264]

(iii) $Co^{I,II,III}$, $Rh^{I,II,III}$

Co^{III} corrinoids form five- and six-coordinate complexes, both of which are diamagnetic. The axial coordination is predominantly σ bonding and better σ donors form stronger bonds with the cobalt atom. Thus a linear relationship between pK_a of 4-substituted pyridines and $\log K_{obs}$ for their $Co^{III}(2,3,14,18\text{-TDC})(CN)$ complexes is observed (equation 35).[265] But, when various types of ligands are compared, the order of stability of the sixth ligand varies as the fifth ligand changes. For example, K_{obs} for $Co^{III}(\text{cobinanmide})(X)(H_2O)$ is $L = CN^- > SO_3^{2-} > NH_3 > OH^- > Im$ for $X = OH_2$, $L = CN^- > SO_3^{2-} \geqslant Im > NH_3 > OH^-$ for $X = CN^-$, and $CN^- > Im > NH_3 > SP_3^{2-} > OH^-$ for $X = Me$.[267]

$$Co^{III}(\text{Con})(X)(OH_2) + L \;\underset{k_{-1}}{\overset{k_1}{\rightleftharpoons}}\; Co^{III}(\text{Con})(X)(L) + H_2O \tag{35}$$

$$K_{obs} = \frac{[Co(\text{Con})(X)(L)]}{[Co(\text{Con})(X)(H_2O)][L]}$$

Con = 2,3,17,18-TDC, X = CN, L = 4-CN, H, Me, NH_2-pyridine

$\log K_{obs} = 0.32 pK_a + 0.50$ in H_2O at 25 °C

If the ligand is ambident, it coordinates preferentially through the less electronegative or more polarizable atom; *i.e.* N in NCO^-, S in SO_3^{2-}, SSO_3^{2-} and NCS^-, Se in $NCSe^-$, NIm in histidine, and N-9 in adenine.

Axial ligands with strong σ-donor character labilize the *trans* ligand independent of its π character (thermodynamic *trans* effect) and K_{obs} for $L = CN^-$ is 2.1 (X = Me), 8 (X = CN^-), 14.1 (X = Bzm) and $\geqslant 14$ (X = OH_2) M^{-1}.

Some ligand substitution rates for cobalamin (X = Bzm) have been reported. Values of k_1 (equation 35) are in the order of $\sim 10^3 \, M^{-1} \, s^{-1}$ for anionic ligands (L = N_3^-, NCO^-, NCS^-, CN^-) $10\text{--}10^2 \, M^{-1} \, s^{-1}$ for neutral ligands (L = NH_3, Im) in buffered solution with a total ionic strength of 0.054 at 25 °C. The aquation rate is highly ligand dependent and varies over a wide range, for instance, $k_{-1} = 1.8$ (NCS^-), 0.95 (NCO^-), 0.7 (NH_3), 0.03 (N_3^-), 6×10^{-4} (Im) and $< 10^{-9}$ (CN^-) s^{-1} (X = Bzm).

Electron-donating *trans* ligands (X) increase the rate of aquation (kinetic *trans* effect), *i.e.* for $L = NH_3$, $k_{-1} = 1.4 \times 10^{-5}$ (X = H_2O), 8.6×10^{-5} (X = Bzm), $\geqslant 3 \times 10^{-1}$ (X = CN^-, SO_3^{2-}, Me^-) s^{-1}. It was pointed out by Pratt[267a] that ligand substitutions of Co^{III}Con are faster than those of $Co^{III}(\text{dimethylglyoximato})_2(L)_2$, $Co^{III}(CN)_4(L)_2$ or $Co^{III}(NH_3)_4(L)_2$ by a factor of $\sim 10^7$ (kinetic *cis* effect).

Axial ligand dissociation is often induced by metal ions such as Ag^I, Hg^{II}, Pd^{II} and Au^{III} for CN^-, Ag^I and Hg^{II} for Bzm, and Cu^{II} for RS^-.

Ethynyl- and ethenyl-Co^{III} corrins form σ and π complexes respectively with Ag^I, and cyano $Co^{III}(\text{Con})$ complexes coordinate metal cyanide *via* a CN bridge, *i.e.* $\{M(CN)_5\}$—CN–Co(Con)CN (M = Fe^{III}, Fe^{II}, Co^{III}).

The pK_a values of Co-bound water molecules were reported to be ~ 6 and ~ 11 for the diaqua and ~ 11 for the cyanoaqua complexes. Ring dehydrogenation increases the pK_a values (equation 36).

$$\{Co^{III}(\text{Con})X(OH_2)\}^{n+} \;\overset{pK_a}{\rightleftharpoons}\; \{Co^{III}(\text{Con})X(OH)\}^{(n-1)+} + H^+ \tag{36}$$

	$\{Co^{III}\text{cobinamide}(OH_2)\}^{2+}$	$\{Co^{III}(2,3,17,18\text{-TDC})(OH_2)_2\}^{2+}$	$\{Co^{III}(ODC)(OH_2)_2\}^{2+}$
pK_{a_1}	6.0	6.2	6.15
pK_{a_2}		11.2	11.50

	$\{Co^{III}\text{cobinamide}(CN)(OH_2)\}^+$	$\{Co^{III}(2,3,17,18\text{-TDC})(CN)(OH_2)\}^+$
pK_a	11.0	11.3

Co^{II} corrinoids show less tendency for axial coordination than the Co^{III} analogues due to an unpaired electron in the metal d_{z^2} orbital ($g_\perp \approx 2.3$, $g_\parallel \approx 2.0$).[254,265] They do not coordinate weak ligands such as NO_3^- or ClO_4^-, even in aprotic solvents,[258] but form five-coordinate complexes with various ligands, *i.e.* alcohol, water, amines, phosphines, halogens, pseudohalogens and cyanide.[267] Two pyridine molecules may ligate to $[Co^{II}Con]^+$ at high pyridine concentrations (since ESR superhyperfine structures consisting of five lines, probably due to two axial nitrogen atoms ($I = 1$), are observed under certain conditions). $Co^{II}(\text{Con})$ complexes are all low-spin compounds irrespective of their coordination state, but hyperfine coupling constants due to the Co nucleus ($I = \frac{7}{2}$, eight lines) decrease in the order four- ($A_\parallel^{Co} \approx 130 \times 10^{-4}\,\text{cm}^{-1}$) > five- ($A_\parallel^{Co} \approx 110 \times 10^{-4}\,\text{cm}^{-1}$) > six-coordinate ($A_\parallel^{Co} \approx 95 \times 10^{-4}\,\text{cm}^{-1}$).

For [Co^{II}(2,3,17,18-TDC)]$^+$, the association constant K_1 was found to increase in the range 10^2–$10^4\,\text{M}^{-1}$ as the pK_a of 4-substituted pyridines increased.[254,265] The K_1 is linearly correlated to the pK_a in the [Co^{II}(ODC)]$^+$–N-donor system (equation 37).[269] Though the K_1 for L = Im fits the equations, aliphatic amines show lower K_1 values than expected from their pK_a values, suggesting some metal–axial ligand π interaction. The steric hindrance of the 2-methyl group lowers K_1 by a factor of $\sim 10^3$.

$$\text{Co}^{II}(\text{Con}) + \text{L} \underset{\text{CH}_2\text{Cl}_2,\ 15\,°\text{C}}{\overset{K_1}{\rightleftharpoons}} \text{Co}^{II}(\text{Con})\text{L} \qquad (37)$$

$\log K_1 = 0.24\,\text{p}K_a + 2.1,\ \log(K_1^X/K_1^H) = 1.6\sigma$

Con = 8,12-diethyl-1,2,3,7,13,17,18,19-octamethyl-ODC

L = substituted pyridines, imidazole, aliphatic amines

The structure of a μ-iodo dimer, [{Co^{II}(cobyrinic acid hexamethyl ester)}$_2$I]I has been determined.[253] The cobalt atoms are displaced from the N$_4$ plane by 0.11 Å and 0.13 Å respectively, and the Co—I—Co group is slightly bent. The dimer is diamagnetic due to strong antiferromagnetic coupling between the two Co atoms. It exists in equilibrium with the monomer in benzene solution.

Co^{II}(Con)L complexes bind molecular oxygen at low temperature.[254,265,267] The ease of O$_2$ addition to five-coordinate cobinamides is in the order L = MeOH > adenine > Bzm > CN$^-$. The ESR hyperfine coupling constants ($A^{Co} = 12$–$18 \times 10^{-4}\,\text{cm}^{-1}$) for the dioxygen adducts are significantly lower than those of Co^{II}(Con)L ($A_\parallel^{Co} \approx 100 \times 10^{-4}\,\text{cm}^{-1}$), and a Co(III)—O$_2^{\overline{\cdot}}$ formulation seems to be appropriate. Detailed analyses of the ESR spectra of crystalline O$_2$B$_{12}$ predict a Co—O—O angle of 111° (105° in O$_2$Co porphyrin); the projection of the O—O bond on to the corrin plane approximately bisects the N^{22}—Co—N^{23} angle and is perpendicular to the Bzm plane.[253]

Co^I corrinoids may coordinate an axial ligand, though, except for Co^{I}ODC, they are strong nucleophiles and react with various electrophiles.

The axial coordination chemistry of Rh corrinoids parallels that of their Co analogues but the former bind axial ligands more tightly than the latter.[264] For example, dicyanorhodibamide remains unchanged at pH 2.7 while the Co analogue undergoes ligand exchange. In 0.1 N KCN solution, cyanocobalamin is converted to dicyanocobalamin, but cyanorhodibalamin is inert.

(iv) Ni^{II}, Pd^{II}, Pt^{II}

There is no solid evidence for the axial ligation of divalent Ni, Pd and Pt corrins.

(v) Cu^{II}

Cu^{II} corrinoids do not coordinate axial ligands.[264] MCD, ESR and ENDOR spectra indicate the Cu atom is in the N$_4$ plane with an unpaired electron in a $d_{x^2-y^2}$ orbital which has little interaction with the corrin π system.

(vi) Zn^{II}, Cd^{II}

The pH-dependent spectral changes of zincobalamins imply coordination of the Bzm group to the Zn atom in neutral and basic solutions.[264] Chloride ion is likely to coordinate to the metal atom in Zn(Con)Cl[270] and Cd(Con)Cl,[262b] since substitution of the counterion from Cl$^-$ to BF$_4^-$ caused a low field shift of the meso-H signals of Zn(Con). Zn(1-MeCon)Cl exhibits two signals corresponding to the C-10 H due to *syn–anti* isomerism of the chloride ion relative to the angular methyl group.[270]

21.1.3.2.7 Reactions at the metal

Metal-centred corrinoid chemistry is represented by the 'chemistry of the Co—C bond'. The reactions which lead to Co—C bond formation are summarized in Scheme 100.[271]

$Co^I(Con)$ complexes, except for $Co^I(ODC)$, are strong nucleophiles and undergo S_N2 reactions with sp^3 carbon attached to a leaving group (Scheme 100, i; X = halogen, tosyl, phosphoryl, epoxy oxygen, *etc.*). Their extremely high nucleophilicity is evident when compared with conventional nucleophiles using the Pearsons constants, $n_{MeI} = \log (k_y/k_{MeOH})$, where k_y and k_{MeOH} are the second order rate constants of the substitution reactions of methyl iodide with nucleophile y and methanol respectively in methanol solution at 25 °C. For example, $n_{MeI} = 14.4$ for $Co^I Con$, 9.92 for PhS^-, 8.72 for Et_3P, 7.42 for I^-, and 6.7 for CN^-. The nucleophilicity of $Co^I(2,3,17,18\text{-}TDC)$ is slightly lower than that of $Co^I(Con)$ while $Co^I(ODC)$ does not react with MeI.[254] The macrocycle with better π-acceptor character reduces the electron density on the Co atom. Though there remains little doubt about the S_N2 mechanism for the reaction of $Co^I Con$ and ordinary alkyl halides, electron transfer mechanisms are most likely for reactions with bridgehead halides.[271] A concerted displacement mechanism has also been proposed for the substitution reactions of vinyl halides with Co^I species which proceed with retention of configuration.

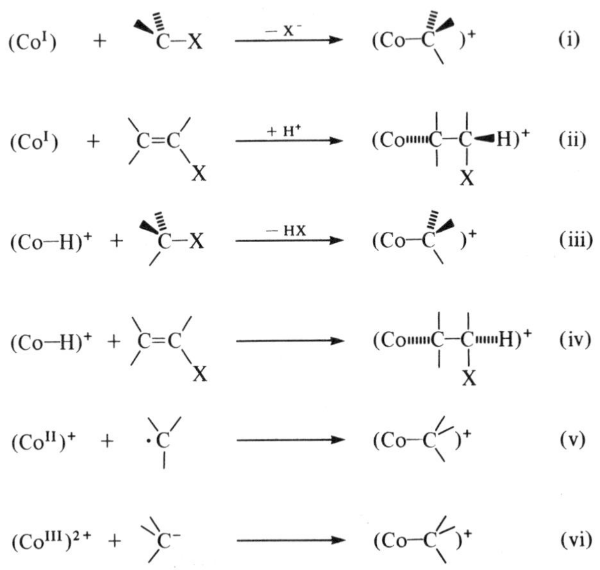

$$(Co^I) \quad + \quad \overset{}{\underset{}{C}}\!-\!X \quad \xrightarrow{\;-X^-\;} \quad (Co\!-\!\overset{}{\underset{}{C}}\,)^+ \qquad (i)$$

$$(Co^I) \quad + \quad \overset{}{\underset{}{C}}\!=\!\overset{}{\underset{X}{C}} \quad \xrightarrow{\;+H^+\;} \quad (Co{\cdots}C\!-\!\overset{}{\underset{X}{C}}\!\blacktriangleleft H)^+ \qquad (ii)$$

$$(Co\!-\!H)^+ \quad + \quad \overset{}{\underset{}{C}}\!-\!X \quad \xrightarrow{\;-HX\;} \quad (Co\!-\!\overset{}{\underset{}{C}}\,)^+ \qquad (iii)$$

$$(Co\!-\!H)^+ \quad + \quad \overset{}{\underset{X}{C}}\!=\!\overset{}{\underset{}{C}} \quad \longrightarrow \quad (Co{\cdots}C\!-\!\overset{}{\underset{X}{C}}{\cdots}H)^+ \qquad (iv)$$

$$(Co^{II})^+ \quad + \quad {\cdot}\overset{}{\underset{}{C}} \quad \longrightarrow \quad (Co\!-\!\overset{}{\underset{}{C}}\,)^+ \qquad (v)$$

$$(Co^{III})^{2+} \quad + \quad \overset{}{\underset{}{C}}^- \quad \longrightarrow \quad (Co\!-\!\overset{}{\underset{}{C}}\,)^+ \qquad (vi)$$

Scheme 100

The β addition of Co^I nucleophiles to a double bond conjugated to π acids, such as CN, CO_2R, Ph, $CH{=}CH_2$, proceeds *via* a π complex to give the *trans* adduct, while direct σ adduct formation is more likely for the *trans* β addition of Co^I species to acetylene (Scheme 100, ii).

$Co^I(Con)$ exists in equilibrium with the hydrido form, $\{HCo(Con)\}^+$ ($pK_a \approx 1.0$), which gradually decomposes to $Co^{II}(Con)$ and H_2. The hydrido complexes are more reactive nucleophiles than are $Co^I(Con)$, and their addition to unsaturated bonds gives *cis* products probably *via* an unsymmetric four-centre mechanism (Scheme 100, iii, iv). The Co—C bond is also formed between $\{Co^{II}(Con)\}^+$ and alkyl radicals (Scheme 100, v), and $\{Co^{III}(Con)\}^{2+}$ and carbanions (Scheme 100, vi).

In the absence of light, the Co—C bond is stable relative to the majority of other metal–carbon bonds. Pratt[267] attributed the exceptional stability of the Co—C bond to the hybridization of metal $4s$ and $4p$ to $3d$ orbitals that enables efficient orbital overlap between cobalt and carbon atoms. According to this theory, alkyl–metal complexes with metal atoms in a $(t_{2g})^6(e_g)^0$ state $[(t_{2g})^3(e_g)^0$ and $(t_{2g})^6(e_g)^2$ states as well] are stable due to the significant mixing of metal $4s$ and $4p$ to e_g $(3d_{x^2-y^2}, d_{z^2})$. The ligand fields of the macrocycle and the carbanion enhance the orbital mixing by bringing the latter two levels closer to the former. However, the apparent stability of alkylcobalt complexes is higher than expected from the Co—C bond energy (84–125 kJ mol^{-1} for primary alkyl–cobalt macrocyles).[272a] Since the recombination rate of a radical pair is very fast ($k_{-1} \approx 10^9\,M^{-1}\,s^{-1}$ for Me$\cdot$ + Co^{II} balamin), the stability of alkyl–Co complexes may be kinetic rather than thermodynamic. Radical scavengers are known to enhance their decomposition. The alkyl–Co complexes become less stable as the alkyl group becomes electron rich, π-conjugative or bulky, as expected from radical stability or the steric hindrance. Interestingly, Me—Co(1-MeCon) complexes exist in a *syn/anti* ratio of 8:92 under thermal or photo equilibrating conditions.[254] Coordination of the sixth ligand sterically destabilizes (increases the F strain) but electronically stabilizes (R—Co^{III} relative to $R^{\ddagger}Co^{II}$) the alkyl—Co bond.[272a]

The Co—C bond can be cleaved either homolytically or heterolytically.[267,272] The photoinduced or thermal fission is homolytic and enhanced by radical acceptors such as mercaptide, Co^{II}, Cr^{II} and Sn^{III}. The radical transfer proceeds with the configuration inversion at the metal-bound carbon atom, and the S_N2 mechanism is considered (reverse process of Scheme 100, v). Electrophiles such as halogen molecules and heavy metals in high oxidation state (Hg^{II}, Pd^{II}, Tl^{III}, Sb^{III}, Bi^{III}, Sn^{IV}, Pb^{IV} and Co^{III}) abstract the alkyl group as an anionic form from the Co complex (reverse process of Scheme 100, vi). For Hg^{II}, the ease of reaction is in the order $R = Me > Et > Pr^i$, and the configuration at the carbon centre is inverted, *i.e.* characteristic of the S_E2 reaction. Alkylcobalt complexes resist acid-catalyzed decomposition but a β-hydroxy group facilitates the alkyl elimination. β elimination is also observed in the cyanide ion-induced cleavage of the Co–adenosyl bond of B_{12}, where CN^- ion promotes the reaction by a *trans* effect (Scheme 101).

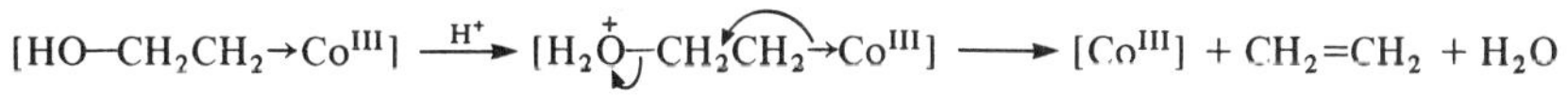

$$[HO\text{—}CH_2CH_2\text{→}Co^{III}] \xrightarrow{H^+} [H_2\overset{+}{O}\text{—}CH_2CH_2\text{→}Co^{III}] \longrightarrow [Co^{III}] + CH_2\text{=}CH_2 + H_2O$$

Scheme 101

Strong nucleophiles such as Co^I complexes and soft nucleophiles such as R_3P, RS^- and RSe^- undergo S_N2 reactions with alkyl–Co complexes (the reverse process of Scheme 100, i), while strong bases such as OH^- cause an $E2$ reaction with β-cyano-, -hydroxy- or -alkoxyethyl-Co complexes (reverse process of Scheme 100, ii). Redox reactions also cause homolytic cleavage of the Co—C bond. For example $[Ir^{IV}Cl_6]^{3-}$ and Au^{III} salts generate alkyl radicals and a Co^{III} species by outer sphere single electron oxidation (Scheme 102).

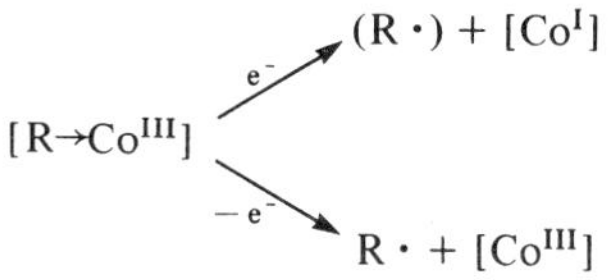

Scheme 102

An interesting and mechanistically unsettled problem of organocobalt chemistry is the 1,2-shift of a β-carboxy or -hydroxy group in the homolytic Co—C bond fission–recombination process. The evidence for cobalt participation in the rearrangement is substantial, but its role is not yet clarified (Scheme 103). The heterolytic process also appears to be viable in the migration of a β-hydroxy or -alkoxy group.

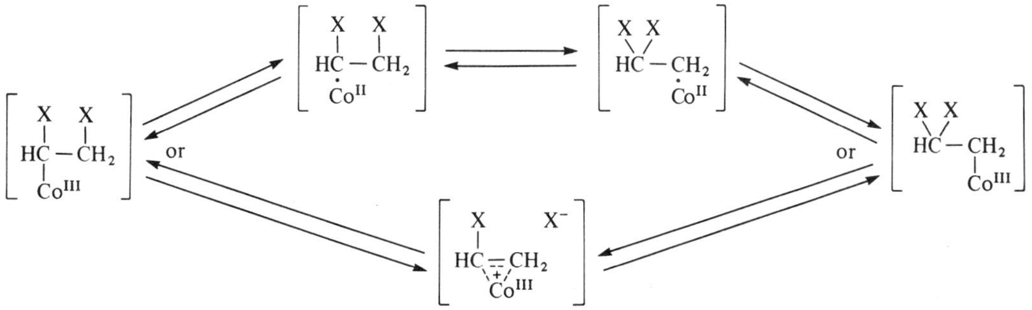

Scheme 103

Co corrinoids play central roles in the two classes of enzymic reactions, *i.e.* methyl transfer mediated by vitamin B_{12} and mutase or isomerase reactions catalyzed by coenzyme B_{12}.[253] Though there remain many ambiguities, the former is considered to be a combination of Scheme 100, i and its reverse process, and the latter to be represented by Scheme 103.

Rh^I corrinoids show reactivity similar to the Co^I analogue, but the alkyl—Rh bonds are more resistant to light than the alkyl—Co bonds.[264]

21.1.3.2.8 Photochemistry

Many $Co^{III}(Con)$ complexes are photosensitive and cleavage of axial bonds to Co is induced by visible light.[254,267] For alkyl, mercapto, sulfinato and selenocyanato Co^{III} corrins, homolytic fission (photoreduction) proceeds, whereas heterolytic dissociation (photoaquation) occurs for amino and cyano complexes. These photoprocesses are considered to be initiated by $\pi \to \pi^*$ excitations.

21.1.3.8.9 Redox chemistry

Aquamanganibalamin appears to be reduced from an Mn^{III} to an Mn^{II} state in 0.1 N KCN solution.[264] $NaBH_4$ reduction of ferribalamin is reported to give a green coloured Fe^I complex.[264] A variety of reductants have been used to reduce Co^{III} corrinoids, but the most common reductant is $NaBH_4$.[267a] Since Co^I complexes are converted to Co^{II} species by protons, Co^{II} corrinoids are conveniently prepared in acidic media, while Co^I corrinoids are obtained in neutral to alkaline solutions by $NaBH_4$ reduction of a Co^{III} precursor. $[Co^{III}(Con)]^{2+}$ is reduced by OH^- ion rapidly to $[Co^{II}(Con)]^+$ but only slowly to $[Co^I(Con)]$. Co^I corrins[267a] and $Co^I(2,3,17,18\text{-}TDC)$[265] are oxidized by O_2 instantaneously to the Co^{II} state and then finally to Co^{III} complexes. The oxidation of $Co^I(ODC)$ is slower than these two, and I_2 and $FeCl_3$ were used to oxidize it to $[Co^{II}(ODC)]^+$.[269,273] N_2O oxidizes Co^I alamin to a Co^{II} species, and I_2 and $[Fe(CN)_6]^{3-}$ to Co^{III}.[267a] Rh^{III} alamin chloride can be reduced by $NaBH_4$, but cyano and dicyano Rh^{III} alamin resist reduction.[264] Like the Co^I analogue, Rh^I alamin is oxidized by CN^- to the $Rh^{III}(CN)_2$ complex.

Electrochemical reduction of $[Ni^{II}(ODC)]^+$ is ligand centred and up to four successive ring reductions are observed (Scheme 104).[247]

$$[Ni^{II}L]^+ \xrightarrow{-0.30\,V} [Ni^{II}L] \xrightarrow{-1.71\,V} [Ni^{II}L]^- \xrightarrow{-2.20\,V} [Ni^{II}L]^{2-} \xrightarrow[\substack{\text{near limit}\\ \text{of detection}}]{} [Ni^{II}L]^{3-}$$

$$g = 2.002 \text{ (THF)}$$

L = (1,19-bis(ethoxycarbonyl)ODC)$^-$

$$[Ni^{II}L] \xleftarrow{-1.17\,V} [Ni^{II}L]^- \xleftarrow{-2.05\,V} [Ni^{II}L]^{2-}$$

L = [1-methyl(ODC)]$^{2-}$

in DMF with a dropping Hg electrode, $E_{1/2}$ *vs.* SCE

Scheme 104

21.1.4 PENTAPYRROLIC MACROCYCLES AND RELATED COMPOUNDS

21.1.4.1 General Properties

Addition of one more pyrrole unit to porphyrins, corroles or phthalocyanines gives pentaphyrin (**38**), sapphyrin (**39**), smaragdyrin (**42**) or superphthalocyanine (**45**) (Figures 16 and 17).

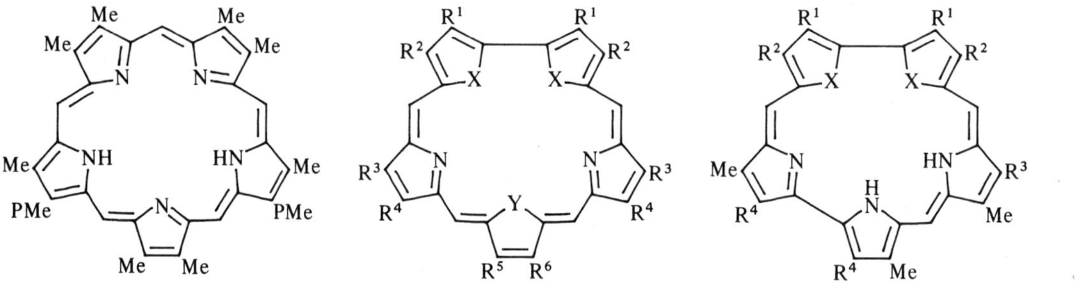

(38)

(39) X = Y = NH
a: $R^1 = R^2 = R^3 = R^4 = R^5 = R^6 = $ Me
b: $R^1 = R^2 = $ H, $R^3 = R^4 = R^5 = R^6 = $ Me
c: $R^1 = R^4 = R^5 = R^6 = $ Me, $R^2 = R^3 = $ Et
d: $R^1 = R^4 = R^5 = $ Me, $R^2 = R^3 = R^6 = $ Et
e: $R^1 = R^4 = R^5 = $ Et, $R^2 = R^3 = R^6 = $ Me
(40) X = O, Y = NH
a: $R^1 = R^2 = $ H, $R^3 = R^4 = R^5 = R^6 = $ Me
b: $R^1 = R^2 = $ H, $R^3 = R^5 = $ Me, $R^4 = R^6 = $ Et
(41) X = NH, Y = S
$R^1 = R^4 = $ Et, $R^2 = R^3 = $ Me, $R^5 = R^6 = $ H

(42) X = NH
a: $R^1 = R^2 = $ H, $R^3 = R^4 = $ Me
b: $R^1 = R^3 = R^4 = $ Et, $R^2 = $ Me
c: $R^1 = $ Et, $R^2 = R^3 = $ Me, $R^4 = $ H
(43) X = O
a: $R^1 = R^2 = $ H, $R^3 = R^4 = $ Et
b: $R^1 = R^2 = R^4 = $ H, $R^3 = $ Me

Figure 16 Pentapyrrolic macrocycles

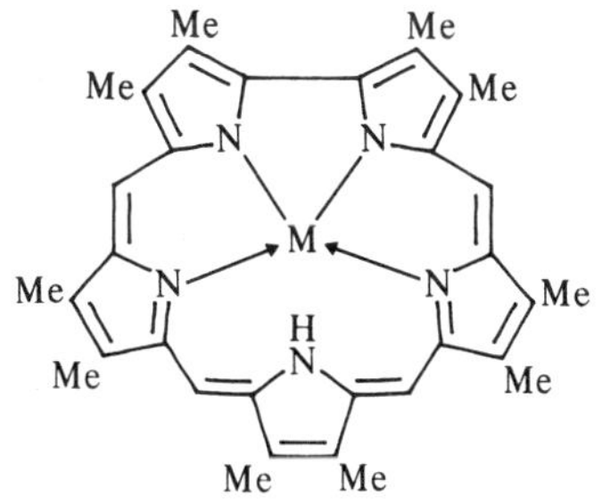

(44) M = Mn^{II}, Fe^{II}, Co^{II}, Ni^{II}, Zn^{II}, Cd^{II}

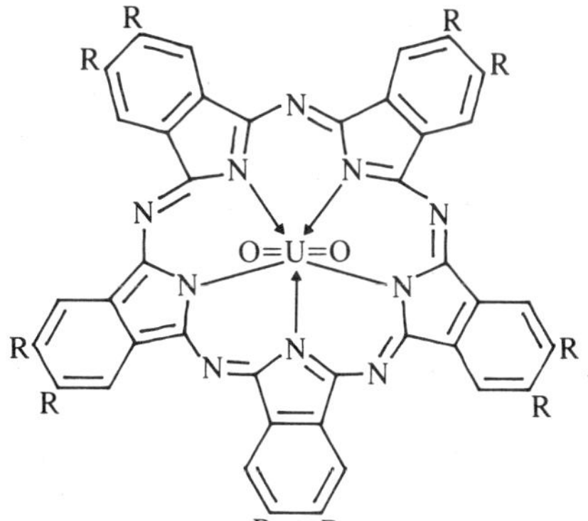

(45) a: R = H
b: R = H and Me
c: R = Me
d: R = Bu^n

Figure 17 Metallopentapyrrolic macrocycles

These pentapyrrolic macrocycles contain a 22-π-electron ring system, and are expected to be the largest members of the [4n + 2] π aromatics. In fact, their spectroscopic characteristics, *i.e.* the intense absorption band in visible region, the large diamagnetic shift of peripheral H and inner NH NMR signals, and relatively strong parent (M$^+$) and dicationic (M^{2+}) mass peaks, indicate the existence of a delocalized cyclic aromatic π system (Table 11).

21.1.4.2 Synthesis

Sapphyrin (**39d**) was the first cyclic pentapyrrole obtained by Woodward *et al.*[274] The acid-catalyzed cyclization of the tetrapyrrole (**46**) gave (**39d**) instead of corrole (**30**) *via* a disproportionation reaction. Accordingly, addition of diformylpyrrole (**47**) to (**46**) leads to (**39c**) (Scheme 105). However, the [3 + 2] route is facile and more satisfactory in general (Scheme 106).[274,275]

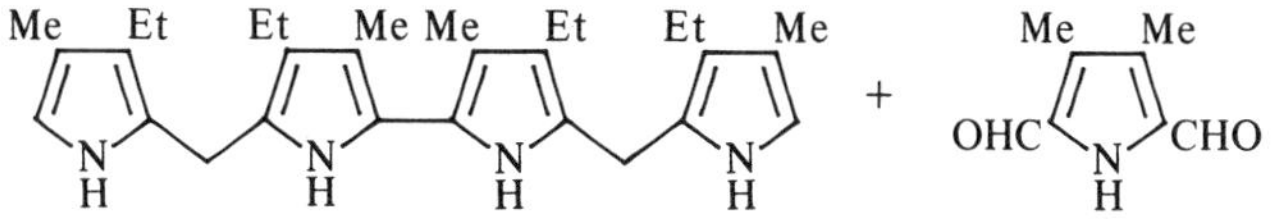

$$\xrightarrow[\text{O}_2 \text{ in air}]{\text{88\% aq. HCO}_2\text{H—MeOH}} \quad (39)(\sim 15\%)$$

Scheme 105

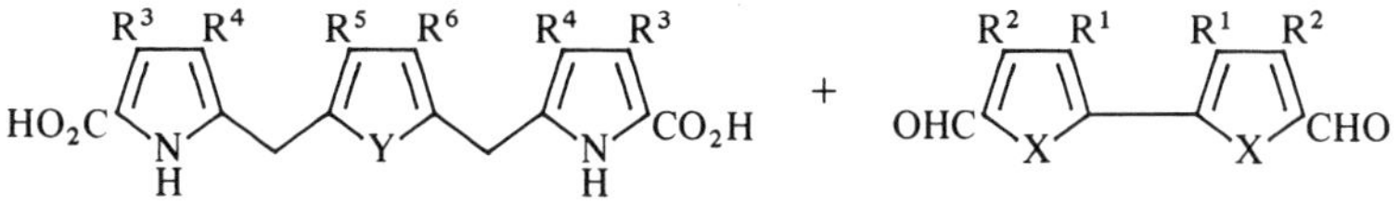

$$\xrightarrow[\substack{\text{or 48\% HBr—AcOH}\\ \text{THF—CHCl}_3\\ \text{O}_2 \text{ in air}}]{\text{TsOH}\cdot\text{H}_2\text{O—EtOH}} \quad \begin{array}{l}(39)(\sim 71\%)\\ (40)(\sim 40\%)\\ (41)(\sim 20\%)\end{array}$$

Scheme 106

Alternatively, the sulfur extrusion method may be useful if the α-pyrrolyl sulfide unit or its analogue (**50**) is more easily available than (**49**).[275]

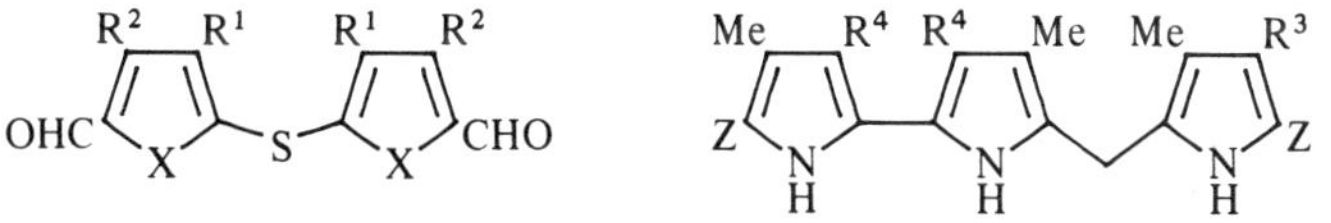

(51) Z = H or CO$_2$H

Similarly, smaragdyrin (**42**)[274] and dioxasmaragdyrin (**43**)[275] are synthesized from the tripyrroles (**51**) and (**49**).

Table 11 ^{1}H NMR Chemical Shifts and Absorption Maxima (Soret Band) of Pentapyrrolic Macrocycles

	NMR solvent	$\delta_{meso\text{-}H}$	$\delta_{\beta\text{-}H}$	δ_{NH}	Vis solvent	Soret λ_{max} (ε)	Ref.
(38)	1% CF$_3$CO$_2$H in CDCl$_3$ (diprotonated)	12.55, 12.46, 12.44			CH$_2$Cl$_2$	458 (5.38)	a
(39a)	CF$_3$CO$_2$H in CDCl$_3$ (diprotonated)	11.70, 11.51		−4.84, −5.0, −5.46	CHCl$_3$ (diprotonated)	457 (5.77)	b
(39e)	CDCl$_3$	10.50, 10.32		−3.9	0.5% HClO$_4$ in MeCOMe (diprotonated)	450 (5.72)	c
(40b)	CDCl$_3$	10.48, 10.45, 10.38, 10.34	10.04, 9.72	−6.55	0.5% HClO$_4$ in MeCOMe (diprotonated)	436 (5.88)	c
	20% CF$_3$CO$_2$H in CDCl$_3$ (diprotonated)	12.23	11.84, 11.27	−5.1, −6.5			
(41)	CDCl$_3$	10.87, 10.31	10.12		CHCl$_3$	468 (5.32)	c
					0.5% HClO$_4$ in MeCOMe (diprotonated)	461 (5.67)	
(42a)					CHCl$_3$	453 (4.92)	d
(43a)	CDCl$_3$	10.56, 10.10	9.86, 9.76, 9.41, 9.37	−4.85	C$_5$H$_5$N	447 (5.28)	c
					HBr–MeCO$_2$H–CHCl$_3$ (diprotonated)	448 (5.26)	
						459 (5.33)	
(44)	Zn complex				CHCl$_3$	462	b
	Co complex				CHCl$_3$	468	
(45a)	C$_6$D$_6$	9.06 ($\delta_{o\text{-}H}$)	7.68 ($\delta_{m\text{-}H}$)		C$_6$H$_6$	424 (4.70)	e
					PhMe	420 (4.73)	
(45d)	C$_6$D$_5$CD$_3$	9.13			PhMe	417 (4.84)	f
					1,2,4-Cl$_3$C$_6$H$_3$	419 (4.86)	

[a] Ref. 277.
[b] Ref. 274.
[c] Ref. 275.
[d] M. M. King, Ph.D. Dissertation, Harvard University, 1970.
[e] T. J. Marks and D. R. Stojakovic, *J. Am. Chem. Soc.*, 1978, **100**, 1695.
[f] Ref. 201.

This strategy is not applicable for pentaphyrin, since the condensation reaction of α,α′-free oligopyrrole and α,α′-diformyldipyrrole invariably results in porphyrin formation.[276] The disproportionation problem can be avoided by using an α,α′-diformyltripyrrole (**52**) and an α,α′-free dipyrrole (**53**) (Scheme 107).[277]

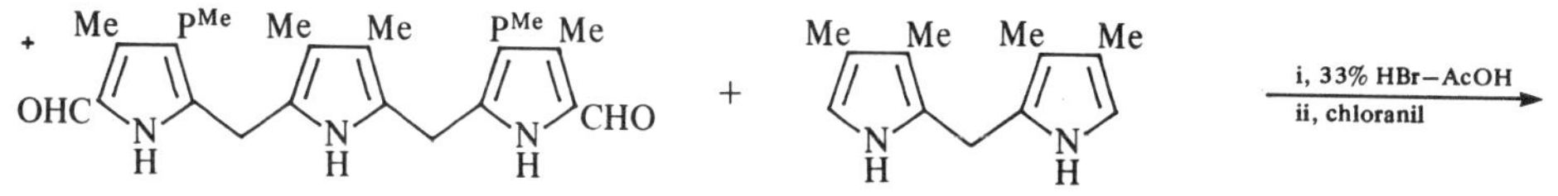

(38)

(52) (53)

Scheme 107

The general phthalocyanine synthesis takes an abnormal course when uranium oxychloride is employed as the template (Scheme 108).[201] The product showed a broadened and bathochromically shifted optical spectrum compared to those of phthalocyanines, and turned out to be the pentameric compound (**45a**) as elucidated by X-ray analysis.

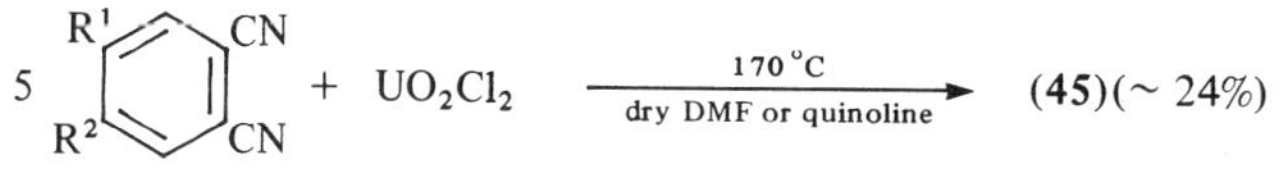

Scheme 108

21.1.4.3 Reactions at the Macrocycle

Like porphyrins, sapphyrins are easily diprotonated in acidic media; the basicity of sapphyrins (**39**)–(**41**) and pentaphyrins (**38**) is reflected in the enhanced $(M + 2)^+$ signal in their mass spectra due to proton capture.[274–276] The less basic smaragdyrins (**42**) and (**43**) exhibit a normal mass spectral pattern.

Electrophilic proton exchange occurs readily for sapphyrin (**39**) and meso deuteration is completed in trifluoroacetic acid and at room temperature overnight.[274] The substitution proceeds at a moderate rate for two of the meso protons but very slowly for the other two in the dioxa analogue (**40**) at 100 °C.[275] These observations suggest a reactivity of sapphyrins intermediate between corroles and porphyrins.

21.1.4.4 Metallation and Demetallation

Sapphyrins (**39**) are metallated with divalent metal acetates in the presence of sodium acetate (Scheme 109).[274] For the Zn^{II} and Co^{II} complexes, structure (**44**) was proposed with the dianionic macrocycle coordinating to the metal ion as a tetradentate ligand. These metallosapphyrins experience ring contraction to porphyrin complexes under mass spectral conditions.

$$(39a) + M^{II}(OCOMe)_2 \xrightarrow[MeOH-CHCl_3]{NaOCOMe} (44)$$

M = Mn, Fe, Co, Ni, Zn, Cd

Scheme 109

Metallation of dioxasapphyrin (**40d**) was reported to be unsuccessful, probably because of the too large ring size.[275] Smaragdyrin (**42a**) decomposed under the metallation conditions.[274]

Metal-free superphthalocyanine has not yet been obtained. The attempted demetallation in acidic media resulted in ring contraction to phthalocyanine or complete hydrolysis to phthalic acid (Scheme 110).[201]

The attempted transmetallation also resulted in metallophthalocyanine formation (Scheme 111).

$$(45) \xrightarrow[72\,^\circ C]{CF_3CO_2H \text{ or } CCl_3CO_2H} \text{phthalocyanine}$$

$$(45) \xrightarrow[\text{at } 25\,^\circ C]{6\text{ N aq. HCl or } 25\%\text{ aq. HF}} \text{phthalic acid}$$

Scheme 110

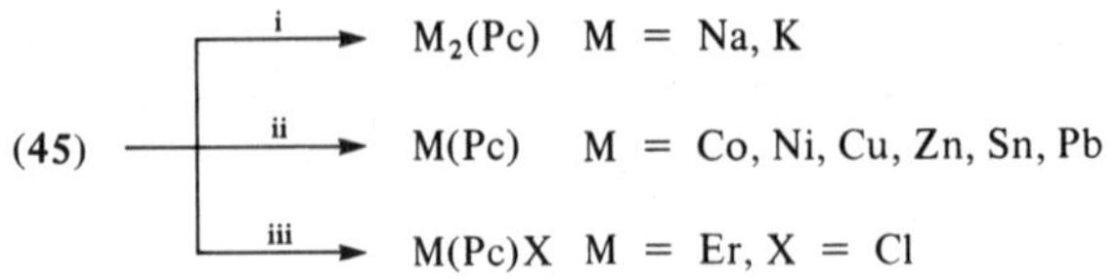

$$i \longrightarrow M_2(Pc) \quad M = Na, K$$

$$(45) \xrightarrow{\quad ii \quad} M(Pc) \quad M = Co, Ni, Cu, Zn, Sn, Pb$$

$$iii \longrightarrow M(Pc)X \quad M = Er, X = Cl$$

i, M, mesitylene, reflux; ii, M^{II} salt, DMF, toluene, quinoline or
1-pentanol, Δ; iii, M^{III} salt, DMF, Δ

Scheme 111

The kinetics of the ring-contraction transmetallation reaction in Scheme 111 ($M = Cu^{II}$) is first
order in both (45) and Cu^{II}, suggesting a mechanism which involves either rate-determining transmetallation followed by ring contraction, or ring opening and closure initiated by metal coordination to meso nitrogens.

This tendency of superphthalocyanine to contract to phthalocyanine may be attributed to the
internal ring strain, the impaired macrocyclic π system, the inadequately large centre hole and the
unstable coordination geometry. In (45a) the ligand ring cannot take a planar conformation but is
buckled severely, and the uranium atom is centred in a distorted pentagonal bipyramidal structure
with an average $U-N_{spc}$ distance of ~ 2.52 Å.

21.1.4.5 Hexaphyrins

The synthesis of hexaphyrin (Figure 18) has been reported.[278]

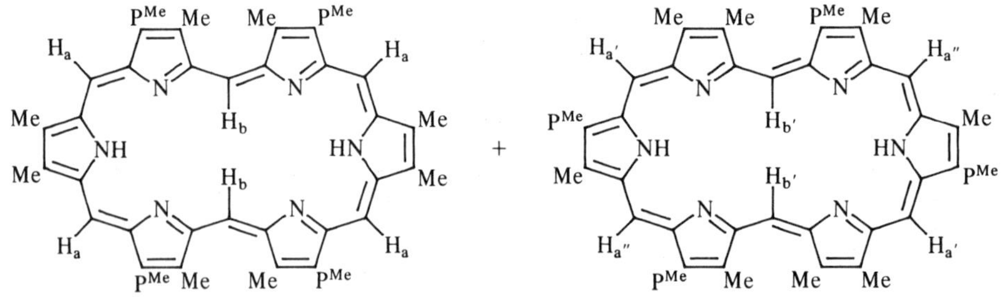

$\delta - 7.54, - 7.40\ (H_b, H_{b'}),\ 12.19,\ 12.33\ (H_{a'}, H_{a''}),\ 12.42\ (H_a)\ (in\ CF_3CO_2H)$

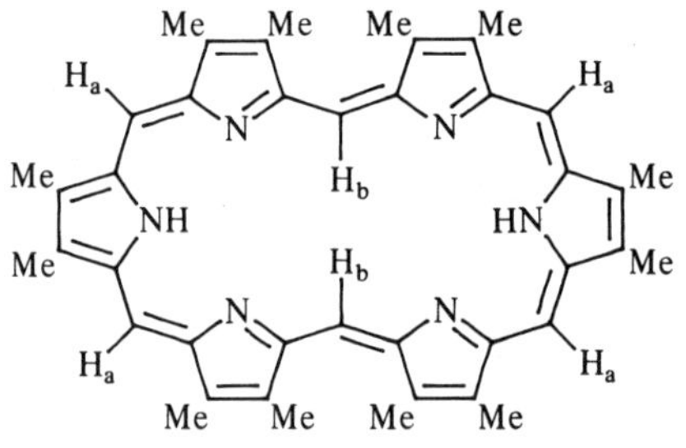

$\delta - 7.3\ (H_b),\ 4.55,\ 4.66\ (Me),\ 12.5\ (H_a)$

Figure 18 Hexaphyrins

ACKNOWLEDGEMENT

This work was supported by the United States National Institute of Health (GM 29198).

21.1.5 REFERENCES

1. P. George, *Chem. Rev.*, 1975, **75**, 85.
2. D. Dolphin (ed.), 'The Porphyrins', Academic, New York, 1978–1979, vols. 1–7.
3. R. J. Abraham, G. E. Hawkes and K. M. Smith, *Tetrahedron Lett.*, 1974, 1483; *J. Chem. Soc., Perkin Trans. 2*, 1974,
   627. H. J. C. Yeh, M. Sato and I. Morishima, *J. Magn. Reson.*, 1977, **26**, 365.

4. M. Gouterman, in ref. 2, vol. 3, chap. 1, p. 1.
5. R. J. Abraham, G. R. Bedford and B. Wright, *Org. Magn. Reson.*, 1983, **21**, 637, and preceding papers.
6. D. F. Evans, *J. Chem. Soc.*, 1959, 2003.
7. F. R. Longo, E. J. Thorne, A. D. Adler and S. Dym, *J. Heterocycl. Chem.*, 1975, **12**, 1305.
8. A. D. Adler, F. R. Longo, J. D. Finarelli, J. Goldmacher, J. Assour and L. Korsakoff, *J. Org. Chem.*, 1967, **32**, 476.
9. D. O. Cheng and E. LeGoff, *Tetrahedron Lett.*, 1977, 1469.
10. J. T. Groves and T. E. Nemo, *J. Am. Chem. Soc.*, 1983, **105**, 6243.
11. K. Rousseau and D. Dolphin, *Tetrahedron Lett.*, 1974, 4251.
12. A. H. Jackson, G. W. Kenner and J. Wass, *J. Chem. Soc., Perkin Trans. 1*, 1972, 1475.
13. B. Evans and K. M. Smith, *Tetrahedron*, 1977, **33**, 629.
14. K. M. Smith (ed.), 'Porphyrins and Metalloporphyrins', Elsevier, Amsterdam, 1975.
15. E. Watanabe, S. Nishimura, H. Ogoshi and Z. Yoshida, *Tetrahedron*, 1975, **31**, 1385.
16. H. J. Callot, J. Fischer and R. Weiss, *J. Am. Chem. Soc.*, 1982, **104**, 1272, and preceding papers, D. Mansuy, I. Morgenstern-Badarau, M. Lange and P. Gans, *Inorg. Chem.*, 1982, **21**, 1427, and preceding papers. P. Batten, A. L. Hamilton, A. W. Johnson, M. Mahendran, D. Ward and T. J. King, *J. Chem. Soc., Perkin Trans. 1*, 1977, 1623, and preceding papers. Y. W. Chan, M. W. Renner and A. L. Balch, *Organometallics*, 1983, **2**, 1888, and preceding papers. T. J. Wisnieff, A. Gold and S. A. Evans, Jr., *J. Am. Chem. Soc.*, 1981, **103**, 5616. K. Ichimura, *Chem. Lett.*, 1976, 641.
17. (a) J. W. Buchler, in ref. 2, vol. 1, chap. 10, p. 389, and references therein. (b) J. W. Buchler, in ref. 14, chap. 5, p. 157.
18. P. Hambright, in ref. 14, chap. 6, p. 233, and references therein.
19. (a) D. Ostfeld and M. Tsutsui, *Acc. Chem. Res.*, 1974, **7**, 52. (b) E. B. Fleischer and F. Dixon, *Bioinorg. Chem.*, 1977, **7**, 129.
20. W. R. Scheidt and C. A. Reed, *Chem. Rev.*, 1981, **81**, 543, and references therein.
21. W. R. Scheidt, in ref. 2, vol. 3, chap. 10, p. 463, and references therein.
22. E. W. Baker, C. B. Storm, G. T. McGrew and A. H. Corwin, *Bioinorg. Chem.*, 1973, **3**, 49. K. M. Kadish and L. R. Shiue, *Inorg. Chem.*, 1982, **21**, 1112, and references therein.
23. P. Dumas and P. Guerin, *Can. J. Chem.*, 1978, **56**, 925.
24. C. J. Carrano and M. Tsutsui, *J. Coord. Chem.*, 1977, **7**, 125.
25. R. G. Ball, K. M. Lee, A. G. Marshall and J. Trotter, *Inorg. Chem.*, 1980, **19**, 1463.
26. P. Cocolios, P. Fournari, R. Guilard, C. LeComte, J. Protas and J. C. Boubel, *J. Chem. Soc., Dalton Trans.*, 1980, 2081.
27. A. Coutsolelos and R. Guilard, *J. Orgnaomet. Chem.*, 1983, **253**, 273.
28. F. Brady, K. Henrick and R. W. Matthews, *J. Organomet. Chem.*, 1981, **210**, 281.
29. K. M. Smith and J. J. Lai, *Tetrahedron Lett.*, 1980, **21**, 433.
30. K. M. Barkigia, J. Fajer, A. D. Adler and G. J. B. Willliams, *Inorg. Chem.*, 1980, **19**, 2057.
31. P. Sayer, M. Gouterman and C. R. Connell, *Acc. Chem. Res.*, 1982, **15**, 73, and references therein.
32. S. Mangani, E. F. Mayer, Jr., D. L. Cullen, M. Tsutsui and C. J. Carrano, *Inorg. Chem.*, 1983, **22**, 400.
33. W. D. Horrocks, Jr. and C. P. Wong, *J. Am. Chem. Soc.*, 1976, **98**, 7157, and preceding papers.
34. J. W. Buchler, H. G. Kapellmann, M. Knoff, K. L. Lay and S. Pfeifer, *Z. Naturforsch., Teil B*, 1983, **38**, 1339.
35. A. Dormond, B. Belkalem and R. Guilard, *Polyhedron*, 1984, **3**, 107.
36. J.-M. Latour, C. J. Boreham and J.-C. Marchon, *J. Organomet. Chem.*, 1980, **190**, C61, and preceding papers. R. Guilard, M. Fontesse, P. Fournari, C. LeComte and J. Protas, *J. Chem. Soc., Chem. Commun.*, 1976, 161. M. Inamo, S. Funahashi and M. Tanaka, *Inorg. Chim. Acta*, 1983, **76**, L93.
37. J. L. Poncet, R. Guilard, P. Friant and J. Goulon, *Polyhedron*, 1983, **2**, 417. P. Richard, J. L. Poncet, J. M. Barbe, R. Guilard, J. Goulon, D. Rinaldi, A. Cartier and P. Tola, *J. Chem. Soc., Dalton Trans.*, 1982, 1451, and preceding papers.
38. C. LeComte, J. Protas, R. Guilard, B. Fliniaux and P. Fournari, *J. Chem. Soc., Dalton Trans.*, 1979, 1306.
39. M. L. H. Green and J. J. E. Moreau, *Inorg. Chim. Acta*, 1978, **31**, L461.
40. P. Richard and R. Guilard, *J. Chem. Soc., Chem. Commun.*, 1983, 1454.
41. D. A. Summerville, R. D. Jones, B. M. Hoffman and F. Basolo, *J. Am. Chem. Soc.*, 1977, **99**, 8195.
42. W. R. Scheidt and C. A. Reed, *Inorg. Chem.*, 1978, **17**, 710.
43. J. R. Budge, R. M. K. Gatehouse, M. C. Nesbit and B. O. West, *J. Chem. Soc., Chem. Commun.*, 1981, 370.
44. J. W. Buchler and C. Dreher, *Z. Naturforsch., Teil B*, 1984, **39**, 222, and references therein.
45. J. T. Groves, W. J. Cruper, Jr., R. C. Haushalter and W. M. Butler, *Inorg. Chem.*, 1982, **21**, 1363, and preceding papers.
46. Y. Matsuda, F. Kubota and Y. Murakami, *Chem. Lett.*, 1977, 1281. F. Basolo, R. D. Jones and D. A. Summerville, *Acta Chem. Scand., Ser. A*, 1978, **32**, 771.
47. T. Imamura, K. Hasegawa and M. Fujimoto, *Chem. Lett.*, 1983, 705.
48. T. Diebold, B. Chevrier and R. Weiss, *Angew. Chem., Int. Ed. Engl.*, 1977, **16**, 788.
49. T. Diebold, M. Schappacher, B. Chevrier and R. Weiss, *J. Chem. Soc., Chem. Commun.*, 1979, 693, and preceding papers.
50. P. A. Loach and M. Calvin, *Biochemistry*, 1963, **2**, 361. M. J. Camenzind, F. J. Hollander and C. L. Hill, *Inorg. Chem.*, 1982, **21**, 4301.
51. L. J. Boucher, *Coord. Chem. Rev.*, 1971, **7**, 289.
52. S. L. Kelly and K. M. Kadish, *Inorg. Chem.*, 1982, **21**, 3631, and preceding papers.
53. B. M. Hoffman, T. Szymanski, T. G. Brown and F. Basolo, *J. Am. Chem. Soc.*, 1978, **100**, 7253, and references therein.
54. T. Watanabe, T. Ama and K. Nakamoto, *Inorg. Chem.*, 1983, **22**, 2470, and preceding papers.
55. W. R. Scheidt, K. Hatano, G. A. Rupprecht and P. L. Piciulo, *Inorg. Chem.*, 1979, **18**, 292, and preceding papers.
56. B. B. Wayland, L. W. Olson and Z. U. Siddiqui, *J. Am. Chem. Soc.*, 1976, **98**, 94.
57. J. T. Landrum, K. Hatano, W. R. Scheidt and C. A. Reed, *J. Am. Chem. Soc.*, 1980, **102**, 6729.
58. E. T. Shimomura, M. A. Phillippi, H. M. Goff, W. F. Scholz and C. A. Reed, *J. Am. Chem. Soc.*, 1981, **103**, 6778.
59. J. A. Smegal, B. C. Schardt and C. L. Hill, *J. Am. Chem. Soc.*, 1983, **105**, 3510, and preceding papers.
60. M. Tsutsui, C. P. Hrung, D. Ostfeld, T. S. Srivastava, D. L. Cullen and E. F. Meyer, Jr., *J. Am. Chem. Soc.*, 1974, **97**, 3952, and preceding papers.
61. K. Tatsumi and R. Hoffman, *Inorg. Chem.*, 1981, **20**, 3771.

62. K. S. Suslick and M. M. Fox, *J. Am. Chem. Soc.*, 1983, **105**, 3507.
63. T. Mincey and T. G. Traylor, *Bioinorg. Chem.*, 1978, **9**, 409.
64. J. P. Collman, J. L. Hoard, N. Kim, G. Lang and C. A. Reed, *J. Am. Chem. Soc.*, 1975, **97**, 2676.
65. D. Dolphin, J. R. Sams, T. B. Tsin and K. L. Wong, *J. Am. Chem. Soc.*, 1976, **98**, 6970.
66. B. B. Wayland, L. F. Mehne and J. Swartz, *J. Am. Chem. Soc.*, 1978, **100**, 2379. W. R. Scheidt and D. K. Geiger, *Inorg. Chem.*, 1982, **21**, 1208, and references therein. C. Goulon-Ginet, J. Goulon, J. P. Battioni, D. Mansuy and J. C. Chottard, *Springer Ser. Chem. Phys.*, 1983, **27**, 349. P. E. Ellis, Jr., R. D. Jones and F. Basolo, *J. Chem. Soc., Chem. Commun.*, 1980, 54.
67. M. Rougee and D. Brault, *Biochemistry*, 1975, **14**, 4100, and preceding papers.
68. T. Mashiko, M. E. Kastner, K. Spartalian, W. R. Scheidt and C. A. Reed, *J. Am. Chem. Soc.*, 1978, **100**, 6354.
69. T. Hashimoto, R. L. Dyer, M. J. Crossley, J. E. Baldwin and F. Basolo, *J. Am. Chem. Soc.*, 1982, **104**, 2101, and preceding papers.
70. J. P. Collman, *Acc. Chem. Res.*, 1977, **10**, 265. D.-H. Chin, G. N. LaMar and A. L. Balch, *J. Am. Chem. Soc.*, 1980, **102**, 4344. T. G. Traylor, *Acc. Chem. Res.*, 1981, **14**, 102.
71. J. H. Wang, *J. Am. Chem. Soc.*, 1958, **80**, 3168.
72. J. P. Collman, J. I. Brauman, T. J. Collins, B. L. Iverson, G. Lang, R. B. Pettman, J. L. Sessler and M. A. Walters, *J. Am. Chem. Soc.*, 1983, **105**, 3038.
73. B. Ward, C.-B. Wang and C. K. Chang, *J. Am. Chem. Soc.*, 1981, **103**, 5236.
74. J. Mispelter, M. Momenteau, D. Lavalette and J.-M. Lhoste, *J. Am. Chem. Soc.*, 1983, **105**, 5165, and preceding papers.
75. (a) M. F. Perutz, *Proc. R. Soc. London, Ser. B*, 1980, **208**, 135, and preceding papers. (b) T. G. Traylor, M. J. Mitchell, S. Tsuchiya, D. H. Campbell, D. V. Stynes and N. Koga, *J. Am. Chem. Soc.*, 1981, **103**, 2450.
76. E. Tsuchida, E. Hasegawa and K. Honda, *Biochem. Biophys. Acta*, 1976, **427**, 520.
77. (a) J. P. Collman, J. I. Brauman, E. Rose and K. S. Suslick, *Proc. Natl. Acad. Sci. USA*, 1978, **75**, 1052. G. B. Jameson, F. S. Molinaro, J. A. Ibers, J. P. Collman, J. I. Brauman, E. Rose and K. S. Suslick, *J. Am. Chem. Soc.*, 1980, **102**, 3224. (b) S. E. V. Phillips and B. P. Schoenborn, *Nature (London)*, 1981, **292**, 81. B. Shaanan, *J. Mol. Biol.*, 1983, **171**, 31.
78. T. G. Traylor, Y. Tatsuno, D. W. Powell and J. B. Cannon, *J. Chem. Soc., Chem. Commun.*, 1977, 732.
79. T. G. Traylor, M. J. Mitchell, J. P. Ciccone and S. Nelson, *J. Am. Chem. Soc.*, 1982, **104**, 4986. I. Tabushi and T. Sasaki, *J. Am. Chem. Soc.*, 1983, **105**, 2901.
80. K. Anzai, K. Hatano, Y. J. Lee and W. R. Scheidt, *Inorg. Chem.*, 1981, **20**, 2337, and references therein.
81. R. Quinn, M. Nappa and J. S. Valentine, *J. Am. Chem. Soc.*, 1982, **104**, 2588.
82. C. A. Reed, T. Mashiko, S. P. Bentley, M. E. Kastner, W. R. Scheidt, K. Spartalian and G. Lang, *J. Am. Chem. Soc.*, 1979, **101**, 2948. H. Masuda, T. Taga, K. Osaki, H. Sugimoto, I. Yoshida and H. Ogoshi, *Inorg. Chem.*, 1980, **19**, 950.
83. M. A. Phillippi, N. Baenziger and H. M. Goff, *Inorg. Chem.*, 1981, **20**, 3904.
84. T. K. Miyamoto, S. Tsuzuki, T. Hasegawa and Y. Sasaki, *Chem. Lett.*, 1983, 1587, and references therein.
85. P. Cocolios, G. Lagrange and R. Guilard, *J. Organomet. Chem.*, 1983, **253**, 65, and references therein.
86. F. A. Walker, M. W. Lo and M. T. Ree, *J. Am. Chem. Soc.*, 1976, **98**, 5552. K. M. Adams, P. G. Rasmussen, W. R. Scheidt and K. Hatano, *Inorg. Chem.*, 1979, **18**, 1892. T. Yoshimura and T. Ozaki, *Bull. Chem. Soc. Jpn.*, 1979, **52**, 2268.
87. D. L. Hickman and H. M. Goff, *Inorg. Chem.*, 1983, **22**, 2787, and references therein.
88. W. R. Scheidt, D. K. Geiger, R. G. Hayes and G. Lang, *J. Am. Chem. Soc.*, 1983, **105**, 2625, and preceding papers.
89. D. Lexa, J. Mispelter and J.-M. Savéant, *J. Am. Chem. Soc.*, 1981, **103**, 6806. C. A. Reed, *Adv. Chem. Ser.*, 1982, **201**, 333, and references therein.
90. D. Mansuy, P. Battioni and J. P. Mahy, *J. Am. Chem. Soc.*, 1982, **104**, 4487, and preceding papers. W. R. Scheidt, D. A. Summerville and I. A. Cohen, *J. Am. Chem. Soc.*, 1976, **98**, 6623, and preceding papers. M. A. Phillippi and H. M. Goff, *J. Am. Chem. Soc.*, 1979, **101**, 7641, and references therein.
91. H. M. Goff and M. A. Phillippi, *J. Am. Chem. Soc.*, 1983, **105**, 7567, and references therein.
92. S. S. Eaton and G. R. Eaton, *Inorg. Chem.*, 1976, **15**, 134; 1977, **16**, 72. H. Ogoshi, H. Sugimoto and Z. Yoshida; *Bull. Chem. Soc. Jpn.*, 1978, **51**, 2369.
93. T. Boschi, G. Bontempelli and G.-A. Mazzocchin, *Inorg. Chim. Acta*, 1979, **37**, 155.
94. R. G. Ball, G. Domazetis, D. Dolphin, B. R. James and J. Trotter, *Inorg. Chem.*, 1981, **20**, 1556. F. Pomposo, D. Carruthers and D. V. Stynes, *Inorg. Chem.*, 1982, **21**, 4245. C. E. Holloway, D. V. Stynes and C. P. J. Vuik, *J. Chem. Soc., Dalton Trans.*, 1982, 95. D. P. Rillema, J. K. Nagle, L. F. Barringer, Jr. and T. J. Meyer, *J. Am. Chem. Soc.*, 1981, **103**, 56.
95. M. Barley, J. Y. Becker, G. Domazetis, D. Dolphin and B. R. James, *J. Chem. Soc., Chem. Commun.*, 1981, 982, and references therein.
96. H. Sugimoto, T. Higashi, M. Mori, M. Nagano, Z. Yoshida and H. Ogoshi, *Bull. Chem. Soc. Jpn.*, 1982, **55**, 822, and preceding papers.
97. J. P. Collman, C. E. Barnes, P. N. Swepston and J. A. Ibers, *J. Am. Chem. Soc.*, 1984, **106**, 3500, and references therein.
98. N. Farrell and A. A. Neves, *Inorg. Chim. Acta*, 1981, **54**, L53.
99. N. Farrell, D. H. Dolphin and B. R. James, *J. Am. Chem. Soc.*, 1978, **100**, 324.
100. A. Antipas, J. W. Buchler, M. Gouterman and P. D. Smith, *J. Am. Chem. Soc.*, 1978, **100**, 3015; 1980, **102**, 198, and preceding papers.
101. J. W. Buchler, W. Kokisch and P. D. Smith, *Struct. Bonding (Berlin)*, 1978, **34**, 79.
102. J. W. Buchler and W. Kokisch, *Angew. Chem., Int. Ed. Engl.*, 1981, **20**, 403.
103. H. Sugimoto, N. Ueda and M. Mori, *Bull. Chem. Soc. Jpn.*, 1981, **54**, 3425.
104. F. A. Walker, D. Beroiz and K. M. Kadish, *J. Am. Chem. Soc.*, 1976, **98**, 3484.
105. R. F. Pasternack and G. R. Parr, *Inorg. Chem.*, 1976, **15**, 3087, and preceding papers.
106. K. R. Ashley and S. Au-Young, *Inorg. Chem.*, 1976, **15**, 1937, and preceding papers.
107. H. Ogoshi, J. Setsune and Z. Yoshida, *J. Organomet. Chem.*, 1980, **185**, 95. J. Setsune, T. Yazawa, H. Ogoshi and Z. Yoshida, *J. Chem. Soc., Perkin Trans. 1*, 1980, 1641, and preceding papers.

108. M. Krishnamurthy, *Inorg. Chim. Acta*, 1977, **25**, 215. K. R. Ashley, S.-B. Shyu and J. G. Leipoldt, *Inorg. Chem.*, 1980, **19**, 1613.
109. B. B. Wayland, B. A. Woods and R. Pierce, *J. Am. Chem. Soc.*, 1982, **104**, 302, and preceding papers. J. Setsune, Z. Yoshida and H. Ogoshi, *J. Chem. Soc., Perkin Trans. 1*, 1982, 983, and preceding papers.
110. B. R. James and D. V. Stynes, *J. Am. Chem. Soc.*, 1972, **94**, 6225.
111. H. Ogoshi, J. Setsune and Z. Yoshida, *J. Organomet. Chem.*, 1978, **159**, 317.
112. H. Sugimoto, N. Ueda and M. Mori, *J. Chem. Soc., Dalton Trans.*, 1982, 1611.
113. R. F. Pasternack, E. G. Spiro and M. Teach, *J. Inorg. Nucl. Chem.*, 1974, **36**, 599.
114. J. W. Buchler, C. Dreher and K.-L. Lay, *Z. Naturforsch., Teil B*, 1982, **37**, 1155, and preceding papers.
115. A. MacCragh, C. B. Storm and W. S. Koski, *J. Am. Chem. Soc.*, 1965, **87**, 1470.
116. D. J. Morano and H. N. Po, *Inorg. Chim. Acta*, 1978, **31**, L421, and references therein.
117. M. E. Jamin and R. T. Iwamoto, *Inorg. Chim. Acta*, 1978, **27**, 135.
118. H. Sugimoto, *J. Chem. Soc., Dalton Trans.*, 1982, 1169.
119. W. R. Scheidt, M. E. Kastner and K. Hatano, *Inorg. Chem.*, 1978, **17**, 706.
120. K. M. Kadish, L. R. Shiue, R. K. Rhodes and L. A. Bottomley, *Inorg. Chem.*, 1981, **20**, 1274.
121. P. F. Rodesiler, E. H. Griffith, P. D. Ellis and E. L. Amma, *J. Chem. Soc., Chem. Commun.*, 1980, 492.
122. J. P. Collman, C. E. Barnes and L. K. Woo, *Proc. Natl. Acad. Sci. USA*, 1983, **80**, 7684.
123. I. A. Cohen, D. Ostfeld and B. Lichtenstein, *J. Am. Chem. Soc.*, 1972, **94**, 4522.
124. P. Cocolios, C. Moise and R. Guilard, *J. Organomet. Chem.*, 1982, **228**, C43.
125. S. Onaka, Y. Kondo, K. Toriumi and T. Ito, *Chem. Lett.*, 1980, 1605.
126. I. Noda, S. Kato, M. Mizuta, N. Yasuoka and N. Kasai, *Angew. Chem., Int. Ed. Engl.*, 1979, **18**, 83, and preceding papers.
127. J. T. Groves and T. E. Nemo, *J. Am. Chem. Soc.*, 1983, **105**, 5786, 6243, and preceding papers; J. T. Groves, and R. S. Myers, *J. Am. Chem. Soc.*, 1983, **105**, 5791.
128. J. R. Lindsay Smith and P. R. Sleath, *J. Chem. Soc., Perkin Trans. 2*, 1982, 1009.
129. J. T. Groves, Y. Watanabe and T. J. McMurry, *J. Am. Chem. Soc.*, 1983, **105**, 4489, and references therein.
130. I. Tabushi and A. Yazaki, *J. Am. Chem. Soc.*, 1981, **103**, 7371, and preceding papers.
131. H. J. Ledon, P. Durbut and F. Varescon, *J. Am. Chem. Soc.*, 1981, **103**, 3601.
132. P. Shannon and T. C. Bruice, *J. Am. Chem. Soc.*, 1981, **103**, 4580.
133. J. T. Groves and T. Takahashi, *J. Am. Chem. Soc.*, 1983, **105**, 2073.
134. A. M. Abeysekera, R. Grigg, J. Trocha-Grimshaw and V. Viswanatha, *J. Chem. Soc., Perkin Trans. 1*, 1977, 36, 1395.
135. D. Lexa and J.-M. Savéant, *J. Am. Chem. Soc.*, 1982, **104**, 3503, and preceding papers.
136. G. Domazetis, B. R. James, B. Tarpey and D. Dolphin, *Am. Chem. Soc. Symp. Ser.* 1981, **152**, 243, and preceding papers.
137. H. J. Callot and F. Metz, *Tetrahedron Lett.*, 1982, **23**, 4321.
138. E. R. Hopf and D. G. Whitten, in ref. 14, chap. 16, p. 667; in ref. 2, vol. 2, chap. 16, p. 161, and references therein.
139. D. Mauserall, in ref. 2, vol. 5, chap. 2, p. 29, and references therein. J. R. Darwent, P. Douglas, A. Harriman, G. Porter and M.-C. Richoux, *Coord. Chem. Rev.*, 1982, **44**, 83.
140. B. C. Chow and I. A. Cohen, *Bioinorg. Chem.*, 1971, **1**, 57. G. W. Sovocool, F. R. Hopf and D. G. Whitten, *J. Am. Chem. Soc.*, 1972, **94**, 4350.
141. S. Inoue and N. Takeda, *Bull. Chem. Soc. Jpn.*, 1977, **50**, 984.
142. M. Perre-Fauvet, A. Gaudemer, P. Boucly and J. Devynck, *J. Organomet. Chem.*, 1976, **120**, 439.
143. C. Cloutour, D. Lafargue, J. A. Richards and J. C. Pommier, *J. Organomet. Chem.*, 1977, **137**, 157.
144. J. H. Fuhrhop, in ref. 14, chap. 14, p. 593. R. H. Felton, in ref. 2, vol. 5, chap. 3, p. 53. D. G. Davies, in ref. 2, vol. 5, chap. 4, p. 127.
145. P. Worthington, P. Hambright, R. F. X. Williams and M. R. Feldman, *Inorg. Nucl. Chem. Lett.*, 1980, **16**, 441, and preceding papers.
146. L. A. Bottomley, L. Olson and K. M. Kadish, *Adv. Chem. Ser.*, 1982, **201**, 279.
147. T. Mashiko, J.-C. Marchon, D. T. Muesser, C. A. Reed, M. E. Kastner and W. R. Scheidt, *J. Am. Chem. Soc.*, 1979, **101**, 3653.
148. P. R. Ortiz de Montellano, K. L. Kunze, H. S. Beilan and C. Wheeler, *Biochemistry*, 1982, **21**, 1331, and preceding papers.
149. A. M. Abeysekera, R. Grigg, J. Trocha-Grimshaw and K. Henrick, *Tetrahedron*, 1980, **36**, 1857, and references therein.
150. H. J. Callot and F. Metz, *J. Chem. Soc., Chem. Commun.*,1982, 947, and preceding papers. P. R. Ortiz de Montellano, K. L. Kunze and O. Augusto, *J. Am. Chem. Soc.*, 1982, **104**, 3545, and preceding papers. D. Mansuy, J. P. Battioni, D. Dupre, E. Sartori and G. Chottard, *J. Am. Chem. Soc.*, 1982, **104**, 6159, and preceding papers.
151. G. M. McLaughlin, *J. Chem. Soc., Perkin Trans. 2*, 1974, 136.
152. C. P. Hrung, M. Tsutsui, D. L. Cullen and E. F. Meyer, Jr., *J. Am. Chem. Soc.*, 1976, **98**, 7878.
153. H. Tori, S. Kitamura, Y. Aoyama and H. Ogoshi, *Chem. Lett.*, 1982, 773.
154. O. P. Anderson, A. B. Kopelove and D. K. Lavallee, *Inorg. Chem.*, 1980, **19**, 2101, and preceding papers.
155. D. E. Goldberg and K. M. Thomas, *J. Am. Chem. Soc.*, 1976, **98**, 913.
156. H. Ogoshi, S. Kitamura, H. Toi and Y. Aoyama, *Chem. Lett.*, 1982, 495.
157. M. Nukui, S. Inoue and Y. Ohkatsu, *Bull. Chem. Soc. Jpn.*, 1983, **56**, 2055, and preceding papers.
158. K. Ichimura, *Bull. Chem. Soc. Jpn.*, 1978, **51**, 1444, and preceding papers. H. J. Callot, *Tetrahedron*, 1979, **35**, 1455, and preceding papers.
159. L. E. Andrews, R. Bonnett, R. J. Ridge and E. H. Appelman, *J. Chem. Soc., Perkin Trans. 1*, 1983, 103.
160. M. J. Broadhurst, R. Grigg and A. W. Johnson, *J. Chem. Soc. (C)*, 1971, 3681. A. Ulman and J. Manassen, *J. Am. Chem. Soc.*, 1975, **97**, 6540.
161. H. J. Callot, T. Tschamber and E. Schaeffer, *J. Am. Chem. Soc.*, 1975, **97**, 6178; *Tetrahedron Lett.*, 1975, 2919.
162. S. H. Strauss, M. E. Silver and J. A. Ibers, *J. Am. Chem. Soc.*, 1983, **105**, 4108.
163. J. C. Gallucci, P. N. Swepston and J. A. Ibers, *Acta Crystallogr., Sect. B*, 1982, **38**, 2134, and preceding papers.
164. A. Ulman, D. Fischer and J. A. Ibers, *J. Heterocycl. Chem.*, 1981, **19**, 409.

165. Y. Harel and J. Manassen, *J . Am. Chem. Soc.*, 1977, **99**, 5817; 1978, **100**, 6228, and references therein.
166. A. M. Stolzenberg, L. O. Spreer and R. H. Holm, *J. Am. Chem. Soc.*, 1980, **102**, 364.
167. M. Strell and T. Urumow, *Liebigs Ann. Chem.*, 1977, 970.
168. U. Eisner and M. J. C. Harding, *J. Chem. Soc.*, 1964, 4089.
169. M. R. Wasielewski, *Tetrahedron Lett.*, 1977, 1373, and references therein.
170. W. Schlesinger, A. H. Corwin and L. J. Sargent, *J. Am. Chem. Soc.*, 1950, **72**, 2867.
171. A. R. Battersby, K. Jones and R. J. Snow, *Angew. Chem., Int. Ed. Engl.*, 1983, **22**, 734.
172. J. R. Miller and G. D. Dorough, *J. Am. Chem. Soc.*, 1952, **74**, 3977.
173. A. Botulinski, J. W. Buchler, J. Ensling, H. Twilfer, H. Billecke, H. Leuken and B. Tonn, *Adv. Chem. Ser.*, 1981, **201**, 253, and references therein.
174. J. W. Buchler, K. L. Lay, Y. J. Lee and W. R. Scheidt, *Angew. Chem., Int. Ed. Engl.*, 1982, **21**, 432.
175. A. Faessler, A. Pfaltz, P. M. Mueller, S. Farooq, C. Kratky, B. Kraeutler and A. Eschenmoser, *Helv. Chim. Acta*, 1982, **65**, 812, and references therein.
176. Yu. K. Grishin, O. A. Subbotin, Yu. A. Ustynyuk, V. N. Kopranenkov, L. S. Goncharova and E. D. Luk'yanets, *J. Struct. Chem. (Engl. Transl.)*, 1979, **20**, 297.
177. J.-H. Fuhrhop, P. Krueger and W. S. Sheldrick, *Liebigs Ann. Chem.*, 1977, 339. J. Engel, A. Gossauer and A. W. Johnson, *J. Chem. Soc., Perkin Trans. 1*, 1978, 871, and preceding papers. S. S. Dvornikov, V. N. Knyukshto, V. A. Kuz'mitskii, A. M. Shul'ga and K. N. Solov'ev, *J. Lumin.*, 1981, **23**, 373.
178. M. Whalley, *J. Chem. Soc.*, 1961, 866, and references therein.
179. Z. Berkovitch-Yellin and D. E. Ellis, *J. Am. Chem. Soc.*, 1981, **103**, 6066, and references therein.
180. Yu. B. Vysotskii, V. A. Kuz'mitskii, K. N. Solv'ev, *J. Struct. Chem. (Engl. Transl.)*, 1981, **22**, 497; *Theor. Chim. Acta*, 9181, **59**, 467, and references therein.
181. J.-H. Fuhrhop and P. Krueger, *Liebigs Ann. Chem.*, 1977, 360.
182. A. M. Shul'ga, G. P. Gurinovitch and I. E. Gurinovitch, *Biofizika*, 1973, **18**, 32.
183. C. J. Schramm and B. M. Hoffman, *Inorg. Chem.*, 1980, **19**, 383. V. N. Kopranenkov, L. S. Goncharova, L. E. Marinina and E. A. Luk'yanets, *Khim. Geterotsikl. Soedin.*, 1982, 1645, and preceding papers.
184. J. A. Clarke, P. J. Dawson, R. Grigg and C. H. Rochester, *J. Chem. Soc., Perkin Trans. 2*, 1973, 414, and preceding papers.
185. A. H. Cook and R. P. Linstead, *J. Chem. Soc.*, 1937, 929. S. S. Iodko, O. L. Kaliya, M. G. Gal'pern, V. N. Kopranenkov, O. L. Lebedev and E. A. Luk'yanets, *Koord. Khim.*, 1982, **8**, 1025.
186. G. E. Ficken and R. P. Linstead, *J. Chem. Soc.*, 1952, 4846.
187. M. E. Baguley, H. France, R. P. Linstead and M. Whalley, *J. Chem. Soc.*, 1955, 3521.
188. V. N. Kopranenkov, L. S. Goncharova and E. A. Luk'yanets, *J. Gen. Chem. USSR (Engl. Transl.)*, 1977, **47**, 1954; 1979, **49**, 1233; *J. Org. Chem. USSR (Engl. Transl.)*, 1979, **15**, 962.
189. B. D. Berezin, O. G. Khelevina, N. D. Gerasimova and P. A. Stuzhin, *Russ. J. Phys. Chem. (Engl. Transl.)*, 1982, **56**, 1699.
190. B. D. Berezin and O. G. Khelevina, *Russ. J. Phys. Chem. (Engl. Transl.)*, 1982, **56**, 538.
191. B. D. Berezin, O. G. Khelevina and G. P. Brin, *Russ. J. Phys. Chem. (Engl. Transl.)*, 1982, **56**, 902.
192. N. S. Hush, *Theor. Chim. Acta*, 1966, **4**, 108, and preceding papers. K. Zielinski, A. I. Vrublevskii, G. N. Sinyakov and G. P. Gurinovich, *Zh. Prikl. Spectrosk.*, 1980, **33**, 114.
193. A. Stanienda, *Z. Naturforsch., Teil B*, 1968, **23**, 145.
194. K. Meier and R. Scheffold, *Helv. Chim. Acta*, 1981, **64**, 1505, and preceding papers.
195. P. A. Barrett, R. P. Linstead, F. G. Rundall and G. A. P. Tuey, *J. Chem. Soc.*, 1940, 1079.
196. F. H. Moser and A. L. Thomas, 'Phthalocyanine Compounds', ACS Monograph No. 157, American Chemical Society, New York, 1963.
197. A. B. P. Lever, *Adv. Inorg. Chem. Radiochem.*, 1965, **7**, 27, and references therein.
198. L. J. Boucher, in 'Coordination Chemistry of Macrocyclic Compounds', ed. G. A. Melson, Plenum, New York, 1979, chap. 7, p. 461, and references therein.
199. K. Kasuga, M. Tsutsui, *Coord. Chem. Rev.*, 1980, **32**, 67, and references therein.
200. A. B. P. Lever, S. R. Pickens, P. C. Minor, S. Licoccia, B. S. Ramaswamy and K. Maqnell, *J. Am. Chem. Soc.*, 1981, **103**, 6800.
201. E. A. Cueller and T. J. Marks, *Inorg. Chem.*, 1981, **20**, 3766, and preceding papers.
202. J. E. Maskasky and M. E. Kenney, *J. Am. Chem. Soc.*, 1973, **95**, 1443, and preceding papers.
203. P. Coppens, L. Li and N. J. Zhu, *J. Am. Chem. Soc.*, 1983, **105**, 6173, and references therein.
204. J. H. Helberger and D. B. Hever, *Liebigs Ann. Chem.*, 1938, **536**, 173, and preceding papers.
205. C. E. Dent, *J. Chem. Soc.*, 1938, 1.
206. J. G. Jones and M. V. Twigg, *Inorg. Chem.*, 1969, **8**, 2018.
207. L. E. Marinina, S. A. Mikhalenko and E. A. Luk'yanets, *J. Gen. Chem. USSR (Engl. Transl.)*, 1973, **43**, 2010, and preceding papers. D. E. Remy, *Tetrahedron Lett.*, 1983, **24**, 1451.
208. J. C. Speakman, *Acta Crystallogr.*, 1953, **6**, 784. J. A. Elvidge, *Chem. Soc.., Spec. Publ.*, 1956, **4**, 28. N. A. Kolesnikov and V. F. Borodkin, *Izv. Vyssh. Uchebn. Zaved., Khim. Khim. Tekhnol.*, 1972, **15**, 880.
209. M. G. Gurevich and K. N. Solv'ev, *Dokl. Akad. Nauk BSSR*, 1961, **5**, 291.
210. M. Hedayatulla, *C. R. Hebd. Seances Acad. Sci., Ser. B*, 1983, **296**, 621.
211. P. N. Moskalev and I. S. Kirin, *Russ. J. Inorg. Chem. (Engl. Transl.)*, 1971, **16**, 57.
212. H. Sugimoto, T. Higashi, A. Maeda, M. Mori, H. Masuda and T. Taga, *J. Chem. Soc., Chem. Commun.*, 1983, 1234, and preceding papers.
213. K. Fischer and M. Hanack, *Chem. Ber.*, 1983, **116**, 1860.
214. M. Hanack, J. Metz and G. Pawlowski, *Chem. Ber.*, 1982, **115**, 2836.
215. Y. Iyechika, K. Yakushi, I. Ikemoto and H. Kuroda, *Acta Crystallogr., Sect. B*, 1982, **38**, 766.
216. M. Gouterman, P. Sayer, E. Shankland and J. P. Smith, *Inorg. Chem.*, 1981, **20**, 87.
217. I. S. Kirin, P. N. Moskalev and V. Ya. Mishin, *J. Gen. Chem. USSR (Engl. Transl.)*, 1967, **37**, 265.
218. P. N. Moskalev, G. N. Shapkin and N. I. Alimova, *Russ. J. Inorg. Chem. (Engl. Transl.)*, 1982, **27**, 794, and preceding

papers. K. Kasuga and M. Tsutsui, *J. Coord. Chem.*, 1980, **10**, 263; 1981, **11**, 177, and references therein. L. G. Tomilova, E. V. Chernykh, V. I. Gavrilov, I. V. Shelepin, V. M. Derkacheva and E. A. Luk'yanets, *J. Gen. Chem. USSR (Engl. Transl.)*, 1982, **52**, 2304.

219. P. N. Moskalev and I. S. Kirin, *Russ. J. Inorg. Chem. (Engl. Transl.)*, 1970, **15**, 7. I. S. Kirin and P. N. Moskalev, *Russ. J. Inorg. Chem. (Engl. Transl.)*, 1971, **16**, 1687.
220. W. Hiller, J. Straehle, W. Kobel and M. Hanack, *Z. Kristallogr.*, 1982, **159**, 173.
221. R. F. Ziolo, C. H. Griffiths and J. M. Troup, *J. Chem. Soc., Dalton Trans.*, 1980, 2300.
222. K. Ukei, *Acta Crystallogr., Sect. B*, 1982, **38**, 1288.
223. Yu. A. Buslaev, A. A. Kuznetsova and L. F. Goryachova, *Izv. Akad. Nauk SSSR, Neorg. Mater.*, 1967, **3**, 1701. G. P. Shaposhnikov, V. F. Borodkin and M. I. Fedorov, *Izv. Vyssh. Uchebn. Zaved., Khim. Khim. Tekhnol.*, 1981, **24**, 1485.
224. Yu. A. Zhukov, B. D. Berezin and I. N. Sokolova, *Izv. Vyssh. Uchebn. Zaved., Khim. Khim. Tekhnol.*, 1970, **13**, 1566.
225. V. F. Borodkin, M. I. Al'yanov, F. P. Snegireva, V. G. Shishkin and A. P. Snegireva, *USSR Pat.*478 016 (*Chem. Abstr.*, 1975, **83**, 147 611x).
226. A. Tamaki, T. Takahashi, I. Sudo and S. Ozaki, *Jpn. Kokai* 75 84 504 (*Chem. Abstr.*, 1976, **84**, 43 811s).
227. A. B. P. Lever, J. P. Wilshire and S. K. Quan, *Inorg. Chem.*, 1981, **20**, 761, and references therein.
228. J. Martinsen, M. Miller, D. Trojan and D. A. Sweigart, *Inorg. Chem.*, 1980, **19**, 2162, and references therein.
229. C. Ercolani, G. Rossi, F. Monacelli and M. Verzino, *Inorg. Chim. Acta*, 1983, **73**, 95, and references therein.
230. N. H. Sabelli and C. A. Melendres, *J. Phys. Chem.*, 1982, **86**, 4342, and preceding papers.
231. F. Pomposo, D. Carruthers and D. V. Stynes, *Inorg. Chem.*, 1982, **21**, 4245, and preceding papers.
232. B. D. Berezin and N. I. Sosnikova, *Dokl. Akad. Nauk SSSR*, 1962, **146**, 604.
233. A. MacCragh and W. S. Koski, *J. Am. Chem. Soc.*, 1965, **87**, 2496.
234. H. Homborg and W. Katz, *Z. Naturforsch., Teil B*, 1978, **33**, 1063.
235. D. R. Prasad and G. Ferraudi, *Inorg. Chem.*, 1982, **21**, 2967, 4241; *J. Phys. Chem.*, 1982, **86**, 4037, and references therein.
236. D. Dolphin, B. R. James, A. J. Murray and J. R. Thornback, *Can. J. Chem.*, 1980, **58**, 1125, and preceding papers. R. Taube, H. Drevs and T. DucHiep, *Z. Chem.*, 1969, **9**, 115. S. Muralidharan, G. Ferraudi and K. Schmatz, *Inorg. Chem.*, 1982, **21**, 2961.
237. A. B. P. Lever, S. Licoccia, K. Magnell, P. C. Minor and B. S. Ramaswamy, *Adv. Chem. Ser.*, 1982, **201**, 237, and references therein.
238. (a) B. F. Anderson, T. J. Bartczak and D. C. Hodgkin, *J. Chem. Soc., Perkin Trans. 2*, 1974, 977. (b) J. M. Dyke, N. S. Hush, M. L. Williams and I. S. Woolsey, *Mol. Phys.*, 1971, **20**, 1149.
239. A. W. Johnson, in ref. 14, chap. 18, p. 729. R. Grigg, in ref. 2, vol. 2, chap. 10, p. 327.
240. J. Engel, A. Gossauer and A. W. Johnson, *J. Chem. Soc., Perkin Trans. 1*, 1978, 871.
241. Y. Murakami, Y. Aoyama and M. Hayashida, *J. Chem. Soc., Chem. Commun.*, 1980, 501.
242. Y. Murakami, Y. Matsuda, K. Sakata, S. Yamada, Y. Tanaka and Y. Aoyama, *Bull. Chem. Soc. Jpn.*, 1981, **54**, 163, and references therein.
243. Y. Matsuda, S. Yamada and Y. Murakami, *Inorg. Chim. Acta*, 1980, **44**, L309, and preceding papers.
244. A. M. Abeysekera, R. Grigg, J. Trocha-Grimshaw and T. J. King, *J. Chem. Soc., Perkin Trans. 1*, 1979, 2184, and preceding papers.
245. P. B. Hitchcock and G. M. McLaughlin, *J. Chem. Soc., Dalton Trans.*, 1976, 1927.
246. N. S. Hush and I. S. Woolsey, *J. Chem. Soc., Dalton Trans.*, 1974, 24, and preceding papers.
247. N. S. Hush and J. M. Dyke, *J. Inorg. Nucl. Chem.*, 1973, **35**, 4341.
248. Y. Matsuda, S. Yamada and Y. Murakami, *Inorg. Chem.*, 1981, **20**, 2239.
249. D. P. Arnold and A. W. Johnson, *J. Chem. Soc., Chem. Commun.*, 1977, 787.
250. A. Eschenmoser, *Pure Appl. Chem.*, 1969, **20**, 1, Q. *Rev., Chem. Soc.*, 1970, **24**, 366, and references therein.
251. E. D. Edmond and D. Crowfoot-Hodgkin, *Helv. Chim. Acta*, 1975, **58**, 641.
252. J. D. Dunitz and E. F. Meyer, Jr., *Helv. Chim. Acta*, 1971, **54**, 77, and preceding papers.
253. D. Dolphin (ed.), 'B$_{12}$', Wiley-Interscience, New York, 1982.
254. Y. Murakami, *Adv. Chem. Ser.*, 1980, **191**, 179, and references therein.
255. N. J. Lewis, R. Nussberger, B. Kraeutler and A. Eschenmoser, *Angew. Chem., Int. Ed. Engl.*, 1983, **22**, 736, and preceding papers.
256. (a) V. Rasetti, K. Hilpert, A. Faessler, A. Pfaltz and A. Eschenmoser, *Angew. Chem., Int. Ed. Engl.*, 1981, **20**, 1058, and preceding papers. (b) B. Kräutler and K. Hilpert, *Angew. Chem., Int. Ed. Engl.*, 1982, **21**, 152.
257. M. Gardiner and A. J. Thomson, *J. Chem. Soc., Dalton Trans.*, 1974, 820, and preceding papers.
258. N. S. Genokhova and T. A. Melent'eva, *J. Gen. Chem. USSR (Engl. Transl.)*, 1979, **49**, 1626, and preceding papers.
259. D. Jauernig, P. Rapp and G. Ruoff, *Hoppe-Seylers Z. Physiol. Chem.*, 1973, **354**, 957.
260. C. Nussbaumer and D. Arigoni, *Angew. Chem., Int. Ed. Engl.*, 1983, **22**, 737.
261. J. Engel and H. H. Inhoffen, *Liebigs Ann. Chem.*, 1977, 767, and preceding papers.
262. (a) V. Rasetti, A. Pfaltz, C. Kratky and A. Eschenmoser, *Proc. Natl. Acad. Sci. USA*, 1981, **78**, 16. (b) A Pfaltz, N. Buehler, R. Neier, K. Hirai and A. Eschenmoser, *Helv. Chim. Acta*, 1977, **60**, 2653.
263. N. J. Lewis, A. Pfaltz and A. Eschenmoser, *Angew. Chem., Int. Ed. Engl.*, 1983, **22**, 735, and preceding papers.
264. V. B. Koppenhagen, in ref. 253, vol. 2, chap. 5, p. 105.
265. Y. Murakami, Y. Aoyama and K. Tokunaga, *J. Am. Chem. Soc.*, 1980, **102**, 6736, and preceding papers.
266. Y. Murakami, K. Sakata, Y. Tanaka and T. Matsuo, *Bull. Chem. Soc. Jpn.*, 1975, **48**, 3622.
267. (a) J. M. Pratt, 'Inorganic Chemistry of Vitamin B$_{12}$', Academic, London, 1972. (b) J. M. Pratt, in ref. 253, vol. 1, chap. 10, p. 326.
268. Y. Murakami, Y. Aoyama and T. Tada, *Bull. Chem. Soc. Jpn.*, 1981, **54**, 2302.
269. Y. Murakami and Y. Aoyama, *Bull. Chem. Soc. Jpn.*, 1976, **49**, 683.
270. A. Fischli and A. Eschenmoser, *Angew. Chem., Int. Ed. Engl.*, 1967, **6**, 866, and preceding papers.
271. K. L. Brown, in ref. 253, vol. 1, chap. 8, p. 245.
272. (a) J. Halpern, in ref. 253, vol. 1, chap. 14, p. 501. (b) H. P. C. Hogenkamp, in ref. 253, vol. 1, chap. 9, p. 295. (c) J. M. Wood, in ref. 253, vol. 2, chap. 6, p. 151.

273. N. S. Hush and I. S. Woolsey, *J. Am. Chem. Soc.*, 1972, **94**, 4107.
274. V. J. Bauer, D. L. J. Clive, D. Dolphin, J. B. Paine, III, F. L. Harris, M. M. King, J. Loder, S.-W. C. Wang and R. B. Woodward, *J. Am. Chem. Soc.*, 1983, **105**, 6429, and references therein.
275. M. J. Broadhurst, R. Grigg and A. W. Johnson, *J. Chem. Soc., Perkin Trans. 1*, 1972, 1124, 2111, and preceding papers.
276. G. Bringmann and B. Franck, *Liebigs Ann. Chem.*, 1982, 1261, 1272, and references therein.
277. H. Rexhausen and A. Gossauer, *J. Chem. Soc., Chem. Commun.*, 1983, 275.
278. A. Gossauer, *Chimia*, 1983, **37**, 341.

21.2

Other Polyaza Macrocycles

NEIL F. CURTIS
Victoria University of Wellington, New Zealand

21.2.1 INTRODUCTION

This section covers the coordination chemistry of aza macrocycles taken as rings of nine or more members, with three or more ring nitrogen atoms, and not included in Chapter 21.1. These ligands have many properties in common with their non-cyclic analogues (Chapter 13) and also have properties in common with other cyclic polydentate ligands (Chapters 21.1 and 21.3). The subject was comprehensively reviewed in 1979,[1-6] and in general references included therein will not be listed.

Systematic names are cumbersome, and a variety of trivial names have been used. Both will be avoided, and the system of abbreviations described in ref. 1 will be used.*

In common with other nitrogen donor ligands, the majority of reported coordination compounds are with later *d* transition elements, or early B subgroup elements, in oxidation states of II or III. Tetraaza macrocycles predominate, with 14 followed by 16 as the most common ring sizes.

---

* Main features: ring size, [*n*]; unsaturation, -ane, -diene; number of ligating atoms, N_4; number of non-ligating atoms (O_2); substituents as prefixed, Me_6, including fused 2,6-pyridine (pyo) and 1,2-benzene (bzo) rings; locants used if required for specificity.

21.2.2 METHODS OF PREPARATION

Compounds of aza macrocycles are prepared by three main types of procedure: (1) conventional organic synthesis of the ligand, (2) metal-ion-promoted reactions, involving condensation of non-cyclic components in the presence of a suitable metal ion, and (3) modification of a compound prepared by methods (1) or (2).

21.2.2.1 Conventional Synthesis

The unsubstituted, saturated aza macrocycles, the family of cyclic secondary amines, $[n]aneN_m$, are generally prepared by the method shown in Scheme 1. Some other types of aza macrocycle, particularly cyclic amides[7] and macrocycles with $N{=}CRCR{=}CRNH$ functions (see below), are also prepared conventionally. The preformed ligands are then reacted with the desired metal ion under appropriate conditions.

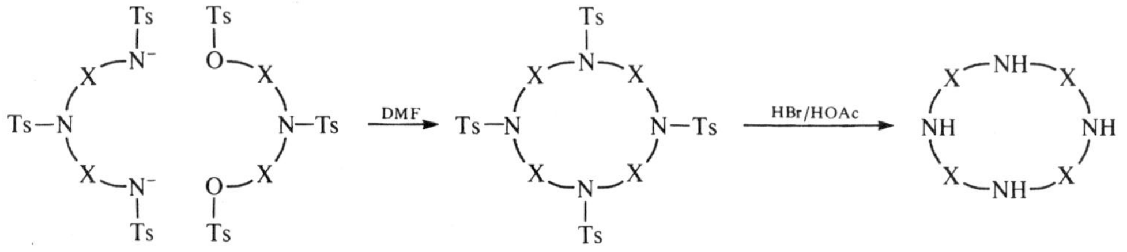

Scheme 1

21.2.2.2 Metal-ion-promoted Synthesis[2,8-11]

A great variety of aza macrocycle complexes have been formed by condensation reactions in the presence of a metal ion, often termed 'template reactions'. The majority of such reactions have imine formation as the ring-closing step. Fourteen- and, to a lesser extent, sixteen-membered tetraaza macrocycles predominate, and nickel(II) and copper(II) are the most widely active metal ions. Only a selection of the more general types of reaction can be described here, and some closely related, but non metal-ion-promoted, reactions will be included for convenience. The reactions are classified according to the nature of the carbonyl and amine reactants.

21.2.2.2.1 Self-condensation of aminocarbonyl compounds

The self-condensation of 2-aminobenzaldehyde in the presence of suitable metal ions to form the cyclic tetramer (**1**), and sometimes the trimer (**2**), is the best-known example of this reaction type (Scheme 2). The cyclization of the β-amino ketone complex (**5**) provides an aliphatic example

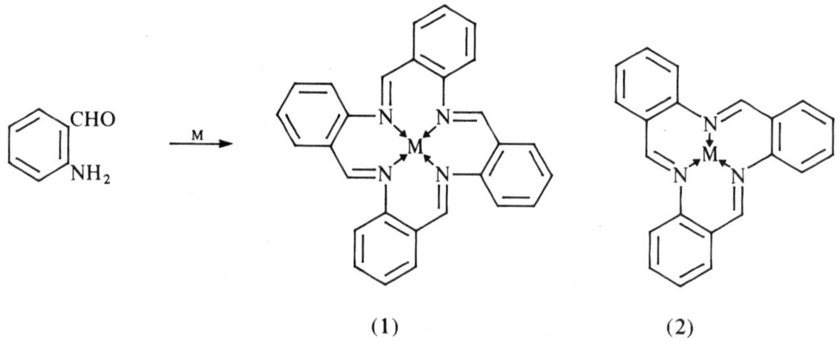

(1)   (2)

Scheme 2

21.2.2.2.2 Reactions of amines with monocarbonyl compounds

Monocarbonyl compounds take part in macrocyclic condensations of two types. The first, represented by the reaction of $[Ni(en)_3]^{2+}$ with acetone, forms isomeric $[14]dieneN_4$ macrocycle complexes (3) and (4) with the amine–imine bridging group linking the amine residues derived from two acetone residues (Scheme 3).[12] $[Ni(en)_2]^{2+}$ reacts with acetone by rapid imine formation, followed by an aldol-type condensation, to give a β-amino ketone complex (5), which can be cyclized by further imine formation. Similar reactions occur for complexes of 1,3- and 1,4-diamines,[13] giving [15]-, [16]- and [18]-dieneN$_4$ macrocycles analogous to (3), and with some tri- and tetra-amines to give [12]eneN$_3$, [13]dieneN$_4$ and [14]eneN$_4$ compounds. Acetone is the most successful carbonyl reactant, so the compounds characteristically have the trimethyl-substituted amine–imine six-membered chelate ring. Similar reactions occur for copper(II) amine complexes.

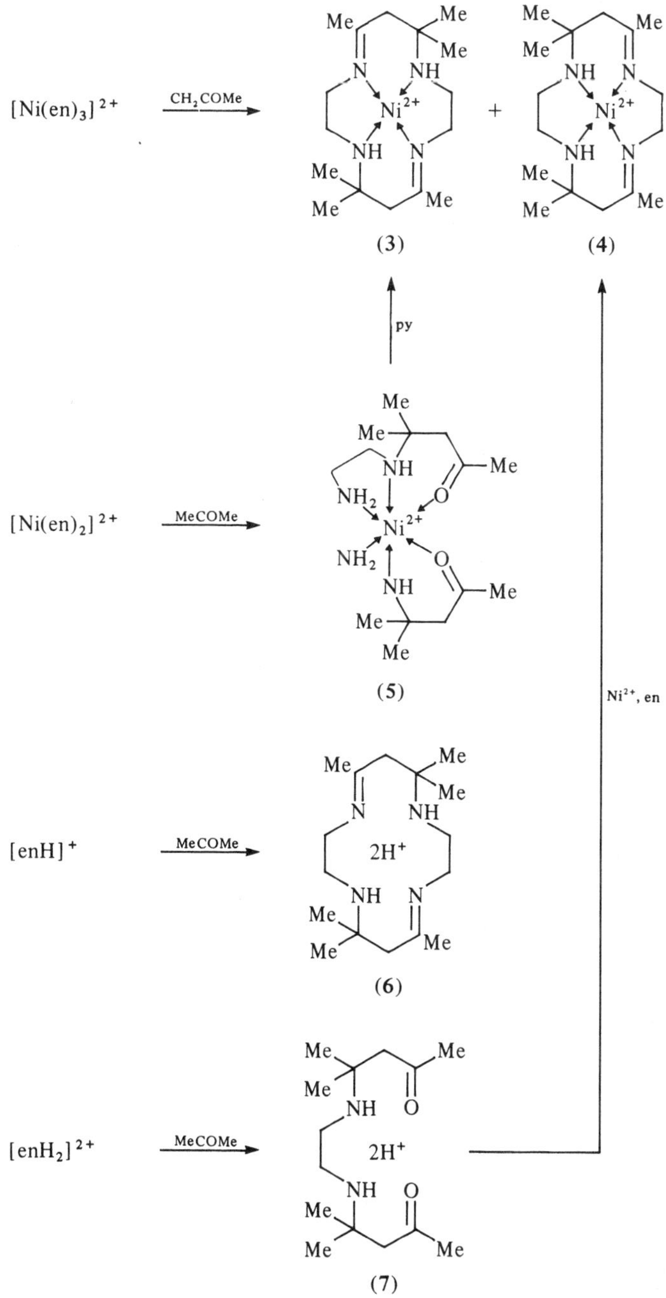

Scheme 3

Although this was one of the earliest reported metal-ion-promoted reactions yielding aza macrocycle complexes, the compounds can often be made without the mediation of metal ions. Thus, salts of the macrocycle (6) are formed rapidly by reaction of [enH]$^+$ with acetone, or with 2-methylpent-3-en-2-one (mesityl oxide), and this reaction has been generalized to the reaction of a variety of α,β-unsaturated carbonyl compounds, yielding [14]4,11-dieneN$_4$ macrocycles with varied substituents on the six-membered chelate rings. The [14]4,14-dieneN$_4$ macrocycle compound (4) has also been prepared indirectly *via* the β-amino ketone salt (7), which is formed by reaction of $[enH_2]^{2+}$ salts with acetone,[14] and which has been used to prepared [15]- and [16]-dieneN macrocycle compounds related to (4), by reaction with the appropriate diamine.

The second type of reaction of a monocarbonyl compound to yield a macrocyclic product is represented by the condensation of $[Ni(en)_2]^{2+}$ or $[Cu(en)_2]^{2+}$ with formaldehyde in the presence of a suitable nucleophile (Scheme 4).[15,16] This reaction is related to the condensations of $[Co(en)_3]^{3+}$ with formaldehyde plus nucleophiles to form clathrochelate compounds (Chapter 21.3), and also to the formation of the $Co^{III}[14]aneN_4(O_2)$ complex (8) by reaction of a bis(ethanediamine)cobalt(III) complex with formaldehyde.[17]

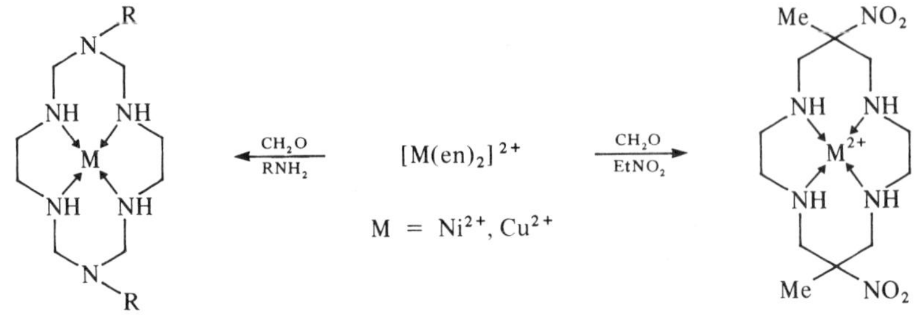

Scheme 4

21.2.2.2.3 *Reactions of amines with dicarbonyl compounds*

Condensations of diamines with dicarbonyl compounds are a fruitful source of imino macrocycles. Condensations can be 2:2 with diamines to form tetraimine macrocycles, or 1:1, with additional heteroatoms present for the amine and/or carbonyl components, to form diimine macrocycles, as for (9) and (10) formed from 1,2-dicarbonyl compounds.

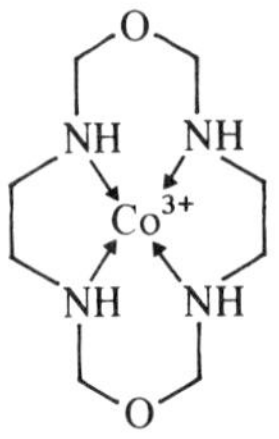

(8) 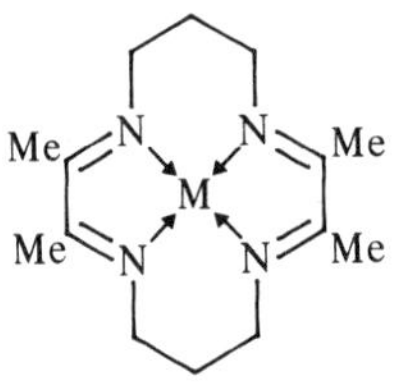   (9) 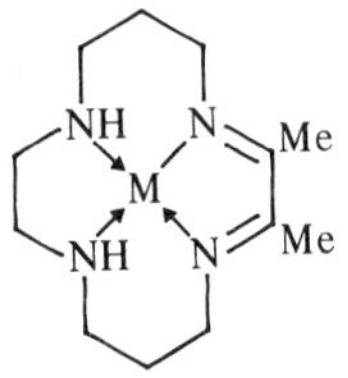   (10)

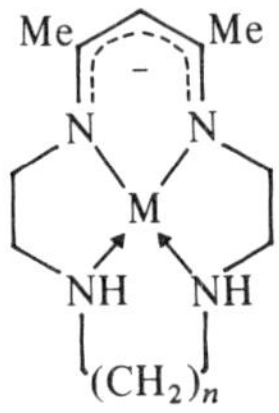

(11) $n = 2$
(12) $n = 3$

1,3-Dicarbonyl compounds form 1,3-diimine macrocycles, which are usually isolated with the deprotonated tautomeric dienylato($1-$) chelate ring, as for (11) and (12) formed from tetramines and pentane-2,4-dione (acetylacetone, acacH). Direct reaction of acacH with ethanediamine complexes of the usual labile metal ions forms the β-iminoenolato complex which will not react further under any conditions devised, apparently because of the low reactivity towards nucleophiles of the oxygen atoms, although the reaction does occur for Pt^{IV} and Au^{III} bisethanediamine compounds (Scheme 5).[18] The reactivity of the oxygen is enhanced with 1,2-diaminobenzene, and benzo and dibenzo macrocycle complexes are readily prepared (Scheme 6). For aliphatic diamines the reactivity is sufficiently enhanced by a variety of substituents at the 2 position of the 1,3-dicarbonyl compound, *e.g.* acyl, bromo, diazophenyl, *etc.* The metal ions can be removed from the macrocycles by a variety

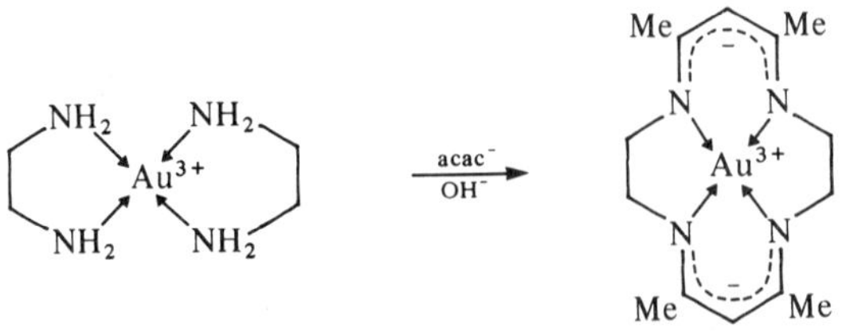

Scheme 5

of reagents: the diacyl compounds, for example, are deacylated and demetallated by HCl in dry ethanol to give the free ligand (Scheme 7). A number of these macrocycles have been prepared in the absence of metal ions, *e.g.* the macrocycles with 1,3-dimethyl- (14) or 2-phenyl-substituted (15) dienylato chelate rings, and the unsubstituted dibenzo macrocycle (16) formed from 1,2-diamino-benzene and propargyl aldehyde.

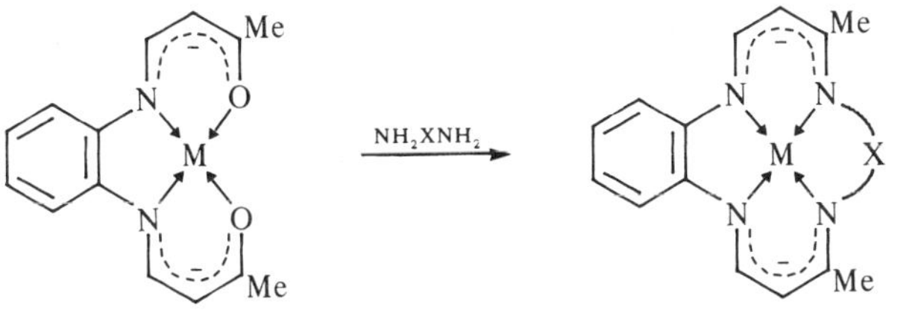

$X = (CH_2)_2, (CH_2)_3, o\text{-}C_6H_4$

Scheme 6

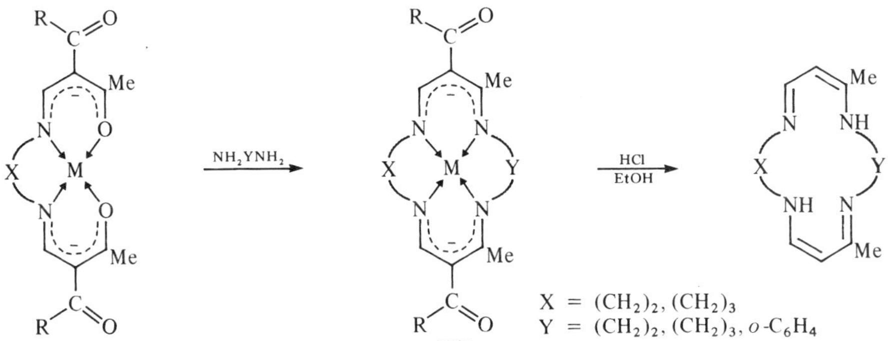

(13)

$X = (CH_2)_2, (CH_2)_3$
$Y = (CH_2)_2, (CH_2)_3, o\text{-}C_6H_4$

Scheme 7

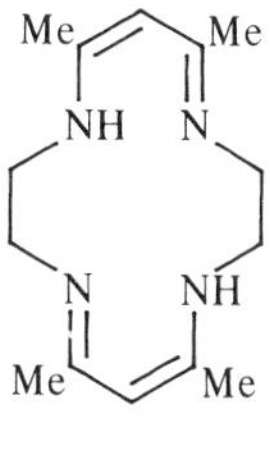

(14) 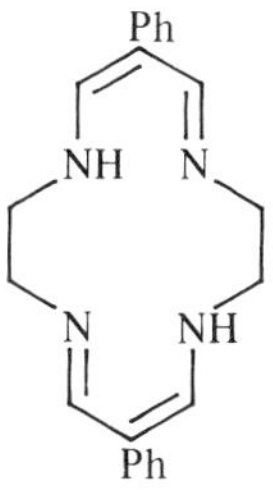   (15) 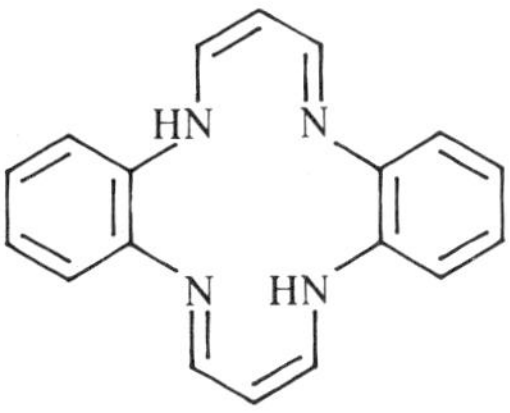   (16)

An extensive family of tetraaza (17) and pentaaza (18) macrocycles with a range of ring sizes are obtained by reactions of 2,6-diacetylpyridine with tri- or tetr-amines in the presence of suitable metal ions, which include as well as the usual M^{II} transition metal ions a variety of non-transition metal ions.[19] Large rings, such as (19), can be formed by 2:2 condensations in the presence of suitable larger metal ions.[20] Reactions of α, ω diamines which have other 'internal' heteroatoms with 2,6-diacetylpyridine (or 2,5-diformylfuran) produce a variety or related mixed heteroatom macrocycles.

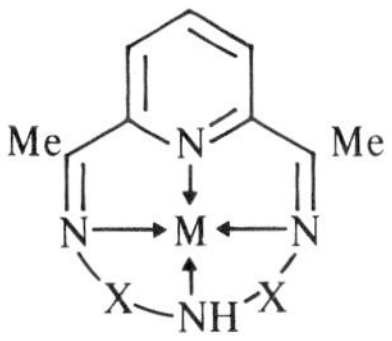

(17) 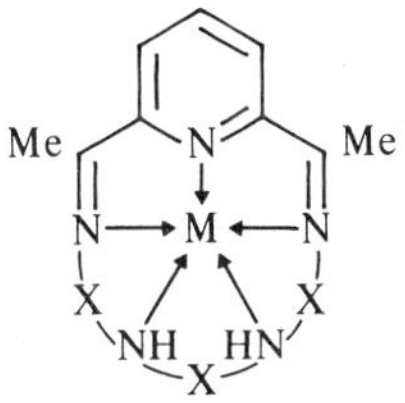   (18) 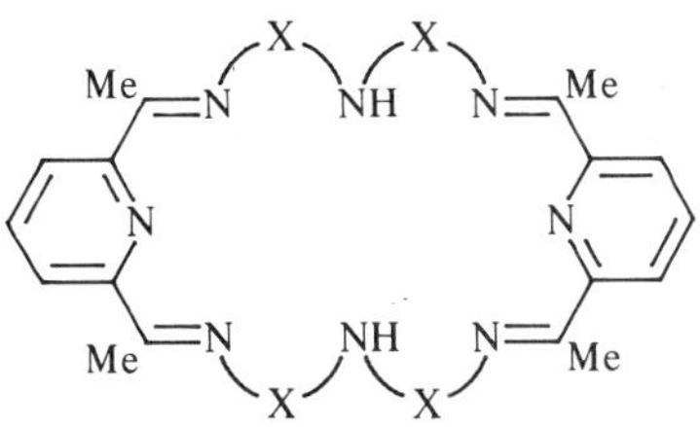   (19)

$X = (CH_2)_2, (CH_2)_3, o\text{-}C_6H_4$, respectively

Other examples of macrocycles formed by 1:1 condensations of diamines with dicarbonyl compounds, with additional donor atoms present for the amine or carbonyl compound, include an extensive family with 2-aminobenzaldimino chelate rings, such as (20), and those formed from (7).

(20)

$$X = (CH_2)_2, (CH_2)_3, (CH_2)_4$$
$$Y = (CH_2)_2, (CH_2)_3, o\text{-}C_6H_4$$

21.2.2.2.4 *Reactions of* cis *diaza compounds with carbonyl compounds*

A variety of macrocyclic complexes which have adjacent nitrogen atoms (cyclic hydrazines, hydrazones or diazines) are formed by condensations of hydrazine, substituted hydrazines or hydrazones with carbonyl compounds. The reactions parallel in diversity those of amines, but are often more facile since the reacting NH_2 groups is generally not coordinated and the electrophile is thus not in competition with the metal ion. The resulting macrocycles may be capable of coordination isomerism, since either of the adjacent nitrogen atoms can act as donor atom.

Condensation of a monocarbonyl compound with a dihydrazone initially yields a macrocycle with a tetraaza six-membered chelate ring, *e.g.* (21; Scheme 8) or (23; Scheme 9), but this can isomerize to give a triaza five-membered chelate ring, as for (27; Scheme 10), where cyclization is by a reaction subsequent to the hydrazone/carbonyl condensation,[21] or for the isomeric pair of compounds (25) and (26) of Scheme 9. Compounds with triaza (28) and tetraaza (22) seven-membered chelate rings have also been prepared.[22] Tetradentate and pentadentate aza macrocycles are formed by condensations of 2,6-diacetylpyridine with hydrazine (29) or with dihydrazines (30).[21]

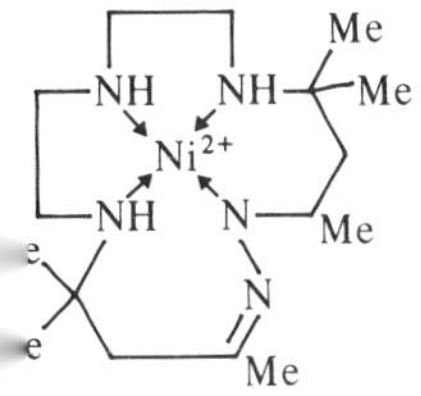

(28) 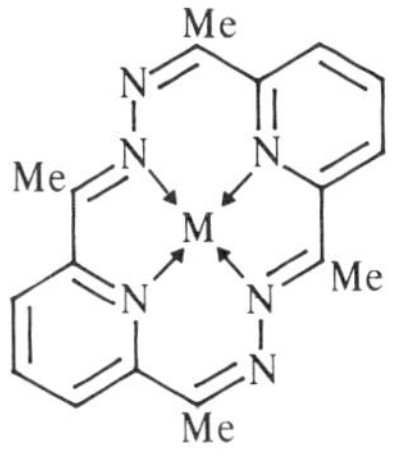   (29) 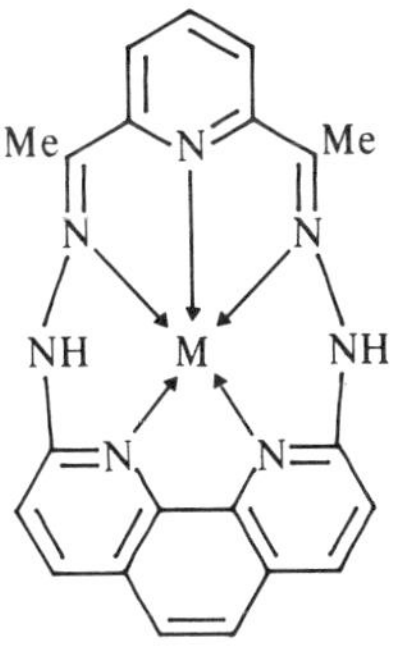   (30) 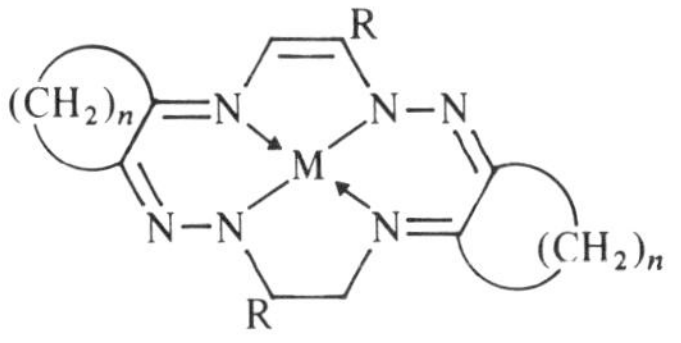   (31)

Dihydrazones of cyclic 1,2-diketones react with ortho esters and similar reagents to form aza macrocycles, which coordinate as for (31).

21.2.2.2.5 Cyclization reactions of oximes

Bora-substituted aza macrocycles are formed by reaction of oxime complexes with boron compounds, *e.g.* (32) and (33) are formed with boron trifluoride.

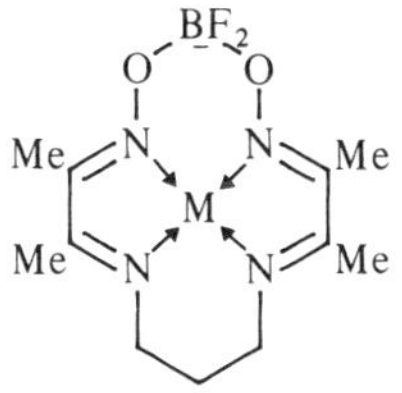

(32) 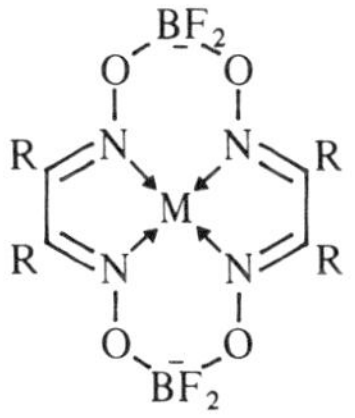   (33)

21.2.3 MODIFICATION OF PREFORMED AZA MACROCYCLE COMPLEXES[2,6]

Complexes of aza macrocycles can be modified by various chemical procedures.

21.2.3.1 Oxidative Dehydrogenation

Oxidative dehydrogenation reactions increase the extent of unsaturation in an aza macrocyclic ligand, usually by converting secondary amines into imine groups, occasionally by introducng $C=C$ double bonds. The reactions tend to be metal-ion specific, and generally proceed by a sequence of one-electron oxidations of the metal ion, which then oxidizes the ligand. In Scheme 11 the Fe^{II} cyclic tetramine complex is oxidized by O_2 in acetonitrile, the Ni^{II} complex is oxidized by dilute HNO_3, forming tetraimine complexes: (34) with α-diimine groups for Fe^{II} and (35) with isolated imine groups for Ni^{II}.

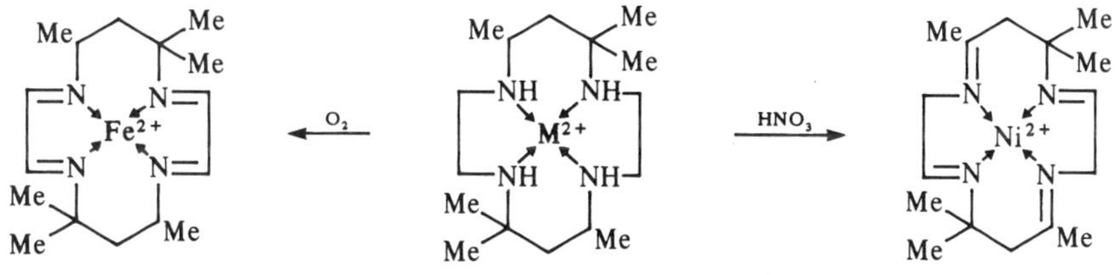

(34)   (35)

Scheme 11

Oxidative dehydrogenation reactions can produce macrocycles with extensive delocalization, *e.g.* Schemes 9 and 12. In Scheme 12 the 15-π-electron radical complex (36) can be further oxidized to the 14-π-electron complex (37), or reduced to the 16-π-electron complex (38) by reversible one-electron redox steps.

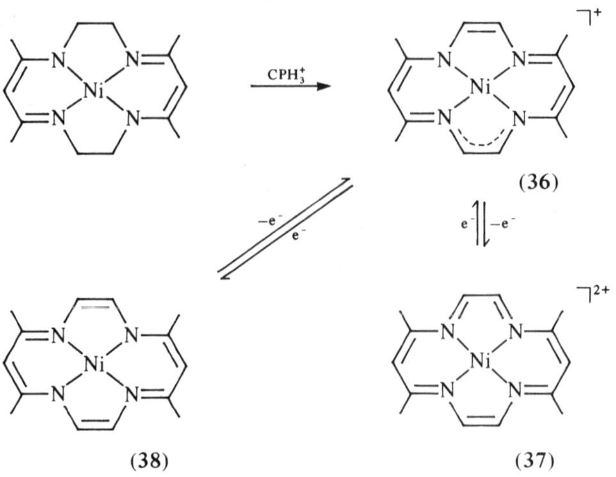

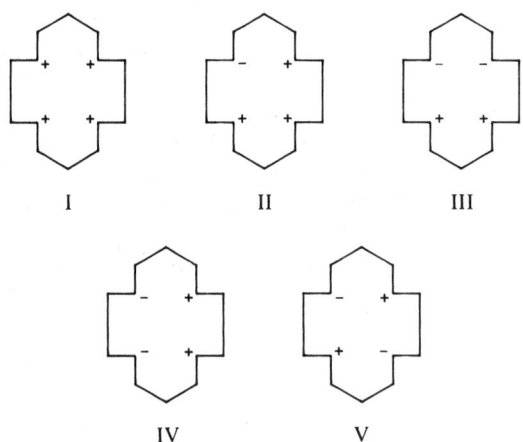

Scheme 12

21.2.3.2 Hydrogenation

Reduction reactions, which usually convert imine to secondary amine functions, are also metal-ion specific and usually most successful for the Ni^{II} complexes. Reductions can be by electrochemical means or by chemical reductants such as $NaBH_4$, $NiAl/OH^-$, H_2/Pt or H_2/Ni. H_3PO_2 is specific for conversion of an α-diimine group to a monoimine. Examples of imine complexes which have been reduced to form cyclic amine complexes include (**1**), (**3**), (**4**), (**9**), (**16**) and (**20**).

21.2.3.3 Substitution Reactions[6]

Coordinated secondary amino groups of aza macrocycles can often be alkylated by standard reagents. Alkylation of $[Ni([14]aneN_4)]^{2+}$ produces $[Ni(N-Me_4[14]aneN_4)]^{2+}$ in the stable nitrogen configuration III (Figure 1), while reaction of Ni^{2+} with preformed $N-Me_4[14]aneN_4$ produces the compound in the thermodynamically and kinetically less stable configuration I. Macrocycles with dienylato(-1) six-membered chelate rings undergo facile nucleophilic substitution reactions at the 'central' carbon atom of the chelate ring (Scheme 12). Alkylation of the carbonyl function of (**13**) produces very reactive 1,3-diimine 'ethylidyne' compounds (**39**) which serve as precursors of a variety of substituted macrocyclic products including a variety of macrobicyclic complexes.[23]

Figure 1 Configurations of coordinated 1,4,8,11-tetraazacyclotetradecane. The + indicates that the NH group is above the plane of the (flattened) macrocycle; the − indicates that it is below

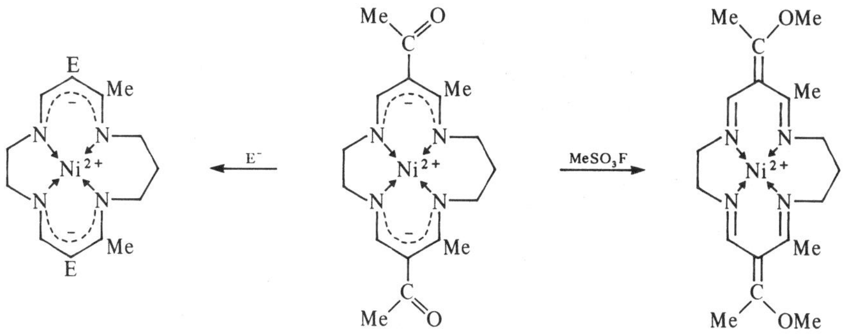

Scheme 13

(39)

21.2.3.4 Addition Reactions

Coordinated imine groups, particularly for (1), react with nucleophiles, *e.g.* with alkoxides to form α-amino ethers (40).

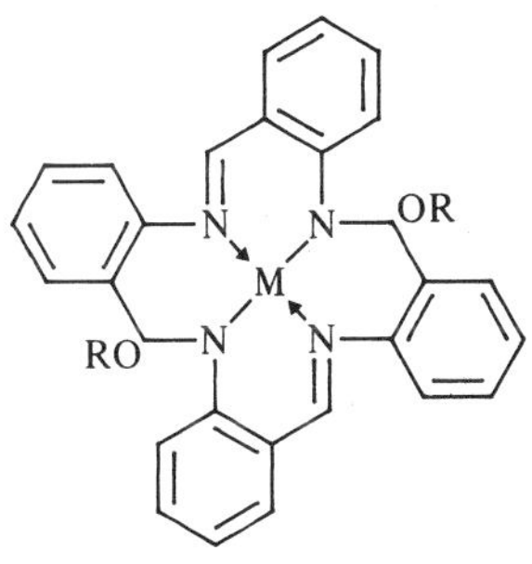

(40)

21.2.4 BINUCLEAR COMPOUNDS

Binuclear compounds of aza macrocycles of four types have been described: (i) compounds of large polydenate rings, *e.g.* (19), which have two metal ions coordinated by different sets of donor atoms, often with a bridging ligand such as azide linking the metal ions;[20] (ii) compounds which have separate aza macrocycles which are covalently linked; (iii) compounds with metal(macrocycle) units linked facially by a bridging group between the metal ions; and (iv) confacial dimers with some degree of metal–metal bonding, *e.g.* of Ru^{II} or Rh^{II} compounds of (Me$_4$-16), with which may be associated some π-type interactions between the rings as indicated by structures with eclipsed macrocycles (*e.g.* for nickel(II) complexes 26 and 36).

21.2.5 PROPERTIES OF AZA MACROCYCLE COORDINATION COMPOUNDS[3-6,24]

The properties of the aza macrocycle compounds generally resemble those of their non-cyclic analogues, and only a few more specifically macrocyclic aspects are described.

21.2.5.1 Structures of Aza Macrocycle Complexes[4,24a]

Tetraaza macrocycles usually coordinate with the N$_4$ donor set approximately coplanar for square planar or *trans* octahedral arrangements. The more flexible macrocycles can also coordinate in 'folded' arrangements, usually to accommodate an additional chelate ligand in an octahedral arrangement, and planar-folded interconversions are facile. In planar coordination some degree of tetrahedral distortion of the donor set is common, and this can become substantial for larger ring sizes, *e.g.* for Ni^{II} complexes of the aminotroponiminato macrocycles (41) the dihedral angle between the NiN$_2$ sets increases from 8.3° to 85.2°, and the average Ni—N distance increases from 186.3 to 195.0 pm as *n* increases from 3 to 6.[25] In some instances the macrocycle configuration orients all nitrogen lone pairs to the same side of the macrocycle, and structures with the metal ion displaced from the N$_4$ plane, usually with square pyramidal coordination geometry, result, *e.g.* for *N*-Me$_4$[14]aneN$_4$, configuration I, or for (*N*-Me$_4$-16).

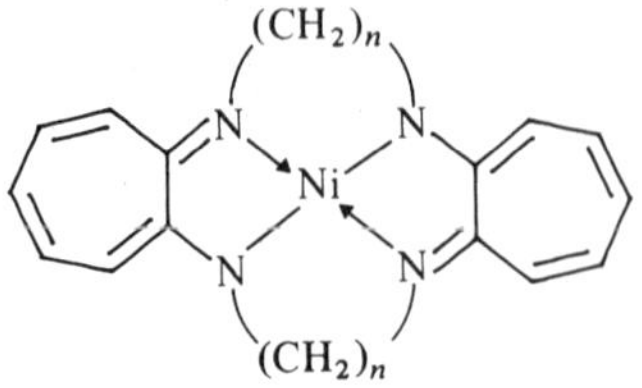

(41)

Cyclic pentamines adopt folded coordination modes in octahedral coordination, with configurational isomerism possible if different chelate segments are present.[26] Pentagonal bipyramidal coordination with planar macrocycles has been observed for tetradentate pyo macrocycles, *e.g.* (18) and (30).[21] Reported hexaaza macrocycles fold to give octahedral coordination arrangements, again with the possibility of configurational isomerism.[27]

Coordinated secondary and tertiary nitrogens become chiral centres, and configurational isomerism arises if two or more of these groups are present, *e.g. meso* and *racemo* isomers of (3), and the configurations I–V for [M([14]aneN$_4$)]$^{n+}$ (Figure 1). These configurations can differ appreciably in internal strain energy, and so III is favoured for planar coordination and V for folded coordination.[28] Interconversion of configurations is base catalyzed, and metastable forms can often be isolated under acid conditions, *e.g.* the reactions in Scheme 14, which permit isolation of [Ni(Me$_6$[14]aneN$_4$)]$^{2+}$ salts in metastable configuration V.[24] For tertiary amines, isomerization of nitrogen configurations requires the breaking of an M—N bond, and the isomers of (*N*-Me$_4$)[14]aneN$_4$ are resistant to interconversion (although it has been reported that the Ni^{II} isomers do interconvert in hot ethylamine).[29]

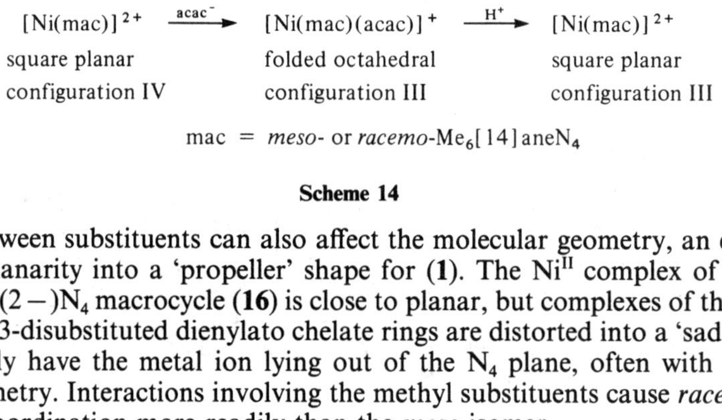

$$[Ni(mac)]^{2+} \xrightarrow{\text{acac}^-} [Ni(mac)(acac)]^+ \xrightarrow{\text{H}^+} [Ni(mac)]^{2+}$$

square planar   folded octahedral   square planar
configuration IV   configuration III   configuration III

mac = *meso*- or *racemo*-Me$_6$[14]aneN$_4$

Scheme 14

Interactions between substituents can also affect the molecular geometry, an example being the distortion from planarity into a 'propeller' shape for (1). The Ni^{II} complex of the unsubstituted bzo$_2$[14]hexaenato(2−)N$_4$ macrocycle (16) is close to planar, but complexes of the macrocycle with 1-substituted or 1,3-disubstituted dienylato chelate rings are distorted into a 'saddle shape' and the complexes typically have the metal ion lying out of the N$_4$ plane, often with square pyramidal coordination geometry. Interactions involving the methyl substituents cause *racemo*-Me$_6$[14]aneN$_4$ to adopt folded coordination more readily than the *meso* isomer.

21.2.5.2 Spectroscopic and Magnetic Properties[5,24,30]

The *d–d* spectra of many aza macrocyclic complexes have been tabulated,[5] and the spectra of the compounds of some metal ions investigated in detail. Only for the tetraaza macrocycles are any specifically macrocyclic effects observed. For *cis*-[M(mac)X$_2$] compounds (mac = [*n*]aneN$_4$) there are changes in the ligand field attributable to the macrocycle, and these can be related to changes in ligand field strengths for linear amines with similar structural features, particularly the numbers and relative linkages of five- and six-membered chelate rings present. For *trans*-[M(mac)X$_2$] compounds of Fe^{II}, Co^{III} and Ni^{II} the equatorial (macrocycle) ligand field strength parameter, $10Dq^{xy}$, is very dependent on the ring size and can have values substantially larger than for related but non-macrocyclic ligands, *e.g.* with Co^{III}: 2750 cm^{-1} for [13]aneN$_4$, decreasing to 2295 cm^{-1} for [16]aneN$_4$, compared with 2530 cm^{-1} for *trans*-bis(ethanediamine).[24,28] These observations have been rationalized by considering the relationship between the optimum size of cation which permits the macrocycle to adopt its minimum-strain *endo* conformation and the 'ideal' M—N distances for the cations. Strain energy minimization calculations indicate that the macrocycle 'hole size' increases by 10–15 pm for each increment in n.[24a,28] If the spectroscopic data are taken to indicate that the macrocycle with $n = 14$ is the optimum size to coordinate Co^{II} with minimum strain, since the $10Dq^{xy}$ value is similar to that of *trans*-bis(ethanediamine), then rings with $n < 14$ and with $n > 14$ will be strained in opposite senses to accommodate the relatively inflexible Co—N distances. If the ring is too small, the inward strain on the donor atoms raises the effective ligand field strength of the equatorial N_4 donor set, $10Dq^{xy}$, by a 'macrocyclic constrictive effect', and conversely for rings larger than optimum, $10Dq^{xy}$ is decreased by an analogous 'dilative effect'.[24]

Alternatively, the maximum ligand field strength can be considered to arise for the least strained system, when the optimum M—N distance and 'hole sizes' are in best agreement. The high values observed with small rings are then taken to arise from the inherently greater ligand field strength of secondary nitrogen donor atoms, compared with primary nitrogen atoms. For non-macrocyclic ligands this effect is usually masked by the greater intra- and inter-ligand repulsions present for secondary amines, which for the small Co^{III} ion in particular prevent most non-macrocyclic ligands from approaching the ideal Co—N distance of about 193 pm. High ligand field strengths for cyclic di- and tri-amines, which cannot 'girdle' the metal ion, support this view.[30]

For many Cu^{II} tetramine complexes a correlation between the enthalpy of formation and the band maximum in the *d–d* spectrum has been observed and attributed to the common presence of tetragonal diaquo species.[31] Values for cyclic amines follow this correlation, both $\Delta H^{\ominus}$ and the band energy having maximum values for [14]aneN$_4$, with deviations for $n = 12$ and $n = 15$ attributable to distortions from tetragonal geometry.

Complexes of several cyclic tetramines with Ni^{II} exist in aqueous solution as an equilibrium between singlet ground state, square planar species and triplet ground state, diaquo species (see Table 2).[32]

[Fe([9]aneN$_3$)$_2$] is low-spin, supporting the assignment of high inherent ligand field strengths to secondary amine donors.[30] [Fe^{II}(mac)X$_2$] complexes of cyclic tetramines–imines lie close to the high-spin–low-spin crossover point. Most compounds with $X = CN^-$ are low-spin and those with $X = MeCN$ are high-spin. Spin equilibria exist for [Fe(*meso*-Me$_6$[14]aneN$_4$)(NCS)$_2$] and [Fe(Me$_6$[14]dieneN$_4$)(phen)]. Square planar compounds of Fe^{II} in unusual $S = 1$ spin state have been reported for macrocycles (Me$_4$-16) and (26), as well as for tetraphenylporphyrins.

21.2.5.3 Reactions Kinetics[3,6]

21.2.5.3.1 Complex formation reactions

The rates of the formation reaction (1) in non-aqueous solvents for a variety of cyclic tetramines are comparable with those for non-cyclic polyamines.[33] In aqueous solution, where protonated amine species predominate, the rates of reactions (2) are slower for the cyclic amines, the effect becoming more pronounced as the protonation number, l, increases. For different metal ions the rates parallel the water exchange rates ($Cu^{2+} > Zn^{2+} > Co^{2+} > Ni^{2+}$), and for any particular ion the rates do not vary systematically with ring size.[34]

$$M^{n+} + [m]aneN_4 \longrightarrow [M([m]aneN_4)]^{n+} \tag{1}$$

$$[M_{aq}]^{n+} + [[m]aneN_4H_l]^{l+} \longrightarrow [M([m]aneN_4)]^{n+} + lH^+ \tag{2}$$

21.2.5.3.2 Acid dissociation reactions

The extreme resistance to acid dissociation of the [14]aneN and [14]dieneN complexes of Ni^{II}, and to a lesser extent Cu^{II}, is one of the features of aza macrocycle coordination chemistry.

These aquation reactions follow the same general mechanism as for non-cyclic amines, even though the rates can be many orders of magnitude less.[35] The rate expression can show acid independent (solvolytic) and/or acid dependent pathways. For secondary amine/imine macrocycles with less than 16 members, reactions with Ni^{2+} are usually first order in $[H^+]$ (cleavage of first M—N bond rate-determining), while for Cu^{2+} they are second order in $[H^+]$ (cleavage of second

M—N bond rate-determining). For complexes of larger macrocycles acid limiting kinetics are often observed. Rates of dissociation are collected in Table 1.

For tetraaza macrocycles, the reaction rate is least for the minimally strained fourteen-membered ring, increasing substantially for larger or smaller ring sizes. Rates vary for configurational isomers, *e.g.* in the extreme case of (tetra-*N*-methyl-)[14]aneN$_4$ nickel^{II}, from acid labile for the configuration I species (formed by reaction of the ligand with Ni^{2+}) to extremely acid resistant for the configuration III species (formed by methylating the [14]aneN$_4$ complex) rate-determining step.

Slow acid dissociation reactions are observed not only for complexes with 'girdling' macrocycles, but also for compounds with *cis* (folded) macrocycles, with cyclic triamines (which must coordinate facially) and even for some cyclic diamines (which must coordinate bidentally). The common feature is the absence of a terminal nitrogen which can initiate an 'unzipping' mechanism. The rate-determining step is associated with a movement of a nitrogen away from a coordination site, with or without concomitant protonation. Because of the restricted conformational freedom of a coordinated macrocycle, this requires a concerted conformational change with high activation energy. Imine groups which make the macrocycle more rigid, and substituents which restrict conformational changes, generally reduce reaction rates.

For the cyclic triamines, Ni^{II} compounds hydrolyze slowly (*e.g.* for [12]aneN$_3$ $k \approx 10^{-6}\,s^{-1}$)[42] but the Cu^{II} compounds react at about the same rate as non-cyclic analogues.[43]

For Ni^{II} and Cu^{II} complexes of [15]aneN$_5$ the reaction is second order in [H$^+$],[44] while for [Cu([18]aneN$_6$)]$^{2+}$ the reaction is third order in [H$^+$], indicating that the step from tetra- to tri-dentate is rate-determining in both cases. The rates for these larger macrocycles are comparable with those of non-cyclic analogues.

Table 1 Rates of Dissociation in Acid Solution at 25 °C

Ligand	Metal		Dissociation rate (s^{-1})	Ref.
[12]aneN$_4$	Cu		2.54[H$^+$] × 10^{-4}	36[a]
[13]aneN$_4$	Ni		2 × 10^{-5}	24[b]
[14]aneN$_4$	Ni	Configuration V	3[H$^+$] × 10^{-6}	37[c]
		Configuration III	*ca.* 10^{-9}	37[c]
Me$_6$[14]aneN$_4$	Cu	C-*meso*	2.5[H$^+$] × 10^{-4}, 4.6[H$^+$] × 10^{-4}	38[d]
		C-*racemo*	(0.52 + 1.42[H$^+$]) × 10^{-7},	38[d]
			(0.115 + 1.7[H$^+$]) × 10^{-7}	
[15]aneN$_4$	Ni		6 × 10^{-5}	24[b]
[16]aneN$_4$	Ni		0.19	24[b]
	Cu		1.1[H$^+$], 1.3, 1.7	39[e,f,h]
[17]aneN$_4$	Cu		0.6[H$^+$], 0.7, 1.3	39[e,f,g]
Me$_6$[14]diene	*trans*-Cu		1.3[H$^+$]^2 × 10^{-7}	40[i,j]
	cis-Cu		1.5[H$^+$]^2 × 10^{-6}	40[i,k]
Me$_6$[15]diene	*trans*-Ni		[H$^+$] × 10^{-5}, 1.4 × 10^{-5}, 0.4	40[e,i,l]
	cis-Ni		1.7[H$^+$] × 10^{-5}	40[i,m]
	cis-Cu		3.3[H$^+$]^2 × 10^{-4}	40[m,n]
Me$_7$[15]diene	Ni		4[H$^+$] × 10^{-7}, 5 × 10^{-7}, 0.13	40[e,l,o]
Me$_6$[16]diene	Cu		7[H$^+$] × 10^{-3}, 9 × 10^{-3}, 1.0	40[e,n,p]
Me$_6$[18]diene	*trans*-Ni		6[H$^+$] × 10^{-4}, 8.5 × 10^{-4}, 1.7	41[e,q,r]
	trans-Cu		3.6[H$^+$] × 10^{-2}, 4.9 × 10^{-2}, 10.2	41[e,q,s]
	cis-Cu		25[H$^+$]	41[s,t]

[a] 1,4,7,10-Tetraazacyclododecane, in 5 M ClO$_4^-$.
[b] 1,4,7,10-Tetraazacyclotridecene. Pseudo-first-order rate constants in 0.3 M HClO$_4$.
[c] 1,4,8,11-Tetraazacyclotetradecane.
[d] 5,5,7,12,12,14-Hexamethyl-1,4,8,11-tetraazacyclotetradecane in 0.1 M NO$_3^-$, consecutive reactions.
[e] Acid limiting kinetics; rate at low [H$^+$], A (M^{-1} s^{-1}), B (M^{-1}) for k_{obs} = A[H$^+$]/(1 + B[H$^+$]).
[f] In 1 M ClO$_4^-$.
[g] 1,4,8,11-Tetraazacyclohexadecane.
[h] 1,4,8,11-Tetraazacycloheptadecane.
[i] In 2 M Cl$^-$.
[j] 5,7,7,12,14,14-Hexamethyl-1,4,8,11-tetraazacyclotetradeca-4,11-diene.
[k] 5,7,7,12,12,14-Hexamethyl-1,4,8,11-tetraazacyclotetradeca-4,14-diene.
[l] 5,5,7,13,15,15-Hexamethyl-1,4,8,12-tetraazacyclopentadeca-4,12-diene.
[m] 5,5,7,13,15,15-Hexamethyl-1,4,8,12-tetraazacyclopentadeca-7,12-diene.
[n] In 1 M Cl$^-$.
[o] 3,5,7,8,13,15,15-Heptamethyl-1,4,8,12-tetraazacyclopentadeca-7,12-diene.
[p] 2,4,4,10,12,12-Hexamethyl-1,5,9,13-tetraazacyclohexadeca-1,9-diene.
[q] 2,4,4,11,13,13-Hexamethyl-1,5,10,14-tetraazacyclohexadeca-1,10-diene.
[r] In 1 M ClO$_4^-$.
[s] In 1 M NO$_3^-$.
[t] 2,4,4,11,11,13-Hexamethyl-1,5,10,14-tetraazacyclohexadeca-1,13-diene.

21.2.5.3.3 Substitution of other ligands

Rates of many substitution reactions of additional ligands of aza macrocyclic complexes have been measured and the effects of ring size, substituents, *etc.* have been established.[6,45] *trans*-$[Co([n]aneN_4)Cl_4)]^+$ complexes provide the most comprehensive example of the effect of ring size: [13] 6.8×10^{-4}, [14] 1.1×10^{-6}, [15] 1.9×10^{-3}, [16] 2.57 (k_1 at 25 °C, s^{-1}), *i.e.* with a minimum rate for [14]aneN$_4$. The rates correlate with calculated strain energies for the complexes, and it has been suggested that relief of this strain energy in the five-coordinate transition state can account for the observed order of rates.[24]

21.2.5.4 Thermodynamics of Complexation of Cyclic Amines[3]

The cyclic amines have very high thermodynamic stability with the ions Co^{II}–Zn^{II}, Cd^{II} and Hg^{II}, for which data have been reported. Reliable values are sparse, and $\Delta H^\ominus$ and $\Delta S^\ominus$ values evaluated from the temperature coefficients of formation constants, K_f, particularly those determined by polarographic techniques, have often been shown to be in appreciable error by later calorimetric studies. The most comprehensive set of values is available for tetramines (Table 2). K_f values are only available for Co^{II}, and for Ni^{II} the very slow achievement of equilibrium has restricted K_f values available. For several amines with Ni^{II} the results are complicated by the presence of an equilibrium between low-spin square planar and high-spin diaquo octahedral forms. For Co^{II}, Ni^{II} (with available data) and Cu^{II}, K_f for the symmetrical $[n]aneN_4$ macrocycles ($n = 12$–16) is a maximum for $n = 13$, while for Zn^{II} and Hg^{II}, K_f decreases with increasing n. For Ni^{II} and Cu^{II} $\Delta H^\ominus$ is a maximum for $n = 14$, attributed to this macrocycle having minimum internal strain in the *endo* conformation for ions with *ca.* 205–215 pm M—N distances. For Zn^{II}, $\Delta H^\ominus$ is approximately constant for $n = 12$–14, somewhat larger for $n = 15$, and much smaller for $n = 16$, attributed to the presence of similar tetrahedral arrangements for $n = 12$–14, diaquo octahedral species for $n = 15$, and probably five-coordinate monoaquo species for $n = 16$. Entropy changes for Cu^{II} (and from available data also for Ni^{II}) and for Hg^{II} are at a minimum for $n = 14$, and for Zn^{II} at $n = 15$. For Zn^{II} and Hg^{II} large variations are probably attributable to changes in the numbers of coordinated water molecules. Isomers of [14]aneN$_4$ with linked 5,5,6,6- and 5,5,5,7-membered chelate rings have appreciably smaller $\Delta H^\ominus$ values with Cu^{II} and Ni^{II}, showing the importance of structural effects. K_f values for Cu^{II} with *meso*-Me$_6$[14]aneN$_4$ differ by *ca.* 10^6 for different nitrogen configurations. *N*-Substituted [14]aneN$_4$ macrocycles have much lower K_f values because the reaction studied yields complexes in the unfavourable nitrogen configuration I. *C*-Alkyl substituents and the presence of a fused pyridine ring also reduce K_f and $\Delta H^\ominus$ values.

The cyclic triamines $[n]aneN_3$ with Ni^{II} and Zn^{II} have K_f values appreciably higher than analogous linear triamines, while for Cu^{II} the values are similar; K_f values decrease with increasing n. Enthalpy changes for formation of complexes of cyclic pentamines $[n]aneN_5$ with Ni^{II} are -67.4 for $n = 18$, -96.2 for $n = 19$ and -81.2 for $n =$ sym-20 (kJ mol^{-1}); *i.e.* the largest value, for coordination of five nitrogens for [19]aneN$_5$, is smaller than the -100.8 kJ mol^{-1} value for coordination of four nitrogens for [14]aneN$_4$.

The enhancement of the formation constants relative to the values for the thermodynamically most favoured, structurally analogous, but non-cyclic amine has been termed the 'macrocycle effect', by analogy with the 'chelate effect'. This enhancement is dependent on an arbitrary choice of comparison non-cyclic amine, but for Ni^{II} and Cu^{II} it is at a maximum for symmetrical [14]aneN$_4$. Unlike the chelate effect, which is largely entropic in origin, the macrocyclic effect has both enthalpic and entropic components. The entropic component is generally attributed to a smaller configurational entropy for the macrocycle compared with the non-macrocyclic ligand. The enthalpic component has been more controversial. An early suggestion was that for the macrocycle the donor atoms were constrained near the required coordination sites, so the ligand was 'prestrained' as compared with the non-macrocycle. However, there is increasing evidence that (gas phase) metal–nitrogen bond formation is inherently more exothermic for secondary nitrogens. This effect is usually masked by additional inter-/intra-ligand repulsions present for non-cyclic secondary amines, but not for their cyclic analogues. Thus, the substitution of primary by secondary nitrogen donor atoms when a non-cyclic and a cyclic amine are compared can account for the more exothermic reaction of the macrocycle. Solvation energy differences between cyclic and non-cyclic amines are also appreciable, and contribute to the greater exothermicity of coordination of the macrocycles.[30]

Table 2 Thermodynamic Data for Complexes of Cyclic Tetramines[a]

n[b]	Co^{II}[c]	Ni^{II}[d]			Cu^{II}[e]			Zn^{II}[f]			Hg^{II}[g]		
	$\log K_f$	$\log K_f$	$-\Delta H^{\ominus}$	$T\Delta S^{\ominus}$	$\log K_f$	$-\Delta H^{\ominus}$	$T\Delta S^{\ominus}$	$\log K_f$	$-\Delta H^{\ominus}$	$T\Delta S^{\ominus}$	$\log K_f$	$-\Delta H^{\ominus}$	$T\Delta S^{\ominus}$
12[h]	13.8	—	47.8	—	24.8	95.0	46.6	16.2*	60.7	31.8	25.5	98.7	46.9
13[i]	14.7	21.9[b]	83.7[j]	—	29.1	122.2	43.9	15.7	64.0	25.6	25.3	103.3	41.1
14[k]	12.7	—	100.8	24.3	27.2	135.6	19.7	15.3	61.9	26.6	23.0	137.7	-6.4
14[m]	—	—	84.4[n]	31.0	—	116.3	—	—	—	—	—	—	—
14[o]	7.6	14.8[p]	53.5	—	22.4	87.5	40.1	10.4	—	—	20.3	—	—
14[q]	—	—	—	—	18.3	—	—	—	—	—	—	—	—
14[r]	—	—	—	—	—	23.9[s]	—	—	—	—	—	—	—
15[t]	—	18.4	75.0	30.0	24.4	110.9	28.4	15.4	69.0	18.9	23.7	103.3	32.0
16[u]	—	13.2	40.6	34.9	20.9	83.7	35.6	15.0	29.7	55.9	—	—	—

[a] kJ mol^{-1} at 25 °C in water. Enthalpy changes determined calorimetrically unless otherwise specified.
[b] Ring size, [n]aneN$_4$.
[c] Ref. 46.
[d] For triplet ground state diaquo species (ref. 47).
[e] Ref. 48.
[f] Ref. 49.
[g] From temperature coefficients of polarographic data (ref. 50).
[h] 1,4,7,10-Tetraazacyclododecane, cyclen.
[i] 1,4,7,10-Tetraazacyclotridecane.
[j] Value(s) for singlet ground state, square planar species: $\Delta H^{\ominus}$ 52.3.
[k] 1,4,8,11-Tetraazacyclotetradecane, cyclam.
[l] As for j: $\log K_f$ 22.3, $\Delta H^{\ominus}$ 78.2, $T\Delta S^{\ominus}$ 49.1
[m] 1,4,7,11-Tetraazacyclotetradecane.
[n] As for j: $\Delta H^{\ominus}$ 60.3.
[o] 1,4,7,10-Tetraazacyclotetradecane.
[p] As for j: $\log K_f$ 14.8, $-\Delta H^{\ominus}$ 36.4, $T\Delta S^{\ominus}$ 48.
[q] 1,4,8,11-Tetramethyl-1,4,8,11-tetraazacyclotetradecane.[3,53]
[r] meso-2,12-Dimethyl-3,7,11,17-tetraazabicyclo[11.3.1]heptadec-1(7),13,15-triene, Me$_2$pyo[14]aneN$_4$.
[s] Ref. 52.
[t] 1,4,8,12-Tetraazacyclopentadecane.
[u] 1,5,9,13-Tetraazacyclohexadecane.

21.2.5.5 Metal Ion Oxidation–Reduction Reactions[6]

The relative stabilites of metal ions in different oxidation states are strongly dependent on ligand structure for the aza macrocycles. The $M^{I/II}$ and $M^{II/III}$ complexes have been studied for Fe–Cu, although the most comprehensive study is for $Ni^{II/III}$ with tetraaza macrocycles, where a wide range of electrode potentials have been observed.[55] For $[n]aneN_4$ ($n = 12$–16) oxidation is easiest for $n = 14$. Replacement of amine by imine donors makes oxidation more difficult, as does the presence of ring substituents, particularly if axially oriented. The bis[dienylato(1 −)] macrocycle complexes (12) are more easily oxidized, and for $n = 14$–16 the potential is not strongly ring size dependent, but is strongly dependent on the electron-withdrawing effect of substituents at the 2 position of the dienylato(1 −) chelate rings.

The ability of aza macrocycles to stabilize particular states has led to the isolation of compounds with metals in unusual oxidation states, *e.g.* Ni^{III}, Fe^{I}, Co^{I}, Ag^{II}, *etc.*

21.2.6 REFERENCES

1. G. A. Melson, in 'Coordination Chemistry of Macrocyclic Ligands', ed. G. A. Melson, Plenum, New York, 1968, p. 1.
2. G. A. Melson, in ref. 1, p. 17.
3. J. D. Lamb, R. M. Izatt, J. J. Christensen and D. J. Etough, in ref. 1, p. 145.
4. N. F. Curtis, in ref. 1, p. 219.
5. F. L. Urbach, in ref. 1, p. 345.
6. J. F. Endicott and B. Durham, in ref. 1, p. 393.
7. H. Sigel and R. B. Martin, *Chem. Rev.*, 1982, **82**, 385.
8. L. F. Lindoy, *Chem. Soc. Rev.*, 1975, **4**, 421; *Q. Rev., Chem. Soc.*, 1971, **25**, 379.
9. M. de S. Healey and A. J. Rest, *Adv. Inorg. Chem. Radiochem.*, 1978, **21**, 1.
10. D. H. Busch, *Helv. Chim. Acta*, 1976, 174; *Recent Chem. Prog.*, 1964, **25**, 107; C. J. Hipp and D. H. Busch, *ACS Monogr.* 1978, **174**, 220.
11. *Inorg. Synth.*, 1978, **18**, 2–52.
12. N. F. Curtis, *Coord. Chem. Rev.*, 1968, **3**, 3.
13. J. W. L. Martin, J. H. Timmons, A. E. Martell and C. J. Willis, *Inorg. Chem.*, 1980, **19**, 2328.
14. N. F. Curtis, *Inorg. Chim. Acta*, 1982, **59**, 171.
15. P. Comba, N. F. Curtis, G. A. Lawrance, A. M. Sargeson, B. W. Skelton and A. H. White, *Inorg. Chem.*, 1986, **25**, 4260.
16. V. L. Goedken and S.-M. Peng, *J. Chem. Soc., Chem. Commun.*, 1973, 62.
17. R. L. Geue, M. R. Snow, J. Springborg, A. J. Hertt, A. M. Sargeson and D. Taylor, *J. Chem. Soc., Chem. Commun.*, 1976, 285.
18. C. H. Park, B. Lee and G. W. Everett, *Inorg. Chem.*, 1982, **21**, 1681.
19. S. M. Nelson, *Pure Appl. Chem.*, 1980, **52**, 2461.
20. J. M. Lehn, *Pure Appl. Chem.*, 1980, **52**, 2431; O. Kahn, *Inorg. Chim. Acta*, 1982, **62**, 3.
21. G. M. Shalhoub, C. A. Reider and G. A. Melson, *Inorg. Chem.*, 1982, **21**, 1998.
22. A. R. Davis, F. W. B. Einstein and A. C. Willis, *Acta Crystallogr., Sect. B*, 1982, **38**, 443; F. W. B. Einstein and T. Jones, *Acta Crystallogr., Sect. C*, 1983, **39**, 872.
23. D. H. Busch, *Pure Appl. Chem.*, 1980, **52**, 2477; J. H. Cameron, M. Kojima, B. Korybut-Daszkiewicz, B. K. Coltrain, T. J. Meade, N. W. Alcock and D. H. Busch, *Inorg. Chem.*, 1987, **26**, 427; N. W. Alcock, W.-K. Lin, A. Jircitano, J. D. Mokren, P. W. R. Corfield, G. Johnson, G. Novotnak, C. Cairns and D. H. Busch, *Inorg. Chem.*, 1987, **26**, 440.
24. D. H. Busch, *Acc. Chem. Res.*, 1978, 392.
24. (a) K. Henrick, P. A. Tasker and L. F. Lindoy, *Prog. Inorg. Chem.*, 1985, **33**, 1.
25. W. M. Davis, M. M. Roberts, A. Zask, K. Nakanishi, T. Nozoe and S. J. Lippard, *J. Am. Chem. Soc.*, 1985, **107**, 3864.
26. G. Bombieri, E. Forsellini, A. Del Pra, C. J. Cooksey, M. Humanes and M. L. Tobe, *Inorg. Chim. Acta*, 1982, **61**, 43.
27. D. J. Roher, G. J. Grant, D. G. van Deveer and M. J. Castillo, *Inorg. Chem.*, 1982, **21**, 1902; Y. Yoshikawa, K. Toriumu, T. Ito and H. Yamatera, *Bull. Chem. Soc. Jpn.*, 1982, **55**, 1422.
28. V. J. Thom, C. C. Fox, J. C. A. Boyens and R. D. Hancock, *J. Am. Chem. Soc.*, 1984, **106**, 5947; E. K. Barefield, A. Bianchi, E. J. Billo, P. J. Connolly, P. Pavletti, J. S. Summers and D. G. Van Derveer, *Inorg. Chem.*, 1986, **25**, 4197.
29. P. Moore, J. Sachinidis and G. R. Willey, *J. Chem. Soc., Chem. Commun.*, 1983, 522.
30. V. J. Thom, J. C. A. Boeyens, G. J. McDougall and R. D. Hancock, *J. Am. Chem. Soc.*, 1984, **106**, 3198.
31. L. Fabbrizzi, P. Paoletti and A. P. B. Lever, *Inorg. Chem.*, 1976, **15**, 1502.
32. R. J. Pell, H. W. Dodgen and J. P. Hunt, *Inorg. Chem.*, 1983, **22**, 529.
33. L. Hertli and T. A. Kaden, *Helv. Chim. Acta*, 1981, **64**, 33.
34. A. P. Leugger, L. Hertli and T. A. Kaden, *Helv. Chim. Acta*, 1978, **61**, 2296.
35. D. W. Margerum, G. R. Caley, D. C. Weatherburn, G. K. Pagenkopf, *ACS Monogr.* 1978, **174**, 1; R. M. Izatt, J. S. Bradshaw, A. S. Nielson, J. D. Lamb and J. J. Christensen, *Chem. Rev.*, 1985, **85**, 271.
36. R. W. Hay and M. P. Pujari, *Inorg. Chim. Acta*, 1985, **100**, L1.
37. E. J. Billo, *Inorg. Chem.*, 1984, **23**, 236.
38. B.-F. Liang and C.-S. Chung, *Inorg. Chem.*, 1981, **20**, 2152; 1983, **22**, 1017 (in $5 M NO_3^-$).
39. R. Bembi, B. K. Bardwarj, R. Singh, R. Singh, K. Taneja and S. Aftab, *Inorg. Chem.*, 1984, **23**, 4153.
40. S. R. Goddard and N. F. Curtis, unpublished results.
41. R. W. Hay and R. Bembi, *Inorg. Chim. Acta*, 1982, **62**, 89.

42. L. J. Murphy and L. J. Zompa, *Inorg. Chem.*, 1979, **18**, 3278.
43. P. G. Graham and D. C. Weatherburn, *Aust. J. Chem.*, 1981, **34**, 291.
44. R. W. Hay, R. Bembi, T. Moodie and P. R. Norman, *J. Chem. Soc., Dalton Trans.*, 1982, 2131.
45. C.-K. Poon, *Coord. Chem. Rev.*, 1973, **10**, 1; C.-K. Poon, T.-C. Lau, C.-L. Wong and Y. P. Kan, *J. Chem. Soc., Dalton Trans.*, 1983, 1641.
46. R. Machida, E. Kimura and M. Kodama, *Inorg. Chem.*, 1983, **22**, 2055.
47. M. Micheloni, P. Paoletti and A. Sabatini, *J. Chem. Soc., Dalton Trans.*, 1983, 1189.
48. M. Micheloni, P. Paoletti, A. Poggi and L. Fabbrizzi, *J. Chem. Soc., Dalton Trans.*, 1982, 61; P. Paoletti, *Pure Appl. Chem.*, 1980, **52**, 2433.
49. A. Anichini, F. Fabbrizzi, P. Paoletti and R. M. Clay, *J. Chem. Soc., Dalton Trans.*, 1978, 577.
50. M. Kodama and E. Kimura, *J. Chem. Soc., Dalton Trans.*, 1976, 2335.
51. M. Micheloni and P. Paoletti, *Inorg. Chim. Acta*, 1980, **43**, 109.
52. L. Fabbrizzi, M. Micheloni and P. Paoletti, *J. Chem. Soc., Dalton Trans.*, 1979, 1581.
53. B. S. Nakani, J. J. B. Welsch and R. D. Hancock, *Inorg. Chem.*, 1983, **22**, 2956.
54. D. H. Busch, D. G. Pillsburg, F. V. Lavocchio, A. M. Tait, V. Hung, S. Jackels, M. C. Rakowski, W. P. Schammel and L. Y. Martin, *ACS Symp. Ser.*, 1977, **38**, 32.

21.3
Multidentate Macrocyclic and Macropolycyclic Ligands

KRISTIN B. MERTES
University of Kansas, Lawrence, KS, USA

and

JEAN-MARIE LEHN
Université Louis Pasteur, Strasbourg, France

21.3.1 INTRODUCTION

Multidentate macrocyclic ligands are cyclic molecules consisting of an organic framework interspersed with heteroatoms which are capable of interacting with a variety of species. Macropolycylic ligands are a three-dimensional extension of macromonocycles, in which more than one macrocycle is incorporated in the same molecule. Macrocyclic and macropolycyclic molecules display unique and exciting chemistries in that they can function as receptors for substrates of widely differing physical and chemical properties and upon complexation can drastically alter these properties. Selective substrate recognition, stable complex formation, transport capabilities and catalysis are examples of the wide ranging properties of these molecules. All of these features, which result from association of two or more species, form what has been coined a supramolecular chemistry, the chemistry of the intermolecular bond, as molecular chemistry is the chemistry of the covalent bond. Thus this chapter uses a generalized view of coordination chemistry which is much broader in scope than the traditional concept of coordination chemistry of metal cations. It is intended to be an overview of the features of multidentate macrocyclic and macropolycyclic ligands. It will not include, however, the naturally occurring porphyrin-related systems and ionophores, nor the polyaza macrocycles, as both these categories are being discussed elsewhere in this volume (Chapters 21.1, 20.2 and 21.2). The coordination chemistry of macrocyclic and macropolycyclic ligands certainly cannot be comprehensive in so few pages, as evidenced by the thousands of papers, hundreds of reviews, numerous patents and countless researchers in this area.

21.3.1.1 Terminology

Useful definitions of terms occuring frequently throughout this chapter are presented below.

(1) A complex is an entity consisting of two or more molecular species which are interacting in such a manner that they are being held together in a physically characterizable structural relationship. It may thus be considered as a supramolecular entity, which implies use of covalent bonds within its components and of intermolecular bonds for holding them together.

(2) Receptor and substrate are terms describing the species involved in complex formation. Throughout the chapter the receptor will refer to the macrocyclic ligand, the substrate to other interacting species. Substrates may be metal or molecular cations, neutral molecules, or atomic or molecular anions. The terms receptor and substrate imply that the complex formed has the well-defined structural and chemical properties of a supermolecule, as in biological receptor–substrate associations. They exclude species formed only in the solid state (clathrates). They are also easily converted and understood in many languages.

(3) Host and guest are terms covering all kinds of intermolecular associations from the well-defined supermolecules to loosely packed solid state inclusion compounds.

(4) Homonuclear refers to multisite binding in which identical substrates are bound.

(5) Heteronuclear refers to multisite binding in which different substrates are bound.

(6) Monotopic refers to receptors which possess only one binding subunit.

(7) Ditopic (or polytopic) receptors are those which possess two (or more) binding subunits.

(8) A cascade complex is one in which an additional substrate may be included between metal cations (or eventually anions) in a ditopic or polytopic ligand.

(9) Macrocyclic and macropolycyclic (cryptate) effects designate the greater thermodynamic stability of macrocyclic ligand complexes compared to nonmacrocyclic analogs.

21.3.2 CLASSIFICATIONS OF LIGANDS, SUBSTRATES AND COMPLEXES

21.3.2.1 Ligands

Receptor capabilities are inescapably correlated with the design of a given ligand. The size, nature and number of substrates to be complexed, as well as the chemical reactivity of the resulting complex and its 'guests', are closely associated with the ligand framework, which is determined by the number and nature of binding sites and overall ligand topology.[1] Ligand geometry, in fact, becomes critical as one progresses from monocyclic to polycyclic systems. General types of macrocyclic ligands in existence are shown in Figure 1. Ligand topologies are drawn graphically in terms of vertices (Z) connected by edges (lines). Often a vertex is a donor atom, but not always. The dimensionality of a macrocycle is defined in terms of the highest number of edges to which a vertex is attached. Hence, only the simple, unappended monocycle is bidimensional, the rest are tridimensional. Specific types of ligands to be discussed in this chapter are defined and common nomenclature is discussed below. Examples of monocyclic and polycyclic ligands are depicted in Figures 2 and 3, respectively, along with references.

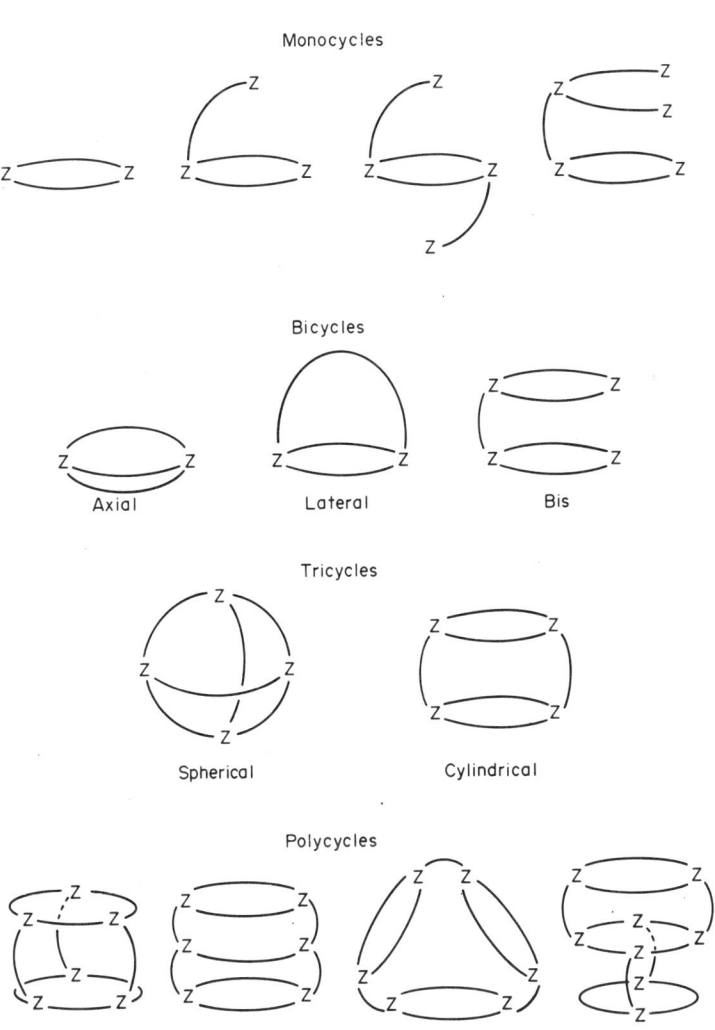

Figure 1 Ligand topology

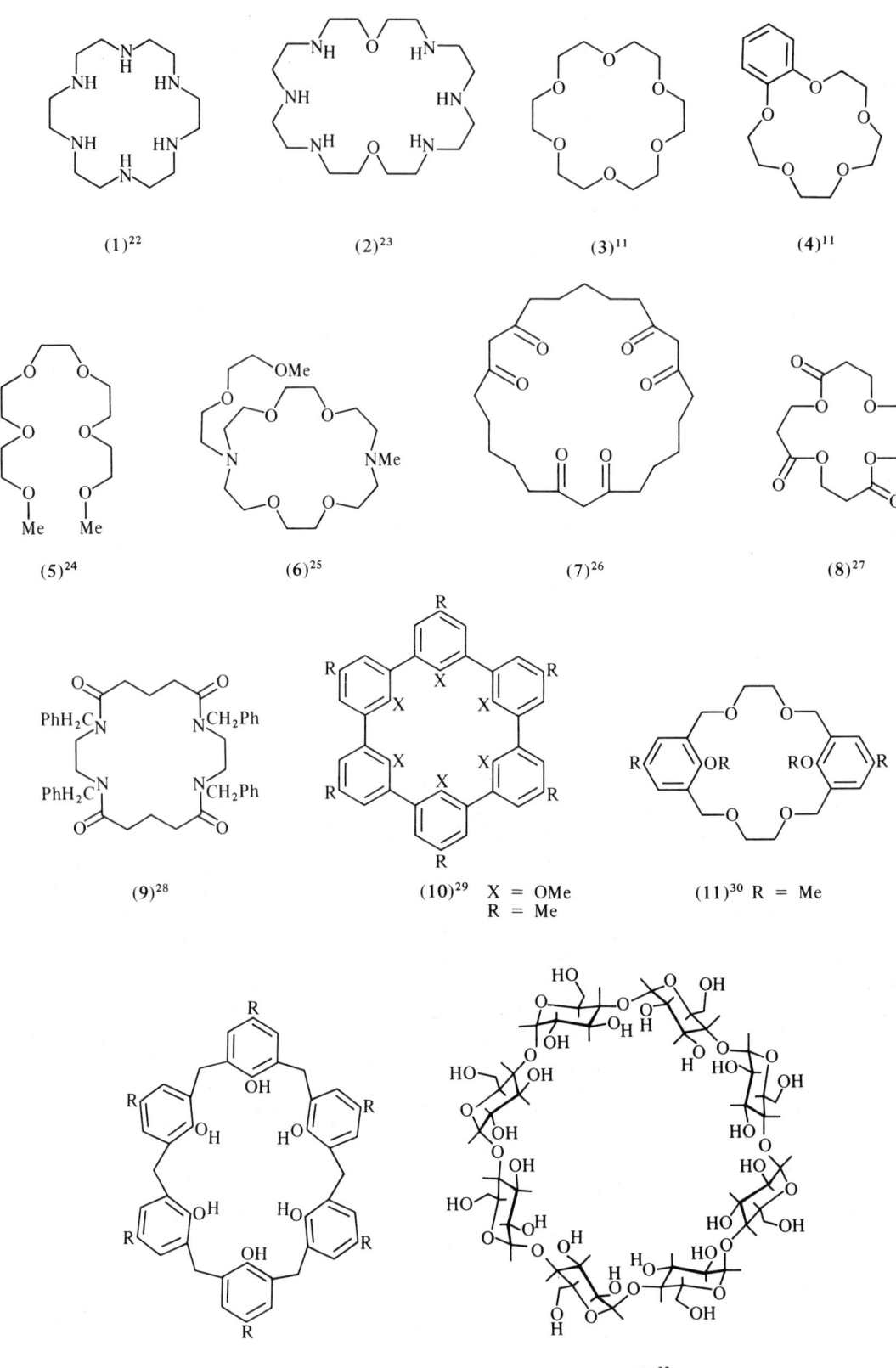

Figure 2   Ligand classification: macromonocycles

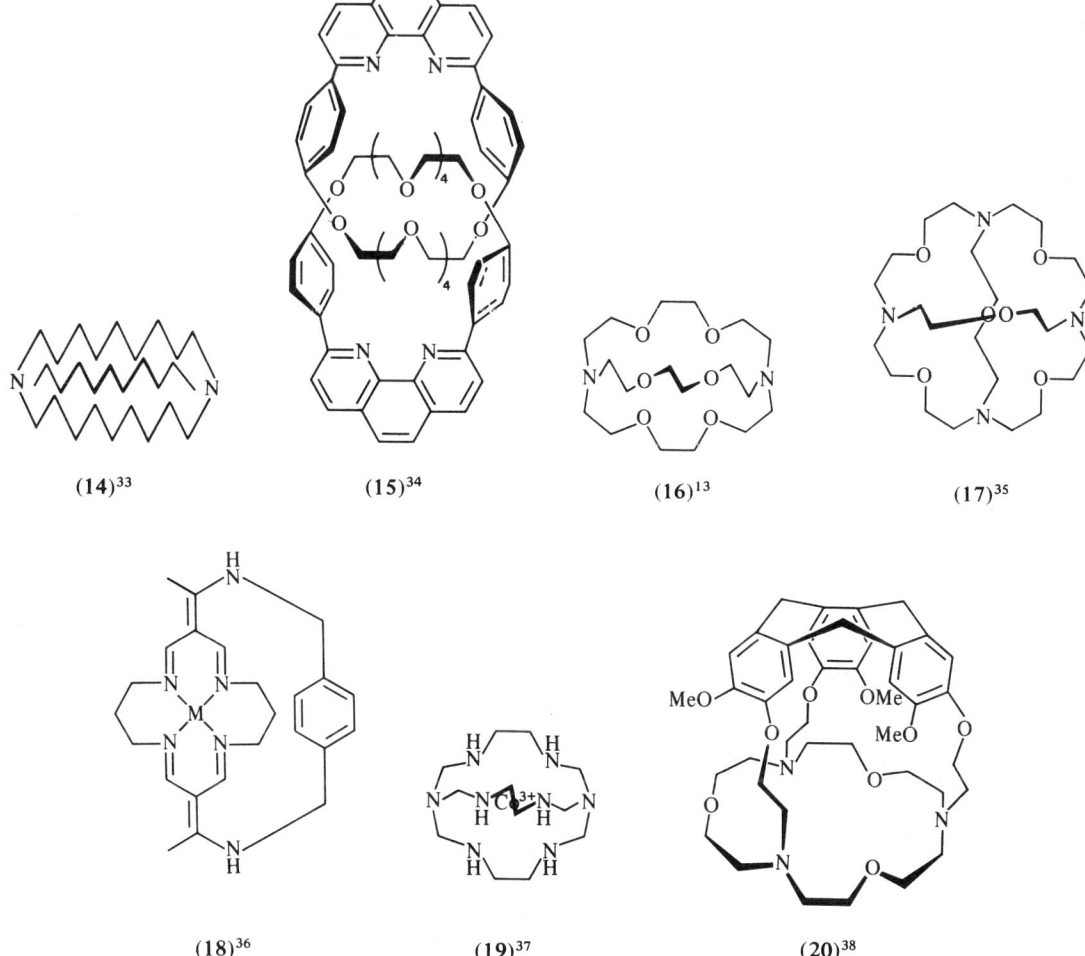

Figure 3 Ligand classification: macropolycycles

21.3.2.1.1 *Ligand classification*

Progressing from macrocyclic to macropolycyclic, the following types of ligands may be schematically distinguished.

(1) Coronands (**1**) and (**2**) are macrocyclic species which contain various heteroatoms as binding sites. The term coronates refers to the complexes. The polyaza macrocycles were the first macrocyclic ligands to be studied.

(2) Crown ethers (**3**) and (**4**) are macrocyclic polyethers. Numerous compounds are possible by modifications which include, but are not limited to, varying ring size, arrangement of donor atoms and addition of aryl groups.

(3) Podands (**5**) are ligands consisting of heteroatoms incorporated into organic chains. This general classification is not included in the scope of this chapter.

(4) Podandocoronands (**6**) refer to macrocyclic ligands with pendant podand chains laterally attached.

(5) Macrocyclic polycarbonyls are cyclic ligands containing carbonyl functionalities. Several categories come under this heading: the macrocyclic oligoketones (**7**), the polylactones (**8**) and the polylactams (**9**).

(6) Spherands (**10**) and hemispherands (**11**) are macrocyclic ligands which consist of arrangements of phenyl groups.

(7) Calixarenes (**12**), from the Greek meaning chalice and arene (incorporation of aromatic rings), are macrocyclic phenol–formaldehyde condensation products.

(8) Cyclodextrins (**13**) are cyclic oligomers of 1,4-glucopyranoside. Since these are naturally occurring molecules, they will not be covered in this chapter.

(9) Catapinands (**14**) are diazabicycloalkanes.

(10) Catenands (**15**) are two separate, but interlocked, macrocyclic ligands.

(11) Cryptands (**16**) and (**17**) are macropolycyclic receptor molecules which provide a cavity for inclusion of a variety of substrates. Cryptate refers to the complexes.

(12) Cyclidenes (**18**) are bicyclic macrocycles which coordinate one metal ion and contain a protected 'void' about the axial site of the metal ion.

(13) Sepulchrates (**19**) are polyaza macrobicycles analogous to the cryptands.

(14) Speleands (**20**) are hollow, macropolycyclic molecules formed by the combination of polar binding units with rigid shaping groups. Speleate refers to the complex.

(15) Cavitands refer to synthetic organic compounds containing rigid cavities at least equal in size to smaller alkali metal ions such as Li^+.

21.3.2.1.2 *Nomenclature*

Common nomenclature has developed over recent years, in particular for coronand and cryptand ligands. The crown-type ligands are commonly named according to the total number of atoms in the macrocyclic ring enclosed in brackets and preceding the classification (crown), followed by the number of oxygens. Hence (**3**) is [18]crown-6. Groups annexed to the ring are added at the beginning of the name, *e.g.* benzo[15]crown-5 would correspond to ligand (**4**). The more general system of nomenclature for coronands, used more for mixed heteroatoms or heteroatoms other than oxygen, is to specify the ring size in brackets followed by the number of each heteroatom. Thus (**1**) is $[18]N_6$ and (**2**) is $[24]N_6O_2$. For bicyclic cryptands, the common practice is to name the ligand by denoting the number of bridging oxygens in each chain in brackets: [2.2.2] for the ligand (**15**). The corresponding monocycle would be the [2.2] under this system, or, better, $[18]N_2O_4$. A hydrocarbon chain in place of one of the bridges would be specified according to the number of carbons, *e.g.* $[2.2.C_8]$. Complexation of a substrate with inclusion into a molecular cavity is often denoted by the symbol $\subset$; for example, potassium ion complexed with the ligand [2.2.2] is written $[K^+ \subset 2.2.2]$.

21.3.2.2 Substrates

Historically, the concept of coordination chemistry was associated with complexation of a metal cation (Lewis acid) by a ligand behaving as a Lewis base. Such was traditionally the case for macrocyclic molecules as ligands. In the early 1970s, however, the concept of coordination chemistry was extended in the area of macrocyclic chemistry to include molecular cations, neutral molecules and anions as substrates. Complexes of all of these species are to be included in the scope of this chapter section. Examples of the types of substrates are discussed below.

21.3.2.2.1 *Metal cations*

Virtually all types of metal ions have been complexed with macrocyclic ligands.[2-7] Complexes of transition metal ions have been studied extensively with tetraaza macrocycles (Chapter 21.2). Porphyrin and porphyrin-related complexes are of course notoriously present in biological systems and have been receiving considerable investigative attention (Chapter 22).[8] Macrocyclic ligands derived from the Schiff base and template-assisted condensation reactions of Curtis and Busch also figure prominently with transition metal ions.[6,7] The chemistry of these ions has been more recently expanded into the realm of polyaza, polynucleating and polycyclic systems.[9] Transition metal complexes with thioether and phosphorus donor macrocycles are also known.[2]

The chemistry of the alkali and alkaline earth metal cations expanded greatly with the discovery of the binding properties of natural macrocyclic antibiotics,[10] of crown ethers in 1967,[11,12] and also with the introduction of the cryptands shortly thereafter.[13] The actinides and lanthanides have not been as extensively studied as the other transition metals, although they form complexes with cryptands as well as other macrocyclic ligands.[14-16] In all of these complexes, circular–spherical recognition of the ligand for the cation is a guiding factor in selective complexation, particularly when the resulting association is primarily electrostatic (alkali and alkaline earth metal ions). Metal cations and their complexes are discussed in greater detail in Section 21.3.5.

21.3.2.2.2 Molecular cations

Molecular cations range from the inorganic ammonium and oxonium ions to complex organic cations.[17-20] Directional influences are crucial to the process of recognition of the macrocycle for the substrate. Stabilizing forces include electrostatic, hydrogen bonding and van der Waals among others. Organic cations as substrates have been the subject of numerous studies, traditionally alkylammonium ions containing NH moieties, where an important factor in complex stability has been found to be the number of hydrogens available for bonding to the macrocycle. Chiral discrimination has also been achieved for organic cations by appropriately designed macrocyclic ligands. Molecular cation complexes will be treated in Section 21.3.6.

21.3.2.2.3 Anions

Binding of anions by organic substrates is, at this time, one of the less developed areas in macrocyclic chemistry, yet anion coordination chemistry[17] is rapidly advancing.[21] These substrates can be divided into several different categories: monatomic and polyatomic inorganic anions, and organic anions. Macrocyclic and macropolycyclic ammonium ions have been found to complex both inorganic and organic anions. Major ligand prerequisites include a large cavity and the possibility of strong electrostatic interactions with the anion. Hydrogen bonding can also provide stabilization. These complexes are discussed in Section 21.3.7.

21.3.2.2.4 Neutral molecules

Macrocyclic receptors have also been found to complex neutral substrates,[21] and complexes of both variable and exact stoichiometries have been prepared. The latter category includes acidic CH— and NH— or polar neutral molecules, such as acetonitrile, nitromethane, benzyl chloride and dimethylmalonitrile. X-Ray data indicate complex formation to be mainly a result of hydrogen bonding and dipole–dipole interactions. Section 21.3.8 contains a more detailed treatment of these complexes.

21.3.2.3 Complexes

Although the ligand topology may govern the type of complex formed, it can not always be used to predict substrate recognition and selectivity. Figure 4 depicts some of the types of complexes available for the ligand systems depicted in Figure 1.

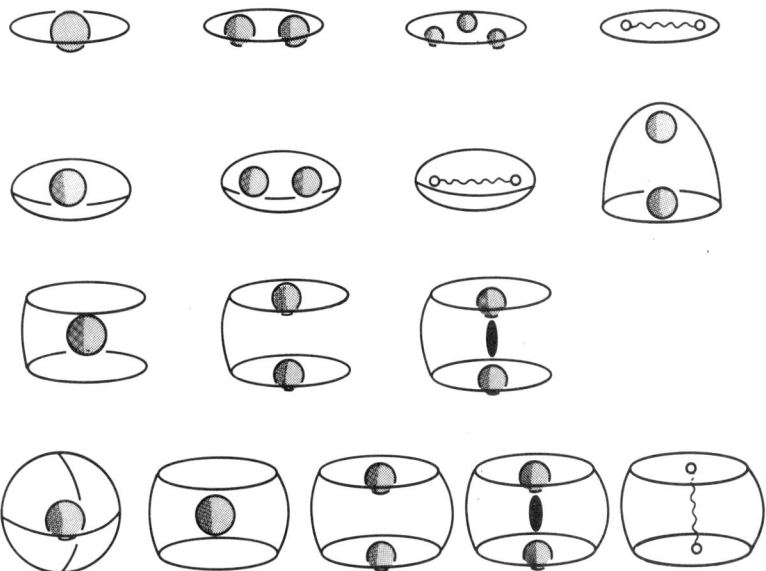

Figure 4 Complexes: ◐ spherical cation or anion; ∘〰∘ difunctional substrate; ▮ bridging substrate in a ditopic receptor

21.3.3 STRATEGIES TO ARCHITECTURAL DESIGN

Knowledge of substrate binding preferences and structure can be efficiently utilized in the synthesis of receptor ligands by introducing site and geometry control into receptor design. Interactions between the macrocyclic ligand and substrates can be fine-tuned by appropriate selection of the binding site and its environment, and overall ligand topology. Specifically to be considered are (1) electronic effects, *e.g.* charge, polarity and polarizability; and (2) structural effects, important from the standpoint of both site and ligand geometry.

21.3.3.1 Electronic Effects

Charge, polarity and polarizability of the binding sites are major influences in complex stability. These factors include: (1) ion-pair interactions for oppositely charged ligand and substrate; (2) ion–dipole and ion–induced dipole interactions for neutral ligands; (3) hydrogen bonding in molecular organic cation and anion substrates; and (4) the hard–soft acid–base (HSAB) rules.[1]

Hard donor atoms, namely nitrogen and oxygen, provide binding sites such as RCO_2^-, RO^-, ROR and R_3N for hard or intermediate acids. Ether oxygens and tertiary amine nitrogens are generally chemically inert, and macrocyclic ligands incorporating these species as binding sites have been the most extensively studied. Nitrogen is found in the naturally occurring porphyrins, corrins and chlorins as part of the heterocyclic pyrrole ring system. Macrocyclic systems containing other forms of this donor atom ranging from ammonium to amine to other heterocycles such as pyridine[39,40] are also known. Porphyrin-related and tetraaza macrocycles are discussed in detail in Chapters 21.1 and 21.2, respectively. Oxygen-containing macrocycles are of several types: ether linkages as in the crown ethers[11,12] and cryptands,[13] polyether esters[41] and oxygens incorporated into heterocyclic rings as in furans.[40,42] Groups with high dipole moments, such as ketones, esters, amides and ureas, are also predicted to behave as good receptor sites, due to the stronger force exerted between substrate and receptor for dipole *versus* induced dipole interactions.[1,21,43] Amide and ester moieties are found in naturally occurring macrocyclic ionophores (Chapter 20.2).[44]

The 'soft' donor atoms, phosphorus, arsenic and sulfur, potentially act as stabilizing influences to more polarizable substrates. These heteroatoms have been less extensively studied than the nitrogen and oxygen analogs. Of the three, the thioether macrocycles are the most common, and complexes of many of the first row as well as the heavier transition metals have been reported.[45]

21.3.3.2 Structural Effects

21.3.3.2.1 Number and arrangement of binding sites

This aspect is crucial for selective complexation of a substrate by a macrocyclic ligand. The influence of site geometry becomes evident when one considers complexation as the replacement of the solvation shell of a given substrate by the ligand donor atoms. Hence, in general, the number of binding sites should be at least equal to the coordination number of the cation with the solvent molecules. The subtle balance between solvation of both ligand and substrate, and complexation, in which the substrate is 'solvated' by the ligand, is critical in maintaining both stability and selectivity (equation 1; where S = solvent molecule and O = ligand donor atom). The influence of solvent effects becomes evident upon examination of the stabilities of the complexes as a function of solvent. For example, for the [*n.n.n*] cryptands, stability constants increase by a factor of about 10^4 from water to methanol.[1] Stabilities are increased even more in ethanol. The choice of solvent as well as accompanying anion may thus influence stabilities to the extent that selectivity changes occur.[1]

$$ (1) $$

The arrangement of binding sites is closely associated with ligand geometry and should also be such as to achieve maximum utilization of the potential ligand–substrate interactions. The simplest geometry for the ligand to recognize selectively is the sphere, *e.g.* metal cations and halide anions. Macromonocyclic molecules maintain a circular cavity, which is suited for 'circular recognition', and these ligands readily complex the spherical cations. The naturally occurring ionophores (Chapter 20.2) and crown ethers are good receptor ligands,[44,46] therefore, for certain metal ions. An increase in dimensionality to macrobicyclic receptors of the cryptand type can result in the generation of a spherical cavity, found to be even more efficient for spherical monatomic ions.[1] Deviations from spherical charge distributions increase the complexity of the recognition pattern, and the influence of the lattice topology becomes even more critical to receptor–substrate stability. Other forms of recognition include tetrahedral, trigonal and linear. Tetrahedral recognition can be achieved by tetrahedral placement of binding sites, *e.g.* the spheroidal ligand (Figure 3; **17**), which readily encapsulates the NH_4^+ ion. Trigonally arranged binding sites allow for recognition of primary alkylammonium ions, which contain trigonally disposed amine hydrogens. Macrotricyclic receptors of appropriate design are also capable of binding polyfunctional substrates such as diammonium or dicarboxylate salts. Examples of recognition patterns are shown in Figure 5.

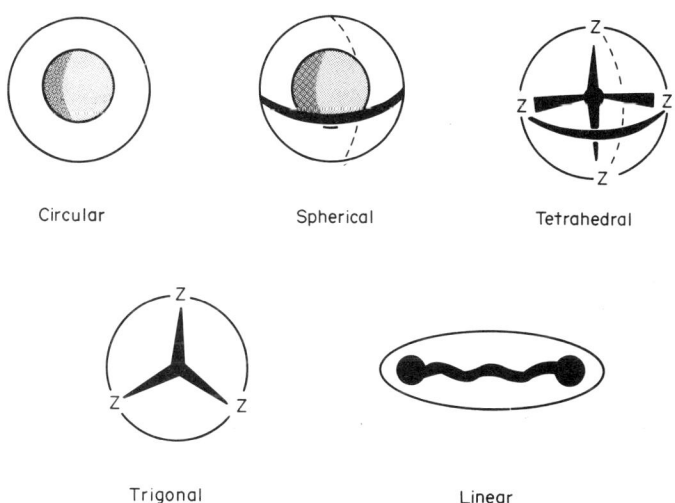

Circular   Spherical   Tetrahedral

Trigonal   Linear

Figure 5 Recognition patterns

21.3.3.2.2 *Conformation*

Complexation can be complicated by the existence of more than one conformation for a given macrocyclic ligand. An example is the bicyclic [*n.n.n*] cryptand, which potentially can exist in *endo–endo*, *endo–exo*, or *exo–exo* conformations (Figure 6).[1,33] The cavity of the *endo–endo* form is more spherical than those of the other conformations. Consideration of ligand conformation is, as in the case of ligand dynamics, an exercise in balancing the desired complexation and exchange properties. Greater complex stability is to be expected when the built-in conformations in the free ligand and complexed ligand are the same. On the other hand, faster exchange rates may result from a stepwise solvation shell replacement, if a change in ligand conformation occurs from uncomplexed to complexed forms.[1] Thus, the goal is to achieve a rigidity/flexibility balance for a given purpose.

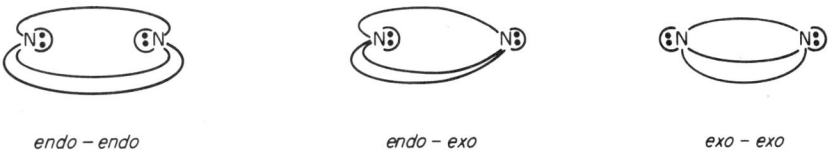

endo – endo   *endo – exo*   *exo – exo*

Figure 6 Conformations of bicyclic [*n.n.n*] cryptands

21.3.3.2.3 *Shaping groups*

The nature and, as a consequence, the selectivity of a particular macrocycle can be further manipulated by the selection of appropriate 'shaping groups' and heteroatoms at the binding site. In general, saturated chains provide greater flexibility, particularly as chelate size (pertaining to the number of atoms between binding sites) and total macrocyclic ring size are increased. Unsaturation results in the imposition of steric constraints on the molecule, to the extent that when donor atoms are connected *via* an aromatic system, flexibility is at a minimum, *e.g.* phthalocyanines. Macrocyclic shaping through the utilization of preselected functional groups is thus a viable part of ligand design. Specific examples of design control are delineated below.

(1) The introduction of aryl groups to the ring tends to stiffen the macrocycle and decrease the donor ability of the oxygens in crown ether macrocycles.[1,21]

(2) Functional group incorporation (*e.g.* amide, ester, thioester, urethane and thiourea) provides polar binding sites and additionally ligand stiffening.[1,21]

A new class of macropolycyclic receptors, called speleands (**20**), epitomizes the use of shaping groups to obtain the desired receptor sites.[38] In these ligands a receptor unit is combined with a rigid shaping unit to form cryptands capable of substrate inclusion. The appeal of these ligands derives from the possibility of combining geometrical and binding interaction requirements into the architectural design of the ligand in order to achieve receptor–substrate compatibility for both bonding and fit.

21.3.3.2.4 *Dimensionality*

Ligand flexibility and shape are also controlled by the dimensionality of the macrocycle. Macrobicyclic ligands are inherently more rigid than their monocyclic analogs. As chain length is increased, however, flexibility returns. Cryptands larger than the [2.2.2] thus display 'plateau' rather than 'peak' selectivity.[1] Peak selectivity refers to the ability of a given ligand to show distinct discrimination of one particular substrate over others, while plateau selectivity denotes a propensity of a given ligand for several substrates over others.

21.3.3.2.5 *Cavity size*

The number of donor atoms in the macrocycle and the imposed degree of rigidity influence the nature of the cavity. While a rigid framework results in a preformed cavity, flexibility allows latent cavity formation. The selectivities observed for the crown ethers and cryptands in the complexation of the alkali and alkaline earth metal ions are closely related to cavity size, although in exceedingly large cavities selectivity may become lost due to a preponderance of flexibility.[1] If a substrate is too small for a given ligand cavity, the resulting complex will be destabilized by substrate–receptor repulsions and ligand deformation. On the other hand, for a substrate that is too large for a macrocycle, destabilized complexes will result due to poor ligand–substrate binding contact or unfavorable ligand deformation in order to achieve binding contact. This influence of cavity size also pertains to cylindrical cavities, for which the length of 'chain' substrates compared to cylindrical macrocyclic cavity size has been found to be distinctly correlated to substrate selectivity and complex stability.[47-49]

21.3.3.2.6 *Chirality*

The incorporation of chiral units within the macrocyclic skeleton is an important route to the design of macrocyclic receptors capable of enantiomeric or chiral substrate recognition. In order to achieve this goal, the receptor must display a fine structure as a result of chiral barriers, which then allows diastereomeric receptor–substrate interactions, in addition to maintaining the desired cavity properties.[18,50,51]

21.3.4 SYNTHETIC PROCEDURES

21.3.4.1 Macromonocycles

Generalized synthetic routes for macromonocycles are presented in equations (2)–(12). A comprehensive review of crown syntheses has been published by Gokel and Korzeniowski.[52] The [18]crown-6 was initially synthesized *via* template assistance using an oligo(ethylene glycol), monoglyme and potassium *t*-butoxide,[11,12] although other efficient methods have since then been reported, *e.g.* equation (2).[52] Benzo derivatives can be obtained *via* the reaction of diphenols with oligo(ethylene glycol) dichlorides (equation 3).[11,12,53] Mixed oxa–thia crowns are obtained from oligo(ethylene glycol) dichloride reactions with dithiols (equation 4).[54-56] Ether–ester and ether–ester–amide macrocycles are derived from acid chlorides and oligo(ethylene glycols) or ethylenediamine (equation 5).[41,57]

$$(2)$$

$$(3)$$

$$(4)$$

$$(5)$$

Polythia macrocycles are achieved as a result of the reaction of an appropriate polythiane with a dibromoalkane. The reactions are sometimes, although not always, aided by metal template assistance (equations 6 and 7).[58-59]

$$(6)$$

$$(7)$$

The phosphorus macrocycles are made *via* template condensation of coordinated polyphosphine ligands and a dibromoalkane (equation 8).[60] A more recently reported method involves a template-assisted single-stage ring closure (equation 9).[61] The arsenic donor macrocycles are synthesized by reacting lithiated polyarsanes with a dichloroalkane (equation 10).[62]

The synthesis of spherands involves ring closures using aryllithiums with Fe(acac)₃. Yields are increased by high dilution techniques (equation 11).[29,63]

Calixarenes can be obtained by base-catalyzed condensation of a *p*-substituted phenol with

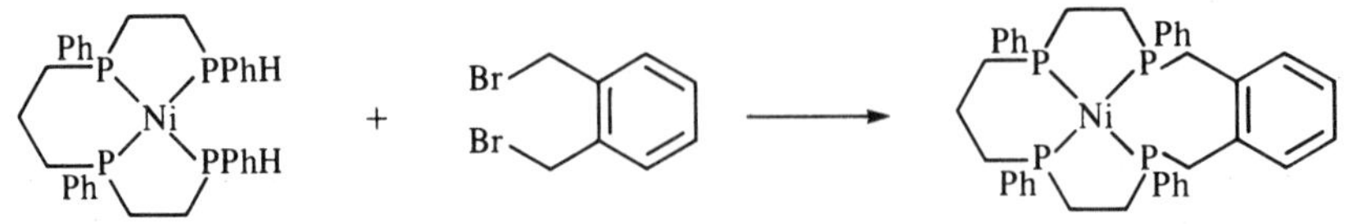

(8)

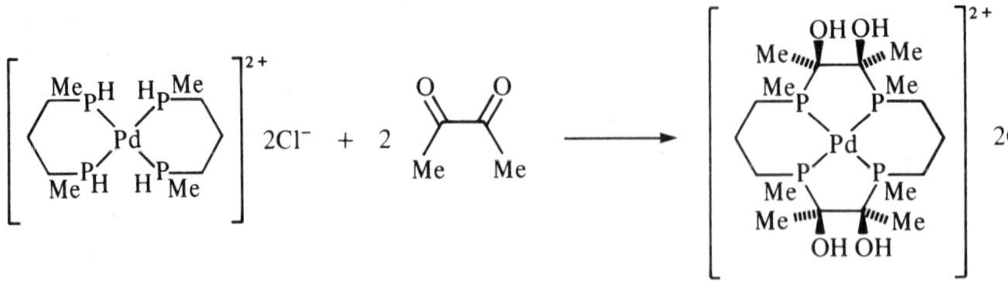

(9)

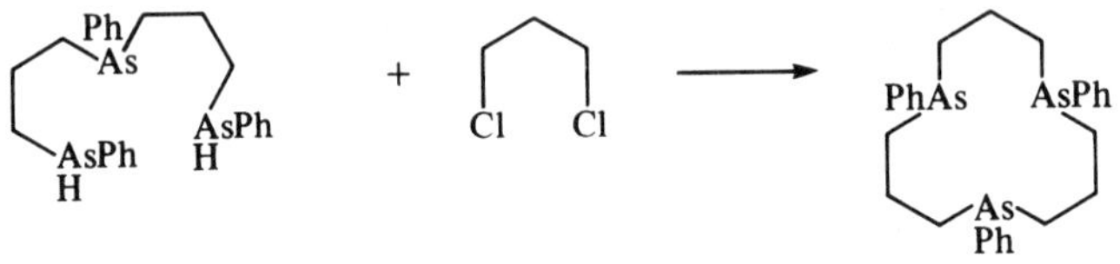

(10)

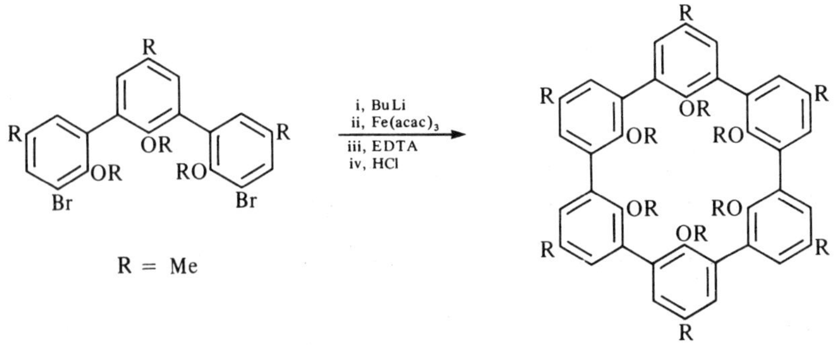

(11)

R = Me

formaldehyde (equation 12).[31,64,65] This method provides good yields of pure products only for the *p*-t-butylphenol. The size of the product macrocycle is dependent on reaction conditions.

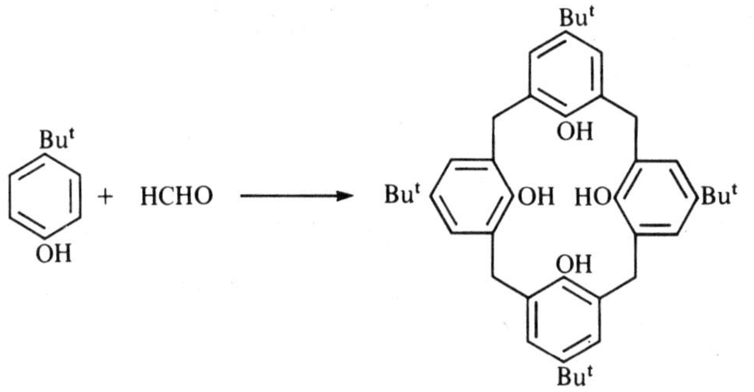

(12)

21.3.4.2 Macrobicycles

The diazapolyoxamacrobicycles are the result of a series of reactions between diamines and acid chlorides (Scheme 1). The product of the first reaction of an oligoether diamine and an oligoether acid chloride gives a diamide monocycle which is then reduced by LiAlH₄ to the saturated monocycle. This compound, when reacted with a second mole of oligoether acid chloride, gives the bicyclic precursor, which is then reduced by diborane.[13] These reactions are normally accomplished using high dilution techniques, which require very gradual mixing of reagents in order to prevent polymerization. Mixed polythia–polyoxa bicyclic cryptands are prepared in the same manner, with the substitution of sulfur for the appropriate oligoether oxygens.[66] Several variants and simplifications of the initial method have been developed over the years.

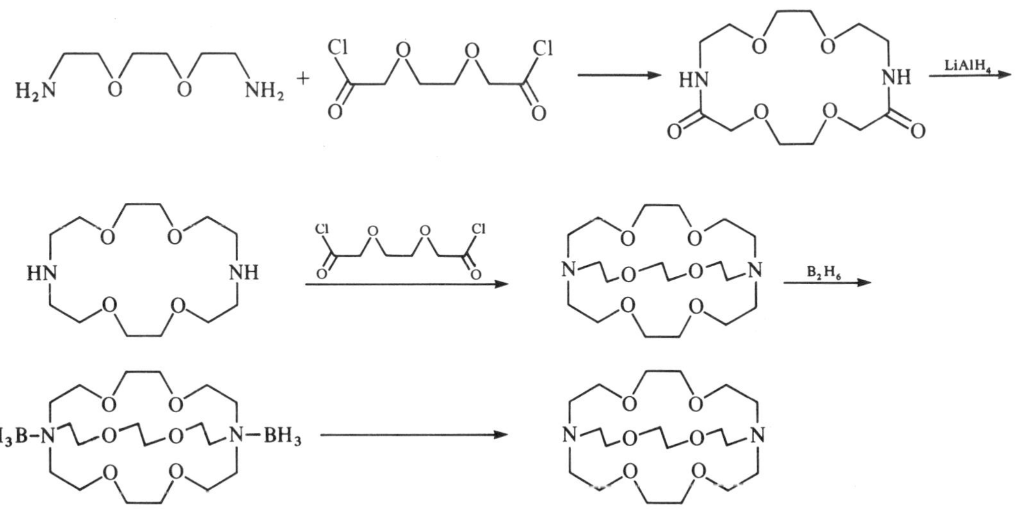

Scheme 1

The lacunar cyclidenes are synthesized from a precursor tetraaza macrocycle containing a pendant vinyl carbon attached to a methoxy group, which readily undergoes addition–elimination reactions with diamines to give the vaulted bridge bicycle.[36]

The sepulchrates (Scheme 2) are synthesized *via* template-assisted condensation of $[Co(en)_3]^{3+}$ with formaldehyde and ammonia under basic conditions.[67]

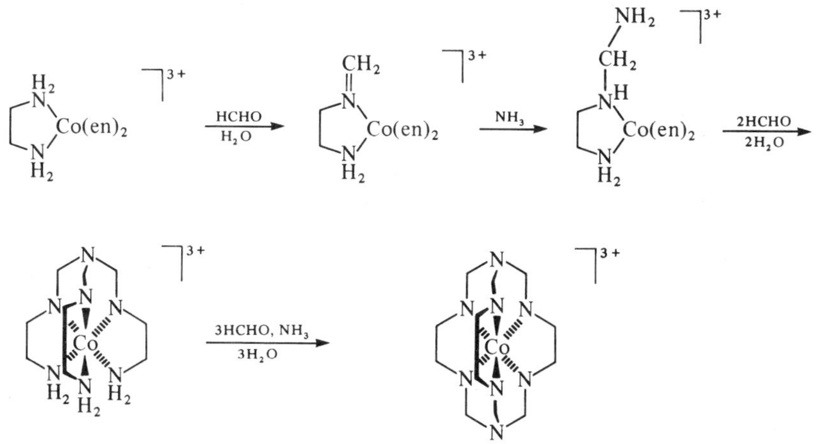

Scheme 2

The catenands are synthesized using the metal ion template effect, whereby a bis complex is formed from an α,α'-disubstituted *o*-phenanthroline. This initial product is treated with a diiodoalkane to effect the ring closures.[34]

21.3.4.3 Macrotricycles

Cylindrical and spherical tricyclic cryptands are made by progressive construction of the framework following the general strategies shown in Schemes 3 and 4.[34,66–69]

21.3.5 METAL CATION COMPLEXES

21.3.5.1 Macromonocycles

Classification of macromonocycles (Figure 2) can be made on the basis of donor atom, *i.e.* nitrogen, oxygen, sulfur, phosphorus, arsenic and mixed donor systems, as well as topicity, *i.e.* mono- *versus* di- or poly-topic. The planar nitrogen donor macrocycles and other tetraaza macro-

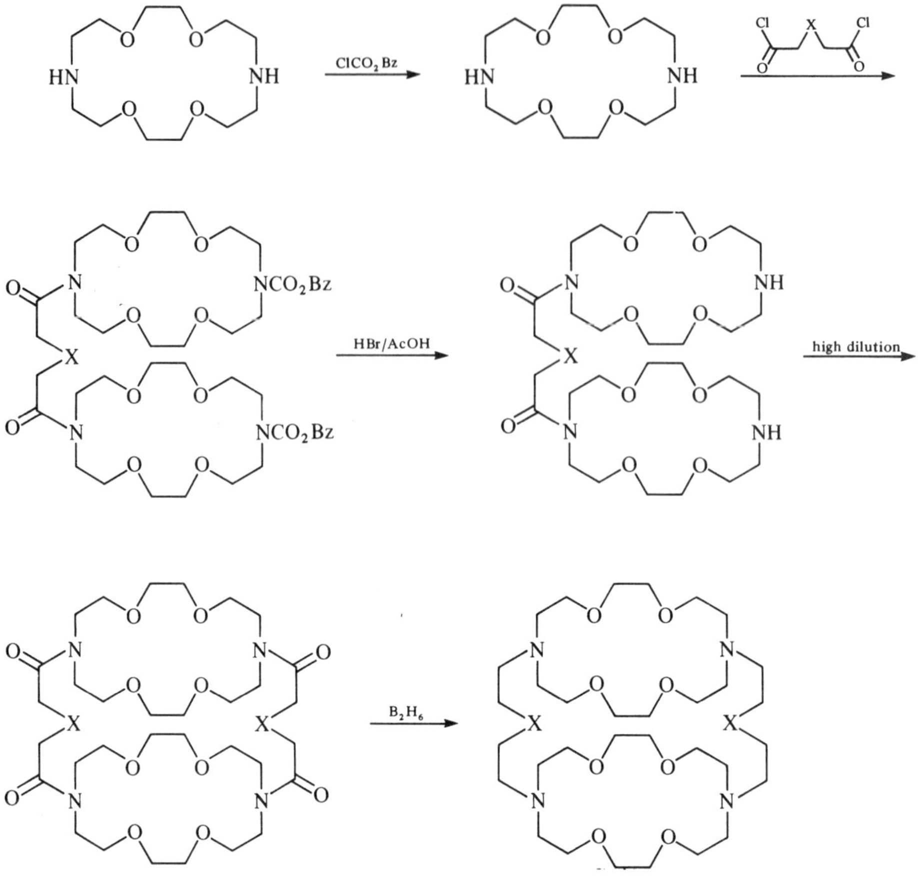

Scheme 3

cycles are covered in Chapters 21.1 and 21.2 respectively, while other mixed donor open polydentate macrocyclic species are covered in Chapter 21.1. Some aspects of mixed donor coronands will be included in the section on crown ethers and related coronands (Section 21.3.5.1.1).

21.3.5.1.1 Coronands and crown ethers

The initial report by Pedersen of the dibenzo[18]crown-6 (21) led to the development of an exciting and extensive area of receptor chemistry.[11] An early discovery concerning the chemistry of these ligands was the observation that they selectively complex alkali and alkaline earth metal ions. The importance of the crowns in separations, solubilizations, ion transport of inorganic salts, anion activation and enantiomeric selection of organic salts is now well recognized. Numerous reviews of the area have appeared, the references to some of which are given.[2-5,50,72-74] In this section some of the salient features of the voluminous chemistry of the crown ethers and coronands in general will be discussed, including metal ion selectivity, thermodynamics and kinetics of complexation, and ion transport. A glimpse at the extent of coronand territory can be seen by the wide variety of ring modifications shown in Figure 7. This figure is by no means an all-inclusive representation, but rather indicates some of the more common (and in a few examples less common) possibilities. References are also provided covering examples of complexes containing each functional group modification, since not all of these systems will be covered in the scope of this section.

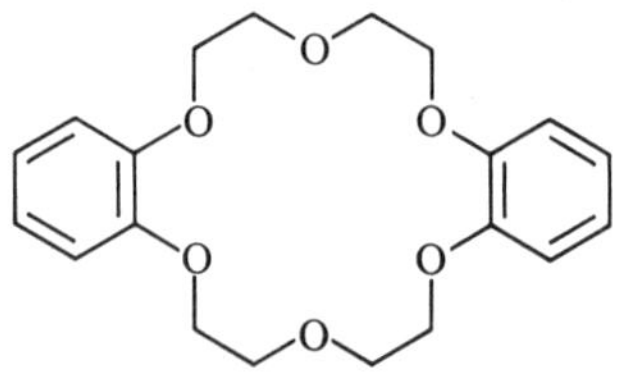

(21)

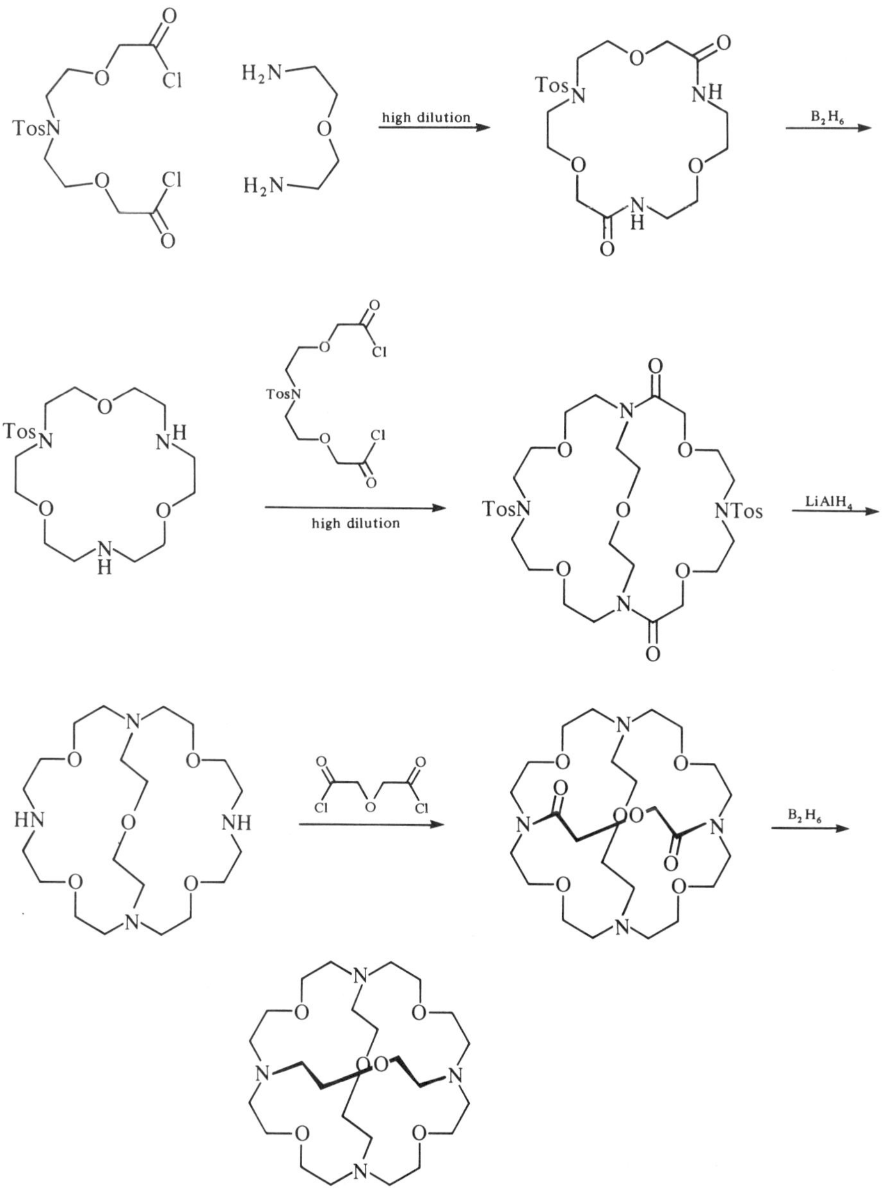

Scheme 4

The observed selectivity of the crown ethers is actually the result of several influences, including cation:ligand radius ratios, the number and nature of donor atoms, entropies and enthalpies of solvation, and ligand conformational preferences.[75] Modified ligands, achieved by variation of the heteroatom or ring substituents, as well as macrocyclic ring size, can greatly influence observed selectivity patterns. Figure 8 shows the relationships between complex stability constants and cation radii for various crown and coronand complexes. A close perusal of the figure indicates that, with the exception of the [21]crown-7, the highest stability constants for each of the polyether crowns are obtained for the K^+ ion. The maximum stability, however, is achieved for the $K^+ \subset$ [18]crown-6, in which the diameter of the central cavity (2.6–3.2 Å) and the ionic diameter of the K^+ ion (2.66 Å) are matched quite well. Furthermore, as can be seen from comparison of the [15]crown-5, [18]crown-6 and [21]crown-7, the discrimination ability of these ligands differs. 'Peak selectivity' is observed for the ideally suited K^+ and [18]crown-6, while 'plateau selectivity' is noted for the 15- and 21-membered crowns. Thus, the 'hole-size selectivity' principle often cited in conjunction with the crowns is not solely responsible for the selectivity patterns of these particular systems.[75]

Cyclic groups affixed to the ring structure have variable influences on complex formation. The addition of cyclohexyl groups does not seem to modify reactivity greatly, probably due to the

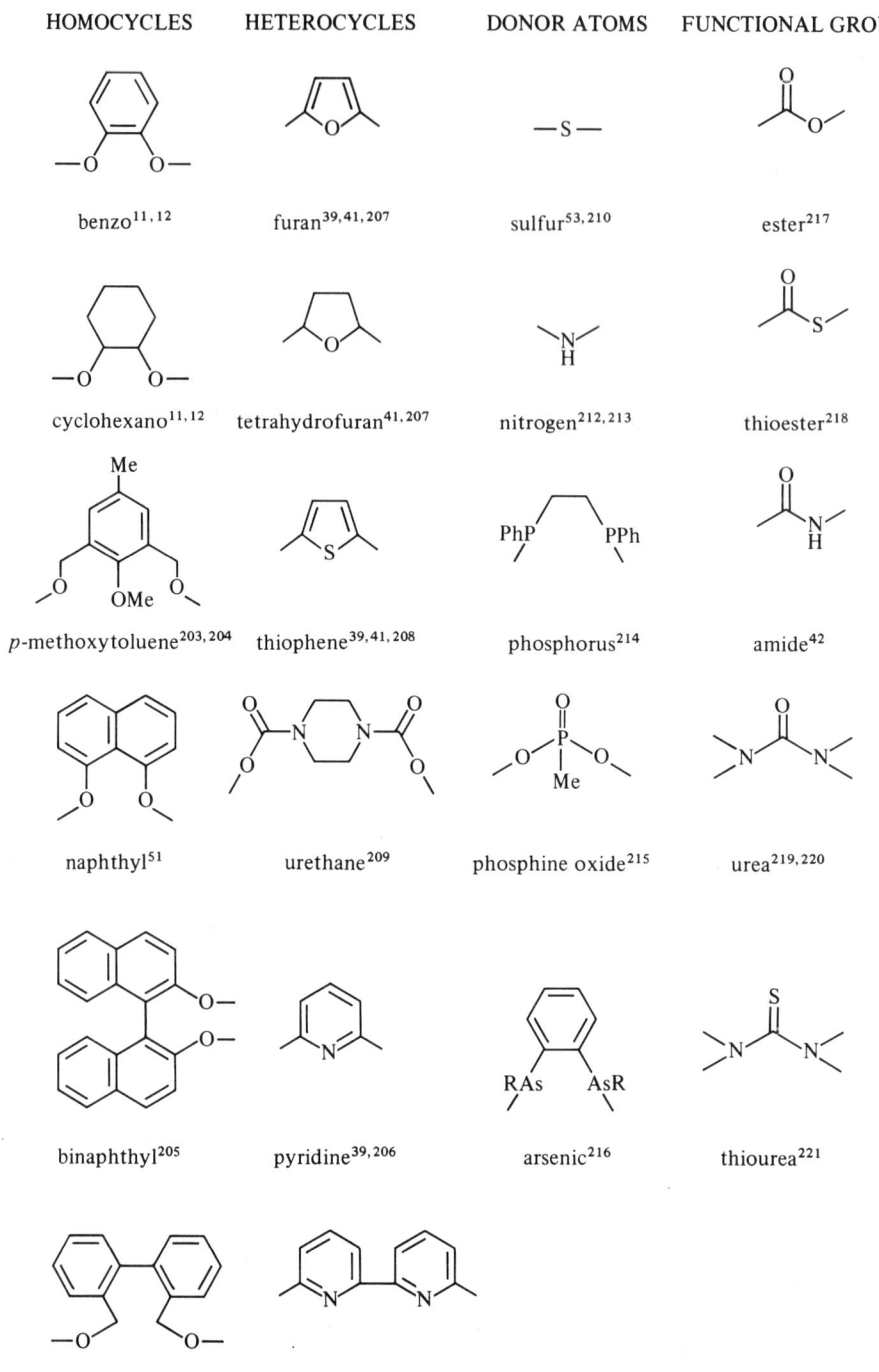

Figure 7 Representative coronand ring modifications

flexibility of the ligand, which makes it not unlike the unsubstituted ring.[76] Benzo substituents, on the other hand, do appear to affect complex stability. Although the observed constants are only slightly lower for the alkali metal ions complexed with dibenzo[18]crown-6 (**21**) compared to the unsubstituted analog, the log K_s (MeOH) for Ba^{2+} decreases by almost one-half from 7.0 to 4.28 for the unsubstituted to dibenzo derivative.[77] Decreased ligand flexibility and oxygen basicity, both of which tend to hamper metal ligand interactions, can be cited as factors contributing to decreased stabilities.[77] Substituents on affixed benzo groups such as NH$_2$, Me, Br, NO$_2$, and CO$_2$H have also been noted to influence selectivity, in some cases even reversing preferences of a given crown for certain ions.[78–80]

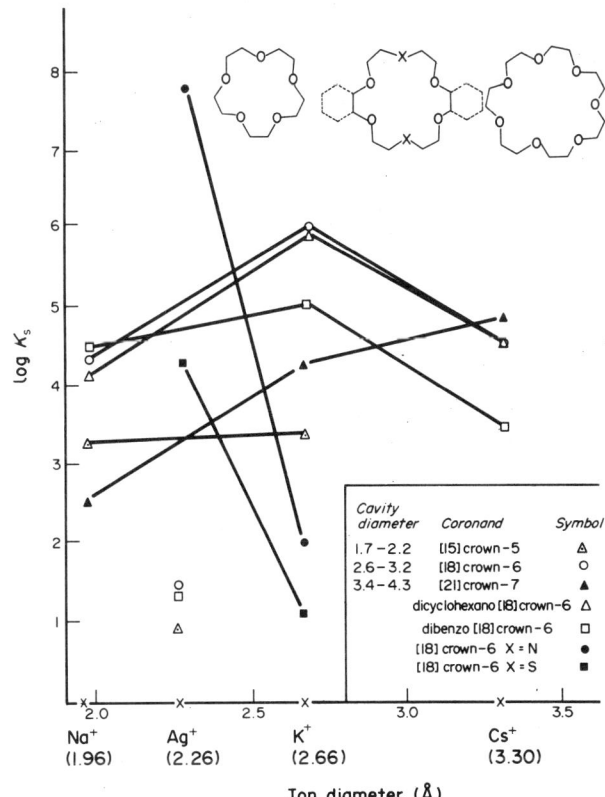

Figure 8 Stability constants for complexes of coronands with metal ions

The incorporation of a 'foreign' donor atom, *i.e.* sulfur or nitrogen, in general, reduces coronand affinity for alkali metal ions, but can in some instances greatly enhance complex stabilities of other metal ions such as Ag^+,[81] as seen in Figure 8. The effect on stabilities of sulfur substitution is not necessarily as straightforward as that of nitrogen, due to the conformational preference of the sulfur, which in most free ligand polyether structures has been shown to be exodentate.[82] Thus, complexes of sulfur-substituted coronands may not be of the inclusion type. A more detailed discussion of this aspect is found in the following section on polythioether crowns.

The thermodynamic origins of the enhanced stabilities of macrocyclic ligands over their acyclic counterparts have been the subject of considerable debate since the term 'macrocyclic effect' was first coined.[83] Comparison of thermodynamic data for the several metal ion complexes of the [18]crown-6 and its acyclic counterpart are shown in Table 1. Enthalpy contributions to stabilization appear strongest for the K^+ complex, while entropic contributions are stronger for the Na^+ complex. Undoubtedly, the factors responsible for the thermodynamics will vary according to ion size, charge, solvation effects and structural preference. Hence, a single definable source of the macrocyclic effect is, in these systems at least, probably nonexistent.

Table 1 Thermodynamic Data ($kJ\,mol^{-1}$) for the [18]crown-6 and Acyclic Analog[a,b]

Ligand	Na^+		K^+		Ba^{2+}	
	ΔH	$T\Delta S$	ΔH	$T\Delta S$	ΔH	$T\Delta S$
$MeO(CH_2CH_2O)_5Me$	−38.2	−32.2	−34.1	−21.2	−23.6	−9.29
[18]crown-6	−33.9	−9.20	−55.3	−20.8	−43.4	−3.47

[a] 25 °C in 99 wt % methanol.
[b] Data from ref. 75.

Kinetic studies, using NMR, of $^{23}Na^+$ ions with dibenzo[18]crown-6 (**21**) in DMF have been reported[84] and indicate a dynamic equilibrium occurring between solvated sodium ions and the uncoordinated macrocycle. In the larger dibenzo[30]crown-10 (**22**), the crown was found to be undergoing a rapid conformation transition prior to complexation.[85] Additional investigations have indicated that complexation kinetics are strongly influenced by conformational aspects of a given macrocycle.[86]

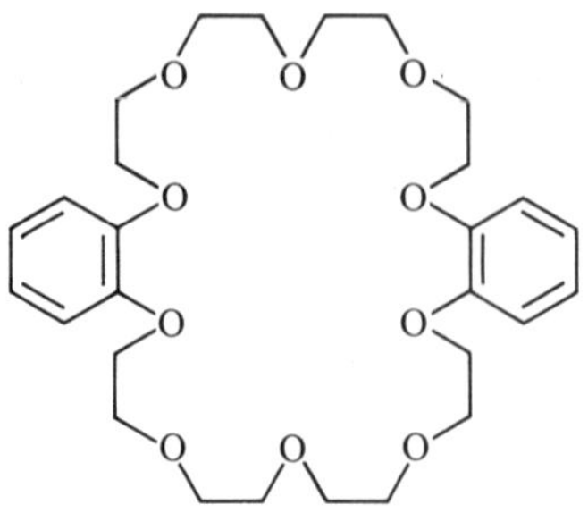

(22)

'Ion transport' has been a catchword in crown chemistry since its inception.[10,87] When complexes of the lipophilic crown ethers were found to be soluble in many organic solvents, applications in phase transfer catalysis were immediately envisioned, whereby the crowns would be capable of solubilizing a wide variety of inorganic salts.[88,89] An extension of this area is the use of macrocyclic ligands to carry cations across hydrophobic membranes.[90,91] The potential of a wide variety of macrocyclic ligands, with variations in cavity size, donor atoms and ring substituents, as cation transport carriers for alkali, alkaline earth and transition metal ions, has been examined.[90,91] In Ag$^+$–M$^+$ binary systems, for example, selectivities were found to be correlated with several factors, including (1) a good cation/macrocycle fit for the smaller (less than 21-membered ring) crowns; (2) the importance of donor atom, *i.e.* substitution of sulfur, amine or pyridine donors to achieve enhanced Ag$^+$ ion selectivity; and (3) substituent effects which help to fine-tune solubility, and electronic and steric influences.[90]

Cation binding in crown ethers can also be influenced by attaching a side arm possessing a donor group (lariat ethers) (23).[92] Enhanced cation binding can be observed by appropriate positioning of the side arm in relation to the macroring. Further control can be achieved by using side arms capable of reduction to anion radicals.[93]

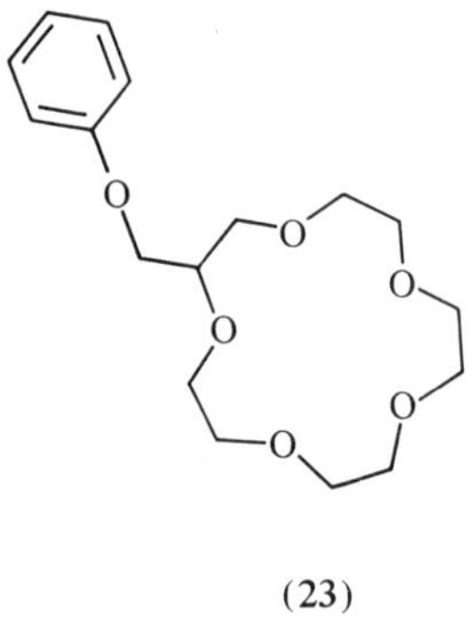

(23)

The formation of cascade complexes by monocyclic coronands and the exceptionally fine control of the chemistry of these systems by minor changes in the macrocyclic framework are exemplified by the series of copper complexes (24)–(26). With only minor donor atom and framework modifications, the magnetic properties of the copper ions are drastically altered,[23] each of the tetraazido complexes displaying different magnetic behavior. In (24) the copper(II) ions are ferromagnetically coupled with a ground state triplet, in (25) the two ions are not coupled, while in (26) the metal ions present strong antiferromagnetic interactions. The coupling behavior is thought to be associated with the orientation of the magnetic orbitals on the metal ions.

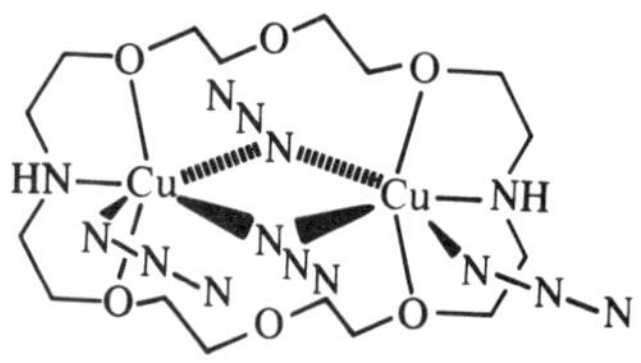

(24) 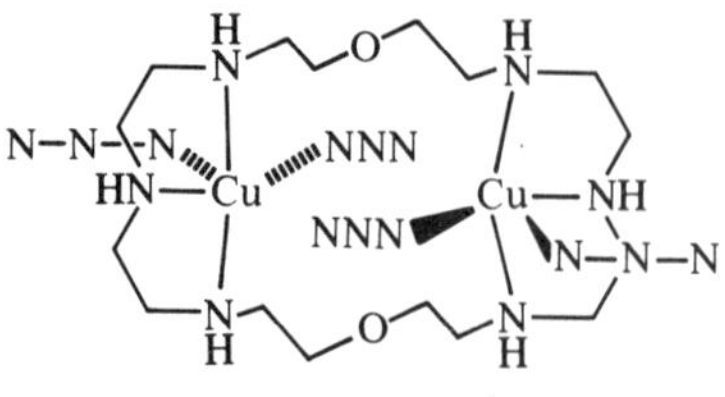   (25) 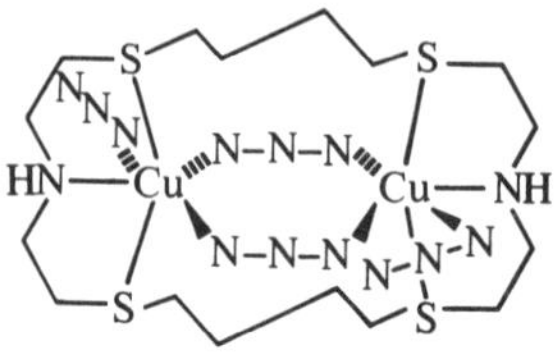   (26)

Binuclear metal ion incorporation can also be achieved by attaching functionalized side chains to the macrocyclic framework (27).[94] In (27) the 'soft' phosphine portion can be thought of as the redox center, while the 'hard' N_2O_2 ring provides a Lewis acid metal ion center.

Higher order polynuclear complexes are accessible *via* polytopic macrocyclic monocycles. The tritopic hexaazamacrocycle $[27]N_6O_3$ was recently shown to bind three copper(II) ions, with two triply bridging μ-OH groups (28).[95]

Organometallic crown ethers have also been synthesized.[96-98] Recently, a crown–cation group was shown to interact with an appended transition metal acyl ligand (29).[99] Complexes of this type have potential applications in Lewis acid-accelerated alkyl migration to coordinated carbonyls.

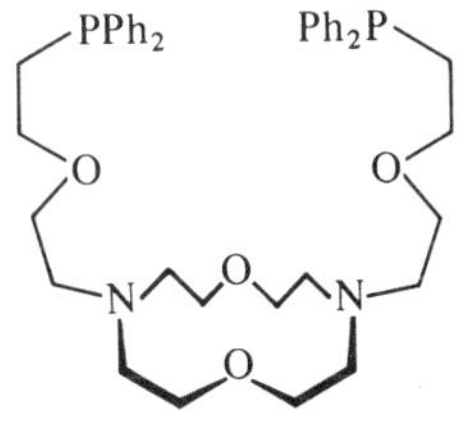

(27) 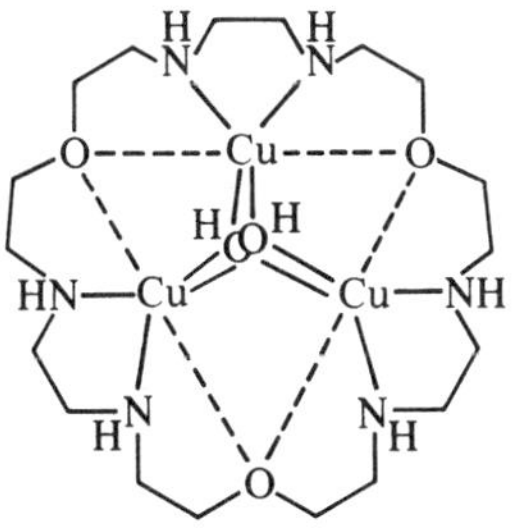 (28) 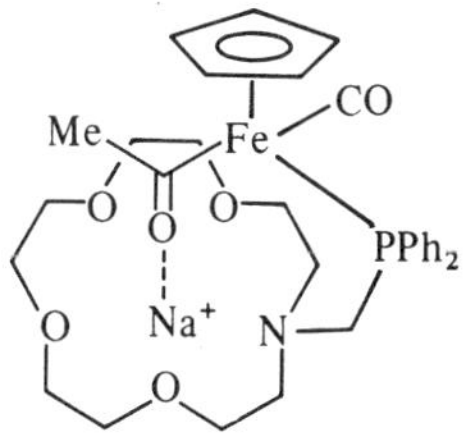 (29)

Multisite crown ethers (30) and (31) are polymacrocycles. These molecules are potentially like cryptands in view of the possibilities for forming inclusion-like species. The photoresponsive crowns provide an excellent example of this aspect, and consist of two crown ethers, as in (30a and 30b), attached *via* a photosensitive azo linkage. This molecule undergoes reversible isomerization (likened to a butterfly motion), shown in equation (13). The *cis* form gives a stable 1:1 cation:ligand complex with the larger alkali cations (actually a 1:2 cation:crown ratio). Concentrations of (30b) in solutions are thus noted to be enhanced by the addition of these cations.[100,101] Other multisite crowns have been prepared from diphenyl- and triphenyl-methane dyes, *e.g.* (31).[102]

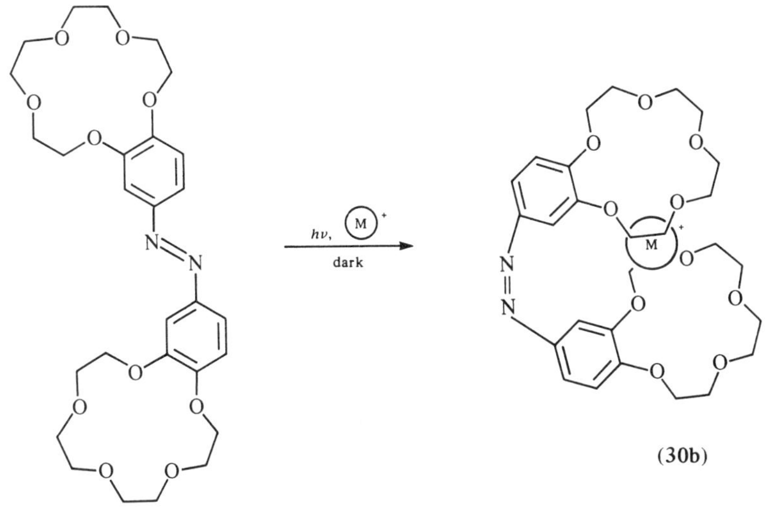

(30a) (30b)

21.3.5.1.2 *Polythia macrocycles*

Thioether analogs of the crowns have been known since the 1930s,[103] but metal ion complexes of these ligands have not been investigated with the intensity found for the crowns and polyaza macrocycles. As anticipated from the 'soft' nature of the heteroatom, the sulfur macrocycles show a preference to bind transition metals rather than alkali and alkaline earth ions. Studies have

(31)

involved copper ions in many instances.[104,105] An important aspect of the chemistry of these ligands pertains to the preference of the free ligand for exodentate conformation, in which the sulfur atom lone pairs are directed outward (32).[106] Thus, structural aspects of complexes of these ligands are not as straightforward as in the other macrocyclic systems. A complexed metal ion could be coordinated in the 'expected' enclosed fashion, as shown crystallographically for several metal ions including the nickel(II) complex (33),[107] or in the exodentate form as found for niobium(IV) and mercury(II) chlorides of (34) and (35), respectively.[108,109] For the trithia[9]crown-3 only endodentate sulfurs were observed, probably the result of ring strain.[110] The hexathia[18]crown-6 analog, on the other hand, possesses both exo- and endo-dentate sulfur atoms.[111] It was found in a comparative study of two conformers of cationic rhodium(I) complexes with [14]S$_4$ and its tetramethyl-substituted analog that the nucleophilicity of the metal ion is influenced by the macrocyclic conformation.[112]

(32)   (33)   (34)   (35)

21.3.5.1.3 Polyaza macrocycles

Of the polyaza macrocycles, the tetraaza group has been the most intensively studied and is dealt with in Chapters 21.1 and 21.2. Fewer than four donor nitrogen atoms result in a ring size so small that metal cations cannot in general be incorporated within the cavity.[45] Pentaaza and higher polyaza macrocycles have begun to appear more frequently, particularly in view of the potential of the larger macrocycles for binding more than one metal ion. Included in this category are ligands derived from Schiff base condensations, particularly 2,6-diacetyl- and diformyl-pyridine (36).[40,113] Schiff base condensations of pyrrole[114] and dipyrromethane[115] moieties have also resulted in macrocycles capable of binuclear coordination as in the octaaza bis(dipyrromethene) macroring (37). The polyaza analogs of the crown ethers are also capable of binuclear metal ion coordination as evidenced by the dicopper(II) complex of [24]N$_8$.[116] The incorporation of two copper(II) ions in polyaza macrocycles holds promise for the design of models for binuclear copper proteins. For example, the dicopper(II) complex of the hexaaza macrocycle (38) has been found to exhibit features distinctly related to superoxide dismutase.[117,118]

The synthesis of pyridine analog to the spherands has also been reported as shown for ligand (39).[119,120] Corey–Pauling–Kotun (CPK) molecular models of cyclosexipyridine molecules indicate a cavity size appropriate for Rb$^+$ and K$^+$ incorporation.[120]

(36)   (37)

(38) X = O, CH$_2$ (39)

21.3.5.1.4 Polyphospha macrocycles

The first tetraphosphine macrocycle was reported in 1975; hence, these molecules are relative newcomers to the macrocyclic circle.[121] The template synthesis of the macrocycle (40) was described in 1977 (equation 8).[60,122] A more rigid dibenzo ligand (41) was reported in 1977 and structurally characterized shortly thereafter.[123,124] By using 1-naphthylmethyl as a P—H protecting group, macrocycles with secondary phosphino groups have also been synthesized.[125] These have potential for conversion into more complex bicyclic or bis-monocyclic systems. The scarcity of reports on polyphospha macrocyclic ligands and their complexes emanates from the fact that the synthetic methods involved are lengthy with poor yields. A template-assisted synthetic route involving a single-stage ring closure has provided the synthesis of a series of tetraphosphorus macrocycles of general type (42; equation 9).[61] The use of protected carbonyl groups in the alkyl side chains of bis(tertiary phosphines) has also been used as another route for template synthesis of 14- to 16-membered tetraphospha macrocycles.[126] The polyphospha macrocycles are capable of complexing a variety of metal ions, notably nickel(II), palladium(II) and platinum(II).[126] Rhodium(I) and molybdenum(0) complexes have also been obtained.[125]

(40)   (41)   (42) R = H, Me

21.3.5.1.5   Polyarsa macrocycles

Macrocyclic polyarsanes (43) have been reported by Ennen and Kauffman.[62] The complexation tendencies of these ligands are as yet unexplored, although they should, due to the 'soft' arsenic donors, form stable transition metal complexes.

(43)

21.3.5.1.6   Spherands

Both spherands (10) and hemispherands (11; Figure 2) complex alkali metal ions, the selectivity and stability of the resulting complexes depending on the size of the ion.[127-129] Owing to structural restraints provided by the rigidly joined phenyl rings, the spherands can be considered to be 'preorganized' so that the binding sites, with their associated lone pairs of electrons, are directed toward the center of the macrocyclic cavity.[127] In complexes of sodium and lithium picrates with (10), for example, stability constants are of the order of 10^{14} to $10^{16}\,l\,mol^{-1}$, whereas the corresponding constants obtained for the dicyclohexano[18]crown-6 were 10^4 to $10^6\,l\,mol^{-1}$ (in $CDCl_3$ saturated with D_2O).[130] Molecular mechanical calculations have correlated well with experimental observations and have been used successfully in predicting complexation tendencies based on ligand conformation.[131]

Hemispherands (11) are by nature only partially 'preorganized' for complexation, depending on their structural characteristics.[30,129,132] Addition of hydrocarbon-based bridges, which essentially converts the monocycles to bicyclic systems, serves to provide a more rigid structural framework as in the cryptands.[132]

21.3.5.1.7   Calixarenes

These ligands, as exemplified by (12; Figure 2), were first reported by Zinke and Ziegler.[31] Synthetic design has since then been modified and improved.[64,65,133-135] The attachment of an aliphatic chain bridging two opposite *para* positions in the calix[4]arenes (*i.e.* those composed of four phenol groups) has been used to fix the 'cone' conformation in these somewhat flexible molecules.[136]

The performance of calixarenes as cation carriers through H_2O–organic solvent–H_2O liquid membranes has also been studied.[137] In basic metal hydroxide solutions, the monodeprotonated phenolate anions complex and transport the cations, while [18]crown-6 does not, under the same conditions. Low water solubility, neutral complex formation and potential coupling of cation transport to reverse proton flux have been cited as desirable transport features inherent in these molecules.[137]

21.3.5.2   Macrobicycles

21.3.5.2.1   Axial macrobicycles

The macrobicyclic cryptands (44) were first synthesized in 1969.[13] As a result of the bridgehead nitrogen atoms, these cryptands can have several conformations, as is the case in diaza macrobicycles in general. It was shown by NMR studies that the three anticipated configurations exist (Figure 6):[138] *exo–exo* or out–out, where the nitrogen lone pairs are directed away from the central cavity; *endo–endo* or in–in, in which the lone pairs are directed inward; and *exo–endo* or out–in. X-Ray structural results indicate the *endo–endo* configuration for the free ligand of [2.2.2] (44d), and the *exo–exo* conformation for the corresponding bis-borohydride derivative (Scheme 1) precursor to the [2.2.2].[139] The metal ion complexes tend to have the *endo–endo* configuration.[140]

(44)

a; $m = n = 0$ [1.1.1]
b; $m = 0, n = 1$ [2.1.1.]
c; $m = 1, n = 0$ [2.2.1]
d; $m = n = 1$ [2.2.2]
e; $m = 1, n = 2$ [3.2.2]
f; $m = 2, n = 1$ [3.3.2]
g; $m = n = 2$ [3.3.3]

The cryptands of (44) are highly selective for recognition of spherical cations, and in particular for alkali and alkaline earth metal ions.[1,13,25,141] Modifications such as the additional incorporation of nitrogen or other heteroatoms, *e.g.* (45) and (46), make these ligands highly suited for transition metal ion complexation. Thus, a variety of ligands of this type can be obtained by varying chain length and/or donor heteroatoms.

(45)

a; X = O, Y = CH_2
b; X = O, Y = S
c; X = S, Y = O
d; X = Y = S
e; X = O, Y = NMe
f; X = NMe, Y = O
g; X = O, Y = NCH_2CH_2Ph

(46)

The increased dimensionality of the tridimensional bicyclic ligands over the monocycles allows for even more efficient spherical selectivity in the cryptands as compared to the two-dimensional, circular recognition of the coronands. The result is enhanced alkali and alkaline earth ion selectivities for ligands of the type (44). Figure 9 shows the relationship of stability constants of the representative cryptands of (44) to ion and cryptand diameters. By comparing these constants to those shown in Figure 8 for the crowns, the exceptional stability and selectivity of the [2.1.1], [2.2.1] and [2.2.2] cryptands becomes apparent. For the macrobicycles the 'hole-size selectivity' principle becomes much more meaningful, therefore, since peak selectivity for these aforementioned cryptands does appear to be closely correlated with the cavity size.[1] The bicyclic topology is largely responsible for these effects, allowing less flexibility for expansion or contraction of the spherical cavity compared to the crowns. Nonetheless, the decreased rigidity of the larger cryptands (> [2.2.2]) tends to 'dilute' selectivity, so that plateau, instead of peak, selection is obtained. X-Ray crystallographic determinations of a variety of these cryptates have confirmed the *endo–endo* conformation of the ligand and the inclusion nature of the complexes.[142] A study of the enthalpies and entropies of formation of alkali and alkaline earth macrobicyclic cryptate complexes has shown that both enthalpy and entropy changes play an important role in selectivity and complex stability.[143] Results indicate that the cryptate effect, *i.e.* the high stability of macrobicyclic complexes as compared to macromonocyclic analogs, is of enthalpic origin.[143]

The number of binding sites and ligand thickness can be used for M^{2+}/M^+ selectivity control for cryptands related to (44).[144] For example, in comparing the [2.2.2] to the [2.2.C$_8$], it is found that the former cryptand shows a selectivity ratio for Ba^{2+}/K^+ of 10^4 whereas for the latter it is 10^{-2}. The loss of binding sites and decreased solvent access to the shielded ion are thus more crucial for the doubly charged Ba^{2+} ion than for K^+. The result is strong destabilization of the dication relative to the monocation for the [2.2.C$_8$].

Transition metal cryptates are also known for (44). A variety of transition metal complexes, including Cu^{2+}, Ni^{2+}, Co^{2+}, Zn^{2+} and Pb^{2+}, with cryptands such as [2.1.1], [2.2.1] and [2.2.2] have been studied.[145-150] In some cases dinuclear complexes are formed for the [2.2], [2.1.1], [2.2.1] and [2.2.2] ligands.[145,146]

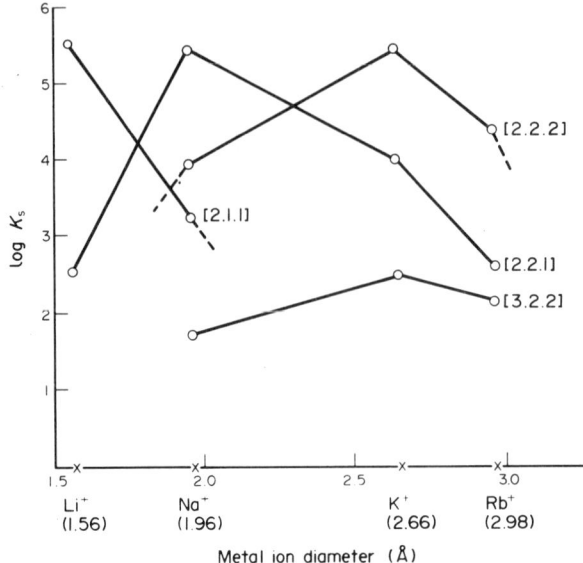

Figure 9 Stability constants for complexes of cryptands with metal ions

Trivalent lanthanides have been examined including lanthanum(III), praseodymium(III), europium(III), gadolinium(III) and ytterbium(III).[14,15,151–154] In an extensive study of the trivalent lanthanides, the bicyclic cryptands were found to be stronger complexing agents than the polyether crowns. In fact, the high degree of kinetic stability supplied to europium(III) and gadolinium(III) by the cryptands renders these ions essentially substitution inert.[15] Studies of the complexation tendencies of uranyl ion with the [2.2] and [2.2.2] have shown that, while in aqueous solution no complex is formed,[16,155] in propylene carbonate 1:2, 1:1 and 2:1 metal:ligand complexes are observed.[16]

Alkali metal anions have also been generated as a result of cryptand stabilization of the corresponding cation. Cryptands were found to enhance the solubility of zerovalent alkali metals in various organic solvents.[156,157] Initially, the solutions apparently contain the cryptate cation and solvated electrons together with free ligand. When more metal is dissolved, metal anions, M^-, are formed.[158] Dye and co-workers have isolated gold-colored crystals of $[Na^+ \subset 2.2.2]Na^-$[159,160] and the crystal structure has been determined.[161,162] Anion clusters such as Sb_7^{3-}, Pb_5^{2-} and Sn_9^{4-} have been isolated as crystalline salts of the [2.2.2] cryptate counterion [2.2.2].[162,163]

As in the coronands, binding site identity plays an important role in the complexation preferences of cryptands. Replacement of oxygens by sulfur or nitrogen in the bridges (45b–d) results in decreased stability and selectivity for alkali metal and alkaline earth cations (as was also noted for the coronands).[66,164] Selectivity is then shifted toward softer metal ions such as Ag^+, Tl^+, Cd^{2+} and Pb^{2+}.

Other modifications to the [n.n.n] cryptands include replacing one or more of the polyether chains by o-phenanthroline and/or bipyridine, as in (46), to obtain photoactive compounds.[165]

An axial macrobicyclic ligand particularly suited for binuclear transition metal ion complexation is (47); it is derived from the linking of two tripodal tren $[N(CH_2CH_2NH_2)_3]$ units.[166] In its 'in–in' form it is capable of incorporating two metal cations including Ag^+, Zn^{2+} Cu^{2+} and Co^{2+}. NMR studies indicate that binuclear complexes are formed as in the route shown in Scheme 5.[9] Both side-to-side (inside the cavity) as well as inside cavity–outside cavity ion exchange are slow. Stability constants for metal ion uptake have been determined by pH metric titration for a variety of metal ions.[9,167] For Cu^{2+} the $\log K_{S1}$ and $\log K_{S2}$ are 16.6 and 11.8, respectively. The formation of a cascade-type complex, by incorporation of bridging species between the two cations, has been indicated for the dicopper(II) complex upon addition of H_2O, CN^- and N_3^-,[166] leading to large decreases in paramagnetism. Chloro and hydroxo bridges have also been reported for these binuclear cryptates. Mononuclear examples indicate that the anions are also coordinated in the cavity and stabilized by hydrogen bonding to the protonated amino groups at the end opposite the copper ion.[168] The bis-cobalt(II) complex gives rise to isolation of the μ-hydroxo-μ-peroxo species (48), as observed by spectral and titration data.[9,166,167] The crystal structure of the bis-tren complex of cobalt(II), $[(tren)Co(O_2,OH)Co(tren)]^{3+}$, indicates bonding to occur in the same fashion as postulated for the cryptate.[169,170]

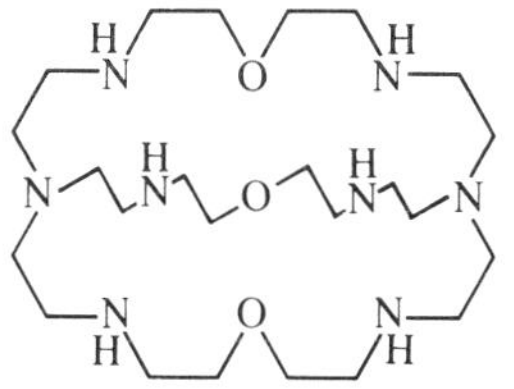

(47) 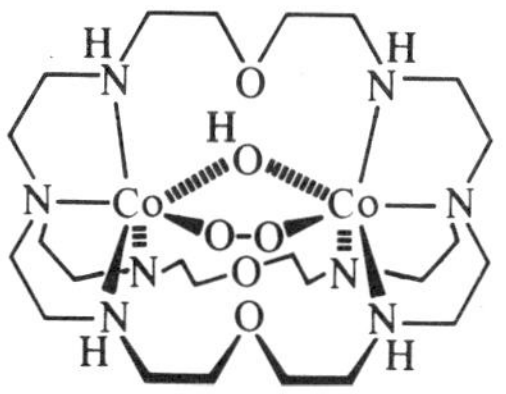   (48)

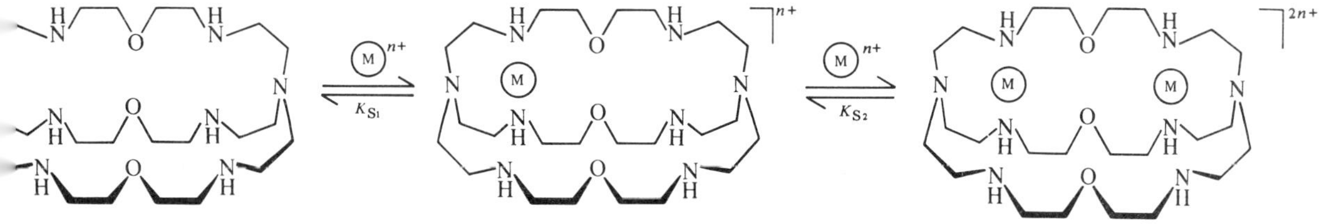

Scheme 5

21.3.5.2.2 Lateral macrobicycles

Although two cations are often observed to complex in a dinuclear fashion in the axial macro-bicycles, as noted in the previous paragraphs, lateral macrobicycles (Figure 1) are clearly designed for incorporation of two metal ions.[9,171,172] These two metal ions are by construction necessarily in chemically different environments, which can greatly affect both chemical and physical properties. For example, in the bis-copper(II) complex of (49b) the two copper ions exhibit greatly different redox potentials (+550 and +70 mV *vs.* NHE in propylene carbonate).[9]

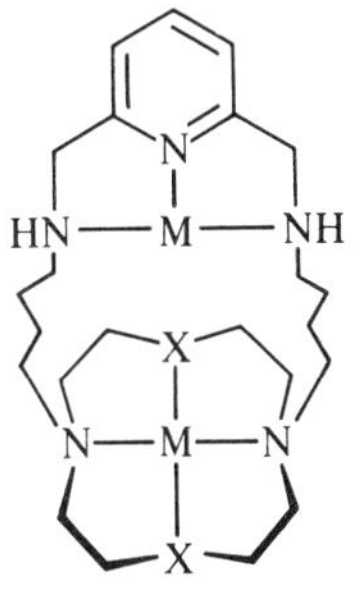

(49) a; X = O
b; X = S
c; X = NH

21.3.5.2.3 Bis macrocycles

These molecules are formed by a single bridge link between two monocycles (Figure 1), and some have already been treated under the section on crown ethers (the multisite crowns). By virtue of possessing two monocycles, these ligands actually are classified as bicycles, however. Both, *syn* and *anti* configurations are possible, as in (30a ↔ 30b; equation 13), which allows for greater flexibility in the complexes. A dicopper complex is shown in (50), which has been reported to undergo reversible CO and O_2 fixation.[173]

21.3.5.2.4 Cyclidenes

The lacunar ligands of Busch and co-workers (18) comprise a series of bicyclic macrocycles called cyclidenes.[36,174-176] These macrocycles typically coordinate one transition metal ion and, due to their

(50)

vaulted structural characteristics, maintain a 'persistent void' accessible to small molecules such as molecular oxygen.[174,175] A ^{13}C NMR study has shown correlations with the chemical shift of the bridgehead carbon and K_{O_2}, the latter increasing with increased bridge length.[177]

21.3.5.2.5 Sepulchrates

Sepulchrates, the polyaza cage macrobicycles analogous to the cryptates, were first synthesized in 1977.[178] The cobalt(III) complex shown in Figure 3 (19) is the octaazasepulchrate analog of the [2.2.2] cryptand (Figure 3; 16), and is commonly written [Co(sep)]$^{3+}$ (sep = sepulchrate).

Cobalt complexes have been the most generally studied, although other metal ions have been incorporated, such as platinum(IV), rhodium(III) and chromium(III).[67] The cobalt(II) complex of (19) is optically stable and kinetically inert (no exchange observed using labelled ^{60}Co^{2+} over a period of 24 hours).[68] Electron transfer studies show that the [Co(sep)]$^{2+/3+}$ couple undergoes electron exchange at a rate 10^5-fold faster than [Co(en)$_3$]$^{2+/3+}$.[178] [Co(sep)]$^{2+}$ also quantitatively reduces O$_2$ to H$_2$O$_2$ *via* a superoxide radical ion O$_2^-$ intermediate.[179]

Cobalt(III) cage complexes can also perform as electron transfer agents in the photoreduction of water.[180,181] Because of the kinetic inertness of the encapsulated cobalt(II) ion, the cobalt(II)/cobalt(III) redox couple can be repeatedly cycled without decomposition. Thus these complexes are potentially useful electron transfer agents, *e.g.* in the photochemical reduction of water, in energy transfer and as relays in photosensitized electron transfer reactions.[180,181] The problem of the short excited-state lifetimes of these complexes can be circumvented by the formation of Co(sep)$^{3+}$ ion pairs, so that the complexes can be used as photosensitizers for cyclic redox processes.[182,183]

Related ligands such as (51) have also been reported,[184] and mixed N and S donor atoms have been used to isolate lower oxidation states, *e.g.* cobalt(I).[185]

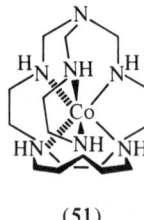

(51)

21.3.5.2.6 Catenands

Two interlocked macrocyclic ligands as in (15) are topologically related to the catenanes, whence the name catenand derives.[34,186,187] These macrocycles complex a variety of metals, presumably in a tetrahedral geometry.[34,186] The stabilizing effect of the catenand topology is evident in the observed redox stability of the nickel(I) complex[186] as well as the reluctance toward demetallation, observed for the copper(I) complexes.[187]

21.3.5.3 Macrotricycles

21.3.5.3.1 Cylindrical macrotricycles

These ligands are obtained by linking two monocycles by two bridges as in (52) and (53). Potentially three cavities are offered for receptor incorporation: the two lateral cavities of the monocycles and the central cavity within the bridge arms. The chemical and physical properties of the resulting species depend on several factors: size of the monocycles, length and nature of the bridges, and nature and number of the binding sites. The ring size of each monocycle is a major determining factor in whether a cation will be included in the monocyclic cavity or sit on top of the ring. Thus, the coordination chemistry of the cation can be manipulated *via* modification of monocyclic ring size.[9] Further control of the chemistry of the system can be maintained by judicious selection of bridges with respect to length and conformational properties. Propitious choices can allow for cation–cation interactions (shortened bridges) or the possible inclusion of an additional substrate (cascade complex), each of which has potential applications in catalysis and the concomitant activation of two or more species held in proximity.

(52) a; Y = O
b; Y = S

(53) a; X = O, Y = O
b; X = CH₂, Y = O
c; X = *o*-phenylene, Y = O
d; X = NH, Y = O
e; X = O, Y = S

The first cylindrical macrotricyclic ligands synthesized were (52a) and 53a–d.[68–70] Cryptands in which the two monocycles are even farther apart as a result of bridging naphthyl, biphenyl and related groups have also been reported.[188,189] The smaller macrocycle (52a) forms complexes with a variety of metal cations, including two silver(I) ions.[69,70,190] Crystal data results for the latter complex indicate both Ag⁺ ions are located slightly out of the plane of the macrocycles (undoubtedly the result of macrocyclic size constraints), but within the central main cavity, with an Ag—Ag distance of 3.88 Å.[191]

The larger-ringed macrocycles of (53a–d) form binuclear complexes with alkali metal, alkaline earth, silver(I) and lead(II) cations.[68,192] The two 18-membered rings are large enough to allow for cations as large as Rb⁺ to be incorporated within their cavity, with a net result of increasing the metal–metal separation. Thus, crystal structure data for the disodium complex of (53a) indicate the sodium ions to be 6.40 Å apart,[193,194] compared to a 3.88 Å separation found for the aforementioned disilver complex of (52a). Heteronuclear complex formation has also been observed, *e.g.* with both Ag⁺ and Pb²⁺ incorporated in the same cryptand.[192]

Stability constants associated with complex formation correspond to two successive steps as shown in Scheme 6.[190] The constants for the formation of 1:1 and 2:1 complexes of the smaller alkali cations are comparable to those of the lone macrocycles [1.1] for complexes of (52). Very stable complexes are formed by the alkaline earth cations with this ligand (log $K_s \approx 7$ in H₂O).[190] The larger macrocycles of (53a–d) form alkali metal and alkaline earth complexes in which the stabilities of the 1:1 complexes are similar to those of the model *N*-methylated monocycle. The stabilities and selectivities of the 2:1 complexes are also similar to those of the 1:1 complexes. In fact, K_{S2} for incorporation of a second Ba²⁺ into (44a) is as high as K_{S1} for the same ligand.[190] Such findings indicate two essentially independent macrocyclic units.

Kinetic processes have been shown by NMR and other spectral data to involve both intra- and inter-molecular exchange for 1:1 complexes of alkaline earth metal ions and (52a) (Scheme 7).[190] The internal exchange between the two monocycles occurs at a faster rate than the intermolecular process, however.

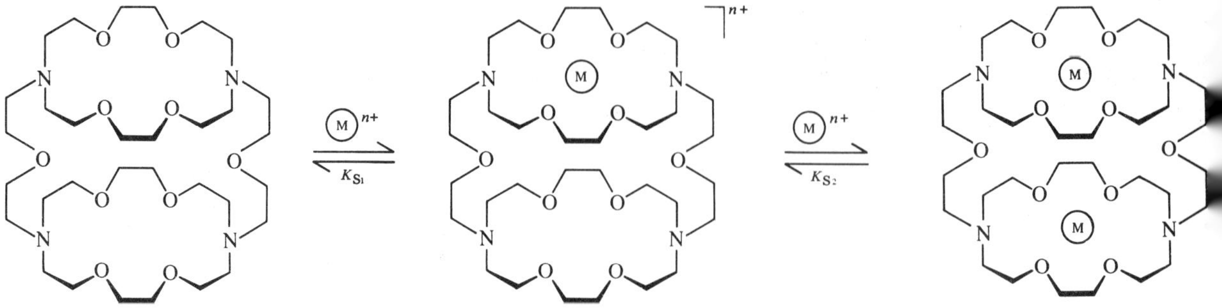

Scheme 6

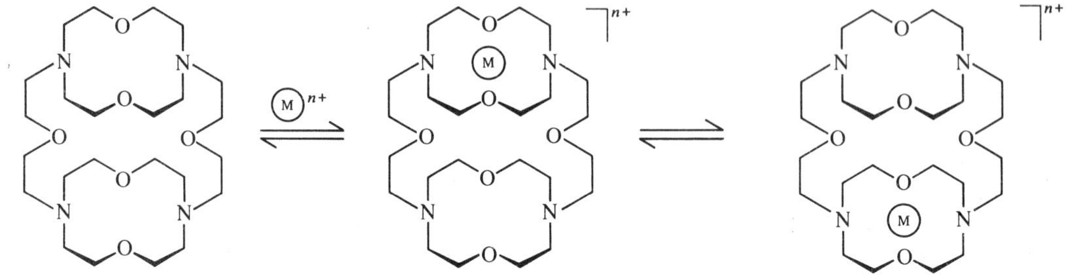

Scheme 7

Replacement of oxygen by sulfur yields the macrotricycles of the type (**52b**) and (**53e**).[9,195] These cryptands readily form dinuclear cryptates with transition metal cations, in particular with copper ion. Bis complexes of copper(II), copper(I) and mixed copper(I)–copper(II) complexes have been obtained.[195] The crystal structure of the bis-copper(II) complex of (**52b**) shows the two copper ions coordinated in a distorted tetragonal pyramid, approximately 0.34 Å out of the N_2S_2 planes of the two monocyclic cavities. A Cu–Cu distance of 5.62 Å is observed.[196] Investigations of the bis-copper complex of (**53e**) indicate highly stable complex formation in aqueous solution for both mono- and di-nuclear complexes.[195] Electrochemical experiments reveal a reversible transfer of two electrons at a positive potential (+445 mV), with little interaction between the two centers. As such, the molecule could possibly be considered as a dielectronic receptor unit and capable of exchanging two electrons at once, potentially of use in electron transfer catalysis.[197] The two copper(II) cations are coupled antiferromagnetically.[198] A study of the electrochemical behavior of a series of mononuclear and dinuclear copper cryptands has been reported.[199]

21.3.5.3.2 Spherical macrotricycles

These macrocycles are particularly suited for spherical recognition as a result of their spherical topology (Figure 1 and **54**).[35] The cavity of (**54a**) is approximately 3.4 Å in diameter and possesses 10 binding sites. The four nitrogen atoms are arranged at tetrahedral sites while the six oxygens are located at octahedral positions. The two different heteroatoms thus generate two polyhedra, the two centers of which coincide.[35,200] These ligands readily recognize spherical alkali metal and alkaline earth cations such as K^+, Rb^+, Cs^+ and Ba^{2+}, forming stable complexes ($\log K_5 = 3.4$, 4.2 and 3.4 for K^+, Rb^+ and Cs^+, respectively, in water). The cations are incorporated into the central cavity and cation exchange is slow. These complexes are one to two orders of magnitude more stable and undergo slower exchange than the macrobicyclic [3.2.2], which has approximately the same cavity size.[1,201]

The spherical macrotricyclic ligands are also particularly suited for tetrahedral recognition by virtue of the location of the nitrogens. This aspect will be discussed in Section 21.3.8.

21.3.5.4 Macropolycycles

Macropolycyclic molecules recognizing more than one type of substrate are well suited as models for metalloenzymes. For example, two porphyrins and two coronands can be alternated in a cyclic framework (**55**).[202] The porphyrin portions readily bind metal cations, while the coronands can

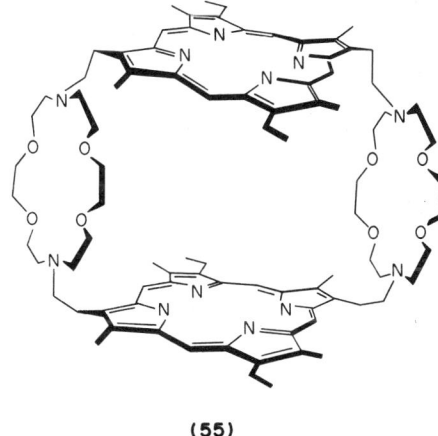

$$(54) \ a; \ X \ = \ O(= 17)$$
$$b; \ X \ = \ CH_2$$
$$c; \ X \ = \ CH_2CH_2$$

complex molecular cations such as diammonium salts. By varying the macrocyclic subunits and overall topology, the chemical and physical interactions between the two types of substrates (*e.g.* metal and molecular cation) can be fine-tuned to achieve the desired metallocatalysis.[202]

(55)

21.3.6 MOLECULAR CATION COMPLEXES

21.3.6.1 Macromonocycles

Primary ammonium ions are probably the most extensively studied of the molecular cations, and the crown ethers are noted for their complexation ability in this area.[11b,17–20,222–224] A primary ammonium ion possesses a trigonal arrangement of hydrogens. Thus, selective complexation can be achieved by a simple crown ether with trigonal recognition capabilities as a result of appropriately placed oxygens, *e.g.* [18]crown-6 (**56**). Hydrogen bonding interactions maintain the primary responsibility for complex stability, which is thus strongly correlated with the number of hydrogens available on the substrate for interactions with the ether oxygens. For example, formation constants for a series of alkylammonium complexes with [18]crown-6 and analogs decrease markedly in the order $NH_4^+ \approx MeNH_3^+ > Me_2NH_2^+ > Me_3NH^+$, as a result of 'central discrimination' of the crown between primary and higher order ammonium salts (Figure 10.)[222,225]

The use of 'lateral discrimination' to enhance complex stability can be accomplished by incorporating ring variations such as appropriate podando appendages. Additional complex stabilization

(56)

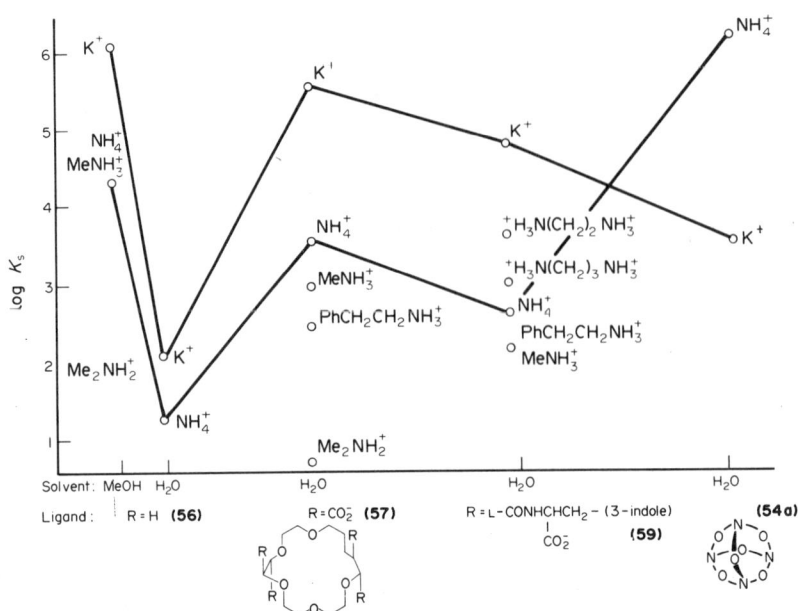

Figure 10 Comparison of stability constants for complexes of [18]crown-6, tetrasubstituted analogs and a spherical cryptand with alkylammonium substrates

is then gained *via* charge transfer, hydrophobic and/or electrostatic interactions between organic residues on the substrate and lateral podando groups, as illustrated in **(57)** and **(58)**.[17] Figure 10 examines complex stability as a function of both substrate and receptor for the [18]crown-6 and two modifications **(56)** and **(57)**, as well as for the tridimensional spherical ligand **(54a)**. The importance of electrostatic interactions is seen by the fact that the tetracarboxylate **(57)** forms some of the most stable complexes reported with both K^+ and RNH_3^+ ($\log K_s$ values in water are 5.5 and 2.9, respectively).[17,222]

(57)   (58) X = CO_2, CO_2H, CO_2Na, $CONHMe$, $CON(NO)Me$

By incorporating diphenylmethane groups into the chiral tartaric acid-derived macrocyclic design, a speleand **(58)** is formed which strongly binds quaternary and primary ammonium ions.[226] The biologically significant neurotransmitter, acetylcholine, and methylviologen, of interest in current photochemical energy storage research, are two of the relevant substrates for this macrocycle.[226] Tryptophanate residues provide additional lipophilic shielding of the carboxylates from solvation, which allows for increased ion-pair interactions with the ammonium salt as shown for the nicotinamide complex **(59)**. These molecular cation complexes are thus stabilized by an onslaught of factors such as electrostatic interactions between the ammonium sites and the carboxylates, hydrogen bonding and hydrophobic effects, among others. The result is a very stable complex, for which $\log K_s$ can exceed 5 in some cases.[226] The synthesis of peptide bonds has been accomplished by crown ethers with thiol appendages according to Scheme 8.[227]

Guanidinium and imidazolium ions are also recognized by the appropriate ligand system.[228-230] These planar ions are surrounded by suitably sized macrocycles which are capable of 'circular recognition' as shown in **(60)**. The size of the [27]crown-9 is structurally ideal for complexation of

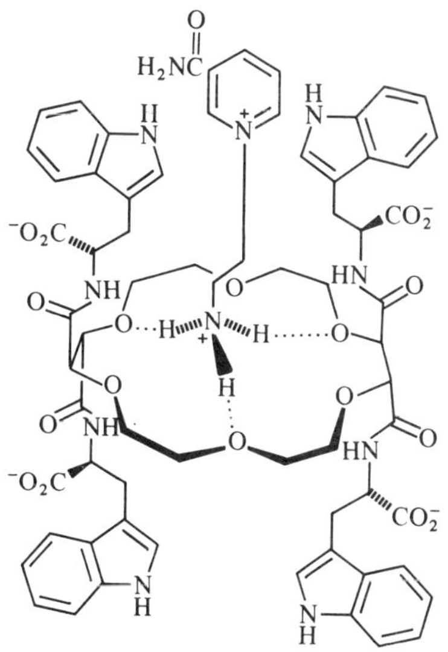

(59)

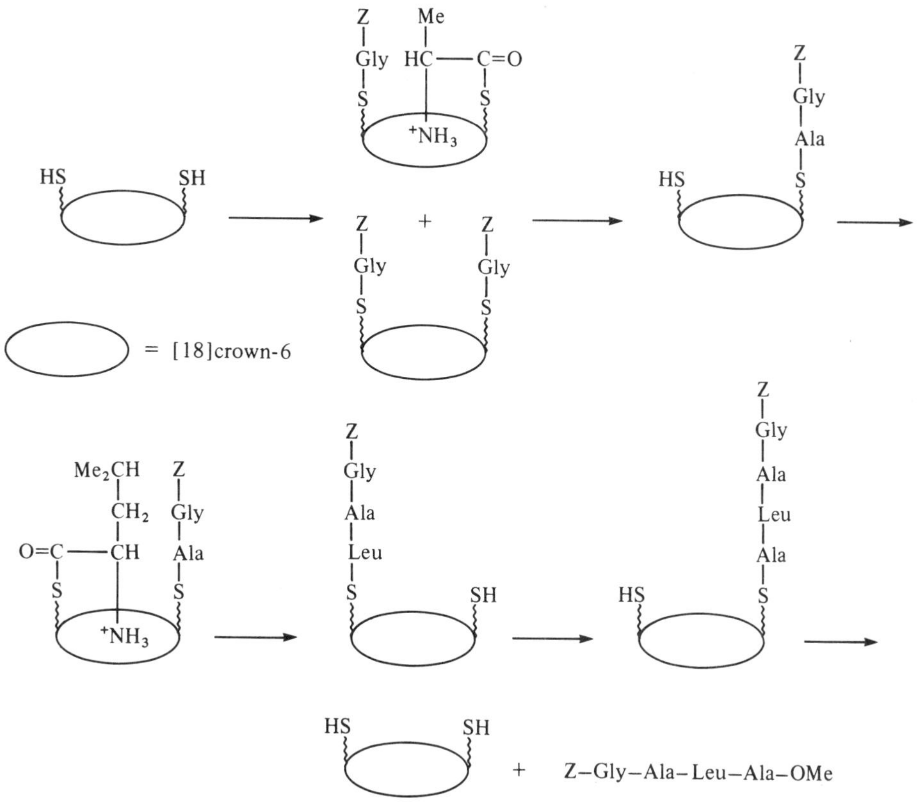

Scheme 8

the guanidinium ion, and carboxylates enhance the complexation ($\log K_s$ in water = 3.8).[229] Again the markedly decreased stabilities of the analogous complexes of substituted guanidines emphasize the importance of maximizing hydrogen bonding interactions in central discrimination.

Chiral recognition has been achieved by combining central recognition and a fine structure which imposes enantiomeric discrimination. The goal is to allow a diastereomeric host/guest complex to

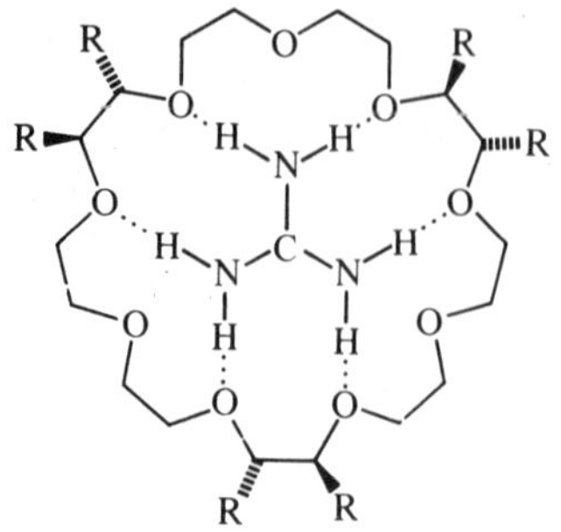

(60) R = $CO_2^-NMe_4^+$

form, in which interactions with the receptor are energetically more favorable for one optical isomer of an enantiomeric pair of substrates, *i.e.* 'chiral recognition'. This is indeed an extensive area of macrocyclic chemistry and numerous reviews have appeared.[18–20,223,224]

A variety of macrocycles with asymmetric centers have been reported, examples of which are shown in (57) to (66). Chiral discrimination has been observed in the study of thiolysis of activated ester bonds with tetracysteinyl[18]crown-6 (67), *e.g.* Gly-L-Phe reacts up to 80 times faster than Gly-D-Phe with this ligand.[231] The chiral macrocyclic ligand (66) is also capable of enantiomeric discrimination, by assisting in the selective reduction of carbonyl compounds with high optical yields.[232,233]

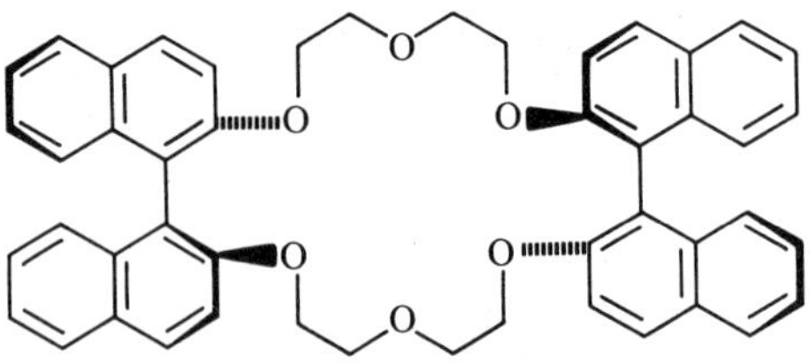

(61)

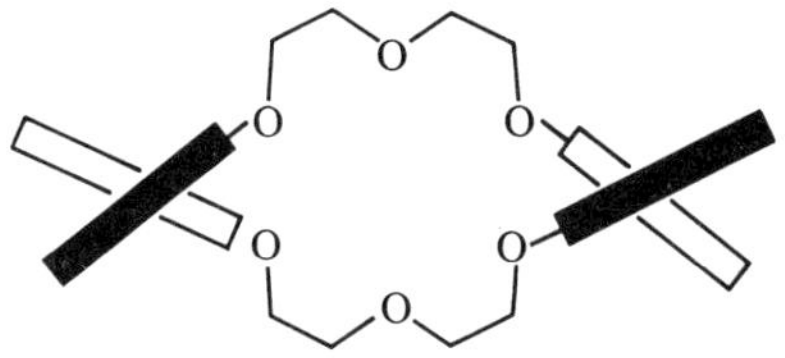

(62) (S, S)

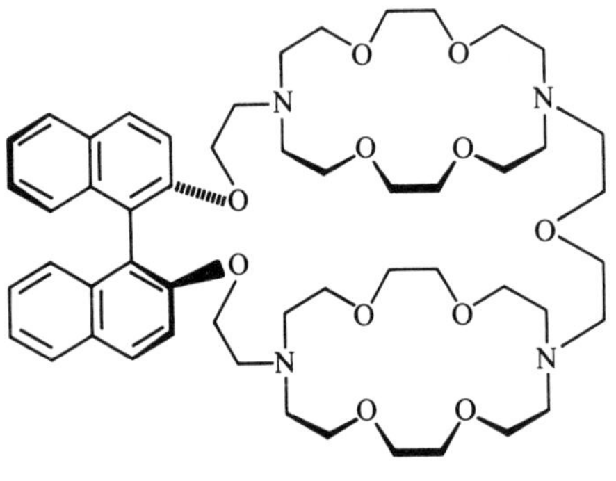

(63)

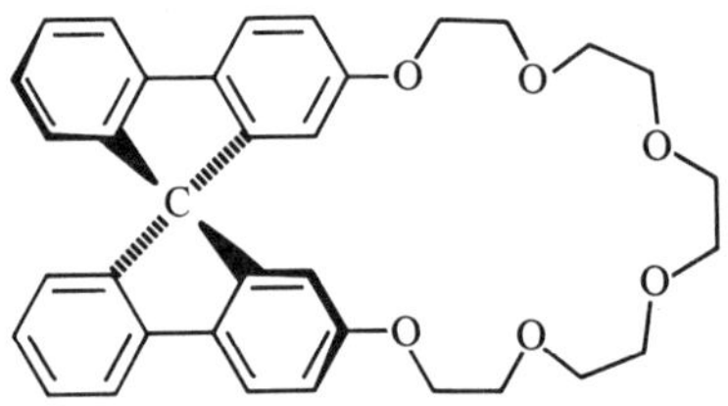

(64)

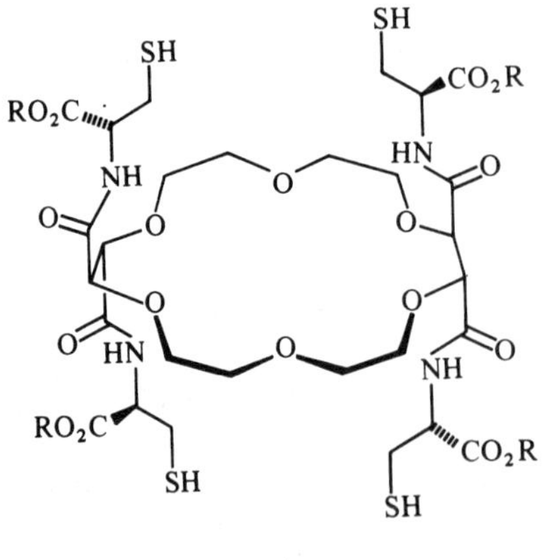

(65)

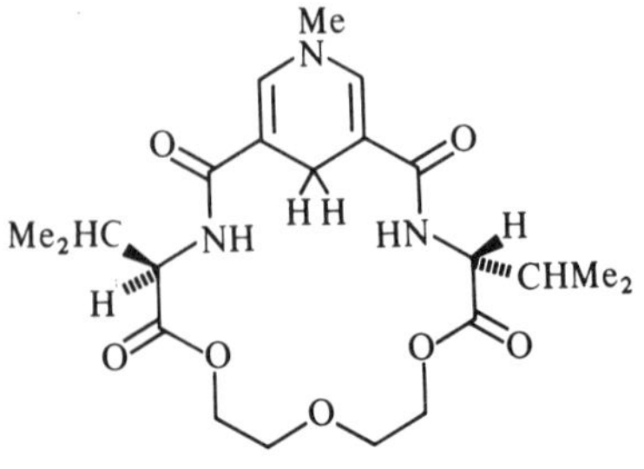

(66)

Attachment of bulky chiral barriers to achieve enantiomeric selection is exemplified by ligands
(**61**),[204] (**63**)[69] and (**64**).[19] Ligands of type (**61**) are especially efficient chiral compounds with 'binap-
thyl hinges'. The orientation of the binaphthyl rings for (**61**) is shown more explicitly in (**62**).[234] With
optically pure (**61**), complete enantiomeric separation of ammonium salt racemates such as (**67**) is
obtained.[235–237] The structure has been determined crystallographically for the less stable D form of
the PF_6^- salt of (**67**) with the (S,S) host (**61**). The results indicate that the receptor, substrate,
counterion and solvent have undergone several steric concessions in order to relieve the strain
imposed on the system by the visit of an unwanted guest.[238]

(**67**)

21.3.6.2 Macropolycycles

The complexation of the ammonium ion by the spherical macrotricyclic ligand (**54a**) exemplifies
the use of appropriate placement of binding sites in order to fine-tune receptor capabilities. This
ligand readily binds the tetrahedral NH_4^+, forming extremely stable complexes (log $K_s = 6$).[35] Stability is maintained by hydrogen bonding between the ammonium hydrogens and the tetrahedrally
placed ligand nitrogens as shown in (**68**). Electrostatic interactions between ligand oxygen atoms
and the positive nitrogen center are also optimized since the oxygens are located so that they bisect
the H—N—H angles.[35,239] The specificity of tetrahedral recognition becomes evident upon noting
that the corresponding potassium complex is 500 times less stable (Figure 11). Complexation
increases the pK_a of the NH_4^+ from 9 to 15. Water is also complexed by this ligand in its diprotonated
form *via* similar interactions.[20]

(**68**)

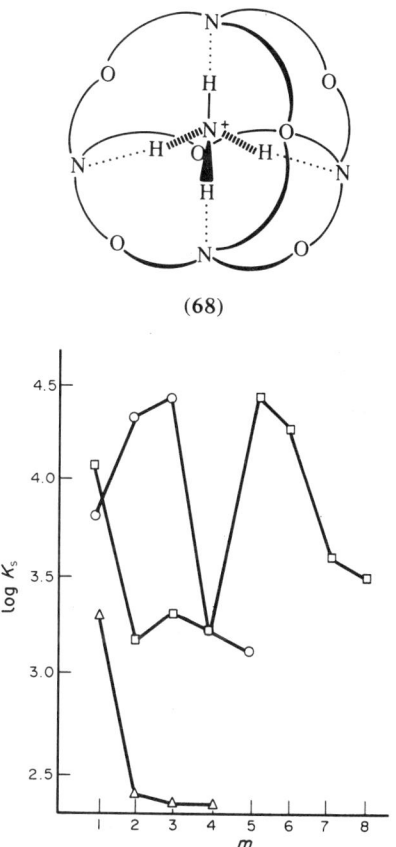

Figure 11 Stability constants for complexes of polyammonium macrocycles with dicarboxylate substrates $^-O_2C(CH_2)_m$-
CO_2^-. △ = [24]N_6·6HCl (**72a**); ○ = [32]N_6·6HCl (**72b**); □ = [38]N_6·6HCl (**72c**)

Cylindrical macrotricycles maintain the capability of binding either two substrates or one substrate, in addition to forming cascade complexes (Figure 3). The ditopic macrocyclic ligand containing two naphthalene bridges (**69**) is very effective in selectively complexing diammonium cations by chain length selection (linear recognition). In fact, while this receptor readily incorporates cadaverine ($^+NH_3(CH_2)_5NH_3^+$), putrescine, with a chain one carbon shorter, is not as readily complexed by this ligand.[189,240,241]

(69)

By combining a receptor unit with a rigid shaping unit, hollow macropolycyclic speleands (Figure 3; **20**) have been synthesized. NMR experiments indicate that (**20**) forms both external and internal complexes with methylammonium cation.[38]

21.3.7 ANION COMPLEXES

21.3.7.1 Introduction

While anion complexation might more appropriately be divided according to the complexity of the anion, the format followed in the preceding two subsections will be maintained for this section, *i.e.* subdivision according to the 'dimensionality' of the macrocycles (mono-, bi- and tri-cyclic systems). A brief overview of anion complexation will, however, be presented in this introductory subsection. The types of anions which undergo complexation with macrocyclic ligands are categorized below.

(1) 'Simple' inorganic ions (halides, NO_3^-, ClO_4^-, SO_4^{2-} and N_3^-), depending on the ion, can be complexed by macromonocycles,[35,242,243] macrobicycles[33,243] and macrotricycles.[35b,244]

(2) More complex organic anions and polyanions (oxalate^{2-}, malonate^{2-}, succinate^{2-}, *etc.*), and phosphates (ATP^{4-}, ADP^{3-}, AMP^{2-}, *etc.*) also form complexes with macrocyclic ligands. Steric constraints imposed by incorporation of the larger, more complex ions into macropolycyclic ligands lead to marked selectivity patterns.[46,242,246]

(3) Metal ion complexes ($[M(CN)_6]^{n-}$; M = Fe, Ru, $n = 3, 4$) can also be incorporated into certain macromonocyclic polyammonium ligands forming 'complexes of complexes', *i.e.* supercomplexes.[242,246,247]

The binding sites for anion complexation are usually ammonium cations forming ^+N—$H \cdots X^-$ hydrogen bonds; however, other sites may be utilized (quaternary ammonium, acidic OH groups, *etc.*). Lewis acid sites, as in anion binding to metal ions (*e.g.* in complex ions like $CuCl_4^{2-}$ or in complexes like nonporphyrins) will not be considered here. The results obtained in recent years on anion binding by organic ligands have begun to build the body of a defined anion coordination chemistry, covering inorganic, organic and biological anionic substrates.[17,21,248]

21.3.7.2 Macromonocycles

Multidentate polyaza and mixed polyaza–polyoxa macrocycles in their polyammonium form complex a variety of anions. Complexation ability is, as anticipated, related to macrocyclic ring size, macrocyclic topology and the relation of these factors to the substrate to be complexed. Polyammonium macromonocycles of various ring sizes and binding subunits complex polycarboxylate anions,[47,242,245] polyphosphate anions[242,249] and also hexacyanometallate anions.[242,244,247] The guest tendencies of organic anions with these macrocycles can be divided for purposes of discussion into those of the 'smaller' macrocycles (< 18-membered rings), compared to larger ligands with more flexible rings. Table 2 shows the relationship of stability constants for several of these ligands with the identity of both ligand and substrate.

Table 2 Stability Constants ($\log K_s$) for Anion Binding

Ligand	Citrate	Succinate	Substrate Malonate	AMP^{2-}	ADP^{3-}	ATP^{4-}
(70a)[a]	1.7	c	c	3.1	3.9	4.0
(70b)[a]	2.4	2.1	0.8	3.1	3.1	3.6
(70c)[a]	3.0	1.0	0.4	2.8	3.0	3.7
(71)[a]	2.4	0.2	0.5	3.2	5.6	6.4
(72a)[b]	4.7	2.4	3.3	3.4	6.5	8.9
(73)[b]	7.6	3.6	3.9	4.1	7.5	8.5
(74)[b]	5.8	2.8	3.8	4.7	7.7	9.1

[a] Determined by polarographic methods from refs. 245 and 249.
[b] Determined by pH metric titration from ref. 242.
[c] Negligible.

From an examination of Table 2, the general trend appears to be that the increased flexibility of the larger macrocycles (72a), (73) and (74) facilitates complexation of the bulkier substrates shown. The greatest stabilities are not always associated with the largest macrocycles, however, and care must also be exercised to make comparisons within the table qualitatively, since the data shown were obtained under different conditions. The complexation studies of dicarboxylate anions by the partially protonated 'smaller' penta- and hexa-amine macrocycles, *e.g.* (70) and (71), indicated that the association constants vary according to macrocyclic ring size, but that the most stable complexes for any given macrocycle were formed with citrate ion.[249] In another report on a more extensive series of tetra-, and penta- and hexa-aza macromonocycles, phosphate and nucleotide association were observed at neutral pH.[249] While the stability constants for the phosphate anion were comparable to those of the dicarboxylate anions (*e.g.* $\log K_s = 2.0$ for 70b), the stabilities of the nucleotide complexes were considerably greater (Table 2). This observation was attributed to possible adenine interactions with the macrocycles.

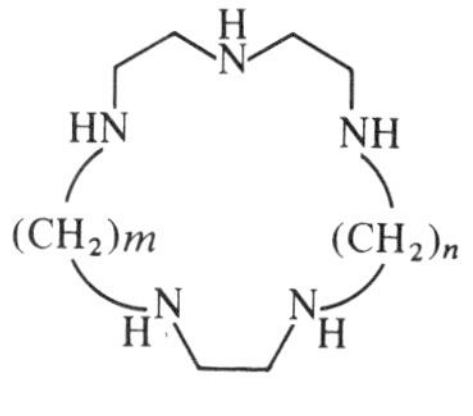

(70) a; $m = n = 2$
　　b; $m = 2, n = 3$
　　c; $m = n = 3$

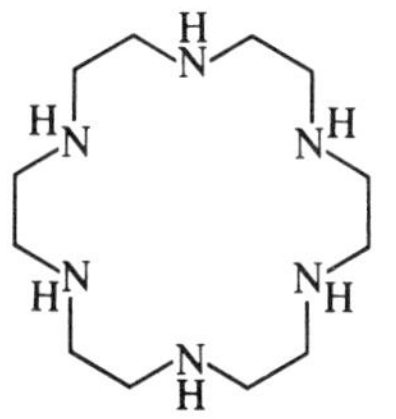

(71)

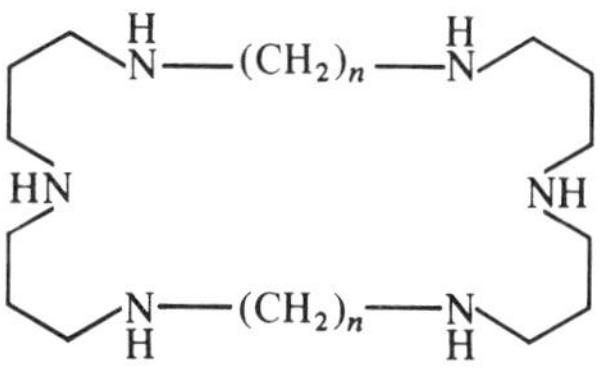

(72) a; $n = 3$
　　b; $n = 7$
　　c; $n = 10$

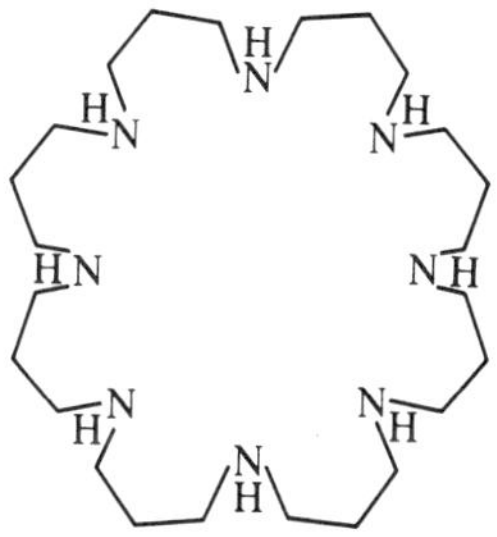

(73)

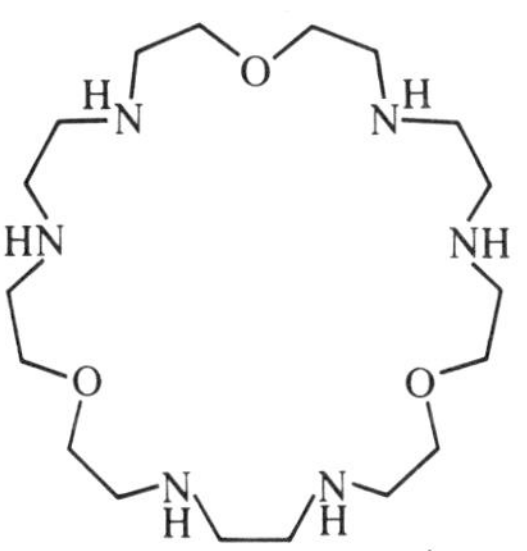

(74)

The ability of larger ring systems to complex organic anions was examined for a number of receptor molecules. Ligands (72a)–(74), varying in ring size from 24 to 32 members, were found to complex dicarboxylates, 1,3,5-benzenetricarboxylate^{3-}, $Co(CN)_6^{3-}$, $Fe(CN)_6^{4-}$, AMP^{2-}, ADP^{3-} and

ATP^{4-}.[242] Results indicated associations to be influenced by macrocyclic and size effects, electrostatic interactions and structural complementarity. Of particular interest in this series is the observed very strong binding of anionic metal ion complexes. The electrochemistry of $M(CN)_6^{3-}/M(CN)_6^{4-}$ (M = Fe, Ru) indicates the 'complexed complex' to be nonlabile and the system to undergo a reversible one-electron exchange. The ability of the macrocycle to 'protect' the reactivity of the complex is evidenced by the hindered photoaquation observed for $Co(CN)_6^{3-} \subset [32]N_8H_8$.[247] The results of competition studies suggest predominantly coulombic interactions between the metal complex and macrocycle for these 1:1 'complexes of complexes'.

In addition to forming stable complexes with ATP, polyammonium macrocyclic ligands have also been shown to catalyze the hydrolysis of ATP.[250] This catalytic behavior is a primary example of the applicability of macrocyclic systems to the design of molecular catalysts. The ability of a series of ligands to catalyze ATP hydrolysis was studied, including monocycles such as (72a), (73) and (74), two tetraaza macrocycles, as well as the bis-tren bicycle (47). For most of the macrocycles, rate enhancements of ATP hydrolysis were observed. The ligand (75) was noted to be especially suited for catalysis of ATP hydrolysis, forming a phosphoramidate intermediate during the process.[250] This intermediate was also observed in the hydrolysis of acetyl phosphate as catalyzed by (75), which led to pyrophosphate formation.[251] In the presence of the biologically relevant metal ions, magnesium and calcium, the phosphorylated macrocycle was found to be stabilized in the ATP reaction, and pyrophosphate formation was also observed.[252]

(75)

Appropriate choice of ditopic polyammonium macrocycles (72b and 72c) can allow for selective linear recognition (Figure 11). These macrocycles contain two binding subunits and are therefore potentially capable of either binding two substrates to form a dinuclear cryptate, or complexing one polyfunctional substrate (Figure 4). At neutral pH, all six amine nitrogens are protonated, and the polyammonium species form strong complexes with a variety of dicarboxylate anions, $^-O_2C(CH_2)_m-CO_2^-$, of varying chain lengths from $m = 0$ to 8. The binding is selective and can be correlated with substrate and receptor hydrocarbon chain lengths. For example, ligand (72a) shows larger stabilities for smaller m, and (72b) exhibits maximum selectivity at succinate ($\log K_s = 4.3$) and glutarate ($\log K_s = 4.4$) ions ($m = 2$ and 3, respectively). By increasing the macrocyclic chain length by three carbons to obtain (72c), selectivity shifts to pimelate ($\log K_s = 4.4$) and suberate ($\log K_s = 4.25$) ions ($m = 5$ and 6, respectively). Thus a macrocyclic chain increase of three is mirrored by more favored selection of substrates three carbons longer, and recognition is again closely associated with the structural design of the ligand.[35]

Additionally, macrocycles containing guanidinium groups (76) have been reported to form inclusion compounds with some anions. Appealing features of the guanidinium group are: (1) the possibility of forming several zwitterionic hydrogen bonds $NH^+ \cdots X$, which renders its chemistry relatively independent of pH changes; and (2) its function in proteins on arginyl residues in maintaining protein conformations and binding and recognizing anionic substrates.[253] These species appear to complex phosphate ion, albeit weakly compared to alkali metal cation complexation.[253]

Heterocyclophane cations (77)[254,255] have also been reported to form inclusion compounds with some anions.

21.3.7.3 Macrobicycles

The protonated *endo–endo* isomers of the diazabicycloalkanes, catapinands, were among the first macrocycles found to be capable of incorporating halide anions (78).[34] Electrostatic and hydrogen bond interactions are undoubtedly the primary driving forces for complex stability. The crystal structure of the chloride catapinate of (78) has confirmed the inclusion nature of the complex and

(76)   (77)

the importance of hydrogen bonding interactions (N—Cl = 3.10Å).[256] As expected, complexation ability is related to cavity size. Chloride ion is more readily encapsulated for the nine-carbon-chained catapinand than bromide ion, and shortening of the alkyl chains effectively precludes enclosure.

(78)

The macrobicyclic ligand (47) derived from two tren units [tren = $N(CH_2CH_2NH_2)_3$] is particularly suited structurally for linear recognition as in (79).[243] The hexaprotonated form contains an ellipsoidal cavity with inverted binding groups, the three positively charged ammonium groups being located about the molecular axis through the two poles of the molecule. The crystal structures of the cryptates [X^- (47)-6H$^+$] formed with $X^- = F^-$, Cl^-, Br^- and N_3^- have been determined.[215] The structures confirm the inclusion nature of the complexes and provide the first sequence of anion coordination geometries: tetrahedral for F^-· octahedral for Cl^- and Br^-, and bipyramidal for N_3^- as shown in (79)–(81). Hydrogen bonds again play an important role in anion complexation. The selectivity sequence is $ClO_4^- < I^- < Br^-$, Cl^-, $NO_3^- \ll N_3^-$, with much stronger binding for polyanions. These findings are indicative of topological discrimination as a result of the shape and size of the cavity, as well as the arrangement of the binding sites.[243,255]

(79)   (80)   (81)

21.3.7.4 Macrotricycles

In 1976, the fully protonated spherical macrotricyclic ligands (54) were observed to exhibit spherical anion recognition as in (82).[35] Such ligands display tetrahedral recognition in the free base form, as discussed in Section 21.3.7.3. The crystal structure of the cryptate (82) indicates that the anion is held in a tetrahedral array of hydrogen bonds.[241] The complex is very stable and highly selective.[257] Other macrotricyclic ligands have also been observed to form inclusion complexes with halide ions.[258] A ^{35}Cl NMR study of a series of the inclusion complexes of chloride ion with tricyclic and bicyclic cryptands has provided comparative information on chemical shifts, exchange processes and chloride quadrupole coupling constants of the cryptates.[259]

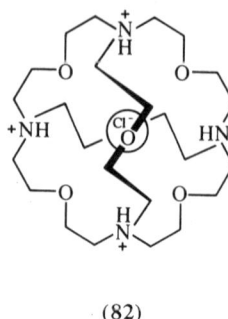

(82)

21.3.8 NEUTRAL MOLECULAR COMPLEXES

Complexes are also formed in certain instances between neutral molecules and macrocyclic receptors. Neutral molecules which form such complexes for the most part contain polar O—H, N—H or C—H bonds, and hydrogen bonding interactions are responsible for the solid state structural characteristics of these complexes. For many of these complexes, stoichiometries range considerably, from 1:1 to 1:6 host:guest, and include a variety of 'odd' ratios such as 3:2, 2:7, *etc.* Structural results for these 'complexes' indicate them not to be of the inclusion type in a majority of cases. Thus, discussion in this subsection will be limited to a general overview. A more complete review of neutral molecule complexation can be obtained elsewhere.[21]

Polar O—H bonds are found in molecules such as water and alcohols. Crystal structures of several crown ether complexes have indicated that the cavity need only be partially filled by water.[260-262] Several host–guest complexes of alcohols with the pyridino crown (83) have been reported.[263] Longer chain and branched alcohols do not in general form crystalline adducts with this ligand.

(83)

Guests with polar N—H bonds include amides, ureas and related sulfur compounds, substituted hydrazines, and aromatic amines.[21,264,265] Complexes with these substrates exhibit a wide variety of stoichiometric and nonstoichiometric associations. A particularly popular association is the 1:2 guest:host relationship as typified by the structure of the benzenesulfonamide complex with [18]crown-6 (84).[266,267] The weak complexes formed between crown ethers and cryptands with certain proteins allow for solubilizing these species in organic solvents.[268]

Compounds with polar C—H bonds, *e.g.* acetonitrile, malonitrile, nitromethane and dimethylformamide, also have a tendency to become the 'guests' of a variety of crown hosts. The 'complexes' that many of these compounds form are not of the inclusion type, and the specific interactions between the host and guests are not always evident. Possibly interstitial trapping of a guest molecule creates a more favorable crystalline lattice along with increased ordering due to interactions between the guest and electronegative heteroatoms in the host.[269] An example is the malonodinitrile structure schematically depicted in (85).

Complexes which contain enforced cavities large enough to incorporate simple molecules or ions are capable of hosting neutral molcules. Included in this category are spherands (Figure 2; 10) and cavitands,[270-272] speleands (Figure 3; 20),[38] water-soluble paracyclophanes (86)[273,274] and cryptopha-

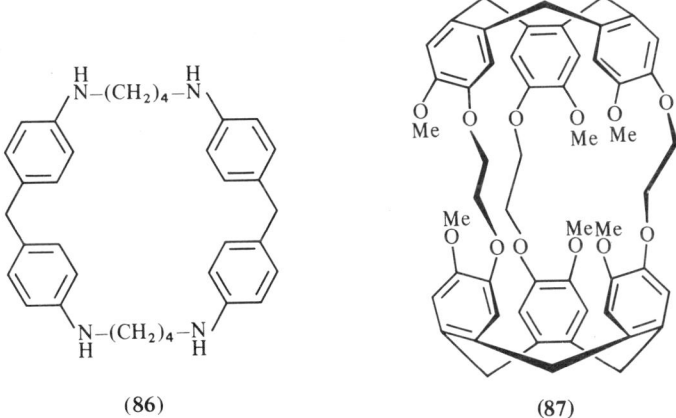

(84)   (85)

nes (87).[275,276] Cavitands can be designed to incorporate only 'slim linear guests' such as CS_2, MeCN and O_2.[271] An even more restrictive framework, termed carcerand, can be obtained from further closing the shell with two cavitands.[272]

(86)   (87)

21.3.9 REFERENCES

1. J.-M. Lehn, *Struct. Bonding (Berlin)*, 1973, **16**, 1.
2. G. A. Melson (ed.), 'Coordination Chemistry of Macrocyclic Compounds', Plenum, New York, 1979.
3. R. M. Izatt and J. J. Christensen (eds.), 'Synthetic Multidentate Macrocyclic Compounds', Academic, New York, 1978.
4. L. F. Lindoy, *Chem. Soc. Rev.* 1975, **4**, 421.
5. R. M. Izatt and J. J. Christensen (eds.), 'Progress in Macrocyclic Chemistry', Wiley, New York, 1979, vol. 1.
6. D. H. Busch, *Acc. Chem. Res.*, 1978, **11**, 392.
7. N. F. Curtis, *Coord. Chem. Rev.*, 1968, **3**, 3.
8. D. Dolphin, 'The Porphyrins', Academic, New York, vols. I–VII.
9. J.-M. Lehn, *Pure Appl. Chem.*, 1980, **52**, 2441; *Science*, 1985, **227**, 4689.
10. B. C. Pressman, *Annu. Rev. Biochem.*, 1976, **45**, 501, and references therein.
11. C. J. Pedersen, *J. Am. Chem. Soc.*, 1967, **89**, 2459, 7017.
12. C. J. Pedersen and H. K. Frensdorff, *Angew. Chem., Int. Ed. Engl.*, 1972, **11**, 16.
13. B. Dietrich, J.-M. Lehn, and J.-P. Sauvage *Tetrahedron Lett.*, 1969, 2885, 2889.
14. J. C. G. Buenzli and D. Wessner, *Coord. Chem. Rev.*, 1984, **60**, 191.
15. M.-C. Almasio, F. Arnaud-Neu and M. J Schwing-Weill, *Helv. Chim. Acta*, 1983, **66**, 1296.
16. M. Brighli, P. Fux, J. Lagrange and P. Lagrange, *Inorg. Chem.*, 1985, **24**, 80.
17. J.-M. Lehn, *Pure Appl. Chem.*, 1978, **50**, 871.
18. D. J. Cram, R. C. Helgeson, L. R. Sousa, J. M. Timko, M. Newcomb, P. Moreau, F. DeJong, G. W. Gokel, D. H. Hoffman, L. A. Domeier, S. C. Peacock, K. Madan and L. Kaplan, *Pure Appl. Chem.*, 1975, **43**, 327.
19. V. Prelog, *Pure Appl. Chem.*, 1978, **50**, 893.
20. J. F. Stoddart, *Chem. Soc. Rev.*, 1979, **8**, 85.
21. F. Vogtle, H. Sieger and W. M. Muller, *Top. Curr. Chem.*, 1981, **98**, 107; and in 'Host Guest Complex Chemistry, Macrocycles', ed. F. Vogtle and G. Weber, Springer, Berlin, 1985, p. 319.
22. M. Kodama, E. Kimura and S. Yamaguchi, *J. Chem. Soc., Dalton Trans.*, 1980, 2536.
23. J. Comarmond, P. Plumeré, J.-M. Lehn, Y. Agnus, R. Louis, R. Weiss, O. Kahn and I. Morgenstern-Badaru, *J. Am. Chem. Soc.*, 1982, **104**, 6330.

24. J. Smid, *Angew. Chem., Int. Ed. Engl.*, 1972, **11**, 112.
25. J.-M. Lehn and J.-P. Sauvage, *J. Am. Chem. Soc.*, 1975, **97**, 6700.
26. I. Tabushi, Y. Kobuke and T. Nishiya, *Tetrahedron Lett.*, 1979, **20**, 3515.
27. A. Shanzer, J. Libman and F. Frolow, *J. Am. Chem. Soc.*, 1981, **103**, 7339.
28. E. Schwartz and A. Shanzer, *J. Chem. Soc., Chem. Commun.*, 1981, 634.
29. D. J. Cram, T. Kaneda, R. C. Helgeson and G. M. Lein, *J. Am. Chem. Soc.*, 1979, **101**, 6752.
30. K. E. Koenig, R. C. Helgeson and D. J. Cram, *J. Am. Chem. Soc.*, 1976, **98**, 4018.
31. A. Zinke and E. Ziegler, *Ber. Dtsch. Chem. Ges. B*, 1944, **72**, 264.
32. M. L. Bender and M. Komiyama, 'Cyclodextrin Chemistry', Springer, Berlin, 1978.
33. C. H. Park and H. E. Simmons, *J. Am. Chem. Soc.*, 1968, **90**, 2431.
34. C. O. Dietrich-Buchecker, J.-M. Kern and J. P. Sauvage, *J. Am. Chem. Soc.*, 1984, **106**, 3043.
35. E. Graf and J.-M. Lehn, *J. Am. Chem. Soc.*, 1975, **97**, 5022; 1976, **98**, 6403.
36. W. P. Schammel, K. B. Mertes, G. G. Christoph and D. H. Busch, *J. Am. Chem. Soc.*, 1979, **101**, 1622.
37. I. I. Creaser, J. MacB. Harrowfield, A. J. Herlt, A. M. Sargeson, J. Springborg, R. J. Geue and M. R. Snoro, *J. Am. Chem. Soc.*, 1977, **99**, 3181.
38. J. Canceill, A. Collet, J. Gabard, F. Kotzyba-Hibert and J.-M. Lehn, *Helv. Chim. Acta*, 1982, **65**, 1894.
39. V. K. Majestic and G. R. Newkome, *Top. Curr. Chem.*, 1982, **106**, 79.
40. G. R. Newkome, J. D. Sauer, J. M. Roper and D. C. Hager, *Chem. Rev.*, 1977, **77**, 513.
41. R. M. Izatt, J. D. Lamb, G. E. Moas, R. E. Asay, J. S. Bradshaw and J. J. Christensen, *J. Am. Chem. Soc.*, 1967, **99**, 2365.
42. Y. Kobuke, K. Hanji, K. Horiguchi, M. Asada, Y. Nakayama and J. Furukawa, *J. Am. Chem. Soc.*, 1976, **98**, 7414.
43. A. Shanzer, J. Libman and F. Frolow, *Acc. Chem. Res.*, 1983, **16**, 60.
44. Y. A. Ovchinnikov, V. T. Ivanov and A. M. Shkrob, 'Membrane Active Complexes', Elsevier, New York, 1974, BBA Library, vol. 12.
45. G. A. Melson, in 'Coordination Chemistry of Macrocyclic Compounds', ed. G. A. Melson, Plenum, New York, 1979, p. 108.
46. D. J. Cram and J. M. Cram, *Acc. Chem. Res.*, 1978, **11**, 8.
47. M. W. Hosseini and J.-M. Lehn, *J. Am. Chem. Soc.*, 1982, **104**, 3525.
48. N. F. Jones, A. Kumar and I. O. Sutherland, *J. Chem. Soc., Chem. Commun.*, 1981, 990.
49. F. Kotzyba-Hibert, J.-M. Lehn and K. Saigo, *J. Am. Chem. Soc.*, 1981, **103**, 4266.
50. F. Vogtle and E. Weber, in 'The Chemistry of the Ether Linkage', ed. S. Patai, Wiley, New York, 1981, Supplement E, Part 1, p. 59.
51. S. T. Jolley, J. S. Bradshaw and R. M. Izatt, *J. Heterocycl. Chem.*, 1982, **19**, 3.
52. G. W. Gokel and S. H. Korzeniowski, in 'Macrocyclic Polyether Synthesis', Springer, Berlin, 1982.
53. C. J. Pedersen, *J. Am. Chem. Soc.*, 1970, **92**, 391.
54. J. S. Bradshaw, J. Y. Hui, B. L Haymore, J. J. Christensen and R. M. Izatt, *J. Heterocycl. Chem.*, 1973, **10**, 1.
55. J. S. Bradshaw, J. Y. Hui, Y. Chan. B. L. Haymore, R. M. Izatt and J. J. Christensen, *J. Heterocycl. Chem.*, 1974, **11**, 45.
56. J. S. Bradshaw, R. A. Reeder, M. D. Thompson, E. D. Flanders, R. L. Carruth, R. M. Izatt and J. J. Christensen, *J. Org. Chem.*, 1976, **41**, 134.
57. J. S. Bradshaw, L. D. Hansen, S. F. Nielsen, M. D. Thompson, R. A. Reeder, R. M. Izatt and J. J. Christensen, *J. Chem. Soc., Chem. Commun.*, 1975, 874.
58. W. Rosen and D. H. Busch, *Chem. Commun.*, 1970, 1041; *J. Am. Chem. Soc.*, 1969, **91**, 4694.
59. M. C. Thompson and D. H. Busch, *J. Am. Chem. Soc.*, 1964, **86**, 3651.
60. T. A. DelDonno and W. Rosen, *J. Am. Chem. Soc.*, 1977, **99**, 8051.
61. R. Bartsch, S. Hietkamp, S. Morton, H. Peters and O. Stelzer, *Inorg. Chem.*, 1983, **22**, 3624.
62. J. Ennen and T. Kauffmann, *Angew. Chem., Int. Ed. Engl.*, 1981, **20**, 118.
63. D. J. Cram, T. Kaneda, G. M. Lein and R. C. Helgeson, *J. Chem. Soc., Chem. Commun.*, 1979, 948.
64. C. D. Gutsche and J. A. Levine, *J. Am. Chem. Soc.*, 1982, **104**, 2652.
65. C. D. Gutsche, B. Dhawan, K. H. No and R. Muthukrishnan, *J. Am. Chem. Soc.*, 1981, **103**, 3782.
66. B. Dietrich, J.-M. Lehn and J.-P. Sauvage, *Chem. Commun.*, 1970, 1055.
67. A. M. Sargeson, *Chem. Br.*, 1979, **15**, 23.
68. J.-M. Lehn, J. Simon and J. Wagner, *Nouv. J. Chim.*, 1977, **1**, 77; *Angew. Chem., Int. Ed. Engl.*, 1973, **12**, 578.
69. J. Cheney, J.-M. Lehn, J.-P. Sauvage and M. E. Stubbs, *J. Chem. Soc., Chem. Commun.*, 1972, 1100.
70. J. Cheney, J. P. Kintzinger and J.-M. Lehn, *Nouv. J. Chim.*, 1978, **2**, 411.
71. B. Dietrich, J.-M. Lehn and J. Simon, *Angew. Chem., Int. Ed. Engl.*, 1974, **13**, 406.
72. M. Hiraoka, in 'Crown Compounds, Their Characteristics and Applications. Studies in Organic Chemistry', Elsevier, Amsterdam, 1982, vol. 12.
73. F. Vogtle, *Top. Curr. Chem.*, 1981, **98**.
74. J. J. Christensen, D. J. Eatough and R. M. Izatt, *Chem. Rev.*, 1974, **74**, 351.
75. G. W. Gokel, D. M. Goll, C. Minganti and L. Echegoyen, *J. Am. Chem. Soc.*, 1983, **105**, 6786.
76. R. M. Izatt, R. E. Terry, B. L. Haymore, L. D. Hanson, J. K. Dalley, A. G. Avondet and J. J. Christensen, *J. Am. Chem. Soc.*, 1976, **98**, 7620.
77. J. D. Lamb, R. M. Izatt, J. J. Christensen and D. J. Eatough, in 'Coordination Chemistry of Macrocyclic Compounds', ed. G. A. Melson, Plenum, New York, 1979, pp. 145–214.
78. E. Shchori and J. Jagur-Grodzinski, *Isr. J. Chem.*, 1973, **11**, 243.
79. R. Ungaro, B. El Hag and J. Smid, *J. Am. Chem. Soc.*, 1976, **98**, 5198.
80. D. Midgley, *Chem. Soc. Rev.*, 1975, **7**, 549.
81. H. K. Frensdorff, *J. Am. Chem. Soc.*, 1971, **93**, 600.
82. N. K. Dalley, J. S. Smith, S. B. Larson, K. L. Metheson, J. J. Christensen and R. M. Izatt, *J. Chem. Soc., Chem. Commun.*, 1975, 84.
83. D. K. Cabbiness and D. W. Margerum, *J. Am. Chem. Soc.*, 1969, **91**, 6540.
84. E. Schori, J. Jagur-Grodzinski, L. Luz and M. Shporer, *J. Am. Chem. Soc.*, 1971, **93**, 7133.

85. P. B. Chock, *Proc. Natl. Acad. Sci. USA*, 1972, **69**, 1939.
86. G. W. Liesegang and E. M. Eyring, in 'Synthetic Multidentate Macrocyclic Ligands', ed. R. M. Izatt and J. J. Christensen, Academic, New York, 1978, chap. 5.
87. B. F. Gisin, R. B. Merrifield and D. C. Tosteson, *J. Am. Chem. Soc.*, 1969, **91**, 2691; D. C. Tosteson, *Fed. Proc.*, 1968, **27**, 1269.
88. C. J. Liotta, in 'Synthetic Multidentate Macrocyclic Compounds', ed. R. M. Izatt and J. J. Christensen, Academic, New York, 1978, chap. 3
89. W. P. Weber and G. W. Gokel, in 'Phase Transfer Catalysis in Organic Chemsitry', Springer, Berlin, 1977.
90. R. M. Izatt, D. V. Dearden, P. R. Brown, J. S. Bradshaw, J. D. Lamb and J. J. Christensen, *J. Am. Chem. Soc.*, 1983, **105**, 1785.
91. J. D. Lamb and J. J. Christensen, in 'Progress in Macrocyclic Chemistry', ed. R. M. Izatt and J. J. Christensen, Wiley, New York, 1981, vol. 2.
92. G. W. Gokel, D. M. Dishong and C. J. Diamond, *J. Chem. Soc., Chem. Commun.*, 1980, 1053.
93. A. Karfer, D. A. Gustowski, L. Echegoyen, V. J. Gatto, R. A. Schultz, T. P. Cleary, C. R. Morgan, D. M. Goli, A. M. Ross and G. W. Gokcl, *J. Am. Chem. Soc.*, 1985 , **107**, 1958.
94. B. A. Boyce, A. Carroy, J.-M. Lehn and D. Parker, *J. Chem. Soc., Chem. Commun.*, 1984, 1546.
95. J. Comarmond, B. Dietrich, J.-M. Lehn and R. Louis, *J. Chem. Soc., Chem. Commun.*, 1985, 75.
96. E. M. Hyde, B. L. Shaw and I. Shepherd, *J. Chem. Soc., Dalton Trans.*, 1978, 1696.
97. B. L. Shaw and I. Shepherd, *J. Chem. Soc., Dalton Trans.*, 1979, 1634.
98. K. J. Odell, E. M. Hyde, B. L. Shaw and I. Shepherd, *J. Organomet. Chem.*, 1979, **168**, 103.
99. S. J. McLain, *J. Am. Chem. Soc.*, 1983, **105**, 6355.
100. S. Shinkai, T. Ogawa, Y. Kusano, O. Manabe, K. Kikukawa, T. Goto and T. Matsuda, *J. Am. Chem. Soc.*, 1982, **104**, 1960.
101. S. Shinkai, T. Nakaji, T. Ogawa, K. Shigematsu and O. Manabe, *J. Am. Chem. Soc.*, 1981, **103**, 111.
102. J. P. Dix and F. Vogtle, *Angew. Chem., Int. Ed. Engl.*, 1978, 857.
103. N. B. Tucker and E. E. Reid, *J. Am. Chem. Soc.*, 1933, **55**, 775.
104. T. E. Jones, D. B. Rorabacher and L. A. Ochrymowycz, *J. Am. Chem. Soc.*, 1975, **97**, 7485.
105. T. E. Jones, L. L. Zimmer, L. L. Diaddario, D. B. Rorabacher and L. A. Ochrymowycz, *J. Am. Chem. Soc.*, 1975, **97**, 7163.
106. R. E. DeSimone and M. D. Glick, *J. Am. Chem. Soc.*, 1976, **98**, 762.
107. P. H. Davis, K. L. White and R. L. Bedford, *Inorg. Chem.*, 1975, **14**, 1753.
108. R. E. DeSimone and M. D. Glick, *J. Am. Chem. Soc.*, 1974, **97**, 942.
109. N. W. Alcock, H. Heron and P. Moore, *J. Chem. Soc., Chem. Commun.*, 1976, 886.
110. R. S. Glass, G. S. Wilson and W. N. Setzer, *J. Am. Chem. Soc.*, 1980, **102**, 5068.
111. J. A. R. Hartman, R. E. Wolf, P. M. Foxman and S. R. Cooper, *J. Am. Chem. Soc.*, 1983, **105**, 131.
112. T. Yoshida, T. Ueda, T. Adachi, K. Yamamota and T. Higuchi, *J. Chem. Soc., Chem. Commun.*, 1985, 1137.
113. D. H. Cook, D. E. Fenton, M. G. B. Drew, A. Rodgers, M. McCanna and S. M. Nelson, *J. Chem. Soc., Dalton Trans.*, 1979, 414.
114. H. Adams, N. A. Bailey, D. E. Fenton, S. Moss and G. Jones, *Inorg. Chim. Acta*, 1984, **83**, L79.
115. F. V. Acholla, F. Takusagawa and K. B. Mertes, *J. Am. Chem. Soc.*, 1985, **107**, 6902.
116. P. K. Coughlin, J. C. Dewan, S. J. Lippard, E.-I. We atanabe and J.-M. Lehn, *J. Am. Chem. Soc.*, 1979, **101**, 265.
117. A. E. Martin and S. J. Lippard, *J. Am. Chem. Soc.*, 1984, **106**, 2579.
118. A. Bianchi, S. Mangani, M. Micheloni, V. Nanini, P. Orioli, P. Paoletti and B. Seghi, *Inorg. Chem.*, 1984, **24**, 1182.
119. G. R. Newkome and H.-W. Lee, *J. Am. Chem. Soc.*, 1983, **105**, 5956.
120. J. L. Toner, *Tetrahedron Lett.*, 1983, **24**, 2707.
121. L. Horner, H. Kunz and P. Walach, *Phosphorus Relat. Group V Elem.*, 1975, **6**, 63.
122. T. A. DelDonno and W. Rosen, *Inorg. Chem.*, 1978, **17**, 3714.
123. E. P. Kyba, C. W. Hudson, M. J. McPhaul and A. M. John, *J. Am. Chem. Soc.*, 1977, **99**, 8053.
124. R. E. Davis, C. W. Hudson and E. P. Kyba, *J. Am. Chem. Soc.*, 1978, **100**, 3642.
125. E. P. Kyba and S.-T. Liu, *Inorg. Chem.*, 1985, **24**, 1613.
126. R. Bartsch, S. Hietkamp, H. Peters and O. Stelzer, *Inorg. Chem.*, 1984, **23**, 3304.
127. D. J. Cram, T. Kaneda, R. C. Helgeson, S. B. Brown, C. B. Knobler, E. Maverick and K. N. Trueblood, *J. Am. Chem. Soc.*, 1985, **107**, 3645.
128. D. J. Cram and G. M. Lein, *J. Am. Chem. Soc.*, 1985, **107**, 3657.
129. K. Koenig, G. M. Lein, P. Stuckler, T. Kaneda and D. J. Cram, *J. Am. Chem. Soc.*, 1979, **101**, 3553.
130. G. M. Lein and D. J. Cram, *J. Chem. Soc., Chem. Commun.*, 1982, 301.
131. P. Kollman, G. Wipff and U. C. Singh, *J. Am. Chem. Soc.*, 1985, **107**, 2212.
132. G. M. Lein and D. J. Cram, *J. Am. Chem. Soc.*, 1985, **107**, 448.
133. C. D. Gutsche and R. Muthukrishnam, *J. Org. Chem.*, 1978, **43**, 4905.
134. H. Kammerer and G. Happel, *Makromol. Chem.*, 1980, **181**, 2049.
135. C. D. Gutsche, *Acc. Chem. Res.*, 1983, **16**, 161; *Top. Curr. Chem.*, 1984, **123**, 1; and in 'Host Guest Complex Chemistry, Macrocycles', ed. F. Vogtle and E. Weber, Springer, Berlin, 1985, pp. 375–421.
136. V. Bohmer, H. Goldmann and W. Vogt, *J. Chem. Soc., Chem. Commun.*, 1985, 667.
137. R. M. Izatt, J. D. Lamb, R. T. Hawkins, P. R. Brown, S. R. Izatt and J. J. Christensen, *J. Am. Chem. Soc.*, 1983, **105**, 1782.
138. E. Simons and C. H. Park, *J. Am. Chem. Soc.*, 1968, **90**, 2428.
139. B. Metz, D. Moras and R. Weiss, *J. Chem. Soc., Perkin Trans.*, 1976, **2**, 423.
140. J.-M. Lehn, in 'Coordination Chemistry of Macrocyclic Compounds', ed. G. A. Melson, Plenum, New York, 1979, pp. 537–602.
141. B. Dietrich, J.-M. Lehn, J.-P. Sauvage and J. Blanzat, *Tetrahedron*, 1973, **29**, 1629.
142. B. Metz, D. Moras and R. Weiss, *Chem. Commun.*, 1971, 445.
143. E. Kauffmann, J.-M. Lehn and J.-P. Sauvage, *Helv. Chim. Acta*, 1976, **59**, 1099.
144. B. Dietrich, J.-M. Lehn and J.-P. Sauvage, *J. Chem. Soc., Chem. Commun.*, 1973, 15.

145. F. Arnaud-Neu, B. Spiess and M.-J. Schwing-Weill, *Helv. Chim. Acta*, 1977, **60**, 2633.
146. B. Spiess, F. Arnaud-Neu and M.-J. Schwing-Weill, *Helv. Chim. Acta*, 1979, **62**, 1531.
147. G. Anderegg, *Helv. Chim. Acta*, 1981, **64**, 1790.
148. B. Spiess, F. Arnaud-Neu and M.-J. Schwing-Weill, *Helv. Chim. Acta*, 1980, **63**, 2287.
149. B. Dietrich, J.-M. Lehn and J.-P. Sauvage, *Tetrahedron*, 1973, **29**, 1647.
150. J. P. Gisselbrecht, F. Peter and M. Gross, *J. Electroanal. Chem.*, 1979, **96**, 81.
151. E. L. Yee, O. A. Gansow and M. J. Weaver, *J. Am. Chem. Soc.*, 1980, **102**, 2278.
152. F. A. Hart, M. B. Hursthouse, K. M. A. Malik and S. Moorhouse, *J. Chem. Soc., Chem. Commun.*, 1978, 549.
153. M. Ciampolini, P. Dapporto and N. Nardi, *J. Chem. Soc., Chem. Commun.*, 1978, 788.
154. B. Spiess, F. Arnaud-Neu and M.-J. Schwing-Weill, *Inorg. Nucl. Chem. Lett.*, 1979, **15**, 13.
155. O. A. Gansow, D. J. Pruett and K. B. Triplett, *J. Am. Chem. Soc.*, 1979, **101**, 4408.
156. M. T. Lok, F. J. Tehan and J. L. Dye, *J. Phys. Chem.*, 1972, **76**, 2975.
157. B. Kaempf, S. Raynal, A. Collet, F. Schue, S. Bodeau and J.-M. Lehn, *Angew. Chem., Int. Ed. Engl.*, 1974, **13**, 611.
158. J. Lacoste, F. Schue, S. Bywater and B. Kaempf, *Polymer Lett.*, 1976, **14**, 201.
159. J. L. Dye, J. M. Ceraso, M. T. Lok, B. L. Barnett and F. J. Tehan, *J. Am. Chem. Soc.*, 1974, **96**, 608.
160. J. L. Dye, C. W. Andrews and S. E. Mathews, *J. Phys. Chem.*, 1975, **79**, 3065.
161. F. J. Tehan, B. L. Barnett and J. L. Dye, *J. Am. Chem. Soc.*, 1974, **96**, 7203.
162. J. D. Corbett and P. A. Edwards, *J. Chem. Soc., Chem. Commun.*, 1975, 984.
163. J. D. Corbett, D. G. Adolphson, D. J. Merryman, P. A. Edwards and F. J. Armates, *J. Am. Chem. Soc.*, 1975, **97**, 6267.
164. J.-M. Lehn and F. Montavon, *Helv. Chim. Acta*, 1976, **59**, 1566; 1978, **61**, 67.
165. J. C. Rodriguez-Ubis, B. Alpha, D. Plancherel and J.-M. Lehn, *Helv. Chim. Acta*, 1984, **67**, 2264.
166. J.-M. Lehn, S. H. Pine, E. I. Watanabe and A. K. Willard, *J. Am. Chem. Soc.*, 1977, **99**, 6766.
167. R. J. Motekaitis, A. E. Martell, J.-M. Lehn and E. I. Watanabe, *Inorg. Chem.*, 1982, **21**, 4253.
168. R. J. Motekaitis, A. E. Martell, B. Dietrich and J.-M. Lehn, *Inorg. Chem.*, 1984, **23**, 1588.
169. G. McLendon and A. E. Martell, *J. Coord. Chem.*, 1975, **4**, 235.
170. M. Zehnder, U. Therwalt and S. Fallab, *Helv. Chim. Acta*, 1976, **59**, 2290.
171. J.-M. Lehn, C. T. Wu and P. Plumeré, *Acta Chim. Sin.*, 1983, 57.
172. J. Comarmond, Ph. D. Thesis, Université Louis Pasteur, Strasbourg, France, 1981.
173. J. E. Bulkowski, P. L. Burk, M. F. Ludmann and J. A. Osborn, *J. Chem. Soc., Chem. Commun.*, 1977, 498.
174. N. Herron, J. H. Cameron, G. L. Neer and D. H. Busch, *J. Am. Chem. Soc.*, 1983, **105**, 298.
175. N. Herron, L. L. Zimmer, J. J. Grzybowski, D. J. Olszansk, S. C. Jackels, R. W. Callahan, J. H. Cameron, G. G. Christoph and D. H. Busch, *J. Am. Chem. Soc.*, 1983, **105**, 6585.
176. N. Herron, M. Y. Chavan and D. H. Busch, *J. Chem. Soc., Dalton Trans.*, 1984, 1491.
177. K. A. Goldsby, T. J. Meade, M. Kojima and D. H. Busch, *Inorg. Chem.*, 1985, **24**, 2588.
178. I. I. Creaser, J. MacB. Harrowfield, A. J. Herlt, A. M. Sargeson, J. Springborg, R. J. Geue and M. R. Snow, *J. Am. Chem. Soc.*, 1977, **99**, 3181.
179. A. Bakac, J. H. Espensen, I. I. Creaser and A. M. Sargeson, *J. Am. Chem. Soc.*, 1983, **105**, 7624.
180. I. I. Creaser, L. R. Gahan, R. J. Geue, A. Launikonis, P. A. Lay, J. D. Lydon, M. G. Mccarthy, A. W.-H. Mau, A. M. Sargeson and W. A. F. Sasse, *Inorg. Chem.*, 1985, **24**, 2671.
181. P. A. Lay, A. W.-H. Mau, W. H. F. Sasse, I. I. Creaser, L. R. Gahan and A. M. Sargeson, *Inorg. Chem.*, 1983, **22**, 2347.
182. F. Pina, M. Ciano, L. Moggi and V. Balzani, *Inorg. Chem.*, 1985, **24**, 844.
183. F. Pina, Q. G. Mulazzani, M. Venturi, M. Ciano and V. Balzani, *Inorg. Chem.*, 1985, **24**, 848.
184. A. Hammershoi and A. M. Sargeson, *Inorg. Chem.*, 1983, **22**, 3554.
185. L. H. Gahan, G. A. Lawrance and A. M. Sargeson, *Inorg. Chem.*, 1984, **23**, 4369.
186. C. O. Dietrich-Buchecker, J.-M. Kern and J.-P. Sauvage, *J. Chem. Soc., Chem. Commun.*, 1985, 760.
187. A. M. Albrecht-Gary, Z. Saad, C. O. Dietrich-Buchecker and J.-P. Sauvage, *J. Am. Chem. Soc.*, 1985, **107**, 3205.
188. I. O. Sutherland, in 'Cyclophanes', ed. P. M. Keehn and S. M. Rosenfeld, Academic, New York, 1983, vol. II, chap. 12.
189. F. Kotzyba-Hibert, J.-M. Lehn and P. Vierling, *Tetrahedron Lett.*, 1980, 941.
190. J.-M. Lehn and M. E. Stubbs, *J. Am. Chem. Soc.*, 1974, **96**, 4011.
191. R. Wiest and R. Weiss, *J. Chem. Soc., Chem. Commun.*, 1973, 678.
192. J.-M. Lehn and J. Simon, *Helv. Chim. Acta*, 1977, **60**, 141.
193. M. Mellinger, J. Fischer and R. Weiss, *Angew. Chem., Int. Ed. Engl.*, 1973, **9**, 771.
194. J. Fischer, M. Mellinger and R. Weiss, *Inorg. Chim. Acta*, 1977, **21**, 259.
195. A. H. Alberts, R. Annunziata and J.-M. Lehn, *J. Am. Chem. Soc.*, 1977, **99**, 8502.
196. R. Louis, Y. Agus and R. Weiss, *J. Am. Chem. Soc.*, 1978, **100**, 3604.
197. J. P. Gisselbrecht, M. Gross, A. H. Alberts, and J.-M. Lehn, *Inorg. Chem.*, 1980, **19**, 1386.
198. O. Kahn, I. Morgenstern-Baderau, J. P. Audiere, J.-M. Lehn and S. A. Sullivan, *J. Am. Chem. Soc.*, 1980, **102**, 5935.
199. J. P. Gisselbrecht and M. Gross, *Adv. Chem. Ser.*, 1982, **201**, 109.
200. E. Graf and J.-M. Lehn, *Helv. Chim. Acta*, 1981, **64**, 1040.
201. J.-M. Lehn, *Acc. Chem. Res.*, 1978, **11**, 49.
202. A. D. Hamilton, J.-M. Lehn and J. L. Sessler, *J. Chem. Soc., Chem. Commun.*, 1984, 311.
203. M. Newcomb and D. J. Cram, *J. Am. Chem. Soc.*, 1975, **97**, 1257.
204. K. E. Koenig, R. C. Helgeson and D. J. Cram, *J. Am. Chem. Soc.*, 1978, **98**, 4018.
205. E. P. Kyba, K. Koga, R. Sousa, M. G. Siegel and D. J. Cram, *J. Am. Chem. Soc.*, 1973, **95**, 1692.
206. E. Weber and F. Vogtle, *Chem. Ber.*, 1976, **109**, 1803.
207. J. M. Timko, R. C. Helgeson, M. Newcomb, G. W. Gokel and D. J. Cram, *J. Am. Chem. Soc.*, 1974, **96**, 7097.
208. F. Vogtle and E. Weber, *Angew. Chem., Int. Ed. Engl.*, 1974, **13**, 149.
209. E. Buhleier, K. Frensch, F. Luppertz and F. Vogtle, *Liebigs Ann. Chem.*, 1978, 1586.
210. G. R. Newkome and D. C. Hager, *J. Am. Chem. Soc.*, 1978, **100**, 5567.
211. C. J. Pedersen, *J. Org. Chem.*, 1971, **36**, 254.

212. J.-M. Lehn and P. Vierling, *Tetrahedron Lett.*, 1980, **21**, 1323.
213. G. W. Gokel and B. J. Garcia, *Tetrahedron Lett.*, 1977, **18**, 317.
214. M. Ciampolini, P. Dapporto, A. Dei and N. Nardi, *Inorg. Chem.*, 1982, **21**, 489.
215. M. Ciampolini, P. Dapporto, N. Nardi and F. Zanobini, *Inorg. Chim. Acta*, 1980, **45**, L239.
216. E. P. Kyba and S.-S. P. Chow, *J. Am. Chem. Soc.*, 1980, **102**, 7012.
217. K. Frensch, G. Oepen and F. Vogtle, *Liebigs Ann. Chem.*, 1979, **85**, 6.
218. J. S. Bradshaw, C. T. Bishop, S. F. Nielsen, R. E. Asay, D. R. K. Masihdas, E. D. Flanders, L. D. Hansen, R. M. Izatt and J. J. Christensen, *J. Chem. Soc., Perkin Trans. 1*, 1976, 2505.
219. R. J. M. Nolte and D. J. Cram, *J. Am. Chem. Soc.*, 1984, **106**, 1416.
220. K. B. Yatsimirskii, E. N. Korol, V. G. Lolovatyi, T. N. Kudrya and G. G. Talanova, *Dokl. Akad. Nauk SSSR*, 1979, **244**, 1359.
221. A. V. Bogatsky, N. G. Lukyanenko and T. I. Kirichenko, *Tetrahedron Lett.*, 1980, **21**, 313.
222. J.-P. Behr, J.-M. Lehn and P. Vierling, *J. Chem. Soc., Chem. Commun.*, 1976, 621; *Helv. Chim. Acta*, 1982, **65**, 1853.
223. D. J. Cram, in 'Application of Biochemical Systems in Organic Chemistry', ed. J. B. Jones, C. J. Sih and C. J. Perlmann, Wiley, New York, 1976, part II, Techniques of Chemistry, vol. X.
224. D. Cram and J. M. Cram, *Science*, 1974, **183**, 803.
225. R. M. Izatt, B. E. Rossiter, J. J. Christensen and B. L. Haymore, *Science*, 1978, **199**, 944.
226. M. Dhaenens, L. Lacombe, J.-M. Lehn and J.-P. Vigneron, *J. Chem. Soc., Chem. Commun.*, 1984, 1097.
227. S. Sasaki, M. Shionoya and K. Koga, *J. Am. Chem. Soc.*, 1985, **107**, 3371.
228. K. Madan and D. J. Cram, *J. Chem. Soc., Chem. Commun.*, 1975, 427.
229. J.-M. Lehn, P. Vierling and R. C. Hayward, *J. Chem. Soc., Chem. Comm.*, 1979, 296.
230. J. W. H. M. Uiterwijk, S. Harkema, J. Geevers and D. N. Reinhoudt, *J. Chem. Soc., Chem. Commun.*, 1982, 200.
231. J.-M. Lehn and C. Sirlin, *J. Chem. Soc., Chem. Commun.*, 1978, 949.
232. J. G. De Vries and R. M. Kellogg, *J. Am. Chem. Soc.*, 1979, **101**, 2759.
233. M. Zinic, B. Bosnic-Kasner and D. Kolbah, *Tetrahedron Lett.*, 1980, **21**, 1365.
234. D. J. Cram and K. N. Trueblood, *Top. Curr. Chem.*, 1981, **98**, 43.
235. L. R. Sousa, D. H. Hoffman, L. Kaplan and D. J. Cram, *J. Am. Chem. Soc.*, 1974, **96**, 7100.
236. G. W. Gokel, J. M. Timko and D. J. Cram, *J. Chem. Soc., Chem. Commun.*, 1975, 394.
237. L. R. Sousa, G. S. Y. Hoffman and D. J. Cram, *J. Am. Chem. Soc.*, 1978, **100**, 4569.
238. I. Goldberg, *J. Am. Chem. Soc.*, 1977, **99**, 6049.
239. B. Metz, J. M. Rosalky and R. Weiss, *J. Chem. Soc., Chem. Commun.*, 1976, 533.
240. J. P. Kintzinger, F. Kotzyba-Hibert, J.-M. Lehn, A. Pagelot and K. Saigo, *J. Chem. Soc., Chem. Commun.*, 1981, 833.
241. C. Pascard, C. Riche, M. Cesario, F. Kotzyba-Hibert and J.-M. Lehn, *J. Chem. Soc., Chem. Commun.*, 1982, 557.
242. B. Dietrich, M. W. Hosseini, J.-M. Lehn and R. B. Sessions, *J. Am. Chem. Soc.*, 1981, **103**, 1282.
243. J.-M. Lehn, E. Sonveaux and A. K. Willard, *J. Am. Chem. Soc.*, 1978, **100**, 4914.
244. F. P. Schmidtchen, *Chem. Ber.*, 1981, **114**, 597.
245. E. Kimura, A. Sakonaka, T. Yatsunami and M. Kodama, *J. Am. Chem. Soc.*, 1981, **103**, 3041.
246. F. Peter, M. Gross, M. W. Hosseini and J.-M. Lehn, *J. Electroanalyt. Chem.*, 1983, **144**, 279.
247. M. F. Manfrin, N. Sabbatini, L. Moggi, V. Balzani, M. W. Hosseini and J.-M. Lehn, *J. Chem. Soc., Chem. Commun.*, 1984, 555.
248. J.-L. Pierre and P. Baret, *Bull. Soc. Chim. Fr., Part 2*, 1983, 367.
249. E. Kimura, M. Kodama and T. Yatsunami, *J. Am. Chem. Soc.*, 1982, **104**, 3182.
250. M. W. Hosseini, J.-M. Lehn and M. P. Mertes, *Helv. Chim. Acta*, 1983, **66**, 2454.
251. M. W. Hosseini and J.-M. Lehn, *J. Chem. Soc., Chem. Commun.*, 1985, 1155.
252. P. G. Yohannes, K. E. Plute, M. P. Mertes and K. B. Mertes, *Inorg. Chem.*, 1987, **26**, 1751.
253. B. Dietrich, T. M. Fyles, J.-M. Lehn, L. G. Pease and D. L. Fyles, *J. Chem. Soc., Chem. Commun.*, 1978, 934.
254. I. Tabushi, I. Sasaki and H. Kuroda, *J. Am. Chem. Soc.*, 1976, **98**, 5727.
255. I. Tabushi, Y. Kuroda and Y. Kimura, *Tetrahedron Lett.*, 1976, 3327.
256. R. A. Bell, G. G. Christoph, F. R. Fronczek and R. E. Marsh, *Science*, 1975, **190**, 151.
257. B. Dietrich, J. Guilhem, J.-M. Lehn, C. Pascard and E. Sonveaux, *Helv. Chim. Acta*, 1984, **67**, 91.
258. F. Vogtle and E. Weber, *Angew. Chem., Int. Ed. Engl.*, 1974, 814.
259. J. P. Kintzinger, J.-M. Lehn, E. Kauffmann, J. L. Dye and A. I. Popov, *J. Am. Chem. Soc.*, 1983, **105**, 7549.
260. I. Goldberg, *Acta Crystallogr., Sect. B*, 1978, **34**, 3387.
261. E. Maverick, P. Seiler, W. B. Schweizer and J. D. Dunitz, *Acta Crystallogr., Sect. B*, 1980, **36**, 615.
262. J. D. Dunitz and P. Seiler, *Acta Crystallogr., Sect. B*, 1974, **30**, 2739.
263. E. Weber and F. Vogtle, *Angew. Chem., Int. Ed. Engl.*, 1980, **19**, 1030.
264. C. J. Pedersen, *J. Org. Chem.*, 1971, **36**, 1690.
265. F. Vogtle and W. M. Muller, *Naturwissenschaften*, 1980, **67**, 255.
266. A. Kochel, J. Kopf, J. Oehler and G. Rudolph, *J. Chem. Soc., Chem. Commun.*, 1978, 595.
267. A. Elbasyouny, H. J. Brugge, K. von Deuten, M. Dickel, A. Knochel, K. U. Koch, J. Kopf, D. Melzer and G. Rudolph, *J. Am. Chem. Soc.*, 1983, 6568.
268. B. Odell and G. Earlam, *J. Chem. Soc., Chem. Commun.*, 1985, 359.
269. F. Vogtle, W. M. Muller and E. Weber, *Chem. Ber.*, 1980, 113, 1130.
270. J. R. Moran, S. Karbach and D. J. Cram, *J. Am. Chem. Soc.*, 1982, **104**, 5826.
271. D. J. Cram, K. D. Stewart, I. Goldberg and K. N. Trueblood, *J. Am. Chem. Soc.*, 1985, **107**, 2574.
272. D. J. Cram, S. Karbach, Y. H. Kim, L. Baczynskyj and G. W. Kalleymeyn, *J. Am. Chem. Soc.*, 1985, **107**, 2575.
273. K. Odashima, A. Itai, Y. Iitaka and K. Koga, *J. Am. Chem. Soc.*, 1980, **102**, 2504.
274. Y. Murakami, J. Kikuchi and Hiroaki Tenma, *J. Chem. Soc., Chem. Commun.*, 1985, 753.
275. J. Gabard and A. Collet, *J. Chem. Soc., Chem. Commun.*, 1981, 1137.
276. J. Canceill, L. Lacombe, and A. Collet, *C. R. Hebd. Seances Acad. Sci., Ser. B*, 1984, **298**, 39.

22

Naturally Occurring Ligands

STUART H. LAURIE
Leicester Polytechnic, UK

22.1 GENERAL CONSIDERATIONS

The naturally occurring ligands to be considered in this chapter are the chelating ligands from biological sources, thus geochemical species (covered in Chapter 64) and the so-called 'biological minerals' are excluded.[1] It is now well established that a number of metal ions are essential for all life forms. Typical compositions of these elements in different systems are shown in Table 1. To be more specific these metal ions are essential to the proper functioning of all living cells. Each organism in fact can be considered as a highly tuned multimetal multiligand system.

The data in Table 1 show that the transition metal elements and zinc are required in minute amounts compared to the other essential elements, hence they are frequently referred to as the *trace elements* or *micronutrients*. Considering the many functions they perform (see Chapter 62) their amounts are quite disproportionate to their roles. It is also clear that it is essential that these minute amounts be delivered to the right compartments, that they be present in the unique forms (oxidation state, coordination geometry) required to carry out their selective functions, and that the organism is able to control the amounts of each of these elements and distinguish between them. All of these requirements can be, and are, achieved by selective *chelation*. Indeed, without chelation as a control any free (*i.e.* hydrated) transition metal ions would indiscriminately bind to a range of biological molecules with concomitant physiological disorders. A good example of this occurring is in Wilson's disease (see Chapter 62) where the excess copper that accumulates becomes lethal. This has been referred to as 'miscoordination'.[2]

Table 1 Typical Concentrations of Metals in Biological
Systems ($g\,kg^{-1}$)

Metal	Man[a]	Soils[b]	Plants[b]
Sodium	1.0	6.3	1.2
Potassium	3.6	14	14
Magnesium	0.6	5	3.2
Calcium	24.3	13.7	18.0
Manganese	<0.01	0.9	0.6
Iron	0.06–0.09	38	0.14
Cobalt	<0.01	<0.01	<0.01
Copper	<0.01	0.02	0.014
Zinc	0.015–0.03	0.05	0.16
Molybdenum	<0.01	<0.01	<0.01

[a] Based on the 'average 70 kg man', data from various sources.
[b] On a dry weight basis, data from ref. 16.

  The possible factors involved in the biological selectivity towards metal ions have been considered by Frausto da Silva and Williams[3] and by Kustin *et al.*[4] In terms of thermodynamic selectivity a useful formalism for the uptake of any metal ion from a multimetal system is the quotient $K_m C_m$, where K_m is a *relative* stability constant and C_m is the concentration of the metal ion. However, as these authors point out,[3] a combination of both thermodynamic and kinetic properties must be considered. An appreciation of kinetic factors is often absent in this field, but must be of prime consideration in chelate exchange reactions and in the final irreversible step of metal ion insertion to form the metalloenzymes.

  The metabolic cycle of an element can be represented by the simple Scheme 1. As will be shown each of these phases involves specific chelation. Passage between the phases is also under the control of chelates. On a more extensive scale an ecological cycle, as shown in Scheme 2, is of significance to both essential and toxic elements. Of particular importance is the interface between any two phases, the interface requiring water for effective transport of elements. Here again, chelation or 'mischelation' can have a serious impact on the ecological cycle of an element.

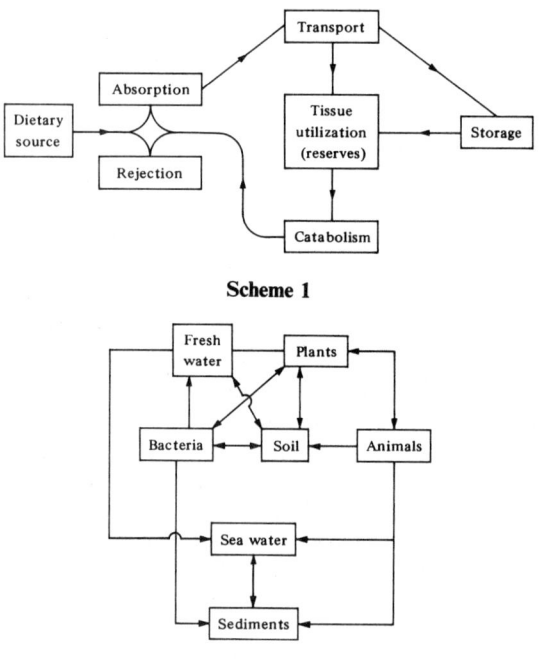

Scheme 1

Scheme 2

  Chelation of the main group IA and IIA metal ions in nature is not extensive, an observation attributable to their relatively weak coordination properties and their large hydration energies. However, when it comes to selecting between similar ions, *e.g.* Na^+ and K^+, then selective chelation appears to offer the most efficient route.

  The evolutionary development of natural chelates has not been extensivelyl considered but could be an instructive field of study. Chelation could have played a role in early protein nucleation[5] and in the development of stereoselectivity in biological molecules.[6] That complex ligands were formed early on in the stages of biological evolution is evident from the discovery of metal–porphyrin complexes (see Section 22.8) in the biolithosphere, *i.e.* the fossil fuels formed from sediments of organic debris. Interestingly the metal ions found in these porphyrins include gallium(III), oxovanadium(IV) and nickel(II), which are not found as such in any known living organism. The concentration of oxovanadium(IV) in some oil shales can reach 1000 p.p.m. The finding of Ga, Mn and Fe porphyrins in lignites has been attributed to chlorophylls originally present in the organic debris but their origin could also have been haem groups from microorganisms.[7] The structures of some of these porphyrins have recently been reported.[8,9] The simulation of the prebiotic synthesis of amino acids was reported some 20 years ago, in a now classical set of experiments, by Urey and Miller. There have been many reports since then on a similar theme; some of the more recent studies have involved metal ions and their complexes, *e.g.* irradiation of mixtures of aqueous ammonia, ammonium chloride and titanium dioxide,[10] and synthesis using metal carboxylates,[11] and mixtures of sodium cyanide and copper(II) sulfate.[12]

The remainder of this chapter will be devoted to the consideration of groups of chelates starting with those involved in the first link of the food chain — the transference of micronutrients from soils to plants.

22.2 CARBOXYLIC AND PHENOLIC ACIDS

22.2.1 Fulvic and Humic Acids

Encapsulated within the pores of soils are aqueous solutions, the so-called *soil solutions*. Radiotracer experiments have shown that there is a rapid, dynamic equilibrium between metal ions in the soil phase and the soil solutions. Hence, the soil solution is regarded as the transport medium for moving the essential nutrients from soils to plants. The soil solutions are highly variable in composition and properties, *e.g.* the pH can vary over the range 2–11. Alkaline soils are notorious for causing metal ion deficiency in plants, principally because many of the micronutrients form insoluble hydroxides and so are biologically unavailable.

The binding of metal ions by soil organic matter is of particular importance in controlling the bioavailability of these metal ions as well as influencing the chemical and physical properties of soils. Extraction methods suggest the binding affinity by soils of metal ions follows the order $Fe^{III} > Al^{III} > Cu^{II} > Mn^{II} > Zn^{II}$. The stronger binding of Cu^{II} relative to Zn^{II} explains why zinc deficiency is more prevalent in high pH soils, the stronger binding of Cu^{II} reducing the amount of insoluble copper hydroxide. The soil organic matter is classified as either humic or non-humic; the latter includes amino acids and carbohydrates, the former comprises fulvic acids, humic acids and humin. Both of the acids are soluble in alkaline solutions and are extracted this way; on acidification the humic acid precipitates. The humic group is also found in natural waters, and in sediments where it accumulates a large number of different cations of both the essential and toxic types. Extremely high geochemical enrichment factors arising from this humic accumulative factor have been reported for the ions VO^{2+} and UO_2^{2+}.[13,14] These high accumulative factors are in proportion to the large concentrations of acidic groups present in both fulvic and humic acids. Typical values are given in Table 2.

Table 2 Typical Compositions of Oxygen Functional Groups in Fulvic and Humic Acids (mmol g^{-1})[166]

Groups	Fulvic acid	Humic acid
Total acidity	10.3	6.7
Carboxylic	8.2	3.6
Phenolic	3.0	3.9
Alcoholic	6.1	2.6
Ketonic and quinonoid	2.7	2.9

Quantification of metal ion binding and the determination of the nature of the binding sites are extremely difficult for humic acid because of its polymeric (molar masses *ca.* 5×10^4) and heterogeneous nature. The results of such attempts are often contradictory.[15,16] Frequently a 1:1 metal: humic acid ratio is assumed for the complexes, a very surprising assumption for such large molecules with high concentrations of carboxylic acids. There is evidence for other stoichiometries, including polynuclear species. Ternary complex formation has also been observed, *e.g.* humic acid–M–phosphate (M = Fe^{III}, Al^{III}), and their equilibrium concentrations in natural water systems have been determined.[17]

The general consensus[15] is that a large fraction of monovalent ions (*e.g.* Na^+, K^+) is condensed on the surfaces of the humic acid molecules, and that most of the divalent metals, except Cu^{II}, Pb^{II} and VO^{2+}, are found as outer-sphere complexes, *i.e.* as the hydrated ions, while the more tightly bound cations are present as partially hydrated inner-sphere complexes. Chelation may not be involved. The two most predominant binding sites are the salicylic and phthalic acid groups. Nitrogen-containing sites, *e.g.* porphyrins, make only a minor contribution, while sulfur-containing sites apparently make no contribution at all. The methods used in establishing metal ion binding are given in the recent text by Stevenson.[18]

The fulvic acids are more soluble and of lower mass than the humic acids. Their greater functional group content means they are also more effective in complexing metal ions, including the toxic ones, and making them available to plants.[19] Even so their metal binding appears just as intractable to quantitative treatment, mainly because they are not well-defined compounds but consist of a collection of macromolecular fractions of different sizes. A recent novel attempt has been made to develop a computer model based on randomized positioning of functional groups and aromatic rings and the deduction of the likely types of metal-binding sites with a statistical estimate of their concentrations.[20] The conclusions are that the major binding sites are, as with humic acid, the salicylate and phthalate groups, with a significant proportion of aromatic carboxyl and phenolic OH groups not participating in forming chelating sites. The deduced metal ion affinity of $Fe^{III} > Cu^{II} > Zn^{II} > Mn^{II} > Ca^{II} > Mg^{II}$ follows that observed experimentally.

22.2.2 Di- and Tri-carboxylic Acids

A wide range of simple aliphatic acids are ubiquitous in soils and plants. Amongst these are a number of di- and tri-carboxylic acids, oxalic (**1**), malonic (**2**), malic (**3**), tartaric (**4**) and citric (**5**), which are implicated as having metal-binding roles.

$$
\begin{array}{ccccc}
\begin{array}{c} CO_2H \\ | \\ CO_2H \end{array}
&
H_2C\begin{array}{c} CO_2H \\ \\ CO_2H \end{array}
&
HOCH\begin{array}{c} CH_2CO_2H \\ \\ CH_2CO_2H \end{array}
&
\begin{array}{c} HOCHCO_2H \\ | \\ HOCHCO_2H \end{array}
&
\begin{array}{c} CH_2CO_2H \\ | \\ HOCCH_2CO_2H \\ | \\ CO_2H \end{array}
\\
(1) & (2) & (3) & (4) & (5)
\end{array}
$$

Plant roots are known to exude a fluid containing a number of amino and carboxylic acids, the amount of the exudate increasing under conditions of metal deficiency. 'Chlorotic' plants, *i.e.* those suffering from iron deficiency, have been found to contain more of the citric and malic acids than their normal green counterparts.[21] Differences in the susceptibilities of plant species to trace metal deficiencies have indeed frequently been attributed to variations in organic acid production.

Much of the work on the identification of carboxylic acid complexation has been pioneered by Tiffen. He was the first to positively identify an iron–citrate complex in plant xylem.[22] Iron–citrate complexes have since been identified in a number of plants. The complex formed in these plant fluids is anionic, and Tiffen has shown that a number of other metal ions (Cr, Cu, Ni, Mn and Zn) are also present as anionic complexes.[23] Although the neutrality of complexes may be considered a prerequisite for metal ion penetration of membranes, this has not been demonstrated with plant roots or with leaf-cell membranes. Involvement of negatively charged citrate complexes of Ni^{II} has been confirmed both for nickel uptake and for translocation in plant species.[24] Trisoxalatochromate(III) anion has been found in the leaf tissues of a plant species.[25]

For animal fluids, amino acid and protein binding appear to dominate for the trace elements. However, for Ca^{2+} and Mg^{2+} citrate binding may well occur. A computer simulation exercise on blood plasma (further details in Section 22.3) suggests approximately 5% of these metal ions are in the form of citrate complexes.[26] The same exercise indicates that any Fe^{III} present would be totally bound in anionic citrate species. Interestingly the greater zinc nutritional value of human milk as opposed to bovine milk is related to zinc being present as labile citrate complexes in the former but as a tightly bound protein complex in the latter.[27]

Unfortunately, inferences as to the nature of the citrate species in the earlier reports suffered from the then inadequate data on the metal speciation. Metal complexation by the hydroxycarboxylates is complicated because of polynuclear speciation. The elucidation of such species is still continuing; some of the more recent data are shown in Table 3. Included in the table are novel heterobinuclear species; those of composition $[MM'(Cit)_2H_{-2}]^{4-}$ are the major species formed in alkaline solutions containing comparable concentrations of citrate and the two metal ions. With Fe^{III} large polymeric ions (molar mass 2.1×10^5) form at pH 7–9.[28]

Table 3 Some Recently Determined Log Formation Constants of Metal–Citric Acid (H_3L) Complexes[a]

Metal ion	Species	Log β	Ref.
Mg^{II}	MH_2L^+	0.6	167
	MHL	1.78	167
	ML^-	3.63	167
Ca^{II}	MH_2L^+	1.0	167
	MHL	2.03	167
	ML^-	3.64	167
Cu^{II}	MHL	9.47	168
	$M_2L_2^{2-}$	14.60	168
	$M_2H_{-1}L_2^{3-}$	10.75, 10.82	168, 169
	$M_2H_{-1}L$	5.07	169
	$M_2H_{-2}L_2^{4-}$	6.00, 5.80	168, 169
Ni^{II}	ML^-	5.30	168
	MHL	8.84	168
	$M_2H_{-2}L_2^{4-}$	−4.71	168
Zn^{II}	ML^-	4.83	168
	MHL	8.43	168
	$M_2H_{-2}L_2^{4-}$	−2.94	168
Mixed species	$CuNiH_{-2}L_2^{4-}$	1.55	168
	$CuZnH_{-2}L_2^{4-}$	1.51	168
	$NiZnH_{-2}L_2^{4-}$	−2.92	168

[a] All data at 25 °C and 0.1 mol dm^{-3} ionic strength.

It has been known for some time that tolerance towards high levels of both essential and toxic metals in a local soil environment is exhibited by species and clones of plants that colonize such sites. Tolerance is generally achieved by a combination of exclusion and poor uptake and translocation. Some species can accumulate large quantities of metals in their leaves and shoots at potentially toxic levels, but without any harmful effects. These *metal-tolerant species* have been used in attempts to reclaim and recolonize metal-contaminated wastelands.[29] More recently such species have attracted the attention of inorganic chemists.[30] There is abundant evidence that the high metal levels are associated with carboxylic acids, particularly with nickel-tolerant species such as *Allysum bertolonii*. The main carboxylic acids implicated are citric, malic and malonic acids (see refs. 30 and 31 and literature cited therein). Complexation of zinc by malic and oxalic acids has been reported in the zinc-tolerant *Agrostis tenuis* and oxalic acid complexation of chromium in the chromium-accumulator species *Leptospermum scoparium*.[31]

Some doubt has been expressed over the complexation of Ni^{II} by carboxylic acids since amino acids, which are also present, form stronger chelates.[32] These authors also point out that donor N ligands are required to ensure the discrimination between Ni^{II} and Co^{II} ions that occurs. Some evidence to reinforce N donor involvement comes from recent studies on Cu^{II} accumulation in the Ni-tolerant genus *Alyssum*, in which ESR spectroscopy confirmed N donor involvement with Cu^{II} and competition between Ni^{II} and Cu^{II} for the same low molecular weight ligands.[33]

Whilst these acids are of minor importance in man, another multifunctional acid, ascorbic acid (vitamin C; **6**), has an apparent role in human iron metabolism as one of the postulated factors aiding iron uptake. Ascorbic acid reduces Fe^{III} to Fe^{II} with probable complexation of the latter. It thus converts Fe^{III} to a more soluble form. Ascorbic acid can also reduce Cu^{II} to Cu^I (equation 1), and there is accumulating evidence of a link between the metabolism of ascorbic acid and that of Cu.[34]

(6)

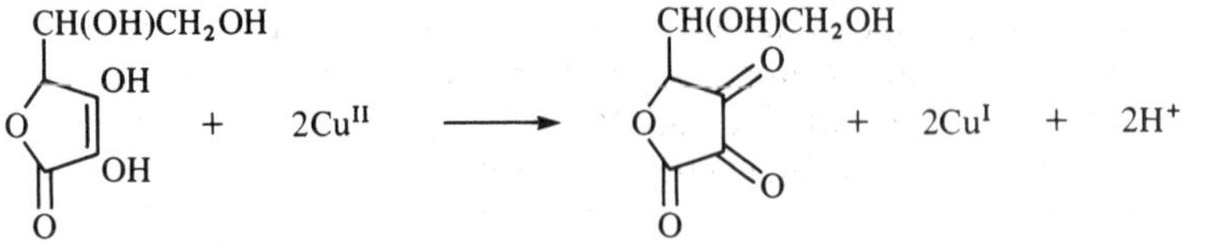

(6)

Another intriguing link between metals and organic acids has been claimed for the ubiquitous lichens, which are known to be able to dissolve minerals from rocks and to absorb the nutrients within their tissues. The basic unit of the lichen acids implicated is a chelating α-hydroxy carboxylic moiety (**7**).

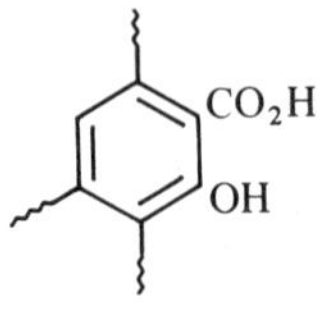

(7)

22.3 AMINO ACIDS

The data in Chapter 20.2 show that the naturally occurring amino acids are strong chelators of the transition metal ions, the presence of the N donor atom considerably enhancing the thermodynamic stabilities relative to those of the carboxylic acids.

Some of these amino acids have been found in soil solutions at concentrations of 10^{-5}–10^{-4} mol dm^{-3} and are considered second only to the carboxylic acids as stabilizers of minerals in soils.[35] The presence of amino acids in plant exudates[36] also raises the possibility of their being involved in the absorption of the transition metals. It is not surprising therefore that evidence has been sought for amino acid complexation of the transition metals in the metal-tolerant species. Increased amino acid levels are sometimes found in these but evidence of any chelation of physiological significance is more difficult to establish. Thus, in the copper-accumulator species *Becium homblei* and *Armeria maritima* the L-proline levels are considerably higher than in other plants, but no evidence has been produced of any physiologically active Cu–proline complex.[30]

In ordinary plants the evidence is for the involvement of anionic complexes of Cu and Ni for uptake and transport. This is generally taken as resulting from citrate rather than amino acid complexation [as neutral bis(amino acidato)metal(II) complexes]. However, this conclusion ignores the presence of aspartic acid, which is often found in exudates in relatively large amounts and would form bis-anionic species. A metal binding role for aspartic acid in xylem sap was proposed in some earlier work.[37]

Related to amino acids is picolinic acid (**8**), which is present at high concentrations in dormant tissues such as plant seeds and whose involvement in the quiescent state may involve iron chelation. Also mugineic acid (**9**) possesses strong Fe^{III} chelating properties. This amino acid derivative has been isolated from the roots of rice plants and has been found to have a strong stimulatory effect on Fe uptake. The crystal structure of a complex of Co^{III} shows that the anion of (**9**) can function as a hexadentate chelate.[38]

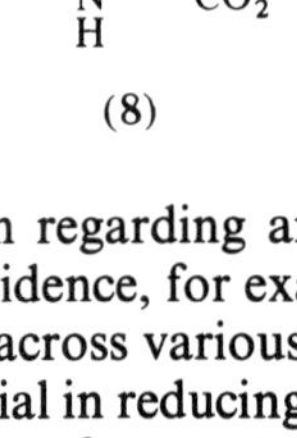

(8) 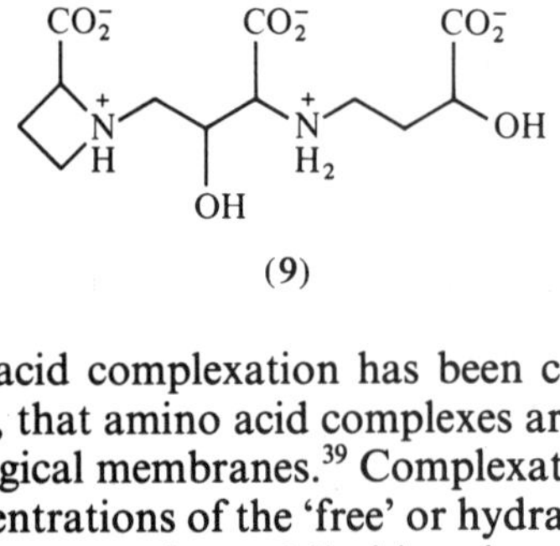   (9)

Most of the attention regarding amino acid complexation has been centred on animal fluids. There is considerable evidence, for example, that amino acid complexes are involved in the (active) transport of metal ions across various biological membranes.[39] Complexation of the trace elements is also considered essential in reducing concentrations of the 'free' or hydrated metal ions and hence preventing the formation of unwanted hydroxy species and limiting the toxicity of the metal ions.

The differences in stability constants of the amino acid complexes (Chapter 20.2, Table 2) can be used to explain the selectivity that enzymes exhibit towards a particular metal ion. Rabbit muscle phosphoglucomutase, for example, can bind Zn^{II} 10^6 times more strongly than Mg^{2+}, and yet *in vivo* it is Mg^{2+} that is selected rather than Zn, even though the concentrations of the two metal ions are similar.[40] The much stronger complexation of Zn^{II} by amino acids in fact reduces the free metal ion concentration to *ca.* 10^{-10} mol dm^{-3}, thus explaining the greater availability of Mg^{II} for incorporation into the enzyme. Similarly, many of the zinc enzymes can bind Cu^{II} equally as well as Zn^{II} but the exclusion of Cu^{II} can be related to its stronger binding by the amino acids so that its free concentration is negligible.

The complexation of copper by the amino acids in human blood plasma has, for several reasons, been the most intensely studied. Of the approximately 5 mg Cu in blood, some 90–95% is tightly and irreversibly bound in the protein caeruloplasmin; most of the remainder is reversibly bound to the protein albumin (see Chapter 20.2 and Section 22.6). Less than 1% has been reported by several groups to be present in the ultrafilterable, low molecular weight fraction which includes all the amino acids and other potential chelating agents.[41] It is frequently stated that this fraction contains amino acid-chelated Cu^{II} which is in rapid equilibrium with the Cu^{II} bound to albumin; the equilibria involved are shown in Scheme 3. Similar complexations have been reported for native Zn^{II} and for extraneous Ni^{II}.[42] Albumin is the postulated transport carrier and the amino acid complexes are of sufficiently small size to be membrane diffusible. It is generally claimed that the equilibria between the various species in Scheme 3 are rapidly established. This has no firm foundation and neither have many of the other assumptions made regarding the nature of the interactions in the equilibria and the physiological significance of the species. The problem lies in the extremely small amount of Cu involved (*ca.* 10^{-6} mol dm^{-3}), which is too low for experimental measurement. Frequently the earlier studies on which many of the conclusions are based involved raising the Cu concentration, and in so doing the very equilibria that were to be measured were altered.[43,44] Neither did these studies demonstrate that the equilibria between the proteins and amino fractions were *rapidly* established, as is often assumed.

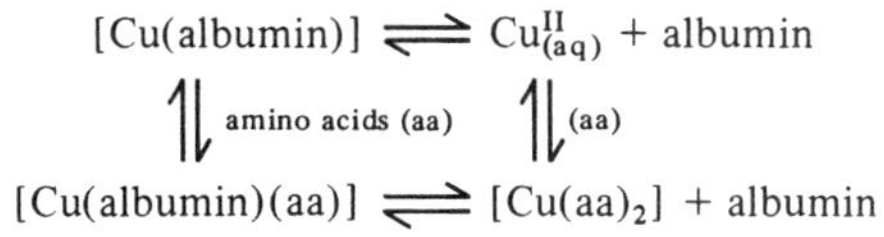

$$[Cu(albumin)] \rightleftharpoons Cu^{II}_{(aq)} + albumin$$

amino acids (aa) ⇅   ⇅ (aa)

$$[Cu(albumin)(aa)] \rightleftharpoons [Cu(aa)_2] + albumin$$

Scheme 3

Credit must be given to these clinicians[44] for their foresight and for demonstrating that histidine (10) is the major amino acid ligand for Cu^{II}. They even suggested the possibility of ternary complexes being present. The same conclusions were later reached by computer simulation of the equilibria in the amino acid pool.[41,45] The results from the more recent computer calculations (involving some 5000 equilibria!) are given in Table 4. Complexation of toxic metal ions can be treated in the same way. The computer simulation also provides valuable information on the likely influence of chelating drugs, *e.g.* in rheumatoid arthritis treatment[46] and on nutritional availability of essential elements, *e.g.* Zn.[47] Several criticisms have been made regarding the assumptions inherent in the computer simulation exercise, *e.g.* of equilibrium conditions prevailing, the stability constants of ternary species and the concentration on blood plasma. However, to date this is the only method available for obtaining speciation information and as more and better data become available so the method will prove its worth.

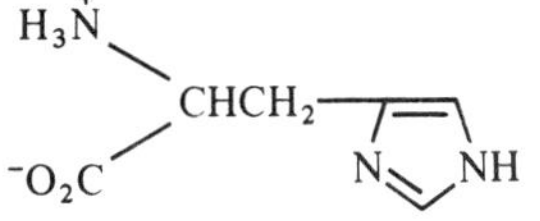

(10)

Table 4 Percentage Occurrence of the Major Complexes (Non-protein) in Blood Plasma from Computer Simulation[a,b,c]

Mn^{II}		Fe^{III}		Cu^{II}		Zn^{II}	
MnHCO$_3^+$	24	Fe(Cit)(OH)$^-$	99	Cu(HisO)(CystO)$^-$	21	Zn(CysO)$_2^{2-}$	40
MnCit$^-$	10	Fe(Cit)(Sal)$^{3-}$	<1	Cu(His)(CystO)	17	Zn(CysO)(HisO)$^-$	24
				Cu(HisO)$_2$	11	Zn(HisO)$^+$	4

[a] Cit, citrate; Sal, salicylate; His, histidine; Cys, cysteine; Cyst, cystine.
[b] Data, except for Zn^{II}, from ref. 41.
[c] Data for Zn^{II} from ref. 170.

One other serious criticism regarding the data on Cu speciation is the neglect of the cysteine present in blood plasma. Cu^{II} and cysteine undergo a facile redox reaction (Chapter 20.2). Since the reaction is irreversible, no quantitative thermodynamic quotient is available for use in the computer calculations. Another assumption often made is that the overwhelming concentration of other amino acids may prevent cysteine coordination and, as a result, stabilize the Cu^{II} state. Recent studies show that this assumption is totally unjustified[48] and so the dilemma still has to be resolved.

Dihydroxyphenylalanine (dopa) is a naturally occurring ambidentate amino acid (Chapter 20.2) of physiological importance (*e.g.* in Parkinson's disease) and a useful therapeutic agent for Mn poisoning.[49] Frequently metal ion involvement has been implicated in the biological roles of dopa, such as in the storage and transport of neurotransmitters derived from dopa. Computer simulation would suggest, however, that dopa is unlikely to be an effective chelator because of overwhelming competition from amino acids, especially histidine.[50] Here again the concentration on blood plasma could be misleading.

22.4 PEPTIDES

A large number of simple and oligopeptides occur naturally: some 150–170 have been identified in human urine, for example. Many of these have important physiological roles and metal ion chelation is often involved.[51] Indeed, in contrast to plants and bacteria, the modulators of metal ion metabolism in higher animals are peptides and proteins. Whether or not a physiological function is involved, the chelation properties of peptides have generated considerable interest because of the versatility of their chelating modes and because of the change in reactivity of the peptide itself (Chapter 20.2). Chelation, for example, can hinder the enzymatic decomposition of some peptides and finds use for this reason in peptide purification procedures. The clinical effects of some peptide-containing medicines (*e.g.* corticotropin, insulin) can be further enhanced by use as their zinc complexes.[52]

22.4.1 Small Peptides

One of the smallest naturally occurring peptides is carnosine (β-Ala-L-His) found in relatively large amounts in various animal tissues. Its exact function is not known. In the kidney an enzyme, carnosinase, hydrolyzes the peptide to its constituent amino acids. Also present in kidney is the highest *in vivo* concentration of cobalt, and Co^{II} complexes with carnosine are known to reversibly bind O_2. The inference is that carnosine (*via* its Co^{II} complex) in kidney may control the O_2 level.[53] Further evidence is still needed for this conclusion. The extra methylene group in the β-Ala moiety considerably alters the chelating properties of carnosine relative to other His-containing dipeptides[54] (Chapter 20.2). This is particularly so with Cu^{II} in aqueous solution where the major species is a dimer formed from the His moiety bridging the two metal centres.[54]

A number of tripeptides of natural origin also have important chelating properties. Gly-L-His-L-Lys, for example, which is found in human blood plasma, is a growth factor whose *in vivo* activity is synergistically enhanced by Cu^{II} ions. An interesting application of X-ray photoelectron spectroscopy was its use in monitoring metal ion levels during the separation of the tripeptide from plasma;[51] the method is non-destructive and requires only very small amounts of material. The method showed the presence of both Cu and Fe in the various fractionation stages with the final pure material containing an equimolar amount of Cu and 0.2 mol of Fe. It is suggested that this co-isolation of metal ions and peptide may be more widespread among growth-modulating peptides and proteins than has been appreciated. The hypothesized physiological function is that the peptide facilitates the uptake of Cu into cells, *i.e.* has a specific transport function for Cu cellular uptake.

From potentiometric and spectroscopic studies it is concluded that the main species at neutral pH is a 1:1 tridentate chelate (11) with a log stability constant of *ca.* −4. The claim[55] that the stability of this species is comparable to that of the Cu(albumin) complex is rather surprising, since for this to occur the involvement of a histidine in the third amino acid position is normally required, and furthermore others have concluded that in blood plasma at least the tripeptide is unlikely to compete against other ligands for the available Cu^{II}.[56,57] To illustrate the point that such conclusions from blood plasma simulations are only applicable to that medium, Pickart and Thaler[58] have shown that in a cell culture medium the tripeptide considerably enhanced Cu uptake into cells and that this was not affected by a 300-fold molar excess of amino acids, including histidine.

The tripeptide amide L-Pro-L-Leu-Glyamide (melanostatin) is a hypothalamic hormone and an effective therapeutic chelating agent. Its chelating properties have been examined by Kozlowski and his co-workers[59] because of their interest in proline-containing peptides (Chapter 20.2). With Cu^{II}

and Ni^{II} a number of pH-dependent species were found. Ultimately, above pH 8 three protons are dissociated to give the 1:1 anionic complex (12).

(11)

(12) M = Cu^{II}, Ni^{II}

The tripeptide which has attracted the most attention from both biochemists and inorganic chemists is glutathione (GSH), γ-L-Glu-L-Cys-Gly (13). The tripeptide is found in numerous cellular systems and is considered an essential constituent of all living cells. It is generally the most abundant intracellular non-protein thiol, its concentration approaching $2-3 \times 10^{-3} \, \text{mol dm}^{-3}$ in erythrocytes. Various biological roles have been assigned to GSH, including a metal detoxification role.[60,61] The Cu^{II} complex has been identified in human red blood cells.[62]

(13)

As with other thiols (*e.g.* cysteine) its oxidation to a disulfide (equation 2) is catalyzed by a number of metal ions, in particular Cu^{II}, Fe^{II}, Co^{II}, Mn^{II} and Mo^{IV}. Conversely, Cu^{I} and Hg_2^{2+} can reduce the disulfide back to GSH.

$$2GSH \longrightarrow GSSG + 2H^+ + 2e^- \qquad (2)$$

GSH has six potential donor atoms, including the peptide linkages. It is not surprising therefore that a number of coordination modes have been reported, as well as protonated and polynuclear species. The emphasis has consequently been very much on determining the species formed in aqueous solution. A review of the available data up to 1979 shows how complex is this area and how little agreement has been obtained.[63] Some representative data from reliable sources are given in Table 5 to illustrate these points. Mixed metal species of the type M—GSH—M′ have also been observed and such species are claimed to increase the concentration of lipophilic complexes.[64] The only cases where the metal binding is known with confidence are the heavy metal ions, Pb^{II}, Hg^{II} and $MeHg^+$, which bind exclusively to the thiol atom as does Cu^{I} in its major bis-ligand species. Ag^{I} forms polymeric species *via* the thiol and terminal amine groups.

Although Cu^{II} oxidizes GSH, attempts have been made to identify complexes between the two, mainly by ^1H NMR line-broadening measurements in earlier work. More recently a potentiometric and spectroscopic study was interpreted as being consistent with the formation of at least three Cu^{II}–GSH and three Cu^{II}–GSSG species.[65] Likewise, with Fe^{III} as reactant a large number of species are apparent, although all involve Fe^{II}.[66] There are no crystal structure reports of any GSH complexes but the structure of a Cu^{II}–GSSG complex, $Na_4[Cu_2(GSSG)_2] \cdot 6H_2O$, has recently been determined.[67] The crystals were isolated from a 1:1 solution of Cu^{II}:GSH at neutral pH. The structure (14) is similar to that proposed as existing in alkaline solution.

In contrast to the speciation work there have only been a modest number of reports on kinetic and mechanistic details. The formation of Zn^{II} complexes of GSH proceeds by two pH-dependent mechanisms involving coordination at the thiol group,[68] although no thiol involvement is apparent in Ni^{II} complex formation.[69] Different behaviour is found again with $MeHg^+$, which binds to GSH by an associative pathway.[63] The Cu^{II}–GSH system is further complicated by the observation that the disulfide hydrolysis in alkaline solution is accelerated by Cu^{II} ions (equations 3–5).

$$GSSG + OH^- \longrightarrow GS^- + GSOH \qquad (3)$$

$$2GSOH \longrightarrow GSH + GSO_2H \qquad (4)$$

$$GS^-/GSH \xrightarrow{Cu^{II}} GSSG + Cu^{I} \qquad (5)$$

Table 5 Log Stability Constants of Cd^{II} and Zn^{II} Glutathione (H_3L) Complexes in Aqueous Media

Metal ion	Equilibrium constant[a]	Log β[b]	Log β[c]
Cd^{II}	$[ML]/[M][L]$	8.59	10.18
	$[ML_2]/[M][L]^2$	13.39	15.00
	$[MHL]/[M][H][L]$	15.19	17.14
	$[ML(HL)]/[M][H][L]^2$	22.81	24.66
	$[M(HL)_2]/[M][H]^2[L]^2$	30.45	32.10
	$[ML(OH)][H]/[ML]$	9.12	—
	$[M_2L]/[M]^2[L]$	13.29	—
	$[ML_2H_{-1}]/[M][L]^2[H]^{-1}$	—	4.50
Zn^{II}	$[ML]/[M][L]$	7.94	8.57
	$[ML_2]/[M][L]^2$	12.41	13.59
	$[ML_2H_2]/[M][L]^2[H]^2$	28.60	30.62
	$[MHL]/[M][H][L]$	14.16	14.76
	$[ML(HL)]/[M][H][L]^2$	21.56	23.27
	$[ML(OH)][H]/[ML]$	8.82	—
	$[MLH_{-1}]/[M][L][H]^{-1}$	—	−0.07
	$[ML_2H_{-1}]/[M][L]^2[H]^{-1}$	—	3.63
	$[M_2L]/[M]^2[L]$	10.62	—

[a] Charges omitted.
[b] Data at 37 °C and 0.15 mol dm^{-3} (KNO_3) ionic strength; from ref. 171.
[c] Data at 25 °C and 3.00 mol dm^{-3} ($NaClO_4$) ionic strength; from ref. 172.

(14)

22.4.2 Large Linear Peptides

The identification, separation and syntheses of such compounds are more difficult than with the smaller di- to tetra-peptides. As a consequence the number of reports and studies of metal interactions have been correspondingly fewer. Nevertheless, a growing number of such naturally occurring peptides with an involvement of metal ions *in vivo* are coming to light. For example, the linear octapeptide angiotensin, L-Asp-L-Arg-L-Val-L-Tyr-L-Ile-L-His-L-Pro-L-Phe, is a potent hypertensive hormone which stimulates the muscles of blood vessels. It also mediates the transport of Mn ions across lipid bilayers[70] and its biological role has been linked with other metal ions, most notably Na$^+$ and Ca^{2+}. An earlier claim that cation binding induces beneficial conformational changes in the peptide has not been substantiated by a more recent study which shows that only minor conformational changes occur upon cation binding.[71]

An oligopeptide of relative molar mass 1600 that binds Fe^{III} and functions to enhance the cellular uptake of Fe has been isolated from mouse cells.[51] The peptide also binds Co^{II} but does not enhance the uptake of this metal. The peptide has been referred to as the siderophore-like growth factor (SGF). There is also a proposed Cr-chelating peptide: this 'glucose tolerant factor' is essential for the utilization of Cr in the proper functioning of insulin in cells.[72] Whilst Zn^{II} is normally associated with hydrolytic enzymes *in vivo*, it appears that it can also bind to polypeptides inducing conformational changes that hinder the cleavage of the peptides by protease enzymes. This interaction is suggested to be of relevance to a number of viral infections and their control.[50] An Fe^{III}-containing pigment, adenochrome, found in the bronchial heart of *Octopus vulgaris*, has been shown to consist of a group of closely related peptides derived from glycine and two unusual Fe-binding amino acids termed adenochromines A and B.[73]

Bleomycin is a glycopeptide, isolated from *Streptomyces verticillus*, which is used as an anticancer drug. It has both metal-binding and DNA binding properties and its biological activity is thought to be related to this bifunctionality.[74] It can cause strand scission of DNA *in vitro* and *in vivo*. The reaction *in vitro* requires the presence of both Fe^{II} and O_2. Co^{III} and Cu^{I} can also enhance the DNA cleavage reaction, but other divalent metal ions, *e.g.* Co^{II}, that can compete with Fe^{II} for the same binding sites in bleomycin suppress the cleavage activity. Metal-binding studies show the bleomycin donor groups to be histidyl groups, a pyrimidine ring, peptide groups and the terminal amine group.[75]

22.4.3 Peptide-based Ionophores

Some 20 years ago it was observed that certain antibiotics could induce the movement of aqueous K^+ ions into the mitochondria of cells, but not that of aqueous Na^+ ions. These antibiotics, many of which are naturally occurring, are termed *ionophores*, *i.e.* neutral molecules which can mediate the transport of the essential groups IA and IIA cations across biological membranes.[76] The essential features of an ionophore are a highly polar interior, a hydrophobic exterior and conformational flexibility. Many are cyclic peptides, the coordination properties of the cyclic molecules are considerably different to those of the linear peptides. These differences are outlined in Chapter 20.2.

The antibiotic studied in most detail is valinomycin (15), a cyclic dodecadepsipeptide ('depsi' indicates an ester linkage) first isolated from *Streptomyces* spp. As might be expected for a 36-membered ring, the molecule can adopt a number of conformations in solution and in the solid phase.[76] Valinomycin exhibits a much stronger affinity for K^+ ions ($\log K_1 = 4.90$ in MeOH at 25 °C) than for Na^+ ions ($\log K_1 = 0.67$ in MeOH at 25 °C). Crystal structure determinations show the 1:1 complexes to contain the non-hydrated cations with the K^+ ion tightly encased in a polar environment of ester carbonyl O atoms, indicated in (15), but the Na^+ gives poorer contacts because of its smaller size. The poorer coordination coupled with the greater hydration energy of Na^+ explains its smaller stability constant. Further molecular details and proposals regarding the biological transport mechanism are given in an excellent review by Fenton.[76]

(15) *metal binding sites

A group of antibiotics isolated from soil fungi, the enniatins,[77] also show ionophoric activity. Enniatin B (16), a cyclohexadepsipeptide, has a greater selectivity for Na^+ over K^+. With the latter ion stoichiometries other than 1:1 are found in solution. These are attributed to stacking interactions between the adjacent disc-like molecules as observed in the solid state.[76] Interestingly, Tl^I,

(16) *metal binding sites

normally a good probe for K^+ ions in biological systems, is found not to complex with enniatin B, presumably because the molecular stacking is inhibited.[78] The cyclodecapeptide antamanide, isolated from mushrooms, can also complex with the group IA cations with a selectivity for Na^+, but apparently has little ionophoric activity. The crystal structure of the Na^+ complex shows only four carbonyl O atom contacts with the metal ion and a fifth coordination site is occupied by a water molecule.[78]

The above ionophores are of the 'carrier' type, whilst the gramicidin ionophores are of the 'channel forming' type. These different classes reflect the two different mechanisms postulated for the membrane diffusion of the IA and IIA cations. Gramicidin A is a linear pentadecapeptide antibiotic, with alternating L- and D-amino acid residues, that dimerizes to form transmembrane channels about 0.5 nm in diameter and 3.2 nm in length.[79] The channels are permeable to the alkali metal cations, but are blocked by Ca^{2+} ions. X-Ray crystallography reveals two metal-binding sites per dimeric channel and changes in the channel size on metal ion binding.[80]

A molybdenum-binding ionophore has been described, namely a peptide secreted by *Bacillus thuringiensis* which can bind MoO_4^{2-} ion. Its biological function is not known.[81] Curiously the secretion of this peptide appears to occur under conditions of iron depletion.

22.4.4 Peptide-based Siderophores

Aerobic and facultative aerobic bacteria produce extremely powerful chelating agents to specifically facilitate the uptake of iron, particularly under conditions of iron deficiency. The availability of Fe to microbes plays an important role in infections. These complexing agents are referred to as *siderophores* (formerly siderochromes). Much of the work in this area has been carried out by Raymond, Neiland and their collaborators.[82,83]

The siderophores contain either hydroxamate or catecholate chelating groups. The former are found in the ferrichromes, which contain three hydroxamate groups covalently bonded to a cyclic hexapeptide framework (**17**; Table 6), and in the ferrioxamines, which also contain three hydroxamate groups, but covalently linked to a linear framework (**18**; Table 7). The linear framework of the ferrioxamines is not a peptide but it does contain units linked by peptide type (amide) bonds; a further amide link between the two terminal groups gives ferrioxamine E (**18**; $n = 5$, no R or R′). The important feature of these siderophores is that the three hydroxamate groups can bind to a single Fe^{III} ion to give a neutral, water soluble, six oxygen-donor octahedral complex (**19**) of high spin configuration which is kinetically labile. The ability of hydroxamates to form strong chelates with Fe^{III} has been known for a long time, and these are discussed in Chapter 15.9.

Table 6 The Known Ferrichromes

(17)

Name	R^1	R^2	R^3	R
Ferrichrome	H	H	H	Me
Ferrichrysin	CH_2OH	CH_2OH	H	Me
Ferricrocin	H	CH_2OH	H	Me
Ferrichrome C	H	Me	H	Me
Ferrichrome A	CH_2OH	CH_2OH	H	$CH=C(Me)CH_2CO_2H$ (*trans*)
Ferrirhodin	CH_2OH	CH_2OH	H	$CH=C(Me)CH_2CH_2OH$ (*cis*)
Ferrirubin	CH_2OH	CH_2OH	H	$CH=C(Me)CH_2CH_2OH$ (*trans*)

Table 7 The Known Ferrioxamines

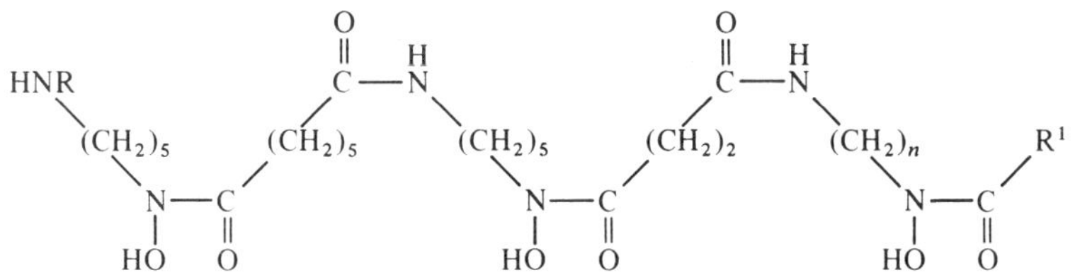

(18)

Name	R	n	R^1
Ferrioxamine B	H	5	Me
Ferrioxamine D₁	MeCO	5	Me
Ferrioxamine G	H	5	HO₂C(CH₂)₂
Ferrioxamine A₁	H	4	HO₂C(CH₂)₂

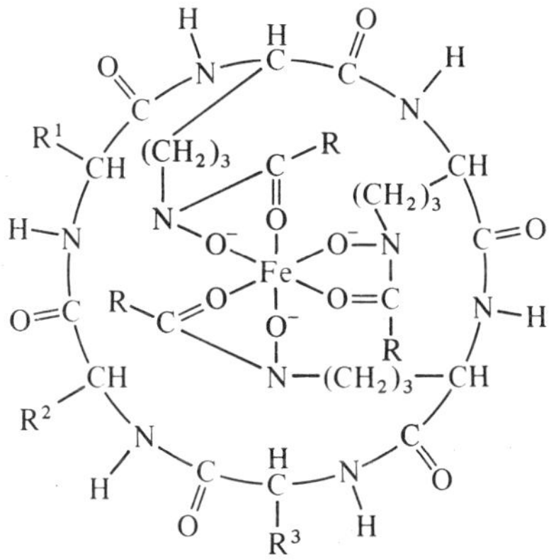

(19)

The stability constants of the Fe^{III} siderophore complexes are some of the largest known, *e.g.* the ferrichrome and ferrioxamine E complexes have log values of the order 29 and 32 respectively as compared to a value of 25 for $Fe(edta)^-$. So strong are these complexes that microbes have been observed to leach iron from stainless steel vessels. Not surprisingly the siderophores also find use in treating cases of iron poisoning and for the elimination of iron from cases of thalassaemia.[84] Complexation of Fe^{II} is considerably weaker than that of Fe^{III} and this is probably utilized for the release of the iron within the cells.

Much of the structural information has come from studies on the Cr^{III} and Co^{III} complexes, which offer advantages in terms of being kinetically inert and having more pronounced *d–d* electronic spectra. The Cr^{III} complexes are isostructural with the Fe^{III} complexes and sufficiently alike in bonding to serve as useful molecular probes. In solution the Cr^{III} complexes of desferriferrichrome and desferriferrochyrin ('desferri' indicating iron-free) form exclusively as the Λ-*cis* isomer, no Δ-*cis* being found although it was predicted. The X-ray structures of the Fe^{III} complexes show the same Λ-*cis* geometry (19).[82] However, the Fe^{III} complex of coprogen is mainly found with the Δ configuration in solution, where coprogen, from the fungus *Neurospora crassa*, closely resembles the ferrichromes with a cyclic trimeric L-ornithine framework.[85] The Al^{III} complexes have also been synthesized and have given useful information in both the solid[86] and solution (¹³C and ¹⁵N NMR) phases.[87]

Enteric bacteria secrete a different type of siderophore, called enterobactin, under conditions of iron deficiency. This is a cyclic depsipeptide derived from L-serine with three covalently linked catechol groups (20). The Fe^{III} complex has the largest known formation constant (*ca.* 10^{52}) of any known Fe^{III} complex.[88] Again, the elucidation of the iron binding site has come from studies on the isostructural Cr^{III} complex. The absolute configuration of the metal site, Δ, is opposite to that of most of the other chiral siderophores, an observation which is also very evident in solution from the circular dichroism spectra. Recently the antipode of enterobactin has been synthesized from D-serine and, as expected, the Fe^{III} complex adopts the Λ configuration.[89] Interestingly, the synthesis of enterobactin has also been achieved using a cyclic organotin compound as a template.[90]

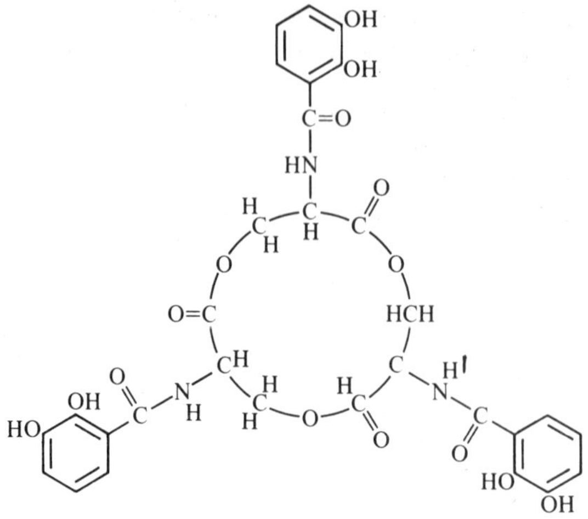

(20)

The mechanism of intracellular iron release from such a strong chelate is of obvious importance. Several proposals have been made but no definite conclusions have been reached as yet. One suggestion is of enzyme cleavage of the peptide chain, *i.e.* by a specific esterase.[91] Another is that protonation of the complex occurs which weakens Fe^{III} binding thus facilitating reduction to, and release of, Fe^{II}. Protonation occurs in mildly acidic aqueous solution making the electrode potential more favourable for reduction. At pH 3 a blue solid forms, [Fe(H$_3$ent)], in which the bonding is suggested as being salicylate-like (equation 6). In methanol solvent further protonation occurs with concomitant reduction of the Fe^{III} by the catechol groups.[92] An interesting parallel to the proposed protonation mechanism is provided by a number of plant species which under conditions of iron deficiency exude from their roots (see Section 22.2.2) both iron reductants and protons. Both enterobactin and ferrichrome have been observed to bind Ca^{2+} ions; any conformational change resulting from this could also weaken the Fe^{III} binding.

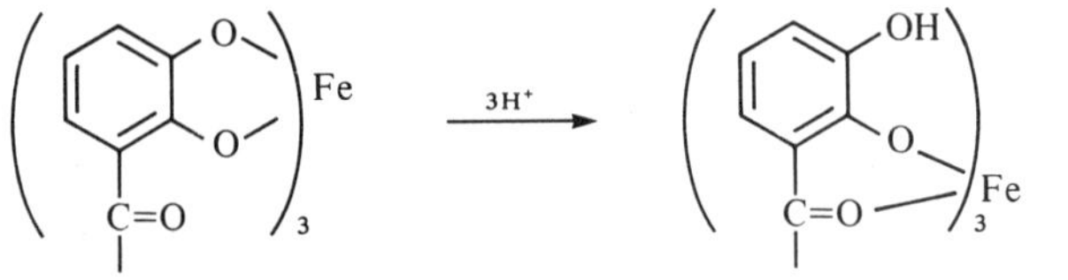

(6)

A further group of siderophores, which are not so well characterized, possess two hydroxamate and one citrate group. These are the aerobactins (21) produced by *Aerobacter aerogenes* under conditions of iron deficiency. Although not strictly peptide based, the molecule contains two peptide-type bonds linking the central citrate moiety. As with the other hydroxamate siderophores the Fe^{III} complex is labile, high-spin and octahedrally coordinated, the donor atoms being from the two hydroxamate groups and carboxyl and hydroxyl O atoms from the citrate. In aqueous solution the configuration is again Λ, but a number of protonated species are in equilibrium with the main Λ-[FeH$_{-1}$L]$^{3-}$ complex. All of these are apparently very effective Fe^{III} chelators. *A. aerogenes* also excretes the stronger iron-binding enterobactin. It is postulated that this is released under conditions of severe iron deficiency, whilst aerobactin is released under less severe conditions.[93]

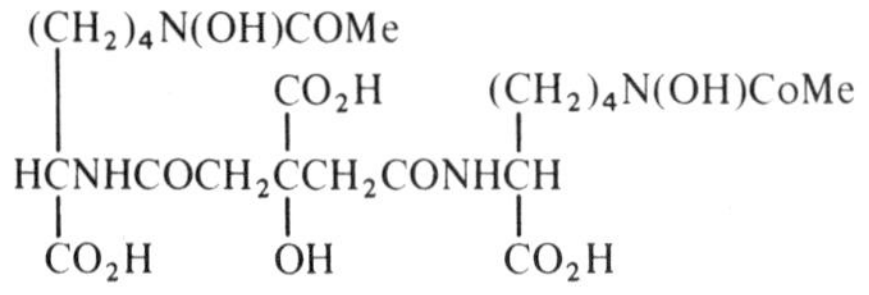

$$\begin{array}{c} (CH_2)_4N(OH)COMe \\ | \\ \qquad\qquad CO_2H \qquad (CH_2)_4N(OH)CoMe \\ \qquad\qquad | \qquad\qquad | \\ H\overset{|}{C}NHCOCH_2\overset{|}{C}CH_2CONH\overset{|}{C}H \\ | \qquad\quad | \qquad\qquad | \\ CO_2H \quad\; OH \qquad\quad CO_2H \end{array}$$

(21)

22.5 NON-PEPTIDE-BASED IONOPHORES

A number of ionophores not based on peptides are also known. Again these are microbial metabolites which can bind alkali metal ions and render them lipophilic. Whether all of these function as ionophores *in vivo* is still to be established in most cases.

Streptomyces spp. produce both peptide- and non-peptide-based ionophores, the latter type comprising the channel-forming amphotericin B and nystatin. Amphotericin B is a polyhydroxylic lactone which can apparently span half the thickness of a lipid bilayer: about 10 molecules are thought to assemble to provide a hydrophilic channel of about 0.5 nm diameter, which allows the passage of small molecules and hydrated metal ions.

A further group with potential as ion carriers are the neutral macrotetrolide actins, with the basic structure (**22**), isolated from *Actinomyces* spp. The various known actins are described in Table 8. In the crystal structure of the K^+ complex of nonactin the metal ion is located within a central cavity bonded by the four furanyl and the four carbonyl O atoms, the ligand adopting a conformation similar to the seam of a tennis ball.[94] This group shows many similarities to valinomycin and the enniatins in terms of their preference for K^+ over Na^+, the conformations of the K^+ complexes, and the dehydration of the K^+ ion occurring in a stepwise fashion.[76] The selectivity of nonactin for K^+ is very much temperature and solvent dependent. The decreased selectivity observed with decreasing temperature is taken as evidence of a carrier mechanism rather than a channel-forming mechanism.[95] The selectivity also virtually disappears in dry organic solvents, indicating again how important is the difference in hydration energies of Na^+ and K^+ in the selectivities exhibited by the ionophores. A useful compilation of some of these stability constants is given by Hughes.[96]

Table 8 The Known Actins

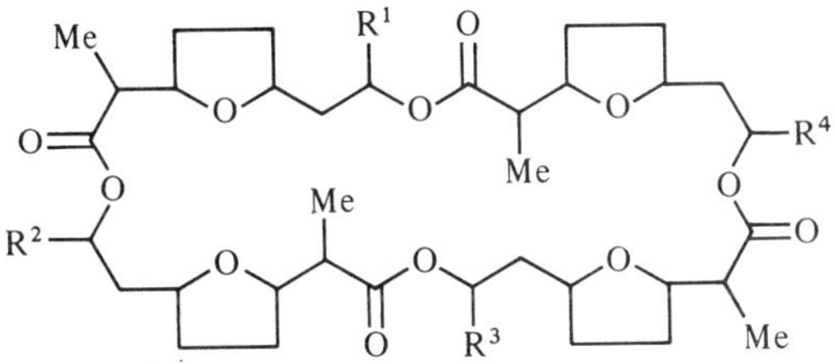

(22)

Name	Substituents
Nonactin	$R^1 = R^2 = R^3 = R^4 = Me$
Monactin	$R^1 = R^2 = R^3 = Me; R^4 = Et$
Dinactin	$R^1 = R^3 = Me; R^2 = R^4 = Et$
Trinactin	$R^1 = Me; R^2 = R^3 = R^4 = Et$
Tetranactin	$R^1 = R^2 = R^3 = R^4 = Et$

Another type of compound with potential ionophoric activity contains a single carboxylic acid group which can neutralize the charge on the metal ion. As the structures of some of these show (**23–28**), they have an abundance of potential O donor atoms.

Both monensin (**24**) and nigericin (**25**) complex Na^+ and K^+ strongly but not selectively. The crystal structures of the Na^+, Tl^+ and Ag^+ complexes all show the metal ion to be in an O-rich cavity. The carboxylate group is not involved however.[97] With the antibiotics (**26**), (**27**) and (**28**) the thermodynamic stabilities (Table 9) are greater for the divalent than for the monovalent metal ions.[98] The conformations adopted in these complexes are very solvent dependent, and the implication of these to the biological transportation of the cations has been discussed.[99]

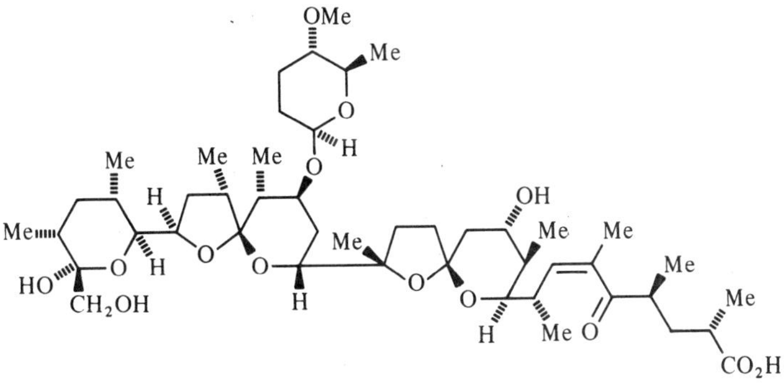

(23)

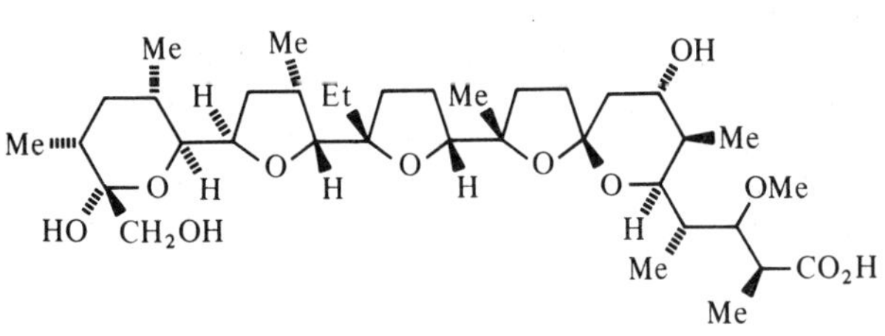

(24)

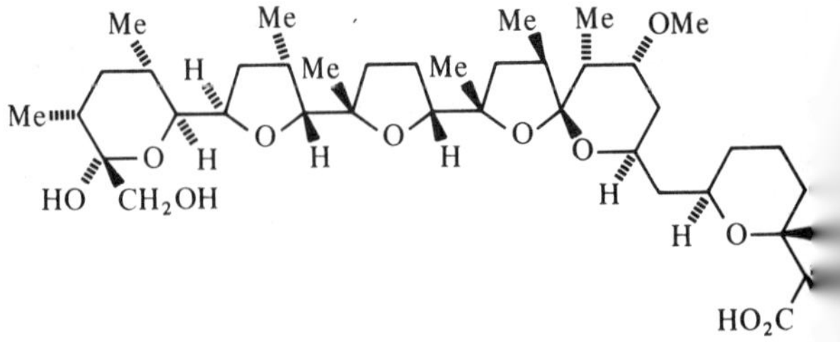

(25)

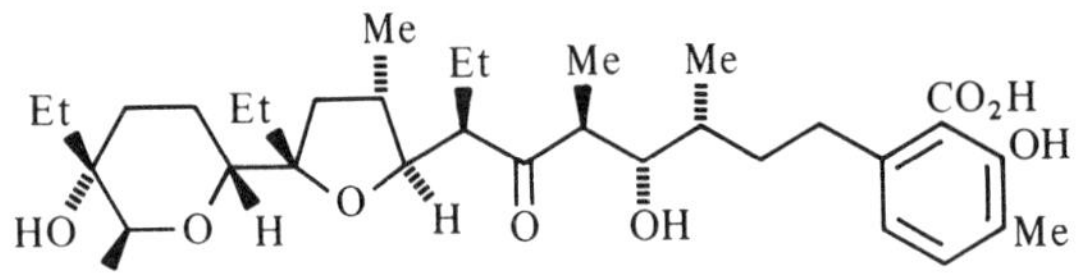

(26)

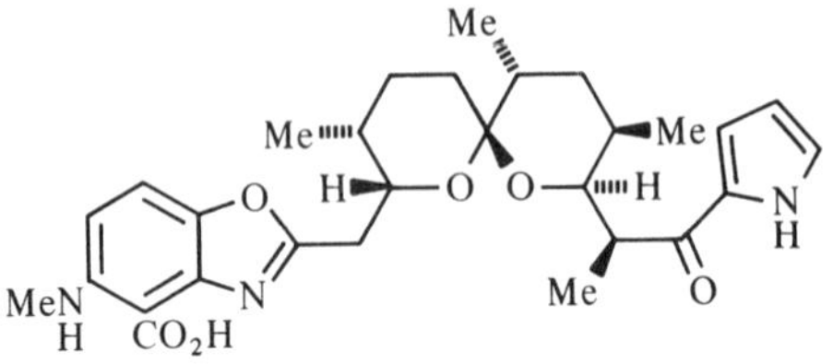

(27)

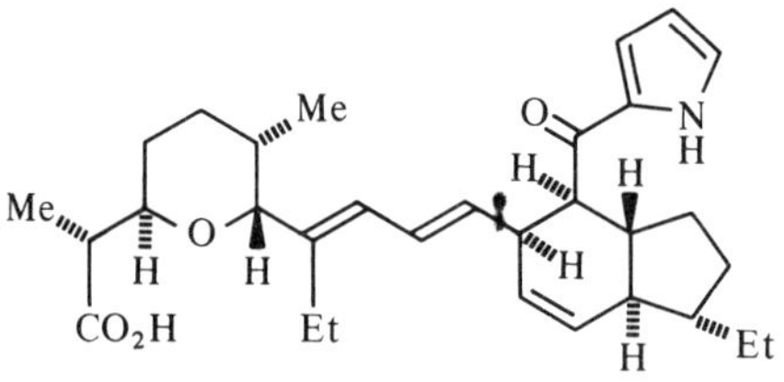

(28)

Table 9 Log Formation Constants of Some Monobasic Ionophores (HL)[a,b]

| | Lasalocide (26) | | Calcimycine (27) | |
Cation	Log β (ML)	Log β (ML$_2$)	Log β (ML)	Log β (ML$_2$)
H$^+$	8.23	—	10.94	—
Na$^+$	2.61	—	3.4	—
K$^+$	3.45	—	2.4	—
Mg^{2+}	4.12	6.07	7.6	15.9
Ca^{2+}	4.88	NF^c	8.2	16.2

[a] Data at 25 °C.
[b] Data from ref. 98.
[c] Species not found.

Naturally occurring compounds with potential ionophoric properties are still being discovered, two recent examples being the antibiotic M139603[100] and the remarkable anticancer substance bryostatin 1.[101]

22.6 PROTEINS

The functional roles of the trace metals inevitably involves the metal ion coordinated to a protein or to a protein prosthetic group, mostly as a metalloenzyme or as a metal-activated enzyme. The difference between these two enzymic forms is given in Chapter 20.2, as are details on the likely coordination sites of protein molecules. The functional roles and their molecular details are covered in Chapter 62. In addition, proteins are involved in the absorption, transportation and storage of the trace metals in biological systems.[102] All of these processes are interrelated (see Scheme 1) and are essential in ensuring the delivery of the right metal ion at the right level to the site of action and in controlling the overall level of a metal in the system (homeostasis). Even the essential trace elements become toxic at high concentrations. There is still much to learn about these processes.

The gut mucosal proteins are implicated in the absorption of the metal ions from the diet, but no molecular details are as yet available. Once absorbed the metal ions are rapidly delivered to the liver by the portal bloodstream, the principal transport form for a number of ions and molecules being the protein albumin.[103] Albumin is a globular protein with a molar mass of *ca.* 66 000 and can coordinate a number of metal ions, including the bulk metal ions Ca^{2+} and Mg^{2+}, and also any extraneous Ni or Hg ions.[104] The protein has a single strong binding site for Cu^{II} and Ni^{II} ions, which has been identified (Chapter 20.2, **19**) as involving the three N-terminal amino acid units. The delivery of Cu from the albumin complex is often considered to involve dissociation *via* the equilibrium with the amino acid pool (see Section 22.3). An attractive alternative[103] is that this could occur at specific albumin receptor sites on membrane surfaces. Zinc resembles copper in terms of its distribution between the protein and amino acid fractions in blood plasma; zinc-albumin is also regarded as the transport form for Zn^{II} ions.[105] The zinc binding site, however, is different to that of Cu^{II}, the evidence pointing to the amino acid residues 37 to 39 (Glu-Asp-His) with binding involving the two carboxyl O atoms and the histidine imidazole N atom.[106] Chelation involving deprotonated peptide N atoms is highly unlikely for Zn^{II} at neutral pH.

Whether specific storage forms are involved for the trace metals is debatable.[107] Such a role has been alluded to for the non-exchangeable protein complexes in serum, *i.e.* caeruloplasmin for Cu, and α_2-macroglobulin for Zn, and to metallothionein protein (Chapter 20.2). However, it has also been argued that there are sufficient deposits within body tissues to provide at least short-term storage of the minute quantities that are required.

With iron, for which there is a greater demand (see Table 1), the transport and storage forms are both proteins and known in some detail. The transport forms are a group of very similar proteins called transferrins. Serum transferrin, a glycoprotein, strongly binds two Fe^{III} ions per molecule, accounting for about 4 mg of the body iron at any one time but turning over 10 times that amount per day. The two Fe^{III}-binding sites are similar in metal-binding affinity[108] (log β values 20.67 and 19.38) and are structurally similar, most likely involving two to three Tyr phenolate O atoms and one or two His imidazole N atoms. Curiously, hydrogen carbonate anion also appears to be essential for iron binding. Release of iron from transferrin probably involves reduction to Fe^{II} for which the apoprotein has a much reduced affinity.

The same storage form, an iron–protein complex, occurs in animals, plants and probably in bacteria. This is the water-soluble complex ferritin, the metal-free form (apoferritin) of which is well characterized from X-ray studies.[109] Ferritin consists of 24 polypeptide subunits arranged in the form of a spherical shell of internal diameter 7.5 nm and external diameter 13.0 nm. The inner cavity is filled with a polymeric hydrated iron(III) oxide phosphate complex which remarkably can contain up to 4500 Fe ions. Whereas most biological systems take great care to prevent Fe^{III} hydrolysis, ferritin is a unique example of the hydrolysis being made use of, albeit under controlled conditions. The protein shell is penetrated by six channels of *ca.* 1 nm diameter through which metal ions may enter or leave. Again, reduction to Fe^{II} appears the most likely method of release of iron from ferritin,[110] the solubility of '$Fe(OH)_2$' being considerably larger than that of '$Fe(OH)_3$'. A second storage form occurs in the cell lysosomes, which resembles ferritin stripped of part of its protein shell. Much less is known concerning the function of this form, called haemosiderin.

In avian egg yolk a different iron storage form is utilized; it is also an important reserve of iron in the eggs of lower vertebrates. This is phosvitin, a medium-sized phosphoglycoprotein that can contain up to 47 Fe^{III} ions. The unusual green colour of this complex is attributed to a tetrahedral $Fe^{III}O_4$ chromophore formed from phosphate O atom bridging of the metal centres. The bridging also creates a linear array of magnetically interacting Fe ions. The phosphate groups are present as phosphoserine residues. Fe^{II} has been shown to be rapidly oxidized by O_2 in the presence of apophosphovitin to give the Fe^{III} phosvitin and the simultaneous release of inorganic phosphate, two phosphate groups remaining per Fe^{III} ion bound.[111] Different iron-storage proteins again have been reported in *Escherichia coli* and *Proteus mirabilis*.[112]

The physiological disorders resulting from the presence of toxic metals invariably arise from their binding to proteins and modifying the protein or enzyme functions. The prime protein coordination sites are the thiol groups of cysteine residues, particularly for the heavier toxic metals, Pb^{II}, Hg^{II} and $MeHg^+$, and sites involving both thiol and imidazole groups. Toxicity can also arise from displacement of the essential metal ions from their protein binding sites with consequent changes in reactivity. The teratogenic effect of Cd^{II}, for example, has been ascribed to its displacement of Zn^{II} in foetal Zn^{II}-containing DNA synthesis enzymes, while As, Co and Pb have been shown to inhibit ferrochelatase activity thereby blocking haem synthesis. Other metals, such as Be, can form stable complexes with certain proteins which provoke an immune response. These are just a few examples of how metal toxicity involves metal–protein interactions.

22.7 NUCLEIC ACIDS AND SUBUNITS

Nucleic acids are the chemical compounds which store and transfer genetic information; they are large linear polymers composed of nucleotide units. The *nucleosides* comprise heterocyclic N bases joined to a pentose sugar molecule and the *nucleotides* are the nucleoside phosphates in which the phosphate moiety is linked to the sugar molecule. The nucleic acids consist of nucleotide units linked together by phosphate diester bonds through the 3' and 5' C atoms of adjacent nucleotides (**29**). *DNA* (deoxyribonucleic acid), found within the nucleus of biological cells, contains deoxyribose (**30**) as its sugar unit and consists of two intertwined polynucleotide chains forming the, now famous, double helix. The polymer sizes are large with molar masses of the order 10^7–10^8. *RNA* (ribonucleic acid) is also found within the nucleus but is mostly found within the cytoplasm of cells. It is a single-stranded structure containing ribose as its sugar unit (an extra OH group at C-2' in **30**).

(29)   (30)

Two types of heterocyclic N bases are found: the pyrimidine bases (**31**), (**32**) and (**33**), and the purine bases (**34**) and (**35**). The pK_a values are given in Table 10 as are the names of the compounds derived from the bases. One other difference between DNA and RNA is that the former is found to contain thymine, while the latter contains uracil. The average cell contains 2–5% of its dry weight as nucleic acids and nucleotides; neither the heterocyclic bases nor the nucleosides occur as such in cells.

(31) Uracil   (32) Thymine   (33) Cytosine   (34) Adenine   (35) Guanine

Table 10   The Nucleic Bases, Their Proton Ionization Constants (pK_a), and the Nomenclature of Their Nucleoside and Nucleotide Derivatives

Base	pK_a[a]	Nucleoside	Nucleotide
Adenine	4.1 (N-1)	Adenosine	Adenosine monophosphate (AMP)
	9.8 (N-9)		
Guanine	3.3 (N-7)	Guanosine	Guanosine monophosphate (GMP)
	9.4 (N-1)		
Cytosine	4.6 (N-3)	Cytidine	Cytidine monophosphate (CMP)
Uracil	9.4 (N-3)	Uridine	Uridine monophosphate (UMP)
Thymine	9.8 (N-3)	Thymidine	Thymidine monophosphate (TMP)

[a] Data at 25 °C, 0.1 mol dm^{-3} ionic strength, from ref. 116; protonation sites identified in parentheses.

The biological functions of the nucleic acids involve the participation of metal ions. In particular K^+ and Mg^{2+} stabilize the active nucleic acid conformations. Mg^{2+} also activates enzymes which are involved in phosphate transfer reactions and in nucleotide transfer. The monomeric nucleotides are also involved in a number of metabolic processes and here again metal ions are implicated. Consequently, there has been considerable attention focussed on the coordination properties of these molecules as a means of understanding the mechanism of the metal ion involvements. A number of reviews are available which cover the studies on metal interactions.[113–117]

22.7.1 Nucleosides and Nucleotides

Since the nucleic bases are not found naturally as free bases, their coordinating properties will only be briefly summarized. It should also be pointed out that two of the potential binding atoms,

N-1 of the pyrimidines and N-9 of the purines, are the same atoms that form the glycoside bonds in the nucleosides and nucleotides, and are therefore not available for coordination in these molecules. For the purines adenine and guanine, either N-9 or N-7 seems to be the preferred donor atom. Chelation is surprisingly rare. There is some evidence that the heavy metals such as Hg^{II} coordinate to ring C atoms.[116] The inability of adenine to form chelates is attributed to the poor basicity of the amine group at C-6. The coordination by thymine and uracil is strongly pH dependent, with O atom donation in acid solutions, N-3 in slightly alkaline solutions and N-1 ($pK_a \approx 14$) in strongly alkaline media. For cytosine the N-3 atom, in contrast to the other pyrimidines (see pK_a values in Table 10), can coordinate at neutral pH and is the preferred donor atom for a number of metal ions.

The nucleosides and nucleotides present a plethora of binding sites, *i.e.* the base N and O atoms, the ribose OH groups and the phosphate O atoms of the nucleotides; the ribose groups are the weakest of these potential donor sites. Most attention has been focussed on the coordinating properties of the adenosine mono-, di- and tri-phosphates (36) (AMP, ADP and ATP respectively) because of their biological importance. The stability constants of some of the complexes formed with these adenosine phosphates are given in Table 11. The species formed are highly pH dependent, and there is also a problem in interpreting potentiometric data because the normal background electrolyte cations, Na^+ and K^+, also form (weak) complexes with ATP. A careful analysis of the data available on the Cu^{II} and Zn^{II} complexation of ATP has recently been published together with measured thermodynamic parameters.[118] The stability constant data, however, do clearly show that the metal affinity follows the order $ATP^{4-} > ADP^{3-} > AMP^{2-}$, and that little variation occurs between the different N bases,[114] all of which points to the phosphate groups as being the major coordination sites, a conclusion also made from one of the earliest applications of NMR (^1H and ^{31}P) to studies on metal complexation.[119] Mg^{2+} and Ca^{2+} were found not to induce any ^1H shifts with ATP and ADP, and only a very weak effect with AMP. ^{31}P NMR established their binding sites as the β- and γ-phosphate groups of ATP, an important finding in view of the known involvement of Mg^{2+}–ATP complexation in biological systems. Similar observations have been made with the guanine nucleotides.

(36)

Table 11 Log Formation Constants of Some 1:1 Divalent Metal Complexes of Adenosine Phosphates[a]

Metal	AMP	ADP	ATP
Mg	1.97	3.17	4.22
Ca	1.85	2.86	3.97
Mn	2.40	4.16	4.78
Co	2.64	4.20	4.66
Ni	2.84	4.50	5.02
Cu	3.18	5.90	6.13
Zn	2.72	4.28	4.85

[a] Data at 25 °C and 0.1 mol dm^{-3} (KNO_3) ionic strength, from ref. 114, ligands in their fully deprotonated forms.

Complexation of Zn^{II} with ATP results in a significant shift of the H-8 resonance which is taken as evidence of Zn binding to the N-7 atom. The effect of coordination on the C(8)–H bond also explains the formation of a metal–C(8) bond with the $MeHg^+$ ion.[120] Multinuclear NMR has been used to identify a number of Al^{III} complexes of ATP in solution, and the results related to the involvement of Al in dialysis dementia disease.[121]

By far the greatest attention has been to the complexing of the transition metal ions, often with surprising results. Thus, Cu^{II} can be shown to bind to N-7 of the adenine moiety of ATP and yet it can considerably enhance the hydrolytic cleavage of the phosphate bonds under the same conditions. Earlier suggestions were that chelate formation occurred *via* the N-7 and phosphate O atoms, giving very large chelate rings. More recent interpretations are that Cu^{II}–phosphate interactions occur *via* an adjacent ATP molecule.[116] In support of this the crystal structures of a number of metal–nucleotide complexes show polymeric-type behaviour.[115] They also show that transition metal–phosphate interactions can occur *via* hydrogen bonding of a coordinated water molecule (outer-sphere association). This is clearly demonstrated in the structures of the compounds $[M(nucleotide)(H_2O)_5]$ (37; $M = Cd^{II}$, Co^{II}, Mn^{II} and Ni^{II}). Curiously, the kinetically labile Cr^{III} and Co^{III} complexes of ATP, *e.g.* $[Cr(H_2O)_4(ATP)]^-$ (38) and $[Co(NH_3)_4ATP]^-$ are found to be good analogues of $Mg(ATP)^{2-}$ in that they can substitute for $Mg(ATP)^{2-}$ *in vitro* in the activation of enzyme reactions.[122] The crystal structure analysis of the Cr^{III} complex (38) shows inner-sphere coordination to the phosphate group but no involvement of the adenine base in coordination. Surprisingly the complex ion is chiral, although the Co^{III} analogue above is not! Of the two chiral isomers only the Δ form was found to activate the phosphoryl transfer reaction of the enzyme pyruvate kinase.[123]

(37)

(38)

Both solution and solid-state behaviour consistently show that stronger metal–base interactions arise with the pyrimidine nucleotides than with the purine nucleotides. The general conclusions[116] are that for the pyrimidine monophosphates the groups IA and IIA metal ions form inner-sphere complexes with the phosphate groups. The $3d$ transition metals do likewise in the absence of significant interactions with the bases. If the latter do arise, then for steric reasons bonding to the phosphate groups must be of the outer-sphere type. For the di- and tri-phosphates metal binding is primarily to the phosphate groups; only Cu^{II} of the $3d$ metal ions shows any significant interaction with the base moiety (N-7 atom).

In the presence of an aromatic ligand such as bipyridyl or *o*-phenanthroline ternary complexes of the type nucleotide—M^{n+}—ligand form in aqueous solution with enhanced thermodynamic stability. The enhanced stability arises from π-stacking interactions between the aromatic ligand and the heterocyclic base.[124] In solution these interactions are evident from induced ^1H NMR shifts and in the solid state the two bases are aligned parallel within van der Waals contact distances. Recently the extent of these stacking interactions for a number of Mn^{II} and Zn^{II} complexes of nucleoside triphosphates have been estimated in terms of intramolecular equilibrium constants.[125,126]

Studies on the kinetic behaviour of nucleoside and nucleotide complexes are less common than those on structural aspects. This arises because of the rapid rates of the formation and dissociation reactions, requiring NMR or temperature-jump relaxation measurements. The number of species that can coexist in solution also hinders interpretation. The earlier kinetic studies have been reviewed by Frey and Stuehr.[127] Two important biological reactions of the nucleotides are phosphoryl and nucleotidyl group transfers. Both reactions are catalytic nucleophilic reactions and they both require the presence of a divalent metal ion, in particular Mg^{2+}. Consequently, one of the main interests has been in understanding the catalytic mechanism of the metal ion involvement. This has mainly involved studies on related non-enzymic reactions.[128]

Biological phosphorylation means in effect the transfer of a PO_3^- group from one nucleophilic atom to another (N or O atoms). For example the phosphorylation of inorganic phosphate requires a divalent metal ion, Ca^{2+} being the most effective (reaction 7). Apart from Mn^{II} the transition metal ions are ineffective in catalyzing reaction (7). The simple dephosphorylation of ATP however (equation 8), is rapidly catalyzed by Cu^{II}; Zn^{II} shows a lesser enhancement, whilst Ni^{II}, Mn^{II}, Mg^{II} and Ca^{II} show very little activity. The mechanisms proposed for these and related reactions have been discussed in reviews by Cooperman[128] and Martin and Mariam.[116]

$$ATP^{4-} + HPO_4^{2-} \longrightarrow ADP^{3-} + HP_2O_7^{3-} \qquad (7)$$

$$ATP^{4-} \longrightarrow ADP^{3-} + PO_4^{3-} \qquad (8)$$

22.7.2 Nucleic Acids

It is now understood that the replication of DNA and virtually every step in the utilization of the genetic code require metal ions in some way. The three fundamental steps of replication, transcription and translation are influenced by metal complexation.[113,129,130]

Mg^{2+}, present in high concentrations within cells, is required in the replication process and appears to function by binding the deoxynucleoside triphosphates to DNA polymerase (a zinc enzyme). Mg^{2+} is also the metal ion essential for the transcription stage. Although other metal ions are also active *in vitro*, *e.g.* Mn^{II}, Co^{II}, they do not give the proper selectivity in terms of the required nucleotides. Heavy divalent metal ions inhibit the process. The final stage of translation of the DNA code into protein synthesis is very much influenced by metal ions, high concentrations of metal ions leading to mistaken identity of amino acid codons. It is not surprising that there have been so many investigations into the affinity of metal ions for DNA and RNA, the nature of the binding sites, and the influence of coordination upon the properties of the nucleic acids.[113,117]

Mg^{II} and Cu^{II} reversibly bind to DNA with stability constants of the same order, however their effects on DNA are very different.[113] Mg^{II} has a stabilizing influence on the double helix structure of DNA in aqueous solution, whilst Cu^{II} has the opposite effect. This difference has been related to their different binding behaviours. Mg^{II} ions coordinate exclusively with the phosphate groups and in so doing reduce coulombic repulsion between adjacent negatively charged phosphate groups. Cu^{II} also appears to bind predominantly to phosphate groups, except in the denatured form of DNA for which additional base interactions are evident both from NMR measurements and from the increased binding strength of the metal ion. The effect of base coordination is to prevent hydrogen-bond formation between base pairs and thus inhibit the reversibility of the DNA unwinding $\rightleftharpoons$ rewinding process. In keeping with these conclusions destabilization also occurs with other metal ions that are known to bind predominantly to base N atoms, *e.g.* Ag^{I} and Hg^{II}. Both of these ions exhibit strong complexation with DNA, however Ag^{I} has a preference for guanine–cytosine base pairs, whilst Hg^{II} prefers adenine–thymine base pairs. Removal of these metal ions regenerates the DNA helix structure but this is not the case with $MeHg^{+}$, which appears to induce some additional cross-linking between polynucleotide strands. Recently, stereospecific binding has been reported between DNA and the chiral isomers of the tris(*o*-phenanthroline) complexes of Zn^{II} and Ru^{III}.[131,132] These should serve as useful molecular probes.

The ordered conformations of RNA molecules *in vivo* also require the presence of divalent metal ions. Again Mg^{II} is the ion of choice. There is also evidence that metal ions are essential in converting inactive forms of transfer RNA (tRNA) into active forms and that the metal ions involved bind to specific sites on the tRNA molecule.[133] Again, as with DNA, Mg^{II} stabilizes and Cu^{II} destabilizes the ordered conformations of RNA molecules.

The reactivity of metal ions is not always the same with DNA and RNA. One reaction that is exclusive to RNA is depolymerization of the polynucleotide structure by the cleavage of the phosphodiester bonds. This depolymerization reaction, as with other RNA hydrolyses, can be induced by metal hydroxides, Zn^{II} being one of the most effective. A simple mechanism is that the Zn^{II} chelates to the phosphate group and the 2′-hydroxyl group of ribose (the 2′-group is absent in DNA). Electron withdrawal by the Zn ion then weakens the phosphodiester linkage. Such a mechanism, however, does not take into account the observed influence of the nature of the adjacent base and the formation of metal-dependent products. Pb^{II} is also an effective catalyst in site-specific depolymerization of tRNA. In this case the metal has been shown to bind to the bases with only weak interactions with phosphate groups. The catalytic action has been interpreted in terms of nucleophilic attack by a metal-bonded hydroxide ion.[134] This may have implications for the mechanisms of other metal ions active in this reaction.

A number of crystal structure determinations have been reported for metal–RNA complexes.[115] The binding sites for some of the metals are summarized in Table 12. One of the surprising results shown in Table 12 is the specific binding of *trans*-[Pt(NH$_3$)$_2$Cl$_2$] to tRNA, since the *cis* isomer was found not to bind. Yet it is the *cis* isomer (trade name cisplatin) which is so effective against cancer tumours not the *trans*, and generally interactions of cisplatin with nucleic acids are implicated. More specifically the weight of evidence[135–137] points to coordination of cisplatin with DNA, not RNA, in the tumour cell preventing replication of the DNA. Spectroscopic studies show that both isomers can bind to DNA bases, in particular guanine base. One conclusion is that the *cis* isomer can form a chelate, *via* Cl ion loss, to the N-7 and O-6 atoms; the *trans* isomer is unable to form such a chelate.[138] Others consider the N-7 (purines) and N-3 (pyrimidines) atoms as likely binding sites.[139] This latter report and more recent ones are in agreement that the interactions are more complex than simple chelate formation.[140,141]

Table 12 Metal Binding Sites in tRNA (phenylalanine) Identified
by X-ray Crystallographic Studies[115]

Metal	Binding sites
Mg[II]	O (phosphates)
Mn[II]	N-7 (guanine no. 20)
Co[II]	N-7 (guanine no. 15)
Hg[II]	O-4 (uracil)
Sm[III], Lu[III]	O (phosphate), O-6 and N-7 (guanine)
trans-[Pt(NH₃)₂Cl₂]	N-7 (guanine) — replaces one Cl⁻
[OsO₃(py)₂]	N-7 (guanine), O-2' and O-3' (ribose)

The bonding of antitumour coordination compounds to DNA is not always necessary. The active tetra(μ-carboxylato)rhodium(III) compounds, for example, show no evidence of binding to DNA or the RNAs but inhibit tumour DNA synthesis, probably by inhibition of enzymes involved in the DNA synthesis.[142]

22.8 PORPHYRINS AND RELATED COMPOUNDS

This group of compounds is widely found in nature as metal complexes; in the chlorophylls, the haem groups of many iron proteins and the corrinoids. They have in common a macrocyclic structure which provides four N donor atoms at the corners of a square plane. Metal coordination to the N atoms results in the displacement of two H⁺ ions. An extremely important feature of these molecules is their extensive π-electron delocalization. The complexation of these and synthetic analogues has been the subject of a number of texts.[143-145] Some of these aspects are also covered by Dolphin (Chapter 21.1), and biological related properties by Hughes (Chapter 62.1).

Mention was made in the introductory section of the occurrence of metal porphyrins in oils and other fossil fuels, thus indicating that the porphyrins were formed early in the evolutionary process. Interestingly, some of the metals found in these porphyrin complexes, Ga, Ni and V, are not found in this form today. Either these originate from early life forms or metal-exchange reactions have occurred. Related to this, the biosynthesis of the naturally occurring porphyrin complexes is known to involve enzyme-catalyzed insertion of the metal ions into the porphyrin rings.

22.8.1 Porphyrins

These are rarely found free but occur as a common prosthetic group of proteins. The first such protein to be extensively studied was the iron porphyrin (Greek *porphyra*, purple) haemoglobin. This and other iron–porphyrin proteins play a vital role in the physiological activity of nearly all forms of life.[146] These forms have the same basic structure (**39**) but differ in the nature of the pyrrole substituents; these are shown for the major porphyrins in Table 13. It has become common practice to refer to all the iron–porphyrin proteins as haem proteins. The function of haemoglobin is, of

(39)

Table 13 A Glossary of Naturally Occurring Iron–Porphyrin (**39**) Structures and Names

R^1	R^2	R^3	R^4	R^5	R^6	R^7	R^8	Name
Me	CH=CH$_2$	Me	CH=CH$_2$	Me	CH$_2$CH$_2$CO$_2$H	CH$_2$CH$_2$CO$_2$H	Me	Protoporphyrin IX (haem)
Me	CH=CH$_2$	Me	CH(OH)Me	Me	CH$_2$CH$_2$CO$_2$H	CH$_2$CH$_2$CO$_2$H	Me	Haem a$_2$
Me	—CHMe–S—	Me	—CHMe–S—	Me	CH$_2$CH$_2$CO$_2$H	CH$_2$CH$_2$CO$_2$H	Me	Haem c
$R^2 = $ —CH(CH$_2$)$_2$CHC(CH$_2$)$_2$CHC(CH$_2$)$_2$CHCMe$_2$ (with OH and Me substituents)				$R^4 = $ CH=CH$_2$, $R^8 = $ CHO Others as above				Haem a
Me	CH$_2$CH$_2$CO$_2$H	Me	CH$_2$CH$_2$CO$_2$H	Me	CH$_2$CH$_2$CO$_2$H	CH$_2$CH$_2$CO$_2$H	Me	Coproporphyrin I
CH$_2$CO$_2$H	CH$_2$CH$_2$CO$_2$H	CH$_2$CO$_2$H	CH$_2$CH$_2$CO$_2$H	CH$_2$CO$_2$H	CH$_2$CH$_2$CO$_2$H	CH$_2$CO$_2$H	CH$_2$CH$_2$CO$_2$H	Uroporphyrin I
Me	CHO	Me	CH=CH$_2$	Me	CH$_2$CH$_2$CO$_2$H	CH$_2$CH$_2$CO$_2$H	Me	Chlorocruoroporphyrin

course, the transportation of O_2 by reversible coordination to Fe^{II}. Thus the prime function of the protoporphyrin IX ligand is to stabilize Fe^{II} against oxidation. This it achieves by inducing a low-spin configuration at the oxyiron centre and by means of its extensive electron delocalization avoids excessive charge accumulation at the iron centre. However, it should be noted that differences in oxygen affinity between haemoglobins from different sources arise from differences in the protein moiety. Also the haem group, which can be isolated from haemoglobin by acid treatment, undergoes the same facile oxidation reaction with O_2 (Scheme 4) as do other Fe^{II} complexes. The conclusion is that the protein plays the predominant role in the reversibility and extent of O_2 binding.

$$2[Fe^{II}haem]_{(aq)} + O_{2(g)} \rightleftharpoons [haem\ Fe^{II}-O_2-Fe^{II}haem] \longleftrightarrow [haem\ Fe^{III}-O_2^{2-}-Fe^{III}haem] \xrightarrow{2H^+}$$

$$2[Fe^{III}haem] + H_2O_2$$

Scheme 4

Respiratory pigments similar to the vertebrate haemoglobins have also been identified in many invertebrates. These vary from small proteins with two Fe–porphyrin units to large molecules containing up to 190 Fe–porphyrin units. Myoglobin, the O_2 storage protein in muscle tissue, is also a small iron–protoporphyrin protein. The crystal structures of this and a number of other porphyrin proteins are now known (Chapter 20.2, Table 11).

The cytochromes, a group of iron–porphyrin proteins found in the cells of all aerobic organisms, are involved in the electron-transfer steps of the reduction of O_2 to H_2O (a four-electron process). The Fe–porphyrin centres (haems a, a2 and c in Table 13, and others) are thus acting as catalytic electron-transfer centres, and as part of this function they exhibit a gradient of electrode potentials along the cytochrome chain. This variation in redox behaviour can be related to different protein environments and to differences in axial ligands (*i.e.* fifth and sixth coordination sites). The iron is not always in the $+2$ oxidation state, for example the enzymes catalase and peroxidase contain Fe^{III}–porphyrin groups. The iron in haemoglobin can also be oxidized to Fe^{III} to give the brown methaemoglobin. Small amounts of this oxidized form are invariably present in blood and, of course, are totally inactive towards O_2 transportation. The iron porphyrins therefore show three types of behaviour: (1) the Fe stays in the $+2$ oxidation state; (2) the Fe oscillates between the $+2$ and $+3$ states; and (3) the Fe remains in the $+3$ state. This diversity in behaviour is often related to properties of the porphyrin ligands, but it is now clear that it is the encompassing protein that is the major determinant of the properties and is particularly so in selecting the appropriate substrates.

One of the distinctive physical properties of the porphyrins is their strong colour. They all strongly absorb at *ca.* 400 nm (the Soret band) and less strongly in the visible region (α and β bands). These latter two bands are of particular diagnostic value. A distinctive red colour is the feature of the one natural Cu porphyrin; this is found in the feathers of certain tropical birds, *Turaco corythaix*, and was first reported in 1880! This Cu complex has no apparent *in vivo* function and its contribution to nature may be just its distinctive colour. The particular porphyrin involved is uroporphyrin III. This same porphyrin together with the isomeric uroporphyrin I are found in the urine of patients suffering from a disorder of porphyrin metabolism known as porphyria. Zn–porphyrin complexes are also occasionally excreted but appear to have no *in vivo* role.

22.8.2 Chlorophylls

These are a group of closely related pigments found in all organisms capable of photosynthesis, their vital function being the efficient absorption of visible radiation and its radiationless transfer. The metal ion involved is Mg^{II}. The most common of the chlorophylls are chlorophyll a (**40**; $R^1 = Me$, $R^2 = CHCH_2$) found in all plants that produce O_2, chlorophyll b (**40**; $R^1 = CHO$, $R^2 = CHCH_2$) found in green algae and some higher plants, and bacteriochlorophyll (**40**; $R^1 = COMe$, $R^2 = Me$, one less double bond in ring II) found in purple photosynthetic bacteria. Other forms are found in brown and red algae. Unique to the chlorophyll molecules is the alicyclic ring adjacent to the pyrrole ring (III), otherwise the other substituents are fairly similar to those in other porphyrins.

From coordination chemistry criteria the choice of Mg^{II} rather than a transition metal ion, which would bind more strongly and be at least as reactive, is rather a surprising one. In fact, the Mg^{II} binding is not strong: it is readily displaced by weak acids and other metal ions. Replacement with Zn^{II} results in the retention of some of the essential properties but no metal ion can exactly duplicate

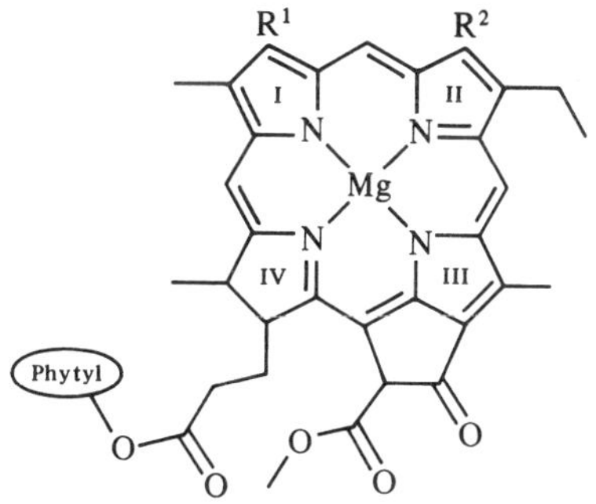

(40)

the behaviour of the Mg^{II} chlorophylls. In nature the Mg^{II} ion is the one of choice and there is no evidence of any other metal ion occurring in the chlorophyll molecule. Possible mechanisms by which the Mg^{II} ion is selected from the biological milieu have been discussed by Frausto da Silva and Williams.[3] The crystal structure of a bacteriochlorophyll–protein complex has been determined. This shows the Mg^{II} ion to be slightly out of the porphyrin plane.[147]

As with the other porphyrins, the chlorophylls have distinctive visible absorption spectra. Since their role is to absorb visible radiation, their absorption spectra are of vital importance. Chlorophyll extracted from plants and dissolved in a polar solvent absorbs near 660 nm; however, in the plant the maximum absorption is nearer 680 nm. The significance of this red shift in terms of molecular aggregation and the relation to photosynthesis has been discussed by Katz[148] but lies outside the scope of this chapter.

22.8.3 Corrinoids

Vitamin B$_{12}$ is a red crystalline, cobalt-containing compound that can be isolated from the liver. It has a functional role in preventing pernicious anaemia and also serves as a coenzyme in hydrogen and methyl transfer reactions (Co appears to be the only metal *in vivo* catalyzing C transfer reactions; O and N transfers are more common). Vitamin B$_{12}$ is also a growth-promoting factor for several microorganisms.

The structure has been established from chemical and X-ray crystallographic studies.[149] The structure of the porphyrin moiety, called a corrin ring, is shown in (41). Its most notable features are the direct link between the pyrrole rings I and IV and the less extensive π-electron delocalization. The corrin ring itself is *non-planar*, although the four N atoms do define a square planar geometry around the central Co^{III} ion. The fifth ligand is an N atom from a benzimidazole group which is covalently linked to pyrrole ring IV. The sixth site of the octahedral structure is occupied by a cyanide ion when the compound is isolated from natural sources; this is an artefact of the isolation procedure. Without a sixth ligand the compound is referred to as *cobalamin*.

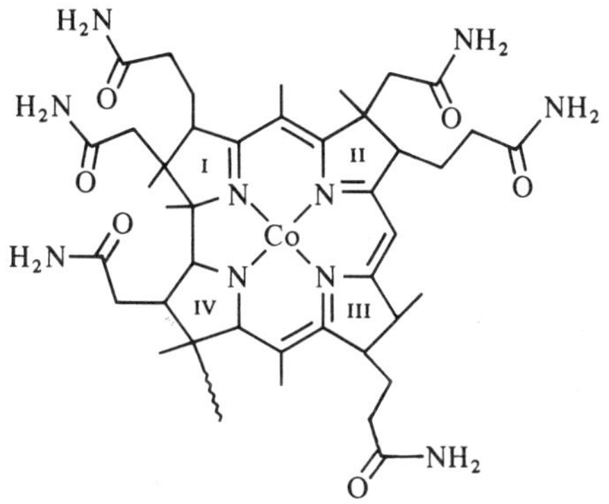

(41)

Although formally regarded as a Co^{III} compound it differs in several important aspects from conventional Co^{III} compounds. Thus, the metal centre is kinetically labile, and the coordinated CN$^-$ can be readily exchanged for other ligands, with rate constants several orders of magnitude larger than those of the more usual inert Co^{III} systems. The reactivity of this sixth ligand position is all important in its coenzyme role. A further unusual feature is the ability of the Co^{III} centre to coordinate a methyl group in an aqueous environment: methylcobalamin is the intermediate in the methyl transfer reactions. In fact the first naturally occurring Co—C bond to be reported was that in adenosylcobalamin (sixth ligand = 5'-deoxyadenosyl; **42**). Normally, Co—C bond formation requires a low oxidation state and stabilizing ligands such as CO. The markedly increased reactivity of the Co in the cobalamins is attributed to the corrin π-electron system and to the relatively small geometric changes that occur.[150]

(42)

The reduction properties of the cobalamins also differ from the normal Co^{III} complexes in that they can be readily reduced to the Co^I state, *e.g.* as in the methylation reaction (9). The electrochemical properties of the cobalamins have recently been reviewed.[151] Several reviews are also available concerning their biological activity, the mechanisms of reactions, and synthetic analogues.[150,152-155]

$$Co^{III}—Me + homocysteine \longrightarrow (Co^I) + methionine \qquad (9)$$
$$methylcobalamin$$

22.9 CARBOHYDRATES

Carbohydrates comprise the many naturally occurring monosaccharides and the oligo- and poly-saccharides derived from them. The oligosaccharides consist of two or more monosaccharide units linked together by glycoside bonds. These bonds are formed by a condensation reaction between a hydroxyl group of one monosaccharide unit and the reducing (aldehydic) group of another monosaccharide: they may be regarded as the carbohydrate equivalent of the peptide bond. Carbohydrate units are also found in nature covalently bonded to lipids, proteins, and of course in the nucleic acids and nucleotides. One of their more important functions is in providing structural rigidity, *e.g.* the plant cellulose and pectic acids. Glycoproteins are often associated with cell membranes.

The larger oligo- and poly-saccharides with their large number of polar hydroxyl groups and their conformational flexibility are potential ligands, particularly for the 'hard' or 'class A' metal ions, *e.g.* Mg^{II}, Ca^{II}, Fe^{III}. This conclusion was also made by Angyal in his review on cation binding of carbohydrates.[156] of the monosaccharides *cis*-inositol (**43**) forms some of the strongest complexes because of its three *syn* axial OH groups. This highlights the requirements of the right conformation and at least three OH groups for effective coordination.

(43)

By far the best method for establishing coordination has been ^1H NMR. Early NMR studies showed, as might be anticipated, that cation binding increases with increasing positive charge. There also seems to be a requirement for cations with radii larger than 80 pm, presumably because of the need to make effective contact with at least three OH groups. Hence, the group IA cations and Mg^{II} form extremely weak complexes, although Ca^{II} forms some of the stronger complexes. The stability constants of the 1:1 complexes are only of the order 1–2 $mol^{-1} dm^3$. Larger values are found with the lanthanide(III) ions. These complexes have proved useful in determining ligand conformations from NMR line-broadening measurements.

The increased cation binding possibilities with the polysaccharides are particularly important for plants where the simplest mode of uptake is by absorption on to the plant and cell walls. This is also considered to be important in removing nutrient cations from clays.[157] Ca^{II} binding is also found to be necessary to stabilize the polysaccharide structures.[158] Deiana and coworkers[159] have established that complex formation with polygalacturonic acids is of the outer-sphere type with non-transition metals but of the inner-sphere type with the transition metal ions Cu^{II} and VO^{2+}. They also found that this and other polyuronic acids reduce Fe^{III} to Fe^{II}. Since these acids are the main components of plant roots and reduction to Fe^{II} is known to be necessary for iron absorption by roots, this observation could be of considerable significance. Pectates, which include polygalacturonic acids, are found in root systems and have been implicated as the major binding sites in roots for the Cu ions that accumulate in the copper-tolerant species *Armeria maritima*.[160]

The highly ionic lipopolysaccharides, present in the outer membrane of Gram-negative bacteria, have significant affinities for metal ions, with the binding not being readily reversible.[161] A more recent quantitative study confirms these observations and shows these membrane components as having ion-exchange properties with particularly strong binding affinities for Mg^{II} and Ca^{II}. The lipopolysaccharides from different sources were found to have metal-binding properties which depended on their gross structures.[162] Other derivatives of polysaccharides which present additional potential donor sites need to be examined to see if they possess enhanced metal coordination properties. A highly important example of this occurring with dramatic effects is the formation of highly insoluble complexes between Zn^{II} ions in the diet and phytic acid. Phytic acid [myoinositol hexakis(dihydrogen phosphate)], a phosphate-substituted monosaccharide, occurs in maize and soya-bean products. Diets high in these materials lead to gross zinc deficiency because of the formation of the insoluble zinc phytates. The results are serious inhibitions of growth, and other physiological disorders.[163] Monosaccharides can also readily form Schiff base complexes;[164] the ligands are tetradentate (**44**; equation 10). Their behaviour closely resembles that of amino acids and peptides in this respect (Chapter 20.2), and perhaps they deserve more attention.

$$(10)$$

* indicates coordination sites

(44)

To finally underline the wide functional roles that metal–carbohydrate coordination may play in nature, complexes of this type have long been used as mineral supplements because of their greater absorption in the digestive system, *e.g.* iron–dextran compositions in treating iron deficiency. More recently, metal pectate compositions (see ref. 165 and literature cited therein) have been advocated for use as mineral supplementation.

22.10 ADDENDUM

Pertinent texts which have been published more recently are: 'Trace Elements in Soils and Plants', ed. A. Kabata-Pendias and H. K. Pendias, CRC Press, Boca-Raton, FL, 1984; 'Biochemistry of the Essential Ultratrace Elements', ed. E. Frieden, Plenum, New York, 1984. Several aspects of the siderophores and their coordination are featured in volume 85 of *Structure and Bonding (Berlin)* (1984). Among the more recent additions to the series *Metal Ions in Biological Systems*, edited by H. Sigel, are included 'Zinc and its Role in Biology and Nutrition', vol. 15 (1983); 'Calcium and its Role in Biology', vol. 17 (1984); 'Circulation of Metals in the Environment', vol. 18 (1984); and 'Antibiotics (Ionophores) and their Complexes', vol. 19 (1985). The proceedings of the second international conference on bioinorganic chemistry, held in Portugal in 1985, have also been published (*Rev. Port. Quim.*, 1985, **27**).

22.11 REFERENCES

1. S. Mann, S. B. Parker and R. J. P. Williams, *J. Inorg. Biochem.*, 1983, **18**, 169.
2. B. Sarkar, in 'An Introduction to Bio-Inorganic Chemistry', ed. D. R. Williams, C. C. Thomas, Springfield, IL, 1976, p. 318.
3. J. J. R. Frausto da Silva and R. J. P. Williams, *Struct. Bonding (Berlin)*, 1976, **29**, 67.
4. K. Kustin, G. C. McLeod, T. R. Gilbert and L. R. Briggs, *Struct. Bonding (Berlin)*, 1983, **53**, 139.

5. R. J. P. Williams, *Chem. Br.*, 1983, **19**, 1009.
6. A. N. Astanina, A. P. Rudenko, M. A. Ismailova, E. Y. Offengenden and H. M. Yakubov, *Inorg. Chim. Acta*, 1983, **79**, 284.
7. R. Bonnett, P. J. Burke and A. Reszka, *J. Chem. Soc., Chem. Commun.*, 1983, 1085.
8. J. M. E. Quirke and J. R. Maxwell, *Tetrahedron*, 1980, **36**, 3453.
9. G. A. Wolff, M. Murray, J. R. Maxwell, B. K. Hunter and J. K. M. Sanders, *J. Chem. Soc., Chem. Commn.*, 1983, 922.
10. H. Reiche and A. J. Bard, *J. Am. Chem. Soc.*, 1979, **101**, 3127.
11. K. Harada and M. Matsuyama, *Biosystems*, 1979, **11**, 47.
12. M. Beck, V. Gaspar and J. Ling, *Magy. Kem. Foly.*, 1979, **85**, 147.
13. M. B. McBride, *Can. J. Soil Sci.*, 1980, **60**, 145.
14. A. Szalay, *Geochim. Cosmochim. Acta*, 1964, **28**, 1605.
15. P. R. Bloom, in 'Chemistry of the Soil Environment', ed. M. Stelly, American Society of Agronomy Special Publication No. 40, Winsconsin, 1981, p. 129.
16. H. J. M. Bowen, 'Trace Elements in Biochemistry', Academic, London, 1966.
17. S. Ramamoorthy and P. G. Manning, *J. Inorg. Nucl. Chem.*, 1973, **35**, 1279.
18. F. J. Stevenson, 'Humus Chemistry', Wiley—Interscience, New York, 1982.
19. M. Schnitzer, *Soil Sci. Soc. Am. Proc.*, 1969, **33**, 75.
20. K. Murray and P. W. Linder, *J. Soil Sci.*, 1983, **34**, 511.
21. J. C. Brown, *Adv. Chem. Ser.*, 1977, **162**, 93.
22. L. O. Tiffen, *Plant Physiol.*, 1966, **41**, 515.
23. L. O. Tiffen, in 'Biological Implications of Metals in the Environment', Proceedings of the 15th Annual Hanford Life Sciences Symposium, Richland, USA, 1975, p. 315.
24. J. Lee, R. D. Reeves, R. R. Brooks and T. Jaffre, *Phytochemistry*, 1977, **16**, 1503.
25. G. L. Lyon, P. J. Peterson and R. R. Brooks, *Planta*, 1969, **88**, 282.
26. P. M. May, P. W. Linder and D. R. Williams, *J. Chem. Soc., Dalton Trans.*, 1977, 588.
27. B. Lonnerdal, A. G. Stanislowski and L. S. Hurley, *J. Inorg. Biochem.*, 1980, **12**, 71.
28. T. G. Spiro and P. Saltman, *Struct. Bonding (Berlin)*, 1969, **6**, 116.
29. G. T. Goodman, *Proc. R. Soc. London, Ser. A*, 1974, **399**, 373.
30. M. E. Farago, *Coord. Chem. Rev.*, 1981, **36**, 155.
31. P. J. Peterson, in 'Metals and Micronutrients: Uptake and Utilization by Plants', ed. D. A. Robb and W. S. Pierpoint, Academic, London, 1983, p. 51.
32. E. R. Still and R. J. P. Williams, *J. Inorg. Biochem.*, 1980, **13**, 35.
33. R. Fiumano, A. Scozzafava and R. Gabrielli, *Inorg. Chim. Acta*, 1983, **79**, 237.
34. E. B. Finley and F. L. Cerklewski, *Am. J. Clin. Nutr.*, 1983, **37**, 553.
35. J. J. Mortvedt, P. M. Giordano and W. L. Lindsay (eds.), 'Micronutrients in Agriculture', Soil Science Society of America, Madison, USA, 1972.
36. A. D. Rovira, *Botanical Rev.*, 1969, **35**, 35.
37. D. G. Hill-Cottingham and C. P. Lloyd-Jones, *Nature (London)*, 1968, **220**, 3889.
38. Y. Mino, T. Ishida, N. Ota, M. Inoue, K. Nomoto, H. Takemoto, H. Tanaka and Y. Sugiura, *J. Am. Chem. Soc.*, 1983, **105**, 4671.
39. D. I. M. Harris and A. Sass-Kortsak, *J. Clin. Invest.*, 1967, **46**, 659.
40. E. J. Peck and W. R. Ray, *J. Biol. Chem.*, 1971, **246**, 1160.
41. P. M. May, P. W. Linder and D. R. Williams, *J. Chem. Soc., Dalton Trans.*, 1977, 588.
42. B. Sarkar, *Met. Ions Biol. Syst.*, 1981, **12**, 233.
43. A. G. Bearn and H. G. Kunkel, *Proc. Soc. Exp. Biol. Med.*, 1954, **85**, 44.
44. P. Z. Neumann and A. Sass-Kortsak, *J. Clin. Invest.*, 1967, **46**, 646.
45. P. S. Hallman, D. D. Perrin and A. M. Watt, *Biochem. J.*, 1971, **121**, 549.
46. P. M. May and D. R. Williams, *Met. Ions Biol. Syst.*, 1981, **12**, 283.
47. T. Alemdaroglu and G. Berthon, *Inorg. Chim. Acta*, 1981, **56**, 115.
48. S. H. Laurie and E. S. Mohammed, *Inorg. Chim. Acta*, 1981, **55**, L63.
49. A. Gergely and T. Kiss, *Met. Ions Biol. Syst.*, 1979, **9**, 143.
50. D. D. Perrin and H. Stunzi, *Met. Ions Biol. Syst.*, 1982, **14**, 207.
51. L. Pickart, in 'Chemistry and Biochemistry of Amino Acids, Peptides and Proteins', ed. B. Weinstein, Dekker, New York, 1982, vol. 6, p. 75.
52. K. Burger, *Met. Ions Biol. Syst.*, 1979, **9**, 213.
53. H. Sigel and R. B. Martin, *Chem. Rev.*, 1974, **74**, 471.
54. I. Sovago, E. Farkas and A. Gergely, *J. Chem. Soc., Dalton Trans.*, 1982, 2159.
55. L. Pickart, *Inorg. Chim. Acta*, 1983, **79**, 305.
56. S.-J. Lau and B. Sarkar, *Biochem. J.*, 1981, **199**, 649.
57. P. M. May, J. Whittaker and D. R. Williams, *Inorg. Chim. Acta*, 1983, **80**, L5.
58. L. Pickart and M. M. Thaler, *Fed. Proc., Fed. Am. Soc. Exp. Biol.*, 1979, **38**, 668.
59. G. Formicka-Kozlowska, H. Kozlowski, M. Bezer, L. D. Pettit, G. Kupryszewski and J. Przybylski, *Inorg. Chim. Acta*, 1981, **56**, 79.
60. L. Flohe, H. Ch. Benohr, H. Sies, H. D. Walker and A. Wendel (eds.), 'Glutathione', Academic, New York, 1974.
61. I. M. Arias and W. B. Jakoby (eds.) 'Glutathione: Metabolism and Function', Raven, New York, 1976.
62. G. Marzullo and A. J. Friedhoff, *Life Sci.*, 1977, **21**, 1559.
63. D. L. Rabenstein, R. Guevremont and C. A. Evans, *Met. Ions Biol. Syst.*, 1979, **9**, 103.
64. E. Friedheim and C. Corvi, *J. Pharm. Pharmacol.*, 1975, **27**, 624.
65. K. Miyoshi, K. Ishizu, H. Tanaka, Y. Sugiura and K. Asada, *Inorg. Chim. Acta*, 1983, **79**, 261.
66. M. Y. Hamed and J. Silver, *Inorg. Chim. Acta*, 1983, **80**, 115.
67. K. Miyoshi, Y. Sugiura, K. Ishizu, Y. Iitaka and H. Nakamura, *J. Am. Chem. Soc.*, 1980, **102**, 6130.
68. L. A. Dominey and K. Kustin, *J. Inorg. Biochem.*, 1983, **18**, 153.

69. J. Letter, Jr. and R. Jordan, *J. Am. Chem. Soc.*, 1975, **97**, 2381.
70. H. Degani and R. E. Lenkinski, *Biochemistry*, 1980, **19**, 3430.
71. R. E. Lenkinski and R. L. Stephens, *J. Inorg. Biochem.*, 1981, **15**, 95.
72. W. Mertz, *Nutr. Rev.*, 1975, **33**, 129.
73. S. Ito, G. Nardi and G. Prota, *J. Chem. Soc., Chem. Commun.*, 1976, 1042.
74. S. K. Carter, S. T. Cook and H. Umezawa (eds.), 'Bleomycin, Status and New Developments', Academic, New York, 1978.
75. J. C. Dabrowiak, *Met. Ions Biol. Syst.*, 1980, **11**, 305.
76. D. E. Fenton, *Chem. Soc. Rev.*, 1977, **6**, 325.
77. Yu. A. Ovchinnikov, V. T. Ivanov and A. M. Shkrob, 'Membrane-Active Complexones', Elsevier, Amsterdam, 1974.
78. M. R. Truter, *Philos. Trans. R. Soc. London, Ser. B*, 1975, **272**, 29.
79. V. T. Ivanov and S. V. Sychev, in 'Structure of Complexes between Biopolymers and Low Molecular Weight Molecules', ed. W. Bartmann and G. Snatzke, Wiley, Chichester, 1982, p. 107.
80. R. E. Koeppe, II, J. M. Berg, K. O. Hodgson and L. Stryer, *Nature (London)*, 1979, **279**, 723.
81. P. A. Ketchum and M. S. Owens, *J. Bacteriol.*, 1975, **122**, 412.
82. K. N. Raymond, *Adv. Chem. Ser.*, 1977, **162**, 33.
83. K. N. Raymond and C. J. Carrano, *Acc. Chem. Res.*, 1979, **12**, 183.
84. R. W. Grady and A. Cerami, *Annu. Rep. Med. Chem.*, 1979, **13**, 219.
85. G. B. Wong, M. J. Kappel, K. N. Raymond, B. Matzanke and G. Winkelmann, *J. Am. Chem. Soc.*, 1983, **105**, 810.
86. D. van der Helm, J. R. Baker, R. A. Loghry and J. D. Ekstrand, *Acta Crystallogr., Sect. B*, 1981, **37**, 323.
87. M. Llinas, D. M. Wilson and M. P. Klein, *J. Am. Chem. Soc.*, 1977, **99**, 6846.
88. W. R. Harris, C. J. Carrano, S. R. Cooper, S. R. Sofen, A. Avdeef, J. V. McArdle and K. N. Raymond, *J. Am. Chem. Soc.*, 1979, **101**, 6097.
89. W. H. Rostetter, T. J. Erickson and M. C. Venuti, *J. Org. Chem.*, 1980, **45**, 5011.
90. A. Shanzer and J. Libman, *J. Chem. Soc., Chem. Commun.*, 1983, 846.
91. I. G. O'Brien, G. B. Cox and F. Gibson, *Biochim. Biophys. Acta*, 1971, **237**, 537.
92. R. C. Hider, A. R. Mohd-Nor, J. Silver, I. E. G. Morrison and L. V. C. Rees, *J. Chem. Soc., Dalton Trans.*, 1981, 609.
93. W. R. Harris, C. J. Carrano and K. N. Raymond, *J. Am. Chem. Soc.*, 1979, **101**, 2722.
94. M. Dobler, *Helv. Chim. Acta*, 1972, **55**, 1371.
95. E. Eyal and G. A. Rechnitz, *Anal. Chem.*, 1971, **43**, 1090.
96. M. N. Hughes, 'The Inorganic Chemistry of Biological Processes', 2nd edn., Wiley, Chichester, 1981, p. 260.
97. W. L. Duax, G. D. Smith and P. D. Strong, *J. Am. Chem. Soc.*, 1980, **102**, 6725.
98. J. Bolte, C. Demuynck, G. Jeminet, J. Juillard and C. Tissier, *Can. J. Chem.*, 1982, **60**, 981.
99. C. C. Chiang and I. C. Paul, *Science*, 1977, **196**, 1441.
100. D. H. Davies, E. W. Snape, P. J. Suter, T. J. King and C. P. Falshaw, *J. Chem. Soc., Chem. Commun.*, 1981, 1073.
101. G. R. Pettit, C. L. Herald, D. L. Doubek and D. L. Herald, *J. Am. Chem. Soc.*, 1982, **104**, 6846.
102. R. R. Crichton and J.-C. Mareschal, in 'Inorganic Biochemistry', ed. H. A. O. Hill, Royal Society of Chemistry Specialist Periodical Report, London, 1982, vol. 3, p. 78.
103. T. Peters, Jr., in 'The Plasma Proteins', ed. F. W. Putnam, 2nd edn., Academic, New York, 1975, vol. 1, p. 133.
104. B. Sarkar, *Life Chem. Rep.*, 1983, **1**, 165.
105. E. L. Giroux and R. I. Henkin, *Biochim. Biophys. Acta*, 1972, **273**, 64.
106. E. L. Giroux and J. Schoun, *J. Inorg. Biochem.*, 1981, **14**, 359.
107. S. H. Laurie, *J. Inherited Metab. Dis.*, 1983, **6**, 9.
108. P. Aisen, A. Leibman and J. Zweir, *J. Biol. Chem.*, 1978, **253**, 1930.
109. S. H. Banyard and D. K. Stammers, *Nature (London)*, 1978, **271**, 282.
110. S. Stefanini, E. Chianeone and E. Antonini, *Biochim. Biophys. Acta*, 1978, **542**, 170.
111. G. Taborsky, *J. Biol. Chem.*, 1980, **255**, 2976.
112. D. P. E. Dickson and S. Rottem, *Eur. J. Biochem.*, 1979, **101**, 291.
113. G. L. Eichhorn, in 'Inorganic Biochemistry', ed. G. L. Eichhorn, Elsevier, Amsterdam, 1973, vol. 2, pp. 1191 and 1210.
114. A. T. Tu and M. J. Heller, *Met. Ions Biol. Syst.*, 1974, **1**, 1.
115. R. W. Gellert and R. Bau, *Met. Ions. Biol. Syst.*, 1979, **8**, 1.
116. R. B. Martin and Y. H. Mariam, ref. 115, p. 57.
117. L. G. Marzilli, *Prog. Inorg. Chem.*, 1977, **23**, 255.
118. G. Arena, R. Cali, V. Cucinotta, S. Musumeci, E. Rizzarelli and S. Sammartano, *J. Chem. Soc., Dalton Trans.*, 1983, 1271.
119. M. Cohn and T. R. Hughes, Jr., *J. Biol. Chem.*, 1962, **237**, 176.
120. E. Buncel, A. R. Norris, W. J. Racz and S. E. Taylor, *J. Chem. Soc., Chem. Commun.*, 1979, 562.
121. S. J. Karlik, E. Tarien, G. A. Elgavish and G. L. Eichhorn, *Inorg. Chem.*, 1983, **22**, 525.
122. D. Dunaway-Mariano and W. Cleland, *Biochemistry*, 1980, **19**, 1506.
123. S. H. McClaugherty and C. M. Grisham, *Inorg. Chem.*, 1982, **21**, 4133.
124. H. Sigel, *Pure Appl. Chem.*, 1983, **55**, 137.
125. P. R. Mitchell, B. Prijs and H. Sigel, *Helv. Chim. Acta*, 1979, **62**, 1723.
126. H. Sigel, B. E. Fischer and E. Farkas, *Inorg. Chem.*, 1983, **22**, 925.
127. C. M. Frey and J. Stuehr, *Met. Ions Biol. Syst.*, 1974, **1**, 52.
128. B. S. Cooperman, *Met. Ions Biol. Syst.*, 1976, **5**, 79.
129. G. L. Eichhorn, in 'Advances in Inorganic Biochemistry', ed. G. L. Eichhorn and L. G. Marzilli, Elsevier, New York, 1981, vol. 3, p. 2.
130. G. L. Eichhorn and L. G. Marzilli, 'Metal Ions in Genetic Information Transfer', Elsevier, New York, 1981.
131. J. K. Barton, J. J. Dannenbey and A. L. Raphael, *J. Am. Chem. Soc.*, 1982, **104**, 4967.
132. A. Yamagishi, *J. Chem. Soc., Chem. Commun.*, 1983, 572.
133. T. Lindahl, A. Adams and J. R. Freoco, *Proc. Natl. Acad. Sci. USA*, 1966, **55**, 941.

134. R. S. Brown, B. E. Hingerty, J. C. Dewan and A. Klug, *Nature (London)*, 1983, **303**, 543.
135. B. Rosenberg, in 'Nucleic Acids and Metal Ion Interactions', ed. T. G. Spiro, Wiley, New York, 1980, p. 1.
136. S. J. Lippard, *ACS Symp. Ser.*, 1983, **209**.
137. A. W. Prestakyo, S. T. Crooke and S. K. Carter (eds.), 'CISPLATIN: Current Status and New Developments', Academic, New York, 1980.
138. J. J. Roberts and A. J. Thomson, *Prog. Nucleic Acid Res.*, 1979, **22**, 71.
139. J. K. Barton and S. J. Lippard, ref. 135, p. 31.
140. U. K. Haring and R. B. Martin, *Inorg. Chim. Acta*, 1983, **78**, 259.
141. A. Pasini, R. Ugo, M. Gullotti, F. Spreafico, S. Filippeschi and A. Velcich, *Inorg. Chim. Acta*, 1983, **79**, 289.
142. P. N. Rao, M. L. Smith, S. Pathak, R. A. Howard and J. L. Bear, *J. Natl. Cancer Inst.*, 1980, **64**, 905.
143. G. L. Eichhorn (ed.), 'Inorganic Biochemistry', Elsevier, Amsterdam, 1973, vol. 2.
144. K. M. Smith (ed.), 'Porphyrins and Metalloporphyrins', Elsevier, New York, 1975.
145. B. D. Berezin, 'Coordination Compounds of Porphyrins and Phthalocyanines', Wiley, Chichester, 1981.
146. A. B. P. Lever and H. B. Gray (eds.), 'Iron Porphyrins', Addison-Wesley, Massachusetts, 1983, parts 1 and 2.
147. R. E. Fenna and B. W. Matthews, *Nature (London)*, 1975, **258**, 573.
148. J. J. Katz, ref. 143, p. 1022.
149. C. Brink-Shoemaker, D. W. J. Cruickshank, D. C. Hodgkin, J. Kamper and D. Pilling, *Proc. R. Soc. London, Ser. A*, 1964, **1**, 278.
150. H. A. O. Hill, ref. 143, p. 1067.
151. D. Lexa and J.-M. Saveant, *Acc. Chem. Res.*, 1983, **16**, 235.
152. J. M. Pratt, 'Inorganic Chemistry of Vitamin B_{12}', Academic, New York, 1972.
153. B. M. Babior (ed.), 'Cobalamin, Biochemistry and Pathophysiology', Wiley-Interscience, New York, 1975.
154. D. Dolphin (ed.), 'B_{12} Chemistry', Wiley-Interscience, New York, 1982, vol. 1.
155. A. I. Scott, *Acc. Chem. Res.*, 1978, **11**, 29.
156. S. J. Angyal, *Chem. Soc. Rev.*, 1980, **9**, 415.
157. S. Ramamoorthy and G. G. Leppard, *J. Theor. Biol.*, 1977, **66**, 527.
158. R. Repaska, *Biochim. Biophys. Acta*, 1958, **30**, 225.
159. C. Gessa, M. L. De Cherchi, A. Dessi, S. Deiana and G. Micera, *Inorg. Chim. Acta*, 1983, **80**, L53.
160. M. E. Farago, W. A. Mullen, M. M. Cole and R. F. Smith, *Environ. Pollut.*, 1980, **A21**, 225.
161. M. Schindler and M. J. Osborn, *Biochemistry*, 1979, **18**, 4425.
162. R. T. Coughlin, S. Tonsager and E. J. McGroarty, *Biochemistry*, 1983, **22**, 2002.
163. N. T. Davies and H. Reid, *Br. J. Nutr.*, 1979, **41**, 579.
164. T. Tsubomura, S. Yano and S. Yoshikawa, *Polyhedron*, 1983, **2**, 123.
165. B. Lakatos, J. Meisel, A. Rockenbauer, P. Simon and L. Korecz, *Inorg. Chim. Acta*, 1983, **79**, 269.
166. M. Schnitzer, *Agrochimica*, 1978, **22**, 216.
167. K. N. Pearce, *Aust. J. Chem.*, 1980, **33**, 1511.
168. P. Amico, P. G. Daniele, G. Ostacoli, G. Arena, E. Rizzarelli and S. Sammartano, *Inorg. Chim. Acta*, 1980, **44**, L219.
169. E. R. Still and P. Wikberg, *Inorg. Chim. Acta*, 1980, **46**, 147.
170. G. Berthon, P. M. May and D. R. Williams, *J. Chem. Soc., Dalton Trans.*, 1978, 1433.
171. D. D. Perrin and A. E. Watt, *Biochim. Biophys. Acta*, 1971, **230**, 96.
172. A. M. Corrie, M. D. Walker and D. R. Williams, *J. Chem. Soc., Dalton Trans.*, 1976, 1012.

14

Phosphorus, Arsenic, Antimony and Bismuth Ligands

CHARLES A. McAULIFFE
University of Manchester Institute of Science and Technology, UK

14.1 INTRODUCTION

Paul Thenard produced trimethylphosphine from methyl chloride and impure calcium phosphide at 180–300 °C in 1847.[1] After the discovery of aliphatic amines, the relationship between amines and phosphines stimulated Hofmann, and then Cahours,[2] to develop this area of chemistry; Michaelis synthesized triphenylphosphine in 1885.[3]

It was F. G. Mann[4] who began the explosion of interest in transition metal phosphine chemistry. A number of key figures should be mentioned: Dwyer,[5] Chatt[4] and Nyholm[6] were paramount. In recent times the general area has itself given rise to important sub-areas, *e.g.* metal hydrides,[7] nitrogen fixation,[8] stability of unusual oxidation states,[6] metal alkyls and aryls,[9] and vast areas of homogeneous (and heterogeneous) catalysis.[10] In this section, the synthesis and stereochemical properties of the ligands containing phosphorus, arsenic, antimony and bismuth are discussed, as well as the nature of the metal–ligand bond.

14.2 SYNTHESIS

There are several excellent works giving details on the preparation of organophosphine ligands[11,12] and some of their physical properties. Until the mid-1970s most of the coordination chemistry was restricted to relatively few ligands, but recently more exotic ligands, *e.g.* phosphorus-containing macrocycles, have been investigated. In this section the main methods for the synthesis of organophosphines are described, and particular attention is paid to the availability of ready starting materials and/or to routes which give good yields. It should be pointed out that the physiological properties of these ligands and their precursors are for the most part unknown, but all should be treated as toxic. The syntheses should always be carried out in fume hoods with adequate ventilation, with thought for fire precautions, and the side products invariably need some chemical treatment before release into the environment.

14.2.1 Phosphines

The most important precursors are those containing a P–halogen bond, R_nPX_{3-n}, or alkali metal derivatives, MPR_2.

14.2.1.1 Halophosphine compounds, R_nPX_{3-n}

The phenyl derivatives $PhPCl_2$, Ph_2PCl and PCl_3 itself are commercially available and a useful review of halophosphines[13] has been published. The reaction of PCl_3 with Grignard or organolithium reagents gives almost solely the tertiary phosphine, R_3P, but with highly branched alkyl groups, *e.g.* cyclohexyl, isopropyl or *t*-butyl, the intermediate R_2PCl or $RPCl_2$ can be isolated under appropriate conditions. Although $PhPCl_2$ and Ph_2PCl can be easily bought, their synthesis is straightforward. Reaction of $PhCl$, PCl_3 and $AlCl_3$, and subsequent reduction by aluminum produces $PhPCl_2$;[14] other (aryl)PCl_2 compounds may be similarly formed, though competing side reactions such as isomerization of the R group and production of (aryl)$_2PCl$ may also occur. Fractional distillation of the products (Ph_2PCl and PCl_3) of the $AlCl_3$-catalyzed disproportionation of $PhPCl_2$ yields Ph_2PCl.[15] An extremely important precursor to a range of bidentate ligands is o-$C_6H_4(Br)PCl_2$, synthesized as in equation (1).

$$ (1) $$

The extremely reactive $MePCl_2$ can best be prepared from PCl_3, $MeCl$ and $AlCl_3$, to form initially $[MePCl_3][AlCl_4]$, which can subsequently be decomposed with $PhPCl_2$ and $POCl_3$.[16] The almost equally reactive Me_2PCl is obtained from the distillation of a mixture of $Me_2P(S)P(S)Me_2$ and $PhPCl_2$, but improved yields are obtained from the two-stage process shown in equation (2).[17]

$$ Me_2P(S)P(S)Me_2 \xrightarrow{SO_2Cl_2} Me_2P(S)Cl \xrightarrow{PBu^n_3} Me_2PCl \ (+ Bu^n_3PS) \qquad (2) $$

The precursor diphosphine disulfide is obtained from $PSCl_3$ and $MeMgI$.[18] Using the appropriate ratios of reactants, dimethylaminochlorophosphines ($Me_2N)_nPCl_{3-n}$ ($n = 1, 2$) are obtained from Me_2NH and PCl_3.[19] These are extremely useful synthetic reagents for the preparation of unsymmetrical multidentate phosphines, since the P—N bonds are easily cleaved by gaseous HCl to give P—Cl bonds.

The ω-chloroalkyldichlorophosphines, $Cl(CH_2)_nPCl_2$, unlike the corresponding chloroalkylarsines, are very difficult to prepare and are rarely used in syntheses. However, $Cl_2PCH_2CH_2PCl_2$ is readily prepared.[20] All of the chlorophosphines are malodorous, moisture sensitive and toxic, and the lower alkyl derivatives are spontaneously inflammable.

14.2.1.2 *Alkali metal derivatives*

The synthesis and uses of alkali metal phosphides have been reviewed.[21] Alkali metal diphenylphosphides are readily obtained by cleavage of a phenyl group from PPh_3 by lithium in tetrahydrofuran (THF),[22] sodium or lithium in liquid ammonia[23] or potassium in dioxane (equation 3).[24] The metal–phenyl product can be destroyed by addition of a calculated amount of Bu^tCl or NH_4X. Usually $MPPh_2$ compounds are prepared *in situ*, but they can be isolated, usually as solvates. These compounds can also be obtained from Ph_2PCl and the alkali metal, or from Ph_2PH and alkyllithium.[25]

$$ 2M + PPh_3 \longrightarrow MPPh_2 + MPh \qquad (3) $$

Tertiary alkyl and cycloaliphatic phosphines do not undergo alkali metal cleavage; however, (aryl)$_{3-n}$(alkyl)$_n$P ($n = 1, 2$) are cleaved[26] by Li/THF to give $LiP(aryl)_{2-n}$(alkyl)$_n$ and aryllithium. Although this appears to be an attractive route to LiP(alkyl)$_2$, it has not been widely exploited. Somewhat anomalously, $PhEt_2P$ and K produce $KPEtPh$ and not $KPEt_2$.[27]

Dimethylphosphide, Me_2P^-, can be easily obtained in ether (Li salt) or ammonia (Na) by alkali metal cleavage of the P—P bond in Me_2PPMe_2, itself being available from desulfurization of $Me_2P(S)P(S)Me_2$ with PBu^n_3 or iron powder.[18,28] This reaction would seem to have general applicability for PR_2^- ($R = Et, Pr^n, Bu^n, \frac{1}{2}MePh$) production, since the corresponding $R_4P_2S_2$ compounds are available from the analogous Grignard route.[29]

Phenyl-substituted diphosphines, $R_2P(CH_2)_nPR_2$, can also be cleaved by Na or Li in liquid ammonia or THF to yield $MPhP(CH_2)_nPPhM$,[30] which are very useful starting materials for the production of unsymmetrical bidentate ligands $PhRP(CH_2)_nPRPh$. However, care should be taken in these reactions as Ph—P cleavage is sometimes accompanied by P—CH_2 fission. For the production of unsymmetrical bidentate phosphines, such as $R_2P(CH_2)_nPPh_2$, stepwise formation of mono- and di-phosphonium salts provides a facile route (see Section 14.2.4).

The alkali metal derivatives are usually coloured deep red and are soluble in ether, THF, dioxane and liquid ammonia (some ammonolyze). They readily hydrolyze and oxidize. They are strong nucleophiles and react with many functional groups;[31] they also slowly ring-open THF under reflux to give $MO(CH_2)_4PR_2$.

14.2.1.3 Primary and secondary phosphines

The coordination chemistry of these ligands does not grow rapidly, but they are also of interest because of their use in the synthesis of multidentate ligands. Phosphine, PH_3, is best obtained by the hydrolysis of aluminum phosphide.[32] The easiest preparations of primary phosphines, RPH_2, are $LiAlH_4$ reductions[1] of RPX_2, $RP(S)X_2$, $RP(O)X_2$, *etc.*, and although $NaPH_2$ is also a useful starting point, it is not as attractive since it initially involves the synthesis and manipulation of PH_3. Lithyl reduction of dimethylmethylphosphonate in dioxane yields $MePH_2$, a spontaneously inflammable gas (b.p. $-14\,°C$).[33] The malodorous liquid, $PhPH_2$, is obtained from $LiAlH_4$ and $PhPCl_2$.[34]

Secondary phosphines, R_2PH, are also readily obtained from $LiAlH_4$ reduction of R_2PX, $R_2P(S)X$, $R_2P(O)X$, *etc.*,[1] but other methods are known, *viz.* hydrolysis of the corresponding phosphide, MPR_2, and alkylation and hydrolysis of $MPRH$. The inflammable liquid Me_2PH (b.p. $20\,°C$) is best prepared[35] from $LiAlH_4$ and $Me_4P_2S_2$; Ph_2PH (b.p. $103\,°C/1\,mmHg$) is obtainable by cleavage of a phenyl group from PPh_3 with Li/THF or $Na/liquid\ NH_3$ and subsequent hydrolysis of the resulting phosphide.[36] Similar treatment of $Ph_2PCH_2CH_2PPh_2$ leads to $PhHPCH_2CH_2PHPh$, but since some Ph_2PH is also formed, a better method is the $LiAlH_4$ reduction of $Ph(Pr^iO)OP\text{-}CH_2CH_2PO(Pr^iO)Ph$.[37]

14.2.1.4 Tertiary phosphines

A large number of tertiary phosphines are known,[1] and are easily prepared by the reaction of PX_3, RPX_2 or R_2PX with Grignard or organolithium reagents to form PR'_3, RPR'_2, R_2PR', respectively, and by reaction of RX with alkali phosphides.[11] Reaction of either $MeMgI$ or $MeLi$[38] with PCl_3 thus gives Me_3P, and since this is a low boiling (37–39 °C) inflammable liquid it is usually stored as $AgI \cdot (PMe_3)_4$,[39] $AgNO_3 \cdot PMe_3$[39] or $InCl_3 \cdot (PMe_3)_2$,[40] from which it is easily recovered. Most other trialkylphosphines are readily obtained by the Grignard route: Et_3P (b.p. 127 °C), Pr_3^nP (b.p. 85–87 °C/24 mmHg), Pr_3^iP (b.p. 81 °C/22 mmHg), Bu_3^nP (b.p. 129 °C/22 mmHg), Cy_3P (m.p. 76–78 °C), although for the last reflux of the $PCl_3/CyMgCl$ mixture, it is necessary to get complete replacement of all the chlorines of PCl_3. The lower alkyl and branch-chained alkylphosphines readily oxidize in air, but the tertiary arylphosphines are air stable; these are also readily obtainable by the Grignard route: PPh_3 (m.p. 79–81 °C), $P(C_6F_5)_3$ (m.p. 115–116 °C) (better prepared from $PhLi + PCl_3$),[41] o-, m-, p-$(ClC_6H_4)_3P$, o-, m-, p-$(tolyl)_3P$ (m.p. 125 °C (o-), 100 °C (m-), 146 °C (p-)).

Although mixed alkylphosphines have had only sparse use as ligands, the alkylarylphosphines (aryl is usually phenyl) have received extensive attention by coordination chemists. They are better donors than the $(aryl)_3P$ and less air-sensitive than the $(alkyl)_3P$ ligands. The usual route to them is either $RMgX$ or RLi with $PhPCl_2$ or Ph_2PCl, or from $MPPh_2$ and RX.[11] The most commonly used are Me_2PPh (b.p. 82 °C/20 mmHg), Et_2PPh (b.p. 108 °C/20 mmHg), $MePPh_2$ (b.p. 84 °C/13 mmHg) and $EtPPh_2$ (b.p. 98 °C/10 mmHg). In recent years the bulky *t*-butyl group has been incorporated into phosphines in order to specially study internal metallation, but Bu_3^tP cannot be synthesized *via* the Grignard route, presumably because of steric crowding in the activated complex, and is made from Bu^tLi and Bu_2^tPCl.[42] A review article is available of the syntheses of optically active tertiary phosphines.[43]

14.2.1.5 Diphosphines

The method of greatest applicability is the reaction of alkali metal phosphides with dihalo-alkanes, -alkenes, *etc.* (equation 4). Generally this reaction proceeds well and in high yield; excess dihaloalkane should be avoided as this leads to the production of phosphonium salts which can contaminate the product.

$$2MPR_2 + X \frown X \longrightarrow R_2P \frown PR_2 + 2MX \qquad (4)$$

The most widely used diphosphine ligands are the bis(diphenylphosphino)alkanes, $Ph_2P(CH_2)_nPPh_2$ ($n = 1–3$), and these are readily obtained from $X(CH_2)_nX$ and $LiPPh_2$ in THF.[43,33] In the case of $n = 2$, the ligand is also obtainable by base-catalyzed addition of PPh_2H to $Ph_2PCH=CH_2$,[45] but since the latter is so much more difficult to prepare than $LiPPh_2$ this method is of interest more as the simplest example of a reaction which is valuable in polydentate phosphine ligand synthesis. The ligands are air-stable crystalline solids:[45,46] $Ph_2PCH_2PPh_2$ (dppm, m.p. 120–121 °C), $Ph_2PCH_2CH_2PPh_2$ (dppe, m.p. 140–142 °C), $Ph_2PCH_2CH_2CH_2PPh_2$ (dpp, m.p. 62 °C); the analogous $n = 4$ ligand is a solid, but for $n = 5$ the compound is a yellow oil.

The *cis*- and *trans*-1,2-bis(diphenylphosphino)ethylene are obtained from the corresponding $ClCH=CHCl$, the substitution being completely stereospecific as shown in equations (5) and (6).[44] The $KOBu^t$-catalyzed addition of PPh_2H to $Ph_2PC\equiv CH$ yields *trans*-$Ph_2PCH=CHPPh_2$.[45]

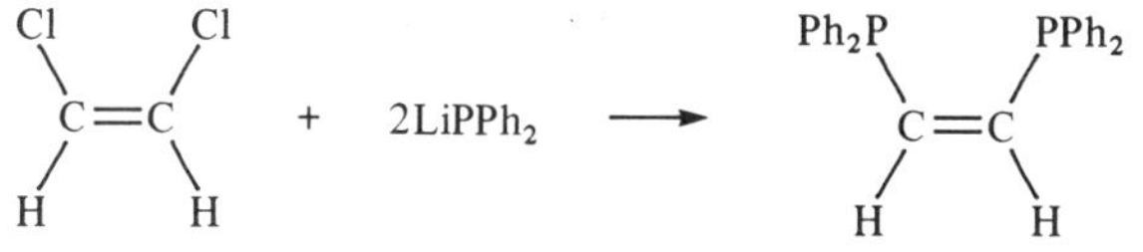

$$\text{(5)}$$

m.p. 125 °C

$$\text{(6)}$$

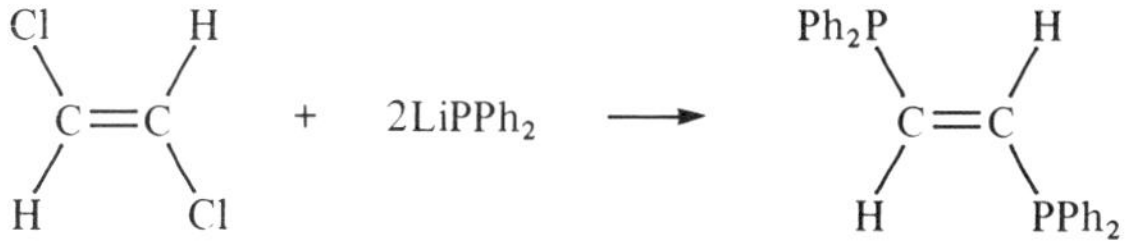

m.p. 126 °C

Although *p*-phenylenebis(diphenylphosphine) is obtained in high yield from $MPPh_2$ and either p-$C_6H_4Br_2$ or p-$ClC_6H_4SO_3Na$,[47] the *o* isomer is produced impure and in low yields from o-$C_6H_4Cl_2$.[48] It can, however, be prepared as shown in equation (7).[49]

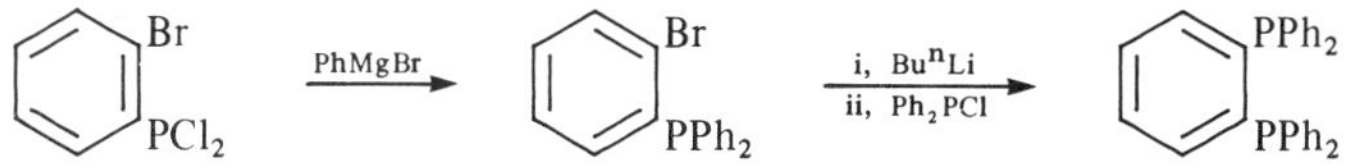

$$\text{(7)}$$

m.p. 185 °C

The analogous tetraflurophenyl analogue (m.p. 108 °C) can be prepared from o-$C_6F_4Br_2$ and Bu^nLi/Ph_2PCl.[50] Bis(diphenylphosphino)acetylene can be prepared from $XMgC\equiv CMgX$ and $2PPh_2Cl$.[51] A range of fluoroalicyclic ligands are available from the PPh_2H addition/elimination to the appropriate chlorofluoroalkene (equation 8).[52]

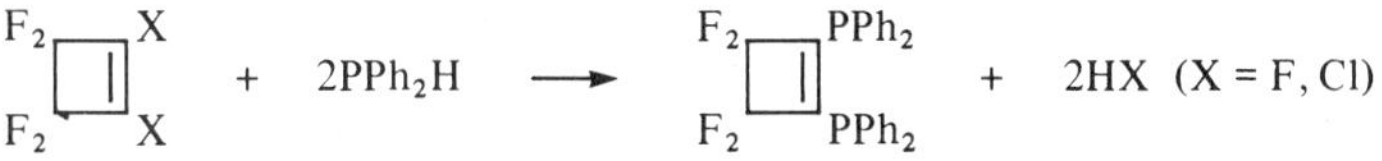

$$\text{(8)}$$

f$_4$fos (m.p. 127 °C)

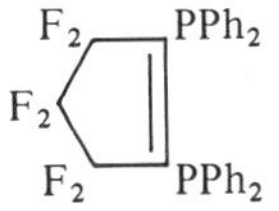

f$_6$fos (m.p. 97 °C)   f$_8$fos (m.p. 135 °C)

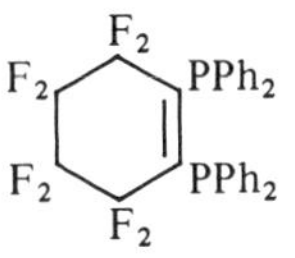

Alkyl-substituted diphosphines have received much less attention than their aryl counterparts, arising undoubtedly from the fact that they are usually very air-sensitive pyrophoric liquids. $Me_2P(CH_2)_nPMe_2$ ($n = 2$, b.p. 65 °C/10 mmHg;[18] $n = 3$, b.p. 62 °C/4 mmHg[28]) and $Et_2P(CH_2)_2PEt_2$ have been prepared from the reaction of MPR_2 and the organic dihalide.[53] The vinyl analogue, *cis*-$Me_2PCH=CHPMe_2$, does not appear to have been prepared but o-$C_6H_4(PMe_2)_2$ (b.p. 98 °C/ 0.3 mmHg) has been obtained in low yield from o-$C_6H_4X_2$ and $NaPMe_2$ in dioxane or liquid ammonia.[48,54] An alternative synthesis, employed for the ethyl-substituted ligand, is not only tedious, but also gives low yields (equation 9).[55]

Recently, unsymmetrical ligands of types $Ph_2PCH_2CH_2PRPh$ and $Ph_2P(CH_2)_nPR_2$ have been prepared. The approach taken by Grim[56] and co-workers is to begin with alkyldiphenylphosphines or dialkylphenylphosphines and to remove a phenyl group with Li in THF (equation 10).

$$Ph_2RP + 2Li \xrightarrow{THF} PhRPLi + PhLi \qquad (10)$$

Reaction with diphenylvinylphosphine followed by hydrolysis gives the required ligand shown in equation (11).

$$PhRPLi + Ph_2PCH{=}CH_2 \longrightarrow PhPCHLiCH_2PRPh \xrightarrow{H_2O} Ph_2PCH_2CH_2PRPh \qquad (11)$$

Briggs et al.[57] have prepared $Ph_2P(CH_2)_6PEtPh$ by a novel synthetic route that can be generalized to prepare a wide range of unsymmetrical bis(phosphines) of the general formula $Ph_2P(CH_2)_nPRPh$ or $Ph_2P(CH_2)_nPR_2$.[58] The method involves the preparation of an unsymmetrical bis(phosphonio) salt, which is subsequently hydrolyzed to the unsymmetrical bis(phosphine oxide), involving specific cleavage of a phenyl group from each phosphorus atom. The bis(phosphine oxide) is then reduced by poly(methylhydrosiloxane) (PMHS) to the corresponding unsymmetrical phosphine in good yield (see equations 12–15).

$$Br(CH_2)_6Br + Ph_3P \xrightarrow{C_6H_6} [Ph_3P(CH_2)_6Br]Br \qquad (12)$$

$$[Ph_3P(CH_2)_6Br]Br + EtPh_2P \xrightarrow{DMF} [Ph_3P(CH_2)_6P(Et)Ph_2]Br_2 \qquad (13)$$

$$[Ph_3P(CH_2)_6P(Et)Ph_2]Br_2 \xrightarrow{NaOH(aq)} Ph_2P(O)(CH_2)_6P(O)EtPh \qquad (14)$$

$$Ph_2P(O)(CH_2)_6P(O)EtPh \xrightarrow{PMHS} Ph_2P(CH_2)_6PEtPh \qquad (15)$$

14.2.1.6 Multidentate phosphines

Chatt[59] pioneered this area by the preparation of a triphosphine (equation 16) and though this method has been extended, e.g. equation (17), it is limited by the difficulty in replacing all halogens[60] and by the availability of suitable haloorganic compounds. Exceptions to this are those ligands containing $o\text{-}C_6H_4PR_2$ or $o\text{-}C_6F_4PPh_2$ moieties, which are readily prepared (equation 18).[61]

$$MeC(CH_2Cl)_3 \xrightarrow{NaPPh_2} MeC(CH_2PPh_2) \qquad (16)$$

$$PhP(CH_2CH_2Br)_2 \xrightarrow[\text{liq. } NH_3]{NaPPh_2} PhP(CH_2CH_2PPh_2)_2 \qquad (17)$$

$$\qquad (18)$$

Triphosphines containing methylene linkages between the donor atoms can be prepared in a number of ways (equation 19).

$$PhPHNa \xrightarrow{ClCH_2CH_2PPh_2} PhHPCH_2CH_2PPh_2 \xrightarrow[\text{ii, } ClCH_2CH_2PPh_2]{i,\ Bu^nLi} PhP(CH_2CH_2PPh_2)_2 \qquad (19)$$

King et al.[45,63] have developed a method using base-catalyzed addition of P—H bonds to vinyl phosphines (equation 20).

This general method can be applied to the introduction of methylphosphine groups via $Me_2P(S)$ or $MeP(S)$ moieties, which are easier to prepare and manipulate than the Me_2P or MeP analogues.

$$Ph_2PCl \xrightarrow{CH_2=CHMgBr} Ph_2PCH=CH_2$$

$$PhPCl_2 \xrightarrow{CH_2=CHMgBr} PhP(CH=CH_2)_2$$

$$\xrightarrow[Ph_2PH/KOBu^t]{PhPH_2/KOBu^t} PhP(CH_2CH_2PPh_2)_2 \qquad (20)$$

The sulfur is subsequently removed by lithyl reduction (equation 21).[64]

$$Me_2P(S)P(S)Me_2 \xrightarrow{Br_2} Me_2P(S)Br \xrightarrow{CH_2=CHMgBr} Me_2P(S)CH=CH_2$$

$$\xrightarrow{PhPH_2, KOBu^t, THF} PhP\{CH_2CH_2P(S)Me_2\}_2 \xrightarrow[dioxane]{LiAlH_4} PhP\{CH_2CH_2PMe_2\}_2 \qquad (21)$$

If $PhPH_2$ is replaced by $MePH_2$, then the completely aliphatic $MeP(CH_2CH_2PMe_2)_2$ is produced. Meek and co-workers[65] have shown that 2,2'-azobis(isobutyronitrile) catalyzes the addition of P—H bonds to vinyl phosphines (equation 22).

$$Ph_2PCH=CH_2 \xrightarrow{PhPH_2} PhP(CH_2CH_2PPh_2)_2 \qquad (22)$$

There are three main types of tetraphosphine ligands: (i) the tetrapodic $C(CH_2PPh_2)_4$, obtainable from $MPPh_2$ and $C(CH_2Cl)_4$;[66] (ii) the essentially C_{3v} symmetry tripod ligands $P(\sim PR_2)_3$. Venanzi's classic, $P(o\text{-}C_6H_4PPh_2)_3$, is obtained from $o\text{-}C_6H_4(PPh_2)Li$ and $\frac{1}{3}PCl_3$,[67] and King's procedure yields $P(CH_2CH_2PPh_2)_3$.[68] The Me_2P analogue of the latter forms only in poor yield from Me_2PH and $P(CH=CH)_3$, and whilst $Me_2P(S)CH=CH_2$ does add to PH_3, the resulting $P(CH_2CH_2P(S)Me_2)_3$ is very difficult to desulfurize. A good route is shown in equation (23).[64]

$$Me_2PCH_2CH_2PH_2 + 2Me_2P(S)CH=CH_2 \longrightarrow Me_2PCH_2CH_2P\{CH_2CH_2P(S)Me_2\}_2$$

$$\xrightarrow{LiAlH_4} (Me_2PCH_2CH_2)_3P \qquad (23)$$

The versatility of King's methods are thus well demonstrated. (iii) Finally there are the so-called facultative or open-chain ligands.[69] The ubiquitous method of King can be used to synthesize 1,1,4,7,10,10-hexaphenyl-1,4,7,10-tetraphosphadecane (equation 24).[45]

$$PhPHCH_2CH_2PHPh + 2PH_2PCH=CH_2 \longrightarrow Ph_2PCH_2CH_2P(Ph)CH_2CH_2P(Ph)CH_2CH_2PPh_2 \quad (24)$$

Meek's free radical-catalyzed reaction of $PhHPCH_2CH_2CH_2PHPh$ with $Ph_2PCH=CH_2$ yields a tetraphosphine with a central trimethylene backbone.[65] As before, dialkylphosphino groups may be incorporated (equation 25).[64]

$$2Me_2P(S)CH=CH_2 + PhHPCH_2CH_2PHPh$$

$$\xrightarrow[THF]{KOBu^t} Me_2P(S)CH_2CH_2P(Ph)CH_2CH_2P(Ph)CH_2CH_2P(S)Me_2$$

$$\xrightarrow[dioxane]{LiAlH_4} Me_2PCH_2CH_2P(Ph)CH_2CH_2P(Ph)CH_2CH_2PMe_2 \qquad (25)$$

Only one pentatertiary phosphine is known,[64] $Me_2PCH_2CH_2P(Ph)CH_2CH_2P(CH_2CH_2PMe_2)_2$, but two hexatertiary phosphine ligands are reported, *viz.* $(R_2PCH_2CH_2)_2PCH_2CH_2P(CH_2CH_2PR_2)_2$ ($R = Ph$,[45] Me[64]). King's group has extended the range of polydentate phosphine ligands to incorporate a range of PH_2, PRH and PR_2 groups.[70] In general these multidentate ligands have physical characteristics reminiscent of similar bidentates, *viz.* the aryl-substituted ligands are air-stable solids, and the alkyl analogues are air-sensitive malodorous ligands or low-melting solids.

14.2.1.7 Macrocyclic phosphines

(i) Introduction

The systematic development of macrocyclic chemistry started in the 1960s with the pioneering work of D. H. Busch. Most macrocyclic ligands have been developed from two biological archetypes. The first, porphyrins, are best exemplified by haemin (protoporphyrin IX). The second group were modelled on valinomycin,[71] which was shown to be capable of transporting metals through biological membranes. Consideration of this system led to the development of macrocyclic polyethers.[72] The substitution of some of the oxygen atoms by phosphorus has led to ligands such as 18-aneP_4O_2 (Table 1). The chemistry of these macrocylic systems is expanding rapidly, but in the main is confined to systems containing N, O and S donor atoms.

Table 1　Structures of Some P-containing Macrocycles

Structure	Ref.	Structure	Ref.	Structure	Ref.
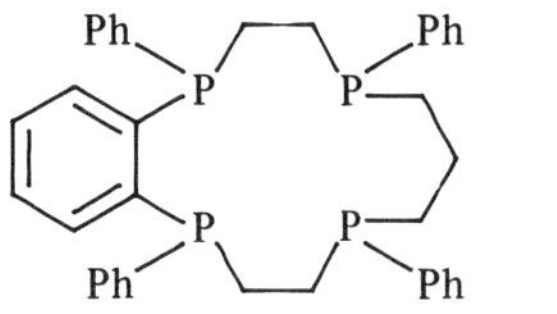	1	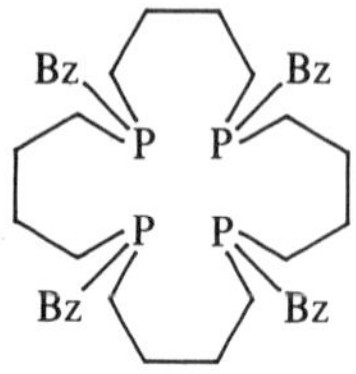	4	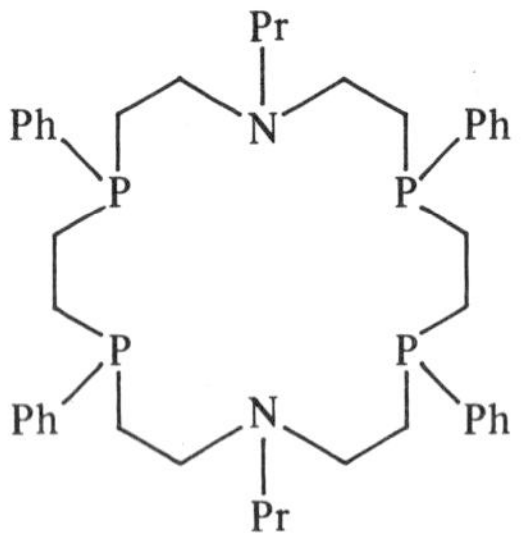	8
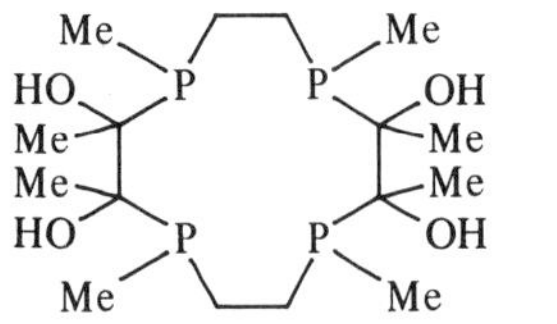	2	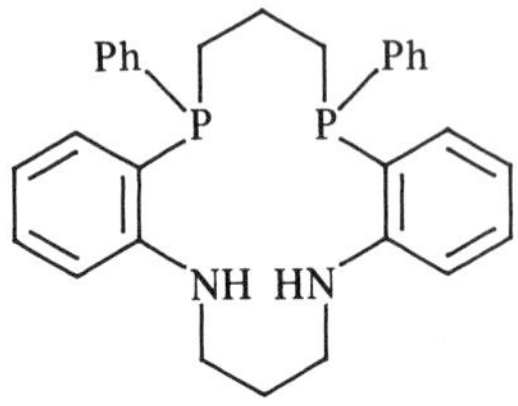	5	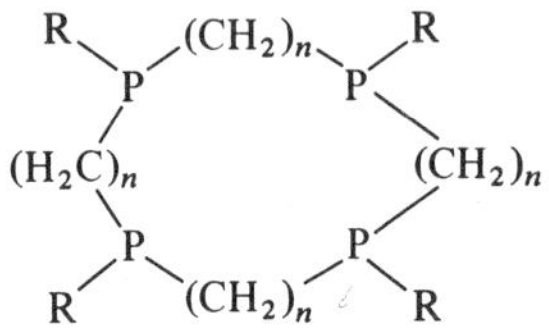	4
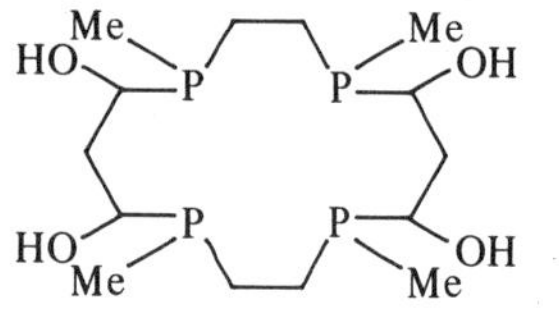	3	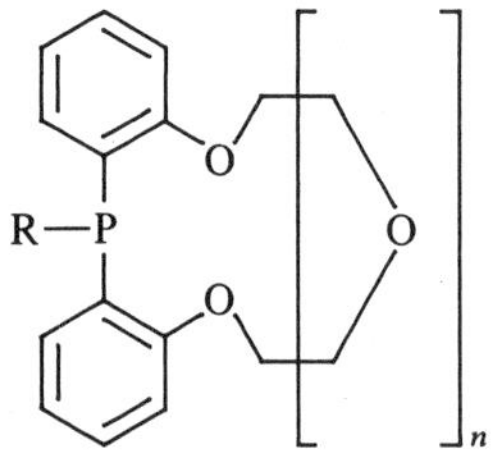	6	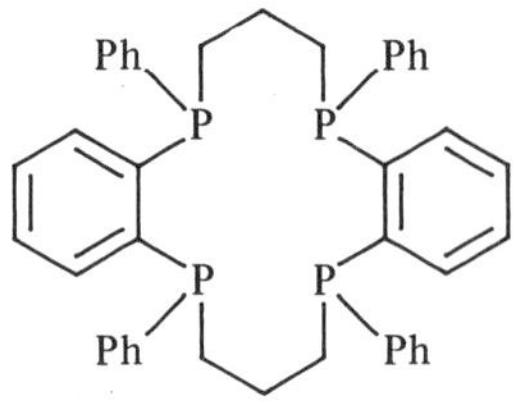	9
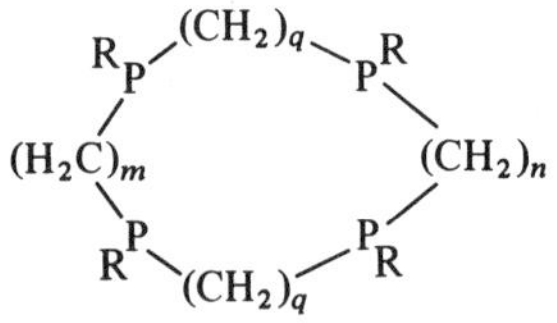	2, 3	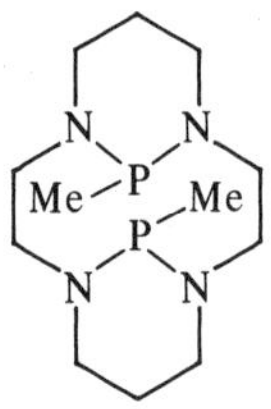	7	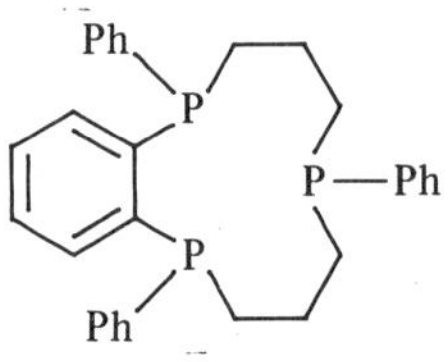	9
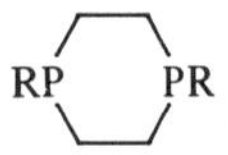	4	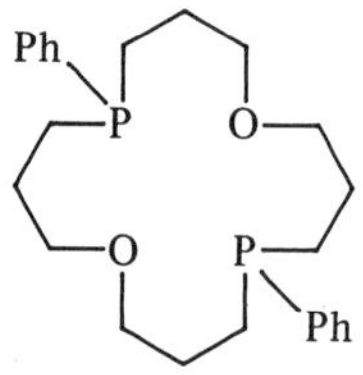	8	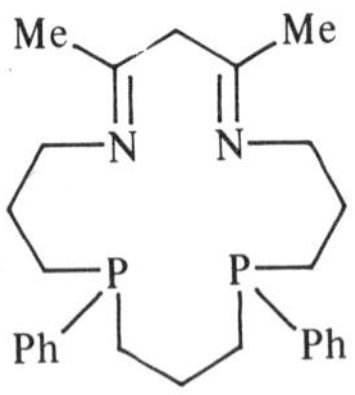	10
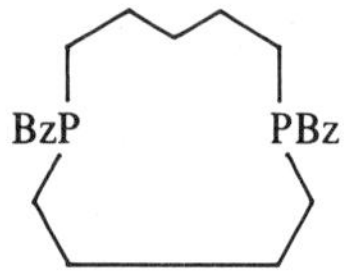	4	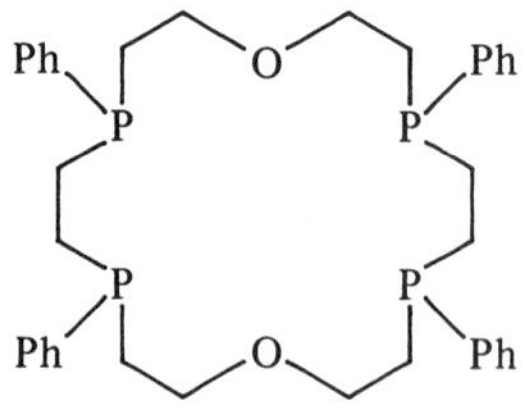	8		
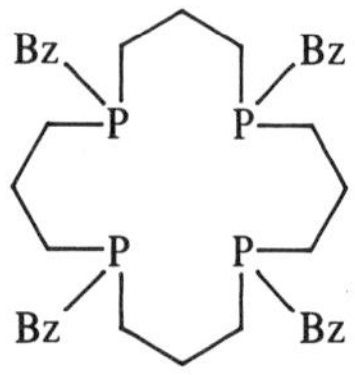	4				
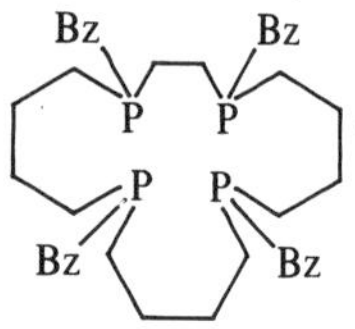	4				

1. W. A. Rosen and T. A. DelDonno, *J. Am. Chem. Soc.*, 1977, **99**, 8051.
2. O. Stelzer, R. Bartsch, S. Hietkamp and H. Peters, *Inorg. Chem.*, 1984, **23**, 3304.
3. O. Stelzer, R. Bartsch, S. Hietkamp, S. Morton and H. Peters, *Inorg. Chem.*, 1983, **22**, 3624.
4. L. Horner, P. Walsch and H. Kunz, *Phosphorus Sulfur*, 1978, **5**, 171.
5. M. K. Cooper, C. W. G. Ansell, K. P. Dancey, P. A. Duckworth, K. Henrick, M. McPartlin and P. A. Tasker, *J. Chem. Soc., Chem. Commun.*, 1985, 439.
6. A. Van Zon, G. J. Torny and J. H. G. Frijns, *Recl. Trav. Chim. Pays-Bas*, 1983, **102**, 326.
7. P. P. Power, M. Viggiano, B. Moezzi and H. Hope, *Inorg. Chem.*, 1984, **23**, 2550.
8. H. J. Cristau, L. Chiche, F. Fallouh, P. Hullot, G. Renard and H. Christol, *Nouv. J. Chim.*, 1984, **8**, 191.
9. E. P. Kyba, R. E. Davis, C. W. Hudson, A. M. John, S. B. Brown, M. J. McPhaul, L. K. Liu and A. C. Glover, *J. Am. Chem. Soc.*, 1981, **103**, 3868 and refs. therein.
10. D. W. Meek, L. G. Scanlon, Y. Y. Tsao, S. C. Cummings and K. Toman, *J. Am. Chem. Soc.*, 1980, **102**, 6849.

In view of the widespread use of tertiary phosphines in coordination chemistry, it is somewhat surprising that phosphorus-containing macrocycles have not evoked more interest. In part this is undoubtedly due to experimental difficulties. The synthesis of P macrocycles is doubly difficult, since the usual difficulties and hazards of phosphorus(III) chemistry are compounded by the complexities and low yields of multistage heterocyclic macrocyclic syntheses. Because of these difficulties, the development of synthetic methods has been slow. Mann's survey[73] of the heterocyclic derivatives of P, As, Sb and Bi did not mention phosphorus macrocycles and recent textbooks[74,75] give them only cursory mention.

For the purposes of this review P, As and Sb macrocycles are defined as carbocyclic systems of ten or more atoms, of which at least two are potential donor atoms and at least one of which is P, As or Sb, *i.e.* (**1**). Since very little has been published on As, Sb or Bi macrocycles they have been included here, but inevitably this survey is mainly concerned with phosphorus compounds. It should be pointed out that cyclic systems of fewer than 12 members are thought to be too small to enclose or encapsulate a metal, but rings of 11 members can show some interesting ligand properties. The figure of 10 members is one chosen arbitrarily for this review. The above definition excludes P-containing macrocycles that have been formed by 'large-bite' bidentate ligands, as in (**2**) and (**3**),[76] and excludes macrocyclic systems where phosphorus is part of a pendent group, *e.g.* (**4**).[77]

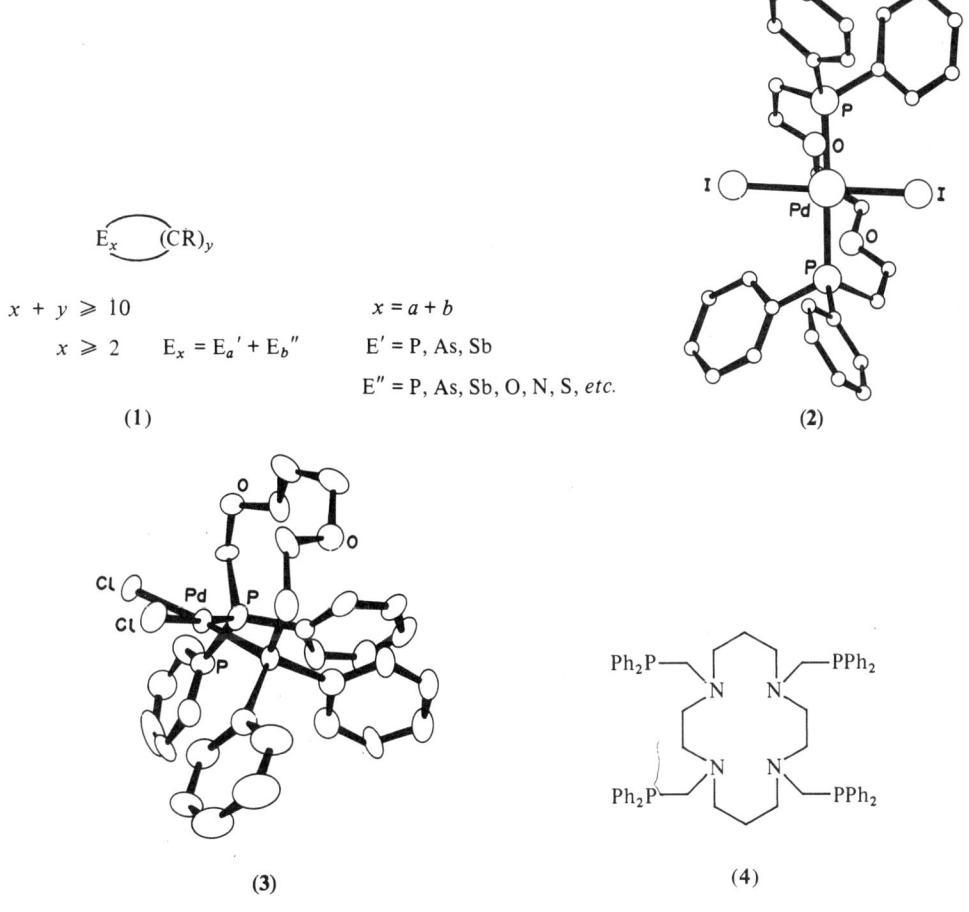

$$E_x \quad (CR)_y$$

$$x + y \geqslant 10 \qquad x = a + b$$

$$x \geqslant 2 \qquad E_x = E_a' + E_b'' \qquad E' = P, As, Sb$$

$$E'' = P, As, Sb, O, N, S, etc.$$

(**1**)

(**2**)

(**3**)

(**4**)

The abbreviated form of nomenclature notation is described in Melson.[78] Two examples here, (5) and (6), should make this clear and other examples are contained in Table 1. Phosphorus is given precedence in this review, although, strictly, oxygen should precede phosphorus.

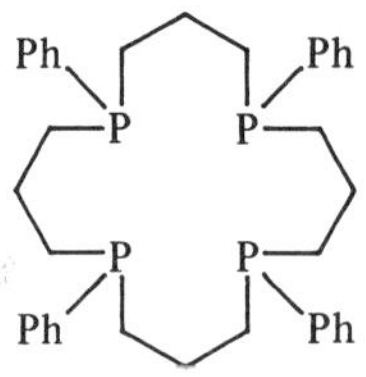

[16]aneP$_4$

(5)

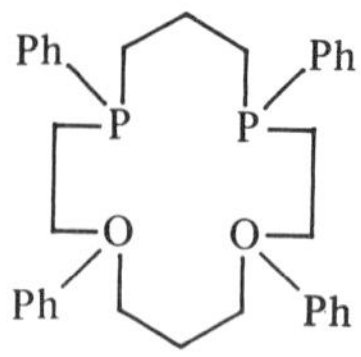

[14]aneP$_2$O$_2$

(6)

(ii) *Formation of P macrocycles*

The synthetic methods can be conveniently divided into those based on conventional (classical) macrocyclic syntheses and those based on template methods. The schools of Ciampolini and Kybe were early, and are still major, contributors to this field.

Ciampolini's group[79] have used the direct method of reacting two moles of dilithio phosphide (ca. 10% w/v in THF) and two moles of bis(dichloroethyl) ether (ca. 25% w/v in THF) at low temperature ($-20\,°C$) to afford a 12% yield of 18-aneP$_4$O$_2$ (equation 26).

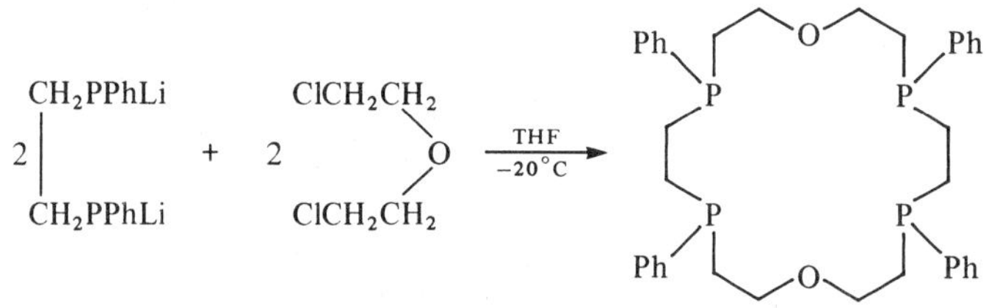

$$(26)$$

The five diastereoisomers are separable by chromatography, Table 1. The ether may be replaced with the corresponding thio compound, producing five isomers of [18]aneP$_4$S$_2$ in ca. 14% yield.[80] The nitrogen-containing analogue necessitated the use of the potassium phosphide.[81]

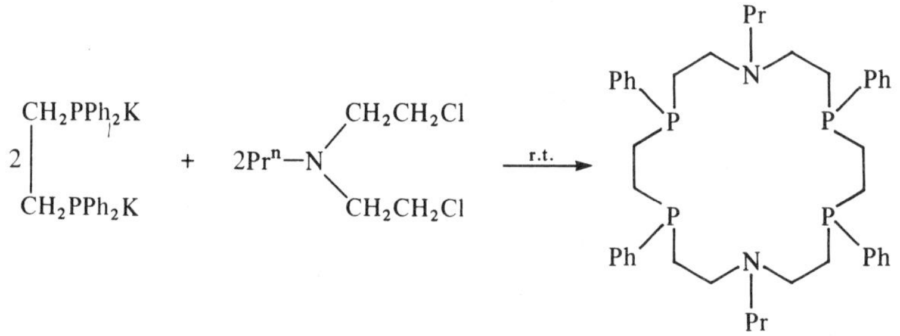

$$(27)$$

These cyclic tertiary phosphines are stable in air and no precautions to exclude it were necessary in the chromatographic separations. Although the yields are low by this method, it has the advantage of simplicity and inexpensive, readily available starting materials. The chromatographic separation of the isomers can be monitored using ^{31}P NMR spectroscopy.

A number of macrocycles containing only one phosphorus atom have been prepared, and they are usually referred to as phosphorus-containing crown ethers rather than as P macrocycles. The 18-crown, or [18]anePO$_5$, was prepared by van Zon (equation 28).[82] This cyclic phosphine is stable in air, even in solution (cf. Ph$_3$P). However, when the non-macrocyclic substituent is Bu^t the phosphorus is oxidized in solution by air in the presence of alumina. This unwelcome discovery was made by van Zon when purifying the material by chromatography on an alumina column.

Russian workers[83,84] have synthesized a series of phosphorus-containing crowns utilizing O—P—O bonds in the macrocycle, i.e. cyclic phosphonate esters (7). One of these is claimed to have been prepared in 40% yield (equation 29).

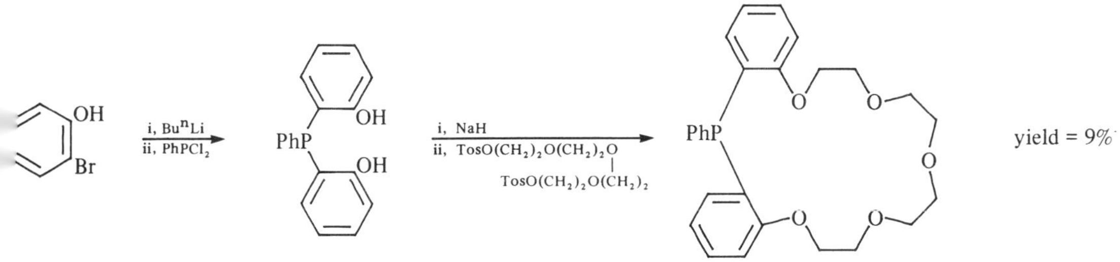

yield = 9% (28)

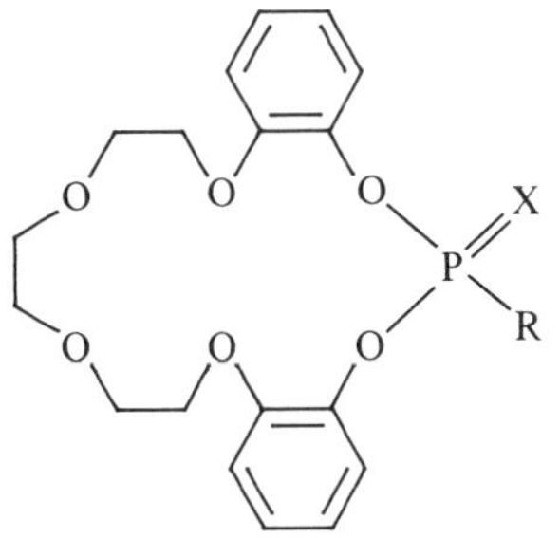

R, X = Me, S; Ph, S; Ph, O; PhO, O

(7)

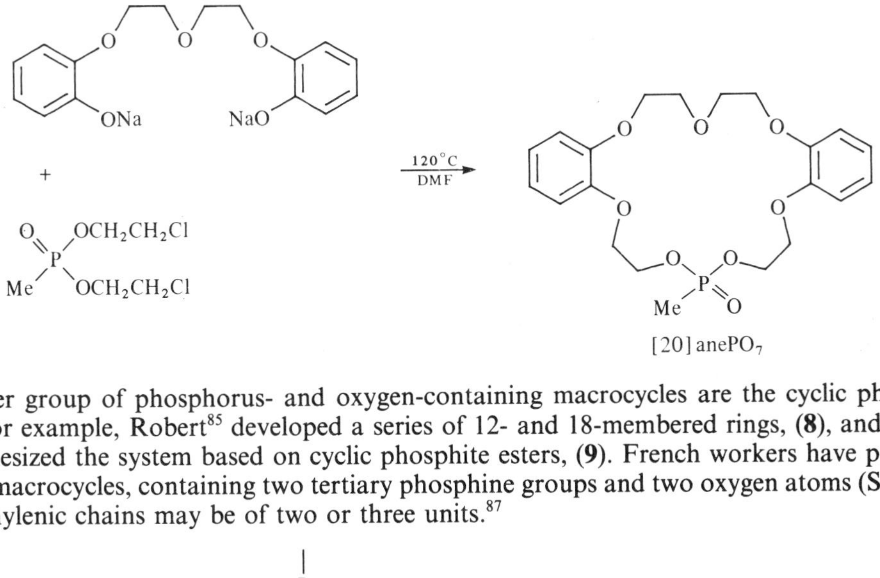

(29)

[20]anePO$_7$

Another group of phosphorus- and oxygen-containing macrocycles are the cyclic phosphorus esters. For example, Robert[85] developed a series of 12- and 18-membered rings, (8), and Picavet[86] has synthesized the system based on cyclic phosphite esters, (9). French workers have prepared a series of macrocycles, containing two tertiary phosphine groups and two oxygen atoms (Scheme 1). The methylenic chains may be of two or three units.[87]

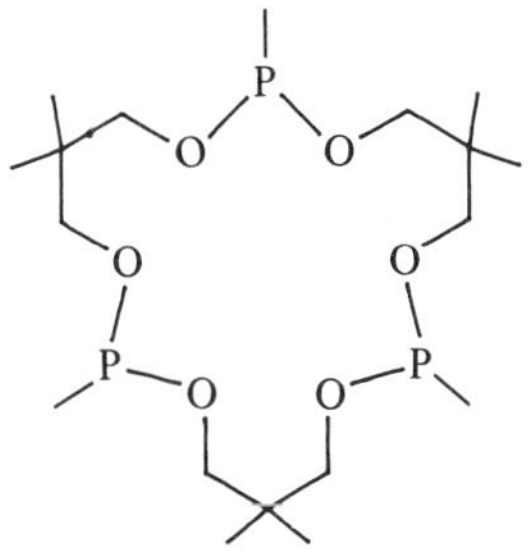

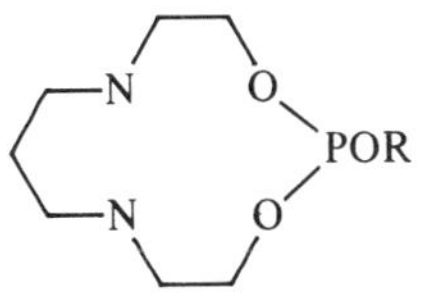

(8) [18]aneP$_3$O$_6$ (9) [12]anePN$_2$O$_2$

Kyba and co-workers employ multistep synthesis and a high dilution technique[88,89] for the final cyclization stage (Scheme 2). By variation of the reactants of the final high dilution stage, various novel macrocycles, *e.g.* [14]aneP$_2$S$_2$ (10), have been synthesized in yields of 30–60%. Kyba also modified this route to produce an 11-membered series of macrocycles (Scheme 3).[90] Kyba *et al.* have used similar techniques to produce a range of arsine macrocycles.

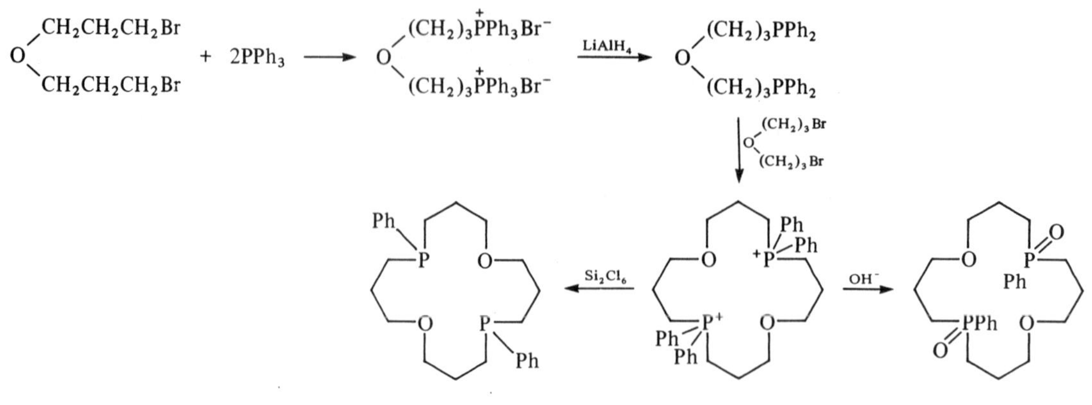

Scheme 1

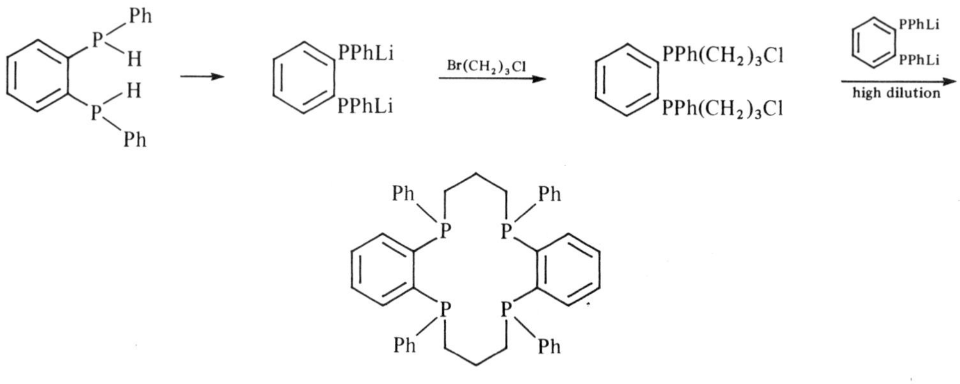

Scheme 2

[14]aneP$_4$

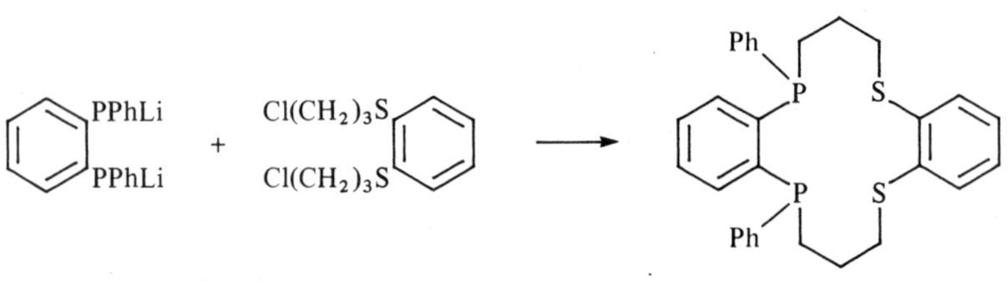

[14]aneP$_2$S$_2$
(10)

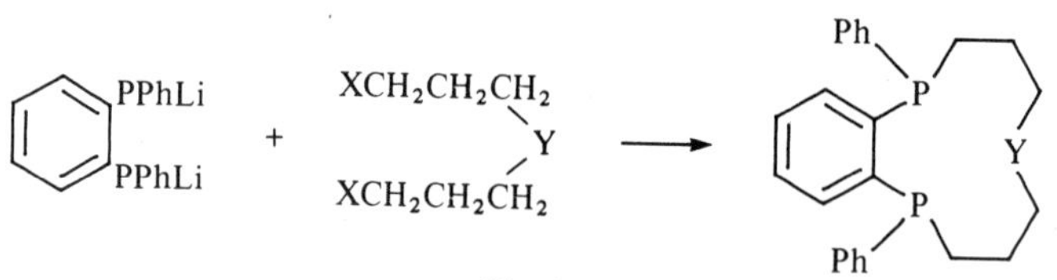

X = Cl, Br
Y = PPh, S, O, NMe, NPh, CH$_2$

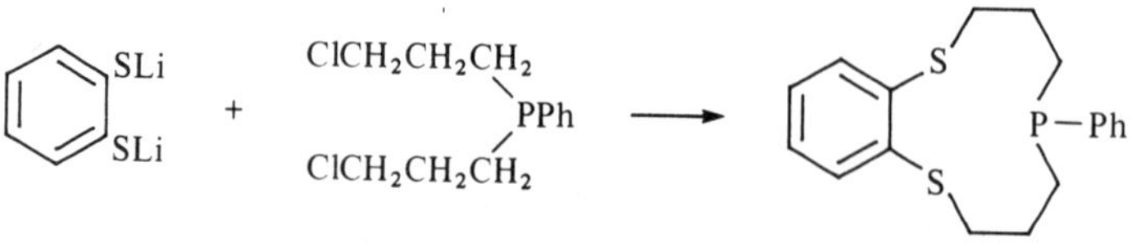

Yield 42%

Scheme 3

The approach of Horner *et al.* is shown in Scheme 4. The macrocyclic phosphonium salts may be converted to phosphine oxides with alkali (yields 30–85%) or to phosphines by reduction using LiAlH$_4$ (yields 80–90%).[91] It was found that dilution did not increase yields.

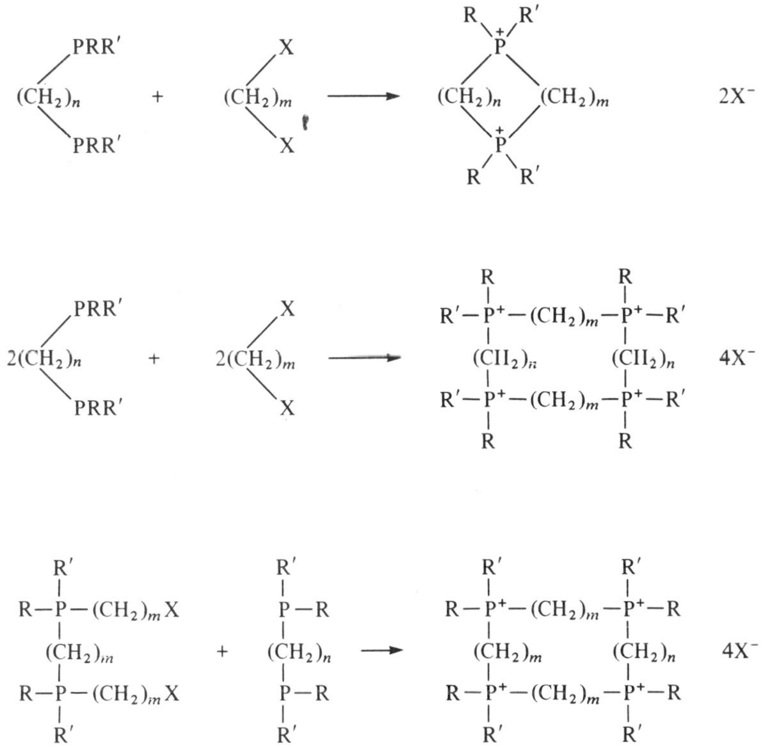

Scheme 4

Berlin and co-workers[92] have also synthesized a large number of phosphonium macrocycles in high yields (up to 96%), *e.g.* [20]aneP$_4$, Scheme 5. For X = Cl, the intermediate diphosphonium salt could be isolated in high yield. Interestingly, attempts to isolate the macrocyclic phosphine oxide by alkaline hydrolysis resulted in ring fission, which was ascribed to the greater stability of the benzylic anion as a leaving group from phosphorus. No report is made of attempts to reduce by LiAlH$_4$. This system is obviously capable of further development.

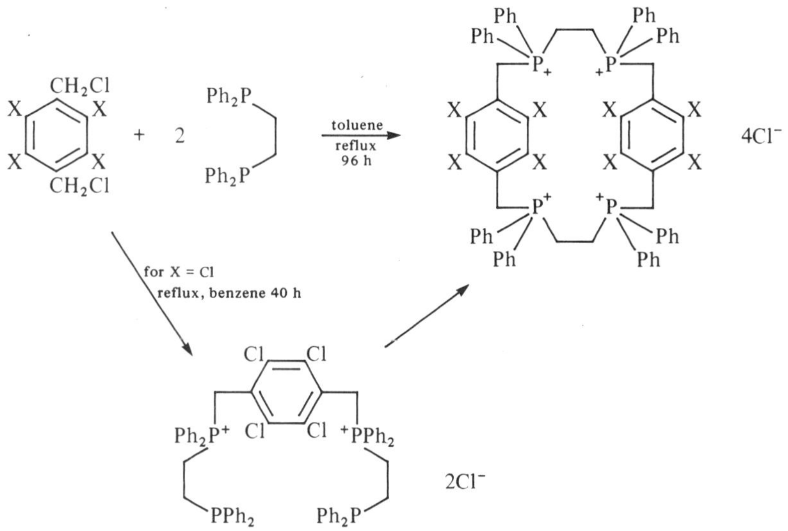

Scheme 5

Template syntheses of P macrocycles are a new area. In fact, a 1978 review[93] of template synthesis made no mention of P macrocycles. Template syntheses have been developed by Stelzer and co-workers.[94] Firstly, two molecules of the bidentate secondary phosphine are complexed with a nickel(II) or palladium(II) salt (Scheme 6) and the resultant secondary phosphine complex is then condensed with a diketone to form the macrocyclic metal complex. Unfortunately, these macrocycles are strong field ligands and no method has yet been devised to remove the metal from the ring. On the other hand, Cooper and co-workers[95] have used a template synthesis to produce a [14]aneP$_2$N$_2$ macrocycle (Scheme 7).

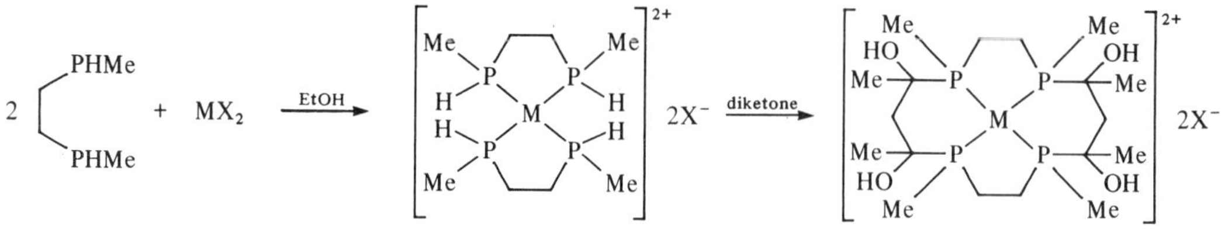

Scheme 6

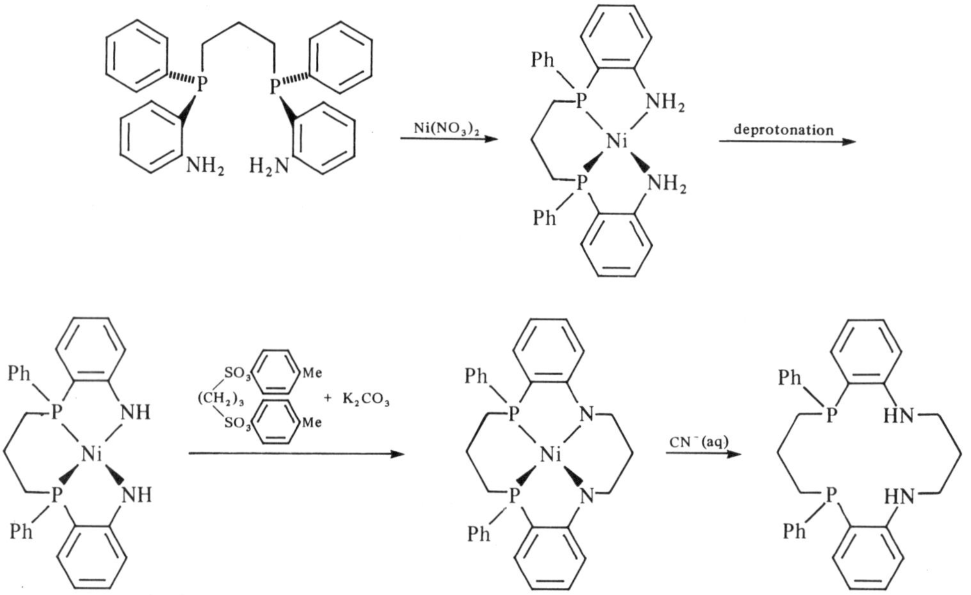

Scheme 7

(iii) The macrocyclic effect

The stability of macrocyclic complexes has been shown to be greater than the comparable open chain complexes. Unfortunately, most quantitative data apply to N and S ligand systems.[96] This increased stability is the net effect of a number of factors including changes in entropy, enthalpy, solvent effects, pH, ring size and conformation (equation 30).[97]

Entropy changes favour such systems by approx. 71.1 J K^{-1} mol^{-1}, ascribed to a large $+ S_{int\,rot}$ and $- S_{solvn}$.

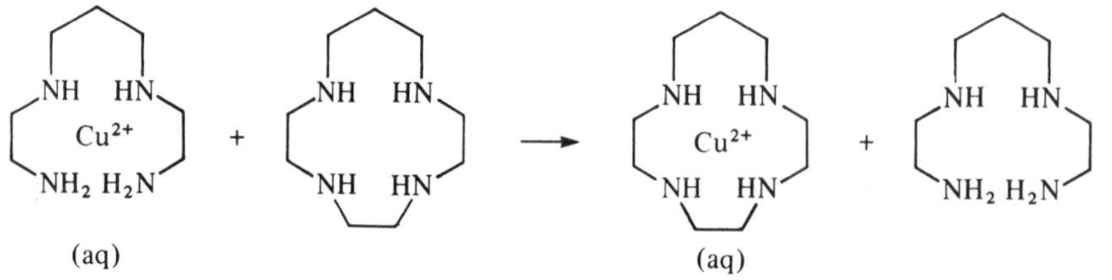

(30)

$$\log K = 5.2 \text{ at } 27\,^\circ\text{C}$$

(iv) Ring size

In macrocycles, ring size is critical. If the ring size is too large, potential ligating atoms will not all be engaged. If the ring is too small, the preferred stereochemical bonding requirements of the transition metal ion will not be met. In both cases partial bonding may occur, with the macrocycle acting as a monodentate or bidentate ligand. Thus, the ring must be of the optimum size for binding a particular metal ion if the macrocyclic effect is to be harnessed. For example, Kyba *et al.* have shown[90] that the [11]aneP$_3$ ring is too small to encapsulate a transition metal ion, which 'sits' above the plane of the P atoms; but these workers showed that [14]aneP$_4$ contained an optimum-sized ring for comfortable metal binding. Finally, it may be pointed out that model-making may not be very helpful in predicting optimum ring size. Different model systems indicate[98] different cavity sizes, and the uncomplexed ligand may adopt a conformation with a greatly reduced cavity due to chain folding.[99] As mentioned above, Stelzer's macrocyclic complexes are stable[94] to treatment with KCN, but Cooper's liberate the macrocycle.[95]

(v) Stereochemistry

The subject of macrocyclic conformation is complex and will be illustrated here by reference to Kyba's systems. These 11-[90] and 14-membered[100] phosphorus macrocycles exist in various conformations. The barrier to inversion for phosphines is approximately 146.4 kJ mol^{-1},[101] but some conformations are remarkably stable. For example, 2,6,10-triphenyl-2,6,10-triphosphabicyclo[9.4.0]pentadeca-11(1),12,14-triene (**11**) has three conformers (a–c in Figure 1). When a xylene solution of (**11**) is refluxed (135 °C, 1 h), only (a) is obtained when crystallized. The phenyl groups at the 2-P and 10-P remain *cis*, although the 6-P does invert, but the crystallization process favours conformer (a) over (b). The equilibrium (a) ⇌ (b) is displaced to the LHS by removal of (a) as it crystallizes out. When (a) is heated above its m.p., a mixture of (a)/(b) in the ratio 1:7 forms. Conformer (c), with 2-P and 6-P phenyl groups *trans*, is not formed. The molecule is L-shaped with the benzo rings at approximate right angles (100°) to the plane of the phosphorus atoms.

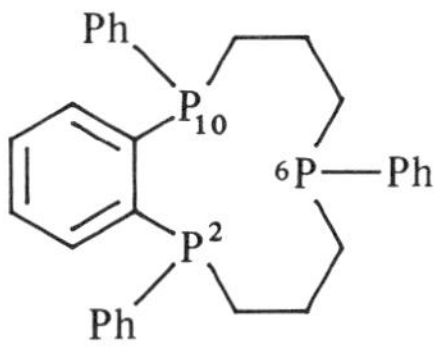

(11)

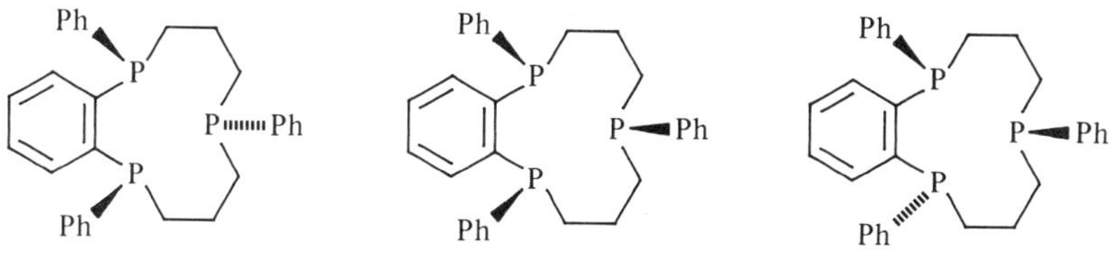

(a)   (b)   (c)

Figure 1

The conformation of the larger ring with four heteroatoms is more complex. The five possible isomers of the benzo-ring phosphino groups in (**12**) are shown schematically in Figure 2. In fact, the synthesis favours (a) (high m.p. 214–6 °C) and (b) (low m.p. 160–4 °C) conformers. In (a) the macrocycle benzo groups are *trans* and in (b) they are *cis*. The structures are shown to good effect in the stereoviews in Figure 3.

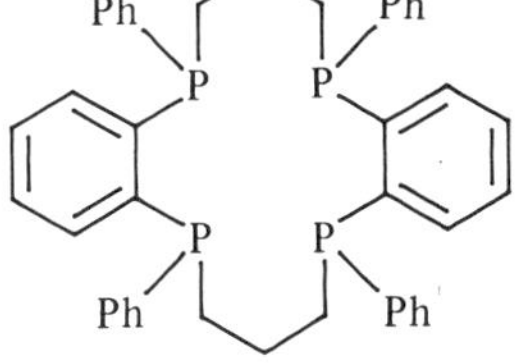

(12)

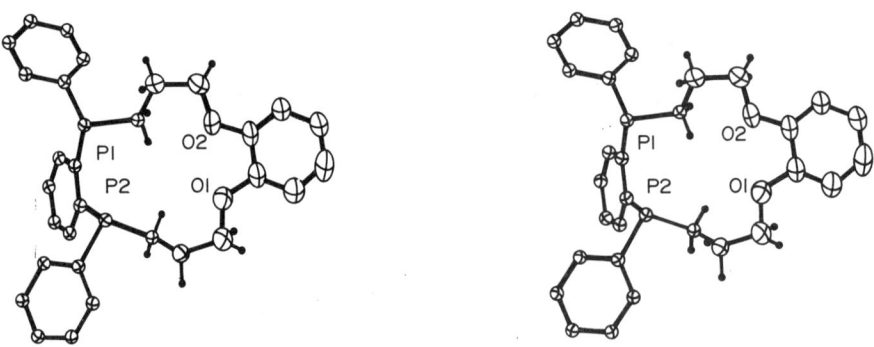

Figure 2

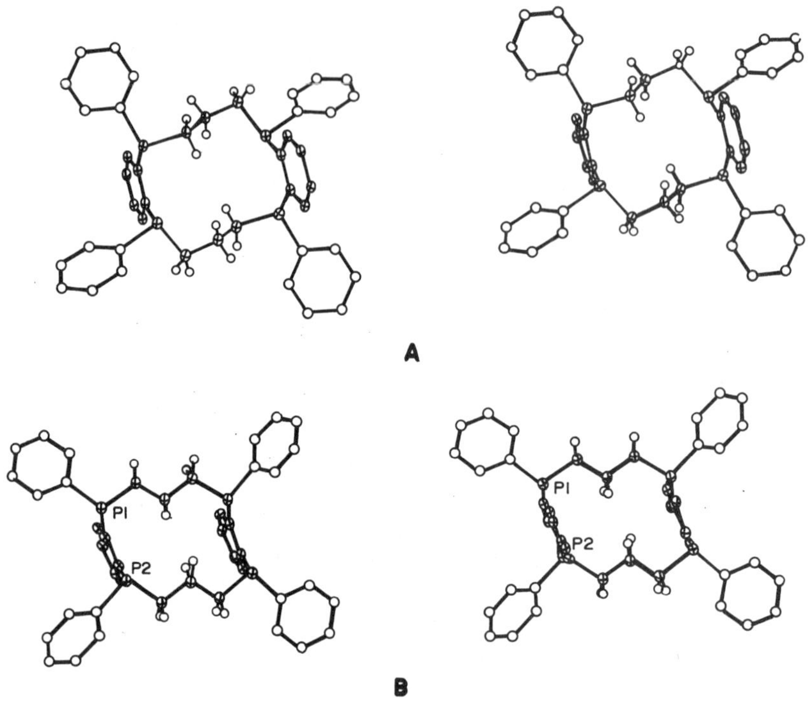

A

B

Figure 3

Replacing two phosphorus atoms in the macrocycle by two sulfur or nitrogen atoms makes slight differences to the *trans* structure as in (a), but when replaced by two oxygen atoms the benzorings tend to line up with the average plane of the macrocyclic ring as shown in the stereoview (Figure 4).

Figure 4

14.2.2 Arsines

14.2.2.1 Haloarsine compounds, R_nAsX_{3-n}

Because arsenic(V) (as arsonic or arsinic acids) is readily reduced to arsenic(III) by sulfur dioxide in concentrated HX,[102,103] this makes chloroarsines much more readily accessible than the corresponding phosphine compounds. This is shown in equation (31),[104]

$$\text{MeI} \xrightarrow{\text{As}_2\text{O}_3/\text{NaOH}} \text{MeAsO}_3\text{H}_2 \xrightarrow{\text{SO}_2/\text{HCl/KI}} \text{MeAsI}_2 \qquad (31)$$

whereas the employment of alkylarsenites (synthesized from $RAsX_2$ + NaOH) instead of As_2O_3/NaOH leads to the formation of $RR'AsX$ compounds (equation 32).[104]

$$MeAsI_2 \xrightarrow{\text{NaOH/MeI}} Me_2AsO_2H \xrightarrow{\text{SO}_2/\text{HCl/KI}} Me_2AsI \qquad (32)$$

In this way PhMeAsI can be prepared from $PhAsCl_2$/NaOH/MeI,[105] but the much less reactive aryl halides do not generally react. Aniline[106] may be used as a precursor for dichlorophenylarsine (equation 33).

$$PhNH_2 \xrightarrow{\text{NaNO}_2/\text{HCl}} PhN_2^+ Cl^- \xrightarrow{\text{As}_2O_3/\text{NaOH}} PhAsO(OH)_2 \xrightarrow{\text{SO}_2/\text{HCl}} PhAsCl_2 \qquad (33)$$

A key precursor to multidentate arsines is *o*-bromophenyldichloroarsine, which may be similarly obtained if the diazotization is conducted in glacial acetic acid and the source of arsenic is $AsCl_3$ (equation 34).[107,108] Conproportionation of $PhAsCl_2$ and PhAsO gives Ph_2AsCl (contaminated with $PhAsCl_2$ and Ph_3As).[106]

$$(34)$$

Also useful in multidentate ligand syntheses are the chloroalkyldichloroarsines, $Cl(CH_2)_nAsCl_2$ (n = 2, 3) (equation 35).[109]

$$Cl(CH_2)_nOH \xrightarrow{\text{As}_2O_3/\text{NaOH}} H_2O_3As(CH_2)_nOH \xrightarrow{\text{SO}_2/\text{HCl}} Cl_2As(CH_2)_nOH \xrightarrow{\text{SOCl}_2} Cl_2As(CH_2)_nCl \qquad (35)$$

Reaction of $Ph_2As(CH_2)_2AsPh_2$ with $AsCl_3$ under pressure gives $Cl_2As(CH_2)_2AsCl_2$, and a valuable intermediate to mixed substituent arsines is obtained as shown in equation (36).[110–113]

$$(36)$$

Another key intermediate is *o*-$C_6H_4(AsCl_2)_2$ (equation 37).[111]

$$(37)$$

14.2.2.2 Alkali metal derivatives

There is a somewhat old, but nonetheless very useful, review of the synthesis and uses of metal arsenides.[114] Triphenylarsine can be cleaved in the same way as triphenylphosphine (Section 14.2.1.2) to form $MAsPh_2$ (M = Na, K).[108,115] Dimethylarsenide, Me_2As^-, is best prepared and used *in situ* from the reaction of sodium with Me_2AsI or with Me_2AsH in THF.[105,116] Reaction of Me_2AsI and Bu^nLi in petroleum ether enables isolation as the lithium salt.[117] Sodium or lithium will cleave phenyl groups from $Ph_2As(CH_2)_nAsPh_2$ in THF or liquid ammonia to form $MPhAs(CH_2)_nAsPhM$, but often products are impure.[118]

14.2.2.3 Tertiary arsines

Trialkylarsines are invariably highly toxic and malodorous; they are readily air-oxidized — indeed, Me_3As spontaneously inflames in air. Mixed arylalkylarsines also tend to be air-sensitive, although Ph_3As is air stable. Like the phosphine analogues, the usual route to tertiary arsines is *via* the Grignard or the lithium reagent and R_nAsCl_{3-n} or RX and $MAsR_2$, although the yields are often poor. The most commonly used ligands are Ph_3As (m.p. 60 °C),[119] Me_3As (b.p. 52 °C),[120] Et_3As (b.p. 48 °C/24 mmHg)[121] and Bu_3^nAs (b.p. 113 °C/10 mmHg).[122] It is also possible to prepare mixed arsines of type $Ph_{3-n}AsR_n$ (R = alkyl; $n = 1, 2$) from RMgX and $Ph_{3-n}AsCl_n$, but Ph_2AsMe (b.p. 112–115 °C/0.35 mmHg)[123] and $PhAsMe_2$ (b.p. 88–90 °C/18 mmHg)[123] are obtainable from the metal arsenide and RX.

14.2.2.4 Diarsines

Ligands such as $Ph_2AsCH_2CH_2AsPh_2$ (m.p. 100 °C)[118] are usually prepared from $MAsPh_2$ and the organohalogen, although elimination and not substitution can occur. Other common ligands prepared in ths way are $Ph_2As(CH_2)_nAsPh_2$ ($n = 1$,[124] 3[118]), p-$C_6H_4(AsPh_2)_2$[125] and *cis*-Ph_2As-*cis*-$Ph_2AsCH{=}CHAsPh_2$ (m.p. 112 °C) (this ligand results from a stereospecific reaction with *cis*-$ClCH{=}CHCl$);[126] *trans*-$Ph_2AsCH{=}CHAsPh_2$ has been prepared similarly.[127]

As for the phenyl-substituted ligands, attempts to synthesize $Me_2AsCH_2CH_2AsMe_2$ from $NaAsMe_2$ and XCH_2CH_2X in THF usually yield $Me_2AsAsMe_2$ and C_2H_4,[126] but there is a report of a low yield of the ligand when the reaction is carried out in liquid ammonia at -78 °C;[128] the reaction in THF can be used to produce $Me_2As(CH_2)_3AsMe_2$. Although the reaction of Ph_2As^- with *cis*-$ClCH{=}CHCl$ is stereospecific, Me_2As^- yields *cis*- and *trans*-$Me_2AsCH{=}CHAsMe_2$ in a ratio of approximately 1:9.[129] The reaction of *trans*-$ClCH{=}CHCl$ with $NaAsMe_2$ yields mainly $Me_2AsAsMe_2$ and a little $Me_2AsCH{=}CHAsMe_2$ (isomer not specified),[130] but the *cis* product may be obtained from the mixture as the $[Fe(ligand)_2Cl_2]FeCl_4$ complex which, upon subsequent reaction with PPh_3, yields the pure *cis* ligand.[131] It may also be obtained by hydroboration of $Me_2AsC{-}CAsMe_2$.[131]

Nyholm exploited *o*-phenylenebis(dimethylarsine) (diars) very successfully. Initially it was produced in a six-stage reaction starting with *o*-nitroaniline,[111] but an easier route is *via* $NaAsMe_2$ and *o*-$C_6H_4Cl_2$[116] (45% yield; b.p. 100 °C/1 mmHg). Feltham and Metzger have also identified such species as Me_2AsH, Me_3As, Me_2PhAs and $MeAs(o$-$C_6H_4AsMe_2)_2$ in the latter reaction products.[131] The tetrafluoro analogue of diars, *o*-$C_6F_4(AsMe_2)_2$, is synthesized from *o*-$C_6F_4Br_2$ by successive treatments with Mg and Me_2AsI.[132] The fluoroalicyclic arsines $f_4$fars and $f_6$fars are obtainable from

f₄fars   f₆fars

Me_2AsH and the perfluorocycloalkene by addition/elimination.[133] Fluoroaliphatic diarsines may be obtained from the addition of tetramethyldiarsine to fluoro-alkenes or -alkynes (equations 38 and 39).[134]

$$F_2C{=}CFX + Me_2AsAsMe_2 \longrightarrow Me_2AsCF_2CFXAsMe_2 \qquad (38)$$

$$FHC{=}CHF + Me_2AsAsMe_2 \longrightarrow Me_2AsCFHCFHAsMe_2 \qquad (39)$$

Mixed aliphatic/aromatic substituted diarsines are available; 1,2-bis(phenylbutylarsino)ethane was first prepared in 1939.[111] Bosnich and Wild were able to separate *meso* and *rac* isomers of $PhMeAsCH_2CH_2AsMePh$ by chromatographing the $PdLCl_2$ complex and, on decomposing by reaction with aqueous KCN, obtained the isomeric ligands.[135] *o*-Phenylenebis(methylphenylarsine) has also been separated into diastereoisomers;[136] some chemistry associated with this ligand and other ligands is shown in Scheme 8.

There has been very little interest in unsymmetrical diarsines, $R_2As{\sim}AsR_2^1$, but they are obtainable from $NaAsR_2$ and $Cl{\sim}AsR_2^1$. An interesting stereospecific reaction is that used (*cis*-$ClCH{=}CHAsPh_2$ and $NaAsMe_2$) to produce *cis*-$Me_2AsCH{=}CHAsPh_2$ (40% yield);[2] this contrasts with the reaction of *cis*-$ClCH{=}CHCl$ and $NaAsMe_2$ which produces a *cis/trans* mixture of $Me_2AsCH{=}CHAsMe_2$).[126]

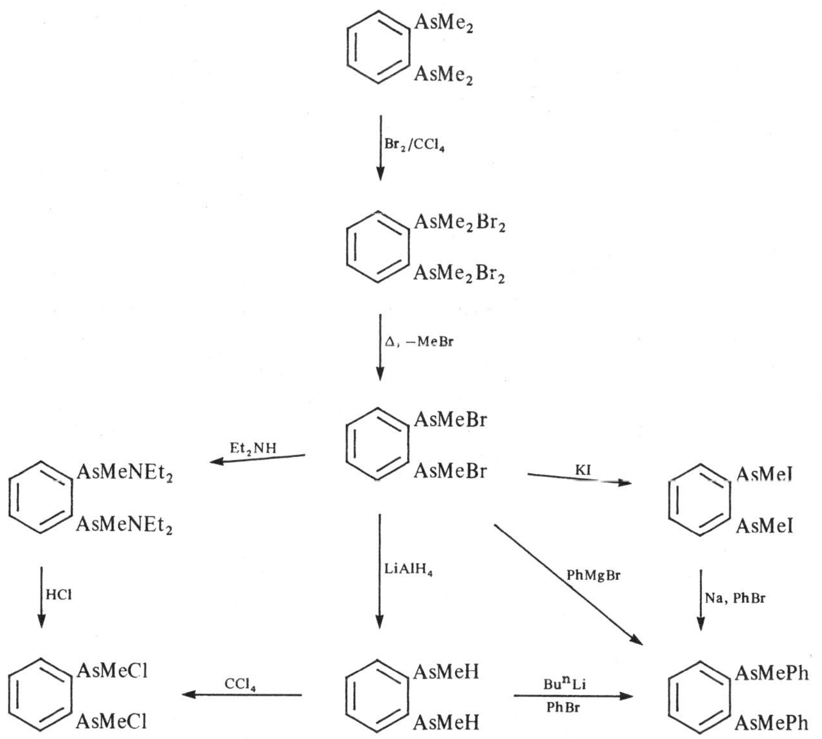

Scheme 8

14.2.2.5 Multidentate arsines

From $Me_2AsCH_2CH_2CH_2MgCl$ and $RAsCl_2$ can be prepared $(Me_2AsCH_2CH_2CH_2)_2AsR$ (R = Me,[137] Ph[138]); *o*-lithiophenyldiphenylarsine and $PhAsI_2$ produce $PhAs(o\text{-}C_6H_4AsPh_2)_2$[139] and $o\text{-}C_6H_4(AsMe_2)_2$.[140] Other triarsines include $MeC(CH_2AsMe_2)_3$[140] and tridentate phosphine–arsines are known.[141] Replacing $RAsX_2$ by $AsCl_3$ allows the preparation of tripodal tetraarsine ligands (equation 40).

$$AsCl_3 \begin{cases} \xrightarrow{3(o\text{-}C_6H_4AsPh_2Li)} As(o\text{-}C_6H_4AsPh_2)_3 \text{ [139]} \\ \xrightarrow{3(o\text{-}C_6H_4AsMe_2Li)} As(o\text{-}C_6H_4AsMe_2)_3 \text{ [142]} \\ \xrightarrow{3ClMgCH_2CH_2CH_2AsMe_2} As(CH_2CH_2CH_2AsMe_2)_3 \text{ [143]} \end{cases} \qquad (40)$$

A range of facultative (open-chain) tetradentate arsines containing both $o\text{-}C_6H_4$[144] and $(CH_2)_3$[145] backbones are known. They can frequently exist in isomeric forms and their separation *via* metal complexation is possible.[144] Reaction of $C(CH_2Br)_4$ with $NaAsPh_2$ yields the tetrapod $C(CH_2AsPh_2)_4$.[146] The hexadentate $o\text{-}C_6H_4As(CH_2CH_2CH_2AsMe_2)_2]_2$ is obtainable from $o\text{-}C_6H_4(AsCl_2)_2$ and $4ClMgCH_2CH_2CH_2AsMe_2$.[147]

14.2.3 Stibines

14.2.3.1 Halostibine compounds, R_nSbX_{3-n}

Alkylstibonic and dialkylstibinic acids have not been well characterized and thus they are not used as starting materials for alkylhalostibine ligands in the same way that the arsenic analogues are. On the other hand, arylstibonic acids have been employed, despite the difficulties in their preparation (yields 20%) and their flocculent nature.

Reaction of Me_3Sb with X_2 gives Me_3SbX_2 (X = Cl, Br) and subsequent heating under reduced pressure gives Me_2SbX (+ MeX). Another route is *via* $Me_2SbSbMe_2$ and SO_2Cl_2, to yield Me_2SbCl.[148] The best method for the preparation of $MeSbCl_2$ appears to be from $PbMe_4$ and $SbCl_3$.[102] Conproportionation of Ph_3Sb and $2SbCl_3$ in ether yields $PhSbCl_2$. Diphenylchlorostibine may be prepared as shown in equation (41) and the valuable intermediate *o*-bromophenyldichlorostibine can be prepared from *o*-bromoaniline *via* the diazonium salt.[151]

$$Ph_3Sb \xrightarrow{HCl/MeOH} Ph_2SbCl \xrightarrow{NaOH} (Ph_2Sb)_2O \xrightarrow{HCl/EtOH} Ph_2SbCl \qquad (41)$$

14.2.3.2 Alkali metal derivatives

Cleavage of a methyl group from Me_3Sb by sodium in liquid ammonia yields $NaSbMe_2$,[148] but a superior route is from Me_2SbBr and the same reagent.[152] Low temperature reaction of R_2SbH and RLi also yields dialkylstibides.[153] Cleavage of $Ph_2Sb(CH_2)_nSbPh_2$ by Na or Li in liquid ammonia gives $MPhSb(CH_2)_nSbPhM$.[154]

14.2.3.3 Tertiary stibines

These are trigonal pyramidal molecules with the CSbC angle 90°, but this angle is sensitive to electronic and steric factors, *e.g.* in $Sb(CF_3)_3$ it is 100° (see Section 28.11.1.1). Their normal method of preparation is *via* the Grignard route,[102,103] but difficulties have been reported in separation of some trialkylstibines from solvents such as ether. A way round this is to oxidize to R_3SbX_2 with halogen and then reduce with zinc or $LiAlH_4$. The most commonly used are either liquids, *e.g.* Me_3Sb (b.p. 80–81 °C) and Et_3Sb (b.p. 158 °C), or solids, *e.g.* Ph_3Sb (m.p. 50 °C). Mixed alkylaryl-stibines are readily prepared from Ar_nSbCl_{3-n} and RMgX.[155] Triphenylstibine is air stable, but the trialkyl derivatives are readily oxidized; Me_3Sb spontaneously flames in air.

14.2.3.4 Distibines

Quite a large number of these are known and they may be prepared from the dihaloalkane and lithium dialkyl- or diphenyl-stibide in THF at low temperature, or from the sodium stibide in liquid ammonia.

$R_2Sb(CH_2)_nSbR_2$	R	n	$Ref.$
	Me	1, 3	152, 156
	Et	1, 4–6	153
	Pr^n	1	148
	Bu^t	4	157
	Cy	3–6	158
	Ph	1, 3–6	154

Reaction of either *cis*- or *trans*-ClCHCHCl and $NaSbPh_2$ gives $Ph_2SbSbPh_2$ rather than $Ph_2SbCHCHSbPh_2$,[159] but Ph_2SbH addition to $Ph_2SbCHCH_2$ does work.[160] From sodium acetylide and Me_2SbBr is obtained $Me_2SbCCSbMe_2$.[161] Asymmetrical distibines of type PhMeSb-$(CH_2)_nSbMePh$ are available *via* PhMeSbNa or by equation (42).[154]

$$Ph_2Sb(CH_2)_nSbPh_2 \xrightarrow[MeCl]{Na/liq.NH_3} PhMeSb(CH_2)_nSbMePh \qquad (42)$$

The antimony analogue of diars, *o*-phenylenebis(dimethylstibine), has been prepared, as has its phenyl analogue (equation 43).[162] Attempts to prepare multidentate stibines have been unsuccessful, probably due to the weakness of the carbon–antimony bond. For example, in the reaction of Bu^nLi with *o*-$C_6H_4(SbPh_2)Br$ the nucleophile attacks the C—Sb as well as the C—Br bond.[162]

(43)

14.2.4 Bismuthines

Little work has been done in this area. Most interest was shown more than 60 years ago.[102,103] The best route to haloorganobismuthines appears to be *via* heating R_3Bi and $BiCl_3$ in an organic solvent, and the phenyl compounds have more recently been isolated.[163]

There is a report[164] of Ph_2Bi^- being formed from Ph_2BiI and $Na/liq.$ NH_3, but instability complicates synthetic utility. Very little is known about tertiary bismuthines, although Ph_3Bi is available.

14.2.5 Mixed Group V Ligands

14.2.5.1 Bidentates

The techniques used to synthesize these ligands are similar to those quoted above. Table 2 contains a list of the most common examples.

Table 2 Some Mixed Group V Bidentate Ligands

Ligand		Ref.
$Ph_2PCH_2CH_2AsPh_2$		159
cis-$Ph_2PCHCHAsPh_2$		159
$trans$-$Ph_2PCHCHAsPh_2$		159
$Ph_2PCCPPh_2$		45
(o-C₆H₄ with PR₂ and AsR₂)	R = Me	165
	R = Ph	166
(o-C₆H₄ with PR₂ and SbR₂)	R = Me	167
	R = Ph	168
(o-C₆H₄ with PEt₂ and AsMe₂)		169
(o-C₆H₄ with AsR₂ and SbR₂)	R = Me	170
	R = Ph	168

14.2.5.2 Multidentate ligands

The same routes for monodonor systems are used for the preparation of mixed ligands. The base-catalyzed addition of As—H bonds (*e.g.* Ph_2AsH) to vinylphosphines yields $PhP(CH_2CH_2AsPh_2)$;[45] $PhPCl_2$ reacts with o-$C_6H_4(AsPh_2)Li$ to form $PhP(CH_2CH_2CH_2AsMe_2)_2$.[171] A large number of tripodal ligands with a varied apical donor are known:

$E(o$-$C_6H_4PPh_2)_3$	E = As[172]	E = Sb[145]	E = Bi[173]
$E(o$-$C_6H_4AsPh_2)_3$	E = P[139]	E = Sb[172]	E = Bi[173]
$E(o$-$C_6H_4AsMe_2)_3$	E = P[174]	E = Sb[175]	E = Bi
$E(CH_2CH_2CH_2AsMe_2)_3$	E = P[176]	E = Sb[177]	E = Bi

For the tripodal ligands which contain o-phenylene linkages, $E(o$-$C_6H_4E'Ph_2)_3$, the yields increase in the order P > As > Sb, and for $E(o$-$C_6H_4E'Me_2)$ as E' = As > Sb > Bi, which has been attributed to steric hindrance and to increasing weakness of the C—E' bond, respectively. Tripodal ligands of C_s symmetry have been produced by Venanzi's group by use of the blocking effect of the $PNMe_2$ group (equation 44).[178]

(44)

14.2.5.3 Other multidentate systems

Routes have been developed to the hitherto unobtainable arsine—alkene ligands $(CH_2{=}CHCH_2CH_2)_n As(CH_2CH_2CH_2AsMe_2)_{3-n}$ ($n = 1$, tasol, but-3-enylbis(3-dimethylarsinopropyl)arsine; $n = 2$, dasdol, 3-dimethylarsinopropylbis(but-3-enyl)arsine) by making use of the difference in reactivity between the Cl—C and As—Cl bonds in the precursor $Cl(CH_2)_m AsCl_2$ ($m = 2, 3$) molecules. Thus, the triarsine obtained by reaction of 2-chloroethyldichloroarsine with the Grignard reagent of 3-chloropropyldimethylarsine yields 2-chloroethylbis(3-dimethylarsinopropyl)arsine, from which tasol is obtainable by subsequent reaction with either the Grignard reagent of vinyl bromide or, preferably, with vinyllithium. Similarly, 3-chloropropyldichloroarsine reacts with the Grignard reagent of 4-chlorobut-1-ene to form 3-chloropropylbis(but-3-enyl)arsine, which on reaction with sodium dimethyl arsenide yields dasdol.[179] There have been an enormous number of mixed Group V/N/S/Se/O/alkene ligand systems synthesized. A good guide to their preparation is available.[12]

14.2.5.4 Arsenic and mixed macrocycles

Kyba and co-workers extended their high dilution synthesis of 11-[180] and 14-membered[181] rings containing phosphorus to incorporate arsenic ligating centres (equation 45). Since the barrier to inversion[182] for arsenic is of the order of $167.4\,kJ\,mol^{-1}$ the [14]aneAs$_4$ macrocycle will exist in five isomeric forms defined by positions of the methyl groups with respect to the mean plane of the ring, as in the case of the phosphorus analogue (see Section 14.2.1.7.V). Three isomers were separated chromatographically as white crystalline needles and their structures assigned on the basis of ^1H and ^{13}C NMR and chromatographic behaviour; these correspond to the phosphorus isomers (a), (b) and (c) shown in Figure 2 (p. 1044).

(45)

The mixed pnictide P_2As_2 macrocycle has also been prepared as shown by equation (46). Kyba's group have also prepared an [11]aneAs$_3$ macrocycle (equation 47).[180] In general it was found[183] that in 11-membered rings the cavities were too small to encapsulate a metal ion. These smaller rings form *fac* or *mer* complexes, whereas the 14-membered rings can encapsulate the metal ion. Kauffmann and Ennen[184] synthesized 12-, 16- and 24-membered arseno macrocycles as shown in Scheme 9, and extended this to prepare a number of large ring arsenothio and arsenothioether macrocycles (Scheme 10).[185] The arsenophosphametallo macrocycle, $Rh_2(CO)_2Cl_2(\mu\text{-dmpa})_2$ **(13)**, may be used as a macrocyclic ligand when arsenic rather than phosphorus are the donor atoms.[186]

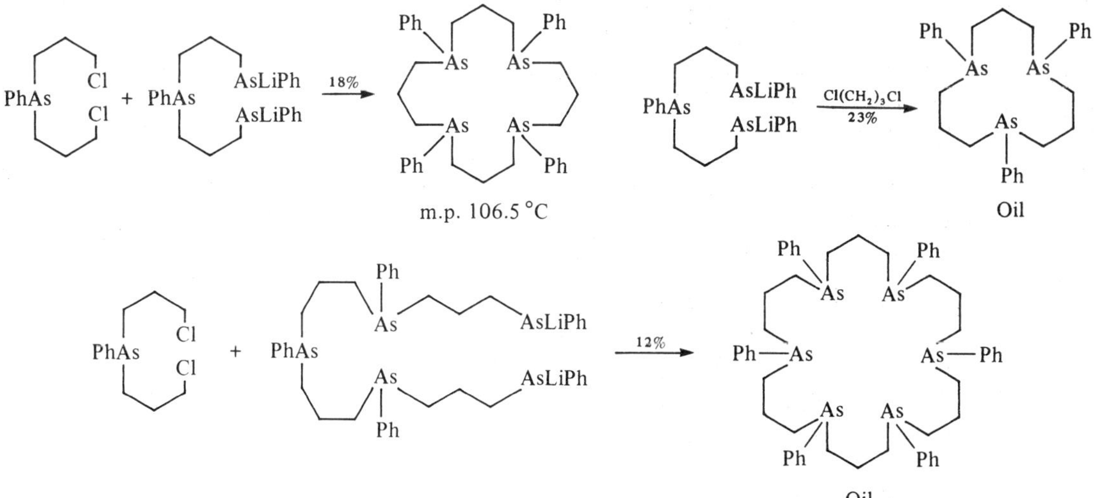

Scheme 9

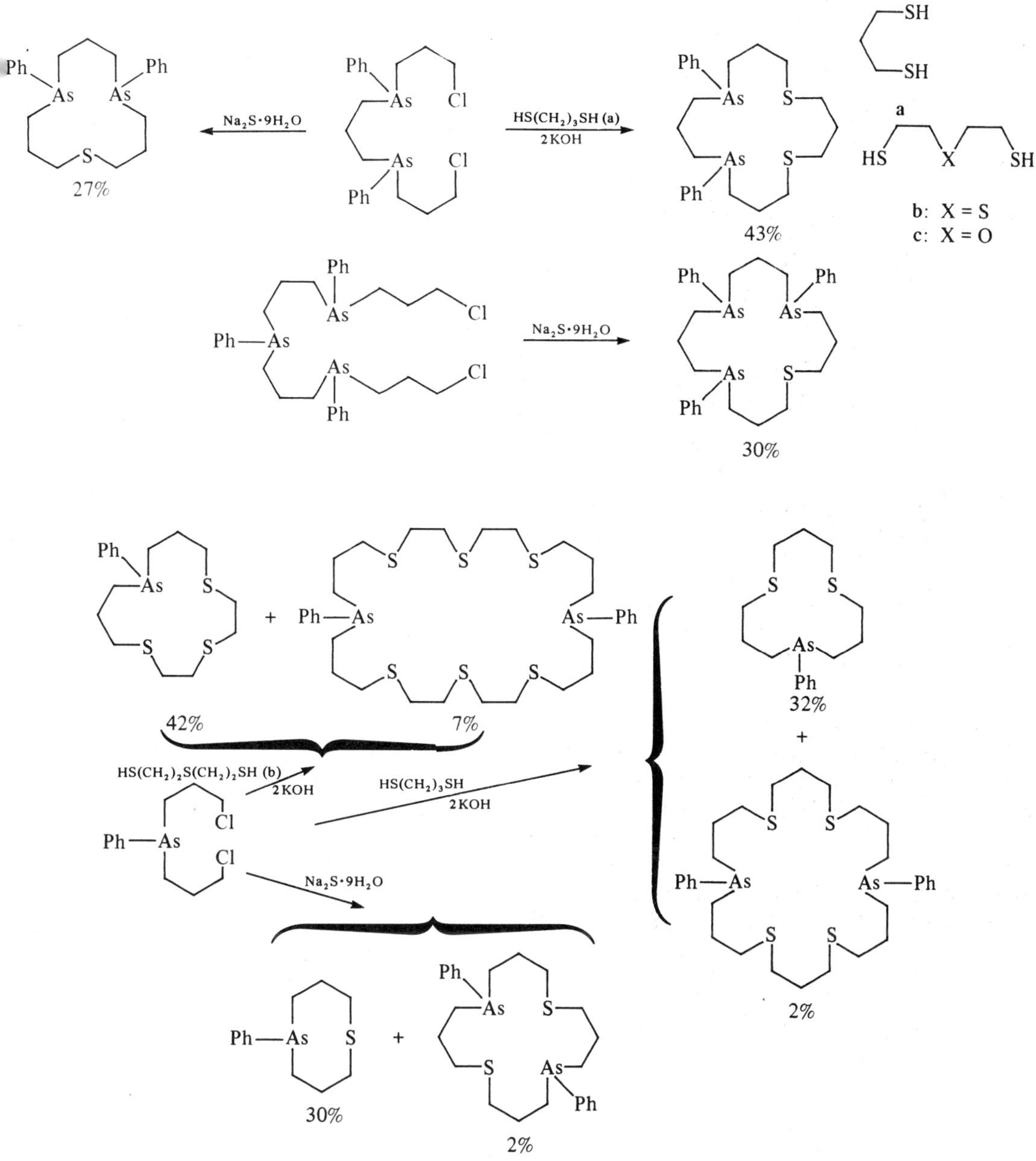

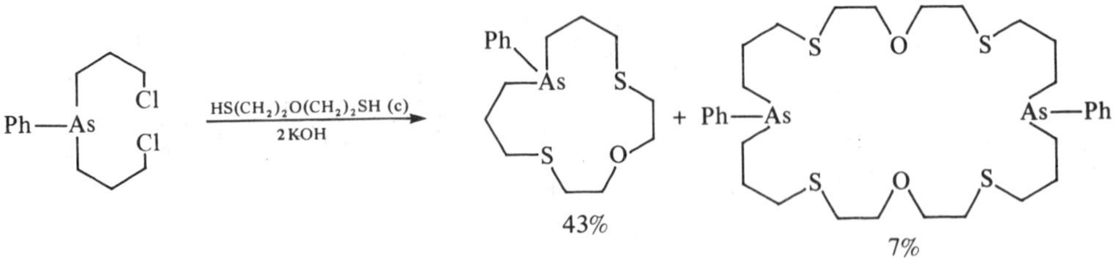

Scheme 10

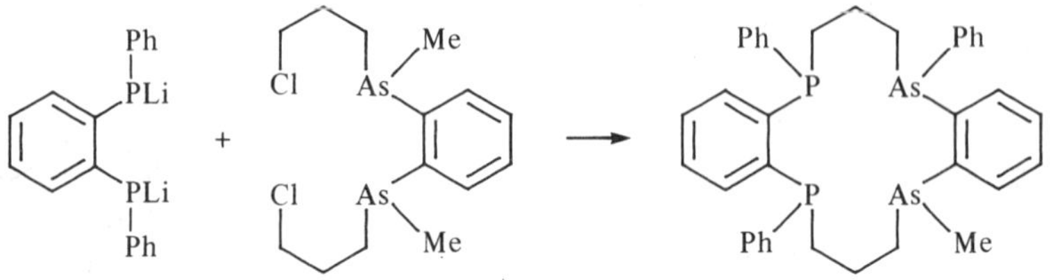

(46)

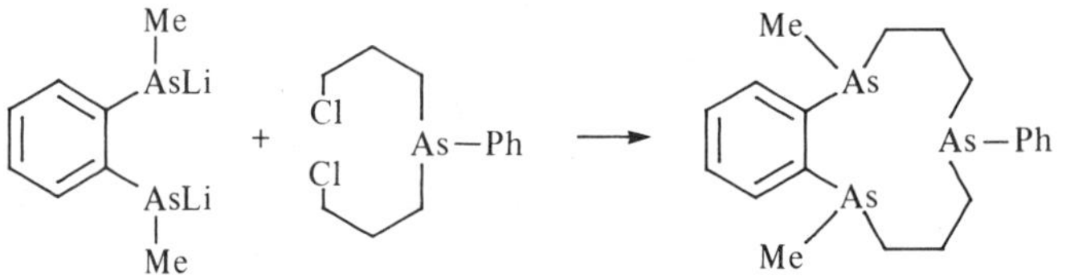

(47)

14.3 THE CONE ANGLE CONCEPT

The ability of heavy Group V ligands and transition metals to form stable L_nM—ER_3 (E = P, As, Sb, Bi; M = transition metal; L = other ligands) is determined by the synergic interplay of their respective donor–acceptor properties, subtly modulated by steric influences. The electronic and steric factors determining the electron availability on the transition metal are determined by the oxidation state, coordination number, orbital geometry and the ligand effects of the other substituents in the coordination sphere. These factors will be discussed later.

Information on the pnictogens, like the chalcogens and halogens, rapidly decreases down the group! Thus, the following discussion will mainly centre on the phosphorus. However, by analogy, the arguments will usually be applicable to arsenic and, with less certainty, to antimony and bismuth.

Tolman[187] introduced the ingenuous phrase *cone angle* to describe in quantifiable terms the bulk of (mainly) phosphine ligands. The term is especially applicable to Group V ligands, because in the free state (coordination number three) they are coordinatively unsaturated, but on bonding to metals increase their coordination number to four, with tetrahedrally disposed bonds. A tetrahedron rotated around an apical axis describes a cone. It is this geometrical fact that makes the term particularly appropriate for pnictide ligands.

The cone angle concept is a useful approach to the understanding of the spatial/steric effects of phosphorus ligands. It is important to appreciate, however, that cone angles are not necessarily the dominant steric/structural effect in ligand behaviour, *i.e.* steric effect and cone angle are not necessarily synonymous. Below are listed some structural features (but not directly affecting electronic effects) dependent on the size, shape and connectivity of the ligand and described in terms other than cone angle.

14.3.1 Chelate Effects

Although cone angles have been given for phosphorus chelate ligands[187] their behaviour is more usefully understood in terms of chelate bite and chain length,[188–190] *e.g.* $R_2PCH_2PR_2$ tend to bridge between two metals, $R_2PCH_2CH_2PR_2$ chelate in *cis* positions, whereas diphosphines with longer chains can *trans* chelate. It is a general observation that the replacement of phosphorus by arsenic or antimony enhances the effect.

Chain length may *inter alia* leave a potential ligand site 'dangling',[191] where the stability of the chelate is determined by its intrinsic strain (equation 48). Chelating ligands' properties may also be described in terms of torsion angles.

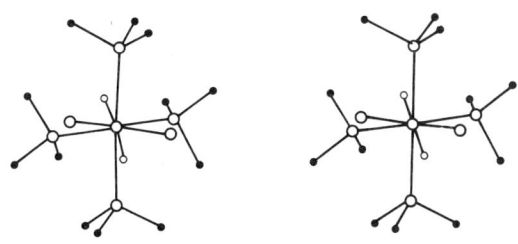

$$CpCoI_2(CO) \ + \ (CH_2)_n \ \longrightarrow \ \overset{PR_2}{\underset{PR_2}{|}} \ \longrightarrow \ Co \ \rightleftharpoons \ \left[\ Co \ \right]^+ \ I^- \quad (48)$$

14.3.2 Polydentate Phosphines

The advantages of polydentate phosphine ligands over isolated phosphorus ligands include: (i) increased nucleophilicity/basicity at the metal centre, (ii) greater control of the stereochemistry and coordination number of the complex, (iii) slower and more controlled intra- and inter-molecular exchange reactions, and (iv) useful bonding information from NMR phosphorus–phosphorus coupling constants and, where applicable, metal–phosphorus coupling constants.

All this is not to say that the steric bulk or cone angle of the phosphine is unimportant. Some authors explain their results in term of cone angles, *e.g.* Sattelberger *et al.*[192] attribute the change of geometry around the tantalum complexes (**14**) and (**15**) to the smaller cone angle of dmpe (107°) compared to PMe₃ (118°): the change is from distorted dodecahedral (**13**) to distorted square antiprismatic (**14**) geometry. The antiprismatic complex has somewhat shorter Ta—P bonds.

Stereoview of $TaCl_2H_2(PMe_3)_4$

$TaCl_2H_2(dmpe)_2$

The importance of connectivity in polydentate ligands has been demonstrated by Dahlenberg.[193] The classic QP ligand of Venanzi, $P(o\text{-}PPh_2C_6H_4)_3$, containing four phosphorus donor atoms, forms a trigonal bipyramidal complex with iron(II), [FeX(QP)]X (X = Cl, Br, I).[194] When flexibility is allowed between the phosphorus donor atoms in $P(CH_2CH_2CH_2PMe_3)_3$ a distorted octahedral complex forms, (16). Further evidence that the overall geometry of the multidentate ligand is of more importance than the geometry of the individual ligating atoms has been shown by Osborn.[195] Thus triangular faces of tetrahedral clusters may be complexed by $CH(PPh_2)_3$, affording complexes of type (17).

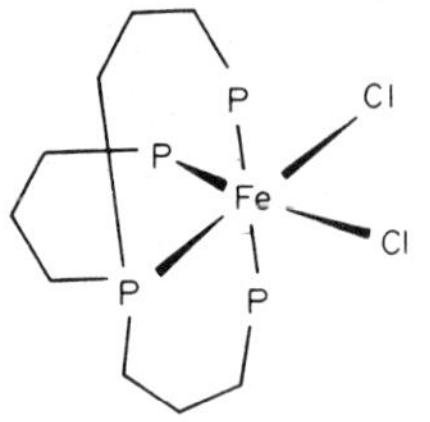

(16) 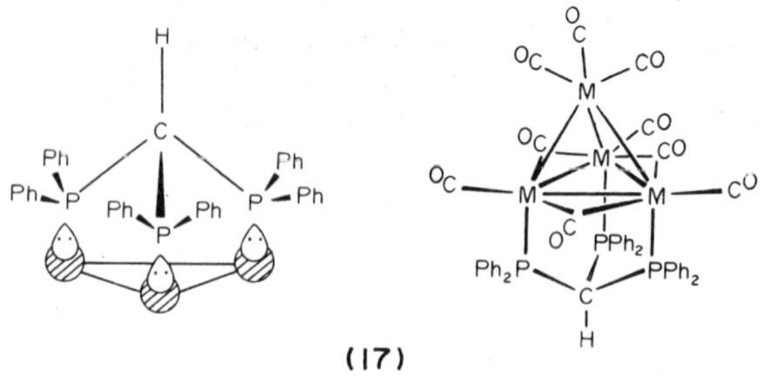　　　　　　　　　　　　　　　　　　　(17)

Another important factor in multidentate ligands is the nature of donor atoms other than phosphorus, *e.g.* Fryzuk's preparation of lanthanide and actinide phosphine complexes using mixed donor ligands such as $Me_2PCH_2SiMe_2NHSiMe_2CH_2PMe_2$ (LH) to prepare $\{[LHf(BH_4)_2](\mu\text{-H})_3\text{-}[(BH_4)HfL]\}$ (18),[196] where the unsymmetrical, complicated geometry is more simply represented as an idealized trigonal bipyramid at one Hf and an octahedral Hf. Macrocyclic ligands are dealt with in Section 14.2.1.7 where it is emphasized that cavity size is probably the dominant factor in determining complex stability for a given group of donor atoms.

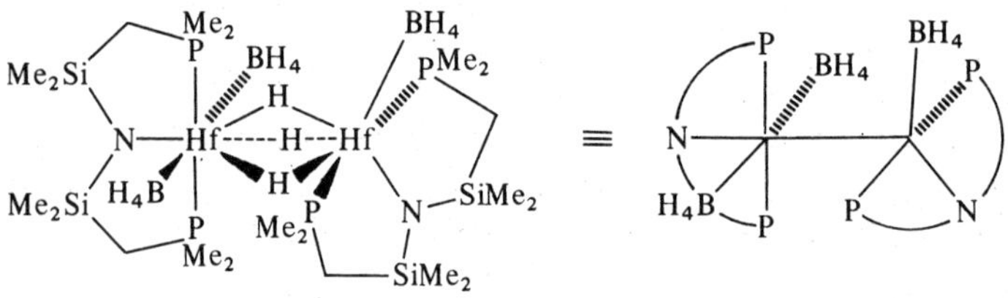

(18)

14.3.3 Solubility Effects

Aside from electronic factors the effect of phosphine substituents on reaction rate need not be steric in origin. In a given solvent system the structure of the phosphine ligand can have an effect on solubility and hence the concentration of a catalyst species. Two examples from different ends of the polar/non-polar range are given here. The solubilization of phosphines in water was achieved by Baird and Smith[197] by attaching a quaternary ammonium group, as in $[Ph_2PCH_2\overset{+}{N}Me_3]I^-$. The donor properties of this ligand appear to differ only slightly from those of PPh_2Me and PPh_3. Enhanced solubility of organometallic complexes in organic solvents was achieved by Chipperfield *et al.*[198] with the long chain phosphines PR_3 (R = $n\text{-}C_{10}H_{21}\text{-}C_{19}H_{39}$) and $P(p\text{-}C_6H_4R)_3$ (R = Et–$n\text{-}C_9H_{19}$). When complexed to Pt, Pd or Rh in catalytic systems the product distribution was related to phosphine chain length.

14.3.4 Chiral Phosphines

The cone angles of optical isomers are, of course, identical, but it has been demonstrated that chiral phosphines in homogenous catalytic systems can determine a reaction pathway to form asymmetric products.[199] Clearly the chirality of the phosphine is the determining factor in these systems, although other essential characteristics of the phosphine ligand's bulk and the π acidity still apply. The work of Ferguson and Alyea's group[200-202] on the cog-like shape of the cone should, eventually, contribute much to the asymmetry problem.

14.3.5 The Definitions of Cone Angle

The cone angle was defined originally by Tolman[203] in 1970 and subsequently amplified[187] in 1977. As defined by Tolman the 'cone angle' was the property of the model of a phosphine. Two lines of work have emerged since which redefine cone angle. Firstly, several groups have measured 'actual' cone angles from X-ray data. Other groups have sought to develop mathematical models which would accurately describe ligand behaviour.

14.3.6 Definitions of Cone Angle Based on Models or on X-Ray Data

14.3.6.1 Tolman's cone angle

Tolman defined the cone angle in terms of a phosphine ligand bonded to nickel. Clearly the M—P bond length will affect the magnitude of the cone angle and so Tolman based his concept on a bond length of 2.28 Å. The simplest case is a symmetrical PR_3 ligand (**19**). In the absence of X-ray crystallographic data the solid angle was estimated by the use of molecular models.

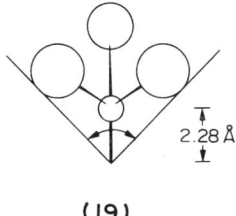

(19)

The concept of cone angles was extended to phosphines of the type PR′R″R‴ by measuring the angle between the M—P axis and the tangent of the van der Waals sphere for each substituent drawn from the neutral centre (**20**) giving a 'half' cone angle. Two-thirds of the sum of these half angles is the cone angle of the unsymmetrical ligand (equation 49).

$$\theta = \frac{2}{3}\sum_{i=1}^{3}\theta_i/2 \qquad (49)$$

This equation may be used to justify using 1/3 cone angle PR_3' to be the contribution of R′ to the cone angle of PR′R″R‴. Thus, if the cone angles of PR_3', PR_3'' and PR_3''' are known that of PR′R″R‴ may be calculated. In complex molecules, if free rotation is possible about the P—R bond, then that conformer is chosen that minimizes θ (or $\theta_i/2$). It is clear from diagram (**19**) that the solid angle θ is strictly geometric and does not depend on the nature of the atoms represented. The use of the concept (Table 3) has been extended to the other heavy Group V elements As, Sb and Bi (Tables 4–6).

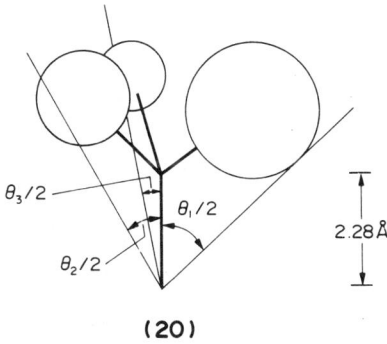

(20)

Table 3 Empirical Formula Index and Cone Angles of Phosphorus Compounds

	Empirical formula	Compound	Cone angle (°)	Ref.
1	Br_3P	Tribromophosphine	131	1
2	Cl_3P	Trichlorophosphine	124	1
3	F_3P	Trifluorophosphine	104	1
4	H_3P	Phosphine	87	1
			91	6
5	CH_5P	Methylphosphine	103	6
6	C_2H_7P	Dimethylphosphine	118	6
7	C_2H_7P	Ethylphosphine	108	6
8	C_3F_9P	Tris(trifluoromethyl) phosphine	137	1
9	$C_3H_9O_3P$	Trimethyl phosphite	107	1
10	C_3H_9P	Trimethylphosphine	118	1
			134	6
11	$C_4H_{11}O_2P$	Dimethyl ethylphosphonite	115	1
12	$C_4H_{11}P$	Diethylphosphine	127	6
13	$C_4H_{11}P$	t-Butylphosphine	116	6
14	$C_6H_6Cl_9O_3P$	Tris(2,2,2-trichloroethyl) phosphite	115	1
15	C_6H_7P	Phenylphosphine	101	1
			106	6
16	$C_6H_9O_3P$	1-Phospha-2,8,9-trioxaadamantane	106	1
17	$C_6H_{11}O_3P$	4-Ethyl-2,6,7-trioxa-1-phosphabicyclo[2.2.2]octane	101	5
18	$C_6H_{12}Cl_3O_3P$	Tris(2-chloroethyl)phosphite	110	1
19	$C_6H_{12}N_3P$	1,3,5-Triaza-7-phosphaadamantane	102	5
20	$C_6H_{12}N_3P$	Tris(aziridinyl)phosphine	108	1
21	$C_6H_{15}O_3P$	Triethyl phosphite	109	1
			110	2
22	$C_6H_{15}P$	Bis(isopropyl)phosphine	135	6
23	$C_6H_{15}P$	Triethylphosphine	132	1
			148	6
24	$C_6H_{16}P_2$	1,2-Bis(dimethylphosphino)ethane	107	1
25	$C_6H_{18}N_3P$	Tris(dimethylamino)phosphine	157	1
26	C_7H_9P	Methylphenylphosphine	118	6
27	C_7H_9P	o-Tolylphosphine	110	6
28	$C_8H_{11}O_2P$	Dimethyl phenylphosphonite	115	1
29	$C_8H_{11}P$	Ethylphenylphosphine	122	6
30	$C_8H_{11}P$	Dimethylphenylphosphine	122	1
			139	6
31	$C_9H_{12}N_3P$	Tris(2-cyanoethyl)phosphine	132	1
32	$C_9H_{21}O_3P$	Triisopropyl phosphite	130	1
33	$C_9H_{21}P$	Tripropylphosphine	132	1
			139	3
34	$C_9H_{21}P$	Tris(isopropyl)phosphine	160	1
35	$C_{10}H_{15}O_2P$	Diethyl phenylphosphonite	116	1
36	$C_{10}H_{15}P$	Diethylphenylphosphine	136	1
37	$C_{10}H_{23}P$	Isobutylbis(isopropyl)phosphine	167	2
38	$C_{10}H_{24}P_2$	1,2-Bis(diethylphosphino)ethane	115	1
39	$C_{12}H_{11}P$	Diphenylphosphine	128	1
40	$C_{12}H_{19}O_2P$	Di-n-propyl phosphonite	120	8
41	$C_{12}H_{19}P$	Bis(isopropyl)phenylphosphine	155	8
42	$C_{12}H_{19}P$	Di-n-propylphenylphospine	141	8
43	$C_{12}H_{23}PSi_2$	Bis(trimethylsilyl)phenylphosphine	175	8
44	$C_{12}H_{25}P$	Dicyclohexylphosphine	142	9
45	$C_{12}H_{27}O_3P$	Tributyl phosphite	112	3
46	$C_{12}H_{27}O_3P$	Tri-t-butyl phosphite	172	1
47	$C_{12}H_{27}P$	Tri-s-butylphosphine	160	1
48	$C_{12}H_{27}P$	Tri-t-butylphosphine	182	1
49	$C_{12}H_{27}P$	Tri(isobutyl)phosphine	143	1
50	$C_{12}H_{27}P$	Tributylphosphine	132	1
51	$C_{13}H_{13}OP$	Methyl diphenylphosphinite	132	1
52	$C_{13}H_{16}P$	Methyldiphenylphosphine	136	1
53	$C_{14}H_{15}OP$	Ethyl diphenylphosphinite	133	1
54	$C_{14}H_{15}P$	Ethyldiphenylphosphine	140	1
55	$C_{14}H_{23}O_2P$	Dibutyl phenylphosphonite	118	3
56	$C_{14}H_{23}P$	Di-t-butylphenylphosphine	170	1
57	$C_{14}H_{23}P$	Di-n-butylphenylphosphine	141	8
58	$C_{15}H_{17}P$	Isopropyldiphenylphosphine	164	7
59	$C_{15}H_{17}P$	Diphenyl-n-propylphosphine	143	8
60	$C_{15}H_{17}OP$	Diphenylpropylphosphinite	140	8

61	$C_{15}H_{19}PSi$	Diphenyltrimethylsilylphosphine	162	8
62	$C_{15}H_{33}P$	Trineopentylphosphine	180	1
63	$C_{16}H_{19}P$	Butyldiphenylphosphine	141	7
64	$C_{16}H_{19}P$	t-Butyldiphenylphosphine	157	1
65	$C_{16}H_{19}OP$	Butyldiphenylphosphinite	145	8
66	$C_{16}H_{20}NP$	Diethylaminodiphenylphosphine	160	8
67	$C_{16}H_{27}P$	Di-n-pentylphenylphosphine	143	8
68	$C_{16}H_{31}P$	t-Butyldicyclohexylphosphine	171	2
69	$C_{17}H_{21}P$	n-Pentyldiphenylphosphine	145	8
70	$C_{18}F_{15}P$	Tris(perfluorophenyl)phosphine	184	1
71	$C_{18}H_{12}Cl_3P$	Tris(p-chlorophenyl)phosphine	145	7
72	$C_{18}H_{12}F_3P$	Tris(p-fluorophenyl)phosphine	145	7
73	$C_{18}H_{12}F_3P$	Tris(m-fluorophenyl)phosphine	145	1
74	$C_{18}H_{15}OP$	Phenyl diphenylphosphinite	139	3
75	$C_{18}H_{15}O_3P$	Triphenyl phosphite	128	1
76	$C_{18}H_{15}P$	Triphenylphosphine	145	1
77	$C_{18}H_{22}P$	Cyclohexyldiphenylphosphine	153	9
78	$C_{18}H_{23}P$	n-Hexyldiphenylphosphine	145	8
79	$C_{18}H_{27}P$	Dicyclohexylphenylphosphine	162	5
80	$C_{18}H_{31}P$	Di-n-hexylphenylphosphine	145	8
81	$C_{18}H_{36}P$	Tricyclohexylphosphine	170	1
82	$C_{21}H_{21}O_3P$	Tri-p-tolyl phosphite	128	1
83	$C_{21}H_{21}O_3P$	Tri-o-tolyl phosphite	141	1
84	$C_{21}H_{21}P$	Tri-o-tolylphosphine	194	1
85	$C_{21}H_{21}P$	Tri-p-tolylphosphine	145	1
86	$C_{21}H_{21}P$	Tribenzylphosphine	165	1
			160	3
87	$C_{21}H_{21}O_3P$	Tris(p-methoxyphenyl)phosphine	145	7
88	$C_{23}H_{45}O_2P$	Dimenthyl isopropylphosphinite	209	4
89	$C_{24}H_{27}O_3P$	Tris(2,6-dimethylphenyl)phosphine	190	1
90	$C_{25}H_{22}P_2$	Bis(diphenylphosphino)methane	121	1
91	$C_{26}H_{24}P_2$	1,2-Bis(diphenylphosphino)ethane	125	1
92	$C_{26}H_{43}O_2P$	Dimenthyl phenylphosphonite	142	2
93	$C_{26}H_{48}P_2$	1,2-Bis(dicyclohexylphosphino)ethane	142	1
94	$C_{27}H_{26}P_2$	1,3-Bis(diphenylphosphino)propane	127	1
95	$C_{27}H_{33}O_3P$	Tris(2-isopropylphenyl) phosphite	148	1
96	$C_{27}H_{33}P$	Trimesitylphosphine	212	1
97	$C_{30}H_{57}O_3P$	Trimenthyl phosphite	140	2
98	$C_{36}H_{27}O_3P$	Tri-o-phenylphenyl phosphite	152	1

1. C. A. Tolman, *Chem. Rev.*, 1977, **77**, 313.
2. H. Schenkluhn, M. Zähres, W. Scheidt and B. Weimann, *Angew. Chem., Int. Ed. Engl.*, 1979, **18**, 401.
3. P. Heimbach, J. Kluth, H. Schenkluhn and B. Weimann, *Angew. Chem., Int. Ed. Engl.*, 1980, **19**, 570.
4. A. Musco and A. Immirzi, *Inorg. Chim. Acta*, 1977, **25**, 41.
5. D. J. Darensbourg and R. L. Kump, *Inorg. Chem.*, 1978, **17**, 2680.
6. J. A. Mosbo, J. T. DeSanto, B. N. Storhoff, P. L. Bock and R. E. Bloss, *Inorg. Chem.*, 1980, **19**, 3086.
7. W. P. Giering, M. N. Golouin, M. M. Rahman and J. E. Belmonte, *Organometallics*, 1985, **4**, 1981.
8. T. Bartik, P. Heimbach and T. Himmler, *J. Organomet. Chem.*, 1984, **276**, 399.
9. M. Basato, *J. Chem. Soc., Dalton Trans.*, 1986, 217.

Table 4 Arsines, Stibines and Bismuthines

ER_3	Cone angle θ (°)		
	E = As	E = Sb	E = Bi
R = alkyl			
EMe_3	121	119	117
EEt_3	134	131	129
EPr_3^n, EBu_3^n	134	131	129
EBu_3^i	145	141	139
EPr_3^i	162	157	155
EBu_3^s	162	157	155
EBz_3	166	162	159
$E(CH_2CMe_3)_3$	181	176	173
EBu_3^t	183	178	175
R = aryl			
EPh_3	147	143	141
$E(o\text{-}tolyl)_3$	195	189	185
$E(mesityl)_3$	212	205	202

Table 5   Miscellaneous Arsines, Stibines and Bismuthines

ER_2R'	Cone angle θ (°)		
	As	Sb	Bi
EMePh?	125	122	120
EMe_2CF_3	127	124	123
EPh_2Me	138	135	133
EPh_2Et	140	137	135
EPh_2Pr^i	152	148	146
EPh_2Bu^t	158	154	152
EBu_2^tPh	171	166	164
$E(CH_2CH_2CN)_3$	134	131	129
$E(CF_3)_3$	139	136	134
EPh_3	185	179	176

Table 6   Chelating Arsines, Stibines and Bismuthines

Ligand	Cone angle θ (°)		
	As	Sb	Bi
$Me_2E(CH_2)_2EMe_2$	110	109	107
$Et_2E(CH_2)_2EEt_2$	118	116	114
$Ph_2ECH_2EPh_2$	124	121	120
$Ph_2E(CH_2)_2EPh_2$	128	125	123
$Ph_2E(CH_2)_3EPh_2$	129	127	125
$Cy_2E(CH_2)_2ECy_2$	144	140	138

[a] The cone angles refer to each coordinating pnictide of the ligand.

14.3.6.2   Mathematical methods for estimating cone angles

Imyanitov[204] has developed a method of estimating cone angles for any ligand provided that the atomic radii, van der Waals radii and the molecular geometry are known. The method involves combination of mathematical and graphical methods. From the geometry of M—AB_n shown in (21) the equation in terms of $\theta/2$ was deduced (equation 50).

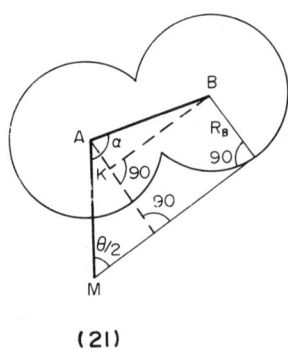

(21)

$$R_B = [r_M + r'_A + 0.33(r_A + r_B)]\sin(\theta/2) - 0.94(r_A + r_B)\cos(\theta/2)$$

$$\frac{\theta}{2} = \arcsin\frac{R_B}{\{[r_M + r'_A + 0.33(r_A + r_B)]^2 + 0.94^2(r_A + r_B)^2\}^{1/2}}$$
$$+ \arctan\frac{0.94(r_A + r_B)}{r_M + r'_A + 0.33(r_A + r_B)} \tag{50}$$

r = covalent radius of subscript atom

R = van der Waals radius of the subscript atom as defined by the geometry of Figure 5

$\dfrac{\theta}{2}$ = half cone angle

For more complex and bulky ligands an accurate geometrical construction was used for measurement (**22a**) to estimate a value for r_s and R_s, where 's' refers to the complex substituent and r,R the sphere enclosing the complex substituent, making it equivalent to the atom B in (**21**). This procedure yields values for phosphorus cone angles systematically 5° higher than that of Tolman,[187] undoubtedly because of the Ni—P distance of 2.23 Å chosen by Imyanitov.

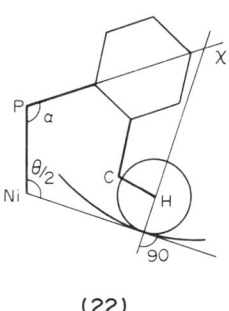

(22)

Bagnall and Xing-fu[205,206] in their study of lanthanide and actinide complexes developed a more sophisticated approach to the analysis of the steric demands of ligands. The emphasis of their work, in contrast to that previously discussed, was on ligands other than phosphines.

First order steric crowding is due to the steric demands of the coordinating atom, *e.g.* the oxygen atom of the phosphine oxide or a chlorine atom on its own. Second order steric crowding is due to the atoms or groups attached to the coordinating atom, *e.g.* the substituents on the phosphorus atom in the phosphine oxide above.

These steric effects are described by two terms: (a) the *cone angle factor* (caf) is the solid angle of the cone comprising the apical centre and the coordinating atom or ligand divided by four, Figure 5(a); (b) the *fan angle* describes the crowding in the plane and is the angle subtended by the atom or group in the various symmetry planes. This is the same as the *ligand profile* of Ferguson *et al.* However, since Bagnall's interest was in non-phosphorus ligands, *e.g.* H_2O or R_2O, the ligand profiles are extreme, and this work will not be further discussed here.

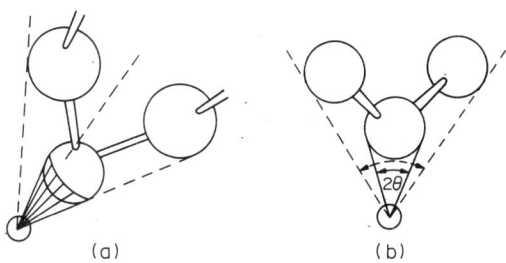

Figure 5 First- and second-order cone angles (a) and fan angles (b). When the ligand is irregular in shape, the fan angles in different planes will not be the same.

The cone angle concept has been further developed by Mingos[207] when applied to metal clusters. Since the function of the phosphine ligand is to stabilize the cluster by covering the surface, the important phosphine parameter is the 'plane of coverage', Figure 6. In the idealized case shown this would be a circle of radius $d/2$. In practice the projection would be cog-like, as demonstrated by Ferguson and Alyea[200] *inter alia* in the context of mononuclear complexes.

In cluster formation the ligand may be viewed from the cluster as a whole rather than the particular metal atom forming the M—P bond, *i.e.* the cluster–phosphine distance that determines

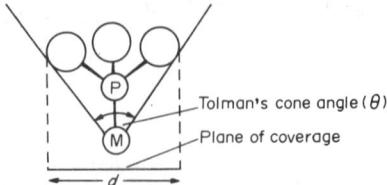

Figure 6   Tolman's cone angle used in mononuclear complexes with the 'plane of coverage'.

the effective coverage of the cluster by the ligand is not the distance from the phosphorus atom of the ligand to the nearest metal atom of the cluster, but the longer distance of the phosphine to cluster as a whole or, at any rate, the mean metal position of that section covered by the ligand, Figure 7. Of necessity the cluster cone angle is less than Tolman's cone angle for any given ligand since, as has been discussed previously, the cone angle is inversely proportional to the M—P bond length for a given ligand.

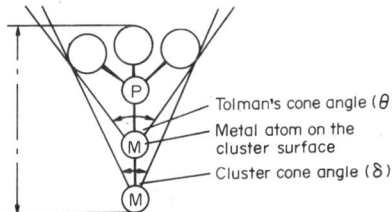

Figure 7   The cluster cone angle

14.3.6.3   Computer-generated cone angles

Mosbo and co-workers[208] have employed semi-empirical molecular orbital calculations, computed using MINDO/3. The heats of formation (ΔH_f) and atom positions for 15 phosphines were calculated using the following bond lengths and angles: P—H = 1.4 Å, P—C = 1.8 Å, C—C(alkyl) = 1.5 Å,   C—C(aryl) = 1.4 Å,   C—H = 1.1 Å;   R—P—R' = 109.5°,   R—C—C'(alkyl) = 109.5°, R—C—R(aryl) = 120°. All the potentially low energy conformations of the ligands were calculated and are given in Table 7. The Hanack diagrams for ethyl, isopropyl and *o*-tolyl substituents are shown in Figure 8.

The MINDO/3 optimized geometries were used in a computer program developed by Mosbo's group. A d(M—P) = 2.28 Å on the vector through P normal to a plane is defined by three points 1.00 Å along each P—R bond, **(22b)**. Rotation (ϕ) through 360° in 1° increments and using van der Waals radii of hydrogen (1.22 Å) and carbon (1.55 Å) enabled the half cone angles $\theta/2$ of the outermost atoms to be computed.

$$R' \quad R'' \quad R$$

|2.28 A
M—

(22b)

The tg_i conformer of H_2P—$CHMe_2$ is shown in Figure 9 and the g_rg_l conformer in Figure 10. Using the ligand profile plots four alternative definitions of the steric factor θ were considered. The best compromise was found by averaging the maximum $\theta/2$ value for the two *gauche* groups corresponding to points c and d in Figure 9 and then averaged with a and b (equation 51).

$$\theta_{\mathrm{III}} = \frac{2}{3}\sum_{i=1}^{3}\left\{\left[\left(\frac{\theta}{2}\right)_{\max}\right]_{av}\right\}_i \tag{51}$$

Table 7 MINDO/3-derived Heats of Formation and Mole Fractions of All Phosphine Complexes

Compound	Unique and contributing conformers	No.	Mole fraction	MINDO/3 cone angles for all conformers (θ)
PH_3			unity	91.2
PH_2Me			unity	103.4
PH_2Et	t	1	0.318	104.2
	g_r; g_l	2	0.682	109.8
PH_2Pr^i	tg_r; tg_l	1	0.648	110.2
	$g_r g_l$	2	0.352	114.3
PH_2Bu^t			unity	116.4
PH_2Ph	s	1	0.856	103.7
	e	2	0.144	121.7
PH_2-o-tol	s_r; s_l	1	0.846	107.2
	e_l	2	0.113	123.3
	e_c	3	0.042	135.3
$PHMe_2$			unity	117.7
$PHEt_2$	t, t	1	0.039	120.4
	t, g_l; g_r, t	2	0.146	124.5
	t, g_r; g_l, t	3	0.207	124.7
	g_l, g_r	4	0.227	128.4
	g_l, g_l; g_r, g_r	5	0.302	128.5
	g_r, g_l	6	0.079	128.5
$PHPr^i_2$	tg_l, tg_r	1	0.083	129.0
	tg_r, tg_l	2	0.055	132.7
	tg_r, tg_r; tg_l, tg_l	3	0.134	133.2
	$tg_r, g_r g_l$; $g_r g_l, tg_l$	4	0.262	135.9
	$tg_l, g_r g_l$; $g_r g_l, tg_r$	5	0.333	136.5
	$g_r g_l, g_r g_l$	6	0.132	138.3
PHMePh	s	1	unity	117.5
	e	2		
PHEtPh	t, s	1	0.290	119.2
	g_l, s	2	0.494	122.8
	g_r, s	3	0.216	121.4
	t, e	4		
	g_l, e	5		
	g_r, e	6		
PMe_3			unity	134.4
PEt_3	t, t, t	1	0.005	139.0
	t, t, g_r; t, t, g_l; t, g_r, t; t, g_l, t; g_r, t, t; g_l, t, t	2	0.113	142.8
	t, g_l, g_r; g_r, t, g_l; g_l, g_r, t	3	0.122	146.7
	t, g_r, g_r; t, g_l, g_l; g_r, t, g_r; g_l, t, g_l; g_r, g_r, t; g_l, g_l, t	4	0.256	146.9
	t, g_r, g_l; g_l, t, g_r; g_r, g_l, t	5	0.085	147.2
	g_r, g_r, g_r; g_l, g_l, g_l	6	0.140	151.0
	g_r, g_l, g_l; g_l, g_r, g_r; g_r, g_r, g_l; g_l, g_l, g_r; g_r, g_l, g_r; g_l, g_r, g_l	7	0.278	151.3
PMe_2Ph	s	1	0.882	137.2
	e	2	0.118	154.7

For ligands with multiple conformations the contribution for each conformer (θ_A for conformer A) and its mole fraction n_A is given by equation (52).

$$\theta = n_A\theta_A + n_B\theta_B + \ldots n_i\theta_i$$

$$n_A = \frac{g_A}{g_A + g_B e^{-\Delta E_{AB}/RT} + \ldots g_i e^{-\Delta E_{Ai}/RT}} \tag{52}$$

E_{Ai} = difference in H_f for conformers A and i

g_A = number of conformers with that unique conformation.

T = 298 K

For those ligands with multiple conformations the weighted average of all conformations was the best compromise value. The results of the calculations are shown in Table 8. In common with all calculated cone angles the main sources of inaccuracies are the values chosen for bond lengths and

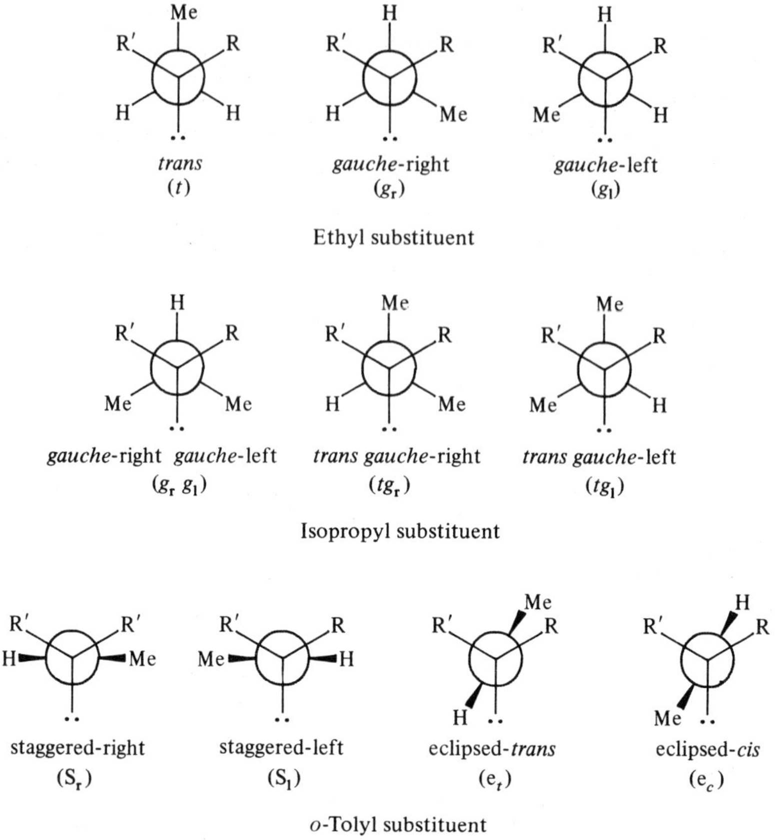

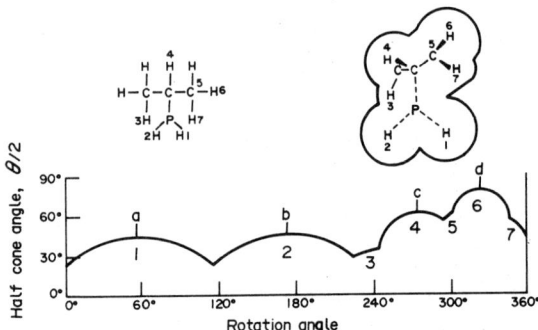

trans
(t)
gauche-right
(g_r)
gauche-left
(g_l)
Ethyl substituent
gauche-right gauche-left
(g_r g_l)
trans gauche-right
(tg_r)
trans gauche-left
(tg_l)
Isopropyl substituent
staggered-right
(S_r)
staggered-left
(S_l)
eclipsed-trans
(e_t)
eclipsed-cis
(e_c)
o-Tolyl substituent

Figure 8 Hanack diagrams (phosphorus is the front atom)

Half cone angle, θ/2
90°
60°
30°
0°
0°
60°
120°
180°
240°
300°
360°
Rotation angle

Figure 9 $tg_l PH_2$-Pr^i: top left, atom number identification; top right, circular trace containing contacted and skeletal atoms; bottom, ligand profile plot

van der Waals radii. Clearly there is considerable scope for improvement in computational methods, when values for the various molecular parameters can be agreed. In these calculations the authors choose to neglect the size of the phosphorus atom in order to emphasize the surface of the ligand.

14.3.6.4 Methods based on X-ray data

Several workers have pointed out that there is a large and growing body of X-ray structural data of transition metal phosphine complexes. Such data can provide 'real' cone angles, which may then be compared to Tolman's cone angles derived from models. Of course this line only applies to crystalline solids. Since the main application of transition metal chemistry is homogeneous catalysis, crystal structure data, though useful, are of limited applicability. Inevitably, cone angles in complexes in solution will be more variable than in crystalline systems.

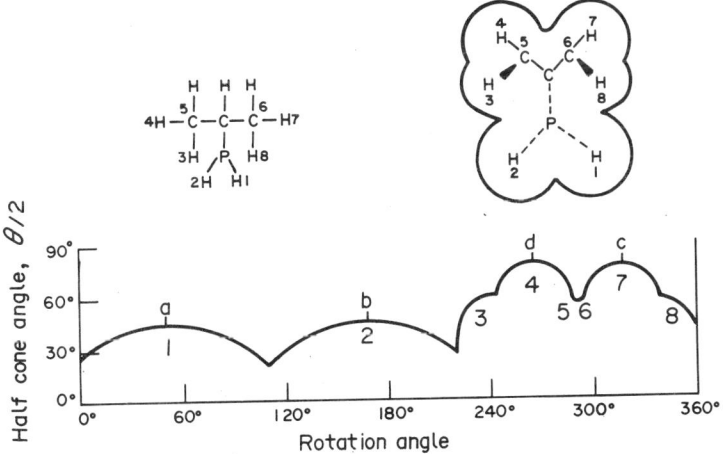

Figure 10 $g_t g_1 PH_2$-Pr^i: top left, atom number identification; top right, circular trace containing contacted and skeletal atoms; bottom, ligand profile plot.

Table 8 Weight-average Cone Angles from MINDO/3-Optimized Geometries and Heats of Formation Data

Compound	θ_{III}
PH_3	91.2
PH_2Me	103.4
PH_2Et	108.0
PH_2Pr^i	111.6
PH_2Bu^t	116.4
PH_2Ph	106.3
PH_2-o-tol	110.2
$PHMe_2$	117.7
$PHEt_2$	126.8
$PHPr^i_2$	135.3
$PHMePh$	117.5
$PHEtPh$	121.5
PMe_3	134.4
PEt_3	148.2
PMe_2Ph	139.3

Ferguson, Alyea and co-workers[200-202] have examined complexes containing bulky tertiary phosphines, and have discussed their results in terms of cone angles. They have, however, been critical of Tolman's approach of treating ligands as if they were solid cones. They have devised a method of graphically displaying the 'cog-like' nature of ligands which also shows the gaps between the ligand moieties—the 'depth of tooth' of the 'conic cog'. These workers have termed these graphs 'ligand profiles' and their data are obtained from X-ray structures. The ligand profile is generated by rotating the tertiary phosphine about the M—P bond axis (angle of rotation is ϕ) and measuring the angle $\theta/2$ formed by the tangent to the van der Waals radius of the 'proudest' hydrogen atom through the central metal atom (Figure 11).

A graph of ϕ *vs* $\theta/2$ is the ligand profile and several examples drawn by Ferguson *et al.* are shown in Figure 12, together with tabulated data (Table 9) from 14 X-ray structural determinations. Ferguson used these profiles to define the cone angle as the maximum value given. As can be seen, there is considerable variation in the cone angle for a given ligand, *e.g.* for PCy_3, from 163° in $[PtI_2(PCy_3)_2]$ to 181° in $[Hg(NO_3)_2(PCy_3)]_2$, *cf.* 179° (from CPK models), 170° (Tolman, 1977), 145° (Immirzi and Musco).[209]

A further refinement of the cone angle concept has been proposed by Oliver.[210] The detailed consideration of the X-ray structure of $[Pt(C_3H_5)(PCy_3)_2]PF_6$ (see stereoview, Figure 13) showed considerable ligand intermeshing, and a radial graphical method of depicting the ligand profile was suggested (Figure 14) (*cf.* Ferguson, Figure 11).

Immirzi and Musco[209] have proposed a modification of Tolman's definition of cone angle to account for the ability of ligands to intermesh when complexed, *e.g.* tricyclohexylphosphine in

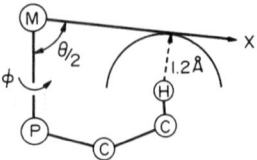

Figure 11   Details of the calculation of the maximum semicone angle, $\theta/2$. The point of contact of the cone-generating vector $M \rightarrow X$ from the metal, with the van der Waals sphere of the hydrogen atom, is coplanar with the metal, phosphorus and hydrogen atoms.

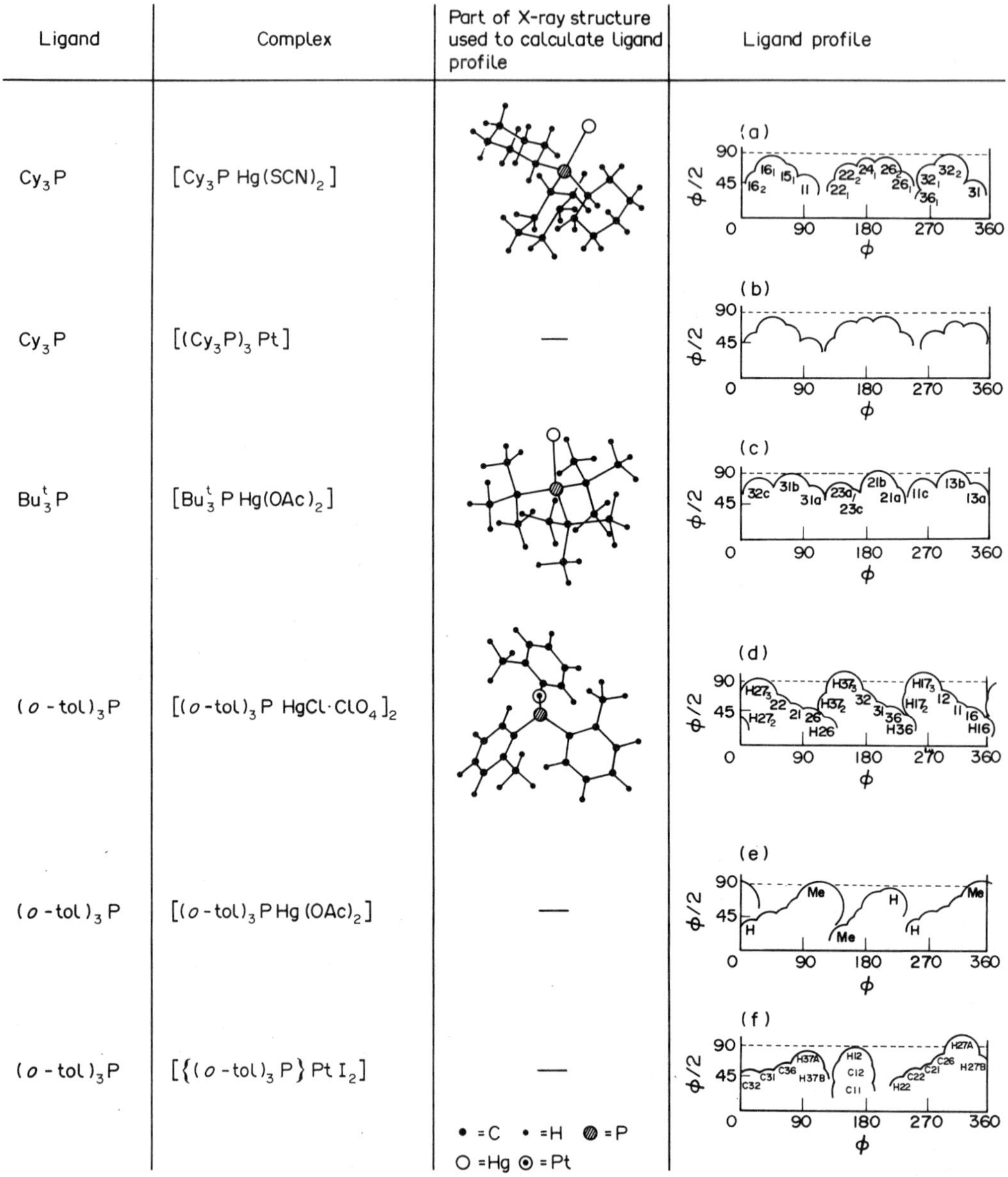

Ligand	Complex	Part of X-ray structure used to calculate ligand profile	Ligand profile
Cy_3P	$[Cy_3P\,Hg(SCN)_2]$		(a)
Cy_3P	$[(Cy_3P)_3\,Pt]$	—	(b)
Bu^t_3P	$[Bu^t_3P\,Hg(OAc)_2]$		(c)
$(o\text{-}tol)_3P$	$[(o\text{-}tol)_3P\,HgCl\cdot ClO_4]_2$		(d)
$(o\text{-}tol)_3P$	$[(o\text{-}tol)_3P\,Hg(OAc)_2]$	—	(e)
$(o\text{-}tol)_3P$	$[\{(o\text{-}tol)_3P\}\,Pt\,I_2]$	—	(f)

● = C   • = H   ◍ = P
○ = Hg   ◉ = Pt

Figure 12   Ligand profiles for (a) Cy_3P in $Cy_3PHg(SCN)_2$; (b) Cy_3P in $(Cy_3P)_3Pt$; (c) Bu^t_3P in $Bu^t_3PHg(OAc)_2$; (d) $(o\text{-}tol)_3P$ in $[(o\text{-}tol)_3PHgCl\cdot ClO_4]_2$; (e) $(o\text{-}tol)_3P$ in $[(o\text{-}tol)_3PHg(OAc)_2]_2$; (f) $(o\text{-}tol)_3P$ in $\{(o\text{-}tol)_3P\}_2Pt]_2$. The ordinate is the maximum semicone angle $\theta/2$. The abscissa is the angle ϕ through which the vector $M \rightarrow X$ (Figure 11) has been rotated about the M—P bond; the origin of ϕ was arbitrarily chosen. The numbers under some of the curves denote the hydrogen and carbon atoms, whose van der Waals spheres define the ligand profile. For (d), (e) and (f) both carbon and hydrogen atoms define the profile; for (a), (b) and (c) only hydrogen atoms are required

$Pt(PCy_3)_3$. They propose that the phosphine be rotated incrementally about the M—P axis (rotation angle ϕ) and a tangent r be drawn for each increment of ϕ (Figure 15). Changes in ϕ change the angle $\theta/2$, termed the *ligand angular encumbrance*. A non-circular cone is generated and at some arbitrary value of r, say X, a section of the cone (or its base) will be a non-circular line.

Table 9 Ferguson and Alyea's Cone Angle Data for Phosphine Ligands

	Maximum semicone angles $\theta/2$ (°)	Cone angle θ (°)	θ from models,[187] (°)	Orientation torsion angle (°)	Comments
$[Cy_3PHg(SCN)_2]_n$	86.7, 87.3, 91.9	177	179	63, 170, 87	Uncrowded
$[Cy_3PHg(NO_3)_2]_2$	89.9, 90.5, 90.9	181	170	65, 172, 95	
$[Cy_3PHg(OAc)_2]_2$	84.5, 88.9, 95.1	179		63, 177, 86	
$(Cy_3P)_2Hg(OAc)_2$	83.0, 86.3, 86.9	171		56, 185, 80	Crowded
	78.5, 85.5, 89.0	170		69, 177, −45	
$(Cy_3P)_2Hg(ClO_4)_2$	79.1, 82.6, 87.2	166		53, 181, −46	
	78.2, 82.9, 86.2	165		56, 182, −45	
$(Cy_3P)_3Pt$	76.3, 84.3, 85.0	164		71, 175, −42	
$(Cy_3P)_2PtI_2$	74.9, 84.3, 85.9	163		63, 175, −44	
$Bu_3^iPHg(OAc)_2$	91.9, 93.9, 95.1	187	182		Uncrowded
$Bu_3^iPHg_2(SCN)_4$	92.9, 94.2, 96.5	189			Crowded
$Bu_3^iPNiBr_3^-$	88.0, 88.2, 88.2	176			Crowded
$[(o\text{-tol})_3PHgCl \cdot ClO_4]_2$	93.6, 101.4, 102.0	198	194	128, 128, 127	Uncrowded
$[(o\text{-tol})_3PHg(OAc)_2]_2$	86.6, 99.7, 99.8	191		32, −125, −121	Crowded
$\{(o\text{-tol})_3P\}_2PtI_2$	82.8, 88.6, 103.8	183		−1, −106, −119	
$\{(o\text{-tol})_3P\}_2IrClCO$	84, 89, 102	183		6, −120, −113	

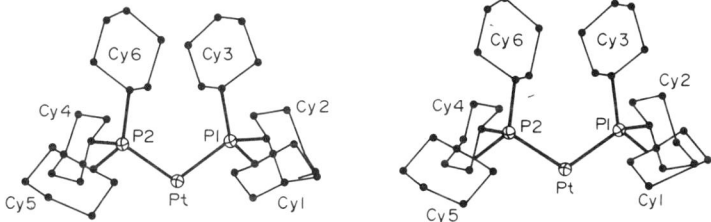

Figure 13 Stereoview of the $Pr(PCy_3)_2$ moiety illustrating the meshing of the bulky ligands

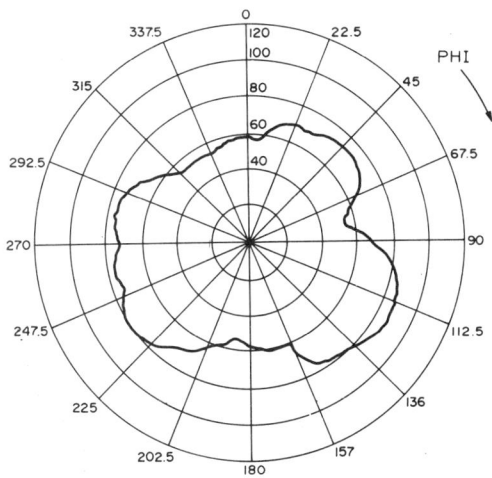

Figure 14 Ligand profile involving P(1)

The solid angle, Ω, of this non-circular cone is given by equation (53).

$$\Omega = \int_{\phi=0}^{2\pi} (1 - \cos\tfrac{1}{2}\theta)\,d\phi \qquad (53)$$

The circular cone solid angle, Ω, that would enclose this irregular cone is given by equation (54).

$$\bar{\Theta} = 2 - \arccos\left(1 - \frac{\Omega}{2\pi}\right) \qquad (54)$$

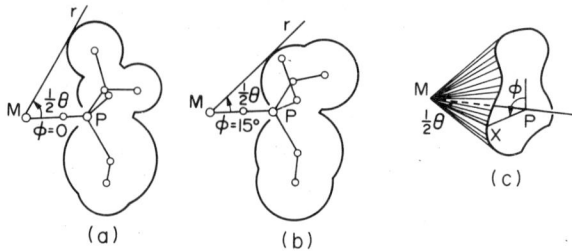

Figure 15 Dependence of the angular encumbrance of a coordinated phosphorus ligand (PEt$_3$ as a working example) on the ligand orientation. Hydrogen atoms are omitted for sake of clarity. In (a) and (b) the curved line indicates the section in the plane of the drawing of the space filling molecular model (carbon radius 1.8 Å). For each orientation ϕ the $\frac{1}{2}\theta$ angle is defined by the straight line r tangent to the model in the plane of the figure. (c) represents the generalized non-circular cone described by r on changing ϕ and $\frac{1}{2}\theta$ angles. X, a point of r at an arbitrary fixed distance from M, describes a closed non-circular line belonging to a sphere centred on M

$\bar{\Theta}$ is the 'actual' cone angle, *i.e.* the value derived from the results of X-ray data. The values calculated by Immirzi and Musco are given in Table 10. It is seen that Tolman's cone angles Θ are systematically greater than those found from crystallographic studies.

Table 10 Solid Cone Angles (Ω) for Some Tertiary Phosphines Based on X-Ray Structures of Metal Complexes*

No.	Ligand	Complex	Ω	$\bar{\Theta}$	Θ
1	PEt$_3$	*trans*-HPdClL$_2$	3.60, 3.72	129.5, 131.8	132
2	PEt$_3$	PtL$_4$	3.09, 3.14	118.9, 119.9	132
3	PMe$_2$Ph	[Ir(CO)$_3$L$_2$]$^+$ClO$_4^-$	3.06, 3.03	118.3, 117.7	122
4	PMe$_2$Ph	*cis*-PdCl$_2$L$_2$	2.90	125.0	122
5	PMe$_2$Ph	PtL$_4$	2.82, 2.81,	113.1, 113.1	122
			2.79, 2.88	112.4, 114.6	
6	PPh$_2$Me	[AuL$_2$]$^+$ PF$_6^-$	3.44	126.2	136
7	PPh$_2$Me	MoH$_4$L$_4$	2.99, 2.99,	116.7, 116.8,	136
			3.05, 3.10	118.1, 119.1	
8	PPh$_2$Me	[IrL$_4$]$^+$ BF$_4^-$	3.31, 3.27	123.4, 122.6	136
9	PPh$_3$	AuCl$_3$L	3.75	132.4	145
10	PPh$_3$	Co(CO)$_2$NOL	3.58	129.0	145
11	PPh$_3$	[CuClL]$_2$	3.57	128.8	145
12	PPh$_3$	Rh(C$_2$H$_4$)(pmcp)L^a	3.54	128.2	145
13	PPh$_3$	*cis*-PtCl(dtt)L$_2^b$	3.82, 3.72	133.8, 131.8	145
14	PPh$_3$	CuClL$_3$	3.70	131.5	145
15	PPh$_3$	Ir(NO)L$_3$	3.59	129.2	145
16	PPh$_3$	[CuL$_3$]$^+$ BF$_4^-$	3.77	132.9	145
17	PPh$_3$	PdL$_4$	3.44, 3.43	126.2, 125.9	145
18	PPh$_3$	RuH$_2$L$_4$	3.31, 3.32,	123.5, 123.7	145
			3.35, 3.38	124.4, 124.9	
19	PPr$^i_3$	IrHL$_2$(C$_4$H$_6$)	3.89, 3.96	135.2, 136.7	160
20	PPr$^i_3$	IrL$_2$(C$_3$H$_5$)	4.02	137.8	160
21	PCy$_3$	PdL$_2$	4.48	146.7	170
22	PCy$_3$	PtH$_2$L$_2$ (monocl.)^c	4.61	149.1	170
23	PCy$_3$	PtH$_2$L$_2$ (tricl.)^c	4.40	145.0	170
24	PCy$_3$	PtL$_3$	4.18	140.6	170
25	PBu$^t_2$Ph	PdL$_2$	4.94	155.3	170
26	PBu$^t_2$Ph	*cis*-PtCl$_2$L$_2$	4.61, 4.50	149.1, 147.0	170
27	PMen$_2$Pr$^{i\,d}$	PdL$_2$	6.09	176.5	209^e

* Hydrogen atom positions have been calculated by assuming C—H bond lengths 1.08 Å, H—C—H angles 109°, local C_{2v} symmetry on each C atom and staggered conformations of C—C bonds for saturated carbons. Methyl groups bonded to phosphorus are assumed staggered with respect to M—P bond. Following Tolman the M—P distance is assumed in all cases 2.30 Å, shifting the metal along M—P if necessary. Van der Waals radii are: $R_P = 1.80$, $R_C = 1.80$, $R_H = 1.17$ Å. When several crystallographically independent ligands are present, all Ω values are listed.
^{a} pmcp = pentamethylcyclopentadienyl.
^{b} dtt = di-*p*-tolyltriazenido.
^{c} Conformational analysis and solid state molecular volumes indicate the triclinic as the more stable form. The Ω value 4.61, surprisingly greater than 4.48 (case 21), is probably due to a strained conformation induced by the crystal packing.
^{d} Men = menthyl.
^{e} Evaluation using Tolman's formula.

Table 11 The Use of Cone Angles in Experimental Systems

Ligands	System studied	Correlation with cone angle	Ref.
$P(OMe)_3$, PMe_3, PBu_3^n, PPh_3, PCy_3, $P(OMe)_2Ph$, $P(OMe)Ph_2$	$CoX(HDMG)_2PR_3$ $X = Cl$, Me	Co—P bond lengths Proton chemical shifts	1
$P(OCH_2)_3CEt$, $PN_3(CH_2)_6$, $PPhMe_2$, $P(OPh)_3$, PBu_3^n, PPh_3, $PPhCy_2$	cis-$Mo(CO)_4(Pip)_2 \xrightarrow{PR_3} cis$-$Mo(CO)_4(PR_3)_2$	Reaction rates	2
PMe_2Ph, $PMePh_2$, $PEtPh_2$, PPh_3, PPr^iPh_2, PBu^iPh_2	$[(solvent)M(COR)] \xrightarrow{PR_3} [R_3PM(COR)]$	Reaction rates	3
$P(OMe)_3$, $P(OPh)_3$, $HPPh_2$, PEt_3, $PMePh_2$, PPh_3, PCy_3	(diagram fac → mer)	Isomerization of products	4
PCy_3, PPh_3, $P(4\text{-}MeOC_6H_4)_3$, PPh_2Et, $PPhEt_2$, PBu_3^n, PEt_3, $P(OPh)_3$, $P(OMe)_3$, $PPh(OMe)_2$	Phosphine and phosphite substitution reactions of metal carbonyls. M = Mn, Fe, Ru	Reaction rates	5
PCy_3, PPh_3, $P(2\text{-}MeC_6H_4)_3$, $P(4\text{-}MeC_6H_4)_3$, PMe_2Ph, $PMePh_2$, PBu_3^n	Reversible phosphine binding to palladium arylazooximes	Reaction rates	6
PPh_3, $PMePh_2$, PMe_2Ph, PBu_3^n	$trans$-$[Co(DH)_2(H_2O)_2]^+ \xrightarrow{PR_3} trans$-$[Co(DH)_2(PR_3)_2]^+$	Reaction rates and ease of reduction (polarographic monitoring)	7
PCy_3, PPh_3, PPh_2Pr^i, $P(4\text{-}MeC_6H_4)_3$, $P(4\text{-}MeOC_6H_4)_3$, PPh_2Me, PBu_3^i, $PPhMe_2$, PBu_3^n	$CpMo(CO)_2(PR_3)COMe \xrightarrow{-CO} CpMo(CO)_2(PR_3)(Me)$	Reaction rates	8
$P(OMe)Ph_2$, $PPhEt_2$, PPh_2Et, PPh_3, $P(4\text{-}MeC_6H_4)_3$, $P(4\text{-}MeOC_6H_4)_3$, PPh_2Pr^i, PPh_2Bu^n, PPh_2Me	$[MeCpMn(CO)_2pyR]^+ \xrightarrow{PR_3} [MeCpMn(CO)_2PR_3]^+$	Reaction rates *via* cyclic voltammetry	9
$P(OMe)_3$, $P(OPh)_3$, $P(OBu^n)_3$, $P(OCy)_3$	$[(\eta\text{-arene})Cr(CO)_2(CX)] \xrightarrow{P(OR)_3} [Cr(CO)_2(CX)\{P(OR)_3\}_3]$ X = S, arene	Reaction rates	10
37 phosphines and phosphites	Nickel-catalyzed cooligomerization reactions of alkenes	Reaction rates and product ratios	11
15 phosphines and phosphites	Attempt to develop NMR techniques to estimate cone angle	NMR	12
$P(OEt)_3$, PMe_2Ph, PPh_3, PBu_3^n, $PMePh_2$, $HPCy_2$, PPh_2Cy	$(CO)_4Mo(\mu\text{-}PEt_2)_2Mo(CO)_4 \underset{}{\overset{PR_3}{\rightleftharpoons}}$ $(CO)_4Mo(\mu\text{-}PEt_2)_2Mo(CO)_3PR_3$	Reaction rates	13

HDMG = monoanion of dimethylglyoxime; Pip = piperidine; DH = dimethylglyoximato; pyR = substituted pyridine.

1. N. Bresciani-Pahor, L. Randaccio and P. J. Toscano, *J. Chem. Soc., Dalton Trans.*, 1982, 1559.
2. D. J. Darensbourg and R. L. Kump, *Inorg. Chem.*, 1978, **17**, 2680.
3. J. D. Cotton and R. D. Markwell, *Inorg. Chim. Acta*, 1982, **83**, 13.
4. U. Kunze, A. Bruns, *J. Organomet. Chem.*, 1985, **292**, 349.
5. A. J. Poë, *Chem. Br.*, 1983, **19**, 997.
6. P. Bandyopadhyay, P. K. Mascharak and A. Chakravorty, *Inorg. Chim. Acta*, 1980, **45**, L219.
7. C. W. Smith, G. W. Vanloon and M. C. Baird, *J. Coord. Chem.*, 1976, **6**, 898.
8. K. W. Barnett and T. G. Pollman, *J. Organomet. Chem.*, 1974, **69**, 413.
9. J. K. Kochi, P. M. Zizelman and C. Amatore, *J. Am. Chem. Soc.*, 1984, **106**, 3771.
10. I. S. Butler and A. A. Ismail, *Inorg. Chem.*, 1986, **25**, 3910.
11. T. Bartik, P. Heimbach and T. Himmler, *J. Organomet. Chem.*, 1984, **276**, 412.
12. L. G. Marzilli and W. C. Trogler, *J. Am. Chem. Soc.*, 1974, **96**, 7589.
13. M. Basato, *J. Chem. Soc., Dalton Trans.*, 1986, 217.

14.3.7   The Use of the Cone Angle Concept

It should be clear by now that the term 'cone angle' is largely a matter of definition and is not an intrinsic property of a molecule comparable with bond lengths. However, the cone angle concept, for all its shortcomings, does have the merit of being a quantifiable variable parameter in chemical studies. Workers in various fields (see Table 11) have used cone angles to attempt to quantify steric factors rather than, as previously, ascribing experimental anomalies to undefined steric factors.

While many workers have used Tolman's steric parameter (the cone angle), not nearly as many have used his electronic parameter. The latter represents the net total of all electronic effects as reflected in the value of the $v(CO)A_1$ vibration (*i.e.* the contributions of σ and π bonding are unresolved) of the $LNi(CO)_3$ (L = phosphine) complex. Tolman further analyzed the contributions of various substituents to v by equation (55).

$$v = 2056.1 + \sum_{i=1}^{3}\chi_i \qquad (55)$$

v = frequency of the A_1 carbonyl mode of $LNi(CO)_3$

χ_i = substituent contribution

The most basic phosphine considered by Tolman was PBu_3^n; this produced an A_1 frequency of $2056.1\,cm^{-1}$ and was made the arbitrary standard for the electronic parameter (Table 12). The electronic parameter, v, and the steric cone angle parameter, θ, when plotted together produce a steric and electronic map (Figure 16).

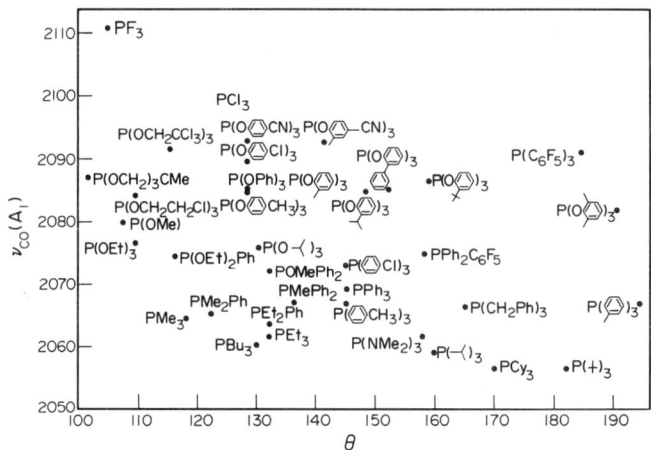

Figure 16   Steric and electronic map

When some other reaction parameter, Z, such as the log of a rate constant, is plotted on to this steric and electronic map on an axis normal to the plane of the paper the comparative contributions of θ and v should become apparent. A purely steric effect will slope north or south (the reader is encouraged to view Figures 26–28 of ref. 187 to appreciate this fully). Weimann and co-workers[211] used Tolman's methodology to show the % steric effect in the oligomerization of butadiene catalyzed by nickel phosphine complexes.

For the reaction of dioxygen with the manganese(II) phosphine complexes see equation (56).

$$Mn(phosphine)X_2 \underset{-O_2}{\overset{O_2}{\rightleftharpoons}} Mn(phosphine)X_2(O_2) \qquad (56)$$

Minten[212] has related the equilibrium constant, K_{O_2}, to the cone angle, θ, and the Strohmeier electronic parameter, v, as shown in Figure 17 for the $Mn(phosphine)Br_2$ complexes in toluene at

Table 12 Substituent Contributions to v for $PX_1X_2X_3$ $v = 2056.1 + \sum_{i=1}^{3} \chi_i$

Substituent χ_i	χ_i	Substituent χ_i	χ_i
Bu^t	0.0	p-C$_6$H$_4$OMe	3.4
Cy	0.1	o-tol, p-tol, Bz	3.5
o-C$_6$H$_4$OMe	0.9	m-tol	3.7
Pr^i	1.0	Ph	4.3
Bu^i	1.2 est	CH=CH$_2$	4.5
Bu	1.4	p-C$_6$H$_4$F	5.0
Et	1.8	p-C$_6$H$_4$Cl	5.6
NMe$_2$	1.9	m-C$_6$H$_4$F	6.0
Pipcridyl	2.0	CH$_2$CH$_2$Cl	6.1^b
Me	2.6	OPr^i	6.3
2,4,6-C$_6$H$_2$Me$_3$	2.7	OBu	6.5
OEt	6.8	(OCH$_2$)$_2$/2	9.8
CH$_2$CH$_2$CN	7.3	O-o-C$_6$H$_4$Bu^t	10.0^b
OMe, OCH$_2$CH=CH$_2$,	7.7	OCH$_2$CH$_2$CN	10.5
OCH$_2$CH$_2$OMe		O-o-tol-p-Cl	11.1
H	8.3	C$_6$F$_5$	11.2
O-2,4-C$_6$H$_3$Me$_2$	9.0	O-o-C$_6$H$_4$Cl	11.4
OCH$_2$CH$_2$Cl, O-o-tol,	9.3	OCH$_2$CCl$_3$	11.9
O-p-tol, O-p-C$_6$H$_4$OMe		O-p-C$_6$H$_4$CN	12.2
O-o-C$_6$H$_4$Pr^i	9.5^b	Cl	14.8
O-o-C$_6$H$_4$Bu^t	9.6^b	F	18.2
OPh	9.7	CF$_3$	19.6

^{a} $v_{CO}(A_1)$ of Ni(CO)$_3$L in CH$_2$Cl$_2$ from ref. 1 unless noted otherwise.
^{b} Previously unpublished value.
^{c} This value for PH$_3$ was estimated from extrapolation of values for PH$_{3-n}$Ph$_n$ ($n = 1$ to 3).

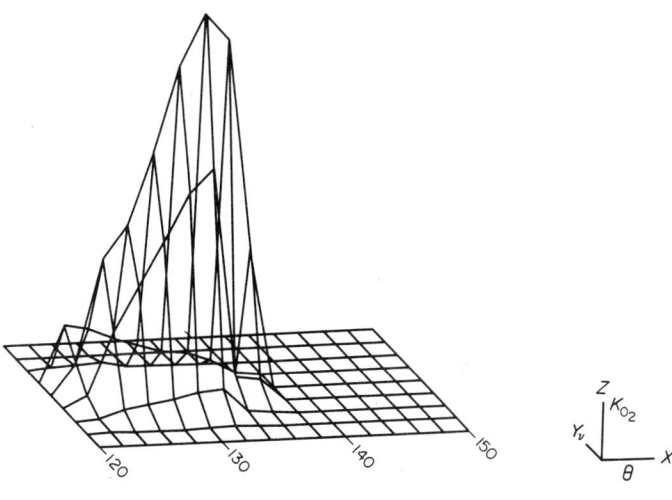

Figure 17 The 3D or spatial diagram of K_{O_2} *vs.* **q** *vs.* v for MnLBr$_2$ complexes in toluene solution at 20°C

20 °C. The graph indicates that no reaction occurs for $\theta > 141°$. This is borne out very well, for no reaction is found for phosphines PPh$_3$, PPr$_3^i$, PCy$_3$ and PBu$_3^i$, all for which $\theta \geqslant 145°$. Whatever the merits of Tolman's cone angle concept, his ideas have stimulated much research, particularly in the field of bulky ligands.

14.4 BONDING IN COMPLEXES OF PNICTIDE LIGANDS

14.4.1 Introduction

A large variety of ligands form bonds with transition metals *via* one or more pnictide atoms. These range from simple ER_3 (E = P, As, Sb, Bi; R = alkyl, aryl, alkoxy, aryloxy, halogen and their combinations/permutations) molecules to ligands of elaborate architecture such as the macrocycles described in Section 14.2.1.7.

Pnictide–transition metal bonds are essentially covalent coordinate, in which the pnictide provides the electrons. However, this simple picture does not account for all the structural data now available. The consensus view is that three factors are involved in the ligand contribution to the M—E bond: (a) σ bonding (Section 14.4.2), (b) π bonding (Section 14.4.3), and (c) steric factors (Section 14.3). The effect of the metal will not be discussed here, except insofar as individual complexes are used as examples.

14.4.2 σ Bonding and the Basicity of Phosphines

The M—P bond is a donor covalent bond with the phosphorus providing both electrons, *i.e.* behaving as a Lewis base. The usual measure of phosphine basicity, $pK_a(H_2O)$, is a measure of Brönsted basicity (proton affinity) and this is not always the same as Lewis basicity. In practice pK_a values are the most widely quoted in the literature, and a selection are given in Table 13. These represent only a fraction of the thousands of phosphines prepared. When pK_a values are not available, estimates often suffice for practical/experimental work and these can be calculated to within one pK_a unit. Kabachnik and Mastryukova[218] have demonstrated that the Hammett equation is applicable to a large number of alkyl/aryl-substituted phosphorus acids by employing a substituent constant, σ^ϕ, for the substituent(s) on phosphorous. Hypophosphorous acid is the reference compound ($\sigma^\phi = 0$ for hydrogen) and the slope is defined as 1 for the acids $RR'PO_2H$. The σ^ϕ are additive for all the substituents on phosphorus, giving the $\Sigma\sigma^\phi$ value. The σ^ϕ values can be used in calculating pK_a values of phosphorus bases. Because the σ^ϕ constants are derived from measurements of phosphorus acids they should, in principle, be better constants to use with phosphorus compounds than the Hammett/Taft parameters based on substituents on carbon.[219] In fact σ^ϕ values give correlations with experimental data superior to those obtained with Hammett σ_p constants.[220]

In general, electron-releasing substituents will increase the electron availability on phosphorus, resulting in a higher pK_a. This has been well illustrated by Goel[213] for a series of *para*-substituted phenyl phosphines (see Table 13). Compared to triphenylphosphine ($pK_a = 2.73$), the dimethyl-amino substitution resulted in a greatly enhanced basicity (*e.g.* X = Cl, $pK_a = 1.03$).

The most extreme effect of substitution in aryl phosphines is reported by Wada *et al.*[214] where electron-releasing OMe groups reinforced the C—P—C angle increase caused by the bulky *ortho*-substituted phenyl groups. They estimated the $pK_a \sim 11.2$ for tris(2,4,6-trimethoxyphenyl)phosphine, the highest for any arylphosphine. However, the highest reported pK_a value is that for PBu_3^t.[213] The increase in bulk of a substituent group on phosphorus will increase the R—P—R angle, which will increase the p character of the lone pair. The stability of the metal–phosphorus σ bond decreases[221] in the series: $PBu_3^t > P(OR)_3 > PR_3 \sim PPh_3 > PH_3 > PF_3 > P(OPh)_3$.

The effect of the basicity of phosphines on ligand substitution and complex stability was demonstrated by Angelici and Ingemanson (equation 57).[222]

$$RNH_2W(CO)_5 + PR_3 \rightleftharpoons R_3PW(CO)_5 + RNH_2 \qquad (57)$$

For a given amine, the equilibrium constant decreased in the order of basicities: $PBu_3^n \sim PCy_3 \geqslant P(OCH_2)_3CEt > P(SCH_2)_3CMe \geqslant P(OBu^n)_3 > PPh_3 > AsPh_3 > SbPh_3 \sim P(OPh_3) > BiPh_3$. Some phosphines can act both as nucleophiles[223] and electrophiles,[224] *e.g.* PCl_3 in equations (58) and (59), respectively.

$$Ni(CO)_4 + 4PCl_3 \longrightarrow Ni(PCl_3)_4 + 4CO \qquad (58)$$

$$Me_3N + PCl_3 \longrightarrow Me_3NPCl_3 \qquad (59)$$

Table 13 Basicities of Tertiary Phosphines

Tertiary phosphine	pK_a	Ref.
PBu_3^t	11.40	213[a]
$P\{2,4,6\text{-}(MeO)_3C_6H_2\}_3$	$\leqslant 11.2$	214[b]
$P\{2,6\text{-}(MeO)_2C_6H_3\}_3$	~ 10.7	214[b]
PCy_3	9.65	213, 215[a]
PEt_3	8.69	215[a]
PMe_3	8.65	215[a]
$P(4\text{-}Me_2NC_6H_4)_3$	8.65	215[a]
PPr_3^n	8.64	215[a]
Me_2PEt	8.62	215[a]
Et_2PMe	8.62	215[a]
PBu_3^u	8.43	215[a]
$P(n\text{-}C_5H_{11})_3$	8.33	215[a]
$P(CH_2CH_2OBu^n)_3$	8.03	215[a]
PBu_3^i	7.97	215[a]
$Cy_2PCH_2CH_2CN$	7.13	215[a]
$PhP\{2,6\text{-}(MeO)_2C_6H_3\}_2$	~ 7.0	214[b]
$P(CH_2CH_2Ph)_3$	6.60	215[a]
$Bu_2^nPCH_2CH_2CN$	6.49	215[a]
Me_2PPh	6.49	215[a]
$Me_2PCH_2CH_2CN$	6.35	215[a]
$(n\text{-}C_8H_{17})_2PCH_2CH_2CN$	6.27	215[a]
Et_2PPh	6.25	215[a]
$4\text{-}Me_2NC_6H_4PEt_2$	5.08	216[c]
$P(4\text{-}MeOC_6H_4)_3$	4.57 (3.15)	213, 216[c]
$P(4\text{-}MeC_6H_4)_3$	3.84	213[a]
$4\text{-}MeOC_6H_4PEt_2$	3.88	217[c]
$(CNCH_2CH_2)_2PEt$	3.80	215[a]
$(CNCH_2CH_2)_2PMe$	3.61	215[a]
$Ph_2P\{2,6\text{-}(MeO)_2C_6H_3\}$	3.5	214[b]
$(PhCH_2CH_2)_2PCH_2CH_2CN$	3.43	215[a]
$P(3\text{-}MeC_6H_4)_3$	3.30	213[a]
$4\text{-}FC_6H_4PEt_2$	3.02	217[c]
$PhP(CH_2CH_2CN)_2$	3.20	215[a]
$4\text{-}Me_2NC_6H_4PPh_2$	2.95	216[c]
$(4\text{-}ClC_6H_4)_3P$	2.86	216[c]
$4\text{-}ClC_6H_4PEt_2$	2.79	217[c]
PPh_3	2.73	213, 215[a]
	(2.30)	216[c]
$4\text{-}BrC_6H_4PEt_2$	2.70	217[c]
Ph_2PPr^n	2.64	217[c]
Ph_2PEt	2.62	217[c]
$(4\text{-}ClC_6H_4)_2PPr^n$	2.59	217[c]
$4\text{-}MeOC_6H_4PPh_2$	2.58	216[c]
$4\text{-}ClC_6H_4PPh_2$	2.18	216[c]
$4\text{-}BrC_6H_4PPh_2$	2.09	216[c]
$(4\text{-}FC_6H_4)_3P$	1.97	213[a]
$P(CH_2CH_2CN)_3$	1.37	215[a]
$P(4\text{-}ClC_6H_4)_3$	1.03	213[a]

[a] pK_a values were determined by recording the half neutralization potential (ΔHNP) of the phosphine in nitromethane, which is linearly related to the $pK_a(H_2O)$, where $pK_a = 10.12 - 0.129(\Delta HNP)$ (see ref. 213 for experimental details).

[b] pK_a values were estimated by comparison of acetone solutions of the phosphine with a range of nitrogen bases of known pK_a by ^1H NMR. The values are approximate.

[c] pK_a values were determined in 80% ethanol with standard HCl and electrochemical monitoring.

14.4.2.1 Gas phase basicities of phosphines

UV photoelectron spectroscopy is a direct method of observing the electronic properties of molecules. In particular, the ionization potentials of phosphines would seem to afford a direct measure of basicity, *i.e.* availability of the electron lone pair. Some surprising results have emerged, *e.g.* the ionization energies of the phosphorus lones pair in the gas phase follow the sequence $PMe_3 > PPhMe_2 > PPh_2Me > PPh_3$.[225] A selection of such ionization energies is given in Table 14. The photoelectron spectroscopy of phosphorus compounds has been reviewed by Bock.[226] Whilst

Table 14 Gas Phase Ionization Energies of Phosphines

Compound	Ionization energy (eV)	Ref.
$P(NMe_2)_3$	7.58	5
PBu_3^t	7.70	1
PPh_3	7.80	2
$P(SPr^i)_3$	7.91	5
I^a	8.22	5
PPh_2Me	8.28	2
II^b	8.3	5
$PPhMe_2$	8.32	2
$P(SMe)_3$	8.33	5
$PHBu_2^t$	8.35	1
$PClBu_2^t$	8.44	1
$PFBu_2^t$	8.50	1
PMe_3	8.63	1
	8.62	2
	8.60	4
$P(NMeCH_2)_3CMe$	9.03	5
$PHMe_2$	9.08	1
	9.10	4
$PClMe_2$	9.19	1
$P(OMe)_3$	9.22	5
PH_2Bu^t	9.32	1
PCl_2Bu^t	9.32	1
$PFMe_2$	9.37	1
PF_2Bu^t	9.63	1
$P(SCH_2)_3CMe$	9.65	5
PH_2Me	9.70	4
	9.72	1
PCl_2Me	9.83	1
$P(OCH_2)_3CH$	9.9	5
$P(OCH_2)_3CMe$	9.95	5
PF_2Me	10.34	1
PH_3	10.58	1
PCl_3	10.59	4
	10.7	1
PF_2H	11.0	4, 3
PF_2I	11.2	3
PF_2Br	11.8	3
$PF_2(NCS)$	11.9	3
$PF_2(CN)$	11.9	3
$PF_2(NCO)$	12.2	3
PF_3	12.23 ($\pm$0.02)	3
	12.3	1
PF_2Cl	12.8	3

$^a I =$ 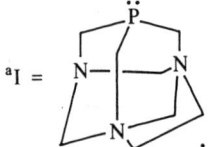 $^b II =$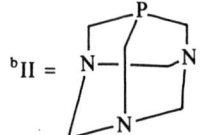

1. M. F. Lappert, J. B. Pedley, B. T. Wilkins, O. Stelzer and E. Unger, *J. Chem. Soc., Dalton Trans.*, 1975, 1207.
2. R. J. Puddephatt, G. M. Bancroft and T. Chan, *Inorg. Chim. Acta*, 1983, **73**, 83.
3. S. Cradock and D. W. H. Rankin, *J. Chem. Soc., Faraday Trans. 2*, 1972, **68**, 940.
4. A. H. Cowley, *Inorg. Chem.*, 1982, **21**, 85.
5. A. H. Cowley, M. Laftman, P. M. Stricklen and J. G. Verkade, *Inorg. Chem.*, 1982, **21**, 543.

there may be good correlations within a series of closely related phosphines between lone pair IP and substituent coefficients from Hammett equations, photoelectron spectroscopy (PES) is not a good technique for determining the relative basicities of different types of phosphines such as PF_3 and PCy_3.

The basicities of phosphines in the liquid phase are dominated by solvation effects in ionizing solvents and the results of measurements shown in Table 13 are clearly the result of gross energy changes of chemical reactions, including solvation energies, whereas those in Table 14 are unencumbered gas phase ionizations. The PES basicities of methyl/phenyl phosphines shown above have been confirmed (in the gas phase) by a study of the equilibrium shown in equation (60), using a pulsed electron beam high ion source pressure mass spectrometer.[219] For the reaction in equation (61) the forward rate, $k_2 \simeq 0.8 \times 10^{-9}$ molecule^{-1} cm^3 s^{-1} and $\Delta G°$ (320 K) = -103.7 kJ mol^{-1}.

$$B^1H^+ + B^2 \rightleftharpoons B^1 + B^2H^+ \tag{60}$$

$$Me_3PH^+ + Me_2PhP \rightleftharpoons Me_3P + Me_2PhPH^+ \tag{61}$$

Since most inorganic chemists are interested in basicities because of their effects on reactions in solution the pK_a values are of practical relevance, but PES does hold great promise for the essential understanding of phosphines. In principle, PES should enable the direct measurement of π and σ effects. Bursten[227] has proposed a ligand additivity model postulating linear contributions from each ligand to the orbital energies. For a d^6 low-spin octahedral complex, $ML_n(PR_3)_{6-n}$, the orbital energy of each d orbital is given by equation (62), where a_{M0} is a constant characteristic of the metal atom in its particular oxidation state, b_{ML} and b_{MP} are constants describing the gross energetic effects of L and PR_3 upon the metal atom, C_{ML} and C_{MP} are constants describing the energetic effect upon the $d\pi$ orbital interacting with L and PR_3, and x_i is equal to the number of ligands, L, with which the $d\pi$ orbital can interact.

$$E_i = a_{M0} + nb_{ML} + (6-n)b_{MP} + x_iC_{ML} + (4-x_i)C_{MP} \tag{62}$$

This can be simplified to equation (63) where $a = a_{M0} + 6b_{MP} + 4C_{MP}$, $b = b_{ML} + b_{MP}$, and $c = C_{ML} - C_{MP}$.

$$E_i = a + bn + cx_i \tag{63}$$

Since the b term represents a combined σ and π influence, whereas the c term is only π influenced, it should be possible, by suitably designed experiments, to separate the σ and π effects on the system studied.[228]

Enthalpy data for the gas phase reaction between ER_3 (E = P, As, Sb) and BX_3 Lewis acids indicate a decrease in the basicity in the order: $PR_3 > AsR_3 > SbR_3$. A similar order was found from the photoelectron spectra of the free ligands.[229]

14.4.3 π Bonding in Pnictide Complexes

The phosphorus–transition metal back bonding is a transfer of charge from the metal to the ligand, and this is described as π bonding because of the symmetry of the orbitals involved. In the case of an octahedral complex, the d orbitals involved in σ bonding with the six ligands are the e_g orbitals, leaving the t_{2g} orbitals non-bonding and directed between the six M—P bonds (Figure 18). The primary P—M is the σ bond formed between a d^2sp^3 metal hybrid orbital and an sp^3 hybrid phosphorus orbital. On coordination, the bonding of the PY_3 changes from pyramidal p^3 to tetrahedral sp^3, which has steric implications discussed in Section 14.3. From various compilations of experimental data, mostly spectroscopic, an empirical π acceptor series can be drawn, and prominent among them are various PY_3 compounds:

π acid series[230]

$NO > CO \sim RNC \sim PF_3 > PCl_3 > PCl_2(OR) > PCl_2R > PBr_2R > PCl(OR)_2 > PClR_2$

$> P(OR)_3 > PR_3 \sim SR_2 > RCN > o$-phenanthroline $> RNH_2 \sim R_2O$

The success of the synergetic ($\sigma + \pi$) approach to the solution of metal carbonyl bonding led to its application to transition metal–phosphorus bonding. Mainly because it is symmetry allowed, chemists for the last two decades have drawn orbital diagrams such as those shown in Figure 19. However, the experimental facts that indicated (M—P)π bonding gave no indication as to the nature of the phosphorus orbital.

Superficially it may seem obvious that empty phosphorus $3d$ orbitals should be the recipients of the metal electron charge surplus, but there have always been those who have argued[231,232] against phosphorus–metal $3d$ bonding on the grounds that the orbitals were energetically unfavourable and too diffuse to have appreciable interaction with the valence orbitals.

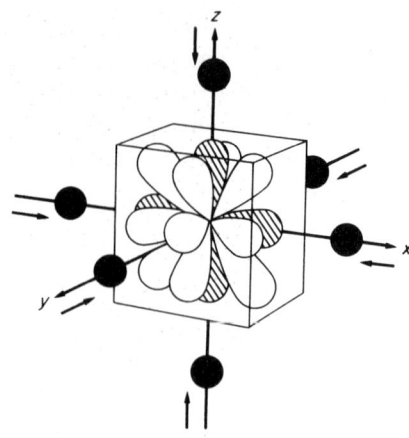

Figure 18 Complete set of d orbitals in an octahedral field. The e_g orbitals are shaded and the t_{2g} orbitals are unshaded. The torus of the d_{z^2} orbital has been omitted for clarity

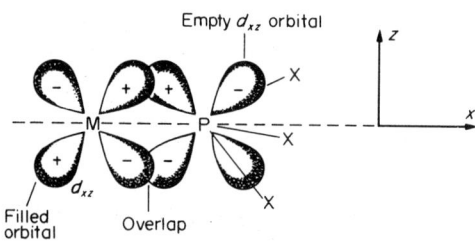

Figure 19 The back-bonding from a filled metal d orbital to an empty phosphorus $3d$ orbital in the PX_3 ligand taking the internuclear axis as the z axis. An exactly similar overlap occurs in the yz plane using the d_{yz} orbitals

Recent quantum mechanical (self consistent multipolar X_α) calculations[233] of the frontier orbitals (HOMO, LUMO) of PH_3, PF_3 and PMe_3 have shown that the LUMO has mostly $3p$ character (Figure 20).

PY_3	$3p$ %	$3d$ %
PH_3	36	23
PF_3	44	23
PMe_3	14	10

Marynick and co-workers[234,235] have calculated (PRDDO calculation) the molecular orbitals of PY_3 complexed to a zerovalent transition metal, $[Cr(NH_3)_5(PH_3)]$ (Figure 21). In this, and other calculations, they find that phosphine ligands are acting as π acceptors, even though no d orbitals are included in the phosphorus base set.

The previous explanation for the superior π acidity of PF_3 and PH_3 was that electronegative substituents lower the energy of the d orbitals on phosphorus, increasing their availability for bonding. The alternative explanation now being proposed is that the highly polar P—F bonds result in low lying σ^* orbitals, which have appropriate symmetry for metal d overlap and are better π acceptors than the higher P—H σ^* orbitals. Further, because the σ bond is highly polar towards F, the σ^* orbital must necessarily be highly polar towards P and thus increases σ^*–metal $d\pi$ overlap.

Ammonia differs from phosphine in that the polarity of the N—H bond results in σ^* polar to H. Further, the phosphine σ^* orbitals are $3s$ and $3p$, which overlap more effectively with the transition metal than amines with $2s$ and $2p$, a point which was argued by Epshtein *et al.* 11 years ago.[232]

Quantum mechanical calculations have had support from the electron transmission spectroscopy of Giordan and co-workers.[236] This technique measures the attachment energy for the formation of

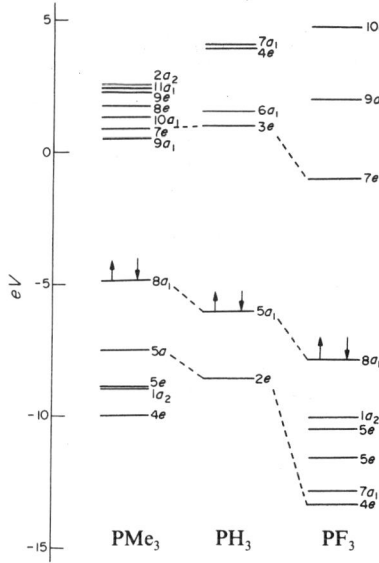

Figure 20 Valence orbital energy diagrams for PMe_3, PH_3 and PF_3

Figure 21 Electron density plot of one of the two π-donating d orbitals in $Cr(NH_3)_5(PH_3)$ without d orbitals on phosphorus. The contour levels are 0.5, 0.4, 0.3, 0.2, 0.1, 0.05, 0.02, 0.01, 0.005, 0.0035 and 0.002 e/au³

a negative ion, *i.e.* it is a direct assessment of the LUMO of the molecule investigated. Qualitative MO theory describing M—P back-bonding indicates that the energy of the unoccupied ligand orbital is paramount in determining bond strength.[327] Giordan and co-workers[236] found that the orbital energies in PR_3 are quite sensitive to substitution, implying that the LUMO orbital is molecular as opposed to being $3d$ in character. If the LUMO is not $3d$ then it strongly suggests that it is a σ^* orbital (what else?). Orpen and Connelly[237] have presented evidence adduced from bond length arguments (Table 15) supporting a σ^* theory but incorporating some $3d$ character from phosphorus, *i.e.* more in accord with Ellis and co-workers than with Marynick. Orpen and Connelly[238] compared pairs of transition metal phosphine and phosphite complexes related by electron transfer reactions, differing only in the total number of electrons. The comparisons are between M—P bonds in very similar environments unaffected by, for example, changes in *trans* ligands. The principal difference is thus in the degree of M—P σ bonding. In each case the d(M—P) increases notably on oxidation, while the P—X distance decreases. In complexes 1, 2 and 3 of Table 15 the loss of electrons from t_{2g} orbitals leads to a direct loss of π back-donation. A consensus is emerging that the nature of the M—P π bonding essentially involves σ^* of the P—X bond rather than the P $3d$ orbital. (It might be thought that there is a consequence to this. Attention was focused on the P atom and its state when considering π back-bonding, *i.e.* substituents on P were thought to influence P through the P—X bond. Now it is the P—X antibonding σ^* orbital that is the focus of attention, and what affects that is not quite the same as what affects the P $3d$ orbital.)

Table 15 Metal–Phosphorus and Phosphorus–Substituent Distances.

	Complex	Charge	Counter ion	M—P^a (Å)	P—X^a (Å)
1	$[Mn(CO)(Ph_2PCH_2CH_2PPh_2)(\eta^5\text{-}C_6H_6Ph)]$	0		2.221(1)	1.849(3) [Ph], 1.864(2) [CH_2]
		1	PF_6	2.339(2)	
2	$[Fe(CO)\{P(OMe)_3\}_2(\eta^4\text{-}C_4Ph_4)]$	0		2.146(1)	1.598(3)
		+1	BF_4	2.262(2)	1.579(6)
3	$[\{(Rh(CO)(PPh_3)\}_2(\eta^5\text{-}\eta'^5\text{-fulvalene})]$	0		2.255(2)	1.844(5)
		+2	PF_6	2.322(4)	1.813(15)
4	$[\{Rh(CO)(PPh_3)\}_2\{\mu\text{-}N(\text{tolyl})NN(\text{tolyl})\}_2]$	0		2.290(5)	1.855(12)
		+1	PF_6	2.335(1)	1.825(5)
5	$[Co(PEt_3)_2(\eta\text{-}C_5H_5)]$	0		2.218(1)	1.846(3)
		+1	BF_4	2.230(1)	1.829(3)
6	$[Fe(\eta^3\text{-}C_8H_{13})\{P(OMe)_3\}_3]$	0		2.138(1)	1.621(1)
		+1	BF_4	2.153(2)	1.600(2)
7	$[Re_2Cl_4(\mu\text{-}Cl)_2(Ph_2PCH_2PPh_2)_2]$	0		2.475(2)	1.828(8) [Ph], 1.843(8) [CH_2]
		+1	$H_2PO_4 H_3PO_4$	2.524(3)	1.82(1) [Ph], 1.84(1) [CH_2]

^{a} M—P and P—X distances are averaged over equivalent bonds.

14.4.3.1 *Trigonal bipyramidal complexes*

The effect of substituents on the π acidity of phosphorus ligands is well exemplified in the series $[F_{3-n}P\{OC(CF_3)_2CN\}_nFe(CO)_4]$, a set of trigonal bipyramidal complexes. In complexes of type $Fe(CO)_4L$ the stronger π acid will occupy equatorial rather than axial sites.[239] The fluxionality of the complex arises by a Berry pseudorotation.[240] If there is no site preference, *i.e.* statistical distribution of isomers, then the axial:equatorial ratio is 2:3. Since the site competition is with CO, equatorial occupation by L implies greater π acidity. In the case of PF_3, substitution of F by OMe results in a lowering of the π acidity and the % equatorial isomer falls from 67% to nil. In contrast, the substitution of F by $OC(CF_3)CN$ results in an increase in equatorial ratio to 100% in the case of complete substitution (Table 16). It may be tempting to ascribe this dramatic change to an increase in steric bulk or cone angle, since $P\{OC(CF_3)_2CN\}_3$ is clearly a bulky ligand (cone angle $\sim 160°$). However, since PBu_3^n is axial in $(OC)_4FePBu_3^n$ the steric requirements of tris(2-cyanohexafluoroisopropyl) phosphite are unlikely to be the cause of the equatorial site preference; rather, it must be due to its π acid strength.

The interplay between σ and π bonding between pnictide ligand and metal has subtle structural consequences. In the trigonal bipyramidal series of complexes $M(CO)_4(ER_3)$ (M = Fe, Ru, Os; E = P, As, Sb; R = alkyl, aryl), in which the ER_3 ligand may occupy an axial or an equatorial site,

Table 16 Isomer Distribution Equilibria in $(CO)_4FeL$ Complexes[a,67]

Complex	% axial	% equatorial	Ref.
$(CO)_4FePF_3$	33	67	1
$(CO)_4FeF_2P(OCH_3)$	100	—	1
$(CO)_4FeFP(OCH_3)_2$	100	—	1
$(CO)_4FePBu_3^{t\ b}$	100	—	2
$(CO)_4FeF_2POC(CF_3)_2CN$	22	—	3
$(CO)_4FeFP\{OC(CF_3)_2CN\}_2$	15	85	3
$(CO)_4FeP\{OC(CF_3)_2CN\}_3$	0	100	3
$(CO)_4FeF_2POCNC_2(CF_3)_4O$	40	60	3

^{a} Ratio of isomers determined by $\nu_{C=O}$ stretch.
^{b} X-Ray structure.

1. C. A. Udovich, R. J. Clark and H. Hass, *Inorg. Chem.*, 1969, **8**, 1066.
2. H. Schumann, J. Pickardt and L. Rosch, *J. Organomet. Chem.*, 1976, **107**, 241.
3. J. K. Ruff and D. P. Bauer, *Inorg. Chem.*, 1983, **22**, 1686.

the complexes are usually the axial isomer even in the case of PBu$_3^t$.[242] Three generalizations have been stated governing the steric properties and σ and π bonding in these complexes.

(i) *Bulky ligands prefer axial sites*[244]

The ligand ER$_3$ in an M(CO)$_4$(ER$_3$) complex comes into closer contact with the carbonyl ligands when it occupies an equatorial as opposed to an axial site, since the equatorial ER$_3$ cannot form a completely staggered structure in relation to the carbonyl ligands. A typical M(CO)$_4$(ER$_3$) structure is shown in the stereoview in Figure 22 of Fe(CO)$_4$(PPh$_3$).[245] This axial preference is shown for ligands as bulky as PBu$_3^t$ ($\theta = 182°$). However, since PBu$_3^t$ is a powerful σ donor the third substitution rule would also apply (see below). In the case of extremely bulky ligands equatorial substitution has been observed, *e.g.* in Fe(CO)$_4${PPh(PPh$_2$)$_2$}, (23).[246] The extremely bulky (Me$_3$Si)$_2$N—P=C(SiMe$_3$)$_2$ also occupies an equatorial site (equation 64). The geometry around iron is slightly distorted trigonal bipyramidal and in order to accommodate the interactions with the substituents the P=C bond is twisted 30.3°; d(P=C) = 1.657 Å.[247]

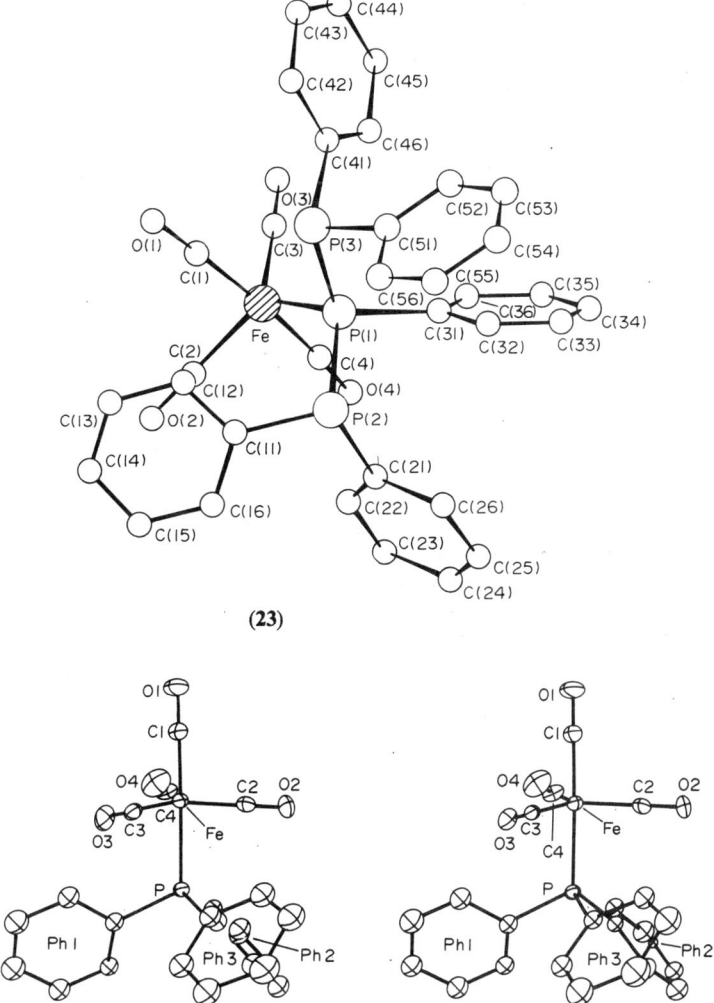

(23)

Figure 22 Stereoscopic view of the Fe(CO)$_4$PPh$_3$ molecule, illustrating the atom numbering scheme. Nonhydrogen atoms are shown as ellipsoids of 30% probability

(ii) *The stronger π acids occupy equatorial sites*[239]

This has already been discussed in relation to π bonding in Fe(CO)$_4$P{OC(CF$_3$)$_2$CN}$_3$ and is confirmed by the fact that the majority of M(CO)$_4$L complexes are axially substituted, since the

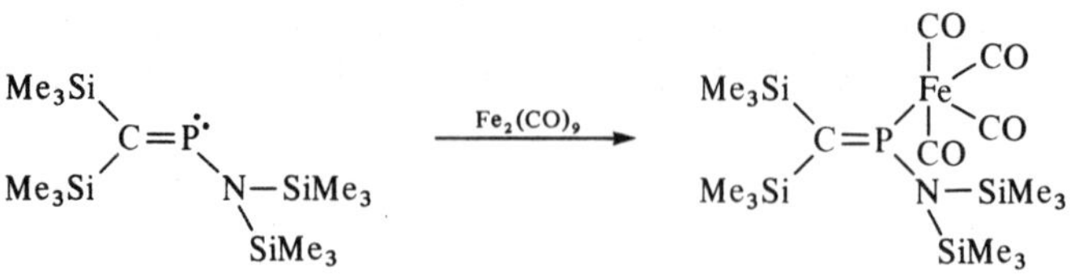

$$d(\text{Fe}-\text{P}) \ 2.208 \ \text{Å}$$
$$\text{Fe}-\text{P}-\text{C} \ 128.0°$$
$$\text{Fe}-\text{P}-\text{N} \ 116.8°$$

carbonyl ligand is a powerful π acid and not easily displaced from axial positions. This is not to say that the complex is rigid in solution. On the contrary, trigonal bipyramidal structures are noted for their lability in solution when exchange between axial and equatorial positions occurs *via* the Berry mechanism.[240]

(iii) *Strong σ donor ligands occupy axial positions in d^8 tbp complexes*[239]

The first $M(CO)_4L$ with L in an equatorial site was prepared by Forbes *et al.*, eq-$Ru(CO)_4(SbPh_3)$ **(24)**,[248] but since $SbPh_3$ is not high on the scale of π acids it was initially inexplicable. Einstein *et al.*[249] prepared a number of $M(CO)_4L$ complexes including ax-$Ru(CO)_4(AsPh_3)$, ax-$Ru(CO)_4(AsMe_3)$ and eq-$Os(CO)_4(SbPh_3)$; all were characterized by X-ray analyses.

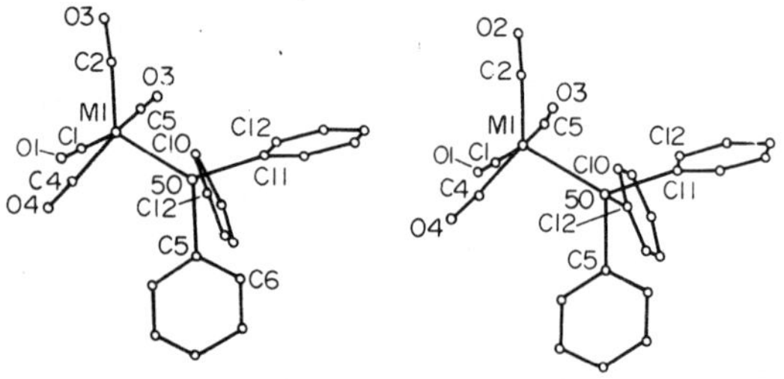

(24)

If this third substitution rule is restated as 'weak σ donor ligands occupy equatorial positions', then in the case of osmium and ruthenium $SbPh_3$ is a weaker σ donor than is CO, as indicated by equatorial substitution $PPh_3 > AsPh_3 > CO > SbPh_3$. The preference for axial substitution for RPh_2Sb ligands was shown *via* X-ray crystallography for $[ax-\{(Ph_3Sb)(CO)_3(Ph)Fe(\mu-SbPh_2)\}Fe(CO)_4]$ **(25)**. Axial substitution was also found for the more basic $M(CO)_4(SbMe_3)$ ($M = Fe,$[250] Ru[249]).

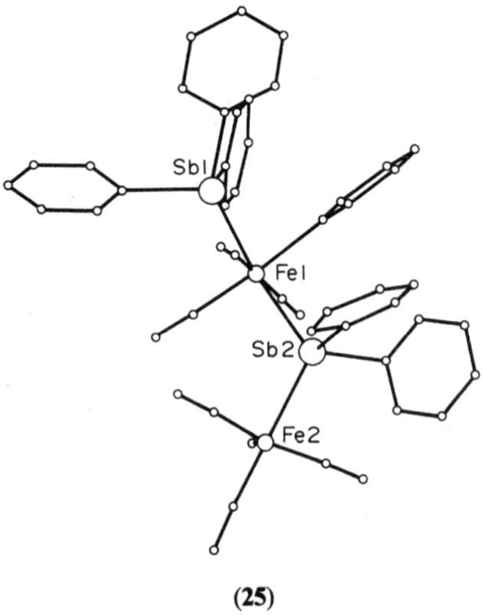

(25)

14.4.3.2 Substituent effects on transition metal–phosphorus bond formation

Metal–phosphine bond formation in general involves a change of phosphorus coordination number from three to four, *i.e.* from pyramidal (p^3) to tetrahedral geometry (sp^3). The importance of substituents and their steric demands lies in their effect on phosphorus geometry, both before and after TM—P bond formation. As discussed in Section 14.3, sterically demanding phosphines interact strongly with other ligands in the coordination sphere resulting in restricted rotation rotamers in some cases, *e.g.* *trans*-RhCl(CO)(PBu$_2^t$Et)$_2$ at $-60\,°$C has a high energy barrier to the eclipsing of a P—Bu^t and an Rh—Cl/CO bond, and rotation about the P—Rh is slow on the NMR timescale and may cease at low temperatures. At $-60\,°$C three sharp NMR patterns result (two A$_2$X and one ABX) corresponding to rotamers (**26a–c**).[251]

(26a)   (26b)   (26c)

To relieve the strain of sterically demanding ligands, a metal often remains coordinatively unsaturated. Copper(I) halides and phosphines form cubane-like metal cluster compounds, $L_m(CuX)_n$.[12] With the bulky trimesitylphosphine, a monomeric two-coordinate [CuBr(Pmes)$_3$] is formed, Br—Cu—P = 173.7°.[252] The d(Cu—P) of 2.193 Å is comparable to that in normal tetrameric complexes, but d(Cu—Br) at 2.225 Å is shorter, no doubt due to the reduced coordination number. Heating crowded complexes can also result in a reduction in coordination number (see equation 65).

$$[Ni\{P(OR)_3\}_4] \xrightarrow{72\,°C} [Ni\{P(OR)_3\}_3] + P(OR)_3 \qquad (65)$$

The ligands used ranged in θ from 109 to 141° and the dissociation constants increased from $\sim 10^{-10}$ to $\sim 10^{-2}$, resulting in the stabilization of nickel(0) 16-electron compounds.[253] Further examples of steric effects occur in the complexes [IrH$_5$(PR$_3$)$_2$] which are labile for R = Me and Et but are robust and easily handled for R$_3$ = Bu$_3^t$, Bu$_2^t$Ph and PBu$_2^t$(p-tolyl).[254]

Shaw and co-workers have shown that large chelating biphosphines are stable with respect to their open-chain components in solution, *i.e.* they are thermodynamically stable. This stability arises from non-bonding interactions between sterically demanding groups which give a favourable conformation and entropy effect. Although one group has a small effect on the ΔH and $T\Delta S$ components of ring formation, several such sterically demanding groups, suitably aligned, have a cumulative effect, *e.g.* the 20-atom ring in (**27**) formed by the reaction (66).[251] The PBu$_3^t$ is the bulkiest of the simple trialkyl phosphines (C—P—C = 109.7°). The [W(CO)$_5$(PBu$_3^t$)] complex has d(W—P) = 2.686 Å, which is the longest W—P bond for this type of complex, *cf.* d(W—P) = 2.516 Å in [W(CO)$_5$(PMe$_3$)], a difference of 0.17 Å, which must be attributed to steric bulk, since both phosphines are strong bases.

(27)

$$\text{Bu}_2^t\text{P(CH}_2)_9\text{PBu}_2^t \;+\; \text{PdCl}_2(\text{PhCN})_2 \;\longrightarrow\; trans\text{-}[\text{Pd}_2\text{Cl}_4\{\text{Bu}_2^t\text{P(CH}_2)_9\text{PBu}_2^t\}] \qquad (66)$$

14.4.4 Soft and Hard Acids and Bases

A further factor influencing transition metal–phosphorus bonding is the SHAB interaction. In general, phosphorus ligands are soft bases and, as such, would be expected to form complexes more readily with transition metals to the right and down the Periodic Table and in lower, rather than higher, oxidation states. Although both Hg^{2+} and Cd^{2+} are considered soft acceptors, slight variations in polarizability lead to structural changes. The SHAB principle has been invoked to explain the 'long' $d(Cd-P)$ of 2.584 Å, in $[Cd(SCN)_2\{P(m\text{-}C_6H_4Me)_3\}]$ (29).[258] Since $d(Hg-P)$ in the analogous complex is 2.43 Å,[259] and since the metals have the same ionic radii, the anomalously long $d(Cd-P)$ has been ascribed to the relatively greater hardness of Cd^{2+} compared to Hg^{2+}. It is significant that the equatorial coordination of Cd is P, N and S and that of the softer Hg is P, S and S.

$$\begin{array}{c}
\text{N} \\
| \\
\text{R}_3\text{P} - \text{Cd} \overset{\displaystyle\nearrow \text{NCS}}{\underset{\displaystyle\searrow \text{SCN}}{}} \\
| \\
\text{S}
\end{array}$$

(29)

Although the SHAB principle is a useful generalization and aids our understanding of many chemical phenomena, it should not inhibit workers from chemical experimentation. On the basis of SHAB theory $Mn^{2+}-PR_3$ bonding is unfavoured and yet an extensive series of MnX_2L[260] and MnX_2L_2[261] (L = tertiary phosphine) complexes has been prepared; even a manganese(III) species has been isolated, $MnI_3(PMe)_3$, which has a *trans* trigonal bipyramidal structure.[262] Moreover, McAuliffe and Levason and co-workers have stabilized nickel(III)[263] and palladium(IV)[264] with phosphine and arsine chelates, *e.g.* $[PdCl_2(diars)_2](ClO_4)_2$ (30), with $d(Pd-As) = 2.454$ Å (av). Such chelating ligands as these have been used to stabilize Cu^{III}, Ag^{III} and the strongly oxidizing Fe^{IV}.[265] The stability of these complexes is thought to arise from a combination of good σ donor properties and small steric requirements; the rigid *o*-phenylene backbone also resists dissociation from the metal.

$$\begin{array}{c}
\text{Me}_2 \quad \text{Cl} \quad \text{Me}_2 \\
\text{As} \qquad | \qquad \text{As} \\
\diagdown \qquad \text{Pd} \qquad \diagup \\
\text{As} \qquad | \qquad \text{As} \\
\text{Me}_2 \quad \text{Cl} \quad \text{Me}_2
\end{array}$$

(30)

In order to overcome the SHAB principle of donor/acceptor incompatibility, a number of groups have developed ligands with both hard and soft ligating centres. Recently Fryzuk and co-workers[266,267] have used ligands such as (31) in which the N atom serves to anchor the hard metal centres, *e.g.* Zr^{IV} and Hf^{IV}, enabling the phosphorus chelating atoms to form a stable complex, *e.g.* $mer\text{-}[ZrCl_3\{N(SiMe_2CH_2PPr_2^i)_2\}]$ (32), $d(Zr-P) = 2.765$ and 2.783 Å, *cf.* $[Zr(dmpe)(CH_2SiMe_3)_4]$, $d(Zr-P) = 2.876$ and 2.972 Å. In the last few years stable phosphine complexes of lanthanides and actinides have also been prepared, and some examples are listed in Table 17.

$$\begin{array}{cc}
\begin{array}{c}
\qquad\quad (CH_2)_n PR_2 \\
\diagup \\
Me_2Si \\
\diagdown \\
\qquad N^- \\
\diagup \\
Me_2Si \\
\diagdown \\
\qquad\quad (CH_2)_n PR_2
\end{array}
&
\begin{array}{c}
Me_2Si \quad\diagup\!\!\!\diagdown\quad PPr_2^i \\
\diagdown \qquad\qquad | \quad Cl \\
\qquad N - Zr - Cl \\
\diagup \qquad\quad Cl \;\; | \\
Me_2Si \quad\diagdown\!\!\!\diagup\quad PPr_2^i
\end{array}
\\[2em]
(31) & (32)
\end{array}$$

The foregoing examples of hard TM/soft phosphine complexes are given to show that such bonding is possible, not that it is preferred. Karsch *et al.*[268] made use of the weak affinity between

Table 17 Actinide/Lanthanide–Phosphorus Interatomic Distances in their Complexes

f-block complex	d(M—P) (Å)	Ref.
$[Yb(\eta^5\text{-}Me_5C_5)_2Cl(Me_2PCH_2PMe_2)]$	2.941	1
$[Yb\{N(SiMe_3)_2\}_2(Me_2PCH_2PMe_2)]$	3.012	2
$[U^{IV}(OPh)_4(Me_2PCH_2CH_2PMe_2)]$	3.104	3
$[U^{III}(\eta^5\text{-}Me_5C_5)_2(Me_2PCH_2CH_2PMe_2)H]$	3.211, 3.092	4
$[U^{IV}(BH_3Me)_4(Me_2PCH_2CH_2PMe_2)]$	3.029, 3.017	5
$[U^{III}(BH_3Me)_3(Me_2PCH_2CH_2PMe_2)_2]$	3.174, 3.085	5
$[U(BH_4)_3(Me_2PCH_2CH_2PMe_2)_2]$	3.051, 3.139	6
$[U(BH_4)_3(2\text{-}Ph_2Ppy)_2]\cdot0.5C_6H_6$	3.162	7

1. A. Zalkin, D. T. Tilley and R. A. Andersen, *Inorg. Chem.*, 1983, **22**, 856.
2. A. Zalkin, D. T. Tilley and R. A. Andersen, *J. Am. Chem. Soc.*, 1982, **104**, 3725.
3. R. A. Andersen, A. Zalkin and P. G. Edwards, *J. Am. Chem. Soc.*, 1981, **103**, 7792.
4. T. J. Marks, M. R. Duttera and P. J. Fagan, *J. Am. Chem. Soc.*, 1982, **104**, 865.
5. A. Zalkin, N. Edelstein, J. Breenan and R. Shinomoto, *Inorg. Chem.*, 1984, **23**, 4143.
6. H. J. Wasserman, D. C. Moody and R. R. Ryan, *J. Chem. Soc., Chem. Commun.*, 1984, 532.
7. H. J. Wasserman, D. C. Moody, R. T. Paine, R. R. Ryan and K. V. Salazar, *J. Chem. Soc., Chem. Commun.*, 1984, 533.

lanthanum(III) and phosphorus(III) to produce a novel bonding system. The ligand, diphenyl-phosphinomethanide anion, tends to form chelate complexes *via* the two phosphine groups. In order to overcome this they used lanthanum(III) to produce an allyl-like system (equation 67).

$$LaCl_3 + 3K[(Ph_2P)_2CH] \xrightarrow{THF} [La\{(Ph_2P)_2CH\}_3] + 3KCl \qquad (67)$$

X-Ray analysis reveals an average $d(La—P)$ of 2.991 Å (**33**). The complex is trigonal planar, with the P—C—P groups at right angles to the plane of the La bonds. There is a large degree of ionic character to the La complex, since the ^{31}P NMR chemical shifts of the P atoms in $[La\{(Ph_2P)_2CH\}_3]$ and $K\{(Ph_2P)_2CH\}$ are nearly identical (C_6D_6/THF, $\delta = -2.85$).

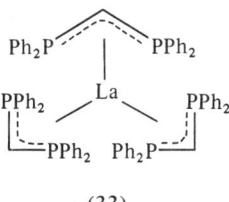

(33)

Although lanthanide phosphine complexes are still comparatively rare, an increasing number are being prepared, *e.g.* Schumann[269] has reported a series (equation 68).

$$Cp_2LnCl + LiPRBu^t \longrightarrow Cp_2LnPRBu^t \qquad (68)$$

(Ln = Tb, Ho, Er, Tm, Yb, La; R = Bu^t, Ph)

14.5 SOME RECENT DEVELOPMENTS

14.5.1 Pnictide Heterocycle Transition Metal Complexes

This discussion will deal mainly with phosphorus heterocycles, but the points will generally be applicable to the emerging chemistry of the lower group members (As, Sb and Bi). Phosphorus heterocycles such as phosphabenzene or the phospholes can form bonds with transition metals in a variety of modes. Because the chemistry of the phospholes is more fully developed, this series will be used to exemplify the area generally. Phospholes are readily synthesized in a 'one pot' process (equation 69).[270] The yield is variable depending on the nature of X and the nitrogen base used, but typically is between 60 and 85%.

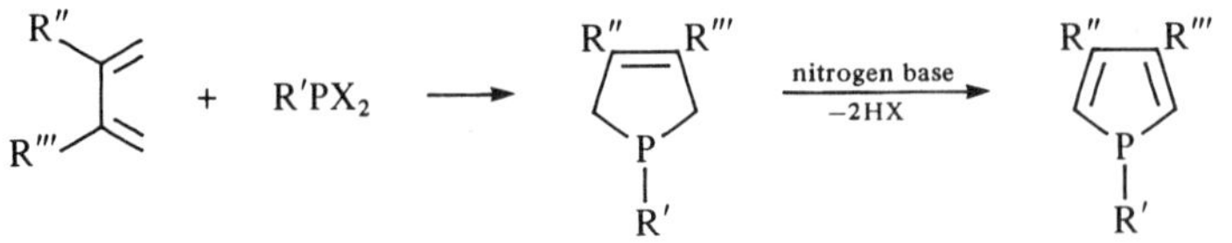

$$\text{(69)}$$

14.5.1.1 Phospholes as two electron donors

Phospholes can behave as simple two electron donors, in the same way as tertiary phosphines, and most of the transition metals have been complexed to phospholes. For example, ruthenium(II) forms a series of complexes $[(Phole)_2 Ru(CO)_2 Cl_2]$ and $[(Phole)_3 Ru(CO)Cl_2]$. The formation of the tris phosphole complex attests to their small size. Because of the ring structure an unusual isomerism has been observed, with the rings either in the basal plane of the square pyramidal complex or normal to the basal plane (Figure 23).

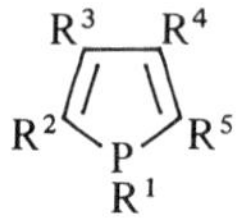

Phole (R = H, alkyl, aryl)[271]

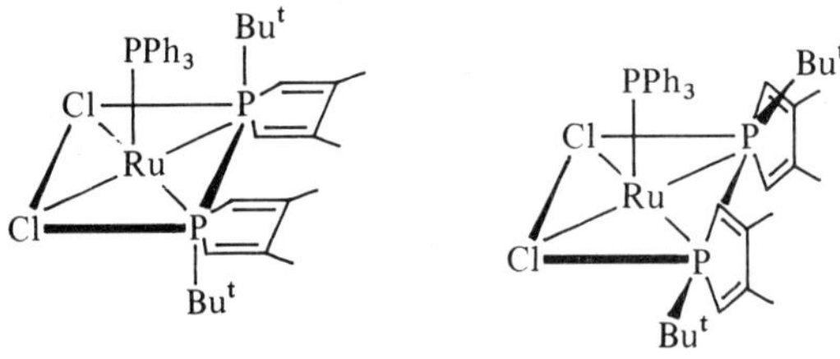

Figure 23

14.5.1.2 Phospholes as two electron donors and four π electron ring donors

The ring π electrons are also available for complexation, enabling the build-up of large complexes as shown in equation (70).

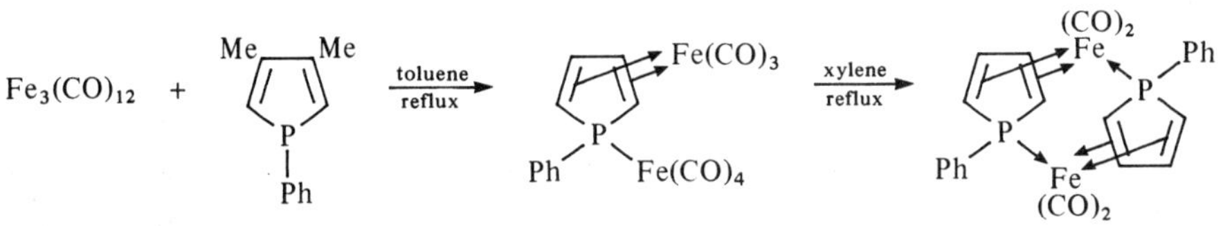

$$\text{(70)}$$

Phospholes with phosphorus(V), *e.g.* the oxide, will form π complexes, *e.g.* (**34**).[274]

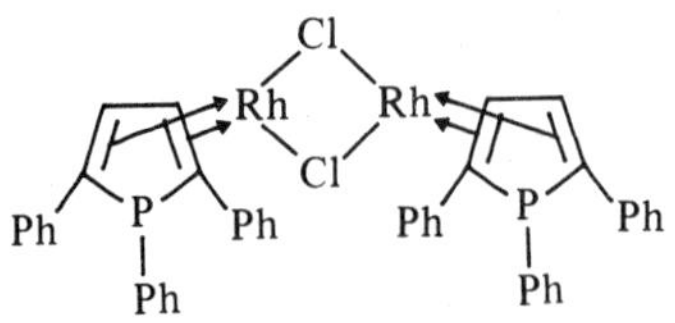

$$\text{(71)}$$

(**34**)

A further bonding mode was utilized by Mathey and co-workers[275] using a localized C=P bond (equation 71).

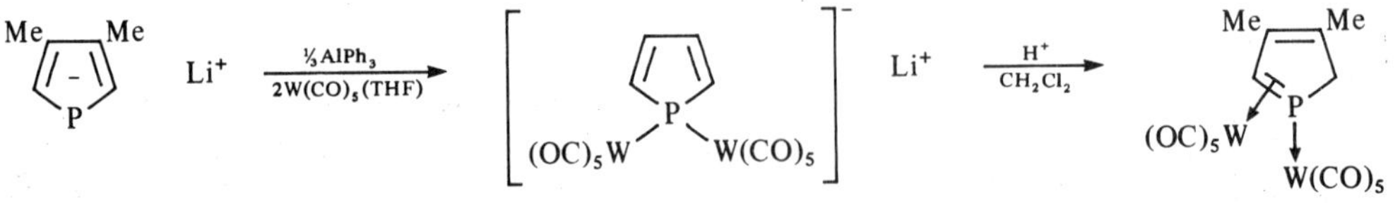

$$d\{W-\eta^2\text{-}(P{=}C)\} = 2.46 \text{ (C)}, \ 2.61 \text{ (P)}; \ d(W-P) = 2.49 \text{ Å}$$

14.5.1.3 Phospholyl complexes

Phospholyl anions are readily generated by R—P cleavage mediated by alkali metal (equation (72)).

$$\text{(72)} \qquad (M = Li, Na, K; \; R = Ph, Me, Bz)$$

The phospholyl anions are resonance structures (equation 73).

$$\text{(73)}$$

If the phospholyl anion is sterically hindered then one electron type ligand complexes are possible, as shown by equation (74).[276]

$$\text{(74)}$$

$$M = CpFe(CO)_2, \; Re(CO)_5, \; CpW(CO)_3$$

The phospholyl anion can also act as a three electron donor, but these complexes are rare (equation 75).[276]

$$\text{hexane} \atop \text{reflux, 24 h} \qquad \text{(75)}$$

(i) Phospholyl anions as five electron donors — the phosphometallocenes

A rich and diverse chemistry has developed based on the phosphorus ring system complexed to transition metals analogous to the metallocenes. Phospholes and phosphametallocenes have recently been reviewed by Mathey *et al.*[277] The phosphacymantrenes and phosphaferrocenes will be used here as representative examples (equation 76). UV photoelectron spectroscopy[278] has shown that the HOMO is not localized on phosphorus, but is mainly on the ring carbons; the lone pair orbital is only the fourth highest occupied orbital. The LUMO is mainly localized on phosphorus in phosphacymantrene, whereas in cymantrene it is on the manganese.

$$+ Mn_2(CO)_{10} \xrightarrow[150\,°C]{xylene} \qquad \text{(76)}$$

$$(R^1 = R^4 = H; \; R^2 = R^3 = Me; \; yield \; 60-80\%)$$

The crystal structure of phosphaferrocene (**35**) shows[279] the C_5 ring to be planar, but the C_4P ring is slightly puckered because of the larger P atom; the P atom is 0.05 Å out of the C plane away from the Fe. (This slight deviation from planarity is typical of all phospholyl rings.) The lone pair is localized on phosphorus, *i.e.* it is not part of the ring system. The availability of the lone pair is demonstrated by such complexes as (**36**).[280]

The phosphorus in the phosphametallocenes is only a very weak nucleophile. Phosphacymantrenes do not react with benzyl bromide in refluxing toluene; however, phosphaferrocenes decompose to form a corresponding phosphonium salt. The complexes undergo a great variety of typical aromatic substitution reactions without affecting the π metal bonding, and some reactions are shown in Scheme 11. It may be realized that many routes are possible for the incorporation of transition metals into organic polymers.

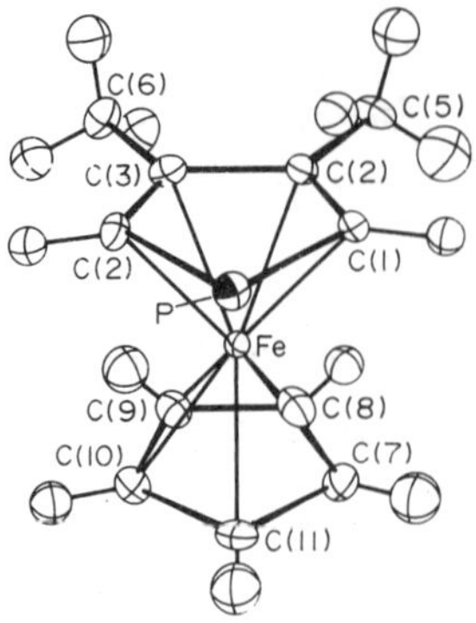

(35)

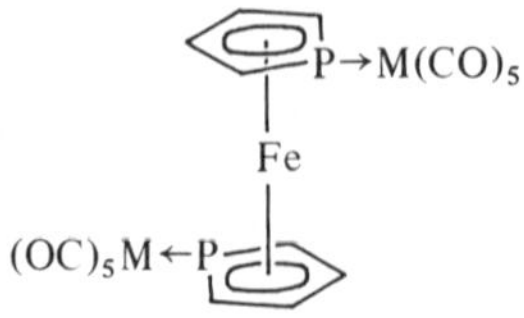

(36) M = Cr, Mo, W

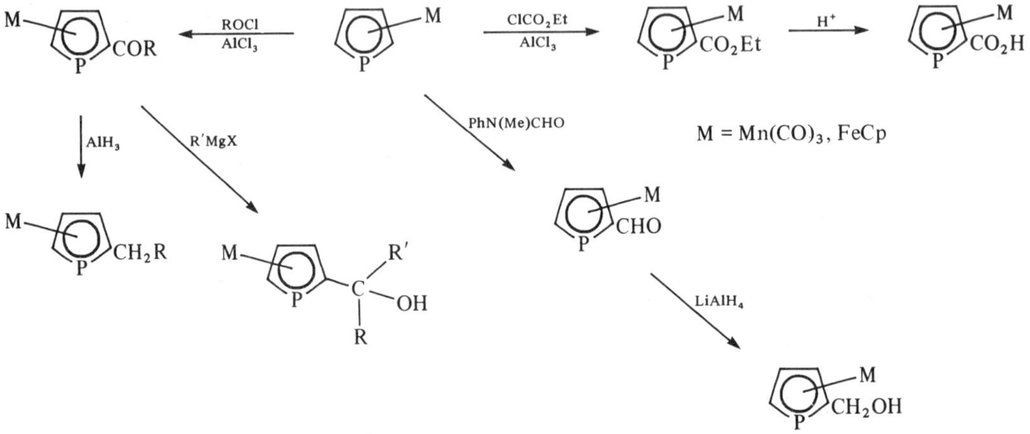

M = Mn(CO)$_3$, FeCp

Scheme 11 Aromatic chemistry of phosphametallocenes

There is a large upfield shift of the ^{31}P NMR of a two electron σ-bonding ligand compared to that of the same ligand acting as a six electron π complex, for example in structures (**37**) and (**38**). This upfield shift is characteristic of cyclic π-bonded systems and is used analytically to detect such systems.

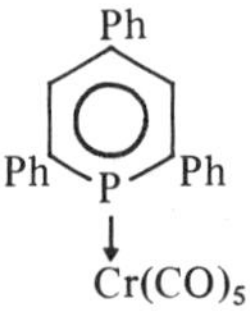

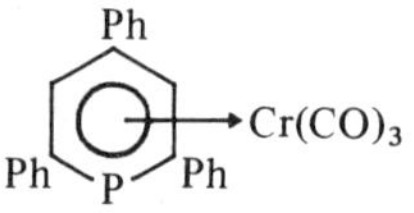

(37) $\delta(^{31}P) = 197.8$ p.p.m.   **(38)** $\delta(^{31}P) = 4.3$ p.p.m.

The unusual phosphole ring C$_2$Fe$_2$P was synthesized by Huttner and co-workers (Scheme 12).[281] In terms of Wade's rules[282] (**b**) is an *arachno* system which can be converted into a *nido* cluster, (**c**), upon decarbonylation. The latter can form directly in the photochemical reaction of (**a**) with alkynes. The five-membered organometallic heterocycle (**c**) is planar (deviation $\leqslant 0.07$ Å) indicating delocalized bonding (structure **39**). Significant developments may be expected from this new class of compounds. New types of heterocyclic phosphorus ligands are continuing to be synthesized, *e.g.* a 1,3,5-triphosphabenzene derivative, 1,1,3,3-tetrakis(dimethylamino)-4-*t*-butyl-1λ^5,3λ^5,5λ^3-tri-phosphabenzene (**40**).

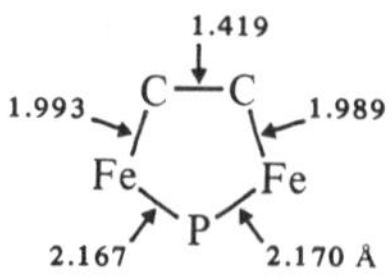

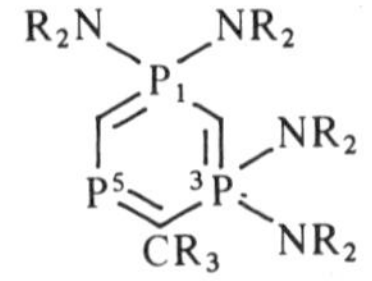

(39)   **(40)** R = Me

14.5.2 Transition Metal Bonding Involving P—P π Bonds and Pnictogen Delocalized Systems

In recent years a number of interesting complexes have been synthesized in which the TM bond to the pnictogen ligand has not been to the pnictogen atom *per se* but to a delocalized π-bonded

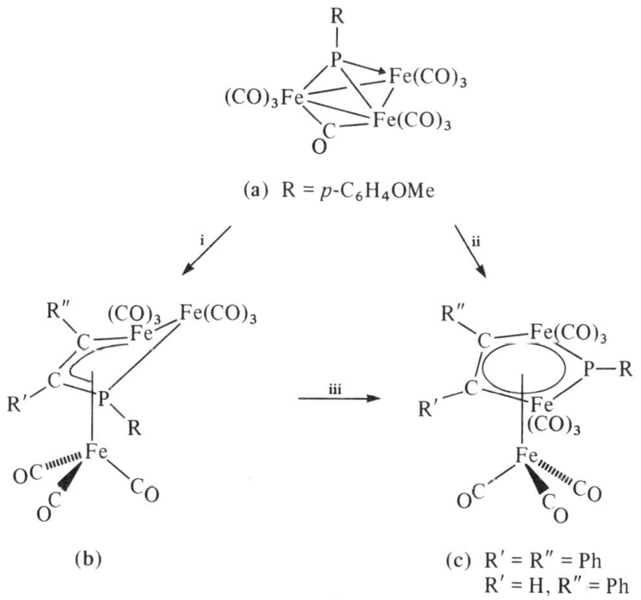

(a) R = p-C_6H_4OMe

i, 40 °C, R'C≡CR" (R' = H, R" = Ph), 71%;
ii, hv, R'C≡CR" (R' = R" = Ph), −CO, 54%; iii, 140 °C, −CO, 90%

(b)

(c) R' = R" = Ph
    R' = H, R" = Ph

Scheme 12

system analogous to metallocenes and the π-allyl complexes. For the purposes of this discussion, they are divided into open-chain ligands, *e.g.* RP=PR, and cyclic ligands, *e.g.* arsenobenzene, C_5H_5As. These are reviewed by Scherer[284] and Mathey.[277]

14.5.2.1 Ligands containing pnictide double bonds

The first complexes to contain the P=P fragment were synthesized in the metal coordination sphere. The parent diphosphene, HP=PH, is unknown in the free state, but is quite stable in (41).[285]

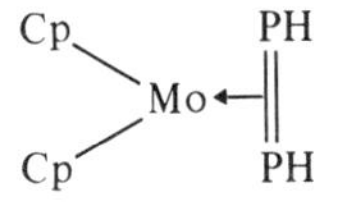

(41)

These complexes may be represented by the canonical forms (42a) and (42b). That the phosphorus lone pairs are not involved in this type of bonding is shown by the existence of other complexes which do not use the lone pairs, *e.g.* (43),[286] (44)[287] and (45).[288]

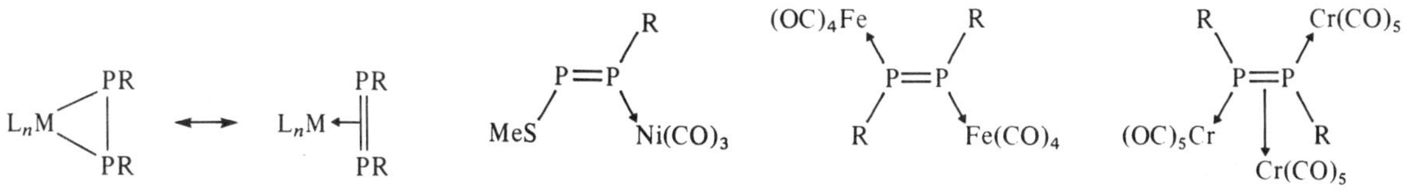

(42a)   (42b)   (43) R = $Bu^t_3C_6H_2$ (44) R = $(Me_3Si)_2CH$ (45) R = Bu^t

Since the synthesis of unsymmetrical RP=PR′ compounds, it has been possible to show with ^{31}P NMR that $^1J_{PP}$ for P=P, *e.g.* 566 Hz for $(Me_3Si)CHP=PR$, is distinct from $^1J_{PP}$ for P—P, usually in the range 200–300 Hz, enabling NMR to be a useful diagnostic tool. Thermolysis of (45) results in a *triangulo* trichromium cluster (46), containing a μ_3-PBu^t bridge resulting from the fission of the P=P bond;[289] this complex represented the first example of an RP=PR ligand acting as a six electron donor in a TM cluster.

The X-ray crystal structures of (47), (48) and (49) show that the angles α and β decrease as the double bond lengthens[290] and the substituent bulk decreases, showing a tendency towards a preference for 90°, *cf.* the alkene bond angle of 120°.[290] A stereoscopic view of bis(2,4,6-tri-t-butyl-phenyl)diphosphine is shown in (50), which illustrates the steric crowding caused by demanding

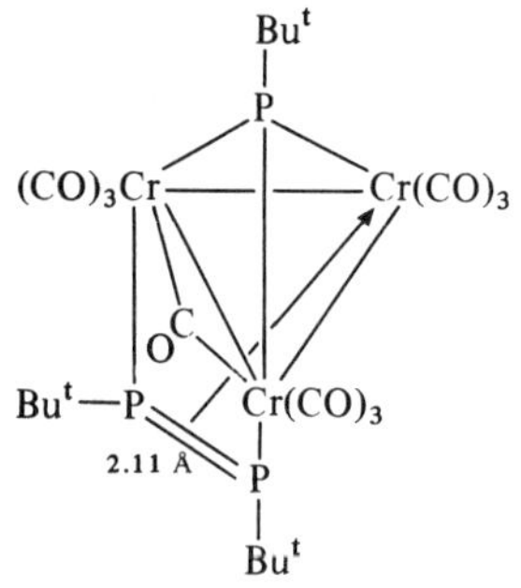

(46)

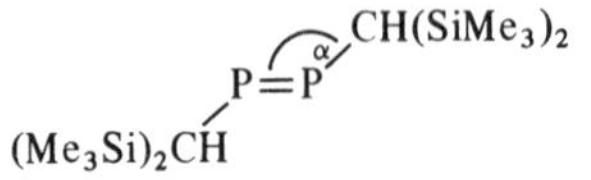

(47) $d(P-P) = 2.014$ Å
$\alpha = 108.2°$

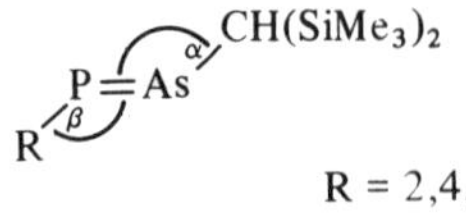

(48) $d(As-P) = 2.125$ Å
$\alpha = 101.4°, \beta = 96.4°$

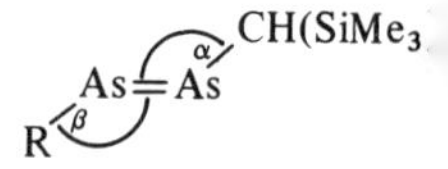

$R = 2,4,6\text{-Bu}_3^tC_6H_2$

(49) $d(As-As) = 2.224$
$\alpha = 99.9°, \beta = 93.6$

substituents, where P—P—C = 102.8° and $d(P—P) = 2.034$ Å.[291] Unlike alkenic sp^2 hybrid bonds, in RP=PR, RP=AsR and RAs=AsR there is a tendency to sp hybridization in going from (47) to (49). This may be the result of an inert s-pair effect when phosphorus is compared to arsenic, where sp hybridization (50a) is preferred to sp^2 (50b). Theoretical studies on the model diphosphenes *trans*-HP=PH and *trans*-MeP=PMe show the presence of relatively low lying LUMOs of P=P π^* character.[292]

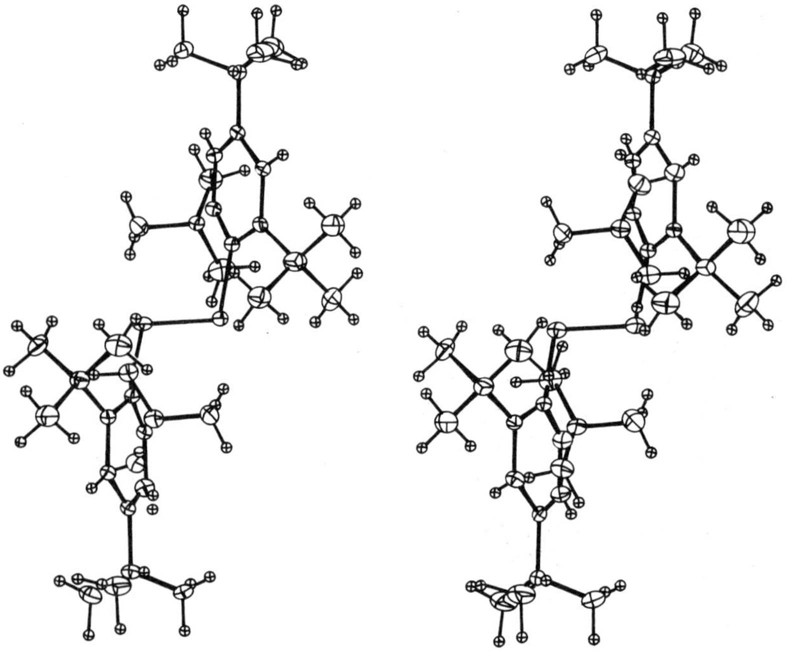

(50)

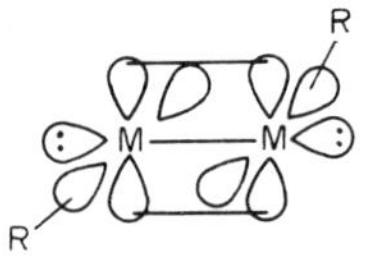

(50a) 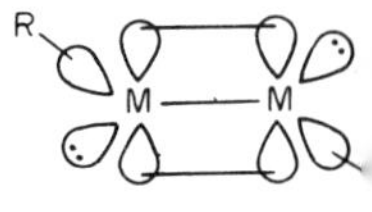   (50b)

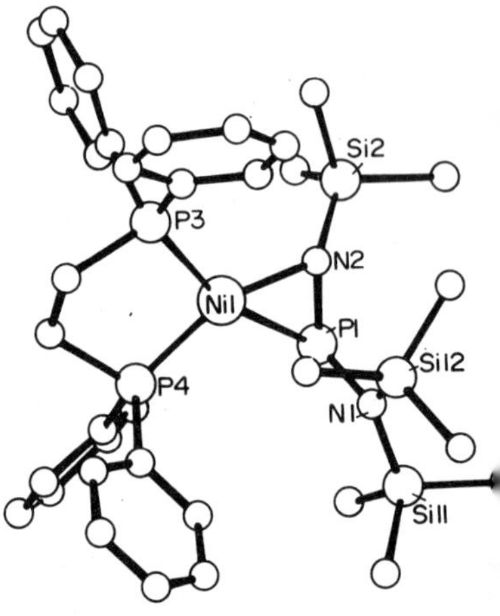

(50c)

14.5.2.2 Ligands containing pnictide triple bonds

Alkene-like E=E ligands occur *via* a number of synthetic routes in pnictide chemistry; those arising, for example, from P_4 are dealt with later. The E≡E ligands are useful in synthesizing clusters. A list of E—E bond lengths from the review by Schere[284] is given in Tables 18 and 19.

Huttner *et al.*[293] synthesized the star-shaped cluster (51; equation 77).

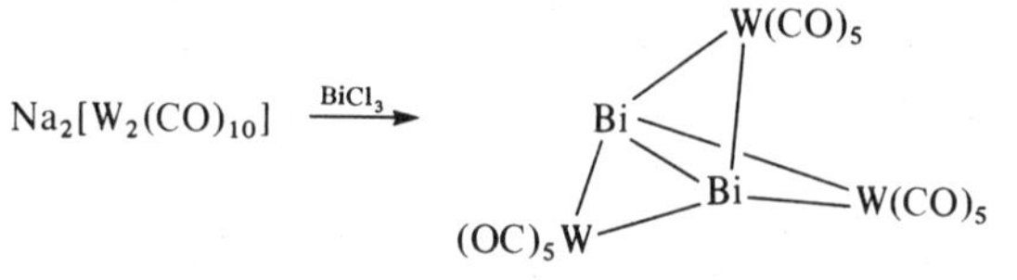

(77)

(51)

Quite elaborate structures have been built up, *e.g.*[294] equation (78), and Schafer and co-workers synthesized (52) while studying the formation of (53; equation 79).[295] Details of the crystal structure of (52) are given in Figure 24, from which it can be seen that $d(P—P)$ is 2.12 Å, *cf.* $d(P—P)$ 2.20–2.25 Å, $d(P=P)$ 2.00–2.05 Å.

Table 18 η^2-R—E=E—R' Ligand Complexes with E—E Bond Lengths

Compound	$^{31}P\{^1H\}$	E—E (Å)	Ref.
η^2-coordination			
$[(PhP)_2[Cr(CO)_5]_3]$	97 (s)	2.125(6)	1
$[(dppe)Pd(PhP)_2[W(CO)_5]_2]$	17.0, ^1J(PP) 344	2.186(6)	2
$[(HP)_2Mo(\eta^5\text{-}C_5H_5)_2]$	203	2.146(3)	3
$[(C_6F_5P)_2Pt(PPh_3)_2]$	− 22 [66d]	2.156(7)	4
$[(Me_3SiP)_2Ni(PEt_3)_2]$	− 62	2.148(2)	5
$[(PhP)_2Pd(dppe)]$	34.3, ^1J(PP) 348	2.121(4)	2
$[(PhP)_2Ni\{\{Ph_2PCC(OSiMe_3)\}_2NMe\}]$	− 4.9	2.140(1)	6
$[\{2,4,6\text{-}(Me_3C)_3C_6H_2OP\}_2\{Fe(CO)_4\}_2]$	233.8, 193.4,^{a} J(PP) 532	2.184(2)	7
$[(Me_3CP)_2\{Fe_2(CO)_6\}]$		2.059(3)	8
$[\{(RP)_2\}_2\{Ni_5(CO)_4Cl\}]$, R = $(Me_3Si)_2$CH	353 (s)	2.085(4), 2.098(4)	9
$[(C_6F_5As)_2\{Fe(CO)_4\}]$		2.388(7)	4
$[(PhAs)_2\{Fe(CO)_4\}]$		2.365(2)	10
$[(PhAs)_2Ni\{\{Ph_2PCC(OSiMe_3)\}_2NMe\}]$		2.372(1)	6
$[\{(Me_3Si)_2CHSb\}_2\{Fe(CO)_4\}]$		2.774(1)	11
$[(PhAs)_2\{Cr(CO)_5\}_3]$		2.371(−)	12
$[(PhAs)_2\{Cr(CO)_5\}_2\{Cr(CO)_4\{P(OMe)_3\}\}]$		2.34(1)	13
$[(PhAs)_2\{Cr(CO)_5\}_2\{\{(MeO)_3P\}_2Pd\}]$		2.366(2)	13
$[(PhAs)_2\{Cr(CO)_5\}_2\{(bipy)Pd\}]$		2.360(6)	14
$[(PhSb)_2\{W(CO)_5\}_3]$		2.706(4)	14
$[(Me_3CSb)_2\{Cr(CO)_5\}_3]$		2.720(3)	15

1. G. Huttner, J. Borm and L. Zsolani, *Angew. Chem., Int. Ed. Engl.*, 1983, **22**, 977.
2. J. Chatt, P. B. Hitchcock, A. Pidcock, C. P. Warrens and K. R. Dixon, *J. Chem. Soc., Chem. Commun.*, 1982, 932.
3. J. C. Green, M. L. H. Green and G. E. Morris, *J. Chem. Soc., Chem. Commun.*, 1974, 212.
4. P. S. Elms, P. Leverett and B. O. West, *Chem. Commun.*, 1971, 747.
5. H. Schäfer, *Z. Naturforsch., Teil B*, 1979, **34**, 1358.
6. D. Fenske and K. Merzweiler, *Angew. Chem., Int. Ed. Engl.*, 1984, **23**, 635.
7. P. P. Power, K. M. Flynn, H. Hope, B. D. Murray and M. M. Olmstead, *J. Am. Chem. Soc.*, 1983, **105**, 7750.
8. D. Woleters and H. Vahrenkamp, *Angew. Chem., Int. Ed. Engl.*, 1983, **22**, 154.
9. P. P. Power and M. M. Olmstead, *J. Am. Chem. Soc.*, 1984, **106**, 1495.
10. E. Weiss and M. Jacob, *J. Organomet. Chem.*, 1978, **153**, 31.
11. A. H. Cowley, N. C. Norman and M. Pakulski, *J. Am. Chem. Soc.*, 1984, **106**, 6844.
12. G. Huttner, H. G. Schmid, A. Frank and O. Orama, *Angew. Chem., Int. Ed. Engl.*, 1976, **15**, 234.
13. G. Huttner and I. Jibril, *Angew. Chem., Int. Ed. Engl.*, 1984, **23**, 740.
14. G. Huttner, U. Weber, B. Sigwarth and O. Scheidsteger, *Angew. Chem., Int. Ed. Engl.*, 1982, **21**, 215.
15. G. Huttner, U. Weber, O. Scheidsteger and L. Zsolnai, *J. Organomet. Chem.*, 1985, **289**, 357.

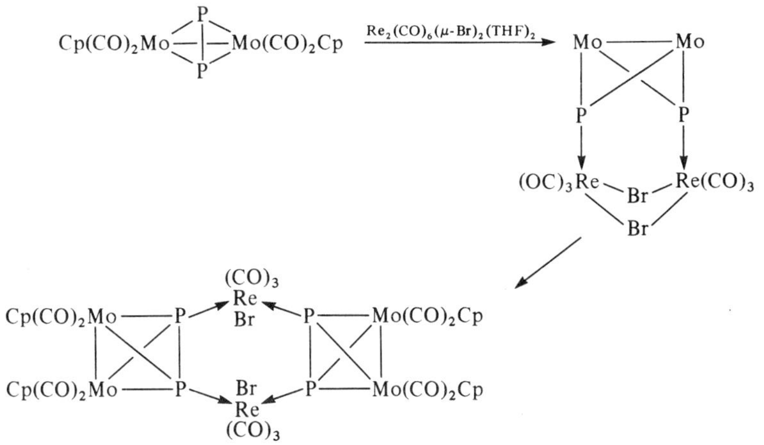

(78)

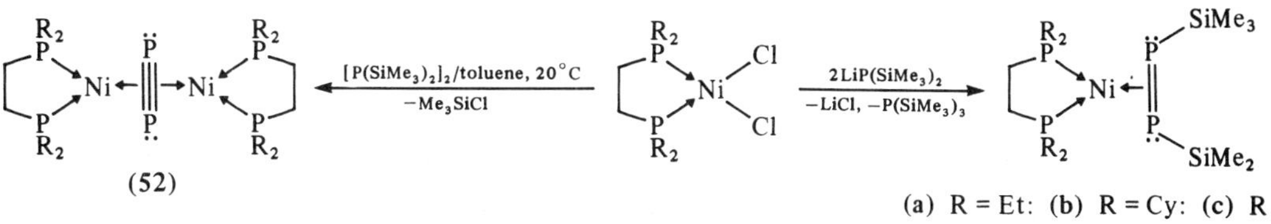

(79)

(a) R = Et: (b) R = Cy: (c) R = Ph

(53)

Table 19 $^{31}P\{^1H\}$ NMR Data (δ Values, J in Hz), Metal–Metal and E—E Bond Lengths (Å) in Complexes with an M_xE_y Framework[a]

Compound[b,c]	$^{31}P\{^1H\}$	M—M	E—E	Ref.
$[Co_2(CO)_5(PPh_3)P_2]$		2.574(3)	2.019(9)	1
$[W_2(OPr^i)_6(py)P_2]$		2.695(1)	2.154(4)	2
$[Cp_2Mo_2(CO)_4P_2]$	−42.9 (s)	3.022(1)	2.079(2)	3
$[\{(Et_2PCH_2CH_2PEt_2)Ni\}_2P_2]^d$	133.0 (q), $^2J(PP)$ 33.0	2.908(3)	2.121(6)	4
$[Co_2(CO)_6P_2\{Cr(CO)_5\}_2]$	146.0 (s)	2.565(3)	2.060(5)	5
$[Cp_2Mo_2(CO)_4P_2\{Re_2(CO)_6(\mu\text{-}Br)_2\}]$	−78.5 (s)	3.077(2)	2.093(8)	6
$[\{Cp_2Mo_2(CO)_4P_2\}_2\{Re_2(CO)_6Br_2\}]$		3.034(2)	2.071(9)	6
$[(Cp^*Mo)_2P_6]$	−315.6 (s)	2.647(1)	2.167(3)–2.175(3)	7
$[Cp^*Mo(CO)P_2\{Cr(CO)_5\}\}]_2$	e	2.905(1)	2.063(5), 2.071(5)	8
$[Co_2(CO)_5(PPh_3)As_2]$		2.594(3)	2.273(3)	9
$[Co_2(CO)_4(PPh_3)_2As_2]$		2.576(3)	2.281(3)	10
$[Cp_2Mo_2(CO)_4As_2]^d$		3.039(2)	2.311(3)	11
$[Cp_2W_2(CO)_4As_2]^d$		3.026(2)	2.326(5)	11
$[Cp^*_2Mn_2(CO)_4As_2]$		4.02	2.225(1)	12
$[\{CpMn(CO)_2\}_4As_2]$			2.445(4)	13
$[W_2(CO)_7I(\mu\text{-}I)As_2]$		3.069(7)	2.305(10)	13
$[\{W(CO)_5\}_3As_2]$			2.279(4)	14
$[Cp_2Mo_2(CO)_4As_2\{Cr(CO)_5\}_2]$		3.064(3)	2.310(3)	13
$[Co_2(CO)_4\{P(OCH_3)_3\}_2As_2\{W(CO)_5\}_2]$		2.59(2)	2.28(1)	15
$[(CpMo)_2As_5]$			2.389(2)–2.762(3)	16
$[\{W(CO)_5\}_3Sb_2]$			2.663(3)	17
$[\{W(CO)_5\}_3Bi_2]$			2.818(3)	18
$[(\mu\text{-}BiCH_3W(CO)_5\}W_2(CO)_8Bi_2]$		3.142(3)	2.795(3)	19

[a] Gas-phase bond length of P_2 for comparison: 1.875 Å (D. E. C. Corbridge, 'The Structural Chemistry of Phosphorus', Elsevier, New York, 1974).
[b] Cp = η^5-C_5H_5.
[c] Cp* = η^5-C_5Me_5.
[d] Two independent molecules in asymmetric unit.
[e] AA'XX' spin system.

1. L. F. Dahl, C. F. Campana, A. Vizi-Orosz, G. Pályi and L. Markó, *Inorg. Chem.*, 1979, **18**, 3054.
2. M. H. Chisholm, K. Folting, J. C. Huffman and J. J. Koh, *Polyhedron*, 1985, **4**, 893.
3. O. J. Scherer, H. Sitzman and G. Wolmershäuser, *J. Organomet. Chem.*, 1984, **268**, C9.
4. H. Schäfer, D. Binder and D. Fenske, *Angew. Chem., Int. Ed. Engl.*, 1985, **24**, 522.
5. G. Huttner, H. Lang and L. Zsolnai, *Angew. Chem., Int. Ed. Engl.*, 1983, **22**, 976.
6. O. J. Scherer, H. Sitzmann and G. Wolmershäuser, *Angew. Chem., Int. Ed. Engl.*, 1984, **23**, 968.
7. O. J. Scherer, H. Sitzmann and G. Wolmershäuser, *Angew. Chem., Int. Ed. Engl.*, 1985, **24**, 924.
8. O. J. Scherer, *Angew. Chem., Int. Ed. Engl.*, 1985, **24**, 924.
9. L. F. Dahl, A. S. Fonst and M. S. Foster, *J. Am. Chem. Soc.*, 1969, **91**, 5633.
10. L. F. Dahl, A. S. Fonst, C. F. Campana and J. D. Sinclair, *Inorg. Chem.*, 1979, **18**, 3047.
11. A. L. Rheingold and P. J. Sullivan, *Organometallics*, 1982, **1**, 1547.
12. M. L. Zeigler, W. A. Herrmann, B. Koumbouris and T. Zahn, *Angew. Chem., Int. Ed. Engl.*, 1984, **23**, 438.
13. G. Huttner, B. Sigwarth, O. Scheidsteger, L. Zsolnai and O. Orama, *Organometallics*, 1985, **4**, 326.
14. G. Huttner, B. Sigwarth, L. Zsolnai and H. Berke, *J. Organomet. Chem.*, 1982, **226**, C5.
15. M. Muller and H. Vahrenkamp, *J. Organomet. Chem.*, 1983, **253**, 95.
16. G. L. Rheingold, M. J. Foley and P. J. Sullivan, *J. Am. Chem. Soc.*, 1982, **104**, 4727.
17. G. Huttner, U. Weber, B. Sigwarth and O. Scheidsteger, *Angew. Chem., Int. Ed. Engl.*, 1982, **21**, 215.
18. G. Huttner, U. Weber and L. Zsolnai, *Z. Naturforsch., Teil B*, 1982, **37**, 707.
19. A. H. Cowley, A. M. Arif, N. C. Norman and M. Pakulski, *J. Am. Chem. Soc.*, 1985, **107**, 1062.

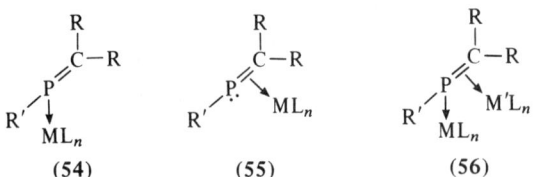

(54)   (55)   (56)

14.5.3 Pnictogen–Element Multiple Bond Ligands

14.5.3.1 E=C type ligands

Methylene phosphines,[296] RP=CR$_2$, can form σ (η^1) and π (η^2) complexes, or both, with transition metals (**54–56**). The σ (η^1) complexes *cis*-$M(CO)_4(mesP=CPh_2)_2$ (M = Cr, Mo, W) have been synthesized, *e.g.* by displacing norbornadiene from $Mo(CO)_4(C_7H_8)$. MO calculations indicate that the HOMO in CH_2=PH is π type, the phosphorus lone pair σ orbital slightly more stable and the π^* LUMO somewhat low-lying.[297] Thus the conditions are met, in theory at any rate, for side-on alkene type π bonding to the transition metal. The first π (η^2) methylene phosphine complex was prepared by Cowley and co-workers (equation 80; Figure 25). The d(P—C) of 1.77 Å is intermediate between a P—C single bond (1.85 Å) and the d(P=C) of an uncoordinated double bond (1.67 Å), *i.e.* this shows typical bond lengthening of π bonds on coordination to TMs (Table 20).

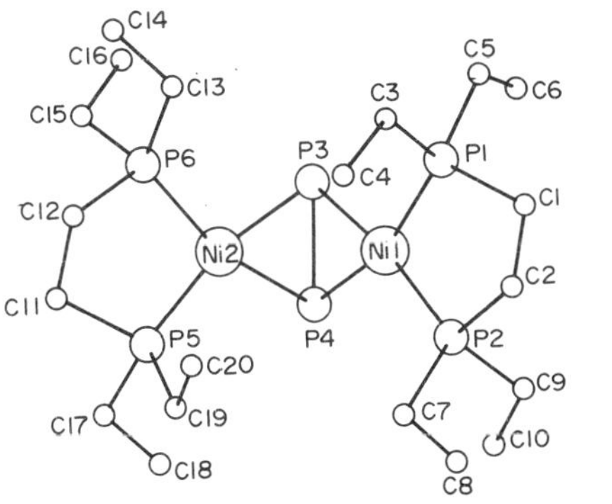

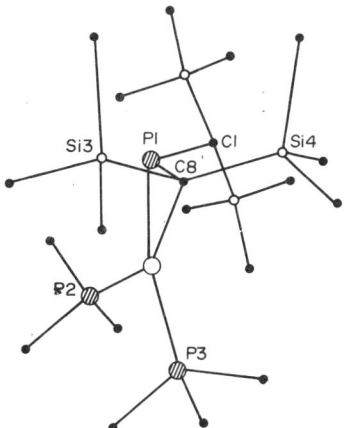

Figure 24 Molecular structure of (**52**) The Ni atoms are surrounded by a nearly planar array of four P atoms and lie 9 pm below the mean plane. Fold angle (Ni1-P3-P4—Ni2-P3-P4) 95.4°. Important distances (pm) and bond angles (°): Ni1—P1 212.5(5), Ni1—P2 213.1(4), Ni1—P3 223.8(5), Ni1—P4 223.1(5), Ni1—Ni2 290.8(3), P3—P4 212.1(6). Ni2—P3 22.3(3), Ni2—P4 223.3(3), Ni2—P5 213.4(4), Ni2—P6 214.9(5), P—C 182–186(1); Ni1—P3—P4 61.5(2), P3—P4—Ni1 61.8(2), P3—Ni1—P4 56.7(2), P1—Ni1—P2 91.5(2), P3—Ni1—P1 101.9(3), P4—Ni1—P2 109.6(2). P3—Ni1—P2 165.7(2), P4—Ni1—P1 158.3(2), Ni1—P3—Ni2 81.1(2), Ni1—P4—Ni2 81.3(2). The distances in molecule (**52**) correspond to the values given above within the standard deviation

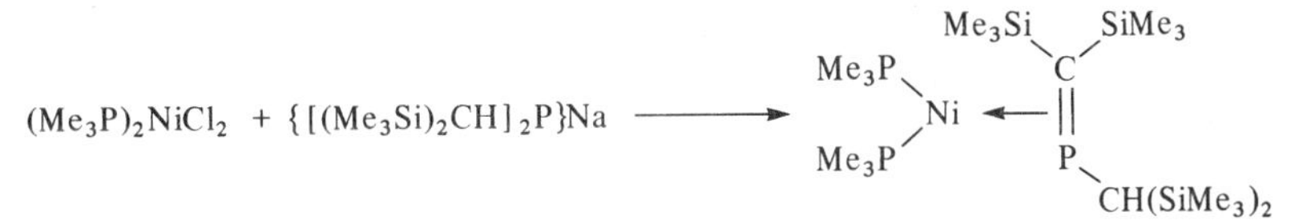

$$(Me_3P)_2NiCl_2 + \{[(Me_3Si)_2CH]_2P\}Na \longrightarrow \begin{matrix} Me_3Si \quad SiMe_3 \\ C \\ Me_3P \\ \searrow \\ Ni \leftarrow \| \\ Me_3P \\ \nearrow \quad P \\ CH(SiMe_3)_2 \end{matrix} \qquad (80)$$

Figure 25 ORTEP view of $(Me_3P)_2Ni[(Me_3Si)_2CPC(H)(SiMe_3)_2]$ showing the atom numbering scheme

Pertinent Bond Lengths (Å) and Bond Angles (°) for $(Me_3P)_2Ni[(Me_3Si)_2CPC(H)(SiMe_3)_2]$

Bond Lengths			
Ni—P(1)	2.239(2)	P(1)—C(8)	1.773(8)
Ni—P(2)	2.195(3)	P(1)—C(1)	1.912(8)
Ni—P(3)	2.202(3)	C(8)—Si(3)	1.894(8)
Ni—C(8)	2.020(8)	C(8)—Si(4)	1.879(8)
Bond Angles			
P(1)—Ni—P(2)	99.9(1)	C(1)—P(1)—C(8)	110.3(4)
P(2)—Ni—P(3)	102.(1)	Ni—P(1)—C(8)	59.1(3)
P(1)—Ni—C(8)	48.9(2)	Ni—C(8)—P(1)	72.0(3)
P(3)—Ni—C(8)	109.9(2)	Si(3)—C(8)—Si(4)	113.8(4)

Table 20 η^2-RP=CR$_2'$ Complexes, Bond Lengths and Angles

Compound	$^{31}P\{^1H\}$	CPC (°)	P=C (°)	Ref.
[{(2,6-dimethylphenyl)P=CPh$_2$}{Ni(bipy)}]	−16.1s	102.6(3)	1.832(6)	1
[{2,4,6-(Me$_3$C)$_3$C$_6$H$_2$P=CH$_2$}{Fe(CO)$_4$}$_2$]	6.5s	102.5(3)	1.737(6)	2
[{(Me$_3$Si)$_2$CHP=C(SiMe$_3$)$_2$}{Ni(PMe$_3$)$_2$}]	23.4^d	110.3(4)	1.773(8)	3
[(H$_2$C(Me)C=(Me)C(H)C=P}{W(CO)$_3$}$_2$]	−31.9s	93.2(8)	1.78(1)	4
[p-MeOC$_6$H$_4$P=CH$_2${Fe$_3$(CO)$_{10}$}]	a	a	1.76(1)	5
[{PhP=CPh(OEt)(μ_3-PPh)}{Fe$_3$(CO)$_9$}]	163.7d, 142.7d, ^2J(PP) 28/23^e	a	1.800(6)	6
[(RPCHPR){Co(CO)$_3$}]	34.4d, ^2J(PH) 7.0	(101.8(4) (PCP))	1.769(7), 1.791(7)	7

R = 2,4,6-(Me$_3$C)$_3$C$_6$H$_2$

^{a} Not given.
^{b} Fe—Fe: 2.666(1) Å.
^{c} Mo—P: 2.174(1) Å.
^{d} X part of an ABX spectrum.
^{e} Diastereomers in solution; μ(P=C).

1. T. A. van der Knaap, L. W. Jenneskens, H. J. Meeuwissen, F. Bickelhaupt, D. Walther, E. Dinjus, E. Uhlig and A. L. Spek, *J. Organomet. Chem.*, 1983, **254**, C33.
2. R. Appel, C. Casser and F. Knoch, *J. Organomet. Chem.*, 1985, **293**, 213.
3. A. H. Cowley, R. A. Jones, C. A. Stewart, A. L. Stuart, J. L. Atwood, W. E. Hunter and H. M. Zhang, *J. Am. Chem. Soc.*, 1983, **105**, 3737.
4. F. Mathey, S. Holand, C. Charrier, J. Fischer and A. Mitschler, *J. Am. Chem. Soc.*, 1984, **106**, 826.
5. G. Huttner, K. Knoll, W. Wasiucionek and L. Zsolnai, *Angew. Chem., Int. Ed. Engl.*, 1984, **23**, 739.
6. A. L. Rheingold, G. D. Williams, G. L. Geoffroy and R. R. Whittle, *J. Am. Chem. Soc.*, 1985, **107**, 729.
7. R. Appel, W. Schuhn and F. Knoch, *Angew. Chem., Int. Ed. Engl.*, 1985, **24**, 420.

Other Group IV analogues R$_2$E=PR (E = Si, Ge, Sn; R = bulky organic group such as mesityl or 2,4,6-tri-*t*-butylphenyl) have recently been prepared,[299] and are interesting potential ligands. Finally, it should be mentioned that the R group itself can be a TM moiety (Figure 26; equation 81).[300]

$$[(\eta^5\text{-}C_5H_5)(CO)_2FeP(SiMe_3)_2] \xrightarrow[-Me_3SiCl]{Bu^tC(O)Cl} [(\eta^5\text{-}C_5H_5)(CO)_2FeP(SiMe_3)\{C(O)Bu^t\}] \longrightarrow$$

$$[(\eta^5\text{-}C_5H_5)(CO)_2FeP=C(OSiMe_3)(Bu^t)] \tag{81}$$

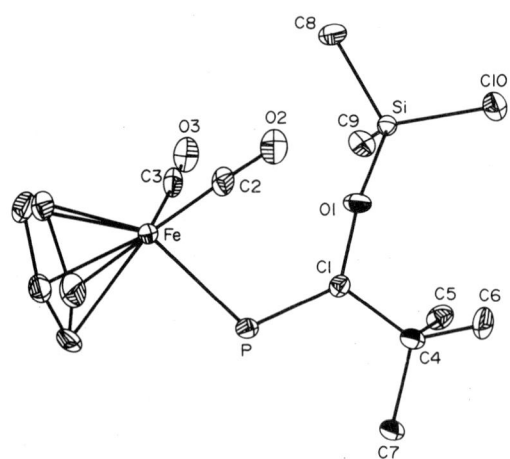

Figure 26 Selected bond lengths (pm) and angles (°): Fe—P 229.8(1), P—C1 170.1(4), C1—O1 136.5(5), C1—C4 152.9(6), Fe—C2 175.3(5), Fe—C3 178.3(5), C2—O2 115.1(6), C3—O3 110.8(7), Fe—C(ring) 211.2(3) to 211.9(3), C2—Fe—P 86.2(1), P—Fe—C3 88.8(2), C2—Fe—C3 96.5(2), Fe—P—C1 113.8(2), P—C1—O1 123.9(3), P—C1—C4 120.5(3), O1—C1—C4 115.3(4), C1—O1—Si 144.7(3)

14.5.3.2 *Phosphorus–nitrogen multiple bond ligands*

Ligands of the type $RP{=}NR^1$ ($2p/3p\pi$ bonding) are known and may form complexes with TMs through P or N or the P=N π bond. The ligand $(Me_2Si)_2NP{=}NSiMe_3$ shows[301] an interesting variation in bonding with nickel, depending upon the other ligands in the coordination sphere, from σ-P bonding to π-NP bonding. If *trans*-N ligands are introduced, an oxidative coupling between the ligand takes place (Scheme 13). The crystal structure of (50c) indicates the side-on P=N coordination; d(P=N) in the free ligand is 1.545 Å[302] and N(1)—P—N(2) is 109°. Upon coordination the d(P=N) lengthens to 1.646 Å and the N—P—N remains unchanged. Complex (50c) is oxidized by Se to afford the unusual P=Se side-on coordination (equation 82).

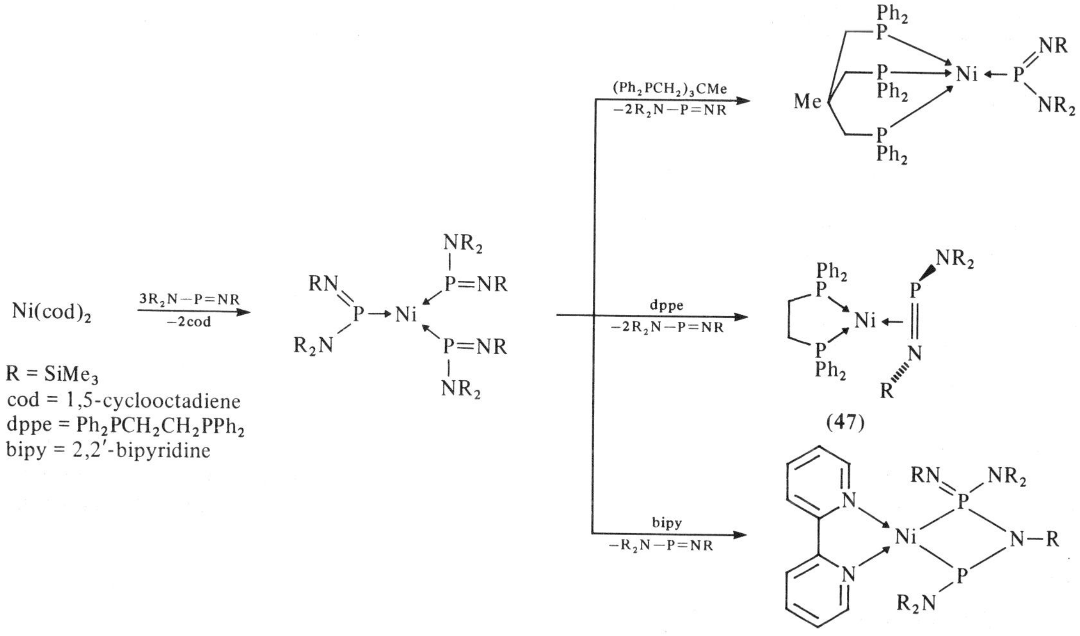

Scheme 13 $\eta^1{-}\eta^2$ change in P=N coordination

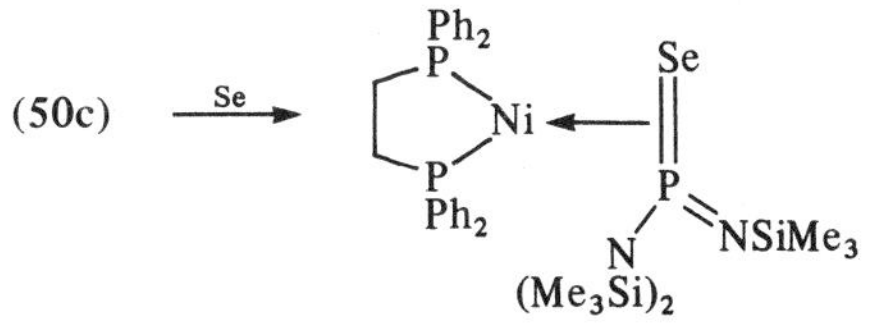

$$(82)$$

The ligand $Me_5C_5P{=}NBu^t$ itself undergoes disproportionation at nickel on complexing, as shown by equation (83).[303] The nickel moiety may be viewed as a substituent of the phosphaalkene, or the phosphaalkene may be viewed as a ligand of nickel.

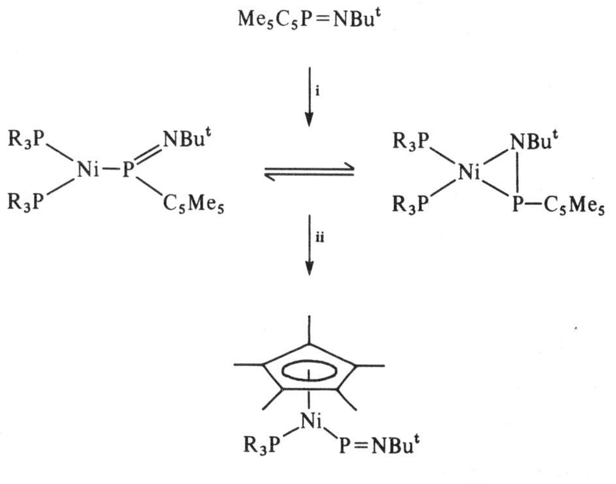

$$(83)$$

i, $(R_3P)_2Ni(cod)$ (1 equiv.), toluene, $-30\,°C$, 2 h; ii, $-20\,°C$, 1 h

14.5.3.3 Miscellaneous phosphorus–element multibond π complexes

Other P=E transition metal π-bonded ligands are not as numerous, but several X-ray crystal structures are available. The three-coordinate P^V ligand $R_2N(RN)PS$ forms the planar platinum(0) complex, (57), where the P=S bond shows bond lengthening on coordination.[304] (The selenium analogue is shown in Section 14.5.3.2.)

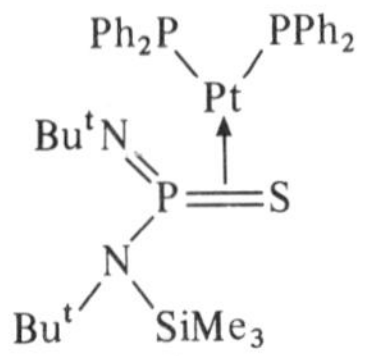

$d(P=S)$ 2.07 Å
$d(Pt-P)$ 2.29 Å
$d(Pt-S)$ 2.38 Å
$d(P=N)$ 1.57 Å
$d(P-N)$ 1.68 Å

(57)

Ligands of type R—P=O are only known in complexes and attempts to isolate them in the free state result in oligomers, *e.g.* equation (84).

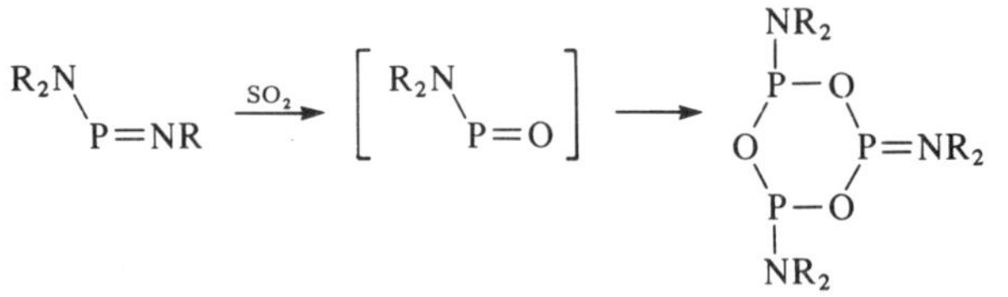

(84)

(R = Pr^i)

But if the complex with chromium carbonyl is first formed, then the P^{III} oxide can be isolated as a bound ligand (equation 85).[305]

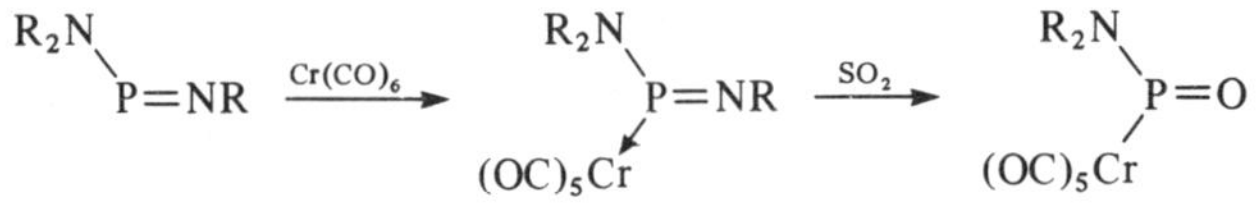

(85)

14.5.4 Pnictogen–Transition Metal Double Bonds

There is a recent review of two-coordinate phosphorus complexes.[306] Malisch *et al.*[307] observed the reversible reaction (86), in which a metal–arsenic(III) double bond is formed, *i.e.* the M—As σ bond is augmented by the arsenic lone pair to form a π bond system (since the cyclopentadienyl coligand is not coplanar with the M=As, the arsenic double bond is isolated). Complex (58) undergoes reactions typical of double bond molecules (Scheme 14). Phosphorus analogues have also been prepared (Scheme 15); the crystal structure of product (c) in Scheme 15 has been solved (59a). The $d(W-P)$ of 2.181 Å is shorter than the predicted $d(W=P)$ of 2.26 Å, and the trigonal planar coordination of phosphorus indicates sp^2 hybridization.[308]

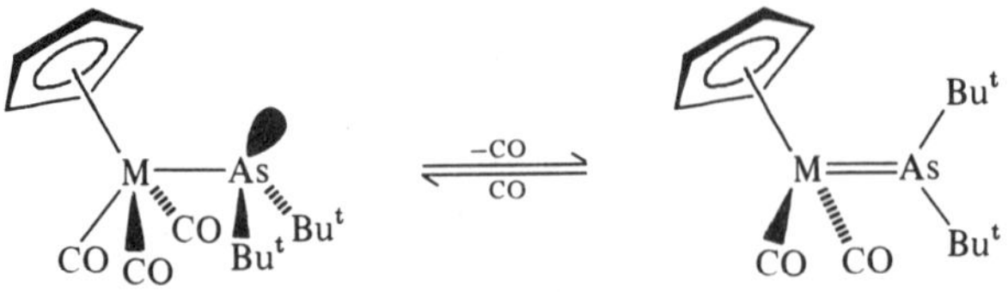

(86)

(a) M = Mo; (b) M = W   (58)

The method has been extended to bimetallic complexes (equation 87) and to butadiene type compounds (equation 88).[309] These dienes may be used in complex formation producing complexes analogous to butadiene, as shown by equation (89). The structure of (59b) and the diene [Cp(CO)$_2$ Mo=P(R)CH=P(R)] (59c) have been proved by X-ray crystallography. A similar series of dienyl complexes have been synthesized by Huttner *et al.*[281] (see Section 14.5.1.3.i).

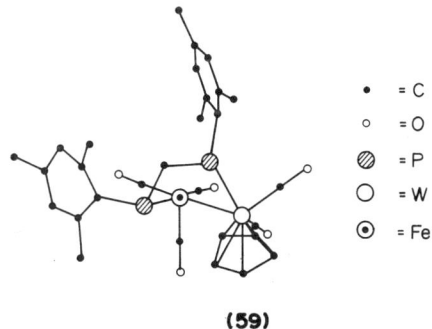

•	= C
○	= O
⊘	= P
◯	= W

(59a)

•	= C
○	= O
⊘	= P
◯	= W
⊙	= Fe

(59)

•	= C
○	= O
⊘	= P
◯	= Mo

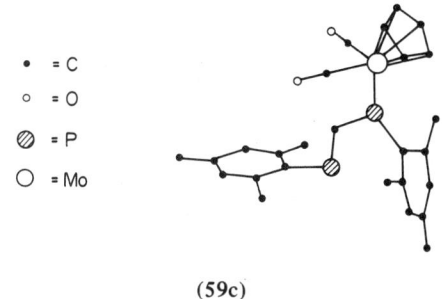

(59c)

An unusual series of complexes was prepared by Baker and co-workers[310] based upon the exceptional properties of the hindered ligand Cy$_2$P$^-$. The complexes are often homoleptic, *e.g.* (Cy$_2$P)$_4$Mo, and involve a P—M σ bond and a second bond from the donation of the lone pair (Figure 27). An interesting phosphabutatriene (60) has been synthesized recently and should be a ligand which provides a good deal of interesting chemistry.[311] Finally, ligands of the type P≡CR are capable of accepting up to six electron donors in complexes such as (61).[312]

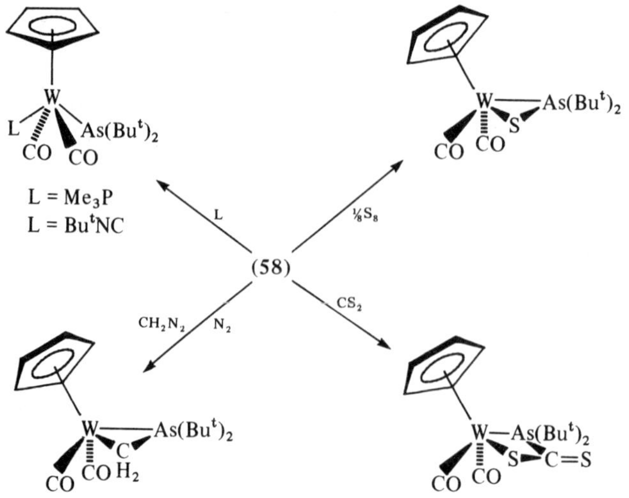

Scheme 14

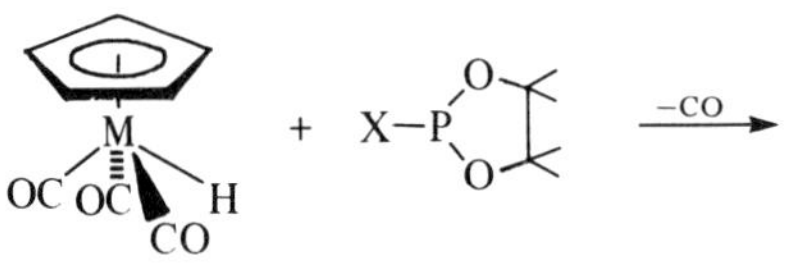

(a) M = Cr, **(b)** M = Mo, **(c)** M = W; **(a)** X = Cl, **(b)** X = NMe$_2$

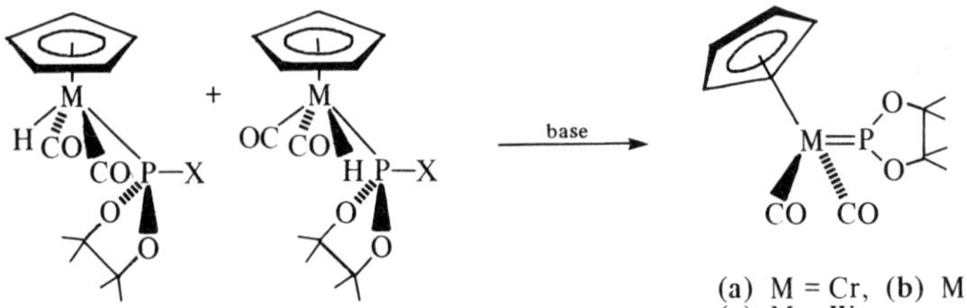

(a) M = Cr, **(b)** M = Mo,
(c) M = W

(a) M = Cr, X = Cl; **(b)** M = Mo, X = Cl; **(c)** M = W, X = Cl;
(d) M = Mo, X = NMe$_2$; **(e)** M = W, X = NMe$_2$

Scheme 15

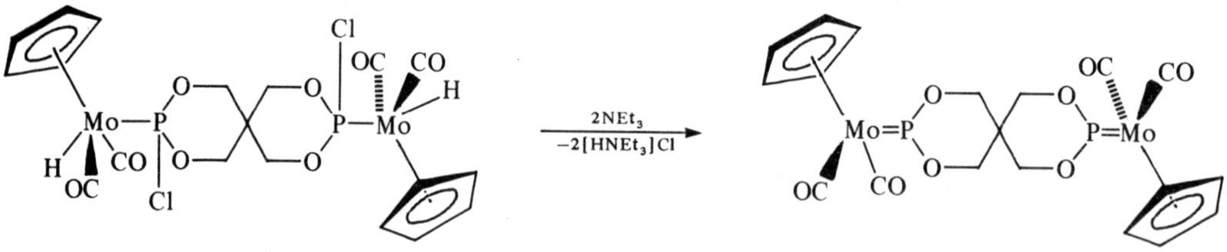

(87)

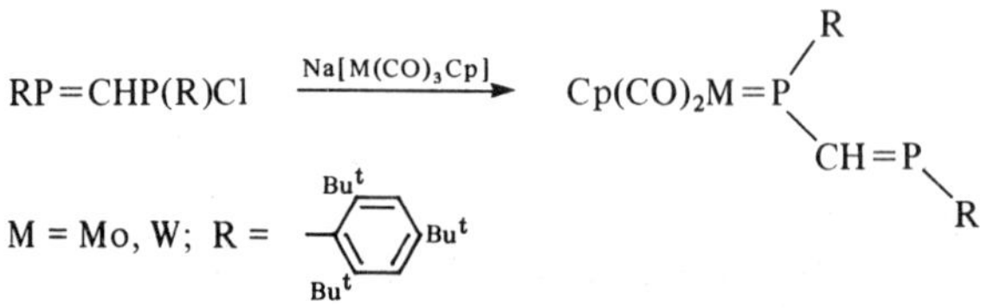

(88)

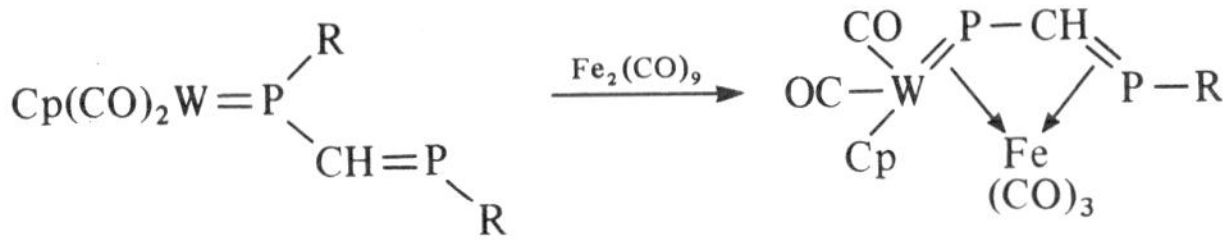

$$Cp(CO)_2W = P \underset{CH=P}{\overset{R}{<}} \xrightarrow{\text{Fe}_2(CO)_9} \quad (89)$$

(59)

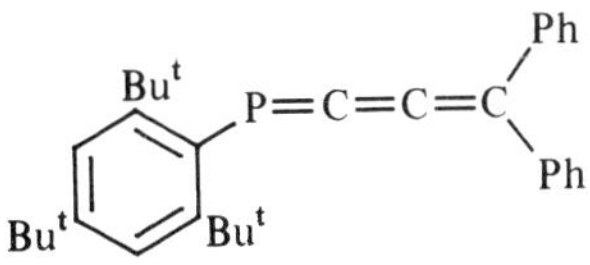

(60)

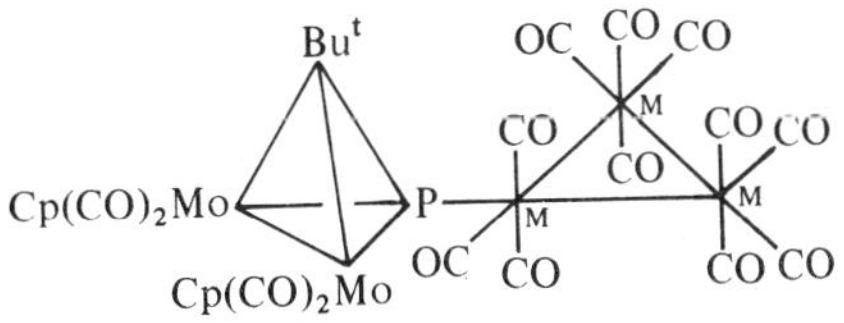

(61) a: M = Os
      b: M = Ru

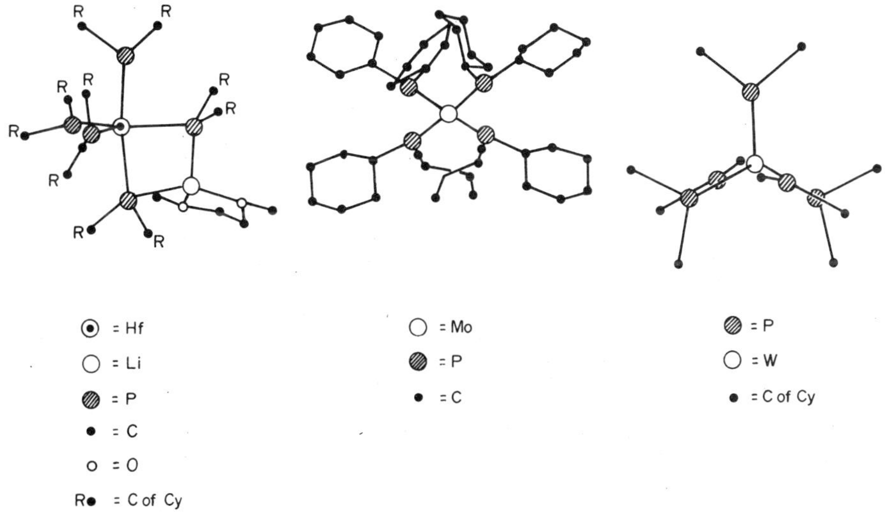

⊙ = Hf	○ = Mo	▨ = P
○ = Li	▨ = P	○ = W
▨ = P	● = C	● = C of Cy
● = C		
o = O		
R● = C of Cy		

Figure 27

14.5.5 Allyl Type Complexes

η^3-1-Phosphaallyliron complexes[313] have been prepared by first stabilizing a secondary vinyl phosphine by complexing with tungsten carbonyl, thus preventing self condensation. The iron phosphaallyl complex then forms, shown by equation (90).

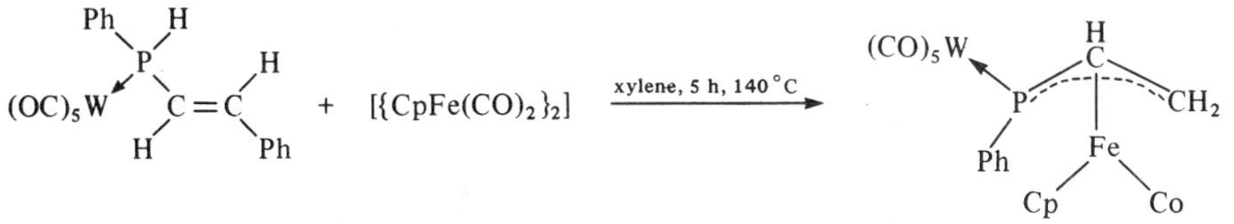

$$\text{(OC)}_5W \quad + \quad [\{CpFe(CO)_2\}_2] \xrightarrow{\text{xylene, 5 h, 140°C}} \quad (90)$$

The phosphinomethanides are not neutral ligands and can be generated in reactions with transition metals, *e.g.* equation (91).[314]

According to an X-ray structure determination (62) can best be described as a square pyramid having two P atoms and two CO ligands at the base and a third carbonyl at the apex, with the alkyl C atom below the basal plane pointing towards the Co centre, and the NMR magnetic equivalence

$$R-P=CH-P\overset{Cl}{\underset{R}{\big\langle}} \ + \ NaCo(CO)_4 \ \xrightarrow{-NaCl} \ R-P=CH-P\overset{Co(CO)_4}{\underset{R}{\big\langle}} \ \xrightarrow{-CO} \ \text{(62)} \tag{91}$$

$$R = 2,4,6\text{-}Bu^t_3C_6H_2$$

of the two aryl P units are in accord with this. [La{(Ph$_2$P)CH}$_3$] (**33**) was discussed in detail in Section 14.4.4.

14.5.6 Phosphorus Ylides as Ligands

The use of phosphorus ylides as ligands was reviewed by Schmidbaur in 1983.[315] These ligands are strongly polarized, R$_3$P$^+$—CR$_2^-$, and thus they bond in the main to transition metals *via* a C—M bond. Although the phosphorus plays a unique role in the properties of such ligands, it is not the province of this article to describe this M—C type of bonding. An example is given of the type of complex formed with lanthanides (equation 92).[316]

$$LnCl_3 \ \xrightarrow{Me_3PCH_2} \ (Me_3PCH_2)_3LnCl_3 \ \xrightarrow[\substack{-3LiCl \\ -3BuH}]{3BuLi} \ \text{(92)}$$

$$Ln = La, \ Pr, \ Nd, \ Sm, \ Gd, \ Ho, \ Er, \ Lu$$

Ylide complexes with P—H bonds can rearrange to phosphine complexes (equation 93).[317]

$$(CO)_5\bar{C}r-\overset{OMe}{\underset{Ph}{\overset{|}{C}}}-\overset{+}{P}Me_2H \ \xrightarrow{\Delta} \ (CO)_5Cr\leftarrow\overset{Me}{\underset{Me}{\overset{|}{P}}}-CH\overset{OMe}{\underset{Ph}{\big\langle}} \tag{93}$$

With ylides carbene complexes usually undergo Wittig alkenation whereby the (OC)$_5$W fragment can formally play the role of the oxygen atom of the aldehyde or ketone (equation 94).[318]

$$(CO)_5W=C\overset{Ph}{\underset{OMe}{\big\langle}} \ \xrightarrow{Ph_3P=CH_2} \ Ph_3P-W(CO)_5 \ + \ \overset{Ph}{\underset{MeO}{\big\rangle}}C=CH_2 \tag{94}$$

Ylides often give unexpected products by rather complicated reactions, *e.g.* Ph$_3$P=CH$_2$ reacts[319] with MeCp(CO)$_3$Mn to give (**63**).

$$MeCpMn\leftarrow P(Ph_2)$$

(**63**)

14.6 P$_4$ AND FRAGMENTS

These unusual ligands, which have no stabilizing organic groups, are formed either by employing white phosphorus, P$_4$, as a ligand or, less commonly, by a phosphine losing its substituents in the course of a reaction. Analogous complexes are known for the heavier pnictides and a few will be mentioned, but the emphasis will be on phosphorus. The P$_4$ group has T_d symmetry and some geometrical facts should be borne in mind when considering the P$_4$ complexes which follow. These geometrical observations are not suggested to have any mechanistic import, merely potentials inherent in the T_d symmetry.

The P_4 tetrahedron may be thought of as half the sites of a cube, or the cube may be thought of as a tetrahedron with capped faces (Figure 28). If the P_4 structure is to be retained then the TM must either form a normal σ bond to phosphorus or form some part of a capped structure; both are known.

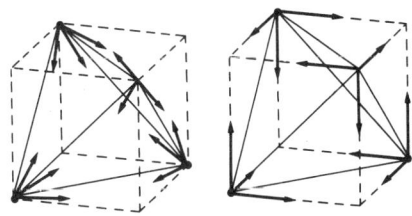

Figure 28

Many cubane type complexes (M_4E_4; E = P, As, Sb, Bi) are known, *e.g.* $Co_4(CO)_{12}Sb_4$[320] has a distorted cubane structure which may be considered as a $Co(CO)_3$ capped Sb_4 tetrahedron with the Sb—Sb expanded to a non-bonding distance, 3.156 Å (Figure 29). Each Sb is trigonal pyramidally bonded to three $Co(CO)_3$ groups; each Co has distorted octahedral geometry with the Sb bonds being *fac*. In forming capped M_4P_4 structures the smaller P atom produces structural strain, causing severe distortion, often to the point where cubic geometry is lost, *e.g.* $Co_4(\eta^5\text{-}Cp)_4P_4$ is a complex[321] in which the cobalt atoms are so close that they must be considered bonded, *i.e.* the ideal cubic M_4P_4 has the M sites pulled out and drawn together (Figure 30).

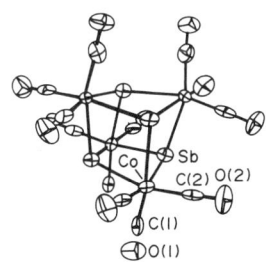

Figure 29   The 50% probability thermal ellipsoids of $Co_4(CO)_{12}Sb_4$. The molecule of idealized T_d-$\bar{4}3m$ geometry has crystallographic site symmetry D_M-$\bar{4}2m$

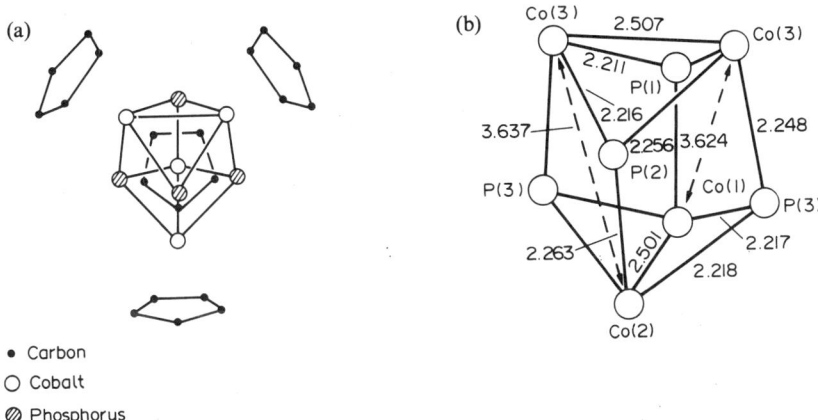

• Carbon
○ Cobalt
◐ Phosphorus

Figure 30   (a) The $[Co_4(\eta^5\text{-}C_5H_5)_4P_4]$ tetramer. (b) The Co_4P_4 core of $[Co_4(\eta^5\text{-}C_5H_5)_4P_4]$ with crystallographic independent distances

14.6.1 η^1- and η^2-P$_4$ Ligands

Sacconi and co-workers prepared (**64**) by reaction (95). The short d(Ni—P) suggests a degree of π bonding.[322] Welsh *et al.*[323] have succeeded in preparing an η^2-P$_4$ complex (Figure 31), in which the η^2-d(P—P) distance lengthens by about 0.25 Å over the P$_4$ free ligand, where d(P—P) = 2.21 Å.

$$\text{(95)}$$

$$\text{(64)}$$

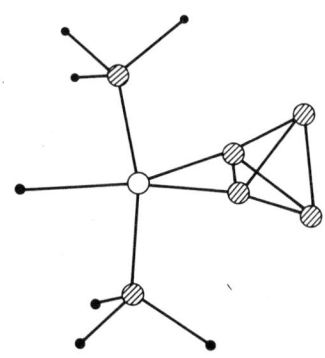

Figure 31 An η^2-P$_4$ complex. Important molecular parameters: Rh(1)—Cl(1), 2.4095 (14); Rh(1)—P(1), 2.3340 (14); Rh(1)—P(2), 2.3312 (14); Rh(1)—P(3), 2.3016 (16); Rh(1)—P(4), 2.2849 (16); P(3)—P(4), 2.4616 (22) Å; P(1)—Rh(1)—Cl(1), 82.65 (5); P(2)—Rh(1)—Cl(1), 83.56 (5); P(3)—Rh(1)—P(4), 64.92 (5)°.

The P$_4$ tetrahedron will, under certain conditions, rearrange to an open chain, as shown by equation (96), *i.e.* the ligand is a chain of six phosphorus atoms, *viz.* Ph$_2$PCH$_2$P(Ph)$_2$P=PP= Ph$_2$PCH$_2$P(Ph$_2$)P=P—P=P(Ph)$_2$CH$_2$PPh$_2$. The double bonds are based on those measurements in the crystal structure of the cobalt(II) complex (Figure 32). The short P—P bonds are P(3)—P(4) and P(5)—P(6) and the longer bond is P(4)—P(5). On reacting the complex with W(CO)$_6$, the tungsten complexes *via* the P—5 atom to give [{Co(Ph$_2$PCH$_2$P(Ph)$_2$P$_4$P(Ph)$_2$CH$_2$P-Ph$_2$)}W(CO)$_5$]BPh$_4$.

$$[\text{Co(H}_2\text{O)}_6](\text{BF}_4)_2 + \text{Ph}_2\text{PCH}_2\text{PPh}_2 + \text{P}_4 \xrightarrow[\text{reflux}]{\text{THF}} [\text{Co}\{\text{Ph}_2\text{PCH}_2\text{P(Ph)}_2\text{P}_4\text{P(Ph)}_2\text{CH}_2\text{PPh}_2\}]\text{BF}_4 \quad \text{(96)}$$

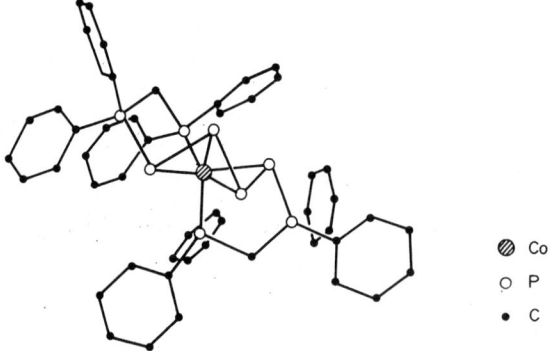

⦻ Co
○ P
● C

Figure 32 Perspective view of the complex cation [Co(Ph$_2$PCH$_2$P(Ph)$_2$P$_4$P(Ph)$_2$CH$_2$PPh$_2$)]$^+$

14.6.2   Phosphorus Ligands Derived from P$_4$ Fragments

The cyclotriphosphorus fragment of P$_4$ is one of the faces of the tetrahedron. Alternatively, a TM may be deemed to have replaced one of the P atoms of the P$_4$ tetrahedra in the case of η^3-P$_3$—M complexes. Typically, the η^3-P_3 complexes are synthesized by reacting a THF solution of white phosphorus with a TM complexed with triphos (triphos is 1,1,1-tris(diphenylphosphino)methyl-ethane, $MeC(CH_2PPh_2)_3$) or other polydentate ligands, *e.g.* [(triphos)Co(η^3-P_3)], Figure 33.[324] The P_3 face of the MP_3 complex tetrahedron is capable of further complex formation to give so-called triple decker complexes, *e.g.* [(triphos)Pd{μ-(η^3-P_3)}Pd(triphos)], Figure 34.[325]

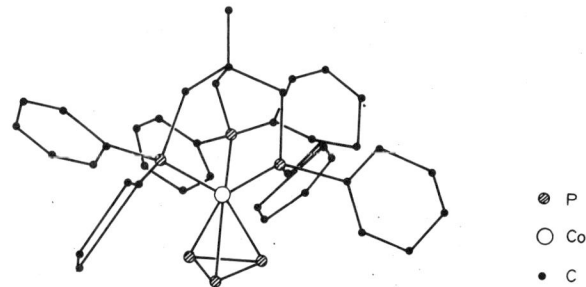

⦻	P
○	Co
●	C

Figure 33 Perspective view of the complex molecule [(triphos)Co(η^3-P_3)]

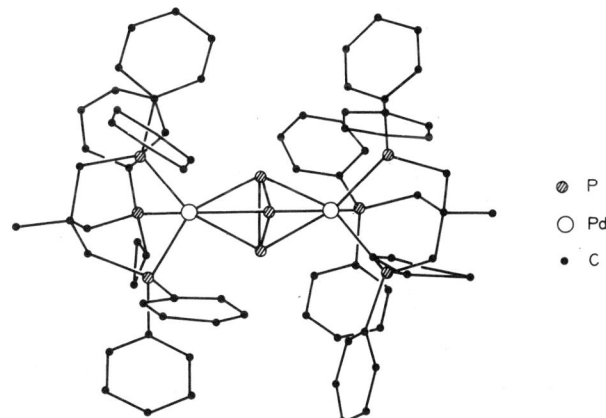

⦻	P
○	Pd
●	C

Figure 34 Perspective view of the [(triphos)Pd(μ-(η^3-P_3))Pd(triphos)]$^+$ cation

14.6.3 P_2 Fragments

Although η^2-P_2 complexes have already been referred to (Section 14.5.2.1), the point may be made in the context of P_4 fragments that M_2P_2 complexes may be viewed as disubstitution products of the P_4 tetrahedron by the TM, *e.g.* the molybdenum complexes (**65**) and (**66**).[326] The P_2 moiety is often formed in reactions of phosphorus-containing compounds by the stripping of constituents, *e.g.* the reaction of $Co(CO)_4^-$ with PCl_3 in THF affords a complex $Co_2(CO)_6P_2$, an air-sensitive red oil which, on refluxing with a benzene solution of PPh_3, gives a crystalline complex

$$Cp(CO)_2Mo{-}|{-}P$$

(**65**) 'monosubstitution'

$$Cp(CO)_2Mo{-}|{-}Mo(CO)_2Cp$$

(**66**) 'disubstitution'

[$Co_2(CO)_5(PPh_3)(P_2)$] (Figure 35; d(P—P) = 2.019, d(Co—P) average = 2.264, d(Co—Co) = 2.574 Å).

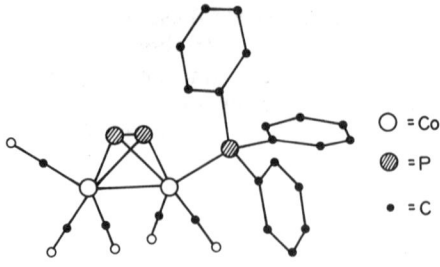

Figure 35 Configuration of the $Co_2(CO)_5(PPh_3)(\mu \cdot P_2)$ molecule showing the atom-labelling scheme

14.6.4 The Naked Phosphorus Ligand, η^9-P and η^6-P_6

For the sake of completeness the ultimate, encapsulated phosphorus ligand in $[Rh_9(CO)_{21}P]^{2-}$ is described.[327] Like $[Co_2(CO)_5(PPh_3)(P_2)]$, $[Rh_9(CO)_{21}P]^{2-}$ is formed by stripping the substituents from a tertiary phosphine ligand in tetraethyleneglycol dimethyl ether solution (equation 97).

$$[Rh(CO)_2(acac)] + PPh_3 \xrightarrow[\text{Cs benzoate}]{CO/H_2} [Rh_9(CO)_{21}P]^{2-} + \text{other products} \qquad (97)$$

The cluster size in such complexes as these is controlled by the size of the encapsulated atom. In this case, the phosphorus is within bonding distance of eight of the rhodium atoms and its position in the eight Rh atoms is displaced towards the ninth — hence it is nine-coordinate. The Rh cluster is a monocapped square antiprism; eight Rh—P distances range from 2.401 to 2.449 Å and one (weak) is at 3.057 Å (the sum of covalent radii is 2.48 Å) (Figure 36). The fragmentation of the P_4 tetrahedron in some cases leads to the formation of larger phosphorus assemblies by recombination. Both P_5 and P_6 rings have been prepared, stabilized by being the 'meat in the sandwich' of triple

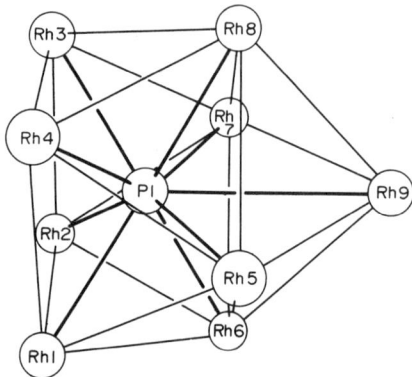

Figure 36 ORTEP diagram of $[Rh_9P(CO)_{21}]^{2-}$ without the carbon monoxide ligands

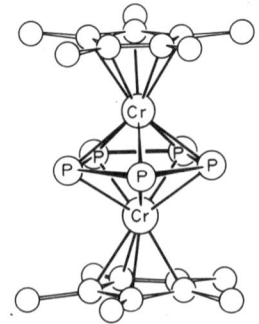

Figure 37 Selected bond lengths (Å) and angles (°): P—P 2.15(2)–2.21(2), P—Cr 2.29(1)–2.32(1), Cr—Cr 2.727(5); P—P—P 107.3(6)–108.9(6). Sum of angles 540, Cr—P—Cr 72.2(3)–72.7(3)

decker sandwich complexes, *e.g.* [(Me$_5$CpCr)-μ-η^5-P$_5$-(Me$_5$CpCr)] (Figure 37)[328] and [(Me$_5$CpMo)-μ-η^6-P$_6$-(Me$_5$CpMo)] (Figure 38).[329] In these complexes the phosphorus rings have a delocalized aromatic character (equation 98).

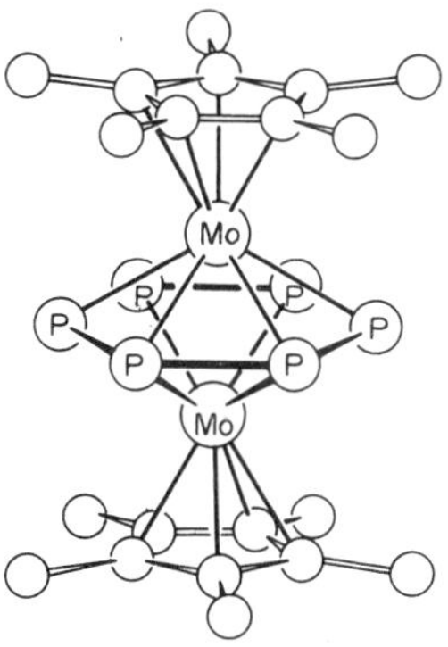

Figure 38 Selected bond lengths (Å) and angles (°): P—P 2.167(3), 2.167(3), 2.175(3), P—Mo 2.541(2), 2.542(2), 2.541(2), mean Mo—C 2.331, Mo—Mo 2.647(1). The methyl C atoms lie, on average, 0.1079 Å above the plane of the five-membered ring (distortion about 4°); P—P—P 119.9(1), 120.1(1), 120.1(1), 120.0(1), Mo—P—Mo 62.7(1), 62.8(1), 62.8(1)

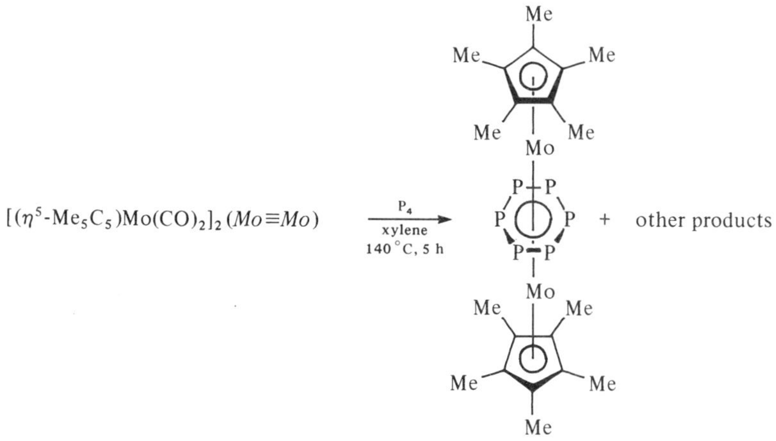

$$[(\eta^5\text{-Me}_5\text{C}_5)\text{Mo(CO)}_2]_2 \ (Mo\equiv Mo) \xrightarrow[\substack{\text{xylene} \\ 140\,^\circ\text{C, 5 h}}]{\text{P}_4} \qquad + \quad \text{other products} \qquad (98)$$

14.7 REFERENCES

1. P. Thenard, *C.R. Hebd. Seances Acad. Sci., Ser. C*, 1847, **25**, 892.
2. A. W. Hoffman and A. Cahours, *Q. J. Chem. Soc.*, 1859, **11**, 56.
3. A. Michaelis and H. V. Soden, *Annalen*, 1885, **229**, 295.
4. J. Chatt and F. G. Mann, *J. Chem. Soc.*, 1938, 1622.
5. F. P. Dwyer and D. M. Stewart, *J. Proc. R. Soc. NSW*, 1949, **83**, 177.
6. D. P. Craig, 'Ronald Sydney Nyholm' *Biogr. Mem. R. Soc.*, 1972, **18**, 445.
7. A. P. Humphries and J. D. Kaesz, *Prog. Inorg. Chem.*, 1979, **25**, 145.
8. J. Chatt, J. R. Dilworth and R. L. Richards, *Chem. Rev.*, 1978, **78**, 589.
9. F. A. Cotton and G. Wilkinson, 'Advanced Inorganic Chemistry', 4th edn., Wiley, New York, 1980.
10. L. H. Pignolet (ed.), 'Homogeneous Catalysis with Metal Phosphine Complexes', Plenum, New York, 1983.
11. L. Maier, in 'Organic Phosphorus Compounds', ed. L. Maier and G. M. Kosolapoff, Wiley, New York, 1972, **1**, 1.
12. C. A. McAuliffe and W. Levason, 'Phosphine, Arsine and Stibine Complexes of the Transition Elements', Elsevier, Amsterdam, 1979.
13. M. Fild and R. Schmutzler, in 'Organic Phosphorus Compounds', ed. L. Maier and G. M. Kosolapoff, Wiley, New York, 1972, vol. 4, p. 75.
14. B. Buchner and L. B. Lockhardt, *Org. Synth.*, 1951, **31**, 88.
15. L. Horner, P. Beck and V. G. Toscano, *Chem. Ber.*, 1961, **94**, 2122.
16. G. W. Parshall, *J. Inorg. Nucl. Chem.*, 1960, **12**, 372.
17. G. W. Parshall, *Inorg. Synth.*, 1974, **15**, 191.
18. S. A. Butler and J. Chatt, *Inorg. Synth.*, 1974, **15**, 185.
19. J. G. Morse, K. Cohn, R. W. Rudolph and R. W. Parry, *Inorg. Synth.*, 1967, **10**, 157.
20. R. A. Henderson, W. Hussain, G. J. Leigh and F. B. Normanton, *Inorg. Synth.*, 1985, **23**, 141.
21. L. Maier, in 'Organic Phosphorus Compounds', ed. L. Maier and G. M. Kosolapoff, Wiley, New York, 1972, vol. 1, p. 289.

22. W. Levason, C. A. McAuliffe, R. C. Barth and S. O. Grim, *Inorg. Synth.*, 1976, **16**, 188.
23. W. Hewertson and H. R. Watson, *J. Chem. Soc.*, 1962, 1490.
24. K. Isslieb and A. Tzschach, *Chem. Ber.*, 1959, **92**, 1118.
25. W. Levason, C. A. McAuliffe and S. G. Murray, *J. Organomet. Chem.*, 1975, **88**, 171.
26. S. O. Grim and R. P. Molenda, *Phosphorus Sulfur*, 1974, **4**, 189.
27. K. Isslieb and R. Voelker, *Chem. Ber.*, 1961, **94**, 392.
28. G. Kordosky, B. R. Cook, J. C. Cloyd and D. W. Meek, *Inorg. Synth.*, 1973, **14**, 14.
29. N. K. Patel and H. J. Harwood, *J. Org. Chem.*, 1967, **32**, 2999.
30. K. Sommer, *Z. Anorg. Allg. Chem.*, 1970, **376**, 37.
31. F. R. Benn, J. C. Briggs and C. A. McAuliffe, *J. Chem. Soc., Dalton Trans.*, 1984, 293.
32. E. Fluck, *Fortschr. Chem. Forsch.*, 1973, **35**, 1.
33. K. D. Crosbie and G. M. Sheldrick, *J. Inorg. Nucl. Chem.*, 1969, **31**, 3684.
34. W. Kuchen and H. Buckwald, *Chem. Ber.*, 1958, **91**, 2296.
35. G. W. Parshall, *Inorg. Synth.*, 1968, **11**, 157.
36. W. Gee, R. A. Shaw and B. C. Smith, *Inorg. Synth.*, 1967, **9**, 19.
37. K. Isslieb and H. Weichmann, *Chem. Ber.*, 1968, **101**, 2197.
38. R. A. Markham, E. A. Dietz and D. R. Martin, *J. Inorg. Nucl. Chem.*, 1973, **35**, 2659.
39. J. G. Evans, P. L. Goggin, R. J. Goodfellow and J. G. Smith, *J. Chem. Soc. (A)*, 1968, 464.
40. A. J. Carty, T. Hinsperger and P. M. Boorman, *Can. J. Chem.*, 1970, **48**, 1959.
41. R. D. W. Kemmitt, D. I. Nichols and R. D. Peacock, *J. Chem. Soc. (A)*, 1968, 2149.
42. H. Hoffmann and P. Schellenbeck, *Chem. Ber.*, 1967, **100**, 692.
43. L. Horner, *Pure Appl. Chem.*, 1964, **9**, 225.
44. A. M. Aguiar and D. Daigle, *J. Am. Chem. Soc.*, 1964, **86**, 2299.
45. R. B. King and P. N. Kapoor, *J. Am. Chem. Soc.*, 1971, **93**, 4158.
46. G. R. Van Hecke and W. de W. Horrocks, *Inorg. Chem.*, 1966, **5**, 1960.
47. W. Schindbauer, *Monatsh. Chem.*, 1965, **96**, 2051.
48. W. Levason, C. A. McAuliffe and S. G. Murray, unpublished observations.
49. L. Horner, P. Beck and V. G. Toscano, *Chem. Ber.*, 1961, **94**, 2122.
50. P. G. Eller and D. W. Meek, *J. Organomet. Chem.*, 1970, **22**, 631.
51. H. Hartmann, C. Beermann and H. Czempik, *Z. Anorg. Allg. Chem.*, 1956, **287**, 261.
52. W. R. Cullen, *Adv. Inorg. Chem. Radiochem.*, 1972, **15**, 323 and refs. therein.
53. C. E. Wymore and J. C. Bailar, *J. Inorg. Nucl. Chem.*, 1960, **14**, 42.
54. L. F. Warren and M. A. Bennett, *J. Am. Chem. Soc.*, 1974, **96**, 3340.
55. F. A. Hart, *J. Chem. Soc.*, 1960, 3324.
56. S. O. Grim, J. Del Gandio, R. P. Molenda, C. A. Tolman and J. P. Jesson, *J. Am. Chem. Soc.*, 1974, **96**, 3416.
57. J. C. Briggs, C. A. McAuliffe, W. E. Hill, D. M. A. Minahan, J. G. Taylor and G. Dyer, *Inorg. Chem.*, 1982, **21**, 4024.
58. F. R. Benn, J. C. Briggs and C. A. McAuliffe, *J. Chem. Soc., Dalton Trans.*, 1984, 293.
59. J. Chatt and F. A. Mart, *J. Chem. Soc.*, 1960, 1378.
60. D. L. Berglund and D. W. Meek, *Inorg. Chem.*, 1969, **8**, 2602.
61. J. G. Hartley, L. M. Venanzi and D. C. Goodall, *J. Chem. Soc.*, 1963, 3930.
62. J. C. Cloyd and D. W. Meek, *Inorg. Chim. Acta*, 1972, **6**, 607.
63. R. B. King, *Acc. Chem. Res.*, 1972, **5**, 177.
64. R. B. King and J. C. Cloyd, *J. Am. Chem. Soc.*, 1975, **97**, 53.
65. T. L. DuBois, W. H. Myers and D. W. Meek, *J. Chem. Soc., Dalton Trans.*, 1975, 1011.
66. J. Ellerman and K. Dorn, *Chem. Ber.*, 1966, **99**, 653.
67. J. G. Hartley and L. M. Venanzi, *J. Chem. Soc.*, 1963, 3930.
68. R. B. King, R. N. Kapoor, M. S. Saran and P. N. Kapoor, *Inorg. Chem.*, 1971, **10**, 1851.
69. C. A. McAuliffe, *Adv. Inorg. Chem. Radiochem.*, 1975, **17**, 165.
70. R. B. King and J. C. Cloyd, *J. Am. Chem. Soc.*, 1975, **97**, 46 and refs. therein.
71. L. Stryer, 'Biochemistry', 2nd edn., Freeman, 1981, p. 875.
72. C. J. Pederson, *J. Am. Chem. Soc.*, 1967, **89**, 7017.
73. F. G. Mann, 'The Heterocyclic Derivatives of Phosphorus, Arsenic, Antimony and Bismuth', Wiley, New York, 1970.
74. G. A. Melson (ed.), 'Coordination Chemistry of Macrocyclic Compounds', Plenum, New York, 1979.
75. L. D. Quin, E. D. Middlemas and N. S. Rao, *J. Org. Chem.*, 1982, **47**, 905.
76. W. E. Hill, J. G. Taylor, C. P. Falshaw, T. J. King, B. Beagley, D. M. Tonge, R. G. Pritchard and C. A. McAuliffe, *J. Chem. Soc., Dalton Trans.*, paper 5/470, in press.
77. H. Hope, M. Viggiano, B. Moezzi and P. P. Power, *Inorg. Chem.*, 1984, **23**, 2550.
78. See reference 64, pp. 6–15.
79. M. Ciampolini, P. Dapporto, A. Dei, N. Nardi and F. Zanobini, *Inorg. Chem.*, 1982, **21**, 489.
80. M. Ciampolini, N. Nardi and F. Zanobini, *Inorg. Chim. Acta*, 1983, **76**, L17.
81. M. Ciampolini, P. Dapporto, A. Dei, N. Nardi and F. Zanobini, *J. Chem. Soc., Dalton Trans.*, 1984, 575.
82. A. Van Zon, J. H. G. Frijns and F. J. Torny, *Recl. Trav. Chim. Pays-Bas*, 1983, **102**, 326.
83. T. N. Kudyra, A. A. Chaikovskaya, Z. Z. Rozhkova and A. M. Pinchuk, *Zh. Obshch. Khim.*, 1982, **52**, 1092.
84. T. N. Kudrya, A. V. Krsanov and A. S. Shtepanek, *Zh. Obshch. Khim.*, 1980, **50**, 2452.
85. J. B. Rober, J. P. Dustasta and J. P. Allbrand, *J. Am. Chem. Soc.*, 1974, **96**, 4584.
86. J. P. Picavet, *Tetrahedron Lett.*, 1977, 1583.
87. H. J. Cristau, L. Chiche, H. Christol, H. Fallou, P. Hullot and G. Renard, *Nouv. J. Chim.*, 1984, **8**, 191.
88. F. Vogtle and G. Wittig, *J. Chem. Educ.*, 1973, **50**, 650.
89. E. P. Kyba, C. N. Clubb, S. B. Larson, V. J. Schueler and R. E. Davis, *J. Am. Chem. Soc.*, 1985, **107**, 2141.
90. E. P. Kyba, A. M. John, B. Brown, C. W. Hudson, M. J. McPhaul, A. Harding, K. Larsen, S. Niedzwiecki and R. E. Davis, *J. Am. Chem. Soc.*, 1980, **102**, 139.
91. L. Horner, H. Kunz and P. Walach, *Phosphorus Sulfur*, 1975, **6**, 63; 1978, **5**, 171.
92. K. D. Berlin, S. D. Venkatermau and M. El-Deek, *Tetrahedron Lett.*, 1976, 3365.

93. M. De Sousa Healy and A. J. Rest, *Adv. Inorg. Chem. Radiochem.*, 1978, **21**, 1.
94. O. Stelzer, R. Bartsch, S. Hietkamp and H. Peters, *Inorg. Chem.*, 1984, **23**, 3304, and refs. therein.
95. M. K. Cooper, C. W. G. Ansell, K. P. Dancey, P. A. Duckworth, K. Hendrick, M. McPartlin and P. A. Tasker, *J. Chem. Soc., Chem. Commun.*, 1985, 437; 1985, 439.
96. D. K. Cabbiness and D. W. Margerum, *J. Am. Chem. Soc.*, 1969, **91**, 6540; 1970, **92**, 2151.
97. D. Munro, *Chem. Br.*, 1977, **13**, 100.
98. N. K. Dalley, in 'Synthetic Multidentate Macrocyclic Systems', ed. R. M. Izatt and J. J. Christensen, Academic, New York, 1978.
99. P. H. Davis, *Inorg. Chem.*, 1975, **14**, 1753.
100. E. P. Kyba, *J. Am. Chem. Soc.*, 1981, **103**, 3868.
101. K. Mislow and R. Baechler, *J. Am. Chem. Soc.*, 1970, **92**, 3090.
102. G. O. Doak and L. D. Freedman, 'Organometallic Compounds of Arsenic, Antimony and Bismuth,' Wiley, New York, 1970.
103. M. Dub, 'Organometallic Compounds', 2nd edn., Springer-Verlag, New York, 1968, vol. 3.
104. I. T. Millar, H. Heaney, D. M. Hcinckcy and W. C. Fernelius, *Inorg. Synth.*, 1960, **6**, 113.
105. R. D. Feltham, A. Kasenally and R. S. Nyholm, *J. Organomet. Chem.*, 1967, **7**, 285.
106. R. L. Barker, E. Booth, W. E. Jones, A. F. Millidge and F. N. Woodward, *Chem. Ind.*, 1949, **68**, 285; 289; 295.
107. E. R. H. Jones and F. G. Mann, *J. Chem. Soc.*, 1955, 4472.
108. W. Levason, C. A. McAuliffe, A. I. Plaza and S. O. Grim, *Inorg. Synth.*, 1976, **16**, 184.
109. G. A. Gough and H. King, *J. Chem. Soc.*, 1928, 2426.
110. K. Sommer, *Z. Anorg. Allg. Chem.*, 1970, **376**, 150.
111. J. Chatt and F. G. Mann, *J. Chem. Soc.*, 1939, 610.
112. B. Bosnich, W. G. Jackson and S. B. Wild, *J. Am. Chem. Soc.*, 1983, **95**, 8269.
113. R. L. Dutta, D. W. Meek and D. H. Busch, *Inorg. Chem.*, 1970, **9**, 1215.
114. G. O. Doak and L. D. Freedman, *Synthesis*, 1974, 328.
115. A. M. Aguair and T. G. Archibald, *J. Org. Chem.*, 1967, **32**, 2627.
116. R. D. Feltham and W. Silverthorn, *Inorg. Synth.*, 1967, **10**, 159.
117. A. Tzschach and R. Nidel, *Z. Chem.*, 1970, **10**, 118.
118. A. Tzschach and W. Lange, *Chem. Ber.*, 1962, **95**, 1360.
119. R. L. Schriner and C. N. Wolf, *Org. Synth.*, 1950, **30**, 95.
120. E. G. Claeys, *J. Organomet. Chem.*, 1966, **5**, 446.
121. M. Durand and J. P. Laurent, *C.R. Hebd. Seances Acad. Sci., Ser. C*, 1965, **261**, 3793.
122. J. Seifter, *J. Am. Chem. Soc.*, 1939, **61**, 530.
123. G. J. Burrows and E. E. Turner, *J. Chem. Soc.*, 1920, 1337.
124. A. M. Aguiar, J. T. Mague, H. J. Aguaia, T. G. Archibald and B. Preien, *J. Org. Chem.*, 1968, **33**, 1681.
125. H. Zorn, H. Schindbauer and D. Hammer, *Monatsh. Chem.*, 1967, **98**, 731.
126. K. K. Chow, W. Levason and C. A. McAuliffe, *J. Chem. Soc., Dalton Trans.*, 1976, 1429.
127. W. Levason and C. A. McAuliffe, *J. Chem. Soc., Dalton Trans.*, 1974, 2238.
128. K. Sommer, *Z. Anorg. Allg. Chem.*, 1970, **377**, 278.
129. R. D. Feltham and H. G. Metzger, *J. Organomet. Chem.*, 1971, **33**, 347; and references therein.
130. J. R. Phillips and J. H. Vis, *Can. J. Chem.*, 1967, **45**, 675.
131. R. D. Feltham and H. G. Metzger, *J. Organomet. Chem.*, 1971, **33**, 347.
132. N. V. Duffy, A. J. Layton, R. S. Nyholm, D. Powell and M. L. Tobe, *Nature (London)*, 1966, **212**, 177.
133. W. R. Cullen, *Adv. Inorg. Chem. Radiochem.*, 1972, **15**, 323.
134. W. R. Cullen, L. D. Hall and J. E. H. Ward, *J. Am. Chem. Soc.*, 1974, **96**, 3422.
135. B. Bosnich and S. B. Wild, *J. Am. Chem. Soc.*, 1970, **92**, 459.
136. K. Henrick and S. B. Wild, *J. Chem. Soc., Dalton Trans.*, 1975, 1506.
137. G. A. Barclay, R. S. Nyholm and R. V. Parish, *J. Chem. Soc.*, 1961, 4433.
138. K. Isslieb and B. Hamman, *Z. Anorg. Allg. Chem.*, 1966, **343**, 196.
139. T. E. W. Howell, S. A. J. Pratt and L. M. Venanzi, *J. Chem. Soc.*, 1961, 3167.
140. R. G. Cunningham, R. S. Nyholm and M. L. Tobe, *J. Chem. Soc.*, 1964, 580.
141. W. E. Hill, J. Dalton and C. A. McAuliffe, *J. Chem. Soc.*, 1973, 143.
142. O. St C. Headley, R. S. Nyholm, C. A. McAuliffe, L. Sindellari, M. L. Tobe and L. M. Venanzi, *Inorg. Chim. Acta*, 1970, **4**, 93.
143. G. A. Barclay and A. K. Barnard, *J. Chem. Soc.*, 1961, 4269.
144. B. Bosnich, S. T. D. Lo and A. E. Sullivan, *Inorg. Chem.*, 1975, **14**, 2305.
145. S. T. Chow and C. A. McAuliffe, *J. Organomet. Chem.*, 1974, **77**, 401.
146. J. Ellerman and K. Dorn, *Chem. Ber.*, 1967, **100**, 1230.
147. G. A. Barclay, C. M. Harris and J. V. Kingston, *Chem. Commun.*, 1968, 965.
148. H. A. Meinema, H. F. Martens and J. G. Noltes, *J. Organomet. Chem.*, 1973, **51**, 223.
149. C. A. McAuliffe, I. E. Niven and R. V. Parish, unpublished observations.
150. K. Isslieb and B. Hamann, *Z. Anorg. Allg. Chem.*, 1966, **343**, 196.
151. B. R. Cook, C. A. McAuliffe and D. W. Meek, *Inorg. Chem.*, 1972, **11**, 2676.
152. Y. Matsumura and R. Okawara, *Inorg. Nucl. Chem. Lett.*, 1969, **5**, 449.
153. K. Isslieb and B. Hamann, *Z. Anorg. Allg. Chem.*, 1965, **339**, 289.
154. S. Sato, Y. Matsumura and R. Okawara, *J. Organomet. Chem.*, 1972, **43**, 333.
155. C. A. McAuliffe, I. E. Niven and R. V. Parish, *Inorg. Chim. Acta*, 1975, **15**, 67.
156. R. J. Dickinson, W. Levason, C. A. McAuliffe and R. V. Parish, *J. Chem. Soc., Chem. Commun.*, 1975, 272.
157. K. Isslieb, B. Hamann and L. Schmidt, *Z. Anorg. Allg. Chem.*, 1965, **339**, 298.
158. K. Isslieb and B. Hamann, *Z. Anorg. Allg. Chem.*, 1964, **332**, 179.
159. K. K. Chow, W. Levason and C. A. McAuliffe, *J. Chem. Soc., Dalton Trans.*, 1976, 1429.
160. A. N. Nesmeyanov, A. E. Borisov and N. V. Novikova, *Dokl. Akad. Nauk SSSR*, 1969, **172**, 1329.
161. H. Hartmann and G. Kuhl, *Z. Anorg. Allg. Chem.*, 1969, **312**, 186.

162. W. Levason, C. A. McAuliffe and S. G. Murray, *J. Organomet. Chem.*, 1975, **88**, 171.
163. J. J. Ventura, *US Pat.* 3347892 (1967) (*Chem. Abstr.*, 1967, **68**, 69129).
164. H. Gilman and H. L. Yablunky, *J. Am. Chem. Soc.*, 1941, **63**, 212.
165. W. Levason and K. G. Smith, *Inorg. Chim. Acta*, 1981, **41**, 133.
166. T. R. Carlton and C. D. Cook, *Inorg. Chem.*, 1971, **10**, 2628.
167. L. R. Gray, S. J. Higgins, W. Levason and M. Webster, *J. Chem. Soc., Dalton Trans.*, 1984, 459.
168. W. Levason and C. A. McAuliffe, *Inorg. Chim. Acta*, 1974, **11**, 33.
169. E. R. H. Jones and F. G. Mann, *J. Chem. Soc.*, 1955, 4472.
170. B. R. Cook, C. A. McAuliffe and D. W. Meek, *Inorg. Chem.*, 1972, **11**, 2676.
171. M. O. Workman, C. A. McAuliffe and D. W. Meek, *J. Coord. Chem.*, 1972, **2**, 137.
172. B. R. Higginson, C. A. McAuliffe and L. M. Venanzi, *Inorg. Chim. Acta*, 1971, **5**, 37.
173. C. A. McAuliffe, D. Phil. Thesis, Oxford University, 1967.
174. E. Grimley and D. W. Meek, *Inorg. Chem.*, 1986, **25**, 2049.
175. L. Baracco and C. A. McAuliffe, *J. Chem. Soc., Dalton Trans.*, 1972, 948.
176. G. S. Benner, W. E. Hatfield and D. W. Meek, *Inorg. Chem.*, 1964, **3**, 1544.
177. C. A. McAuliffe and D. W. Meek, *Inorg. Chim. Acta*, 1971, **5**, 270.
178. J. W. Dawson and L. M. Venanzi, *J. Chem. Soc. (A)*, 1971, 2897.
179. F. M. Ashmawy, F. R. Benn, C. A. McAuliffe and D. G. Watson, *J. Organomet. Chem.*, 1985, **287**, 65.
180. E. P. Kyba and S.-S. P. Chou, *J. Chem. Soc., Chem. Commun.*, 1980, 449.
181. E. P. Kyba and S.-S. P. Chou, *J. Org. Chem.*, 1981, **46**, 860.
182. K. Mislow, R. C. Braechler, J. P. Casey, R. J. Cook and G. H. J. Senkler, *J. Am. Chem. Soc.*, 1972, **94**, 2859.
183. E. P. Kyba, R. E. Davis, S.-T. Liu, K. A. Hassett and S. B. Larson, *Inorg. Chem.*, 1985, **24**, 4629.
184. T. Kauffmann and J. Ennen, *Angew. Chem., Int. Ed. Engl.*, 1981, **20**, 118.
185. T. Kauffmann and J. Ennen, *Chem. Ber.*, 1985, **118**, 2703, 2714.
186. A. L. Balch, L. A. Fossett, M. M. Olmstead, D. E. Oram and P. E. Reedy, *J. Am. Chem. Soc.*, 1985, **107**, 5272.
187. C. A. Tolman, *Chem. Rev.*, 1977, **77**, 313.
188. B. L. Shaw, A. Pryde and B. Weeks, *J. Chem. Soc., Dalton Trans.*, 1976, 322.
189. R. J. Puddephatt, *Chem. Soc. Rev.*, 1983, **12**, 99.
190. J. H. Espenson and S. Muraldiharan, *J. Am. Chem. Soc.*, 1984, **106**, 8104.
191. T. B. Brill, Q. B. Bao, S. J. Landon, A. L. Rheingold and T. M. Haller, *Inorg. Chem.*, 1985, **24**, 900.
192. A. P. Sattelberger, M. L. Keutkens, W. L. Elcesser and J. C. Hoffman, *Inorg. Chem.*, 1984, **23**, 1718.
193. L. Dahlengburg, *Inorg. Chim. Acta*, 1985, **104**, 51.
194. M. T. Halfpenny, J. G. Hartley and L. M. Venanzi, *J. Chem. Soc. (A)*, 1967, 627.
195. J. A. Osborn, *Angew. Chem., Int. Ed. Engl.*, 1980, **19**, 1024.
196. M. D. Fryzuk, S. J. Rettig, A. Westerhaus and H. D. Williams, *Inorg. Chem.*, 1985, **24**, 4316.
197. M. C. Baird and R. T. Smith, *Inorg. Chim. Acta*, 1982, **62**, 135.
198. J. R. Chipperfield, S. Franks and F. R. Hartley, in 'Catalytic Aspects of Metal–Phosphine Chemistry', ed. E. C. Alyea and D. W. Meek, Am. Chem. Soc. Special Publication, 1982, chap. 16.
199. B. Bosnich and N. K. Roberts, chap. 21, 'Asymmetric Catalytic Hydrogenation', in ref. 198.
200. G. Ferguson, E. C. Alyea, S. A. Dias and R. J. Restivo, *Inorg. Chem.*, 1977, **16**, 2329.
201. G. Ferguson, E. C. Alyea and M. Khan, *Inorg. Chem.*, 1978, **17**, 2965.
202. G. Ferguson, E. C. Alyea, S. A. Dias and M. Pavez, *Inorg. Chim. Acta*, 1979, **37**, 45.
203. C. A. Tolman, *J. Am. Chem. Soc.*, 1970, **92**, 2953.
204. N. S. Imyanitov, *Koord. Khim.*, 1985, **11**, 1041; 1181.
205. K. W. Bagnall and L. Xing-Fu, *J. Chem. Soc., Dalton Trans.*, 1982, 1365.
206. K. W. Bagnall, *J. Less-Common Met.*, 1984, **98**, 309.
207. D. M. P. Mingos, *Inorg. Chem.*, 1982, **21**, 464.
208. J. A. Mosbo, J. T. DeSanto, B. N. Storhoff, P. L. Block and R. E. Bloss, *Inorg. Chem.*, 1980, **19**, 3086.
209. A. Immirzi and A. Musco, *Inorg. Chim. Acta*, 1977, **25**, L41.
210. J. D. Oliver and J. D. Smith, *Inorg. Chem.*, 1978, **17**, 2585.
211. P. Heimbach, J. Kluth, H. Schenkluhn and B. Weimann, *Angew. Chem., Int. Ed. Engl.*, 1980, **19**, 569.
212. K. L. Minten, Ph.D. Thesis, University of Manchester, 1983.
213. R. G. Goel and T. Allman, *Can. J. Chem.*, 1982, **60**, 716.
214. M. Wada, S. Higashizaki and A. Tsuboi, *J. Chem. Res. (S)*, 1985, 38.
215. C. A. Streuli and W. A. Henderson, *J. Am. Chem. Soc.*, 1960, **82**, 5791; C. A. Streuli, *Anal. Chem.*, 1960, **32**, 985.
216. H. Goetz and A. Sidhu, *Liebigs Ann. Chem.*, 1965, **682**, 71.
217. H. Goetz and S. Domin, *Liebigs Ann. Chem.*, 1967, **704**, 1.
218. M. I. Kabachnik and T. A. Mastryukova, *Russ. Chem. Rev. (Engl. Transl.)*, 1969, **38**, 795.
219. J. R. Chipperfield, chap. 7 in 'Advances in Linear Free Energy Relationships', ed. N. B. Chapman and J. Shorter, Plenum, London, 1972, pp. 355–361.
220. W. H. Thompson and C. T. Sears, *Inorg. Chem.*, 1977, **16**, 769.
221. N. N. Greenwood and A. Earnshaw, 'Chemistry of the Elements', Pergamon, Oxford, 1984, p. 566.
222. R. J. Angelici and C. M. Ingemanson, *Inorg. Chem.*, 1969, **8**, 83.
223. W. C. Smith, *Inorg. Synth.*, 1960, **6**, 20.
224. R. H. Holmes and R. P. Wagner, *Inorg. Chem.*, 1963, **2**, 384.
225. R. J. Puddephatt, G. M. Bancroft and T. Chan, *Inorg. Chim. Acta*, 1983, **73**, 83.
226. H. Bock, *Pure Appl. Chem.*, 1975, **77**, 313.
227. B. E. Bursten, *J. Am. Chem. Soc.*, 1982, **104**, 1299.
228. B. E. Bursten, D. J. Darensbourg, G. E. Kellog and D. L. Lichtenberger, *Inorg. Chem.*, 1984, **23**, 4361.
229. T. P. Edbiew and J. W. Rabalais, *Inorg. Chem.*, 1974, **13**, 308.
230. J. E. Huheey, 'Inorganic Chemistry', 3rd edn., Harper and Row, New York, 1983, p. 436.
231. L. M. Venanzi, *Chem. Br.*, 1968, **4**, 162.
232. D. A. Bochvar, N. P. Gambaryan and L. M. Epshtein, *Russ. Chem. Rev. (Engl. Transl.)*, 1976, **45**, 660.

233. D. E. Ellis, W. C. Trogler, S.-X. Xiao and Z. Berkovitch-Yellin, *J. Am. Chem. Soc.*, 1983, **105**, 7033.
234. D. S. Marynick, *J. Am. Chem. Soc.*, 1984, **106**, 4064.
235. D. S. Marynick, S. Askari and D. F. Nickerson, *Inorg. Chem.* 1985, **24**, 868.
236. J. C. Giordan, J. A. Tossell and J. H. Moore, *Inorg. Chem.*, 1985, **24**, 1100.
237. J. K. Burdett, 'Molecular Shapes', Wiley-Interscience, New York, 1980, p. 205.
238. A. G. Orpen and N. G. Connelly, *J. Chem. Soc., Chem. Commun.*, 1985, 1310.
239. R. Hoffmann and A. R. Rossi, *Inorg. Chem.*, 1975, **14**, 365.
240. R. S. Berry, *J. Chem. Phys.*, 1960, **32**, 933.
241. C. A. Udovich, R. J. Clark and H. Haas, *Inorg. Chem.*, 1969, **8**, 1066.
242. H. Schumann, J. Pickardt and L. Rosch, *J. Organomet. Chem.*, 1976, **107**, 241.
243. J. K. Ruff and D. P. Bauer, *Inorg. Chem.*, 1983, **22**, 1686.
244. T. L. Brown and D. L. Lichtenberger, *J. Am. Chem. Soc.*, 1977, **99**, 8187.
245. P. E. Riley and R. E. Davis, *Inorg. Chem.*, 1980, **19**, 159.
246. O. Stelzer, S. Morton and W. S. Sheldrick, *Z. Anorg. Allg. Chem.*, 1981, **475**, 232.
247. R. H. Neilson, R. J. Thoma, I. Vickovic and W. H. Watson, *Organometallics*, 1984, **3**, 1132.
248. E. J. Forbes, D. L. Jones, K. Paxton and T. A. Hamor, *J. Chem. Soc., Dalton Trans.*, 1979, 879.
249. F. W. B. Einstein, R. K. Pomeroy and L. R. Martin, *Inorg. Chem.*, 1985, **24**, 2777.
250. J. J. Legendre, C. Girard and M. Huber, *Bull. Soc. Chim. Fr.*, 1971, 1998.
251. B. L. Shaw, *J. Organomet. Chem.*, 1980, **200**, 307.
252. E. C. Alyea, G. Ferguson, J. Malito and B. L. Ruhl, *Inorg. Chem.*, 1985, **24**, 3720.
253. C. A. Tolman and W. C. Seidel, *Ann. N.Y. Acad. Sci.*, 1983, **415**, 201.
254. H. D. Empsall, E. M. Hyde, E. Mentzer, B. L. Shaw and M. F. Uttley, *J. Chem. Soc., Dalton Trans.*, 1976, 2069.
255. K. Mislow, J. F. Blount and C. A. Maryanoff, *Tetrahedron Lett.*, 1975, 913.
256. H. Schumann, L. Rosch and J. Pickardt, *Z. Anorg. Allg. Chem.*, 1976, **426**, 66.
257. F. A. Cotton, D. J. Darensbourg and B. W. S. Kolthammer, *Inorg. Chem.*, 1981, **20**, 4440.
258. R. G. Goel, W. P. Henry, M. J. Oliver and A. L. Beauchamp, *Inorg. Chem.*, 1981, **20**, 2924.
259. A. L. Beauchamp and C. Cagnon, *Acta Crystallogr., Sect. B*, 1979, **35**, 166.
260. C. A. McAuliffe, H. F. Al-Khateeb, D. S. Barratt, J. C. Briggs, A. Challita, A. Hosseiny, M. G. Little, A. G. Mackie and K. Minten, *J. Chem. Soc., Dalton Trans.*, 1983, 2147.
261. C. G. Howard, G. Wilkinson, M. Thornton-Pett and M. B. Hursthouse, *J. Chem. Soc., Dalton Trans.*, 1983, 2025.
262. B. Beagley, C. A. McAuliffe, K. Minten and R. G. Pritchard, *J. Chem. Soc., Chem. Commun.*, 1984, 658.
263. M. J. Crook, W. Levason and C. A. McAuliffe, *Inorg. Chem.*, 1978, **17**, 325.
264. W. Levason, L. R. Gray, S. J. Higgins and M. Webster, *J. Chem. Soc., Dalton Trans.*, 1984, 459.
265. L. F. Warren and M. A. Bennett, *Inorg. Chem.*, 1976, **15**, 3126.
266. M. D. Fryzuk, A. Carter and A. Westerhaus, *Inorg. Chem.*, 1985, **24**, 642.
267. M. D. Fryzuk, H. D. Williams and S. J. Rettig, *Inorg. Chem.*, 1983, **22**, 863.
268. H. H. Karsch, G. Muller and A. Appelt, *Angew. Chem., Int. Ed. Engl.*, 1986, **25**, 823.
269. H. Schumann, *Angew. Chem., Int. Ed. Engl.*, 1984, **23**, 474.
270. F. Mathey, A. Breque and P. Savignac, *Synthesis*, 1981, 983.
271. L. M. Wilkes, J. H. Nelson, L. B. McCusker, K. Seff and F. Mathey, *Inorg. Chem.*, 1983, **22**, 2476.
272. F. Mathey and G. Muller, *J. Organomet. Chem.*, 1977, **136**, 241.
273. C. C. Santini, J. Fischer, F. Mathey and A. Mitschler, *Inorg. Chem.*, 1981, **20**, 2848.
274. D. G. Holah, A. N. Hughes and B. Hui, *Can. J. Chem.*, 1972, **50**, 3714.
275. F. Mathey, P. Lemoine, M. Gross, P. Braunstein and B. Deschamps, *J. Organomet. Chem.*, 1985, **295**, 189.
276. E. W. Abel and C. Towers, *J. Chem. Soc., Dalton Trans.*, 1979, 814.
277. F. Mathey, J. Fischer and J. H. Nelson, *Struct. Bonding (Berlin)*, 1983, **55**, 153.
278. F. Mathey, C. Guimon and G. Pfister-Guillouzo, *Nouv. J. Chim.*, 1979, **3**, 725.
279. F. Mathey, R. Wiest, B. Rees and A. Mitschler, *Inorg. Chem.*, 1981, **20**, 2966.
280. F. Mathey, P. Lemoine, M. Gross, P. Braunstein and B. Deschamps, *J. Organomet. Chem.*, 1985, **295**, 189.
281. G. Huttner, K. Knoll and O. Orama, *Angew. Chem., Int. Ed. Engl.*, 1984, **23**, 976.
282. K. Wade, *Adv. Inorg. Chem. Radiochem.*, 1976, **18**, 1.
283. E. Fluck, G. Becker, B. Neumiller, G. Knebel, G. Heckmann and H. Riffel, *Angew. Chem., Int Ed. Engl.*, 1986, **25**, 1002.
284. O. J. Schere, *Angew. Chem., Int Ed. Engl.*, 1985, **24**, 924.
285. E. Cannillo, A. Coda, K. Pront and J. C. Daren, *Acta Crystallogr., Sect. B*, 1977, **33**, 2608.
286. A. H. Cowley, J. E. Kilduff, J. G. Lasch, N. C. Norman, M. Pakulski, F. Ando and T. C. Wright, *J. Am. Chem. Soc.*, 1983, **105**, 2085.
287. P. P. Power, K. M. Flynn and M. M. Olstead, *J. Am. Chem. Soc.*, 1983, **105**, 2085.
288. A. H. Cowley, J. E. Kilduff, S. K. Mehotra, N. C. Norman and M. Pakulski, *J. Chem. Soc., Chem. Commun.*, 1983, 528.
289. G. Huttner, J. Borm and L. Zsolnai, *Angew. Chem., Int Ed. Engl.*, 1985, **24**, 1069.
290. A. H. Cowley, J. E. Kilduff, J. G. Lasch, S. K. Mehotra, N. C. Norman, M. Pakulski, B. R. Whittlesey, J. L. Atwood and W. E. Hunter, *Inorg. Chem.*, 1984, **23**, 2582.
291. M. Yoshifuji, I. Shima and N. Inamoto, *J. Am. Chem. Soc.*, 1981, **103**, 4587.
292. A. H. Cowley, J. G. Lee and J. E. Boggs, *Inorg. Chim. Acta*, 1983, **77**, L61; V. Galasso, *Chem. Phys.*, 1984, **83**, 407; D. Gonbeau and G. Pfister-Guillouzo, *J. Electron. Spectrosc. Relat. Phenom.*, 1984, **33**, 279.
293. G. Huttner and H. Jungmann, *Angew. Chem., Int. Ed. Engl.*, 1979, **18**, 953.
294. O. J. Scherer, H. Sitzmann and G. Wolmershauser, *Angew. Chem., Int. Ed. Engl.*, 1984, **23**, 968.
295. H. Schafter, D. Binder and D. Fenske, *Angew. Chem., Int. Ed. Engl.*, 1985, **24**, 522.
296. R. Appel, F. Knoll and I. Ruppert, *Angew. Chem., Int. Ed. Engl.*, 1981, **20**, 731.
297. W. W. Schoeller and E. Niecke, *J. Chem. Soc., Chem. Commun.*, 1982, 569.
298. A. H. Cowley, J. L. Atwood, R. A. Jones, C. A. Stewart, W. E. Hunter and H. M. Zhang, *J. Am. Chem. Soc.*, 1983, **105**, 3737.

299. J. Stage, J. Escudie, C. Couret and M. Andrianarison, *J. Am. Chem. Soc.,* 1987, **109**, 386.
300. L. Weber, K. Reizig, R. Boese and M. Polk, *Angew. Chem., Int. Ed. Engl.*, 1985, **24**, 604.
301. O. J. Scherer, R. Walter and W. S. Sheldrick, *Angew. Chem., Int. Ed. Engl.*, 1985, **24**, 525.
302. S. Pohl, *Chem. Ber.*, 1979, **112**, 3159.
303. E. Niecke and D. Gudat, *J. Chem. Soc., Chem. Commun.*, 1987, 10.
304. O. J. Scherer and H. Jungmann, *Angew. Chem., Int. Ed. Engl.*, 1979, **18**, 953.
305. G. Henckel, E. Niecke, M. Engelmann, H. Zorn and B. Krebs, *Angew. Chem., Int Ed. Engl.*, 1980, **19**, 710.
306. H. Schmidbaur, *Angew. Chem., Int. Ed. Engl.*, 1983, **22**, 907.
307. W. Malisch, S. Himmel and M. Luksza, *Angew. Chem., Int. Ed. Engl.*, 1983, **22**, 416.
308. W. Malisch, E. Gross, K. Jorg, K. Fiederling, A. Gottlein and R. Boese, *Angew. Chem., Int. Ed. Engl.*, 1984, **23**, 738.
309. W. Malisch, H. H. Karsch, H. U. Reisacher, B. Huber, G. Muller and K. Jorg, *Angew. Chem., Int. Ed. Engl.*, 1986, **25**, 455.
310. R. T. Baker, P. J. Krusic, T. H. Tulip, J. C. Calebrese and S. S. Wreford, *J. Am. Chem. Soc.*, 1983, **105**, 6763.
311. G. Markl, H. Sejpka, S. Dietl, B. Nuber and M. L. Zieger, *Angew. Chem., Int. Ed. Engl.*, 1986, **25**, 1003.
312. J. F. Nixon, R. Bertsch, P. B. Hitchcock and M. F. Meidine, *J. Organomet. Chem.*, 1984, **266**, C41.
313. F. Mathey, F. Mercier and J. Fischer, *Angew. Chem., Int. Ed. Engl.*, 1986, **25**, 357.
314. R. Appel, W. Schunn and F. Knoch, *Angew. Chem., Int. Ed. Engl.*, 1985, **24**, 420.
315. H. Schmidbaur, *Angew. Chem., Int. Ed. Engl.*, 1983, **22**, 907.
316. H. Schumann and S. Hohmann, *Chem.-Ztg.*, 1976, **100**, 336.
317. E. O. Fischer, F. R. Kreissl, C. G. Kreiter and H. Fischer, *Chem. Ber.*, 1973, **106**, 1262.
318. C. P. Casey and T. J. Burkhardt, *J. Am. Chem. Soc.*, 1972, **94**, 6543.
319. M. L. Ziegler, R. Korswagen, R. Alt and D. Speth, *Angew. Chem., Int. Ed. Engl.*, 1981, **20**, 1049.
320. L. F. Dahl and A. S. Foust, *J. Am. Chem. Soc.*, 1970, **92**, 7337.
321. L. F. Dahl and G. L. Simon, *J. Am. Chem. Soc.*, 1973, **95**, 2175.
322. L. Sacconi, P. Dapporto and S. Midollini, *Angew. Chem., Int. Ed. Engl.*, 1979, **18**, 469.
323. A. J. Welch, W. E. Lindsell and K. J. McCullough, *J. Am. Chem. Soc.*, 1983, **105**, 4487.
324. L. Sacconi, C. A. Ghilardi, S. Midollini and A. Orlandini, *Inorg. Chem.*, 1980, **19**, 301.
325. L. Sacconi, P. Dapporto, P. Stoppioni and F. Zanobini, *Inorg. Chem.*, 1981, **20**, 3834.
326. O. J. Scherer, H. Sitzmann and G. Wolmershauser, *Angew. Chem., Int. Ed. Engl.*, 1985, **24**, 351.
327. J. L. Vidal, W. E. Walker, R. L. Pruett and R. C. Schoenings, *Inorg. Chem.*, 1979, **18**, 129.
328. O. J. Scherer, J. Schwalb, G. Wolmershauser, W. Kain and R. Gorss, *Angew. Chem., Int. Ed. Engl.*, 1986, **25**, 363.
329. O. J. Scherer, H. Sitzmann and G. Wolmershauser, *Angew. Chem., Int. Ed. Engl.*, 1985, **24**, 351.

Subject Index

Absolute configuration
  amine metal complexes, 26
Acetaldehyde
  oximes
    metal complexes, 270
Acetamide
  metal complexes, 490
Acetamide, *N*-methyl-
  metal complexes, 494
Acetic acid
  metal complexes, 436, 438
  organometallic ligands, 442
Acetic acid, (carboxymethyl)iminobis(ethylene-
      imino)tetra-
  metal complexes
    equilibrium data, 788
    stability, 787
Acetic acid, *trans*-1,2-cyclohexylenediiminotetra-
  metal complexes, 786
Acetic acid, ethylenebis[(carboxymethyl)imino-
      ethyleneimino]tetra-
  metal complexes
    equilibrium data, 788
    stability, 787
Acetic acid, ethylenebis(oxyethyleneimino)tetra-
  metal complexes
    stability, 786
Acetic acid, ethylenediaminetetra-
  acid–base equilibria, 779
  complexes
    composition, 783
    coordination numbers, 783
    solid state structure, 783
  cyclic derivatives
    complexes, 785
  heteroatom derivatives
    metal complexes, 786
  homologs
    binuclear complexes, 784
    complexes, 784
    complexes, stability constants, 784
    protonated complexes, 784
  metal complexes
    equilibrium constants, 784
    formation, 781
    hydrolytic, 783
    stability constants, 782
  methyl derivatives
    complexes, 785
  phenyl derivatives
    complexes, 785
  synthesis, 778
Acetic acid, *N*-(cyanomethyl)iminodi-
  synthesis, 779
Acetic acid, ethylenenitrilo[(hydroxyethyl)nitrilo]tri-
  iron(III) complexes, 788
Acetic acid, hexamethylenediaminetetra-
  synthesis, 779
Acetic acid, iminodi-
  metal complexes, 788
Acetic acid, nitrilotri-

metal complexes
  stability, 789
  stability constants, 789
  synthesis, 778
Acetic acid, octamethylenediaminetetra-
  synthesis, 779
Acetic acid, pentamethylenediaminetetra-
  metal complexes
    equilibrium constants, 784
Acetic acid, tetramethylenediaminetetra-
  metal complexes
    equilibrium constants, 784
  synthesis, 779
Acetic acid, trimethylenediaminetetra-
  metal complexes
    equilibrium constants, 784
  synthesis, 779
Acetohydroxamic acid
  metal complexes
    spectroscopy, 506
Acetohydroxamic acid, *N*-phenyl-
  metal complexes, 507
  as metal precipitants, 506
Acetone
  oximes
    metal complexes, 270, 798
Acetone, acetyl-
  dianion
    metal complexes, 369
  hydrogen bonds, 365
  ions
    metal complexes, 369
  metal complexes
    alcoholysis, 379
    C-η^1, 367
    C-η^3, σ-bonded, 367
    charge transfer complexes, 386
    dienediolate ligands, 369
    electrochemical synthesis, 375
    standard enthalpies of formation, 366
  reactions
    with metal surfaces, 395
  reaction with amines
    aza macrocycles from, 902
  tautomers
    reactions with iron(III), 367
  trivalent metal complexes
    polymorphs, 371
Acetone, dibenzoylacetyl-
  metal complexes, 399
Acetone, hexafluoroacetyl-
  metal complexes
    decomposition, 385
Acetonitrile, ethylenediaminetetra-
  synthesis, 778
Acetonitriles
  ligand exchange reactions, 261
Acetophenone, *o*-mercapto-
  metal complexes, 654
Acetylene, bis(diphenylphosphino)-, 993
Acid phosphatases, 773

Formula Index